Differentiation	Integration
$(cu)' = cu'$ (c constant)	$\int uv'\,dx = uv - \int u'v\,dx$
$(u + v)' = u' + v'$	$\int x^n\,dx = \frac{x^{n+1}}{n+1} + c \quad (n \neq -1)$
$(uv)' = u'v + v'u$	$\int \frac{1}{x}\,dx = \ln\lvert x\rvert + c$
$\left(\frac{u}{v}\right)' = \frac{u'v - v'u}{v^2}$	$\int e^{ax}\,dx = \frac{1}{a}e^{ax} + c$
$\frac{du}{dx} = \frac{du}{dy}\cdot\frac{dy}{dx}$ (Chain rule)	$\int \sin x\,dx = -\cos x + c$
	$\int \cos x\,dx = \sin x + c$
	$\int \tan x\,dx = -\ln\lvert\cos x\rvert + c$
$(x^n)' = nx^{n-1}$	$\int \cot x\,dx = \ln\lvert\sin x\rvert + c$
$(e^x)' = e^x$	$\int \sec x\,dx = \ln\lvert\sec x + \tan x\rvert + c$
$(a^x)' = a^x \ln a$	$\int \csc x\,dx = \ln\lvert\csc x - \cot x\rvert + c$
$(\sin x)' = \cos x$	$\int \frac{dx}{x^2 + a^2} = \frac{1}{a}\arctan\frac{x}{a} + c$
$(\cos x)' = -\sin x$	$\int \frac{dx}{\sqrt{a^2 - x^2}} = \arcsin\frac{x}{a} + c$
$(\tan x)' = \sec^2 x$	$\int \frac{dx}{\sqrt{x^2 + a^2}} = \sinh^{-1}\frac{x}{a} + c$
$(\cot x)' = -\csc^2 x$	$\int \frac{dx}{\sqrt{x^2 - a^2}} = \cosh^{-1}\frac{x}{a} + c$
$(\sinh x)' = \cosh x$	$\int \sin^2 x\,dx = \frac{1}{2}x - \frac{1}{4}\sin 2x + c$
$(\cosh x)' = \sinh x$	$\int \cos^2 x\,dx = \frac{1}{2}x + \frac{1}{4}\sin 2x + c$
$(\ln x)' = \frac{1}{x}$	$\int \tan^2 x\,dx = \tan x - x + c$
$(\log_a x)' = \frac{\log_a e}{x}$	$\int \cot^2 x\,dx = -\cot x - x + c$
$(\arcsin x)' = \frac{1}{\sqrt{1 - x^2}}$	$\int \ln x\,dx = x\ln x - x + c$
$(\arccos x)' = -\frac{1}{\sqrt{1 - x^2}}$	$\int e^{ax}\sin bx\,dx = \frac{e^{ax}}{a^2 + b^2}(a\sin bx - b\cos bx) + c$
$(\arctan x)' = \frac{1}{1 + x^2}$	$\int e^{ax}\cos bx\,dx = \frac{e^{ax}}{a^2 + b^2}(a\cos bx + b\sin bx) + c$
$(\text{arc cot}\ x)' = -\frac{1}{1 + x^2}$	

ADVANCED ENGINEERING MATHEMATICS

SEVENTH EDITION

ADVANCED ENGINEERING MATHEMATICS

ERWIN KREYSZIG
Professor of Mathematics
Ohio State University
Columbus, Ohio

JOHN WILEY & SONS, INC.
New York Chichester Brisbane Toronto Singapore

Publisher: Wayne Anderson
Mathematics Editors: Barbara Holland, Robert Macek
Developmental Editor: Joan Carrafiello
Marketing Manager: Susan Elbe
Production Supervised by: Suzanne Ingrao, Lucille Buonocore, Hudson River Studio
Designer: Edward A. Burke, Hudson River Studio
Manufacturing Manager: Lorraine Fumoso
Illustrations: John Balbalis

This book was set in Times Roman by General Graphic Services

Library of Congress Cataloging in Publication Data

Kreyszig, Erwin.
Advanced engineering mathematics / Erwin Kreyszig. — 7th ed.
p. cm.
Includes index.
1. Mathematical physics. 2. Engineering mathematics. I. Title.
QA401.K7 1993
510′.2462—dc20 92-6989
CIP

Printed in Singapore

10 9 8 7 6 5 4 3

Preface

Purpose of the Book

This book introduces students of engineering, physics, mathematics, and computer science to those areas of mathematics which, from a modern point of view, are most important in connection with practical problems.

The content and character of mathematics needed in applications are changing rapidly. Linear algebra—especially matrices—and numerical methods for computers are of increasing importance. Statistics and graph theory play more prominent roles. Real analysis (ordinary and partial differential equations) and complex analysis remain indispensable. The material in this book is arranged accordingly, in seven independent parts (see also the diagram on the next page):

A Ordinary Differential Equations (Chaps. 1–6)
B Linear Algebra, Vector Calculus (Chaps. 7–9)
C Fourier Analysis and Partial Differential Equations (Chaps. 10, 11)
D Complex Analysis (Chaps. 12–17)
E Numerical Methods (Chaps. 18–20)
F Optimization, Graphs (Chaps. 21, 22)
G Probability and Statistics (Chaps. 23, 24)

This is followed by

References (Appendix 1)
Answers to Problems (Appendix 2)
Auxiliary Material (Appendix 3 and inside of covers)
Additional Proofs (Appendix 4)
Tables of Functions (Appendix 5).

This book has helped to pave the way for the present development and will prepare students for the present situation and the future by a modern approach to the areas listed above and the ideas—some of them computer related—that are presently causing basic changes: Many methods have become obsolete. New ideas are emphasized, for instance, stability, error estimation, and structural problems of algorithms, to mention just a few. Trends are driven by supply and demand: supply of powerful new mathematical and computational methods and of enormous computer capacities, demand to solve problems of growing complexity and size, arising from more

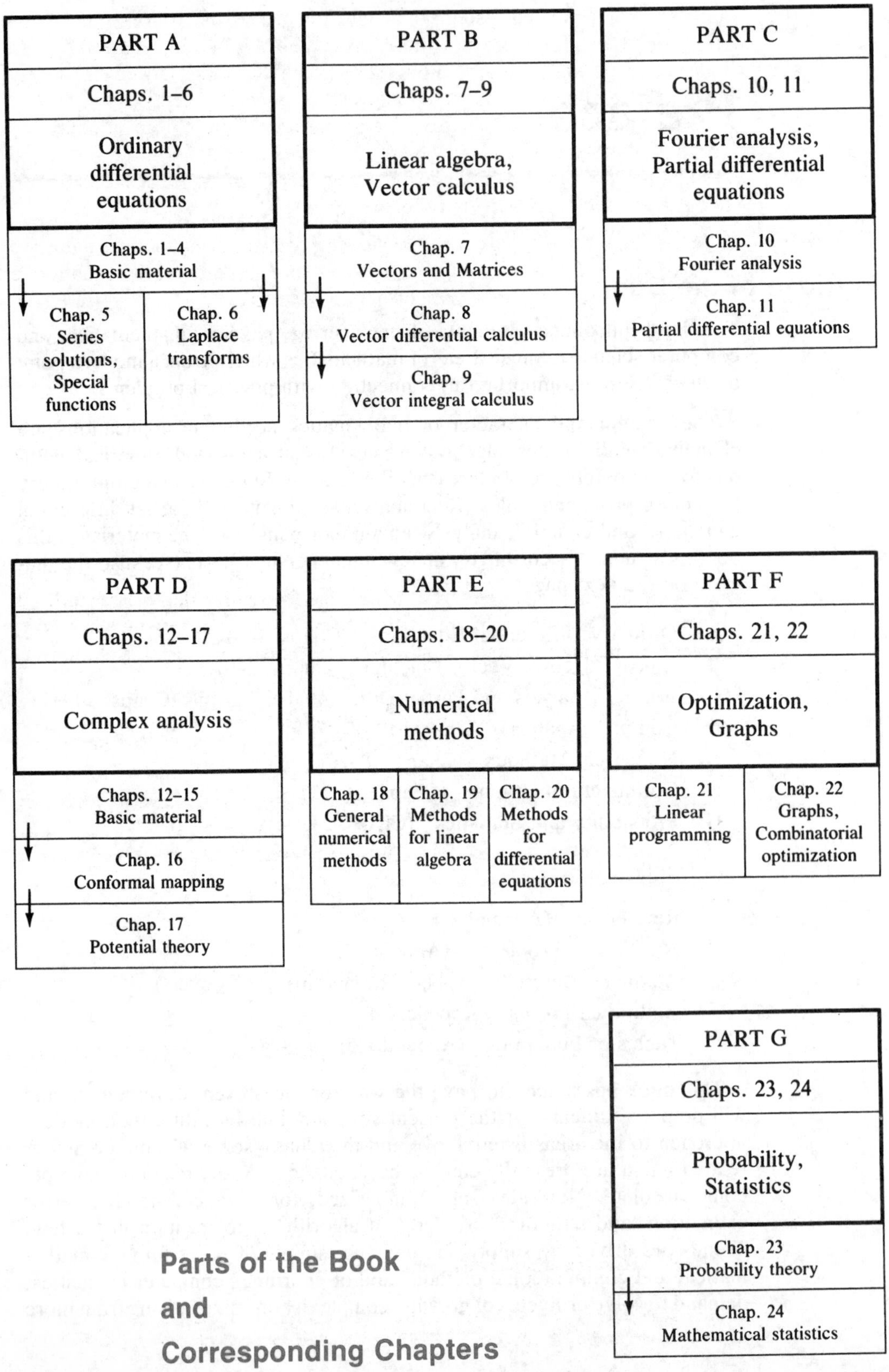

PART A
Chaps. 1–6
Ordinary differential equations
Chaps. 1–4
Basic material
Chap. 5
Series solutions, Special functions
Chap. 6
Laplace transforms
PART B
Chaps. 7–9
Linear algebra, Vector calculus
Chap. 7
Vectors and Matrices
Chap. 8
Vector differential calculus
Chap. 9
Vector integral calculus
PART C
Chaps. 10, 11
Fourier analysis, Partial differential equations
Chap. 10
Fourier analysis
Chap. 11
Partial differential equations
PART D
Chaps. 12–17
Complex analysis
Chaps. 12–15
Basic material
Chap. 16
Conformal mapping
Chap. 17
Potential theory
PART E
Chaps. 18–20
Numerical methods
Chap. 18
General numerical methods
Chap. 19
Methods for linear algebra
Chap. 20
Methods for differential equations
PART F
Chaps. 21, 22
Optimization, Graphs
Chap. 21
Linear programming
Chap. 22
Graphs, Combinatorial optimization
PART G
Chaps. 23, 24
Probability, Statistics
Chap. 23
Probability theory
Chap. 24
Mathematical statistics

Parts of the Book and Corresponding Chapters

and more sophisticated systems or production processes, from extreme physical conditions (for example, those in space travel), from materials with unusual properties (plastics, alloys, superconductors, etc.), or from entirely new tasks in computer vision, robotics, and other new fields.

The general trend seems clear. Details are more difficult to predict. Accordingly, students need solid knowledge of basic principles, methods, and results, and a clear perception of what engineering mathematics is all about, in all three phases of solving problems:

Modeling: Translating given physical or other information and data into mathematical form, into a mathematical *model* (a differential equation, a system of equations, or some other expression).

Solving: Obtaining the solution by selecting and applying suitable mathematical methods, and in most cases doing numerical work on a computer. This is the main task of this book.

Interpreting: Understanding the meaning and the implications of the mathematical solution for the original problem in terms of physics—or wherever the problem comes from.

It would make no sense to overload students with all kinds of little things that might be of occasional use. Instead, it is important that students become familiar with ways to think mathematically, recognize the need for applying mathematical methods to engineering problems, realize that mathematics is a systematic science built on relatively few basic concepts and involving powerful unifying principles, and get a firm grasp for the interrelation between theory, computing, and experiment.

The rapid ongoing developments just sketched* have led to many changes and new features in the present edition of this book, causing it to differ very substantially from previous editions.

In particular, many sections have been rewritten in a more detailed and leisurely fashion to make it a simpler book.

This has also led to a still better balance between applications, algorithmic ideas, worked-out examples, and theory.

The Four Big Changes in This Edition

1. **INSTRUCTOR'S MANUAL TRIPLED IN CONTENT**

2. **PROBLEM SETS CHANGED,**
   more closely related to the fully worked-out examples in the text.

3. **DIFFERENTIAL EQUATIONS REORGANIZED:**
   ***n*th Order Equations Enlarged to a Separate Chapter**
   Systems Greatly Extended and Modernized

4. **LINEAR ALGEBRA COMPLETELY REORGANIZED:**
   Vectors and Matrices (Chap. 7)
   Vector Algebra and Differential Calculus in R^3 (Chap. 8)
   Vector Integral Calculus in R^3 (Chap. 9)

---

*For some related recent ideas, see also "Revitalizing Undergraduate Mathematics" in *Notices Amer. Math. Soc.* **38** (1991), 545–559, or "Refreshing Curricula" in *I E E E Spectrum* (March 1992), 31–35.

Further Changes and New Features in Chapters

Ordinary Differential Equations (Chaps. 1–6)

- **First order** (Chap. 1). Integrating factors presented more systematically (Sec. 1.6); Riccati and Clairaut equations added (Sec. 1.7); mixture problems added (Sec. 1.7, etc.).
- **Second order** (Chap. 2). Streamlined by reordering—the whole theory is now in subsequent sections (Secs. 2.7, 2.8), followed by the two main methods for particular solutions (Secs. 2.9, 2.10) and the basic applications of forced vibrations (Secs. 2.11, 2.12).
- ***n*th order** (Chap. 3). Separated from the material on second-order equations and put into a chapter of their own, with an extension of the material; the presentation parallels that of Chap. 2 as much as possible.
- **Systems** (Chap. 4). Completely rewritten and extended, with a systematic use of 2×2 matrices (which are reviewed in Sec. 4.0).
- **Frobenius method** (Chap. 5). Simpler examples; more on Bessel functions (Sec. 5.6); more on eigenfunction expansions (Sec. 5.9).
- **Laplace transform** (Chap. 6). Transfer function added (Sec. 6.2); Laguerre equation added (Sec. 6.5); more on discontinuous inputs and convolution techniques (Sec. 6.6); partial fractions improved (Sec. 6.7).

Linear Algebra, Vector Calculus (Chaps. 7–9)

- **Vectors and matrices in R^n** now come first (Chap. 7), followed by
- **Vector algebra, geometry, and differential calculus in R^3** (Chap. 8, the old Chap. 6 shortened and combined with the old Chap. 8), followed by
- **Vector integral calculus** (Chap. 9, the old Chap. 9, with path independence moved to the front, to Sec. 9.2).

This new arrangement will provide a much better flow of the material.

Fourier Analysis and Partial Differential Equations (Chaps. 10, 11)

- **Fourier series and integrals** (Chap. 10). New section on the complex Fourier series (Sec. 10.6); new discussion of the amplitude spectrum of the Fourier integral and its physical significance (Secs. 10.9, 10.11).
- **Partial differential equations** (Chap. 11). Extended discussion of d'Alembert's solution (Sec. 11.4); more boundary value problems (Sec. 11.5, etc.); material taken from the problem sets into the text, to give the student more help.

Complex Analysis (Chaps. 12–17)

- **Complex numbers** (Sec. 12.1) now introduced with algebraic and geometric aspects carefully disentangled.

- **Series** (Chap. 14). Review sections combined into one section (Sec. 14.1); uniform convergence of *general* series made optional (Sec. 14.6).
- **Mappings** (Chaps. 16, 17). Simplified discussion of some of the more involved problems.

Numerical Methods (Chaps. 18–20)

- **Computer-related aspects and algorithms** still more strongly emphasized.
- **Modernization** and simplified discussions throughout the three chapters; more details on stability (Sec. 18.1, etc.); better general discussion of interpolation errors (Sec. 18.3); more on splines (Sec. 18.4) and convergence improvement by shifting (Sec. 19.8).

Appendices

- **Appendix 1** (References) updated.
- **Appendix 4** contains the optional proofs that were scattered throughout the text in the last edition.

Suggestions for Courses: A Four-Semester Sequence

The material may be taken in sequence and is suitable for four consecutive semester courses, meeting 3–5 hours a week:

First semester.	Ordinarily differential equations (Chaps. 1–6)
Second semester.	Linear algebra and vector analysis (Chaps. 7–9)
Third semester.	Complex analysis (Chaps. 12–17)
Fourth semester.	Numerical methods (Chaps. 18–20)

For the remaining chapters, see below. Possible interchanges are obvious; for instance, numerical methods could precede complex analysis, etc.

Suggestions for Courses: Independent One-Semester Courses

The book is also suitable for various independent one-semester courses meeting 3 hours a week; for example:

Introduction to ordinary differential equations (Chaps. 1–3)
Laplace transform (Chap. 6)
Vector algebra and calculus (Chaps. 8, 9)
Matrices and linear systems of equations (Chap. 7)
Fourier series and partial differential equations (Chaps. 10, 11, Secs. 20.4–20.7)
Introduction to complex analysis (Chaps. 12–15)
Numerical analysis (Chaps. 18, 20)
Numerical linear algebra (Chap. 19)
Optimization (Chaps. 21, 22)
Graphs and combinatorial optimization (Chap. 22)
Probability and statistics (Chaps. 23, 24)

General Features of This Edition

The selection, arrangement, and presentation of the material has been made with greatest care, based on past and present teaching, research, and consulting experience. Some major features of the book are these:

The book is **self-contained,** except for a few clearly marked places where a proof would be beyond the level of a book of the present type and a reference is given instead. Hiding difficulties or oversimplifying would be of no real help to students.

The presentation is **detailed,** to avoid irritating readers by frequent references to details in other books.

The examples are **simple,** to make the book teachable—why choose complicated examples when simple ones are as instructive or even better?

The notations are **modern and standard,** to help students read articles in journals or other *modern* books and understand other mathematically oriented courses.

The chapters are largely **independent,** providing flexibility in teaching special courses (see above).

Acknowledgments

I am indebted to many of my former teachers, colleagues, and students who directly or indirectly helped me in preparing this book, in particular, the present edition of it. Various parts of the manuscript were distributed to my classes in mimeographed form and returned to me with suggestions for improvement. Discussions with engineers and mathematicians (as well as written comments) were of great help to me; I want to mention particularly Professors S. L. Campbell, J. T. Cargo, P. L. Chambré, V. F. Connolly, A. Cronheim, J. Delany, J. W. Dettman, D. Dicker, D. Ellis, W. Fox, R. G. Helsel, V. W. Howe, W. N. Huff, J. Keener, E. C. Klipple, V. Komkow, H. Kuhn, G. Lamb, H. B. Mann, I. Marx, K. Millet, J. D. Moore, W. D. Munroe, J. N. Ong, Jr., P. J. Pritchard, H.-W. Pu, W. O. Ray, P. V. Reichelderfer, J. T. Scheick, H. A. Smith, J. P. Spencer, J. Todd, H. Unz, A. L. Villone, H. J. Weiss, A. Wilansky, C. H. Wilcox, L. Zia, A. D. Ziebur, all from this country, Professors H. S. M. Coxeter and R. Vaillancourt and Mr. H. Kreyszig (whose computer expertise was of great help in Chaps. 18–20) from Canada, and Professors H. Florian, M. Kracht, H. Unger, H. Wielandt, all from Europe. I can offer here only an inadequate acknowledgment of my appreciation.

Furthermore, I wish to thank John Wiley and Sons (see the list on p. IV), Mr. and Mrs. E. A. Burke of Hudson River Studio, and General Graphic Services for their effective cooperation and great care in preparing this edition.

Suggestions of many readers were evaluated in preparing the present edition. Any further comment and suggestion for improvement of the book will be gratefully received.

ERWIN KREYSZIG

Contents

PART A

ORDINARY DIFFERENTIAL EQUATIONS

Differential equations are of fundamental importance in engineering mathematics because many physical laws and relations appear mathematically in the form of such equations. In the present part, which consists of six chapters, we shall consider various physical and geometrical problems that lead to differential equations, and we shall explain the most important standard methods for solving such equations.

We shall pay particular attention to the derivation of differential equations from given physical situations. This transition from the physical problem to a corresponding "mathematical model" is called **modeling.** It is of great practical importance to the engineer and physicist, and will be illustrated by typical examples.

Differential equations are particularly convenient for modern computers. Corresponding **NUMERICAL METHODS** for obtaining approximate solutions of differential equations will be explained in Chap. 20, which is independent of the other chapters of Part E on numerical methods, so that Secs. 20.1 and 20.2 can be studied directly after Chap. 1, and Sec. 20.3 directly after Chap. 2.

Chapter 1

First-Order Differential Equations

In the present chapter we begin our program of studying ordinary differential equations and their applications by considering the simplest of these equations. These are called differential equations of the *first* order since they involve only the *first* derivative of the unknown function. In this chapter and the following chapters, one of our main goals is to equip the student with methods for solving differential equations, concentrating on those which are of practical importance.

Prerequisite for this chapter: integral calculus.
Sections that may be omitted in a shorter course: 1.8–1.11.
References: Appendix 1, Part A.
Answers to Problems: Appendix 2.

1.1 Basic Concepts and Ideas

An **ordinary differential equation** is an equation that contains one or several derivatives of an unknown function, which we call $y(x)$ and which we want to determine from the equation; the equation may also contain y itself as well as given functions and constants. For example,

$$y' = \cos x, \tag{1}$$

$$y'' + 4y = 0, \tag{2}$$

$$x^2y'''y' + 2e^xy'' = (x^2 + 2)y^2 \tag{3}$$

are ordinary differential equations. The word ***"ordinary"*** distinguishes them from ***partial*** differential equations, in which the unknown function depends on two or more variables, so that they are more complicated and will be considered later (in Chap. 11).

Differential equations arise in many engineering and other applications as mathematical models of various physical and other systems. The simplest of them can be solved by remembering elementary calculus.

For example, if a population (of humans, animals, bacteria, etc.) grows at a rate $y' = dy/dx$ (x = time) equal to the population $y(x)$ present, the population model is $y' = y$, a differential equation. If we remember from calculus that $y = e^x$ (or more generally $y = ce^x$) has the property that $y' = y$, we have obtained a solution of our problem.

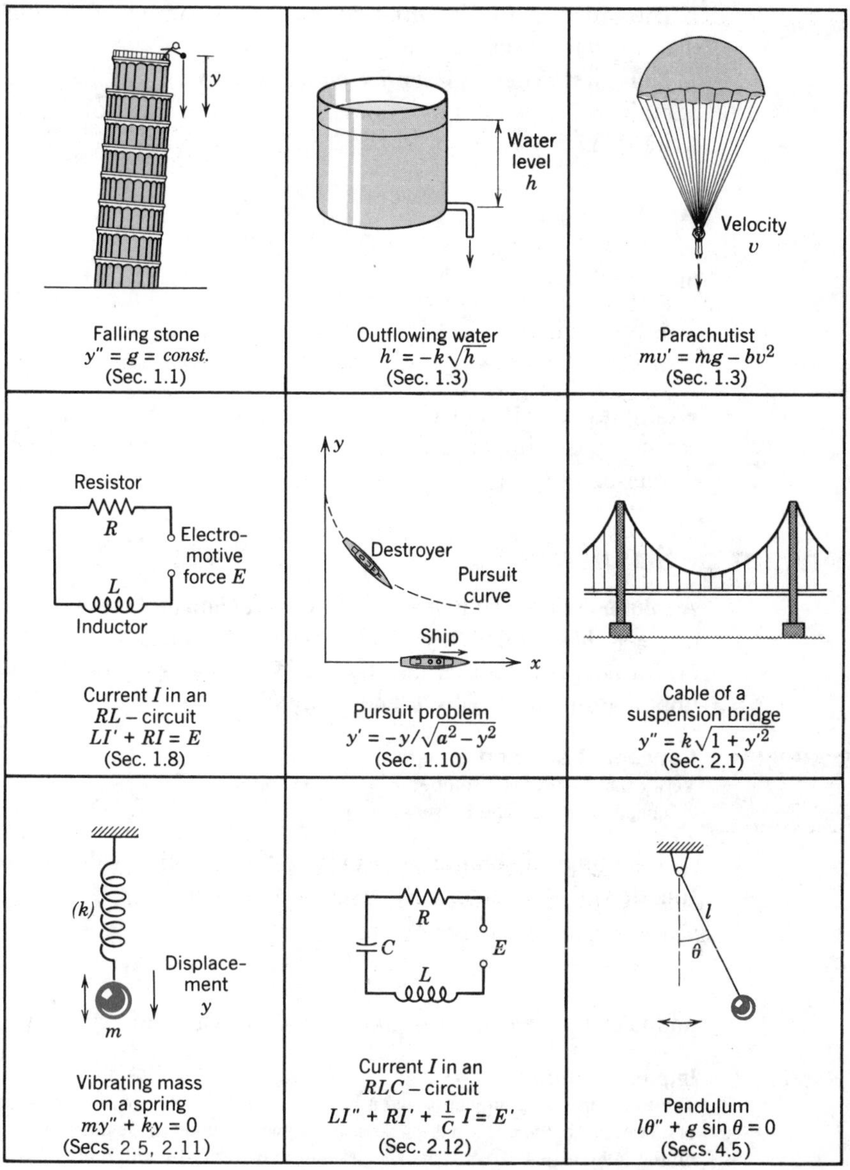

Fig. 1. Some applications of differential equations

As another example, if we drop a stone, then its acceleration $y'' = d^2y/dx^2$ (x = time, as before) is equal to the acceleration of gravity g (a constant). Hence the model of this problem of "free fall" is $y'' = g$, in good approximation, since the air resistance will not matter too much in this case. By integration we get the velocity $y' = dy/dx = gx + v_0$, where v_0 is the initial velocity with which the motion started (e.g., $v_0 = 0$). Integrating once more, we get the distance traveled $y = \frac{1}{2}gx^2 + v_0x + y_0$, where y_0 is the distance from 0 at the beginning (e.g., $y_0 = 0$).

More complicated models, such as those illustrated in Fig. 1 and in many other figures in this book, need for their solution and discussion more refined methods, which we shall discuss systematically. This begins with a classification of the differential equations according to their "order":

The **order** of a differential equation is the order of the highest derivative that appears in the equation.

Thus **first-order differential equations,** to be considered in this chapter, contain only y' and may contain y and given functions of x; hence we can write them

$$F(x, y, y') = 0 \tag{4}$$

or sometimes

$$y' = f(x, y).$$

Examples are (1) and $y' = y$ just considered. Equations (2) and (3) are of second and third order, respectively; such higher order equations will be discussed in Chaps. 2–6.

Concept of Solution

A **solution** of a given first-order differential equation (4) on some open interval[1] $a < x < b$ is a function $y = h(x)$ that has a derivative $y' = h'(x)$ and satisfies (4) for all x in that interval; that is, (4) becomes an identity if we replace the unknown function y by h and y' by h'.

EXAMPLE 1 **Concept of solution**

Verify that $y = x^2$ is a solution of $xy' = 2y$ for all x.

Indeed, $y' = 2x$, and by substitution, $xy' = x(2x) = 2y = 2x^2$, an identity in x. ∎

Sometimes a solution of a differential equation will appear as an implicit function, that is, implicitly given in the form

$$H(x, y) = 0,$$

and is called an *implicit solution*, in contrast to an *explicit solution* $y = h(x)$.

EXAMPLE 2 **Implicit solution**

The function y of x implicitly given by $x^2 + y^2 - 1 = 0 \quad (y > 0)$, representing a semicircle of unit radius in the upper half-plane, is an implicit solution of the differential equation $yy' = -x$, on the interval $-1 < x < 1$, as the student may verify by differentiation. ∎

We next observe that a differential equation may (and in general, will) have many solutions. This should not really surprise us because we know from calculus that integration introduces arbitrary constants.

EXAMPLE 3 Our equation (1) was $y' = \cos x$ and can be solved by calculus. Integration gives sine curves $y = \sin x + c$ with arbitrary c. Each c gives one of them, and these are all possible solutions, as we know from calculus. Figure 2 shows some of them, for $c = -2, -1, 0, 1, 2, 3, 4$. ∎

---

[1]By definition, the concept of **interval** includes as special cases $a < x < \infty$ and $-\infty < x < b$ as well as the whole x-axis $-\infty < x < \infty$. All our intervals are **open,** that is, their endpoints are not regarded as points belonging to the interval.

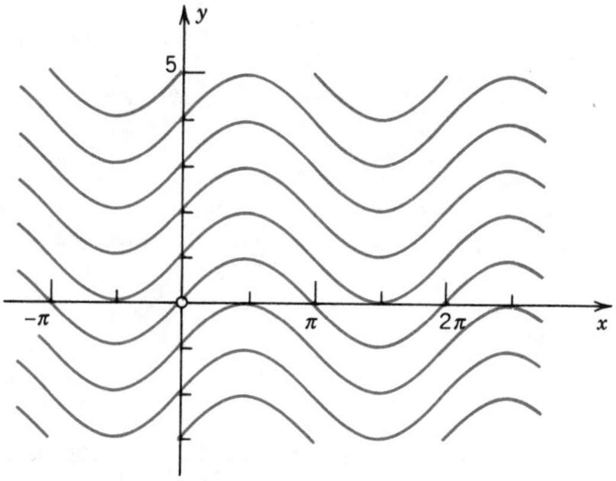

Fig. 2. Solutions of $y' = \cos x$

This example, simple as it is, is typical of most equations of first order. It illustrates that all solutions are represented by a single formula involving an arbitrary constant c. It is customary to call such a function involving an arbitrary[2] constant a **general solution** of a first-order differential equation. And if we choose a specific c (e.g., $c = 2$ or 0 or $-5/3$, etc.), we obtain what is called a **particular solution** of that equation.

Thus, $y = \sin x + c$ is a general solution of $y' = \cos x$, and $y = \sin x$, $y = \sin x - 2$, $y = \sin x + 0.75$, etc. are particular solutions.

In the following sections we shall develop various methods for obtaining general solutions of first-order equations. For a given equation, a general solution obtained by such a method is unique, except for notation, and will then be called ***the*** general solution of that differential equation.

COMMENT **Singular solutions**

A differential equation may sometimes have an additional solution that cannot be obtained from the general solution and is then called a **singular solution**. This is not of great engineering interest, and we mention it merely for completeness. For example,

$$y'^2 - xy' + y = 0 \tag{5}$$

has the general solution $y = cx - c^2$, as the student may verify by differentiation and substitution. This represents a family of straight lines, one line for each c. These are the particular solutions shown in Fig. 3. Substitution also shows that the parabola $y = x^2/4$ in Fig. 3 is also a solution. This is a singular solution of (5) because we cannot obtain it from $y = cx - c^2$ by choosing a suitable c.

We shall see that the conditions under which a given differential equation has solutions are fairly general. But we should note that there are simple equations that do not have solutions at all, and others that do not have a general solution. For example, the equation $y'^2 = -1$ does not have a solution for real y. (Why?) The equation $|y'| + |y| = 0$ does not have a general solution, because its only solution is $y \equiv 0$.

[2]The range of the constant may have to be restricted in some cases to avoid imaginary expressions or other degeneracies.

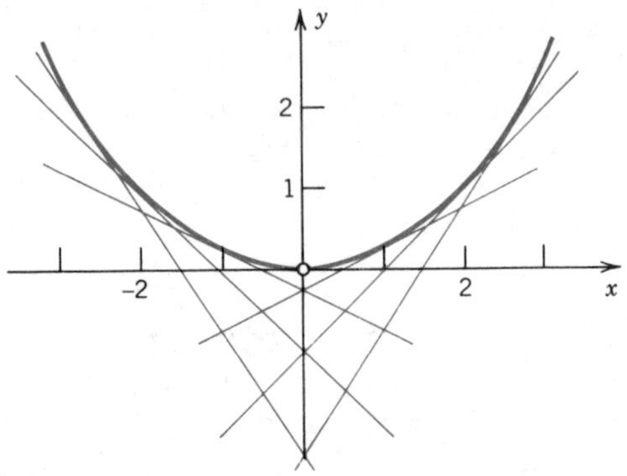

Fig. 3. Singular solution (representing a parabola) and particular solutions of equation (5)

Applications. Modeling. Initial Value Problems

Differential equations are of great importance in engineering, because many physical laws and relations appear mathematically in the form of differential equations.

To begin with, let us consider a basic physical application that will illustrate the typical steps of **modeling,** that is, the steps that lead from the physical situation (*physical system*) to a mathematical formulation (*mathematical model*) and solution, and to the physical interpretation of the result. This may be the easiest way of obtaining a first idea of the nature and purpose of differential equations and their applications.

EXAMPLE 4 **Radioactivity, exponential decay**

Experiments show that a radioactive substance decomposes at a rate proportional to the amount present. Starting with a given amount of substance, say, 2 grams, at a certain time, say, $t = 0$, what can be said about the amount available at a later time?

Solution. 1st Step. Setting up a mathematical model (a differential equation) of the physical process. We denote by $y(t)$ the amount of substance still present at time t. The rate of change is dy/dt. According to the physical law governing the process of radiation, dy/dt is proportional to y:

$$\frac{dy}{dt} = ky. \tag{6}$$

Hence y is the unknown function, depending on t. The constant k is a definite physical constant whose numerical value is known for various radioactive substances. (For example, in the case of radium ${}_{88}\mathrm{Ra}^{226}$ we have $k \approx -1.4 \cdot 10^{-11}\ \mathrm{sec}^{-1}$.) Clearly, since the amount of substance is positive and decreases with time, dy/dt is negative, and so is k. We see that the physical process under consideration is described mathematically by an ordinary differential equation of the first order. Hence this equation is the mathematical model of that physical process. ***Whenever a physical law involves a rate of change of a function, such as velocity, acceleration, etc., it will lead to a differential equation. For this reason differential equations occur frequently in physics and engineering.***

2nd Step. Solving the differential equation. We do not yet know any methods of solution, but calculus will help us here. Indeed, Eq. (6) tells us that if there is a solution $y(t)$, its derivative must be proportional to y. Now we remember from calculus that exponential functions have this property. By differentiation and substitution we see that a solution for all t is e^{kt} since $(e^{kt})' = ke^{kt}$, or more generally,

$$y(t) = ce^{kt} \tag{7}$$

with any constant c because $y'(t) = cke^{kt} = ky(t)$. Since c is arbitrary, (7) is a **general solution** of (6), by definition.

3rd Step. Determination of a particular solution from an initial condition. Clearly, our physical process behaves uniquely. Hence we should be able to get from (7) a unique particular solution. Now the amount of substance at some time t will depend on the initial amount $y = 2$ grams at time $t = 0$, or, written as a formula,

$$y(0) = 2. \tag{8}$$

This is called an ***initial condition.*** We use it to find c in (7):

$$y(0) = ce^0 = 2, \qquad \text{thus} \qquad c = 2.$$

With this c, Eq. (7) gives as the answer the particular solution

$$y(t) = 2e^{kt} \qquad \text{(Fig. 4).} \tag{9}$$

Thus the amount of radioactive substance shows exponential decay (exponential decrease with time). This agrees with physical experiments.

4th Step. Checking. From (9) we have

$$\frac{dy}{dt} = 2ke^{kt} = ky \qquad \text{and} \qquad y(0) = 2e^0 = 2.$$

We see that the function (9) satisfies the equation (6) as well as the initial condition (8).

The student should never forget to carry out this important final step, which shows whether the function is (or is not) the solution of the problem. ▮

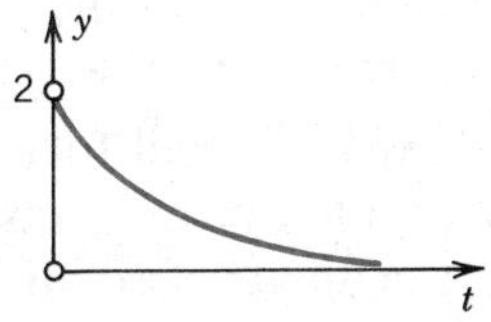

Fig. 4. Radioactivity (exponential decay)

A differential equation together with an initial condition, as in our example, is called an **initial value problem.** With x as the independent variable (instead of t) it is of the form

$$y' = f(x, y), \qquad y(x_0) = y_0 \tag{10}$$

where x_0 and y_0 are given values. (In our example, $x_0 = t_0 = 0$ and $y_0 = y(0) = 2$.) The initial condition $y(x_0) = y_0$ is used to determine a value of c in the general solution.

Let us show next that geometrical problems also lead to differential equations and initial value problems.

EXAMPLE 5 **A geometrical application**

Find the curve through the point (1, 1) in the xy-plane having at each of its points the slope $-y/x$.

Solution. The function giving the desired curve must be a solution of the differential equation

$$y' = -\frac{y}{x}. \tag{11}$$

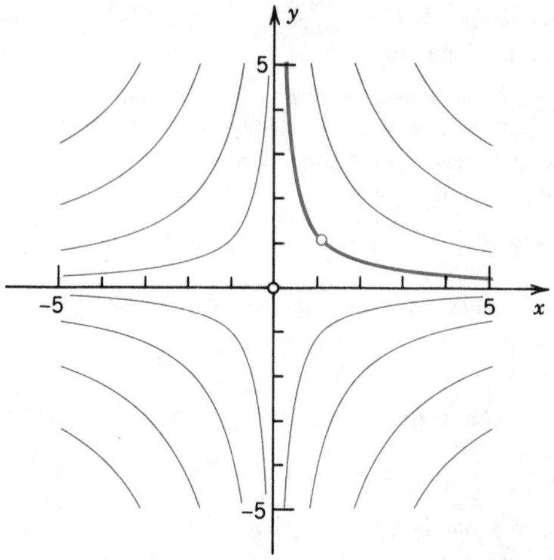

Fig. 5. Solutions of $y' = -y/x$ (hyperbolas)

We shall soon learn how to solve such an equation. Meanwhile the student may verify that the general solution of (11) is (Fig. 5)

$$y = \frac{c}{x} \qquad (c \text{ arbitrary}). \tag{12}$$

Since we are looking for the curve that passes through (1, 1), we must have $y = 1$ when $x = 1$. This initial condition $y(1) = 1$ gives $c = 1$ in (12) and as the answer the particular solution $y = 1/x$. ■

Many more examples will follow throughout the chapter; so we should without delay turn to our main task, the explanation and application of the most important methods of solution, beginning in the next section.

Problem Set 1.1

State the order of the differential equation and verify that the given function is a solution.

1. $y' + 4y = 8x, \quad y = ce^{-4x} + 2x - \frac{1}{2}$

2. $y'' + 9y = 0, \quad y = A\cos 3x + B\sin 3x$

3. $y' - 0.5y = 1, \quad y = ce^{0.5x} - 2$

4. $y''' = 6, \quad y = x^3 + ax^2 + bx + c$

5. $y' + y\tan x = 0, \quad y = c\cos x$

6. $y'' - 2y' + 2y = 0, \quad y = e^x(A\cos x + B\sin x)$

Solve the following differential equations.

7. $y' = e^{-2x}$ **8.** $y' = xe^{x^2}$ **9.** $y' = -\cos\frac{1}{2}x$ **10.** $y''' = 48x$

Verify that the given function is a solution of the corresponding differential equation and determine c so that the resulting particular solution satisfies the given initial condition.

11. $y' + y = 1, \quad y = ce^{-x} + 1, \quad y = 2.5$ when $x = 0$

12. $y' = 2xy, \quad y = ce^{x^2}, \quad y = 4$ when $x = 1$

13. $xy' = 2y$, $y = cx^2$, $y = 12$ when $x = 2$

14. $yy' = x$, $y^2 - x^2 = c$, $y(0) = 1$

15. $y' = y \cot x$, $y = c \sin x$, $y(-\pi/2) = 2$

16. $yy' + x = 0$, $x^2 + y^2 = c$, $y(\sqrt{2}) = \sqrt{2}$

Find a first-order differential equation involving both y and y' for which the given function is a solution.

17. $y = x^2$ **18.** $y = x^3 - 4$ **19.** $y = \tan x$ **20.** $x^2 + 9y^2 = 9$

Applications, Modeling

21. **(Falling body)** Experiments show that if a body falls in vacuum due to the action of gravity, then its acceleration is constant (equal to $g = 9.80$ meters/sec^2 = 32.1 ft/sec^2; this is called the *acceleration of gravity*). State this law as a differential equation for $y(t)$, the distance fallen as a function of time t (already mentioned in the text), and solve it to get the familiar law

$$y(t) = \tfrac{1}{2}gt^2.$$

(In practice, this also applies to the free fall in air if we can neglect the air resistance, for instance, if we drop a stone or an iron ball.)

22. **(Falling body, general initial conditions)** If in Prob. 21 the body starts at $t = 0$ from initial position $y = y_0$ with initial velocity $v = v_0$, show that then the solution is

$$y(t) = \tfrac{1}{2}gt^2 + v_0 t + y_0.$$

23. **(Airplane takeoff)** An airplane taking off from a landing field has a run of 1.8 kilometers. If the plane starts with speed 5 meters/sec, moves with constant acceleration, and makes the run in 40 sec, with what speed does it take off?

24. In Prob. 23, if you want to reduce the take-off speed to 250 km/hr, to what value can you reduce the constant acceleration, the other data being as before?

25. **(Exponential growth)** We know from the text that $y' = y$ with solution $y(x) = ce^x$ governs the growth of a population if the growth rate $y' = dy/dx$ equals the population $y(x)$ present (x = time). (a) What is the particular solution satisfying $y(0) = 3$? (b) What initial amount $y(0)$ is necessary to get $y = 100$ after $x = 2$ [hours]?

26. **(Exponential growth)** If in a culture of yeast the rate of growth $y'(x)$ is proportional to the population present at time x, say, $y' = ky$, verify that $y(x) = ce^{kx}$. If y doubles in 1 day, how much can be expected after 1 week at the same rate of growth? After 2 weeks?

27. **(Malthus's law)** The law in Prob. 26 (growth rate proportional to the population present) is called *Malthus's law*.[3] For the United States, observed values of $y(t) = y_0 e^{kt}$, in millions, are as follows.

t	0	30	60	90	120	150	180	190
Year	1800	1830	1860	1890	1920	1950	1980	1990
Population	5.3	13	31	63	105	150	230	250

Use the first two columns for determining y_0 and k. Then calculate values for 1860, 1890, $\cdots$, 1990 and compare them with the observed values. Comment.

[3]THOMAS ROBERT MALTHUS (1766—1834), English social scientist, one of the leaders in classical national economy.

28. **(Exponential decay; atmospheric pressure)** Observations show that the rate of change of the atmospheric pressure y with altitude x is proportional to the pressure. Assuming that the pressure at 6000 meters (about 18,000 ft) is half its value y_0 at sea level, find the formula for the pressure at any height.

29. **(Half-life)** The *half-life* of a radioactive substance is the time in which half of a given amount will disappear. What is the half-life of $_{88}Ra^{226}$ (in years) in Example 4?

30. **(Interest rates)** Let $y(x)$ be the investment resulting from a deposit y_0 after x years at an interest rate r. Show that

$$y(x) = y_0[1 + r]^x \qquad \text{(interest compounded annually)}$$

$$y(x) = y_0[1 + (r/4)]^{4x} \qquad \text{(interest compounded quarterly)}$$

$$y(x) = y_0[1 + (r/365)]^{365x} \qquad \text{(interest compounded daily).}$$

Now recall from calculus that $[1 + (1/n)]^n \to e$ as $n \to \infty$, hence $[1 + (r/n)]^{nx} \to e^{rx}$, which gives

$$y(x) = y_0 e^{rx} \qquad \text{(interest compounded continuously).}$$

What differential equation does the last function satisfy? Let $y_0 = \$1000.00$ and $r = 8\%$. Compute $y(1)$ and $y(5)$ from each of the four formulas and confirm that there is not much difference between daily and continuous compounding.

1.2 Separable Differential Equations

Many first-order differential equations can be reduced to the form

$$g(y)y' = f(x) \tag{1}$$

by algebraic manipulations. Since $y' = dy/dx$, we find it convenient to write

$$g(y)\,dy = f(x)\,dx, \tag{2}$$

but we keep in mind that this is merely another way of writing (1). Such an equation is called a **separable equation,** because in (2) the variables x and y are *separated* so that x appears only on the right and y appears only on the left.

To solve (1), we integrate on both sides with respect to x, obtaining

$$\int g(y)\frac{dy}{dx}\,dx = \int f(x)\,dx + c.$$

Now on the left we can switch to y as the variable of integration. By calculus, $(dy/dx)\,dx = dy$, so that we get

$$\int g(y)\,dy = \int f(x)\,dx + c. \tag{3}$$

If we assume that f and g are continuous functions, the integrals in (3) will exist, and by evaluating these integrals we obtain the general solution of (1).

EXAMPLE 1 Solve the differential equation

$$9yy' + 4x = 0.$$

Solution. By separating variables we have

$$9y\,dy = -4x\,dx.$$

By integrating on both sides we obtain the general solution

$$\frac{9}{2}y^2 = -2x^2 + \tilde{c}, \qquad \text{thus} \qquad \frac{x^2}{9} + \frac{y^2}{4} = c \qquad \left(c = \frac{\tilde{c}}{18}\right).$$

The solution represents a family of ellipses. Can you sketch some of them? ▮

EXAMPLE 2 Solve the differential equation

$$y' = 1 + y^2.$$

Solution. By separating variables and integrating we obtain

$$\frac{dy}{1 + y^2} = dx, \qquad \arctan y = x + c, \qquad y = \tan(x + c).$$

It is of great importance to introduce the constant of integration immediately when the integration is carried out. ▮

EXAMPLE 3 **Initial value problem**

Solve the initial value problem

$$y' + 5x^4y^2 = 0, \qquad y(0) = 1.$$

Solution. Separating variables and integrating, we obtain

$$\frac{dy}{y^2} = -5x^4\,dx, \qquad -\frac{1}{y} = -x^5 + c, \qquad y = \frac{1}{x^5 - c}.$$

From this and the initial condition we find

$$y(0) = \frac{1}{-c} = 1, \qquad c = -1, \qquad \text{thus} \qquad y = \frac{1}{x^5 + 1}.$$

Checking. Check the result.

$$y' + 5x^4y^2 = -\frac{5x^4}{(x^5 + 1)^2} + 5x^4\frac{1}{(x^5 + 1)^2} = 0, \qquad y(0) = \frac{1}{1} = 1.$$ ▮

EXAMPLE 4 **Initial value problem**

Solve

$$y' = \frac{x}{y}, \qquad y(1) = 3.$$

Solution. Separation, integration, and the use of the initial condition gives

$$y\,dy = x\,dx, \qquad \tfrac{1}{2}y^2 = \tfrac{1}{2}x^2 + c \qquad \tfrac{1}{2}\cdot 3^2 = \tfrac{1}{2}\cdot 1^2 + c, \qquad c = 4.$$

Hence the answer is the upper branch of the hyperbola $y^2 - x^2 = 2c = 8$. (Why not the lower?) ▮

EXAMPLE 5 **Initial value problem**

Solve the initial value problem

$$y' = ky, \qquad y(0) = 2.$$

Solution. By separation of variables and integration,

$$\frac{dy}{y} = k, \qquad \ln|y| = kx + \tilde{c}$$

[because by calculus, $(\ln|y|)' = y'/y$; indeed, when $y > 0$, then $(\ln|y|)' = (\ln y)' = y'/y$; and when $y < 0$, then $-y > 0$ and $(\ln|y|)' = (\ln(-y))' = -y'/(-y) = y'/y$]. Taking exponentials and noting that $e^{a+b} = e^a e^b$, we get

$$|y| = e^{kx+\tilde{c}} = e^{kx}e^{\tilde{c}}, \qquad \text{thus} \qquad y = ce^{kx},$$

where $c = +e^{\tilde{c}}$ when $y > 0$ and $c = -e^{\tilde{c}}$ when $y < 0$, and we can also admit $c = 0$ (giving $y \equiv 0$). This is the general solution. From the initial condition $y(0) = ce^0 = 2$ we have $c = 2$ and obtain as the answer the particular solution $y = 2e^{kx}$, in agreement with Example 4 in Sec. 1.1. ∎

EXAMPLE 6 **Bell-shaped curves**

Solve the initial value problem

$$y' = -2xy, \qquad y(0) = 1.$$

Solution. Separating variables, integrating, and taking exponents, we obtain

$$\frac{dy}{y} = -2x\,dx, \qquad \ln|y| = -x^2 + \tilde{c}, \qquad |y| = e^{-x^2+\tilde{c}}.$$

Setting $e^{\tilde{c}} = +c$ when $y > 0$, and $e^{\tilde{c}} = -c$ when $y < 0$, and admitting also $c = 0$ (which gives the solution $y \equiv 0$), we get the general solution

$$y = ce^{-x^2},$$

representing so-called **"bell-shaped curves,"** which play a role in heat conduction (Sec. 11.6) and in probability and statistics (Sec. 23.7 and Chap. 24). Figure 6 shows some of them for $c > 0$.

The student may show that the particular solution of our initial value problem is

$$y = e^{-x^2}.$$ ∎

Many physical and engineering problems have as their model a differential equation that is separable. We show this in the next section by discussing a careful selection of some typical problems of this nature.

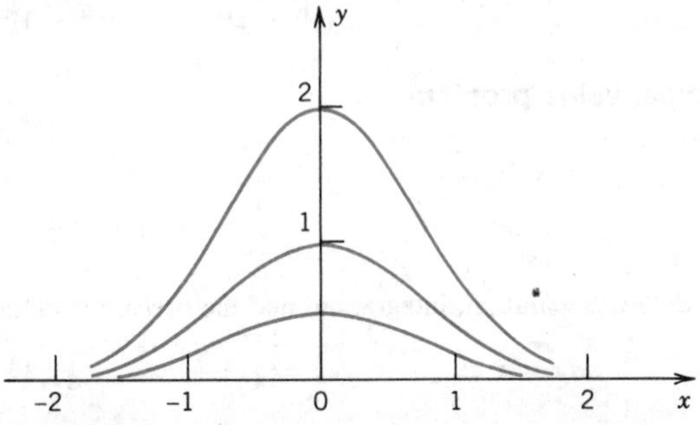

Fig. 6. Solutions of $y' = -2xy$ ("bell-shaped curves") in the upper half-plane ($y > 0$)

Problem Set 1.2

1. Why is it important to introduce the constant of integration immediately when the integration is performed?

Find a general solution. Check your answer by substitution.

2. $y' + (x + 1)y^3 = 0$

3. $y' = 3(y + 1)$

4. $y' + \csc y = 0$

5. $y' = (1 + x)(1 + y^2)$

6. $yy' = \frac{1}{2}\sin^2 \omega x \quad (\omega \neq 0)$

7. $y' \sin 2x = y \cos 2x$

8. $y' = \cos x \tan y$

9. $y' = y \tanh x$

10. $y' + y^2 = 1$

11. $y' = e^{2x} \cos^2 y$

12. $y' = y^2 \sin x$

13. $y' = y/(x \ln x)$

14. $y' = x^2y^2 - 2y^2 + x^2 - 2$

15. $y' = \sqrt{1 - y^2}$

Solve the following initial value problems. Check your answer. (L and R are constant.)

16. $y' = x^3 e^{-y}, \quad y(2) = 0$

17. $yy' + x = 0, \quad y(0) = -2$

18. $y' = 2e^x y^3, \quad y(0) = 0.5$

19. $y' \cosh^2 x + \sin^2 y = 0, \quad y(0) = \frac{1}{4}\pi$

20. $y' = 4\sqrt{y + 1}\cos 2x, \quad y(\frac{1}{4}\pi) = -1$

21. $y' = (1 - x)/y, \quad y(1) = 1$

22. $dr/dt = -4tr, \quad r(0) = 8.2$

23. $v(dv/dt) = g = const, \quad v(t_0) = v_0$

24. $e^x y' = 2(x + 3)y^3, \quad y(0) = \frac{1}{4}$

25. $dr \sin\theta = r\cos\theta\, d\theta, \quad r(\frac{1}{2}\pi) = -0.3$

26. $(x^2 + 1)^{1/2} y' = xy^3, \quad y(0) = 2$

27. $L(dI/dt) + RI = 0, \quad I(0) = I_0$

28. An initial value problem is usually solved by first determining the general solution of the equation and then the particular solution. Using (3), show that the particular solution of (1) satisfying the initial condition $y(x_0) = y_0$ can also be obtained directly from

$$\int_{y_0}^{y} g(y^*)\, dy^* = \int_{x_0}^{x} f(x^*)\, dx^*.$$

Using the formula in Prob. 28, solve:

29. Problem 17.

30. Problem 19.

1.3 Modeling: Separable Equations

Modeling means setting up mathematical models of physical or other systems. In this section we consider some of the many systems that can be modeled in terms of a separable differential equation.

EXAMPLE 1 **Radiocarbon dating**

If a fossilized bone contains 25% of the original amount of radioactive carbon ${}_6C^{14}$, what is its age?

Idea of the method to be used. In the atmosphere, the ratio of radioactive carbon ${}_6C^{14}$ and ordinary carbon ${}_6C^{12}$ is constant, and the same holds for *living* organisms. When an organism dies, the absorption of ${}_6C^{14}$ by breathing and eating terminates. Hence one can estimate the age of a fossil by comparing the carbon ratio in the fossil with that in the atmosphere. This is W. Libby's idea of radiocarbon dating (Nobel Prize for chemistry, 1960). The half-life of ${}_6C^{14}$ is 5730 years (*CRC Handbook of Chemistry and Physics*, 54th ed., 1973, p. B251).

Solution. As in Example 4 of Sec. 1.1, the mathematical model of the process of radioactive decay is

$$y' = ky, \qquad \text{solution} \qquad y(t) = y_0 e^{kt}.$$

Here, y_0 is the initial amount of $_6C^{14}$. By definition, the **half-life** (5730 years) is the time after which the amount of radioactive substance ($_6C^{14}$) has decreased to half its original value. Thus,

$$(1) \qquad y_0 e^{k \cdot 5730} = \tfrac{1}{2} y_0, \qquad e^{5730k} = \tfrac{1}{2}, \qquad k = \frac{\ln 1/2}{5730} = -0.000\,121.$$

The time after which 25% of the original amount of $_6C^{14}$ is still present can now be computed from

$$(2) \qquad y_0 e^{-0.000\,121t} = \tfrac{1}{4} y_0, \qquad t = \frac{\ln 1/4}{-0.000\,121} = 11\,460 \text{ [years]}.$$

Hence the mathematical answer is that the bone has an age of 11 460 years. Actually, the experimental determination of the half-life of $_6C^{14}$ involves an error of about 40 years. Also, a comparison with other methods shows that radiocarbon dating tends to give values that are too small, perhaps because the ratio of $_6C^{14}$ to $_6C^{12}$ may have changed over long periods of time. Hence 12 000 or 13 000 years is probably a more realistic answer to our present problem.

Have you noticed that the answer is twice the half-life? Is this just by chance? ■

EXAMPLE 2 **Newton's law of cooling**[4]

A copper ball is heated to a temperature of 100°C. Then at time $t = 0$ it is placed in water that is maintained at a temperature of 30°C. At the end of 3 minutes the temperature of the ball is reduced to 70°C. Find the time at which the temperature of the ball is reduced to 31°C.

Physical information. Experiments show that the time rate of change dT/dt of the temperature T of the ball is proportional to the difference between T and the temperature T_0 of the surrounding medium (**Newton's law of cooling**). Also, in copper, heat flows so rapidly that at any time the temperature is practically the same at all points of the ball.

Solution. 1st Step. Modeling. All we have to do is to write Newton's law as an equation; denoting the (unknown) constant of proportionality by k, we have

$$(3) \qquad \frac{dT}{dt} = k(T - 30).$$

2nd Step. General solution. Separation and integration leads to the general solution of (3),

$$\frac{dT}{T - 30} = k\,dt, \qquad \ln|T - 30| = kt + \tilde{c}, \qquad T(t) = ce^{kt} + 30.$$

3rd Step. Particular solution. The given initial condition $T(0) = 100$ yields

$$T(0) = ce^0 + 30 = 100, \qquad \text{thus} \qquad T(t) = 70e^{kt} + 30.$$

4th Step. Determination of k and of the answer. The given $T(3) = 70$ yields

$$T(3) = 70e^{k \cdot 3} + 30 = 70, \qquad \text{thus} \qquad k = \frac{1}{3} \ln \frac{70 - 30}{70} = -0.1865.$$

---

[4]Sir ISAAC NEWTON (1642—1727), great English physicist and mathematician, became a professor at Cambridge in 1669 and Master of the Mint in 1699. He and the German mathematician and philosopher GOTTFRIED WILHELM LEIBNIZ (1646—1716) invented (independently) the differential and integral calculus. Newton discovered many basic physical laws and created the method of investigating physical problems by means of calculus. His *Philosophiae naturalis principia mathematica* (*Mathematical Principles of Natural Philosophy*, 1687) contain the development of classical mechanics. His work is of greatest importance to both mathematics and physics.

This negative k makes our dT/dt negative [see (3)], as it should because T must decrease. We now get the answer from

$$T(t) = 70e^{-0.1865t} + 30 = 31, \qquad -0.1865t = \ln(1/70), \qquad t = 22.78.$$

Thus the answer is that the ball reaches the temperature of 31°C after about 23 minutes.

5th Step. Checking. Check that the answer satisfies (3) and all given conditions. ■

EXAMPLE 3 **Flow of water through a hole (Torricelli's law[5])**

A cylindrical tank 1.50 meters high stands on its circular base of diameter 1.00 meter and is initially filled with water. At the bottom of the tank there is a hole of diameter 1.00 cm, which is opened at some instant, so that the water starts draining under the influence of gravity (Fig. 7). Find the height $h(t)$ of the water in the tank at any time t. Find the times at which the tank is one-half full, one-quarter full, and empty.

Physical information. Experiments show that water issues from a hole with velocity

$$v(t) = 0.600\sqrt{2gh(t)} \qquad \text{(4)} \qquad \textbf{(Torricelli's law}^5\textbf{)}$$

where t is time, $h(t)$ the instantaneous height of the water above the hole, $g = 980 \text{ cm/sec}^2 = 32.17 \text{ ft/sec}^2$ the acceleration of gravity at the surface of the earth, and 0.600 a contraction factor[6] (because the stream has a smaller cross section than the hole). Equation (4) looks reasonable because $\sqrt{2gh}$ is the speed a body gains if it falls a distance h (under negligibly small air resistance).

Solution. 1st Step. Modeling. We shall see that the mathematical model of our physical problem will be a differential equation. The idea is to relate the decrease of the water level $h(t)$ to the outflow. If A is the area of the hole, then the volume ΔV of water flowing out with velocity v during a short time interval Δt is

$$\Delta V = Av\Delta t.$$

This causes a decrease Δh of the water level h in the tank, which must be such that the corresponding volume $B\Delta h$ (B = the cross-sectional area of the tank) equals ΔV; thus

$$B\Delta h = -\Delta V = -Av\Delta t,$$

where the minus sign appears since the water level decreases. We now divide by $B\Delta t$, substitute v from (4), and let $\Delta t \to 0$. Then, using $g = 980$ [cm/sec^2], we get

$$\frac{dh}{dt} = -\frac{A}{B}v = -26.56\frac{A}{B}\sqrt{h}. \qquad \text{(5)}$$

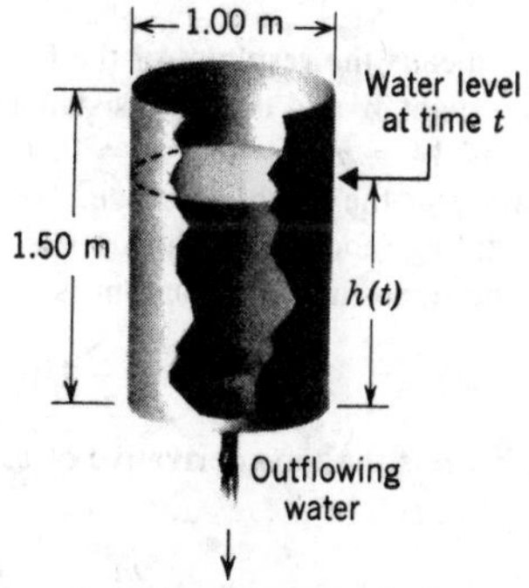

Fig. 7. Tank in Example 3

[5]EVANGELISTA TORRICELLI (1608—1647), Italian physicist and mathematician, pupil and later successor of GALILEO GALILEI (1564—1642) at Florence.

[6]Suggested by J. C. BORDA in 1766.

From the given data, $A/B = 0.500^2\pi/50.0^2\pi = 0.000\ 1000$, so that our model is the differential equation

$$\frac{dh}{dt} = -0.002656h^{1/2}. \tag{6}$$

2nd Step. General solution. By separation and integration we get

$$\frac{dh}{h^{1/2}} = -0.002656\,dt, \qquad 2h^{1/2} = -0.002656t + \tilde{c}.$$

Writing $c = \tilde{c}/2$, we have the general solution

$$h(t) = (c - 0.001328t)^2. \tag{7}$$

3rd Step. Particular solution. The initial condition is $h(0) = 150$ [cm] and gives from (7) the value $h(0) = c^2 = 150$, $c = 12.25$ and thus the particular solution

$$h(t) = (12.25 - 0.001328t)^2. \tag{7*}$$

4th Step. Answers are obtained if we solve (7*) algebraically for t, getting

$$t = \frac{12.25 - \sqrt{h}}{0.001328} = 9224 - 753\sqrt{h}.$$

Hence the tank will be half full ($h = 75.0$), one-quarter full ($h = 37.5$), and empty after $t = 2703$ sec $= 45$ min, 77 min, and 154 min, respectively.

5th Step. Checking. Check the result. ∎

EXAMPLE 4 **Skydiver**

Suppose that a skydiver falls from rest toward the earth and the parachute opens at an instant, call it $t = 0$, when the skydiver's speed is $v(0) = v_0 = 10.0$ meters/sec. Find the speed $v(t)$ of the skydiver at any later time t. Does $v(t)$ increase indefinitely?

Physical assumptions and laws. Suppose that the weight of the man plus the equipment is $W = 712$ nt (read "newtons"; about 160 lb), the air resistance U is proportional to v^2, say, $U = bv^2$ nt, where the constant of proportionality b depends mainly on the parachute, and we assume that $b = 30.0$ nt $\cdot$ sec^2/meter^2 = 30.0 kg/meter.

Solution. 1st Step. Modeling. We set up the mathematical model (the differential equation) of the problem. **Newton's second law** is

$$\text{Mass} \times \text{Acceleration} = \text{Force}$$

where "force" means the resultant of the forces acting on the skydiver at any instant. These forces are the weight W and the air resistance U.

The weight is $W = mg$, where $g = 9.80$ meters/sec^2 (about 32.2 ft/sec^2) is the acceleration of gravity at the earth's surface. Hence the mass of the man plus the equipment is $m = W/g = 72.7$ kg (about 5.00 slugs). The air resistance U acts upward (against the direction of the motion), so that the resultant is

$$W - U = mg - bv^2.$$

The acceleration a is the time derivative of v, that is, $a = dv/dt$. By Newton's second law,

$$m\frac{dv}{dt} = mg - bv^2. \tag{8}$$

This is the differential equation of our problem.

2nd Step. General solution. Division of (8) by m gives

$$\frac{dv}{dt} = -\frac{b}{m}(v^2 - k^2), \qquad k^2 = \frac{mg}{b}. \tag{8*}$$

By separation and integration

$$\int \frac{dv}{v^2 - k^2} = -\int \frac{b}{m}\,dt = -\frac{b}{m}t + \tilde{c}. \tag{9}$$

The rest of the step is integration technique. We represent the integrand in terms of partial fractions (as explained in most calculus books under "Integration techniques"). We get

$$\frac{1}{v^2 - k^2} = \frac{1}{2k}\left(\frac{1}{v - k} - \frac{1}{v + k}\right),$$

as can be readily verified, and we can now integrate; using (9), we then obtain

$$\int \frac{dv}{v^2 - k^2} = \frac{1}{2k}\left[\ln|v - k| - \ln|v + k|\right] = \frac{1}{2k}\ln\left|\frac{v - k}{v + k}\right| = -\frac{b}{m}t + \tilde{c}.$$

Multiplying by $2k$ and taking exponents, we get

$$\frac{v - k}{v + k} = ce^{-pt}, \qquad p = \frac{2kb}{m}, \tag{10}$$

where $c = +e^{2k\tilde{c}}$ or $-e^{2k\tilde{c}}$ when the fraction is positive or negative, respectively, and $c = 0$ also gives a solution ($v = k = const$), as can be seen from (8*). Solving algebraically for v, we obtain

$$v(t) = k\,\frac{1 + ce^{-pt}}{1 - ce^{-pt}}. \tag{11}$$

We see that $v(t) \to k$ as $t \to \infty$; that is, $v(t)$ does not increase indefinitely but approaches a limit, k. It is interesting to note that this limit is independent of the initial condition $v(0) = v_0$.

3rd Step. Particular solution. From the initial condition $v(0) = v_0$ and (10) we immediately have

$$c = \frac{v_0 - k}{v_0 + k}.$$

Equation (11) with this c is the particular solution we are looking for.

4th Step. Computations. Note that we have not yet made use of the given numerical values. Very often it is a good principle to hunt for general formulas first and to substitute numerical data later. In this way one often gets a better idea of what is going on and, moreover, saves work if one wants to obtain various solutions corresponding to different sets of data. In the present case we now compute the limiting speed (see before)

$$k = \sqrt{\frac{mg}{b}} = \sqrt{\frac{W}{b}} = \sqrt{\frac{712}{30.0}} = 4.87 \text{ [meters/sec]},$$

which the skydiver will reach after some time (actually fairly soon, since we are dealing with exponential functions that go to zero rather fast; see Fig. 8 on p. 18), the value of c for $v_0 = 10.0$ meters/sec, obtaining $c = (v_0 - k)/(v_0 + k) = 0.345$, and the constant in the exponent of (11),

$$p = \frac{2kb}{m} = \frac{2 \cdot 4.87 \cdot 30.0}{72.7} = 4.02 \text{ [sec}^{-1}\text{]}.$$

This altogether yields the result (Fig. 8)

$$v(t) = 4.87\,\frac{1 + 0.345e^{-4.02t}}{1 - 0.345e^{-4.02t}}. \tag{12}$$

5th Step. Checking. Check the result. ∎

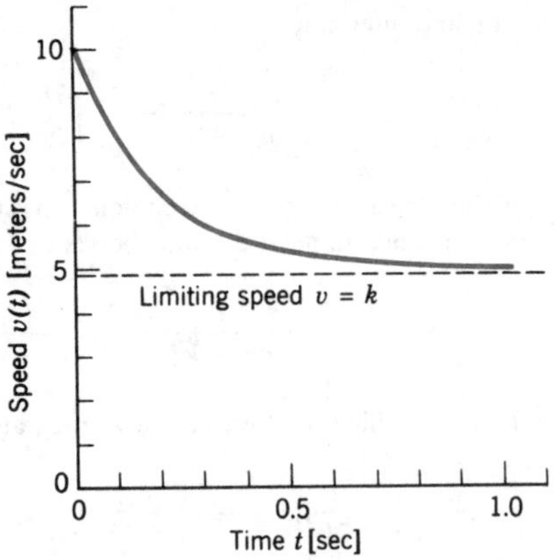

Fig. 8. Speed $v(t)$ of the skydiver in Example 4

Problem Set 1.3

1. Show that a ball thrown vertically upward with initial velocity v_0 takes twice as much time to return as to reach the highest point. Find the velocity upon return. (Air resistance is assumed negligible.)

2. **(Linear accelerator)** Linear accelerators are used in physics for accelerating charged particles. Suppose that an alpha particle enters an accelerator and undergoes a constant acceleration which increases the speed of the particle from 10^3 meters/sec to 10^4 meters/sec in 10^{-3} sec. Find the acceleration a and the distance traveled during this period of 10^{-3} sec.

3. **(Radiocarbon dating)** What should be the ${}_6C^{14}$ content of a bone that is claimed to be 2000 years old?

4. Can you think of a very short argument to get the answer in Example 1 (practically without calculation)? After what time will 12.5% of ${}_6C^{14}$ be left?

5. **(Exponential decay, half-life)** What percentage of a radioactive substance will still be present at time $H/2$, where H is the half-life of the substance? At time $2H$?

6. **(Newton's law of cooling)** A thermometer, reading 10°C, is brought into a room whose temperature is 18°C. One minute later the thermometer reading is 14°C. How long does it take until the reading is practically 18°C, say, 17.9°C?

7. **(Evaporation)** Experiments show that a wet porous substance in the open air loses its moisture at a rate proportional to the moisture content. If a sheet hung in the wind loses half its moisture during the first hour, when will it be practically dry, say, when will it have lost 99.9% of its moisture, weather conditions remaining the same?

8. **(Evaporation)** Suppose that a mothball loses volume by evaporation at a rate proportional to its instantaneous area. If the diameter of the ball decreases from 2 cm to 1 cm in 3 months, how long will it take until the ball has practically gone, say, until its diameter is 1 mm?

9. **(Exponential decay) Lambert's law of absorption**[7] states that the absorption of light in a very thin transparent layer is proportional to the thickness of the layer and to the amount incident on that layer. Formulate this in terms of a differential equation and solve it.

---

[7]JOHANN HEINRICH LAMBERT (1728—1777), German physicist and mathematician, known for his contributions to cartography and astronomy.

10. **(Torricelli's law)** In Example 3 it is claimed that the speed a body gains in free fall from rest through height h (without air resistance) is $\sqrt{2gh}$. Derive this.

11. If the diameter of the hole in Example 3 is doubled and all other data are left as before, how will the answer change?

12. **(Torricelli's law)** Suppose that the tank in Example 3 is hemispherical, of radius R, is initially full of water, and has an outlet of 5 cm^2 cross-sectional area at the bottom. The outlet is opened at some instant. Find the time it takes to empty the tank (a) for any given R, (b) for $R = 1$ meter. *Hint.* You can use (5) (why?), where B is the cross-sectional area of the tank at height $h(t)$, so that B now depends on h.

13. The time to empty the tank in Example 3 is greater than twice the time at which the tank is one-half full. Is this physically understandable?

14. **(Skydiver)** In Example 4, the speed of the skydiver decreases. What happens if the initial speed is less than k?

15. To what height of free fall does the limiting speed (4.87 meters/sec) in Example 4 correspond?

16. How do the equation and solution in Example 4 change if we assume the air resistance to be proportional to v (instead of v^2), say, $U = \tilde{b}v$ where we assume that $\tilde{b} = 7.3$ kg/sec? Is this model still reasonable?

17. **(Boyle–Mariotte's law for ideal gases[8])** Experiments show that for a gas at low pressure p (and constant temperature) the rate of change of the volume $V(p)$ equals $-V/p$. Solve the corresponding differential equation.

18. **(Flywheel)** A flywheel of moment of inertia I is rotating with a relatively small constant angular speed ω_0 (radians/sec). At some instant, call it $t = 0$, the power is shut off and the motion starts slowing down because of friction. Let the friction torque be proportional to $\sqrt{\omega}$, where $\omega(t)$ is the angular speed. Using **Newton's second law** (Moment of inertia $\times$ Angular acceleration $=$ Torque), find $\omega(t)$ and the times t_1 at which the wheel is rotating with $\omega_0/2$ and t_2 when it comes to rest.

19. **(Sugar inversion)** Experiments show that the rate of inversion of cane sugar in dilute solution is proportional to the concentration $y(t)$ of unaltered sugar. Let the concentration be 1/100 at $t = 0$ and 1/300 at $t = 4$ hours. Find $y(t)$.

20. **(Rope)** If we wind a rope around a rough cylinder which is fixed on the ground, we need only a small force at one end to resist a large force at the other. Let S be the force in the rope. Experiments show that the change ΔS of S in a small portion of the rope is proportional to S and to the small angle $\Delta\phi$ in Fig. 9. Show that the differential equation for S is $dS/d\phi = \mu S$. If $\mu = 0.2$ radian^{-1}, how many times must the rope be snubbed around the cylinder in order that a man holding one end of the rope can resist a force one thousand times greater than he can exert?

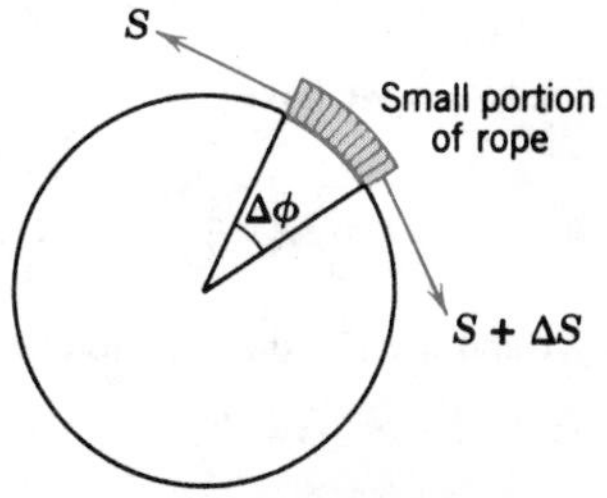

Fig. 9. Problem 20

---

[8]ROBERT BOYLE (1627—1691), English physicist and chemist, one of the founders of the Royal Society; EDMÉ MARIOTTE (about 1620—1684). French physicist and prior of a monastery near Dijon.

1.4 Reduction to Separable Form *Optional*

Certain first-order differential equations are not separable but can be made separable by a simple change of variables. This holds for equations of the form[9]

$$y' = g\left(\frac{y}{x}\right) \tag{1}$$

where g is any given function of y/x, for example $(y/x)^3$, $\sin(y/x)$, etc. The form of the equation suggests that we set

$$\frac{y}{x} = u.$$

Multiply this by x to get $y = xu$. Product differentiation now gives

$$y' = u + xu' \qquad \text{where} \quad u' = \frac{du}{dx}. \tag{2}$$

This is the left side of (1), and the right side is $g(y/x) = g(u)$. Together,

$$u + xu' = g(u).$$

Now we may separate the variables u and x, finding

$$\frac{du}{g(u) - u} = \frac{dx}{x}.$$

If we integrate on both sides and in the result replace u again by y/x, we obtain the general solution of (1), as is explained in the following examples.

EXAMPLE 1 Solve

$$2xyy' - y^2 + x^2 = 0.$$

Solution. Dividing by x^2, we have

$$2\frac{y}{x}y' - \left(\frac{y}{x}\right)^2 + 1 = 0.$$

If we set $u = y/x$ and use (2), the equation becomes

---

[9]These equations are sometimes called **homogeneous equations.** We shall not use this terminology but reserve the term "homogeneous" for a much more important purpose (see Sec. 1.7).

$$2u(u + u'x) - u^2 + 1 = 0, \qquad \text{thus} \qquad 2xuu' + u^2 + 1 = 0.$$

Separating variables, we find

$$\frac{2u\,du}{1 + u^2} = -\frac{dx}{x}.$$

By integration

$$\ln(1 + u^2) = -\ln|x| + c^*, \qquad \text{thus} \qquad 1 + u^2 = \frac{c}{x}.$$

Replacing u by y/x, we finally obtain the family of circles

$$x^2 + y^2 = cx, \qquad \text{thus} \qquad \left(x - \frac{c}{2}\right)^2 + y^2 = \frac{c^2}{4}.$$

■

EXAMPLE 2 **An initial value problem**

Solve the initial value problem

$$y' = \frac{y}{x} + \frac{2x^3 \cos x^2}{y}, \qquad y(\sqrt{\pi}) = 0.$$

Solution. We set $u = y/x$. Then $y = xu$, $y' = xu' + u$, and the equation becomes

$$xu' + u = u + \frac{2x^2 \cos x^2}{u}.$$

We simplify algebraically and then integrate:

$$uu' = 2x \cos x^2, \qquad \tfrac{1}{2}u^2 = \sin x^2 + c.$$

Since $u = y/x$, this gives

$$y = ux = x\sqrt{2 \sin x^2 + 2c}.$$

Since $\sin \pi = 0$, the initial condition yields $c = 0$. Hence the answer is

$$y = x\sqrt{2 \sin x^2}.$$

■

Sometimes the form of a given differential equation suggests other simple substitutions, as the following example illustrates.

EXAMPLE 3 Solve

$$(2x - 4y + 5)y' + x - 2y + 3 = 0. \tag{3}$$

Solution. We set $x - 2y = v$. Then $y' = \frac{1}{2}(1 - v')$ and the equation takes the form

$$(2v + 5)v' = 4v + 11.$$

Separating variables and integrating, we find

$$\left(1 - \frac{1}{4v + 11}\right) dv = 2\,dx \qquad \text{and} \qquad v - \frac{1}{4}\ln|4v + 11| = 2x + c^*.$$

Since $v = x - 2y$, this may be written

$$4x + 8y + \ln|4x - 8y + 11| = c.$$

■

Further simple substitutions are illustrated by the equations in Probs. 17–27. Some of those substitutions can be found systematically, as is shown in Ref. [A6], pp. 19–20, listed in Appendix 1. Instead of spending more time on that technique of limited applicability, we turn in the next sections to much more fundamental methods of solution.

Problem Set 1.4

Find the general solution of the following equations.

1. $xy' = x + y$

2. $xy' = 2x + 2y$

3. $xyy' = \frac{1}{2}(y^2 + x^2)$

4. $x^2y' = y^2 + 5xy + 4x^2$

5. $x^2y' = y^2 + xy + x^2$

6. $(xy' - y)\cos(2y/x) = -3x^4$

7. $x^2y' = y^2 + xy$

8. $xy' = x\sec(y/x) + y$

9. $y' = \dfrac{y + x}{y - x}$

10. $y' = \dfrac{y - x}{y + x}$

Solve the following initial value problems.

11. $xy' = x + y, \quad y(1) = -7.4$

12. $xyy' = 2y^2 + 4x^2, \quad y(2) = 4$

13. $xy' = y + x^5e^x/4y^3, \quad y(1) = 0$

14. $xy' = y^2 + y, \quad y(4) = 2$

15. $yy' = x^3 + y^2/x, \quad y(2) = 6$

16. $xy' = y + x^2\sec(y/x), \quad y(1) = \pi/2$

Using the indicated transformation, find the general solution.

17. $y' = (y + x)^2 \quad (y + x = v)$

18. $y' = \tan(x + y) - 1 \quad (x + y = v)$

19. $2x^2yy' = \tan(x^2y^2) - 2xy^2 \quad (x^2y^2 = z)$

20. $y' = (x + e^y - 1)e^{-y} \quad (x + e^y = w)$

21. $y' = \dfrac{y - x + 1}{y - x + 5} \quad (y - x = v)$

22. $y' = \dfrac{1 - 2y - 4x}{1 + y + 2x} \quad (y + 2x = v)$

23. Consider $y' = f(ax + by + k)$, where f is continuous. If $b = 0$, the solution is immediate. (Why?) If $b \neq 0$, show that one obtains a separable equation by using $u(x) = ax + by + k$ as a new dependent variable.

Using Prob. 23, find the general solution of the following equations.

24. $y' = (x + y - 7)^2$

25. $y' = 2y + 8x$

26. $y' = (6y - y^2 - 8)^{1/2}$
(*Hint.* $u = y - 3$)

27. $y' = (5y + 2)^4$

28. Find the curve $y(x)$ that passes through $(1, 1/2)$ and is such that at each point (x, y) the intercept of the tangent on the y-axis is equal to $2xy^2$.

29. Show that a straight line through the origin intersects all solution curves of a given differential equation $y' = g(y/x)$ at the same angle.

30. The positions of four battle ships on the ocean are such that the ships form the vertices of a square of length l. At some instant each ship fires a missile that directs its motion steadily toward the missile on its right. Assuming that the four missiles fly horizontally and with the same speed, find the path of each.

1.5 Exact Differential Equations

We remember from calculus that if a function $u(x, y)$ has continuous partial derivatives, its total or exact differential is

$$du = \frac{\partial u}{\partial x} dx + \frac{\partial u}{\partial y} dy.$$

From this it follows that if $u(x, y) = c = const$, then $du = 0$.

For example, if $u = x + x^2y^3 = c$, then

$$du = (1 + 2xy^3)\, dx + 3x^2y^2\, dy = 0$$

or

$$y' = -\frac{1 + 2xy^3}{3x^2y^2},$$

a differential equation that we can solve by going backward. This idea gives a powerful solution method, as follows.

A first-order differential equation of the form

$$M(x, y)\, dx + N(x, y)\, dy = 0 \tag{1}$$

is called **exact** if its left side is the total or exact differential

$$du = \frac{\partial u}{\partial x} dx + \frac{\partial u}{\partial y} dy \tag{2}$$

of some function $u(x, y)$. Then the differential equation (1) can be written

$$du = 0.$$

By integration we immediately obtain the general solution of (1) in the form

$$u(x, y) = c. \tag{3}$$

Comparing (1) and (2), we see that (1) is exact if there is some function $u(x, y)$ such that

$$\text{(a)} \quad \frac{\partial u}{\partial x} = M, \qquad \text{(b)} \quad \frac{\partial u}{\partial y} = N. \tag{4}$$

Suppose that M and N are defined and have continuous first partial derivatives in a region in the xy-plane whose boundary is a closed curve having no self-intersections. Then from (4) (see Appendix 3.2 for notation)

$$\frac{\partial M}{\partial y} = \frac{\partial^2 u}{\partial y\, \partial x},$$

$$\frac{\partial N}{\partial x} = \frac{\partial^2 u}{\partial x\, \partial y}.$$

By the assumption of continuity the two second derivatives are equal. Thus

$$\frac{\partial M}{\partial y} = \frac{\partial N}{\partial x}. \tag{5}$$

This condition is not only necessary but also sufficient[10] for $M\,dx + N\,dy$ to be an exact differential.

If (1) is exact, the function $u(x, y)$ can be found by guessing or in the following systematic way. From (4a) we have by integration with respect to x

$$u = \int M\,dx + k(y); \tag{6}$$

in this integration, y is to be regarded as a constant, and $k(y)$ plays the role of a "constant" of integration. To determine $k(y)$, we derive $\partial u/\partial y$ from (6), use (4b) to get dk/dy, and integrate dk/dy to get k.

Formula (6) was obtained from (4a). Instead of (4a) we may equally well use (4b). Then instead of (6) we first have

$$u = \int N\,dy + l(x). \tag{6*}$$

To determine $l(x)$ we derive $\partial u/\partial x$ from (6*), use (4a) to get dl/dx, and integrate.

EXAMPLE 1 **An exact equation**

Solve

$$(x^3 + 3xy^2)\,dx + (3x^2y + y^3)\,dy = 0. \tag{7}$$

Solution. 1st Step. Test for exactness. Our equation is of the form (1) with

$$M = x^3 + 3xy^2, \qquad N = 3x^2y + y^3. \qquad \text{Thus} \qquad \frac{\partial M}{\partial y} = 6xy, \qquad \frac{\partial N}{\partial x} = 6xy.$$

From this and (5) we see that (7) is exact.

---

[10]We shall prove this fact at another occasion (Theorem 3 in Sec. 9.2); the proof can also be found in some books on elementary calculus; see Ref. [12] in Appendix 1.

2nd Step. Implicit solution. From (6) we obtain

$$u = \int M\,dx + k(y) = \int (x^3 + 3xy^2)\,dx + k(y) = \frac{1}{4}x^4 + \frac{3}{2}x^2y^2 + k(y). \tag{8}$$

To find $k(y)$, we differentiate this formula with respect to y and use formula (4b), obtaining

$$\frac{\partial u}{\partial y} = 3x^2y + \frac{dk}{dy} = N = 3x^2y + y^3.$$

Hence $dk/dy = y^3$, so that $k = (y^4/4) + \tilde{c}$. Inserting this into (8) we get the answer

$$u(x, y) = \tfrac{1}{4}(x^4 + 6x^2y^2 + y^4) = c. \tag{9}$$

3rd Step. Checking. CAUTION! Note well that the present method gives the solution in implicit form, $u(x, y) = c = const$, not in explicit form, $y = f(x)$. For ***checking,*** we can differentiate $u(x, y) = c$ implicitly and see whether this leads to $dy/dx = -M/N$ or $M\,dx + N\,dy = 0$, the given equation.

In the present case, differentiating (9) implicitly with respect to x, we obtain

$$\tfrac{1}{4}(4x^3 + 12xy^2 + 12x^2yy' + 4y^3y') = 0.$$

Collecting terms, we see that this equals $M + Ny' = 0$ with M and N as in (7); thus $M\,dx + N\,dy = 0$. This completes the check. ∎

EXAMPLE 2 **An initial value problem**

Solve the initial value problem

$$(\sin x \cosh y)\,dx - (\cos x \sinh y)\,dy = 0, \quad y(0) = 0. \tag{10}$$

Solution. The student may verify that the equation is exact. From (6) we obtain

$$u = \int \sin x \cosh y\,dx + k(y) = -\cos x \cosh y + k(y).$$

From this, $\partial u/\partial y = -\cos x \sinh y + dk/dy$. Hence $dk/dy = 0$, and $k = const$. The general solution is $u = const$, that is, $\cos x \cosh y = c$. The initial condition gives $\cos 0 \cosh 0 = 1 = c$. Hence the answer is $\cos x \cosh y = 1$.

Checking. $(\cos x \cosh y)' = -\sin x \cosh y + \cos x (\sinh y)y' = 0$, which gives (10). Also, $\cos 0 \cosh 0 = 1$ shows that the answer satisfies the initial condition. ∎

EXAMPLE 3 **WARNING! Breakdown in the case of nonexactness**

Consider the equation

$$y\,dx - x\,dy = 0.$$

We see that $M = y$, $N = -x$, hence $\partial M/\partial y = 1$ but $\partial N/\partial x = -1$. Hence the equation is not exact. Let us show that in such a case, the present method does not work. From (6),

$$u = \int M\,dx + k(y) = xy + k(y).$$

From this,

$$\frac{\partial u}{\partial y} = x + k'(y).$$

This should equal $N = -x$ But this is impossible, since $k(y)$ can depend only on y. Try (6*): it will also fail. Solve the equation by another method we have discussed. ∎

Problem Set 1.5

Given $u(x, y)$, find the exact differential equation $du = 0$. What kind of curves are the solution curves $u(x, y) = const$?

1. $x^2 + y^2 = c$
2. $u = y/x^2$
3. $u = (x - a)(b - y)$
4. $u = \cos(x^2 - y^2)$
5. $u = \exp(xy^2)$
6. $u = \sin xy$
7. $u = \ln(x^2y^2)$
8. $u = \tan(x^2 + 4y^2)$
9. $u = (y - x + 1)^2$
10. $u = \cosh(x^3 - y^2)$

Show that the following differential equations are exact and solve them.

11. $y\,dx + x\,dy = 0$
12. $x\,dx + 9y\,dy = 0$
13. $y^3\,dx + 3xy^2\,dy = 0$
14. $ye^x\,dx + [e^x + (y + 1)e^y]\,dy = 0$
15. $e^{-\theta}\,dr - re^{-\theta}\,d\theta = 0$
16. $\frac{1}{4}e^{4\theta}\,dr + re^{4\theta}\,d\theta = 0$
17. $\cosh x \cos y\,dx = \sinh x \sin y\,dy$
18. $e^x(\cos y\,dx - \sin y\,dy) = 0$
19. $(2x + e^y)\,dx + xe^y\,dy = 0$
20. $(\cot y + x^2)\,dx = x\,\mathrm{cosec}^2\,y\,dy$

Are the following equations exact? Solve the initial value problems.

21. $x\,dy + y^2\,dx = 0, \quad y(1) = 0.2$
22. $4\,dx + x^{-1}\,dy = 0, \quad y(1) = -8$
23. $(y - 1)\,dx + (x - 3)\,dy = 0, \quad y(0) = 2/3$
24. $(x - 1)\,dx + (y + 1)\,dy = 0, \quad y(1) = 0$
25. $e^{y/x}(-y\,dx + x\,dy)/x^2 = 0, \quad y(-2) = -2$
26. $(2xy\,dx + dy)e^{x^2} = 0, \quad y(0) = 2$
27. $2xy\,dy = (x^2 + y^2)\,dx, \quad y(1) = 2$
28. $\cos \pi x \cos 2\pi y\,dx = 2 \sin \pi x \sin 2\pi y\,dy, \quad y(3/2) = 1/2$
29. $\sinh x\,dx + y^{-1} \cosh x\,dy = 0, \quad y(0) = \pi$
30. $2 \sin \omega y\,dx + \omega \cos \omega y\,dy = 0, \quad y(0) = \pi/2\omega$

31. Solve the equation in Example 3.
32. If an equation is separable, show that it is exact. Is the converse true?
33. Under what conditions is $(ax + by)\,dx + (kx + ly)\,dy = 0$ exact? (Here, a, b, k, l are constants.) Solve the exact equation.
34. Under what conditions is $[f(x) + g(y)]\,dx + [h(x) + p(y)]\,dy = 0$ exact?
35. Under what conditions is $f(x, y)\,dx + g(x)h(y)\,dy = 0$ exact?

To see that a differential equation can sometimes be solved by several methods, solve (a) by the present method, (b) by separation or inspection:

36. $xy' + y + 4 = 0$
37. $2x\,dx + x^{-2}(x\,dy - y\,dx) = 0$
38. $b^2x\,dx + a^2y\,dy = 0$
39. $3x^{-4}y\,dx = x^{-3}\,dy$

40. Can you figure out what the solution curves in Example 1 look like? *Hint.* Set $x = s + t$, $y = s - t$.

1.6 Integrating Factors

The idea of the method in this section is quite simple. We sometimes have an equation

$$(1) \qquad P(x, y)\,dx + Q(x, y)\,dy = 0$$

that is not exact, but if we multiply it by a suitable function $F(x, y)$, the new equation

$$(2) \qquad FP\,dx + FQ\,dy = 0$$

is exact, so that it can be solved by the method of Sec. 1.5. The function $F(x, y)$ is then called an **integrating factor** of (1).

Let us first consider some simple examples and then see how we can obtain integrating factors in a systematic fashion.

EXAMPLE 1 **Integrating factors**

Show that the differential equation

$$(3) \qquad y\,dx - x\,dy = 0$$

is not exact, but has an integrating factor, namely, $F = 1/x^2$, and solve the new equation.

Solution. The equation is of the form (1) with $P = y$ and $Q = -x$. Since $\partial P/\partial y = 1$ but $\partial Q/\partial x = -1$, it is not exact. Multiplying it by $F = 1/x^2$, we get the exact equation (verify exactness!)

$$FP\,dx + FQ\,dy = \frac{y\,dx - x\,dy}{x^2} = -d\left(\frac{y}{x}\right) = 0. \qquad \text{Solution} \qquad \frac{y}{x} = c.$$

These are straight lines $y = cx$ through the origin.

Other integrating factors of (3) are $1/y^2$, $1/xy$, $1/(x^2 + y^2)$ because

$$(4) \qquad \frac{y\,dx - x\,dy}{y^2} = d\left(\frac{x}{y}\right), \quad \frac{y\,dx - x\,dy}{xy} = d\left(\ln\frac{x}{y}\right), \quad \frac{y\,dx - x\,dy}{x^2 + y^2} = -d\left(\arctan\frac{y}{x}\right). \quad ■$$

EXAMPLE 2 **Integrating factor**

Verify that $F(x) = x^3$ is an integrating factor of

$$2\sin(y^2)\,dx + xy\cos(y^2)\,dy = 0$$

and then find the general solution.

Solution. Multiplication by $F = x^3$ gives the new equation

$$2x^3\sin(y^2)\,dx + x^4y\cos(y^2)\,dy = 0.$$

This equation is exact because

$$\frac{\partial}{\partial y}[2x^3\sin(y^2)] = 4x^3y\cos(y^2) = \frac{\partial}{\partial x}[x^4y\cos(y^2)].$$

We solve it by the method of Sec. 1.5 (or by inspection) to get $x^4\sin(y^2) = c = const.$

Checking. Verify the solution by differentiating it implicitly with respect to x. ■

How to Find Integrating Factors

In simpler cases, integrating factors may be found by inspection or perhaps after some trials [keeping (4) in mind]. In the general case, the idea is this: Equation (2) is $M\,dx + N\,dy = 0$ with $M = FP$, $N = FQ$, and is exact by the definition of an integrating factor. Hence the exactness criterion $\partial M/\partial y = \partial N/\partial x$ (Eq. (5) in Sec. 1.5) now is

$$\frac{\partial}{\partial y}(FP) = \frac{\partial}{\partial x}(FQ), \tag{5}$$

that is, $F_yP + FP_y = F_xQ + FQ_x$ (subscripts denoting partial derivatives). In the general case, this would be complicated and useless. So we follow the ***Golden Rule:*** If you cannot solve your problem, try to solve a simpler one—the result may be useful (and may also help you later on). Hence we look for an integrating factor depending only on ***one*** variable; fortunately, in many practical cases, there are such factors, as we shall see. Thus, let $F = F(x)$. Then $F_y = 0$ and $F_x = F' = dF/dx$, so that (5) becomes

$$FP_y = F'Q + FQ_x.$$

Dividing by FQ and reshuffling terms, we have

$$\frac{1}{F}\frac{dF}{dx} = \frac{1}{Q}\left(\frac{\partial P}{\partial y} - \frac{\partial Q}{\partial x}\right). \tag{6}$$

This proves

Theorem 1 **[Integrating factor $F(x)$]**

If (1) *is such that the right side of* (6), *call it R, depends only on x, then* (1) *has an integrating factor $F = F(x)$, which is obtained by integrating* (6) *and taking exponents on both sides,*

$$F(x) = \exp \int R(x)\,dx. \tag{7}$$

Similarly, if $F = F(y)$, then instead of (6) we get

$$\frac{1}{F}\frac{dF}{dy} = \frac{1}{P}\left(\frac{\partial Q}{\partial x} - \frac{\partial P}{\partial y}\right) \tag{8}$$

and have the companion

Theorem 2 **[Integrating factor $F(y)$]**

If (1) *is such that the right side $\tilde{R}$ of* (8) *depends only on y, then* (1) *has an integrating factor $F = F(y)$, which is obtained from* (8) *in the form*

$$F(y) = \exp \int \tilde{R}(y)\,dy. \tag{9}$$

EXAMPLE 3 **Integrating factor $F(x)$**

Solve Example 2 by Theorem 1.

Solution. We have $P = 2 \sin (y^2)$, $Q = xy \cos (y^2)$, hence in (6) on the right,

$$R = \frac{1}{xy \cos (y^2)} [4y \cos (y^2) - y \cos (y^2)] = \frac{3}{x}$$

and thus $F(x) = \exp \int (3/x)\, dx = x^3$, as in Example 2, and so on. ■

EXAMPLE 4 **Application of Theorems 1 and 2**

Solve the initial value problem

$$2xy\, dx + (4y + 3x^2)\, dy = 0, \qquad y(0.2) = -1.5.$$

Solution. Here, $P = 2xy$, $Q = 4y + 3x^2$, the equation is not exact, the right side of (6) depends on both x *and* y (verify!), but the right side of (8) is

$$\tilde{R} = \frac{1}{2xy}(6x - 2x) = \frac{2}{y}. \qquad \text{Thus} \qquad F(y) = y^2$$

is an integrating factor by (9). Multiplication by y^2 gives the exact equation

$$2xy^3\, dx + (4y^3 + 3x^2y^2)\, dy = 0,$$

which we can write as

$$4y^3\, dy + (2xy^3\, dx + 3x^2y^2\, dy) = 0$$

and solve by inspection or by the method in Sec. 1.5 to get $y^4 + x^2y^3 = c$; from this we obtain $y^4 + x^2y^3 = 4.9275$ by the initial condition. ■

Problem Set 1.6

1. Verify (4).
2. Verify exactness in Example 1 by the usual test.
3. Verify the solution in Example 2, as indicated.
4. Give the details of the derivation of (8).
5. Verify that Theorem 1 cannot be used to solve Example 4.
6. Verify that y, xy^3, and x^2y^5 are integrating factors of $y\, dx + 2x\, dy = 0$ and solve.

Verify that the given function F is an integrating factor and solve the initial value problem:

7. $2y\, dx + x\, dy = 0$, $y(0.5) = 8$, $F = x$
8. $3y\, dx + 2x\, dy = 0$, $y(-1) = 1.4$, $F = x^2y$
9. $(1 + xy)\, dx + x^2\, dy = 0$, $y(1) = 0$, $F = e^{xy}$
10. $dx + (x + y + 1)\, dy = 0$, $y(2.5) = 0.5$, $F = e^y$
11. $(2x^{-1}y - 3)\, dx + (3 - 2y^{-1}x)\, dy = 0$, $y(1) = -1$, $F = x^2y^2$
12. $y\, dx + [y + \tan (x + y)]\, dy = 0$, $y(0) = \pi/2$, $F = \cos (x + y)$
13. $y\, dx + [\coth (x - y) - y]\, dy = 0$, $y(3) = 3$, $F = \sinh (x - y)$
14. $(2xe^x - y^2)\, dx + 2y\, dy = 0$, $y(0) = \sqrt{2}$, $F = e^{-x}$

Find an integrating factor F and solve (using inspection or Theorems 1 and 2):

15. $2 \cos \pi y \, dx = \pi \sin \pi y \, dy$

16. $y \cos x \, dx + 3 \sin x \, dy = 0$

17. $(2y + xy) \, dx + 2x \, dy = 0$

18. $2y \, dx + 3x \, dy = 0$

19. $(1 + 2x^2 + 4xy) \, dx + 2 \, dy = 0$

20. $2x \, dx = [3y^2 + (x^2 - y^3) \tan y] \, dy$

21. $(y + 1) \, dx - (x + 1) \, dy = 0$

22. $5 \, dx - e^{y-x} \, dy = 0$

23. $ay \, dx + bx \, dy = 0$

24. $(3xe^y + 2y) \, dx + (x^2 e^y + x) \, dy = 0$

In each case find conditions such that F is an integrating factor of (1). *Hint.* Assume $F(P \, dx + Q \, dy) = 0$ to be exact and apply (5) in Sec. 1.5.

25. $F = x^a$ **26.** $F = y^b$ **27.** $F = x^a y^b$ **28.** $F = e^y$

29. Using Prob. 27, derive the integrating factor in Prob. 11.

30. (Checking) Checking of solutions is always important. In connection with the present method it is particularly essential since one may have to exclude the function $y(x)$ given by $F(x, y) = 0$. To see this, consider $(xy)^{-1} \, dy - x^{-2} \, dx = 0$; show that an integrating factor is $F = y$ and leads to $d(y/x) = 0$, hence $y = cx$, where c is arbitrary, but $F = y = 0$ is not a solution of the original equation.

1.7 Linear Differential Equations

A first-order differential equation is said to be **linear** if it can be written

$$y' + p(x)y = r(x). \tag{1}$$

The characteristic feature of this equation is that it is linear in y and y', whereas p and r on the right may be *any* given functions of x.

If the right side $r(x)$ is zero for all x in the interval in which we consider the equation (written $r(x) \equiv 0$), the equation is said to be **homogeneous;** otherwise it is said to be **nonhomogeneous.**

Let us find a formula for the general solution of (1) in some interval I, assuming that p and r are continuous in I. For the homogeneous equation

$$y' = p(x)y = 0 \tag{2}$$

this is very simple. Indeed, by separating variables we have

$$\frac{dy}{y} = -p(x) \, dx, \qquad \text{thus} \qquad \ln |y| = -\int p(x) \, dx + c^*$$

and by taking exponents on both sides

$$y(x) = ce^{-\int p(x) \, dx} \qquad (c = \pm e^{c^*} \quad \text{when} \quad y \gtrless 0); \tag{3}$$

here we may also take $c = 0$ and obtain the *trivial solution* $y \equiv 0$.

The nonhomogeneous equation (1) will now be solved. It turns out that it has the pleasant property of possessing an integrating factor depending only on x. Indeed, we first write (1) as

$$(py - r)\,dx + dy = 0.$$

This is $P\,dx + Q\,dy = 0$, where $P = py - r$ and $Q = 1$. Hence (6) in Sec. 1.6 becomes simply

$$\frac{1}{F}\frac{dF}{dx} = p(x).$$

Since this depends only on x, Eq. (1) has an integrating factor $F(x)$, which we obtain directly by integration and exponentiation [as in (7), Sec. 1.6]:

$$F(x) = e^{\int p\,dx}.$$

Multiplication of (1) by this F and observing the product rule of differentiation gives

$$e^{\int p\,dx}(y' + py) = (e^{\int p\,dx}y)' = e^{\int p\,dx}r.$$

We now integrate with respect to x,

$$e^{\int p\,dx}y = \int e^{\int p\,dx}r\,dx + c$$

and solve for y, abbreviating $\int p\,dx$ by h,

$$\text{(4)} \qquad y(x) = e^{-h}\left[\int e^{h}r\,dx + c\right], \qquad h = \int p(x)\,dx.$$

This represents the general solution of (1) in the form of an integral.[11] (The choice of the value of the constant of integration in $\int p\,dx$ does not matter; see Prob. 2.)

EXAMPLE 1 Solve the linear differential equation

$$y' - y = e^{2x}.$$

Solution. Here

$$p = -1, \qquad r = e^{2x}, \qquad h = \int p\,dx = -x$$

and from (4) we obtain the general solution

$$y(x) = e^{x}\left[\int e^{-x}e^{2x}\,dx + c\right] = e^{x}[e^{x} + c] = ce^{x} + e^{2x}.$$

---

[11]If the integral cannot be integrated by the usual methods of calculus (as often happens in practice), we may have to use a numerical method for integrals (Sec. 18.5) or for the differential equation itself (Secs. 20.1, 20.2).

Alternatively, we may multiply the given equation by $e^h = e^{-x}$, finding

$$(y' - y)e^{-x} = (ye^{-x})' = e^{2x}e^{-x} = e^x$$

and integrate on both sides, obtaining the same result as before:

$$ye^{-x} = e^x + c, \qquad \text{hence} \qquad y = e^{2x} + ce^x.$$

EXAMPLE 2 Mixing problem

The tank in Fig. 10 contains 200 gal of water in which are dissolved 40 lb of salt. 5 gal of brine, each containing 2 lb of dissolved salt, run into the tank per minute and the mixture, kept uniform by stirring, runs out at the same rate. Find the amount of salt $y(t)$ in the tank at any time t.

Solution. 1st Step. Modeling. The time rate of change $y' = dy/dt$ of $y(t)$ equals the inflow $5 \times 2 = 10$ [lb/min] of salt minus the outflow. The outflow [lb/min] is $(5/200) \times y(t) = 0.025y(t)$ because $y(t)$ is the total amount of salt in the tank and 5 gal/200 gal is the fraction of volume that flows out per minute. Alternatively, $y(t)$ is the total amount of salt, hence $y(t)/200$ is the amount of salt per gallon, and 5 gal/min flow out. Thus the model is $y' = 10 - 0.0025y$, that is, the initial value problem

$$y' + 0.025y = 10, \qquad y(0) = 40.$$

2nd Step. Solving the equation. In (4), with t instead of x, we have $p = 0.025$, $h = 0.025t$, $r = 10$ and get the general solution

$$\begin{aligned} y(t) &= e^{-0.025t}\left[\int e^{0.025t} \cdot 10\,dt + c\right] \\ &= e^{-0.025t}\left[\frac{10}{0.025}e^{0.025t} + c\right] \\ &= ce^{-0.025t} + 400. \end{aligned}$$

The initial condition $y(0) = c + 400 = 40$ gives $c = -360$ and as the answer the particular solution

$$y(t) = 400 - 360e^{-0.025t} \text{ [lb]}.$$

We see that the amount of salt in the tank increases monotone. Can you explain this physically?

Of course, (4) may not be needed in simple cases such as Examples 1 or 2, but will be useful for more complicated equations.

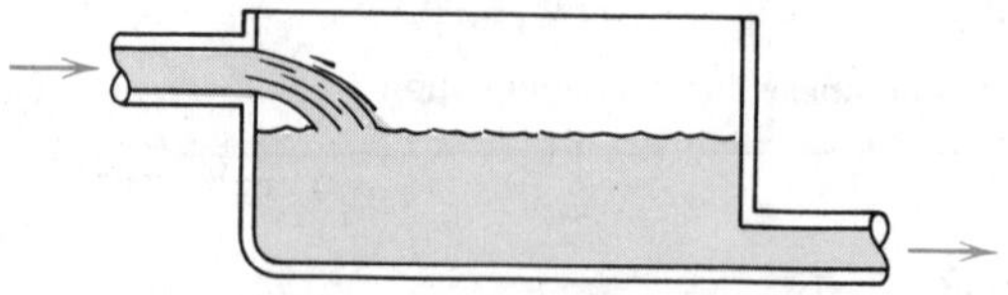

Fig. 10. Tank in Example 2

EXAMPLE 3 Solve

$$y' + 2y = e^x(3 \sin 2x + 2 \cos 2x).$$

Solution. Here $p = 2$, $h = 2x$, so that (4) gives

$$\begin{aligned} y &= e^{-2x}\left[\int e^{2x}e^x(3 \sin 2x + 2 \cos 2x)\,dx + c\right] \\ &= e^{-2x}[e^{3x} \sin 2x + c] \\ &= ce^{-2x} + e^x \sin 2x. \end{aligned}$$

EXAMPLE 4 Initial value problem

Solve the initial value problem

$$y' + y \tan x = \sin 2x, \qquad y(0) = 1.$$

Solution. Here $p = \tan x$, $r = \sin 2x = 2 \sin x \cos x$, and

$$\int p\,dx = \int \tan x\,dx = \ln |\sec x|.$$

From this we see that in (4),

$$e^h = \sec x, \qquad e^{-h} = \cos x, \qquad e^h r = (\sec x)(2 \sin x \cos x) = 2 \sin x,$$

and the general solution of our equation is

$$y(x) = \cos x \left[2 \int \sin x\,dx + c\right] = c \cos x - 2 \cos^2 x.$$

From this and the initial condition, $1 = c \cdot 1 - 2 \cdot 1^2$, thus $c = 3$ and the solution of our initial value problem is

$$y = 3 \cos x - 2 \cos^2 x.$$

■

Reduction to Linear Form. Bernoulli Equation

Certain nonlinear differential equations can be reduced to linear form, as we shall illustrate in the problem set. The practically most famous of these is the ***Bernoulli equation***[12] (see also Probs. 37, 38, etc.)

$$(5) \qquad y' + p(x)y = g(x)y^a \qquad (a \text{ any real number}).$$

If $a = 0$ or $a = 1$, the equation is linear. Otherwise it is nonlinear. Then we set

$$u(x) = [y(x)]^{1-a}.$$

We differentiate this and substitute y' from (5), obtaining

$$u' = (1 - a)y^{-a}y' = (1 - a)y^{-a}(gy^a - py)$$
$$= (1 - a)(g - py^{1-a}),$$

where $y^{1-a} = u$ on the right, so that we get the linear equation

$$(6) \qquad u' + (1 - a)pu = (1 - a)g.$$

---

[12]JAKOB BERNOULLI (1654—1705), Swiss mathematician, professor at Basel, also known for his contributions to elasticity theory and mathematical probability. The method for solving Bernoulli's equation was discovered by Leibniz in 1696. Jakob Bernoulli's students include his nephew NIKLAUS BERNOULLI (1687—1759), who contributed to probability theory and infinite series, and his youngest brother JOHANN BERNOULLI (1667—1748), who had profound influence on the development of calculus, became Jakob's successor at Basel, and had among his students GABRIEL CRAMER (see Sec. 7.9) and LEONHARD EULER (see Sec. 2.6). His son DANIEL BERNOULLI (1700—1782) is known for his basic work in fluid flow and the kinetic theory of gases.

EXAMPLE 5 Bernoulli equation. Verhulst equation. Logistic population model

Solve the special Bernoulli equation, called the **Verhulst equation:**

$$y' - Ay = -By^2 \qquad (A, B \text{ positive constants}). \tag{7}$$

Solution. Here, $a = 2$, so that $u = y^{-1}$, and by differentiation and substitution of y' from (7),

$$u' = -y^{-2}y' = -y^{-2}(-By^2 + Ay) = B - Ay^{-1},$$

that is,

$$u' + Au = B.$$

From (4) with $p = A$, hence $h = Ax$, and $r = B$ we obtain

$$u = e^{-Ax}\left[\int Be^{Ax}\,dx + c\right] = e^{-Ax}\left[\frac{B}{A}e^{Ax} + c\right] = ce^{-Ax} + \frac{B}{A}.$$

This gives the general solution of (7),

$$y = \frac{1}{u} = \frac{1}{(B/A) + ce^{-Ax}}, \tag{8}$$

and directly from (7) we see that $y \equiv 0$ is also a solution.

Equation (8) is called the **logistic law** of population growth, where x is time. For $B = 0$ it gives exponential growth $y = (1/c)e^{Ax}$ (**Malthus's law,** Prob. 27 in Sec. 1.1). $-By^2$ in (7) is a "braking term," preventing the population from growing without bound. Indeed, (8) shows that initially small populations $[0 < y(0) < A/B]$ increase monotone to A/B, whereas initially large populations $[y(0) > A/B]$ decrease monotone to the same limit A/B (Fig. 11).

The logistic law has useful applications to human populations (see Prob. 54) and animal populations (see C. W. Clark, *Mathematical Bioeconomics*, New York, Wiley, 1976). ∎

Input and Output

Linear differential equations (1) have various applications, as we shall illustrate in the problem set as well as in the next section. Then the independent variable x will often be time; the function $r(x)$ on the right side of (1) may represent a force, and the solution $y(x)$ a displacement, a current, or some other variable physical quantity. In engineering mathematics $r(x)$ is frequently called the **input,** and $y(x)$ is called the **output** or *response to the input* (and the initial conditions). For instance, in electrical engineering the differential equation may govern the behavior of an electric circuit and the output $y(x)$ is obtained as the solution of that equation corresponding to the input $r(x)$. We shall discuss this idea in terms of typical examples in the next section (and for second-order equations in Secs. 2.5, 2.11, 2.12).

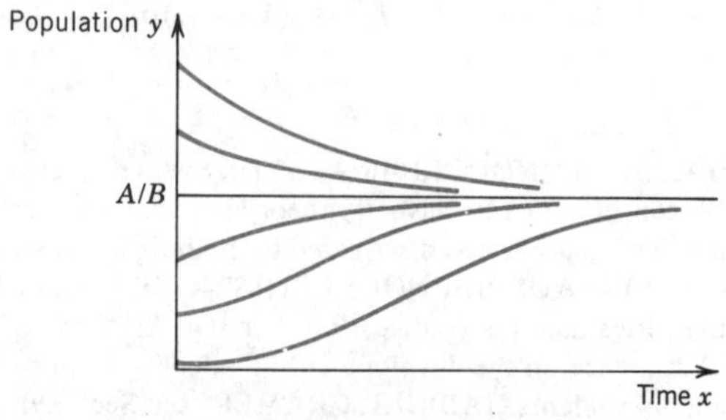

Fig. 11. Logistic population model. Curves (8) in Example 5

Problem Set 1.7

1. Show that $e^{-\ln x} = 1/x$ (but not $-x$) and $e^{-\ln(\sec x)} = \cos x$.
2. Show that the choice of the value of the constant of integration in $\int p\,dx$ [see (4)] does not matter (so that we may choose it to be zero).
3. What is the limit of $y(t)$ as $t \to \infty$ in Example 2? Is it physically reasonable?

Find the general solutions of the following differential equations.

4. $y' + y = 5$
5. $y' - 4y = 0.8$
6. $y' + 2xy = 0$
7. $y' + 2y = 6e^x$
8. $y' - 2y = 2 - 4x$
9. $y' - 4y = 2x - 4x^2$
10. $y' = (y - 1)\cot x$
11. $xy' + 2y = 9x$
12. $y' \tan x = 2y$
13. $y' + 3y = e^{-3x}$
14. $y' + 2y = \cos x$
15. $xy' + 2y = 4e^{x^2}$
16. $(x^2 - 1)y' = xy$
17. $xy' - 2y = x^3e^x$
18. $x^2y' + 2xy = \sinh 3x$

Solve the following initial value problems.

19. $y' + 3y = 12, \quad y(0) = 6$
20. $y' = y\cot x, \quad y(\tfrac{1}{2}\pi) = 2$
21. $y' + y = (x + 1)^2, \quad y(0) = 3$
22. $y' + x^3y = 4x^3, \quad y(0) = -1$
23. $y' + 2xy = 4x, \quad y(0) = 3$
24. $xy' = (1 + x)y, \quad y(1) = 3e$
25. $y' \coth 2x = 2y - 2, \quad y(0) = 0$
26. $y' = 2y/x + x^2e^x, \quad y(2) = 0$
27. $y' + ky = e^{-kx}, \quad y(0) = 0.7$
28. $y' + 3x^2y = xe^{-x^3}, \quad y(0) = -1$

General properties of linear differential equations. The *linear* differential equations (1) and (2) have certain important properties. In the next two chapters we shall see that the same is true for *linear* differential equations of higher order. This fact is quite important, since we can use it to obtain new solutions from given ones. Indeed, prove and illustrate with an example that the **homogeneous equation** (2) has the following properties.

29. $y \equiv 0$ is a solution of (2), called the ***trivial solution***.
30. If y_1 is a solution of (2), then $y = cy_1$ (c any constant) is a solution of (2).
31. If y_1 and y_2 are solutions of (2), then their sum $y = y_1 + y_2$ is a solution of (2).

Prove and illustrate with an example that the **nonhomogeneous equation (1)** has the following properties.

32. If y_1 is a solution of (1) and y_2 is a solution of (2), then $y = y_1 + y_2$ is a solution of (1).
33. The difference $y = y_1 - y_2$ of two solutions y_1 and y_2 of (1) is a solution of (2).
34. If y_1 is a solution of (1), then $y = cy_1$ is a solution of $y' + py = cr$.
35. If y_1 is a solution of $y_1' + py_1 = r_1$ and y_2 is a solution of $y_2' + py_2 = r_2$ (with the same p), then $y = y_1 + y_2$ is a solution of $y' + py = r_1 + r_2$.
36. If $p(x)$ and $r(x)$ in (1) are constant, say, $p(x) = p_0$ and $r(x) = r_0$, then (1) can be solved by separating variables and the result will agree with that obtained from (4).

Reduction of nonlinear differential equations to linear form. Applying suitable transformations of variables, reduce to linear form and solve the following equations. *Hint.* Some are Bernoulli equations; some become linear if one takes y as the independent variable and x as the unknown function.

37. $y' + y = y^2$
38. $y' + y = x/y$

Reduction (*continued*)

39. $y' \cos y + x \sin y = 2x \quad (\sin y = z)$

40. $y' - 1 = e^{-y} \sin x$

41. $(e^y + x)y' = 1$

42. $y'(\sinh 3y - 2xy) = y^2$

43. $3y' + y = (1 - 2x)y^4$

44. $2xy' = 10x^3y^5 + y$

45. $2xyy' + (x - 1)y^2 = x^2e^x$

46. $y' \cos y + 2x \sin y = 2x$

Some applications (More in the next section)

47. How long will it take $y(t)$ in Example 2 to practically reach the limit, say, the value 399.9 lb? First guess.

48. Show that if in Example 2 we double the influx (but make no further changes), the model is $y' = 20 - [5/(200 + 5t)]y$, $y(0) = 40$. Solve this initial value problem.

49. If in Example 2 we replace the inflowing brine after 10 minutes by pure water (still flowing at 5 gal/min), how long will it take to get the tank practically salt-free, say, to decrease $y(t)$ to 0.01 lb? First guess.

50. **(Motion of a boat)** Two persons are riding in a motorboat, the combined weight of the persons and the boat being 4900 nt (about 1100 lb). Suppose that the motor exerts a constant force of 200 nt (about 45 lb) and the resistance R of the water is proportional to the speed v, say, $R = kv$ nt, where $k = 10$ nt · sec/meter. Set up the differential equation for $v(t)$, using Newton's second law

$$\text{Mass} \times \text{Acceleration} = \text{Force}.$$

Find $v(t)$ satisfying $v(0) = 0$. Find the maximum speed v_∞ at which the boat will travel (practically after a sufficiently long time). If the boat starts from rest, how long will it take to reach $0.9v_\infty$ and what distance does the boat travel during that time?

51. **(Newton's law of cooling)** Solve the differential equation in Example 2 of Sec. 1.3 by our present method, assuming the initial temperature of the ball to be $T(0) = T_0$.

52. **Hormone secretion** can be modeled by

$$y' = a - b\cos\frac{2\pi t}{24} - ky.$$

Here, t is time [in hours, with $t = 0$ suitably chosen, e.g., 8:00 A.M.], $y(t)$ is the amount of a certain hormone in the blood, a is the average secretion rate, $b\cos(\pi t/12)$ models the daily 24-hr secretion cycle, and ky models the removal rate of the hormone from the blood. Find the solution when $a = b = k = 1$ and $y(0) = 2$.

53. **(Logistic population model)** Show that (8) with $0 < y(0) < A/B$ grows monotone and with $y(0) > A/B$ decreases monotone.

54. **(United States)** For the United States, Verhulst predicted in 1845 the values $A = 0.03$ and $B = 1.6 \cdot 10^{-4}$, where x is measured in years and $y(x)$ in millions. Find the particular solution (8) satisfying $y(0) = 5.3$ (corresponding to the year 1800) and compare the values of this solution with some actual (rounded) values:

1800	1830	1860	1890	1920	1950	1980	1990
5.3	13	31	63	105	150	230	250

55. Show that the curves (8) have a point of inflection if $y(x) = A/2B$ [Use (7).]

Riccati and Clairaut equations

56. A **Riccati equation**[13] is of the form $y' + p(x)y = g(x)y^2 + h(x)$. Verify that the Riccati equation $y' = x^3(y - x)^2 + x^{-1}y$ has the solution $y = x$ and reduce it to a Bernoulli equation by the substitution $w = y - x$ and solve it.

57. Show that the general Riccati equation in Prob. 56 (which is a Bernoulli equation when $h \equiv 0$) can be reduced to a Bernoulli equation if one knows a solution $y = v$, by setting $w = y - v$.

58. A **Clairaut equation**[14] is of the form $y = xy' + g(y')$. Solve the Clairaut equation $y = xy' + 1/y'$. *Hint.* Differentiate the equation with respect to x.

59. Show that the general Clairaut equation in Prob. 58, with arbitrary $g(s)$ has as solutions a family of straight lines $y = cx + g(c)$ and a singular solution determined by $g'(s) = -x$, where $s = y'$. (Those lines are tangents to the latter.) *Hint.* Differentiate the equation with respect to x, as in Prob. 58.

60. Show that the straight lines, whose segment between the positive x-axis and y-axis has constant length 1, are solutions of the Clairaut equation $y = xy' - y'/\sqrt{1 + y'^2}$, whose singular solution is the **astroid** $x^{2/3} + y^{2/3} = 1$. Make a sketch.

1.8 Modeling: Electric Circuits

Recall from Secs. 1.1 and 1.3 that **modeling** means setting up mathematical models of physical or other systems. In the present section we shall model electric circuits. Their models will be linear differential equations. Although of particular interest to students of electrical engineering, computer engineering, etc., our discussion will be profitable to ***all*** students, because modeling skills can be acquired most successfully by considering practical problems from *various* fields. To help everybody, we first explain the basic concepts needed.

The simplest electric circuit is a series circuit in which we have a source of electric energy (**electromotive force**) such as a generator or a battery, and a resistor, which uses energy, for example an electric light bulb (Fig. 12). If we close the switch, a current I will flow through the resistor, and this will cause a **voltage drop,** that is, the electric potential at the two ends of the resistor will be different; this potential difference or voltage drop can be measured by a voltmeter. Experiments show that the following law holds.

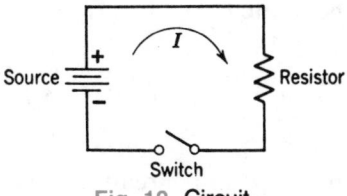

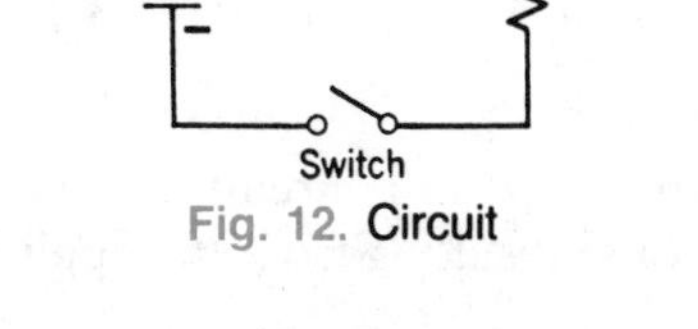

Fig. 12. Circuit

[13]JACOPO FRANCESCO RICCATI (1676—1754), Italian mathematician, who introduced his equation in 1723.

[14]ALEXIS CLAUDE CLAIRAUT (1713—1765), French mathematician, also known by his work in geodesy and astronomy.

The **voltage drop E_R across a resistor** *is proportional to the instantaneous current I,* say,

$$E_R = RI \qquad \text{(1)} \qquad \textbf{(Ohm's law)}$$

where the constant of proportionality R is called the **resistance** of the resistor. The current I is measured in *amperes,* the resistance R in *ohms,* and the voltage E_R in *volts.*[15]

The other two important elements in more complicated circuits are *inductors* and *capacitors.* An inductor opposes a change in current, having an inertia effect in electricity similar to that of mass in mechanics; we shall consider this analogy later (Sec. 2.12). Experiments yield the following law.

The **voltage drop E_L across an inductor** *is proportional to the instantaneous time rate of change of the current I,* say,

$$E_L = L\frac{dI}{dt} \qquad \text{(2)}$$

where the constant of proportionality L is called the **inductance** of the inductor and is measured in *henrys;* time t is measured in seconds.

A capacitor is an element which stores energy. Experiments yield the following law.

The **voltage drop E_C across a capacitor** *is proportional to the instantaneous electric charge Q on the capacitor,* say,

$$E_C = \frac{1}{C}Q \qquad \text{(3*)}$$

where C is called the **capacitance** and is measured in *farads;* the charge Q is measured in *coulombs.* Since

$$I(t) = \frac{dQ}{dt} \qquad \text{(3')}$$

this may be written

$$E_C = \frac{1}{C}\int_{t_0}^{t} I(t^*)\,dt^*. \qquad \text{(3)}$$

The current $I(t)$ in a circuit may be determined by solving the equation (or equations) resulting from the application of the following physical law.

[15]These and the subsequent units are named after ANDRÉ MARIE AMPÈRE (1775—1836), French physicist; GEORG SIMON OHM (1789—1854), German physicist; ALESSANDRO VOLTA (1745—1827), Italian physicist; JOSEPH HENRY (1797—1878), American physicist; MICHAEL FARADAY (1791—1867), English physicist; and CHARLES AUGUSTIN DE COULOMB (1736—1806), French physicist and engineer.

Kirchhoff's voltage law (KVL)[16]

The algebraic sum of all the instantaneous voltage drops around any closed loop is zero, or the voltage impressed on a closed loop is equal to the sum of the voltage drops in the rest of the loop.

EXAMPLE 1 ***RL*-circuit**

Model the "RL-circuit" in Fig. 13 and solve the resulting equation for (A) a constant electromotive force, (B) a periodic electromotive force.

Solution. 1st Step. Modeling. By (1) the voltage drop across the resistor is RI. By (2) the voltage drop across the inductor is $L\,dI/dt$. By KVL the sum of the two voltage drops must equal the electromotive force $E(t)$; thus

$$L\frac{dI}{dt} + RI = E(t). \tag{4}$$

2nd Step. Solution of the equation. We get the general solution from (4) in Sec. 1.7, which was derived for an equation $y' + py = r$, with y' having coefficient 1; thus we must first divide (4) by L,

$$\frac{dI}{dt} + \frac{R}{L}I = \frac{E}{L}.$$

Then (4), Sec. 1.7, with $x = t$, $y = I$, $p = R/L$, and $r = E/L$ gives

$$I(t) = e^{-\alpha t}\left[\int e^{\alpha t}\frac{E}{L}\,dt + c\right], \qquad \alpha = \frac{R}{L}. \tag{5}$$

3rd Step. Case A. Constant electromotive force $E = E_0$. Since $\int e^{\alpha t}\,dt = e^{\alpha t}/\alpha$, for constant $E = E_0$, Eq. (5) gives the solution

$$I(t) = e^{-\alpha t}\left[\frac{E_0}{L}\cdot\frac{L}{R}e^{\alpha t} + c\right] = \frac{E_0}{R} + ce^{-\alpha t}. \tag{5*}$$

The last term goes to zero as $t \to \infty$; practically, after some time the current will be constant, equal to E_0/R, the value it would have immediately (by Ohm's law) had we no inductor in the circuit, and we see that this limit is independent of the initial value $I(0)$. Figure 14 shows the particular solution for which $I(0) = 0$, namely, by (5*)

$$I(t) = \frac{E_0}{R}(1 - e^{-\alpha t}) = \frac{E_0}{R}(1 - e^{-t/\tau_L}) \tag{5**}$$

where $\tau_L = L/R$ $(= 1/\alpha)$ is called the **inductive time constant** of the circuit (see Prob. 3).

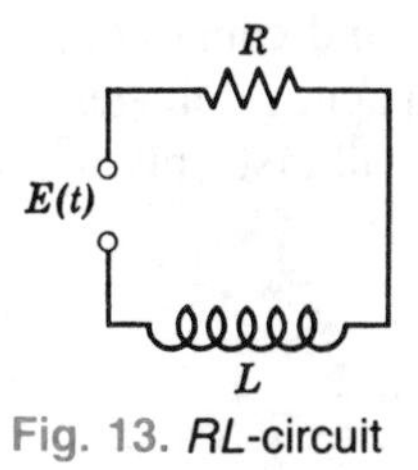

Fig. 13. *RL*-circuit

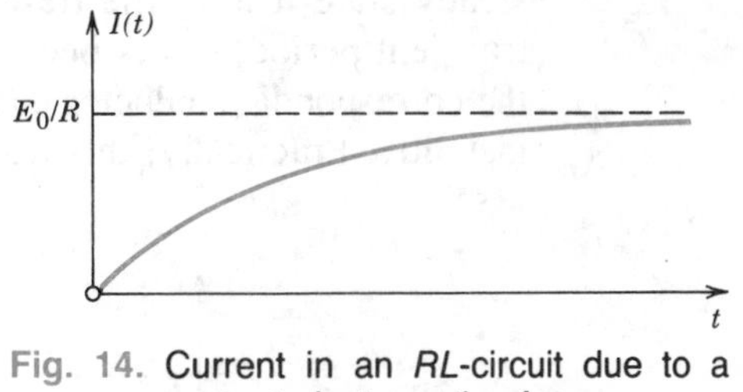

Fig. 14. Current in an *RL*-circuit due to a constant electromotive force

[16]GUSTAV ROBERT KIRCHHOFF (1824—1887), German physicist. Later we shall also need **Kirchhoff's current law (KCL):**

At any point of a circuit, the sum of the inflowing currents is equal to the sum of the outflowing currents.

4th Step. Case B. Periodic electromotive force $E(t) = E_0 \sin \omega t$. For this $E(t)$, Eq. (5) is

$$I(t) = e^{-\alpha t}\left[\frac{E_0}{L}\int e^{\alpha t}\sin \omega t\, dt + c\right] \qquad (\alpha = R/L).$$

Integration by parts yields

$$I(t) = ce^{-(R/L)t} + \frac{E_0}{R^2 + \omega^2 L^2}(R \sin \omega t - \omega L \cos \omega t).$$

This may be written [see (14) in Appendix 3]

$$(6) \qquad I(t) = ce^{-(R/L)t} + \frac{E_0}{\sqrt{R^2 + \omega^2 L^2}} \sin(\omega t - \delta), \qquad \delta = \arctan \frac{\omega L}{R}.$$

The exponential term will approach zero as t approaches infinity. This means that after some time the current $I(t)$ will execute practically harmonic oscillations. (See Fig. 15.) Figure 16 shows the phase angle δ as a function of $\omega L/R$. If $L = 0$, then $\delta = 0$, and the oscillations of $I(t)$ are in phase with those of $E(t)$.

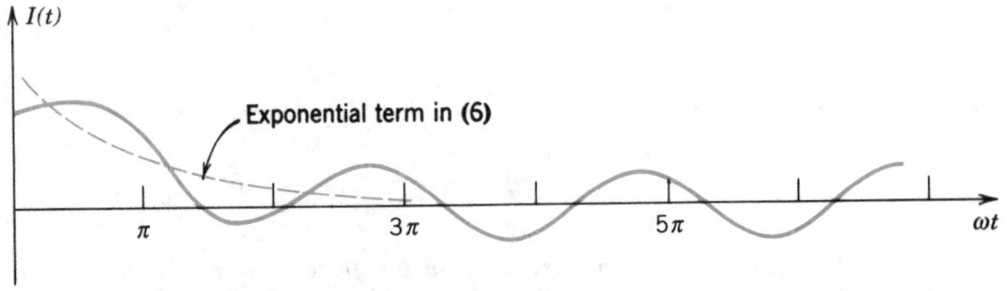

Fig. 15. Current in an RL-circuit due to a sinusoidal electromotive force, as given by (6) (with $\delta = \pi/4$)

An electrical (or dynamical) system is said to be in the **steady state** when the variables describing its behavior are periodic functions of time or constant, and it is said to be in the **transient state** (or *unsteady state*) when it is not in the steady state. The corresponding variables are called *steady-state functions* and *transient functions,* respectively.

In Example 1, Case A, the function E_0/R is a steady-state function or **steady-state solution** of (4), and in Case B the steady-state solution is represented by the last term in (6). Before the circuit (practically) reaches the steady state it is in the transient state. It is clear that such an interim or transient period occurs because inductors and capacitors store energy, and the corresponding inductor currents and capacitor voltages cannot be changed instantly. Practically, this transient state will last only a short time.

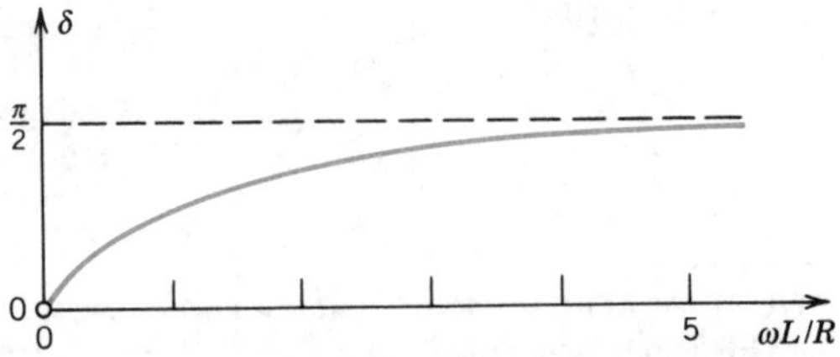

Fig. 16. Phase angle α in (6) as a function of $\omega L/R$

EXAMPLE 2 ***RC*-circuit**

Model the "RC-circuit" in Fig. 17 and find the current in the circuit for the two cases of the electromotive force $E(t)$ considered in Example 1.

Solution. 1st Step. Modeling. From (1), (3), and KVL we get

$$RI + \frac{1}{C}\int I\,dt = E(t). \tag{7}$$

To get rid of the integral we differentiate the equations with respect to t, finding

$$R\frac{dI}{dt} + \frac{1}{C}I = \frac{dE}{dt}. \tag{8}$$

2nd Step. Solution of the equation. Dividing (8) by R, we get from (4), Sec. 1.7, the general solution

$$I(t) = e^{-t/(RC)}\left(\frac{1}{R}\int e^{t/(RC)}\frac{dE}{dt}\,dt + c\right). \tag{9}$$

3rd Step. Case A. Constant electromotive force. If $E = const$, then $dE/dt = 0$ and (9) is simply

$$I(t) = ce^{-t/(RC)} = ce^{-t/\tau_C} \qquad \text{(Fig. 18)} \tag{10}$$

where $\tau_C = RC$ is called the **capacitive time constant** of the circuit.

4th Step. Case B. Periodic electromotive force $\boldsymbol{E(t) = E_0 \sin \omega t}$. For this $E(t)$ we have

$$\frac{dE}{dt} = \omega E_0 \cos \omega t.$$

By inserting this in (9) and integrating by parts we find

$$\begin{aligned} I(t) &= ce^{-t/(RC)} + \frac{\omega E_0 C}{1 + (\omega RC)^2}(\cos \omega t + \omega RC \sin \omega t) \\ &= ce^{-t/(RC)} + \frac{\omega E_0 C}{\sqrt{1 + (\omega RC)^2}} \sin(\omega t - \delta), \end{aligned} \tag{11}$$

where $\tan \delta = -1/(\omega RC)$. The first term decreases steadily as t increases, and the last term represents the steady-state current, which is sinusoidal. The graph of $I(t)$ is similar to that in Fig. 15. ∎

"***RLC*-circuits**" containing all three kinds of components lead to second-order differential equations and will be considered in Sec. 2.12; electrical networks leading to systems of differential equations, will be discussed in Chap. 4.

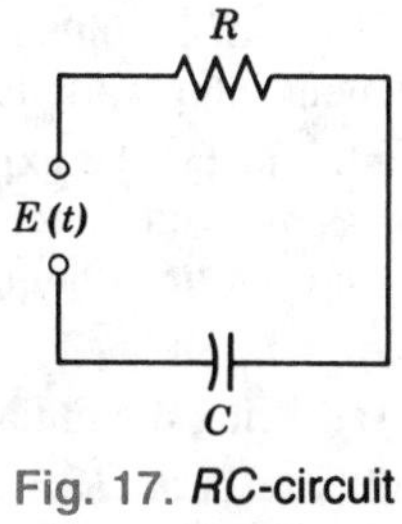

Fig. 17. RC-circuit

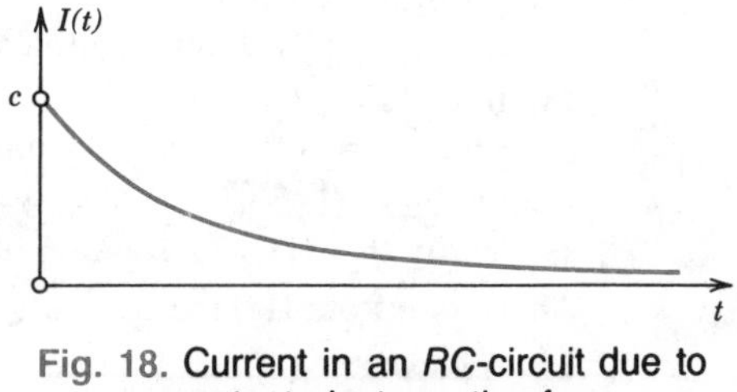

Fig. 18. Current in an RC-circuit due to a constant electromotive force

Problem Set 1.8

RL-Circuits

1. Derive (5*) from (5).

2. Obtain from (5*) the solutions satisfying $I(0) = 0.5E_0/R,\ E_0/R,\ 2E_0/R$. Sketch the corresponding curves.

3. Show that the inductive time constant $\tau_L = L/R$ is the time t at which the current (5**) reaches about 63% of its final value.

4. In Example 1, Case A, let $R = 20$ ohms, $L = 0.03$ millihenry $(3 \cdot 10^{-5}$ henry) and $I(0) = 0$. Find the time when the current reaches 99.9% of its final value.

5. Will the time in Prob. 4 decrease or increase if we change $I(0) = 0$ to some positive value less than E_0/R?

6. What L should we choose in an RL-circuit with $R = 100$ ohms for (5**) to reach 99.9% of its final value at $t = 0.01$ sec?

7. What L should we choose in (4) with $E = E_0 = 100$ volts and $R = 1000$ ohms if we want the current to grow from 0 to 25% of its final value within 10^{-4} sec?

8. In Example 1, case A, let $R = 100$ ohms, $L = 2.5$ henrys, $E_0 = 110$ volts, and $I(0) = 0$. Find the time constant and the time necessary for the current to rise from 0 to 0.6 ampere.

9. Derive (6) from (4) and check it by substitution.

10. Derive the steady-state solution in (6) by substituting $I_p = A \cos \omega t + B \sin \omega t$ into (4) with $E(t) = E_0 \sin \omega t$ and determining A and B by equating the cosine and sine terms in the resulting equation. (This avoids integration by parts.)

11. For what initial condition does (6) yield the steady-state solution?

12. How does the phase angle δ in (6) depend on L? Is this physically understandable?

13. Solve (4) with $E(t) = e^{-t}$ when (a) $R \neq L$, (b) $R = L$.

14. Solve (4) with electromotive force $E(t) = t$.

15. Find the current in the RL-circuit [see (4)] with $R = 1$ ohm, $L = 1$ henry, and $E(t) = 1$ if $0 \leqq t \leqq 3$ sec, $E(t) = 0$ if $t > 3$ sec, and $I(0) = \frac{1}{2}$ ampere. *Hint.* Use that $I(t)$ must be continuous at $t = 3$.

RC-Circuits

16. **(Discharge of a capacitor)** Show that (7) can also be written

$$R\frac{dQ}{dt} + \frac{1}{C}Q = E(t). \tag{12}$$

Solve this equation with $E(t) = 0$, assuming $Q(0) = Q_0$. Find the time when the capacitor has lost 99.9% of its initial charge.

17. Write the general solution of (12) in the form of an integral and derive from it formula (9) by differentiation and subsequent integration by parts.

18. In (12), let $R = 20$ ohms, $C = 0.01$ farad, and $E(t)$ be exponentially decaying, say, $E(t) = 60e^{-2t}$ volts. Assuming $Q(0) = 0$, find and sketch $Q(t)$. Also determine the time when $Q(t)$ reaches a maximum and that maximum charge.

19. Verify that (11) is a solution of (8) with $E = E_0 \sin \omega t$.

20. Obtain from (11) the particular solution satisfying the initial condition $I(0) = 0$.

21. A capacitor ($C = 0.1$ farad) in series with a resistor ($R = 200$ ohms) is charged from a source ($E_0 = 12$ volts); see Fig. 17 with $E(t) = E_0$. Find the voltage $V(t)$ on the capacitor, assuming that at $t = 0$ the capacitor is completely uncharged.

22. Find the current $I(t)$ in the RC-circuit shown in Fig. 17, assuming that $E = 100$ volts, $C = 0.25$ farad, R is variable according to $R = (100 - t)$ ohms when $0 \leqq t \leqq 100$ sec, $R = 0$ when $t > 100$ sec, and $I(0) = 1$ ampere.

23. Solve (12) when $R = 500$ ohms, $C = 10^{-3}$ farad, and $E(t) = 1 - e^{-t}$ volt.

Find the steady-state solution of (12) when $R = 50$ ohms, $C = 0.04$ farad, and $E(t)$ equals:

24. $110 \cos 314t$

25. $50e^{-t} + 1012 \sin \pi t$

26. Show that if the initial charge in the capacitor in Fig. 17 is $Q(0)$, the initial current in the RC-circuit is $I(0) = E(0)/R - Q(0)/RC$.

27. Show that if $E(t)$ in Fig. 17 has a jump of magnitude J at $t = a$, then the current $I(t)$ in the RC-circuit has a jump of magnitude J/R at $t = a$.

Using the result of Prob. 27, find the current $I(t)$ in the RC-circuit in Fig. 17, assuming $R = 1$ ohm, $C = 1$ farad, zero initial charge on the capacitor, and

28. $E = t$ when $0 < t < 1$ and $E = 1$ when $t > 1$

29. $E = t + 1$ when $0 < t < a$ and $E = 0$ when $t > a$

30. $E = t + 1$ when $0 < t < a$ and $E = a + 1$ when $t > a$

1.9 Orthogonal Trajectories of Curves *Optional*

As another interesting application, we shall see in this section how to use differential equations for finding curves that intersect given curves at right angles,[17] a task that arises rather often in practice. The new curves are then called the **orthogonal trajectories** of the given curves (and conversely). Here, "orthogonal" is another word for "perpendicular." Figure 19 shows an example.

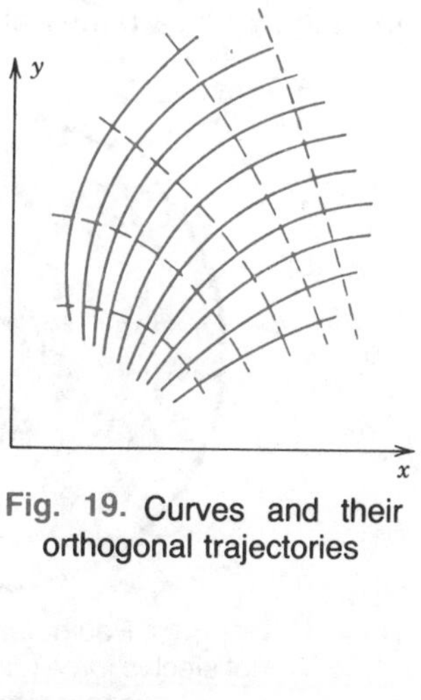

Fig. 19. Curves and their orthogonal trajectories

[17] Remember that the **angle of intersection** of two curves is defined to be the angle between the tangents of the curves at the point of intersection.

For instance, the meridians and parallels on the globe are orthogonal trajectories of each other, and so are the curves of steepest descent and the contour lines on a map. In an electric field, the curves of electric force are the orthogonal trajectories of the equipotential lines (= curves of constant voltage) and conversely (for an example, see Fig. 20). Other important examples arise in fluid flow, heat conduction, and other fields of physics.

Our idea of finding the orthogonal trajectories of given curves is to represent these curves by the general solution of a differential equation $y' = f(x, y)$, then to replace the slope y' by its negative reciprocal $-1/y'$ (the condition for perpendicularity, as we recall from analytic geometry), and finally to solve the new differential equation $-1/y' = f(x, y)$. The details are as follows.

Family of curves. If for each fixed value of c the equation

$$F(x, y, c) = 0 \tag{1}$$

represents a curve in the xy-plane and if for variable c it represents infinitely many curves, then the totality of these curves is called a **one-parameter family of curves,** and c is called the *parameter* of the family.

EXAMPLE 1 **Families of curves**

The equation

$$F(x, y, c) = x + y + c = 0 \tag{2}$$

represents a family of parallel straight lines; each line corresponds to precisely one value of the parameter c. The equation

$$F(x, y, c) = x^2 + y^2 - c^2 = 0 \tag{3}$$

represents a family of concentric circles of radius c with center at the origin. ∎

Now many one-parameter families can be given by the general solution of a differential equation, which contains an arbitrary parameter c, as we know. Accordingly, given a family of curves, the first step of our method is

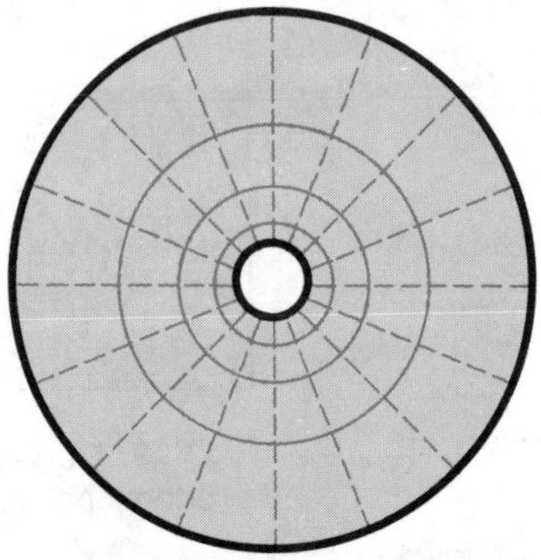

Fig. 20. Equipotential lines and lines of electric force (dashed) between two concentric cylinders

to find a differential equation of it, usually by differentiating (1). Example 2 shows three families, the third one illustrating how to get rid of c if it does not go.

EXAMPLE 2 Differential equations of families. Elimination of parameter

Differentiating (2) gives as the differential equation of those straight lines

$$y' = -1.$$

Differentiating (3) gives $2x + 2yy' = 0$, so that the differential equation of those circles is

$$y' = -x/y.$$

Differentiating the family of parabolas

$$y = cx^2 \tag{4}$$

we first get $y' = 2cx$. We must get rid of c. Now (4) gives $c = y/x^2$. By substitution, $y' = 2(y/x^2)x$, that is,

$$y' = 2y/x. \tag{5}$$

Faster: solve (4) for c, that is, $c = yx^{-2}$, and differentiate implicitly with respect to x; thus, $0 = y'x^{-2} - 2yx^{-3}$, which gives (5). ■

Determination of Orthogonal Trajectories

1st Step. Given a family (1), find its differential equation in the form

$$y' = f(x, y). \tag{6}$$

2nd Step. Find the orthogonal trajectories by solving their differential equation

$$y' = -\frac{1}{f(x, y)}. \tag{7}$$

Proof of Method. From (6) we see that a given curve that passes through a point P: (x_0, y_0) has at P the slope $f(x_0, y_0)$. The slope of the orthogonal trajectory through P should at P be the negative reciprocal of $f(x_0, y_0)$, that is, $-1/f(x_0, y_0)$ because this is the condition for the tangents of the two curves at P to be perpendicular. By this, (6) implies (7). ■

EXAMPLE 3 Orthogonal trajectories

Find the orthogonal trajectories of the parabolas in Example 2.

Solution. From (5) we see that the differential equation (7) of the orthogonal trajectories is

$$y' = -\frac{1}{2y/x} = -\frac{x}{2y}.$$

By separating variables and integrating we find that the orthogonal trajectories are the ellipses

$$\frac{x^2}{2} + y^2 = c^* \qquad \text{(Fig. 21, p. 46).}$$ ■

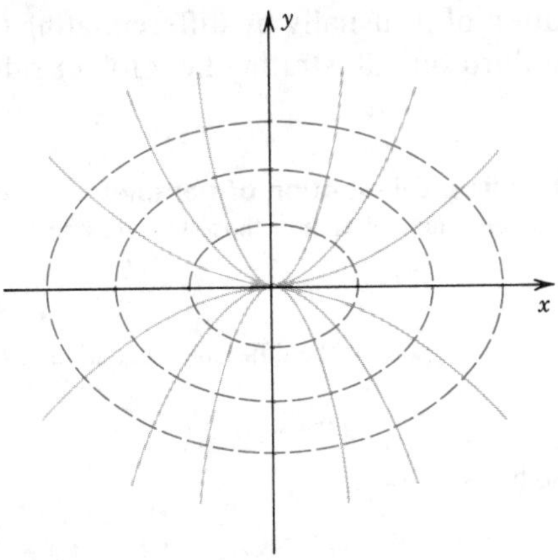

Fig. 21. Parabolas and their orthogonal trajectories in Example 3

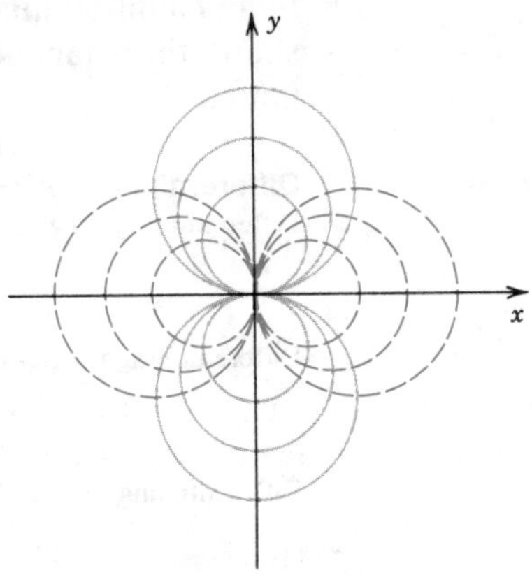

Fig. 22. Circles and their orthogonal trajectories (dashed) in Example 4

EXAMPLE 4 **Orthogonal trajectories**

Find the orthogonal trajectories of the circles

$$x^2 + (y - c)^2 = c^2 \qquad \text{(Fig. 22).} \tag{8}$$

Solution. 1st Step. Differential equation of (8). Differentiation of (8) gives $2x + 2(y - c)y' = 0$, still containing c. So we solve (8), which is $x^2 + y^2 - 2cy = 0$, for $2c$, obtaining

$$2c = y^{-1}(x^2 + y^2) = y^{-1}x^2 + y.$$

Implicit differentiation with respect to x gives

$$0 = -\frac{x^2}{y^2}y' + \frac{2x}{y} + y'.$$

Solving this algebraically for y', we get the differential equation of (8),

$$y' = \frac{2xy}{x^2 - y^2}. \tag{9}$$

2nd Step. Orthogonal trajectories. From (9) we have the differential equation of the orthogonal trajectories of (8):

$$y' = -\frac{x^2 - y^2}{2xy}, \qquad \text{thus} \qquad 2xyy' - y^2 + x^2 = 0.$$

Its general solution (obtained in Example 1 of Sec. 1.4) gives as the trajectories of (8) the circles (Fig. 22)

$$(x - \tilde{c})^2 + y^2 = \tilde{c}^2.$$ ∎

Problem Set 1.9

Find the differential equation (6) of the following families.

1. $xy = c$ **2.** $e^{3x}y = c$ **3.** $y = c \sin 2x$

4. $y = cxe^x$ **5.** $y = e^{cx^2}$ **6.** $y = 1/(1 + ce^x)$

7. $y = cx^2 + x^2e^x$ **8.** $c^2x^2 + y^2 = c^2$ **9.** $y^2 = ce^{-2x} + x - \frac{1}{2}$

Using differential equations, find the orthogonal trajectories of the following curves. Graph some of the curves and the trajectories.

10. $x^2 - y^2 = c^2$ **11.** $x^2 + y^2 = c^2$ **12.** $y = c/x^2$

13. $y = \ln|x| + c$ **14.** $xy = c$ **15.** $y = \sqrt{x + c}$

16. $y = cx^3$ **17.** $x = ce^{-y^2}$ **18.** $x^2 + 2y^2 = c$

19. $x^2 + 4y^2 = c$ **20.** $(x - c)^2 + y^2 = c^2$ **21.** $y = ce^{8x}$

Applications

22. (Temperature field) If the **isotherms** (= curves of constant temperature) in a body are $T(x, y) = 2x^2 + y^2 = const$, what are their orthogonal trajectories (the curves along which heat will flow in regions free of heat sources or sinks and filled with homogeneous material)?

23. (Electric field) In the electric field between two concentric cylinders (Fig. 20) the **equipotential lines** (= curves of constant potential) are circles given by $U(x, y) = x^2 + y^2 = const$ [volts]. Use our present method to get their trajectories (the curves of electric force).

24. (Electric field) Experiments show that the electric lines of force of two opposite charges of the same strength at $(-1, 0)$ and $(1, 0)$ are the circles through $(-1, 0)$ and $(1, 0)$. Show that these circles can be represented by the equation $x^2 + (y - c)^2 = 1 + c^2$. Show that the equipotential lines (orthogonal trajectories) are the circles $(x + c^*)^2 + y^2 = c^{*2} - 1$ (dashed in Fig. 23).

25. (Fluid flow) If the **streamlines** of the flow (= paths of the particles of the fluid) in the channel in Fig. 24 are $\Psi(x, y) = xy = const$, what are their orthogonal trajectories (called **equipotential lines,** for reasons explained in Sec. 17.4)?

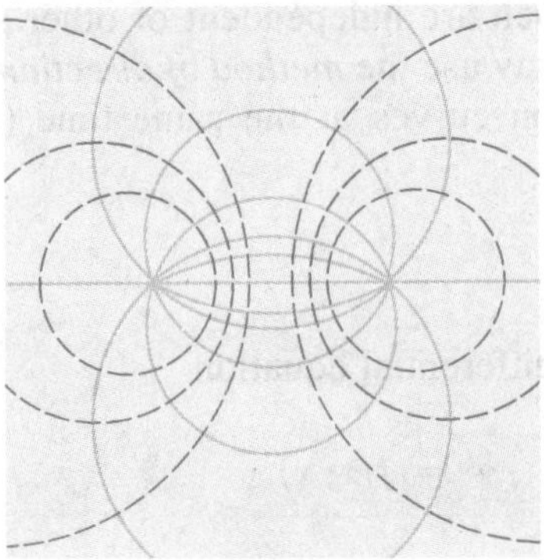

Fig. 23. Electric field in Problem 24

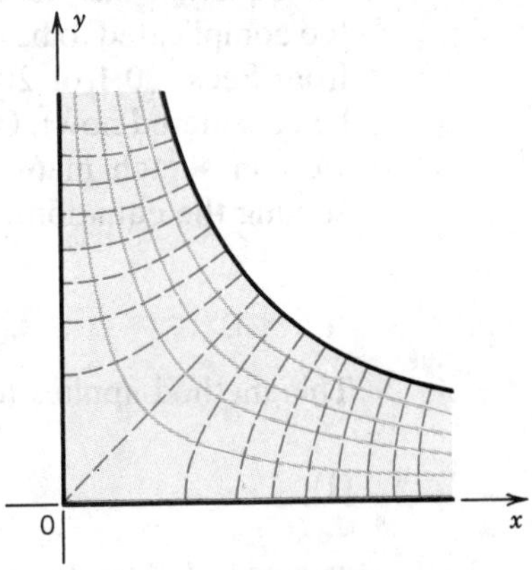

Fig. 24. Flow around a corner in Problem 25

Other forms of the differential equations. Isogonal trajectories

26. Show that (7) can be written as $dx/dy = -f(x, y)$. Use it to get the orthogonal trajectories of the curves $y = x + ce^{-x}$.

27. Show that the orthogonal trajectories of a family $g(x, y) = c$ can be obtained from the following differential equation and use it to solve Prob. 25:

$$\frac{dy}{dx} = \frac{\partial g/\partial y}{\partial g/\partial x}.$$

28. **(Cauchy–Riemann equations).** Show that for a family $u(x, y) = c = const$ the orthogonal trajectories $v(x, y) = c^* = const$ can be obtained from the following so-called *Cauchy–Riemann equations* (which are basic in complex analysis in Chap. 12):

$$\frac{\partial u}{\partial x} = \frac{\partial v}{\partial y}, \qquad \frac{\partial u}{\partial y} = -\frac{\partial v}{\partial x}.$$

29. Find the orthogonal trajectories of $e^x \cos y = c$ by the Cauchy–Riemann equations.

30. **Isogonal trajectories** of a given family of curves are curves that intersect the given curves at a constant angle θ. Show that at each point the slopes m_1 and m_2 of the tangents to the corresponding curves satisfy the relation

$$\frac{m_2 - m_1}{1 + m_1 m_2} = \tan\theta = const.$$

Using this, find the curves that intersect the circles $x^2 + y^2 = c$ at 45°.

1.10 Approximate Solutions: Direction Fields, Iteration

Approximate solutions of a differential equation are of practical interest if the equation has no explicit exact solution formula[18] or if that formula is too complicated to be of practical value. Then we can use a **numerical method** from Secs. 20.1 or 20.2 (which are independent of other sections and can be considered now). Or we may use the *method of direction fields*, by which we can sketch many solution curves at the same time (without actually solving the equation).

Method of Direction Fields

This method applies to any differential equation

$$y' = f(x, y). \tag{1}$$

The idea is simply this: y' is the slope of the unknown solution curves. If such a curve passes through a point P: (x_0, y_0), its slope at P must be $f(x_0, y_0)$, as we can see directly from (1). Hence we could draw at various points **"lineal elements,"** meaning short segments indicating the tangent directions of solution curves as determined by (1), and then fit solution curves through this field of tangent directions.

It is better and more economical if we first draw curves of constant slope $f(x, y) = const$ (so these are not yet solution curves!); these are also called **isoclines** (meaning curves of equal inclinations). Then, as the second step,

[18]Reference [A7] in Appendix 1 includes more than 1500 important differential equations and their solutions, systematically arranged and with numerous references to original literature.

we draw along each isocline $f(x, y) = k = const$ many lineal elements of slope k; this we do for one isocline after another. What we get is called a **direction field** of (1). In it, as the third step, we can now sketch approximate solution curves of (1), guided by the tangent directions as given by those lineal elements.

It suffices to illustrate the method by a simple equation that can be solved exactly, so that we get a feeling for the accuracy of the method.

EXAMPLE 1 **Isoclines, direction field**

Graph the direction field of the first-order differential equation

(2) $$y' = xy$$

and an approximation to the solution curve through the point (1, 2). Compare with the exact solution.

Solution. The isoclines are the equilateral hyperbolas $xy = k$ together with the two coordinate axes. We graph some of them. Then we draw lineal elements by sliding a triangle along a fixed ruler. The result is shown in Fig. 25, which also shows an approximation to the solution curve passing through the point (1, 2).

By separating variables, $y = ce^{x^2/2}$. The initial condition is $y(1) = 2$. Hence $2 = ce^{1/2}$, and the exact solution is

$$y = 2e^{(x^2-1)/2}.$$ ∎

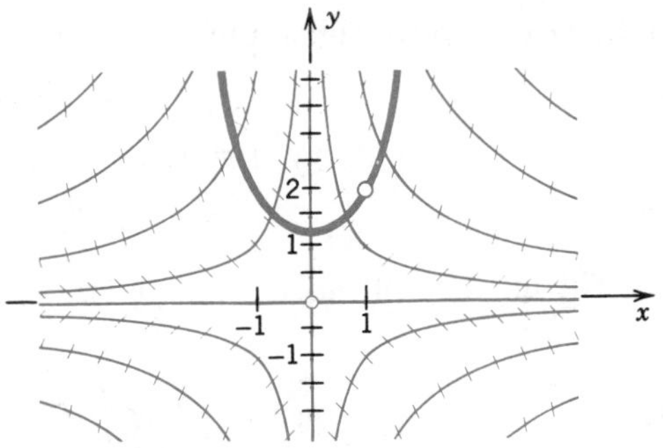

Fig. 25. Direction field of the differential equation (2)

Picard's Iteration Method[19]

Picard's method gives approximate solutions of an initial value problem

(3) $$y' = f(x, y), \qquad y(x_0) = y_0$$

which is assumed to have a unique solution in some open interval on the x-axis containing x_0. The method is not very practical because it involves integrations. Nevertheless, we discuss it here for two reasons:

1. Picard's method is the basis of Picard's existence and uniqueness theorems to be discussed in the next section.

2. Picard's method illustrates the idea of **iteration methods,** in which one does several steps of calculation or computation by the same rule but with

[19]EMILE PICARD (1856—1941), French mathematician, professor in Paris since 1881, also known for his important contributions to complex analysis (see Sec. 14.8 for his famous theorem).

changing data (usually with the data obtained in the previous step, as we shall see). Iteration methods are used quite often in applied mathematics, particularly in numerical work (see Chaps. 18–20).

The idea of Picard's method is simple. By integration we see that (3) may be written in the form

$$y(x) = y_0 + \int_{x_0}^{x} f[t, y(t)]\, dt \tag{4}$$

where t denotes the variable of integration. In fact, when $x = x_0$ the integral is zero and $y = y_0$, so that (4) satisfies the initial condition in (3); furthermore, by differentiating (4) we obtain the differential equation in (3).

To find approximations to the solution $y(x)$ of (4) we proceed as follows. We substitute the crude approximation $y = y_0 = const$ on the right; this yields the presumably better approximation

$$y_1(x) = y_0 + \int_{x_0}^{x} f(t, y_0)\, dt.$$

In the next step we substitute the function $y_1(x)$ in the same way to get

$$y_2(x) = y_0 + \int_{x_0}^{x} f[t, y_1(t)]\, dt,$$

etc. The nth step of this iteration gives an approximating function

$$y_n(x) = y_0 + \int_{x_0}^{x} f[t, y_{n-1}(t)]\, dt. \tag{5}$$

In this way we obtain a sequence of approximations

$$y_1(x), \quad y_2(x), \cdots, \quad y_n(x), \cdots,$$

and we shall see in the next section that the conditions under which this sequence converges to the solution $y(x)$ of (3) are fairly general.

Let us illustrate Picard's method for an equation that we can readily solve exactly, so that we may compare the approximations with the exact solution.

EXAMPLE 2 Picard iteration

Find approximate solutions to the initial value problem

$$y' = 1 + y^2, \qquad y(0) = 0.$$

Solution. In this case, $x_0 = 0$, $y_0 = 0$, $f(x, y) = 1 + y^2$, and (5) becomes

$$y_n(x) = \int_0^x [1 + y_{n-1}^2(t)]\, dt = x + \int_0^x y_{n-1}^2(t)\, dt.$$

Starting from $y_0 = 0$, we thus obtain (see Fig. 26)

$$y_1(x) = x + \int_0^x 0\, dt = x$$

$$y_2(x) = x + \int_0^x t^2\, dt = x + \tfrac{1}{3}x^3$$

$$y_3(x) = x + \int_0^x \left(t + \frac{t^3}{3}\right)^2 dt = x + \tfrac{1}{3}x^3 + \tfrac{2}{15}x^5 + \tfrac{1}{63}x^7$$

etc. Of course, we can obtain the exact solution of our present problem by separating variables (see Example 2 in Sec. 1.2), finding

$$y(x) = \tan x = x + \tfrac{1}{3}x^3 + \tfrac{2}{15}x^5 + \tfrac{17}{315}x^7 + \cdots \qquad \left(-\frac{\pi}{2} < x < \frac{\pi}{2}\right). \tag{6}$$

The first three terms of $y_3(x)$ and the series in (6) are the same. The series in (6) converges for $|x| < \pi/2$, and all we may expect is that our sequence $y_1, y_2, \cdots$ converges to a function that is the solution of our problem for $|x| < \pi/2$. This illustrates that the study of convergence is of practical importance. ∎

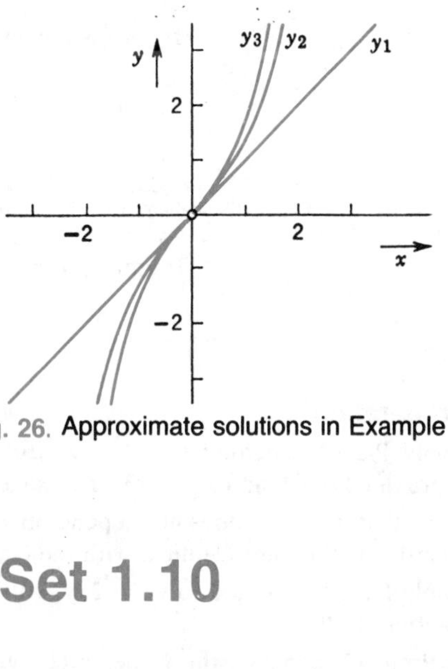

Fig. 26. Approximate solutions in Example 2

Problem Set 1.10

Direction Fields

In each case draw a good direction field. Sketch several approximate solution curves. Then solve the equation analytically and compare, to get a feeling for the accuracy of the present method.

1. $y' = 2y/x$ **2.** $y' = -x/y$ **3.** $y' = -xy$

4. $y' = x + y$ **5.** $9yy' + x = 0$ **6.** $y' = y^2$

7. (Skydiver) Sketch the direction field of the differential equation

$$\frac{dv}{dt} = 10 - 0.4v^2$$

and draw from it the following conclusions. The isoclines are horizontal straight lines. The isocline $v = 5$ is at the same time a solution curve. All solution curves in the upper half-plane ($v > 0$) seem to approach the line $v = 5$ as $t \to \infty$; they are monotone increasing if $0 < v(0) < 5$ and monotone decreasing if $v(0) > 5$. Sketch the solution curves for which $v(0) = 0$ and $v(0) = 7$ and compare with the exact solutions obtained from (11), Sec. 1.3.

8. **(Verhulst population model)** Draw the direction field of the differential equation in Example 5 of Sec. 1.7, with $A = 0.03$ and $B = 1.6 \cdot 10^{-4}$ and use it to discuss the general behavior of solutions corresponding to initial conditions greater and smaller than 187.5.

9. **(Pursuit problem. Tractrix)** In Fig. 27, the destroyer D (the *pursuer*) *pursues* the ship S (the *target*), that is, moves in the direction of S at all times. Assume that S moves along the x-axis and the distance a from D to S is constant. Show that $y' = -y/\sqrt{a^2 - y^2}$. Sketch a direction field (for $a = 1$ [nautical mile]) and the solution satisfying $y(0) = 1$. (This curve is called a *tractrix*, from Latin *trahere*, meaning "to pull.") Separating variables, show that

$$x = -\int y^{-1}\sqrt{a^2 - y^2}\, dy = -\sqrt{a^2 - y^2} + a \ln |y^{-1}(a + \sqrt{a^2 - y^2})| + c.$$

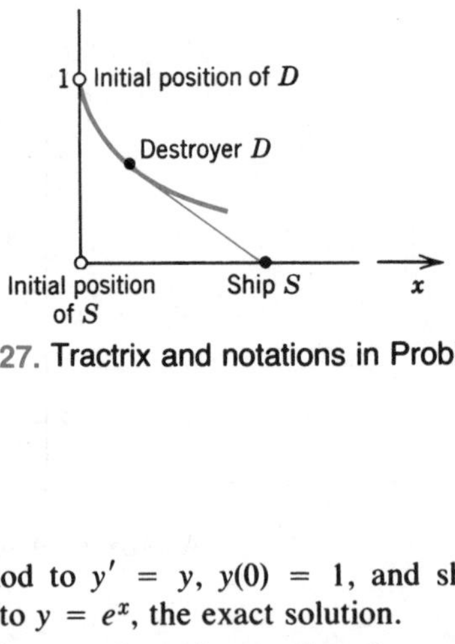

Fig. 27. Tractrix and notations in Problem 9

Picard Iteration

10. Apply Picard's method to $y' = y$, $y(0) = 1$, and show that the successive approximations tend to $y = e^x$, the exact solution.

11. Show that if f in (3) does not depend on y, then the approximations obtained by Picard's method are identical with the exact solution. Why?

12. Apply Picard's method to $y' = 2xy$, $y(0) = 1$. Sketch y_1, y_2, y_3, and the exact solution for $0 \leqq x \leqq 2$.

13. In Prob. 12, compute the values $y_1(1)$, $y_2(1)$, $y_3(1)$ and compare them with the exact value $y(1) = e = 2.718 \cdots$.

Apply Picard's method to the following initial value problems. Determine also the exact solution. Compare.

14. $y' = xy, \quad y(0) = 1$

15. $y' = 2y, \quad y(0) = 1$

16. $y' = x + y, \quad y(0) = 0$

17. $y' = x + y, \quad y(0) = -1$

18. $y' = y^2, \quad y(0) = 1$

19. $y' = xy + 2x - x^3, \quad y(0) = 0$

20. $y' - xy = 1, \quad y(0) = 1$

1.11 Existence and Uniqueness of Solutions

So far, for the differential equations considered there existed a general solution, and for an **initial value problem,** say,

$$y' = f(x, y), \qquad y(x_0) = y_0 \tag{1}$$

consisting of a differential equation and an initial condition $y(x_0) = y_0$, we got a unique particular solution. However, this was just one of three possibilities illustrated by the following examples:

The initial value problem

$$|y'| + |y| = 0, \qquad y(0) = 1$$

has no solution because $y \equiv 0$ is the only solution of the differential equation. (Why?) The initial value problem

$$y' = x, \qquad y(0) = 1$$

has precisely one solution, namely, $y = \frac{1}{2}x^2 + 1$. The initial value problem

$$xy' = y - 1, \qquad y(0) = 1$$

has infinitely many solutions, namely, $y = 1 + cx$ where c is an arbitrary constant. From these three examples we see that an initial value problem may have no solutions, precisely one solution, or more than one solution. This leads to the following two fundamental questions.

Problem of existence. *Under what conditions does an initial value problem of the form* (1) *have at least one solution?*

Problem of uniqueness. *Under what conditions does that problem have at most one solution?*

Theorems that state such conditions are called **existence theorems** and **uniqueness theorems**, respectively.

Of course, our three examples are so simple that we can find the answer to those two questions by inspection, without using any theorems. However, it is clear that in more complicated cases—for example, when the equation cannot be solved by elementary methods—existence and uniqueness theorems may be of considerable practical importance. Even when you are sure that your physical or other system behaves uniquely, once in a while your model may be oversimplified and may not give a faithful picture of the reality. So make sure that your model has a unique solution before you try to compute that solution. The following two theorems take care of almost all conceivable practical cases.

The first theorem says that if $f(x, y)$ in (1) is continuous in some region of the xy-plane containing the point (x_0, y_0) (corresponding to the given initial condition), then the problem (1) has at least one solution. If, moreover, the partial derivative $\partial f/\partial y$ exists and is continuous in that region, then the problem (1) has precisely one solution. This solution can then be obtained by Picard's iteration method. Let us formulate these statements in a precise way.

Theorem 1 **Existence theorem**

If $f(x, y)$ is continuous at all points (x, y) in some rectangle (Fig. 28)

$$R: \qquad |x - x_0| < a, \qquad |y - y_0| < b$$

and bounded[20] *in R, say,*

$$(2) \qquad |f(x, y)| \leqq K \qquad \textit{for all } (x, y) \textit{ in } R,$$

then the initial value problem (1) *has at least one solution $y(x)$. This solution is defined at least for all x in the interval $|x - x_0| < \alpha$ where α is the smaller of the two numbers a and b/K.*

Theorem 2 **Uniqueness theorem**

If $f(x, y)$ and $\partial f/\partial y$ are continuous for all (x, y) in that rectangle R and bounded, say,

$$(3) \qquad (a)\ \ |f| \leqq K, \qquad (b)\ \ \left|\frac{\partial f}{\partial y}\right| \leqq M \qquad \textit{for all } (x, y) \textit{ in } R,$$

then the initial value problem (1) *has at most one solution $y(x)$. Hence, by Theorem* 1, *it has precisely one solution. This solution is defined at least for all x in that interval $|x - x_0| < \alpha$. It can be obtained by Picard's method, that is, the sequence $y_0, y_1, \cdots, y_n, \cdots$, where*

$$y_n(x) = y_0 + \int_{x_0}^{x} f[t, y_{n-1}(t)]\, dt, \qquad n = 1, 2, \cdots,$$

converges to that solution $y(x)$.

Since the proofs of these theorems require familiarity with uniformly convergent series and other concepts to be considered later in this book, we shall not present these proofs at this time but refer the student to Ref. [A6] in Appendix 1. However, we want to include some remarks and examples that may be helpful for a good understanding of the two theorems.

---

[20] A function $f(x, y)$ is said to be **bounded** when (x, y) varies in a region in the xy-plane, if there is a number K such that $|f(x, y)| \leqq K$ when (x, y) is in that region. For example, $f = x^2 + y^2$ is bounded, with $K = 2$ if $|x| < 1$ and $|y| < 1$. The function $f = \tan(x + y)$ is not bounded for $|x + y| < \pi/2$.

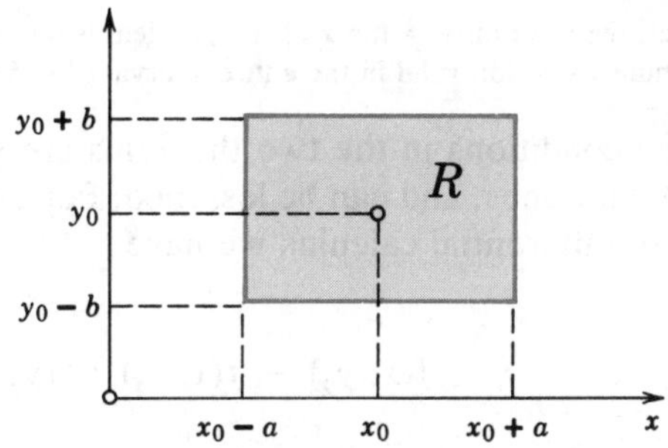

Fig. 28. Rectangle R in the existence and uniqueness theorems

Since $y' = f(x, y)$, the condition (2) implies that $|y'| \leqq K$, that is, the slope of any solution curve $y(x)$ in R is at least $-K$ and at most K. Hence a solution curve that passes through the point (x_0, y_0) must lie in the colored region in Fig. 29 bounded by the lines l_1 and l_2 whose slopes are $-K$ and K, respectively. Depending on the form of R, two different cases may arise. In the first case, shown in Fig. 29*a*, we have $b/K \geqq a$ and therefore $\alpha = a$ in the existence theorem, which then asserts that the solution exists for all x between $x_0 - a$ and $x_0 + a$. In the second case, shown in Fig. 29*b*, we have $b/K < a$. Therefore $\alpha = b/K$, and all we can conclude from the theorems is that the solution exists for all x between $x_0 - b/K$ and $x_0 + b/K$. For larger or smaller x's the solution curve may leave the rectangle R, and since nothing is assumed about f outside R, nothing can be concluded about the solution for those larger or smaller x's; that is, for such x's the solution may or may not exist—we don't know.

Let us illustrate our discussion by a simple example.

EXAMPLE 1 Consider the problem

$$y' = 1 + y^2, \qquad y(0) = 0$$

(see Example 2 in Sec. 1.10) and take R: $|x| < 5$, $|y| < 3$. Then $a = 5$, $b = 3$ and

$$|f| = |1 + y^2| \leqq K = 10, \qquad |\partial f/\partial y| = 2|y| \leqq M = 6, \qquad \alpha = b/K = 0.3 < a.$$

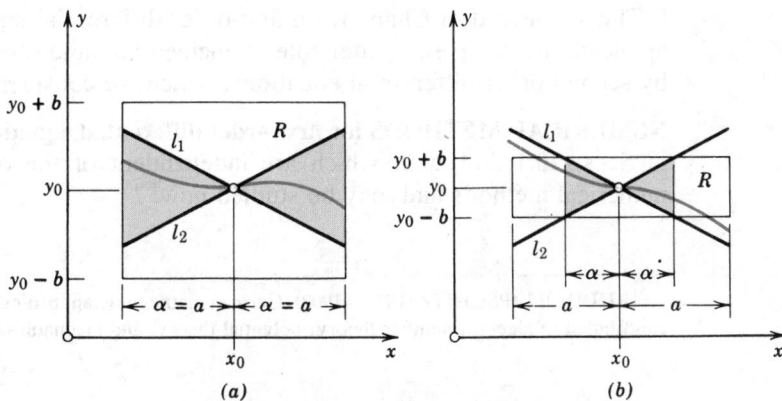

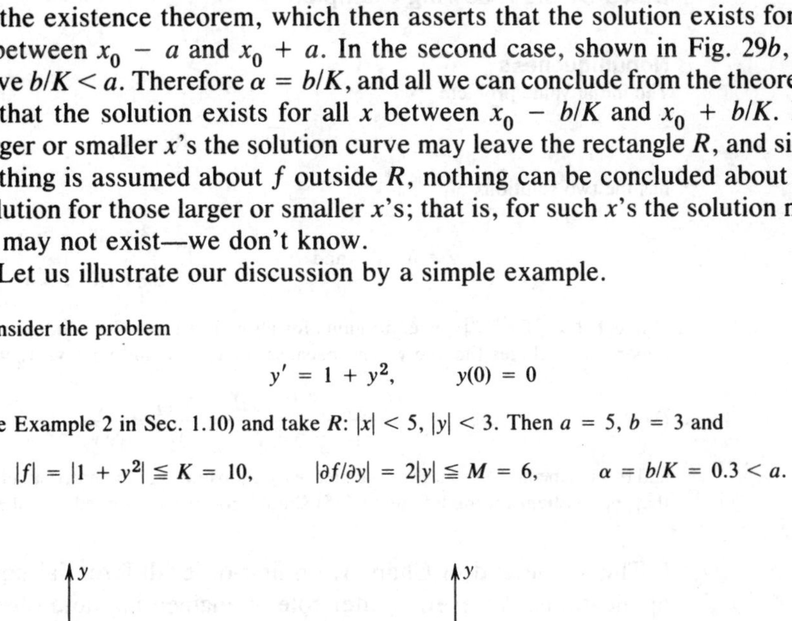

Fig. 29. The condition (2) of the existence theorem. (*a*) First case. (*b*) Second case

In fact, the solution $y = \tan x$ of the problem is discontinuous at $x = \pm\pi/2$, and there is no continuous solution valid in the entire interval $|x| < 5$ from which we started. ∎

The conditions in the two theorems are sufficient conditions rather than necessary ones, and can be lessened. For example, by the mean value theorem of differential calculus we have

$$f(x, y_2) - f(x, y_1) = (y_2 - y_1) \left.\frac{\partial f}{\partial y}\right|_{y=\tilde{y}}$$

where (x, y_1) and (x, y_2) are assumed to be in R, and $\tilde{y}$ is a suitable value between y_1 and y_2. From this and (3*b*) it follows that

$$|f(x, y_2) - f(x, y_1)| \leqq M|y_2 - y_1|, \tag{4}$$

and it can be shown that (3*b*) may be replaced by the weaker condition (4), which is known as a **Lipschitz condition.**[21] However, continuity of $f(x, y)$ is not enough to guarantee the *uniqueness* of the solution. This may be illustrated by the following example.

EXAMPLE 2 **Nonuniqueness**

The initial value problem

$$y' = \sqrt{|y|}, \qquad y(0) = 0$$

has the two solutions

$$y \equiv 0 \qquad \text{and} \qquad y^* = \begin{cases} x^2/4 & \text{if } x \geqq 0 \\ -x^2/4 & \text{if } x < 0 \end{cases}$$

although $f(x, y) = \sqrt{|y|}$ is continuous for all y. The Lipschitz condition (4) is violated in any region that includes the line $y = 0$, because for $y_1 = 0$ and positive y_2 we have

$$\frac{|f(x, y_2) - f(x, y_1)|}{|y_2 - y_1|} = \frac{\sqrt{y_2}}{y_2} = \frac{1}{\sqrt{y_2}}, \qquad (\sqrt{y_2} > 0) \tag{5}$$

and this can be made as large as we please by choosing y_2 sufficiently small, whereas (4) requires that the quotient on the left side of (5) should not exceed a fixed constant M. ∎

This is the end of Chap. 1, on first-order differential equations and their applications. An even greater role in engineering (and elsewhere) is played by second-order differential equations, which we consider in Chap. 2.

NUMERICAL METHODS for first-order differential equations are presented in Secs. 20.1 and 20.2, which are independent of the other sections on numerical methods and may be studied now.

---

[21] RUDOLF LIPSCHITZ (1832—1903), German mathematician, professor at Bonn, who also contributed to algebra, number theory, potential theory, and mechanics.

Problem Set 1.11

1. Find all solutions of the initial value problem $xy' = 3y$, $y(0) = 0$. Sketch some of them. Does your answer contradict our present theorems?
2. Show that the initial value problem $xy' = 2y$, $y(0) = 1$ has no solution. Does this contradict our present existence theorem?
3. Find all solutions of the initial value problem $y' = 2\sqrt{y}$, $y(1) = 0$. Which of them do we obtain by Picard's method if we start from $y_0 = 0$?
4. Consider the equation $(x^2 - x)y' = (2x - 1)y$ and find all initial conditions $y(x_0) = y_0$ such that the resulting initial value problem has (a) no solution, (b) more than one solution, (c) precisely one solution. Does your answer contradict our present theorems?
5. Perform the same task as in Prob. 4, for the equation $xy' = 2y$.
6. For what initial conditions does the problem $(x^3 - x)y' = (3x^2 - 1)y$, $y(x_0) = y_0$, have infinitely many solutions? Sketch some solution curves in the xy-plane.
7. Determine the largest set S in the xy-plane such that through each point of S there passes one and only one solution curve of $x\,dy = y\,dx$.
8. Find all solutions of $y' = x|y|$.
9. Show that if $y' = f(x, y)$ satisfies the assumptions of the present theorems in a rectangle R and y_1 and y_2 are two solutions of the equation whose curves lie in R, then these curves cannot have a point in common (unless they are identical).
10. If the assumptions of the present theorems are satisfied not merely in a rectangle R but even in a vertical strip given by $|x - x_0| < a$, show that then the solution of (1) exists for all x in the interval $|x - x_0| < a$.
11. It is worth noting that the solution of an initial value problem may exist in an interval larger than that in the theorem, the latter being $|x - x_0| < \alpha$ with α defined in the theorem. This is illustrated by Example 1 in the text. Consider that example. Find α in terms of general a and b. What is the maximum possible α that we can achieve by a suitable choice of a and b?
12. Find the largest α for which the present theorems guarantee the existence of the problem $y' = y^2$, $y(1) = 1$. For what x does the solution actually exist?
13. Perform the same task as in Prob. 12, for $y' - y^2 = 4$, $y(0) = 0$.
14. Show that $f(x, y)$ in Prob. 13 satisfies a Lipschitz condition in R (see Fig. 28).
15. Find all solutions of $y' = 4\sqrt{y}$, $y(2) = 0$. Does $4\sqrt{y}$ satisfy a Lipschitz condition?
16. Show that $f(x, y) = |\sin y| + x$ satisfies a Lipschitz condition (4), with $M = 1$, on the whole xy-plane, but $\partial f/\partial y$ does not exist when $y = 0$.
17. Does $f(x, y) = |x| + |y|$ satisfy a Lipschitz condition in the xy-plane? Does $\partial f/\partial y$ exist?
18. Does $f(x, y) = x\,|y|$ in Prob. 8 satisfy a Lipschitz condition in a rectangle? Does $\partial f/\partial y$ exist?
19. **(Linear differential equation)** Write $y' + p(x)y = r(x)$ in the form (1). If p and r are continuous for all x such that $|x - x_0| \leqq a$, show that $f(x, y)$ in this equation satisfies the conditions of our present existence and uniqueness theorems, so that a corresponding initial value problem has a unique solution. [This also follows directly from (4) in Sec. 1.7, so that for the *linear* differential equation, we do not need these theorems.]
20. Show that in Prob. 19 a Lipschitz condition holds.

Review Questions and Problems for Chapter 1

1. What is the difference between an ordinary and a partial differential equation?
2. How many arbitrary constants does a general solution of a first-order differential equation contain? How many conditions are needed to determine these constants?
3. What do we mean by a particular solution of a first-order differential equation? By a singular solution?
4. What gives us the right to talk about *the* general solution of a first-order differential equation?
5. Does every first-order differential equation have a solution? A general solution? (Give reason.)
6. What do we mean by exponential growth and exponential decay? What is the corresponding differential equation? In what applications does it occur?
7. Why do electric circuits lead to differential equations?
8. Make a list of some of the problems in mechanics in the text and problem sets.
9. What are orthogonal trajectories? Why do they lead to differential equations? In what practical problems do they occur?
10. What is a direction field? What is it good for?

Find the general solution, using one of the methods discussed in this chapter.

11. $(2xe^{2xy} + \cos y)y' + 2ye^{2xy} = 0$
12. $xy' = (y - x)^3 + y$
13. $y' = 2x(y + x^2 - 1)$
14. $y' = x^3y^2 + xy$
15. $y' = (1 + x)(1 + y^2)$
16. $x^2y' + 2xy = \sinh 3x$
17. $y' + xy = xy^{-1}$
18. $y' + y \cos x = \cos x$
19. $2xyy' + 3y^2 + 4x = 0$
20. $(1 + xy)y' + y^2 = 0$
21. $2x \tan y \, dx + \sec^2 y \, dy = 0$
22. $(x^2 + 1)y' + y^2 + 1 = 0$
23. $y' - x^{-1}y + x^{-1}(x - y)^3 = 0$
24. $(1 + x^2) \, dy = -2xy \, dx$
25. $y' = (y - x)/(y + x)$

Solve the following initial value problems.

26. $y' = (1 - x)/(1 + y), \quad y(1) = 0$
27. $4x \, dx + 9y \, dy = 0, \quad y(3) = 0$
28. $3x^{-1}y^2 \, dx - 2y \, dy = 0, \quad y(4) = 8$
29. $x^2y' + 2xy - x + 1 = 0, \quad y(1) = 0$
30. $xy' - 3y = x^4(e^x + \cos x) - 2x^2, \quad y(\pi) = \pi^3e^\pi + 2\pi^2$
31. $y' - y \cot x = 2x - x^2 \cot x, \quad y(\tfrac{1}{2}\pi) = \tfrac{1}{4}\pi^2 + 1$
32. $xy' + y = x^2y^2, \quad y(0.5) = 0.5$
33. $(2x + e^y) \, dx + xe^y \, dy = 0, \quad y(1) = 0$
34. $y' + \csc y = 0, \quad y(4.1) = -6.3$
35. $(2x + y^4)y' = y, \quad y(24) = 2$

36. Apply Picard's method to $y' = xy$, $y(0) = 2$. Sketch y_1, y_2, y_3, and the exact solution for $-2 \leqq x \leqq 2$. Compute $y_1(1)$, $y_2(1)$, $y_3(1)$ and compare the values with the exact value $y(1) = 2\sqrt{e} \approx 3.297$.

37. Perform the same tasks as in Prob. 36 for the problem $y' = 2xy$, $y(0) = 1$. [Here, $y(1) = e = 2.718 \cdots$.]

Find the orthogonal trajectories. Sketch some of the curves and their trajectories.

38. $y = 1 - cx^2$ **39.** $(x - 1)(y - 1) = c$ **40.** $2x^2 + 3y^2 = c$

41. Uranium ${}_{92}U^{232}$ is radioactive, with half-life $H = 74$ years. What percentage will disappear within 1 year? Within 10 years? Within 37 years?

42. Show that the half-life H of a radioactive substance can be determined from two measurements $y_1 = y(t_1)$ and $y_2 = y(t_2)$ of the amounts present at times t_1 and t_2 by the formula $H = (t_2 - t_1)(\ln 2)/\ln(y_1/y_2)$.

43. If the growth rate of a culture of bacteria is proportional to the number of bacteria present and after 1 day is 1.1 times the original number, within what interval of time will the number of bacteria (a) double, (b) triple?

44. A metal bar whose temperature is 25°C is placed in boiling water. How long does it take to heat the bar to 99°C, if the temperature of the bar after 1 min of heating is 26°C? First guess, then calculate.

45. **(Evaporation)** Two liquids are boiling in a vessel. It is found that the ratio of the quantities of each passing off as vapor at any instant is proportional to the ratio of the quantities x and y still in the liquid state. Show that $y' = ky/x$ and solve this differential equation.

46. If in an RL-circuit, $R = 12$ ohms, $L = 0.02$ millihenry, $E = E_0 = const$, and $I(0) = 0$, how long does it take the current $I(t)$ to reach practically (theoretically 99.9% of) its final value? Guess first.

47. If $L = 10$ in Prob. 46, what R should we choose for $I(t)$ to reach practically (theoretically 99% of) its final value after 1 sec? Guess first.

48. **(Heart pacemaker)** Figure 30 shows a heart pacemaker consisting of a capacitor of capacitance C, a battery of voltage E_0, and a switch that is periodically moved from A (charging period $t_1 < t < t_2$ of the capacitor) to B (discharging period $t_2 < t < t_3$ during which the capacitor sends an electric stimulus to the heart, which acts as a resistor of resistance R). Find the current during the discharging period, assuming that the charging period is such that the charge on the capacitor is 99% of its maximum possible value.

49. **(Friction)** If a body slides on a surface, it experiences friction F (a force against the direction of motion). Experiments show that $|F| = \mu|N|$ (*Coulomb's*[22] *law of kinetic friction without lubrication*), where N is the normal force (force that holds the two surfaces together; see Fig. 31) and the constant of proportionality μ is called the *coefficient of kinetic friction*. In Fig. 31, assume that the body weighs 45 nt (about 10 lb; see front cover for conversion), $\mu = 0.20$

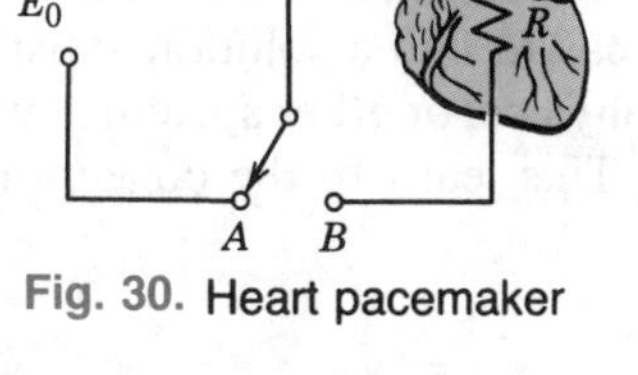

Fig. 30. Heart pacemaker

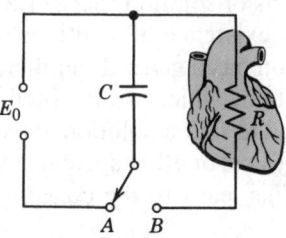

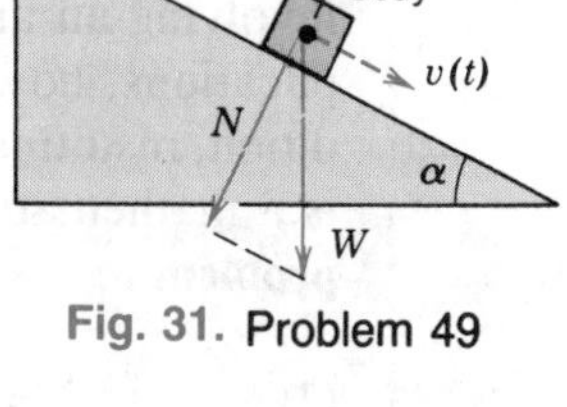

Fig. 31. Problem 49

[22] See footnote 15 in Sec. 1.8.

(corresponding to steel on steel), $\alpha = 30°$, the slide is 10 meters long, the initial velocity is zero, and air resistance is negligible. Find the velocity of the body at the end of the slide.

50. If the isotherms in a body of homogeneous material are $T(x, y) = y - x^3 = const$, what are the curves of heat flow?

51. The **law of mass action** states that under constant temperature the velocity of a chemical reaction is proportional to the product of the concentrations of the reacting substances. A bimolecular reaction $A + B \rightarrow M$ combines a moles per liter of a substance A and b moles of a substance B. If $y(t)$ is the number of moles per liter that have reacted after time t, the rate of reaction is $dy/dt = k(a - y)(b - y)$. Solve this equation, assuming that $a \neq b$.

52. In a room containing 10,000 cubic feet of air, 200 cubic feet of fresh air and 200 cubic feet of polluted air flow in per minute and the mixture (made practically uniform by circulating fans) is exhausted at the rate of 400 cubic feet per minute. Find the amount of fresh air $y(t)$ [cubic feet] at any time, assuming that $y(0) = 0$.

53. The solution curve of $dr/d\theta + (a^2/r) \sin 2\theta = 0$, $r^2(0) = a^2$, $a \neq 0$, is called a **lemniscate.** Find and sketch it. (r and θ are polar coordinates.)

54. Find all curves such that for every tangent, the segment between the coordinate axes is bisected by the point of tangency.

55. Find all curves such that the tangent at each point (x, y) intersects the x-axis at $(x - 1, 0)$.

Summary of Chapter 1 First-Order Differential Equations

This chapter concerns **first-order differential equations** and their applications; these are equations

(1) $F(x, y, y') = 0$ or in explicit form $y' = f(x, y)$

involving the derivative y' of an unknown function y, given functions of x, and, perhaps, y itself. We started with the basic concepts (Sec. 1.1), then considered methods of solution (Secs. 1.2–1.9), and, finally, ideas on approximation and existence of solutions (Secs. 1.10, 1.11).

Ordinarily such an equation has a **general solution,** that is, a solution involving an arbitrary constant, which we denoted by c. In most applications, however, one has to find a solution satisfying a given condition, resulting from the physical or other system of which the equation is a mathematical model. This leads to the concept of an **initial value problem**

(2) $y' = f(x, y)$, $y(x_0) = y_0$ (x_0, y_0 given numbers)

in which the **initial condition** $y(x_0) = y_0$ is used to determine a **particular solution,** that is, a solution obtained from the general solution by specifying a value of c. Geometrically, a general solution represents a family of curves, and each particular solution corresponds to a curve of this family.

Perhaps the simplest equations are the **separable equations,** those we can put into the form $g(y)\,dy = f(x)\,dx$, so that we can solve them by integrating on both sides (Secs. 1.2, 1.3). This method can be extended to certain other equations by first applying suitable substitutions (Sec. 1.4).

An **exact equation**

$$M(x, y)\,dx + N(x, y)\,dy = 0$$

is one for which $M\,dx + N\,dy$ is an exact differential

$$du = \frac{\partial u}{\partial x}\,dx + \frac{\partial u}{\partial y}\,dy,$$

so that we get the implicit solution $u(x, y) = c$ (Sec. 1.5). This method extends to nonexact equations that can be made exact by multiplying them by some function F, called an **integrating factor** (Sec. 1.6).

Linear equations (Sec. 1.7)

$$(3) \qquad y' + p(x)y = r(x)$$

are of that type. For them an integrating factor is $F(x) = \exp\left(\int p(x)\,dx\right)$ and leads to the solution formula (4), Sec. 1.7. Certain nonlinear equations can be reduced to linear form by substituting new variables. This holds for the **Bernoulli equation** $y' + p(x)y = g(x)y^a$ (Sec. 1.7).

Applications are included at various places. Sections entirely devoted to applications are 1.3 on separable equations, 1.8 on linear equations applied to **electrical circuits,** and 1.9 on **orthogonal trajectories,** that is, curves that intersect given curves at right angles.

Direction fields (Sec. 1.10) help in sketching families of solution curves, for instance, in order to gain an impression of their general behavior.

Picard's iteration method (Sec. 1.10) gives approximate solutions of initial value problems by iteration. It is important as the theoretical basis of **Picard's existence and uniqueness theorems** (Sec. 1.11).

Chapter 2

Second-Order Linear Differential Equations

The ordinary differential equations may be divided into two large classes, namely, **linear equations** and **nonlinear equations.** Whereas nonlinear equations are difficult in general, linear equations are much simpler because properties of their solutions can be characterized in a general way and standard methods are available for solving many of these equations.

In this chapter we consider second-order linear differential equations, **homogeneous** equations in Secs. **2.1–2.7** and **nonhomogeneous** equations in Secs. **2.8–2.13.** There are two main reasons for concentrating on equations of *second* order. First, they have important applications in mechanics (Secs. 2.5, 2.11) and in electric circuit theory (Sec. 2.12), which make them more important than higher order linear equations. Second, their theory is typical of that of linear differential equations of any order (but with simpler formulas than in higher order cases), so that the transition to higher order equations (in the next chapter) involves only very few new ideas (although an increase in technical difficulties).

Numerical methods for solving second-order differential equations are included in Sec. 20.3, which is independent of the other sections in Chap. 20.

(Legendre's, Bessel's, and the hypergeometric equations will be considered in Chap. 5.)

Prerequisite for this chapter: Chap. 1, in particular Sec. 1.7.
Sections that may be omitted in a shorter course: 2.4, 2.7, 2.10, 2.12, 2.13.
References: Appendix 1, Part A.
Answers to problems: Appendix 2.

2.1 Homogeneous Linear Equations

The student has already met with linear differential equations of the first order (Secs. 1.7, 1.8), and we shall now define and consider linear differential equations of the second order.

Linear Differential Equation of the Second Order

A second-order differential equation is called **linear** if it can be written

$$y'' + p(x)y' + q(x)y = r(x) \tag{1}$$

and **nonlinear** if it cannot be written in this form.

The characteristic feature of Eq. (1) is that it is linear in the unknown function y and its derivatives, whereas p and q as well as r on the right may be any given functions of x. If the first term is, say, $f(x)y''$, we have to divide by $f(x)$ to get the "standard form" (1), with y'' as the first term, which is practical.

If $r(x) \equiv 0$ (that is, $r(x) = 0$ for all x considered), then (1) becomes simply

$$y'' + p(x)y' + q(x)y = 0 \tag{2}$$

and is called **homogeneous.** If $r(x) \not\equiv 0$, then (1) is called **nonhomogeneous.** This is similar to Sec. 1.7.

The functions p and q in (1) and (2) are called the **coefficients** of the equations.

An example of a nonhomogeneous linear differential equation is

$$y'' + 4y = e^{-x} \sin x.$$

An example of a homogeneous linear equation is

$$(1 - x^2)y'' - 2xy' + 6y = 0.$$

Examples of nonlinear differential equations are

$$x(y''y + y'^2) + 2y'y = 0$$

and

$$y'' = \sqrt{y'^2 + 1}.$$

We shall always suppose that x varies on some open interval I, and all our assumptions and statements will refer to such an I, which we need not specify in each case. (Recall from footnote 1 in Sec. 1.1 that I may be the entire x-axis.)

A **solution** of a second-order (linear or nonlinear) differential equation on some open interval $a < x < b$ is a function $y = h(x)$ that has derivatives $y' = h'(x)$ and $y'' = h''(x)$ and satisfies that differential equation for all x in that interval; that is, the equation becomes an identity if we replace the unknown function y and its derivatives by h and its corresponding derivatives.

Second-order linear differential equations have many basic applications, as we shall see. Some of them are very simple, their solutions being familiar functions from calculus. Others are more involved, their solutions being higher functions (for instance, Bessel functions) occurring in engineering problems.

Homogeneous Equations: Superposition or Linearity Principle

We now begin our discussion of second-order homogeneous linear differential equations (Secs. 2.1–2.7). Nonhomogeneous equations follow in Secs. 2.8–2.13.

EXAMPLE 1 **Solutions of a homogeneous linear differential equation**

$y = e^x$ and $y = e^{-x}$ are solutions of the homogeneous linear differential equation

$$y'' - y = 0$$

for all x because for $y = e^x$ we get $(e^x)'' - e^x = e^x - e^x = 0$ and similarly for $y = e^{-x}$, as the student should verify.

We can even go an important step further. We can multiply e^x and e^{-x} by different constants, say, -3 and 8 (or any other numbers) and then take the sum

$$y = -3e^x + 8e^{-x}$$

and verify that this is another solution of our homogeneous equation for all x because

$$(-3e^x + 8e^{-x})'' - (-3e^x + 8e^{-x}) = -3e^x + 8e^{-x} - (-3e^x + 8e^{-x}) = 0.$$ ∎

This example illustrates the very important fact that for a ***homogeneous linear*** equation (2), we can always obtain new solutions from known solutions by multiplication by constants and by addition. Of course, this is of great advantage because in this way we can get further solutions from given ones. Now from y_1 $(= e^x)$ and y_2 $(= e^{-x})$ we got a function of the form

$$y = c_1y_1 + c_2y_2 \qquad (c_1, c_2 \text{ arbitrary constants}). \tag{3}$$

This is called a **linear combination** of y_1 and y_2. Using this concept, we can now formulate the result suggested by our example, often called the **superposition principle** or **linearity principle.**

Theorem 1 **Fundamental Theorem 1 for the homogenous equation (2)**

For a homogeneous linear differential equation (2), *any linear combination of two solutions on an open interval I is again a solution of* (2) *on I. In particular, for such an equation, sums and constant multiples of solutions are again solutions.*

Proof. Let y_1 and y_2 be solutions of (2) on I. Then by substituting $y = c_1y_1 + c_2y_2$ and its derivatives into (2), and using the familiar rule $(c_1y_1 + c_2y_2)' = c_1y_1' + c_2y_2'$, etc., we get

$$\begin{aligned} y'' + py' + qy &= (c_1y_1 + c_2y_2)'' + p(c_1y_1 + c_2y_2)' + q(c_1y_1 + c_2y_2) \\ &= c_1y_1'' + c_2y_2'' + p(c_1y_1' + c_2y_2') + q(c_1y_1 + c_2y_2) \\ &= c_1(y_1'' + py_1' + qy_1) + c_2(y_2'' + py_2' + qy_2) = 0, \end{aligned}$$

since in the last line, $(\cdots) = 0$ because y_1 and y_2 are solutions, by assumption. This shows that y is a solution of (2) on I. ∎

Caution! Always remember this highly important theorem, but don't forget that it ***does not hold*** for *nonhomogeneous linear equations* or *nonlinear equations*, as the following two examples illustrate.

EXAMPLE 2 A nonhomogeneous linear differential equation

Substitution shows that the functions $y = 1 + \cos x$ and $y = 1 + \sin x$ are solutions of the nonhomogeneous linear differential equation

$$y'' + y = 1,$$

but the following functions are *not* solutions of this differential equation:

$$2(1 + \cos x) \qquad \text{and} \qquad (1 + \cos x) + (1 + \sin x).$$

■

EXAMPLE 3 A nonlinear differential equation

Substitution shows that the functions $y = x^2$ and $y = 1$ are solutions of the nonlinear differential equation

$$y''y - xy' = 0,$$

but the following functions are *not* solutions of this differential equation:

$$-x^2 \qquad \text{and} \qquad x^2 + 1.$$

■

Initial Value Problem. General Solution. Basis

For a *first-order* differential equation, a general solution involved one arbitrary constant c, and in an initial value problem we used one initial condition $y(x_0) = y_0$ to get a particular solution in which c had a definite value. The idea of a general solution was to get all possible solutions, and we know that for *linear* equations (Sec. 1.7) we succeeded (because those have no singular solutions). We now extend this idea to second-order equations:

For second-order homogeneous linear equations (2), a general solution will be of the form

$$y = c_1y_1 + c_2y_2, \tag{4}$$

a linear combination of two solutions involving two arbitrary constants c_1, c_2. An **initial value problem** now consists of Eq. (2) and two **initial conditions**

$$y(x_0) = K_0, \qquad y'(x_0) = K_1, \tag{5}$$

prescribing values K_0 and K_1 of the solution and its derivative (slope of the curve) at the same given x_0 in the open interval considered. We shall use (5) to get from (4) a particular solution of (2), in which c_1 and c_2 have definite values.

Let us illustrate this by a simple example, which will also help us to see that we have to impose a condition on y_1 and y_2 in (4).

EXAMPLE 4 **Initial value problem**

Solve the initial value problem

$$y'' - y = 0, \qquad y(0) = 5, \qquad y'(0) = 3.$$

Solution. ***1st Step.*** e^x and e^{-x} are solutions (by Example 1), and and we take

$$y = c_1e^x + c_2e^{-x}.$$

(This will turn out to be a general solution as defined below.)

2nd Step. From the initial conditions, since $y' = c_1e^x - c_2e^{-x}$, we get

$$y(0) = c_1 + c_2 = 5$$

$$y'(0) = c_1 - c_2 = 3.$$

Hence $c_1 = 4$, $c_2 = 1$. *Answer.* $y = 4e^x + e^{-x}$.

Observation. Had we taken $y_1 = e^x$ and $y_2 = le^x$ and thus got

$$y = c_1e^x + c_2le^x = (c_1 + c_2l)e^x = y',$$

our solution would not have been general enough to satisfy the two initial conditions and solve the problem. Why? Well, our present y_1 and y_2 are proportional, $y_1/y_2 = 1/l$, whereas before they were not, $y_1/y_2 = e^x/e^{-x} = e^{2x}$. This is the point. It motivates the following definition as well as its importance in connection with initial value problems. ∎

Definition (General solution, basis, particular solution)

A **general solution** of an equation (2) on an open interval[1] I is a solution (4) with y_1 and y_2 not proportional solutions of (2) on I and c_1, c_2 arbitrary[2] constants. These y_1, y_2 are then called a **basis** (or **fundamental system**) of (2) on I.

A **particular solution** of (2) on I is obtained if we assign specific values to c_1 and c_2 in (4). ∎

As usual, y_1 and y_2 are called *proportional* on I if[3]

$$\text{(6)} \qquad \text{(a)}\ \ y_1 = ky_2 \qquad \text{or} \qquad \text{(b)}\ \ y_2 = ly_1$$

holds for all x on I, where k and l are numbers, zero or not.

Actually, we can also formulate our definition of a basis in terms of "linear independence." We call two functions $y_1(x)$ and $y_2(x)$ **linearly independent** on an interval I where they are defined if

$$\text{(7)} \qquad k_1y_1(x) + k_2y_2(x) = 0 \quad \text{on } I \text{ implies} \quad k_1 = 0, \ \ k_2 = 0,$$

and we call them **linearly dependent** on I if this equation also holds for some

[1]See Sec. 1.1.

[2]The range of the constants may have to be restricted in some cases to avoid imaginary expressions or other degeneracies.

[3]If $k \neq 0$ in (a), it implies (b) with $l = 1/k$, but not when $k = 0$, that is, $y_1 = 0$.

constants k_1, k_2 not both zero. Then, if $k_1 \neq 0$ or $k_2 \neq 0$, we can divide and solve, obtaining

$$y_1 = -\frac{k_2}{k_1} y_2 \quad \text{or} \quad y_2 = -\frac{k_1}{k_2} y_1.$$

Hence then y_1 and y_2 are proportional, whereas in the case of linear independence, they are not proportional. We thus have the following

Definition of a basis (Reformulated)

A **basis of solutions** of (2) on an interval I is a pair y_1, y_2 of linearly independent solutions of (2) on I. ∎

EXAMPLE 5 **Basis, general solution, particular solution**

e^x and e^{-x} in Example 4 form a basis of the differential equation $y'' - y = 0$ for all x. Hence a general solution is $y = c_1 e^x + c_2 e^{-x}$. The answer in Example 4 is a particular solution of the equation. ∎

EXAMPLE 6 **Basis, general solution**

The student may verify that $y_1 = \cos x$ and $y_2 = \sin x$ are solutions of the differential equation

$$y'' + y = 0.$$

Now $\cos x$ and $\sin x$ are not proportional, $y_1/y_2 = \cot x \neq const.$ Hence they form a basis of our equation for all x, and a general solution is

$$y = c_1 \cos x + c_2 \sin x.$$ ∎

In practice, one mainly uses a general solution to get particular solutions from it, by imposing two initial conditions (5), because it is the *particular solution* that describes the unique behavior of a given physical or other system. Being interested in first gaining experience in that practical task, we give the underlying theory afterwards (in Sec. 2.7); for the time being, it suffices to know the following. If the coefficients p and q of (1) and the function r are continuous on some interval I, then (1) always has a general solution on I, from which one obtains the solution of any initial value problem (1), (5) on I, which is unique. Also, (1) does not have **singular solutions** (i.e., solutions not obtainable from a general solution).

Problem Set 2.1

Important general properties of homogeneous and nonhomogeneous linear differential equations. Prove the following statements, which refer to any fixed open interval I, and illustrate them with examples. Here we assume that $r(x) \not\equiv 0$ in (1).

1. $y \equiv 0$ is a solution of (2) (called the "*trivial solution*") but not of (1).

2. The sum of two solutions of (1) is **not** a solution of (1).

3. A multiple $y = cy_1$ of a solution of (1) is **not** a solution of (1), unless $c = 1$.

4. The sum $y = y_1 + y_2$ of a solution y_1 of (1) and y_2 of (2) is a solution of (1).

5. The difference $y = y_1 - y_2$ of two solutions of (1) is a solution of (2).

Second-order differential equations reducible to the first order

Problems 6–19 illustrate that certain second-order differential equations can be reduced to the first order.

6. If in a second-order equation the dependent variable y does not appear explicitly, the equation is of the form $F(x, y', y'') = 0$. Show that by setting $y' = z$ we obtain a first-order differential equation in z and from its solution the solution of the original equation by integration.

Reduce to the first order and solve:

7. $xy'' = 2y'$ **8.** $y'' = y'$ **9.** $y'' = 2y' \coth 2x$
10. $y'' + 9y' = 0$ **11.** $xy'' + y' = y'^2$ **12.** $xy'' + y' = 0$

13. Another type of equation reducible to first order is $F(y, y', y'') = 0$, in which the independent variable x does not appear explicitly. Using the chain rule, show that $y'' = (dz/dy)z$, where $z = y'$, so that we obtain a first-order equation with y as the *independent* variable.

Reduce to first order and solve:

14. $yy'' + y'^2 = 0$ **15.** $y'' + e^{2y}y'^3 = 0$ **16.** $y'' + y'^2 = 0$
17. $y'' + y'^3 \cos y = 0$ **18.** $yy'' - y'^2 = 0$ **19.** $y'' + (1 + y^{-1})y'^2 = 0$

20. A particle moves on a straight line so that the product of its velocity and acceleration is constant, say, 2 meters²/sec³. At time $t = 0$ its displacement from the origin is 5 meters and its velocity is zero. Find its position and velocity when $t = 9$ sec.

21. Find the curve in the xy-plane which passes through the point (1, 1), intersects the line $y = x$ at a right angle, and satisfies $xy'' + 2y' = 0$.

22. **(Hanging cable)** It can be shown that the curve $y(x)$ of an inextensible flexible homogeneous cable hanging between two fixed points is obtained by solving $y'' = k\sqrt{1 + y'^2}$, where the constant k depends on the weight. This curve is called a *catenary* (from Latin *catena* = the chain). Find and graph $y(x)$, assuming $k = 1$ and those fixed points are $(-1, 0)$ and $(1, 0)$ in a vertical xy-plane.

Initial value problems

Verify that the given functions form a basis of solutions of the given equation and solve the given initial value problem.

23. $y'' - 9y = 0,\quad y(0) = 2,\quad y'(0) = 0;\quad e^{3x}, e^{-3x}$
24. $y'' + 4y = 0,\quad y(0) = -5,\quad y'(0) = 2;\quad \cos 2x, \sin 2x$
25. $y'' - 2y' + y = 0,\quad y(0) = 4,\quad y'(0) = 3;\quad e^x, xe^x$
26. $y'' + \omega^2 y = 0\ (\omega \neq 0),\quad y(0) = -1.5,\quad y'(0) = 0;\quad \cos \omega x, \sin \omega x$
27. $y'' - 4y' + 3y = 0,\quad y(0) = -1,\quad y'(0) = -5;\quad e^x, e^{3x}$
28. $x^2y'' - 1.5xy' - 1.5y = 0,\quad y(1) = 1,\quad y'(1) = -4;\quad x^{-1/2}, x^3$
29. $x^2y'' - 3xy' + 4y = 0,\quad y(1) = 2,\quad y'(1) = 5;\quad x^2, x^2 \ln x$
30. $y'' - 2y' + 10y = 0,\quad y(0) = 1,\quad y'(0) = 10;\quad e^x \cos 3x, e^x \sin 3x$

Are the following functions linearly dependent or independent on the given interval?

31. $x + 1, x - 1 \quad (0 < x < 1)$

32. $0, \cosh x$, any interval

33. $\sin 2x, \sin x \cos x$, any interval

34. $\ln x, \ln x^2 \quad (x > 1)$

35. $|x|x, x^2 \quad (0 < x < 1)$

36. $|x|x, x^2 \quad (-1 < x < 1)$

37. $|\cos x|, \cos x \quad (0 < x < \pi)$

38. $x^2 - 3, -3x^2 + 9 \quad (x < 0)$

39. $\cosh x, \cosh 2x$, any interval

40. $1, e^{-2x} \quad (x > 0)$

2.2 Homogeneous Equations with Constant Coefficients

In this section and the next one, we show how to solve homogeneous linear equations

$$y'' + ay' + by = 0 \tag{1}$$

whose coefficients a and b are constant. These equations have important applications, especially in connection with mechanical and electrical vibrations, as we shall see in Secs. 2.5, 2.11 and 2.12.

To solve (1), we remember from Sec. 1.7 that a *first-order* linear differential equation $y' + ky = 0$ with constant coefficient k has an exponential function as solution, $y = e^{-kx}$. This gives us the idea to try as a solution of (1) the function

$$y = e^{\lambda x}. \tag{2}$$

Substituting (2) and the derivatives

$$y' = \lambda e^{\lambda x} \qquad \text{and} \qquad y'' = \lambda^2 e^{\lambda x}$$

into our equation (1), we obtain

$$(\lambda^2 + a\lambda + b)e^{\lambda x} = 0.$$

Hence (2) is a solution of (1), if λ is a solution of the quadratic equation

$$\lambda^2 + a\lambda + b = 0. \tag{3}$$

This equaton is called the **characteristic equation** (or *auxiliary equation*) of (1). Its roots are

$$\lambda_1 = \tfrac{1}{2}(-a + \sqrt{a^2 - 4b}), \qquad \lambda_2 = \tfrac{1}{2}(-a - \sqrt{a^2 - 4b}). \tag{4}$$

Our derivation shows that the functions

$$y_1 = e^{\lambda_1 x} \quad \text{and} \quad y_2 = e^{\lambda_2 x} \tag{5}$$

are solutions of (1). The student may check this by substituting (5) into (1).

Directly from (4) we see that, depending on the sign of the discriminant $a^2 - 4b$, we obtain

(Case I) *two real roots if* $a^2 - 4b > 0$,
(Case II) *a real double root if* $a^2 - 4b = 0$,
(Case III) *complex conjugate roots if* $a^2 - 4b < 0$.

We discuss Cases I and II now and Case III in the next section.

Case I. Two Distinct Real Roots λ_1 and λ_2

In this case,

$$y_1 = e^{\lambda_1 x} \quad \text{and} \quad y_2 = e^{\lambda_2 x}$$

constitute a basis of solutions of (1) on any interval (because y_1/y_2 is not constant; see Sec. 2.1). The corresponding general solution is

$$y = c_1 e^{\lambda_1 x} + c_2 e^{\lambda_2 x}. \tag{6}$$

EXAMPLE 1 General solution in the case of distinct real roots

We can now solve $y'' - y = 0$ (in Example 1 of Sec. 2.1) in a systematic fashion. The characteristic equation is $\lambda^2 - 1 = 0$. Its roots are $\lambda_1 = 1$ and $\lambda_2 = -1$. Hence a basis is e^x and e^{-x} and, as before, gives the general solution

$$y = c_1 e^x + c_2 e^{-x}.$$

■

EXAMPLE 2 An initial value problem in the case of distinct real roots

Solve the initial value problem

$$y'' + y' - 2y = 0, \qquad y(0) = 4, \qquad y'(0) = -5.$$

Solution. 1st Step. General solution. The characteristic equation is $\lambda^2 + \lambda - 2 = 0$. Its roots are

$$\lambda_1 = \tfrac{1}{2}(-1 + \sqrt{9}) = 1 \quad \text{and} \quad \lambda_2 = \tfrac{1}{2}(-1 - \sqrt{9}) = -2,$$

so that we obtain the general solution

$$y = c_1 e^x + c_2 e^{-2x}.$$

2nd Step. Particular solution. Since $y'(x) = c_1 e^x - 2c_2 e^{-2x}$, we obtain from the general solution and the initial conditions

$$y(0) = c_1 + c_2 = 4,$$
$$y'(0) = c_1 - 2c_2 = -5.$$

Hence $c_1 = 1$ and $c_2 = 3$. *Answer.* $y = e^x + 3e^{-2x}$.

■

Case II. Real Double Root $\lambda = -a/2$

When the discriminant $a^2 - 4b = 0$, then (4) gives only one root $\lambda = \lambda_1 = \lambda_2 = -a/2$ and we get at first only one solution

$$y_1 = e^{-(a/2)x}$$

To find a second solution, needed for a basis, we use the "*method of reduction of order.*"[4] That is, we set

$$y_2 = uy_1$$

and try to determine the function u such that y_2 becomes a solution of (1). For this, we substitute $y_2 = uy_1$ and its derivatives

$$y_2' = u'y_1 + uy_1'$$

$$y_2'' = u''y_1 + 2u'y_1' + uy_1''$$

into (1). This gives

$$(u''y_1 + 2u'y_1' + uy_1'') + a(u'y_1 + uy_1') + buy_1 = 0.$$

Collecting terms, we obtain

$$u''y_1 + u'(2y_1' + ay_1) + u(y_1'' + ay_1' + by_1) = 0$$

The expression in the last parentheses is zero, since y_1 is a solution of (1). The expression in the first parentheses is zero, too, since

$$2y_1' = -ae^{-ax/2} = -ay_1.$$

We are thus left with $u''y_1 = 0$. Hence $u'' = 0$. By two integrations, $u = c_1x + c_2$. To get a second independent solution $y_2 = uy_1$, we can simply take $u = x$. Then $y_2 = xy_1$. Since these solutions are not proportional, they form a basis. Our result is that in the case of a double root of (3) a basis of solutions of (1) on any interval is

$$e^{-ax/2}, \qquad xe^{-ax/2}.$$

The corresponding general solution is

$$y = (c_1 + c_2x)e^{-ax/2}. \tag{7}$$

Warning. If λ is a *simple* root of (4), then $(c_1 + c_2x)e^{\lambda x}$ is ***not*** a solution of (1).

---

[4]Actually, we use a special case of it. We discuss the general method in Sec. 2.7.

EXAMPLE 3 General solution in the case of a double root

Solve

$$y'' + 8y' + 16y = 0.$$

Solution. The characteristic equation has the double root $\lambda = -4$. Hence a basis is

$$e^{-4x} \quad \text{and} \quad xe^{-4x}$$

and the corresponding general solution is

$$y = (c_1 + c_2 x)e^{-4x}.$$

EXAMPLE 4 An initial value problem in the case of a double root

Solve the initial value problem

$$y'' - 4y' + 4y = 0, \qquad y(0) = 3, \qquad y'(0) = 1.$$

Solution. A general solution of the differential equation is

$$y(x) = (c_1 + c_2 x)e^{2x}.$$

By differentiation we obtain

$$y'(x) = c_2 e^{2x} + 2(c_1 + c_2 x)e^{2x}.$$

From this and the initial conditions it follows that

$$y(0) = c_1 = 3, \qquad y'(0) = c_2 + 2c_1 = 1.$$

Hence $c_1 = 3$, $c_2 = -5$, and the answer is

$$y = (3 - 5x)e^{2x}.$$

The remaining Case III of **complex conjugate roots** of the characteristic equation (4) will be discussed in the next section.

Problem Set 2.2

Find a general solution of the following differential equations.

1. $y'' + 3y' + 2y = 0$ **2.** $y'' - 9y = 0$ **3.** $y'' + 10y' + 25y = 0$
4. $y'' + 4y' = 0$ **5.** $y'' - 6y' + 9y = 0$ **6.** $y'' - 2y' + 0.75y = 0$

Find the differential equation (1) for which the given functions form a basis of solutions.

7. e^{2x}, e^{-3x} **8.** $e^{-\pi x}, xe^{-\pi x}$ **9.** $e^{-2x}, e^{-x/2}$
10. e^{kx}, e^{-kx} **11.** $\cosh x, \sinh x$ **12.** $1, e^{-kx}$

13. Verify directly that in the case of a double root, $xe^{\lambda x}$ with $\lambda = -a/2$ is a solution of (1).

14. Verify that $y = e^{-3x}$ is a solution of $y'' + 5y' + 6y = 0$, but $y = xe^{-3x}$ is not. Explain.

15. Show that a and b in (1) can be expressed in terms of λ_1 and λ_2 by the formulas $a = -\lambda_1 - \lambda_2$, $b = \lambda_1\lambda_2$.

16. Solve $y'' + 3y' = 0$ (a) by the present method, (b) by reduction to the first order.

Solve the following initial value problems.

17. $y'' - 16y = 0, \quad y(0) = 1, \quad y'(0) = 20$

18. $y'' - 4y' + 4y = 0, \quad y(0) = 0, \quad y'(0) = -3$

19. $y'' + 6y' + 9y = 0, \quad y(0) = -4, \quad y'(0) = 14$

20. $y'' + 3.7y' = 0, \quad y(-2) = 4, \quad y'(-2) = 0$

21. $y'' + 2.2y' + 0.4y = 0, \quad y(0) = 3.3, \quad y'(0) = -1.2$

22. $y'' - 25y = 0, \quad y(0) = 0, \quad y'(0) = 10$

23. $4y'' - 4y' - 3y = 0, \quad y(-2) = e, \quad y'(-2) = -\frac{1}{2}e$

24. $5y'' + 16y' + 12.8y = 0, \quad y(0) = 0, \quad y'(0) = -2.3$

25. Different bases lead to the same solution. To illustrate this, solve $y'' - 9y = 0$, $y(0) = 4$, $y'(0) = -6$, using (a) e^{3x}, e^{-3x}, (b) $\cosh 3x$, $\sinh 3x$.

2.3 Case of Complex Roots. Complex Exponential Function

For homogeneous linear differential equations with constant coefficients

$$y'' + ay' + by = 0 \tag{1}$$

we now discuss the remaining case that the characteristic equation

$$\lambda^2 + a\lambda + b = 0 \tag{2}$$

has roots

$$\lambda_1 = -\tfrac{1}{2}a + \tfrac{1}{2}\sqrt{a^2 - 4b}, \qquad \lambda_2 = -\tfrac{1}{2}a - \tfrac{1}{2}\sqrt{a^2 - 4b} \tag{3}$$

which are complex. Equation (3) shows that this happens if the discriminant $a^2 - 4b$ is negative. This is Case III of the last section. In this case it is practical to pull out $\sqrt{-1} = i$ from the root, pull $1/2 = \sqrt{1/4}$ under the root, and write

$$\lambda_1 = -\tfrac{1}{2}a + i\omega, \qquad \lambda_2 = -\tfrac{1}{2}a - i\omega, \tag{4}$$

where $\omega = \sqrt{b - \frac{1}{4}a^2}$. We see that $e^{\lambda_1 x}$ and $e^{\lambda_2 x}$ now are *complex* solutions of (1). But we claim that in this Case III, a basis of *real* solutions of (1) on any interval is

$$y_1 = e^{-ax/2}\cos \omega x, \qquad y_2 = e^{-ax/2}\sin \omega x. \tag{5}$$

Indeed, that these are solutions follows by differentiation and substitution. Also, $y_2/y_1 = \tan \omega x$ is not constant, since $\omega \neq 0$, so that y_1 and y_2 are not proportional. The corresponding general solution is

$$y = e^{-ax/2}(A\cos \omega x + B\sin \omega x). \tag{6}$$

EXAMPLE 1 **Complex roots. General solution (6)**

Find a general solution of the equation

$$y'' - 2y' + 10y = 0.$$

Solution. The characteristic equation $\lambda^2 - 2\lambda + 10 = 0$ has the complex conjugate roots

$$\lambda_1 = 1 + \sqrt{1 - 10} = 1 + 3i, \qquad \lambda_2 = 1 - 3i.$$

(Verify!) This yields the basis (5),

$$y_1 = e^x \cos 3x, \qquad y_2 = e^x \sin 3x$$

and the corresponding general solution (6),

$$y = e^x(A \cos 3x + B \sin 3x).$$ ■

Complex Exponential Function

Case III (complex roots) has been settled, and nothing is left to prove. We just want to see how one can get the *idea* that y_1 and y_2 might be solutions in this case. We show that this follows from the complex exponential function.

The **complex exponential function** e^z of a complex variable[5] $z = s + it$ is defined by

$$e^z = e^{s+it} = e^s(\cos t + i \sin t). \tag{7}$$

For real $z = s$, this becomes the familiar real exponential function e^s of calculus because then $\cos t = \cos 0 = 1$ and $\sin t = \sin 0 = 0$. Also, e^z has properties quite similar to those of the real exponential function; in particular, one can show that it is differentiable and that it satisfies $e^{z_1+z_2} = e^{z_1}e^{z_2}$. (Proofs in Sec. 12.6.) This may suffice to motivate the definition (7).

In (7) we now take $z = \lambda_1 x$ with λ_1 as in (4), that is,

$$z = s + it = \lambda_1 x = -\tfrac{1}{2}ax + i\omega x.$$

Then (7) gives

$$e^{-(a/2)x + i\omega x} = e^{-(a/2)x}(\cos \omega x + i \sin \omega x).$$

Similarly, since $\sin(-\alpha) = -\sin \alpha$, we get for $e^{\lambda_2 x}$ [see (4)]

$$e^{-(a/2)x - i\omega x} = e^{-(a/2)x}(\cos \omega x - i \sin \omega x).$$

Adding these two formulas and dividing the sum by 2, we get on the right y_1 as given in (5). Similarly, by subtracting those formulas and dividing the result by $2i$ we get on the right y_2 as given in (5). From Fundamental Theorem 1 in Sec. 2.1 it follows that y_1 and y_2 are again solutions. This confirms that (6) is a general solution of (1) in Case III (complex roots). ■

[5] We write $z = s + it$ instead of $z = x + iy$ since x and y are used as variable and unknown function in our equations.

For later use we note that (7) with $s = 0$ gives the so-called **Euler formula**

$$e^{it} = \cos t + i \sin t. \tag{8}$$

Further Examples. Summary of Cases I–III

EXAMPLE 2 Initial value problem (Complex roots)

Solve the initial value problem

$$y'' + 2y' + 5y = 0, \qquad y(0) = 1, \qquad y'(0) = 5.$$

Solution. 1st Step. General solution. The characteristic equation $\lambda^2 + 2\lambda + 5 = 0$ has the complex roots $-1 \pm \sqrt{1 - 5} = -1 \pm 2i$. Hence (6) becomes

$$y(x) = e^{-x}(A \cos 2x + B \sin 2x).$$

2nd Step. Particular solution. The first initial condition gives $y(0) = A = 1$. The derivative of the general solution is

$$y'(x) = e^{-x}(-A \cos 2x - B \sin 2x - 2A \sin 2x + 2B \cos 2x),$$

and the second initial condition gives (since $\sin 0 = 0$)

$$y'(0) = -A + 2B = -1 + 2B = 5.$$

Hence $B = 3$. *Answer.* $y = e^{-x}(\cos 2x + 3 \sin 2x)$. ∎

EXAMPLE 3 A general solution of the equation

$$y'' + \omega^2 y = 0 \qquad (\omega \text{ constant, not zero})$$

is

$$y = A \cos \omega x + B \sin \omega x.$$

With $\omega = 1$ this confirms Example 6 in Sec. 2.1. ∎

This completes the discussion of all three cases, and we may sum it up:

Case	Roots of (2)	Basis of (1)	General Solution of (1)
I	Distinct real λ_1, λ_2	$e^{\lambda_1 x}, e^{\lambda_2 x}$	$y = c_1 e^{\lambda_1 x} + c_2 e^{\lambda_2 x}$
II	Real double root $\lambda = -\frac{1}{2}a$	$e^{-ax/2}, xe^{-ax/2}$	$y = (c_1 + c_2 x)e^{-ax/2}$
III	Complex conjugate $\lambda_1 = -\frac{1}{2}a + i\omega$, $\lambda_2 = -\frac{1}{2}a - i\omega$	$e^{-ax/2} \cos \omega x$ $e^{-ax/2} \sin \omega x$	$y = e^{-ax/2}(A \cos \omega x + B \sin \omega x)$

It is very interesting that in applications to mechanical systems or electrical circuits, these three cases correspond to three different forms of motion or flows of current, respectively. We shall discuss this basic relation between theory and practice in detail in Sec. 2.5 (and again in Sec. 2.12).

Boundary Value Problems

Applications sometimes also lead to conditions of the type

$$(9) \qquad y(P_1) = k_1, \qquad y(P_2) = k_2.$$

These are known as **boundary conditions,** since they refer to the endpoints P_1, P_2 (*boundary points* P_1, P_2) of an interval I on which the equation (1) is considered. Equation (1) and conditions (9) together constitute what is known as a **boundary value problem.** We confine ourselves to the discussion of a typical example.

EXAMPLE 4 **Boundary value problem**

Solve the boundary value problem

$$y'' + y = 0, \qquad y(0) = 3, \qquad y(\pi) = -3.$$

Solution. 1st Step. A basis is $y_1 = \cos x$, $y_2 = \sin x$. The corresponding general solution is

$$y(x) = c_1 \cos x + c_2 \sin x.$$

2nd Step. The left boundary condition gives $y(0) = c_1 = 3$. The right boundary condition gives $y(\pi) = c_1 \cos \pi + c_2 \cdot 0 = -3$. Now $\cos \pi = -1$ and $c_1 = 3$, so that this equation holds, and we see that it yields no condition for c_2. Hence a solution of the problem is

$$y = 3 \cos x + c_2 \sin x.$$

Here c_2 is still arbitrary. This is a surprise. Of course, the reason is that $\sin x$ is zero at 0 and π. The reader may conclude and prove (Prob. 30) that the solution of a boundary value problem (1), (9) is unique if and only if no solution $y \neq 0$ of (1) satisfies $y(P_1) = y(P_2) = 0$. ∎

Problem Set 2.3

Verify that the following functions are solutions of the given differential equation and obtain from them a real-valued general solution of the form (6).

1. $y = c_1 e^{3ix} + c_2 e^{-3ix}$, $y'' + 9y = 0$

2. $y = c_1 e^{(-1+3i)x} + c_2 e^{(-1-3i)x}$, $y'' + 2y' + 10y = 0$

3. $y = c_1 e^{-(\alpha+i)x} + c_2 e^{-(\alpha-i)x}$, $y'' + 2\alpha y' + (\alpha^2 + 1)y = 0$

4. $y = c_1 e^{(-5+2i)x} + c_2 e^{(-5-2i)x}$, $y'' + 10y' + 29y = 0$

5. $y = c_1 e^{-(\alpha-i\omega)x} + c_2 e^{-(\alpha+i\omega)x}$, $y'' + 2\alpha y' + (\alpha^2 + \omega^2)y = 0$

General Solution. State whether the given equation corresponds to Case I, Case II, or Case III and find a general solution involving real-valued functions.

6. $y'' + 25y = 0$

7. $y'' - 25y = 0$

8. $y'' - 8y' + 16y = 0$

9. $y'' + 6y' + 9y = 0$

10. $y'' + y' + 0.25y = 0$

11. $y'' + 2y' = 0$

12. $8y'' - 2y' - y = 0$

13. $10y'' + 6y' + 10.9y = 0$

14. $2y'' + 10y' + 25y = 0$

15. $y'' + 2y' + (\omega^2 + 1)y = 0$

Initial Value Problems. Solve the following initial value problems.

16. $y'' - 9y = 0, \quad y(0) = 5, \quad y'(0) = 9$

17. $y'' + 9y = 0, \quad y(\pi) = -2, \quad y'(\pi) = 3$

18. $y'' + 2y' + 2y = 0, \quad y(0) = 1, \quad y'(0) = 0$

19. $y'' - 4y' + 4y = 0, \quad y(0) = 3, \quad y'(0) = 10$

20. $y'' - 6y' + 18y = 0, \quad y(0) = 0, \quad y'(0) = 6$

21. $y'' + 20y' + 100y = 0, \quad y(0.1) = 3.2/e \approx 1.177, \quad y'(0.1) = -30/e \approx -11.04$

22. $10y'' + 2y' + 0.1y = 0, \quad y(10) = -40/e \approx -14.72, \quad y'(10) = 0$

23. $2y'' + y' - y = 0, \quad y(4) = e^2 - e^{-4} \approx 7.371, \quad y'(4) = \frac{1}{2}e^2 + e^{-4} \approx 3.713$

24. $y'' + 4y' + 4.25y = 0, \quad y(0) = 1, \quad y'(0) = -2$

Boundary Value Problems. Solve the following boundary value problems.

25. $y'' - 16y = 0, \quad y(0) = 5, \quad y(\frac{1}{4}) = 5e$

26. $y'' - 9y = 0, \quad y(-4) = y(4) = \cosh 12$

27. $y'' - 2y' = 0, \quad y(0) = -1, \quad y(\frac{1}{2}) = e - 2$

28. $y'' + 4y' + 5y = 0, \quad y(\frac{1}{2}\pi) = 14e^{-\pi} \approx 0.6050, \quad y(\frac{3}{2}\pi) = -14e^{-3\pi} \approx -0.0011$

29. $y'' - 2y' + 2y = 0, \quad y(0) = -3, \quad y(\frac{1}{2}\pi) = 0$

30. Show that the solution of a boundary value problem (1), (9) is unique if and only if no solution $y \neq 0$ of (1) satisfies $y(P_1) = y(P_2) = 0$.

2.4 Differential Operators. *Optional*

This section gives an introduction to differential operators. It will be used only once and for a minor purpose (in Sec. 3.2), so that it can be omitted without interrupting the flow of ideas.

By an **operator** we mean a transformation that transforms a function into another function. Operators and corresponding techniques, which are called **operational methods,** play an increasing role in engineering mathematics.

Differentiation suggests an operator as follows. Let D denote differentiation with respect to x, that is, write

$$Dy = y'.$$

D is an operator; it transforms y (assumed differentiable) into its derivative y'. For example,

$$D(x^2) = 2x, \quad D(\sin x) = \cos x.$$

Applying D twice, we obtain the second derivative $D(Dy) = Dy' = y''$. We simply write $D(Dy) = D^2y$, so that

$$Dy = y', \qquad D^2y = y'', \qquad D^3y = y''', \cdots .$$

More generally,

(1) $$L = P(D) = D^2 + aD + b$$

is called a **second-order differential operator.** Here a and b are constant. P suggests "polynomial." L suggests "linear" (explained below). When L is applied to a function y (assumed twice differentiable), it produces

(2) $$L[y] = (D^2 + aD + b)y = y'' + ay' + by.$$

L is a ***linear*** operator. By definition this means that we have

$$L[\alpha y + \beta w] = \alpha L[y] + \beta L[w]$$

for any constants α and β and any (twice differentiable) functions y and w.

The homogeneous linear differential equation $y'' + ay' + by = 0$ may now be simply written

(3) $$L[y] = P(D)[y] = 0.$$

For example,

(4) $$L[y] = (D^2 + D - 6)y = y'' + y' - 6y = 0.$$

Since

$$D[e^{\lambda x}] = \lambda e^{\lambda x}, \qquad D^2[e^{\lambda x}] = \lambda^2 e^{\lambda x},$$

we have from (2) and (3)

(5) $$P(D)[e^{\lambda x}] = (\lambda^2 + a\lambda + b)e^{\lambda x} = P(\lambda)e^{\lambda x} = 0.$$

This confirms our result of Sec. 2.2 that $e^{\lambda x}$ is a solution of (3) if and only if λ is a solution of the characteristic equation $P(\lambda) = 0$. If $P(\lambda)$ has two different roots, we obtain a basis. If $P(\lambda)$ has a double root, we need a second independent solution. To obtain that solution, we differentiate

$$P(D)[e^{\lambda x}] = P(\lambda)e^{\lambda x}$$

[see (5)] on both sides with respect to λ and interchange differentiation with respect to λ and x, obtaining

$$P(D)[xe^{\lambda x}] = P'(\lambda)e^{\lambda x} + P(\lambda)xe^{\lambda x}$$

where $P' = dP/d\lambda$. For a double root, $P(\lambda) = P'(\lambda) = 0$, so that we have $P(D)[xe^{\lambda x}] = 0$. Hence $xe^{\lambda x}$ is the desired second solution. This agrees with Sec. 2.2.

$P(\lambda)$ is a polynomial in λ, in the usual sense of algebra. If we replace λ by D, then we obtain the "operator polynomial" $P(D)$. The point of this **"operational calculus"** is that $P(D)$ can be treated just like an algebraic quantity. In particular, we can factor it.

EXAMPLE 1 Factorization, solution of a differential equation

Factor $P(D) = D^2 + D - 6$ and solve $P(D)[y] = 0$.

Solution. $D^2 + D - 6 = (D + 3)(D - 2)$. Now $(D - 2)y = y' - 2y$ by definition. Hence

$$(D + 3)(D - 2)y = (D + 3)[y' - 2y] = y'' - 2y' + 3y' - 6y$$
$$= y'' + y' - 6y.$$

Hence our factorization is "permissible," that is, yields the correct result. Solutions of $(D + 3)y = 0$ and $(D - 2)y = 0$ are $y_1 = e^{-3x}$ and $y_2 = e^{2x}$. This is a basis of $P(D)[y] = 0$ on any inerval. The student should verify that our method in Sec. 2.2 gives the same result. This is not unexpected, since we factored $P(D)$ in the same way as we factor the characteristic polynomial $P(\lambda) = \lambda^2 + \lambda - 6$. ∎

Without going into details at this time, we want to mention that operational methods can also be used for operators $M = D^2 + fD + g$ with *variable* coefficients $f(x)$ and $g(x)$ but are more difficult and require more care. For example, $xD \neq Dx$ because

$$xDy = xy' \quad \text{but} \quad Dxy = (xy)' = y + xy'.$$

If operational calculus were limited to the simple situations illustrated in this section, it would perhaps not be worth mentioning. Actually, the power of the linear operator approach comes out in more complicated engineering problems, as we shall see in Chap. 6.

Problem Set 2.4

In each case apply the given operator to each of the given functions.

1. $D + 3;\quad x^2 + 6x - 2,\quad 9e^{-3x},\quad \sin \pi x + 2 \cos \pi x$
2. $D^2 - 2D;\quad xe^x,\quad \sinh 2x,\quad e^{2x} + 5$
3. $(D + 4)(D - 1);\quad e^{-4x},\quad xe^{-4x},\quad e^x$
4. $(D - 5)^2;\quad 5x + \cosh 5x,\quad e^{5x},\quad xe^{5x}$

Find a general solution of the following equations.

5. $(D^2 + 2D + 2)y = 0$
6. $(4D^2 + 4D + 1)y = 0$
7. $(4D^2 - 12D + 9)y = 0$
8. $(D^2 + 6D + 12)y = 0$
9. $(\pi^2 D^2 - 4\pi D + 4)y = 0$
10. $(4D^2 + 4D + 17)y = 0$
11. $(10D^2 + 12D + 3.6)y = 0$
12. $(D^2 + 2kD + k^2 + 3)y = 0$

Solve the following initial value problems.

13. $(D^2 + 4D + 5)y = 0,\quad y(0) = 0,\quad y'(0) = -3$
14. $(D^2 + 5D + 6)y = 0,\quad y(0) = 2,\quad y'(0) = -3$
15. $(D^2 - 2D + \pi^2 + 1)y = 0,\quad y(0) = 1,\quad y'(0) = 1 - \pi$
16. $(D^2 - 0.1D - 3.8)y = 0,\quad y(0) = -3.9,\quad y'(0) = 7.8$
17. $(9D^2 + 6D + 1)y = 0,\quad y(-3) = 10e \approx 27.18,\quad y'(-3) = -\frac{19}{3}e \approx -17.22$
18. $(D^2 - 0.2D + 100.01)y = 0,\quad y(0) = 0,\quad y'(0) = 40$
19. $(D + 1)^2 y = 0,\quad y(0) = 1,\quad y'(0) = -2$
20. Prove that the operator L in (2) is linear.

2.5 Modeling: Free Oscillations (Mass–Spring System)

Homogeneous linear differential equations with constant coefficients have basic engineering applications. In this section we consider an important application from mechanics (a vibrating mass on an elastic spring). We model the system (i.e., set up its mathematical equation), solve it, and discuss the types of motion, which—interestingly enough—will correspond to Cases I–III in Secs. 2.2 and 2.3. Incidentally, our mechanical system has a complete analog in electric circuits, as we shall discover later (in Sec. 2.12).

Setting up the Model

We take an ordinary spring that resists compression as well as extension and suspend it vertically from a fixed support (Fig. 32). At the lower end of the spring we attach a body of mass m. We assume m to be so large that we may disregard the mass of the spring. If we pull the body down a certain distance and then release it, it undergoes a motion. We assume that the body moves strictly vertically.

We want to determine the motion of our mechanical system. For this purpose we consider the forces[6] acting on the body during the motion. This will lead to a differential equation, whose solution $y(t)$ will give the displacement of the mass as a function of time t.

We choose the *downward direction* as the positive direction and thus regard downward forces as positive and upward forces as negative.

The most obvious force acting on the body is the *attraction of gravity*

$$F_1 = mg \tag{1}$$

where m is the mass of the body and g (= 980 cm/sec^2) is the acceleration of gravity.

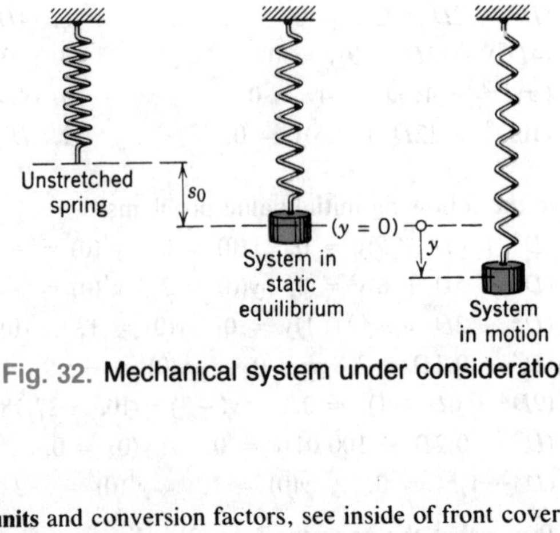

Fig. 32. Mechanical system under consideration

[6]For **systems of units** and conversion factors, see inside of front cover.

We next consider the *spring force* F_2 acting on the body. Experiments show that within reasonable limits, its magnitude is proportional to the change in the length of the spring. Its direction is upward if the spring is extended and downward if the spring is compressed. Thus,

$$F_2 = -ks \qquad \textbf{(Hooke's}^7\textbf{ law)}, \tag{2}$$

where s is the vertical displacement of the body (recall that the upper end of the spring is fixed), the constant of proportionality k is called the *spring modulus*, and the minus sign makes F_2 negative (upward) for positive s (extension of the spring) and positive (downward) for negative s (compression of the spring).

If $s = 1$, then $F_2 = -k$. The stiffer the spring, the larger k.

When the body is at rest (motionless), gravitational force and spring force are in equilibrium, their resultant is the zero force,

$$F_1 + F_2 = mg - ks_0 = 0 \tag{3}$$

where s_0 is the extension of the spring corresponding to this position, which is called the *static equilibrium position*.

We denote by $y = y(t)$ [t time] the displacement of the body from the static equilibrium position ($y = 0$), with the positive direction downward (Fig. 32). This displacement causes an additional force $-ky$ on the body, by Hooke's law. Hence the resultant of the forces on the body at position $y(t)$ is [see (3)]

$$F_1 + F_2 - ky = -ky. \tag{4}$$

Undamped System: Equation and Solution

If the damping of the system is so small that it can be disregarded, then (4) is the resultant of *all* the forces acting on the body. The differential equation will now be obtained by the use of **Newton's second law**

$$\text{Mass} \times \text{Acceleration} = \text{Force}$$

where *force* means the resultant of the forces acting on the body at any instant. In our case, the acceleration is $y'' = d^2y/dt^2$ and that resultant is given by (4). Thus

$$my'' = -ky.$$

Hence the motion of our system is governed by the linear differential equation with constant coefficients

$$\boxed{my'' + ky = 0.} \tag{5}$$

[7]ROBERT HOOKE (1635—1703), English physicist, a forerunner of Newton with respect to the law of gravitation.

By the method in Sec. 2.3 (see Example 3) we get the general solution

$$(6) \qquad y(t) = A\cos\omega_0 t + B\sin\omega_0 t \qquad \omega_0 = \sqrt{k/m}.$$

The corresponding motion is called a **harmonic oscillation.** Figure 33 shows typical forms of (6) corresponding to some positive initial displacement $y(0)$ [which determines $A = y(0)$ in (6)] and different initial velocities $y'(0)$ [each of which determines a value of B in (6), since $y'(0) = \omega_0 B$].

By applying the addition formula for the cosine, the student may verify that (6) can be written [see also (13) in Appendix 3]

$$(6^*) \qquad y(t) = C\cos(\omega_0 t - \delta) \qquad \left(C = \sqrt{A^2 + B^2}, \quad \tan\delta = \frac{B}{A}\right).$$

Since the period of the trigonometric functions in (6) is $2\pi/\omega_0$, the body executes $\omega_0/2\pi$ cycles per second. The quantity $\omega_0/2\pi$ is called the **frequency** of the oscillation and is measured in cycles per second. Another name for cycles/sec is hertz (Hz).[8]

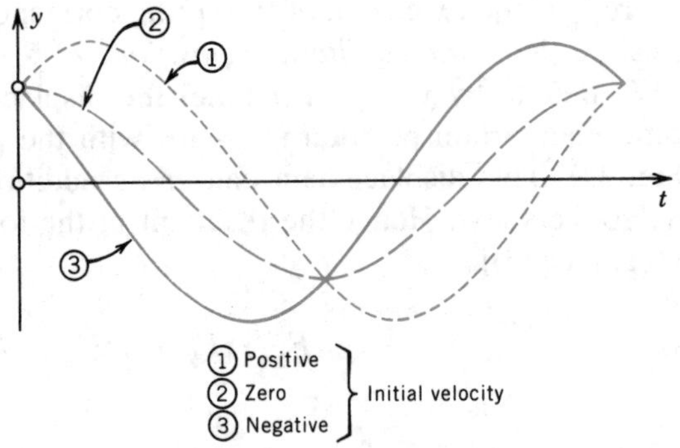

Fig. 33. Harmonic oscillations

EXAMPLE 1 **Undamped system. Harmonic oscillations**

If an iron ball of weight $W = 89.00$ nt (about 20 lb) stretches a spring 10.00 cm (about 4 inches), how many cycles per minute will this mass–spring system execute? What will its motion be if we pull down the weight an additional 15.00 cm (about 6 inches)?

Solution. From (2) we obtain the value $k = 89.00/0.1000 = 890.0$ [nt/meter]. The mass is $m = W/g = 89.00/9.8000 = 9.082$ [kg]. This gives the frequency

$$\omega_0/2\pi = \sqrt{890.0/9.082}/2\pi = 9.899/2\pi = 1.576 \text{ Hz}$$

or 94.5 cycles per minute. From (6) and the initial conditions, $y(0) = A = 0.1500$ [meter] and $y'(0) = \omega_0 B = 0$. Hence the motion is

$$y(t) = 0.1500\cos 9.899t \text{ [meters]} \qquad \text{or} \qquad 0.492\cos 9.899t \text{ [ft]}.$$

If you have a chance of experimenting with a mass–spring system, don't miss it. You will be surprised about the good agreement between theory and experiment, usually within a fraction of one percent if you measure carefully.

[8]HEINRICH HERTZ (1857—1894), German physicist, who discovered electromagnetic waves and made important contributions to electrodynamics.

Damped System: Equation and Solutions

If we connect the mass to a dashpot (Fig. 34), then we have to take the corresponding viscous damping into account. The corresponding damping force has the direction opposite to the instantaneous motion, and we assume that it is proportional to the velocity $y' = dy/dt$ of the body. This is generally a good approximation, at least for small velocities. Thus the damping force is of the form

$$F_3 = -cy'.$$

Let us show that the *damping constant* c is positive. If y' is positive, the body moves downward (in the positive y-direction) and $-cy'$ must be an upward force, that is, by agreement, $-cy' < 0$, which implies $c > 0$. For negative y' the body moves upward and $-cy'$ must represent a downward force, that is, $-cy' > 0$, which implies $c > 0$.

The resultant of the forces acting on the body is now [see (4)]

$$F_1 + F_2 + F_3 = -ky - cy'.$$

Hence, by Newton's second law,

$$my'' = -ky - cy',$$

and we see that the motion of the damped mechanical system is governed by the linear differential equation with constant coefficients

$$my'' + cy' + ky = 0. \qquad (7)$$

The corresponding characteristic equation is

$$\lambda^2 + \frac{c}{m}\lambda + \frac{k}{m} = 0.$$

The roots are

$$\lambda_{1,2} = -\frac{c}{2m} \pm \frac{1}{2m}\sqrt{c^2 - 4mk}.$$

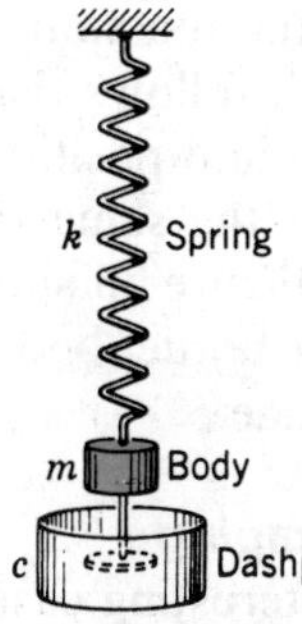

Fig. 34 Damped system

Using the short notations

$$\alpha = \frac{c}{2m} \quad \text{and} \quad \beta = \frac{1}{2m}\sqrt{c^2 - 4mk}, \tag{8}$$

we can write

$$\lambda_1 = -\alpha + \beta \quad \text{and} \quad \lambda_2 = -\alpha - \beta.$$

The form of the solution of (7) will depend on the damping, and, as in Secs. 2.2 and 2.3, we now have the following three cases:

Case I.	$c^2 > 4mk$.	*Distinct real roots* λ_1, λ_2.	***(Overdamping)***
Case II.	$c^2 = 4mk$.	*A real double root.*	***(Critical damping)***
Case III.	$c^2 < 4mk$.	*Complex conjugate roots.*	***(Underdamping)***

Let us discuss these three cases separately.

Case I. Overdamping

If the damping constant c is so large that $c^2 > 4mk$, then λ_1 and λ_2 are distinct real roots, and the general solution of (7) is

$$y(t) = c_1 e^{-(\alpha-\beta)t} + c_2 e^{-(\alpha+\beta)t}. \tag{9}$$

We see that in this case the body does not oscillate. For $t > 0$ both exponents in (9) are negative because $\alpha > 0$, $\beta > 0$, and $\beta^2 = \alpha^2 - k/m < \alpha^2$. Hence both terms in (9) approach zero as t approaches infinity. Practically speaking, after a sufficiently long time the mass will be at rest at the static equilibrium position ($y = 0$). This is understandable since the damping takes energy from the system and there is no external force that keeps the motion going. Figure 35 shows (9) for some typical initial conditions.

Case II. Critical damping

If $c^2 = 4mk$, then $\beta = 0$, $\lambda_1 = \lambda_2 = -\alpha$, and the general solution is

$$y(t) = (c_1 + c_2 t)e^{-\alpha t}. \tag{10}$$

Since the exponential function is never zero and $c_1 + c_2 t$ can have at most one positive zero, it follows that the motion can have at most one passage through the equilibrium position ($y = 0$). If the initial conditions are such that c_1 and c_2 have the same sign, there is no such passage at all. This is similar to Case I. Figure 36 shows typical forms of (10).

Case II marks the border between nonoscillatory motions and oscillations; this explains its name.

Case III. Underdamping

This is the most interesting case. If the damping constant c is so small that $c^2 < 4mk$, then β in (8) is pure imaginary, say,

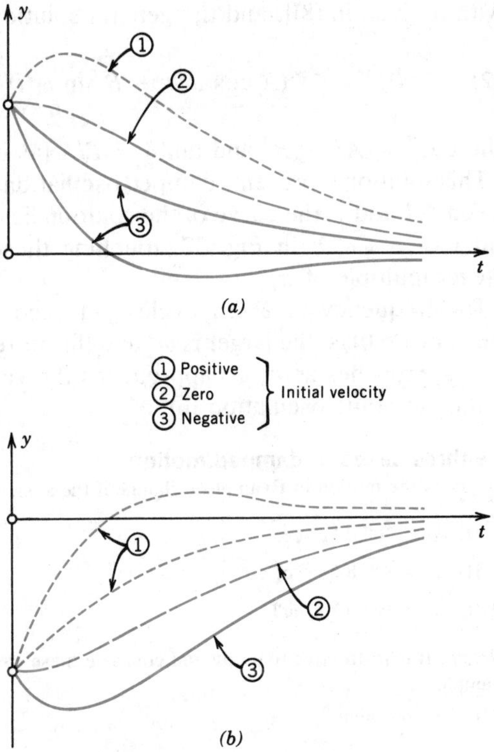

Fig. 35. Typical motions in the overdamped case
(a) Positive initial displacement
(b) Negative initial displacement

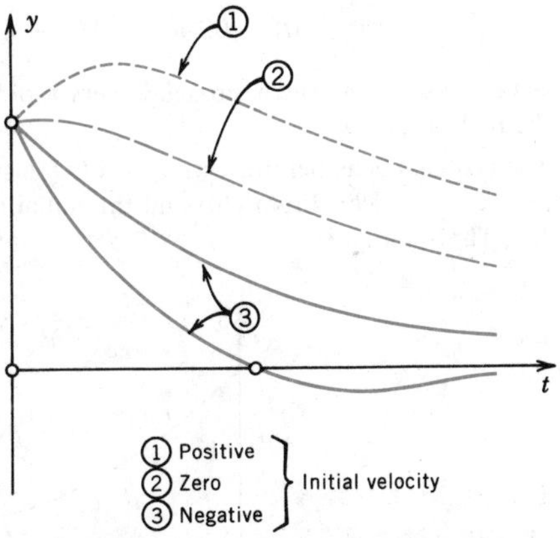

Fig. 36. Critical damping

$$(11) \quad \beta = i\omega^* \quad \text{where} \quad \omega^* = \frac{1}{2m}\sqrt{4mk - c^2} = \sqrt{\frac{k}{m} - \frac{c^2}{4m^2}} \quad (> 0).$$

(We write ω^* to reserve ω for Sec. 2.11.) The roots of the characteristic equation are complex conjugate,

$$\lambda_1 = -\alpha + i\omega^*, \qquad \lambda_2 = -\alpha - i\omega^*$$

[with α given in (8)], and the general solution is

$$y(t) = e^{-\alpha t}(A \cos \omega^* t + B \sin \omega^* t) = Ce^{-\alpha t} \cos (\omega^* t - \delta) \tag{12}$$

where $C^2 = A^2 + B^2$ and $\tan \delta = B/A$ [as in (6*)].

This solution represents damped oscillations. Since $\cos (\omega^* t - \delta)$ varies between -1 and 1, the curve of the solution lies between the curves $y = Ce^{-\alpha t}$ and $y = -Ce^{-\alpha t}$ in Fig. 37, touching these curves when $\omega^* t - \delta$ is an integer multiple of π.

The frequency is $\omega^*/2\pi$ cycles per second. From (11) we see that the smaller c (> 0) is, the larger is ω^* and the more rapid the oscillations become. As c approaches zero, ω^* approaches the value $\omega_0 = \sqrt{k/m}$ corresponding to the harmonic oscillation (6).

EXAMPLE 2 The three cases of damped motion

How does the motion in Example 1 change if the system has damping given by

(I) $c = 200.0$ kg/sec,

(II) $c = 179.8$ kg/sec,

(III) $c = 100.0$ kg/sec?

Solution. It is instructive to study and compare these cases with the behavior of the system in Example 1.

(I) The problem is

$$9.082y'' + 200.0y' + 890.0y = 0, \qquad y(0) = 0.1500 \text{ [meter]}, \qquad y'(0) = 0.$$

The characteristic equation has the roots $\lambda_{1,2} = -\alpha \pm \beta = -11.01 \pm 4.822$, thus $\lambda_1 = -6.190$, $\lambda_2 = -15.83$. From (9) and the initial conditions, $c_1 + c_2 = 0.1500$, $\lambda_1 c_1 + \lambda_2 c_2 = 0$. The solution is

$$y(t) = 0.2463e^{-6.190t} - 0.0963e^{-15.83t}.$$

It approaches 0 as $t \to \infty$. The approach is very rapid. After a few seconds, it is practically 0; that is, the body is at rest.

(II) The problem is as before, with $c = 179.8$ instead of 200. Since $c^2 = 4mk$, we get the double root $\lambda = -9.899$. From (10) and the initial conditions, $c_1 = 0.1500$, $c_2 + \lambda c_1 = 0$, $c_2 = 1.485$. The solution is

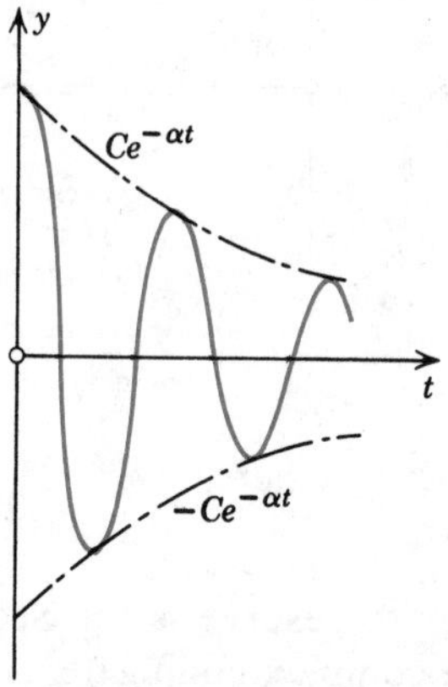

Fig. 37. Damped oscillation in Case III

$$y(t) = (0.150 + 1.485t)e^{-9.899t}.$$

It decreases rapidly to zero.

(III) The problem is as before, with $c = 100$. The roots are complex conjugate, $\lambda_{1,2} = -\alpha \pm i\omega^* = -5.506 \pm 8.227i$. From (12) and the initial conditions, $A = 0.1500$, $-\alpha A + \omega^* B = 0$ or $B = 0.1004$. This gives the solution

$$y(t) = e^{-5.506t}(0.1500 \cos 8.227t + 0.1004 \sin 8.227t).$$

These damped oscillations are slower than the harmonic oscillations in Example 1 by about 17%. ■

This whole section concerned **free motions** of mass–spring systems. These are governed by homogeneous differential equations, as we have seen. **Forced motions** under the influence of a "driving force" lead to nonhomogeneous equations and will be studied in Sec. 2.11, after we have learned how to solve such equations.

Problem Set 2.5

Harmonic oscillations (Undamped motion)

1. Show that the harmonic oscillation (6) starting from initial displacement y_0 with initial velocity v_0 is $y(t) = y_0 \cos \omega_0 t + (v_0/\omega_0) \sin \omega_0 t$ and represent this in the form (6*).
2. To find the modulus k of a spring S, take a body B, determine its weight W, attach B to S (whose upper end is fixed), let B come to rest, and measure the stretch s_0 of S. If $W = 69.77$ nt [= 15.68 lb] and $s_0 = 0.441$ meter, what is k and how many cycles per minute will B execute when allowed to vibrate?
3. How does the frequency of the harmonic oscillation change if we (i) double the mass, (ii) take a stiffer spring, (iii) change the initial conditions?
4. If a spring is such that a weight of 40 nt (about 9 lb) would stretch it 4 cm, what would the frequency of the corresponding harmonic oscillation be? The period?
5. Show that the frequency of a harmonic oscillation of a body on a spring is $(\sqrt{g/s_0})/2\pi$, so that the period is $2\pi\sqrt{s_0/g}$, where s_0 is the elongation in Fig. 32.
6. A 20.0 nt weight (about 4.5 lb) stretches a spring 9.8 cm, and we pull the weight on the spring down 5 cm from rest and give it an upward velocity of 30.0 cm/sec. Find the resulting motion $y(t)$, assuming no damping.
7. If a body hangs on a spring of modulus $k_1 = 3$, which in turn hangs on a spring of modulus $k_2 = 7$, what is the modulus k of this combination of springs?
8. What are the frequencies of vibration of a mass $m = 5$ kg (i) on a spring with modulus $k_1 = 10$ nt/m, (ii) on a spring with $k_2 = 20$ nt/m, (iii) on the two springs in parallel? See Fig. 38.

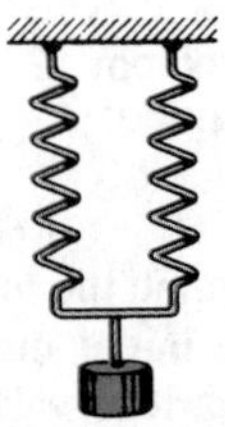

Fig. 38. Problem 8

9. **(Pendulum)** Determine the frequency of oscillation of the pendulum of length L in Fig. 39. Neglect air resistance and the weight of the rod. Assume that θ is small enough that $\sin\theta \approx \theta$.

10. A clock has a 1-meter pendulum. The clock ticks once for each time the pendulum completes a swing, returning to its original position. How many times a minute does the clock tick?

11. Suppose that the system in Fig. 40 consists of a pendulum as in Prob. 9 and two springs with constants k_1 and k_2 attached to the vibrating body and two vertical walls such that $\theta = 0$ remains the position of static equilibrium and $\theta(t)$ remains small during the motion. Find the period T.

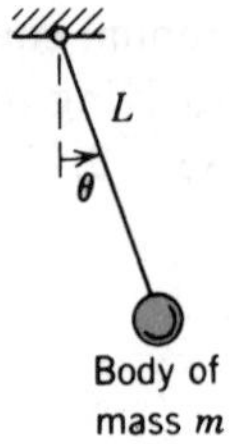

Fig. 39. Pendulum

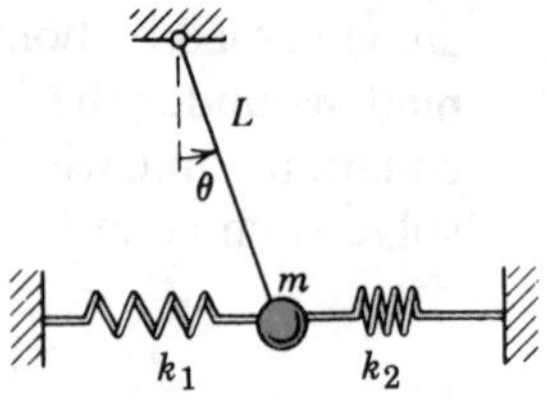

Fig. 40. Problem 11

12. **(Flat spring)** Our present equation $my'' + k_0y = 0$ also governs the (undamped) vibrations of a body attached to a flat spring (of negligible mass) whose other end is horizontally clamped (Fig. 41); here k_0 is the spring constant in Hooke's law $F = -k_0s$. What is the motion if the body weighs 4 nt (about 0.9 lb), the system has its equilibrium 1 cm below the horizontal line, and we let it start from this position with downward initial velocity 20 cm/sec? When will the body reach its highest position for the first time?

13. **(Torsional vibrations)** Undamped torsional vibrations (rotations back and forth) of a wheel attached to an elastic thin rod or wire (Fig. 42) are governed by the equation

$$I_0\theta'' + K\theta = 0,$$

where θ is the angle measured from the state of equilibrium, I_0 the polar moment of inertia of the wheel about its center, and K the torsional stiffness of the rod. Solve the equation for $K/I_0 = 13.69\ \text{sec}^{-2}$, initial angle 15° (= 0.2618 rad) and initial angular velocity 10° sec^{-1} (= 0.1745 rad · sec^{-1}).

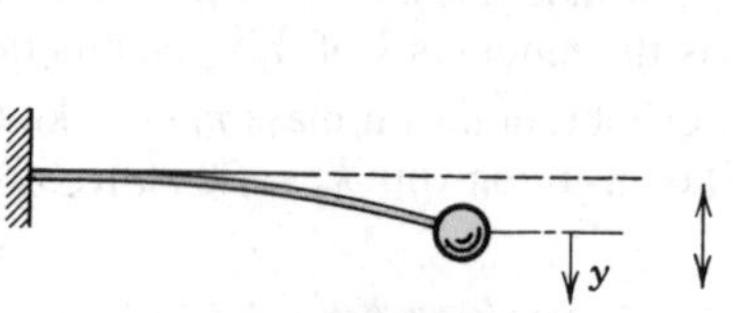

Fig. 41. Problem 12
(Flat spring)

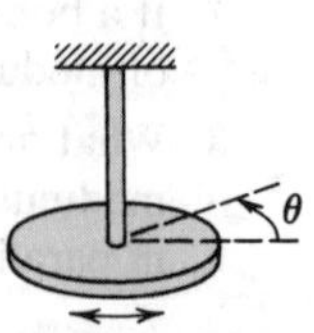

Fig. 42. Problem 13
(Torsional vibrations)

14. **(Damping)** Determine the motion $y(t)$ of the mechanical system described by (7) corresponding to initial displacement 1, initial velocity zero, mass 1, spring modulus 1, and various values of the damping constant, say, $c = 0, 1, 2, 10$.

Case I. Overdamped motion

15. Show that for (9) to satisfy initial conditions $y(0) = y_0$ and $v(0) = v_0$ we must have $c_1 = [(1 + \alpha/\beta)y_0 + v_0/\beta]/2$ and $c_2 = [(1 - \alpha/\beta)y_0 - v_0/\beta]/2$.

16. Show that in the overdamped case, the body can pass through $y = 0$ at most once (Fig. 35).

17. In Prob. 16 find conditions for c_1 and c_2 such that the body does not pass through $y = 0$ at all.

18. Show that an overdamped motion with zero initial displacement cannot pass through $y = 0$.

Case II. Critical damping

19. Find the critical motion (10) that starts from y_0 with initial velocity v_0.

20. Under what conditions does (10) have a maximum or minimum at some instant $t > 0$?

21. Represent the maximum or minimum amplitude in Prob. 20 in terms of the initial values y_0 and v_0.

Case III. Underdamped motion (Damped oscillation)

22. Find and graph the three damped oscillations of the form

$$y = e^{-t}(A \cos t + B \sin t) = Ce^{-t} \cos (t - \delta)$$

starting from $y = 1$ with initial velocity $-1, 0, 1$, respectively.

23. Show that the damped oscillation satisfying the initial conditions $y(0) = y_0$, $v(0) = v_0$ is

$$y = e^{-\alpha t}[y_0 \cos \omega^* t + \omega^{*-1}(v_0 + \alpha y_0) \sin \omega^* t].$$

24. Show that the frequency $\omega^*/2\pi$ of the underdamped motion decreases as the damping increases.

25. Show that for small damping, $\omega^* \approx \omega_0 [1 - (c^2/8mk)]$.

26. For what c (in terms of m and k) is $\omega^*/\omega_0 = 99\%$? 95%? Calculate (a) exactly, (b) by the formula in Prob. 25.

27. Determine the values of t corresponding to the maxima and minima of the oscillation $y(t) = e^{-t} \sin t$. Check your result by graphing $y(t)$.

28. Show that the maxima and minima of an underdamped motion occur at equidistant values of t, the distance between two consecutive maxima being $2\pi/\omega^*$.

29. Consider an underdamped motion of a body of mass $m = 2$ kg. If the time between two consecutive maxima is 3 sec and the maximum amplitude decreases to $\frac{1}{2}$ its initial value after 20 cycles, what is the damping constant of the system?

30. **(Logarithmic decrement)** Prove that the ratio of two consecutive maximum amplitudes of a damped oscillation (12) is constant, the natural logarithm of this ratio being $\Delta = 2\pi\alpha/\omega^*$. ($\Delta$ is called the *logarithmic decrement* of the oscillation.) Find Δ in the case of $y = e^{-t} \cos t$ and determine the values of t corresponding to the maxima and minima.

2.6 Euler–Cauchy Equation

Constant-coefficient equations are solvable without integration, as we have seen. Similarly, the so-called **Euler–Cauchy equation**[9]

$$x^2y'' + axy' + by = 0 \qquad (a, b \text{ constant}) \tag{1}$$

can also be solved by purely algebraic manipulations. Indeed, substituting

$$y = x^m \tag{2}$$

and its derivatives into the differential equation (1), we find

$$x^2m(m - 1)x^{m-2} + axmx^{m-1} + bx^m = 0.$$

Omitting x^m, which is not zero if $x \neq 0$, we obtain the auxiliary equation

$$m^2 + (a - 1)m + b = 0. \tag{3}$$

Case I. Distinct real roots. If the roots m_1, m_2 of (3) are real and distinct, then

$$y_1(x) = x^{m_1} \quad \text{and} \quad y_2(x) = x^{m_2}$$

constitute a basis of solutions of the differential equation (1) for all x for which these functions are defined. The corresponding general solution is

$$y = c_1x^{m_1} + c_2x^{m_2} \qquad (c_1, c_2 \text{ arbitrary}). \tag{4}$$

EXAMPLE 1 **General solution in the case of different real roots**

Solve the Euler–Cauchy equation

$$x^2y'' - 2.5xy' - 2.0y = 0.$$

---

[9]LEONHARD EULER (1707—1783) was an enormously creative Swiss mathematician. He studied in Basel under JOHANN BERNOULLI and in 1727 became a professor of physics (and later of mathematics) in St. Petersburg, Russia. In 1741 he went to Berlin as a member of the Berlin Academy. In 1766 he returned to St. Petersburg. He contributed to almost all branches of mathematics and its applications to physical problems, even after he became totally blind in 1771; we mention his fundamental work in differential and difference equations, Fourier and other infinite series, special functions, complex analysis, the calculus of variations, mechanics, and hydrodynamics. He is the exponent of a very rapid expansion of analysis. (So far, the first seventy (!) volumes of his Collected Works have appeared.)

This expansion of mathematics was followed by a period characterized by greater rigor, dominated by the great French mathematician AUGUSTIN-LOUIS CAUCHY (1789—1857), the father of modern analysis. Cauchy studied and taught mainly in Paris. He is the creator of complex analysis and exercised a great influence on the theory of infinite series and ordinary and partial differential equations (see the Index to this book for some of these contributions). He is also known for his work in elasticity and optics. Cauchy published nearly 800 mathematical research papers, many of them of basic importance.

Solution. The auxiliary equation is

$$m^2 - 3.5m - 2.0 = 0.$$

The roots are $m_1 = -0.5$ and $m_2 = 4$. Hence a basis of real solutions for all positive x is

$$y_1 = \frac{1}{\sqrt{x}}, \qquad y_2 = x^4$$

and the corresponding general solution for all those x is

$$y = \frac{c_1}{\sqrt{x}} + c_2 x^4$$

Case II. Double root. If (3) has a double root $m = \frac{1}{2}(1 - a)$, we get a first solution

$$y_1 = x^{(1-a)/2} \tag{5}$$

and a second solution y_2 by the method of reduction of order (as in Sec. 2.2). Thus, substituting $y_2 = uy_1$ and its derivatives into (1), we obtain

$$x^2(u''y_1 + 2u'y_1' + uy_1'') + ax(u'y_1 + uy_1') + buy_1 = 0.$$

Reshuffling terms gives

$$u''x^2y_1 + u'x(2xy_1' + ay_1) + u(x^2y_1'' + axy_1' + by_1) = 0. \tag{6}$$

The last expression $(\cdot\ \cdot\ \cdot)$ is zero since y_1 is a solution of (1). From (5) we get in (6)

$$2xy_1' + ay_1 = (1 - a)x^{(1-a)/2} + ax^{(1-a)/2} = x^{(1-a)/2} = y_1.$$

This reduces (6) to $(u''x^2 + u'x)y_1 = 0$. We divide this by y_1 $(\neq 0)$, separate variables, and integrate. Then for $x > 0$ we get

$$\frac{u''}{u'} = -\frac{1}{x}, \qquad \ln|u'| = -\ln x, \qquad u' = \frac{1}{x}, \qquad u = \ln x.$$

Thus $y_2 = y_1 \ln x$, which is not proportional to y_1. Hence in the case of a double root of (3), a basis of (1) for all positive x is

$$y_1 = x^m, \qquad y_2 = x^m \ln x \qquad\qquad m = \tfrac{1}{2}(1 - a) \tag{7*}$$

and gives the general solution

$$y = (c_1 + c_2 \ln x)x^{(1-a)/2} \qquad (c_1, c_2 \text{ arbitrary}). \tag{7}$$

EXAMPLE 2 **General solution in the case of a double root**

Solve

$$x^2y'' - 3xy' + 4y = 0.$$

Solution. The auxiliary equation has the double root $m = 2$. Hence a basis of real solutions for all positive x is x^2, $x^2 \ln x$, and the corresponding general solution is

$$y = (c_1 + c_2 \ln x)x^2.$$

Case III. Complex conjugate roots. If the roots m_1 and m_2 of (3) are complex, they are conjugate, say $m_1 = \mu + i\nu$, $m_2 = \mu - i\nu$. We claim that in this case, a basis of solutions of (1) for all positive x is

$$y_1 = x^\mu \cos(\nu \ln x), \qquad y_2 = x^\mu \sin(\nu \ln x). \tag{8}$$

Indeed, these functions are not proportional, and they are solutions of (1), as follows by differentiation and substitution. The corresponding general solution is

$$y = x^\mu [A \cos(\nu \ln x) + B \sin(\nu \ln x)]. \tag{9}$$

This proves everything and settles the case.

Another question is how we got the *idea* that (8) might be solutions. To answer it, we state that the formula $x^k = (e^{\ln x})^k = e^{k \ln x}$ extends from real to complex $k = i\nu$ and, together with (7) in Sec. 2.3 (with $s = 0$) gives

$$x^{i\nu} = e^{i\nu \ln x} = \cos(\nu \ln x) + i \sin(\nu \ln x),$$

$$x^{-i\nu} = e^{-i\nu \ln x} = \cos(\nu \ln x) - i \sin(\nu \ln x).$$

Now multiply by x^μ and add and subtract. This gives $2y_1$ and $2iy_2$, respectively. From this, dividing by 2 and $2i$, we have (8).

EXAMPLE 3 General solution in the case of complex conjugate roots

Solve

$$x^2y'' + 7xy' + 13y = 0.$$

Solution. The auxiliary equation (3) is $m^2 + 6m + 13 = 0$. The roots of this equation are $m_{1,2} = -3 \pm \sqrt{9 - 13} = -3 \pm 2i$. By (9) the answer is

$$y = x^{-3}[A \cos(2 \ln x) + B \sin(2 \ln x)].$$

Euler–Cauchy equations occur in certain applications, and we illustrate this by a simple example from electrostatics.

EXAMPLE 4 Electric potential field between two concentric spheres

Find the electrostatic potential $v = v(r)$ between two concentric spheres of radii $r_1 = 4$ cm and $r_2 = 8$ cm kept at potentials $v_1 = 110$ volts and $v_2 = 0$, respectively.

Physical information. $v(r)$ is a solution of $rv'' + 2v' = 0$, where $v' = dv/dr$.

Solution. The auxiliary equation $m^2 + m = 0$ has the roots 0 and -1. This gives the general solution $v(r) = c_1 + c_2/r$. From the "boundary conditions" (the potentials on the spheres),

$$v(8) = c_1 + c_2/8 = 0, \qquad v(4) = c_1 + c_2/4 = 110.$$

By subtraction, $c_2/8 = 110$, $c_2 = 880$, thus $c_1 = -110$. *Answer.* $v(r) = -110 + 880/r$ volts.

Problem Set 2.6

1. Verify directly by substitution that y_2 in (7*) is a solution of (1) if (3) has a double root, but $x^{m_1} \ln x$ and $x^{m_2} \ln x$ are **not** solutions of (1) if the roots m_1 and m_2 of (3) are different.

Find a general solution of the following differential equations.

2. $x^2y'' - 6y = 0$
3. $xy'' + 4y' = 0$
4. $x^2y'' - 2xy' + 2y = 0$
5. $(x^2D^2 + 9xD + 16)y = 0$
6. $x^2y'' + xy' - y = 0$
7. $(x^2D^2 + 3xD + 1)y = 0$
8. $(x^2D^2 - 1.5xD + 1)y = 0$
9. $x^2y'' + 6.2xy' + 6.76y = 0$
10. $x^2y'' + 3xy' + 5y = 0$
11. $(x^2D^2 + xD + 1)y = 0$
12. $(x^2D^2 - 3xD + 20)y = 0$
13. $(4x^2D^2 + 8xD - 15)y = 0$

Solve the following initial value problems.

14. $x^2y'' - 4xy' + 4y = 0, \quad y(1) = 4, \quad y'(1) = 13$
15. $(4x^2D^2 + 4xD - 1)y = 0, \quad y(4) = 2, \quad y'(4) = -0.25$
16. $(x^2D^2 - 5xD + 8)y = 0, \quad y(1) = 5, \quad y'(1) = 18$
17. $(x^2D^2 - xD + 2)y = 0, \quad y(1) = -1, \quad y'(1) = -1$
18. $10x^2y'' + 46xy' + 32.4y = 0, \quad y(1) = 0, \quad y'(1) = 2$
19. $(x^2D^2 + xD - 0.01)y = 0, \quad y(1) = 1, \quad y'(1) = 0.1$

20. (Potential between two spheres) Find the potential in Example 4 if $r_1 = 2$ cm, $r_2 = 20$ cm and the potentials on the spheres are $v_1 = 220$ volts and $v_2 = 130$ volts.

21. (Equations with constant coefficients) Setting $x = e^t$ $(x > 0)$, transform the Euler–Cauchy equation (1) into the equation $\ddot{y} + (a - 1)\dot{y} + by = 0$, whose coefficients are constant. Here, dots denote derivatives with respect to t.

22. Transform the equation in Prob. 21 back into (1).

23. Show that if we apply the transformation in Prob. 21, then (2) yields an expression of the form (2), Sec. 2.2, and (7) yields an expression of the form (7), Sec. 2.2, except for notation.

Reduce to the form (1) and solve:

24. $2(3z + 1)^2y'' + 21(3z + 1)y' + 18y = 0$
25. $(z - 2)^2y'' + 5(z - 2)y' + 3y = 0$

2.7 Existence and Uniqueness Theory. Wronskian

In this section we give a general theory for homogeneous linear equations

$$(1) \qquad y'' + p(x)y' + q(x)y = 0$$

with continuous, but otherwise arbitrary variable coefficients p and q. This will concern the existence of a general solution

$$y = c_1y_1 + c_2y_2 \tag{2}$$

of (1) as well as initial value problems consisting of (1) and two initial conditions

$$y(x_0) = K_0, \qquad y'(x_0) = K_1 \tag{3}$$

with given x_0, K_0, and K_1.

(Clearly, no such theory was needed for constant-coefficient or Euler–Cauchy equations because everything came out explicitly from our calculations.)

Central to our present discussion is the following theorem.

Theorem 1 **Existence and Uniqueness Theorem for Initial Value Problems**

If $p(x)$ and $q(x)$ are continuous functions on some open interval I and x_0 is in I, then the initial value problem (1), (3) *has a unique solution $y(x)$ on the interval I.*

The proof of existence uses the same prerequisites as those of the existence theorem in Sec. 1.11 and will not be presented here; it can be found in Ref. [A6] listed in Appendix 1. Uniqueness proofs are usually simpler than existence proofs. But in the present case, even the uniqueness proof is long, and we give it as an additional proof in Appendix 4.

Linear Independence of Solutions. Wronskian

Theorem 1 will imply very important properties of general solutions (2) of (1). As we know, these are made up of a **basis** y_1, y_2, that is, of a pair of linearly independent solutions. Remember from Sec. 2.1 that we call y_1 and y_2 **linearly independent** on an interval I if

$$k_1y_1(x) + k_2y_2(x) = 0 \quad \text{on } I \qquad \text{implies} \qquad k_1 = 0, k_2 = 0;$$

and we call y_1 and y_2 **linearly dependent** on I if this equation also holds for k_1, k_2 not both zero. In this case, and only in this case, y_1 and y_2 are proportional on I, that is (see Sec. 2.1),

$$\text{(a)} \quad y_1 = ky_2 \qquad \text{or} \qquad \text{(b)} \quad y_2 = ly_1. \tag{4}$$

For our discussion, the following criterion of linear independence and dependence of solutions will be helpful. This criterion uses the so-called *Wronski determinant*[10] or, briefly, the **Wronskian,** of two solutions y_1 and y_2 of (1), defined by

---

[10]Introduced by I. M. HÖNE (1778—1853), Polish mathematician, who changed his name to Wrónski. Second-order determinants should be familiar from elementary calculus; otherwise, consult the beginning of Sec. 7.8, which is independent of the other sections in Chap. 7.

$$W(y_1, y_2) = \begin{vmatrix} y_1 & y_2 \\ y_1' & y_2' \end{vmatrix} = y_1y_2' - y_2y_1'. \tag{5}$$

Theorem 2 **(Linear dependence and independence of solutions)**
Suppose that (1) *has continuous coefficients* $p(x)$ *and* $q(x)$ *on an open interval* I. *Then two solutions* y_1 *and* y_2 *of* (1) *on* I *are linearly dependent on* I *if and only if their Wronskian* W *is zero at some* x_0 *in* I. *Furthermore, if* $W = 0$ *for* $x = x_0$, *then* $W \equiv 0$ *on* I; *hence if there is an* x_1 *in* I *at which* $W \neq 0$, then y_1, y_2 *are linearly independent on* I.

Proof. **(a)** If y_1 and y_2 are linearly dependent on I, then (4a) or (4b) holds on I and gives for (4a)

$$W(y_1, y_2) = W(ky_2, y_2) = \begin{vmatrix} ky_2 & y_2 \\ ky_2' & y_2' \end{vmatrix} = ky_2y_2' - y_2ky_2' \equiv 0;$$

similarly when (4b) holds.

(b) Conversely, we assume that $W(y_1, y_2) = 0$ for some $x = x_0$ in I and show that then y_1, y_2 are linearly dependent. We consider the linear system of equations

$$\begin{aligned} k_1y_1(x_0) + k_2y_2(x_0) &= 0 \\ k_1y_1'(x_0) + k_2y_2'(x_0) &= 0 \end{aligned} \tag{6}$$

in the unknowns k_1, k_2. Now this system is homogeneous and its determinant is just the Wronskian $W[y_1(x_0), y_2(x_0)]$, which is zero by assumption. Hence the system has a solution k_1, k_2 where k_1 and k_2 are not both zero (see Theorem 2, Sec. 7.9). Using these numbers k_1, k_2, we introduce the function

$$y(x) = k_1y_1(x) + k_2y_2(x).$$

By Fundamental Theorem 1 in Sec. 2.1 the function $y(x)$ is a solution of (1) on I. From (6) we see that it satisfies the initial conditions $y(x_0) = 0$, $y'(x_0) = 0$. Now another solution of (1) satisfying the same initial conditions is $y^* \equiv 0$. Since p and q are continuous, Theorem 1 applies and guarantees uniqueness, that is, $y \equiv y^*$, written out,

$$k_1y_1 + k_2y_2 \equiv 0$$

on I. Now since k_1 and k_2 are not both zero, this means linear dependence of y_1, y_2 on I.

(c) We prove the last statement of the theorem. If $W = 0$ at an x_0 in I, we have linear dependence of y_1, y_2 on I by part (b), hence $W \equiv 0$ by part (a) of this proof. Hence $W \neq 0$ at an x_1 in I cannot happen in the case of linear dependence, so that $W \neq 0$ at x_1 implies linear independence. ∎

EXAMPLE 1 Application of Theorem 2

Show that $y_1 = \cos \omega x$, $y_2 = \sin \omega x$ form a basis of solutions of $y'' + \omega^2 y = 0$, $\omega \neq 0$, on any interval.

Solution. Substitution shows that they are solutions, and linear independence follows from Theorem 2, since

$$W(\cos \omega x, \sin \omega x) = \begin{vmatrix} \cos \omega x & \sin \omega x \\ -\omega \sin \omega x & \omega \cos \omega x \end{vmatrix} = \omega(\cos^2 \omega x + \sin^2 \omega x) = \omega.$$ ∎

EXAMPLE 2 Application of Theorem 2

Show that $y = (c_1 + c_2 x)e^x$ is a general solution of $y'' - 2y' + y = 0$ on any interval.

Solution. Substitution shows that $y_1 = e^x$ and $y_2 = xe^x$ are solutions, and Theorem 2 implies linear independence, since

$$W(e^x, xe^x) = \begin{vmatrix} e^x & xe^x \\ e^x & (x + 1)e^x \end{vmatrix} = (x + 1)e^{2x} - xe^{2x} = e^{2x} \neq 0.$$ ∎

A General Solution of (1) Includes All Solutions

We prove this in two steps, showing first that general solutions always exist:

Theorem 3 (Existence of a general solution)

If the coefficients $p(x)$ and $q(x)$ of (1) *are continuous on some open interval I, then* (1) *has a general solution on I.*

Proof. By Theorem 1, equation (1) has a solution $y_1(x)$ on I satisfying the initial conditions

$$y_1(x_0) = 1, \qquad y_1'(x_0) = 0$$

and a solution $y_2(x)$ on I satisfying the initial conditions

$$y_2(x_0) = 0, \qquad y_2'(x_0) = 1.$$

From this we see that the Wronskian $W(y_1, y_2)$ has at x_0 the value 1. Hence y_1, y_2 are linearly independent on I, by Theorem 2; they form a basis of solutions of (1) on I, and $y = c_1 y_1 + c_2 y_2$ with arbitrary c_1, c_2 is a general solution of (1) on I. ∎

We now reach the final goal of this section by proving that a general solution of (1) is as general as it can be, namely, it includes *all* solutions of (1):

Theorem 4 (General solution)

Suppose that (1) *has continuous coefficients $p(x)$ and $q(x)$ on some open interval I. Then every solution $y = Y(x)$ of* (1) *on I is of the form*

$$Y(x) = C_1 y_1(x) + C_2 y_2(x), \tag{7}$$

where y_1, y_2 form a basis of solutions of (1) *on I and C_1, C_2 are suitable constants.*

Hence (1) *does not have* **singular solutions** (*i.e., solutions not obtainable from a general solution*).

Proof. By Theorem 3, our equation has a general solution

$$y(x) = c_1y_1(x) + c_2y_2(x) \tag{8}$$

on I. We have to find suitable values of c_1, c_2 such that $y(x) = Y(x)$ on I. We choose any fixed x_0 in I and show first that we can find c_1, c_2 such that

$$y(x_0) = Y(x_0), \qquad y'(x_0) = Y'(x_0)$$

written out

$$\begin{aligned} c_1y_1(x_0) + c_2y_2(x_0) &= Y(x_0), \\ c_1y_1'(x_0) + c_2y_2'(x_0) &= Y'(x_0). \end{aligned} \tag{9}$$

In fact, this is a system of linear equations in the unknowns c_1, c_2. Its determinant is the Wronskian of y_1 and y_2 at $x = x_0$. Since (8) is a general solution, y_1 and y_2 are linearly independent on I, and from Theorem 2 it follows that their Wronskian is not zero. Hence the system has a unique solution $c_1 = C_1$, $c_2 = C_2$ which can be obtained by Cramer's rule (Sec. 7.8). By using these constants we obtain from (8) the particular solution

$$y^*(x) = C_1y_1(x) + C_2y_2(x).$$

Since C_1, C_2 are solutions of (9), from (9) we now see that

$$y^*(x_0) = Y(x_0), \qquad y^{*\prime}(x_0) = Y'(x_0).$$

From this and the uniqueness theorem (Theorem 1) we conclude that y^* and Y must be equal on I, and the proof is complete. ∎

Reduction of Order: How to Obtain a Second Solution

In attempting to find a basis of solutions, one can often get one solution by guessing or by some method. Typical cases were considered in Sec. 2.2 for the constant coefficient equation and in the last section for the Euler–Cauchy equation, and we show that those were just particular cases of a general method, the **method of reduction of order**[11] applicable to any equation (1) as follows. Let y_1 be a solution of (1) on some interval I. Substitute $y_2 = uy_1$ and its derivatives $y_2' = u'y_1 + uy_1'$ and $y_2'' = u''y_1 + 2u'y_1' + uy_1''$ into (1) and collect terms, obtaining

$$u''y_1 + u'(2y_1' + py_1) + u(y_1'' + py_1' + qy_1) = 0.$$

[11]Credited to the great mathematician JOSEPH LOUIS LAGRANGE (1736—1813), who was born in Turin, of French extraction, got his first professorship when he was 19 (at the Military Academy of Turin), became director of the mathematical section of the Berlin Academy in 1766, and moved to Paris in 1787. His important major work was in the calculus of variations, celestial mechanics, general mechanics (*Mécanique analytique*, Paris, 1788), differential equations, approximation theory, algebra and number theory.

Since y_1 is a solution of (1), the expression in the last parentheses is zero. We divide the remaining formula by y_1 and write $u' = U$. Then $u'' = U'$ and we have

$$U' + \left(\frac{2y_1'}{y_1} + p\right)U = 0.$$

We now separate variables and integrate, choosing the constant of integration to be zero (since we need no arbitrary constant). This gives

$$\ln |U| = -2 \ln |y_1| - \int p \, dx$$

and by taking exponents

$$U = \frac{1}{y_1^2} e^{-\int p \, dx}. \tag{10}$$

Here $U = u'$. Hence the desired second solution is $y_2 = uy_1 = y_1 \int U \, dx$. Since $y_2/y_1 = u = \int U \, dx$ cannot be a constant (why?), we see that y_1 and y_2 form a basis. ■

We have already applied this method in Secs. 2.2 and 2.6, but we add another typical example.

EXAMPLE 3 **Reduction of order**

Inspection shows that

$$(x^2 - 1)y'' - 2xy' + 2y = 0$$

has $y_1 = x$ as a first solution. Find another independent solution.

Solution. We set $y_2 = uy_1$ and use (10). ***Caution!*** It is crucial that we first write the equation in standard form,

$$y'' - \frac{2x}{x^2 - 1} y' + \frac{2}{x^2 - 1} y = 0,$$

because (10) was derived under this assumption. Then in (10),

$$-\int p \, dx = \int \frac{2x}{x^2 - 1} \, dx = \ln |x^2 - 1|.$$

Thus

$$U = x^{-2}(x^2 - 1) = 1 - x^{-2} \qquad \text{and} \qquad u = \int U \, dx = x + x^{-1}.$$

Answer. $y_2 = uy_1 = (x + x^{-1})x = x^2 + 1$. Check by substitution. ■

Problem Set 2.7

Find the Wronskian of the following bases (that we have used before), thereby verifying Theorem 2 for any interval. (In Probs. 4–6, assume $x > 0$.)

1. $e^{\lambda_1 x}, e^{\lambda_2 x}, \quad \lambda_1 \neq \lambda_2$

2. $e^{\lambda x}, xe^{\lambda x}$

3. $e^{-\alpha x/2} \cos \omega x, e^{-\alpha x/2} \sin \omega x$

4. $x^{m_1}, x^{m_2}, \quad m_1 \neq m_2$

5. $x^{\mu} \cos (\nu \ln x), x^{\mu} \sin (\nu \ln x)$

6. $x^m, x^m \ln x$

Find a second-order homogeneous linear differential equation for which the given functions are solutions. Find the Wronskian and use it to verify Theorem 2.

7. e^x, xe^x **8.** x, $x \ln x$ **9.** $e^x \cos x$, $e^x \sin x$

10. $\cosh kx$, $\sinh kx$ **11.** $\cos \pi x$, $\sin \pi x$ **12.** $\sqrt{x}$, $1/\sqrt{x}$

13. 1, x^3 **14.** 1, e^{-2x} **15.** $x^{1/2}$, $x^{3/2}$

16. Suppose that (1) has continuous coefficients on I. Show that two solutions of (1) on I that are zero at the same point in I cannot form a basis of solutions of (1) on I.

17. Suppose that (1) has continuous coefficients on I. Show that two solutions of (1) on I that have maxima or minima at the same point in I cannot form a basis of solutions of (1) on I.

18. Suppose that y_1, y_2 constitute a basis of solutions of a differential equation satisfying the assumptions of Theorem 2. Show that $z_1 = a_{11}y_1 + a_{12}y_2$, $z_2 = a_{21}y_1 + a_{22}y_2$ is a basis of that equation on the interval I if and only if the determinant of the coefficients a_{jk} is not zero.

19. Illustrate Prob. 18 with $y_1 = e^x$, $y_2 = e^{-x}$, $z_1 = \cosh x$, $z_2 = \sinh x$.

20. **(Euler–Cauchy equation)** Show that $x^2y'' - 4xy' + 6y = 0$ (Sec. 2.6) has $y_1 = x^2$, $y_2 = x^3$ as a basis of solutions for all x. Show that $W(x^2, x^3) = 0$ at $x = 0$. Does this contradict Theorem 2?

Reduction of order. Show that the given function y_1 is a solution of the given equation. Using the method of reduction of order, find y_2 such that y_1, y_2 form a basis. ***Caution!*** First write the equation in standard form.

21. $(x + 1)^2y'' - 2(x + 1)y' + 2y = 0, \quad y_1 = x + 1$

22. $(x - 1)y'' - xy' + y = 0, \quad y_1 = e^x$

23. $(1 - x)^2y'' - 4(1 - x)y' + 2y = 0, \quad y_1 = (1 - x)^{-1}, \quad x \neq 1$

24. $x^2y'' + xy' + (x^2 - \tfrac{1}{4})y = 0, \quad y_1 = x^{-1/2}\cos x, \quad x > 0$

25. $xy'' + 2y' + xy = 0, \quad y_1 = x^{-1}\sin x, \quad x \neq 0$

2.8 Nonhomogeneous Equations

***Beginning in this section, we turn from* homogeneous *to* nonhomogeneous linear equations**

$$y'' + p(x)y' + q(x)y = r(x) \tag{1}$$

where $r(x) \not\equiv 0$. How can we solve such an equation? Before we consider methods, let us first explore what we actually need for proceeding from the corresponding homogeneous equation

$$y'' + p(x)y' + q(x)y = 0 \tag{2}$$

to the nonhomogeneous equation (1). The key that relates (1) to (2) and gives us a plan for solving (1) is the following theorem.

Theorem 1 **[Relations between solutions of (1) and (2)]**

(a) *The difference of two solutions of* (1) *on some open interval I is a solution of* (2) *on I.*

(b) *The sum of a solution of* (1) *on I and a solution of* (2) *on I is a solution of* (1) *on I.*

Proof. **(a)** Denote the left side of (1) by $L[y]$. Let y and $\tilde{y}$ be any solutions of (1) on I. Then $L[y] = r(x)$, $L[\tilde{y}] = r(x)$, and since we have $(y - \tilde{y})' = y' - \tilde{y}'$, etc., we obtain the first assertion,

$$L[y - \tilde{y}] = L[y] - L[\tilde{y}] = r(x) - r(x) \equiv 0.$$

(b) Similarly, for y as before and any solution y^* of (2) on I,

$$L[y + y^*] = L[y] + L[y^*] = r(x) + 0 = r(x).$$ ∎

This situation suggests the following concepts.

Definition (General solution, particular solution)

A **general solution** of the nonhomogeneous equation (1) on some open interval I is a solution of the form

$$y(x) = y_h(x) + y_p(x), \tag{3}$$

where $y_h(x) = c_1y_1(x) + c_2y_2(x)$ is a general solution of the homogeneous equation (2) on I and $y_p(x)$ is any solution of (1) on I containing no arbitrary constants.

A **particular solution** of (1) on I is a solution obtained from (3) by assigning specific values to the arbitrary constants c_1 and c_2 in $y_h(x)$. ∎

If the coefficients of (1) and $r(x)$ are continuous functions on I, then (1) has a general solution on I because $y_h(x)$ exists on I by Theorem 3, Sec. 2.7, and the existence of $y_p(x)$ will be shown in Sec. 2.10. Also, an initial value problem for (1) has a unique solution on I. This follows from Theorem 1, Sec. 2.7, once the existence of $y_p(x)$ has been established. Indeed, if initial conditions

$$y(x_0) = K_0, \qquad y'(x_0) = K_1$$

are given and a y_p has been determined, by that theorem there exists a unique solution $\tilde{y}$ of the homogeneous equation (2) on I satisfying

$$\tilde{y}(x_0) = K_0 - y_p(x_0), \qquad \tilde{y}'(x_0) = K_1 - y_p'(x_0),$$

and $y = \tilde{y} + y_p$ is the unique solution of (1) on I satisfying the given initial conditions.

Furthermore, justifying the terminology, we now prove that a general solution of (1) includes *all* solutions of (1); hence the situation is the same as for the homogeneous equation:

Theorem 2 **(General solution)**
Suppose that the coefficients and $r(x)$ in (1) are continuous on some open interval I. Then every solution of (1) on I is obtained by assigning suitable values to the arbitrary constants in a general solution (3) of (1) on I.

Proof. Let $\tilde{y}(x)$ be any solution of (1) on I. Let (3) be any general solution of (1) on I; this solution exists because of our continuity assumption. Theorem 1(a) implies that the difference $Y(x) = \tilde{y}(x) - y_p(x)$ is a solution of the *homogeneous* equation (2). By Theorem 4 in Sec. 2.7, this solution $Y(x)$ is obtained from $y_h(x)$ by assigning suitable values to the arbitrary constants c_1, c_2. From this and $\tilde{y}(x) = Y(x) + y_p(x)$, the statement follows. ∎

Practical conclusion
To solve the nonhomogeneous equation (1) or an initial value problem for (1), we have to solve the homogeneous equation (2) and find any particular solution y_p of (1). Methods for this and applications will be the subject for the remaining sections of Chap. 2, which contain various examples, so that at present, with methods not yet available, we merely illustrate the basic technique and our notation by a simple example.

EXAMPLE 1 **Initial value problem for a nonhomogeneous equation**
Solve the initial value problem

$$y'' - 4y' + 3y = 10e^{-2x}, \qquad y(0) = 1, \qquad y'(0) = -3.$$

Solution. 1st Step. General solution of the homogeneous equation. The characteristic equation $\lambda^2 - 4\lambda + 3 = 0$ has the roots 1 and 3. This gives as a general solution of the homogeneous equation

$$y_h = c_1e^x + c_2e^{3x}.$$

2nd Step. Particular solution of the nonhomogeneous equation. Since e^{-2x} has derivatives e^{-2x} times some constants, we try

$$y_p = Ce^{-2x}.$$

Then $y_p' = -2Ce^{-2x}$, $y_p'' = 4Ce^{-2x}$. Substitution gives

$$4Ce^{-2x} - 4(-2Ce^{-2x}) + 3Ce^{-2x} = 10e^{-2x}.$$

Hence $4C + 8C + 3C = 10$, $C = \frac{2}{3}$, and a general solution of the nonhomogeneous equation is

$$y = y_h + y_p = c_1e^x + c_2e^{3x} + \tfrac{2}{3}e^{-2x}. \tag{4}$$

3rd Step. Particular solution satisfying the initial conditions. By differentiation,

$$y'(x) = c_1e^x + 3c_2e^{3x} - \tfrac{4}{3}e^{-2x}. \tag{5}$$

From (4) and (5) and the initial conditions,

$$y(0) = c_1 + c_2 + \tfrac{2}{3} = 1$$

$$y'(0) = c_1 + 3c_2 - \tfrac{4}{3} = -3.$$

This gives $c_1 = 4/3$, $c_2 = -1$. *Answer.* $y = \frac{4}{3}e^x - e^{3x} + \frac{2}{3}e^{-2x}$. ∎

Solution methods follow in the next sections.

Problem Set 2.8

In each case verify that $y_p(x)$ is a solution of the given differential equation and find a general solution.

1. $y'' - y = 3e^{2x}, \quad y_p = e^{2x}$
2. $y'' - y' - 2y = -4x, \quad y_p = 2x - 1$
3. $y'' + y = -3 \sin 2x, \quad y_p = \sin 2x$
4. $y'' - 2y' + y = 12e^x/x^3, \quad y_p = 6e^x/x$
5. $(D^2 + 3D - 4)y = 5e^x, \quad y_p = xe^x$
6. $(D^2 + 3D - 4)y = -6.8 \sin x, \quad y_p = \sin x + 0.6 \cos x$
7. $(x^2D^2 - 2xD + 2)y = 5x^3 \cos x, \quad y_p = -5x \cos x$
8. $(x^2D^2 - 4xD + 6)y = 42/x^4, \quad y_p = 1/x^4$
9. $(4x^2D^2 + 1)y = (1 - x^2) \cos 0.5x, \quad y_p = \cos 0.5x$
10. $(x^2D^2 - 3xD + 3)y = 3 \ln x - 4, \quad y_p = \ln x$

Verify that y_p is a solution of the given equation and solve the given initial value problem.

11. $y'' - 6y' + 9y = 2e^{3x}, \quad y(0) = 0, \quad y'(0) = 1; \quad y_p = x^2e^{3x}$
12. $8y'' - 6y' + y = 6e^x + 3x - 16, \quad y(0) = 7, \quad y'(0) = 6.5;$
$y_p = 2e^x + 2x + 2$
13. $y'' + 4y = -12 \sin 2x, \quad y(0) = 1, \quad y'(0) = 3; \quad y_p = 3x \cos 2x$
14. $(D^2 - 4D + 3)y = 10e^{-2x}, \quad y(0) = y'(0) = 5/3; \quad y_p = 2e^{-2x}/3$
15. $(D^2 - 2D + 1)y = e^x \sin x, \quad y(0) = 1, y'(0) = 0; \quad y_p = -e^x \sin x$
16. $(D^2 + 4D + 4)y = e^{-2x}/x^2, \quad y(1) = 1/e^2, \quad y'(1) = -2/e^2;$
$y_p = -e^{-2x} \ln x$

17. Show that if y_1 is a solution of (1) with $r = r_1$ and y_2 is a solution of (1) with $r = r_2$, then $y = y_1 + y_2$ is a solution of (1) with $r = r_1 + r_2$.

18. As an illustration of Prob. 17, find a general solution of the differential equation $y'' + 3y' - 4y = 5e^x - 6.8 \sin x$. *Hint.* Use Probs. 5 and 6.

19. Verify that $y_1 = e^x$ is a solution of $y'' + y = 2e^x$ and $y_2 = x \sin x$ is a solution of $y'' + y = 2 \cos x$. Using Prob. 17, find a general solution of the equation $y'' + y = 2e^x + 2 \cos x$.

20. To illustrate that the choice of y_p in (3) is immaterial, show that $y_{p1} = -\cos x$ and $y_{p2} = e^x - \cos x$ are particular solutions of $y'' - y = 2 \cos x$, find the corresponding general solutions (3), and show that one of them can be expressed in terms of the other.

2.9 Solution by Undetermined Coefficients

A general solution of a nonhomogeneous linear equation is a sum of the form

$$y = y_h + y_p,$$

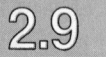

where y_h is a general solution of the corresponding homogeneous equation and y_p is any particular solution of the nonhomogeneous equation. This has just been shown. Hence our main task is to discuss methods for finding such y_p. There is a general method for this which always works and which we shall consider in the next section. There also is a much simpler special method of practical interest which we discuss now. It is called the **method of undetermined coefficients** and applies to equations

$$y'' + ay' + by = r(x) \tag{1}$$

with constant coefficients and special right sides $r(x)$, namely, exponential functions, polynomials, cosines, sines, or sums or products of such functions. These $r(x)$ have derivatives of a form similar to $r(x)$ itself. This gives the key idea: Choose for y_p a form similar to that of $r(x)$ and involving unknown coefficients to be determined by substituting that choice for y_p into (1). Example 1 in the last section illustrates this for an exponential function; the undetermined coefficient was C. The rules of the method are as follows.

Rules for the Method of Undetermined Coefficients

(A) Basic Rule. *If $r(x)$ in* (1) *is one of the functions in the first column in Table* 2.1, *choose the corresponding function y_p in the second column and determine its undetermined coefficients by substituting y_p and its derivatives into* (1).

(B) Modification Rule. *If a term in your choice for y_p happens to be a solution of the homogeneous equation corresponding to* (1), *then multiply your choice of y_p by x (or by x^2 if this solution corresponds to a double root of the characteristic equation of the homogeneous equation).*

(C) Sum Rule. *If $r(x)$ is a sum of functions listed in several lines of Table* 2.1, *first column, then choose for y_p the sum of the functions in the corresponding lines of the second column.*

The Basic Rule tells us what to do in general. The Modification Rule takes care of difficulties that occur in the case indicated. Accordingly, we always have to solve the homogeneous equation first. The Sum Rule is obtained if we note that the sum of two solutions of (1) with $r = r_1$ and $r = r_2$, respectively, is a solution of (1) with $r = r_1 + r_2$. (Verify!)

The method corrects itself in the sense that a false choice of y_p or one with too few terms will lead to a contradiction, usually indicating the necessary correction, and a choice of too many terms will give a correct result, with superfluous coefficients coming out zero.

Table 2.1
Method of Undetermined Coefficients

Term in $r(x)$	Choice for y_p
$ke^{\gamma x}$	$Ce^{\gamma x}$
$kx^n \ (n = 0, 1, \cdots)$	$K_n x^n + K_{n-1}x^{n-1} + \cdots + K_1 x + K_0$
$k \cos \omega x$ $k \sin \omega x$	$K \cos \omega x + M \sin \omega x$
$ke^{\alpha x} \cos \omega x$ $ke^{\alpha x} \sin \omega x$	$e^{\alpha x}(K \cos \omega x + M \sin \omega x)$

EXAMPLE 1 Application of Rule (A)

Solve the nonhomogeneous equation

(2) $$y'' + 4y = 8x^2.$$

Solution. Table 2.1 suggests the choice

$$y_p = K_2x^2 + K_1x + K_0. \qquad \text{Then} \qquad y_p'' = 2K_2.$$

Substitution gives

$$2K_2 + 4(K_2x^2 + K_1x + K_0) = 8x^2.$$

Equating the coefficients of x^2, x, and x^0 on both sides, we have $4K_2 = 8$, $4K_1 = 0$, $2K_2 + 4K_0 = 0$. Thus $K_2 = 2$, $K_1 = 0$, $K_0 = -1$. Hence $y_p = 2x^2 - 1$, and a general solution of (2) is

$$y = y_h + y_p = A \cos 2x + B \sin 2x + 2x^2 - 1.$$

Note well that although $r(x) = 8x^2$, a trial $y_p = K_2x^2$ would fail. Try it. Can you see why it fails? ∎

EXAMPLE 2 Modification Rule (B) in the case of a simple root

Solve

(3) $$y'' - 3y' + 2y = e^x.$$

Solution. The characteristic equation $\lambda^2 - 3\lambda + 2 = 0$ has the roots 1 and 2. Hence $y_h = c_1e^x + c_2e^{2x}$. Ordinarily, our choice would be $y_p = Ce^x$. But we see that e^x is a solution of the homogeneous equation corresponding to a simple root (namely, 1). Hence Rule (B) suggests

$$y_p = Cxe^x. \qquad \text{We need} \qquad y_p' = C(e^x + xe^x), \qquad y_p'' = C(2e^x + xe^x).$$

Substitution gives

$$C(2 + x)e^x - 3C(1 + x)e^x + 2Cxe^x = e^x.$$

The xe^x-terms cancel out, and $-Ce^x = e^x$ remains. Hence $C = -1$. A general solution is

$$y = c_1e^x + c_2e^{2x} - xe^x.$$

Check it! Try $y_p = Ce^x$ to convince yourself that it does not work. ∎

EXAMPLE 3 Modification Rule (B) (double root) and Sum Rule (C)

Solve the initial value problem

(4) $$y'' - 2y' + y = (D - 1)^2y = e^x + x, \qquad y(0) = 1, \qquad y'(0) = 0.$$

Solution. The characteristic equation has the double root $\lambda = 1$. Hence $y_h = (c_1 + c_2x)e^x$. We determine a particular solution y_p. By Table 2.1, the term x indicates a particular solution choice

$$K_1x + K_0.$$

Since 1 is a double root of the characteristic equation $(\lambda - 1)^2 = 0$, by the Modification Rule the term e^x calls for the particular solution

$$Cx^2e^x \qquad \text{(instead of } Ce^x\text{)}.$$

Together,

$$y_p = K_1x + K_0 + Cx^2e^x.$$

Substituting this into (4) and simplifying, we obtain

$$y_p'' - 2y_p' + y_p = 2Ce^x + K_1x - 2K_1 + K_0 = e^x + x.$$

Hence $C = \frac{1}{2}$, $K_1 = 1$, $K_0 = 2$, and a general solution of (4) is

$$y = y_h + y_p = (c_1 + c_2x)e^x + \tfrac{1}{2}x^2e^x + x + 2.$$

To take care of the initial conditions, we also need

$$y' = (c_1 + c_2 + c_2x)e^x + (x + \tfrac{1}{2}x^2)e^x + 1.$$

Hence

$$y(0) = c_1 + 2 = 1, \qquad c_1 = -1,$$
$$y'(0) = c_1 + c_2 + 1 = 0, \qquad c_2 = 0.$$

Answer. $y = -e^x + \frac{1}{2}x^2e^x + x + 2.$ ■

EXAMPLE 4 Another application of Sum Rule (C)

Solve

$$\text{(5)} \qquad y'' + 2y' + 5y = 16e^x + \sin 2x.$$

Solution. The characteristic equation $\lambda^2 + 2\lambda + 5 = 0$ has the complex roots $-1 + 2i$ and $-1 - 2i$. Hence $y_h = e^{-x}(A \cos 2x + B \sin 2x)$. By Table 2.1 and differentiation,

$$y_p = Ce^x + K \cos 2x + M \sin 2x,$$
$$y_p' = Ce^x - 2K \sin 2x + 2M \cos 2x,$$
$$y_p'' = Ce^x - 4K \cos 2x - 4M \sin 2x.$$

We substitute this into (5) and collect terms, finding

$$8Ce^x + (-4K + 4M + 5K) \cos 2x + (-4M - 4K + 5M) \sin 2x = 16e^x + \sin 2x.$$

Hence $8C = 16$, $K + 4M = 0$, $-4K + M = 1$, so that $C = 2$, $K = -4/17$, $M = 1/17$. The answer is

$$y = e^{-x}(A \cos 2x + B \sin 2x) + 2e^x - \tfrac{4}{17} \cos 2x + \tfrac{1}{17} \sin 2x.$$ ■

Problem Set 2.9

Find a general solution of the following differential equations.

1. $y'' + y = 3x^2$
2. $y'' - 4y = e^{2x}$
3. $y'' + 6y' + 9y = 18 \cos 3x$
4. $y'' + 4y' + 4y = 9 \cosh x$
5. $y'' - y' - 2y = e^x + x$
6. $y'' - 2y' + 2y = 2e^x \cos x$
7. $y'' + 2y' + 5y = 5x^2 + 4x + 2$
8. $3y'' + 10y' + 3y = x^2 + \sin x$
9. $(D^2 - 4D + 3)y = 2 \sin x - 4 \cos x$
10. $(D^2 + 5D + 6)y = 9x^4 - x$
11. $(D^2 - 2D + 1)y = 2e^x$
12. $(D^2 - 4D + 3)y = 8e^{-3x} + e^{3x}$
13. $(D^2 + 9)y = \cos 3x$
14. $(D^2 + 5D)y = 1 + x + e^x$
15. $(D^2 - 4)y = 2 \sinh 2x + x$
16. $(D^2 + D - 6)y = 52 \cos 2x$

Solve the following initial value problems.

17. $y'' - y' - 2y = 3e^{2x}$, $y(0) = 0$, $y'(0) = -2$

18. $y'' - y = x$, $y(2) = e^2 - 2 \approx 5.389$, $y'(2) = e^2 - 1 \approx 6.389$

19. $y'' + y' - 2y = 14 + 2x - 2x^2$, $y(0) = 0$, $y'(0) = 0$

20. $y'' - 4y' + 3y = 4e^{3x}$, $y(0) = -1$, $y'(0) = 3$

21. $y'' - y' - 2y = 10 \sin x$, $y(\frac{1}{2}\pi) = -3$, $y'(\frac{1}{2}\pi) = -1$

22. $y'' + 4y' + 4y = 4 \cos x + 3 \sin x$, $y(0) = 1$, $y'(0) = 0$

23. $y'' + y' - 2y = -6 \sin 2x - 18 \cos 2x$, $y(0) = 2$, $y'(0) = 2$

First-Order Equations. The method of undetermined coefficients can also be applied to certain first-order linear differential equations, and may sometimes be simpler than the usual method (Sec. 1.7). Using both methods, solve:

24. $y' - y = x^5$

25. $y' + 2y = \cos 2x$

2.10 Solution by Variation of Parameters

The method in the last section is simple and has important engineering applications (as we shall see in the next sections), but it applies only to constant-coefficient equations with special right sides $r(x)$. In this section we discuss the so-called method of variation of parameters,[12] which is completely general; that is, it applies to equations

$$y'' + p(x)y' + q(x)y = r(x) \tag{1}$$

with arbitrary variable function p, q, and r that are continuous on some interval I. The method gives a particular solution y_p of (1) on I in the form

$$y_p(x) = -y_1 \int \frac{y_2 r}{W}\, dx + y_2 \int \frac{y_1 r}{W}\, dx, \tag{2}$$

where y_1, y_2 form a basis of solutions of the homogeneous equation

$$y'' + p(x)y' + q(x)y = 0 \tag{3}$$

corresponding to (1) and

$$W = y_1 y_2' - y_1' y_2 \tag{4}$$

is the Wronskian of y_1, y_2 (see Sec. 2.7).

In practice, this method is much more complicated than our previous method, because of the integrations in (2). Let us first see an example to which our previous method does not apply (as the answer will show).

[12] Credited to Lagrange (see footnote 11 in Sec. 2.7).

EXAMPLE 1 Solve the differential equation

$$y'' + y = \sec x.$$

Solution. A basis of solutions of the homogeneous equation on any interval is

$$y_1 = \cos x, \qquad y_2 = \sin x.$$

This gives the Wronskian

$$W(y_1, y_2) = \cos x \cos x - (-\sin x) \sin x = 1.$$

Hence from (2), choosing the constants of integration to be zero, we get the particular solution

$$\begin{aligned} y_p &= -\cos x \int \sin x \sec x \, dx + \sin x \int \cos x \sec x \, dx \\ &= \cos x \ln |\cos x| + x \sin x \end{aligned}$$

of the given equation and from this the general solution

$$y = y_h + y_p = [c_1 + \ln |\cos x|] \cos x + (c_2 + x) \sin x.$$

This is the answer. Had we included two arbitrary constants of integration $-c_1$, c_2, we would have obtained in (2) the additional $c_1 \cos x + c_2 \sin x = c_1 y_1 + c_2 y_2$, that is, a general solution of the given equation directly from (2). This will always be the case. ∎

Idea of the Method. Derivation of (2)

What was Lagrange's idea? Where does the name of the method come from? How can we get (2)? Where do we use the continuity assumption?

The continuity of p and q implies that the homogeneous equation (3) has a general solution

$$y_h(x) = c_1 y_1(x) + c_2 y_2(x)$$

on I, by Theorem 3 in Sec. 2.7. The method of variation of parameters involves replacing the constants c_1 and c_2 (here regarded as "parameters" in y_h) by functions $u(x)$ and $v(x)$ to be determined so that the resulting function

$$y_p(x) = u(x)y_1(x) + v(x)y_2(x) \tag{5}$$

is a particular solution of (1) on I. By differentiating (5) we obtain

$$y_p' = u'y_1 + uy_1' + v'y_2 + vy_2'.$$

Now (5) contains *two* functions u and v, but the requirement that y_p satisfy (1) imposes only *one* condition on u and v. Hence it seems plausible that we may impose a second arbitrary condition. Indeed, our further calculation will show that we can determine u and v such that y_p satisfies (1) and u and v satisfy as a second condition the relationship

$$u'y_1 + v'y_2 = 0. \tag{6}$$

This reduces the expression for y_p' to the form

$$y_p' = uy_1' + vy_2'. \tag{7}$$

By differentiating this function we have

$$y_p'' = u'y_1' + uy_1'' + v'y_2' + vy_2''. \tag{8}$$

Substituting (5), (7), and (8) into (1) and collecting terms containing u and terms containing v, we readily obtain

$$u(y_1'' + py_1' + qy_1) + v(y_2'' + py_2' + qy_2) + u'y_1' + v'y_2' = r.$$

Since y_1 and y_2 are solutions of the homogeneous equation (2), this reduces to

$$u'y_1' + v'y_2' = r.$$

Equation (6) is

$$u'y_1 + v'y_2 = 0.$$

This is a linear system of two algebraic equations for the unknown functions u' and v'. The solution is obtained by Cramer's rule (Sec. 7.8) or as follows. Multiply the first equation by $-y_2$ and the second by y_2' and add to get

$$u'(y_1y_2' - y_2y_1') = -y_2r, \qquad \text{thus} \qquad u'W = -y_2r,$$

where W is the Wronskian (4) of y_1, y_2. Now multiply the first equation by y_1 and the second by $-y_1'$ and add to get

$$v'(y_1y_2' - y_2y_1') = y_1r, \qquad \text{thus} \qquad v'W = y_1r.$$

Division by $W \neq 0$ (y_1, y_2 form a basis, hence $W \neq 0$ by Theorem 2 in Sec. 2.7!) gives.

$$u' = -\frac{y_2r}{W}, \qquad v' = \frac{y_1r}{W}. \tag{9}$$

By integration,

$$u = -\int \frac{y_2r}{W}\,dx, \qquad v = \int \frac{y_1r}{W}\,dx.$$

These integrals exist because $r(x)$ is continuous. Substituting them into (5), we obtain (2). This completes the derivation. ∎

CAUTION! Before applying (2), make sure that your equation is written in the standard form (1), with y'' as the first term; divide by $f(x)$ if it starts with $f(x)y''$.

Problem Set 2.10

Find a general solution of the following equations.

1. $y'' - 2y' + y = x^{3/2}e^x$

2. $y'' + 4y = 2 \sec 2x$

3. $y'' - 2y' + y = 12e^x/x^3$

4. $y'' + y = \csc x$

5. $y'' + 4y' + 4y = e^{-2x}/x^2$

6. $y'' - 4y' + 5y = 2e^{2x}/\sin x$

7. $y'' - 2y' + y = 35x^{3/2}e^x + x^2$

8. $y'' - 4y' + 4y = (3x^2 + 2)e^x$

9. $y'' - 2y' + y = e^x \sin x$

10. $y'' + 6y' + 9y = 8e^{-3x}/(x^2 + 1)$

11. $(D^2 - 4D + 4)y = 6x^{-4}e^{2x}$

12. $(D^2 + 9)y = \sec 3x$

13. $(D^2 - 2D + 1)y = e^x/x^3$

14. $(D^2 + 2D + 2)y = e^{-x}/\cos^3 x$

Nonhomogeneous Euler–Cauchy Equations. Find a general solution of the following equations. ***Caution!*** First divide the equation by the coefficient of y'' to get the standard form (1).

15. $(x^2D^2 - 4xD + 6)y = 42/x^4$

16. $(x^2D^2 - 2)y = 9x^2$

17. $(x^2D^2 - 2xD + 2)y = 5x^3 \cos x$

18. $(xD^2 - D)y = (3 + x)x^2e^x$

19. $(xD^2 - D)y = x^2e^x$

20. $(x^2D^2 - 4xD + 6)y = -7x^4 \sin x$

21. $x^2y'' - 2xy' + 2y = 24/x^2$

22. $4x^2y'' + 4xy' - y = 12/x$

23. $x^2y'' - 4xy' + 6y = 1/x^4$

24. $x^2y'' - 2xy' + 2y = x^4$

25. **(Comparison of methods)** Whenever the method of undetermined coefficients (Sec. 2.9) is applicable, it should be used because it is simpler than the present method. To illustrate this fact, solve by both methods

$$y'' + 4y' + 3y = 65 \cos 2x.$$

2.11 Modeling: Forced Oscillations. Resonance

Free motions of the mass–spring system in Fig. 43 are motions in the absence of external forces, and they are governed by the homogeneous differential equation

$$my'' + cy' + ky = 0 \qquad (1) \qquad \text{(Sec. 2.5).}$$

Here, y is the displacement of the body from rest, m the mass of the body,

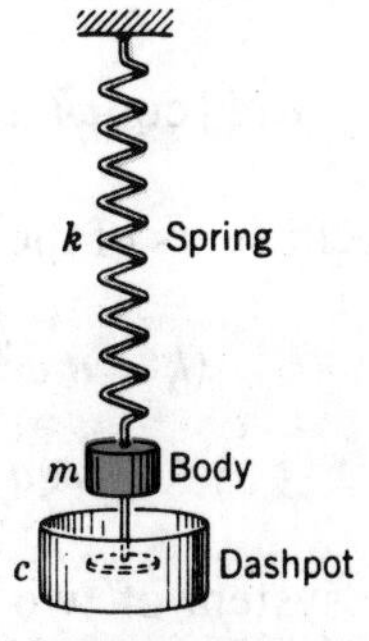

Fig. 43. Mass on a spring

my'' the force of inertia, cy' the damping force, and ky the spring force. **Forced motions** are obtained if we let an external force $r(t)$ act on the body. To get the model, we simply have to add our new force $r(t)$ to those forces; this gives the nonhomogeneous differential equation

$$my'' + cy' + ky = r(t).$$

$r(t)$ is called the **input** or **driving force,** and a corresponding solution is called an **output** or a **response** *of the system to the driving force*. (See also Sec. 1.7.)

Of particular interest are *periodic inputs*, and we shall consider a sinusoidal force, say,

$$r(t) = F_0 \cos \omega t \qquad (F_0 > 0, \ \omega > 0).$$

Then we have the equation

$$my'' + cy' + ky = F_0 \cos \omega t, \tag{2}$$

whose solution will familiarize us with further interesting facts fundamental in engineering mathematics, in particular with resonance.

Solving the Equation

A general solution of (2) is the sum of a general solution y_h of (1), which we know from Sec. 2.5, and a particular solution y_p of (2), which we can best determine by the method of undetermined coefficients (Sec. 2.9). Accordingly, we start from

$$y_p(t) = a \cos \omega t + b \sin \omega t. \tag{3}$$

By differentiating this function we have

$$\begin{aligned} y_p' &= -\omega a \sin \omega t + \omega b \cos \omega t, \\ y_p'' &= -\omega^2 a \cos \omega t - \omega^2 b \sin \omega t. \end{aligned}$$

We substitute these expressions into (2) and collect the cosine and sine terms:

$$[(k - m\omega^2)a + \omega c b] \cos \omega t + [-\omega c a + (k - m\omega^2)b] \sin \omega t = F_0 \cos \omega t.$$

Equating the coefficients of the cosine and sine terms on both sides we have

$$\begin{aligned} (k - m\omega^2)a + \qquad \omega c b \quad &= F_0 \\ -\omega c a \quad + (k - m\omega^2)b &= 0. \end{aligned} \tag{4}$$

This is a linear system of two algebraic equations in the two unknowns a and b. The solution is obtained in the usual way by elimination or by Cramer's rule (if necessary, see Sec. 7.8). We find

$$a = F_0 \frac{k - m\omega^2}{(k - m\omega^2)^2 + \omega^2 c^2}, \qquad b = F_0 \frac{\omega c}{(k - m\omega^2)^2 + \omega^2 c^2}$$

provided the denominator is not zero. If we set $\sqrt{k/m} = \omega_0$ (> 0) as in Sec. 2.5, this becomes

$$a = F_0 \frac{m(\omega_0^2 - \omega^2)}{m^2(\omega_0^2 - \omega^2)^2 + \omega^2 c^2}, \qquad b = F_0 \frac{\omega c}{m^2(\omega_0^2 - \omega^2)^2 + \omega^2 c^2}. \tag{5}$$

We thus obtain the general solution of (2) in the form

$$y(t) = y_h(t) + y_p(t), \tag{6}$$

where y_h is a general solution of (1) and y_p is given by (3) with coefficients (5).

Discussion of Types of Solutions

We shall now discuss the behavior of the mechanical system, distinguishing between the two cases $c = 0$ (no damping) and $c > 0$ (damping). These cases will correspond to two different types of output.

Case 1. Undamped forced oscillations

If there is no damping, then $c = 0$. We first assume that $\omega^2 \neq \omega_0^2$ (where $\omega_0^2 = k/m$, as in Sec. 2.5). This is essential. We then obtain from (3) and (5) (where $b = 0$ because $c = 0$)

$$y_p(t) = \frac{F_0}{m(\omega_0^2 - \omega^2)} \cos \omega t = \frac{F_0}{k[1 - (\omega/\omega_0)^2]} \cos \omega t. \tag{7}$$

From this and (6*) in Sec. 2.5 we have the general solution

$$y(t) = C \cos (\omega_0 t - \delta) + \frac{F_0}{m(\omega_0^2 - \omega^2)} \cos \omega t. \tag{8}$$

*This output represents a superposition of two harmonic oscillations; the frequencies are the "**natural frequency**" $\omega_0/2\pi$ [cycles/sec] of the system (that is, the frequency of the free undamped motion) and the frequency $\omega/2\pi$ of the input.*

From (7) we see that the maximum amplitude of y_p is

$$a_0 = \frac{F_0}{k}\rho \qquad \text{where} \qquad \rho = \frac{1}{1 - (\omega/\omega_0)^2}. \tag{9}$$

a_0 depends on ω and ω_0. If $\omega \to \omega_0$, then ρ and a_0 tend to infinity. This phenomenon of excitation of large oscillations by matching input and natural frequencies ($\omega = \omega_0$) is known as **resonance,** and is of basic importance in the study of vibrating systems (see below). The quantity ρ is called the *resonance factor* (Fig. 44). From (9) we see that $\rho/k = a_0/F_0$ is the ratio of the amplitudes of the function y_p and of the input.

In the case of resonance, equation (2) becomes

$$y'' + \omega_0^2 y = \frac{F_0}{m} \cos \omega_0 t. \tag{10}$$

From the Modification Rule in Sec. 2.9 we conclude that a particular solution of (10) is of the form

$$y_p(t) = t(a \cos \omega_0 t + b \sin \omega_0 t).$$

By substituting this into (10) we find $a = 0$, $b = F_0/2m\omega_0$, and (Fig. 45)

$$y_p(t) = \frac{F_0}{2m\omega_0} t \sin \omega_0 t. \tag{11}$$

We see that y_p becomes larger and larger. In practice, this means that systems with very little damping may undergo large vibrations that can destroy the system; we shall return to this practical aspect of resonance later in this section.

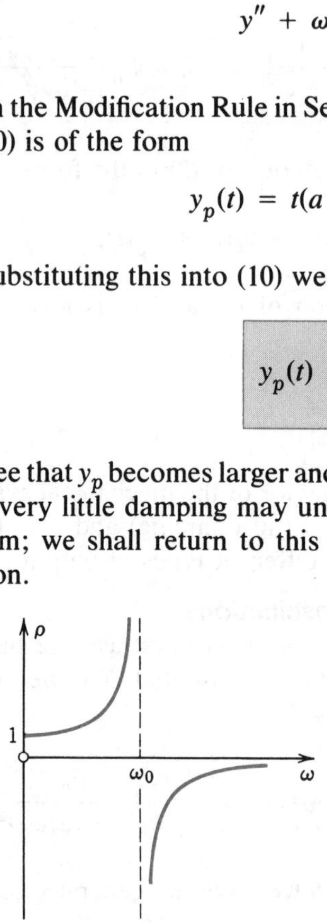

Fig. 44. Resonance factor $\rho(\omega)$

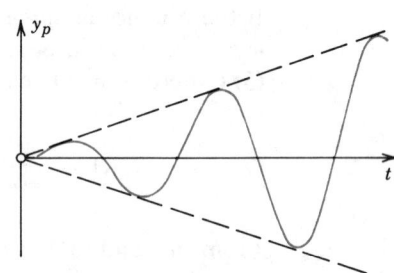

Fig. 45. Particular solution in the case of resonance

Another interesting and highly important type of oscillation is obtained when ω is close to ω_0. Take, for example, the particular solution [see (8)]

$$y(t) = \frac{F_0}{m(\omega_0^2 - \omega^2)} (\cos \omega t - \cos \omega_0 t) \qquad (\omega \neq \omega_0) \tag{12}$$

corresponding to the initial conditions $y(0) = 0$, $y'(0) = 0$. This may be written [see (12) in Appendix 3]

$$y(t) = \frac{2F_0}{m(\omega_0^2 - \omega^2)} \sin \frac{\omega_0 + \omega}{2} t \sin \frac{\omega_0 - \omega}{2} t.$$

Since ω is close to ω_0, the difference $\omega_0 - \omega$ is small, so that the period of the last sine function is large, and we obtain an oscillation of the type shown in Fig. 46. This is what musicians are listening to when they ***tune*** their instruments.

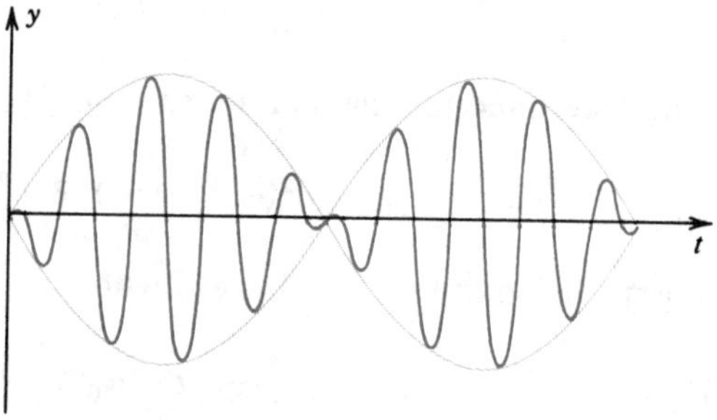

Fig. 46. Forced undamped oscillation when the difference of the input and natural frequencies is small (**"beats"**)

Case 2. Damped forced oscillations

If there is damping, then $c > 0$, and we know from Sec. 2.5 that the general solution y_h of (1) is

$$y_h(t) = e^{-\alpha t}(A \cos \omega^* t + B \sin \omega^* t) \qquad \left(\alpha = \frac{c}{2m} > 0\right)$$

and this solution approaches zero as t approaches infinity (practically, after a sufficiently long time); that is, the general solution (6) of (2) now represents the **transient solution** and tends to the **steady-state solution** y_p. *Hence, after a sufficiently long time, the output corresponding to a purely sinusoidal input will practically be a harmonic oscillation whose frequency is that of the input. This is what happens in practice, because no physical system is completely undamped.*

Whereas in the undamped case the amplitude of y_p approaches infinity as ω approaches ω_0, this will not happen in the damped case; *in this case the amplitude will always be finite,* but may have a maximum for some ω, depending on c. This may be called **practical resonance.** It is of great importance because it shows that some input may excite oscillations with such a large amplitude that the system can be destroyed. Such cases happened in practice, in particular in earlier times when less was known about resonance. Machines, cars, ships, airplanes, and bridges are vibrating mechanical systems, and it is sometimes rather difficult to find constructions that are completely free of undesired resonance effects.

Amplitude of y_p

To study the amplitude of y_p as a function of ω, we write (3) in the form

$$y_p(t) = C^* \cos(\omega t - \eta) \tag{13}$$

where, according to (5),

$$
(14) \qquad C^*(\omega) = \sqrt{a^2 + b^2} = \frac{F_0}{\sqrt{m^2(\omega_0^2 - \omega^2)^2 + \omega^2 c^2}}, \qquad \tan \eta = \frac{b}{a} = \frac{\omega c}{m(\omega_0^2 - \omega^2)}.
$$

Let us determine the maximum of $C^*(\omega)$. By setting $dC^*/d\omega = 0$ we find

$$
[-2m^2(\omega_0^2 - \omega^2) + c^2]\,\omega = 0.
$$

The expression in brackets is zero when

$$
(15) \qquad c^2 = 2m^2(\omega_0^2 - \omega^2).
$$

For sufficiently large damping ($c^2 > 2m^2\omega_0^2 = 2mk$) equation (15) has no real solution, and C^* decreases in a monotone way as ω increases (Fig. 47). If $c^2 \leqq 2mk$, equation (15) has a real solution $\omega = \omega_{max}$, which increases as c decreases and approaches ω_0 as c approaches zero. The amplitude $C^*(\omega)$ has a maximum at $\omega = \omega_{max}$, and by inserting $\omega = \omega_{max}$ into (14) we find

$$
(16) \qquad C^*(\omega_{max}) = \frac{2mF_0}{c\sqrt{4m^2\omega_0^2 - c^2}}.
$$

We see that $C^*(\omega_{max})$ is finite when $c > 0$. Since $dC^*(\omega_{max})/dc < 0$ when $c^2 < 2mk$, the value of $C^*(\omega_{max})$ increases as c ($\leqq \sqrt{2mk}$) decreases and approaches infinity as c approaches zero, in agreement with our result in Case 1. Figure 47 shows the **amplification** C^*/F_0 (ratio of the amplitudes of output and input) as a function of ω for $m = 1$, $k = 1$, and various values of the damping constant c.

The angle η in (14) is called the **phase angle** or **phase lag** (Fig. 48) because it measures the lag of the output behind the input. If $\omega < \omega_0$, then $\eta < \pi/2$; if $\omega = \omega_0$, then $\eta = \pi/2$, and if $\omega > \omega_0$, then $\eta > \pi/2$.

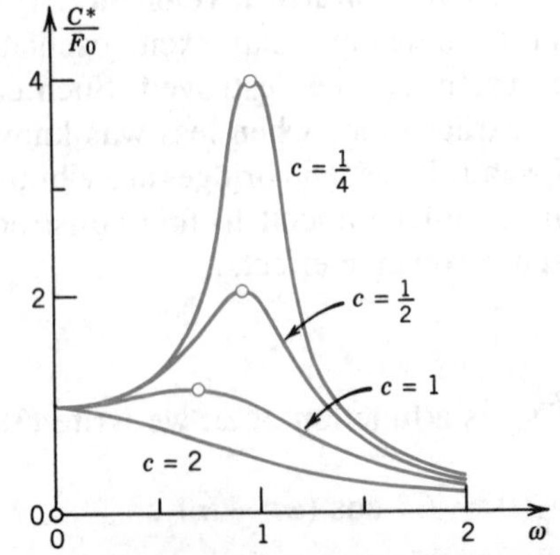

Fig. 47. Amplification C^*/F_0 as a function of ω for $m = 1$, $k = 1$, and various values of the damping constant c

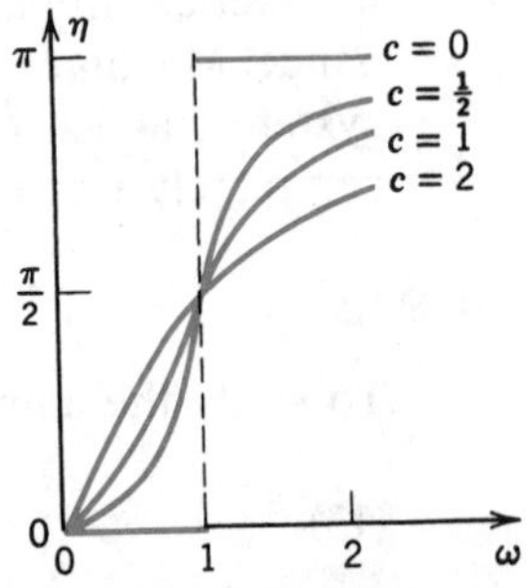

Fig. 48. Phase lag η as a function of ω for $m = 1$, $k = 1$, and various values of the damping constant c

Problem Set 2.11

Find the steady-state oscillations of the vibrating systems governed by the following equations.

1. $(D^2 + 4)y = 15 \sin t$

2. $y'' + y = \cos 2t$

3. $y'' + 3y' + 2y = 40 \sin t$

4. $2y'' + 2y' + 3y = \cos 3t - 5 \sin t$

5. $y'' + 5y' + 6y = 6.29 \sin \frac{1}{2}t$

6. $(D^2 + D + 1)y = \cos t + 13 \cos 2t$

7. $(D^2 + 2D + 4)y = \sin 0.2t$

8. $(3D^2 + D + 2)y = 2 \cos t - 52 \sin 2t$

Find the transient motions of the vibrating systems governed by the following equations.

9. $y'' + 25y = 48 \sin t$

10. $y'' + 2y' + 5y = -\sin t$

11. $y'' + 2y' + 2y = \cos t$

12. $y'' + 2y' + y = 50 \sin 3t$

13. $(D^2 + \frac{1}{2}D + 2)y = 5 \cos t$

14. $(D^2 + 4D + 5)y = 37.7 \sin 4t$

15. $(D^2 + 1)y = \cos \omega t, \ \omega^2 \neq 1$

16. $(D^2 + 4D + 20)y = \sin t + \frac{4}{19} \cos t$

In each case, the given differential equation is the mathematical model of a vibrating system. Find the motion of the system corresponding to the given initial displacement and initial velocity.

17. $y'' + 9y = 8 \sin t, \quad y(0) = 1, \quad y'(0) = 1$

18. $y'' + 4y' + 20y = 23 \sin t - 15 \cos t, \quad y(0) = 0, \quad y(0) = -1$

19. $(D^2 + \omega_0^2)y = \cos \omega t, \quad y(0) = y_0, \quad y(0) = v_0, \quad \omega^2 \neq \omega_0^2$

20. $(D^2 + 4)y = \sin t + \frac{1}{3} \sin 3t + \frac{1}{5} \sin 5t, \quad y(0) = 1, \quad y(0) = \frac{3}{35}$

21. $(D^2 + D + 0.25)y = 2 \cos t - \frac{3}{2} \sin t - 2 \cos 2t + 3.75 \sin 2t,$
$y(0) = 0, \quad y'(0) = 1.5$

22. (Gun barrel) Solve

$$y'' + y = \begin{cases} 1 - t^2/\pi^2 & \text{if } 0 \leqq t \leqq \pi \\ 0 & \text{if } t > \pi \end{cases} \qquad y(0) = y'(0) = 0.$$

This may be interpreted as an undamped system on which a force F acts during some interval of time (see Fig. 49), for instance, the force acting on a gun barrel when a shell is fired, the barrel being braked by heavy springs (and then damped by a dashpot, which we disregard for simplicity). *Hint.* At $t = \pi$ both y and y' must be continuous.

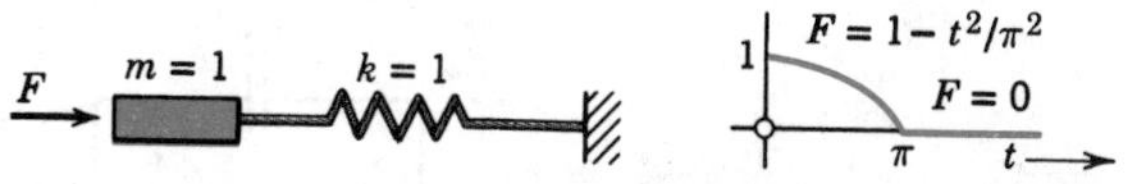

Fig. 49. Problem 22

23. In (12) let ω approach ω_0. Show that this leads to a solution of the form (11).

24. For what initial conditions $y(0) = y_0$, $y'(0) = v_0$ does the solution of Prob. 15 represent an oscillation whose frequency equals the frequency of the input?

25. Solve the initial value problem $y'' + y = \cos \omega t$, $\omega^2 \neq 1$, $y(0) = 0$, $y'(0) = 0$. Graph the maximum amplitude as a function of ω. Show that the solution can be written

$$y(t) = \frac{2}{1 - \omega^2} \sin \left[\frac{1}{2}(1 + \omega)t\right] \sin \left[\frac{1}{2}(1 - \omega)t\right].$$

Sketch graphs of $y(t)$ with $\omega = 0.5, 0.9, 1.1, 2$.

2.12 Modeling of Electric Circuits

The last section was devoted to the study of a mechanical system that is of great practical interest. We shall now consider a similarly important *electrical system*, which may be regarded as a basic building block in electrical networks. This consideration will also provide a striking example of the important fact that ***entirely different physical systems may correspond to the same mathematical model***—in the present case, to the same differential equation—so that they can be treated and solved by the same methods. This is an impressive demonstration of the ***unifying power*** of mathematics.

Indeed, we shall obtain a *correspondence between mechanical and electrical systems that is not merely qualitative but strictly quantitative* in the sense that to a given mechanical system we can construct an electric circuit whose current will give the exact values of the displacement in the mechanical system when suitable scale factors are introduced.

The practical importance of such an ***analogy between mechanical and electrical systems*** is almost obvious. The analogy may be used for constructing an "electrical model" of a given mechanical system; in many cases this will be an essential simplification, because electric circuits are easy to assemble and currents and voltages are easy to measure, whereas the construction of a mechanical model may be complicated and expensive, and the measurement of displacements will be time-consuming and relatively inaccurate.

Setting up the Model

We consider the *RLC*-circuit in Fig. 50, in which an Ohm's resistor of resistance R [ohms], an inductor of inductance L [henrys], and a capacitor of capacitance C [farads] are connected in series to a source of electromotive force $E(t)$ [volts], where t is time. The equation for the current $I(t)$ [amperes]

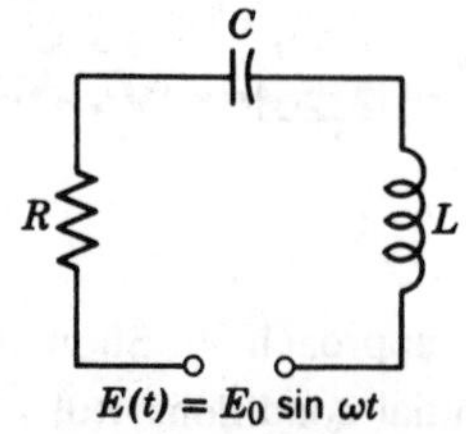

Fig. 50. *RLC*-circuit

in the *RLC*-circuit is obtained by considering the three voltage drops

$E_L = LI'$ across the inductor,

$E_R = RI$ across the resistor (*Ohm's law*), and

$E_C = \frac{1}{C}\int I(t)\,dt$ across the capacitor.

Their sum equals the electromotive force $E(t)$. This is **Kirchhoff's voltage law** (Sec. 1.8), the analog of Newton's second law (Sec. 2.5) for mechanical systems. For a sinusoidal $E(t) = E_0 \sin \omega t$ (E_0 constant), this law gives

$$(1') \qquad LI' + RI + \frac{1}{C}\int I\,dt = E(t) = E_0 \sin \omega t.$$

This process of modeling is the same as that in Sec. 1.8. Indeed, if we add $E_L = LI'$ to the equation (7) in Sec. 1.8 for the *RC*-circuit, we obtain our present equation (1′) for the *RLC*-circuit.

To get rid of the integral in (1′), we differentiate with respect to t, obtaining

$$(1) \qquad LI'' + RI' + \frac{1}{C}I = E_0\omega \cos \omega t.$$

This is of the same form as (2), *Sec.* 2.11. *Hence our RLC-circuit is the electrical analog of the mechanical system in Sec.* 2.11. *The corresponding analogy of electrical and mechanical quantities is shown in Table* 2.2.

Table 2.2
Analogy of Electrical and Mechanical Quantities in (1), This Section, and (2), Sec. 2.11

Electrical System	Mechanical System
Inductance L	Mass m
Resistance R	Damping constant c
Reciprocal $1/C$ of capacitance	Spring modulus k
Derivative $E_0\omega \cos \omega t$ of electromotive force	Driving force $F_0 \cos \omega t$
Current $I(t)$	Displacement $y(t)$

Comment. Recalling from Sec. 1.8 that $I = Q'$, we have $I' = Q''$ and $Q = \int I\,dt$ in (1′). Hence we obtain from (1′) as the differential equation for the charge Q on the capacitor

$$(1'') \qquad LQ'' + RQ' + \frac{1}{C}Q = E_0 \sin \omega t.$$

In most practical problems, the current $I(t)$ is more important than $Q(t)$, and for this reason we shall concentrate on (1) rather than on (1″).

Solving Equation (1), Discussion of Solution

To obtain a particular solution of (1) we may proceed as in Sec. 2.11. By substituting

$$I_p(t) = a \cos \omega t + b \sin \omega t \tag{2}$$

into (1) we obtain

$$a = \frac{-E_0 S}{R^2 + S^2}, \qquad b = \frac{E_0 R}{R^2 + S^2} \tag{3}$$

where S is the so-called **reactance**, given by the expression

$$S = \omega L - \frac{1}{\omega C}. \tag{4}$$

In any practical case, $R \neq 0$, so that the denominator in (3) is not zero. The result is that (2), with a and b given by (3), is a particular solution of (1).

Using (3), we may write I_p in the form

$$I_p(t) = I_0 \sin(\omega t - \theta) \tag{5}$$

where [see (14) in Appendix 3]

$$I_0 = \sqrt{a^2 + b^2} = \frac{E_0}{\sqrt{R^2 + S^2}}, \qquad \tan \theta = -\frac{a}{b} = \frac{S}{R}.$$

The quantity $\sqrt{R^2 + S^2}$ is called the **impedance**. Our formula shows that the impedance equals the ratio E_0/I_0, which is somewhat analogous to $E/I = R$ (Ohm's law).

A general solution of the homogeneous equation corresponding to (1) is

$$I_h = c_1 e^{\lambda_1 t} + c_2 e^{\lambda_2 t},$$

where λ_1 and λ_2 are the roots of the characteristic equation

$$\lambda^2 + \frac{R}{L}\lambda + \frac{1}{LC} = 0,$$

which we can write in the form $\lambda_1 = -\alpha + \beta$ and $\lambda_2 = -\alpha - \beta$, where

$$\alpha = \frac{R}{2L}, \qquad \beta = \frac{1}{2L}\sqrt{R^2 - \frac{4L}{C}}.$$

As in Sec. 2.11 we conclude that if $R > 0$ (which, of course, is true in any practical case), the general solution $I_h(t)$ of the homogeneous equation approaches zero as t approaches infinity (practically: after a sufficiently long time). Hence, the transient current $I = I_h + I_p$ tends to the steady-state current I_p, and *after some time* ***the output will practically be a harmonic oscillation,*** *which is given by* (5) *and whose frequency is that of the input.*

EXAMPLE 1 ***RLC*-circuit**

Find the current $I(t)$ in an RLC-circuit with $R = 100$ ohms, $L = 0.1$ henry, $C = 10^{-3}$ farad, which is connected to a source of voltage $E(t) = 155 \sin 377t$ (hence 60 Hz = 60 cycles/sec), assuming zero charge and current when $t = 0$.

Solution. 1st Step. General solution. Equation (1) is

$$0.1I'' + 100I' + 1000I = 155 \cdot 377 \cos 377t.$$

We calculate the reactance $S = 37.7 - 1/0.377 = 35.0$ and the steady-state current

$$I_p(t) = a \cos 377t + b \sin 377t$$

where

$$a = \frac{-155 \cdot 35.0}{100^2 + 35^2} = -0.484, \qquad b = \frac{155 \cdot 100}{100^2 + 35^2} = 1.380.$$

Then we solve the characteristic equation

$$0.1\lambda^2 + 100\lambda + 1000 = 0.$$

The roots are $\lambda_1 = -10$ and $\lambda_2 = -990$. Hence the general solution is

$$I(t) = c_1 e^{-10t} + c_2 e^{-990t} - 0.484 \cos 377t + 1.380 \sin 377t. \tag{6}$$

2nd Step. Particular solution. We determine c_1 and c_2 from the initial conditions $Q(0) = 0$ and $I(0) = 0$. The second condition gives

$$I(0) = c_1 + c_2 - 0.484 = 0. \tag{7}$$

How to use $Q(0) = 0$? Solving (1′) algebraically for I', we have

$$I' = \frac{1}{L}\left[E(t) - RI(t) - \frac{1}{C}Q(t)\right] \tag{8}$$

since $\int I\,dt = Q$. Here $E(0) = 0$, $I(0) = 0$, and $Q(0) = 0$, so that $I'(0) = 0$. Differentiating (6), we thus obtain

$$I'(0) = -10c_1 - 990c_2 + 1.380 \cdot 377 = 0. \tag{9}$$

The solution of (7) and (9) is $c_1 = -0.042$, $c_2 = 0.526$. From (6) we thus have the answer

$$I(t) = -0.042e^{-10t} + 0.526e^{-990t} - 0.484 \cos 377t + 1.380 \sin 377t.$$

The first two terms will die out rapidly, and after a very short time the current will practically execute harmonic oscillations of frequency 60 Hz = 60 cycles/sec, which is the frequency of the impressed voltage.

Note that by (5) we can write the steady-state current in the form

$$I_p(t) = 1.463 \sin(377t - 0.34).$$

■

Problem Set 2.12

RLC-circuits

1. Derive (3) in two ways, namely, (a) directly by substituting (2) into (1), (b) from (5) in Sec. 2.11, using Table 2.2 and taking $E_0\omega$ instead of F_0.

2. It was claimed in the text that if $R > 0$, then the transient current approaches $I_p(t)$ as $t \to \infty$. How can this be proved?

3. **(Types of damping)** What are the conditions for an *RLC*-circuit to be overdamped (Case I), critically damped (Case II), and underdamped (Case III)? In particular, what is the critical resistance R_{crit} (the analog of the critical damping constant $2\sqrt{mk}$)?

4. **(Tuning)** In tuning a radio to a station we turn a knob on the radio that changes C (or perhaps L) in an *RLC*-circuit (Fig. 51) so that the amplitude of the steady-state current becomes maximum. For what C will this be the case?

Find the steady-state current in the *RLC*-circuit in Fig. 51, where

5. $R = 4$ ohms, $L = 1$ henry, $C = 2 \cdot 10^{-4}$ farad, $E = 220$ volts

6. $R = 10$ ohms, $L = 5$ henrys, $C = 10^{-2}$ farad, $E = 87\frac{2}{9} \sin 3t$ volts

7. $R = 20$ ohms, $L = 10$ henrys, $C = 10^{-3}$ farad, $E = 100 \cos t$ volts

Find the transient current in the *RLC*-circuit in Fig. 51, where

8. $R = 20$ ohms, $L = 5$ henrys, $C = 10^{-2}$ farad, $E = 85 \sin 4t$ volts

9. $R = 200$ ohms, $L = 100$ henrys, $C = 0.005$ farad, $E = 500 \sin t$ volts

10. $R = 16$ ohms, $L = 8$ henrys, $C = \frac{1}{8}$ farad, $E = 100 \cos 2t$ volts

Using (8), find the current in the *RLC*-circuit in Fig. 51, assuming zero initial current and charge, and

11. $R = 80$ ohms, $L = 20$ henrys, $C = 0.01$ farad, $E = 100$ volts

12. $R = 160$ ohms, $L = 20$ henrys, $C = 0.002$ farad, $E = 481 \sin 10t$ volts

13. $R = 6$ ohms, $L = 1$ henry, $C = 0.04$ farad, $E = 24 \cos 5t$ volts

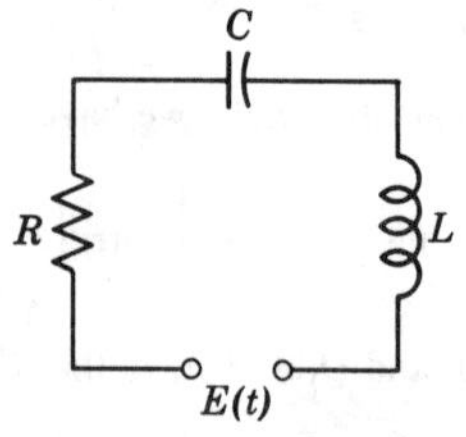

Fig. 51. *RLC*-circuit

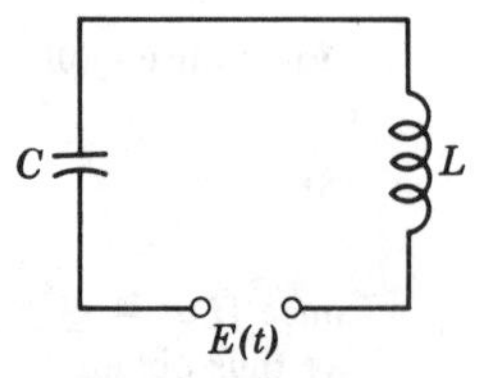

Fig. 52. *LC*-circuit

LC-circuits

(In practice, these are *RLC*-circuits with negligibly small *R*.)

14. Find the natural frequency (= frequency of free oscillations) of an *LC*-circuit (a) directly, (b) from Sec. 2.5 by means of Table 2.2.

Find the current $I(t)$ in the *LC*-circuit in Fig. 52, where

15. $L = 0.4$ henry, $C = 0.1$ farad, $E = 110 \sin \omega t$ volts, $\omega^2 \neq 25$

16. $L = 0.2$ henry, $C = 0.05$ farad, $E = 100$ volts

17. $L = 2.5$ henrys, $C = 10^{-3}$ farad, $E = 10t^2$ volts

Find the current $I(t)$ in the LC-circuit in Fig. 52, assuming zero initial current and charge, and

18. $L = 10$ henrys, $C = 0.004$ farad, $E = 250$ volts

19. $L = 1$ henry, $C = 0.25$ farad, $E = 30 \sin t$ volts

20. $L = 10$ henrys, $C = \frac{1}{90}$ farad, $E = 10 \cos 2t$ volts

21. $L = 10$ henrys, $C = 0.1$ farad, $E = 10t$ volts

22. Show that if $E(t)$ in Fig. 52 has a jump of magnitude J at $t = a$, then $I'(t)$ has a jump of magnitude J/L at $t = a$, whereas $I(t)$ is continuous at $t = a$.

Using the result of Prob. 22, find the current $I(t)$ in the LC-circuit in Fig. 52, assuming $L = 1$ henry, $C = 1$ farad, zero initial current and charge, and

23. $E = 1$ when $0 < t < 1$ and $E = 0$ when $t > 1$

24. $E = t$ when $0 < t < a$ and $E = a$ when $t > a$

25. $E = 1 - e^{-t}$ when $0 < t < \pi$ and $E = 0$ when $t > \pi$

2.13 Complex Method for Particular Solutions. *Optional*

Engineers like to solve equations such as

$$LI'' + RI' + \frac{1}{C}I = E_0\omega \cos \omega t \qquad \text{(Sec. 2.12)} \tag{1}$$

by an elegant complex method, using that $\cos \omega t$ is the real part of $e^{i\omega t}$ (see Sec. 2.3), finding a particular solution I_p of the resulting complex equation

$$LI'' + RI' + \frac{1}{C}I = E_0\omega e^{i\omega t} \qquad (i = \sqrt{-1}) \tag{2}$$

and finally taking the real part $\tilde{I}_p$ of I_p as a particular solution of the given real equation (1). To find I_p, we substitute (using $i^2 = -1$)

$$I_p = Ke^{i\omega t}, \qquad I_p' = i\omega Ke^{i\omega t}, \qquad I_p'' = -\omega^2 Ke^{i\omega t}$$

into (2). This gives

$$\left(-\omega^2 L + i\omega R + \frac{1}{C}\right) Ke^{i\omega t} = E_0\omega e^{i\omega t}.$$

Dividing by $\omega e^{i\omega t}$ on both sides, solving for K, and using the reactance $S = \omega L - 1/\omega C$ [(4), Sec. 2.12], we obtain

$$K = \frac{E_0}{-\left(\omega L - \dfrac{1}{\omega C}\right) + iR} = \frac{E_0}{-S + iR} = \frac{-E_0(S + iR)}{S^2 + R^2}, \tag{3}$$

where the last equality follows by multiplication of numerator and denominator by $-S - iR$. Thus the complex particular solution of (2) is

$$I_p = Ke^{i\omega t} = \frac{-E_0}{S^2 + R^2}(S + iR)(\cos \omega t + i \sin \omega t).$$

The real part is (use again $i^2 = -1$)

$$\tilde{I}_p = \frac{-E_0}{S^2 + R^2}(S \cos \omega t - R \sin \omega t). \tag{4}$$

This agrees with (2), (3) in Sec. 2.12.

EXAMPLE 1 Complex method

Solve by the complex method:

$$I'' + I' + 3I = 5 \cos t.$$

Solution. The corresponding complex differential equation is

$$I'' + I' + 3I = 5e^{it}.$$

We substitute

$$I_p = Ke^{it}, \qquad I'_p = iKe^{it}, \qquad I''_p = -Ke^{it}$$

into the complex equation. This gives

$$(-1 + i + 3)Ke^{it} = 5e^{it}.$$

Solving for K, we obtain

$$K = \frac{5}{2 + i} = \frac{5(2 - i)}{(2 + i)(2 - i)} = \frac{10 - 5i}{5} = 2 - i.$$

From this we get the complex solution

$$I_p = (2 - i)e^{it} = (2 - i)(\cos t + i \sin t).$$

The real part is

$$\tilde{I}_p = 2 \cos t + \sin t.$$

The student should verify that this is indeed a solution of the given equation. ∎

Formula (3) suggests to introduce the so-called **complex impedance**

$$Z = R + iS = R + i\left(\omega L - \frac{1}{\omega C}\right). \tag{5}$$

Then we can write (3) in the form

$$K = \frac{E_0}{iZ}. \tag{3*}$$

We see that the real part of Z is R, the imaginary part is the reactance S, and the absolute value is the impedance

$$|Z| = \sqrt{R^2 + S^2} \qquad \text{(Sec. 2.12).}$$

Thus (Fig. 53)

$$Z = |Z|e^{i\theta} \qquad \text{where} \qquad \tan\theta = \frac{S}{R}. \tag{6}$$

Consequently, the solution I_p of (2) can now be written (use $1/i = -i$)

$$I_p = \frac{E_0}{iZ} e^{i\omega t} = -i \frac{E_0}{|Z|} e^{i(\omega t - \theta)}.$$

Its real part is

$$\tilde{I}_p = \frac{E_0}{|Z|} \sin(\omega t - \theta) = \frac{E_0}{\sqrt{R^2 + S^2}} \sin(\omega t - \theta), \tag{7}$$

and this solution of the given equation (1) agrees with (5) in Sec. 2.12, fully justifying our present complex method. ∎

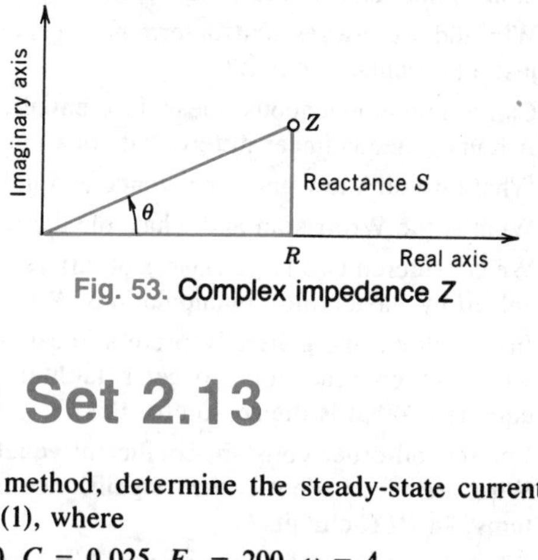

Fig. 53. Complex impedance Z

Problem Set 2.13

Using the complex method, determine the steady-state current $I_p(t)$ in the *RLC*-circuit governed by (1), where

1. $L = 30, R = 50, C = 0.025, E_0 = 200, \omega = 4$
2. $L = 4, R = 20, C = 0.5, E_0 = 10, \omega = 10$
3. $L = 2, R = 4, C = \frac{1}{8}, E_0 = 10, \omega = 5$
4. $R = 50, L = 25, C = 0.01, E_0 = 500, \omega = 3$
5. $R = 20, L = 10, C = 0.05, E_0 = 5, \omega = 2$

6. Find the complex impedance Z and the reactance in Prob. 1.
7. Verify I_p in Example 1 of Sec. 2.12 by the complex method.

Using the complex method, find the steady-state output of the following equations.

8. $y'' + 5y' + \frac{1}{2}y = 25 \cos 10t$
9. $y'' + y' + 4y = 8 \sin 2t$
10. $y'' + 0.5y' + 2y = 5 \cos t$
11. $y'' + 2y' + 2y = \cos t$
12. $y'' - y' - 2y = \sin t$
13. $y'' + 4y' + 3y = 65 \cos 2t$
14. $y'' + 3y' + 2y = 20 \sin t$
15. $y'' + 2y' + y = 50 \sin 3t$

Review Questions and Problems for Chapter 2

1. What is the superposition principle? Does it hold for nonlinear equations? For nonhomogeneous linear equations? For homogeneous linear equations?
2. How many arbitrary constants does a general solution of a second-order nonhomogeneous linear equation involve? How many additional conditions does one need to determine them?
3. How would you practically determine whether two solutions of a differential equation are linearly independent? Why is this important and relevant in this chapter?
4. Does it make sense to talk about linear dependence of two functions at a single point? Explain.
5. If we know two solutions of a nonhomogeneous linear differential equation on the same interval, can we find from them a particular solution of the corresponding homogeneous equation? A general solution of the latter?
6. What is the difference between an initial value problem and a boundary value problem? Why did we make no such distinction in the case of a first-order equation?
7. What is a particular solution? Why are particular solutions generally more common as final answers to practical problems than general solutions?
8. Why did we always first determine a general solution, even when we needed just a particular solution?
9. Can a nonhomogeneous linear differential equation have the trivial solution 0? A homogeneous linear differential equation?
10. What do you know about existence and uniqueness of solutions?
11. What is the Wronskian and what role did it play in our discussions?
12. We considered two large classes of differential equations that can essentially be solved by "algebraic" manipulations. What were they?
13. In modeling, one generally prefers linear over nonlinear differential equations whenever one can hope to get a faithful picture of the reality from a linear equation. What is the reason for this?
14. For second-order constant-coefficient equations, we distinguished three cases. What are they? What is their significance in connection with mass–spring systems? In *RLC*-circuits?
15. What do we mean by "resonance"? Where and under what conditions does it occur?

Find a general solution of the following differential equations.

16. $y'' - 2y' + y = e^x \sin x$

17. $y'' + y' - 2y = 3e^x$

18. $x^2y'' + xy' - y = 4$

19. $x^2y'' - 0.4xy' + 0.49y = 0$

20. $x^2y'' - xy' + 2y = 0$

21. $y'' - 4y' + 4y = e^{2x}/x$

22. $y'' + y' + y = \cos x + 13 \cos 2x$

23. $y'' + 2y' + y = e^{-x} \cos x$

24. $y'' + 9y = \sec 3x + 18x - 36$

25. $x^2y'' - 4xy' + 6y = x^4 \sin x$

26. $(D^2 - D - 2)y = 4 \sin x$

27. $(D^2 + 4D + 3)y = 2 \cos x + \sin x$

28. $(x^2D^2 - 5xD + \frac{82}{9})y = 0$

29. $(D^2 + 4D + 4)y = e^{-2x}/x^2$

30. $(x^2D^2 + xD - 1)y = x^3e^x$

Solve the following initial value problems.

31. $y'' - 2y' + y = 2x^2 - 8x + 4, \quad y(0) = 0.3, \quad y'(0) = 0.3$

32. $(D^2 + 2D + 10)y = 10x^2 + 4x + 2, \quad y(0) = 1, \quad y'(0) = -1$

33. $4x^2y'' + 12xy' + 3y = 0, \quad y(4) = \frac{3}{4}, \quad y'(4) = -\frac{5}{32}$

34. $(D^2 + 4)y = 8e^{-2x} + 4x^2 + 2, \quad y(0) = 2, \quad y'(0) = 2$

35. $(x^2D^2 + xD - 1)y = 16x^3, \quad y(1) = -1, \quad y'(1) = 11$

36. $y'' - 0.2y' + 100.01y = 0, \quad y(0) = 1, \quad y'(0) = -9.9$

37. $y'' + 2\alpha y' + (\alpha^2 + \pi^2)y = 0, \quad y(0) = 3, \quad y'(0) = -3\alpha$

38. $(D^2 + 2D + 5)y = 16 \sin x, \quad y(0) = -0.6, \quad y'(0) = 0.2$

39. $y'' + 2y' + 2y = 0, \quad y(\frac{1}{2}\pi) = 0, \quad y'(\frac{1}{2}\pi) = -2e^{-\pi/2}$

40. $y'' + 4y' + (4 + \omega^2)y = 0, \quad y(0) = 1, \quad y'(0) = \omega - 2$

41. $y'' + 2y' + 2y = 0 \quad y(0) = 1, \quad y'(0) = -1$

42. $(x^2D^2 - 2)y = 3x^2, \quad y(1) = 0, \quad y'(1) = 0$

43. $(D^2 - 4D + 3)y = 2 \sin x - 4 \cos x, \quad y(\frac{1}{4}\pi) = 1/\sqrt{2}, \quad y'(\frac{1}{4}\pi) = 1/\sqrt{2}$

44. $(4D^2 - 4D + 65)y = 64e^{x/2} + 65x - 4, \quad y(0) = 1, \quad y'(0) = 5.5$

45. $(D^2 + 0.5D - 0.5)y = 3 \cos x + \sin x + e^x, \quad y(0) = 0, \quad y'(0) = 1.5$

46. Find the steady-state current in the RLC-circuit in Fig. 54, assuming that $L = 1$ henry, $R = 2000$ ohms, $C = 4 \cdot 10^{-3}$ farad, and $E(t) = 110 \sin 415t$ (66 cycles/sec).

47. Find a general solution of the homogeneous equation corresponding to the equation in Prob. 46.

48. Find the current in the RLC-circuit in Fig. 54 when $R = 20$ ohms, $L = 0.1$ henry, $C = 1.5625 \cdot 10^{-3}$ farad, $E(t) = 160t$ volts if $0 < t < 0.01$, $E(t) = 1.6$ volts if $t > 0.01$ sec, assuming that $I(0) = 0$, $I'(0) = 0$.

49. Find the steady-state current in the RLC-circuit in Fig. 54 when $R = 50$ ohms, $L = 30$ henrys, $C = 0.025$ farad, $E(t) = 200 \sin 4t$ volts.

50. Find the motion of the mass–spring system in Fig. 55 with mass 0.125 kg, damping 0, spring constant 1.125 kg/sec^2, and driving force $\cos t - 4 \sin t$ nt, assuming zero initial displacement and velocity. For what frequency of the driving force would you get resonance?

51. Find the steady-state solution of the system in Fig. 55 when $m = 1$, $c = 2$, $k = 6$ and the driving force is $\sin 2t + 2 \cos 2t$.

52. In Fig. 55, let $m = 1$, $c = 4$, $k = 24$ and $r(t) = 10 \cos \omega t$. Determine ω such that you get the steady-state vibration of maximum possible amplitude. Determine this amplitude. Then find the general solution with this ω and check whether the results are in agreement.

53. In Prob. 51, find the solution corresponding to initial displacement 1 and initial velocity 0.

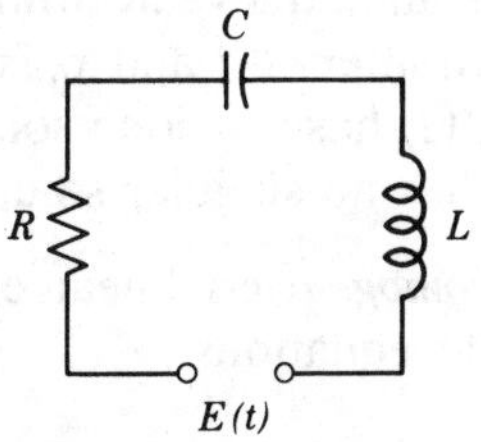

Fig. 54. RLC-circuit

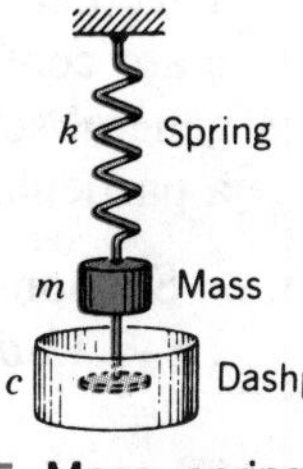

Fig. 55. Mass–spring system

54. A cylindrical buoy 60 cm in diameter stands in water with its axis vertical (Fig. 56). When depressed slightly and released, it is found that the period of vibration is 2 sec. Determine the weight of the buoy. *Hint.* By **Archimedes' principle**, the buoyancy force equals the weight of the water displaced by the body (assumed partly or totally submerged).

55. Solve $y'' + 6y' + 8y = 40 \sin 2x$ by undetermined coefficients and by variation of parameters; comment on the amount of work in both methods.

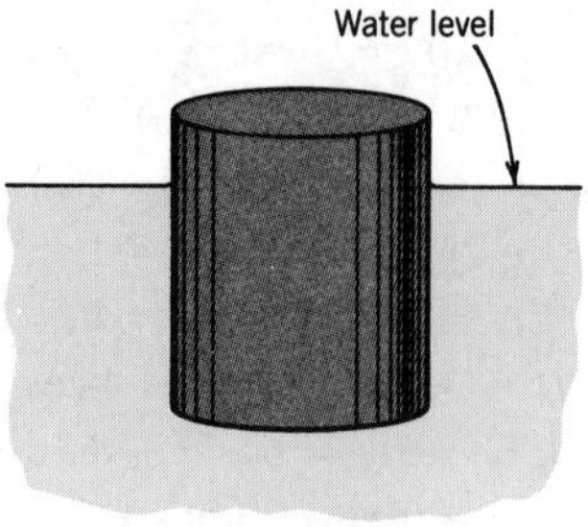

Fig. 56. Buoy

Summary of Chapter 2 Second-Order Differential Equations

A *first-order* homogeneous linear equation is $y' + p(x)y = 0$ (Sec. 1.7). A **second-order homogeneous linear equation** is an equation that can be written

$$y'' + p(x)y' + q(x)y = 0. \tag{1}$$

It has the very important property that a linear combination of solutions is again a solution (**superposition principle** or **linearity principle**, Sec. 2.1). Two linearly independent solutions y_1, y_2 of (1) form a **basis** of solutions and $y = c_1y_1 + c_2y_2$ with arbitrary constants c_1, c_2 a **general solution**. From it we obtain a **particular solution** if we specify numerical values of c_1 and c_2, for instance, by imposing two **initial conditions** (Sec. 2.1)

$$y(x_0) = K_0, \quad y'(x_0) = K_1 \qquad (x_0, K_0, K_1 \text{ given numbers}). \tag{2}$$

Together, (1) and (2) constitute an **initial value problem** for (1). If p and q are continuous on some open interval I and x_0 is in I, then (1) has a general solution on I, and (1), (2) has a unique solution on I (which is a particular solution; thus (1) has no singular solutions).

Sections 2.1–2.7 concern ***homogeneous*** linear equations and Secs. 2.8–2.13 ***nonhomogeneous*** linear equations

$$y'' + p(x)y' + q(x)y = r(x) \qquad r(x) \not\equiv 0. \tag{3}$$

A **general solution** of (3) is of the form

$$y = y_h + y_p,$$

where y_h is a general solution of the corresponding homogeneous equation (1) and y_p is a particular solution of (3). Hence the additional practical problem in solving (3) is the determination of such a y_p. For this we give a general method, called the method of **variation of parameters** (Sec. 2.10), a simpler special method of **undetermined coefficients**, valid for constant p, q and special r (powers of x, sine, cosine, etc.; Sec. 2.9), and a complex method for sinusoidal driving forces (Sec. 2.13).

If $p(x)$ and $q(x)$ in (1) or (3) are *variable*, the solutions will in general be higher functions. We study the most important ones of them in Chap. 5. If $p(x)$ and $q(x)$ are ***constant***, we write $p(x) = a$, $q(x) = b$ and obtain from (1) an equation

$$y'' + ay' + by = 0. \tag{4}$$

This equation can be solved by substituting $y = e^{\lambda x}$. Then λ is a root of the **characteristic equation**

$$\lambda^2 + a\lambda + b = 0.$$

Hence there are three cases (Sec. 2.2):

Case	Type of Roots	General Solution
I	Distinct real λ_1, λ_2	$y = c_1e^{\lambda_1 x} + c_2e^{\lambda_2 x}$
II	Double $-\frac{1}{2}a$	$y = (c_1 + c_2x)e^{-ax/2}$
III	Complex $-\frac{1}{2}a \pm i\omega$	$y = e^{-ax/2}(A \cos \omega x + B \sin \omega x)$

Equation (4) and the nonhomogeneous equation

$$y'' + ay' + by = r(x) \qquad (a, b \text{ constant}) \tag{5}$$

have important applications in mechanics (Secs. 2.5, 2.11) and electrical engineering (Sec. 2.12), which are fundamental in the study of vibrations and resonance.

Another large class of equations also solvable by an "algebraic" method consists of the **Euler–Cauchy equations** (Sec. 2.6)

$$x^2y'' + axy' + by = 0.$$

If we substitute $y = x^m$, we can determine m from the auxiliary equation

$$m^2 + (a - 1)m + b = 0$$

(where in the case of a double root $m = (1 - a)/2$, a basis is x^m, $x^m \ln x$).

Chapter 3

Higher Order Linear Differential Equations

In this chapter we show that the concepts and the methods for solving second-order linear differential equations in Chap. 2 extend in a straightforward fashion to linear differential equations of higher order. No really new ideas are needed in that extension. The correspondence between sections is roughly as follows (and should be used to make study of Chap. 3 easier):

Section 3.1 extends 2.1 and 2.7.
3.2 extends 2.2 and 2.3.
3.3 extends 2.8.
3.4 extends 2.9.
3.5 extends 2.10.

Some new features are as follows:

(i) In Secs. 3.1 and 3.5 the more prominent role of the Wronskian.

(ii) In Sec. 3.2 the larger number of possibilities of roots (instead of only three cases in Secs. 2.2 and 2.3).

(iii) In Sec. 3.5 the interesting extension of Lagrange's proof.

Prerequisite for this chapter: Chap. 2
References: Appendix 1, Part A
Answers to problems: Appendix 2

3.1 Homogeneous Linear Equations

Recall from Sec. 1.1 that an ordinary differential equation of **nth order** is an equation in which the nth derivative $y^{(n)} = d^ny/dx^n$ of the unknown function $y(x)$ is the highest occurring derivative. Thus the equation is of the form

$$F(x, y, y', \cdots, y^{(n)}) = 0,$$

with the understanding that lower order derivatives of y or y itself may or may not occur in the equation.

The equation is called **linear** if it can be written

$$y^{(n)} + p_{n-1}(x)y^{(n-1)} + \cdots + p_1(x)y' + p_0(x)y = r(x) \qquad \textbf{(1)}$$

where r on the right and the **coefficients** $p_0, p_1, \cdots, p_{n-1}$ are any given function of x. Any nth order differential equation that cannot be written in the form (1) is called **nonlinear**.

As in Sec. 2.1 for $n = 2$, the **"standard form"** (1), with $y^{(n)}$ as the first term, is practical. (Divide an equation by $f(x)$ if its first term is $f(x)y^{(n)}$.)

If $r(x) \equiv 0$, equation (1) becomes

$$y^{(n)} + p_{n-1}(x)y^{(n-1)} + \cdots + p_1(x)y' + p_0(x)y = 0 \tag{2}$$

and is called **homogeneous**. If $r(x)$ is not identically zero, the equation is called **nonhomogeneous**. This is as in Chap. 2.

Solution. General Solution. Linear Independence

A **solution** of an nth order (linear or nonlinear) differential equation on some open interval I is a function $y = h(x)$ that is defined and n times differentiable on I and is such that the equation becomes an identity if we replace the unknown function y and its derivatives in the equation by h and its corresponding derivatives.

We now turn to the homogeneous equation (2) and begin with the following.

Theorem 1 **(Superposition principle or linearity principle)**
*For the **homogeneous** linear differential equation* (2), *sums and constant multiples of solutions on some open interval I are again solutions of* (2) *on I.*

The proof is a simple generalization of that in Sec. 2.1 and is left to the student. And we repeat our ***warning*** that the theorem ***does not hold*** for the nonhomogeneous equation (1) or for a nonlinear equation.

Our further discussion parallels and extends that for second-order equations in Chap. 2. So we define next a general solution of (2), which will need an extension of linear independence from two to n functions, a concept of great general importance, far beyond our present purpose.

Definition (General solution, basis, particular solution)
A **general solution** of (2) on an open interval I is a solution of (2) on I of the form

$$y(x) = c_1y_1(x) + \cdots + c_ny_n(x) \qquad (c_1, \cdots, c_n \text{ arbitrary}^1) \tag{3}$$

where $y_1, \cdots, y_n$ is a **basis** (or **fundamental system**) of solutions of (2) on I; that is, these solutions are linearly independent on I, as defined below.

A **particular solution** of (2) on I is obtained if we assign specific values to the n constants $c_1, \cdots, c_n$ in (3). ∎

[1]See footnote 2 in Sec. 2.1.

Definition (Linear independence and dependence)
n functions $y_1(x), \cdots, y_n(x)$ are called **linearly independent** *on some interval* I where they are defined if the equation

$$k_1y_1(x) + \cdots + k_ny_n(x) = 0 \quad \text{on } I \tag{4}$$

implies that all $k_1, \cdots, k_n$ are zero. These functions are called **linearly dependent** on I if this equation also holds on I for some $k_1, \cdots, k_n$ not all zero.

If and only if $y_1, \cdots, y_n$ are linearly dependent on I, we can express (at least) one of these functions on I as a **"linear combination"** of the other $n - 1$ functions, that is, as a sum of those functions, each multiplied by a constant (zero or not). This motivates the term "linearly dependent." For instance, if (4) holds with $k_1 \neq 0$, we can divide by k_1 and express y_1 as the linear combination

$$y_1 = -\frac{1}{k_1}(k_2y_2 + \cdots + k_ny_n).$$

Note that when $n = 2$, these concepts reduce to those defined in Sec. 2.1.

EXAMPLE 1 Linear dependence

Show that the functions $y_1 = x$, $y_2 = 3x$, $y_3 = x^2$ are linearly dependent on any interval.

Solution. $y_2 = 3y_1 + 0y_3$.

EXAMPLE 2 Linear Independence

Show that $y_1 = x$, $y_2 = x^2$, $y_3 = x^3$ are linearly independent on any interval, for instance, on $-1 \leqq x \leqq 2$.

Solution. Equation (4) is $k_1x + k_2x^2 + k_3x^3 = 0$. Taking $x = -1, 1, 2$, we get

$$-k_1 + k_2 - k_3 = 0, \qquad k_1 + k_2 + k_3 = 0, \qquad 2k_1 + 4k_2 + 8k_3 = 0,$$

respectively, which implies $k_1 = 0$, $k_2 = 0$, $k_3 = 0$, that is, linear independence.

This calculation was not too pleasant and illustrates the need for a better method of testing linear independence, at least for solutions. We shall get to this soon.

EXAMPLE 3 General solution, basis

Solve the fourth-order differential equation

$$y^{\text{IV}} - 5y'' + 4y = 0.$$

Solution. As in Sec. 2.2, we try $y = e^{\lambda x}$. Then substitution and omission of the common (nonzero) factor $e^{\lambda x}$ gives the characteristic equation

$$\lambda^4 - 5\lambda^2 + 4 = 0,$$

which is a quadratic equation in $\mu = \lambda^2$,

$$\mu^2 - 5\mu + 4 = 0.$$

The roots are $\mu = 1$ and $\mu = 4$. Hence $\lambda = -2, -1, 1, 2$, which gives four solutions, so that a general solution on any interval is

$$y = c_1e^{-2x} + c_2e^{-x} + c_3e^{x} + c_4e^{2x}$$

provided those solutions are linearly independent. This is true, but will be shown later.

Initial Value Problem, Existence and Uniqueness

An **initial value problem** for equation (2) consists of (2) and n **initial conditions**

$$(5)\quad y(x_0) = K_0, \quad y'(x_0) = K_1, \quad \cdots, \quad y^{(n-1)}(x_0) = K_{n-1},$$

where x_0 is some fixed point in the interval I considered.

In extension of Theorem 1 in Sec. 2.7 we now have the following.

Theorem 2 **Existence and Uniqueness Theorem for Initial Value Problems**

If $p_0(x), \cdots, p_{n-1}(x)$ are continuous functions on some open interval I and x_0 is in I, then the initial value problem (2), (5) *has a unique solution $y(x)$ on the interval I.*

Existence is proved in Ref. [A6] in Appendix 1, and uniqueness can be proved by a slight generalization of the uniqueness proof at the beginning of Appendix 4.

EXAMPLE 4 **Initial value problem for a third-order Euler–Cauchy equation**

Solve the initial value problem

$$x^3y''' - 3x^2y'' + 6xy' - 6y = 0, \qquad y(1) = 2, \qquad y'(1) = 1, \qquad y''(1) = -4$$

on any open interval I on the positive x-axis containing $x = 1$.

Solution. 1st Step. General solution. As in Sec. 2.6 we try $y = x^m$. Differentiation and substitution gives

$$m(m-1)(m-2)x^m - 3m(m-1)x^m + 6mx^m - 6x^m = 0.$$

Ordering terms and dropping the factor x^m, we obtain

$$m^3 - 6m^2 + 11m - 6 = 0.$$

If we can guess the root $m = 1$, we can divide and find as the other roots $m = 2$ and $m = 3$. (Without guessing, for orders higher than four, one has to use a numerical root-finding method, such as Newton's; see Sec. 18.2.) The corresponding solutions x, x^2, x^3 are linearly independent on I (see Example 2). Hence a general solution on I is

$$y = c_1x + c_2x^2 + c_3x^3.$$

(Our interval I does not include 0, where the coefficients of our equation in standard form [the given form divided by x^3] are not continuous, but we see that, actually, y is a general solution on any interval.)

2nd Step. Particular solution. We now also need the derivatives

$$y' = c_1 + 2c_2x + 3c_3x^2, \qquad y'' = 2c_2 + 6c_3x.$$

From this and y and the initial conditions we get

$$\begin{aligned} y(1) &= c_1 + c_2 + c_3 = 2 \\ y'(1) &= c_1 + 2c_2 + 3c_3 = 1 \\ y''(1) &= 2c_2 + 6c_3 = -4. \end{aligned}$$

By elimination or Cramer's rule (Sec. 7.8) we obtain $c_1 = 2$, $c_2 = 1$, $c_3 = -1$. *Answer.* $y = 2x + x^2 - x^3$. ∎

Linear Independence of Solutions. Wronskian

We have seen that it would be good to have a practical criterion for checking linear independence of solutions. Fortunately, the criterion involving the Wronskian (Theorem 2 in Sec. 2.7) extends to nth order. It uses the **Wronskian** W of n solutions defined as the nth order determinant

$$W(y_1, \cdots, y_n) = \begin{vmatrix} y_1 & y_2 & \cdots & y_n \\ y_1' & y_2' & \cdots & y_n' \\ \cdot & \cdot & \cdots & \cdot \\ y_1^{(n-1)} & y_2^{(n-1)} & \cdots & y_n^{(n-1)} \end{vmatrix} \tag{6}$$

and can be stated as follows.

Theorem 3 **(Linear dependence and independence of solutions)**
Suppose that the coefficients $p_0(x), \cdots, p_{n-1}(x)$ of (2) are continuous on some open interval I. Then n solutions $y_1, \cdots, y_n$ of (2) on I are linearly dependent on I if and only if their Wronskian is zero for some $x = x_0$ in I. Furthermore, if $W = 0$ for $x = x_0$, then $W \equiv 0$ on I; hence if there is an x_1 in I at which $W \neq 0$, then $y_1, \cdots, y_n$ are linearly independent on I.

Proof. **(a)** Let $y_1, \cdots, y_n$ be linearly dependent on I. Then there are constants $k_1, \cdots, k_n$, not all zero, such that

$$k_1y_1 + \cdots + k_ny_n = 0 \tag{7}$$

for all x in I, and by $n - 1$ differentiations of this identity,

$$\begin{aligned} k_1y_1' + \cdots + k_ny_n' &= 0 \\ &\vdots \\ k_1y_1^{(n-1)} + \cdots + k_ny_n^{(n-1)} &= 0. \end{aligned} \tag{8}$$

(7), (8) is a homogeneous linear system of algebraic equations with a nontrivial solution $k_1, \cdots, k_n$, so that its coefficient determinant must be zero for every x on I, by Cramer's theorem (Sec. 7.9). But that determinant is the Wronskian W, as we see, so that $W = 0$ for every x on I.

(b) Conversely, let $W = 0$ for an x_0 in I. Then the system (7), (8) with $x = x_0$ has a solution $\tilde{k}_1, \cdots, \tilde{k}_n$, not all zero, by that same theorem. With these constants we define the solution $\tilde{y} = \tilde{k}_1y_1 + \cdots + \tilde{k}_ny_n$ of (2). By (7), (8) it satisfies the initial conditions $\tilde{y}(x_0) = 0, \cdots, \tilde{y}^{(n-1)}(x_0) = 0$. But another solution also satisfying these conditions is $y \equiv 0$. Hence $\tilde{y} \equiv y$ on I by Theorem 2; that is, (7) holds identically on I, which means linear dependence of $y_1, \cdots, y_n$.

(c) If $W = 0$ at an x_0 in I, we have linear dependence by (b), hence $W \equiv 0$ by (a), so that $W \neq 0$ at any x_1 implies linear independence of the solutions $y_1, \cdots, y_n$. ∎

EXAMPLE 5 Basis, Wronskian

We can now prove that in Example 3 we do have a basis. In evaluating W, pull out the exponentials columnwise. In the result, subtract column 1 from columns 2, 3, 4 and expand by row 1. In the resulting third-order determinant, subtract column 1 from column 2 and expand the result by row 2:

$$W = \begin{vmatrix} e^{-2x} & e^{-x} & e^{x} & e^{2x} \\ -2e^{-2x} & -e^{-x} & e^{x} & 2e^{2x} \\ 4e^{-2x} & e^{-x} & e^{x} & 4e^{2x} \\ -8e^{-2x} & -e^{-x} & e^{x} & 8e^{2x} \end{vmatrix} = \begin{vmatrix} 1 & 1 & 1 & 1 \\ -2 & -1 & 1 & 2 \\ 4 & 1 & 1 & 4 \\ -8 & -1 & 1 & 8 \end{vmatrix} = \begin{vmatrix} 1 & 3 & 4 \\ -3 & -3 & 0 \\ 7 & 9 & 16 \end{vmatrix} = 72.$$

A General Solution of (2) Includes All Solutions

We first show that general solutions always exist. Indeed, Theorem 3 in Sec. 2.7 extends as follows.

Theorem 4 (Existence of a general solution)

If the coefficients $p_0(x), \cdots, p_{n-1}(x)$ of (2) are continuous on some open interval I, then (2) has a general solution on I.

Proof. We choose any fixed x_0 in I. By Theorem 2, equation (2) has n solutions $y_1, \cdots, y_n$, where y_j satisfies initial conditions (5) with $K_j = 1$ and all other K's equal to zero. Their Wronskian at x_0 equals 1; for instance, when $n = 3$,

$$W(y_1(x_0), y_2(x_0), y_3(x_0)) = \begin{vmatrix} y_1(x_0) & y_2(x_0) & y_3(x_0) \\ y_1'(x_0) & y_2'(x_0) & y_3'(x_0) \\ y_1''(x_0) & y_2''(x_0) & y_3''(x_0) \end{vmatrix} = \begin{vmatrix} 1 & 0 & 0 \\ 0 & 1 & 0 \\ 0 & 0 & 1 \end{vmatrix} = 1.$$

Hence these solutions are linearly independent on I, by Theorem 3; they form a basis on I, and $y = c_1y_1 + \cdots + c_ny_n$ with arbitrary constants $c_1, \cdots, c_n$ is a general solution of (2) on I.

We can now prove the basic property that from a general solution of (2) every solution of (2) can be obtained by choosing suitable values of the arbitrary constants. Hence an nth order ***linear*** differential equation has no **singular solutions**, that is, solutions that cannot be obtained from a general solution.

Theorem 5 (General solution)

Suppose that (2) has continuous coefficients $p_0(x), \cdots, p_{n-1}(x)$ on some open interval I. Then every solution $y = Y(x)$ of (2) on I is of the form

$$Y(x) = C_1y_1(x) + \cdots + C_ny_n(x), \tag{9}$$

where $y_1, \cdots, y_n$ is a basis of solutions of (2) on I and $C_1, \cdots, C_n$ are suitable constants.

Proof. Let $y = c_1y_1 + \cdots + c_ny_n$ be a general solution of (2) on I and choose any fixed x_0 in I. We show that we can find values of $c_1, \cdots, c_n$ for which y and its first $n - 1$ derivatives agree with Y and its corresponding derivatives at x_0. Written out, this means that for $x = x_0$ we should have

$$
\begin{aligned}
c_1y_1 &+ \cdots + c_ny_n &&= Y \\
c_1y_1' &+ \cdots + c_ny_n' &&= Y' \\
&\vdots \\
c_1y_1^{(n-1)} &+ \cdots + c_ny_n^{(n-1)} &&= Y^{(n-1)}.
\end{aligned}
\tag{10}
$$

But this is a linear system of equations in the unknowns $c_1, \cdots, c_n$. Its coefficient determinant is the Wronskian of $y_1, \cdots, y_n$ at $x = x_0$, which is not zero by Theorem 3 because $y_1, \cdots, y_n$ are linearly independent on I (they form a basis!). Hence (10) has a unique solution $c_1 = C_1, \cdots, c_n = C_n$ (by Cramer's theorem, Sec. 7.9). With these values we get from our general solution the particular solution

$$y^*(x) = C_1y_1(x) + \cdots + C_ny_n(x)$$

on I. From (10) we see that y^* agrees with Y at x_0, and the same holds for the first $n - 1$ derivatives of y^* and Y. That is, y^* and Y satisfy at x_0 the same initial conditions. From the uniqueness theorem (Theorem 2) it now follows that $y^* \equiv Y$ on I, and the theorem is proved. ■

This completes our theory of the homogeneous linear equation (2), and in the next section we begin with solution methods.

Problem Set 3.1

Important general properties of homogeneous and nonhomogeneous linear differential equations. Prove the following statements, which refer to any fixed open interval I. Here we assume that $r(x) \not\equiv 0$ in (1).

1. The "**trivial solution**" $y(x) \equiv 0$ is a solution of (2) but not of (1).
2. The sum of a solution of (1) and a solution of (2) is a solution of (1).
3. The difference of two solutions of (1) is a solution of (2).
4. The sum of two solutions of (1) is ***not*** a solution of (1).
5. A multiple cy of a solution y of (1) with $c \neq 1$ is ***not*** a solution of (1).

Show that the given functions form a basis of solutions of the given differential equations on any open interval, verifying linear independence by Theorem 3. (In Prob. 7 assume $x > 0$.)

6. $1, x, x^2, x^3, \quad y^{\text{IV}} = 0$
7. $1, x^2, x^4, \quad x^2y''' - 3xy'' + 3y' = 0$
8. $e^{-x}, xe^{-x}, x^2e^{-x}, \quad y''' + 3y'' + 3y' + y = 0$
9. $e^{-x}, xe^{-x}, x^3e^{-x}, x^2e^{-x}, \quad (D^4 + 4D^3 + 6D^2 + 4D + 1)y = 0$

10. x, x^2, e^x, $[(x^2 - 2x + 2)D^3 - x^2D^2 + 2xD - 2]y = 0$
11. $\cos x, \sin x, e^{-x}$, $(D^3 + D^2 + D + 1)y = 0$
12. $e^x, e^{-x}, \cos x, \sin x$, $y^{IV} - y = 0$
13. $e^x \cos x, e^x \sin x, e^{-x} \cos x, e^{-x} \sin x$, $(D^4 + 4)y = 0$
14. $\cos x, \sin x, \cos 2x, \sin 2x$, $(D^4 + 5D^2 + 4)y = 0$
15. $\cosh x, \sinh x, \cos x, \sin x$, $y^{IV} - y = 0$

Verify that the given functions are solutions of the given differential equation. Show their linear independence by Theorem 3. Solve the given initial value problem.

16. $y''' - y' = 0$, $y(0) = 6$, $y'(0) = -4$, $y''(0) = 2$; $1, e^{-x}, e^x$
17. $y^{IV} = 0$, $y(0) = 0$, $y'(0) = 6$, $y''(0) = 0$, $y'''(0) = -48$; $1, x, x^2, x^3$
18. $y''' - 6y'' + 11y' - 6y = 0$, $y(0) = 4$, $y'(0) = 11$, $y''(0) = 17$; e^x, e^{2x}, e^{3x}
19. $xy''' + 3y'' = 0$, $y(1) = 4$, $y'(1) = -8$, $y''(1) = 10$; $1, x, x^{-1}$
20. $(x + 1)y''' - y'' - (x + 1)y' + y = 0$, $y(0) = 2$, $y'(0) = 1$, $y''(0) = 1$; $e^x, e^{-x}, x + 1$

Linear Independence and Dependence

Since these concepts are of *general* importance, far beyond our present study, we add a few more problems on them.

Are the following functions linearly independent or dependent on the positive x-axis?

21. $1, x, x^2$
22. $x, x + 1, x + 2$
23. $(x - 1)^2, (x + 1)^2, x$
24. $\cos^2 x, \sin^2 x, -2$
25. $\sin x, \sin 2x, \sin 3x$
26. $\ln x, \ln x^2, (\ln x)^2$
27. $x, 1/x, 1$
28. $\cosh^2 x \sinh^2 x, 1$
29. $(x - 1)^2, (x + 1)^2, x$
30. $\cos x, \sin x, 1$
31. $\cos^2 x, \sin^2 x, \cos 2x$
32. $\cos x, \cosh x, e^x$

33. Find an interval I on which the three functions x^3, $|x|^3$, and 1 are linearly independent and a subinterval of I on which the functions are linearly dependent. What fact does this illustrate?
34. If a set of p functions is linearly dependent on an interval I, prove that a set of n ($\geqq p$) functions on I containing that set is linearly dependent on I.
35. If n functions are linearly dependent on an interval I, prove that they are also linearly dependent on any subinterval of I.
36. If $y \equiv 0$ is a function of a set of functions on an interval I, prove that the set is linearly dependent on I.
37. Verify the calculations in Example 5.
38. Show that the functions of a basis of an equation (2) with continuous coefficients cannot all have a maximum or minimum at the same point.
39. Show that the functions in Prob. 38 cannot all be zero at the same point.
40. Verify that e^x, e^{-x}, x form a basis of

$$xy''' - y'' - xy' + y = 0$$

on any interval. Show that $W(e^x, e^{-x}, x) = 0$ at $x = 0$. Does this contradict Theorem 3?

3.2 Homogeneous Equations with Constant Coefficients

We now turn to nth order homogeneous linear differential equations with *constant* coefficients and write these equations in the form

$$y^{(n)} + a_{n-1}y^{(n-1)} + \cdots + a_1 y' + a_0 y = 0. \tag{1}$$

The idea of solution is the same as for $n = 2$. Indeed, by substitution of $y = e^{\lambda x}$ and its derivatives we obtain the **characteristic equation**

$$\lambda^n + a_{n-1}\lambda^{n-1} + \cdots + a_1\lambda + a_0 = 0 \tag{2}$$

of (1). To get solutions of (1), we have to determine the roots of (2), which will be difficult in practice and will have to be done by a numerical method, unless one can guess some roots or find them by trial and error. We shall discuss possible cases (which combine and extend those in Secs. 2.2 and 2.3) and illustrate them by typical examples.

Real Different Roots

If (2) has n real unequal roots $\lambda_1, \cdots, \lambda_n$, then the n solutions

$$y_1 = e^{\lambda_1 x}, \quad \cdots, \quad y_n = e^{\lambda_n x} \tag{3}$$

constitute a basis for all x, and the corresponding general solution of (1) is

$$y = c_1 e^{\lambda_1 x} + \cdots + c_n e^{\lambda_n x}. \tag{4}$$

Indeed, the solutions in (3) are linearly independent, as we shall see after the example.

EXAMPLE 1 Distinct real roots

Solve the differential equation

$$y''' - 2y'' - y' + 2y = 0.$$

Solution. The roots of the characteristic equation

$$\lambda^3 - 2\lambda^2 - \lambda + 2 = 0$$

are -1, 1 and 2, and the corresponding general solution (4) is

$$y = c_1 e^{-x} + c_2 e^x + c_3 e^{2x}.$$

The student should verify linear independence by using the Wronskian. ∎

Students familiar with nth order determinants may verify that by pulling out all exponentials from the columns, the Wronskian of $e^{\lambda_1 x}, \cdots, e^{\lambda_n x}$ becomes

$$W = \begin{vmatrix} e^{\lambda_1 x} & e^{\lambda_2 x} & \cdots & e^{\lambda_n x} \\ \lambda_1 e^{\lambda_1 x} & \lambda_2 e^{\lambda_2 x} & \cdots & \lambda_n e^{\lambda_n x} \\ \lambda_1^2 e^{\lambda_1 x} & \lambda_2^2 e^{\lambda_2 x} & \cdots & \lambda_n^2 e^{\lambda_n x} \\ \cdot & \cdot & \cdots & \cdot \\ \lambda_1^{n-1} e^{\lambda_1 x} & \lambda_2^{n-1} e^{\lambda_2 x} & \cdots & \lambda_n^{n-1} e^{\lambda_n x} \end{vmatrix}$$

(5)

$$= e^{(\lambda_1 + \cdots + \lambda_n)x} \begin{vmatrix} 1 & 1 & \cdots & 1 \\ \lambda_1 & \lambda_2 & \cdots & \lambda_n \\ \lambda_1^2 & \lambda_2^2 & \cdots & \lambda_n^2 \\ \cdot & \cdot & \cdots & \cdot \\ \lambda_1^{n-1} & \lambda_2^{n-1} & \cdots & \lambda_n^{n-1} \end{vmatrix}.$$

The exponential function is never zero. Hence $W = 0$ if and only if the determinant on the right is zero. This is a so-called **Vandermonde** or **Cauchy determinant.**[2] It can be shown that it equals

(6) $$(-1)^{n(n-1)/2} V$$

where V is the product of all factors $\lambda_j - \lambda_k$ with $j < k$ ($\leqq n$); for instance, when $n = 3$ we get $-V = -(\lambda_1 - \lambda_2)(\lambda_1 - \lambda_3)(\lambda_2 - \lambda_3)$. This shows that the Wronskian is not zero if and only if all the n roots of (2) are different and thus gives the following.

Theorem 1 (Basis)

Solutions $y_1 = e^{\lambda_1 x}, \cdots, y_n = e^{\lambda_n x}$ of (1) *(with any real or complex λ_j's) form a basis of solutions of* (1) *if and only if all n roots of* (2) *are different.*

Actually, Theorem 1 is an important special case of our more general result obtained from (5) and (6):

Theorem 2 (Linear independence)

Any number of solutions $y_1 = e^{\lambda_1 x}, \cdots, y_m = e^{\lambda_m x}$ of (1) *are linearly independent on an open interval I if and only if $\lambda_1, \cdots, \lambda_m$ are all different.*

---

[2] ALEXANDRE-THÉOPHILE VANDERMONDE (1735—1796), French mathematician, who worked on solution of equations by determinants. For CAUCHY, see footnote 9 in Sec. 2.6.

Complex Simple Roots

If complex roots occur, they must occur in conjugate pairs since the coefficients of (1) are real. Thus, if $\lambda = \gamma + i\omega$ is a simple root of (2), so is the conjugate $\bar{\lambda} = \gamma - i\omega$, and two corresponding linearly independent solutions are (as in Sec. 2.3, except for notation)

$$y_1 = e^{\gamma x} \cos \omega x, \qquad y_2 = e^{\gamma x} \sin \omega x.$$

EXAMPLE 2 **Complex conjugate simple roots**

Solve the initial value problem

$$y''' - 2y'' + 2y' = 0, \qquad y(0) = 0.5, \qquad y'(0) = -1, \qquad y''(0) = 2.$$

Solution. A root of $\lambda^3 - 2\lambda^2 + 2\lambda = 0$ is $\lambda_1 = 0$. A corresponding solution is $y_1 = e^{0x} = 1$. Division by λ gives

$$\lambda^2 - 2\lambda + 2 = 0.$$

The roots are $\lambda_2 = 1 + i$ and $\lambda_3 = 1 - i$. Corresponding solutions are $y_2 = e^x \cos x$ and $y_3 = e^x \sin x$. The corresponding general solution and its derivatives are

$$y = c_1 + e^x[A \cos x + B \sin x],$$

$$y' = e^x[(A + B) \cos x + (B - A) \sin x],$$

$$y'' = e^x[2B \cos x - 2A \sin x],$$

as follows by differentiation. From this and the initial conditions we obtain

$$y(0) = c_1 + A = 0.5, \qquad y'(0) = A + B = -1, \qquad y''(0) = 2B = 2.$$

Hence $B = 1$, $A = -2$, $c_1 = 2.5$. The answer is

$$y = 2.5 + e^x(-2 \cos x + \sin x).$$

■

Real Multiple Roots

If a real **double root** occurs, say, $\lambda_1 = \lambda_2$, then $y_1 = y_2$ in (3) and we take y_1 and $y_2 = xy_1$ as two linearly independent solutions corresponding to this root; this is as in Sec. 2.2.

If a **triple root** occurs, say, $\lambda_1 = \lambda_2 = \lambda_3$, then $y_1 = y_2 = y_3$ in (3) and three linearly independent solutions corresponding to this root are

$$y_1, \qquad xy_1, \qquad x^2y_1. \tag{7}$$

More generally, *if λ is a* **root of order *m***, *then m corresponding linearly independent solutions are*

$$e^{\lambda x}, \qquad xe^{\lambda x}, \qquad \cdots, \qquad x^{m-1}e^{\lambda x}. \tag{8}$$

Linear independence of these functions on any open interval follows from that of $1, x, \cdots, x^{m-1}$, which in turn follows from Theorem 3 in Sec. 3.1 and the fact that these are solutions of $y^{(m)} = 0$ with a nonzero Wronskian W. How does one get (8)? We show this after the example.

EXAMPLE 3 Real double and triple roots

Solve the differential equation

$$y^{V} - 3y^{IV} + 3y''' - y'' = 0.$$

Solution. The characteristic equation

$$\lambda^5 - 3\lambda^4 + 3\lambda^3 - \lambda^2 = 0$$

has the roots $\lambda_1 = \lambda_2 = 0$ and $\lambda_3 = \lambda_4 = \lambda_5 = 1$, and the answer is

$$y = c_1 + c_2 x + (c_3 + c_4 x + c_5 x^2)e^x. \tag{9}$$

Can you solve the equation by setting $y'' = z$? ∎

As promised, we now show how (8) is obtained [and that these functions are solutions of (1) in the present case]. To simplify formulas a little, we use operator notation (see Sec. 2.4), writing the left side of (1) as

$$L[y] = [D^n + a_{n-1}D^{n-1} + \cdots + a_0]y.$$

For $y = e^{\lambda x}$ we can perform the indicated differentiations and get

$$L[e^{\lambda x}] = (\lambda^n + a_{n-1}\lambda^{n-1} + \cdots + a_0)e^{\lambda x}.$$

Let λ_1 be an mth order root of the polynomial on the right, and let $\lambda_{m+1}, \cdots, \lambda_n$ be the other roots, all different from λ_1, when $m < n$. In product form we then have

$$L[e^{\lambda x}] = (\lambda - \lambda_1)^m h(\lambda)e^{\lambda x}$$

with $h(\lambda) = 1$ if $m = n$ or $h(\lambda) = (\lambda - \lambda_{m+1}) \cdots (\lambda - \lambda_n)$ if $m < n$. Now comes the key idea: We differentiate on both sides with respect to λ,

$$\frac{\partial}{\partial \lambda} L[e^{\lambda x}] = m(\lambda - \lambda_1)^{m-1}h(\lambda)e^{\lambda x} + (\lambda - \lambda_1)^m \frac{\partial}{\partial \lambda}[h(\lambda)e^{\lambda x}]. \tag{10}$$

Differentiations with respect to x and λ are independent, so we can interchange their order on the left:

$$\frac{\partial}{\partial \lambda} L[e^{\lambda x}] = L\left[\frac{\partial}{\partial \lambda} e^{\lambda x}\right] = L[xe^{\lambda x}]. \tag{11}$$

Now the right side of (10) is zero for $\lambda = \lambda_1$ because of the factors $\lambda - \lambda_1$ (and $m \geqq 2$). Hence (11) shows that $xe^{\lambda_1 x}$ is a solution of (1).

We can repeat this step and produce $x^2e^{\lambda_1 x}, \cdots, x^{m-1}e^{\lambda_1 x}$ by another $n - 2$ such differentiations with respect to λ. Going one step further would no longer produce zero on the right because the lowest power of $\lambda - \lambda_1$ would then be $(\lambda - \lambda_1)^0$, multiplied by $m!h(\lambda)$ and $h(\lambda_1) \neq 0$ because $h(\lambda)$ has no factors $\lambda - \lambda_1$; so we get *precisely* the solutions in (8). ∎

Complex Multiple Roots

In this case, real solutions are obtained as for complex simple roots above. Consequently, if $\lambda = \gamma + i\omega$ is a **complex double root**, so is the conjugate $\bar{\lambda} = \gamma - i\omega$. Corresponding linearly independent solutions are

$$(12) \qquad e^{\gamma x}\cos\omega x \quad e^{\gamma x}\sin\omega x, \qquad xe^{\gamma x}\cos\omega x, \qquad xe^{\gamma x}\sin\omega x.$$

The first two of these result from $e^{\lambda x}$ and $e^{\bar{\lambda} x}$ as before, and the second two from $xe^{\lambda x}$ and $xe^{\bar{\lambda} x}$ in the same fashion.

For *complex triple roots* (which hardly ever occur in applications), one would obtain two more solutions $x^2e^{\gamma x}\cos\omega x$, $x^2e^{\gamma x}\sin\omega x$, and so on.

EXAMPLE 4 Complex double roots

Solve

$$y^{(7)} + 18y^{(5)} + 81y''' = 0.$$

Solution. The characteristic equation

$$\begin{aligned}\lambda^7 + 18\lambda^5 + 81\lambda^3 &= \lambda^3(\lambda^4 + 18\lambda^2 + 81)\\ &= \lambda^3(\lambda^2 + 9)^2\\ &= \lambda^3[(\lambda + 3i)(\lambda - 3i)]^2 = 0\end{aligned}$$

has a triple root 0 and double roots $-3i$ and $3i$. hence, by (12) with $\gamma = 0$ and $\omega = 3$, a general solution is

$$y = c_1 + c_2x + c_3x^2 + A_1\cos 3x + B_1\sin 3x + x(A_2\cos 3x + B_2\sin 3x). \qquad \blacksquare$$

Problem Set 3.2

Find an equation (1) for which the given functions form a basis.

1. e^x, e^{2x}, e^{3x}

2. $e^x, e^{-x}, \cos x, \sin x$

3. e^x, xe^x, x^2e^x

4. $\cos x, \sin x, x\cos x, x\sin x$

5. $e^{-x}, xe^{-x}, e^x, xe^x$

6. $e^{-2x}, e^{-x}, e^x, e^{2x}, 1$

7. $1, x, \cos 2x, \sin 2x$

8. $\cosh x, \sinh x, x\cosh x, x\sinh x$

Find a general solution.

9. $y''' - y' = 0$

10. $y^{IV} + 4y = 0$

11. $y''' - y'' - y' + y = 0$

12. $y''' + 6y'' + 11y' + 6y = 0$

13. $y^{IV} + 2y'' + y = 0$

14. $y''' + y'' - y' - y = 0$

15. $y^{IV} - 81y = 0$

16. $y^{IV} - 29y'' + 100y = 0$

Solve the following initial value problems.

17. $y''' = 0,\quad y(2) = 12,\quad y'(2) = 16,\quad y''(2) = 8$

18. $y''' - 3y'' + 3y' - y = 0,\quad y(0) = 2,\quad y'(0) = 2,\quad y''(0) = 10$

19. $y''' - y'' - y' + y = 0,\quad y(0) = 2,\quad y'(0) = 1,\quad y''(0) = 0$

20. $y^{IV} + 3y'' - 4y = 0,\quad y(0) = 0,\quad y'(0) = -10,\quad y''(0) = 0,$ $y'''(0) = 40$

21. $y''' - 2y'' - y' + 2y = 0, \quad y(0) = 3, \quad y'(0) = 0, \quad y''(0) = 3$
22. $y''' + 3y'' + 3y' + y = 0, \quad y(0) = 12, \quad y'(0) = -12, \quad y''(0) = 6$

Reduction of order extends from second-order (see Sec. 2.7) to higher order equations. It is particularly simple in the case of constant coefficients, since then we simply have to divide the characteristic equation by $\lambda - \lambda_1$, where $y_1 = e^{\lambda_1 x}$ is a known solution. Reduce and solve the following equations, using the given y_1.

23. $y''' - 2y'' - 5y' + 6y = 0, y_1 = e^x$
24. $y''' + 4y'' + 6y' + 4y = 0, y_1 = e^{-2x}$
25. $y''' - 2y'' + 4y' - 8y = 0, y_1 = e^{2x}$
26. $4y''' + 8y'' + 5y' + y = 0, y_1 = e^{-x}$
27. $y''' - 2y'' - 7.25y' - 3y = 0, y_1 = e^{4x}$
28. $y''' - y'' + 4y' - 4y = 0, y_1 = e^x$

29. Reduction of order is more complicated for an equation with *variable* coefficients $y''' + p_2(x)y'' + p_1(x)y' + p_0(x)y = 0$. If $y_1(x)$ is a solution of it, show that another solution is $y_2(x) = u(x)y_1(x)$ with $u(x) = \int z(x)\,dx$ and z obtained from

$$y_1 z'' + (3y_1' + p_2 y_1)z' + (3y_1'' + 2p_2 y_1' + p_1 y_1)z = 0.$$

30. Solve $x^3y''' - 3x^2y'' + (6 - x^2)xy' - (6 - x^2)y = 0$ by reducing the order, using $y_1 = x$.

31. (Euler–Cauchy equation) The *Euler–Cauchy equation of the third order* is

$$x^3y''' + ax^2y'' + bxy' + cy = 0.$$

Generalizing the method of Sec. 2.6, show that $y = x^m$ is a solution of the equation if and only if m is a root of the auxiliary equation

$$m^3 + (a - 3)m^2 + (b - a + 2)m + c = 0.$$

Solve:

32. $x^2y''' + 3xy'' - 3y' = 0$
33. $x^3y''' + 2x^2y'' - 4xy' + 4y = 0$
34. $x^3y''' + x^2y'' - 2xy' + 2y = 0$
35. $x^3y''' + 5x^2y'' + 2xy' - 2y = 0$

3.3 Nonhomogeneous Equations

From homogeneous linear equations we shall now turn to ***nonhomogeneous*** linear differential equations of nth order, which we write in the standard form

$$y^{(n)} + p_{n-1}(x)y^{(n-1)} + \cdots + p_1(x)y' + p_0(x)y = r(x) \tag{1}$$

with $y^{(n)} = d^ny/dx^n$ as the first term, which is practical. Here, $r(x) \not\equiv 0$. In studying (1) we also need the corresponding homogeneous equation

$$y^{(n)} + p_{n-1}(x)y^{(n-1)} + \cdots + p_1(x)y' + p_0(x)y = 0 \tag{2}$$

just as in Sec. 2.8 for the case $n = 2$. The theory of (1) will result from that of (2) by means of the relations between solutions:

Theorem 1 **[Relations between solutions of (1) and (2)]**

(a) *The difference of two solutions of* (1) *on some open interval I is a solution of* (2) *on I.*

(b) *The sum of a solution of* (1) *on I and a solution of* (2) *on I is a solution of* (1) *on I.*

The proof of this theorem is the same as that for the special case $n = 2$ (Theorem 1 in Sec. 2.8); even the formulas remain the same if we write (1) in the form $L[y] = r(x)$.

Together with the theory of the homogeneous equation in Sec. 3.1, this theorem suggests the following concepts of general and particular solutions, generalizing those for $n = 2$.

General Solution

Definition (General solution, particular solution)

A **general solution** of the nonhomogeneous equation (1) on some open interval I is a solution of the form

$$y(x) = y_h(x) + y_p(x), \tag{3}$$

where $y_h(x) = c_1y_1(x) + \cdots + c_ny_n(x)$ is a general solution of the homogeneous equation (2) on I and $y_p(x)$ is any solution of (1) on I containing no arbitrary constants.

A **particular solution** of (1) on I is a solution obtained from (3) by assigning specific values to the arbitrary constants $c_1, \cdots, c_n$ in $y_h(x)$. ∎

As for the homogeneous equation we can prove that (1) has a general solution, which includes all solutions, so that (1) has no singular solutions:

Theorem 2 **(General solution)**

If the coefficients $p_0(x), \cdots, p_{n-1}(x)$ of (1) *and $r(x)$ are continuous on some open interval I, then* (1) *has a general solution on I, and every solution of* (1) *on I is obtained by assigning suitable values to the constants in that general solution.*

Proof. **(a)** $y_h(x)$ in (3) exists by Theorem 4 in Sec. 3.1, and the existence of $y_p(x)$ will be shown in Sec. 3.5, where we shall construct $y_p(x)$ by the method of variation of parameters.

(b) Let $\tilde{y}(x)$ be any solution of (1) on I and (3) any general solution of (1) on I. Then Theorem 1(a) implies that $Y(x) = \tilde{y}(x) - y_p(x)$ is a solution of (2). By Theorem 5 in Sec. 3.1, this $Y(x)$ can be obtained from $y_h(x)$ by assigning suitable values to the arbitrary constants $c_1, \cdots, c_n$ in $y_h(x)$. From this and $\tilde{y}(x) = Y(x) + y_p(x)$ the second statement of the theorem follows. ∎

Initial Value Problem

An **initial value problem** for (1) consists of (1) and n initial conditions

$$y(x_0) = K_0, \quad y'(x_0) = K_1, \quad \cdots, \quad y^{(n-1)}(x_0) = K_{n-1} \tag{4}$$

[as for (2)] and has a unique solution:

Theorem 3 Existence and Uniqueness Theorem for Initial Value Problems

If the coefficients of (1) *and* $r(x)$ *are continuous on some open interval* I *and* x_0 *is in* I, *then the initial value problem* (1), (4) *has a unique solution on* I.

Proof. Choose any general solution (3) of (1) on I, which exists by Theorem 2. Then Theorem 2 in Sec. 3.1 implies that the initial value problem for the homogeneous equation (2) with initial conditions

$$y(x_0) = K_0 - y_p(x_0), \quad \cdots, \quad y^{(n-1)}(x_0) = K_{n-1} - y_p^{(n-1)}(x_0)$$

has a unique solution $y^*(x)$ on I. Hence $y(x) = y^*(x) + y_p(x)$ is a solution of (1) on I, which satisfies the initial conditions (4); that is, $y(x)$ is the desired solution, and the theorem is proved. ∎

Our discussion shows that to solve (1) or initial value problems for (1), we need methods for obtaining particular solutions y_p of (1). Such methods will be considered and applied to typical examples in the next two sections.

3.4 Method of Undetermined Coefficients

As for second-order linear equations, the method of undetermined coefficients gives particular solutions y_p of the constant-coefficient equation

$$y^{(n)} + a_{n-1}y^{(n-1)} + \cdots + a_1y' + a_0y = r(x). \tag{1}$$

In this method, the range of application (to functions $r(x)$ whose derivatives have forms similar to that of $r(x)$ itself) and the technical details of calculations and computations remain the same as for $n = 2$ in Sec. 2.9.

The only small difference concerns the Modification Rule and comes from the fact that whereas for $n = 2$ the characteristic equation of the homogeneous equation can have simple or double roots only, the characteristic equation of the present homogeneous equation

$$y^{(n)} + a_{n-1}y^{(n-1)} + \cdots + a_1y' + a_0y = 0 \tag{2}$$

can have multiple roots of larger orders m ($\leqq n$). This gives:

Rules for the Method of Undetermined Coefficients

(A) Basic Rule as in Sec. 2.9, with Table 3.1 (which is the same as Table 2.1 in Sec. 2.9).

(B) Modification Rule. *If a term in your choice for* y_p *is a solution of the homogeneous equation* (2), *then multiply* $y_p(x)$ *by* x^k, *where k is the smallest positive integer such that no term of* $x^k y_p(x)$ *is a solution of* (2).

(C) Sum Rule as in Sec. 2.9.

Thus for an initial value problem we have to do three steps:

1st Step. Find a general solution of the homogeneous equation (2).

2nd Step. Check whether the Modification Rule must be applied, and then determine a particular solution $y_p(x)$ of (1).

3rd Step. Find the particular solution of (1) that satisfies the given initial conditions.

Table 3.1
Method of Undetermined Coefficients

Term in $r(x)$	Choice for y_p
$ke^{\gamma x}$	$Ce^{\gamma x}$
$kx^n \ (n = 0, 1, \cdots)$	$C_n x^n + C_{n-1}x^{n-1} + \cdots + C_1 x + C_0$
$k \cos \omega x$ $k \sin \omega x$	$K \cos \omega x + M \sin \omega x$
$ke^{\alpha x} \cos \omega x$ $ke^{\alpha x} \sin \omega x$	$e^{\alpha x}(K \cos \omega x + M \sin \omega x)$

EXAMPLE 1

Solve

$$y^{\text{IV}} - y = 4.5e^{-2x}.$$

Solution. 1st Step. The characteristic equation $\lambda^4 - 1 = 0$ has the roots ± 1 and $\pm i$. Hence a general solution is

$$y_h = c_1 e^x + c_2 e^{-x} + c_2 e^{ix} + c_4 e^{-ix}$$

or

$$y_h = c_1 e^x + c_2 e^{-x} + A \cos x + B \sin x.$$

2nd Step. The Modification Rule is not needed. From $y_p = Ce^{-2x}$ we get by substitution

$$(-2)^4 Ce^{-2x} - Ce^{-2x} = 4.5e^{-2x}.$$

This gives $C = 0.3$. *Answer:*

$$y = y_h + y_p = c_1 e^x + c_2 e^{-x} + \text{A} \cos x + B \sin x + 0.3e^{-2x}.$$ ∎

EXAMPLE 2 Modification Rule

Consider

$$y''' - 3y'' + 3y' - y = 30e^x. \tag{3}$$

Find a general solution.

Solution. 1st Step. The characteristic equation $\lambda^3 - 3\lambda^2 + 3\lambda - 1 = 0$ has a triple root, $\lambda = 1$. Hence a general solution is

$$y_h = c_1e^x + c_2xe^x + c_3x^2e^x.$$

2nd Step. If we try $y_p = Ce^x$, we get $C - 3C + 3C - C = 30$, which has no solution. Try Cxe^x and Cx^2e^x. The Modification Rule calls for

$$y_p = Cx^3e^x.$$

Then

$$y_p' = C(x^3 + 3x^2)e^x,$$

$$y_p'' = C(x^3 + 6x^2 + 6x)e^x,$$

$$y_p''' = C(x^3 + 9x^2 + 18x + 6)e^x.$$

Substitution of these expressions into (3) and omission of the common factor e^x gives

$$(x^3 + 9x^2 + 18x + 6)C - 3(x^3 + 6x^2 + 6x)C + 3(x^3 + 3x^2)C - x^3C = 30.$$

The linear, quadratic, and cubic terms drop out, and $6C = 30$. Hence $C = 5$. *Answer*:

$$y = y_h + y_p = (c_1 + c_2x + c_3x^2)e^x + 5x^3e^x.$$ ∎

EXAMPLE 3 **An initial value problem**

Solve the initial problem

$$y''' - 2y'' - y' + 2y = 2x^2 - 6x + 4, \qquad y(0) = 5, \qquad y'(0) = -5, \qquad y''(0) = 1.$$

Solution. A general solution of the homogeneous equation is $y_h = c_1e^{-x} + c_2e^x + c_3e^{2x}$; see Example 1 in Sec. 3.2. We need a solution y_p. The form of the right side suggests that we try

$$y_p = Kx^2 + Mx + N. \qquad \text{Then} \qquad y_p' = 2Kx + M, \qquad y_p'' = 2K, \qquad y_p''' = 0.$$

Substitution into the equation yields

$$-2 \cdot 2K - (2Kx + M) + 2(Kx^2 + Mx + N) = 2x^2 - 6x + 4.$$

By equating like powers,

$$2Kx^2 = 2x^2, \qquad (-2K + 2M)x = -6x, \qquad -4K - M + 2N = 4.$$

Hence $K = 1$, $M = -2$, $N = 3$. This gives the general solution

$$y = y_h + y_p = c_1e^{-x} + c_2e^x + c_3e^{2x} + x^2 - 2x + 3.$$

For determining the constants from the initial conditions we also need the derivatives

$$y' = -c_1e^{-x} + c_2e^x + 2c_3e^{2x} + 2x - 2,$$

$$y'' = c_1e^{-x} + c_2e^x + 4c_3e^{2x} + 2.$$

Setting $x = 0$ and using the initial conditions, we obtain

$$\text{(a)} \qquad y(0) = c_1 + c_2 + c_3 + 3 = 5,$$

$$\text{(b)} \qquad y'(0) = -c_1 + c_2 + 2c_3 - 2 = -5,$$

$$\text{(c)} \qquad y''(0) = c_1 + c_2 + 4c_3 + 2 = 1.$$

(a) plus (b) gives $2c_2 + 3c_3 = -1$, and (c) minus (a) gives $3c_3 = -3$. Hence $c_3 = -1$, $c_2 = 1$, and $c_1 = 2$ from (a). The answer is

$$y = 2e^{-x} + e^x - e^{2x} + x^2 - 2x + 3.$$ ∎

Problem Set 3.4

Find a general solution.

1. $y''' + 2y'' - y' - 2y = 12e^{2x}$
2. $y'' + 3y' - 4y = -13.6 \sin x$
3. $y''' - y' = 10 \cos 2x$
4. $y^{\text{IV}} - 5y'' + 4y = 80e^{3x}$
5. $y''' - 3y'' + 3y' - y = 12e^{x}$
6. $y''' + 2y'' - y' - 2y = 1 - 4x^3$
7. $y''' + 3y'' + 3y' + y = 16e^{x} + x + 3$
8. $y''' + 5y'' + 7y' - 13y = -48 \cos x - 36 \sin x$

Solve the following initial value problems.

9. $y^{\text{IV}} - 5y'' + 4y = 10 \cos x$, $y(0) = 2$, $y'(0) = 0$, $y''(0) = 0$, $y'''(0) = 0$
10. $y''' - y'' - 4y' + 4y = 6e^{-x}$, $y(0) = 2$, $y'(0) = 3$, $y''(0) = -1$
11. $y^{\text{IV}} + y''' - 2y'' = -4x^2 + 18$, $y(1) = -3/2$, $y'(1) = -10/3$, $y''(1) = -2$, $y'''(1) = 6$
12. $y''' - 4y' = 10 \cos x + 5 \sin x$, $y(0) = 3$, $y'(0) = -2$, $y''(0) = -1$
13. $y''' + y'' - 2y = 2x^2 + 2x$, $y(0) = -1$, $y'(0) = 0$, $y''(0) = -4$
14. $y''' + 3y'' - y' - 3y = 21e^{x/2} + 3x + 1$, $y(0) = -7$, $y'(0) = -4$, $y''(0) = -9$
15. $y^{\text{IV}} + 10y'' + 9y = 2 \sinh x$, $y(0) = 0$, $y'(0) = 4.1$, $y''(0) = 0$, $y'''(0) = -27.9$

3.5 Method of Variation of Parameters

The **method of variation of parameters** is a method for finding particular solutions y_p of nth order nonhomogeneous linear differential equations

$$y^{(n)} + p_{n-1}(x)y^{(n-1)} + \cdots + p_1(x)y' + p_0(x)y = r(x). \tag{1}$$

It applies to any equation (1) with continuous coefficients and right side $r(x)$ on some open interval I, but is more complicated than the special method in the last section. (For second-order equations it was explained in Sec. 2.10.)

The method gives a particular solution y_p of (1) on I in the form

$$\begin{aligned} y_p(x) &= y_1(x) \int \frac{W_1(x)}{W(x)} r(x)\,dx + y_2(x) \int \frac{W_2(x)}{W(x)} r(x)\,dx \\ &\quad + \cdots + y_n(x) \int \frac{W_n(x)}{W(x)} r(x)\,dx. \end{aligned} \tag{2}$$

Here, $y_1, \cdots, y_n$ is a basis of solutions of the homogeneous equation

$$y^{(n)} + p_{n-1}(x)y^{(n-1)} + \cdots + p_1(x)y' + p_0(x)y = 0 \tag{3}$$

on I, with Wronskian W, and W_j ($j = 1, \cdots, n$) is obtained from W by replacing the jth column of W by the column $[0 \quad 0 \quad \cdots \quad 0 \quad 1]$.

Thus, when $n = 2$,

$$W = \begin{vmatrix} y_1 & y_2 \\ y_1' & y_2' \end{vmatrix}, \qquad W_1 = \begin{vmatrix} 0 & y_2 \\ 1 & y_2' \end{vmatrix} = -y_2, \qquad W_2 = \begin{vmatrix} y_1 & 0 \\ y_1' & 1 \end{vmatrix} = y_1,$$

and we see that (2) becomes identical with (2) in Sec. 2.10.

Also, the idea of proof in Sec. 2.10 extends to arbitrary n, as follows. We write (3) as $L[y] = 0$. In a general solution of (3),

$$y = c_1y_1 + \cdots + c_ny_n$$

we replace the constants (the "parameters") by functions $u_1(x), \cdots, u_n(x)$ to be determined so that

$$y_p = u_1y_1 + \cdots + u_ny_n \tag{4}$$

becomes a solution of (1) on I. This is one condition on n arbitrary functions u_j, and it seems plausible that we may impose $n - 1$ more conditions. To simplify calculations, we choose the latter conditions so that in $y_p', y_p'', \cdots$ we get rid of as many derivatives of the u_j as possible. Thus, from (4),

$$y_p' = (u_1y_1' + \cdots + u_ny_n') + (u_1'y_1 + \cdots + u_n'y_n),$$

and we choose as the first of the $n - 1$ conditions

$$u_1'y_1 + \cdots + u_n'y_n = 0. \tag{5/1}$$

Differentiating what is left, we get

$$y_p'' = (u_1y_1'' + \cdots + u_ny_n'') + (u_1'y_1' + \cdots + u_n'y_n')$$

and we impose as the second condition

$$u_1'y_1' + \cdots + u_n'y_n' = 0, \tag{5/2}$$

and so on until we reach

$$y_p^{(n-1)} = (u_1 y_1^{(n-1)} + \cdots + u_n y_n^{(n-1)}) + (u_1' y_1^{(n-2)} + \cdots + u_n' y_n^{(n-2)})$$

and impose as the last of the $n - 1$ conditions

$$(5/n-1) \qquad u_1' y_1^{(n-2)} + \cdots + u_n' y_n^{(n-2)} = 0.$$

The expressions for the derivatives, as reduced by these conditions, are

$$(6) \qquad y_p^{(j)} = u_1 y_1^{(j)} + \cdots + u_n y_n^{(j)}, \quad j = 1, \cdots, n-1.$$

Differentiating the last of these gives

$$(7) \quad y_p^{(n)} = (u_1 y_1^{(n)} + \cdots + u_n y_n^{(n)}) + (u_1' y_1^{(n-1)} + \cdots + u_n' y_n^{(n-1)}).$$

As the nth condition we require y_p to be a solution of (1); by substituting (6), (5), and (4) into (1) this gives

$$\begin{aligned}(u_1 y_1^{(n)} + \cdots + u_n y_n^{(n)}) + (u_1' y_1^{(n-1)} + \cdots + u_n' y_n^{(n-1)})&\\ (8) \qquad + p_{n-1}(u_1 y_1^{(n-1)} + \cdots + u_n y_n^{(n-1)})&\\ + \cdots + p_0(u_1 y_1 + \cdots + u_n y_n) &= r(x).\end{aligned}$$

Collecting the terms in u_1, then in u_2, etc., gives $u_1 L[y_1] = 0$, then $u_2 L[y_2] = 0$, etc., because $y_1, \cdots, y_n$ are solutions of (2). This reduces (8) to

$$(8^*) \qquad u_1' y_1^{(n-1)} + \cdots + u_n' y_n^{(n-1)} = r.$$

The conditions (5/1), (5/2), $\cdots$, ($5/n - 1$), (8*) form a system of n equations for the unknown functions $u_1', \cdots, u_n'$:

$$\begin{aligned} y_1 u_1' + \cdots + y_n u_n' &= 0\\ y_1' u_1' + \cdots + y_n' u_n' &= 0\\ &\vdots\\ y_1^{(n-2)} u_1' + \cdots + y_n^{(n-2)} u_n' &= 0\\ y_1^{(n-1)} u_1' + \cdots + y_n^{(n-1)} u_n' &= r.\end{aligned}$$

The coefficient determinant of the system is the Wronskian W, which is not zero since $y_1, \cdots, y_n$ is a basis of solutions of (2). Cramer's rule (Sec. 7.9) gives for $u_1', \cdots, u_n'$ the integrands in (2), and integration and substitution into (4) yields (2) and completes the derivation. ∎

EXAMPLE 1 Variation of parameters

Solve the nonhomogeneous Euler–Cauchy equation

$$x^3y''' - 3x^2y'' + 6xy' - 6y = x^4 \ln x \qquad (x > 0).$$

Solution. 1st Step. General solution. Substitution of $y = x^m$ and the derivatives and deletion of the factor x^m gives

$$m(m-1)(m-2) - 3m(m-1) + 6m - 6 = 0.$$

The roots are 1, 2, 3 and give as a basis of the homogeneous equation

$$y_1 = x, \qquad y_2 = x^2, \qquad y_3 = x^3.$$

2nd Step. Determinants needed in **(2).** These are

$$W = \begin{vmatrix} x & x^2 & x^3 \\ 1 & 2x & 3x^2 \\ 0 & 2 & 6x \end{vmatrix} = 2x^3, \qquad W_1 = \begin{vmatrix} 0 & x^2 & x^3 \\ 0 & 2x & 3x^2 \\ 1 & 2 & 6x \end{vmatrix} = x^4,$$

$$W_2 = \begin{vmatrix} x & 0 & x^3 \\ 1 & 0 & 3x^2 \\ 0 & 1 & 6x \end{vmatrix} = -2x^3, \qquad W_3 = \begin{vmatrix} x & x^2 & 0 \\ 1 & 2x & 0 \\ 0 & 2 & 1 \end{vmatrix} = x^2.$$

3rd Step. Integration. In (2) we also need the right side $r(x)$ of our equation in standard form, which we obtain by dividing the given equation by x^3 (the coefficient of y'''); this gives $r(x) = (x^4 \ln x)/x^3 = x \ln x$. From this and (2),

$$y_p = x \int \frac{x}{2} x \ln x \, dx - x^2 \int x \ln x \, dx + x^3 \int \frac{1}{2x} x \ln x \, dx$$

$$= \frac{x}{2}\left(\frac{x^3}{3} \ln x - \frac{x^3}{9}\right) - x^2\left(\frac{x^2}{2} \ln x - \frac{x^2}{4}\right) + \frac{x^3}{2}(x \ln x - x).$$

Simplification gives the answer

$$y_p = \frac{x^4}{6}\left(\ln x - \frac{11}{6}\right).$$

■

Problem Set 3.5

Find a general solution.

1. $y''' - 3y' + 3y' - y = x^{1/2}e^x$
2. $y''' - 6y'' + 12y' - 8y = \sqrt{2x}\, e^{2x}$
3. $y''' - 6y'' + 11y' - 6y = e^{2x} \sin x$
4. $y''' - y' = \cosh x$
5. $y''' + y' = \sec x$
6. $x^3y''' - 3x^2y'' + 6xy' - 6y = x^4 \sinh x$
7. $xy''' + 3y'' = e^x$
8. $x^3y''' + x^2y'' - 2xy' + 2y = x^3 \ln x$
9. $x^3y''' + x^2y'' - 2xy' + 2y = x^{-2}$
10. $4x^3y''' + 3xy' - 3y = 4x^{11/2}$

Review Questions and Problems for Chapter 3

1. What is the superposition or linearity principle? For what nth order equations does it hold?
2. List some other basic theorems that extend from second-order to nth order differential equations.
3. Under what (sufficient) conditions do those theorems hold?
4. How is linear independence and dependence of n functions defined? Why are these concepts important in the present chapter?
5. What form does a general solution of a homogeneous differential equation have? Of a nonhomogeneous equation?
6. What does an initial value problem for an nth order differential equation look like?
7. What is the Wronskian? What role did it play in this chapter?
8. Describe the idea and some details of the method of variation of parameters from memory, as best as you remember.
9. Discuss the advantages and disadvantages of our two methods for finding particular solutions.
10. What is the Modification Rule and when did we need it?

Find a general solution.

11. $y''' - y'' - y' + y = 0$
12. $y^{IV} + 20y'' + 64y = 0$
13. $x^2y''' + 3xy'' + 0.75y' = 0$
14. $y^{IV} + 4y = 0$
15. $(x^2D^3 - 3xD^2 + 3D)y = 0$
16. $(D^3 - 2D^2 - D + 2)y = 8x^3$
17. $(D^4 - 1)y = 30e^{2x} - x^2$
18. $(D^3 + 3D^2 + 3D + 1)y = 4 \sin x$
19. $(D^4 - 5D^2 + 4)y = 40 \cos 2x$
20. $(D^3 - D)y = \sinh x$

Solve the following initial value problems.

21. $y''' - y' = 0, \quad y(0) = 3.4, \quad y'(0) = -5.4, \quad y''(0) = -0.6$
22. $y^{IV} + 3y'' - 4y = 0, \quad y(0) = 0, \quad y'(0) = -20, \quad y''(0) = 0, \quad y'''(0) = 80$
23. $y''' - y'' - y' + y = 0, \quad y(0) = 2, \quad y'(0) = 1, \quad y''(0) = 0$
24. $4x^2y''' + 12xy'' + 3y' = 0, \quad y(1) = 0, \quad y'(1) = 1.5, \quad y''(1) = -1.75$
25. $(D^3 - 3D + 2)y = 0, \quad y(0) = 1, \quad y'(0) = 2, \quad y''(0) = 3$
26. $(D^3 - D)y = \cosh x, \quad y(0) = 5, \quad y'(0) = -1.25, \quad y''(0) = 3$
27. $(x^3D^3 - 3x^2D^2 + 6xD - 6)y = 12x^{-1}, \quad y(1) = 2.5, \quad y'(1) = 6.5,$ $y''(1) = 5$
28. $(D^3 + 3D^2 + 3D + 1) = 4 \sin x, \quad y(0) = 0, \quad y'(0) = -2, \quad y''(0) = 4$
29. $(D^3 - 3D^2 + 3D - 1)y = 0, \quad y(0) = 2, \quad y'(0) = 2, \quad y''(0) = 10$
30. $(x^3D^3 + x^2D^2 - 2xD + 2)y = x^3 \ln x, \quad y(1) = 25/32, \quad y'(1) = 47/32,$ $y''(1) = 21/16$

Summary of Chapter 3 Higher Order Linear Differential Equations

Compare with the similar Summary of Chap. 2 (the case $n = 2$).

An ***n*th order homogeneous linear differential equation** is an equation that can be written in the "*standard form*" (Sec. 3.1)

$$(1) \qquad y^{(n)} + p_{n-1}(x)y^{(n-1)} + \cdots + p_1(x)y' + p_0(x)y = 0$$

with $y^{(n)} = d^n y/dx^n$ as the first term. It has the very important property that a linear combination of solutions is again a solution (**superposition principle** or **linearity principle**, Sec. 3.1). A **basis** of solutions of (1) consists of n ***linearly independent*** solutions $y_1, \cdots, y_n$ of (1). The corresponding **general solution** is a linear combination

$$y = c_1y_1 + \cdots + c_ny_n \qquad (c_1, \cdots, c_n \text{ arbitrary constants}).$$

From it we obtain a particular solution if we choose n numbers for $c_1, \cdots, c_n$, for instance, by imposing n **initial conditions**

$$(2) \quad y(x_0) = K_0, \qquad y'(x_0) = K_1, \qquad \cdots, \qquad y^{(n-1)}(x_0) = K_{n-1}$$

($x_0, K_0, \cdots, K_{n-1}$ given). Equations (1), (2) constitute an **initial value problem** for (1). If $p_0, \cdots, p_{n-1}$ are continuous on some open interval I and x_0 is in I, then (1) has a general solution and (1), (2) has a unique solution on I, which is a particular solution. Hence (1) and the **nonhomogeneous linear differential equation**

$$(3) \qquad y^{(n)} + p_{n-1}(x)y^{(n-1)} + \cdots + p_1(x)y' + p_0(x)y = r(x)$$

(with continuous $r(x)$ on I) have no singular solutions. A **general solution** of (3) is of the form

$$y = y_h + y_p \qquad \text{(Sec. 3.3)},$$

where y_h is a general solution of (1) and y_p a particular solution of (3). For two methods of determining y_p, see Secs. 3.4 and 3.5.

In the case of constant coefficients we write (3) as

$$y^{(n)} + a_{n-1}y^{(n-1)} + \cdots + a_1y' + a_0y = r(x)$$

and similarly for (1), which can then be solved in terms of exponential and functions and sine and cosine (Sec. 3.2). In contrast, variable coefficient equations (1) generally have higher functions as solutions, as we show in Chap. 5.

Chapter 4

Systems of Differential Equations. Phase Plane, Stability

Systems of differential equations occur in various practical problems (see Secs. 4.1 and 4.5), and their theory (outlined in Sec. 4.2) includes that of single equations. Linear systems (Secs. 4.3, 4.4, 4.6) are best treated by the use of matrices and vectors, of which, however, only modest knowledge will be needed here (see Sec. 4.0). In addition to *solving* linear systems of differential equations (Secs. 4.3, 4.6), there is the powerful method of discussing the general behavior of solutions in the **phase plane** (Sec. 4.4). This also relates to **stability,** a concept that is of increasing importance in engineering science (for instance, in control theory). It means that, roughly speaking, small changes of a physical system at some instant cause only small changes in the behavior of the system at all later times. Phase plane methods can be extended to nonlinear systems (Sec. 4.5), for which they are particularly important.

Prerequisites for this chapter: Secs. 1.7–1.11, Chap. 2.
References: Appendix 1, Part A.
Answers to problems: Appendix 2.

4.0 Introduction: Vectors, Matrices

In discussing linear systems of differential equations we shall use matrices and vectors, in order to simplify formulas and thereby to clarify ideas. This requires knowledge of some rather elementary facts probably at the disposal of most students, who therefore may immediately go to Sec. 4.1 and use the present section ***for reference*** as needed. (The full treatment of matrices in Chap. 7 will ***not*** be needed in this chapter.)

Most of our linear systems will consist of two equations in two unknown functions $y_1(t)$, $y_2(t)$ of the form

$$(1)\qquad \begin{aligned} y_1' &= a_{11}y_1 + a_{12}y_2, \\ y_2' &= a_{21}y_1 + a_{22}y_2, \end{aligned} \qquad \text{for example,} \qquad \begin{aligned} y_1' &= -5y_1 + 2y_2 \\ y_2' &= 13y_1 + \tfrac{1}{2}y_2 \end{aligned}$$

(probably with an additional *known* function in each equation on the right).

In Sec. 4.2 we shall briefly discuss systems of n equations in n unknown functions $y_1(t), \cdots, y_n(t)$ of the form

$$
(2) \qquad
\begin{aligned}
y_1' &= a_{11}y_1 + a_{12}y_2 + \cdots + a_{1n}y_n \\
y_2' &= a_{21}y_1 + a_{22}y_2 + \cdots + a_{2n}y_n \\
&\cdots\cdots\cdots\cdots\cdots\cdots\cdots\cdots \\
y_n' &= a_{n1}y_1 + a_{n2}y_2 + \cdots + a_{nn}y_n.
\end{aligned}
$$

Some Definitions and Terms

In (1) the (constant or variable) coefficients form a **2 × 2 matrix A**, that is, an array

$$
(3) \quad \mathbf{A} = [a_{jk}] = \begin{bmatrix} a_{11} & a_{12} \\ a_{21} & a_{22} \end{bmatrix}, \qquad \text{for example,} \qquad \mathbf{A} = \begin{bmatrix} -5 & 2 \\ 13 & \frac{1}{2} \end{bmatrix}.
$$

Similarly, the coefficients in (2) form an ***n* × *n* matrix**

$$
(4) \qquad \mathbf{A} = [a_{jk}] = \begin{bmatrix} a_{11} & a_{12} & \cdots & a_{1n} \\ a_{21} & a_{22} & \cdots & a_{2n} \\ \cdot & \cdot & \cdots & \cdot \\ a_{n1} & a_{n2} & \cdots & a_{nn} \end{bmatrix}.
$$

The $a_{11}, a_{12}, \cdots$ are called **entries**, the horizontal lines **rows** or *row vectors*, and the vertical lines **columns** or *column vectors*. Thus, in (3) the first row is $[a_{11} \quad a_{12}]$, the second row is $[a_{21} \quad a_{22}]$ and the first and second columns are

$$
\begin{bmatrix} a_{11} \\ a_{21} \end{bmatrix} \quad \text{and} \quad \begin{bmatrix} a_{12} \\ a_{22} \end{bmatrix}.
$$

In the "***double subscript notation***" for entries, the first subscript denotes the *row* and the second the *column* in which the entry stands. Similarly in (4). The **main diagonal** is the diagonal $a_{11} \quad a_{22} \quad \cdots \quad a_{nn}$ in (4), hence $a_{11} \quad a_{22}$ in (3).

We shall also need vectors. A **column vector x** with n **components** $x_1, \cdots, x_n$ is

$$
\mathbf{x} = \begin{bmatrix} x_1 \\ x_2 \\ \vdots \\ x_n \end{bmatrix}, \qquad \text{thus if } n = 2, \qquad \mathbf{x} = \begin{bmatrix} x_1 \\ x_2 \end{bmatrix}.
$$

Similarly, a **row vector** $\mathbf{v}$ is of the form

$$\mathbf{v} = [v_1 \quad \cdots \quad v_n], \qquad \text{thus if } n = 2, \qquad \mathbf{v} = [v_1 \quad v_2].$$

Calculations with Matrices and Vectors

Equality. Two matrices are *equal* if and only if they are both $n \times n$ and corresponding entries are equal. Thus for $n = 2$,

$$\mathbf{A} = \begin{bmatrix} a_{11} & a_{12} \\ a_{21} & a_{22} \end{bmatrix} = \mathbf{B} = \begin{bmatrix} b_{11} & b_{12} \\ b_{21} & b_{22} \end{bmatrix}$$

if and only if

$$\begin{aligned} a_{11} &= b_{11}, & a_{12} &= b_{12} \\ a_{21} &= b_{21}, & a_{22} &= b_{22}. \end{aligned}$$

Two vectors are *equal* if and only if they both have n components and corresponding components are equal. Thus for $n = 2$,

$$\mathbf{v} = \begin{bmatrix} v_1 \\ v_2 \end{bmatrix} = \mathbf{x} = \begin{bmatrix} x_1 \\ x_2 \end{bmatrix} \quad \text{if and only if} \quad \begin{aligned} v_1 &= x_1 \\ v_2 &= x_2. \end{aligned}$$

Addition is performed by adding corresponding entries (or components); here, matrices must both be $n \times n$, and vectors must both have the same number of components. Thus for $n = 2$,

$$\text{(5)} \quad \mathbf{A} + \mathbf{B} \begin{bmatrix} a_{11} + b_{11} & a_{12} + b_{12} \\ a_{21} + b_{21} & a_{22} + b_{22} \end{bmatrix}, \qquad \mathbf{v} + \mathbf{x} = \begin{bmatrix} v_1 + x_1 \\ v_2 + x_2 \end{bmatrix}.$$

Scalar multiplication (multiplication by a number c) is performed by multiplying each entry (or component) by c. For example, if

$$\mathbf{A} = \begin{bmatrix} 9 & 3 \\ -2 & 0 \end{bmatrix}, \quad \text{then} \quad -7\mathbf{A} = \begin{bmatrix} -63 & -21 \\ 14 & 0 \end{bmatrix}.$$

If

$$\mathbf{v} = \begin{bmatrix} 0.4 \\ -13 \end{bmatrix}, \quad \text{then} \quad 10\mathbf{v} = \begin{bmatrix} 4 \\ -130 \end{bmatrix}.$$

Matrix multiplication. The product $\mathbf{C} = \mathbf{AB}$ (in this order) of two $n \times n$ matrices $\mathbf{A} = [a_{jk}]$ and $\mathbf{B} = [b_{jk}]$ is the $n \times n$ matrix $\mathbf{C} = [c_{jk}]$ with entries

$$c_{jk} = \sum_{m=1}^{n} a_{jm} b_{mk} \qquad \begin{array}{l} j = 1, \cdots, n \\ k = 1, \cdots, n, \end{array} \tag{6}$$

that is, multiply each entry in the jth *row* of **A** by the corresponding entry in the kth *column* of **B** and then add these n products. One says briefly that this is a "multiplication of rows into columns." For example,

$$\begin{bmatrix} 9 & 3 \\ -2 & 0 \end{bmatrix} \begin{bmatrix} 1 & -4 \\ 2 & 5 \end{bmatrix} = \begin{bmatrix} 9 \cdot 1 + 3 \cdot 2 & 9 \cdot (-4) + 3 \cdot 5 \\ -2 \cdot 1 + 0 \cdot 2 & (-2) \cdot (-4) + 0 \cdot 5 \end{bmatrix}$$

$$= \begin{bmatrix} 15 & -21 \\ -2 & 8 \end{bmatrix}.$$

Caution! Matrix multiplication is ***not commutative***, $\mathbf{AB} \neq \mathbf{BA}$ in general. In our example,

$$\begin{bmatrix} 1 & -4 \\ 2 & 5 \end{bmatrix} \begin{bmatrix} 9 & 3 \\ -2 & 0 \end{bmatrix} = \begin{bmatrix} 1 \cdot 9 + (-4) \cdot (-2) & 1 \cdot 3 + (-4) \cdot 0 \\ 2 \cdot 9 + 5 \cdot (-2) & 2 \cdot 3 + 5 \cdot 0 \end{bmatrix}$$

$$= \begin{bmatrix} 17 & 3 \\ 8 & 6 \end{bmatrix}.$$

Multiplication of an $n \times n$ matrix **A** by a vector **x** with n components is defined by the same rule: $\mathbf{v} = \mathbf{Ax}$ is the vector with the n components

$$v_j = \sum_{m=1}^{n} a_{jm} x_m \qquad j = 1, \cdots, n.$$

For example,

$$\begin{bmatrix} 12 & 7 \\ -8 & 3 \end{bmatrix} \begin{bmatrix} x_1 \\ x_2 \end{bmatrix} = \begin{bmatrix} 12x_1 + 7x_2 \\ -8x_1 + 3x_2 \end{bmatrix}.$$

Differentiation. The *derivative* of a matrix (or vector) with variable entries (or components) is obtained by differentiating each entry (or component). Thus, if

$$\mathbf{y}(t) = \begin{bmatrix} y_1(t) \\ y_2(t) \end{bmatrix} = \begin{bmatrix} e^{-2t} \\ \sin t \end{bmatrix}, \quad \text{then} \quad \mathbf{y}'(t) = \begin{bmatrix} y_1'(t) \\ y_2'(t) \end{bmatrix} = \begin{bmatrix} -2e^{-2t} \\ \cos t \end{bmatrix}$$

Transposition is the operation of writing columns as rows and conversely and is indicated by T. Thus the transpose $\mathbf{A}^{\mathsf{T}}$ of the 3×3 matrix

$$\mathbf{A} = \begin{bmatrix} a_{11} & a_{12} & a_{13} \\ a_{21} & a_{22} & a_{23} \\ a_{31} & a_{32} & a_{33} \end{bmatrix} = \begin{bmatrix} 4 & 0 & -8 \\ 1 & -6 & 7 \\ 9 & 9 & 5 \end{bmatrix}$$

is

$$\mathbf{A}^{\mathsf{T}} = \begin{bmatrix} a_{11} & a_{21} & a_{31} \\ a_{12} & a_{22} & a_{32} \\ a_{13} & a_{23} & a_{33} \end{bmatrix} = \begin{bmatrix} 4 & 1 & 9 \\ 0 & -6 & 9 \\ -8 & 7 & 5 \end{bmatrix}.$$

The transpose of a column vector, say,

$$\mathbf{v} = \begin{bmatrix} v_1 \\ v_2 \end{bmatrix}, \quad \text{is a row vector,} \quad \mathbf{v}^{\mathsf{T}} = [v_1 \quad v_2],$$

and conversely.

REMARK Using matrix multiplication and differentiation, we can now write (1) as

$$\text{(7)} \quad \mathbf{y}' = \begin{bmatrix} y_1' \\ y_2' \end{bmatrix} = \mathbf{Ay} = \begin{bmatrix} a_{11} & a_{12} \\ a_{21} & a_{22} \end{bmatrix} \begin{bmatrix} y_1 \\ y_2 \end{bmatrix}, \text{ e.g., } \mathbf{y}' = \begin{bmatrix} -5 & 2 \\ 13 & \frac{1}{2} \end{bmatrix} \begin{bmatrix} y_1 \\ y_2 \end{bmatrix}$$

and similarly for (2) by means of an $n \times n$ matrix $\mathbf{A}$ and a column vector $\mathbf{y}$ with n components, namely, $\mathbf{y}' = \mathbf{Ay}$. The vector equation (7) is equivalent to two equations for the components, and these are precisely the two equations in (1).

Linear independence. r given vectors $\mathbf{v}^{(1)}, \cdots, \mathbf{v}^{(r)}$ with n components are called a *linearly independent set* or, more briefly, **linearly independent,** if

$$\text{(8)} \qquad c_1\mathbf{v}^{(1)} + \cdots + c_r\mathbf{v}^{(r)} = \mathbf{0}$$

implies that all scalars $c_1, \cdots, c_r$ must be zero; here, $\mathbf{0}$ denotes the **zero vector**, whose n components are all zero. If (8) also holds for scalars not all zero, then these vectors are called a *linearly dependent set* or, more briefly, **linearly dependent**.

Inverse of a matrix. The $n \times n$ **unit matrix I** is the $n \times n$ matrix with main diagonal $1, 1, \cdots, 1$ and all other entries zero. If for a given $n \times n$ matrix $\mathbf{A}$ there is an $n \times n$ matrix $\mathbf{B}$ such that $\mathbf{AB} = \mathbf{BA} = \mathbf{I}$, then $\mathbf{A}$ is called **nonsingular** and $\mathbf{B}$ is called the **inverse** of $\mathbf{A}$ and is denoted by $\mathbf{A}^{-1}$; thus

$$\text{(9)} \qquad \mathbf{AA}^{-1} = \mathbf{A}^{-1}\mathbf{A} = \mathbf{I}.$$

If $\mathbf{A}$ has no inverse, it is called **singular**. For $n = 2$,

$$\mathbf{A}^{-1} = \frac{1}{\det \mathbf{A}} \begin{bmatrix} a_{22} & -a_{12} \\ -a_{21} & a_{11} \end{bmatrix}, \tag{10}$$

where the **determinant** of $\mathbf{A}$ is

$$\det \mathbf{A} = \begin{vmatrix} a_{11} & a_{12} \\ a_{21} & a_{22} \end{vmatrix} = a_{11}a_{22} - a_{12}a_{21}. \tag{11}$$

(For general n, see Sec. 7.8, but this will not be needed in this chapter.)

Eigenvalues, Eigenvectors

Let $\mathbf{A} = [a_{jk}]$ be a given matrix. Consider the equation

$$\mathbf{Ax} = \lambda\mathbf{x} \tag{12}$$

where λ is a scalar (a real or complex number) to be determined and $\mathbf{x}$ is a vector to be determined. Now for every λ a solution is $\mathbf{x} = \mathbf{0}$. A scalar λ such that (12) holds for some vector $\mathbf{x} \neq \mathbf{0}$ is called an **eigenvalue** of $\mathbf{A}$, and this vector is called an **eigenvector** of $\mathbf{A}$ corresponding to this eigenvalue λ. We can write (12) as $\mathbf{Ax} - \lambda\mathbf{x} = \mathbf{0}$ or

$$(\mathbf{A} - \lambda\mathbf{I})\mathbf{x} = \mathbf{0}. \tag{13}$$

These are n linear algebraic equations in the n unknowns $x_1, \cdots, x_n$ (the components of $\mathbf{x}$), and for these equations to have a solution $\mathbf{x} \neq \mathbf{0}$, their coefficient matrix $\mathbf{A} - \lambda\mathbf{I}$ must be singular, which is proved as a basic fact in linear algebra (Sec. 7.6). In this chapter we need this only for $n = 2$. Then (13) is

$$\begin{bmatrix} a_{11} - \lambda & a_{12} \\ a_{21} & a_{22} - \lambda \end{bmatrix} \begin{bmatrix} x_1 \\ x_2 \end{bmatrix} = \begin{bmatrix} 0 \\ 0 \end{bmatrix}; \tag{14}$$

in components,

$$\begin{aligned} (a_{11} - \lambda)x_1 + \quad a_{12}x_2 &= 0 \\ a_{21}x_1 \quad + (a_{22} - \lambda)x_2 &= 0. \end{aligned} \tag{14*}$$

Now $\mathbf{A} - \lambda\mathbf{I}$ is singular if and only if its determinant $\det(\mathbf{A} - \lambda\mathbf{I})$, called the **characteristic determinant** of $\mathbf{A}$ (also for general n), is zero:

$$\det(\mathbf{A} - \lambda\mathbf{I}) = \begin{vmatrix} a_{11} - \lambda & a_{12} \\ a_{21} & a_{22} - \lambda \end{vmatrix}$$

$$(15) \qquad = (a_{11} - \lambda)(a_{22} - \lambda) - a_{12}a_{21}$$

$$= \lambda^2 - (a_{11} + a_{22})\lambda + a_{11}a_{22} - a_{12}a_{21} = 0.$$

This quadratic equation is called the **characteristic equation** of **A**. Its solutions are the eigenvalues λ_1 and λ_2 of **A**. First determine these, then from (14*) with $\lambda = \lambda_1$ an eigenvector $\mathbf{x}^{(1)}$ of **A** corresponding to λ_1 and then from (14*) with $\lambda = \lambda_2$ an eigenvector $\mathbf{x}^{(2)}$ of **A** corresponding to λ_2. Note that if **x** is an eigenvector of **A**, so is $k\mathbf{x}$ for any $k \neq 0$.

EXAMPLE 1 Eigenvalue problem

Find the eigenvalues and eigenvectors of

$$(16) \qquad \mathbf{A} = \begin{bmatrix} -5 & 2 \\ 2 & -2 \end{bmatrix}.$$

Solution. The characteristic equation is

$$\det(\mathbf{A} - \lambda\mathbf{I}) = \begin{vmatrix} -5 - \lambda & 2 \\ 2 & -2 - \lambda \end{vmatrix} = \lambda^2 + 7\lambda + 6 = 0.$$

Its solutions are $\lambda_1 = -1$ and $\lambda_2 = -6$. For $\lambda = \lambda_1 = -1$ we have from (14*)

$$-4x_1 + 2x_2 = 0$$

$$2x_1 - x_2 = 0.$$

A solution of the first equation is $x_1 = 1$, $x_2 = 2$. The second equation gives the same. (Why?) Hence an eigenvector corresponding to $\lambda_1 = -1$ is

$$(17) \qquad \mathbf{x}^{(1)} = \begin{bmatrix} 1 \\ 2 \end{bmatrix}. \qquad \text{Similarly,} \qquad \mathbf{x}^{(2)} = \begin{bmatrix} 2 \\ -1 \end{bmatrix}$$

is an eigenvector of **A** corresponding to $\lambda_2 = -6$, as obtained from (14*) with $\lambda = \lambda_2$. ∎

4.1 Introductory Examples

We first illustrate with a few typical examples that systems of differential equations have various applications, and that a higher order equation can always be reduced to a first-order system. This accounts for the practical importance of these systems.

Vibrating masses on springs

From Sec. 2.5 we know that the undamped free motions of a mass on an elastic spring are governed by $my'' + ky = 0$ or

$$my'' = -ky$$

where $y = y(t)$ is the displacement of the mass. By the same arguments, for the two masses on the two springs in Fig. 57 we obtain the linear homogeneous system

$$\begin{aligned} m_1 y_1'' &= -k_1 y_1 + k_2(y_2 - y_1) \\ m_2 y_2'' &= -k_2(y_2 - y_1) \end{aligned} \tag{1}$$

for the unknown displacements $y_1 = y_1(t)$ of the first mass m_1 and $y_2 = y_2(t)$ of the second mass m_2. The forces acting on the first mass give the first equation, and the forces acting on the second mass give the second equation. Now $m_1 = m_2 = 1$, $k_1 = 3$, and $k_2 = 2$ in Fig. 57 so that by ordering (1) we obtain

$$\begin{aligned} y_1'' &= -5y_1 + 2y_2 \\ y_2'' &= 2y_1 - 2y_2 \end{aligned} \tag{2*}$$

or, written as a single vector equation,

$$\mathbf{y}'' = \begin{bmatrix} y_1'' \\ y_2'' \end{bmatrix} = \begin{bmatrix} -5 & 2 \\ 2 & -2 \end{bmatrix} \begin{bmatrix} y_1 \\ y_2 \end{bmatrix}. \tag{2}$$

As for a single equation, we try an exponential function of t,

$$\mathbf{y} = \mathbf{x}e^{\omega t}. \qquad \text{Then} \qquad \mathbf{y}'' = \omega^2 \mathbf{x} e^{\omega t} = \mathbf{A}\mathbf{x}e^{\omega t}.$$

Then, writing $\omega^2 = \lambda$ and dividing by $e^{\omega t}$, we get

$$\mathbf{A}\mathbf{x} = \lambda \mathbf{x}.$$

Eigenvalues and eigenvectors are (see Example 1, Sec. 5.0)

$$\lambda_1 = -1, \qquad \mathbf{x}^{(1)} = \begin{bmatrix} 1 \\ 2 \end{bmatrix} \qquad \text{and} \qquad \lambda_2 = -6, \qquad \mathbf{x}^{(2)} = \begin{bmatrix} 2 \\ -1 \end{bmatrix}.$$

Since $\omega = \sqrt{\lambda}$ and $\sqrt{-1} = \pm i$ and $\sqrt{-6} = \pm i\sqrt{6}$, we get

$$\mathbf{y} = \mathbf{x}^{(1)}(c_1 e^{it} + c_2 e^{-it}) + \mathbf{x}^{(2)}(c_3 e^{i\sqrt{6}t} + c_4 e^{-i\sqrt{6}t})$$

or, by (7) in Sec. 2.3,

$$\mathbf{y} = a_1 \mathbf{x}^{(1)} \cos t + b_1 \mathbf{x}^{(1)} \sin t + a_2 \mathbf{x}^{(2)} \cos \sqrt{6}t + b_2 \mathbf{x}^{(2)} \sin \sqrt{6}t$$

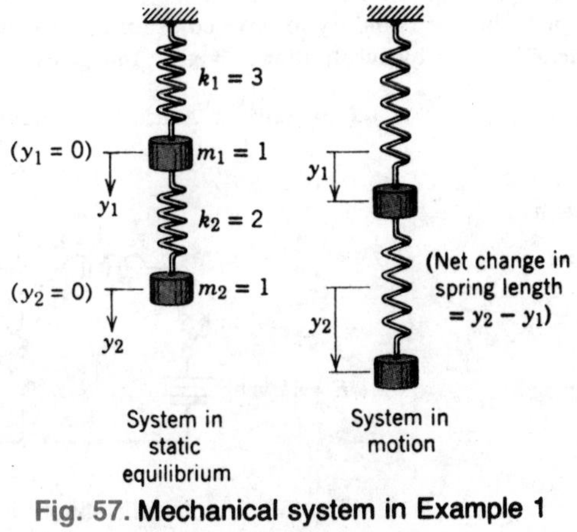

Fig. 57. Mechanical system in Example 1

where $a_1 = c_1 + c_2$, $b_1 = i(c_1 - c_2)$, $a_2 = c_3 + c_4$, $b_2 = i(c_3 - c_4)$. These four arbitrary constants can be specified by four initial conditions. In components, this solution is

$$y_1 = a_1 \cos t + b_1 \sin t + 2a_2 \cos \sqrt{6}t + 2b_2 \sin \sqrt{6}t$$

$$y_2 = 2a_1 \cos t + 2b_1 \sin t - a_2 \cos \sqrt{6}t - b_2 \sin \sqrt{6}t.$$ ∎

EXAMPLE 2 **Model of an electrical network**

Find the currents $I_1(t)$ and $I_2(t)$ in the network shown in Fig. 58, assuming that all charges and currents are zero when the switch is closed at $t = 0$.

Solution. 1st Step. Setting up the mathematical model. The mathematical model of this network is obtained from Kirchhoff's voltage law, as in Secs. 1.8 and 2.12 (where we considered single circuits). The left loop yields

$$I_1' + 4(I_1 - I_2) = 12$$

or

$$I_1' = -4I_1 + 4I_2 + 12, \tag{3a}$$

where I_1 is the current in the left loop and I_2 the current in the right loop, and $4(I_1 - I_2)$ is the voltage drop across the resistor because I_1 and I_2 flow through the resistor in opposite directions. Similarly, for the right loop we obtain

$$6I_2 + 4(I_2 - I_1) + 4\int I_2\,dt = 0$$

or, by differentiation and division by 10,

$$I_2' - 0.4I_1' + 0.4I_2 = 0.$$

Replacing $-0.4I_1'$ by using (3a) and ordering, we obtain

$$I_2' = -1.6I_1 + 1.2I_2 + 4.8. \tag{3b}$$

In matrix form, (3) is (we write **J** since **I** is the unit matrix)

$$\mathbf{J}' = \mathbf{AJ} + \mathbf{g}, \quad \text{where} \quad \mathbf{J} = \begin{bmatrix} I_1 \\ I_2 \end{bmatrix}, \quad \mathbf{A} = \begin{bmatrix} -4.0 & 4.0 \\ -1.6 & 1.2 \end{bmatrix}, \quad \mathbf{g} = \begin{bmatrix} 12.0 \\ 4.8 \end{bmatrix}. \tag{4}$$

2nd Step. Solving (4). This is a nonhomogeneous linear system. No method of solution is yet available, but we could try to proceed as for a single equation, solving first the homogeneous system $\mathbf{J}' = \mathbf{AJ}$ by substituting $\mathbf{J} = \mathbf{x}e^{\lambda t}$; this gives

$$\mathbf{J}' = \lambda\mathbf{x}e^{\lambda t} = \mathbf{A}\mathbf{x}e^{\lambda t}, \qquad \text{thus} \qquad \mathbf{Ax} = \lambda\mathbf{x}.$$

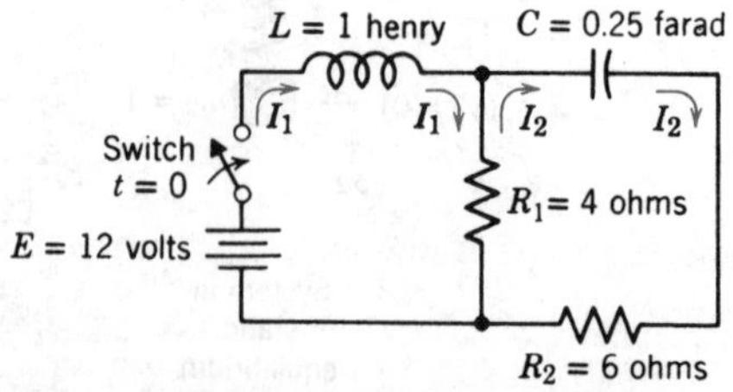

Fig. 58. Electrical network in Example 2

The characteristic equation

$$\begin{vmatrix} -4-\lambda & 4 \\ -1.6 & 1.2-\lambda \end{vmatrix} = \lambda^2 + 2.8\lambda + 1.6 = 0$$

has the solutions $\lambda_1 = -2$, $\lambda_2 = -0.8$. For $\lambda_1 = -2$ we get

$$(-4+2)x_1 + 4x_2 = 0, \quad \text{say,} \quad x_1 = 2, \ x_2 = 1, \quad \text{thus} \quad \mathbf{x}^{(1)} = \begin{bmatrix} 2 \\ 1 \end{bmatrix}.$$

For $\lambda_2 = -0.8$ we get

$$(-4+0.8)x_1 + 4x_2 = 0, \quad \text{say,} \quad x_1 = 1, \ x_2 = 0.8, \quad \text{thus} \quad \mathbf{x}^{(2)} = \begin{bmatrix} 1 \\ 0.8 \end{bmatrix}.$$

Hence a "general solution" of the homogeneous system is

$$\mathbf{J}_h = c_1\mathbf{x}^{(1)}e^{-2t} + c_2\mathbf{x}^{(2)}e^{-0.8t}.$$

For a particular solution of (4), since $\mathbf{g}$ is constant, we could try a constant vector $\mathbf{J}_p{}^\mathsf{T} = \mathbf{a}^\mathsf{T} = [a_1 \quad a_2]$. Then $\mathbf{J}_p' = \mathbf{0}$, and substitution gives $\mathbf{Aa} + \mathbf{g} = \mathbf{0}$; in components,

$$\begin{aligned} -4.0a_1 + 4.0a_2 + 12.0 &= 0 \\ -1.6a_1 + 1.2a_2 + \ \ 4.8 &= 0. \end{aligned}$$

The solution is $a_1 = 3$, $a_2 = 0$; thus $\mathbf{a}^\mathsf{T} = [3 \quad 0]$. Hence

(5) $$\mathbf{J} = \mathbf{J}_h + \mathbf{J}_p = c_1\mathbf{x}^{(1)}e^{-2t} + c_2\mathbf{x}^{(2)}e^{-0.8t} + \mathbf{a};$$

in components,

$$\begin{aligned} I_1 &= 2c_1e^{-2t} + c_2e^{-0.8t} + 3 \\ I_2 &= c_1e^{-2t} + 0.8c_2e^{-0.8t}. \end{aligned}$$

The initial conditions give

$$\begin{aligned} I_1(0) &= 2c_1 + c_2 + 3 = 0 \\ I_2(0) &= c_1 + 0.8c_2 = 0. \end{aligned}$$

Hence $c_1 = -4$ and $c_2 = 5$. We see that $I_1(t) \to 3$ amperes as $t \to \infty$, whereas $I_2(t) \to 0$. Why was this to be expected physically? ∎

Basic applications account for one reason for the importance of systems of differential equations. Another reason results from the fact that an nth order differential equation

(6) $$y^{(n)} = F(t, y, y', \cdots, y^{(n-1)})$$

can always be reduced to a system of n first-order differential equations simply by setting

(7) $$y_1 = y, \quad y_2 = y', \quad y_3 = y'', \quad \cdots, \quad y_n = y^{(n-1)}.$$

We then immediately obtain the system

(8a)
$$
\begin{aligned}
y_1' &= y_2 \\
y_2' &= y_3 \\
&\vdots \\
y_{n-1}' &= y_n
\end{aligned}
$$

and from (6)

(8b)
$$y_n' = F(t, y_1, y_2, \cdots, y_n).$$

This reduction is also a reason why one generally concentrates on *first-order* systems. To gain confidence, let us consider an old friend of ours:

EXAMPLE 3 **Mass on a spring**

For

$$y'' + \frac{c}{m} y' + \frac{k}{m} y = 0 \qquad \text{(Sec. 2.5)}$$

the system (8) is linear and homogeneous,

$$
\begin{aligned}
y_1' &= y_2 \\
y_2' &= -\frac{k}{m} y_1 - \frac{c}{m} y_2.
\end{aligned}
$$

Setting $\mathbf{y}^\mathsf{T} = [y_1 \quad y_2]$, we get in matrix form

$$\mathbf{y}' = \begin{bmatrix} 0 & 1 \\ -\dfrac{k}{m} & -\dfrac{c}{m} \end{bmatrix} \mathbf{y}.$$

The characteristic equation is

$$\det(\mathbf{A} - \lambda\mathbf{I}) = \begin{vmatrix} -\lambda & 1 \\ -\dfrac{k}{m} & -\dfrac{c}{m} - \lambda \end{vmatrix} = \lambda^2 + \frac{c}{m}\lambda + \frac{k}{m} = 0$$

and agrees with the characteristic equation in Sec. 2.5. For a simple illustrative computation, assume that $m = 1$, $c = 2$, and $k = 0.75$. Then $\lambda^2 + 2\lambda + 0.75 = 0$, and this gives the eigenvalues $\lambda_1 = -0.5$, $\lambda_2 = -1.5$, and eigenvectors are $\mathbf{x}^{(1)\mathsf{T}} = [2 \quad -1]$ (from $0.5x_1 + x_2 = 0$) and $\mathbf{x}^{(2)\mathsf{T}} = [1 \quad -1.5]$ (from $1.5x_1 + x_2 = 0$). Hence we get the solution

$$\mathbf{y} = c_1 \begin{bmatrix} 2 \\ -1 \end{bmatrix} e^{-0.5t} + c_2 \begin{bmatrix} 1 \\ -1.5 \end{bmatrix} e^{-1.5t}.$$

The first component of this vector equation is the expected solution

$$y = y_1 = 2c_1 e^{-0.5t} + c_2 e^{-1.5t}$$

and $y_2 = y_1'$ comes out as it should,

$$y_2 = -c_1 e^{-0.5t} - 1.5c_2 e^{-1.5t} = y_1'. \qquad \blacksquare$$

Similarly, many systems of differential equations of higher order can be reduced to first-order systems. For instance, in Example 1 we can obtain such a system by setting $y_3 = y_1'$, $y_4 = y_2'$; then $y_3' = y_1''$ and $y_4' = y_2''$, so that by (2*),

$$
\begin{aligned}
y_1' &= y_3 \\
y_2' &= y_4 \\
y_3' &= -5y_1 + 2y_2 \\
y_4' &= 2y_1 - 2y_2.
\end{aligned} \tag{9}
$$

Problem Set 4.1

1. Show the details of the calculations in Example 1 that lead from the complex general solution to the real general solution.

In Example 1, find the particular solution satisfying the following initial conditions.

2. $y_1(0) = 1$, $y_2(0) = 2$, initial velocities zero

3. Initial displacements zero, $y_1'(0) = 2$, $y_2'(0) = -1$

4. $y_1(0) = 3$, $y_1'(0) = 0$, $y_2(0) = 1$, $y_2'(0) = 0$

5. $y_1(0) = -2$, $y_1'(0) = 8\sqrt{6}$, $y_2(0) = -4$, $y_2'(0) = -4\sqrt{6}$

In Example 2, find the currents when

6. $I_1(0) = 28$, $I_2(0) = 14$

7. $I_1(0) = 9$, $I_2(0) = 0$

4.2 Basic Concepts and Theory

In this section we discuss some basic concepts and facts that are similar to those for single equations.

The first-order systems in the last section were special cases of the more general system

$$
\begin{aligned}
y_1' &= f_1(t, y_1, \cdots, y_n) \\
y_2' &= f_2(t, y_1, \cdots, y_n) \\
&\cdots \\
y_n' &= f_n(t, y_1, \cdots, y_n)
\end{aligned} \tag{1}
$$

which includes almost all cases of practical interest. When $n = 1$, this becomes $y_1' = f_1(t, y_1)$ or, simply, $y' = f(t, y)$, well known to us from Chap. 1.

A **solution** of (1) on some interval $a < t < b$ is a set of n differentiable functions

$$y_1 = \phi_1(t), \cdots, y_n = \phi_n(t)$$

on $a < t < b$ that satisfy (1) throughout this interval.

An **initial value problem** for (1) consists of (1) and n given initial conditions

$$y_1(t_0) = K_1, \quad y_2(t_0) = K_2, \quad \cdots, \quad y_n(t_0) = K_n \tag{2}$$

where t_0 is a specified value of t in the interval considered and $K_1, \cdots, K_n$ are given numbers. Sufficient conditions for the existence and uniqueness of a solution of an initial value problem (1), (2) are stated in the following theorem, which extends the theorems in Sec. 1.11 for a single equation. (For a proof, see Ref. [A4].)

Theorem 1 (Existence and uniqueness theorem)

Let $f_1, \cdots, f_n$ in (1) be continuous functions having continuous partial derivatives $\partial f_1/\partial y_1, \cdots, \partial f_1/\partial y_n, \cdots, \partial f_n/\partial y_n$ in some domain R of $ty_1y_2 \cdots y_n$-space containing the point $(t_0, K_1, \cdots, K_n)$. Then (1) has a solution on some interval $t_0 - \alpha < t < t_0 + \alpha$ satisfying (2), and this solution is unique.

Extending the notion of a linear differential equation, we call (1) a **linear system** if it is linear in $y_1 \cdots, y_n$; that is, if it can be written

$$\begin{aligned} y_1' &= a_{11}(t)y_1 + \cdots + a_{1n}(t)y_n + g_1(t) \\ &\;\;\vdots \\ y_n' &= a_{n1}(t)y_1 + \cdots + a_{nn}(t)y_n + g_n(t). \end{aligned} \tag{3*}$$

In vector form, this becomes

$$\mathbf{y}' = \mathbf{A}\mathbf{y} + \mathbf{g} \tag{3}$$

where $\mathbf{A} = \begin{bmatrix} a_{11} & \cdots & a_{1n} \\ \cdot & \cdots & \cdot \\ a_{n1} & \cdots & a_{nn} \end{bmatrix}, \quad \mathbf{y} = \begin{bmatrix} y_1 \\ \vdots \\ y_n \end{bmatrix}, \quad \mathbf{g} = \begin{bmatrix} g_1 \\ \vdots \\ g_n \end{bmatrix}.$

This system is called **homogeneous** if $\mathbf{g} = \mathbf{0}$, so that it is

$$\mathbf{y}' = \mathbf{A}\mathbf{y}. \tag{4}$$

If $\mathbf{g} \neq \mathbf{0}$, then (3) is called **nonhomogeneous**. Example 1 in the previous section is homogeneous and Example 2 nonhomogeneous.

For a linear system (3) the existence and uniqueness theorem is simpler:

Theorem 2 **(Existence and uniqueness in the linear case)**
Let the a_{jk}*'s and* g_j*'s in* (3) *be continuous functions of* t *on an open interval* $\alpha < t < \beta$ *containing the point* $t = t_0$. *Then* (3) *has a solution* $y(t)$ *on this interval satisfying* (2), *and this solution is unique.*

As for a single homogeneous linear equation we have

Theorem 3 **(Superposition principle or linearity principle)**
If $y^{(1)}$ *and* $y^{(2)}$ *are solutions of the* ***homogeneous linear*** *system* (4) *on some interval, so is any linear combination* $y = c_1 y^{(1)} + c_2 y^{(2)}$.

Proof. Differentiating and using (4), we obtain

$$\begin{aligned} \mathbf{y}' &= [c_1\mathbf{y}^{(1)} + c_2\mathbf{y}^{(2)}]' \\ &= c_1\mathbf{y}^{(1)\prime} + c_2\mathbf{y}^{(2)\prime} \\ &= c_1\mathbf{A}\mathbf{y}^{(1)} + c_2\mathbf{A}\mathbf{y}^{(2)} \\ &= \mathbf{A}(c_1\mathbf{y}^{(1)} + c_2\mathbf{y}^{(2)}) = \mathbf{A}\mathbf{y}. \end{aligned}$$ ∎

The general theory of linear systems is quite similar to that of a single linear equation in Secs. 2.1, 2.7, and 2.8. To see this, we shall explain the most basic concepts and facts, referring to more advanced texts for proofs, such as Ref. [A4].

By a **basis** of solutions of the homogeneous system (4) on some interval J we mean a linearly independent set of n solutions $\mathbf{y}^{(1)}, \cdots, \mathbf{y}^{(n)}$ of (4) on that interval, and we call a corresponding linear combination

$$\mathbf{y} = c_1\mathbf{y}^{(1)} \cdots + c_n\mathbf{y}^{(n)} \qquad (c_1, \cdots, c_n \text{ arbitrary}) \tag{5}$$

a **general solution** of (4) on J, because this $\mathbf{y}$ includes every solution of (4) on J, as can be shown. If the $a_{jk}(t)$ in (4) are continuous on J, then (4) has a basis of solutions on J.

Furthermore, the **Wronskian** of $\mathbf{y}^{(1)}, \cdots, \mathbf{y}^{(n)}$ is the determinant

$$W(\mathbf{y}^{(1)}, \cdots, \mathbf{y}^{(n)}) = \begin{vmatrix} y_1^{(1)} & y_1^{(2)} & \cdots & y_1^{(n)} \\ y_2^{(1)} & y_2^{(2)} & \cdots & y_2^{(n)} \\ \cdot & \cdot & \cdots & \cdot \\ y_n^{(1)} & y_n^{(2)} & \cdots & y_n^{(n)} \end{vmatrix} \tag{6}$$

whose columns are these solutions, each written in terms of components. These solutions form a basis on J if and only if W is not zero at any t_1 in this interval. W either is identically zero or is nowhere zero in J.

Note that the Wronskian of two solutions y and z of a single second-order equation is (see Sec. 2.8)

$$W(y, z) = \begin{vmatrix} y & z \\ y' & z' \end{vmatrix}.$$

Now if we write the equation as a first-order system, we must set $y_1 = y$, $y_2 = y'$ and $z_1 = z$, $z_2 = z'$ (see Sec. 4.1). Then $W(y, z)$ takes the form (6) with $n = 2$ (except for the notation). This motivates our definition in (6).

We shall now turn to **methods of solution**, beginning with homogeneous linear systems with constant coefficients in the next section.

4.3 Homogeneous Linear Systems with Constant Coefficients

We continue our discussion of homogeneous linear systems

$$\mathbf{y}' = \mathbf{A}\mathbf{y}, \tag{1}$$

now assuming that the $n \times n$ matrix $\mathbf{A} = [a_{jk}]$ is constant, that is, its entries do not depend on t. We wish to solve (1). For this we remember that a single equation $y' = ky$ has the solution $y = Ce^{kt}$. Accordingly, we try

$$\mathbf{y} = \mathbf{x}e^{\lambda t}. \tag{2}$$

By substitution into (1) we get

$$\mathbf{y}' = \lambda\mathbf{x}e^{\lambda t} = \mathbf{A}\mathbf{y} = \mathbf{A}\mathbf{x}e^{\lambda t}.$$

Dividing by $e^{\lambda t}$, we are left with the eigenvalue problem

$$\mathbf{A}\mathbf{x} = \lambda\mathbf{x}. \tag{3}$$

Thus the nontrivial solutions of (1) are of the form (2), where λ is an eigenvalue of $\mathbf{A}$ and $\mathbf{x}$ is a corresponding eigenvector.

Let us further assume that $\mathbf{A}$ has a basis of n eigenvectors $\mathbf{x}^{(1)}, \cdots, \mathbf{x}^{(n)}$ corresponding to eigenvalues $\lambda_1, \cdots, \lambda_n$ (which may all be different or some of which—or all—may be equal). Then the corresponding solutions (2) are

$$\mathbf{y}^{(1)} = \mathbf{x}^{(1)}e^{\lambda_1 t}, \qquad \cdots, \qquad \mathbf{y}^{(n)} = \mathbf{x}^{(n)}e^{\lambda_n t} \tag{4}$$

and have the Wronskian

$$W(\mathbf{y}^{(1)}, \cdots, \mathbf{y}^{(n)}) = \begin{vmatrix} x_1^{(1)}e^{\lambda_1 t} & \cdots & x_1^{(n)}e^{\lambda_n t} \\ x_2^{(1)}e^{\lambda_1 t} & \cdots & x_2^{(n)}e^{\lambda_n t} \\ \cdot & \cdots & \cdot \\ x_n^{(1)}e^{\lambda_1 t} & \cdots & x_n^{(n)}e^{\lambda_n t} \end{vmatrix}$$

$$= e^{\lambda_1 t + \cdots + \lambda_n t} \begin{vmatrix} x_1^{(1)} & \cdots & x_1^{(n)} \\ x_2^{(1)} & \cdots & x_2^{(n)} \\ \cdot & \cdots & \cdot \\ x_n^{(1)} & \cdots & x_n^{(n)} \end{vmatrix}.$$

On the right, the exponential function is never zero, and the determinant is not zero either because its columns are the linearly independent eigenvectors that form a basis. This proves

Theorem 1 **(General solution)**

If the constant matrix **A** *in the system* (1) *has a linearly independent set of n eigenvectors,*[1] *then the corresponding solutions* $\mathbf{y}^{(1)}, \cdots, \mathbf{y}^{(n)}$ *in* (4) *form a basis of solutions of* (1), *and the corresponding general solution is*

$$\mathbf{y} = c_1\mathbf{x}^{(1)}e^{\lambda_1 t} + \cdots + c_n\mathbf{x}^{(n)}e^{\lambda_n t}. \tag{5}$$

Let us see how to solve a system (1) of two equations in practice, and how to visualize solution curves in the y_1y_2-plane, also known as **paths** or **trajectories**. This plane is called **phase plane**.

EXAMPLE 1 **Node**

Find the general solution of the homogeneous linear system

$$\mathbf{y}' = \mathbf{A}\mathbf{y} = \begin{bmatrix} -3 & 1 \\ 1 & -3 \end{bmatrix}\mathbf{y}, \qquad \text{thus} \qquad \begin{aligned} y_1' &= -3y_1 + y_2 \\ y_2' &= y_1 - 3y_2. \end{aligned} \tag{6}$$

Solution. By substituting $\mathbf{y} = \mathbf{x}e^{\lambda t}$ and $\mathbf{y}' = \lambda\mathbf{x}e^{\lambda t}$ and dropping the exponential function we get $\mathbf{A}\mathbf{x} = \lambda\mathbf{x}$. The characteristic equation is

$$\det(\mathbf{A} - \lambda\mathbf{I}) = \begin{vmatrix} -3-\lambda & 1 \\ 1 & -3-\lambda \end{vmatrix} = \lambda^2 + 6\lambda + 8 = 0.$$

This gives the eigenvalues $\lambda_1 = -2$ and $\lambda_2 = -4$. Eigenvectors are then obtained from

$$(-3-\lambda)x_1 + x_2 = 0.$$

For $\lambda_1 = -2$ this is $-x_1 + x_2 = 0$, so that we can take $\mathbf{x}^{(1)\mathsf{T}} = [1 \quad 1]$. For $\lambda_2 = -4$ this becomes $x_1 + x_2 = 0$, and an eigenvector is $\mathbf{x}^{(2)\mathsf{T}} = [1 \quad -1]$. Hence this gives the general solution

[1]This holds if **A** is symmetric ($a_{jk} = a_{kj}$) or if it has *n different* eigenvalues. (See Theorems 2 and 4 in Sec. 7.14.)

$$\mathbf{y} = \begin{bmatrix} y_1 \\ y_2 \end{bmatrix} = c_1\mathbf{y}^{(1)} + c_2\mathbf{y}^{(2)} = c_1\begin{bmatrix} 1 \\ 1 \end{bmatrix}e^{-2t} + c_2\begin{bmatrix} 1 \\ -1 \end{bmatrix}e^{-4t}.$$

For every choice of c_1, c_2 we get some path. When $c_2 = 0$ and $c_1 > 0$, we get the straight ray $y_1 = y_2$ in the first quadrant in Fig. 59, the arrow indicating the direction of increasing t. When $c_2 = 0$ and $c_1 < 0$, we get the ray $y_1 = y_2$ in the third quadrant. Similarly, for $c_1 = 0$ and $c_2 > 0$ or $c_2 < 0$ we get the rays of $y_1 = -y_2$ in the fourth and second quadrants, respectively. When $c_1 \neq 0$ and $c_2 \neq 0$ we get a curve whose tangent direction at the origin is that of $\mathbf{x}^{(1)}$, because e^{-4t} goes to zero faster than e^{-2t} as t increases. For instance, $c_1 = c_2 = 1$ gives

$$y_1 = e^{-2t} + e^{-4t}, \qquad y_2 = e^{-2t} - e^{-4t},$$

so that $y_1 \approx y_2$ when t is large, that is, near the origin. Thus every path has at 0 the same limiting direction, except for the one pair of paths for which $y_1 = -y_2$. A point at which the solutions show this behavior is called a **node**. Thus 0 is a node of our system (6). More precisely, 0 is called an **improper node**, as opposed to a **proper node**, which is a point at which there is a solution curve in every possible direction. For instance, the system

$$\text{(7)} \qquad \mathbf{y}' = \begin{bmatrix} 1 & 0 \\ 0 & 1 \end{bmatrix}\mathbf{y}, \qquad \text{thus} \qquad \begin{aligned} y_1' &= y_1 \\ y_2' &= y_2 \end{aligned}$$

has a proper node at 0 (see Fig. 60) because the general solution is

$$\mathbf{y} = c_1\begin{bmatrix} 1 \\ 0 \end{bmatrix}e^t + c_2\begin{bmatrix} 0 \\ 1 \end{bmatrix}e^t \qquad \text{or} \qquad \begin{aligned} y_1 &= c_1e^t \\ y_2 &= c_2e^t \end{aligned} \qquad \text{or} \qquad c_1y_2 = c_2y_1.$$

■

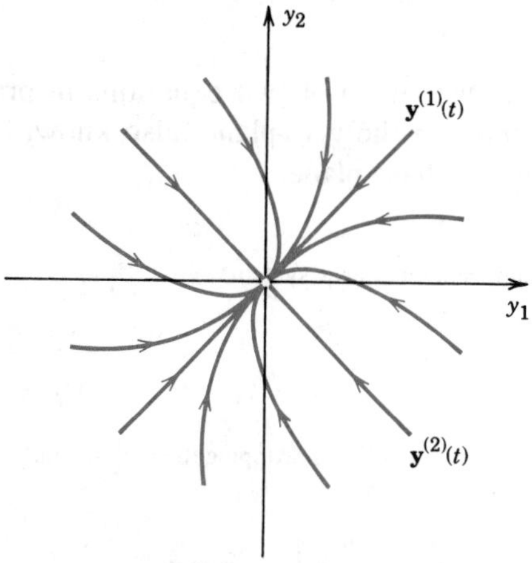

Fig. 59. Solutions of the system (6), improper node at the origin

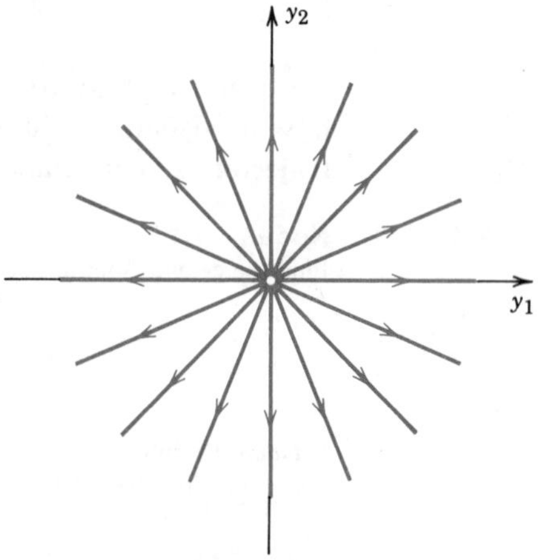

Fig. 60. Solutions of the system (7), proper node at the origin

EXAMPLE 2 Saddle point. Initial value problem

Find the general solution of the homogeneous linear system

$$\text{(8)} \qquad \mathbf{y}' = \mathbf{A}\mathbf{y} = \begin{bmatrix} 2 & -4 \\ 1 & -3 \end{bmatrix}\mathbf{y}, \qquad \text{thus} \qquad \begin{aligned} y_1' &= 2y_1 - 4y_2 \\ y_2' &= y_1 - 3y_2 \end{aligned}$$

and from it the particular solution satisfying the initial conditions $y_1(0) = 3$, $y_2(0) = 0$; in vector form, $\mathbf{y}(0)^{\mathsf{T}} = [3 \quad 0]$.

Solution. The systems in Example 1 had real eigenvalues of the same sign, and this led to nodes. Our present system will have real eigenvalues of opposite signs, and this will lead to a so-called **saddle point** (again at 0), characterized by two incoming paths, two outgoing paths, and all other paths bypassing the point. Indeed, the characteristic equation is

$$\det(\mathbf{A} - \lambda \mathbf{I}) = \begin{vmatrix} 2-\lambda & -4 \\ 1 & -3-\lambda \end{vmatrix} = \lambda^2 + \lambda - 2 = (\lambda - 1)(\lambda + 2) = 0.$$

It gives the eigenvalues $\lambda_1 = 1$ and $\lambda_2 = -2$. Eigenvectors are obtained from the equation $(2 - \lambda)x_1 - 4x_2 = 0$. For $\lambda_1 = 1$ this gives $x_1 = 4x_2$, and we can take $\mathbf{x}^{(1)\mathsf{T}} = [4 \quad 1]$. For $\lambda_2 = -2$ we get $4x_1 = 4x_2$ and can take $\mathbf{x}^{(2)\mathsf{T}} = [1 \quad 1]$. We thus obtain the general solution

$$\mathbf{y} = \begin{bmatrix} y_1 \\ y_2 \end{bmatrix} = c_1\mathbf{y}^{(1)} + c_2\mathbf{y}^{(2)} = c_1\begin{bmatrix} 4 \\ 1 \end{bmatrix} e^t + c_2\begin{bmatrix} 1 \\ 1 \end{bmatrix} e^{-2t}.$$

This represents the paths shown in Fig. 61. We can discuss them by first choosing $c_2 = 0$, which gives the two outgoing straight paths (for $c_1 > 0$ and $c_1 < 0$), then choosing $c_1 = 0$ to get the two incoming straight paths, and, third, making more general choices of c_1 and c_2 to get paths bypassing the saddle point 0. Arrows indicate directions of increasing t, as before.

Finally, from the initial conditions $y_1(0) = 3$, $y_2(0) = 0$ we obtain

$$\mathbf{y}(0) = c_1\begin{bmatrix} 4 \\ 1 \end{bmatrix} + c_2\begin{bmatrix} 1 \\ 1 \end{bmatrix} = \begin{bmatrix} 3 \\ 0 \end{bmatrix}, \quad \text{thus} \quad \begin{aligned} 4c_1 + c_2 &= 3 \\ c_1 + c_2 &= 0 \end{aligned}$$

giving $c_1 = 1$, $c_2 = -1$. Hence the desired particular solution is

$$\mathbf{y} = \begin{bmatrix} 4 \\ 1 \end{bmatrix} e^t - \begin{bmatrix} 1 \\ 1 \end{bmatrix} e^{-2t} \quad \text{or} \quad \begin{aligned} y_1 &= 4e^t - e^{-2t} \\ y_2 &= e^t - e^{-2t}. \end{aligned}$$

A simpler system having a saddle point (at 0) is

(9) $$\mathbf{y}' = \begin{bmatrix} 1 & 0 \\ 0 & -1 \end{bmatrix}\mathbf{y}, \quad \text{thus} \quad \begin{aligned} y_1' &= y_1 \\ y_2' &= -y_2. \end{aligned}$$

It has the general solution

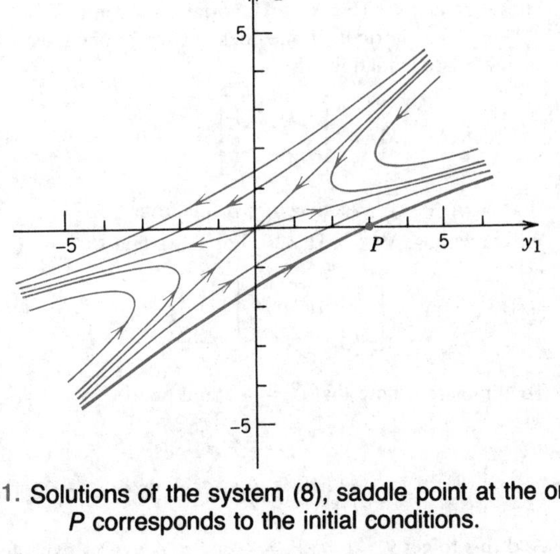

Fig. 61. Solutions of the system (8), saddle point at the origin. P corresponds to the initial conditions.

$$\mathbf{y} = c_1 \begin{bmatrix} 1 \\ 0 \end{bmatrix} e^t + c_2 \begin{bmatrix} 0 \\ 1 \end{bmatrix} e^{-t} \quad \text{or} \quad \begin{aligned} y_1 &= c_1 e^t \\ y_2 &= c_2 e^{-t} \end{aligned} \quad \text{or} \quad y_1 y_2 = const.$$

This is a family of hyperbolas (and the coordinate axes); see Fig. 62. ∎

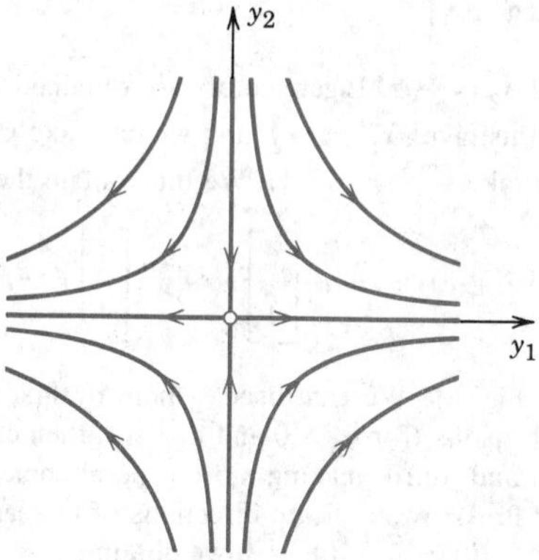

Fig. 62. Solutions of the system (9), saddle point at the origin

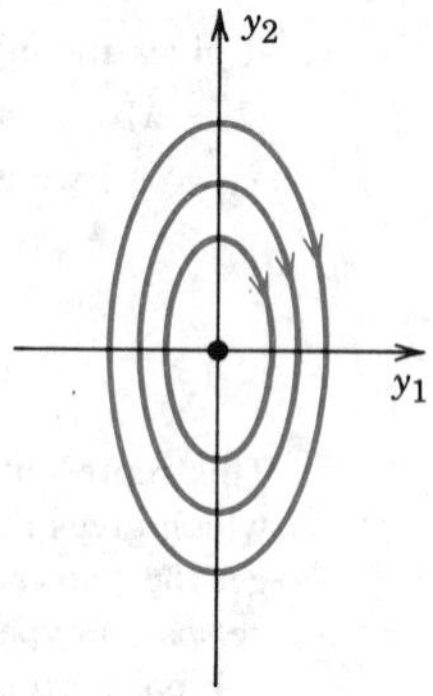

Fig. 63. Solutions of the system (10), center at the origin

EXAMPLE 3 Center

We show next that pure imaginary eigenvalues lead to a **center**, that is, a point P such that the paths are closed curves around P. Find the general solution of the system

$$(10) \qquad \mathbf{y}' = \begin{bmatrix} 0 & 1 \\ -4 & 0 \end{bmatrix} \mathbf{y}, \qquad \text{thus} \qquad \begin{aligned} y_1' &= y_2 \\ y_2' &= -4y_1. \end{aligned}$$

Solution. We write the given equations as $y_1' = y_2$ and $4y_1 = -y_2'$. Clearly the product of the left sides must equal the product of the right sides,

$$4y_1 y_1' = -y_2 y_2'. \qquad \text{By integration,} \qquad 2y_1^2 + \tfrac{1}{2}y_2^2 = const.$$

These are ellipses (Fig. 63). The origin is a center.

This was a shortcut. If one does not see it, one must proceed systematically as follows. The characteristic equation is

$$\det(\mathbf{A} - \lambda\mathbf{I}) = \begin{bmatrix} -\lambda & 1 \\ -4 & -\lambda \end{bmatrix} = \lambda^2 + 4 = 0. \qquad \text{Eigenvalues } \lambda_1 = 2i, \ \lambda_2 = -2i.$$

Here, $i = \sqrt{-1}$. Eigenvectors result from $-\lambda x_1 + x_2 = 0$. For λ_1 this is $-2ix_1 + x_2 = 0$. We can choose $\mathbf{x}^{(1)\mathsf{T}} = [1 \quad 2i]$. For λ_2 we find $\mathbf{x}^{(2)\mathsf{T}} = [1 \quad -2i]$. This gives the general solution

$$(11) \qquad \mathbf{y} = c_1 \begin{bmatrix} 1 \\ 2i \end{bmatrix} e^{2it} + c_2 \begin{bmatrix} 1 \\ -2i \end{bmatrix} e^{-2it}, \qquad \text{thus} \qquad \begin{aligned} y_1 &= c_1 e^{2it} + c_2 e^{-2it} \\ y_2 &= 2ic_1 e^{2it} - 2ic_2 e^{-2it}. \end{aligned}$$

To eliminate t, note that $i^2 = -1$ and take

$$\begin{aligned} y_1^2 &= c_1^2 e^{4it} + 2c_1c_2 + c_2^2 e^{-4it} \\ (\tfrac{1}{2}y_2)^2 &= -c_1^2 e^{4it} + 2c_1c_2 - c_2^2 e^{-4it}. \end{aligned}$$

Add this to get $y_1^2 + \frac{1}{4}y_2^2 = 4c_1c_2 = const$, as expected.

The solution (11) is complex, but we can readily obtain a *real* basis and a *real* general solution by applying (7), Sec. 2.3, with $s = 0$ and $\pm 2t$ instead of t,

$$e^{2it} = \cos 2t + i \sin 2t, \qquad e^{-2it} = \cos 2t - i \sin 2t.$$

Collecting the real and imaginary parts, we thus obtain in (11)

$$\text{(11*)} \qquad \begin{bmatrix} 1 \\ 2i \end{bmatrix} e^{2it} = \begin{bmatrix} 1 \\ 2i \end{bmatrix} (\cos 2t + i \sin 2t) = \begin{bmatrix} \cos 2t \\ -2 \sin 2t \end{bmatrix} + i \begin{bmatrix} \sin 2t \\ 2 \cos 2t \end{bmatrix}$$

and similarly

$$\begin{bmatrix} 1 \\ -2i \end{bmatrix} e^{-2it} = \begin{bmatrix} 1 \\ -2i \end{bmatrix} (\cos 2t - i \sin 2t) = \begin{bmatrix} \cos 2t \\ -2 \sin 2t \end{bmatrix} - i \begin{bmatrix} \sin 2t \\ 2 \cos 2t \end{bmatrix}.$$

Substitution into (10) shows that the real part and the imaginary part in (11*),

$$\mathbf{u} = \begin{bmatrix} u_1 \\ u_2 \end{bmatrix} = \begin{bmatrix} \cos 2t \\ -2 \sin 2t \end{bmatrix} \quad \text{and} \quad \mathbf{v} = \begin{bmatrix} v_1 \\ v_2 \end{bmatrix} = \begin{bmatrix} \sin 2t \\ 2 \cos 2t \end{bmatrix},$$

are solutions. These real solutions form a basis because their Wronskian is not zero,

$$W(\mathbf{u}, \mathbf{v}) = \begin{vmatrix} \cos 2t & \sin 2t \\ -2 \sin 2t & 2 \cos 2t \end{vmatrix} = 2 \cos^2 2t + 2 \sin^2 2t = 2.$$

Hence a real general solution of (10) is

$$\text{(12)} \qquad \mathbf{y} = \begin{bmatrix} y_1 \\ y_2 \end{bmatrix} = A\mathbf{u} + B\mathbf{v} = A \begin{bmatrix} \cos 2t \\ -2 \sin 2t \end{bmatrix} + B \begin{bmatrix} \sin 2t \\ 2 \cos 2t \end{bmatrix}.$$

This represents the same ellipses as before because by calculation and simplification we find

$$\text{(13)} \qquad y_1^2 + \tfrac{1}{4}y_2^2 = (A^2 + B^2)(\cos^2 2t + \sin^2 2t) = A^2 + B^2 = \mathit{const.}$$

∎

EXAMPLE 4 Spiral point

Find the general solution of

$$\text{(14)} \qquad \mathbf{y}' = \begin{bmatrix} -1 & 1 \\ -1 & -1 \end{bmatrix} \mathbf{y}, \quad \text{thus} \quad \begin{aligned} y_1' &= -y_1 + y_2 \\ y_2' &= -y_1 - y_2. \end{aligned}$$

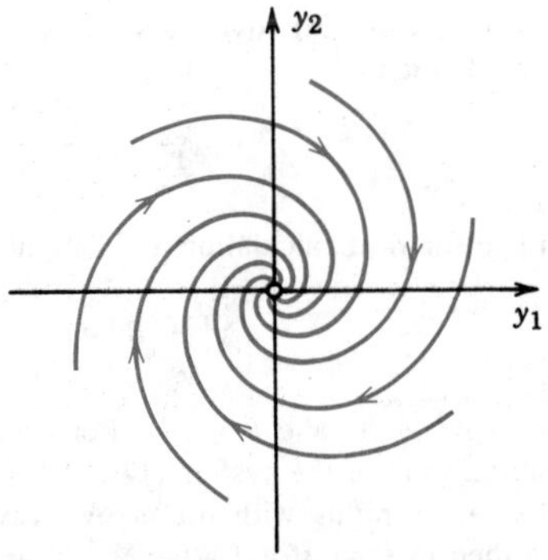

Fig. 64. Solutions of the system (14), spiral point at the origin

Solution. This example will show that complex (not pure imaginary) eigenvalues lead to a **spiral point** (at 0) defined to be a point P about which the solutions spiral, approaching P as $t \to \infty$; see Fig. 64 (or tracing these spirals in the opposite sense). Indeed, the characteristic equation

$$\det(\mathbf{A} - \mathbf{I}) = \begin{vmatrix} -1-\lambda & 1 \\ -1 & -1-\lambda \end{vmatrix} = \lambda^2 + 2\lambda + 2 = 0$$

gives the eigenvalues $\lambda_1 = -1 + i$ and $\lambda_2 = -1 - i$. Eigenvectors are obtained from $(-1 - \lambda)x_1 + x_2 = 0$. For λ_1 this becomes $-ix_1 + x_2 = 0$, and we can take as an eigenvector $\mathbf{x}^{(1)\mathsf{T}} = [1 \quad i]$. For λ_2 we get $ix_1 + x_2 = 0$ and can take $\mathbf{x}^{(2)} = [1 \quad -i]$. Hence the (complex) general solution of (14) is

$$\mathbf{y} = c_1 \begin{bmatrix} 1 \\ i \end{bmatrix} e^{(-1+i)t} + c_2 \begin{bmatrix} 1 \\ -i \end{bmatrix} e^{(-1-i)t}. \tag{15}$$

We derive from this a *real* general solution. This is similar to the previous example. In (15) we have

$$\begin{bmatrix} 1 \\ i \end{bmatrix} e^{(-1+i)t} = \begin{bmatrix} e^{-t}(\cos t + i \sin t) \\ ie^{-t}(\cos t + i \sin t) \end{bmatrix} = \begin{bmatrix} e^{-t} \cos t \\ -e^{-t} \sin t \end{bmatrix} + i \begin{bmatrix} e^{-t} \sin t \\ e^{-t} \cos t \end{bmatrix}$$

and

$$\begin{bmatrix} 1 \\ -i \end{bmatrix} e^{(-1-i)t} = \begin{bmatrix} e^{-t}(\cos t - i \sin t) \\ -ie^{-t}(\cos t - i \sin t) \end{bmatrix} = \begin{bmatrix} e^{-t} \cos t \\ -e^{-t} \sin t \end{bmatrix} - i \begin{bmatrix} e^{-t} \sin t \\ e^{-t} \cos t \end{bmatrix}.$$

The real and imaginary parts on the right are real solutions of (14)—call them **u** and **v**—as can be seen by substitution. They form a basis because their Wronskian is not zero,

$$W(\mathbf{u}, \mathbf{v}) = \begin{vmatrix} e^{-t} \cos t & e^{-t} \sin t \\ -e^{-t} \sin t & e^{-t} \cos t \end{vmatrix} = e^{-2t}(\cos^2 t + \sin^2 t) = e^{-2t}.$$

The corresponding real general solution is

$$\mathbf{y} = \begin{bmatrix} y_1 \\ y_2 \end{bmatrix} = A\mathbf{u} + B\mathbf{v} = A \begin{bmatrix} e^{-t} \cos t \\ -e^{-t} \sin t \end{bmatrix} + B \begin{bmatrix} e^{-t} \sin t \\ e^{-t} \cos t \end{bmatrix}; \tag{16}$$

in components,

$$y_1 = e^{-t}(A \cos t + B \sin t), \qquad y_2 = e^{-t}(B \cos t - A \sin t). \tag{16*}$$

It represents spirals (see Fig. 64). To see this, we introduce the usual polar coordinates r, θ in the y_1y_2-plane defined by

$$r^2 = y_1^2 + y_2^2, \qquad \tan \theta = y_2/y_1.$$

Then by straightforward calculation and simplification of the result we obtain from (16*)

$$r^2 = (A^2 + B^2)e^{-2t}, \qquad \text{thus} \qquad r = c_0 e^{\theta} \tag{17}$$

where $c_0 = \sqrt{A^2 + B^2}$ and $\theta = -t$. For each c_0 this represents a spiral, as claimed. The origin is a spiral point of the system (14).

To get the same spirals with the arrows reversed, all we have to do is to replace t by $-t$, because then instead of e^{-t} a factor e^{+t} appears, causing each spiral to be traced in the sense away from the origin because e^{+t} increases with t, whereas e^{-t} decreases (to zero) with increasing t, causing the spirals to approach the origin. ∎

No Basis of Eigenvectors Available

If $\mathbf{A}$ does not have a basis of eigenvectors, we get fewer linearly independent solutions of (1) and ask how we could get a basis of solutions in such a case. Suppose μ is a *double eigenvalue* of $\mathbf{A}$ [i.e., the product representation of $\det(\mathbf{A} - \lambda\mathbf{I})$ has a factor $(\lambda - \mu)^2$] for which there is only one eigenvector $\mathbf{x}$ (instead of two linearly independent ones), so that we get only one solution $\mathbf{y}^{(1)} = \mathbf{x}e^{\mu t}$. Then we can obtain a second independent solution of (1) by substituting

$$\mathbf{y}^{(2)} = \mathbf{x}te^{\mu t} + \mathbf{u}e^{\mu t} \tag{18}$$

into (1). (The $\mathbf{x}t$-term alone, resembling what we did in Sec. 2.2 in the case of a double root, would not be enough. Try it.) This gives

$$\mathbf{y}^{(2)\prime} = \mathbf{x}e^{\mu t} + \mu\mathbf{x}te^{\mu t} + \mu\mathbf{u}e^{\mu t} = \mathbf{A}\mathbf{y}^{(2)} = \mathbf{A}\mathbf{x}te^{\mu t} + \mathbf{A}\mathbf{u}e^{\mu t}.$$

Since $\mu\mathbf{x} = \mathbf{A}\mathbf{x}$, two terms cancel, and division by $e^{\mu t}$ gives

$$\mathbf{x} + \mu\mathbf{u} = \mathbf{A}\mathbf{u}, \qquad \text{thus} \qquad (\mathbf{A} - \mu\mathbf{I})\mathbf{u} = \mathbf{x}. \tag{19}$$

Although $\det(\mathbf{A} - \mu\mathbf{I}) = 0$, this can always be solved for $\mathbf{u}$, as can be shown.

EXAMPLE 5 No basis of eigenvectors available

Find a general solution of

$$\mathbf{y}' = \mathbf{A}\mathbf{y} = \begin{bmatrix} 4 & 1 \\ -1 & 2 \end{bmatrix}\mathbf{y}.$$

Solution. The characteristic equation is

$$\det(\mathbf{A} - \lambda\mathbf{I}) = \begin{bmatrix} 4-\lambda & 1 \\ -1 & 2-\lambda \end{bmatrix} = \lambda^2 - 6\lambda + 9 = (\lambda - 3)^2 = 0.$$

It has a double root $\lambda = 3$. Eigenvectors are obtained from $(4 - \lambda)x_1 + x_2 = 0$, thus from $x_1 + x_2 = 0$, say, $\mathbf{x}^{(1)\mathsf{T}} = [1 \quad -1]$ and multiples of this (which do not help). Now (19) is

$$(\mathbf{A} - 3\mathbf{I})\mathbf{u} = \begin{bmatrix} 1 & 1 \\ -1 & -1 \end{bmatrix}\mathbf{u} = \begin{bmatrix} 1 \\ -1 \end{bmatrix}, \qquad \text{thus} \qquad \begin{aligned} u_1 + u_2 &= 1 \\ -u_1 - u_2 &= -1 \end{aligned}$$

and we can take simply $\mathbf{u}^{\mathsf{T}} = [0 \quad 1]$. This gives the answer (Fig. 65, p. 174)

$$\mathbf{y} = c_1\mathbf{y}^{(1)} + c_2\mathbf{y}^{(2)} = c_1\begin{bmatrix} 1 \\ -1 \end{bmatrix}e^{3t} + c_2\left(\begin{bmatrix} 1 \\ -1 \end{bmatrix}t + \begin{bmatrix} 0 \\ 1 \end{bmatrix}\right)e^{3t}.$$

■

If $\mathbf{A}$ has a triple eigenvalue μ and only a single linearly independent eigenvector $\mathbf{x}$ corresponding to it, we get a second solution (18) with $\mathbf{u}$ satisfying (19), as just discussed, and a third of the form

$$\mathbf{y}^{(3)} = \tfrac{1}{2}\mathbf{x}t^2e^{\mu t} + \mathbf{u}te^{\mu t} + \mathbf{v}e^{\mu t} \tag{20}$$

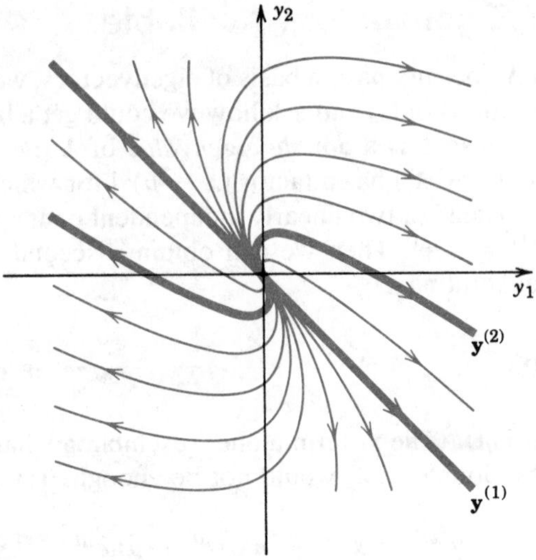

Fig. 65. Improper node in Example 5, p. 173

with **u** satisfying (19) and **v** determined from

$$(\mathbf{A} - \mu\mathbf{I})\mathbf{v} = \mathbf{u}, \tag{21}$$

which can always be solved.

We finally mention that if **A** has a triple eigenvalue and two linearly independent eigenvectors $\mathbf{x}^{(1)}$, $\mathbf{x}^{(2)}$ corresponding to it, then three linearly independent solutions are

$$\mathbf{y}^{(1)} = \mathbf{x}^{(1)}e^{\mu t}, \qquad \mathbf{y}^{(2)} = \mathbf{x}^{(2)}e^{\mu t}, \qquad \mathbf{y}^{(3)} = \mathbf{x}te^{\mu t} + \mathbf{u}e^{\mu t} \tag{22}$$

where **x** is a linear combination of $\mathbf{x}^{(1)}$ and $\mathbf{x}^{(2)}$ such that

$$(\mathbf{A} - \mu\mathbf{I})\mathbf{u} = \mathbf{x} \tag{23}$$

is solvable for **u**.

Problem Set 4.3

Find a real general solution of the following systems of differential equations.

1. $y_1' = y_2$, $y_2' = y_1$

2. $y_1' = 4y_2$, $y_2' = 4y_1$

3. $y_1' = -2y_2$, $y_2' = 2y_1$

4. $y_1' = 5y_2$, $y_2' = -5y_1$

5. $y_1' = 2y_1 + 3y_2$, $y_2' = \frac{1}{3}y_1 + 2y_2$

6. $y_1' = y_1 + 4y_2$, $y_2' = y_1 + y_2$

7. $y_1' = -4y_1 - 6y_2$, $y_2' = y_1 + y_2$

8. $y_1' = -2y_1 + 2y_2$, $y_2' = -2y_1 - 2y_2$

9. $y_1' = y_1 - y_2$, $y_2' = y_1 + y_2$

Solve the following initial value problems.

10. $y_1' = -y_2$
$y_2' = -y_1$
$y_1(0) = 3,\ y_2(0) = 1$

11. $y_1' = 2y_2$
$y_2' = 2y_1$
$y_1(0) = -9,\ y_2(0) = 15$

12. $y_1' = 6y_1 + 9y_2$
$y_2' = y_1 + 6y_2$
$y_1(0) = -3,\ y_2(0) = -3$

13. $y_1' = 2y_1 + 4y_2$
$y_2' = y_1 + 2y_2$
$y_1(0) = -4,\ y_2(0) = -4$

14. $y_1' = 5y_1 + y_2$
$y_2' = y_1 + 5y_2$
$y_1(0) = -3,\ y_2(0) = 7$

15. $y_1' = -y_1 + 4y_2$
$y_2' = 3y_1 - 2y_2$
$y_1(0) = 3,\ y_2(0) = 4$

Find a general solution of the following systems by converting them to a single higher order equation.

16. The system in Prob. 1

17. The system in Example 4

18. Tank T_1 in Fig. 66 contains initially 100 gal of pure water. Tank T_2 contains initially 100 gal of water in which 150 lb of salt are dissolved. Liquid circulates through the tanks at a constant rate of 2 gal/min, and the mixture is kept uniform by stirring. Show that the amounts of salt $y_1(t)$ lb and $y_2(t)$ lb in T_1 and T_2, respectively, are obtained from the following system and find the solution.

$$y_1' = \text{inflow/min} - \text{outflow/min} = \frac{2}{100}y_2 - \frac{2}{100}y_1 \qquad \text{(Tank } T_1\text{)}$$

$$y_2' = \text{inflow/min} - \text{outflow/min} = \frac{2}{100}y_1 - \frac{2}{100}y_2 \qquad \text{(Tank } T_2\text{).}$$

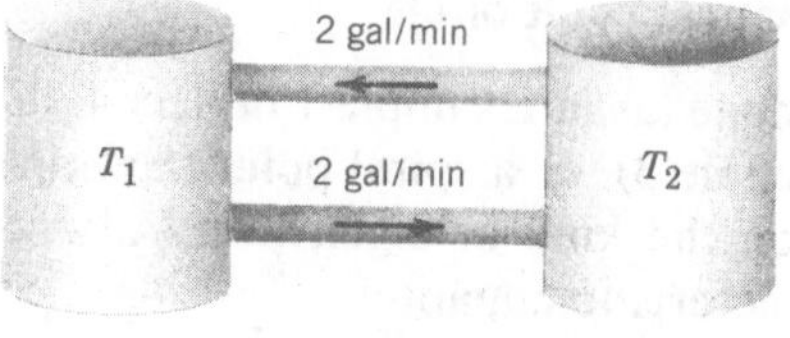

Fig. 66. Tanks in Problem 18

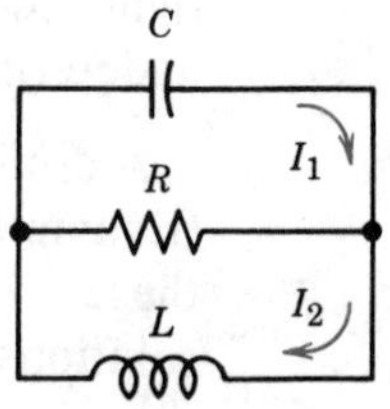

Fig. 67. Network in Problem 19

19. Show that a model for the currents $I_1(t)$ and $I_2(t)$ in the network in Fig. 67 is

$$\frac{1}{C}\int I_1\,dt + R(I_1 - I_2) = 0, \qquad LI_2' + R(I_2 - I_1) = 0.$$

Find a general solution, assuming that $R = 3$ ohms, $L = 4$ henrys, and $C = 1/12$ farad.

20. Find the matrix **A** of the system in Fig. 67 for any R, L, C. When will **A** have real eigenvalues? Complex conjugate eigenvalues?

4.4 Phase Plane, Critical Points, Stability

In this section we give a systematic discussion of the **phase plane method**. This is the method of exhibiting solutions of systems

$$\text{(1)} \qquad \mathbf{y}' = \mathbf{A}\mathbf{y} = \begin{bmatrix} a_{11} & a_{12} \\ a_{21} & a_{22} \end{bmatrix} \mathbf{y} \qquad (\det \mathbf{A} \neq 0),$$

thus,

$$y_1' = a_{11}y_1 + a_{12}y_2$$
$$y_2' = a_{21}y_1 + a_{22}y_2,$$

as **paths** $y_1 = y_1(t)$, $y_2 = y_2(t)$ in the **phase plane**, the y_1y_2-plane, as we have just illustrated by examples in the last section. The phase plane has recently become of great importance, along with advances in computer graphics, for instance, in connection with stability, as we shall see. Its name comes from a special case in physics (where it is the y-(mv)-plane, m = mass, $v = y'$ = velocity), but is now generally used for the y_1y_2-plane.

To begin our discussion, we see that from (1) we get

$$\text{(2)} \qquad \frac{dy_2}{dy_1} = \frac{dy_2/dt}{dy_1/dt} = \frac{y_2'}{y_1'} = \frac{a_{21}y_1 + a_{22}y_2}{a_{11}y_1 + a_{12}y_2}.$$

This associates with every point P: (y_1, y_2) a unique tangent direction dy_2/dy_1 of the path passing though P, except for the point $P = P_0$: $(0, 0)$, where the right side of (2) becomes 0/0. This point P_0, at which dy_2/dy_1 becomes undetermined, is called a **critical point** of (1).

A critical point can be a node (as in Example 1 in Sec. 4.3), a saddle point (Example 2), a center (Example 3), or a spiral point (Example 4). Which of these will occur depends on the kind of eigenvalues of **A**, which are the solutions λ_1, λ_2 of the characteristic equation

$$\text{(3)} \qquad \begin{aligned} \det(\mathbf{A} - \lambda\mathbf{I}) &= \begin{vmatrix} a_{11} - \lambda & a_{12} \\ a_{21} & a_{22} - \lambda \end{vmatrix} \\ &= \lambda^2 - (a_{11} + a_{22})\lambda + \det \mathbf{A} = 0. \end{aligned}$$

Using the standard notation

$$\text{(4)} \qquad p = a_{11} + a_{22}, \qquad q = \det \mathbf{A} = a_{11}a_{22} - a_{12}a_{21}, \qquad \Delta = p^2 - 4q,$$

from the right side of (3) and the product representation we have

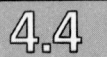

(5) $\lambda^2 - p\lambda + q = (\lambda - \lambda_1)(\lambda - \lambda_2) = \lambda^2 - (\lambda_1 + \lambda_2)\lambda + \lambda_1\lambda_2.$

Hence p is the sum of the eigenvalues, q the product, and Δ the discriminant. The critical point P_0 of (1) is a:

(6)
(a) *node* if $q > 0$ and $\Delta \geqq 0$,
(b) *saddle point* if $q < 0$,
(c) *center* if $p = 0$ and $q > 0$,
(d) *spiral point* if $p \neq 0$ and $\Delta < 0$.

We shall indicate below how (6) is obtained. This classifies critical points P_0 in terms of the geometric shapes of paths near P_0. Another classification is in terms of stability. Stability concepts are important in many applications. They are suggested by physics, where **stability** means, roughly speaking, that a small change (small disturbance) of a physical system at some instant changes the behavior of the system only slightly at all future times t. For critical points, the following definitions are appropriate.

P_0 is called a **stable**[2] **critical point** of (1) if, roughly speaking, all paths of (1) that at some instant are sufficiently close to P_0 remain close to P_0 at all future times; precisely: if for every disk D_ϵ of radius $\epsilon > 0$ with center P_0 there is a disk D_δ of radius $\delta > 0$ with center P_0 such that every path of (1) that has a point P_1 (corresponding to $t = t_1$, say) in D_δ has all its points corresponding to $t \geqq t_1$ in D_ϵ. See Fig. 68.

P_0 is called a **stable and attractive critical point**[3] of (1) if P_0 is stable and every path that has a point in D_δ approaches P_0 as $t \to \infty$. See Fig. 69.

P_0 is called **unstable** if P_0 is not stable.

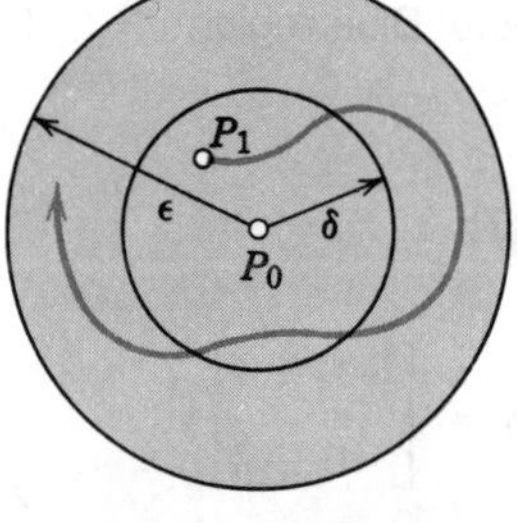

Fig. 68. Stable critical point P_0 of (1) (The path initiating at P_1 stays in the disk of radius ϵ.)

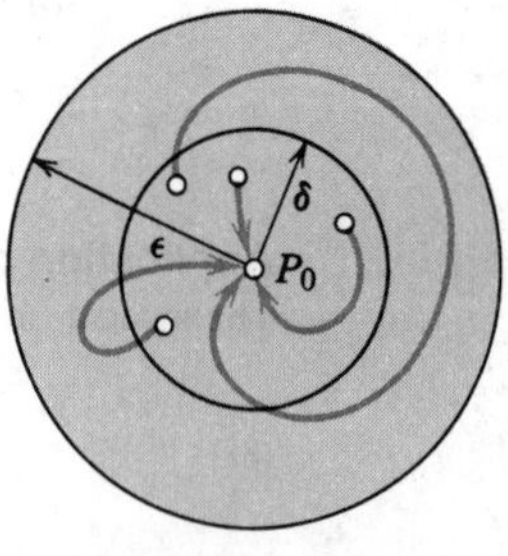

Fig. 69. Stable and attractive critical point P_0 of (1)

---

[2]More precisely: **stable in the sense of Liapunov**. There exist other definitions of stability, but Liapunov's definition—the only we shall consider—is probably the most useful one.

[3]Or an **asymptotically stable critical point**.

A critical point P_0 is:

(7)	(a)	*stable and attractive* if $p < 0$ and $q > 0$,
	(b)	*stable* if $p \leqq 0$ and $q > 0$,
	(c)	*unstable* if $p > 0$ or $q < 0$.

The criteria in (6) and (7) are summarized in the **stability chart** in Fig. 70. In this chart, the region of instability is blue.

We indicate how these criteria are obtained. If $q = \lambda_1\lambda_2 > 0$, both eigenvalues are positive or both negative or complex conjugates. If also $p = \lambda_1 + \lambda_2 < 0$, both are negative or have a negative real part. Hence P_0 is stable and attractive. The reasoning for the other two lines in (7) is similar.

If $\Delta < 0$, the eigenvalues are complex conjugates, say, $\lambda_1 = \alpha + i\beta$ and $\lambda_2 = \alpha - i\beta$. If also $p = \lambda_1 + \lambda_2 = 2\alpha < 0$, this gives a spiral point that is stable and attractive.

If $p = 0$, then $\lambda_2 = -\lambda_1$ and $q = \lambda_1\lambda_2 = -\lambda_1^2$. If also $q > 0$, then $\lambda_1^2 = -q < 0$, so that λ_1, and thus λ_2, must be pure imaginary. This gives periodic solutions, their paths being closed curves around P_0, which is a center.

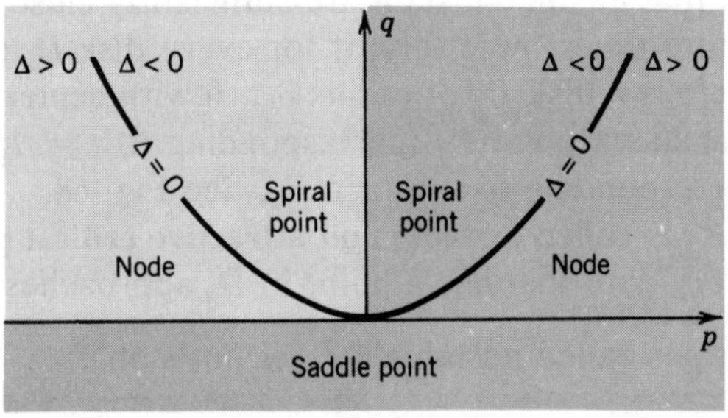

Fig. 70. Stability chart of the system (1) with p, q, Δ defined in (4). Stable and attractive: The second quadrant without the q-axis. Stability also on the positive q-axis (which corresponds to centers). Unstable: Blue region

EXAMPLE 1 Application of the criteria (6) and (7)

The system (6) in the previous section is

$$\text{(8)} \qquad \mathbf{y}' = \mathbf{A}\mathbf{y} = \begin{bmatrix} -3 & 1 \\ 1 & -3 \end{bmatrix} \mathbf{y}.$$

We see that $p = a_{11} + a_{22} = -6$, $q = \det \mathbf{A} = 8$, and $\Delta = (-6)^2 - 4 \cdot 8 = 4$. Hence, by (6a), the critical point at the origin is a node, which, by (7a), is stable and attractive. This agrees with our previous result. The other examples in Sec. 4.3 can be discussed similarly. ■

EXAMPLE 2 Free motions of a mass on a spring

What kind of critical point does the model of a mass on an elastic spring have?

Solution. The equation is [see (7) in Sec. 2.5]

$$y'' + \frac{c}{m}y' + \frac{k}{m}y = 0.$$

To get a system, we set $y_1 = y$, $y_2 = y'$ (see Sec. 4.1), obtaining

$$\mathbf{y}' = \begin{bmatrix} 0 & 1 \\ -\dfrac{k}{m} & -\dfrac{c}{m} \end{bmatrix} \mathbf{y}, \qquad \text{thus} \qquad \begin{aligned} y_1' &= y_2 \\ y_2' &= -\frac{k}{m} y_1 - \frac{c}{m} y_2. \end{aligned}$$

For our criteria (6) and (7) we need $p = -c/m$, $q = k/m$, and $\Delta = (c/m)^2 - 4k/m$. This gives the following results.

No damping. $c = 0$, $p = 0$, $q > 0$, a center.
Underdamping. $c^2 < 4mk$, $p < 0$, $q > 0$, $\Delta < 0$, a stable and attractive spiral point.
Critical damping. $c^2 = 4mk$, $p < 0$, $q > 0$, $\Delta = 0$, a stable and attractive node.
Overdamping. $c^2 > 4mk$, $p < 0$, $q > 0$, $\Delta > 0$, a stable and attractive node. ■

Problem Set 4.4

Find a general solution and determine the type and stability of the critical point.

1. $y_1' = 3y_1 - y_2$, $y_2' = -y_1 + 3y_2$

2. $y_1' = -4y_2$, $y_2' = y_1 - 5y_2$

3. $y_1' = -3y_1 + y_2$, $y_2' = -4y_1 + 2y_2$

4. $y_1' = -2y_2$, $y_2' = 8y_1$

5. $y_1' = y_1 + 3y_2$, $y_2' = 4y_1 + 2y_2$

6. $y_1' = y_1 - 4y_2$, $y_2' = -3y_1 + 2y_2$

7. $y_1' = -2y_1 + 2y_2$, $y_2' = -2y_1 - 2y_2$

8. $y_1' = 4y_1 + 5y_2$, $y_2' = -5y_1 - 4y_2$

9. $y_1' = 6y_1 + 9y_2$, $y_2' = y_1 + 6y_2$

10. In (8), let t be time. What happens to the critical point if you introduce $\tau = -t$ as a new independent variable? What does this transformation mean mechanically?

11. Discuss the critical points in (7), (8), (9) in Sec. 4.3 by applying (6) and (7) of the present section.

12. What happens to the center of (10) in Sec. 4.3 if you replace **A** by $\mathbf{A} + k\mathbf{I}$ (where k may represent the effect of errors of measurement of the main diagonal entries)?

13. Let **A** be the matrix of the system in Prob. 12. Is there a k for which you will get a system $\mathbf{y}' = (\mathbf{A} + k\mathbf{I})\mathbf{y}$ with a node?

What kind of curves are the paths of the following equations in the phase plane?

14. $y'' + \frac{1}{4}y = 0$

15. $y'' - k^2y = 0$

16. $y'' + 2y' = 0$

17. (Harmonic oscillations) Solve $y'' + \omega_0{}^2y = 0$ and indicate the kind of paths in the phase plane.

18. (Damped oscillations) Consider the damped oscillation $y(t) = e^{-t}\sin t$. Plot some points of the corresponding path in the phase plane to obtain the impression that this must be a spiral.

19. (Forced oscillations) What paths in the phase plane correspond to steady-state forced oscillations of a damped system (Sec. 2.11) under a sinuosidal driving force?

20. (Free damped motions) From Sec. 2.5 we know that there are three types of free motions of a damped system. Using the figures of solution curves in Sec. 2.5, give a rough qualitative description and sketch of the corresponding paths in the phase plane. (Assume zero initial velocity, for simplicity.)

4.5 Phase Plane Methods for Nonlinear Systems

Phase plane methods become particularly valuable for systems whose solution by analytic methods is difficult or impossible, as is the case for many practically important first-order ***nonlinear*** systems

$$\mathbf{y}' = \mathbf{f}(\mathbf{y}), \qquad \text{thus} \qquad \begin{aligned} y_1' &= f_1(y_1, y_2) \\ y_2' &= f_2(y_1, y_2), \end{aligned} \tag{1}$$

to which we now extend the methods just discussed for linear systems. We assume that (1) is an **autonomous system**, that is, the independent variable t does not occur explicitly. We shall see that the extended methods give a characterization of various general properties of solutions, without actually solving the system. As before, we shall exhibit whole families of solutions. This generally is an advantage over numerical methods, which give only one (approximate) solution at a time (albeit with much greater accuracy).

As before we call the y_1y_2-plane the *phase plane*, a solution curve of (1) in the phase plane a *path* (or *trajectory*), and a point P_0: (y_1, y_2) at which both $f_1(y_1, y_2) = 0$ and $f_2(y_1, y_2) = 0$ a **critical point**. If (1) has several critical points, we discuss one after the other. Each time we move the point P_0: (a, b) to be discussed to the origin. This can be done by a translation $\tilde{y}_1 = y_1 - a$, $\tilde{y}_2 = y_2 - b$. Thus we can assume that P_0 is the origin (0, 0), and we continue to write y_1, y_2 (instead of $\tilde{y}_1, \tilde{y}_2$), for simplicity. We also assume that P_0 is ***isolated***, that is, it is the only critical point within a (sufficiently small) circular disk with center at the origin.

How can we determine the type and stability property of a critical point P_0: (0, 0)? In most practical cases this can be done by **linearization**, that is, by investigating a certain *linear* system, as follows, assuming that f_1 and f_2 in (1) are continuous and have continuous partial derivatives in a neighborhood of P_0. Since P_0 is a critical point, we have $f_1(0, 0) = 0$ and $f_2(0, 0) = 0$. Hence f_1 and f_2 have no constant terms. Their linear terms we write explicitly. Then (1) becomes

$$\mathbf{y}' = \mathbf{A}\mathbf{y} + \mathbf{h}(\mathbf{y}), \qquad \text{thus} \qquad \begin{aligned} y_1' &= a_{11}y_1 + a_{12}y_2 + h_1(y_1, y_2) \\ y_2' &= a_{21}y_1 + a_{22}y_2 + h_2(y_1, y_2). \end{aligned} \tag{2}$$

Here $\mathbf{A}$ is constant (independent of t) because (1) is autonomous. One can prove that if $\det \mathbf{A} \neq 0$, then the type and stability of P_0 is the same as that of the critical point (0, 0) of the *linear system* obtained by **linearization**, that is, by dropping $\mathbf{h}(\mathbf{y})$ from (2):

$$\mathbf{y}' = \mathbf{A}\mathbf{y}, \qquad \text{thus} \qquad \begin{aligned} y_1' &= a_{11}y_1 + a_{12}y_2 \\ y_2' &= a_{21}y_1 + a_{22}y_2. \end{aligned} \tag{3}$$

Indeed, the above assumption on the derivatives implies that h_1 and h_2 are small near P_0. ***Two exceptions*** occur: if the eigenvalues of **A** are equal or pure imaginary, in addition to the type of critical points of the *linear* system, the nonlinear system may have a spiral point. For proofs, see Ref. [A4], pp. 375–388.

EXAMPLE 1 **Free undamped pendulum. Linearization**

Figure 71 shows a pendulum consisting of a body of mass m (the bob) and a rod of length L. Determine the locations and types of the critical points. Assume that the mass of the rod and air resistance are negligible.

Solution. 1st Step. Setting up the mathematical model. Let θ denote the angular displacement, measured counterclockwise from the equilibrium position. The weight of the bob is mg (g the acceleration of gravity). It causes a restoring force $mg \sin \theta$ tangent to the curve of motion (circular arc) of the bob. By Newton's second law, at each instant this force is balanced by the force of acceleration $mL\theta''$, where $L\theta''$ is the acceleration; hence the resultant of these two forces is zero, and we obtain as the mathematical model

$$mL\theta'' + mg \sin \theta = 0.$$

Dividing this by mL, we have

$$\theta'' + k \sin \theta = 0 \qquad \left(k = \frac{g}{L}\right). \tag{4}$$

When θ is very small, we could approximate $\sin \theta$ rather accurately by θ and obtain as an *approximate* solution $A \cos \sqrt{k}t + B \sin \sqrt{k}t$, but the *exact* solution for any θ is not an elementary function.

2nd Step. Discussion of critical points by linearization. To obtain a system of equations, we set $\theta = y_1$, $\theta' = y_2$. Then from (4),

$$\begin{aligned} y_1' &= f_1(y_1, y_2) = y_2 \\ y_2' &= f_2(y_1, y_2) = -k \sin y_1. \end{aligned} \tag{4*}$$

The right sides are both zero when $y_2 = 0$ and $\sin y_1 = 0$. This gives the infinitely many critical points $(n\pi, 0)$, where $n = 0, \pm 1, \pm 2, \cdots$. We consider $(0, 0)$. Since the Maclaurin series is

$$\sin y_1 = y_1 - \tfrac{1}{6}y_1^3 + - \cdots \approx y_1,$$

the linearized system is

$$\mathbf{y}' = \mathbf{A}\mathbf{y} = \begin{bmatrix} 0 & 1 \\ -k & 0 \end{bmatrix} \mathbf{y}, \qquad \text{thus} \qquad \begin{aligned} y_1' &= y_2 \\ y_2' &= -ky_1. \end{aligned}$$

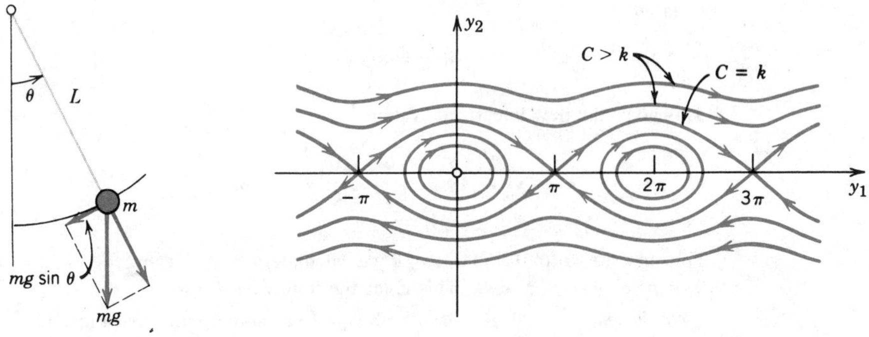

(a) Pendulum *(b)* Solution curves $y_2(y_1)$ of (4) in the phase plane

Fig. 71. Examples 1 and 2
(C will be explained in Example 3.)

In our criteria (6), (7) in Sec. 4.4 we need $p = a_{11} + a_{22} = 0$, $q = \det \mathbf{A} = k\ (> 0)$, and $\Delta = p^2 - 4q = -4k$. From this and (6c), in Sec. 4.4 we conclude that (0, 0) is a center, which is always stable. Since $\sin \theta = \sin y_1$ is periodic, with period 2π, the critical points $(n\pi, 0)$, $n = \pm 2, \pm 4, \cdots$, are all centers.

We now consider the critical point $(\pi, 0)$, setting $\theta - \pi = y_1$ and $(\theta - \pi)' = \theta' = y_2$, so that in (4),

$$\sin \theta = \sin (y_1 + \pi) = -\sin y_1 = -y_1 + \tfrac{1}{6}y_1^3 - + \cdots \approx -y_1$$

and the linearized system is now

$$\mathbf{y}' = \mathbf{A}\mathbf{y} = \begin{bmatrix} 0 & 1 \\ k & 0 \end{bmatrix} \mathbf{y}, \qquad \text{thus} \qquad \begin{aligned} y_1' &= y_2 \\ y_2' &= ky_1. \end{aligned}$$

We see that $p = 0$, $q = -k\ (< 0)$, and $\Delta = -4q = 4k$, so that (6b) in Sec. 4.4 gives a saddle point, which is always unstable. By periodicity, the critical points $(n\pi, 0)$, $n = \pm 1, \pm 3, \cdots$, are all saddle points. These results agree with the impression we get from Fig. 71. ∎

EXAMPLE 2 Linearization of the damped pendulum equation

To gain further experience in investigating critical points, as another practically important case, let us see how Example 1 changes when we add a damping term $c\theta'$ (damping proportional to the angular velocity) to equation (4), so that it becomes

$$\theta'' + c\theta' + k \sin \theta = 0, \tag{5}$$

where $k > 0$ and $c \geqq 0$. Setting $\theta = y_1$, $\theta' = y_2$, as before, we obtain the system

$$\begin{aligned} y_1' &= y_2 \\ y_2' &= -k \sin y_1 - cy_2. \end{aligned}$$

We see that the critical points have the same locations as before, namely, (0, 0), $(\pm\pi, 0)$, $(\pm 2\pi, 0)$, $\cdots$. We consider (0, 0). Linearizing $\sin y_1 \approx y_1$ as in Example 1, we get the linearized system

$$\mathbf{y}' = \mathbf{A}\mathbf{y} = \begin{bmatrix} 0 & 1 \\ -k & -c \end{bmatrix} \mathbf{y}, \qquad \text{thus} \qquad \begin{aligned} y_1' &= y_2 \\ y_2' &= -ky_1 - cy_2. \end{aligned} \tag{6}$$

This is identical with the system in Example 2 of Sec. 4.4, except for the (positive!) factor m (and except for the physical meaning of y_1); hence for $c = 0$ (no damping) we have a center (see Fig. 71), for small damping we have a spiral point (see Fig. 72), and so on.

We now consider the critical point $(\pi, 0)$. We set $\theta - \pi = y_1$, $(\theta - \pi)' = \theta' = y_2$ and linearize

$$\sin \theta = \sin (y_1 + \pi) = -\sin y_1 \approx -y_1.$$

This gives the new linearized system

$$\mathbf{y}' = \mathbf{A}\mathbf{y} = \begin{bmatrix} 0 & 1 \\ k & -c \end{bmatrix} \mathbf{y}, \qquad \text{thus} \qquad \begin{aligned} y_1' &= y_2 \\ y_2' &= ky_1 - cy_2. \end{aligned} \tag{7}$$

For our criteria (6), (7) in Sec. 4.4 we calculate $p = a_{11} + a_{22} = -c$, $q = \det \mathbf{A} = -k$, and $\Delta = p^2 - 4q = c^2 + 4k$. This gives the following results.

No damping. $c = 0$, $p = 0$, $q < 0$, $\Delta > 0$, a saddle point. See Fig. 71.

Damping. $c > 0$, $p < 0$, $q < 0$, $\Delta > 0$, a saddle point. See Fig. 72.

Since $\sin y_1$ is periodic with period 2π, the critical points $(\pm 2\pi, 0)$, $(\pm 4\pi, 0)$, $\cdots$ are of the same type as (0, 0), and the critical points $(-\pi, 0)$, $(\pm 3\pi, 0)$, $\cdots$ are of the same type as $(\pi, 0)$, so that our task is finished.

Figure 72 shows the paths in the case of damping. What we see agrees with our physical intuition. Indeed, damping means loss of energy; hence instead of the closed paths of periodic solutions in Fig. 71 we now have paths spiraling around one of the critical points (0, 0), $(\pm 2\pi, 0)$, · · · . Even the wavy paths corresponding to whirly motions eventually spiral around one of these points; furthermore, there are no more paths that connect critical points (as there were in the undamped case).

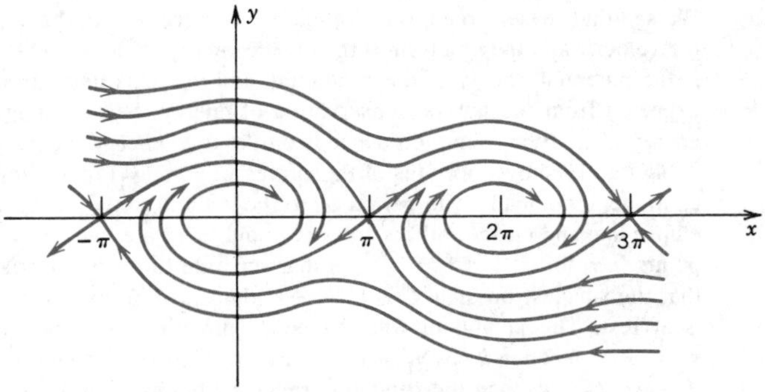

Fig. 72. Paths in the phase plane for the damped pendulum

Transformation to a First-Order Equation in the Phase Plane

Another phase plane method is based on the idea of transforming a second-order autonomous equation

$$F(y, y', y'') = 0$$

to first order by taking $y = y_1$ as the independent variable, setting $y' = y_2$ and transforming y'' by the chain rule,

$$y'' = y_2' = \frac{dy_2}{dt} = \frac{dy_2}{dy_1}\frac{dy_1}{dt} = \frac{dy_2}{dy_1}y_2.$$

Then the equation becomes of first order,

$$F\left(y_1, y_2, \frac{dy_2}{dy_1}y_2\right) = 0 \tag{8}$$

and can sometimes be solved or treated by direction fields. We illustrate this for the equation in Example 1 and shall gain much more insight into the behavior of solutions.

EXAMPLE 3 **An equation (8) for the free undamped pendulum**

If in (4) we set $\theta = y_1$, $\theta' = y_2$ (the angular velocity) and use

$$\theta'' = \frac{dy_2}{dt} = \frac{dy_2}{dy_1}\frac{dy_1}{dt} = \frac{dy_2}{dy_1}y_2,$$

we get from (4)

$$\frac{dy_2}{dy_1}y_2 = -k \sin y_1.$$

We can now separate variables, $y_2\,dy_2 = -k \sin y_1\,dy_1$, and integrate,

(9) $$\tfrac{1}{2}y_2^2 = k \cos y_1 + C \qquad (C \text{ constant}).$$

Multiplying this by mL^2, we get

$$\tfrac{1}{2}m(Ly_2)^2 - mL^2k \cos y_1 = mL^2C.$$

We see that these three terms are energies. Indeed, y_2 is the angular velocity, so that Ly_2 is the velocity and the first term is the kinetic energy. The second term (including the minus sign) is the potential energy of the pendulum, and mL^2C is its total energy, which is constant, as expected from the law of conservation of energy, because there is no damping (no loss of energy). The type of motion depends on the total energy, hence on C, as follows.

Figure 71*b* shows portions of the curves $y_2 = y_2(y_1)$ for various values of C. These graphs continue periodically with period 2π to the left and to the right. We see that some of them are ellipse-like and closed, others are wavy, and there are two curves (passing through the saddle points $(n\pi, 0)$, $n = \pm 1, \pm 3, \cdots$) that separate those two types of curves. From (9) we see that the smallest possible C is $C = -k$; then $y_2 = 0$, and $\cos y_1 = 1$, so that the pendulum is at rest. The pendulum will change its direction of motion if there are points at which $y_2 = \theta' = 0$. Then $k \cos y_1 + C = 0$ by (9). If $y_1 = \pi$, then $\cos y_1 = -1$ and $C = k$. Hence if $-k < C < k$, then the pendulum reverses its direction for a $|y_1| = |\theta| < \pi$, and for these values of C with $|C| < k$ the pendulum oscillates. This corresponds to the closed curves in the figure. However, if $C > k$, then $y_2 = 0$ is impossible and the pendulum makes a whirly motion which appears as a wavy curve in the y_1y_2-plane. Finally, the value $C = k$ corresponds to the two "separating curves" in Fig. 71b connecting the saddle points. ∎

The phase plane method of deriving a single first-order equation (8) may be of practical interest not only when (8) can be solved (as in Example 3) but also when solution is not possible and we have to utilize direction fields (Sec. 1.10). We illustrate this with a very famous example:

EXAMPLE 4 Self-sustained oscillations, van der Pol equation

There are physical systems such that for small oscillations, energy is fed into the system, whereas for large oscillations, energy is taken from the system. In other words, large oscillations will be damped, whereas for small oscillations there is "negative damping" (feeding of energy into the system). For physical reasons we expect such a system to approach a periodic behavior, which will thus appear as a closed curve in the phase plane, called a **limit cycle**. A differential equation describing such vibrations is the famous **van der Pol equation**[4]

(10) $$\boxed{y'' - \mu(1 - y^2)y' + y = 0} \qquad (\mu > 0, \text{ constant}).$$

It occurs in the study of electrical circuits containing vacuum tubes. For $\mu = 0$ this equation becomes $y'' + y = 0$ and we obtain harmonic oscillations. Let $\mu > 0$. The damping term has the coefficient $-\mu(1 - y^2)$. This is negative for small oscillations, namely, $y^2 < 1$, so that we have "negative damping," is zero for $y^2 = 1$ (no damping) and is positive if $y^2 > 1$ (positive damping, loss of energy). If μ is small, we expect a limit cycle that is almost a circle because then our equation differs but little from $y'' + y = 0$. If μ is large, the limit cycle will probably look different.

Setting $y = y_1$, $y' = y_2$ and using $y'' = (dy_2/dy_1)y_2$ as in (8), we have from (10)

(11) $$\frac{dy_2}{dy_1}y_2 - \mu(1 - y_1^2)y_2 + y_1 = 0.$$

The isoclines in the y_1y_2-plane (the phase plane) are the curves $dy_2/dy_1 = k = const$, that is,

$$\frac{dy_2}{dy_1} = \mu(1 - y_1^2) - \frac{y_1}{y_2} = k.$$

[4]BALTHASAR VAN DER POL (1889—1959). Dutch physicist.

We can solve this equation for y_2 in terms of y_1,

$$y_2 = \frac{y_1}{\mu(1 - y_1^2) - k} \qquad \text{(Figs. 73, 74).}$$

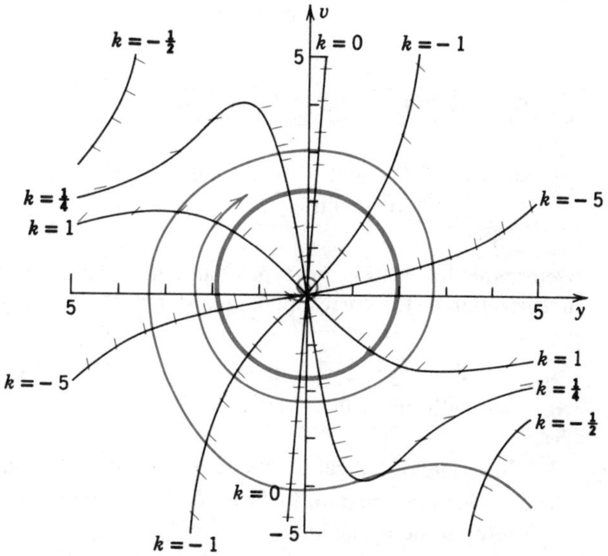

Fig. 73. Lineal element diagram for the van der Pol equation with $\mu = 0.1$ in the phase plane, showing also the limit cycle and two solution curves

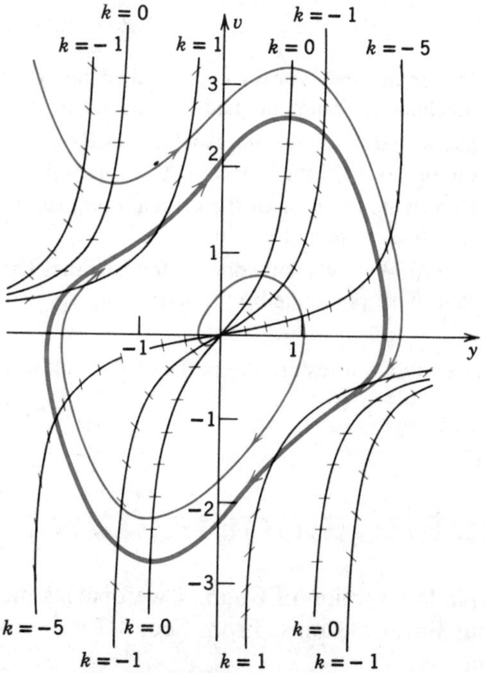

Fig. 74. Lineal element diagram for the van der Pol equation with $\mu = 1$ in the phase plane, showing also the limit cycle and two solution curves approaching it

These curves are a bit complicated. Figure 73 shows some of them for a small $\mu = 0.1$ as well as the limit cycle (almost a circle) and two solution curves approaching the limit cycle, one from the outside and one from the inside. The latter is a narrow spiral and only the initial part of it is shown in the figure. For larger μ the situation changes and the limit cycle no longer resembles a circle. Figure 74 illustrates this for $\mu = 1$. Note that the approach of solution curves to the limit cycle is much more rapid than for $\mu = 0.1$. ▮

Problem Set 4.5

1. Show that $y'' - y + y^2 = 0$ has the critical points (0, 0) and (1, 0) and determine their type and stability.

Determine the location and type of all critical points of the following equations. (Use linearization of the corresponding systems.)

2. $y'' + y - y^3 = 0$

3. $y'' - 9y + y^3 = 0$

4. $y'' + 4y - 5y^3 + y^5 = 0$

5. $y'' + \cos y = 0$

6. $y'' - 4y + y^3 = 0$

7. What type of critical points do we obtain in Example 1 if $k < 0$?

8. Convert the equation in Prob. 6 to a system, solve it in the phase plane, and sketch some of the paths.

9. To what state (position, speed, direction of motion) of the oscillating pendulum do the four points of intersection of a closed curve with the axes in Fig. 71 correspond? The point of intersection of a wavy curve with the y_2-axis?

10. In Prob. 8, add a linear damping term to get

$$y'' + y' - 4y + y^3 = 0.$$

Determine the type of each critical point by linearization. First guess (using mechanical arguments and the answer to Prob. 8), then calculate.

11. Show that if $\mu \to 0$, the isoclines in Example 4 approach straight lines through the origin, of slope $-1/k$. Why is this to be expected?

12. Determine the type of the critical point (0, 0) of the van der Pol equation when $\mu > 0$, $\mu = 0$, and $\mu < 0$.

13. Figure 73 shows that some of the isoclines have a maximum or minimum. Show that these points lie on the hyperbola $y_1 y_2 = -1/2\mu$.

What kind of curves are the paths of the following equations in the phase plane?

14. $y'' + y'^2 = 0$

15. $yy'' + y'^2 = 0$

4.6 Nonhomogeneous Linear Systems

In this last section of Chap. 4 we discuss methods for solving nonhomogeneous linear systems. From Sec. 4.2 we recall that such a system is of the form

$$\mathbf{y}' = \mathbf{A}\mathbf{y} + \mathbf{g} \qquad (1)$$

with the vector $\mathbf{g}(t)$ not identically zero. We assume $\mathbf{g}(t)$ and the $n \times n$ matrix $\mathbf{A}(t)$ to be continuous on some interval J of the t-axis. From a general solution $\mathbf{y}^{(h)}(t)$ of the homogeneous system $\mathbf{y}' = \mathbf{A}\mathbf{y}$ on J and a **particular solution** $\mathbf{y}^{(p)}(t)$ of (1) on J [i.e., a solution of (1) containing no arbitrary constants], we get a solution of (1),

$$\mathbf{y} = \mathbf{y}^{(h)} + \mathbf{y}^{(p)}, \tag{2}$$

called a **general solution** of (1) on J, because it includes every solution of (1) on J, as can be shown. Having studied homogeneous linear systems in Secs. 4.1–4.4, our present task will be to explain methods for obtaining particular solutions of (1). We discuss the method of undetermined coefficients and the method of the variation of parameters; these have counterparts for a single equation, as we know from Secs. 2.9 and 2.10. As a third method of solution we also consider the reduction to diagonal form.

Method of Undetermined Coefficients

This method is suitable if the components of $\mathbf{g}$ are integer powers of t, exponential functions or sines and cosines. We explain this in terms of an example.

EXAMPLE 1 Method of undetermined coefficients

Find a general solution of the nonhomogeneous linear system

$$\mathbf{y}' = \mathbf{A}\mathbf{y} + \mathbf{g} = \begin{bmatrix} 2 & -4 \\ 1 & -3 \end{bmatrix}\mathbf{y} + \begin{bmatrix} 2t^2 + 10t \\ t^2 + 9t + 3 \end{bmatrix}. \tag{3}$$

Solution. The form of $\mathbf{g}$ suggests to assume $\mathbf{y}^{(p)}$ in the form

$$\mathbf{y}^{(p)} = \mathbf{u} + \mathbf{v}t + \mathbf{w}t^2$$

and to determine the vectors $\mathbf{u}$, $\mathbf{v}$, and $\mathbf{w}$. By substitution,

$$\mathbf{y}^{(p)\prime} = \mathbf{v} + 2\mathbf{w}t = \mathbf{A}\mathbf{u} + \mathbf{A}\mathbf{v}t + \mathbf{A}\mathbf{w}t^2 + \mathbf{g}.$$

In terms of components, this gives

$$\begin{bmatrix} v_1 \\ v_2 \end{bmatrix} + \begin{bmatrix} 2w_1t \\ 2w_2t \end{bmatrix} = \begin{bmatrix} 2u_1 - 4u_2 \\ u_1 - 3u_2 \end{bmatrix} + \begin{bmatrix} 2v_1 - 4v_2 \\ v_1 - 3v_2 \end{bmatrix}t + \begin{bmatrix} 2w_1 - 4w_2 \\ w_1 - 3w_2 \end{bmatrix}t^2 + \begin{bmatrix} 2t^2 + 10t \\ t^2 + 9t + 3 \end{bmatrix}.$$

Equating the t^2-terms on both sides, we get

$$0 = 2w_1 - 4w_2 + 2, \quad 0 = w_1 - 3w_2 + 1, \quad \text{thus} \quad w_1 = -1, \; w_2 = 0.$$

From the t-terms we get

$$2w_1 = 2v_1 - 4v_2 + 10, \quad 2w_2 = v_1 - 3v_2 + 9, \quad \text{thus} \quad v_1 = 0, \; v_2 = 3.$$

From the constant terms we finally get

$$v_1 = 2u_1 - 4u_2, \quad v_2 = u_1 - 3u_2 + 3, \quad \text{thus} \quad u_1 = 0, \; u_2 = 0.$$

From this and the general solution of the homogeneous system in Example 2 of Sec. 4.3 we obtain the answer

$$\mathbf{y} = \mathbf{y}^{(h)} + \mathbf{y}^{(p)} = c_1 \begin{bmatrix} 4 \\ 1 \end{bmatrix} e^{t} + c_2 \begin{bmatrix} 1 \\ 1 \end{bmatrix} e^{-2t} + \begin{bmatrix} -t^2 \\ 3t \end{bmatrix}.$$

A **modification** is necessary if a term $\mathbf{g}$ involves $e^{\lambda t}$ with λ being an eigenvalue of $\mathbf{A}$. Then instead of assuming in $\mathbf{y}^{(p)}$ a term $\mathbf{u}e^{\lambda t}$, we must start from $\mathbf{u}te^{\lambda t} + \mathbf{v}e^{\lambda t}$. (The first of these two terms is the analog of the modification in Sec. 2.9, but it would not be sufficient here. Try it out.)

EXAMPLE 2 Modification of the undetermined-coefficient method

Find a general solution of

$$(4) \qquad \mathbf{y}' = \mathbf{A}\mathbf{y} + \mathbf{g} = \begin{bmatrix} -3 & 1 \\ 1 & -3 \end{bmatrix} \mathbf{y} + \begin{bmatrix} -6 \\ 2 \end{bmatrix} e^{-2t}.$$

Solution. A general equation of the homogeneous system is (see Example 1 in Sec. 4.3)

$$(5) \qquad \mathbf{y}^{(h)} = c_1 \begin{bmatrix} 1 \\ 1 \end{bmatrix} e^{-2t} + c_2 \begin{bmatrix} 1 \\ -1 \end{bmatrix} e^{-4t}.$$

Since $\lambda = -2$ is an eigenvalue of $\mathbf{A}$, we must assume $\mathbf{y}^{(p)} = \mathbf{u}te^{-2t} + \mathbf{v}e^{-2t}$ (rather than $\mathbf{u}e^{-2t}$). By substitution,

$$\mathbf{y}^{(p)\prime} = \mathbf{u}e^{-2t} - 2\mathbf{u}te^{-2t} - 2\mathbf{v}e^{-2t} = \mathbf{A}\mathbf{u}te^{-2t} + \mathbf{A}\mathbf{v}e^{-2t} + \mathbf{g}.$$

Equating the te^{-2t}-terms on both sides, we have $-2\mathbf{u} = \mathbf{A}\mathbf{u}$. Hence $\mathbf{u}$ is an eigenvector of $\mathbf{A}$ corresponding to $\lambda = -2$; thus [see (5)] $\mathbf{u}^{\mathsf{T}} = a[1 \quad 1]$. Equating the other terms gives

$$\mathbf{u} - 2\mathbf{v} = \mathbf{A}\mathbf{v} + \begin{bmatrix} -6 \\ 2 \end{bmatrix} \qquad \text{or} \qquad (\mathbf{A} + 2\mathbf{I})\mathbf{v} = \mathbf{u} - \begin{bmatrix} -6 \\ 2 \end{bmatrix} = a \begin{bmatrix} 1 \\ 1 \end{bmatrix} - \begin{bmatrix} -6 \\ 2 \end{bmatrix};$$

in components,

$$-v_1 + v_2 = a + 6$$

$$v_1 - v_2 = a - 2.$$

Hence $a = -2$ (to have a solution) and then $v_2 = v_1 + 4$, say, $v_1 = k$ and $v_2 = k + 4$; thus

$$\mathbf{v} = k \begin{bmatrix} 1 \\ 1 \end{bmatrix} + \begin{bmatrix} 0 \\ 4 \end{bmatrix}$$

and we can simply choose $k = 0$. This gives the answer

$$(6) \qquad \mathbf{y} = \mathbf{y}^{(h)} + \mathbf{y}^{(p)} = c_1 \begin{bmatrix} 1 \\ 1 \end{bmatrix} e^{-2t} + c_2 \begin{bmatrix} 1 \\ -1 \end{bmatrix} e^{-4t} - 2 \begin{bmatrix} 1 \\ 1 \end{bmatrix} te^{-2t} + \begin{bmatrix} 0 \\ 4 \end{bmatrix} e^{-2t}.$$

For other k we get other $\mathbf{v}$; for instance, $k = -2$ gives $\mathbf{v}^{\mathsf{T}} = [-2 \quad 2]$, so that the answer becomes

$$(6^*) \qquad \mathbf{y} = c_1 \begin{bmatrix} 1 \\ 1 \end{bmatrix} e^{-2t} + c_2 \begin{bmatrix} 1 \\ -1 \end{bmatrix} e^{-4t} - 2 \begin{bmatrix} 1 \\ 1 \end{bmatrix} te^{-2t} + \begin{bmatrix} -2 \\ 2 \end{bmatrix} e^{-2t},$$

and so on.

Method of Variation of Parameters

This method can be applied to nonhomogeneous linear systems

$$\mathbf{y}' = \mathbf{A}(t)\mathbf{y} + \mathbf{g}(t) \tag{7}$$

with variable $\mathbf{A} = \mathbf{A}(t)$ and general $\mathbf{g}(t)$. It yields a particular solution $\mathbf{y}^{(p)}$ of (7) on some interval J of the t-axis if a general solution

$$\mathbf{y}^{(h)} = c_1\mathbf{y}^{(1)} + \cdots + c_n\mathbf{y}^{(n)} \tag{8}$$

of the homogeneous system on J is known. In components, (8) is

$$\mathbf{y}^{(h)} = \begin{bmatrix} c_1y_1^{(1)} + \cdots + c_ny_1^{(n)} \\ \vdots \\ c_1y_n^{(1)} + \cdots + c_ny_n^{(n)} \end{bmatrix} = \begin{bmatrix} y_1^{(1)} & \cdots & y_1^{(n)} \\ & \vdots & \\ y_n^{(1)} & \cdots & y_n^{(n)} \end{bmatrix} \begin{bmatrix} c_1 \\ \vdots \\ c_n \end{bmatrix} = \mathbf{Y}(t)\mathbf{c},$$

where $\mathbf{Y}(t)$ is the matrix with columns $\mathbf{y}^{(1)}, \cdots, \mathbf{y}^{(n)}$, the vectors of the basis in (8), and $\mathbf{c}^{\mathsf{T}} = [c_1 \quad \cdots \quad c_n]$ is constant. As in Sec. 2.10, the method consists of replacing the constant vector $\mathbf{c}$ by a variable vector $\mathbf{u}(t)$,

$$\mathbf{y}^{(p)} = \mathbf{Y}(t)\mathbf{u}(t) \tag{9}$$

and determining $\mathbf{u}(t)$ by substituting $\mathbf{y}^{(p)}$ into (7). This gives

$$\mathbf{Y}'\mathbf{u} + \mathbf{Y}\mathbf{u}' = \mathbf{A}\mathbf{Y}\mathbf{u} + \mathbf{g}. \tag{10}$$

Now

$$\mathbf{y}^{(1)\prime} = \mathbf{A}\mathbf{y}^{(1)}, \qquad \mathbf{y}^{(2)\prime} = \mathbf{A}\mathbf{y}^{(2)}, \qquad \cdots, \qquad \mathbf{y}^{(n)\prime} = \mathbf{A}\mathbf{y}^{(n)}$$

since these are solutions of the homogeneous system. These n vector equations can be written as a single matrix equation $\mathbf{Y}' = \mathbf{A}\mathbf{Y}$. Hence (10) reduces to $\mathbf{Y}\mathbf{u}' = \mathbf{g}$. Now $\mathbf{Y}$ is nonsingular because the n solutions of a basis are linearly independent. Hence $\mathbf{Y}^{-1}$ exists and we can solve

$$\mathbf{Y}\mathbf{u}' = \mathbf{g} \qquad \text{to get} \qquad \mathbf{u}' = \mathbf{Y}^{-1}\mathbf{g}.$$

Integrating from some t_0 in J to a variable t, we obtain

$$\mathbf{u}(t) = \int_{t_0}^{t} \mathbf{Y}^{-1}(\tilde{t})\mathbf{g}(\tilde{t})\, d\tilde{t} + \mathbf{C}.$$

By leaving the constant vector $\mathbf{C}$ general we get a general solution

$$\mathbf{y} = \mathbf{Y}\mathbf{u} = \mathbf{Y}\mathbf{C} + \mathbf{Y}\int_{t_0}^{t} \mathbf{Y}^{-1}(\tilde{t})\mathbf{g}(\tilde{t})\, d\tilde{t}, \tag{11}$$

and for $\mathbf{C} = \mathbf{0}$ this is a particular solution (9) of (7).

EXAMPLE 3 **Solution by the method of variation of parameters**

For the system (4) in Example 2 we have from (5) and (4)

$$\text{(12)} \qquad \mathbf{Y} = [\mathbf{y}^{(1)} \quad \mathbf{y}^{(2)}] = \begin{bmatrix} e^{-2t} & e^{-4t} \\ e^{-2t} & -e^{-4t} \end{bmatrix}, \qquad \mathbf{g} = \begin{bmatrix} -6 \\ 2 \end{bmatrix} e^{-2t}.$$

From this and (10) in Sec. 4.0 we obtain the inverse

$$\mathbf{Y}^{-1} = \frac{1}{-2e^{-6t}} \begin{bmatrix} -e^{-4t} & -e^{-4t} \\ -e^{-2t} & e^{-2t} \end{bmatrix} = \frac{1}{2} \begin{bmatrix} e^{2t} & e^{2t} \\ e^{4t} & -e^{4t} \end{bmatrix}.$$

We multiply this by $\mathbf{g}$, obtaining

$$\mathbf{Y}^{-1}\mathbf{g} = \frac{1}{2} \begin{bmatrix} e^{2t} & e^{2t} \\ e^{4t} & -e^{4t} \end{bmatrix} \begin{bmatrix} -6e^{-2t} \\ 2e^{-2t} \end{bmatrix} = \frac{1}{2} \begin{bmatrix} -4 \\ -8e^{2t} \end{bmatrix} = \begin{bmatrix} -2 \\ -4e^{2t} \end{bmatrix}.$$

We now integrate. Choosing the constant vector of integration to be the zero vector, we get

$$\mathbf{u}(t) = \int_0^t \begin{bmatrix} -2 \\ -4e^{2\tilde{t}} \end{bmatrix} d\tilde{t} = \begin{bmatrix} -2t \\ -2e^{2t} + 2 \end{bmatrix}.$$

From this and (12) we obtain

$$\mathbf{Yu} = \begin{bmatrix} e^{-2t} & e^{-4t} \\ e^{-2t} & -e^{-4t} \end{bmatrix} \begin{bmatrix} -2t \\ -2e^{2t} + 2 \end{bmatrix} = \begin{bmatrix} -2te^{-2t} - 2e^{-2t} + 2e^{-4t} \\ -2te^{-2t} + 2e^{-2t} - 2e^{-4t} \end{bmatrix}$$

$$= \begin{bmatrix} -2te^{-2t} - 2e^{-2t} \\ -2te^{-2t} + 2e^{-2t} \end{bmatrix} + \begin{bmatrix} 2 \\ -2 \end{bmatrix} e^{-4t}.$$

The last term is a solution of the homogeneous system, so that we can absorb it into $\mathbf{y}^{(h)}$ and obtain as a general solution of the system (4)

$$\text{(13)} \qquad \mathbf{y} = c_1 \begin{bmatrix} 1 \\ 1 \end{bmatrix} e^{-2t} + c_2 \begin{bmatrix} 1 \\ -1 \end{bmatrix} e^{-4t} - 2 \begin{bmatrix} 1 \\ 1 \end{bmatrix} te^{-2t} + \begin{bmatrix} -2 \\ 2 \end{bmatrix} e^{-2t},$$

in agreement with (6*). ∎

Method of Diagonalization

The idea of this method is to "decouple" the n equations of a linear system, so that each equation contains only one of the unknown functions $y_1, \cdots, y_n$ and thus can be solved independently of the other equations. This works for systems

$$\text{(14)} \qquad \boxed{\mathbf{y}' = \mathbf{Ay} + \mathbf{g}(t)}$$

for which $\mathbf{A}$ has a basis of eigenvectors $\mathbf{x}^{(1)}, \cdots, \mathbf{x}^{(n)}$. It can be shown that then

$$\text{(15)} \qquad \boxed{\mathbf{D} = \mathbf{X}^{-1}\mathbf{AX}}$$

is a diagonal matrix with the eigenvalues $\lambda_1, \cdots, \lambda_n$ of $\mathbf{A}$ on the main diagonal; here $\mathbf{X}$ is the $n \times n$ matrix with columns $\mathbf{x}^{(1)}, \cdots, \mathbf{x}^{(n)}$. (Proof in

Sec. 7.14.) Note that $\mathbf{X}^{-1}$ exists because these columns are linearly independent. For instance, in Example 2,

$$\mathbf{X} = \begin{bmatrix} 1 & 1 \\ 1 & -1 \end{bmatrix},$$

(16)
$$\mathbf{D} = \begin{bmatrix} \frac{1}{2} & \frac{1}{2} \\ \frac{1}{2} & -\frac{1}{2} \end{bmatrix} \begin{bmatrix} -3 & 1 \\ 1 & -3 \end{bmatrix} \begin{bmatrix} 1 & 1 \\ 1 & -1 \end{bmatrix} = \begin{bmatrix} -2 & 0 \\ 0 & -4 \end{bmatrix},$$

as the reader may verify.

To apply diagonalization to (14), we define the new unknown function

(17) (a) $\mathbf{z} = \mathbf{X}^{-1}\mathbf{y}$. Then (b) $\mathbf{y} = \mathbf{X}\mathbf{z}$.

Substituting this into (14), we have (note that $\mathbf{X}$ is constant!)

$$\mathbf{X}\mathbf{z}' = \mathbf{A}\mathbf{X}\mathbf{z} + \mathbf{g}.$$

We multiply this by $\mathbf{X}^{-1}$ from the left, obtaining

$$\mathbf{z}' = \mathbf{X}^{-1}\mathbf{A}\mathbf{X}\mathbf{z} + \mathbf{h} \qquad \text{where} \qquad \mathbf{h} = \mathbf{X}^{-1}\mathbf{g}.$$

Because of (15) we can write this

(18) $\mathbf{z}' = \mathbf{D}\mathbf{z} + \mathbf{h}$; in components, $z_j' = \lambda_j z_j + h_j$,

where $j = 1, \cdots, n$. We can now solve each of these n linear equations as in Sec. 1.7, to get

(19)
$$z_j(t) = c_j e^{\lambda_j t} + e^{\lambda_j t}\int e^{-\lambda_j t} h_j(t)\, dt.$$

These are the components of $\mathbf{z}(t)$, and from them we obtain the answer $\mathbf{y} = \mathbf{X}\mathbf{z}$ by (17b).

EXAMPLE 4 **Method of diagonalization**

For the system (4) in Example 2 we get from (5) and (4) [see also (16)]

$$\mathbf{h} = \mathbf{X}^{-1}\mathbf{g} = \begin{bmatrix} \frac{1}{2} & \frac{1}{2} \\ \frac{1}{2} & -\frac{1}{2} \end{bmatrix} \begin{bmatrix} -6e^{-2t} \\ 2e^{-2t} \end{bmatrix} = \begin{bmatrix} -2e^{-2t} \\ -4e^{-2t} \end{bmatrix}.$$

Since the eigenvalues are $\lambda_1 = -2$ and $\lambda_2 = -4$, the diagonalized system is

$$\mathbf{z}' = \begin{bmatrix} -2 & 0 \\ 0 & -4 \end{bmatrix}\mathbf{z} + \mathbf{h}, \qquad \text{thus} \qquad \begin{aligned} z_1' &= -2z_1 - 2e^{-2t} \\ z_2' &= -4z_2 - 4e^{-2t}. \end{aligned}$$

From (19) we obtain the solutions

$$z_1 = c_1e^{-2t} - 2te^{-2t}, \qquad z_2 = c_2e^{-4t} - 2e^{-2t}.$$

From this and (17b) we get the answer

$$\mathbf{y} = \mathbf{Xz} = \begin{bmatrix} 1 & 1 \\ 1 & -1 \end{bmatrix} \begin{bmatrix} c_1e^{-2t} - 2te^{-2t} \\ c_2e^{-4t} - 2e^{-2t} \end{bmatrix} = \begin{bmatrix} c_1e^{-2t} - 2te^{-2t} + c_2e^{-4t} - 2e^{-2t} \\ c_1e^{-2t} - 2te^{-2t} - c_2e^{-4t} + 2e^{-2t} \end{bmatrix}.$$

This is identical with (13) and (6*). ∎

Problem Set 4.6

Find a general solution of the following nonhomogeneous linear systems.

1. $y_1' = y_2 + 6e^{2t}$
$y_2' = y_1 - 3e^{2t}$

2. $y_1' = 4y_2$
$y_2' = 4y_1 + 2 - 16t^2$

3. $y_1' = -3y_1 + y_2 + 3\cos t$
$y_2' = y_1 - 3y_2 - 2\cos t - 3\sin t$

4. $y_1' = 4y_1 - 8y_2 + 2\cosh t$
$y_2' = 2y_1 - 6y_2 + \cosh t + 2\sinh t$

5. $y_1' = y_2 - 5\sin t$
$y_2' = -4y_1 + 17\cos t$

6. $y_1' = -3y_1 - 4y_2 + 5e^t$
$y_2' = 5y_1 + 6y_2 - 6e^t$

Solve the following initial value problems.

7. $y_1' = 5y_2 + 23$
$y_2' = -5y_1 + 15t$
$y_1(0) = 1, \quad y_2(0) = -2$

8. $y_1' = y_1 + 4y_2 - t^2 + 6t$
$y_2' = y_1 + y_2 - t^2 + t - 1$
$y_1(0) = 2, \quad y_2(0) = -1$

9. $y_1' = 4y_1 + 8y_2 + 2\cos t - 16\sin t$
$y_2' = 6y_1 + 2y_2 + \cos t - 14\sin t$
$y_1(0) = 0, \quad y_2(0) = 1.75$

10. $y_1' = -y_2 + \cos t - \sin t$
$y_2' = -y_1 + \cos t + \sin t$
$y_1(0) = -1, \quad y_2(0) = -6$

11. $y_1' = 5y_1 + 4y_2 - 5t^2 + 6t + 25$
$y_2' = y_1 + 2y_2 - t^2 + 2t + 4$
$y_1(0) = 0, \quad y_2(0) = 0$

12. $y_1' = 2y_1 + 3y_2 + 2e^{2t}$
$y_2' = y_1 + 4y_2 + 3e^{2t}$
$y_1(0) = -2/3, \quad y_2(0) = 1/3$

Find a general solution (currents) of the network in Fig. 75, where:

13. $R_1 = 1$ ohm, $R_2 = 4$ ohms, $L = 0.5$ henry, $C = 1$ farad, $E = 100$ volts.

14. $E = 220 \sin t$ volts, the other data as in Prob. 13.

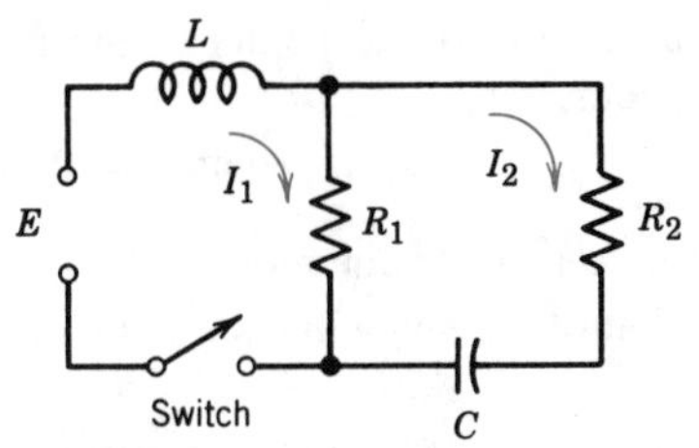

Fig. 75. Problems 13–15

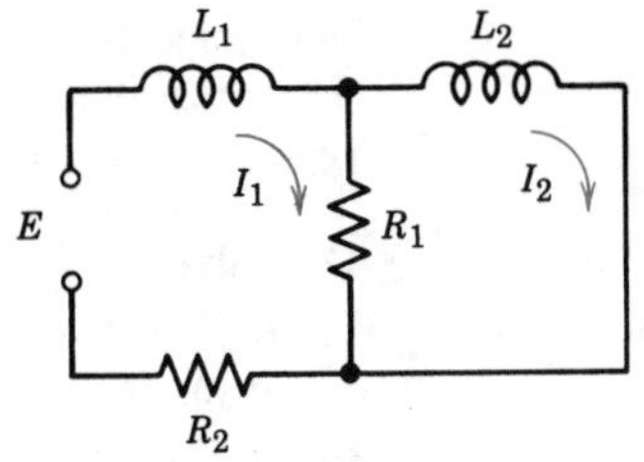

Fig. 76. Problem 16

15. In Prob. 13, find the particular solution corresponding to zero currents and charges when $t = 0$.

16. Find the currents in Fig. 76, where $R_1 = 1$ ohm, $R_2 = 1.4$ ohms, $L_1 = 0.8$ henry, $L_2 = 1$ henry, $E = 100$ volts, and $I_1(0) = I_2(0) = 0$.

Review Questions and Problems for Chapter 4

1. Without looking in the text, write down a general first-order linear system of n differential equations in n unknown functions (i) in matrix form and (ii) written out.

2. Without looking into the text, convert the second-order equation $y'' + p(t)y' + q(t)y = 0$ into a system of first-order equations. Give an example.

3. What is the Wronskian of two solutions and what role did it play in this chapter?

4. What are eigenvalues and eigenvectors of a matrix? Why were they important in this chapter?

5. What is a basis of eigenvectors? How did we use it in a basis of solutions? What can we do if a matrix has no such basis?

6. What do we mean by the phase plane? By a path in the phase plane? By a critical point of a system?

7. What do the paths of a system look like near a saddle point? Near a spiral point? Near a center? Near a node?

8. What do we mean by linearization of a system and what is the purpose of it?

9. What are self-sustained oscillations? Write down a corresponding differential equation (as simple as possible).

10. List and explain methods of obtaining particular solutions of nonhomogeneous linear systems.

Find a general solution of the following systems and determine the type and stability of the critical points.

11. $y_1' = 2y_1 + 4y_2$
$y_2' = 3y_1 + y_2$

12. $y_1' = y_1 + 4y_2$
$y_2' = y_1 + y_2$

13. $y_1' = 4y_1 - 2y_2$
$y_2' = 13y_1 - 6y_2$

14. $y_1' = y_1 + y_2$
$y_2' = -6y_1 - 4y_2$

Determine the location and type of all critical points of the following equations. (Linearize the corresponding systems.)

15. $y'' - y + y^3 = 0$

16. $y'' + \tan y = 0$

17. Solve Prob. 11 by conversion to a single equation.

18. If $\mathbf{y}' = \mathbf{A}\mathbf{y}$ has a saddle point at (0, 0), show that $\mathbf{y}' = \mathbf{A}^2\mathbf{y}$ has an unstable node at (0, 0).

19. If $\mathbf{A}$ in $\mathbf{y}' = \mathbf{A}\mathbf{y}$ has the eigenvalues -5 and 3, what type of critical point is (0, 0)?

20. Find the currents in Fig. 77, assuming that $R = 1$ ohm, $L = 1.25$ henrys, $C = 0.2$ farad, and $I_1(0) = I_2(0) = 1$ amp.

21. Find the currents in Fig. 78, where $R = 2.5$ ohms, $L = 1$ henry, $C = 0.04$ farad, $E(t) = 169 \sin t$ volts, and $I_1(0) = I_2(0) = 0$.

22. Find the currents in Fig. 78, where $R = 1$ ohm, $L = 10$ henrys, $C = 1$ farad, $E = 100$ volts, and $I_1(0) = I_2(0) = 0$.

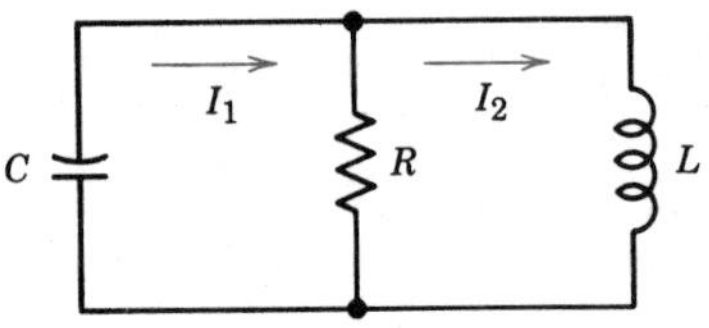

Fig. 77. Network in Problem 20

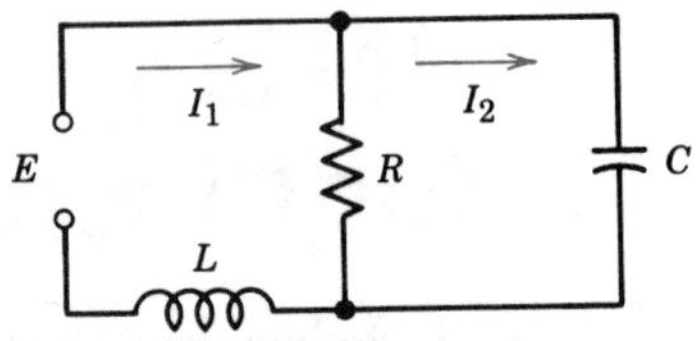

Fig. 78. Network in Problems 21, 22

23. Tank T_1 in Fig. 79 contains initially 100 gal of pure water. Tank T_2 contains initially 100 gal of water in which 90 lb of salt are dissolved. Liquid is pumped through the system as indicated, and the mixtures are kept uniform by stirring. Find the amounts of salt $y_1(t)$ and $y_2(t)$ in T_1 and T_2, respectively.

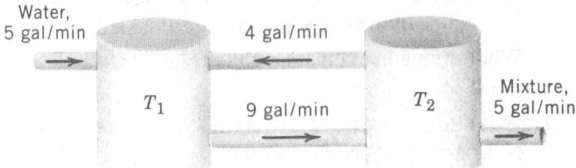

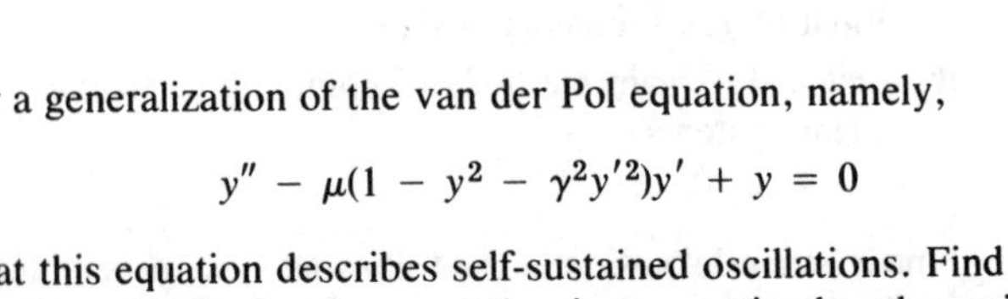

Fig. 79. Tanks in Problem 23

24. Consider a generalization of the van der Pol equation, namely,

$$y'' - \mu(1 - y^2 - \gamma^2 y'^2)y' + y = 0 \qquad (\gamma \text{ constant}).$$

Show that this equation describes self-sustained oscillations. Find the equation of the isoclines. Verify that for $\gamma = 1$ the given equation has the periodic solution $y = \sin t$.

25. **(Rayleigh equation)** Show that the so-called *Rayleigh equation*

$$Y'' - \mu(1 - \tfrac{1}{3}Y'^2)Y' + Y = 0 \qquad (\mu > 0)$$

also describes self-sustained oscillations and that by differentiating it and setting $y = Y'$ one obtains the van der Pol equation.

26. (Duffing equation) The Duffing equation is

$$y'' + \omega_0{}^2 y + \beta y^3 = 0$$

where usually $|\beta|$ is small, thus characterizing a small deviation of the restoring force from linearity. $\beta > 0$ and $\beta < 0$ are called the cases of a *hard* and a *soft spring*, respectively. Find the equation of the solution curves in the phase plane. (For $\beta > 0$ all these curves are closed.)

Find a general solution of the following nonhomogeneous systems.

27. $y_1' = 2y_1 + 3y_2 - 2e^{-t}$
$y_2' = -y_1 - 2y_2$

28. $y_1' = 2y_1 + 2y_2 + e^t$
$y_2' = -2y_1 - 3y_2 + e^t$

29. Solve the initial value problem

$$\begin{aligned} y_1' &= -y_1 + 4y_2 + 6e^t - t + 11, \\ y_2' &= 3y_1 - 2y_2 - 6e^t + 3t - 6, \end{aligned} \qquad y_1(0) = 4, \quad y_2(0) = 0.$$

30. Find a general solution of $y_1' = y_2 + t$, $y_2' = -y_1$ by applying each of the three methods discussed in Sec. 4.6.

Summary of Chapter 4 Systems of Differential Equations, Phase Plane, Stability

Whereas single electric circuits or single mass–spring systems are governed by single differential equations as their mathematical model (Chap. 2), networks consisting of several circuits, mechanical systems of several masses and springs and other problems of engineering interest lead to **systems of differential equations** (Sec. 4.1) in which we simultaneously deal with several unknown functions, representing the currents in the various circuits, the displacements of those masses, etc. Of central interest are first-order systems (Sec. 4.2)

$$\begin{aligned} y_1' &= f_1(t, y_1, \cdots, y_n) \\ &\vdots \\ y_n' &= f_n(t, y_1, \cdots, y_n), \end{aligned}$$

to which higher order equations and systems can be reduced. In this summary we let $n = 2$, for simplicity; thus

$$\text{(1)} \qquad \begin{aligned} y_1' &= f_1(t, y_1, y_2) \\ y_2' &= f_2(t, y_1, y_2). \end{aligned}$$

(Corresponding formulas for general n and some theoretical ideas and concepts are included in Secs. 4.2 and 4.3.) A **linear system** is of the form

$$
(2') \qquad \begin{aligned} y_1' &= a_{11}y_1 + a_{12}y_2 + g_1 \\ y_2' &= a_{21}y_1 + a_{22}y_2 + g_2; \end{aligned}
$$

in vector form

$$
(2) \quad \mathbf{y}' = \mathbf{A}\mathbf{y} + \mathbf{g}, \text{ where } \mathbf{A} = \begin{bmatrix} a_{11} & a_{12} \\ a_{21} & a_{22} \end{bmatrix}, \ \mathbf{y} = \begin{bmatrix} y_1 \\ y_2 \end{bmatrix}, \ \mathbf{g} = \begin{bmatrix} g_1 \\ g_2 \end{bmatrix}.
$$

A **homogeneous linear system** is of the form

$$
(3) \qquad \mathbf{y}' = \mathbf{A}\mathbf{y}, \qquad \text{written out} \qquad \begin{aligned} y_1' &= a_{11}y_1 + a_{12}y_2 \\ y_2' &= a_{21}y_1 + a_{22}y_2. \end{aligned}
$$

Systems (3) with constant $a_{11}, \cdots, a_{22}$ have solutions $\mathbf{y} = \mathbf{x}e^{\lambda t} \not\equiv \mathbf{0}$ with λ being the solutions of the quadratic equation (Sec. 4.3)

$$
\begin{vmatrix} a_{11} - \lambda & a_{12} \\ a_{21} & a_{22} - \lambda \end{vmatrix} = (a_{11} - \lambda)(a_{22} - \lambda) - a_{12}a_{21} = 0
$$

and vector $\mathbf{x} \neq \mathbf{0}$ with components x_1, x_2 determined up to a multiplicative constant by

$$
(a_{11} - \lambda)x_1 + a_{12}x_2 = 0
$$

(2×2 matrices needed here are explained in Sec. 4.0.)

The y_1y_2-plane is called the **phase plane**. A **path** (also called a **trajectory**) is the curve of a solution $y_1(t)$, $y_2(t)$ in the phase plane. A **critical point** P: (y_1, y_2) of a system (1) [or (2)] is one at which the right sides of the system are both zero. Depending on the behavior of the paths near P we call P a **node, saddle point, center,** or **spiral point**, and critical points are also classified in terms of **stability** (Secs. 4.3, 4.4). For nonlinear systems this is done by linearization (Sec. 4.5). Phase plane methods are primarily applied to **autonomous systems**, that is, systems in which t does not occur explicitly. They are qualitative and reveal a surprisingly large amount of information, particularly on nonlinear equations and systems that cannot be solved analytically. Applications to the famous pendulum and van der Pol equations are shown in Sec. 4.5.

Methods for solving nonhomogeneous linear systems are discussed in Sec. 4.6.

Chapter 5

Series Solutions of Differential Equations. Special Functions

In Chap. 2 we saw that a homogeneous linear differential equation whose coefficients are *constant* can be solved by algebraic methods, and the solutions are elementary functions known from calculus. However, if those coefficients are not constant but depend on x, the situation is more complicated and the solutions may be nonelementary functions. **Legendre's equation** (Sec. 5.3), the **hypergeometric equation** (Sec. 5.4), and **Bessel's equation** (Sec. 5.5) are of this type. Since these and other equations and their solutions play an important role in applied mathematics, we devote a whole chapter to two standard methods of solution and their applications: the **power series method** (Secs. 5.1, 5.2), which yields solutions in the form of power series, and an extension of it, called the **Frobenius method** (Sec. 5.4).

This will also give the student a chance to become familiar with methods and techniques on **special functions,** meaning "higher transcendental functions" (of great practical importance, but not considered in calculus courses), and some of their properties. This will include **Sturm–Liouville theory** (Secs. 5.8, 5.9) based on **orthogonality of functions,** an idea whose significance to mathematical physics and its engineering applications can hardly be overestimated.

COMMENT. This chapter can also be studied directly after Chap. 2 because it uses no material from Chaps. 3 and 4.

Prerequisite for this chapter: Chap. 2.
Sections that may be omitted in a shorter course: 5.2, 5.6–5.9.
References: Appendix 1, Part A.
Answers to problems: Appendix 2.

5.1 Power Series Method

The **power series method** is the standard basic method for solving linear differential equations with ***variable*** coefficients. It gives solutions in the form of power series; this explains the name. These series can be used for computing values of solutions, for exploring their properties, and for deriving other kinds of representations of those solutions, as we shall see. In this section we begin by explaining and illustrating the basic idea of the method.

Power Series

We first remember from calculus that a **power series**[1] (in powers of $x - x_0$) is an infinite series of the form

$$(1)\qquad \sum_{m=0}^{\infty} a_m(x - x_0)^m = a_0 + a_1(x - x_0) + a_2(x - x_0)^2 + \cdots .$$

$a_0, a_1, a_2, \cdots$ are constants, called the **coefficients** of the series. x_0 is a constant, called the **center** of the series, and x is a variable.

If in particular $x_0 = 0$, we obtain a *power series in powers of* x

$$(2)\qquad \sum_{m=0}^{\infty} a_m x^m = a_0 + a_1 x + a_2 x^2 + a_3 x^3 + \cdots .$$

We shall assume in this section that all variables and constants are real.

Familiar examples of power series are the Maclaurin series

$$\frac{1}{1-x} = \sum_{m=0}^{\infty} x^m = 1 + x + x^2 + \cdots \qquad (|x| < 1, \textit{ geometric series}),$$

$$e^x = \sum_{m=0}^{\infty} \frac{x^m}{m!} = 1 + x + \frac{x^2}{2!} + \frac{x^3}{3!} + \cdots ,$$

$$\cos x = \sum_{m=0}^{\infty} \frac{(-1)^m x^{2m}}{(2m)!} = 1 - \frac{x^2}{2!} + \frac{x^4}{4!} - + \cdots ,$$

$$\sin x = \sum_{m=0}^{\infty} \frac{(-1)^m x^{2m+1}}{(2m+1)!} = x - \frac{x^3}{3!} + \frac{x^5}{5!} - + \cdots .$$

Idea of the Power Series Method

The idea of the power series method for solving differential equations is simple and natural. We begin by describing the practical procedure and illustrate it for simple equations whose solutions we know, so that we can see what is going on. The mathematical justification of the method follows in the next section.

For a given equation

$$y'' + p(x)y' + q(x)y = 0$$

[1]The term "power series" alone usually refers to a series of the form (1), but does not include series of negative powers of x such as $a_0 + a_1x^{-1} + a_2x^{-2} + \cdots$ or series involving fractional powers of x. Note that in (1) we write, for convenience, $(x - x_0)^0 = 1$, even when $x = x_0$.

We use m as the summation letter, reserving n as a standard notation for the parameters in the Legendre and Bessel equations for integer values.

we first represent $p(x)$ and $q(x)$ by power series in powers of x (or of $x - x_0$ if solutions in powers of $x - x_0$ are wanted). Often, $p(x)$ and $q(x)$ are polynomials, and then nothing needs to be done in this first step. Next we assume a solution in the form of a power series with unknown coefficients,

$$y = \sum_{m=0}^{\infty} a_m x^m = a_0 + a_1 x + a_2 x^2 + a_3 x^3 + \cdots \tag{3}$$

and insert this series and the series obtained by termwise differentiation,

$$\text{(4)}\quad \begin{aligned} &\text{(a)}\quad y' = \sum_{m=1}^{\infty} m a_m x^{m-1} = a_1 + 2a_2 x + 3a_3 x^2 + \cdots \\ &\text{(b)}\quad y'' = \sum_{m=2}^{\infty} m(m-1) a_m x^{m-2} = 2a_2 + 3 \cdot 2a_3 x + 4 \cdot 3a_4 x^2 + \cdots \end{aligned}$$

into the equation. Then we collect like powers of x and equate the sum of the coefficients of each occurring power of x to zero, starting with the constant terms, the terms containing x, the terms containing x^2, etc. This gives relations from which we can determine the unknown coefficients in (3) successively.

We illustrate this for some simple equations that can also be solved by elementary methods.

EXAMPLE 1 Solve

$$y' - y = 0.$$

Solution. In the first step, we insert (3) and (4a) into the equation:

$$(a_1 + 2a_2 x + 3a_3 x^2 + \cdots) - (a_0 + a_1 x + a_2 x^2 + \cdots) = 0.$$

Then we collect like powers of x, finding

$$(a_1 - a_0) + (2a_2 - a_1)x + (3a_3 - a_2)x^2 + \cdots = 0.$$

Equating the coefficient of each power of x to zero, we have

$$a_1 - a_0 = 0, \qquad 2a_2 - a_1 = 0, \qquad 3a_3 - a_2 = 0, \cdots .$$

Solving these equations, we may express $a_1, a_2, \cdots$ in terms of a_0, which remains arbitrary:

$$a_1 = a_0, \qquad a_2 = \frac{a_1}{2} = \frac{a_0}{2!}, \qquad a_3 = \frac{a_2}{3} = \frac{a_0}{3!}, \cdots .$$

With these coefficients the series (3) becomes

$$y = a_0 + a_0 x + \frac{a_0}{2!} x^2 + \frac{a_0}{3!} x^3 + \cdots ,$$

and we see that we have obtained the familiar general solution

$$y = a_0 \left(1 + x + \frac{x^2}{2!} + \frac{x^3}{3!} + \cdots\right) = a_0 e^x. \quad \blacksquare$$

EXAMPLE 2 Solve

$$y' = 2xy.$$

Solution. We insert (3) and (4a) into the equation:

$$a_1 + 2a_2x + 3a_3x^2 + \cdots = 2x(a_0 + a_1x + a_2x^2 + \cdots).$$

We must perform the multiplication by $2x$ on the right and can write the resulting equation in the convenient form

$$\begin{aligned} &a_1 + 2a_2x + 3a_3x^2 + 4a_4x^3 + 5a_5x^4 + 6a_6x^5 + \cdots \\ = \quad &2a_0x + 2a_1x^2 + 2a_2x^3 + 2a_3x^4 + 2a_4x^5 + \cdots. \end{aligned}$$

From this we see that

$$a_1 = 0, \qquad 2a_2 = 2a_0, \qquad 3a_3 = 2a_1, \qquad 4a_4 = 2a_2, \qquad 5a_5 = 2a_3, \qquad \cdots.$$

Hence $a_3 = 0$, $a_5 = 0$, $\cdots$ and for the coefficients with even subscripts,

$$a_2 = a_0, \qquad a_4 = \frac{a_2}{2} = \frac{a_0}{2!}, \qquad a_6 = \frac{a_4}{3} = \frac{a_0}{3!}, \quad \cdots;$$

a_0 remains arbitrary. With these coefficients the series (3) becomes

$$y = a_0\left(1 + x^2 + \frac{x^4}{2!} + \frac{x^6}{3!} + \frac{x^8}{4!} + \cdots\right) = a_0e^{x^2}.$$

Confirm this solution by the method of separating variables. ∎

EXAMPLE 3 Solve

$$y'' + y = 0.$$

Solution. By inserting (3) and (4b) into the equation we obtain

$$(2a_2 + 3 \cdot 2a_3x + 4 \cdot 3a_4x^2 + \cdots) + (a_0 + a_1x + a_2x^2 + \cdots) = 0.$$

Collecting like powers of x, we find

$$(2a_2 + a_0) + (3 \cdot 2a_3 + a_1)x + (4 \cdot 3a_4 + a_2)x^2 + \cdots = 0.$$

Equating the coefficient of each power of x to zero, we have

$$\begin{aligned} 2a_2 + a_0 &= 0 && \text{coefficient of } x^0 \\ 3 \cdot 2a_3 + a_1 &= 0 && \text{coefficient of } x^1 \\ 4 \cdot 3a_4 + a_2 &= 0 && \text{coefficient of } x^2 \end{aligned}$$

etc. Solving these equations, we see that $a_2, a_4, \cdots$ may be expressed in terms of a_0; and $a_3, a_5, \cdots$ may be expressed in terms of a_1:

$$a_2 = -\frac{a_0}{2!}, \qquad a_3 = -\frac{a_1}{3!}, \qquad a_4 = -\frac{a_2}{4 \cdot 3} = \frac{a_0}{4!}, \quad \cdots;$$

a_0 and a_1 are arbitrary. With these coefficients the series (3) becomes

$$y = a_0 + a_1x - \frac{a_0}{2!}x^2 - \frac{a_1}{3!}x^3 + \frac{a_0}{4!}x^4 + \frac{a_1}{5!}x^5 + \cdots.$$

This can be written

$$y = a_0\left(1 - \frac{x^2}{2!} + \frac{x^4}{4!} - + \cdots\right) + a_1\left(x - \frac{x^3}{3!} + \frac{x^5}{5!} - + \cdots\right)$$

and we recognize the familiar general solution

$$y = a_0 \cos x + a_1 \sin x.$$ ∎

Do we need the power series method for these and similar equations? Of course not; this was just to explain the method. So what happens if we apply the method to an equation not of the kind considered in Chap. 2, even an innocent-looking one such as $y'' + xy = 0$ ("Airy's equation")? We most likely end up with new functions given by power series. And if such an equation and its solutions are of practical (or theoretical) interest, they are given names and are thoroughly investigated. This is what happened to Legendre's, Bessel's, and Gauss's hypergeometric equations, to mention just the most prominent ones. However, before we can discuss these highly important equations, we must first explain the power series method (and an extension of it) in more detail.

Problem Set 5.1

Apply the power series method to the following differential equations.

1. $y' = 2y$ **2.** $y' + y = 0$ **3.** $y' = ky$

4. $(1 - x)y' = y$ **5.** $(x + 1)y' = 3y$ **6.** $(1 + x)y' + y = 0$

7. $y' + 2xy = 0$ **8.** $y' = 3x^2y$ **9.** $y'' - y = 0$

10. $y'' + 4y = 0$ **11.** $y'' - y' = 0$ **12.** $y'' - 9y = 0$

(More problems of this type are included in the next problem set.)

5.2 Theory of the Power Series Method

In the last section we saw that the power series method gives solutions of differential equations in the form of power series. In this section we first review a few relevant facts on power series from calculus, then list the operations on power series needed in the method (differentiation, addition, multiplication, etc.), and finally say a word about the existence of power series solutions.

Basic Concepts

Recall from calculus that a **power series** is an infinite series of the form

$$\sum_{m=0}^{\infty} a_m(x - x_0)^m = a_0 + a_1(x - x_0) + a_2(x - x_0)^2 + \cdots . \tag{1}$$

As before, we assume the variable x, the **center** x_0, and the **coefficients** $a_0, a_1, \cdots$ to be real. The ***n*th partial sum** of (1) is

$$s_n(x) = a_0 + a_1(x - x_0) + a_2(x - x_0)^2 + \cdots + a_n(x - x_0)^n, \tag{2}$$

where $n = 0, 1, \cdots$. Clearly, if we omit the terms of s_n from (1), the remaining expression is

$$R_n(x) = a_{n+1}(x - x_0)^{n+1} + a_{n+2}(x - x_0)^{n+2} + \cdots. \tag{3}$$

This expression is called the **remainder** *of* (1) *after the term* $a_n(x - x_0)^n$.

For example, in the case of the geometric series

$$1 + x + x^2 + \cdots + x^n + \cdots$$

we have

$$s_0 = 1, \qquad R_0 = x + x^2 + x^3 + \cdots,$$

$$s_1 = 1 + x, \qquad R_1 = x^2 + x^3 + x^4 + \cdots,$$

$$s_2 = 1 + x + x^2, \qquad R_2 = x^3 + x^4 + x^5 + \cdots, \qquad \text{etc.}$$

In this way we have now associated with (1) the sequence of the partial sums $s_0(x), s_1(x), s_2(x), \cdots$. If for some $x = x_1$ this sequence converges, say,

$$\lim_{n\to\infty} s_n(x_1) = s(x_1),$$

then the series (1) is called **convergent** *at* $x = x_1$, the number $s(x_1)$ is called the **value** or *sum* of (1) at x_1, and we write

$$s(x_1) = \sum_{m=0}^{\infty} a_m(x_1 - x_0)^m.$$

Then we have for every n,

$$s(x_1) = s_n(x_1) + R_n(x_1). \tag{4}$$

If that sequence diverges at $x = x_1$, the series (1) is called **divergent** at $x = x_1$.

In the case of convergence, for any positive ϵ there is an N (depending on ϵ) such that, by (4),

$$|R_n(x_1)| = |s(x_1) - s_n(x_1)| < \epsilon \qquad \text{for all } n > N. \tag{5}$$

Geometrically, this means that all $s_n(x_1)$ with $n > N$ lie between $s - \epsilon$ and $s + \epsilon$ (Fig. 80). Practically, this means that in the case of convergence we

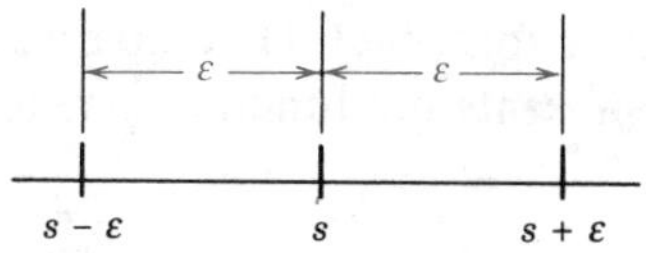

Fig. 80. Inequality (5)

can approximate the sum $s(x_1)$ of (1) at x_1 by $s_n(x_1)$ as accurately as we please, by taking n large enough.

Convergence Interval. Radius of Convergence

1. The series (1) always converges at $x = x_0$, because then all its terms except for the first, a_0, are zero. In exceptional cases this may be the only x for which (1) converges. Such a series is of no practical interest.

2. If there are further values of x for which the series converges, these values form an interval, called the **convergence interval.** If this interval is finite, it has the midpoint x_0, so that it is of the form

$$|x - x_0| < R \qquad \text{(Fig. 81)} \tag{6}$$

and the series (1) converges for all x such that $|x - x_0| < R$ and diverges for all x such that $|x - x_0| > R$. The number R is called the **radius[2] of convergence** of (1). It can be obtained from either of the formulas

$$\text{(7)} \quad \text{(a)} \quad R = 1 \Big/ \lim_{m\to\infty} \sqrt[m]{|a_m|} \qquad \text{(b)} \quad R = 1 \Big/ \lim_{m\to\infty} \left|\frac{a_{m+1}}{a_m}\right|$$

provided these limits exist and are not zero. [If they are infinite, then (1) converges only at the center x_0.]

3. The convergence interval may sometimes be infinite, that is, (1) converges for all x. For instance, if the limit in (7a) or (7b) is zero, this case occurs. One then writes $R = \infty$, for convenience. (Proofs of all these facts can be found in Sec. 14.2.)

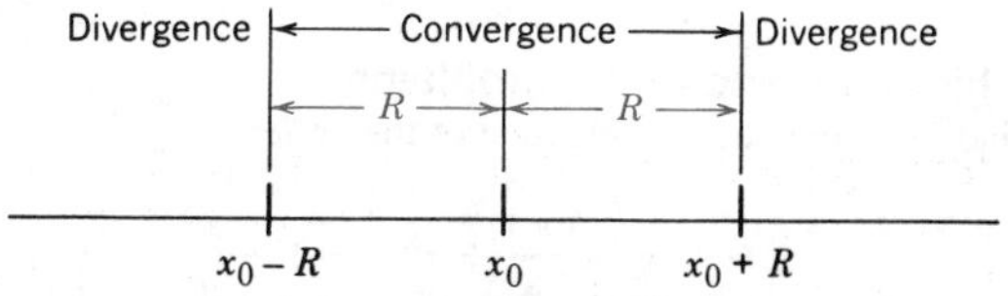

Fig. 81. Convergence interval (6) of a power series with center x_0

[2]Because for a *complex* power series, one has convergence in an open disk of *radius R*. No general statement about convergence or divergence can be made for $x - x_0 = R$ or $-R$.

For each x for which (1) converges, it has a certain value $s(x)$. We say that (1) **represents** the function $s(x)$ in the convergence interval and write

$$s(x) = \sum_{m=0}^{\infty} a_m(x - x_0)^m \qquad (|x - x_0| < R).$$

Let us illustrate the three possibilities by typical examples.

EXAMPLE 1 The useless Case 1 of convergence only at the center

In the case of the series

$$\sum_{m=0}^{\infty} m!\,x^m = 1 + x + 2x^2 + 6x^3 + \cdots$$

we have $a_m = m!$, and in (7b),

$$\frac{a_{m+1}}{a_m} = \frac{(m+1)!}{m!} = m + 1 \to \infty \qquad \text{as } m \to \infty.$$

Thus this series converges only at the center $x = 0$. Such a series is useless. ∎

EXAMPLE 2 Case 2. The geometric series

For the **geometric series** we have

$$\frac{1}{1-x} = \sum_{m=0}^{\infty} x^m = 1 + x + x^2 + \cdots \qquad (|x| < 1).$$

In fact, $a_m = 1$ for all m, and from (7) we obtain $R = 1$, that is, the geometric series converges and represents $1/(1 - x)$ when $|x| < 1$. ∎

EXAMPLE 3 Case 3

In the case of the series

$$e^x = \sum_{m=0}^{\infty} \frac{x^m}{m!} = 1 + x + \frac{x^2}{2!} + \cdots$$

we have $a_m = 1/m!$. Hence in (7b),

$$\frac{a_{m+1}}{a_m} = \frac{1/(m+1)!}{1/m!} = \frac{1}{m+1} \to 0 \qquad \text{as } m \to \infty,$$

so that the series converges for all x. ∎

EXAMPLE 4 A hint for some of the problems

Find the radius of convergence of the series

$$\sum_{m=0}^{\infty} \frac{(-1)^m}{8^m} x^{3m} = 1 - \frac{x^3}{8} + \frac{x^6}{64} - \frac{x^9}{512} + - \cdots.$$

Solution. This is a series in powers of $t = x^3$ with coefficients $a_m = (-1)^m/8^m$, so that in (7b),

$$\left|\frac{a_{m+1}}{a_m}\right| = \frac{8^m}{8^{m+1}} = \frac{1}{8}.$$

Thus $R = 8$. Hence the series converges for $|t| < 8$, that is, $|x| < 2$. ∎

Operations on Power Series

In the power series method we differentiate, add, and multiply power series. These three operations are permissible, in the sense explained in what follows. We also list a condition about the vanishing of all coefficients of a power series, which is a basic tool of the power series method. (Proofs can be found in Sec. 14.3.)

Termwise differentiation

A power series may be differentiated term by term. More precisely: if

$$y(x) = \sum_{m=0}^{\infty} a_m(x - x_0)^m$$

converges for $|x - x_0| < R$, where $R > 0$, then the series obtained by differentiating term by term also converges for those x and represents the derivative y' of y for those x, that is,

$$y'(x) = \sum_{m=1}^{\infty} ma_m(x - x_0)^{m-1} \qquad (|x - x_0| < R).$$

Similarly,

$$y''(x) = \sum_{m=2}^{\infty} m(m - 1)a_m(x - x_0)^{m-2} \qquad (|x - x_0| < R),$$

and so on.

Termwise addition

Two power series may be added term by term. More precisely: if the series

$$\text{(8)} \qquad \sum_{m=0}^{\infty} a_m(x - x_0)^m \qquad \text{and} \qquad \sum_{m=0}^{\infty} b_m(x - x_0)^m$$

have positive radii of convergence and their sums are $f(x)$ and $g(x)$, then the series

$$\sum_{m=0}^{\infty} (a_m + b_m)(x - x_0)^m$$

converges and represents $f(x) + g(x)$ for each x that lies in the interior of the convergence interval of each of the given series.

Termwise multiplication

Two power series may be multiplied term by term. More precisely: Suppose that the series (8) have positive radii of convergence and let $f(x)$ and $g(x)$

be their sums. Then the series obtained by multiplying each term of the first series by each term of the second series and collecting like powers of $x - x_0$, that is,

$$\sum_{m=0}^{\infty} (a_0 b_m + a_1 b_{m-1} + \cdots + a_m b_0)(x - x_0)^m$$

$$= a_0 b_0 + (a_0 b_1 + a_1 b_0)(x - x_0) + (a_0 b_2 + a_1 b_1 + a_2 b_0)(x - x_0)^2 + \cdots$$

converges and represents $f(x)g(x)$ for each x in the interior of the convergence interval of each of the given series.

Vanishing of all coefficients

If a power series has a positive radius of convergence and a sum that is identically zero throughout its interval of convergence, then each coefficient of the series is zero.

Shifting summation indices

This is a technicality best explained in terms of typical examples. An index of summation is a dummy and can be changed; for example,

$$\sum_{m=1}^{\infty} \frac{3^m m^2}{m!} = \sum_{k=1}^{\infty} \frac{3^k k^2}{k!} = 3 + 18 + \frac{81}{2} + \cdots .$$

We can "shift" an index of summation; for example, if we set $m = s + 2$, then $s = m - 2$, and

$$\sum_{m=2}^{\infty} m(m - 1)a_m x^{m-2} = \sum_{s=0}^{\infty} (s + 2)(s + 1)a_{s+2} x^s$$

$$= 2a_2 + 6a_3 x + 12a_4 x^2 + \cdots .$$

This is needed, for instance, in writing the sum of two series as a single series. For instance, write

$$x^2 \sum_{m=2}^{\infty} m(m - 1)a_m x^{m-2} + 2 \sum_{m=1}^{\infty} m a_m x^{m-1}$$

$$= x^2(2a_2 + 6a_3 x + 12a_4 x^2 + \cdots) + 2(a_1 + 2a_2 x + 3a_3 x^2 + \cdots)$$

as a single series by first taking x^2 inside the summation, obtaining

$$\sum_{m=2}^{\infty} m(m - 1)x^m + \sum_{m=1}^{\infty} 2m a_m x^{m-1},$$

and then setting $m = s$ and $m - 1 = s$, respectively, obtaining

$$\sum_{s=2}^{\infty} s(s-1)a_s x^s + \sum_{s=0}^{\infty} 2(s+1)a_{s+1}x^s,$$

where $s = 2$ can be replaced by $s = 0$ (why?), so that we have

$$\sum_{s=0}^{\infty} [s(s-1)a_s + 2(s+1)a_{s+1}]x^s$$
$$= 2a_1 + 4a_2x + (2a_2 + 6a_3)x^2 + (6a_3 + 8a_4)x^3 + \cdots.$$

Existence of Power Series Solutions. Real Analytic Functions

The properties of power series just discussed form the foundation of the power series method. The remaining question is whether an equation has power series solutions at all. An answer is simple: If the coefficients p and q and the function r on the right side of

$$y'' + p(x)y' + q(x)y = r(x) \tag{9}$$

have power series representations, then (9) has power series solutions. The same is true if $\tilde{h}$, $\tilde{p}$, $\tilde{q}$, and $\tilde{r}$ in

$$\tilde{h}(x)y'' + \tilde{p}(x)y' + \tilde{q}(x)y = \tilde{r}(x) \tag{10}$$

have power series representations and $\tilde{h}(x_0) \neq 0$ (x_0 the center of the series). Almost all equations in practice have polynomials as coefficients (thus terminating power series), so that (when $r(x) \equiv 0$ or is a power series, too) those conditions are satisfied, except perhaps $\tilde{h}(x_0) \neq 0$, a fact that will keep us busy in later sections.

To formulate all of this in a precise and simple way, we use the following concept (which is of general interest).

Definition of real analytic function

A real function $f(x)$ is called **analytic** *at a point* $x = x_0$ if it can be represented by a power series in powers of $x - x_0$ with radius of convergence $R > 0$. ∎

With this we have

Theorem 1 (Existence of power series solutions)

If p, q, and r in (9) are analytic at $x = x_0$, then every solution of (9) is analytic at $x = x_0$ and can thus be represented by a power series in powers of $x - x_0$ with radius of convergence[3] $R > 0$.

Hence the same is true if $\tilde{h}$, $\tilde{p}$, $\tilde{q}$, and $\tilde{r}$ in (10) are analytic at $x = x_0$ and $\tilde{h}(x_0) \neq 0$.

---

[3] R is at least equal to the distance between the point $x = x_0$ and that point (or those points) closest to $x = x_0$ at which one of the functions p, q, r, *as functions of a complex variable*, is not analytic. (Note that that point may not lie on the x-axis, but somewhere in the complex plane.)

The proof of this theorem requires advanced methods of complex analysis and can be found in Ref. [A6] in Appendix 1.

Problem Set 5.2

Apply the power series method to the following differential equations.

1. $xy' = 3y + 3$
2. $(x - 3)y' - xy = 0$
3. $y' = 2xy$
4. $(1 - x^4)y' = 4x^3y$
5. $(x + 1)y' - (2x + 3)y = 0$
6. $y'' - y = x$
7. $y'' - 3y' + 2y = 0$
8. $y'' - 4xy' + (4x^2 - 2)y = 0$
9. $(1 - x^2)y'' - 2xy' + 2y = 0$
10. $y'' - xy' + y = 0$

11. Show that $y' = (y/x) + 1$ cannot be solved for y as a power series in x. Solve this equation for y as a power series in powers of $x - 1$. (*Hint.* Introduce $t = x - 1$ as a new independent variable and solve the resulting equation for y as a power series in t.) Compare the result with that obtained by the appropriate elementary method.

Solve for y as a power series in powers of $x - 1$:

12. $y' = ky$
13. $y'' + y = 0$
14. $y'' - y = 0$

Radius of convergence. Find the radius of convergence of the following series.

15. $\sum_{m=0}^{\infty} \frac{x^m}{3^m}$
16. $\sum_{m=0}^{\infty} (-1)^m x^{2m}$
17. $\sum_{m=0}^{\infty} \frac{x^{2m}}{(2m)!}$

18. $\sum_{m=2}^{\infty} \frac{m(m-1)}{3^m} x^m$
19. $\sum_{m=0}^{\infty} \frac{(-1)^m}{k^m} x^{2m}$
20. $\sum_{m=0}^{\infty} \frac{(-1)^m x^{2m+1}}{(2m+1)!}$

21. $\sum_{m=0}^{\infty} \left(\frac{7}{5}\right)^m x^{2m}$
22. $\sum_{m=0}^{\infty} \frac{(-1)^m}{5^m} (x-5)^m$
23. $\sum_{m=0}^{\infty} m^{2m} x^m$

24. $\sum_{m=1}^{\infty} \frac{1}{3^m m^2} (x+1)^m$
25. $\sum_{m=0}^{\infty} \frac{1}{2^m} (x-1)^{2m}$
26. $\sum_{m=0}^{\infty} \frac{(3m)!}{(m!)^3} x^m$

27. (Shift of summation index) Show that

$$\sum_{m=2}^{\infty} m(m-1)a_m x^{m-2} = \sum_{j=1}^{\infty} (j+1)ja_{j+1}x^{j-1} = \sum_{s=0}^{\infty} (s+2)(s+1)a_{s+2}x^s.$$

Shift of summation index. Shift the index so that the power under the summation sign is x^m. Check your result by writing the first few terms explicitly. Also determine the radius of convergence.

28. $\sum_{n=1}^{\infty} \frac{(-1)^{n+1}}{3n} x^{n+2}$
29. $\sum_{s=2}^{\infty} \frac{s(s+1)}{s^2+1} x^{s-1}$
30. $\sum_{k=3}^{\infty} \frac{(-1)^{k+1}}{6^k} x^{k-3}$

5.3 Legendre's Equation. Legendre Polynomials $P_n(x)$

The examples and problems considered so far were supposed to let us gain confidence in the power series method and to obtain skill in handling the method practically, but, of course, those equations could be solved by other methods (to permit comparison of results). We now consider the first "big" equation of physics, **Legendre's differential equation**[4]

$$(1 - x^2)y'' - 2xy' + n(n + 1)y = 0 \tag{1}$$

which arises in numerous problems, mostly in those exhibiting spherical symmetry (for instance, in electrostatics—take a quick look at the first few lines of Sec. 11.12). The parameter n in (1) is a given real number. The solutions of (1) are called **Legendre functions.** These are new functions; they belong to what is called "**special functions,**" meaning higher functions, as opposed to sine, cosine, logarithm, etc., known from calculus. (Further special functions occur in the next sections.)

We see that the coefficients of (1) are analytic at $x = 0$ (they are very special polynomials) and $\tilde{h}(x) = 1 - x^2 \neq 0$ at $x = 0$. Hence, by Theorem 1 of Sec. 5.2, every solution of (1) is some power series

$$y = \sum_{m=0}^{\infty} a_m x^m. \tag{2}$$

Substituting (2) and its derivatives into (1) and denoting the constant $n(n + 1)$ by k, we obtain

$$(1 - x^2) \sum_{m=2}^{\infty} m(m - 1)a_m x^{m-2} - 2x \sum_{m=1}^{\infty} ma_m x^{m-1} + k \sum_{m=0}^{\infty} a_m x^m = 0.$$

By writing the first expression as two separate series, we have the equation

$$\begin{aligned} \sum_{m=2}^{\infty} m(m - 1)a_m x^{m-2} &- \sum_{m=2}^{\infty} m(m - 1)a_m x^m \\ &- 2 \sum_{m=1}^{\infty} ma_m x^m + k \sum_{m=0}^{\infty} a_m x^m = 0, \end{aligned} \tag{1*}$$

[4]ADRIEN MARIE LEGENDRE (1752—1833), French mathematician, who became a professor in Paris in 1775 and made important contributions to special functions, elliptic integrals, number theory, and the calculus of variations. His book *Éléments de géométrie* (1794) became very famous and had 12 editions in less than 30 years.

Formulas on Legendre functions are contained in Refs. [1], [6], and [10].

written out

$$
\begin{aligned}
&2\cdot 1a_2 + 3\cdot 2a_3x + 4\cdot 3a_4x^2 + \cdots + (s+2)(s+1)a_{s+2}x^s + \cdots \\
&\qquad\qquad - 2\cdot 1a_2x^2 - \cdots \qquad\qquad - s(s-1)a_sx^s - \cdots \\
&\qquad - 2\cdot 1a_1x - 2\cdot 2a_2x^2 - \cdots \qquad\qquad - 2sa_sx^s - \cdots \\
&+ ka_0 + \quad ka_1x + \quad ka_2x^2 + \cdots \qquad\qquad + ka_sx^s + \cdots = 0.
\end{aligned}
$$

Since this must be an identity in x if (2) is to be a solution of (1), the sum of the coefficients of each power of x must be zero; since $k = n(n+1)$, this gives

$$
\text{(3a)}\qquad
\begin{aligned}
2a_2 + n(n+1)a_0 &= 0, &\quad &\text{coefficients of } x^0\\
6a_3 + [-2 + n(n+1)]a_1 &= 0, &\quad &\text{coefficients of } x^1
\end{aligned}
$$

and in general, when $s = 2, 3, \cdots$,

$$
\text{(3b)}\qquad (s+2)(s+1)a_{s+2} + [-s(s-1) - 2s + n(n+1)]a_s = 0.
$$

Now the expression in brackets $[\cdots]$ can be written $(n-s)(n+s+1)$, as the student may readily verify. We thus obtain from (3)

$$
\text{(4)}\qquad a_{s+2} = -\frac{(n-s)(n+s+1)}{(s+2)(s+1)}\,a_s \qquad (s = 0, 1, \cdots).
$$

This is called a **recurrence relation** or **recursion formula.** It gives each coefficient in terms of the second one preceding it, except for a_0 and a_1, which are left as arbitrary constants. We find successively

$$
\begin{aligned}
a_2 &= -\frac{n(n+1)}{2!}\,a_0 &\qquad a_3 &= -\frac{(n-1)(n+2)}{3!}\,a_1\\
a_4 &= -\frac{(n-2)(n+3)}{4\cdot 3}\,a_2 &\qquad a_5 &= -\frac{(n-3)(n+4)}{5\cdot 4}\,a_3\\
&= \frac{(n-2)n(n+1)(n+3)}{4!}\,a_0 &\qquad &= \frac{(n-3)(n-1)(n+2)(n+4)}{5!}\,a_1
\end{aligned}
$$

etc. By inserting these values for the coefficients into (2) we obtain

$$
\text{(5)}\qquad y(x) = a_0y_1(x) + a_1y_2(x)
$$

where

$$
\text{(6)}\qquad y_1(x) = 1 - \frac{n(n+1)}{2!}x^2 + \frac{(n-2)n(n+1)(n+3)}{4!}x^4 - + \cdots
$$

and

$$(7)\quad y_2(x) = x - \frac{(n-1)(n+2)}{3!}x^3 + \frac{(n-3)(n-1)(n+2)(n+4)}{5!}x^5 - + \cdots.$$

These series converge for $|x| < 1$ (see Prob. 5; or they may terminate, see below). Since (6) contains even powers of x only, while (7) contains odd powers of x only, the ratio y_1/y_2 is not a constant, so that y_1 and y_2 are not proportional and are thus linearly independent solutions. Hence (5) is a general solution of (1) on the interval $-1 < x < 1$.

When n is a nonnegative integer, then y_1 or y_2 reduces to a polynomial (and the other of the two solutions becomes a polynomial multiplied by $\ln [(1 + x)/(1 - x)]$, as can be shown). We illustrate this for the simplest cases.

EXAMPLE 1 **Legendre functions for $n = 0$**

When $n = 0$, then (6) gives $y_1(x) = 1$, the simplest "Legendre polynomial" (see below) and (7) gives

$$y_2(x) = x + \frac{2}{3!}x^3 + \frac{(-3)(-1)\cdot 2\cdot 4}{5!}x^5 + \cdots = x + \frac{x^3}{3} + \frac{x^5}{5} + \cdots = \frac{1}{2}\ln\frac{1+x}{1-x}.$$

The student may verify this by solving Legendre's equation $(1 - x^2)z' - 2xz = 0$, $z = y'$, by separating variables. ∎

EXAMPLE 2 **Legendre functions for $n = 1$**

When $n = 1$, then (7) gives $y_2(x) = x$, the next "Legendre polynomial," and (6) gives

$$\begin{aligned} y_1(x) &= 1 - \frac{x^2}{1} - \frac{x^4}{3} - \frac{x^6}{5} - \cdots \\ &= 1 - x\left(x + \frac{x^3}{3} + \frac{x^5}{5} + \cdots\right) = 1 - \frac{1}{2}x\ln\frac{1+x}{1-x}. \end{aligned}$$ ∎

Legendre Polynomials

In many applications the parameter n in Legendre's equation will be a nonnegative integer. Then the right side of (4) is zero when $s = n$, and, therefore, $a_{n+2} = 0$, $a_{n+4} = 0$, $a_{n+6} = 0, \cdots$. Hence, if n is even, $y_1(x)$ reduces to a polynomial of degree n. If n is odd, the same is true for $y_2(x)$. These polynomials, multiplied by some constants, are called **Legendre polynomials.** Since they are of great practical importance, let us consider them in more detail. For this purpose we solve (4) for a_s, obtaining

$$(8)\qquad a_s = -\frac{(s+2)(s+1)}{(n-s)(n+s+1)}a_{s+2} \qquad (s \leqq n-2)$$

and may then express all the nonvanishing coefficients in terms of the coefficient a_n of the highest power of x of the polynomial. The coefficient a_n is at first still arbitrary. It is customary to choose $a_n = 1$ when $n = 0$ and

$$(9)\qquad a_n = \frac{(2n)!}{2^n(n!)^2} = \frac{1\cdot 3\cdot 5\cdots(2n-1)}{n!}, \qquad n = 1, 2, \cdots.$$

The reason is that for this choice of a_n all those polynomials will have the value 1 when $x = 1$; this follows from (14) in Prob. 10. We then obtain from (8) and (9)

$$
\begin{aligned}
a_{n-2} &= -\frac{n(n-1)}{2(2n-1)} a_n = -\frac{n(n-1)(2n)!}{2(2n-1)2^n(n!)^2} \\
&= -\frac{n(n-1)2n(2n-1)(2n-2)!}{2(2n-1)2^n n(n-1)!\, n(n-1)(n-2)!},
\end{aligned}
\tag{9*}
$$

that is,

$$a_{n-2} = -\frac{(2n-2)!}{2^n(n-1)!\,(n-2)!}.$$

Similarly,

$$a_{n-4} = -\frac{(n-2)(n-3)}{4(2n-3)} a_{n-2} = \frac{(2n-4)!}{2^n 2!\,(n-2)!\,(n-4)!}$$

etc., and in general, when $n - 2m \geqq 0$,

$$a_{n-2m} = (-1)^m \frac{(2n-2m)!}{2^n m!\,(n-m)!\,(n-2m)!}. \tag{10}$$

The resulting solution of Legendre's differential equation (1) is called the **Legendre polynomial** *of degree n* and is denoted by $P_n(x)$. From (10) we obtain

$$
\begin{aligned}
P_n(x) &= \sum_{m=0}^{M} (-1)^m \frac{(2n-2m)!}{2^n m!\,(n-m)!\,(n-2m)!} x^{n-2m} \\
&= \frac{(2n)!}{2^n(n!)^2} x^n - \frac{(2n-2)!}{2^n 1!\,(n-1)!\,(n-2)!} x^{n-2} + - \cdots,
\end{aligned}
\tag{11}
$$

where $M = n/2$ or $(n-1)/2$, whichever is an integer. In particular (Fig. 82),

$$
\begin{aligned}
&P_0(x) = 1, && P_1(x) = x, \\
&P_2(x) = \tfrac{1}{2}(3x^2 - 1), && P_3(x) = \tfrac{1}{2}(5x^3 - 3x), \\
&P_4(x) = \tfrac{1}{8}(35x^4 - 30x^2 + 3), && P_5(x) = \tfrac{1}{8}(63x^5 - 70x^3 + 15x),
\end{aligned}
\tag{11'}
$$

etc.

The so-called ***orthogonality*** of the Legendre polynomials will be considered in Secs. 5.8 and 5.9.

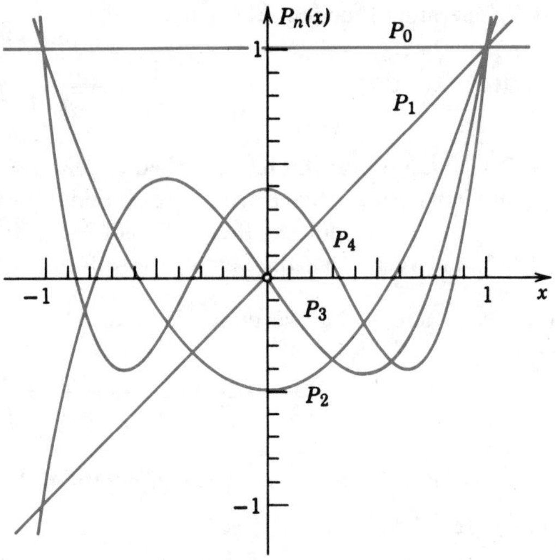

Fig. 82. Legendre polynomials

Problem Set 5.3

1. Using (11′), verify by substitution that $P_0, \cdots, P_5$ satisfy Legendre's equation.
2. Find and graph $P_6(x)$.
3. Derive (11′) from (11).
4. Show that we can get (3) from (1*) more quickly if we write $m - 2 = s$ in the first sum in (1*) and $m = s$ in the other sums, obtaining

$$\sum_{s=0}^{\infty} \{(s + 2)(s + 1)a_{s+2} - [s(s - 1) + 2s - k]a_s\}x^s = 0.$$

5. Show that for any n for which the series (6) or (7) do not reduce to a polynomial, these series have radius of convergence 1.
6. Solve (1) with $n = 0$ as indicated in Example 1.
7. Using (11), show that $P_n(-x) = (-1)^n P_n(x)$ and $P_n'(-x) = (-1)^{n+1}P_n'(x)$.
8. **(Rodrigues's formula[5])** Applying the binomial theorem to $(x^2 - 1)^n$, differentiating n times term by term, and comparing with (11), show that

$$P_n(x) = \frac{1}{2^n n!}\frac{d^n}{dx^n}[(x^2 - 1)^n] \qquad (12)$$

(*Rodrigues's formula*).

9. Using (12) and integrating n times by parts, show that

$$\int_{-1}^{1} P_n^2(x)\,dx = \frac{2}{2n + 1} \qquad (13)$$

$(n = 0, 1, \cdots)$.

[5]OLINDE RODRIGUES (1794—1851), French mathematician and economist.

10. (Generating function) Show that

$$\frac{1}{\sqrt{1 - 2xu + u^2}} = \sum_{n=0}^{\infty} P_n(x)u^n. \tag{14}$$

The function on the left is called a *generating function* of the Legendre polynomials. *Hint.* Start from the binomial expansion of $1/\sqrt{1 - v}$, then set $v = 2xu - u^2$, multiply the powers of $2xu - u^2$ out, collect all the terms involving u^n, and verify that the sum of these terms is $P_n(x)u^n$.

11. Let A_1 and A_2 be two points in space (Fig. 83, $r_2 > 0$). Using (14), show that

$$\frac{1}{r} = \frac{1}{\sqrt{r_1^2 + r_2^2 - 2r_1r_2 \cos \theta}} = \frac{1}{r_2} \sum_{m=0}^{\infty} P_m(\cos \theta) \left(\frac{r_1}{r_2}\right)^m.$$

This formula has applications in potential theory.

Using (14), show that

12. $P_n(1) = 1$

13. $P_n(-1) = (-1)^n$

14. $P_{2n+1}(0) = 0$

15. $P_{2n}(0) = (-1)^n \cdot 1 \cdot 3 \cdots (2n - 1)/2 \cdot 4 \cdots (2n)$.

16. (Bonnet's recursion[6]) Differentiating (14) with respect to u, using (14) in the resulting formula, and comparing coefficients of u^n, obtain the *Bonnet recursion*

$$(n + 1)P_{n+1}(x) = (2n + 1)xP_n(x) - nP_{n-1}(x), \qquad n = 1, 2, \cdots. \tag{15}$$

17. (Computation) Formula (15) is useful for computations, the loss of significant figures being small (except at zeros). Using (15), compute $P_2(2.6)$ and $P_3(2.6)$.

18. Using (15) and (11′), find P_6.

19. (Associated Legendre functions) Consider

$$(1 - x^2)y'' - 2xy' + \left[n(n + 1) - \frac{m^2}{1 - x^2}\right] y = 0.$$

Substituting $y(x) = (1 - x^2)^{m/2}u(x)$, show that u satisfies

$$(1 - x^2)u'' - 2(m + 1)xu' + [n(n + 1) - m(m + 1)]u = 0. \tag{16}$$

Starting from (1) and differentiating it m times, show that a solution of (16) is

$$u = \frac{d^mP_n}{dx^m}.$$

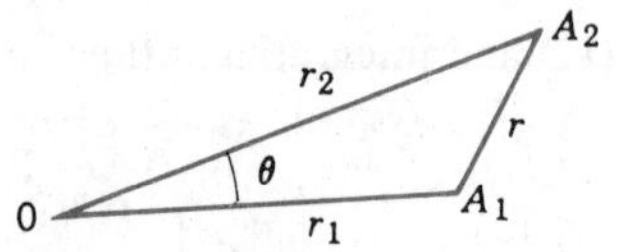

Fig. 83. Problem 11

[6]OSSIAN BONNET (1819—1892), French mathematician, whose main work was in the differential geometry of surfaces.

The corresponding $y(x)$ is denoted by $P_n{}^m(x)$. It is called an *associated Legendre function* and plays a role in quantum physics. Thus

$$P_n{}^m(x) = (1 - x^2)^{m/2} \frac{d^m P_n}{dx^m}.$$

20. Find $P_1{}^1(x)$, $P_2{}^1(x)$, $P_2{}^2(x)$, $P_4{}^2(x)$.

5.4 Frobenius Method

Just as the power series method, the **Frobenius**[7] **method** is a method for solving linear differential equations with variable coefficients; however, the Frobenius method applies to more general equations for which the power series method no longer works, some of them basic (Bessel's equation, in particular; see Secs. 5.5–5.7). Thus the Frobenius method has great practical importance.

We begin by explaining to what kind of equations the Frobenius method applies, as follows.

A point $x = x_0$ at which the coefficients p and q of an equation

$$y'' + p(x)y' + q(x)y = 0$$

are analytic (see Sec. 5.3) is called a **regular point** of the equation. Similarly, a **regular point** $x = x_0$ of

$$\tilde{h}(x)y'' + \tilde{p}(x)y' + \tilde{q}(x)y = 0$$

is one at which $\tilde{h}$, $\tilde{p}$, and $\tilde{q}$ are analytic and $\tilde{h}(x_0) \neq 0$ (so that by dividing by $\tilde{h}$ we can get the standard form with y'' as the first term and *analytic* $p = \tilde{p}/\tilde{h}$ and $q = \tilde{q}/\tilde{h}$). A point that is not a regular point is called a **singular point** of the equation.

For example, the famous Bessel equation (see Sec. 5.5)

$$x^2y'' + xy' + (x^2 - \nu^2)y = 0 \qquad (\nu \text{ a given number})$$

has $x = 0$ as a singular point, as we see. (Divide by x^2 if that helps.)

Now if x_0 is a regular point of an equation, the power series method takes care of the equation, as we know, and gives power series solutions in powers of $x - x_0$. However, if x_0 is a singular point, this is no longer the case, simply because the equation may not have a power series solution in powers of $x - x_0$. Fortunately, the behavior of the coefficients of an equation at a singular point is often "not too bad," but is such that the following theorem applies.

---

[7]GEORG FROBENIUS (1849—1917), German mathematician, who also made important contributions to the theory of matrices and groups.

Theorem 1 **(Frobenius method)**

Any differential equation of the form

$$y'' + \frac{b(x)}{x} y' + \frac{c(x)}{x^2} y = 0, \tag{1}$$

where the functions $b(x)$ *and* $c(x)$ *are analytic at* $x = 0$*, has at least one solution that can be represented in the form*

$$y(x) = x^r \sum_{m=0}^{\infty} a_m x^m = x^r(a_0 + a_1 x + a_2 x^2 + \cdots) \qquad (a_0 \neq 0) \tag{2}$$

where the exponent r may be any (real or complex) number (and r is chosen so that $a_0 \neq 0$*).*[8]

The equation also has a second solution (such that these two solutions are linearly independent) that may be similar to (2) *(with a different r and different coefficients) or may contain a logarithmic term* (details in Theorem 2, below).

For example, Bessel's equation

$$y'' + \frac{1}{x} y' + \left(\frac{x^2 - \nu^2}{x^2}\right) y = 0$$

is of the form (1) with $b(x) = 1$ and $c(x) = x^2 - \nu^2$ analytic at $x = 0$, so that the theorem applies.

The point is that in (2) we have a power series times a single power of x whose exponent r is not restricted to be a nonnegative integer. (The latter restriction would make the whole expression a power series, by definition; see footnote 1 in Sec. 5.1.)

The proof of the theorem requires advanced methods of complex analysis and can be found in Ref. [A6] listed in Appendix 1.

The subsequent method for solving (1) is based on this theorem and is called the **Frobenius method** or **extended power series method.**

Indicial Equation, Indicating the Form of Solutions

To solve (1), we write it in the somewhat more convenient form

$$x^2 y'' + x b(x) y' + c(x) y = 0. \tag{1'}$$

We first expand $b(x)$ and $c(x)$ in power series,

---

[8]In this theorem, we may replace x by $x - x_0$, where x_0 is any number. Note that the condition $a_0 \neq 0$ is no restriction of generality; it simply means that we factor out the highest possible power of x.

The singular point of (1) at $x = 0$ is sometimes called a **regular singular point,** a term, confusing to the student, which we shall not use.

$$b(x) = b_0 + b_1x + b_2x^2 + \cdots, \qquad c(x) + c_0 + c_1x + c_2x^2 + \cdots.$$

Then we differentiate (2) term by term, finding

$$y'(x) = \sum_{m=0}^{\infty} (m + r)a_m x^{m+r-1} = x^{r-1}[ra_0 + (r + 1)a_1x + \cdots],$$

$$\text{(2*)} \quad y''(x) = \sum_{m=0}^{\infty} (m + r)(m + r - 1)a_m x^{m+r-2}$$

$$= x^{r-2}[r(r - 1)a_0 + (r + 1)ra_1x + \cdots].$$

By inserting all these series into (1′) we readily obtain

$$\text{(3)} \quad \begin{aligned} x^r[r(r - 1)a_0 + \cdots] + (b_0 + b_1x + \cdots)x^r(ra_0 + \cdots) \\ + (c_0 + c_1x + \cdots)x^r(a_0 + a_1x + \cdots) = 0. \end{aligned}$$

We now equate the sum of the coefficients of each power of x to zero, as before. This yields a system of equations involving the unknown coefficients a_m. The smallest power is x^r, and the corresponding equation is

$$[r(r - 1) + b_0r + c_0]a_0 = 0.$$

Since by assumption $a_0 \neq 0$, the expression in the brackets must be zero. This gives

$$\text{(4)} \quad \boxed{r(r - 1) + b_0r + c_0 = 0.}$$

This important quadratic equation is called the **indicial equation** of the differential equation (1). Its role is as follows.

Our method will yield a basis of solutions. One of the two solutions will always be of the form (2), where r is a root of (4). The form of the other solution will be indicated by the indicial equation; depending on the roots, there are three possible cases, as we shall see. Two of them are expected by the Euler–Cauchy equation (Sec. 2.6). The third one is perhaps unexpected, but can also be seen from a simple equation. To see what to expect, let us first show examples.

EXAMPLE 1 **Case 1. Distinct roots not differing by an integer**

Substituting $y = x^r$ into the Euler–Cauchy equation

$$x^2y'' - \tfrac{1}{2}xy' + \tfrac{1}{2}y = 0$$

we get the auxiliary equation, which is the indicial equation,

$$r(r - 1) - \tfrac{1}{2}r + \tfrac{1}{2} = 0$$

with roots $r_1 = 1$ and $r_2 = \frac{1}{2}$. A basis is $y_1 = x$, $y_2 = \sqrt{x}$. ∎

EXAMPLE 2 Case 2. Double root

For the Euler–Cauchy equation

$$x^2y'' - xy' + y = 0$$

the indicial equation

$$r(r - 1) - r + 1 = (r - 1)^2 = 0$$

has a double root $r = r_1 = r_2 = 1$ and gives the basis $y_1 = x$, $y_2 = x \ln x$ (see Sec. 2.6). ■

EXAMPLE 3A

Case 3. Roots differing by an integer, no logarithmic term

For the Euler–Cauchy equation

$$x^2y'' + xy' - y = 0$$

the indicial equation

$$r(r - 1) + r - 1 = 0$$

has roots $r_1 = 1$ and $r_2 = -1$ and gives the basis $y_1 = x$, $y_2 = 1/x$. ■

EXAMPLE 3B

Case 3. Roots differing by an integer, logarithmic term

For the equation

$$(x^2 - x)y'' - xy' + y = 0$$

we can get the indicial equation by first casting the given equation into the form (1),

$$y'' + \frac{1}{x}\left(-\frac{x^2}{x^2 - x}\right)y' + \frac{1}{x^2}\left(\frac{x^2}{x^2 - x}\right)y = 0,$$

which shows that in (1),

$$b(x) = -c(x) = -\frac{x^2}{x^2 - x} = \frac{x}{1 - x} = x(1 + x + \cdots), \qquad \text{thus} \quad b_0 = c_0 = 0.$$

Hence (4) is $r(r - 1) = 0$ and its roots $r_1 = 1$, $r_2 = 0$ differ by an integer. The student may verify by substitution that a basis of solutions for all positive x is

$$y_1 = x, \qquad y_2 = x \ln x + 1.$$

A derivation of this follows in Example 7. ■

For a general equation (1) we need infinite series for solutions, not just a single term (as in the examples) or a few terms; and Theorem 2 now gives the general form of a basis of solutions in each case.[9]

Theorem 2 (Frobenius method. Basis of solutions)

Suppose that the differential equation (1) *satisfies the assumptions in Theorem* 1. *Let r_1 and r_2 be the roots of the indicial equation* (4). *Then we have the following three cases.*

---

[9] A general theory of convergence of the occurring series will not be presented here, but in each individual case convergence may be tested in the usual way.

Case 1. Distinct roots not differing by an integer.[10] *A basis is*

$$y_1(x) = x^{r_1}(a_0 + a_1x + a_2x^2 + \cdots) \tag{5}$$

and

$$y_2(x) = x^{r_2}(A_0 + A_1x + A_2x^2 + \cdots) \tag{6}$$

with coefficients obtained successively from (3) *with* $r = r_1$ *and* $r = r_2$, *respectively.*

Case 2. Double root $r_1 = r_2 = r$. *A basis is*

$$y_1(x) = x^r(a_0 + a_1x + a_2x^2 + \cdots) \qquad [r = \tfrac{1}{2}(1 - b_0)] \tag{7}$$

(*of the same general form as before*) *and*

$$y_2(x) = y_1(x)\ln x + x^r(A_1x + A_2x^2 + \cdots) \qquad (x > 0). \tag{8}$$

Case 3. Roots differing by an integer. *A basis is*

$$y_1(x) = x^{r_1}(a_0 + a_1x + a_2x^2 + \cdots) \tag{9}$$

(*of the same general form as before*) *and*

$$y_2(x) = ky_1(x)\ln x + x^{r_2}(A_0 + A_1x + A_2x^2 + \cdots), \tag{10}$$

where the roots are so denoted that $r_1 - r_2 > 0$ *and* k *may turn out to be zero.*

Proofs are given in Appendix 4. Note that in Case 2 we ***must*** have a logarithm, whereas in Case 3 we may or may not.

Typical Applications

Technically, the Frobenius method is similar to the power series method, once the roots of the indicial equation have been determined. However, (5)–(10) merely indicate the general form of a basis, and Examples 6 and 7 (below) show that one often has more convenient ways of determining solutions.

EXAMPLE 4 Euler–Cauchy equation

For the Euler–Cauchy equation

$$x^2y'' + b_0xy' + c_0y = 0 \qquad (b_0, c_0 \text{ constant})$$

[10] Note that this case includes complex conjugate roots r_1 and $r_2 = \bar{r}_1$, because we have $r_1 - r_2 = r_1 - \bar{r}_1 = 2i \operatorname{Im} r_1$, which is imaginary, hence cannot be a *real* integer.

substitution of $y = x^r$ gives the auxiliary equation

$$r(r-1) + b_0 r + c_0 = 0,$$

which is the indicial equation [and $y = x^r$ is a very special form of (2)!]. For different roots r_1, r_2 we get a basis $y_1 = x^{r_1}$, $y_2 = x^{r_2}$, and for a double root r we get a basis x^r, $x^r \ln x$. Accordingly, for this simple equation, Case 3 without a logarithmic term plays no extra role, and Case 3 with a logarithmic term is not possible. ∎

EXAMPLE 5 An equation that leads to Case 1

Solve the differential equation

$$y'' + \frac{1}{2x}y' + \frac{1}{4x}y = 0.$$

Solution. This equation satisfies the assumption of Theorems 1 and 2, so that the Frobenius method applies. We write the equation more conveniently as

$$4xy'' + 2y' + y = 0.$$

Substitution of (2) and of its derivatives (2*) gives

$$4\sum_{m=0}^{\infty}(m+r)(m+r-1)a_m x^{m+r-1} + 2\sum_{m=0}^{\infty}(m+r)a_m x^{m+r-1} + \sum_{m=0}^{\infty} a_m x^{m+r} = 0. \tag{11}$$

If you feel more comfortable by writing this out, go ahead:

$$\begin{aligned} &4r(r-1)a_0x^{r-1} + 4(r+1)ra_1x^r + 4(r+2)(r+1)a_2x^{r+1} + \cdots \\ &+ 2ra_0x^{r-1} + 2(r+1)\,a_1x^r + \qquad 2(r+2)\,a_2x^{r+1} + \cdots \\ &+ \qquad\qquad a_0x^r + \qquad\qquad a_1x^{r+1} + \cdots = 0. \end{aligned}$$

By equating the sum of the coefficients of x^{r-1} to zero we obtain the indicial equation

$$4r(r-1) + 2r = 0, \qquad \text{thus} \qquad r^2 - \tfrac{1}{2}r = 0.$$

The roots are $r_1 = \frac{1}{2}$ and $r_2 = 0$. This is Case 1.

By equating the sum of the coefficients of x^{r+s} in (11) to zero we obtain (take $m + r - 1 = r + s$, thus $m = s + 1$ in the first two series and $m = s$ in the last series)

$$4(s+r+1)(s+r)a_{s+1} + 2(s+r+1)a_{s+1} + a_s = 0.$$

By simplification we find that this can be written

$$4(s+r+1)(s+r+\tfrac{1}{2})a_{s+1} + a_s = 0.$$

We solve this for a_{s+1} in terms of a_s:

$$a_{s+1} = -\frac{a_s}{(2s+2r+2)(2s+2r+1)} \qquad (s = 0, 1, \cdots). \tag{12}$$

First solution. We determine a first solution $y_1(x)$ corresponding to $r_1 = \frac{1}{2}$. For $r = r_1$, formula (12) becomes

$$a_{s+1} = -\frac{a_s}{(2s+3)(2s+2)} \qquad (s = 0, 1, \cdots).$$

From this we get successively

$$a_1 = -\frac{a_0}{3\cdot 2}, \qquad a_2 = -\frac{a_1}{5\cdot 4}, \qquad a_3 = -\frac{a_2}{7\cdot 6}, \qquad \text{etc.}$$

In many practical situations an explicit formula for a_m will be rather complicated. Here it is simple: by successive substitution we get

$$a_1 = -\frac{a_0}{3!}, \qquad a_2 = \frac{a_0}{5!}, \qquad a_3 = -\frac{a_0}{7!}, \qquad \cdots$$

and in general, taking $a_0 = 1$,

$$a_m = \frac{(-1)^m}{(2m+1)!} \qquad (m = 0, 1, \cdots).$$

Hence the first solution is

$$y_1(x) = x^{1/2} \sum_{m=0}^{\infty} \frac{(-1)^m}{(2m+1)!} x^m = \sqrt{x}\left(1 - \frac{1}{6}x + \frac{1}{120}x^2 - + \cdots\right).$$

Second solution. If you recognize y_1 as a familiar function, apply reduction of order (see Sec. 2.7). If not, start from (6) with $r_2 = 0$. For $r = r_2 = 0$, formula (12) [with A_{s+1} and A_s instead of a_{s+1} and a_s] becomes

$$A_{s+1} = -\frac{A_s}{(2s+2)(2s+1)} \qquad (s = 0, 1, \cdots).$$

From this we get successively

$$A_1 = -\frac{A_0}{2 \cdot 1}, \qquad A_2 = -\frac{A_1}{4 \cdot 3}, \qquad A_3 = -\frac{A_2}{6 \cdot 5},$$

and by successive substitution we have

$$A_1 = -\frac{A_0}{2!}, \qquad A_2 = \frac{A_0}{4!}, \qquad A_3 = -\frac{A_0}{6!}, \qquad \cdots$$

and in general, taking $A_0 = 1$,

$$A_m = \frac{(-1)^m}{(2m)!}.$$

Hence the second solution, of the form (6) with $r_2 = 0$, is

$$y_2(x) = \sum_{m=0}^{\infty} \frac{(-1)^m}{(2m)!} x^m = 1 - \tfrac{1}{2}x + \tfrac{1}{24}x^2 - + \cdots.$$

y_1 and y_2 are linearly independent on the positive x-axis since their quotient is not constant. Hence these solutions form a basis for all $x > 0$. ∎

EXAMPLE 6 Illustration of Case 2 (Double root)

Solve the differential equation

$$x(x-1)y'' + (3x-1)y' + y = 0. \tag{13}$$

(This is a special hypergeometric differential equation, as we shall see in the problem set.)

Solution. Writing this equation in the standard form (1), we see that it satisfies the assumptions in Theorem 1. By inserting (2) and its derivatives (2*) into (13) we obtain

$$\begin{aligned}\sum_{m=0}^{\infty} (m+r)(m+r-1)a_m x^{m+r} - \sum_{m=0}^{\infty} (m+r)(m+r-1)a_m x^{m+r-1} \\ + 3\sum_{m=0}^{\infty} (m+r)a_m x^{m+r} - \sum_{m=0}^{\infty} (m+r)a_m x^{m+r-1} + \sum_{m=0}^{\infty} a_m x^{m+r} = 0.\end{aligned} \tag{14}$$

The smallest power is x^{r-1}; by equating the sum of its coefficients to zero we have

$$[-r(r-1)-r]a_0 = 0, \qquad \text{thus} \qquad r^2 = 0.$$

Hence this indicial equation has the double root $r = 0$.

First solution. We insert this value into (14) and equate the sum of the coefficients of the power x^s to zero, finding

$$s(s-1)a_s - (s+1)sa_{s+1} + 3sa_s - (s+1)a_{s+1} + a_s = 0,$$

thus $a_{s+1} = a_s$. Hence $a_0 = a_1 = a_2 = \cdots$, and by choosing $a_0 = 1$ we obtain the solution

$$y_1(x) = \sum_{m=0}^{\infty} x^m = \frac{1}{1-x}.$$

Second solution. We get a second independent solution y_2 by the method of reduction of order (Sec. 2.7), substituting $y_2 = uy_1$ and its derivatives into the equation. This leads to (10), Sec. 2.7, which we shall use in this example, instead of starting from scratch (as we shall do in the next example). In (10) of Sec. 2.7 we have $p = (3x-1)/(x^2-x)$, the coefficient of y' in (13) ***in standard form.*** By partial fractions,

$$-\int p\,dx = -\int \frac{3x-1}{x(x-1)}\,dx = -\int\left(\frac{2}{x-1}+\frac{1}{x}\right)dx = -2\ln(x-1) - \ln x.$$

Hence (10), Sec. 2.7, becomes

$$u' = U = y_1^{-2}e^{-\int p\,dx} = \frac{(x-1)^2}{(x-1)^2 x}, \qquad u = \ln x, \qquad y_2 = uy_1 = \frac{\ln x}{1-x}.$$

y_1 and y_2 are linearly independent and thus form a basis on the interval $0 < x < 1$ as well as on $1 < x < \infty$. ∎

EXAMPLE 7 Case 3, second solution with logarithmic term

Solve

$$(x^2 - x)y'' - xy' + y = 0. \tag{15}$$

Solution. The answer was given in Example 3B, and we now show how to get it. Substituting (2) and (2*) into (15), we have

$$(x^2-x)\sum_{m=0}^{\infty}(m+r)(m+r-1)a_m x^{m+r-2} - x\sum_{m=0}^{\infty}(m+r)a_m x^{m+r-1} + \sum_{m=0}^{\infty} a_m x^{m+r} = 0.$$

We now take x^2, x and x inside the summations and collect all terms with power x^{m+r} and simplify algebraically, obtaining

$$\sum_{m=0}^{\infty}(m+r-1)^2 a_m x^{m+r} - \sum_{m=0}^{\infty}(m+r)(m+r-1)a_m x^{m+r-1} = 0.$$

In the first series we set $m = s$ and in the second $m = s+1$, thus $s = m-1$. Then

$$\sum_{s=0}^{\infty}(s+r-1)^2 a_s x^{s+r} - \sum_{s=-1}^{\infty}(s+r+1)(s+r)a_{s+1}x^{s+r} = 0. \tag{16}$$

The lowest power is x^{r-1} (take $s = -1$ in the second series) and gives the indicial equation

$$r(r-1) = 0.$$

The roots are $r_1 = 1$ and $r_2 = 0$. They differ by an integer. This is Case 3.

First solution. From (16) with $r = r_1 = 1$ we have

$$\sum_{s=0}^{\infty} [s^2 a_s - (s + 2)(s + 1)a_{s+1}]x^{s+1} = 0.$$

This gives the recurrence relation

$$a_{s+1} = \frac{s^2}{(s + 2)(s + 1)} a_s \qquad (s = 0, 1, \cdots).$$

Hence $a_1 = 0$, $a_2 = 0$, $\cdots$ successively. Taking $a_0 = 1$, we get as a first solution $y_1 = x^{r_1}a_0 = x$.

Second solution. Applying reduction of order (Sec. 2.7), we substitute $y_2 = y_1 u = xu$, $y_2' = xu' + u$ and $y_2'' = xu'' + 2u'$ into the equation, obtaining

$$(x^2 - x)(xu'' + 2u') - x(xu' + u) + xu = 0.$$

xu drops out. Division by x and simplification gives

$$(x^2 - x)u'' + (x - 2)u' = 0.$$

From this, using partial fractions and integrating, we get

$$\frac{u''}{u'} = -\frac{x - 2}{x^2 - x} = -\frac{2}{x} + \frac{1}{x - 1}, \qquad \ln u' = \ln \frac{x - 1}{x^2}.$$

Taking exponents and integrating (again taking the integration constant zero), we obtain

$$u' = \frac{x - 1}{x^2} = \frac{1}{x} - \frac{1}{x^2}, \qquad u = \ln x + \frac{1}{x}, \qquad y_2 = xu = x \ln x + 1.$$

y_1 and y_2 are linearly independent, and y_2 has a logarithmic term. ∎

The Frobenius method solves the **hypergeometric equation,** whose solutions include many known functions as special cases (see the problem set). In the next section we use the method for solving Bessel's equation.

Problem Set 5.4

Find a basis of solutions of the following differential equations. Try to identify the series obtained by the Frobenius method as expansions of known functions.

1. $xy'' + 2y' + xy = 0$

2. $xy'' + 2y' + 4xy = 0$

3. $x^2y'' + 6xy' + (6 - x^2)y = 0$

4. $16x^2y'' + 3y = 0$

5. $x^2y'' + 6xy' + (x^2 + 6)y = 0$

6. $(x + 2)^2y'' + (x + 2)y' - y = 0$

7. $2x^2y'' + xy' - 3y = 0$

8. $x^2y'' + x^3y' + (x^2 - 2)y = 0$

9. $x^2y'' + 4xy' + (x^2 + 2)y = 0$

10. $(x + 1)^2y'' + (x + 1)y' - y = 0$

11. $xy'' + (1 - 2x)y' + (x - 1)y = 0$

12. $xy'' + (2 - 2x)y' + (x - 2)y = 0$

13. $xy'' + 3y' + 4x^3y = 0$

14. $x^2y'' + xy' + (x^2 - \frac{1}{4})y = 0$

15. $(x - 1)^2y'' + (x - 1)y' - 4y = 0$

16. $xy'' + y' - xy = 0$

17. $(1 + x)x^2y'' - (1 + 2x)xy' + (1 + 2x)y = 0$

18. $(1 + \frac{1}{2}x)x^2y'' - (1 + x)xy' + (1 + x)y = 0$

19. $2x(x - 1)y'' - (4x^2 - 3x + 1)y' + (2x^2 - x + 2)y = 0$

20. $(x^2 - 1)x^2y'' - (x^2 + 1)xy' + (x^2 + 1)y = 0$

Hypergeometric equation, hypergeometric series, hypergeometric functions

21. Gauss's hypergeometric differential equation[11] is

$$x(1 - x)y'' + [c - (a + b + 1)x]y' - aby = 0 \tag{17}$$

where a, b, c are constants. Show that the corresponding indicial equation has the roots $r_1 = 0$ and $r_2 = 1 - c$. Show that for $r_1 = 0$ the Frobenius method yields

$$y_1(x) = 1 + \frac{ab}{1!\,c}x + \frac{a(a+1)b(b+1)}{2!\,c(c+1)}x^2 + \frac{a(a+1)(a+2)b(b+1)(b+2)}{3!\,c(c+1)(c+2)}x^3 + \cdots \tag{18}$$

where $c \neq 0, -1, -2, \cdots$. This series is called the **hypergeometric series.** Its sum $y_1(x)$ is denoted by $F(a, b, c; x)$ and is called the **hypergeometric function.**

Using (18), prove:

22. The series (18) converges for $|x| < 1$.

23. $F(1, b, b; x) = 1 + x + x^2 + \cdots$, the geometric series.

24. If a or b is a negative integer, (18) reduces to a polynomial.

25. $\dfrac{dF(a, b, c; x)}{dx} = \dfrac{ab}{c}F(a + 1, b + 1, c + 1; x)$,

$\dfrac{d^2F(a, b, c; x)}{dx^2} = \dfrac{a(a + 1)b(b + 1)}{c(c + 1)}F(a + 2, b + 2, c + 2; x)$, etc.

Many elementary functions are special cases of $F(a, b, c; x)$. Prove

26. $\dfrac{1}{1 - x} = F(1, 1, 1; x) = F(1, b, b; x) = F(a, 1, a; x)$

27. $(1 + x)^n = F(-n, b, b; -x)$, $\quad (1 - x)^n = 1 - nxF(1 - n, 1, 2; x)$

28. $\arctan x = xF(\tfrac{1}{2}, 1, \tfrac{3}{2}; -x^2)$, $\quad \arcsin x = xF(\tfrac{1}{2}, \tfrac{1}{2}, \tfrac{3}{2}; x^2)$

29. $\ln(1 + x) = xF(1, 1, 2; -x)$, $\quad \ln\dfrac{1 + x}{1 - x} = 2xF(\tfrac{1}{2}, 1, \tfrac{3}{2}; x^2)$

30. (Second solution) Show that for $r_2 = 1 - c$ in Prob. 21 the Frobenius method yields the following solution (where $c \neq 2, 3, 4, \cdots$):

$$y_2(x) = x^{1-c}\left(1 + \frac{(a - c + 1)(b - c + 1)}{1!(-c + 2)}x + \frac{(a - c + 1)(a - c + 2)(b - c + 1)(b - c + 2)}{2!(-c + 2)(-c + 3)}x^2 + \cdots\right) \tag{19}$$

31. Show that in Prob. 30,

$$y_2(x) = x^{1-c}F(a - c + 1, b - c + 1, 2 - c; x).$$

---

[11] CARL FRIEDRICH GAUSS (1777—1855), great German mathematician. He already made the first of his great discoveries as a student at Helmstedt and Göttingen. In 1807 he became a professor and director of the Observatory at Göttingen. His work was of basic importance in algebra, number theory, differential equations, differential geometry, non-Euclidean geometry, complex analysis, numerical analysis, astronomy, geodesy, electromagnetism, and theoretical mechanics. He also paved the way for a general and systematic use of complex numbers.

32. Show that if $c \neq 0, \pm 1, \pm 2, \cdots$, the functions (18) and (19) constitute a basis of solutions of (17).

33. Consider the differential equation

$$(t^2 + At + B)\ddot{y} + (Ct + D)\dot{y} + Ky = 0 \tag{20}$$

where A, B, C, D, K are constants, $\dot{y} = dy/dt$, and $t^2 + At + B$ has distinct zeros t_1 and t_2. Show that by introducing the new independent variable

$$x = \frac{t - t_1}{t_2 - t_1}$$

the equation (20) becomes the hypergeometric equation, where the parameters are related by $Ct_1 + D = -c(t_2 - t_1)$, $C = a + b + 1$, $K = ab$.

Solve the following equations in terms of hypergeometric functions.

34. $x(1 - x)y'' + (3 - 5x)y' - 4y = 0$

35. $8x(1 - x)y'' + (4 - 14x)y' - y = 0$

36. $x(1 - x)y'' + (\frac{1}{2} + 2x)y' - 2y = 0$

37. $5x(1 - x)y'' + (4 - x)y' + y = 0$

38. $4x(1 - x)y'' + (1 - 10x)y' - 2y = 0$

39. $4x(1 - x)y'' + y' + 8y = 0$

40. $2x(1 - x)y'' + (7 + 2x)y' - 2y = 0$

41. $4x(1 - x)y'' + (2 + 12x)y' - 15y = 0$

42. $3x(1 - x)y'' + (1 - 5x)y' + y = 0$

43. $4(t^2 - 3t + 2)\ddot{y} - 2\dot{y} + y = 0$

44. $2(t^2 - 5t + 6)\ddot{y} + (2t - 3)\dot{y} - 8y = 0$

45. $3t(1 + t)\ddot{y} + t\dot{y} - y = 0$

5.5 Bessel's Equation. Bessel Functions $J_\nu(x)$

In this section we use the Frobenius method for solving one of the most important equations in applied mathematics, **Bessel's differential equation**[12]

$$x^2y'' + xy' + (x^2 - \nu^2)y = 0 \tag{1}$$

or, in standard form,

$$y'' + \frac{1}{x}y' + \left(1 - \frac{\nu^2}{x^2}\right)y = 0. \tag{1*}$$

[12]FRIEDRICH WILHELM BESSEL (1784—1846), German astronomer and mathematician, started out as an apprentice of a trade company, studying astronomy on his own in his spare time, later became an assistant at a small private observatory, and finally director of the new Königsberg observatory. His paper on the Bessel functions (dated 1824) appeared in 1826.

Formulas are contained in Refs. [1], [6], [10], and the standard treatise [A9].

Here, the parameter ν is a given nonnegative real number. This equation appears in problems on vibrations, electric fields, heat conduction, fluid flow, etc., in most cases when the problem shows cylindrical symmetry (just as problems exhibiting *spherical* symmetry may lead to Legendre's equation). If you wish, take a quick look at Sec. 11.10, where vibrations of a circular membrane lead to Bessel's equation.

Formula (1*) shows that Bessel's equation is of the type characterized in Theorem 1, Sec. 5.4. Hence we can solve it by the Frobenius method (Sec. 5.4), substituting a series of the form

$$y(x) = \sum_{m=0}^{\infty} a_m x^{m+r} \qquad (a_0 \neq 0). \tag{2}$$

with undetermined coefficients and its derivatives into Bessel's equation (1). This gives

$$\sum_{m=0}^{\infty} (m+r)(m+r-1)a_m x^{m+r} + \sum_{m=0}^{\infty} (m+r)a_m x^{m+r}$$
$$+ \sum_{m=0}^{\infty} a_m x^{m+r+2} - \nu^2 \sum_{m=0}^{\infty} a_m x^{m+r} = 0.$$

We equate the sum of the coefficients of x^{s+r} to zero. Note that this power x^{s+r} corresponds to $m = s$ in the first, second, and fourth series, and to $m = s - 2$ in the third series. Hence for $s = 0$ and $s = 1$, the third series does not contribute since $m \geqq 0$. For $s = 2, 3, \cdots$ all four series contribute, so that we get a general formula for all these s. We find

(3)
$$\text{(a)} \quad r(r-1)a_0 + ra_0 - \nu^2 a_0 = 0 \qquad (s = 0)$$
$$\text{(b)} \quad (r+1)ra_1 + (r+1)a_1 - \nu^2 a_1 = 0 \qquad (s = 1)$$
$$\text{(c)} \quad (s+r)(s+r-1)a_s + (s+r)a_s + a_{s-2} - \nu^2 a_s = 0 \qquad (s = 2, 3, \cdots).$$

From (3a) we obtain the **indicial equation**

$$(r+\nu)(r-\nu) = 0. \tag{4}$$

The roots are $r_1 = \nu \ (\geqq 0)$ and $r_2 = -\nu$.

Coefficient Recurrence in the Case of $r = r_1 = \nu$

For $r = r_1 = \nu$, equation (3b) yields $a_1 = 0$. Equation (3c) may be written

$$(s + r + \nu)(s + r - \nu)a_s + a_{s-2} = 0,$$

and for $r = \nu$ this takes the form

$$(s + 2\nu)sa_s + a_{s-2} = 0. \tag{5}$$

Since $a_1 = 0$ and $\nu \geqq 0$, it follows that $a_3 = 0, a_5 = 0, \cdots$, successively. If we set $s = 2m$ in (5), we obtain for the other coefficients

$$a_{2m} = -\frac{1}{2^2 m(\nu + m)} a_{2m-2}, \qquad m = 1, 2, \cdots, \tag{6}$$

and may determine these coefficients $a_2, a_4, \cdots$ successively. This gives

$$a_2 = -\frac{a_0}{2^2(\nu + 1)}$$

$$a_4 = -\frac{a_2}{2^2 2(\nu + 2)} = \frac{a_0}{2^4 2!\,(\nu + 1)(\nu + 2)}$$

and so on, and in general

$$a_{2m} = \frac{(-1)^m a_0}{2^{2m} m!\,(\nu + 1)(\nu + 2) \cdots (\nu + m)}, \qquad m = 1, 2, \cdots. \tag{7}$$

Bessel Function $J_n(x)$ for Integer $\nu = n$

Integer values of ν are denoted by n. This is standard. For $\nu = n$ the relation (7) becomes

$$a_{2m} = \frac{(-1)^m a_0}{2^{2m} m!\,(n + 1)(n + 2) \cdots (n + m)}, \qquad m = 1, 2, \cdots. \tag{8}$$

a_0 is still arbitrary, so that the series (2) with these coefficients would contain this arbitrary factor a_0, a highly impractical situation for developing formulas or computing values of this new function. Accordingly, we have to make a choice: $a_0 = 1$ would be possible, but more practical is

$$a_0 = \frac{1}{2^n n!} \tag{9}$$

because then $n!\,(n + 1) \cdots (n + m) = (m + n)!$ in (8), so that

$$a_{2m} = \frac{(-1)^m}{2^{2m+n} m!(n + m)!}, \qquad m = 1, 2, \cdots. \tag{10}$$

With these coefficients and $r_1 = \nu = n$ we get from (2) a particular solution of (1), denoted by $J_n(x)$, given by

$$J_n(x) = x^n \sum_{m=0}^{\infty} \frac{(-1)^m x^{2m}}{2^{2m+n} m!\,(n + m)!} \tag{11}$$

and called the **Bessel function of the first kind** *of order n*. This series converges for all x, as the ratio test shows, and in fact converges very rapidly because of the factorials in the denominator.

EXAMPLE 1 **Bessel functions $J_0(x)$ and $J_1(x)$**

For $n = 0$ we obtain from (11) the Bessel function of order 0

$$J_0(x) = \sum_{m=0}^{\infty} \frac{(-1)^m x^{2m}}{2^{2m}(m!)^2} = 1 - \frac{x^2}{2^2(1!)^2} + \frac{x^4}{2^4(2!)^2} - \frac{x^6}{2^6(3!)^2} + - \cdots, \tag{12}$$

which looks similar to a cosine (Fig. 84). For $n = 1$ we obtain the Bessel function of order 1

$$J_1(x) = \sum_{m=0}^{\infty} \frac{(-1)^m x^{2m+1}}{2^{2m+1} m!\,(m+1)!} = \frac{x}{2} - \frac{x^3}{2^3 1!\,2!} + \frac{x^5}{2^5 2!\,3!} - \frac{x^7}{2^7 3!\,4!} + - \cdots, \tag{13}$$

which looks similar to a sine (Fig. 84). But the zeros of these functions are not completely regularly spaced (see also Table A1 in Appendix 5) and the height of the "waves" decreases with increasing x. Heuristically, n^2/x^2 in (1*) is zero (if $n = 0$) or small in absolute value for large x, and so is y'/x, so that then, Bessel's equation comes close to $y'' + y = 0$, the equation of $\cos x$ and $\sin x$; also, y'/x acts as a "damping term," in part responsible for the decrease in height. One can show that for large x,

$$J_n(x) \approx \sqrt{\frac{2}{\pi x}} \cos\left(x - \frac{n\pi}{2} - \frac{\pi}{4}\right). \tag{14}$$

∎

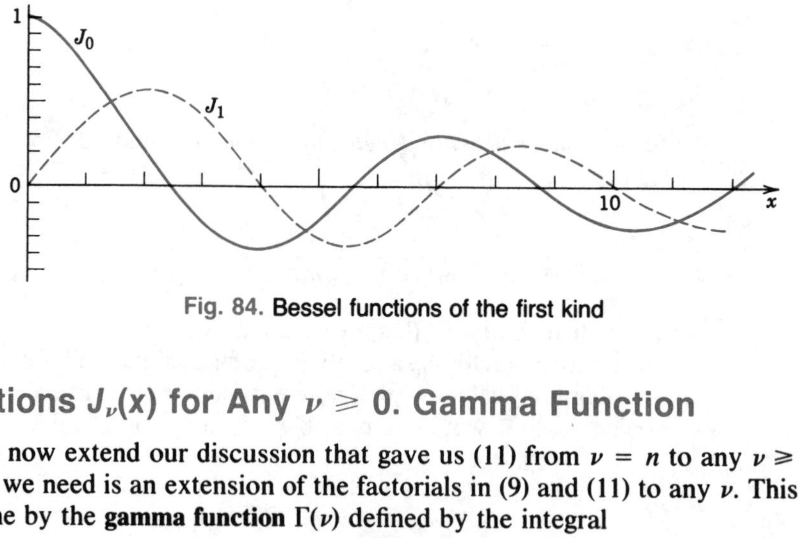

Fig. 84. Bessel functions of the first kind

Bessel Functions $J_\nu(x)$ for Any $\nu \geqq 0$. Gamma Function

We now extend our discussion that gave us (11) from $\nu = n$ to any $\nu \geqq 0$. All we need is an extension of the factorials in (9) and (11) to any ν. This is done by the **gamma function** $\Gamma(\nu)$ defined by the integral

$$\Gamma(\nu) = \int_0^\infty e^{-t} t^{\nu-1}\, dt \qquad (\nu > 0). \tag{15}$$

By integration by parts we obtain

$$\Gamma(\nu + 1) = \int_0^\infty e^{-t} t^{\nu}\, dt = -e^{-t} t^{\nu} \Big|_0^\infty + \nu \int_0^\infty e^{-t} t^{\nu-1}\, dt.$$

The first expression on the right is zero, and the integral on the right is $\Gamma(\nu)$. This yields the basic relation

$$\Gamma(\nu + 1) = \nu\Gamma(\nu). \tag{16}$$

Since

$$\Gamma(1) = \int_0^\infty e^{-t}\, dt = 1,$$

we conclude from (16) that

$$\Gamma(2) = \Gamma(1) = 1!, \qquad \Gamma(3) = 2\Gamma(2) = 2!, \cdots$$

and in general

$$\Gamma(n + 1) = n! \qquad (n = 0, 1, \cdots). \tag{17}$$

This shows that the gamma function does in fact generalize the factorial function known from calculus.

Now in (9) we had $a_0 = 1/2^n n!$, which is $1/2^n\Gamma(n + 1)$ by (17) and suggests to choose, for any ν,

$$a_0 = \frac{1}{2^\nu \Gamma(\nu + 1)}. \tag{18}$$

Then (7) becomes

$$a_{2m} = \frac{(-1)^m}{2^{2m+\nu} m!\, (\nu + 1)(\nu + 2) \cdots (\nu + m)\Gamma(\nu + 1)}.$$

But (16) gives in the denominator

$$(\nu + 1)\Gamma(\nu + 1) = \Gamma(\nu + 2), \qquad (\nu + 2)\Gamma(\nu + 2) = \Gamma(\nu + 3)$$

and so on, so that

$$(\nu + 1)(\nu + 2) \cdots (\nu + m)\Gamma(\nu + 1) = \Gamma(\nu + m + 1).$$

Hence the coefficients are

$$a_{2m} = \frac{(-1)^m}{2^{2m+\nu} m!\, \Gamma(\nu + m + 1)}. \tag{19}$$

With these coefficients and $r = r_1 = \nu$ we get from (2) a particular solution of (1), denoted by $J_\nu(x)$, given by

$$J_\nu(x) = x^\nu \sum_{m=0}^{\infty} \frac{(-1)^m x^{2m}}{2^{2m+\nu} m!\, \Gamma(\nu + m + 1)} \tag{20}$$

and called the **Bessel function of the first kind of order ν.** This series converges for all x, as one can verify by the ratio test. (More formulas for the gamma function are listed in Appendix 3, but were not needed here.)

Solution $J_{-\nu}$ of the Bessel Equation

So far we have one solution, J_ν, and our further work will be motivated by the goal of obtaining a general solution of the Bessel equation. For this we must derive a second independent solution. If ν is not an integer, this will be easy. If ν is an integer n, this will need more effort, as we show in Sec. 5.7. The details are as follows.

Replacing ν by $-\nu$ in (20), we have

$$J_{-\nu}(x) = x^{-\nu} \sum_{m=0}^{\infty} \frac{(-1)^m x^{2m}}{2^{2m-\nu} m!\, \Gamma(m - \nu + 1)} . \tag{21}$$

Since Bessel's equation involves ν^2, the functions J_ν and $J_{-\nu}$ are solutions of the equation for the same ν. If ν is not an integer, they are linearly independent, because the first term in (20) and the first term in (21) are finite nonzero multiples of x^ν and $x^{-\nu}$, respectively. This yields the following result.

Theorem 1 (General solution of Bessel's equation)
If ν is not an integer, a general solution of Bessel's equation for all $x \neq 0$ is

$$y(x) = c_1 J_\nu(x) + c_2 J_{-\nu}(x). \tag{22}$$

But if ν is an integer, then (22) is not a general solution. Indeed, in this case the two solutions in (22) become linearly dependent, as follows.

Theorem 2 (Linear dependence of Bessel functions J_n and J_{-n})
For integer $\nu = n$ the Bessel functions $J_n(x)$ and $J_{-n}(x)$ are linearly dependent, because

$$J_{-n}(x) = (-1)^n J_n(x) \qquad (n = 1, 2, \cdots). \tag{23}$$

Proof. We use (21) and let ν approach a positive integer n. Then the gamma functions in the coefficients of the first n terms become infinite (see Fig. 527 in Appendix 3), the coefficients become zero, and the summation starts with $m = n$. Since in this case $\Gamma(m - n + 1) = (m - n)!$ by (17), we obtain

$$J_{-n}(x) = \sum_{m=n}^{\infty} \frac{(-1)^m x^{2m-n}}{2^{2m-n} m!\, (m - n)!} = \sum_{s=0}^{\infty} \frac{(-1)^{n+s} x^{2s+n}}{2^{2s+n} (n + s)!\, s!}$$

where $m = n + s$ and $s = m - n$. From (11) we see that the last series represents $(-1)^n J_n(x)$. This completes the proof. ∎

A general solution of the Bessel equation with integer $\nu = n$ will be obtained in Sec. 5.7.

Problem Set 5.5

1. Show that the series in (11) converge for all x.
2. The series (11)–(13) converge very rapidly (why?), so that they are useful in computations. For illustration, find out how many terms of (12) one needs to compute $J_0(1)$ with an error less than 1 unit of the 5th decimal place. (*Hint.* Use the Leibniz test in Appendix 3.) How many terms would you need to compute ln 2 from the Maclaurin series of ln $(1 + x)$ with the same accuracy?
3. Show that $J_n(x)$ for even n is an even function and for odd n is an odd function.
4. Show that for small $|x|$ we have $J_0(x) \approx 1 - 0.25x^2$. Using this formula, compute $J_0(x)$ for $x = 0.1, 0.2, \cdots, 1.0$ and determine the relative error by comparing with Table A1 in Appendix 5.
5. **(Behavior for large x)** It can be shown that for large x,

$$\begin{aligned} J_{2n}(x) &\approx (-1)^n(\pi x)^{-1/2}(\cos x + \sin x) \\ J_{2n+1}(x) &\approx (-1)^{n+1}(\pi x)^{-1/2}(\cos x - \sin x). \end{aligned} \tag{24}$$

Using (24), sketch $J_0(x)$ for large x, compute approximate values of the first five positive zeros of $J_0(x)$, and compare them with the more accurate values 2.405, 5.520, 8.654, 11.792, 14.931.

6. Using (12) and the Leibniz test in Appendix 3, can you think of an argument why $2 < x_0 < \sqrt{8}$, where $x_0 \approx 2.405$ is the smallest positive zero of $J_0(x)$?
7. Using (24), compute approximate values of the first four positive zeros of $J_1(x)$ and determine the relative error, using the more exact values 3.832, 7.016, 10.173, 13.324.
8. Using (12) and (13), show that $J_0'(x) = -J_1(x)$.
9. Referring to Prob. 8, does Fig. 84 give the impression that $J_1(x) = 0$ when $J_0(x)$ has a horizontal tangent?
10. Using (12) and (13), show that $J_1'(x) = J_0(x) - \dfrac{1}{x} J_1(x)$.

Differential equations reducible to Bessel's equation

Various differential equations can be reduced to Bessel's equation. To see this, use the indicated substitutions and find a general solution in terms of J_ν and $J_{-\nu}$, or indicate why these functions do not give a general solution. (***More such equations follow in Problem Set* 5.7.**)

11. $x^2y'' + xy' + (x^2 - \frac{1}{9})y = 0$
12. $x^2y'' + xy' + (x^2 - 16)y = 0$
13. $4x^2y'' + 4xy' + (100x^2 - 9)y = 0 \quad (5x = z)$
14. $4x^2y'' + 4xy' + (x - \frac{1}{36})y = 0 \quad (\sqrt{x} = z)$
15. $9x^2y'' + 9xy' + (36x^4 - 16)y = 0 \quad (x^2 = z)$
16. $xy'' + 2y' + xy = 0 \quad (y = u/\sqrt{x})$
17. $xy'' + 5y' + xy = 0 \quad (y = u/x^2)$
18. $xy'' - 5y' + xy = 0 \quad (y = x^3u)$
19. $81x^2y'' + 27xy' + (9x^{2/3} + 8)y = 0 \quad (y = x^{1/3}u, \ x^{1/3} = z)$
20. $x^2y'' + \frac{1}{2}xy' + \frac{1}{16}(x^{1/2} + \frac{15}{16})y = 0 \quad (y = x^{1/4}u, \ x^{1/4} = z)$

5.6 Further Properties of $J_\nu(x)$

Reference [A9] in Appendix 1 shows that Bessel functions $J_\nu(x)$ satisfy an incredibly large number of relationships. We discuss four most basic ones, because of their importance and to show how properties of special functions can be discovered from their series (a fact that was also illustrated by the proof of Theorem 2 in Sec. 5.5).

Multiplying (20), Sec. 5.5, by x^ν, we have

$$x^\nu J_\nu(x) = \sum_{m=0}^{\infty} \frac{(-1)^m x^{2m+2\nu}}{2^{2m+\nu} m!\, \Gamma(\nu + m + 1)}.$$

Differentiating this and then using (16), Sec. 5.5, we obtain

$$\begin{aligned}(x^\nu J_\nu)' &= \sum_{m=0}^{\infty} \frac{(-1)^m 2(m + \nu) x^{2m+2\nu-1}}{2^{2m+\nu} m!\, \Gamma(\nu + m + 1)} \\ &= x^\nu x^{\nu-1} \sum_{m=0}^{\infty} \frac{(-1)^m x^{2m}}{2^{2m+\nu-1} m!\, \Gamma(\nu + m)}.\end{aligned}$$

From (20) in Sec. 5.5 we see that the right side is $x^\nu J_{\nu-1}(x)$. This proves

$$\frac{d}{dx}[x^\nu J_\nu(x)] = x^\nu J_{\nu-1}(x). \tag{1}$$

Similarly, from (20) in Sec. 5.5, multiplied by $x^{-\nu}$, we get by differentiation and an index shift

$$\begin{aligned}(x^{-\nu} J_\nu)' &= \sum_{m=1}^{\infty} \frac{(-1)^m x^{2m-1}}{2^{2m+\nu-1}(m - 1)!\, \Gamma(\nu + m + 1)} \\ &= \sum_{s=0}^{\infty} \frac{(-1)^{s+1} x^{2s+1}}{2^{2s+\nu+1} s!\, \Gamma(\nu + s + 2)}\end{aligned}$$

where $m = s + 1$. By (20), Sec. 5.5, the expression on the right is $-x^{-\nu} J_{\nu+1}(x)$. This proves

$$\frac{d}{dx}[x^{-\nu} J_\nu(x)] = -x^{-\nu} J_{\nu+1}(x). \tag{2}$$

Writing out formulas (1) and (2) and multiplying (2) by $x^{2\nu}$, we get

$$\nu x^{\nu-1} J_\nu + x^\nu J_\nu' = x^\nu J_{\nu-1} \tag{1*}$$

and

(2*) $$-\nu x^{\nu-1}J_\nu + x^\nu J_\nu' = -x^\nu J_{\nu+1}.$$

Substracting (2*) from (1*) and dividing the result by x^ν, we obtain the first recurrence relation

(3) $$J_{\nu-1}(x) + J_{\nu+1}(x) = \frac{2\nu}{x} J_\nu(x).$$

Adding (1*) and (2*) and dividing the result by x^ν, we obtain the second recurrence relation

(4) $$J_{\nu-1}(x) - J_{\nu+1}(x) = 2J_\nu'(x).$$

These relations are of considerable practical importance, for instance, in numerical work, where (3) can be used to express Bessel functions of high orders in terms of Bessel functions of low orders for computing tables.

EXAMPLE 1 Computation of $J_3(x)$

Find a formula for computing $J_3(1)$ and $J_3(2)$ from Table A1 in Appendix 5.

Solution. Formula (3) with $\nu = 1$ and $\nu = 2$ gives

$$J_2(x) = \frac{2}{x} J_1(x) - J_0(x),$$

$$J_3(x) = \frac{4}{x} J_2(x) - J_1(x) = \left(\frac{8}{x^2} - 1\right) J_1(x) - \frac{4}{x} J_0(x).$$

Hence $J_3(1) = 7J_1(1) - 4J_0(1) = 0.0199$ (actually, 0.0196, the discrepancy being due to round-off), and $J_3(2) = J_1(2) - 2J_0(2) = 0.1289$ (correct to 4 digits). ∎

Formulas (1) and (2) are useful for integrals involving Bessel functions.

EXAMPLE 2 Integral involving a Bessel function

Integrate

$$I = \int_1^2 x^{-3} J_4(x)\, dx.$$

Solution. From (2) with $\nu = 3$ and Example 1 we get

$$I = -x^{-3}J_3(x)\Big|_1^2 = -\tfrac{1}{8}J_3(2) + J_3(1) = 0.0038$$

[0.0035 if we use the above 4D value 0.0196 of $J_3(1)$]. ∎

EXAMPLE 3 Integral involving a Bessel function

Express the following integral in terms of Bessel functions.

$$\int x^{-2} J_2(x)\, dx.$$

Solution. Integrating by parts and using (1) with $\nu = 2$, we get

$$\int x^{-2}J_2\, dx = \int x^{-4}(x^2J_2)\, dx = -\frac{x^{-3}}{3}x^2J_2 + \int \frac{x^{-3}}{3}x^2J_1\, dx = -\frac{J_2}{3x} + \frac{1}{3}\int x^{-1}J_1\, dx.$$

On the right, by another integration by parts and use of (1) with $\nu = 1$, we obtain

$$\int x^{-1}J_1\,dx = \int x^{-2}(xJ_1)\,dx = -x^{-1}xJ_1 + \int x^{-1}xJ_0\,dx.$$

Together,

$$\int x^{-2}J_2(x)\,dx = -\frac{J_2(x)}{3x} - \frac{J_1(x)}{3} + \frac{1}{3}\int J_0(x)\,dx.$$

The remaining integral cannot be evaluated in finite form but has been tabulated; see Ref. [A9] in Appendix 1. ∎

$J_\nu(x)$ with $\nu = \pm\frac{1}{2}, \pm\frac{3}{2}, \pm\frac{5}{2}, \cdots$ Are Elementary

It often happens that in special cases, higher functions become functions known from calculus. We study this for $J_\nu(x)$.

When $\nu = \frac{1}{2}$, then (20), Sec. 5.5, is

$$J_{1/2}(x) = \sqrt{x}\sum_{m=0}^{\infty}\frac{(-1)^m x^{2m}}{2^{2m+1/2}m!\,\Gamma(m+\frac{3}{2})} = \sqrt{\frac{2}{x}}\sum_{m=0}^{\infty}\frac{(-1)^m x^{2m+1}}{2^{2m+1}m!\,\Gamma(m+\frac{3}{2})}.$$

Now we use without proof that

$$\Gamma(\tfrac{1}{2}) = \sqrt{\pi}. \tag{5}$$

From this and (16), Sec. 5.5, we get in the denominator

$$\Gamma(m+\tfrac{3}{2}) = (m+\tfrac{1}{2})(m-\tfrac{1}{2})\cdots\tfrac{3}{2}\cdot\tfrac{1}{2}\Gamma(\tfrac{1}{2})$$
$$= 2^{-(m+1)}(2m+1)(2m-1)\cdots 3\cdot 1\sqrt{\pi}.$$

In the denominator we further have

$$2^{2m+1}m! = 2^{2m+1}m(m-1)\cdots 2\cdot 1 = 2^{m+1}2m(2m-2)\cdots 4\cdot 2.$$

Together the denominator becomes $(2m+1)!\sqrt{\pi}$, so that

$$J_{1/2}(x) = \sqrt{\frac{2}{\pi x}}\sum_{m=0}^{\infty}\frac{(-1)^m x^{2m+1}}{(2m+1)!}.$$

This series is the familiar Maclaurin series of $\sin x$. Thus

$$J_{1/2}(x) = \sqrt{\frac{2}{\pi x}}\sin x. \tag{6}$$

By differentiation and by (1) with $\nu = \frac{1}{2}$ we get from this

$$[\sqrt{x}J_{1/2}(x)]' = \sqrt{\frac{2}{\pi}}\cos x = x^{1/2}J_{-1/2}(x)$$

and our next result is

(7)
$$J_{-1/2}(x) = \sqrt{\frac{2}{\pi x}} \cos x.$$

From this and (3) we thus have the following interesting result.

Theorem 1 **(Elementary Bessel functions)**
Bessel functions J_ν of orders $\nu = \pm\frac{1}{2}, \pm\frac{3}{2}, \pm\frac{5}{2}, \cdots$ are elementary; they can be expressed by finitely many cosines and sines and powers of x.

EXAMPLE 4 **Further elementary Bessel functions**

From (3), (6), and (7) we get

$$J_{3/2}(x) = \frac{1}{x} J_{1/2}(x) - J_{-1/2}(x) = \sqrt{\frac{2}{\pi x}} \left(\frac{\sin x}{x} - \cos x\right)$$

$$J_{-3/2}(x) = -\frac{1}{x} J_{-1/2}(x) - J_{1/2}(x) = -\sqrt{\frac{2}{\pi x}} \left(\frac{\cos x}{x} + \sin x\right)$$

and so on. ∎

We hope that our study has not only helped us to become acquainted with Bessel functions, but has also convinced us that series can be quite useful in obtaining various properties of the corresponding functions.

Problem Set 5.6

Using (1)–(4), show that

1. $J_0'(x) = -J_1(x)$

2. $J_1'(x) = J_0(x) - x^{-1}J_1(x)$

3. $J_2'(x) = \frac{1}{2}[J_1(x) - J_3(x)]$

4. $J_2'(x) = (1 - 4x^{-2})J_1(x) + 2x^{-1}J_0(x)$

5. Using Table A1 in Appendix 5 and (3), compute $J_2(x)$ for the values $x = 0, 0.1, 0.2, \cdots, 1.0$.

6. Compute $J_3(x)$ for the values $x = 2.0, 2.2, 2.4, 2.6, 2.8$ from (3) and Table A1 in Appendix 5.

7. Derive Bessel's equation from (1) and (2).

8. **(Interlacing of zeros)** Using (1), (2), and Rolle's theorem, show that between two consecutive zeros of $J_0(x)$ there is precisely one zero of $J_1(x)$.

9. Show that between any two consecutive positive zeros of $J_n(x)$ there is precisely one zero of $J_{n+1}(x)$.

Integrals involving Bessel functions can often be evaluated or at least simplified by the use of (1)–(4). Show that

10. $\int x^\nu J_{\nu-1}(x)\,dx = x^\nu J_\nu(x) + c$

11. $\int x^{-\nu} J_{\nu+1}(x)\,dx = -x^{-\nu} J_\nu(x) + c$

12. $\int J_{\nu+1}(x)\,dx = \int J_{\nu-1}(x)\,dx - 2J_\nu(x)$

Using the formulas in Probs. 10–12 and, if necessary, integration by parts, evaluate

13. $\int J_3(x)\,dx$

14. $\int x^3 J_0(x)\,dx$

15. $\int J_5(x)\,dx$

16. Derive the formulas in Example 4 of the text.

17. **(Gamma function)** Using (5) and $\sqrt{\pi} = 1.772\,454$, compute $\Gamma(1.5)$, $\Gamma(2.5)$, and $\Gamma(3.5)$.

18. Compute $\Gamma(4.6)$ from Table A2 in Appendix 5.

Elimination of first derivative

19. Substitute $y(x) = u(x)v(x)$ into $y'' + p(x)y' + q(x)y = 0$ and show that for obtaining a second-order differential equation for u not containing u', we must take

$$v(x) = \exp\left(-\tfrac{1}{2}\int p(x)\,dx\right).$$

20. Show that for the Bessel equation the substitution in Prob. 19 is $y = ux^{-1/2}$ and gives

$$x^2u'' + (x^2 + \tfrac{1}{4} - \nu^2)u = 0. \tag{8}$$

Solve this equation with $\nu = \frac{1}{2}$ and compare the result with (6) and (7). Comment.

5.7 Bessel Functions of the Second Kind

What is left is the task of getting a general solution of Bessel's equation for integer $\nu = n$; remember from Sec. 5.5 that for noninteger ν we already have a basis J_ν, $J_{-\nu}$, but for $\nu = n$ these two solutions become linearly dependent, so that we need a second independent solution, which will be denoted by Y_n.

$n = 0$: Bessel Function of the Second Kind $Y_0(x)$

We first consider the case $n = 0$. Then Bessel's equation can be written

$$xy'' + y' + xy = 0, \tag{1}$$

and the indicial equation (4), Sec. 5.5, has a double root $r = 0$. This is Case 2 in Sec. 5.4. In this case we first have only one solution, $J_0(x)$. From (8) in Sec. 5.4 we see that the desired second solution must be of the form

$$y_2(x) = J_0(x)\ln x + \sum_{m=1}^{\infty} A_m x^m. \tag{2}$$

We substitute y_2 and its derivatives

$$y_2' = J_0' \ln x + \frac{J_0}{x} + \sum_{m=1}^{\infty} mA_m x^{m-1}$$

$$y_2'' = J_0'' \ln x + \frac{2J_0'}{x} - \frac{J_0}{x^2} + \sum_{m=1}^{\infty} m(m-1)A_m x^{m-2}$$

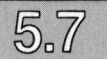

into (1). Then the logarithmic terms disappear because J_0 is a solution of (1), the other two terms containing J_0 cancel, and we are left with

$$2J_0' + \sum_{m=1}^{\infty} m(m-1)A_m x^{m-1} + \sum_{m=1}^{\infty} mA_m x^{m-1} + \sum_{m=1}^{\infty} A_m x^{m+1} = 0.$$

From (12) in Sec. 5.5 we obtain the power series of J_0' in the form

$$J_0'(x) = \sum_{m=1}^{\infty} \frac{(-1)^m 2m x^{2m-1}}{2^{2m}(m!)^2} = \sum_{m=1}^{\infty} \frac{(-1)^m x^{2m-1}}{2^{2m-1} m!\,(m-1)!}.$$

By inserting this series we have

$$\sum_{m=1}^{\infty} \frac{(-1)^m x^{2m-1}}{2^{2m-2} m!\,(m-1)!} + \sum_{m=1}^{\infty} m^2 A_m x^{m-1} + \sum_{m=1}^{\infty} A_m x^{m+1} = 0.$$

We first show that the A_m with odd subscripts are all zero. The coefficient of the power x^0 is A_1, and so, $A_1 = 0$. By equating the sum of the coefficients of the power x^{2s} to zero we have

$$(2s+1)^2 A_{2s+1} + A_{2s-1} = 0, \qquad s = 1, 2, \cdots.$$

Since $A_1 = 0$, we thus obtain $A_3 = 0$, $A_5 = 0$, $\cdots$, successively.

We now equate the sum of the coefficients of x^{2s+1} to zero. This gives

$$-1 + 4A_2 = 0 \quad \text{or} \quad A_2 = \tfrac{1}{4} \qquad (s = 0)$$

and for the other values $s = 1, 2, \cdots$

$$\frac{(-1)^{s+1}}{2^{2s}(s+1)!\,s!} + (2s+2)^2 A_{2s+2} + A_{2s} = 0.$$

For $s = 1$ this yields

$$\tfrac{1}{8} + 16A_4 + A_2 = 0 \quad \text{or} \quad A_4 = -\tfrac{3}{128}$$

and in general

$$\text{(3)} \qquad A_{2m} = \frac{(-1)^{m-1}}{2^{2m}(m!)^2}\left(1 + \frac{1}{2} + \frac{1}{3} + \cdots + \frac{1}{m}\right), \qquad m = 1, 2, \cdots.$$

Using the short notation

$$\text{(4)} \qquad h_m = 1 + \frac{1}{2} + \cdots + \frac{1}{m}$$

and inserting (3) and $A_1 = A_3 = \cdots = 0$ into (2), we obtain the result

(5)
$$y_2(x) = J_0(x) \ln x + \sum_{m=1}^{\infty} \frac{(-1)^{m-1}h_m}{2^{2m}(m!)^2} x^{2m}$$
$$= J_0(x) \ln x + \tfrac{1}{4}x^2 - \tfrac{3}{128}x^4 + - \cdots .$$

Since J_0 and y_2 are linearly independent functions, they form a basis of (1). Of course, another basis is obtained if we replace y_2 by an independent particular solution of the form $a(y_2 + bJ_0)$ where a ($\neq 0$) and b are constants. It is customary to choose $a = 2/\pi$ and $b = \gamma - \ln 2$, where the number $\gamma = 0.577\,215\,664\,90 \cdots$ is the so-called **Euler constant,** which is defined as the limit of

$$1 + \frac{1}{2} + \cdots + \frac{1}{s} - \ln s$$

as s approaches infinity. The standard particular solution thus obtained is called the **Bessel function of the second kind** *of order zero* (Fig. 85) or **Neumann's function**[13] *of order zero* and is denoted by $Y_0(x)$. Thus [see (4)]

(6)
$$Y_0(x) = \frac{2}{\pi}\left[J_0(x)\left(\ln \frac{x}{2} + \gamma\right) + \sum_{m=1}^{\infty} \frac{(-1)^{m-1}h_m}{2^{2m}(m!)^2} x^{2m}\right].$$

For small $x > 0$ the function $Y_0(x)$ behaves about like $\ln x$ (see Fig. 85; why?), and $Y_0(x) \to -\infty$ as $x \to 0$.

Bessel Functions of the Second Kind $Y_n(x)$

For $\nu = n = 1, 2, \cdots$ a second solution can be obtained by manipulations similar to those for $n = 0$, starting from (10), Sec. 5.4. It turns out that in these cases the solution also contains a logarithmic term.

The situation is not yet completely satisfactory, because the second solution is defined differently, depending on whether the order ν is an integer or not. To provide uniformity of formalism and numerical tabulation, it is desirable to adopt a form of the second solution that is valid for all values of the order. This is the reason for introducing a standard second solution $Y_\nu(x)$ defined for all ν by the formula

(7)
(a) $$Y_\nu(x) = \frac{1}{\sin \nu\pi}[J_\nu(x) \cos \nu\pi - J_{-\nu}(x)]$$

(b) $$Y_n(x) = \lim_{\nu \to n} Y_\nu(x).$$

[13] CARL NEUMANN (1832—1925), German mathematician and physicist, became a professor at Leipzig in 1868. His work on potential theory sparked the development in the field of integral equations by VITO VOLTERRA (1860—1940) of Rome, ERIC IVAR FREDHOLM (1866—1927) of Stockholm, whose famous 1901–1903 papers were a sensation to the mathematical world of his time, and DAVID HILBERT (1862—1943) of Göttingen (see the footnote in Sec. 7.15).

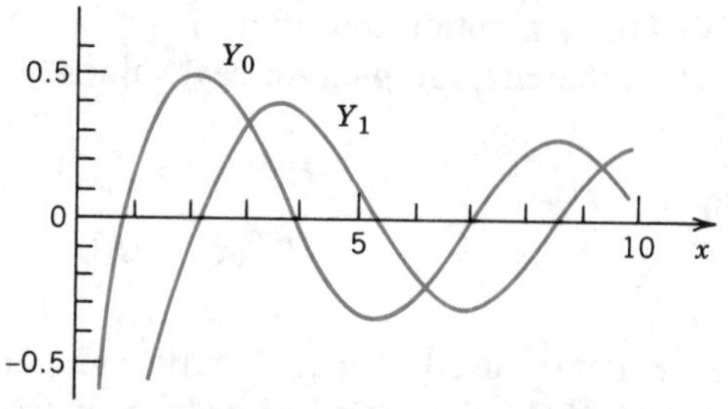

Fig. 85. Bessel functions of the second kind.
(For a small table, see Appendix 5.)

This function is called the **Bessel function of the second kind** *of order* ν or **Neumann's function**[14] *of order* ν. Figure 85 shows $Y_0(x)$ and $Y_1(x)$.

We discuss the linear independence of J_ν and Y_ν.

For noninteger order ν, the function $Y_\nu(x)$ is evidently a solution of Bessel's equation because $J_\nu(x)$ and $J_{-\nu}(x)$ are solutions of that equation. Since for those ν the solutions J_ν and $J_{-\nu}$ are linearly independent and Y_ν involves $J_{-\nu}$, the functions J_ν and Y_ν are linearly independent. Furthermore, it can be shown that the limit in (7b) exists and Y_n is a solution of Bessel's equation for integer order; see Ref. [A9] in Appendix 1. We shall see that the series development of $Y_n(x)$ contains a logarithmic term. Hence $J_n(x)$ and $Y_n(x)$ are linearly independent solutions of Bessel's equation. The series development of $Y_n(x)$ can be obtained if we insert the series (20) and (21), Sec. 5.5, for $J_\nu(x)$ and $J_{-\nu}(x)$ into (7a) and then let ν approach n; for details see Ref. [A9]; the result is

(8)
$$Y_n(x) = \frac{2}{\pi} J_n(x)\left(\ln\frac{x}{2} + \gamma\right) + \frac{x^n}{\pi}\sum_{m=0}^{\infty}\frac{(-1)^{m-1}(h_m + h_{m+n})}{2^{2m+n}m!\,(m+n)!}x^{2m}$$
$$- \frac{x^{-n}}{\pi}\sum_{m=0}^{n-1}\frac{(n-m-1)!}{2^{2m-n}m!}x^{2m}$$

where $x > 0$, $n = 0, 1, \cdots$, and

$$h_0 = 0, \qquad h_s = 1 + \frac{1}{2} + \frac{1}{3} + \cdots + \frac{1}{s} \qquad (s = 1, 2, \cdots),$$

and when $n = 0$ the last sum in (8) is to be replaced by 0. For $n = 0$ the representation (8) takes the form (6). Furthermore, it can be shown that

$$Y_{-n}(x) = (-1)^n Y_n(x).$$

We may formulate our main result as follows.

Theorem 1 **(General solution of Bessel's equation)**
A general solution of Bessel's equation for all values of ν *is*

(9)
$$y(x) = C_1 J_\nu(x) + C_2 Y_\nu(x).$$

---

[14] See footnote 13. The solutions $Y_\nu(x)$ are sometimes denoted by $N_\nu(x)$; in Ref. [A9] they are called **Weber's functions;** Euler's constant in (6) is often denoted by C or $\ln\gamma$.

We finally mention that there is a practical need for solutions of Bessel's equation that are complex for real values of x. For this purpose the solutions

$$(10)\qquad \begin{aligned} H_\nu^{(1)}(x) &= J_\nu(x) + iY_\nu(x) \\ H_\nu^{(2)}(x) &= J_\nu(x) - iY_\nu(x) \end{aligned}$$

are frequently used. These linearly independent functions are called **Bessel functions of the third kind** *of order* ν or *first and second* **Hankel functions**[15] *of order* ν.

This finishes our discussion on Bessel functions, except for their "orthogonality," which we explain in Sec. 5.8. Applications to vibrations follow in Sec. 11.10.

Problem Set 5.7

Some further differential equations reducible to Bessel's equation (See also Sec. 5.5.) Using the indicated substitutions, reduce the following equations to Bessel's differential equation and find a general solution in terms of Bessel functions.

1. $x^2y'' + xy' + (x^2 - 4)y = 0$

2. $x^2y'' + xy' + (\lambda^2x^2 - \nu^2)y = 0 \qquad (\lambda x = z)$

3. $xy'' + y' + \frac{1}{4}y = 0 \qquad (\sqrt{x} = z)$

4. $4x^2y'' + 4xy' + (x - n^2)y = 0 \qquad (\sqrt{x} = z)$

5. $x^2y'' + xy' + (4x^4 - \frac{1}{4})y = 0 \qquad (x^2 = z)$

6. $xy'' + (1 + 2n)y' + xy = 0 \qquad (y = x^{-n}u)$

7. $x^2y'' - 3xy' + 4(x^4 - 3)y = 0 \qquad (y = x^2u,\ x^2 = z)$

8. $x^2y'' + (1 - 2\nu)xy' + \nu^2(x^{2\nu} + 1 - \nu^2)y = 0 \qquad (y = x^\nu u,\ x^\nu = z)$

9. $x^2y'' + \frac{1}{4}(x + \frac{3}{4})y = 0 \qquad (y = u\sqrt{x},\ \sqrt{x} = z)$

10. $y'' + xy = 0 \qquad (y = u\sqrt{x},\ \frac{2}{3}x^{3/2} = z)$

11. $y'' + x^2y = 0 \qquad (y = u\sqrt{x},\ \frac{1}{2}x^2 = z)$

12. $y'' + k^2xy = 0 \qquad (y = u\sqrt{x},\ \frac{2}{3}kx^{3/2} = z)$

13. $y'' + k^2x^2y = 0 \qquad (y = u\sqrt{x},\ \frac{1}{2}kx^2 = z)$

14. $y'' + k^2x^4y = 0 \qquad (y = u\sqrt{x},\ \frac{1}{3}kx^3 = z)$

15. Show that for small $x > 0$ we have $Y_0(x) \approx 2\pi^{-1}(\ln \frac{1}{2}x + \gamma)$. Using this formula, compute an approximate value of the smallest positive zero of $Y_0(x)$ and compare it with the more accurate value 0.9.

16. It can be shown that for large x,

$$(11)\qquad Y_n(x) \approx \sqrt{2/(\pi x)}\,\sin\left(x - \tfrac{1}{2}n\pi - \tfrac{1}{4}\pi\right).$$

Using (11), sketch $Y_0(x)$ and $Y_1(x)$ for $0 < x \leqq 15$. Using (11), compute approximate values of the first three positive zeros of $Y_0(x)$ and compare these values with the more accurate values 0.89, 3.96, and 7.09.

17. Show that the Hankel functions (10) constitute a basis of solutions of Bessel's equation for any ν.

---

[15] HERMANN HANKEL (1839—1873), German mathematician.

Modified Bessel functions

18. The function $I_\nu(x) = i^{-\nu}J_\nu(ix)$, $i = \sqrt{-1}$, is called the *modified Bessel function of the first kind of order* ν. Show that $I_\nu(x)$ is a solution of the differential equation

$$x^2y'' + xy' - (x^2 + \nu^2)y = 0 \tag{12}$$

and has the representation

$$I_\nu(x) = \sum_{m=0}^{\infty} \frac{x^{2m+\nu}}{2^{2m+\nu}m!\,\Gamma(m + \nu + 1)}. \tag{13}$$

19. Show that $I_\nu(x)$ is real for all real x (and real ν), $I_\nu(x) \neq 0$ for all real $x \neq 0$, and $I_{-n}(x) = I_n(x)$, where n is any integer.

20. Show that another solution of the differential equation (12) is the so-called *modified Bessel function of the third kind* (sometimes: *of the second kind*)

$$K_\nu(x) = \frac{\pi}{2 \sin \nu\pi}[I_{-\nu}(x) - I_\nu(x)]. \tag{14}$$

5.8 Sturm–Liouville Problems. Orthogonality

Sturm and Liouville[16] have shown that Bessel's, Legendre's, and other equations of engineering significance can be considered from a common viewpoint, with the outcome that one obtains families of solutions that are very convenient for series representations of given functions as they occur in mechanics, heat conduction, electricity, and other physical applications. We begin with two motivating examples, Bessel's and Legendre's equations.

Bessel's equation (with the independent variable written as $\tilde{x}$ and $\dot{y} = dy/d\tilde{x}$, $\ddot{y} = d^2y/d\tilde{x}^2$)

$$\tilde{x}^2\ddot{y} + \tilde{x}\dot{y} + (\tilde{x}^2 - n^2)y = 0$$

can be transformed by setting $\tilde{x} = kx$; then $\dot{y} = y'/k$, $\ddot{y} = y''/k^2$ and we get (k^2 and k dropping out in the first two terms)

$$x^2y'' + xy' + (k^2x^2 - n^2)y = 0$$

and by division by x,

$$[xy']' + \left(-\frac{n^2}{x} + \lambda x\right)y = 0 \qquad (\lambda = k^2).$$

---

[16]JACQUES CHARLES FRANÇOIS STURM (1803—1855), was born and studied in Switzerland and then moved to Paris, where he later became the successor of Poisson in the chair of mechanics at the Sorbonne.

JOSEPH LIOUVILLE (1809—1882), French mathematician and professor in Paris, contributed to various fields in mathematics and is particularly known by his important work in complex analysis (Liouville's theorem; Sec. 13.6), special functions, differential geometry, and number theory.

Similarly, Legendre's equation (Sec. 5.3)

$$(1 - x^2)y'' - 2xy' + n(n + 1)y = 0$$

can be written

$$[(1 - x^2)y']' + \lambda y = 0 \qquad [\lambda = n(n + 1)].$$

Both equations obtained are of the form

$$[r(x)y']' + [q(x) + \lambda p(x)]y = 0, \tag{1}$$

called a **Sturm–Liouville equation,** and there are other equations that can be cast into the form (1). The simplest equation (1) is

$$y'' + \lambda y = 0$$

(with $r = p = 1$ and $q = 0$).

Equation (1) is considered on some given interval $a \leqq x \leqq b$, and we assume continuity of p, q, r, r' as well as

$$p(x) > 0$$

on this interval. At the endpoints (boundary points) a and b we impose **boundary conditions**

$$\text{(2)} \qquad \begin{aligned} &\text{(a)} \quad k_1 y(a) + k_2 y'(a) = 0 \\ &\text{(b)} \quad l_1 y(b) + l_2 y'(b) = 0 \end{aligned}$$

with given constants k_1, k_2, not both zero, and given constants l_1, l_2, not both zero. The **boundary value problem** consisting of (1) and (2) is called a **Sturm–Liouville problem.**

Clearly, $y \equiv 0$ is always a solution of the problem, but is of no practical use. What we want to find is solutions of (1) satisfying (2) without being identically zero. We call such a solution $y(x)$—if it exists—an **eigenfunction** and a number λ for which an eigenfunction exists, an **eigenvalue** of the problem.

EXAMPLE 1 Vibrating elastic string

Find the eigenvalues and eigenfunctions of the Sturm–Liouville problem

$$\text{(3)} \qquad \text{(a)} \quad y'' + \lambda y = 0 \qquad \text{(b)} \quad y(0) = 0, \; y(\pi) = 0.$$

This problem arises, for instance, if an elastic string (a violin string, for example) is stretched a little and then fixed at its ends $x = 0$ and $x = \pi$ and allowed to vibrate. Then $y(x)$ is the "space function" of the deflection $u(x, t)$ of the string, assumed in the form $u(x, t) = y(x)w(t)$, where t is time. (This model will be discussed in Secs. 11.2–11.4.)

Solution. For negative $\lambda = -\nu^2$ a general solution of the equation is

$$y(x) = c_1 e^{\nu x} + c_2 e^{-\nu x}.$$

From (3b) we obtain $c_1 = c_2 = 0$ and $y \equiv 0$, so that $y \equiv 0$, which is not an eigenfunction. For $\lambda = 0$ the situation is similar. For positive $\lambda = \nu^2$ a general solution is

$$y(x) = A \cos \nu x + B \sin \nu x.$$

From the first boundary condition we obtain $y(0) = A = 0$. The second boundary condition then yields

$$y(\pi) = B \sin \nu\pi = 0, \qquad \text{thus} \qquad \nu = 0, \pm 1, \pm 2, \cdots.$$

For $\nu = 0$ we have $y \equiv 0$. For $\lambda = \nu^2 = 1, 4, 9, 16, \cdots$, taking $B = 1$, we obtain

$$y(x) = \sin \nu x \qquad (\nu = 1, 2, \cdots).$$

Hence the eigenvalues of the problem are $\lambda = \nu^2$, where $\nu = 1, 2, \cdots$; and corresponding eigenfunctions are $y(x) = \sin \nu x$, where $\nu = 1, 2, \cdots$. ∎

Existence of eigenvalues

Eigenvalues of a Sturm–Liouville problem (1), (2), even infinitely many, exist under rather general conditions on p, q, r in (1). (Sufficient are the conditions in Theorem 1, below, together with $p(x) > 0$ and $r(x) > 0$ on $a < x < b$. Proofs are complicated; see Refs. [A1] or [A6] listed in Appendix 1.)

Reality of eigenvalues

Furthermore, if p, q, r, and r' in (1) are real-valued and continuous on the interval $a \leqq x \leqq b$ and p is positive throughout that interval (or negative throughout that interval), then all the eigenvalues of the Sturm–Liouville problem (1), (2) are real. (A proof is included in Appendix 4.)

This is what the engineer would expect because eigenvalues are often related to frequencies, energies, or other physical quantities that must be real.

Orthogonality

Eigenfunctions of Sturm–Liouville problems have remarkable general properties—above all, orthogonality, which is defined as follows.

Definition of orthogonality

Functions $y_1, y_2, \cdots$ defined on some interval $a \leqq x \leqq b$ are called **orthogonal** on $a \leqq x \leqq b$ with respect to a **weight function** $p(x) > 0$ if

$$\int_a^b p(x) y_m(x) y_n(x)\, dx = 0 \qquad \text{for } m \neq n. \tag{4}$$

The **norm** $\|y_m\|$ of y_m is defined by

$$\|y_m\| = \sqrt{\int_a^b p(x) y_m^2(x)\, dx}. \tag{5}$$

The functions are called **orthonormal** on $a \leqq x \leqq b$ if they are orthogonal on $a \leqq x \leqq b$ and all have norm 1.

For "orthogonal with respect to $p(x) = 1$" we simply say "orthogonal." Thus, functions $y_1, y_2, \cdots$ are **orthogonal** on some interval $a \leqq x \leqq b$ if

$$\text{(4')} \qquad \int_a^b y_m(x) y_n(x)\, dx = 0 \qquad \text{for } m \neq n.$$

The **norm** $\|y_m\|$ of y_m is then defined by

$$\text{(5')} \qquad \|y_m\| = \sqrt{\int_a^b y_m^2(x)\, dx}.$$

And the functions are called **orthonormal** on $a \leqq x \leqq b$ if they are orthogonal there and all have norm (5′) equal to 1. ∎

EXAMPLE 2 Orthogonal set, orthonormal set

The functions $y_m(x) = \sin mx$, $m = 1, 2, \cdots$ form an orthogonal set on the interval $-\pi \leqq x \leqq \pi$, because for $m \neq n$ we obtain [see (11) in Appendix 3]

$$\int_{-\pi}^{\pi} y_m(x) y_n(x)\, dx = \int_{-\pi}^{\pi} \sin mx \sin nx\, dx = \frac{1}{2}\int_{-\pi}^{\pi} \cos (m - n)x\, dx - \frac{1}{2}\int_{-\pi}^{\pi} \cos (m + n)x\, dx = 0.$$

The norm $\|y_m\|$ equals $\sqrt{\pi}$, because

$$\|y_m\|^2 = \int_{-\pi}^{\pi} \sin^2 mx\, dx = \pi \qquad (m = 1, 2, \cdots).$$

Hence the corresponding orthonormal set, obtained by division by the norm, is

$$\frac{\sin x}{\sqrt{\pi}}, \qquad \frac{\sin 2x}{\sqrt{\pi}}, \qquad \frac{\sin 3x}{\sqrt{\pi}}, \qquad \cdots.$$ ∎

Orthogonality of Eigenfunctions

Theorem 1 (Orthogonality of eigenfunctions)

Suppose that the functions p, q, r, and r' in the Sturm–Liouville equation (1) *are real-valued and continuous and $p(x) > 0$ on the interval $a \leqq x \leqq b$. Let $y_m(x)$ and $y_n(x)$ be eigenfunctions of the Sturm–Liouville problem* (1), (2) *that correspond to different eigenvalues λ_m and λ_n, respectively. Then y_m, y_n are orthogonal on that interval with respect to the weight function p.*

If $r(a) = 0$, then (2a) *can be dropped from the problem. If $r(b) = 0$, then* (2b) *can be dropped.* [It is then required that y and y' remain bounded at such a point, and the problem is called **singular**, as opposed to a **regular problem** in which (2) is used.]

If $r(a) = r(b)$, then (2) *can be replaced by the* "**periodic boundary conditions**"

$$\text{(6)} \qquad y(a) = y(b), \qquad y'(a) = y'(b).$$

Remark. The boundary value problem consisting of the Sturm-Liouville equation (1) and the periodic boundary conditions (6) is called a **periodic Sturm-Liouville problem.**

Proof of Theorem 1. By assumption, y_m satisfies

$$(ry_m')' + (q + \lambda_m p)y_m = 0,$$

and y_n satisfies

$$(ry_n')' + (q + \lambda_n p)y_n = 0.$$

Multiplying the first equation by y_n, the second by $-y_m$ and adding, we get

$$\begin{aligned}(\lambda_m - \lambda_n)py_my_n &= y_m(ry_n')' - y_n(ry_m')' \\ &= [(ry_n')y_m - (ry_m')y_n]'\end{aligned}$$

where the last equality can be readily verified by performing the indicated differentiation of the last expression in brackets. This expression is continuous on $a \leqq x \leqq b$ since r and r' are continuous by assumption and y_m, y_n are solutions of (1). Integrating over x from a to b, we thus obtain

$$(7) \qquad (\lambda_m - \lambda_n)\int_a^b py_my_n\,dx = \Big[r(y_n'y_m - y_m'y_n)\Big]_a^b .$$

The expression on the right equals

$$(8) \qquad \begin{aligned}&r(b)[y_n'(b)y_m(b) - y_m'(b)y_n(b)] \\ &-r(a)[y_n'(a)y_m(a) - y_m'(a)y_n(a)].\end{aligned}$$

We now have to consider several cases depending on whether r vanishes or does not vanish at a or b.

Case 1. If $r(a) = 0$ and $r(b) = 0$, then the expression in (8) is zero. Hence the expression on the left side of (7) must be zero, because y_m, y_m', y_n, y_n' on the right remain continuous at a and b (by assumption) and λ_m and λ_n are distinct. We thus obtain the desired orthogonality

$$(9) \qquad \int_a^b p(x)y_m(x)y_n(x)\,dx = 0 \qquad (m \neq n)$$

without the use of the boundary conditions (2).

Case 2. Let $r(b) = 0$, but $r(a) \neq 0$. Then the first line in (8) is zero. We consider the remaining expression in (8). From (2a) we have

$$k_1y_n(a) + k_2y_n'(a) = 0,$$

$$k_1y_m(a) + k_2y_m'(a) = 0.$$

Let $k_2 \neq 0$. Then by multiplying the first equation by $y_m(a)$ and the last by $-y_n(a)$ and adding, we have

$$k_2[y_n'(a)y_m(a) - y_m'(a)y_n(a)] = 0.$$

Since $k_2 \neq 0$, the expression in brackets must be zero. This expression is identical with that in the last line of (8). Hence (8) is zero, and from (7) we obtain (9) as before. If $k_2 = 0$, then by assumption $k_1 \neq 0$, and the argument of proof is similar.

Case 3. If $r(a) = 0$, but $r(b) \neq 0$, the proof is similar to that in Case 2, but instead of (2a) we now have to use (2b).

Case 4. If $r(a) \neq 0$ and $r(b) \neq 0$, we have to use both boundary conditions (2) and proceed as in Cases 2 and 3.

Case 5. Let $r(a) = r(b)$. Then (8) takes the form

$$r(b)[y_n'(b)y_m(b) - y_m'(b)y_n(b) - y_n'(a)y_m(a) + y_m'(a)y_n(a)].$$

We may use (2) as before and conclude that the expression in brackets is zero. However, we immediately see that this also follows from (6), so that we may replace (2) by (6). Hence (7) yields (9), as before. This completes the proof of Theorem 1. ∎

EXAMPLE 3 Vibrating elastic string

The differential equation in Example 1 is of the form (1) where $r = 1$, $q = 0$, and $p = 1$. From Theorem 1 it follows that the eigenfunctions are orthogonal on the interval $0 \leqq x \leqq \pi$. ∎

EXAMPLE 4 Another application of Theorem 1

For the periodic Sturm–Liouville problem

$$y'' + \lambda y = 0, \qquad y(\pi) = y(-\pi), \qquad y'(\pi) = y'(-\pi)$$

we obtain from the general solution $y = A \cos kx + B \sin kx$, where $k = \sqrt{\lambda}$, and the boundary conditions the two equations

$$A \cos k\pi + B \sin k\pi = A \cos(-k\pi) + B \sin(-k\pi)$$

$$-kA \sin k\pi + kB \cos k\pi = -kA \sin(-k\pi) + kB \cos(-k\pi).$$

Since $\cos(-\alpha) = \cos \alpha$ and $\sin(-\alpha) = -\sin \alpha$, this gives

$$\sin k\pi = 0, \qquad \lambda = k^2 = n^2 = 0, 1, 4, 9, \cdots.$$

Hence the eigenfunctions are

$$1, \qquad \cos x, \qquad \sin x, \qquad \cos 2x, \qquad \sin 2x, \qquad \cdots.$$

By Theorem 1, any two of these belonging to different eigenvalues are orthogonal on the interval $-\pi \leqq x \leqq \pi$ (note that $p(x) = 1$ for the present equation), and the orthogonality of $\cos mx$ and $\sin mx$ for the same m follows by integration,

$$\int_{-\pi}^{\pi} \cos mx \sin mx \, dx = \frac{1}{2}\int_{-\pi}^{\pi} \sin 2mx \, dx = 0.$$

For the norms we get $\|1\| = \sqrt{2\pi}$ and $\sqrt{\pi}$ for all the others, as the student may verify by integrating 1, $\cos^2 x$, $\sin^2 x$, etc., from $-\pi$ to π. Accordingly, an orthonormal set of eigenfunctions of our present problem is

$$\frac{1}{\sqrt{2\pi}}, \quad \frac{\cos x}{\sqrt{\pi}}, \quad \frac{\sin x}{\sqrt{\pi}}, \quad \frac{\cos 2x}{\sqrt{\pi}}, \quad \frac{\sin 2x}{\sqrt{\pi}}, \quad \cdots .$$

EXAMPLE 5 Orthogonality of Legendre polynomials

Legendre's equation is a Sturm–Liouville equation (see the beginning of this section)

$$[(1 - x^2)y']' + \lambda y = 0, \qquad \lambda = n(n + 1),$$

with $r = 1 - x^2$, $q = 0$, and $p = 1$. Since $r(-1) = r(1) = 0$, we need no boundary conditions, but have a ***singular Sturm–Liouville problem*** on the interval $-1 \leqq x \leqq 1$. We know that for $n = 0, 1, \cdots$, hence $\lambda = 0, 1 \cdot 2, 2 \cdot 3, \cdots$, the Legendre polynomials $P_n(x)$ are solutions of the problem. Hence these are the eigenfunctions. From Theorem 1 it follows that they are orthogonal on that interval, that is,

$$\int_{-1}^{1} P_m(x)P_n(x)\, dx = 0 \qquad (m \neq n). \tag{10}$$

The norm is (see Prob. 9 in Sec. 5.3)

$$\|P_m\| = \sqrt{\int_{-1}^{1} P_m(x)^2\, dx} = \sqrt{\frac{2}{2m + 1}} \qquad (m = 0, 1, \cdots). \tag{11}$$

EXAMPLE 6 Orthogonality of Bessel functions $J_n(x)$

The Bessel function $J_n(\tilde{x})$ with fixed integer $n \geqq 0$ satisfies Bessel's equation (Sec. 5.5)

$$\tilde{x}^2\ddot{J}_n(x) + \tilde{x}\dot{J}_n(\tilde{x}) + (\tilde{x}^2 - n^2)J_n(\tilde{x}) = 0,$$

where $\dot{J}_n = dJ_n/d\tilde{x}$, $\ddot{J}_n = d^2J_n/d\tilde{x}^2$. At the beginning of this section we transformed this equation, by setting $\tilde{x} = kx$, into a Sturm–Liouville equation

$$[xJ_n'(kx)]' + \left(-\frac{n^2}{x} + k^2x\right)J_n(kx) = 0$$

with $p(x) = x$, $q(x) = -n^2/x$, $r(x) = x$, and parameter $\lambda = k^2$. Since $r(0) = 0$, Theorem 1 implies orthogonality on an interval $0 \leqq x \leqq R$ of those solutions $J_n(kx)$ that are zero at $x = R$, that is,

$$J_n(kR) = 0 \qquad (n \text{ fixed}).$$

[Note that $q(x) = -n^2/x$ is discontinuous at 0, but this does not affect the proof of Theorem 1.] It can be shown (see Ref. [A9]) that $J_n(\tilde{x})$ has infinitely many real zeros, say, $\tilde{x} = \alpha_{1n} < \alpha_{2n} < \cdots$ (see Fig. 84 in Sec. 5.5 for $n = 0$ and 1). Hence we must have

$$kR = \alpha_{mn}, \qquad \text{thus} \qquad k = k_{mn} = \alpha_{mn}/R \qquad (m = 1, 2, \cdots). \tag{12}$$

This proves

Theorem 2 (Orthogonality of Bessel functions)

For each fixed nonnegative integer n the Bessel functions $J_n(k_{1n}x)$, $J_n(k_{2n}x)$, $J_n(k_{3n}x)$, $\cdots$, with k_{mn} as in (12), form an orthogonal set on the interval $0 \leqq x \leqq R$ with respect to the weight $p(x) = x$, that is,

$$\int_{0}^{R} xJ_n(k_{mn}x)J_n(k_{jn}x)\, dx = 0 \qquad (j \neq m). \tag{13}$$

Hence we have obtained infinitely many orthogonal sets, each corresponding to one of the fixed values n.

Problem Set 5.8

1. Carry out the details of the proof of Theorem 1 in Cases 3 and 4.
2. **(Normalization of eigenfunctions)** Show that if $y = y_0$ is an eigenfunction of (1), (2) corresponding to some eigenvalue $\lambda = \lambda_0$, then $y = \alpha y_0$ ($\alpha \neq 0$, arbitrary) is an eigenfunction of (1), (2) corresponding to λ_0. (Note that this property can be used to "normalize" eigenfunctions, that is, to obtain eigenfunctions of norm 1.)

Find the eigenvalues and eigenfunctions of the following Sturm–Liouville problems. In Probs. 3–9 also verify orthogonality by direct calculation.

3. $y'' + \lambda y = 0, \quad y(0) = 0, \ y'(1) = 0$
4. $y'' + \lambda y = 0, \quad y(0) = 0, \ y(L) = 0$
5. $y'' + \lambda y = 0, \quad y(0) = 0, \ y'(L) = 0$
6. $y'' + \lambda y = 0, \quad y'(0) = 0, \ y(L) = 0$
7. $y'' + \lambda y = 0, \quad y'(0) = 0, \ y'(\pi) = 0$
8. $y'' + \lambda y = 0, \quad y(0) = y(2\pi), \ y'(0) = y'(2\pi)$
9. $y'' + \lambda y = 0, \quad y'(0) = 0, \ y'(L) = 0$
10. $(xy')' + \lambda x^{-1}y = 0, \quad y(1) = 0, \ y(e) = 0.$ *Hint.* Set $x = e^t$.
11. $(xy')' + \lambda x^{-1}y = 0, \quad y(1) = 0, \ y'(e) = 0$
12. $(e^{2x}y')' + e^{2x}(\lambda + 1)y = 0, \quad y(0) = 0, \ y(\pi) = 0$ *Hint.* Set $y = e^{-x}u$.
13. $(x^{-1}y')' + (\lambda + 1)x^{-3}y = 0, \quad y(1) = 0, \ y(e) = 0$

14. Show that the eigenvalues of the Sturm–Liouville problem $y'' + \lambda y = 0$, $y(0) = 0$, $y(1) + y'(1) = 0$ are obtained as solutions of the equation $\tan k = -k$, where $k = \sqrt{\lambda}$. Show graphically that this equation has infinitely many solutions $k = k_n$ and the eigenfunctions are $y_n = \sin k_n x$ ($k_n \neq 0$). Show that the positive k_n are of the form $k_n = \frac{1}{2}(2n + 1)\pi + \delta_n$, where δ_n is positive and small and $\delta_n \to 0$ as $n \to \infty$. Compute k_0 and k_1 (by Newton's method, Sec. 18.2).
15. Verify by direct calculation that $P_0(x)$, $P_1(x)$, $P_2(x)$ form an orthogonal set on $-1 \leqq x \leqq 1$ (with $p(x) = 1$) and find the corresponding orthonormal set, also by direct calculation.
16. Determine constants $a_0, b_0, \cdots, c_2$ so that $y_0 = a_0$, $y_1 = b_0 + b_1 x$, $y_2 = c_0 + c_1 x + c_2 x^2$ form an orthonormal set on $-1 \leqq x \leqq 1$ (with $p(x) = 1$). Compare the result with that of Prob. 15 and comment.
17. Show that if the functions $y_0(x), y_1(x), \cdots$ form an orthogonal set on an interval $a \leqq x \leqq b$ (with $p(x) = 1$), then the functions $y_0(ct + k), y_1(ct + k), \cdots$, $c > 0$, form an orthogonal set on the interval $(a - k)/c \leqq t \leqq (b - k)/c$.
18. Using Prob. 17, derive the orthogonality of $1, \cos \pi x, \sin \pi x, \cos 2\pi x, \sin 2\pi x, \cdots$ on $-1 \leqq x \leqq 1$ ($p(x) = 1$) from that in Example 4 of the text.

Verify that the given functions are orthogonal on the given interval with respect to the given $p(x)$ and have the indicated norm.

19. $L_0(x) = 1$, $L_1(x) = 1 - x$, $L_2(x) = 1 - 2x + \frac{1}{2}x^2$, $0 \leqq x < \infty$, $p(x) = e^{-x}$, norm 1
20. $T_0(x) = 1$, $T_1(x) = x$, $T_2(x) = 2x^2 - 1$, $-1 \leqq x \leqq 1$, $p(x) = (1 - x^2)^{-1/2}$, $\|T_0\| = \sqrt{\pi}$, $\|T_1\| = \|T_2\| = \sqrt{\pi/2}$. *Hint.* Set $x = \cos \theta$.

5.9 Eigenfunction Expansions

Why are orthogonal sets of eigenfunctions important? What are they good for? The answer is that they yield series developments of given functions in a simple fashion. This includes the famous "Fourier series" and other "eigenfunction expansions" arising from Sturm–Liouville problems, the daily bread of the physicist and engineer in heat conduction, fluid flow, solid mechanics, numerical work, and so on.

To explain this, we begin by introducing some practical notation. We denote the integral (4), Sec. 5.8, in the definition of orthogonality and orthonormality by (y_m, y_n). This is standard; thus, for an orthonormal set with respect to weight $p(x) > 0$ on $a \leqq x \leqq b$,

$$(1) \qquad (y_m, y_n) = \int_a^b p(x) y_m(x) y_n(x)\, dx = \begin{cases} 0 & \text{if} \quad m \neq n \\ 1 & \text{if} \quad m = n \end{cases}$$

where $m = 0, 1, 2, \cdots$ and $n = 0, 1, 2, \cdots$. Even more briefly,

$$(1^*) \qquad (y_m, y_n) = \delta_{mn} \qquad \text{where} \qquad \delta_{mn} = \begin{cases} 0 & \text{if} \quad m \neq n \\ 1 & \text{if} \quad m = n \end{cases}$$

is called the **Kronecker delta.**[17] For the norm $\|y_m\|$ of y_m we can now write

$$(2) \qquad \|y_m\| = \sqrt{(y_m, y_m)} = \sqrt{\int_a^b p(x) y_m^2(x)\, dx}.$$

Orthogonal Series, Eigenfunction Expansions

Let $y_0, y_1, \cdots$ be an orthogonal set with respect to weight $p(x)$ on an interval $a \leqq x \leqq b$. Let $f(x)$ be a given function that can be represented in terms of the y_m by a convergent series

$$(3) \qquad f(x) = \sum_{m=0}^{\infty} a_m y_m(x) = a_0 y_0(x) + a_1 y_1(x) + \cdots.$$

This is called an **orthogonal expansion** or **generalized Fourier series** or, if the y_m are eigenfunctions of a Sturm–Liouville problem, an **eigenfunction expansion.** [We use m in (3) because we need n later as a fixed index of Bessel functions.]

The point is that because of the orthogonality we get the unknown coefficients $a_0, a_1, \cdots$ in a simple fashion; these are called the **Fourier constants**

[17] LEOPOLD KRONECKER (1823—1891), German mathematician at Berlin, who made important contributions to algebra, group theory, and number theory.

of $f(x)$ with respect to $y_0, y_1, \cdots$. In fact, if we multiply both sides of (3) by $p(x)y_n(x)$ (n fixed) and then integrate over $a \leqq x \leqq b$, we obtain, assuming that term-by-term integration is permissible,[18]

$$(f, y_n) = \int_a^b pfy_n\, dx = \int_a^b p\left(\sum_{m=0}^{\infty} a_m y_m\right) y_n\, dx = \sum_{m=0}^{\infty} a_m (y_m, y_n).$$

Crucial now is that because of the orthogonality, all the integrals (y_m, y_n) on the right are zero, except when $m = n$; then $(y_n, y_n) = \|y_n\|^2$, so that the whole formula reduces to

$$(f, y_n) = a_n \|y_n\|^2 . \tag{3*}$$

Writing m for n, to be in agreement with the notation in (3), we get the desired formula for the Fourier constants

$$a_m = \frac{(f, y_m)}{\|y_m\|^2} = \frac{1}{\|y_m\|^2}\int_a^b p(x)f(x)y_m(x)\, dx \qquad (m = 0, 1, \cdots). \tag{4}$$

EXAMPLE 1 **Fourier series**

The Sturm–Liouville problem in Example 4 of Sec. 5.8 gave an orthogonal set 1, $\cos x$, $\sin x$, $\cos 2x$, $\sin 2x$, $\cdots$ on the interval $-\pi \leqq x \leqq \pi$ with $p(x) = 1$. Hence a corresponding eigenfunction expansion can be written

$$f(x) = a_0 + \sum_{m=1}^{\infty} (a_m \cos mx + b_m \sin mx). \tag{5}$$

This is called the **Fourier series** of $f(x)$. Fourier series are by far the most important eigenfunction expansions (so important to the engineer that we shall devote two chapters, 10 and 11, to them and their applications). The coefficients in (5) are called the **Fourier coefficients** of $f(x)$. From (4) and the norms $\sqrt{2\pi}$ and $\sqrt{\pi}$ (as given in Example 4, Sec. 5.8) we get for these Fourier coefficients the so-called **Euler formulas**

$$\begin{aligned} a_0 &= \frac{1}{2\pi}\int_{-\pi}^{\pi} f(x)\, dx \\ a_m &= \frac{1}{\pi}\int_{-\pi}^{\pi} f(x)\cos mx\, dx \qquad (m = 1, 2, \cdots) \\ b_m &= \frac{1}{\pi}\int_{-\pi}^{\pi} f(x)\sin mx\, dx \qquad (m = 1, 2, \cdots). \end{aligned} \tag{6}$$

For instance, for the "**periodic square wave**" in Fig. 86, given by

$$f(x) = \begin{cases} -1 & \text{if} \quad -\pi < x < 0 \\ 1 & \text{if} \quad 0 < x < \pi \end{cases} \qquad \text{and} \qquad f(x + 2\pi) = f(x)$$

---

[18]This is justified, for instance, in the case of uniform convergence (Theorem 3, Sec. 14.6).

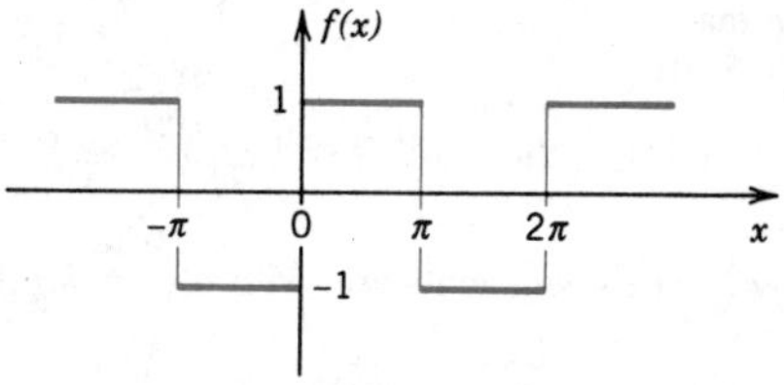

Fig. 86. Periodic square wave in Example 1

we get from (6) the values $a_0 = 0$ and

$$a_m = \frac{1}{\pi}\left[\int_{-\pi}^{0} (-1)\cos mx\,dx + \int_{0}^{\pi} 1\cdot\cos mx\,dx\right] = 0,$$

$$b_m = \frac{1}{\pi}\left[\int_{-\pi}^{0} (-1)\sin mx\,dx + \int_{0}^{\pi} 1\cdot\sin mx\,dx\right]$$

$$= \frac{1}{\pi}\left[\frac{\cos mx}{m}\bigg|_{-\pi}^{0} - \frac{\cos mx}{m}\bigg|_{0}^{\pi}\right]$$

$$= \frac{1}{\pi m}[1 - 2\cos m\pi + 1] = \begin{cases} 4/\pi m & \text{if } m = 1, 3, \cdots, \\ 0 & \text{if } m = 2, 4, \cdots. \end{cases}$$

Hence the Fourier series of the periodic square wave is

$$f(x) = \frac{4}{\pi}\left(\sin x + \frac{1}{3}\sin 3x + \frac{1}{5}\sin 5x + \cdots\right).$$

Many further Fourier series will be discussed in Chap. 10. ∎

EXAMPLE 2 Fourier–Legendre series

This is an eigenfunction expansion

$$f(x) = \sum_{m=0}^{\infty} a_m P_m(x) = a_0P_0 + a_1P_1(x) + \cdots = a_0 + a_1x + a_2\left(\frac{3}{2}x^2 - \frac{1}{2}\right) + \cdots$$

in terms of Legendre polynomials (Sec. 5.3), the eigenfunctions of the Sturm–Liouville problem in Example 5 of Sec. 5.8 on the interval $-1 \leqq x \leqq 1$. We have $p(x) = 1$ for Legendre's equation, and (4) gives, because of (11) in Sec. 5.8,

$$a_m = \frac{2m+1}{2}\int_{-1}^{1} f(x)P_m(x)\,dx, \qquad m = 0, 1, \cdots. \tag{7}$$

∎

EXAMPLE 3 Fourier–Bessel series

In Example 6 of Sec. 5.8 we have obtained infinitely many orthogonal sets of Bessel functions, one for each of $J_0, J_1, J_2, \cdots$. The orthogonality is on an interval $0 \leqq x \leqq R$ with any fixed positive R and with respect to weight x. The set for J_n is $J_n(k_{1n}x), J_n(k_{2n}x), J_n(k_{3n}x), \cdots$, where n is fixed, and k_{mn} is given in (12), Sec. 5.8. The corresponding Fourier–Bessel series is

$$f(x) = \sum_{m=1}^{\infty} a_m J_n(k_{mn}x) = a_1J_n(k_{1n}x) + a_2J_n(k_{2n}x) + a_3J_n(k_{3n}x) + \cdots.$$

We claim that

$$\|J_n(k_{mn}x)\|^2 = \int_0^R xJ_n{}^2(k_{mn}x)\,dx = \frac{R^2}{2}J_{n+1}^2(k_{mn}R), \tag{8}$$

so that by (4) the coefficients are (with $\alpha_{mn} = k_{mn}R$)

$$a_m = \frac{2}{R^2J_{n+1}^2(\alpha_{mn})}\int_0^R xf(x)J_n(k_{mn}x)\,dx, \qquad m = 1, 2, \cdots. \tag{9}$$

Proof of (8). Bessel's equation in Sturm–Liouville form is (see Example 6 in Sec. 5.8)

$$[xJ_n'(kx)]' + \left(-\frac{n^2}{x} + k^2x\right)J_n(kx) = 0.$$

Multiplication by $2xJ_n'(kx)$ gives an equation that can be written

$$\{[xJ_n'(kx)]^2\}' + (k^2x^2 - n^2)\{J_n{}^2(kx)\}' = 0,$$

as can be verified by performing the indicated differentiations $\{\cdots\}'$ in the result. Integration over x from 0 to R now gives

$$[xJ_n'(kx)]^2\Big|_0^R = -\int_0^R (k^2x^2 - n^2)\{J_n{}^2(kx)\}'\,dx. \tag{10}$$

From (2) in Sec. 5.6, writing $\bar{x}$ and n instead of x and ν, and $\dot{J}_n(\bar{x}) = dJ_n/d\bar{x}$, we obtain

$$-n\bar{x}^{-n-1}J_n(\bar{x}) + \bar{x}^{-n}\dot{J}_n(\bar{x}) = -\bar{x}^{-n}J_{n+1}(\bar{x}).$$

Taking the first term to the right, multiplying the equation by $\bar{x}^{n+1}$, and setting $\bar{x} = kx$, we get

$$kxJ_n'(kx)\frac{1}{k} = nJ_n(kx) - kxJ_{n+1}(kx),$$

where the prime denotes the derivative with respect to x. Hence the left side of (10) equals

$$\Big[[nJ_n(kx) - kxJ_{n+1}(kx)]^2\Big]_{x=0}^R.$$

Now if $k = k_{mn} = \alpha_{mn}/R$, then $J_n(kR) = J_n(\alpha_{mn}) = 0$, and since $J_n(0) = 0$ [$n = 1, 2, \cdots$; see (11), Sec. 5.5], that left side reduces to

$$k_{mn}{}^2R^2J_{n+1}^2(k_{mn}R). \tag{11}$$

If we integrate by parts on the right side of (10), the right side becomes

$$-\Big[(k^2x^2 - n^2)J_n{}^2(kx)\Big]_0^R + 2k^2\int_0^R xJ_n{}^2(kx)\,dx.$$

For $k = k_{mn}$ the first expression is zero at $x = R$. It is also zero at $x = 0$ because $k^2x^2 - n^2 = 0$ if $n = x = 0$, and $J_n(kx) = 0$ if $x = 0$ and $n = 1, 2, \cdots$. From this and (11) we readily obtain (8), and the proof is complete. ∎

Completeness of Orthonormal Sets

In practice, one uses only orthonormal sets that consist of "sufficiently many" functions, so that one can represent large classes of functions—certainly all continuous functions on an interval $a \leqq x \leqq b$—by a generalized Fourier series (3). These orthonormal sets are called *"complete"* (in the set of functions considered; definition below). For instance, the orthonormal set in Example 1 is complete in the set of continuous[19] functions on the interval $-\pi \leqq x \leqq \pi$, and so are the sets of Legendre polynomials and Bessel functions in Examples 2 and 3 on their respective intervals.

In this connection, convergence is *convergence in the norm* (also called *mean-square convergence* or *mean convergence*); that is, a sequence of functions f_n is called **convergent** with the limit f if

$$\lim_{k\to\infty} \|f_k - f\| = 0, \tag{12*}$$

written out by (2) (where we can drop the square root)

$$\lim_{k\to\infty} \int_a^b p(x)[f_k(x) - f(x)]^2\, dx = 0. \tag{12}$$

Accordingly, (3) converges and represents f if

$$\lim_{k\to\infty} \int_a^b p(x)[s_k(x) - f(x)]^2\, dx = 0 \tag{13}$$

where s_k is the kth partial sum of (3),

$$s_k(x) = \sum_{m=0}^{k} a_m y_m(x). \tag{14}$$

By definition, an orthonormal set $y_0, y_1, \cdots$ on an interval $a \leqq x \leqq b$ is **complete** *in a set of functions* S defined on $a \leqq x \leqq b$ if we can approximate every f belonging to S arbitrarily closely by a linear combination $a_0y_0 + a_1y_1 + \cdots + a_ky_k$, that is, technically, if for every $\epsilon > 0$ we can find constants $a_0, \cdots, a_k$ (with k large enough) such that

$$\|f - (a_0y_0 + \cdots + a_ky_k)\| < \epsilon. \tag{15}$$

An interesting and basic consequence of the integral in (13) is obtained as follows. Performing the square and using (14), we first have

---

[19]Actually, piecewise continuous and much more general functions, but a full discussion of completeness would need prerequisites not required in this book. See Ref. [9], Secs. 3.4–3.7 listed in Appendix 1. (There "complete sets" are called "total sets," a more modern term.)

$$\int_a^b p(x)[s_k(x) - f(x)]^2\,dx = \int_a^b p s_k^{\,2}\,dx - 2\int_a^b p f s_k\,dx + \int_a^b p f^2\,dx$$

$$= \int_a^b p\left[\sum_{m=0}^{k} a_m y_m\right]^2 dx - 2\sum_{m=0}^{k} a_m \int_a^b p f y_m\,dx + \int_a^b p f^2\,dx.$$

The first integral on the right equals $a_m^{\,2}$ because $\int p y_m y_l\,dx = 0$ for $m \neq l$, and $\int p y_m^{\,2}\,dx = 1$. In the second sum on the right, the integral equals a_m, by (4) with $\|y_m\|^2 = 1$. Hence the right side reduces to

$$-\sum_{m=0}^{k} a_m^{\,2} + \int_a^b p f^2\,dx.$$

This is nonnegative because in the previous formula the integrand on the left and thus the integral on the left are nonnegative. This proves the important **Bessel's inequality**

$$\sum_{m=0}^{k} a_m^{\,2} \leqq \|f\|^2 = \int_a^b p(x) f(x)^2\,dx \qquad (k = 1, 2, \cdots). \tag{16}$$

Here we can let $k \to \infty$, because the left sides form a monotone increasing sequence that is bounded by the right side, so that we have convergence by the familiar Theorem 1 in Appendix 3 (see A3.3). Hence

$$\sum_{m=0}^{\infty} a_m^{\,2} \leqq \|f\|^2. \tag{17}$$

Furthermore, if $y_0, y_1, \cdots$ is complete in a set of functions S, then (13) holds for every f belonging to S. By (15) this implies equality in (16) with $k \to \infty$; hence in the case of completeness the so-called **Parseval's equality**

$$\sum_{m=0}^{\infty} a_m^{\,2} = \|f\|^2 = \int_a^b p(x) f(x)^2\,dx \tag{18}$$

holds for every f in S.

As a consequence of (18) we prove that in the case of *completeness* there is no function orthogonal to *every* function of the orthonormal set, with the trivial exception of a function of zero norm:

Theorem 1 (Completeness)

Let $y_0, y_1, \cdots$ be a complete orthonormal set on $a \leqq x \leqq b$ in a set of functions S. Then if a function f belongs to S and is orthogonal to every y_m, it must have norm zero. In particular, if f is continuous, it must be identically zero.

Proof. From the orthogonality assumption we see that the left side of (18) must be zero. This proves the first statement. If f is continuous, then $\|f\| = 0$ implies $f(x) \equiv 0$, as can be seen directly from (2), with f instead of y_m. ∎

EXAMPLE 4 Fourier series

The orthonormal set in Example 1 is complete in the set of continuous functions on $-\pi \leqq x \leqq \pi$. Verify directly that $f(x) \equiv 0$ is the only continuous function orthogonal to all the functions of that set.

Solution. Let f be any continuous function. By the orthogonality (we can omit $\sqrt{2\pi}$ and $\sqrt{\pi}$),

$$\int_{-\pi}^{\pi} 1 \cdot f(x)\, dx = 0, \qquad \int_{-\pi}^{\pi} f(x) \cos mx\, dx = 0, \qquad \int_{-\pi}^{\pi} f(x) \sin mx\, dx = 0.$$

Hence $a_m = 0$ and $b_m = 0$ in (6) for all m, so that (3) reduces to $f(x) \equiv 0$. ∎

This is the end of Chap. 5 on the power series and Frobenius methods, which are indispensible in solving linear differential equations with variable coefficients, some of the most important of which we have discussed and solved. We have also seen that the latter equations are important sources of special functions having orthogonality properties that make them suitable for orthogonal series representations of given functions.

Problem Set 5.9

Legendre polynomials. Represent the following polynomials in terms of Legendre polynomials.

1. $5x^3 + x$

2. $10x^3 - 3x^2 - 5x - 1$

3. $1, x, x^2, x^3$

4. $35x^4 + 15x^3 - 30x^2 - 15x + 3$

In each case, obtain the first few terms of the expansion of $f(x)$ in terms of Legendre polynomials and graph the first three partial sums.

5. $f(x) = \begin{cases} 0 & \text{if } -1 < x < 0 \\ x & \text{if } \ \ 0 < x < 1 \end{cases}$

6. $f(x) = \begin{cases} 0 & \text{if } -1 < x < 0 \\ 1 & \text{if } \ \ 0 < x < 1 \end{cases}$

7. $f(x) = |x| \ \text{ if } -1 < x < 1$

8. $f(x) = e^x \ \text{ if } -1 < x < 1$

9. Show that the functions $P_n(\cos\theta)$, $n = 0, 1, \cdots$, form an orthogonal set on the interval $0 \leqq \theta \leqq \pi$ with respect to the weight function $\sin\theta$.

Chebyshev polynomials.[20] The functions

$$T_n(x) = \cos\,(n \arccos x), \quad U_n(x) = \frac{\sin\,[(n+1) \arccos x]}{\sqrt{1 - x^2}} \qquad (n = 0, 1, \cdots)$$

are called *Chebyshev polynomials of the first and second kind,* respectively.

[20]PAFNUTI CHEBYSHEV (1821—1894), Russian mathematician, is known by his work in approximation theory and the theory of numbers. Another transliteration of the name is TCHEBICHEF.

10. Show that

$$T_0 = 1, \qquad T_1(x) = x, \qquad T_2(x) = 2x^2 - 1, \qquad T_3(x) = 4x^3 - 3x,$$

$$U_0 = 1, \qquad U_1(x) = 2x, \qquad U_2(x) = 4x^2 - 1, \qquad U_3(x) = 8x^3 - 4x.$$

11. Show that the Chebyshev polynomials $T_n(x)$ are orthogonal on the interval $-1 \leqq x \leqq 1$ with respect to the weight function $p(x) = 1/\sqrt{1 - x^2}$. *Hint.* To evaluate the integral, set arc cos $x = \theta$.

12. Show that $T_n(x)$ is a solution of the differential equation

$$(1 - x^2)T_n'' - xT_n' + n^2T_n = 0.$$

Laguerre polynomials.[21] The functions

$$L_0 = 1, \qquad L_n(x) = \frac{e^x}{n!} \frac{d^n(x^n e^{-x})}{dx^n}, \qquad n = 1, 2, \cdots$$

are called *Laguerre polynomials*.

13. Show that

$$L_1(x) = 1 - x, \quad L_2(x) = 1 - 2x + x^2/2, \quad L_3(x) = 1 - 3x + 3x^2/2 - x^3/6.$$

14. Verify by direct integration that L_0, $L_1(x)$, $L_2(x)$ are orthogonal on the positive axis $0 \leqq x < \infty$ with respect to the weight function $p(x) = e^{-x}$.

15. Prove that the set of all Laguerre polynomials is orthogonal on $0 \leqq x < \infty$ with respect to the weight function $p(x) = e^{-x}$.

16. Show that

$$L_n(x) = \sum_{m=0}^{n} \frac{(-1)^m}{m!} \binom{n}{m} x^m = 1 - nx + \frac{n(n-1)}{4} x^2 - + \cdots + \frac{(-1)^n}{n!} x^n.$$

17. $L_n(x)$ satisfies Laguerre's differential equation $xy'' + (1 - x)y' + ny = 0$. Verify this fact for $n = 0, 1, 2, 3$.

Hermite polynomials[22]. The functions

$$He_0 = 1, \qquad He_n(x) = (-1)^n e^{x^2/2} \frac{d^n}{dx^n} (e^{-x^2/2}), \qquad n = 1, 2, \cdots$$

are called *Hermite polynomials*.

Remark. As is true for many special functions, the literature contains more than one notation, and one sometimes defines as Hermite polynomials the functions

$$H_0^* = 1, \qquad H_n^*(x) = (-1)^n e^{x^2} \frac{d^n e^{-x^2}}{dx^n}.$$

This differs from our definition, which is preferably used in applications.

---

[21] EDMOND LAGUERRE (1834—1886), French mathematician, who did research work in geometry and the theory of infinite series.

[22] CHARLES HERMITE (1822—1901), French mathematician, is known by his work in algebra and number theory. The great HENRI POINCARÉ (1854—1912) was one of his students.

18. Show that

$$He_1(x) = x, \ He_2(x) = x^2 - 1, \ He_3(x) = x^3 - 3x, \ He_4(x) = x^4 - 6x^2 + 3.$$

19. Show that the Hermite polynomials are related to the coefficients of the Maclaurin series

$$e^{tx - t^2/2} = \sum_{n=0}^{\infty} a_n(x) t^n$$

by the formula $He_n(x) = n!a_n(x)$. *Hint*. Note that $tx - . \ 2 = x^2/2 - (x - t)^2/2$. (The exponential function is called the **generating function** of the He_n.)

20. Show that the Hermite polynomials satisfy the relation

$$He_{n+1}(x) = xHe_n(x) - He_n'(x).$$

21. Differentiating the generating function in Prob. 19 with respect to x, show that

$$He_n'(x) = nHe_{n-1}(x).$$

Using this and the formula in Prob. 20 (with n replaced by $n - 1$), prove that $He_n(x)$ satisfies the differential equation

$$y'' - xy' + ny = 0.$$

22. Using the differential equation in Prob. 21, show that $w = e^{-x^2/4}He_n(x)$ is a solution of **Weber's equation**[23]

$$w'' + (n + \tfrac{1}{2} - \tfrac{1}{4}x^2)w = 0 \qquad (n = 0, 1, \cdots).$$

23. Show that the Hermite polynomials are orthogonal on the x-axis $-\infty < x < \infty$ with respect to the weight function $p(x) = e^{-x^2/2}$.

Bessel functions

24. Sketch $J_0(\lambda_{10}x)$, $J_0(\lambda_{20}x)$, $J_0(\lambda_{30}x)$, and $J_0(\lambda_{40}x)$ for $R = 1$ in the interval $0 \leqq x \leqq 1$. (Use Table A1 in Appendix 5.)

Develop the following functions $f(x)$ $(0 < x < R)$ in a **Fourier–Bessel series**

$$f(x) = a_1J_0(\lambda_{10}x) + a_2J_0(\lambda_{20}x) + a_3J_0(\lambda_{30}x) + \cdots$$

and sketch the first few partial sums.

25. $f(x) = 1$. *Hint*. Use (1), Sec. 5.6.

26. $f(x) = \begin{cases} 1 & \text{if} \quad 0 < x < R/2 \\ 0 & \text{if} \quad R/2 < x < R \end{cases}$

27. $f(x) = \begin{cases} k & \text{if} \quad 0 < x < a \\ 0 & \text{if} \quad a < x < R \end{cases}$

28. $f(x) = \begin{cases} 0 & \text{if} \quad 0 < x < R/2 \\ k & \text{if} \quad R/2 < x < R \end{cases}$

29. $f(x) = 1 - x^2$ $(R = 1)$. *Hint*. Use (1), Sec. 5.6, and integration by parts.[24]

30. $f(x) = R^2 - x^2$

31. $f(x) = x^2$

32. $f(x) = x^4$

---

[23]HEINRICH WEBER (1842—1913), German mathematician.

[24]This problem will be needed in Example 1 of Sec. 11.10.

33. Show that $f(x) = x^n$ $(0 < x < 1, n = 0, 1, \cdots)$ can be represented by the Fourier–Bessel series

$$x^n = \frac{2J_n(\alpha_{1n}x)}{\alpha_{1n}J_{n+1}(\alpha_{1n})} + \frac{2J_n(\alpha_{2n}x)}{\alpha_{2n}J_{n+1}(\alpha_{2n})} + \cdots.$$

34. Find a representation of $f(x) = x^n$ $(0 < x < R, n = 0, 1, \cdots)$ similar to that in Prob. 33.

35. Represent $f(x) = x^3$ $(0 < x < 2)$ by a Fourier–Bessel series involving J_3.

Review Questions and Problems for Chapter 5

1. What is a power series? Can it contain negative powers? Fractional powers? How would you test for convergence?
2. We always took $x_0 = 0$ and considered power series in powers of x. Why was this no essential restriction of generality?
3. Can a power series solution reduce to a polynomial? To a constant?
4. Why did we generalize the power series method? Give an equation to which the Frobenius method applies whereas the power series method does not.
5. Can a solution in the Frobenius method reduce to a power series?
6. What is the indicial equation? How did it arise? What is it good for?
7. Can the indicial equation have complex roots?
8. What do we mean by saying that a function is analytic at some point? Why did this concept arise in this chapter?
9. For which of the two equations, Legendre's and Bessel's equations, did we need the Frobenius method?
10. In the Frobenius method, we had three cases. What were they? Which of them did the Euler–Cauchy equation reveal and which not?
11. What is a Sturm–Liouville problem? Give an example.
12. What is an eigenfunction expansion? How did it arise in Sturm–Liouville problems?
13. Why is orthogonality of eigenfunctions important?
14. What do you remember about Fourier–Legendre series?
15. What do you know about the orthogonality of Bessel functions?

Using one of the two methods discussed in this chapter, find a basis of solutions of the following equations. Try to identify the series obtained as expansions of known functions.

16. $y'' - 4y = 0$
17. $x(1 - x)y'' + 2(1 - 2x)y' - 2y = 0$
18. $(x - 1)y'' - xy' + y = 0$
19. $xy'' + (1 - 2x)y' + (x - 1)y = 0$
20. $(x - 1)^2y'' + (x - 1)y' - y = 0$
21. $2x(x - 1)y'' - (x + 1)y' + y = 0$
22. $xy'' + 2y' + xy = 0$

23. $x(x + 1)^2y'' + (1 - x^2)y' + (x - 1)y = 0$
24. $x^2y'' - 5xy' + 9y = 0$
25. $x^2(x + 1)^2y'' - (5x^2 + 8x + 3)xy' + (9x^2 + 11x + 4)y = 0$

Using the indicated substitutions, solve, in terms of Bessel functions:

26. $x^2y'' + (x^2 + \frac{1}{4})y = 0 \quad (y = u\sqrt{x})$
27. $xy'' + 3y' + xy = 0 \quad (y = u/x)$
28. $x^2y'' + xy' + 4(x^4 - \nu^2)y = 0 \quad (x^2 = z)$
29. $xy'' - y' + xy = 0 \quad (y = xu)$
30. $y'' - xy = 0 \quad (y = u\sqrt{x}, \frac{2}{3}ix^{3/2} = z; i = \sqrt{-1}$; **Airy's equation**)

Show orthogonality on the given interval and determine the corresponding orthonormal set of functions.

31. $1, \cos \pi x, \cos 2\pi x, \cos 3\pi x, \cdots, \quad -1 \leqq x \leqq 1$
32. $1, \cos \omega x, \cos 2\omega x, \cos 3\omega x, \cdots, \quad 0 \leqq x \leqq 2\pi/\omega$
33. $1, \cos 4mx, \sin 4mx, m = 1, 2, \cdots, \quad 0 \leqq x \leqq \frac{1}{2}\pi$
34. $1, \cos(\pi x/L), \sin(\pi x/L), \cos(2\pi x/L), \sin(2\pi x/L), \cdots, \quad 0 \leqq x \leqq 2L$
35. $P_0(\frac{1}{2}x), P_1(\frac{1}{2}x), P_2(\frac{1}{2}x), \quad -2 \leqq x \leqq 2$; see Sec. 5.3.

Find the eigenvalues and eigenfunctions of the following problems.

36. $y'' + \lambda y = 0, \quad y(0) = 0, \quad y(\frac{1}{2}\pi) = 0$
37. $(xy')' + \lambda y/x = 0 \quad (\lambda > 0), \quad y(1) = 0, \quad y(e^{\pi}) = 0.$ *Hint.* Set $x = e^t$.
38. $y'' + \lambda y = 0, \quad y(0) = y(2L), \quad y'(0) = y'(2L)$
39. $(xy')' + \lambda y/x = 0 \quad (\lambda > 0), \quad y(e^{1/2}) = 0, \quad y(e^{3/2}) = 0$
40. $x^2y'' + xy' + (\lambda^2x^2 - 1)y = 0, \quad y(0) = 0, \quad y(1) = 0$

Expand in terms of Legendre polynomials:

41. $5x^3 - 3x^2 - x - 1$
42. $1, x, x^2, x^4$
43. $70x^4 - 45x^2 - 3$
44. $231(x^6 + 1)$
45. $(x - 3)^3$

Summary of Chapter 5
Series Solutions. Special Functions

The **power series method** is a general method for solving linear differential equations

$$(1) \qquad y'' + p(x)y' + q(x)y = r(x)$$

with variable $p(x)$, $q(x)$, and $r(x)$; it also applies to higher order equations. It gives solutions in the form of power series; this motivates the name. In this method, one substitutes a power series (with any center x_0, e.g., $x_0 = 0$)

$$(2) \qquad y(x) = a_0 + a_1(x - x_0) + a_2(x - x_0)^2 + \cdots$$

into (1), similarly $y'(x) = a_1 + 2a_2(x - x_0) + \cdots$ and $y''(x)$. In this way, one determines the undetermined coefficients a_m in (2), as explained in Sec. 5.1 and other sections. This gives solutions y represented by power series. If $p(x)$, $q(x)$, and $r(x)$ are **analytic** at $x = x_0$ (Sec. 5.2), then (1) has solutions of this form. The same holds if $\tilde{h}(x)$, $\tilde{p}(x)$, $\tilde{q}(x)$, and $\tilde{r}(x)$ in

$$\tilde{h}(x)y'' + \tilde{p}(x)y' + \tilde{q}(x)y = \tilde{r}(x)$$

are analytic at $x = x_0$ and $\tilde{h}(x_0) \neq 0$.

The **Frobenius method** (Sec. 5.4) extends the power series method to equations

$$y'' + \frac{a(x)}{x - x_0} y' + \frac{b(x)}{(x - x_0)^2} y = 0 \tag{3}$$

whose coefficients are **singular** (i.e., not analytic) at $x = x_0$, but are not "too bad," namely, such that $a(x)$ and $b(x)$ are analytic at $x = x_0$. Then (3) has at least one solution of the form

$$y(x) = x^r[a_0 + a_1(x - x_0) + a_2(x - x_0)^2 + \cdots], \tag{4}$$

where r can be any real number or even a complex number and is determined by substituting (4) into (3); this also gives the a_m's. A second independent solution may be of a similar form (with different r and a_m's) or may involve a logarithmic term.

"Special functions" is a common name for higher functions, as opposed to the usual functions of calculus. They arise from (1) or (3) and get a special name and notation if they are important in applications. Of this kind, and particularly useful to the engineer and physicist, are **Legendre's equation** and the Legendre polynomials $P_0(x)$, $P_1(x)$, $P_2(x)$, $\cdots$ (Sec. 5.3), the **hypergeometric equation** and the hypergeometric functions $F(a, b, c; x)$ (Sec. 5.4), and **Bessel's equation** and the Bessel functions J_ν and Y_ν (Secs. 5.5–5.7). Indeed, second-order linear differential equations are one of the two main sources of such "higher functions." [Other special functions arise from nonelementary integrals, such as those listed in Appendix 3. The gamma function (Sec. 5.5) is of that type.]

Modeling involving differential equations in most cases leads to initial value (Chap. 2) or boundary value problems. Many of the latter can be written in the form of **Sturm–Liouville problems** (Sec. 5.8). These are **eigenvalue problems** involving a parameter λ, which in applications may be related to frequencies, energies, or other physical quantities. Solutions of these problems, called *eigenfunctions,* have many general properties in common, notably the highly important **orthogonality** (Sec. 5.8). This leads to **eigenfunction expansions** (Sec. 5.9), such as those involving cosine and sine ("*Fourier series*"), Legendre polynomials, or Bessel functions.

Chapter 6

Laplace Transforms

The Laplace transform method solves differential equations and corresponding initial and boundary value problems. The process of solution consists of three main steps:

1st step. The given "hard" problem is transformed into a "simple" equation **(subsidiary equation).**

2nd step. The subsidiary equation is solved by purely algebraic manipulations.

3rd step. The solution of the subsidiary equation is transformed back to obtain the solution of the given problem.

In this way Laplace transforms reduce the problem of solving a differential equation to an algebraic problem. The third step is made easier by tables, whose role is similar to that of integral tables in integration. (These tables are also useful in the first step.) Such a table is included at the end of the chapter.

This switching from operations of calculus to *algebraic* operations on transforms is called **operational calculus,** a very important area of applied mathematics, and the Laplace transform method is practically the most important method for this purpose. (For another such method, the Fourier transform method, see Sec. 10.11.) Indeed, Laplace transforms have numerous engineering applications. They are particularly useful in problems where the (mechanical or electrical) driving force has discontinuities, is impulsive or is periodic but not merely a sine or cosine function. Another advantage is that the method solves problems directly. Indeed, initial value problems are solved without first determining a general solution. Similarly, nonhomogeneous equations are solved without first solving the corresponding homogeneous equation.

In this chapter we consider Laplace transforms from a practical point of view and illustrate their use by important engineering problems, many of them related to *ordinary* differential equations.

Partial differential equations can also be treated by Laplace transforms, as we show in Sec. 11.13.

Section 6.9 contains a list of general formulas and Sec. 6.10 a list of transforms $F(s)$ and corresponding functions $f(t)$.

Prerequisite for this chapter: Chap. 2.
Sections that may be omitted in a very short course: 6.5–6.7.
References: Appendix 1, Part A.
Answers to problems: Appendix 2.

6.1 Laplace Transform. Inverse Transform. Linearity

Let $f(t)$ be a given function that is defined for all $t \geqq 0$. We multiply $f(t)$ by e^{-st} and integrate with respect to t from zero to infinity. Then, if the resulting integral exists (that is, has some finite value), it is a function of s, say, $F(s)$:

$$F(s) = \int_0^\infty e^{-st} f(t)\, dt.$$

This function $F(s)$ of the variable s is called the **Laplace transform**[1] of the original function $f(t)$, and will be denoted by $\mathcal{L}(f)$. Thus

$$F(s) = \mathcal{L}(f) = \int_0^\infty e^{-st} f(t)\, dt. \tag{1}$$

So remember: the given function depends on t and the new function (its transform) depends on s.

The operation just described, which yields $F(s)$ from a given $f(t)$, is also called the **Laplace transform.**

Furthermore, the original function $f(t)$ in (1) is called the *inverse transform* or **inverse** of $F(s)$ and will be denoted by $\mathcal{L}^{-1}(F)$; that is, we shall write

$$f(t) = \mathcal{L}^{-1}(F).$$

Notation

Original functions are denoted by lowercase letters and their transforms by the same letters in capitals, so that $F(s)$ denotes the transform of $f(t)$, and $Y(s)$ denotes the transform of $y(t)$, and so on.

EXAMPLE 1 Laplace transform

Let $f(t) = 1$ when $t \geqq 0$. Find $F(s)$.

Solution. From (1) we obtain by integration

$$\mathcal{L}(f) = \mathcal{L}(1) = \int_0^\infty e^{-st}\, dt = -\frac{1}{s} e^{-st} \Big|_0^\infty ;$$

---

[1] PIERRE SIMON MARQUIS DE LAPLACE (1749—1827), great French mathematician, was professor in Paris. He developed the foundation of potential theory and made important contributions to celestial mechanics, astronomy in general, special functions, and probability theory. Napoléon Bonaparte was his student for a year. For Laplace's interesting political involvements, see Ref. [2], p. 260, listed in Appendix 1.

The powerful practical Laplace transform techniques were developed over a century later by the English electrical engineer OLIVER HEAVISIDE (1850—1925) and were often called "Heaviside calculus."

hence, when $s > 0$,

$$\mathcal{L}(1) = \frac{1}{s}.$$

Our notation in the first line on the right is convenient, but we should say a word about it. The interval of integration in (1) is infinite. Such an integral is called an **improper integral** and, by definition, is evaluated according to the rule

$$\int_0^\infty e^{-st} f(t)\, dt = \lim_{T\to\infty} \int_0^T e^{-st} f(t)\, dt.$$

Hence our convenient notation means

$$\int_0^\infty e^{-st}\, dt = \lim_{T\to\infty} \left[-\frac{1}{s} e^{-st} \right]_0^T = \lim_{T\to\infty} \left[-\frac{1}{s} e^{-sT} + \frac{1}{s} e^0 \right] = \frac{1}{s} \qquad (s > 0).$$

We shall use that notation throughout this chapter. ■

EXAMPLE 2 **Laplace transform**

Let $f(t) = e^{at}$ when $t \geqq 0$, where a is a constant. Find $\mathcal{L}(f)$.

Solution. Again by (1),

$$\mathcal{L}(e^{at}) = \int_0^\infty e^{-st} e^{at}\, dt = \frac{1}{a - s} e^{-(s-a)t} \Big|_0^\infty ;$$

hence, when $s - a > 0$,

$$\mathcal{L}(e^{at}) = \frac{1}{s - a}.$$ ■

Must we go on in this fashion and obtain the transform of one function after another directly from the definition? The answer is no. And the reason is that the Laplace transform has many general properties that are helpful for that purpose. Above all, the Laplace transform is a "linear operation," just as differentiation and integration. By this we mean the following.

Theorem 1 **(Linearity of the Laplace transform)**

The Laplace transform is a linear operation; that is, for any functions $f(t)$ and $g(t)$ whose Laplace transforms exist and any constants a and b,

$$\mathcal{L}\{af(t) + bg(t)\} = a\mathcal{L}\{f(t)\} + b\mathcal{L}\{g(t)\}.$$

Proof. By the definition,

$$\begin{aligned} \mathcal{L}\{af(t) + bg(t)\} &= \int_0^\infty e^{-st}[af(t) + bg(t)]\, dt \\ &= a\int_0^\infty e^{-st} f(t)\, dt + b\int_0^\infty e^{-st} g(t)\, dt \\ &= a\mathcal{L}\{f(t)\} + b\mathcal{L}\{g(t)\}. \end{aligned}$$ ■

EXAMPLE 3 An application of Theorem 1

Let $f(t) = \cosh at = (e^{at} + e^{-at})/2$. Find $\mathcal{L}(f)$.

Solution. From Theorem 1 and Example 2 we obtain

$$\mathcal{L}(\cosh at) = \frac{1}{2}\mathcal{L}(e^{at}) + \frac{1}{2}\mathcal{L}(e^{-at}) = \frac{1}{2}\left(\frac{1}{s-a} + \frac{1}{s+a}\right);$$

that is, when $s > a\ (\geqq 0)$,

$$\mathcal{L}(\cosh at) = \frac{s}{s^2 - a^2}.$$

■

EXAMPLE 4 A simple partial fraction reduction

Let $F(s) = \dfrac{1}{(s-a)(s-b)}$, $a \neq b$. Find $\mathcal{L}^{-1}(F)$.

Solution. The inverse of a linear transformation is linear (see Prob. 40). By partial fraction reduction we thus obtain from Example 2

$$\mathcal{L}^{-1}(F) = \mathcal{L}^{-1}\left\{\frac{1}{a-b}\left(\frac{1}{s-a} - \frac{1}{s-b}\right)\right\}$$

$$= \frac{1}{a-b}\left[\mathcal{L}^{-1}\left(\frac{1}{s-a}\right) - \mathcal{L}^{-1}\left(\frac{1}{s-b}\right)\right] = \frac{1}{a-b}(e^{at} - e^{bt}).$$

This proves formula 11 in the table in Sec. 6.10 near the end of the chapter. ■

EXAMPLE 5 Derivation of formula 12 in the table in Sec. 6.10

Let $F(s) = \dfrac{s}{(s-a)(s-b)}$, $a \neq b$. Find $\mathcal{L}^{-1}(F)$.

Solution. Using the idea of Example 4, we obtain

$$\mathcal{L}^{-1}\left\{\frac{s}{(s-a)(s-b)}\right\} = \mathcal{L}^{-1}\left\{\frac{1}{a-b}\left(\frac{a}{s-a} - \frac{b}{s-b}\right)\right\} = \frac{1}{a-b}(ae^{at} - be^{bt}).$$ ■

A short list of some important elementary functions and their Laplace transforms is given in Table 6.1, and a more extensive list in Sec. 6.10. See also the references in Appendix 1, Part A.

Once we know the transforms in Table 6.1, nearly all the transforms we shall need can be obtained through the use of some simple general theorems which we consider in the next sections.

Formulas 1, 2, and 3 in Table 6.1 are special cases of formula 4. Formula 4 is proved by induction as follows. It is true for $n = 0$ because of Example 1 and $0! = 1$. We now make the induction hypothesis that it hold for any positive integer n. From (1) we get by integration by parts

$$\mathcal{L}(t^{n+1}) = \int_0^\infty e^{-st}t^{n+1}\,dt = -\frac{1}{s}e^{-st}t^{n+1}\Big|_0^\infty + \frac{(n+1)}{s}\int_0^\infty e^{-st}t^n\,dt.$$

The integral-free part is zero at $t = 0$ and for $t \to \infty$. The right side equals $(n+1)\mathcal{L}(t^n)/s$. From this and the induction hypothesis we obtain

$$\mathcal{L}(t^{n+1}) = \frac{n+1}{s}\mathcal{L}(t^n) = \frac{(n+1)n!}{s \cdot s^n} = \frac{(n+1)!}{s^{n+1}}.$$

This proves formula 4.

Table 6.1
Some Functions $f(t)$ and Their Laplace Transforms $\mathcal{L}(f)$

	$f(t)$	$\mathcal{L}(f)$		$f(t)$	$\mathcal{L}(f)$
1	1	$1/s$	6	e^{at}	$\dfrac{1}{s-a}$
2	t	$1/s^2$	7	$\cos \omega t$	$\dfrac{s}{s^2+\omega^2}$
3	t^2	$2!/s^3$	8	$\sin \omega t$	$\dfrac{\omega}{s^2+\omega^2}$
4	t^n $(n = 0, 1, \cdots)$	$\dfrac{n!}{s^{n+1}}$	9	$\cosh at$	$\dfrac{s}{s^2-a^2}$
5	t^a (a positive)	$\dfrac{\Gamma(a+1)}{s^{a+1}}$	10	$\sinh at$	$\dfrac{a}{s^2-a^2}$

$\Gamma(a + 1)$ in formula 5 is the so-called *gamma function* [(15) in Sec. 5.5 or (24) in Appendix 3], and we get formula 5 from (1), setting $st = x$:

$$\mathcal{L}(t^a) = \int_0^\infty e^{-st} t^a \, dt = \int_0^\infty e^{-x} \left(\frac{x}{s}\right)^a \frac{dx}{s} \qquad (s > 0)$$

$$= \frac{1}{s^{a+1}} \int_0^\infty e^{-x} x^a \, dx = \frac{\Gamma(a+1)}{s^{a+1}}$$

because the last integral is precisely that defining $\Gamma(a + 1)$. Note further that $\Gamma(n + 1) = n!$ when n is a nonnegative integer, so that formula 4 also follows from formula 5.

Formula 6 was proved in Example 2. To prove the formulas 7 and 8, we set $a = i\omega$ in formula 6. Then

$$\mathcal{L}(e^{i\omega t}) = \frac{1}{s - i\omega} = \frac{s + i\omega}{(s - i\omega)(s + i\omega)}$$

$$= \frac{s + i\omega}{s^2 + \omega^2} = \frac{s}{s^2 + \omega^2} + i \frac{\omega}{s^2 + \omega^2} .$$

On the other hand, by Theorem 1 and $e^{i\omega t} = \cos \omega t + i \sin \omega t$ (Sec. 2.3),

$$\mathcal{L}(e^{i\omega t}) = \mathcal{L}(\cos \omega t + i \sin \omega t) = \mathcal{L}(\cos \omega t) + i\mathcal{L}(\sin \omega t).$$

Equating the real and imaginary parts of these two equations, we obtain the formulas 7 and 8. Avoiding working in complex, the reader may prove formulas 7 and 8 by using the defining integrals (1) and integration by parts. Another derivation of these formulas in real will be given in the next section.

Formula 9 was proved in Example 3, and formula 10 can be proved in a similar manner. ∎

Existence of Laplace Transforms

In conclusion of this introductory section we should say something about the existence of the Laplace transform. Roughly and intuitively speaking, the situation is as follows. For a fixed s the integral in (1) will exist if the whole integrand $e^{-st}f(t)$ goes to zero fast enough as $t \to \infty$, say, at least like an exponential function with a negative exponent. This motivates the inequality (2) in the subsequent existence theorem. $f(t)$ need not be continuous. This is of practical importance since discontinuous inputs (driving forces) are just those for which the Laplace transform method becomes particularly useful. It suffices to require that $f(t)$ be piecewise continuous on every finite interval in the range $t \geqq 0$.

By definition, a function $f(t)$ is **piecewise continuous** on a finite interval $a \leqq t \leqq b$ if $f(t)$ is defined on that interval and is such that the interval can be subdivided into finitely many intervals, in each of which $f(t)$ is continuous and has finite limits as t approaches either endpoint of the interval of subdivision from the interior.

It follows from this definition that finite jumps are the only discontinuities that a piecewise continuous function may have; these are known as *ordinary discontinuities*. Figure 87 shows an example. Clearly, the class of piecewise continuous functions includes every continuous function.

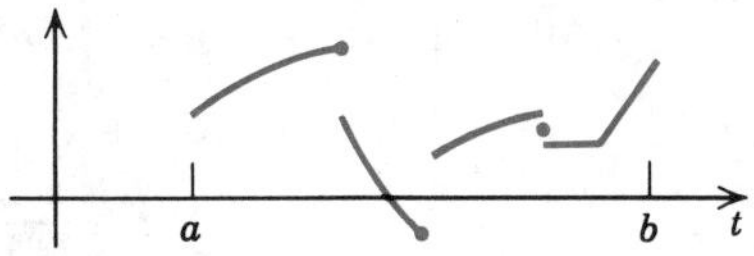

Fig. 87. Example of a piecewise continuous function $f(t)$ (The dots mark the function values at the jumps.)

Theorem 2 **(Existence theorem for Laplace transforms)**

Let $f(t)$ be a function that is piecewise continuous on every finite interval in the range $t \geqq 0$ and satisfies

$$|f(t)| \leqq Me^{\gamma t} \qquad \text{for all } t \geqq 0 \tag{2}$$

and for some constants γ and M. Then the Laplace transform of $f(t)$ exists for all $s > \gamma$.

Proof. Since $f(t)$ is piecewise continuous, $e^{-st}f(t)$ is integrable over any finite interval on the t-axis. From (2), assuming that $s > \gamma$, we obtain

$$|\mathcal{L}(f)| = \left| \int_0^\infty e^{-st} f(t)\, dt \right| \leqq \int_0^\infty |f(t)| e^{-st}\, dt \leqq \int_0^\infty Me^{\gamma t} e^{-st}\, dt = \frac{M}{s - \gamma}$$

where the condition $s > \gamma$ was needed for the existence of the last integral. This completes the proof. ∎

The conditions in Theorem 2 are sufficient for most applications, and it is easy to find out whether a given function satisfies an inequality of the form (2). For example,

$$\text{(3)} \qquad \cosh t < e^t, \qquad t^n < n!\, e^t \quad (n = 0, 1, \cdots) \qquad \text{for all } t > 0,$$

and any function that is bounded in absolute value for all $t \geqq 0$, such as the sine and cosine functions of a real variable, satisfies that condition. An example of a function that does not satisfy a relation of the form (2) is the exponential function e^{t^2}, because, no matter how large we choose M and γ in (2),

$$e^{t^2} > Me^{\gamma t} \qquad \text{for all } t > t_0$$

where t_0 is a sufficiently large number, depending on M and γ.

It should be noted that the conditions in Theorem 2 are sufficient rather than necessary. For example, the function $1/\sqrt{t}$ is infinite at $t = 0$, but its transform exists; in fact, from the definition and $\Gamma(\frac{1}{2}) = \sqrt{\pi}$ [see (30) in Appendix 3] we obtain

$$\mathcal{L}(t^{-1/2}) = \int_0^\infty e^{-st}t^{-1/2}\,dt = \frac{1}{\sqrt{s}}\int_0^\infty e^{-x}x^{-1/2}\,dx = \frac{1}{\sqrt{s}}\,\Gamma(\tfrac{1}{2}) = \sqrt{\frac{\pi}{s}}\,.$$

Uniqueness. If the Laplace transform of a given function exists, it is uniquely determined. Conversely, it can be shown that if two functions (both defined on the positive real axis) have the same transform, these functions cannot differ over an interval of positive length, although they may differ at various isolated points (see Ref. [A10] in Appendix 1). Since this is of no importance in applications, we may say that the inverse of a given transform is essentially unique. In particular, if two *continuous* functions have the same transform, they are completely identical. Of course, this *is* of practical importance. Why? (Remember the introduction to the chapter.)

Problem Set 6.1

Find the Laplace transforms of the following functions. (a, b, T, ω, and θ are constants.)

1. $3t + 4$ **2.** $at + b$ **3.** $t^2 + at + b$ **4.** $(a + bt)^2$

5. $\cos(\omega t + \theta)$ **6.** $\sin(\omega t + \theta)$ **7.** $\sin(2n\pi t/T)$ **8.** $\sin^2 t$

9. $\cos^2 \omega t$ **10.** $\sin^2 \omega t$ **11.** $\cos^2 t$ **12.** $-5\cos 0.4t$

13. e^{at+b} **14.** $\cosh^2 3t$ **15.** $\sinh^2 2t$ **16.** $\sin t \cos t$

17.
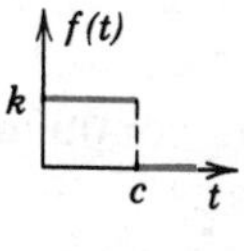

18.
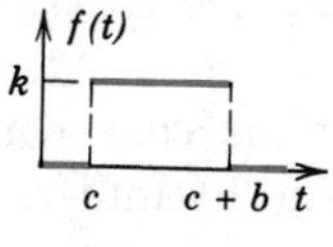

19.
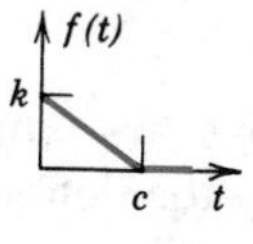

20.
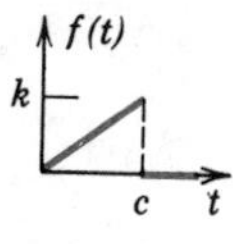

21. Obtain formula 6 in Table 6.1 from formulas 9 and 10.

22. Obtain formula 10 in Table 6.1 from formula 6.

23. Derive formulas 7 and 8 in Table 6.1 by integration by parts.

24. Obtain the answer to Prob. 20 from the answers to Probs. 17 and 19.

25. Using $\cos x = \cosh ix$ and $\sin x = -i \sinh ix$, derive formulas 7 and 8 in Table 6.1 from formulas 9 and 10.

26. Using Prob. 17, find $\mathcal{L}(f)$, where $f(t) = 0$ if $t \leqq 4$, $f(t) = 1$ if $t > 4$.

Find $f(t)$ if $F(s) = \mathcal{L}(f)$ is as follows. (a, b, etc. are constants.)

27. $\dfrac{5}{s+3}$ **28.** $\dfrac{s-9}{s^2-9}$ **29.** $\dfrac{1}{s^2+25}$ **30.** $\dfrac{4}{(s+1)(s+2)}$

31. $\dfrac{1}{s^4}$ **32.** $\dfrac{s+1}{s^2+1}$ **33.** $\dfrac{1}{s(s+1)}$ **34.** $\dfrac{n\pi L}{L^2s^2+n^2\pi^2}$

35. $\dfrac{9}{s^2+3s}$ **36.** $\dfrac{4(s+1)}{s^2-16}$ **37.** $\dfrac{2}{s}+\dfrac{1}{s+2}$ **38.** $\dfrac{0.25}{s-5}-\dfrac{0.4}{s^3}$

39. Prove (3).

40. **(Linearity of the inverse Laplace transform)** Show that $\mathcal{L}^{-1}$ is linear. *Hint.* Use that $\mathcal{L}$ is linear.

6.2 Transforms of Derivatives and Integrals

In this section we discuss and apply the most crucial property of the Laplace transform, namely, that, roughly speaking, differentiation of functions corresponds to the multiplication of transforms by s, and integration of functions corresponds to the division of transforms by s. Hence ***the Laplace transform replaces operations of calculus by operations of algebra on transforms.*** This, in a nutshell, is Laplace's basic idea, for which we should admire him.

Our program for this section is as follows. Theorem 1 concerns the differentiation of $f(t)$, Theorem 2 the extension to higher derivatives, and Theorem 3 the integration of $f(t)$. We also include examples as well as a first application to a differential equation.

Theorem 1 **[Laplace transform of the derivative of $f(t)$]**

Suppose that $f(t)$ is continuous for all $t \geqq 0$, satisfies (2), *Sec.* 6.1, *for some γ and M, and has a derivative $f'(t)$ that is piecewise continuous on every finite interval in the range $t \geqq 0$. Then the Laplace transform of the derivative $f'(t)$ exists when $s > \gamma$, and*

$$\mathcal{L}(f') = s\mathcal{L}(f) - f(0) \qquad (s > \gamma). \tag{1}$$

Proof. We first consider the case when $f'(t)$ is continuous for all $t \geqq 0$. Then, by the definition and by integrating by parts,

$$\mathcal{L}(f') = \int_0^\infty e^{-st} f'(t)\, dt = [e^{-st} f(t)] \Big|_0^\infty + s \int_0^\infty e^{-st} f(t)\, dt.$$

Since f satisfies (2), Sec. 6.1, the integrated portion on the right is zero at the upper limit when $s > \gamma$, and at the lower limit it is $-f(0)$. The last integral is $\mathcal{L}(f)$, the existence for $s > \gamma$ being a consequence of Theorem 2 in Sec. 6.1. This proves that the expression on the right exists when $s > \gamma$, and is equal to $-f(0) + s\mathcal{L}(f)$. Consequently, $\mathcal{L}(f')$ exists when $s > \gamma$, and (1) holds.

If the derivative $f'(t)$ is merely piecewise continuous, the proof is quite similar; in this case, the range of integration in the original integral must be broken up into parts such that f' is continuous in each such part. ∎

REMARK This theorem may be extended to piecewise continuous functions $f(t)$, but in place of (1) we then obtain the formula (1*) in Prob. 40 at the end of the current section.

By applying (1) to the second derivative $f''(t)$ we obtain

$$\begin{aligned} \mathcal{L}(f'') &= s\mathcal{L}(f') - f'(0) \\ &= s[s\mathcal{L}(f) - f(0)] - f'(0); \end{aligned}$$

that is,

$$\mathcal{L}(f'') = s^2\mathcal{L}(f) - sf(0) - f'(0). \tag{2}$$

Similarly,

$$\mathcal{L}(f''') = s^3\mathcal{L}(f) - s^2 f(0) - sf'(0) - f''(0), \tag{3}$$

etc. By induction we thus obtain the following extension of Theorem 1.

Theorem 2 **(Laplace transform of the derivative of any order n)**

Let $f(t)$ and its derivatives $f'(t), f''(t), \cdots, f^{(n-1)}(t)$ be continuous functions for all $t \geqq 0$, satisfying (2), Sec. 6.1, for some γ and M, and let the derivative $f^{(n)}(t)$ be piecewise continuous on every finite interval in the range $t \geqq 0$. Then the Laplace transform of $f^{(n)}(t)$ exists when $s > \gamma$, and is given by

$$\mathcal{L}(f^{(n)}) = s^n\mathcal{L}(f) - s^{n-1}f(0) - s^{n-2}f'(0) - \cdots - f^{(n-1)}(0). \tag{4}$$

EXAMPLE 1 Let $f(t) = t^2$. Find $\mathcal{L}(f)$.

Solution. Since $f(0) = 0$, $f'(0) = 0$, $f''(t) = 2$, and $\mathcal{L}(2) = 2\mathcal{L}(1) = 2/s$, we obtain from (2)

$$\mathcal{L}(f'') = \mathcal{L}(2) = \frac{2}{s} = s^2\mathcal{L}(f), \qquad \text{hence} \qquad \mathcal{L}(t^2) = \frac{2}{s^3},$$

in agreement with Table 6.1. The example is typical: it illustrates that in general there are several ways of obtaining the transforms of given functions. ∎

EXAMPLE 2 Derive the Laplace transforms of $\cos \omega t$ and $\sin \omega t$.

Solution. Let $f(t) = \cos \omega t$. Then $f''(t) = -\omega^2 \cos \omega t = -\omega^2 f(t)$. Also $f(0) = 1$, $f'(0) = 0$. From this and (2),

$$-\omega^2 \mathcal{L}(f) = \mathcal{L}(f'') = s^2\mathcal{L}(f) - s, \qquad \text{hence} \qquad \mathcal{L}(f) = \mathcal{L}(\cos \omega t) = \frac{s}{s^2 + \omega^2}.$$

Similarly for $g(t) = \sin \omega t$. Then $g(0) = 0$, $g'(0) = \omega$, and

$$-\omega^2 \mathcal{L}(g) = \mathcal{L}(g'') = s^2\mathcal{L}(g) - \omega, \qquad \text{hence} \qquad \mathcal{L}(g) = \mathcal{L}(\sin \omega t) = \frac{\omega}{s^2 + \omega^2}.$$ ∎

EXAMPLE 3 Let $f(t) = \sin^2 t$. Find $\mathcal{L}(f)$.

Solution. We have $f(0) = 0$, $f'(t) = 2 \sin t \cos t = \sin 2t$, and (1) gives

$$\mathcal{L}(\sin 2t) = \frac{2}{s^2 + 4} = s\mathcal{L}(f) \qquad \text{or} \qquad \mathcal{L}(\sin^2 t) = \frac{2}{s(s^2 + 4)}.$$ ∎

EXAMPLE 4 Let $f(t) = t \sin \omega t$. Find $\mathcal{L}(f)$.

Solution. We have $f(0) = 0$ and

$$f'(t) = \sin \omega t + \omega t \cos \omega t, \qquad f'(0) = 0,$$

$$f''(t) = 2\omega \cos \omega t - \omega^2 t \sin \omega t$$

$$= 2\omega \cos \omega t - \omega^2 f(t),$$

so that by (2),

$$\mathcal{L}(f'') = 2\omega\mathcal{L}(\cos \omega t) - \omega^2\mathcal{L}(f) = s^2\mathcal{L}(f).$$

Using the formula for the Laplace transform of $\cos \omega t$, we thus obtain

$$(s^2 + \omega^2)\mathcal{L}(f) = 2\omega\mathcal{L}(\cos \omega t) = \frac{2\omega s}{s^2 + \omega^2}.$$

Hence the result is

$$\mathcal{L}(t \sin \omega t) = \frac{2\omega s}{(s^2 + \omega^2)^2}.$$ ∎

Differential Equations, Initial Value Problems

We consider an initial value problem

$$(5) \qquad y'' + ay' + by = r(t), \qquad y(0) = K_0, \qquad y'(0) = K_1$$

with constant a and b. Here $r(t)$ is the **input** (driving force) applied to the system and $y(t)$ is the **output** (response of the system; see also Sec. 1.7). In Laplace's method we do three steps:

1st Step. We transform (5) by means of (1) and (2), writing $Y = \mathcal{L}(y)$ and $R = \mathcal{L}(r)$. This gives

$$[s^2 Y - sy(0) - y'(0)] + a[sY - y(0)] + bY = R(s)$$

and is called the **subsidiary equation.** Collecting Y-terms, we have

$$(s^2 + as + b)Y = (s + a)y(0) + y'(0) + R(s)$$

2nd Step. Division by $s^2 + as + b$ and use of the so-called **transfer function**[2]

$$Q(s) = \frac{1}{s^2 + as + b} \tag{6}$$

gives as the solution of the subsidiary equation

$$Y(s) = [(s + a)y(0) + y'(0)]Q(s) + R(s)Q(s) \tag{7}$$

If $y(0) = y'(0) = 0$, this is simply $Y = RQ$; thus Q is the quotient

$$Q = \frac{Y}{R} = \frac{\mathcal{L}(\text{output})}{\mathcal{L}(\text{input})}$$

and this explains the name of Q. Note that Q depends only on a and b, but neither on $r(t)$ nor on the initial conditions.

3rd Step. We reduce (7) (usually by partial fractions, as in integral calculus) to a sum of terms whose inverses can be found from the table, so that the solution $y(t) = \mathcal{L}^{-1}(Y)$ of (5) is obtained.

EXAMPLE 5 **Initial value problem**

Solve

$$y'' - y = t, \qquad y(0) = 1, \qquad y'(0) = 1.$$

Solution. 1st Step. From (2) and Table 6.1 we get the subsidiary equation

$$s^2Y - sy(0) - y'(0) - Y = 1/s^2, \qquad \text{thus,} \qquad (s^2 - 1)Y = s + 1 + 1/s^2.$$

2nd Step. The transfer function is $Q = 1/(s^2 - 1)$, and (7) becomes

$$Y = (s + 1)Q + \frac{1}{s^2}Q = \frac{s + 1}{s^2 - 1} + \frac{1}{s^2(s^2 - 1)}$$

$$= \frac{1}{s - 1} + \left(\frac{1}{s^2 - 1} - \frac{1}{s^2}\right).$$

3rd Step. From this expression for Y and Table 6.1 we obtain the solution

$$y(t) = \mathcal{L}^{-1}(Y) = \mathcal{L}^{-1}\left\{\frac{1}{s - 1}\right\} + \mathcal{L}^{-1}\left\{\frac{1}{s^2 - 1}\right\} - \mathcal{L}^{-1}\left\{\frac{1}{s^2}\right\} = e^t + \sinh t - t.$$

The diagram summarizes our approach. ∎

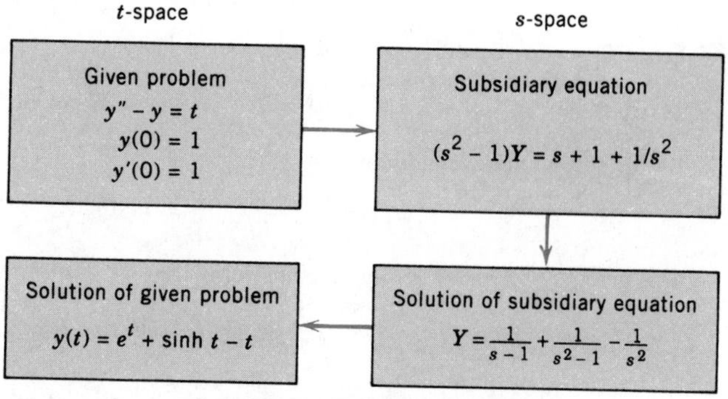

Laplace transform method

[2]Often denoted by H, but we need H much more frequently for other purposes.

In practice, instead of justifying the use of formulas and theorems in this method, one simply checks at the end whether $y(t)$ satisfies the given equation and initial conditions.

Gains in the method, as compared to that in Chap. 2, and illustrated by the example, are as follows:

1. No determination of a general solution of the homogeneous equation.
2. No determination of values for arbitrary constants in a general solution.

Shifted data problems is a short name for initial value problems in which the initial conditions refer to some later instant instead of $t = 0$. We explain the idea of solving such a problem by the Laplace transform in terms of a simple example.

EXAMPLE 6 Shifted data problem

Solve the initial value problem

$$y'' + y = 2t, \qquad y(\tfrac{1}{4}\pi) = \tfrac{1}{2}\pi, \qquad y'(\tfrac{1}{4}\pi) = 2 - \sqrt{2}$$

Solution. We see by inspection that $y = A \cos t + B \sin t + 2t$ is a general solution, and we know how we would have to go from here to take care of the initial conditions. What we want to learn is how the Laplace transform can proceed although $y(0)$ and $y'(0)$ are unknown.

1st Step. Setting up the subsidiary equation. From (2) and Table 6.1 (Sec. 6.1) we obtain

$$s^2Y - sy(0) - y'(0) + Y = 2/s^2.$$

2nd Step. Solution of the subsidiary equation. Solving algebraically and using partial fractions, we get

$$\begin{aligned} Y &= \frac{2}{(s^2+1)s^2} + y(0)\frac{s}{s^2+1} + y'(0)\frac{1}{s^2+1} \\ &= 2\left(\frac{1}{s^2} - \frac{1}{s^2+1}\right) + y(0)\frac{s}{s^2+1} + y'(0)\frac{1}{s^2+1}. \end{aligned}$$

3rd Step. Solution of the given problem. From Table 6.1 we obtain $y = \mathcal{L}^{-1}(Y)$ in the form

$$\begin{aligned} y(t) &= 2t + y(0)\cos t + [y'(0) - 2]\sin t \\ &= 2t + A\cos t + B\sin t, \end{aligned}$$

(where $A = y(0)$ and $B = y'(0) - 2$, but this is of no further interest). The first initial condition gives

$$y(\tfrac{1}{4}\pi) = \tfrac{1}{2}\pi + A/\sqrt{2} + B/\sqrt{2} = \tfrac{1}{2}\pi,$$

hence $B = -A$. By differentiation,

$$y'(t) = 2 - A\sin t + B\cos t.$$

By the second initial condition,

$$y'(\tfrac{1}{4}\pi) = 2 - A/\sqrt{2} + B/\sqrt{2} = 2 - \sqrt{2}.$$

This gives $A = 1$, $B = -1$ and the answer

$$y(t) = \cos t - \sin t + 2t. \qquad \blacksquare$$

Laplace Transform of the Integral of a Function

Since differentiation and integration are inverse processes, and since, roughly speaking, differentiation of a function corresponds to the multiplication of its transform by s, we expect integration of a function to correspond to division of its transform by s, because division is the inverse operation of multiplication:

Theorem 3 [Integration of $f(t)$]

If $f(t)$ is piecewise continuous and satisfies an inequality of the form (2), *Sec.* 6.1, *then*

$$\mathcal{L}\left\{\int_0^t f(\tau)\,d\tau\right\} = \frac{1}{s}\,\mathcal{L}\{f(t)\} \qquad (s > 0,\ s > \gamma). \tag{8}$$

Proof. Suppose that $f(t)$ is piecewise continuous and satisfies (2), Sec. 6.1, for some γ and M. Clearly, if (2) holds for some negative γ, it also holds for positive γ, and we may assume that γ is positive. Then the integral

$$g(t) = \int_0^t f(\tau)\,d\tau$$

is continuous, and by using (2) in Sec. 6.1 we obtain

$$|g(t)| \leqq \int_0^t |f(\tau)|\,d\tau \leqq M\int_0^t e^{\gamma\tau}\,d\tau = \frac{M}{\gamma}(e^{\gamma t} - 1) \leqq \frac{M}{\gamma}e^{\gamma t} \quad (\gamma > 0).$$

This shows that $g(t)$ also satisfies an inequality of the form (2), Sec. 6.1. Also, $g'(t) = f(t)$, except for points at which $f(t)$ is discontinuous. Hence $g'(t)$ is piecewise continuous on each finite interval, and, by Theorem 1,

$$\mathcal{L}\{f(t)\} = \mathcal{L}\{g'(t)\} = s\mathcal{L}\{g(t)\} - g(0) \qquad (s > \gamma).$$

Here, clearly, $g(0) = 0$, so that $\mathcal{L}(f) = s\mathcal{L}(g)$. This implies (8). ∎

Equation (8) has a useful companion, which we obtain by writing $\mathcal{L}\{f(t)\} = F(s)$, interchanging the two sides and taking the inverse transform on both sides. Then

$$\mathcal{L}^{-1}\left\{\frac{1}{s}F(s)\right\} = \int_0^t f(\tau)\,d\tau. \tag{9}$$

EXAMPLE 7 **An application of Theorem 3**

Let $\mathcal{L}(f) = \dfrac{1}{s(s^2 + \omega^2)}$. Find $f(t)$.

Solution. From Table 6.1 in Sec. 6.1 we have

$$\mathcal{L}^{-1}\left(\frac{1}{s^2 + \omega^2}\right) = \frac{1}{\omega} \sin \omega t.$$

From this and Theorem 3 we obtain the answer

$$\mathcal{L}^{-1}\left\{\frac{1}{s}\left(\frac{1}{s^2 + \omega^2}\right)\right\} = \frac{1}{\omega}\int_0^t \sin \omega\tau \, d\tau = \frac{1}{\omega^2}(1 - \cos \omega t).$$

This proves formula 19 in the table in Sec. 6.10. ∎

EXAMPLE 8 **Another application of Theorem 3**

Let $\mathcal{L}(f) = \dfrac{1}{s^2(s^2 + \omega^2)}$. Find $f(t)$.

Solution. Applying Theorem 3 to the answer in Example 7, we obtain the desired formula

$$\mathcal{L}^{-1}\left\{\frac{1}{s^2}\left(\frac{1}{s^2 + \omega^2}\right)\right\} = \frac{1}{\omega^2}\int_0^t (1 - \cos \omega\tau)\, d\tau = \frac{1}{\omega^2}\left(t - \frac{\sin \omega t}{\omega}\right).$$

This proves formula 20 in the table in Sec. 6.10. ∎

Problem Set 6.2

Using (1) or (2), find the transform $\mathcal{L}(f)$ of the given function $f(t)$.

1. $\cos^2 t$ **2.** $\sin^2 \omega t$ **3.** te^{at} **4.** $t \cos t$
5. $\cosh^2 2t$ **6.** $\sinh^2 2t$ **7.** $\cos^2 3t$ **8.** $\cos^2 \pi t$

Further transforms. Using Theorems 1 and 2, derive the following transforms that occur in applications (in connection with resonance, etc.).

9. $\mathcal{L}(t \cos \omega t) = \dfrac{s^2 - \omega^2}{(s^2 + \omega^2)^2}$ **10.** $\mathcal{L}(t \sin \omega t) = \dfrac{2\omega s}{(s^2 + \omega^2)^2}$

11. $\mathcal{L}(t \cosh at) = \dfrac{s^2 + a^2}{(s^2 - a^2)^2}$ **12.** $\mathcal{L}(t \sinh at) = \dfrac{2as}{(s^2 - a^2)^2}$

Using the formulas in Probs. 9 and 10, show that

13. $\mathcal{L}^{-1}\left(\dfrac{1}{(s^2 + \omega^2)^2}\right) = \dfrac{1}{2\omega^3}(\sin \omega t - \omega t \cos \omega t)$

14. $\mathcal{L}^{-1}\left(\dfrac{s^2}{(s^2 + \omega^2)^2}\right) = \dfrac{1}{2\omega}(\sin \omega t + \omega t \cos \omega t)$

Application of Theorem 3. Find $f(t)$ if $\mathcal{L}(f)$ equals

15. $\dfrac{3}{s^2 + s}$ **16.** $\dfrac{4}{s^3 - 4s}$ **17.** $\dfrac{4}{s^3 + 4s}$ **18.** $\dfrac{1}{s^2 + as}$

19. $\dfrac{8}{s^4 - 4s^2}$ **20.** $\dfrac{1}{s}\left(\dfrac{s - a}{s + a}\right)$ **21.** $\dfrac{1}{s^4 - 2s^3}$ **22.** $\dfrac{1}{s^2}\left(\dfrac{s - a}{s + a}\right)$

23. $\dfrac{2s - \pi}{s^3(s - \pi)}$ **24.** $\dfrac{1}{s^2}\left(\dfrac{s - 2}{s^2 + 4}\right)$ **25.** $\dfrac{1}{s^2}\left(\dfrac{s + 1}{s^2 + 1}\right)$ **26.** $\dfrac{1}{s^4(s^2 + \pi^2)}$

Initial value problems. Using Laplace transforms, solve:

27. $y'' + 4y = 0, \quad y(0) = 2, \quad y'(0) = -8$

28. $4y'' + \pi^2 y = 0, \quad y(0) = 0, \quad y'(0) = 1$

29. $y'' + \omega^2 y = 0, \quad y(0) = A, \quad y'(0) = B, \quad (\omega$ real, not zero)

30. $y'' + 2y' - 8y = 0, \quad y(0) = 0, \quad y'(0) = 6$

31. $y'' + 5y' + 6y = 0, \quad y(0) = 0, \quad y'(0) = 1$

32. $y'' - 4y' + 3y = 2t - \frac{8}{3}, \quad y(0) = 0, \quad y'(0) = -\frac{16}{3}$

33. $y'' + 25y = t, \quad y(0) = 1, \quad y'(0) = 0.04$

34. $y'' + 4y = 1 - 2t, \quad y(0) = 0, \quad y'(0) = 0$

35. $y'' + ky' - 2k^2 y = 0, \quad y(0) = 2, \quad y'(0) = 2k$

36. $y'' + \pi^2 y = t^3, \quad y(0) = 6/\pi^4, \quad y'(0) = 0$

Derivation by different methods is possible for various formulas, and is typical of Laplace transforms. Problems 37–39 illustrate this point.

37. Using (1), derive $\mathcal{L}(\sin \omega t)$ from $\mathcal{L}(\cos \omega t)$.

38. Find $\mathcal{L}(\cos^2 t)$ (a) by the use of the result of Example 3, (b) by the method used in Example 3, (c) by expressing $\cos^2 t$ in terms of $\cos 2t$.

39. Check the answer to Prob. 5 by writing $\cosh 2t$ in terms of exponential functions.

40. **(Extension of Theorem 1)** Show that if $f(t)$ is continuous, except for an ordinary discontinuity (finite jump) at $t = a\ (> 0)$, the other conditions remaining the same as in Theorem 1, then (see Fig. 88)

$$\mathcal{L}(f') = s\mathcal{L}(f) - f(0) - [f(a + 0) - f(a - 0)]e^{-as}. \tag{1*}$$

Using (1*), find the Laplace transform of $f(t) = t$ if $0 < t < 1$, $f(t) = 1$ if $1 < t < 2$, $f(t) = 0$ otherwise.

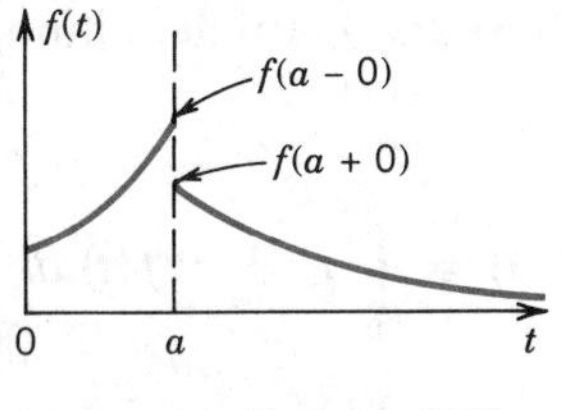

Fig. 88. Formula (1*)

6.3 s-Shifting, t-Shifting, Unit Step Function

What state have we reached and what is our next goal? We know that the Laplace transform is linear (Theorem 1, Sec. 6.1), that differentiation of $f(t)$ roughly corresponds to the multiplication of $\mathcal{L}(f)$ by s (Theorems 1 and 2, Sec. 6.2), and that this property is essential in solving differential equations. A first illustration of the technique is given in Example 5, Sec. 6.2. But there the solution may easily be found by the usual methods.

To provide applications such that the Laplace transform can show its real power, we first have to derive some further general properties of it. Two very important properties concern the shifting on the s-axis and the shifting on the t-axis, as expressed in the two shifting theorems (Theorems 1 and 2 of this section).

s-Shifting: Replacing s by $s - a$ in $F(s)$

Theorem 1 (First shifting theorem; s-shifting)
If $f(t)$ has the transform $F(s)$ where $s > \gamma$, then $e^{at}f(t)$ has the transform $F(s - a)$ where $s - a > \gamma$; thus, if $\mathcal{L}\{f(t)\} = F(s)$, then

$$\mathcal{L}\{e^{at}f(t)\} = F(s - a). \tag{1}$$

*Hence if we know the transform $F(s)$ of $f(t)$, we get the transform of $e^{at}f(t)$ by "**shifting on the s-axis**"* [*i.e., by replacing s with $s - a$, to get $F(s - a)$*]. See Fig. 89.

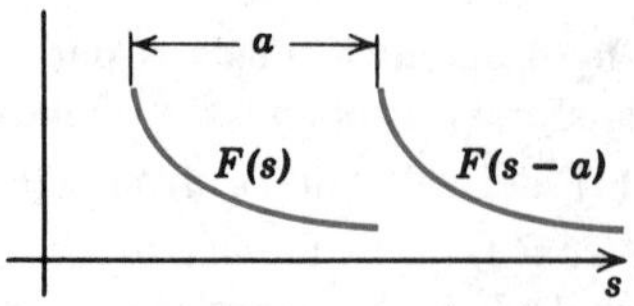

Fig. 89. First shifting theorem, shifting on the s-axis (In this figure, $a > 0$.)

Remark. By taking the inverse transform on both sides and interchanging the left and right sides we obtain from (1)

$$\mathcal{L}^{-1}\{F(s - a)\} = e^{at}f(t). \tag{1*}$$

Proof of Theorem 1. By definition, $F(s) = \int_0^\infty e^{-st}f(t)\,dt$ and, therefore,

$$F(s - a) = \int_0^\infty e^{-(s-a)t}f(t)\,dt = \int_0^\infty e^{-st}[e^{at}f(t)]\,dt = \mathcal{L}\{e^{at}f(t)\}. \quad \blacksquare$$

EXAMPLE 1 s-shifting by means of Theorem 1

By applying Theorem 1 to the formulas 4, 7, and 8 in Table 6.1 we obtain the following results.

$f(t)$	$\mathcal{L}(f)$
$e^{at}t^n$	$\dfrac{n!}{(s-a)^{n+1}}$
$e^{at}\cos\omega t$	$\dfrac{s-a}{(s-a)^2+\omega^2}$
$e^{at}\sin\omega t$	$\dfrac{\omega}{(s-a)^2+\omega^2}$

This proves formulas 8, 9, 17, 18 in the table in Sec. 6.10. ■

EXAMPLE 2 Damped free vibrations

An iron ball of mass $m = 2$ is attached at the lower end of an elastic spring whose upper end is fixed, the spring modulus being $k = 10$. Let $y(t)$ be the displacement of the ball from the position of static equilibrium. Determine the free vibrations of the ball, starting from the initial position $y(0) = 2$ with the initial velocity $y'(0) = -4$, assuming that there is damping proportional to the velocity, the damping constant being $c = 4$.

Solution. The motion is described by the solution $y(t)$ of the initial value problem

$$y'' + 2y' + 5y = 0, \qquad y(0) = 2, \qquad y'(0) = -4,$$

see (7), Sec. 2.5. Using (1) and (2) in Sec. 6.2, we obtain the subsidiary equation

$$s^2Y - 2s + 4 + 2(sY - 2) + 5Y = 0.$$

The solution is

$$Y(s) = \frac{2s}{(s+1)^2 + 2^2} = 2\,\frac{s+1}{(s+1)^2 + 2^2} - \frac{2}{(s+1)^2 + 2^2}.$$

Now

$$\mathcal{L}^{-1}\left(\frac{s}{s^2 + 2^2}\right) = \cos 2t, \qquad \mathcal{L}^{-1}\left(\frac{2}{s^2 + 2^2}\right) = \sin 2t.$$

From this and Theorem 1 we obtain the expected type of solution

$$y(t) = \mathcal{L}^{-1}(Y) = e^{-t}(2\cos 2t - \sin 2t).$$ ∎

EXAMPLE 3 **Nonhomogeneous equation**

Solve the initial value problem

$$y'' - 2y' + y = e^t + t, \qquad y(0) = 1, \qquad y'(0) = 0.$$

Solution. From (1) and (2) in Sec. 6.2 and the initial conditions we obtain the subsidiary equation

$$(s^2Y - s) - 2(sY - 1) + Y = \frac{1}{s-1} + \frac{1}{s^2}.$$

Collecting Y-terms gives

$$(s^2 - 2s + 1)Y = (s-1)^2Y = s - 2 + \frac{1}{s-1} + \frac{1}{s^2}.$$

Hence by division,

$$Y = \frac{s-2}{(s-1)^2} + \frac{1}{(s-1)^3} + \frac{1}{s^2(s-1)^2}.$$

We now apply the first shifting theorem. The first term is

$$\frac{s-2}{(s-1)^2} = \frac{s-1}{(s-1)^2} - \frac{1}{(s-1)^2} = \frac{1}{s-1} - \frac{1}{(s-1)^2}; \qquad \text{inverse:} \qquad e^t - te^t.$$

The inverse of the second term is $t^2e^t/2$, from Table 6.1 and by the first shifting theorem. In terms of partial fractions, the last term is

$$\frac{1}{s^2(s-1)^2} = \frac{A}{(s-1)^2} + \frac{B}{s-1} + \frac{C}{s^2} + \frac{D}{s}.$$

Multiplication by the common denominator gives

$$1 = As^2 + Bs^2(s-1) + C(s-1)^2 + Ds(s-1)^2.$$

For $s = 0$ this gives $C = 1$. For $s = 1$ it gives $A = 1$. Equating the sum of the s^3-terms to zero gives $D + B = 0$, $D = -B$. Equating the sum of the s-terms to zero gives $-2C + D = 0$, $D = 2$, and $B = -D = -2$. The sum of the partial fractions now is

$$\frac{1}{(s-1)^2} + \frac{-2}{s-1} + \frac{1}{s^2} + \frac{2}{s}; \qquad \text{inverse:} \qquad te^t - 2e^t + t + 2.$$

The inverse results from Table 6.1 and the first shifting theorem. Collecting terms, we get

$$y(t) = \mathcal{L}^{-1}(Y) = e^t - te^t + \tfrac{1}{2}t^2e^t + (t-2)e^t + t + 2 = -e^t + \tfrac{1}{2}t^2e^t + t + 2.$$

This agrees with Example 3 in Sec. 2.9. ∎

t-Shifting: Replacing t by $t - a$ in $f(t)$

The first shifting theorem (Theorem 1) concerns shifting on the s-axis: the replacement of s in $F(s)$ by $s - a$ corresponds to the multiplication of the original function $f(t)$ by e^{at}. We shall now state the second shifting theorem (Theorem 2), which concerns shifting on the t-axis: the replacement of t in $f(t)$ by $t - a$ corresponds roughly to the multiplication of the transform $F(s)$ by e^{-as}; the precise formulation of this is as follows.

Theorem 2 **(Second shifting theorem; t-shifting)**
If $f(t)$ has the transform $F(s)$, then the function

$$\tilde{f}(t) = \begin{cases} 0 & \text{if } t < a \\ f(t-a) & \text{if } t > a \end{cases} \tag{2}$$

with arbitrary $a \geqq 0$ has the transform $e^{-as}F(s)$. Hence if we know that transform $F(s)$ of $f(t)$, we get the transform of the function (2), *whose variable has been shifted* (**"shifting on the t-axis"**), *by multiplying $F(s)$ by e^{-as}.* (Proof and applications follow below.)

Unit Step Function $u(t - a)$

By definition, $u(t - a)$ is 0 for $t < a$, has a jump of size 1 at $t = a$ (where we can leave it undefined) and is 1 for $t > a$:

$$u(t-a) = \begin{cases} 0 & \text{if } t < a \\ 1 & \text{if } t > a \end{cases} \qquad (a \geqq 0). \tag{3}$$

Figure 90 shows the special case $u(t)$, which has the jump at zero, and Fig. 91 the general case $u(t - a)$ for an arbitrary positive a. The unit step function is also called the **Heaviside function.**[3]

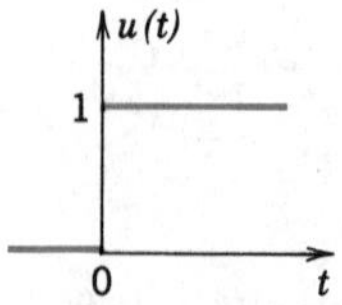

Fig. 90. Unit step function $u(t)$

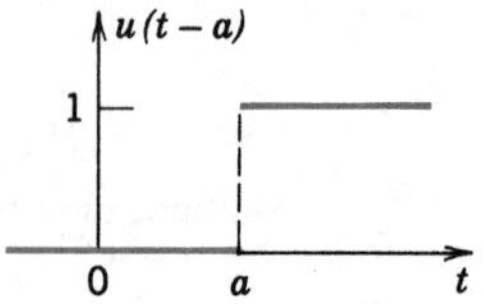

Fig. 91. Unit step function $u(t - a)$

[3]See footnote 1 in Sec. 6.1.

The unit step function $u(t - a)$ is a basic building block of various functions, as we shall see, and it greatly increases the usefulness of Laplace transform methods. At present we can use it to write $\tilde{f}(t)$ in (2) in the form $f(t - a)u(t - a)$, that is,

$$f(t - a)u(t - a) = \begin{cases} 0 & \text{if } t < a \\ f(t - a) & \text{if } t > a. \end{cases} \tag{2*}$$

This is the graph of $f(t)$ for $t > 0$, but shifted a units to the right.

Figure 92 shows an example. This is the cosine curve $f(t) = \cos t$ for $t > 0$ (Fig. 92a) and the curve $f(t - 2)u(t - 2) = \cos (t - 2)\, u(t - 2)$ obtained by shifting it 2 units to the right. For $t < a = 2$ this function is zero because $u(t - 2)$ has this property.

Using (2*), we can now reformulate

Theorem 2 **(Second shifting theorem; *t*-shifting)**

If $\mathcal{L}\{f(t)\} = F(s)$, *then*

$$\mathcal{L}\{f(t - a)u(t - a)\} = e^{-as}F(s). \tag{4}$$

Remark. Taking the inverse transform on both sides of (4) and interchanging sides, we obtain the companion formula

$$\mathcal{L}^{-1}\{e^{-as}F(s)\} = f(t - a)u(t - a). \tag{4*}$$

Proof of Theorem 2. From the definition we have

$$e^{-as}F(s) = e^{-as}\int_0^\infty e^{-s\tau}f(\tau)\, d\tau = \int_0^\infty e^{-s(\tau+a)}f(\tau)\, d\tau.$$

Substituting $\tau + a = t$ in the integral, we obtain

$$e^{-as}F(s) = \int_a^\infty e^{-st}f(t - a)\, dt.$$

We can write this as an integral from 0 to ∞ if we make sure that the integrand

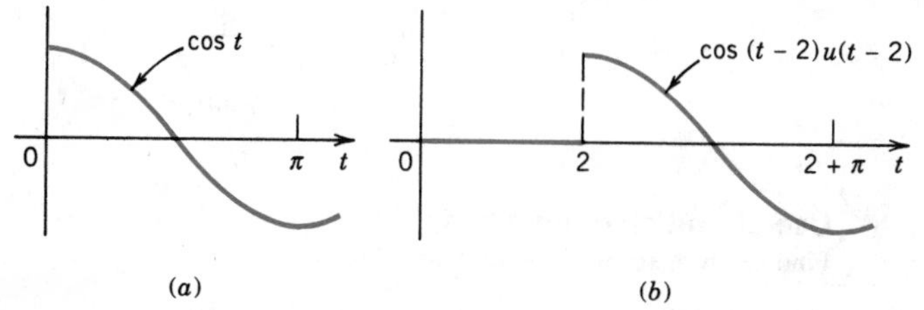

Fig. 92. $f(t) = \cos t$ and $f(t - 2)u(t - 2) = \cos (t - 2)\, u(t - 2)$

is zero for all t from 0 to a. We may easily accomplish this by multiplying the present integrand by the step function $u(t - a)$, thereby obtaining (4) and completing the proof:

$$e^{-as}F(s) = \int_0^\infty e^{-st} f(t - a)u(t - a)\, dt = \mathcal{L}\{f(t - a)u(t - a)\}. \qquad \blacksquare$$

It is fair to say that we are already approaching the stage where we can attack problems for which the Laplace transform method is preferable to the usual method, as the examples in the next section will illustrate. In this connection we need the transform of the unit step function $u(t - a)$,

$$\text{(5)} \qquad \mathcal{L}\{u(t - a)\} = \frac{e^{-as}}{s} \qquad (s > 0).$$

This formula follows directly from the definition because

$$\mathcal{L}\{u(t - a)\} = \int_0^\infty e^{-st} u(t - a)\, dt$$

$$= \int_0^a e^{-st} 0\, dt + \int_a^\infty e^{-st} 1\, dt = -\frac{1}{s} e^{-st} \Big|_a^\infty .$$

Let us consider two simple examples. Further applications follow in the problem set and in the next sections.

EXAMPLE 4 Application of Theorem 2

Find the inverse transform of e^{-3s}/s^3.

Solution. Since $\mathcal{L}^{-1}(1/s^3) = t^2/2$ (see Table 6.1 in Sec. 6.1), Theorem 2 gives (Fig. 93)

$$\mathcal{L}^{-1}(e^{-3s}/s^3) = \tfrac{1}{2}(t - 3)^2 u(t - 3) = \begin{cases} 0 & \text{if } t < 3 \\ \tfrac{1}{2}(t - 3)^2 & \text{if } t > 3. \end{cases} \qquad \blacksquare$$

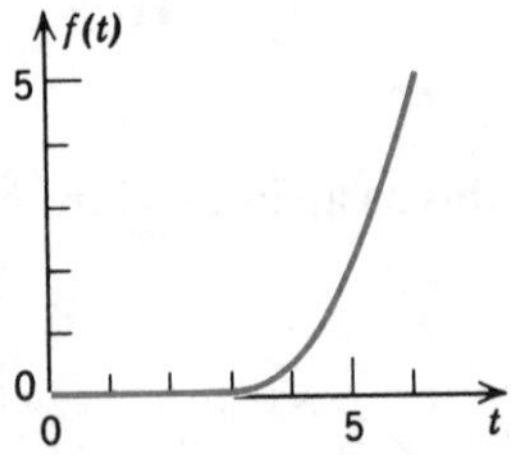

Fig. 93. Example 4

EXAMPLE 5 Use of unit step functions

Find the transform of the function (Fig. 94)

$$f(t) = \begin{cases} 2 & \text{if } 0 < t < \pi \\ 0 & \text{if } \pi < t < 2\pi \\ \sin t & \text{if } \quad t > 2\pi. \end{cases}$$

Solution. 1st Step. We write $f(t)$ in terms of step functions. For $0 < t < \pi$, we take $2u(t)$. For $t > \pi$ we want 0, so we must subtract the step function $2u(t - \pi)$ with step at π. Then we have $2u(t) - 2u(t - \pi) = 0$ when $t > \pi$. This is fine until we reach 2π where we want $\sin t$ to come in; so we add $u(t - 2\pi) \sin t$. Together,

$$f(t) = 2u(t) - 2u(t - \pi) + u(t - 2\pi) \sin t.$$

2nd Step. The last term equals $u(t - 2\pi) \sin (t - 2\pi)$ because of the periodicity, so that (5), (4), and Table 6.1 (Sec. 6.1) give

$$\mathcal{L}(f) = \frac{2}{s} - \frac{2e^{-\pi s}}{s} + \frac{e^{-2\pi s}}{s^2 + 1}. \quad ■$$

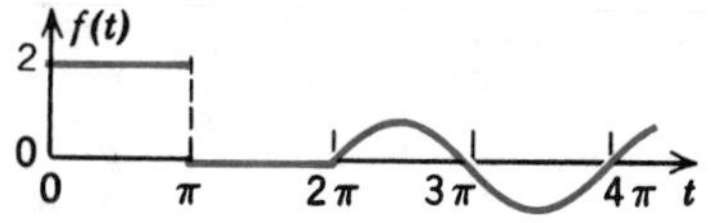

Fig. 94. Example 5

EXAMPLE 6 Find the inverse Laplace transform $f(t)$ of

$$F(s) = \frac{2}{s^2} - \frac{2e^{-2s}}{s^2} - \frac{4e^{-2s}}{s} + \frac{se^{-\pi s}}{s^2 + 1}.$$

Solution. From Table 6.1 (Sec. 6.1) and the Theorem 2,

$$f(t) = 2t - 2(t - 2)u(t - 2) - 4u(t - 2) + \cos (t - \pi)\, u(t - \pi)$$

$$= 2t - 2tu(t - 2) - \cos t\, u(t - \pi) = \begin{cases} 2t & \text{if } 0 < t < 2 \\ 0 & \text{if } 2 < t < \pi \\ -\cos t & \text{if } \quad t > \pi. \end{cases} \quad ■$$

Problem Set 6.3

Applications of the First Shifting Theorem

Find the Laplace transforms of the following functions.

1. $4.5te^{3.5t}$ **2.** t^2e^{-2t} **3.** $e^t \sin t$ **4.** $e^{-t} \cos t$

5. $e^{-t} \sin (\omega t + \theta)$ **6.** $e^{-\alpha t} \sin \omega t$

7. $2t^3e^{-t/2}$ **8.** $e^{-2t}(2 \cos 3t - \sin 3t)$

9. $e^{-\alpha t}(A \cos \beta t + B \sin \beta t)$ **10.** $e^{-t}(a_0 + a_1 t + \cdots + a_n t^n)$

Find $f(t)$ if $\mathcal{L}(f)$ equals

11. $\dfrac{\pi}{(s + \pi)^2}$ **12.** $\dfrac{1}{(s + \frac{1}{2})^3}$ **13.** $\dfrac{s - 2}{s^2 - 4s + 5}$ **14.** $\dfrac{1}{s^2 + 2s + 5}$

15. $\dfrac{s}{(s + 3)^2 + 1}$ **16.** $\dfrac{as + b}{(s + c)^2 + \omega^2}$ **17.** $\dfrac{6}{s^2 - 4s - 5}$ **18.** $\dfrac{2}{s^2 + s + \frac{1}{2}}$

Representing the hyperbolic functions in terms of exponential functions and applying the first shifting theorem, show that

19. $\mathcal{L}(\cosh at \cos at) = \dfrac{s^3}{s^4 + 4a^4}$

20. $\mathcal{L}(\cosh at \sin at) = \dfrac{a(s^2 + 2a^2)}{s^4 + 4a^4}$

21. $\mathcal{L}(\sinh at \cos at) = \dfrac{a(s^2 - 2a^2)}{s^4 + 4a^4}$

22. $\mathcal{L}(\sinh at \sin at) = \dfrac{2a^2 s}{s^4 + 4a^4}$

Unit step function. Represent the following functions in terms of unit step functions and find their Laplace transforms.

23.

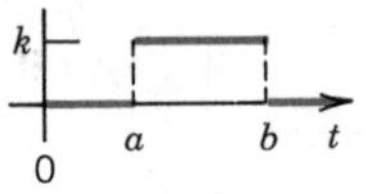

24.

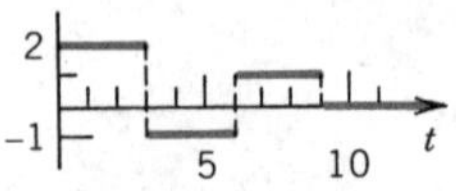

25.

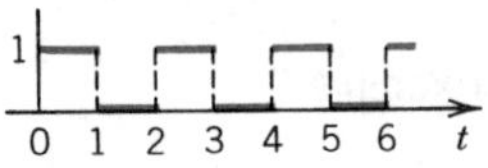

26.

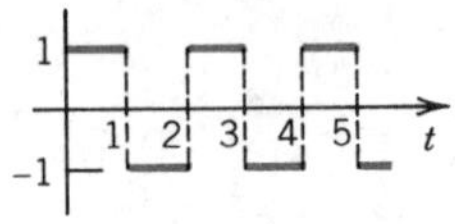

Applications of the Second Shifting Theorem

Sketch the following functions and find their Laplace transforms.

27. $(t - 1)u(t - 1)$ **28.** $tu(t - 1)$ **29.** $(t - 1)^2 u(t - 1)$ **30.** $t^2 u(t - 1)$

31. $e^t u(t - \frac{1}{2})$ **32.** $u(t - 1)\cosh t$ **33.** $u(t - \pi)\cos t$ **34.** $u(t - \frac{1}{2}\pi)\sin t$

In each case sketch the given function, which is assumed to be zero outside the given interval, and find its Laplace transform.

35. $t \quad (0 < t < 1)$

36. $t \quad (0 < t < 2)$

37. $e^t \quad (0 < t < 1)$

38. $t^2 \quad (0 < t < 3)$

39. $t \quad (0 < t < a)$

40. $\sin t \quad (2\pi < t < 4\pi)$

41. $2\cos \pi t \quad (1 < t < 2)$

42. $1 - e^{-t} \quad (0 < t < \pi)$

Find and sketch the inverse Laplace transforms of the following functions.

43. e^{-3s}/s^2

44. $e^{-\pi s}/(s^2 + 2s + 2)$

45. $3(e^{-4s} - e^{-s})/s$

46. $(e^{-2s} - e^{-4s})/(s - 2)$

47. $se^{-\pi s}/(s^2 + 4)$

48. $e^{-s}/(s^2 + \omega^2)$

49. e^{-s}/s^4

50. $(1 - e^{-\pi s})/(s^2 + 4)$

Initial value problems. Using Laplace transforms, solve:

51. $y'' + 2y' + 2y = 0, \quad y(0) = 0, \quad y'(0) = 1$

52. $y'' - 4y' + 5y = 0. \quad y(0) = 1, \quad y'(0) = 2$

53. $4y'' - 4y' + 37y = 0, \quad y(0) = 3, \quad y'(0) = 1.5$

54. $9y'' - 6y' + y = 0, \quad y(0) = 3, \quad y'(0) = 1$

55. $y'' + 4y' + 13y = 145 \cos 2t, \quad y(0) = 9, \quad y'(0) = 19$

56. $y'' + 2y' - 8y = -256t^3, \quad y(0) = 15, \quad y'(0) = 36$

57. $y'' + 6y' + 8y = -e^{-3t} + 3e^{-5t}$, $y(0) = 4$, $y'(0) = -14$

58. $y'' + 2y' + 5y = 9 \cosh 2t + 4 \sinh 2t$, $y(0) = 1$, $y'(0) = 2$

59. $y'' + y = r(t)$, $r(t) = t$ if $0 < t < 1$ and 0 if $t > 1$, $y(0) = y'(0) = 0$

60. $y'' - 5y' + 6y = r(t)$, $r(t) = 4e^t$ if $0 < t < 2$ and 0 if $t > 2$, $y(0) = 1$, $y'(0) = -2$

61. $y'' + 9y = r(t)$, $r(t) = 8 \sin t$ if $0 < t < \pi$ and 0 if $t > \pi$, $y(0) = 0$, $y'(0) = 4$

62. $y'' + 3y' + 2y = r(t)$, $r(t) = 4t$ if $0 < t < 1$ and 8 if $t > 1$, $y(0) = y'(0) = 0$

63. $y'' + 4y = r(t)$, $r(t) = 3 \sin t$ if $0 < t < \pi$ and $-3 \sin t$ if $t > \pi$, $y(0) = 0$, $y'(0) = 3$

64. $y'' + 2y' + 2y = r(t)$, $r(t) = t$ if $0 < t < 1$ and $2t^2$ if $t > 1$, $y(0) = \frac{1}{2}$, $y'(0) = -\frac{1}{2}$

65. $y'' + y' - 2y = r(t)$, $r(t) = 3 \sin t - \cos t$ if $0 < t < 2\pi$ and $3 \sin 2t - \cos 2t$ if $t > 2\pi$, $y(0) = 1$, $y'(0) = 0$

Models of electric circuits

66. A capacitor of capacitance C is charged so that its potential is V_0. At $t = 0$ the switch in Fig. 95 is closed and the capacitor starts to discharge through the resistor of resistance R. Using Laplace transforms, find the charge $q(t)$ on the capacitor.

67. Find the current $i(t)$ in the circuit in Fig. 96, assuming that no current flows when $t \leqq 0$ and the switch is closed at $t = 0$.

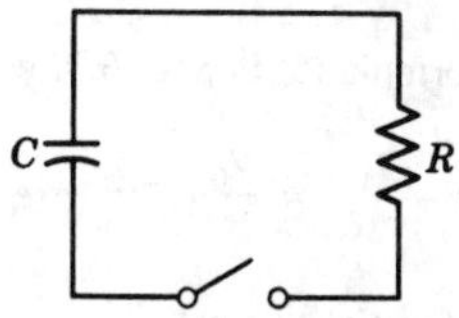

Fig. 95. Problem 66

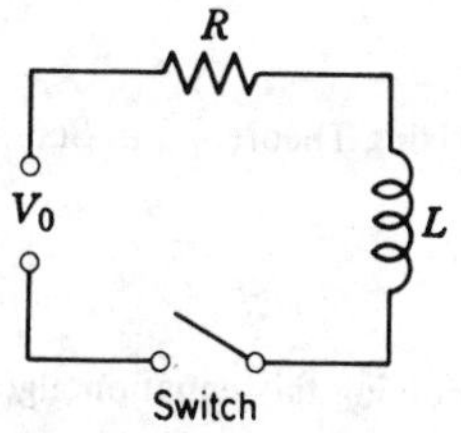

Fig. 96. Problem 67

Find the current $i(t)$ in the LC-circuit shown in Fig. 97, assuming $L = 1$ henry, $C = 1$ farad, zero initial current and charge on the capacitor, and $v(t)$ as follows.

68. $v = 1$ if $0 < t < a$ and 0 otherwise

69. $v = t$ if $0 < t < 1$ and $v = 1$ if $t > 1$

70. $v = 1 - e^{-t}$ if $0 < t < \pi$ and 0 otherwise

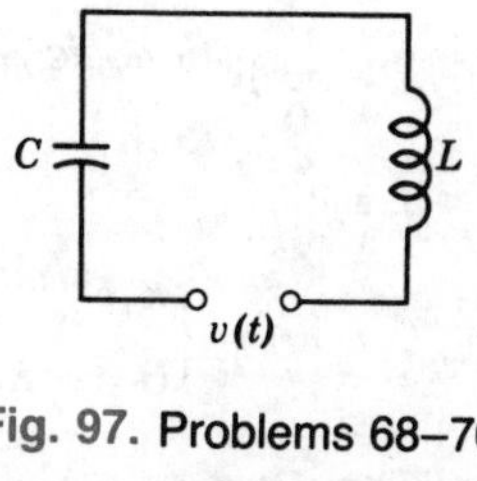

Fig. 97. Problems 68–70

6.4 Further Applications. Dirac's Delta Function

In this section we consider some further applications and then introduce Dirac's delta function, which greatly increases the usefulness of the Laplace transform in connection with impulsive inputs.

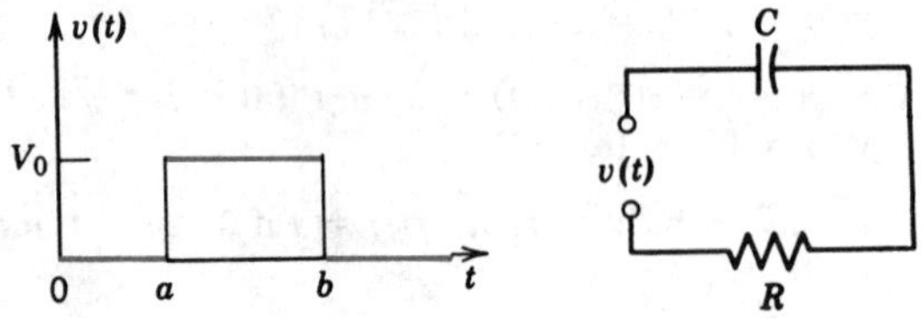

Fig. 98. Example 1

EXAMPLE 1 **Response of an *RC*-circuit to a single square wave**

Find the current $i(t)$ in the circuit in Fig. 98 if a single square wave with voltage V_0 is applied. The circuit is assumed to be quiescent before the square wave is applied.

Solution. The equation of the circuit is (see Sec. 1.8)

$$Ri(t) + \frac{q(t)}{C} = Ri(t) + \frac{1}{C}\int_0^t i(\tau)\, d\tau = v(t)$$

where $v(t)$ can be represented in terms of two unit step functions:

$$v(t) = V_0[u(t-a) - u(t-b)].$$

Using Theorem 3 in Sec. 6.2 and formula (5) in Sec. 6.3, we obtain the subsidiary equation

$$RI(s) + \frac{I(s)}{sC} = \frac{V_0}{s}[e^{-as} - e^{-bs}].$$

Solving this equation algebraically for $I(s)$, we get

$$I(s) = F(s)(e^{-as} - e^{-bs}) \qquad \text{where} \qquad F(s) = \frac{V_0/R}{s + 1/RC}.$$

From Table 6.1 in Sec. 6.1 we have

$$\mathcal{L}^{-1}(F) = \frac{V_0}{R} e^{-t/(RC)}.$$

Hence Theorem 2 in Sec. 6.3 yields the solution (Fig. 99)

$$i(t) = \mathcal{L}^{-1}(I) = \mathcal{L}^{-1}\{e^{-as}F(s)\} - \mathcal{L}^{-1}\{e^{-bs}F(s)\}$$

$$= \frac{V_0}{R}[e^{-(t-a)/(RC)}u(t-a) - e^{-(t-b)/(RC)}u(t-b)];$$

that is, $i = 0$ if $t < a$, and

$$i(t) = \begin{cases} K_1 e^{-t/(RC)} & \text{if } a < t < b \\ (K_1 - K_2)e^{-t/(RC)} & \text{if } t > b \end{cases}$$

where $K_1 = V_0 e^{a/(RC)}/R$ and $K_2 = V_0 e^{b/(RC)}/R$. ∎

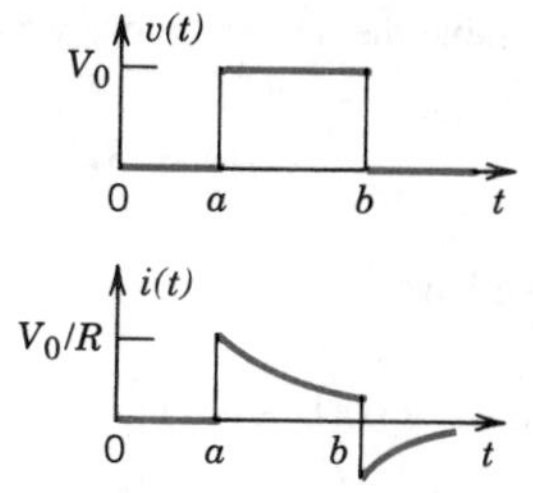

Fig. 99. Voltage and current in Example 1

EXAMPLE 2 Response of an undamped system to a single square wave

Solve the initial value problem

$$y'' + 2y = r(t), \qquad y(0) = 0, \qquad y'(0) = 0,$$

where $r(t) = 1$ if $0 < t < 1$ and 0 otherwise (Fig. 100).

Solution. From Theorem 2 in Sec. 6.2, the initial conditions and (3) and (5) in Sec. 6.3 we obtain the subsidiary equation

$$s^2Y + 2Y = \frac{1}{s} - \frac{e^{-s}}{s}.$$

The solution is

$$Y(s) = \frac{1}{s(s^2+2)} - \frac{e^{-s}}{s(s^2+2)}.$$

On the right we use partial fractions:

$$\frac{1}{s(s^2+2)} = \frac{1}{2}\left(\frac{1}{s} - \frac{s}{s^2+2}\right).$$

Hence by Table 6.1 (Sec. 6.1) and Theorem 2 in Sec. 6.3,

$$y(t) = \tfrac{1}{2}[1 - \cos\sqrt{2}\,t] - \tfrac{1}{2}[1 - \cos\sqrt{2}(t-1)]u(t-1),$$

that is,

$$y(t) = \begin{cases} \frac{1}{2} - \frac{1}{2}\cos\sqrt{2}\,t & \text{if } 0 \leqq t < 1 \\ \frac{1}{2}\cos\sqrt{2}(t-1) - \frac{1}{2}\cos\sqrt{2}\,t & \text{if } \quad t > 1. \end{cases}$$

We see that $y(t)$ represents a composite of harmonic oscillations. ∎

EXAMPLE 3 Response of a damped vibrating system to a single square wave

Determine the response of the damped vibrating system corresponding to

$$y'' + 3y' + 2y = r(t), \qquad y(0) = 0, \qquad y'(0) = 0$$

with $r(t)$ as in Example 2 (Fig. 100).

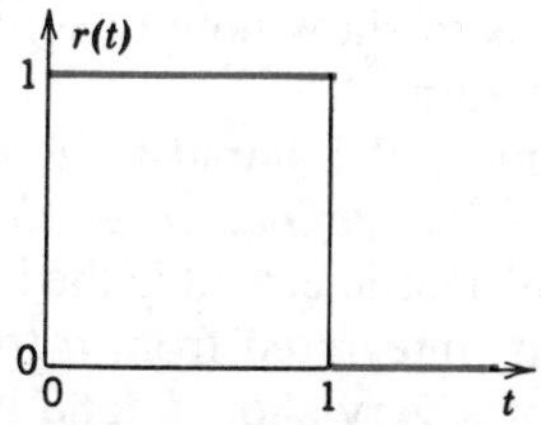

Fig. 100. Input in Examples 2 and 3

Solution. By the same method as before we obtain the subsidiary equation

$$s^2Y + 3sY + 2Y = \frac{1}{s}(1 - e^{-s}).$$

Solving for Y, we have

$$Y(s) = F(s)(1 - e^{-s}) \qquad \text{where} \qquad F(s) = \frac{1}{s(s+1)(s+2)}.$$

In terms of partial fractions,

$$F(s) = \frac{1/2}{s} - \frac{1}{s+1} + \frac{1/2}{s+2}.$$

Hence by Table 6.1 (Sec. 6.1),

$$f(t) = \mathcal{L}^{-1}(F) = \tfrac{1}{2} - e^{-t} + \tfrac{1}{2}e^{-2t}.$$

Therefore, by Theorem 2 in Sec. 6.3 we have

$$\mathcal{L}^{-1}\{e^{-s}F(s)\} = f(t-1)u(t-1) = \begin{cases} 0 & (0 \leqq t < 1) \\ \tfrac{1}{2} - e^{-(t-1)} + \tfrac{1}{2}e^{-2(t-1)} & (t > 1). \end{cases}$$

This yields the solution (Fig. 101)

$$y(t) = \mathcal{L}^{-1}(Y) = f(t) - f(t-1)u(t-1) = \begin{cases} \tfrac{1}{2} - e^{-t} + \tfrac{1}{2}e^{-2t} & (0 \leqq t < 1) \\ K_1e^{-t} - K_2e^{-2t} & (t > 1) \end{cases}$$

where $K_1 = e - 1$ and $K_2 = (e^2 - 1)/2$. ∎

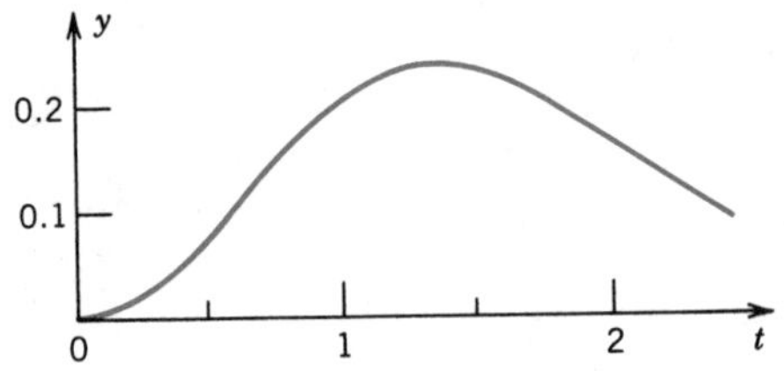

Fig. 101. Output in Example 3

Short Impulses. Dirac's Delta Function

Phenomena of an impulsive nature, such as the action of very large forces (or voltages) over very short intervals of time, are of great practical interest, since they arise in various applications. This situation occurs, for instance, when a tennis ball is hit, a system is given a blow by a hammer, an airplane makes a "hard" landing, a ship is hit by a high single wave, and so on. Our present goal is to show how to solve problems involving short impulses by Laplace transforms.

In mechanics, the **impulse** of a force $f(t)$ over a time interval, say, $a \leqq t \leqq a + k$ *is defined to be the integral of* $f(t)$ from a to $a + k$. The analog for an electric circuit is the integral of the electromotive force applied to the circuit, integrated from a to $a + k$. Of particular practical interest is the case of a very short k (and its limit $k \to 0$), that is, the impulse of a force acting only for an instant. To handle the case, we consider the function

$$f_k(t) = \begin{cases} 1/k & \text{if } a \leqq t \leqq a + k \\ 0 & \text{otherwise} \end{cases} \qquad \text{(Fig. 102).} \tag{1}$$

Its impulse I_k is 1, since the integral evidently gives the area of the rectangle in Fig. 102

$$I_k = \int_0^\infty f_k(t)\, dt = \int_a^{a+k} \frac{1}{k}\, dt = 1. \tag{2}$$

We can represent $f_k(t)$ in terms of two unit step functions (Sec. 6.3), namely,

$$f_k(t) = \frac{1}{k}[u(t - a) - u(t - (a + k))].$$

From (5) in Sec. 6.3 we obtain the Laplace transform

$$\mathcal{L}\{f_k(t)\} = \frac{1}{ks}\,[e^{-as} - e^{-(a+k)s}] = e^{-as}\,\frac{1 - e^{-ks}}{ks}\,. \tag{3}$$

The limit of $f_k(t)$ as $k \to 0$ is denoted by

$$\delta(t - a)$$

and is called the **Dirac delta function**[4] (sometimes the **unit impulse function**). The quotient in (3) has the limit 1 as $k \to 0$, as follows by l'Hôpital's rule (differentiate the numerator and also the denominator with respect to k). Thus

$$\mathcal{L}\{\delta(t - a)\} = e^{-as}. \tag{4}$$

We note that $\delta(t - a)$ is not a function in the ordinary sense as used in calculus, but a so-called *"generalized function,"*[5] because (1) and (2) with $k \to 0$ imply

$$\delta(t - a) = \begin{cases} \infty & \text{if } t = a \\ 0 & \text{otherwise} \end{cases} \qquad \text{and} \qquad \int_0^\infty \delta(t - a)\, dt = 1,$$

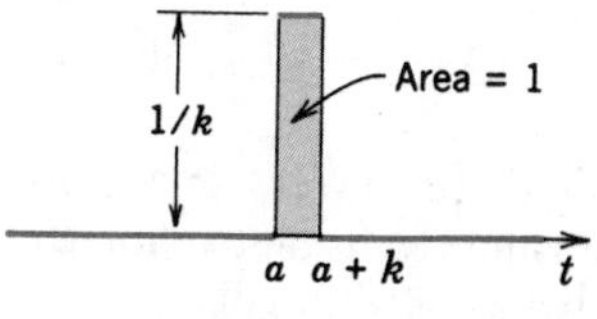

Fig. 102. The function $f_k(t)$ in (1)

[4]PAUL DIRAC (1902—1984), English physicist, was awarded the Nobel Prize [jointly with ERWIN SCHRÖDINGER (1887—1961)] in 1933 for his work in quantum mechanics.

[5]Or *"distribution."* A systematic theory of generalized functions was created in 1936 by the Russian mathematician, SERGEI L'VOVICH SOBOLEV (1908—1989) and in 1945, under wider aspects, by the French mathematician, LAURENT SCHWARTZ (born 1915).

but an ordinary function which is everywhere 0 except at a single point must have the integral 0. Nevertheless, in impulse problems it is convenient to operate on $\delta(t - a)$ as though it were an ordinary function.

EXAMPLE 4 Response of a damped vibrating system to a unit impulse

Determine the response of the damped mass–spring system (see Sec. 2.11) governed by

$$y'' + 3y' + 2y = \delta(t - a), \qquad y(0) = 0, \qquad y'(0) = 0.$$

Thus the system is initially at rest and at time $t = a$ is suddenly given a sharp hammerblow.

Solution. By (4) we obtain the subsidiary equation

$$s^2Y + 3sY + 2Y = e^{-as}.$$

Solving for Y, we have

$$Y(s) = F(s)e^{-as}, \qquad \text{where} \qquad F(s) = \frac{1}{(s+1)(s+2)} = \frac{1}{s+1} - \frac{1}{s+2}.$$

Taking the inverse transform, we obtain

$$f(t) = \mathcal{L}^{-1}(F) = e^{-t} - e^{-2t}.$$

Hence by the second shifting theorem (Sec. 6.3) we have

$$y(t) = \mathcal{L}^{-1}\{e^{-as}F(s)\} = f(t - a)u(t - a) = \begin{cases} 0 & \text{if } 0 \leqq t < a \\ e^{-(t-a)} - e^{-2(t-a)} & \text{if } t > a. \end{cases}$$

Figure 103 shows this solution for $a = 1$. The reader may compare this with the output in Example 3 (Fig. 101) and comment. ∎

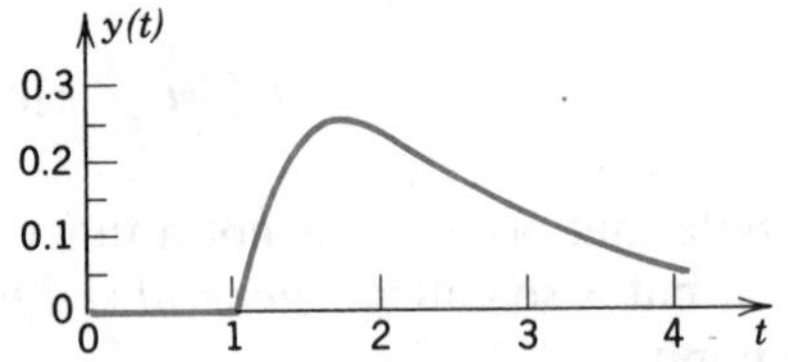

Fig. 103. Output in Example 4 with $a = 1$

Problem Set 6.4

***RC*-circuit.** Find the current in the RC-circuit in Fig. 104 with $R = 100$ ohms, $C = 0.1$ farad, and electromotive force $v(t)$ [volts] as follows. Assume that the circuit is quiescent before $v(t)$ is applied.

1. $v(t) = 100$ if $1 < t < 2$ and 0 otherwise.

2. $v(t) = 10000$ if $1 < t < 1.01$ and 0 otherwise

3. $v(t) = e^{-t}$ if $t > 2$ and 0 otherwise

4. $v(t) = 50(t - 3)$ if $t > 3$ and 0 otherwise

5. $v(t) = 200t$ if $0 < t < 1$ and 0 if $t > 1$

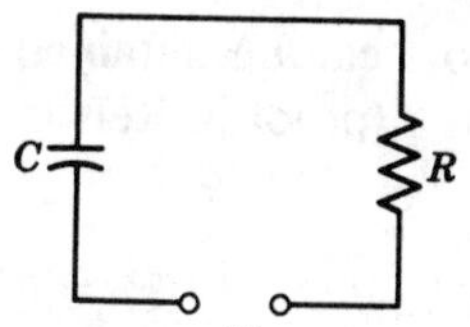

Fig. 104. *RC*-circuit in Probs. 1–5

Effect of the delta function on vibrating systems and in other initial value problems. Solve:

6. $y'' + 4y = \delta(t - \pi), \quad y(0) = 2, \quad y'(0) = 0$

7. $y'' + 9y = \delta(t - 1), \quad y(0) = 0, \quad y'(0) = 0$

8. $y'' - y = \sin t + \delta(t - \frac{1}{2}\pi), \quad y(0) = 3.0, \quad y'(0) = -3.5$

9. $y'' + y = \delta(t - \pi) - \delta(t - 2\pi), \quad y(0) = 0, \quad y'(0) = 1$

10. $y'' - y = -2 \sin t + \delta(t - 1), \quad y(0) = 0, \quad y'(0) = 2$

11. $y'' + 2y' + 2y = \delta(t - 2\pi), \quad y(0) = 1, \quad y'(0) = -1$

12. $y'' + 3y' + 2y = \delta(t - 4), \quad y(0) = y'(0) = 0$

13. $y'' + 4y' + 5y = \delta(t - 1), \quad y(0) = 0, \quad y'(0) = 3$

14. $y'' + 2y' - 3y = \delta(t - 2) + \delta(t - 3), \quad y(0) = 1, \quad y'(0) = 0$

15. $y'' + 2y' - 3y = -8e^{-t} - \delta(t - \frac{1}{2}), \quad y(0) = 3, \quad y'(0) = -5$

16. $y'' + 4y' + 4y = 1 + t + \delta(t - 1), \quad y(0) = 0, \quad y'(0) = 2.25$

17. $y'' + 2y' + 5y = 8e^{t} + \delta(t - 1), \quad y(0) = 2, \quad y'(0) = 0$

18. $y'' + 5y' + 6y = u(t - 1) + \delta(t - 2), \quad y(0) = 0, \quad y'(0) = 1$

19. $y'' + 2y' + 5y = 25t - \delta(t - \pi), \quad y(0) = -2, \quad y'(0) = 5$

20. $y'' - y' - 2y = -4t^2 + \delta(t - 2), \quad y(0) = 5, \quad y'(0) = -1$

6.5 Differentiation and Integration of Transforms

The number of methods for obtaining transforms or inverse transforms and their application in solving differential equations is surprisingly large. They include direct integration (Sec. 6.1), the use of linearity (Sec. 6.1), shifting (Sec. 6.3), and differentiation or integration of original functions $f(t)$ (Sec. 6.2). But this is not all: in this section we consider differentiation and integration of *transforms* $F(s)$ and find out the corresponding operations for original functions $f(t)$.

Differentiation of Transforms

It can be shown that if $f(t)$ satisfies the conditions of the existence theorem in Sec. 6.1, then the derivative of its transform

$$F(s) = \mathcal{L}(f) = \int_0^\infty e^{-st} f(t)\, dt$$

with respect to s can be obtained by differentiating under the integral sign with respect to s (proof in Ref. [5] listed in Appendix 1); thus

$$F'(s) = -\int_0^\infty e^{-st}[tf(t)]\,dt.$$

Consequently, if $\mathcal{L}(f) = F(s)$, then

$$\mathcal{L}\{tf(t)\} = -F'(s); \tag{1}$$

differentiation of the transform of a function corresponds to the multiplication of the function by $-t$. Equivalently,

$$\mathcal{L}^{-1}\{F'(s)\} = -tf(t). \tag{1*}$$

This property enables us to get new transforms from given ones, as we show next.

EXAMPLE 1 Differentiation of transforms

We shall derive the following three formulas (formulas 21–23 in the table in Sec. 6.10):

	$\mathcal{L}(f)$	$f(t)$
(2)	$\dfrac{1}{(s^2+\beta^2)^2}$	$\dfrac{1}{2\beta^3}(\sin\beta t - \beta t\cos\beta t)$
(3)	$\dfrac{s}{(s^2+\beta^2)^2}$	$\dfrac{t}{2\beta}\sin\beta t$
(4)	$\dfrac{s^2}{(s^2+\beta^2)^2}$	$\dfrac{1}{2\beta}(\sin\beta t + \beta t\cos\beta t)$

Solution. From (1) and formula 8 (with $\omega = \beta$) in Table 6.1, Sec. 6.1, we obtain

$$\mathcal{L}(t\sin\beta t) = \frac{2\beta s}{(s^2+\beta^2)^2}.$$

By dividing by 2β we obtain (3). From (1) and formula 7 (with $\omega = \beta$) in Table 6.1 we find

$$\mathcal{L}(t\cos\beta t) = -\frac{(s^2+\beta^2) - 2s^2}{(s^2+\beta^2)^2} = \frac{s^2-\beta^2}{(s^2+\beta^2)^2}. \tag{5}$$

From this and formula 8 (with $\omega = \beta$) in Table 6.1 we have

$$\mathcal{L}\left(t\cos\beta t \pm \frac{1}{\beta}\sin\beta t\right) = \frac{s^2-\beta^2}{(s^2+\beta^2)^2} \pm \frac{1}{s^2+\beta^2}.$$

On the right we now take the common denominator. Then for the plus sign the numerator is $s^2 - \beta^2 + s^2 + \beta^2 = 2s^2$, so that (4) follows. For the minus sign the numerator takes the form $s^2 - \beta^2 - s^2 - \beta^2 = -2\beta^2$, and we obtain (2). ∎

Integration of Transforms

Similarly, if $f(t)$ satisfies the conditions of the existence theorem in Sec. 6.1 and the limit of $f(t)/t$, as t approaches 0 from the right, exists, then

$$\text{(6)} \qquad \mathcal{L}\left\{\frac{f(t)}{t}\right\} = \int_s^\infty F(\tilde{s})\, d\tilde{s} \qquad (s > \gamma);$$

in this manner, *integration of the transform of a function $f(t)$ corresponds to the division of $f(t)$ by t.* Equivalently,

$$\text{(6*)} \qquad \mathcal{L}^{-1}\left\{\int_s^\infty F(\tilde{s})\, d\tilde{s}\right\} = \frac{f(t)}{t}.$$

In fact, from the definition it follows that

$$\int_s^\infty F(\tilde{s})\, d\tilde{s} = \int_s^\infty \left[\int_0^\infty e^{-\tilde{s}t} f(t)\, dt\right] d\tilde{s},$$

and it can be shown (see Ref. [5] in Appendix 1) that under the above assumptions we may reverse the order of integration, that is,

$$\int_s^\infty F(\tilde{s})\, d\tilde{s} = \int_0^\infty \left[\int_s^\infty e^{-\tilde{s}t} f(t)\, d\tilde{s}\right] dt = \int_0^\infty f(t) \left[\int_s^\infty e^{-\tilde{s}t}\, d\tilde{s}\right] dt.$$

The integral over $\tilde{s}$ on the right equals e^{-st}/t when $s > \gamma$, and, therefore,

$$\int_s^\infty F(\tilde{s})\, d\tilde{s} = \int_0^\infty e^{-st} \frac{f(t)}{t}\, dt = \mathcal{L}\left\{\frac{f(t)}{t}\right\} \qquad (s > \gamma). \quad \blacksquare$$

EXAMPLE 2 Integration of transforms

Find the inverse transform of the function $\ln\left(1 + \frac{\omega^2}{s^2}\right)$.

Solution. By differentiation,

$$-\frac{d}{ds} \ln\left(1 + \frac{\omega^2}{s^2}\right) = \frac{2\omega^2}{s(s^2 + \omega^2)} = \frac{2}{s} - 2\frac{s}{s^2 + \omega^2},$$

where the last equality can be readily verified by direct calculation. This is our present $F(s)$. It is the derivative of the given function (times -1), so that the latter is the integral of $F(s)$ from s to ∞. From Table 6.1 in Sec. 6.1 we obtain

$$f(t) = \mathcal{L}^{-1}(F) = \mathcal{L}^{-1}\left\{\frac{2}{s} - 2\frac{s}{s^2 + \omega^2}\right\} = 2 - 2\cos \omega t.$$

This function satisfies the conditions under which (6) holds. Therefore,

$$\mathcal{L}^{-1}\left\{\ln\left(1 + \frac{\omega^2}{s^2}\right)\right\} = \mathcal{L}^{-1}\left\{\int_s^\infty F(\tilde{s})\, d\tilde{s}\right\} = \frac{f(t)}{t}.$$

Our result is

$$\mathcal{L}^{-1}\left\{\ln\left(1 + \frac{\omega^2}{s^2}\right)\right\} = \frac{2}{t}(1 - \cos \omega t).$$

This proves formula 42 in the table in Sec. 6.10. ∎

EXAMPLE 3 Integration of transforms

Reasoning as in Example 2, we obtain (see formula 43 in the table in Sec. 6.10)

$$\mathcal{L}^{-1}\left\{\ln\left(1 - \frac{a^2}{s^2}\right)\right\} = \frac{2}{t}(1 - \cosh at).$$ ∎

Differential Equations with Variable Coefficients

From (1) and (1), (2) in Sec. 6.2 and by differentiating we have

$$\text{(7)} \quad \mathcal{L}(ty') = -\frac{d}{ds}[sY - y(0)] = -Y - sY',$$

$$\text{(8)} \quad \mathcal{L}(ty'') = -\frac{d}{ds}[s^2Y - sy(0) - y'(0)] = -2sY - s^2Y' + y(0).$$

Hence if a differential equation has coefficients such as $at + b$, we get a first-order differential equation for Y, which is sometimes simpler than the given equation. But if the latter has coefficients $at^2 + bt + c$, we get, by two applications of (1), a second-order differential equation for Y, and this shows that the Laplace transform method works well only for very special equations with variable coefficients. We illustrate it for an important equation in the following example.

EXAMPLE 4 Laguerre's differential equation, Laguerre polynomials

Laguerre's differential equation is (see also Problem Set 5.9)

$$\text{(9)} \qquad ty'' + (1 - t)y' + ny = 0.$$

We determine a solution of (9) with $n = 0, 1, 2, \cdots$. From (7)–(9) we get

$$-2sY - s^2Y' + y(0) + sY - y(0) - (-Y - sY') + nY = 0.$$

Simplification gives

$$(s - s^2)Y' + (n + 1 - s)Y = 0.$$

Separating variables, using partial fractions, and integrating (with the constant of integration taken zero), we get

$$\text{(10*)} \qquad \frac{dY}{Y} = \left(\frac{n}{s-1} - \frac{n+1}{s}\right) ds \qquad \text{and} \qquad Y = \frac{(s-1)^n}{s^{n+1}}.$$

We write $l_n = \mathcal{L}^{-1}(Y)$ and show that

$$\text{(10)} \qquad l_0 = 1, \qquad l_n(t) = \frac{e^t}{n!}\frac{d^n}{dt^n}(t^n e^{-t}), \qquad n = 1, 2, \cdots.$$

These are polynomials because the exponential terms cancel if we perform the indicated differentiations. They are called **Laguerre polynomials** and are usually denoted by L_n (but we conform to our convention of reserving capital letters for transforms). We prove (10). By Table 6.1 and the first shifting theorem,

$$\mathcal{L}(t^n e^{-t}) = \frac{n!}{(s+1)^{n+1}}.$$

By (4) in Sec. 6.2,

$$\mathcal{L}\{(t^n e^{-t})^{(n)}\} = \frac{n!s^n}{(s+1)^{n+1}}.$$

Now make another shift and divide by $n!$ to get [see (10) and (10*)]

$$\mathcal{L}(l_n) = \frac{(s-1)^n}{s^{n+1}} = Y. \qquad \blacksquare$$

Problem Set 6.5

Using (1), find the Laplace transform of

1. $2t \cos 2t$ **2.** $4te^{2t}$ **3.** $t^2 e^t$ **4.** $t \cosh t$
5. $t \sinh 3t$ **6.** $t^2 \sinh 2t$ **7.** $t \sin \omega t$ **8.** $t^2 \sin \omega t$
9. $t^2 \cos t$ **10.** $t^2 \cos \omega t$ **11.** $\frac{1}{2}te^{-2t} \sin t$ **12.** $te^{-2t} \sin \omega t$

Using (6) or (1), find $f(t)$ if $\mathcal{L}(f)$ equals

13. $\dfrac{4}{(s+1)^2}$ **14.** $\dfrac{8s}{(s^2+4)^2}$ **15.** $\dfrac{4s}{(s^2-4)^2}$ **16.** $\dfrac{2}{(s-a)^3}$
17. $\dfrac{s}{(s^2+1)^2}$ **18.** $\ln \dfrac{s+a}{s+b}$ **19.** $\ln \dfrac{s}{s-1}$ **20.** $\ln \dfrac{s^2+1}{(s-1)^2}$
21. arc cot (s/ω) **22.** arc cot $(s+1)$

23. Solve Probs. 2 and 3 by using the first shifting theorem.
24. Find $\mathcal{L}(t^n e^{at})$ by repeated application of (1), choosing $f(t) = e^{at}$.
25. Carry out the derivation in Example 3.

6.6 Convolution. Integral Equations

Another important general property of the Laplace transform has to do with products of transforms. It often happens that we are given two transforms $F(s)$ and $G(s)$ whose inverses $f(t)$ and $g(t)$ we know, and we would like to calculate the inverse of the product $H(s) = F(s)G(s)$ from those known inverses $f(t)$ and $g(t)$. This inverse $h(t)$ is written $(f * g)(t)$, which is a standard notation, and is called the **convolution** of f and g. How can we find h from f and g? This is stated in the following theorem. Since the situation and task just described arise quite often in applications, this theorem is of considerable practical importance.

Theorem 1 **(Convolution theorem)**

Let $f(t)$ and $g(t)$ satisfy the hypothesis of the existence theorem (Sec. 6.1) Then the product of their transforms $F(s) = \mathcal{L}(f)$ and $G(s) = \mathcal{L}(g)$ is the transform $H(s) = \mathcal{L}(h)$ of the **convolution** *$h(t)$ of $f(t)$ and $g(t)$, written $\mathcal{L}(f * g)(t)$ and defined by*

$$h(t) = (f * g)(t) = \int_0^t f(\tau)g(t - \tau)\, d\tau. \tag{1}$$

(The proof follows after Example 3.)

EXAMPLE 1 **Convolution**

Using convolution, find the inverse $h(t)$ of

$$H(s) = \frac{1}{(s^2 + 1)^2} = \frac{1}{s^2 + 1} \cdot \frac{1}{s^2 + 1}.$$

Solution. We know that each factor on the right has the inverse $\sin t$. Hence by the convolution theorem and by (11), Appendix 3.1, we get

$$\begin{aligned} h(t) = \mathcal{L}^{-1}(H) &= \sin t * \sin t \\ &= \int_0^t \sin \tau \sin (t - \tau)\, d\tau \\ &= \frac{1}{2}\int_0^t -\cos t\, d\tau + \frac{1}{2}\int_0^t \cos (2\tau - t)\, d\tau \\ &= -\tfrac{1}{2}t \cos t + \tfrac{1}{2} \sin t. \end{aligned}$$

■

EXAMPLE 2 **Convolution**

$1/s^2$ has the inverse t and $1/s$ has the inverse 1, and the convolution theorem confirms that $1/s^3 = (1/s^2)(1/s)$ has the inverse

$$t * 1 = \int_0^t \tau \cdot 1\, d\tau = \frac{t^2}{2}.$$

■

EXAMPLE 3 **Convolution**

Let $H(s) = 1/[s^2(s - a)]$. Find $h(t)$.

Solution. From Table 6.1 (Sec. 6.1) we know that

$$\mathcal{L}\left\{\frac{1}{s^2}\right\} = t, \qquad \mathcal{L}^{-1}\left\{\frac{1}{s - a}\right\} = e^{at}.$$

Using the convolution theorem and integrating by parts, we get the answer

$$\begin{aligned} h(t) = t * e^{at} &= \int_0^t \tau e^{a(t-\tau)}\, d\tau = e^{at}\int_0^t \tau e^{-a\tau}\, d\tau \\ &= \frac{1}{a^2}(e^{at} - at - 1). \end{aligned}$$

■

Proof of Theorem 1. By the definition of $G(s)$ and the second shifting theorem, for each fixed τ ($\tau \geqq 0$) we have

$$
\begin{aligned}
e^{-s\tau}G(s) &= \mathcal{L}\{g(t-\tau)u(t-\tau)\} \\
&= \int_0^\infty e^{-st}g(t-\tau)u(t-\tau)\,dt \\
&= \int_\tau^\infty e^{-st}g(t-\tau)\,dt
\end{aligned}
$$

where $s > \gamma$. From this and the definition of $F(s)$ we obtain

$$F(s)G(s) = \int_0^\infty e^{-s\tau}f(\tau)G(s)\,d\tau = \int_0^\infty f(\tau)\int_\tau^\infty e^{-st}g(t-\tau)\,dt\,d\tau$$

where $s > \gamma$. Here we integrate over t from τ to ∞ and then over τ from 0 to ∞; this corresponds to the colored wedge-shaped region extending to infinity in the $t\tau$-plane shown in Fig. 105. Our assumptions on f and g are such that the order of integration can be reversed. (A proof requiring the knowledge of uniform convergence is included in Ref. [A3] listed in Appendix 1.) We then integrate first over τ from 0 to t (see Fig. 105) and then over t from 0 to ∞; thus

$$
\begin{aligned}
F(s)G(s) &= \int_0^\infty e^{-st}\int_0^t f(\tau)g(t-\tau)\,d\tau\,dt \\
&= \int_0^\infty e^{-st}h(t)\,dt = \mathcal{L}(h)
\end{aligned}
$$

where h is given by (1). This completes the proof. ∎

Using the definition, the reader may show that the convolution $f * g$ has the properties

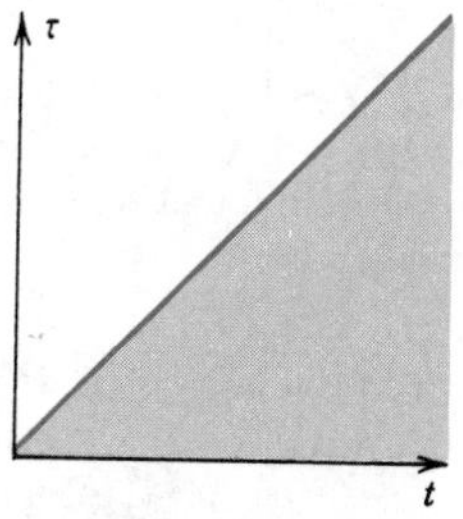

Fig. 105. Region of integration in the $t\tau$-plane in the proof of Theorem 1

$$f * g = g * f \qquad \text{(commutative law)}$$

$$f * (g_1 + g_2) = f * g_1 + f * g_2 \qquad \text{(distributive law)}$$

$$(f * g) * v = f * (g * v) \qquad \text{(associative law)}$$

$$f * 0 = 0 * f = 0,$$

just as for numbers. But $f * 1 \neq f$ in general, as Example 2 shows. Another unusual property is that $(f * f)(t) \geqq 0$ may not hold, as we can see from Example 1.

Very useful applications of convolution occur in a natural way in the solution of differential equations, as we shall now discuss.

Differential Equations

From Sec. 6.2 we recall that the subsidiary equation of the differential equation

$$y'' + ay' + by = r(t) \tag{2}$$

has the solution

$$Y(s) = [(s + a)y(0) + y'(0)]Q(s) + R(s)Q(s) \tag{3}$$

with $R(s) = \mathcal{L}(r)$ and $Q(s) = 1/(s^2 + as + b)$ the transfer function. Hence for the solution $y(t)$ of (2) satisfying $y(0) = y'(0) = 0$ we have $Y = RQ$ in (3) and obtain from the convolution theorem the integral representation

$$y(t) = \int_0^t q(t - \tau)r(\tau)\,d\tau, \qquad q(t) = \mathcal{L}^{-1}(Q). \tag{4}$$

EXAMPLE 4 **Response of an undamped system to a single square wave**

We reconsider the model in Sec. 6.4, Example 2,

$$y'' + 2y = r(t), \qquad r(t) = 1 \text{ if } 0 < t < 1 \text{ and 0 otherwise,} \qquad y(0) = y'(0) = 0.$$

We solve it by the convolution technique in order to see how it works for ***inputs that act for some time only***.

Solution. We have $Q(s) = 1/(s^2 + 2)$, hence $q(t) = (\sin \sqrt{2}t)/\sqrt{2}$. Now we must be careful and remember that $r(t) = 1$ if $0 < t < 1$ but 0 if $t > 1$. Accordingly, in (4) we integrate from 0 to t when $t < 1$ but from 0 to 1 only when $t > 1$. Hence for $t < 1$ (we integrate over τ, so that the chain rule gives a -1),

$$y(t) = \frac{1}{\sqrt{2}}\int_0^t \sin \sqrt{2}(t - \tau)\,d\tau = \frac{1}{2}\cos \sqrt{2}(t - \tau)\Big|_0^t = \frac{1}{2}(1 - \cos \sqrt{2}t)$$

and for $t > 1$,

$$y(t) = \frac{1}{\sqrt{2}}\int_0^1 \sin \sqrt{2}(t - \tau)\,d\tau = \frac{1}{2}[\cos \sqrt{2}(t - 1) - \cos \sqrt{2}t].$$

■

Integral Equations

Convolution also helps in solving certain **integral equations,** that is, equations in which the unknown function $y(t)$ appears under the integral (and perhaps also outside of it). This concerns only very special ones (those whose integral is of the form of a convolution), so that it suffices to consider a typical example and a handful of problems, but we do this because integral equations are practically important and often difficult to solve.

EXAMPLE 5 Integral equation

Solve the integral equation

$$y(t) = t + \int_0^t y(\tau) \sin(t - \tau)\, d\tau.$$

Solution. 1st Step. Equation in terms of convolution. We see that the given equation can be written

$$y = t + y * \sin t.$$

2nd Step. Application of the convolution theorem. We write $Y = \mathcal{L}(y)$. By the convolution theorem,

$$Y(s) = \frac{1}{s^2} + Y(s)\frac{1}{s^2+1}.$$

Solving for $Y(s)$, we obtain

$$Y(s) = \frac{s^2+1}{s^4} = \frac{1}{s^2} + \frac{1}{s^4}.$$

3rd Step. Taking the inverse transform. This gives the solution

$$y(t) = t + \tfrac{1}{6}t^3.$$

The reader may check this by substitution and evaluating the integral by repeated integration by parts (which will need patience). ■

Problem Set 6.6

Find the following convolutions. [*Hint.* In Probs. 7–10 use (11) in Appendix 3.]

1. $1 * 1$ **2.** $t * t^2$ **3.** $e^t * e^t$

4. $e^{kt} * e^{-kt}$ **5.** $t^2 * t^2$ **6.** $t * e^{at}$

7. $\sin \omega t * \cos \omega t$ **8.** $\sin \omega t * \sin \omega t$ **9.** $\sin t * \sin 2t$

10. $\cos \omega t * \cos \omega t$ **11.** $u(t - \pi) * \cos t$ **12.** $u(t - 1) * t$

Application of the Convolution Theorem. Find $h(t)$ by the convolution theorem if $H(s) = \mathcal{L}\{h(t)\}$ equals

13. $\dfrac{1}{(s-1)^2}$ **14.** $\dfrac{1}{s(s-1)}$ **15.** $\dfrac{1}{s^2(s-3)}$

16. $\dfrac{1}{s(s-2)^2}$ **17.** $\dfrac{1}{s(s^2+\omega^2)}$ **18.** $\dfrac{1}{s^2(s^2+\omega^2)}$

19. $\dfrac{s}{(s^2 + \omega^2)^2}$ 20. $\dfrac{s^2}{(s^2 + 4)^2}$ 21. $\dfrac{s^2 - \omega^2}{(s^2 + \omega^2)^2}$

22. $\dfrac{s^2 + a^2}{(s^2 - a^2)^2}$ 23. $\dfrac{1}{(s + 1)(s + 2)}$ 24. $\dfrac{1}{(s^2 + \omega^2)^2}$

General properties of convolution

25. Prove the commutative law $f * g = g * f$.
26. Prove the associative law $(f * g) * v = f * (g * v)$.
27. Prove the distributive law $f * (g_1 + g_2) = f * g_1 + f * g_2$.
28. **(Dirac's delta function)** Using the convolution theorem and treating Dirac's delta function (Sec. 6.4) as though it were an ordinary function, show that

$$(\delta * f)(t) = f(t).$$

29. Derive the formula in Prob. 28 by using f_k with $a = 0$ (Sec. 6.4) and applying the mean value theorem for integrals.
30. Using the convolution theorem, prove by induction that

$$\mathcal{L}^{-1}\left\{\frac{1}{(s - a)^n}\right\} = \frac{1}{(n - 1)!} t^{n-1}e^{at}.$$

Application of convolution to initial value problems. Using the convolution theorem, solve:

31. $y'' + y = \sin 3t, \quad y(0) = 0, \quad y'(0) = 0$
32. $y'' + y = \sin t, \quad y(0) = 0, \quad y'(0) = 0$
33. $y'' + y = t, \quad y(0) = 0, \quad y'(0) = 0$
34. $y'' + 3y' + 2y = e^{-t}, \quad y(0) = 0, \quad y'(0) = 0$
35. $y'' + 25y = 5.2e^{-t}, \quad y(0) = 1.2, \quad y'(0) = -10.2$
36. $y'' + 2y = r(t)$, $r(t) = 1$ if $0 < t < 1$ and 0 if $t > 1$; $y(0) = 0$, $y'(0) = 0$
37. The gun barrel problem (Prob. 22 in Sec. 2.11).
38. $y'' + 3y' + 2y = r(t)$, $r(t) = 1$ if $0 < t < t_0$ and 0 if $t > t_0$; $y(0) = 0$, $y'(0) = 0$
39. $y'' + 4y = u(t - 1), \quad y(0) = 0, \quad y'(0) = 0$
40. $y'' + 3y' + 2y = 1 - u(t - 1), \quad y(0) = 0, \quad y'(0) = 1$
41. $y'' + 9y = r(t)$, $r(t) = 8 \sin t$ if $0 < t < \pi$ and 0 if $t > \pi$; $y(0) = 0$, $y'(0) = 4$
42. $y'' - 5y' + 6y = r(t)$, $r(t) = 4e^t$ if $0 < t < 2$ and 0 if $t > 2$; $y(0) = 1$, $y'(0) = -2$
43. $y'' + 4y = r(t)$, $r(t) = 3 \sin t$ if $0 < t < \pi$ and $-3 \sin t$ if $t > \pi$; $y(0) = 0$, $y'(0) = 3$
44. $y'' + 3y' + 2y = r(t)$, $r(t) = 4t$ if $0 < t < 1$ and 8 if $t > 1$; $y(0) = y'(0) = 0$

Integral equations

Using Laplace transforms, solve:

45. $y(t) = 1 + \displaystyle\int_0^t y(\tau)\, d\tau$ 46. $y(t) = \sin 2t + \displaystyle\int_0^t y(\tau) \sin 2(t - \tau)\, d\tau$

47. $y(t) = 1 - \int_0^t (t - \tau)y(\tau)\, d\tau$

48. $y(t) = \sin t + \int_0^t y(\tau) \sin (t - \tau)\, d\tau$

49. $y(t) = te^t - 2e^t \int_0^t e^{-\tau}y(\tau)\, d\tau$

50. $y(t) = t + e^t - \int_0^t y(\tau) \cosh (t - \tau)\, d\tau$

6.7 Partial Fractions. Systems of Differential Equations

We have seen that partial fractions are needed to obtain the solution $y(t) = \mathcal{L}^{-1}(Y)$ of a problem from the solution $Y(s)$ of the subsidiary equation, because Y usually comes out as a quotient of two polynomials,

$$Y(s) = \frac{F(s)}{G(s)},$$

and for a partial fraction P, the inverse $\mathcal{L}^{-1}(P)$ is easy to get from a table and the first shifting theorem.

You may skip this section. Use your favorite method for partial fractions from calculus and any shortcuts you see. Call on this section for help only when you get stuck. We develop matters systematically in this order:

(Case 1) Unrepeated factor $s - a$.
(Case 2) Repeated factor $(s - a)^m$.
(Case 3) Complex factors $(s - a)(s - \bar{a})$.
(Case 4) Repeated complex factors $[(s - a)(s - \bar{a})]^2$.

Higher powers $[(s - a)(s - \bar{a})]^m$ are left aside because they are of minor practical interest. We discuss each case along with an example, and give the proofs for all four cases at the end. The formulas in this section are often called the **Heaviside expansion formulas.**

General assumption

$F(s)$ and $G(s)$ have real coefficients and no common factors. The degree of $F(s)$ is lower than that of $G(s)$.

Case 1. Unrepeated Factor $s - a$

To this factor corresponds in $Y = F/G$ a fraction

$$\frac{A}{s - a}. \tag{1a}$$

From Table 6.1 in Sec. 6.1 we see that its inverse transform is

$$Ae^{at}. \tag{1b}$$

We prove below that the constant A is given by

$$A = \lim_{s \to a} \frac{(s - a)F(s)}{G(s)} \qquad \text{or by} \qquad A = \frac{F(a)}{G'(a)}. \tag{1c}$$

EXAMPLE 1 **Unrepeated factors**

Find the inverse transform of

$$Y(s) = \frac{F(s)}{G(s)} = \frac{s + 1}{s^3 + s^2 - 6s}$$

Solution. Since $G(s) = s(s - 2)(s + 3)$ has three distinct linear factors, Y has the representation

$$Y(s) = \frac{A_1}{s} + \frac{A_2}{s - 2} + \frac{A_3}{s + 3}.$$

We determine A_1, A_2, A_3. Multiplication by $G(s)$ gives

$$s + 1 = (s - 2)(s + 3)A_1 + s(s + 3)A_2 + s(s - 2)A_3.$$

Taking $s = 0$, $s = 2$, $s = -3$, we obtain

$$1 = -2 \cdot 3A_1, \qquad 3 = 2 \cdot 5A_2, \qquad -2 = -3(-5)A_3.$$

Hence $A_1 = -1/6$, $A_2 = 3/10$, $A_3 = -2/15$. The answer is

$$\mathcal{L}^{-1}(Y) = -\tfrac{1}{6} + \tfrac{3}{10}e^{2t} - \tfrac{2}{15}e^{-3t}.$$ ∎

Case 2. Repeated Factor $(s - a)^m$

To this factor corresponds in $Y = F/G$ a sum of m fractions

$$\frac{A_m}{(s - a)^m} + \frac{A_{m-1}}{(s - a)^{m-1}} + \cdots + \frac{A_1}{s - a}. \tag{2a}$$

From Table 6.1 and the first shifting theorem (Sec. 6.3) it follows that its inverse transform is

$$e^{at}\left(A_m \frac{t^{m-1}}{(m - 1)!} + A_{m-1}\frac{t^{m-2}}{(m - 2)!} + \cdots + A_2 \frac{t}{1!} + A_1\right). \tag{2b}$$

We prove below that the constant A_m is given by

$$A_m = \lim_{s \to a} \frac{(s - a)^m F(s)}{G(s)} \tag{2c}$$

and the other constants are given by

$$\text{(2d)} \quad A_k = \frac{1}{(m-k)!} \lim_{s \to a} \frac{d^{m-k}}{ds^{m-k}} \left[\frac{(s-a)^m F(s)}{G(s)} \right], \qquad k = 1, \cdots, m-1.$$

EXAMPLE 2 **Repeated factor. An initial value problem**

Solve the initial value problem

$$y'' - 3y' + 2y = 4t \qquad y(0) = 1, \qquad y'(0) = -1.$$

Solution. From Table 6.1, Sec. 6.1, and (1) and (2) in Sec. 6.2 we get the subsidiary equation

$$s^2 Y - s + 1 - 3(sY - 1) + 2Y = \frac{4}{s^2}.$$

Collecting the terms in Y on the left and the others on the right, we have

$$(s^2 - 3s + 2)Y = \frac{4}{s^2} + s - 4 = \frac{4 + s^3 - 4s^2}{s^2}.$$

Since $s^2 - 3s + 2 = 0$ has the solutions 2 and 1, this gives the partial fraction representation

$$Y(s) = \frac{F(s)}{G(s)} = \frac{s^3 - 4s^2 + 4}{s^2(s-2)(s-1)} = \frac{A_2}{s^2} + \frac{A_1}{s} + \frac{B}{s-2} + \frac{C}{s-1}.$$

We first determine the constants A_1 and A_2 in the partial fractions corresponding to the root $a = 0$. From (2c) with $m = 2$,

$$A_2 = \lim_{s \to 0} \frac{s^2(s^3 - 4s^2 + 4)}{s^2(s^2 - 3s + 2)} = \left. \frac{s^3 - 4s^2 + 4}{s^2 - 3s + 2} \right|_{s=0} = 2.$$

From (2d) with $m = 2$ and $k = 1$,

$$A_1 = \lim_{s \to 0} \frac{d}{ds} \left(\frac{s^3 - 4s^2 + 4}{s^2 - 3s + 2} \right) = 3.$$

We determine B and C. From the first formula in (1c),

$$B = \left. \frac{F(s)}{s^2(s-1)} \right|_{s=2} = -1, \qquad C = \left. \frac{F(s)}{s^2(s-2)} \right|_{s=1} = -1.$$

From this we obtain the answer

$$y(t) = \mathcal{L}^{-1}(Y) = \mathcal{L}^{-1} \left\{ \frac{2}{s^2} + \frac{3}{s} - \frac{1}{s-2} - \frac{1}{s-1} \right\} = 2t + 3 - e^{2t} - e^t. \qquad \blacksquare$$

Case 3. Unrepeated Complex Factors $(s - a)(s - \bar{a})$

Here $a = \alpha + i\beta$, say, and $\bar{a} = \alpha - i\beta$ is the complex conjugate of a. We could work in complex, but it is preferable to stay in real. Then to

$$\begin{aligned}(s - a)(s - \bar{a}) &= s^2 - (a + \bar{a})s + a\bar{a} \\ &= s^2 - 2\alpha s + \alpha^2 + \beta^2 \\ &= (s - \alpha)^2 + \beta^2\end{aligned}$$

corresponds the partial fraction

(3a) $$\frac{As + B}{(s-\alpha)^2 + \beta^2}, \qquad \text{which we can write} \qquad \frac{A(s-\alpha) + \alpha A + B}{(s-\alpha)^2 + \beta^2}.$$

From Table 6.1 and the first shifting theorem we obtain the inverse transform

(3b) $$e^{\alpha t}\left(A \cos \beta t + \frac{\alpha A + B}{\beta} \sin \beta t\right).$$

We prove below that A is the imaginary part and $(\alpha A + B)/\beta$ the real part of

(3c) $$Q_a = \frac{1}{\beta} \lim_{s \to a} \frac{[(s-\alpha)^2 + \beta^2]F(s)}{G(s)}.$$

EXAMPLE 3A

Unrepeated complex and real factors. Solution by inspection

Solve the initial value problem

$$y^{\mathrm{IV}} - k^4 y = 0, \qquad y(0) = y'(0) = y''(0) = 0, \qquad y'''(0) = 1 \qquad (k \neq 0).$$

Solution. By Theorem 2 in Sec. 6.2, the subsidiary equation is

$$s^4 Y - 1 - k^4 Y = 0, \qquad \text{thus,} \qquad (s^4 - k^4)Y = 1.$$

By inspection,

$$Y(s) = \frac{1}{(s^2 + k^2)(s^2 - k^2)} = \frac{1}{2k^2}\left(\frac{1}{s^2 - k^2} - \frac{1}{s^2 + k^2}\right).$$

From this and Table 6.1,

$$y(t) = \mathcal{L}^{-1}\left\{\frac{1}{s^4 - k^4}\right\} = \frac{1}{2k^3}(\sinh kt - \sin kt).$$

This also proves formula 27 in the table in Sec. 6.10.

Our example illustrates an important point. If one does not see shortcuts, one can still get ***help from a systematic partial fraction reduction:***

$$Y(s) = \frac{1}{(s^2 + k^2)(s - k)(s + k)} = \frac{A_1 s + A_2}{s^2 + k^2} + \frac{B}{s - k} + \frac{C}{s + k}.$$

In (3c) we have $a = ik$. Hence $\alpha = 0$ and $\beta = k$. Thus,

$$Q_a = \frac{1}{k} \cdot \left.\frac{1}{(s - k)(s + k)}\right|_{s=ik} = \left.\frac{1}{k(s^2 - k^2)}\right|_{s=ik} = -\frac{1}{2k^3}.$$

This is real. Hence $(\alpha A + B)/\beta = Q_a$ and $A = 0$. From (3b) we now obtain the inverse transform

$$-\frac{1}{2k^3} \sin kt.$$

Also, by (1c),

$$B = \left.\frac{1}{(s^2 + k^2)(s + k)}\right|_{s=k} = \frac{1}{4k^3}, \qquad C = \left.\frac{1}{(s^2 + k^2)(s - k)}\right|_{s=-k} = -\frac{1}{4k^3}.$$

Together, this gives as the answer an expression that agrees with that above:

$$-\frac{1}{2k^3} \sin kt + \frac{1}{4k^3}(e^{kt} - e^{-kt}) = -\frac{1}{2k^3} \sin kt + \frac{1}{2k^3} \sinh kt.$$

■

EXAMPLE 3B

Forced vibrations, no resonance, unrepeated complex factors

Solve the initial value problem

$$my'' + ky = K_0 \sin pt, \qquad y(0) = 0, \qquad y'(0) = 0 \qquad (k/m \neq p^2).$$

Solution. From Sec. 2.11 we know that this equation is the mathematical model of forced oscillations of a body of mass m attached at the lower end of an elastic spring whose upper end is fixed (Fig. 106). k is the spring modulus, and $K_0 \sin pt$ is the driving force or input. Setting $\omega_0 = \sqrt{k/m}$, we may write the equation in the form

$$y'' + \omega_0^2 y = K \sin pt \qquad \left(K = \frac{K_0}{m}\right).$$

The subsidiary equation is

$$s^2 Y + \omega_0^2 Y = K \frac{p}{s^2 + p^2}.$$

The solution is

$$Y(s) = \frac{Kp}{(s^2 + \omega_0^2)(s^2 + p^2)} \qquad (\omega_0^2 \neq p^2).$$

By (3a) the partial fraction form of $Y(s)$ is

$$Y(s) = \frac{Kp}{(s^2 + \omega_0^2)(s^2 + p^2)} = \frac{As + B}{s^2 + \omega_0^2} + \frac{Ms + N}{s^2 + p^2}.$$

We first determine A and B in the fraction corresponding to $a_1 = \alpha_1 + i\beta_1 = i\omega_0$. In (3c) we thus have $\alpha = \alpha_1 = 0$, $\beta = \beta_1 = \omega_0$, and $F(s) = Kp$, so that

$$Q_{a_1} = \frac{1}{\omega_0} \lim_{s \to a_1} \frac{(s^2 + \omega_0^2)Kp}{(s^2 + \omega_0^2)(s^2 + p^2)} = \left.\frac{Kp}{s^2 + p^2}\right|_{s = a_1} = \frac{Kp}{p^2 - \omega_0^2}.$$

Since this Q_{a_1} is real, $A = 0$ in (3b), and Q_{a_1} is the other coefficient, $(\alpha A + B)/\beta$ in (3b), with $\beta = \omega_0$. Hence (3b) gives the inverse

$$\frac{Kp}{\omega_0(p^2 - \omega_0^2)} \sin \omega_0 t.$$

We determine M and N in the last fraction, which corresponds to $a_2 = \alpha_2 + i\beta_2 = ip$. In (3c) we now have

$$Q_{a_2} = \left.\frac{Kp}{s^2 + \omega_0^2}\right|_{s = a_2} = \frac{Kp}{\omega_0^2 - p^2},$$

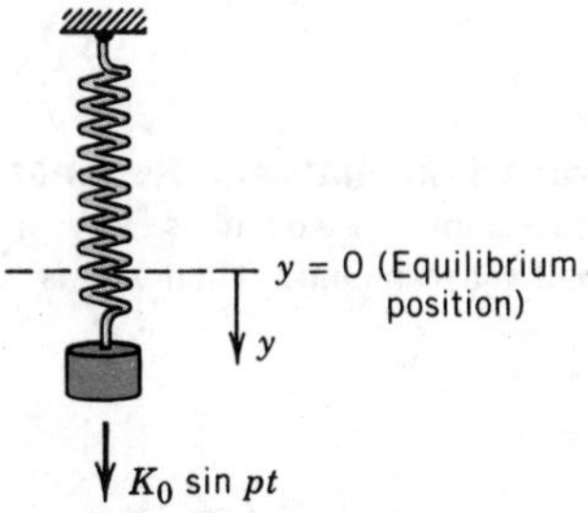

Fig. 106. Vibrating system. Forced oscillations

and (3b) now gives the inverse

$$\frac{Kp}{p(\omega_0^2 - p^2)} \sin pt.$$

By adding these two inverses we obtain the desired solution

$$y(t) = \mathcal{L}^{-1}(Y) = \frac{K}{p^2 - \omega_0^2}\left(\frac{p}{\omega_0} \sin \omega_0 t - \sin pt\right).$$

The case of resonance follows in Example 4. ■

Case 4. Repeated Complex Factors $[(s - a)(s - \bar{a})]^2$

To these factors correspond in $Y = F/G$ the partial fractions

(4a)
$$\frac{As + B}{[(s - \alpha)^2 + \beta^2]^2} + \frac{Cs + D}{(s - \alpha)^2 + \beta^2}.$$

The second fraction is as in (3a). For the first fraction we can use (3) and (2) in Sec. 6.5 and the first shifting theorem. Since we can write $As + B = A(s - \alpha) + (\alpha A + B)$, this gives as the inverse transform of both fractions

(4b)
$$e^{\alpha t}\left[\frac{A}{2\beta} t \sin \beta t + \frac{\alpha A + B}{2\beta^3}(\sin \beta t - \beta t \cos \beta t)\right]$$
$$+ e^{\alpha t}\left[C \cos \beta t + \frac{\alpha C + D}{\beta} \sin \beta t\right].$$

We prove below that the constants in (4b) are given by the formulas

(4c)
$$A = \operatorname{Im} R_a/\beta, \qquad \alpha A + B = \operatorname{Re} R_a,$$
$$C = (A - \operatorname{Re} S_a)/2\beta^2, \qquad \alpha C + D = \operatorname{Im} S_a/2\beta$$

where Re denotes the real part and Im the imaginary part,

$$R_a = \lim_{s \to a} R(s), \qquad S_a = \lim_{s \to a} R'(s)$$

and

$$R(s) = \frac{[(s - \alpha)^2 + \beta^2]^2 F(s)}{G(s)}.$$

EXAMPLE 4 Forced oscillations. Resonance. Repeated complex factors

In Example 3B we had $\omega_0^2 = k/m \neq p^2$ and got no resonance. We now assume $\omega_0^2 = p^2$ and shall get resonance. Then $Y(s)$ is (see Example 3B)

$$Y(s) = \frac{K\omega_0}{(s^2 + \omega_0^2)^2}.$$

The denominator has the double roots $a = \alpha + i\beta = i\omega_0$ and $\bar{a} = -i\omega_0$, that is, $\alpha = 0$ and $\beta = \omega_0$. This denominator cancels out in $R(s)$ in (4c) and we simply have

$$R(s) = K\omega_0, \qquad R_a = K\omega_0, \qquad S_a = 0.$$

This gives $\alpha A + B = K\omega_0$ in (4c), whereas the other expressions in (4c) are zero. Hence (4b) yields the solution

$$y(t) = \mathcal{L}^{-1}(Y) = \frac{K}{2\omega_0^2}(\sin \omega_0 t - \omega_0 t \cos \omega_0 t).$$

The last term grows beyond bound, indicating resonance, as in Sec. 2.11. ∎

Proofs of Formulas (1c), (2c), (2d), (3c), (4c)

In these proofs, $W(s)$ always denotes the sum of the other partial fractions of $Y(s) = F(s)/G(s)$.

Proof of (1c) *in Case* 1

In this case,

$$Y(s) = \frac{F(s)}{G(s)} = \frac{A}{s-a} + W(s).$$

Multiplication by $s - a$ gives

$$\frac{(s-a)F(s)}{G(s)} = A + (s-a)W(s).$$

The first formula in (1c) now follows by letting $s \to a$ and noting that $(s - a)\, W(s) \to 0$, since $W(s)$ has no factor that could cancel $s - a$. The second formula in (1c) is obtained by writing

$$\frac{(s-a)F(s)}{G(s)} = \frac{F(s)}{G(s)/(s-a)}$$

and letting $s \to a$. Then $F(s) \to F(a)$ in the numerator, while the denominator appears as an indeterminate of the form 0/0; evaluating it in the usual way, we get the denominator of the second formula in (1c):

$$\lim_{s\to a} \frac{G(s)}{s-a} = \lim_{s\to a} \frac{G'(s)}{(s-a)'} = G'(a). \qquad ∎$$

Proof of (2c) *and* (2d) *in Case* 2

In this case,

$$Y(s) = \frac{F(s)}{G(s)} = \frac{A_m}{(s-a)^m} + \frac{A_{m-1}}{(s-a)^{m-1}} + \cdots + \frac{A_1}{s-a} + W(s).$$

Multiplication by $(s - a)^m$ gives

$$\frac{(s-a)^m F(s)}{G(s)} = A_m + (s-a)A_{m-1} + (s-a)^2 A_{m-2} + \cdots + (s-a)^m W(s).$$

Letting $s \to a$, we obtain (2c). By differentiation,

$$\frac{d}{ds}\left[\frac{(s-a)^m F(s)}{G(s)}\right] = A_{m-1} + \text{further terms containing factors } s - a.$$

Letting $s \to a$, we obtain A_{m-1} as given by (2d) with $k = m - 1$. Differentiating again, we obtain A_{m-2}, and so on. This proves (2d). ∎

Proof of (3c) in Case 3

We simply write in this proof $p(s) = (s - a)(s - \bar{a}) = (s - \alpha)^2 + \beta^2$. Then by (3a)

$$Y(s) = \frac{F(s)}{G(s)} = \frac{As + B}{p(s)} + W(s) \qquad (A, B \text{ real}),$$

We multiply this by $p(s)$,

$$\frac{p(s)F(s)}{G(s)} = As + B + p(s)W(s).$$

We now let $s \to a$. Since $p(a) = 0$ and $W(s)$ has no factor that could cancel $s - a$, and since $a = \alpha + i\beta$, we get

$$Q_a = \lim_{s \to a} \frac{p(s)F(s)}{G(s)} = Aa + B = (\alpha A + B) + i\beta A.$$

Dividing by β and separating the real and the imaginary parts on both sides, we obtain the statement involving (3c). ∎

Proof of (4c) in Case 4

Writing $p(s)$ as before, from (4a) we have

$$Y(s) = \frac{F(s)}{G(s)} = \frac{As + B}{p^2(s)} + \frac{Cs + D}{p(s)} + W(s).$$

Multiplying this by $p^2(s)$, we get

$$R(s) = \frac{p^2(s)F(s)}{G(s)} = As + B + p(s)(Cs + D) + p^2(s)W(s).$$

We let $s \to a$. Then we get, since $p(a) = 0$,

$$R_a = \lim_{s \to a} R(s) = Aa + B = (\alpha A + B) + i\beta A.$$

Separating the real and the imaginary parts, we obtain the first two formulas in (4c). By differentiation,

$$R'(s) = A + p'(s)(Cs + D) + \text{terms containing } p(s).$$

Here $p(s) = (s - \alpha)^2 + \beta^2$ gives $p'(s) = 2(s - \alpha)$. We now let $s \to a$. Since $p'(a) = 2(a - \alpha) = 2i\beta$, we then obtain

$$S_a = \lim_{s \to a} R'(s) = A + 2i\beta[C(\alpha + i\beta) + D]$$

$$= (A - 2\beta^2 C) + 2i\beta(\alpha C + D).$$

Separation of the real and the imaginary parts gives the other two formulas in (4c). ∎

Systems of Differential Equations

The Laplace transform may also be used for solving systems of differential equations, a task that also requires partial fractions. We explain the method in terms of a typical example.

EXAMPLE 5 Model of two masses on springs

The mechanical system in Fig. 107 consists of two masses on three springs and is governed by the differential equations

$$\begin{aligned} y_1'' &= -ky_1 + k(y_2 - y_1) \\ y_2'' &= -k(y_2 - y_1) - ky_2, \end{aligned} \tag{5}$$

where k is the spring modulus of each of the three springs, y_1 and y_2 are the displacements of the masses from their positions of static equilibrium; the masses of the springs and the damping are neglected. The derivation of (5) is similar to that of the differential equations in Sec. 4.1.

We shall determine the solution corresponding to the initial conditions $y_1(0) = 1$, $y_2(0) = 1$, $y_1'(0) = \sqrt{3k}$, $y_2'(0) = -\sqrt{3k}$. Let $Y_1 = \mathcal{L}(y_1)$ and $Y_2 = \mathcal{L}(y_2)$. Then from (2) in Sec. 6.2 and the initial conditions we obtain the subsidiary equations

$$s^2Y_1 - s - \sqrt{3k} = -kY_1 + k(Y_2 - Y_1)$$

$$s^2Y_2 - s + \sqrt{3k} = -k(Y_2 - Y_1) - kY_2.$$

This system of linear algebraic equations in the unknowns Y_1 and Y_2 may be written

$$\begin{aligned} (s^2 + 2k)Y_1 - \quad kY_2 \quad &= s + \sqrt{3k} \\ -kY_1 \quad + (s^2 + 2k)Y_2 &= s - \sqrt{3k}. \end{aligned}$$

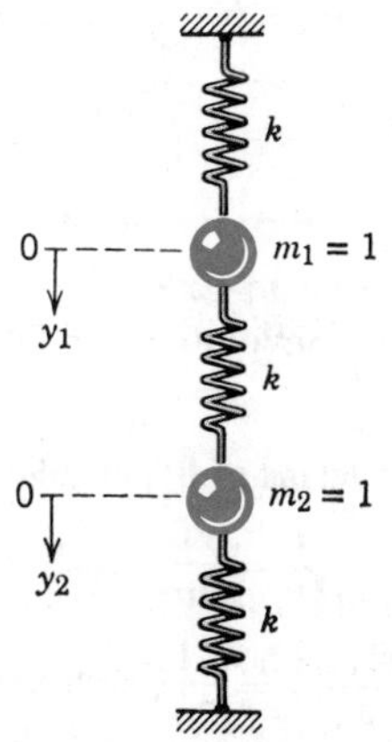

Fig. 107. Example 5

Cramer's rule (Sec. 7.9) or elimination yields the solution

$$Y_1 = \frac{(s + \sqrt{3k})(s^2 + 2k) + k(s - \sqrt{3k})}{(s^2 + 2k)^2 - k^2}$$

$$Y_2 = \frac{(s^2 + 2k)(s - \sqrt{3k}) + k(s + \sqrt{3k})}{(s^2 + 2k)^2 - k^2}.$$

The representations in terms of partial fractions are

$$Y_1 = \frac{s}{s^2 + k} + \frac{\sqrt{3k}}{s^2 + 3k}, \qquad Y_2 = \frac{s}{s^2 + k} - \frac{\sqrt{3k}}{s^2 + 3k}.$$

Hence the solution of our initial value problem is

$$y_1(t) = \mathcal{L}^{-1}(Y_1) = \cos \sqrt{k}t + \sin \sqrt{3k}t$$

$$y_2(t) = \mathcal{L}^{-1}(Y_2) = \cos \sqrt{k}t - \sin \sqrt{3k}t.$$

We see that the motion of each mass is harmonic (the system is undamped!), being the superposition of a "slow" and a "rapid" oscillation. ∎

Problem Set 6.7

Using partial fractions, find $f(t)$ if $\mathcal{L}(f)$ equals

1. $\dfrac{1}{(s - 4)(s - 1)}$ **2.** $\dfrac{s - 3}{s^2 - 1}$ **3.** $\dfrac{3s}{s^2 + 2s - 8}$

4. $\dfrac{s + 12}{s^2 + 4s}$ **5.** $\dfrac{s + 13}{s^2 + 2s + 10}$ **6.** $\dfrac{-s + 4.5}{s^2 + 2.25}$

7. $\dfrac{s^2 - 6s + 4}{s^3 - 3s^2 + 2s}$ **8.** $\dfrac{s}{(s - 2)^3}$ **9.** $\dfrac{10 - 4s}{(s - 2)^2}$

10. $\dfrac{s}{s^2 + 2s + 2}$ **11.** $\dfrac{s^2 + s - 2}{(s + 1)^3}$ **12.** $\dfrac{s^2 + 2s}{(s^2 + 2s + 2)^2}$

13. $\dfrac{-2s^3 + 26s}{s^4 - 10s^2 + 9}$ **14.** $\dfrac{s + 1}{s^2 + 4s + 13}$ **15.** $\dfrac{s^3 + 3s^2 - s - 3}{(s^2 + 2s + 5)^2}$

16. $\dfrac{6s^2 - 26s + 26}{s^3 - 6s^2 + 11s - 6}$ **17.** $\dfrac{s^3 + 6s^2 + 14s}{(s + 2)^4}$ **18.** $\dfrac{s^4 + 3(s + 1)^3}{s^4(s + 1)^3}$

19. $\dfrac{2s^2 - 3s}{(s - 2)(s - 1)^2}$ **20.** $\dfrac{s^3 - 7s^2 + 14s - 9}{(s - 1)^2(s - 2)^3}$

21. Solve Prob. 1 by convolution.

22. Check the result in Example 1 by (i) working backward, (ii) using Theorem 3, Sec. 6.2, (iii) convolution methods.

Some inverses in terms of hyperbolic functions. Show that

23. $\mathcal{L}^{-1}\left\{\dfrac{1}{s^4 + 4a^4}\right\} = \dfrac{1}{4a^3}(\cosh at \sin at - \sinh at \cos at)$

24. $\mathcal{L}^{-1}\left\{\dfrac{s}{s^4 + 4a^4}\right\} = \dfrac{1}{2a^2} \sinh at \sin at$

25. $\mathcal{L}^{-1}\left\{\dfrac{s^2}{s^4 + 4a^4}\right\} = \dfrac{1}{2a}(\cosh at \sin at + \sinh at \cos at)$

26. $\mathcal{L}^{-1}\left\{\dfrac{s^3}{s^4 + 4a^4}\right\} = \cosh at \cos at$

Systems of differential equations. Solve the following initial value problem by means of Laplace transforms.

27. $y_1' = -y_2, \quad y_2' = y_1, \qquad y_1(0) = 1, \quad y_2(0) = 0$

28. $y_1' + y_2 = 2\cos t, \quad y_1 + y_2' = 0, \qquad y_1(0) = 0, \quad y_2(0) = 1$

29. $y_1' = -y_1 + y_2, \quad y_2' = -y_1 - y_2, \qquad y_1(0) = 1, \quad y_2(0) = 0$

30. $y_1' = 6y_1 + 9y_2, \quad y_2' = y_1 + 6y_2, \qquad y_1(0) = -3, \quad y_2(0) = -3$

31. $y_1' = 2y_1 + 4y_2, \quad y_2' = y_1 + 2y_2, \qquad y_1(0) = -4, \quad y_2(0) = -4$

32. $y_1' = -y_1 + 4y_2, \quad y_2' = 3y_1 - 2y_2, \qquad y_1(0) = 3, \quad y_2(0) = 4$

33. $y_1' = 2y_1 - 4y_2, \quad y_2' = y_1 - 3y_2, \qquad y_1(0) = 3, \quad y_2(0) = 0$

34. $y_1' = 5y_1 + y_2, \quad y_2' = y_1 + 5y_2, \qquad y_1(0) = -3, \quad y_2(0) = 7$

35. $y_1' = -2y_1 + 3y_2, \quad y_2' = 4y_1 - y_2, \qquad y_1(0) = 4, \quad y_2(0) = 3$

36. $y_1'' + y_2 = -5\cos 2t, \quad y_2'' + y_1 = 5\cos 2t,$
$y_1(0) = 1, \quad y_1'(0) = 1, \quad y_2(0) = -1, \quad y_2'(0) = 1$

37. $y_1'' = y_1 + 3y_2, \quad y_2'' = 4y_1 - 4e^t,$
$y_1(0) = 2, \quad y_1'(0) = 3, \quad y_2(0) = 1, \quad y_2'(0) = 2$

38. $y_1'' = -5y_1 + 2y_2, \quad y_2'' = 2y_1 - 2y_2,$
$y_1(0) = 3, \quad y_2(0) = 1, \quad y_1'(0) = y_2'(0) = 0$

39. $y_1' + y_2' = 2\sinh t, \quad y_2' + y_3' = e^t, \quad y_3' + y_1' = 2e^t + e^{-t},$
$y_1(0) = 1, \quad y_2(0) = 1, \quad y_3(0) = 0$

40. $2y_1' - y_2' - y_3' = 0, \quad y_1' + y_2' = 4t + 2, \quad y_2' + y_3 = t^2 + 2,$
$y_1(0) = y_2(0) = y_3(0) = 0$

6.8 Periodic Functions. Further Applications

Periodic functions appear in many practical problems, and in most cases they are more complicated than just single cosine or sine functions. This justifies the topic of the present section, which is a systematic approach to the transformation of periodic functions. The text and the problem set also include further applications.

Let $f(t)$ be a function that is defined for all positive t and has the period p (> 0), that is,

$$f(t + p) = f(t) \qquad \text{for all } t > 0.$$

If $f(t)$ is piecewise continuous on an interval of length p, then its Laplace transform exists, and we can write the integral from zero to infinity as the series of integrals over successive periods:

$$\mathcal{L}(f) = \int_0^\infty e^{-st} f(t)\, dt = \int_0^p e^{-st} f\, dt + \int_p^{2p} e^{-st} f\, dt + \int_{2p}^{3p} e^{-st} f\, dt + \cdots .$$

If we substitute $t = \tau + p$ in the second integral, $t = \tau + 2p$ in the third integral, $\cdots$, $t = \tau + (n - 1)p$ in the nth integral, $\cdots$, then the new limits in every integral are 0 and p. Since

$$f(\tau + p) = f(\tau), \qquad f(\tau + 2p) = f(\tau),$$

etc., we thus obtain

$$\mathcal{L}(f) = \int_0^p e^{-s\tau} f(\tau)\, d\tau + \int_0^p e^{-s(\tau+p)} f(\tau)\, d\tau + \int_0^p e^{-s(\tau+2p)} f(\tau)\, d\tau + \cdots .$$

The factors that do not depend on τ can be taken out from under the integral signs; this gives

$$\mathcal{L}(f) = [1 + e^{-sp} + e^{-2sp} + \cdots] \int_0^p e^{-s\tau} f(\tau)\, d\tau.$$

The series in brackets $[\cdots]$ is a geometric series whose sum is $1/(1 - e^{-ps})$. This establishes the following result.

Theorem 1 (Transform of periodic functions)

The Laplace transform of a piecewise continuous periodic function $f(t)$ with period p is

$$\mathcal{L}(f) = \frac{1}{1 - e^{-ps}} \int_0^p e^{-st} f(t)\, dt \qquad (s > 0). \tag{1}$$

EXAMPLE 1 Periodic square wave

Find the transform of the square wave shown in Fig. 108.

Solution. We use (1). Since $p = 2a$, we obtain by direct integration and simplification

$$\mathcal{L}(f) = \frac{1}{1 - e^{-2as}} \left(\int_0^a k e^{-st}\, dt + \int_a^{2a} (-k) e^{-st}\, dt \right)$$

$$= \frac{k}{s} \frac{1 - 2e^{-as} + e^{-2as}}{(1 + e^{-as})(1 - e^{-as})} = \frac{k}{s} \left(\frac{1 - e^{-as}}{1 + e^{-as}} \right)$$

$$= \frac{k}{s} \frac{e^{-as/2}(e^{as/2} - e^{-as/2})}{e^{-as/2}(e^{as/2} + e^{-as/2})} = \frac{k}{s} \frac{2 \sinh (as/2)}{2 \cosh (as/2)} .$$

Hence the result is

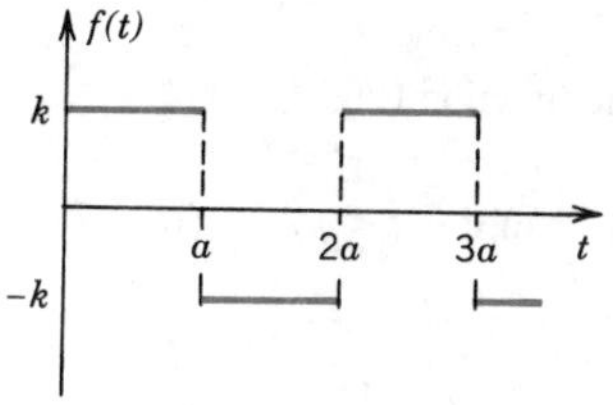

Fig. 108. Example 1

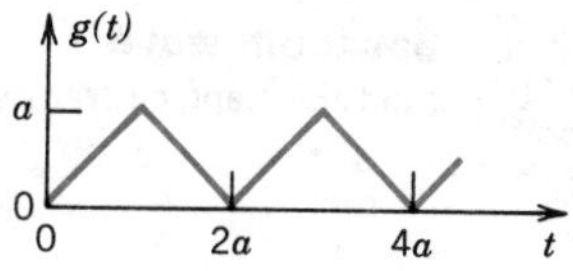

Fig. 109. Example 2

$$\mathcal{L}(f) = \frac{k}{s} \tanh \frac{as}{2}.$$

We can also obtain a less elegant but often more useful form of the result if we write

$$\mathcal{L}(f) = \frac{k}{s}\left(\frac{1 - e^{-as}}{1 + e^{-as}}\right) = \frac{k}{s}\left(1 - \frac{2e^{-as}}{1 + e^{-as}}\right) = \frac{k}{s}\left(1 - \frac{2}{e^{as} + 1}\right). \quad \blacksquare$$

EXAMPLE 2 **Periodic triangular wave**

Find the transform of the periodic function shown in Fig. 109.

Solution. We see that $g(t)$ is the integral of the function $f(t)$ with $k = 1$ in Example 1. Hence, by Theorem 3 in Sec. 6.2,

$$\mathcal{L}(g) = \frac{1}{s}\mathcal{L}(f) = \frac{1}{s^2} \tanh \frac{as}{2}. \quad \blacksquare$$

EXAMPLE 3 **Half-wave rectifier**

Find the transform of the following function $f(t)$ with period $p = 2\pi/\omega$:

$$f(t) = \begin{cases} \sin \omega t & \text{if} \quad 0 < t < \pi/\omega, \\ 0 & \text{if} \quad \pi/\omega < t < 2\pi/\omega. \end{cases}$$

Note that this function is the half-wave rectification of $\sin \omega t$ (Fig. 110).

Solution. From (1) we obtain

$$\mathcal{L}(f) = \frac{1}{1 - e^{-2\pi s/\omega}} \int_0^{\pi/\omega} e^{-st} \sin \omega t \, dt.$$

Using $1 - e^{-2\pi s/\omega} = (1 + e^{-\pi s/\omega})(1 - e^{-\pi s/\omega})$ and integrating by parts or noting that the integral is the imaginary part of the integral

$$\int_0^{\pi/\omega} e^{(-s+i\omega)t}\, dt = \frac{1}{-s + i\omega} e^{(-s+i\omega)t} \Big|_0^{\pi/\omega} = \frac{-s - i\omega}{s^2 + \omega^2}(-e^{-s\pi/\omega} - 1)$$

we obtain the result

$$\mathcal{L}(f) = \frac{\omega(1 + e^{-\pi s/\omega})}{(s^2 + \omega^2)(1 - e^{-2\pi s/\omega})} = \frac{\omega}{(s^2 + \omega^2)(1 - e^{-\pi s/\omega})}. \quad \blacksquare$$

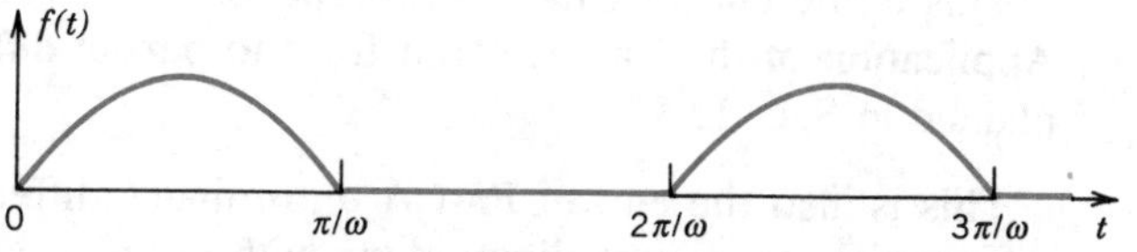

Fig. 110. Half-wave rectification of sin ωt

EXAMPLE 4 **Sawtooth wave**

Find the Laplace transform of the function (Fig. 111)

$$f(t) = \frac{k}{p} t \qquad \text{if } 0 < t < p, \qquad f(t + p) = f(t).$$

Solution. Integration by parts yields

$$\int_0^p e^{-st} t \, dt = -\frac{t}{s} e^{-st} \Big|_0^p + \frac{1}{s} \int_0^p e^{-st} \, dt$$

$$= -\frac{p}{s} e^{-sp} - \frac{1}{s^2}(e^{-sp} - 1),$$

and from (1) we thus obtain the result

$$\mathcal{L}(f) = \frac{k}{ps^2} - \frac{ke^{-ps}}{s(1 - e^{-ps})} \qquad (s > 0).$$ ∎

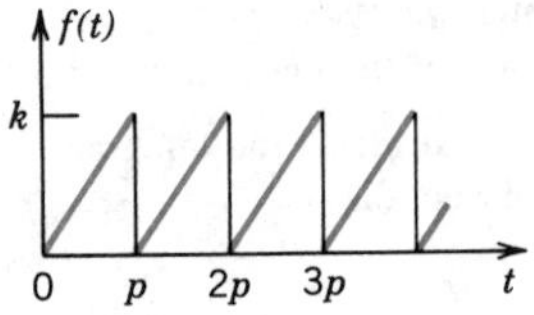

Fig. 111. Sawtooth wave

EXAMPLE 5 **Staircase function**

Find the Laplace transform of the staircase function (Fig. 112)

$$g(t) = kn \qquad [np < t < (n + 1)p, \quad n = 0, 1, 2, \cdots].$$

Solution. Since $g(t)$ is the difference of the functions $h(t) = kt/p$ (whose transform is k/ps^2) and $f(t)$ in Example 4, we obtain

$$\mathcal{L}(g) = \mathcal{L}(h) - \mathcal{L}(f) = \frac{ke^{-ps}}{s(1 - e^{-ps})} \qquad (s > 0).$$ ∎

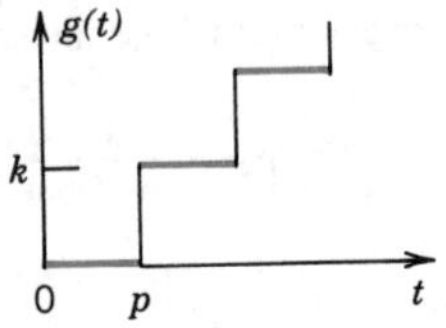

Fig. 112. Staircase function

This is the end of Chap. 6 (except for the tables in Secs. 6.9 and 6.10). Application of the Laplace transform to **partial differential equations** is explained in Sec. 11.13.

This is also the end of Part A on ordinary differential equations. Partial differential equations follow, along with Fourier series, in Chaps. 10 and 11 and **numerical methods** for both ordinary and partial differential equations in Chap. 20.

Problem Set 6.8

Laplace transforms of periodic functions. Sketch the following functions, which are assumed to have the period 2π, and find their transforms.

1. $f(t) = \pi - t \quad (0 < t < 2\pi)$

2. $f(t) = t \quad (0 < t < 2\pi)$

3. $f(t) = 4\pi^2 - t^2 \quad (0 < t < 2\pi)$

4. $f(t) = t^2 \quad (0 < t < 2\pi)$

5. $f(t) = e^t \quad (0 < t < 2\pi)$

6. $f(t) = \sin \frac{1}{2}t \quad (0 < t < 2\pi)$

7. $f(t) = \begin{cases} t & \text{if } 0 < t < \pi \\ 0 & \text{if } \pi < t < 2\pi \end{cases}$

8. $f(t) = \begin{cases} 1 & \text{if } 0 < t < \pi \\ -1 & \text{if } \pi < t < 2\pi \end{cases}$

9. $f(t) = \begin{cases} t & \text{if } 0 < t < \pi \\ \pi - t & \text{if } \pi < t < 2\pi \end{cases}$

10. $f(t) = \begin{cases} 0 & \text{if } 0 < t < \pi \\ t - \pi & \text{if } \pi < t < 2\pi \end{cases}$

11. How can the answer to Prob. 9 be obtained from the answers to Probs. 7 and 10?

12. Solve Prob. 10 by applying the second shifting theorem (Sec. 6.3) to Prob. 7.

13. Apply Theorem 1 to the function $f(t) = 1$, which is periodic with any period p.

Half-wave and full-wave rectifiers

14. Find the Laplace transform of the half-wave rectification of $-\sin \omega t$ (Fig. 113).

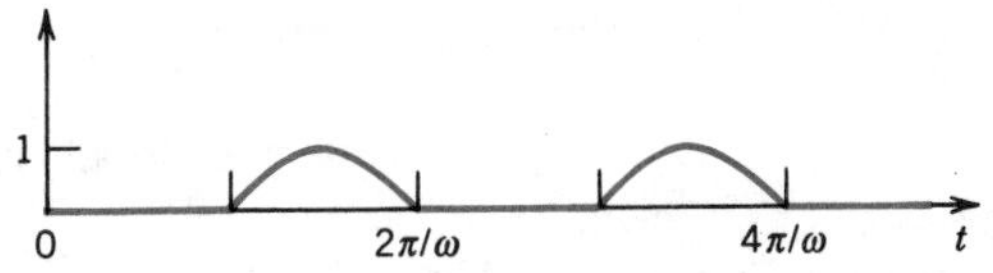

Fig. 113. Problem 14

15. Find the Laplace transform of the full-wave rectification of $\sin \omega t$ (Fig. 114).

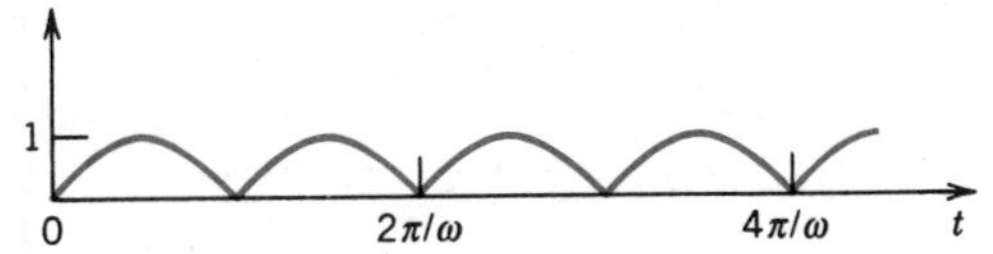

Fig. 114. Problem 15

16. Find the Laplace transform of the full-wave rectification $|\cos \omega t|$ of $\cos \omega t$.

17. Solve Prob. 14 by applying Theorem 2, Sec. 6.3, to the result of Example 3.

18. Check the answer to Prob. 15 by using the results of Example 3 and Prob. 14.

Models of electric circuits

19. Using Laplace transforms, show that the current $i(t)$ in the RLC-circuit in Fig. 115 (constant electromotive force V_0, zero initial current and charge) is

$$i(t) = \begin{cases} (K/\omega^*)e^{-\alpha t} \sin \omega^* t & \text{if } \omega^{*2} > 0 \\ Kte^{-\alpha t} & \text{if } \omega^{*2} = 0 \\ (K/\beta)e^{-\alpha t} \sinh \beta t & \text{if } \omega^{*2} = -\beta^2 < 0; \end{cases}$$

here $K = V_0/L$, $\alpha = R/2L$, $\omega^{*2} = (1/LC) - \alpha^2$.

20. Find the current in the circuit in Prob. 19, assuming that the electromotive force applied at $t = 0$ is $V_0 \sin pt$, and the current and charge at $t = 0$ are zero.

21. Find the current in the *RLC*-circuit in Fig. 115 if a battery of electromotive force V_0 is connected to the circuit at $t = 0$ and short-circuited at $t = a$. Assume that the initial current and charge are zero, and ω^{*2}, as defined in Prob. 19, is positive.

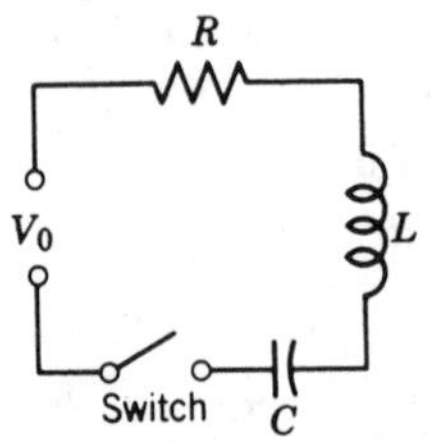

Fig. 115. Problem 19

22. Find the steady-state current in the *RL*-circuit in Fig. 116.

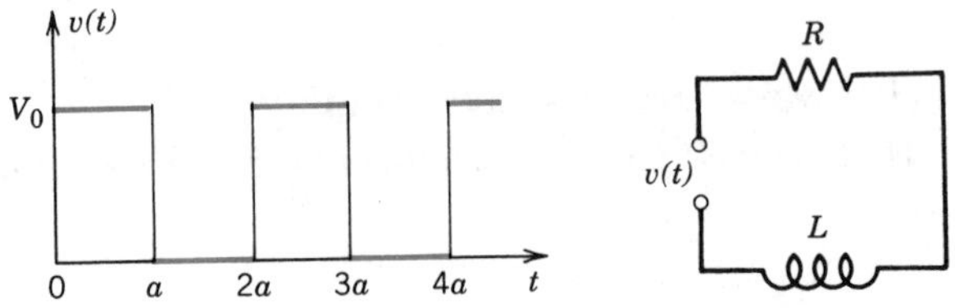

Fig. 116. Problem 22

23. Solve Prob. 22 without the use of Laplace transforms.

24. Find the current $i(t)$ in the circuit in Fig. 117, assuming that $i(0) = 0$.

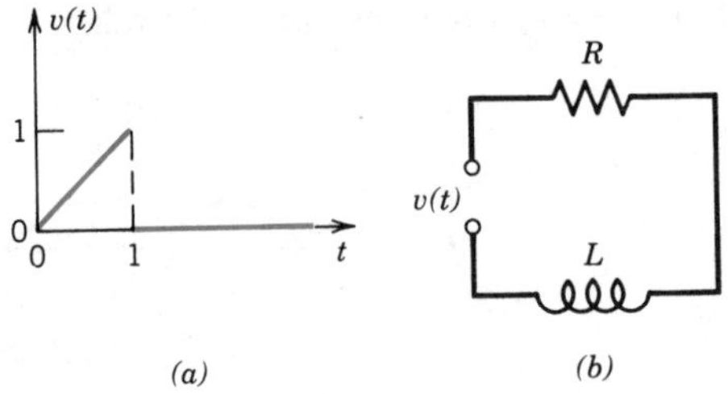

Fig. 117. Problem 24

25. Find the steady-state current in the circuit in Fig. 117*b*, if $v(t) = t$ when $0 < t < 1$ and $v(t + 1) = v(t)$ as shown in Fig. 118.

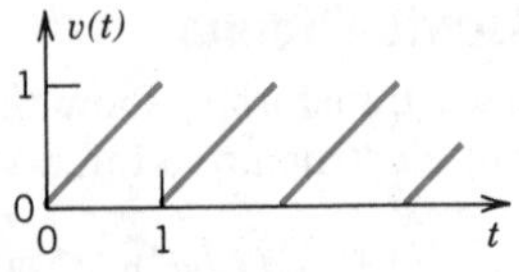

Fig. 118. Problem 25

26. Find the ramp-wave response of the *RC*-circuit in Fig. 119, assuming that the circuit is quiescent at $t = 0$.

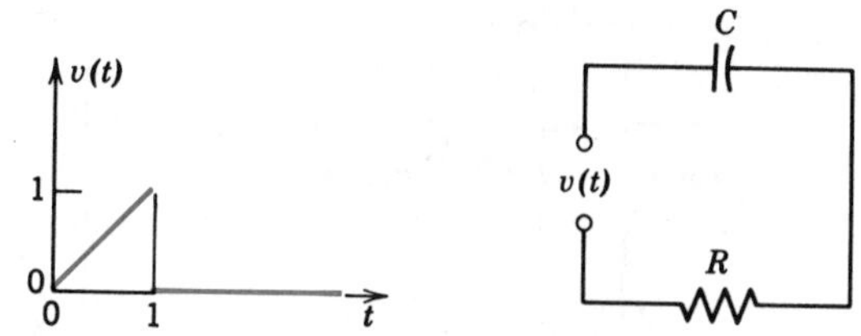

Fig. 119. Problem 26

27. Solve Prob. 26 without the use of Laplace transforms. Explain the reason for the jump of $i(t)$ at $t = 1$.

28. Solve Prob. 26 by means of Laplace transforms, starting from the equation of the form $Ri' + (1/C)i = v'$.

29. Find the current $i(t)$ in the RC-circuit in Fig. 119 when $v(t) = \sin \omega t$ $(0 < t < \pi/\omega)$, $v(t) = 0$ $(t > \pi/\omega)$, and $i(0) = 0$.

30. Steady current is flowing in the circuit in Fig. 120 with the switch closed. At $t = 0$ the switch is opened. Find the current $i(t)$.

31. A capacitor ($C = 1$ farad) is charged to the potential $V_0 = 100$ volts and discharged starting at $t = 0$ by closing the switch in Fig. 121. Find the current in the circuit and the charge on the capacitor.

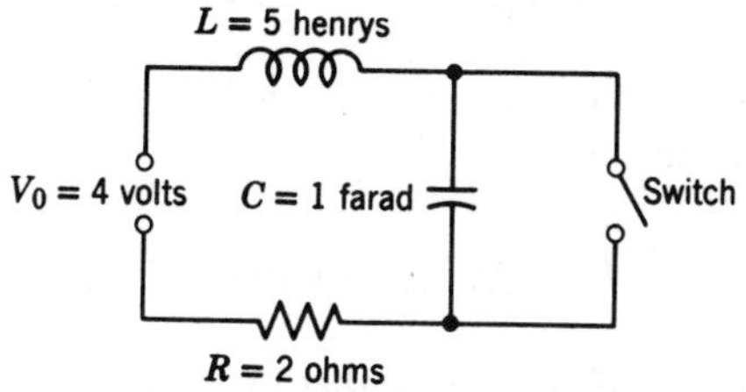

Fig. 120. Problem 30

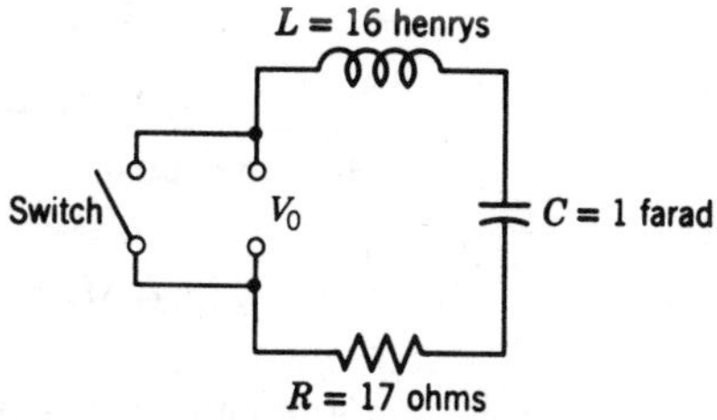

Fig. 121. Problem 31

Models of mechanical systems

32. (Automatic pressure control) Figure 122 on p. 316 shows a system for automatic control of pressure. $y(t)$ is the displacement, where $y = 0$ corresponds to the equilibrium position due to a given constant pressure. We make the following assumptions. The damping of the system is proportional to the velocity y'. For $t < 0$, the system is at rest. At $t = 0$ the pressure is suddenly increased in the form of a unit step function. Show that the corresponding differential equation is

$$my'' + cy' + ky = Pu(t)$$

(m = effective mass of the moving parts, c = damping constant, k = spring modulus, P = force due to the increase of pressure at $t = 0$). Using Laplace transforms, show that if $c^2 < 4mk$, then (Fig. 123)

$$y(t) = \frac{P}{k}\left[1 - e^{-\alpha t}\sqrt{1 + (\alpha/\omega^*)^2}\cos(\omega^* t + \theta)\right]$$

where $\alpha = c/2m$, $\omega^* = \sqrt{(k/m) - \alpha^2}$ (> 0), $\tan \theta = -\alpha/\omega^*$.

33. Solve Prob. 32 when $c^2 > 4mk$ and when $c^2 = 4mk$.

34. (Flywheels) Two flywheels (moments of inertia M_1 and M_2) are connected by an elastic shaft (moment of inertia negligible), and are rotating with constant angular

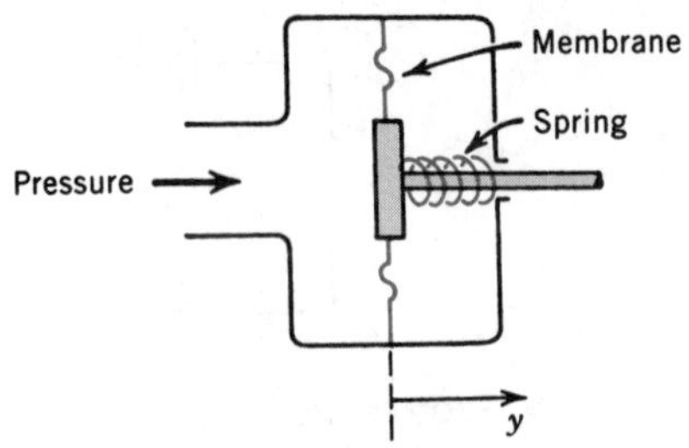

Fig. 122. Problem 32

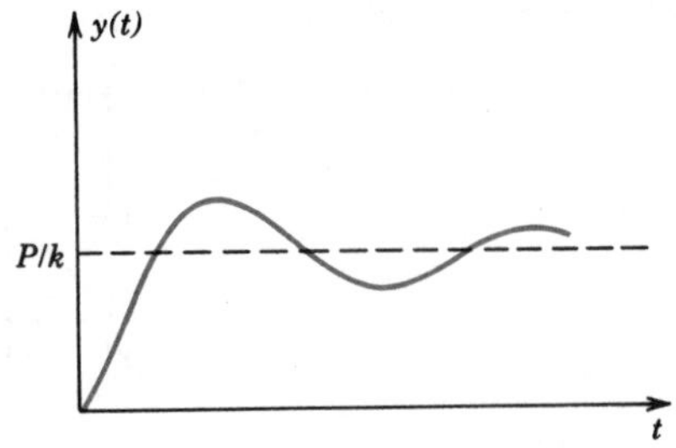

Fig. 123. Displacement $y(t)$ in Problem 32

velocity ω. At $t = 0$ a constant retarding couple P is applied to the first wheel. Find the subsequent angular velocity $v(t)$ of the other wheel.

35. Same conditions as in Prob. 34, but the retarding couple is applied during the interval $0 < t < 1$ only. Find $v(t)$.

Model of two circuits coupled by inductance. The circuits in Fig. 124 are coupled by mutual inductance M, and at $t = 0$ the currents $i_1(t)$ and $i_2(t)$ are zero.

36. Applying Kirchhoff's voltage law (Sec. 1.8), show that the differential equations for the currents are

$$L_1 i_1' + R_1 i_1 = M i_2' + V_0 u(t), \qquad L_2 i_2' + R_2 i_2 = M i_1'$$

where $u(t)$ is the unit step function. Setting $\mathcal{L}(i_1) = I_1(s)$ and $\mathcal{L}(i_2) = I_2(s)$, show that the subsidiary equations are

$$L_1 s I_1 + R_1 I_1 = M s I_2 + \frac{V_0}{s}, \qquad L_2 s I_2 + R_2 I_2 = M s I_1.$$

Assuming that $A \equiv L_1 L_2 - M^2 > 0$, show that the expression for I_2, obtained by solving these algebraic equations, may be written

$$I_2 = \frac{K}{(s + \alpha)^2 + \omega^{*2}}$$

where $K = A^{-1} V_0 M$, $\alpha = (2A)^{-1}(R_1 L_2 + R_2 L_1)$, $\omega^{*2} = A^{-1} R_1 R_2 - \alpha^2$. Show that $i_2(t) = \mathcal{L}^{-1}(I_2)$ is of the same form as $i(t)$ in Prob. 19, where the constants K, α, and ω^* are now those defined in the present problem.

37. Show that if $A = 0$ in Prob. 36, then $i_2(t) = K_0 e^{-at}$ where $K_0 = BV_0 M$, $a = BR_1 R_2$, $B = (R_1 L_2 + R_2 L_1)^{-1}$.

38. Suppose that in the first circuit in Fig. 124 the switch is closed and a steady current V_0/R_1 is flowing in it. At $t = 0$ the switch is opened. Find the secondary current $i_2(t)$. *Hint.* Note that $i_1(0) = V_0/R_1$ and $i_1 = 0$ when $t > 0$.

39. Find $i_1(t)$ in Prob. 38, assuming that $\omega^{*2} > 0$.

40. Verify the details of the calculations in Example 3.

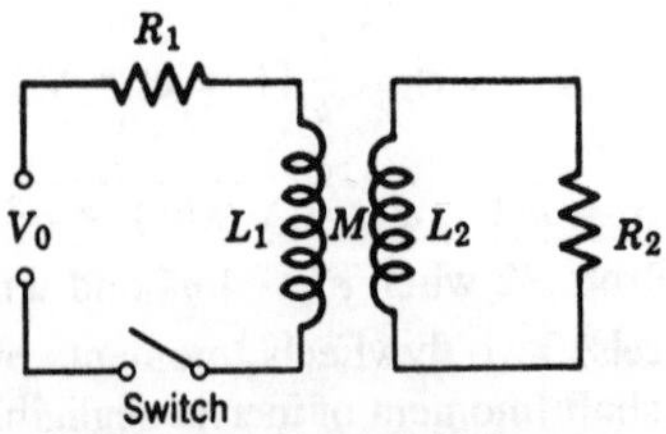

Fig. 124. Problems 36–39

6.9 Laplace Transform: General Formulas

Formula	Name, Comments	Sec.
$F(s) = \mathcal{L}\{f(t)\} = \int_0^\infty e^{-st} f(t)\,dt$ $f(t) = \mathcal{L}^{-1}\{F(s)\}$	Definition of Transform Inverse Transform	6.1
$\mathcal{L}\{af(t) + bg(t)\} = a\mathcal{L}\{f(t)\} + b\mathcal{L}\{g(t)\}$	Linearity	6.1
$\mathcal{L}(f') = s\mathcal{L}(f) - f(0)$ $\mathcal{L}(f'') = s^2\mathcal{L}(f) - sf(0) - f'(0)$ $\mathcal{L}(f^{(n)} = s^n\mathcal{L}(f) - s^{n-1}f(0) - \cdots - f^{(n-1)}(0)$ $\mathcal{L}\left\{\int_0^t f(\tau)\,d\tau\right\} = \frac{1}{s}\mathcal{L}(f)$	Differentiation of Function Integration of Function	6.2
$\mathcal{L}\{e^{at}f(t)\} = F(s - a)$ $\mathcal{L}^{-1}\{F(s - a)\} = e^{at}f(t)$	s-Shifting (1st Shifting Theorem)	6.3
$\mathcal{L}\{f(t - a)u(t - a)\} = e^{-as}F(s)$ $\mathcal{L}^{-1}\{e^{-as}F(s)\} = f(t - a)u(t - a)$	t-Shifting (2nd Shifting Theorem)	6.3
$\mathcal{L}\{tf(t)\} = -F'(s)$ $\mathcal{L}\left\{\frac{f(t)}{t}\right\} = \int_s^\infty F(\tilde{s})\,d\tilde{s}$	Differentiation of Transform Integration of Transform	6.5
$(f * g)(t) = \int_0^t f(\tau)g(t - \tau)\,d\tau$ $= \int_0^t f(t - \tau)g(\tau)\,d\tau$ $\mathcal{L}(f * g) = \mathcal{L}(f)\mathcal{L}(g)$	Convolution	6.6
$\mathcal{L}(f) = \frac{1}{1 - e^{-ps}} \int_0^p e^{-st} f(t)\,dt$	f Periodic with Period p	6.8

6.10 Table of Laplace Transforms

For more extensive tables, see Refs.* [A5] *and* [A8] *in Appendix 1.

	$F(s) = \mathcal{L}\{f(t)\}$	$f(t)$	Sec.
1	$1/s$	1	6.1
2	$1/s^2$	t	
3	$1/s^n, \quad (n = 1, 2, \cdots)$	$t^{n-1}/(n-1)!$	
4	$1/\sqrt{s}$	$1/\sqrt{\pi t}$	
5	$1/s^{3/2}$	$2\sqrt{t/\pi}$	
6	$1/s^a \quad (a > 0)$	$t^{a-1}/\Gamma(a)$	
7	$\dfrac{1}{s-a}$	e^{at}	6.1
8	$\dfrac{1}{(s-a)^2}$	te^{at}	6.3
9	$\dfrac{1}{(s-a)^n} \quad (n = 1, 2, \cdots)$	$\dfrac{1}{(n-1)!}t^{n-1}e^{at}$	
10	$\dfrac{1}{(s-a)^k} \quad (k > 0)$	$\dfrac{1}{\Gamma(k)}t^{k-1}e^{at}$	
11	$\dfrac{1}{(s-a)(s-b)} \quad (a \neq b)$	$\dfrac{1}{(a-b)}(e^{at} - e^{bt})$	6.1
12	$\dfrac{s}{(s-a)(s-b)} \quad (a \neq b)$	$\dfrac{1}{(a-b)}(ae^{at} - be^{bt})$	
13	$\dfrac{1}{s^2+\omega^2}$	$\dfrac{1}{\omega}\sin \omega t$	6.1
14	$\dfrac{s}{s^2+\omega^2}$	$\cos \omega t$	
15	$\dfrac{1}{s^2-a^2}$	$\dfrac{1}{a}\sinh at$	
16	$\dfrac{s}{s^2-a^2}$	$\cosh at$	
17	$\dfrac{1}{(s-a)^2+\omega^2}$	$\dfrac{1}{\omega}e^{at}\sin \omega t$	6.3
18	$\dfrac{s-a}{(s-a)^2+\omega^2}$	$e^{at}\cos \omega t$	
19	$\dfrac{1}{s(s^2+\omega^2)}$	$\dfrac{1}{\omega^2}(1 - \cos \omega t)$	6.2
20	$\dfrac{1}{s^2(s^2+\omega^2)}$	$\dfrac{1}{\omega^3}(\omega t - \sin \omega t)$	
21	$\dfrac{1}{(s^2+\omega^2)^2}$	$\dfrac{1}{2\omega^3}(\sin \omega t - \omega t \cos \omega t)$	

Table of Laplace Transforms (*continued*)

	$F(s) = \mathcal{L}\{f(t)\}$	$f(t)$	Sec.
22	$\dfrac{s}{(s^2+\omega^2)^2}$	$\dfrac{t}{2\omega}\sin\omega t$	6.5
23	$\dfrac{s^2}{(s^2+\omega^2)^2}$	$\dfrac{1}{2\omega}(\sin\omega t + \omega t\cos\omega t)$	
24	$\dfrac{s}{(s^2+a^2)(s^2+b^2)}$ $(a^2 \neq b^2)$	$\dfrac{1}{b^2-a^2}(\cos at - \cos bt)$	
25	$\dfrac{1}{s^4+4k^4}$	$\dfrac{1}{4k^3}(\sin kt\cosh kt - \cos kt\sinh kt)$	6.7
26	$\dfrac{s}{s^4+4k^4}$	$\dfrac{1}{2k^2}\sin kt\sinh kt$	
27	$\dfrac{1}{s^4-k^4}$	$\dfrac{1}{2k^3}(\sinh kt - \sin kt)$	
28	$\dfrac{s}{s^4-k^4}$	$\dfrac{1}{2k^2}(\cosh kt - \cos kt)$	
29	$\sqrt{s-a} - \sqrt{s-b}$	$\dfrac{1}{2\sqrt{\pi t^3}}(e^{bt} - e^{at})$	
30	$\dfrac{1}{\sqrt{s+a}\sqrt{s+b}}$	$e^{-(a+b)t/2}I_0\left(\dfrac{a-b}{2}t\right)$	5.7
31	$\dfrac{1}{\sqrt{s^2+a^2}}$	$J_0(at)$	5.5
32	$\dfrac{s}{(s-a)^{3/2}}$	$\dfrac{1}{\sqrt{\pi t}}e^{at}(1+2at)$	
33	$\dfrac{1}{(s^2-a^2)^k}$ $(k>0)$	$\dfrac{\sqrt{\pi}}{\Gamma(k)}\left(\dfrac{t}{2a}\right)^{k-1/2} I_{k-1/2}(at)$	5.7
34	e^{-as}/s	$u(t-a)$	6.3
35	e^{-as}	$\delta(t-a)$	6.4
36	$\dfrac{1}{s}e^{-k/s}$	$J_0(2\sqrt{kt})$	5.5
37	$\dfrac{1}{\sqrt{s}}e^{-k/s}$	$\dfrac{1}{\sqrt{\pi t}}\cos 2\sqrt{kt}$	
38	$\dfrac{1}{s^{3/2}}e^{k/s}$	$\dfrac{1}{\sqrt{\pi k}}\sinh 2\sqrt{kt}$	
39	$e^{-k\sqrt{s}}$ $(k>0)$	$\dfrac{k}{2\sqrt{\pi t^3}}e^{-k^2/4t}$	
40	$\dfrac{1}{s}\ln s$	$-\ln t - \gamma$ $(\gamma \approx 0.5772)$	5.7

(continued)

Table of Laplace Transforms (*continued*)

	$F(s) = \mathcal{L}\{f(t)\}$	$f(t)$	Sec.
41	$\ln \dfrac{s-a}{s-b}$	$\dfrac{1}{t}(e^{bt} - e^{at})$	6.5
42	$\ln \dfrac{s^2+\omega^2}{s^2}$	$\dfrac{2}{t}(1 - \cos \omega t)$	
43	$\ln \dfrac{s^2-a^2}{s^2}$	$\dfrac{2}{t}(1 - \cosh at)$	
44	$\arctan \dfrac{\omega}{s}$	$\dfrac{1}{t}\sin \omega t$	App. 3
45	$\dfrac{1}{s}\,\text{arc cot}\, s$	$\text{Si}(t)$	

Review Questions and Problems for Chapter 6

1. What do we mean by saying that the Laplace transform is a *linear* operation? Why is this practically important?
2. Does every continuous function have a Laplace transform? (Give a reason or counterexample.)
3. Can a discontinuous function have a Laplace transform? (Give a reason for your answer.)
4. Why is it of practical importance that two continuous functions that are different also have different Laplace transforms?
5. What is the purpose of the Laplace transform method? What are its advantages over the classical method?
6. For what kind of problems would you prefer the Laplace transform over the classical method?
7. What do we mean by the subsidiary equation?
8. State the formula for the Laplace transform of the nth derivative of a function $f(t)$ from memory.
9. What do you know about differentiation and integration of transforms?
10. If you know $f(t) = \mathcal{L}^{-1}\{F(s)\}$, how would you find $\mathcal{L}^{-1}\{F(s)/s^2\}$?
11. What is the difference in the shifting by the first shifting theorem and by the second shifting theorem?
12. Show that $e^{-as}\mathcal{L}\{f(t+a)\} = \mathcal{L}\{f(t)u(t-a)\}$. (Use a shifting theorem.)
13. Is $\mathcal{L}\{f(t)g(t)\} = \mathcal{L}\{f(t)\}\mathcal{L}\{g(t)\}$? Or what?
14. Is $\int_0^t f(\tau)g(t-\tau)\,d\tau = \int_0^t f(t-\tau)g(\tau)\,d\tau$? (Give a reason for your answer.)
15. Obtain the Laplace transform of t^n from Theorem 2 in Sec. 6.2.

Find the Laplace transform of the given function.

16. $e^{-2t}\cos 3t$
17. $e^{-t}\sin 2\pi t$
18. $tu(t-2)$
19. $\cosh^2 t$
20. $\sinh^2 2t$
21. $\sin t + t\cos t$
22. $t * e^{-2t}$
23. $\cos^2 t$
24. $t^3u(t-1)$
25. $t^{-1}\sin t$

Find the inverse Laplace transform of the given function.

26. $\dfrac{5s+3}{s^2+4}$
27. $\dfrac{s-2}{s^2+2s+10}$
28. $\dfrac{1-2s^3}{s^5}$
29. $\dfrac{1}{s^2+3s}$
30. $\dfrac{s^2+2}{s^3+4s}$
31. $\dfrac{8}{s^4-2s^3}$
32. $\dfrac{\omega\cos\theta + s\sin\theta}{s^2+\omega^2}$
33. $\dfrac{s^2-6s+4}{s^3-3s^2+2s}$
34. $\dfrac{3s^2-2s-1}{(s-3)(s^2+1)}$
35. $\dfrac{s^4-7s^3+13s^2+4s-12}{s^2(s-3)(s^2-3s+2)}$

Using Laplace transforms, find $y(t)$ satisfying the given equation and conditions.

36. $y'' + y = \delta(t-2), \quad y(0) = 2.5, \ y'(0) = 0$
37. $y'' + 9y = 1.8u(t-3), \quad y(0) = y'(0) = 0$
38. $y'' + y = 0$ if $0 < t < 1$ and $t - 1$ if $t > 1$; $\quad y(0) = y'(0) = 0$
39. $y'' + y = 0$ if $0 < t < 2\pi$ and $e^{-(t-2\pi)}$ if $t > 2\pi$; $\quad y(0) = y'(0) = 0$
40. $y'' + 3y' + 2y = u(t-2), \quad y(0) = y'(0) = 0$
41. $y'' - 2y' + 2y = 2e^t\cos t, \quad y(0) = 1.5, \ y'(0) = 1.5$
42. $y'' + 2y' + 5y = t - 0.04\,\delta(t-\pi), \quad y(0) = -0.08, \ y'(0) = 0.2$
43. $y'' - 4y' + 4y = (3t^2+2)e^t, \quad y(0) = 20, \ y'(0) = 34$
44. $y'' + 4y' + 3y = t - 1$ if $1 < t < 2$ and 0 otherwise; $\quad y(0) = 2,$ $y'(0) = -2$
45. $y'' + 4y = \delta(t-\pi) - \delta(t-2\pi), \quad y(0) = 0, \ y'(0) = 2$
46. $y(t) = 1 - \sinh t + \displaystyle\int_0^t (1+\tau)y(t-\tau)\,d\tau$
47. $y(t) = \cosh 3t - 3e^{3t}\displaystyle\int_0^t y(\tau)e^{-3\tau}\,d\tau$
48. $y(t) = t^2 + \frac{1}{60}t^6 - \displaystyle\int_0^t y(\tau)(t-\tau)^3\,d\tau$

Models of mass–spring systems, circuits, networks

Solve the following problems by Laplace transforms.

49. Show that the model of the system in Fig. 125 on p. 322 (no friction, no damping) is

$$m_1y_1'' = -k_1y_1 + k_2(y_2 - y_1)$$
$$m_2y_2'' = -k_2(y_2 - y_1) - k_3y_2.$$

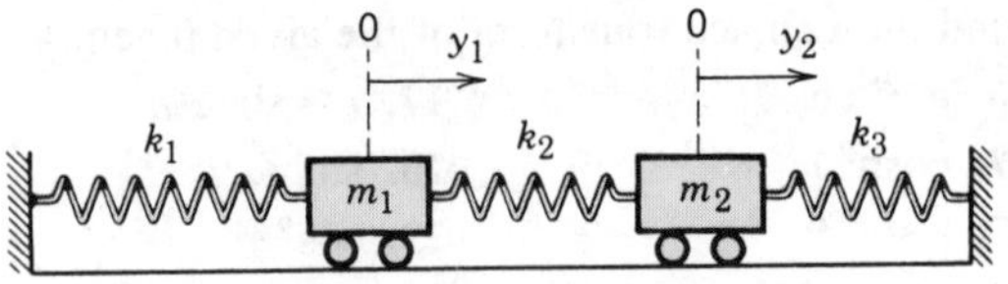

Fig. 125. System in Problems 49–52

50. In Prob. 49, let $m_1 = m_2 = 10$ kg, $k_1 = k_3 = 20$ kg/sec^2, $k_2 = 40$ kg/sec^2. Find the solution satisfying the initial conditions $y_1(0) = y_2(0) = 0$, $y_1'(0) = 1$ meter/sec, $y_2'(0) = -1$ meter/sec.

51. Solve Prob. 50, assuming that $y_1(0) = y_2(0) = 1$ meter, $y_1'(0) = y_2'(0) = 0$, all the other data being as before. Compare the solution (frequency, type of motion) with that in Prob. 50.

52. Solve Prob. 50 by classical methods.

53. Find the current $i(t)$ in the RC-circuit in Fig. 126, where $R = 10$ ohms, $C = 0.1$ farad, $e(t) = 10t$ volts if $0 < t < 4$, $e(t) = 40$ volts if $t > 4$, and the initial charge on the capacitor is 0.

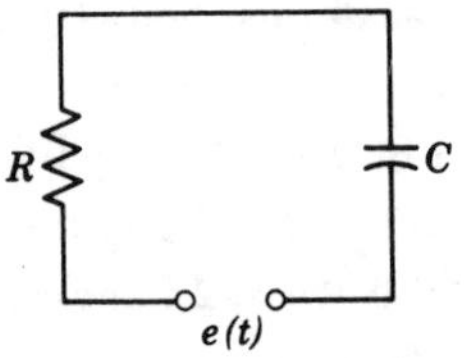

Fig. 126. *RC*-circuit

54. Find the charge $q(t)$ and the current $i(t)$ in the LC-circuit in Fig. 127, assuming $L = 1$ henry, $C = 1$ farad, $e(t) = 1 - e^{-t}$ if $0 < t < \pi$, $e(t) = 0$ if $t > \pi$, and zero initial current and charge.

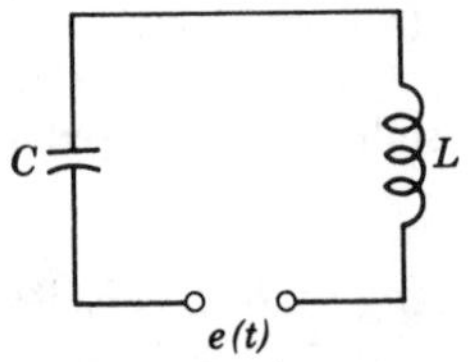

Fig. 127. *LC*-circuit

55. Find the current $i(t)$ in the RLC-circuit in Fig. 128, where $R = 160$ ohms, $L = 20$ henrys, $C = 0.002$ farad, $e(t) = 37 \sin 10t$ volts, assuming zero initial current and charge.

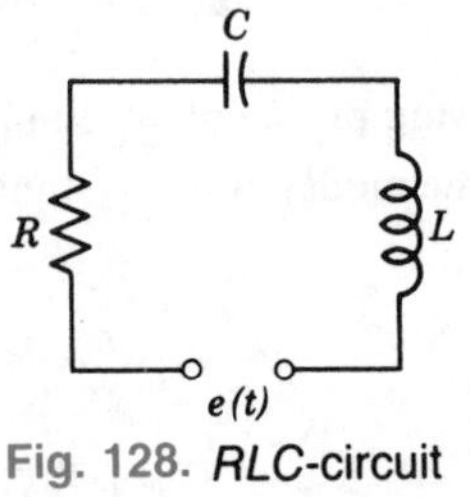

Fig. 128. *RLC*-circuit

56. Show that, by Kirchhoff's voltage law (Sec. 1.8), the currents in the network in Fig. 129 are obtained from the system

$$Li_1' + R(i_1 - i_2) = e(t)$$

$$R(i_2' - i_1') + \frac{1}{C} i_2 = 0$$

57. Solve the system in Prob. 56, where $R = 10$ ohms, $L = 20$ henrys, $C = 0.05$ farad, $e = 20$ volts, $i_1(0) = 0$, $i_2(0) = 2$ amperes.

58. Show that in Prob. 57, we have $i_1(t) \to 2$, $i_2(t) \to 0$ as $t \to \infty$. Can you conclude this directly from the network?

59. Solve the system in Prob. 56, where $R = 0.8$ ohm, $L = 1$ henry, $C = 0.25$ farad, $e(t) = \frac{4}{5}t + \frac{21}{25}$ volt, $i(0) = 1$ ampere, $i_2(0) = -3.8$ ampere.

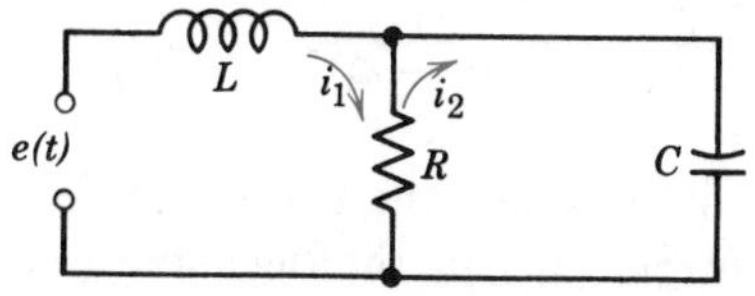

Fig. 129. Network in Problems 56–59

60. Set up the model of the network in Fig. 130 and find the solution, assuming that all charges and currents are 0 when the switch is closed at $t = 0$. Find the limits of $i_1(t)$ and $i_2(t)$ as $t \to \infty$, (i) from the solution, (ii) directly from the given network.

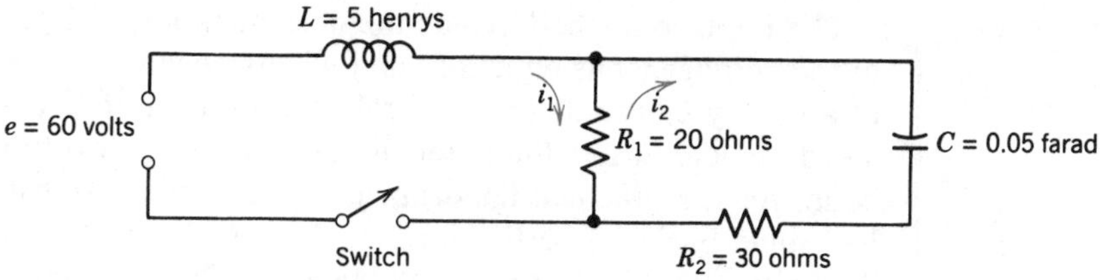

Fig. 130. Network in Problem 60

Summary of Chapter 6
Laplace Transforms

The main purpose of Laplace transforms is the solution of differential equations and systems of such equations, as well as corresponding initial value problems. The **Laplace transform** $F(s) = \mathcal{L}(f)$ of a function $f(t)$ is defined by (Sec. 6.1)

$$F(s) = \mathcal{L}(f) = \int_0^{\infty} e^{-st} f(t)\, dt. \qquad (1)$$

This definition is motivated by the property that the differentiation of f with respect to t corresponds to the multiplication of the transform F by s; precisely (Sec. 6.2):

$$\begin{aligned} \mathcal{L}(f') &= s\mathcal{L}(f) - f(0), \\ \mathcal{L}(f'') &= s^2\mathcal{L}(f) - sf(0) - f'(0), \end{aligned} \tag{2}$$

etc. Hence by taking the transform of a given differential equation

$$y'' + ay' + by = r(t) \tag{3}$$

and writing $\mathcal{L}(y) = Y(s)$, we obtain the **subsidiary equation**

$$(s^2 + as + b)Y = \mathcal{L}(r) + sf(0) + f'(0) + af(0). \tag{4}$$

Here, in obtaining the transform $\mathcal{L}(r)$ we can get help from the small table in Sec. 6.1 or the big table in Sec. 6.10. In the second step we solve the subsidiary equation *algebraically* for $Y(s)$. In the third step we determine the inverse transform $y(t) = \mathcal{L}^{-1}(Y)$, that is, the solution of the problem. This is generally the hardest step, and in it we may again use one of those two tables. $Y(s)$ will often be a rational function, so that we can obtain the inverse $\mathcal{L}^{-1}(Y)$ by partial fraction reduction (Sec. 6.7) if we see no simpler way.

The Laplace method avoids the determination of a general solution of the homogeneous equation, and we also need not determine values of arbitrary constants in a general solution from initial conditions; instead, we can insert the latter directly into (4). Two further facts account for the practical importance of the Laplace transform. First, it has some basic properties and resulting techniques that simplify the determination of transforms and inverses. The most important of these properties are listed in Sec. 6.9, together with references to the corresponding sections. More on the use of unit step functions and Dirac's delta can be found in Secs. 6.3 and 6.4, and on convolution in Sec. 6.6. Second, due to these properties, the present method is particularly suitable for handling right sides $r(t)$ given by different expressions over different intervals of time, for instance, when $r(t)$ is a square wave or an impulse or of a form such as $r(t) = \cos t$ if $0 \leqq t \leqq 4\pi$ and 0 elsewhere.

The application of Laplace transforms to *partial* differential equations follows in Sec. 11.13.

PART B

LINEAR ALGEBRA, VECTOR CALCULUS

Two main factors have affected the development of engineering mathematics during the past decades, namely, the extensive application of computers to engineering problems and the increasing use of linear algebra and linear analysis, for instance, in handling large-scale problems in systems analysis.

The first chapter of the present part is devoted to **linear algebra,** consisting of the theory and application of vectors and matrices in connection with the solution of linear systems of equations, eigenvalue problems, and so on.

Numerical methods in linear algebra are presented in Chap. 19, which is independent of the other chapters in Part E on numerical methods. Thus Chap. 19 can be studied immediately after Chap. 7, if desired.

The last two chapters of the present part are devoted to **linear analysis,** usually called **vector calculus.** Chapter 8 concerns vector *differential* calculus (vector fields, curves, velocity, directional derivative, gradient, divergence, curl), and Chap. 9 covers vector *integral* calculus (line, surface, and triple integrals and their transformation by the integral theorems of Green, Gauss, and Stokes).

Chapter 7

Linear Algebra: Matrices, Vectors, Determinants

Linear algebra includes the theory and application of linear systems of equations (briefly called linear systems), linear transformations, and eigenvalue problems, as they arise, for instance, from electrical networks, frameworks in mechanics, curve fitting and other optimization problems, processes in statistics, systems of differential equations, and so on.

Linear algebra makes systematic use of **vectors** and **matrices** (Sec. 7.1) and, to a lesser extent, **determinants** (Sec. 7.8); and the study of properties of matrices is by itself a central task of linear algebra.

A matrix is a rectangular array of numbers. Matrices occur in various problems, for instance, as arrays of coefficients of equations (Sec. 7.4). Matrices (and vectors) are useful because they enable us to consider an array of many numbers as a single object, denote it by a single symbol, and perform calculations with these symbols in a very compact form. The "mathematical shorthand" thus obtained is very elegant and powerful and is suitable for various practical problems. It entered applied mathematics more than 60 years ago and is of increasing importance in various fields.

This chapter has three big parts:

Calculation with matrices, Secs. 7.1–7.3
Systems of linear equations, Secs. 7.4–7.9
Eigenvalue problems, Secs. 7.10–7.14

and a (more abstract) optional section (7.15) on vector and inner product spaces and linear transformations.

Thus we first introduce matrices and vectors and related concepts (Sec. 7.1) and define the algebraic operations for matrices (Secs. 7.2, 7.3). Next we consider linear systems—solution by Gauss elimination in Sec. 7.4, existence of solutions in Sec. 7.6, determinants and Cramer's rule in Secs. 7.8 and 7.9. Then we study eigenvalue problems in general (Secs. 7.10, 7.11) and for important special real matrices (Sec. 7.12) and complex matrices (Sec. 7.13). Finally, we discuss the diagonalization of matrices and the reduction of quadratic forms to principal axes (Sec. 7.14). Other important concepts in this chapter are the rank of a matrix (Secs. 7.5, 7.9) and the inverse of a matrix (Sec. 7.7). Applications of matrices to practical problems are shown throughout the chapter.

NUMERICAL METHODS in Chap. 19 can be studied immediately after the corresponding material in the present chapter.

Prerequisite for this chapter: None.
Sections that may be omitted in a shorter course: 7.12–7.15.
References: Appendix 1, Part B.
Answers to problems: Appendix 2.

7.1 Basic Concepts

The first three sections of this chapter introduce the basic concepts and rules of matrix and vector algebra. The main application to linear systems of equations begins in Sec. 7.4.

A **matrix** is a rectangular array of numbers (or functions) enclosed in brackets. These numbers (or functions) are called *entries* or *elements* of the matrix. For example,

$$(1)\qquad \begin{bmatrix} 2 & 0.4 & 8 \\ 5 & -32 & 0 \end{bmatrix}, \quad \begin{bmatrix} 6 \\ 1 \end{bmatrix}, \quad [a_1 \quad a_2 \quad a_3], \quad \begin{bmatrix} a & b \\ c & d \end{bmatrix}, \quad \begin{bmatrix} e^x & 3x \\ e^{2x} & x^2 \end{bmatrix}$$

are matrices. The first has two "*rows*" (horizontal lines) and three "*columns*" (vertical lines). The second consists of a single column, and we call it a *column vector*. The third consists of a single row, and we call it a *row vector*. The last two are *square matrices*, that is, each has as many rows as columns (two in this case).

Matrices are practical in many applications. For example, in a system of equations such as

$$\begin{aligned} 5x - 2y + z &= 0 \\ 3x + 4z &= 0 \end{aligned}$$

the coefficients of the unknowns x, y, z are the entries of the *coefficient matrix*, call it $\mathbf{A}$,

$$\mathbf{A} = \begin{bmatrix} 5 & -2 & 1 \\ 3 & 0 & 4 \end{bmatrix},$$

which displays these coefficients in the pattern of the equations. Sales figures for three products I, II, III in a store on Monday (M), Tuesday (T), · · · may for each week be arranged in a matrix

$$\mathbf{A} = \begin{array}{c} \begin{array}{cccccc} \text{M} & \text{T} & \text{W} & \text{Th} & \text{F} & \text{S} \end{array} \\ \begin{bmatrix} 40 & 33 & 81 & 0 & 21 & 47 \\ 0 & 12 & 78 & 50 & 50 & 96 \\ 10 & 0 & 0 & 27 & 43 & 78 \end{bmatrix} \end{array} \begin{array}{c} \\ \text{I} \\ \text{II} \\ \text{III} \end{array}$$

and if the company has ten stores, we can set up ten such matrices, one for each store; then by adding corresponding entries of these matrices we can get a matrix showing the total sales of each product on each day. Can you think of other data for which matrices are feasible? For instance, in transportation or storage problems? Or in recording phone calls, or in listing distances in a network of roads?

General Notations and Concepts

Our discussion suggests the following. We denote matrices by capital boldface letters **A, B, C,** $\cdots$, or by writing the general entry in brackets; thus, $\mathbf{A} = [a_{jk}]$, and so on. By an ***m* × *n* matrix** (read "*m* by *n* matrix") we mean a matrix with *m* rows, also called **row vectors,** and *n* columns, also called **column vectors** of the matrix. Thus, an $m \times n$ matrix **A** is of the form

$$\mathbf{A} = [a_{jk}] = \begin{bmatrix} a_{11} & a_{12} & \cdots & a_{1n} \\ a_{21} & a_{22} & \cdots & a_{2n} \\ \cdot & \cdot & \cdots & \cdot \\ a_{m1} & a_{m2} & \cdots & a_{mn} \end{bmatrix}. \tag{2}$$

Hence the matrices in (1) are 2×3, 2×1, 1×3, 2×2, and 2×2.

In the **double-subscript notation** *for the entries, the first subscript always denotes the* **row** *and the second the* **column** *in which the given entry stands.* Thus a_{23} is the entry in the second row and third column.

If $m = n$, we call **A** an $n \times n$ **square matrix.** Then its diagonal containing the entries $a_{11}, a_{22}, \cdots, a_{nn}$ is called the **main diagonal** or *principal diagonal* of **A.** Thus the last two matrices in (1) are square. Square matrices are particularly important, as we shall see.

A **submatrix** of an $m \times n$ matrix **A** is a matrix obtained by omitting some rows or columns (or both) from **A.** For convenience, this includes **A** itself (as the matrix obtained by omitting no rows or columns of **A**).

EXAMPLE 1 Submatrices of a matrix

The 2×3 matrix

$$\begin{bmatrix} a_{11} & a_{12} & a_{13} \\ a_{21} & a_{22} & a_{23} \end{bmatrix}$$

contains three 2×2 submatrices, namely,

$$\begin{bmatrix} a_{11} & a_{12} \\ a_{21} & a_{22} \end{bmatrix}, \quad \begin{bmatrix} a_{11} & a_{13} \\ a_{21} & a_{23} \end{bmatrix}, \quad \begin{bmatrix} a_{12} & a_{13} \\ a_{22} & a_{23} \end{bmatrix},$$

two 1×3 submatrices (the two row vectors), three 2×1 submatrices (the column vectors), six 1×2 submatrices, namely,

$$[a_{11} \quad a_{12}], \quad [a_{11} \quad a_{13}], \quad [a_{12} \quad a_{13}],$$

$$[a_{21} \quad a_{22}], \quad [a_{21} \quad a_{23}], \quad [a_{22} \quad a_{23}],$$

and six 1×1 submatrices, $[a_{11}], [a_{12}], \cdots, [a_{23}]$. ∎

Vectors

A **vector** is a matrix that has only one row—then we call it a **row vector**—or only one column—then we call it a **column vector.** In both cases we call its entries **components** and denote the vector by a ***lowercase*** boldface letter such as **a, b,** $\cdots$, or by its general component in brackets, $\mathbf{a} = [a_j]$, and so on. Thus

$$\mathbf{a} = [a_1 \quad a_2 \cdots \quad a_n]$$

is a row vector, and

$$\mathbf{b} = \begin{bmatrix} b_1 \\ b_2 \\ \cdot \\ \cdot \\ \cdot \\ b_m \end{bmatrix}$$

is a column vector. It will depend on our purpose as to which of the two is more practical, but we often want to switch from one type of vector to the other. We can do this by "**transposition,**" which is indicated by ^{T}; thus, if

$$\mathbf{b} = \begin{bmatrix} 4 \\ 0 \\ -7 \end{bmatrix}, \qquad \text{then} \qquad \mathbf{b}^{\mathsf{T}} = [4 \quad 0 \quad -7];$$

Conversely, if

$$\mathbf{a} = [5 \quad 3 \quad \tfrac{1}{2}], \qquad \text{then} \qquad \mathbf{a}^{\mathsf{T}} = \begin{bmatrix} 5 \\ 3 \\ \tfrac{1}{2} \end{bmatrix}.$$

Transposition

It is practical to define **transposition** for any matrix. The **transpose** $\mathbf{A}^{\mathsf{T}}$ of an $m \times n$ matrix $\mathbf{A} = [a_{jk}]$ as given in (2) is the $n \times m$ matrix that has the first *row* of **A** as its first *column,* the second *row* of **A** as its second *column* and so on. Thus the transpose of **A** in (2) is

$$\text{(3)} \qquad \mathbf{A}^{\mathsf{T}} = [a_{kj}] = \begin{bmatrix} a_{11} & a_{21} & \cdots & a_{m1} \\ a_{12} & a_{22} & \cdots & a_{m2} \\ \cdot & \cdot & \cdots & \cdot \\ a_{1n} & a_{2n} & \cdots & a_{mn} \end{bmatrix}.$$

EXAMPLE 2 Transposition of a matrix

If

$$\mathbf{A} = \begin{bmatrix} 5 & -8 & 1 \\ 4 & 0 & 0 \end{bmatrix}, \qquad \text{then} \qquad \mathbf{A}^{\mathsf{T}} = \begin{bmatrix} 5 & 4 \\ -8 & 0 \\ 1 & 0 \end{bmatrix}.$$ ∎

Symmetric matrices and **skew-symmetric matrices** are square matrices whose transpose equals the matrix or minus the matrix, respectively:

$$\mathbf{A}^{\mathsf{T}} = \mathbf{A} \quad \text{(symmetric matrix)}, \qquad \mathbf{A}^{\mathsf{T}} = -\mathbf{A} \quad \text{(skew-symmetric matrix)}.$$

These matrices are quite important, and we shall use them often in this chapter.

Rules of matrix calculation follow in the next section and problems at the end of it.

7.2 Matrix Addition, Scalar Multiplication

What makes matrices and vectors really useful is the fact that we can calculate with them almost as easily as with numbers. Indeed, practical applications suggested the rules of addition and multiplication by scalars (numbers), which we now introduce. (Multiplication of matrices by matrices follows in the next section.)

We say briefly that two matrices have the **same size** if they are both $m \times n$, for instance, both 3×4. We begin by defining equality.

Definition. Equality of matrices

Two matrices $\mathbf{A} = [a_{jk}]$ and $\mathbf{B} = [b_{jk}]$ are equal, written $\mathbf{A} = \mathbf{B}$, if and only if they have the same size and the corresponding entries are equal, that is, $a_{11} = b_{11}$, $a_{12} = b_{12}$, and so on.

EXAMPLE 1 Equality of matrices

The definition implies that

$$\mathbf{A} = \begin{bmatrix} a_{11} & a_{12} \\ a_{21} & a_{22} \end{bmatrix} = \mathbf{B} = \begin{bmatrix} 4 & 0 \\ 3 & -1 \end{bmatrix} \quad \text{if and only if} \quad \begin{array}{l} a_{11} = 4, \ a_{12} = 0, \\ a_{21} = 3, \ a_{22} = -1. \end{array}$$

A cannot be equal to, say, a 2 × 3 matrix. A column vector cannot be equal to a row vector, by the very definition of equality. ∎

We shall now define two algebraic operations, called *matrix addition* and *scalar multiplication,* which turn out to be practical and very useful in applications, as we shall see later in this chapter.

Definition. Addition of matrices

Addition is defined only for matrices $\mathbf{A} = [a_{jk}]$ and $\mathbf{B} = [b_{jk}]$ of the same and their **sum,** written $\mathbf{A} + \mathbf{B}$, is then obtained by adding the corresponding entries. Matrices of different sizes cannot be added.

As a special case, the **sum a + b** of two row vectors or two column vectors, which must have the same number of components, is obtained by adding the corresponding components.

EXAMPLE 2 **Addition of matrices and vectors**

If $\mathbf{A} = \begin{bmatrix} -4 & 6 & 3 \\ 0 & 1 & 2 \end{bmatrix}$ and $\mathbf{B} = \begin{bmatrix} 5 & -1 & 0 \\ 3 & 1 & 0 \end{bmatrix}$, then $\mathbf{A} + \mathbf{B} = \begin{bmatrix} 1 & 5 & 3 \\ 3 & 2 & 2 \end{bmatrix}$.

Our present **A** and $\mathbf{A}^\mathsf{T}$ cannot be added. **B** in Example 1 and the present **A** cannot be added. If $\mathbf{a} = [5 \quad 7 \quad 2]$ and $\mathbf{b} = [-6 \quad 2 \quad 0]$, then $\mathbf{a} + \mathbf{b} = [-1 \quad 9 \quad 2]$. ∎

Definition. Scalar multiplication (Multiplication by a number)

The **product** of any $m \times n$ matrix $\mathbf{A} = [a_{jk}]$ and any scalar c (number c), written $c\mathbf{A}$, is the $m \times n$ matrix $c\mathbf{A} = [ca_{jk}]$ obtained by multiplying each entry in **A** by c.

Here $(-1)\mathbf{A}$ is simply written $-\mathbf{A}$ and is called the **negative** of **A.** Similarly, $(-k)\mathbf{A}$ is written $-k\mathbf{A}$. Also, $\mathbf{A} + (-\mathbf{B})$ is written $\mathbf{A} - \mathbf{B}$ and is called the **difference** of **A** and **B** (which must have the same size!).

EXAMPLE 3 **Scalar multiplication**

If

$$\mathbf{A} = \begin{bmatrix} 2.7 & -1.8 \\ 0 & 0.9 \\ 9.0 & -4.5 \end{bmatrix},$$

then

$$-\mathbf{A} = \begin{bmatrix} -2.7 & 1.8 \\ 0 & -0.9 \\ -9.0 & 4.5 \end{bmatrix}, \qquad \frac{10}{9}\mathbf{A} = \begin{bmatrix} 3 & -2 \\ 0 & 1 \\ 10 & -5 \end{bmatrix}, \qquad 0\mathbf{A} = \begin{bmatrix} 0 & 0 \\ 0 & 0 \\ 0 & 0 \end{bmatrix}.$$ ∎

An $m \times n$ **zero matrix** is an $m \times n$ matrix with all entries zero. It is denoted by **0.** The last matrix in Example 3 is the 3×2 zero matrix.

From the definition we see that matrix addition enjoys properties quite similar to those of the addition of real numbers; namely, for matrices of the same size we have

$$
(1)\qquad
\begin{aligned}
&\text{(a)}\quad \mathbf{A} + \mathbf{B} = \mathbf{B} + \mathbf{A} \\
&\text{(b)}\quad (\mathbf{U} + \mathbf{V}) + \mathbf{W} = \mathbf{U} + (\mathbf{V} + \mathbf{W}) \qquad (\text{written } \mathbf{U} + \mathbf{V} + \mathbf{W}) \\
&\text{(c)}\quad \mathbf{A} + \mathbf{0} = \mathbf{A} \\
&\text{(d)}\quad \mathbf{A} + (-\mathbf{A}) = \mathbf{0}.
\end{aligned}
$$

Furthermore, from the definitions of matrix addition and scalar multiplication we also obtain

$$
(2)\qquad
\begin{aligned}
&\text{(a)}\quad c(\mathbf{A} + \mathbf{B}) = c\mathbf{A} + c\mathbf{B} \\
&\text{(b)}\quad (c + k)\mathbf{A} = c\mathbf{A} + k\mathbf{A} \\
&\text{(c)}\quad c(k\mathbf{A}) = (ck)\mathbf{A} \qquad (\text{written } ck\mathbf{A}) \\
&\text{(d)}\quad 1\mathbf{A} = \mathbf{A}.
\end{aligned}
$$

For the transpose (Sec. 7.1) of a sum of two $m \times n$ matrices we have

$$
(3)\qquad (\mathbf{A} + \mathbf{B})^{\mathsf{T}} = \mathbf{A}^{\mathsf{T}} + \mathbf{B}^{\mathsf{T}},
$$

as the reader may prove; also,

$$
(4)\qquad (c\mathbf{A})^{\mathsf{T}} = c\mathbf{A}^{\mathsf{T}}.
$$

One more algebraic operation, the multiplication of matrices by matrices, follows in the next section. Then we shall be ready for applications.

Problem Set 7.1–7.2

Let $\mathbf{A} = \begin{bmatrix} 3 & 2 \\ 4 & 1 \end{bmatrix}$, $\mathbf{B} = \begin{bmatrix} 0 & -2 \\ 4 & 5 \end{bmatrix}$, $\mathbf{C} = \begin{bmatrix} 3 & 0 & 2 \\ 4 & 0 & 1 \end{bmatrix}$, $\mathbf{D} = \begin{bmatrix} 6 & 1 & -5 \\ 5 & -2 & 13 \end{bmatrix}$.

Find the following expressions or give reasons why they are undefined.

1. $\mathbf{A} + \mathbf{B}$, $\mathbf{B} + \mathbf{A}$
2. $4\mathbf{A}$, $-3\mathbf{C}$, $3\mathbf{A} - 3\mathbf{B}$, $3(\mathbf{A} - \mathbf{B})$
3. $2\mathbf{C} + 2\mathbf{D}$, $2(\mathbf{C} + \mathbf{D})$
4. $\mathbf{A} + \mathbf{B} + \mathbf{C}$, $\mathbf{C} - \mathbf{D}$
5. $\mathbf{A} - \mathbf{C}$, $\mathbf{A} + 0\mathbf{C}$, $\mathbf{C} + 0\mathbf{A}$
6. $\mathbf{A} + \mathbf{A}^{\mathsf{T}}$, $(\mathbf{A} + \mathbf{B})^{\mathsf{T}}$, $\mathbf{A}^{\mathsf{T}} + \mathbf{B}^{\mathsf{T}}$, $(\mathbf{A}^{\mathsf{T}})^{\mathsf{T}}$
7. $4\mathbf{B} + 8\mathbf{B}^{\mathsf{T}}$, $4(\mathbf{B} + 2\mathbf{B}^{\mathsf{T}})$
8. $(2\mathbf{C})^{\mathsf{T}}$, $2\mathbf{C}^{\mathsf{T}}$, $\mathbf{C} + \mathbf{C}^{\mathsf{T}}$, $\mathbf{C}^{\mathsf{T}} - 2\mathbf{D}^{\mathsf{T}}$

Let $\mathbf{K} = \begin{bmatrix} 4 & 1 & 0 \\ 1 & 3 & 2 \\ 0 & 2 & 5 \end{bmatrix}$, $\mathbf{L} = \begin{bmatrix} 0 & 2 & -8 \\ -2 & 0 & 6 \\ 8 & -6 & 0 \end{bmatrix}$, $\mathbf{a} = \begin{bmatrix} 2 \\ 1 \\ 4 \end{bmatrix}$, $\mathbf{b} = \begin{bmatrix} 9 \\ 0 \\ 5 \end{bmatrix}$.

Find the following expressions or give reasons why they are undefined.

9. $\mathbf{K} + \mathbf{L}$, $\mathbf{K} - \mathbf{L}$
10. $3(\mathbf{a} - 4\mathbf{b})$, $3\mathbf{a} - 12\mathbf{b}$, $\mathbf{K} + \mathbf{a}$, $\mathbf{a} + \mathbf{a}^{\mathsf{T}}$
11. $\mathbf{K} - \mathbf{K}^{\mathsf{T}}$, $\mathbf{L} + \mathbf{L}^{\mathsf{T}}$, $\mathbf{a}^{\mathsf{T}} + \mathbf{b}^{\mathsf{T}}$
12. $3\mathbf{K} + 4\mathbf{L}$, $6\mathbf{K} + 8\mathbf{L}$
13. $\mathbf{K} + \mathbf{K}^{\mathsf{T}} + \mathbf{L} - \mathbf{L}^{\mathsf{T}}$
14. $6\mathbf{a}^{\mathsf{T}} - 9\mathbf{b}^{\mathsf{T}}$, $3(2\mathbf{a} - 3\mathbf{b})^{\mathsf{T}}$, $3(3\mathbf{b}^{\mathsf{T}} - 2\mathbf{a}^{\mathsf{T}})$

Symmetric and skew-symmetric matrices

15. Show that **K** is symmetric and **L** is skew-symmetric.

16. Show that for a symmetric matrix $\mathbf{A} = [a_{jk}]$ we have $a_{jk} = a_{kj}$.

17. Show that if $\mathbf{A} = [a_{jk}]$ is skew-symmetric, then $a_{jk} = -a_{kj}$, in particular, $a_{jj} = 0$.

18. Write **A** (in Probs. 1–8) as the sum of a symmetric and a skew-symmetric matrix.

19. Write **B** as the sum of a symmetric and a skew-symmetric matrix.

20. Show that if **A** is any square matrix, then $\mathbf{S} = \frac{1}{2}(\mathbf{A} + \mathbf{A}^\mathsf{T})$ is symmetric, $\mathbf{T} = \frac{1}{2}(\mathbf{A} - \mathbf{A}^\mathsf{T})$ is skew-symmetric, and $\mathbf{A} = \mathbf{S} + \mathbf{T}$.

21. Prove (3) and (4) in Sec. 7.2 as well as $(\mathbf{A}^\mathsf{T})^\mathsf{T} = \mathbf{A}$.

Use of matrices in modeling networks. Matrices have various engineering applications, as we shall see. For instance, they may be used to characterize connections (in electrical networks, in nets of roads connecting cities, in production processes, etc.), as follows.

22. **(Nodal incidence matrix)** Figure 131 shows an electrical network having 6 *branches* (connections) and 4 *nodes* (points where two or more branches come together). One node is the *reference node* (grounded node, whose voltage is zero). We number the other nodes and number and direct the branches. This we do arbitrarily. The network can now be described by a matrix $\mathbf{A} = [a_{jk}]$, where

$$a_{jk} = \begin{cases} +1 & \text{if branch } k \text{ leaves node } (j) \\ -1 & \text{if branch } k \text{ enters node } (j) \\ 0 & \text{if branch } k \text{ does not touch node } (j) \, . \end{cases}$$

A is called the *nodal incidence matrix* of the network. Show that for the network in Fig. 131, **A** has the given form.

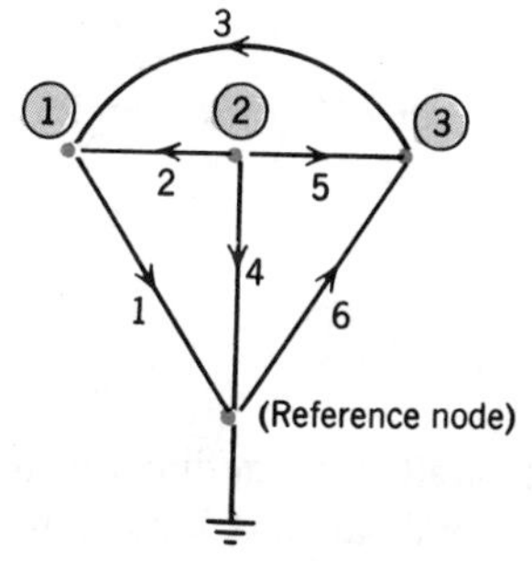

Branch	1	2	3	4	5	6
Node (1)	1	−1	−1	0	0	0
Node (2)	0	1	0	1	1	0
Node (3)	0	0	1	0	−1	−1

Fig. 131. Network and nodal incidence matrix in Prob. 22

23. Find the nodal incidence matrix of the electrical network in Fig. 132A.

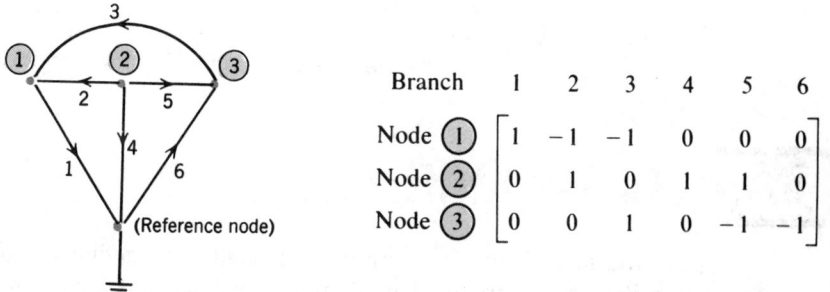

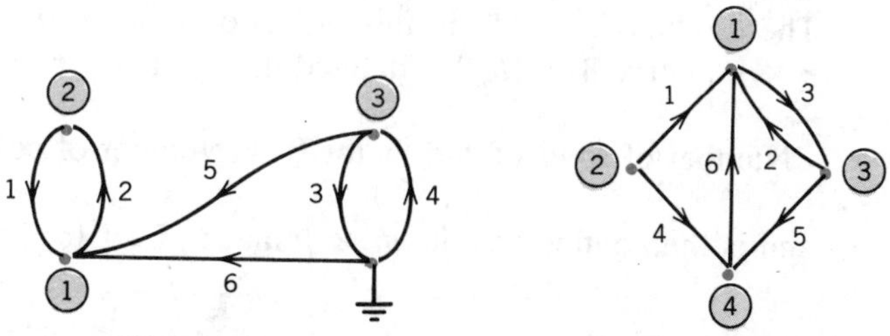

(A) Problem 23 **(B) Problem 24**

Fig. 132. Electrical network and net of one-way streets

24. Methods of electrical network analysis have applications in other fields, too. Determine the analog of the nodal incidence matrix for the net of one-way streets (directions as indicated by the arrows) shown in Fig. 132B.

Sketch the network whose nodal incidence matrix is

25. $\begin{bmatrix} 0 & 0 & -1 \\ 1 & 1 & 0 \\ 0 & -1 & 0 \end{bmatrix}$ **26.** $\begin{bmatrix} 1 & -1 & 0 & 0 \\ -1 & 1 & 1 & -1 \\ 0 & 0 & 0 & 1 \end{bmatrix}$ **27.** $\begin{bmatrix} -1 & 1 & 0 & 0 \\ 0 & -1 & 1 & 0 \\ 0 & 0 & -1 & 1 \end{bmatrix}$

28. **(Mesh incidence matrix)** A network can also be characterized by the *mesh incidence matrix* $\mathbf{M} = [m_{jk}]$, where

$$m_{jk} = \begin{cases} +1 \text{ if branch } k \text{ is in mesh } \boxed{j} \text{ and has the same orientation} \\ -1 \text{ if branch } k \text{ is in mesh } \boxed{j} \text{ and has the opposite orientation} \\ 0 \text{ if branch } k \text{ is not in mesh } \boxed{j} \end{cases}$$

where a *mesh* is a loop with no branch in its interior (or in its exterior). Here, the meshes are numbered and directed (oriented) in an arbitrary fashion. Show that for the network in Fig. 133, the matrix **M** has the given form.

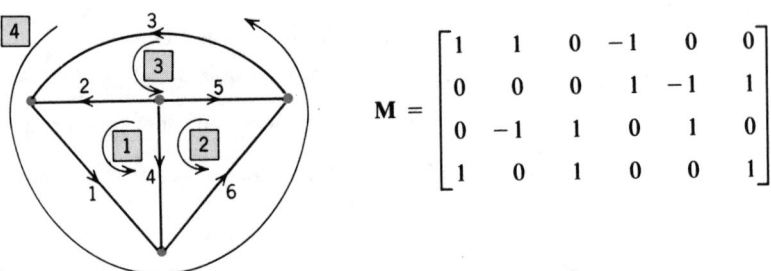

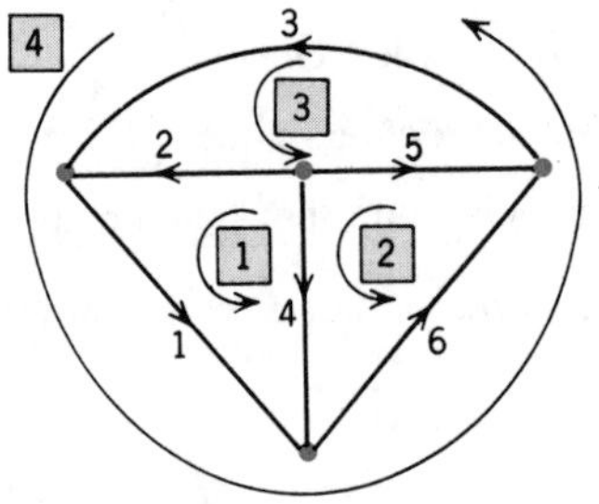

$$\mathbf{M} = \begin{bmatrix} 1 & 1 & 0 & -1 & 0 & 0 \\ 0 & 0 & 0 & 1 & -1 & 1 \\ 0 & -1 & 1 & 0 & 1 & 0 \\ 1 & 0 & 1 & 0 & 0 & 1 \end{bmatrix}$$

Fig. 133. Network and matrix **M** in Problem 28

7.3 Matrix Multiplication

As the last algebraic operation we shall now define the multiplication of matrices by matrices. This definition will at first look somewhat artificial, but afterward it will be fully motivated by the use of matrices in linear transformations, by which this multiplication is suggested.

Definition. Multiplication of a matrix by a matrix

The product $\mathbf{C} = \mathbf{AB}$ (in this order) of an $m \times n$ matrix $\mathbf{A} = [a_{jk}]$ and an $r \times p$ matrix $\mathbf{B} = [b_{jk}]$ is defined if and only if $r = n$, that is,

Number of rows of 2nd factor **B** = Number of columns of 1st factor **A**,

and is then defined as the $m \times p$ matrix $\mathbf{C} = [c_{jk}]$ with entries

$$c_{jk} = \sum_{l=1}^{n} a_{jl}b_{lk} = a_{j1}b_{1k} + a_{j2}b_{2k} + \cdots + a_{jn}b_{nk} \tag{1}$$

(where $j = 1, \cdots, m$ and $k = 1, \cdots, p$); that is, multiply each entry in the jth *row* of **A** by the corresponding entry in the kth *column* of **B** and then add these n products. One says briefly that this is a ***"multiplication of rows into columns."*** Figure 134 illustrates this.

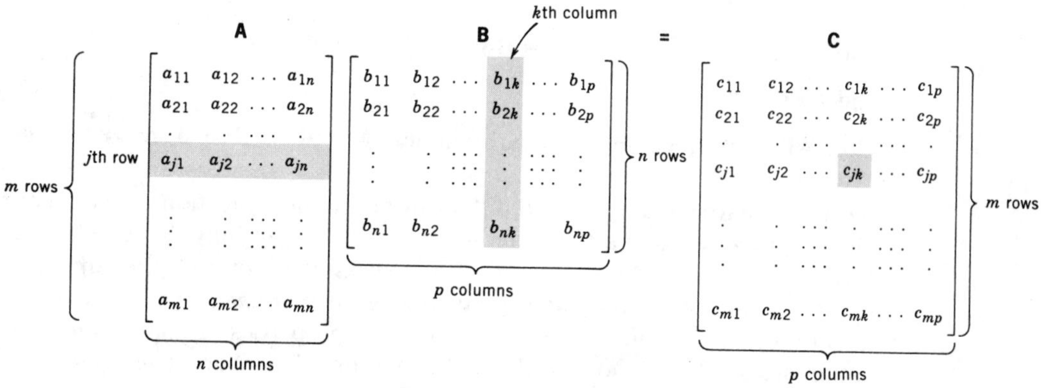

Fig. 134. Matrix multiplication **AB** = **C**

Examples. Properties of Matrix Multiplication

EXAMPLE 1 **Matrix multiplication**

$$\mathbf{AB} = \begin{bmatrix} 4 & 3 \\ 7 & 2 \\ 9 & 0 \end{bmatrix} \begin{bmatrix} 2 & 5 \\ 1 & 6 \end{bmatrix} = \begin{bmatrix} 4\cdot 2 + 3\cdot 1 & 4\cdot 5 + 3\cdot 6 \\ 7\cdot 2 + 2\cdot 1 & 7\cdot 5 + 2\cdot 6 \\ 9\cdot 2 + 0\cdot 1 & 9\cdot 5 + 0\cdot 6 \end{bmatrix} = \begin{bmatrix} 11 & 38 \\ 16 & 47 \\ 18 & 45 \end{bmatrix}.$$

Here **A** is 3×2 and **B** is 2×2, so that **AB** comes out 3×2, whereas **BA** is not defined. ∎

EXAMPLE 2 **Multiplication of a matrix and a vector**

$$\begin{bmatrix} 4 & 2 \\ 1 & 8 \end{bmatrix} \begin{bmatrix} 3 \\ 5 \end{bmatrix} = \begin{bmatrix} 12 + 10 \\ 3 + 40 \end{bmatrix} = \begin{bmatrix} 22 \\ 43 \end{bmatrix} \quad \text{whereas} \quad \begin{bmatrix} 3 \\ 5 \end{bmatrix} \begin{bmatrix} 4 & 2 \\ 1 & 8 \end{bmatrix} \quad \text{is undefined.}$$ ∎

EXAMPLE 3 **Products of row and column vectors**

$$\begin{bmatrix} 3 & 6 & 1 \end{bmatrix} \begin{bmatrix} 1 \\ 2 \\ 4 \end{bmatrix} = [19], \qquad \begin{bmatrix} 1 \\ 2 \\ 4 \end{bmatrix} \begin{bmatrix} 3 & 6 & 1 \end{bmatrix} = \begin{bmatrix} 3 & 6 & 1 \\ 6 & 12 & 2 \\ 12 & 24 & 4 \end{bmatrix}.$$ ∎

EXAMPLE 4 **CAUTION! Matrix multiplication is not commutative, AB ≠ BA in general**

This is illustrated by Examples 2 and 3, but also holds for square matrices; for instance,

$$\begin{bmatrix} 9 & 3 \\ -2 & 0 \end{bmatrix} \begin{bmatrix} 1 & -4 \\ 2 & 5 \end{bmatrix} = \begin{bmatrix} 9\cdot 1 + 3\cdot 2 & 9\cdot(-4) + 3\cdot 5 \\ -2\cdot 1 + 0\cdot 2 & (-2)\cdot(-4) + 0\cdot 5 \end{bmatrix} = \begin{bmatrix} 15 & -21 \\ -2 & 8 \end{bmatrix}$$

whereas

$$\begin{bmatrix} 1 & -4 \\ 2 & 5 \end{bmatrix} \begin{bmatrix} 9 & 3 \\ -2 & 0 \end{bmatrix} = \begin{bmatrix} 1\cdot 9 + (-4)\cdot(-2) & 1\cdot 3 + (-4)\cdot 0 \\ 2\cdot 9 + 5\cdot(-2) & 2\cdot 3 + 5\cdot 0 \end{bmatrix} = \begin{bmatrix} 17 & 3 \\ 8 & 6 \end{bmatrix}.$$ ∎

EXAMPLE 5 AB = 0 does not necessarily imply A = 0 or B = 0 or BA = 0

$$\begin{bmatrix} 1 & 1 \\ 2 & 2 \end{bmatrix}\begin{bmatrix} -1 & 1 \\ 1 & -1 \end{bmatrix} = \begin{bmatrix} 0 & 0 \\ 0 & 0 \end{bmatrix}, \qquad \begin{bmatrix} -1 & 1 \\ 1 & -1 \end{bmatrix}\begin{bmatrix} 1 & 1 \\ 2 & 2 \end{bmatrix} = \begin{bmatrix} 1 & 1 \\ -1 & -1 \end{bmatrix}. \quad \blacksquare$$

We have thus discovered the two properties

$$\text{(2a)} \qquad \mathbf{AB} \neq \mathbf{BA} \qquad \text{in general}$$

and

$$\text{(2b)} \quad \mathbf{AB} = \mathbf{0} \text{ does not necessarily imply } \mathbf{A} = \mathbf{0} \text{ or } \mathbf{B} = \mathbf{0} \text{ or } \mathbf{BA} = \mathbf{0},$$

by which matrix multiplication differs from the multiplication of numbers. Hence, always observe the order of factors very carefully! To emphasize this, we say that in **AB**, the matrix **B** is ***premultiplied,*** or ***multiplied from the left,*** by **A**, and **A** is ***postmultiplied,*** or ***multiplied from the right,*** by **B**. More about (2b) will be said in Sec. 7.7. The other properties of matrix multiplication are similar to those of the multiplication of numbers, namely,

$$\text{(2)} \qquad \begin{aligned} &\text{(c)} \quad (k\mathbf{A})\mathbf{B} = k(\mathbf{AB}) = \mathbf{A}(k\mathbf{B}) && \textit{written } k\mathbf{AB} \textit{ or } \mathbf{A}k\mathbf{B} \\ &\text{(d)} \quad \mathbf{A}(\mathbf{BC}) = (\mathbf{AB})\mathbf{C} && \textit{written } \mathbf{ABC} \\ &\text{(e)} \quad (\mathbf{A} + \mathbf{B})\mathbf{C} = \mathbf{AC} + \mathbf{BC} \\ &\text{(f)} \quad \mathbf{C}(\mathbf{A} + \mathbf{B}) = \mathbf{CA} + \mathbf{CB} \end{aligned}$$

provided **A**, **B**, and **C** are such that the expressions on the left are defined; here, k is any scalar.

Special Matrices

Certain kinds of matrices will occur quite frequently in our further work, and we now list the most important ones of them.

Triangular matrices

A square matrix whose entries above the main diagonal are all zero is called a *lower triangular matrix*. Similarly, an *upper triangular matrix* is a square matrix whose entries below the main diagonal are all zero. For instance,

$$\mathbf{T}_1 = \begin{bmatrix} 1 & 0 & 0 \\ -2 & 3 & 0 \\ 5 & 0 & 2 \end{bmatrix} \qquad \text{and} \qquad \mathbf{T}_2 = \begin{bmatrix} 1 & 6 & -1 \\ 0 & 2 & 3 \\ 0 & 0 & 4 \end{bmatrix}$$

are lower and upper triangular, respectively. An entry *on* the main diagonal of a triangular matrix may be zero or not.

Diagonal matrices

A square matrix $\mathbf{A} = [a_{jk}]$ whose entries above *and* below the main diagonal are all zero, that is, $a_{jk} = 0$ for all $j \neq k$, is called a **diagonal matrix.** For example,

$$\begin{bmatrix} 2 & 0 & 0 \\ 0 & 0 & 0 \\ 0 & 0 & -4 \end{bmatrix} \quad \text{and} \quad \begin{bmatrix} 2 & 0 & 0 \\ 0 & 2 & 0 \\ 0 & 0 & 2 \end{bmatrix}$$

are diagonal matrices.

A diagonal matrix whose entries on the main diagonal are all equal is called a **scalar matrix.** Thus a scalar matrix is of the form

$$\mathbf{S} = \begin{bmatrix} c & 0 & \cdots & 0 \\ 0 & c & \cdots & \cdot \\ \cdot & \cdot & \cdots & \cdot \\ 0 & 0 & \cdots & c \end{bmatrix}$$

where c is any number. The name comes from the fact that an $n \times n$ scalar matrix $\mathbf{S}$ commutes with any $n \times n$ matrix $\mathbf{A}$, and the multiplication by $\mathbf{S}$ has the same effect as the multiplication by a scalar,

$$\text{(3)} \qquad \mathbf{AS} = \mathbf{SA} = c\mathbf{A}.$$

In particular, a scalar matrix whose entries on the main diagonal are all 1 is called a **unit matrix** and is denoted by $\mathbf{I}_n$ or simply by $\mathbf{I}$. For $\mathbf{I}$, formula (3) becomes

$$\text{(4)} \qquad \mathbf{AI} = \mathbf{IA} = \mathbf{A}.$$

For example, the 3×3 unit matrix is

$$\mathbf{I} = \begin{bmatrix} 1 & 0 & 0 \\ 0 & 1 & 0 \\ 0 & 0 & 1 \end{bmatrix}.$$

Transpose of a Product

The transpose (see Sec. 7.1) *of a product equals the product of the transposed factors, taken in* ***reverse*** *order,*

$$\text{(5)} \qquad (\mathbf{AB})^{\mathsf{T}} = \mathbf{B}^{\mathsf{T}}\mathbf{A}^{\mathsf{T}}.$$

The proof of the useful formula (5) follows from the definition of matrix multiplication and is left to the student.

EXAMPLE 6 Transposition of a product

Formula (5) is illustrated by

$$(\mathbf{AB})^{\mathsf{T}} = \left(\begin{bmatrix} 4 & 9 \\ 0 & 2 \\ 1 & 6 \end{bmatrix}\begin{bmatrix} 3 & 7 \\ 2 & 8 \end{bmatrix}\right)^{\mathsf{T}} = \begin{bmatrix} 30 & 100 \\ 4 & 16 \\ 15 & 55 \end{bmatrix}^{\mathsf{T}} = \begin{bmatrix} 30 & 4 & 15 \\ 100 & 16 & 55 \end{bmatrix},$$

$$\mathbf{B}^{\mathsf{T}}\mathbf{A}^{\mathsf{T}} = \begin{bmatrix} 3 & 2 \\ 7 & 8 \end{bmatrix}\begin{bmatrix} 4 & 0 & 1 \\ 9 & 2 & 6 \end{bmatrix} = \begin{bmatrix} 30 & 4 & 15 \\ 100 & 16 & 55 \end{bmatrix}.$$

■

Inner Product of Vectors

This is just a special case of our definition of matrix multiplication, which occurs frequently, so that it pays to give it a special name and notation, as follows.

If **a** and **b** are column vectors with n components, then $\mathbf{a}^{\mathsf{T}}$ is a row vector, and matrix multiplication of these vectors gives a 1×1 matrix, thus a real number, which is called the **inner product** or **dot product** of **a** and **b** and is denoted by $\mathbf{a} \cdot \mathbf{b}$; thus

$$\text{(6)} \qquad \mathbf{a} \cdot \mathbf{b} = \mathbf{a}^{\mathsf{T}}\mathbf{b} = [a_1 \cdots a_n]\begin{bmatrix} b_1 \\ \vdots \\ b_n \end{bmatrix} = \sum_{l=1}^{n} a_l b_l = a_1 b_1 + \cdots + a_n b_n.$$

Inner products have interesting applications in mechanics and geometry, as we shall see in Sec. 8.2. At present we shall use them to express matrix products in a condensed form, which is often quite useful.

Product in Terms of Row and Column Vectors

Matrix multiplication is a multiplication of rows into columns, as we know, and we can thus write (1) in terms of inner products. Indeed, every entry of $\mathbf{C} = \mathbf{AB}$ is an inner product,

$$c_{11} = \mathbf{a}_1 \cdot \mathbf{b}_1 = (\text{first row of } \mathbf{A}) \cdot (\text{first column of } \mathbf{B})$$

$$c_{12} = \mathbf{a}_1 \cdot \mathbf{b}_2 = (\text{first row of } \mathbf{A}) \cdot (\text{second column of } \mathbf{B})$$

and so on, the general term being

$$\text{(7)} \qquad \boxed{c_{jk} = \mathbf{a}_j \cdot \mathbf{b}_k = (j\text{th row of } \mathbf{A}) \cdot (k\text{th column of } \mathbf{B}).}$$

Accordingly, if we write **A** in terms of its row vectors,

$$\mathbf{A} = \begin{bmatrix} \mathbf{a}_1 \\ \mathbf{a}_2 \\ \vdots \\ \mathbf{a}_m \end{bmatrix}, \quad \text{where} \quad \begin{aligned} \mathbf{a}_1 &= [a_{11} \quad a_{12} \quad \cdots \quad a_{1n}] \\ \mathbf{a}_2 &= [a_{21} \quad a_{22} \quad \cdots \quad a_{2n}] \\ &\vdots \\ \mathbf{a}_n &= [a_{m1} \quad a_{m2} \quad \cdots \quad a_{mn}] \end{aligned}$$

and **B** in terms of its column vectors, $\mathbf{B} = [\mathbf{b}_1 \quad \mathbf{b}_2 \quad \cdots \quad \mathbf{b}_p]$, where

$$\mathbf{b}_1 = \begin{bmatrix} b_{11} \\ b_{21} \\ \vdots \\ b_{n1} \end{bmatrix}, \quad \mathbf{b}_2 = \begin{bmatrix} b_{12} \\ b_{22} \\ \vdots \\ b_{n2} \end{bmatrix}, \quad \cdots, \quad \mathbf{b}_p = \begin{bmatrix} b_{1p} \\ b_{2p} \\ \vdots \\ b_{np} \end{bmatrix}$$

we see from (1) or (7) that the product $\mathbf{C} = \mathbf{AB}$ can be written

$$\text{(8)} \qquad \mathbf{C} = \mathbf{AB} = \begin{bmatrix} \mathbf{a}_1 \cdot \mathbf{b}_1 & \mathbf{a}_1 \cdot \mathbf{b}_2 & \cdots & \mathbf{a}_1 \cdot \mathbf{b}_p \\ \mathbf{a}_2 \cdot \mathbf{b}_1 & \mathbf{a}_2 \cdot \mathbf{b}_2 & \cdots & \mathbf{a}_2 \cdot \mathbf{b}_p \\ \cdot & \cdot & \cdots & \cdot \\ \mathbf{a}_m \cdot \mathbf{b}_1 & \mathbf{a}_m \cdot \mathbf{b}_2 & \cdots & \mathbf{a}_m \cdot \mathbf{b}_p \end{bmatrix}.$$

This idea sometimes helps in applications to see more clearly what is going on.

Furthermore, $\mathbf{Ab}_1$ is a column vector

$$\mathbf{Ab}_1 = \begin{bmatrix} a_{11} & \cdots & a_{1n} \\ \cdot & \cdots & \cdot \\ \cdot & \cdots & \cdot \\ a_{m1} & \cdots & a_{mn} \end{bmatrix} \begin{bmatrix} b_{11} \\ \cdot \\ \cdot \\ \cdot \\ b_{n1} \end{bmatrix} = \begin{bmatrix} \mathbf{a}_1 \cdot \mathbf{b}_1 \\ \cdot \\ \cdot \\ \cdot \\ \mathbf{a}_m \cdot \mathbf{b}_1 \end{bmatrix}$$

and (8) shows that this is the first column of **AB**. Similarly for the other columns of **AB**, so that we can write

$$\text{(9)} \qquad \mathbf{AB} = [\mathbf{Ab}_1 \quad \mathbf{Ab}_2 \quad \cdots \quad \mathbf{Ab}_p],$$

a formula that is often useful (for instance, in Sec. 7.14).

EXAMPLE 7 Product in terms of row and column vectors

Writing a 2 × 2 matrix **A** in terms of row vectors, say,

$$\mathbf{A} = \begin{bmatrix} a_{11} & a_{12} \\ a_{21} & a_{22} \end{bmatrix} = \begin{bmatrix} \mathbf{a}_1 \\ \mathbf{a}_2 \end{bmatrix} \quad \text{where} \quad \begin{matrix} \mathbf{a}_1 = [a_{11} \quad a_{12}] \\ \mathbf{a}_2 = [a_{21} \quad a_{22}] \end{matrix}$$

and a 2 × 2 matrix **B** in terms of column vectors, say,

$$\mathbf{B} = \begin{bmatrix} b_{11} & b_{12} \\ b_{21} & b_{22} \end{bmatrix} = [\mathbf{b}_1 \quad \mathbf{b}_2] \quad \text{where} \quad \mathbf{b}_1 = \begin{bmatrix} b_{11} \\ b_{21} \end{bmatrix}, \quad \mathbf{b}_2 = \begin{bmatrix} b_{12} \\ b_{22} \end{bmatrix}.$$

we see that (8) takes the form

$$\mathbf{AB} = \begin{bmatrix} \mathbf{a}_1 \\ \mathbf{a}_2 \end{bmatrix} [\mathbf{b}_1 \quad \mathbf{b}_2] = \begin{bmatrix} \mathbf{a}_1 \cdot \mathbf{b}_1 & \mathbf{a}_1 \cdot \mathbf{b}_2 \\ \mathbf{a}_2 \cdot \mathbf{b}_1 & \mathbf{a}_2 \cdot \mathbf{b}_2 \end{bmatrix} = \begin{bmatrix} a_{11}b_{11} + a_{12}b_{21} & a_{11}b_{12} + a_{12}b_{22} \\ a_{21}b_{11} + a_{22}b_{21} & a_{21}b_{12} + a_{22}b_{22} \end{bmatrix}.$$

Also, $\mathbf{AB} = [\mathbf{Ab}_1 \quad \mathbf{Ab}_2]$ by (9). ∎

Motivation of Matrix Multiplication

Matrix multiplication may look somewhat strange at first sight, but there is a good reason for such an "unnatural" definition, which comes from the use of matrices in connection with "linear transformations." To see this, we consider three coordinate systems in the plane, which we denote as the w_1w_2-system, the x_1x_2-system, and the y_1y_2-system, and we assume that these systems are related by transformations

$$(10) \qquad \begin{aligned} y_1 &= a_{11}x_1 + a_{12}x_2 \\ y_2 &= a_{21}x_1 + a_{22}x_2 \end{aligned}$$

and

$$(11) \qquad \begin{aligned} x_1 &= b_{11}w_1 + b_{12}w_2 \\ x_2 &= b_{21}w_1 + b_{22}w_2 \end{aligned}$$

which are (special) *linear transformations*. By substituting (11) into (10) we see that the y_1y_2-coordinates can be obtained directly from the w_1w_2-coordinates by a single linear transformation of the form

$$(12) \qquad \begin{aligned} y_1 &= c_{11}w_1 + c_{12}w_2 \\ y_2 &= c_{21}w_1 + c_{22}w_2. \end{aligned}$$

Now this substitution gives

$$y_1 = a_{11}(b_{11}w_1 + b_{12}w_2) + a_{12}(b_{21}w_1 + b_{22}w_2)$$

$$y_2 = a_{21}(b_{11}w_1 + b_{12}w_2) + a_{22}(b_{21}w_1 + b_{22}w_2).$$

Comparing this with (12), we see that we must have

$$c_{11} = a_{11}b_{11} + a_{12}b_{21} \qquad c_{12} = a_{11}b_{12} + a_{12}b_{22}$$

$$c_{21} = a_{21}b_{11} + a_{22}b_{21} \qquad c_{22} = a_{21}b_{12} + a_{22}b_{22}$$

or briefly

$$c_{jk} = a_{j1}b_{1k} + a_{j2}b_{2k} = \sum_{i=1}^{2} a_{ji}b_{ik} \qquad j, k = 1, 2. \tag{13}$$

This is (1) with $m = n = p = 2$.

What does our calculation show? Essentially two things. First, matrix multiplication is defined in such a way that linear transformations can be written in compact form, using matrices; in our case, (10) becomes

$$\text{(10*)} \quad \mathbf{y} = \mathbf{Ax} \quad \text{where} \quad \mathbf{y} = \begin{bmatrix} y_1 \\ y_2 \end{bmatrix}, \quad \mathbf{A} = \begin{bmatrix} a_{11} & a_{12} \\ a_{21} & a_{22} \end{bmatrix}, \quad \mathbf{x} = \begin{bmatrix} x_1 \\ x_2 \end{bmatrix}$$

and (11) becomes

$$\text{(11*)} \quad \mathbf{x} = \mathbf{Bw} \quad \text{where} \quad \mathbf{x} = \begin{bmatrix} x_1 \\ x_2 \end{bmatrix}, \quad \mathbf{B} = \begin{bmatrix} b_{11} & b_{12} \\ b_{21} & b_{22} \end{bmatrix}, \quad \mathbf{w} = \begin{bmatrix} w_1 \\ w_2 \end{bmatrix}.$$

Second, if we substitute linear transformations into each other, we can obtain the coefficient matrix **C** of the composite transformation (the transformation obtained by the substitution) simply by multiplying the coefficient matrices **A** and **B** of the given transformations, in the right order suggested by the substitution; from (10*), (11*), and (12) we get

$$\mathbf{y} = \mathbf{Ax} = \mathbf{A(Bw)} = \mathbf{ABw} = \mathbf{Cw}, \qquad \text{where} \qquad \mathbf{C} = \mathbf{AB}.$$

For higher dimensions the idea and the result are exactly the same; only the number of variables changes. We then have m variables $y_1, \cdots, y_m$ and n variables $x_1, \cdots, x_n$ and p variables $w_1, \cdots, w_p$. The matrix **A** is $m \times n$, the matrix **B** is $n \times p$, and **C** is $m \times p$, as in Fig. 134. And the requirement that **C** be the product **AB** leads to formula (1) in its general form. This completely motivates the definition of matrix multiplication.

We shall say more about (general) linear transformations and related matrices in Sec. 7.15, after we have gained more experience with matrices by considering linear systems of equations, beginning in the next section.

An Application of Matrix Multiplication

EXAMPLE 8 **Stochastic matrix. Markov process**

Suppose that the 1993 state of land use in a city of 50 square miles of (nonvacant) area is

I	(Residentially used)	30%
II	(Commercially used)	20%
III	(Industrially used)	50%.

Find the states in 1998 and 2003, assuming that the transition probabilities for 5-year intervals are given by the following matrix $\mathbf{A} = [a_{jk}]$.

	To I	To II	To III
From I	0.8	0.1	0.1
From II	0.1	0.7	0.2
From III	0	0.1	0.9

Remark. A square matrix with nonnegative entries and row sums all equal to 1 is called a **stochastic matrix.** Thus **A** is a stochastic matrix. A stochastic process for which the probability of entering a certain state depends only on the *last* state occupied (and on the matrix governing the process) is called a **Markov process.**[1] Thus our example concerns a Markov process.

Solution. From the matrix **A** and the 1993 state we can compute the 1998 state

I	(Residential)	$0.8 \cdot 30 + 0.1 \cdot 20 + 0 \cdot 50 = 26$ [%]
II	(Commercial)	$0.1 \cdot 30 + 0.7 \cdot 20 + 0.1 \cdot 50 = 22$ [%]
III	(Industrial)	$0.1 \cdot 30 + 0.2 \cdot 20 + 0.9 \cdot 50 = 52$ [%].

The sum is 100%, as it should be. We write this in matrix form. Let the column vector **x** denote the 1993 state; thus, $\mathbf{x}^\mathsf{T} = [30 \quad 20 \quad 50]$. Let **y** denote the 1998 state. Then

$$\mathbf{y}^\mathsf{T} = \mathbf{x}^\mathsf{T}\mathbf{A} = [30 \quad 20 \quad 50]\begin{bmatrix} 0.8 & 0.1 & 0.1 \\ 0.1 & 0.7 & 0.2 \\ 0 & 0.1 & 0.9 \end{bmatrix} = [26 \quad 22 \quad 52].$$

Similarly, for the vector **z** of the 2003 state we get, as the reader may verify.

$$\mathbf{z}^\mathsf{T} = \mathbf{y}^\mathsf{T}\mathbf{A} = (\mathbf{x}^\mathsf{T}\mathbf{A})\mathbf{A} = \mathbf{x}^\mathsf{T}\mathbf{A}^2 = [23 \quad 23.2 \quad 53.8].$$

Answer. In 1998, the residential area will be 26% (13 square miles), the commercial 22% (11 square miles) and the industrial 52% (26 square miles). For 2003, the corresponding figures are 23%, 23.2%, 53.8%. ■

This is the end of the first portion of Chap. 7, in which we have defined the rules of matrix and vector algebra. We are now ready for applications, beginning in the next section.

---

[1] ANDREI ANDREJEVITCH MARKOV (1856—1922), Russian mathematician, known for his work in probability theory.

Problem Set 7.3

Let $\mathbf{a} = \begin{bmatrix} 3 \\ 1 \\ 4 \end{bmatrix}$, $\mathbf{B} = \begin{bmatrix} 0 & 1 \\ 0 & -2 \\ 2 & 3 \end{bmatrix}$, $\mathbf{C} = \begin{bmatrix} 1 & 0 & -1 \\ 2 & 3 & 0 \\ 0 & 3 & 4 \end{bmatrix}$, $\mathbf{d} = [1 \quad 0 \quad 2]$.

Find those of the following expressions that are defined.

1. $\mathbf{CB}$, $\mathbf{B}^\mathsf{T}\mathbf{C}^\mathsf{T}$, $\mathbf{BC}^\mathsf{T}$

2. $\mathbf{C}^2$, $\mathbf{C}^3$, $\mathbf{CC}^\mathsf{T}$, $\mathbf{C}^\mathsf{T}\mathbf{C}$

3. $\mathbf{Ca}$, $\mathbf{Cd}^\mathsf{T}$, $\mathbf{C}^\mathsf{T}\mathbf{d}^\mathsf{T}$

4. $\mathbf{B}^\mathsf{T}\mathbf{a}$, $\mathbf{Bd}$, $\mathbf{dB}$, $\mathbf{ad}$

5. $\mathbf{B}^\mathsf{T}\mathbf{C}$, $\mathbf{B}^\mathsf{T}\mathbf{B}$

6. $\mathbf{BB}^\mathsf{T}$, $\mathbf{BB}^\mathsf{T}\mathbf{C}$, $\mathbf{BB}^\mathsf{T}\mathbf{a}$

7. $\mathbf{a}^\mathsf{T}\mathbf{a}$, $\mathbf{a}^\mathsf{T}\mathbf{Ca}$, $\mathbf{dCd}^\mathsf{T}$

8. $\mathbf{dd}^\mathsf{T}$, $\mathbf{d}^\mathsf{T}\mathbf{d}$, $\mathbf{adB}$, $\mathbf{adBB}^\mathsf{T}$

9. Prove (5).

10. Find real 2×2 matrices (as many as you can) whose square is $\mathbf{I}$, the unit matrix.

11. Find a 2×2 matrix $\mathbf{A} \neq \mathbf{0}$ such that $\mathbf{A}^2 = \mathbf{0}$.

12. Find two 2×2 matrices $\mathbf{A}$, $\mathbf{B}$ such that $(\mathbf{A} + \mathbf{B})^2 \neq \mathbf{A}^2 + 2\mathbf{AB} + \mathbf{B}^2$.

13. **(Idempotent matrix)** A matrix $\mathbf{A}$ is said to be *idempotent* if $\mathbf{A}^2 = \mathbf{A}$. Give examples of idempotent matrices, different from the zero or unit matrix.

14. Show that $\mathbf{AA}^\mathsf{T}$ is symmetric.

15. Find all real square matrices that are both symmetric and skew-symmetric.

16. Show that the product of symmetric matrices $\mathbf{A}$, $\mathbf{B}$ is symmetric if and only if $\mathbf{A}$ and $\mathbf{B}$ commute, $\mathbf{AB} = \mathbf{BA}$.

Special linear transformations were used in the text to motivate matrix multiplication, and we add some problems of practical interest. (Linear transformations in general follow in Sec. 7.15.)

17. **(Rotation)** Show that the linear transformation $\mathbf{y} = \mathbf{Ax}$ with matrix

$$\mathbf{A} = \begin{bmatrix} \cos\theta & -\sin\theta \\ \sin\theta & \cos\theta \end{bmatrix} \quad \text{and} \quad \mathbf{x} = \begin{bmatrix} x_1 \\ x_2 \end{bmatrix}, \quad \mathbf{y} = \begin{bmatrix} y_1 \\ y_2 \end{bmatrix}$$

is a counterclockwise rotation of the Cartesian x_1x_2-coordinate system in the plane about the origin, where θ is the angle of rotation.

18. Show that in Prob. 17,

$$\mathbf{A}^n = \begin{bmatrix} \cos n\theta & -\sin n\theta \\ \sin n\theta & \cos n\theta \end{bmatrix}.$$

What does this result mean geometrically?

19. **(Computer graphics)** To visualize a three-dimensional object with plane faces (e.g., a cube), we may store the position vectors of the vertices with respect to a suitable $x_1x_2x_3$-coordinate system (and a list of the connecting edges) and then obtain a two-dimensional image on a video screen by projecting the object onto a coordinate plane, for instance, onto the x_1x_2-plane by setting $x_3 = 0$. To change the appearance of the image, we can impose a linear transformation on the position vectors stored. Show that a diagonal matrix $\mathbf{D}$ with main diagonal entries 3, 1, $\frac{1}{2}$ gives from an $\mathbf{x} = [x_j]$ the new position vector $\mathbf{y} = \mathbf{Dx}$, where $y_1 = 3x_1$ (stretch in the x_1-direction by a factor 3), $y_2 = x_2$ (unchanged), $y_3 = \frac{1}{2}x_3$ (contraction in the x_3-direction). What effect would a scalar matrix have?

20. (Rotations in space in computer graphics) What effect would the following matrices have in the situation described in Prob. 19?

$$\begin{bmatrix} 1 & 0 & 0 \\ 0 & \cos\theta & -\sin\theta \\ 0 & \sin\theta & \cos\theta \end{bmatrix}, \quad \begin{bmatrix} \cos\varphi & 0 & -\sin\varphi \\ 0 & 1 & 0 \\ \sin\varphi & 0 & \cos\varphi \end{bmatrix}, \quad \begin{bmatrix} \cos\psi & -\sin\psi & 0 \\ \sin\psi & \cos\psi & 0 \\ 0 & 0 & 1 \end{bmatrix}$$

21. (Assignment problem) Contractors C_1, C_2, C_3 bid for jobs J_1, J_2, J_3 as the cost matrix (in 100 000-dollar units) shows. What assignment minimizes the total cost (a) under no condition? (b) Under the condition that each contractor be assigned to only one job?

	J_1	J_2	J_3
C_1	24	4	10
C_2	18	6	12
C_3	16	8	8

Cost matrix for Problem 21

$$\mathbf{A} = [a_{jk}] = \begin{bmatrix} 40 & 28 & 20 \\ 34 & 26 & 14 \\ 36 & 30 & 20 \end{bmatrix}$$

Matrix A for Problem 22

22. If worker W_j can do job J_k in a_{jk} hours, as shown by the matrix **A**, and each worker should do one job only, which assignment would minimize the total time?

23. (Markov process) For the Markov process with transition matrix $\mathbf{A} = [a_{jk}]$, whose entries are $a_{11} = a_{12} = 0.5$, $a_{21} = 0.2$, $a_{22} = 0.8$, and initial state $[0.7 \quad 0.7]^\mathsf{T}$, compute the next 3 states.

24. In a production process, let N mean "no trouble" and T "trouble." Let the transition probabilities from one day to the next be 0.8 for $N \to N$, hence 0.2 for $N \to T$, and 0.5 for $T \to N$, hence 0.5 for $T \to T$. If today there is no trouble, what is the probability of trouble 2 days after today? 3 days after today?

7.4 Linear Systems of Equations. Gauss Elimination

The most important practical use of matrices is in the solution of linear systems of equations, which appear frequently as models of various problems, for instance, in frameworks, electrical networks, traffic flow, production and consumption, assignment of jobs to workers, population growth, statistics, numerical methods for differential equations (Chap. 20), and many others. We begin in this section with an important solution method, the Gauss elimination, and discuss general properties of solutions in the next sections.

Linear systems. A *linear system of m equations in n unknowns* $x_1, \cdots, x_n$ is a set of equations of the form

$$\begin{aligned} a_{11}x_1 + \cdots + a_{1n}x_n &= b_1 \\ a_{21}x_1 + \cdots + a_{2n}x_n &= b_2 \\ \cdots\cdots\cdots&\cdots\cdots\cdots \\ a_{m1}x_1 + \cdots + a_{mn}x_n &= b_m \end{aligned} \tag{1}$$

Thus, a system of two equations in three unknown is

$$\begin{aligned} a_{11}x_1 + a_{12}x_2 + a_{13}x_3 &= b_1 \\ a_{21}x_1 + a_{22}x_2 + a_{23}x_3 &= b_2, \end{aligned} \qquad \text{for example,} \qquad \begin{aligned} 5x_1 + 2x_2 - x_3 &= 4 \\ x_1 - 4x_2 + 3x_3 &= 6. \end{aligned}$$

The a_{jk} are given numbers, which are called the **coefficients** of the system. The b_i are also given numbers. If the b_i are all zero, then (1) is called a **homogeneous system.** If at least one b_i is not zero, then (1) is called a **nonhomogenous system.**

A **solution** of (1) is a set of numbers $x_1, \cdots, x_n$ that satisfy all the m equations. A **solution vector** of (1) is a vector **x** whose components constitute a solution of (1). If the system (1) is homogeneous, it has at least the **trivial solution** $x_1 = 0, \cdots, x_n = 0$.

Coefficient Matrix, Augmented Matrix

From the definition of matrix multiplication we see that the m equations of (1) may be written as a single vector equation

$$\mathbf{Ax} = \mathbf{b} \tag{2}$$

where the **coefficient matrix** $\mathbf{A} = [a_{jk}]$ is the $m \times n$ matrix

$$\mathbf{A} = \begin{bmatrix} a_{11} & a_{12} & \cdots & a_{1n} \\ a_{21} & a_{22} & \cdots & a_{2n} \\ \cdot & \cdot & \cdots & \cdot \\ a_{m1} & a_{m2} & \cdots & a_{mn} \end{bmatrix}, \quad \text{and} \quad \mathbf{x} = \begin{bmatrix} x_1 \\ \cdot \\ \cdot \\ \cdot \\ x_n \end{bmatrix} \quad \text{and} \quad \mathbf{b} = \begin{bmatrix} b_1 \\ \cdot \\ \cdot \\ \cdot \\ b_m \end{bmatrix}$$

are column vectors. We assume that the coefficients a_{jk} are not all zero, so that **A** is not a zero matrix. Note that **x** has n components, whereas **b** has m components. The matrix

$$\tilde{\mathbf{A}} = \begin{bmatrix} a_{11} & \cdots & a_{1n} & b_1 \\ \cdot & \cdots & \cdot & \cdot \\ \cdot & \cdots & \cdot & \cdot \\ a_{m1} & \cdots & a_{mn} & b_m \end{bmatrix}$$

is called the **augmented matrix** of the system (1). We see that $\tilde{\mathbf{A}}$ is obtained by augmenting **A** by the column **b**. The matrix $\tilde{\mathbf{A}}$ determines the system (1) completely, because it contains all the given numbers appearing in (1).

EXAMPLE 1 Geometric interpretation. Existence of solutions

If $m = n = 2$, we have two equations in two unknowns x_1, x_2

$$a_{11}x_1 + a_{12}x_2 = b_1$$
$$a_{21}x_1 + a_{22}x_2 = b_2.$$

If we interpret x_1, x_2 as coordinates in the x_1x_2-plane, then each of the two equations represents a straight line, and (x_1, x_2) is a solution if and only if the point P with coordinates x_1, x_2 lies on both lines. Hence there are three possible cases:

(a) No solution if the lines are parallel.
(b) Precisely one solution if they intersect.
(c) Infinitely many solutions if they coincide.

For instance,

Case (a)	Case (b)	Case (c)
$x + y = 1$	$x + y = 1$	$x + y = 1$
$x + y = 0$	$x - y = 0$	$2x + 2y = 2$

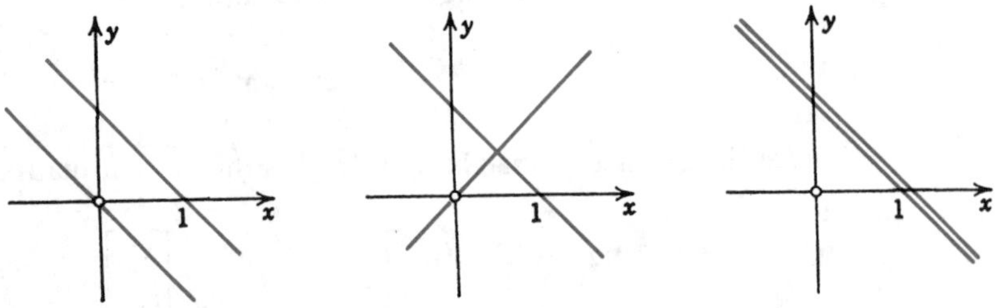

If the system is homogeneous, Case (*a*) cannot happen, because then those two straight lines pass through the origin, whose coordinates 0, 0 constitute the trivial solution. The reader may consider three equations in three unknowns as representations of three planes in space and discuss the various possible cases in a similar fashion. ∎

Our simple example illustrates that a system (1) may not always have a solution, and relevant problems are as follows. Does a given system (1) have a solution? Under what conditions does it have precisely one solution? If it has more than one solution, how can we characterize the set of all solutions? How can we obtain the solutions? We discuss the last question first and the others in Sec. 7.6.

Gauss Elimination

The Gauss elimination is a standard method for solving linear systems. This is a systematic process of elimination, a method of great importance that works in practice and is reasonable with respect to computing time and storage demand (two aspects we shall consider in Sec. 19.1 on numerical methods). We first explain the method by some typical examples. Since a linear system is completely determined by its augmented matrix, the process of elimination can be performed by merely considering the matrices. To see this correspondence, we shall write systems of equations and augmented matrices side by side.

EXAMPLE 2 Gauss elimination. Electrical network

Solve the linear system

$$\begin{aligned} x_1 - x_2 + x_3 &= 0 \\ -x_1 + x_2 - x_3 &= 0 \\ 10x_2 + 25x_3 &= 90 \\ 20x_1 + 10x_2 &= 80. \end{aligned}$$

Derivation from the circuit in Fig. 135 (Optional). This is the system for the unknown currents $x_1 = i_1, x_2 = i_2, x_3 = i_3$ in the electrical network in Fig. 135. To obtain it, we label the currents as shown, choosing directions arbitrarily; if a current will come out negative, this will simply mean that the current flows against the direction of our arrow. The current entering each battery will be the same as the current leaving it. The equations for the currents result from Kirchhoff's laws:

Kirchhoff's current law (KCL). *At any point of a circuit, the sum of the inflowing currents equals the sum of the outflowing currents.*

Kirchhoff's voltage law (KVL). *In any closed loop, the sum of all voltage drops equals the impressed electromotive force.*

Node P gives the first equation, node Q the second, the right loop the third, and the left loop the fourth, as indicated in the figure.

Solution by Gauss's method. This system is so simple that we could almost solve it by inspection. This is not the point. The point is to perform a systematic elimination—the Gauss elimination—which will work in general, also for large systems. It is a reduction to "*triangular form*" from which we shall then readily obtain the values of the unknowns by "*back substitution.*"

We write the system and its augmented matrix side by side:

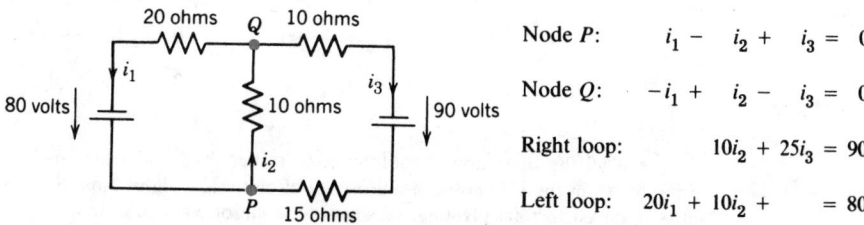

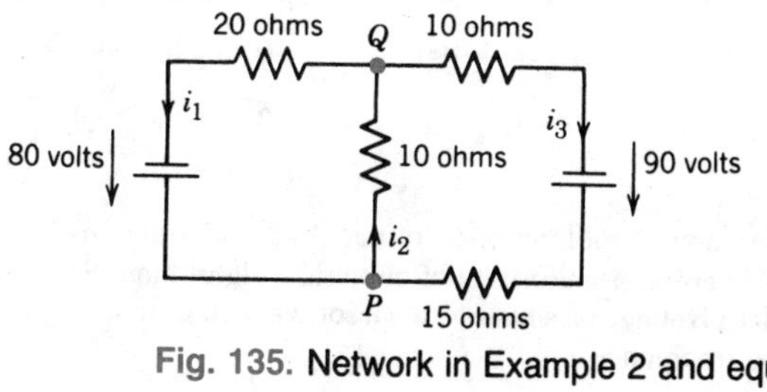

Node P: $i_1 - i_2 + i_3 = 0$

Node Q: $-i_1 + i_2 - i_3 = 0$

Right loop: $10i_2 + 25i_3 = 90$

Left loop: $20i_1 + 10i_2 + \quad = 80$

Fig. 135. Network in Example 2 and equations for the currents

Equations	Augmented Matrix $\tilde{\mathbf{A}}$
Pivot → $\boxed{x_1} - x_2 + x_3 = 0$ $-x_1 + x_2 - x_3 = 0$ Eliminate → $10x_2 + 25x_3 = 90$ $20x_1 + 10x_2 = 80$	$\begin{bmatrix} 1 & -1 & 1 & 0 \\ -1 & 1 & -1 & 0 \\ 0 & 10 & 25 & 90 \\ 20 & 10 & 0 & 80 \end{bmatrix}$

First Step. Elimination of x_1

Call the first equation the **pivot equation** and its x_1-term the **pivot** in this step, and use this equation to eliminate x_1 (get rid of x_1) in the other equations. For this, do these operations:

Subtract -1 times the pivot equation from the second equation.[2]

Subtract 20 times the pivot equation from the fourth equation.

This corresponds to row operations on the augmented matrix, which we indicate behind the *new* matrix in (3). The result is

$$
(3)\qquad
\begin{aligned}
x_1 - x_2 + x_3 &= 0\\
0 &= 0\\
10x_2 + 25x_3 &= 90\\
30x_2 - 20x_3 &= 80
\end{aligned}
\qquad
\begin{bmatrix} 1 & -1 & 1 & 0 \\ 0 & 0 & 0 & 0 \\ 0 & 10 & 25 & 90 \\ 0 & 30 & -20 & 80 \end{bmatrix}
\begin{matrix} \\ \text{Row 2 + Row 1} \\ \\ \text{Row 4 - 20 Row 1} \end{matrix}
$$

Second Step. Elimination of x_2

The first equation, which has just served as the pivot equation, remains untouched. We want to take the (new!) second equation as the next pivot equation. Since it contains no x_2-term (needed as the next pivot)—in fact, it is $0 = 0$—first we have to change the order of equations (and corresponding rows of the new matrix) to get a nonzero pivot. We put the second equation ($0 = 0$) at the end and move the third and fourth equations one place up; this is called **partial pivoting.**[3] We get

$$
\begin{aligned}
x_1 - x_2 + x_3 &= 0\\
\text{Pivot} \rightarrow \boxed{10x_2} + 25x_3 &= 90\\
\text{Eliminate} \rightarrow \boxed{30x_2} - 20x_3 &= 80\\
0 &= 0
\end{aligned}
\qquad
\begin{bmatrix} 1 & -1 & 1 & 0 \\ 0 & 10 & 25 & 90 \\ 0 & 30 & -20 & 80 \\ 0 & 0 & 0 & 0 \end{bmatrix}
$$

To eliminate x_2, do:

Subtract 3 times the pivot equation from the third equation.

The result is

$$
(4)\qquad
\begin{aligned}
x_1 - x_2 + x_3 &= 0\\
10x_2 + 25x_3 &= 90\\
-95x_3 &= -190\\
0 &= 0
\end{aligned}
\qquad
\begin{bmatrix} 1 & -1 & 1 & 0 \\ 0 & 10 & 25 & 90 \\ 0 & 0 & -95 & -190 \\ 0 & 0 & 0 & 0 \end{bmatrix}
\begin{matrix} \\ \\ \text{Row 3 - 3 Row 2} \\ \\ \end{matrix}
$$

---

[2]To call all the operations "subtractions" rather than "subtractions" and "additions" is preferable from the viewpoint of uniformity of numerical algorithms. See also Sec. 19.1.

[3]As opposed to **total pivoting,** in which also the order of the unknowns is changed. Total pivoting is hardly used in practice.

Back Substitution. Determination of x_3, x_2, x_1
Working backward from the last to the first equation of this "triangular" system (4), we can now readily find x_3, then x_2 and then x_1:

$$\begin{aligned} -95x_3 &= -190, & x_3 &= i_3 = 2 \text{ [amperes]}, \\ 10x_2 + 25x_3 &= 90, & x_2 &= \tfrac{1}{10}(90 - 25x_3) = i_2 = 4 \text{ [amperes]}, \\ x_1 - x_2 + x_3 &= 0, & x_1 &= x_2 - x_3 = i_1 = 2 \text{ [amperes]}. \end{aligned}$$

This is the answer to our problem. The solution is unique. ∎

A system (1) is called **overdetermined** if it has more equations than unknowns, as in Example 2, **determined** if $m = n$, as in Example 1, and **underdetermined** if (1) has fewer equations than unknowns. An underdetermined system always has solutions, whereas in the other two cases, solutions may or may not exist. (Details follow in Sec. 7.6.) We want to illustrate next that the Gauss elimination applies to any system, no matter whether it has many solutions, a unique solution, or no solutions.

EXAMPLE 3 Gauss elimination for an underdetermined system

Solve the linear system of three equations in four unknowns

$$(5)\quad \begin{aligned} 3.0x_1 + 2.0x_2 + 2.0x_3 - 5.0x_4 &= 8.0 \\ 0.6x_1 + 1.5x_2 + 1.5x_3 - 5.4x_4 &= 2.7 \\ 1.2x_1 - 0.3x_2 - 0.3x_3 + 2.4x_4 &= 2.1 \end{aligned} \qquad \begin{bmatrix} 3.0 & 2.0 & 2.0 & -5.0 & 8.0 \\ 0.6 & 1.5 & 1.5 & -5.4 & 2.7 \\ 1.2 & -0.3 & -0.3 & 2.4 & 2.1 \end{bmatrix}$$

Solution. As in the previous example, we circle pivots and box terms to be eliminated.

First Step. Elimination of x_1 from the second and third equations by subtracting

$0.6/3.0 = 0.2$ times the first equation from the second equation,

$1.2/3.0 = 0.4$ times the first equation from the third equation.

This gives a new system of equations

$$(6)\quad \begin{aligned} 3.0x_1 + 2.0x_2 + 2.0x_3 - 5.0x_4 &= 8.0 \\ 1.1x_2 + 1.1x_3 - 4.4x_4 &= 1.1 \\ -1.1x_2 - 1.1x_3 + 4.4x_4 &= -1.1 \end{aligned} \qquad \begin{bmatrix} 3.0 & 2.0 & 2.0 & -5.0 & 8.0 \\ 0 & 1.1 & 1.1 & -4.4 & 1.1 \\ 0 & -1.1 & -1.1 & 4.4 & -1.1 \end{bmatrix}$$

and we circle the pivot to be used in the next step.

Second Step. Elimination of x_2 from the third equation of (6) by subtracting

$-1.1/1.1 = -1$ times the second equation from the third equation.

This gives

$$(7)\quad \begin{aligned} 3.0x_1 + 2.0x_2 + 2.0x_3 - 5.0x_4 &= 8.0 \\ 1.1x_2 + 1.1x_3 - 4.4x_4 &= 1.1 \\ 0 &= 0 \end{aligned} \qquad \begin{bmatrix} 3.0 & 2.0 & 2.0 & -5.0 & 8.0 \\ 0 & 1.1 & 1.1 & -4.4 & 1.1 \\ 0 & 0 & 0 & 0 & 0 \end{bmatrix}$$

Back Substitution. From the second equation, $x_2 = 1 - x_3 + 4x_4$. From this and the first equation, $x_1 = 2 - x_4$. Since x_3 and x_4 remain arbitrary, we have infinitely many solutions; if we choose a value of x_3 and a value of x_4, then the corresponding values of x_1 and x_2 are uniquely determined. ∎

EXAMPLE 4 Gauss elimination if a unique solution exists

Solve the system

$$\begin{aligned} -x_1 + x_2 + 2x_3 &= 2 \\ 3x_1 - x_2 + x_3 &= 6 \\ -x_1 + 3x_2 + 4x_3 &= 4 \end{aligned} \qquad \begin{bmatrix} -1 & 1 & 2 & 2 \\ 3 & -1 & 1 & 6 \\ -1 & 3 & 4 & 4 \end{bmatrix}$$

First Step. Elimination of x_1 from the second and third equations gives

$$\begin{aligned} -x_1 + x_2 + 2x_3 &= 2 \\ 2x_2 + 7x_3 &= 12 \\ 2x_2 + 2x_3 &= 2 \end{aligned} \qquad \begin{bmatrix} -1 & 1 & 2 & 2 \\ 0 & 2 & 7 & 12 \\ 0 & 2 & 2 & 2 \end{bmatrix} \begin{matrix} \\ \text{Row 2 + 3 Row 1} \\ \text{Row 3 - Row 1} \end{matrix}$$

Second Step. Elimination of x_2 from the third equation gives

$$\begin{aligned} -x_1 + x_2 + 2x_3 &= 2 \\ 2x_2 + 7x_3 &= 12 \\ -5x_3 &= -10 \end{aligned} \qquad \begin{bmatrix} -1 & 1 & 2 & 2 \\ 0 & 2 & 7 & 12 \\ 0 & 0 & -5 & -10 \end{bmatrix} \begin{matrix} \\ \\ \text{Row 3 - Row 2} \end{matrix}$$

Back Substitution. Beginning with the last equation, we obtain successively $x_3 = 2$, $x_2 = -1$, $x_1 = 1$. We see that the system has a unique solution. ∎

EXAMPLE 5 Gauss elimination if no solution exists

What will happen if we apply the Gauss elimination to a linear system that has no solution? The answer is that in this case the method will show this fact by producing a contradiction. For instance, consider

$$\begin{aligned} \boxed{3x_1} + 2x_2 + x_3 &= 3 \\ \boxed{2x_1} + x_2 + x_3 &= 0 \\ \boxed{6x_1} + 2x_2 + 4x_3 &= 6 \end{aligned} \qquad \begin{bmatrix} 3 & 2 & 1 & 3 \\ 2 & 1 & 1 & 0 \\ 6 & 2 & 4 & 6 \end{bmatrix}$$

First Step. Elimination of x_1 from the second and third equations by subtracting

2/3 times the first equation from the second equation,

6/3 = 2 times the first equation from the third equation.

This gives

$$\begin{aligned} 3x_1 + 2x_2 + x_3 &= 3 \\ \boxed{-\tfrac{1}{3}x_2} + \tfrac{1}{3}x_3 &= -2 \\ \boxed{-2x_2} + 2x_3 &= 0 \end{aligned} \qquad \begin{bmatrix} 3 & 2 & 1 & 3 \\ 0 & -\frac{1}{3} & \frac{1}{3} & -2 \\ 0 & -2 & 2 & 0 \end{bmatrix}$$

Second Step. Elimination of x_2 from the third equation gives

$$\begin{aligned} 3x_1 + 2x_2 + x_3 &= 3 \\ -\tfrac{1}{3}x_2 + \tfrac{1}{3}x_3 &= -2 \\ 0 &= 12 \end{aligned} \qquad \begin{bmatrix} 3 & 2 & 1 & 3 \\ 0 & -\frac{1}{3} & \frac{1}{3} & -2 \\ 0 & 0 & 0 & 12 \end{bmatrix}$$

This shows that the system has no solution. ∎

The form of the system and of the matrix in the last step of the Gauss elimination is called the **echelon form.** Thus in Example 5 the echelon forms of the coefficient matrix and the augmented matrix are

$$\begin{bmatrix} 3 & 2 & 1 \\ 0 & -\frac{1}{3} & \frac{1}{3} \\ 0 & 0 & 0 \end{bmatrix} \quad \text{and} \quad \begin{bmatrix} 3 & 2 & 1 & 3 \\ 0 & -\frac{1}{3} & \frac{1}{3} & -2 \\ 0 & 0 & 0 & 12 \end{bmatrix}.$$

At the end of the Gauss elimination (before the back substitution) the reduced system will have the form

$$\begin{aligned} a_{11}x_1 + a_{12}x_2 + \cdots\cdots + a_{1n}x_n &= b_1 \\ c_{22}x_2 + \cdots\cdots + c_{2n}x_n &= b_2{}^* \\ &\vdots \\ k_{rr}x_r + \cdots + k_{rn}x_n &= \tilde{b}_r \\ 0 &= \tilde{b}_{r+1} \\ &\vdots \\ 0 &= \tilde{b}_m \end{aligned} \tag{8}$$

where $r \leqq m$ (and $a_{11} \neq 0$, $c_{22} \neq 0, \cdots, k_{rr} \neq 0$). From this we see that with respect to solutions of this system (8), there are three possible cases:

(a) No solution if $r < m$ and one of the numbers $\tilde{b}_{r+1}, \cdots, \tilde{b}_m$ is not zero. This is illustrated by Example 5, where $r = 2 < m = 3$ and $\tilde{b}_{r+1} = \tilde{b}_3 = 12$.

(b) Precisely one solution if $r = n$ and $\tilde{b}_{r+1}, \cdots, \tilde{b}_m$, if present, are zero. This solution is obtained by solving the nth equation of (8) for x_n, then the $(n - 1)$th equation for x_{n-1}, and so on up the line. See Example 2, where $r = n = 3$ and $m = 4$.

(c) Infinitely many solutions if $r < n$ and $\tilde{b}_{r+1}, \cdots, \tilde{b}_m$, if present, are zero. Then any of these solutions is obtained by choosing values at pleasure for the unknowns $x_{r+1}, \cdots, x_n$, solving the rth equation for x_r, then the $(r - 1)$th equation for x_{r-1}, and so on up the line. Example 3 illustrates this case.

Elementary Row Operations

To justify the Gauss elimination as a method of solving linear systems, we first introduce two related concepts.

Elementary operations for equations

Interchange of two equations
Multiplication of an equation by a nonzero constant
Addition of a constant multiple of one equation to another equation.

To these correspond the following

Elementary row operations for matrices

Interchange of two rows
Multiplication of a row by a nonzero constant
Addition of a constant multiple of one row to another row.

The Gauss elimination consists of the use of the third of these operations[4] (for getting zeros) and of the first (in pivoting).

Now call a linear system S_1 **row-equivalent** to a linear system S_2 if S_1 can be obtained from S_2 by (finitely many!) elementary row operations. Clearly, the system produced by the Gauss elimination at the end is row-equivalent to the original system to be solved. Hence the desired justification of the Gauss elimination as a solution method now follows from the subsequent theorem, which implies that the Gauss elimination gives all solutions of the *original* system.

Theorem 1 (Row-equivalent systems)

Row-equivalent linear systems have the same sets of solutions.

Proof. The interchange of two equations does not alter the solution set. Neither does the multiplication of an equation by a (nonzero!) constant c, because multiplication of the new equation by $1/c$ produces the original equation. Similarly for the addition of an equation E_1 to an equation E_2, since by adding $-E_1$ (the equation obtained from E_1 by multiplying E_1 by -1) to the equation resulting from the addition we get back the original equation. ∎

This justifies the Gauss elimination. Numerical aspects of it are discussed in Sec. 19.1 (which is independent of other sections on numerical methods) and popular variants of it (Doolittle's, Crout's, and Cholesky's methods) in Sec. 19.2.

Problem Set 7.4

Solve the following linear systems by the Gauss elimination.

1. $2x + 3y = 4$
$3x + 2y = -4$

2. $3x + 2y = -17$
$10x + y = 0$

3. $-x + 2y = 4$
$3x + 4y = 38$

[4] In the Gauss elimination we said "subtraction of a constant multiple" (rather than "addition"), as being more suggestive in getting zeros; of course, this is a mere matter of language.

4. $$\begin{aligned} x + 2y - 8z &= 0 \\ 2x - 3y + 5z &= 0 \\ 3x + 2y - 12z &= 0 \end{aligned}$$

5. $$\begin{aligned} 3x - y + z &= -2 \\ x + 5y + 2z &= 6 \\ 2x + 3y + z &= 0 \end{aligned}$$

6. $$\begin{aligned} 7x - y - 2z &= 0 \\ 9x - y - 3z &= 0 \\ 2x + 4y - 7z &= 0 \end{aligned}$$

7. $$\begin{aligned} x + y + z &= -1 \\ 4y + 6z &= 6 \\ y + z &= 1 \end{aligned}$$

8. $$\begin{aligned} 5x + 3y &= 22 \\ -4x + 7y &= 20 \\ 9x - 2y &= 15 \end{aligned}$$

9. $$\begin{aligned} 4y + 3z &= 13 \\ x - 2y + z &= 3 \\ 3x + 5y &= 11 \end{aligned}$$

10. $$\begin{aligned} 7x - 4y - 2z &= -6 \\ 16x + 2y + z &= 3 \end{aligned}$$

11. $$\begin{aligned} x - 3y + 2z &= 2 \\ 5x - 15y + 7z &= 10 \end{aligned}$$

12. $$\begin{aligned} 3x - 3y - 7z &= -4 \\ x - y + 2z &= 3 \end{aligned}$$

13. $$\begin{aligned} 3w - 6x - y - z &= 0 \\ w - 2x + 5y - 3z &= 0 \\ 2w - 4x + 3y - z &= 3 \end{aligned}$$

14. $$\begin{aligned} 4w + 3x - 9y + z &= 1 \\ -w + 2x - 13y + 3z &= 3 \\ 3w - x + 8y - 2z &= -2 \end{aligned}$$

15. $$\begin{aligned} w + x + y &= 6 \\ -3w - 17x + y + 2z &= 2 \\ 4w - 17x + 8y - 5z &= 2 \\ -5x - 2y + z &= 2 \end{aligned}$$

16. $$\begin{aligned} w - x + 3y - 3z &= 3 \\ -5w + 2x - 5y + 4z &= -5 \\ -3w - 4x + 7y - 2z &= 7 \\ 2w + 3x + y - 11z &= 1 \end{aligned}$$

Models of electrical networks

Using Kirchhoff's laws (see Example 2), find the currents in the following networks.

17.

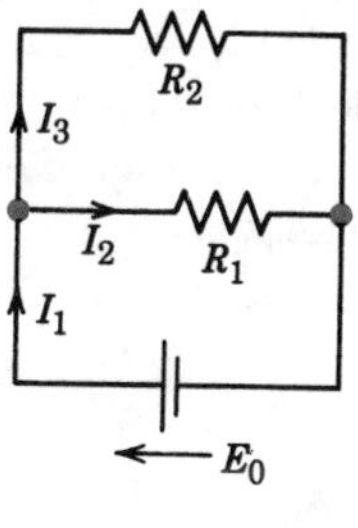

18.

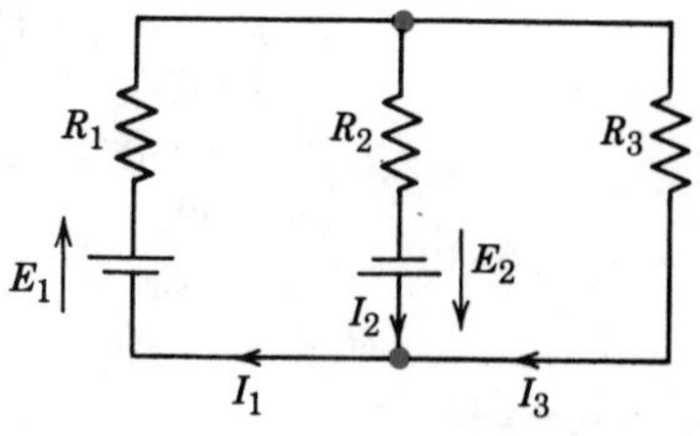

19.

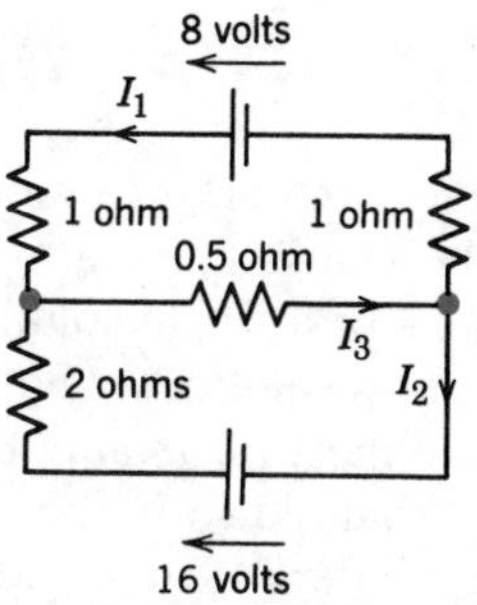

20.

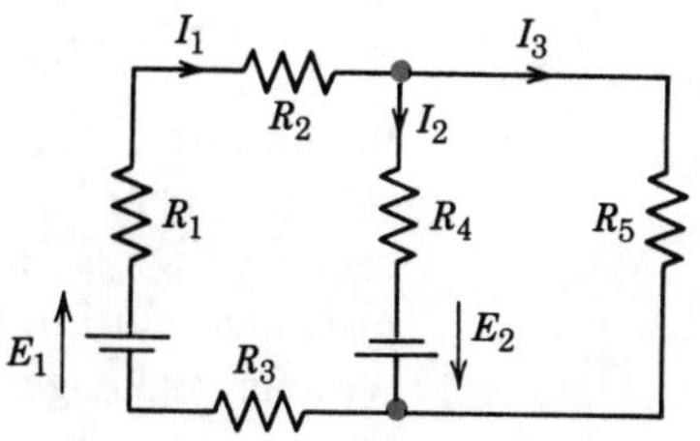

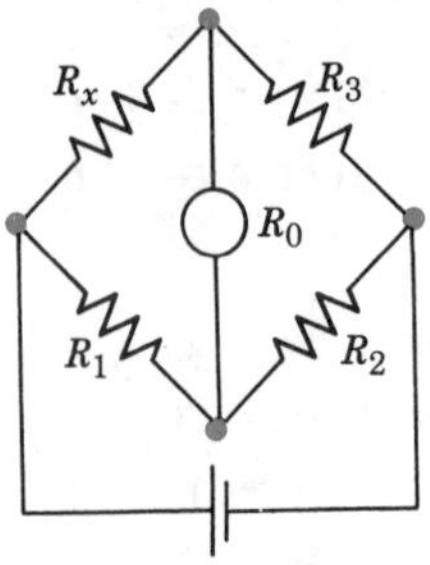

Problem 21
Wheatstone bridge

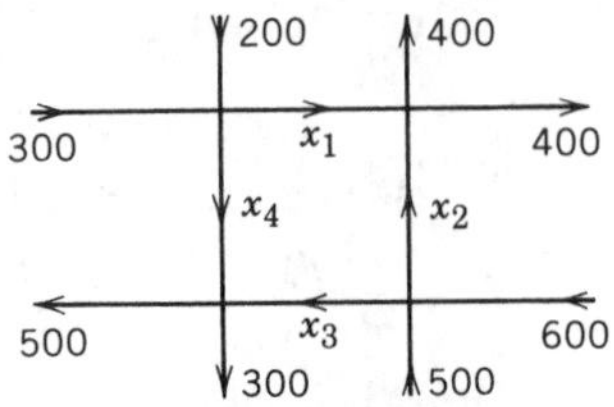

Problem 22
Net of one-way streets

21. **(Wheatstone bridge)** Show that if $R_x/R_3 = R_1/R_2$ in the figure, then $I = 0$. (R_0 is the resistance of the instrument by which I is measured.)
22. **(Traffic flow)** Methods of electrical circuit analysis have applications to other fields. For instance, applying the analog of Kirchhoff's current law, find the traffic flow (cars per hour) in the net of one-way streets (in the directions indicated by the arrows) shown in the figure. Is the solution unique?
23. **(Models of markets)** Determine the equilibrium solution ($D_1 = S_1$, $D_2 = S_2$) of the two-commodity market with linear model

$$D_1 = 40 - 2P_1 - P_2, \qquad S_1 = 4P_1 - P_2 + 4$$

$$D_2 = 5P_1 - 2P_2 + 16, \qquad S_2 = 3P_2 - 4$$

where D, S, P mean demand, supply, price, and the subscripts 1 and 2 refer to the first and second commodity, respectively.

24. **(Equivalence relation)** By definition, an *equivalence relation* on a set is a relation satisfying three conditions:
(1) Each element A of the set is equivalent to itself.
(2) If A is equivalent to B, then B is equivalent to A.
(3) If A is equivalent to B and B is equivalent to C, then A is equivalent to C.
For instance, equality is an equivalence relation on the set of real numbers. Show that row equivalence satisfies these three conditions.

7.5 Linear Independence. Vector Space. Rank of a Matrix

In the last section we explained the most important practical method for solving linear systems of equations, the Gauss elimination. We have also seen that a system may have no solutions or a single solution or more than just one solution (and then infinitely many solutions). So we ask whether we can make general statements about these *problems of existence and uniqueness*. The answer is yes, and we shall do so in the next section. For this we shall need the concepts of linear independence and rank, which we now introduce; these are of great general importance, far beyond our present discussion.

Linear Independence and Dependence of Vectors

Given any set of m vectors[5] $\mathbf{a}_{(1)}, \cdots, \mathbf{a}_{(m)}$ (with the same number of components), a **linear combination** of these vectors is an expression of the form

$$c_1\mathbf{a}_{(1)} + \cdots + c_m\mathbf{a}_{(m)}$$

where $c_1, \cdots, c_m$ are any scalars.[6] Now consider the equation

$$c_1\mathbf{a}_{(1)} + c_2\mathbf{a}_{(2)} + \cdots + c_m\mathbf{a}_{(m)} = \mathbf{0}. \tag{1}$$

Clearly, this holds if we choose all c's zero, because then it becomes $\mathbf{0} = \mathbf{0}$. If this is the only m-tuple of scalars for which (1) holds, then our vectors $\mathbf{a}_{(1)}, \cdots, \mathbf{a}_{(m)}$ are said to form a *linearly independent set* or, more briefly, we call them **linearly independent.** Otherwise, if (1) also holds with scalars not all zero, we call these vectors **linearly dependent,** because then we can express (at least) one of them as a linear combination of the others; for instance, if (1) holds with, say, $c_1 \neq 0$, we can solve (1) for $\mathbf{a}_{(1)}$:

$$\mathbf{a}_{(1)} = k_2\mathbf{a}_{(2)} + \cdots + k_m\mathbf{a}_{(m)} \qquad \text{where } k_j = -c_j/c_1$$

(and some or even all k's may be zero).

EXAMPLE 1 **Linear independence and dependence**

The three vectors

$$\mathbf{a}_{(1)} = [\;3 \quad 0 \quad 2 \quad 2]$$

$$\mathbf{a}_{(2)} = [-6 \quad 42 \quad 24 \quad 54]$$

$$\mathbf{a}_{(3)} = [\;21 \quad -21 \quad 0 \quad -15]$$

are linearly dependent because

$$6\mathbf{a}_{(1)} - \tfrac{1}{2}\mathbf{a}_{(2)} - \mathbf{a}_{(3)} = \mathbf{0}.$$

Although this is easily checked (do it!), it is not so easy to discover; however, a method for finding out about linear independence and dependence follows below.

The first two of the three vectors are linearly independent because $c_1\mathbf{a}_{(1)} + c_2\mathbf{a}_{(2)} = \mathbf{0}$ implies $c_2 = 0$ (from the second components) and then $c_1 = 0$ (from any other pair of components). ∎

Vector Space, Dimension, Basis

Given m vectors $\mathbf{a}_{(1)}, \cdots, \mathbf{a}_{(m)}$ with n components each, as before, we can form the set V of all linear combinations of these vectors. V is called the **span** of these m vectors.

V is a **vector space.**[7] By definition, this means that V is a set of vectors with the two algebraic operations of *addition* and *scalar multiplication* defined for these vectors such that the following holds.

---

[5] Write simply $\mathbf{a}_1, \cdots, \mathbf{a}_m$ if you wish, but keep in mind that these are *vectors*, not vector *components*.

[6] In this section, scalars will be *real* numbers.

[7] Here, we give just what we need in the next section. General vector spaces follow in Sec. 7.15.

1. The sum $\mathbf{a} + \mathbf{b}$ of any vectors $\mathbf{a}$ and $\mathbf{b}$ in V is also in V and the product $k\mathbf{a}$ of any vector $\mathbf{a}$ in V and scalar k is also in V.

2. For all vectors and scalars we have the familiar rules

$$\mathbf{a} + \mathbf{b} = \mathbf{b} + \mathbf{a}$$

$$(\mathbf{a} + \mathbf{b}) + \mathbf{c} = \mathbf{a} + (\mathbf{b} + \mathbf{c}) \qquad (\text{written } \mathbf{a} + \mathbf{b} + \mathbf{c})$$

$$\mathbf{a} + \mathbf{0} = \mathbf{a}$$

$$\mathbf{a} + (-\mathbf{a}) = \mathbf{0}$$

as well as

$$k(\mathbf{a} + \mathbf{b}) = k\mathbf{a} + k\mathbf{b}$$

$$(k + \ell)\mathbf{a} = k\mathbf{a} + \ell\mathbf{a}$$

$$k(\ell\mathbf{a}) = (k\ell)\mathbf{a} \qquad (\text{written } k\ell\mathbf{a})$$

$$1\mathbf{a} = \mathbf{a}.$$

(For our vectors, these rules follow from (1) and (2) in Sec. 7.2—after all, vectors are special matrices!)

The maximum number of linearly independent vectors in V is called the **dimension** of V and is denoted by dim V.

Clearly, if those given m vectors are linearly independent, then dim $V = m$; and if they are linearly dependent, then dim $V < m$.

A linearly independent set in V consisting of a maximum possible number of vectors in V is called a **basis** for V. Thus the number of vectors of a basis for V equals dim V.

EXAMPLE 2 Vector space, Dimension, Basis

The span of the three vectors in Example 1 is a vector space of dimension 2, and a basis is $\mathbf{a}_{(1)}$, $\mathbf{a}_{(2)}$, for instance, or $\mathbf{a}_{(1)}$, $\mathbf{a}_{(3)}$, etc. ∎

By the **real *n*-dimensional vector space R^n** we mean the space of all vectors with n *real* numbers as components ("*real vectors*") and real numbers as scalars. This is a standard name and notation. Hence each such vector is an ordered n-tuple of real numbers, as we know.

Thus for $n = 3$ we get R^3 consisting of ordered triples ("vectors in 3-space"), and for $n = 2$ we get R^2 consisting of ordered pairs ("vectors in the plane"). In Chaps. 8 and 9 we shall see that these special cases provide wide areas for applications of geometry, mechanics, and calculus which are of basic importance to the engineer and physicist.

Rank of a Matrix

The maximum number of linearly independent row vectors of a matrix $\mathbf{A} = [a_{jk}]$ is called the **rank** of $\mathbf{A}$ and is denoted by

$$\text{rank } \mathbf{A}.$$

EXAMPLE 3 Rank

The matrix

$$\mathbf{A} = \begin{bmatrix} 3 & 0 & 2 & 2 \\ -6 & 42 & 24 & 54 \\ 21 & -21 & 0 & -15 \end{bmatrix} \tag{2}$$

has rank 2, because Example 1 shows that the first two row vectors are linearly independent, whereas all three row vectors are linearly dependent. ∎

Note further that rank $\mathbf{A} = 0$ if and only if $\mathbf{A} = \mathbf{0}$. This follows directly from the definition.

In our proposed discussion of the existence and uniqueness of solutions of systems of linear equations we shall need the following very important

Theorem 1 (Rank in terms of column vectors)

The rank of a matrix **A** *equals the maximum number of linearly independent* ***column*** *vectors of* **A**.

Hence **A** *and its transpose* $\mathbf{A}^\mathsf{T}$ *have the same rank.*

Proof. Let $\mathbf{A} = [a_{jk}]$ and let rank $\mathbf{A} = r$. Then, by definition, **A** has a linearly independent set of r row vectors, call them $\mathbf{v}_{(1)}, \cdots, \mathbf{v}_{(r)}$, and all row vectors $\mathbf{a}_{(1)}, \cdots, \mathbf{a}_{(m)}$ of **A** are linear combinations of those independent ones, say,

$$\begin{aligned} \mathbf{a}_{(1)} &= c_{11}\mathbf{v}_{(1)} + c_{12}\mathbf{v}_{(2)} + \cdots + c_{1r}\mathbf{v}_{(r)} \\ \mathbf{a}_{(2)} &= c_{21}\mathbf{v}_{(1)} + c_{22}\mathbf{v}_{(2)} + \cdots + c_{2r}\mathbf{v}_{(r)} \\ &\vdots \\ \mathbf{a}_{(m)} &= c_{m1}\mathbf{v}_{(1)} + c_{m2}\mathbf{v}_{(2)} + \cdots + c_{mr}\mathbf{v}_{(r)} \end{aligned}$$

These are vector equations. Each of them is equivalent to n equations for corresponding components. Denoting the components of $\mathbf{v}_{(1)}$ by $v_{11}, \cdots, v_{1n}$, the components of $\mathbf{v}_{(2)}$ by $v_{21}, \cdots, v_{2n}$, etc., and similarly for the vectors on the left, we thus have

$$\begin{aligned} a_{1k} &= c_{11}v_{1k} + c_{12}v_{2k} + \cdots + c_{1r}v_{rk} \\ a_{2k} &= c_{21}v_{1k} + c_{22}v_{2k} + \cdots + c_{2r}v_{rk} \\ &\vdots \\ a_{mk} &= c_{m1}v_{1k} + c_{m2}v_{2k} + \cdots + c_{mr}v_{rk} \end{aligned}$$

where $k = 1, \cdots, n$. This can be written

$$\begin{bmatrix} a_{1k} \\ a_{2k} \\ \vdots \\ a_{mk} \end{bmatrix} = v_{1k}\begin{bmatrix} c_{11} \\ c_{21} \\ \vdots \\ c_{m1} \end{bmatrix} + v_{2k}\begin{bmatrix} c_{12} \\ c_{22} \\ \vdots \\ c_{m2} \end{bmatrix} + \cdots + v_{rk}\begin{bmatrix} c_{1r} \\ c_{2r} \\ \vdots \\ c_{mr} \end{bmatrix}$$

where $k = 1, \cdots, n$. The vector on the left is the kth column vector of **A**. Hence the equation shows that each column vector of **A** is a linear combination of the r vectors on the right. Hence the maximum number of linearly independent column vectors of **A** cannot exceed r, which is the maximum number of linearly independent row vectors of **A**, by the definition of rank.

Now the same conclusion applies to the transpose $\mathbf{A}^{\mathsf{T}}$ of **A**. Since the row vectors of $\mathbf{A}^{\mathsf{T}}$ are the column vectors of **A**, and the column vectors of $\mathbf{A}^{\mathsf{T}}$ are the row vectors of **A**, that conclusion means that the maximum number of linearly independent row vectors of **A** (which is r) cannot exceed the maximum number of linearly independent column vectors of **A**. Hence that number must equal r, and the proof is complete. ∎

EXAMPLE 4 Illustration of Theorem 1

What does Theorem 1 mean with respect to our matrix **A** in (2)? Since we have rank $\mathbf{A} = 2$, the column vectors should contain two linearly independent ones, and the other two should be linear combinations of them. Indeed, the first two column vectors are linearly independent, and

$$\begin{bmatrix} 2 \\ 24 \\ 0 \end{bmatrix} = \frac{2}{3}\begin{bmatrix} 3 \\ -6 \\ 21 \end{bmatrix} + \frac{2}{3}\begin{bmatrix} 0 \\ 42 \\ -21 \end{bmatrix} \quad \text{and} \quad \begin{bmatrix} 2 \\ 54 \\ -15 \end{bmatrix} = \frac{2}{3}\begin{bmatrix} 3 \\ -6 \\ 21 \end{bmatrix} + \frac{29}{21}\begin{bmatrix} 0 \\ 42 \\ -21 \end{bmatrix}.$$

This is easy to verify but not so easy to see. Imagining that $\mathbf{A}^{\mathsf{T}}$ were given, we realize that the determination of a rank by a direct application of the definition is not the proper way, unless a matrix is sufficiently simple. This suggests asking whether we can "simplify" (transform) a matrix without altering its rank. The answer is yes, as we show next. ∎

The span of the row vectors of a matrix **A** is called the **row space** of **A** and the span of the columns the **column space** of **A**. From this and Theorem 1 we have

Theorem 2 (Row space and column space)

The row space and the column space of a matrix **A** *have the same dimension, equal to* rank **A**.

Invariance of Rank Under Elementary Row Operations

We claim that *elementary row operations* (Sec. 7.4) *do not alter the rank of a matrix* **A**.

For the first operation (interchange of two row vectors) this is clear. The second operation (multiplication of a row vector by a nonzero constant) does not alter the rank either, since it does not alter the maximum number of linearly independent row vectors. Finally, the third operation is the addition of c times a row vector $\mathbf{a}_{(j)}$, say, to another row vector, say, $\mathbf{a}_{(i)}$. This produces a matrix that differs from **A** only in the ith row vector, which is of the form $\mathbf{a}_{(i)} + c\mathbf{a}_{(j)}$, a linear combination of the row vectors $\mathbf{a}_{(i)}$ and $\mathbf{a}_{(j)}$, so that the number of linearly independent row vectors remains the same. Hence the new matrix has the same rank as **A**. Remembering from the previous section that *row-equivalent matrices* are those that can be obtained from each other by finitely many elementary row operations, our result is

Theorem 3 (Row-equivalent matrices)

Row-equivalent matrices have the same rank.

This theorem tells us what we can do to determine the rank of a matrix **A**, namely, we can reduce **A** to echelon form (Sec. 7.4), using the technique of the Gauss elimination, because this leaves the rank unchanged, by Theorem 2, and from the echelon form we can recognize the rank directly.

EXAMPLE 5 Determination of rank

For the matrix in Example 3 we obtain successively

$$\mathbf{A} = \begin{bmatrix} 3 & 0 & 2 & 2 \\ -6 & 42 & 24 & 54 \\ 21 & -21 & 0 & -15 \end{bmatrix} \quad \text{(given)}$$

$$\begin{bmatrix} 3 & 0 & 2 & 2 \\ 0 & 42 & 28 & 58 \\ 0 & -21 & -14 & -29 \end{bmatrix} \quad \begin{matrix} \\ \text{Row 2} + 2 \text{ Row 1} \\ \text{Row 3} - 7 \text{ Row 1} \end{matrix}$$

$$\begin{bmatrix} 3 & 0 & 2 & 2 \\ 0 & 42 & 28 & 58 \\ 0 & 0 & 0 & 0 \end{bmatrix} \quad \begin{matrix} \\ \\ \text{Row 3} + \tfrac{1}{2} \text{ Row 2} \end{matrix}$$

The last matrix is in echelon form. From the row vectors and Theorem 3 we see immediately that rank $\mathbf{A} \leqq 2$, and rank $\mathbf{A} = 2$ by Theorem 1, since the first two column vectors are certainly linearly independent. ■

This method of determining rank has practical applications in connection with the determination of linear dependence and independence of vectors. The key to this is the following theorem, which results immediately from the definition of rank.

Theorem 4 (Linear dependence and independence)

p vectors $\mathbf{x}_{(1)}, \cdots, \mathbf{x}_{(p)}$ (with n components each) are linearly independent if the matrix with row vectors $\mathbf{x}_{(1)}, \cdots, \mathbf{x}_{(p)}$ has rank p; they are linearly dependent if that rank is less than p.

Since each of those p vectors has n components, that matrix, call it **A**, has p rows and n columns; and if $n < p$, then by Theorem 1 we must have rank $\mathbf{A} \leqq n < p$, so that Theorem 4 yields the following result, which one should keep in mind.

Theorem 5 *p vectors with $n < p$ components are always linearly dependent.*

For instance, three or more vectors in the plane are linearly dependent. Similarly, four or more vectors in space are linearly dependent.

By the definition of dimension, we also have

Theorem 6 *The vector space R^n consisting of all vectors with n components has dimension n.*

Basic applications of rank follow in the next section.

Problem Set 7.5

Vector spaces. Is the given set of vectors a vector space? (Give a reason.) If your answer is yes, determine the dimension and find a basis.

1. All vectors $[v_1 \;\; v_2 \;\; v_3]^\mathsf{T}$ in R^3 such that $v_1 + 2v_2 = 0$.

2. All vectors in R^4 such that $v_1 + v_2 = 0$, $v_3 + v_4 = 0$.

3. All vectors in R^3 satisfying $v_1 + v_2 + v_3 = 1$.

4. All vectors in R^2 such that $v_1 + v_2 = k$ (= *const*).

5. All real numbers.

6. All vectors in R^5 such that $v_1 + v_2 = 0$, $v_1 + 2v_2 + v_3 - v_4 = 0$.

7. All ordered quintuples of positive real numbers.

8. All vectors in R^4 such that $v_1 = 0$, $v_2 + v_3 + v_4 \geqq 0$.

9. All vectors in R^n with the first $n - 1$ components zero.

10. **(Subspace)** A nonempty subset W of a vector space V is called a *subspace* of V if W is itself a vector space with respect to the algebraic operations defined in V. Give examples of one- and two-dimensional subspaces of R^3.

Rank. Find the rank by inspection or by the method in Example 5.

11. $\begin{bmatrix} 7 & 7 \\ \frac{1}{2} & -\frac{1}{2} \\ 0 & 0 \end{bmatrix}$ **12.** $\begin{bmatrix} 3 & 2 & -9 \\ -6 & -4 & 18 \\ 12 & 8 & -36 \end{bmatrix}$ **13.** $\begin{bmatrix} 2 & 0 & 9 & 2 \\ 1 & 4 & 6 & 0 \\ 3 & 5 & 7 & 1 \end{bmatrix}$

14. $\begin{bmatrix} 0 & 1 & 0 \\ 1 & 0 & 0 \\ 0 & 0 & 1 \end{bmatrix}$ **15.** $\begin{bmatrix} a & b & c \\ b & a & c \end{bmatrix}$, $a \neq \pm b$ **16.** $\begin{bmatrix} 8 & 1 & 3 & 6 \\ 0 & 3 & 2 & 2 \\ -8 & -1 & -3 & 4 \end{bmatrix}$

17. $\begin{bmatrix} 3 & 1 & 4 \\ 0 & 5 & 8 \\ -3 & 4 & 4 \\ 1 & 2 & 4 \end{bmatrix}$ **18.** $\begin{bmatrix} 6 & 0 & 2 \\ 0 & 8 & 3 \\ 2 & 7 & 5 \\ 5 & 5 & 0 \end{bmatrix}$ **19.** $\begin{bmatrix} 1 & 0 & 3 & 0 \\ 0 & 2 & 4 & 6 \\ 3 & 0 & 5 & 1 \\ 2 & 3 & 0 & 1 \end{bmatrix}$

20. Show by an example that rank $\mathbf{A}$ = rank $\mathbf{B}$ does *not* imply rank $\mathbf{A}^2$ = rank $\mathbf{B}^2$.

21. Show that rank $\mathbf{B}^\mathsf{T}\mathbf{A}^\mathsf{T}$ = rank $\mathbf{AB}$.

22. Prove that if the row vectors of an $n \times n$ matrix $\mathbf{A}$ are linearly independent, then so are the column vectors of $\mathbf{A}$ (and vice versa).

23. Prove that if $\mathbf{A}$ is not square, then either the row vectors or the column vectors of $\mathbf{A}$ are linearly dependent.

24. Prove that row-equivalent matrices have the same row space.

25. Find a basis of the row space and of the column space of the matrix in Prob. 11.

26. Do the same task as in Prob. 25 for the matrix in Prob. 17.

Linear independence. State whether the given vectors are linearly independent or dependent.

27. [1 5 3], [2 4 6], [3 9 11]

28. [4 3 9], [0 0 0], [1 $\frac{3}{2}$ $\frac{2}{3}$]

29. [1 0], [1 2], [3 4]

30. [1 1 0], [0 1 1], [1 0 1]

31. [1 2 3], [4 5 6], [7 8 9]

32. [3 −1 4], [6 7 5], [9 6 9]

33. [9 0 9], [0 6 6], [3 3 0]

34. [2 1 0 6], [1 9 9 0]

35. [3 0 2 4 5], [7 2 6 1 0], [1 2 2 −7 −10]

7.6 Linear Systems: General Properties of Solutions

Using the concept of the rank of a matrix, as defined in the last section, we can now settle the issue of existence and uniqueness of solutions of linear systems. The central theorem (which the student should memorize!) is as follows. (For illustrative examples, see Sec. 7.4.)

Theorem 1 Fundamental Theorem for linear systems

(a) *A linear system of m equations*

$$\begin{aligned} a_{11}x_1 + a_{12}x_2 + \cdots + a_{1n}x_n &= b_1 \\ a_{21}x_1 + a_{22}x_2 + \cdots + a_{2n}x_n &= b_2 \\ \cdots\cdots\cdots\cdots\cdots\cdots\cdots\cdots\cdots \\ a_{m1}x_1 + a_{m2}x_2 + \cdots + a_{mn}x_n &= b_m \end{aligned} \tag{1}$$

in n unknowns $x_1, \cdots, x_n$ has solutions if and only if the coefficient matrix **A** *and the augmented matrix* $\tilde{\mathbf{A}}$, *that is,*

$$\mathbf{A} = \begin{bmatrix} a_{11} & \cdots & a_{1n} \\ \cdot & \cdots & \cdot \\ \cdot & \cdots & \cdot \\ a_{m1} & \cdots & a_{mn} \end{bmatrix} \quad \text{and} \quad \tilde{\mathbf{A}} = \begin{bmatrix} a_{11} & \cdots & a_{1n} & b_1 \\ \cdot & \cdots & \cdot & \cdot \\ \cdot & \cdots & \cdot & \cdot \\ a_{m1} & \cdots & a_{mn} & b_m \end{bmatrix},$$

have the same rank.

(b) *If this rank r equals n, the system* (1) *has precisely one solution.*

(c) *If $r < n$, the system* (1) *has infinitely many solutions, all of which are obtained by determining r suitable unknowns (whose submatrix of coefficients must have rank r) in terms of the remaining $n - r$ unknowns, to which arbitrary values can be assigned.*

(d) *If solutions exist, they can all be obtained by the Gauss elimination* (see Sec. 7.4). (This elimination may be started without first looking at the ranks of **A** and $\tilde{\mathbf{A}}$, since it will automatically reveal whether or not solutions exist; see, for instance, Example 5 in Sec. 7.4.)

Proof. **(a)** We can write the system (1) in the form

(1) $$\mathbf{Ax} = \mathbf{b}$$

or in terms of the column vectors $\mathbf{c}_{(1)}, \cdots, \mathbf{c}_{(n)}$ of **A**:

(2) $$\mathbf{c}_{(1)}x_1 + \mathbf{c}_{(2)}x_2 + \cdots + \mathbf{c}_{(n)}x_n = \mathbf{b}.$$

Since $\tilde{\mathbf{A}}$ is obtained by attaching to **A** the additional column **b**, Theorem 1 in Sec. 7.5 implies that rank $\tilde{\mathbf{A}}$ equals rank **A** or rank $\mathbf{A} + 1$. Now if (1) has a solution **x**, then (2) shows that **b** must be a linear combination of those column vectors. Hence rank $\tilde{\mathbf{A}}$ cannot exceed rank **A**, so that we must have rank $\tilde{\mathbf{A}}$ = rank **A**.

Conversely, if rank $\tilde{\mathbf{A}}$ = rank **A**, then **b** must be a linear combination of the column vectors of **A**, say,

$$\mathbf{b} = \alpha_1\mathbf{c}_{(1)} + \cdots + \alpha_n\mathbf{c}_{(n)}$$

since otherwise rank $\tilde{\mathbf{A}}$ = rank $\mathbf{A} + 1$. But this means that (1) has a solution, namely, $x_1 = \alpha_1, \cdots, x_n = \alpha_n$.

(b) If rank $\mathbf{A} = r = n$, then the set $C = \{\mathbf{c}_{(1)}, \cdots, \mathbf{c}_{(n)}\}$ is linearly independent, by Theorem 1 in Sec. 7.5. It follows that then the representation (2) of **b** is unique because

$$\mathbf{c}_{(1)}x_1 + \cdots + \mathbf{c}_{(n)}x_n = \mathbf{c}_{(1)}\tilde{x}_1 + \cdots + \mathbf{c}_{(n)}\tilde{x}_n$$

would imply

$$(x_1 - \tilde{x}_1)\mathbf{c}_{(1)} + \cdots + (x_n - \tilde{x}_n)\mathbf{c}_{(n)} = \mathbf{0}$$

and $x_1 - \tilde{x}_1 = 0, \cdots, x_n - \tilde{x}_n = 0$ by the linear independence. Hence the scalars $x_1, \cdots, x_n$ in (2) are uniquely determined, that is, the solution of (1) is unique.

(c) If rank **A** = rank $\tilde{\mathbf{A}} = r < n$, by Theorem 1, Sec. 7.5, there is a linearly independent set K of r column vectors of **A** such that the other $n - r$ column vectors of **A** are linear combinations of those vectors. We renumber the columns and unknowns, denoting the renumbered quantities by $\hat{}$, so that $\{\hat{\mathbf{c}}_{(1)}, \cdots, \hat{\mathbf{c}}_{(r)}\}$ is that linearly independent set K. Then (2) becomes

$$\hat{\mathbf{c}}_{(1)}\hat{x}_1 + \cdots + \hat{\mathbf{c}}_{(n)}\hat{x}_n = \mathbf{b},$$

$\hat{\mathbf{c}}_{(r+1)}, \cdots, \hat{\mathbf{c}}_{(n)}$ are linear combinations of the vectors of K, and so are the vectors $\hat{x}_{r+1}\hat{\mathbf{c}}_{(r+1)}, \cdots, \hat{x}_n\hat{\mathbf{c}}_{(n)}$. Expressing these vectors in terms of the vectors of K and collecting terms, we can thus write the system in the form

(3) $$\hat{\mathbf{c}}_{(1)}y_1 + \cdots + \hat{\mathbf{c}}_{(r)}y_r = \mathbf{b}$$

with $y_j = \hat{x}_j + \beta_j$, where β_j results from the terms $\hat{x}_{r+1}\hat{\mathbf{c}}_{(r+1)}, \cdots, \hat{x}_n\hat{\mathbf{c}}_{(n)}$; here, $j = 1, \cdots, r$. Since the system has a solution, there are $y_1, \cdots, y_r$ satisfying (3). These scalars are unique since K is linearly independent. Choosing $\hat{x}_{r+1}, \cdots, \hat{x}_n$ fixes the β_j and corresponding $\hat{x}_j = y_j - \beta_j$, where $j = 1, \cdots, r$.

(d) This was proved in Sec. 7.5 and is restated here as a reminder. ∎

The theorem is illustrated by the examples in Sec. 7.4: in Example 3 we have rank $\mathbf{A}$ = rank $\tilde{\mathbf{A}} = 2 < n = 4$ and can choose x_3 and x_4 arbitrarily; in Example 4 there is a unique solution since rank $\mathbf{A}$ = rank $\tilde{\mathbf{A}} = n = 3$; and in Example 5 there is no solution, since rank $\mathbf{A} = 2 <$ rank $\tilde{\mathbf{A}} = 3$.

The Homogeneous System

The system (1) is called **homogeneous** if all the b_j's on the right side are zero. Otherwise it is called **nonhomogeneous.** (See also Sec. 7.4.) From the Fundamental Theorem we readily obtain the following results.

Theorem 2 (Homogeneous system)
A homogeneous linear system

(4) $$\begin{aligned} a_{11}x_1 + a_{12}x_2 + \cdots + a_{1n}x_n &= 0 \\ a_{21}x_1 + a_{22}x_2 + \cdots + a_{2n}x_n &= 0 \\ \cdots\cdots\cdots\cdots\cdots\cdots\cdots\cdots\cdots \\ a_{m1}x_1 + a_{m2}x_2 + \cdots + a_{mn}x_n &= 0 \end{aligned}$$

always has the **trivial solution** $x_1 = 0, \cdots, x_n = 0$. *Nontrivial solutions exist if and only if* rank $\mathbf{A} < n$. *If* rank $\mathbf{A} = r < n$, *these solutions, together with* $\mathbf{x} = \mathbf{0}$, *form a vector space of dimension* $n - r$ *(see Sec. 7.5). In particular, if* $\mathbf{x}_{(1)}$ *and* $\mathbf{x}_{(2)}$ *are solution vectors of* (4), *then* $\mathbf{x} = c_1\mathbf{x}_{(1)} + c_2\mathbf{x}_{(2)}$, *where* c_1 *and* c_2 *are any scalars, is a solution vector of* (4). (*This* ***does not hold*** *for nonhomogeneous systems.*)

Proof. The first proposition is obvious and is in agreement with the fact that for a homogeneous system the matrix of the coefficients and the augmented matrix have the same rank. The solution vectors form a vector space because if $\mathbf{x}_{(1)}$ and $\mathbf{x}_{(2)}$ are any of them, then $\mathbf{Ax}_{(1)} = \mathbf{0}$, $\mathbf{Ax}_{(2)} = \mathbf{0}$, and this implies $\mathbf{A}(\mathbf{x}_{(1)} + \mathbf{x}_{(2)}) = \mathbf{Ax}_{(1)} + \mathbf{Ax}_{(2)} = \mathbf{0}$ as well as $\mathbf{A}(c\mathbf{x}_{(1)}) = c\mathbf{Ax}_{(1)} = \mathbf{0}$, where c is arbitrary. If rank $\mathbf{A} = r < n$, the Fundamental Theorem implies that we can choose $n - r$ suitable unknowns, call them $x_{r+1}, \cdots, x_n$, in an arbitrary fashion, and every solution is obtained in this way. It follows that a **basis of solutions** (that is, a basis for the vector space of these solutions) is $\mathbf{y}_{(1)}, \cdots, \mathbf{y}_{(n-r)}$, where the solution vector $\mathbf{y}_{(j)}$, $j = 1, \cdots, n - r$, is obtained by choosing $x_{r+j} = 1$ and the other $x_{r+1}, \cdots, x_n$ zero; the corresponding $x_1, \cdots, x_r$ are then determined. This proves that the vector space of all solutions has dimension $n - r$ and completes the proof. ∎

We mention that the vector space of all solutions of (4) is called the **null space** of the coefficient matrix **A**, because if we multiply any **x** in this null space by **A** we get **0**. The dimension of the null space is called the **nullity** of **A**. In terms of these concepts, Theorem 2 states that

$$\text{rank } \mathbf{A} + \text{nullity } \mathbf{A} = n \tag{5}$$

where n is the number of unknowns (number of columns of **A**). If we have rank $\mathbf{A} = n$, then nullity $\mathbf{A} = 0$, so that the system has only the trivial solution. If rank $\mathbf{A} = r < n$, then nullity $\mathbf{A} = n - r > 0$, so that we have nontrivial solutions which, together with **0**, form a vector space of dimension $n - r > 0$.

Note that rank $\mathbf{A} \leqq m$ in (4), by the definition, so that rank $\mathbf{A} < n$ when $m < n$. By Theorem 2 this proves the following theorem, which is of considerable practical importance.

Theorem 3 **(System with fewer equations than unknowns)**
A homogeneous system of linear equations with fewer equations than unknowns always has nontrivial solutions.

The Nonhomogeneous System

If a nonhomogeneous system of linear equations has solutions, their totality can be characterized as follows.

Theorem 4 **(Nonhomogeneous system)**
If a nonhomogeneous linear system of equations of the form (1) *has solutions, then all these solutions are of the form*

$$\mathbf{x} = \mathbf{x}_0 + \mathbf{x}_h$$

where $\mathbf{x}_0$ *is any fixed solution of* (1) *and* $\mathbf{x}_h$ *runs through all the solutions of the corresponding homogeneous system* (4).

Proof. Let **x** be any given solution of (1) and $\mathbf{x}_0$ an arbitrarily chosen solution of (1). Then $\mathbf{Ax} = \mathbf{b}$, $\mathbf{Ax}_0 = \mathbf{b}$ and, therefore,

$$\mathbf{A}(\mathbf{x} - \mathbf{x}_0) = \mathbf{Ax} - \mathbf{Ax}_0 = \mathbf{0}.$$

This shows that the difference $\mathbf{x} - \mathbf{x}_0$ of any solution **x** of (1) and any fixed solution $\mathbf{x}_0$ of (1) is a solution of (4), say, $\mathbf{x}_h$. Hence all solutions of (1) are obtained by letting $\mathbf{x}_h$ run through all the solutions of the homogeneous system (4), and the proof is complete. ∎

7.7 Inverse of a Matrix

In this section we consider exclusively* square *matrices.

The **inverse** of an $n \times n$ matrix $\mathbf{A} = [a_{jk}]$ is denoted by $\mathbf{A}^{-1}$ and is an $n \times n$ matrix such that

$$\mathbf{A}\mathbf{A}^{-1} = \mathbf{A}^{-1}\mathbf{A} = \mathbf{I}, \tag{1}$$

where $\mathbf{I}$ is the $n \times n$ unit matrix (see Sec. 7.3).

If $\mathbf{A}$ has an inverse, then $\mathbf{A}$ is called a **nonsingular matrix.** If $\mathbf{A}$ has no inverse, then $\mathbf{A}$ is called a **singular matrix.**

If $\mathbf{A}$ *has an inverse, the inverse is unique.*

Indeed, if both $\mathbf{B}$ and $\mathbf{C}$ are inverses of $\mathbf{A}$, then $\mathbf{AB} = \mathbf{I}$ and $\mathbf{CA} = \mathbf{I}$, so that we obtain the uniqueness from

$$\mathbf{B} = \mathbf{IB} = (\mathbf{CA})\mathbf{B} = \mathbf{C}(\mathbf{AB}) = \mathbf{CI} = \mathbf{C}.$$

We prove next that $\mathbf{A}$ has an inverse (is nonsingular) if and only if it has maximum possible rank n. The proof will also show that $\mathbf{Ax} = \mathbf{b}$ implies $\mathbf{x} = \mathbf{A}^{-1}\mathbf{b}$ provided $\mathbf{A}^{-1}$ exists, and thus give a motivation for the inverse as well as a relation to linear systems.[8]

Theorem 1 (Existence of the inverse)

The inverse $\mathbf{A}^{-1}$ *of an* $n \times n$ *matrix* $\mathbf{A}$ *exists if and only if* rank $\mathbf{A} = n$. *Hence* $\mathbf{A}$ *is nonsingular if* rank $\mathbf{A} = n$, *and is singular if* rank $\mathbf{A} < n$.

Proof. Consider the linear system

$$\mathbf{Ax} = \mathbf{b} \tag{2}$$

with the given matrix $\mathbf{A}$ as coefficient matrix. If the inverse exists, then multiplication from the left on both sides gives by (1)

$$\mathbf{A}^{-1}\mathbf{Ax} = \mathbf{x} = \mathbf{A}^{-1}\mathbf{b}.$$

This shows that (2) has a unique solution $\mathbf{x}$, so that $\mathbf{A}$ must have rank n by the Fundamental Theorem in the last section.

Conversely, let rank $\mathbf{A} = n$. Then by the same theorem, the system (2) has a unique solution $\mathbf{x}$ for any $\mathbf{b}$, and the Gauss elimination (in Sec. 7.4) shows that its components x_j are linear combinations of those of $\mathbf{b}$, so that we can write

$$\mathbf{x} = \mathbf{Bb}. \tag{3}$$

[8] But *not* a method of solving $\mathbf{Ax} = \mathbf{b}$ ***numerically,*** because the Gauss elimination (Sec. 7.4) requires fewer computations.

Substitution into (2) gives

$$\mathbf{Ax} = \mathbf{A(Bb)} = \mathbf{(AB)b} = \mathbf{Cb} = \mathbf{b} \qquad (\mathbf{C} = \mathbf{AB})$$

for any **b.** Hence $\mathbf{C} = \mathbf{AB} = \mathbf{I}$, the unit matrix. Similarly, if we substitute (2) into (3) we get

$$\mathbf{x} = \mathbf{Bb} = \mathbf{B(Ax)} = \mathbf{(BA)x}$$

for any **x** (and $\mathbf{b} = \mathbf{Ax}$). Hence $\mathbf{BA} = \mathbf{I}$. Together, $\mathbf{B} = \mathbf{A}^{-1}$ exists. ∎

Determination of the Inverse

We want to show that for practically determining the inverse $\mathbf{A}^{-1}$ of a nonsingular $n \times n$ matrix **A** we can use the Gauss elimination (Sec. 7.4), actually, a variant of it, called the **Gauss–Jordan elimination.**[9] Our idea is as follows. Using **A**, we form the n systems $\mathbf{Ax}_{(1)} = \mathbf{e}_{(1)}, \cdots, \mathbf{Ax}_{(n)} = \mathbf{e}_{(n)}$, where $\mathbf{e}_{(j)}$ has the jth component 1 and the other components 0. Introducing the $n \times n$ matrices $\mathbf{X} = [\mathbf{x}_{(1)} \cdots \mathbf{x}_{(n)}]$ and $\mathbf{I} = [\mathbf{e}_{(1)} \cdots \mathbf{e}_{(n)}]$, we combine the n systems into a single matrix equation $\mathbf{AX} = \mathbf{I}$ and the n augmented matrices $[\mathbf{A} \ \ \mathbf{e}_{(1)}], \cdots, [\mathbf{A} \ \ \mathbf{e}_{(n)}]$ into a single augmented matrix $\tilde{\mathbf{A}} = [\mathbf{A} \ \ \mathbf{I}]$. Now $\mathbf{AX} = \mathbf{I}$ implies $\mathbf{X} = \mathbf{A}^{-1}\mathbf{I} = \mathbf{A}^{-1}$, and to solve $\mathbf{AX} = \mathbf{I}$ for **X** we can apply the Gauss elimination to $\tilde{\mathbf{A}} = [\mathbf{A} \ \ \mathbf{I}]$ to get $[\mathbf{U} \ \ \mathbf{H}]$, where **U** is upper triangular, since the Gauss elimination triangularizes systems. The Gauss–Jordan elimination now operates on $[\mathbf{U} \ \ \mathbf{H}]$ and, by eliminating the entries in **U** above the main diagonal, reduces it to $[\mathbf{I} \ \ \mathbf{K}]$, the augmented matrix of $\mathbf{IX} = \mathbf{A}^{-1}$. Hence we must have $\mathbf{K} = \mathbf{A}^{-1}$ and can thus read off $\mathbf{A}^{-1}$ at the end.

(A formula for the entries of $\mathbf{A}^{-1}$ in terms of those of **A** follows in Sec. 7.9, in connection with determinants.)

EXAMPLE 1 **Inverse of a matrix. Gauss–Jordan elimination**

Find the inverse $\mathbf{A}^{-1}$ of

$$\mathbf{A} = \begin{bmatrix} -1 & 1 & 2 \\ 3 & -1 & 1 \\ -1 & 3 & 4 \end{bmatrix}$$

---

[9]WILHELM JORDAN (1842—1899), German mathematician and geodesist. [See *American Mathematical Monthly* **94** (1987), 130–142.]

We do ***not recommend*** it as a method for solving systems of linear equations, since the number of operations in addition to those of the Gauss elimination is larger than that for back substitution, which the Gauss–Jordan elimination avoids. See also Sec. 19.1.

Solution. **We apply the Gauss elimination (Sec. 7.4) to**

$$[\mathbf{A} \quad \mathbf{I}] = \left[\begin{array}{rrr|rrr} -1 & 1 & 2 & 1 & 0 & 0 \\ 3 & -1 & 1 & 0 & 1 & 0 \\ -1 & 3 & 4 & 0 & 0 & 1 \end{array}\right]$$

$$\left[\begin{array}{rrr|rrr} -1 & 1 & 2 & 1 & 0 & 0 \\ 0 & 2 & 7 & 3 & 1 & 0 \\ 0 & 2 & 2 & -1 & 0 & 1 \end{array}\right] \quad \begin{array}{l} \\ \text{Row 2 + 3 Row 1} \\ \text{Row 3 − Row 1} \end{array}$$

$$\left[\begin{array}{rrr|rrr} -1 & 1 & 2 & 1 & 0 & 0 \\ 0 & 2 & 7 & 3 & 1 & 0 \\ 0 & 0 & -5 & -4 & -1 & 1 \end{array}\right] \quad \begin{array}{l} \\ \\ \text{Row 3 − Row 2} \end{array}$$

This is [U H] as produced by the Gauss elimination, and U agrees with Example 4 in Sec. 7.4. Now follow the additional Gauss–Jordan steps, reducing U to I, that is, to diagonal form with entries 1 on the main diagonal.

$$\left[\begin{array}{rrr|rrr} 1 & -1 & -2 & -1 & 0 & 0 \\ 0 & 1 & 3.5 & 1.5 & 0.5 & 0 \\ 0 & 0 & 1 & 0.8 & 0.2 & -0.2 \end{array}\right] \quad \begin{array}{l} -\text{Row 1} \\ 0.5\ \text{Row 2} \\ -0.2\ \text{Row 3} \end{array}$$

$$\left[\begin{array}{rrr|rrr} 1 & -1 & 0 & 0.6 & 0.4 & -0.4 \\ 0 & 1 & 0 & -1.3 & -0.2 & 0.7 \\ 0 & 0 & 1 & 0.8 & 0.2 & -0.2 \end{array}\right] \quad \begin{array}{l} \text{Row 1 + 2 Row 3} \\ \text{Row 2 − 3.5 Row 3} \\ \\ \end{array}$$

$$\left[\begin{array}{rrr|rrr} 1 & 0 & 0 & -0.7 & 0.2 & 0.3 \\ 0 & 1 & 0 & -1.3 & -0.2 & 0.7 \\ 0 & 0 & 1 & 0.8 & 0.2 & -0.2 \end{array}\right] \quad \begin{array}{l} \text{Row 1 + Row 2} \\ \\ \\ \end{array}$$

The last three columns constitute $\mathbf{A}^{-1}$. Check:

$$\begin{bmatrix} -1 & 1 & 2 \\ 3 & -1 & 1 \\ -1 & 3 & 4 \end{bmatrix} \begin{bmatrix} -0.7 & 0.2 & 0.3 \\ -1.3 & -0.2 & 0.7 \\ 0.8 & 0.2 & -0.2 \end{bmatrix} = \begin{bmatrix} 1 & 0 & 0 \\ 0 & 1 & 0 \\ 0 & 0 & 1 \end{bmatrix}.$$

Hence $\mathbf{A}\mathbf{A}^{-1} = \mathbf{I}$. Similarly, $\mathbf{A}^{-1}\mathbf{A} = \mathbf{I}$. ∎

Some Useful Formulas for Inverses

For a nonsingular 2 × 2 matrix we obtain

$$\textbf{(4)} \qquad \mathbf{A} = \begin{bmatrix} a_{11} & a_{12} \\ a_{21} & a_{22} \end{bmatrix}, \qquad \mathbf{A}^{-1} = \frac{1}{\det \mathbf{A}} \begin{bmatrix} a_{22} & -a_{12} \\ -a_{21} & a_{11} \end{bmatrix},$$

where det $\mathbf{A} = a_{11}a_{22} - a_{12}a_{21}$ and will be discussed in the next section. Indeed, one can readily verify that (1) holds.

Similarly, for a nonsingular diagonal matrix we simply have

$$(5)\quad \mathbf{A} = \begin{bmatrix} a_{11} & \cdots & 0 \\ \cdot & \cdots & \cdot \\ \cdot & \cdots & \cdot \\ 0 & \cdots & a_{nn} \end{bmatrix}, \quad \mathbf{A}^{-1} = \begin{bmatrix} 1/a_{11} & \cdots & 0 \\ \cdot & \cdots & \cdot \\ \cdot & \cdots & \cdot \\ 0 & \cdots & 1/a_{nn} \end{bmatrix};$$

the entries of $\mathbf{A}^{-1}$ on the main diagonal are the reciprocals of those of $\mathbf{A}$.

EXAMPLE 2 Inverse of a 2 × 2 matrix

$$\mathbf{A} = \begin{bmatrix} 3 & 1 \\ 2 & 4 \end{bmatrix}, \qquad \mathbf{A}^{-1} = \frac{1}{10}\begin{bmatrix} 4 & -1 \\ -2 & 3 \end{bmatrix} = \begin{bmatrix} 0.4 & -0.1 \\ -0.2 & 0.3 \end{bmatrix}$$ ■

EXAMPLE 3 Inverse of a diagonal matrix

$$\mathbf{A} = \begin{bmatrix} -0.5 & 0 & 0 \\ 0 & 4 & 0 \\ 0 & 0 & 1 \end{bmatrix}, \qquad \mathbf{A}^{-1} = \begin{bmatrix} -2 & 0 & 0 \\ 0 & 0.25 & 0 \\ 0 & 0 & 1 \end{bmatrix}$$ ■

The inverse of the inverse is the given matrix $\mathbf{A}$:

$$(6)\qquad (\mathbf{A}^{-1})^{-1} = \mathbf{A}.$$

The simple proof is left to the reader (Prob. 16).

Inverse of a product. The inverse of a product $\mathbf{AC}$ can be calculated by inverting each factor separately and multiplying the results *in reverse order:*

$$(7)\qquad \boxed{(\mathbf{AC})^{-1} = \mathbf{C}^{-1}\mathbf{A}^{-1}.}$$

To prove (7), we start from (1), with $\mathbf{A}$ replaced by $\mathbf{AC}$, that is,

$$\mathbf{AC}(\mathbf{AC})^{-1} = \mathbf{I}.$$

Multiplying this by $\mathbf{A}^{-1}$ from the left and using $\mathbf{A}^{-1}\mathbf{A} = \mathbf{I}$, we obtain

$$\mathbf{C}(\mathbf{AC})^{-1} = \mathbf{A}^{-1}.$$

If we multiply this by $\mathbf{C}^{-1}$ from the left, the result follows.

Of course, (7) may be generalized to products of more than two matrices; by induction we obtain

$$(8)\qquad (\mathbf{AC}\cdots\mathbf{PQ})^{-1} = \mathbf{Q}^{-1}\mathbf{P}^{-1}\cdots\mathbf{C}^{-1}\mathbf{A}^{-1}.$$

Vanishing of Matrix Products. Cancellation Law

We can now obtain more information about the strange fact that for matrix multiplication, the cancellation law is not true, in general; that is, $\mathbf{AB} = \mathbf{0}$ does not necessarily imply that $\mathbf{A} = \mathbf{0}$ or $\mathbf{B} = \mathbf{0}$ (as for numbers), and it does also not imply that $\mathbf{BA} = \mathbf{0}$. These facts were stated in Sec. 7.3 and illustrated with

$$\begin{bmatrix} 1 & 1 \\ 2 & 2 \end{bmatrix} \begin{bmatrix} -1 & 1 \\ 1 & -1 \end{bmatrix} = \begin{bmatrix} 0 & 0 \\ 0 & 0 \end{bmatrix},$$

$$\begin{bmatrix} -1 & 1 \\ 1 & -1 \end{bmatrix} \begin{bmatrix} 1 & 1 \\ 2 & 2 \end{bmatrix} = \begin{bmatrix} 1 & 1 \\ -1 & -1 \end{bmatrix}.$$

Each of these two matrices has rank less than $n = 2$. This is typical, and the situation changes when $n \times n$ matrices have rank n:

Theorem 2 (Cancellation law)

Let $\mathbf{A}, \mathbf{B}, \mathbf{C}$ *be* $n \times n$ *matrices. Then:*

(a) *If* rank $\mathbf{A} = n$ *and* $\mathbf{AB} = \mathbf{AC}$, *then* $\mathbf{B} = \mathbf{C}$.

(b) *If* rank $\mathbf{A} = n$, *then* $\mathbf{AB} = \mathbf{0}$ *implies* $\mathbf{B} = \mathbf{0}$. *Hence if* $\mathbf{AB} = \mathbf{0}$, *but* $\mathbf{A} \neq \mathbf{0}$ *as well as* $\mathbf{B} \neq \mathbf{0}$, *then* rank $\mathbf{A} < n$ *and* rank $\mathbf{B} < n$.

(c) *If* $\mathbf{A}$ *is singular, so are* $\mathbf{AB}$ *and* $\mathbf{BA}$.

Proof. **(a)** Premultiply $\mathbf{AB} = \mathbf{AC}$ on both sides by $\mathbf{A}^{-1}$, which exists by Theorem 1.

(b) Premultiply $\mathbf{AB} = \mathbf{0}$ on both sides by $\mathbf{A}^{-1}$.

(c_1) Rank $\mathbf{A} < n$ by Theorem 1. Hence $\mathbf{Ax} = \mathbf{0}$ has nontrivial solutions, by Theorem 2 in Sec. 7.6. Multiplication gives $\mathbf{BAx} = \mathbf{0}$. Hence those solutions also satisfy $\mathbf{BAx} = \mathbf{0}$. Hence rank $(\mathbf{BA}) < n$ by Theorem 2 in Sec. 7.6, and $\mathbf{BA}$ is singular by Theorem 1.

(c_2) $\mathbf{A}^{\mathsf{T}}$ is singular by Theorem 1 in Sec. 7.5. Hence $\mathbf{B}^{\mathsf{T}}\mathbf{A}^{\mathsf{T}}$ is singular by part (c_1). But $\mathbf{B}^{\mathsf{T}}\mathbf{A}^{\mathsf{T}} = (\mathbf{AB})^{\mathsf{T}}$; see Sec. 7.3. Hence $\mathbf{AB}$ is singular by Theorem 1 in Sec. 7.5. ∎

Problem Set 7.7

Find the inverse and check the result, or state that the inverse does not exist, giving a reason.

1. $\begin{bmatrix} 1 & 2 \\ 3 & 4 \end{bmatrix}$ **2.** $\begin{bmatrix} \cos\theta & \sin\theta \\ -\sin\theta & \cos\theta \end{bmatrix}$ **3.** $\begin{bmatrix} 0.6 & 0.8 \\ 0.8 & -0.6 \end{bmatrix}$

Find the inverse and check the result, or state that the inverse does not exist, giving a reason.

4. $\begin{bmatrix} -1 & 2 & 2 \\ 2 & -1 & 2 \\ 2 & 2 & -1 \end{bmatrix}$ **5.** $\begin{bmatrix} 6 & -2 & \frac{1}{2} \\ 1 & 5 & 2 \\ -8 & 24 & 7 \end{bmatrix}$ **6.** $\begin{bmatrix} 0 & 0 & 5 \\ 0 & -1 & 0 \\ 3 & 0 & 0 \end{bmatrix}$

7. $\begin{bmatrix} 0.5 & 0 & -0.5 \\ -0.1 & 0.2 & 0.3 \\ 0.5 & 0 & -1.5 \end{bmatrix}$ **8.** $\begin{bmatrix} 7 & 9 & 11 \\ 8 & -8 & 5 \\ 4 & 60 & 29 \end{bmatrix}$ **9.** $\begin{bmatrix} 1 & 0 & 0 \\ \frac{1}{2} & 1 & 0 \\ 1 & 5 & 2 \end{bmatrix}$

10. $\begin{bmatrix} 0 & 2 & 0 \\ 4 & 0 & 0 \\ 0 & 0 & 5 \end{bmatrix}$ **11.** $\begin{bmatrix} 1 & 4 & 8 \\ 0 & 5 & 2 \\ 0 & 0 & 10 \end{bmatrix}$ **12.** $\begin{bmatrix} 10 & 0 & 0 \\ 0 & 9 & 17 \\ 0 & 4 & 8 \end{bmatrix}$

13. $\begin{bmatrix} 3 & -1 & 1 \\ -15 & 6 & -5 \\ 5 & -2 & 2 \end{bmatrix}$ **14.** $\begin{bmatrix} 19 & 2 & -9 \\ -4 & -1 & 2 \\ -2 & 0 & 1 \end{bmatrix}$ **15.** $\begin{bmatrix} 371 & -76 & -40 \\ 36 & -7 & -4 \\ -176 & 36 & 19 \end{bmatrix}$

16. Prove (6).

17. Verify (4) and (5) by showing that $\mathbf{AA}^{-1} = \mathbf{A}^{-1}\mathbf{A} = \mathbf{I}$.

18. Show that $(\mathbf{A}^{-1})^{\mathsf{T}} = (\mathbf{A}^{\mathsf{T}})^{-1}$.

19. Show that the inverse of a nonsingular symmetric matrix is symmetric.

20. Show that $(\mathbf{A}^2)^{-1} = (\mathbf{A}^{-1})^2$. Find $(\mathbf{A}^2)^{-1}$ for the matrix in Prob. 14.

7.8 Determinants

Determinants were first defined for solving linear systems and, although ***impractical in computations,***[10] they have important engineering applications in eigenvalue problems (Sec. 7.10), differential equations (Chaps. 2, 4), vector algebra (vector products, scalar triple products, Sec. 8.3), and so on. They can be introduced in several equivalent ways, and our definition is particularly practical in connection with those systems.

An *nth-order determinant* is an expression associated with an $n \times n$ (hence square!) matrix $\mathbf{A} = [a_{jk}]$, as we now explain, beginning with $n = 2$.

Second-Order Determinants

A **determinant of second order** is denoted and defined by

$$D = \det \mathbf{A} = \begin{vmatrix} a_{11} & a_{12} \\ a_{21} & a_{22} \end{vmatrix} = a_{11}a_{22} - a_{12}a_{21}. \tag{1}$$

[10] In numerical work, use a method from Secs. 7.4, 19.1–19.3; ***do not use*** **Cramer's rule** (see Sec. 7.9).

So here we have ***bars*** (whereas a matrix has *brackets*). For example,

$$\begin{vmatrix} 4 & 3 \\ 2 & 5 \end{vmatrix} = 4 \cdot 5 - 3 \cdot 2 = 14.$$

This definition is suggested by systems

$$\text{(2)} \qquad \begin{aligned} &\text{(a)} \quad a_{11}x_1 + a_{12}x_2 = b_1 \\ &\text{(b)} \quad a_{21}x_1 + a_{22}x_2 = b_2 \end{aligned}$$

whose solution can be written $x_1 = D_1/D$, $x_2 = D_2/D$ with D as in (1) and

$$D_1 = \begin{vmatrix} b_1 & a_{12} \\ b_2 & a_{22} \end{vmatrix} = b_1a_{22} - a_{12}b_2,$$

$$D_2 = \begin{vmatrix} a_{11} & b_1 \\ a_{21} & b_2 \end{vmatrix} = a_{11}b_2 - b_1a_{21},$$

provided $D \neq 0$; this is called **Cramer's rule.**[11] It follows by the usual elimination. Indeed, to eliminate x_2, multiply (2a) by a_{22} and (2b) by $-a_{12}$ and add, finding

$$(a_{11}a_{22} - a_{12}a_{21})x_1 = b_1a_{22} - a_{12}b_2, \qquad \text{thus} \qquad Dx_1 = D_1.$$

To eliminate x_1, multiply (2a) by $-a_{21}$ and (2b) by a_{11} and add, finding

$$(a_{11}a_{22} - a_{12}a_{21})x_2 = a_{11}b_2 - b_1a_{21}, \qquad \text{thus} \qquad Dx_2 = D_2.$$

Now divide by D (if $D \neq 0$) to get x_1 and x_2.

EXAMPLE 1 Use of second-order determinants

If

$$\begin{aligned} 4x_1 + 3x_2 &= 12 \\ 2x_1 + 5x_2 &= -8, \end{aligned}$$

then

$$D = \begin{vmatrix} 4 & 3 \\ 2 & 5 \end{vmatrix} = 14, \qquad D_1 = \begin{vmatrix} 12 & 3 \\ -8 & 5 \end{vmatrix} = 84, \qquad D_2 = \begin{vmatrix} 4 & 12 \\ 2 & -8 \end{vmatrix} = -56,$$

so that $x_1 = 84/14 = 6$ and $x_2 = -56/14 = -4$. ∎

---

[11]GABRIEL CRAMER (1704—1752). Swiss mathematician.

If the system (2) is homogeneous ($b_1 = b_2 = 0$) and $D \neq 0$, it has only the trivial solution $x_1 = x_2 = 0$, and if $D = 0$, it also has nontrivial solutions.

Third-Order Determinants

A **determinant of third order** can be defined by

$$(3)\quad D = \begin{vmatrix} a_{11} & a_{12} & a_{13} \\ a_{21} & a_{22} & a_{23} \\ a_{31} & a_{32} & a_{33} \end{vmatrix} = a_{11}\begin{vmatrix} a_{22} & a_{23} \\ a_{32} & a_{33} \end{vmatrix} - a_{21}\begin{vmatrix} a_{12} & a_{13} \\ a_{32} & a_{33} \end{vmatrix} + a_{31}\begin{vmatrix} a_{12} & a_{13} \\ a_{22} & a_{23} \end{vmatrix}.$$

Note the following. The signs on the right are $+\ -\ +$. Each of the three terms on the right is an entry in the first column of D times its "**minor,**" that is, the second-order determinant obtained by deleting from D the row and column of that entry (thus for a_{11} delete the first row and first column, etc.).

If we write out the minors, we get

$$(4)\quad \begin{aligned} D = {} & a_{11}a_{22}a_{33} - a_{11}a_{32}a_{23} + a_{21}a_{32}a_{13} \\ & - a_{21}a_{12}a_{33} + a_{31}a_{12}a_{23} - a_{31}a_{22}a_{13}. \end{aligned}$$

For linear systems of three equations in three unknowns

$$(5)\quad \begin{aligned} a_{11}x_1 + a_{12}x_2 + a_{13}x_3 &= b_1 \\ a_{21}x_1 + a_{22}x_2 + a_{23}x_3 &= b_2 \\ a_{31}x_1 + a_{32}x_2 + a_{33}x_3 &= b_3 \end{aligned}$$

Cramer's rule is

$$(6)\quad x_1 = \frac{D_1}{D}, \qquad x_2 = \frac{D_2}{D}, \qquad x_3 = \frac{D_3}{D} \qquad (D \neq 0)$$

with the "*determinant of the system*" D given by (3) and

$$D_1 = \begin{vmatrix} b_1 & a_{12} & a_{13} \\ b_2 & a_{22} & a_{23} \\ b_3 & a_{32} & a_{33} \end{vmatrix}, \qquad D_2 = \begin{vmatrix} a_{11} & b_1 & a_{13} \\ a_{21} & b_2 & a_{23} \\ a_{31} & b_3 & a_{33} \end{vmatrix}, \qquad D_3 = \begin{vmatrix} a_{11} & a_{12} & b_1 \\ a_{21} & a_{22} & b_2 \\ a_{31} & a_{32} & b_3 \end{vmatrix}.$$

This could be derived by elimination similarly as above, but, instead, we shall obtain Cramer's rule for general n in the next section.

Determinant of Any Order n

A **determinant of order n** is a scalar associated with an $n \times n$ matrix $\mathbf{A} = [a_{jk}]$, which is written

$$D = \det \mathbf{A} = \begin{vmatrix} a_{11} & a_{12} & \cdots & a_{1n} \\ a_{21} & a_{22} & \cdots & a_{2n} \\ \cdot & \cdot & \cdots & \cdot \\ \cdot & \cdot & \cdots & \cdot \\ a_{n1} & a_{n2} & \cdots & a_{nn} \end{vmatrix} \tag{7}$$

and is defined for $n = 1$ by

$$D = a_{11} \tag{8}$$

and for $n \geqq 2$ by

$$D = a_{j1}C_{j1} + a_{j2}C_{j2} + \cdots + a_{jn}C_{jn} \qquad (j = 1, 2, \cdots, \text{ or } n) \tag{9a}$$

or

$$D = a_{1k}C_{1k} + a_{2k}C_{2k} + \cdots + a_{nk}C_{nk} \qquad (k = 1, 2, \cdots, \text{ or } n) \tag{9b}$$

where

$$C_{jk} = (-1)^{j+k}M_{jk}$$

and M_{jk} is a determinant of order $n - 1$, namely, the determinant of the submatrix of **A** obtained from **A** by deleting the row and column of the entry a_{jk} (the jth row and the kth column). ∎

In this way, D is defined in terms of n determinants of order $n - 1$, each of which is, in turn, defined in terms of $n - 1$ determinants of order $n - 2$, and so on; we finally arrive at second-order determinants, in which those submatrices consist of single entries whose determinant is defined by (8).

From the definition it follows that *we may* **expand** *D by any row or column,* that is, choose in (9) the entries in any row or column, similarly when expanding the C_{jk}'s in (9), and so on.

This definition is unambiguous, that is, yields the same value for D no matter which columns or rows we choose. A proof is given in Appendix 4.

Terms used in connection with determinants are taken from matrices: in D we have n^2 **entries** or *elements* a_{jk}, also n **rows** and n **columns,** a **main diagonal** or *principal diagonal* on which $a_{11}, a_{22}, \cdots, a_{nn}$ stand. Two names are new:

M_{jk} is called the **minor** *of a_{jk} in D,* and C_{jk} the **cofactor** *of a_{jk} in D.*

For later use we note that (9) may also be written in terms of minors

$$D = \sum_{k=1}^{n} (-1)^{j+k} a_{jk} M_{jk} \qquad (j = 1, 2, \cdots, \text{ or } n) \tag{10a}$$

$$D = \sum_{j=1}^{n} (-1)^{j+k} a_{jk} M_{jk} \qquad (k = 1, 2, \cdots, \text{ or } n). \tag{10b}$$

EXAMPLE 2 **Determinant of second order**

For

$$D = \det \mathbf{A} = \begin{vmatrix} a_{11} & a_{12} \\ a_{21} & a_{22} \end{vmatrix}$$

formula (9) gives four possibilities of expanding it, namely, by

$$\begin{aligned} &\text{the first row:} && D = a_{11}a_{22} + a_{12}(-a_{21}), \\ &\text{the second row:} && D = a_{21}(-a_{12}) + a_{22}a_{11}, \\ &\text{the first column:} && D = a_{11}a_{22} + a_{21}(-a_{12}), \\ &\text{the second column:} && D = a_{12}(-a_{21}) + a_{22}a_{11}. \end{aligned}$$

They all give the same value $D = a_{11}a_{22} - a_{12}a_{21}$, which agrees with (1). ■

EXAMPLE 3 **Minors and cofactors of a third-order determinant**

In the third-order determinant

$$\begin{vmatrix} a_{11} & a_{12} & a_{13} \\ a_{21} & a_{22} & a_{23} \\ a_{31} & a_{32} & a_{33} \end{vmatrix}$$

the minors are

$$M_{11} = \begin{vmatrix} a_{22} & a_{23} \\ a_{32} & a_{33} \end{vmatrix}, \quad M_{12} = \begin{vmatrix} a_{21} & a_{23} \\ a_{31} & a_{33} \end{vmatrix}, \quad M_{13} = \begin{vmatrix} a_{21} & a_{22} \\ a_{31} & a_{32} \end{vmatrix},$$

$$M_{21} = \begin{vmatrix} a_{12} & a_{13} \\ a_{32} & a_{33} \end{vmatrix}, \quad M_{22} = \begin{vmatrix} a_{11} & a_{13} \\ a_{31} & a_{33} \end{vmatrix}, \quad M_{23} = \begin{vmatrix} a_{11} & a_{12} \\ a_{31} & a_{32} \end{vmatrix},$$

$$M_{31} = \begin{vmatrix} a_{12} & a_{13} \\ a_{22} & a_{23} \end{vmatrix}, \quad M_{32} = \begin{vmatrix} a_{11} & a_{13} \\ a_{21} & a_{23} \end{vmatrix}, \quad M_{33} = \begin{vmatrix} a_{11} & a_{12} \\ a_{21} & a_{22} \end{vmatrix},$$

and the cofactors are

$$C_{11} = +M_{11}, \quad C_{12} = -M_{12}, \quad C_{13} = +M_{13},$$

$$C_{21} = -M_{21}, \quad C_{22} = +M_{22}, \quad C_{23} = -M_{23},$$

$$C_{31} = +M_{31}, \quad C_{32} = -M_{32}, \quad C_{33} = +M_{33}.$$

Hence the signs form a **checkerboard pattern:**

$$\begin{matrix} + & - & + \\ - & + & - \\ + & - & + \end{matrix}$$

EXAMPLE 4 A third-order determinant

Let

$$D = \begin{vmatrix} 1 & 3 & 0 \\ 2 & 6 & 4 \\ -1 & 0 & 2 \end{vmatrix}$$

The expansion by the first row is

$$D = 1\begin{vmatrix} 6 & 4 \\ 0 & 2 \end{vmatrix} - 3\begin{vmatrix} 2 & 4 \\ -1 & 2 \end{vmatrix} = 1(12 - 0) - 3(4 + 4) = -12.$$

The expansion by the third column is

$$D = 0\begin{vmatrix} 2 & 6 \\ -1 & 0 \end{vmatrix} - 4\begin{vmatrix} 1 & 3 \\ -1 & 0 \end{vmatrix} + 2\begin{vmatrix} 1 & 3 \\ 2 & 6 \end{vmatrix} = 0 - 12 + 0 = -12,$$

etc. ∎

EXAMPLE 5 Determinant of a triangular matrix

The determinant of any triangular matrix equals the product of all the entries on the main diagonal. To see this, expand by rows if the matrix is lower triangular, and by columns if it is upper triangular. For instance,

$$\begin{vmatrix} -3 & 0 & 0 \\ 6 & 4 & 0 \\ -1 & 2 & 5 \end{vmatrix} = -3\begin{vmatrix} 4 & 0 \\ 2 & 5 \end{vmatrix} = -3 \cdot 4 \cdot 5 = -60.$$ ∎

General Properties of Determinants

From our definition we may now readily obtain the most important properties of determinants, as follows.

Since the same value is obtained whether we expand a determinant by any row or any column, we have

Theorem 1 (Transposition)

The value of a determinant is not altered if its rows are written as columns, in the same order.

EXAMPLE 6 Transposition

$$\begin{vmatrix} 1 & 3 & 0 \\ 2 & 6 & 4 \\ -1 & 0 & 2 \end{vmatrix} = \begin{vmatrix} 1 & 2 & -1 \\ 3 & 6 & 0 \\ 0 & 4 & 2 \end{vmatrix} = -12.$$ ∎

Theorem 2 **(Multiplication by a constant)**
If all the entries in one row (or one column) of a determinant are multiplied by the same factor k, the value of the new determinant is k times the value of the given determinant.

Proof. Expand the determinant by that row (or column) whose entries are multiplied by k. ∎

Caution! det $k\mathbf{A} = k^n$ det $\mathbf{A}$ (not k det $\mathbf{A}$). Explain why.

EXAMPLE 7 **Application of Theorem 2**

$$\begin{vmatrix} 1 & 3 & 0 \\ 2 & 6 & 4 \\ -1 & 0 & 2 \end{vmatrix} = 2\begin{vmatrix} 1 & 3 & 0 \\ 1 & 3 & 2 \\ -1 & 0 & 2 \end{vmatrix} = 6\begin{vmatrix} 1 & 1 & 0 \\ 1 & 1 & 2 \\ -1 & 0 & 2 \end{vmatrix} = 12\begin{vmatrix} 1 & 1 & 0 \\ 1 & 1 & 1 \\ -1 & 0 & 1 \end{vmatrix} = -12$$ ∎

From Theorem 2, with $k = 0$, or directly by expanding, we obtain

Theorem 3 *If all the entries in a row (or a column) of a determinant are zero, the value of the determinant is zero.*

Theorem 4 *If each entry in a row (or a column) of a determinant is expressed as a binomial, the determinant can be written as the sum of two determinants.*

Proof. Expand the determinant by the row (or column) whose entries are binomials. ∎

EXAMPLE 8 **Illustration of Theorem 4**

$$\begin{vmatrix} a_1 + d_1 & b_1 & c_1 \\ a_2 + d_2 & b_2 & c_2 \\ a_3 + d_3 & b_3 & c_3 \end{vmatrix} = \begin{vmatrix} a_1 & b_1 & c_1 \\ a_2 & b_2 & c_2 \\ a_3 & b_3 & c_3 \end{vmatrix} + \begin{vmatrix} d_1 & b_1 & c_1 \\ d_2 & b_2 & c_2 \\ d_3 & b_3 & c_3 \end{vmatrix}$$ ∎

Theorem 5 **(Interchange of rows or columns)**
If any two rows (or two columns) of a determinant are interchanged, the value of the determinant is multiplied by −1.

Proof. The proof is by induction. We see that the theorem holds for determinants of order $n = 2$. Assuming that it holds for determinants of order $n - 1$, we will show that it holds for determinants of order n.

Let D be of order n and E obtained from D by interchanging two rows. Expand D and E by a row that is not one of those interchanged, call it the jth row. Then, by (10a),

$$(11) \qquad D = \sum_{k=1}^{n} (-1)^{j+k} a_{jk} M_{jk}, \qquad E = \sum_{k=1}^{n} (-1)^{j+k} a_{jk} N_{jk}$$

where N_{jk} is obtained from the minor M_{jk} of a_{jk} in D by interchanging two rows. Since these minors are of order $n - 1$, the induction hypothesis applies and gives $N_{jk} = -M_{jk}$. Hence $E = -D$ by (11). This proves the statement for *rows*. To get it for *columns*, apply Theorem 1. ∎

EXAMPLE 9 Interchange of two rows

$$\begin{vmatrix} 2 & 6 & 4 \\ 1 & 3 & 0 \\ -1 & 0 & 2 \end{vmatrix} = -\begin{vmatrix} 1 & 3 & 0 \\ 2 & 6 & 4 \\ -1 & 0 & 2 \end{vmatrix} = 12$$ ∎

Theorem 6 (Proportional rows or columns)

If corresponding entries in two rows (or two columns) of a determinant are proportional, the value of the determinant is zero.

Proof. Let the entries in the ith and jth rows of D be proportional, say, $a_{ik} = ca_{jk}$, $k = 1, \cdots, n$. If $c = 0$, then $D = 0$. Now let $c \neq 0$. By Theorem 2,

$$D = cB$$

where the ith and jth rows of B are identical. Interchange these rows. Then B goes over into $-B$, by Theorem 5. On the other hand, since the rows are identical, the new determinant is still B. Thus $B = -B$, $B = 0$, and $D = 0$. ∎

EXAMPLE 10 Proportional rows

$$\begin{vmatrix} 3 & 6 & -4 \\ 1 & -1 & 3 \\ -6 & -12 & 8 \end{vmatrix} = 0$$ ∎

Theorem 7 (Addition of a row or column)

The value of a determinant is left unchanged if the entries in a row (or column) are altered by adding to them any constant multiple of the corresponding entries in any other row (or column, respectively).

Proof. Apply Theorem 4 to the determinant that results from the given addition. This yields a sum of two determinants; one is the original determinant and the other contains two proportional rows. According to Theorem 6, the second determinant is zero, and the proof is complete. ∎

Theorem 7 shows that we can evaluate a determinant by first creating zeros as in the Gauss elimination (Sec. 7.4), a method that can be programmed easily.[12] We explain it in terms of an example.

---

[12]In specific cases, selecting rows or columns by inspection may save work (e.g., in pocket computations), an art which old-fashioned texts emphasize to this day.

EXAMPLE 11 Evaluation of a determinant by reduction to "triangular form"

Explanations of computations, such as "Row 2 − 2 Row 1," always refer to the preceding determinant; they are placed behind the row where the result goes.

$$D = \begin{vmatrix} 2 & 0 & -4 & 6 \\ 4 & 5 & 1 & 0 \\ 0 & 2 & 6 & -1 \\ -3 & 8 & 9 & 1 \end{vmatrix}$$

$$= \begin{vmatrix} 2 & 0 & -4 & 6 \\ 0 & 5 & 9 & -12 \\ 0 & 2 & 6 & -1 \\ 0 & 8 & 3 & 10 \end{vmatrix} \quad \begin{matrix} \\ \text{Row 2} - 2 \text{ Row 1} \\ \\ \text{Row 4} + 1.5 \text{ Row 1} \end{matrix}$$

$$= \begin{vmatrix} 2 & 0 & -4 & 6 \\ 0 & 5 & 9 & -12 \\ 0 & 0 & 2.4 & 3.8 \\ 0 & 0 & -11.4 & 29.2 \end{vmatrix} \quad \begin{matrix} \\ \\ \text{Row 3} - 0.4 \text{ Row 2} \\ \text{Row 4} - 1.6 \text{ Row 2} \end{matrix}$$

$$= \begin{vmatrix} 2 & 0 & -4 & 6 \\ 0 & 5 & 9 & -12 \\ 0 & 0 & 2.4 & 3.8 \\ 0 & 0 & 0 & 47.25 \end{vmatrix} \quad \begin{matrix} \\ \\ \\ \text{Row 4} + 4.75 \text{ Row 3} \end{matrix}$$

$$= 2 \times 5 \times 2.4 \times 47.25 = 1134.$$

In work with pencil and paper, one writes down lower order determinants when they appear, instead of carrying along zeros,

$$D = \begin{vmatrix} 2 & 0 & -4 & 6 \\ 4 & 5 & 1 & 0 \\ 0 & 2 & 6 & -1 \\ -3 & 8 & 9 & 1 \end{vmatrix} = \cdots = 2 \begin{vmatrix} 5 & 9 & -12 \\ 2 & 6 & -1 \\ 8 & 3 & 10 \end{vmatrix} = \cdots = 10 \begin{vmatrix} 2.4 & 3.8 \\ -11.4 & 29.2 \end{vmatrix} = 1134. \quad \blacksquare$$

For determinants of products of matrices, there is a very useful formula that has various applications. A proof of this formula will be given in the next section.

Theorem 8 (Determinant of a product of matrices)

For any $n \times n$ matrices **A** *and* **B**,

$$\boxed{\det(\mathbf{AB}) = \det(\mathbf{BA}) = \det \mathbf{A} \det \mathbf{B}.} \tag{12}$$

EXAMPLE 12 Illustration of Theorem 8

$$\begin{vmatrix} 2 & 4 & 3 \\ 6 & 10 & 14 \\ 4 & 7 & 9 \end{vmatrix} \begin{vmatrix} 4 & 0 & 5 \\ -2 & 1 & -1 \\ 3 & 0 & 4 \end{vmatrix} = \begin{vmatrix} 9 & 4 & 18 \\ 46 & 10 & 76 \\ 29 & 7 & 49 \end{vmatrix} \qquad \blacksquare$$

If the entries of a square matrix are scalars (numbers), so is the determinant. If they are functions, so is the determinant, and in this case, one occasionally needs the following theorem, which can be obtained by the product rule.

Theorem 9 (Derivative of a determinant)

The derivative D' of a determinant D of order n whose entries are differentiable functions can be written

$$D' = D_{(1)} + D_{(2)} + \cdots + D_{(n)} \tag{13}$$

where $D_{(j)}$ is obtained from D by differentiating the entries in the jth row.

EXAMPLE 13 Derivative of a third-order determinant

$$\frac{d}{dx}\begin{vmatrix} f & g & h \\ p & q & r \\ u & v & w \end{vmatrix} = \begin{vmatrix} f' & g' & h' \\ p & q & r \\ u & v & w \end{vmatrix} + \begin{vmatrix} f & g & h \\ p' & q' & r' \\ u & v & w \end{vmatrix} + \begin{vmatrix} f & g & h \\ p & q & r \\ u' & v' & w' \end{vmatrix} \qquad \blacksquare$$

Problem Set 7.8

Evaluate

1. $\begin{vmatrix} 17 & 9 \\ -4 & 13 \end{vmatrix}$

2. $\begin{vmatrix} \cos n\theta & \sin n\theta \\ -\sin n\theta & \cos n\theta \end{vmatrix}$

3. $\begin{vmatrix} 4.3 & 0.7 \\ 0.8 & -9.2 \end{vmatrix}$

4. $\begin{vmatrix} 1.0 & 0.2 & 1.6 \\ 3.0 & 0.6 & 1.2 \\ 2.0 & 0.8 & 0.4 \end{vmatrix}$

5. $\begin{vmatrix} 5 & 1 & 8 \\ 15 & 3 & 6 \\ 10 & 4 & 2 \end{vmatrix}$

6. $\begin{vmatrix} 4 & 6 & 5 \\ 0 & 1 & -7 \\ 0 & 0 & 6 \end{vmatrix}$

7. $\begin{vmatrix} 16 & 22 & 4 \\ 4 & -3 & 2 \\ 12 & 25 & 2 \end{vmatrix}$

8. $\begin{vmatrix} a & b & c \\ c & a & b \\ b & c & a \end{vmatrix}$

9. $\begin{vmatrix} 1.1 & 8.7 & 3.6 \\ 0 & 9.1 & -1.7 \\ 0 & 0 & 4.5 \end{vmatrix}$

10. $\begin{vmatrix} 2 & 8 & 0 & 0 \\ 9 & -4 & 0 & 0 \\ 0 & 0 & 7 & 1 \\ 0 & 0 & 6 & -2 \end{vmatrix}$

11. $\begin{vmatrix} 3 & 2 & 0 & 0 \\ 6 & 8 & 0 & 0 \\ 0 & 0 & 4 & 7 \\ 0 & 0 & 2 & 5 \end{vmatrix}$

12. $\begin{vmatrix} -6 & 4 & 5 & 6 \\ 2 & 7 & 2 & 1 \\ -1 & 7 & 2 & 4 \\ -7 & 4 & 5 & 7 \end{vmatrix}$

Evaluate

13. $\begin{vmatrix} 4 & 3 & 9 & 9 \\ -8 & 3 & 5 & -4 \\ -8 & 0 & -2 & -8 \\ -16 & 6 & 14 & -5 \end{vmatrix}$ **14.** $\begin{vmatrix} 4 & 3 & 0 & 0 \\ -8 & 1 & 2 & 0 \\ 0 & -7 & 3 & -6 \\ 0 & 0 & 5 & -5 \end{vmatrix}$ **15.** $\begin{vmatrix} 12 & 6 & 1 & 11 \\ 4 & 4 & 1 & 4 \\ 7 & 4 & 3 & 7 \\ 8 & 2 & 3 & 9 \end{vmatrix}$

16. Show that det $(k\mathbf{A}) = k^n$ det **A** (*not* k det **A**), where **A** is any $n \times n$ matrix.

17. Write the product of the determinants in Probs. 5 and 6 as a determinant.

18. Do the same task as in Prob. 17, taking the determinants in reverse order.

19. Verify that the answer to Prob. 11 equals the product of the determinants of the 2×2 submatrices containing no zero entries. Explain why.

20. Show that the straight line through two points P_1: (x_1, y_1) and P_2: (x_2, y_2) in the xy-plane is given by formula (a) (below), and derive from (a) the familiar formula (b).

$$\text{(a)} \quad \begin{vmatrix} x & y & 1 \\ x_1 & y_1 & 1 \\ x_2 & y_2 & 1 \end{vmatrix} = 0 \qquad \text{(b)} \quad \frac{x - x_1}{x_1 - x_2} = \frac{y - y_1}{y_1 - y_2}$$

7.9 Rank in Terms of Determinants. Cramer's Rule

In this section we first show that the rank of a matrix **A** (defined as the maximum number of linearly independent row or column vectors of **A**; see Sec. 7.5) can also be characterized in terms of determinants. This remarkable property is often used for *defining* rank. We formulate this as follows, assuming rank $\mathbf{A} > 0$ (since rank $\mathbf{A} = 0$ if and only if $\mathbf{A} = \mathbf{0}$; see Sec. 7.5).

Theorem 1 **(Rank in terms of determinants)**

An $m \times n$ matrix $\mathbf{A} = [a_{jk}]$ has rank $r \geqq 1$ if and only if **A** *has an $r \times r$ submatrix with nonzero determinant, whereas the determinant of every square submatrix with $r + 1$ or more rows that* **A** *has (or does not have!) is zero.*

In particular, if **A** *is a square matrix,* **A** *is nonsingular, so that the inverse $\mathbf{A}^{-1}$ of* **A** *exists, if and only if*

$$\det \mathbf{A} \neq 0.$$

Proof. The key lies in the fact that elementary row operations (Sec. 7.4), which do not alter the rank (by Theorem 3 in Sec. 7.5), also do not alter the property of a determinant of being zero or not zero, since the determinant is multiplied by

(*i*) -1 if we interchange two rows (Theorem 5, Sec. 7.8),
(*ii*) $c \neq 0$ if we multiply a row by $c \neq 0$ (Theorem 2, Sec. 7.8),
(*iii*) 1 if we add a multiple of a row to another row (Theorem 7, Sec. 7.8).

Let $\hat{\mathbf{A}}$ denote the echelon form of $\mathbf{A}$ (see Sec. 7.4). $\hat{\mathbf{A}}$ has r nonzero row vectors (which are the first r row vectors) if and only if rank $\mathbf{A} = r$. Let $\hat{\mathbf{R}}$ be the $r \times r$ submatrix of $\hat{\mathbf{A}}$ consisting of the r^2 entries that are simultaneously in the first r rows and the first r columns of $\hat{\mathbf{A}}$. Since $\hat{\mathbf{R}}$ is triangular and has all diagonal entries different from zero, det $\hat{\mathbf{R}} \neq 0$. Since $\hat{\mathbf{R}}$ is obtained from the corresponding $r \times r$ submatrix $\mathbf{R}$ of $\mathbf{A}$ by elementary operations, we have det $\mathbf{R} \neq 0$. Similarly, det $\mathbf{S} = 0$ for a square submatrix $\mathbf{S}$ of $r + 1$ or more rows possibly contained in $\mathbf{A}$, since the corresponding submatrix $\hat{\mathbf{S}}$ of $\hat{\mathbf{A}}$ must contain a row of zeros, so that det $\hat{\mathbf{S}} = 0$ by Theorem 3 in Sec. 7.8. This proves the assertion of the theorem for an $m \times n$ matrix.

If $\mathbf{A}$ is square, say, an $n \times n$ matrix, the statement just proved implies that rank $\mathbf{A} = n$ if and only if $\mathbf{A}$ has an $n \times n$ submatrix with a nonzero determinant; but this submatrix is $\mathbf{A}$ itself, so that det $\mathbf{A} \neq 0$. ∎

Using this theorem, we shall now derive Cramer's rule, which gives solutions of linear systems as quotients of determinants. ***Cramer's rule is not practical in computations*** (for which the methods in Secs. 7.4 and 19.1–19.3 are suitable), but is of ***theoretical interest*** in differential equations (Sec. 3.5) and other theories that have engineering applications.

Theorem 2 **Cramer's Theorem (Solution of linear systems by determinants)**

(a) *If the determinant* $D = \det \mathbf{A}$ *of a linear system of n equations*

$$\begin{aligned} a_{11}x_1 + a_{12}x_2 + \cdots + a_{1n}x_n &= b_1 \\ a_{21}x_1 + a_{22}x_2 + \cdots + a_{2n}x_n &= b_2 \\ \cdots\cdots\cdots\cdots\cdots\cdots\cdots\cdots \\ a_{n1}x_1 + a_{n2}x_2 + \cdots + a_{nn}x_n &= b_n \end{aligned} \tag{1}$$

in the same number of unknowns $x_1, \cdots, x_n$ *is not zero, the system has precisely one solution. This solution is given by the formulas*

$$x_1 = \frac{D_1}{D}, \quad x_2 = \frac{D_2}{D}, \cdots, \quad x_n = \frac{D_n}{D} \tag{2}$$

(Cramer's rule)

where D_k *is the determinant obtained from D by replacing in D the kth column by the column with the entries* $b_1, \cdots, b_n$.

(b) *Hence if* (1) *is* ***homogeneous*** *and* $D \neq 0$, *it has only the trivial solution* $x_1 = 0, x_2 = 0, \cdots, x_n = 0$. *If* $D = 0$, *the homogeneous system also has nontrivial solutions.*

Proof. From Theorem 1 and the Fundamental Theorem in Sec. 7.6 it follows that (1) has a unique solution, because if

$$D = \det \mathbf{A} = \begin{vmatrix} a_{11} & \cdots & a_{1n} \\ \cdot & \cdots & \cdot \\ & \cdots & \\ \cdot & \cdots & \cdot \\ a_{n1} & \cdots & a_{nn} \end{vmatrix} \neq 0, \tag{3}$$

then rank $\mathbf{A} = n$. We prove (2). Expanding D by the kth column, we obtain

$$D = a_{1k}C_{1k} + a_{2k}C_{2k} + \cdots + a_{nk}C_{nk}, \tag{4}$$

where C_{ik} is the cofactor of the entry a_{ik} in D. If we replace the entries in the kth column of D by any other numbers, we obtain a new determinant, say, $\tilde{D}$. Clearly, its expansion by the kth column will be of the form (4), with $a_{1k}, \cdots, a_{nk}$ replaced by those new numbers and the cofactors C_{ik} as before. In particular, if we choose as new numbers the entries $a_{1l}, \cdots, a_{nl}$ in the lth column of D (where $l \neq k$), then the expansion of the resulting determinant $\tilde{D}$ becomes

$$a_{1l}C_{1k} + a_{2l}C_{2k} + \cdots + a_{nl}C_{nk} = 0 \qquad (l \neq k) \tag{5}$$

because $\tilde{D}$ has two identical columns and is zero (by Theorem 6 in Sec. 7.8). If we multiply the first equation in (1) by C_{1k}, the second by $C_{2k}, \cdots$, the last by C_{nk} and add the resulting equations, we first obtain

$$C_{1k}(a_{11}x_1 + \cdots + a_{1n}x_n) + \cdots + C_{nk}(a_{n1}x_1 + \cdots + a_{nn}x_n)$$
$$= b_1C_{1k} + \cdots + b_nC_{nk}.$$

Collecting terms with the same x, we can write the left side as

$$x_1(a_{11}C_{1k} + \cdots + a_{n1}C_{nk}) + \cdots + x_n(a_{1n}C_{1k} + \cdots + a_{nn}C_{nk}).$$

From this we see that x_k is multiplied by

$$a_{1k}C_{1k} + \cdots + a_{nk}C_{nk},$$

which equals D by (4), and x_l is multiplied by

$$a_{1l}C_{1k} + \cdots + a_{nl}C_{nk},$$

which is zero by (5) when $l \neq k$. Hence the left side equals x_kD, and we thus have

$$x_kD = b_1C_{1k} + \cdots + b_nC_{nk}.$$

The right side is D_k (as defined in the theorem) expanded by its kth column. Division by D ($\neq 0$) gives (2).

If (1) is homogeneous and $D \neq 0$, then each D_k has a column of zeros, so that $D_k = 0$ by Theorem 3 in Sec. 7.8, and (2) gives the trivial solution.

Finally, if (1) is homogeneous and $D = 0$, then rank $\mathbf{A} < n$ by Theorem 1, so that nontrivial solutions exist by Theorem 2 in Sec. 7.6. ∎

An example is included in Sec. 7.8 (Example 1).

As an important consequence of Cramer's theorem, we may now express the entries of the inverse of a matrix as follows.

Theorem 3 **(Inverse of a matrix)**

The inverse of a nonsingular $n \times n$ *matrix* $\mathbf{A} = [a_{jk}]$ *is given by*

$$\text{(6)} \qquad \mathbf{A}^{-1} = \frac{1}{\det \mathbf{A}} [A_{jk}]^{\mathsf{T}} = \frac{1}{\det \mathbf{A}} \begin{bmatrix} A_{11} & A_{21} & \cdots & A_{n1} \\ A_{12} & A_{22} & \cdots & A_{n2} \\ \cdot & \cdot & \cdots & \cdot \\ A_{1n} & A_{2n} & \cdots & A_{nn} \end{bmatrix},$$

where A_{jk} *is the cofactor of* a_{jk} *in* det $\mathbf{A}$ (see Sec. 7.8). *Note well that in* $\mathbf{A}^{-1}$, *the cofactor* A_{jk} *occupies the same place as* a_{kj} *(not* a_{jk}*) does in* $\mathbf{A}$.

Proof. We denote the right side of (6) by $\mathbf{B}$ and show that $\mathbf{BA} = \mathbf{I}$. We write

$$\text{(7)} \qquad \mathbf{BA} = \mathbf{G} = [g_{kl}].$$

Here, by the definition of matrix multiplication, and by the form of the entries of $\mathbf{B}$ as given in (6),

$$\text{(8)} \qquad g_{kl} = \sum_{s=1}^{n} \frac{A_{sk}}{\det \mathbf{A}} a_{sl} = \frac{1}{\det \mathbf{A}} (a_{1l}A_{1k} + \cdots + a_{nl}A_{nk}).$$

Now (3) and (4) (with C_{jk} written in our present notation A_{jk}) show that the sum $(\cdots)$ on the right is $D = \det \mathbf{A}$ when $l = k$ and zero when $l \neq k$. Hence

$$g_{kk} = \frac{1}{\det \mathbf{A}} \det \mathbf{A} = 1, \qquad g_{kl} = 0 \quad (l \neq k),$$

so that $\mathbf{G} = [g_{kl}] = \mathbf{BA} = \mathbf{I}$ in (7). Similarly $\mathbf{AB} = \mathbf{I}$. Hence $\mathbf{B} = \mathbf{A}^{-1}$. ∎

The explicit formula (6) is often useful in theoretical studies, as opposed to *methods of actually **computing** inverses* (see Sec. 7.7).

EXAMPLE 1 Illustration of Theorem 3

Using (6), find the inverse of

$$\mathbf{A} = \begin{bmatrix} -1 & 1 & 2 \\ 3 & -1 & 1 \\ -1 & 3 & 4 \end{bmatrix}.$$

Solution. We get det $\mathbf{A} = -1(-7) - 13 + 2 \cdot 8 = 10$, and in (6),

$$A_{11} = \begin{vmatrix} -1 & 1 \\ 3 & 4 \end{vmatrix} = -7, \qquad A_{21} = -\begin{vmatrix} 1 & 2 \\ 3 & 4 \end{vmatrix} = 2, \qquad A_{31} = \begin{vmatrix} 1 & 2 \\ -1 & 1 \end{vmatrix} = 3,$$

$$A_{12} = -\begin{vmatrix} 3 & 1 \\ -1 & 4 \end{vmatrix} = -13, \qquad A_{22} = \begin{vmatrix} -1 & 2 \\ -1 & 4 \end{vmatrix} = -2, \qquad A_{32} = -\begin{vmatrix} -1 & 2 \\ 3 & 1 \end{vmatrix} = 7,$$

$$A_{13} = \begin{vmatrix} 3 & -1 \\ -1 & 3 \end{vmatrix} = 8, \qquad A_{23} = -\begin{vmatrix} -1 & 1 \\ -1 & 3 \end{vmatrix} = 2, \qquad A_{33} = \begin{vmatrix} -1 & 1 \\ 3 & -1 \end{vmatrix} = -2,$$

so that by (6), in agreement with Example 1 in Sec. 7.7,

$$\mathbf{A}^{-1} = \begin{bmatrix} -0.7 & 0.2 & 0.3 \\ -1.3 & -0.2 & 0.7 \\ 0.8 & 0.2 & -0.2 \end{bmatrix}.$$

∎

Using Theorem 1, we may now also prove the theorem on determinants of matrix products (Theorem 8 in Sec. 7.8), which we first restate.

Theorem 4 (Determinant of a product of matrices)

For any $n \times n$ matrices **A** *and* **B**,

$$\det(\mathbf{AB}) = \det(\mathbf{BA}) = \det \mathbf{A} \det \mathbf{B}. \tag{9}$$

Proof. If **A** is singular, so is **AB** by Theorem 2(c) in Sec. 7.7. Hence we have det $\mathbf{A} = 0$, det $(\mathbf{AB}) = 0$ by Theorem 1, and (9) is $0 = 0$, which holds.

Let **A** be nonsingular. Then we can reduce **A** to a diagonal matrix $\hat{\mathbf{A}} = [\hat{a}_{jk}]$ by Gauss–Jordan steps (Sec. 7.7). Under these operations, det **A** retains its value, by Theorem 7 in Sec. 7.8, except perhaps for a sign reversal if we have to interchange two rows to get a nonzero pivot (see Theorem 5 in Sec. 7.8). But the same operations reduce **AB** to $\hat{\mathbf{A}}\mathbf{B}$ with the same effect on det (**AB**). Hence it remains to prove (9) for $\hat{\mathbf{A}}\mathbf{B}$; written out,

$$\hat{\mathbf{A}}\mathbf{B} = \begin{bmatrix} \hat{a}_{11} & 0 & \cdots & 0 \\ 0 & \hat{a}_{22} & \cdots & 0 \\ & & \ddots & \\ 0 & 0 & \cdots & \hat{a}_{nn} \end{bmatrix} \begin{bmatrix} b_{11} & b_{12} & \cdots & b_{1n} \\ b_{21} & b_{22} & \cdots & b_{2n} \\ & & \vdots & \\ b_{n1} & b_{n2} & \cdots & b_{nn} \end{bmatrix}$$

$$= \begin{bmatrix} \hat{a}_{11}b_{11} & \hat{a}_{11}b_{12} & \cdots & \hat{a}_{11}b_{1n} \\ \hat{a}_{22}b_{21} & \hat{a}_{22}b_{22} & \cdots & \hat{a}_{22}b_{2n} \\ \vdots & & & \vdots \\ \hat{a}_{nn}b_{n1} & \hat{a}_{nn}b_{n2} & \cdots & \hat{a}_{nn}b_{nn} \end{bmatrix}.$$

We now take the determinant det $(\hat{\mathbf{A}}\mathbf{B})$. On the right we can take out a factor $\hat{a}_{11}$ from the first row, $\hat{a}_{22}$ from the second, $\cdots$, $\hat{a}_{nn}$ from the nth. But this product $\hat{a}_{11}\hat{a}_{22}\cdots\hat{a}_{nn}$ equals det $\hat{\mathbf{A}}$, since $\hat{\mathbf{A}}$ is diagonal. The remaining determinant is det $\mathbf{B}$, and (9) is proved. ∎

This completes our discussion of linear systems (Secs. 7.4–7.9). (For numerical methods, see Secs. 19.1–19.4, which are independent of other sections on numerical methods.) Beginning with Sec. 7.10, we turn to *eigenvalue problems,* whose importance in engineering and physics can hardly be overestimated.

Problem Set 7.9

Using Theorem 1, find the rank of the following matrices.

1. $\begin{bmatrix} -1 & 3 \\ 3 & -9 \end{bmatrix}$ **2.** $\begin{bmatrix} 30 & -70 & 50 \\ -36 & 84 & -60 \end{bmatrix}$ **3.** $\begin{bmatrix} 3 & 6 & 12 & 10 \\ 2 & 4 & 8 & 7 \end{bmatrix}$

4. $\begin{bmatrix} 0.4 & 2.0 \\ 3.2 & 1.6 \\ 0 & 1.1 \end{bmatrix}$ **5.** $\begin{bmatrix} 0 & 5 & 2 \\ 3 & 0 & -1 \\ 7 & 9 & 0 \end{bmatrix}$ **6.** $\begin{bmatrix} 21 & -3 & 17 & 13 \\ 46 & 11 & 52 & 14 \\ 33 & 48 & 71 & -23 \end{bmatrix}$

Using Theorem 3, find the inverse. Check your answer.

7. $\begin{bmatrix} 9 & 5 \\ 25 & 14 \end{bmatrix}$ **8.** $\begin{bmatrix} 0.8 & 0.6 \\ 0.6 & -0.8 \end{bmatrix}$ **9.** $\begin{bmatrix} \cos 3\theta & \sin 3\theta \\ -\sin 3\theta & \cos 3\theta \end{bmatrix}$

10. $\begin{bmatrix} -3 & 1 & -1 \\ 15 & -6 & 5 \\ -5 & 2 & -2 \end{bmatrix}$ **11.** $\begin{bmatrix} 1 & 0 & 0 \\ -2 & 1 & 0 \\ 3 & -4 & 1 \end{bmatrix}$ **12.** $\begin{bmatrix} 2 & 5 & 4 \\ 0 & 1 & 8 \\ 0 & 0 & 10 \end{bmatrix}$

13. $\begin{bmatrix} 0 & -0.4 & 0.2 \\ 0.1 & 0.1 & -0.1 \\ -0.2 & 0.4 & 0 \end{bmatrix}$ **14.** $\begin{bmatrix} 0 & 1 & 0 \\ 1 & 0 & 0 \\ 0 & 0 & 1 \end{bmatrix}$ **15.** $\begin{bmatrix} 19 & 2 & -9 \\ -4 & -1 & 2 \\ -2 & 0 & 1 \end{bmatrix}$

Solve by Cramer's rule and by the Gauss elimination:

16. $\begin{aligned} 4x - y &= 3 \\ -2x + 5y &= 21 \end{aligned}$

17. $\begin{aligned} -x + 3y - 2z &= 7 \\ 3x \qquad + 3z &= -3 \\ 2x + y + 2z &= -1 \end{aligned}$

18. $\begin{aligned} 2x + 5y + 3z &= 1 \\ -x + 2y + z &= 2 \\ x + y + z &= 0 \end{aligned}$

19. Using A^{-1} as given by (6), show that $AA^{-1} = I$.

20. Obtain (4) and (5) in Sec. 7.7 from the present Theorem 3.

21. Show that the product of two $n \times n$ matrices is singular if and only if at least one of the two matrices is singular.

Geometrical applications. Using Cramer's theorem, part (b), show:

22. The **plane through three points** (x_1, y_1, z_1), (x_2, y_2, z_2), (x_3, y_3, z_3) in space is given by the formula below.

23. The **circle through three points** (x_1, y_1), (x_2, y_2), (x_3, y_3) in the plane is given by the formula below.

$$\begin{vmatrix} x & y & z & 1 \\ x_1 & y_1 & z_1 & 1 \\ x_2 & y_2 & z_2 & 1 \\ x_3 & y_3 & z_3 & 1 \end{vmatrix} = 0 \qquad \begin{vmatrix} x^2 + y^2 & x & y & 1 \\ x_1^2 + y_1^2 & x_1 & y_1 & 1 \\ x_2^2 + y_2^2 & x_2 & y_2 & 1 \\ x_3^2 + y_3^2 & x_3 & y_3 & 1 \end{vmatrix} = 0$$

Problem 22 **Problem 23**

24. Find the plane through (1, 1, 1), (5, 0, 5), (3, 2, 6).

25. Find the circle through (2, 6), (6, 4), (7, 1).

7.10 Eigenvalues, Eigenvectors

From the standpoint of engineering applications, eigenvalue problems are among the most important problems in connection with matrices, and the student should follow our present discussion with particular attention. We first define the basic concepts and explain them in terms of typical examples. Then we shall turn to practical applications.

Let $\mathbf{A} = [a_{jk}]$ be a given $n \times n$ matrix and consider the vector equation

$$\mathbf{Ax} = \lambda \mathbf{x} \tag{1}$$

where λ is a number.

It is clear that the zero vector $\mathbf{x} = \mathbf{0}$ is a solution of (1) for any value of λ. A value of λ for which (1) has a solution $\mathbf{x} \neq \mathbf{0}$ is called an **eigenvalue**[13] or **characteristic value** (or *latent root*) of the matrix **A**. The corresponding solutions $\mathbf{x} \neq \mathbf{0}$ of (1) are called **eigenvectors** or **characteristic vectors of A** corresponding to that eigenvalue λ. The set of the eigenvalues is called the **spectrum** of **A**. The largest of the absolute values of the eigenvalues of **A** is called the **spectral radius** of **A**.

The set of all eigenvectors corresponding to an eigenvalue of **A**, together with **0**, forms a vector space (Sec. 7.5), called the **eigenspace** of **A** corresponding to this eigenvalue.

The problem of determining the eigenvalues and eigenvectors of a matrix is called an *eigenvalue problem.*[14] Problems of this type occur in connection with physical and technical applications, as we shall see.

[13] German: *Eigenwert: "eigen"* means "proper"; *"wert"* means "value."

[14] More precisely: an *algebraic* eigenvalue problem, because there are other eigenvalue problems involving a differential equation (see Secs. 5.8 and 11.3) or an integral equation.

How to Find Eigenvalues and Eigenvectors

The following example illustrates all steps.

EXAMPLE 1 Determination of eigenvalues and eigenvectors

Find the eigenvalues and eigenvectors of the matrix

$$\mathbf{A} = \begin{bmatrix} -5 & 2 \\ 2 & -2 \end{bmatrix}.$$

Solution. **(a)** ***Eigenvalues.*** These must be determined ***first.*** Equation (1) is

$$\mathbf{Ax} = \begin{bmatrix} -5 & 2 \\ 2 & -2 \end{bmatrix} \begin{bmatrix} x_1 \\ x_2 \end{bmatrix} = \lambda \begin{bmatrix} x_1 \\ x_2 \end{bmatrix};$$

written out in components,

$$\begin{aligned} -5x_1 + 2x_2 &= \lambda x_1 \\ 2x_1 - 2x_2 &= \lambda x_2. \end{aligned}$$

Transferring the terms on the right to the left, we get

$$\begin{aligned} (-5 - \lambda)x_1 + \quad 2x_2 \quad &= 0 \\ 2x_1 \quad + (-2 - \lambda)x_2 &= 0. \end{aligned} \tag{2*}$$

This can be written in matrix notation

$$(\mathbf{A} - \lambda\mathbf{I})\mathbf{x} = \mathbf{0}. \tag{3*}$$

[Indeed, (1) is $\mathbf{Ax} - \lambda\mathbf{x} = \mathbf{0}$ or $\mathbf{Ax} - \lambda\mathbf{Ix} = 0$, which gives (3).] We see that this is a *homogeneous* linear system. By Cramer's Theorem in Sec. 7.9 it has a nontrivial solution $\mathbf{x} \neq \mathbf{0}$ (an eigenvector of $\mathbf{A}$ we are looking for) if and only if its coefficient determinant is zero,

$$\begin{aligned} D(\lambda) = \det(\mathbf{A} - \lambda\mathbf{I}) &= \begin{vmatrix} -5 - \lambda & 2 \\ 2 & -2 - \lambda \end{vmatrix} \\ &= (-5 - \lambda)(-2 - \lambda) - 4 = \lambda^2 + 7\lambda + 6 = 0. \end{aligned} \tag{4*}$$

We call $D(\lambda)$ the **characteristic determinant** or, if expanded, the **characteristic polynomial**, and $D(\lambda) = 0$ the **characteristic equation** of $\mathbf{A}$. The solutions of this quadratic equation are $\lambda_1 = -1$ and $\lambda_2 = -6$. These are the eigenvalues of $\mathbf{A}$.

***(b_1) Eigenvector of* A *corresponding to* λ_1.** This vector is obtained from (2*) with $\lambda = \lambda_1 = -1$, that is,

$$\begin{aligned} -4x_1 + 2x_2 &= 0 \\ 2x_1 - x_2 &= 0. \end{aligned}$$

A solution is x_1 arbitrary, $x_2 = 2x_1$. If we choose $x_1 = 1$, then $x_2 = 2$, and an eigenvector of $\mathbf{A}$ corresponding to $\lambda_1 = -1$ is

$$\mathbf{x}_1 = \begin{bmatrix} 1 \\ 2 \end{bmatrix}.$$

We can easily check this:

$$\mathbf{Ax}_1 = \begin{bmatrix} -5 & 2 \\ 2 & -2 \end{bmatrix} \begin{bmatrix} 1 \\ 2 \end{bmatrix} = \begin{bmatrix} -1 \\ -2 \end{bmatrix} = (-1)\mathbf{x}_1 = \lambda_1\mathbf{x}_1.$$

***(b_2) Eigenvector of* A *corresponding to* λ_2.** For $\lambda = \lambda_2 = -6$, equation (2*) becomes

$$x_1 + 2x_2 = 0$$
$$2x_1 + 4x_2 = 0.$$

A solution is $x_2 = -x_1/2$. If we choose $x_1 = 2$, we get $x_2 = -1$, and an eigenvector of A corresponding to $\lambda_2 = -6$ is

$$\mathbf{x}_2 = \begin{bmatrix} 2 \\ -1 \end{bmatrix}.$$

Check this. ■

This example illustrates the general case as follows. Equation (1) written in components is

$$\begin{aligned} a_{11}x_1 + \cdots + a_{1n}x_n &= \lambda x_1 \\ a_{21}x_1 + \cdots + a_{2n}x_n &= \lambda x_2 \\ &\cdots\cdots \\ a_{n1}x_1 + \cdots + a_{nn}x_n &= \lambda x_n. \end{aligned}$$

Transferring the terms on the right side to the left side, we have

$$\text{(2)} \quad \begin{array}{ccccccc} (a_{11} - \lambda)x_1 & + & a_{12}x_2 & + \cdots + & a_{1n}x_n & = 0 \\ a_{21}x_1 & + & (a_{22} - \lambda)x_2 & + \cdots + & a_{2n}x_n & = 0 \\ \cdots & & \cdots & & \cdots & \\ a_{n1}x_1 & + & a_{n2}x_2 & + \cdots + & (a_{nn} - \lambda)x_n & = 0. \end{array}$$

In matrix notation,

$$\text{(3)} \quad \boxed{(\mathbf{A} - \lambda\mathbf{I})\mathbf{x} = \mathbf{0}.}$$

By Cramer's Theorem in Sec. 7.9, this homogeneous linear system of equations has a nontrivial solution if and only if the corresponding determinant of the coefficients is zero:

$$\text{(4)} \quad D(\lambda) = \det(\mathbf{A} - \lambda\mathbf{I}) = \begin{vmatrix} a_{11} - \lambda & a_{12} & \cdots & a_{1n} \\ a_{21} & a_{22} - \lambda & \cdots & a_{2n} \\ \cdot & \cdot & \cdots & \cdot \\ a_{n1} & a_{n2} & \cdots & a_{nn} - \lambda \end{vmatrix} = 0.$$

$D(\lambda)$ is called the **characteristic determinant,** and (4) is called the **characteristic equation** corresponding to the matrix **A.** By developing $D(\lambda)$ we obtain a polynomial of nth degree in λ. This is called the **characteristic polynomial** corresponding to **A.**

This proves the following important theorem.

Theorem 1 **(Eigenvalues)**
The eigenvalues of a square matrix **A** *are the roots of the corresponding characteristic equation* (4).

Hence an $n \times n$ *matrix has at least one eigenvalue and at most* n *numerically different eigenvalues.*

For larger n, the actual computation of eigenvalues will in general require the use of Newton's method (Sec. 18.2) or another numerical approximation method in Secs. 19.7–19.10. Sometimes it may also help to observe that the product and sum of the eigenvalues are the constant term and $(-1)^{n-1}$ times the coefficient of the second highest term, respectively, of the characteristic polynomial (why?). Once an eigenvalue λ_1 has been found, one may divide the characteristic polynomial by $\lambda - \lambda_1$.

The eigen*values* must be determined first. Once these are known, corresponding eigen*vectors* are obtained from the system (2), for instance, by the Gauss elimination, where λ is the eigenvalue for which an eigenvector is wanted. This is what we did in Example 1 and shall do again in the examples below.

Theorem 2 **(Eigenvectors)**
If **x** *is an eigenvector of a matrix* **A** *corresponding to an eigenvalue* λ, *so is* $k\mathbf{x}$ *with any* $k \neq 0$.

Proof. $\mathbf{Ax} = \lambda\mathbf{x}$ implies $k(\mathbf{Ax}) = \mathbf{A}(k\mathbf{x}) = \lambda(k\mathbf{x})$. ∎

Examples 2 and 3 will illustrate that an $n \times n$ matrix may have n linearly independent eigenvectors,[15] or it may have fewer than n. In Example 4 we shall see that a *real* matrix may have *complex* eigenvalues and eigenvectors.

EXAMPLE 2 **Multiple eigenvalues**
Find the eigenvalues and eigenvectors of the matrix

$$\mathbf{A} = \begin{bmatrix} -2 & 2 & -3 \\ 2 & 1 & -6 \\ -1 & -2 & 0 \end{bmatrix}.$$

Solution. For our matrix, the characteristic determinant gives the characteristic equation

$$-\lambda^3 - \lambda^2 + 21\lambda + 45 = 0.$$

The roots (eigenvalues of **A**) are $\lambda_1 = 5$, $\lambda_2 = \lambda_3 = -3$. To find eigenvectors, we apply the Gauss elimination (Sec. 7.4) to the system $(\mathbf{A} - \lambda\mathbf{I})\mathbf{x} = \mathbf{0}$, first with $\lambda = 5$ and then with $\lambda = -3$. We find that the vector

[15] A property that will play a role in Sec. 7.14.

$$\mathbf{x}_1 = \begin{bmatrix} 1 \\ 2 \\ -1 \end{bmatrix}$$

is an eigenvector of $\mathbf{A}$ corresponding to the eigenvalue 5, and the vectors

$$\mathbf{x}_2 = \begin{bmatrix} -2 \\ 1 \\ 0 \end{bmatrix} \quad \text{and} \quad \mathbf{x}_3 = \begin{bmatrix} 3 \\ 0 \\ 1 \end{bmatrix}$$

are two linearly independent eigenvectors of $\mathbf{A}$ corresponding to the eigenvalue -3. This agrees with the fact that, for $\lambda = -3$, the matrix $\mathbf{A} - \lambda\mathbf{I}$ has rank 1 and so, by Theorem 2 in Sec. 7.6, a basis of solutions of the corresponding system (2) with $\lambda = -3$,

$$\begin{aligned} x_1 + 2x_2 - 3x_3 &= 0 \\ 2x_1 + 4x_2 - 6x_3 &= 0 \\ -x_1 - 2x_2 + 3x_3 &= 0 \end{aligned}$$

consists of two linearly independent vectors. ∎

If an eigenvalue λ of a matrix $\mathbf{A}$ is a root of order M_λ of the characteristic polynomial of $\mathbf{A}$, then M_λ is called the **algebraic multiplicity** of λ, as opposed to the **geometric multiplicity** m_λ of λ, which is defined to be the number of linearly independent eigenvectors corresponding to λ, thus, the dimension of the corresponding eigenspace. Since the characteristic polynomial has degree n, the sum of all algebraic multiplicities equals n. In Example 2, for $\lambda = -3$ we have $m_\lambda = M_\lambda = 2$. In general, $m_\lambda \leqq M_\lambda$, as can be shown. We convince ourselves that $m_\lambda < M_\lambda$ is possible:

EXAMPLE 3 Algebraic and geometric multiplicity

The characteristic equation of the matrix

$$\mathbf{A} = \begin{bmatrix} 0 & 1 \\ 0 & 0 \end{bmatrix} \quad \text{is} \quad \det(\mathbf{A} - \lambda\mathbf{I}) = \begin{vmatrix} -\lambda & 1 \\ 0 & -\lambda \end{vmatrix} = \lambda^2 = 0.$$

Hence $\lambda = 0$ is an eigenvalue of algebraic multiplicity 2. But its geometric multiplicity is only 1, since eigenvectors result from $-0x_1 + x_2 = 0$, hence $x_2 = 0$, in the form $[x_1 \quad 0]^\mathsf{T}$. ∎

EXAMPLE 4 Real matrices with complex eigenvalues and eigenvectors

Since real polynomials may have complex roots (which then occur in conjugate pairs), a real matrix may have complex eigenvalues and eigenvectors. For instance, the characteristic equation of the skew-symmetric matrix

$$\mathbf{A} = \begin{bmatrix} 0 & 1 \\ -1 & 0 \end{bmatrix} \quad \text{is} \quad \det(\mathbf{A} - \lambda\mathbf{I}) = \begin{vmatrix} -\lambda & 1 \\ -1 & -\lambda \end{vmatrix} = \lambda^2 + 1 = 0$$

and gives the eigenvalues $\lambda_1 = i\ (= \sqrt{-1})$, $\lambda_2 = -i$. Eigenvectors are obtained from $-ix_1 + x_2 = 0$ and $ix_1 + x_2 = 0$, respectively, and we can choose $x_1 = 1$ to get

$$\begin{bmatrix} 1 \\ i \end{bmatrix} \quad \text{and} \quad \begin{bmatrix} 1 \\ -i \end{bmatrix}.$$

The reader may show that, more generally, these vectors are eigenvectors of the matrix

$$\mathbf{A} = \begin{bmatrix} a & b \\ -b & a \end{bmatrix} \qquad (a, b \text{ real})$$

and that $\mathbf{A}$ has the eigenvalues $a + ib$ and $a - ib$. ∎

Having gained a first impression of matrix eigenvalue problems, in the next section we illustrate their importance by some typical applications.

Problem Set 7.10

Find the eigenvalues and eigenvectors of the following matrices.

1. $\begin{bmatrix} 1 & 0 \\ 0 & -4 \end{bmatrix}$ **2.** $\begin{bmatrix} 1 & 2 \\ 0 & 3 \end{bmatrix}$ **3.** $\begin{bmatrix} 2 & 0 \\ 4 & 5 \end{bmatrix}$ **4.** $\begin{bmatrix} 3 & 4 \\ 4 & -3 \end{bmatrix}$

5. $\begin{bmatrix} 0 & 0 \\ 0 & 0 \end{bmatrix}$ **6.** $\begin{bmatrix} 1 & 0 \\ 0 & 1 \end{bmatrix}$ **7.** $\begin{bmatrix} 0 & 1 \\ 1 & 0 \end{bmatrix}$ **8.** $\begin{bmatrix} 14 & -10 \\ 5 & -1 \end{bmatrix}$

9. $\begin{bmatrix} -5 & 2 \\ -9 & 6 \end{bmatrix}$ **10.** $\begin{bmatrix} 0 & 2 \\ -2 & 0 \end{bmatrix}$ **11.** $\begin{bmatrix} 0 & 1 \\ -1 & 0 \end{bmatrix}$ **12.** $\begin{bmatrix} 0.8 & -0.6 \\ 0.6 & 0.8 \end{bmatrix}$

13. $\begin{bmatrix} 4 & 0 & 0 \\ 0 & 8 & 0 \\ 0 & 0 & 6 \end{bmatrix}$ **14.** $\begin{bmatrix} 3 & 0 & 0 \\ 5 & 4 & 0 \\ 3 & 6 & 1 \end{bmatrix}$ **15.** $\begin{bmatrix} -1 & 1 & 0 \\ 1 & -1 & 0 \\ 0 & 0 & 0 \end{bmatrix}$

16. $\begin{bmatrix} 2 & -2 & 3 \\ -2 & -1 & 6 \\ 1 & 2 & 0 \end{bmatrix}$ **17.** $\begin{bmatrix} 6 & 10 & 6 \\ 0 & 8 & 12 \\ 0 & 0 & 2 \end{bmatrix}$ **18.** $\begin{bmatrix} 32 & -24 & -8 \\ 16 & -11 & -4 \\ 72 & -57 & -18 \end{bmatrix}$

19. $\begin{bmatrix} 8 & 0 & 3 \\ 2 & 2 & 1 \\ 2 & 0 & 3 \end{bmatrix}$ **20.** $\begin{bmatrix} 5 & 0 & -15 \\ -3 & -4 & 9 \\ 5 & 0 & -15 \end{bmatrix}$ **21.** $\begin{bmatrix} 1 & 2 & -3 \\ 2 & 4 & -6 \\ -1 & -2 & 3 \end{bmatrix}$

Some general properties of the spectrum. Let $\lambda_1, \cdots, \lambda_n$ be the eigenvalues of a given matrix $\mathbf{A} = [a_{jk}]$. In each case prove the proposition and illustrate it with an example.

22. **(Trace)** The so-called *trace* of $\mathbf{A}$, given by trace $\mathbf{A} = a_{11} + a_{22} + \cdots + a_{nn}$, is equal to $\lambda_1 + \cdots + \lambda_n$. The constant term of $D(\lambda)$ equals det $\mathbf{A}$.

23. If $\mathbf{A}$ is real, the eigenvalues are real or complex conjugates in pairs.

24. **(Inverse)** The inverse $\mathbf{A}^{-1}$ exists if and only if $\lambda_j \neq 0$ $(j = 1, \cdots, n)$.

25. The inverse $\mathbf{A}^{-1}$ has the eigenvalues $1/\lambda_1, \cdots, 1/\lambda_n$.

26. **(Triangular matrix)** If $\mathbf{A}$ is triangular, the entries on the main diagonal are the eigenvalues of $\mathbf{A}$.

27. **("Spectral shift")** The matrix $\mathbf{A} - k\mathbf{I}$ has the eigenvalues $\lambda_1 - k, \cdots, \lambda_n - k$.

28. The matrix $k\mathbf{A}$ has the eigenvalues $k\lambda_1, \cdots, k\lambda_n$.

29. The matrix $\mathbf{A}^m$ (m a nonnegative integer) has the eigenvalues $\lambda_1^m, \cdots, \lambda_n^m$.

30. (Spectral mapping theorem) The matrix

$$k_m\mathbf{A}^m + k_{m-1}\mathbf{A}^{m-1} + \cdots + k_1\mathbf{A} + k_0\mathbf{I},$$

which is called a **polynomial matrix,** has the eigenvalues

$$k_m\lambda_j^{\,m} + k_{m-1}\lambda_j^{\,m-1} + \cdots + k_1\lambda_j + k_0 \quad (j = 1, \cdots, n).$$

(This proposition is called the *spectral mapping theorem for polynomial matrices.*) The eigenvectors of that matrix are the same as those of **A**.

7.11 Some Applications of Eigenvalue Problems

In this section we discuss a few typical examples from the range of applications of matrix eigenvalue problems, which is incredibly large. Chapter 4 shows matrix eigenvalue problems related to differential equations governing mechanical systems and electrical networks. To keep our present discussion independent, for students not familiar with Chap. 4 we include a typical application of that kind as our last example.

EXAMPLE 1 Stretching of an elastic membrane

An elastic membrane in the x_1x_2-plane with boundary circle $x_1^2 + x_2^2 = 1$ (Fig. 136) is stretched so that a point P: (x_1, x_2) goes over into the point Q: (y_1, y_2) given by

$$(1) \qquad \mathbf{y} = \begin{bmatrix} y_1 \\ y_2 \end{bmatrix} = \mathbf{A}\mathbf{x} = \begin{bmatrix} 5 & 3 \\ 3 & 5 \end{bmatrix} \begin{bmatrix} x_1 \\ x_2 \end{bmatrix}; \qquad \text{in components,} \qquad \begin{aligned} y_1 &= 5x_1 + 3x_2 \\ y_2 &= 3x_1 + 5x_2. \end{aligned}$$

Find the "*principal directions,*" that is, directions of the position vector **x** of P for which the direction of the position vector **y** of Q is the same or exactly opposite. What shape does the boundary circle take under this deformation?

Solution. We are looking for vectors **x** such that $\mathbf{y} = \lambda\mathbf{x}$. Since $\mathbf{y} = \mathbf{A}\mathbf{x}$, this gives $\mathbf{A}\mathbf{x} = \lambda\mathbf{x}$, an equation of the form (1), an eigenvalue problem. In components, $\mathbf{A}\mathbf{x} = \lambda\mathbf{x}$ is

$$(2) \qquad \begin{aligned} 5x_1 + 3x_2 &= \lambda x_1 \\ 3x_1 + 5x_2 &= \lambda x_2 \end{aligned} \qquad \text{or} \qquad \begin{aligned} (5-\lambda)x_1 + \qquad 3x_2 &= 0 \\ 3x_1 + (5-\lambda)\,x_2 &= 0. \end{aligned}$$

The characteristic equation is

$$(3) \qquad \begin{vmatrix} 5-\lambda & 3 \\ 3 & 5-\lambda \end{vmatrix} = (5-\lambda)^2 - 9 = 0.$$

Its solutions are $\lambda_1 = 8$ and $\lambda_2 = 2$. These are the eigenvalues of our problem. For $\lambda = \lambda_1 = 8$, our system (2) becomes

$$\begin{aligned} -3x_1 + 3x_2 &= 0, \\ 3x_1 - 3x_2 &= 0. \end{aligned} \qquad \begin{aligned} &\text{Solution } x_2 = x_1, \quad x_1 \text{ arbitrary,} \\ &\text{for instance, } x_1 = x_2 = 1. \end{aligned}$$

For $\lambda_2 = 2$, our system (2) becomes

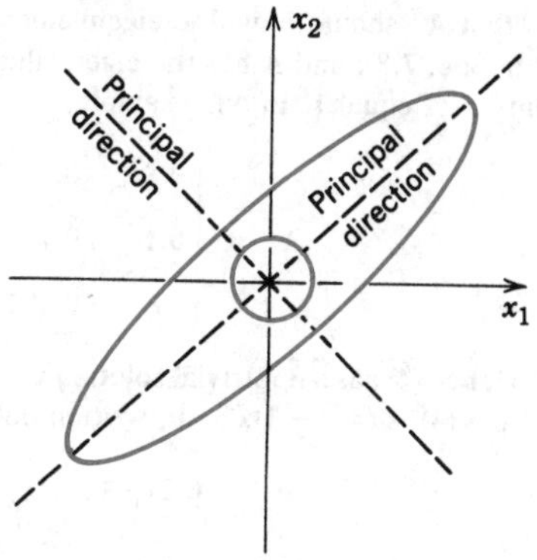

Fig. 136. **Undeformed and deformed membrane in Example 1**

$$\begin{aligned} 3x_1 + 3x_2 &= 0, \\ 3x_1 + 3x_2 &= 0. \end{aligned} \quad \Bigg| \quad \begin{aligned} &\text{Solution } x_2 = -x_1, \quad x_1 \text{ arbitrary,} \\ &\text{for instance, } x_1 = 1, x_2 = -1. \end{aligned}$$

We thus obtain as eigenvectors of **A**, for instance,

$$\begin{bmatrix} 1 \\ 1 \end{bmatrix} \text{ corresponding to } \lambda_1; \qquad \begin{bmatrix} 1 \\ -1 \end{bmatrix} \text{ corresponding to } \lambda_2;$$

(or a nonzero scalar multiple of these). These vectors make 45° and 135° angles with the positive x_1-direction. They give the principal directions, the answer to our problem. The eigenvalues show that in the principal directions the membrane is stretched by factors 8 and 2, respectively; see Fig. 136.

Accordingly, if we choose the principal directions as directions of a new Cartesian u_1u_2-coordinate system, say, with the positive u_1-semiaxis in the first quadrant and the positive u_2-semiaxis in the second quadrant of the x_1x_2-system, and if we set

$$u_1 = r\cos\phi, \qquad u_2 = r\sin\phi,$$

then a boundary point of the unstretched circular membrane has coordinates cos ϕ, sin ϕ. Hence, after the stretch we have

$$z_1 = 8\cos\phi, \qquad z_2 = 2\sin\phi.$$

Since $\cos^2\phi + \sin^2\phi = 1$, this shows that the deformed boundary is an ellipse (Fig. 136)

$$\frac{z_1^2}{8^2} + \frac{z_2^2}{2^2} = 1$$

with principal semiaxes 8 and 2 in the principal directions. ∎

EXAMPLE 2 **Eigenvalue problems arising from Markov processes**

As another application, let us show that Markov processes also lead to eigenvalue problems. To see this, let us determine the limit state of the land-use succession in Example 8, Sec. 7.3.

Solution. We recall that Example 8 in Sec. 7.3 concerns a *Markov process* and that such a transition process is governed by a **stochastic matrix** $\mathbf{A} = [a_{jk}]$, that is, a square matrix with nonnegative entries a_{jk} (giving transition probabilities) and all row sums equal to 1. Furthermore, state **y** (a column vector) is obtained from state **x** according to $\mathbf{y}^\mathsf{T} = \mathbf{x}^\mathsf{T}\mathbf{A}$, equivalently, $\mathbf{y} = \mathbf{A}^\mathsf{T}\mathbf{x}$. A limit is reached if states remain unchanged, if $\mathbf{x}^\mathsf{T} = \mathbf{x}^\mathsf{T}\mathbf{A}$ or

$$\mathbf{A}^\mathsf{T}\mathbf{x} = \mathbf{x}. \tag{4}$$

This means that $\mathbf{A}^T$ should have the eigenvalue 1. But $\mathbf{A}^T$ has the same eigenvalues as $\mathbf{A}$ (by Theorem 1 in Sec. 7.8); and $\mathbf{A}$ has the eigenvalue 1, with eigenvector $\mathbf{v}^T = [1 \cdots 1]$, because the row sums of $\mathbf{A}$ equal 1. In our example,

$$\mathbf{Av} = \begin{bmatrix} 0.8 & 0.1 & 0.1 \\ 0.1 & 0.7 & 0.2 \\ 0 & 0.1 & 0.9 \end{bmatrix} \begin{bmatrix} 1 \\ 1 \\ 1 \end{bmatrix} = \begin{bmatrix} 1 \\ 1 \\ 1 \end{bmatrix}$$

as claimed. Hence (4) has a nontrivial solution $\mathbf{x} \neq \mathbf{0}$, which is an eigenvector of $\mathbf{A}^T$ corresponding to $\lambda = 1$. Now (4) is $(\mathbf{A}^T - \mathbf{I})\mathbf{x} = \mathbf{0}$; written out,

$$\begin{aligned} -0.2x_1 + 0.1x_2 \qquad\quad &= 0 \\ 0.1x_1 - 0.3x_2 + 0.1x_3 &= 0 \\ 0.1x_1 + 0.2x_2 - 0.1x_3 &= 0. \end{aligned}$$

A solution is $x^T = [12.5 \quad 25 \quad 62.5]$. *Answer.* Assuming that the probabilities remain the same as time progresses, we see that the states tend to 12.5% residentially, 25% commercially, and 62.5% industrially used area. ∎

EXAMPLE 3 **Eigenvalue problems arising from population models. Leslie model**

The Leslie model describes age-specified population growth, as follows. Let the oldest age attained by the females in some animal population be 6 years. Divide the population into three age classes of 2 years each. Let the "*Leslie matrix*" be

$$\mathbf{L} = [l_{jk}] = \begin{bmatrix} 0 & 2.3 & 0.4 \\ 0.6 & 0 & 0 \\ 0 & 0.3 & 0 \end{bmatrix} \tag{5}$$

where l_{1k} is the average number of daughters born to a single female during the time she is in age class k, and $l_{j,j-1}$ ($j = 2, 3$) is the fraction of females in age class $j - 1$ that will survive and pass into class j. (a) What is the number of females in each class after 2, 4, 6 years if each class initially consists of 500 females? (b) For what initial distribution will the number of females in each class change by the same proportion? What is this rate of change?

Solution. **(a)** Initially, $\mathbf{x}_{(0)}^T = [500 \quad 500 \quad 500]$. After 2 years,

$$\mathbf{x}_{(2)} = \mathbf{Lx}_{(0)} = \begin{bmatrix} 0 & 2.3 & 0.4 \\ 0.6 & 0 & 0 \\ 0 & 0.3 & 0 \end{bmatrix} \begin{bmatrix} 500 \\ 500 \\ 500 \end{bmatrix} = \begin{bmatrix} 1350 \\ 300 \\ 150 \end{bmatrix}.$$

Similarly, after 4 years we have $\mathbf{x}_{(4)}^T = (\mathbf{Lx}_{(2)})^T = [750 \quad 810 \quad 90]$ and after 6 years we have $\mathbf{x}_{(6)}^T = (\mathbf{Lx}_{(4)})^T = [1899 \quad 450 \quad 243]$.

(b) Proportional change means that we are looking for a distribution vector $\mathbf{x}$ such that $\mathbf{Lx} = \lambda\mathbf{x}$, where λ is the rate of change (growth if $\lambda > 1$, decrease if $\lambda < 1$). The characteristic equation is

$$\det(\mathbf{L} - \lambda\mathbf{I}) = -\lambda^3 - 0.6(-2.3\lambda - 0.3 \cdot 0.4) = -\lambda^3 + 1.38\lambda + 0.072 = 0.$$

A positive root is found to be (for instance, by Newton's method, Sec. 18.2) $\lambda = 1.2$. A corresponding eigenvector can be determined from $0.6x_1 - 1.2x_2 = 0$, $0.3x_2 - 1.2x_3 = 0$. Thus, $\mathbf{x}^T = [1 \quad 0.5 \quad 0.125]$. To get an initial population of 1500, as before, we multiply $\mathbf{x}^T$ by 923. *Answer.* 923 females in class 1, 462 in class 2, 115 in class 3. Growth rate 1.2. ∎

EXAMPLE 4 Vibrating system of two masses on two springs (Fig. 57 in Sec. 4.1)

Mass–spring systems involving several masses and springs can be treated as eigenvalue problems. For instance, the mechanical system in Fig. 57 (Sec. 4.1) is governed by the differential equations

$$(6)\qquad \begin{aligned} y_1'' &= -5y_1 + 2y_2 \\ y_2'' &= 2y_1 - 2y_2 \end{aligned}$$

where y_1 and y_2 are the displacements of the masses from rest, as shown in the figure, and primes denote derivatives with respect to time t. In vector form, this becomes

$$(7)\qquad \mathbf{y}'' = \begin{bmatrix} y_1'' \\ y_2'' \end{bmatrix} = \mathbf{A}\mathbf{y} = \begin{bmatrix} -5 & 2 \\ 2 & -2 \end{bmatrix} \begin{bmatrix} y_1 \\ y_2 \end{bmatrix}.$$

We try a vector solution of the form

$$(8)\qquad \mathbf{y} = \mathbf{x}e^{\omega t}.$$

This is suggested by a mechanical system of a single mass on a spring (Sec. 2.5), whose motion is given by exponential functions (and sines and cosines). Substitution into (7) gives

$$\omega^2 \mathbf{x} e^{\omega t} = \mathbf{A}\mathbf{x}e^{\omega t}.$$

Dividing by $e^{\omega t}$ and writing $\omega^2 = \lambda$, we see that our mechanical system leads to the eigenvalue problem

$$(9)\qquad \mathbf{A}\mathbf{x} = \lambda \mathbf{x} \qquad \text{where } \lambda = \omega^2.$$

From Example 1 in Sec. 7.10 we see that $\mathbf{A}$ has the eigenvalues $\lambda_1 = -1$; consequently, $\omega = \sqrt{-1} = \pm i$, and $\lambda_2 = -6$, thus $\omega = \sqrt{-6} = \pm i\sqrt{6}$, and corresponding eigenvectors

$$(10)\qquad \mathbf{x}_1 = \begin{bmatrix} 1 \\ 2 \end{bmatrix} \quad \text{and} \quad \mathbf{x}_2 = \begin{bmatrix} 2 \\ -1 \end{bmatrix}.$$

From (8) we thus obtain the four complex solutions [see (7), Sec. 2.3]

$$\begin{aligned} \mathbf{x}_1 e^{\pm it} &= \mathbf{x}_1(\cos t \pm i \sin t), \\ \mathbf{x}_2 e^{\pm i\sqrt{6}t} &= \mathbf{x}_2(\cos \sqrt{6}\,t \pm i \sin \sqrt{6}\,t), \end{aligned}$$

and by addition and subtraction (see Sec. 2.3) we get the four real solutions

$$\mathbf{x}_1 \cos t, \qquad \mathbf{x}_1 \sin t, \qquad \mathbf{x}_2 \cos \sqrt{6}\,t, \qquad \mathbf{x}_2 \sin \sqrt{6}\,t.$$

A general solution is obtained by taking a linear combination of these,

$$\mathbf{y} = \mathbf{x}_1(a_1 \cos t + b_1 \sin t) + \mathbf{x}_2(a_2 \cos \sqrt{6}\,t + b_2 \sin \sqrt{6}\,t)$$

with arbitrary constants a_1, b_1, a_2, b_2 (to which values can be assigned by prescribing initial displacement and initial velocity of each of the two masses). By (10), the components of $\mathbf{y}$ are

$$\begin{aligned} y_1 &= a_1 \cos t + b_1 \sin t + 2a_2 \cos \sqrt{6}\,t + 2b_2 \sin \sqrt{6}\,t \\ y_2 &= 2a_1 \cos t + 2b_1 \sin t - a_2 \cos \sqrt{6}\,t - b_2 \sin \sqrt{6}\,t. \end{aligned}$$

These functions describe harmonic oscillations of the two masses. ∎

Problem Set 7.11

Find the principal directions and corresponding factors of extension or contraction of the elastic deformation $\mathbf{y} = \mathbf{Ax}$, where $\mathbf{A}$ equals

1. $\begin{bmatrix} 3 & \sqrt{2} \\ \sqrt{2} & 2 \end{bmatrix}$ **2.** $\begin{bmatrix} 3/2 & 1/\sqrt{2} \\ 1/\sqrt{2} & 1 \end{bmatrix}$ **3.** $\begin{bmatrix} 1 & 0.2 \\ 0.2 & 1 \end{bmatrix}$

4. $\begin{bmatrix} 2.00 & 1.75 \\ 2.00 & 2.25 \end{bmatrix}$ **5.** $\begin{bmatrix} 2 & 1 \\ 1 & 2 \end{bmatrix}$ **6.** $\begin{bmatrix} 3.50 & 1.00 \\ 0.75 & 2.50 \end{bmatrix}$

Find limit states of the Markov processes governed by the following stochastic matrices.

7. $\begin{bmatrix} 0.3 & 0.7 \\ 0.5 & 0.5 \end{bmatrix}$ **8.** $\begin{bmatrix} 0.50 & 0.25 & 0.25 \\ 0.25 & 0.50 & 0.25 \\ 0.25 & 0.25 & 0.50 \end{bmatrix}$ **9.** $\begin{bmatrix} 0.6 & 0.4 & 0 \\ 0.1 & 0.1 & 0.8 \\ 0 & 1.0 & 0 \end{bmatrix}$

Find the growth rate in the Leslie model with Leslie matrix

10. $\begin{bmatrix} 0 & 5.2 & 2.125 \\ 0.4 & 0 & 0 \\ 0 & 0.3 & 0 \end{bmatrix}$ **11.** $\begin{bmatrix} 0 & 8 & 0 \\ 0.5 & 0 & 0 \\ 0 & 0.2 & 0 \end{bmatrix}$ **12.** $\begin{bmatrix} 0 & 4.5 & 2.5 \\ 0.2 & 0 & 0 \\ 0 & 0.2 & 0 \end{bmatrix}$

13. **(Leontief[16] input–output model)** Suppose that three industries are interrelated so that their outputs are used as inputs by themselves, according to the 3×3 **consumption matrix**

$$\mathbf{A} = [a_{jk}] = \begin{bmatrix} 0.2 & 0.5 & 0 \\ 0.6 & 0 & 0.4 \\ 0.2 & 0.5 & 0.6 \end{bmatrix}$$

where a_{jk} is the fraction of the output of industry k consumed (purchased) by industry j. Let p_j be the price charged by industry j for its total output. A problem is to find prices so that for each industry, total expenditures equal total income. Show that this leads to $\mathbf{Ap} = \mathbf{p}$, where $\mathbf{p} = [p_1 \quad p_2 \quad p_3]^\mathsf{T}$, and find a solution $\mathbf{p}$ with nonnegative p_1, p_2, p_3.

14. Show that a consumption matrix as considered in Prob. 13 must have column sums 1 and always has the eigenvalue 1.

15. **(Open Leontief input–output model)** If not the whole output is consumed by the industries themselves (as in Prob. 13), then instead of $\mathbf{Ax} = \mathbf{x}$ we have $\mathbf{x} - \mathbf{Ax} = \mathbf{y}$, where $\mathbf{x} = [x_1 \quad x_2 \quad x_3]^\mathsf{T}$ is produced, $\mathbf{Ax}$ is consumed by the industries, and, thus, $\mathbf{y}$ is the net production available for other consumers. Find for what production $\mathbf{x}$ a given $\mathbf{y} = [0.1 \quad 0.3 \quad 0.1]^\mathsf{T}$ can be achieved if the consumption matrix is

$$\mathbf{A} = \begin{bmatrix} 0.1 & 0.4 & 0.2 \\ 0.5 & 0 & 0.1 \\ 0.1 & 0.4 & 0.4 \end{bmatrix}.$$

[16] WASSILY LEONTIEF (born 1906). American economist. For his work he was awarded the Nobel Prize in 1973.

16. **(Perron–Frobenius theorem)** Show that a Leslie matrix **L** with positive l_{12}, l_{13}, l_{21}, l_{32} has a positive eigenvalue. *Hint.* Use Probs. 22, 23 in Sec. 7.10. (This is a special case of the famous *Perron–Frobenius theorem* in Sec. 19.7, which is difficult to prove in its general form.)

For **differential equations** and related matrix eigenvalue problems, see Chap. 5.

7.12 Symmetric, Skew-Symmetric, and Orthogonal Matrices

We consider three classes of real square matrices that occur quite frequently in applications. These are defined as follows.

Definitions of symmetric, skew-symmetric, and orthogonal matrices

A *real* square matrix $\mathbf{A} = [a_{jk}]$ is called

symmetric if transposition leaves it unchanged,

$$\mathbf{A}^\mathsf{T} = \mathbf{A}, \qquad \text{thus} \qquad a_{kj} = a_{jk}, \tag{1}$$

skew-symmetric if transposition gives the negative of **A**,

$$\mathbf{A}^\mathsf{T} = -\mathbf{A}, \qquad \text{thus} \qquad a_{kj} = -a_{jk}, \tag{2}$$

orthogonal if transposition gives the inverse of **A**,

$$\mathbf{A}^\mathsf{T} = \mathbf{A}^{-1}. \tag{3}$$

EXAMPLE 1 **Symmetric, skew-symmetric, and orthogonal matrices**

The matrices

$$\begin{bmatrix} -3 & 1 & 5 \\ 1 & 0 & -2 \\ 5 & -2 & 4 \end{bmatrix}, \qquad \begin{bmatrix} 0 & 9 & -12 \\ -9 & 0 & 20 \\ 12 & -20 & 0 \end{bmatrix}, \qquad \begin{bmatrix} \frac{2}{3} & \frac{1}{3} & \frac{2}{3} \\ -\frac{2}{3} & \frac{2}{3} & \frac{1}{3} \\ \frac{1}{3} & \frac{2}{3} & -\frac{2}{3} \end{bmatrix}$$

are symmetric, skew-symmetric, and orthogonal, respectively, as the student should verify. Every skew-symmetric matrix has all main diagonal entries zero. (Can you prove this?) ∎

Any real square matrix **A** may be written as the sum of a symmetric matrix **R** and a skew-symmetric matrix **S**, where

$$\mathbf{R} = \tfrac{1}{2}(\mathbf{A} + \mathbf{A}^\mathsf{T}) \qquad \text{and} \qquad \mathbf{S} = \tfrac{1}{2}(\mathbf{A} - \mathbf{A}^\mathsf{T}). \tag{4}$$

EXAMPLE 2 Illustration of formula (4)

$$\mathbf{A} = \begin{bmatrix} 3 & -4 & -1 \\ 6 & 0 & -1 \\ -3 & 13 & -4 \end{bmatrix} = \mathbf{R} + \mathbf{S} = \begin{bmatrix} 3 & 1 & -2 \\ 1 & 0 & 6 \\ -2 & 6 & -4 \end{bmatrix} + \begin{bmatrix} 0 & -5 & 1 \\ 5 & 0 & -7 \\ -1 & 7 & 0 \end{bmatrix}$$ ∎

Theorem 1 (Eigenvalues of symmetric and skew-symmetric matrices)

(a) *The eigenvalues of a symmetric matrix are real.*

(b) *The eigenvalues of a skew-symmetric matrix are pure imaginary or zero.* (Proofs see in the next section.)

EXAMPLE 3 Eigenvalues of symmetric and skew-symmetric matrices

The matrices in (1) and (7) of Sec. 7.11 are symmetric and have real eigenvalues. The skew-symmetric matrix in Example 1 has the eigenvalues 0, $-25i$, and $25i$. (Verify this.) The matrix

$$\begin{bmatrix} 3 & 4 \\ 1 & 3 \end{bmatrix}$$

has the real eigenvalues 1 and 5 and is not symmetric. Does this contradict Theorem 1? ∎

Orthogonal Transformations and Matrices

Orthogonal transformations are transformations

(5) $$\mathbf{y} = \mathbf{A}\mathbf{x} \qquad \text{with } \mathbf{A} \text{ an orthogonal matrix.}$$

With each vector $\mathbf{x}$ in R^n such a transformation assigns a vector $\mathbf{y}$ in R^n. For instance, the plane rotation through an angle θ

(6) $$\mathbf{y} = \begin{bmatrix} y_1 \\ y_2 \end{bmatrix} = \begin{bmatrix} \cos\theta & -\sin\theta \\ \sin\theta & \cos\theta \end{bmatrix} \begin{bmatrix} x_1 \\ x_2 \end{bmatrix}$$

is an orthogonal transformation, and one can show that any orthogonal transformation in the plane or in three-dimensional space is a rotation (possibly combined with a reflection in a straight line or a plane, respectively).

The following property of orthogonal transformations is the main reason for the importance of orthogonal matrices.

Theorem 2 (Invariance of inner product)

An orthogonal transformation preserves the value of the inner product of vectors (see Sec. 7.3)

(7) $$\mathbf{a}\cdot\mathbf{b} = \mathbf{a}^{\mathsf{T}}\mathbf{b},$$

hence also the **length** *or* **norm** *of a vector in R^n given by*

(8) $$\|\mathbf{a}\| = \sqrt{\mathbf{a}\cdot\mathbf{a}} = \sqrt{\mathbf{a}^{\mathsf{T}}\mathbf{a}}.$$

Proof. Let $\mathbf{u} = \mathbf{Aa}$ and $\mathbf{v} = \mathbf{Ab}$, where $\mathbf{A}$ is orthogonal. We must show that $\mathbf{u}\cdot\mathbf{v} = \mathbf{a}\cdot\mathbf{b}$. Now (5) in Sec. 7.3 gives $\mathbf{u}^T = (\mathbf{Aa})^T = \mathbf{a}^T\mathbf{A}^T$. Also, $\mathbf{A}^T\mathbf{A} = \mathbf{A}^{-1}\mathbf{A} = \mathbf{I}$ by (3). Hence

$$(9) \qquad \mathbf{u}\cdot\mathbf{v} = \mathbf{u}^T\mathbf{v} = (\mathbf{Aa})^T\mathbf{Ab} = \mathbf{a}^T\mathbf{A}^T\mathbf{Ab} = \mathbf{a}^T\mathbf{Ib} = \mathbf{a}^T\mathbf{b} = \mathbf{a}\cdot\mathbf{b}.$$ ∎

Orthogonal matrices have further interesting properties, as follows.

Theorem 3 (Orthonormality of column and row vectors)

A real square matrix is orthogonal if and only if its column vectors $\mathbf{a}_1, \cdots, \mathbf{a}_n$ *(and also its row vectors) form an* **orthonormal system,** *that is,*

$$(10) \qquad \mathbf{a}_j\cdot\mathbf{a}_k = \mathbf{a}_j^T\mathbf{a}_k = \begin{cases} 0 & \text{if } j \neq k \\ 1 & \text{if } j = k. \end{cases}$$

Proof. **(a)** Let $\mathbf{A}$ be orthogonal. Then $\mathbf{A}^{-1}\mathbf{A} = \mathbf{A}^T\mathbf{A} = \mathbf{I}$, in terms of column vectors,

$$(11) \quad \mathbf{A}^{-1}\mathbf{A} = \mathbf{A}^T\mathbf{A} = \begin{bmatrix} \mathbf{a}_1^T \\ \vdots \\ \mathbf{a}_n^T \end{bmatrix} [\mathbf{a}_1 \cdots \mathbf{a}_n] = \begin{bmatrix} \mathbf{a}_1^T\mathbf{a}_1 & \mathbf{a}_1^T\mathbf{a}_2 & \cdots & \mathbf{a}_1^T\mathbf{a}_n \\ \mathbf{a}_2^T\mathbf{a}_1 & \mathbf{a}_2^T\mathbf{a}_2 & \cdots & \mathbf{a}_2^T\mathbf{a}_n \\ \cdot & \cdot & \cdots & \cdot \\ \mathbf{a}_n^T\mathbf{a}_1 & \mathbf{a}_n^T\mathbf{a}_2 & \cdots & \mathbf{a}_n^T\mathbf{a}_n \end{bmatrix} = \mathbf{I},$$

where the last equality implies (10), by the definition of the $n \times n$ unit matrix $\mathbf{I}$. From (3) it follows that the inverse of an orthogonal matrix is orthogonal (see Prob. 22), and the column vectors of $\mathbf{A}^{-1}$ $(= \mathbf{A}^T)$ are the row vectors of $\mathbf{A}$; hence the row vectors of $\mathbf{A}$ also form an orthonormal system.

(b) Conversely, if the column vectors of $\mathbf{A}$ satisfy (10), the off-diagonal entries in the big matrix in (11) are 0 and the diagonal entries are 1. Hence $\mathbf{A}^T\mathbf{A} = \mathbf{I}$, as (11) shows. Similarly, $\mathbf{AA}^T = \mathbf{I}$. This implies $\mathbf{A}^T = \mathbf{A}^{-1}$ because also $\mathbf{A}^{-1}\mathbf{A} = \mathbf{AA}^{-1} = \mathbf{I}$ and the inverse is unique. Hence $\mathbf{A}$ is orthogonal. Similarly when the row vectors of $\mathbf{A}$ form an orthonormal system, by what has been said at the end of part **(a)**. ∎

Theorem 4 (Determinant of an orthogonal matrix)

The determinant of an orthogonal matrix has the value $+1$ *or* -1.

Proof. This follows from $\det \mathbf{AB} = \det \mathbf{A} \det \mathbf{B}$ and $\det \mathbf{A}^T = \det \mathbf{A}$ (Theorems 1 and 8 in Sec. 7.8). Indeed, if $\mathbf{A}$ is orthogonal, then

$$1 = \det \mathbf{I} = \det (\mathbf{AA}^{-1}) = \det (\mathbf{AA}^T) = \det \mathbf{A} \det \mathbf{A}^T = (\det \mathbf{A})^2.$$ ∎

EXAMPLE 4 Illustration of Theorems 3 and 4

The last matrix in Example 1 and the matrix in (6) illustrate Theorems 3 and 4, their determinants being -1 and $+1$, as the student should verify. ∎

Theorem 5 (Eigenvalues of an orthogonal matrix)
The eigenvalues of an orthogonal matrix **A** *are real or complex conjugates in pairs and have absolute value* 1.

Proof. The first part of the statement holds for any real matrix **A** because its characteristic polynomial has real coefficients, so that its zeros (the eigenvalues of **A**) must be as indicated. The claim that $|\lambda| = 1$ will be proved in the next section. ∎

EXAMPLE 5 Eigenvalues of an orthogonal matrix

The orthogonal matrix in Example 1 has the characteristic equation

$$-\lambda^3 + \tfrac{2}{3}\lambda^2 + \tfrac{2}{3}\lambda - 1 = 0.$$

Now one of the eigenvalues must be real (why?), hence $+1$ or -1. Trying, we find -1. Division by $\lambda + 1$ gives $\lambda^2 - 5\lambda/3 + 1 = 0$ and the two eigenvalues $(5 + i\sqrt{11})/6$ and $(5 - i\sqrt{11})/6$. Verify all of this. ∎

Problem Set 7.12

Write the following matrices as the sum of a symmetric and a skew-symmetric matrix.

1. $\begin{bmatrix} 4 & -3 \\ 7 & -1 \end{bmatrix}$ 2. $\begin{bmatrix} -2 & -2 & 3 \\ 4 & 0 & -5 \\ -1 & -3 & 5 \end{bmatrix}$ 3. $\begin{bmatrix} 0 & -1 & 4 \\ 9 & 1 & -7 \\ -10 & 11 & -1 \end{bmatrix}$

4. Show that the main diagonal entries of a skew-symmetric matrix are all zero.

Are the following matrices symmetric? Skew-symmetric? Orthogonal? Find their eigenvalues (thereby illustrating Theorems 1 and 5).

5. $\begin{bmatrix} 6 & 8 \\ 8 & -6 \end{bmatrix}$ 6. $\begin{bmatrix} 0.8 & 0.6 \\ -0.6 & 0.8 \end{bmatrix}$ 7. $\begin{bmatrix} 0 & 3 \\ -3 & 0 \end{bmatrix}$

8. $\begin{bmatrix} 0 & 1 & -1 \\ -1 & 0 & 1 \\ 1 & -1 & 0 \end{bmatrix}$ 9. $\begin{bmatrix} 0 & 0 & 1 \\ 0 & 1 & 0 \\ -1 & 0 & 0 \end{bmatrix}$ 10. $\begin{bmatrix} \cos\theta & -\sin\theta & 0 \\ \sin\theta & \cos\theta & 0 \\ 0 & 0 & 1 \end{bmatrix}$

11. $\begin{bmatrix} 1 & 1 & 1 \\ 1 & 1 & 1 \\ 1 & 1 & 1 \end{bmatrix}$ 12. $\begin{bmatrix} 0.50 & 0.25 & 0.25 \\ 0.25 & 0.50 & 0.25 \\ 0.25 & 0.25 & 0.50 \end{bmatrix}$ 13. $\begin{bmatrix} 0 & 18 & -24 \\ -18 & 0 & 40 \\ 24 & -40 & 0 \end{bmatrix}$

14. **(Symmetric matrix)** Prove that eigenvectors of a symmetric matrix corresponding to different eigenvalues are orthogonal. Give an example.
15. Find a real matrix that has real eigenvalues but is not symmetric. Does this contradict Theorem 1?
16. Show that (6) is an orthogonal transformation. Verify that Theorem 3 holds. Find the inverse transformation.
17. Let $\mathbf{v}^T = [4 \quad 2]$, $\mathbf{x}^T = [-2 \quad 1]$, $\mathbf{w} = \mathbf{Av}$, $\mathbf{y} = \mathbf{Ax}$ with **A** given in (6). Find $|\mathbf{v}|$, $|\mathbf{x}|$, $|\mathbf{w}|$, $|\mathbf{y}|$. Which theorem do the results illustrate?
18. Find **A** such that $\mathbf{y} = \mathbf{Ax}$ is a counterclockwise rotation through 30° in the plane.

19. Interpret the transformation $\mathbf{y} = \mathbf{Ax}$ geometrically, where $\mathbf{A}$ is the matrix in Prob. 10 and the components of $\mathbf{x}$ and $\mathbf{y}$ are Cartesian coordinates.
20. Find a 2×2 matrix that is both orthogonal and skew-symmetric. Find its eigenvalues.
21. Does there exist an orthogonal skew-symmetric 3×3 matrix? An orthogonal symmetric 3×3 matrix? (Give a reason.)
22. Show that the inverse of an orthogonal matrix is orthogonal.
23. Show that the product of two orthogonal $n \times n$ matrices is orthogonal.
24. Is the sum of two orthogonal matrices orthogonal?
25. Show that the inverse of a nonsingular skew-symmetric matrix is skew-symmetric.

7.13 Hermitian, Skew-Hermitian, and Unitary Matrices

We shall now introduce three classes of complex square matrices that generalize the three classes of real matrices just considered and have important applications, for instance, in quantum mechanics.

In this connection we use the standard notation

$$\overline{\mathbf{A}} = [\overline{a}_{jk}]$$

for the matrix obtained from $\mathbf{A} = [a_{jk}]$ by replacing each entry by its complex conjugate, and we also use the notation

$$\overline{\mathbf{A}}^{\mathsf{T}} = [\overline{a}_{kj}]$$

for the conjugate transpose. For example, if

$$\mathbf{A} = \begin{bmatrix} 3+4i & -5i \\ -7 & 6-2i \end{bmatrix}, \quad \text{then} \quad \overline{\mathbf{A}}^{\mathsf{T}} = \begin{bmatrix} 3-4i & -7 \\ 5i & 6+2i \end{bmatrix}.$$

Definitions of Hermitian,[17] Skew-Hermitian, and unitary matrices

A square matrix $\mathbf{A} = [a_{jk}]$ is called

$$\begin{array}{llll} \textbf{Hermitian if} & \overline{\mathbf{A}}^{\mathsf{T}} = \mathbf{A}, & \text{that is,} & \overline{a}_{kj} = a_{jk} \\ \textbf{skew-Hermitian if} & \overline{\mathbf{A}}^{\mathsf{T}} = -\mathbf{A}, & \text{that is,} & \overline{a}_{kj} = -a_{jk} \\ \textbf{unitary if} & \overline{\mathbf{A}}^{\mathsf{T}} = \mathbf{A}^{-1}. & & \end{array}$$

From these definitions we see the following. If $\mathbf{A}$ is Hermitian, the entries on the main diagonal must satisfy $\overline{a}_{jj} = a_{jj}$, that is, they are real. Similarly, if $\mathbf{A}$ is skew-Hermitian, then $\overline{a}_{jj} = -a_{jj}$ or, if we set $a_{jj} = \alpha + i\beta$, this becomes $\alpha - i\beta = -(\alpha + i\beta)$, so that $\alpha = 0$ and a_{jj} is pure imaginary or 0.

[17] See footnote 22 in Problem Set 5.9.

EXAMPLE 1 Hermitian, skew-Hermitian, and unitary matrices

The matrices

$$\mathbf{A} = \begin{bmatrix} 4 & 1-3i \\ 1+3i & 7 \end{bmatrix}, \quad \mathbf{B} = \begin{bmatrix} 3i & 2+i \\ -2+i & -i \end{bmatrix}, \quad \mathbf{C} = \begin{bmatrix} \frac{1}{2}i & \frac{1}{2}\sqrt{3} \\ \frac{1}{2}\sqrt{3} & \frac{1}{2}i \end{bmatrix}$$

are Hermitian, skew-Hermitian, and unitary, respectively, as the reader may verify. ∎

If a Hermitian matrix is real, then $\overline{\mathbf{A}}^{\mathsf{T}} = \mathbf{A}^{\mathsf{T}} = \mathbf{A}$. Hence a real Hermitian matrix is a symmetric matrix (Sec. 7.12).

Similarly, if a skew-Hermitian matrix is real, then $\overline{\mathbf{A}}^{\mathsf{T}} = \mathbf{A}^{\mathsf{T}} = -\mathbf{A}$. Hence a real skew-Hermitian matrix is a skew-symmetric matrix.

Finally, if a unitary matrix is real, then $\overline{\mathbf{A}}^{\mathsf{T}} = \mathbf{A}^{\mathsf{T}} = \mathbf{A}^{-1}$. Hence a real unitary matrix is an orthogonal matrix.

This shows that *Hermitian, skew-Hermitian, and unitary matrices generalize symmetric, skew-symmetric, and orthogonal matrices, respectively.*

Eigenvalues

It is quite remarkable and in part accounts for the importance of the matrices under consideration that their spectra (their sets of eigenvalues; see Sec. 7.10) can be characterized in a general way as follows (see Fig. 137).

Theorem 1 (Eigenvalues)

(a) *The eigenvalues of a Hermitian matrix* (*and thus of a symmetric matrix*) *are real.*

(b) *The eigenvalues of a skew-Hermitian matrix* (*and thus of a skew-symmetric matrix*) *are pure imaginary or zero.*

(c) *The eigenvalues of a unitary matrix* (*and thus of an orthogonal matrix*) *have absolute value* 1.

Proof. Let λ be an eigenvalue of $\mathbf{A}$ and $\mathbf{x}$ a corresponding eigenvector. Then

$$\mathbf{Ax} = \lambda\mathbf{x}. \tag{1}$$

(a) Let $\mathbf{A}$ be Hermitian. Multiplying (1) by $\overline{\mathbf{x}}^{\mathsf{T}}$ from the left, we obtain

$$\overline{\mathbf{x}}^{\mathsf{T}}\mathbf{Ax} = \overline{\mathbf{x}}^{\mathsf{T}}\lambda\mathbf{x} = \lambda\overline{\mathbf{x}}^{\mathsf{T}}\mathbf{x}.$$

Now $\overline{\mathbf{x}}^{\mathsf{T}}\mathbf{x} = \bar{x}_1 x_1 + \cdots + \bar{x}_n x_n = |x_1|^2 + \cdots + |x_n|^2$ is real, and is not 0 since $\mathbf{x} \neq \mathbf{0}$. Hence we may divide to get

$$\lambda = \frac{\overline{\mathbf{x}}^{\mathsf{T}}\mathbf{Ax}}{\overline{\mathbf{x}}^{\mathsf{T}}\mathbf{x}}. \tag{2}$$

We see that λ is real if the numerator is real. We prove that the numerator is real by showing that it is equal to its complex conjugate, using $\overline{\mathbf{A}}^{\mathsf{T}} = \mathbf{A}$ or $\overline{\mathbf{A}} = \mathbf{A}^{\mathsf{T}}$ and (5) in Sec. 7.3. Indeed, beginning with the application of a transposition, which has no effect on a number (the numerator), we get

$$\overline{\mathbf{x}}^{\mathsf{T}}\mathbf{Ax} = (\overline{\mathbf{x}}^{\mathsf{T}}\mathbf{Ax})^{\mathsf{T}} = \mathbf{x}^{\mathsf{T}}\mathbf{A}^{\mathsf{T}}\overline{\mathbf{x}} = \mathbf{x}^{\mathsf{T}}\overline{\mathbf{A}}\overline{\mathbf{x}} = \overline{(\overline{\mathbf{x}}^{\mathsf{T}}\mathbf{Ax})}. \tag{3}$$

From this and (2), whose denominator is real, we see that λ is real.

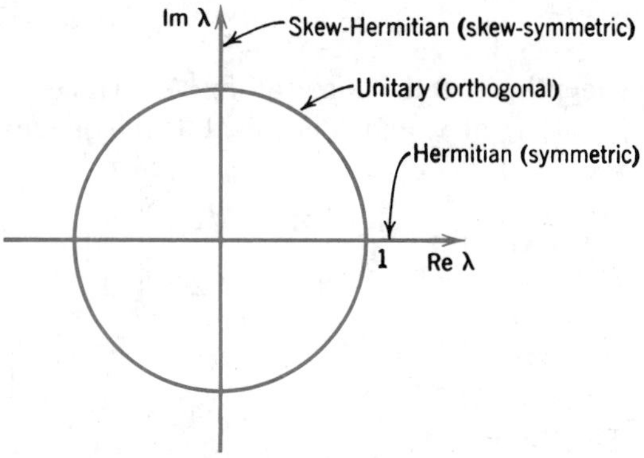

Fig. 137. Location of the eigenvalues of Hermitian, skew-Hermitian, and unitary matrices in the complex λ-plane

(b) If $\mathbf{A}$ is skew-Hermitian, then $\overline{\mathbf{A}}^{\mathsf{T}} = -\mathbf{A}$, thus $\overline{\mathbf{A}} = -\mathbf{A}^{\mathsf{T}}$, so that we get a minus sign in (3),

$$\overline{\mathbf{x}}^{\mathsf{T}}\mathbf{A}\mathbf{x} = -(\overline{\overline{\mathbf{x}}^{\mathsf{T}}\mathbf{A}\mathbf{x}}). \tag{4}$$

So this is a complex number $c = a + ib$ that equals minus its conjugate $\overline{c} = a - ib$, that is, $a + ib = -(a - ib)$. Hence $a = 0$, so that c is pure imaginary or zero, and division by the real $\overline{\mathbf{x}}^{\mathsf{T}}\mathbf{x}$ in (2) gives a pure imaginary λ or $\lambda = 0$.

(c) Let $\mathbf{A}$ be unitary. We take (1) and its conjugate transpose,

$$\mathbf{A}\mathbf{x} = \lambda\mathbf{x} \qquad \text{and} \qquad (\overline{\mathbf{A}\mathbf{x}})^{\mathsf{T}} = (\overline{\lambda\mathbf{x}})^{\mathsf{T}} = \overline{\lambda}\overline{\mathbf{x}}^{\mathsf{T}}$$

and multiply the two left sides and the two right sides,

$$(\overline{\mathbf{A}\mathbf{x}})^{\mathsf{T}}\mathbf{A}\mathbf{x} = \overline{\lambda}\lambda\overline{\mathbf{x}}^{\mathsf{T}}\mathbf{x} = |\lambda|^2\overline{\mathbf{x}}^{\mathsf{T}}\mathbf{x}.$$

But $\mathbf{A}$ is unitary, $\overline{\mathbf{A}}^{\mathsf{T}} = \mathbf{A}^{-1}$, so that on the left we obtain

$$(\overline{\mathbf{A}\mathbf{x}})^{\mathsf{T}}\mathbf{A}\mathbf{x} = \overline{\mathbf{x}}^{\mathsf{T}}\overline{\mathbf{A}}^{\mathsf{T}}\mathbf{A}\mathbf{x} = \overline{\mathbf{x}}^{\mathsf{T}}\mathbf{A}^{-1}\mathbf{A}\mathbf{x} = \overline{\mathbf{x}}^{\mathsf{T}}\mathbf{I}\mathbf{x} = \overline{\mathbf{x}}^{\mathsf{T}}\mathbf{x}.$$

Together, $\overline{\mathbf{x}}^{\mathsf{T}}\mathbf{x} = |\lambda|^2\overline{\mathbf{x}}^{\mathsf{T}}\mathbf{x}$. Now divide by $\overline{\mathbf{x}}^{\mathsf{T}}\mathbf{x}$ ($\neq 0$) to get $|\lambda|^2 = 1$, hence $|\lambda| = 1$.

This proves our present theorem as well as Theorems 1 and 5 in the previous section. ∎

EXAMPLE 2 **Illustration of Theorem 1**

For the matrices in Example 1 we find by direct calculation

Matrix	Characteristic Equation	Eigenvalues
A Hermitian	$\lambda^2 - 11\lambda + 18 = 0$	$9, \ 2$
B Skew-Hermitian	$\lambda^2 - 2i\lambda + 8 = 0$	$4i, \ -2i$
C Unitary	$\lambda^2 - i\lambda - 1 = 0$	$\frac{1}{2}\sqrt{3} + \frac{1}{2}i, \ -\frac{1}{2}\sqrt{3} + \frac{1}{2}i$

and $|\pm\frac{1}{2}\sqrt{3} + \frac{1}{2}i|^2 = \frac{3}{4} + \frac{1}{4} = 1.$ ∎

Forms

We mention that the numerator $\bar{\mathbf{x}}^T\mathbf{A}\mathbf{x}$ in (2) is called a **form** in the components $x_1, \cdots, x_n$ of **x**, and **A** is called its *coefficient matrix*. When $n = 2$, we get

$$\bar{\mathbf{x}}^T\mathbf{A}\mathbf{x} = [\bar{x}_1 \quad \bar{x}_2]\begin{bmatrix} a_{11} & a_{12} \\ a_{21} & a_{22} \end{bmatrix}\begin{bmatrix} x_1 \\ x_2 \end{bmatrix} = [\bar{x}_1 \quad \bar{x}_2]\begin{bmatrix} a_{11}x_1 + a_{12}x_2 \\ a_{21}x_1 + a_{22}x_2 \end{bmatrix}$$

$$= \begin{cases} a_{11}\bar{x}_1x_1 + a_{12}\bar{x}_1x_2 \\ + a_{21}\bar{x}_2x_1 + a_{22}\bar{x}_2x_2. \end{cases}$$

Similarly for general n,

$$\text{(5)} \qquad \begin{aligned} \bar{\mathbf{x}}^T\mathbf{A}\mathbf{x} = \sum_{j=1}^{n}\sum_{k=1}^{n} a_{jk}\bar{x}_jx_k = \quad & a_{11}\bar{x}_1x_1 + \cdots + a_{1n}\bar{x}_1x_n \\ & + a_{21}\bar{x}_2x_1 + \cdots + a_{2n}\bar{x}_2x_n \\ & + \cdots\cdots\cdots\cdots\cdots\cdots \\ & + a_{n1}\bar{x}_nx_1 + \cdots + a_{nn}\bar{x}_nx_n. \end{aligned}$$

If **x** and **A** are real, then (5) becomes

$$\text{(6)} \qquad \begin{aligned} \mathbf{x}^T\mathbf{A}\mathbf{x} = \sum_{j=1}^{n}\sum_{k=1}^{n} a_{jk}x_jx_k = \quad & a_{11}x_1^2 + a_{12}x_1x_2 + \cdots + a_{1n}x_1x_n \\ & + a_{21}x_2x_1 + a_{22}x_2^2 + \cdots + a_{2n}x_2x_n \\ & + \cdots\cdots\cdots\cdots\cdots\cdots\cdots\cdots \\ & + a_{n1}x_nx_1 + a_{n2}x_nx_2 + \cdots + a_{nn}x_n^2 \end{aligned}$$

and is called a **quadratic form.** Without restriction we may then assume the coefficient matrix to be ***symmetric,*** because we can take off-diagonal terms together in pairs and then write the result as a sum of two equal terms, as the next example (Example 3) illustrates. Quadratic forms occur in physics and geometry, for instance, in connection with conic sections (ellipses $x_1^2/a^2 + x_2^2/b^2 = 1$, etc.) and quadratic surfaces. (Their "transformation to principal axes" will be discussed in the next section.)

EXAMPLE 3 Quadratic form. Symmetric coefficient matrix C

Let

$$\mathbf{x}^T\mathbf{A}\mathbf{x} = [x_1 \quad x_2]\begin{bmatrix} 3 & 4 \\ 6 & 2 \end{bmatrix}\begin{bmatrix} x_1 \\ x_2 \end{bmatrix} = 3x_1^2 + 4x_1x_2 + 6x_2x_1 + 2x_2^2 = 3x_1^2 + 10x_1x_2 + 2x_2^2.$$

Here $4 + 6 = 10 = 5 + 5$. From the corresponding ***symmetric*** matrix $\mathbf{C} = [c_{jk}]$, where $c_{jk} = \frac{1}{2}(a_{jk} + a_{jk})$, thus $c_{11} = 3$, $c_{12} = c_{21} = 5$, $c_{22} = 2$, we get the same result

$$\mathbf{x}^T\mathbf{C}\mathbf{x} = [x_1 \quad x_2]\begin{bmatrix} 3 & 5 \\ 5 & 2 \end{bmatrix}\begin{bmatrix} x_1 \\ x_2 \end{bmatrix} = 3x_1^2 + 5x_1x_2 + 5x_2x_1 + 2x_2^2 = 3x_1^2 + 10x_1x_2 + 2x_2^2. \quad \blacksquare$$

If the matrix **A** in (5) is Hermitian or skew-Hermitian, the form (5) is called a **Hermitian form** or **skew-Hermitian form,** respectively. These forms have the following property, which accounts for their importance in physics.

Theorem 1* **(Hermitian and skew-Hermitian forms)**

For every choice of the vector **x** *the value of a Hermitian form is real, and the value of a skew-Hermitian form is pure imaginary or* 0.

Proof. In proving (3) and (4), we made no use of the fact that **x** was an eigenvector, and the proofs remain valid for any vectors (and Hermitian or skew-Hermitian matrices). From this, our present theorem follows. ∎

EXAMPLE 4 **Hermitian form**

If

$$\mathbf{A} = \begin{bmatrix} 3 & 2-i \\ 2+i & 4 \end{bmatrix} \quad \text{and} \quad \mathbf{x} = \begin{bmatrix} 1+i \\ 2i \end{bmatrix},$$

then

$$\bar{\mathbf{x}}^{\mathsf{T}}\mathbf{A}\mathbf{x} = [1-i \quad -2i]\begin{bmatrix} 3 & 2-i \\ 2+i & 4 \end{bmatrix}\begin{bmatrix} 1+i \\ 2i \end{bmatrix}$$

$$= [1-i \quad -2i]\begin{bmatrix} 3(1+i)+(2-i)2i \\ (2+i)(1+i)+4\cdot 2i \end{bmatrix} = 34.$$ ∎

Properties of Unitary Matrices. Complex Vector Space C^n

We now extend our discussion of orthogonal matrices in Sec. 7.12 to unitary matrices. Instead of the real vector space R^n of all real vectors with n components and real numbers as scalars, we now use the **complex vector space C^n** of all complex vectors with n complex numbers as components and complex numbers as scalars. For such complex vectors, the **inner product** is defined by

$$\mathbf{a}\cdot\mathbf{b} = \bar{\mathbf{a}}^{\mathsf{T}}\mathbf{b} \tag{7}$$

and the **length** or **norm** of a vector by

$$\|\mathbf{a}\| = \sqrt{\mathbf{a}\cdot\mathbf{a}} = \sqrt{\bar{\mathbf{a}}^{\mathsf{T}}\mathbf{a}} = \sqrt{\bar{a}_1 a_1 + \cdots + \bar{a}_n a_n} = \sqrt{|a_1|^2 + \cdots + |a_n|^2}. \tag{8}$$

Note that for *real* vectors this reduces to the inner product as defined in Sec. 7.3.

Theorem 2 **(Invariance of inner product)**

A **unitary transformation,** *that is,* $\mathbf{y} = \mathbf{Ax}$ *with a unitary matrix* **A,** *preserves the value of the inner product* (7), *hence also the norm* (8).

Proof. The proof is the same as that of Theorem 2 in Sec. 7.12, which the theorem generalizes; in the analog of (9), Sec. 7.12, we now have bars,

$$\mathbf{u}\cdot\mathbf{v} = \bar{\mathbf{u}}^{\mathsf{T}}\mathbf{v} = (\bar{\mathbf{A}}\bar{\mathbf{a}})^{\mathsf{T}}\mathbf{A}\mathbf{b} = \bar{\mathbf{a}}^{\mathsf{T}}\bar{\mathbf{A}}^{\mathsf{T}}\mathbf{A}\mathbf{b} = \bar{\mathbf{a}}^{\mathsf{T}}\mathbf{I}\mathbf{b} = \bar{\mathbf{a}}^{\mathsf{T}}\mathbf{b} = \mathbf{a}\cdot\mathbf{b}.$$ ∎

The complex analog of an *orthonormal system* of real vectors (see Sec. 7.12) is a **unitary system,** defined by

$$\mathbf{a}_j \cdot \mathbf{a}_k = \bar{\mathbf{a}}_j^{\mathsf{T}} \mathbf{a}_k = \begin{cases} 0 & \text{if } j \neq k \\ 1 & \text{if } j = k, \end{cases} \tag{9}$$

and the extension of Theorem 3, Sec. 7.12, to complex is as follows.

Theorem 3 **(Unitary systems of column and row vectors)**
A square matrix is unitary if and only if its column vectors (and also its row vectors) form a unitary system.

Proof. The proof is the same as that of Theorem 3 in Sec. 7.12, except for the bars required by the definitions $\bar{\mathbf{A}}^{\mathsf{T}} = \mathbf{A}^{-1}$ and (7) and (9). ∎

Theorem 4 **(Determinant of a unitary matrix)**
The determinant of a unitary matrix has absolute value 1.

Proof. Similarly as in Sec. 7.12 we get

$$1 = \det \mathbf{A}\mathbf{A}^{-1} = \det (\mathbf{A}\bar{\mathbf{A}}^{\mathsf{T}}) = \det \mathbf{A} \det \bar{\mathbf{A}}^{\mathsf{T}} = \det \mathbf{A} \det \bar{\mathbf{A}}$$
$$= \det \mathbf{A}\, \overline{\det \mathbf{A}} = |\det \mathbf{A}|^2.$$

Hence $|\det \mathbf{A}| = 1$ (where det **A** may now be complex). ∎

EXAMPLE 5 **Unitary matrix illustrating Theorems 2–4**

For the vectors $\mathbf{a}^{\mathsf{T}} = [1 \quad i]$ and $\mathbf{b}^{\mathsf{T}} = [3i \quad 2+i]$ we get $\bar{\mathbf{a}}^{\mathsf{T}}\mathbf{b} = 3i - i(2+i) = 1 + i$, and with

$$\mathbf{A} = \begin{bmatrix} 0.6i & 0.8 \\ 0.8 & 0.6i \end{bmatrix} \quad \text{also} \quad \mathbf{A}\mathbf{a} = \begin{bmatrix} 1.4i \\ 0.2 \end{bmatrix} \quad \text{and} \quad \mathbf{A}\mathbf{b} = \begin{bmatrix} -0.2 + 0.8i \\ -0.6 + 3.6i \end{bmatrix},$$

as one can readily verify. This gives $(\overline{\mathbf{A}\mathbf{a}})^{\mathsf{T}}\mathbf{A}\mathbf{b} = 1 + i$, illustrating Theorem 2. The matrix is unitary. Its columns form a unitary system, and so do the rows, as we see. Also, $\det \mathbf{A} = -1$. ∎

Problem Set 7.13

1. Verify the eigenvalues in Example 2.

In Examples 1 and 2, find eigenvectors of

2. The matrix **A** **3.** The matrix **B** **4.** The matrix **C**

Indicate whether the following matrices are Hermitian, skew-Hermitian, or unitary and find their eigenvalues (thereby verifying Theorem 1) and eigenvectors.

5. $\begin{bmatrix} 0 & i \\ i & 0 \end{bmatrix}$ **6.** $\begin{bmatrix} 0 & 2i \\ -2i & 0 \end{bmatrix}$ **7.** $\begin{bmatrix} 4 & i \\ -i & 2 \end{bmatrix}$

8. $\begin{bmatrix} 1/\sqrt{3} & i\sqrt{2/3} \\ -i\sqrt{2/3} & -1/\sqrt{3} \end{bmatrix}$ **9.** $\begin{bmatrix} 1/\sqrt{2} & i/\sqrt{2} \\ -i/\sqrt{2} & -1/\sqrt{2} \end{bmatrix}$ **10.** $\begin{bmatrix} i & 0 & 0 \\ 0 & 0 & i \\ 0 & i & 0 \end{bmatrix}$

11. Show that the product of two $n \times n$ unitary matrices is unitary.

12. Show that the inverse of a unitary matrix is unitary. Verify this for the matrix in Prob. 10.

13. Verify Theorems 3 and 4 for the matrix in Prob. 9.

14. Show that any square matrix may be written as the sum of a Hermitian matrix and a skew-Hermitian matrix.

15. **(Normal matrix)** By definition, a *normal matrix* is a square matrix that commutes with its conjugate transpose,

$$\mathbf{A}\bar{\mathbf{A}}^{\mathsf{T}} = \bar{\mathbf{A}}^{\mathsf{T}}\mathbf{A}.$$

Show that Hermitian, skew-Hermitian, and unitary matrices are normal.

Quadratic forms. Find a symmetric matrix $\mathbf{C}$ such that $Q = \mathbf{x}^{\mathsf{T}}\mathbf{C}\mathbf{x}$, where Q equals

16. $x_1^2 - 4x_1x_2 + 7x_2^2$

17. $(x_1 - 3x_2)^2$

18. $(x_1 + x_2 + x_3)^2$

19. $-3x_1^2 + 4x_1x_2 - x_2^2 + 2x_1x_3 - 5x_3^2$

20. $(x_1 - x_2 + 2x_3 - 2x_4)^2$

21. $(x_1 + x_2)^2 + (x_3 + x_4)^2$

22. **(Definiteness)** A real quadratic form $Q = \mathbf{x}^{\mathsf{T}}\mathbf{C}\mathbf{x}$ and its symmetric matrix $\mathbf{C} = [c_{jk}]$ are said to be **positive definite** if $Q > 0$ for all $[x_1 \cdots x_n] \neq [0 \cdots 0]$. A necessary and sufficient condition for positive definiteness is that all the determinants

$$C_1 = c_{11}, \quad C_2 = \begin{vmatrix} c_{11} & c_{12} \\ c_{21} & c_{22} \end{vmatrix}, \quad C_3 = \begin{vmatrix} c_{11} & c_{12} & c_{13} \\ c_{21} & c_{22} & c_{23} \\ c_{31} & c_{32} & c_{33} \end{vmatrix}, \quad \cdots, \quad C_n = \det \mathbf{C}$$

are positive (see Ref. [B2], vol. 1, p. 306). Show that the form in Prob. 16 is positive definite, whereas that in Example 3 is not positive definite.

Hermitian and skew-Hermitian forms. Is $\mathbf{A}$ Hermitian or skew-Hermitian? Find $\bar{\mathbf{x}}^{\mathsf{T}}\mathbf{A}\mathbf{x}$.

23. $\mathbf{A} = \begin{bmatrix} 0 & i \\ -i & 0 \end{bmatrix}, \quad \mathbf{x} = \begin{bmatrix} 1 \\ i \end{bmatrix}$

24. $\mathbf{A} = \begin{bmatrix} 2 & 1+i \\ 1-i & 1 \end{bmatrix}, \quad \mathbf{x} = \begin{bmatrix} 1 \\ 2 \end{bmatrix}$

25. $\mathbf{A} = \begin{bmatrix} i & 1 \\ -1 & 2i \end{bmatrix}, \quad \mathbf{x} = \begin{bmatrix} 1 \\ i \end{bmatrix}$

26. $\mathbf{A} = \begin{bmatrix} a & b+ic \\ b-ic & k \end{bmatrix}, \quad \mathbf{x} = \begin{bmatrix} x_1 \\ x_2 \end{bmatrix}$

27. $\mathbf{A} = \begin{bmatrix} 0 & i & 0 \\ -i & 1 & -2i \\ 0 & 2i & 2 \end{bmatrix}, \quad \mathbf{x} = \begin{bmatrix} i \\ 1 \\ -i \end{bmatrix}$

28. $\mathbf{A} = \begin{bmatrix} i & 1+i & 2 \\ -1+i & -3i & 3+i \\ -2 & -3+i & 0 \end{bmatrix}, \quad \mathbf{x} = \begin{bmatrix} i \\ 1 \\ -i \end{bmatrix}$

Is **A** Hermitian or skew-Hermitian? Find $\bar{\mathbf{x}}^T\mathbf{A}\mathbf{x}$.

29. $\mathbf{A} = \begin{bmatrix} 3 & -i & 0 \\ i & 0 & 2i \\ 0 & -2i & 4 \end{bmatrix}, \quad \mathbf{x} = \begin{bmatrix} x_1 \\ x_2 \\ x_3 \end{bmatrix}$

30. $\mathbf{A} = \begin{bmatrix} 2i & 0 & 4 \\ 0 & i & 5-i \\ -4 & -5-i & 4i \end{bmatrix}, \quad \mathbf{x} = \begin{bmatrix} 0 \\ 2i \\ -3 \end{bmatrix}$

7.14 Properties of Eigenvectors. Diagonalization

In our discussion of eigenvalue problems so far we have emphasized properties of eigen*values*. We now turn to eigen*vectors* and their properties. Eigenvectors of an $n \times n$ matrix **A** may (or may not!) form a basis for R^n or C^n (see Sec. 7.13), and if they do, we can use them for "*diagonalizing*" **A**, that is, for transforming it into diagonal form with the eigenvalues on the main diagonal. These are the key issues in this section.

We begin with a concept of central interest in eigenvalue problems:

Similarity of Matrices

An $n \times n$ matrix $\hat{\mathbf{A}}$ is called **similar** to an $n \times n$ matrix **A** if

$$\hat{\mathbf{A}} = \mathbf{T}^{-1}\mathbf{A}\mathbf{T} \tag{1}$$

for some (nonsingular!) $n \times n$ matrix **T**. This transformation, which gives $\hat{\mathbf{A}}$ from **A**, is called a **similarity transformation.**

Similarity transformations are important since they preserve eigenvalues:

Theorem 1 (Eigenvalues and eigenvectors of similar matrices)

If $\hat{\mathbf{A}}$ is similar to **A**, *then $\hat{\mathbf{A}}$ has the same eigenvalues as* **A**.

Furthermore, if **x** *is an eigenvector of* **A**, *then* $\mathbf{y} = \mathbf{T}^{-1}\mathbf{x}$ *is an eigenvector of $\hat{\mathbf{A}}$ corresponding to the same eigenvalue.*

Proof. From $\mathbf{A}\mathbf{x} = \lambda\mathbf{x}$ (λ an eigenvalue, $\mathbf{x} \neq \mathbf{0}$) we get $\mathbf{T}^{-1}\mathbf{A}\mathbf{x} = \lambda\mathbf{T}^{-1}\mathbf{x}$. Now $\mathbf{I} = \mathbf{T}\mathbf{T}^{-1}$, so that

$$\mathbf{T}^{-1}\mathbf{A}\mathbf{x} = \mathbf{T}^{-1}\mathbf{A}\mathbf{I}\mathbf{x} = \mathbf{T}^{-1}\mathbf{A}\mathbf{T}\mathbf{T}^{-1}\mathbf{x} = \hat{\mathbf{A}}(\mathbf{T}^{-1}\mathbf{x}) = \lambda\mathbf{T}^{-1}\mathbf{x}.$$

Hence λ is an eigenvalue of $\hat{\mathbf{A}}$ and $\mathbf{T}^{-1}\mathbf{x}$ a corresponding eigenvector, because $\mathbf{T}^{-1}\mathbf{x} = \mathbf{0}$ would give $\mathbf{x} = \mathbf{I}\mathbf{x} = \mathbf{T}\mathbf{T}^{-1}\mathbf{x} = \mathbf{T}\mathbf{0} = \mathbf{0}$, contradicting $\mathbf{x} \neq \mathbf{0}$. ∎

Properties of Eigenvectors

The next theorem is of interest in itself and of help in connection with bases of eigenvectors.

Theorem 2 **(Linear independence of eigenvectors)**
*Let $\lambda_1, \lambda_2, \cdots, \lambda_k$ be **distinct** eigenvalues of an $n \times n$ matrix. Then corresponding eigenvectors $\mathbf{x}_1, \mathbf{x}_2, \cdots, \mathbf{x}_k$ form a linearly independent set.*

Proof. Suppose that the conclusion is false. Let r be the largest integer such that $\{\mathbf{x}_1, \cdots, \mathbf{x}_r\}$ is a linearly independent set. Then $r < k$ and the set $\{\mathbf{x}_1, \cdots, \mathbf{x}_{r+1}\}$ is linearly dependent. Thus there are scalars $c_1, \cdots, c_{r+1}$, not all zero, such that

$$c_1\mathbf{x}_1 + \cdots + c_{r+1}\mathbf{x}_{r+1} = \mathbf{0} \tag{2}$$

(see Sec. 7.5). Multiplying both sides by $\mathbf{A}$ and using $\mathbf{A}\mathbf{x}_j = \lambda_j\mathbf{x}_j$, we obtain

$$c_1\lambda_1\mathbf{x}_1 + \cdots + c_{r+1}\lambda_{r+1}\mathbf{x}_{r+1} = \mathbf{0}. \tag{3}$$

To get rid of the last term, we subtract λ_{r+1} times (2) from this, obtaining

$$c_1(\lambda_1 - \lambda_{r+1})\mathbf{x}_1 + \cdots + c_r(\lambda_r - \lambda_{r+1})\mathbf{x}_r = \mathbf{0}.$$

Here $c_1(\lambda_1 - \lambda_{r+1}) = 0, \cdots, c_r(\lambda_r - \lambda_{r+1}) = 0$ since $\{x_1, \cdots, x_r\}$ is linearly independent. Hence $c_1 = \cdots = c_r = 0$, since all the eigenvalues are distinct. But with this, (2) reduces to $c_{r+1}\mathbf{x}_{r+1} = \mathbf{0}$, hence $c_{r+1} = 0$, since $\mathbf{x}_{r+1} \neq \mathbf{0}$ (an eigenvector!). This contradicts the fact that not all scalars in (2) are zero. Hence the conclusion of the theorem must hold. ∎

This theorem immediately implies the following.

Theorem 3 **(Basis of eigenvectors)**
*If an $n \times n$ matrix $\mathbf{A}$ has n **distinct** eigenvalues, then $\mathbf{A}$ has a basis of eigenvectors for C^n (or R^n).*

EXAMPLE 1 **Basis of eigenvectors**

The matrix

$$\mathbf{A} = \begin{bmatrix} 5 & 3 \\ 3 & 5 \end{bmatrix} \quad \text{has a basis of eigenvectors} \quad \begin{bmatrix} 1 \\ 1 \end{bmatrix}, \begin{bmatrix} 1 \\ -1 \end{bmatrix}$$

corresponding to the eigenvalues $\lambda_1 = 8$, $\lambda_2 = 2$. (See Example 1 in Sec. 7.11.) ∎

EXAMPLE 2 **Basis when not all eigenvalues are distinct. Nonexistence of basis**

Even if not all n eigenvalues are different, a matrix $\mathbf{A}$ may still provide a basis of eigenvectors for C^n or R^n. This is illustrated by Example 2 in Sec. 7.10, where $n = 3$. On the other hand, $\mathbf{A}$ may not have enough linearly independent eigenvectors to make up a basis. For instance, the matrix in Example 3, Sec. 7.10,

$$\mathbf{A} = \begin{bmatrix} 0 & 1 \\ 0 & 0 \end{bmatrix} \quad \text{has only one eigenvector} \quad \begin{bmatrix} k \\ 0 \end{bmatrix},$$

where k is arbitrary, not zero. Hence $\mathbf{A}$ does not provide a basis of eigenvectors for R^2. ∎

Actually, bases of eigenvectors exist under much more general conditions than those given in Theorem 3, and for the matrices in the previous section we can even choose a unitary system of eigenvectors, as follows.

Theorem 4 (Basis of eigenvectors)
A Hermitian, skew-Hermitian, or unitary matrix has a basis of eigenvectors for C^n *that is a unitary system* (see Sec. 7.13). *A symmetric matrix has an orthonormal basis of eigenvectors for* R^n. (Proof see Ref. [B2], vol. 1, pp. 270–272.)

EXAMPLE 3 Orthonormal basis of eigenvectors
The matrix in Example 1 is symmetric, and an orthonormal basis of eigenvectors is $[1/\sqrt{2} \quad 1/\sqrt{2}]^{\mathsf{T}}, [1/\sqrt{2} \quad -1/\sqrt{2}]^{\mathsf{T}}$. ∎

A basis of eigenvectors of a matrix **A** is of great advantage if we are interested in a transformation $\mathbf{y} = \mathbf{Ax}$ because then we can represent any $\mathbf{x}$ uniquely as

$$\mathbf{x} = c_1\mathbf{x}_1 + c_2\mathbf{x}_2 + \cdots + c_n\mathbf{x}_n$$

in terms of such a basis $\mathbf{x}_1, \cdots, \mathbf{x}_n$, and if these eigenvectors of **A** correspond to (not necessarily distinct) eigenvalues $\lambda_1, \cdots, \lambda_n$ of **A**, then we get

$$\begin{aligned} \mathbf{y} = \mathbf{Ax} &= \mathbf{A}(c_1\mathbf{x}_1 + \cdots + c_n\mathbf{x}_n) \\ &= c_1\mathbf{Ax}_1 + \cdots + c_n\mathbf{Ax}_n \qquad (4) \\ &= c_1\lambda_1\mathbf{x}_1 + \cdots + c_n\lambda_n\mathbf{x}_n. \end{aligned}$$

This shows the advantage: we have decomposed the complicated action of **A** on arbitrary vectors **x** into a sum of simple actions (multiplication by scalars) on the eigenvectors of **A**.

Diagonalization

Bases of eigenvectors also play a central role in the diagonalization of an $n \times n$ matrix **A**, as the following theorem explains.

Theorem 5 (Diagonalization of a matrix)
If an $n \times n$ *matrix* **A** *has a basis of eigenvectors, then*

$$\mathbf{D} = \mathbf{X}^{-1}\mathbf{AX} \qquad (5)$$

is diagonal, with the eigenvalues of **A** *as the entries on the main diagonal. Here* **X** *is the matrix with these eigenvectors as column vectors. Also,*

$$\mathbf{D}^m = \mathbf{X}^{-1}\mathbf{A}^m\mathbf{X}. \qquad (5^*)$$

Proof. Let $\mathbf{x}_1, \cdots, \mathbf{x}_n$ form a basis of eigenvectors of $\mathbf{A}$ for C^n (or R^n) corresponding to the eigenvalues $\lambda_1, \cdots, \lambda_n$, respectively, of $\mathbf{A}$. Then $\mathbf{X} = [\mathbf{x}_1 \cdots \mathbf{x}_n]$ has rank n, by Theorem 1 in Sec. 7.5. Hence $\mathbf{X}^{-1}$ exists by Theorem 1 in Sec. 7.7. Now (9) in Sec. 7.3 and $\mathbf{A}\mathbf{x}_j = \lambda_j\mathbf{x}_j$ give

$$\mathbf{AX} = \mathbf{A}[\mathbf{x}_1 \quad \cdots \quad \mathbf{x}_n] = [\mathbf{A}\mathbf{x}_1 \quad \cdots \quad \mathbf{A}\mathbf{x}_n] = [\lambda_1\mathbf{x}_1 \quad \cdots \quad \lambda_n\mathbf{x}_n].$$

Together, $\mathbf{AX} = \mathbf{XD}$. This we multiply on both sides by $\mathbf{X}^{-1}$ from the left to get $\mathbf{X}^{-1}\mathbf{AX} = \mathbf{X}^{-1}\mathbf{XD} = \mathbf{D}$, which is (5). Also, (5*) follows by noting that

$$\mathbf{D}^2 = \mathbf{DD} = \mathbf{X}^{-1}\mathbf{AXX}^{-1}\mathbf{AX} = \mathbf{X}^{-1}\mathbf{AAX} = \mathbf{X}^{-1}\mathbf{A}^2\mathbf{X}, \qquad \text{etc.} \quad ∎$$

EXAMPLE 4 Diagonalization

Calculation as in the examples in Sec. 7.10, etc. shows that the matrix

$$\mathbf{A} = \begin{bmatrix} 5 & 4 \\ 1 & 2 \end{bmatrix} \quad \text{has eigenvectors} \quad \begin{bmatrix} 4 \\ 1 \end{bmatrix} \quad \text{and} \quad \begin{bmatrix} 1 \\ -1 \end{bmatrix}. \qquad \text{Hence} \quad \mathbf{X} = \begin{bmatrix} 4 & 1 \\ 1 & -1 \end{bmatrix}$$

and, using (4) in Sec. 7.7, we obtain

$$\mathbf{X}^{-1}\mathbf{AX} = \frac{1}{-5}\begin{bmatrix} -1 & -1 \\ -1 & 4 \end{bmatrix}\begin{bmatrix} 5 & 4 \\ 1 & 2 \end{bmatrix}\begin{bmatrix} 4 & 1 \\ 1 & -1 \end{bmatrix} = \begin{bmatrix} 0.2 & 0.2 \\ 0.2 & -0.8 \end{bmatrix}\begin{bmatrix} 24 & 1 \\ 6 & -1 \end{bmatrix} = \begin{bmatrix} 6 & 0 \\ 0 & 1 \end{bmatrix}.$$

The student may show that an interchange of the columns of $\mathbf{X}$ results in an interchange of the eigenvalues 6 and 1 in the diagonal matrix. ∎

EXAMPLE 5 Diagonalization

Diagonalize

$$\mathbf{A} = \begin{bmatrix} 7.3 & 0.2 & -3.7 \\ -11.5 & 1.0 & 5.5 \\ 17.7 & 1.8 & -9.3 \end{bmatrix}.$$

Solution. The characteristic determinant gives the characteristic equation $-\lambda^3 - \lambda^2 + 12\lambda = 0$. The roots (eigenvalues of $\mathbf{A}$) are $\lambda_1 = 3$, $\lambda_2 = -4$, $\lambda_3 = 0$. By the Gauss elimination applied to $(\mathbf{A} - \lambda\mathbf{I})\mathbf{x} = \mathbf{0}$ with $\lambda = \lambda_1, \lambda_2, \lambda_3$ we find eigenvectors and then $\mathbf{X}^{-1}$ by the Gauss–Jordan elimination (Sec. 7.7, Example 1). The results are

$$\begin{bmatrix} -1 \\ 3 \\ -1 \end{bmatrix}, \quad \begin{bmatrix} 1 \\ -1 \\ 3 \end{bmatrix}, \quad \begin{bmatrix} 2 \\ 1 \\ 4 \end{bmatrix}, \qquad \mathbf{X} = \begin{bmatrix} -1 & 1 & 2 \\ 3 & -1 & 1 \\ -1 & 3 & 4 \end{bmatrix}, \quad \mathbf{X}^{-1} = \begin{bmatrix} -0.7 & 0.2 & 0.3 \\ -1.3 & -0.2 & 0.7 \\ 0.8 & 0.2 & -0.2 \end{bmatrix}.$$

Calculating $\mathbf{AX}$ and multiplying by $\mathbf{X}^{-1}$ from the left, we thus obtain

$$\mathbf{D} = \mathbf{X}^{-1}\mathbf{AX} = \begin{bmatrix} -0.7 & 0.2 & 0.3 \\ -1.3 & -0.2 & 0.7 \\ 0.8 & 0.2 & -0.2 \end{bmatrix}\begin{bmatrix} -3 & -4 & 0 \\ 9 & 4 & 0 \\ -3 & -12 & 0 \end{bmatrix} = \begin{bmatrix} 3 & 0 & 0 \\ 0 & -4 & 0 \\ 0 & 0 & 0 \end{bmatrix}. \quad ∎$$

Transformation of Forms to Principal Axes

This is an important practical task related to the diagonalization of matrices. We explain the idea for quadratic forms (see Sec. 7.13)

$$Q = \mathbf{x}^{\mathsf{T}}\mathbf{A}\mathbf{x}. \tag{6}$$

Without restriction we can assume that **A** is real ***symmetric*** (see Sec. 7.13). Then **A** has an orthonormal basis of n eigenvectors, by Theorem 4. Hence the matrix **X** with these vectors as column vectors is orthogonal, so that $\mathbf{X}^{-1} = \mathbf{X}^{\mathsf{T}}$. From (5) we thus have $\mathbf{A} = \mathbf{X}\mathbf{D}\mathbf{X}^{-1} = \mathbf{X}\mathbf{D}\mathbf{X}^{\mathsf{T}}$. Substitution into (6) gives

$$Q = \mathbf{x}^{\mathsf{T}}\mathbf{X}\mathbf{D}\mathbf{X}^{\mathsf{T}}\mathbf{x}.$$

If we set $\mathbf{X}^{\mathsf{T}}\mathbf{x} = \mathbf{y}$, then, since $\mathbf{X}^{\mathsf{T}} = \mathbf{X}^{-1}$, we get

$$\mathbf{x} = \mathbf{X}\mathbf{y} \tag{7}$$

and Q becomes simply

$$Q = \mathbf{y}^{\mathsf{T}}\mathbf{D}\mathbf{y} = \lambda_1 y_1^2 + \lambda_2 y_2^2 + \cdots + \lambda_n y_n^2. \tag{8}$$

This proves

Theorem 6 (Principal axes theorem)

The substitution (7) *transforms a quadratic form*

$$Q = \mathbf{x}^{\mathsf{T}}\mathbf{A}\mathbf{x} = \sum_{j=1}^{n}\sum_{k=1}^{n} a_{jk}x_j x_k$$

to the principal axes form (8), *where* $\lambda_1, \cdots, \lambda_n$ *are the* (*not necessarily distinct*) *eigenvalues of the* (*symmetric!*) *matrix* **A**, *and* **X** *is an orthogonal matrix with corresponding eigenvectors* $\mathbf{x}_1, \cdots, \mathbf{x}_n$, *respectively, as column vectors.*

EXAMPLE 6 Transformation to principal axes. Conic sections

Find out what type of conic section the following quadratic form represents and transform it to principal axes:

$$Q = 17x_1^2 - 30x_1x_2 + 17x_2^2 = 128.$$

Solution. We have $Q = \mathbf{x}^{\mathsf{T}}\mathbf{A}\mathbf{x}$, where

$$\mathbf{A} = \begin{bmatrix} 17 & -15 \\ -15 & 17 \end{bmatrix}, \qquad \mathbf{x} = \begin{bmatrix} x_1 \\ x_2 \end{bmatrix}.$$

This gives the characteristic equation $(17 - \lambda)^2 - 15^2 = 0$. It has the roots $\lambda_1 = 2, \lambda_2 = 32$. Hence (8) becomes

$$Q = 2y_1^2 + 32y_2^2.$$

We see that $Q = 128$ represents the ellipse $2y_1^2 + 32y_2^2 = 128$, that is,

$$\frac{y_1^2}{8^2} + \frac{y_2^2}{2^2} = 1.$$

If we want to know the direction of the principal axes in the x_1x_2-coordinates, we have to determine normalized eigenvectors from $(\mathbf{A} - \lambda\mathbf{I})\mathbf{x} = \mathbf{0}$ with $\lambda = \lambda_1 = 2$ and $\lambda = \lambda_2 = 32$ and then use (7). We get

$$\begin{bmatrix} 1/\sqrt{2} \\ 1/\sqrt{2} \end{bmatrix} \quad \text{and} \quad \begin{bmatrix} -1/\sqrt{2} \\ 1/\sqrt{2} \end{bmatrix},$$

hence

$$\mathbf{x} = \mathbf{X}\mathbf{y} = \begin{bmatrix} 1/\sqrt{2} & -1/\sqrt{2} \\ 1/\sqrt{2} & 1/\sqrt{2} \end{bmatrix} \begin{bmatrix} y_1 \\ y_2 \end{bmatrix}, \qquad \begin{aligned} x_1 &= y_1/\sqrt{2} - y_2/\sqrt{2} \\ x_2 &= y_1/\sqrt{2} + y_2/\sqrt{2}. \end{aligned}$$

This is a 45° rotation. Our results agree with those in Sec. 7.11, Example 1, except for the notations. See also Fig. 136 in that example. ∎

Problem Set 7.14

Similarity transformations

Find $\hat{\mathbf{A}} = \mathbf{T}^{-1}\mathbf{A}\mathbf{T}$. Find the eigenvalues of $\hat{\mathbf{A}}$ and $\mathbf{A}$ and verify that they are the same. Find corresponding eigenvectors $\mathbf{y}$ of $\hat{\mathbf{A}}$, compute $\mathbf{x} = \mathbf{T}\mathbf{y}$, and verify that they are eigenvectors of $\mathbf{A}$.

1. $\mathbf{A} = \begin{bmatrix} -4 & 0 \\ 0 & 2 \end{bmatrix}, \mathbf{T} = \begin{bmatrix} 4 & -2 \\ -3 & 1 \end{bmatrix}$

2. $\mathbf{A} = \begin{bmatrix} 6 & 8 \\ 8 & -6 \end{bmatrix}, \mathbf{T} = \begin{bmatrix} 10 & 4 \\ 4 & 2 \end{bmatrix}$

3. $\mathbf{A} = \begin{bmatrix} 4 & -2 \\ 1 & 1 \end{bmatrix}, \mathbf{T} = \begin{bmatrix} 5 & 2 \\ 2 & 1 \end{bmatrix}$

4. $\mathbf{A} = \begin{bmatrix} -2 & 0 \\ -4 & 2 \end{bmatrix}, \mathbf{T} = \begin{bmatrix} 10 & 3 \\ 0 & 1 \end{bmatrix}$

5. $\mathbf{A} = \begin{bmatrix} 5 & 10 \\ 4 & -1 \end{bmatrix}, \mathbf{T} = \begin{bmatrix} 5 & 1 \\ 2 & -1 \end{bmatrix}$

6. $\mathbf{A} = \begin{bmatrix} 2 & 1 \\ -2 & -1 \end{bmatrix}, \mathbf{T} = \begin{bmatrix} 7 & -5 \\ 10 & -7 \end{bmatrix}$

7. $\mathbf{A} = \begin{bmatrix} 8 & 0 & 3 \\ 2 & 2 & 1 \\ 2 & 0 & 3 \end{bmatrix}, \mathbf{T} = \begin{bmatrix} 0 & 1 & 0 \\ 1 & 0 & 0 \\ 0 & 0 & 1 \end{bmatrix}$

8. $\mathbf{A} = \begin{bmatrix} 5 & 0 & -15 \\ -3 & -4 & 9 \\ 5 & 0 & -15 \end{bmatrix}, \mathbf{T} = \begin{bmatrix} 2 & 0 & 3 \\ 0 & 1 & 0 \\ 3 & 0 & 5 \end{bmatrix}$

Traces of similar matrices

The sum of the entries on the main diagonal of an $n \times n$ matrix $\mathbf{A} = [a_{jk}]$ is called the **trace** of $\mathbf{A}$; thus trace $\mathbf{A} = a_{11} + a_{22} + \cdots + a_{nn}$.

9. Show that trace $\mathbf{AB} = \sum_{i=1}^{n} \sum_{l=1}^{n} a_{il}b_{li} =$ trace $\mathbf{BA}$, where $\mathbf{A} = [a_{jk}]$ and $\mathbf{B} = [b_{jk}]$ are $n \times n$ matrices.

10. Using Prob. 9, show that similar matrices have equal traces.

11. **(Sum of the eigenvalues)** Show that trace **A** equals the sum of the eigenvalues of **A**, each counted as often as its algebraic multiplicity indicates.

12. Using Prob. 11, show that trace $\hat{\mathbf{A}}$ = trace **A** when $\hat{\mathbf{A}}$ is similar to **A**.

Find a basis of eigenvectors that form a unitary system (or a real orthogonal system).

13. $\begin{bmatrix} 0 & i \\ -i & 0 \end{bmatrix}$ **14.** $\begin{bmatrix} 0 & 2i \\ 2i & 0 \end{bmatrix}$ **15.** $\begin{bmatrix} 0.8i & 0.6i \\ 0.6i & -0.8i \end{bmatrix}$

16. $\begin{bmatrix} 1 & 1+i \\ 1-i & 1 \end{bmatrix}$ **17.** $\begin{bmatrix} i & 1 \\ -1 & i \end{bmatrix}$ **18.** $\begin{bmatrix} \frac{1}{2}i & \frac{1}{2}\sqrt{3} \\ \frac{1}{2}\sqrt{3} & \frac{1}{2}i \end{bmatrix}$

Diagonalization

Find a basis of eigenvectors and diagonalize:

19. $\begin{bmatrix} 0 & 16 \\ 4 & 0 \end{bmatrix}$ **20.** $\begin{bmatrix} 3 & 3 \\ 0 & 0 \end{bmatrix}$ **21.** $\begin{bmatrix} 0 & 1 \\ -1 & 0 \end{bmatrix}$

22. $\begin{bmatrix} 3 & 4 \\ 1 & 3 \end{bmatrix}$ **23.** $\begin{bmatrix} 1 & 0 & 1 \\ 0 & 3 & 2 \\ 0 & 0 & 2 \end{bmatrix}$ **24.** $\begin{bmatrix} 5 & 0 & -6 \\ 2 & 1 & -4 \\ 3 & 0 & -4 \end{bmatrix}$

Transformation of quadratic forms to principal axes. Conic sections

Find out what type of conic section (or pair of straight lines) is represented by the given quadratic form. Transform it to principal axes. Express $\mathbf{x}^\mathsf{T} = [x_1 \quad x_2]$ in terms of the new coordinate vector $\mathbf{y}^\mathsf{T} = [y_1 \quad y_2]$, as in Example 6.

25. $x_1^2 + 24x_1x_2 - 6x_2^2 = 5$

26. $2x_1^2 + 2\sqrt{3}x_1x_2 + 4x_2^2 = 5$

27. $3x_1^2 + 4\sqrt{3}x_1x_2 + 7x_2^2 = 9$

28. $-3x_1^2 + 8x_1x_2 + 3x_2^2 = 0$

29. $x_1^2 + 6x_1x_2 + 9x_2^2 = 10$

30. $6x_1^2 + 16x_1x_2 - 6x_2^2 = 10$

7.15 Vector Spaces, Inner Product Spaces, Linear Transformations *Optional*

From Sec. 7.5 we recall that the **real vector space** R^n is the set of all real vectors with n components (thus, each such vector is an ordered n-tuple of real numbers), with the two algebraic operations of vector addition and multiplication by scalars (real numbers). Similarly, taking ordered n-tuples of *complex* numbers as vectors and *complex* numbers as scalars, we obtain the **complex vector space** C^n (see Sec. 7.13).

There are other sets of practical interest (sets of matrices, functions, transformations, etc.) for which an addition and a scalar multiplication can be defined in a natural way. The desire to treat such sets as "vector spaces" suggests we create from the "**concrete model**" R^n the "**abstract concept**" of a "real vector space" V by taking the most basic properties of R^n as axioms by which V is defined, properties without which one would not be able to create a useful and applicable theory of those more general situations. Selecting good axioms is not easy, but needs experience, sometimes gained only over a long period of time. In the present case, the following system of axioms turned out to be useful; note that each axiom expresses a simple property of R^n, or of R^3, as a matter of fact.

Definition of a real vector space

A nonempty set V of elements **a, b,** $\cdots$ is called a *real vector space* (or *real linear space*), and these elements are called **vectors,**[18] if in V there are defined two algebraic operations (called *vector addition* and *scalar multiplication*) as follows.

I. Vector addition associates with every pair of vectors **a** and **b** of V a unique vector of V, called the *sum* of **a** and **b** and denoted by $\mathbf{a} + \mathbf{b}$, such that the following axioms are satisfied.

I.1 *Commutativity.* For any two vectors **a** and **b** of V,

$$\mathbf{a} + \mathbf{b} = \mathbf{b} + \mathbf{a}.$$

I.2 *Associativity.* For any three vectors **u, v, w** of V,

$$(\mathbf{u} + \mathbf{v}) + \mathbf{w} = \mathbf{u} + (\mathbf{v} + \mathbf{w}) \qquad (\text{written } \mathbf{u} + \mathbf{v} + \mathbf{w}).$$

I.3 There is a unique vector in V, called the *zero vector* and denoted by **0**, such that for every **a** in V,

$$\mathbf{a} + \mathbf{0} = \mathbf{a}.$$

I.4 For every **a** in V there is a unique vector in V that is denoted by $-\mathbf{a}$ and is such that

$$\mathbf{a} + (-\mathbf{a}) = \mathbf{0}.$$

II. Scalar multiplication. The real numbers are called **scalars.** Scalar multiplication associates with every **a** in V and every scalar c a unique vector of V, called the *product* of c and **a** and denoted by $c\mathbf{a}$ (or $\mathbf{a}c$) such that the following axioms are satisfied.

II.1 *Distributivity.* For every scalar c and vectors **a** and **b** in V,

$$c(\mathbf{a} + \mathbf{b}) = c\mathbf{a} + c\mathbf{b}.$$

II.2 *Distributivity.* For all scalars c and k and every **a** in V,

$$(c + k)\mathbf{a} = c\mathbf{a} + k\mathbf{a}.$$

[18] Regardless of what they actually are; this convention causes no confusion because in any specific case the nature of those elements is clear from the context.

II.3 *Associativity.* For all scalars c and k and every $\mathbf{a}$ in V,

$$c(k\mathbf{a}) = (ck)\mathbf{a} \qquad \text{(written } ck\mathbf{a}\text{).}$$

II.4 For every $\mathbf{a}$ in V,

$$1\mathbf{a} = \mathbf{a}. \qquad \blacksquare$$

A **complex vector space** is obtained if, instead of real numbers, we take complex numbers as scalars.

Basic Concepts Related to Vector Space

These concepts are defined as in Sec. 7.5.

A **linear combination** of vectors $\mathbf{a}_{(1)}, \cdots, \mathbf{a}_{(m)}$ in a vector space V is an expression

$$c_1\mathbf{a}_{(1)} + \cdots + c_m\mathbf{a}_{(m)} \qquad (c_1, \cdots, c_m \text{ any scalars).}$$

These vectors form a **linearly independent set** (briefly, they are called **linearly independent**) if

$$(1) \qquad\qquad c_1\mathbf{a}_{(1)} + \cdots + c_m\mathbf{a}_{(m)} = \mathbf{0}$$

implies that $c_1 = 0, \cdots, c_m = 0$; otherwise they are called **linearly dependent**. Note that (1) with $m = 1$ is $c\mathbf{a} = \mathbf{0}$ and shows that a single vector $\mathbf{a}$ is linearly independent if and only if $\mathbf{a} \neq \mathbf{0}$.

V has **dimension** n, or is **n-dimensional**, if it contains a linearly independent set of n vectors, called a **basis** for V, whereas any set of more than n vectors in V is linearly dependent. Then every vector in V can be uniquely written as a linear combination of the basis vectors.

EXAMPLE 1   **Vector space of matrices**

The real 2×2 matrices form a four-dimensional real vector space. A basis is

$$\mathbf{B}_{11} = \begin{bmatrix} 1 & 0 \\ 0 & 0 \end{bmatrix}, \quad \mathbf{B}_{12} = \begin{bmatrix} 0 & 1 \\ 0 & 0 \end{bmatrix}, \quad \mathbf{B}_{21} = \begin{bmatrix} 0 & 0 \\ 1 & 0 \end{bmatrix}, \quad \mathbf{B}_{22} = \begin{bmatrix} 0 & 0 \\ 0 & 1 \end{bmatrix}$$

because any $\mathbf{A} = [a_{jk}] = a_{11}\mathbf{B}_{11} + a_{12}\mathbf{B}_{12} + a_{21}\mathbf{B}_{21} + a_{22}\mathbf{B}_{22}$ in a unique fashion. Similarly, the real $m \times n$ matrices with fixed m and n form an mn-dimensional vector space. What is the dimension of the vector space of all skew-symmetric 3×3 matrices? Can you find a basis? $\blacksquare$

EXAMPLE 2   **Polynomials**

The set of all constant, linear and quadratic polynomials in x together forms a vector space under the usual addition and multiplication by a real number, since these two operations give polynomials of degree not exceeding 2, and the axioms in our definition follow by direct calculation. This space has dimension 3. A basis is $\{1, \ x, \ x^2\}$. $\blacksquare$

EXAMPLE 3   **Second-order homogeneous linear differential equations**

The solutions of such an equation on a fixed interval $a < x < b$ form a vector space under the usual addition and multiplication by a number since these two operations give again such a solution, by Fundamental Theorem 1 in Sec. 2.1, and I.1 to II.4 follow by direct calculation. Do the solutions of a nonhomogeneous linear differential equation form a vector space? $\blacksquare$

If a vector space V contains a linearly independent set of n vectors for every n, no matter how large, then V is called **infinite dimensional,** as opposed to a *finite dimensional* (n-dimensional) vector space as defined before. An example is the space of all continuous functions on some interval $[a, b]$ of the x-axis, as we mention without proof.

Inner Product Spaces

From Sec. 7.3 we know that for vectors $\mathbf{a}$ and $\mathbf{b}$ in R^n we can define an inner product $\mathbf{a}\cdot\mathbf{b} = \mathbf{a}^T\mathbf{b}$. This definition can be extended to general real vector spaces by taking basic properties of $\mathbf{a}\cdot\mathbf{b}$ as axioms for an "abstract" inner product, denoted by $(\mathbf{a}, \mathbf{b})$.

Definition of a real inner product space

A real vector space V is called a *real inner product space* (or *real pre-Hilbert*[19] *space*) if it has the following property. With every pair of vectors $\mathbf{a}$ and $\mathbf{b}$ in V there is associated a real number, which is denoted by $(\mathbf{a}, \mathbf{b})$ and is called the **inner product** of $\mathbf{a}$ and $\mathbf{b}$, such that the following axioms are satisfied.

I. For all scalars q_1 and q_2 and all vectors $\mathbf{a}, \mathbf{b}, \mathbf{c}$ in V,

$$(q_1\mathbf{a} + q_2\mathbf{b}, \mathbf{c}) = q_1(\mathbf{a}, \mathbf{c}) + q_2(\mathbf{b}, \mathbf{c}) \qquad (\textit{Linearity}).$$

II. For all vectors $\mathbf{a}$ and $\mathbf{b}$ in V,

$$(\mathbf{a}, \mathbf{b}) = (\mathbf{b}, \mathbf{a}) \qquad (\textit{Symmetry}).$$

III. For every $\mathbf{a}$ in V,

$$\left.\begin{aligned} (\mathbf{a}, \mathbf{a}) \geqq 0, \quad \text{and} \\ (\mathbf{a}, \mathbf{a}) = 0 \quad \text{if and only if} \quad \mathbf{a} = \mathbf{0} \end{aligned}\right\} (\textit{Positive-definiteness}).$$

Vectors whose inner product is zero are called **orthogonal.**

The *length* or **norm** of a vector in V is now defined by

$$\|\mathbf{a}\| = \sqrt{(\mathbf{a}, \mathbf{a})} \qquad (\geqq 0), \tag{2}$$

This generalizes (8) in Sec. 7.12.

A vector of norm 1 is called a **unit vector.**

[19]DAVID HILBERT (1862—1943), great German mathematician, taught at Königsberg and Göttingen and was the creator of the famous Göttingen mathematical school. He is known for his basic work in algebra, the calculus of variations, integral equations, functional analysis, and mathematical logic. His "Foundations of Geometry" helped the axiomatic method to gain general recognition. His famous 23 problems (presented in 1900 at the International Congress of Mathematicians in Paris) considerably influenced the development of modern mathematics.

If V is finite dimensional, it is actually a so-called *Hilbert space;* see Ref. [9], p. 73, listed in Appendix 1.

From these axioms and from (2) one can derive the basic inequality

$$|(\mathbf{a}, \mathbf{b})| \leqq \|\mathbf{a}\| \, \|\mathbf{b}\| \qquad (3)$$

(*Schwarz*[20] *inequality*),

from this

$$\|\mathbf{a} + \mathbf{b}\| \leqq \|\mathbf{a}\| + \|\mathbf{b}\| \qquad (4)$$

(*Triangle inequality*),

and by a simple direct calculation

$$\|\mathbf{a} + \mathbf{b}\|^2 + \|\mathbf{a} - \mathbf{b}\|^2 = 2(\|\mathbf{a}\|^2 + \|\mathbf{b}\|^2) \qquad (5)$$

(*Parallelogram equality*).

EXAMPLE 4 *n*-dimensional Euclidean space

R^n with the inner product of Sec. 7.3,

$$(\mathbf{a}, \mathbf{b}) = \mathbf{a}^\mathsf{T}\mathbf{b} = a_1 b_1 + \cdots + a_n b_n, \qquad (6)$$

is called *n-dimensional Euclidean space* and denoted by E^n or again simply by R^n. Axioms I–III hold, as direct calculation shows. Equation (2) gives the "**Euclidean norm**" (8) of Sec. 7.12,

$$\|\mathbf{a}\| = \sqrt{(\mathbf{a}, \mathbf{a})} = \sqrt{\mathbf{a}^\mathsf{T}\mathbf{a}} = \sqrt{a_1^2 + \cdots + a_n^2}. \qquad (7)$$

■

EXAMPLE 5 An inner product for functions

The set of all real-valued continuous functions $f(x), g(x), \cdots$ on a given interval $\alpha \leqq x \leqq \beta$ forms a real vector space under the usual addition of functions and multiplication by scalars (real numbers). On this space we can define an inner product by the integral

$$(f, g) = \int_\alpha^\beta f(x)g(x)\, dx. \qquad (8)$$

Axioms I–III can be verified by direct calculation. Equation (2) gives the norm

$$\|f\| = \sqrt{(f, f)} = \sqrt{\int_\alpha^\beta f(x)^2\, dx}. \qquad (9)$$

■

Our examples give a first impression of the great generality of the abstract concepts of vector spaces and inner product spaces. Further details belong to more advanced courses (on functional analysis, meaning abstract modern analysis; see Ref. [9] listed in Appendix 1) and cannot be discussed here. Instead we now take up a related topic where matrices play a central role.

---

[20] HERMANN AMANDUS SCHWARZ (1843—1921). German mathematician, successor of Weierstrass at Berlin, known by his work in complex analysis (conformal mapping), differential geometry, and the calculus of variations (minimal surfaces).

Linear Transformations

Let X and Y be any vector spaces. To each vector $\mathbf{x}$ in X we assign a unique vector $\mathbf{y}$ in Y. Then we say that a **mapping** (or **transformation** or **operator**) of X into Y is given. Such a mapping is denoted by a capital letter, say F. The vector $\mathbf{y}$ in Y assigned to a vector $\mathbf{x}$ in X is called the **image** of $\mathbf{x}$ and is denoted by $F(\mathbf{x})$ [or $F\mathbf{x}$, without parentheses].

F is called a **linear mapping** or **linear transformation** if for all vectors $\mathbf{v}$ and $\mathbf{x}$ in X and scalars c,

$$\textbf{(10)} \qquad \begin{aligned} F(\mathbf{v} + \mathbf{x}) &= F(\mathbf{v}) + F(\mathbf{x}) \\ F(c\mathbf{x}) &= cF(\mathbf{x}). \end{aligned}$$

Linear transformation of space R^n into space R^m

From now on we let $X = R^n$ and $Y = R^m$. Then any real $m \times n$ matrix $\mathbf{A} = [a_{jk}]$ gives a transformation of R^n into R^m,

$$(11) \qquad \mathbf{y} = \mathbf{Ax}.$$

Since $\mathbf{A}(\mathbf{u} + \mathbf{x}) = \mathbf{Au} + \mathbf{Ax}$ and $\mathbf{A}(c\mathbf{x}) = c\mathbf{Ax}$, this transformation is linear.

We show that, conversely, every linear transformation F of R^n into R^m can be given in terms of an $m \times n$ matrix $\mathbf{A}$, after a basis for R^n and a basis for R^m have been chosen. This can be proved as follows.

Let $\mathbf{e}_{(1)}, \cdots, \mathbf{e}_{(n)}$ be any basis for R^n. Then every $\mathbf{x}$ in R^n has a unique representation

$$\mathbf{x} = x_1\mathbf{e}_{(1)} + \cdots + x_n\mathbf{e}_{(n)}.$$

Since F is linear, this implies for the image $F(\mathbf{x})$:

$$F(\mathbf{x}) = F(x_1\mathbf{e}_{(1)} + \cdots + x_n\mathbf{e}_{(n)}) = x_1F(\mathbf{e}_{(1)}) + \cdots + x_nF(\mathbf{e}_{(n)}).$$

Hence F is uniquely determined by the images of the vectors of a basis for R^n. We now choose for R^n the **"standard basis"**

$$(12) \qquad \mathbf{e}_{(1)} = \begin{bmatrix} 1 \\ 0 \\ \vdots \\ 0 \end{bmatrix}, \quad \mathbf{e}_{(2)} = \begin{bmatrix} 0 \\ 1 \\ \vdots \\ 0 \end{bmatrix}, \quad \cdots, \quad \mathbf{e}_{(n)} = \begin{bmatrix} 0 \\ 0 \\ \vdots \\ 1 \end{bmatrix}$$

where $\mathbf{e}_{(j)}$ has its jth component equal to 1 and all others 0. We can now determine an $m \times n$ matrix $\mathbf{A} = [a_{jk}]$ such that for every $\mathbf{x}$ in R^n and image $\mathbf{y} = F(\mathbf{x})$ in R^m,

$$\mathbf{y} = F(\mathbf{x}) = \mathbf{Ax}.$$

Indeed, from the image $\mathbf{y}^{(1)} = F(\mathbf{e}_{(1)})$ of $\mathbf{e}_{(1)}$ we get the condition

$$\mathbf{y}^{(1)} = \begin{bmatrix} y_1^{(1)} \\ y_2^{(1)} \\ \vdots \\ y_m^{(1)} \end{bmatrix} = \begin{bmatrix} a_{11} & \cdots & a_{1n} \\ a_{21} & \cdots & a_{2n} \\ \vdots & & \vdots \\ a_{m1} & \cdots & a_{mn} \end{bmatrix} \begin{bmatrix} 1 \\ 0 \\ \vdots \\ 0 \end{bmatrix}.$$

from which we can determine the first column of **A**, namely $a_{11} = y_1^{(1)}$, $a_{21} = y_2^{(1)}, \cdots, a_{m1} = y_m^{(1)}$. Similarly, from the image of $\mathbf{e}_{(2)}$ we get the second column of **A**, and so on. This completes the proof. ∎

We say that **A represents** F, or *is a representation of* F, with respect to the bases for R^n and R^m. Quite generally, the purpose of a "*representation*" is the replacement of one object of study by another object whose properties are more readily apparent.

The standard basis (12) for R^3 is usually written $\mathbf{e}_{(1)} = \mathbf{i}$, $\mathbf{e}_{(2)} = \mathbf{j}$, $\mathbf{e}_{(3)} = \mathbf{k}$; thus

$$\text{(13)} \qquad \mathbf{i} = \begin{bmatrix} 1 \\ 0 \\ 0 \end{bmatrix}, \qquad \mathbf{j} = \begin{bmatrix} 0 \\ 1 \\ 0 \end{bmatrix}, \qquad \mathbf{k} = \begin{bmatrix} 0 \\ 0 \\ 1 \end{bmatrix}.$$

EXAMPLE 6 Linear transformations

Interpreted as transformations of Cartesian coordinates in the plane, the matrices

$$\begin{bmatrix} 0 & 1 \\ 1 & 0 \end{bmatrix}, \quad \begin{bmatrix} 1 & 0 \\ 0 & -1 \end{bmatrix}, \quad \begin{bmatrix} -1 & 0 \\ 0 & -1 \end{bmatrix}, \quad \begin{bmatrix} a & 0 \\ 0 & 1 \end{bmatrix}$$

represent a reflection in the line $x_2 = x_1$, a reflection in the x_1-axis, a reflection in the origin, and a stretch (when $a > 1$, or a contraction when $0 < a < 1$) in the x_1-direction, respectively. ∎

EXAMPLE 7 Linear transformations

Our discussion preceding Example 6 is simpler than it may look at first sight. To see this, find **A** representing the linear transformation that maps (x_1, x_2) onto $(2x_1 - 5x_2, 3x_1 + 4x_2)$.

Solution. Since

$$\begin{bmatrix} 1 \\ 0 \end{bmatrix} \text{ and } \begin{bmatrix} 0 \\ 1 \end{bmatrix} \text{ are mapped onto } \begin{bmatrix} 2 \\ 3 \end{bmatrix} \text{ and } \begin{bmatrix} -5 \\ 4 \end{bmatrix},$$

respectively, we obtain, according to our discussion,

$$\mathbf{A} = \begin{bmatrix} 2 & -5 \\ 3 & 4 \end{bmatrix}.$$

We check this, finding

$$\begin{bmatrix} y_1 \\ y_2 \end{bmatrix} = \begin{bmatrix} 2 & -5 \\ 3 & 4 \end{bmatrix} \begin{bmatrix} x_1 \\ x_2 \end{bmatrix} = \begin{bmatrix} 2x_1 - 5x_2 \\ 3x_1 + 4x_2 \end{bmatrix}.$$

The reader may obtain **A** also at once by writing the given transformation in the form

$$y_1 = 2x_1 - 5x_2$$
$$y_2 = 3x_1 + 4x_2.$$ ∎

If **A** in (11) is square, $n \times n$, then (11) maps R^n into R^n. If this **A** is nonsingular, so that $\mathbf{A}^{-1}$ exists (see Sec. 7.7), then multiplication of (11) by $\mathbf{A}^{-1}$ from the left and use of $\mathbf{A}^{-1}\mathbf{A} = \mathbf{I}$ gives the **inverse transformation**

$$\mathbf{x} = \mathbf{A}^{-1}\mathbf{y}. \tag{14}$$

It maps every $\mathbf{y} = \mathbf{y}_0$ onto that $\mathbf{x}$ which by (11) is mapped onto $\mathbf{y}_0$. *The inverse of a linear transformation is itself linear,* because it is given by a matrix, as (14) shows.

This is the end of Chap. 7 on linear algebra, which was concerned with algebraic operations on matrices and vectors and applications to systems of linear equations and eigenvalue problems. The next chapter is devoted to the application of differential calculus to vector functions in 3-space, a field of basic importance in engineering and physics.

Problem Set 7.15

Vector spaces, bases

Is the given set (taken with the usual addition and scalar multiplication) a vector space or not? (Give reason.) If your answer is yes, determine the dimension and find a basis. (See Sec. 7.5 for similar problems.)

1. All polynomials in x, of degree not exceeding 4.

2. All symmetric real 3×3 matrices.

3. All skew-symmetric real 2×2 matrices.

4. All vectors in R^3 satisfying $v_1 - 2v_2 + v_3 = 0$.

5. All real 4×4 matrices with positive entries.

6. All real 2×3 matrices with the first row any multiple of $[1 \quad 0 \quad 2]$.

7. All functions $f(x) = (ax + b)e^{-x}$ with arbitrary constant a and b.

8. All ordered quintuples of nonnegative real numbers.

Linear independence, bases. (See Sec. 7.5 for further problems.)

9. If a subset S_0 of a set S is linearly dependent, show that S is itself linearly dependent.

10. Show that a subset of a linearly independent set is itself linearly independent.

11. Find three different bases for R^2.

12. **(Uniqueness)** Show that the representation $\mathbf{v} = c_1\mathbf{a}_{(1)} + \cdots + c_n\mathbf{a}_{(n)}$ of any given vector $\mathbf{v}$ in an n-dimensional vector space V in terms of a basis $\mathbf{a}_{(1)}, \cdots, \mathbf{a}_{(n)}$ for V is unique.

Inner product, orthogonality

Find the Euclidean norm of the vectors:

13. $[1 \quad 3 \quad -1]^T$ **14.** $[2 \quad 4 \quad 1 \quad 3]^T$ **15.** $[3 \quad 0 \quad 0 \quad 4]^T$

16. $[4 \quad 2 \quad 0]^T$ **17.** $[5 \quad 1 \quad 0 \quad 6]^T$ **18.** $[\frac{1}{2} \quad 3 \quad \frac{1}{4} \quad 2]^T$

Using the Euclidean norm, verify:

19. The Schwarz inequality for the vectors in Probs. 13 and 16.

20. The Schwarz inequality for the vectors in Probs. 14 and 18.

21. The triangle inequality for the vectors in Probs. 15 and 17.

22. The parallelogram equality for the vectors $[2 \quad 1]^T$ and $[1 \quad 3]^T$. Graph the parallelogram with these sides and explain the geometric meaning of this equality.

23. Find all vectors $\mathbf{v}$ orthogonal to $\mathbf{a} = [1 \quad 2 \quad 0]^T$. Do they form a vector space?

24. Using (6), find all unit vectors $\mathbf{v} = [v_1 \quad v_2]$ orthogonal to $[3 \quad -4]$.

Linear transformations

Find the inverse transformation:

25. $$\begin{aligned} y_1 &= 3x_1 - x_2 \\ y_2 &= -5x_1 + 2x_2 \end{aligned}$$

26. $$\begin{aligned} y_1 &= 4x_1 + 3x_2 \\ y_2 &= 3x_1 + 2x_2 \end{aligned}$$

27. $$\begin{aligned} y_1 &= 2x_1 + 4x_2 + x_3 \\ y_2 &= x_1 + 2x_2 + x_3 \\ y_3 &= 3x_1 + 4x_2 + 2x_3 \end{aligned}$$

28. $$\begin{aligned} y_1 &= x_1 + 3x_3 \\ y_2 &= 2x_2 + x_3 \\ y_3 &= 3x_1 + x_2 + 10x_3 \end{aligned}$$

29. $$\begin{aligned} y_1 &= 0.2x_1 - 0.1x_2 \\ y_2 &= -0.2x_2 + 0.1x_3 \\ y_3 &= 0.1x_1 + 0.1x_3 \end{aligned}$$

30. $$\begin{aligned} y_1 &= 3x_1 - x_2 + x_3 \\ y_2 &= -15x_1 + 6x_2 - 5x_3 \\ y_3 &= 5x_1 - 2x_2 + 2x_3 \end{aligned}$$

Review Questions and Problems for Chapter 7

1. Let $\mathbf{A}$ be a 20×20 matrix and $\mathbf{B}$ a 20×10 matrix. Indicate whether or not the following expressions are defined: $\mathbf{A} + \mathbf{B}$, $\mathbf{AB}$, $\mathbf{A}^T\mathbf{B}$, $\mathbf{AB}^T$, $\mathbf{B}^T\mathbf{A}$, $\mathbf{A}^2$, $\mathbf{AA}^T$, $\mathbf{B}^2$, $\mathbf{BB}^T$, $\mathbf{B}^T\mathbf{BA}$. (Give reasons.)

2. What properties of matrix multiplication are "unusual" (i.e., differ from those for the multiplication of numbers)?

3. How can the rank of a matrix be given in terms of row vectors? Column vectors? Determinants?

4. What do you know about the existence and number of solutions of a nonhomogeneous system of linear equations? A homogeneous system?

5. What is the Gauss elimination good for? What is its basic idea? Why is this elimination generally better than Cramer's rule? What does pivoting mean?

6. What is the inverse of a matrix? When does it exist? How would you practically determine it?

7. Give simple examples of linear systems of equation without solutions. With a unique solution. With more than one solution.
8. Write down the formulas for $(\mathbf{AB})^{\mathsf{T}}$ and $(\mathbf{AB})^{-1}$ from memory. Work an example.
9. Can the row vectors of an 8×6 matrix be linearly independent?
10. What are symmetric, skew-symmetric, and orthogonal matrices? Hermitian, skew-Hermitian, and unitary matrices?
11. Show that the symmetric 4×4 matrices form a vector space. What is its dimension? Find a basis.
12. What is the nullity of a matrix **A**? The row space of **A**? The column space of **A**?
13. State the definitions of an eigenvalue and an eigenvector of a matrix **A**. What is the spectrum of **A**? The characteristic equation?
14. What do you know about the eigenvalues of the matrices in Prob. 10?
15. Do there exist square matrices without eigenvalues? Can a real matrix have complex eigenvalues? Does a real 3×3 matrix always have a real eigenvalue?

Let $\mathbf{a} = \begin{bmatrix} 2 \\ 0 \\ -1 \end{bmatrix}$, $\mathbf{b} = \begin{bmatrix} 6 \\ -4 \\ 3 \end{bmatrix}$, $\mathbf{C} = \begin{bmatrix} 0 & 0 & 5 \\ 0 & 4 & -2 \\ 1 & 3 & 2 \end{bmatrix}$, $\mathbf{D} = \begin{bmatrix} 6 & -2 & 2 \\ 8 & 3 & 2 \\ 1 & 5 & -9 \end{bmatrix}$. Find

16. $8\mathbf{C} - 5\mathbf{D}$
17. $\mathbf{a} + 4\mathbf{b}$
18. $\mathbf{Ca}$, $\mathbf{a}^{\mathsf{T}}\mathbf{C}$
19. $\mathbf{Cb}$
20. $\mathbf{CD}$, $\mathbf{D}^{\mathsf{T}}\mathbf{C}^{\mathsf{T}}$
21. $\mathbf{C}^{-1}$
22. rank $\mathbf{D}$
23. rank $(\mathbf{b}^{\mathsf{T}}\mathbf{D})$
24. $\mathbf{C} + \mathbf{C}^{\mathsf{T}}$
25. $\mathbf{CC}^{\mathsf{T}}$
26. $\mathbf{DCab}^{\mathsf{T}}$
27. $\mathbf{a} + \mathbf{Db}$
28. $\mathbf{C}^4$
29. $(\mathbf{C} - \mathbf{D})\,\mathbf{a}$
30. $\mathbf{a}^{\mathsf{T}}\mathbf{b}$, $\mathbf{ab}^{\mathsf{T}}$
31. $\mathbf{aa}^{\mathsf{T}}$, $\mathbf{a}^{\mathsf{T}}\mathbf{a}$

Solve the following systems of equations or indicate that no solutions exist.

32. $\begin{aligned} 5x + 3y - 3z &= -1 \\ 3x + 2y - 2z &= -1 \\ 2x - y + 2z &= 8 \end{aligned}$

33. $\begin{aligned} 2y - z &= -1 \\ x + 3z &= 11 \\ 2x - 4y + 2z &= 6 \end{aligned}$

34. $\begin{aligned} 7x - 4y - 2z &= -6 \\ 16x + 2y + z &= 3 \end{aligned}$

35. $\begin{aligned} 6x + 4y &= 4 \\ 8x - 6z &= 7 \\ -8y - 2z &= -1 \end{aligned}$

36. $\begin{aligned} 4y + 7z &= -13 \\ 5x - 3y + 4z &= -23 \\ -x + 2y - 8z &= 29 \end{aligned}$

37. $\begin{aligned} 4x + 2y - 6z &= -6 \\ 5x + 3y - 8z &= -9 \end{aligned}$

38. $\begin{aligned} 3x - 2y &= -2 \\ 5x + 4y &= 26 \\ x - 3y &= 8 \end{aligned}$

39. $\begin{aligned} 3x + 4y + 6z &= 1 \\ -2x + 8y - 4z &= 2 \\ 4x - 8y + 8z &= -2 \end{aligned}$

40. $\begin{aligned} 4x - y + z &= 0 \\ x + 2y - z &= 0 \\ 3x + y + 5z &= 0 \end{aligned}$

Determine the rank of the following matrices.

41. $\begin{bmatrix} 9 & 1 \\ 0 & 4 \\ 2 & 6 \end{bmatrix}$

42. $\begin{bmatrix} 3 & 1 & 4 \\ 0 & 5 & 8 \\ -3 & 4 & 4 \end{bmatrix}$

43. $\begin{bmatrix} 8 & 1 & 3 & 6 \\ 0 & 3 & 2 & 2 \\ -8 & -1 & -3 & 4 \end{bmatrix}$

Find the inverse or indicate that it does not exist. Check your result.

44. $\begin{bmatrix} 0.3 & 0.1 \\ -0.4 & 0.2 \end{bmatrix}$

45. $\begin{bmatrix} 0 & 5 \\ 1 & 8 \end{bmatrix}$

46. $\begin{bmatrix} 0.6 & 0.3 \\ -1.6 & -0.8 \end{bmatrix}$

Find the inverse or indicate that it does not exist. Check your result.

47. $\begin{bmatrix} 2 & 4 & 1 \\ 1 & 2 & 1 \\ 3 & 4 & 2 \end{bmatrix}$ **48.** $\begin{bmatrix} -1 & 1 & 2 \\ 3 & -1 & 1 \\ -1 & 3 & 4 \end{bmatrix}$ **49.** $\begin{bmatrix} 0.2 & 0.1 & 0.2 \\ 0 & 0.5 & 0.4 \\ 0 & 0 & 0.1 \end{bmatrix}$

Find the eigenvalues and eigenvectors:

50. $\begin{bmatrix} 0 & 0 \\ 0 & -1 \end{bmatrix}$ **51.** $\begin{bmatrix} 1 & 0 \\ 2 & -1 \end{bmatrix}$ **52.** $\begin{bmatrix} 1.4 & 0.5 \\ -1.0 & -0.1 \end{bmatrix}$

53. $\begin{bmatrix} 15 & 0 & -15 \\ -3 & 6 & 9 \\ 5 & 0 & -5 \end{bmatrix}$ **54.** $\begin{bmatrix} 3 & 1 & 4 \\ 0 & 2 & 6 \\ 0 & 0 & 5 \end{bmatrix}$ **55.** $\begin{bmatrix} 1 & 2 & 4 \\ -2 & -4 & 2 \\ 2 & 4 & 3 \end{bmatrix}$

56. $\begin{bmatrix} 0 & 1 & 0 \\ 1 & 0 & 0 \\ 0 & 0 & 1 \end{bmatrix}$ **57.** $\begin{bmatrix} 0 & 2 & 0 \\ 3 & -2 & 3 \\ 0 & 3 & 0 \end{bmatrix}$ **58.** $\begin{bmatrix} 0 & 2 & 0 \\ 0 & 0 & 2 \\ 2 & 0 & 0 \end{bmatrix}$

Find a basis of eigenvectors and diagonalize:

59. $\begin{bmatrix} 0 & 4 \\ 9 & 0 \end{bmatrix}$ **60.** $\begin{bmatrix} 0 & 0 & 1 \\ 0 & 2 & 0 \\ 3 & 0 & 0 \end{bmatrix}$ **61.** $\begin{bmatrix} -2 & 0 & 0 \\ -2 & 0 & 0 \\ 0 & 0 & 2 \end{bmatrix}$

What kind of conic section is represented by the given quadratic form? Transform it to principal axes. Express $[x_1 \quad x_2]^\mathsf{T}$ in terms of the new coordinates.

62. $10x_1^2 - 9x_1x_2 + \frac{25}{4}x_2^2 = 13$ **63.** $7x_1^2 + 48x_1x_2 - 7x_2^2 = 25$

64. $4x_1x_2 + 3x_2^2 = 1$ **65.** $801x_1^2 - 600x_1x_2 + 1396x_2^2 = 169$

Find $\bar{\mathbf{x}}^\mathsf{T}\mathbf{A}\mathbf{x}$ where

66. $\mathbf{A} = \begin{bmatrix} 5i & 2+i \\ -2+i & i \end{bmatrix}, \quad \mathbf{x} = \begin{bmatrix} 2i \\ 3 \end{bmatrix}$

67. $\mathbf{A} = \begin{bmatrix} 2i & 4 \\ -4 & 0 \end{bmatrix}, \quad \mathbf{x} = \begin{bmatrix} x_1 \\ x_2 \end{bmatrix}$

68. $\mathbf{A} = \begin{bmatrix} 1 & i & 0 \\ -i & 0 & 3 \\ 0 & 3 & 2 \end{bmatrix}, \quad \mathbf{x} = \begin{bmatrix} x_1 \\ x_2 \\ x_3 \end{bmatrix}$ **69.** $\mathbf{A} = \begin{bmatrix} 1 & -i & 2i \\ i & 1 & 0 \\ -2i & 0 & 1 \end{bmatrix}, \quad \mathbf{x} = \begin{bmatrix} 0 \\ 1 \\ i \end{bmatrix}$

70. **(Pauli spin matrices)** Find the eigenvalues and eigenvectors of the so-called *Pauli spin matrices* and show that $\mathbf{S}_x\mathbf{S}_y = i\mathbf{S}_z$, $\mathbf{S}_y\mathbf{S}_x = -i\mathbf{S}_z$, $\mathbf{S}_x^2 = \mathbf{S}_y^2 = \mathbf{S}_z^2 = \mathbf{I}$, where

$$\mathbf{S}_x = \begin{bmatrix} 0 & 1 \\ 1 & 0 \end{bmatrix}, \qquad \mathbf{S}_y = \begin{bmatrix} 0 & -i \\ i & 0 \end{bmatrix}, \qquad \mathbf{S}_z = \begin{bmatrix} 1 & 0 \\ 0 & -1 \end{bmatrix}.$$

Networks. Find the currents in the following networks.

71.

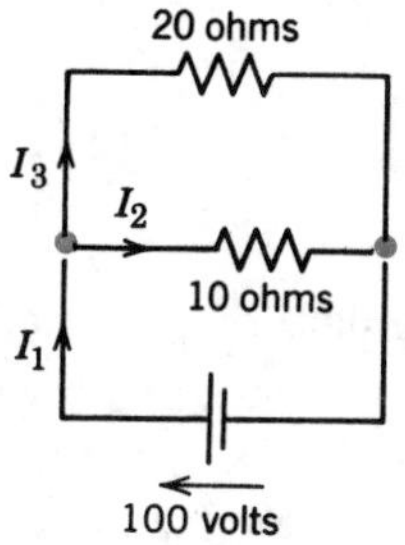

72.

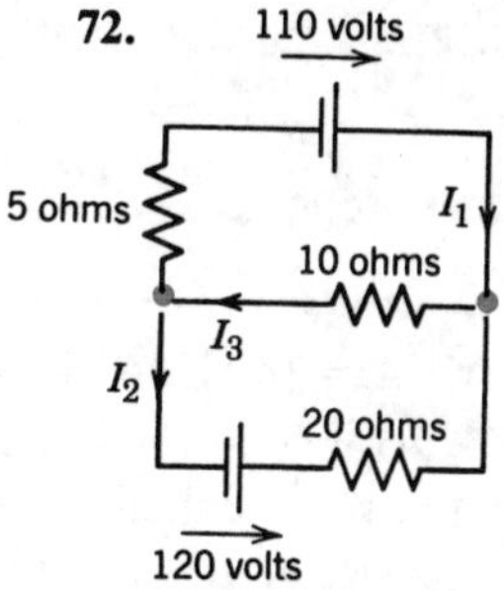

73.

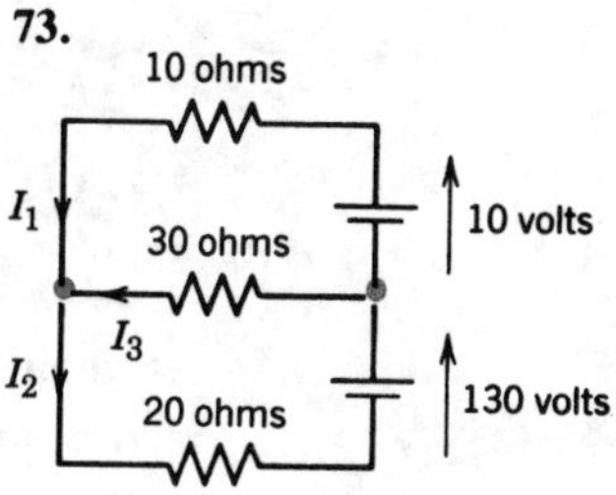

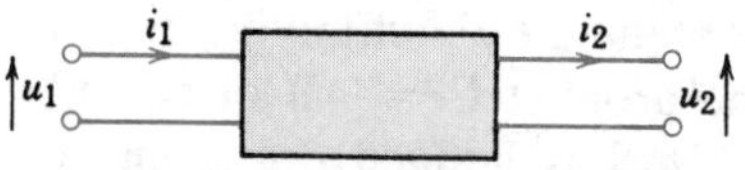

Fig. 138. Four-terminal network

Four-terminal networks. Assume that the input current i_1 and voltage u_1 of the four-terminal network in Fig. 138 are related to the output current i_2 and voltage u_2 according to

$$\mathbf{v}_1 = \mathbf{T}\mathbf{v}_2, \qquad \text{where} \qquad \mathbf{v}_1 = \begin{bmatrix} u_1 \\ i_1 \end{bmatrix}, \quad \mathbf{T} = \begin{bmatrix} t_{11} & t_{12} \\ t_{21} & t_{22} \end{bmatrix}, \quad \mathbf{v}_2 = \begin{bmatrix} u_2 \\ i_2 \end{bmatrix}$$

and where $\mathbf{T}$ is called the transmission matrix of the network. Verify the form of $\mathbf{T}$:

74.

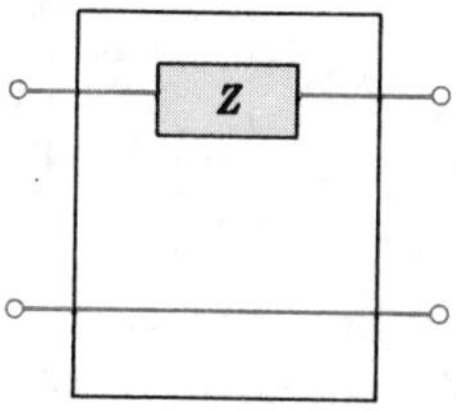

$$\mathbf{T} = \begin{bmatrix} 1 & Z \\ 0 & 1 \end{bmatrix}$$

75.

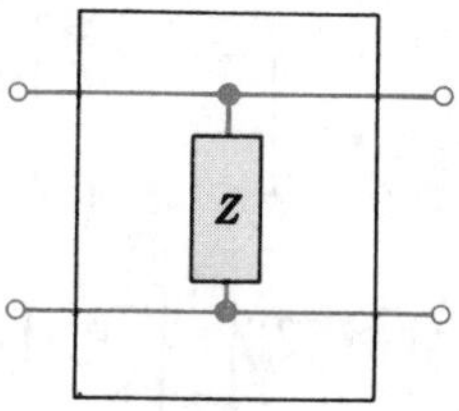

$$\mathbf{T} = \begin{bmatrix} 1 & 0 \\ 1/Z & 1 \end{bmatrix}$$

76.

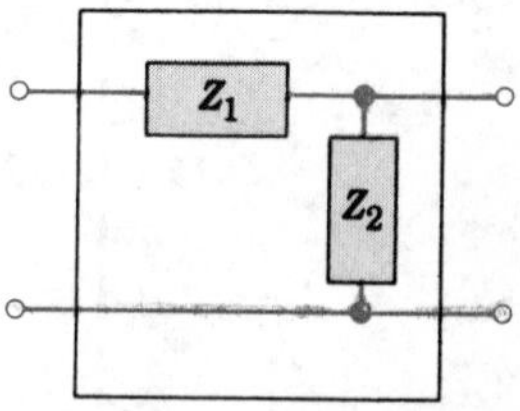

$$\mathbf{T} = \begin{bmatrix} 1 + Z_1/Z_2 & Z_1 \\ 1/Z_2 & 1 \end{bmatrix}$$

77. Show that for the networks in cascade in Fig. 139 we have $\mathbf{v}_1 = \mathbf{T}\mathbf{v}_2$ with $\mathbf{T} = \mathbf{T}_1\mathbf{T}_2$ and $\mathbf{v}_1$ and $\mathbf{v}_2$ as before.

78. Use Prob. 77 to get the matrix in Prob. 76 from those in Probs. 74 and 75.

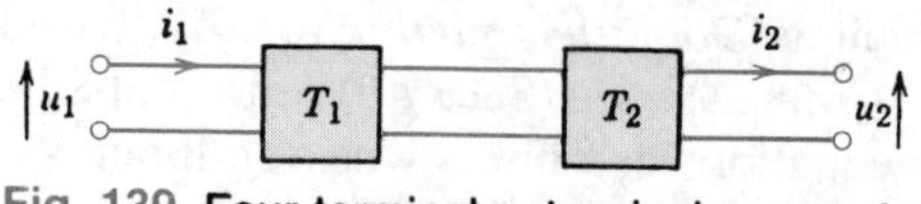

Fig. 139. Four-terminal networks in cascade

Summary of Chapter 7 Linear Algebra: Matrices, Vectors, Determinants

An $m \times n$ **matrix** $\mathbf{A} = [a_{jk}]$ is a rectangular array of numbers ("entries" or "elements") arranged in m horizontal rows and n vertical columns. If $m = n$, the matrix is called **square.** A $1 \times n$ matrix is called a **row vector** and an $m \times 1$ matrix a **column vector** (see Sec. 7.1).

The **sum** $\mathbf{A} + \mathbf{B}$ of matrices of the same **size** (i.e., both $m \times n$) is obtained by adding corresponding entries. The **product** of $\mathbf{A}$ by a scalar c is obtained by multiplying each a_{jk} by c (see Sec. 7.2).

The **product** $\mathbf{C} = \mathbf{AB}$ of an $m \times n$ matrix $\mathbf{A}$ by an $r \times p$ matrix $\mathbf{B} = [b_{jk}]$ is defined only when $r = n$, and is the $m \times p$ matrix $\mathbf{C} = [c_{jk}]$ with entries

$$c_{jk} = a_{j1}b_{1k} + a_{j2}b_{2k} + \cdots + a_{jn}b_{nk} \qquad \text{(row } j \text{ of } \mathbf{A} \text{ times column } k \text{ of } \mathbf{B}\text{).} \tag{1}$$

This multiplication is motivated by the composition of **linear transformations** (Secs. 7.3, 7.15). It is associative, but is ***not commutative:*** if $\mathbf{AB}$ is defined, $\mathbf{BA}$ may not be defined, but even if $\mathbf{BA}$ is defined, $\mathbf{AB} \neq \mathbf{BA}$ in general. Also $\mathbf{AB} = \mathbf{0}$ may not imply $\mathbf{A} = \mathbf{0}$ or $\mathbf{B} = \mathbf{0}$ or $\mathbf{BA} = \mathbf{0}$ (Secs. 7.3, 7.7):

$$\begin{bmatrix} 1 & 1 \\ 2 & 2 \end{bmatrix}\begin{bmatrix} -1 & 1 \\ 1 & -1 \end{bmatrix} = \begin{bmatrix} 0 & 0 \\ 0 & 0 \end{bmatrix},$$

$$\begin{bmatrix} -1 & 1 \\ 1 & -1 \end{bmatrix}\begin{bmatrix} 1 & 1 \\ 2 & 2 \end{bmatrix} = \begin{bmatrix} 1 & 1 \\ -1 & -1 \end{bmatrix}$$

$$[1 \quad 2]\begin{bmatrix} 3 \\ 4 \end{bmatrix} = [11], \qquad \begin{bmatrix} 3 \\ 4 \end{bmatrix}[1 \quad 2] = \begin{bmatrix} 3 & 6 \\ 4 & 8 \end{bmatrix}.$$

A main application of matrices concerns **linear systems of equations**

$$\mathbf{Ax} = \mathbf{b} \qquad \text{(Sec. 7.4)} \tag{2}$$

(m equations in n unknowns $x_1, \cdots, x_n$; $\mathbf{b}$ given). The most important method of solution is the **Gauss elimination** (Sec. 7.4), which reduces the system to "triangular" form by *elementary row operations,* which leave the set of solutions unchanged. (Numerical aspects and variants, such as *Doolittle's method* are discussed in Secs. 19.1 and 19.2.)

Cramer's rule (Sec. 7.9) represents the unknowns in a system (2) of n equations in n unknowns as quotients of determinants; for numerical work it is impractical. **Determinants** (Sec. 7.8) have decreased in importance, but will retain their place in eigenvalue problems, elementary geometry, etc.

The **inverse** $\mathbf{A}^{-1}$ of a square matrix satisfies $\mathbf{AA}^{-1} = \mathbf{A}^{-1}\mathbf{A} = \mathbf{I}$. It exists if and only if det $\mathbf{A} \neq 0$. It can be computed by the *Gauss–Jordan elimination*. See Secs. 7.7, 7.9.

The **rank** r of a matrix $\mathbf{A}$ is the maximum number of linearly indepedent rows or columns of $\mathbf{A}$, equivalently, the number of rows of the largest square submatrix of $\mathbf{A}$ with nonzero determinant (Secs. 7.5, 7.9). The system (2) has solutions if and only if rank $\mathbf{A}$ = rank $[\mathbf{A} \;\; \mathbf{b}]$, $[\mathbf{A} \;\; \mathbf{b}]$ the *augmented matrix* (Fundamental Theorem, Sec. 7.6). The *homogeneous system*

$$\mathbf{Ax} = \mathbf{0} \tag{3}$$

has solutions $\mathbf{x} \neq \mathbf{0}$ ("nontrivial solutions") if and only if rank $\mathbf{A} < n$, in the case $m = n$ equivalently if and only if det $\mathbf{A} = 0$ (Secs. 7.6, 7.9).

Hence the system of n equations in n unknowns

$$\mathbf{Ax} = \lambda\mathbf{x} \qquad \text{or} \qquad (\mathbf{A} - \lambda\mathbf{I})\mathbf{x} = \mathbf{0} \tag{4}$$

has solutions $\mathbf{x} \neq \mathbf{0}$ if and only if λ is a root of the *characteristic equation*

$$\det(\mathbf{A} - \lambda\mathbf{I})\mathbf{x} = 0 \qquad \text{(Sec. 7.10).} \tag{5}$$

Such a (real or complex) number λ is called an **eigenvalue** of $\mathbf{A}$ and that solution $\mathbf{x} \neq \mathbf{0}$ an **eigenvector** of $\mathbf{A}$ corresponding to this λ. Equation (4) is called an (algebraic) **eigenvalue problem** (Sec. 7.10). Eigenvalue problems are of great importance in physics and engineering (Sec. 7.11), and they also have applications in economics and statistics. They are basic in solving systems of differential equations (Chap. 4).

The *transpose* $\mathbf{A}^\mathsf{T}$ of a matrix $\mathbf{A} = [a_{jk}]$ is $\mathbf{A}^\mathsf{T} = [a_{kj}]$; rows become columns and conversely (Sec. 7.1). For a product, $(\mathbf{AB})^\mathsf{T} = \mathbf{B}^\mathsf{T}\mathbf{A}^\mathsf{T}$ (Sec. 7.3). The *complex conjugate* of $\mathbf{A}$ is $\overline{\mathbf{A}} = [\overline{a}_{jk}]$. Six classes of square matrices of practical importance arise from this: a real matrix $\mathbf{A}$ is called *real* **symmetric** if $\mathbf{A}^\mathsf{T} = \mathbf{A}$, *real* **skew-symmetric** if $\mathbf{A}^\mathsf{T} = -\mathbf{A}$, **orthogonal** if $\mathbf{A}^\mathsf{T} = \mathbf{A}^{-1}$ (Secs. 7.2, 7.12); a complex matrix is called **Hermitian** if $\overline{\mathbf{A}}^\mathsf{T} = \mathbf{A}$, **skew-Hermitian** if $\overline{\mathbf{A}}^\mathsf{T} = -\mathbf{A}$, and **unitary** if $\overline{\mathbf{A}}^\mathsf{T} = \mathbf{A}^{-1}$ (Sec. 7.13). The eigenvalues of Hermitian (and real-symmetric) matrices are real; those of skew-Hermitian (and real skew-symmetric) are pure imaginary or 0; those of unitary (and orthogonal) matrices have absolute value 1 (Sec. 7.13).

The diagonalization of matrices and the transformation of quadratic forms to principal axes are discussed in Sec. 7.14.

General vector spaces and inner product spaces are discussed in Sec. 7.15. For R^n and C^n, see also Secs. 7.5 and 7.13.

Chapter 8

Vector Differential Calculus. Grad, Div, Curl

In this chapter we shall be concerned with vectors and vector functions in 3-space and the application of differential calculus to them. This leads to useful applications in physics and engineering, notably in connection with forces and velocities of motions. For vectors and vector functions in 3-space we have the advantage that we can give *geometrical* interpretations of concepts and relations, in addition to algebraic aspects. This begins with the geometric characterization of vectors in terms of directed segments in space. We shall see that vectors provide a shorthand that simplifies many calculations considerably and helps to visualize physical and geometrical quantities and relations between them. For all these reasons, vector methods are used extensively in applied mathematics. Important to the engineer is the great impact of these methods on the study of physical phenomena, such as elasticity, fluid flow, heat flow, electrostatics, and waves in solids and fluids, which the engineer must understand as the basis of the design and construction of systems such as airplanes, laser generators, thermodynamical systems, or robots.

Sections 8.1–8.3 concentrate on basic *algebraic* operations with vectors in 3-space. Vector calculus, consisting of **vector differential calculus** (this chapter) and **vector integral calculus** (Chap. 9) begins with a discussion of vector functions, which represent vector fields (Sec. 8.4) and have various physical and geometrical applications. Then we extend the basic concepts of differential calculus to vector functions in a simple and natural fashion. In Secs. 8.5–8.7 we show that vector functions are useful in studying curves and their applications as paths of moving bodies in mechanics (Sec. 8.6). Physically and geometrically important concepts in connection with scalar and vector fields are the gradient (Sec. 8.9), divergence (Sec. 8.10), and curl (Sec. 8.11). (Integral theorems involving these concepts will be considered in Chap. 9.) In the (optional) last section (Sec. 8.12) we show how the gradient, divergence, and curl as well as the Laplacian can be transformed into general curvilinear coordinates.

We shall keep this chapter independent of Chap. 7.[1]

[1]Readers familiar with Chap. 7 will notice that our present approach is in harmony with that in Chap. 7. The restriction to two and three dimensions will provide for a richer theory with basic physical, engineering, and geometrical applications.

Prerequisites for this chapter: In Sec. 8.3 we shall make elementary use of second- and third-order determinants.
Sections that may be omitted in a shorter course: 8.6–8.8, 8.12.
References: Appendix 1, Part B.
Answers to problems: Appendix 2.

8.1 Vector Algebra in 2-Space and 3-Space

In geometry and physics and its engineering applications we use two kinds of quantities, scalars and vectors. A **scalar** is a quantity that is determined by its magnitude, its number of units measured on a suitable scale. For instance, length, temperature, and voltage are scalars.

A **vector** is a quantity that is determined by both its magnitude and its direction; thus it is an **arrow** or **directed line segment.** For instance, a force is a vector, and so is a velocity, giving the speed and direction of motion (Fig. 140).

We denote vectors by lowercase boldface letters[2] **a, b, v,** etc.

A vector (arrow) has a tail, called its **initial point,** and a tip, called its **terminal point.** For instance, in Fig. 141, the triangle is translated (displaced without rotation); the initial point P of the vector **a** is the original position of a point and the terminal point Q is its position after the translation.

The length of a vector **a** (length of the arrow) is also called the **norm** (or Euclidean norm) of **a** and is denoted by $|\mathbf{a}|$.

A vector of length 1 is called a **unit vector.**

By definition, two vectors **a** and **b** are **equal,** written, $\mathbf{a} = \mathbf{b}$, if they have the same length and the same direction (Fig. 142, p. 430). Hence a vector can be arbitrarily translated, that is, its initial point can be chosen arbitrarily. This definition is practical in connection with forces and other applications.

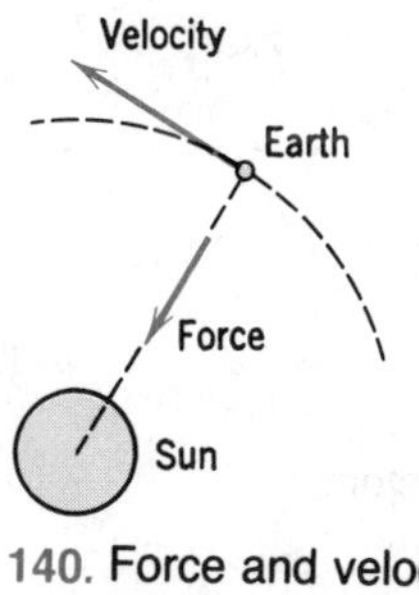

Fig. 140. Force and velocity

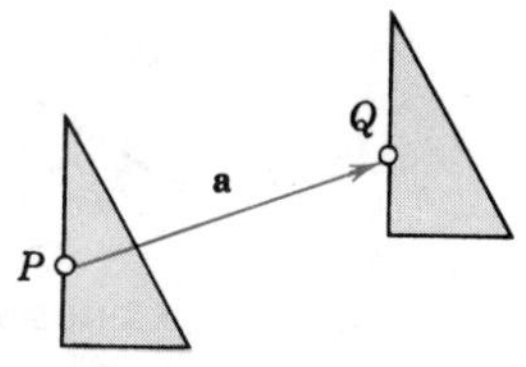

Fig. 141. Translation

[2]This is customary in printed work; in handwritten work one may characterize vectors by arrows, for example, $\vec{a}$ (in place of **a**), $\vec{b}$, etc.

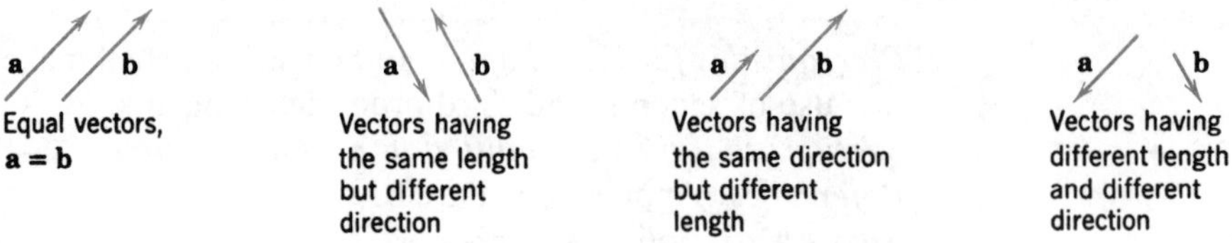

Fig. 142. Vectors

Components of a Vector

If we choose an xyz **Cartesian coordinate system**[3] in space (Fig. 143), that is, a usual rectangular coordinate system with the same scale of measurement on the three mutually perpendicular coordinate axes, and a given vector $\mathbf{a}$ has initial point P: (x_1, y_1, z_1) and terminal point Q: (x_2, y_2, z_2), then the three numbers (Fig. 144)

$$a_1 = x_2 - x_1, \qquad a_2 = y_2 - y_1, \qquad a_3 = z_2 - z_1 \tag{1}$$

are called the **components** of the vector $\mathbf{a}$ with respect to that coordinate system, and we write simply

$$\mathbf{a} = [a_1, a_2, a_3].$$

Length in terms of components. By definition, the length $|\mathbf{a}|$ of a vector $\mathbf{a}$ is the distance between its initial point P and terminal point Q, and from the Pythagorean theorem and (1) we see that

$$|\mathbf{a}| = \sqrt{a_1^2 + a_2^2 + a_3^2}. \tag{2}$$

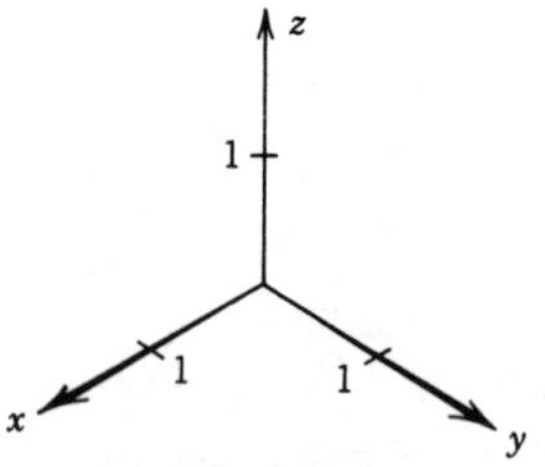

Fig. 143. Cartesian coordinate system

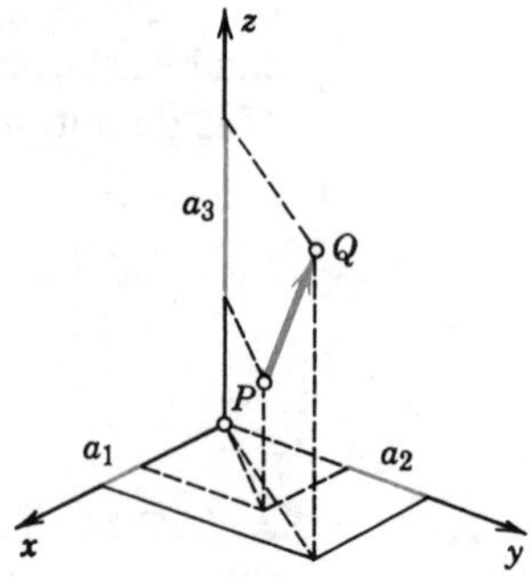

Fig. 144. Components of a vector

---

[3]Named after the French philosopher and mathematician RENATUS CARTESIUS, latinized for RENÉ DESCARTES (1596—1650), who invented analytic geometry. His basic work *Géométrie* appeared in 1637, as an appendix to his *Discours de la méthode*.

EXAMPLE 1 Components and length of a vector

The vector **a** with initial point P: (3, 1, 4) and terminal point Q: (1, −2, 4) has the components

$$a_1 = 1 - 3 = -2, \qquad a_2 = -2 - 1 = -3, \qquad a_3 = 4 - 4 = 0.$$

Hence $\mathbf{a} = [-2, -3, 0]$, so that (2) gives the length

$$|\mathbf{a}| = \sqrt{(-2)^2 + (-3)^2 + 0^2} = \sqrt{13}.$$

If we choose (−1, 5, 8) as the initial point of **a**, the corresponding terminal point is (−3, 2, 8). If we choose the origin (0, 0, 0) as the initial point of **a**, the corresponding terminal point is (−2, −3, 0); its coordinates equal the components of **a**. ∎

Position vector. A Cartesian coordinate system being given, we can determine each point **A:** (x, y, z) by its **position vector r** $= [x, y, z]$, whose initial point is the origin and whose terminal point is A (see Fig. 145). This can be seen directly from (1).

Vectors as ordered triples of numbers. If we translate a vector **a**, with initial point P and terminal point Q, then corresponding coordinates of P and Q change by the same amount, so that the differences in (1) remain unchanged. This proves

Theorem 1 (Vectors as ordered triples of real numbers)

A fixed Cartesian coordinate system being given, each vector is uniquely determined by its ordered triple of corresponding components. Conversely, to each ordered triple of real numbers (a_1, a_2, a_3) there corresponds precisely one vector $\mathbf{a} = [a_1, a_2, a_3]$, *with* (0, 0, 0) *corresponding to the* **zero vector 0**, *which has length* 0 *and no direction.*

Hence a vector equation $\mathbf{a} = \mathbf{b}$ is equivalent to the three equations $a_1 = b_1$, $a_2 = b_2$, $a_3 = b_3$ for the components.

We see that from our "geometrical" definition of vectors as arrows we have arrived at an "algebraic" characterization by Theorem 1; we could have started from the latter[4] and reversed our process. This shows that the two approaches are equivalent.

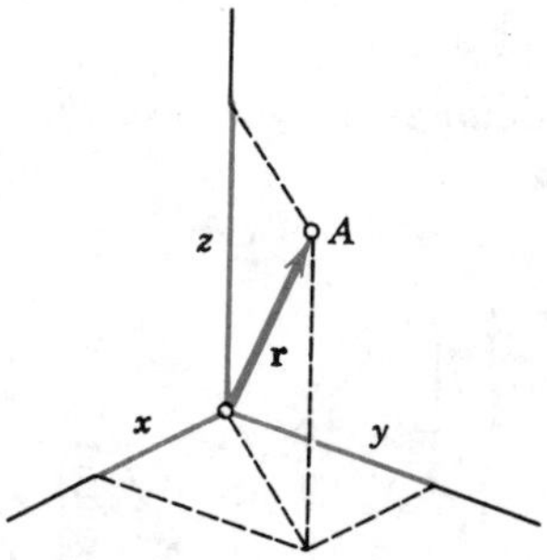

Fig. 145. Position vector **r** of a point A: (x, y, z)

[4]This is in agreement with Chap. 7 (which we shall not use here).

Vector Addition, Scalar Multiplication

Applications have suggested algebraic calculations with vectors that are practically useful and almost as simple as calculations with numbers.

Definition 1. Addition of vectors

The **sum** $\mathbf{a} + \mathbf{b}$ of two vectors $\mathbf{a} = [a_1, a_2, a_3]$ and $\mathbf{b} = [b_1, b_2, b_3]$ is obtained by adding the corresponding components,

$$\mathbf{a} + \mathbf{b} = [a_1 + b_1,\quad a_2 + b_2,\quad a_3 + b_3]. \tag{3}$$

Geometrically, place the vectors as in Fig. 146 (the initial point of **b** at the terminal point of **a**); then $\mathbf{a} + \mathbf{b}$ is the vector drawn from the initial point of **a** to the terminal point of **b**. ∎

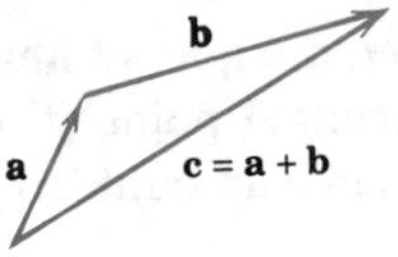

Fig. 146. Vector addition

Figure 147 illustrates that for forces, this addition is the parallelogram law by which we obtain the resultant of two forces in mechanics.

Figure 148 illustrates (for the plane) that the "algebraic" way and the "geometric" way of vector addition amount to the same thing.

Basic properties of vector addition, as obtained from the definition, are (see also Figs. 149, 150)

$$\text{(4)}\quad \begin{aligned} &\text{(a)} && \mathbf{a} + \mathbf{b} = \mathbf{b} + \mathbf{a} && \textit{(commutativity)} \\ &\text{(b)} && (\mathbf{u} + \mathbf{v}) + \mathbf{w} = \mathbf{u} + (\mathbf{v} + \mathbf{w}) && \textit{(associativity)} \\ &\text{(c)} && \mathbf{a} + \mathbf{0} = \mathbf{0} + \mathbf{a} = \mathbf{a} && \\ &\text{(d)} && \mathbf{a} + (-\mathbf{a}) = \mathbf{0} && \end{aligned}$$

where $-\mathbf{a}$ denotes the vector having the length $|\mathbf{a}|$ and the direction opposite to that of **a**.

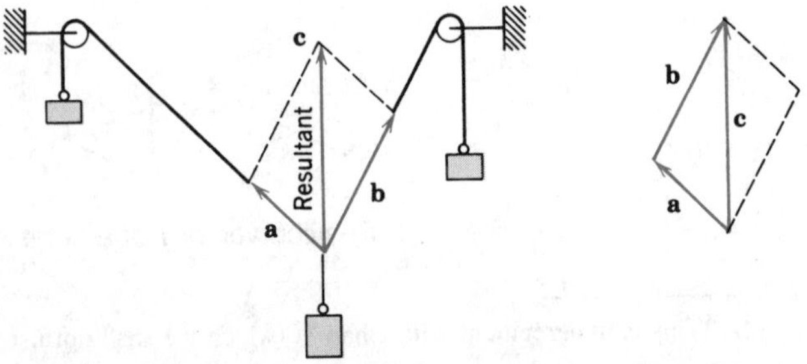

Fig. 147. Resultant of two forces (parallelogram law)

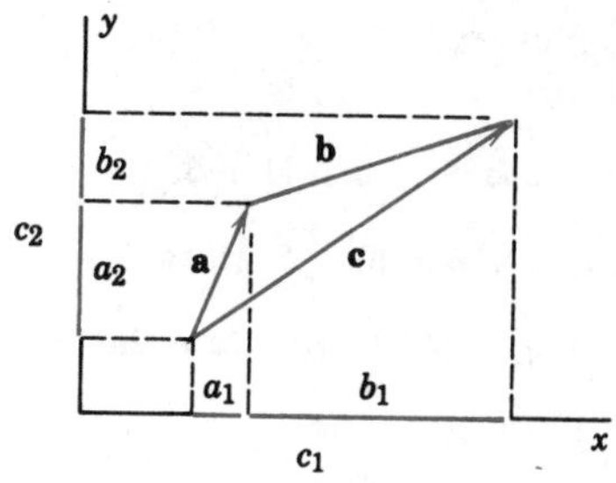

Fig. 148. Vector addition

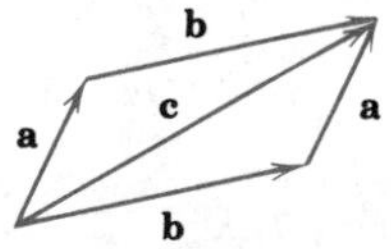

Fig. 149. Commutativity of vector addition

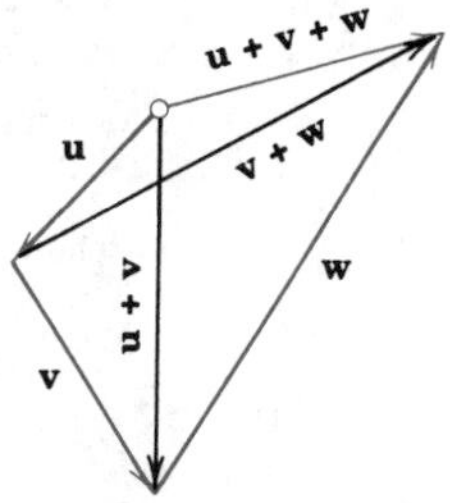

Fig. 150. Associativity of vector addition

In (4b) we may simply write **u** + **v** + **w**, and similarly for sums of more than three vectors. Instead of **a** + **a** we also write 2**a**, and so on. This (and the notation −**a** used before) suggests that we define the second algebraic operation for vectors, the multiplication of a vector by a scalar, as follows.

Definition 2. Scalar multiplication (Multiplication by a number)
The product $c\mathbf{a}$ of any vector $\mathbf{a} = [a_1, a_2, a_3]$ and any scalar c (real number c) is the vector

$$c\mathbf{a} = [ca_1, ca_2, ca_3] \tag{5}$$

obtained by multiplying each component of **a** by c.

Geometrically, if $\mathbf{a} \neq \mathbf{0}$, then $c\mathbf{a}$ with $c > 0$ has the direction of **a** and with $c < 0$ the direction opposite to **a**. In any case, the length of $c\mathbf{a}$ is $|c\mathbf{a}| = |c||\mathbf{a}|$, and $c\mathbf{a} = \mathbf{0}$ if $\mathbf{a} = \mathbf{0}$ or $c = 0$ (or both). (See Fig. 151.) ∎

Basic properties, as obtained from Definitions 1 and 2, are

$$\text{(6)}\qquad \begin{aligned} &\text{(a)} && c(\mathbf{a} + \mathbf{b}) = c\mathbf{a} + c\mathbf{b} \\ &\text{(b)} && (c + k)\mathbf{a} = c\mathbf{a} + k\mathbf{a} \\ &\text{(c)} && c(k\mathbf{a}) = (ck)\mathbf{a} \qquad \text{(written } ck\mathbf{a}\text{)} \\ &\text{(d)} && 1\mathbf{a} = \mathbf{a}. \end{aligned}$$

The student may prove that (4) and (6) imply that for any **a**,

$$\text{(7)}\qquad \begin{aligned} &\text{(a)} && 0\mathbf{a} = \mathbf{0} \\ &\text{(b)} && (-1)\mathbf{a} = -\mathbf{a}. \end{aligned}$$

Instead of $\mathbf{b} + (-\mathbf{a})$ we simply write $\mathbf{b} - \mathbf{a}$ (Fig. 152).

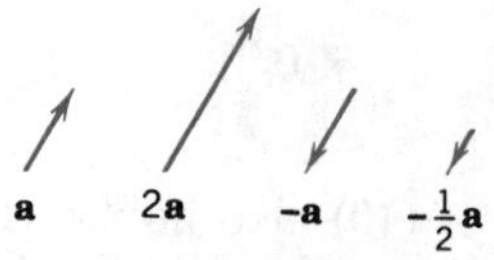

Fig. 151. Multiplication of vectors by scalars (numbers)

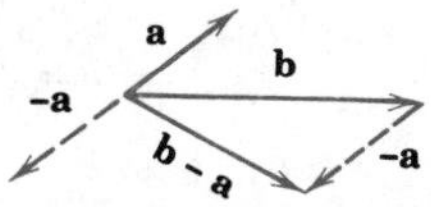

Fig. 152. Difference of vectors

EXAMPLE 2 Vector addition. Multiplication by scalars

With respect to a given coordinate system, let

$$\mathbf{a} = [4, 0, 1] \quad \text{and} \quad \mathbf{b} = [2, -5, \tfrac{1}{3}].$$

Then $-\mathbf{a} = [-4, 0, -1]$, $7\mathbf{a} = [28, 0, 7]$, $\mathbf{a} + \mathbf{b} = [6, -5, \tfrac{4}{3}]$, and

$$2(\mathbf{a} - \mathbf{b}) = 2[2, 5, \tfrac{2}{3}] = [4, 10, \tfrac{4}{3}] = 2\mathbf{a} - 2\mathbf{b}.$$

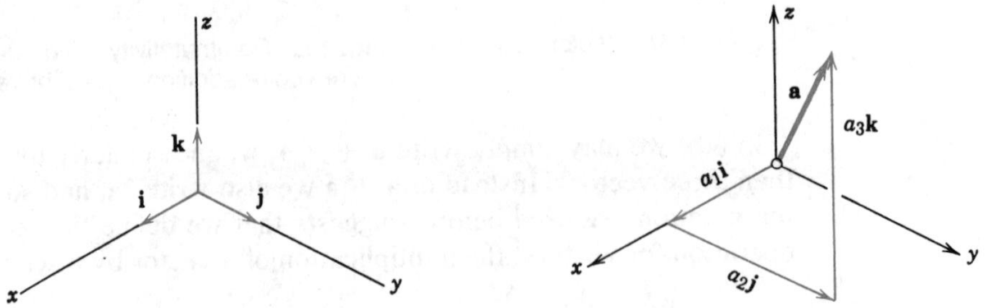

Fig. 153. The unit vectors **i, j, k** and the representation (8)

Unit vectors i, j, k. Another popular representation of vectors is

$$\mathbf{a} = [a_1, a_2, a_3] = a_1\mathbf{i} + a_2\mathbf{j} + a_3\mathbf{k} \tag{8}$$

where **i, j, k** are the unit vectors in the positive directions of the axes of a Cartesian coordinate system (Fig. 153); hence

$$\mathbf{i} = [1, 0, 0], \qquad \mathbf{j} = [0, 1, 0], \qquad \mathbf{k} = [0, 0, 1] \tag{9}$$

and the right side of (8) is a sum of three vectors parallel to the three axes.

Vector Space R^3 (*Optional*)

We mention that the set of all vectors considered are said to form the *real three-dimensional vector space* R^3 with the two *algebraic operations* of vector addition and scalar multiplication. To explain "dimension," we need the concept of linear independence. By a **linear combination** of given vectors $\mathbf{a}_{(1)}, \mathbf{a}_{(2)}, \cdots, \mathbf{a}_{(m)}$ we mean an expression of the form

$$c_1\mathbf{a}_{(1)} + c_2\mathbf{a}_{(2)} + \cdots + c_m\mathbf{a}_{(m)},$$

where $c_1, \cdots, c_m$ are any scalars; clearly, this is again a vector. The given vectors are called a *linearly independent set* or, briefly, **linearly independent** if and only if the only solution of

$$c_1\mathbf{a}_{(1)} + \cdots + c_m\mathbf{a}_{(m)} = \mathbf{0} \qquad (c_1, \cdots, c_m \text{ any scalars}) \tag{10}$$

is $c_1 = c_2 = \cdots = c_m = 0$. If (10) also holds with scalars not all zero, then those m vectors are called **linearly dependent.** Now it can be shown that 4 or more vectors in R^3 are always linearly dependent, but R^3 contains linearly independent sets of 3 vectors, so that the maximum possible number of vectors in a linearly independent set in R^3 is 3. This number 3 is called the **dimension** of R^3, and any such set is called a **basis** for R^3. Particularly useful is the "standard basis" (9). Any basis being given, every vector in R^3 can be written as a linear combination of the basis vectors in a unique fashion. Representation (8) is an example of this.

Vector space R^3 is a model of a *general vector space,* as discussed in Sec. 7.15 but not needed in the present chapter.

Problem Set 8.1

Find the components of the vector **v** with given initial point P: (x_1, y_1, z_1) and terminal point Q: (x_2, y_2, z_2). Find $|\mathbf{v}|$.

1. P: (1, 0, 0), Q: (4, 2, 0)

2. P: (6, −1, 0), Q: (3, 3, 0)

3. P: (4, 0, −1), Q: (1, 0, 2)

4. P: (0, 0, 0), Q: (a, b, c)

5. P: (−1, −1, −1), Q: (3, 0, 0)

6. P: (8, 6, 1), Q: (−8, 6, 1)

7. P: (−1, 7, 5), Q: (−1, 7, 5)

8. P: (−3, −2, 1), Q: (3, 2, −1)

In each case, the components v_1, v_2, v_3 of a vector **v** and a particular initial point P are given. Find the corresponding terminal point and the length (norm) of **v**.

9. 1, −1, 0, P: (2, 1, 0)

10. 0, 0, 1, P: (−3, 2, 0)

11. 6, 2, 1, P: (−6, −2, −1)

12. 2, 2, 2, P: (4, 4, 0)

13. 0, 0, 0, P: (1, −1, −2)

14. 1, 2, 3, P: (0, 0, 0)

15. 2, −4, 6, P: (4, −2, 6)

16. $\frac{1}{2}$, 1, $\frac{3}{2}$, P: $(-\frac{1}{2}, 1, \frac{1}{2})$

Let $\mathbf{a} = [2, -1, 3] = 2\mathbf{i} - \mathbf{j} + 3\mathbf{k}$, $\mathbf{b} = [1, 1, -1] = \mathbf{i} + \mathbf{j} - \mathbf{k}$, $\mathbf{c} = [0, 0, 4] = 4\mathbf{k}$. Find

17. $\mathbf{a} + \mathbf{b}$, $\mathbf{b} + \mathbf{a}$

18. $-\mathbf{a}$, $2\mathbf{a}$, $0.5\mathbf{a}$

19. $3\mathbf{a} - 2\mathbf{b} + 4\mathbf{c}$

20. $2\mathbf{a} + 2\mathbf{b}$, $2(\mathbf{a} + \mathbf{b})$

21. $3\mathbf{b} - 6\mathbf{c}$, $3(\mathbf{b} - 2\mathbf{c})$

22. $(\mathbf{a} + \mathbf{b}) + \mathbf{c}$, $\mathbf{a} + (\mathbf{b} + \mathbf{c})$

23. $|\mathbf{a} + \mathbf{b}|$, $|\mathbf{a}| + |\mathbf{b}|$

24. $(1/|\mathbf{b}|)\mathbf{b}$, $(1/|\mathbf{c}|)\mathbf{c}$

25. What laws do Probs. 17, 20, and 22 illustrate?

Forces. In each case, find the resultant (its components) and its magnitude.

26. $\mathbf{p} = [1, 1, 1]$, $\mathbf{q} = [1, -2, 3]$, $\mathbf{u} = [-2, 3, 1]$

27. $\mathbf{p} = [1, 3, -1]$, $\mathbf{q} = [5, 0, -2]$, $\mathbf{u} = [0, -1, 3]$

28. $\mathbf{p} = [3, 2, 1]$, $\mathbf{q} = [4, -1, -1]$, $\mathbf{u} = [-7, -1, 0]$

29. $\mathbf{p} = [3, 2, 1]$, $\mathbf{q} = [4, 1, 1]$, $\mathbf{u} = [7, 1, 0]$

30. $\mathbf{p} = [-1, 0, 2]$, $\mathbf{q} = [0, 4, 13]$, $\mathbf{u} = [1, 2, 3]$, $\mathbf{v} = [3, -8, 0]$

31. Determine a force **p** such that **p**, $\mathbf{q} = 3\mathbf{j} - 4\mathbf{k}$, and $\mathbf{u} = \mathbf{i} - \mathbf{j}$ are in equilibrium.

32. Determine three forces **p**, **q**, **u** in the direction of the coordinate axes such that **p**, **q**, **u**, $\mathbf{v} = 3\mathbf{i} - 2\mathbf{j} + \mathbf{k}$, and $\mathbf{w} = -2\mathbf{k}$ are in equilibrium. Are **p**, **q**, **u** uniquely determined?

33. If $|\mathbf{p}| = 1$ and $|\mathbf{q}| = 2$, what can be said about the magnitude and direction of the resultant?

Geometrical applications. Using vectors, prove the following statements.

34. The line that passes through the midpoints of adjacent sides of a rectangle divides one of the diagonals in the ratio 1:3.

35. The diagonals of a parallelogram bisect each other.

36. The four space diagonals of a parallelepiped meet and bisect each other.

37. The line that passes through one vertex of a parallelogram and the midpoint of an opposite side divides one of the diagonals in the ratio 1:2.

38. The sum of the vectors drawn from the center of a regular polygon to its vertices is the zero vector.

8.2 Inner Product (Dot Product)

We shall now define a multiplication of two vectors that gives a scalar as the product and is suggested by various applications.

Definition. Inner product (dot product) of vectors

The **inner product** or **dot product** $\mathbf{a} \cdot \mathbf{b}$ (read "**a** dot **b**") of two vectors $\mathbf{a} = [a_1, a_2, a_3]$ and $\mathbf{b} = [b_1, b_2, b_3]$ is defined as

(1)
$$\begin{aligned} \mathbf{a} \cdot \mathbf{b} &= |\mathbf{a}||\mathbf{b}| \cos \gamma \qquad && \text{if } \mathbf{a} \neq \mathbf{0}, \mathbf{b} \neq \mathbf{0} \\ \mathbf{a} \cdot \mathbf{b} &= 0 && \text{if } \mathbf{a} = \mathbf{0} \text{ or } \mathbf{b} = \mathbf{0} \end{aligned}$$

where γ, $0 \leqq \gamma \leqq \pi$, is the angle between **a** and **b** (measured when the vectors have their initial point coinciding, as in Fig. 154). In components,

(2)
$$\mathbf{a} \cdot \mathbf{b} = a_1 b_1 + a_2 b_2 + a_3 b_3.$$

For the derivation of (2) from (1), see below. ∎

Since the cosine in (1) may be positive, zero, or negative, so may be the inner product (Fig. 154). The case that the inner product is zero is of great practical interest and suggests the following.

A vector **a** is called **orthogonal** *to a vector* **b** if $\mathbf{a} \cdot \mathbf{b} = 0$. Then **b** is also orthogonal to **a** and we call these vectors **orthogonal vectors.** Clearly, the zero vector is orthogonal to every vector. For nonzero vectors we have $\mathbf{a} \cdot \mathbf{b} = 0$ if and only if $\cos \gamma = 0$, thus $\gamma = \pi/2$ (90°). This proves the important

Theorem 1 (Orthogonality)

The inner product of two nonzero vectors is zero if and only if these vectors are perpendicular.

Length in terms of inner product. Equation (1) with $\mathbf{b} = \mathbf{a}$ gives $\mathbf{a} \cdot \mathbf{a} = |\mathbf{a}|^2$, so that we have

(3)
$$|\mathbf{a}| = \sqrt{\mathbf{a} \cdot \mathbf{a}}.$$

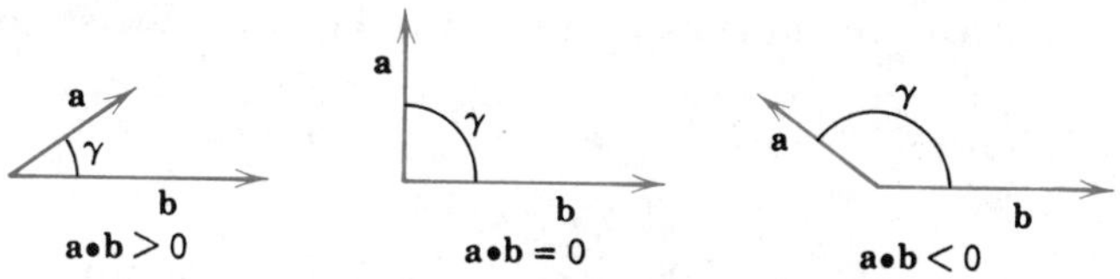

Fig. 154. Angle between vectors and value of inner product

From (3) and (1) we obtain for the angle γ between two nonzero vectors

$$\cos\gamma = \frac{\mathbf{a}\cdot\mathbf{b}}{|\mathbf{a}||\mathbf{b}|} = \frac{\mathbf{a}\cdot\mathbf{b}}{\sqrt{\mathbf{a}\cdot\mathbf{a}}\sqrt{\mathbf{b}\cdot\mathbf{b}}}. \tag{4}$$

EXAMPLE 1 Inner product. Angle between vectors

Find the inner product and the lengths of $\mathbf{a} = [1, 2, 0]$ and $\mathbf{b} = [3, -2, 1]$ as well as the angle between these vectors.

Solution. $\mathbf{a}\cdot\mathbf{b} = 1\cdot 3 + 2\cdot(-2) + 0\cdot 1 = -1$, $|\mathbf{a}| = \sqrt{\mathbf{a}\cdot\mathbf{a}} = \sqrt{5}$, $|\mathbf{b}| = \sqrt{\mathbf{b}\cdot\mathbf{b}} = \sqrt{14}$, and (4) gives the angle

$$\gamma = \text{arc cos}\,\frac{\mathbf{a}\cdot\mathbf{b}}{|\mathbf{a}||\mathbf{b}|} = \text{arc cos}\,(-0.11952) = 1.69061 = 96.865°.$$

Can you sketch these vectors and convince yourself that they make an angle greater than 90°, so that the inner product comes out negative? ∎

From the definition we also see that the inner product has the following properties. For any vectors **a**, **b**, **c** and scalars q_1, q_2,

(5)

(a) $[q_1\mathbf{a} + q_2\mathbf{b}]\cdot\mathbf{c} = q_1\mathbf{a}\cdot\mathbf{c} + q_2\mathbf{b}\cdot\mathbf{c}$ *(Linearity)*

(b) $\mathbf{a}\cdot\mathbf{b} = \mathbf{b}\cdot\mathbf{a}$ *(Symmetry)*

(c) $\left.\begin{array}{c}\mathbf{a}\cdot\mathbf{a} \geqq 0\\ \mathbf{a}\cdot\mathbf{a} = 0 \text{ if and only if } \mathbf{a} = \mathbf{0}\end{array}\right\}$ *(Positive-definiteness).*

Hence *dot multiplication is commutative* [see (5b)] *and is distributive with respect to vector addition;* in fact, from (5a) with $q_1 = 1$ and $q_2 = 1$ we have

(5a*) $(\mathbf{a} + \mathbf{b})\cdot\mathbf{c} = \mathbf{a}\cdot\mathbf{c} + \mathbf{b}\cdot\mathbf{c}$ *(Distributivity).*

Furthermore, from (1) and $|\cos\gamma| \leqq 1$ we see that

(6) $|\mathbf{a}\cdot\mathbf{b}| \leqq |\mathbf{a}||\mathbf{b}|$ *(Schwarz[5] inequality).*

Using this and (3), the reader may prove

(7) $|\mathbf{a} + \mathbf{b}| \leqq |\mathbf{a}| + |\mathbf{b}|$ **(Triangle inequality).**

---

[5]See the footnote in Sec. 7.15.

A simple direct calculation with inner products shows that

$$|\mathbf{a} + \mathbf{b}|^2 + |\mathbf{a} - \mathbf{b}|^2 = 2(|\mathbf{a}|^2 + |\mathbf{b}|^2) \qquad (8) \quad \textit{(Parallelogram equality)}.$$

Derivation of (2) from (1). Using (8) in Sec. 8.1, we can write

$$\mathbf{a} = a_1\mathbf{i} + a_2\mathbf{j} + a_3\mathbf{k} \quad \text{and} \quad \mathbf{b} = b_1\mathbf{i} + b_2\mathbf{j} + b_3\mathbf{k}.$$

Since **i**, **j**, and **k** are unit vectors, we have from (3)

$$\mathbf{i}\cdot\mathbf{i} = 1, \quad \mathbf{j}\cdot\mathbf{j} = 1, \quad \mathbf{k}\cdot\mathbf{k} = 1,$$

and since they are orthogonal (because the coordinate axes are perpendicular), Theorem 1 gives

$$\mathbf{i}\cdot\mathbf{j} = 0, \quad \mathbf{j}\cdot\mathbf{k} = 0, \quad \mathbf{k}\cdot\mathbf{i} = 0,$$

Hence if we substitute those representations of **a** and **b** into $\mathbf{a}\cdot\mathbf{b}$ and use (5a*) and (5b), we first have a sum of nine inner products,

$$\mathbf{a}\cdot\mathbf{b} = a_1b_1\mathbf{i}\cdot\mathbf{i} + a_1b_2\mathbf{i}\cdot\mathbf{j} + \cdots + a_3b_3\mathbf{k}\cdot\mathbf{k}.$$

Since six of these products are zero, we obtain (2). ∎

Applications of Inner Products

Typical applications of inner products are shown in the following examples and in Problem Set 8.2.

EXAMPLE 2 Work done by a force as inner product

Consider a body on which a constant force **a** acts. Let the body be given a displacement **d**. Then the work done by **a** in the displacement is defined as

$$W = |\mathbf{a}||\mathbf{d}| \cos\alpha = \mathbf{a}\cdot\mathbf{d}, \qquad (9)$$

that is, magnitude $|\mathbf{a}|$ of the force times length $|\mathbf{d}|$ of the displacement times the cosine of the angle α between **a** and **d** (Fig. 155). If $\alpha < 90°$, as in Fig. 155, then $W > 0$. If **a** and **d** are orthogonal, then the work is zero (why?). If $\alpha > 90°$, then $W < 0$, which means that in the displacement one has to do work against the force. ∎

EXAMPLE 3 Component of a force in a given direction

What force in the rope in Fig. 156 will hold a car of 5000 lb in equilibrium if the ramp makes an angle of 25° with the horizontal?

Solution. Introducing coordinates as shown, the weight is $\mathbf{a} = [0, -5000]$ because this force points downward, in the negative y-direction. We now have to represent **a** as a sum (resultant) of two forces, $\mathbf{a} = \mathbf{c} + \mathbf{p}$, where **c** is the force the car exerts on the ramp, which is of no interest to us, and **p** is parallel to the rope, of magnitude (see Fig. 156)

$$|\mathbf{p}| = |\mathbf{a}| \cos\gamma = 5000 \cos 65° = 2113 \text{ [lb]}$$

and direction of the unit vector **u** opposite to the direction of the rope; here $\gamma = 90° - 25° =$

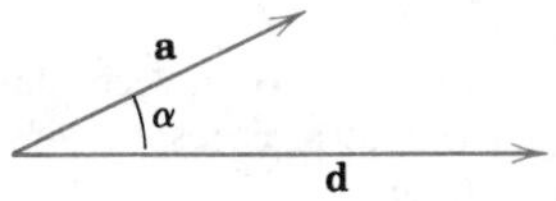

Fig. 155. Work done by a force

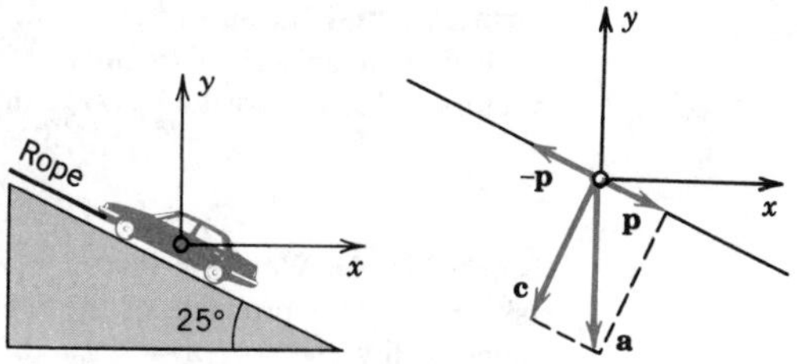

Fig. 156. Example 3

65° is the angle between **a** and **p**. Now a vector in the direction of the rope is

$$\mathbf{b} = [-1, \tan 25°] = [-1, 0.46631], \qquad \text{thus} \qquad |\mathbf{b}| = 1.10338,$$

so that

$$\mathbf{u} = -\frac{\mathbf{b}}{|\mathbf{b}|} = [0.90631, \ -0.42262].$$

Since $|\mathbf{u}| = 1$ and $\cos\gamma > 0$, we see that we can also write our result as

$$|\mathbf{p}| = (|\mathbf{a}| \cos \gamma)|\mathbf{u}| = \mathbf{a} \cdot \mathbf{u} = -\frac{\mathbf{a} \cdot \mathbf{b}}{|\mathbf{b}|} = \frac{5000 \cdot 0.46631}{1.10338} = 2113 \text{ [lb]}.$$

Answer. About 2100 lb. ∎

Example 3 is typical of applications in which one uses the concept of the **component** or **projection** *of a vector* **a** *in the direction of a vector* **b** ($\neq \mathbf{0}$), defined by (see Fig. 157)

$$p = |\mathbf{a}| \cos \gamma. \tag{10}$$

Thus p is the length of the orthogonal projection of **a** on a straight line l parallel to **b**, taken with the plus sign if $p\mathbf{b}$ has the direction of **b** and with the minus sign if $p\mathbf{b}$ has the direction opposite to **b**; see Fig. 157. If **b** is a unit vector, then (10) gives $p = |\mathbf{a}| \cos \gamma \, |\mathbf{b}|$; thus

$$p = \mathbf{a} \cdot \mathbf{b} \qquad (|\mathbf{b}| = 1) \tag{11}$$

and if not, then

$$p = \frac{\mathbf{a} \cdot \mathbf{b}}{|\mathbf{b}|} \qquad (\mathbf{b} \neq \mathbf{0}). \tag{12}$$

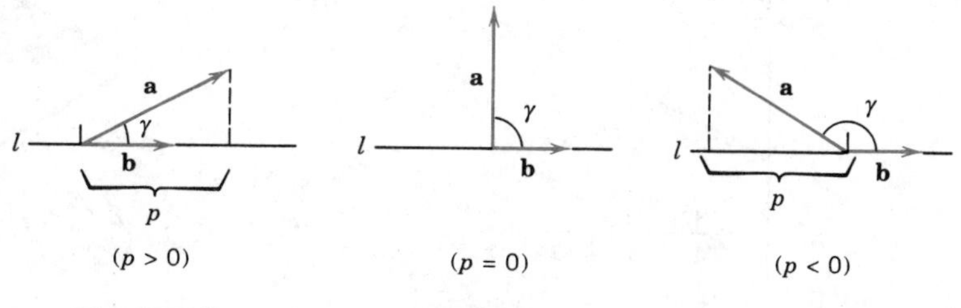

Fig. 157. Component of a vector **a** in the direction of a vector **b**

EXAMPLE 4 **Orthonormal basis**

By definition, an *orthonormal basis* for 3-space is a basis {**a**, **b**, **c**} consisting of orthogonal unit vectors. It has the great advantage that the determination of the coefficients in representations

$$\mathbf{v} = l_1\mathbf{a} + l_2\mathbf{b} + l_3\mathbf{c} \qquad (\mathbf{v} \text{ a given vector})$$

is very simple. We claim that $l_1 = \mathbf{a}\cdot\mathbf{v}$, $l_2 = \mathbf{b}\cdot\mathbf{v}$, $l_3 = \mathbf{c}\cdot\mathbf{v}$. Indeed, this follows simply by taking the inner products of the representation with **a**, **b**, **c**, respectively, and using the orthonormality, $\mathbf{a}\cdot\mathbf{v} = l_1\mathbf{a}\cdot\mathbf{a} + l_2\mathbf{a}\cdot\mathbf{b} + l_3\mathbf{a}\cdot\mathbf{c} = l_1$, etc.

For example, the unit vectors **i**, **j**, **k** in (8), Sec. 8.1, associated with a Cartesian coordinate system form an orthonormal basis. ∎

EXAMPLE 5 **Orthogonal straight lines in the plane**

Find the straight line L_1 through the point P: (1, 3) in the xy-plane and perpendicular to the line L_2: $x - 2y + 2 = 0$; see Fig. 158.

Solution. Any straight line L_1 in the xy-plane can be written $a_1x + a_2y = c$ or, by (2), $\mathbf{a}\cdot\mathbf{r} = c$, where $\mathbf{a} = [a_1, a_2] \neq \mathbf{0}$ and $\mathbf{r} = [x, y]$. Now the line L_1^* through the origin and parallel to L_1 is $\mathbf{a}\cdot\mathbf{r} = 0$. Hence, by Theorem 1, the vector **a** is perpendicular to **r** and, therefore, perpendicular to L_1^* and also to L_1 because L_1 and L_1^* are parallel. **a** is called a **normal vector** to L_1 (and to L_1^*).

Now a normal vector to the given line $x - 2y + 2 = 0$ is $\mathbf{b} = [1, -2]$, so that L_1 is perpendicular to L_2 if $\mathbf{b}\cdot\mathbf{a} = a_1 - 2a_2 = 0$, for instance, if $\mathbf{a} = [2, 1]$. Hence L_1 is given by $2x + y = c$ and passes through P: (1, 3) when $2 \cdot 1 + 3 = c = 5$.

Answer. $y = -2x + 5$ (Fig. 158). ∎

EXAMPLE 6 **Normal vector to a plane**

Find a unit vector perpendicular to the plane $4x + 2y + 4z = -7$.

Solution. Using (2), we may write any plane in space as

$$\mathbf{a}\cdot\mathbf{r} = a_1x + a_2y + a_3z = c \tag{13}$$

where $\mathbf{a} = [a_1, a_2, a_3] \neq \mathbf{0}$ and $\mathbf{r} = [x, y, z]$. The unit vector in the direction of **a** is (Fig. 159)

$$\mathbf{n} = \frac{1}{|\mathbf{a}|}\mathbf{a}.$$

Dividing by $|\mathbf{a}|$, we obtain from (13)

$$\mathbf{n}\cdot\mathbf{r} = p \qquad \text{where} \qquad p = \frac{c}{|\mathbf{a}|}. \tag{14}$$

From (11) we see that p is the projection of **r** in the direction of **n**, and this projection has the same constant value $c/|\mathbf{a}|$ for the position vector **r** of any point in the plane. Clearly this holds if and only if **n** is perpendicular to the plane. **n** is called a **unit normal vector** to the plane (the other being $-\mathbf{n}$). Furthermore, from this and the definition of projection it follows that $|p|$ is

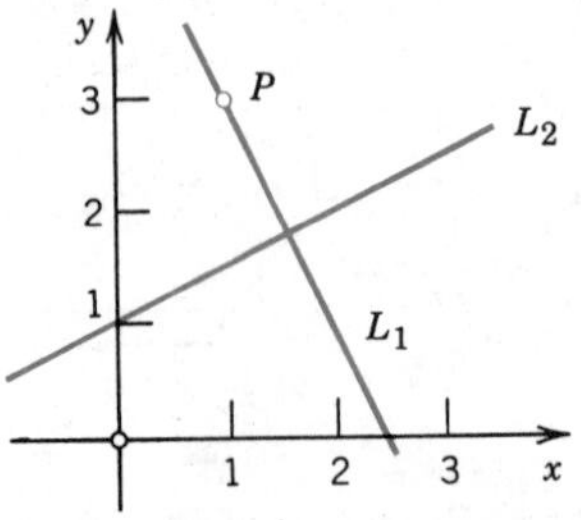

Fig. 158. Example 5

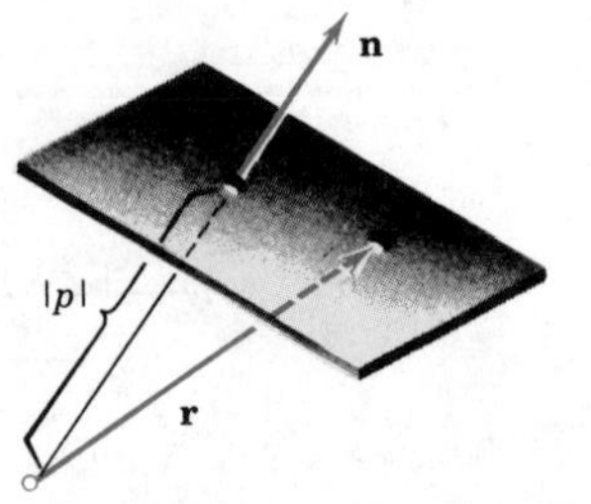

Fig. 159. Normal vector to a plane

the distance of the plane from the origin. Representation (14) is called **Hesse's[6] normal form** of a plane. In our case, $\mathbf{a} = [4, 2, 4]$, $c = -7$, $|\mathbf{a}| = 6$, $\mathbf{n} = \frac{1}{6}\mathbf{a} = [\frac{2}{3}, \frac{1}{3}, \frac{2}{3}]$, and the plane has the distance 7/6 from the origin. ∎

We mention that R^3 with the inner product as just defined is a special case of a general inner product space as defined in Sec. 7.15 (but not needed here).

Problem Set 8.2

Let $\mathbf{a} = [2, 1, 3]$, $\mathbf{b} = [1, 0, -4]$, $\mathbf{c} = [3, -1, 2]$. Find

1. $\mathbf{a} \cdot \mathbf{b}$, $\mathbf{b} \cdot \mathbf{a}$
2. $|\mathbf{a}|$, $|\mathbf{b}|$, $|\mathbf{c}|$
3. $(\mathbf{a} + \mathbf{b}) \cdot \mathbf{c}$, $\mathbf{a} \cdot \mathbf{c} + \mathbf{b} \cdot \mathbf{c}$
4. $3\mathbf{a} \cdot 2\mathbf{c}$, $6(\mathbf{a} \cdot \mathbf{c})$
5. $|\mathbf{a} + \mathbf{b} + \mathbf{c}|$
6. $(\mathbf{a} - \mathbf{c}) \cdot \mathbf{b}$, $\mathbf{a} \cdot \mathbf{b} - \mathbf{c} \cdot \mathbf{b}$
7. $\mathbf{a} \cdot (\mathbf{b} - \mathbf{c})$, $\mathbf{a} \cdot (\mathbf{c} - \mathbf{b})$
8. $|\mathbf{a} + \mathbf{c}|$, $|\mathbf{a}| + |\mathbf{c}|$
9. $|\mathbf{a} - 2\mathbf{b}|$, $|2\mathbf{b} - \mathbf{a}|$
10. $\mathbf{a} \cdot (\mathbf{b} + \mathbf{c})$, $(\mathbf{a} \cdot \mathbf{b})\mathbf{c}$
11. $|\mathbf{a} - \mathbf{c}|$, $|\mathbf{a}| - |\mathbf{c}|$
12. $\mathbf{a} \cdot \mathbf{b} + \mathbf{b} \cdot \mathbf{c} + \mathbf{c} \cdot \mathbf{a}$

13. What laws do Probs. 1, 3, and 6 illustrate?

14. If **a** is given and $\mathbf{a} \cdot \mathbf{b} = \mathbf{a} \cdot \mathbf{c}$, can we conclude that $\mathbf{b} = \mathbf{c}$?

15. Find $\mathbf{a} \cdot \mathbf{b}$ where $\mathbf{a} = 2\mathbf{i}$ and $\mathbf{b} = \mathbf{i} + \mathbf{j}$, $\mathbf{b} = \mathbf{j}$, $\mathbf{b} = -\mathbf{i} + \mathbf{j}$, and sketch a figure similar to Fig. 157.

Work. Find the work done by a force **p** acting on a body if the body is displaced from a point A to a point B along the straight segment AB, where

16. $\mathbf{p} = [2, 1, 0]$, A: (0, 0, 0), B: (0, 1, 0)

17. $\mathbf{p} = [1, 2, 0]$, A: (4, −7, 3), B: (4, −7, 8)

18. $\mathbf{p} = [1, 1, 1]$, A: (1, −1, 2), B: (2, 1, 3)

19. $\mathbf{p} = [3, -2, 4]$, A: (8, −2, −3), B: (−2, 0, 6)

20. Show that the work done by the resultant of two constant forces **p** and **q** in the displacement of a body from a point A to a point B along the straight segment $\overrightarrow{AB}$ is the same as the sum of the works done by each of the forces in this displacement.

21. Why is the work in Prob. 17 equal to zero?

Orthogonality of vectors

22. Show that $\mathbf{a} = [1, -1, 2]$, $\mathbf{b} = [0, 4, 2]$, $\mathbf{c} = [-10, -2, 4]$ are orthogonal.

23. For what values of a_1 are $\mathbf{a} = [a_1, 2, 0]$ and $\mathbf{b} = [3, 4, -1]$ orthogonal?

24. Find all unit vectors $\mathbf{a} = [a_1, a_2]$ orthogonal to $[2, -5]$.

25. Show that the straight lines $3x + 5y = 1$ and $10x - 6y = 7$ are orthogonal.

26. Find an orthonormal basis {**a**, **b**, **c**} in three-dimensional space, where $\mathbf{b} = q_1[3, 4, 0]$, $\mathbf{c} = q_2[4, -3, 0]$, and q_1, q_2 are suitable scalars.

27. Find a unit normal vector to the plane $2x + y - 2z = 5$.

28. Find all vectors **v** orthogonal to $\mathbf{a} = [1, 2, 0]$. Do they form a vector space?

29. For what c are the planes $x + 2y + 3z = 6$ and $x + cy + z = 2$ orthogonal?

30. Using vectors, show that if the diagonals of a rectangle are orthogonal, the rectangle must be a square.

---

[6]LUDWIG OTTO HESSE (1811—1874), German mathematician, who contributed to the theory of curves and surfaces.

Angle between vectors. Let $\mathbf{a} = [1, 1, 1]$, $\mathbf{b} = [-1, 1, 0]$, $\mathbf{c} = [3, 0, 1]$. Find the cosine of the angle between the following vectors.

31. $\mathbf{a}, \mathbf{b}$ **32.** $\mathbf{a}, \mathbf{b} + \mathbf{c}$ **33.** $\mathbf{a} + \mathbf{b}, \mathbf{c}$ **34.** $\mathbf{a} + \mathbf{c}, \mathbf{a} - \mathbf{c}$

35. Find the angle between the straight lines $4x - y = 2$ and $x + 4y = 3$.

36. Find the angle between the straight lines $x + y = 1$ and $2x - 3y = 0$.

37. Find the angle between the planes $x + 2y + z = 2$ and $2x - y + 3z = -4$.

38. Find the angle between the planes $x + y + z = 3$ and $x - y = 4$.

39. Find the angles of the triangle with vertices A: $(0, 0, 0)$, B: $(1, 2, 3)$, C: $(4, -1, 3)$.

40. Find the angles of the triangle with vertices A: $(2, 1, 5)$, B: $(4, 1, 5)$, C: $(2, 3, 5)$.

41. Deduce the law of cosines by using vectors **a**, **b**, and $\mathbf{a} - \mathbf{b}$.

42. Let $\mathbf{a} = \cos\alpha\, \mathbf{i} + \sin\alpha\, \mathbf{j}$ and $\mathbf{b} = \cos\beta\, \mathbf{i} + \sin\beta\, \mathbf{j}$, where $0 \leqq \alpha \leqq \beta \leqq 2\pi$. Show that **a** and **b** are unit vectors. Use (2) to obtain the trigonometric identity for $\cos(\beta - \alpha)$.

Component in the direction of a vector. In each case find the component of **a** in the direction of **b**.

43. $\mathbf{a} = [1, 1, 2]$, $\mathbf{b} = [0, 0, 6]$

44. $\mathbf{a} = [3, 0, -2]$, $\mathbf{b} = [1, 1, 1]$

45. $\mathbf{a} = [0, 3, -4]$, $\mathbf{b} = [0, 4, 3]$

46. $\mathbf{a} = [-2, -5, 6]$, $\mathbf{b} = [2, 1, 2]$

47. $\mathbf{a} = [2, 3, 0]$, $\mathbf{b} = [-2, -3, 0]$

48. $\mathbf{a} = [7, -5, 3]$, $\mathbf{b} = [0.8, 0, 0.6]$

49. Prove the triangle inequality (7).

50. Prove the parallelogram equality (8).

8.3 Vector Product (Cross Product)

Dot multiplication gives a scalar as the product of two vectors (Sec. 8.2), but there are other applications that suggest another multiplication, "*cross multiplication*," that again gives a vector as the product of two vectors in three-space.

Definition. Vector product (Cross product)

The **vector product (cross product)** $\mathbf{a} \times \mathbf{b}$ of two vectors $\mathbf{a} = [a_1, a_2, a_3]$ and $\mathbf{b} = [b_1, b_2, b_3]$ is a vector

$$\mathbf{v} = \mathbf{a} \times \mathbf{b}$$

as follows. If **a** and **b** have the same or opposite direction or if one of these vectors is the zero vector, then $\mathbf{v} = \mathbf{a} \times \mathbf{b} = \mathbf{0}$.

In any other case, $\mathbf{v} = \mathbf{a} \times \mathbf{b}$ has the length

$$|\mathbf{v}| = |\mathbf{a}||\mathbf{b}| \sin\gamma, \tag{1}$$

which is the area of the parallelogram in Fig. 160 with **a** and **b** as adjacent sides. The direction of $\mathbf{v} = \mathbf{a} \times \mathbf{b}$ is ***perpendicular to both* a *and* b** and such that **a, b, v,** in this order, form a *right-handed triple* as in Fig. 160 (explanation below). ∎

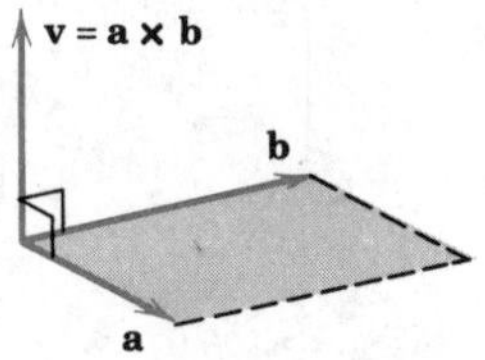

Fig. 160. Vector product

In components, $\mathbf{v} = [v_1, v_2, v_3] = \mathbf{a} \times \mathbf{b}$ is

$$(2) \qquad v_1 = a_2b_3 - a_3b_2, \quad v_2 = a_3b_1 - a_1b_3, \quad v_3 = a_1b_2 - a_2b_1.$$

Here we assume that the Cartesian coordinate system is *right-handed* (explanation below), whereas for a *left-handed* system these three components must be multiplied by -1. Formula (2) is derived in Appendix 4.

Explanations. First, a **right-handed triple** of vectors **a, b, v** is one in which the vectors, in the order given, assume the same sort of orientation as the thumb, index finger, and middle finger of the right hand when these are held as shown in Fig. 161. We may also say that if **a** is rotated into the direction of **b** through the angle α ($< \pi$), then **v** advances in the same direction as a right-handed screw would if turned in the same way (Fig. 162).

Second, a Cartesian coordinate system is called **right-handed** if the corresponding unit vectors **i, j, k** in the positive directions of the axes (see Sec. 8.1) form a right-handed triple as in Fig. 163*a* on p. 444. The system is called *left-handed* if the sense of **k** is reversed, as in Fig. 163*b*. In applications, we prefer right-handed systems.

EXAMPLE 1 **Vector product**

For the vector product $\mathbf{v} = \mathbf{a} \times \mathbf{b}$ of $\mathbf{a} = [1, \ 1, \ 0]$ and $\mathbf{b} = [3, \ 0, \ 0]$ in right-handed coordinates we obtain from (2)

$$v_1 = 0, \qquad v_2 = 0, \qquad v_3 = 1 \cdot 0 - 1 \cdot 3 = -3.$$

To check the result in this simple case, sketch **a, b,** and **v.** Can you see that two vectors in the xy-plane must always have their vector product parallel to the z-axis (or equal to the zero vector)? ∎

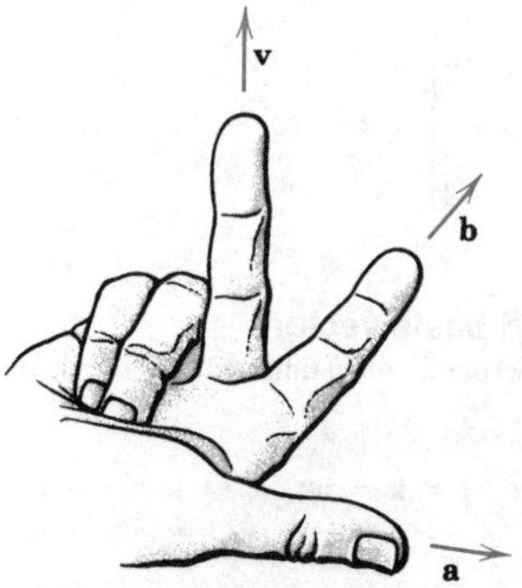

Fig. 161. Right-handed triple of vectors **a, b, v**

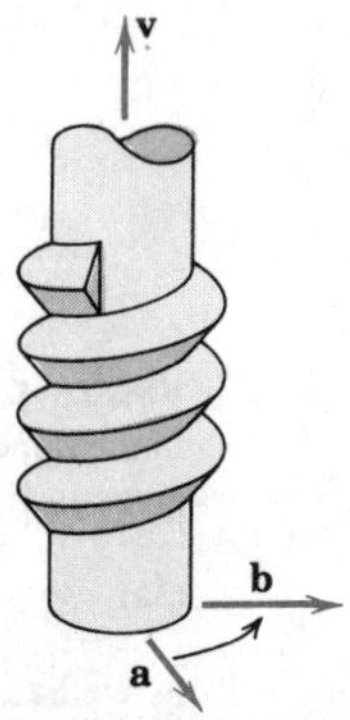

Fig. 162. Right-handed screw

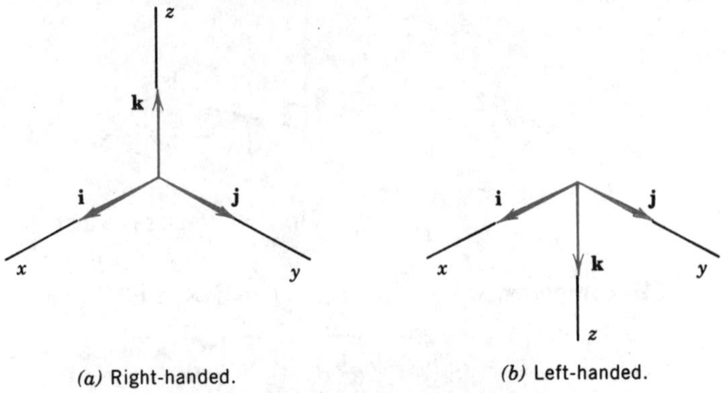

Fig. 163. The two types of Cartesian coordinate systems

How to memorize formula (2) on p. 443. Students familiar with second-order and third-order determinants will notice that in (2),

$$(2^*)\qquad v_1 = \begin{vmatrix} a_2 & a_3 \\ b_2 & b_3 \end{vmatrix}, \qquad v_2 = \begin{vmatrix} a_3 & a_1 \\ b_3 & b_1 \end{vmatrix}, \qquad v_3 = \begin{vmatrix} a_1 & a_2 \\ b_1 & b_2 \end{vmatrix},$$

so that $\mathbf{v} = [v_1, v_2, v_3] = v_1\mathbf{i} + v_2\mathbf{j} + v_3\mathbf{k}$ becomes the expansion of the symbolical third-order determinant

$$(2^{**})\qquad \mathbf{a} \times \mathbf{b} = \begin{vmatrix} \mathbf{i} & \mathbf{j} & \mathbf{k} \\ a_1 & a_2 & a_3 \\ b_1 & b_2 & b_3 \end{vmatrix}$$

by the first row. (We call it "symbolical" because the first row consists of vectors rather than numbers.) In left-handed coordinates these determinants must be multiplied by -1.

EXAMPLE 2 An application of (2)**

With respect to a right-handed Cartesian coordinate system, let $\mathbf{a} = [4, 0, -1]$ and $\mathbf{b} = [-2, 1, 3]$. Then

$$\mathbf{a} \times \mathbf{b} = \begin{vmatrix} \mathbf{i} & \mathbf{j} & \mathbf{k} \\ 4 & 0 & -1 \\ -2 & 1 & 3 \end{vmatrix} = \mathbf{i} - 10\mathbf{j} + 4\mathbf{k} = [1, -10, 4].$$ ∎

EXAMPLE 3 Vector products of the standard basis vectors

Since **i**, **j**, **k** are orthogonal (mutually perpendicular) unit vectors, the definition of vector product gives in right-handed coordinates

$$(3a)\qquad \mathbf{i} \times \mathbf{j} = \mathbf{k}, \qquad \mathbf{j} \times \mathbf{k} = \mathbf{i}, \qquad \mathbf{k} \times \mathbf{i} = \mathbf{j}$$

as well as

$$(3b)\qquad \mathbf{j} \times \mathbf{i} = -\mathbf{k}, \qquad \mathbf{k} \times \mathbf{j} = -\mathbf{i}, \qquad \mathbf{i} \times \mathbf{k} = -\mathbf{j}.$$ ∎

General Properties of Vector Products

Cross multiplication has the property that

$$(4)\qquad (k\mathbf{a}) \times \mathbf{b} = k(\mathbf{a} \times \mathbf{b}) = \mathbf{a} \times (k\mathbf{b}) \qquad \text{for every scalar } k.$$

It is distributive with respect to vector addition, that is,

$$(5)\qquad \begin{aligned} &\text{(a)} \quad \mathbf{a} \times (\mathbf{b} + \mathbf{c}) = (\mathbf{a} \times \mathbf{b}) + (\mathbf{a} \times \mathbf{c}), \\ &\text{(b)} \quad (\mathbf{a} + \mathbf{b}) \times \mathbf{c} = (\mathbf{a} \times \mathbf{c}) + (\mathbf{b} \times \mathbf{c}). \end{aligned}$$

It is ***not commutative*** but **anticommutative,** that is,

$$(6)\qquad \mathbf{b} \times \mathbf{a} = -(\mathbf{a} \times \mathbf{b}) \qquad \text{(Fig. 164).}$$

It is ***not associative,*** that is,

$$(7)\qquad \mathbf{a} \times (\mathbf{b} \times \mathbf{c}) \neq (\mathbf{a} \times \mathbf{b}) \times \mathbf{c} \qquad \text{in general,}$$

so that the parentheses cannot be omitted.

Proofs. (4) follows directly from the definition. In (5a), formula (2*) gives for the first component on the left

$$\begin{aligned} \begin{vmatrix} a_2 & a_3 \\ b_2 + c_2 & b_3 + c_3 \end{vmatrix} &= a_2(b_3 + c_3) - a_3(b_2 + c_2) \\ &= (a_2b_3 - a_3b_2) + (a_2c_3 - a_3c_2) \\ &= \begin{vmatrix} a_2 & a_3 \\ b_2 & b_3 \end{vmatrix} + \begin{vmatrix} a_2 & a_3 \\ c_2 & c_3 \end{vmatrix}, \end{aligned}$$

and, by (2*), the sum of the two determinants is the first component of $(\mathbf{a} \times \mathbf{b}) + (\mathbf{a} \times \mathbf{c})$, the right side of (5a). For the other components in (5a) and in (5b), equality follows by the same idea.

Anticommutativity (6) follows from (2**) by noting that the interchange of rows 2 and 3 multiplies the determinant by -1. We can confirm this geometrically if we set $\mathbf{a} \times \mathbf{b} = \mathbf{v}$ and $\mathbf{b} \times \mathbf{a} = \mathbf{w}$; then $|\mathbf{v}| = |\mathbf{w}|$ by (1), and for $\mathbf{b}$, $\mathbf{a}$, $\mathbf{w}$ to form a *right-handed* triple, we must have $\mathbf{w} = -\mathbf{v}$.

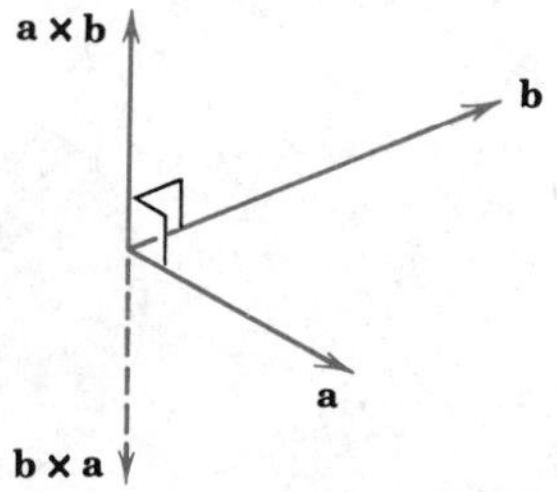

Fig. 164. Anticommutativity of cross multiplication

Finally, $\mathbf{i} \times (\mathbf{i} \times \mathbf{j}) = \mathbf{i} \times \mathbf{k} = -\mathbf{j}$, whereas $(\mathbf{i} \times \mathbf{i}) \times \mathbf{j} = \mathbf{0} \times \mathbf{j} = \mathbf{0}$ (see Example 3), which illustrates (7). ▮

Typical Applications of Vector Products

EXAMPLE 4 Moment of a force

In mechanics the moment m of a force $\mathbf{p}$ about a point Q is defined as the product $m = |\mathbf{p}|d$, where d is the (perpendicular) distance between Q and the line of action L of $\mathbf{p}$ (Fig. 165). If $\mathbf{r}$ is the vector from Q to any point A on L, then $d = |\mathbf{r}| \sin \gamma$ (Fig. 165) and

$$m = |\mathbf{r}|\,|\mathbf{p}| \sin \gamma.$$

Since γ is the angle between $\mathbf{r}$ and $\mathbf{p}$,

$$m = |\mathbf{r} \times \mathbf{p}|,$$

as follows from (1). The vector,

$$\mathbf{m} = \mathbf{r} \times \mathbf{p} \tag{8}$$

is called the **moment vector or vector moment** of $\mathbf{p}$ about Q. Its magnitude is m, and its direction is that of the axis of the rotation about Q which $\mathbf{p}$ has the tendency to produce. ▮

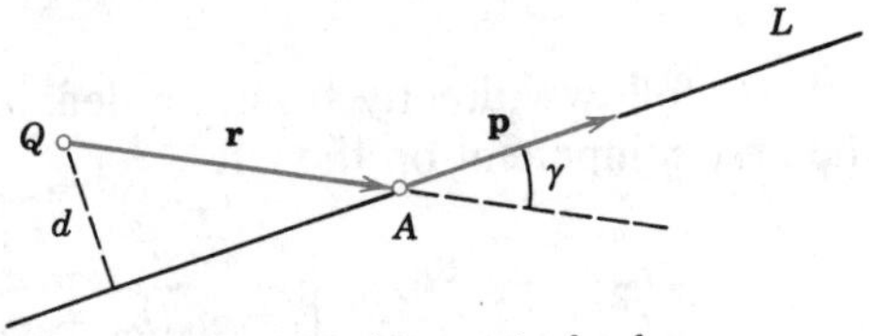

Fig. 165. Moment of a force

EXAMPLE 5 Moment of a force

Find the moment of the force $\mathbf{p}$ in Fig. 166 about the center of the wheel.

Solution. Introducing coordinates as shown in Fig. 166, we have

$$\mathbf{p} = [1000 \cos 30°,\ \ 1000 \sin 30°,\ \ 0] = [866,\ \ 500,\ \ 0], \qquad \mathbf{r} = [0,\ \ -1.5,\ \ 0],$$

so that (8) and (2**) give

$$\mathbf{m} = \mathbf{r} \times \mathbf{p} = \begin{vmatrix} \mathbf{i} & \mathbf{j} & \mathbf{k} \\ 0 & -1.5 & 0 \\ 866 & 500 & 0 \end{vmatrix} = 0\mathbf{i} - 0\mathbf{j} + \begin{vmatrix} 0 & -1.5 \\ 866 & 500 \end{vmatrix}\mathbf{k} = [0, 0, 1299].$$

This moment vector is normal (perpendicular) to the plane of the wheel; hence it has the direction of the axis of rotation about the center of the wheel that the force has the tendency to produce. ▮

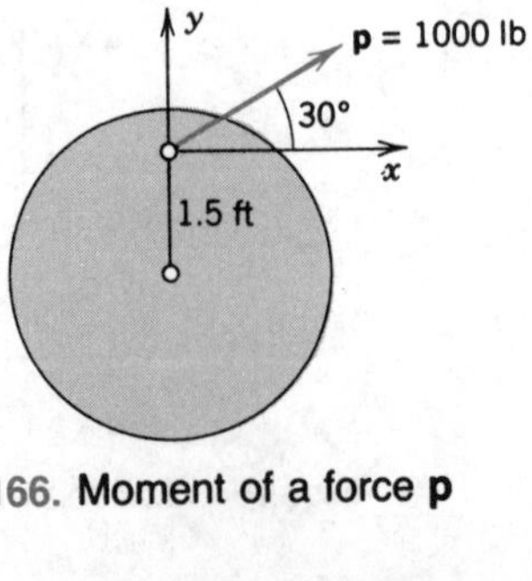

Fig. 166. Moment of a force p

EXAMPLE 6 Velocity of a rotating body

A rotation of a rigid body B in space can be simply and uniquely described by a vector **w** as follows. The direction of **w** is that of the axis of rotation and such that the rotation appears clockwise if one looks from the initial point of **w** to its terminal point. The length of **w** is equal to the **angular speed** ω (> 0) of the rotation, that is, the linear (or tangential) speed of a point of B divided by its distance from the axis of rotation.

Let P be any point of B and d its distance from the axis. Then P has the speed ωd. Let **r** be the position vector of P referred to a coordinate system with origin 0 on the axis of rotation. Then $d = |\mathbf{r}| \sin \gamma$ where γ is the angle between **w** and **r**. Therefore,

$$\omega d = |\mathbf{w}|\,|\mathbf{r}| \sin \gamma = |\mathbf{w} \times \mathbf{r}|.$$

From this and the definition of vector product we see that the velocity vector **v** of P can be represented in the form (Fig. 167)

$$\mathbf{v} = \mathbf{w} \times \mathbf{r}. \tag{9}$$

This simple formula is useful for determining **v** at any point of B. ∎

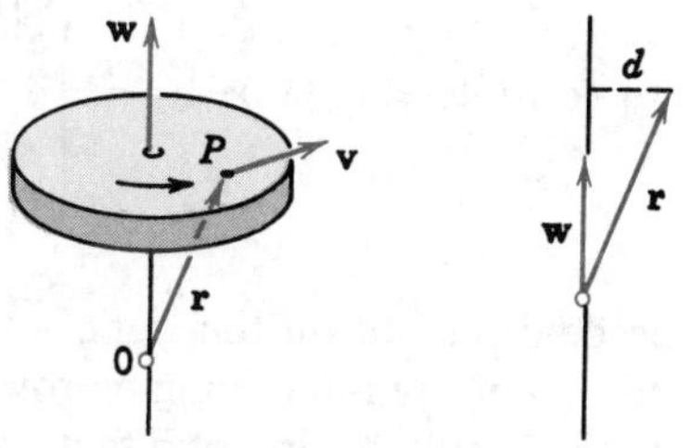

Fig. 167. Rotation of a rigid body

Scalar Triple Product

The **scalar triple product** or **mixed triple product** of three vectors

$$\mathbf{a} = [a_1, a_2, a_3], \qquad \mathbf{b} = [b_1, b_2, b_3], \qquad \mathbf{c} = [c_1, c_2, c_3]$$

is denoted by $(\mathbf{a} \;\; \mathbf{b} \;\; \mathbf{c})$ and is defined by

$$(\mathbf{a} \;\; \mathbf{b} \;\; \mathbf{c}) = \mathbf{a} \cdot (\mathbf{b} \times \mathbf{c}).$$

Now from the dot product in components [formula (2) in Sec. 8.2] and from (2*), denoting $\mathbf{b} \times \mathbf{c}$ by $\mathbf{v} = [v_1, v_2, v_3]$, we first obtain

$$\mathbf{a} \cdot (\mathbf{b} \times \mathbf{c}) = a_1 v_1 + a_2 v_2 + a_3 v_3$$

$$= a_1 \begin{vmatrix} b_2 & b_3 \\ c_2 & c_3 \end{vmatrix} + a_2 \begin{vmatrix} b_3 & b_1 \\ c_3 & c_1 \end{vmatrix} + a_3 \begin{vmatrix} b_1 & b_2 \\ c_1 & c_2 \end{vmatrix}.$$

The expression on the right is the development of a third-order determinant, say, by its first row; thus

(10) $$(\mathbf{a}\ \mathbf{b}\ \mathbf{c}) = \mathbf{a}\cdot(\mathbf{b}\times\mathbf{c}) = \begin{vmatrix} a_1 & a_2 & a_3 \\ b_1 & b_2 & b_3 \\ c_1 & c_2 & c_3 \end{vmatrix}.$$

We also have for any scalar k

(11) $$(k\mathbf{a}\ \mathbf{b}\ \mathbf{c}) = k(\mathbf{a}\ \mathbf{b}\ \mathbf{c})$$

because the multiplication of a row of a determinant by k multiplies the value of the determinant by k. Furthermore,

(12) $$\mathbf{a}\cdot(\mathbf{b}\times\mathbf{c}) = (\mathbf{a}\times\mathbf{b})\cdot\mathbf{c}$$

because dot multiplication is commutative, so that

$$(\mathbf{a}\times\mathbf{b})\cdot\mathbf{c} = \mathbf{c}\cdot(\mathbf{a}\times\mathbf{b}) = \begin{vmatrix} c_1 & c_2 & c_3 \\ a_1 & a_2 & a_3 \\ b_1 & b_2 & b_3 \end{vmatrix} = \begin{vmatrix} a_1 & a_2 & a_3 \\ b_1 & b_2 & b_3 \\ c_1 & c_2 & c_3 \end{vmatrix},$$

where the determinant on the right is that in (10) and is obtained from the other determinant by interchanging rows 1 and 2 and by interchanging rows 2 and 3 in the result, giving two factors -1, and $(-1)(-1) = 1$.

Geometric interpretation. The absolute value of the scalar triple product (10) is the volume of the parallelepiped with **a, b, c** as edge vectors (Fig. 168), because, by (1) in Sec. 8.2,

$$|\mathbf{a}\cdot(\mathbf{b}\times\mathbf{c})| = |\mathbf{a}|\,|\mathbf{b}\times\mathbf{c}|\cos\beta \qquad \text{(Fig. 168)}$$

where $|\mathbf{a}|\,|\cos\beta|$ is the height h and, by (1), the base, the parallelogram with sides **b** and **c**, has area $|\mathbf{b}\times\mathbf{c}|$.

EXAMPLE 7 Tetrahedron

A tetrahedron is determined by three edge vectors **a, b, c** as indicated in Fig. 169.

Find the volume of the tetrahedron with $\mathbf{a} = [2, 0, 3]$, $\mathbf{b} = [0, 6, 2]$, $\mathbf{c} = [3, 3, 0]$ as edge vectors.

Solution. The volume V of the parallelepiped with these vectors as edge vectors is the absolute value of the scalar triple product

$$(\mathbf{a}\ \mathbf{b}\ \mathbf{c}) = \begin{vmatrix} 2 & 0 & 3 \\ 0 & 6 & 2 \\ 3 & 3 & 0 \end{vmatrix} = 2\begin{vmatrix} 6 & 2 \\ 3 & 0 \end{vmatrix} + 3\begin{vmatrix} 0 & 6 \\ 3 & 3 \end{vmatrix} = -12 - 54 = -66,$$

that is, $V = 66$. The minus sign indicates that **a, b, c,** in this order, form a left-handed triple if the Cartesian coordinate system is right-handed (and a right-handed triple if this system is left-handed). The volume of the tetrahedron is $\frac{1}{6}$ of that of the parallelepiped, hence 11.

Can you sketch the tetrahedron, choosing the origin as the common initial point of the three vectors? What are the coordinates of the four vertices? ▮

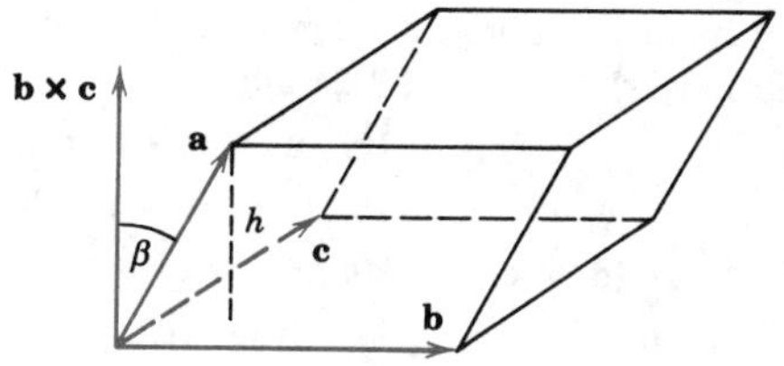

Fig. 168. Geometrical interpretation of a scalar triple product

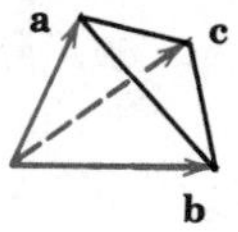

Fig. 169. Tetrahedron

Linear independence of three vectors can be tested by scalar triple products, as follows.

We call a given set of vectors $\mathbf{a}_{(1)}, \cdots, \mathbf{a}_{(m)}$ **linearly independent** if the only scalars $c_1, \cdots, c_m$ for which the vector equation

$$c_1\mathbf{a}_{(1)} + c_2\mathbf{a}_{(2)} + \cdots + c_m\mathbf{a}_{(m)} = \mathbf{0}$$

is satisfied are $c_1 = 0, c_2 = 0, \cdots, c_m = 0$. Otherwise, that is, if that equation also holds for an m-tuple of scalars not all zero, we call that set of vectors **linearly dependent.**

Now three vectors, if we let their initial point coincide, form a linearly independent set if and only if they do not lie in the same plane (or on the same line). The interpretation of a scalar triple product as a volume thus gives the following criterion.

Theorem 1 **(Linear independence of three vectors)**

Three vectors form a linearly independent set if and only if their scalar triple product is not zero.

The scalar triple product is the most important "repeated product." Several others that one needs occasionally are included in the problem set.

This is the end of vector *algebra* (in 3-space and in the plane). Vector *calculus* (differentiation) begins in the next section.

Problem Set 8.3

With respect to a right-handed Cartesian coordinate system, let $\mathbf{a} = [1, 1, 0]$, $\mathbf{b} = [-1, 2, 0]$, $\mathbf{c} = [2, 3, 1]$, $\mathbf{d} = [5, -7, 2]$. Find

1. $\mathbf{a} \times \mathbf{b}, \mathbf{b} \times \mathbf{a}$

2. $\mathbf{b} \times \mathbf{c}, |\mathbf{b} \times \mathbf{c}|, |\mathbf{c} \times \mathbf{b}|$

3. $\mathbf{a} \times \mathbf{c}, |\mathbf{a} \times \mathbf{c}|, \mathbf{a} \cdot \mathbf{c}$

4. $2\mathbf{a} \times 3\mathbf{b}, 3\mathbf{a} \times 2\mathbf{b}, 6\mathbf{a} \times \mathbf{b}$

5. $(\mathbf{a} + \mathbf{b}) \times \mathbf{c}, \mathbf{a} \times \mathbf{c} + \mathbf{b} \times \mathbf{c}$

6. $(\mathbf{a} + \mathbf{b}) \times \mathbf{c}, \mathbf{a} \times (\mathbf{b} + \mathbf{c})$

7. $(\mathbf{a} + \mathbf{c}) \times \mathbf{c}, \mathbf{a} \times \mathbf{c}$

8. $(\mathbf{a} - \mathbf{b}) \times (\mathbf{a} - \mathbf{c}), (\mathbf{a} - \mathbf{b}) \times (\mathbf{c} - \mathbf{a})$

9. $\mathbf{a} \times \mathbf{c} + \mathbf{c} \times \mathbf{a}$

10. $(\mathbf{a} + \mathbf{b}) \times (\mathbf{b} + \mathbf{a})$

11. $(3\mathbf{a} - 6\mathbf{b}) \times \mathbf{c}, 3\mathbf{c} \times (2\mathbf{b} - \mathbf{a})$

12. $\mathbf{a} \times (3\mathbf{b} - 5\mathbf{c}), 3\mathbf{a} \times \mathbf{b} - 5\mathbf{a} \times \mathbf{c}$

13. $(\mathbf{a} \cdot \mathbf{b})\mathbf{c}, (\mathbf{a} \times \mathbf{b}) \cdot \mathbf{c}$

14. $(\mathbf{a} \times \mathbf{b}) \times \mathbf{c}, \mathbf{a} \times (\mathbf{b} \times \mathbf{c})$

With respect to a right-handed Cartesian coordinate system, let **a** = [1, 1, 0], **b** = [−1, 2, 0], **c** = [2, 3, 1], **d** = [5, −7, 2]. Find

15. (**i** **j** **k**), (**i** **k** **j**)

16. (**b** × **c**) •**d**, **b**•(**c** × **d**)

17. (**a** × **b**) × (**c** × **d**)

18. **b** × (**b** × **c**), (**b** × **b**) × **c**

19. (**a** × **b**)•(**c** × **d**), (**b** × **a**)•(**d** × **c**)

20. [(**a** × **b**) × **c**]•**d**

21. (**a** − **b** **b** − **c** **c**), (**a** **b** **c**)

22. (**b** **c** **d**), (**c** **b** **d**)

23. (2**a** 3**b** 5**d**), 30(**a** **b** **d**)

24. (**a** + **b** **b** + **c** **c** + **d**)

25. What properties of cross multiplication do Probs. 1, 4, 5, and 14 illustrate?

Applications of vector product

Moment of a force. A force **p** acts on a line through a point A. Find the moment vector **m** of **p** about a point Q, where the force, the point A, and the point Q are

26. [1, −2, 0], (1, 1, 1), (2, −1, 3)

27. [0, 0, 1], (0, 0, 0), (0, 0, 5)

28. [3, −1, 2], (0, −1, 4), (3, 0, 2)

29. [2, 4, 1], (4, 2, −1), (0, 1, 2)

30. [1, −2, 3], (4, 3, 1), (6, −1, 7)

31. [3, 0, −6], (1, 8, 1), (4, 6, −1)

Area of a parallelogram. Find the area if the vertices in the xy-plane are as follows. (*Hint.* First obtain two adjacent edge vectors.)

32. (−1, 2), (2, 0), (4, 3), (7, 1)

33. (0, 0), (4, 1), (2, 3), (6, 4)

34. (4, 4), (−1, 9), (6, 6), (1, 11)

35. (2, −3), (1, 1), (5, −6), (4, −2)

Area of a triangle. Find the area if the vertices are

36. (0, 0, 2), (0, 1, 2), (1, 1, 2)

37. (6, −1, 3), (6, 1, 1), (3, 3, 3)

38. (1, 3, 0), (0, 2, 5), (−1, 0, 2)

39. (2, 2, 2), (5, 2, 4), (−2, 4, −1)

Plane. Find the plane through the three given points. (*Hint.* Find two vectors in the plane; their cross product then gives a normal vector. See also Example 6 in Sec. 8.2.)

40. (2, 0, 0), (0, 2, 0), (0, 0, 2)

41. (1, 2, $\frac{1}{4}$), (4, 2, −2), (0, 8, 4)

42. (4, $\frac{5}{3}$, 3), (1, 2, 1), (7, −1, 4)

43. (1, 6, 1), (9, 1, −31), (−5, −2, 25)

Application of scalar triple product

Volume of a parallelepiped. Find the volume from the given edges.

44. **i**, 2**j**, −3**k**

45. **i** + **j**, **i** − **j**, **i** + 2**j** + 4**k**

46. 2**j** + **k**, **i** − **j**, −**j** + 4**k**

47. 2**i** − 6**k**, **j** + **k**, **i** + **j**

Volume of a tetrahedron. Find the volume if the vertices are

48. (2, 2, 2), (3, 2, 2), (2, 3, 2), (2, 2, 3)

49. (0, 0, 0), (1, 0, 0), (0, 1, 0), (0, 0, 1)

50. (2, 1, 1), (3, 3, 4), (0, 1, 5), (1, 2, 2)

51. (0, 1, 2), (5, 5, 6), (1, 2, 1), (3, 3, 1)

Linear independence. Are the following vectors linearly independent?

52. [4, −8, 3], [7, 2, 5], [10, −52, 8]

53. [1, 1, 0], [0, 1, 1], [1, 0, 1]

54. [7, 3, −2], [5, 9, 8], [2, −7, 3]

55. [3, 5, −7], [−1, 37, −43], [8, −6, 4]

Additional formulas

56. Using $\sin^2 \gamma = 1 - \cos^2 \gamma$ in (1), show that

$$|\mathbf{a} \times \mathbf{b}| = \sqrt{(\mathbf{a}\cdot\mathbf{a})(\mathbf{b}\cdot\mathbf{b}) - (\mathbf{a}\cdot\mathbf{b})^2}. \tag{13}$$

57. Using Theorem 5, Sec. 7.8, show that

$$(14) \qquad \begin{aligned} (\mathbf{a}\ \ \mathbf{b}\ \ \mathbf{c}) &= -(\mathbf{b}\ \ \mathbf{a}\ \ \mathbf{c}) = -(\mathbf{a}\ \ \mathbf{c}\ \ \mathbf{b}) \\ (\mathbf{a}\ \ \mathbf{b}\ \ \mathbf{c}) &= (\mathbf{b}\ \ \mathbf{c}\ \ \mathbf{a}) = (\mathbf{c}\ \ \mathbf{a}\ \ \mathbf{b}). \end{aligned}$$

58. Show that

$$(15) \qquad \mathbf{b} \times (\mathbf{c} \times \mathbf{d}) = (\mathbf{b}\cdot\mathbf{d})\mathbf{c} - (\mathbf{b}\cdot\mathbf{c})\mathbf{d}.$$

Hint. Choose special right-handed Cartesian coordinates such that $\mathbf{d} = [d_1, 0, 0]$ and $\mathbf{c} = [c_1, c_2, 0]$. Then verify that each side equals $[-b_2c_2d_1, \ b_1c_2d_1, \ 0]$. Then give a reason why the two sides must thus be equal in any Cartesian coordinate system.

59. Using (15), show that

$$(16) \qquad (\mathbf{a} \times \mathbf{b}) \times (\mathbf{c} \times \mathbf{d}) = (\mathbf{a}\ \ \mathbf{b}\ \ \mathbf{d})\mathbf{c} - (\mathbf{a}\ \ \mathbf{b}\ \ \mathbf{c})\mathbf{d}.$$

60. **(Identity of Lagrange)** Show that (15) implies the *identity of Lagrange*

$$(17) \qquad (\mathbf{a} \times \mathbf{b})\cdot(\mathbf{c} \times \mathbf{d}) = (\mathbf{a}\cdot\mathbf{c})(\mathbf{b}\cdot\mathbf{d}) - (\mathbf{a}\cdot\mathbf{d})(\mathbf{b}\cdot\mathbf{c}).$$

8.4 Vector and Scalar Functions and Fields. Derivatives

This is the beginning of vector calculus, which involves two kinds of functions, **vector functions,** whose values are vectors

$$\mathbf{v} = \mathbf{v}(P) = [v_1(P), v_2(P), v_3(P)]$$

depending on the points P in space, and **scalar functions,** whose values are scalars

$$f = f(P)$$

depending on P. In applications, the domain of definition for such a function is a region of space or a surface in space or a curve in space. We say that a vector function defines a **vector field** in that region (or on that surface or curve). Examples are shown in Figs. 170–173. Similarly, a scalar function defines a **scalar field** in a region or on a surface or a curve. Examples are the temperature field in a body and the pressure field of the air in the earth's atmosphere. Vector and scalar functions may also depend on time t or on further parameters.

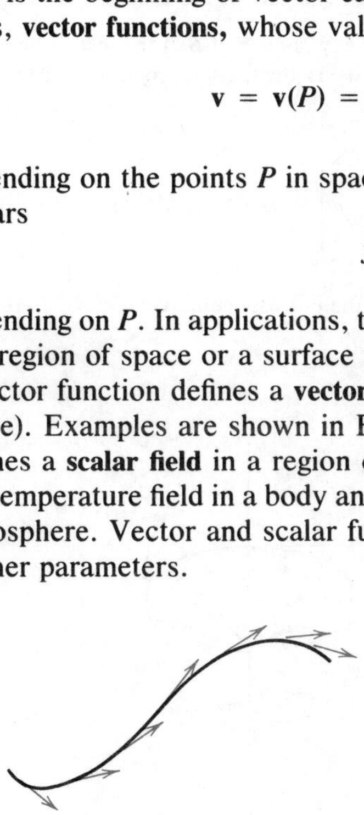

Fig. 170. Field of tangent vectors of a curve

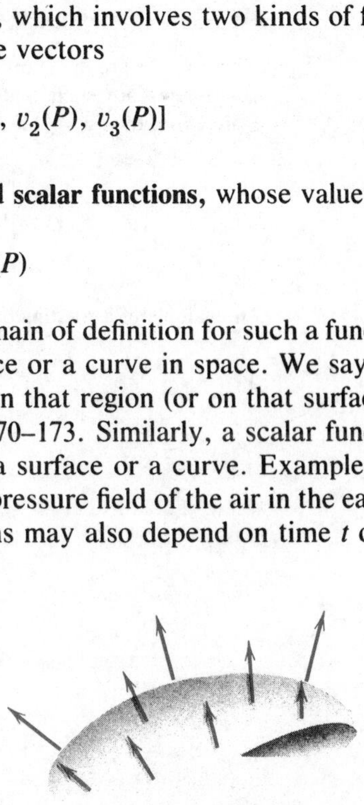

Fig. 171. Field of normal vectors of a surface

Comment on Notation. If we introduce Cartesian coordinates x, y, z, then instead of $\mathbf{v}(P)$ and $f(P)$ we can also write

$$\mathbf{v}(x, y, z) = [v_1(x, y, z), \quad v_2(x, y, z), \quad v_3(x, y, z)]$$

and $f(x, y, z)$, but we keep in mind that a vector or scalar field that has a geometrical or physical meaning should depend only on the points P where it is defined but not on the particular choice of Cartesian coordinates.

EXAMPLE 1 Scalar function (Euclidean distance in space)

The distance $f(P)$ of any point P from a fixed point P_0 in space is a scalar function whose domain of definition is the whole space. $f(P)$ defines a scalar field in space. If we introduce a Cartesian coordinate system and P_0 has the coordinates x_0, y_0, z_0, then f is given by the well-known formula

$$f(P) = f(x, y, z) = \sqrt{(x - x_0)^2 + (y - y_0)^2 + (z - z_0)^2}$$

where x, y, z are the coordinates of P. If we replace the given Cartesian coordinate system by another such system, then the values of the coordinates of P and P_0 will in general change, but $f(P)$ will have the same value as before. Hence $f(P)$ is a scalar function. The direction cosines of the line through P and P_0 are not scalars because their values will depend on the choice of the coordinate system. ∎

EXAMPLE 2 Vector field (Velocity field)

At any instant the velocity vectors $\mathbf{v}(P)$ of a rotating body B constitute a vector field, the so-called **velocity field** of the rotation. If we introduce a Cartesian coordinate system having the origin on the axis of rotation, then (see Example 6 in Sec. 8.3)

$$\mathbf{v}(x, y, z) = \mathbf{w} \times \mathbf{r} = \mathbf{w} \times [x, y, z] = \mathbf{w} \times (x\mathbf{i} + y\mathbf{j} + z\mathbf{k}), \tag{1}$$

where x, y, z are the coordinates of any point P of B at the instant under consideration. If the coordinates are such that the z-axis is the axis of rotation and $\mathbf{w}$ points in the positive z-direction, then $\mathbf{w} = \omega\mathbf{k}$ and

$$\mathbf{v} = \begin{vmatrix} \mathbf{i} & \mathbf{j} & \mathbf{k} \\ 0 & 0 & \omega \\ x & y & z \end{vmatrix} = \omega[-y, x, 0] = \omega(-y\mathbf{i} + x\mathbf{j}).$$

An example of a rotating body and the corresponding velocity field are shown in Fig. 172. ∎

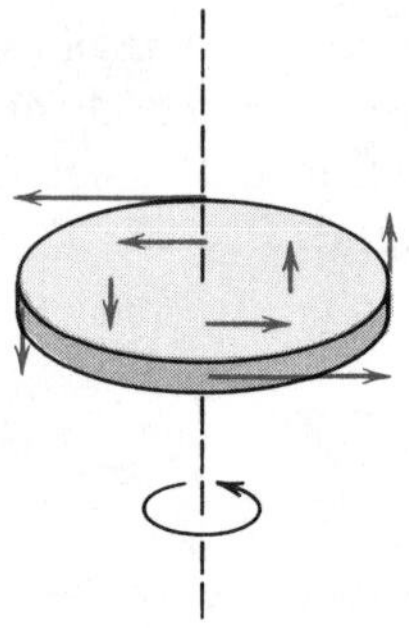

Fig. 172. Velocity field of a rotating body

EXAMPLE 3 **Vector field (Field of force)**

Let a particle A of mass M be fixed at a point P_0 and let a particle B of mass m be free to take up various positions P in space. Then A attracts B. According to **Newton's law of gravitation** the corresponding gravitational force **p** is directed from P to P_0, and its magnitude is proportional to $1/r^2$ where r is the distance between P and P_0, say,

$$(2) \qquad |\mathbf{p}| = \frac{c}{r^2}, \qquad c = GMm,$$

where G ($= 6.67 \cdot 10^{-8}$ cm^3/gm $\cdot$ sec^2) is the gravitational constant. Hence **p** defines a vector field in space. If we introduce Cartesian coordinates such that P_0 has the coordinates x_0, y_0, z_0 and P has the coordinates x, y, z, then by the Pythagorean theorem,

$$r = \sqrt{(x - x_0)^2 + (y - y_0)^2 + (z - z_0)^2} \qquad (\geqq 0).$$

Assuming that $r > 0$ and introducing the vector

$$\mathbf{r} = [x - x_0, y - y_0, z - z_0] = (x - x_0)\mathbf{i} + (y - y_0)\mathbf{j} + (z - z_0)\mathbf{k},$$

we have $|\mathbf{r}| = r$, and $(-1/r)\mathbf{r}$ is a unit vector in the direction of **p**; the minus sign indicates that **p** is directed from P to P_0 (Fig. 173). From this and (2) we obtain

$$(3) \qquad \mathbf{p} = |\mathbf{p}|\left(-\frac{1}{r}\mathbf{r}\right) = -\frac{c}{r^3}\mathbf{r} = -c\,\frac{x - x_0}{r^3}\,\mathbf{i} - c\,\frac{y - y_0}{r^3}\,\mathbf{j} - c\,\frac{z - z_0}{r^3}\,\mathbf{k}.$$

This vector function describes the gravitational force acting on B. ■

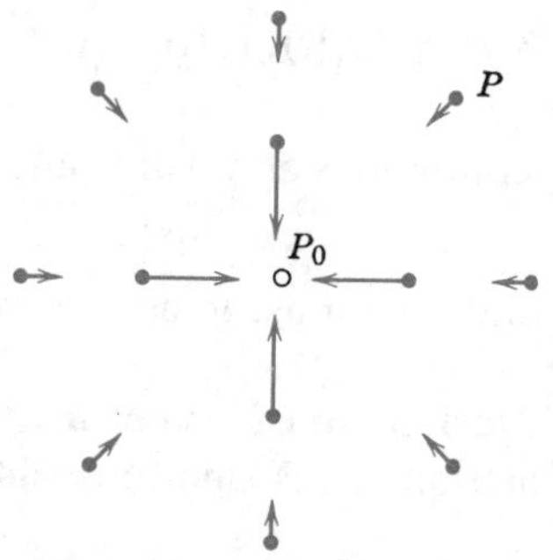

Fig. 173. Gravitational field in Example 3

Vector Calculus

We show next that the basic concepts of calculus, such as convergence, continuity, and differentiability, can be defined for vector functions in a simple and natural way. Most important here is the derivative.

Convergence. An infinite sequence of vectors $\mathbf{a}_{(n)}$, $n = 1, 2, \cdots$, is said to **converge** if there is a vector **a** such that

$$(4) \qquad \lim_{n\to\infty} |\mathbf{a}_{(n)} - \mathbf{a}| = 0.$$

a is called the **limit vector** of that sequence, and we write

$$(5) \qquad \lim_{n\to\infty} \mathbf{a}_{(n)} = \mathbf{a}.$$

Cartesian coordinates being given, this sequence of vectors converges to **a** if and only if the three sequences of components of the vectors converge to the corresponding components of **a**. We leave the simple proof to the student.

Similarly, a vector function $\mathbf{v}(t)$ of a real variable t is said to have the **limit** $\boldsymbol{l}$ as t approaches t_0, if $\mathbf{v}(t)$ is defined in some *neighborhood*[7] of t_0 (possibly except at t_0) and

$$\lim_{t \to t_0} |\mathbf{v}(t) - \boldsymbol{l}| = 0. \tag{6}$$

Then we write

$$\lim_{t \to t_0} \mathbf{v}(t) = \boldsymbol{l}. \tag{7}$$

Continuity. A vector function $\mathbf{v}(t)$ is said to be **continuous** at $t = t_0$ if it is defined in some neighborhood of t_0 and

$$\lim_{t \to t_0} \mathbf{v}(t) = \mathbf{v}(t_0). \tag{8}$$

If we introduce a Cartesian coordinate system, we may write

$$\mathbf{v}(t) = [v_1(t), v_2(t), v_3(t)] = v_1(t)\mathbf{i} + v_2(t)\mathbf{j} + v_3(t)\mathbf{k}.$$

Then $\mathbf{v}(t)$ is continuous at t_0 if and only if its three components are continuous at t_0.

We now state the most important of these definitions.

Definition. Derivative of a vector function

A vector function v(t) is said to be **differentiable** at a point t if the limit

$$\mathbf{v}'(t) = \lim_{\Delta t \to 0} \frac{\mathbf{v}(t + \Delta t) - \mathbf{v}(t)}{\Delta t} \tag{9}$$

exists. The vector $\mathbf{v}'(t)$ is called the **derivative** of $\mathbf{v}(t)$. See Fig. 174. (The curve in this figure is the locus of the heads of the arrows representing **v** for values of the independent variable in some interval containing t and $t + \Delta t$.) ∎

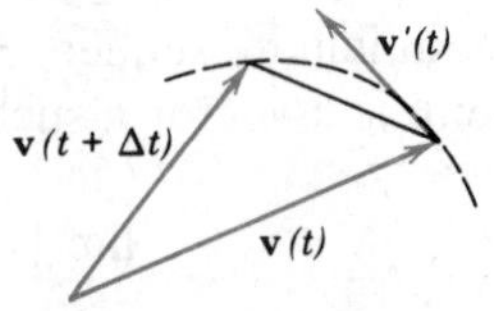

Fig. 174. Derivative of a vector function

[7]That is, in some interval (segment) on the t-axis containing t_0 as an interior point (not as an endpoint). In mathematics (as opposed to everyday language), a point (here, t_0) is always regarded as a point of any of its neighborhoods, by definition; this turns out to be practical.

In terms of components with respect to a given Cartesian coordinate system, $\mathbf{v}(t)$ is differentiable at a point t if and only if its three components are differentiable at t, and then *the derivative* $\mathbf{v}'(t)$ *is obtained by differentiating each component separately,*

$$\mathbf{v}'(t) = [v_1'(t), \quad v_2'(t), \quad v_3'(t)]. \tag{10}$$

It follows that the familiar rules of differentiation yield corresponding rules for differentiating vector functions, for example,

$$(c\mathbf{v})' = c\mathbf{v}' \qquad (c \text{ constant}),$$

$$(\mathbf{u} + \mathbf{v})' = \mathbf{u}' + \mathbf{v}'$$

and in particular

$$(\mathbf{u} \bullet \mathbf{v})' = \mathbf{u}' \bullet \mathbf{v} + \mathbf{u} \bullet \mathbf{v}' \tag{11}$$

$$(\mathbf{u} \times \mathbf{v})' = \mathbf{u}' \times \mathbf{v} + \mathbf{u} \times \mathbf{v}' \tag{12}$$

$$(\mathbf{u} \quad \mathbf{v} \quad \mathbf{w})' = (\mathbf{u}' \quad \mathbf{v} \quad \mathbf{w}) + (\mathbf{u} \quad \mathbf{v}' \quad \mathbf{w}) + (\mathbf{u} \quad \mathbf{v} \quad \mathbf{w}'). \tag{13}$$

The simple proofs are left to the reader. In (12), the order of the vectors must be carefully observed because cross multiplication is not commutative.

EXAMPLE 4 Derivative of a vector function of constant length

Let $\mathbf{v}(t)$ be a vector function whose length is constant, say, $|\mathbf{v}(t)| = c$. Then $|\mathbf{v}|^2 = \mathbf{v} \bullet \mathbf{v} = c^2$, and $(\mathbf{v} \bullet \mathbf{v})' = 2\mathbf{v} \bullet \mathbf{v}' = 0$, by differentiation [see (11)]. This yields the following result. *The derivative of a vector function* $\mathbf{v}(t)$ *of constant length is either the zero vector or is perpendicular to* $\mathbf{v}(t)$. ∎

Partial Derivatives of a Vector Function

From our present discussion we see that partial differentiation of vector functions depending on two or more variables can be introduced as follows. Suppose that the components of a vector function

$$\mathbf{v} = [v_1, \quad v_2, \quad v_3] = v_1\mathbf{i} + v_2\mathbf{j} + v_3\mathbf{k}$$

are differentiable functions of n variables $t_1, \cdots, t_n$. Then the **partial derivative** of $\mathbf{v}$ with respect to t_l is denoted by $\partial\mathbf{v}/\partial t_l$ and is defined as the vector function

$$\frac{\partial \mathbf{v}}{\partial t_l} = \frac{\partial v_1}{\partial t_l}\mathbf{i} + \frac{\partial v_2}{\partial t_l}\mathbf{j} + \frac{\partial v_3}{\partial t_l}\mathbf{k}.$$

Similarly,

$$\frac{\partial^2 \mathbf{v}}{\partial t_l \partial t_m} = \frac{\partial^2 v_1}{\partial t_l \partial t_m}\mathbf{i} + \frac{\partial^2 v_2}{\partial t_l \partial t_m}\mathbf{j} + \frac{\partial^2 v_3}{\partial t_l \partial t_m}\mathbf{k},$$

and so on.

EXAMPLE 5 Partial derivatives

Let $\mathbf{r}(t_1, t_2) = a \cos t_1\, \mathbf{i} + a \sin t_1\, \mathbf{j} + t_2\mathbf{k}$. Then

$$\frac{\partial \mathbf{r}}{\partial t_1} = -a \sin t_1\, \mathbf{i} + a \cos t_1\, \mathbf{j}, \qquad \frac{\partial \mathbf{r}}{\partial t_2} = \mathbf{k}.$$

Various physical and geometrical applications of derivatives of vector functions will be discussed in the next sections and in Chap. 9.

Problem Set 8.4

Scalar fields

Level curves. Determine the *level curves* $T(x, y) = const$ (curves of constant temperature or *isotherms*) of the temperature fields given by the following functions. Sketch some of these isotherms.

1. $T = x^2 - y^2$ **2.** $T = \ln(x^2 + y^2)$ **3.** $T = 3x - 4y$

4. $T = x^2 - y^2 + 4y$ **5.** $T = xy$ **6.** $T = \arctan(y/x)$

Consider the scalar field (pressure field) determined by $f(x, y) = x^2 + 4y^2$. Find:

7. The pressure at the points $(-1, 2)$, $(3, 8)$, and $(0, 5)$.

8. The level curves (curves of constant pressure or *isobars*). (Graph some of them.)

9. The pressure at the points of the parabola $y = 2x^2$.

10. The region in which $1 \leqq f(x, y) \leqq 4$.

Level surfaces. Find the *level surfaces* $f(x, y, z) = const$ of the scalar fields in space given by the following functions.

11. $f = x + y + z$ **12.** $f = x^2 + y^2 + z^2$ **13.** $f = x^2 + y^2$

14. $f = x^2 + 9y^2 + 25z^2$ **15.** $f = z - \sqrt{x^2 + y^2}$ **16.** $f = x^2 + y^2 - z$

Consider the pressure field in space given by $f(x, y, z) = 3x^2 + y^2 + z^2$. Find:

17. The level surfaces.

18. A formula for the pressure on the surface $xyz = 1$.

19. The level curves in the plane $z = 2$. (Graph some of them.)

20. The region in which $3 \leqq f(x, y, z) \leqq 12$.

Vector fields

Sketch figures (similar to Fig. 173) of the vector fields in the xy-plane given by the following vector functions.

21. $\mathbf{v} = \mathbf{i}$ **22.** $\mathbf{v} = \mathbf{i} - \mathbf{j}$

23. $\mathbf{v} = 2\mathbf{i} + 3x\mathbf{j}$ **24.** $\mathbf{v} = x\mathbf{i} + y\mathbf{j}$

25. $\mathbf{v} = x^2\mathbf{i} + y^2\mathbf{j}$ **26.** $\mathbf{v} = (x - y)\mathbf{i} + (x + y)\mathbf{j}$

Find and sketch the curves on which $\mathbf{v}$ has constant length and the curves on which $\mathbf{v}$ has constant direction, where

27. $\mathbf{v} = x\mathbf{i} + y\mathbf{j}$ **28.** $\mathbf{v} = x^2\mathbf{i} - y^2\mathbf{j}$ **29.** $\mathbf{v} = y\mathbf{i} + x^2\mathbf{j}$

Derivatives

Find the first and second derivatives of the following vector functions and the norms (lengths) of these derivatives.

30. $\mathbf{a} + \mathbf{b}t$ **31.** $[t, \ t^2, \ 0]$ **32.** $[\cos t, \ \sin t, \ 0]$

33. $[3\cos t, \ 2\sin t, \ 0]$ **34.** $[t, \ t^2, \ t^3]$ **35.** $[3\cos t, \ 3\sin t, \ 4t]$

36. $[\cos t, \ 4\sin t, \ t]$ **37.** $[\cos t, \ \sin 2t, \ 0]$ **38.** $[e^{-t}\cos t, e^{-t}\sin t, 0]$

Partial derivatives. Find the first partial derivatives with respect to x, y, z:

39. $[x, \ y, \ z]$ **40.** $[2, \ x^3y^2z, \ 3x]$

41. $[xy, \ yz, \ 0]$ **42.** $[e^x \sin y, \ e^x \cos y, \ e^z]$

43. $[y^2, \ z^2, \ x^2]$ **44.** $\sin xyz\,(\mathbf{i} + \mathbf{j})$

Derivatives of products. For $\mathbf{u} = [t^2, \ 0, \ -3t]$, $\mathbf{v} = [0, \ -4t^3, t + 1]$, and $\mathbf{w} = [3t^2, \ t, \ -t^3]$ verify

45. Formula (11) **46.** Formula (12) **47.** Formula (13)

48. Using (2), Sec. 8.2, prove (11). Prove (12).

49. Derive (13) from (11) and (12).

50. Show that $\left(\frac{1}{|\mathbf{v}|}\mathbf{v}\right)' = (\mathbf{v}\bullet\mathbf{v})^{-3/2}\,[\mathbf{v}'(\mathbf{v}\bullet\mathbf{v}) - \mathbf{v}(\mathbf{v}\bullet\mathbf{v}')]$.

8.5 Curves. Tangents. Arc Length

As an important application of vector calculus, let us now consider some basic facts about curves in space. The student will know that curves occur in various problems in calculus as well as in physics, for example, as paths of moving bodies. We mention that the study of curves and surfaces in space by means of calculus is an important branch of mathematics that is called **differential geometry.**

A Cartesian coordinate system being given, we may represent a curve C by a vector function (see Fig. 175 on p. 458)

$$\mathbf{r}(t) = [x(t), \ y(t), \ z(t)] = x(t)\mathbf{i} + y(t)\mathbf{j} + z(t)\mathbf{k}; \tag{1}$$

to each value t_0 of the real variable t there corresponds a point of C having the position vector $\mathbf{r}(t_0)$, that is, the coordinates $x(t_0)$, $y(t_0)$, $z(t_0)$.

A representation of the form (1) is called a **parametric representation** of the curve C, and t is called the *parameter* of this representation.

The sense of increasing values of t is called the **positive sense** on C, which we can indicate by the tip of an arrow as in Fig. 175, and the sense of decreasing t the *negative sense* on C. We say that in this way the function $\mathbf{r}(t)$ defines an **orientation** of C. Obviously, there are two ways of orienting a curve.

Parametric representations are useful in many applications, for instance, in mechanics, where t may be time, and the positive sense is the sense in which a body moves along a curve.

Another type of representation of a curve C in space is

$$y = f(x), \qquad z = g(x); \tag{2}$$

here, $y = f(x)$ is the projection of C into the xy-plane, and $z = g(x)$ is the projection of C into the xz-plane.

Finally, a third type is the representation of a curve in space as the intersection of two surfaces

$$F(x, y, z) = 0 \quad \text{and} \quad G(x, y, z) = 0. \tag{3}$$

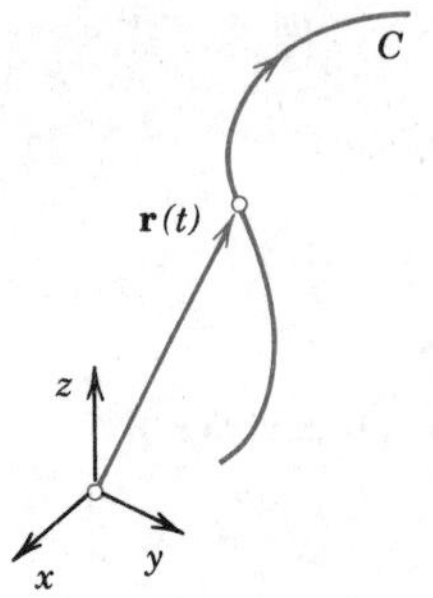

Fig. 175. Parametric representation of a curve

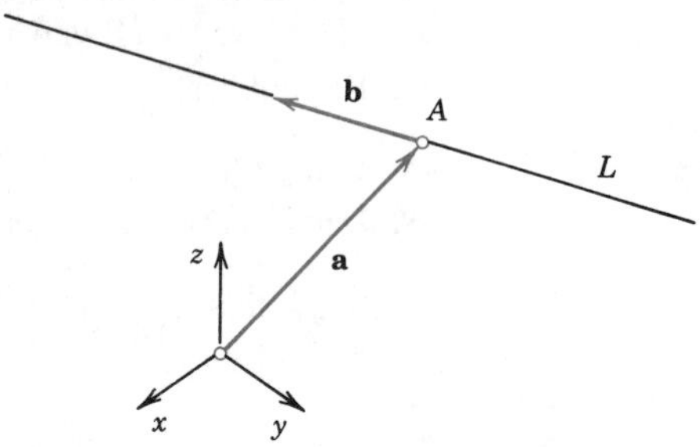

Fig. 176. Parametric representation of a straight line

EXAMPLE 1 **Straight line**

A straight line L through a point A with position vector **a** in the direction of a constant vector **b** (see Fig. 176) can be represented in the form

$$\mathbf{r}(t) = \mathbf{a} + t\mathbf{b} = [a_1 + tb_1, \;\; a_2 + tb_2, \;\; a_3 + tb_3]. \tag{4}$$

If **b** is a unit vector, its components are the **direction cosines** of L, and in this case, $|t|$ measures the distance of the points of L from A. For instance, the straight line in the xy-plane through A: (3, 2) having slope 1 is

$$\mathbf{r}(t) = [3, \;\; 2, \;\; 0] + t[1, \;\; 1, \;\; 0] = [3 + t, \;\; 2 + t, \;\; 0].$$

∎

EXAMPLE 2 **Ellipse, circle**

The vector function

$$\mathbf{r}(t) = [a \cos t, \;\; b \sin t, \;\; 0] = a \cos t \, \mathbf{i} + b \sin t \, \mathbf{j} \tag{5}$$

represents an ellipse in the xy-plane with center at the origin and principal axes in the direction of the x and y axes. In fact, since $\cos^2 t + \sin^2 t = 1$, we obtain from (5)

$$\frac{x^2}{a^2} + \frac{y^2}{b^2} = 1, \qquad z = 0.$$

If $b = a$, then (5) represents a *circle* of radius a. ∎

EXAMPLE 3 **Circular helix**

The twisted curve C represented by the vector function

$$\mathbf{r}(t) = [a \cos t, \;\; a \sin t, \;\; ct] = a \cos t \, \mathbf{i} + a \sin t \, \mathbf{j} + ct\mathbf{k} \qquad (c \neq 0) \tag{6}$$

is called a *circular helix*. It lies on the cylinder $x^2 + y^2 = a^2$. If $c > 0$, the helix is shaped like a right-handed screw (Fig. 177). If $c < 0$, it looks like a left-handed screw (Fig. 178). If $c = 0$, then (6) is a circle. ∎

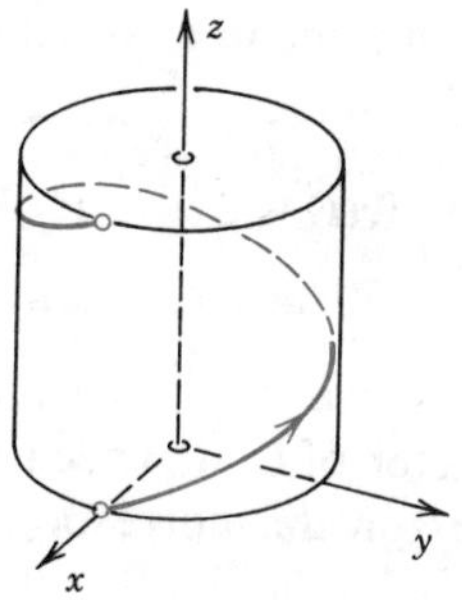

Fig. 177. Right-handed circular helix

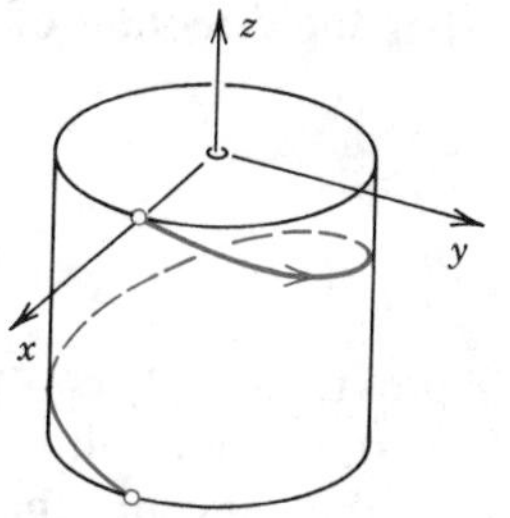

Fig. 178. Left-handed circular helix

A **plane curve** is a curve that lies in a plane in space. A curve that is not plane is called a **twisted curve.** For example, the helix is twisted.

A **simple curve** is a curve without **multiple points,** that is, without points at which the curve intersects or touches itself. Circle and helix are simple. Figure 179 shows curves that are not simple.

An **arc** of a curve is the portion between any two points of the curve. For simplicity, we say "curve" for curves as well as for arcs.

Fig. 179. Curves with multiple points

Comment on parameter. A curve C may be given by various vector functions. If C is given by (1) and we set $t = h(t^*)$, we obtain a new vector function $\tilde{\mathbf{r}}(t^*)$ representing C. In mechanics, when t is time, this means that we change the motion of a body in time without changing its path. *Example*. For the parabola $\mathbf{r}(t) = [t, \quad t^2]$, by setting $t = -2t^*$, we get the new representation $\tilde{\mathbf{r}}(t^*) = \mathbf{r}(-2t^*) = [-2t^*, \quad 4t^{*2}]$. ∎

The next idea is the approximation of a curve by straight lines, leading to tangents and to a definition of length. Tangents are straight lines touching a curve, as follows.

Tangent to a Curve

The **tangent** to a curve C at a point P of C is the limiting position of a straight line L through P and a point Q of C as Q approaches P along C. See Fig. 180.

If C is given by $\mathbf{r}(t)$, with P and Q corresponding to t and $t + \Delta t$, respectively, then the vector

$$\frac{1}{\Delta t}\,[\mathbf{r}(t + \Delta t) - \mathbf{r}(t)]$$

has the direction of L and in the limit becomes the derivative

(7) $$\mathbf{r}'(t) = \lim_{\Delta t \to 0} \frac{1}{\Delta t} [\mathbf{r}(t + \Delta t) - \mathbf{r}(t)],$$

provided $\mathbf{r}(t)$ is differentiable, as we shall assume from now on. If $\mathbf{r}'(t) \neq \mathbf{0}$, we call $\mathbf{r}'(t)$ a **tangent vector** of C at P because it has the direction of the tangent; the same is true for the corresponding unit vector, the **unit tangent vector** (see Fig. 180)

(8) $$\mathbf{u} = \frac{1}{|\mathbf{r}'|} \mathbf{r}'.$$

Note that both $\mathbf{r}'$ and $\mathbf{u}$ point in the direction of increasing t; so their sense depends on the orientation of C.

It is now easy to see that the **tangent** to C at P is given by

(9) $$\mathbf{q}(w) = \mathbf{r} + w\mathbf{r}' \qquad \text{(Fig. 181)},$$

the sum of the position vector $\mathbf{r}$ of P and a multiple of the tangent vector $\mathbf{r}'$ of C at P, both vectors depending on P; here the real variable w is the parameter in (9).

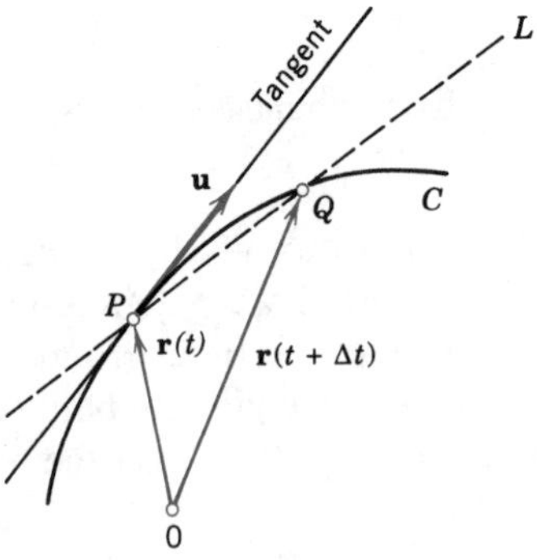

Fig. 180. Tangent to a curve

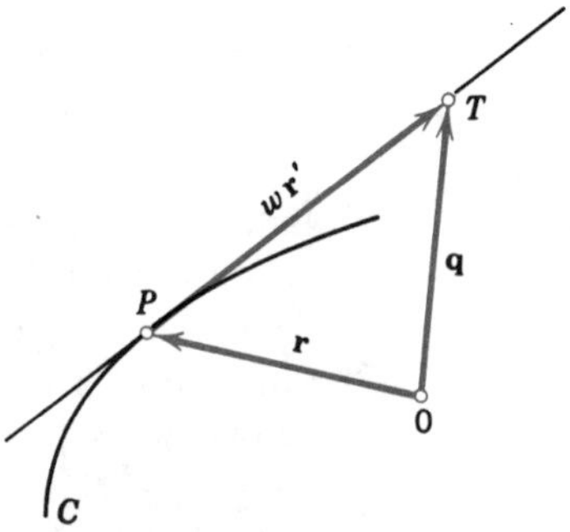

Fig. 181. Formula (9) for the tangent to a curve

EXAMPLE 4 Tangent to an ellipse

Find the tangent to the ellipse $\frac{1}{4}x^2 + y^2 = 1$ at P: $(\sqrt{2}, 1/\sqrt{2})$.

Solution. $\mathbf{r}(t) = 2 \cos t \, \mathbf{i} + \sin t \, \mathbf{j}$, hence $\mathbf{r}'(t) = -2 \sin t \, \mathbf{i} + \cos t \, \mathbf{j}$, and P corresponds to $t = \pi/4$, since $2 \cos (\pi/4) = \sqrt{2}$ and $\sin (\pi/4) = 1/\sqrt{2}$. Thus $\mathbf{r}'(\pi/4) = [-\sqrt{2}, 1/\sqrt{2}]$. *Answer:*

$$\mathbf{q}(w) = [\sqrt{2}, \ 1/\sqrt{2}] + w[-\sqrt{2}, \ 1/\sqrt{2}] = \sqrt{2}(1 - w)\mathbf{i} + (1/\sqrt{2})(1 + w)\mathbf{j}.$$

To check the result, sketch the ellipse and the tangent. ∎

Length of a Curve

We are now ready to define the length l of a curve C as the limit of the lengths of broken lines of n chords as in Fig. 182 (where $n = 5$) with larger and larger n, as follows. Let $\mathbf{r}(t)$, $a \leqq t \leqq b$, represent C. For each $n = 1, 2, \cdots$ we subdivide ("partition") the interval $a \leqq t \leqq b$ by points

$$t_0 \; (= a), \quad t_1, \quad \cdots, \quad t_{n-1}, \quad t_n \; (= b),$$

where $t_0 < t_1 < \cdots < t_n$. This gives a broken line of chords with endpoints $\mathbf{r}(t_0), \cdots, \mathbf{r}(t_n)$. This we do arbitrarily but so that the greatest $|\Delta t_m| = |t_m - t_{m-1}|$ approaches 0 as $n \to \infty$. The lengths $l_1, l_2, \cdots$ of these lines of chords can be obtained from the Pythagorean theorem. If $\mathbf{r}(t)$ has a continuous derivative $\mathbf{r}'(t)$, it can be shown that the sequence $l_1, l_2, \cdots$ has a limit l, which is independent of the particular choice of the representation of C and of the choice of subdivisions, and is given by the integral

$$l = \int_a^b \sqrt{\mathbf{r}' \cdot \mathbf{r}'} \, dt \qquad \left(\mathbf{r}' = \frac{d\mathbf{r}}{dt}\right). \tag{10}$$

This is called the **length** of C, and C is called **rectifiable.** The proof is similar to that for plane curves given in calculus (see Ref. [12]) and can be found in Ref. [B5] listed in Appendix 1. The practical evaluation of the integral (10) will be difficult in general; some simple cases are given in the problem set.

Fig. 182. Length of a curve

Arc Length s of a Curve

The length l of a curve C is a constant, a positive number, but if we replace the fixed upper limit b in (10) with a variable upper limit t, the integral becomes a function of t, commonly denoted by $s(t)$ and called the *arc length function* or, simply, the **arc length** of C, given by

$$s(t) = \int_a^t \sqrt{\mathbf{r}' \cdot \mathbf{r}'} \, d\tilde{t} \qquad \left(\mathbf{r}' = \frac{d\mathbf{r}}{d\tilde{t}}\right). \tag{11}$$

(Here, $\tilde{t}$ is the variable of integration because t is used in the upper limit.) Geometrically, $s(t_0)$ with some $t_0 > a$ is the length of the arc of C between the points with parametric values a and t_0. The choice of a (the point $s = 0$) is arbitrary; changing a means changing s by a constant.

Linear element ds. If we differentiate (11) and square, we have

$$\left(\frac{ds}{dt}\right)^2 = \frac{d\mathbf{r}}{dt}\bullet\frac{d\mathbf{r}}{dt} = |\mathbf{r}'(t)|^2 = \left(\frac{dx}{dt}\right)^2 + \left(\frac{dy}{dt}\right)^2 + \left(\frac{dz}{dt}\right)^2. \tag{12}$$

It is customary to write

$$d\mathbf{r} = [dx,\ dy,\ dz] = dx\,\mathbf{i} + dy\,\mathbf{j} + dz\,\mathbf{k} \tag{13*}$$

and

$$ds^2 = d\mathbf{r}\bullet d\mathbf{r} = dx^2 + dy^2 + dz^2. \tag{13}$$

ds is called the **linear element** of C.

Arc length s as parameter in representations of curves. The use of the special $t = s$ (the arc length) in representations (1) of curves simplifies various formulas. A first instance of this occurs for the unit tangent vector, namely,

$$\mathbf{u}(s) = \mathbf{r}'(s), \tag{14}$$

as follows from (12) with $t = s$ because $ds/ds = 1$. Even greater simplifications occur in curvature and torsion (Sec. 8.7).

EXAMPLE 5 Circular helix. Circle. Arc length as parameter

For the helix in Example 3,

$$\mathbf{r}(t) = [a\cos t,\quad a\sin t,\quad ct] = a\cos t\,\mathbf{i} + a\sin t\,\mathbf{j} + ct\mathbf{k},$$

we get

$$\mathbf{r}'(t) = [-a\sin t,\quad a\cos t,\quad c] = -a\sin t\,\mathbf{i} + a\cos t\,\mathbf{j} + c\mathbf{k},$$

From this, $\mathbf{r}'\bullet\mathbf{r}' = a^2 + c^2$, so that (11) gives

$$s = \int_0^t \sqrt{a^2 + c^2}\,d\tilde{t} = t\sqrt{a^2 + c^2}.$$

Hence $t = s/\sqrt{a^2 + c^2}$, and a formula for the helix with the arc length s as parameter is

$$\mathbf{r}^*(s) = \mathbf{r}\left(\frac{s}{\sqrt{a^2 + c^2}}\right) = a\cos\frac{s}{\sqrt{a^2 + c^2}}\,\mathbf{i} + a\sin\frac{s}{\sqrt{a^2 + c^2}}\,\mathbf{j} + \frac{cs}{\sqrt{a^2 + c^2}}\,\mathbf{k}.$$

Setting $c = 0$, we have $t = s/a$ and obtain for a circle of radius a the representation

$$\mathbf{r}\left(\frac{s}{a}\right) = \left[a\cos\frac{s}{a},\quad a\sin\frac{s}{a}\right] = a\cos\frac{s}{a}\,\mathbf{i} + a\sin\frac{s}{a}\,\mathbf{j}.$$

The circle is oriented in the counterclockwise sense, which corresponds to increasing values of s. Setting $s = -s^*$ and using $\cos(-\alpha) = \cos\alpha$ and $\sin(-\alpha) = -\sin\alpha$, we obtain

$$\mathbf{r}\left(-\frac{s^*}{a}\right) = \left[a\cos\frac{s^*}{a},\quad -a\sin\frac{s^*}{a}\right] = a\cos\frac{s^*}{a}\,\mathbf{i} - a\sin\frac{s^*}{a}\,\mathbf{j};$$

we have $ds/ds^* = -1 < 0$, and the circle is now oriented clockwise. ∎

Problem Set 8.5

Representations of curves

Straight lines. Find a parametric representation of the straight lines through a point A in the direction of a vector $\mathbf{b}$, where

1. A: $(1, 1, 0)$, $\mathbf{b} = \mathbf{k}$

2. A: $(2, 3, 1)$, $\mathbf{b} = [1, -1, 2]$

3. A: $(-3, 1, -2)$, $\mathbf{b} = [3, 0, -1]$

4. A: $(3, 4, 1)$, $\mathbf{b} = 3\mathbf{i} + 5\mathbf{j} - \mathbf{k}$

Find a parametric representation of the straight line through the points A and B, where

5. A: $(0, 0, 0)$, B: $(1, 1, 1)$

6. A: $(1, 4, 2)$, B: $(1, 4, -2)$

7. A: $(3, -1, 5)$, B: $(5, -5, 5)$

8. A: $(0, 2, 0)$, B: $(1, 5, -2)$

9. A: $(1, 4, -2)$, B: $(2, 2, 3)$

10. A: $(9, 3, 8)$, B: $(0, 3, 10\frac{1}{2})$

Find a parametric representation of the straight line represented by

11. $y = x, \quad z = 0$

12. $x = y, \quad y = z$

13. $x + y + z = 1, \quad y - z = 0$

14. $x + y - z = 2, \quad 2x - 5y + z = 3$

General curves. What curves are represented by the following parametric representations?

15. $[t, \quad 1/t, \quad 0]$

16. $[0, \quad t^2, \quad t]$

17. $\cos t\,\mathbf{j} + (2 + 2\sin t)\mathbf{k}$

18. $[\cos 3t, \quad \sin 3t, \quad -t]$

19. $\cosh t\,\mathbf{i} + 4\sinh t\,\mathbf{j}$

20. $4\mathbf{i} + (5 + 3\cos t)\mathbf{j} + (2 + 3\sin t)\mathbf{k}$

Represent the following curves in parametric form.

21. $x^2 + y^2 = 1, \quad z = 0$

22. $(x + 2)^2 + (y - 2)^2 = 4, \quad z = 6$

23. $\frac{1}{4}x^2 + \frac{1}{16}y^2 = 1, \quad z = 1$

24. $y^2 + 25z^2 = 25, \quad x = -2$

25. $4x^2 - 9y^2 = 36, \quad z = 0$

26. $x^2 + y^2 = 4, \quad z = 3 \arctan(y/x)$

27. $(x - 1)^2 + 4(y + 2)^2 = 4, \quad z = \frac{1}{2}$

28. $(y - 5)(z + 5) = 1, \quad x = 1$

29. Set $t = -t^*$ in $\mathbf{r}(t) = \cos t\,\mathbf{i} + \sin t\,\mathbf{j}$ and show that the sense of increasing t^* is the clockwise sense on the circle.

30. If in $\mathbf{r}(t) = (1 - 4t)\mathbf{i} + (3 + 2t)\mathbf{j}$ you set $t = e^{\tau}$, does the resulting representation represent the entire line?

Tangent vectors, tangents

Given a curve C: $\mathbf{r}(t)$, find (a) a tangent vector $\mathbf{r}'(t)$ and the corresponding unit tangent vector $\mathbf{u}(t)$, (b) $\mathbf{r}'$ and $\mathbf{u}$ at the given point P, and (c) the tangent at P.

31. $\mathbf{r}(t) = t\,\mathbf{i} + t^2\mathbf{j}, \quad P: (1, 1, 0)$

32. $\mathbf{r}(t) = \cos t\,\mathbf{i} + \sin t\,\mathbf{j}, \quad P: (1/\sqrt{2}, 1/\sqrt{2})$

33. $\mathbf{r}(t) = t\mathbf{i} + t^3\mathbf{j}, \quad P: (2, 8, 0)$

34. $\mathbf{r}(t) = \cos t\,\mathbf{i} + \sin t\,\mathbf{j} + t\mathbf{k}, \quad P: (1, 0, 0)$

35. $\mathbf{r}(t) = 2\cos t\,\mathbf{i} + \sin t\,\mathbf{j}, \quad P: (\sqrt{2}, 1/\sqrt{2}, 0)$

36. $\mathbf{r}(t) = \cosh t\,\mathbf{i} + \sinh t\,\mathbf{j}, \quad P: (\frac{5}{3}, \frac{4}{3}, 0)$

37. $\mathbf{r}(t) = 3\cos t\,\mathbf{i} + 3\sin t\,\mathbf{j} + 4t\mathbf{k}, \quad P: (0, 3, 2\pi)$

38. $\mathbf{r}(t) = t\mathbf{i} + t^2\mathbf{j} + t^3\mathbf{k}, \quad P: (1, 1, 1)$

Length of a curve

Find the lengths of the following curves. Sketch the curves.

39. **Catenary** $\mathbf{r}(t) = t\mathbf{i} + \cosh t\,\mathbf{j}$ from $t = 0$ to $t = 1$

40. **Circular helix** $\mathbf{r}(t) = a \cos t\,\mathbf{i} + a \sin t\,\mathbf{j} + ct\mathbf{k}$ from $(a, 0, 0)$ to $(a, 0, 2\pi c)$

41. **Semicubical parabola** $\mathbf{r}(t) = t\mathbf{i} + t^{3/2}\mathbf{j}$ from $(0, 0, 0)$ to $(4, 8, 0)$

42. Four-cusped **hypocycloid** $\mathbf{r}(t) = a \cos^3 t\,\mathbf{i} + a \sin^3 t\,\mathbf{j}$, total length

43. If a plane curve is represented in the form $y = f(x)$, $z = 0$, using (10) show that its length between $x = a$ and $x = b$ is

$$l = \int_a^b \sqrt{1 + y'^2}\, dx.$$

44. Show that if a plane curve is represented in polar coordinates $\rho = \sqrt{x^2 + y^2}$ and $\theta = \arctan (y/x)$, then $ds^2 = \rho^2\, d\theta^2 + d\rho^2$ and

$$l = \int_\alpha^\beta \sqrt{\rho^2 + \rho'^2}\, d\theta \qquad (\rho' = d\rho/d\theta).$$

Using this formula and (10) in Appendix A3.1, find the total length of the **cardioid** $\rho = a(1 - \cos \theta)$. Sketch this curve.

8.6 Velocity and Acceleration

Curves as just discussed from the viewpoint of geometry also play an important role in mechanics as paths of moving bodies. To see this, we consider a path C (a curve) given by $\mathbf{r}(t)$, where t now is ***time,*** and we show that the derivative $\mathbf{r}'(t)$, interpreted geometrically in the last section, is also basic in mechanics, and so is the second derivative $\mathbf{r}''(t)$.

From Sec. 8.5 we know that the vector

$$\mathbf{v} = \mathbf{r}' = \frac{d\mathbf{r}}{dt} \tag{1}$$

is tangent to C and, therefore, points in the instantaneous direction of motion of the moving body P. From (12) in Sec. 8.5 we see that this vector has the length

$$|\mathbf{v}| = \sqrt{\mathbf{r}' \cdot \mathbf{r}'} = \frac{ds}{dt},$$

where s is the arc length, which measures the distance of P from a fixed point ($s = 0$) on C along the curve. Hence ds/dt is the **speed** of P. The vector $\mathbf{v}$ is, therefore, called the **velocity vector**[8] of the motion.

The derivative of the velocity vector is called the **acceleration vector** and will be denoted by $\mathbf{a}$; thus

$$\mathbf{a}(t) = \mathbf{v}'(t) = \mathbf{r}''(t). \tag{2}$$

[8]When no confusion is likely to arise, the speed $|\mathbf{v}|$ is also called the **velocity**.

EXAMPLE 1 Centripetal acceleration

The vector function

$$\mathbf{r}(t) = R \cos \omega t \,\mathbf{i} + R \sin \omega t \,\mathbf{j} \qquad (\omega > 0)$$

represents a circle C of radius R with center at the origin of the xy-plane and describes a motion of a particle P in the counterclockwise sense. The velocity vector

$$\mathbf{v} = \mathbf{r}' = -R\omega \sin \omega t \,\mathbf{i} + R\omega \cos \omega t \,\mathbf{j}$$

(see Fig. 183) is tangent to C, and its magnitude, the speed

$$|\mathbf{v}| = \sqrt{\mathbf{r}' \cdot \mathbf{r}'} = R\omega$$

is constant. The **angular speed** (speed divided by the distance R from the center) is equal to ω. The acceleration vector is

$$\mathbf{a} = \mathbf{v}' = -R\omega^2 \cos \omega t \,\mathbf{i} - R\omega^2 \sin \omega t \,\mathbf{j} = -\omega^2 \mathbf{r}. \tag{3}$$

We see that there is an acceleration of constant magnitude $|\mathbf{a}| = \omega^2|\mathbf{r}| = \omega^2 R$ toward the origin. $\mathbf{a}$ is called the **centripetal acceleration.** It results from the fact that the velocity vector is changing direction at a constant rate. The **centripetal force** is $m\mathbf{a}$, where m is the mass of P. The opposite vector $-m\mathbf{a}$ is called the **centrifugal force,** and the two forces are in equilibrium at each instant of the motion. ∎

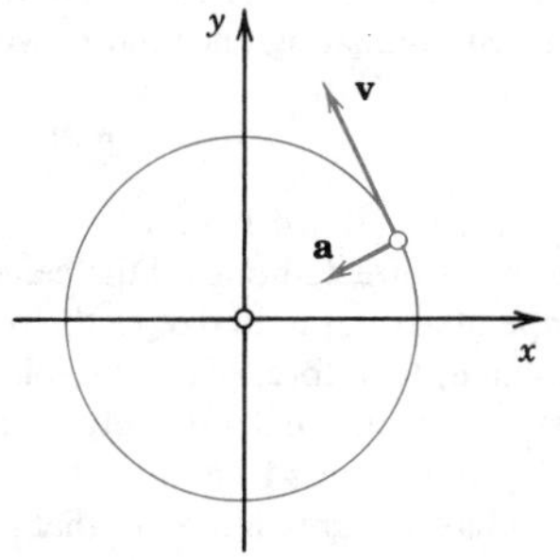

Fig. 183. Centripetal acceleration

Tangential Acceleration and Normal Acceleration

It is clear that $\mathbf{a}$ is the time rate of change of $\mathbf{v}$. In Example 1 we have $|\mathbf{v}| = const$, but $|\mathbf{a}| \neq 0$. This illustrates that the magnitude of $\mathbf{a}$ is not in general the rate of change of $|\mathbf{v}|$. The reason is that, in general, $\mathbf{a}$ is not tangent to the path C. In fact, by applying the chain rule to (1) we have

$$\mathbf{v}(t) = \frac{d\mathbf{r}}{dt} = \frac{d\mathbf{r}}{ds}\frac{ds}{dt} = \mathbf{u}(s)\frac{ds}{dt},$$

where $\mathbf{u}(s) = d\mathbf{r}/ds$ is the unit tangent vector of C (Sec. 8.5), and, by differentiating this again,

$$\mathbf{a}(t) = \frac{d\mathbf{v}}{dt} = \frac{d}{dt}\left(\mathbf{u}(s)\frac{ds}{dt}\right) = \frac{d\mathbf{u}}{ds}\left(\frac{ds}{dt}\right)^2 + \mathbf{u}(s)\frac{d^2s}{dt^2}. \tag{4}$$

Now $\mathbf{u}(s)$ is tangent to C and of constant length (one), so that $d\mathbf{u}/ds$ is perpendicular to $\mathbf{u}(s)$ (recall Example 4 in Sec. 8.4). Hence (4) is a decomposition of the acceleration vector into its normal component $(d\mathbf{u}/ds)(ds/dt)^2$, called the **normal acceleration,** and its tangential component $\mathbf{u}(s)d^2s/dt^2$, called the **tangential acceleration.** (Examples below.) From this we see that if and only if the normal acceleration is zero, $|\mathbf{a}|$ equals the time rate of change of $|\mathbf{v}| = ds/dt$ (except for the sign), because then we have $|\mathbf{a}| = |\mathbf{u}(s)|\,|d^2s/dt^2| = |d^2s/dt^2|$ from (4).

EXAMPLE 2 **Coriolis acceleration**[9]

A particle P moves on a disk toward the edge, the position vector being

$$\mathbf{r}(t) = t\mathbf{b} \tag{5}$$

where $\mathbf{b}$ is a unit vector, rotating together with the disk with constant angular speed ω in the counterclockwise sense (Fig. 184). Find the acceleration $\mathbf{a}$ of P.

Solution. Because of the rotation, $\mathbf{b}$ is of the form

$$\mathbf{b}(t) = \cos \omega t\ \mathbf{i} + \sin \omega t\ \mathbf{j}. \tag{6}$$

Differentiating (5) with respect to t, we obtain the velocity

$$\mathbf{v} = \mathbf{r}' = \mathbf{b} + t\mathbf{b}'. \tag{7}$$

Obviously $\mathbf{b}$ is the velocity of P relative to the disk, and $t\mathbf{b}'$ is the additional velocity due to the rotation. Differentiating once more, we obtain the acceleration

$$\mathbf{a} = \mathbf{v}' = 2\mathbf{b}' + t\mathbf{b}''. \tag{8}$$

In the last term of (8) we have $\mathbf{b}'' = -\omega^2\mathbf{b}$, as follows by differentiating (6). Hence this acceleration $t\mathbf{b}''$ is directed toward the center of the disk, and from Example 1 we see that this is the centripetal acceleration due to the rotation. In fact, the distance of P from the center is equal to t, which, therefore, plays the role of R in Example 1.

The most interesting and probably unexpected term in (8) is $2\mathbf{b}'$, the so-called **Coriolis acceleration,** which results from the interaction of the rotation of the disk and the motion of P on the disk. It has the direction of $\mathbf{b}'$; that is, it is tangential to the edge of the disk and, referred to the fixed xy-coordinate system, it points in the direction of the rotation. If P is a person of mass m_0 walking on the disk according to (5), then P will feel a force $-2m_0\mathbf{b}'$ in the opposite direction, that is, against the sense of the rotation. ∎

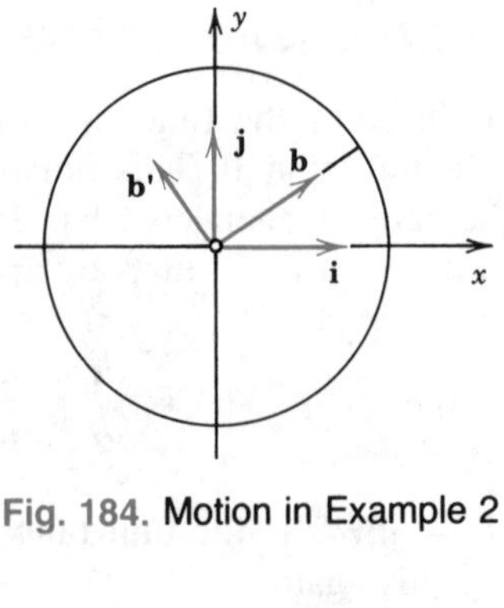

Fig. 184. Motion in Example 2

[9]GUSTAVE GASPARD CORIOLIS (1792—1843), French engineer, who did research in mechanics.

EXAMPLE 3 **Superposition of two rotations, Coriolis acceleration**

Find the acceleration of a projectile P moving along a "meridian" M of a rotating sphere (for instance, the surface of our earth) with constant speed relative to the sphere, which also rotates with constant angular speed.

Solution. The motion of P on M can be described in the form

$$\mathbf{r}(t) = R\cos\gamma t\,\mathbf{b} + R\sin\gamma t\,\mathbf{k} \tag{9}$$

where R is the radius of the sphere, γ (> 0) the angular speed of P on M, $\mathbf{b}$ a horizontal unit vector in the plane of M (Fig. 185), and $\mathbf{k}$ the unit vector in the positive z-direction. Since $\mathbf{b}$ rotates together with the sphere, it is of the form

$$\mathbf{b} = \cos\omega t\,\mathbf{i} + \sin\omega t\,\mathbf{j} \tag{10}$$

where ω (> 0) is the angular speed of the sphere and $\mathbf{i}$ and $\mathbf{j}$ are the unit vectors in the positive x- and y-directions, which are fixed in space. By differentiating (9) we obtain the velocity

$$\mathbf{v} = \mathbf{r}' = R\cos\gamma t\,\mathbf{b}' - \gamma R\sin\gamma t\,\mathbf{b} + \gamma R\cos\gamma t\,\mathbf{k}. \tag{11}$$

By differentiating this again with respect to t we obtain the acceleration

$$\mathbf{a} = \mathbf{v}' = R\cos\gamma t\,\mathbf{b}'' - 2\gamma R\sin\gamma t\,\mathbf{b}' - \gamma^2 R\cos\gamma t\,\mathbf{b} - \gamma^2 R\sin\gamma t\,\mathbf{k}, \tag{12}$$

where, by (10),

$$\mathbf{b}' = -\omega\sin\omega t\,\mathbf{i} + \omega\cos\omega t\,\mathbf{j},$$

$$\mathbf{b}'' = -\,\omega^2\cos\omega t\,\mathbf{i} - \omega^2\sin\omega t\,\mathbf{j} = -\omega^2\mathbf{b}.$$

From (9) we see that the sum of the last two terms in (12) is equal to $-\gamma^2\mathbf{r}$, and (12) becomes

$$\mathbf{a} = -\omega^2 R\cos\gamma t\,\mathbf{b} - 2\gamma R\sin\gamma t\,\mathbf{b}' - \gamma^2\mathbf{r}. \tag{13}$$

The first term on the right is the centripetal acceleration caused by the rotation of the sphere, and the last term is the centripetal acceleration resulting from the rotation of P on M. The second term is the **Coriolis acceleration**

$$\mathbf{a}_c = -2\gamma R\sin\gamma t\,\mathbf{b}'. \tag{14}$$

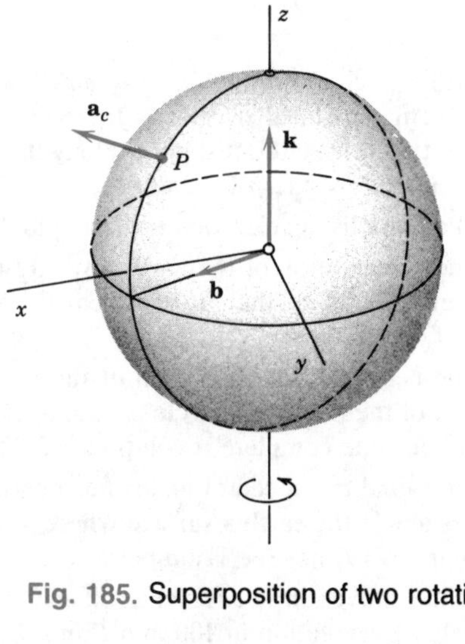

Fig. 185. Superposition of two rotations

On the "Northern Hemisphere," $\sin \gamma t > 0$ [see (9)] and because of the minus sign, $\mathbf{a}_c$ is directed opposite to $\mathbf{b}'$, that is, tangential to the surface of the sphere, perpendicular to M, and opposite to the rotation of the sphere. Its magnitude $2\gamma R\,|\sin \gamma t|\,\omega$ is maximum at the "North Pole" and zero at the equator. If the projectile P has mass m_0, it experiences a force $-m_0\mathbf{a}_c$, opposite to $m_0\mathbf{a}_c$; this is quite similar to the force in Example 2. This force tends to let the projectile deviate from the path M to the right. On the "Southern Hemisphere," $\sin \gamma t < 0$ and that force acts in the opposite direction, tending to let the projectile deviate from M to the left. This effect can be observed in connection with missiles, rockets, and shells. The flow of air toward an area of low pressure also shows those deviations. ∎

Problem Set 8.6

Let $\mathbf{r}(t)$ be the position vector of a moving particle where t ($\geqq 0$) is time. Describe the geometric shape of the path and find the velocity vector, the speed, and the acceleration vector.

1. $\mathbf{r}(t) = 3t\mathbf{i}$

2. $\mathbf{r}(t) = 4t\mathbf{i} - 4t\mathbf{j} + 2t\mathbf{k}$

3. $\mathbf{r}(t) = 4t^2\mathbf{k}$

4. $\mathbf{r}(t) = (1 + t^3)\mathbf{i} + 2t^3\mathbf{j} + (2 - t^3)\mathbf{k}$

5. $\mathbf{r}(t) = 3\cos 2t\,\mathbf{i} + 3\sin 2t\,\mathbf{j}$

6. $\mathbf{r}(t) = \cos t\,\mathbf{i} + \sin t\,\mathbf{j} + t\mathbf{k}$

7. $\mathbf{r}(t) = e^t\mathbf{i} + e^{-t}\mathbf{j}$

8. $\mathbf{r}(t) = \cos t^2\,\mathbf{i} + \sin t^2\,\mathbf{j}$

9. $\mathbf{r}(t) = \sin t\,\mathbf{j}$

10. $\mathbf{r}(t) = 3\cos 2t\,\mathbf{i} + 2\sin 3t\,\mathbf{j}$

11. Find the motion for which the acceleration vector is constant.

12. Obtain (3) from $\mathbf{v} = \mathbf{w} \times \mathbf{r}$ [(9), Sec. 8.3] by differentiation.

13. Find (a) the velocity vector $\mathbf{v}(t)$ and the acceleration vector $\mathbf{a}(t)$ of the motion $\mathbf{r}(t) = \sin t\,\mathbf{j}$, (b) the points at which $\mathbf{v}(t) = \mathbf{0}$, and (c) the points at which $|\mathbf{a}(t)|$ is maximum.

14. Find the tangential acceleration and the normal acceleration of the motion given by $\mathbf{r}(t) = t\mathbf{i} - t^2\mathbf{j}$.

15. Let a motion be given by $\mathbf{r}(t) = \cos t\,\mathbf{i} + 2\sin t\,\mathbf{j}$. Find the tangential acceleration.

16. In Prob. 15, find the points of maximum and minimum speed and acceleration. (Guess first.)

17. **(Cycloid)** Sketch $\mathbf{r}(t) = (R\sin \omega t + \omega Rt)\mathbf{i} + (R\cos \omega t + R)\mathbf{j}$, taking $R = 1$ and $\omega = 1$. This so-called *cycloid* is the path of a point on the rim of a wheel of radius R that rolls without slipping along the x-axis. Find $\mathbf{v}$ and $\mathbf{a}$ at the maximum and minimum y-values of the curve.

18. Find the Coriolis acceleration in Example 2 with (5) replaced by $\mathbf{r} = t^2\mathbf{b}$.

19. Find the acceleration of the earth toward the sun from (3) and the fact that the earth revolves about the sun in a nearly circular orbit with an almost constant speed of 30 km/sec.

20. Find the centripetal acceleration of the moon toward the earth, assuming that the orbit of the moon is a circle of radius 239,000 miles $= 3.85 \cdot 10^8$ meters and the time for one complete revolution is 27.3 days $= 2.36 \cdot 10^6$ sec.

21. **(Satellite)** Find the speed of an artificial earth satellite traveling at an altitude of 80 miles above the earth's surface where $g = 31$ ft/sec^2. (The radius of the earth is 3960 miles.) *Hint.* See Example 1.

22. A satellite moves in a circular orbit 450 miles above the earth's surface and completes 1 revolution in 100 min. Find the acceleration of gravity at the orbit from these data and from the radius of the earth (3960 miles).

8.7 Curvature and Torsion of a Curve *Optional*

Together with Sec. 8.5, this section gives the foundations of the theory of space curves, but we shall not need the material included here, so that we can leave this section optional.

The **curvature** $\kappa(s)$ of a curve C, represented by $\mathbf{r}(s)$ with arc length s as parameter, is defined by

$$\kappa(s) = |\mathbf{u}'(s)| = |\mathbf{r}''(s)| \qquad (' = d/ds). \tag{1}$$

Here $\mathbf{u}(s) = \mathbf{r}'(s)$ is the unit tangent vector of C (Sec. 8.5), and we have to assume that $\mathbf{r}(s)$ is twice differentiable, so that $\mathbf{r}''(s)$ exists.

Since κ is the length of the rate of change of the unit tangent vector of C with s, we see that κ measures the deviation of C from the tangent at each point of C. Examples are given below and in the problem set. For general parameters t, the formula for the curvature is complicated (Prob. 2).

If $\kappa(s) \neq 0$ (hence > 0), the unit vector $\mathbf{p}(s)$ in the direction of $\mathbf{u}'(s)$ is

$$\mathbf{p} = \frac{1}{\kappa}\mathbf{u}' \qquad (\kappa > 0) \tag{2}$$

and is called the **unit principal normal vector** of C. From Example 4 in Sec. 8.4 we see that $\mathbf{p}$ is perpendicular to $\mathbf{u}$. The vector

$$\mathbf{b} = \mathbf{u} \times \mathbf{p} \qquad (\kappa > 0) \tag{3}$$

is called the **unit binormal vector** of C. From the definition of a vector product it follows that $\mathbf{u}$, $\mathbf{p}$, $\mathbf{b}$ constitute a right-handed triple of orthogonal unit vectors (Sec. 8.3). This triple is called the **trihedron** of C at the point under consideration (Fig. 186). The three straight lines through that point in the

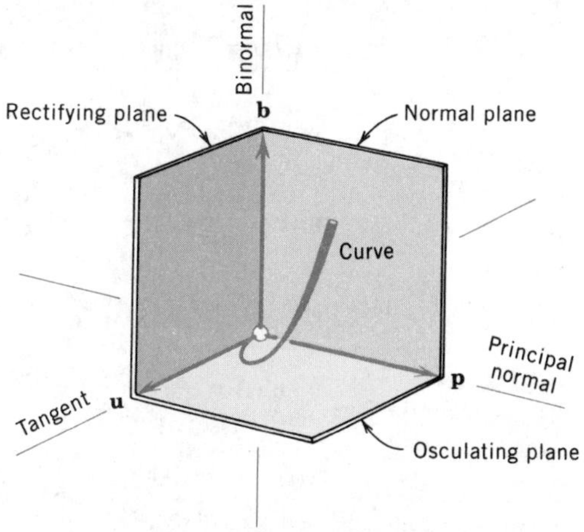

Fig. 186. Trihedron

directions of **u, p, b** are called the **tangent,** the **principal normal,** and the **binormal** of C. Figure 186 also shows the names of the three planes spanned by each pair of those vectors.

Torsion of a Curve

To introduce the torsion $\tau(s)$ of a curve C represented by $\mathbf{r}(s)$, we consider the derivative $\mathbf{b}' = d\mathbf{b}/ds$ of the unit binormal vector $\mathbf{b}$ of C. Here we assume that $\mathbf{r}(s)$ is three times differentiable, so that $\mathbf{b}'$ exists. If $\mathbf{b}'(s) \neq \mathbf{0}$, it is perpendicular to $\mathbf{b}$ (see Example 4 in Sec. 8.4). We show that $\mathbf{b}'$ is also perpendicular to $\mathbf{u}$. In fact, by differentiating $\mathbf{b} \cdot \mathbf{u} = 0$ we have $\mathbf{b}' \cdot \mathbf{u} + \mathbf{b} \cdot \mathbf{u}' = 0$; hence $\mathbf{b}' \cdot \mathbf{u} = 0$ because $\mathbf{b} \cdot \mathbf{u}' = (\mathbf{u} \times \mathbf{p}) \cdot \kappa \mathbf{p} = 0$ by (3) and (2). Consequently, $\mathbf{b}'$ is of the form $\mathbf{b}' = \alpha \mathbf{p}$ where α is a scalar. It is customary to set $\alpha = -\tau$. Then

$$\mathbf{b}' = -\tau \mathbf{p} \qquad (\kappa > 0). \tag{4}$$

The scalar function τ is called the **torsion** of C. Taking the dot product by $\mathbf{p}$ on both sides of (4), we obtain

$$\tau(s) = -\mathbf{p}(s) \cdot \mathbf{b}'(s). \tag{5}$$

The concepts just introduced are basic in the theory and application of curves. Let us illustrate them by a typical example. Further applications are given in the problem set.

EXAMPLE 1 Circular helix. Circle

In the case of the circular helix (6) in Sec. 8.5 we obtain the arc length $s = t\sqrt{a^2 + c^2}$; see Example 5 in Sec. 8.5. Hence we may represent the helix in the form

$$\mathbf{r}(s) = a \cos \frac{s}{K}\,\mathbf{i} + a \sin \frac{s}{K}\,\mathbf{j} + c\,\frac{s}{K}\,\mathbf{k} \qquad \text{where} \qquad K = \sqrt{a^2 + c^2}.$$

It follows that

$$\mathbf{u}(s) = \mathbf{r}'(s) = -\frac{a}{K} \sin \frac{s}{K}\,\mathbf{i} + \frac{a}{K} \cos \frac{s}{K}\,\mathbf{j} + \frac{c}{K}\,\mathbf{k}$$

$$\mathbf{r}''(s) = -\frac{a}{K^2} \cos \frac{s}{K}\,\mathbf{i} - \frac{a}{K^2} \sin \frac{s}{K}\,\mathbf{j}$$

$$\kappa = |\mathbf{r}''| = \sqrt{\mathbf{r}'' \cdot \mathbf{r}''} = \frac{a}{K^2} = \frac{a}{a^2 + c^2}$$

$$\mathbf{p}(s) = \frac{1}{\kappa(s)}\,\mathbf{r}''(s) = -\cos \frac{s}{K}\,\mathbf{i} - \sin \frac{s}{K}\,\mathbf{j}$$

$$\mathbf{b}(s) = \mathbf{u}(s) \times \mathbf{p}(s) = \frac{c}{K} \sin \frac{s}{K}\,\mathbf{i} - \frac{c}{K} \cos \frac{s}{K}\,\mathbf{j} + \frac{a}{K}\,\mathbf{k}$$

$$\mathbf{b}'(s) = \frac{c}{K^2} \cos \frac{s}{K}\,\mathbf{i} + \frac{c}{K^2} \sin \frac{s}{K}\,\mathbf{j}$$

$$\tau(s) = -\mathbf{p}(s) \cdot \mathbf{b}'(s) = \frac{c}{K^2} = \frac{c}{a^2 + c^2} \cdot$$

Hence the circular helix has constant curvature and torsion. If $c > 0$ (right-handed helix, see Fig. 177 in Sec. 8.5), then $\tau > 0$, and if $c < 0$ (left-handed helix, Fig. 178), then $\tau < 0$.

If $c = 0$, we get a circle of radius a, and our formulas then yield $\kappa = 1/a$ (thus the curvature is the reciprocal of the radius) and $\tau = 0$; also, $\mathbf{b}(s) = \mathbf{k}$ is constant, namely, perpendicular to the plane of the circle (the xy-plane). ∎

Since **u, p,** and **b** are linearly independent vectors, we may represent any vector in space as a linear combination of these vectors. Hence if the derivatives $\mathbf{u}'$, $\mathbf{p}'$, and $\mathbf{b}'$ exist, they may be represented in this fashion. The corresponding formulas are the so-called **Frenet formulas**[10]

$$\text{(6)}\qquad \begin{aligned} &\text{(a)}\quad \mathbf{u}' = \kappa\mathbf{p} \\ &\text{(b)}\quad \mathbf{p}' = -\kappa\mathbf{u} + \tau\mathbf{b} \\ &\text{(c)}\quad \mathbf{b}' = -\tau\mathbf{p} \end{aligned}$$

Formula (6a) follows from (2), and (6c) is identical to (4). The derivation of (6b) is not very difficult either and is left to the reader (Prob. 16).

A curve is uniquely determined (except for its position in space) if we prescribe its curvature κ (> 0) and torsion τ as continuous functions of arc length s. (Proof in Ref. [B5] listed in Appendix 1.) For this reason, one calls $\kappa = \kappa(s)$ and $\tau = \tau(s)$ the **natural equations** of a curve. This also shows why curvature and torsion are basic in the differential geometry of space curves.

This is the end of our discussion of vector functions of a single variable and their application in geometry and mechanics.

Problem Set 8.7

1. Show that the curvature of a circle of radius a equals $1/a$.
2. Using (1), show that if a curve is represented by $\mathbf{r}(t)$, where t is any parameter, then its curvature is

$$\text{(1')}\qquad \kappa(t) = \frac{\sqrt{(\mathbf{r}'\bullet\mathbf{r}')(\mathbf{r}''\bullet\mathbf{r}'') - (\mathbf{r}'\bullet\mathbf{r}'')^2}}{(\mathbf{r}'\bullet\mathbf{r}')^{3/2}}.$$

3. Using (1′), show that for a curve $y = y(x)$ in the xy-plane,

$$\text{(1'')}\qquad \kappa(x) = |y''|/(1 + y'^2)^{3/2} \qquad (y' = dy/dx, \text{ etc.}).$$

Indicate what kind of curve is represented and find its curvature, using (1′) or (1″).

4. $\mathbf{r} = a \cos t\, \mathbf{i} + b \sin t\, \mathbf{j}$
5. $y = x^2$
6. $xy = 1$
7. $\mathbf{r} = a \cos t\, \mathbf{i} + a \sin t\, \mathbf{j} + ct\mathbf{k}$
8. $\mathbf{r} = \cosh t\, \mathbf{i} + \sinh t\, \mathbf{j}$
9. $\mathbf{r} = t\mathbf{i} + t^{3/2}\mathbf{j}$ $(t \geqq 0)$

[10] JEAN-FRÉDÉRIC FRENET (1816—1900), French mathematician.

Torsion of a curve

10. Using (3) and (5), show that the torsion $\tau(s)$ of a curve C: $\mathbf{r}(s)$ is

$$(5') \qquad \tau(s) = (\mathbf{u}\ \ \mathbf{p}\ \ \mathbf{p}') \qquad (\kappa > 0).$$

11. Using (2), show that (5′) may be written ($' = d/ds$, etc.)

$$(5'') \qquad \tau(s) = (\mathbf{r}'\ \ \mathbf{r}''\ \ \mathbf{r}''')/\kappa^2 \qquad (\kappa > 0).$$

12. Show that if a curve C is represented by $\mathbf{r}(t)$, where t is any parameter, then (5″) becomes ($' = d/dt$, etc.)

$$(5''') \qquad \tau(t) = \frac{(\mathbf{r}'\ \ \mathbf{r}''\ \ \mathbf{r}''')}{(\mathbf{r}'\cdot\mathbf{r}')(\mathbf{r}''\cdot\mathbf{r}'') - (\mathbf{r}'\cdot\mathbf{r}'')^2} \qquad (\kappa > 0).$$

13. Show that the torsion of a plane curve (with $\kappa > 0$) is identically zero.

14. Find the torsion of the helix $\mathbf{r}(t) = a \cos t\,\mathbf{i} + a \sin t\,\mathbf{j} + ct\mathbf{k}$. Use (5‴). Compare the result with Example 1.

15. Find the torsion of the curve C: $\mathbf{r}(t) = t\mathbf{i} + t^2\mathbf{j} + t^3\mathbf{k}$ (which looks similar to the curve in Fig. 186).

16. Prove the Frenet formula (6b). *Hint.* Start by differentiating $\mathbf{p} = \mathbf{b} \times \mathbf{u}$.

8.8 Review from Calculus in Several Variables. *Optional*

From vector functions of a single variable we now proceed to vector functions of several variables, beginning with a review from calculus. ***The student should go on to the next section, consulting this material only when needed.*** (We include this short section to keep the book reasonably self-contained. For partial derivatives see Appendix 3.)

Chain Rules

Theorem 1 **(Chain rule)**

Let $w = f(x, y, z)$ be continuous and have continuous first partial derivatives in a domain[11] *D in xyz-space. Let $x = x(u, v)$, $y = y(u, v)$, $z = z(u, v)$ be functions that are continuous and have first partial derivatives in a domain B in the uv-plane, where B is such that for every point (u, v) in B, the corresponding point $[x(u, v), y(u, v), z(u, v)]$ lies in D. Then the function*

$$w = f(x(u, v),\ y(u, v),\ z(u, v))$$

[11] A **domain** D is an open connected point set, where "connected" means that any two points of D can be joined by a broken line of finitely many linear segments all of whose points belong to D, and "open" means that every point P of D has a neighborhood (a little ball with center P) all of whose points belong to D. For example, the interior of a cube or of an ellipsoid (the solid without the boundary surface) is a domain.

is defined in B, has first partial derivatives with respect to u and v in B, and

(1)
$$\frac{\partial w}{\partial u} = \frac{\partial w}{\partial x}\frac{\partial x}{\partial u} + \frac{\partial w}{\partial y}\frac{\partial y}{\partial u} + \frac{\partial w}{\partial z}\frac{\partial z}{\partial u},$$
$$\frac{\partial w}{\partial v} = \frac{\partial w}{\partial x}\frac{\partial x}{\partial v} + \frac{\partial w}{\partial y}\frac{\partial y}{\partial v} + \frac{\partial w}{\partial z}\frac{\partial z}{\partial v}.$$

Special cases of practical interest

If $w = f(x, y)$ and $x = x(u, v)$, $y = y(u, v)$ as before, then (1) becomes

(2)
$$\frac{\partial w}{\partial u} = \frac{\partial w}{\partial x}\frac{\partial x}{\partial u} + \frac{\partial w}{\partial y}\frac{\partial y}{\partial u},$$
$$\frac{\partial w}{\partial v} = \frac{\partial w}{\partial x}\frac{\partial x}{\partial v} + \frac{\partial w}{\partial y}\frac{\partial y}{\partial v}.$$

If $w = f(x, y, z)$ and $x = x(t)$, $y = y(t)$, $z = z(t)$, then (1) gives

(3)
$$\frac{dw}{dt} = \frac{\partial w}{\partial x}\frac{dx}{dt} + \frac{\partial w}{\partial y}\frac{dy}{dt} + \frac{\partial w}{\partial z}\frac{dz}{dt}.$$

If $w = f(x, y)$ and $x = x(t)$, $y = y(t)$, then (3) reduces to

(4)
$$\frac{dw}{dt} = \frac{\partial w}{\partial x}\frac{dx}{dt} + \frac{\partial w}{\partial y}\frac{dy}{dt}.$$

Finally, the simplest case $w = f(x)$, $x = x(t)$ gives

(5)
$$\frac{dw}{dt} = \frac{dw}{dx}\frac{dx}{dt}.$$

EXAMPLE 1 Chain rule

If $w = x^2 - y^2$ and we define polar coordinates r, θ by $x = r\cos\theta$, $y = r\sin\theta$, then (2) gives

$$\frac{\partial w}{\partial r} = 2x\cos\theta - 2y\sin\theta = 2r\cos^2\theta - 2r\sin^2\theta = 2r\cos 2\theta,$$

$$\frac{\partial w}{\partial \theta} = 2x(-r\sin\theta) - 2y(r\cos\theta)$$

$$= -2r^2\cos\theta\sin\theta - 2r^2\sin\theta\cos\theta = -2r^2\sin 2\theta.$$

∎

Mean Value Theorems

Theorem 2 **(Mean value theorem)**

Let $f(x, y, z)$ be continuous and have continuous first partial derivatives in a domain D in xyz-space. Let P_0: (x_0, y_0, z_0) and P: $(x_0 + h, y_0 + k, z_0 + l)$ be points in D such that the straight line segment P_0P joining these points lies entirely in D. Then

$$f(x_0 + h, y_0 + k, z_0 + l) - f(x_0, y_0, z_0) = h\frac{\partial f}{\partial x} + k\frac{\partial f}{\partial y} + l\frac{\partial f}{\partial z},$$

the partial derivatives being evaluated at a suitable point of that segment.

Special cases

For a function $f(x, y)$ of two variables (satisfying assumptions as in the theorem), formula (6) reduces to (Fig. 187)

$$f(x_0 + h, y_0 + k) - f(x_0, y_0) = h\frac{\partial f}{\partial x} + k\frac{\partial f}{\partial y}, \tag{7}$$

and for a function $f(x)$ of a single variable, (6) becomes

$$f(x_0 + h) - f(x_0) = h\frac{df}{dx}, \tag{8}$$

where in (8), the domain D is a segment of the x-axis and the derivative is taken at a suitable point between x_0 and $x_0 + h$.

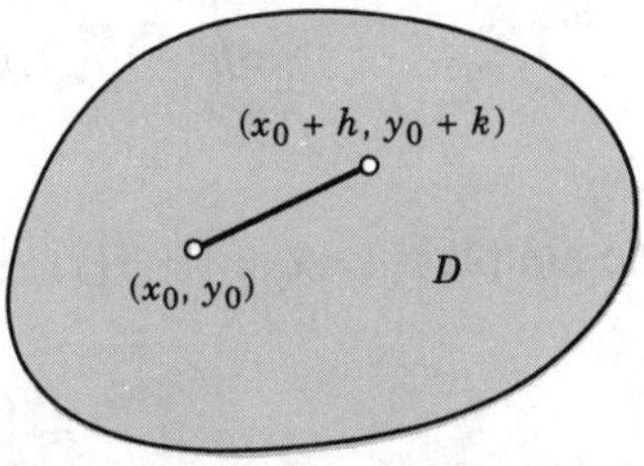

Fig. 187. Mean value theorem for a function of two variables [Formula (7)]

Problem Set 8.8

Find dw/dt by (3) or (4) and check the result by substitution and differentiation, where

1. $w = \sqrt{x^2 + y^2}, \quad x = e^t, \quad y = \sin t$

2. $w = x/y, \quad x = g(t), \quad y = h(t)$

3. $w = x^y, \quad x = \cos t, \quad y = \sin t$

4. $w = xy + yz + zx, \quad x = t, \quad y = \cos t, \quad z = \sin t$

5. $w = (x + y)/(y + z), \quad x = e^t, \quad y = e^{2t}, \quad z = e^{-t}$

Find $\partial w/\partial u$ and $\partial w/\partial v$, where

6. $w = x^2 + y^2, \quad x = u + v, \quad y = u - v$

7. $w = xy, \quad x = e^u \cos v, \; y = e^u \sin v$

8. $w = x^4 - 4x^2y^2 + y^4, \quad x = uv, \quad y = u/v$

9. $w = \frac{1}{2}(x^2 + y^2 + z^2)^{-1}, \quad x = u^2 + v^2, \quad y = u^2 - v^2, \quad z = 2uv$

10. (Partial derivatives on a surface) Let $w = f(x, y, z)$, and let $z = g(x, y)$ represent a surface S in space. Then on S, the function becomes

$$\tilde{w}(x, y) = f[x, y, g(x, y)].$$

Show that its partial derivatives are obtained from

$$\frac{\partial \tilde{w}}{\partial x} = \frac{\partial f}{\partial x} + \frac{\partial f}{\partial z}\frac{\partial g}{\partial x}, \qquad \frac{\partial \tilde{w}}{\partial y} = \frac{\partial f}{\partial y} + \frac{\partial f}{\partial z}\frac{\partial g}{\partial y} \qquad [z = g(x, y)].$$

Apply this to $f = x^3 + y^3 + z^2$, $g = x^2 + y^2$ and check by substitution and direct differentiation.

8.9 Gradient of a Scalar Field. Directional Derivative

We shall see that in applications there often occur vector fields which have the advantage that they can be obtained from scalar fields, which, of course, are easier to handle. This relation between the two types of fields is accomplished by the "gradient," which therefore is of great importance.

Gradient. For a given scalar function $f(x, y, z)$ the *gradient* grad f of f is the vector function defined by

$$\text{grad}\, f = \frac{\partial f}{\partial x}\mathbf{i} + \frac{\partial f}{\partial y}\mathbf{j} + \frac{\partial f}{\partial z}\mathbf{k}. \tag{1*}$$

Here we must assume that f is differentiable. It has become popular, particularly with physicists and engineers, to introduce the differential operator

$$\nabla = \frac{\partial}{\partial x}\mathbf{i} + \frac{\partial}{\partial y}\mathbf{j} + \frac{\partial}{\partial z}\mathbf{k} \tag{2}$$

(read **nabla** or *del*) and to write

$$\text{grad}\, f = \nabla f = \frac{\partial f}{\partial x}\mathbf{i} + \frac{\partial f}{\partial y}\mathbf{j} + \frac{\partial f}{\partial z}\mathbf{k}. \tag{1}$$

For instance, the gradient of $f(x, y, z) = 2x + yz - 3y^2$ is grad $f = \nabla f = 2\mathbf{i} + (z - 6y)\mathbf{j} + y\mathbf{k}$.

We show later that grad f is a vector; that is, although it is defined in (1) in terms of components, it has a length and direction that is independent of the particular choice of Cartesian coordinates. But first we explore how the gradient is related to the rate of change of f in various directions. In the directions of the three coordinate axes, this rate is given by the partial derivatives, as we know from calculus. The idea of extending this to arbitrary directions seems natural and leads to the concept of directional derivative.

Directional Derivative

The rate of change of f at any point P in any fixed direction given by a vector $\mathbf{b}$ is defined as in calculus; we denote it by $D_{\mathbf{b}}f$ or df/ds, call it the **directional derivative** *of f at P in the direction of* $\mathbf{b}$, and define it by (see Fig. 188)

$$D_{\mathbf{b}}f = \frac{df}{ds} = \lim_{s \to 0} \frac{f(Q) - f(P)}{s} \qquad (s = \text{distance between } P \text{ and } Q), \tag{3}$$

where Q is a variable point on the ray C in the direction of $\mathbf{b}$ as in Fig. 188.

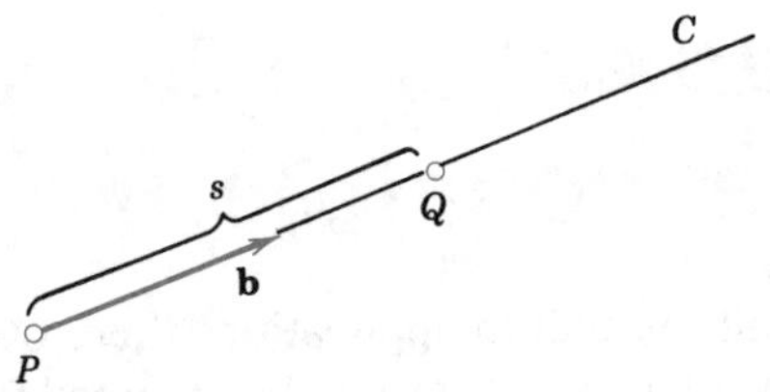

Fig. 188. Directional derivative

The next idea is to use Cartesian xyz-coordinates and for $\mathbf{b}$ a *unit vector*. Then the ray C is given by

$$\mathbf{r}(s) = x(s)\mathbf{i} + y(s)\mathbf{j} + z(s)\mathbf{k} = \mathbf{p}_0 + s\mathbf{b} \qquad (s \geqq 0, \quad |\mathbf{b}| = 1) \tag{4}$$

($\mathbf{p}_0$ the position vector of P) and (3) now shows that $D_{\mathbf{b}}f = df/ds$ is the derivative of the function $f(x(s), y(s), z(s))$ with respect to the arc length s of C. Hence, assuming that f has continuous partial derivatives and applying the chain rule [formula (3) in the previous section], we obtain

$$D_{\mathbf{b}}f = \frac{df}{ds} = \frac{\partial f}{\partial x}x' + \frac{\partial f}{\partial y}y' + \frac{\partial f}{\partial z}z' \tag{5}$$

where primes denote derivatives with respect to s (which are taken at $s = 0$). But here, $\mathbf{r}' = x'\mathbf{i} + y'\mathbf{j} + z'\mathbf{k} = \mathbf{b}$ by (4), and we see that (5) is simply the inner product of $\mathbf{b}$ and grad f [see (2), Sec. 8.2]; thus

$$D_{\mathbf{b}}f = \frac{df}{ds} = \mathbf{b} \cdot \operatorname{grad} f \qquad (|\mathbf{b}| = 1). \tag{6}$$

Attention! If the direction is given by a vector **a** of any length ($\neq 0$), then

$$(6') \qquad D_{\mathbf{a}}f = \frac{df}{ds} = \frac{1}{|\mathbf{a}|}\,\mathbf{a}\cdot\operatorname{grad} f.$$

EXAMPLE 1 **Gradient. Directional derivative**

Find the directional derivative of $f(x, y, z) = 2x^2 + 3y^2 + z^2$ at the point P: (2, 1, 3) in the direction of the vector $\mathbf{a} = \mathbf{i} - 2\mathbf{k}$.

Solution. We obtain

$$\operatorname{grad} f = 4x\mathbf{i} + 6y\mathbf{j} + 2z\mathbf{k}, \qquad \text{and at } P, \qquad \operatorname{grad} f = 8\mathbf{i} + 6\mathbf{j} + 6\mathbf{k}.$$

From this and (6′),

$$D_{\mathbf{a}}f = \frac{1}{\sqrt{5}}(\mathbf{i} - 2\mathbf{k})\cdot(8\mathbf{i} + 6\mathbf{j} + 6\mathbf{k}) = \frac{1\cdot 8 - 2\cdot 6}{\sqrt{5}} = -\frac{4}{\sqrt{5}} \approx -1.789.$$

The minus sign indicates that f decreases at P in the direction of **a**. ∎

Gradient Characterizes Maximum Increase

Theorem 1 **(Gradient, maximum increase)**

Let $f(P) = f(x, y, z)$ be a scalar function having continuous first partial derivatives. Then grad *f exists and its length and direction are independent of the particular choice of Cartesian coordinates in space. If at a point P the gradient of f is not the zero vector, it has the direction of maximum increase of f at P.*

Proof. From (6) and the definition of inner product [(1) in Sec. 8.2] we have

$$(7) \qquad D_{\mathbf{b}}f = |\mathbf{b}|\,|\operatorname{grad} f| \cos\gamma = |\operatorname{grad} f| \cos\gamma,$$

where γ is the angle between **b** and grad f. Now f is a scalar function. Hence its value at a point P depends on P but not on the particular choice of coordinates. The same holds for the arc length s of the ray C (see Fig. 188), hence also for $D_{\mathbf{b}}f$. Now (7) shows that $D_{\mathbf{b}}f$ is maximum when $\cos\gamma = 1$, $\gamma = 0$, and then $D_{\mathbf{b}}f = |\operatorname{grad} f|$. It follows that the length and direction of grad f are independent of the coordinates. Since $\gamma = 0$ if and only if **b** has the direction of grad f, the latter is the direction of maximum increase of f at P, provided grad $f \neq \mathbf{0}$ at P. ∎

Gradient as Surface Normal Vector

Another basic use of the gradient results in connection with surfaces S in space given by

$$(8) \qquad f(x, y, z) = c = const,$$

as follows. We recall that a curve C in space can be given by

$$(9) \qquad \mathbf{r}(t) = x(t)\mathbf{i} + y(t)\mathbf{j} + z(t)\mathbf{k} \qquad \text{[see (1), Sec. 8.5].}$$

Now if we want C to lie on S, its components must satisfy (8); thus

$$f(x(t),\ y(t),\ z(t)) = c. \tag{10}$$

A tangent vector of C is [see (7), Sec. 8.5]

$$\mathbf{r}'(t) = x'(t)\mathbf{i} + y'(t)\mathbf{j} + z'(t)\mathbf{k}.$$

If C lies on S, this vector is tangent to S. At a fixed point P on S, these tangent vectors of all curves on S through P will generally form a plane, called the **tangent plane** of S at P (Fig. 189). Its normal (the straight line through P and perpendicular to the tangent plane) is called the **surface normal** of S at P, and a vector parallel to it is called a **surface normal vector** of S at P. Now if we differentiate (10) with respect to t, we get by the chain rule

$$\frac{\partial f}{\partial x}x' + \frac{\partial f}{\partial y}y' + \frac{\partial f}{\partial z}z' = (\text{grad } f)\cdot\mathbf{r}' = 0, \tag{11}$$

which means orthogonality of grad f and all the vectors $\mathbf{r}'$ in the tangent plane. Our result is (see Fig. 189)

Theorem 2 (Gradient as surface normal vector)

Let f be a differentiable scalar function that represents a surface S: $f(x, y, z) = c = const$, as shown in (8). Then if the gradient of f at a point P of S is not the zero vector, it is a normal vector of S at P.

Comment. The surfaces given by (8) with various values of c are called the **level surfaces** of the scalar function f.

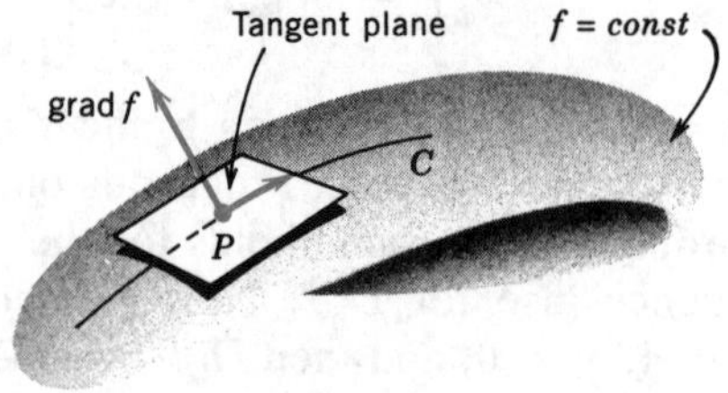

Fig. 189. Gradient as surface normal vector

EXAMPLE 2 Gradient as surface normal vector

Find a unit normal vector $\mathbf{n}$ of the cone of revolution $z^2 = 4(x^2 + y^2)$ at the point P: (1, 0, 2).

Solution. The cone is the level surface $f = 0$ of $f(x, y, z) = 4(x^2 + y^2) - z^2$. Thus

$$\text{grad } f = 8x\mathbf{i} + 8y\mathbf{j} - 2z\mathbf{k} \quad \text{and at } P, \quad \text{grad } f = 8\mathbf{i} - 4\mathbf{k}.$$

Hence, by Theorem 2, a unit normal vector of the cone at P is

$$\mathbf{n} = \frac{1}{|\text{grad } f|}\text{grad } f = \frac{2}{\sqrt{5}}\mathbf{i} - \frac{1}{\sqrt{5}}\mathbf{k} \qquad \text{(Fig. 190)}$$

and the other one is $-\mathbf{n}$. ∎

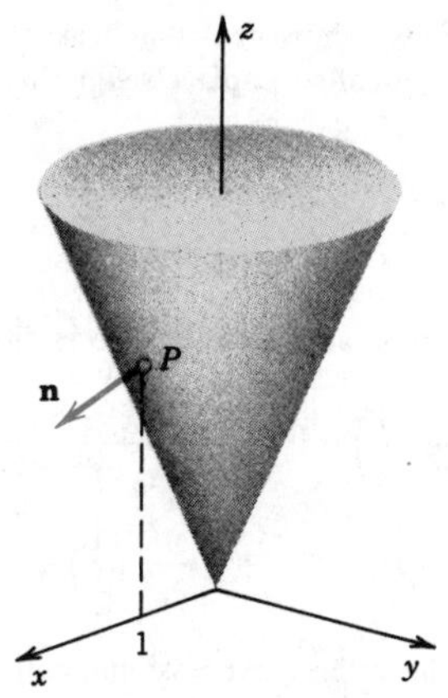

Fig. 190. Cone and unit normal vector **n**

Vector Fields That Are Gradients of a Scalar Field ("Potential")

Scalar fields are generally easier to handle than vector fields. Some vector functions **v**(P) have the advantage that they can be obtained as the gradient of a scalar function $f(P)$. The latter is then called a *potential function* or **potential** of **v**(P), and such a **v**(P) is called **conservative** because in such a vector field, energy is conserved; that is, no energy is lost (or gained) in displacing a body (or a charge in the case of an electrical field) from a point P to another point in the field and back to P. We show this in Sec. 9.2. To obtain a first impression of this method of handling vector fields, let us consider an important example.

EXAMPLE 3 Gravitational field. Laplace's equation

In Example 3 of Sec. 8.4 we saw that, by Newton's law of gravitation, the force of attraction between two particles is

$$\mathbf{p} = -\frac{c}{r^3}\mathbf{r} = -c\left(\frac{x-x_0}{r^3}\mathbf{i} + \frac{y-y_0}{r^3}\mathbf{j} + \frac{z-z_0}{r^3}\mathbf{k}\right), \tag{12}$$

where

$$r = \sqrt{(x-x_0)^2 + (y-y_0)^2 + (z-z_0)^2}$$

is the distance between the two particles and c is a constant. Observing that

$$\frac{\partial}{\partial x}\left(\frac{1}{r}\right) = -\frac{2(x-x_0)}{2[(x-x_0)^2 + (y-y_0)^2 + (z-z_0)^2]^{3/2}} = -\frac{x-x_0}{r^3} \tag{13a}$$

and similarly

$$\frac{\partial}{\partial y}\left(\frac{1}{r}\right) = -\frac{y-y_0}{r^3}, \qquad \frac{\partial}{\partial z}\left(\frac{1}{r}\right) = -\frac{z-z_0}{r^3}, \tag{13b}$$

we see that **p** is the gradient of the scalar function

$$f(x, y, z) = \frac{c}{r} \qquad (r > 0);$$

thus, f is a potential of that gravitational field (and further potentials are $f + k$ with constant k).

Furthermore, we show next that f satisfies the most important partial differential equation of physics, the so-called **Laplace's equation**

$$\frac{\partial^2 f}{\partial x^2} + \frac{\partial^2 f}{\partial y^2} + \frac{\partial^2 f}{\partial z^2} = 0. \tag{14}$$

Indeed, to see this, all we have to do is differentiate (13),

$$\frac{\partial^2}{\partial x^2}\left(\frac{1}{r}\right) = -\frac{1}{r^3} + \frac{3(x - x_0)^2}{r^5}, \qquad \frac{\partial^2}{\partial y^2}\left(\frac{1}{r}\right) = -\frac{1}{r^3} + \frac{3(y - y_0)^2}{r^5},$$

$$\frac{\partial^2}{\partial z^2}\left(\frac{1}{r}\right) = -\frac{1}{r^3} + \frac{3(z - z_0)^2}{r^5},$$

and then add these three expressions, noting that the sum of the right sides is zero, so that we obtain (14).

In (14), the expression on the left is called the **Laplacian** of f and is denoted by $\nabla^2 f$ or Δf. The differential operator

$$\nabla^2 = \Delta = \frac{\partial^2}{\partial x^2} + \frac{\partial^2}{\partial y^2} + \frac{\partial^2}{\partial z^2}$$

(read "nabla squared" or "delta") is called the **Laplace operator.** Using this operator, we may write (14) in the form

$$\nabla^2 f = 0.$$

It can be shown that the field of force produced by any distribution of masses is given by a vector function that is the gradient of a scalar function f, and f satisfies (14) in any region of space that is free of matter.

There are other laws in physics that are of the same form as Newton's law of gravitation. For example, in electrostatics the force of attraction (or repulsion) between two particles of opposite (or like) charges Q_1 and Q_2 is

$$\mathbf{p} = \frac{k}{r^3}\mathbf{r} \qquad \text{(Coulomb's law[12])}$$

where $k = Q_1 Q_2 / 4\pi\epsilon$, and ϵ is the dielectric constant. Hence $\mathbf{p}$ is the gradient of the potential $f = -k/r$, and f satisfies (14) when $r > 0$.

Laplace's equation will be discussed in detail in Chaps. 11 and 17. ∎

Problem Set 8.9

Gradient. (a) Find the gradient ∇f. (b) Graph some level curves $f = const$ and indicate ∇f by arrows at some points of these curves.

1. $f = xy$ **2.** $f = x^2 + y^2$ **3.** $f = x^2 - y^2$

4. $f = \frac{1}{4}x^2 + y^2$ **5.** $f = x/y$ **6.** $f = \text{arc tan}\,(y/x)$

[12] CHARLES AUGUSTIN DE COULOMB (1736—1806), French physicist and engineer. Coulomb's law was derived by him from his own very precise measurements.

Find the gradient ∇f, where f equals

7. $e^x \sin y$ **8.** $x^2 + 3y^2 - 2z^2$ **9.** $\frac{1}{2} \ln (x^2 + y^2)$

10. $\sin x \sinh y$ **11.** $(x^2 + y^2 + z^2)^{-1/2}$ **12.** $yz + zx + xy$

13. $z/(x^2 + y^2)$ **14.** $e^{x^2-y^2} \sin 2xy$ **15.** $(y^2 - z^2)/(y^2 + z^2)$

Find a potential f of the vector field given by

16. $x\mathbf{i} + y\mathbf{j} - \mathbf{k}$ **17.** $\mathbf{i} - \mathbf{j} + \mathbf{k}$

18. $6x\mathbf{i} - 4y\mathbf{j} + 2z\mathbf{k}$ **19.** $e^{xy}(y\mathbf{i} + x\mathbf{j})$

20. $(x + y)(\mathbf{i} + \mathbf{j})$ **21.** $(x\mathbf{i} + y\mathbf{j})/(x^2 + y^2)$

22. $(x^2 + y^2 + z^2)^{-3/2} (x\mathbf{i} + y\mathbf{j} + z\mathbf{k})$ **23.** $y^{-1}\mathbf{i} - xy^{-2}\mathbf{j} + z\mathbf{k}$

Formulas for the gradient. Granted sufficient differentiability, show that

24. $\nabla(fg) = f\nabla g + g\nabla f$ **25.** $\nabla(f^n) = nf^{n-1}\nabla f$

26. $\nabla(f/g) = (1/g^2)(g\nabla f - f\nabla g)$ **27.** $\nabla^2(fg) = g\nabla^2 f + 2\nabla f \cdot \nabla g + f\nabla^2 g$

28. Solve Probs. 14 and 15 by the formulas in Probs. 24 and 26.

Directional derivative. Find the directional derivative of f at P in the direction of $\mathbf{a}$, where

29. $f = x - y$, P: (4, 5), $\mathbf{a} = 2\mathbf{i} + \mathbf{j}$

30. $f = x^2 - y^2$, P: (2, 3), $\mathbf{a} = \mathbf{i} + \mathbf{j}$

31. $f = x^2 + y^2$, P: (1, 2), $\mathbf{a} = \mathbf{i} - \mathbf{j}$

32. $f = x^2 + y^2 + z^2$, P: (1, 2, 3), $\mathbf{a} = \mathbf{i} + \mathbf{j}$

33. $f = x^2 + y^2 + z^2$, P: (2, 2, 2), $\mathbf{a} = \mathbf{i} + 2\mathbf{j} - 3\mathbf{k}$

34. $f = 1/\sqrt{x^2 + y^2 + z^2}$, P: (3, 0, 4), $\mathbf{a} = \mathbf{i} + \mathbf{j} + \mathbf{k}$

35. $f = e^x \cos y$, P: $(2, \pi, 0)$, $\mathbf{a} = 2\mathbf{i} + 3\mathbf{k}$

36. $f = xy + 10$, P: $(1, 1, -\frac{1}{2})$, $\mathbf{a} = \mathbf{i} - \mathbf{j} + 7\mathbf{k}$

37. $f = xyz$, P: $(-1, 1, 3)$, $\mathbf{a} = \mathbf{i} - 2\mathbf{j} + 2\mathbf{k}$

38. $f = \ln (x^2 + y^2)$, P: (4, 0), $\mathbf{a} = \mathbf{i} - \mathbf{j}$

Some applications. Find a unit normal vector to the given curve in the xy-plane or surface in space at the given point P.

39. $y = \frac{4}{3}x - \frac{2}{3}$, P: (2, 2) **40.** $y = 2x^2$, P: (2, 8)

41. $x^2 + y^2 = 100$, P: (6, 8) **42.** $x^2 - y^2 = 1$, P: $(2, \sqrt{3})$

43. $4x^2 + 9y^2 = 36$, P: (0, 2) **44.** $ax + by + cz + d = 0$, any P

45. $z = \sqrt{x^2 + y^2}$, P: (3, 4, 5) **46.** $z = xy$, P: $(2, -1, -2)$

47. **(Heat flow)** In a temperature field, heat flows in the direction of maximum decrease of temperature T. Find this direction at P: (2, 1) when $T = x^3 - 3xy^2$.

48. Repeat Prob. 47, when $T = 4x^2 + y^2 - 5z^2$ and P: $(\frac{1}{4}, -2, \frac{1}{2})$.

49. If on a mountain, the elevation above sea level is $z(x, y) = 1500 - 3x^2 - 5y^2$ [meters], what is the direction of steepest ascent at P: $(-0.2, 0.1)$?

50. **(Electric field)** If the potential between two concentric cylinders is $V(x, y) = 110 + 30 \ln (x^2 + y^2)$ [volts], what is the direction of the electric force (the gradient) at P: (2, 5)? Show that $\nabla^2 V = 0$.

8.10 Divergence of a Vector Field

Vector calculus owes much of its importance in engineering and physics to the gradient, divergence, and curl. Having discussed the gradient, we turn next to the divergence. The curl follows in Sec. 8.11.

Let $\mathbf{v}(x, y, z)$ be a differentiable vector function, where x, y, z are Cartesian coordinates, and let v_1, v_2, v_3 be the components of $\mathbf{v}$. Then the function

$$\text{div } \mathbf{v} = \frac{\partial v_1}{\partial x} + \frac{\partial v_2}{\partial y} + \frac{\partial v_3}{\partial z} \tag{1}$$

is called the **divergence** *of* $\mathbf{v}$ or the *divergence of the vector field defined by* $\mathbf{v}$. Another common notation for the divergence of $\mathbf{v}$ is $\nabla \cdot \mathbf{v}$,

$$\begin{aligned} \text{div } \mathbf{v} = \nabla \cdot \mathbf{v} &= \left(\frac{\partial}{\partial x} \mathbf{i} + \frac{\partial}{\partial y} \mathbf{j} + \frac{\partial}{\partial z} \mathbf{k} \right) \cdot (v_1 \mathbf{i} + v_2 \mathbf{j} + v_3 \mathbf{k}) \\ &= \frac{\partial v_1}{\partial x} + \frac{\partial v_2}{\partial y} + \frac{\partial v_3}{\partial z}, \end{aligned}$$

with the understanding that the "product" $(\partial/\partial x)v_1$ in the dot product means the partial derivative $\partial v_1/\partial x$, etc. This is a convenient notation, but nothing more. Note that $\nabla \cdot \mathbf{v}$ means the scalar div $\mathbf{v}$, whereas ∇f means the vector grad f defined in Sec. 8.9.

For example, if

$$\mathbf{v} = 3xz\mathbf{i} + 2xy\mathbf{j} - yz^2\mathbf{k}, \qquad \text{then} \qquad \text{div } \mathbf{v} = 3z + 2x - 2yz.$$

We shall see below that the divergence has an important physical meaning. Clearly the values of a function that characterize a physical or geometrical property must be independent of the particular choice of coordinates; that is, those values must be invariant with respect to coordinate transformations.

Theorem 1 **(Invariance of the divergence)**

The values of div $\mathbf{v}$ *depend only on the points in space (and, of course, on* $\mathbf{v}$*) but not on the particular choice of the coordinates in* (1), *so that with respect to other Cartesian coordinates* x^*, y^*, z^* *and corresponding components* v_1^*, v_2^*, v_3^* *of* $\mathbf{v}$ *the function* div $\mathbf{v}$ *is given by*

$$\text{div } \mathbf{v} = \frac{\partial v_1^*}{\partial x^*} + \frac{\partial v_2^*}{\partial y^*} + \frac{\partial v_3^*}{\partial z^*}. \tag{2}$$

A first proof is given in Appendix 4 and a second proof (requiring integrals) in Sec. 9.8. Presently, let us turn to the more immediate practical task of getting a feeling for the significance of the divergence.

If $f(x, y, z)$ is a twice differentiable scalar function, then

$$\text{grad } f = \frac{\partial f}{\partial x}\mathbf{i} + \frac{\partial f}{\partial y}\mathbf{j} + \frac{\partial f}{\partial z}\mathbf{k}$$

and by (1),

$$\text{div (grad } f) = \frac{\partial^2 f}{\partial x^2} + \frac{\partial^2 f}{\partial y^2} + \frac{\partial^2 f}{\partial z^2}.$$

The expression on the right is the Laplacian of f (see Sec. 8.9). Thus

$$\text{div (grad } f) = \nabla^2 f. \qquad (3)$$

EXAMPLE 1 Gravitational force

The gravitational force $\mathbf{p}$ in Example 3, Sec. 8.9, is the gradient of the scalar function $f(x, y, z) = c/r$, which satisfies Laplace's equation $\nabla^2 f = 0$. According to (3), this means that div $\mathbf{p} = 0$ $(r > 0)$. ∎

The following example, taken from hydrodynamics, shows the physical significance of the divergence of a vector field (and more will be added in Sec. 9.8 when the so-called divergence theorem of Gauss will be available).

EXAMPLE 2 Motion of a compressible fluid. Meaning of the divergence.

We consider the motion of a fluid in a region R having no **sources** or **sinks** in R, that is, no points at which fluid is produced or disappears. The concept of **fluid state** is meant to cover also gases and vapors. Fluids in the restricted sense, or liquids, have a small compressibility which can be neglected in many problems. Gases and vapors have a large compressibility; that is, their density ρ (= mass per unit volume) depends on the coordinates x, y, z in space (and may depend on time t). We assume that our fluid is compressible.

We consider the flow through a small rectangular box W of dimensions[13] Δx, Δy, Δz with edges parallel to the coordinate axes (Fig. 191). W has the volume $\Delta V = \Delta x\,\Delta y\,\Delta z$. Let

$$\mathbf{v} = [v_1, v_2, v_3] = v_1\mathbf{i} + v_2\mathbf{j} + v_3\mathbf{k}$$

be the velocity vector of the motion. We set

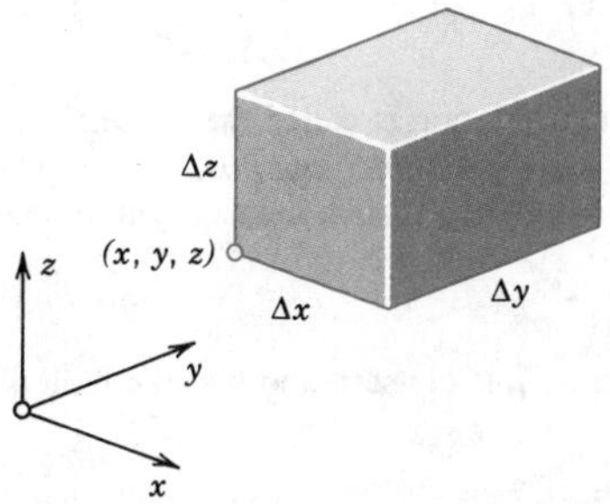

Fig. 191. Physical interpretation of the divergence

[13]It is a standard usage to indicate small quantities by Δ; this has, of course, nothing to do with the Laplacian.

$$\mathbf{u} = \rho\mathbf{v} = [u_1, u_2, u_3] = u_1\mathbf{i} + u_2\mathbf{j} + u_3\mathbf{k} \tag{4}$$

and assume that **u** and **v** are continuously differentiable vector functions of x, y, z and t. Let us calculate the change in the mass included in W by considering the **flux** across the boundary, that is, the total loss of mass leaving W per unit time. Consider the flow through the left face of W, whose area is $\Delta x\, \Delta z$. The components v_1 and v_3 of **v** are parallel to that face and contribute nothing to this flow. hence the mass of fluid entering through that face during a short time interval Δt is given approximately by

$$(\rho v_2)_y\, \Delta x\, \Delta z\, \Delta t = (u_2)_y\, \Delta x\, \Delta z\, \Delta t,$$

where the subscript y indicates that this expression refers to the left face. The mass of fluid leaving the box W through the opposite face during the same time interval is approximately $(u_2)_{y+\Delta y}\, \Delta x\, \Delta z\, \Delta t$, where the subscript $y + \Delta y$ indicates that this expression refers to the right face. The difference

$$\Delta u_2\, \Delta x\, \Delta z\, \Delta t = \frac{\Delta u_2}{\Delta y}\, \Delta V\, \Delta t \qquad [\Delta u_2 = (u_2)_{y+\Delta y} - (u_2)_y]$$

is the approximate loss of mass. Two similar expressions are obtained by considering the other two pairs of parallel faces of W. If we add these three expressions, we find that the total loss of mass in W during the time inerval Δt is approximately

$$\left(\frac{\Delta u_1}{\Delta x} + \frac{\Delta u_2}{\Delta y} + \frac{\Delta u_3}{\Delta z}\right) \Delta V\, \Delta t,$$

where

$$\Delta u_1 = (u_1)_{x+\Delta x} - (u_1)_x \qquad \text{and} \qquad \Delta u_3 = (u_3)_{z+\Delta z} - (u_3)_z.$$

This loss of mass in W is caused by the time rate of change of the density and is thus equal to

$$-\frac{\partial \rho}{\partial t}\, \Delta V\, \Delta t.$$

If we equate both expressions, divide the resulting equation by $\Delta V\, \Delta t$, and let Δx, Δy, Δz and Δt approach zero, then we obtain

$$\operatorname{div} \mathbf{u} = \operatorname{div} (\rho\mathbf{v}) = -\frac{\partial \rho}{\partial t}$$

or

$$\frac{\partial \rho}{\partial t} + \operatorname{div} (\rho\mathbf{v}) = 0. \tag{5}$$

This important relation is called the *condition for the conservation of mass* or the **continuity equation** *of a compressible fluid flow.*

If the flow is **steady**, that is, independent of time, then $\partial\rho/\partial t = 0$ and the continuity equation is

$$\operatorname{div} (\rho\mathbf{v}) = 0. \tag{6}$$

If the density ρ is constant, so that the fluid is incompressible, then equation (6) becomes

$$\operatorname{div} \mathbf{v} = 0. \tag{7}$$

This relation is known as the **condition of incompressibility.** It expresses the fact that the balance of outflow and inflow for a given volume element is zero at any time. Clearly, the assumption that the flow has no sources or sinks in R is essential to our argument. ∎

Comment. The **divergence theorem** of Gauss, an integral theorem involving the divergence, follows in the next chapter (Sec. 9.7).

Problem Set 8.10

Find the divergence of the following vector functions.

1. $x\mathbf{i} + y\mathbf{j} + z\mathbf{k}$

2. $yz\mathbf{i} + zx\mathbf{j} + xy\mathbf{k}$

3. $y^2e^z\mathbf{i} + x^2z^2\mathbf{k}$

4. $xyz(x\mathbf{i} + y\mathbf{j} + z\mathbf{k})$

5. $\cos x \cosh y\,\mathbf{i} + \sin x \sinh y\,\mathbf{j}$

6. $e^{-xy}\mathbf{i} + e^{-yz}\mathbf{j} + e^{-zx}\mathbf{k}$

7. $x^2y\mathbf{i} + x^2y\mathbf{j} + y^2z\mathbf{k}$

8. $(x^2 + y^2 + z^2)^{-3/2}(x\mathbf{i} + y\mathbf{j} + z\mathbf{k})$

9. $v_1(y, z)\mathbf{i} + v_2(x, z)\mathbf{j}$

10. $\sin xy\,(\mathbf{i} + \mathbf{j}) - z\cos xy\,\mathbf{k}$

Find $\nabla^2 f$ by (3). Check the result by direct differentiation.

11. $f = x^2 + 4y^2 + 9z^2$

12. $f = (x - y)/(x + y)$

13. $f = \arctan(y/x)$

14. $f = e^{xyz}$

15. $f = xz/y$

16. $f = \sin x \cosh y$

Find grad (div **v**), where **v** equals

17. $x^3\mathbf{i} + y^2\mathbf{j} + z\mathbf{k}$

18. $z\sinh x\,\mathbf{i} + y\cosh z\,\mathbf{k}$

19. $x^2\mathbf{i} + y^2\mathbf{j} - 4z^2\mathbf{k}$

20. $e^x\mathbf{i} + e^y\mathbf{j} + e^z\mathbf{k}$

Formulas for the divergence. Show that

21. $\operatorname{div}(k\mathbf{v}) = k \operatorname{div}\mathbf{v}$ (k constant)

22. $\operatorname{div}(f\mathbf{v}) = f\operatorname{div}\mathbf{v} + \mathbf{v}\cdot\nabla f$

23. $\operatorname{div}(f\nabla g) = f\nabla^2 g + \nabla f\cdot\nabla g$

24. $\operatorname{div}(f\nabla g) - \operatorname{div}(g\nabla f) = f\nabla^2 g - g\nabla^2 f$

25. Verify the formula in Prob. 22 when $f = e^{xyz}$ and $\mathbf{v} = ax\mathbf{i} + by\mathbf{j} + cz\mathbf{k}$.

26. Obtain the answer to Prob. 8 from the formula in Prob. 22.

27. Verify the formula in Prob. 23 for $f = x^2 - y^2$ and $g = e^{x+y}$.

28. **(Rotational flow)** The velocity vector $\mathbf{v}(x, y, z)$ of an incompressible fluid rotating in a cylindrical vessel is of the form $\mathbf{v} = \mathbf{w} \times \mathbf{r}$, where **w** is the (constant) rotation vector; see Example 6 in Sec. 8.3. Show that $\operatorname{div}\mathbf{v} = 0$. Is this plausible from our present Example 2?

29. **(Flow)** Consider a steady flow whose velocity vector is $\mathbf{v} = y\mathbf{i}$. Show that it has the following properties. The flow is incompressible. The particles that at time $t = 0$ are in the cube bounded by the planes $x = 0$, $x = 1$, $y = 0$, $y = 1$, $z = 0$, $z = 1$ occupy at $t = 1$ the volume 1.

30. **(Flow)** Consider a steady flow having the velocity $\mathbf{v} = x\mathbf{i}$. Show that the individual particles have the position vectors $\mathbf{r}(t) = c_1e^t\mathbf{i} + c_2\mathbf{j} + c_3\mathbf{k}$ where c_1, c_2, c_3 are constants, the flow is compressible, and e is the volume occupied at $t = 1$ by the particles that at $t = 0$ fill the cube in Prob. 29.

8.11 Curl of a Vector Field

Gradient (Sec. 8.9), divergence (Sec. 8.10), and curl are basic in connection with fields. We now define and discuss the curl.

Let x, y, z be right-handed Cartesian coordinates (Sec. 8.3), and let

$$\mathbf{v}(x, y, z) = v_1\mathbf{i} + v_2\mathbf{j} + v_3\mathbf{k}$$

be a differentiable vector function. Then the function

$$\text{(1)} \qquad \operatorname{curl} \mathbf{v} = \nabla \times \mathbf{v} = \begin{vmatrix} \mathbf{i} & \mathbf{j} & \mathbf{k} \\ \dfrac{\partial}{\partial x} & \dfrac{\partial}{\partial y} & \dfrac{\partial}{\partial z} \\ v_1 & v_2 & v_3 \end{vmatrix}$$

$$= \left(\frac{\partial v_3}{\partial y} - \frac{\partial v_2}{\partial z}\right)\mathbf{i} + \left(\frac{\partial v_1}{\partial z} - \frac{\partial v_3}{\partial x}\right)\mathbf{j} + \left(\frac{\partial v_2}{\partial x} - \frac{\partial v_1}{\partial y}\right)\mathbf{k}$$

is called the **curl** *of the vector function* **v** or the *curl of the vector field defined by* **v**. For a left-handed Cartesian coordinate system, the determinant in (1) has a minus sign in front, in agreement with the remark following (2**) in Sec. 8.3.

Instead of curl **v** the notation rot **v** (suggested by "rotation"; see Example 2, below) is also used.

EXAMPLE 1 Curl of a vector function

With respect to right-handed Cartesian coordinates, let $\mathbf{v} = yz\mathbf{i} + 3zx\mathbf{j} + z\mathbf{k}$. Then (1) gives

$$\operatorname{curl} \mathbf{v} = \begin{vmatrix} \mathbf{i} & \mathbf{j} & \mathbf{k} \\ \partial/\partial x & \partial/\partial y & \partial/\partial z \\ yz & 3zx & z \end{vmatrix} = -3x\mathbf{i} + y\mathbf{j} + (3z - z)\mathbf{k} = -3x\mathbf{i} + y\mathbf{j} + 2z\mathbf{k}.$$ ∎

The curl plays an important role in many applications. Let us illustrate this with a typical basic example. (We shall say more about the role and nature of the curl in Sec. 9.9.)

EXAMPLE 2 Rotation of a rigid body. Relation to the curl

We have seen in Example 6, Sec. 8.3, that a rotation of a rigid body B about a fixed axis in space can be described by a vector **w** of magnitude ω in the direction of the axis of rotation, where ω (> 0) is the angular speed of the rotation, and **w** is directed so that the rotation appears clockwise if we look in the direction of **w**. According to (9), Sec. 8.3, the velocity field of the rotation can be represented in the form

$$\mathbf{v} = \mathbf{w} \times \mathbf{r}$$

where **r** is the position vector of a moving point with respect to a Cartesian coordinate system having the origin on the axis of rotation. Let us choose right-handed Cartesian coordinates such that $\mathbf{w} = \omega\mathbf{k}$; that is, the axis of rotation is the z-axis. Then (see Example 2 in Sec. 8.4)

$$\mathbf{v} = \mathbf{w} \times \mathbf{r} = -\omega y\mathbf{i} + \omega x\mathbf{j}$$

and, therefore,

$$\operatorname{curl} \mathbf{v} = \begin{vmatrix} \mathbf{i} & \mathbf{j} & \mathbf{k} \\ \dfrac{\partial}{\partial x} & \dfrac{\partial}{\partial y} & \dfrac{\partial}{\partial z} \\ -\omega y & \omega x & 0 \end{vmatrix} = 2\omega\mathbf{k},$$

that is,

$$\operatorname{curl} \mathbf{v} = 2\mathbf{w}. \tag{2}$$

Hence, in the case of a rotation of a rigid body, the curl of the velocity field has the direction of the axis of rotation, and its magnitude equals twice the angular speed ω *of the rotation.*
Note that our result does not depend on the particular choice of the Cartesian coordinate system in space. ∎

For any twice continuously differentiable scalar function f,

$$\operatorname{curl} (\operatorname{grad} f) = \mathbf{0}, \tag{3}$$

as can easily be verified by direct calculation. *Hence if a vector function is the gradient of a scalar function, its curl is the zero vector.* Since the curl characterizes the rotation in a field, we also say more briefly that *gradient fields describing a motion are* ***irrotational.*** (If such a field occurs in some other connection, not as a velocity field, it is usually called ***conservative;*** see Sec. 8.9.)

EXAMPLE 3 The gravitational field in Example 3, Sec. 8.9, has curl $\mathbf{p} = \mathbf{0}$. The field in Example 2 of the present section is not irrotational. A similar velocity field is obtained by stirring coffee in a cup. ∎

The curl is defined in (1) in terms of coordinates, but if it is supposed to have a physical or geometrical significance, it should not depend on the choice of these coordinates. This is true, as follows.

Theorem 1 **(Invariance of the curl)**
The length and direction of curl $\mathbf{v}$ *are independent of the particular choice of Cartesian coordinate systems in space.* (Proof in Appendix 4.)

Problem Set 8.11

Find curl $\mathbf{v}$ where, with respect to right-handed Cartesian coordinates, $\mathbf{v}$ equals

1. $y\mathbf{i} + 2x\mathbf{j}$
2. $z\mathbf{i} + x\mathbf{j} + y\mathbf{k}$
3. $v_1(x)\mathbf{i} + v_2(y)\mathbf{j} + v_3(z)\mathbf{k}$
4. $\sin y\,\mathbf{i} + \cos x\,\mathbf{j}$
5. $(x^2 + y^2 + z^2)^{-3/2}(x\mathbf{i} + y\mathbf{j} + z\mathbf{k})$
6. $y^2\mathbf{i} + z^2\mathbf{j} + x^2\mathbf{k}$
7. $e^x(\cos z\,\mathbf{j} + \sin z\,\mathbf{k})$
8. $xyz(x\mathbf{i} + y\mathbf{j} + z\mathbf{k})$
9. $\ln (x^2 + y^2)\,\mathbf{i} + \arctan (y/x)\,\mathbf{j}$
10. $e^{xyz}(\mathbf{i} + \mathbf{j} + \mathbf{k})$

Fluid motion. In each case the velocity vector **v** of a steady fluid motion is given. Find curl **v**. Is the motion incompressible? Find the paths of the particles.

11. $\mathbf{v} = z^2\mathbf{j}$

12. $\mathbf{v} = y^3\mathbf{i}$

13. $\mathbf{v} = y\mathbf{i} - x\mathbf{j}$

14. $\mathbf{v} = -y^2\mathbf{i} + 2\mathbf{j}$

15. $\mathbf{v} = x\mathbf{i} + y\mathbf{j}$

16. $\mathbf{v} = -\frac{1}{4}y\mathbf{i} + 4x\mathbf{j}$

Formulas for curl, div, etc. Show that, granted sufficient differentiability,

17. $\text{curl}\,(\mathbf{u} + \mathbf{v}) = \text{curl}\,\mathbf{u} + \text{curl}\,\mathbf{v}$

18. $\text{div}\,(\text{curl}\,\mathbf{v}) = 0$

19. $\text{curl}\,(f\mathbf{v}) = \text{grad}\,f \times \mathbf{v} + f\,\text{curl}\,\mathbf{v}$

20. $\text{curl}\,(\text{grad}\,f) = \mathbf{0}$

21. $\text{div}\,(\mathbf{u} \times \mathbf{v}) = \mathbf{v} \cdot \text{curl}\,\mathbf{u} - \mathbf{u} \cdot \text{curl}\,\mathbf{v}$

22. $\text{div}\,(g\nabla f \times f\nabla g) = 0$

With respect to right-handed Cartesian coordinates, let $\mathbf{u} = z\mathbf{i} + x\mathbf{j} + y\mathbf{k}$, $\mathbf{v} = xy\mathbf{i} + yz\mathbf{j} + zx\mathbf{k}$ and $f = xyz$. Find the following expressions. If one of the formulas in Probs. 17–22 applies, use it to check your result.

23. $\text{curl}\,(\mathbf{u} + \mathbf{v})$

24. $\text{curl}\,(f\mathbf{u})$

25. $\text{div}\,(\mathbf{u} \times \mathbf{v})$

26. $\text{curl}\,(f\mathbf{v})$

27. $\mathbf{u} \cdot \text{curl}\,\mathbf{v}$, $\mathbf{u} \times \text{curl}\,\mathbf{v}$

28. $\text{curl}\,(\mathbf{u} \times \mathbf{v})$, $\text{curl}\,(\mathbf{v} \times \mathbf{u})$

29. $\mathbf{v} \times \text{curl}\,\mathbf{u}$

30. $\mathbf{v} \cdot \text{curl}\,\mathbf{v}$, $\mathbf{v} \times \text{curl}\,\mathbf{v}$

8.12 Grad, Div, Curl in Curvilinear Coordinates. *Optional*

In applications we often use coordinates other than Cartesian coordinates, for instance, when a problem involves cylindrical or spherical symmetry. (Examples will be shown in Chap. 11.) Then we may have to transform the gradient, divergence, or curl into these new coordinates. We show how this can be done and give explicit formulas for the cases of cylindrical and spherical coordinates, the two most important coordinate systems in space (other than Cartesian).

To simplify formulas, we write Cartesian coordinates x, y, z as

$$x = x_1, \qquad y = x_2, \qquad z = x_3.$$

Cylindrical coordinates r, θ, z are given by (Fig. 192*c*)

$$x_1 = r\cos\theta, \qquad x_2 = r\sin\theta, \qquad x_3 = z. \tag{1}$$

Spherical coordinates r, θ, ϕ are given by (Fig. 192*d*)[14]

$$x_1 = r\cos\theta\sin\phi, \qquad x_2 = r\sin\theta\sin\phi, \qquad x_3 = r\cos\phi. \tag{2}$$

[14] This is the notation used in calculus and many other books. It is logical since in it, θ plays the same role as in polar coordinates. *Caution!* Some books interchange the roles of θ and ϕ.

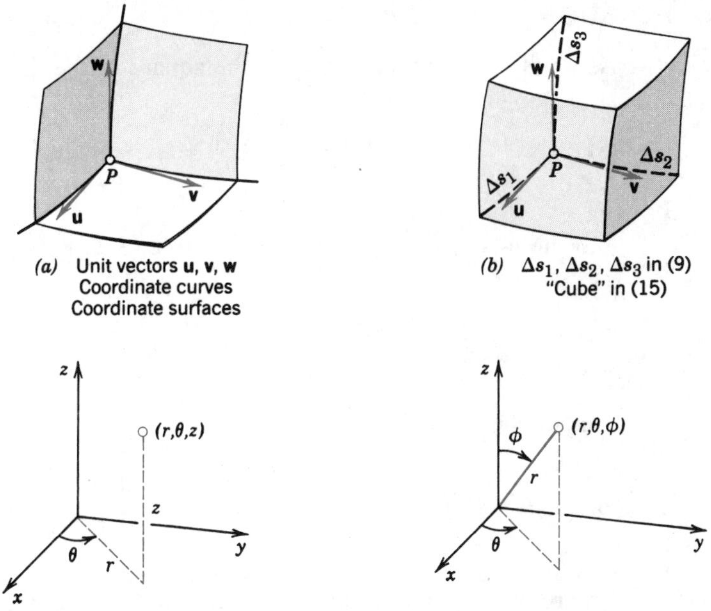

(a) Unit vectors **u**, **v**, **w** Coordinate curves Coordinate surfaces

(b) $\Delta s_1, \Delta s_2, \Delta s_3$ in (9) "Cube" in (15)

(c) Cylindrical coordinates defined by (1)

(d) Spherical coordinates defined by (2)

Fig. 192. Curvilinear coordinates

Equations (1) and (2) and similar equations for other coordinates q_1, q_2, q_3 in space are of the form

$$(3)\quad x_1 = x_1(q_1, q_2, q_3), \qquad x_2 = x_2(q_1, q_2, q_3), \qquad x_3 = x_3(q_1, q_2, q_3).$$

We assume that every point P is determined by a unique triple of coordinates q_1, q_2, q_3, so that we can solve (3) in the form

$$(4)\quad q_1 = q_1(x_1, x_2, x_3), \qquad q_2 = q_2(x_1, x_2, x_3), \qquad q_3 = q_3(x_1, x_2, x_3).$$

Through P there pass three surfaces $q_1 = const$, $q_2 = const$, $q_3 = const$, called **coordinate surfaces,** which are curved in general (Fig. 192*a*). Their three pairwise intersections are called **coordinate curves** through P. This motivates the name **"curvilinear coordinates"** of the q-coordinates. The coordinate systems used in practice suggest that we assume the q-coordinates to be **orthogonal,** that is, the coordinate curves through P are mutually perpendicular (their tangents at P make right angles, see Fig. 192*a*). For instance, cylindrical and spherical coordinates have this property (except when $r = 0$). We further assume the functions in (3) and (4) to be differentiable.

We transform the linear element, the gradient, the divergence, jointly with the latter the Laplacian, and finally the curl.

Transformation of the Linear Element

As a first task, let us transform the square of the linear element

$$ds^2 = \sum_{i=1}^{3} dx_i^2 = dx_1^2 + dx_2^2 + dx_3^2 \tag{5}$$ (Sec. 8.5)

into the q-coordinates. By the chain rule (Sec. 8.8) we have

$$dx_i = \sum_{j=1}^{3} \frac{\partial x_i}{\partial q_j} dq_j.$$

This implies

$$ds^2 = \sum_{i=1}^{3} dx_i^2 = \sum_{i=1}^{3} dx_i\, dx_i = \sum_{i=1}^{3} \left[\sum_{j=1}^{3} \frac{\partial x_i}{\partial q_j} dq_j \sum_{k=1}^{3} \frac{\partial x_i}{\partial q_k} dq_k \right] \tag{6}$$

(where we used a different summation letter, k, in the last sum because the two summations are independent). Since the x-coordinates and the q-coordinates are orthogonal, we have in (5) only squares, and similarly in (6). Consequently, we can write (6)

$$ds^2 = h_1^2\, dq_1^2 + h_2^2\, dq_2^2 + h_3^2\, dq_3^2 \tag{7}$$

and determine h_1^2, h_2^2, h_3^2 by collecting the coefficients of dq_1^2, dq_2^2, dq_3^2 in (6). Here, dq_1^2 is obtained in (6) by taking $j = 1$ and $k = 1$; hence h_1^2 has three terms because i goes from 1 to 3. Similarly for h_2^2, h_3^2:

$$h_1^2 = \sum_{i=1}^{3} \left(\frac{\partial x_i}{\partial q_1}\right)^2, \qquad h_2^2 = \sum_{i=1}^{3} \left(\frac{\partial x_i}{\partial q_2}\right)^2, \qquad h_3^2 = \sum_{i=1}^{3} \left(\frac{\partial x_i}{\partial q_3}\right)^2. \tag{8}$$

EXAMPLE 1 Linear element in cylindrical coordinates

In our present q-notation, cylindrical coordinates $q_1 = r$, $q_2 = \theta$, $q_3 = z$ are defined by (1) in the form

$$x_1 = q_1 \cos q_2, \qquad x_2 = q_1 \sin q_2, \qquad x_3 = q_3.$$

By differentiation we have from (8)

$$h_1^2 = \cos^2 q_2 + \sin^2 q_2 = 1$$

$$h_2^2 = (-q_1 \sin q_2)^2 + (q_1 \cos q_2)^2 = q_1^2$$

$$h_3^2 = 1.$$

Hence, by (7) the square of the linear element in cylindrical coordinates is

$$ds^2 = dq_1^2 + q_1^2\, dq_2^2 + dq_3^2,$$

that is,

$$ds^2 = dr^2 + r^2\, d\theta^2 + dz^2.$$

For $z = const$ we have $dz = 0$ and obtain the familiar $ds^2 = dr^2 + r^2\, d\theta^2$ in **polar coordinates** in the plane. ∎

EXAMPLE 2 Linear element in spherical coordinates

Spherical coordinates $q_1 = r$, $q_2 = \theta$, $q_3 = \phi$ are defined by [see (2)]

$$x_1 = q_1 \cos q_2 \sin q_3, \qquad x_2 = q_1 \sin q_2 \sin q_3, \qquad x_3 = q_1 \cos q_3.$$

Hence (8) gives

$$h_1^2 = (\cos q_2 \sin q_3)^2 + (\sin q_2 \sin q_3)^2 + \cos^2 q_3 = 1$$

$$h_2^2 = (-q_1 \sin q_2 \sin q_3)^2 + (q_1 \cos q_2 \sin q_3)^2 + 0 = q_1^2 \sin^2 q_3$$

$$h_3^2 = (q_1 \cos q_2 \cos q_3)^2 + (q_1 \sin q_2 \cos q_3)^2 + (-q_1 \sin q_3)^2 = q_1^2.$$

This and (7) give the square of the linear element in spherical coordinates

$$ds^2 = dq_1^2 + q_1^2 \sin^2 q_3\, dq_2^2 + q_1^2\, dq_3^2,$$

that is,

$$ds^2 = dr^2 + r^2 \sin^2 \phi\, d\theta^2 + r^2\, d\phi^2.$$ ∎

Transformation of the Gradient

For the x-coordinates we have [see (13*), Sec. 8.5, where $d\mathbf{x}$ is denoted $d\mathbf{r}$]

$$d\mathbf{x} = dx_1\, \mathbf{i} + dx_2\, \mathbf{j} + dx_3\, \mathbf{k}$$

where **i, j, k** are unit vectors in the directions of the coordinate axes, and

$$ds^2 = d\mathbf{x} \cdot d\mathbf{x} = dx_1^2 + dx_2^2 + dx_3^2.$$

Similarly, for the q-coordinates we can write

$$d\mathbf{q} = h_1\, dq_1\, \mathbf{u} + h_2\, dq_2\, \mathbf{v} + h_3\, dq_3\, \mathbf{w}$$

where **u, v, w** are orthogonal unit vectors in the directions of the coordinate curves at P (Fig. 192a; see before) and we obtain, in agreement with (7),

$$ds^2 = d\mathbf{q} \cdot d\mathbf{q} = h_1^2\, dq_1^2 + h_2^2\, dq_2^2 + h_3^2\, dq_3^2.$$

If we now let q_1 increase by Δq_1, then we move in the **u**-direction by $\Delta s_1 = h_1 \Delta q_1$ (see Fig. 192b). In the limit this gives $\partial q_1/\partial s_1 = 1/h_1$. Similarly for the other two directions. Together,

$$\text{(9)} \qquad \frac{\partial q_1}{\partial s_1} = \frac{1}{h_1}, \qquad \frac{\partial q_2}{\partial s_2} = \frac{1}{h_2}, \qquad \frac{\partial q_3}{\partial s_3} = \frac{1}{h_3}.$$

From this we now get the components of the gradient ∇f of a differentiable scalar function $f(q_1, q_2, q_3)$ as the projections in the **u**-, **v**-, **w**-directions; and on the other hand these are the directional derivatives $\partial f/\partial s_i$ in these three directions. Thus by (9),

$$\frac{\partial f}{\partial s_1} = \frac{\partial f}{\partial q_1}\frac{\partial q_1}{\partial s_1} = \frac{\partial f}{\partial q_1}\frac{1}{h_1}$$

and similarly for the other two components. This gives the following expression for the **gradient** ∇f in orthogonal curvilinear coordinates q_1, q_2, q_3 [with h_1, h_2, h_3 given by (8) and **u, v, w** shown in Fig. 192*a*]:

$$\text{(10)} \qquad \operatorname{grad} f = \nabla f = \frac{1}{h_1}\frac{\partial f}{\partial q_1}\mathbf{u} + \frac{1}{h_2}\frac{\partial f}{\partial q_2}\mathbf{v} + \frac{1}{h_3}\frac{\partial f}{\partial q_3}\mathbf{w}.$$

EXAMPLE 3 **Gradient**

In the cylindrical coordinates r, θ, z given by (1) we have (see Example 1)

$$\text{(11)} \qquad h_1 = h_r = 1, \qquad h_2 = h_\theta = q_1 = r, \qquad h_3 = h_z = 1$$

and (10) with $q_1 = r, q_2 = \theta, q_3 = z$ gives the gradient

$$\text{(12)} \qquad \operatorname{grad} f = \nabla f = \frac{\partial f}{\partial r}\mathbf{u} + \frac{1}{r}\frac{\partial f}{\partial \theta}\mathbf{v} + \frac{\partial f}{\partial z}\mathbf{w}.$$

In the spherical coordinates, $r\ \theta, \phi$ given by (2) we have (see Example 2)

$$\text{(13)} \qquad h_1 = h_r = 1, \qquad h_2 = h_\theta = q_1 \sin q_3 = r \sin\phi, \qquad h_3 = h_\phi = q_1 = r$$

and (10) with $q_1 = r, q_2 = \theta, q_3 = \phi$ gives the gradient

$$\text{(14)} \qquad \operatorname{grad} f = \nabla f = \frac{\partial f}{\partial r}\mathbf{u} + \frac{1}{r\sin\phi}\frac{\partial f}{\partial \theta}\mathbf{v} + \frac{1}{r}\frac{\partial f}{\partial \phi}\mathbf{w}.$$

Transformation of the Divergence and the Laplacian

For the **divergence** we get in orthogonal curvilinear coordinates q_1, q_2, q_3 [with h_1, h_2, h_3 given by (8)]

$$\text{(15)} \qquad \operatorname{div} \mathbf{F} = \nabla \cdot \mathbf{F} = \frac{1}{h_1 h_2 h_3}\left[\frac{\partial}{\partial q_1}(h_2 h_3 F_1) + \frac{\partial}{\partial q_2}(h_3 h_1 F_2) + \frac{\partial}{\partial q_3}(h_1 h_2 F_3)\right].$$

This follows as in Example 2 of Sec. 8.10 by considering the flux through the faces of the "cube" in Fig. 192*b*, the divergence being the limit of the flux divided by the volume $h_1\Delta q_1 h_2 \Delta q_2 h_3 \Delta q_3$; here $h_2 \Delta q_2 h_3 \Delta q_3$ is the area of the faces $q_1 = const$ and $q_1 + h_1\Delta q_1 = const$, so that $\partial(h_2 h_3 F_1)/\partial q_1$ results from the contribution of these faces to the flux. The other two terms in (15) correspond to the other two pairs of faces of the "cube."

EXAMPLE 4 **Divergence**

In the cylindrical coordinates given by (1) we obtain from (11) and (15) with $q_1 = r$, $q_2 = \theta$, $q_3 = z$ the divergence

$$\nabla \cdot \mathbf{F} = \frac{1}{r}\left[\frac{\partial}{\partial r}(rF_1) + \frac{\partial F_2}{\partial \theta} + \frac{\partial}{\partial z}(rF_3)\right],$$

that is,

$$\text{div}\, \mathbf{F} = \nabla \cdot \mathbf{F} = \frac{1}{r}\frac{\partial}{\partial r}(rF_1) + \frac{1}{r}\frac{\partial F_2}{\partial \theta} + \frac{\partial F_3}{\partial z}. \tag{16}$$

In the spherical coordinates given by (2) we obtain from (13) and (15) with $q_1 = r$, $q_2 = \theta$, $q_3 = \phi$ the divergence

$$\nabla \cdot \mathbf{F} = \frac{1}{r^2 \sin \phi}\left[\frac{\partial}{\partial r}(r^2 \sin \phi\, F_1) + \frac{\partial}{\partial \theta}(rF_2) + \frac{\partial}{\partial \phi}(r \sin \phi\, F_3)\right],$$

that is,

$$\text{div}\, \mathbf{F} = \nabla \cdot \mathbf{F} = \frac{1}{r^2}\frac{\partial}{\partial r}(r^2F_1) + \frac{1}{r \sin \phi}\frac{\partial F_2}{\partial \theta} + \frac{1}{r \sin \phi}\frac{\partial}{\partial \phi}(\sin \phi\, F_3). \tag{17}$$

Since $\nabla^2 f = \nabla \cdot \nabla f = \text{div}\,(\text{grad}\, f)$ (see Sec. 8.10), we also obtain from (15) with $\mathbf{F} = \nabla f$ and ∇f given by (10) the **Laplacian** in orthogonal curvilinear coordinates q_1, q_2, q_3

$$\nabla^2 f = \frac{1}{h_1h_2h_3}\left[\frac{\partial}{\partial q_1}\left(\frac{h_2h_3}{h_1}\frac{\partial f}{\partial q_1}\right) + \frac{\partial}{\partial q_2}\left(\frac{h_3h_1}{h_2}\frac{\partial f}{\partial q_2}\right) + \frac{\partial}{\partial q_3}\left(\frac{h_1h_2}{h_3}\frac{\partial f}{\partial q_3}\right)\right] \tag{18}$$

with h_1, h_2, h_3 given by (8).

EXAMPLE 5 **Laplacian**

In the cylindrical coordinates $q_1 = r$, $q_2 = \theta$, $q_3 = z$ given by (1) we obtain from (11) and (18) the Laplacian

$$\nabla^2 f = \frac{1}{r}\left[\frac{\partial}{\partial r}\left(r\frac{\partial f}{\partial r}\right) + \frac{\partial}{\partial \theta}\left(\frac{1}{r}\frac{\partial f}{\partial \theta}\right) + \frac{\partial}{\partial z}\left(r\frac{\partial f}{\partial z}\right)\right];$$

hence

$$\nabla^2 f = \frac{1}{r}\frac{\partial}{\partial r}\left(r\frac{\partial f}{\partial r}\right) + \frac{1}{r^2}\frac{\partial^2 f}{\partial \theta^2} + \frac{\partial^2 f}{\partial z^2},$$

or by performing the indicated product differentiation,

$$\nabla^2 f = \frac{\partial^2 f}{\partial r^2} + \frac{1}{r}\frac{\partial f}{\partial r} + \frac{1}{r^2}\frac{\partial^2 f}{\partial \theta^2} + \frac{\partial^2 f}{\partial z^2}. \tag{19}$$

In the spherical coordinates $q_1 = r$, $q_2 = \theta$, $q_3 = \phi$ given by (2) we obtain from (13) and (18) the Laplacian

$$\nabla^2 f = \frac{1}{r^2 \sin\phi}\left[\frac{\partial}{\partial r}\left(r^2 \sin\phi \frac{\partial f}{\partial r}\right) + \frac{\partial}{\partial \theta}\left(\frac{1}{\sin\phi}\frac{\partial f}{\partial \theta}\right) + \frac{\partial}{\partial \phi}\left(\sin\phi \frac{\partial f}{\partial \phi}\right)\right];$$

hence

$$\nabla^2 f = \frac{1}{r^2}\frac{\partial}{\partial r}\left(r^2 \frac{\partial f}{\partial r}\right) + \frac{1}{r^2 \sin^2\phi}\frac{\partial^2 f}{\partial \theta^2} + \frac{1}{r^2 \sin\phi}\frac{\partial}{\partial \phi}\left(\sin\phi \frac{\partial f}{\partial \phi}\right),$$

or by performing the indicated product differentiations,

$$\nabla^2 f = \frac{\partial^2 f}{\partial r^2} + \frac{2}{r}\frac{\partial f}{\partial r} + \frac{1}{r^2 \sin^2\phi}\frac{\partial^2 f}{\partial \theta^2} + \frac{1}{r^2}\frac{\partial^2 f}{\partial \phi^2} + \frac{\cot\phi}{r^2}\frac{\partial f}{\partial \phi}. \tag{20}$$

■

Transformation of the Curl

For curl $\mathbf{F} = \nabla \times \mathbf{F}$ the general formula in *right-handed* orthogonal curvilinear coordinates q_1, q_2, q_3 is the following. (Proof in Ref. [7] listed in Appendix 1.)

$$\operatorname{curl} \mathbf{F} = \nabla \times \mathbf{F} = \frac{1}{h_1 h_2 h_3}\begin{vmatrix} h_1\mathbf{u} & h_2\mathbf{v} & h_3\mathbf{w} \\ \dfrac{\partial}{\partial q_1} & \dfrac{\partial}{\partial q_2} & \dfrac{\partial}{\partial q_3} \\ h_1F_1 & h_2F_2 & h_3F_3 \end{vmatrix}. \tag{21}$$

From this and (11) or (13) the curl can now readily be obtained in cylindrical or spherical coordinates.

Because of the practical importance of cylindrical and spherical coordinates we have used them for illustration in our examples, but we want to emphasize that our formulas are *general* and can be used in any orthogonal coordinate system.

Review Questions and Problems for Chapter 8

1. Give examples of quantities that are vectors. Give examples of vector fields.
2. Why do the components of a vector remain unchanged if we change the initial point of the vector?
3. What are typical applications that motivate the inner product and the vector product?
4. In what case will the work done by a constant force in a displacement be zero?

5. What are right-handed and left-handed coordinates? Where in this chapter was this distinction essential?
6. The vector product has two unusual properties. Can you remember them?
7. Vector formulas in two and three dimensions often look the same. Explain this in terms of representations of a circle and a sphere. Can you think of other examples?
8. What do we mean by orthogonality? Why is this concept important, and in what connection?
9. What is wrong with $\mathbf{a} \times \mathbf{b} \times \mathbf{c}$? With $\mathbf{a} \cdot \mathbf{b} \cdot \mathbf{c}$? With $(\mathbf{a} \cdot \mathbf{b}) \times \mathbf{c}$?
10. How many different bases does R^3 have?
11. What is an orthonormal basis? What is the advantage of using it?
12. Without looking in the text, state the definition of the derivative of a vector function $\mathbf{v}(t)$ and of the first partial derivatives of a vector function $\mathbf{v}(x, y, z)$. What is $\mathbf{v}'(t)$ geometrically?
13. If $\mathbf{r}(t)$ represents a curve, what is $\mathbf{r}'(t)$ in mechanics when t is time?
14. Can a body have constant speed and still have a varying velocity vector?

15. What do we mean by the directional derivative?
16. What are the directional derivatives of a function $f(x, y, z)$ in the x-, y-, and z-directions? In the negative x-, y-, and z-directions?
17. Without looking in the text, write down the definitions of grad, div, and curl.
18. What properties of a scalar function do the direction and length of the gradient characterize?
19. If f is a scalar function and $\mathbf{v}$ a vector function, both sufficiently often differentiable, which of the following expressions make sense: grad f, f grad f, $\mathbf{v} \cdot$grad f, div f, div $(f\mathbf{v})$, curl $(f\mathbf{v})$ curl f, f curl $\mathbf{v}$?
20. Can you remember the role of the divergence in connection with fluid motions? The role of the curl in connection with rotations?

Let $\mathbf{a} = [2, -1, 5]$, $\mathbf{b} = [1, -2, 3]$, $\mathbf{c} = [2, 2, -1]$, $\mathbf{d} = [3, 0, 2]$. Find the following expressions. (In connection with vector products assume the coordinate system to be right-handed.)

21. $\mathbf{b} + \mathbf{c} - \mathbf{a}$
22. $\mathbf{a} + \mathbf{b} + \mathbf{c} + \mathbf{d}$
23. $4\mathbf{a} + 8\mathbf{b}$, $4(2\mathbf{b} + \mathbf{a})$
24. $5\mathbf{a} + 2\mathbf{b}$, $|5\mathbf{a} + 2\mathbf{b}|$
25. $|\mathbf{c} - \mathbf{b}|$, $|\mathbf{c}| - |\mathbf{b}|$
26. $(1/|\mathbf{c}|)\mathbf{c}$, $(1/|\mathbf{d}|)\mathbf{d}$
27. $\mathbf{a} \cdot \mathbf{b}$, $\mathbf{b} \cdot \mathbf{a}$, $\mathbf{c} \cdot \mathbf{d}$, $\mathbf{d} \cdot \mathbf{c}$
28. $8\mathbf{c} \cdot \mathbf{d}$, $4\mathbf{d} \cdot 2\mathbf{c}$
29. $\mathbf{a} \times \mathbf{b}$, $\mathbf{b} \times \mathbf{a}$
30. $\mathbf{a} \times \mathbf{d}$, $|\mathbf{a} \times \mathbf{d}|$
31. $(\mathbf{b} \quad -\mathbf{b} \quad \mathbf{d})$, $(\mathbf{b} \quad \mathbf{b} + \mathbf{d} \quad \mathbf{d})$
32. $(\mathbf{b} \quad \mathbf{c} \quad \mathbf{d})$, $(\mathbf{c} \quad \mathbf{d} \quad \mathbf{b})$
33. $(\mathbf{c} \quad \mathbf{b} \quad \mathbf{a})$, $(\mathbf{b} \quad \mathbf{c} \quad 2\mathbf{a})$
34. $\mathbf{d} \times (\mathbf{c} \times \mathbf{b})$, $(\mathbf{d} \times \mathbf{c}) \times \mathbf{b}$

35. Find a force $\mathbf{p}$ such that the resultant of $\mathbf{p}$, $\mathbf{q} = [1, -2, 2]$ and $\mathbf{u} = [2, -3, 4]$ is the zero vector.
36. Find the angle between the forces $[2, 2]$ and $[3, 1]$. Make a figure.
37. Find the angle between the forces $[2, 1, 3]$ and $[4, 4, 0]$.

38. Find the angles α, β, γ between the force $\mathbf{p} = [3, 6, 8]$ and the coordinate axes. Verify that $\cos^2 \alpha + \cos^2 \beta + \cos^2 \gamma = 1$. Prove this in general.

Find the work done by a force $\mathbf{p}$ acting on a body if the body is displaced from a point P to a point Q along the straight segment PQ, where P, Q, and $\mathbf{p}$ are

39. $(1, 1), (4, 3), [3, 8]$

40. $(6, 8), (8, 6), [14, -2]$

41. $(-2, 2), (0, 10), [8, -2]$

42. $(-8, 4, 0), (8, 2, 9), [4, 9, -2]$

43. $(3, 3, 8), (4, 6, 1), [3, 4, 6]$

44. $(1, 5, 9), (-1, 1, 1), [-2, -4, -8]$

Find the component of a force $\mathbf{p}$ in the direction of a vector $\mathbf{b}$, where

45. $\mathbf{p} = [1, 2], \mathbf{b} = [4, 0]$

46. $\mathbf{p} = [-5, 3], \mathbf{b} = [8, 6]$

47. $\mathbf{p} = [4, 0, -2], \mathbf{b} = [1, -3, 5]$

48. $\mathbf{p} = [6, 2, -\frac{1}{2}], \mathbf{b} = [3, 1, 40]$

49. In what case is the component of $\mathbf{a}$ in the direction of $\mathbf{b}$ equal to the component of $\mathbf{b}$ in the direction of $\mathbf{a}$?

Find the moment vector $\mathbf{m}$ of a force $\mathbf{p}$ about a point Q if $\mathbf{p}$ acts on a line through a point A, where $\mathbf{p}$, A, Q are

50. $[0, 10, 0], (2, 3, 0), (0, 0, 0)$

51. $[4, 3, 0], (6, 5, 0), (9, 3, 0)$

52. $[0, 2, -2], (0, 7, 1), (0, 2, 8)$

53. $[1, 4, 9], (3, 5, 7), (2, 0, 1)$

54. $[2, 1, -10], (4, 7, 1), (3, 3, 6)$

55. $[5, 4, 3], (3, 4, 5), (-1, 1, 3)$

Find the volume of the parallelepiped with edge vectors

56. $[4, 3, 2], [1, 0, -1], [0, -2, -3]$

57. $[7, 6, 4], [1, 9, 3], [8, 15, 6]$

Find the volume of the tetrahedron with vertices

58. $(2, 1, 8), (3, 2, 9), (2, 1, 4), (3, 3, 10)$

59. $(-1, 0, 1), (4, 4, 5), (0, 1, 0), (2, 2, 0)$

Find a representation of the plane through the points

60. $(-1, 0, 2), (8, 6, 4), (1, 3, 5)$

61. $(1, 1, 1), (5, 0, -5), (3, 2, 0)$

Let $f = 4xyz$, $g = x^4 + y^4 + z$, $\mathbf{v} = x^2\mathbf{i} + (y - z)^2\mathbf{j} + xy\mathbf{k}$, $\mathbf{w} = (x + y)^2\mathbf{i} + z^2\mathbf{j} + 2yz\mathbf{k}$. (Assume the coordinate system to be right-handed whenever this is essential.) Find

62. grad f at $(5, 7, -2)$

63. grad g at $(4, -1, 3)$

64. $\nabla^2 f$, $\nabla^2 g$

65. div $\mathbf{v}$ at $(3, 1, 5)$

66. $\mathbf{w} \cdot \nabla g$

67. grad (div $\mathbf{w}$)$\cdot\mathbf{v}$

68. $\mathbf{v} \times \mathbf{w}$, $\mathbf{w} \times \mathbf{v}$

69. $\nabla f \cdot \nabla g$, $\nabla g \cdot \nabla f$

70. curl $\mathbf{v}$, curl $\mathbf{w}$

71. div (curl $(\mathbf{v} + \mathbf{w})$)

72. curl $(\mathbf{v} \times \mathbf{k})$

73. curl (grad g)$\cdot\mathbf{v}$

74. curl $(f(\mathbf{i} + \mathbf{j}))$

75. $D_{\mathbf{v}} f$ at $(1, 1, 1)$

76. $D_{\mathbf{w}} g$ at $(3, 0, 2)$

77. $D_{\mathbf{v}}(\mathbf{v} \cdot \mathbf{k})$ at $(4, 3, 0)$

Indicate the kind of curve represented by $\mathbf{r}(t)$, where t is time, and find the velocity vector, speed, and acceleration vector at P, where

78. $\mathbf{r} = (1 + \cos t)\mathbf{i} + 9 \sin t\, \mathbf{j}$, $P: (1, -9)$

79. $\mathbf{r} = t\mathbf{i} + t^{-1}\mathbf{j}$, $P: (2, \frac{1}{2})$

80. $\mathbf{r} = \cos 2t\, \mathbf{i} + \sin 2t\, \mathbf{j} + t\mathbf{k}$, $P: (1, 0, \pi)$

Summary of Chapter 8 Vector Differential Calculus. Grad, Div, Curl

All vectors of the form

(1) $$\mathbf{a} = [a_1, a_2, a_3] = a_1\mathbf{i} + a_2\mathbf{j} + a_3\mathbf{k} \qquad \text{(Sec. 8.1)}$$

constitute the real vector space R^3 with *vector addition* defined by

(2) $$[a_1, a_2, a_3] + [b_1, b_2, b_3] = [a_1 + b_1, a_2 + b_2, a_3 + b_3]$$

and *scalar multiplication* defined by

(3) $$c[a_1, a_2, a_3] = [ca_1, ca_2, ca_3] \qquad \text{(Sec. 8.1)},$$

c being a scalar (a real number). For instance, the *resultant* of forces **a** and **b** is **a** + **b**.

The **inner product** or **dot product** of two vectors is defined by

(4) $$\mathbf{a} \cdot \mathbf{b} = |\mathbf{a}||\mathbf{b}| \cos \gamma = a_1 b_1 + a_2 b_2 + a_3 b_3 \qquad \text{(Sec. 8.2)}$$

(γ the angle between **a** and **b**). This gives for the **norm** or **length** of **a**

(5) $$|\mathbf{a}| = \sqrt{\mathbf{a} \cdot \mathbf{a}} = \sqrt{a_1^2 + a_2^2 + a_3^2}$$

as well as a formula for γ. If $\mathbf{a} \cdot \mathbf{b} = 0$, we call **a** and **b** **orthogonal.** The dot product is suggested by the *work* $W = \mathbf{p} \cdot \mathbf{d}$ done by a force **p** in a displacement **d**.

The **vector product** or **cross product** $\mathbf{v} = \mathbf{a} \times \mathbf{b}$ is a vector of length

(6) $$|\mathbf{a} \times \mathbf{b}| = |\mathbf{a}|\,|\mathbf{b}| \sin\gamma \qquad \text{(Sec. 8.3)}$$

and perpendicular to both **a** and **b** such that **a**, **b**, **v** form a *right-handed* triple. In terms of components with respect to right-handed coordinates,

(7) $$\mathbf{a} \times \mathbf{b} = \begin{vmatrix} \mathbf{i} & \mathbf{j} & \mathbf{k} \\ a_1 & a_2 & a_3 \\ b_1 & b_2 & b_3 \end{vmatrix} \qquad \text{(Sec. 8.3)}.$$

The vector product is suggested, for instance, by moments of forces or by rotations. Caution! This multiplication is *anti*commutative, $\mathbf{a} \times \mathbf{b} = -\mathbf{b} \times \mathbf{a}$, and is *not* associative.

Most important among repeated products is the scalar triple product

$$(\mathbf{a}\ \ \mathbf{b}\ \ \mathbf{c}) = \mathbf{a} \cdot (\mathbf{b} \times \mathbf{c}) = (\mathbf{a} \times \mathbf{b}) \cdot \mathbf{c}. \tag{8}$$

Sections 8.4–8.12 concern the extension of differential calculus to **vector functions.** If they depend on a single variable, t, they are of the form

$$\mathbf{v}(t) = [v_1(t),\ v_2(t),\ v_3(t)] = v_1(t)\mathbf{i} + v_2(t)\mathbf{j} + v_3(t)\mathbf{k}. \tag{9}$$

Then the derivative is

$$\mathbf{v}' = \frac{d\mathbf{v}}{dt} = \lim_{\Delta t \to 0} \frac{\mathbf{v}(t + \Delta t) - \mathbf{v}(t)}{\Delta t} \qquad \text{(Sec. 8.4).} \tag{10}$$

Formula (10) looks quite similar to the familiar formula in calculus. It implies that in terms of components,

$$\mathbf{v}' = [v_1',\ v_2',\ v_3'] = v_1'\mathbf{i} + v_2'\mathbf{j} + v_3'\mathbf{k}.$$

Differentiation rules are as in calculus. They imply (Sec. 8.4)

$$(\mathbf{u} \cdot \mathbf{v})' = \mathbf{u}' \cdot \mathbf{v} + \mathbf{u} \cdot \mathbf{v}', \qquad (\mathbf{u} \times \mathbf{v})' = \mathbf{u}' \times \mathbf{v} + \mathbf{u} \times \mathbf{v}'.$$

A vector function of a single variable t, usually denoted by $\mathbf{r}(t)$, can be used to represent a curve C in space. Then $\mathbf{r}(t)$ associates with each $t = t_0$ in some interval $a < t < b$ the point of C with position vector $\mathbf{r}(t_0)$. The derivative $\mathbf{r}'(t)$ is a **tangent vector** of C (Sec. 8.5). If we choose the **arc length** s (Sec. 8.5) as a parameter, then we get the *unit tangent vector* $\mathbf{r}'(s)$. The length of the derivative of this vector is the **curvature** $\kappa(s) = |\mathbf{r}''(s)|$; see Sec. 8.7.

In mechanics, $\mathbf{r}(t)$ may represent the path of a moving body, where t is time. Then $\mathbf{v}(t) = \mathbf{r}'(t)$ is the **velocity vector,** its length is the *speed,* and its derivative $\mathbf{a}(t) = \mathbf{v}'(t) = \mathbf{r}''(t)$ is the **acceleration vector** (see Sec. 8.6).

A vector function of Cartesian coordinates x, y, z in space, say,

$$\begin{aligned} \mathbf{v}(x, y, z) &= [v_1(x, y, z),\ \ v_2(x, y, z),\ \ v_3(x, y, z)] \\ &= v_1(x, y, z)\mathbf{i} + v_2(x, y, z)\mathbf{j} + v_3(x, y, z)\mathbf{k}, \end{aligned} \tag{11}$$

defines a **vector field** in the region in space in which $\mathbf{v}$ is defined;

with every point P_0: (x_0, y_0, z_0) in that region, **v** associates a vector $\mathbf{v}(P_0) = \mathbf{v}(x_0, y_0, z_0)$. Partial derivatives of **v** are obtained by taking partial derivatives of components; for instance,

$$\frac{\partial \mathbf{v}}{\partial x} = \left[\frac{\partial v_1}{\partial x}, \frac{\partial v_2}{\partial x}, \frac{\partial v_3}{\partial x}\right] = \frac{\partial v_1}{\partial x}\mathbf{i} + \frac{\partial v_2}{\partial x}\mathbf{j} + \frac{\partial v_3}{\partial x}\mathbf{k}.$$

Section 8.8 on the chain rule and a few other facts about functions of several variables prepared us for Secs. 8.9–8.11, in which we discussed the **gradient** of a scalar function f (Sec. 8.9),

$$\text{(12)} \qquad \operatorname{grad} f = \nabla f = \left[\frac{\partial f}{\partial x}, \frac{\partial f}{\partial y}, \frac{\partial f}{\partial z}\right],$$

the **divergence** of a vector function **v** (Sec. 8.10),

$$\text{(13)} \qquad \operatorname{div} \mathbf{v} = \nabla \cdot \mathbf{v} = \frac{\partial v_1}{\partial x} + \frac{\partial v_2}{\partial y} + \frac{\partial v_3}{\partial z},$$

and the **curl** of **v** (Sec. 8.11),

$$\text{(14)} \qquad \operatorname{curl} \mathbf{v} = \nabla \times \mathbf{v} = \begin{vmatrix} \mathbf{i} & \mathbf{j} & \mathbf{k} \\ \dfrac{\partial}{\partial x} & \dfrac{\partial}{\partial y} & \dfrac{\partial}{\partial z} \\ v_1 & v_2 & v_3 \end{vmatrix}$$

(with a minus sign in front of the determinant for left-handed coordinates). Some basic formulas for grad, div, curl are (Secs. 8.9–8.11)

$$\text{(15)} \qquad \nabla(fg) = f\nabla g + g\nabla f, \qquad \nabla(f/g) = (1/g^2)(g\nabla f - f\nabla g)$$

$$\text{(16)} \qquad \operatorname{div}(f\mathbf{v}) = f \operatorname{div} \mathbf{v} + \mathbf{v} \cdot \nabla f, \qquad \operatorname{div}(f\nabla g) = f\nabla^2 g + \nabla f \cdot \nabla g$$

$$\text{(17)} \qquad \nabla^2 f = \operatorname{div}(\nabla f), \qquad \nabla^2(fg) = g\nabla^2 f + 2\nabla f \cdot \nabla g + f\nabla^2 g$$

$$\text{(18)} \qquad \begin{aligned} \operatorname{curl}(f\mathbf{v}) &= \nabla f \times \mathbf{v} + f \operatorname{curl} \mathbf{v} \\ \operatorname{div}(\mathbf{u} \times \mathbf{v}) &= \mathbf{v} \cdot \operatorname{curl} \mathbf{u} - \mathbf{u} \cdot \operatorname{curl} \mathbf{v} \end{aligned}$$

$$\text{(19)} \qquad \operatorname{curl}(\nabla f) = \mathbf{0}, \qquad \operatorname{div}(\operatorname{curl} \mathbf{v}) = 0.$$

The form of grad, div, curl and the Laplacian in general rectangular coordinates is shown in Sec. 8.12.

The **directional derivative** of f in the direction of a ***unit*** vector **b** is

$$\text{(20)} \qquad D_{\mathbf{b}} f = \frac{df}{ds} = \mathbf{b} \cdot \nabla f \qquad \text{(Sec. 8.9).}$$

Chapter 9

Vector Integral Calculus. Integral Theorems

In this chapter we shall define line integrals and surface integrals and consider some important applications of such integrals, which occur frequently in connection with physical and engineering problems. We shall see that a line integral is a natural generalization of a definite integral, and a surface integral is a generalization of a double integral.

Line integrals can be transformed into double integrals (Sec. 9.4) or into surface integrals (Sec. 9.9), and conversely. Triple integrals can be transformed into surface integrals (Sec. 9.7), and conversely. These transformations are of great practical importance. The corresponding formulas of Gauss, Green, and Stokes serve as powerful tools in many applications as well as in theoretical problems. We shall see that they also lead to a better understanding of the physical meaning of the divergence and the curl of a vector function.

Prerequisites for this chapter: elementary integral calculus and Chap. 8.
Sections that may be omitted in a shorter course: 9.3, 9.5, 9.8.
References: Appendix 1, Part B.
Answers to problems: Appendix 2.

9.1 Line Integrals

The concept of a line integral is a simple and natural generalization of the concept of a definite integral

$$\int_a^b f(x)\,dx. \tag{1}$$

In (1) we integrate along the x-axis from a to b and the integrand f is a function defined at each point between a and b. In the case of a line integral we shall integrate along a curve C in space (or in the plane) and the integrand will be a function defined at each point of C. (Hence *curve integral* would be a better name, but *line integral* is the standard term.) The curve C is called the **path of integration.** (For a discussion of curves, see Sec. 8.5.)

We call C a **smooth curve** if C has a representation

$$\mathbf{r}(t) = [x(t), y(t), z(t)] = x(t)\mathbf{i} + y(t)\mathbf{j} + z(t)\mathbf{k} \qquad (a \leqq t \leqq b) \tag{2}$$

such that $\mathbf{r}(t)$ has a continuous derivative $\mathbf{r}'(t) = d\mathbf{r}/dt$ that is nowhere the zero vector. Geometrically: the curve C has a unique tangent at each of its points whose direction varies continuously as we traverse C. We call A: $\mathbf{r}(a)$ the *initial point* and B: $\mathbf{r}(b)$ the *terminal point* of C, and we say that C is now *oriented*, calling the direction from A to B the *positive direction* along C and indicating it by an arrow (as in Fig. 193*a*). The points A and B may coincide (as in Fig. 193*b*); then C is called a *closed curve.*

General assumption

In this book, every path of integration of a line integral is assumed to be **piecewise smooth;** *that is, it consists of finitely many smooth curves.*

Definition and Evaluation of Line Integrals

A **line integral** of a vector function $\mathbf{F}(\mathbf{r})$ over a curve C is defined by

$$\int_C \mathbf{F}(\mathbf{r}) \cdot d\mathbf{r} = \int_a^b \mathbf{F}(\mathbf{r}(t)) \cdot \frac{d\mathbf{r}}{dt}\, dt \qquad (3) \qquad \text{[see (2)]}$$

(see Sec. 8.2 for the dot product). In terms of components, with $d\mathbf{r} = [dx,\ dy,\ dz]$ as in Sec. 8.5 and $' = d/dt$, formula (3) becomes

$$\begin{aligned} \int_C \mathbf{F}(\mathbf{r}) \cdot d\mathbf{r} &= \int_C (F_1\, dx + F_2\, dy + F_3\, dz) \\ &= \int_a^b (F_1 x' + F_2 y' + F_3 z')\, dt. \end{aligned} \qquad (3')$$

If the path of integration C in (3) is a *closed* curve, then instead of

$$\int_C \quad \text{we also write} \quad \oint_C .$$

$\mathbf{F} \cdot d\mathbf{r}/dt$ with $t = s$ (the arc length of C) is the tangential component of $\mathbf{F}$, and this integral arises naturally in mechanics, where it gives the work done by a force $\mathbf{F}$ in a displacement along C (details and examples below). We may thus call the line integral (3) the **work integral.** Other forms of the line

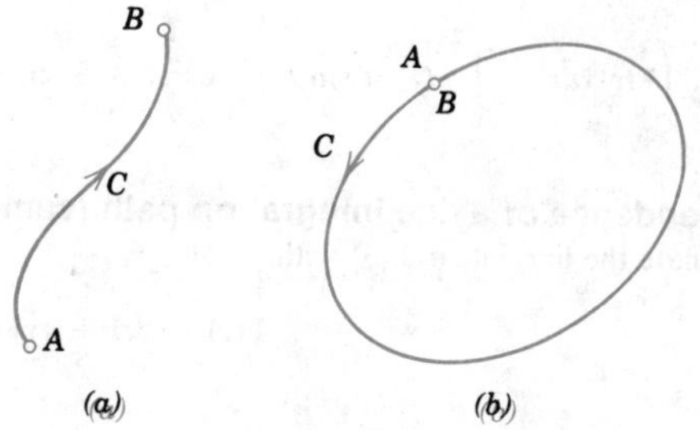

Fig. 193. Oriented curve

integral will be discussed later in this section. We see that the integral in (3) on the right is a definite integral taken over the interval $a \leqq t \leqq b$ on the t-axis in the ***positive*** direction (the direction of increasing t). This definite integral exists for continuous **F** and piecewise smooth C, because this makes $\mathbf{F} \cdot \mathbf{r}'$ piecewise continuous.

EXAMPLE 1 Evaluation of a line integral in the plane

Find the value of the line integral (3) when $\mathbf{F}(\mathbf{r}) = -y\mathbf{i} + xy\mathbf{j}$ and C is the circular arc in Fig. 194 from A to B.

Solution. We may represent C by

$$\mathbf{r}(t) = \cos t\,\mathbf{i} + \sin t\,\mathbf{j} \qquad (0 \leqq t \leqq \pi/2).$$

Thus $x(t) = \cos t$, $y(t) = \sin t$, so that

$$\mathbf{F}(\mathbf{r}(t)) = -y(t)\mathbf{i} + x(t)y(t)\mathbf{j} = -\sin t\,\mathbf{i} + \cos t \sin t\,\mathbf{j}.$$

By differentiation,

$$\mathbf{r}'(t) = -\sin t\,\mathbf{i} + \cos t\,\mathbf{j}$$

so that

$$\int_C \mathbf{F}(\mathbf{r}) \cdot d\mathbf{r} = \int_0^{\pi/2} (-\sin t\,\mathbf{i} + \cos t \sin t\,\mathbf{j}) \cdot (-\sin t\,\mathbf{i} + \cos t\,\mathbf{j})\,dt$$

$$= \int_0^{\pi/2} (\sin^2 t + \cos^2 t \sin t)\,dt = \frac{\pi}{4} + \frac{1}{3} \approx 1.119$$

[use (10) in Appendix 3.1; set $\cos t = u$ in the second term]. ∎

Two important questions now arise. Does this value depend on the particular choice of a representation of the circular arc C? The answer is no; see Theorem 1 below. Does this value change if we integrate from the same A to the same B as before but along another path? The answer is yes, in general; see Example 3.

EXAMPLE 2 Line integral in space

To see that the method of computing line integrals in space is the same as that in the plane just considered, find the value of (3) when $\mathbf{F}(\mathbf{r}) = z\mathbf{i} + x\mathbf{j} + y\mathbf{k}$ and C is the helix (Fig. 195)

$$\text{(4)} \qquad \mathbf{r}(t) = \cos t\,\mathbf{i} + \sin t\,\mathbf{j} + 3t\mathbf{k} \qquad (0 \leqq t \leqq 2\pi).$$

Solution. From (4) we have $x(t) = \cos t$, $y(t) = \sin t$, $z(t) = 3t$. Thus

$$\mathbf{F}(\mathbf{r}(t)) \cdot \mathbf{r}'(t) = (3t\mathbf{i} + \cos t\,\mathbf{j} + \sin t\,\mathbf{k}) \cdot (-\sin t\,\mathbf{i} + \cos t\,\mathbf{j} + 3\mathbf{k}).$$

The dot product is $3t(-\sin t) + \cos^2 t + 3 \sin t$. Hence (3) gives

$$\int_C \mathbf{F}(\mathbf{r}) \cdot d\mathbf{r} = \int_0^{2\pi} (-3t \sin t + \cos^2 t + 3 \sin t)\,dt = 6\pi + \pi + 0 = 7\pi \approx 21.99.$$ ∎

EXAMPLE 3 Dependence of a line integral on path (same endpoints)

Evaluate the line integral (3) with

$$\mathbf{F}(\mathbf{r}) = 5z\mathbf{i} + xy\mathbf{j} + x^2z\mathbf{k}$$

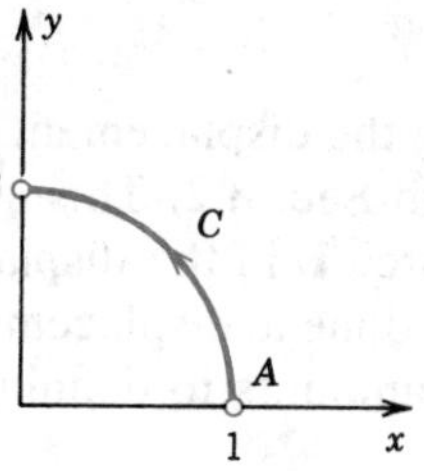

Fig. 194. Example 1

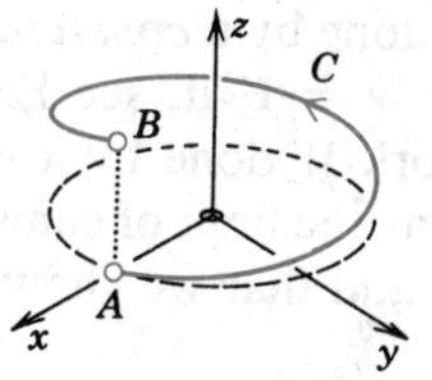

Fig. 195. Example 2

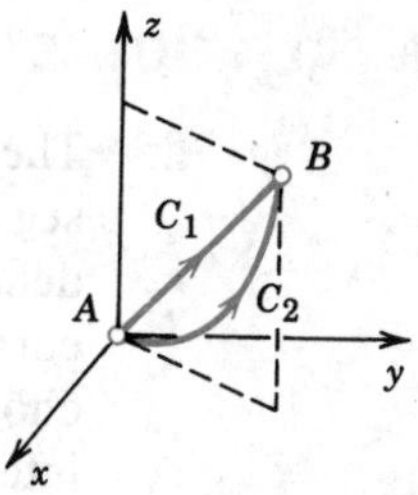

Fig. 196. Example 3

along two different paths with the same initial point A: (0, 0, 0) and the same terminal point B: (1, 1, 1), namely (Fig. 196),

(a) C_1: the straight-line segment $\mathbf{r}_1(t) = t\mathbf{i} + t\mathbf{j} + t\mathbf{k}, \quad 0 \leqq t \leqq 1$, and

(b) C_2: the parabolic arc $\mathbf{r}_2(t) = t\mathbf{i} + t\mathbf{j} + t^2\mathbf{k}, \quad 0 \leqq t \leqq 1$.

Solution. **(a)** By substituting $\mathbf{r}_1$ into $\mathbf{F}$ we obtain

$$\mathbf{F}(\mathbf{r}_1(t)) = 5t\mathbf{i} + t^2\mathbf{j} + t^3\mathbf{k}.$$

We also need

$$\mathbf{r}_1' = \mathbf{i} + \mathbf{j} + \mathbf{k}.$$

Hence the integral over C_1 is

$$\begin{aligned}\int_{C_1} \mathbf{F}(\mathbf{r}) \cdot d\mathbf{r} &= \int_0^1 \mathbf{F}(\mathbf{r}_1(t)) \cdot \mathbf{r}_1'(t)\,dt \\ &= \int_0^1 (5t\mathbf{i} + t^2\mathbf{j} + t^3\mathbf{k}) \cdot (\mathbf{i} + \mathbf{j} + \mathbf{k})\,dt \\ &= \int_0^1 (5t + t^2 + t^3)\,dt \\ &= \frac{5}{2} + \frac{1}{3} + \frac{1}{4} = 3\frac{1}{12}.\end{aligned}$$

(b) Similarly, by substituting $\mathbf{r}_2$ into $\mathbf{F}$ and calculating $\mathbf{r}_2'$ we obtain for the integral over the path C_2

$$\begin{aligned}\int_{C_2} \mathbf{F}(\mathbf{r}) \cdot d\mathbf{r} &= \int_0^1 \mathbf{F}(\mathbf{r}_2(t)) \cdot \mathbf{r}_2'(t)\,dt \\ &= \int_0^1 (5t^2 + t^2 + 2t^5)\,dt \\ &= \frac{5}{3} + \frac{1}{3} + \frac{2}{6} = 2\frac{1}{3}.\end{aligned}$$

The two results are different, although the endpoints are the same. This shows that **the value of a line integral (3) will in general depend not only on F and on the endpoints A, B of the path but also on the path along which we integrate from A to B.**

Can we find conditions that guarantee independence? This is a basic question in connection with physical applications. The answer is yes, as we show in the next section. ∎

Motivation of the Line Integral (3): Work by a Force

The work W done by a *constant* force $\mathbf{F}$ in the displacement along a *straight* segment $\mathbf{d}$ is $W = \mathbf{F} \cdot \mathbf{d}$; see Example 2 in Sec. 8.2. This suggests that we define the work W done by a *variable* force $\mathbf{F}$ in the displacement along a curve C: $\mathbf{r}(t)$ as the limit of sums of works done in displacements along small chords of C, and that we show that this amounts to defining W by the line integral (3).

For this we choose points $t_0 (= a) < t_1 < \cdots < t_n (= b)$. Then the work ΔW_m done by $\mathbf{F}(\mathbf{r}(t_m))$ in the straight displacement from $\mathbf{r}(t_m)$ to $\mathbf{r}(t_{m+1})$ is

$$\Delta W_m = \mathbf{F}(\mathbf{r}(t_m)) \cdot [\mathbf{r}(t_{m+1}) - \mathbf{r}(t_m)] \approx \mathbf{F}(\mathbf{r}(t_m)) \cdot \mathbf{r}'(t_m)\Delta t_m \quad (\Delta t_m = t_{m+1} - t_m).$$

The sum of these n works is $W_n = \Delta W_0 + \cdots + \Delta W_{n-1}$. If we choose points and consider W_n for every n arbitrarily but so that the greatest Δt_m approaches zero as $n \to \infty$, then the limit of W_n as $n \to \infty$ exists and is the line integral (3) since $\mathbf{F}$ is continuous and C is piecewise smooth, which makes $\mathbf{r}'(t)$ continuous, except at finitely many points where C may have corners or cusps.

EXAMPLE 4 **Work done by a variable force**

If $\mathbf{F}$ in Example 1 is a force, the work done by $\mathbf{F}$ in the displacement along the quarter-circle is 1.119, measured in suitable units, say, newton-meters (nt·m, also called joules, abbreviation J; see also front cover). Similarly in Examples 2 and 3. ∎

EXAMPLE 5 **Work done equals the gain in kinetic energy**

Let $\mathbf{F}$ be a force, so that (3) is work. Let t be time, so that $d\mathbf{r}/dt = \mathbf{v}$, velocity. Then we can write (3) as

$$W = \int_C \mathbf{F} \cdot d\mathbf{r} = \int_a^b \mathbf{F}(\mathbf{r}(t)) \cdot \mathbf{v}(t)\, dt. \tag{5}$$

Now by Newton's second law (force = mass × acceleration),

$$\mathbf{F} = m\mathbf{r}''(t) = m\mathbf{v}'(t),$$

where m is the mass of the body displaced. Substitution into (5) gives [see (11), Sec. 8.4]

$$W = \int_a^b m\mathbf{v}' \cdot \mathbf{v}\, dt = \int_a^b m \left(\frac{\mathbf{v} \cdot \mathbf{v}}{2}\right)' dt = \frac{m}{2} |\mathbf{v}|^2 \Big|_{t=a}^{t=b}.$$

On the right, $m|\mathbf{v}|^2/2$ is the kinetic energy. Hence *the work done equals the gain in kinetic energy*. This is a basic law in mechanics. ∎

Other Forms of Line Integrals

$$\int_C F_1\, dx, \qquad \int_C F_2\, dy, \qquad \int_C F_3\, dz \tag{6}$$

are special cases of (3) when $\mathbf{F} = F_1\mathbf{i}$ or $F_2\mathbf{j}$ or $F_3\mathbf{k}$, respectively. Another form is

(7) $$\int_C f(\mathbf{r})\,dt = \int_a^b f(\mathbf{r}(t))\,dt$$

with C as in (2). But this definition can also be regarded as a special case of (3), with $\mathbf{F} = F_1\mathbf{i}$ and $F_1 = f/(dx/dt)$, so that $f = F_1x'$, as in (3'). The evaluation of (7) is similar to that in the previous examples.

EXAMPLE 6 A line integral of the form (7)

Find the value of (7) when $f = (x^2 + y^2 + z^2)^2$ and C is the helix (4) in Example 2.

Solution. From $\mathbf{r}(t) = \cos t\,\mathbf{i} + \sin t\,\mathbf{j} + 3t\mathbf{k}$ we see that on C,

$$(x^2 + y^2 + z^2)^2 = [\cos^2 t + \sin^2 t + (3t)^2]^2 = (1 + 9t^2)^2.$$

This gives

$$\int_C f(\mathbf{r})\,dt = \int_0^{2\pi} (1 + 9t^2)^2\,dt = 2\pi + 6(2\pi)^3 + \frac{81}{5}(2\pi)^5 \approx 160\ 135.$$ ■

General Properties of the Line Integral (3)

From familiar properties of integrals in calculus we obtain corresponding formulas for line integrals (3):

(8a) $$\int_C k\mathbf{F}\cdot d\mathbf{r} = k\int_C \mathbf{F}\cdot d\mathbf{r} \qquad (k \text{ constant})$$

(8b) $$\int_C (\mathbf{F} + \mathbf{G})\cdot d\mathbf{r} = \int_C \mathbf{F}\cdot d\mathbf{r} + \int_C \mathbf{G}\cdot d\mathbf{r}$$

(8c) $$\int_C \mathbf{F}\cdot d\mathbf{r} = \int_{C_1} \mathbf{F}\cdot d\mathbf{r} + \int_{C_2} \mathbf{F}\cdot d\mathbf{r} \qquad \text{(Fig. 197)}$$

where in (8c) the path C is subdivided into two arcs C_1 and C_2 that have the same orientation as C (Fig. 197). In (8b) the orientation of C is the same in all three integrals. If the sense of integration along C is reversed, the value of the integral is multiplied by -1.

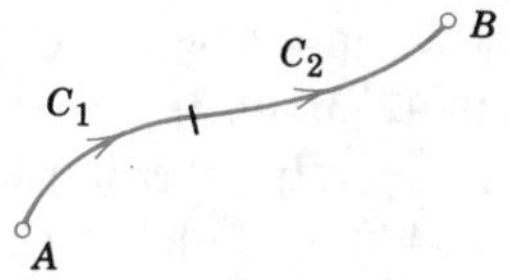

Fig. 197. Formula (8c)

If a line integral (3) is supposed to represent physical quantities, such as work, the choice of one or another representation of a given curve C should not be essential, as long as the positive directions are the same in both cases. This is what we now show.

Theorem 1 **(Direction-preserving transformations of parameter)**
Any representations of C that give the same positive direction on C also yield the same value of the line integral (3).

Proof. We represent C in (3) using another parameter t^* given by a function $t = \phi(t^*)$ that has a positive derivative and is such that $a^* \leqq t^* \leqq b^*$ corresponds to $a \leqq t \leqq b$. Then, writing $\mathbf{r}(\phi(t^*)) = \mathbf{r}^*(t^*)$ and using the chain rule, we have $dt^* = (dt^*/dt)\, dt$ and thus

$$\int_C \mathbf{F}(\mathbf{r}^*) \bullet d\mathbf{r}^* = \int_{a^*}^{b^*} \left[\mathbf{F}(\mathbf{r}^*(t^*)) \bullet \frac{d\mathbf{r}^*}{dt^*} \right] dt^*$$

$$= \int_{a^*}^{b^*} \mathbf{F}(\mathbf{r}(\phi(t^*))) \bullet \frac{d\mathbf{r}}{dt} \frac{dt}{dt^*}\, dt^*$$

$$= \int_a^b \mathbf{F}(\mathbf{r}(t)) \bullet \frac{d\mathbf{r}}{dt}\, dt = \int_C \mathbf{F}(\mathbf{r}) \bullet d\mathbf{r}. \qquad \blacksquare$$

Problem Set 9.1

Compute $\int_C \mathbf{F}(\mathbf{r}) \bullet d\mathbf{r}$, where

1. $\mathbf{F} = y^2\mathbf{i} - x^2\mathbf{j}$, C: the straight-line segment from (0, 0) to (1, 2)
2. $\mathbf{F}$ as in Prob. 1, C: $y = 2x^2$, $0 \leqq x \leqq 1$
3. $\mathbf{F} = xy\mathbf{i} + (y - x)^2\mathbf{j}$, C: $xy = 1$, $1 \leqq x \leqq 3$
4. $\mathbf{F} = xy^2\mathbf{i} + x^5y\mathbf{j}$, C: $\cosh t\,\mathbf{i} + \sinh t\,\mathbf{j}$, $0 \leqq t \leqq 5$
5. $\mathbf{F} = 3x^4\mathbf{i} + 3y^6\mathbf{j}$, C: from (2, 0) to (−2, 0) counterclockwise along $x^2 + y^2 = 4$
6. $\mathbf{F} = \exp(y^{2/3})\,\mathbf{i} - \exp(x^{3/2})\,\mathbf{j}$, C: $\mathbf{r} = t\mathbf{i} + t^{3/2}\mathbf{j}$ from (0, 0) to (0, 1)
7. $\mathbf{F} = z\mathbf{i} + x\mathbf{j} + y\mathbf{k}$, C: $\mathbf{r} = \cos t\,\mathbf{i} + \sin t\,\mathbf{j} + t\mathbf{k}$ from (1, 0, 0) to (1, 0, 4π)
8. $\mathbf{F} = x^5\mathbf{i} + x^4y\mathbf{j} + (z^2x + 2zy)\mathbf{k}$, C as in Prob. 7
9. $\mathbf{F} = x\mathbf{i} - z\mathbf{j} + 2y\mathbf{k}$, C: around the triangle from vertex (0, 0, 0) to vertex (1, 1, 0) to vertex (1, 1, 1) and back to (0, 0, 0)
10. $\mathbf{F} = 2x(5y + 2)\mathbf{i} + 12y\mathbf{j}$, C: clockwise around the rectangle with vertices (0, 0), (2, 0), (2, 3), (0, 3)
11. $\mathbf{F} = \frac{1}{4}xy^3\mathbf{i} + x^2y^{-1}\mathbf{j} - (yz \ln y)\mathbf{k}$, C: $\mathbf{r} = t\mathbf{i} + e^t\mathbf{j} + \cosh t\,\mathbf{k}$, $1 \leqq t \leqq 3$
12. $\mathbf{F} = e^x\mathbf{i} + e^{4y/x}\mathbf{j} + e^{2z/y}\mathbf{k}$, C: $\mathbf{r} = t\mathbf{i} + t^2\mathbf{j} + t^3\mathbf{k}$, $0 \leqq t \leqq 1$

Find the work done by the force $\mathbf{p} = 4xy\mathbf{i} - 8y\mathbf{j} + 2\mathbf{k}$ in the displacement:
13. Along the straight line $y = 2x$, $z = 2x$ from (0, 0, 0) to (3, 6, 6).
14. Along the parabola $y = 2x^2/3$, $z = 0$ from (0, 0, 0) to (3, 6, 0).
15. Counterclockwise around the circle $x^2 + y^2 = 4$, $z = 0$.
16. Along the hyperbola $x^2 - y^2 = 1$, $z = 0$ from (1, 0, 0) to (2, $\sqrt{3}$, 0).

17. Along the ellipse $x^2 + 4y^2 = 4$, $z = 0$ counterclockwise from $(0, -1, 0)$ to $(0, 1, 0)$.

18. Along the helix in Example 2 from $(1, 0, 0)$ to $(1, 0, 6\pi)$.

Orienting C so that the sense of integration becomes the positive sense on C, evaluate $\int_C (3x^2 + 3y^2)\, ds$:

19. Over the path $y = 3x$ from $(0, 0)$ to $(2, 6)$.

20. Over the y-axis from $(0, 0)$ to $(0, 1)$, then parallel to the x-axis from $(0, 1)$ to $(1, 1)$.

21. Over the x-axis from $(0, 0)$ to $(1, 0)$, then parallel to the y-axis from $(1, 0)$ to $(1, 1)$.

22. Clockwise along the circle $x^2 + y^2 = 1$ from $(0, 1)$ to $(1, 0)$.

23. Counterclockwise along the circle $x^2 + y^2 = 1$ from $(1, 0)$ to $(0, 1)$.

24. **(Invariance)** Verify Theorem 1 for $\mathbf{F} = y^2\mathbf{i} + x^2\mathbf{k}$, C: $\mathbf{r}(t) = t\mathbf{i} + t^2\mathbf{j} + t^3\mathbf{k}$, $1 \leqq t \leqq 2$, and $t = t^{*2}$.

Estimation of line integrals

25. Let $\mathbf{F}$ be a vector function defined at all points of a curve C, and suppose that $|\mathbf{F}|$ is bounded, say $|\mathbf{F}| \leqq M$ on C, where M is some positive number. Show that

$$\left| \int_C \mathbf{F} \cdot d\mathbf{r} \right| \leqq ML \qquad (L = \text{length of } C). \tag{9}$$

26. Using (9), find an upper bound for the absolute value of the work W done by the force $\mathbf{F} = -x^3\mathbf{i} - y\mathbf{j}$ in the displacement along the straight line from $(0, 0, 0)$ to $(1, 1, 0)$. Find W by integration and compare the results.

9.2 Line Integrals Independent of Path

The value of a line integral

$$\int_C \mathbf{F}(\mathbf{r}) \cdot d\mathbf{r} = \int_C (F_1\, dx + F_2\, dy + F_3\, dz) \tag{1}$$

over a path C from a point A to a point B in general depends not only on A and B but also on the path C along which we integrate. This was shown in Example 3 of the last section. It raises the question of conditions for independence of path, so that we get the same value in integrating from A to B along any path C. This is of great practical importance. For instance, in mechanics, independence of path may mean that we have to do the same amount of work regardless of the path to the mountaintop, be it short and steep or long and gentle, or that we gain back the work done in extending an elastic spring when we release it. Not all forces are of this type—think of swimming in a big whirlpool.

We define a line integral (1) to be **independent of path in a domain D in space** if for every pair of endpoints A, B in D the integral (1) has the same value for all paths in D that start at A and end at B.

A very practical criterion for path independence is the following. (For the gradient, see Sec. 8.9.)

Theorem 1 **(Independence of path)**

A line integral (1) *with continuous* F_1, F_2, F_3 *in a domain D in space is independent of path in D if and only if* **F** *is the gradient of some function f in D,*

$$\mathbf{F} = \operatorname{grad} f; \tag{2}$$

in components,

$$F_1 = \frac{\partial f}{\partial x}, \qquad F_2 = \frac{\partial f}{\partial y}, \qquad F_3 = \frac{\partial f}{\partial z}. \tag{2'}$$

EXAMPLE 1 **Independence of path**

Show that the integral

$$\int_C \mathbf{F} \cdot d\mathbf{r} = \int_C (2x\,dx + 2y\,dy + 4z\,dz)$$

is independent of path in any domain in space and find its value if C has the initial point A: (0, 0, 0) and terminal point B: (2, 2, 2).

Solution. By inspection we find that

$$\mathbf{F} = 2x\mathbf{i} + 2y\mathbf{j} + 4z\mathbf{k} = \operatorname{grad} f, \qquad \text{where} \qquad f = x^2 + y^2 + 2z^2.$$

(If **F** is more complicated, proceed by integration, as in Example 2, below.) Theorem 1 now implies independence of path. To find the value of the integral, we can choose the convenient straight path

$$C\colon\ \mathbf{r}(t) = t(\mathbf{i} + \mathbf{j} + \mathbf{k}), \qquad 0 \leqq t \leqq 2,$$

and get $\mathbf{r}'(t) = \mathbf{i} + \mathbf{j} + \mathbf{k}$; thus $\mathbf{F} \cdot \mathbf{r}' = 2t + 2t + 4t = 8t$ and from this

$$\int_C (2x\,dx + 2y\,dy + 4z\,dz) = \int_0^2 \mathbf{F} \cdot \mathbf{r}'\,dt = \int_0^2 8t\,dt = 16.$$

Better methods of solution follow below. ∎

Proof of Theorem 1. **(a)** Let (2) hold for some f in D and let C be any path in D from any point A to any point B, given by

$$\mathbf{r}(t) = x(t)\mathbf{i} + y(t)\mathbf{j} + z(t)\mathbf{k}, \qquad a \leqq t \leqq b.$$

Then from (2′), the chain rule (Sec. 8.8) and (3′) in Sec. 9.1 we get

$$\int_A^B (F_1\,dx + F_2\,dy + F_3\,dz) = \int_A^B \left(\frac{\partial f}{\partial x}\,dx + \frac{\partial f}{\partial y}\,dy + \frac{\partial f}{\partial z}\,dz\right)$$

$$= \int_a^b \frac{df}{dt}\,dt = f[x(t), y(t), z(t)]\Big|_{t=a}^{t=b}$$

$$= f(B) - f(A).$$

This shows that the value of the integral is simply the difference of the values of f at the two endpoints of C and is, therefore, independent of the path C.

(b) The more complicated proof of the converse, that independence of path implies (2) for some f, is given in Appendix 4. ∎

The last formula in part (a) of the proof,

$$\text{(3)} \qquad \int_A^B (F_1\,dx + F_2\,dy + F_3\,dz) = f(B) - f(A) \qquad [\mathbf{F} = \text{grad}\, f]$$

is the analog of the usual formula

$$\int_a^b g(x)\,dx = G(x)\Big|_a^b = G(b) - G(a) \qquad [G'(x) = g(x)]$$

for evaluating definite integrals in calculus and should be applied whenever a line integral is independent of path.

Potential theory relates to our present discussion if we remember from Sec. 8.9 that f is called a *potential* of $\mathbf{F} = \text{grad}\, f$. Thus the integral (1) is independent of path in D if and only if $\mathbf{F}$ is the gradient of a potential in D.

EXAMPLE 2 **Independence of path. Determination of a potential**

Evaluate the integral

$$I = \int_C (3x^2\,dx + 2yz\,dy + y^2\,dz)$$

from A: $(0, 1, 2)$ to B: $(1, -1, 7)$ by showing that $\mathbf{F}$ has a potential and applying (3).

Solution. If $\mathbf{F}$ has a potential f, we should have

$$f_x = F_1 = 3x^2, \qquad f_y = F_2 = 2yz, \qquad f_z = F_3 = y^2.$$

We show that we can satisfy these conditions. By integration and differentiation,

$$f = x^3 + g(y, z), \qquad f_y = g_y = 2yz, \qquad g = y^2 z + h(z),$$

$$f_z = y^2 + h' = y^2, \qquad h' = 0 \qquad h = 0, \text{ say}.$$

This gives $f(x, y, z) = x^3 + y^2 z$ and by (3),

$$I = f(1, -1, 7) - f(0, 1, 2)$$

$$= 1 + 7 - (0 + 2) = 6.$$

∎

Integration Around Closed Curves and Independence of Path

The simple idea that two paths with common endpoints (Fig. 198) make up a single closed curve gives almost immediately

Theorem 2 (Independence of path)
The integral (1) is independent of path in a domain D if and only if its value around every closed path in D is zero.

Proof. If we have independence of path, integration from A to B along C_1 and along C_2 in Fig. 198 gives the same value. Now C_1 and C_2 together make up a closed curve C, and if we integrate from A along C_1 to B as before, but then in the opposite sense along C_2 back to A (so that this integral is multiplied by -1), the sum of the two integrals is zero, but this is the integral around the closed curve C.

Conversely, assume that the integral around any closed path C in D is zero. Given any points A and B in D, choose any closed curve C in D passing through A and B, so that C is cut into C_1 and C_2 as in Fig. 198. Since the value of (1) around C is zero, it follows that the integrals over C_1 and C_2, both taken from A to B, must have the same value, and the theorem is proved. ∎

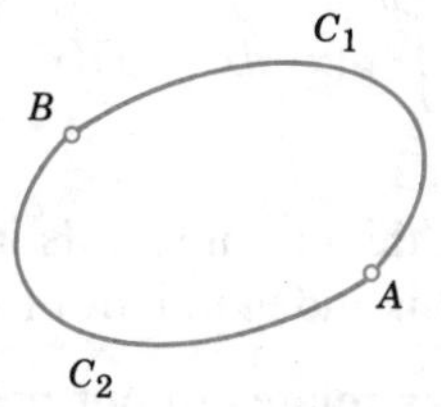

Fig. 198. Proof of Theorem 2

Work. Recall from the last section that in mechanics, the integral (1) represents the work done by a force **F** in the displacement of a particle along C. Then Theorem 2 states that work is independent of path if and only if it is zero for displacement around any closed path. Furthermore, Theorem 1 tells us that this happens if and only if **F** is the gradient of a potential. In this case, **F** and the vector field defined by **F** are called **conservative,** because in this case mechanical energy is conserved, that is, no work is done in the displacement from a point A and back to A. Similarly for the displacement of an electrical charge (an electron, for instance) in an electrostatic field.

Physically, the kinetic energy of a body can be interpreted as the ability of the body to do work by virtue of its motion, and if the body moves in a conservative field of force, after the completion of a round-trip the body will return to its initial position with the same kinetic energy it had originally. For instance, the gravitational force is conservative; if we throw a ball vertically up, it will (if we assume air resistance to be negligible) return to our hand with the same kinetic energy it had when it left our hand.

Friction, air resistance, or water resistance always act against the direction of motion, tending to diminish the total mechanical energy of a system (usually converting it into heat or mechanical energy of the surrounding medium, or both), and if in the motion of a body these forces are so large that they can no longer be neglected, then the resultant **F** of the forces acting on the body is no longer conservative. Quite generally, a physical system is called **conservative** if all the forces acting in it are conservative; otherwise it is called **nonconservative** or **dissipative.**

Exactness and Independence of Path

A third idea of practical interest is to relate independence of path to the exactness of the **differential form**

$$F_1\,dx + F_2\,dy + F_3\,dz \tag{4}$$

under the integral sign in (1). The form (4) is called **exact** or *an exact differential* in a domain D in space if it is the differential

$$df = \frac{\partial f}{\partial x}\,dx + \frac{\partial f}{\partial y}\,dy + \frac{\partial f}{\partial z}\,dz$$

of a differentiable function $f(x, y, z)$ everywhere in D, that is, if we have

$$F_1\,dx + F_2\,dy + F_3\,dz = df.$$

Comparing these two formulas, we see that the form (4) is exact if and only if there is a differentiable function $f(x, y, z)$ in D such that

$$F_1 = \frac{\partial f}{\partial x}, \qquad F_2 = \frac{\partial f}{\partial y}, \qquad F_3 = \frac{\partial f}{\partial z} \tag{5'}$$

everywhere in D; in vectorial form,

$$\mathbf{F} = \operatorname{grad} f. \tag{5}$$

Hence, by Theorem 1, *the integral* (1) *is independent of path in D if and only if the differential form* (4) *is exact in D*. This becomes of practical importance because there is a useful exactness criterion involving the following concept.

A domain D is called **simply connected** if every closed curve in D can be continuously shrunk to any point in D without leaving D.

For example, the interior of a sphere or a cube, the interior of a sphere with finitely many points removed, and the domain between two concentric spheres are simply connected, while the interior of a torus (a doughnut; see Fig. 222 in Sec. 9.6) and the interior of a cube with one space diagonal removed are not simply connected.

The criterion for independence of path based on exactness is then as follows.

Theorem 3 **(Criterion for exactness and independence of path)**
Let F_1, F_2, F_3 in the line integral (1),

$$\int_C \mathbf{F}(\mathbf{r}) \cdot d\mathbf{r} = \int_C (F_1\,dx + F_2\,dy + F_3\,dz),$$

be continuous and have continuous first partial derivatives in a domain D in space. Then:

(a) *If* (1) *is independent of path in D—and thus the differential form* (4) *under the integral sign is exact—then in D,*

$$\text{curl}\,\mathbf{F} = \mathbf{0}; \tag{6}$$

in components (see Sec. 8.11)

$$\frac{\partial F_3}{\partial y} = \frac{\partial F_2}{\partial z}, \qquad \frac{\partial F_1}{\partial z} = \frac{\partial F_3}{\partial x}, \qquad \frac{\partial F_2}{\partial x} = \frac{\partial F_1}{\partial y}. \tag{6'}$$

(b) *If* (6′) *holds in D and D is simply connected, then* (1) *is independent of path in D.*

Proof. **(a)** If (1) is independent of path in D, then $\mathbf{F} = \text{grad}\, f$ by (2) and

$$\text{curl}\,\mathbf{F} = \text{curl}\,(\text{grad}\, f) = \mathbf{0}$$

[see (3) in Sec. 8.11], so that (6) holds.

(b) The proof of the converse requires "Stokes's theorem" and is given in Sec. 9.9. ∎

Comment For a line integral in the plane

$$\int_C \mathbf{F}(\mathbf{r}) \cdot d\mathbf{r} = \int_C (F_1\,dx + F_2\,dy),$$

curl $\mathbf{F}$ has just one component and (6′) reduces to the single relation

$$\frac{\partial F_2}{\partial x} = \frac{\partial F_1}{\partial y}. \tag{6''}$$

EXAMPLE 3 **Exactness and independence of path. Determination of a potential**

Using (6′), show that the differential form under the integral sign of

$$I = \int_C [2xyz^2\,dx + (x^2z^2 + z\cos yz)\,dy + (2x^2yz + y\cos yz)\,dz]$$

is exact, so that we have independence of path in any domain, and find the value of I from A: (0, 0, 1) to B: (1, $\pi/4$, 2).

Solution. Exactness follows from (6′), which gives

$$(F_3)_y = 2x^2z + \cos yz - yz \sin yz = (F_2)_z$$

$$(F_1)_z = 4xyz = (F_3)_x$$

$$(F_2)_x = 2xz^2 = (F_1)_y.$$

To find f, we integrate F_2 (which is "long," so that we save work) and then differentiate to compare with F_1 and F_3,

$$f = \int F_2\, dy = \int (x^2z^2 + z \cos yz)\, dy = x^2z^2y + \sin yz + g(x, z)$$

$$f_x = 2xz^2y + g_x = F_1 = 2xyz^2, \qquad g_x = 0, \qquad g = h(z)$$

$$f_z = 2x^2zy + y \cos yz + h' = F_3 = 2x^2zy + y \cos z, \qquad h' = 0,$$

so that, taking $h = 0$, we have

$$f(x, y, z) = x^2yz^2 + \sin yz.$$

From this and (3) we get

$$I = f(1, \pi/4, z) - f(0, 0, 1) = \pi + \sin \tfrac{1}{2}\pi - 0 = \pi + 1.$$ ∎

The assumption in Theorem 3 that D be simply connected is essential and cannot be omitted. This can be seen from the following example.

EXAMPLE 4 **On the assumption of simple connectedness in Theorem 3**

Let

$$F_1 = -\frac{y}{x^2 + y^2}, \qquad F_2 = \frac{x}{x^2 + y^2}, \qquad F_3 = 0.$$

Differentiation shows that (6′) is satisfied in any domain of the xy-plane not containing the origin, for example, in the domain D: $\frac{1}{2} < \sqrt{x^2 + y^2} < \frac{3}{2}$ shown in Fig. 199. Indeed, F_1 and F_2 do not depend on z, and $F_3 = 0$, so that the first two relations in (6′) are trivially true, and the third is verified by differentiation:

$$\frac{\partial F_2}{\partial x} = \frac{x^2 + y^2 - x \cdot 2x}{(x^2 + y^2)^2} = \frac{y^2 - x^2}{(x^2 + y^2)^2},$$

$$\frac{\partial F_1}{\partial y} = -\frac{x^2 + y^2 - y \cdot 2y}{(x^2 + y^2)^2} = \frac{y^2 - x^2}{(x^2 + y^2)^2}.$$

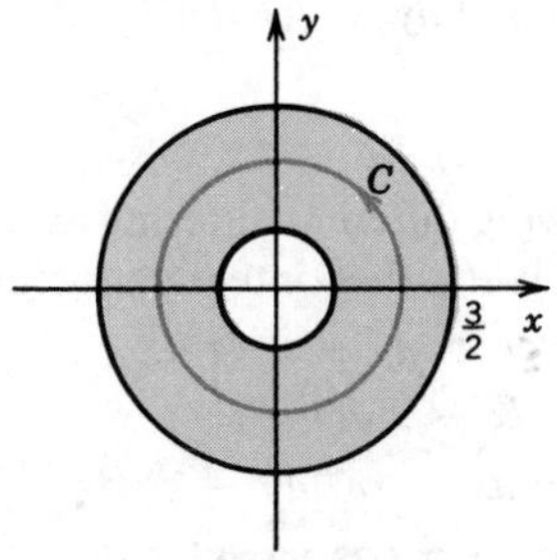

Fig. 199. Example 4

Clearly, D in Fig. 199 is not simply connected. If the integral

$$I = \int_C (F_1\,dx + F_2\,dy) = \int_C \frac{-y\,dx + x\,dy}{x^2 + y^2}$$

were independent of path in D, then $I = 0$ on any closed curve in D, for example, on the circle $x^2 + y^2 = 1$. But setting $x = r\cos\theta$, $y = r\sin\theta$, and noting that the circle is represented by $r = 1$, we have

$$x = \cos\theta, \qquad dx = -\sin\theta\,d\theta, \qquad y = \sin\theta, \qquad dy = \cos\theta\,d\theta,$$

so that $-y\,dx + x\,dy = d\theta$ and counterclockwise integration gives

$$I = \int_0^{2\pi} \frac{d\theta}{1} = 2\pi.$$

Since D is not simply connected, we cannot apply Theorem 3 and conclude that I is independent of path in D. ∎

Problem Set 9.2

Show that the form under the integral sign is exact and evaluate

1. $\displaystyle\int_{(0,0,0)}^{(1,1,1)} (4\,dx - 6y\,dy + 5\,dz)$

2. $\displaystyle\int_{(0,-1,-1)}^{(1,0,1)} (e^x\,dx - e^z\,dz)$

3. $\displaystyle\int_{(0,0,0)}^{(2,2,2)} (2x\,dx + 3y^2z\,dy + y^3\,dz)$

4. $\displaystyle\int_{(1,1,0)}^{(1,4,3)} (\cosh^2 z\,dy + y\sinh 2z\,dz)$

5. $\displaystyle\int_{(1,2,3)}^{(0,0,1)} (3x^2y^2z^4\,dx + 2x^3yz^4\,dy + 4x^3y^2z^3\,dz)$

6. $\displaystyle\int_{(2,1,0)}^{(0,1,2)} (ze^{xz}\,dx - dy + xe^{xz}\,dz)$

7. $\displaystyle\int_{(0,2,0)}^{(1,\pi,0)} \sin xy\,(y\,dx + x\,dy)$

8. $\displaystyle\int_{(0,1,2)}^{(0,2,3)} e^{y^2}(2yz^3\,dy + 3z^2\,dz)$

Are the following differential forms exact? In the case of exactness, find f such that the form equals df; then integrate the exact form from A: $(0, 0, 0)$ to B: (a, b, c).

9. $xyz^2\,dx + \frac{1}{2}x^2z^2\,dy + x^2yz\,dz$

10. $\sin x\,dy + \sin y\,dx$

11. $\sinh xz\,(z\,dx - x\,dz)$

12. $e^y(-2\sin 2x\,dx + \cos 2x\,dy)$

13. $(e^y - ze^x)\,dx + xe^y\,dy - e^x\,dz$

14. $\sin 2yz\,(z\,dy + y\,dz)$

15. $y\cos xy\,dx + x\cos xy\,dy - dz$

16. $\sinh x\cos z\,dx + \cosh x\sin z\,dz$

17. $xe^{2z}\,dx + x^2e^{2z}\,dz$

18. $e^z\,dx + e^y\,dy + e^x\,dz$

19. $ye^z\,dy - ze^y\,dz$

20. $e^{xyz}(yz\,dx + xz\,dy + xy\,dz)$

9.3 From Calculus: Double Integrals *Optional*

Students familiar with double integrals from calculus should go on to the next section, skipping the present review (included to make the book reasonably self-contained).

In a definite integral (1), Sec. 9.1, we integrate a function $f(x)$ over an interval (a segment) of the x-axis. In a double integral we integrate a function $f(x, y)$, called the *integrand,* over a closed bounded[1] region R in the xy-plane, whose boundary curve has a unique tangent at each point, but may have finitely many cusps (such as the vertices of a triangle or rectangle).

The definition of the double integral is quite similar to that of the definite integral. We subdivide the region R by drawing parallels to the x and y axes (Fig. 200). We number the rectangles that are within R from 1 to n. In each such rectangle we choose a point, say, (x_k, y_k) in the kth rectangle, and then we form the sum

$$J_n = \sum_{k=1}^{n} f(x_k, y_k)\,\Delta A_k$$

where ΔA_k is the area of the kth rectangle. This we do for larger and larger positive integers n in a completely independent manner but so that the length of the maximum diagonal of the rectangles approaches zero as n approaches infinity. In this fashion we obtain a sequence of real numbers $J_{n_1}, J_{n_2}, \cdots$. Assuming that $f(x, y)$ is continuous in R and R is bounded by finitely many smooth curves (see Sec. 9.1), one can show[2] that this sequence converges and its limit is independent of the choice of subdivisions and corresponding points (x_k, y_k). This limit is called the **double integral** *of* $f(x, y)$ *over the region* R, and is denoted by

$$\iint_R f(x, y)\,dx\,dy \qquad \text{or} \qquad \iint_R f(x, y)\,dA.$$

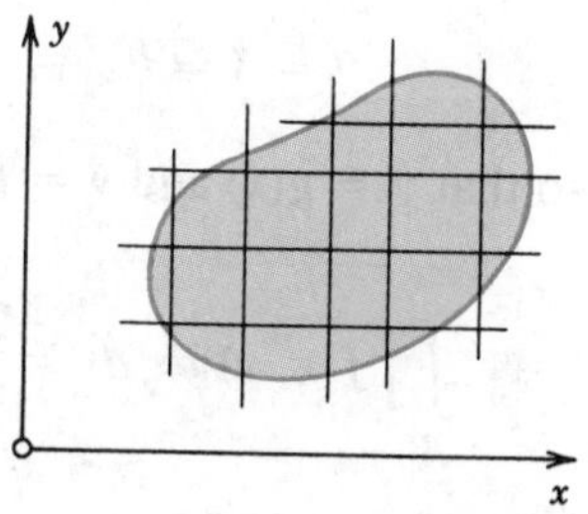

Fig. 200. Subdivision of R

[1] "Closed" means that the boundary is part of the region, and "bounded" means that the region can be enclosed in a circle of sufficiently large radius.

[2] See Ref. [5] in Appendix 1.

From the definition it follows that double integrals have properties quite similar to those of definite integrals. Indeed, for any function f and g of (x, y), defined and continuous in a region R,

$$\iint_R kf\,dx\,dy = k\iint_R f\,dx\,dy \qquad (k \text{ constant})$$

$$\text{(1)} \qquad \iint_R (f+g)\,dx\,dy = \iint_R f\,dx\,dy + \iint_R g\,dx\,dy$$

$$\iint_R f\,dx\,dy = \iint_{R_1} f\,dx\,dy + \iint_{R_2} f\,dx\,dy \qquad \text{(see Fig. 201).}$$

Furthermore, there exists at least one point (x_0, y_0) in R such that we have

$$\text{(2)} \qquad \iint_R f(x, y)\,dx\,dy = f(x_0, y_0)A,$$

where A is the area of R; this is called the **mean value theorem** *for double integrals.*

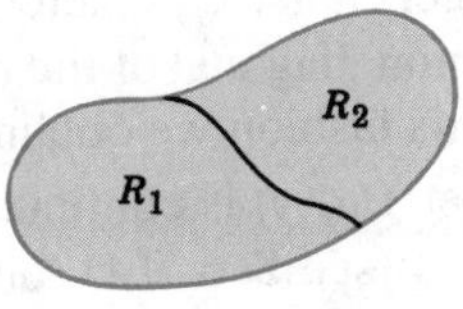

Fig. 201. Formula (1)

Evaluation of Double Integrals

Double integrals over a region R may be evaluated by two successive integrations as follows. Suppose that R can be described by inequalities of the form

$$a \leqq x \leqq b, \qquad g(x) \leqq y \leqq h(x)$$

(Fig. 202), so that $y = g(x)$ and $y = h(x)$ represent the boundary of R. Then

$$\text{(3)} \qquad \iint_R f(x, y)\,dx\,dy = \int_a^b \left[\int_{g(x)}^{h(x)} f(x, y)\,dy\right] dx.$$

We first integrate the inner integral

$$\int_{g(x)}^{h(x)} f(x, y)\,dy.$$

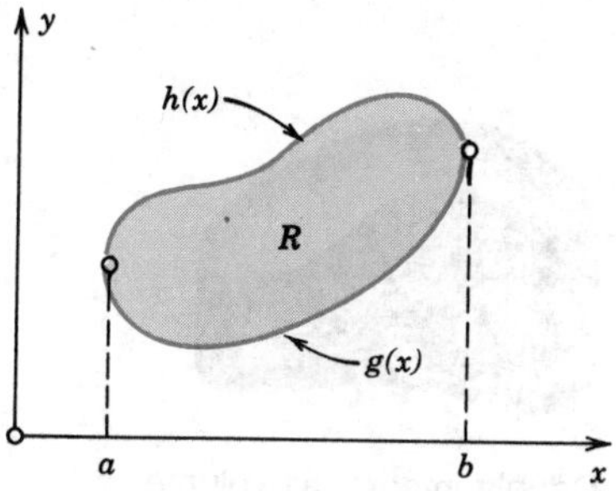

Fig. 202. Evaluation of a double integral

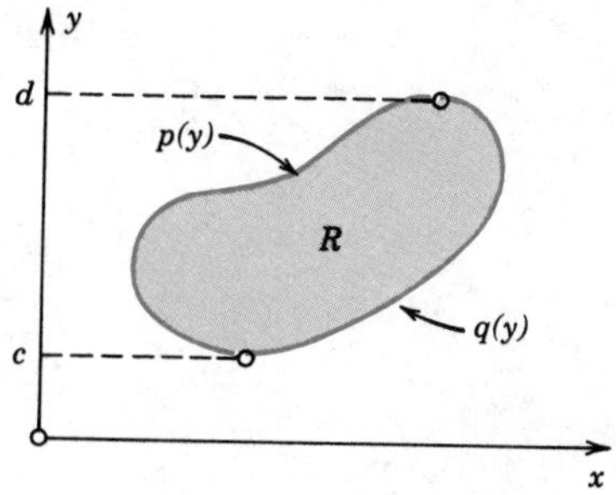

Fig. 203. Evaluation of a double integral

In this integration we keep x fixed, that is, we regard x as a constant. The result of this integration will be a function of x, say, $F(x)$. Integrating $F(x)$ over x from a to b we then obtain the value of the double integral in (3).

Similarly, if R can be described by inequalities of the form

$$c \leqq y \leqq d, \qquad p(y) \leqq x \leqq q(y)$$

(Fig. 203), then we obtain

$$\iint_R f(x, y)\, dx\, dy = \int_c^d \left[\int_{p(y)}^{q(y)} f(x, y)\, dx \right] dy; \tag{4}$$

we now integrate first over x (treating y as a constant) and then the resulting function of y from c to d.

If R cannot be represented by those inequalities, but can be subdivided into finitely many portions that have that property, we may integrate $f(x, y)$ over each portion separately and add the results; this will give us the value of the double integral of $f(x, y)$ over that region R.

Applications of Double Integrals

Double integrals have various geometrical and physical applications. For example, the **area** A of a region R in the xy-plane is given by the double integral

$$A = \iint_R dx\, dy.$$

The **volume** V beneath the surface $z = f(x, y)$ (> 0) and above a region R in the xy-plane is (Fig. 204)

$$V = \iint_R f(x, y)\, dx\, dy,$$

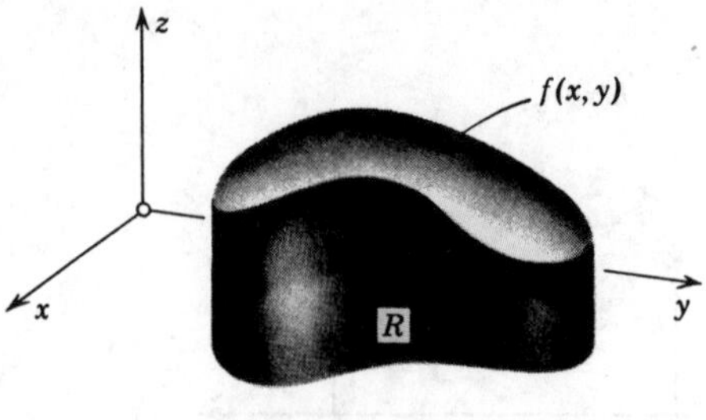

Fig. 204. Double integral as volume

because the term $f(x_k, y_k)\,\Delta A_k$ in J_n at the beginning of this section represents the volume of a rectangular parallelepiped with base ΔA_k and altitude $f(x_k, y_k)$.

Let $f(x, y)$ be the density (= mass per unit area) of a distribution of mass in the xy-plane. Then the total mass M in R is

$$M = \iint_R f(x, y)\, dx\, dy,$$

the **center of gravity** of the mass in R has the coordinates $\bar{x}, \bar{y}$, where

$$\bar{x} = \frac{1}{M}\iint_R x f(x, y)\, dx\, dy \quad \text{and} \quad \bar{y} = \frac{1}{M}\iint_R y f(x, y)\, dx\, dy,$$

the **moments of inertia** I_x and I_y of the mass in R about the x- and y-axes, respectively, are

$$I_x = \iint_R y^2 f(x, y)\, dx\, dy, \qquad I_y = \iint_R x^2 f(x, y)\, dx\, dy,$$

and the **polar moment of inertia** I_0 about the origin of the mass in R is

$$I_0 = I_x + I_y = \iint_R (x^2 + y^2) f(x, y)\, dx\, dy.$$

EXAMPLE 1 **Center of gravity. Moments of inertia**

Let $f(x, y) = 1$ be the density of mass in the region R: $0 \leqq y \leqq \sqrt{1 - x^2}$, $0 \leqq x \leqq 1$ (Fig. 205). Find the center of gravity and the moments of inertia I_x, I_y, and I_0.

Solution. The total mass in R is

$$M = \iint_R dx\, dy = \int_0^1 \left[\int_0^{\sqrt{1-x^2}} dy\right] dx = \int_0^1 \sqrt{1 - x^2}\, dx = \int_0^{\pi/2} \cos^2\theta\, d\theta = \frac{\pi}{4}$$

($x = \sin\theta$), which is the area of R. The coordinates of the center of gravity are

$$\bar{x} = \frac{4}{\pi}\iint_R x\, dx\, dy = \frac{4}{\pi}\int_0^1 \left[\int_0^{\sqrt{1-x^2}} x\, dy\right] dx = \frac{4}{\pi}\int_0^1 x\sqrt{1 - x^2}\, dx = -\frac{4}{\pi}\int_1^0 z^2\, dz = \frac{4}{3\pi}$$

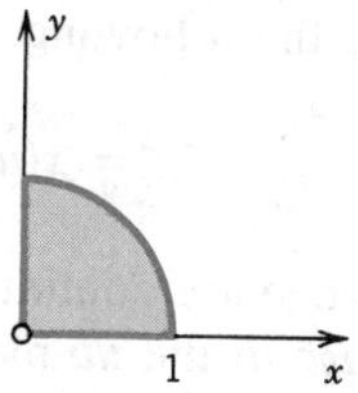

Fig. 205. Example 1

($\sqrt{1 - x^2} = z$), and $\bar{y} = \bar{x}$, for reasons of symmetry. Furthermore,

$$I_x = \iint_R y^2\,dx\,dy = \int_0^1 \left[\int_0^{\sqrt{1-x^2}} y^2\,dy\right] dx = \frac{1}{3}\int_0^1 (\sqrt{1 - x^2})^3\,dx$$

$$= \frac{1}{3}\int_0^{\pi/2} \cos^4\theta\,d\theta = \frac{\pi}{16}, \qquad I_y = \frac{\pi}{16}, \qquad I_0 = I_x + I_y = \frac{\pi}{8} \approx 0.3927.$$

These integrations will become much simpler if we first make a suitable change of variables. This is what we show next. ■

Change of Variables in Double Integrals

Practical problems will often require a change of the variables of integration in double integrals. Recall from calculus that for a definite integral the formula for the change from x to u is

$$\text{(5)} \qquad \int_a^b f(x)\,dx = \int_\alpha^\beta f(x(u))\,\frac{dx}{du}\,du.$$

Here we assume that $x = x(u)$ is continuous and has a continuous derivative in some interval $\alpha \leqq u \leqq \beta$ such that $x(\alpha) = a$, $x(\beta) = b$ [or $x(\alpha) = b$, $x(\beta) = a$] and $x(u)$ varies between a and b when u varies between α and β.

The formula for a change of variables in double integrals from x, y to u, v is

$$\textbf{(6)} \qquad \iint_R f(x, y)\,dx\,dy = \iint_{R^*} f(x(u, v), y(u, v)) \left|\frac{\partial(x, y)}{\partial(u, v)}\right| du\,dv;$$

that is, the integrand is expressed in terms of u and v, and $dx\,dy$ is replaced by $du\,dv$ times the absolute value of the **Jacobian**[3]

$$J = \frac{\partial(x, y)}{\partial(u, v)} = \begin{vmatrix} \dfrac{\partial x}{\partial u} & \dfrac{\partial x}{\partial v} \\[2ex] \dfrac{\partial y}{\partial u} & \dfrac{\partial y}{\partial v} \end{vmatrix}.$$

---

[3]Named after the German mathematician CARL GUSTAV JACOB JACOBI (1804—1851), professor at Königsberg and Berlin, who made important contributions to elliptic functions, partial differential equations, mechanics, astronomy, and the calculus of variations.

Here we assume the following. The functions

$$x = x(u, v), \qquad y = y(u, v)$$

effecting the change are continuous and have continuous partial derivatives in some region R^* in the uv-plane such that the point (x, y) corresponding to any (u, v) in R^* lies in R and, conversely, to every (x, y) in R there corresponds one and only one (u, v) in R^*; furthermore, the Jacobian J is either positive throughout R^* or negative throughout R^*. For a proof, see Ref. [5] in Appendix 1.

Of particular practical interest are **polar coordinates** r and θ, which can be introduced by setting

$$x = r\cos\theta, \qquad y = r\sin\theta.$$

Then

$$J = \frac{\partial(x, y)}{\partial(r, \theta)} = \begin{vmatrix} \cos\theta & -r\sin\theta \\ \sin\theta & r\cos\theta \end{vmatrix} = r,$$

and

$$\iint_R f(x, y)\,dx\,dy = \iint_{R^*} f(r\cos\theta, r\sin\theta)r\,dr\,d\theta, \tag{7}$$

where R^* is the region in the $r\theta$-plane corresponding to R in the xy-plane.

EXAMPLE 2 Double integral in polar coordinates

Using (7), we obtain for I_x in Example 1

$$I_x = \iint_R y^2\,dx\,dy = \int_0^{\pi/2}\int_0^1 r^2\sin^2\theta\, r\,dr\,d\theta = \int_0^{\pi/2}\sin^2\theta\,d\theta\int_0^1 r^3\,dr = \frac{\pi}{4}\cdot\frac{1}{4} = \frac{\pi}{16}.$$ ∎

EXAMPLE 3 Change of variables in a double integral

Evaluate the double integral

$$\iint_R (x^2 + y^2)\,dx\,dy$$

where R is the square in Fig. 206.

Solution. The shape of R suggests the transformation $x + y = u$, $x - y = v$. Then $x = \frac{1}{2}(u + v)$, $y = \frac{1}{2}(u - v)$, the Jacobian is

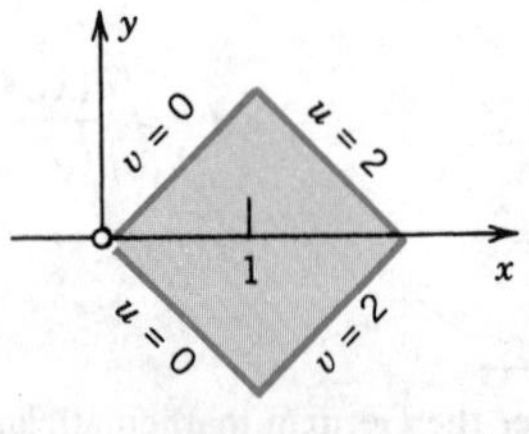

Fig. 206. Region in Example 3

$$J = \frac{\partial(x, y)}{\partial(u, v)} = \begin{vmatrix} \frac{1}{2} & \frac{1}{2} \\ \frac{1}{2} & -\frac{1}{2} \end{vmatrix} = -\tfrac{1}{2},$$

R corresponds to the square $0 \leqq u \leqq 2$, $0 \leqq v \leqq 2$, and, therefore,

$$\iint_R (x^2 + y^2)\, dx\, dy = \int_0^2 \int_0^2 \tfrac{1}{2}(u^2 + v^2)\tfrac{1}{2}\, du\, dv = \tfrac{8}{3}.$$

This is the end of our review on double integrals. These integrals will be needed in this chapter.

Problem Set 9.3

Describe the region of integration and evaluate:

1. $\int_0^1 \int_x^{2x} (2 + x^2 + y^2)\, dy\, dx$
2. $\int_0^{\pi/2} \int_{-1}^{1} x^2 y^2\, dx\, dy$
3. $\int_0^1 \int_{x^2}^{x} (1 - xy)\, dy\, dx$
4. $\int_0^2 \int_0^x e^{x+y}\, dy\, dx$
5. $\int_0^{\pi} \int_0^{\sin x} y\, dy\, dx$
6. $\int_0^1 \int_y^{y^2+1} x^2 y\, dx\, dy$
7. $\int_0^2 \int_{\sinh x^2}^{\cosh x^2} x\, dy\, dx$
8. $\int_0^{\pi/2} \int_0^y \frac{\sin y}{y}\, dx\, dy$
9. $\int_0^{\pi/2} \int_0^{\cos y} x^2 \sin y\, dx\, dy$
10. $\int_1^2 \int_{-x}^{x} e^y \cosh x\, dy\, dx$

Volume. Find the volume of the following regions in space.

11. The region beneath $z = x^2 + y^2$ and above the square with vertices (0, 0), (1, 0), (1, 1), (0, 1) in the xy-plane
12. The region beneath the plane $z = 6x - y + 12$ and above the rectangle with vertices (0, 0), (2, 0), (2, 6), (0, 6) in the xy-plane
13. The first octant section cut from the region inside the cylinder $x^2 + z^2 = a^2$ by the planes $y = 0$, $z = 0$, $x = y$
14. The tetrahedron cut from the first octant by the plane $3x + 4y + 2z = 12$
15. The first octant region bounded by the coordinate planes and the surfaces $y = 1 - x^2$, $z = 1 - x^2$

Use of polar coordinates. Using polar coordinates, evaluate $\iint_R f(x, y)\, dx\, dy$, where

16. $f = 2(x + y)$, R: $x^2 + y^2 \leqq 9$, $x \geqq 0$
17. $f = \cos(x^2 + y^2)$, R: $x^2 + y^2 \leqq \pi/2$, $x \geqq 0$
18. $f = x^2 y - xy^2 + 3$, R: $x^2 + y^2 \leqq a^2$
19. $f = e^{-x^2-y^2}$, R: the annulus bounded by $x^2 + y^2 = 1$ and $x^2 + y^2 = 4$
20. $f = (x + y)^2 + 2x - 2y$, R: $x^2 + y^2 \leqq a^2$, $x \geqq 0$, $-x \leqq y \leqq x$

Jacobian. Find the Jacobian and give a geometrical reason for your result.

21. Translation $x = u + a,\ y = v + b$

22. Expansion $x = au,\ y = bv$ (where $a > 1,\ b > 1$)

23. Rotation $x = u \cos \phi - v \sin \phi,\ y = u \sin \phi + v \cos \phi$

Center of gravity. Find the coordinates $\bar{x}, \bar{y}$ of the center of gravity of a mass of density $f(x, y) = 1$ in a region R, where R is

24. The rectangle $0 \leqq x \leqq 2,\ 0 \leqq y \leqq 4$

25. The triangle with vertices $(0, 0), (b, 0), (b, h)$

26. The region $x^2 + y^2 \leqq a^2$ in the first quadrant

Moments of inertia. Find the moments of inertia I_x, I_y, I_0 of a mass of density $f(x, y) = 1$ in a region R shown in the following figures (which the engineer is likely to need in applications).

27.

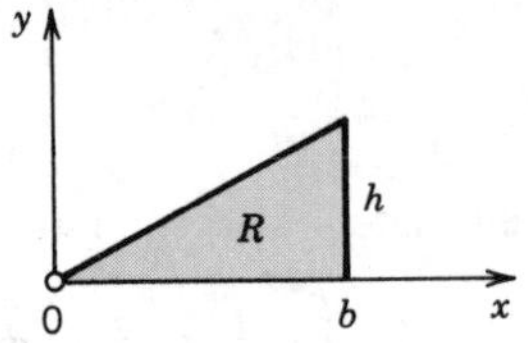

28.

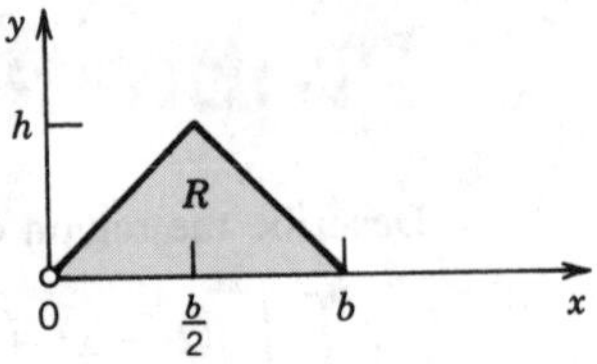

29.

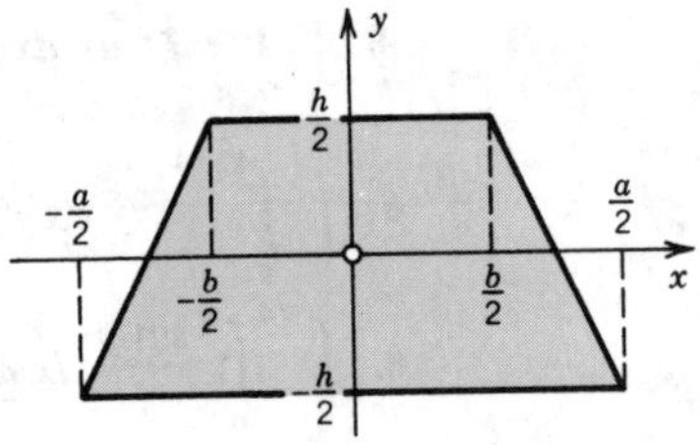

30.

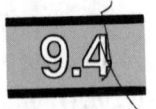

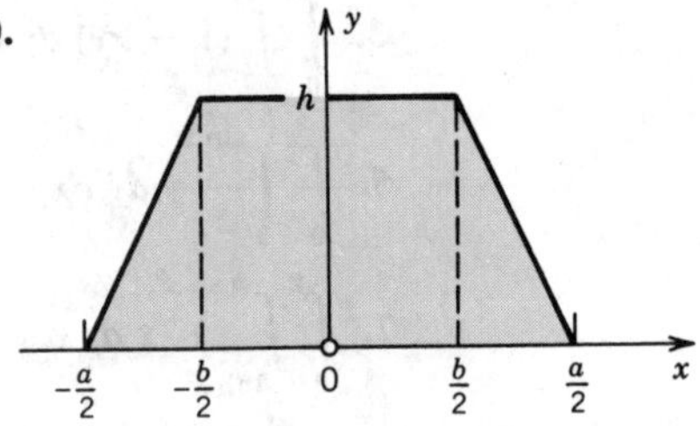

9.4 Green's Theorem in the Plane

Double integrals over a plane region may be transformed into line integrals over the boundary of the region and conversely. This is of practical interest because it may help to make the evaluation of an integral easier. It also helps in the theory whenever one wants to switch from one kind of integral to the other. The transformation can be done by the following theorem.

Theorem 1 **Green's theorem in the plane**
(Transformation between double integrals and line integrals)

Let R be a closed bounded region in the xy-plane whose boundary C consists of finitely many smooth curves. Let $F_1(x, y)$ and $F_2(x, y)$ be functions that are continuous and have continuous partial derivatives $\partial F_1/\partial y$ and $\partial F_2/\partial x$ everywhere in some domain containing R. Then

$$(1) \qquad \iint_R \left(\frac{\partial F_2}{\partial x} - \frac{\partial F_1}{\partial y}\right) dx\, dy = \oint_C (F_1\, dx + F_2\, dy);$$

here we integrate along the entire boundary C of R such that R is on the left as we advance in the direction of integration[4] (see Fig. 207). (Proof below.)

Comment. Formula (1) can be written in vectorial form

$$\text{(1}') \qquad \iint_R (\text{curl } \mathbf{F}) \cdot \mathbf{k} \, dx \, dy = \oint_C \mathbf{F} \cdot d\mathbf{r} \qquad (\mathbf{F} = F_1 \mathbf{i} + F_2 \mathbf{j}).$$

This follows from (1), Sec. 8.11, which shows that the third component of curl **F** is $\partial F_2/\partial x - \partial F_1/\partial y$.

EXAMPLE 1 Verification of Green's theorem in the plane

Green's theorem in the plane will be quite important in our further work. Before proving it, let us get used to it by verifying it for $F_1 = y^2 - 7y$, $F_2 = 2xy + 2x$ and C the circle $x^2 + y^2 = 1$.

Solution. In (1) on the left we get

$$\iint_R \left(\frac{\partial F_2}{\partial x} - \frac{\partial F_1}{\partial y}\right) dx\, dy = \iint_R [(2y + 2) - (2y - 7)] \, dx\, dy = 9 \iint_R dx\, dy = 9\pi$$

since the circular disk R has area π. On the right in (1) we represent C (oriented counterclockwise!) by

$$\mathbf{r}(t) = [\cos t, \sin t]. \qquad \text{Then} \qquad \mathbf{r}'(t) = [-\sin t, \cos t].$$

Then on C we have

$$F_1 = \sin^2 t - 7 \sin t, \qquad F_2 = 2 \cos t \sin t + 2 \cos t.$$

Hence the integral in (1) on the right becomes

$$\oint_C (F_1 x' + F_2 y')\, dt = \int_0^{2\pi} [(\sin^2 t - 7 \sin t)(-\sin t) + 2(\cos t \sin t + \cos t)(\cos t)]\, dt$$

$$= 0 + 7\pi + 0 + 2\pi = 9\pi.$$

This verifies the theorem. ∎

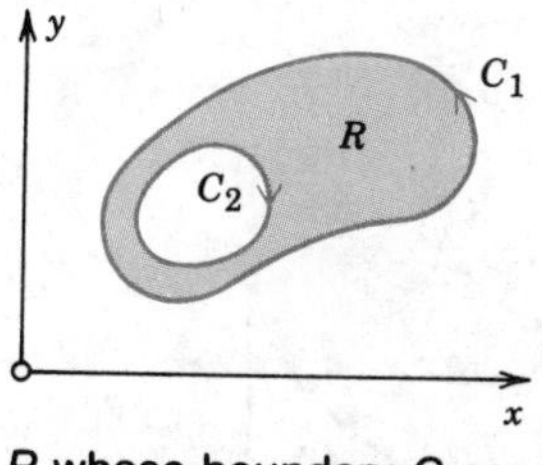

Fig. 207. Region R whose boundary C consists of two parts; C_1 is traversed counterclockwise, while C_2 is traversed clockwise

[4]GEORGE GREEN (1793—1841), English mathematician, who was self-educated, started out as a baker, and at his death was fellow of Caius College, Cambridge. His work concerned potential theory in connection with electricity and magnetism, vibrations, waves, and elasticity theory. It remained almost unknown, even in England, until after his death.

Proof of Green's theorem. We first prove Green's theorem for a *special region* R that can be represented in both the forms

$$a \leqq x \leqq b, \qquad u(x) \leqq y \leqq v(x) \qquad \text{(Fig. 208)}$$

and

$$c \leqq y \leqq d, \qquad p(y) \leqq x \leqq q(y) \qquad \text{(Fig. 209).}$$

Using (3) in Sec. 9.3, we obtain

$$\iint_R \frac{\partial F_1}{\partial y}\, dx\, dy = \int_a^b \left[\int_{u(x)}^{v(x)} \frac{\partial F_1}{\partial y}\, dy \right] dx. \tag{2}$$

We integrate the inner integral:

$$\int_{u(x)}^{v(x)} \frac{\partial F_1}{\partial y}\, dy = F_1(x, y) \Big|_{y=u(x)}^{y=v(x)} = F_1[x, v(x)] - F_1[x, u(x)].$$

By inserting this into (2) we find

$$\begin{aligned} \iint_R \frac{\partial F_1}{\partial y}\, dx\, dy &= \int_a^b F_1[x, v(x)]\, dx - \int_a^b F_1[x, u(x)]\, dx \\ &= -\int_a^b F_1[x, u(x)]\, dx - \int_b^a F_1[x, v(x)]\, dx. \end{aligned}$$

Since $y = u(x)$ represents the curve C^* (Fig. 208) and $y = v(x)$ represents C^{**}, the integrals on the right may be written as line integrals over C^* and C^{**} (oriented as shown in Fig. 208); therefore,

$$\begin{aligned} \iint_R \frac{\partial F_1}{\partial y}\, dx\, dy &= -\int_{C^*} F_1(x, y)\, dx - \int_{C^{**}} F_1(x, y)\, dx \\ &= -\oint_C F_1(x, y)\, dx. \end{aligned} \tag{3}$$

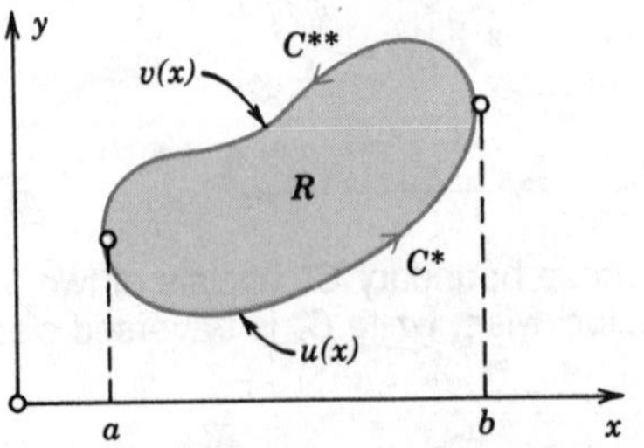

Fig. 208. Example of a special region

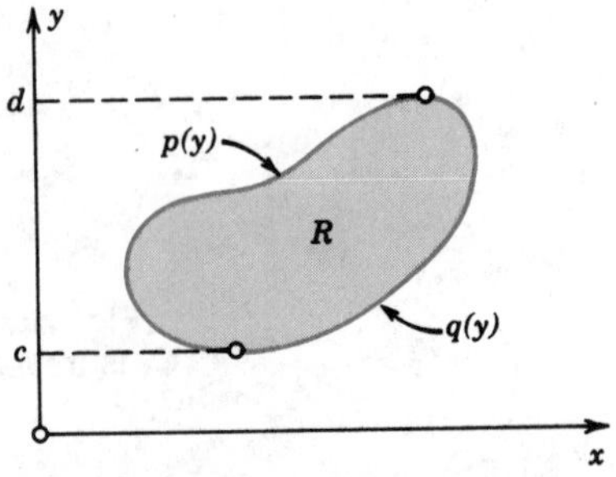

Fig. 209. Example of a special region

If portions of C are segments parallel to the y-axis (such as $\tilde{C}$ and $\tilde{\tilde{C}}$ in Fig. 210), the result is the same as before, because the integrals over these portions are zero and may be added to the integrals over C^* and C^{**} to obtain the integral over the whole boundary C in (3). Similarly, using (4), Sec. 9.3, we obtain (see Fig. 209)

$$\begin{aligned}\iint_R \frac{\partial F_2}{\partial x}\,dx\,dy &= \int_c^d \left[\int_{p(y)}^{q(y)} \frac{\partial F_2}{\partial x}\,dx\right] dy \\ &= \int_c^d F_2[q(y), y]\,dy + \int_d^c F_2[p(y), y]\,dy \\ &= \oint_C F_2(x, y)\,dy.\end{aligned}$$

From this and (3), the formula (1) follows, and the theorem is proved for special regions.

We now prove the theorem for a region R that itself is not a special region but can be subdivided into finitely many special regions (Fig. 211). In this case we apply the theorem to each subregion and then add the results; the left-hand members add up to the integral over R while the right-hand members add up to the line integral over C plus integrals over the curves introduced for subdividing R. Each of the latter integrals occurs twice, taken once in each direction. Hence these two integrals cancel each other, and we are left with the line integral over C.

The proof thus far covers all regions that are of interest in practical problems. To prove the theorem for the most general region R satisfying the conditions in the theorem, we must approximate R by a region of the type just considered and then use a limiting process. For details see Ref. [5] in Appendix 1. ∎

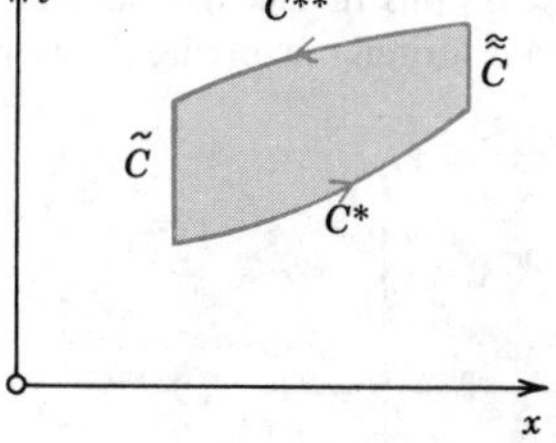

Fig. 210. Proof of Green's theorem

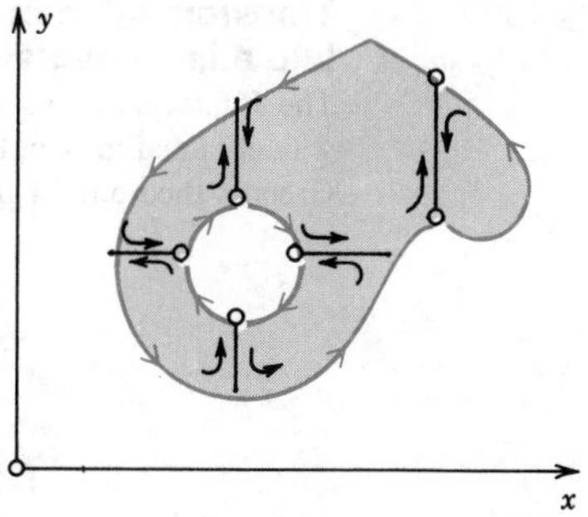

Fig. 211. Proof of Green's theorem

Further Applications of Green's Theorem

EXAMPLE 2 Area of a plane region as a line integral over the boundary

In (1), let $F_1 = 0$ and $F_2 = x$. Then

$$\iint_R dx\, dy = \oint_C x\, dy.$$

The integral on the left is the area A of R. Similarly, let $F_1 = -y$ and $F_2 = 0$; then from (1),

$$A = \iint_R dx\, dy = -\oint_C y\, dx.$$

By adding both formulas we obtain

$$A = \frac{1}{2}\oint_C (x\, dy - y\, dx), \tag{4}$$

where we integrate as indicated in Green's theorem. This interesting formula expresses the area of R in terms of a line integral over the boundary. It has various applications; for instance, the theory of certain **planimeters** (instruments for measuring area) is based on it.

For an **ellipse** $x^2/a^2 + y^2/b^2 = 1$ or $x = a \cos t$, $y = b \sin t$ we get $x' = -a \sin t$, $y' = b \cos t$; thus from (4) we obtain the familiar result

$$A = \frac{1}{2}\int_0^{2\pi} (xy' - yx')\, dt = \frac{1}{2}\int_0^{2\pi} [ab \cos^2 t - (-ab \sin^2 t)]\, dt = \pi ab.$$ ∎

EXAMPLE 3 Area of a plane region in polar coordinates

Let r and θ be polar coordinates defined by $x = r \cos \theta$, $y = r \sin \theta$. Then

$$dx = \cos \theta\, dr - r \sin \theta\, d\theta, \qquad dy = \sin \theta\, dr + r \cos \theta\, d\theta,$$

and (4) becomes a formula that is well known from calculus, namely,

$$A = \frac{1}{2}\oint_C r^2\, d\theta. \tag{5}$$

As an application of (5), we consider the **cardioid** $r = a(1 - \cos \theta)$, where $0 \leqq \theta \leqq 2\pi$ (Fig. 212). We find

$$A = \frac{a^2}{2}\int_0^{2\pi} (1 - \cos \theta)^2\, d\theta = \frac{3\pi}{2} a^2.$$ ∎

EXAMPLE 4 Transformation of a double integral of the Laplacian of a function into a line integral of its normal derivative

The Laplacian plays an important role in physics and engineering. A first impression of this was obtained in Sec. 8.9, and we shall discuss this further in Chap. 11. At present, let us use Green's theorem for deriving a basic integral formula involving the Laplacian.

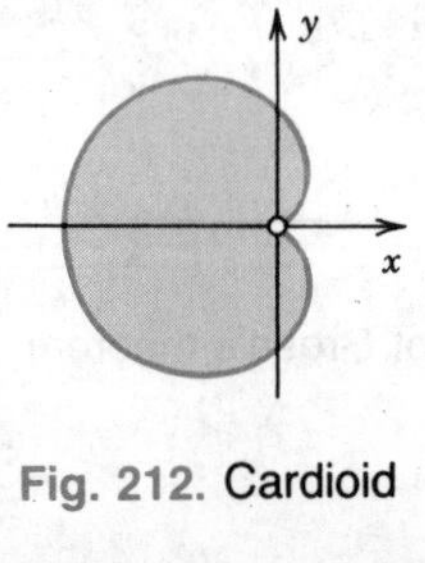

Fig. 212. Cardioid

We take a function $w(x, y)$ that is continuous and has continuous first and second partial derivatives in a domain of the xy-plane containing a region R of the type indicated in Green's theorem. We set $F_1 = -\partial w/\partial y$ and $F_2 = \partial w/\partial x$. Then $\partial F_1/\partial y$ and $\partial F_2/\partial x$ are continuous in R, and

$$\frac{\partial F_2}{\partial x} - \frac{\partial F_1}{\partial y} = \frac{\partial^2 w}{\partial x^2} + \frac{\partial^2 w}{\partial y^2} = \nabla^2 w, \tag{6}$$

the Laplacian of w (see Sec. 8.9). Furthermore, using those expressions for F_1 and F_2, we get

$$\oint_C (F_1\,dx + F_2\,dy) = \oint_C \left(F_1 \frac{dx}{ds} + F_2 \frac{dy}{ds}\right) ds = \oint_C \left(-\frac{\partial w}{\partial y}\frac{dx}{ds} + \frac{\partial w}{\partial x}\frac{dy}{ds}\right) ds, \tag{7}$$

where s is the arc length of C, and C is oriented as shown in Fig. 213. The integrand of the last integral may be written as the dot product of the vectors

$$\operatorname{grad} w = \frac{\partial w}{\partial x}\mathbf{i} + \frac{\partial w}{\partial y}\mathbf{j} \qquad \text{and} \qquad \mathbf{n} = \frac{dy}{ds}\mathbf{i} - \frac{dx}{ds}\mathbf{j};$$

that is,

$$-\frac{\partial w}{\partial y}\frac{dx}{ds} + \frac{\partial w}{\partial x}\frac{dy}{ds} = (\operatorname{grad} w)\cdot\mathbf{n}. \tag{8}$$

The vector $\mathbf{n}$ is a unit normal vector to C, because the vector

$$\mathbf{r}'(s) = \frac{d\mathbf{r}}{ds} = \frac{dx}{ds}\mathbf{i} + \frac{dy}{ds}\mathbf{j} \qquad \text{(Sec. 8.5)}$$

is the unit tangent vector to C, and $\mathbf{r}'\cdot\mathbf{n} = 0$. Furthermore, it is not difficult to see that $\mathbf{n}$ is directed to the *exterior* of C. From this and (6) in Sec. 8.9, it follows that the expression on the right side of (8) is the derivative of w in the direction of the outward normal to C. Denoting this directional derivative by $\partial w/\partial n$ and taking (6), (7), and (8) into account, we obtain from Green's theorem the desired integral fomula

$$\iint_R \nabla^2 w\,dx\,dy = \oint_C \frac{\partial w}{\partial n}\,ds. \tag{9}$$

For instance, $w = x^2 - y^2$ satisfies Laplace's equation $\nabla^2 w = 0$. Hence its normal derivative integrated over a closed curve must give 0. Can you verify this directly by integration, say, for the square $0 \leq x \leq 1$, $0 \leq y \leq 1$? ∎

Green's theorem in the plane may facilitate the evaluation of integrals and can be used in both directions, depending on the kind of integral that is simpler in a concrete case. This is illustrated further in the problem set below. Moreover, and perhaps more fundamentally, Green's theorem will be the essential tool in the proof of Stokes's theorem in Sec. 9.9.

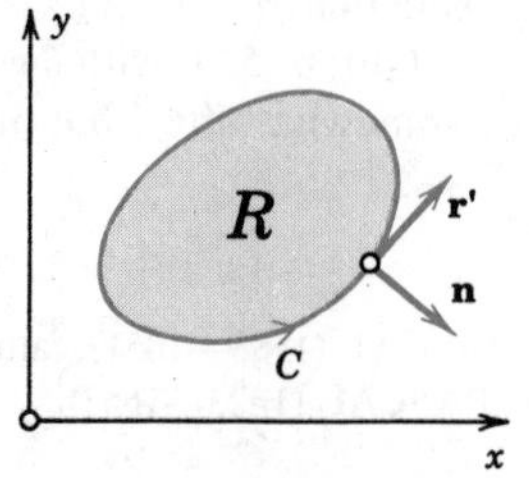

Fig. 213. Example 4

Problem Set 9.4

Using Green's theorem, evaluate the line integral $\oint_C \mathbf{F}(\mathbf{r}) \cdot d\mathbf{r}$ counterclockwise around the boundary C of the region R, where

1. $\mathbf{F} = y\mathbf{i} + 4x\mathbf{j}$, R the square $0 \leqq x \leqq a$, $0 \leqq y \leqq a$

2. $\mathbf{F} = x \sin y\, \mathbf{i} - y \sin x\, \mathbf{j}$, R the rectangle $0 \leqq x \leqq \pi$, $0 \leqq y \leqq \pi/2$

3. $\mathbf{F} = -y^3\mathbf{i} + x^3\mathbf{j}$, R the circular disk $x^2 + y^2 \leqq 4$

4. $\mathbf{F} = (2x - y)\mathbf{i} + (x + 3y)\mathbf{j}$, R the elliptical region $x^2 + 4y^2 \leqq 4$

5. $\mathbf{F} = 2xy\mathbf{i} + (e^x + x^2)\mathbf{j}$, R the triangle with vertices $(0, 0)$, $(1, 0)$, $(1, 1)$

6. $\mathbf{F} = x \ln y\, \mathbf{i} + ye^x\mathbf{j}$, R the rectangle $0 \leqq x \leqq 3$, $1 \leqq y \leqq 2$

7. $\mathbf{F} = \tan 0.1x\, \mathbf{i} + x^5y\mathbf{j}$, R: $x^2 + y^2 \leqq 25$, $y \geqq 0$

8. $\mathbf{F} = y^3\mathbf{i} + (x^3 + 3xy^2)\mathbf{j}$, R: $x^2 \leqq y \leqq x$

9. $\mathbf{F} = (x^2 + y^2)\mathbf{i} + (x^2 - y^2)\mathbf{j}$, R: $0 \leqq y \leqq 1/x$, $1 \leqq x \leqq 3$

10. $\mathbf{F} = x^2y^2\mathbf{i} - (x/y^2)\mathbf{j}$, R: $1 \leqq x^2 + y^2 \leqq 4$, $x \geqq 0$, $y \geqq x$

11. $\mathbf{F} = e^{x+y}\mathbf{i} + e^{x-y}\mathbf{j}$, R the triangle $x \leqq y \leqq 2x$, $x \leqq 1$

12. $\mathbf{F} = x \cos y\, \mathbf{i} + x^2 \sin y\, \mathbf{j}$, R: $1 + x^2 \leqq y \leqq 2$, $x \geqq 0$

13. $\mathbf{F} = (xy^2 + \cosh x)\mathbf{i} - x^2y\mathbf{j}$, R: $x \geqq 0$, $0 \leqq y \leqq 1 - x^2$

14. $\mathbf{F} = x \cosh y\, \mathbf{i} + x^2 \sinh y\, \mathbf{j}$, R: $x^2 \leqq y \leqq x$

15. $\mathbf{F} = (x^3 - 2y^3)\mathbf{i} + (x^3 + 2y^3)\mathbf{j}$, R: $x^2 + y^2 \leqq a^2$, $x \geqq 0$, $y \geqq 0$

Area. Using (4) or (5), find the area of the following regions.

16. The region under one arch of the **cycloid** $\mathbf{r} = a(t - \sin t)\mathbf{i} + a(1 - \cos t)\mathbf{j}$, $0 \leqq t \leqq 2\pi$

17. The region in the first quadrant within the **cardioid** (see Example 3)

18. The region in the first quadrant under the arc of the **limaçon** (snail of Pascal[5]) $r = 1 + 2 \cos \theta$, $0 \leqq \theta \leqq \pi/2$

Integral of the normal derivative. Using (9), evaluate $\oint_C \frac{\partial w}{\partial n}\, ds$ counterclockwise over the boundary curve C of the region R, where

19. $w = e^x + e^y$, R the square $0 \leqq x \leqq 2$, $0 \leqq y \leqq 2$

20. $w = 2x^2 + y^2$, R: $x^2 \leqq y \leqq x + 2$

21. $w = (x + 3y)^2 + 3x$, R: $x^2 + y^2 \leqq 16$, $x \leqq 0$, $y \geqq 0$

22. $w = e^x \cos y + x^3 - 3xy^2$, R the triangle with vertices $(1, 1)$, $(2, -1)$, $(4, 2)$

23. $w = \ln(x^2 + y^2) + xy^3$, R: $1 \leqq y \leqq 2 - x^2$, $x \geqq 0$

24. $w = x^5y + xy^5$, R: $x^2 + y^2 \leqq 1$, $y \geqq 0$

25. $w = x^3y + 2e^y$, R the triangle with vertices $(0, 0)$, $(1, 0)$, $(0, 1)$

26. **(Laplace's equation)** If $w(x, y)$ satisfies Laplace's equation $\nabla^2 w = 0$ in a region R, show that (10), p. 529, with $\partial w/\partial n$ defined as in Example 4 holds. *Hint.* Model your work somewhat after that of Example 4.

---

[5]ETIENNE PASCAL (1588—1651), father of the famous French mathematician and philosopher BLAISE PASCAL (1623—1662).

(10)
$$\iint_R \left[\left(\frac{\partial w}{\partial x}\right)^2 + \left(\frac{\partial w}{\partial y}\right)^2\right] dx\,dy = \oint_C w \frac{\partial w}{\partial n}\, ds.$$

27. Show that $w = 2e^x \cos y$ satisfies Laplace's equation and, using (10), integrate $w(\partial w/\partial n)$ counterclockwise around the boundary C of the square $0 \leqq x \leqq 2$, $0 \leqq y \leqq 2$.

Other forms of Green's theorem

28. Show that (1) may be written in the form (11), below, where $\mathbf{n}$ is the outward unit normal vector to the curve C (Fig. 213) and s is the arc length of C. *Hint.* Introduce $\mathbf{F} = F_2\mathbf{i} - F_1\mathbf{j}$.

(11)
$$\iint_R \operatorname{div} \mathbf{F}\, dx\,dy = \oint_C \mathbf{F}\bullet\mathbf{n}\, ds.$$

29. Verify (11) when $\mathbf{F} = 7x\mathbf{i} - 3y\mathbf{j}$ and C is the circle $x^2 + y^2 = 4$.

30. Show that (1) may be written in the form (12), below, where $\mathbf{k}$ is a unit vector perpendicular to the xy-plane, $\mathbf{r}'$ is the unit tangent vector to C, and s is the arc length of C.

(12)
$$\iint_R (\operatorname{curl} \mathbf{F})\bullet\mathbf{k}\, dx\,dy = \oint_C \mathbf{F}\bullet\mathbf{r}'\, ds.$$

9.5 Surfaces for Surface Integrals

Having introduced line and double integrals over regions in the plane, we turn next to surface integrals, in which we integrate over surfaces in space (a sphere, a portion of a cylinder, etc.). Hence we must first see how we can represent surfaces. We do this now and then discuss surface normals, which will also be needed in surface integrals.

Representations of Surfaces

We say briefly "surface," also for a *portion* of a surface, just as we said "curve" for an *arc* of a curve, for simplicity.

Representations of a surface S in xyz-space are

(1)
$$z = f(x, y) \qquad \text{or} \qquad g(x, y, z) = 0.$$

For example, $z = +\sqrt{a^2 - x^2 - y^2}$ or $x^2 + y^2 + z^2 - a^2 = 0$ $(z \geqq 0)$ represents a hemisphere of radius a and center 0.

Now for *curves* C in line integrals, it was more practical and gave greater flexibility to use a *parametric* representation $\mathbf{r} = \mathbf{r}(t)$, where $a \leqq t \leqq b$. This is a mapping of the interval $a \leqq t \leqq b$, located on the t-axis, onto the curve

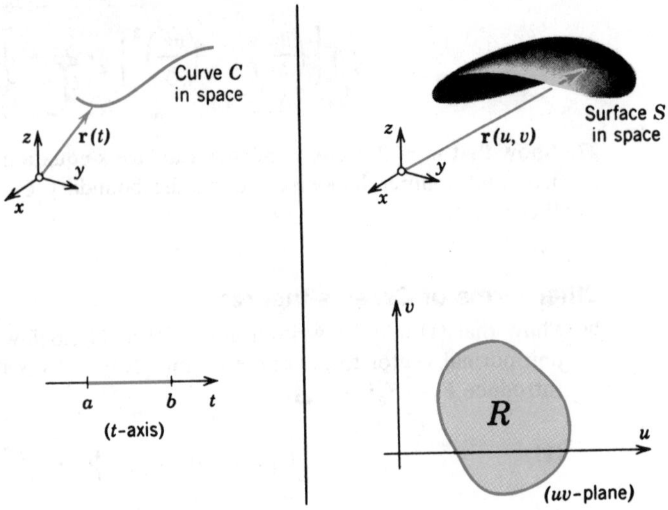

Fig. 214. Parametric representations of a curve and a surface

C in xyz-space. It maps every t in that interval onto the point of C with position vector $\mathbf{r}(t)$. See Fig. 214.

Similarly, for surfaces S in surface integrals, it will often be more practical to use a *parametric* representation. Surfaces are *two*-dimensional. Hence we need *two* parameters, which we call u and v. Thus a **parametric representation** of a surface S in space is of the form

$$\mathbf{r}(u, v) = x(u, v)\mathbf{i} + y(u, v)\mathbf{j} + z(u, v)\mathbf{k} \qquad (u, v) \text{ in } R \tag{2}$$

where R is some region in the uv-plane. This mapping (2) maps every point (u, v) in R onto the point of S with position vector $\mathbf{r}(u, v)$. See Fig. 214.

EXAMPLE 1 Parametric representation of a cylinder

The circular cylinder $x^2 + y^2 = a^2$, $-1 \leqq z \leqq 1$, has radius a, height 2, and the z-axis as axis. A parametric representation is

$$\mathbf{r}(u, v) = a \cos u\, \mathbf{i} + a \sin u\, \mathbf{j} + v\mathbf{k} \qquad \text{(Fig. 215)}$$

where the parameters u, v vary in the rectangle R in the uv-plane given by the inequalities $0 \leqq u \leqq 2\pi$, $-1 \leqq v \leqq 1$. The components of $\mathbf{r}(u, v)$ are

$$x = a \cos u, \qquad y = a \sin u, \qquad z = v.$$

The curves $v = const$ are parallel circles. The curves $u = const$ are vertical straight lines. The point P in Fig. 215 corresponds to $u = \pi/3 = 60°$, $v = 0.7$. ∎

EXAMPLE 2 Parametric representation of a sphere

A sphere $x^2 + y^2 + z^2 = a^2$ can be represented in the form

$$\mathbf{r}(u, v) = a \cos v \cos u\, \mathbf{i} + a \cos v \sin u\, \mathbf{j} + a \sin v\, \mathbf{k} \tag{3}$$

where the parameters u, v vary in the rectangle R in the uv-plane given by the inequalities $0 \leqq u \leqq 2\pi$, $-\pi/2 \leqq v \leqq \pi/2$. The components of $\mathbf{r}$ are

$$x = a \cos v \cos u, \qquad y = a \cos v \sin u, \qquad z = a \sin v.$$

The curves $u = const$ and $v = const$ are the "meridians" and "parallels" on S (see Fig. 216).

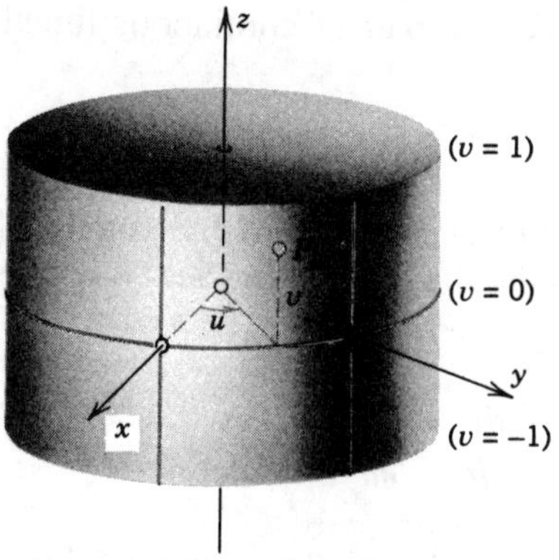

Fig. 215. Parametric representation of a cylinder

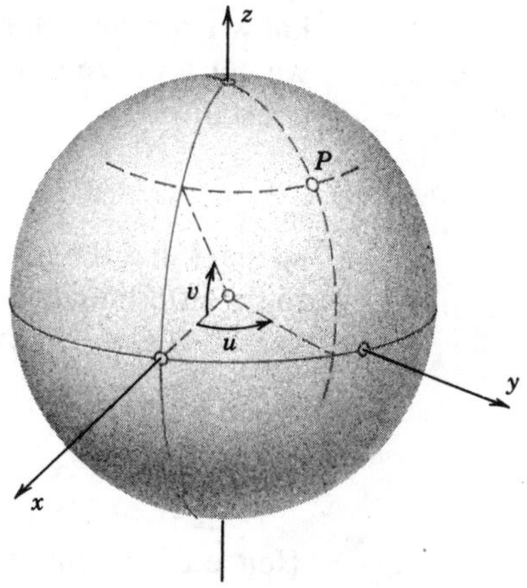

Fig. 216. Parametric representation of a sphere

This representation is used in ***geography*** *for measuring the latitude and longitude of points on the globe.*

Another parametric representation of the sphere also used in mathematics is

$$\mathbf{r}(u, v) = a \cos u \sin v\, \mathbf{i} + a \sin u \sin v\, \mathbf{j} + a \cos v\, \mathbf{k} \tag{3*}$$

where $0 \leqq u \leqq 2\pi$, $0 \leqq v \leqq \pi$. ■

EXAMPLE 3 **Parametric representation of a cone**

A circular cone $z = \sqrt{x^2 + y^2}$, $0 \leqq z \leqq H$ can be represented by

$$\mathbf{r}(u, v) = u \cos v\, \mathbf{i} + u \sin v\, \mathbf{j} + u\mathbf{k}$$

where u, v vary in the rectangle R: $0 \leqq u \leqq H$, $0 \leqq v \leqq 2\pi$. The components of $\mathbf{r}(u, v)$ are

$$x = u \cos v, \qquad y = u \sin v, \qquad z = u$$

and we can check that $x^2 + y^2 = z^2$, as it should be. What are the curves $u = const$ and $v = const$? Make a sketch. ■

Tangent Plane and Surface Normal

Before defining surface integrals we go one more step and introduce surface normal vectors, which we shall need. A **normal vector** of a surface S at a point P is a vector perpendicular to the **tangent plane** of S at P (Fig. 217), the plane containing all the tangent vectors of curves on S through P, as we

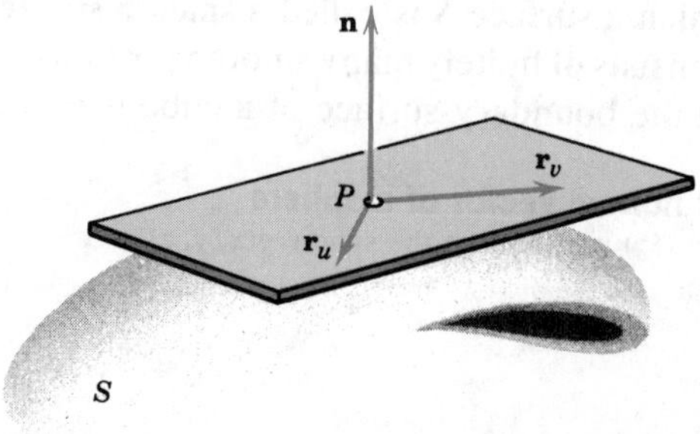

Fig. 217. Tangent plane and normal vector

know from Sec. 8.9. Since S is given by $\mathbf{r} = \mathbf{r}(u, v)$ in (2), the idea is that we get a curve C on S by taking a pair of continuous functions (not both constant)

$$u = u(t), \qquad v = v(t),$$

so that C has the position vector $\tilde{\mathbf{r}}(t) = \mathbf{r}(u(t), v(t))$. Assuming these functions to be differentiable and applying the chain rule (Sec. 8.8), we get a tangent vector of C given by

$$\tilde{\mathbf{r}}'(t) = \frac{d\tilde{\mathbf{r}}}{dt} = \frac{\partial \mathbf{r}}{\partial u} u' + \frac{\partial \mathbf{r}}{\partial v} v'.$$

Hence the partial derivatives $\mathbf{r}_u$ and $\mathbf{r}_v$ at P are tangential to S at P, and we assume that they are linearly independent, so that they span the tangent plane of S at P. Then their vector product gives a normal vector $\mathbf{N}$ of S at P,

$$\mathbf{N} = \mathbf{r}_u \times \mathbf{r}_v \neq \mathbf{0}. \tag{4}$$

The corresponding unit normal vector $\mathbf{n}$ of S at P is (Fig. 217, p. 531)

$$\mathbf{n} = \frac{1}{|\mathbf{N}|}\mathbf{N} = \frac{1}{|\mathbf{r}_u \times \mathbf{r}_v|}\mathbf{r}_u \times \mathbf{r}_v. \tag{5}$$

Also, if S is represented by $g(x, y, z) = 0$, then, by Theorem 2 in Sec. 8.9,

$$\mathbf{n} = \frac{1}{|\text{grad } g|} \text{grad } g. \tag{5*}$$

We can summarize as follows:

Theorem 1 **(Tangent plane and surface normal)**

If a surface S is given by (2) with continuous $\mathbf{r}_u = \partial\mathbf{r}/\partial u$ and $\mathbf{r}_v = \partial\mathbf{r}/\partial v$ satisfying (4) at every point of S, then S has at every point P a unique tangent plane passing through P and spanned by $\mathbf{r}_u$ and $\mathbf{r}_v$, and a unique normal whose direction depends continuously on the points of S.

A unit normal vector $\mathbf{n}$ of S is given by (5). (See Fig. 217.)

Such a surface S is called a **smooth surface.** We call S **piecewise smooth** if it consists of finitely many smooth portions. For instance, a sphere is smooth, and the boundary surface of a cube is piecewise smooth.

EXAMPLE 4 **Unit normal vector of a sphere**

From (5*) we find that the sphere $g(x, y, z) = x^2 + y^2 + z^2 - a^2 = 0$ has the unit normal vector

$$\mathbf{n}(x, y, z) = \frac{x}{a}\mathbf{i} + \frac{y}{a}\mathbf{j} + \frac{z}{a}\mathbf{k}.$$

∎

EXAMPLE 5 Unit normal vector of a cone

At the apex of the cone $g(x, y, z) = -z + \sqrt{x^2 + y^2} = 0$ in Example 3, the unit normal vector **n** becomes undetermined because from (5*) we get

$$\mathbf{n} = \frac{1}{\sqrt{2}}\left(\frac{x}{\sqrt{x^2 + y^2}}\mathbf{i} + \frac{y}{\sqrt{x^2 + y^2}}\mathbf{j} - \mathbf{k}\right).$$

What we have learned about surfaces in the present section will be applied to discuss **surface integrals** in the next section.

Problem Set 9.5

Parametric representations and surface normal. As a preparation for surface integrals, familiarize yourself with parametric representations of surfaces by deriving a representation of the form (1), by stating what the **parameter curves** (curves $u = const$ and $v = const$) on the surface are and by finding the normal vector $\mathbf{N} = \mathbf{r}_u \times \mathbf{r}_v$ of the surface.

1. xy-plane $\mathbf{r} = u\mathbf{i} + v\mathbf{j}$

2. xy-plane in polar coordinates $\mathbf{r} = u \cos v\, \mathbf{i} + u \sin v\, \mathbf{j}$

3. Elliptic cylinder $\mathbf{r} = 4 \cos u\, \mathbf{i} + 2 \sin u\, \mathbf{j} + v\mathbf{k}$

4. Helicoid $\mathbf{r} = u \cos v\, \mathbf{i} + u \sin v\, \mathbf{j} + v\mathbf{k}$

5. Cone $\mathbf{r} = u \cos v\, \mathbf{i} + u \sin v\, \mathbf{j} + cu\mathbf{k}$

6. Ellipsoid $\mathbf{r} = a \cos v \cos u\, \mathbf{i} + b \cos v \sin u\, \mathbf{j} + c \sin v\, \mathbf{k}$

7. Paraboloid of revolution $\mathbf{r} = u \cos v\, \mathbf{i} + u \sin v\, \mathbf{j} + u^2\mathbf{k}$

8. Ellipsoid of revolution $\mathbf{r} = \cos v \cos u\, \mathbf{i} + \cos v \sin u\, \mathbf{j} + 2 \sin v\, \mathbf{k}$

9. Elliptic paraboloid $\mathbf{r} = au \cos v\, \mathbf{i} + bu \sin v\, \mathbf{j} + u^2\mathbf{k}$

10. Hyperbolic paraboloid $\mathbf{r} = au \cosh v\, \mathbf{i} + bu \sinh v\, \mathbf{j} + u^2\mathbf{k}$

11. Hyperboloid $\mathbf{r} = a \sinh u \cos v\, \mathbf{i} + b \sinh u \sin v\, \mathbf{j} + c \cosh u\, \mathbf{k}$

12. Catenoid $\mathbf{r} = v \cos u\, \mathbf{i} + v \sin u\, \mathbf{j} + \cosh^{-1} v\, \mathbf{k}$

Derivation of parametric representations of surfaces. Find a parametric representation of the following surfaces. (The answer gives *one* such representation and there are many others.) Find a normal vector.

13. Plane $y = z$

14. yz-plane

15. Plane $x + y + z = 1$

16. Sphere $(x - 1)^2 + y^2 + (z + 1)^2 = 4$

17. Hyperbolic cylinder $x^2 - y^2 = 1$

18. Elliptic cylinder $x^2 + 4z^2 = 4$

19. Paraboloid $z = 16(x^2 + y^2)$

20. Elliptic cone $z = \sqrt{x^2 + 9y^2}$

21. Find a parametric representation of the paraboloid in Prob. 29 such that $\mathbf{N}(0, 0) \neq \mathbf{0}$. Find **N**.

22. Find the points in Probs. 2 and 20 at which (4) does not hold and indicate whether this is because of the shape of the surface or the choice of the representation.

23. Show that a representation $z = f(x, y)$ can be written ($f_u = \partial f/\partial u$, etc.)

$$\mathbf{r}(u, v) = u\mathbf{i} + v\mathbf{j} + f(u, v)\mathbf{k}, \quad \text{and} \quad \mathbf{N} = -f_u\mathbf{i} - f_v\mathbf{j} + \mathbf{k}. \tag{6}$$

24. Write $x = h(y, z)$ in parametric form and find a normal vector.

25. (Orthogonal parameters on a surface) Show that the parameter curves $u = const$ and $v = const$ on a surface $\mathbf{r} = \mathbf{r}(u, v)$ intersect at right angles if and only if $\mathbf{r}_u \cdot \mathbf{r}_v = 0$.

Surface normal. Using (5*), find a unit normal vector of the surfaces given by

26. $x^2 + 4y^2 + 16z^2 = 64$
27. $2x + 3y - 5z = 7$
28. $4x^2 + 4y^2 - z^2 = 0$
29. $x^2 + y^2 + z^2 = 9$
30. $3x^2 + 3y^2 - z = 0$
31. $x^2 + y^2 = a^2$
32. $z - x^2 + y^2 = 0$
33. $z - xy = 0$

Tangent plane. The tangent plane $T(P)$ as such will be of lesser importance in our further work, but one should know how to represent it. Show that:

34. If S: $\mathbf{r}(u, v)$, then $T(P)$: $(\mathbf{r}^* - \mathbf{r} \;\; \mathbf{r}_u \;\; \mathbf{r}_v) = 0$ (see Sec. 8.3).

35. If S: $\mathbf{r}(u, v)$, then $T(P)$: $\mathbf{r}^*(p, q) = \mathbf{r}(P) + p\mathbf{r}_u(P) + q\mathbf{r}_v(P)$.

36. If S: $g(x, y, z) = 0$, then $T(P)$: $[\mathbf{r}^* - \mathbf{r}(P)] \cdot \nabla g(P) = 0$.
If S: $z = f(x, y)$, then $T(P)$: $z^* - z = (x^* - x)f_x(P) + (y^* - y)f_y(P)$.

Find a representation of the tangent plane of the following surfaces at P_0: (x_0, y_0, z_0).

37. $y^2 + z^2 = 4$, P_0: $(3, -\sqrt{2}, \sqrt{2})$
38. $z^2 = x^2 + y^2$, P_0: $(1, 0, 1)$
39. $x^2 + y^2 + z^2 = 25$, P_0: $(3, 4, 0)$
40. $3x^2 + 2y^2 + z^2 = 20$, P_0: $(1, 2, 3)$

9.6 Surface Integrals

To define a surface integral, we take a surface S, given by a parametric representation as just discussed,

$$\mathbf{r}(u, v) = x(u, v)\mathbf{i} + y(u, v)\mathbf{j} + z(u, v)\mathbf{k} \qquad (u, v) \text{ in } R, \tag{1}$$

and assume that S is piecewise smooth, so that S has a normal vector

$$\mathbf{N} = \mathbf{r}_u \times \mathbf{r}_v \qquad \text{and unit normal vector} \qquad \mathbf{n} = \frac{1}{|\mathbf{N}|}\mathbf{N} \tag{2}$$

(see Sec. 9.5) at every point (except perhaps for some edges or cusps, as for a cube or cone). We can now define a **surface integral** of a vector function $\mathbf{F}$ over S by

$$\iint_S \mathbf{F} \cdot \mathbf{n} \, dA = \iint_R \mathbf{F}[\mathbf{r}(u, v)] \cdot \mathbf{N}(u, v) \, du \, dv. \tag{3}$$

$\mathbf{F} \cdot \mathbf{n}$ is the normal component of $\mathbf{F}$, and this integral arises naturally in flow problems where it gives the *flux* across S (= mass of fluid crossing S per unit time; see Sec. 8.10) when $\mathbf{F} = \rho\mathbf{v}$, where ρ is the density of the fluid and $\mathbf{v}$ the velocity vector of the flow (example below). We may thus call the surface integral (3) the **flux integral.**

Other forms of surface integrals will be discussed on pp. 539–542.

We see that the integral in (3) on the right is a double integral (Sec. 9.3) over the region R in the uv-plane corresponding to S and exists for continuous $\mathbf{F}$ and piecewise smooth S, because this makes $\mathbf{F}\bullet\mathbf{N}$ piecewise continuous. Note also that the integrand is a scalar, not a vector, since we take dot products.

In (3) we have $\mathbf{N}\,du\,dv = \mathbf{n}|\mathbf{N}|\,du\,dv$ by (5) in Sec. 9.5. Now, by the definition of vector product, $|\mathbf{N}| = |\mathbf{r}_u \times \mathbf{r}_v|$ is the area of the parallelogram with sides $\mathbf{r}_u$ and $\mathbf{r}_v$. Hence $|\mathbf{N}|\,du\,dv$ is the element of area dA of S, so that

$$\mathbf{n}\,dA = \mathbf{N}\,du\,dv; \tag{3*}$$

this motivates (3).

Setting $\mathbf{F} = F_1\mathbf{i} + F_2\mathbf{j} + F_3\mathbf{k}$, $\mathbf{n} = \cos\alpha\,\mathbf{i} + \cos\beta\,\mathbf{j} + \cos\gamma\,\mathbf{k}$ and $\mathbf{N} = N_1\mathbf{i} + N_2\mathbf{j} + N_3\mathbf{k}$, we claim that (3) can be written

$$\begin{aligned}\iint_S \mathbf{F}\bullet\mathbf{n}\,dA &= \iint_S (F_1\cos\alpha + F_2\cos\beta + F_3\cos\gamma)\,dA \\ &= \iint_R (F_1N_1 + F_2N_2 + F_3N_3)\,du\,dv.\end{aligned} \tag{4}$$

Indeed, both right sides follow by (2), Sec. 8.2, the first from $\mathbf{F}\bullet\mathbf{n}$ and the second from $\mathbf{F}\bullet\mathbf{N}$. Here, α, β, γ are the angles between $\mathbf{n}$ and the positive directions of the coordinate axes, because we obtain the dot product $\mathbf{n}\bullet\mathbf{i} = \cos\alpha$ and, on the other hand, from (4) in Sec. 8.2 the same result, $\cos\alpha = \mathbf{n}\bullet\mathbf{i}/|\mathbf{n}||\mathbf{i}| = \mathbf{n}\bullet\mathbf{i}$. Similarly for β and γ. ∎

EXAMPLE 1 **Flux through a surface**

Compute the flux of water through the parabolic cylinder S: $y = x^2$, $0 \leqq x \leqq 2$, $0 \leqq z \leqq 3$ (Fig. 218, p. 536) if the velocity vector is $\mathbf{F} = y\mathbf{i} + 2\mathbf{j} + xz\mathbf{k}$, speed being measured in meters/sec.

Solution. Writing $x = u$ and $z = v$, we have $y = x^2 = u^2$. Hence a representation of S is

$$S: \quad \mathbf{r} = u\mathbf{i} + u^2\mathbf{j} + v\mathbf{k} \qquad (0 \leqq u \leqq 2,\ 0 \leqq v \leqq 3).$$

From this,

$$\mathbf{r}_u = \mathbf{i} + 2u\mathbf{j}$$

$$\mathbf{r}_v = \mathbf{k}$$

$$\mathbf{N} = \mathbf{r}_u \times \mathbf{r}_v = 2u\mathbf{i} - \mathbf{j}.$$

On S,

$$\mathbf{F} = u^2\mathbf{i} + 2\mathbf{j} + uv\mathbf{k}.$$

Hence

$$\mathbf{F}\bullet\mathbf{N} = u^2 \cdot 2u + 2(-1) = 2u^3 - 2.$$

By integration we thus get from (3)

$$\iint_S \mathbf{F}\bullet\mathbf{n}\,dA = \int_0^3\int_0^2 (2u^3 - 2)\,du\,dv = 3\int_0^2 (2u^3 - 2)\,du = 12\ [\text{meters}^3/\text{sec}].$$

Since water has the density $\rho = 1$ gram/cm^3 = 1 kg/liter = 1000 kg/meter^3, the answer is 12 000 kg/sec. ∎

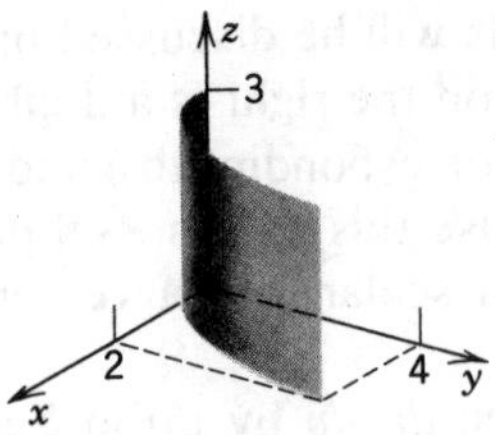

Fig. 218. Surface S in Example 1

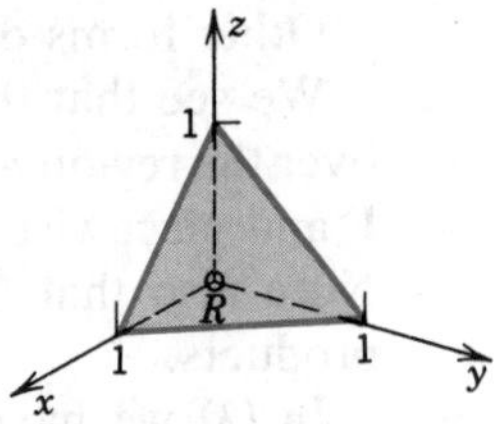

Fig. 219. Portion of a plane in Example 2

EXAMPLE 2 **Surface integral**

Evaluate (3) when $\mathbf{F} = x^2\mathbf{i} + 3y^2\mathbf{k}$ and S is the portion of the plane $x + y + z = 1$ in the first octant (Fig. 219).

Solution. Writing $x = u$ and $y = v$, we have $z = 1 - x - y = 1 - u - v$. Hence we can represent the plane $x + y + z = 1$ in the form

$$\mathbf{r}(u, v) = u\mathbf{i} + v\mathbf{j} + (1 - u - v)\mathbf{k}.$$

We obtain the first-octant portion S of this plane by restricting $x = u$ and $y = v$ to the projection R of S in the xy-plane. R is the triangle bounded by the two coordinate axes and the straight line $x + y = 1$; thus $0 \leqq x \leqq 1 - y$, $0 \leqq y \leqq 1$.

Differentiating $\mathbf{r}(u, v)$, we calculate

$$\mathbf{N} = \mathbf{r}_u \times \mathbf{r}_v = (\mathbf{i} - \mathbf{k}) \times (\mathbf{j} - \mathbf{k}) = \mathbf{i} + \mathbf{j} + \mathbf{k}.$$

Hence $\mathbf{F} \cdot \mathbf{N} = u^2 + 3v^2$, so that (3) gives

$$\iint_S \mathbf{F} \cdot \mathbf{n}\, dA = \iint_R (u^2 + 3v^2)\, du\, dv = \int_0^1 \int_0^{1-v} (u^2 + 3v^2)\, du\, dv$$

$$= \int_0^1 \left[\frac{1}{3}(1 - v)^3 + 3v^2(1 - v)\right] dv = \frac{1}{3}. \qquad ■$$

From (3) or (4) we see that the value of the integral depends on the choice of the unit normal vector $\mathbf{n}$. (Instead of $\mathbf{n}$ we could choose $-\mathbf{n}$.) We express this by saying that such an integral is an *integral over an* **oriented surface** S, that is, over a surface S on which we have chosen one of the two possible unit normal vectors in a continuous fashion. (For a piecewise smooth surface, this needs some further discussion, which we give below.) If we change the orientation of S, which means that we replace $\mathbf{n}$ by $-\mathbf{n}$, then each component of $\mathbf{n}$ in (4) is multiplied by -1, so that we have

Theorem 1 **(Change of orientation)**

The replacement of $\mathbf{n}$ *by* $-\mathbf{n}$ (*hence of* $\mathbf{N}$ *by* $-\mathbf{N}$) *corresponds to the multiplication of the integral in* (3) *or* (4) *by* -1.

How to effect such a change of $\mathbf{N}$ in practice if S is given in the form (1)? The simplest way is to interchange u and v, because then $\mathbf{r}_u$ becomes $\mathbf{r}_v$ and conversely, so that $\mathbf{N} = \mathbf{r}_u \times \mathbf{r}_v$ becomes $\mathbf{r}_v \times \mathbf{r}_u = -\mathbf{r}_u \times \mathbf{r}_v = -\mathbf{N}$, as wanted. Let us illustrate this.

EXAMPLE 3 **Change of orientation**

In Example 1 we had $\mathbf{F} = y\mathbf{i} + 2\mathbf{j} + xz\mathbf{k}$ and S: $y = x^2$, where $0 \leqq x \leqq 2$, $0 \leqq z \leqq 3$. By setting $x = u$ and $z = v$ we got the representation

$$\mathbf{r} = u\mathbf{i} + u^2\mathbf{j} + v\mathbf{k} \qquad (0 \leqq u \leqq 2,\ 0 \leqq v \leqq 3)$$

and $+12$ as the value of the integral.

Let us verify that if we interchange u and v, so that

$$\mathbf{r} = v\mathbf{i} + v^2\mathbf{j} + u\mathbf{k} \qquad (0 \leqq u \leqq 3,\ 0 \leqq v \leqq 2),$$

the integral will be -12. Indeed, by straightforward calculation we get

$$\mathbf{N} = \mathbf{r}_u \times \mathbf{r}_v = \mathbf{k} \times (\mathbf{i} + 2v\mathbf{j}) = \mathbf{j} - 2v\mathbf{i},$$

$$\mathbf{F} = v^2\mathbf{i} + 2\mathbf{j} + uv\mathbf{k} \qquad (\text{on } S)$$

and from this,

$$\iint_R \mathbf{F}\cdot\mathbf{N}\,du\,dv = \int_0^2\int_0^3 (-2v^3 + 2)\,du\,dv = \int_0^2 (-6v^3 + 6)\,dv = -12.$$

More about orientation. We consider first a ***smooth*** surface. If S is smooth and P any of its points, we may choose a unit normal vector $\mathbf{n}$ of S at P. The direction of $\mathbf{n}$ is then called the *positive normal direction* of S at P. Obviously there are two possibilities in choosing $\mathbf{n}$.

A smooth surface S is said to be **orientable** if the positive normal direction, when given at an arbitrary point P_0 of S, can be continued in a unique and continuous way to the entire surface.

Essential in practice is the fact that *a sufficiently small portion of a smooth surface is always orientable.* From a theoretical point of view it is interesting that this may not hold in the large. There are nonorientable surfaces. A well-known example of such a surface is the **Möbius strip**[6] shown in Fig. 220. When a normal vector, which is given at P_0, is displaced continuously along the curve C in Fig. 220, the resulting normal vector upon returning to P_0 is opposite to the original vector at P_0. A model of a Möbius strip can be made by taking a long rectangular piece of paper, making a half-twist and sticking the shorter sides together so that the two points A and the two points B in Fig. 220 coincide.

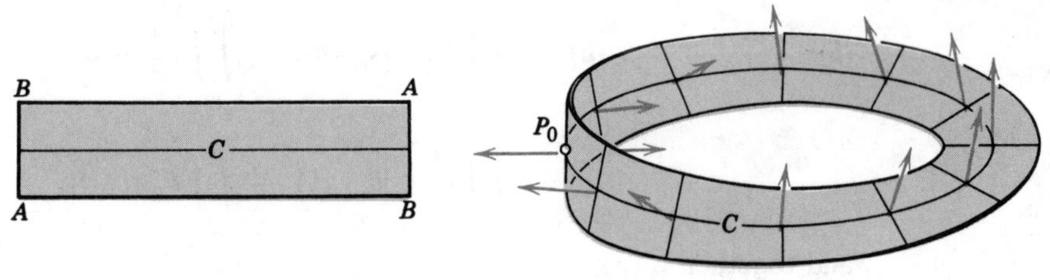

Fig. 220. Möbius strip

[6]AUGUST FERDINAND MÖBIUS (1790—1868), German mathematician, student of Gauss, professor of astronomy at Leipzig, known for his important work in the theory of surfaces, projective geometry, and mechanics. He also contributed to number theory.

If the boundary of an orientable smooth surface S is a simple closed curve C, then we may associate with each of the two possible orientations of S an orientation of C, as shown in Fig. 221a. Using this simple idea, we may now readily extend the concept of orientation to piecewise smooth surfaces as follows.

A ***piecewise smooth*** surface S is called **orientable** if we can orient each smooth piece of S in such a manner that along each curve C^* which is a common boundary of two pieces S_1 and S_2 the positive direction of C^* relative to S_1 is opposite to the positive direction of C^* relative to S_2.

Figure 221b illustrates the situation for a surface consisting of two smooth pieces.

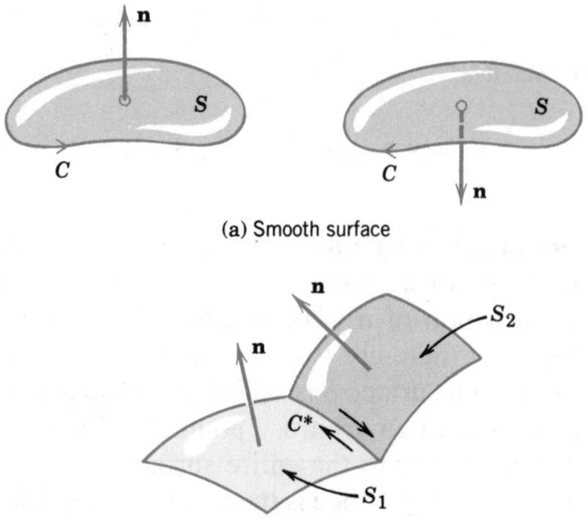

(a) Smooth surface

(b) Piecewise smooth surface

Fig. 221. Orientation of a surface

Another notation for the integrals in (4). After this discussion of orientation we can now explain another way of writing (4). It is customary to write in (4) also

$$\text{(5)}\quad \begin{aligned} &\text{(a)} \quad \iint_S F_1 \cos\alpha \, dA = \iint_S F_1 \, dy\, dz \\ &\text{(b)} \quad \iint_S F_2 \cos\beta \, dA = \iint_S F_2 \, dz\, dx \\ &\text{(c)} \quad \iint_S F_3 \cos\gamma \, dA = \iint_S F_3 \, dx\, dy \end{aligned}$$

and together

$$\text{(6)} \quad \iint_S \mathbf{F} \cdot \mathbf{n} \, dA = \iint_S (F_1 \, dy\, dz + F_2 \, dz\, dx + F_3 \, dx\, dy).$$

This is an analog of (3′) in Sec. 9.1 for line integrals. We can use these formulas for evaluating surface integrals by converting them to double integrals over plane regions, but must carefully take into account the orientation of S (the choice of $\mathbf{n}$). We explain this for (5c). If the surface S is given by $z = h(x, y)$ with (x, y) varying in a region R in the xy-plane, and if S is oriented so that $\cos\gamma > 0$, then

$$\text{(5c}')\qquad \iint_S F_3 \cos\gamma \, dA = +\iint_R F_3[x, y, h(x, y)] \, dx \, dy,$$

but if $\cos\gamma < 0$, then

$$\text{(5c}'')\qquad \iint_S F_3 \cos\gamma \, dA = -\iint_R F_3[x, y, h(x, y)] \, dx \, dy.$$

This follows by noting that the element of area $dx\,dy$ in the xy-plane is the projection $|\cos\gamma|\,dA$ of the element of area dA of S, and we have $\cos\gamma = +|\cos\gamma|$ in (5c′), where $\cos\gamma > 0$, but $\cos\gamma = -|\cos\gamma|$ in (5c″), where $\cos\gamma < 0$. Similarly for (5a), (5b) and in (6). ∎

EXAMPLE 4 An application of (6)

Verify the result in Example 1 by (6).

Solution. In Example 1 we have $F_1 = y$, $F_2 = 2$, $F_3 = xz$ and S: $y = x^2$ with $0 \leqq x \leqq 2$ (thus $0 \leqq y \leqq 4$) and $0 \leqq z \leqq 3$. The surface normal is parallel to the xy-plane, thus $\cos\gamma = 0$, and $\mathbf{n}$ points in the positive x-direction, thus $\cos\alpha > 0$, and somewhat down in the negative y-direction, thus $\cos\beta < 0$, causing a minus sign in (5b). (Indeed, $\mathbf{N} = 2u\mathbf{i} - \mathbf{j} = 2x\mathbf{i} - \mathbf{j}$.) We thus obtain from (6)

$$\begin{aligned}\iint_S \mathbf{F}\cdot\mathbf{n} \, dA &= \iint_R (y \, dy \, dz - 2 \, dz \, dx) \\ &= \int_0^3\int_0^4 y \, dy \, dz - \int_0^2\int_0^3 2 \, dz \, dx \\ &= 3 \cdot \frac{y^2}{2}\bigg|_0^4 - 2 \cdot 2 \cdot 3 = 24 - 12 = 12.\end{aligned}$$

This confirms the result in Example 1. ∎

Integrals over Nonoriented Surfaces

Another type of surface integral is

$$\text{(7)}\qquad \iint_S G(\mathbf{r}) \, dA = \iint_R G[\mathbf{r}(u, v)]|\mathbf{N}(u, v)| \, du \, dv.$$

Here $dA = |\mathbf{N}| \, du \, dv = |\mathbf{r}_u \times \mathbf{r}_v| \, du \, dv$ is the element of area of the surface S represented by (1) [see before (3*)] and we disregard the orientation.

As for applications, if $G(\mathbf{r})$ is the mass density of S, then (7) is the total mass of S. If $G = 1$, then (7) gives the **area** $A(S)$ of S,

$$\text{(8)}\qquad A(S) = \iint_S dA = \iint_R |\mathbf{r}_u \times \mathbf{r}_v| \, du \, dv.$$

EXAMPLE 5 **Area of a sphere**

A sphere of radius a can be represented by (3), Sec. 9.5, that is,

$$\mathbf{r}(u, v) = a \cos v \cos u\, \mathbf{i} + a \cos v \sin u\, \mathbf{j} + a \sin v\, \mathbf{k},$$

where $0 \leqq u \leqq 2\pi$, $-\pi/2 \leqq v \leqq \pi/2$. By direct calculation we obtain (verify!)

$$\mathbf{r}_u \times \mathbf{r}_v = a^2 \cos^2 v \cos u\, \mathbf{i} + a^2 \cos^2 v \sin u\, \mathbf{j} + a^2 \cos v \sin v\, \mathbf{k}.$$

Hence

$$|\mathbf{r}_u \times \mathbf{r}_v| = a^2[\cos^4 v \cos^2 u + \cos^4 v \sin^2 u + \cos^2 v \sin^2 v]^{1/2} = a^2 |\cos v|.$$

With this, (8) gives the familiar formula

$$A(S) = a^2 \int_{-\pi/2}^{\pi/2} \int_0^{2\pi} |\cos v|\, du\, dv = 2\pi a^2 \int_{-\pi/2}^{\pi/2} \cos v\, dv = 4\pi a^2.$$

■

EXAMPLE 6 **Representation and area of a torus surface (doughnut)**

A *torus surface* S is obtained by rotating a circle C about a straight line L in space so that C does not intersect or touch L but its plane always passes through L. If L is the z-axis and C has radius b and its center has distance a from L, as in Fig. 222, then S can be represented by

$$\mathbf{r}(u, v) = (a + b \cos v) \cos u\, \mathbf{i} + (a + b \cos v) \sin u\, \mathbf{j} + b \sin v\, \mathbf{k}.$$

Thus

$$\mathbf{r}_u = -(a + b \cos v) \sin u\, \mathbf{i} + (a + b \cos v) \cos u\, \mathbf{j}$$

$$\mathbf{r}_v = -b \sin v \cos u\, \mathbf{i} - b \sin v \sin u\, \mathbf{j} + b \cos v\, \mathbf{k}$$

$$\mathbf{r}_u \times \mathbf{r}_v = b(a + b \cos v)[\cos u \cos v\, \mathbf{i} + \sin u \cos v\, \mathbf{j} + \sin v\, \mathbf{k}].$$

Hence $|\mathbf{r}_u \times \mathbf{r}_v| = b(a + b \cos v)$, and (8) gives the total area of the torus

$$A(S) = \int_0^{2\pi} \int_0^{2\pi} b(a + b \cos v)\, du\, dv = 4\pi^2 ab. \tag{9}$$

■

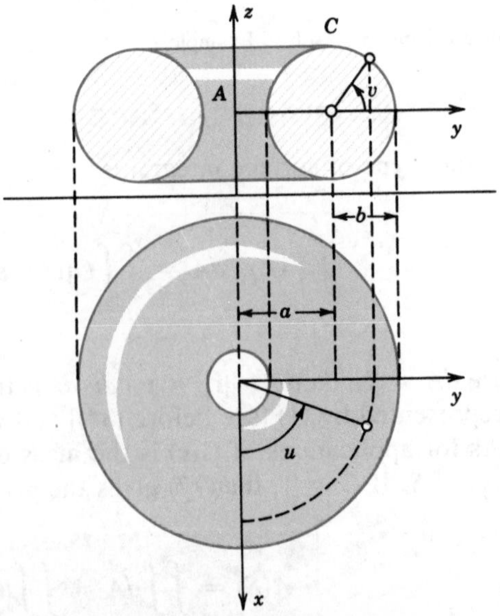

Fig. 222. Torus in Example 6

EXAMPLE 7 **Moment of inertia**

Find the moment of inertia I of a homogeneous spherical lamina S: $x^2 + y^2 + z^2 = a^2$ of mass M about the z-axis.

Solution. If a mass is distributed over a surface S and $\mu(x, y, z)$ is the density of the mass (= mass per unit area), then the moment of inertia I of the mass with respect to a given axis L is defined by the surface integral

$$I = \iint_S \mu D^2 \, dA \tag{10}$$

where $D(x, y, z)$ is the distance of the point (x, y, z) from L. Since, in the present example, μ is constant and S has the area $A = 4\pi a^2$, we have

$$\mu = \frac{M}{A} = \frac{M}{4\pi a^2}.$$

Using for S the representation (3) in Sec. 9.5,

$$\mathbf{r}(u, v) = a \cos v \cos u \,\mathbf{i} + a \cos v \sin u \,\mathbf{j} + a \sin v \,\mathbf{k},$$

we get for the square of the distance of a point (x, y, z) from the z-axis the expression $D^2 = x^2 + y^2 = a^2 \cos^2 v$. Also, by straightforward calculation,

$$dA = |\mathbf{N}| \, du \, dv = |\mathbf{r}_u \times \mathbf{r}_v| \, du \, dv = a^2 \cos v \, du \, dv$$

(verify this!). Hence we obtain the result

$$I = \iint_S \mu D^2 \, dA = \frac{M}{4\pi a^2} \int_{-\pi/2}^{\pi/2} \int_0^{2\pi} a^4 \cos^3 v \, du \, dv = \frac{Ma^2}{2} \int_{-\pi/2}^{\pi/2} \cos^3 v \, dv = \frac{2Ma^2}{3}. \quad \blacksquare$$

If S is given by $z = f(x, y)$, then setting $u = x$, $v = y$, $\mathbf{r} = [u, v, f]$ gives

$$|\mathbf{N}| = |\mathbf{r}_u \times \mathbf{r}_v| = |[1, 0, f_u] \times [0, 1, f_v]|$$
$$= |[-f_u, -f_v, 1]| = \sqrt{1 + f_u^2 + f_v^2}$$

and, since $f_u = f_x$, $f_v = f_y$, formula (7) becomes

$$\iint_S G(\mathbf{r}) \, dA = \iint_{R^*} G[x, y, f(x, y)] \sqrt{1 + \left(\frac{\partial f}{\partial x}\right)^2 + \left(\frac{\partial f}{\partial y}\right)^2} \, dx \, dy \tag{11}$$

where R^* is the projection of S into the xy-plane (Fig. 223) and the normal vector $\mathbf{N}$ on S points *up*. If it points *down*, the integral on the right is preceded by a minus sign.

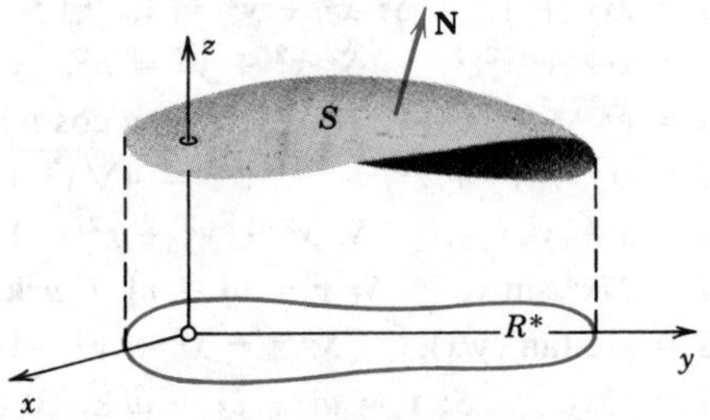

Fig. 223. Formula (11)

From (11) with $G = 1$ we obtain for the **area** $A(S)$ of S: $z = f(x, y)$ the formula

$$A(S) = \iint_{\bar{S}} \sqrt{1 + \left(\frac{\partial f}{\partial x}\right)^2 + \left(\frac{\partial f}{\partial y}\right)^2}\, dx\, dy \tag{12}$$

where $\bar{S}$ is the projection of S into the xy-plane, as before.

Problem Set 9.6

Flux integrals of the form (3). Evaluate $\iint_S \mathbf{F} \cdot \mathbf{n}\, dA$, where

1. $\mathbf{F} = 2x\mathbf{i} + 2y\mathbf{j}$, $\quad S$: $z = 2x + 3y$, $0 \leqq x \leqq 2$, $-1 \leqq y \leqq 1$
2. $\mathbf{F} = x^2\mathbf{i} + e^y\mathbf{j} + \mathbf{k}$, $\quad S$: $x + y + z = 1$, $x \geqq 0$, $y \geqq 0$, $z \geqq 0$
3. $\mathbf{F} = e^y\mathbf{i} - e^z\mathbf{j} + e^x\mathbf{k}$, $\quad S$: $x^2 + y^2 = 9$, $x \geqq 0$, $y \geqq 0$, $0 \leqq z \leqq 2$
4. $\mathbf{F} = \cosh yz\, \mathbf{i} + y^4\mathbf{k}$, $\quad S$: $y^2 + z^2 = 1$, $-5 \leqq x \leqq 5$, $z \geqq 0$
5. $\mathbf{F} = \mathbf{i} + 2\mathbf{j} + 7\mathbf{k}$, $\quad S$: $z = 4(x^2 + y^2)$, $z \leqq 4$
6. $\mathbf{F} = y^2\mathbf{i} + x^4\mathbf{j} + z^2\mathbf{k}$, $\quad S$: $z = x^2 + y^2$, $y \geqq 0$, $z = 9$
7. $\mathbf{F} = z^3(\mathbf{i} - \mathbf{k})$, $\quad S$: $\mathbf{r} = u \cos v\, \mathbf{i} + u \sin v\, \mathbf{j} + u\mathbf{k}$, $0 \leqq u \leqq 5$
8. $\mathbf{F} = (x - z)\mathbf{i} + (y - x)\mathbf{j} + (z - y)\mathbf{k}$, $\quad S$ as in Prob. 7
9. $\mathbf{F} = y^3\mathbf{i} + x^3\mathbf{j} + z^3\mathbf{k}$, $\quad S$: $x^2 + 4y^2 = 4$, $x \geqq 0$, $y \geqq 0$, $0 \leqq z \leqq h$
10. $\mathbf{F} = \tan xy\, \mathbf{i} + x^2y\mathbf{j} - z\mathbf{k}$, $\quad S$: $9y^2 + z^2 = 9$, $1 \leqq x \leqq 4$
11. $\mathbf{F} = \mathbf{i} + x^2\mathbf{j} + xyz\mathbf{k}$, $\quad S$: $z = xy$, $0 \leqq x \leqq y$, $0 \leqq y \leqq 1$
12. $\mathbf{F} = \mathbf{i} + \mathbf{j} + \mathbf{k}$, $\quad S$: $x^2 + y^2 + 4z^2 = 4$, $z \geqq 0$
13. $\mathbf{F} = z\mathbf{i} - xz\mathbf{j} + y\mathbf{k}$, $\quad S$: $x^2 + 9y^2 + 4z^2 = 36$, $x \geqq 0$, $y \geqq 0$, $z \geqq 0$
14. $\mathbf{F} = \cosh x\, \mathbf{i} + \sinh y\, \mathbf{k}$, $\quad S$: $z = x + y^2$, $0 \leqq y \leqq x$, $0 \leqq x \leqq 1$
15. $\mathbf{F} = 2xy\mathbf{i} + x^2\mathbf{j}$, $\quad S$: $\mathbf{r} = \cosh u\, \mathbf{i} + \sinh u\, \mathbf{j} + v\mathbf{k}$, $0 \leqq u \leqq 2$, $-3 \leqq v \leqq 3$

Surface integrals of the form (7). Using (7) or (11), evaluate $\iint_S G(\mathbf{r})\, dA$, where

16. $G = \cos x + \sin y$, $\quad S$: the portion of $x + y + z = 1$ in the first octant
17. $G = 3(x + y + z)$, $\quad S$: $z = x + y$, $0 \leqq y \leqq x$, $0 \leqq x \leqq 1$
18. $G = 2xy + 1$, $\quad S$: $x^2 + y^2 = 4$, $|z| \leqq 1$
19. $G = xe^y + x^2z^2$, $\quad S$: $x^2 + y^2 = a^2$, $y \geqq 0$, $0 \leqq z \leqq h$
20. $G = e^{x^2+y^2} + x^2 - z$, $\quad S$: $\mathbf{r} = u \cos v\, \mathbf{i} + u \sin v\, \mathbf{j} + au\mathbf{k}$, $0 \leqq u \leqq 1$
21. $G = (x^2 + y^2 + z^2)^2$, $\quad S$: $z = 4\sqrt{x^2 + y^2}$, $y \geqq 0$, $0 \leqq z \leqq 4$
22. $G = x + y + z$, $\quad S$: $x^2 + y^2 + z^2 = 1$, $y \geqq 0$, $z \geqq 0$
23. $G = 27x^3 \sin y$, $\quad S$: $\mathbf{r} = u\mathbf{i} + v\mathbf{j} + u^3\mathbf{k}$, $0 \leqq u \leqq 1$, $0 \leqq v \leqq \pi$
24. $G = \arctan(y/x)$, $\quad S$: $z = x^2 + y^2$, $1 \leqq z \leqq 4$, $x \geqq 0$, $y \geqq 0$
25. $G = 15xy$, $\quad S$: $\mathbf{r} = u\mathbf{i} + v\mathbf{j} + uv\mathbf{k}$, $0 \leqq u \leqq 1$, $0 \leqq v \leqq 1$

Center of gravity. Moments of inertia

26. **(Center of gravity)** Justify the following formulas for the mass M and the center of gravity $(\bar{x}, \bar{y}, \bar{z})$ of a lamina S of density (mass per unit area) $\sigma(x, y, z)$ in space:

$$M = \iint_S \sigma\, dA, \quad \bar{x} = \frac{1}{M}\iint_S x\sigma\, dA, \quad \bar{y} = \frac{1}{M}\iint_S y\sigma\, dA, \quad \bar{z} = \frac{1}{M}\iint_S z\sigma\, dA.$$

27. **(Moments of inertia)** Justify the following formulas for the moments of inertia of the lamina in Prob. 26 about the x-, y- and z-axes, respectively:

$$I_x = \iint_S (y^2 + z^2)\sigma\, dA, \quad I_y = \iint_S (x^2 + z^2)\sigma\, dA, \quad I_z = \iint_S (x^2 + y^2)\sigma\, dA.$$

28. Find a formula for the moment of inertia of the lamina in Prob. 26 about the line $y = x$, $z = 0$.

Find the moment of inertia of a lamina S of density 1 about an axis A, where

29. S: $x^2 + y^2 = 1$, $0 \leqq z \leqq h$, A: the z-axis

30. S as in Prob. 29, A: the line $z = h/2$ in the xz-plane

31. S: $x^2 + y^2 = z^2$, $0 \leqq z \leqq h$, A: the z-axis

32. **(Steiner's theorem[7])** If I_A is the moment of inertia of a mass distribution of total mass M with respect to an axis A through the center of gravity, show that its moment of inertia I_B with respect to an axis B, which is parallel to A and has the distance k from it, is

$$I_B = I_A + k^2 M.$$

33. Using Steiner's theorem, find the moment of inertia of S in Prob. 30 about the x-axis.

34. Construct a paper model of a Möbius strip. What happens if you cut it along the curve C in Fig. 220?

First fundamental form of a surface. Given a surface S: $\mathbf{r}(u, v)$, the corresponding quadratic differential form

$$ds^2 = E\, du^2 + 2F\, du\, dv + G\, dv^2 \tag{13}$$

with coefficients[8]

$$E = \mathbf{r}_u \cdot \mathbf{r}_u, \qquad F = \mathbf{r}_u \cdot \mathbf{r}_v, \qquad G = \mathbf{r}_v \cdot \mathbf{r}_v \tag{14}$$

is called the **first fundamental form** of S. It is basic in the theory of surfaces, since with its help, one can determine lengths (Prob. 36, below), angles (Prob. 37), and areas (Prob. 38) on S, as the following problems illustrate.

---

[7]JACOB STEINER (1796—1863), Swiss geometer, born in a small village, learned to write only at age 14, became a pupil of Pestalozzi at 18, later studied at Heidelberg and Berlin and, finally, because of his outstanding research, was appointed professor at Berlin University.

[8]E, F, G are standard notations; of course, they have nothing to do with the functions F and G occurring at some places in this chapter.

35. From (13), Sec. 8.5, we have $ds^2 = d\mathbf{r} \cdot d\mathbf{r}$. Show that for a curve C: $u = u(t)$, $v = v(t)$, $a \leqq t \leqq b$, on S this yields (13) and (14).

36. Show that C in Prob. 35 has length

$$(15) \qquad l = \int_a^b \sqrt{\mathbf{r}'(t) \cdot \mathbf{r}'(t)}\, dt = \int_a^b \sqrt{Eu'^2 + 2Fu'v' + Gv'^2}\, dt.$$

37. Show that the angle γ between two intersecting curves C_1: $u = g(t)$, $v = h(t)$ and C_2: $u = p(t)$, $v = q(t)$ on S: $\mathbf{r}(u, v)$ is obtained from

$$(16) \qquad \cos \gamma = \frac{\mathbf{a} \cdot \mathbf{b}}{|\mathbf{a}|\ |\mathbf{b}|}$$

w' 're $\mathbf{a} = \mathbf{r}_u g' + \mathbf{r}_v h'$ and $\mathbf{b} = \mathbf{r}_u p' + \mathbf{r}_v q'$ are tangent vectors of C_1 and C_2.

38. Show that

$$(17) \qquad |\mathbf{N}|^2 = |\mathbf{r}_u \times \mathbf{r}_v|^2 = EG - F^2,$$

so that formula (8) for the area $A(S)$ of S becomes

$$(18) \qquad A(S) = \iint_S dA = \iint_R |\mathbf{N}|\, du\, dv = \iint_R \sqrt{EG - F^2}\, du\, dv.$$

39. (Polar coordinates) Show that for polar coordinates u $(= r)$ and v $(= \theta)$ defined by $x = u \cos v$, $y = u \sin v$ we have $E = 1$, $F = 0$, $G = u^2$, so that

$$ds^2 = du^2 + u^2\, dv^2 = dr^2 + r^2\, d\theta^2$$

and calculate from this and (18) the area of a disk of radius a.

40. (Theorem of Pappus[9]) Show that the area A of the torus in Example 6 can be obtained by the *theorem of Pappus,* which states that the area of a surface of revolution equals the product of the length of a meridian C and the length of the path of the center of gravity of C when C is rotated through the angle 2π.

9.7 Triple Integrals. Divergence Theorem of Gauss

In this section we first discuss triple integrals. Then we obtain the first "big" integral theorem, which transforms surface integrals into triple integrals. It is called **Gauss's divergence theorem** because it involves the divergence of a vector function (see Sec. 8.10, which the student may wish to review).

The triple integral is a generalization of the double integral introduced in Sec. 9.3. For defining this integral we consider a function $f(x, y, z)$ defined in a bounded closed[10] region T in space. We subdivide this three-dimensional region T by planes parallel to the three coordinate planes. Then we number the boxes of subdivision (parallelepipeds) inside T from 1 to n. In each box

[9]PAPPUS OF ALEXANDRIA (about 300), Greek mathematician. The theorem is also called *Guldin's theorem,* after the Austrian mathematician HABAKUK GULDIN (1577—1643), professor at Graz and Vienna.

[10]Explained in footnote 1, Sec. 9.3 (with "sphere" instead of "circle").

we choose an arbitrary point, say, (x_k, y_k, z_k) in box k, and form the sum

$$J_n = \sum_{k=1}^{n} f(x_k, y_k, z_k)\,\Delta V_k$$

where ΔV_k is the volume of box k. This we do for larger and larger positive integers n in an arbitrary manner, but so that the lengths of the edges of the largest box of subdivision approach zero as n approaches infinity. In this way we obtain a sequence of real numbers $J_{n_1}, J_{n_2}, \cdots$. We assume that $f(x, y, z)$ is continuous in a domain containing T and T is bounded by finitely many *smooth surfaces* (see Sec. 9.5). Then it can be shown (see Ref. [5] in Appendix 1) that the sequence converges to a limit that is independent of the choice of subdivisions and corresponding points (x_k, y_k, z_k). This limit is called the **triple integral** *of* $f(x, y, z)$ *over the region* T, and is denoted by

$$\iiint_T f(x, y, z)\,dx\,dy\,dz \qquad \text{or} \qquad \iiint_T f(x, y, z)\,dV.$$

Triple integrals can be evaluated by three successive integrations. This is similar to the evaluation of double integrals by two successive integrations, as discussed in Sec. 9.3. An example is shown below (Example 1).

Divergence Theorem of Gauss

We shall now show that triple integrals can be transformed into surface integrals over the boundary surface of a region in space and conversely. This is of practical interest, since in many cases, one of the two kinds of integral is simpler than the other; and it also helps in establishing fundamental equations in fluid flow, heat conduction, etc., as we shall see. The transformation is done by the so-called *divergence theorem* (below), which involves the divergence of a vector function **F**,

$$\text{(1)} \qquad \operatorname{div} \mathbf{F} = \frac{\partial F_1}{\partial x} + \frac{\partial F_2}{\partial y} + \frac{\partial F_3}{\partial z} \qquad \text{(Sec. 8.10).}$$

Theorem 1 Divergence theorem of Gauss
(Transformation between volume integrals and surface integrals)

Let T be a closed[11] *bounded region in space whose boundary is a piecewise smooth*[12] *orientable surface S. Let $\mathbf{F}(x, y, z)$ be a vector function that is continuous and has continuous first partial derivatives in some domain containing T. Then*

$$\textbf{(2)} \qquad \iiint_T \operatorname{div} \mathbf{F}\,dV = \iint_S \mathbf{F} \cdot \mathbf{n}\,dA$$

where **n** *is the* ***outer*** *unit normal vector of S.* (Proof on p. 546.)

[11]"Closed" means that the boundary surface S is part of the region.
[12]See Sec. 9.5.

Comment. If we write $\mathbf{F}$ and the outer unit normal vector $\mathbf{n}$ in components,

$$\mathbf{F} = F_1\mathbf{i} + F_2\mathbf{j} + F_3\mathbf{k} \quad \text{and} \quad \mathbf{n} = \cos\alpha\,\mathbf{i} + \cos\beta\,\mathbf{j} + \cos\gamma\,\mathbf{k},$$

so that α, β, and γ are the angles between $\mathbf{n}$ and the positive x-, y-, and z-axes, respectively, formula (2) takes the form

$$\iiint_T \left(\frac{\partial F_1}{\partial x} + \frac{\partial F_2}{\partial y} + \frac{\partial F_3}{\partial z}\right) dx\,dy\,dz = \iint_S (F_1 \cos\alpha + F_2 \cos\beta + F_3 \cos\gamma)\, dA. \tag{3*}$$

Because of (6) in the last section this may be written

$$\iiint_T \left(\frac{\partial F_1}{\partial x} + \frac{\partial F_2}{\partial y} + \frac{\partial F_3}{\partial z}\right) dx\,dy\,dz = \iint_S (F_1\, dy\,dz + F_2\, dz\,dx + F_3\, dx\,dy). \tag{3}$$

EXAMPLE 1 **Evaluation of a surface integral by the divergence theorem**

Before we prove the divergence theorem, let us show a typical application. By transforming to a triple integral, evaluate

$$I = \iint_S (x^3\, dy\,dz + x^2 y\, dz\,dx + x^2 z\, dx\,dy)$$

where S is the closed surface consisting of the cylinder $x^2 + y^2 = a^2$ $(0 \leqq z \leqq b)$ and the circular disks $z = 0$ and $z = b$ $(x^2 + y^2 \leqq a^2)$.

Solution. In (3) we now have

$$F_1 = x^3, \quad F_2 = x^2 y, \quad F_3 = x^2 z, \qquad \text{hence} \qquad \operatorname{div} \mathbf{F} = 3x^2 + x^2 + x^2 = 5x^2.$$

Introducing polar coordinates r, θ defined by $x = r\cos\theta$, $y = r\sin\theta$ (thus, cylindrical coordinates r, θ, z), we have $dx\,dy\,dz = r\,dr\,d\theta\,dz$, and we obtain

$$I = \iiint_T 5x^2\, dx\,dy\,dz = 5 \int_{z=0}^{b} \int_{r=0}^{a} \int_{\theta=0}^{2\pi} r^2 \cos^2\theta\; r\, dr\, d\theta\, dz$$

$$= 5b \int_0^a \int_0^{2\pi} r^3 \cos^2\theta\, dr\, d\theta = 5b\,\frac{a^4}{4} \int_0^{2\pi} \cos^2\theta\, d\theta = \frac{5}{4}\pi b a^4. \qquad \blacksquare$$

Proof of the divergence theorem. Clearly, (3*) is true if the integrals of each component on both sides of (3*) are equal, that is,

$$\iiint_T \frac{\partial F_1}{\partial x}\, dx\,dy\,dz = \iint_S F_1 \cos\alpha\, dA, \tag{4}$$

$$\iiint_T \frac{\partial F_2}{\partial y}\, dx\, dy\, dz = \iint_S F_2 \cos\beta\, dA, \tag{5}$$

$$\iiint_T \frac{\partial F_3}{\partial z}\, dx\, dy\, dz = \iint_S F_3 \cos\gamma\, dA. \tag{6}$$

We first prove (6) for a *special region* T that is bounded by a piecewise smooth orientable surface S and has the property that any straight line parallel to any one of the coordinate axes and intersecting T has at most *one* segment (or a single point) in common with T. This implies that T can be represented in the form

$$g(x, y) \leqq z \leqq h(x, y) \tag{7}$$

where (x, y) varies in the orthogonal projection R of T in the xy-plane. Clearly, $z = g(x, y)$ represents the "bottom" S_2 of S (Fig. 224), whereas $z = h(x, y)$ represents the "top" S_1 of S, and there may be a remaining vertical portion S_3 of S. (The portion S_3 may degenerate into a curve, as for a sphere.)

To prove (6), we use (7). Since $\mathbf{F}$ is continuously differentiable in some domain containing T, we have

$$\iiint_T \frac{\partial F_3}{\partial z}\, dx\, dy\, dz = \iint_R \left[\int_{g(x,y)}^{h(x,y)} \frac{\partial F_3}{\partial z}\, dz\right] dx\, dy. \tag{8}$$

We integrate the inner integral:

$$\int_g^h \frac{\partial F_3}{\partial z}\, dz = F_3[x, y, h(x, y)] - F_3[x, y, g(x, y)].$$

Hence the left side of (8) equals

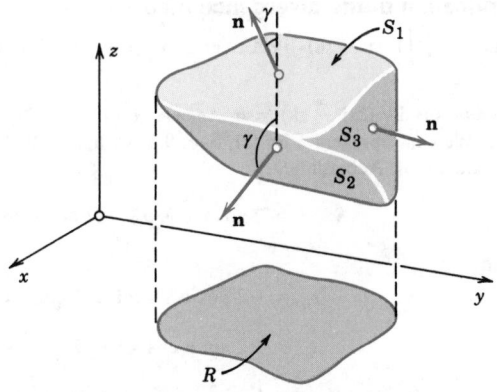

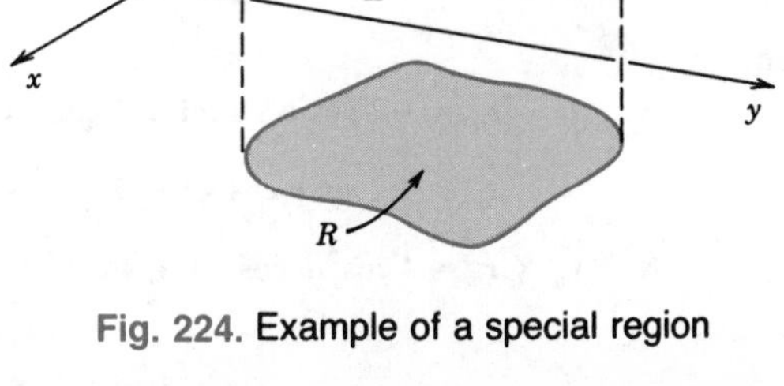

Fig. 224. Example of a special region

$$(9) \qquad \iint_R F_3[x, y, h(x, y)]\, dx\, dy - \iint_R F_3[x, y, g(x, y)]\, dx\, dy.$$

But the same result is also obtained by evaluating the right side of (6), that is [see also (3)],

$$\iint_S F_3 \cos \gamma\, dA = \iint_S F_3\, dx\, dy$$

$$= + \iint_R F_3[x, y, h(x, y)]\, dx\, dy - \iint_R F_3[x, y, g(x, y)]\, dx\, dy,$$

where the first integral gets a plus sign because $\cos \gamma > 0$ on S_1 in Fig. 224 [as in (5c′), Sec. 9.6], and the second integral gets a minus sign because $\cos \gamma < 0$ on S_2 [as in (5c″), Sec. 9.6, with g instead of h]. This proves (6).

The relations (4) and (5) now follow by merely relabeling the variables and using the fact that, by assumption, T has representations similar to (7), namely,

$$\tilde{g}(y, z) \leqq x \leqq \tilde{h}(y, z) \quad \text{and} \quad \tilde{\tilde{g}}(z, x) \leqq y \leqq \tilde{\tilde{h}}(z, x).$$

This establishes the divergence theorem for special regions.

For any region T that can be subdivided into *finitely many* special regions by means of auxiliary surfaces, the theorem follows by adding the result for each part separately; this procedure is analogous to that in the proof of Green's theorem in Sec. 9.4. The surface integrals over the auxiliary surfaces cancel in pairs, and the sum of the remaining surface integrals is the surface integral over the whole boundary S of T; the volume integrals over the parts of T add up to the volume integral over T.

The divergence theorem is now proved for any bounded region that is of interest in practical problems. The extension to the most general region T of the type characterized in the theorem would require a certain limit process; this is similar to the situation in the case of Green's theorem in Sec. 9.4. ∎

EXAMPLE 2 Verification of the divergence theorem

Evaluate $\iint_S (7x\mathbf{i} - z\mathbf{k})\cdot\mathbf{n}\, dA$ over S: $x^2 + y^2 + z^2 = 4$ (a) by (2), (b) directly.

Solution. (a) div $(7x\mathbf{i} - z\mathbf{k}) = 6$. *Answer.* $6 \cdot (4/3)\pi \cdot 2^3 = 64\pi$.

(b) We can represent S by (3), Sec. 9.5 (with $a = 2$), and we shall use $\mathbf{n}\, dA = \mathbf{N}\, du\, dv$ [see (3*), Sec. 9.6]. Accordingly,

$$S\colon \quad \mathbf{r} = 2 \cos v \cos u\, \mathbf{i} + 2 \cos v \sin u\, \mathbf{j} + 2 \sin v\, \mathbf{k}.$$

Then

$$\mathbf{r}_u = -2 \cos v \sin u\, \mathbf{i} + 2 \cos v \cos u\, \mathbf{j}$$

$$\mathbf{r}_v = -2 \sin v \cos u\, \mathbf{i} - 2 \sin v \sin u\, \mathbf{j} + 2 \cos v\, \mathbf{k}$$

$$\mathbf{N} = \mathbf{r}_u \times \mathbf{r}_v = 4 \cos^2 v \cos u\, \mathbf{i} + 4 \cos^2 v \sin u\, \mathbf{j} + 4 \cos v \sin v\, \mathbf{k}.$$

From the representation of S we see that on S,

$$7x\mathbf{i} - z\mathbf{k} = 14 \cos v \cos u\, \mathbf{i} - 2 \sin v\, \mathbf{k}.$$

From this and our expression for $\mathbf{N}$ we thus obtain

$$\begin{aligned}(7x\mathbf{i} - z\mathbf{k})\bullet\mathbf{N} &= (14 \cos v \cos u)(4 \cos^2 v \cos u) - (2 \sin v)(4 \cos v \sin v) \\ &= 56 \cos^2 u \cos^3 v - 8 \sin^2 v \cos v \\ &= 56(\tfrac{1}{2} + \tfrac{1}{2} \cos 2u)(\tfrac{1}{4} \cos 3v + \tfrac{3}{4} \cos v) - 8 \sin^2 v \cos v.\end{aligned}$$

We integrate this over u from 0 to 2π and over v from $-\pi/2$ to $\pi/2$, since this corresponds to the entire sphere. This gives

$$56(\pi + 0)\left[\frac{1}{4}\frac{\sin 3v}{3}\Bigg|_{-\pi/2}^{\pi/2} + \frac{3}{4}\sin v\Bigg|_{-\pi/2}^{\pi/2}\right] - 8 \cdot 2\pi \frac{\sin^3 v}{3}\Bigg|_{-\pi/2}^{\pi/2}$$

$$= 56\pi\left[\frac{1}{12}(-1 - 1) + \frac{3}{4}(1 - (-1))\right] - \frac{16\pi}{3}(1 - (-1)) = 64\pi. \quad ■$$

Further applications of the divergence theorem follow in the problem set and in the next section. The examples in the next section shed further light on the nature of the divergence and its applications.

Problem Set 9.7

Application of triple integrals

Find the total **mass** of a mass distribution of density σ in a region T in space, where

1. $\sigma = x^2y^2z^2$, T the cube $|x| \leqq 1$, $|y| \leqq 1$, $|z| \leqq 1$

2. $\sigma = x + y + z$, T the box $0 \leqq x \leqq 1$, $0 \leqq y \leqq 2$, $0 \leqq z \leqq 3$

3. $\sigma = 2(x^2 + y^2)$, T the cylinder $x^2 + y^2 \leqq 4$, $0 \leqq z \leqq 6$

4. $\sigma = 1 + y + z^2$, T the cylinder $y^2 + z^2 \leqq 9$, $1 \leqq x \leqq 3$

5. $\sigma = xy$, T the tetrahedron with vertices (0, 0, 0), (1, 0, 0), (0, 1, 0), (0, 0, 1)

6. $\sigma = 1 - xy$, T as in Prob. 5

7. $\sigma = \sin x \cos y + 1$, T the box $0 \leqq x \leqq 2\pi$, $0 \leqq y \leqq \pi$, $0 \leqq z \leqq 4$

8. $\sigma = 2z$, T the region in the first octant bounded by $y = 1 - x^2$ and $z = x$

9. $\sigma = x^2y^2$, T: $0 \leqq x \leqq 1$, $1 - x \leqq y \leqq 1$, $1 \leqq z \leqq 2$

10. $\sigma = 1/yz$, T: $0 \leqq x \leqq 2$, $e^{-x} \leqq y \leqq 1$, $e^{-x} \leqq z \leqq 1$

Moment of inertia. Find the moment of inertia $I_x = \iiint_T (y^2 + z^2)\, dx\, dy\, dz$ of a mass of density 1 in T about the x-axis, where T is

11. The cube $0 \leqq x \leqq 1$, $0 \leqq y \leqq 1$, $0 \leqq z \leqq 1$

12. The box $0 \leqq x \leqq a$, $-b/2 \leqq y \leqq b/2$, $-c/2 \leqq z \leqq c/2$

13. The cylinder $y^2 + z^2 \leqq a^2$, $0 \leqq x \leqq h$

14. The cone $y^2 + z^2 \leqq x^2$, $0 \leqq x \leqq h$

15. The ball $x^2 + y^2 + z^2 \leqq a^2$

Application of the divergence theorem. Evaluate the surface integral $\iint_S \mathbf{F}\cdot\mathbf{n}\, dA$ by the divergence theorem, where

16. $\mathbf{F} = e^y\mathbf{j}$, S the surface of the box $0 \leqq x \leqq 3$, $0 \leqq y \leqq 2$, $0 \leqq z \leqq 1$

17. $\mathbf{F} = x^3\mathbf{i} + z^3\mathbf{k}$, S the surface of the cube in Prob. 1

18. $\mathbf{F} = x^3\mathbf{i} + y^2z\mathbf{j} + z^2x\mathbf{k}$, S the surface of the box in Prob. 2

19. $\mathbf{F} = e^x\mathbf{i} - ye^x\mathbf{j} + 3z\mathbf{k}$, S the surface of $x^2 + y^2 \leqq a^2$, $|z| \leqq h$

20. $\mathbf{F} = x^3z^2\mathbf{i} + y^3z^2\mathbf{j} + x^2z\mathbf{k}$, S as in Prob. 19

21. $\mathbf{F} = y^2\mathbf{i} + z^2\mathbf{j} + x^2z\mathbf{k}$, S the surface of $x^2 + y^2 \leqq 4$, $x \geqq 0$, $y \geqq 0$, $|z| \leqq 1$

22. $\mathbf{F} = 10y\mathbf{j} + z^3\mathbf{k}$, S the surface of $0 \leqq x \leqq 6$, $0 \leqq y \leqq 1$, $0 \leqq z \leqq y$

23. $\mathbf{F} = x\mathbf{i} + x^2y\mathbf{j} - x^2z\mathbf{k}$, S the surface of the tetrahedron in Prob. 5

24. $\mathbf{F} = x^3z\mathbf{i} - xz^2\mathbf{j} + 3\mathbf{k}$, S the surface of $x^2 + y^2 \leqq 4z^2$, $0 \leqq z \leqq 1$

25. $\mathbf{F} = x^2\mathbf{i} - (2x - 1)y\mathbf{j} + 4z\mathbf{k}$, S the surface of $x^2 + y^2 \leqq z^2$, $0 \leqq z \leqq 2$

26. $\mathbf{F} = (x + z)\mathbf{i} + (y + z)\mathbf{j} + (x + y)\mathbf{k}$, S: $x^2 + y^2 + z^2 = 4$, $z \geqq 0$

27. $\mathbf{F} = x^3\mathbf{i} + y^3\mathbf{j} + z^3\mathbf{k}$, S: $x^2 + y^2 + z^2 = 4$

28. $\mathbf{F} = x^3\mathbf{i} + y^3\mathbf{j} + 3z(2 - x^2 - y^2)\mathbf{k}$, S: $9x^2 + y^2 + 9z^2 = 9$

29. $\mathbf{F} = \sin^2 x\,\mathbf{i} - z(1 + \sin 2x)\mathbf{k}$, S the surface of $x^2 + y^2 \leqq z$, $z \leqq 1/2$

30. $\mathbf{F} = x^3\mathbf{i} + y^3\mathbf{j} + z^3\mathbf{k}$, S as in Prob. 29

9.8 Further Applications of the Divergence Theorem

The divergence theorem has various applications and consequences. We illustrate some of them in the following examples. Here, the occurring regions and functions are assumed to satisfy the conditions under which the divergence theorem is valid, and in each case, $\mathbf{n}$ is the ***outer*** unit normal vector of the boundary surface of the region, as before.

EXAMPLE 1 Representation of the divergence independent of coordinates

Dividing both sides of (2) in Sec. 9.7 by the volume $V(T)$ of the region T, we obtain

$$\frac{1}{V(T)}\iiint_T \operatorname{div}\mathbf{F}\, dV = \frac{1}{V(T)}\iint_{S(T)} \mathbf{F}\cdot\mathbf{n}\, dA \tag{1}$$

where $S(T)$ is the boundary surface of T. The basic properties of the triple integral are essentially the same as those of the double integral considered in Sec. 9.3. In particular the **mean value theorem** *for triple integrals* asserts that for any continuous function $f(x, y, z)$ in the region T under consideration there is a point Q: (x_0, y_0, z_0) in T such that

$$\iiint_T f(x, y, z)\, dV = f(x_0, y_0, z_0)V(T).$$

Setting $f = \operatorname{div}\mathbf{F}$ and using (1), we have

$$\frac{1}{V(T)}\iiint_T \operatorname{div}\mathbf{F}\, dV = \operatorname{div}\mathbf{F}(x_0, y_0, z_0). \tag{2}$$

Let P: (x_1, y_1, z_1) be any fixed point in T, and let T shrink down onto P, so that the maximum distance $d(T)$ of the points of T from P approaches zero. Then Q must approach P, and from (1) and (2) it follows that the divergence of $\mathbf{F}$ at P is

$$\text{(3)} \qquad \operatorname{div} \mathbf{F}(x_1, y_1, z_1) = \lim_{d(T)\to 0} \frac{1}{V(T)} \iint_{S(T)} \mathbf{F}\cdot\mathbf{n}\, dA.$$

This formula is sometimes used as a ***definition*** of the divergence. While the definition of the divergence in Sec. 8.10 involves coordinates, formula (3) is independent of coordinates. Hence from (3) it follows immediately that ***the divergence is independent of the particular choice of Cartesian coordinates.***

EXAMPLE 2 **Physical interpretation of the divergence**

From the divergence theorem we may obtain an intuitive interpretation of the divergence of a vector. For this purpose we consider the flow of an incompressible fluid (see Sec. 8.10) of constant density $\rho = 1$ which is **steady,** that is, does not vary with time. Such a flow is determined by the field of its velocity vector $\mathbf{v}(P)$ at any point P.

Let S be the boundary surface of a region T in space, and let $\mathbf{n}$ be the outer unit normal vector of S. The mass of fluid that flows through a small portion ΔS of S of area ΔA per unit time from the interior of S to the exterior is equal to $\mathbf{v}\cdot\mathbf{n}\ \Delta A$, where[13] $\mathbf{v}\cdot\mathbf{n}$ is the normal component of $\mathbf{v}$ in the direction of $\mathbf{n}$, taken at a suitable point of ΔS. Consequently, the total mass of fluid that flows across S from T to the outside per unit of time is given by the surface integral

$$\iint_S \mathbf{v}\cdot\mathbf{n}\, dA.$$

Hence this integral represents the total flow out of T, and the integral

$$\text{(4)} \qquad \frac{1}{V} \iint_S \mathbf{v}\cdot\mathbf{n}\, dA$$

where V is the volume of T, represents the average flow out of T. Since the flow is steady and the fluid is incompressible, the amount of fluid flowing outward must be continuously supplied. Hence, if the value of the integral (4) is different from zero, there must be **sources** (*positive sources and negative sources, called* **sinks**) in T, that is, points where fluid is produced or disappears.

If we let T shrink down to a fixed point P in T, we obtain from (4) the **source intensity** at P given by the right side of (3) with $\mathbf{F}\cdot\mathbf{n}$ replaced by $\mathbf{v}\cdot\mathbf{n}$. From this and (3) it follows that ***the divergence of the velocity vector* v *of a steady incompressible flow is the source intensity of the flow at the corresponding point.*** There are no sources in T if and only if $\operatorname{div} \mathbf{v} \equiv 0$; in this case,

$$\iint_S \mathbf{v}\cdot\mathbf{n}\, dA = 0$$

for any closed surface S in T.

EXAMPLE 3 **Modeling of heat flow. Heat equation**

We know that in a body heat will flow in the direction of decreasing temperature. Physical experiments show that the rate of flow is proportional to the gradient of the temperature. This means that the velocity $\mathbf{v}$ of the heat flow in a body is of the form

$$\text{(5)} \qquad \mathbf{v} = -K \operatorname{grad} U$$

[13]Note that $\mathbf{v}\cdot\mathbf{n}$ may be negative at a certain point, which means that fluid *enters* the interior of S at such a point.

where $U(x, y, z, t)$ is temperature, t is time and K is called the *thermal conductivity* of the body; in ordinary physical circumstances K is a constant. Using this information, set up the mathematical model of heat flow, the so-called **heat equation.**

Solution. Let T be a region in the body and let S be its boundary surface. Then the amount of heat leaving T per unit of time is

$$\iint_S \mathbf{v} \cdot \mathbf{n} \, dA,$$

where $\mathbf{v} \cdot \mathbf{n}$ is the component of $\mathbf{v}$ in the direction of the outer unit normal vector $\mathbf{n}$ of S. This expression is obtained in a fashion similar to that in the preceding example. From (5) and the divergence theorem we obtain [see (3), Sec. 8.10]

$$\text{(6)} \qquad \iint_S \mathbf{v} \cdot \mathbf{n} \, dA = -K \iiint_T \operatorname{div} (\operatorname{grad} U) \, dx\, dy\, dz = -K \iiint_T \nabla^2 U \, dx\, dy\, dz$$

where $\nabla^2 U = U_{xx} + U_{yy} + U_{zz}$ is the Laplacian of U.

On the other hand, the total amount of heat H in T is

$$H = \iiint_T \sigma\rho U \, dx\, dy\, dz$$

where the constant σ is the specific heat of the material of the body and ρ is the density (= mass per unit volume) of the material. Hence the time rate of decrease of H is

$$-\frac{\partial H}{\partial t} = -\iiint_T \sigma\rho \frac{\partial U}{\partial t} \, dx\, dy\, dz$$

and this must be equal to the above amount of heat leaving T; from (6) we thus have

$$-\iiint_T \sigma\rho \frac{\partial U}{\partial t} \, dx\, dy\, dz = -K \iiint_T \nabla^2 U \, dx\, dy\, dz$$

or

$$\iiint_T \left(\sigma\rho \frac{\partial U}{\partial t} - K \nabla^2 U \right) dx\, dy\, dz = 0.$$

Since this holds for any region T in the body, the integrand (if continuous) must be zero everywhere; that is,

$$\text{(7)} \qquad \boxed{\frac{\partial U}{\partial t} = c^2 \nabla^2 U} \qquad c^2 = \frac{K}{\sigma\rho},$$

where c^2 is called the *thermal diffusivity* of the material. This partial differential equation is called the **heat equation;** it is fundamental for heat conduction. Methods for solving problems in heat conduction will be considered in Chap. 11.

If the heat flow does not depend on time t (***"steady-state flow"***), then $\partial U/\partial t = 0$ and the heat equation (7) reduces to Laplace's equation $\nabla^2 U = 0$; see below. ■

Potential Theory. Harmonic Functions

The theory of solutions of Laplace's equation

$$\text{(8)} \qquad \boxed{\nabla^2 f = \frac{\partial^2 f}{\partial x^2} + \frac{\partial^2 f}{\partial y^2} + \frac{\partial^2 f}{\partial z^2} = 0}$$

is called **potential theory.** A solution of (8) that has ***continuous*** second-order partial derivatives is called a **harmonic function.**[14] We shall consider details of potential theory in Chaps. 11 and 17. At present let us show that the divergence theorem is very useful in potential theory.

EXAMPLE 4 **A basic property of solutions of Laplace's equation**

Consider the formula in the divergence theorem:

$$\iiint_T \operatorname{div} \mathbf{F}\, dV = \iint_S \mathbf{F} \cdot \mathbf{n}\, dA. \tag{8*}$$

Assume that $\mathbf{F}$ is the gradient of a scalar function, say, $\mathbf{F} = \operatorname{grad} f$. Then [see (3) in Sec. 8.10]

$$\operatorname{div} \mathbf{F} = \operatorname{div} (\operatorname{grad} f) = \nabla^2 f.$$

Furthermore,

$$\mathbf{F} \cdot \mathbf{n} = \mathbf{n} \cdot \operatorname{grad} f,$$

and from (6) in Sec. 8.9 we see that the right side is the directional derivative of f in the outer normal direction of S. If we denote this derivative by $\partial f/\partial n$, formula (8*) becomes

$$\iiint_T \nabla^2 f\, dV = \iint_S \frac{\partial f}{\partial n}\, dA. \tag{9}$$

Obviously this is the three-dimensional analog of the formula (9) in Sec. 9.4.

Taking into account the assumptions under which the divergence theorem holds, we immediately obtain from (9) the following result.

Theorem 1 **(A basic property of harmonic functions)**

Let $f(x, y, z)$ be a harmonic function in some domain D. Then the integral of the normal derivative of the function f over any piecewise smooth[15] *closed orientable surface S in D whose entire interior belongs to D is zero.* ∎

EXAMPLE 5 **Green's theorems**

Let f and g be scalar functions such that $\mathbf{F} = f \operatorname{grad} g$ satisfies the assumptions of the divergence theorem in some region T. Then, by Prob. 23 in Problem Set 8.10,

$$\operatorname{div} \mathbf{F} = \operatorname{div} (f \operatorname{grad} g) = f\nabla^2 g + \operatorname{grad} f \cdot \operatorname{grad} g.$$

Furthermore, since f is a scalar function,

$$\mathbf{F} \cdot \mathbf{n} = \mathbf{n} \cdot (f \operatorname{grad} g) = f(\mathbf{n} \cdot \operatorname{grad} g).$$

The expression $\mathbf{n} \cdot \operatorname{grad} g$ is the directional derivative of g in the direction of the outer normal vector $\mathbf{n}$ of the surface S in the divergence theorem. If we denote this derivative by $\partial g/\partial n$, the formula in the divergence theorem becomes **"Green's first formula"**

$$\iiint_T (f\nabla^2 g + \operatorname{grad} f \cdot \operatorname{grad} g)\, dV = \iint_S f \frac{\partial g}{\partial n}\, dA. \tag{10}$$

---

[14] This continuity requirement should not be deleted from the definition of a harmonic function, as some older books do.

[15] See Sec. 9.5.

Formula (10) together with the assumptions is known as the *first form of Green's theorem.*

Interchanging f and g we obtain a similar formula. Subtracting this formula from (10) we find

$$\iiint_T (f\nabla^2 g - g\nabla^2 f)\, dV = \iint_S \left(f\frac{\partial g}{\partial n} - g\frac{\partial f}{\partial n}\right) dA. \tag{11}$$

This formula is called **Green's second formula** or (together with the assumptions) the *second form of Green's theorem.* ∎

EXAMPLE 6 Uniqueness of solutions of Laplace's equation

Suppose that f satisfies the assumptions in Theorem 1 and is zero everywhere on a piecewise smooth closed orientable surface S in D whose entire interior T belongs to D. Then, setting $g = f$ in (10), we get

$$\iiint_T \operatorname{grad} f \bullet \operatorname{grad} f\, dV = \iiint_T |\operatorname{grad} f|^2\, dV = 0.$$

Since by assumption $|\operatorname{grad} f|$ is continuous in T and on S and is nonnegative, it must be zero everywhere in T. Hence $f_x = f_y = f_z = 0$, and f is constant in T and, because of continuity, equal to its value 0 on S. This proves

Theorem 2 (Harmonic functions)

If a function $f(x, y, z)$ is harmonic in some domain D and is zero at every point of a piecewise smooth closed orientable surface S in D whose entire interior T belongs to D, then f is identically zero in T.

This theorem has an important consequence. Let f_1 and f_2 be functions that satisfy the assumptions of Theorem 1 and take on the same values on S. Then their difference $f_1 - f_2$ satisfies those assumptions and has the value 0 everywhere on S. Hence, Theorem 2 implies that $f_1 - f_2 = 0$ throughout T, and we have the following result.

Theorem 3 (Uniqueness theorem for Laplace's equation)

Let T be a region that satisfies the assumptions of the divergence theorem, and let $f(x, y, z)$ be a harmonic function in a domain D that contains T and its boundary surface S. Then f is uniquely determined in T by its values on S.

The problem of determining a solution u of a partial differential equation in a region T such that u assumes given values on the boundary surface S of T is called the **Dirichlet problem.**[16] We may thus reformulate Theorem 3 as follows.

Theorem 3* (Uniqueness theorem for the Dirichlet problem)

If the assumptions in Theorem 3 are satisfied and the Dirichlet problem for the Laplace equation has a solution in T, then this solution is unique.

These theorems demonstrate the importance of Gauss's theorem in potential theory. ∎

[16] PETER GUSTAV LEJEUNE DIRICHLET (1805—1859), German mathematician, studied in Paris under Cauchy and others and succeeded Gauss at Göttingen in 1855. He became known by his important research on Fourier series (he knew Fourier personally) and in number theory.

Problem Set 9.8

1. Verify Theorem 1 for $f = 2z^2 - x^2 - y^2$ and S the surface of the box $0 \leqq x \leqq 1, 0 \leqq y \leqq 2, 0 \leqq z \leqq 4$.
2. Verify Theorem 1 for $f = x^2 - y^2$ and the surface of the cylinder $x^2 + y^2 \leqq 4$, $0 \leqq z \leqq 1$.

Let f and g be harmonic functions in some domain D that contains a region T and its boundary surface S such that T satisfies the assumptions in the divergence theorem. Show that then:

3. $\displaystyle\iint_S g \frac{\partial g}{\partial n}\, dA = \iiint_T |\text{grad } g|^2\, dV.$
4. If $\partial g/\partial n = 0$ on S, then g is constant in T.
5. $\displaystyle\iint_S \left(f \frac{\partial g}{\partial n} - g \frac{\partial f}{\partial n}\right) dA = 0.$
6. If $\partial f/\partial n = \partial g/\partial n$ on S, then $f = g + c$ in T, where c is a constant.

7. **(Laplacian)** Show that the Laplacian can be represented independently of all coordinate systems in the form

$$\nabla^2 f = \lim_{d(T)\to 0} \frac{1}{V(T)} \iint_{S(T)} \frac{\partial f}{\partial n}\, dA$$

where $d(T)$ is the maximum distance of the points of a region T bounded by $S(T)$ from the point at which the Laplacian is evaluated and $V(T)$ is the volume of T. *Hint.* Put $\mathbf{F} = \text{grad } f$ in (3) and use (6), Sec. 8.9, with $\mathbf{b} = \mathbf{n}$, the outer unit normal vector of S.

8. **(Volume)** Show that a region T with boundary surface S has the volume

$$V = \frac{1}{3} \iint_S r \cos \phi\, dA$$

where r is the distance of a variable point P: (x, y, z) on S from the origin O and ϕ is the angle between the directed line OP and the outer normal of S at P. *Hint.* Use (2), Sec. 9.7, with $\mathbf{F} = x\mathbf{i} + y\mathbf{j} + z\mathbf{k}$.

9. Find the volume of a ball of radius a by means of the formula in Prob. 8.
10. **(Volume)** Show that a region T with boundary surface S has the volume

$$V = \iint_S x\, dy\, dz$$

$$= \iint_S y\, dz\, dx = \iint_S z\, dx\, dy$$

$$= \frac{1}{3} \iint_S (x\, dy\, dz + y\, dz\, dx + z\, dx\, dy).$$

9.9 Stokes's Theorem

Having seen the great usefulness of Gauss's theorem (Secs. 9.7, 9.8), we now turn to the second "big" theorem in this chapter, *Stokes's theorem,* which transforms line integrals into surface integrals and conversely. Hence this theorem generalizes Green's theorem of Sec. 9.4. It involves the curl,

$$\text{curl}\, \mathbf{F} = \begin{vmatrix} \mathbf{i} & \mathbf{j} & \mathbf{k} \\ \partial/\partial x & \partial/\partial y & \partial/\partial z \\ F_1 & F_2 & F_3 \end{vmatrix} \qquad \text{(see Sec. 8.11).} \tag{1}$$

Stokes's theorem[17]

(Transformation between surface integrals and line integrals)

Let S be a piecewise smooth[18] *oriented surface in space and let the boundary of S be a piecewise smooth*[18] *simple closed curve C. Let* $\mathbf{F}(x, y, z)$ *be a continuous vector function that has continuous first partial derivatives in a domain in space containing S. Then*

$$\iint_S (\text{curl}\, \mathbf{F}) \cdot \mathbf{n}\, dA = \oint_C \mathbf{F} \cdot \mathbf{r}'(s)\, ds \tag{2}$$

where $\mathbf{n}$ *is a unit normal vector of S and, depending on* $\mathbf{n}$, *the integration around C is taken in the sense shown in Fig. 225; also,* $\mathbf{r}' = d\mathbf{r}/ds$ *is the unit tangent vector and s the arc length of C.* (Proof on p. 558.)

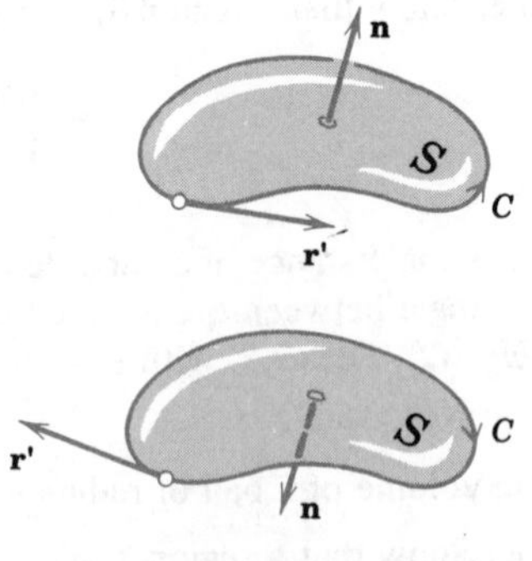

Fig. 225. Stokes's theorem

[17]Sir GEORGE GABRIEL STOKES (1819—1903), Irish mathematician and physicist, who became a professor in Cambridge in 1849. He is also known for his important contribution to the theory of infinite series and to viscous flow (Navier–Stokes equations), geodesy, and optics.

[18]"Piecewise smooth" is defined in Secs. 9.1 and 9.5.

Comment. From (3) in Sec. 9.6, with curl **F** instead of **F**, we get (2) in terms of components [see (1)]

$$\iint_R \left[\left(\frac{\partial F_3}{\partial y} - \frac{\partial F_2}{\partial z}\right) N_1 + \left(\frac{\partial F_1}{\partial z} - \frac{\partial F_3}{\partial x}\right) N_2 + \left(\frac{\partial F_2}{\partial x} - \frac{\partial F_1}{\partial y}\right) N_3\right] du\, dv \tag{3}$$

$$= \oint_{\bar{C}} (F_1\, dx + F_2\, dy + F_3\, dz)$$

where R is the region with boundary curve $\bar{C}$ in the uv-plane corresponding to S represented by $\mathbf{r}(u, v)$, and $\mathbf{N} = [N_1, N_2, N_3] = \mathbf{r}_u \times \mathbf{r}_v$.

EXAMPLE 1 Verification of Stokes's theorem

Before proving Stokes's theorem, let us get used to it by verifying it for

$$\mathbf{F} = y\mathbf{i} + z\mathbf{j} + x\mathbf{k}$$

and S the paraboloid

$$z = f(x, y) = 1 - (x^2 + y^2), \quad z \geqq 0$$

in Fig. 226.

Solution. C: $\mathbf{r}(s) = \cos s\,\mathbf{i} + \sin s\,\mathbf{j}$, $\mathbf{r}'(s) = -\sin s\,\mathbf{i} + \cos s\,\mathbf{j}$, hence

$$\oint_C \mathbf{F}\cdot d\mathbf{r} = \int_0^{2\pi} [(\sin s)(-\sin s) + 0 + 0]\, ds = -\pi.$$

On the other hand we now consider the integral over S in (2) on the left, using (3), Sec. 9.6, with curl **F** instead of **F** and $u = x$, $v = y$. We need (verify this)

$$\text{curl}\,\mathbf{F} = -\mathbf{i} - \mathbf{j} - \mathbf{k}, \qquad \mathbf{N} = -f_x\mathbf{i} - f_y\mathbf{j} + \mathbf{k} = 2x\mathbf{i} + 2y\mathbf{j} + \mathbf{k}$$

and calculate from this $(\text{curl}\,\mathbf{F})\cdot\mathbf{N} = -2x - 2y - 1$. Now S corresponds to $x^2 + y^2 \leqq 1$. Using polar coordinates r, θ, given by $x = r\cos\theta$, $y = r\sin\theta$, we thus get from (3), Sec. 9.6,

$$\iint_S (\text{curl}\,\mathbf{F})\cdot\mathbf{n}\, dA = \int_0^{2\pi}\int_0^1 (-2r\cos\theta - 2r\sin\theta - 1)r\, dr\, d\theta$$

$$= \int_0^{2\pi} [\tfrac{2}{3}(-\cos\theta - \sin\theta) - \tfrac{1}{2}]\, d\theta = -\pi.$$

Note that **N** is an upper normal vector of S (why?) and C is oriented counterclockwise, as required in the theorem. ∎

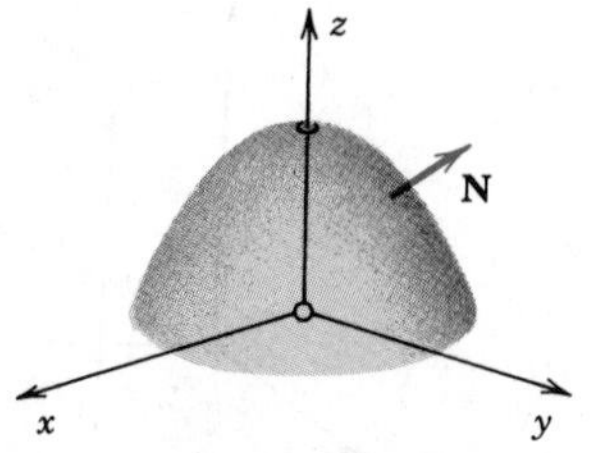

Fig. 226. Surface S in Example 1

Proof of Stokes's theorem. Obviously, (2) holds if the integrals of each component on both sides of (3) are equal, that is,

$$\iint_R \left(\frac{\partial F_1}{\partial z} N_2 - \frac{\partial F_1}{\partial y} N_3\right) du\, dv = \oint_{\bar{C}} F_1\, dx \tag{4}$$

$$\iint_R \left(-\frac{\partial F_2}{\partial z} N_1 + \frac{\partial F_2}{\partial x} N_3\right) du\, dv = \oint_{\bar{C}} F_2\, dy \tag{5}$$

$$\iint_R \left(\frac{\partial F_3}{\partial y} N_1 - \frac{\partial F_3}{\partial x} N_2\right) du\, dv = \oint_{\bar{C}} F_3\, dz. \tag{6}$$

We prove this first for a surface S that can be represented simultaneously in the forms

$$\text{(7)} \qquad \text{(a)}\quad z = f(x, y), \qquad \text{(b)}\quad y = g(x, z), \qquad \text{(c)}\quad x = h(y, z).$$

We prove (4), using (7a). Setting $u = x$, $v = y$, we have from (7a)

$$\mathbf{r}(u, v) = \mathbf{r}(x, y) = x\mathbf{i} + y\mathbf{j} + f\mathbf{k}$$

and in (3), Sec. 9.6, by direct calculation

$$\mathbf{N} = \mathbf{r}_u \times \mathbf{r}_v = \mathbf{r}_x \times \mathbf{r}_y = -f_x\mathbf{i} - f_y\mathbf{j} + \mathbf{k}.$$

Note that $\mathbf{N}$ is an *upper* normal vector of S, since it has a *positive* z-component. Also, $R = S^*$, the projection of S into the xy-plane, with boundary curve $\bar{C} = C^*$ (Fig. 227). Hence the left side of (4) is

$$\iint_{S^*} \left[\frac{\partial F_1}{\partial z}(-f_y) - \frac{\partial F_1}{\partial y}\right] dx\, dy. \tag{8}$$

We now consider the right side of (4). We transform this line integral over $\bar{C} = C^*$ into a double integral over S^* by applying Green's theorem [formula (1) in Sec. 9.4 with $F_2 = 0$]. This gives

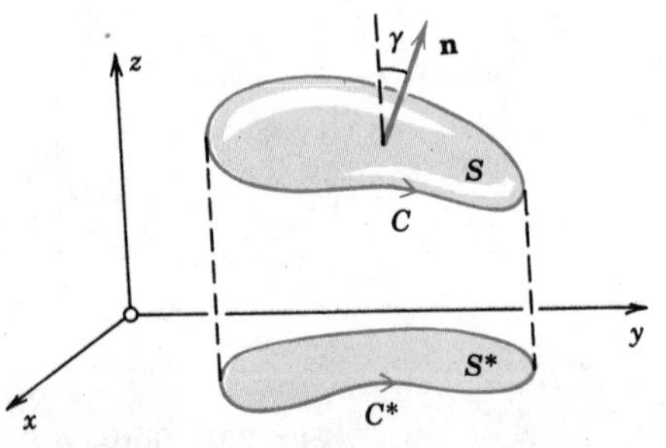

Fig. 227. Proof of Stokes's theorem

$$\oint_{C^*} F_1\, dx = \iint_{S^*} -\frac{\partial F_1}{\partial y}\, dx\, dy.$$

Here, $F_1 = F_1[x, y, f(x, y)]$. Hence by the chain rule (see also Prob. 10 in Problem Set 8.8),

$$-\frac{\partial F_1[x, y, f(x, y)]}{\partial y} = -\frac{\partial F_1(x, y, z)}{\partial y} - \frac{\partial F_1(x, y, z)}{\partial z}\frac{\partial f}{\partial y} \qquad [z = f(x, y)].$$

We see that the right side of this equals the integrand in (8). This proves (4). Relations (5) and (6) follow in the same way if we use (7b) and (7c), respectively. By addition we obtain (3). This proves Stokes's theorem for a surface S that can be represented simultaneously in the forms (7a), (7b), (7c).

As in the proof of the divergence theorem, our result may be immediately extended to a surface S that can be decomposed into finitely many pieces, each of which is of the kind just considered. This covers most of the cases of practical interest. The proof in the case of a most general surface S satisfying the assumptions of the theorem would require a limit process; this is similar to the situation in the case of Green's theorem in Sec. 9.4. ∎

EXAMPLE 2 Green's theorem in the plane as a special case of Stokes's theorem

Let $\mathbf{F} = F_1\mathbf{i} + F_2\mathbf{j}$ be a vector function that is continuously differentiable in a domain in the xy-plane containing a simply connected bounded closed region S whose boundary C is a piecewise smooth simple closed curve. Then, according to (1),

$$(\operatorname{curl} \mathbf{F})\cdot\mathbf{n} = (\operatorname{curl} \mathbf{F})\cdot\mathbf{k} = \frac{\partial F_2}{\partial x} - \frac{\partial F_1}{\partial y}.$$

Hence the formula in Stokes's theorem now takes the form

$$\iint_S \left(\frac{\partial F_2}{\partial x} - \frac{\partial F_1}{\partial y}\right) dA = \oint_C (F_1\, dx + F_2\, dy).$$

This shows that Green's theorem in the plane (Sec. 9.4) is a special case of Stokes's theorem. ∎

EXAMPLE 3 Evaluation of a line integral by Stokes's theorem

Evaluate $\int_C \mathbf{F}\cdot\mathbf{r}'\, ds$, where C is the circle $x^2 + y^2 = 4$, $z = -3$, oriented counterclockwise as seen by a person standing at the origin, and, with respect to right-handed Cartesian coordinates,

$$\mathbf{F} = y\mathbf{i} + xz^3\mathbf{j} - zy^3\mathbf{k}.$$

Solution. As a surface S bounded by C we can take the plane circular disk $x^2 + y^2 \leqq 4$ in the plane $z = -3$. Then $\mathbf{n}$ in Stokes's theorem points in the positive z-direction; thus $\mathbf{n} = \mathbf{k}$. Hence $(\operatorname{curl} \mathbf{F})\cdot\mathbf{n}$ is simply the component of curl $\mathbf{F}$ in the positive z-direction. Since $\mathbf{F}$ with $z = -3$ has the components $F_1 = y$, $F_2 = -27x$, $F_3 = 3y^3$, we thus obtain

$$(\operatorname{curl} \mathbf{F})\cdot\mathbf{n} = \frac{\partial F_2}{\partial x} - \frac{\partial F_1}{\partial y} = -27 - 1 = -28.$$

Hence the integral over S in Stokes's theorem equals -28 times the area 4π of the disk S. This yields the answer $-28 \cdot 4\pi = -112\pi \approx -352$.

The student may confirm the result by direct calculation, which involves somewhat more work. ∎

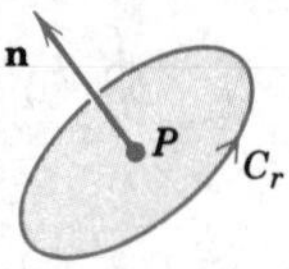

Fig. 228. Example 4

EXAMPLE 4 Physical interpretation of the curl. Circulation

Let S_r be a circular disk of radius r and center P bounded by the circle C_r (Fig. 228), and let $\mathbf{F}(Q) \equiv \mathbf{F}(x, y, z)$ be a continuously differentiable vector function in a domain containing S_r. Then by Stokes's theorem and the mean value theorem for surface integrals,

$$\oint_{C_r} \mathbf{F} \cdot \mathbf{r}' \, ds = \iint_{S_r} (\text{curl } \mathbf{F}) \cdot \mathbf{n} \, dA = (\text{curl } \mathbf{F}) \cdot \mathbf{n}(P^*) A_r$$

where A_r is the area of S_r and P^* is a suitable point of S_r. This may be written in the form

$$(\text{curl } \mathbf{F}) \cdot \mathbf{n}(P^*) = \frac{1}{A_r} \oint_{C_r} \mathbf{F} \cdot \mathbf{r}' \, ds.$$

In the case of a fluid motion with velocity vector $\mathbf{F} = \mathbf{v}$, the integral

$$\oint_{C_r} \mathbf{v} \cdot \mathbf{r}' \, ds$$

is called the **circulation** of the flow around C_r; it measures the extent to which the corresponding fluid motion is a rotation around the circle C_r. If we now let r approach zero, we find

$$(\text{curl } \mathbf{v}) \cdot \mathbf{n}(P) = \lim_{r \to 0} \frac{1}{A_r} \oint_{C} \mathbf{v} \cdot \mathbf{r}' \, ds; \tag{9}$$

that is, the component of the curl in the positive normal direction can be regarded as the **specific circulation** (circulation per unit area) of the flow in the surface at the corresponding point. ∎

EXAMPLE 5 Work done in the displacement around a closed curve

Find the work done by the force $\mathbf{F} = 2xy^3 \sin z \, \mathbf{i} + 3x^2y^2 \sin z \, \mathbf{j} + x^2y^3 \cos z \, \mathbf{k}$ in the displacement around the curve of intersection of the paraboloid $z = x^2 + y^2$ and the cylinder $(x - 1)^2 + y^2 = 1$.

Solution. This work is given by the line integral in Stokes's theorem. Now $\mathbf{F} = \text{grad } f$, where $f = x^2y^3 \sin z$ and curl grad $f = \mathbf{0}$ [see (3) in Sec. 8.11], so that $(\text{curl } \mathbf{F}) \cdot \mathbf{n} = 0$ and the work is 0 by Stokes's theorem. ∎

Stokes's theorem applied to path independence. We have seen and emphasized in Sec. 9.2 that the value of a line integral generally depends not only on the function to be integrated and on the two endpoints A and B of the path of integration C, but also on the particular choice of a path from A to B. In Theorem 3 of Sec. 9.2 we proved that if a line integral

$$\int_C \mathbf{F}(\mathbf{r}) \cdot d\mathbf{r} = \int_C (F_1 \, dx + F_2 \, dy + F_3 \, dz) \tag{10}$$

(involving continuous F_1, F_2, F_3 that have continuous first partial derivatives) is independent of path in a domain D, then

$$\text{curl } \mathbf{F} = \mathbf{0}. \tag{11}$$

We claimed that the converse is also true, that is, if (11) holds, then (10) is independent of path in D, provided D is simply connected (see Sec. 9.2).

The proof needs Stokes's theorem and can now be given as follows. Let C be any simple closed path in D. Since D is simply connected, we can find a surface S in D bounded by C. Stokes's theorem is applicable and gives

$$\oint_C (F_1\,dx + F_2\,dy + F_3\,dz) = \oint_C \mathbf{F}\cdot\mathbf{r}'\,ds = \iint_S (\text{curl }\mathbf{F})\cdot\mathbf{n}\,dA$$

for proper direction on C and normal vector $\mathbf{n}$ on S. Now, regardless of the choice of C, the integral over S is zero because, by (11), its integrand is identically zero. From this and Theorem 2 in Sec. 9.2 it follows that the integral (10) is independent of path in D. This completes the proof. ∎

Problem Set 9.9

Evaluate $\iint_S (\text{curl }\mathbf{F})\cdot\mathbf{n}\,dA$ by direct integration, where

1. $\mathbf{F} = 2z^2\mathbf{i} + 3x\mathbf{j}$, S the square $0 \leqq x \leqq 1$, $0 \leqq y \leqq 1$, $z = 1$
2. $\mathbf{F} = 5\cos y\,\mathbf{i} + \cosh z\,\mathbf{j} + x\mathbf{k}$, S as in Prob. 1
3. $\mathbf{F} = 2z^2\mathbf{i} + 8x\mathbf{j}$, S the rectangle $0 \leqq x \leqq 1$, $0 \leqq y \leqq 1$, $z = y$
4. $\mathbf{F} = y^3\mathbf{i} - x^3\mathbf{j}$, S the circular disk $x^2 + y^2 \leqq 1$, $z = 0$
5. $\mathbf{F} = -e^y\mathbf{i} + e^z\mathbf{j} + e^x\mathbf{k}$, S the quadrangle $0 \leqq x \leqq 1, 0 \leqq y \leqq 1, z = x + y$
6. $\mathbf{F} = z^2\mathbf{i} + x^2\mathbf{j} + y^2\mathbf{k}$, S: $z^2 = x^2 + y^2$, $y \geqq 0$, $0 \leqq z \leqq 2$
7. $\mathbf{F} = e^{2z}\mathbf{i} + e^z \sin y\,\mathbf{j} + e^z \cos y\,\mathbf{k}$, S: $0 \leqq x \leqq 2$, $0 \leqq y \leqq 1$, $z = y^2$
8. $\mathbf{F} = 2x\cos z\,\mathbf{k}$, S: $x^2 + y^2 = 4$, $y \geqq 0$, $0 \leqq z \leqq \pi/2$
9. Verify Stokes's theorem for $\mathbf{F}$ and S in Prob. 1.
10. Verify Stokes's theorem for $\mathbf{F}$ and S in Prob. 4.

Evaluate $\oint_C \mathbf{F}\cdot\mathbf{r}'(s)\,ds$ (clockwise as seen by a person standing at the origin) by Stokes's theorem, where, with respect to right-handed Cartesian coordinates,

11. $\mathbf{F} = -3y\mathbf{i} + 3x\mathbf{j} + z\mathbf{k}$, C the circle $x^2 + y^2 = 4$, $z = 1$
12. $\mathbf{F} = y\mathbf{i} + xz^3\mathbf{j} - zy^3\mathbf{k}$, C the circle $x^2 + y^2 = a^2$, $z = b\ (> 0)$
13. $\mathbf{F} = -2z\mathbf{i} + x\mathbf{j} - x\mathbf{k}$, C the ellipse $x^2 + y^2 = 1$, $z = y + 1$
14. $\mathbf{F} = y\mathbf{i} + z^2\mathbf{j} + x^3\mathbf{k}$, C as in Prob. 13
15. $\mathbf{F} = xyz\mathbf{j}$, C the boundary of the triangle with vertices (1, 0, 0), (0, 1, 0), (0, 0, 1)
16. $\mathbf{F} = z\mathbf{i} + x\mathbf{j} + y\mathbf{k}$, C as in Prob. 15
17. $\mathbf{F} = y^2\mathbf{i} + x^2\mathbf{k}$, C as in Prob. 15
18. $\mathbf{F} = x^2\mathbf{i} + y^2\mathbf{j} + z^2\mathbf{k}$, C the intersection of $x^2 + y^2 + z^2 = a^2$ and $z = y^2$
19. $\mathbf{F} = 2y\mathbf{i} + z\mathbf{j} + 3y\mathbf{k}$, C the circle $x^2 + y^2 + z^2 = 6z$, $z = x + 3$

20. Evaluate $\int_C \mathbf{F}\cdot\mathbf{r}'\,ds$, $\mathbf{F} = (x^2 + y^2)^{-1}(-y\mathbf{i} + x\mathbf{j})$, C: $x^2 + y^2 = 1$, $z = 0$, oriented clockwise. Note that Stokes's theorem cannot be applied. Why?

Review Questions and Problems for Chapter 9

1. Describe from memory how you can evaluate a line integral. What were our assumptions on the path of integration?
2. How can work done by a variable force be expressed as a line integral?
3. How can a double integral be transformed into a line integral?
4. Answer the same question as in Prob. 3 for a surface integral.
5. What do we mean by orienting a surface S? How many possibilities of doing this at a given point of S do we have?
6. Explain the role of orientation for surface integrals.
7. What is a smooth surface? A piecewise smooth surface? Give examples. Why did we need these concepts?
8. Why is independence of path significant in physical applications?
9. State the definition of curl. Explain its significance in this chapter.
10. How did we use Green's theorem in the plane to prove Stokes's theorem?
11. How did we use Stokes's theorem in connection with independence of path?
12. State the divergence theorem from memory and explain an application of it.
13. Our line and surface integrals were made up in terms of a *vector* function **F**, but their integrands are actually *scalar* functions. How did we accomplish this? Why did we proceed in this fashion (which at first sight looks somewhat like a detour)?
14. State Laplace's equation. Where in physics is it important?
15. What is a harmonic function? What facts about harmonic functions did we discover in this chapter?

Evaluate $\int_C \mathbf{F}(\mathbf{r}) \cdot d\mathbf{r}$, with **F** and C as given, by the method that seems most suitable (direct integration, use of exactness, or Green's or Stokes's theorems). Recall that if **F** is a force, the integral gives the work done in a displacement.

16. $\mathbf{F} = 2x^2\mathbf{i} - 4y^2\mathbf{j}$, C the straight line segment from (1, 1) to (3, 3)
17. $\mathbf{F} = x^2\mathbf{i} + y^2\mathbf{j}$, C: $y = 1 - x^2$ from $x = -1$ to $x = 1$
18. $\mathbf{F} = xy\mathbf{i} + z^2\mathbf{j}$, C: $y = x^2$, $z = x$ from (1, 1, 1) to (2, 4, 2)
19. $\mathbf{F} = -y\mathbf{i} + x\mathbf{j} + z\mathbf{k}$, C: $\mathbf{r} = \cosh t\,\mathbf{i} + \sinh t\,\mathbf{j} + e^{2t}\mathbf{k}$, $0 \leqq t \leqq 1/2$
20. $\mathbf{F} = y \sin x\,\mathbf{i} + x \cos y\,\mathbf{j}$, C the boundary of $-1 \leqq x \leqq 1$, $0 \leqq y \leqq 1$
21. $\mathbf{F} = 5(z\mathbf{i} + x\mathbf{j} + y\mathbf{k})$, C the ellipse $25x^2 + y^2 = 25$, $z = 0$
22. $\mathbf{F} = e^{2y}(z^3\mathbf{i} + 2xz^3\mathbf{j} + 3xz^2\mathbf{k})$, C: $x^2 + y^2 = 4$, $z = -x$
23. $\mathbf{F} = 2xz\mathbf{i} - z^2\mathbf{j} + (x^2 - 2yz)\mathbf{k}$, C: $x^2 + 9y^2 = 9$, $z = x^2$
24. $\mathbf{F} = (y - 3z)\mathbf{i} + (4y^3 + x)\mathbf{j} - 3x\mathbf{k}$, C: $\mathbf{r} = \cos t\,\mathbf{i} + \sin t\,\mathbf{j} + t\mathbf{k}$, $0 \leqq t \leqq \pi$
25. $\mathbf{F} = 2yz\mathbf{j} + (e^z + y^2)\mathbf{k}$, C: $y = x$, $z = x^2$ from (0, 0, 0) to (2, 2, 4)
26. $\mathbf{F} = \cosh y\,\mathbf{i} + xy^2\mathbf{j} + z\mathbf{k}$, C: $\mathbf{r} = t\mathbf{i} + 3t\mathbf{j} + t^2\mathbf{k}$, $0 \leqq t \leqq 1$
27. $\mathbf{F} = z^2\mathbf{i} + x^3\mathbf{j} + y^2\mathbf{k}$, C: $x^2 + y^2 = 4$, $x + y + z = 0$
28. $\mathbf{F} = 4(z^2\mathbf{i} + x^2\mathbf{j} + y^2\mathbf{k})$, C the boundary of $0 \leqq x \leqq 1$, $0 \leqq y \leqq 1$, $z = 1$
29. $\mathbf{F} = x^3\mathbf{i} + e^{3y}\mathbf{j} - 3e^{-3z}\mathbf{k}$, C: $9x^2 + y^2 + z^2 = 9$, $x = z$
30. $\mathbf{F} = y\mathbf{i} + (x + z)\mathbf{j} + x^5\mathbf{k}$, C: $x^2 + y^2 = 1$, $z = x$, $x \geqq 0$, $y \geqq 0$
31. $\mathbf{F} = 3x^2e^{-3z}(\mathbf{i} - x\mathbf{k})$, C: $y = x^2$, $z = x^3$ from (0, 0, 0) to (2, 4, 8)
32. $\mathbf{F} = y^3\mathbf{i} - z\mathbf{j} + x\mathbf{k}$, C as in Prob. 31
33. $\mathbf{F} = \sin \pi x\,\mathbf{i} + z\mathbf{j}$, C the boundary of the triangle with vertices (0, 0, 0), (1, 0, 0), (1, 1, 0)

34. $\mathbf{F} = -y^2e^x\mathbf{i} + 4xy^3\mathbf{j}$, C as in Prob. 33

35. $\mathbf{F} = ye^{xy}\mathbf{i} + xe^{xy}\mathbf{j} + \sinh z\,\mathbf{k}$, C as in Prob. 30

Evaluate the following double integrals.

36. $\int_{-\pi}^{\pi}\int_{-1}^{1} xy\,dx\,dy$

37. $\int_{0}^{2}\int_{0}^{x} e^{x+y}\,dy\,dx$

38. $\int_{0}^{\pi/2}\int_{-1}^{1} x^2y^2\,dx\,dy$

39. $\int_{0}^{1}\int_{x^2}^{x} (1 - xy)\,dy\,dx$

40. $\int_{0}^{3}\int_{-y}^{y} (x^2 + y^2)\,dx\,dy$

41. $\int_{-1}^{2}\int_{x^2}^{x+2} (x + y)\,dy\,dx$

Find the coordinates $\bar{x}$, $\bar{y}$ of the center of gravity of a mass of density $f(x, y)$ in a region R, where

42. $f = 1$, R: $0 \leqq y \leqq 1 - x^2$

43. $f = xy$, R: $0 \leqq y \leqq x, 0 \leqq x \leqq 1$

44. $f = 1$, R: $x^2 + y^2 \leqq a^2, y \geqq 0$

45. $f = x^2 + y^2$, R: $x^2 + y^2 \leqq a^2$, $x \geqq 0, y \geqq 0$

Represent the following surfaces in parametric form $\mathbf{r}(u, v)$ and find a normal vector.

46. $4x - 3y + z = 5$

47. $x^2 + y^2 = 9$

48. $x^2 + y^2 + z^2 = 4$

49. $x^2 + y^2 - z^2 = 0$

50. $x^2 - y^2 = 1$

51. $x^2 + 4y^2 - 4z = 0$

Evaluate $\iint_S \mathbf{F}\cdot\mathbf{n}\,dA$, where

52. $\mathbf{F} = 2x^2\mathbf{i} + 4y\mathbf{j}$, S: $z = 1 - x - y$, $x \geqq 0$, $y \geqq 0$, $z \geqq 0$

53. $\mathbf{F} = y\mathbf{i} - x\mathbf{j}$, S: $3x + 2y + z = 1$, $x \geqq 0$, $y \geqq 0$, $z \geqq 0$

54. $\mathbf{F} = y^3\mathbf{i} + x^3\mathbf{j} + 3z^2\mathbf{k}$, S: $z = x^2 + y^2$, $z \leqq 4$

55. $\mathbf{F} = x\mathbf{i} - y\mathbf{j} + x^2y^2\mathbf{k}$, S: $z = e^{xy}$, $-1 \leqq x \leqq 1$, $-1 \leqq y \leqq 1$

56. $\mathbf{F} = x\mathbf{i} + xy\mathbf{j} + e^{yz}\mathbf{k}$, S: $x^2 + y^2 = 1$, $0 \leqq z \leqq h$

57. $\mathbf{F} = z\mathbf{i} - z^4\mathbf{k}$, S: $x^2 + 4y^2 + z^2 = 4$, $x \geqq 0$, $y \geqq 0$, $z \geqq 0$

58. $\mathbf{F} = 3z^2x\mathbf{i} - (x^2 + 1)y\mathbf{j} + (x^2 - z^2)z\mathbf{k}$, S: $x^2 + 4y^2 + z^2 = 4$

59. $\mathbf{F} = \sin^2 x\,\mathbf{i} - y\sin 2x\,\mathbf{j} + 5z\mathbf{k}$, S the surface of $|x| \leqq 1$, $|y| \leqq 1$, $|z| \leqq 1$

60. $\mathbf{F} = 3x^2\mathbf{i} - (6xy + 4yz)\mathbf{j} + 2z(z - 1)\mathbf{k}$, S: $x^2 + y^2 + z^2 = 9$

Summary of Chapter 9 Vector Integral Calculus. Integral Theorems

Chapter 8 extended *differential* calculus to vectors, that is, to vector functions $\mathbf{v}(x, y, z)$ or $\mathbf{v}(t)$, and Chap. 9 extended *integral* calculus to vector functions. This involved *line integrals* (Sec. 9.1), *double integrals* (Sec. 9.3), *surface integrals* (Sec. 9.6), and *triple integrals* (Sec. 9.7) and the three "big" theorems for transforming these integrals into one another, the theorems by Green (Sec. 9.4), Gauss (Sec. 9.7), and Stokes (Sec. 9.9).

The analog of the definite integral of calculus is the **line integral**

$$(1)\quad \int_C \mathbf{F}(\mathbf{r})\cdot d\mathbf{r} = \int_C (F_1\,dx + F_2\,dy + F_3\,dz) = \int_a^b \mathbf{F}(\mathbf{r}(t))\cdot\frac{d\mathbf{r}}{dt}\,dt$$

where C: $\mathbf{r}(t) = x(t)\mathbf{i} + y(t)\mathbf{j} + z(t)\mathbf{k}$ $(a \leqq t \leqq b)$ is a curve in space (or in the plane). See Sec. 9.1. Physically, (1) may represent the work done by a (variable) force in a displacement. Other kinds of line integrals and their applications are also discussed in Sec. 9.1.

Independence of path of a line integral in a region D means that the integral of a given function over any path C with endpoints P and Q has the same value for all paths from P to Q that lie in D; here P and Q are fixed. An integral (1) is independent of path in D if and only if the differential form $F_1\,dx + F_2\,dy + F_3\,dz$ is exact in D (Sec. 9.2). Also, if curl $\mathbf{F} = \mathbf{0}$, where $\mathbf{F} = [F_1, F_2, F_3]$, has continuous first partial derivatives in a *simply connected* domain D, then the integral (1) is independent of path in D (Sec. 9.2).

Double integrals over a region R in the plane (Sec. 9.3) can be transformed into line integrals over the boundary C of R (and conversely) by **Green's theorem in the plane** (Sec. 9.4)

$$(2)\quad \iint_R \left(\frac{\partial F_2}{\partial x} - \frac{\partial F_1}{\partial y}\right) dx\,dy = \oint_C (F_1\,dx + F_2\,dy).$$

Other forms of this are given in Sec. 9.4. This theorem is also needed as the essential tool in proving Stokes's theorem (see below), and is a special case of the latter.

Triple integrals (Sec. 9.7) taken over a region T in space can be transformed into surface integrals over the boundary surface S of T (and conversely) by the **divergence theorem of Gauss** (Sec. 9.7)

$$(3)\quad \iiint_T \operatorname{div} \mathbf{F}\,dV = \iint_S \mathbf{F}\cdot\mathbf{n}\,dA$$

Among other things, this theorem implies **Green's formulas** (Sec. 9.8)

$$(4)\quad \iiint_T (f\nabla^2 g + \nabla f\cdot\nabla g)\,dV = \iint_S f\frac{\partial g}{\partial n}\,dA,$$

$$(5)\quad \iiint_T (f\nabla^2 g - g\nabla^2 f)\,dV = \iint_S \left(f\frac{\partial g}{\partial n} - g\frac{\partial f}{\partial n}\right) dA.$$

Surface integrals over a surface S with boundary curve C can be transformed into line integrals over C (and conversely) by **Stokes's theorem** (Sec. 9.9)

$$(6)\quad \iint_S (\operatorname{curl} \mathbf{F})\cdot\mathbf{n}\,dA = \oint_C \mathbf{F}\cdot\mathbf{r}'(s)\,ds.$$

PART C

FOURIER ANALYSIS AND PARTIAL DIFFERENTIAL EQUATIONS

Periodic phenomena occur quite frequently in physics and its engineering applications, and it is an important practical problem to represent the corresponding periodic functions in terms of simple periodic functions such as sine and cosine. This leads to **Fourier series,** whose terms are sine and cosine functions. Their introduction by Fourier (after work by Euler and Daniel Bernoulli) was one of the most important events in the development of applied mathematics. Chapter 10 is primarily concerned with Fourier series. The corresponding ideas and techniques can also be extended to nonperiodic phenomena. This leads to **Fourier integrals** and **Fourier transforms** (Secs. 10.9–10.11), and a common name for the whole area is **Fourier analysis.**

Chapter 11 is concerned with the most important **partial differential equations** of physics and engineering. In this area, Fourier analysis has its most important applications, as a basic tool in solving boundary and initial value problems in mechanics, heat flow, electrostatics, and other fields.

Chapter 10

Fourier Series, Integrals, and Transforms

Fourier series[1] (Sec. 10.2) are series of cosine and sine terms and arise in the important practical task of representing general periodic functions. They constitute a very important tool in solving problems that involve ordinary and partial differential equations.

In the present chapter we discuss basic concepts, facts, and techniques in connection with Fourier series. Illustrative examples and some important engineering applications will be included. Further applications follow in the next chapter on partial differential equations and initial and boundary value problems.

The *theory* of Fourier series is rather complicated, but the *application* of these series is simple. Fourier series are, in a certain sense, more universal than Taylor series, because many *discontinuous* periodic functions of practical interest can be developed in Fourier series, but, of course, do not have Taylor series representations.

The last three sections of this chapter concern **Fourier integrals** and **Fourier transforms,** which extend the ideas and techniques of Fourier series to nonperiodic functions defined for all x. Corresponding applications to partial differential equations will be considered in the next chapter (Sec. 11.14).

Prerequisite for this chapter: elementary integral calculus.
Sections that may be omitted in a shorter course: 10.6–10.11.
References: Appendix 1, Part C.
Answers to problems: Appendix 2.

10.1 Periodic Functions. Trigonometric Series

A function $f(x)$ is called **periodic** if it is defined for all real x and if there is some positive number p such that

$$f(x + p) = f(x) \qquad \text{for all } x. \tag{1}$$

[1]JEAN-BAPTISTE JOSEPH FOURIER (1768—1830), French physicist and mathematician, lived and taught in Paris, accompanied Napoleon to Egypt and was later made prefect of Grenoble. He utilized Fourier series in his main work *Théorie analytique de la chaleur* (*Analytic Theory of Heat,* Paris 1822) in which he developed the theory of heat conduction (heat equation, see Sec. 11.5). These new series became a most important tool in mathematical physics and also had considerable influence on the further development of mathematics itself; see Ref. [9] in Appendix 1.

This number p is called a **period** of $f(x)$. The graph of such a function is obtained by periodic repetition of its graph in any interval of length p (Fig. 229). Periodic phenomena and functions occur in many applications.

Familiar periodic functions are the sine and cosine functions, and we note that the function $f = c = const$ is also a periodic function in the sense of the definition, because it satisfies (1) for every positive p. Examples of functions that are *not* periodic are x, x^2, x^3, e^x, and $\ln x$, to mention just a few.[2]

From (1) we have $f(x + 2p) = f[(x + p) + p] = f(x + p) = f(x)$, etc., and for any integer n,

$$f(x + np) = f(x) \qquad \text{for all } x. \tag{2}$$

Hence $2p, 3p, 4p, \cdots$ are also periods of $f(x)$. Furthermore, if $f(x)$ and $g(x)$ have period p, then the function

$$h(x) = af(x) + bg(x) \qquad (a, b \text{ constant})$$

also has the period p.

Our problem in the first few sections of this chapter will be the representation of various functions of period $p = 2\pi$ in terms of the simple functions

$$1, \quad \cos x, \quad \sin x, \quad \cos 2x, \quad \sin 2x, \cdots, \quad \cos nx, \quad \sin nx, \cdots, \tag{3}$$

which have the period 2π (Fig. 230, p. 568). The series that will arise in this connection will be of the form

$$a_0 + a_1 \cos x + b_1 \sin x + a_2 \cos 2x + b_2 \sin 2x + \cdots, \tag{4}$$

where $a_0, a_1, a_2, \cdots, b_1, b_2, \cdots$ are real constants. Such a series is called a **trigonometric series,** and the a_n and b_n are called the **coefficients** of the series. Using the summation sign,[3] we may write this series

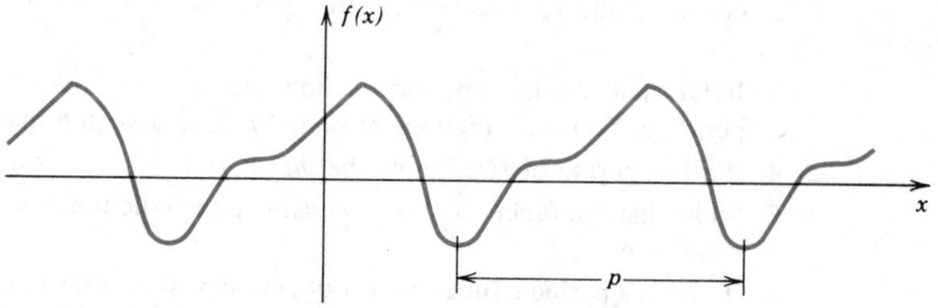

Fig. 229. Periodic function

[2] If a periodic function $f(x)$ has a smallest period p (> 0), this is often called the **primitive period** of $f(x)$. For example, the primitive period of $\sin x$ is 2π, and the primitive period of $\sin 2x$ is π. A periodic function without primitive period is $f = const$.

[3] And inserting parentheses; from a convergent series this gives again a convergent series with the same sum, as can be proved.

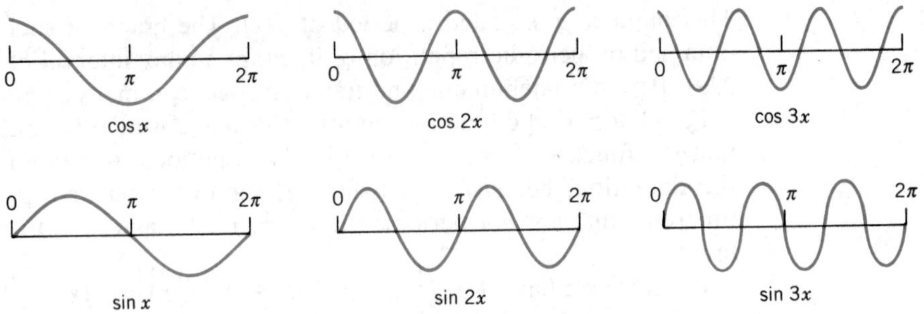

Fig. 230. Cosine and sine functions having the period 2π

$$a_0 + \sum_{n=1}^{\infty} (a_n \cos nx + b_n \sin nx). \tag{4}$$

The set of functions (3) from which we have made up the series (4) is often called the **trigonometric system,** to have a short name for it.

We see that each term of the series (4) has the period 2π. Hence *if the series* (4) *converges, its sum will be a function of period* 2π.

Periodic functions that occur in practical problems are often rather complicated, and it is desirable to represent them in terms of simple periodic functions. We shall see that almost any periodic function $f(x)$ of period 2π that appears in applications—for example, in connection with vibrations—can be represented by a trigonometric series (which will then be called the *Fourier series* of f).

Problem Set 10.1

Find the smallest positive period p of the following functions.

1. $\cos x$, $\sin x$, $\cos 2x$, $\sin 2x$, $\cos \pi x$, $\sin \pi x$, $\cos 2\pi x$, $\sin 2\pi x$

2. $\cos nx$, $\sin nx$, $\cos \dfrac{2\pi x}{k}$, $\sin \dfrac{2\pi x}{k}$, $\cos \dfrac{2\pi nx}{k}$, $\sin \dfrac{2\pi nx}{k}$

3. If $f(x)$ and $g(x)$ have period p, show that $h = af + bg$ (a, b constant) has the period p. Thus all functions of period p form a vector space.

4. If p is a period of $f(x)$, show that np, $n = 2, 3, \cdots$, is a period of $f(x)$.

5. Show that the function $f(x) = const$ is a periodic function of period p for every positive p.

6. If $f(x)$ is a periodic function of x of period p, show that $f(ax)$, $a \neq 0$, is a periodic function of x of period p/a, and $f(x/b)$, $b \neq 0$, is a periodic function of x of period bp. Verify these results for $f(x) = \cos x$, $a = b = 2$.

Sketch the following functions $f(x)$, which are assumed to be periodic of period 2π and, for $-\pi < x < \pi$, are given by the formulas

7. $f(x) = x$

8. $f(x) = x^2$

9. $f(x) = e^{|x|}$

10. $f(x) = |x|$

11. $f(x) = \begin{cases} x^2 & \text{if } -\pi < x < 0 \\ 0 & \text{if } 0 < x < \pi \end{cases}$

12. $f(x) = \begin{cases} 1 & \text{if } -\pi < x < 0 \\ 1 - x/\pi & \text{if } 0 < x < \pi \end{cases}$

13. $f(x) = \begin{cases} \pi + x & \text{if } -\pi < x < 0 \\ \pi - x & \text{if } 0 < x < \pi \end{cases}$

14. $f(x) = \begin{cases} 1 & \text{if } -\pi < x < 0 \\ \cos x/2 & \text{if } 0 < x < \pi \end{cases}$

15. $f(x) = \begin{cases} x & \text{if } -\pi < x < 0 \\ \pi - x & \text{if } 0 < x < \pi \end{cases}$

16. $f(x) = \begin{cases} 0 & \text{if } -\pi < x < 0 \\ \sin x & \text{if } 0 < x < \pi \end{cases}$

Evaluate the following integrals where $n = 0, 1, 2, \cdots$. (These are typical examples of integrals that will be needed in our further work.)

17. $\int_0^{\pi} \sin nx \, dx$

18. $\int_{-\pi/2}^{0} \cos nx \, dx$

19. $\int_{-\pi/2}^{\pi/2} x \cos nx \, dx$

20. $\int_{-\pi}^{\pi} x \sin nx \, dx$

21. $\int_{-\pi}^{0} e^x \sin nx \, dx$

22. $\int_{-\pi/2}^{\pi/2} x \sin nx \, dx$

23. $\int_0^{\pi} e^x \cos nx \, dx$

24. $\int_{-\pi}^{\pi} x^2 \cos nx \, dx$

25. $\int_0^{\pi} x \sin nx \, dx$

10.2 Fourier Series

Fourier series arise from the practical task of representing a given periodic function $f(x)$ in terms of cosine and sine functions. These series are trigonometric series (Sec. 10.1) whose coefficients are determined from $f(x)$ by certain formulas [the "Euler formulas" (6), below], which we shall derive first. Afterwards we shall take a look at the theory of Fourier series.

Euler Formulas for the Fourier Coefficients

Let us assume that $f(x)$ is a periodic function of period 2π that can be **represented** by a trigonometric series,

$$f(x) = a_0 + \sum_{n=1}^{\infty} (a_n \cos nx + b_n \sin nx); \tag{1}$$

that is, we assume that this series converges and has $f(x)$ as its sum. Given such a function $f(x)$, we want to determine the coefficients a_n and b_n of the corresponding series (1).

We determine a_0. Integrating on both sides of (1) from $-\pi$ to π, we get

$$\int_{-\pi}^{\pi} f(x) \, dx = \int_{-\pi}^{\pi} \left[a_0 + \sum_{n=1}^{\infty} (a_n \cos nx + b_n \sin nx) \right] dx.$$

If term-by-term integration of the series is allowed,[4] we obtain

$$\int_{-\pi}^{\pi} f(x)\,dx = a_0 \int_{-\pi}^{\pi} dx + \sum_{n=1}^{\infty} \left(a_n \int_{-\pi}^{\pi} \cos nx\,dx + b_n \int_{-\pi}^{\pi} \sin nx\,dx \right).$$

The first term on the right equals $2\pi a_0$. All the other integrals on the right are zero, as can be readily seen by integration. Hence our first result is

$$a_0 = \frac{1}{2\pi} \int_{-\pi}^{\pi} f(x)\,dx. \tag{2}$$

We now determine $a_1, a_2, \cdots$ by a similar procedure. We multiply (1) by $\cos mx$, where m is any fixed positive integer, and integrate from $-\pi$ to π:

$$\int_{-\pi}^{\pi} f(x) \cos mx\,dx = \int_{-\pi}^{\pi} \left[a_0 + \sum_{n=1}^{\infty} (a_n \cos nx + b_n \sin nx) \right] \cos mx\,dx. \tag{3}$$

Integrating term by term, we see that the right side becomes

$$a_0 \int_{-\pi}^{\pi} \cos mx\,dx + \sum_{n=1}^{\infty} \left[a_n \int_{-\pi}^{\pi} \cos nx \cos mx\,dx + b_n \int_{-\pi}^{\pi} \sin nx \cos mx\,dx \right].$$

The first integral is zero. By applying (11) in Appendix 3.1 we obtain

$$\int_{-\pi}^{\pi} \cos nx \cos mx\,dx = \frac{1}{2} \int_{-\pi}^{\pi} \cos (n+m)x\,dx + \frac{1}{2} \int_{-\pi}^{\pi} \cos (n-m)x\,dx,$$

$$\int_{-\pi}^{\pi} \sin nx \cos mx\,dx = \frac{1}{2} \int_{-\pi}^{\pi} \sin (n+m)x\,dx + \frac{1}{2} \int_{-\pi}^{\pi} \sin (n-m)x\,dx.$$

Integration shows that the four terms on the right are zero, except for the last term in the first line, which equals π when $n = m$. Since in (3) this term is multiplied by a_m, the right side in (3) equals $a_m \pi$. Our second result is

$$a_m = \frac{1}{\pi} \int_{-\pi}^{\pi} f(x) \cos mx\,dx, \qquad m = 1, 2, \cdots. \tag{4}$$

We finally determine $b_1, b_2, \cdots$ in (1). If we multiply (1) by $\sin mx$, where m is any fixed positive integer, and then integrate from $-\pi$ to π, we have

$$\int_{-\pi}^{\pi} f(x) \sin mx\,dx = \int_{-\pi}^{\pi} \left[a_0 + \sum_{n=1}^{\infty} (a_n \cos nx + b_n \sin nx) \right] \sin mx\,dx. \tag{5}$$

[4]This is justified, for instance, in the case of uniform convergence (see Theorem 3 in Sec. 14.6).

Integrating term by term, we see that the right side becomes

$$a_0 \int_{-\pi}^{\pi} \sin mx\,dx + \sum_{n=1}^{\infty} \left[a_n \int_{-\pi}^{\pi} \cos nx \sin mx\,dx + b_n \int_{-\pi}^{\pi} \sin nx \sin mx\,dx \right].$$

The first integral is zero. The next integral is of the kind considered before, and is zero for all $n = 1, 2, \cdots$. For the last integral we obtain

$$\int_{-\pi}^{\pi} \sin nx \sin mx\,dx = \frac{1}{2}\int_{-\pi}^{\pi} \cos (n - m)x\,dx - \frac{1}{2}\int_{-\pi}^{\pi} \cos (n + m)x\,dx.$$

The last term is zero. The first term on the right is zero when $n \neq m$ and is π when $n = m$. Since in (5) this term is multiplied by b_m, the right side in (5) is equal to $b_m \pi$, and our last result is

$$b_m = \frac{1}{\pi}\int_{-\pi}^{\pi} f(x) \sin mx\,dx, \qquad m = 1, 2, \cdots .$$

Writing n in place of m, we altogether have the so-called **Euler[5] formulas**

(6)

$$\text{(a)} \quad a_0 = \frac{1}{2\pi}\int_{-\pi}^{\pi} f(x)\,dx$$

$$\text{(b)} \quad a_n = \frac{1}{\pi}\int_{-\pi}^{\pi} f(x) \cos nx\,dx \qquad n = 1, 2, \cdots,$$

$$\text{(c)} \quad b_n = \frac{1}{\pi}\int_{-\pi}^{\pi} f(x) \sin nx\,dx \qquad n = 1, 2, \cdots.$$

These numbers given by (6) are called the **Fourier coefficients** of $f(x)$. The trigonometric series

$$(7) \qquad a_0 + \sum_{n=1}^{\infty} (a_n \cos nx + b_n \sin nx)$$

with coefficients given in (6) is called the **Fourier series** of $f(x)$ (regardless of convergence—we discuss this on p. 574).

EXAMPLE 1 **Square wave**

Find the Fourier coefficients of the periodic function $f(x)$ in Fig. 231a, p. 573. The formula is

$$f(x) = \begin{cases} -k & \text{if} \quad -\pi < x < 0 \\ k & \text{if} \quad 0 < x < \pi \end{cases} \qquad \text{and} \qquad f(x + 2\pi) = f(x).$$

[5]See footnote 9 in Sec. 2.6.

Functions of this kind occur as external forces acting on mechanical systems, electromotive forces in electric circuits, etc. (The value of $f(x)$ at a single point does not affect the integral; hence we can leave $f(x)$ undefined at $x = 0$ and $x = \pm\pi$.

Solution. From (6a) we obtain $a_0 = 0$. This can also be seen without integration, since the area under the curve of $f(x)$ between $-\pi$ and π is zero. From (6b),

$$a_n = \frac{1}{\pi}\int_{-\pi}^{\pi} f(x)\cos nx\,dx = \frac{1}{\pi}\left[\int_{-\pi}^{0}(-k)\cos nx\,dx + \int_{0}^{\pi} k\cos nx\,dx\right]$$

$$= \frac{1}{\pi}\left[-k\frac{\sin nx}{n}\bigg|_{-\pi}^{0} + k\frac{\sin nx}{n}\bigg|_{0}^{\pi}\right] = 0$$

because $\sin nx = 0$ at $-\pi$, 0, and π for all $n = 1, 2, \cdots$. Similarly, from (6c) we obtain

$$b_n = \frac{1}{\pi}\int_{-\pi}^{\pi} f(x)\sin nx\,dx = \frac{1}{\pi}\left[\int_{-\pi}^{0}(-k)\sin nx\,dx + \int_{0}^{\pi} k\sin nx\,dx\right]$$

$$= \frac{1}{\pi}\left[k\frac{\cos nx}{n}\bigg|_{-\pi}^{0} - k\frac{\cos nx}{n}\bigg|_{0}^{\pi}\right].$$

Since $\cos(-\alpha) = \cos\alpha$ and $\cos 0 = 1$, this yields

$$b_n = \frac{k}{n\pi}[\cos 0 - \cos(-n\pi) - \cos n\pi + \cos 0] = \frac{2k}{n\pi}(1 - \cos n\pi).$$

Now, $\cos\pi = -1$, $\cos 2\pi = 1$, $\cos 3\pi = -1$, etc.; in general,

$$\cos n\pi = \begin{cases} -1 & \text{for odd } n,\\ 1 & \text{for even } n,\end{cases} \qquad \text{and thus} \qquad 1 - \cos n\pi = \begin{cases} 2 & \text{for odd } n,\\ 0 & \text{for even } n.\end{cases}$$

Hence the Fourier coefficients b_n of our function are

$$b_1 = \frac{4k}{\pi}, \qquad b_2 = 0, \qquad b_3 = \frac{4k}{3\pi}, \qquad b_4 = 0, \qquad b_5 = \frac{4k}{5\pi}, \cdots,$$

and since the a_n are zero, the Fourier series of $f(x)$ is

(8)
$$\frac{4k}{\pi}\left(\sin x + \frac{1}{3}\sin 3x + \frac{1}{5}\sin 5x + \cdots\right).$$

The partial sums are

$$S_1 = \frac{4k}{\pi}\sin x, \qquad S_2 = \frac{4k}{\pi}\left(\sin x + \frac{1}{3}\sin 3x\right), \qquad \text{etc.,}$$

and their graphs in Fig. 231 seem to indicate that the series is convergent and has the sum $f(x)$, the given function. We notice that at $x = 0$ and $x = \pi$, the points of discontinuity of $f(x)$, all partial sums have the value zero, the arithmetic mean of the values $-k$ and k of our function.

Furthermore, assuming that $f(x)$ is the sum of the series and setting $x = \pi/2$, we have

$$f\left(\frac{\pi}{2}\right) = k = \frac{4k}{\pi}\left(1 - \frac{1}{3} + \frac{1}{5} - + \cdots\right),$$

thus

$$1 - \frac{1}{3} + \frac{1}{5} - \frac{1}{7} + - \cdots = \frac{\pi}{4}.$$

This is a famous result by Leibniz (obtained in 1673 from geometrical considerations). It illustrates that the values of various series with constant terms can be obtained by evaluating Fourier series at specific points. ∎

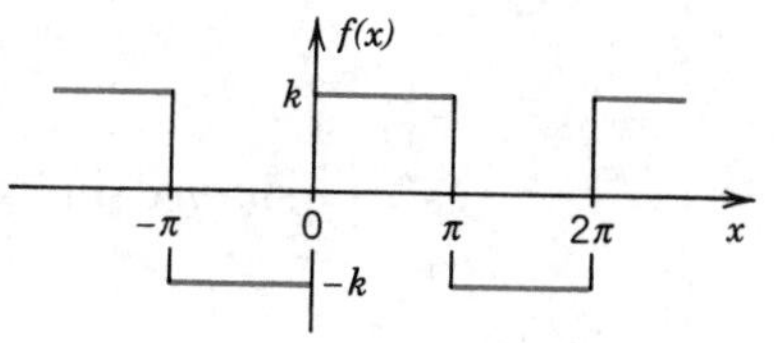

(a) The given function *f(x)* (Periodic square wave)

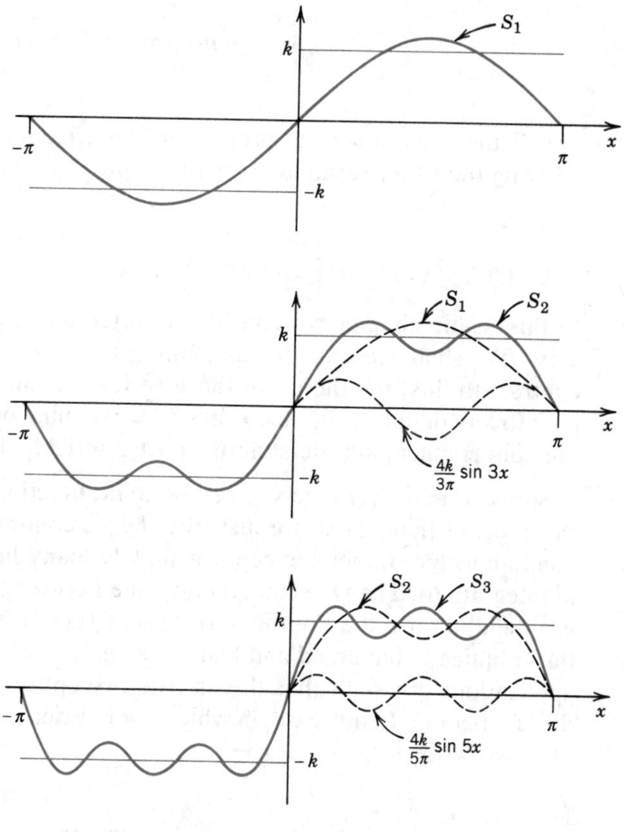

(b) The first three partial sums of the corresponding Fourier series

Fig. 231. Example 1

Orthogonality of the Trigonometric System

The trigonometric system (3), Sec. 10.1,

$$1, \quad \cos x, \quad \sin x, \quad \cos 2x, \quad \sin 2x, \quad \cdots, \quad \cos nx, \quad \sin nx, \quad \cdots$$

is **orthogonal** *on the interval* $-\pi \leqq x \leqq \pi$ (hence on any interval of length 2π, because of periodicity). By definition, this means that the integral of the product of any two different of these functions over that interval is zero; in formulas, for any integers m and $n \neq m$ we have

$$\int_{-\pi}^{\pi} \cos mx \cos nx \, dx = 0 \qquad (m \neq n)$$

and

$$\int_{-\pi}^{\pi} \sin mx \sin nx \, dx = 0 \qquad (m \neq n)$$

and for any integers m and n (including $m = n$) we have

$$\int_{-\pi}^{\pi} \cos mx \sin nx \, dx = 0.$$

This is the most important property of the trigonometric system, the key in deriving the Euler formulas (where we proved this orthogonality).

Convergence and Sum of Fourier Series

In this whole chapter we consider Fourier series from a practical point of view. We shall see that the application of these series is rather simple. In contrast to this, the theory of these series is complicated, and we shall not go into any details of it. Accordingly, we confine ourselves to a theorem on the convergence and the sum of Fourier series, which we present next.

Suppose that $f(x)$ is any given periodic function of period 2π for which the integrals in (6) exist; for instance, $f(x)$ is continuous or merely piecewise continuous (continuous except for finitely many finite jumps in the interval of integration). Then we can compute the Fourier coefficients (6) of $f(x)$ and use them to form the Fourier series (7) of $f(x)$. It would be nice if the series thus obtained converged and had the sum $f(x)$. Most functions appearing in applications are such that this is true (except at jumps of $f(x)$, which we discuss below). In this case, in which the Fourier series of $f(x)$ does represent $f(x)$, we write

$$f(x) = a_0 + \sum_{n=1}^{\infty} (a_n \cos nx + b_n \sin nx)$$

with an equality sign. If the Fourier series of $f(x)$ does not have the sum $f(x)$ or does not converge, one still writes

$$f(x) \sim a_0 + \sum_{n=1}^{\infty} (a_n \cos nx + b_n \sin nx)$$

with a tilde $\sim$, which indicates that the trigonometric series on the right has the Fourier coefficients of $f(x)$ as its coefficients, so it is the Fourier series of $f(x)$.

The class of functions that can be represented by Fourier series is surprisingly large and general. Corresponding sufficient conditions covering almost any conceivable application are as follows.

Theorem 1 **(Representation by a Fourier series)**

If a periodic function $f(x)$ with period 2π is piecewise continuous[6] in the interval $-\pi \leqq x \leqq \pi$ and has a left-hand derivative and right-hand derivative[7] at each point of that interval, then the Fourier series (7) of $f(x)$ [with coefficients (6)] is convergent. Its sum is $f(x)$, except at a point x_0 at which $f(x)$ is discontinuous and the sum of the series is the average of the left- and right-hand limits[7] of $f(x)$ at x_0.

Proof of convergence in Theorem 1 for a continuous function $f(x)$ having continuous first and second derivatives. Integrating (6b) by parts, we get

$$a_n = \frac{1}{\pi}\int_{-\pi}^{\pi} f(x)\cos nx\,dx = \frac{f(x)\sin nx}{n\pi}\bigg|_{-\pi}^{\pi} - \frac{1}{n\pi}\int_{-\pi}^{\pi} f'(x)\sin nx\,dx.$$

The first term on the right is zero. Another integration by parts gives

$$a_n = \frac{f'(x)\cos nx}{n^2\pi}\bigg|_{-\pi}^{\pi} - \frac{1}{n^2\pi}\int_{-\pi}^{\pi} f''(x)\cos nx\,dx.$$

The first term on the right is zero because of the periodicity and continuity of $f'(x)$. Since f'' is continuous in the interval of integration, we have

$$|f''(x)| < M$$

for an appropriate constant M. Furthermore, $|\cos nx| \leqq 1$. It follows that

$$|a_n| = \frac{1}{n^2\pi}\left|\int_{-\pi}^{\pi} f''(x)\cos nx\,dx\right| < \frac{1}{n^2\pi}\int_{-\pi}^{\pi} M\,dx = \frac{2M}{n^2}.$$

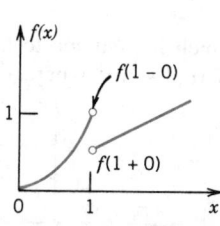

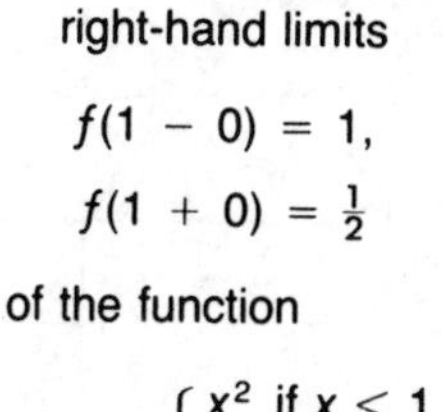

Fig. 232. Left- and right-hand limits

$f(1-0) = 1,$

$f(1+0) = \frac{1}{2}$

of the function

$$f(x) = \begin{cases} x^2 & \text{if } x < 1 \\ x/2 & \text{if } x > 1 \end{cases}$$

[6]Definition in Sec. 6.1.

[7]The **left-hand limit** of $f(x)$ at x_0 is defined as the limit of $f(x)$ as x approaches x_0 from the left and is frequently denoted by $f(x_0 - 0)$. Thus

$$f(x_0 - 0) = \lim_{h\to 0} f(x_0 - h) \quad \text{as } h \to 0 \text{ through positive values.}$$

The **right-hand limit** is denoted by $f(x_0 + 0)$ and

$$f(x_0 + 0) = \lim_{h\to 0} f(x_0 + h) \quad \text{as } h \to 0 \text{ through positive values.}$$

The **left-** and **right-hand derivatives** of $f(x)$ at x_0 are defined as the limits of

$$\frac{f(x_0 - h) - f(x_0 - 0)}{-h} \quad \text{and} \quad \frac{f(x_0 + h) - f(x_0 + 0)}{h},$$

respectively, as $h \to 0$ through positive values. Of course if $f(x)$ is continuous at x_0 the last term in both numerators is simply $f(x_0)$.

Similarly, $|b_n| < 2M/n^2$ for all n. Hence the absolute value of each term of the Fourier series of $f(x)$ is at most equal to the corresponding term of the series

$$|a_0| + 2M\left(1 + 1 + \frac{1}{2^2} + \frac{1}{2^2} + \frac{1}{3^2} + \frac{1}{3^2} + \cdots\right)$$

which is convergent. Hence that Fourier series converges and the proof is complete. (Readers already familiar with uniform convergence will see that, by the Weierstrass test in Sec. 14.6, under our present assumptions the Fourier series converges uniformly, and our derivation of (6) by integrating term by term is then justified by Theorem 3 of Sec. 14.6.)

The proof of convergence in the case of a piecewise continuous function $f(x)$ and the proof that under the assumptions in the theorem the Fourier series (7) with coefficients (6) represents $f(x)$ are substantially more complicated; see, for instance, Ref. [C14]. ■

EXAMPLE 2 Convergence at a jump as indicated in Theorem 1

The square wave in Example 1 has a jump at $x = 0$. Its left-hand limit there is $-k$ and its right-hand limit is k (Fig. 231), so that the average of these limits is 0. The Fourier series (8) of the square wave does indeed converge to this value when $x = 0$ because then all its terms are 0. Similarly for the other jumps. This is in agreement with Theorem 1. ■

Summary. A Fourier series of a given function $f(x)$ of period 2π is a series of the form (7) with coefficients given by the Euler formulas (6). Theorem 1 gives conditions that are sufficient for this series to converge and at each x to have the value $f(x)$, except at discontinuities of $f(x)$, where the series equals the arithmetic mean of the left-hand and right-hand limits of $f(x)$ at that point.

Problem Set 10.2

Find the Fourier series of the function $f(x)$, which is assumed to have the period 2π, and plot accurate graphs of the first three partial sums[8] where $f(x)$ equals

1.

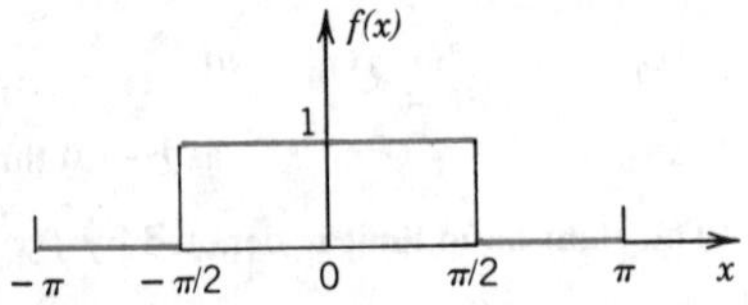

2.

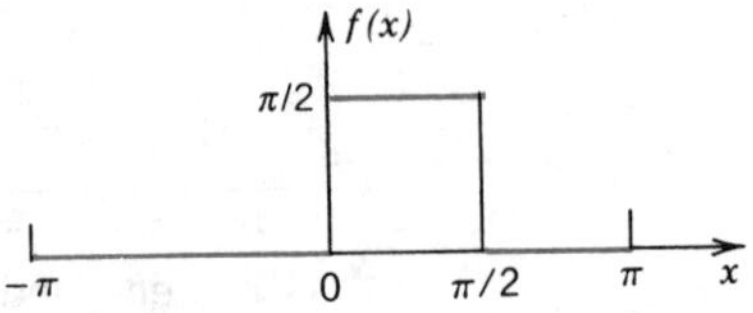

3.

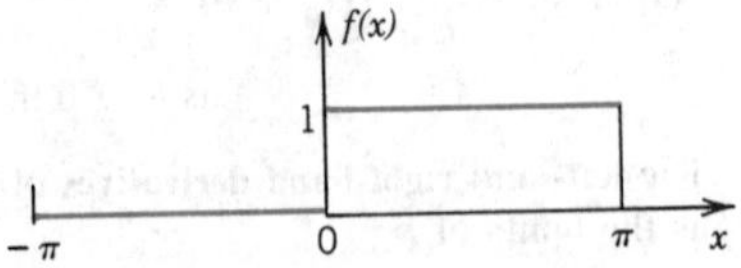

4.

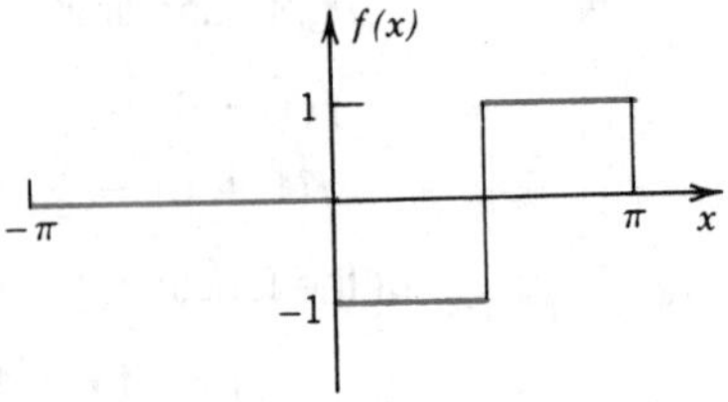

---

[8]That is, $a_0 + \sum_{n=1}^{N} (a_n \cos nx + b_n \sin nx)$ for $N = 1, 2, 3$.

5. $f(x) = \begin{cases} 1 & \text{if } -\pi/2 < x < \pi/2 \\ -1 & \text{if } \pi/2 < x < 3\pi/2 \end{cases}$

6. $f(x) = \begin{cases} -1 & \text{if } 0 < x < \pi/2 \\ 0 & \text{if } \pi/2 < x < 2\pi \end{cases}$

7. $f(x) = x \ (-\pi < x < \pi)$

8. $f(x) = x \ (0 < x < 2\pi)$

9. $f(x) = x^2 \ (-\pi < x < \pi)$

10. $f(x) = x^2 \ (0 < x < 2\pi)$

11. $f(x) = x^3 \ (-\pi < x < \pi)$

12. $f(x) = x + |x| \ (-\pi < x < \pi)$

13. $f(x) = \begin{cases} x & \text{if } -\pi/2 < x < \pi/2 \\ 0 & \text{if } \pi/2 < x < 3\pi/2 \end{cases}$

14. $f(x) = \begin{cases} 0 & \text{if } -\pi < x < 0 \\ x & \text{if } 0 < x < \pi \end{cases}$

15. $f(x) = \begin{cases} x & \text{if } -\pi/2 < x < \pi/2 \\ \pi - x & \text{if } \pi/2 < x < 3\pi/2 \end{cases}$

16. $f(x) = \begin{cases} x^2 & \text{if } -\pi/2 < x < \pi/2 \\ \pi^2/4 & \text{if } \pi/2 < x < 3\pi/2 \end{cases}$

17. Verify the last statement in Theorem 1 concerning the discontinuities for the function in Prob. 1.

18. Obtain the Fourier series in Prob. 3 from that in Prob. 1.

19. Show that if $f(x)$ has the Fourier coefficients a_n, b_n and $g(x)$ has the Fourier coefficients $a_n{}^*$, $b_n{}^*$, then $kf(x) + lg(x)$ has the Fourier coefficients $ka_n + la_n{}^*$, $kb_n + lb_n{}^*$.

20. Using Prob. 19, find the Fourier series in Prob. 2 from those in Probs. 3 and 4.

10.3 Functions of Any Period $p = 2L$

The functions considered so far had period 2π, whereas most periodic functions in applications will have other periods. But we show that the transition from functions of period $p = 2\pi$ to those of period[9] $p = 2L$ is quite simple, basically a stretch of scale on the axis.

If a function $f(x)$ of period $p = 2L$ has a **Fourier series,** we claim that this series is

$$f(x) = a_0 + \sum_{n=1}^{\infty} \left(a_n \cos \frac{n\pi}{L} x + b_n \sin \frac{n\pi}{L} x \right) \tag{1}$$

with the **Fourier coefficients** of $f(x)$ given by the **Euler formulas**

$$\text{(2)} \quad \begin{aligned} &\text{(a)} \quad a_0 = \frac{1}{2L} \int_{-L}^{L} f(x)\, dx \\ &\text{(b)} \quad a_n = \frac{1}{L} \int_{-L}^{L} f(x) \cos \frac{n\pi x}{L}\, dx \qquad n = 1, 2, \cdots, \\ &\text{(c)} \quad b_n = \frac{1}{L} \int_{-L}^{L} f(x) \sin \frac{n\pi x}{L}\, dx \qquad n = 1, 2, \cdots. \end{aligned}$$

[9]This notation is practical since in applications, L will be the length of a vibrating string (Sec. 11.2), of a rod in heat conduction (Sec. 11.5), etc.

Proof. The idea is to derive this from Sec. 10.2 by a change of scale. We set $v = \pi x/L$, hence $x = Lv/\pi$. Then $x = \pm L$ corresponds to $v = \pm\pi$. Thus f, regarded as a function of v which we call $g(v)$,

$$f(x) = g(v),$$

has period 2π. Accordingly, by (7) and (6), Sec. 10.2, with v instead of x, this 2π-periodic function $g(v)$ has the Fourier series

$$g(v) = a_0 + \sum_{n=1}^{\infty} (a_n \cos nv + b_n \sin nv) \tag{3}$$

with coefficients

$$a_0 = \frac{1}{2\pi}\int_{-\pi}^{\pi} g(v)\, dv$$

$$a_n = \frac{1}{\pi}\int_{-\pi}^{\pi} g(v) \cos nv\, dv \tag{4}$$

$$b_n = \frac{1}{\pi}\int_{-\pi}^{\pi} g(v) \sin nv\, dv.$$

Since $v = \pi x/L$ and $g(v) = f(x)$, formula (3) gives (1). In (4) we introduce $x = Lv/\pi$ as variable of integration. Then the limits of integration $v = \pm\pi$ become $x = \pm L$. Also, $v = \pi x/L$ implies $dv = \pi\, dx/L$. Thus $dv/2\pi = dx/2L$ in a_0. Similarly, $dv/\pi = dx/L$ in a_n and b_n. Hence (4) gives (2). ∎

The interval of integration in (2) may be replaced by any interval of length $p = 2L$, for example, by the interval $0 \leqq x \leqq 2L$.

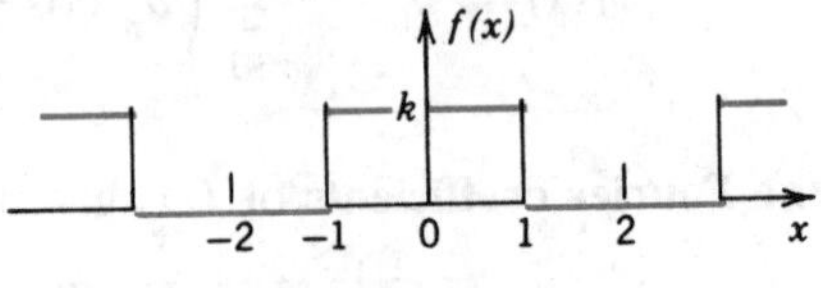

Fig. 233. Example 1

EXAMPLE 1 Periodic square wave

Find the Fourier series of the function (see Fig. 233)

$$f(x) = \begin{cases} 0 & \text{if } -2 < x < -1 \\ k & \text{if } -1 < x < 1 \\ 0 & \text{if } 1 < x < 2 \end{cases} \qquad p = 2L = 4, \quad L = 2.$$

Solution. From (2a) and (2b) we obtain

$$a_0 = \frac{1}{4}\int_{-2}^{2} f(x)\, dx = \frac{1}{4}\int_{-1}^{1} k\, dx = \frac{k}{2},$$

$$a_n = \frac{1}{2} \int_{-2}^{2} f(x) \cos \frac{n\pi x}{2}\, dx = \frac{1}{2} \int_{-1}^{1} k \cos \frac{n\pi x}{2}\, dx = \frac{2k}{n\pi} \sin \frac{n\pi}{2}.$$

Thus $a_n = 0$ if n is even and

$$a_n = 2k/n\pi \quad \text{if} \quad n = 1, 5, 9, \cdots, \qquad a_n = -2k/n\pi \quad \text{if} \quad n = 3, 7, 11, \cdots.$$

From (2c) we find that $b_n = 0$ for $n = 1, 2, \cdots$. Hence the result is

$$f(x) = \frac{k}{2} + \frac{2k}{\pi} \left(\cos \frac{\pi}{2}x - \frac{1}{3} \cos \frac{3\pi}{2}x + \frac{1}{5} \cos \frac{5\pi}{2}x - + \cdots \right). \qquad \blacksquare$$

EXAMPLE 2   **Half-wave rectifier**

A sinusoidal voltage $E \sin \omega t$, where t is time, is passed through a half-wave rectifier that clips the negative portion of the wave (Fig. 234). Find the Fourier series of the resulting periodic function

$$u(t) = \begin{cases} 0 & \text{if} \quad -L < t < 0, \\ E \sin \omega t & \text{if} \quad 0 < t < L \end{cases} \qquad p = 2L = \frac{2\pi}{\omega}, \quad L = \frac{\pi}{\omega}.$$

Solution. Since $u = 0$ when $-L < t < 0$, we obtain from (2a), with t instead of x,

$$a_0 = \frac{\omega}{2\pi} \int_{0}^{\pi/\omega} E \sin \omega t\, dt = \frac{E}{\pi}$$

and from (2b), by using formula (11) in Appendix 3.1 with $x = \omega t$ and $y = n\omega t$,

$$a_n = \frac{\omega}{\pi} \int_{0}^{\pi/\omega} E \sin \omega t \cos n\omega t\, dt = \frac{\omega E}{2\pi} \int_{0}^{\pi/\omega} [\sin (1 + n)\omega t + \sin (1 - n)\omega t]\, dt.$$

If $n = 1$, the integral on the right is zero, and if $n = 2, 3, \cdots$, we readily obtain

$$a_n = \frac{\omega E}{2\pi} \left[- \frac{\cos (1 + n)\omega t}{(1 + n)\omega} - \frac{\cos (1 - n)\omega t}{(1 - n)\omega} \right]_{0}^{\pi/\omega}$$

$$= \frac{E}{2\pi} \left(\frac{-\cos (1 + n)\pi + 1}{1 + n} + \frac{-\cos (1 - n)\pi + 1}{1 - n} \right).$$

If n is odd, this is equal to zero, and for even n we have

$$a_n = \frac{E}{2\pi} \left(\frac{2}{1 + n} + \frac{2}{1 - n} \right) = - \frac{2E}{(n - 1)(n + 1)\pi} \qquad (n = 2, 4, \cdots).$$

In a similar fashion we find from (2c) that $b_1 = E/2$ and $b_n = 0$ for $n = 2, 3, \cdots$. Consequently,

$$u(t) = \frac{E}{\pi} + \frac{E}{2} \sin \omega t - \frac{2E}{\pi} \left(\frac{1}{1 \cdot 3} \cos 2\omega t + \frac{1}{3 \cdot 5} \cos 4\omega t + \cdots \right). \qquad \blacksquare$$

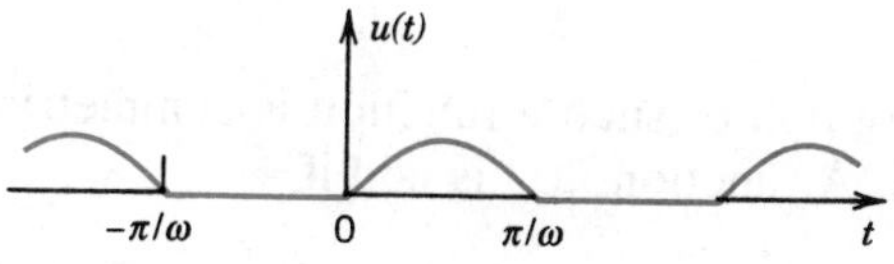

Fig. 234. Half-wave rectifier

Problem Set 10.3

Find the Fourier series of the periodic function $f(x)$, of period $p = 2L$, and sketch $f(x)$ and the first three partial sums.

1. $f(x) = -1 \quad (-1 < x < 0), \quad f(x) = 1 \quad (0 < x < 1), \quad p = 2L = 2$

2. $f(x) = 0 \quad (-1 < x < 1), \quad f(x) = 1 \quad (1 < x < 3), \quad p = 2L = 4$

3. $f(x) = 0 \quad (-2 < x < 0), \quad f(x) = 2 \quad (0 < x < 2), \quad p = 2L = 4$

4. $f(x) = x \quad (-1 < x < 1), \quad p = 2L = 2$

5. $f(x) = 1 - x^2 \quad (-1 < x < 1), \quad p = 2L = 2$

6. $f(x) = 2|x| \quad (-2 < x < 2), \quad p = 2L = 4$

7. $f(x) = 0, \quad (-1 < x < 0), \quad f(x) = x \quad (0 < x < 1), \quad p = 2L = 2$

8. $f(x) = x \quad (0 < x < 1), \quad f(x) = 1 - x \quad (1 < x < 2), \quad p = 2L = 2$

9. $f(x) = -1 \quad (-1 < x < 0), \quad f(x) = 2x \quad (0 < x < 1), \quad p = 2L = 2$

10. $f(x) = \frac{1}{2} + x \quad (-\frac{1}{2} < x < 0), \quad f(x) = \frac{1}{2} - x \quad (0 < x < \frac{1}{2}), \quad p = 2L = 1$

11. $f(x) = 3x^2 \quad (-1 < x < 1), \quad p = 2L = 2$

12. $f(x) = \pi x^3/2 \quad (-1 < x < 1), \quad p = 2L = 2$

13. $f(x) = \pi \sin \pi x \quad (0 < x < 1), \quad p = 2L = 1$

14. $f(x) = x^2/4 \quad (0 < x < 2), \quad p = 2L = 2$

15. Obtain the Fourier series in Prob. 1 directly from that in Example 1, Sec. 10.2.

16. Obtain the Fourier series in Prob. 11 directly from that in Prob. 9, Sec. 10.2.

17. Obtain the Fourier series in Prob. 3 directly from that in Example 1.

18. Find the Fourier series of the periodic function that is obtained by passing the voltage $v(t) = k \cos 100\pi t$ through a half-wave rectifier.

19. Show that each term in (1) has the period $p = 2L$.

20. Show that in (2) the interval of integration may be replaced by any interval of length $p = 2L$.

10.4 Even and Odd Functions

The function in Example 1 of the last section was odd and had only sine terms in its Fourier series, no cosine terms. This is typical. In fact, unnecessary work (and corresponding sources of errors) in determining Fourier coefficients can be avoided if a function is odd or even.

We first remember that a function $y = g(x)$ is **even** if

$$g(-x) = g(x) \qquad \text{for all } x.$$

The graph of such a function is symmetric with respect to the y-axis (Fig. 235). A function $h(x)$ is **odd** if

$$h(-x) = -h(x) \qquad \text{for all } x.$$

(See Fig. 236.) The function $\cos nx$ is even, while $\sin nx$ is odd.

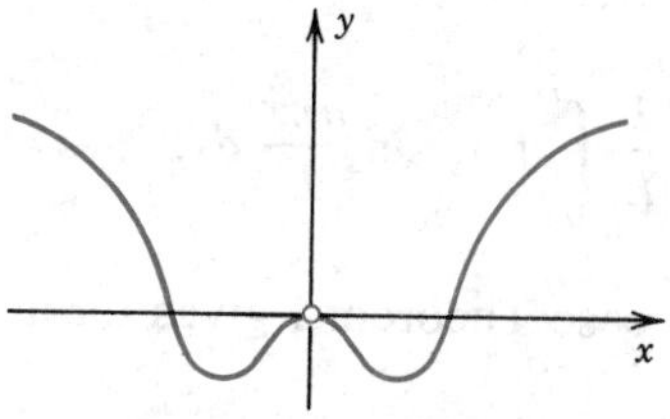

Fig. 235. Even function

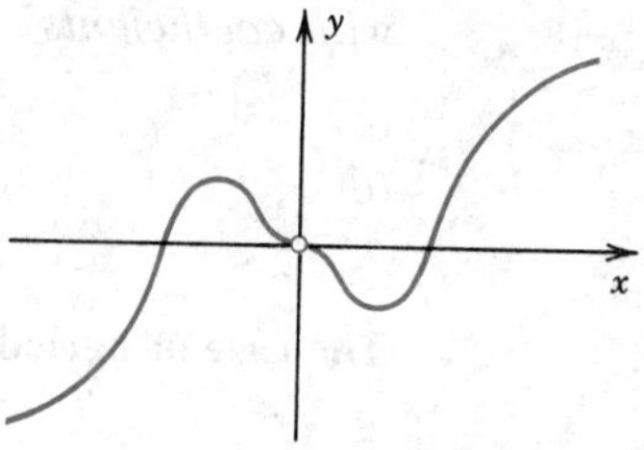

Fig. 236. Odd function

If $g(x)$ is an even function, then

$$\int_{-L}^{L} g(x)\,dx = 2\int_{0}^{L} g(x)\,dx \qquad (g \text{ even}). \tag{1}$$

If $h(x)$ is an odd function, then

$$\int_{-L}^{L} h(x)\,dx = 0 \qquad (h \text{ odd}). \tag{2}$$

Formulas (1) and (2) are obvious from the graphs of g and h, and we leave the formal proofs to the student.

The product $q = gh$ of an even function g and an odd function h is odd, because

$$q(-x) = g(-x)h(-x) = g(x)[-h(x)] = -q(x).$$

Hence if $f(x)$ is even, then the integrand $f \sin(n\pi x/L)$ in (2c), Sec. 10.3, is odd, and $b_n = 0$. Similarly, if $f(x)$ is odd, then $f \cos(n\pi x/L)$ in (2b), Sec. 10.3, is odd, and $a_n = 0$. From this and (1) we obtain the following theorem.

Theorem 1 (Fourier series of even and odd functions)
The Fourier series of an even function of period $2L$ is a **"Fourier cosine series"**

$$f(x) = a_0 + \sum_{n=1}^{\infty} a_n \cos\frac{n\pi}{L}x \qquad (f \text{ even}) \tag{3}$$

with coefficients

$$a_0 = \frac{1}{L}\int_0^L f(x)\,dx, \qquad a_n = \frac{2}{L}\int_0^L f(x)\cos\frac{n\pi x}{L}\,dx, \qquad n = 1, 2, \cdots. \tag{4}$$

The Fourier series of an odd function of period $2L$ is a **"Fourier sine series"**

$$f(x) = \sum_{n=1}^{\infty} b_n \sin\frac{n\pi}{L}x \qquad (f \text{ odd}) \tag{5}$$

with coefficients

$$(6) \qquad b_n = \frac{2}{L}\int_0^L f(x) \sin \frac{n\pi x}{L}\, dx. \qquad \blacksquare$$

The case of period 2π. In this case Theorem 1 gives for an even function

$$(3^*) \qquad f(x) = a_0 + \sum_{n=1}^{\infty} a_n \cos nx \qquad (f \text{ even})$$

with coefficients

$$(4^*) \qquad a_0 = \frac{1}{\pi}\int_0^{\pi} f(x)\, dx, \qquad a_n = \frac{2}{\pi}\int_0^{\pi} f(x) \cos nx\, dx, \qquad n = 1, 2, \cdots$$

and for an odd function

$$(5^*) \qquad f(x) = \sum_{n=1}^{\infty} b_n \sin nx \qquad (f \text{ odd})$$

with coefficients

$$(6^*) \qquad b_n = \frac{2}{\pi}\int_0^{\pi} f(x) \sin nx\, dx, \qquad n = 1, 2, \cdots .$$

For instance, $f(x)$ in Example 1, Sec. 10.2, is odd and, therefore, is represented by a Fourier sine series.

Further simplifications result from the following property (already mentioned in Prob. 19 of Problem Set 10.2):

Theorem 2 (Sum of functions)

The Fourier coefficients of a sum $f_1 + f_2$ are the sums of the corresponding Fourier coefficients of f_1 and f_2.

The Fourier coefficients of cf are c times the corresponding Fourier coefficients of f.

EXAMPLE 1 Rectangular pulse

The function $f^*(x)$ in Fig. 237 is the sum of the function $f(x)$ in Example 1 of Sec. 10.2 and the constant k. Hence, from that example and Theorem 2 we conclude that

$$f^*(x) = k + \frac{4k}{\pi}\left(\sin x + \frac{1}{3}\sin 3x + \frac{1}{5}\sin 5x + \cdots\right). \qquad \blacksquare$$

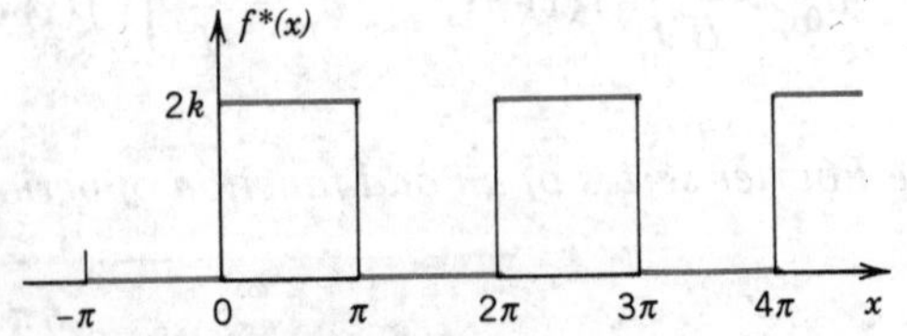

Fig. 237. Example 1

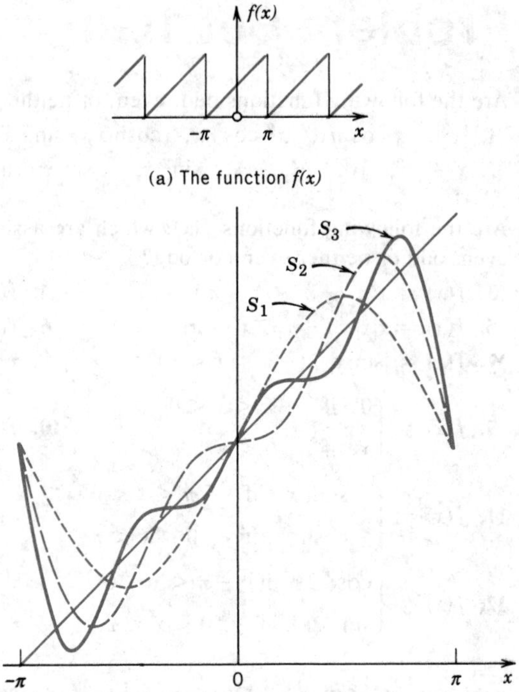

(a) The function $f(x)$

(b) Partial sums $S_n(x)$

Fig. 238. Example 2

EXAMPLE 2 **Saw tooth wave**

Find the Fourier series of the function (Fig. 238a)

$$f(x) = x + \pi \quad \text{if} \quad -\pi < x < \pi \qquad \text{and} \qquad f(x + 2\pi) = f(x).$$

Solution. We may write

$$f = f_1 + f_2$$

where

$$f_1 = x \qquad \text{and} \qquad f_2 = \pi.$$

The Fourier coefficients of f_2 are zero, except for the first one (the constant term), which is π. Hence, by Theorem 2, the Fourier coefficients a_n, b_n are those of f_1, except for a_0, which is π. Since f_1 is odd, $a_n = 0$ for $n = 1, 2, \cdots$, and

$$b_n = \frac{2}{\pi}\int_0^{\pi} f_1(x) \sin nx\, dx = \frac{2}{\pi}\int_0^{\pi} x \sin nx\, dx.$$

Integrating by parts we obtain

$$b_n = \frac{2}{\pi}\left[\frac{-x \cos nx}{n}\Bigg|_0^{\pi} + \frac{1}{n}\int_0^{\pi} \cos nx\, dx\right] = -\frac{2}{n} \cos n\pi.$$

Hence $b_1 = 2$, $b_2 = -2/2$, $b_3 = 2/3$, $b_4 = -2/4$, $\cdots$, and the Fourier series of $f(x)$ is

$$f(x) = \pi + 2\left(\sin x - \frac{1}{2}\sin 2x + \frac{1}{3}\sin 3x - + \cdots\right).$$ ■

Problem Set 10.4

Are the following functions odd, even, or neither odd nor even?

1. $|x^3|$, $x \cos nx$, $x^2 \cos nx$, $\cosh x$, $\sinh x$, $\sin x + \cos x$, $x|x|$

2. $x + x^2$, $|x|$, e^x, e^{x^2}, $\sin^2 x$, $x \sin x$, $\ln x$, $x \cos x$, $e^{-|x|}$

Are the following functions $f(x)$, which are assumed to be periodic, of period 2π, even, odd or neither even nor odd?

3. $f(x) = x^3 \quad (-\pi < x < \pi)$

4. $f(x) = x^4 \quad (0 < x < 2\pi)$

5. $f(x) = x|x| \quad (-\pi < x < \pi)$

6. $f(x) = e^{-|x|} \quad (-\pi < x < \pi)$

7. $f(x) = |\sin x| \quad (-\pi < x < \pi)$

8. $x^3 - x$

9. $f(x) = \begin{cases} 0 & \text{if } -\pi < x < 0 \\ x & \text{if } 0 < x < \pi \end{cases}$

10. $f(x) = \begin{cases} 0 & \text{if } 1 < x < 2\pi - 1 \\ x & \text{if } -1 < x < 1 \end{cases}$

11. $f(x) = \begin{cases} \sinh x & \text{if } -\pi < x < 0 \\ -\cosh x & \text{if } 0 < x < \pi \end{cases}$

12. $f(x) = \begin{cases} \cos^2 2x & \text{if } -\pi < x < 0 \\ \sin^2 2x & \text{if } 0 < x < \pi \end{cases}$

Represent the following functions as the sum of an even and an odd function.

13. $1/(1 - x)$ **14.** $1/(1 - x)^2$ **15.** e^{kx} **16.** $x/(x + 1)$

17. Prove Theorem 2.

18. Find all functions that are both even and odd.

19. Show that the familiar identity $\sin^3 x = \frac{3}{4}\sin x - \frac{1}{4}\sin 3x$ can be interpreted as a Fourier series expansion, and the same holds for the identity $\cos^3 x = \frac{3}{4}\cos x + \frac{1}{4}\cos 3x$.

Prove:

20. The sum and the product of even functions are even functions.

21. The sum of odd functions is odd. The product of two odd functions is even.

22. If $f(x)$ is odd, then $|f(x)|$ and $f^2(x)$ are even functions.

23. If $f(x)$ is even, then $|f(x)|$, $f^2(x)$, and $f^3(x)$ are even functions.

24. If $g(x)$ is defined for all x, then the function $p(x) = [g(x) + g(-x)]/2$ is even and the function $q(x) = [g(x) - g(-x)]/2$ is odd.

Find the Fourier series of the following functions, which are assumed to have the period 2π. *Hint.* Use that some of these functions are even or odd.

25. $f(x) = \begin{cases} k & \text{if } -\pi/2 < x < \pi/2 \\ 0 & \text{if } \pi/2 < x < 3\pi/2 \end{cases}$

26. $f(x) = \begin{cases} x & \text{if } 0 < x < \pi \\ \pi - x & \text{if } \pi < x < 2\pi \end{cases}$

27. $f(x) = \begin{cases} x & \text{if } -\pi/2 < x < \pi/2 \\ \pi - x & \text{if } \pi/2 < x < 3\pi/2 \end{cases}$

28. $f(x) = \begin{cases} -x & \text{if } -\pi < x < 0 \\ x & \text{if } 0 < x < \pi \end{cases}$

29. $f(x) = \begin{cases} x^2 & \text{if } -\pi/2 < x < \pi/2 \\ \pi^2/4 & \text{if } \pi/2 < x < 3\pi/2 \end{cases}$

30. $f(x) = \begin{cases} -x^2 & \text{if } -\pi < x < 0 \\ x^2 & \text{if } 0 < x < \pi \end{cases}$

31. $f(x) = x^2/4 \quad (-\pi < x < \pi)$

32. $f(x) = x(\pi^2 - x^2) \quad (-\pi < x < \pi)$

Show that

33. $1 - \frac{1}{3} + \frac{1}{5} - \frac{1}{7} + - \cdots = \frac{\pi}{4}$ (Use Prob. 25.)

34. $1 + \frac{1}{4} + \frac{1}{9} + \frac{1}{16} + \frac{1}{25} + \cdots = \frac{\pi^2}{6}$ (Use Prob. 31.)

35. $1 - \frac{1}{4} + \frac{1}{9} - \frac{1}{16} + - \cdots = \frac{\pi^2}{12}$ (Use Prob. 31.)

10.5 Half-Range Expansions

In various applications there is a practical need to use Fourier series in connection with functions $f(x)$ that are given on some interval only, say, $0 \leqq x \leqq L$, as in Fig. 239a. Typical cases follow in the next chapter (Secs. 11.3 and 11.5). We could extend $f(x)$ periodically with period L and then represent the extended function by a Fourier series, which in general would involve both cosine *and* sine terms. We can do better and always get a cosine series by first extending $f(x)$ from $0 \leqq x \leqq L$ as an *even* function on the range (the interval) $-L \leqq x \leqq L$, as in Fig. 239b, and then extend this new function as a periodic function of period $2L$ and, since it is even, represent it by a Fourier cosine series. Or we can extend $f(x)$ from $0 \leqq x \leqq L$ as an

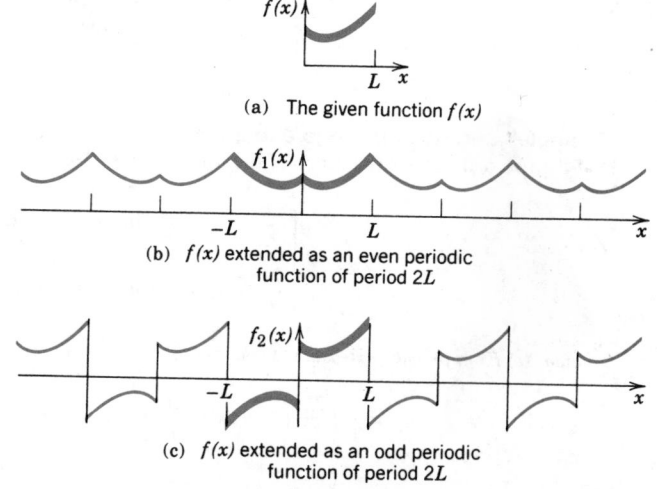

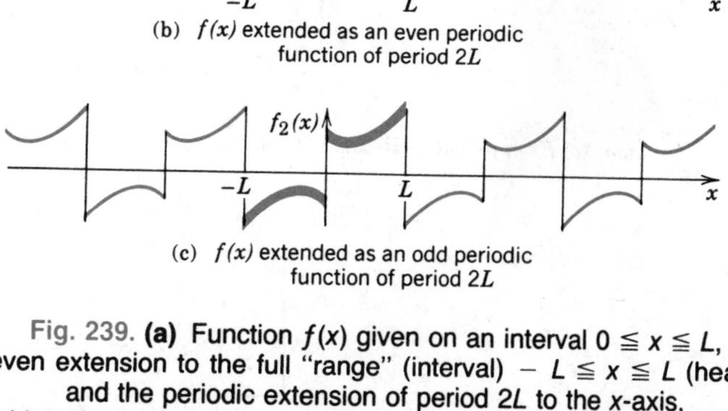

Fig. 239. **(a)** Function $f(x)$ given on an interval $0 \leqq x \leqq L$,
(b) its even extension to the full "range" (interval) $-L \leqq x \leqq L$ (heavy curve) and the periodic extension of period $2L$ to the x-axis,
(c) its odd extension to $-L \leqq x \leqq L$ (heavy curve) and the periodic extension of period $2L$ to the x-axis

odd function on $-L \leqq x \leqq L$, as in Fig. 239c, and then extend this new function as a periodic function of period $2L$ and, since it is odd, represent it by a Fourier sine series. These two series are called the two **half-range expansions** of the function $f(x)$, which is given only on "half the range" (half the periodicity interval of these series). The form of these series is given in Sec. 10.4. The cosine half-range expansion is [see (3), (4), Sec. 10.4]

$$f(x) = a_0 + \sum_{n=1}^{\infty} a_n \cos \frac{n\pi}{L} x \tag{1}$$

where

$$a_0 = \frac{1}{L} \int_0^L f(x)\, dx, \qquad a_n = \frac{2}{L} \int_0^L f(x) \cos \frac{n\pi x}{L}\, dx, \qquad n = 1, 2, \cdots. \tag{2}$$

The sine half-range expansion is [see (5), (6), Sec. 10.4]

$$f(x) = \sum_{n=1}^{\infty} b_n \sin \frac{n\pi}{L} x \tag{3}$$

where

$$b_n = \frac{2}{L} \int_0^L f(x) \sin \frac{n\pi x}{L}\, dx, \qquad n = 1, 2, \cdots. \tag{4}$$

EXAMPLE 1 **"Triangle" and its half-range expansions**

Find the two half-range expansions of the function (Fig. 240)

$$f(x) = \begin{cases} \dfrac{2k}{L} x & \text{if } 0 < x < \dfrac{L}{2} \\[2mm] \dfrac{2k}{L}(L - x) & \text{if } \dfrac{L}{2} < x < L. \end{cases}$$

Solution. ***(a) Even periodic extension.*** From (4), Sec. 10.4, we obtain

$$a_0 = \frac{1}{L}\left[\frac{2k}{L}\int_0^{L/2} x\, dx + \frac{2k}{L}\int_{L/2}^{L} (L - x)\, dx\right] = \frac{k}{2},$$

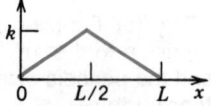

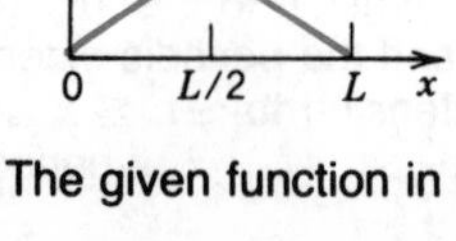

Fig. 240. The given function in Example 1

$$a_n = \frac{2}{L}\left[\frac{2k}{L}\int_0^{L/2} x \cos\frac{n\pi}{L}x\,dx + \frac{2k}{L}\int_{L/2}^{L}(L - x)\cos\frac{n\pi}{L}x\,dx\right].$$

Now by integration by parts,

$$\int_0^{L/2} x\cos\frac{n\pi}{L}x\,dx = \frac{Lx}{n\pi}\sin\frac{n\pi}{L}x\,\Big|_0^{L/2} - \frac{L}{n\pi}\int_0^{L/2}\sin\frac{n\pi}{L}x\,dx$$

$$= \frac{L^2}{2n\pi}\sin\frac{n\pi}{2} + \frac{L^2}{n^2\pi^2}\left(\cos\frac{n\pi}{2} - 1\right).$$

Similarly,

$$\int_{L/2}^{L}(L - x)\cos\frac{n\pi}{L}x\,dx = -\frac{L^2}{2n\pi}\sin\frac{n\pi}{2} - \frac{L^2}{n^2\pi^2}\left(\cos n\pi - \cos\frac{n\pi}{2}\right).$$

By inserting these two results we obtain

$$a_n = \frac{4k}{n^2\pi^2}\left(2\cos\frac{n\pi}{2} - \cos n\pi - 1\right).$$

Thus,

$$a_2 = -16k/2^2\pi^2, \qquad a_6 = -16k/6^2\pi^2, \qquad a_{10} = -16k/10^2\pi^2, \cdots,$$

and $a_n = 0$ if $n \neq 2, 6, 10, 14, \cdots$. Hence the first half-range expansion of $f(x)$ is

$$f(x) = \frac{k}{2} - \frac{16k}{\pi^2}\left(\frac{1}{2^2}\cos\frac{2\pi}{L}x + \frac{1}{6^2}\cos\frac{6\pi}{L}x + \cdots\right).$$

This series represents the even periodic extension of the given function $f(x)$, of period $2L$, shown in Fig. 241a.

(b) Odd periodic extension. Similarly, from (6), Sec. 10.4, we obtain

$$b_n = \frac{8k}{n^2\pi^2}\sin\frac{n\pi}{2}. \tag{5}$$

Hence the other half-range expansion of $f(x)$ is

$$f(x) = \frac{8k}{\pi^2}\left(\frac{1}{1^2}\sin\frac{\pi}{L}x - \frac{1}{3^2}\sin\frac{3\pi}{L}x + \frac{1}{5^2}\sin\frac{5\pi}{L}x - + \cdots\right).$$

This series represents the odd periodic extension of $f(x)$, of period $2L$, shown in Fig. 241b. ■

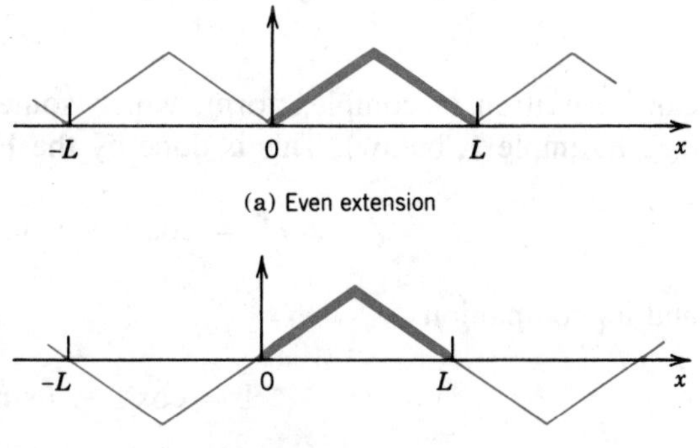

(a) Even extension

(b) Odd extension

Fig. 241. Periodic extensions of $f(x)$ in Example 1

Problem Set 10.5

Represent the following functions $f(x)$ by a Fourier sine series and sketch the corresponding periodic extension of $f(x)$.

1. $f(x) = k \quad (0 < x < L)$

2. $f(x) = kx \quad (0 < x < L)$

3. $f(x) = x^2 \quad (0 < x < L)$

4. $f(x) = 1 - (2/L)x \quad (0 < x < L)$

5. $f(x) = L - x \quad (0 < x < L)$

6. $f(x) = x^3 \quad (0 < x < L)$

7. $f(x) = \begin{cases} x & \text{if} \quad 0 < x < \pi/2 \\ \pi/2 & \text{if} \quad \pi/2 < x < \pi \end{cases}$

8. $f(x) = \begin{cases} \pi/2 & \text{if} \quad 0 < x < \pi/2 \\ \pi - x & \text{if} \quad \pi/2 < x < \pi \end{cases}$

9. $f(x) = \begin{cases} x & \text{if} \quad 0 < x < L/2 \\ L - x & \text{if} \quad L/2 < x < L \end{cases}$

10. $f(x) = \begin{cases} x & \text{if} \quad 0 < x < L/2 \\ L/2 & \text{if} \quad L/2 < x < L \end{cases}$

Represent the following functions $f(x)$ by a Fourier cosine series and sketch the corresponding periodic extension of $f(x)$.

11. $f(x) = x \quad (0 < x < L)$

12. $f(x) = 1 \quad (0 < x < L)$

13. $f(x) = x^2 \quad (0 < x < L)$

14. $f(x) = \sin^2 3x \quad (0 < x < \pi)$

15. $f(x) = \begin{cases} 0 & \text{if} \quad 0 < x < L/2 \\ 1 & \text{if} \quad L/2 < x < L \end{cases}$

16. $f(x) = \begin{cases} 1 & \text{if} \quad 0 < x < L/2 \\ 0 & \text{if} \quad L/2 < x < L \end{cases}$

17. $f(x) = e^x \quad (0 < x < L)$

18. $f(x) = x^3 \quad (0 < x < L)$

19. $f(x) = \sin \dfrac{\pi x}{L} \quad (0 < x < L)$

20. $f(x) = \dfrac{\pi}{4} \sin \dfrac{\pi x}{2L} \quad (0 < x < L)$

10.6 Complex Fourier Series *Optional*

In this optional section we show that the Fourier series

$$f(x) = a_0 + \sum_{n=1}^{\infty} (a_n \cos nx + b_n \sin nx) \tag{1}$$

can be written in complex form, which sometimes simplifies calculations (see Example 1, below). This is done by the Euler formula (8), Sec. 2.3,

$$e^{it} = \cos t + i \sin t$$

and its companion

$$e^{-it} = \cos t - i \sin t$$

[obtained from $\cos(-t) = \cos t$, $\sin(-t) = -\sin t$] with $t = nx$, that is,

$$e^{inx} = \cos nx + i \sin nx, \tag{2}$$

$$e^{-inx} = \cos nx - i \sin nx. \tag{3}$$

By addition and division by 2 we get

$$\cos nx = \frac{1}{2}(e^{inx} + e^{-inx}). \tag{4}$$

Subtraction and division by $2i$ gives

$$\sin nx = \frac{1}{2i}(e^{inx} - e^{-inx}). \tag{5}$$

From this, using $1/i = -i$, we have in (1)

$$\begin{aligned} a_n \cos nx + b_n \sin nx &= \frac{1}{2} a_n(e^{inx} + e^{-inx}) + \frac{1}{2i} b_n(e^{inx} - e^{-inx}) \\ &= \tfrac{1}{2}(a_n - ib_n)e^{inx} + \tfrac{1}{2}(a_n + ib_n)e^{-inx}. \end{aligned}$$

With this, (1) becomes

$$f(x) = c_0 + \sum_{n=1}^{\infty} (c_n e^{inx} + k_n e^{-inx}) \tag{6}$$

where $c_0 = a_0$, and by (1)–(3) and the Euler formulas (6), Sec. 10.2,

$$\begin{aligned} c_n &= \frac{1}{2}(a_n - ib_n) = \frac{1}{2\pi}\int_{-\pi}^{\pi} f(x)e^{-inx}\,dx, \\ k_n &= \frac{1}{2}(a_n + ib_n) = \frac{1}{2\pi}\int_{-\pi}^{\pi} f(x)e^{inx}\,dx, \end{aligned} \qquad n = 1, 2, \cdots. \tag{7}$$

Finally, if we introduce the notation $k_n = c_{-n}$, we obtain from (6) and (7)

$$\begin{aligned} f(x) &= \sum_{n=-\infty}^{\infty} c_n e^{inx}, \\ c_n &= \frac{1}{2\pi}\int_{-\pi}^{\pi} f(x)e^{-inx}\,dx, \end{aligned} \qquad n = 0, \pm 1, \pm 2, \cdots. \tag{8}$$

This is the so-called **complex form of the Fourier series** or, more briefly, the **complex Fourier series**, of $f(x)$, and the c_n are called the **complex Fourier coefficients** of $f(x)$.

It is interesting that (8) can be derived independently as follows. Multiplication of the series in (8) by e^{-imx} with fixed integer m and termwise integration from $-\pi$ to π (allowed, for instance, in the case of uniform convergence) gives

$$\int_{-\pi}^{\pi} f(x)e^{-imx}\,dx = \sum_{n=-\infty}^{\infty} c_n \int_{-\pi}^{\pi} e^{i(n-m)x}\,dx$$

When $n = m$, the integrand is $e^0 = 1$, and the integral equals 2π. This gives

$$\int_{-\pi}^{\pi} f(x)e^{-imx}\,dx = 2\pi c_m, \tag{9}$$

provided the other integrals are zero, which is true by (5),

$$\int_{-\pi}^{\pi} e^{i(n-m)x}\,dx = \frac{1}{i(n-m)}\left(e^{i(n-m)\pi} - e^{-i(n-m)\pi}\right)$$

$$= \frac{1}{i(n-m)}\, 2i \sin (n-m)\pi = 0.$$

Now writing n for m in (9) gives the coefficient formula in (8). ■

For a function of period $2L$, our reasoning gives the **complex Fourier series**

$$f(x) = \sum_{n=-\infty}^{\infty} c_n e^{in\pi x/L}, \qquad c_n = \frac{1}{2L}\int_{-L}^{L} f(x)e^{-in\pi x/L}\,dx. \tag{10}$$

EXAMPLE 1 Complex Fourier series

Find the complex Fourier series of $f(x) = e^x$ if $-\pi < x < \pi$ and $f(x + 2\pi) = f(x)$ and obtain from it the usual Fourier series.

Solution. By (8),

$$c_n = \frac{1}{2\pi}\int_{-\pi}^{\pi} e^x e^{-inx}\,dx = \frac{1}{2\pi}\frac{1}{1-in}\, e^{(1-in)x}\Big|_{-\pi}^{\pi}.$$

Multiplication of numerator and denominator by $1 + in$ and the use of $e^{in\pi} = e^{-in\pi} = (-1)^n$ gives

$$c_n = \frac{1}{2\pi}\frac{1+in}{1+n^2}(-1)^n(e^{\pi} - e^{-\pi}).$$

The last factor $(\cdots)$ is $2 \sinh \pi$, so that the complex Fourier series is

$$e^x = \frac{\sinh \pi}{\pi}\sum_{n=-\infty}^{\infty} (-1)^n \frac{1+in}{1+n^2} e^{inx} \qquad (-\pi < x < \pi). \tag{11}$$

Here, by (2),

$$(1 + in)(\cos nx + i \sin nx) = (\cos nx - n \sin nx) + i(n \cos nx + \sin nx).$$

The corresponding term with $-n$ instead of n is (note that $\cos(-nx) = \cos nx$ and $\sin(-nx) = -\sin nx$)

$$(1 - in)(\cos nx - i \sin nx) = (\cos nx - n \sin nx) - i(n \cos nx + \sin nx).$$

The imaginary parts cancel if we add the two terms, so that their sum is

$$2(\cos nx - n \sin nx), \qquad n = 1, 2, \cdots.$$

For $n = 0$ we get 1 (not 2) because there is only one term. Hence the real Fourier series is

$$(12) \qquad e^x = \frac{2 \sinh \pi}{\pi}\left[\frac{1}{2} - \frac{1}{1+1^2}(\cos x - \sin x) + \frac{1}{1+2^2}(\cos 2x - 2 \sin 2x) - + \cdots\right]$$

where $-\pi < x < \pi$. ∎

Problem Set 10.6

1. Show that the complex Fourier coefficients of an odd function are pure imaginary and those of an even function are real.
2. Show that $a_0 = c_0, \quad a_n = c_n + c_{-n}, \quad b_n = i(c_n - c_{-n}), \quad n = 1, 2, \cdots.$
3. Find the complex Fourier series of $f(x) = -1$ if $-\pi < x < 0$, $f(x) = 1$ if $0 < x < \pi$.
4. Convert the Fourier series in Prob. 3 to real form.
5. Find the complex Fourier series of $f(x) = x \quad (-\pi < x < \pi)$.
6. Find the complex Fourier series of $f(x) = 0$ if $-\pi < x < 0$, $f(x) = 1$ if $0 < x < \pi$.
7. Find the complex Fourier series of $f(x) = x \quad (0 < x < 2\pi)$.
8. Convert the Fourier series in Prob. 7 to real form.
9. Find the complex Fourier series of $f(x) = x^2 \quad (-\pi < x < \pi)$.
10. Convert the Fourier series in Prob. 9 to real form.

10.7 Forced Oscillations

Fourier series have important applications in differential equations. We show this for a basic problem involving an ordinary differential equation. (Numerous applications to partial differential equations follow in Chap. 11.)

From Sec. 2.11 we know that forced oscillations of a body of mass m on a spring of modulus k are governed by the equation

$$(1) \qquad my'' + cy' + ky = r(t),$$

where $y = y(t)$ is the displacement from rest, c the damping constant, and $r(t)$ the external force depending on time t. Figure 242 on p. 592 shows the model and Fig. 243 its electrical analog, an RLC-circuit governed by

$$(1^*) \qquad LI'' + RI' + \frac{1}{C}I = E'(t)$$

(see Sec. 2.12).

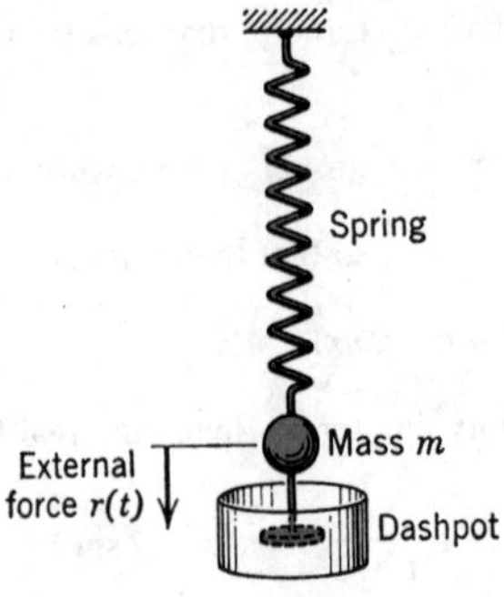

Fig. 242. Vibrating system under consideration

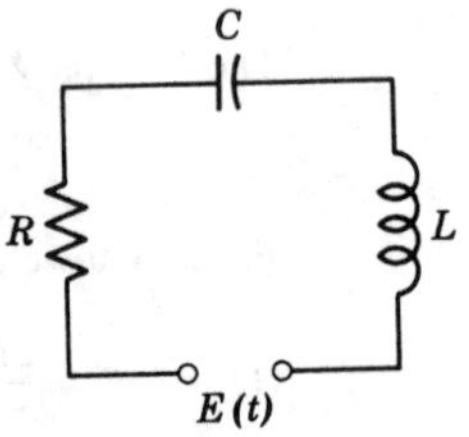

Fig. 243. Electrical analog of the system in Fig. 242 (*RLC*-circuit)

We consider (1). If $r(t)$ is a sine or cosine function and if there is damping ($c > 0$), then the steady-state solution is a harmonic oscillation with frequency equal to that of $r(t)$. However, if $r(t)$ is not a pure sine or cosine function but is any other periodic function, then the steady-state solution will be a superposition of harmonic oscillations with frequencies equal to that of $r(t)$ and integer multiples of the latter. And if one of these frequencies is close to the resonant frequency of the vibrating system (see Sec. 2.11), then the corresponding oscillation may be the dominant part of the response of the system to the external force. This is what the use of Fourier series will show us; of course, this is quite surprising to an observer unfamiliar with the underlying theory, which is highly important in the study of vibrating systems and resonance. Let us discuss the whole situation in terms of a typical example.

EXAMPLE 1 Forced oscillations under a nonsinusoidal periodic driving force

In (1), let $m = 1$ (gm), $c = 0.02$ (gm/sec), and $k = 25$ (gm/sec^2), so that (1) becomes

$$y'' + 0.02y' + 25y = r(t) \tag{2}$$

where $r(t)$ is measured in gm · cm/sec^2. Let (Fig. 244)

$$r(t) = \begin{cases} t + \dfrac{\pi}{2} & \text{if} \quad -\pi < t < 0, \\ -t + \dfrac{\pi}{2} & \text{if} \quad 0 < t < \pi, \end{cases} \qquad r(t + 2\pi) = r(t).$$

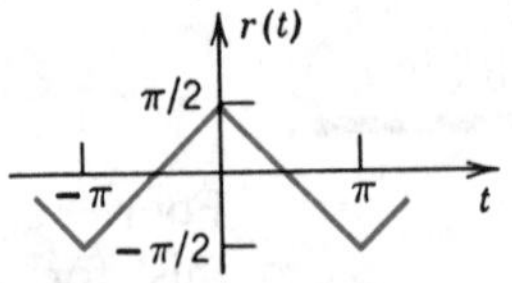

Fig. 244. Force in Example 1

Find the steady-state solution $y(t)$.

Solution. We represent $r(t)$ by a Fourier series, finding

$$r(t) = \frac{4}{\pi}\left(\cos t + \frac{1}{3^2}\cos 3t + \frac{1}{5^2}\cos 5t + \cdots\right) \tag{3}$$

(take $\pi/2$ minus the answer to Prob. 11 in Problem Set 10.5 with $L = \pi$). Then we consider the differential equation

$$y'' + 0.02y' + 25y = \frac{4}{n^2\pi}\cos nt \qquad (n = 1, 3, \cdots) \tag{4}$$

whose right side is a single term of the series (3). From Sec. 2.11 we know that the steady-state solution $y_n(t)$ of (4) is of the form

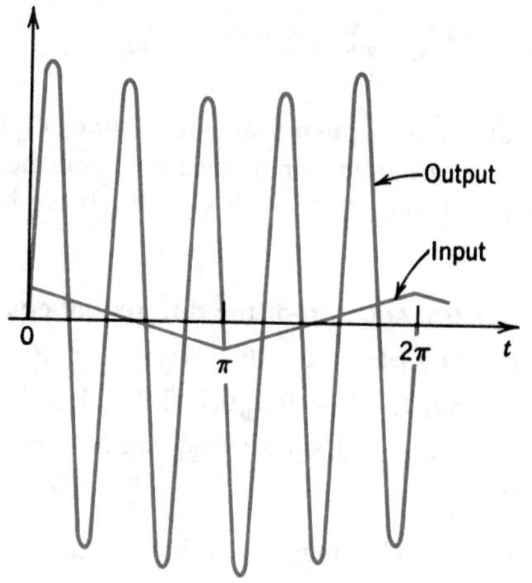

Fig. 245. Input and steady-state output in Example 1

(5) $$y_n = A_n \cos nt + B_n \sin nt.$$

By substituting this into (4) we find that

(6) $$A_n = \frac{4(25 - n^2)}{n^2 \pi D}, \qquad B_n = \frac{0.08}{n\pi D}, \qquad \text{where} \qquad D = (25 - n^2)^2 + (0.02n)^2.$$

Since the differential equation (2) is linear, we may expect the steady-state solution to be

(7) $$y = y_1 + y_3 + y_5 + \cdots$$

where y_n is given by (5) and (6). In fact, this follows readily by substituting (7) into (2) and using the Fourier series of $r(t)$, provided that termwise differentiation of (7) is permissible. (Readers already familiar with the notion of uniform convergence [Sec. 14.6] may prove that (7) may be differentiated term by term.)

From (6) we find that the amplitude of (5) is

$$C_n = \sqrt{A_n^{\,2} + B_n^{\,2}} = \frac{4}{n^2 \pi \sqrt{D}}.$$

Numerical values are

$$C_1 = 0.0530$$

$$C_3 = 0.0088$$

$$C_5 = 0.5100$$

$$C_7 = 0.0011$$

$$C_9 = 0.0003.$$

For $n = 5$ the quantity D is very small, the denominator of C_5 is small, and C_5 is so large that y_5 is the dominating term in (7). This implies that the steady-state motion is almost a harmonic oscillation whose frequency equals five times that of the exciting force (Fig. 245). ∎

The application of Fourier series to more general vibrating systems, heat conduction, and other problems follows in Chap. 11.

Problem Set 10.7

1. What would happen to the amplitudes C_n in Example 1 (and, accordingly, to the form of the vibrations) if we changed the spring constant to the value 9? If we took a stiffer spring with $k = 49$? If we increased the damping?

Find a general solution of the differential equation $y'' + \omega^2 y = r(t)$, where

2. $r(t) = \cos \alpha t + \cos \beta t \qquad (\omega^2 \neq \alpha^2, \beta^2)$

3. $r(t) = \sin t, \quad \omega = 0.5, 0.7, 0.9, 1.1, 1.5, 2.0, 10.0$

4. $r(t) = \sin t + \frac{1}{9}\sin 3t + \frac{1}{25}\sin 5t, \quad \omega = 0.5, 0.9, 1.1, 2, 2.9, 3.1, 4, 4.9, 5.1, 6, 8$

5. $r(t) = \sum_{n=1}^{N} a_n \cos nt, \qquad |\omega| \neq 1, 2, \cdots, N$

6. $r(t) = \dfrac{\pi}{4}|\sin t|$ when $-\pi < t < \pi$ and $r(t + 2\pi) = r(t)$, $|\omega| \neq 0, 2, 4, \cdots$

7. $r(t) = \begin{cases} t + \pi & \text{if } -\pi < t < 0 \\ -t + \pi & \text{if } 0 < t < \pi \end{cases}$ and $r(t + 2\pi) = r(t)$, $|\omega| \neq 0, 1, 3, \cdots$

8. $r(t) = \begin{cases} t & \text{if } -\pi/2 < t < \pi/2 \\ \pi - t & \text{if } \pi/2 < t < 3\pi/2 \end{cases}$ and $r(t + 2\pi) = r(t)$, $|\omega| \neq 1, 3, 5, \cdots$

Find the steady-state oscillation corresponding to $y'' + cy' + y = r(t)$, where $c > 0$ and

9. $r(t) = a_n \cos nt$

10. $r(t) = \sum_{n=1}^{N} b_n \sin nt$

11. $r(t) = \sin 3t$

12. $r(t) = \begin{cases} \pi t/4 & \text{if } -\pi/2 < t < \pi/2 \\ \pi(\pi - t)/4 & \text{if } \pi/2 < t < 3\pi/2 \end{cases}$ and $r(t + 2\pi) = r(t)$

13. $r(t) = \dfrac{t}{12}(\pi^2 - t^2)$ if $-\pi < t < \pi$ and $r(t + 2\pi) = r(t)$

14. ***(RLC-circuit)*** Find the steady-state current $I(t)$ in the *RLC*-circuit in Fig. 243, where $R = 100$ ohms, $L = 10$ henrys, $C = 10^{-2}$ farad,

$$E(t) = \begin{cases} 100(\pi t + t^2) & \text{if } -\pi < t < 0 \\ 100(\pi t - t^2) & \text{if } 0 < t < \pi \end{cases} \quad \text{and} \quad E(t + 2\pi) = E(t).$$

Proceed as follows. Develop $E(t)$ in a Fourier series. $I(t)$ will appear in the form of a trigonometric series. Find the general formulas for the coefficients of this series. Compute numerical values of the first few coefficients. Graph the sum of the first few terms of that series.

15. Same task as in Prob. 14 with R, L, C as before, and $E(t) = 200t(\pi^2 - t^2)$ volts if $-\pi < t < \pi$ and $E(t + 2\pi) = E(t)$.

10.8 Approximation by Trigonometric Polynomials

A main field of application of Fourier series is in differential equations, as we have said. Another area of practical interest in which Fourier series play a major role is the approximation of functions by simpler functions, known as **approximation theory,** as we shall now explain.

Let $f(x)$ be a function of period 2π that can be represented by a Fourier series. Then the Nth partial sum of this series is an approximation to $f(x)$:

$$f(x) \approx a_0 + \sum_{n=1}^{N} (a_n \cos nx + b_n \sin nx). \tag{1}$$

It is natural to ask whether (1) is the "best" approximation to f by a **trigonometric polynomial** *of degree* N (N fixed), that is, a function of the form

$$F(x) = \alpha_0 + \sum_{n=1}^{N} (\alpha_n \cos nx + \beta_n \sin nx), \tag{2}$$

where "best" means that the "error" of the approximation is minimum.

Of course, we must first define what we mean by the error E of such an approximation. We want to choose a definition that measures the goodness of agreement between f and F *on the whole interval* $-\pi \leqq x \leqq \pi$. Obviously, the maximum of $|f - F|$ is not suitable for that purpose: in Fig. 246, the function F is a good approximation to f, but $|f - F|$ is large near x_0. We choose

$$E = \int_{-\pi}^{\pi} (f - F)^2 \, dx. \tag{3}$$

This is called the **total square error** of F relative to the function f on the interval $-\pi \leqq x \leqq \pi$. Clearly, $E \geqq 0$.

N being fixed, we want to determine the coefficients in (2) such that E is minimum. Since $(f - F)^2 = f^2 - 2fF + F^2$, we have

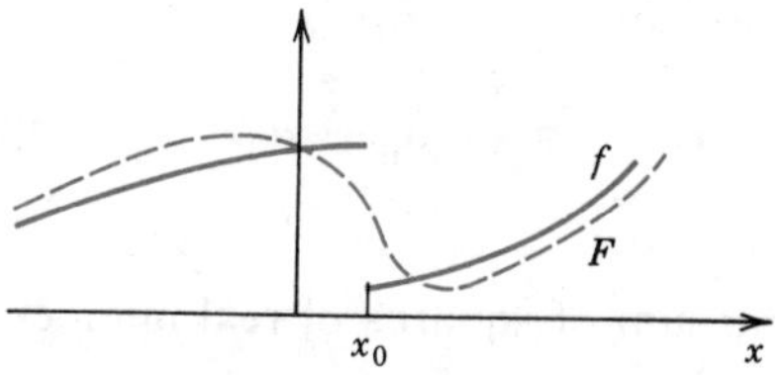

Fig. 246. Error of approximation

$$E = \int_{-\pi}^{\pi} f^2\,dx - 2\int_{-\pi}^{\pi} fF\,dx + \int_{-\pi}^{\pi} F^2\,dx. \tag{4}$$

By inserting (2) into the last integral and evaluating the occurring integrals as in Sec. 10.2 we see that all terms $\cos^2 nx$ and $\sin^2 nx$ ($n \geqq 1$) have integral π and all "mixed terms" $(\cos nx)(\sin mx)$ have integral zero. Thus

$$\int_{-\pi}^{\pi} F^2\,dx = \pi(2\alpha_0^2 + \alpha_1^2 + \cdots + \alpha_N^2 + \beta_1^2 + \cdots + \beta_N^2).$$

By inserting (2) into the second integral in (4) we see that the occurring integrals are those in the Euler formulas (6), Sec. 10.2, and we thus obtain

$$\int_{-\pi}^{\pi} fF\,dx = \pi(2\alpha_0 a_0 + \alpha_1 a_1 + \cdots + \alpha_N a_N + \beta_1 b_1 + \cdots + \beta_N b_N).$$

With these expressions (4) becomes

$$\begin{aligned} E &= \int_{-\pi}^{\pi} f^2\,dx \\ &\quad - 2\pi\left[2\alpha_0 a_0 + \sum_{n=1}^{N} (\alpha_n a_n + \beta_n b_n)\right] \\ &\quad + \pi\left[2\alpha_0^2 + \sum_{n=1}^{N} (\alpha_n^2 + \beta_n^2)\right]. \end{aligned} \tag{5}$$

If we take $\alpha_n = a_n$ and $\beta_n = b_n$ in (2), then from (5) we see that the square error corresponding to this choice of the coefficients of F is given by

$$E^* = \int_{-\pi}^{\pi} f^2\,dx - \pi\left[2a_0^2 + \sum_{n=1}^{N} (a_n^2 + b_n^2)\right]. \tag{6}$$

By subtracting (6) from (5) we obtain

$$E - E^* = \pi\left\{2(\alpha_0 - a_0)^2 + \sum_{n=1}^{N} [(\alpha_n - a_n)^2 + (\beta_n - b_n)^2]\right\}.$$

Since the sum of squares of real numbers on the right cannot be negative,

$$E - E^* \geqq 0, \qquad \text{thus} \qquad E \geqq E^*,$$

and $E = E^*$ if and only if $\alpha_0 = a_0, \cdots, \beta_N = b_N$. This proves

Theorem 1 (Minimum square error)

The total square error of F in (2) (with fixed N) relative to f on the interval $-\pi \leqq x \leqq \pi$ is minimum if and only if the coefficients of F in (2) are the Fourier coefficients of f. This minimum value is given by (6).

From (6) we see that E^* cannot increase as N increases, but may decrease. Hence, *with increasing N, the partial sums of the Fourier series of f yield better and better approximations to f,* considered from the viewpoint of the square error.

Since $E^* \geqq 0$ and (6) holds for every N, we obtain from (6) the important **Bessel inequality**[10]

$$2a_0^2 + \sum_{n=1}^{\infty} (a_n^2 + b_n^2) \leqq \frac{1}{\pi} \int_{-\pi}^{\pi} f(x)^2 \, dx \tag{7}$$

for the Fourier coefficients of any function f for which the integral on the right exists.

It can be shown (see [C14]) that for such a function f, **Parseval's theorem** holds, that is, formula (7) holds with the equality sign, so that it becomes **"Parseval's identity"**[11]

$$2a_0^2 + \sum_{n=1}^{\infty} (a_n^2 + b_n^2) = \frac{1}{\pi} \int_{-\pi}^{\pi} f(x)^2 \, dx. \tag{8}$$

EXAMPLE 1 Square error for the sawtooth wave

Compute the total square error of F with $N = 3$ relative to

$$f(x) = x + \pi \quad (-\pi < x < \pi) \qquad \text{(Fig. 238a, Sec. 10.4)}$$

on the interval $-\pi \leqq x \leqq \pi$.

Solution. $F(x) = \pi + 2 \sin x - \sin 2x + \frac{2}{3} \sin 3x$ by Example 2, Sec. 10.4. From this and (6),

$$E^* = \int_{-\pi}^{\pi} (x + \pi)^2 \, dx - \pi[2\pi^2 + 2^2 + 1^2 + (\tfrac{2}{3})^2],$$

hence

$$E^* = \tfrac{8}{3}\pi^3 - \pi(2\pi^2 + \tfrac{49}{9}) \approx 3.567.$$

$F = S_3$ is shown in Fig. 238b, and although $|f(x) - F(x)|$ is large at $x = \pm\pi$ (how large?), where f is discontinuous, F approximates f quite well on the whole interval. ▮

This brings to an end our discussion of Fourier series, which has emphasized the practical aspects of these series, as needed in applications. In the last four sections of this chapter we show how ideas and techniques in Fourier series can be extended to nonperiodic functions.

---

[10] See footnote 12 in Sec. 5.5.

[11] MARC ANTOINE PARSEVAL (1755—1836), French mathematician. A physical interpretation of the identity follows in the next section.

Problem Set 10.8

1. Let $f(x) = -1$ if $-\pi < x < 0$, $f(x) = 1$ if $0 < x < \pi$, and periodic with 2π. Find the function $F(x)$ of the form (2) for which the total square error (3) is minimum.

2. Compute the minimum square error in Prob. 1 for $N = 1, 3, 5, 7$. What is the smallest N such that $E^* \leqq 0.2$?

3. Show that the minimum square error (6) is a monotone decreasing function of N.

In each case, find the function $F(x)$ of the form (2) for which the total square error E on the interval $-\pi \leqq x \leqq \pi$ is minimum and compute this minimum value for $N = 1, 2, \cdots, 5$, where, for $-\pi < x < \pi$,

4. $f(x) = |x|$

5. $f(x) = x$

6. $f(x) = x(\pi^2 - x^2)/12$

7. $f(x) = x^2$

8. $f(x) = -\pi - x$ if $-\pi < x < -\pi/2$, $f(x) = x$ if $-\pi/2 < x < \pi/2$, $f(x) = \pi - x$ if $\pi/2 < x < \pi$

9. $f(x) = x$ if $-\pi/2 < x < \pi/2$, $f(x) = 0$ elsewhere in $-\pi < x < \pi$

10. Compare the rapidity of decrease of the square error for the discontinuous function in Prob. 5 and the continuous function in Prob. 7 and comment.

Using Parseval's identity, show that

11. $1 + \dfrac{1}{3^4} + \dfrac{1}{5^4} + \dfrac{1}{7^4} + \cdots = \dfrac{\pi^4}{96}$ (Use Prob. 27 in Sec. 10.4.)

12. $1 + \dfrac{1}{9} + \dfrac{1}{25} + \cdots = \dfrac{\pi^2}{8}$ (Use Prob. 5 in Sec. 10.2.)

13. $1 + \dfrac{1}{2^4} + \dfrac{1}{3^4} + \dfrac{1}{4^4} + \cdots = \dfrac{\pi^4}{90}$ (Use Prob. 9 in Sec. 10.2.)

14. $\displaystyle\int_{-\pi}^{\pi} \cos^4 x \, dx = \frac{3\pi}{4}$

15. $\displaystyle\int_{-\pi}^{\pi} \cos^6 x \, dx = \frac{5\pi}{8}$

10.9 Fourier Integrals

Fourier series are powerful tools in treating various problems involving *periodic* functions. Section 10.7 contained a first illustration of this, and various further applications follow in Chap. 11. Since, of course, many practical problems involve **nonperiodic functions,** we ask what can be done to extend the method of Fourier series to such functions. This is our goal in this section. In Example 1 we begin with a special function $f_L(x)$ of period $2L$ and see what happens to its Fourier series if we let $L \to \infty$. Then we consider the Fourier series of an arbitrary function f_L of period $2L$ and again let $L \to \infty$. This will motivate and suggest the main result of this section, an integral representation in Theorem 1 (p. 601).

EXAMPLE 1 **Square wave**

Consider the periodic square wave $f_L(x)$ of period $2L > 2$ given by

$$f_L(x) = \begin{cases} 0 & \text{if} \quad -L < x < -1 \\ 1 & \text{if} \quad -1 < x < \ \ 1 \\ 0 & \text{if} \quad \ \ 1 < x < \ \ L. \end{cases}$$

The left part of Fig. 247 shows this function for $2L = 4, 8, 16$ as well as the nonperiodic function

$$f(x) = \lim_{L\to\infty} f_L(x) = \begin{cases} 1 & \text{if } -1 < x < 1 \\ 0 & \text{otherwise} \end{cases}$$

which we obtain from f_L if we let $L \to \infty$.

We now explore what happens to the Fourier coefficients of f_L as L increases. Since f_L is even, $b_n = 0$ for all n. For a_n the Euler formulas (2), Sec. 10.3, give

$$a_0 = \frac{1}{2L}\int_{-1}^{1} dx = \frac{1}{L}, \quad a_n = \frac{1}{L}\int_{-1}^{1} \cos\frac{n\pi x}{L}\, dx = \frac{2}{L}\int_{0}^{1} \cos\frac{n\pi x}{L}\, dx = \frac{2}{L}\,\frac{\sin(n\pi/L)}{n\pi/L}.$$

This sequence of Fourier coefficients is called the **amplitude spectrum** of f_L because $|a_n|$ is the maximum amplitude of the wave $a_n \cos(n\pi x/L)$. Figure 247 shows this spectrum for the periods $2L = 4, 8, 16$. We see that for increasing L the amplitudes become more and more dense on the positive w_n-axis, where $w_n = n\pi/L$. Indeed, for $2L = 4, 8, 16$ we have 1, 3, 7 amplitudes per "half-wave" of the function $(2 \sin w_n)/Lw_n$ (dashed in the figure). Hence for $2L = 2^k$ we have $2^{k-1} - 1$ amplitudes per half-wave, so that these amplitudes will eventually be everywhere dense on the positive w_n-axis (and will decrease to zero). ∎

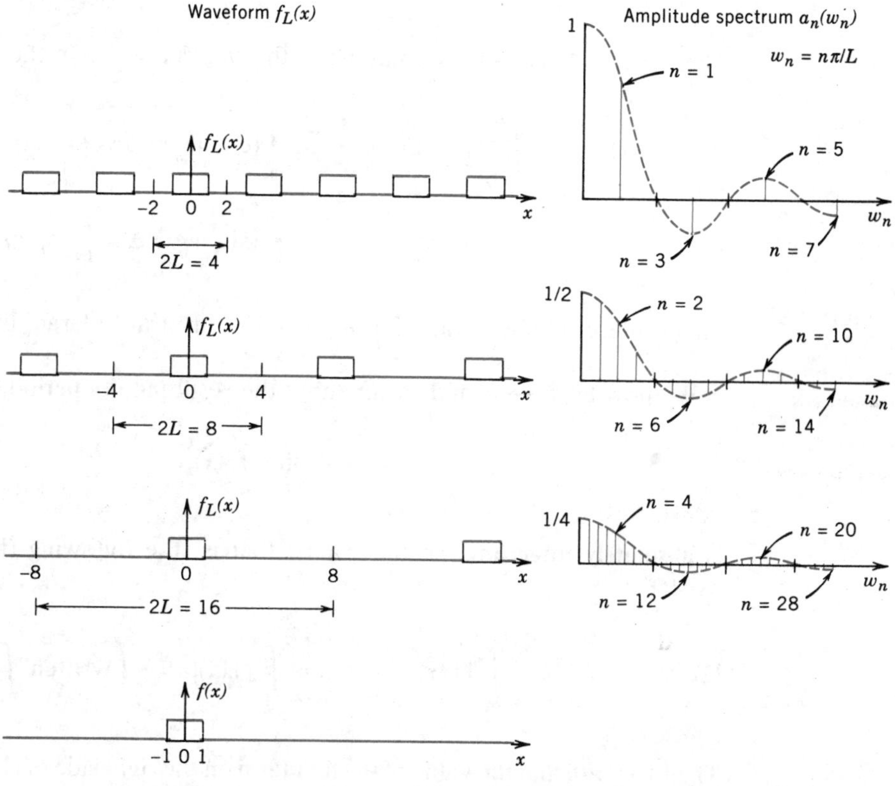

Fig. 247 Waveforms and amplitude spectra in Example 1

We now consider any periodic function $f_L(x)$ of period $2L$ that can be represented by a Fourier series

$$f_L(x) = a_0 + \sum_{n=1}^{\infty} (a_n \cos w_n x + b_n \sin w_n x), \qquad w_n = \frac{n\pi}{L},$$

and find out what happens if we let $L \to \infty$. Together with our example, the present calculation will suggest that we should expect an integral (instead of a series) involving $\cos wx$ and $\sin wx$ with w no longer restricted to integer multiples $w = w_n = n\pi/L$ of π/L but taking *all* values, and we shall also see what form such an integral might have.

If we insert a_n and b_n from the Euler formulas (2), Sec. 10.3, and denote the variable of integration by v, the Fourier series of $f_L(x)$ becomes

$$f_L(x) = \frac{1}{2L}\int_{-L}^{L} f_L(v)\,dv + \frac{1}{L}\sum_{n=1}^{\infty}\left[\cos w_n x \int_{-L}^{L} f_L(v)\cos w_n v\,dv + \sin w_n x \int_{-L}^{L} f_L(v)\sin w_n v\,dv\right].$$

We now set

$$\Delta w = w_{n+1} - w_n = \frac{(n+1)\pi}{L} - \frac{n\pi}{L} = \frac{\pi}{L}.$$

Then $1/L = \Delta w/\pi$, and we may write the Fourier series in the form

$$(1) \quad f_L(x) = \frac{1}{2L}\int_{-L}^{L} f_L(v)\,dv + \frac{1}{\pi}\sum_{n=1}^{\infty}\left[(\cos w_n x)\,\Delta w \int_{-L}^{L} f_L(v)\cos w_n v\,dv + (\sin w_n x)\,\Delta w \int_{-L}^{L} f_L(v)\sin w_n v\,dv\right].$$

This representation is valid for any fixed L, arbitrarily large, but finite.

We now let $L \to \infty$ and assume that the resulting nonperiodic function

$$f(x) = \lim_{L\to\infty} f_L(x)$$

is **absolutely integrable** on the x-axis; that is, the following (finite!) limits exist:

$$(2) \quad \lim_{a\to-\infty}\int_a^0 |f(x)|\,dx + \lim_{b\to\infty}\int_0^b |f(x)|\,dx \quad \left(\text{written } \int_{-\infty}^{\infty} |f(x)|\,dx\right).$$

Then $1/L \to 0$, and the value of the first term on the right side of (1) approaches zero. Also $\Delta w = \pi/L \to 0$ and it seems ***plausible*** that the infinite series in

(1) becomes an integral from 0 to ∞, which represents $f(x)$, namely,

$$f(x) = \frac{1}{\pi}\int_0^\infty \left[\cos wx \int_{-\infty}^{\infty} f(v)\cos wv\, dv + \sin wx \int_{-\infty}^{\infty} f(v)\sin wv\, dv\right] dw. \tag{3}$$

If we introduce the notations

$$A(w) = \frac{1}{\pi}\int_{-\infty}^{\infty} f(v)\cos wv\, dv, \qquad B(w) = \frac{1}{\pi}\int_{-\infty}^{\infty} f(v)\sin wv\, dv, \tag{4}$$

we can write this in the form

$$f(x) = \int_0^\infty [A(w)\cos wx + B(w)\sin wx]\, dw. \tag{5}$$

This is a representation of $f(x)$ by a so-called **Fourier integral.**

It is clear that our naive approach merely ***suggests*** the representation (5), but by no means establishes it; in fact, the limit of the series in (1) as Δw approaches zero is not the definition of the integral (3). Sufficient conditions for the validity of (5) are as follows.

Theorem 1 (Fourier integral)

If $f(x)$ is piecewise continuous (see Sec. 6.1) *in every finite interval and has a right-hand derivative and a left-hand derivative at every point* (see Sec. 10.2) *and if the integral* (2) *exists, then $f(x)$ can be represented by a Fourier integral* (5). *At a point where $f(x)$ is discontinuous the value of the Fourier integral equals the average of the left- and right-hand limits of $f(x)$ at that point* (see Sec. 10.2). (Proof in Ref. [C14]; see Appendix 1.)

The main use of the Fourier integral is in solving differential equations, as we shall see in Sec. 11.14. However, we can also use the Fourier integral in integration and in discussing functions defined by integrals, as the next examples illustrate.

EXAMPLE 2 Single pulse, sine integral

Find the Fourier integral representation of the function (Fig. 248)

$$f(x) = \begin{cases} 1 & \text{if} \quad |x| < 1, \\ 0 & \text{if} \quad |x| > 1. \end{cases}$$

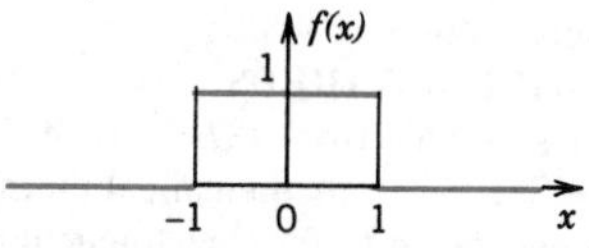

Fig. 248. Example 2

Solution. From (4) we obtain

$$A(w) = \frac{1}{\pi}\int_{-\infty}^{\infty} f(v)\cos wv\, dv = \frac{1}{\pi}\int_{-1}^{1}\cos wv\, dv = \left.\frac{\sin wv}{\pi w}\right|_{-1}^{1} = \frac{2\sin w}{\pi w},$$

$$B(w) = \frac{1}{\pi}\int_{-1}^{1}\sin wv\, dv = 0,$$

and (5) gives the answer

$$f(x) = \frac{2}{\pi}\int_{0}^{\infty}\frac{\cos wx \sin w}{w}\, dw. \tag{6}$$

The average of the left- and right-hand limits of $f(x)$ at $x = 1$ is equal to $(1 + 0)/2$, that is, $1/2$. Furthermore, from (6) and Theorem 1 we obtain

$$\int_{0}^{\infty}\frac{\cos wx \sin w}{w}\, dw = \begin{cases} \pi/2 & \text{if} \quad 0 \leqq x < 1, \\ \pi/4 & \text{if} \quad x = 1, \\ 0 & \text{if} \quad x > 1. \end{cases} \tag{7*}$$

We mention that this integral is called **Dirichlet's discontinuous factor.**[12] Let us consider the case $x = 0$, which is of particular interest. If $x = 0$, then

$$\int_{0}^{\infty}\frac{\sin w}{w}\, dw = \frac{\pi}{2}. \tag{7}$$

We see that this integral is the limit of the so-called **sine integral**

$$\mathrm{Si}(z) = \int_{0}^{z}\frac{\sin w}{w}\, dw \tag{8}$$

as $z \to \infty$ (z real). The graph of Si(z) is shown in Fig. 249.

In the case of a Fourier series the graphs of the partial sums are approximation curves of the curve of the periodic function represented by the series. Similarly, in the case of the Fourier integral (5), approximations are obtained by replacing ∞ by numbers a. Hence the integral

$$\int_{0}^{a}\frac{\cos wx \sin w}{w}\, dw \tag{9}$$

approximates the integral in (6) and therefore $f(x)$; see Fig. 250.

Figure 250 shows oscillations near the points of discontinuity of $f(x)$. We might expect that these oscillations disappear as a approaches infinity, but this is not true; with increasing a, they are shifted closer to the points $x = \pm 1$. This unexpected behavior, which also occurs in connection with Fourier series, is known as the **Gibbs phenomenon.**[13] It can be explained by representing (9) in terms of the sine integral as follows. Using (11) in Appendix 3.1, we have

$$\frac{2}{\pi}\int_{0}^{a}\frac{\cos wx \sin w}{w}\, dw = \frac{1}{\pi}\int_{0}^{a}\frac{\sin (w + wx)}{w}\, dw + \frac{1}{\pi}\int_{0}^{a}\frac{\sin (w - wx)}{w}\, dw.$$

---

[12]See footnote 16 in Sec. 9.8.

[13]JOSIAH WILLARD GIBBS (1839—1903), American mathematician, professor of mathematical physics at Yale from 1871, one of the founders of vector calculus [another being O. Heaviside (see Sec. 6.1)], mathematical thermodynamics, and statistical mechanics. His work was of great importance to the development of mathematical physics.

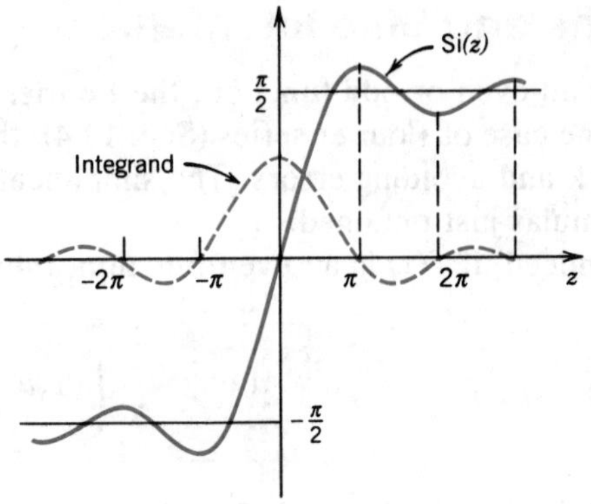

Fig. 249. Sine integral Si(z)

In the first integral on the right we set $w + wx = t$. Then $dw/w = dt/t$, and $0 \leqq w \leqq a$ corresponds to $0 \leqq t \leqq (x + 1)a$. In the last integral we set $w - wx = -t$. Then $dw/w = dt/t$, and $0 \leqq w \leqq a$ corresponds to $0 \leqq t \leqq (x - 1)a$. Since $\sin(-t) = -\sin t$, we thus obtain

$$\frac{2}{\pi}\int_0^a \frac{\cos wx \sin w}{w}\,dw = \frac{1}{\pi}\int_0^{(x+1)a} \frac{\sin t}{t}\,dt - \frac{1}{\pi}\int_0^{(x-1)a} \frac{\sin t}{t}\,dt.$$

From this and (8) we see that our integral equals

$$\frac{1}{\pi}\operatorname{Si}(a[x + 1]) - \frac{1}{\pi}\operatorname{Si}(a[x - 1]),$$

and the oscillations in Fig. 250 result from those in Fig. 249. The increase of a amounts to a transformation of the scale on the axis and causes the shift of the oscillations. ∎

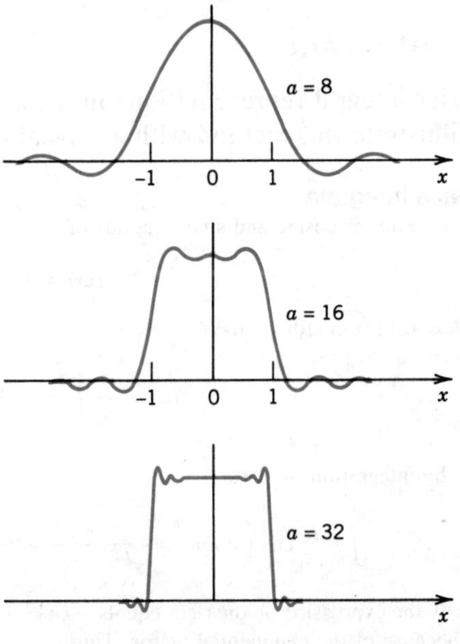

Fig. 250. The integral (9) for a = 8, 16, and 32

Fourier Cosine and Sine Integrals

For an even or odd function, the Fourier integral becomes simpler. Just as in the case of Fourier series (Sec. 10.4), this is of practical interest in saving work and avoiding errors. The simplifications follow immediately from our formulas just obtained.

Indeed, if $f(x)$ is an even function, then $B(w) = 0$ in (4) and

$$A(w) = \frac{2}{\pi}\int_0^\infty f(v) \cos wv \, dv, \tag{10}$$

and the Fourier integral (5) reduces to the so-called **Fourier cosine integral**

$$f(x) = \int_0^\infty A(w) \cos wx \, dw \qquad (f \text{ even}). \tag{11}$$

Similarly, if $f(x)$ is odd, then in (4) we have $A(w) = 0$ and

$$B(w) = \frac{2}{\pi}\int_0^\infty f(v) \sin wv \, dv, \tag{12}$$

and the Fourier integral (5) reduces to the so-called **Fourier sine integral**

$$f(x) = \int_0^\infty B(w) \sin wx \, dw \qquad (f \text{ odd}). \tag{13}$$

Evaluation of Integrals

Fourier integral representations may also be used for evaluating integrals. We illustrate this method with a typical example.

EXAMPLE 3 Laplace integrals

Find the Fourier cosine and sine integrals of

$$f(x) = e^{-kx} \qquad (x > 0,\ k > 0).$$

Solution. (a) From (10) we have

$$A(w) = \frac{2}{\pi}\int_0^\infty e^{-kv} \cos wv \, dv.$$

Now, by integration by parts,

$$\int e^{-kv} \cos wv \, dv = -\frac{k}{k^2 + w^2} e^{-kv}\left(-\frac{w}{k} \sin wv + \cos wv\right).$$

If $v = 0$, the expression on the right equals $-k/(k^2 + w^2)$; if v approaches infinity, it approaches zero because of the exponential factor. Thus

$$A(w) = \frac{2k/\pi}{k^2 + w^2}. \tag{14}$$

By substituting this into (11) we thus obtain the Fourier cosine integral representation

$$f(x) = e^{-kx} = \frac{2k}{\pi}\int_0^\infty \frac{\cos wx}{k^2 + w^2}\,dw \qquad (x > 0,\ k > 0).$$

From this representation we see that

$$\int_0^\infty \frac{\cos wx}{k^2 + w^2}\,dw = \frac{\pi}{2k}e^{-kx} \qquad (x > 0,\ k > 0). \tag{15}$$

(b) Similarly, from (12) we have

$$B(w) = \frac{2}{\pi}\int_0^\infty e^{-kv}\sin wv\,dv.$$

By integration by parts,

$$\int e^{-kv}\sin wv\,dv = -\frac{w}{k^2 + w^2}e^{-kv}\left(\frac{k}{w}\sin wv + \cos wv\right).$$

This equals $-w/(k^2 + w^2)$ if $v = 0$, and approaches 0 as $v \to \infty$. Thus

$$B(w) = \frac{2w/\pi}{k^2 + w^2}. \tag{16}$$

From (13) we thus obtain the Fourier sine integral representation

$$f(x) = e^{-kx} = \frac{2}{\pi}\int_0^\infty \frac{w \sin wx}{k^2 + w^2}\,dw.$$

From this we see that

$$\int_0^\infty \frac{w \sin wx}{k^2 + w^2}\,dw = \frac{\pi}{2}e^{-kx} \qquad (x > 0,\ k > 0). \tag{17}$$

The integrals (15) and (17) are the so-called **Laplace integrals.** ∎

Problem Set 10.9

Using the Fourier integral representation, show that

1. $\displaystyle\int_0^\infty \frac{\cos xw + w \sin xw}{1 + w^2}\,dw = \begin{cases} 0 & \text{if } x < 0 \\ \pi/2 & \text{if } x = 0 \\ \pi e^{-x} & \text{if } x > 0 \end{cases}$ [Use (5).]

2. $\displaystyle\int_0^\infty \frac{w^3 \sin xw}{w^4 + 4}\,dw = \frac{\pi}{2}e^{-x}\cos x \quad \text{if } x > 0$ [Use (13).]

3. $\displaystyle\int_0^\infty \frac{\cos xw}{1 + w^2}\,dw = \frac{\pi}{2}e^{-x} \quad \text{if } x > 0$ [Use (11).]

4. $\displaystyle\int_0^\infty \frac{\sin w \cos xw}{w}\,dw = \begin{cases} \pi/2 & \text{if } 0 \leqq x < 1 \\ \pi/4 & \text{if } x = 1 \\ 0 & \text{if } x > 1 \end{cases}$ [Use (11).]

5. $\displaystyle\int_0^\infty \frac{\sin \pi w \sin xw}{1 - w^2}\, dw = \begin{cases} \frac{\pi}{2} \sin x & \text{if } 0 \leqq x \leqq \pi \\ 0 & \text{if } x > \pi \end{cases}$ [Use (13).]

6. $\displaystyle\int_0^\infty \frac{1 - \cos \pi w}{w} \sin xw\, dw = \begin{cases} \frac{\pi}{2} & \text{if } 0 < x < \pi \\ 0 & \text{if } x > \pi \end{cases}$ [Use (13).]

7. $\displaystyle\int_0^\infty \frac{\cos (\pi w/2) \cos xw}{1 - w^2}\, dw = \begin{cases} \frac{\pi}{2} \cos x & \text{if } |x| < \frac{\pi}{2} \\ 0 & \text{if } |x| > \frac{\pi}{2} \end{cases}$ [Use (11).]

Represent the following functions $f(x)$ in the form (11).

8. $f(x) = \begin{cases} 1 & \text{if } 0 < x < 1 \\ 0 & \text{if } x > 1 \end{cases}$

9. $f(x) = \begin{cases} x^2 & \text{if } 0 < x < 1 \\ 0 & \text{if } x > 1 \end{cases}$

10. $f(x) = \begin{cases} x/2 & \text{if } 0 < x < 1 \\ 1 - x/2 & \text{if } 1 < x < 2 \\ 0 & \text{if } x > 2 \end{cases}$

11. $f(x) = \begin{cases} x & \text{if } 0 < x < a \\ 0 & \text{if } x > a \end{cases}$

12. $f(x) = \begin{cases} a^2 - x^2 & \text{if } 0 < x < a \\ 0 & \text{if } x > a \end{cases}$

13. $f(x) = \dfrac{1}{1 + x^2}$ [see (15)]

14. $f(x) = e^{-x} + e^{-2x} \quad (x > 0)$

If $f(x)$ has the representation (11), show that

15. $\displaystyle f(ax) = \frac{1}{a} \int_0^\infty A\left(\frac{w}{a}\right) \cos xw\, dw \quad (a > 0)$

16. $\displaystyle xf(x) = \int_0^\infty B^*(w) \sin xw\, dw, \quad B^* = -\frac{dA}{dw}, \quad A \text{ as in (10)}$

17. $\displaystyle x^2 f(x) = \int_0^\infty A^*(w) \cos xw\, dw, \quad A^* = -\frac{d^2A}{dw^2}$

18. Solve Prob. 9 by applying the formula in Prob. 17 to the result of Prob. 8.

19. Verify the formula in Prob. 16 for $f(x) = 1$ if $0 < x < a$ and $f(x) = 0$ if $x > a$.

20. Show that $f(x) = 1$ $(0 < x < \infty)$ cannot be represented by a Fourier integral.

10.10 Fourier Cosine and Sine Transforms

An **integral transform** is a transformation that produces from given functions new functions which depend on a different variable and appear in the form of an integral. These transformations are of interest mainly as tools in solving ordinary differential equations, partial differential equations, and integral equations, and they often also help in handling and applying special functions. The **Laplace transform** (Chap. 6) is of this kind and is by far the most important integral transform in engineering. From the viewpoint of applications, the next in order of importance are perhaps the **Fourier transforms,** although these are somewhat more difficult to handle than the Laplace transform. We shall see that they can be obtained from the Fourier integral representations in Sec. 10.9. In this section we consider two of them, called the *Fourier cosine* and *sine transforms,* which are real, and in the next section a third one, which is complex.

Fourier Cosine Transforms

For an ***even*** function $f(x)$, the Fourier integral is the Fourier cosine integral

$$\text{(1) (a)}\quad f(x) = \int_0^\infty A(w)\cos wx\,dw, \quad \text{where} \quad \text{(b)}\quad A(w) = \frac{2}{\pi}\int_0^\infty f(v)\cos wv\,dv$$

[see (10), (11), Sec. 10.9]. We now set $A(w) = \sqrt{2/\pi}\,\hat{f}_c(w)$, where c suggests "cosine." Then from (1b), writing $v = x$, we have

$$\hat{f}_c(w) = \sqrt{\frac{2}{\pi}}\int_0^\infty f(x)\cos wx\,dx \tag{2}$$

and from (1a),

$$f(x) = \sqrt{\frac{2}{\pi}}\int_0^\infty \hat{f}_c(w)\cos wx\,dw. \tag{3}$$

Attention! In (2) we integrate with respect to x and in (3) with respect to w. Formula (2) gives from $f(x)$ a new function $\hat{f}_c(w)$, called the **Fourier cosine transform** of $f(x)$. Formula (3) gives us back $f(x)$ from $\hat{f}_c(w)$, and we therefore call $f(x)$ the **inverse Fourier cosine transform** of $\hat{f}_c(w)$.

The process of obtaining the transform $\hat{f}_c$ from a given f is also called the **Fourier cosine transform** or the *Fourier cosine transform method.*

Fourier Sine Transforms

Similarly, for an ***odd*** function $f(x)$, the Fourier integral is the Fourier sine integral [see (12), (13), Sec. 10.9]

$$\text{(4) (a)}\quad f(x) = \int_0^\infty B(w)\sin wx\,dw, \quad \text{where} \quad \text{(b)}\quad B(w) = \frac{2}{\pi}\int_0^\infty f(v)\sin wv\,dv.$$

We now set $B(w) = \sqrt{2/\pi}\,\hat{f}_s(w)$, where s suggests "sine." Then from (4b), writing $v = x$, we have

(5)
$$\hat{f}_s(w) = \sqrt{\frac{2}{\pi}} \int_0^\infty f(x) \sin wx \, dx,$$

called the **Fourier sine transform** of $f(x)$, and from (4a)

(6)
$$f(x) = \sqrt{\frac{2}{\pi}} \int_0^\infty \hat{f}_s(w) \sin wx \, dw,$$

called the **inverse Fourier sine transform** of $\hat{f}_s(w)$. The process of obtaining $\hat{f}_s(w)$ from $f(x)$ is also called the **Fourier sine transform** or the *Fourier sine transform method.*

Other notations are

$$\mathcal{F}_c(f) = \hat{f}_c, \qquad \mathcal{F}_s(f) = \hat{f}_s$$

and $\mathcal{F}_c^{-1}$ and $\mathcal{F}_s^{-1}$ for the inverses of $\mathcal{F}_c$ and $\mathcal{F}_s$, respectively.

EXAMPLE 1 Fourier cosine and Fourier sine transforms

Find the Fourier cosine and sine transforms of the function

$$f(x) = \begin{cases} k & \text{if } 0 < x < a \\ 0 & \text{if } x > a. \end{cases}$$

Solution. From the definitions (2) and (5) we obtain by integration

$$\hat{f}_c(w) = \sqrt{\frac{2}{\pi}}\, k \int_0^a \cos wx \, dx = \sqrt{\frac{2}{\pi}}\, k \left[\frac{\sin aw}{w}\right],$$

$$\hat{f}_s(w) = \sqrt{\frac{2}{\pi}}\, k \int_0^a \sin wx \, dx = \sqrt{\frac{2}{\pi}}\, k \left[\frac{1 - \cos aw}{w}\right].$$

This agrees with formulas 1 in the first two tables in Sec. 10.12 (where $k = 1$).
Note that for $f(x) = k = const$ $(0 < x < \infty)$, these transforms do not exist. (Why?) ∎

EXAMPLE 2 Fourier cosine transform of the exponential function

Find $\mathcal{F}_c(e^{-x})$.

Solution. By integration by parts and recursion,

$$\mathcal{F}_c(e^{-x}) = \sqrt{\frac{2}{\pi}} \int_0^\infty e^{-x} \cos wx \, dx = \sqrt{\frac{2}{\pi}} \frac{e^{-x}}{1 + w^2}(-\cos wx + w \sin wx) \Bigg|_0^\infty = \frac{\sqrt{2/\pi}}{1 + w^2}.$$

This agrees with formula 3 in Table 1, Sec. 10.12, with $a = 1$. ∎

What have we done in order to introduce the two integral transforms under consideration? Actually not much: We have changed the notations A and B to get a "symmetric" distribution of the constant $2/\pi$ in the original formulas (10)–(13), Sec. 10.9. This redistribution is a standard convenience, but is not essential. One could do without it.

What have we gained? We show next that these transforms have operational properties that permit them to convert differentiations into algebraic operations (just as the Laplace transform does). This is the key to their application in solving differential equations.

Linearity, Transforms of Derivatives

If $f(x)$ is absolutely integrable (see Sec. 10.9) on the positive x-axis and piecewise continuous (see Sec. 6.1) on every finite interval, then the Fourier cosine and sine transforms of f exist.

Furthermore, for a function $af(x) + bg(x)$ we have from (2)

$$\mathcal{F}_c(af + bg) = \sqrt{\frac{2}{\pi}} \int_0^\infty [af(x) + bg(x)] \cos wx \, dx$$

$$= a \sqrt{\frac{2}{\pi}} \int_0^\infty f(x) \cos wx \, dx + b \sqrt{\frac{2}{\pi}} \int_0^\infty g(x) \cos wx \, dx.$$

The right side is $a\mathcal{F}_c(f) + b\mathcal{F}_c(g)$. Similarly for $\mathcal{F}_s$, by (5). This shows that the Fourier cosine and sine transforms are **linear operations,**

$$\text{(7)} \qquad \begin{aligned} &\text{(a)} \quad \mathcal{F}_c(af + bg) = a\mathcal{F}_c(f) + b\mathcal{F}_c(g), \\ &\text{(b)} \quad \mathcal{F}_s(af + bg) = a\mathcal{F}_s(f) + b\mathcal{F}_s(g). \end{aligned}$$

Theorem 1 (Cosine and sine transforms of derivatives)

Let $f(x)$ be continuous and absolutely integrable on the x-axis, let $f'(x)$ be piecewise continuous on each finite interval, and let $f(x) \to 0$ as $x \to \infty$. Then

$$\text{(8)} \qquad \begin{aligned} &\textbf{(a)} \quad \mathcal{F}_c\{f'(x)\} = w\mathcal{F}_s\{f(x)\} - \sqrt{\frac{2}{\pi}} f(0), \\ &\textbf{(b)} \quad \mathcal{F}_s\{f'(x)\} = -w\mathcal{F}_c\{f(x)\}. \end{aligned}$$

Proof. This follows from the definitions by integration by parts, namely,

$$\mathcal{F}_c\{f'(x)\} = \sqrt{\frac{2}{\pi}} \int_0^\infty f'(x) \cos wx \, dx$$

$$= \sqrt{\frac{2}{\pi}} \left[f(x) \cos wx \Big|_0^\infty + w \int_0^\infty f(x) \sin wx \, dx \right]$$

$$= -\sqrt{\frac{2}{\pi}} f(0) + w\mathcal{F}_s\{f(x)\};$$

similarly,

$$\begin{aligned}\mathcal{F}_s\{f'(x)\} &= \sqrt{\frac{2}{\pi}}\int_0^\infty f'(x)\sin wx\,dx \\ &= \sqrt{\frac{2}{\pi}}\left[f(x)\sin wx\,\Big|_0^\infty - w\int_0^\infty f(x)\cos wx\,dx\right] \\ &= 0 - w\mathcal{F}_c\{f(x)\}.\end{aligned}$$

Formula (8a) with f' instead of f gives

$$\mathcal{F}_c\{f''(x)\} = w\mathcal{F}_s\{f'(x)\} - \sqrt{\frac{2}{\pi}}\,f'(0);$$

hence by (8b),

$$\mathcal{F}_c\{f''(x)\} = -w^2\mathcal{F}_c\{f(x)\} - \sqrt{\frac{2}{\pi}}\,f'(0). \tag{9a}$$

Similarly,

$$\mathcal{F}_s\{f''(x)\} = -w^2\mathcal{F}_s\{f(x)\} + \sqrt{\frac{2}{\pi}}\,wf(0). \tag{9b}$$

An application of (9) to differential equations will be given in Sec. 11.14. For the time being, we show how (9) can be used to derive transforms.

EXAMPLE 3 **An application of the operational formula (9)**

Find the Fourier cosine transform of $f(x) = e^{-ax}$, where $a > 0$.

Solution. By differentiation, $(e^{-ax})'' = a^2e^{-ax}$; thus $a^2f(x) = f''(x)$. From this and (9a),

$$a^2\mathcal{F}_c(f) = \mathcal{F}_c(f'') = -w^2\mathcal{F}_c(f) - \sqrt{\frac{2}{\pi}}\,f'(0) = -w^2\mathcal{F}_c(f) + a\sqrt{\frac{2}{\pi}}.$$

Hence $(a^2 + w^2)\mathcal{F}_c(f) = a\sqrt{2/\pi}$. The answer is (see Table I, Sec. 10.12)

$$\mathcal{F}_c(e^{-ax}) = \sqrt{\frac{2}{\pi}}\left(\frac{a}{a^2 + w^2}\right) \qquad (a > 0).$$

Tables of Fourier cosine and sine transforms are included in Sec. 10.12. For more extensive tables, see Ref. [C4] in Appendix 1.

Problem Set 10.10

1. Let $f(x) = -1$ if $0 < x < 1$, $f(x) = 1$ if $1 < x < 2$. Find $\hat{f}_c(w)$.
2. Derive $f(x)$ in Prob. 1 from the answer to Prob. 1. *Hint.* Use Prob. 4 in Sec. 10.9.

3. Find the Fourier cosine transform of $f(x) = x$ if $0 < x < a$, $f(x) = 0$ if $x > a$.
4. Find $\mathcal{F}_s(e^{-ax})$, $a > 0$, by integration.
5. Derive formula 3 in Table I of Sec. 10.12 by integration.
6. Obtain the answer to Prob. 4 from (9b).
7. Obtain $\mathcal{F}_c(1/(1 + x^2))$ from Prob. 3 in Sec. 10.9.
8. Obtain the inverse Fourier cosine transform of e^{-w}.
9. Find the Fourier sine transform of $f(x) = x^2$ if $0 < x < 1$, $f(x) = 0$ if $x > 1$.
10. Find the Fourier cosine transform of the function in Prob. 9.

11. Obtain $\mathcal{F}_s(x^{-1} - x^{-1} \cos \pi x)$. *Hint.* Use Prob. 6, Sec. 10.9, with w and x interchanged.
12. Obtain formula 10 in Table I of Sec. 10.12 with $a = 1$ from Example 2 in Sec. 10.9.
13. Find $\mathcal{F}_c\{(\cos \pi x/2)/(1 - x^2)\}$. *Hint.* Use Prob. 7, Sec. 10.9.
14. Using (8b), obtain $\mathcal{F}_s(xe^{-x^2/2})$ from a suitable formula in Table I, Sec. 10.12.
15. Find $\mathcal{F}_s(e^{-x})$ from (8a) and formula 3 of Table I, Sec. 10.12.
16. Using $\Gamma(\frac{1}{2}) = \sqrt{\pi}$, obtain formula 2 in Table II, Sec. 10.12, from formula 4 in that table.
17. Let $f(x) = x^3/(x^4 + 4)$. Find $\hat{f}_s(w)$ for $w > 0$. *Hint.* Use Prob. 2 in Sec. 10.9.
18. Show that $f(x) = 1$ has no Fourier cosine or sine transform.
19. Do the Fourier cosine and sine transforms of $f(x) = e^x$ exist?
20. Does the Fourier cosine transform of $(\cos x)/x$ exist? Of $(\sin x)/x$?

10.11 Fourier Transform

The previous section concerned two transforms obtained from the Fourier cosine and sine integrals in Sec. 10.9. We now consider a third transform, the *Fourier transform,* which is obtained from the Fourier integral in complex form. (For a motivation of this transform, see the beginning of Sec. 10.10.) We therefore consider first the complex form of the Fourier integral.

Complex Form of the Fourier Integral

The (real) Fourier integral is [see (4), (5), Sec. 10.9]

$$f(x) = \int_0^\infty [A(w) \cos wx + B(w) \sin wx]\, dw$$

where

$$A(w) = \frac{1}{\pi}\int_{-\infty}^{\infty} f(v) \cos wv\, dv, \qquad B(w) = \frac{1}{\pi}\int_{-\infty}^{\infty} f(v) \sin wv\, dv.$$

Substituting A and B into the integral for f, we have

$$f(x) = \frac{1}{\pi}\int_0^{\infty}\int_{-\infty}^{\infty} f(v)[\cos wv \cos wx + \sin wv \sin wx]\, dv\, dw.$$

By the addition formula for the cosine [(6) in Appendix 3.1] the expression in the brackets [· · ·] equals cos $(wv - wx)$ or, since the cosine is even, cos $(wx - wv)$, so that we get

$$f(x) = \frac{1}{\pi}\int_0^{\infty}\left[\int_{-\infty}^{\infty} f(v)\cos(wx - wv)\, dv\right] dw. \tag{1*}$$

The integral in brackets is an *even* function of w, call it $F(w)$, because cos $(wx - wv)$ is an even function of w, the function f does not depend on w, and we integrate with respect to v (not w). Hence the integral of $F(w)$ from $w = 0$ to ∞ is 1/2 times the integral of $F(w)$ from $-\infty$ to ∞. Thus

$$f(x) = \frac{1}{2\pi}\int_{-\infty}^{\infty}\left[\int_{-\infty}^{\infty} f(v)\cos(wx - wv)\, dv\right] dw. \tag{1}$$

We claim that the integral of the form (1) with sin instead of cos is zero:

$$\frac{1}{2\pi}\int_{-\infty}^{\infty}\left[\int_{-\infty}^{\infty} f(v)\sin(wx - wv)\, dv\right] dw = 0. \tag{2}$$

This is true since sin $(wx - wv)$ is an odd function of w, which makes the integral in brackets an odd function of w, call it $G(w)$, so that the integral of $G(w)$ from $-\infty$ to ∞ is zero, as claimed. We now use the **Euler formula**

$$e^{it} = \cos t + i \sin t \tag{3}$$

for the complex exponential function [formula (8) in Sec. 2.3]. Setting $t = wx - wv$ and adding (1) and i times (2), we obtain

$$f(x) = \frac{1}{2\pi}\int_{-\infty}^{\infty}\int_{-\infty}^{\infty} f(v)e^{iw(x-v)}\, dv\, dw \qquad (i = \sqrt{-1}). \tag{4}$$

This is called the **complex Fourier integral.**

It is now a short step from this to the Fourier transform, our present goal.

Fourier Transforms

Writing the exponential function in (4) as a product of exponential functions, we have

$$f(x) = \frac{1}{\sqrt{2\pi}}\int_{-\infty}^{\infty}\left[\frac{1}{\sqrt{2\pi}}\int_{-\infty}^{\infty} f(v)e^{-iwv}\, dv\right] e^{iwx}\, dw. \tag{5}$$

The expression in brackets is a function of w, is denoted by $\hat{f}(w)$, and is called the **Fourier transform** of f; writing $v = x$, we have

$$\hat{f}(w) = \frac{1}{\sqrt{2\pi}} \int_{-\infty}^{\infty} f(x)e^{-iwx}\, dx. \tag{6}$$

With this, (5) becomes

$$f(x) = \frac{1}{\sqrt{2\pi}} \int_{-\infty}^{\infty} \hat{f}(w)e^{iwx}\, dw \tag{7}$$

and is called the **inverse Fourier transform** of $\hat{f}(w)$.

Another notation is $\mathcal{F}(f) = \hat{f}(w)$ and $\mathcal{F}^{-1}$ for the inverse.

The process of obtaining the Fourier transform $\mathcal{F}(f) = \hat{f}$ from a given f is also called the **Fourier transform** or the *Fourier transform method.*

Existence. The following two conditions are sufficient for the existence of the Fourier transform (6) of a function $f(x)$ defined on the x-axis, as we mention without proof.

1. $f(x)$ is piecewise continuous on every finite interval.
2. $f(x)$ is absolutely integrable on the x-axis.

(For the definitions of piecewise continuity and absolute integrability see Secs. 6.1 and 10.9, respectively.)

EXAMPLE 1 **Fourier transform**

Find the Fourier transform of $f(x) = k$ if $0 < x < a$ and $f(x) = 0$ otherwise.

Solution. From (6) by integration,

$$\hat{f}(w) = \frac{1}{\sqrt{2\pi}} \int_0^a ke^{-iwx}\, dx = \frac{k}{\sqrt{2\pi}} \left(\frac{e^{-iwa} - 1}{-iw} \right) = \frac{k(1 - e^{-iaw})}{iw\sqrt{2\pi}}.$$

This shows that the Fourier transform will in general be a complex-valued function. ∎

EXAMPLE 2 **Fourier transform**

Find the Fourier transform of e^{-ax^2}, where $a > 0$.

Solution. We use the definition, complete the square in the exponent, and pull out the exponential factor that contains no x:

$$\begin{aligned}
\mathcal{F}(e^{-ax^2}) &= \frac{1}{\sqrt{2\pi}} \int_{-\infty}^{\infty} \exp\,[-ax^2 - iwx]\, dx \\
&= \frac{1}{\sqrt{2\pi}} \int_{-\infty}^{\infty} \exp \left[-\left(\sqrt{a}x + \frac{iw}{2\sqrt{a}} \right)^2 + \left(\frac{iw}{2\sqrt{a}} \right)^2 \right] dx \\
&= \frac{1}{\sqrt{2\pi}} \exp\left(-\frac{w^2}{4a} \right) \int_{-\infty}^{\infty} \exp \left[-\left(\sqrt{a}x + \frac{iw}{2\sqrt{a}} \right)^2 \right] dx.
\end{aligned}$$

We denote the integral by I and show that it equals $\sqrt{\pi/a}$. For this we use $\sqrt{a}x + iw/2\sqrt{a} = v$ as a new variable of integration. Then $dx = dv/\sqrt{a}$, so that

$$I = \frac{1}{\sqrt{a}} \int_{-\infty}^{\infty} e^{-v^2}\, dv.$$

We now get the result by the following trick. We square the integral, convert it to a double integral, and use polar coordinates $r = \sqrt{u^2 + v^2}$ and θ. Since $du\, dv = r\, dr\, d\theta$, we get

$$I^2 = \frac{1}{a} \int_{-\infty}^{\infty} e^{-u^2}\, du \int_{-\infty}^{\infty} e^{-v^2}\, dv = \frac{1}{a} \int_{-\infty}^{\infty} \int_{-\infty}^{\infty} e^{-(u^2+v^2)}\, du\, dv$$

$$= \frac{1}{a} \int_0^{2\pi} \int_0^{\infty} e^{-r^2}\, r\, dr\, d\theta = \frac{2\pi}{a} \left(-\frac{1}{2} e^{-r^2}\right)\Big|_0^{\infty} = \frac{\pi}{a}.$$

Hence $I = \sqrt{\pi/a}$. From this and the first formula in this solution,

$$\mathcal{F}(e^{-ax^2}) = \frac{1}{\sqrt{2\pi}} \exp\left(-\frac{w^2}{4a}\right) \sqrt{\frac{\pi}{a}} = \frac{1}{\sqrt{2a}} e^{-w^2/4a}.$$

This agrees with formula 9 in Table III, Sec. 10.12. ∎

Physical Interpretation: Spectrum

The nature of the representation (7) of $f(x)$ becomes clear if we think of it as a superposition of sinusoidal oscillations of all possible frequencies, called a **spectral representation.** This name is suggested by optics, where light is such a superposition of colors (frequencies). In (7), the **"spectral density"** $\hat{f}(w)$ measures the intensity of $f(x)$ in the frequency interval between w and $w + \Delta w$ (Δw small, fixed). We claim that in connection with vibrations, the integral

$$\int_{-\infty}^{\infty} |\hat{f}(w)|^2\, dw$$

can be interpreted as the **total energy** of the physical system; hence an integral of $|\hat{f}(w)|^2$ from a to b gives the contribution of the frequencies w between a and b to the total energy.

To make this plausible, we begin with a mechanical system giving a single frequency, namely, the harmonic oscillator (mass on a spring, Sec. 2.5)

$$my'' + ky = 0,$$

denoting time t by x. Multiplication by y' and integration gives

$$\tfrac{1}{2}mv^2 + \tfrac{1}{2}ky^2 = E_0 = const,$$

where $v = y'$ is the velocity, the first term is the kinetic energy, the second the potential energy, and E_0 the total energy of the system. Now a general solution is [use (4), (5), Sec. 10.6]

$$y = a_1 \cos w_0 x + b_1 \sin w_0 x = c_1 e^{iw_0 x} + c_{-1} e^{-iw_0 x}, \quad w_0^2 = k/m,$$

where $c_1 = (a_1 - ib_1)/2$, $c_{-1} = \bar{c}_1 = (a_1 + ib_1)/2$. Since $mw_0^2 = k$ and $(iw_0)^2 = -w_0^2$, we get by straightforward calculation and simplification

$$E_0 = \tfrac{1}{2}m(c_1 iw_0 e^{iw_0 x} - c_{-1} iw_0 e^{-iw_0 x})^2 + \tfrac{1}{2}k(c_1 e^{iw_0 x} + c_{-1} e^{-iw_0 x})^2$$
$$= 2kc_1 c_{-1} = 2k|c_1|^2.$$

Hence *the energy is proportional to the square of the amplitude* $|c_1|$.

As the next step, if a more complicated system leads to a periodic solution $y = f(x)$ that can be represented by a Fourier series, then instead of the single energy term $|c_1|^2$ we get a series of squares $|c_n|^2$ of Fourier coefficients c_n given by (8), Sec. 10.6. In this case we have a "**discrete spectrum**" (or "**point spectrum**") consisting of countably many isolated frequencies (infinitely many, in general), the corresponding $|c_n|^2$ being the contributions to the total energy.

Finally, a system whose solution can be represented by a Fourier integral (7) leads to the above integral for the energy, as is plausible from the cases just discussed.

Linearity. Fourier Transform of Derivatives

New transforms can be obtained from given ones by

Theorem 1 **(Linearity of the Fourier transform)**

The Fourier transform is a linear operation; that is, for any functions f(x) and g(x) whose Fourier transforms exist and any constants a and b,

$$\mathcal{F}(af + bg) = a\mathcal{F}(f) + b\mathcal{F}(g). \tag{8}$$

Proof. This is true since integration is a linear operation, so that (6) gives

$$\mathcal{F}\{af(x) + bg(x)\} = \frac{1}{\sqrt{2\pi}} \int_{-\infty}^{\infty} [af(x) + bg(x)]e^{-iwx}\,dx$$

$$= a\frac{1}{\sqrt{2\pi}} \int_{-\infty}^{\infty} f(x)e^{-iwx}\,dx + b\frac{1}{\sqrt{2\pi}} \int_{-\infty}^{\infty} g(x)e^{-iwx}\,dx$$

$$= a\mathcal{F}\{f(x)\} + b\mathcal{F}\{g(x)\}.$$ ∎

In the application of the Fourier transform to differential equations, the key property is that differentiation of functions corresponds to multiplication of transforms by iw:

Theorem 2 [Fourier transform of the derivative of $f(x)$]

Let $f(x)$ be continuous on the x-axis and $f(x) \to 0$ as $|x| \to \infty$. Furthermore, let $f'(x)$ be absolutely integrable on the x-axis. Then

$$\mathcal{F}\{f'(x)\} = iw\mathcal{F}\{f(x)\}. \tag{9}$$

Proof. Integrating by parts and using $f(x) \to 0$ as $|x| \to \infty$, we obtain

$$\begin{aligned}\mathcal{F}\{f'(x)\} &= \frac{1}{\sqrt{2\pi}} \int_{-\infty}^{\infty} f'(x)e^{-iwx}\, dx \\ &= \frac{1}{\sqrt{2\pi}}\left[f(x)e^{-iwx} \Big|_{-\infty}^{\infty} - (-iw) \int_{-\infty}^{\infty} f(x)e^{-iwx}\, dx \right] \\ &= 0 + iw\mathcal{F}\{f(x)\}.\end{aligned}$$

Two successive applications of (9) give

$$\mathcal{F}(f'') = iw\mathcal{F}(f') = (iw)^2\mathcal{F}(f).$$

Since $(iw)^2 = -w^2$, we have for the transform of the second derivative of f

$$\mathcal{F}\{f''(x)\} = -w^2\mathcal{F}\{f(x)\}. \tag{10}$$

Similarly for higher derivatives.

An application of (10) to differential equations will be given in Sec. 11.14. For the time being we show how (9) can be used to derive transforms.

EXAMPLE 3 An application of the operational formula (9)

Find the Fourier transform of xe^{-x^2} from Table III, Sec. 10.12.

Solution. We use (9). By formula 9 in Table III,

$$\begin{aligned}\mathcal{F}(xe^{-x^2}) &= \mathcal{F}\left\{-\frac{1}{2}(e^{-x^2})'\right\} = -\frac{1}{2}\mathcal{F}\left\{(e^{-x^2})'\right\} \\ &= -\frac{1}{2}iw\mathcal{F}(e^{-x^2}) = -\frac{1}{2}iw\frac{1}{\sqrt{2}}e^{-w^2/4} = -\frac{iw}{2\sqrt{2}}e^{-w^2/4}.\end{aligned}$$

Convolution

The **convolution** $f * g$ of functions f and g is defined by

$$h(x) = (f * g)(x) = \int_{-\infty}^{\infty} f(p)g(x-p)\, dp = \int_{-\infty}^{\infty} f(x-p)g(p)\, dp. \tag{11}$$

The purpose is the same as in the case of Laplace transforms (Sec. 6.6): the convolution of functions corresponds to the multiplication of their Fourier transforms (except for a factor $\sqrt{2\pi}$):

Theorem 3 (Convolution theorem)

Suppose that $f(x)$ and $g(x)$ are piecewise continuous, bounded, and absolutely integrable on the x-axis. Then

$$\mathcal{F}(f * g) = \sqrt{2\pi}\,\mathcal{F}(f)\,\mathcal{F}(g). \tag{12}$$

Proof. By the definition and an interchange of the order of integration,

$$\mathcal{F}(f * g) = \frac{1}{\sqrt{2\pi}} \int_{-\infty}^{\infty} \int_{-\infty}^{\infty} f(p)g(x - p)e^{-iwx}\, dp\, dx$$

$$= \frac{1}{\sqrt{2\pi}} \int_{-\infty}^{\infty} \int_{-\infty}^{\infty} f(p)g(x - p)e^{-iwx}\, dx\, dp.$$

Instead of x we now take $x - p = q$ as a new variable of integration. Then $x = p + q$ and

$$\mathcal{F}(f * g) = \frac{1}{\sqrt{2\pi}} \int_{-\infty}^{\infty} \int_{-\infty}^{\infty} f(p)g(q)e^{-iw(p+q)}\, dq\, dp$$

$$= \frac{1}{\sqrt{2\pi}} \int_{-\infty}^{\infty} f(p)e^{-iwp}\, dp \int_{-\infty}^{\infty} g(q)e^{-iwq}\, dq$$

$$= \sqrt{2\pi}\,\mathcal{F}(f)\,\mathcal{F}(g). \qquad \blacksquare$$

By taking the inverse Fourier transform on both sides of (12), writing $\hat{f} = \mathcal{F}(f)$ and $\hat{g} = \mathcal{F}(g)$ as before, and noting that $\sqrt{2\pi}$ and $1/\sqrt{2\pi}$ cancel each other, we obtain

$$(f * g)(x) = \int_{-\infty}^{\infty} \hat{f}(w)\hat{g}(w)e^{iwx}\, dw, \tag{13}$$

a formula that will help us in solving partial differential equations (Sec. 11.14).

A **table** of Fourier transforms is included in the next section. For more extensive tables, see Ref. [C4] in Appendix 1.

This is the end of Chap. 10 on Fourier series, Fourier integrals, and Fourier transforms. The introduction of Fourier series (and Fourier integrals) was one of the greatest advances ever made in mathematical physics and its engineering applications, because Fourier series (and Fourier integrals) are probably the most important tool in solving boundary value problems. This will be explained in the next chapter.

Problem Set 10.11

Find the Fourier transforms of the following functions $f(x)$ (without using Table III, Sec. 10.12).

1. $f(x) = \begin{cases} e^{-x} & \text{if } x > 0 \\ 0 & \text{if } x < 0 \end{cases}$

2. $f(x) = \begin{cases} e^{x} & \text{if } x < 0 \\ 0 & \text{if } x > 0 \end{cases}$

3. $f(x) = \begin{cases} e^{2ix} & \text{if } -1 < x < 1 \\ 0 & \text{otherwise} \end{cases}$

4. $f(x) = \begin{cases} e^{x} & \text{if } -1 < x < 1 \\ 0 & \text{otherwise} \end{cases}$

5. $f(x) = \begin{cases} x & \text{if } 0 < x < a \\ 0 & \text{otherwise} \end{cases}$

6. $f(x) = \begin{cases} xe^{-x} & \text{if } x > 0 \\ 0 & \text{if } x < 0 \end{cases}$

7. $f(x) = \begin{cases} -1 & \text{if } -1 < x < 0 \\ 1 & \text{if } \quad 0 < x < 1 \\ 0 & \text{otherwise} \end{cases}$

8. $f(x) = \begin{cases} e^{x} & \text{if } x < 0 \\ e^{-x} & \text{if } x > 0 \end{cases}$

9. Derive formula 1 in Table III, Sec. 10.12.

10. Using the answer $\hat{f}(w)$ to Prob. 9, obtain $f(x)$ from (7). *Hint.* Use (7*) in Sec. 10.9.

11. Obtain formula 1 in Table III, Sec. 10.12, from formula 2 in that table.

12. Solve Prob. 6 by (9), using the answer to Prob. 1.

13. **(Shifting)** Show that if $f(x)$ has a Fourier transform, so does $f(x - a)$, and

$$\mathcal{F}\{f(x - a)\} = e^{-iwa}\mathcal{F}\{f(x)\}.$$

14. Solve Prob. 6 by convolution. *Hint.* Show that $xe^{-x} = e^{-x} * e^{-x}$ $(x > 0)$.

15. Using Prob. 13, obtain formula 1 in Table III, Sec. 10.12, from formula 2 (with $c = 3b$).

16. Using the answer to Prob. 8, write down the Fourier integral representation of $f(x)$ and convert it to a Fourier cosine integral (see Sec. 10.9).

17. **(Shifting on the w-axis)** Show that if $\hat{f}(w)$ is the Fourier transform of $f(x)$, then $\hat{f}(w - a)$ is the Fourier transform of $e^{iax}f(x)$.

18. Using Prob. 17, obtain formula 7 in Table III, Sec. 10.12, from formula 1 in that table.

19. Using Prob. 17, obtain formula 8 in Table III, Sec. 10.12, from formula 2 in that table.

20. Verify formula 3 in Table III, Sec. 10.12, with $a = 1$. *Hint.* Use (15) in Sec. 10.9 and (3) in this section.

10.12 Tables of Transforms

For more extensive tables, see Ref.* [C4] *in Appendix 1.

Table I. Fourier Cosine Transforms

See (2) in Sec. 10.10.

	$f(x)$	$\hat{f}_c(w) = \mathcal{F}_c(f)$	
1	$\begin{cases} 1 & \text{if } 0 < x < a \\ 0 & \text{otherwise} \end{cases}$	$\sqrt{\frac{2}{\pi}} \frac{\sin aw}{w}$	
2	$x^{a-1} \quad (0 < a < 1)$	$\sqrt{\frac{2}{\pi}} \frac{\Gamma(a)}{w^a} \cos \frac{aw}{2}$	($\Gamma(a)$ see Appendix 3.1.)
3	$e^{-ax} \quad (a > 0)$	$\sqrt{\frac{2}{\pi}} \left(\frac{a}{a^2 + w^2} \right)$	
4	$e^{-x^2/2}$	$e^{-w^2/2}$	
5	$e^{-ax^2} \quad (a > 0)$	$\frac{1}{\sqrt{2a}} e^{-w^2/4a}$	
6	$x^n e^{-ax} \quad (a > 0)$	$\sqrt{\frac{2}{\pi}} \frac{n!}{(a^2 + w^2)^{n+1}} \operatorname{Re} (a + iw)^{n+1}$	Re = Real part
7	$\begin{cases} \cos x & \text{if } 0 < x < a \\ 0 & \text{otherwise} \end{cases}$	$\frac{1}{\sqrt{2\pi}} \left[\frac{\sin a(1 - w)}{1 - w} + \frac{\sin a(1 + w)}{1 + w} \right]$	
8	$\cos ax^2 \quad (a > 0)$	$\frac{1}{\sqrt{2a}} \cos \left(\frac{w^2}{4a} - \frac{\pi}{4} \right)$	
9	$\sin ax^2 \quad (a > 0)$	$\frac{1}{\sqrt{2a}} \cos \left(\frac{w^2}{4a} + \frac{\pi}{4} \right)$	
10	$\frac{\sin ax}{x} \quad (a > 0)$	$\sqrt{\frac{\pi}{2}} u(a - w)$	(See Sec. 6.3.)
11	$\frac{e^{-x} \sin x}{x}$	$\frac{1}{\sqrt{2\pi}} \arctan \frac{2}{w^2}$	
12	$J_0(ax) \quad (a > 0)$	$\sqrt{\frac{2}{\pi}} \frac{u(a - w)}{\sqrt{a^2 - w^2}}$	(See Secs. 5.5, 6.3.)

Table II. Fourier Sine Transforms

See (5) in Sec. 10.10.

	$f(x)$	$\hat{f}_s(w) = \mathcal{F}_s(f)$
1	$\begin{cases} 1 & \text{if } 0 < x < a \\ 0 & \text{otherwise} \end{cases}$	$\sqrt{\frac{2}{\pi}}\left[\frac{1 - \cos aw}{w}\right]$
2	$1/\sqrt{x}$	$1/\sqrt{w}$
3	$1/x^{3/2}$	$2\sqrt{w}$
4	x^{a-1} $(0 < a < 1)$	$\sqrt{\frac{2}{\pi}}\frac{\Gamma(a)}{w^a}\sin\frac{a\pi}{2}$ ($\Gamma(a)$ see Appendix 3.1.)
5	e^{-x}	$\sqrt{\frac{2}{\pi}}\left(\frac{w}{1 + w^2}\right)$
6	e^{-ax}/x $(a > 0)$	$\sqrt{\frac{2}{\pi}}\arctan\frac{w}{a}$
7	$x^n e^{-ax}$ $(a > 0)$	$\sqrt{\frac{2}{\pi}}\frac{n!}{(a^2 + w^2)^{n+1}}\operatorname{Im}(a + iw)^{n+1}$ Im = Imaginary part
8	$xe^{-x^2/2}$	$we^{-w^2/2}$
9	xe^{-ax^2} $(a > 0)$	$\frac{w}{(2a)^{3/2}}e^{-w^2/4a}$
10	$\begin{cases} \sin x & \text{if } 0 < x < a \\ 0 & \text{otherwise} \end{cases}$	$\frac{1}{\sqrt{2\pi}}\left[\frac{\sin a(1 - w)}{1 - w} - \frac{\sin a(1 + w)}{1 + w}\right]$
11	$\frac{\cos ax}{x}$ $(a > 0)$	$\sqrt{\frac{\pi}{2}}u(w - a)$ (See Sec. 6.3.)
12	$\arctan\frac{2a}{x}$ $(a > 0)$	$\sqrt{2\pi}\frac{\sinh aw}{w}e^{-aw}$

Table III. Fourier Transforms

See (6) in Sec. 10.11.

	$f(x)$	$\hat{f}(w) = \mathcal{F}(f)$
1	$\begin{cases} 1 & \text{if } -b < x < b \\ 0 & \text{otherwise} \end{cases}$	$\sqrt{\frac{2}{\pi}}\,\frac{\sin bw}{w}$
2	$\begin{cases} 1 & \text{if } b < x < c \\ 0 & \text{otherwise} \end{cases}$	$\frac{e^{-ibw} - e^{-icw}}{iw\sqrt{2\pi}}$
3	$\frac{1}{x^2 + a^2} \quad (a > 0)$	$\sqrt{\frac{\pi}{2}}\,\frac{e^{-a\lvert w\rvert}}{a}$
4	$\begin{cases} x & \text{if } 0 < x < b \\ 2x - a & \text{if } b < x < 2b \\ 0 & \text{otherwise} \end{cases}$	$\frac{-1 + 2e^{ibw} - e^{-2ibw}}{\sqrt{2\pi}\, w^2}$
5	$\begin{cases} e^{-ax} & \text{if } x > 0 \\ 0 & \text{otherwise} \end{cases} \quad (a > 0)$	$\frac{1}{\sqrt{2\pi}(a + iw)}$
6	$\begin{cases} e^{ax} & \text{if } b < x < c \\ 0 & \text{otherwise} \end{cases}$	$\frac{e^{(a-iw)c} - e^{(a-iw)b}}{\sqrt{2\pi}(a - iw)}$
7	$\begin{cases} e^{iax} & \text{if } -b < x < b \\ 0 & \text{otherwise} \end{cases}$	$\sqrt{\frac{2}{\pi}}\,\frac{\sin b(w - a)}{w - a}$
8	$\begin{cases} e^{iax} & \text{if } b < x < c \\ 0 & \text{otherwise} \end{cases}$	$\frac{i}{\sqrt{2\pi}}\,\frac{e^{ib(a-w)} - e^{ic(a-w)}}{a - w}$
9	$e^{-ax^2} \quad (a > 0)$	$\frac{1}{\sqrt{2a}}\, e^{-w^2/4a}$
10	$\frac{\sin ax}{x} \quad (a > 0)$	$\sqrt{\frac{\pi}{2}}$ if $\lvert w\rvert < a$; 0 if $\lvert w\rvert > a$

Review Questions and Problems for Chapter 10

1. What do we mean by a trigonometric series? By a Fourier series?
2. What do we mean by orthogonality? What role did it play in the derivation of the Euler formulas?
3. How did we accomplish the transition from a function of period 2π to a function having an arbitrary period?
4. What is an odd periodic function? To what form does its Fourier series reduce?
5. Answer the same questions as in Prob. 4 for an even periodic function.
6. What do we mean by half-range expansions?
7. Can a discontinuous function have a Fourier series? A Taylor series?
8. If the Fourier series of a function $f(x)$ has both cosine and sine terms, what is the sum of the cosine terms expressed in terms of $f(x)$? The sum of the sine terms?
9. What is piecewise continuity? In what connection did it occur in this chapter?
10. What is a trigonometric polynomial? Why did we consider it?
11. What is the mean square error? Why did we consider it?
12. What is a complex Fourier series? How is it related to the usual real form of a Fourier series?
13. What is the basic difference in the vibrations of a mass–spring system governed by $my'' + cy' + ky = r(t)$ with an arbitrary periodic driving force as opposed to a pure sinusoidal driving force?
14. What is the Fourier transform? How did we obtain it from the Fourier integral?
15. Does every continuous function have a Fourier cosine or sine transform? Can a discontinuous function have a Fourier cosine or sine transform?

Find the Fourier series of the given functions that are assumed to have period 2π.

16. $f(x) = \begin{cases} k & \text{if } -\pi/2 < x < \pi/2 \\ -k & \text{if } \pi/2 < x < 3\pi/2 \end{cases}$

17. $f(x) = \begin{cases} 0 & \text{if } 0 < x < \pi \\ 1 & \text{if } \pi < x < 2\pi \end{cases}$

18. $f(x) = \begin{cases} 1 & \text{if } -\pi < x < -\pi/2 \\ -1 & \text{if } -\pi/2 < x < 0 \\ 0 & \text{if } 0 < x < \pi \end{cases}$

19. $f(x) = \begin{cases} -1 & \text{if } -\pi < x < -\pi/2 \\ 0 & \text{if } -\pi/2 < x < \pi/2 \\ 1 & \text{if } \pi/2 < x < \pi \end{cases}$

20. $f(x) = \frac{1}{2}(\pi - x) \quad (0 < x < 2\pi)$

21. $f(x) = |x| \quad (-\pi < x < \pi)$

22. $f(x) = \frac{1}{2}kx|x| \quad (-\pi < x < \pi)$

23. $f(x) = -x/2 \quad (-\pi < x < \pi)$

24. $f(x) = x^4 \quad (-\pi < x < \pi)$

25. $f(x) = \pi - 2|x| \quad (-\pi < x < \pi)$

26. $f(x) = \begin{cases} 0 & \text{if } -\pi < x < 0 \\ \pi x & \text{if } 0 < x < \pi \end{cases}$

27. $f(x) = \begin{cases} x^2/2 & \text{if } -\pi/2 < x < \pi/2 \\ \pi^2/8 & \text{if } \pi/2 < x < 3\pi/2 \end{cases}$

Find the Fourier series of the given functions that are assumed to have period 2π.

28. $f(x) = \begin{cases} 0 & \text{if } -\pi < x < 0 \\ x^2 & \text{if } \quad 0 < x < \pi \end{cases}$

29. $f(x) = \begin{cases} \pi x + x^2 & \text{if } -\pi < x < 0 \\ \pi x - x^2 & \text{if } \quad 0 < x < \pi \end{cases}$

30. $f(x) = \begin{cases} x/a & \text{if } -a < x < a\ (< \pi) \\ \dfrac{\pi - x}{\pi - a} & \text{if } \quad a < x < 2\pi - a \end{cases}$

Find the Fourier series of the following functions that are assumed to have period $p = 2L$.

31. $f(x) = 1 \quad (-1 < x < 0), \quad f(x) = -1 \quad (0 < x < 1), \qquad p = 2L = 2$

32. $f(x) = 1 \quad (-1 < x < 1), \quad f(x) = 0 \quad (1 < x < 3), \qquad p = 2L = 4$

33. $f(x) = x \quad (-2 < x < 2), \qquad p = 2L = 4$

34. $f(x) = 0 \quad (-1 < x < 0), \quad f(x) = x \quad (0 < x < 1), \qquad p = 2L = 2$

35. $f(x) = -x \quad (-1 < x < 0), \quad f(x) = 0 \quad (0 < x < 1), \qquad p = 2L = 2$

36. $f(x) = x \quad (-4 < x < 4), \qquad p = 2L = 8$

37. $f(x) = -1 \quad (-1 < x < 0), \quad f(x) = 2x \quad (0 < x < 1), \qquad p = 2L = 2$

38. $f(x) = 1 - x^2 \quad (0 < x < 2), \qquad p = 2L = 2$

39. $f(x) = x - x^2 \quad (-1 < x < 1), \qquad p = 2L = 2$

40. $f(x) = 1 \quad (-2 < x < 0), \quad f(x) = e^{-x} \quad (0 < x < 2), \qquad p = 2L = 4$

Find the sum of

41. $1 - \frac{1}{3} + \frac{1}{5} - \frac{1}{7} + - \cdots$ (Use Prob. 17.)

42. $1 + \frac{1}{9} + \frac{1}{25} + \frac{1}{47} + \cdots$ (Use Prob. 21.)

43. $1 - 3^{-3} + 5^{-3} - 7^{-3} + - \cdots$ (Use Prob. 29.)

Using Parseval's identity, show that

44. $1 + \dfrac{1}{2^2} + \dfrac{1}{3^2} + \dfrac{1}{4^2} + \cdots = \dfrac{\pi^2}{6}$ (Use Prob. 23.)

45. $1 + \dfrac{1}{3^4} + \dfrac{1}{5^4} + \dfrac{1}{7^4} + \cdots = \dfrac{\pi^4}{96}$ (Use Prob. 25.)

46. $1 + \dfrac{1}{3^6} + \dfrac{1}{5^6} + \dfrac{1}{7^6} + \cdots = \dfrac{\pi^6}{960}$ (Use Prob. 29.)

Compute the first six minimum square errors E^* corresponding to the first few partial sums:

47. In Prob. 21.

48. In Prob. 29.

Find a general solution of $y'' + \omega^2 y = r(t)$, where

49. $r(t) = t^2/4 \quad (-\pi < t < \pi), \quad r(t + 2\pi) = r(t), \qquad |\omega| \neq 0, 1, 2, \cdots$

50. $r(t) = t(\pi^2 - t^2)/12 \quad (-\pi < t < \pi), \quad r(t + 2\pi) = r(t), \qquad |\omega| \neq 1, 2, \cdots$

Summary of Chapter 10 Fourier Series, Integrals, and Transforms

A **trigonometric series** (Sec. 10.1) is a series of the form

$$a_0 + \sum_{n=1}^{\infty} (a_n \cos nx + b_n \sin nx). \tag{1}$$

The **Fourier series** of a given periodic function $f(x)$ of period 2π is a trigonometric series (1) whose coefficients are the **Fourier coefficients** of $f(x)$ given by the **Euler formulas** (Sec. 10.2)

$$\begin{gathered} a_0 = \frac{1}{2\pi}\int_{-\pi}^{\pi} f(x)\,dx, \qquad a_n = \frac{1}{\pi}\int_{-\pi}^{\pi} f(x)\cos nx\,dx, \\ b_n = \frac{1}{\pi}\int_{-\pi}^{\pi} f(x)\sin nx\,dx. \end{gathered} \tag{2}$$

For a function $f(x)$ of any period $p = 2L$ the **Fourier series** is (Sec. 10.3)

$$f(x) = a_0 + \sum_{n=1}^{\infty}\left(a_n \cos\frac{n\pi}{L}x + b_n \sin\frac{n\pi}{L}x\right) \tag{3}$$

with the Fourier coefficients of $f(x)$ given by the **Euler formulas**

$$\begin{gathered} a_0 = \frac{1}{2L}\int_{-L}^{L} f(x)\,dx, \qquad a_n = \frac{1}{L}\int_{-L}^{L} f(x)\cos\frac{n\pi x}{L}\,dx \\ b_n = \frac{1}{L}\int_{-L}^{L} f(x)\sin\frac{n\pi x}{L}\,dx \end{gathered} \tag{4}$$

These series are fundamental in connection with periodic phenomena, particularly in models involving differential equations (Sec. 10.7 and Chap. 11). If $f(x)$ is even $[f(-x) = f(x)]$ or odd $[f(-x) = -f(x)]$, they reduce to **Fourier cosine** or **Fourier sine series,** respectively (Sec. 10.4).

A function $f(x)$ given on an interval $0 \leqq x \leqq L$ can be developed in a Fourier cosine or sine series of period $2L$; these are called **half-range expansions** of $f(x)$; see Sec. 10.5.

The set of cosine and sine functions in (3) is called the **trigonometric system.** Its most basic property is its **orthogonality** on an interval of length $2L$, that is, for all integers m and $n \neq m$ we have

$$\int_{-L}^{L} \cos\frac{m\pi x}{L}\cos\frac{n\pi x}{L}\,dx = 0, \qquad \int_{-L}^{L} \sin\frac{m\pi x}{L}\sin\frac{n\pi x}{L}\,dx = 0$$

and for all integers m and n,

$$\int_{-L}^{L} \cos\frac{m\pi x}{L}\sin\frac{n\pi x}{L}\,dx = 0.$$

This orthogonality was crucial in deriving the Euler formulas for the Fourier coefficients.

Partial sums of Fourier series minimize the square error (Sec. 10.8).

Ideas and techniques of Fourier series extend to nonperiodic functions $f(x)$ defined on the entire real line; this leads to the **Fourier integral**

$$f(x) = \int_0^{\infty} [A(w)\cos wx + B(w)\sin wx]\,dw \tag{5}$$

(Sec. 10.9), where

$$A(w) = \frac{1}{\pi}\int_{-\infty}^{\infty} f(v)\cos wv\,dv, \qquad B(w) = \frac{1}{\pi}\int_{-\infty}^{\infty} f(v)\sin wv\,dv \tag{6}$$

or, in complex form (Sec. 10.11),

$$f(x) = \frac{1}{\sqrt{2\pi}}\int_{-\infty}^{\infty} \hat{f}(w)e^{iwx}\,dw \qquad (i = \sqrt{-1}), \tag{7}$$

where

$$\hat{f}(w) = \frac{1}{\sqrt{2\pi}}\int_{-\infty}^{\infty} f(x)e^{-iwx}\,dx. \tag{8}$$

Formula (8) transforms $f(x)$ into its **Fourier transform** $\hat{f}(w)$.

Related to this are the **Fourier cosine transform**

$$\hat{f}_c(w) = \sqrt{\frac{2}{\pi}}\int_0^{\infty} f(x)\cos wx\,dx \tag{9}$$

and the **Fourier sine transform** (Sec. 10.10)

$$\hat{f}_s(w) = \sqrt{\frac{2}{\pi}}\int_0^{\infty} f(x)\sin wx\,dx. \tag{10}$$

Chapter 11

Partial Differential Equations

Partial differential equations arise in connection with various physical and geometrical problems when the functions involved depend on two or more independent variables. It is fair to say that only the simplest physical systems can be modeled by *ordinary* differential equations, whereas most problems in fluid and solid mechanics (dynamics, elasticity), heat transfer, electromagnetic theory, quantum mechanics, and other areas of physics lead to ***partial*** differential equations. Indeed, the range of application of the latter is enormous, compared to that of ordinary differential equations. The independent variables involved may be time and one or several coordinates in space. The present chapter is devoted to some of the most important partial differential equations occurring in engineering applications. We derive these equations as models of physical systems and consider methods for solving **initial** and **boundary value problems,** that is, methods for obtaining solutions of those equations corresponding to the given physical situations.

In Sec. 11.1 we define the notion of a solution of a partial differential equation. Sections 11.2–11.4 concern the one-dimensional wave equation, governing the motion of a vibrating string. The heat equation is considered in Secs. 11.5 and 11.6, the two-dimensional wave equation (vibrating membranes) in Secs. 11.7–11.10, and Laplace's equation in Secs. 11.11 and 11.12.

In Secs. 11.13 and 11.14 we see that partial differential equations can also be solved by Laplace transform or Fourier transform methods.

Numerical methods for partial differential equations are presented in Secs. 20.4–20.7.

Prerequisites for this chapter: ordinary linear differential equations (Chap. 2) and Fourier series (Chap. 10).

Sections that may be omitted in a shorter course: 11.6, 11.9, 11.10, 11.13, 11.14.

References: Appendix 1, Part C.

Answers to problems: Appendix 2.

11.1 Basic Concepts

An equation involving one or more partial derivatives of an (unknown) function of two or more independent variables is called a **partial differential equation.** The order of the highest derivative is called the **order** of the equation.

Just as in the case of an ordinary differential equation, we say that a partial differential equation is **linear** if it is of the first degree in the dependent variable (the unknown function) and its partial derivatives. If each term of such an equation contains either the dependent variable or one of its derivatives, the equation is said to be **homogeneous;** otherwise it is said to be **nonhomogeneous.**

EXAMPLE 1 **Important linear partial differential equations of the second order**

(1) $$\frac{\partial^2 u}{\partial t^2} = c^2 \frac{\partial^2 u}{\partial x^2}$$ ***One-dimensional wave equation***

(2) $$\frac{\partial u}{\partial t} = c^2 \frac{\partial^2 u}{\partial x^2}$$ ***One-dimensional heat equation***

(3) $$\frac{\partial^2 u}{\partial x^2} + \frac{\partial^2 u}{\partial y^2} = 0$$ ***Two-dimensional Laplace equation***

(4) $$\frac{\partial^2 u}{\partial x^2} + \frac{\partial^2 u}{\partial y^2} = f(x, y)$$ ***Two-dimensional Poisson equation***

(5) $$\frac{\partial^2 u}{\partial x^2} + \frac{\partial^2 u}{\partial y^2} + \frac{\partial^2 u}{\partial z^2} = 0$$ ***Three-dimensional Laplace equation***

Here c is a constant, t is time, and x, y, z are Cartesian coordinates. Equation (4) (with $f(x, y) \not\equiv 0$) is nonhomogeneous, while the other equations are homogeneous. ∎

A **solution** *of a partial differential equation in some region R of the space of the independent variables* is a function that has all the partial derivatives appearing in the equation in some domain containing R and satisfies the equation everywhere in R. (Often one merely requires that that function is continuous on the boundary of R, has those derivatives in the interior of R, and satisfies the equation in the interior of R.)

In general, the totality of solutions of a partial differential equation is very large. For example, the functions

(6) $$u = x^2 - y^2, \qquad u = e^x \cos y, \qquad u = \ln (x^2 + y^2),$$

which are entirely different from each other, are solutions of (3), as the student may verify. We shall see later that the unique solution of a partial differential equation corresponding to a given physical problem will be obtained by the use of additional conditions arising from the problem, for instance, the condition that the solution u assume given values on the boundary of the region considered (**"boundary conditions"**), or, when time t is one of the variables, that u (or $u_t = \partial u/\partial t$ or both) are prescribed at $t = 0$ (**"initial conditions"**).

We know that if an *ordinary* differential equation is linear and homogeneous, then from known solutions we can obtain further solutions by superposition. For a homogeneous linear *partial* differential equation the situation is quite similar. In fact, the following theorem holds.

Theorem 1 Fundamental Theorem (Superposition or linearity principle)

If u_1 and u_2 are any solutions of a linear homogeneous partial differential equation in some region R, then

$$u = c_1u_1 + c_2u_2,$$

where c_1 and c_2 are any constants, is also a solution of that equation in R.

The proof of this important theorem is simple and quite similar to that of Theorem 1 in Sec. 2.1 and is left to the student.

Verification of solutions in Probs. 2–23 proceeds as for ordinary differential equations. Problems 24–35 concern partial differential equations that can be solved like ordinary ones; to help the student with them, we consider two typical examples.

EXAMPLE 2 Find a solution $u(x, y)$ of the partial differential equation $u_{xx} - u = 0$.

Solution. Since no y-derivatives occur, we can solve this like $u'' - u = 0$. In Sec. 2.2 we would have obtained $u = Ae^x + Be^{-x}$ with constant A and B. Here A and B may be functions of y, so that the answer is

$$u(x, y) = A(y)e^x + B(y)e^{-x}$$

with arbitrary functions A and B, so that we have a great variety of solutions. Check the result by differentiation. ∎

EXAMPLE 3 Solve the partial differential equation $u_{xy} = -u_x$.

Solution. Setting $u_x = p$, we have $p_y = -p$, $p_y/p = -1$, $\ln p = -y + \tilde{c}(x)$, $p = c(x)e^{-y}$ and by integration with respect to x,

$$u(x, y) = f(x)e^{-y} + g(y) \qquad \text{where} \qquad f(x) = \int c(x)\,dx;$$

here, $f(x)$ and $g(y)$ are arbitrary. ∎

Problem Set 11.1

1. Prove Fundamental Theorem 1 for second-order differential equations in two and three independent variables.
2. Verify that the functions (6) are solutions of (3).

Verify that the following functions are solutions of Laplace's equation.

3. $u = 2xy$
4. $u = x^3 - 3xy^2$
5. $u = x^4 - 6x^2y^2 + y^4$
6. $u = e^x \sin y$
7. $u = \sin x \sinh y$
8. $u = \arctan (y/x)$

Verify that the following functions are solutions of the wave equation (1) for a suitable value of c.

9. $u = x^2 + 4t^2$
10. $u = x^3 + 3xt^2$
11. $u = \sin 2ct \sin 2x$
12. $u = \cos 4t \sin x$
13. $u = \cos ct \sin x$
14. $u = \sin \omega ct \sin \omega x$

Verify that the following functions are solutions of the heat equation (2) for a suitable value of c.

15. $u = e^{-t} \cos x$ **16.** $u = e^{-2t} \cos x$ **17.** $u = e^{-t} \sin 3x$

18. $u = e^{-4t} \cos \omega x$ **19.** $u = e^{-16t} \cos 2x$ **20.** $u = e^{-\omega^2 c^2 t} \sin \omega x$

21. Show that $u = 1/\sqrt{x^2 + y^2 + z^2}$ is a solution of Laplace's equation (5).

22. Verify that $u(x, y) = a \ln (x^2 + y^2) + b$ satisfies Laplace's equation (3) and determine a and b so that u satisfies the boundary conditions $u = 0$ on the circle $x^2 + y^2 = 1$ and $u = 5$ on the circle $x^2 + y^2 = 9$.

23. Show that $u(x, t) = v(x + ct) + w(x - ct)$ is a solution of the wave equation (1); here, v and w are any twice differentiable functions.

Partial differential equations solvable as ordinary differential equations

If an equation involves derivatives with respect to one variable only, we can solve it like an ordinary differential equation, treating the other variable (or variables) as parameters. Find solutions $u(x, y)$ of

24. $u_x = 0$ **25.** $u_y = 0$ **26.** $u_{xx} + 4u = 0$

27. $u_{xx} = 0$ **28.** $u_y + 2yu = 0$ **29.** $u_x = 2xyu$

Setting $u_x = p$, solve

30. $u_{xy} = u_x$ **31.** $u_{xy} = 0$ **32.** $u_{xyy} + u_x = 0$

Solve the following systems of partial differential equations.

33. $u_x = 0, \quad u_y = 0$ **34.** $u_{xx} = 0, \quad u_{yy} = 0$ **35.** $u_{xx} = 0, \quad u_{xy} = 0$

11.2 Modeling: Vibrating String, Wave Equation

As a first important partial differential equation, let us derive the equation governing small transverse vibrations of an elastic string, such as a violin string. We stretch the string to length L and fix it at the ends. We then distort it and at some instant, say, $t = 0$, we release it and allow it to vibrate. The problem is to determine the vibrations of the string, that is, to find its deflection $u(x, t)$ at any point x and at any time $t > 0$; see Fig. 251.

When deriving a differential equation corresponding to a given physical problem, we usually have to make simplifying assumptions to ensure that the resulting equation does not become too complicated. We know this important fact from our study of ordinary differential equations, and for partial differential equations the situation is similar.

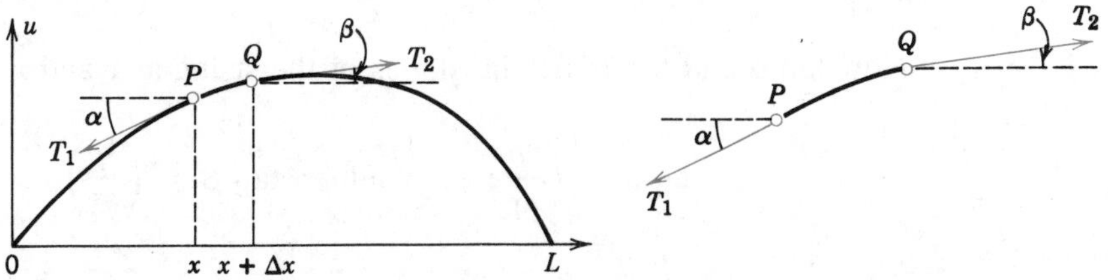

Fig. 251. Deflected string at fixed time t

Physical Assumptions. We assume the following:

1. *The mass of the string per unit length is constant ("homogeneous string"). The string is perfectly elastic and does not offer any resistance to bending.*
2. *The tension caused by stretching the string before fixing it at the ends is so large that the action of the gravitational force on the string can be neglected.*
3. *The string performs small transverse motions in a vertical plane; that is, every particle of the string moves strictly vertically and so that the deflection and the slope at every point of the string always remain small in absolute value.*

Under these assumptions we may expect that the solution $u(x, t)$ of the differential equation to be obtained will reasonably well describe small vibrations of the physical "nonidealized" string of small homogeneous mass under large tension.

Differential Equation from Forces. To obtain the differential equation we consider the forces acting on a small portion of the string (Fig. 251). Since the string does not offer resistance to bending, the tension is tangential to the curve of the string at each point. Let T_1 and T_2 be the tension at the endpoints P and Q of that portion. Since there is no motion in horizontal direction, the horizontal components of the tension must be constant. Using the notation shown in Fig. 251 we thus obtain

$$T_1 \cos \alpha = T_2 \cos \beta = T = const. \tag{1}$$

In vertical direction we have two forces, namely, the vertical components $-T_1 \sin \alpha$ and $T_2 \sin \beta$ of T_1 and T_2; here the minus sign appears because that component at P is directed downward. By Newton's second law the resultant of these two forces is equal to the mass $\rho \, \Delta x$ of the portion times the acceleration $\partial^2 u/\partial t^2$, evaluated at some point between x and $x + \Delta x$; here ρ is the mass of the undeflected string per unit length, and Δx is the length of the portion of the undeflected string. Hence

$$T_2 \sin \beta - T_1 \sin \alpha = \rho \, \Delta x \frac{\partial^2 u}{\partial t^2} .$$

Using (1), we can divide this by $T_2 \cos \beta = T_1 \cos \alpha = T$, obtaining

$$\frac{T_2 \sin \beta}{T_2 \cos \beta} - \frac{T_1 \sin \alpha}{T_1 \cos \alpha} = \tan \beta - \tan \alpha = \frac{\rho \, \Delta x}{T} \frac{\partial^2 u}{\partial t^2} . \tag{2}$$

Now $\tan \alpha$ and $\tan \beta$ are the slopes of the string at x and $x + \Delta x$:

$$\tan \alpha = \left(\frac{\partial u}{\partial x}\right)_x \qquad \text{and} \qquad \tan \beta = \left(\frac{\partial u}{\partial x}\right)_{x+\Delta x} .$$

Here we have to write *partial* derivatives because u also depends on t. Dividing (2) by Δx, we thus have

$$\frac{1}{\Delta x}\left[\left(\frac{\partial u}{\partial x}\right)_{x+\Delta x} - \left(\frac{\partial u}{\partial x}\right)_{x}\right] = \frac{\rho}{T}\frac{\partial^2 u}{\partial t^2}.$$

If we let Δx approach zero, we obtain the linear partial differential equation

(3)
$$\frac{\partial^2 u}{\partial t^2} = c^2 \frac{\partial^2 u}{\partial x^2} \qquad c^2 = \frac{T}{\rho}.$$

This is the so-called **one-dimensional wave equation,** which governs our problem. We see that it is homogeneous and of the second order. The notation c^2 (instead of c) for the physical constant T/ρ has been chosen to indicate that this constant is positive. "One-dimensional" indicates that the equation involves only one space variable, x. Solutions will be obtained in the next section.

11.3 Separation of Variables, Use of Fourier Series

In the last section we showed that the vibrations of an elastic string, such as a violin string, are governed by the **one-dimensional wave equation**

(1)
$$\frac{\partial^2 u}{\partial t^2} = c^2 \frac{\partial^2 u}{\partial x^2},$$

where $u(x, t)$ is the deflection of the string. To find out how the string moves, we solve this equation; more precisely, we determine a solution u of (1) that also satisfies the conditions imposed by the physical system. Since the string is fixed at the ends $x = 0$ and $x = L$, we have the two **boundary conditions**

(2)
$$u(0, t) = 0, \qquad u(L, t) = 0 \qquad \text{for all } t.$$

The form of the motion of the string will depend on the initial deflection (deflection at $t = 0$) and on the initial velocity (velocity at $t = 0$). Denoting the initial deflection by $f(x)$ and the initial velocity by $g(x)$, we thus obtain the two **initial conditions**

(3)
$$u(x, 0) = f(x)$$

and

(4)
$$\left.\frac{\partial u}{\partial t}\right|_{t=0} = g(x).$$

Our problem is now to find a solution of (1) satisfying the conditions (2)–(4). We shall proceed step by step, as follows.

First Step. By applying the so-called **method of separating variables** or *product method,* we shall obtain two *ordinary* differential equations.

Second Step. We shall determine solutions of those two equations that satisfy the boundary conditions.

Third Step. Using **Fourier series,** we shall compose those solutions, in order to get a solution of the wave equation (1) that also satisfies the given initial conditions.

First Step. Two Ordinary Differential Equations

In the **method of separating variables,** or *product method*, we determine solutions of the wave equation (1) of the form

$$u(x, t) = F(x)G(t) \tag{5}$$

which are a product of two functions, each depending only on one of the variables x and t. We shall see later that this method has various other applications in engineering mathematics. By differentiating (5) we obtain

$$\frac{\partial^2 u}{\partial t^2} = F\ddot{G} \qquad \text{and} \qquad \frac{\partial^2 u}{\partial x^2} = F''G,$$

where dots denote derivatives with respect to t and primes derivatives with respect to x. By inserting this into our differential equation (1) we have

$$F\ddot{G} = c^2 F''G.$$

Dividing by c^2FG, we find

$$\frac{\ddot{G}}{c^2G} = \frac{F''}{F}.$$

We now claim that both sides must be constant. Indeed, the expression on the left depends only on t and that on the right only on x, and if they were variable, then changing t or x would affect only the left or the right side, respectively, leaving the other side unaltered. Thus,

$$\frac{\ddot{G}}{c^2G} = \frac{F''}{F} = k.$$

This yields immediately two ordinary linear differential equations, namely,

$$F'' - kF = 0 \tag{6}$$

and

$$\ddot{G} - c^2kG = 0. \tag{7}$$

Here, k is still arbitrary.

Second Step. Satisfying the Boundary Conditions

We shall now determine solutions F and G of (6) and (7) so that $u = FG$ satisfies the boundary conditions (2), that is,

$$u(0, t) = F(0)G(t) = 0, \qquad u(L, t) = F(L)G(t) = 0 \qquad \text{for all } t.$$

Solving (6). If $G \equiv 0$, then $u \equiv 0$, which is of no interest. Thus $G \not\equiv 0$ and then

$$\text{(8)} \qquad \text{(a)} \quad F(0) = 0, \qquad \text{(b)} \quad F(L) = 0.$$

For $k = 0$ the general solution of (6) is $F = ax + b$, and from (8) we obtain $a = b = 0$. Hence $F \equiv 0$, which is of no interest because then $u \equiv 0$. For positive $k = \mu^2$ the general solution of (6) is

$$F = Ae^{\mu x} + Be^{-\mu x},$$

and from (8) we obtain $F \equiv 0$, as before. Hence we are left with the possibility of choosing k negative, say, $k = -p^2$. Then equation (6) takes the form

$$F'' + p^2F = 0.$$

Its general solution is

$$F(x) = A \cos px + B \sin px.$$

From this and (8) we have

$$F(0) = A = 0 \qquad \text{and then} \qquad F(L) = B \sin pL = 0.$$

We must take $B \neq 0$ since otherwise $F \equiv 0$. Hence $\sin pL = 0$. Thus

$$\text{(9)} \qquad pL = n\pi, \qquad \text{so that} \qquad p = \frac{n\pi}{L} \qquad (n \text{ integer}).$$

Setting $B = 1$, we thus obtain infinitely many solutions $F(x) = F_n(x)$, where

$$\text{(10)} \qquad F_n(x) = \sin \frac{n\pi}{L} x \qquad (n = 1, 2, \cdots).$$

These solutions satisfy (8). [For negative integer n we obtain essentially the same solutions, except for a minus sign, because $\sin(-\alpha) = -\sin \alpha$.]

Solving (7). The constant k is now restricted to the values $k = -p^2 = -(n\pi/L)^2$, resulting from (9). For these k, equation (7) becomes

$$\ddot{G} + \lambda_n^{\,2}G = 0 \qquad \text{where} \qquad \lambda_n = \frac{cn\pi}{L}.$$

A general solution is

$$G_n(t) = B_n \cos \lambda_n t + B_n^* \sin \lambda_n t.$$

Hence the functions $u_n(x, t) = F_n(x)G_n(t)$, written out

$$u_n(x, t) = (B_n \cos \lambda_n t + B_n^* \sin \lambda_n t) \sin \frac{n\pi}{L} x \qquad (n = 1, 2, \cdots), \tag{11}$$

are solutions of (1), satisfying the boundary conditions (2). These functions are called the **eigenfunctions,** or *characteristic functions,* and the values $\lambda_n = cn\pi/L$ are called the **eigenvalues,** or *characteristic values,* of the vibrating string. The set $\{\lambda_1, \lambda_2, \cdots\}$ is called the **spectrum.**

Discussion of eigenfunctions. We see that each u_n represents a **harmonic** motion having the frequency $\lambda_n/2\pi = cn/2L$ cycles per unit time. This motion is called the nth **normal mode** of the string. The first normal mode is known as the *fundamental mode* ($n = 1$), and the others are known as *overtones;* musically they give the octave, octave plus fifth, etc. Since in (11)

$$\sin \frac{n\pi x}{L} = 0 \quad \text{at} \quad x = \frac{L}{n}, \frac{2L}{n}, \cdots, \frac{n-1}{n} L,$$

the nth normal mode has $n - 1$ so-called **nodes,** that is, points of the string that do not move (in addition to the fixed endpoints; see Fig. 252).

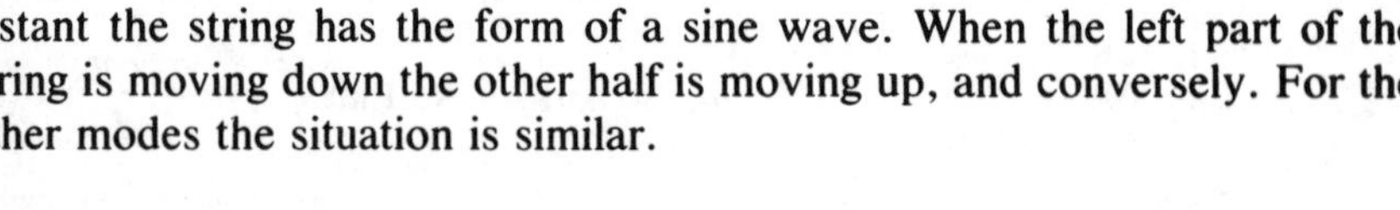

Fig. 252. Normal modes of the vibrating string

Figure 253 shows the second normal mode for various values of t. At any instant the string has the form of a sine wave. When the left part of the string is moving down the other half is moving up, and conversely. For the other modes the situation is similar.

Tuning is done by changing the tension T, and our above formula for the frequency $\lambda_n/2\pi = cn/2L$ of u_n with $c = T/\rho$ [see (3), Sec. 11.2] confirms that effect because it shows that the frequency is proportional to the tension. T cannot be increased indefinitely, but can you see what to do to get a string with a high fundamental mode? (Think of both L and ρ.)

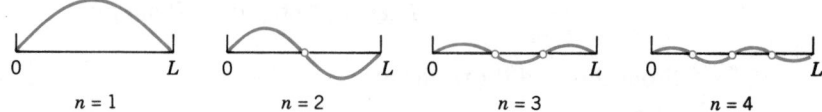

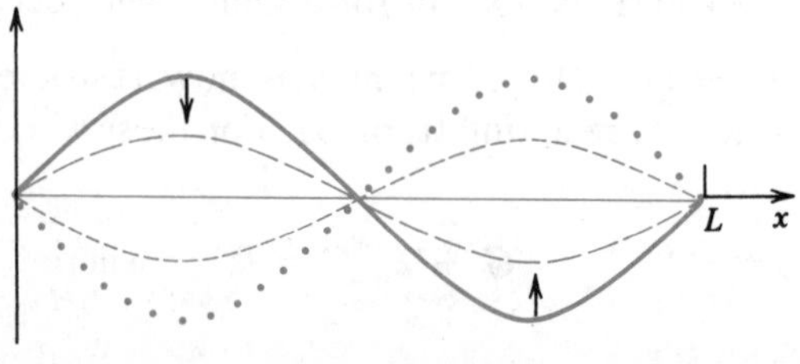

Fig. 253. Second normal mode for various values of t

Third Step. Solution of the Entire Problem

Clearly, a single solution $u_n(x, t)$ will in general not satisfy the initial conditions (3) and (4). Now, since the equation (1) is linear and homogeneous, it follows from Fundamental Theorem 1 in Sec. 11.1 that the sum of finitely many solutions u_n is a solution of (1). To obtain a solution that satisfies (3) and (4), we consider the infinite series (with $\lambda_n = cn\pi/L$ as before)

$$\textbf{(12)} \quad u(x, t) = \sum_{n=1}^{\infty} u_n(x, t) = \sum_{n=1}^{\infty} (B_n \cos \lambda_n t + B_n^* \sin \lambda_n t) \sin \frac{n\pi}{L} x.$$

Satisfying initial condition (3) (given initial displacement). From (12) and (3) we obtain

$$(13) \qquad u(x, 0) = \sum_{n=1}^{\infty} B_n \sin \frac{n\pi}{L} x = f(x).$$

Hence we must choose the B_n's so that $u(x, 0)$ becomes the Fourier sine series of $f(x)$; thus, by (4) in Sec. 10.5,

$$\textbf{(14)} \qquad B_n = \frac{2}{L} \int_0^L f(x) \sin \frac{n\pi x}{L}\, dx, \qquad n = 1, 2, \cdots .$$

Satisfying initial condition (4) (given initial velocity). Similarly, by differentiating (12) with respect to t and using (4), we obtain

$$\left.\frac{\partial u}{\partial t}\right|_{t=0} = \left[\sum_{n=1}^{\infty} (-B_n\lambda_n \sin \lambda_n t + B_n^*\lambda_n \cos \lambda_n t) \sin \frac{n\pi x}{L}\right]_{t=0}$$

$$= \sum_{n=1}^{\infty} B_n^*\lambda_n \sin \frac{n\pi x}{L} = g(x).$$

Hence we must choose the B_n^*'s so that for $t = 0$ the derivative $\partial u/\partial t$ becomes the Fourier sine series of $g(x)$; thus, by (4) in Sec. 10.5,

$$B_n^*\lambda_n = \frac{2}{L} \int_0^L g(x) \sin \frac{n\pi x}{L}\, dx$$

or, since $\lambda_n = cn\pi/L$,

$$\textbf{(15)} \qquad B_n^* = \frac{2}{cn\pi} \int_0^L g(x) \sin \frac{n\pi x}{L}\, dx, \qquad n = 1, 2, \cdots .$$

Result. Our discussion shows that $u(x, t)$ given by (12) with coefficients (14) and (15) is a solution of (1) that satisfies all the conditions (2), (3), (4) of our problem, provided the series (12) converges and so do the series obtained by differentiating (12) twice termwise with respect to x and t and have the sums $\partial^2 u/\partial x^2$ and $\partial^2 u/\partial t^2$, respectively, which are continuous.

Solution (12) established. According to our discussion, the solution (12) is at first a purely formal expression, but we shall now establish it. For the sake of simplicity we consider only the case when the initial velocity $g(x)$ is identically zero. Then the B_n^* are zero, and (12) reduces to

$$u(x, t) = \sum_{n=1}^{\infty} B_n \cos \lambda_n t \sin \frac{n\pi x}{L}, \qquad \lambda_n = \frac{cn\pi}{L}. \tag{16}$$

It is possible to ***sum this series,*** that is, to write the result in a closed or finite form. For this purpose we use the formula [see (11), Appendix 3.1]

$$\cos \frac{cn\pi}{L} t \sin \frac{n\pi}{L} x = \frac{1}{2}\left[\sin\left\{\frac{n\pi}{L}(x - ct)\right\} + \sin\left\{\frac{n\pi}{L}(x + ct)\right\}\right].$$

Consequently, we may write (16) in the form

$$u(x, t) = \frac{1}{2}\sum_{n=1}^{\infty} B_n \sin\left\{\frac{n\pi}{L}(x - ct)\right\} + \frac{1}{2}\sum_{n=1}^{\infty} B_n \sin\left\{\frac{n\pi}{L}(x + ct)\right\}.$$

These two series are those obtained by substituting $x - ct$ and $x + ct$, respectively, for the variable x in the Fourier sine series (13) for $f(x)$. Thus

$$u(x, t) = \tfrac{1}{2}[f^*(x - ct) + f^*(x + ct)] \tag{17}$$

where f^* is the odd periodic extension of f with the period $2L$ (Fig. 254). Since the initial deflection $f(x)$ is continuous on the interval $0 \leqq x \leqq L$ and zero at the endpoints, it follows from (17) that $u(x, t)$ is a continuous function of both variables x and t for all values of the variables. By differentiating (17) we see that $u(x, t)$ is a solution of (1), provided $f(x)$ is twice differentiable on the interval $0 < x < L$, and has one-sided second derivatives at $x = 0$ and $x = L$, which are zero. Under these conditions $u(x, t)$ is established as a solution of (1), satisfying (2)–(4). ∎

Fig. 254. Odd periodic extension of $f(x)$

If $f'(x)$ and $f''(x)$ are merely piecewise continuous (see Sec. 6.1), or if those one-sided derivatives are not zero, then for each t there will be finitely many values of x at which the second derivatives of u appearing in (1) do not exist. Except at these points the wave equation will still be satisfied, and we may then regard $u(x, t)$ as a **"generalized solution,"** as it is called, that is, as a solution in a broader sense. For instance, a triangular initial deflection as in Example 1 (below) leads to a generalized solution.

Representation (17) has an interesting physical interpretation, as follows. The graph of $f^*(x - ct)$ is obtained from the graph of $f^*(x)$ by shifting the latter ct units to the right (Fig. 255). This means that $f^*(x - ct)$ $(c > 0)$

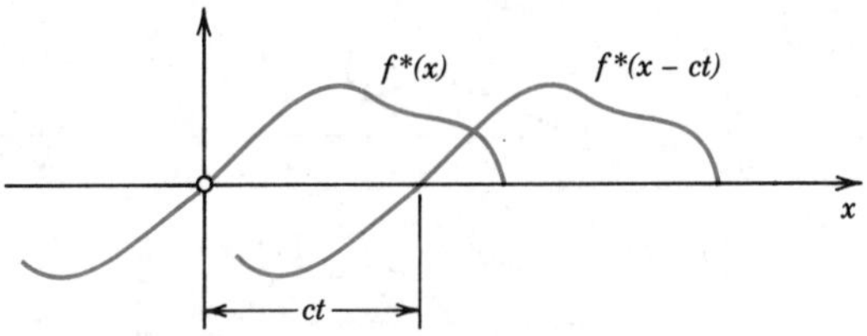

Fig. 255. Interpretation of (17)

represents a wave that is traveling to the right as t increases. Similarly, $f^*(x + ct)$ represents a wave that is traveling to the left, and $u(x, t)$ is the superposition of these two waves.

EXAMPLE 1 **Vibrating string if the initial deflection is triangular**

Find the solution of the wave equation (1) corresponding to the triangular initial deflection

$$f(x) = \begin{cases} \dfrac{2k}{L}x & \text{if } 0 < x < \dfrac{L}{2} \\ \dfrac{2k}{L}(L - x) & \text{if } \dfrac{L}{2} < x < L \end{cases}$$

and initial velocity zero. (Figure 256 on p. 638 shows $f(x) = u(x, 0)$ at the top.)

Solution. Since $g(x) \equiv 0$, we have $B_n^* = 0$ in (12), and from Example 1 in Sec. 10.5 we see that the B_n are given by (5), Sec. 10.5. Thus (12) takes the form

$$u(x, t) = \frac{8k}{\pi^2}\left[\frac{1}{1^2}\sin\frac{\pi}{L}x\cos\frac{\pi c}{L}t - \frac{1}{3^2}\sin\frac{3\pi}{L}x\cos\frac{3\pi c}{L}t + - \cdots\right].$$

For plotting the graph of the solution we may use $u(x, 0) = f(x)$ and the above interpretation of the two functions in the representation (17). This leads to the graph shown in Fig. 256. ∎

Problem Set 11.3

Find the deflection $u(x, t)$ of the vibrating string (length $L = \pi$, ends fixed, and $c^2 = T/\rho = 1$) corresponding to zero initial velocity and initial deflection:

1. $0.02 \sin x$ **2.** $k \sin 3x$ **3.** $k(\sin x - \sin 2x)$

4.

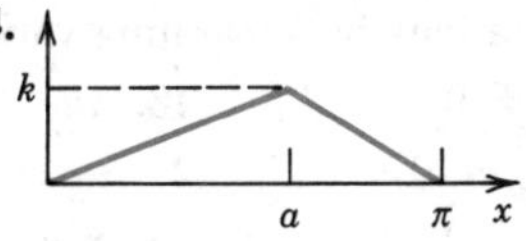

5.

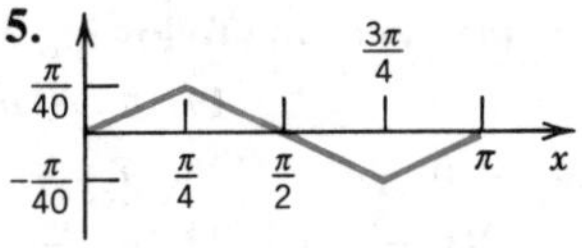

6.

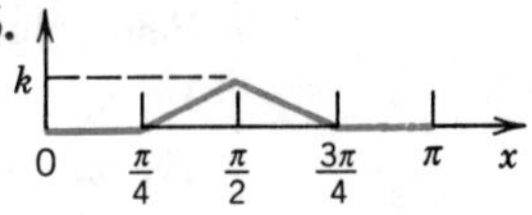

7. $k(\pi x - x^2)$ **8.** $k(\pi^2 x - x^3)$ **9.** $k[(\frac{1}{2}\pi)^4 - (x - \frac{1}{2}\pi)^4]$

Find the deflection $u(x, t)$ of the vibrating string (length $L = \pi$, ends fixed, $c^2 = 1$) if the initial deflection $f(x)$ and the initial velocity $g(x)$ are

10. $f = 0$, $g(x) = 0.1 \sin 2x$ **11.** $f(x) = 0.1 \sin x$, $g(x) = -0.2 \sin x$

12. $f = 0$, $g(x) = 0.01x$ if $0 \leqq x \leqq \frac{1}{2}\pi$, $g(x) = 0.01(\pi - x)$ if $\frac{1}{2}\pi < x \leqq \pi$

13. How does doubling the tension affect the pitch of the fundamental tone of a string?

14. How does the frequency of the fundamental mode of the vibrating string depend on the length of the string, the tension, and the mass per unit length?

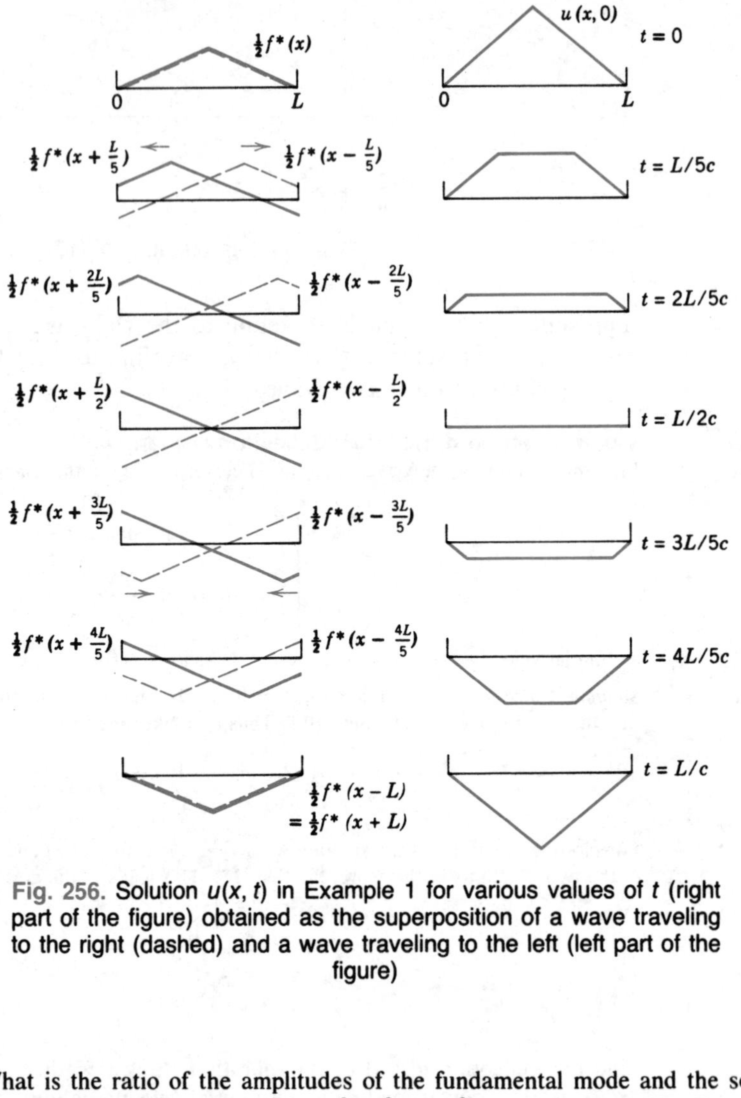

Fig. 256. Solution $u(x, t)$ in Example 1 for various values of t (right part of the figure) obtained as the superposition of a wave traveling to the right (dashed) and a wave traveling to the left (left part of the figure)

15. What is the ratio of the amplitudes of the fundamental mode and the second overtone in Prob. 7? The ratio $a_1^2/(a_1^2 + a_2^2 + \cdots)$? *Hint.* Use Parseval's identity in Sec. 10.8.

Find solutions $u(x, y)$ of the following equations by separating variables.

16. $u_x + u_y = 0$ **17.** $u_x - u_y = 0$ **18.** $xu_x - yu_y = 0$

19. $yu_x - xu_y = 0$ **20.** $u_{xx} + u_{yy} = 0$ **21.** $u_x - yu_y = 0$

22. $u_x + u_y = 2(x + y)u$ **23.** $u_{xy} - u = 0$ **24.** $x^2u_{xy} + 3y^2u = 0$

Forced vibrations of an elastic string

25. Show that forced vibrations of an elastic string are governed by

$$u_{tt} = c^2u_{xx} + \frac{P}{\rho}, \tag{18}$$

where $P(x, t)$ is the external force per unit length acting perpendicular to the string.

26. Assume the external force to be sinusoidal, say, $P = A\rho \sin \omega t$. Show that

$$P/\rho = A \sin \omega t = \sum_{n=1}^{\infty} k_n(t) \sin \frac{n\pi x}{L}$$

where $k_n(t) = (2A/n\pi)(1 - \cos n\pi) \sin \omega t$; consequently $k_n = 0$ (n even), and $k_n = (4A/n\pi) \sin \omega t$ (n odd). Furthermore, show that substitution of

$$u(x, t) = \sum_{n=1}^{\infty} G_n(t) \sin \frac{n\pi x}{L} \quad \text{into (1) gives} \quad \ddot{G}_n + \lambda_n{}^2 G_n = 0, \quad \lambda_n = \frac{cn\pi}{L}.$$

27. Show that by substituting u and P/ρ from Prob. 26 into (18) we obtain

$$\ddot{G}_n + \lambda_n{}^2 G_n = \frac{2A}{n\pi}(1 - \cos n\pi) \sin \omega t, \quad \lambda_n = \frac{cn\pi}{L}.$$

Show that if $\lambda_n{}^2 \neq \omega^2$, the solution is

$$G_n(t) = B_n \cos \lambda_n t + B_n{}^* \sin \lambda_n t + \frac{2A(1 - \cos n\pi)}{n\pi(\lambda_n{}^2 - \omega^2)} \sin \omega t.$$

28. Determine B_n and $B_n{}^*$ in Prob. 27 so that u satisfies the initial conditions $u(x, 0) = f(x)$, $u_t(x, 0) = 0$.

29. Show that in the case of resonance ($\lambda_n = \omega$),

$$G_n(t) = B_n \cos \omega t + B_n{}^* \sin \omega t - \frac{A}{n\pi\omega}(1 - \cos n\pi)t \cos \omega t.$$

30. Show that a problem (1)–(4) with more complicated boundary conditions, say, $u(0, t) = 0$, $u(L, t) = h(t)$, can be reduced to a problem for a new function v satisfying conditions $v(0, t) = v(L, t) = 0$, $v(x, 0) = f_1(x)$, $v_t(x, 0) = g_1(x)$ but a nonhomogeneous wave equation. *Hint.* Set $u = v + w$ and determine w suitably.

11.4 D'Alembert's Solution of the Wave Equation

It is interesting to note that the solution (17), Sec. 11.3, of the wave equation

$$\text{(1)} \qquad \frac{\partial^2 u}{\partial t^2} = c^2 \frac{\partial^2 u}{\partial x^2}, \qquad c^2 = \frac{T}{\rho},$$

can be immediately obtained by transforming (1) in a suitable way, namely, by introducing the new independent variables[1]

$$\text{(2)} \qquad v = x + ct, \qquad z = x - ct.$$

[1]We mention that the general theory of partial differential equations provides a systematic way for finding this transformation that will simplify the equation. See Ref. [C9] in Appendix 1.

Then u becomes a function of v and z, and the derivatives in (1) can be expressed in terms of derivatives with respect to v and z by the use of the chain rule in Sec. 8.8. Denoting partial derivatives by subscripts, we see from (2) that $v_x = 1$ and $z_x = 1$. For simplicity let us denote $u(x, t)$, as a function of v and z, by the same letter u. Then

$$u_x = u_v v_x + u_z z_x = u_v + u_z.$$

Applying the chain rule to the right side and using $v_x = 1$ and $z_x = 1$ we find

$$u_{xx} = (u_v + u_z)_x = (u_v + u_z)_v v_x + (u_v + u_z)_z z_x = u_{vv} + 2u_{vz} + u_{zz}.$$

We transform the other derivative in (1) by the same procedure, finding

$$u_{tt} = c^2(u_{vv} - 2u_{vz} + u_{zz}).$$

By inserting these two results in (1) we get (see footnote 2 in Appendix 3.2)

(3)
$$u_{vz} \equiv \frac{\partial^2 u}{\partial z \, \partial v} = 0.$$

Obviously, the point of the present method is that the resulting equation (3) can be readily solved by two successive integrations. In fact, integrating with respect to z, we find

$$\frac{\partial u}{\partial v} = h(v)$$

where $h(v)$ is an arbitrary function of v. Integration with respect to v gives

$$u = \int h(v) \, dv + \psi(z)$$

where $\psi(z)$ is an arbitrary function of z. Since the integral is a function of v, say, $\phi(v)$, the solution u is of the form $u = \phi(v) + \psi(z)$. Because of (2),

(4)
$$u(x, t) = \phi(x + ct) + \psi(x - ct).$$

This is known as **d'Alembert's solution**[2] of the wave equation (1).

This derivation was much more elegant than the method in Sec. 11.3, but d'Alembert's method is special, whereas the use of Fourier series applies to various equations, as we shall see.

---

[2]JEAN LE ROND D'ALEMBERT (1717—1783), French mathematician, who is also known for his important work in mechanics.

D'Alembert's solution satisfying the initial conditions. These are

$$u(x, 0) = f(x) \tag{5}$$

$$u_t(x, 0) = g(x) \tag{6}$$

as in Sec. 11.3. By differentiating (4) we have

$$u_t(x, t) = c\phi'(x + ct) - c\psi'(x - ct), \tag{7}$$

where primes denote derivatives with respect to the *entire* arguments $x + ct$ and $x - ct$, respectively. From (4)–(7) we have

$$u(x, 0) = \phi(x) + \psi(x) = f(x), \tag{8}$$

$$u_t(x, 0) = c\phi'(x) - c\psi'(x) = g(x). \tag{9}$$

Dividing (9) by c and integrating with respect to x, we obtain

$$\phi(x) - \psi(x) = k(x_0) + \frac{1}{c}\int_{x_0}^{x} g(s)\, ds, \qquad k(x_0) = \phi(x_0) + \psi(x_0). \tag{10}$$

If we add this to (8), then ψ drops out and division by 2 gives

$$\phi(x) = \frac{1}{2} f(x) + \frac{1}{2c}\int_{x_0}^{x} g(s)\, ds + \frac{1}{2} k(x_0). \tag{11}$$

Similarly, subtraction of (10) from (8) and division by 2 gives

$$\psi(x) = \frac{1}{2} f(x) - \frac{1}{2c}\int_{x_0}^{x} g(s)\, ds - \frac{1}{2} k(x_0). \tag{12}$$

In (11) we replace x by $x + ct$; we then get an integral from x_0 to $x + ct$. In (12) we replace x by $x - ct$ and get minus an integral from x_0 to $x - ct$ or plus an integral from $x - ct$ to x_0. Addition gives exactly the expression needed in (4); thus, our final result is

$$u(x, t) = \frac{1}{2}[f(x + ct) + f(x - ct)] + \frac{1}{2c}\int_{x-ct}^{x+ct} g(s)\, ds. \tag{13}$$

If the initial velocity is zero, we see that this reduces to

$$u(x, t) = \tfrac{1}{2}[f(x + ct) + f(x - ct)], \tag{14}$$

in agreement with (17) in Sec. 11.3. The student may show that because of the boundary conditions (2) in that section the function f must be odd and must have period $2L$.

Our result shows that the two initial conditions and the boundary conditions determine the solution uniquely.

The solution of the wave equation by the Laplace and Fourier transform methods will be shown in Secs. 11.13 and 11.14.

Problem Set 11.4

Using (14), sketch a figure (of the type in Fig. 256, Sec. 11.3) of the deflection $u(x, t)$ of a vibrating string (length $L = 1$, ends fixed, $c = 1$) starting with initial velocity zero and the following initial deflection $f(x)$, where k is small, say, $k = 0.01$.

1. $f(x) = kx(1 - x)$ **2.** $f(x) = k \sin 2\pi x$ **3.** $f(x) = k(x - x^3)$

4. $f(x) = k(x^2 - x^4)$ **5.** $f(x) = k \sin^2 \pi x$ **6.** $f(x) = k(x^3 - x^5)$

7. Show that c is the speed of the two waves given by (4).

8. If a steel wire 2 meters in length weighs 0.8 nt (about 0.18 lb) and is stretched by a tensile force of 200 nt (about 45 lb), what is the corresponding speed c of transverse waves?

9. What are the frequencies of the eigenfunctions in Prob. 8?

10. Solve the equation of a string of length L

$$u_{tt} = c^2 u_{xx} - \gamma^2 u$$

moving in an elastic medium ($\gamma^2 = const$, proportional to the elasticity coefficient of the medium), fixed at the ends and subject to initial displacement $f(x)$ and initial velocity zero.

11. Show that, because of the boundary condition (2) in Sec. 11.3, the function f in (14) of the present section must be odd and of period $2L$.

Using the indicated transformations, solve the following equations.

12. $u_{xy} - u_{yy} = 0 \quad (v = x, z = x + y)$

13. $xu_{xy} = yu_{yy} + u_y \quad (v = x, z = xy)$

14. $u_{xx} - 2u_{xy} + u_{yy} = 0 \quad (v = x, z = x + y)$

15. $u_{xx} + 2u_{xy} + u_{yy} = 0 \quad (v = x, z = x - y)$

16. $u_{xx} + u_{xy} - 2u_{yy} = 0 \quad (v = x + y, z = 2x - y)$

17. $u_{xx} - 4u_{xy} + 3u_{yy} = 0 \quad (v = x + y, z = 3x + y)$

Types and normal forms of linear partial differential equations. An equation of the form

$$Au_{xx} + 2Bu_{xy} + Cu_{yy} = F(x, y, u, u_x, u_y) \tag{15}$$

is said to be **elliptic** if $AC - B^2 > 0$, **parabolic** if $AC - B^2 = 0$, and **hyperbolic** if $AC - B^2 < 0$. [Here A, B, C may be functions of x and y, and the type of (15) may be different in different parts of the xy-plane.]

18. Show that

Laplace's equation $u_{xx} + u_{yy} = 0$ is elliptic,
the **heat equation** $u_t = c^2 u_{xx}$ is parabolic,
the **wave equation** $u_{tt} = c^2 u_{xx}$ is hyperbolic,
the **Tricomi equation** $yu_{xx} + u_{yy} = 0$ is of mixed type (elliptic in the upper half-plane and hyperbolic in the lower half-plane).

19. If the equation (15) is *hyperbolic,* it can be transformed to the *normal form* $u_{vz} = F^*(v, z, u, u_v, u_z)$ by setting $v = \Phi(x, y)$, $z = \Psi(x, y)$, where $\Phi = const$ and $\Psi = const$ are the solutions $y = y(x)$ of the equation $Ay'^2 - 2By' + C = 0$ (see Ref. [C9]). Show that in the case of the wave equation (1),

$$\Phi = x + ct, \qquad \Psi = x - ct.$$

20. If (15) is *parabolic,* the substitution $v = x$, $z = \Psi(x, y)$, with Ψ defined as in Prob. 19, reduces it to the *normal form* $u_{vv} = F^*(v, z, u, u_v, u_z)$. Verify this result for the equation $u_{xx} + 2u_{xy} + u_{yy} = 0$.

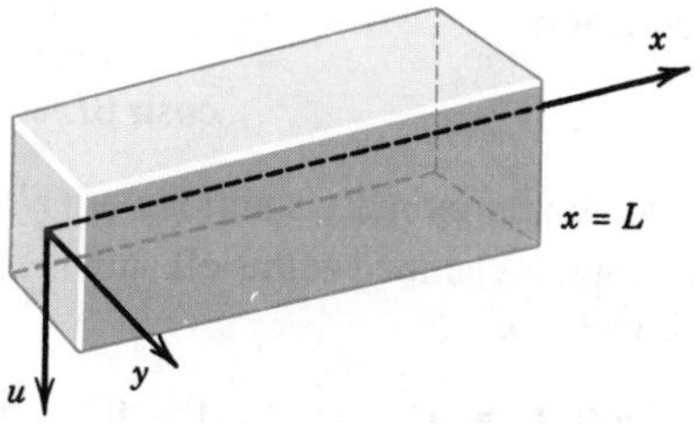

Fig. 257. Undeformed beam in Problem 21

Vibrations of a beam. It can be shown that the small free vertical vibrations of a uniform beam (Fig. 257) are governed by the fourth-order equation

$$\frac{\partial^2 u}{\partial t^2} + c^2 \frac{\partial^4 u}{\partial x^4} = 0 \qquad \text{(Ref. [C9].)} \tag{16}$$

where $c^2 = EI/\rho A$ (E = Young's modulus of elasticity, I = moment of inertia of the cross section with respect to the y-axis in the figure, ρ = density, A = cross-sectional area).

21. Substituting $u = F(x)G(y)$ into (16) and separating variables, show that

$$F^{(4)}/F = -\ddot{G}/c^2 G = \beta^4 = const,$$

$$F(x) = A \cos \beta x + B \sin \beta x + C \cosh \beta x + D \sinh \beta x,$$

$$G(t) = a \cos c\beta^2 t + b \sin c\beta^2 t.$$

22. Find solutions $u_n = F_n(x)G_n(t)$ of (16) corresponding to zero initial velocity and satisfying the boundary conditions (see Fig. 258)
$u(0, t) = 0$, $u(L, t) = 0$ (ends simply supported for all times t),
$u_{xx}(0, t) = 0$, $u_{xx}(L, t) = 0$ (zero moments, hence zero curvature, at the ends).

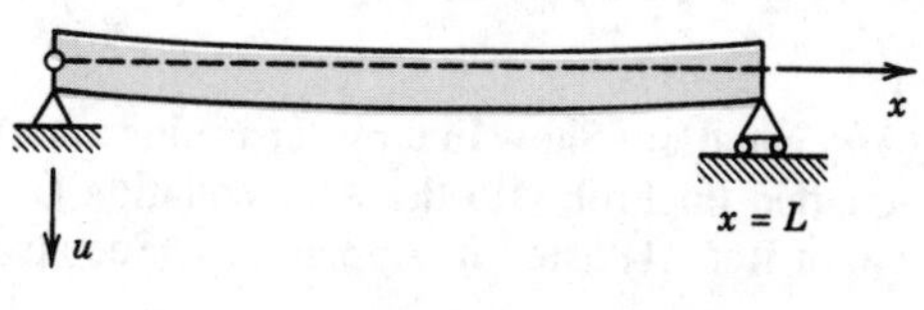

Fig. 258. Beam in Problem 22

23. Find the solution of (16) that satisfies the conditions in Prob. 22 and the initial condition $u(x, 0) = f(x) = x(L - x)$.

24. Compare the results of Prob. 23 and Prob. 7, Sec. 11.3. What is the basic difference between the frequencies of the normal modes of the vibrating string and the vibrating beam?

25. What are the boundary conditions if the beam is clamped at both ends? (See Fig. 259.)

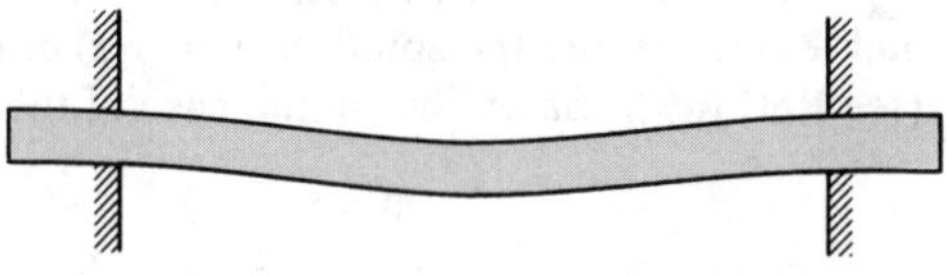

Fig. 259. Beam in Problem 25

26. Show that $F(x)$ in Prob. 21 satisfies the conditions in Prob. 25 if βL is a root of the equation

$$\cosh \beta L \cos \beta L = 1. \tag{17}$$

27. Determine approximate solutions of (17).

28. If the beam is clamped at the left and free at the other end (Fig. 260), the boundary conditions are

$$u(0, t) = 0, \quad u_x(0, t) = 0, \quad u_{xx}(L, t) = 0, \quad u_{xxx}(L, t) = 0.$$

Show that $F(x)$ in Prob. 21 satisfies these conditions if βL is a root of the equation

$$\cosh \beta L \cos \beta L = -1. \tag{18}$$

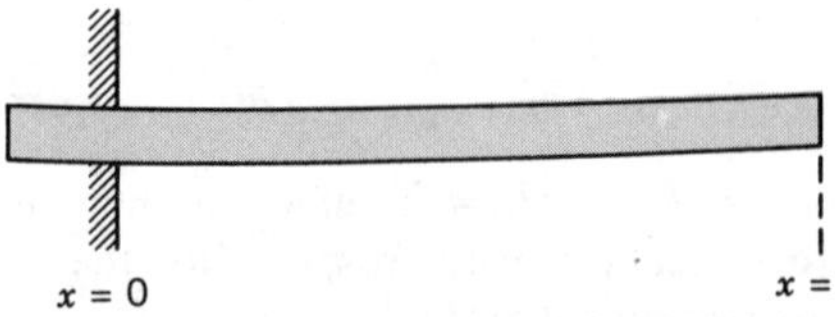

Fig. 260. Beam in Problem 28

29. Find approximate solutions of (18).

30. Longitudinal vibrations of an elastic bar or rod in the direction of the x-axis are governed by the wave equation $u_{tt} = c^2 u_{xx}$, $c^2 = E/\rho$ (see Tolstov [C11], p. 275). If the rod is fastened at one end, $x = 0$, and free at the other, $x = L$, we have $u(0, t) = 0$ and $u_x(L, t) = 0$ (because the force at the free end is zero). Show that the motion corresponding to initial displacement $u(x, 0) = f(x)$ and initial velocity zero is

$$u = \sum_{n=0}^{\infty} A_n \sin p_n x \cos p_n ct, \quad A_n = \frac{2}{L}\int_0^L f(x) \sin p_n x \, dx, \quad p_n = \frac{(2n + 1)\pi}{2L}.$$

31. (Airy equation) Show that by separating variables we can obtain from the **Tricomi equation** (in Prob. 18) the Airy equation $G'' - yG = 0$. (For solutions, see p. 446 of Ref. [1] listed in Appendix 1. See also Review Prob. 30 for Chap. 5.)

11.5 Heat Equation: Solution by Fourier Series

From the wave equation we now turn to the next "big" equation, the **heat equation**

$$\frac{\partial u}{\partial t} = c^2 \nabla^2 u, \qquad c^2 = \frac{K}{\sigma \rho},$$

which gives the temperature $u(x, y, z, t)$ in a body of homogeneous material. Here c^2 is the thermal diffusivity, K the thermal conductivity, σ the specific heat, and ρ the density of the material of the body. $\nabla^2 u$ is the Laplacian of u, and with respect to Cartesian coordinates x, y, z,

$$\nabla^2 u = \frac{\partial^2 u}{\partial x^2} + \frac{\partial^2 u}{\partial y^2} + \frac{\partial^2 u}{\partial z^2}.$$

(The heat equation was derived in Sec. 9.8.)

As an important application, let us first consider the temperature in a long thin bar or wire of constant cross section and homogeneous material, which is oriented along the x-axis (Fig. 261) and is perfectly insulated laterally, so that heat flows in the x-direction only. Then u depends only on x and time t, and the heat equation becomes the **one-dimensional heat equation**

$$\frac{\partial u}{\partial t} = c^2 \frac{\partial^2 u}{\partial x^2}. \tag{1}$$

This seems to differ only very little from the wave equation, which has a term u_{tt} instead of u_t, but we shall see that this will make the solutions of (1) behave quite differently from those of the wave equation.

We shall solve (1) for some important types of boundary and initial conditions. We begin with the case in which the ends $x = 0$ and $x = L$ of the bar are kept at temperature zero, so that we have the **boundary conditions**

$$u(0, t) = 0, \qquad u(L, t) = 0 \qquad \text{for all } t, \tag{2}$$

and the initial temperature in the bar is $f(x)$, so that we have the **initial condition**

$$u(x, 0) = f(x) \qquad [f(x) \text{ given}]. \tag{3}$$

Fig. 261. Bar under consideration

We shall determine a solution $u(x, t)$ of (1) satisfying (2) and (3)—one initial condition will be enough, as opposed to two initial conditions for the wave equation. Technically, our method will parallel that for the wave equation in Sec. 11.3: an application of separation of variables, followed by the use of Fourier series. The student may find a step-by-step comparison worthwhile.

First Step. Two Ordinary Differential Equations. Substitution of

$$u(x, t) = F(x)G(t) \tag{4}$$

into (1) gives $F\dot{G} = c^2F''G$ with $\dot{G} = dG/dt$ and $F'' = d^2F/dx^2$. To separate the variables, we divide by c^2FG, obtaining

$$\frac{\dot{G}}{c^2G} = \frac{F''}{F}. \tag{5}$$

The left side depends only on t and the right side only on x, so that both sides must be equal to a constant k (as in Sec. 11.3). The student may show that for $k \geqq 0$ the only solution $u = FG$ that satisfies (2) is $u \equiv 0$. For negative $k = -p^2$ we have from (5)

$$\frac{\dot{G}}{c^2G} = \frac{F''}{F} = -p^2.$$

We see that this yields the two linear ordinary differential equations

$$F'' + p^2F = 0 \tag{6}$$

and

$$\dot{G} + c^2p^2G = 0. \tag{7}$$

Second Step. Satisfying the Boundary Conditions. We first solve (6). A general solution is

$$F(x) = A\cos px + B\sin px. \tag{8}$$

From the boundary conditions (2) it follows that

$$u(0, t) = F(0)G(t) = 0 \quad \text{and} \quad u(L, t) = F(L)G(t) = 0.$$

Since $G \equiv 0$ would give $u \equiv 0$, we require $F(0) = 0$, $F(L) = 0$ and get $F(0) = A = 0$ by (8) and then $F(L) = B\sin pL = 0$, with $B \neq 0$ (to avoid $F \equiv 0$); thus,

$$\sin pL = 0, \qquad \text{hence} \qquad p = \frac{n\pi}{L}, \qquad n = 1, 2, \cdots.$$

Setting $B = 1$, we thus obtain the following solutions of (6) satisfying (2):

$$F_n(x) = \sin \frac{n\pi x}{L}, \qquad n = 1, 2, \cdots.$$

(As in Sec. 11.3, we need not consider negative integral values of n.)

We now solve the differential equation (7), which for $p = n\pi/L$, as just obtained, is

$$\dot{G} + \lambda_n^2 G = 0 \qquad \text{where} \qquad \lambda_n = \frac{cn\pi}{L}.$$

It has the general solution

$$G_n(t) = B_n e^{-\lambda_n^2 t}, \qquad n = 1, 2, \cdots,$$

where B_n is a constant. Hence the functions

$$\text{(9)} \qquad u_n(x, t) = F_n(x)G_n(t) = B_n \sin \frac{n\pi x}{L} e^{-\lambda_n^2 t} \qquad (n = 1, 2, \cdots)$$

are solutions of the heat equation (1), satisfying (2). These are the **eigenfunctions** of the problem, corresponding to the **eigenvalues** $\lambda_n = cn\pi/L$.

Third Step. Solution of the Entire Problem. So far we have solutions (9) of (1) satisfying the boundary conditions (2). To obtain a solution also satisfying the initial condition (3), we consider a series of these eigenfunctions,

$$\text{(10)} \qquad u(x, t) = \sum_{n=1}^{\infty} u_n(x, t) = \sum_{n=1}^{\infty} B_n \sin \frac{n\pi x}{L} e^{-\lambda_n^2 t} \qquad \left(\lambda_n = \frac{cn\pi}{L}\right).$$

From this and (3) we have

$$u(x, 0) = \sum_{n=1}^{\infty} B_n \sin \frac{n\pi x}{L} = f(x).$$

Hence for (10) to satisfy (3), the B_n's must be the coefficients of the Fourier sine series, as given by (4) in Sec. 10.5; thus

$$\text{(11)} \qquad B_n = \frac{2}{L} \int_0^L f(x) \sin \frac{n\pi x}{L} \, dx \qquad (n = 1, 2, \cdots).$$

The solution of our problem can be established, assuming that $f(x)$ is piecewise continuous on the interval $0 \leqq x \leqq L$ (see Sec. 6.1), and has one-sided derivatives[3] at all interior points of that interval; that is, under these

[3]See footnote 7 in Sec. 10.2.

assumptions the series (10) with coefficients (11) is the solution of our physical problem. The proof, which requires the knowledge of uniform convergence of series, will be given at a later occasion (Probs. 23, 24 in Problem Set 14.6).

Because of the exponential factor all the terms in (10) approach zero as t approaches infinity. The rate of decay increases with n.

EXAMPLE 1 Sinusoidal initial temperature

Find the temperature $u(x, t)$ in a laterally insulated copper bar 80 cm long if the initial temperature is $100 \sin(\pi x/80)$ °C and the ends are kept at 0°C. How long will it take for the maximum temperature in the bar to drop to 50°C? First guess, then calculate. Physical data for copper: density 8.92 gm/cm³, specific heat 0.092 cal/gm °C, thermal conductivity 0.95 cal/cm sec °C.

Solution. The initial condition gives

$$u(x, 0) = \sum_{n=1}^{\infty} B_n \sin \frac{n\pi x}{80} = f(x) = 100 \sin \frac{\pi x}{80}.$$

Hence, by inspection or from (10) we get $B_1 = 100$, $B_2 = B_3 = \cdots = 0$. In (10) we need $\lambda_1^2 = c^2\pi^2/L^2$, where $c^2 = K/\sigma\rho = 0.95/0.092 \cdot 8.92 = 1.158$ [cm²/sec]. Hence we obtain $\lambda_1^2 = 1.158 \cdot 9.870/6400 = 0.001785$ [sec⁻¹]. The solution (10) is

$$u(x, t) = 100 \sin \frac{\pi x}{80} e^{-0.001785t}.$$

Also, $100e^{-0.001785t} = 50$ when $t = (\ln 0.5)/(-0.001785) = 388$ [seconds] ≈ 6.5 [minutes]. ∎

EXAMPLE 2 Speed of decay

Solve the problem in Example 1 when the initial temperature is $100 \sin(3\pi x/80)$ °C and the other data are as before.

Solution. In (10), instead of $n = 1$ we now have $n = 3$, and $\lambda_3^2 = 3^2\lambda_1^2 = 9 \cdot 0.001785 = 0.01607$, so that the solution now is

$$u(x, t) = 100 \sin \frac{3\pi x}{80} e^{-0.01607t}.$$

Hence the maximum temperature drops to 50°C in $t = (\ln 0.5)/(-0.01607) \approx 43$ [seconds], which is much faster (9 times as fast as in Example 1).

Had we chosen a bigger n, the decay would have been still faster, and in a sum or series of such terms, each term has its own rate of decay, and terms with large n are practically 0 after a very short time. Our next example is of this type, and the curve in Fig. 262 corresponding to $t = 0.5$ looks almost like a sine curve; that is, it is practically the graph of the first term of the solution. ∎

EXAMPLE 3 "Triangular" initial temperature in a bar

Find the temperature in a laterally insulated bar of length L whose ends are kept at temperature 0, assuming that the initial temperature is

$$f(x) = \begin{cases} x & \text{if } 0 < x < L/2, \\ L - x & \text{if } L/2 < x < L. \end{cases}$$

(The uppermost part of Fig. 262 shows this function for the special $L = \pi$.)

Solution. From (11) we get

$$B_n = \frac{2}{L}\left(\int_0^{L/2} x \sin \frac{n\pi x}{L}\, dx + \int_{L/2}^{L} (L - x) \sin \frac{n\pi x}{L}\, dx\right). \tag{11*}$$

Integration yields $B_n = 0$ if n is even and

$$B_n = \frac{4L}{n^2\pi^2} \qquad (n = 1, 5, 9, \cdots),$$

$$B_n = -\frac{4L}{n^2\pi^2} \qquad (n = 3, 7, 11, \cdots)$$

(see also Example 1 in Sec. 10.5 with $k = L/2$). Hence the solution is

$$u(x, t) = \frac{4L}{\pi^2}\left[\sin\frac{\pi x}{L}\exp\left[-\left(\frac{c\pi}{L}\right)^2 t\right] - \frac{1}{9}\sin\frac{3\pi x}{L}\exp\left[-\left(\frac{3c\pi}{L}\right)^2 t\right] + - \cdots\right].$$

Figure 262 shows that the temperature decreases with increasing t, because of the heat loss due to the cooling of the ends.

The student may compare Fig. 262 and Fig. 256 in Sec. 11.3 and comment. ■

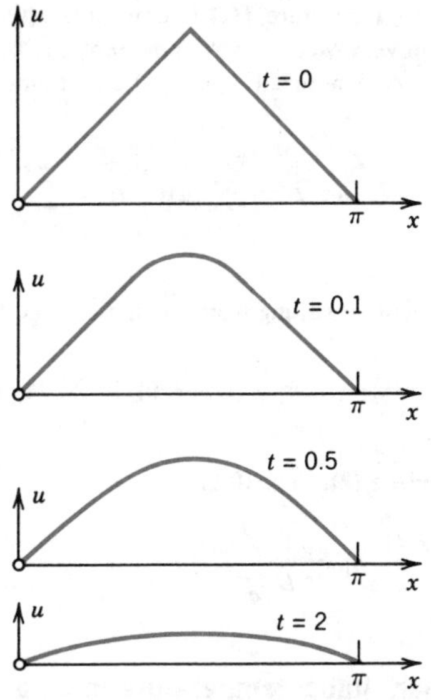

Fig. 262. Solution of Example 3 for $L = \pi$, $c = 1$ and various values of t

EXAMPLE 4 Bar with insulated ends. Eigenvalue 0

Find a solution formula of (1), (3) with (2) replaced by the condition that both ends of the bar are insulated.

Solution. Physical experiments show that the rate of heat flow is proportional to the gradient of the temperature. Hence if the ends $x = 0$ and $x = L$ of the bar are insulated, so that no heat can flow through the ends, we have the boundary conditions

$$(2^*) \qquad u_x(0, t) = 0, \qquad u_x(L, t) = 0 \qquad \text{for all } t.$$

Since $u(x, t) = F(x)G(t)$, this gives

$$u_x(0, t) = F'(0)G(t) = 0, \qquad u_x(L, t) = F'(L)G(t) = 0.$$

Differentiating (8), we have $F'(x) = -Ap \sin px + Bp \cos px$, so that

$$F'(0) = Bp = 0 \qquad \text{and then} \qquad F'(L) = -Ap \sin pL = 0.$$

The second of these conditions gives

$$p = p_n = \frac{n\pi}{L}, \qquad n = 0, 1, 2, \cdots.$$

From this and (8) with $A = 1$ and $B = 0$ we get

$$F_n(x) = \cos\frac{n\pi x}{L}, \qquad n = 0, 1, 2, \cdots.$$

With G_n as before this yields the eigenfunctions

$$(12) \qquad u_n(x, t) = F_n(x)G_n(t) = A_n \cos\frac{n\pi x}{L} e^{-\lambda_n^2 t} \qquad (n = 0, 1, \cdots)$$

corresponding to the eigenvalues $\lambda_n = cn\pi/L$. The latter are as before, but we now have the additional eigenvalue $\lambda_0 = 0$ and eigenfunction $u_0 = const$, which is the solution of the problem if the initial temperature $f(x)$ is constant. This shows the remarkable fact that a separation constant can very well be zero, and zero can be an eigenvalue.

Furthermore, whereas (9) gave a Fourier sine series, we now get from (12) a Fourier cosine series

$$(13) \qquad u(x, t) = \sum_{n=0}^{\infty} u_n(x, t) = \sum_{n=0}^{\infty} A_n \cos\frac{n\pi x}{L} e^{-\lambda_n^2 t} \qquad \left(\lambda_n = \frac{cn\pi}{L}\right)$$

with coefficients resulting from the initial condition (3),

$$u(x, 0) = \sum_{n=0}^{\infty} A_n \cos\frac{n\pi x}{L} = f(x),$$

in the form [see (2), Sec. 10.5]

$$(14) \qquad A_0 = \frac{1}{L}\int_0^L f(x)\,dx, \qquad A_n = \frac{2}{L}\int_0^L f(x)\cos\frac{n\pi x}{L}\,dx, \qquad n = 1, 2, \cdots. \quad ■$$

EXAMPLE 5 **"Triangular" initial temperature in a bar with insulated ends**

Find the temperature in the bar in Example 3, assuming that the ends are insulated (instead of kept at temperature 0).

Solution. For the triangular initial temperature, (14) gives $A_0 = L/4$ and (see also Example 1 in Sec. 10.5 with $k = L/2$)

$$A_n = \frac{2}{L}\left[\int_0^{L/2} x\cos\frac{n\pi x}{L}\,dx + \int_{L/2}^{L} (L - x)\cos\frac{n\pi x}{L}\,dx\right]$$

$$= \frac{2L}{n^2\pi^2}\left(2\cos\frac{n\pi}{2} - \cos n\pi - 1\right).$$

Hence the solution (13) is

$$u(x, t) = \frac{L}{4} - \frac{8L}{\pi^2}\left\{\frac{1}{2^2}\cos\frac{2\pi x}{L}\exp\left[-\left(\frac{2c\pi}{L}\right)^2 t\right] + \frac{1}{6^2}\cos\frac{6\pi x}{L}\exp\left[-\left(\frac{6c\pi}{L}\right)^2 t\right] + \cdots\right\}.$$

We see that the terms decrease with increasing t, and $u \to L/4$, the mean value of the initial temperature. This is plausible because no heat can escape from this totally insulated bar. In contrast, the cooling of the ends in Example 3 led to heat loss and $u \to 0$, the temperature at which the ends were kept. ■

Steady-State Two-Dimensional Heat Flow

The two-dimensional heat equation is (see the beginning of this section)

$$\frac{\partial u}{\partial t} = c^2 \nabla^2 u = c^2 \left(\frac{\partial^2 u}{\partial x^2} + \frac{\partial^2 u}{\partial y^2} \right).$$

If the heat flow is **steady** (that is, time independent), then $\partial u/\partial t = 0$, and the heat equation reduces to **Laplace's equation**[4]

$$\nabla^2 u = \frac{\partial^2 u}{\partial x^2} + \frac{\partial^2 u}{\partial y^2} = 0. \tag{15}$$

A heat problem then consists of this equation to be considered in some region R of the xy-plane and a given boundary condition on the boundary curve of R. This is called a **boundary value problem.** One calls it:

Dirichlet problem if u is prescribed on C,
Neumann problem if the normal derivative $u_n = \partial u/\partial n$ is prescribed on C,
Mixed problem if u is prescribed on a portion of C and u_n on the rest of C.

Dirichlet Problem in a Rectangle R (Fig. 263). We consider a Dirichlet problem for (15) in a rectangle R, assuming that the temperature $u(x, y)$ equals a given function $f(x)$ on the upper side and 0 on the other three sides of the rectangle.

We solve this problem by separating variables. Substitution of

$$u(x, y) = F(x)G(y)$$

into (15) and division by FG gives

$$\frac{1}{F} \cdot \frac{d^2F}{dx^2} = -\frac{1}{G} \cdot \frac{d^2G}{dy^2} = -k.$$

From this and the left and right boundary conditions,

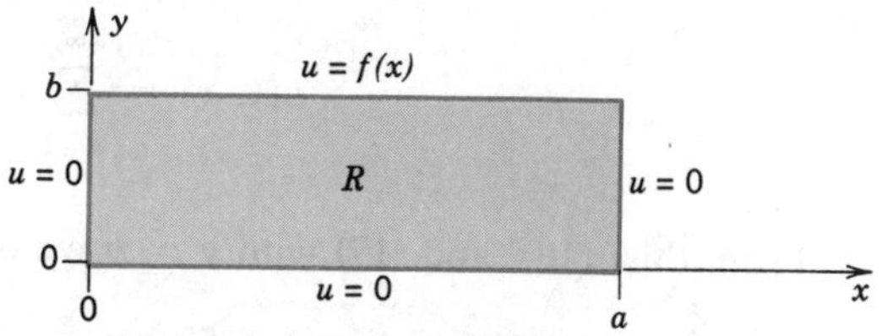

Fig. 263. Rectangle R and given boundary values

[4]This very important equation appeared in Sec. 9.8 and will be discussed further in Secs. 11.9, 11.11, 11.12, 12.5, and in Chap. 17.

$$\frac{d^2F}{dx^2} + kF = 0, \qquad F(0) = 0, \quad F(a) = 0.$$

This gives $k = (n\pi/a)^2$ and corresponding nonzero solutions

$$F(x) = F_n(x) = \sin\frac{n\pi}{a}x, \qquad n = 1, 2, \cdots. \tag{16}$$

The equation for G then becomes

$$\frac{d^2G}{dy^2} - \left(\frac{n\pi}{a}\right)^2 G = 0.$$

Solutions are

$$G(y) = G_n(y) = A_n e^{n\pi y/a} + B_n e^{-n\pi y/a}.$$

Now the boundary condition $u = 0$ on the lower side of R implies that $G_n(0) = 0$; that is, $G_n(0) = A_n + B_n = 0$ or $B_n = -A_n$. This gives

$$G_n(y) = A_n(e^{n\pi y/a} - e^{-n\pi y/a}) = 2A_n \sinh\frac{n\pi y}{a}.$$

From this and (16), writing $2A_n = A_n^*$, we obtain as the **eigenfunctions** of our problem

$$u_n(x, y) = F_n(x)G_n(y) = A_n^* \sin\frac{n\pi x}{a}\sinh\frac{n\pi y}{a}. \tag{17}$$

These satisfy the boundary condition $u = 0$ on the left, right, and lower sides.

To get a solution also satisfying the boundary condition

$$u(x, b) = f(x) \tag{18}$$

on the upper side, we consider the infinite series

$$u(x, y) = \sum_{n=1}^{\infty} u_n(x, y).$$

From this, (18), and (17) with $y = b$ we obtain

$$u(x, b) = f(x) = \sum_{n=1}^{\infty} A_n^* \sin\frac{n\pi x}{a}\sinh\frac{n\pi b}{a}.$$

We can write this in the form

$$u(x, b) = \sum_{n=1}^{\infty} \left(A_n^* \sinh \frac{n\pi b}{a}\right) \sin \frac{n\pi x}{a}.$$

This shows that the expressions in the parentheses must be the Fourier coefficients b_n of $f(x)$; that is, by (4) in Sec. 10.5,

$$b_n = A_n^* \sinh \frac{n\pi b}{a} = \frac{2}{a} \int_0^a f(x) \sin \frac{n\pi x}{a}\, dx.$$

From this and (17) we see that the solution of our problem is

$$u(x, y) = \sum_{n=1}^{\infty} A_n^* \sin \frac{n\pi x}{a} \sinh \frac{n\pi y}{a} \tag{19}$$

where

$$A_n^* = \frac{2}{a \sinh (n\pi b/a)} \int_0^a f(x) \sin \frac{n\pi x}{a}\, dx. \tag{20}$$

This solution, obtained formally without regard to the convergence and the sums of the series for u, u_{xx}, and u_{yy}, can be established when f and f' are continuous and f'' is piecewise continuous on the interval $0 \leqq x \leqq a$. The proof is somewhat involved and relies on uniform convergence; it can be found in Ref. [C2] listed in Appendix 1.

Electrostatic Potential. Elastic Membrane

The Laplace equation (15) also governs the electrostatic potential of electrical charges in any region that is free of these charges. Thus our steady-state heat problem can also be interpreted as an electrostatic potential problem, so that (19), (20) is the potential in the rectangle R when the upper side of R is at potential $f(x)$ and the other three sides are grounded.

Actually, in the steady-state case, the two-dimensional wave equation (to be considered in Secs. 11.7, 11.8) also reduces to (15), and (19), (20) then is the displacement of a rectangular elastic membrane (rubber sheet, drumhead) that is fixed along its boundary, with three sides lying in the xy-plane and the fourth side given the displacement $f(x)$.

This is another impressive demonstration of the ***unifying power*** of mathematics, illustrating that entirely different physical systems may have the same mathematical model and can thus be treated by the same mathematical methods.

Problem Set 11.5

1. Sketch u_1, u_2, u_3 [see (9), with $B_n = 1$, $c = 1$, $L = \pi$] as functions of x for the values $t = 0, 1, 2, 3$. Compare the behavior of these functions.
2. How does the rate of decay of (9) for fixed n depend on the specific heat, the density, and the thermal conductivity of the material?
3. If the first eigenfunction (9) of the bar decreases to half its value within 10 seconds, what is the value of the diffusivity?

Find the temperature $u(x, t)$ in a bar of silver (length 10 cm, constant cross section of area 1 cm², density 10.6 gm/cm³, thermal conductivity 1.04 cal/cm sec °C, specific heat 0.056 cal/gm °C) that is perfectly insulated laterally, whose ends are kept at temperature 0°C and whose initial temperature (in °C) is $f(x)$, where

4. $f(x) = \sin 0.4\pi x$
5. $f(x) = k \sin 0.1\pi x$
6. $f(x) = x$ if $0 < x < 5$ and 0 otherwise
7. $f(x) = 5 - |x - 5|$
8. $f(x) = 0.1x(100 - x^2)$
9. $f(x) = 0.01x(10 - x)$
10. $f(x) = x$ if $0 < x < 2.5$, $f(x) = 2.5$ if $2.5 < x < 7.5$, $f(x) = 10 - x$ if $7.5 < x < 10$

11. Suppose that a bar satisfies the assumptions in the text and that its ends are kept at different constant temperatures $u(0, t) = U_1$ and $u(L, t) = U_2$. Find the temperature $u_I(x)$ in the bar after a long time (theoretically: as $t \to \infty$).
12. In Prob. 11, let the initial temperature be $u(x, 0) = f(x)$. Show that the temperature for any time $t > 0$ is $u(x, t) = u_I(x) + u_{II}(x, t)$ with u_I as before and

$$u_{II} = \sum_{n=1}^{\infty} B_n \sin \frac{n\pi x}{L} e^{-(cn\pi/L)^2 t},$$

where

$$B_n = \frac{2}{L}\int_0^L [f(x) - u_I(x)] \sin \frac{n\pi x}{L}\, dx$$

$$= \frac{2}{L}\int_0^L f(x) \sin \frac{n\pi x}{L}\, dx + \frac{2}{n\pi}[(-1)^n U_2 - U_1].$$

13. **(Insulated ends, adiabatic boundary conditions)** Find the temperature $u(x, t)$ in a bar of length L that is perfectly insulated, also at the ends at $x = 0$ and $x = L$, assuming that $u(x, 0) = f(x)$. *Physical information:* The flux of heat through the faces at the ends is proportional to the values of $\partial u/\partial x$ there. Show that this situation corresponds to the conditions

$$u_x(0, t) = 0, \qquad u_x(L, t) = 0, \qquad u(x, 0) = f(x).$$

Show that the method of separating variables yields the solution

$$u(x, t) = A_0 + \sum_{n=1}^{\infty} A_n \cos \frac{n\pi x}{L} e^{-(cn\pi/L)^2 t}$$

where, by (2) in Sec. 10.5,

$$A_0 = \frac{1}{L}\int_0^L f(x)\, dx, \qquad A_n = \frac{2}{L}\int_0^L f(x) \cos \frac{n\pi x}{L}\, dx, \qquad n = 1, 2, \cdots.$$

Note that $u \to A_0$ as $t \to \infty$. Does this agree with your physical intuition?

14. Find the temperature in the bar in Prob. 13 if the left end is kept at temperature zero, the right end is perfectly insulated, and the initial temperature is $U_0 = const.$

Find the temperature in the bar in Prob. 13 if $L = \pi$, $c = 1$, and

15. $f(x) = 1$

16. $f(x) = x$

17. $f(x) = 0.5 \cos 2x$

18. $f(x) = x^2$

19. $f(x) = x$ if $0 < x < \frac{1}{2}\pi$, $f(x) = \pi - x$ if $\frac{1}{2}\pi < x < \pi$

20. $f(x) = 1$ if $0 < x < \frac{1}{2}\pi$, $f(x) = 0$ if $\frac{1}{2}\pi < x < \pi$

21. $f(x) = x$ if $0 < x < \frac{1}{2}\pi$, $f(x) = 0$ if $\frac{1}{2}\pi < x < \pi$

22. Consider the bar in Probs. 4–10. Assume that the ends are kept at 100°C for a long time. Then at some instant, say, at $t = 0$, the temperature at $x = L$ is suddenly changed to 0°C and kept at this value, while the temperature at $x = 0$ is kept at 100°C. What are the temperatures in the middle of the bar at $t = 1$, 2, 3, 10, 50 sec?

23. **(Radiation at end of bar)** Consider a laterally insulated bar of length π and such that $c = 1$ in (1), whose left end is kept at 0°C and whose right end radiates freely into air of constant temperature 0°C. *Physical information:* The **"radiation boundary condition"** is

$$-u_x(\pi, t) = k[u(\pi, t) - u_0],$$

where $u_0 = 0$ is the temperature of the surrounding air and k is a constant, say, $k = 1$ for simplicity. Show that a solution satisfying these boundary conditions is $u(x, t) = \sin px\, e^{-p^2 t}$, where p is a solution of $\tan p\pi = -p$. Show graphically that this equation has infinitely many positive solutions $p_1, p_2, p_3, \cdots$, where $p_n > n - \frac{1}{2}$ and $\lim_{n\to\infty} (p_n - n + \frac{1}{2}) = 0$.

24. **(Nonhomogeneous heat equation)** Consider the problem consisting of

$$u_t - c^2 u_{xx} = Ne^{-\alpha x}$$

and conditions (2), (3). Here the term on the right may represent loss of heat due to radioactive decay in the bar. Show that this problem may be reduced to a problem for the homogeneous equation by setting $u(x, t) = v(x, t) + w(x)$ and determining $w(x)$ so that v satisfies the homogeneous equation and the conditions $v(0, t) = v(L, t) = 0$, $v(x, 0) = f(x) - w(x)$.

25. **(Radiation)** If the bar in the text is free to radiate into the surrounding medium kept at temperature zero, the equation becomes

$$v_t = c^2 v_{xx} - \beta v.$$

Show that this equation can be reduced to the form (1) by setting $v(x, t) = u(x, t)w(t)$.

26. Consider $v_t = c^2 v_{xx} - v$ $(0 < x < L, t > 0)$, $v(0, t) = 0$, $v(L, t) = 0$, $v(x, 0) = f(x)$, where the term $-v$ corresponds to heat transfer to the surrounding medium kept at temperature zero. Reduce the equation by setting $v(x, t) = u(x, t)w(t)$ with w such that u is given by (10), (11).

27. **(Heat flux)** What is the heat flux $\phi(t) = -Ku_x(0, t)$ across $x = 0$ for the solution (10)? Note that $\phi(t) \to 0$ as $t \to \infty$. Is this physically understandable?

28. Solve (1), (2), (3) with $L = \pi$ and $f(x) = U_0 = const$ if $0 < x < \pi/2$ and $f(x) = 0$ if $\pi/2 < x < \pi$.

29. If the bar in Prob. 28 consists of two iron parts ($c^2 = 0.16$ Cgs unit) of initial temperatures 20°C and 0°C brought into perfect contact at $t = 0$, what is the approximate temperature at their common face at $t = 10, 20, 30$ seconds? (Use only the first term of the solution.)

30. **(Bar with heat generation)** If heat is generated at a constant rate throughout a bar of length $L = \pi$ with initial temperature $f(x)$ and ends at $x = 0$ and $x = \pi$ kept at temperature zero, the heat equation is $u_t = c^2 u_{xx} + H$ with constant $H > 0$. Solve this problem. *Hint.* Set $u = v - Hx(x - \pi)/2c^2$.

Two-Dimensional Problems

31. **(Laplace's equation)** Find the potential in the square $0 \leqq x \leqq 2$, $0 \leqq y \leqq 2$ if the upper side is kept at the potential $\sin \frac{1}{2}\pi x$ and the other sides are kept at zero.

32. Find the potential in the rectangle $0 \leqq x \leqq 20$, $0 \leqq y \leqq 40$ whose upper side is kept at potential 220 V and whose other sides are grounded.

33. **(Heat flow in a plate)** The faces of a thin square plate (Fig. 264, where $a = 24$) are perfectly insulated. The upper side is kept at temperature 20°C and the other sides are kept at 0°C. Find the steady-state temperature $u(x, y)$ in the plate.

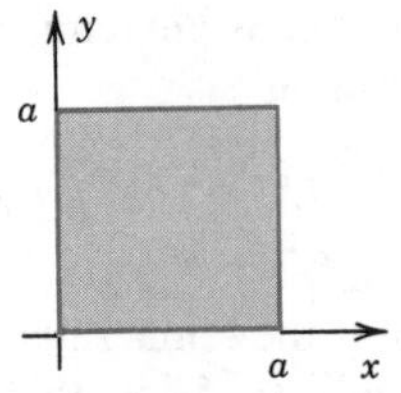

Fig. 264. Square plate

34. Find solutions u of the two-dimensional heat equation

$$u_t = c^2(u_{xx} + u_{yy})$$

in the thin plate in Fig. 264 with $a = \pi$ such that $u = 0$ on the vertical sides, assuming that the faces and the horizontal sides of the plate are perfectly insulated.

35. Find formulas similar to (19), (20), for the temperature distribution in the rectangle R considered in the text when the lower side of R is kept at temperature $f(x)$ and the three other sides are kept at 0.

36. Find the steady-state temperature in the plate in Prob. 33 if the lower side is kept at U_0 °C, the upper at U_1 °C, and the other two sides at 0°C. *Hint.* Split the problem into two problems in which the boundary temperature is zero on three sides for each problem.

37. **(Mixed boundary value problem)** Find the steady-state temperature in the plate in Prob. 33 with the upper and lower sides perfectly insulated, the left side kept at 0°C, and the right side at $f(y)$ °C.

38. **(Radiation)** Find steady-state temperatures in the rectangle in Fig. 263 with the upper and left sides perfectly insulated and the right side radiating into a medium at zero temperature according to $u_x(a, y) + hu(a, y) = 0$, $h > 0$ constant. (You will get many solutions because no condition on the lower side is given.)

39. **(Infinite region)** Find the steady-state temperature $u(x, y)$ in the strip $0 < x < \pi$, $y > 0$ with the vertical sides perfectly insulated and the lower side kept at temperature $f(x)$. (Assume that $|u|$ is bounded.)

40. **(Neumann problem)** Solve $\nabla^2 u = 0$ in the rectangle in Fig. 263 subject to the Neumann conditions $u_y(x, 0) = f(x)$ and $u_n = 0$ on the other sides.

11.6 Heat Equation: Solution by Fourier Integrals

Our discussion in the last section extends to infinite bars, with Fourier series replaced by Fourier integrals (Sec. 10.9); such bars are good models of very long bars or wires (such as a wire 300 ft long). Specifically, we shall find solutions of the heat equation

$$\frac{\partial u}{\partial t} = c^2 \frac{\partial^2 u}{\partial x^2} \tag{1}$$

in a bar that extends to infinity on both sides (and is laterally insulated, as before). Then we do not have boundary conditions, but only the initial condition

$$u(x, 0) = f(x) \qquad (-\infty < x < \infty) \tag{2}$$

where $f(x)$ is the given initial temperature of the bar.

To solve our present problem, we start as in the last section, substituting $u(x, t) = F(x)G(t)$ into (1). This gives the two ordinary differential equations

$$F'' + p^2F = 0 \qquad \text{[see (6), Sec. 11.5]} \tag{3}$$

and

$$\dot{G} + c^2p^2G = 0 \qquad \text{[see (7), Sec. 11.5].} \tag{4}$$

Solutions are

$$F(x) = A \cos px + B \sin px \quad \text{and} \quad G(t) = e^{-c^2p^2t},$$

respectively; here A and B are any constants. Hence a solution of (1) is

$$u(x, t; p) = FG = (A \cos px + B \sin px)e^{-c^2p^2t}. \tag{5}$$

[As in the last section, we had to choose the separation constant k negative, $k = -p^2$, because positive values of k would lead to an increasing exponential function in (5), which has no physical meaning.]

Any series of functions (5), found in the usual manner by taking p as multiples of a fixed number, would lead to a function that is periodic in x when $t = 0$. However, since $f(x)$ in (2) is not assumed to be periodic, it is natural to use **Fourier integrals** instead of Fourier series. Also, A and B in (5) are arbitrary and we may regard them as functions of p, writing $A = A(p)$ and $B = B(p)$. Now, since the heat equation (1) is linear and homogeneous, the function

$$u(x, t) = \int_0^\infty u(x, t; p)\, dp = \int_0^\infty [A(p) \cos px + B(p) \sin px]e^{-c^2p^2t}\, dp \tag{6}$$

is then a solution of (1), provided this integral exists and can be differentiated twice with respect to x and once with respect to t.

Determination of $A(p)$ and $B(p)$ from the initial condition. From (6) and (2) we get

$$u(x, 0) = \int_0^\infty [A(p) \cos px + B(p) \sin px] \, dp = f(x). \tag{7}$$

This gives $A(p)$ and $B(p)$ in terms of $f(x)$; indeed, from (4) and (5) in Sec. 10.9 we have

$$A(p) = \frac{1}{\pi} \int_{-\infty}^{\infty} f(v) \cos pv \, dv, \qquad B(p) = \frac{1}{\pi} \int_{-\infty}^{\infty} f(v) \sin pv \, dv. \tag{8}$$

And by (1*), Sec. 10.11, our Fourier integral (7) with these $A(p)$ and $B(p)$ can be written

$$u(x, 0) = \frac{1}{\pi} \int_0^\infty \left[\int_{-\infty}^{\infty} f(v) \cos (px - pv) \, dv \right] dp.$$

From this $u(x, 0)$ we simply get $u(x, t)$ by putting the exponential function into the integral [compare (7) and (6)]:

$$u(x, t) = \frac{1}{\pi} \int_0^\infty \left[\int_{-\infty}^{\infty} f(v) \cos (px - pv) \, e^{-c^2p^2t} \, dv \right] dp.$$

Assuming that we may invert the order of integration, we obtain

$$u(x, t) = \frac{1}{\pi} \int_{-\infty}^{\infty} f(v) \left[\int_0^\infty e^{-c^2p^2t} \cos (px - pv) \, dp \right] dv. \tag{9}$$

Then we can evaluate the inner integral by the formula

$$\int_0^\infty e^{-s^2} \cos 2bs \, ds = \frac{\sqrt{\pi}}{2} e^{-b^2}. \tag{10}$$

[A derivation of (10) is given in Problem Set 15.4 (Prob. 14).] This takes the form of our inner integral if we choose $p = s/(c\sqrt{t})$ as a new variable of integration and set

$$b = \frac{x - v}{2c\sqrt{t}}.$$

Then $2bs = (x - v)p$ and $ds = c\sqrt{t} \, dp$, so that (10) becomes

$$\int_0^\infty e^{-c^2p^2t} \cos(px - pv)\, dp = \frac{\sqrt{\pi}}{2c\sqrt{t}} \exp\left\{-\frac{(x-v)^2}{4c^2t}\right\}.$$

By inserting this result into (9) we obtain the representation

$$u(x, t) = \frac{1}{2c\sqrt{\pi t}} \int_{-\infty}^{\infty} f(v) \exp\left\{-\frac{(x-v)^2}{4c^2t}\right\} dv. \tag{11}$$

Taking $z = (v - x)/(2c\sqrt{t})$ as a variable of integration, we get the alternative form

$$u(x, t) = \frac{1}{\sqrt{\pi}} \int_{-\infty}^{\infty} f(x + 2cz\sqrt{t})\, e^{-z^2}\, dz. \tag{12}$$

If $f(x)$ is bounded for all values of x and integrable in every finite interval, it can be shown (see Ref. [C12]) that the function (11) or (12) satisfies (1) and (2). Hence this function is the required solution in the present case.

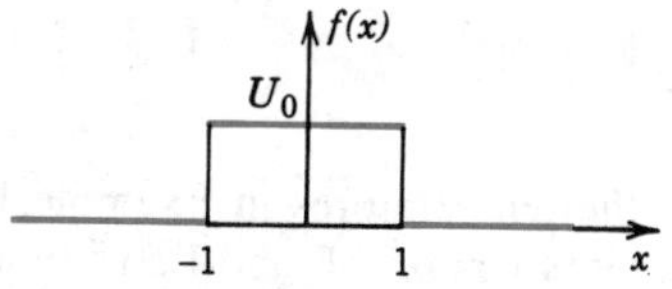

Fig. 265. Initial temperature in Example 1

EXAMPLE 1 Temperature in an infinite bar

Find the temperature in the infinite bar if the initial temperature is (Fig. 265)

$$f(x) = \begin{cases} U_0 = const & \text{if } |x| < 1, \\ 0 & \text{if } |x| > 1. \end{cases}$$

Solution. From (11) we have

$$u(x, t) = \frac{U_0}{2c\sqrt{\pi t}} \int_{-1}^{1} \exp\left\{-\frac{(x-v)^2}{4c^2t}\right\} dv.$$

If we introduce the above variable of integration z, then the integration over v from -1 to 1 corresponds to the integration over z from $(-1 - x)/2c\sqrt{t}$ to $(1 - x)/2c\sqrt{t}$, and

$$u(x, t) = \frac{U_0}{\sqrt{\pi}} \int_{-(1+x)/2c\sqrt{t}}^{(1-x)/2c\sqrt{t}} e^{-z^2}\, dz \qquad (t > 0). \tag{13}$$

We mention that this integral is not an elementary function, but can easily be expressed in terms of the error function, whose values have been tabulated. (Table A4 in Appendix 5 contains a few values; larger tables are listed in Ref. [1] in Appendix 1. See also Probs. 16–25, below.) Figure 266 on p. 660 shows $u(x, t)$ for $U_0 = 100°C$, $c^2 = 1\ cm^2/sec$, and several values of t. ∎

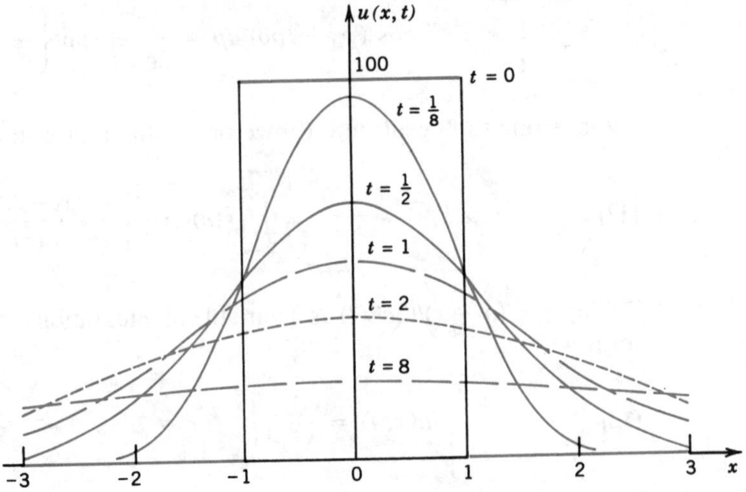

Fig. 266. Solution $u(x, t)$ to Example 1 for $U_0 = 100°C$, $c^2 = 1\ \text{cm}^2/\text{sec}$, and several values of t

Problem Set 11.6

1. Sketch the temperatures in Example 1 (with $U_0 = 100°C$ and $c^2 = 1\ \text{cm}^2/\text{sec}$) at the points $x = 0.5$, 1, and 1.5 as functions of t. Do the results agree with your physical intuition?

2. If $f(x) = 1$ when $x > 0$ and $f(x) = 0$ when $x < 0$, show that (12) becomes

$$u(x, t) = \frac{1}{\sqrt{\pi}} \int_{-x/2c\sqrt{t}}^{\infty} e^{-z^2}\, dz \qquad (t > 0).$$

3. Show that for $x = 0$ the solution in Prob. 2 is independent of t. Can this result be expected for physical reasons?

4. If the bar is semi-infinite, extending from 0 to ∞, the end at $x = 0$ is held at temperature 0, and the initial temperature is $f(x)$, show that the temperature in the bar is

$$(14) \qquad u(x, t) = \frac{1}{\sqrt{\pi}} \left[\int_{-x/\tau}^{\infty} f(x + \tau w) e^{-w^2}\, dw - \int_{x/\tau}^{\infty} f(-x + \tau w) e^{-w^2}\, dw \right],$$

where $\tau = 2c\sqrt{t}$.

5. Obtain (14) from (11) by assuming that $f(v)$ in (11) is odd.

6. If $f(x) = 1$ in Prob. 4, show that

$$u(x, t) = \frac{2}{\sqrt{\pi}} \int_0^{x/\tau} e^{-w^2}\, dw \qquad (t > 0).$$

7. What form does (14) take if $f(x) = 1$ when $a < x < b$ (where $a > 0$) and $f(x) = 0$ otherwise?

8. Show that the result of Prob. 6 can be obtained from (11) or (12) by using $f(x) = 1$ if $x > 0$, and $f(x) = -1$ if $x < 0$. What is the reason?

9. Show that in Prob. 6 the times required for any two points to reach the same temperature are proportional to the squares of their distances from the boundary at $x = 0$.

10. **(Normal distribution)** Introducing $w = z\sqrt{2}$ as a new variable of integration, show that (12) becomes

$$u(x, t) = \frac{1}{\sqrt{2\pi}} \int_{-\infty}^{\infty} f(x + cw\sqrt{2t})e^{-w^2/2}\, dw.$$

Students familiar with probability theory will notice that this involves the density $e^{-w^2/2}/\sqrt{2\pi}$ of the normal distribution (see Sec. 23.7).

Using (6), obtain the solution of (1) (in integral form) subject to the initial condition $u(x, 0) = f(x)$, where

11. $f(x) = e^{-|x|}$. [Use Example 3, Sec. 10.9.]

12. $f(x) = 1/(1 + x^2)$. [Use (15), Sec. 10.9.]

13. $f(x) = 1$ if $|x| < 1$ and 0 otherwise. [Use Example 2, Sec. 10.9.]

14. $f(x) = (\sin x)/x$. [Use Prob. 4, Sec. 10.9.]

15. Verify by integration that u in Prob. 14 satisfies the initial condition.

Error function. The error function is defined by the integral

$$\operatorname{erf} x = \frac{2}{\sqrt{\pi}} \int_0^x e^{-w^2}\, dw,$$

which is important in engineering mathematics. To become familiar with this function, the student may solve the following problems. [A few formulas are included in Appendix 3.1; see (35)–(37).]

16. Sketch the integrand of erf x (the so-called **bell-shaped curve**).

17. Show that erf x is odd.

18. Show that (13) may be written

$$u(x, t) = \frac{U_0}{2}\left[\operatorname{erf}\frac{1 - x}{2c\sqrt{t}} + \operatorname{erf}\frac{1 + x}{2c\sqrt{t}}\right] \qquad (t > 0).$$

19. Show that in Prob. 6 we have $u(x, t) = \operatorname{erf}(x/2c\sqrt{t})$.

20. It can be shown that $\operatorname{erf}(\infty) = 1$. Using this, show that in Prob. 2,

$$u(x, t) = \tfrac{1}{2} + \tfrac{1}{2}\operatorname{erf}(x/2c\sqrt{t}).$$

21. Show that $\displaystyle\int_a^b e^{-w^2}\, dw = \frac{\sqrt{\pi}}{2}(\operatorname{erf} b - \operatorname{erf} a), \qquad \int_{-b}^b e^{-w^2}\, dw = \sqrt{\pi}\operatorname{erf} b.$

22. Obtain the Maclaurin series of erf x by term-by-term integration of that of the integrand.

23. Compute erf x for $x = 0(0.1)0.5$ from Prob. 22 and compare with the 3D-values 0.000, 0.112, 0.223, 0.329, 0.428, 0.520.

24. Show that $\displaystyle\operatorname{erf} x = \sqrt{\frac{2}{\pi}} \int_0^{x\sqrt{2}} e^{-s^2/2}\, ds.$

25. **(Plane quadrant)** Conclude from Prob. 19 that the temperature in the quadrant $x > 0$, $y > 0$ with initial temperature 1, lateral insulation, and boundary kept at zero temperature is $u(x, y, t) = \operatorname{erf}(x/2c\sqrt{t}) \operatorname{erf}(y/2c\sqrt{t})$.

11.7 Modeling: Membrane, Two-Dimensional Wave Equation

As another basic vibrational problem, let us consider the motion of a stretched elastic membrane, such as a drumhead. This is the two-dimensional analog of the vibrating string problem; indeed, the modeling will be similar to that in Sec. 11.2.

Physical Assumptions. We assume the following:

1. *The mass of the membrane per unit area is constant ("homogeneous membrane"). The membrane is perfectly flexible and offers no resistance to bending.*
2. *The membrane is stretched and then fixed along its entire boundary in the xy-plane. The tension per unit length T caused by stretching the membrane is the same at all points and in all directions and does not change during the motion.*
3. *The deflection $u(x, y, t)$ of the membrane during the motion is small compared to the size of the membrane, and all angles of inclination are small.*

Although one cannot realize these assumptions exactly, they hold relatively accurately for small transverse vibrations of a thin elastic membrane, so that we obtain a good model, for instance, of a drumhead, as we shall see.

Derivation of Equation. We obtain the differential equation that governs the motion of the membrane by considering the forces acting on a small portion of the membrane in Fig. 267. Since the deflections of the membrane and the

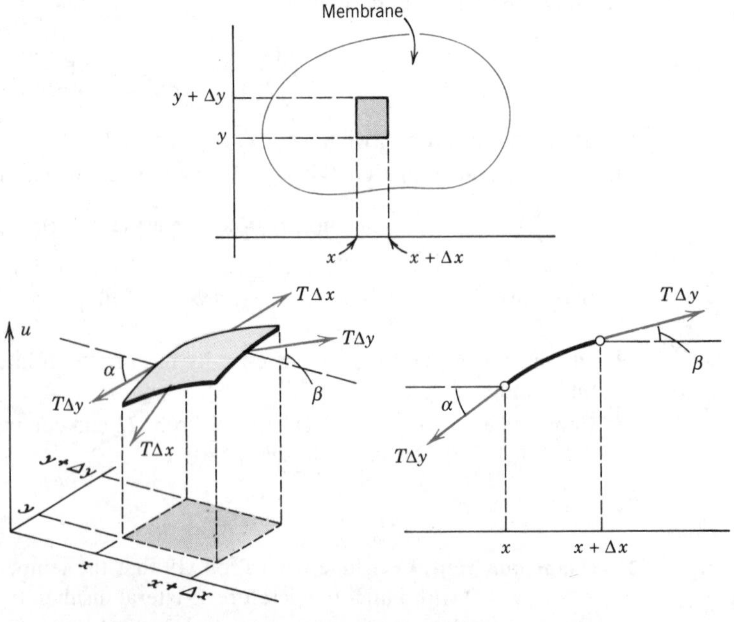

Fig. 267. Vibrating membrane

angles of inclination are small, the sides of the portion are approximately equal to Δx and Δy. The tension T is the force per unit length. Hence the forces acting on the sides of the portion are approximately $T\,\Delta x$ and $T\,\Delta y$. Since the membrane is perfectly flexible, these forces are tangent to the membrane.

Horizontal Components of the Forces. We first consider the horizontal components of the forces, which are obtained by multiplying the forces by the cosines of the angles of inclination. Since these angles are small, their cosines are close to 1. Hence the horizontal components of the forces at opposite sides are approximately equal. Therefore, the motion of the particles of the membrane in a horizontal direction will be negligibly small. From this we conclude that we may regard the motion of the membrane as transversal; that is, each particle moves vertically.

Vertical Components of the Forces. These components along the right side and the left side are[5] (Fig. 267)

$$T\,\Delta y \sin\beta \qquad \text{and} \qquad -T\,\Delta y \sin\alpha,$$

respectively; here the minus sign appears because the force on the left side is directed downward. Since the angles are small, we may replace their sines by their tangents. Hence the resultant of those two vertical components is

$$\begin{aligned} T\,\Delta y(\sin\beta - \sin\alpha) &\approx T\,\Delta y\,(\tan\beta - \tan\alpha) \\ &= T\,\Delta y\,[u_x(x + \Delta x, y_1) - u_x(x, y_2)] \end{aligned} \tag{1}$$

where subscripts x denote partial derivatives and y_1 and y_2 are values between y and $y + \Delta y$. Similarly, the resultant of the vertical components of the forces acting on the other two sides of the portion is

$$T\,\Delta x\,[u_y(x_1, y + \Delta y) - u_y(x_2, y)] \tag{2}$$

where x_1 and x_2 are values between x and $x + \Delta x$.

Newton's Second Law Gives the Equation. By Newton's second law (see Sec. 2.5), the sum of the forces given by (1) and (2) is equal to the mass $\rho\,\Delta A$ of that small portion times the acceleration $\partial^2 u/\partial t^2$; here ρ is the mass of the undeflected membrane per unit area and $\Delta A = \Delta x\,\Delta y$ is the area of that portion when it is undeflected. Thus

$$\begin{aligned} \rho\,\Delta x\,\Delta y\,\frac{\partial^2 u}{\partial t^2} &= T\,\Delta y\,[u_x(x + \Delta x, y_1) - u_x(x, y_2)] \\ &\quad + T\,\Delta x\,[u_y(x_1, y + \Delta y) - u_y(x_2, y)] \end{aligned}$$

where the derivative on the left is evaluated at some suitable point $(\tilde{x}, \tilde{y})$ corresponding to that portion. Division by $\rho\,\Delta x\,\Delta y$ yields

[5]Note that the angle of inclination varies along the sides, and α and β represent values of that angle at a suitable point of the sides under consideration.

$$\frac{\partial^2 u}{\partial t^2} = \frac{T}{\rho}\left[\frac{u_x(x + \Delta x, y_1) - u_x(x, y_2)}{\Delta x} + \frac{u_y(x_1, y + \Delta y) - u_y(x_2, y)}{\Delta y}\right].$$

If we let Δx and Δy approach zero, we obtain the partial differential equation

$$\text{(3)} \qquad \frac{\partial^2 u}{\partial t^2} = c^2\left(\frac{\partial^2 u}{\partial x^2} + \frac{\partial^2 u}{\partial y^2}\right), \qquad c^2 = \frac{T}{\rho}.$$

This equation is called the **two-dimensional wave equation.** The expression in parentheses is the Laplacian $\nabla^2 u$ of u (Sec. 11.5), and (3) can be written

$$\text{(3')} \qquad \frac{\partial^2 u}{\partial t^2} = c^2 \nabla^2 u.$$

Solutions will be obtained and discussed in the next section.

11.8 Rectangular Membrane. Use of Double Fourier Series

To solve the problem of a vibrating membrane, we have to determine a solution $u(x, y, t)$ of the two-dimensional wave equation

$$\text{(1)} \qquad \frac{\partial^2 u}{\partial t^2} = c^2\left(\frac{\partial^2 u}{\partial x^2} + \frac{\partial^2 u}{\partial y^2}\right)$$

that satisfies the boundary condition

$$\text{(2)} \qquad u = 0 \qquad \text{on the boundary of the membrane for all } t \geqq 0$$

and the two initial conditions

$$\text{(3)} \qquad u(x, y, 0) = f(x, y) \qquad \text{[given initial displacement } f(x, y)]$$

and

$$\text{(4)} \qquad \left.\frac{\partial u}{\partial t}\right|_{t=0} = g(x, y) \qquad \text{[given initial velocity } g(x, y)].$$

$u(x, y, t)$ gives the displacement of the point (x, y) of the membrane from rest ($u = 0$) at time t. We see that the conditions (2)–(4) are similar to those for the vibrating string.

As a first important case, let us consider the rectangular membrane R shown in Fig. 268.

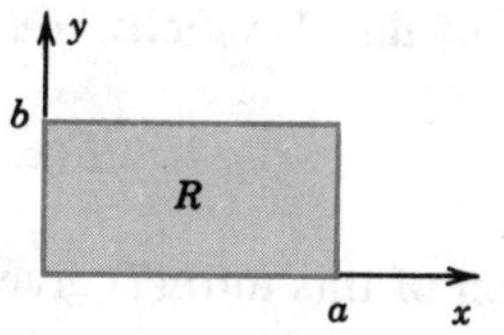

Fig. 268. Rectangular membrane

First Step. Three Ordinary Differential Equations

By applying the method of separation of variables we first determine solutions of (1) that satisfy the boundary condition (2). We start from

$$u(x, y, t) = F(x, y)G(t). \tag{5}$$

By substituting this into the wave equation (1) we have

$$F\ddot{G} = c^2(F_{xx}G + F_{yy}G)$$

where subscripts denote partial derivatives and dots denote derivatives with respect to t. To separate the variables, we divide both sides by c^2FG:

$$\frac{\ddot{G}}{c^2G} = \frac{1}{F}(F_{xx} + F_{yy}).$$

Since the left side depends only on t, whereas the right side is independent of t, both sides must equal a constant. By a little investigation we see that only negative values of that constant will lead to solutions that satisfy (2) without being identically zero; this is similar to Sec. 11.3. Denoting that negative constant by $-\nu^2$, we have

$$\frac{\ddot{G}}{c^2G} = \frac{1}{F}(F_{xx} + F_{yy}) = -\nu^2.$$

This gives two equations: for the **"time function"** $G(t)$ we have the ordinary differential equation

$$\ddot{G} + \lambda^2 G = 0 \qquad \text{where } \lambda = c\nu, \tag{6}$$

and for the **"amplitude function"** $F(x, y)$ the partial differential equation

$$F_{xx} + F_{yy} + \nu^2 F = 0, \tag{7}$$

known as the two-dimensional **Helmholtz[6] equation.**

[6]HERMANN VON HELMHOLTZ (1821—1894), German physicist, known for his basic work in thermodynamics, fluid flow, and acoustics.

Separation of the Helmholtz equation is achieved if we set

$$F(x, y) = H(x)Q(y). \tag{8}$$

Substitution of this into (7) gives

$$\frac{d^2H}{dx^2}Q = -\left(H\frac{d^2Q}{dy^2} + \nu^2 HQ\right).$$

To separate the variables, we divide both sides by HQ, finding

$$\frac{1}{H}\frac{d^2H}{dx^2} = -\frac{1}{Q}\left(\frac{d^2Q}{dy^2} + \nu^2 Q\right).$$

Both sides must equal a constant, by the usual argument. This constant must be negative, say, $-k^2$, because only negative values will lead to solutions that satisfy (2) without being identically zero. Thus

$$\frac{1}{H}\frac{d^2H}{dx^2} = -\frac{1}{Q}\left(\frac{d^2Q}{dy^2} + \nu^2 Q\right) = -k^2.$$

This yields two ordinary linear differential equations for H and Q:

$$\frac{d^2H}{dx^2} + k^2H = 0 \tag{9}$$

and

$$\frac{d^2Q}{dy^2} + p^2Q = 0 \qquad \text{where} \quad p^2 = \nu^2 - k^2. \tag{10}$$

Second Step. Satisfying the Boundary Conditions

The general solutions of (9) and (10) are

$$H(x) = A\cos kx + B\sin kx \qquad \text{and} \qquad Q(y) = C\cos py + D\sin py$$

where A, B, C, and D are constants. From (5) and (2) it follows that the function $F = HQ$ must be zero on the boundary, which corresponds to $x = 0$, $x = a$, $y = 0$, and $y = b$; see Fig. 268. This yields the conditions

$$H(0) = 0, \qquad H(a) = 0, \qquad Q(0) = 0, \qquad Q(b) = 0.$$

Therefore, $H(0) = A = 0$, and then

$$H(a) = B\sin ka = 0.$$

We must take $B \neq 0$, since otherwise $H \equiv 0$ and $F \equiv 0$. Hence $\sin ka = 0$ or $ka = m\pi$, that is,

$$k = \frac{m\pi}{a} \qquad (m \text{ integer}).$$

In precisely the same fashion we conclude that $C = 0$ and p must be restricted to the values $p = n\pi/b$ where n is an integer. We thus obtain the solutions

$$H_m(x) = \sin\frac{m\pi x}{a} \quad \text{and} \quad Q_n(y) = \sin\frac{n\pi y}{b}, \quad \begin{array}{l} m = 1, 2, \cdots, \\ n = 1, 2, \cdots. \end{array}$$

(As in the case of the vibrating string, it is not necessary to consider $m, n = -1, -2, \cdots$ since the corresponding solutions are essentially the same as for positive m and n, except for a factor -1.) Hence the functions

$$F_{mn}(x, y) = H_m(x)Q_n(y) = \sin\frac{m\pi x}{a}\sin\frac{n\pi y}{b}, \quad \begin{array}{l} m = 1, 2, \cdots, \\ n = 1, 2, \cdots, \end{array} \tag{11}$$

are solutions of (7) that are zero on the boundary of our membrane.

Eigenfunctions and Eigenvalues. Having taken care of (7), we turn to (6). Since $p^2 = \nu^2 - k^2$ in (10) and $\lambda = c\nu$ in (6), we have

$$\lambda = c\sqrt{k^2 + p^2}.$$

Hence to $k = m\pi/a$ and $p = n\pi/b$ there corresponds the value

$$\lambda = \lambda_{mn} = c\pi\sqrt{\frac{m^2}{a^2} + \frac{n^2}{b^2}}, \quad \begin{array}{l} m = 1, 2, \cdots, \\ n = 1, 2, \cdots, \end{array} \tag{12}$$

in the differential equation (6). The corresponding general solution of (6) is

$$G_{mn}(t) = B_{mn}\cos\lambda_{mn}t + B^*_{mn}\sin\lambda_{mn}t.$$

It follows that the functions $u_{mn}(x, y, t) = F_{mn}(x, y)G_{mn}(t)$, written out

$$u_{mn}(x, y, t) = (B_{mn}\cos\lambda_{mn}t + B^*_{mn}\sin\lambda_{mn}t)\sin\frac{m\pi x}{a}\sin\frac{n\pi y}{b} \tag{13}$$

with λ_{mn} according to (12), are solutions of the wave equation (1) that are zero on the boundary of the rectangular membrane in Fig. 268. These functions are called the **eigenfunctions** or *characteristic functions,* and the numbers λ_{mn} are called the **eigenvalues** or *characteristic values* of the vibrating membrane. The frequency of u_{mn} is $\lambda_{mn}/2\pi$.

Discussion of Eigenfunctions. It is very interesting to note that, depending on a and b, several functions F_{mn} may correspond to the same eigenvalue. Physically this means that there may exist vibrations having the same frequency but entirely different **nodal lines** (curves of points on the membrane that do not move). Let us illustrate this by the following example.

EXAMPLE 1 Eigenvalues and eigenfunctions of the square membrane

Consider the square membrane for which $a = b = 1$. From (12) we obtain the eigenvalues

$$\lambda_{mn} = c\pi\sqrt{m^2 + n^2}. \tag{14}$$

Hence

$$\lambda_{mn} = \lambda_{nm}$$

but for $m \neq n$ the corresponding functions

$$F_{mn} = \sin m\pi x \sin n\pi y \qquad \text{and} \qquad F_{nm} = \sin n\pi x \sin m\pi y$$

are certainly different. For example, to $\lambda_{12} = \lambda_{21} = c\pi\sqrt{5}$ there correspond the two functions

$$F_{12} = \sin \pi x \sin 2\pi y \qquad \text{and} \qquad F_{21} = \sin 2\pi x \sin \pi y.$$

Hence the corresponding solutions

$$u_{12} = (B_{12} \cos c\pi\sqrt{5}t + B_{12}^* \sin c\pi\sqrt{5}t)F_{12}$$

and

$$u_{21} = (B_{21} \cos c\pi\sqrt{5}t + B_{21}^* \sin c\pi\sqrt{5}t)F_{21}$$

have the nodal lines $y = \frac{1}{2}$ and $x = \frac{1}{2}$, respectively (see Fig. 269). Taking $B_{12} = 1$ and $B_{12}^* = B_{21}^* = 0$, we obtain

$$u_{12} + u_{21} = \cos c\pi\sqrt{5}t\,(F_{12} + B_{21}F_{21}) \tag{15}$$

which represents another vibration corresponding to the eigenvalue $c\pi\sqrt{5}$. The nodal line of this function is the solution of the equation

$$F_{12} + B_{21}F_{21} = \sin \pi x \sin 2\pi y + B_{21} \sin 2\pi x \sin \pi y = 0$$

or, since $\sin 2\alpha = 2 \sin \alpha \cos \alpha$,

$$\sin \pi x \sin \pi y\,(\cos \pi y + B_{21} \cos \pi x) = 0. \tag{16}$$

This solution depends on the value of B_{21} (see Fig. 270).

From (14) we see that even more than two functions may correspond to the same numerical value of λ_{mn}. For example, the four functions F_{18}, F_{81}, F_{47}, and F_{74} correspond to the value $\lambda_{18} = \lambda_{81} = \lambda_{47} = \lambda_{74} = c\pi\sqrt{65}$, because

$$1^1 + 8^2 = 4^2 + 7^2 = 65.$$

This happens because 65 can be expressed as the sum of two squares of positive integers in several ways. According to a theorem by Gauss, this is the case for every sum of two squares among whose prime factors there are at least two different ones of the form $4n + 1$ where n is a positive integer. In our case, $65 = 5 \cdot 13 = (4 + 1)(12 + 1)$. ∎

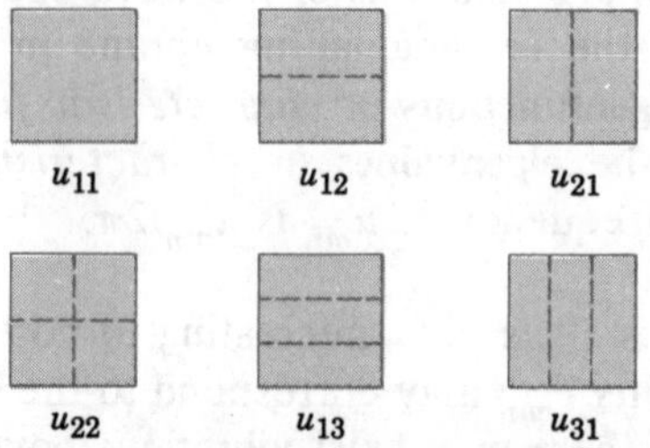

Fig. 269. Nodal lines of the solutions u_{11}, u_{12}, u_{21}, u_{22}, u_{13}, u_{31} in the case of the square membrane

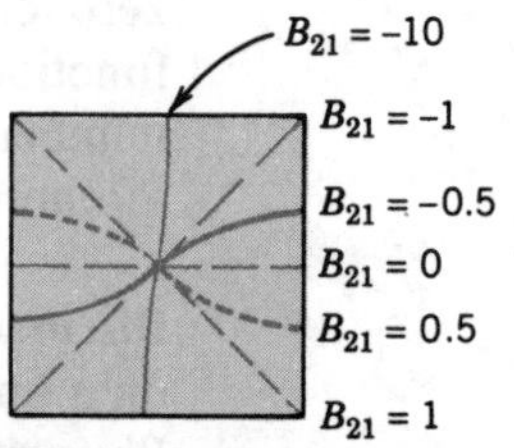

Fig. 270. Nodal lines of the solution (15) for some values of B_{21}

Third Step. Solution of the Entire Problem

To obtain the solution that also satisfies the initial conditions (3) and (4), we proceed similarly as in Sec. 11.3. We consider the double series[7]

$$
\begin{aligned}
u(x, y, t) &= \sum_{m=1}^{\infty} \sum_{n=1}^{\infty} u_{mn}(x, y, t) \\
&= \sum_{m=1}^{\infty} \sum_{n=1}^{\infty} (B_{mn} \cos \lambda_{mn} t + B^*_{mn} \sin \lambda_{mn} t) \sin \frac{m\pi x}{a} \sin \frac{n\pi y}{b} .
\end{aligned}
\tag{17}
$$

From this and (3) we obtain

$$
u(x, y, 0) = \sum_{m=1}^{\infty} \sum_{n=1}^{\infty} B_{mn} \sin \frac{m\pi x}{a} \sin \frac{n\pi y}{b} = f(x, y). \tag{18}
$$

This series is called a **double Fourier series.** Suppose that $f(x, y)$ can be developed in such a series.[8] Then the Fourier coefficients B_{mn} of $f(x, y)$ in (18) may be determined as follows. Setting

$$
K_m(y) = \sum_{n=1}^{\infty} B_{mn} \sin \frac{n\pi y}{b} \tag{19}
$$

we can write (18) in the form

$$
f(x, y) = \sum_{m=1}^{\infty} K_m(y) \sin \frac{m\pi x}{a} .
$$

For fixed y this is the Fourier sine series of $f(x, y)$, considered as a function of x. From (4) in Sec. 10.5 we see that the coefficients of this expansion are

$$
K_m(y) = \frac{2}{a} \int_0^a f(x, y) \sin \frac{m\pi x}{a} \, dx. \tag{20}
$$

Furthermore, (19) is the Fourier sine series of $K_m(y)$, and from (4) in Sec. 10.5 it follows that the coefficients are

$$
B_{mn} = \frac{2}{b} \int_0^b K_m(y) \sin \frac{n\pi y}{b} \, dy.
$$

From this and (20) we obtain the **generalized Euler formula**

$$
B_{mn} = \frac{4}{ab} \int_0^b \int_0^a f(x, y) \sin \frac{m\pi x}{a} \sin \frac{n\pi y}{b} \, dx \, dy \qquad \begin{aligned} m &= 1, 2, \cdots , \\ n &= 1, 2, \cdots . \end{aligned} \tag{21}
$$

for the Fourier coefficients of $f(x, y)$ in the double Fourier series (18).

---

[7]We shall not consider the problems of convergence and uniqueness.

[8]Sufficient conditions: f, $\partial f/\partial x$, $\partial f/\partial y$, $\partial^2 f/\partial x \, \partial y$ continuous in the rectangle R under consideration.

The B_{mn} in (17) are now determined in terms of $f(x, y)$. To determine the B^*_{mn}, we differentiate (17) termwise with respect to t; using (4), we obtain

$$\left.\frac{\partial u}{\partial t}\right|_{t=0} = \sum_{m=1}^{\infty} \sum_{n=1}^{\infty} B^*_{mn}\lambda_{mn} \sin \frac{m\pi x}{a} \sin \frac{n\pi y}{b} = g(x, y).$$

Suppose that $g(x, y)$ can be developed in this double Fourier series. Then, proceeding as before, we find

$$\text{(22)} \quad B^*_{mn} = \frac{4}{ab\lambda_{mn}} \int_0^b \int_0^a g(x, y) \sin \frac{m\pi x}{a} \sin \frac{n\pi y}{b}\, dx\, dy \qquad \begin{aligned} m &= 1, 2, \cdots, \\ n &= 1, 2, \cdots. \end{aligned}$$

The result is that, for (17) to satisfy the initial conditions, the coefficients B_{mn} and B^*_{mn} must be chosen according to (21) and (22).

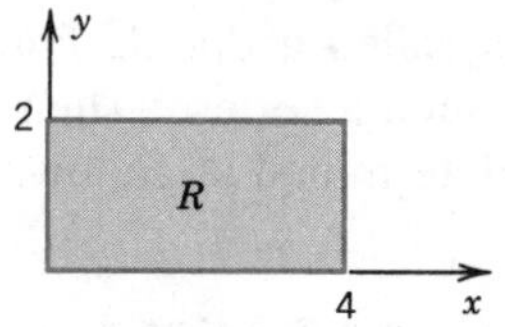

Fig. 271. Membrane in Example 2

EXAMPLE 2 Vibrations of a rectangular membrane

Find the vibrations of a rectangular membrane of sides $a = 4$ ft and $b = 2$ ft (Fig. 271) if the tension is 12.5 lb/ft, the density is 2.5 slugs/ft^2 (as for light rubber), the initial velocity is 0, and the initial displacement is

$$\text{(23)} \qquad f(x, y) = 0.1(4x - x^2)(2y - y^2) \text{ ft}.$$

Solution. $c^2 = T/\rho = 12.5/2.5 = 5$ [ft^2/sec^2]. Also, $B^*_{mn} = 0$ from (22). From (21) and (23),

$$\begin{aligned} B_{mn} &= \frac{4}{4 \cdot 2} \int_0^2 \int_0^4 0.1(4x - x^2)(2y - y^2) \sin \frac{m\pi x}{4} \sin \frac{n\pi y}{2}\, dx\, dy \\ &= \frac{1}{20} \int_0^4 (4x - x^2) \sin \frac{m\pi x}{4}\, dx \int_0^2 (2y - y^2) \sin \frac{n\pi y}{2}\, dy. \end{aligned}$$

Two integrations by parts give for the first integral on the right

$$\frac{128}{m^3\pi^3}[1 - (-1)^m] = \frac{256}{m^3\pi^3} \qquad (m \text{ odd})$$

and for the second integral

$$\frac{16}{n^3\pi^3}[1 - (-1)^n] = \frac{32}{n^3\pi^3} \qquad (n \text{ odd}).$$

For even m or n we get 0. Together with the factor 1/20 we thus have $B_{mn} = 0$ if m or n is even and

$$B_{mn} = \frac{256 \cdot 32}{20m^3n^3\pi^6} \approx \frac{0.426\ 050}{m^3n^3} \qquad (m \text{ and } n \text{ both odd}).$$

From this and (17) we obtain the answer

$$u(x, y, t) = 0.426\,050 \sum_{m,n \text{ odd}} \sum \frac{1}{m^3 n^3} \cos\left(\frac{5\pi}{4}\sqrt{m^2 + 4n^2}\right) t \sin\frac{m\pi x}{4} \sin\frac{n\pi y}{2}$$

$$(24) \qquad = 0.426\,050 \left(\cos\frac{5\pi\sqrt{5}}{4} t \sin\frac{\pi x}{4} \sin\frac{\pi y}{2} + \frac{1}{27} \cos\frac{5\pi\sqrt{37}}{4} t \sin\frac{\pi x}{4} \sin\frac{3\pi y}{2} \right.$$

$$\left. + \frac{1}{27} \cos\frac{5\pi\sqrt{13}}{4} t \sin\frac{3\pi x}{4} \sin\frac{\pi y}{2} + \frac{1}{729} \cos\frac{5\pi\sqrt{45}}{4} t \sin\frac{3\pi x}{4} \sin\frac{3\pi y}{2} + \cdots \right).$$

To discuss this solution, we note that the first term is very similar to the initial shape of the membrane, has no nodal lines, and is by far the dominating term because the coefficients of the next terms are much smaller. The second term has two horizontal nodal lines ($y = 2/3$, $4/3$), the third term two vertical ones ($x = 4/3, 8/3$), the fourth term two horizontal and two vertical ones, and so on. ■

Problem Set 11.8

1. How does the frequency of a solution (13) change if the tension of the membrane is increased?

2. Determine and sketch the nodal lines of the solutions (13) with $m = 1, 2, 3, 4$ and $n = 1, 2, 3, 4$ in the case $a = b = 1$.

3. Same task as in Prob. 2, when $a = 3$ and $b = 1$.

4. Find further eigenvalues of the square membrane with side 1 such that four different eigenfunctions correspond to each such eigenvalue.

5. Find eigenvalues of the rectangular membrane of sides $a = 2$, $b = 1$ such that two or more different eigenfunctions correspond to each such eigenvalue.

6. Show that, among all rectangular membranes of the same area $A = ab$ and the same c, the square membrane is that for which u_{11} [see (13)] has the lowest frequency.

7. Find a similar result as in Prob. 6 for the frequency of a solution (13) with arbitrary fixed m and n.

8. Using integration by parts, verify the calculation of B_{mn} in Example 2.

9. B_{mn} in Example 2 is a product of two integrals. Find out to what functions these integrals correspond and verify their values.

Double Fourier series. Represent $f(x, y)$ by a double Fourier series of the form (18), where $0 < x < \pi$, $0 < y < \pi$.

10. $f(x, y) = 1$

11. $f(x, y) = x$

12. $f(x, y) = \begin{cases} 1 & \text{if} \quad 0 < x < \pi/2 \\ 0 & \text{if} \quad \pi/2 < x < \pi \end{cases}$

13. $f(x, y) = \begin{cases} 1 & \text{if } \pi/2 < x, y < \pi \\ 0 & \text{otherwise} \end{cases}$

Represent the following functions $f(x, y)$ $(0 < x < a, 0 < y < b)$ by a double Fourier series of the form (18).

14. $f = k$

15. $f = 0.25\,xy$

16. $f = 0.125(x + y)$

17. $f = (x + 1)(y + 1)$

18. $f = xy(a - x)(b - y)$

19. $f = xy(a^2 - x^2)(b^2 - y^2)$

Find the deflection $u(x, y, t)$ of the square membrane with $a = b = 1$ and $c = 1$ if the initial velocity is zero and the initial deflection is $f(x, y)$, where

20. $f(x, y) = 0.1 \sin \pi x \sin \pi y$

21. $f(x, y) = k \sin \pi x \sin 2\pi y$

22. $f(x, y) = 0.01xy(1 - x)(1 - y)$

23. $f(x, y) = k \sin 3\pi x \sin 4\pi y$

24. $f(x, y) = k \sin^2 \pi x \sin^2 \pi y$

25. (Forced vibrations of a membrane) Show that the forced vibrations of a membrane are governed by the equation

$$u_{tt} = c^2\nabla^2 u + P/\rho$$

where $P(x, y, t)$ is the external force per unit area acting normal to the xy-plane.

11.9 Laplacian in Polar Coordinates

In connection with boundary value problems for partial differential equations, it is a general principle to use coordinates with respect to which the boundary of the region under consideration is given by simple formulas. In the next section we shall discuss circular membranes (drumheads). Then the usual polar coordinates r and θ, defined by

$$x = r \cos \theta, \qquad y = r \sin \theta,$$

will be appropriate, because they give the boundary of the membrane by the simple equation $r = const.$ Their use requires the transformation of the Laplacian

$$\nabla^2 u = \frac{\partial^2 u}{\partial x^2} + \frac{\partial^2 u}{\partial y^2}$$

in the wave equation into these new coordinates. This we do now once and for all.

Transformations of differential expressions from one coordinate system into another are frequently needed in practice, and the student should follow our present discussion with great attention.

We denote partial derivatives by subscripts and $u(x, y, t)$ as a function of r, θ, t by the same letter u, for simplicity. As in Sec. 11.4 we use the chain rule (Sec. 8.8), obtaining

$$u_x = u_r r_x + u_\theta \theta_x.$$

Another differentiation with respect to x and application of the product rule gives

$$\begin{aligned} u_{xx} &= (u_r r_x)_x + (u_\theta \theta_x)_x \\ &= (u_r)_x r_x + u_r r_{xx} + (u_\theta)_x \theta_x + u_\theta \theta_{xx}. \end{aligned} \tag{1}$$

Now, by applying the chain rule again, we find

$$(u_r)_x = u_{rr}r_x + u_{r\theta}\theta_x \qquad \text{and} \qquad (u_\theta)_x = u_{\theta r}r_x + u_{\theta\theta}\theta_x.$$

To determine the partial derivatives r_x and θ_x, we have to differentiate

$$r = \sqrt{x^2 + y^2} \qquad \text{and} \qquad \theta = \arctan\frac{y}{x},$$

finding

$$r_x = \frac{x}{\sqrt{x^2 + y^2}} = \frac{x}{r}, \qquad \theta_x = \frac{1}{1 + (y/x)^2}\left(-\frac{y}{x^2}\right) = -\frac{y}{r^2}.$$

Differentiating these two formulas again, we obtain

$$r_{xx} = \frac{r - xr_x}{r^2} = \frac{1}{r} - \frac{x^2}{r^3} = \frac{y^2}{r^3}, \qquad \theta_{xx} = -y\left(-\frac{2}{r^3}\right)r_x = \frac{2xy}{r^4}.$$

We substitute all these expressions into (1). Assuming continuity of the first and second partial derivatives, we have $u_{r\theta} = u_{\theta r}$, and by simplifying,

$$u_{xx} = \frac{x^2}{r^2}u_{rr} - 2\frac{xy}{r^3}u_{r\theta} + \frac{y^2}{r^4}u_{\theta\theta} + \frac{y^2}{r^3}u_r + 2\frac{xy}{r^4}u_\theta. \tag{2}$$

In a similar fashion it follows that

$$u_{yy} = \frac{y^2}{r^2}u_{rr} + 2\frac{xy}{r^3}u_{r\theta} + \frac{x^2}{r^4}u_{\theta\theta} + \frac{x^2}{r^3}u_r - 2\frac{xy}{r^4}u_\theta. \tag{3}$$

By adding (2) and (3) we see that the Laplacian of u in polar coordinates is

$$\nabla^2 u = \frac{\partial^2 u}{\partial r^2} + \frac{1}{r}\frac{\partial u}{\partial r} + \frac{1}{r^2}\frac{\partial^2 u}{\partial \theta^2}. \tag{4}$$

In the next section we shall apply this formula to the study of the vibrations of a drumhead (circular membrane).

Problem Set 11.9

1. Perform the details of the calculations that lead to (2) and (3).
2. Transform (4) back into Cartesian coordinates.
3. Show that (4) may be written

$$\nabla^2 u = \frac{1}{r}\frac{\partial}{\partial r}\left(r\frac{\partial u}{\partial r}\right) + \frac{1}{r^2}\frac{\partial^2 u}{\partial \theta^2}.$$

4. If u is independent of θ, then (4) reduces to $\nabla^2 u = u_{rr} + u_r/r$. Derive this result directly from the Laplacian in Cartesian coordinates by assuming that u is independent of θ.
5. Show that the only solution of $\nabla^2 u = 0$ depending only on $r = \sqrt{x^2 + y^2}$ is $u = a \ln r + b$.

6. Show that $u_n = r^n \cos n\theta$, $u_n = r^n \sin n\theta$, $n = 0, 1, \cdots$, are solutions of $\nabla^2 u = 0$ with $\nabla^2 u$ given by (4).

7. Assuming that termwise differentiation is permissible, show that a solution of the Laplace equation in the disk $R < 1$ satisfying the boundary condition $u(R, \theta) = f(\theta)$ (f given) is

$$u(r, \theta) = a_0 + \sum_{n=1}^{\infty} \left[a_n \left(\frac{r}{R}\right)^n \cos n\theta + b_n \left(\frac{r}{R}\right)^n \sin n\theta \right]$$

where a_n, b_n are the Fourier coefficients of f (see Sec. 10.3).

Electrostatic potential. Steady-state heat problems. The electrostatic potential u satisfies Laplace's equation $\nabla^2 u = 0$ in any region free of charges. Also, the heat equation $u_t = c^2\nabla^2 u$ (see Sec. 11.5) reduces to Laplace's equation if the temperature u is independent of time t (**"steady-state case"**). Find the electrostatic potential (equivalently: the steady-state temperature distribution) in the disk $r < 1$ corresponding to the following boundary values.

8. $u(\theta) = 10 \cos^2 \theta$

9. $u(\theta) = 40 \sin^3 \theta$

10. $u(\theta) = \begin{cases} 110 & \text{if } -\pi/2 < \theta < \pi/2 \\ 0 & \text{if } \pi/2 < \theta < 3\pi/2 \end{cases}$

11. $u(\theta) = \begin{cases} -100 & \text{if } -\pi < \theta < 0 \\ 100 & \text{if } 0 < \theta < \pi \end{cases}$

12. $u(\theta) = \begin{cases} \theta & \text{if } -\pi/2 < \theta < \pi/2 \\ \pi - \theta & \text{if } \pi/2 < \theta < 3\pi/2 \end{cases}$

13. $u(\theta) = \begin{cases} \theta & \text{if } -\pi/2 < \theta < \pi/2 \\ 0 & \text{if } \pi/2 < \theta < 3\pi/2 \end{cases}$

14. $u(\theta) = \theta^2 \quad (-\pi < \theta < \pi)$

15. $u(\theta) = |\theta| \quad (-\pi < \theta < \pi)$

16. Find a formula for the potential u on the x-axis in Prob. 15. Use the first four terms of this series for computing u at $x = -0.75, -0.5, -0.25, 0, 0.25, 0.5, 0.75$ (two decimals).

17. Find a formula for the potential u on the y-axis in Prob. 15.

18. Find the electrostatic potential in the semidisk $r < 1$, $0 < \theta < \pi$, which is equal to $110\theta(\pi - \theta)$ on the semicircle $r = 1$ and 0 on the segment $-1 < x < 1$.

19. Find the steady-state temperature u in a semicircular thin plate $r < a$, $0 < \theta < \pi$, if the semicircle $r = a$ is kept at constant temperature u_0 and the bounding segment $-a < x < a$ is kept at $u = 0$. (Use separation of variables.)

20. **(Laplacian in cylindrical coordinates)** Show that the Laplacian in cylindrical coordinates r, θ, z defined by $x = r \cos \theta$, $y = r \sin \theta$, $z = z$ is

$$\nabla^2 u = u_{rr} + \frac{1}{r} u_r + \frac{1}{r^2} u_{\theta\theta} + u_{zz}.$$

Express $\nabla^2 u = u_{xx} + u_{yy}$ in terms of the coordinates x^*, y^* given by

21. $x^* = ax + b$, $y^* = cy + d$

22. $x^* = x + y$, $y^* = x - y$

23. $x^* = x^2$, $y^* = y^2$

24. $x^* = 1/x$, $y^* = 1/y$

25. $x^* = x \cos \alpha - y \sin \alpha$, $y^* = x \sin \alpha + y \cos \alpha$ (**"*rotation*** through α")

26. **(Neumann problem)** Show that the solution of the Neumann problem $\nabla^2 u = 0$ if $r < R$, $u_n(R, \theta) = f(\theta)$ (n the outer normal) is

$$u(r, \theta) = A_0 + \sum_{n=1}^{\infty} r^n (A_n \cos n\theta + B_n \sin n\theta)$$

with arbitrary A_0 and

$$A_n = \frac{2}{\pi n R^{n-1}} \int_{-\pi}^{\pi} f(\theta) \cos n\theta \, d\theta, \qquad B_n = \frac{2}{\pi n R^{n-1}} \int_{-\pi}^{\pi} f(\theta) \sin n\theta \, d\theta.$$

27. Show that (9), Sec. 9.4, imposes on $f(\theta)$ in Prob. 26 the condition

$$\int_{-\pi}^{\pi} f(\theta) \, d\theta = 0,$$

usually called a "*compatibility condition.*"

28. **(Neumann problem)** Solve $\nabla^2 u = 0$ in the annulus $1 < r < 3$ if $u_r(1, \theta) = \sin \theta$, $u_r(3, \theta) = 0$.

11.10 Circular Membrane. Use of Fourier–Bessel Series

Circular membranes occur in drums, pumps, microphones, telephones, and so on, and this accounts for their great importance in engineering. Whenever a circular membrane is plane and its material is elastic, but offers no resistance to bending (this excludes thin metallic membranes!), its vibrations are governed by the two-dimensional wave equation (3′), Sec. 11.7, which we now write in polar coordinates defined by $x = r \cos \theta$, $y = r \sin \theta$ in the form [see (4) in the last section]

$$\frac{\partial^2 u}{\partial t^2} = c^2 \left(\frac{\partial^2 u}{\partial r^2} + \frac{1}{r} \frac{\partial u}{\partial r} + \frac{1}{r^2} \frac{\partial^2 u}{\partial \theta^2} \right).$$

Figure 272 shows our membrane of radius R, for which we shall determine solutions $u(r, t)$ that are radially symmetric,[9] that is, do not depend on θ. Then the wave equation reduces to

$$\text{(1)} \qquad \frac{\partial^2 u}{\partial t^2} = c^2 \left(\frac{\partial^2 u}{\partial r^2} + \frac{1}{r} \frac{\partial u}{\partial r} \right).$$

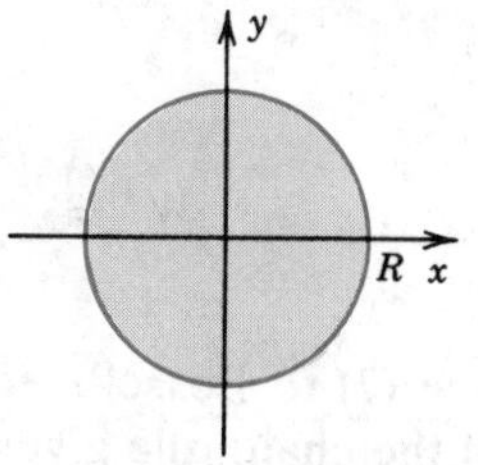

Fig. 272. Circular membrane

[9] For solutions depending on θ, see the problem set.

Boundary and Initial Conditions. Since the membrane is fixed along its boundary $r = R$, we have the boundary condition

$$u(R, t) = 0 \qquad \text{for all } t \geqq 0. \tag{2}$$

Solutions not depending on θ will occur if the initial conditions do not depend on θ, that is, if they are of the form

$$u(r, 0) = f(r) \qquad \text{[initial deflection } f(r)\text{]} \tag{3}$$

and

$$\left.\frac{\partial u}{\partial t}\right|_{t=0} = g(r) \qquad \text{[initial velocity } g(r)\text{].} \tag{4}$$

First Step. Ordinary Differential Equations. Bessel's Equation

Separating variables, we first determine solutions

$$u(r, t) = W(r)G(t) \tag{5}$$

satisfying the boundary condition (2). Substituting (5) and its derivatives into (1) and dividing the result by c^2WG, we get

$$\frac{\ddot{G}}{c^2G} = \frac{1}{W}\left(W'' + \frac{1}{r}W'\right)$$

where dots denote derivatives with respect to t and primes denote derivatives with respect to r. The expressions on both sides must be equal to a constant, and this constant must be negative, say, $-k^2$, in order to obtain solutions that satisfy the boundary condition without being identically zero. Thus,

$$\frac{\ddot{G}}{c^2G} = \frac{1}{W}\left(W'' + \frac{1}{r}W'\right) = -k^2.$$

This yields the two ordinary linear differential equations

$$\boxed{\ddot{G} + \lambda^2 G = 0} \qquad \text{where } \lambda = ck \tag{6}$$

and

$$\boxed{W'' + \frac{1}{r}W' + k^2W = 0.} \tag{7}$$

We can reduce (7) to Bessel's equation (Sec. 5.5) if we set $s = kr$. Then $1/r = k/s$ and the chain rule gives

$$W' = \frac{dW}{dr} = \frac{dW}{ds}\frac{ds}{dr} = \frac{dW}{ds}k \qquad \text{and} \qquad W'' = \frac{d^2W}{ds^2}k^2.$$

By substituting this into (7) and omitting the common factor k^2 we obtain

$$\frac{d^2W}{ds^2} + \frac{1}{s}\frac{dW}{ds} + W = 0. \tag{7*}$$

This is **Bessel's equation** (1), Sec. 5.5, with parameter $\nu = 0$.

Second Step. Satisfying the Boundary Condition

Solutions of (7*) are the Bessel functions J_0 and Y_0 of the first and second kind (see Secs. 5.5, 5.7), but Y_0 becomes infinite at 0, so that we cannot use it because the deflection of the membrane must always remain finite. This leaves us with

$$W(r) = J_0(s) = J_0(kr) \qquad (s = kr). \tag{8}$$

Now on the boundary $r = R$ we get $W(R) = J_0(kR) = 0$ from (2) (because $G \equiv 0$ would imply $u \equiv 0$). We can satisfy this condition because J_0 has (infinitely many) positive zeros, $s = \alpha_1, \alpha_2, \cdots$ (see Fig. 273), with numerical values

$$\alpha_1 = 2.4048, \quad \alpha_2 = 5.5201, \quad \alpha_3 = 8.6537, \quad \alpha_4 = 11.7915, \quad \alpha_5 = 14.9309,$$

and so on (see Ref. [1], Appendix 1, for more extensive tables). These are irregularly spaced, as we see. Equation (8) now implies

$$kR = \alpha_m \quad \text{thus} \quad k = k_m = \frac{\alpha_m}{R}, \qquad m = 1, 2, \cdots. \tag{9}$$

Hence the functions

$$W_m(r) = J_0(k_m r) = J_0\left(\frac{\alpha_m}{R}r\right), \qquad m = 1, 2, \cdots, \tag{10}$$

are solutions of (7) that vanish at $r = R$.

Eigenfunctions and Eigenvalues. For W_m in (10), a corresponding general solution of (6) with $\lambda = \lambda_m = ck_m$ is

$$G_m(t) = a_m \cos \lambda_m t + b_m \sin \lambda_m t.$$

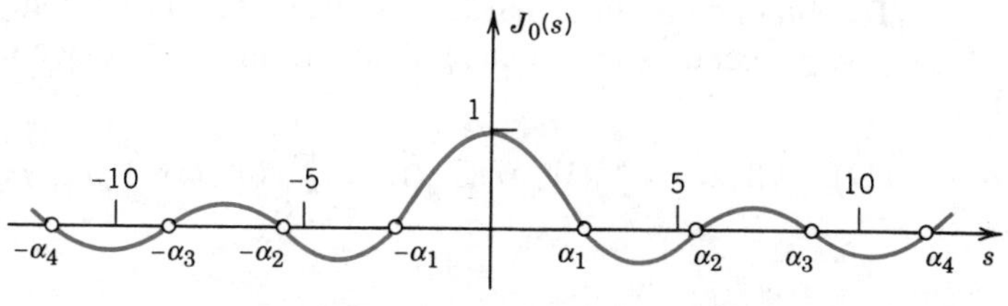

Fig. 273. Bessel function $J_0(s)$

Hence the functions

$$u_m(r, t) = W_m(r)G_m(t) = (a_m \cos \lambda_m t + b_m \sin \lambda_m t)J_0(k_m r) \tag{11}$$

where $m = 1, 2, \cdots$, are solutions of the wave equation (1), satisfying the boundary condition (2). These are the **eigenfunctions** of our problem, and the corresponding **eigenvalues** are λ_m.

The vibration of the membrane corresponding to u_m is called the mth **normal mode;** it has the frequency $\lambda_m/2\pi$ cycles per unit time. Since the zeros of the Bessel function J_0 are not regularly spaced on the axis (in contrast to the zeros of the sine functions appearing in the case of the vibrating string), the sound of a drum is entirely different from that of a violin. The forms of the normal modes can easily be obtained from Fig. 273 and are shown in Fig. 274. For $m = 1$, all the points of the membrane move up (or down) at the same time. For $m = 2$, the situation is as follows. The function

$$W_2(r) = J_0\left(\frac{\alpha_2}{R} r\right)$$

is zero for $\alpha_2 r/R = \alpha_1$, thus $r = \alpha_1 R/\alpha_2$. The circle $r = \alpha_1 R/\alpha_2$ is, therefore, a nodal line, and when at some instant the central part of the membrane moves up, the outer part ($r > \alpha_1 R/\alpha_2$) moves down, and conversely. The solution $u_m(r, t)$ has $m - 1$ nodal lines, which are circles (Fig. 274).

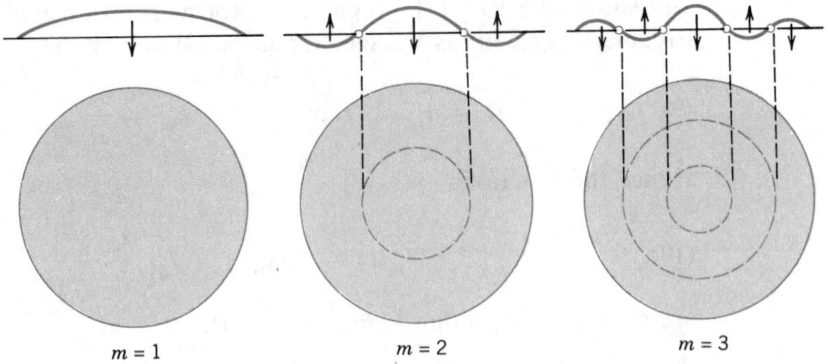

Fig. 274. Normal modes of the circular membrane in the case of vibrations independent of the angle

Third Step. Solution of the Entire Problem

To obtain a solution that also satisfies the initial conditions (3) and (4), we may proceed as in the case of the string, that is, we consider the series[10]

$$u(r, t) = \sum_{m=1}^{\infty} W_m(r)G_m(t) = \sum_{m=1}^{\infty} (a_m \cos \lambda_m t + b_m \sin \lambda_m t)J_0\left(\frac{\alpha_m}{R} r\right). \tag{12}$$

[10]We shall not consider the problems of convergence and uniqueness.

Setting $t = 0$ and using (3), we obtain

$$u(r, 0) = \sum_{m=1}^{\infty} a_m J_0\left(\frac{\alpha_m}{R} r\right) = f(r). \tag{13}$$

Thus for (12) to satisfy (3), the constants a_m must be the coefficients of the **Fourier–Bessel series** (13) that represents $f(r)$ in terms of $J_0(\alpha_m r/R)$; that is [see (9) in Sec. 5.9 with $n = 0$],

(14) $$a_m = \frac{2}{R^2 J_1^{\,2}(\alpha_m)} \int_0^R r f(r) J_0\left(\frac{\alpha_m}{R} r\right) dr \qquad (m = 1, 2, \cdots).$$

Differentiability of $f(r)$ in the interval $0 \leqq r \leqq R$ is sufficient for the existence of the development (13); see Ref. [A9]. The coefficients b_m in (12) can be determined from (4) in a similar fashion. To obtain numerical values of a_m and b_m, we may apply one of the usual methods of approximate integration, using tables of J_0 and J_1. Sometimes numerical integration can be avoided, as the following example illustrates.

EXAMPLE 1 **Vibrations of a circular membrane**

Find the vibrations of a circular drumhead of radius 1 ft and density 2 slugs/ft^2 if the tension is 8 lb/ft, the initial velocity is 0, and the initial displacement is

$$f(r) = 1 - r^2 \text{ [ft]}.$$

Solution. $c^2 = T/\rho = 8/2 = 4$ [ft^2/sec^2]. Also $b_m = 0$, since the initial velocity is 0. From (14) and Prob. 29 in Sec. 5.9, since $R = 1$, we obtain

$$a_m = \frac{2}{J_1^{\,2}(\alpha_m)} \int_0^1 r(1 - r^2) J_0(\alpha_m r)\, dr = \frac{4J_2(\alpha_m)}{\alpha_m^{\,2} J_1^{\,2}(\alpha_m)} = \frac{8}{\alpha_m^{\,3} J_1(\alpha_m)},$$

where the last equality follows from (3), Sec. 5.6, with $\nu = 1$, that is,

$$J_2(\alpha_m) = \frac{2}{\alpha_m} J_1(\alpha_m) - J_0(\alpha_m) = \frac{2}{\alpha_m} J_1(\alpha_m).$$

Table 9.5 on p. 409 of [1] gives α_m and $J_0'(\alpha_m)$. From this we get $J_1(\alpha_m) = -J_0'(\alpha_m)$ by (2), Sec. 5.6, with $\nu = 0$, and compute the a_m's:

m	α_m	$J_1(\alpha_m)$	$J_2(\alpha_m)$	a_m
1	2.40483	0.51915	0.43176	1.10801
2	5.52008	−0.34026	−0.12328	−0.13978
3	8.65373	0.27145	0.06274	0.04548
4	11.79153	−0.23246	−0.03943	−0.02099
5	14.93092	0.20655	0.02767	0.01164
6	18.07106	−0.18773	−0.02078	−0.00722
7	21.21164	0.17327	0.01634	0.00484
8	24.35247	−0.16170	−0.01328	−0.00343
9	27.49348	0.15218	0.01107	0.00253
10	30.63461	−0.14417	−0.00941	−0.00193

Thus

$$f(r) = 1.108J_0(2.4048r) - 0.140J_0(5.5201r) + 0.045J_0(8.6537r) - \cdots .$$

We see that the coefficients decrease relatively slowly. The sum of the explicitly given coefficients in the table is 0.99915. The sum of *all* the coefficients should be 1. (Why?) Hence by the Leibniz test in Appendix 3.3 the partial sum of those terms gives about three correct decimals of the amplitude $f(r)$.

Since

$$\lambda_m = ck_m = c\alpha_m/R = 2\alpha_m,$$

from (12) we thus obtain the solution (with r measured in feet and t in seconds)

$$u(r, t) = 1.108J_0(2.4048r) \cos 4.8097t - 0.140J_0(5.5201r) \cos 11.0402t$$
$$+ 0.045J_0(8.6537r) \cos 17.3075t - \cdots .$$

In Fig. 274, $m = 1$ gives an idea of the motion of the first term of our series, $m = 2$ of the second term, and $m = 3$ of the third term, so that we can "see" our result about as well as for a violin string in Sec. 11.3. ∎

Problem Set 11.10

1. What is the reason for using polar coordinates in the present section?
2. If the tension of the membrane is increased, how does the frequency of each normal mode (11) change?
3. A small drum should have a higher fundamental mode than a large drum, tension and density being the same. How can this be concluded from our formulas?
4. Using the numerical value of α_1, find a formula for the fundamental frequency of a circular drum.
5. Find a formula for the tension T required to produce a desired fundamental note of a circular drum.
6. Determine numerical values of the radii of the nodal lines of u_2 and u_3 [see (11)] when $R = 1$.
7. Sketch a figure similar to Fig. 274, for u_4, u_5, u_6.
8. Same task as in Prob. 6, for u_4, u_5, u_6.
9. Is it possible that, for fixed c and R, two or more functions u_m [see (11)] that have different nodal lines correspond to the same eigenvalue?
10. Why is $a_1 + a_2 + \cdots = 1$ in Example 1? Compute the first five partial sums of this series.
11. Verify the computation of the a_m in Example 1.
12. Using the values in the text, verify that $\alpha_m \approx (m - \frac{1}{4})\pi$ for $m = 1, \cdots, 5$.
13. Show that for (12) to satisfy (4),

$$(15) \qquad b_m = \frac{2}{c\alpha_m R J_1^2(\alpha_m)} \int_0^R r g(r) J_0(\alpha_m r/R)\, dr, \qquad m = 1, 2, \cdots .$$

14. Find the deflection $u(r, t)$ of the circular membrane of radius $R = 1$ if $c = 1$, the initial velocity is zero, and the initial deflection is $f(r) = k(1 - r^4)$. *Hint.* Remember Probs. 25–32, Sec. 5.9.

15. Solve Prob. 14 when $f(r) = k(1 - r^2)$, the other data being as before.

Vibrations of a Circular Membrane Depending on r and θ

16. Show that substitution of $u = F(r, \theta)G(t)$ into the wave equation

$$u_{tt} = c^2\left(u_{rr} + \frac{1}{r}u_r + \frac{1}{r^2}u_{\theta\theta}\right) \tag{16}$$

leads to

$$\ddot{G} + \lambda^2 G = 0, \qquad \text{where } \lambda = ck, \tag{17}$$

$$F_{rr} + \frac{1}{r}F_r + \frac{1}{r^2}F_{\theta\theta} + k^2F = 0. \tag{18}$$

17. Show that substitution of $F = W(r)Q(\theta)$ in (18) yields

$$Q'' + n^2Q = 0, \tag{19}$$

$$r^2W'' + rW' + (k^2r^2 - n^2)W = 0. \tag{20}$$

18. Show that $Q(\theta)$ must be periodic with period 2π, and, therefore, $n = 0, 1, \cdots$ in (19) and (20). Show that this yields the solutions $Q_n = \cos n\theta$, $Q_n^* = \sin n\theta$, $W_n = J_n(kr)$, $n = 0, 1, \cdots$.

19. Show that the boundary condition

$$u(R, \theta, t) = 0 \tag{21}$$

leads to $k = k_{mn} = \alpha_{mn}/R$, where $s = \alpha_{mn}$ is the mth positive zero of $J_n(s)$.

20. Show that solutions of (16) that satisfy (21) are (Fig. 275)

$$\begin{aligned} u_{mn} &= (A_{mn}\cos ck_{mn}t + B_{mn}\sin ck_{mn}t)J_n(k_{mn}r)\cos n\theta, \\ u_{mn}^* &= (A_{mn}^*\cos ck_{mn}t + B_{mn}^*\sin ck_{mn}t)J_n(k_{mn}r)\sin n\theta. \end{aligned} \tag{22}$$

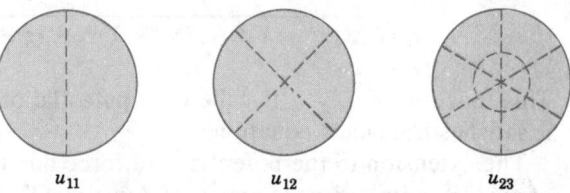

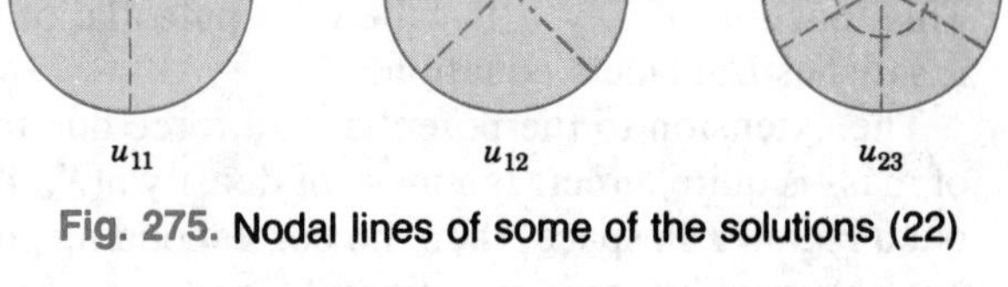

Fig. 275. Nodal lines of some of the solutions (22)

21. Show that the initial condition $u_t(r, \theta, 0) = 0$ leads to $B_{mn} = 0$, $B_{mn}^* = 0$ in (22).

22. Show that $u_{m0}^* \equiv 0$ and u_{m0} is identical with (11) in the current section.

23. Show that u_{11} represents the fundamental mode of a semicircular membrane and find the corresponding frequency when $c^2 = 1$ and $R = 1$.

11.11 Laplace's Equation. Potential

One of the most important partial differential equations in physics is **Laplace's equation**

$$\nabla^2 u = 0. \tag{1}$$

Here $\nabla^2 u$ is the Laplacian of u. In Cartesian coordinates x, y, z in space,

$$\nabla^2 u = \frac{\partial^2 u}{\partial x^2} + \frac{\partial^2 u}{\partial y^2} + \frac{\partial^2 u}{\partial z^2}. \tag{2}$$

The theory of the solutions of Laplace's equation is called **potential theory.** Solutions of (1) that have ***continuous*** second partial derivatives are called **harmonic functions.**

We have met this equation before, in three dimensions (Secs. 8.9, 9.8) as well as in two dimensions (Sec. 11.5), and we shall now consider it in cylindrical and in spherical coordinates, along with typical boundary value problems. Let us first say a few words about its importance by mentioning some of the main areas where it plays a basic role.

Areas of Application

Gravitation is governed by Laplace's equation, as follows. In Example 3, Sec. 8.9, we have seen that if a particle A of mass M is fixed at a point (X, Y, Z) and another particle B of mass m is at a point (x, y, z), then A attracts B, the gravitational force being the gradient of the scalar function

$$u(x, y, z) = \frac{c}{r}, \qquad c = GMm = const,$$

$$r = \sqrt{(x - X)^2 + (y - Y)^2 + (z - Z)^2} \qquad (> 0).$$

This function of x, y, z is called the **potential** of the gravitational field, and it satisfies Laplace's equation.

The extension to the potential and force due to a continuous distribution of mass is quite direct. If a mass of density $\rho(X, Y, Z)$ is distributed throughout a region T in space, then the corresponding potential u at a point (x, y, z) not occupied by mass is defined to be

$$u(x, y, z) = k \iiint_T \frac{\rho(X, Y, Z)}{r}\, dX\, dY\, dZ \qquad (k > 0), \tag{3}$$

with r as given by the previous formula. Since $1/r$ $(r > 0)$ is a solution of (1), that is, $\nabla^2(1/r) = 0$, and ρ does not depend on x, y, z, we obtain

$$\nabla^2 u = k \iiint_T \rho \nabla^2\left(\frac{1}{r}\right) dX\,dY\,dZ = 0;$$

that is, the gravitational potential defined by (3) satisfies Laplace's equation at any point that is not occupied by matter.

Electrostatics. The electrical force of attraction or repulsion between charged particles is governed by *Coulomb's law* (see Sec. 8.9), which is of the same mathematical form as Newton's law of gravitation. From this it follows that the field created by a distribution of electrical charges can be described mathematically by a potential function that satisfies Laplace's equation at any point not occupied by charges.

Heat Flow is governed by the heat equation

$$u_t = c^2 \nabla^2 u,$$

which in the steady-state case ($u_t = 0$) reduces to Laplace's equation, as we know from Sec. 11.5 for two-dimensional problems, and in three dimensions, this is similar.

Hydrodynamics. Laplace's equation also appears in connection with steady-state incompressible fluid flow, as will be shown in Chap. 17 (where we shall also see that ***two-dimensional*** problems can be solved by methods of complex analysis).

Laplacian in Cylindrical and Spherical Coordinates

In most applications leading to Laplace's equation, it is required to solve a **boundary value problem,** that is, to determine the solution of (1) satisfying given boundary conditions on the boundary surface S of the region T in which the equation is considered. This is called:

(I) First boundary value problem or **Dirichlet problem** if u is prescribed on S.

(II) Second boundary value problem or **Neumann problem** if the normal derivative $u_n = \partial u/\partial n$ is prescribed on S.

(III) Third or **mixed boundary value problem** if u is prescribed on a portion of S and u_n on the remaining portion of S.

It is then necessary to introduce coordinates in space such that S is given by simple formulas. For this we must transform the Laplacian into other coordinates. This is the same in principle as for two dimensions in Sec. 11.9.

Indeed, for **cylindrical coordinates** r, θ, z, which are related to Cartesian coordinates by (see Fig. 276 on p. 684)

$$(4) \qquad x = r\cos\theta, \qquad y = r\sin\theta, \qquad z = z,$$

we immediately get $\nabla^2 u$ by adding u_{zz} to (4), Sec. 11.9, obtaining

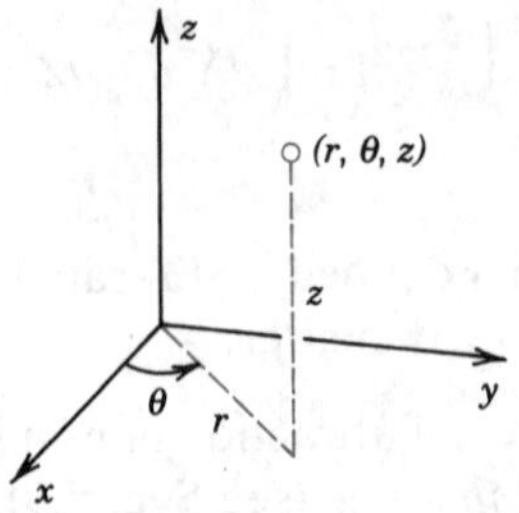

Fig. 276. Cylindrical coordinates

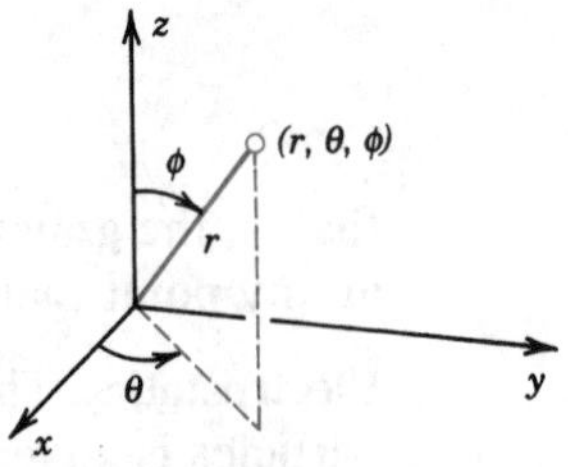

Fig. 277. Spherical coordinates

$$\nabla^2 u = \frac{\partial^2 u}{\partial r^2} + \frac{1}{r}\frac{\partial u}{\partial r} + \frac{1}{r^2}\frac{\partial^2 u}{\partial \theta^2} + \frac{\partial^2 u}{\partial z^2}. \tag{5}$$

Equally important in practice are **spherical coordinates** r, θ, ϕ, which are related to Cartesian coordinates by (see Fig. 277)[11]

$$x = r\cos\theta\sin\phi, \qquad y = r\sin\theta\sin\phi, \qquad z = r\cos\phi. \tag{6}$$

The Laplacian of a function u in spherical coordinates is

$$\nabla^2 u = \frac{\partial^2 u}{\partial r^2} + \frac{2}{r}\frac{\partial u}{\partial r} + \frac{1}{r^2}\frac{\partial^2 u}{\partial \phi^2} + \frac{\cot\phi}{r^2}\frac{\partial u}{\partial \phi} + \frac{1}{r^2\sin^2\phi}\frac{\partial^2 u}{\partial \theta^2}. \tag{7}$$

This may also be written

$$\nabla^2 u = \frac{1}{r^2}\left[\frac{\partial}{\partial r}\left(r^2\frac{\partial u}{\partial r}\right) + \frac{1}{\sin\phi}\frac{\partial}{\partial \phi}\left(\sin\phi\frac{\partial u}{\partial \phi}\right) + \frac{1}{\sin^2\phi}\frac{\partial^2 u}{\partial \theta^2}\right]. \tag{7'}$$

This formula can be derived in a manner similar to that in Sec. 11.9; the details are left as an exercise for the student.

In the next section we show that the separation of variables in spherical coordinates leads to *Legendre's equation,* which we discussed in Sec. 5.3.

Problem Set 11.11

1. Verify that $u = c/r$ satisfies Laplace's equation in spherical coordinates.
2. Show that the only solution of Laplace's equation depending only on $r = \sqrt{x^2 + y^2 + z^2}$ is $u = c/r + k$; here c and k are constants.
3. Determine c and k in Prob. 2 such that u represents the electrostatic potential between two concentric spheres of radii $r_1 = 2$ cm and $r_2 = 4$ cm kept at the potentials $U_1 = 110$ volts and $U_2 = 70$ volts, respectively.

---

[11] Equation (6) is used in calculus and extends the familiar notation for polar coordinates. Unfortunately, some books use θ and ϕ interchanged, an extension of the notation $x = r\cos\phi$, $y = r\sin\phi$ (used in some European countries).

4. Show that the only solution of the two-dimensional Laplace equation depending only on $r = \sqrt{x^2 + y^2}$ is $u = c \ln r + k$.

5. Find the electrostatic potential between two coaxial cylinders of radii $r_1 = 2$ cm and $r_2 = 4$ cm kept at the potentials $U_1 = 110$ volts and $U_2 = 70$ volts, respectively. Sketch and compare the equipotential lines in Probs. 3 and 5. Comment.

6. Substituting $u(r)$, r as in Prob. 2, into $u_{xx} + u_{yy} + u_{zz} = 0$, verify that $u'' + 2u'/r = 0$, in agreement with (7).

7. Derive (7) from $\nabla^2 u$ in Cartesian coordinates.

8. Verify (5) by transforming $\nabla^2 u$ back into Cartesian coordinates.

Show that the following functions $u = f(x, y)$ satisfy the Laplace equation and sketch some of the equipotential lines $u = const.$

9. xy **10.** $x^2 - y^2$ **11.** $x^3 - 3xy^2$ **12.** $x/(x^2 + y^2)$

13. $y/(x^2 + y^2)$ **14.** $(x + 1)(y - 1)$ **15.** $(x^2 - y^2)/(x^2 + y^2)^2$

Find the steady-state (time-independent) temperature distribution:

16. Between two parallel plates $x = x_0$ and $x = x_1$ kept at the temperatures u_0 and u_1, respectively.

17. Between two coaxial circular cylinders of radii r_0 and r_1 kept at the temperatures u_0 and u_1, respectively.

18. Between two concentric spheres of radii r_0 and r_1 kept at the temperatures u_0 and u_1, respectively.

19. If the surface of the ball $r^2 = x^2 + y^2 + z^2 \leqq R^2$ is kept at temperature zero and the initial temperature in the ball is $f(r)$, show that the temperature $u(r, t)$ in the ball is the solution of $u_t = c^2\left(u_{rr} + \frac{2}{r} u_r\right)$, satisfying the conditions $u(R, t) = 0$, $u(r, 0) = f(r)$.

20. Show that by setting $v = ru$ the formulas in Prob. 19 take the form $v_t = c^2 v_{rr}$, $v(R, t) = 0$, $v(r, 0) = rf(r)$. Include the condition $v(0, t) = 0$ (which holds because u must be bounded at $r = 0$), and solve the resulting problem by separating variables.

21. **(Helmholtz equation)** Show that the substitution of $u = U(x, y, z)e^{-i\omega t}$ $(i = \sqrt{-1})$ into the three-dimensional wave equation $u_{tt} = c^2 \nabla^2 u$ yields the so-called three-dimensional *Helmholtz equation* (see also Sec. 11.8)

$$\nabla^2 U + k^2 U = 0, \qquad k = \omega/c.$$

22. Let r, θ, ϕ be spherical coordinates. If $u(r, \theta, \phi)$ satisfies $\nabla^2 u = 0$, show that $v(r, \theta, \phi) = r^{-1}u(r^{-1}, \theta, \phi)$ satisfies $\nabla^2 v = 0$.

23. If $u(r, \theta)$ satisfies $\nabla^2 u = 0$, show that $v(r, \theta) = u(r^{-1}, \theta)$ satisfies $\nabla^2 v = 0$. (r and θ are polar coordinates.)

24. Solve the boundary value problem $\nabla^2 u = 0$ $(0 < x < \pi, y > 0)$, $u(0, y) = 0$, $u(\pi, y) = 0$, $u(x, 0) = 1$, $|u(x, y)|$ bounded. (Use separation of variables.)

25. It can be shown that the solution in Prob. 24 can be written in the form $u = (2/\pi) \arctan (\sin x / \sinh y)$. Verify directly that this u satisfies all the conditions in Prob. 24.

11.12 Laplacian in Spherical Coordinates. Legendre's Equation

Let us consider a typical boundary value problem that involves Laplace's equation in spherical coordinates. Suppose that a sphere S of radius R is kept at a fixed distribution of electric potential

$$u(R, \theta, \phi) = f(\phi) \tag{1}$$

where r, θ, ϕ are the spherical coordinates defined in the last section, with the origin at the center of S, and $f(\phi)$ is a given function. (Figure 278 on p. 689 shows a special example.) We wish to find the potential u at all points in space, which is assumed to be free of further charges. Since the potential on S is independent of θ, so is the potential in space. Thus $\partial^2 u/\partial\theta^2 = 0$ in (7′) of the last section, so that Laplace's equation reduces to

$$\frac{\partial}{\partial r}\left(r^2 \frac{\partial u}{\partial r}\right) + \frac{1}{\sin\phi}\frac{\partial}{\partial \phi}\left(\sin\phi \frac{\partial u}{\partial \phi}\right) = 0. \tag{2}$$

Furthermore, at infinity the potential will be zero; that is, we must have

$$\lim_{r\to\infty} u(r, \phi) = 0. \tag{3}$$

Thus we have to solve the boundary value problem for (2) with boundary condition (1) and condition at infinity (3). Substituting

$$u(r, \phi) = G(r)H(\phi)$$

into (2) and dividing the resulting equation by the function GH, we obtain

$$\frac{1}{G}\frac{d}{dr}\left(r^2 \frac{dG}{dr}\right) = -\frac{1}{H\sin\phi}\frac{d}{d\phi}\left(\sin\phi \frac{dH}{d\phi}\right)$$

and have the variables separated. By the usual argument, the two sides of this equation must be equal to a constant, say, k, so that

$$\frac{1}{\sin\phi}\frac{d}{d\phi}\left(\sin\phi \frac{dH}{d\phi}\right) + kH = 0 \tag{4}$$

and

$$\frac{1}{G}\frac{d}{dr}\left(r^2 \frac{dG}{dr}\right) = k.$$

The last equation may be written $(r^2G')' = kG$ or

$$r^2G'' + 2rG' - kG = 0.$$

This is an Euler–Cauchy equation. From Sec. 2.6 we know that it has solutions $G = r^\alpha$. These will have a particularly simple form if we change our notation and write $n(n + 1)$ for k. Then

$$r^2G'' + 2rG' - n(n + 1)G = 0 \tag{5}$$

where n is still arbitrary. By substituting $G = r^\alpha$ into (5) we have

$$[\alpha(\alpha - 1) + 2\alpha - n(n + 1)]r^\alpha = 0.$$

The zeros of the expression in brackets are $\alpha = n$ and $\alpha = -n - 1$. Hence we obtain the solutions

$$G_n(r) = r^n \qquad \text{and} \qquad G_n^*(r) = \frac{1}{r^{n+1}}\,. \tag{6}$$

Transformation of Equation (4). Setting $\cos\phi = w$, we get $\sin^2\phi = 1 - w^2$ and

$$\frac{d}{d\phi} = \frac{d}{dw}\frac{dw}{d\phi} = -\sin\phi\,\frac{d}{dw}\,.$$

Consequently, (4) with $k = n(n + 1)$ takes the form

$$\frac{d}{dw}\left[(1 - w^2)\frac{dH}{dw}\right] + n(n + 1)H = 0. \tag{7}$$

This is **Legendre's equation** (see Sec. 5.3), written out

$$(1 - w^2)\frac{d^2H}{dw^2} - 2w\frac{dH}{dw} + n(n + 1)H = 0. \tag{7$'$}$$

Solution Using a Fourier–Legendre Series

For integer[12] $n = 0, 1, \cdots$, the Legendre polynomials

$$H = P_n(w) = P_n(\cos\phi), \qquad n = 0, 1, \cdots,$$

are solutions of Legendre's equation (7). We thus obtain the following two sequences of solutions $u = GH$ of Laplace's equation (2):

$$u_n(r, \phi) = A_n r^n P_n(\cos\phi), \qquad u_n^*(r, \phi) = \frac{B_n}{r^{n+1}}P_n(\cos\phi) \tag{8*}$$

where $n = 0, 1, \cdots$, and A_n and B_n are constants.

[12]So far, n was arbitrary since k was arbitrary. It can be shown that the restriction of n to integers is necessary to make the solution of (7) continuous, together with its derivative of the first order, in the interval $-1 \leqq w \leqq 1$ or $0 \leqq \phi \leqq \pi$.

Solution of the Interior Problem. This means a solution of (2) *inside* the sphere and satisfying (1). For this we consider the series[13]

$$u(r, \phi) = \sum_{n=0}^{\infty} A_n r^n P_n(\cos \phi). \tag{8}$$

For (8) to satisfy (1) we must have

$$u(R, \phi) = \sum_{n=0}^{\infty} A_n R^n P_n(\cos \phi) = f(\phi); \tag{9}$$

that is, (9) must be the **Fourier–Legendre series** of $f(\phi)$. From (7) in Sec. 5.9 we get the coefficients

$$A_n R^n = \frac{2n+1}{2} \int_{-1}^{1} \tilde{f}(w) P_n(w)\, dw$$

where $\tilde{f}(w)$ denotes $f(\phi)$ as a function of $w = \cos \phi$. Since we have $dw = -\sin \phi\, d\phi$, and the limits of integration -1 and 1 correspond to $\phi = \pi$ and $\phi = 0$, respectively, we also obtain

$$A_n = \frac{2n+1}{2R^n} \int_0^{\pi} f(\phi) P_n(\cos \phi) \sin \phi\, d\phi, \qquad n = 0, 1, \cdots. \tag{10}$$

Thus the series (8) with coefficients (10) is the solution of our problem for points inside the sphere.

Solution of the Exterior Problem. Outside the sphere we cannot use the functions $u_n(r, \phi)$ because these functions do not satisfy (3), but we may use the functions $u_n^*(r, \phi)$, which satisfy (3), and proceed as before. This leads to the solution

$$u(r, \phi) = \sum_{n=0}^{\infty} \frac{B_n}{r^{n+1}} P_n(\cos \phi) \qquad (r \geqq R) \tag{11}$$

with coefficients

$$B_n = \frac{2n+1}{2} R^{n+1} \int_0^{\pi} f(\phi) P_n(\cos \phi) \sin \phi\, d\phi. \tag{12}$$

[13]Convergence will not be considered. It can be shown that if $f(\phi)$ and $f'(\phi)$ are piecewise continuous in the interval $0 \leqq \phi \leqq \pi$, the series (8) with coefficients (10) can be differentiated termwise twice with respect to r and with respect to ϕ and the resulting series converge and represent $\partial^2 u/\partial r^2$ and $\partial^2 u/\partial \phi^2$, respectively. Hence the series (8) with coefficients (10) is then the solution of our problem inside the sphere.

EXAMPLE 1 Spherical capacitor

Find the potential inside and outside a spherical capacitor consisting of two metallic hemispheres of radius 1 ft separated by a small slit for reasons of insulation, if the upper hemisphere is kept at 110 volts and the lower is grounded (Fig. 278).

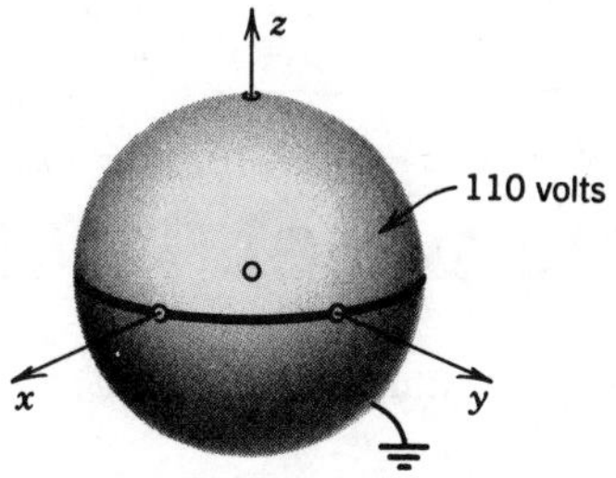

Fig. 278. Spherical capacitor in Example 1

Solution. The given boundary condition is (recall Fig. 277 in Sec. 11.11)

$$f(\phi) = \begin{cases} 110 & \text{if} \quad 0 \leqq \phi < \pi/2 \\ 0 & \text{if} \quad \pi/2 < \phi \leqq \pi. \end{cases}$$

Since $R = 1$, we thus obtain from (10)

$$A_n = \frac{2n+1}{2} \cdot 110 \int_0^{\pi/2} P_n(\cos \phi) \sin \phi \, d\phi.$$

We set $w = \cos \phi$. Then $P_n(\cos \phi) \sin \phi \, d\phi = -P_n(w)\, dw$ and we integrate from 1 to 0. We get rid of the minus by integrating from 0 to 1. Then from (11) in Sec. 5.3,

$$A_n = 55(2n+1) \sum_{m=0}^{M} (-1)^m \frac{(2n-2m)!}{2^n m!(n-m)!(n-2m)!} \int_0^1 w^{n-2m}\, dw$$

where $M = n/2$ for even n and $M = (n-1)/2$ for odd n. The integral equals $1/(n-2m+1)$. Thus

$$A_n = \frac{55(2n+1)}{2^n} \sum_{m=0}^{M} (-1)^m \frac{(2n-2m)!}{m!(n-m)!(n-2m+1)!}. \tag{13}$$

Taking $n = 0$, we get $A_0 = 55$ (since $0! = 1$). For $n = 1, 2, 3, \cdots$ we get

$$A_1 = \frac{165}{2} \cdot \frac{2!}{0!1!2!} = \frac{165}{2},$$

$$A_2 = \frac{275}{4} \left(\frac{4!}{0!2!3!} - \frac{2!}{1!1!1!} \right) = 0,$$

$$A_3 = \frac{385}{8} \left(\frac{6!}{0!3!4!} - \frac{4!}{1!2!2!} \right) = -\frac{385}{8},$$

and so on. Hence the potential (8) inside the sphere is (since $P_0 = 1$)

$$u(r, \phi) = 55 + \frac{165}{2} r P_1(\cos \phi) - \frac{385}{8} r^3 P_3(\cos \phi) + \cdots$$

with $P_1, P_3, \cdots$ given by (11′), Sec. 5.3. Since $R = 1$, we see from (10) and (12) in the present section that $B_n = A_n$, and (11) thus gives the potential outside the sphere

$$u(r, \phi) = \frac{55}{r} + \frac{165}{2r^2} P_1(\cos \phi) - \frac{385}{8r^4} P_3(\cos \phi) + \cdots.$$

Partial sums of these series can now be used for computing approximate values of the potential. Also, it is interesting to see that far away from the sphere the potential is approximately that of a point charge, namely, $55/r$. (Compare with Example 3 in Sec. 8.9.) ▮

Problem Set 11.12

1. Verify by substitution that $u_n(r, \phi)$ and $u_n^*(r, \phi)$, $n = 0, 1, 2$, in (8*) are solutions of (2).
2. Find the surfaces on which the functions u_1, u_2, u_3 are zero.
3. Sketch the functions $P_n(\cos \phi)$ for $n = 0, 1, 2$, [see (11′), Sec. 5.3].
4. Sketch the functions $P_3(\cos \phi)$ and $P_4(\cos \phi)$.

Let r, θ, ϕ be the spherical coordinates used in the text. Find the potential in the interior of the sphere $R = 1$, assuming that there are no charges in the interior and the potential on the surface is $f(\phi)$, where

5. $f(\phi) = 1$
6. $f(\phi) = \cos \phi$
7. $f(\phi) = \cos 2\phi$
8. $f(\phi) = 1 - \cos^2 \phi$
9. $f(\phi) = \cos^3 \phi$
10. $f(\phi) = \cos 3\phi + 3 \cos \phi$
11. $f(\phi) = 10 \cos^3 \phi - 3 \cos^2 \phi - 5 \cos \phi - 1$

12. Show that in Prob. 5, the potential exterior to the sphere is the same as that of a point charge at the origin.
13. Sketch the intersections of the equipotential surfaces in Prob. 6 with the xz-plane.
14. Find the potential exterior to the sphere in Probs. 5–11.
15. Derive the values of A_0, A_1, A_2, A_3 in Example 1 from (13).
16. In Example 1, sketch the sum of the three explicitly given terms for $r = 1$ and see how well this sum approximates the given boundary function.
17. Find the temperature in a homogeneous ball of radius 1 if its lower boundary hemisphere is kept at 0°C and its upper at 20°C.
18. Show that $P'_{n+1}(x) - P'_{n-1}(x) = (2n + 1)P_n(x)$. (Use Prob. 8, Sec. 5.3.)
19. Show that $\int_0^1 P_n(x)\, dx = [P_{n-1}(0) - P_{n+1}(0)]/(2n + 1)$. (Use Prob. 18 and Prob. 12, Sec. 5.3.) Using this, verify A_1, A_2, A_3 in Example 1 and compute A_5.

20. **(Transmission line equations)** Consider a long cable or telephone wire (Fig. 279) that is imperfectly insulated so that leaks occur along the entire length of the cable. The source S of the current $i(x, t)$ in the cable is at $x = 0$, the receiving end T at $x = l$. The current flows from S to T, through the load, and returns to the ground. Let the constants R, L, C, and G denote the resistance, inductance,

capacitance to ground, and conductance to ground, respectively, of the cable per unit length. Show that

$$-\frac{\partial u}{\partial x} = Ri + L\frac{\partial i}{\partial t} \qquad \textbf{(First transmission line equation)}$$

where $u(x, t)$ is the potential in the cable. *Hint.* Apply Kirchhoff's voltage law to a small portion of the cable between x and $x + \Delta x$ (difference of the potentials at x and $x + \Delta x$ = resistive drop + inductive drop).

21. Show that for the cable in Prob. 20,

$$-\frac{\partial i}{\partial x} = Gu + C\frac{\partial u}{\partial t} \qquad \textbf{(Second transmission line equation).}$$

Hint. Use Kirchhoff's current law (difference of the currents at x and $x + \Delta x$ = loss due to leakage to ground + capacitive loss).

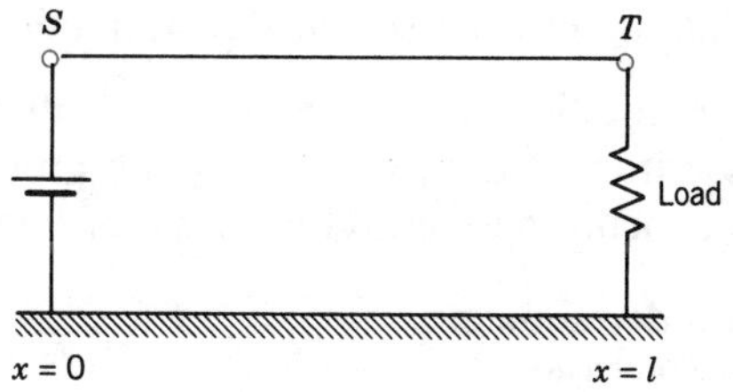

Fig. 279. Transmission line

22. Show that elimination of i or u from the transmission line equations leads to

$$u_{xx} = LCu_{tt} + (RC + GL)u_t + RGu,$$

$$i_{xx} = LCi_{tt} + (RC + GL)i_t + RGi.$$

23. (Telegraph equations) For a submarine cable, G is negligible and the frequencies are low. Show that this leads to the so-called *submarine cable equations* or **telegraph equations**

$$u_{xx} = RCu_t, \qquad i_{xx} = RCi_t.$$

24. Find the potential in a submarine cable with ends ($x = 0$, $x = l$) grounded and initial voltage distribution $U_0 = const.$

25. (High-frequency line equations) Show that in the case of alternating currents of high frequencies the equations in Prob. 22 can be approximated by the so-called **high-frequency line equations**

$$u_{xx} = LCu_{tt}, \qquad i_{xx} = LCi_{tt}.$$

Solve the first of them, assuming that the initial potential is $U_0 \sin(\pi x/l)$, $u_t(x, 0) = 0$ and $u = 0$ at the ends $x = 0$ and $x = l$ for all t.

11.13 Solution by Laplace Transforms

Readers familiar with Chap. 6 may wonder whether Laplace transforms can also be used for solving *partial* differential equations. This answer is yes, particularly if one of the independent variables ranges over the positive axis. The steps of solution are similar to those in Chap. 6. For an equation in two variables they are as follows.

1. Take the Laplace transform with respect to one of the two variables, usually t. This gives an *ordinary differential equation* for the transform of the unknown function. This is so since the derivatives of this function with respect to the other variable slip into the transformed equation. The latter also incorporates the given boundary and initial conditions.

2. Solving that ordinary differential equation, obtain the transform of the unknown function.

3. Taking the inverse transform, obtain the solution of the given problem.

If the coefficients of the given equation do not depend on t, the use of Laplace transforms will simplify the problem.

We explain the method in terms of two typical examples.

EXAMPLE 1 **A first-order equation**

Solve the problem

$$\frac{\partial w}{\partial x} + x\frac{\partial w}{\partial t} = 0, \quad w(x, 0) = 0, \quad w(0, t) = t \qquad (t \geqq 0). \tag{1}$$

We write w since we need u to denote the unit step function (Sec. 6.3).

Solution. We take the Laplace transform of (1) ***with respect to t.*** By (1) in Sec. 6.2,

$$\mathcal{L}\left\{\frac{\partial w}{\partial x}\right\} + x[s\mathcal{L}\{w\} - w(x, 0)] = 0. \tag{2}$$

Here $w(x, 0) = 0$. In the first term we assume that we may interchange integration and differentiation:

$$\mathcal{L}\left\{\frac{\partial w}{\partial x}\right\} = \int_0^\infty e^{-st}\frac{\partial w}{\partial x}\,dt = \frac{\partial}{\partial x}\int_0^\infty e^{-st}w(x, t)\,dt = \frac{\partial}{\partial x}\mathcal{L}\{w(x, t)\}. \tag{3}$$

Writing $W(x, s) = \mathcal{L}\{w(x, t)\}$, we thus obtain from (2)

$$\frac{\partial W}{\partial x} + xsW = 0.$$

This may be regarded as an *ordinary* differential equation with x as the independent variable, since derivatives with respect to s do not occur in the equation. The general solution is

$$W(x, s) = c(s)e^{-sx^2/2} \qquad \text{(Sec. 1.7).}$$

Since $\mathcal{L}\{t\} = 1/s^2$, the condition $w(0, t) = t$ yields $W(0, s) = 1/s^2$, that is,

$$W(0, s) = c(s) = 1/s^2.$$

Hence

$$W(x, s) = \frac{1}{s^2}e^{-sx^2/2}.$$

Now $\mathcal{L}^{-1}\{1/s^2\} = t$, so that the second shifting theorem (Sec. 6.3) with $a = x^2/2$ gives

(4) $$w(x, t) = \left(t - \frac{x^2}{2}\right) u(t - \tfrac{1}{2}x^2) = \begin{cases} 0 & \text{if } t < x^2/2 \\ t - \tfrac{1}{2}x^2 & \text{if } t > x^2/2. \end{cases}$$

Since we proceeded formally, we have to verify that (4) satisfies (1). We leave this to the student. ∎

EXAMPLE 2 Semi-infinite string

Find the displacement $w(x, t)$ of an elastic string subject to the following conditions.

(*i*) The string is initially at rest on the x-axis from $x = 0$ to ∞ (*"semi-infinite string"*).

(*ii*) For time $t > 0$ the left end of the string is moved in a given fashion (Fig. 280)

$$w(0, t) = f(t) = \begin{cases} \sin t & \text{if } 0 \leqq t \leqq 2\pi \\ 0 & \text{otherwise} \end{cases}$$

(*iii*) Furthermore

$$\lim_{x\to\infty} w(x, t) = 0 \quad \text{for } t \geqq 0.$$

Of course there is no infinite string, but our model describes a long string or rope (of negligible weight) with its right end fixed far out on the x-axis. (We again write w since we need u to denote the unit step function.)

Solution. We have to solve the wave equation (Sec. 11.2)

(5) $$\frac{\partial^2 w}{\partial t^2} = c^2 \frac{\partial^2 w}{\partial x^2}, \qquad c^2 = \frac{T}{\rho},$$

for positive x and t, subject to the "boundary conditions"

(6) $$w(0, t) = f(t), \qquad \lim_{x\to\infty} w(x, t) = 0 \qquad (t \geqq 0)$$

with f as given above, and the initial conditions

(7) $$w(x, 0) = 0,$$

(8) $$\left.\frac{\partial w}{\partial t}\right|_{t=0} = 0.$$

We take the Laplace transform ***with respect to t***. By (2) in Sec. 6.2,

$$\mathcal{L}\left\{\frac{\partial^2 w}{\partial t^2}\right\} = s^2\mathcal{L}\{w\} - sw(x, 0) - \left.\frac{\partial w}{\partial t}\right|_{t=0} = c^2\mathcal{L}\left\{\frac{\partial^2 w}{\partial x^2}\right\}.$$

Two terms drop out, by (7) and (8). On the right we assume that we may interchange integration and differentiation:

$$\mathcal{L}\left\{\frac{\partial^2 w}{\partial x^2}\right\} = \int_0^\infty e^{-st}\frac{\partial^2 w}{\partial x^2}\,dt = \frac{\partial^2}{\partial x^2}\int_0^\infty e^{-st} w(x, t)\,dt = \frac{\partial^2}{\partial x^2}\mathcal{L}\{w(x, t)\}.$$

Writing $W(x, s) = \mathcal{L}\{w(x, t)\}$, we thus obtain

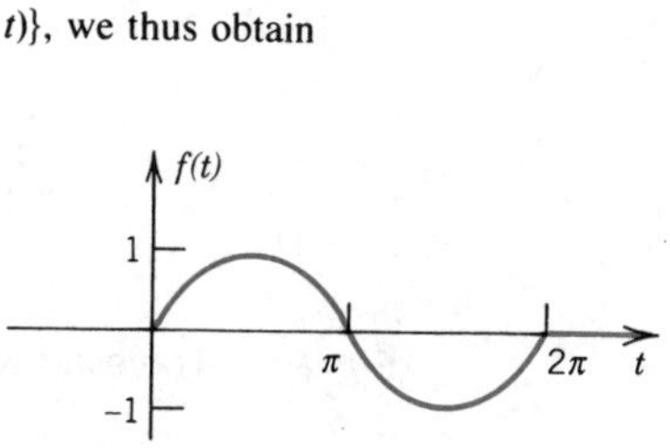

Fig. 280. Motion of the left end of the string in Example 2 as a function of time t

$$s^2 W = c^2 \frac{\partial^2 W}{\partial x^2}, \qquad \text{thus} \qquad \frac{\partial^2 W}{\partial x^2} - \frac{s^2}{c^2} W = 0.$$

Since this equation contains only a derivative with respect to x, it may be regarded as an ordinary differential equation for $W(x, s)$ considered as a function of x. A general solution is

$$W(x, s) = A(s)e^{sx/c} + B(s)e^{-sx/c}. \tag{9}$$

From (6) we obtain, writing $F(s) = \mathcal{L}\{f(t)\}$,

$$W(0, s) = \mathcal{L}\{w(0, t)\} = \mathcal{L}\{f(t)\} = F(s)$$

and, assuming that the order of integrating with respect to t and taking the limit as $x \to \infty$ can be interchanged,

$$\lim_{x\to\infty} W(x, s) = \lim_{x\to\infty} \int_0^\infty e^{-st} w(x, t)\, dt = \int_0^\infty e^{-st} \lim_{x\to\infty} w(x, t)\, dt = 0.$$

This implies $A(s) = 0$ in (9) because $c > 0$, so that for every fixed positive s the function $e^{sx/c}$ increases as x increases. Note that we may assume $s > 0$ since a Laplace transform generally exists for *all* s greater than some fixed γ (Sec. 6.2). Hence we have

$$W(0, s) = B(s) = F(s),$$

so that (9) becomes

$$W(x, s) = F(s)e^{-sx/c}.$$

From the second shifting theorem (Sec. 6.3) with $a = x/c$ we obtain the inverse transform

$$w(x, t) = f\left(t - \frac{x}{c}\right) u\left(t - \frac{x}{c}\right) \qquad \text{(Fig. 281)}, \tag{10}$$

that is,

$$w(x, t) = \sin\left(t - \frac{x}{c}\right) \quad \text{if} \quad \frac{x}{c} < t < \frac{x}{c} + 2\pi \quad \text{or} \quad ct > x > (t - 2\pi)c$$

and zero otherwise. This is a single sine wave traveling to the right with speed c. Note that a point x remains at rest until $t = x/c$, the time needed to reach that x if one starts at $t = 0$ (start of the motion of the left end) and travels with speed c. The result agrees with our physical intuition. Since we proceeded formally, we must verify that (10) satisfies the given conditions. We leave this to the student. ∎

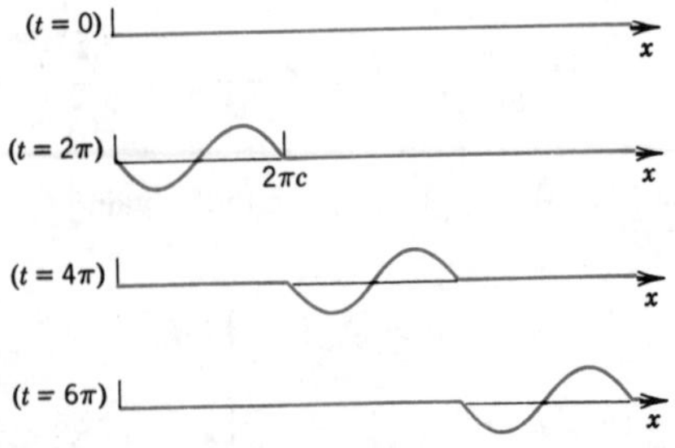

Fig. 281. Traveling wave in Example 2

Problem Set 11.13

1. Sketch a figure similar to Fig. 281 if $c = 1$ and f is "triangular" as in Example 1, Sec. 11.3, with $k = L/2 = 1$.
2. How does the speed of the wave in Example 2 depend on the tension and the mass of the string?
3. Verify the solution in Example 2. What traveling wave do we obtain in Example 2 if we impose a (nonterminating) sinusoidal motion of the left end starting at $t = 0$?

Solve by Laplace transforms:

4. $\dfrac{\partial u}{\partial x} + 2x\dfrac{\partial u}{\partial t} = 2x,\ u(x, 0) = 1,\ u(0, t) = 1$
5. $x\dfrac{\partial u}{\partial x} + \dfrac{\partial u}{\partial t} = xt,\ u(x, 0) = 0$ if $x \geqq 0$, $u(0, t) = 0$ if $t \geqq 0$.
6. Solve Prob. 5 by another method.

Find the temperature $w(x, t)$ in a semi-infinite laterally insulated bar extending from $x = 0$ along the x-axis to ∞, assuming that the initial temperature is 0, $w(x, t) \to 0$ as $x \to \infty$ for every fixed $t \geqq 0$, and $w(0, t) = f(t)$. Proceed as follows.

7. Set up the model and show that the Laplace transform leads to

$$sW(x, s) = c^2\frac{\partial^2 W}{\partial x^2}, \qquad W = \mathcal{L}\{w\},$$

and

$$W(x, s) = F(s)e^{-\sqrt{s}x/c}, \qquad F = \mathcal{L}\{f\}.$$

8. Applying the convolution theorem in Prob. 7, show that

$$w(x, t) = \frac{x}{2c\sqrt{\pi}}\int_0^t f(t - \tau)\tau^{-3/2}e^{-x^2/4c^2\tau}\,d\tau.$$

9. Let $w(0, t) = f(t) = u(t)$ (Sec. 6.3). Denote the corresponding w, W, and F by w_0, W_0, and F_0. Show that then in Prob. 8,

$$w_0(x, t) = \frac{x}{2c\sqrt{\pi}}\int_0^t \tau^{-3/2}e^{-x^2/4c^2\tau}\,d\tau = 1 - \operatorname{erf}\left(\frac{x}{2c\sqrt{t}}\right)$$

with the error function erf as defined in Problem Set 11.6.

10. **(Duhamel's formula**[14]**)** Show that in Prob. 9,

$$W_0(x, s) = \frac{1}{s}e^{-\sqrt{s}x/c}$$

and the convolution theorem gives *Duhamel's formula*

$$w(x, t) = \int_0^t f(t - \tau)\frac{\partial w_0}{\partial \tau}\,d\tau.$$

[14]JEAN MARIE CONSTANT DUHAMEL (1797—1872), French mathematician.

11.14 Solution by Fourier Transforms

Partial differential equations may also be solved by other operational methods. In this section we explain this for Fourier transforms. Indeed, if initial or boundary data are given on the positive half-axis, then Fourier cosine or sine transforms (Sec. 10.10) may be appropriate, and if they are given on the entire axis, we may use Fourier transforms (Sec. 10.11). We discuss these methods in terms of typical applications.

EXAMPLE 1 A heat problem on the x-axis

Find the temperature $u(x, t)$ in a laterally insulated homogeneous bar of constant cross section extending from $x = -\infty$ to ∞, for time $t > 0$, assuming that the given initial temperature is

$$u(x, 0) = f(x) \qquad (-\infty < x < \infty), \tag{1}$$

and for all $t \geqq 0$ the solution and its x derivative satisfy

$$u(x, t) \to 0, \qquad u_x(x, t) \to 0 \qquad \text{as } |x| \to \infty. \tag{2}$$

As a particular case, find $u(x, t)$ when

$$f(x) = U_0 = const \quad \text{if } |x| < 1 \qquad \text{and} \qquad f(x) = 0 \quad \text{if } |x| > 1. \tag{3}$$

Solution. We have to solve the **heat equation**

$$u_t = c^2 u_{xx} \tag{4}$$

subject to conditions (1), (2). Our strategy is to take the Fourier transform with respect to x and then solve the resulting ordinary differential equation in t. The details are as follows.

Let $\hat{u} = \mathcal{F}(u)$ denote the Fourier transform of u, ***regarded as a function of x.*** From (10) in Sec. 10.11 we see that (4) gives

$$\mathcal{F}(u_t) = c^2\mathcal{F}(u_{xx}) = c^2(-w^2)\mathcal{F}(u) = -c^2w^2\hat{u}.$$

On the left, assuming that we may interchange the order of differentiation and integration,

$$\mathcal{F}(u_t) = \frac{1}{\sqrt{2\pi}}\int_{-\infty}^{\infty} u_t e^{-iwx}\,dx = \frac{1}{\sqrt{2\pi}}\frac{\partial}{\partial t}\int_{-\infty}^{\infty} u e^{-iwx}\,dx = \frac{\partial \hat{u}}{\partial t}.$$

Thus

$$\frac{\partial \hat{u}}{\partial t} = -c^2w^2\hat{u}.$$

Since this equation involves only a derivative with respect to t but none with respect to w, this is a first-order ***ordinary*** differential equation with t as the independent variable and w as a parameter. By separating variables (Sec. 1.2) we get the general solution

$$\hat{u}(w, t) = C(w)e^{-c^2w^2t},$$

with the arbitrary "constant" $C(w)$ depending on the parameter w. The initial condition (1) gives $\hat{u}(w, 0) = C(w) = \hat{f}(w) = \mathcal{F}(f)$. Our intermediate result is

$$\hat{u}(w, t) = \hat{f}(w)e^{-c^2w^2t}.$$

The inversion formula (7), Sec. 10.11, now gives the solution

$$u(x, t) = \frac{1}{\sqrt{2\pi}}\int_{-\infty}^{\infty} \hat{f}(w)e^{-c^2w^2t}e^{iwx}\,dw. \tag{5}$$

In this we may insert the Fourier transform

$$\hat{f}(w) = \frac{1}{\sqrt{2\pi}} \int_{-\infty}^{\infty} f(v)e^{-iwv}\, dv.$$

Assuming that we may invert the order of integration, we then obtain

$$u(x, t) = \frac{1}{2\pi} \int_{-\infty}^{\infty} f(v) \left[\int_{-\infty}^{\infty} e^{-c^2w^2t} e^{i(wx - wv)}\, dw \right] dv.$$

By the Euler formula (3), Sec. 10.11, the integrand of the inner integral equals

$$e^{-c^2w^2t} \cos (wx - wv) + ie^{-c^2w^2t} \sin (wx - wv).$$

This shows that its imaginary part is an odd function of w, so that the integral[15] of this part is 0, and the real part is even, so that its integral is twice the integral from 0 to ∞:

$$u(x, t) = \frac{1}{\pi} \int_{-\infty}^{\infty} f(v) \left[\int_{0}^{\infty} e^{-c^2w^2t} \cos (wx - wv)\, dw \right] dv.$$

This agrees with (9), Sec. 11.6, and leads to the further formulas (11) and (13) in Sec. 11.6. ■

EXAMPLE 2 **Problem in Example 1 solved by the method of convolution**

Solve the heat problem in Example 1 by the method of convolution.

Solution. The beginning is as before and leads to (5), that is,

$$u(x, t) = \frac{1}{\sqrt{2\pi}} \int_{-\infty}^{\infty} \hat{f}(w)e^{-c^2w^2t} e^{iwx}\, dw. \tag{5}$$

Now comes the crucial idea. We recognize that this is of the form (13) in Sec. 10.11, that is,

$$u(x, t) = (f * g)(x) = \int_{-\infty}^{\infty} \hat{f}(w)\hat{g}(w)e^{iwx}\, dw \tag{6}$$

where

$$\hat{g}(w) = \frac{1}{\sqrt{2\pi}} e^{-c^2w^2t}. \tag{7}$$

Since, by the definition of convolution [(11), Sec. 10.11],

$$(f * g)(x) = \int_{-\infty}^{\infty} f(p)g(x - p)\, dp, \tag{8}$$

as our next and last step we must determine the inverse Fourier transform g of $\hat{g}$. For this we can use formula 9 in Table III of Sec. 10.12 (which was derived in Example 2 of Sec. 10.11),

$$\mathcal{F}(e^{-ax^2}) = \frac{1}{\sqrt{2a}} e^{-w^2/4a}$$

with a suitable a. With $c^2t = 1/4a$ or $a = 1/4c^2t$, using (7) we obtain

$$\mathcal{F}(e^{-x^2/4c^2t}) = \sqrt{2c^2t}\, e^{-c^2w^2t} = \sqrt{2c^2t}\,\sqrt{2\pi}\, \hat{g}(w).$$

Hence $\hat{g}$ has the inverse

[15] Actually, the principal part of the integral; see Sec. 15.3.

$$\frac{1}{\sqrt{2c^2t}\sqrt{2\pi}}e^{-x^2/4c^2t}.$$

Replacing x with $x - p$ and substituting this into (8) we finally have

$$u(x, t) = (f * g)(x) = \frac{1}{2c\sqrt{\pi t}}\int_{-\infty}^{\infty} f(p)\exp\left\{-\frac{(x-p)^2}{4c^2t}\right\}dp. \tag{9}$$

This solution formula of our problem agrees with (11) in Sec. 11.6. We wrote $(f * g)(x)$, without indicating the parameter t with respect to which we did not integrate. ∎

EXAMPLE 3 Fourier sine transform applied to the heat equation

Find a solution formula in Example 1, assuming that the bar extends from 0 to ∞, the initial temperature is

$$u(x, 0) = f(x) \qquad (0 \leqq x < \infty),$$

and at the left end we have the boundary condition

$$u(0, t) = 0 \qquad (t \geqq 0).$$

Solution. Instead of the Fourier transform we may now apply the Fourier sine transform (Sec. 10.10), since x varies from 0 to ∞. Proceeding as in Example 1, we obtain from the heat equation and (9b), Sec. 10.10, since $f(0) = u(0, 0) = 0$,

$$\mathcal{F}_s(u_t) = \frac{\partial \hat{u}_s}{\partial t} = c^2\mathcal{F}_s(u_{xx}) = -c^2w^2\mathcal{F}_s(u) = -c^2w^2\hat{u}_s(w, t).$$

The solution of this first-order ordinary differential equation is

$$\hat{u}_s(w, t) = C(w)e^{-c^2w^2t}.$$

From the initial condition $u(x, 0) = f(x)$ we have $\hat{u}_s(w, 0) = \hat{f}_s(w) = C(w)$. Hence

$$\hat{u}_s(w, t) = \hat{f}_s(w)e^{-c^2w^2t}.$$

Taking the inverse sine transform and substituting

$$\hat{f}_s(w) = \sqrt{\frac{2}{\pi}}\int_0^{\infty} f(p)\sin wp\,dp,$$

we obtain the desired solution formula

$$u(x, t) = \frac{2}{\pi}\int_0^{\infty}\int_0^{\infty} f(p)\sin wp\; e^{-c^2w^2t}\sin wx\,dp\,dw. \tag{10}$$

∎

EXAMPLE 4 Wave equation on an infinite interval. D'Alembert's solution

Solve the wave equation

$$u_{tt} = c^2u_{xx} \qquad (-\infty < x < \infty,\ t > 0)$$

subject to the conditions

$$u(x, 0) = f(x) \qquad \text{(given initial deflection)}$$

$$u_t(x, 0) = 0 \qquad \text{(initial speed zero)}$$

$$u \to 0, \quad u_x \to 0 \qquad \text{as } |x| \to \infty \text{ for all } t,$$

where f and g are assumed to have a Fourier transform.

Solution. We transform with respect to x, writing $\hat{u} = \mathcal{F}(u)$. Using (10), Sec. 10.11, we get from the wave equation

$$\mathcal{F}(u_{tt}) = \hat{u}_{tt} = c^2\mathcal{F}(u_{xx}) = -c^2w^2\hat{u}.$$

Thus

$$\hat{u}_{tt} + c^2w^2\hat{u} = 0.$$

Since no w-derivatives occur, this is a second-order ***ordinary*** differential equation whose coefficient c^2w^2 is "constant" (that is, independent of t). A general solution is

$$u(w, t) = A(w) \cos cwt + B(w) \sin cwt.$$

For $t = 0$, since $\mathcal{F}\{u(x, 0)\} = \hat{u}(w, 0)$, from the initial conditions we obtain

$$\hat{u}(w, 0) = A(w) = \hat{f}(w)$$

$$\hat{u}_t(w, 0) = cwB(w) = 0.$$

Thus,

$$\hat{u}(w, t) = \hat{f}(w) \cos cwt.$$

We want u. This needs an idea. The idea is to express the cosine in terms of exponential functions and then apply the shifting formula

(11) $$\mathcal{F}\{f(x - a)\} = e^{-iwa}\mathcal{F}\{f(x)\},$$

which follows directly from the definition of the Fourier transformation by setting $x - a = p$, $x = p + a$, $dx = dp$, so that

$$\mathcal{F}\{f(x - a)\} = \frac{1}{\sqrt{2\pi}} \int_{-\infty}^{\infty} f(x - a)e^{-iwx}\, dx = \frac{1}{\sqrt{2\pi}} \int_{-\infty}^{\infty} f(p)e^{-iw(p+a)}\, dp$$

$$= \frac{1}{\sqrt{2\pi}} e^{-iwa} \int_{-\infty}^{\infty} f(p)e^{-iwp}\, dp.$$

Thus

$$\hat{f}(w) \cos cwt = \tfrac{1}{2} \hat{f}(w)[e^{icwt} + e^{-icwt}]$$

has the inverse Fourier transform

(12) $$u(x, t) = \tfrac{1}{2}[f(x - ct) + f(x + ct)].$$

This is **d'Alembert's solution** (14), Sec. 11.4. ∎

This is the end of Chap. 11, in which we concentrated on the most important partial differential equations in physics and engineering. This is also the end of Part C on Fourier analysis and partial differential equations.

We have seen that these equations have various basic engineering applications. For this reason they are the subject of many ongoing research projects.

Numerical methods for partial differential equations follow in Secs. 20.4–20.7, which are independent of the other sections in Part E on numerical methods.

In the next part (Part D, Chaps. 12–17) we again turn to an area of different nature, **complex analysis,** which is also highly important to the engineer, as our examples and problems will show.

Review Questions and Problems for Chapter 11

1. By what physical law did we obtain the one-dimensional wave equation?
2. What conditions did we have in the vibrating string problem?
3. What are the eigenfunctions and their frequencies of the vibrating string?
4. How did we obtain d'Alembert's solution of the vibrating string problem?
5. Why did we discuss Fourier series solutions of the wave equation, although d'Alembert's solution was obtained much more quickly?
6. What form does the heat equation have and what additional condition did we consider in heat problems?
7. What is the superposition principle and to what equations does it apply?
8. What are elliptic, parabolic, and hyperbolic equations? Give an example for each of these.
9. What are the eigenfunctions of a quadratic membrane? What do you know about their frequencies?
10. What are the eigenfunctions of a circular membrane? How do the frequencies of these functions differ in principle from those of the vibrating string?
11. In separating the wave equation we got only trigonometric functions, whereas in separating the heat equation we also got exponential functions. What was the reason?
12. Certain simpler partial differential equations can be solved by methods for ordinary differential equations. Explain; give examples.
13. Why did Legendre and Bessel functions appear in this chapter? Where in this book did these functions first occur?
14. What is the error function and in what connection did it occur?
15. Why could we use Fourier series, although the functions of physical interest were not periodic, in general?
16. How many initial conditions can be given in the case of the wave equation? In the case of the heat equation?
17. Verify that $u = x^4 - 6x^2y^2 + y^4$ and $u = \sin x \sinh y$ are solutions of Laplace's equation.
18. Verify that $u = (x^2 - y^2)/(x^2 + y^2)^2$ and $u = 2xy/(x^2 + y^2)^2$ are solutions of Laplace's equation.
19. For what reasons did the Fourier integral occur in this chapter?
20. In Chap. 6, the subsidiary equation was an *algebraic* equation. Why do we obtain an ordinary *differential* equation in solving a partial differential equation by a transform method (an operational method)?

Solve:

21. $u_{yy} + 16u = 0$
22. $u_{xy} + u_y + x + y + 1 = 0$
23. $u_{xx} + u_x - 2u = 10$
24. $u_{xx} + u = 0,\ u(0, y) = f(y),\ u_x(0, y) = g(y)$
25. $u_{yy} + u_y = 0,\ u(x, 0) = f(x),\ u_y(x, 0) = g(x)$
26. Find all solutions $u(x, y) = F(x)G(y)$ of Laplace's equation in two variables.

Using the indicated transformations, solve the following equations.

27. $u_{xy} = u_{xx}$ $\quad (v = y,\ z = x + y)$

28. $yu_{xy} = xu_{xx} + u_x \quad (v = y,\ z = xy)$
29. $u_{xx} - 2u_{xy} + u_{yy} = 0 \quad (v = y,\ z = x + y)$
30. $u_{xx} = u_{yy} \quad (v = y + x,\ z = y - x)$
31. $u_{yy} + u_{xy} - 2u_{xx} = 0 \quad (v = x + y,\ z = 2y - x)$
32. $u_{xx} + 2u_{xy} + u_{yy} = 0 \quad (v = x,\ z = x - y)$
33. $2u_{xx} + 5u_{xy} + 2u_{yy} = 0 \quad (v = 2x - y,\ z = 2y - x)$

Find the motion of the vibrating string of length π and $c^2 = T/\rho = 1$ starting with initial velocity 0 and deflection

34. $f(x) = \sin x - \frac{1}{2} \sin 2x$
35. $f(x) = -0.1 \sin 3x$
36. $f(x) = \pi/2 - |x - \pi/2|$
37. $f(x) = \sin^3 x$
38. $f(x) = x$ if $0 < x < \pi/3$, $\quad f(x) = (\pi - x)/2$ if $\pi/3 < x < \pi$

Find the temperature $u(x, t)$ in a bar of length $L = \pi$ and $c = 1$ that is perfectly insulated, also at the ends $x = 0$ and $x = \pi$, if the initial temperature $u(x, 0) = f(x)$ is

39. $f(x) = k = const$
40. $f(x) = \sin x$
41. $f(x) = \cos^2 x$
42. $f(x) = 1 - x/\pi$

Find the temperature distribution in a laterally insulated thin copper bar ($c^2 = K/\sigma\rho = 1.158$ cm²/sec), 50 cm long and of constant cross section whose endpoints at $x = 0$ and $x = 50$ are kept at 0°C and whose initial temperature is

43. $f(x) = \sin(\pi x/50)$
44. $f(x) = 100 \sin(\pi x/25)$
45. $f(x) = \sin^3(\pi x/10)$
46. $f(x) = x(50 - x)$
47. $f(x) = x$ if $0 < x < 25$, $\quad f(x) = 50 - x$ if $25 < x < 50$

Recall from Sec. 11.5 that for adiabatic boundary conditions the temperature $u(x, t)$ in a laterally insulated bar of length L is

$$u(x, t) = A_0 + \sum_{n=1}^{\infty} A_n \cos \frac{n\pi x}{L} \exp\left[-\left(\frac{cn\pi}{L}\right)^2 t\right].$$

Assuming $L = \pi$ and $c = 1$, find from this the solution satisfying the initial condition

48. $f(x) = 30x^2$
49. $f(x) = 95 \cos 2x$
50. $f(x) = 4x$ if $0 < x < \pi/2$, $\quad f(x) = 4(\pi - x)$ if $\pi/2 < x < \pi$

51. Suppose that the faces of the thin plate given by $0 \leqq x \leqq \pi$, $0 \leqq y \leqq \pi$ are perfectly insulated, the edges are kept at zero temperature, and the initial temperature is $u(x, y, 0) = f(x, y)$. By applying the method of separating variables to the two-dimensional heat equation $u_t = c^2\nabla^2 u$, show that the temperature in the plate is

$$u(x, y, t) = \sum_{m=1}^{\infty} \sum_{n=1}^{\infty} B_{mn} \sin mx \sin ny\ e^{-c^2(m^2+n^2)t}$$

where

$$B_{mn} = \frac{4}{\pi^2} \int_0^{\pi} \int_0^{\pi} f(x, y) \sin mx \sin ny\ dx\ dy.$$

52. Find the temperature in the plate of Prob. 51, if $f(x, y) = x(\pi - x)y(\pi - y)$.

Show that the following membranes of area 1 with $c^2 = 1$ have the frequencies of the fundamental mode as given (4-decimal values). Compare.

53. Circle: $\alpha_1/2\sqrt{\pi} = 0.6784$
54. Square: $1/\sqrt{2} = 0.7071$
55. Quadrant of circle: $\alpha_{12}/4\sqrt{\pi} = 0.7244$ ($\alpha_{12} = 5.13562$ = first positive zero of J_2)
56. Semicircle: $3.832/\sqrt{8\pi} = 0.7644$
57. Rectangle (sides 1:2): $\sqrt{5/8} = 0.7906$

58. Find the deflection $u(x, y, t)$ of the square membrane corresponding to $0 \leqq x \leqq 1, 0 \leqq y \leqq 1$ with $c = 1$, initial deflection $f(x, y) = kx(1 - x^2)y(1 - y^2)$, and initial velocity zero.

59. Find the electrostatic potential between two concentric spheres of radii 2 cm and 10 cm kept at potentials 110 volts and 0 volt, respectively.

60. Solve $x^2u_{xx} + 2xyu_{xy} + y^2u_{yy} = 0$ by setting $v = y/x$, $z = y$.

Summary of Chapter 11 Partial Differential Equations

Whereas *ordinary* differential equations (Chaps. 1–6) appear as models of simpler problems involving a single independent variable, problems involving two or more independent variables (space variables, or time t and one or several space variables) lead to *partial* differential equations. Hence the importance of these equations to the engineer and physicist can hardly be overestimated.

In this chapter we were mainly concerned with the most important partial differential equations of physics and engineering, namely:

(1) $u_{tt} = c^2u_{xx}$ **One-dimensional wave equation** (Secs. 11.2–11.4)

(2) $u_{tt} = c^2(u_{xx} + u_{yy})$ **Two-dimensional wave equation** (Secs. 11.7–11.10)

(3) $u_t = c^2u_{xx}$ **One-dimensional heat equation** (Secs. 11.5, 11.6)

(4) $\nabla^2u = u_{xx} + u_{yy} = 0$ **Two-dimensional Laplace equation** (Secs. 11.5, 11.9)

(5) $\nabla^2u = u_{xx} + u_{yy} + u_{zz} = 0$ **Three-dimensional Laplace equation** (Secs. 11.11, 11.12).

(1) and (2) are hyperbolic, (3) is parabolic, (4) and (5) are elliptic. (See Problem Set 11.4.)

In practice, one is interested in obtaining the solution of such an equation in a given region satisfying given additional conditions, such as **initial conditions** (conditions at time $t = 0$) or **boundary conditions** (prescribed values of the solution u or some of its derivatives on the boundary surface S, or boundary curve C, of the region) or both. For (1) and (2) one prescribes two initial conditions (initial displacement and initial velocity). For (3) one prescribes the initial temperature distribution. For (4) or (5) one prescribes a boundary condition and calls the resulting problem (see Sec. 11.5)

Dirichlet problem if u is prescribed on S,
Neumann problem if $u_n = \partial u/\partial n$ is prescribed on S,
Mixed problem if u is prescribed on one part of S and u_n on the other.

A general method for solving such problems is the method of **separating variables** or **product method,** in which one assumes solutions in the form of products of functions each depending on one variable only. Thus equation (1) is solved by setting (Sec. 11.3)

$$u(x, t) = F(x)G(t);$$

similarly for (3) in Sec. 11.5. Substitution into the given equation yields *ordinary* differential equations for F and G, and from these one gets infinitely many solutions $F = F_n$ and $G = G_n$ such that the corresponding functions

$$u_n(x, t) = F_n(x)G_n(t)$$

are solutions of the partial differential equations satisfying the given boundary conditions. These are the **eigenfunctions** of the problem, and the corresponding **eigenvalues** determine the frequency of the vibration (or the rapidity of the decrease of temperature in the case of the heat equation, etc.). To satisfy also the initial condition (or conditions), one must consider infinite series of the u_n, whose coefficients turn out to be the Fourier coefficients of the functions f and g representing the given initial conditions (Secs. 11.3, 11.5). Hence **Fourier series** (and *Fourier integrals*) are of basic importance here (Secs. 11.3, 11.5, 11.6, 11.8).

Steady-state problems are those in which the solution does not depend on time t. For these, the heat equation $u_t = c^2\nabla^2 u$ becomes the Laplace equation (Sec. 11.5).

Before solving an initial or boundary value problem, one often transforms the equation into coordinates in which the boundary of the region considered is given by simple formulas. Thus in polar coordinates given by $x = r \cos \theta$, $y = r \sin \theta$, the **Laplacian** becomes (Sec. 11.9)

$$\nabla^2 u = u_{rr} + \frac{1}{r} u_r + \frac{1}{r^2} u_{\theta\theta}, \tag{6}$$

and for spherical coordinates see Sec. 11.11. If one now separates the variables, one gets **Bessel's equation** from (2) and (6) (vibrating circular membrane, Sec. 11.10) and **Legendre's equation** from (5) transformed into spherical coordinates (Sec. 11.12).

Operational methods (Laplace transforms, Fourier transforms) are helpful in solving partial differential equations in infinite regions (Secs. 11.13, 11.14), particularly if the coefficients of the equation are constant or depend on one variable only.

PART D

COMPLEX ANALYSIS

Many engineering problems may be treated and solved by methods involving complex numbers and complex functions. Roughly, there are two kinds of such problems. The first kind consists of "elementary problems" for which some acquaintance with complex numbers is sufficient. For instance, many applications to electric circuits or mechanical vibrating systems are of that kind.

The second kind consists of more advanced problems for which we must be familiar with the theory of complex analytic functions—"**complex function theory**" or "**complex analysis,**" for short—and with its powerful and elegant methods. Interesting problems in heat conduction, fluid flow, and electrostatics are of that kind.

We devote the next six chapters (Chaps. 12–17) to complex analysis and its applications. We shall see that the importance of complex analytic functions in engineering mathematics has the following three main roots.

1. *The real and imaginary parts of an analytic function are solutions of Laplace's equation in two independent variables. Consequently, two-dimensional potential problems can be treated by methods developed for analytic functions.*
2. *Many complicated real and complex integrals that occur in applications can be evaluated by methods of complex integration.*
3. *Most higher functions in engineering mathematics are analytic functions, and their study for complex values of the independent variable leads to a much deeper understanding of their properties.*

Chapter 12

Complex Numbers. Complex Analytic Functions

Complex numbers and the complex plane are discussed in Secs. 12.1–12.3. Complex analysis is concerned with complex analytic functions, as defined in Sec. 12.4. In Sec. 12.5 we explain a check for analyticity based on the so-called Cauchy–Riemann equations. The latter are of basic importance. They are related to Laplace's equation. In the remaining sections of Chap. 12 we study the most important elementary complex functions (exponential function, trigonometric functions, etc.), which generalize familiar real functions known from calculus. In Sec. 12.9 we discuss mappings given by these functions. (In Chaps. 16 and 17 we extend this discussion and show applications to potential problems.)

Prerequisites for this chapter: elementary calculus.
References: Appendix 1, Part D.
Answers to problems: Appendix 2.

12.1 Complex Numbers. Complex Plane

Equations without *real* solutions, such as $x^2 = -1$ or $x^2 - 10x + 40 = 0$, were observed early in history and led to the introduction of complex numbers.[1]

By definition, a **complex number** z is an ordered pair (x, y) of real numbers x and y, written

$$z = (x, y).$$

x is called the **real part** and y the **imaginary part** of z, written

$$x = \text{Re } z, \qquad y = \text{Im } z.$$

[1]First to use complex numbers for this purpose was the Italian mathematician GIROLAMO CARDANO (1501—1576), who found the formula for solving cubic equations. The term "complex number" was introduced by the great German mathematician CARL FRIEDRICH GAUSS (see the footnote in Sec. 5.4), who also paved the way for a general use of complex numbers.

By definition, two complex numbers are **equal** if and only if their real parts are equal and their imaginary parts are equal.

(0, 1) is called the **imaginary unit** and is denoted by

$$i = (0, 1). \tag{1}$$

Addition and multiplication of complex numbers will now be defined so that our complex number system "extends" the real number system. We define **addition** of $z_1 = (x_1, y_1)$ and $z_2 = (x_2, y_2)$ by[2]

$$z_1 + z_2 = (x_1, y_1) + (x_2, y_2) = (x_1 + x_2, \quad y_1 + y_2) \tag{2*}$$

and **multiplication** by

$$z_1 z_2 = (x_1, y_1)(x_2, y_2) = (x_1 x_2 - y_1 y_2, \quad x_1 y_2 + x_2 y_1). \tag{3*}$$

Then

$$(x_1, 0) + (x_2, 0) = (x_1 + x_2, 0),$$

$$(x_1, 0)(x_2, 0) = (x_1 x_2, 0),$$

as for real numbers x_1, x_2. Hence the complex numbers **extend** the reals, as wanted, and we can write

$$x = (x, 0).$$

From this and (1) and (3*) for any real y,

$$iy = (0, 1)(y, 0) = (0, y).$$

Since $z = (x, y) = (x, 0) + (0, y)$ by (2*), we thus have

$$z = x + iy. \tag{4}$$

If $x = 0$, then $z = iy$ and is called **pure imaginary.** Also, (1) and (3*) give

$$i^2 = -1 \tag{5}$$

because $i^2 = ii = (0, 1)(0, 1) = (-1, 0) = -1$.

The notation (4) for complex numbers is exclusively used in practice.[3] For **addition** it gives [see (2*)]

$$(x_1 + iy_1) + (x_2 + iy_2) = (x_1 + x_2) + i(y_1 + y_2). \tag{2}$$

---

[2]Students familiar with vectors see that this is *vector addition,* whereas this multiplication has no analog in the usual vector algebra.

[3]Electrical engineers often use j to reserve i for the current.

For **multiplication** it gives the following very simple recipe. Multiply each term by each term and use $i^2 = -1$ when it occurs [see (3*)]:

$$(3) \qquad \begin{aligned}(x_1 + iy_1)(x_2 + iy_2) &= x_1x_2 + ix_1y_2 + iy_1x_2 + i^2y_1y_2 \\ &= (x_1x_2 - y_1y_2) + i(x_1y_2 + x_2y_1).\end{aligned}$$

This agrees with (3*).

EXAMPLE 1 **Real part, imaginary part, sum and product of complex numbers**

Let $z_1 = 8 + 3i$ and $z_2 = 9 - 2i$. Then $\text{Re}\, z_1 = 8$, $\text{Im}\, z_1 = 3$, $\text{Re}\, z_2 = 9$, $\text{Im}\, z_2 = -2$ and

$$z_1 + z_2 = (8 + 3i) + (9 - 2i) = 17 + i,$$

$$z_1z_2 = (8 + 3i)(9 - 2i) = 72 + 6 + i(-16 + 27) = 78 + 11i.$$ ■

Subtraction and division are defined as the inverse operations of addition and multiplication. Thus the **difference** $z = z_1 - z_2$ is the complex number z for which $z_1 = z + z_2$. Hence by (2),

$$(6) \qquad z_1 - z_2 = (x_1 - x_2) + i(y_1 - y_2).$$

The **quotient** $z = z_1/z_2$ $(z_2 \neq 0)$ is the complex number z for which $z_1 = zz_2$. We show after the example that

$$(7^*) \qquad z = \frac{z_1}{z_2} = x + iy, \qquad x = \frac{x_1x_2 + y_1y_2}{x_2^2 + y_2^2}, \qquad y = \frac{x_2y_1 - x_1y_2}{x_2^2 + y_2^2}.$$

The ***practical rule*** to get this is by multiplying numerator and denominator of z_1/z_2 by $x_2 - iy_2$ and simplifying:

$$(7) \qquad z = \frac{x_1 + iy_1}{x_2 + iy_2} = \frac{(x_1 + iy_1)(x_2 - iy_2)}{(x_2 + iy_2)(x_2 - iy_2)} = \frac{x_1x_2 + y_1y_2}{x_2^2 + y_2^2} + i\frac{x_2y_1 - x_1y_2}{x_2^2 + y_2^2}.$$

EXAMPLE 2 **Difference and quotient of complex numbers**

For $z_1 = 8 + 3i$ and $z_2 = 9 - 2i$ we get

$$z_1 - z_2 = (8 + 3i) - (9 - 2i) = -1 + 5i$$

$$\frac{z_1}{z_2} = \frac{8 + 3i}{9 - 2i} = \frac{(8 + 3i)(9 + 2i)}{(9 - 2i)(9 + 2i)} = \frac{66 + 43i}{81 + 4} = \frac{66}{85} + \frac{43}{85}i.$$

Check the division by multiplication to get $8 + 3i$. ■

Proof of (7*). We have $z_1 = x_1 + iy_1 = zz_2 = (x + iy)(x_2 + iy_2)$. By the definition of equality, the real parts and the imaginary parts on both sides must be equal:

$$x_1 = x_2x - y_2y$$

$$y_1 = y_2x + x_2y.$$

This is a linear system of two equations in the unknowns x and y. Under our assumption $z_2 \neq 0$ (thus $x_2^2 + y_2^2 \neq 0$), we obtain the unique solution (7*) by elimination. ∎

Properties of addition and multiplication are the same as for *real* numbers, from which they follow (here, $-z = -x - iy$):

$$
\begin{aligned}
&\left.\begin{aligned} z_1 + z_2 &= z_2 + z_1 \\ z_1 z_2 &= z_2 z_1 \end{aligned}\right\} && \textit{(Commutative laws)} \\
&\left.\begin{aligned} (z_1 + z_2) + z_3 &= z_1 + (z_2 + z_3) \\ (z_1 z_2) z_3 &= z_1 (z_2 z_3) \end{aligned}\right\} && \textit{(Associative laws)} \\
&z_1(z_2 + z_3) = z_1 z_2 + z_1 z_3 && \textit{(Distributive law)} \\
&0 + z = z + 0 = z \\
&z + (-z) = (-z) + z = 0 \\
&z \cdot 1 = z
\end{aligned}
\tag{8}
$$

Complex Plane

This was algebra. Now comes geometry: the geometrical representation of complex numbers as points in the plane. This is of great practical importance. The idea is quite simple and natural. We choose two perpendicular coordinate axes, the horizontal x-axis, called the **real axis,** and the vertical y-axis, called the **imaginary axis.** On both axes we choose the same unit of length (Fig. 282). This is called a **Cartesian coordinate system.**[4] We now plot a given complex number $z = (x, y) = x + iy$ as the point P with coordinates x, y. The xy-plane in which the complex numbers are represented in this way is called the **complex plane.**[5] Figure 283 shows an example.

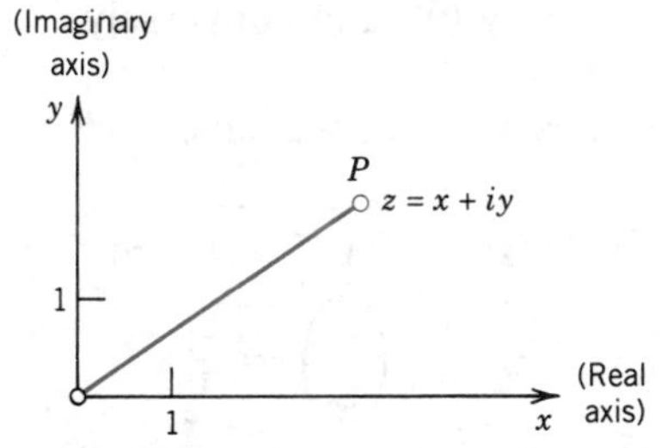

Fig. 282. The complex plane

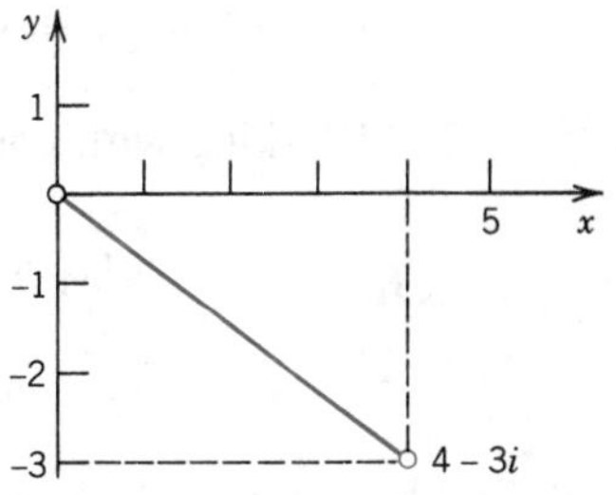

Fig. 283. The number $4 - 3i$ in the complex plane

[4]See footnote 3 in Sec. 8.1.

[5]Sometimes called the **Argand diagram,** after the French mathematician JEAN ROBERT ARGAND (1768—1822), born in Geneva and later librarian in Paris. His paper on the complex plane appeared in 1806, nine years after a similar memoir by the Norwegian mathematician CASPAR WESSEL (1745—1818), a surveyor of the Danish Academy of Science.

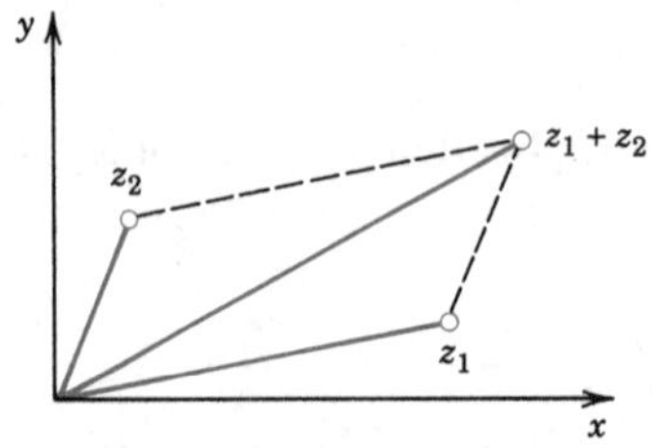

Fig. 284. Addition of complex numbers

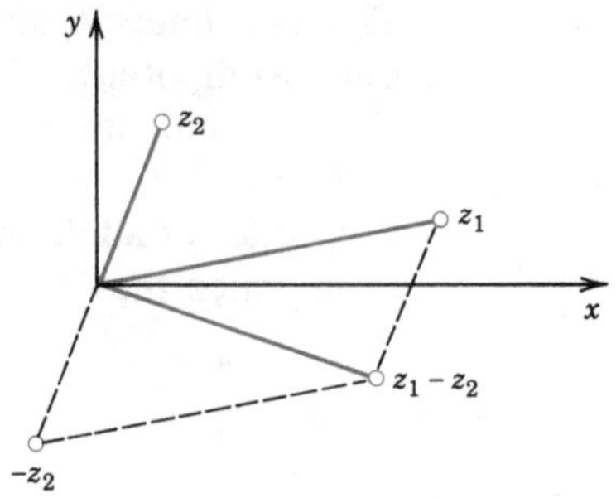

Fig. 285. Subtraction of complex numbers

Instead of saying "the point represented by z in the complex plane" we say briefly and simply ***"the point z in the complex plane."*** This will cause no misunderstandings.

Addition and subtraction can now be visualized as illustrated in Figs. 284 and 285.

Complex Conjugate Numbers. The **complex conjugate** $\bar{z}$ of a complex number

$$z = x + iy \qquad \text{is defined by} \qquad \bar{z} = x - iy$$

and is obtained geometrically by reflecting the point z in the real axis. Figure 286 shows this for $z = 5 + 2i$ and its conjugate $\bar{z} = 5 - 2i$.

Conjugates are useful since $z\bar{z} = x^2 + y^2$ is real, a property we used in the above division (for z_2 instead of z). Moreover, addition and subtraction yields $z + \bar{z} = 2x$, $z - \bar{z} = 2iy$, so that we can express the real part and the imaginary part of z by the important formulas

$$\text{(9)} \qquad \operatorname{Re} z = x = \frac{1}{2}(z + \bar{z}), \qquad \operatorname{Im} z = y = \frac{1}{2i}(z - \bar{z}).$$

If z is real, $z = x$, then $\bar{z} = z$ by (9), and conversely.

Working with conjugates is easy, since we have

$$\text{(10)} \qquad \begin{aligned} &\overline{(z_1 + z_2)} = \bar{z}_1 + \bar{z}_2, \qquad \overline{(z_1 - z_2)} = \bar{z}_1 - \bar{z}_2, \\ &\overline{(z_1 z_2)} = \bar{z}_1 \bar{z}_2, \qquad \overline{\left(\frac{z_1}{z_2}\right)} = \frac{\bar{z}_1}{\bar{z}_2}. \end{aligned}$$

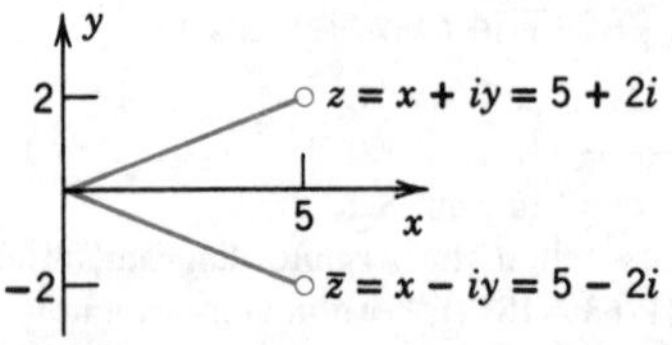

Fig. 286. Complex conjugate numbers

EXAMPLE 3 Illustration of (9) and (10)

Let $z_1 = 4 + 3i$ and $z_2 = 2 + 5i$. Then by (1),

$$\text{Im } z_1 = \frac{1}{2i}[(4 + 3i) - (4 - 3i)] = \frac{3i + 3i}{2i} = 3.$$

Also, the multiplication formula in (10) is verified by

$$\overline{(z_1 z_2)} = \overline{(4 + 3i)(2 + 5i)} = \overline{-7 + 26i} = -7 - 26i,$$

$$\bar{z}_1 \bar{z}_2 = (4 - 3i)(2 - 5i) = -7 - 26i.$$

Problem Set 12.1

1. (Powers of the imaginary unit) Show that

$$\text{(11)} \qquad \begin{aligned} &i^2 = -1, \quad i^3 = -i, \quad i^4 = 1, \quad i^5 = i, \cdots \\ &\frac{1}{i} = -i, \quad \frac{1}{i^2} = -1, \quad \frac{1}{i^3} = i, \cdots, \end{aligned}$$

Let $z_1 = 4 - 5i$ and $z_2 = 2 + 3i$. Find (in the form $x + iy$)

2. $z_1 z_2$ **3.** $(z_1 + z_2)^2$ **4.** $1/z_2$ **5.** z_2/z_1

6. $3z_1 - 6z_2$ **7.** $0.2z_1^{\ 3}$ **8.** $z_1/(z_1 + z_2)$ **9.** $338/z_2^{\ 2}$

Find

10. $\text{Re } \dfrac{1}{1 + i}$ **11.** $\text{Im } \dfrac{3 + 4i}{7 - i}$ **12.** $\text{Re } \dfrac{(2 - 3i)^2}{2 + 3i}$ **13.** $\text{Im } \dfrac{z}{\bar{z}}$

14. $(0.3 + 0.4i)^4$ **15.** $\text{Re } z^2$, $(\text{Re } z)^2$ **16.** $\text{Im } z^3$, $(\text{Im } z)^3$ **17.** $(1 + i)^8$

18. Show that z is pure imaginary if and only if $\bar{z} = -z$.

19. Verify the formulas in (10) for $z_1 = 31 - 34i$ and $z_2 = 2 - 5i$.

20. If the product of two complex numbers is zero, show that at least one factor must be zero.

12.2 Polar Form of Complex Numbers. Powers and Roots

We can substantially increase the usefulness of the complex plane and gain further insight into the nature of complex numbers if in addition to the xy-coordinates we also employ the usual polar coordinates r, θ defined by

$$\text{(1)} \qquad x = r\cos\theta, \qquad y = r\sin\theta.$$

Then $z = x + iy$ takes the so-called **polar form**

$$\text{(2)} \qquad z = r(\cos\theta + i\sin\theta).$$

r is called the **absolute value** or **modulus** of z and is denoted by $|z|$. Hence

$$|z| = r = \sqrt{x^2 + y^2} = \sqrt{z\bar{z}}. \tag{3}$$

Geometrically, $|z|$ is the distance of the point z from the origin (Fig. 287). Similarly, $|z_1 - z_2|$ is the distance between z_1 and z_2 (Fig. 288).

θ is called the **argument** of z and is denoted by arg z. Thus (Fig. 287)

$$\theta = \arg z = \arctan \frac{y}{x} \qquad (z \neq 0) \tag{4}$$

Geometrically, θ is the directed angle from the positive x-axis to OP in Fig. 287. Here, as in calculus, *all angles are measured in* ***radians*** *and positive in the counterclockwise sense.*

For $z = 0$ this angle θ is undefined. (Why?) For given $z \neq 0$ it is determined only up to integer multiples of 2π. The value of θ that lies in the interval $-\pi < \theta \leqq \pi$ is called the **principal value** of the argument of z ($\neq 0$) and is denoted by Arg z. Thus $\theta =$ Arg z satisfies by definition

$$-\pi < \operatorname{Arg} z \leqq \pi$$

EXAMPLE 1 Polar form of complex numbers. Principal value

Let $z = 1 + i$ (Fig. 289). Then

$$z = \sqrt{2}\left(\cos\frac{\pi}{4} + i \sin\frac{\pi}{4}\right), \qquad |z| = \sqrt{2}, \qquad \arg z = \frac{\pi}{4} \pm 2n\pi \qquad (n = 0, 1, \cdots).$$

The principal value of the argument is Arg $z = \pi/4$. Other values are $-7\pi/4$, $9\pi/4$, etc. ∎

EXAMPLE 2 Polar form of complex numbers. Principal value

Let $z = 3 + 3\sqrt{3}\, i$. Then $z = 6\left(\cos\frac{\pi}{3} + i \sin\frac{\pi}{3}\right)$, the absolute value of z is $|z| = 6$, and the principal value of arg z is Arg $z = \pi/3$. ∎

Caution! In using (4), we must pay attention to the quadrant in which z lies, since $\tan\theta$ has period π, so that the arguments of z and $-z$ have the same tangent. *Example:* for $\theta_1 = \arg(1 + i)$ and $\theta_2 = \arg(-1 - i)$ we have $\tan\theta_1 = \tan\theta_2 = 1$.

Triangle inequality

For any complex numbers we have the important **triangle inequality**

$$|z_1 + z_2| \leqq |z_1| + |z_2| \qquad \text{(Fig. 290)} \tag{5}$$

which we shall use quite frequently. This inequality follows by noting that the three points 0, z_1, and $z_1 + z_2$ are the vertices of a triangle (Fig. 290) with sides $|z_1|$, $|z_2|$, and $|z_1 + z_2|$, and one side cannot exceed the sum of the other two sides. A formal proof is left to the reader (Prob. 39).

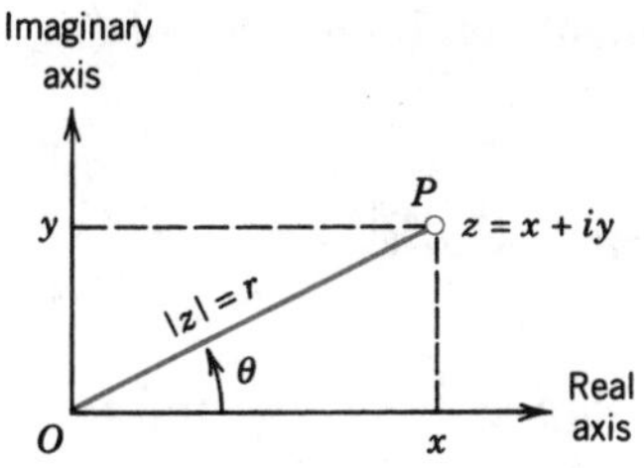

Fig. 287. Complex plane, polar form of a complex number

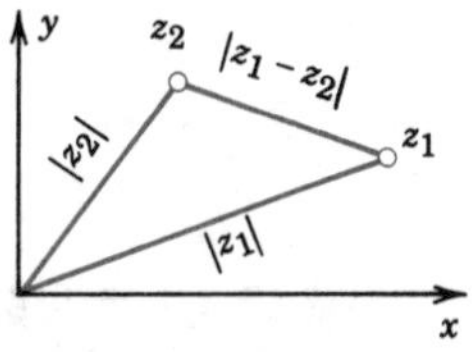

Fig. 288. Distance between two points in the complex plane

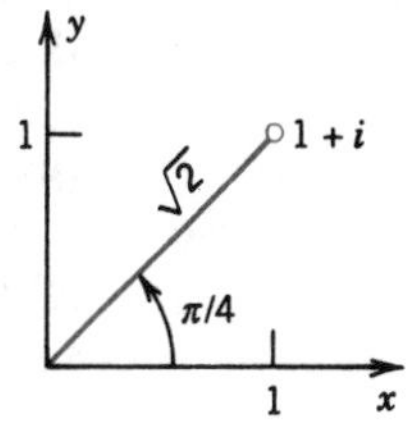

Fig. 289. Example 1

EXAMPLE 3 Triangle inequality

If $z_1 = 1 + i$ and $z_2 = -2 + 3i$, then (sketch a figure!)

$$|z_1 + z_2| = |-1 + 4i| = \sqrt{17} = 4.123 < \sqrt{2} + \sqrt{13} = 5.020.$$

Generalized triangle inequality. By induction we obtain from (5) for any sum

$$|z_1 + z_2 + \cdots + z_n| \leqq |z_1| + |z_2| + \cdots + |z_n|; \tag{6}$$

that is, *the absolute value of a sum cannot exceed the sum of the absolute values of the terms.*

Multiplication and Division in Polar Form

This will give us a better understanding of multiplication and division. Let

$$z_1 = r_1(\cos\theta_1 + i\sin\theta_1) \quad \text{and} \quad z_2 = r_2(\cos\theta_2 + i\sin\theta_2).$$

Multiplication. By (3), Sec. 12.1, the product is at first

$$z_1 z_2 = r_1 r_2[(\cos\theta_1\cos\theta_2 - \sin\theta_1\sin\theta_2) + i(\sin\theta_1\cos\theta_2 + \cos\theta_1\sin\theta_2)].$$

The addition rules for the sine and cosine [(6) in Appendix 3.1] now yield

$$z_1 z_2 = r_1 r_2[\cos(\theta_1 + \theta_2) + i\sin(\theta_1 + \theta_2)]. \tag{7}$$

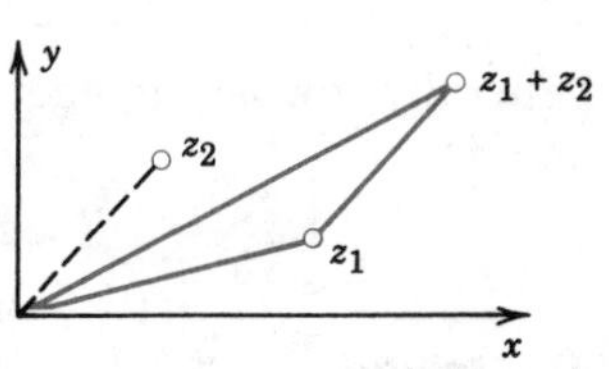

Fig. 290. Triangle inequality[6]

[6]The triangle degenerates if z_1 and z_2 lie on the same straight line through the origin.

Taking absolute values and arguments on both sides of (7), we thus obtain the important rules

$$|z_1 z_2| = |z_1||z_2| \tag{8}$$

and

$$\arg(z_1 z_2) = \arg z_1 + \arg z_2 \qquad \text{(up to multiples of } 2\pi\text{)}. \tag{9}$$

Division. The quotient $z = z_1/z_2$ is the number z satisfying $zz_2 = z_1$. Hence $|zz_2| = |z|\,|z_2| = |z_1|$, $\arg(zz_2) = \arg z + \arg z_2 = \arg z_1$. This yields

$$\left|\frac{z_1}{z_2}\right| = \frac{|z_1|}{|z_2|} \qquad (z_2 \neq 0) \tag{10}$$

and

$$\arg \frac{z_1}{z_2} = \arg z_1 - \arg z_2 \qquad \text{(up to multiples of } 2\pi\text{)}. \tag{11}$$

By combining these two formulas (10) and (11) we also have

$$\frac{z_1}{z_2} = \frac{r_1}{r_2}[\cos(\theta_1 - \theta_2) + i \sin(\theta_1 - \theta_2)]. \tag{12}$$

EXAMPLE 4 **Illustration of formulas (8)–(11)**

Let $z_1 = -2 + 2i$ and $z_2 = 3i$. Then $z_1 z_2 = -6 - 6i$, $z_1/z_2 = 2/3 + (2/3)i$. Hence

$$|z_1 z_2| = 6\sqrt{2} = 3\sqrt{8} = |z_1||z_2|, \qquad |z_1/z_2| = 2\sqrt{2}/3 = |z_1|/|z_2|,$$

and for the arguments we obtain $\operatorname{Arg} z_1 = 3\pi/4$, $\operatorname{Arg} z_2 = \pi/2$,

$$\operatorname{Arg} z_1 z_2 = -\frac{3\pi}{4} = \operatorname{Arg} z_1 + \operatorname{Arg} z_2 - 2\pi,$$

$$\operatorname{Arg}(z_1/z_2) = \frac{\pi}{4} = \operatorname{Arg} z_1 - \operatorname{Arg} z_2. \qquad \blacksquare$$

Integer powers of z. From (7) and (12) we have

$$z^2 = r^2(\cos 2\theta + i \sin 2\theta),$$

$$z^{-2} = r^{-2}[\cos(-2\theta) + i \sin(-2\theta)]$$

and, more generally, for any integer n,

$$z^n = r^n(\cos n\theta + i \sin n\theta). \tag{13}$$

EXAMPLE 5 **Formula of De Moivre**

For $|z| = r = 1$, formula (13) yields the so-called **formula of De Moivre**[7]

$$(\cos\theta + i \sin\theta)^n = \cos n\theta + i \sin n\theta. \tag{13*}$$

[7] ABRAHAM DE MOIVRE (1667—1754), French mathematician, who introduced imaginary quantities in trigonometry and contributed to probability theory (see Sec. 23.7).

This formula is useful for expressing $\cos n\theta$ and $\sin n\theta$ in terms of $\cos\theta$ and $\sin\theta$. For instance, if $n = 2$ and we take the real and imaginary parts on both sides of (13*), we get the familiar formulas

$$\cos 2\theta = \cos^2\theta - \sin^2\theta, \qquad \sin 2\theta = 2\cos\theta\sin\theta.$$

This shows that *complex* methods often simplify the derivation of *real* formulas. ∎

Roots

If $z = w^n$ $(n = 1, 2, \cdots)$, then to each value of w there corresponds one value of z. We shall immediately see that, conversely, to a given $z \neq 0$ there correspond precisely n distinct values of w. Each of these values is called an **nth root** of z, and we write

$$w = \sqrt[n]{z}. \tag{14}$$

Hence this symbol is *multivalued,* namely, *n-valued,* in contrast to the usual conventions made in *real* calculus. The n values of $\sqrt[n]{z}$ can easily be obtained as follows. In terms of polar forms for z and

$$w = R(\cos\phi + i\sin\phi),$$

the equation $w^n = z$ becomes

$$w^n = R^n(\cos n\phi + i\sin n\phi) = z = r(\cos\theta + i\sin\theta).$$

By equating the absolute values on both sides we have

$$R^n = r, \qquad \text{thus} \qquad R = \sqrt[n]{r}$$

where the root is real positive and thus uniquely determined. By equating the arguments we obtain

$$n\phi = \theta + 2k\pi, \qquad \text{thus} \qquad \phi = \frac{\theta}{n} + \frac{2k\pi}{n}$$

where k is an integer. For $k = 0, 1, \cdots, n - 1$ we get n *distinct* values of w. Further integers of k would give values already obtained. For instance, $k = n$ gives $2k\pi/n = 2\pi$, hence the w corresponding to $k = 0$, etc. Consequently, $\sqrt[n]{z}$, for $z \neq 0$, has the n distinct values

$$\sqrt[n]{z} = \sqrt[n]{r}\left(\cos\frac{\theta + 2k\pi}{n} + i\sin\frac{\theta + 2k\pi}{n}\right) \tag{15}$$

where $k = 0, 1, \cdots, n - 1$. These n values lie on a circle of radius $\sqrt[n]{r}$ with center at the origin and constitute the vertices of a regular polygon of n sides.

The value of $\sqrt[n]{z}$ obtained by taking the principal value of arg z and $k = 0$ in (15) is called the **principal value** *of* $w = \sqrt[n]{z}$.

EXAMPLE 6 Square root

From (15) it follows that $w = \sqrt{z}$ has the two values

$$w_1 = \sqrt{r}\left(\cos\frac{\theta}{2} + i\sin\frac{\theta}{2}\right) \tag{16a}$$

and

$$w_2 = \sqrt{r}\left[\cos\left(\frac{\theta}{2} + \pi\right) + i\sin\left(\frac{\theta}{2} + \pi\right)\right] = -w_1 \tag{16b}$$

which lie symmetric with respect to the origin. For instance, the square root of $4i$ has the values

$$\sqrt{4i} = \pm 2\left(\cos\frac{\pi}{4} + i\sin\frac{\pi}{4}\right) = \pm(\sqrt{2} + i\sqrt{2}).$$

From (16) we can obtain the much more practical formula

$$\sqrt{z} = \pm\left[\sqrt{\tfrac{1}{2}(|z| + x)} + (\text{sign } y)i\sqrt{\tfrac{1}{2}(|z| - x)}\right] \tag{17}$$

where sign $y = 1$ if $y \geqq 0$, sign $y = -1$ if $y < 0$, and all square roots of positive numbers are taken with the positive sign. This follows from (16) if we use the trigonometric identities

$$\cos\tfrac{1}{2}\theta = \sqrt{\tfrac{1}{2}(1 + \cos\theta)}, \qquad \sin\tfrac{1}{2}\theta = \sqrt{\tfrac{1}{2}(1 - \cos\theta)},$$

multiply them by $\sqrt{r}$,

$$\sqrt{r}\cos\tfrac{1}{2}\theta = \sqrt{\tfrac{1}{2}(r + r\cos\theta)}, \qquad \sqrt{r}\sin\tfrac{1}{2}\theta = \sqrt{\tfrac{1}{2}(r - r\cos\theta)},$$

use $r\cos\theta = x$, and finally choose the sign of Im $\sqrt{z}$ so that sign [(Re $\sqrt{z}$)(Im $\sqrt{z}$)] = sign y (why?). ∎

EXAMPLE 7 Complex quadratic equation

Solve $z^2 - (5 + i)z + 8 + i = 0$.

Solution.

$$z = \tfrac{1}{2}(5 + i) \pm \sqrt{\tfrac{1}{4}(5 + i)^2 - 8 - i} = \tfrac{1}{2}(5 + i) \pm \sqrt{-2 + \tfrac{3}{2}i}$$

$$= \tfrac{1}{2}(5 + i) \pm \left[\sqrt{\tfrac{1}{2}(\tfrac{5}{2} + (-2))} + i\sqrt{\tfrac{1}{2}(\tfrac{5}{2} - (-2))}\right]$$

$$= \tfrac{1}{2}(5 + i) \pm [\tfrac{1}{2} + \tfrac{3}{2}i] = \begin{cases} 3 + 2i \\ 2 - i. \end{cases}$$ ∎

The circle of radius 1 and center 0 (appearing in the next example and often later on) is called the **unit circle.**

EXAMPLE 8 *n*th root of unity. Unit circle

Solve the equation $z^n = 1$.

Solution. From (15) we obtain

$$\sqrt[n]{1} = \cos\frac{2k\pi}{n} + i\sin\frac{2k\pi}{n}, \qquad k = 0, 1, \cdots, n - 1. \tag{18}$$

If ω denotes the value corresponding to $k = 1$, then the n values of $\sqrt[n]{1}$ can be written as $1, \omega, \omega^2, \cdots, \omega^{n-1}$. These values are the vertices of a regular polygon of n sides inscribed in the **unit circle** (the circle of radius 1 with center 0), with one vertex at the point 1. Each of these n values is called an **nth root of unity.** For instance, $\sqrt[3]{1} = 1, -\frac{1}{2} \pm \frac{1}{2}\sqrt{3}\,i$ (Fig. 291) and $\sqrt[4]{1} = 1, i, -1, -i$ (Fig. 292). Figure 293 shows $\sqrt[5]{1}$.

If w_1 is any nth root of an *arbitrary* complex number z, then the n values of $\sqrt[n]{z}$ are

$$w_1, \quad w_1\omega, \quad w_1\omega^2, \quad \cdots, \quad w_1\omega^{n-1}$$

since multiplying w_1 by ω^k corresponds to increasing the argument of w_1 by $2k\pi/n$. ∎

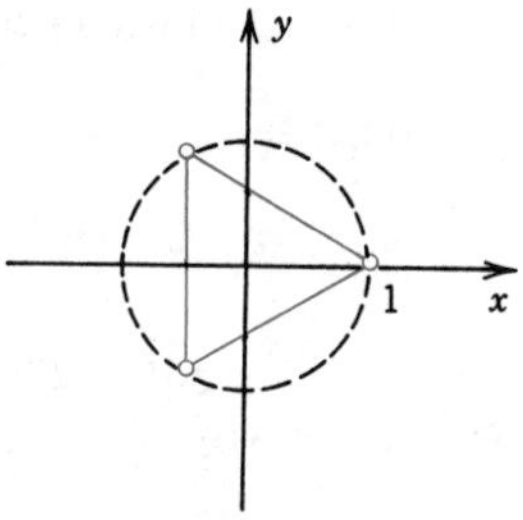

Fig. 291. $\sqrt[3]{1}$

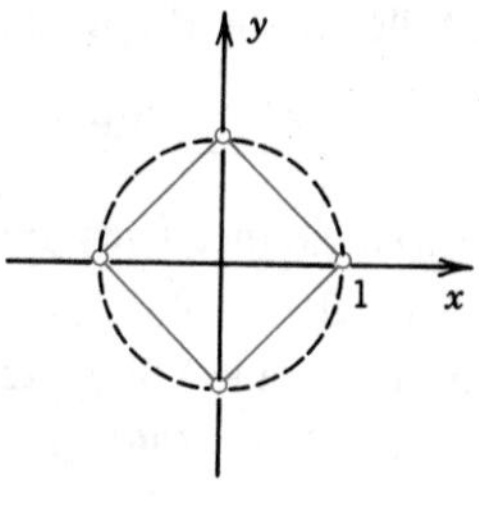

Fig. 292. $\sqrt[4]{1}$

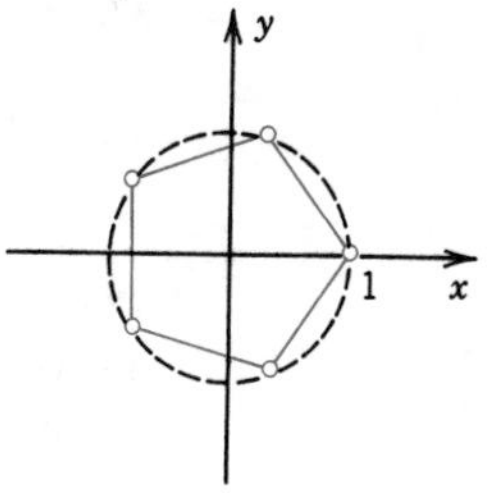

Fig. 293. $\sqrt[5]{1}$

The student should do the problems related to the polar representation with particular care, since we shall need this representation quite often in our work.

Problem Set 12.2

1. **(Multiplication by i)** Show that multiplication of a complex number by i corresponds to a **counterclockwise rotation** of the corresponding vector through the angle $\pi/2$.

Find

2. $|-0.2i|$
3. $|1.5 + 2i|$
4. $|z|^4$, $|z^4|$
5. $|\cos\theta + i\sin\theta|$
6. $\left|\dfrac{\bar{z}}{z}\right|$
7. $\left|\dfrac{5+7i}{7-5i}\right|$
8. $\left|\dfrac{z+1}{z-1}\right|$
9. $\left|\dfrac{(1+i)^6}{i^3(1+4i)^2}\right|$

Represent in polar form:

10. $2i$, $-2i$
11. $1 + i$
12. -3
13. $6 + 8i$
14. $\dfrac{1+i}{1-i}$
15. $\dfrac{i\sqrt{2}}{4+4i}$
16. $\dfrac{3\sqrt{2}+2i}{-\sqrt{2}-2i/3}$
17. $\dfrac{2+3i}{5+4i}$

Determine the principal value of the arguments of

18. $-6 - 6i$
19. $-10 - i$
20. $-\pi$
21. $2 + 2i$

Represent in the form $x + iy$:

22. $4(\cos\frac{1}{3}\pi + i\sin\frac{1}{3}\pi)$
23. $2\sqrt{2}(\cos\frac{3}{4}\pi + i\sin\frac{3}{4}\pi)$
24. $10(\cos 0.4 + i\sin 0.4)$
25. $\cos(-1.8) + i\sin(-1.8)$

Find all values of the following roots and plot them in the complex plane.

26. $\sqrt{i}$
27. $\sqrt{-8i}$
28. $\sqrt{-7-24i}$
29. $\sqrt[8]{1}$
30. $\sqrt[4]{-7+24i}$
31. $\sqrt[4]{-1}$
32. $\sqrt[5]{-1}$
33. $\sqrt[3]{1+i}$

Solve the equations:

34. $z^2 + z + 1 - i = 0$
35. $z^2 - (5+i)z + 8 + i = 0$
36. $z^4 - 3(1+2i)z^2 - 8 + 6i = 0$

37. Prove the following useful inequalities, which we shall need occasionally:

$$|\text{Re } z| \leqq |z|, \qquad |\text{Im } z| \leqq |z|. \tag{19}$$

38. Verify the triangle inequality for $z_1 = 4 + 5i$, $z_2 = -2 + 1.5i$.

39. Prove the triangle inequality.

40. **(Parallelogram equality)** Show that $|z_1 + z_2|^2 + |z_1 - z_2|^2 = 2(|z_1|^2 + |z_2|^2)$. This is called the *parallelogram equality*. Can you see why?

12.3 Curves and Regions in the Complex Plane

In this section we consider some important curves and regions and some related concepts we shall frequently need. This will also help us to become more familiar with the complex plane.

Circles and disks. The distance between two points z and a is $|z - a|$. Hence a **circle** C of radius ρ and center at a (Fig. 294) can be given by

$$|z - a| = \rho. \tag{1}$$

In particular, the **unit circle,** that is, the circle of radius 1 and center at the origin $a = 0$, is

$$|z| = 1 \qquad \text{(Fig. 295).}$$

Furthermore, the inequality

$$|z - a| < \rho \tag{2}$$

holds for every point z inside C; that is, (2) represents the interior of C. Such a region is called a **circular disk,** or, more precisely, an *open* circular disk, in contrast to the *closed* circular disk

$$|z - a| \leqq \rho,$$

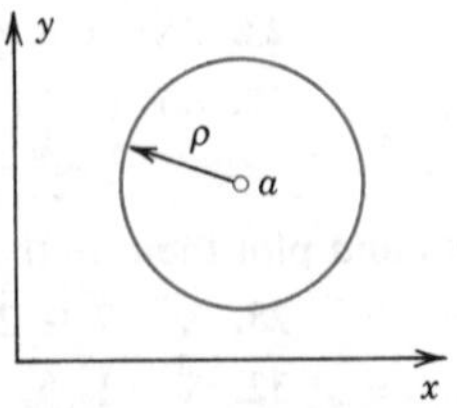

Fig. 294. Circle in the complex plane

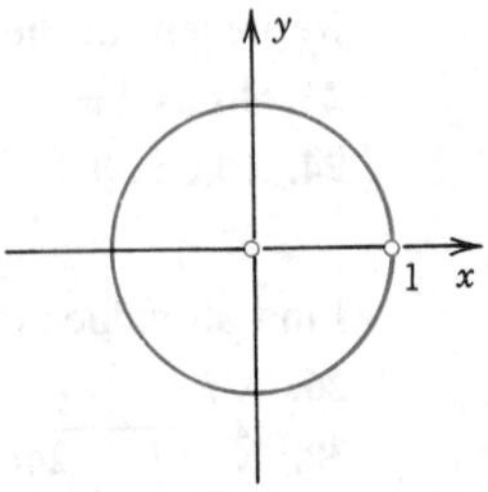

Fig. 295. Unit circle

which consists of the interior of C and C itself. The open circular disk (2) is also called a **neighborhood** of the point a. Obviously, a has infinitely many such neighborhoods, each corresponding to a certain value of ρ (> 0), and a is a point of each such neighborhood.[8]

Similarly, the inequality

$$|z - a| > \rho$$

represents the exterior of the circle C. Furthermore, the region between two concentric circles of radii ρ_1 and ρ_2 ($> \rho_1$) can be given by

$$\rho_1 < |z - a| < \rho_2, \tag{3}$$

where a is the center of the circles. Such a region is called an *open circular ring* or *open* **annulus** (Fig. 296).

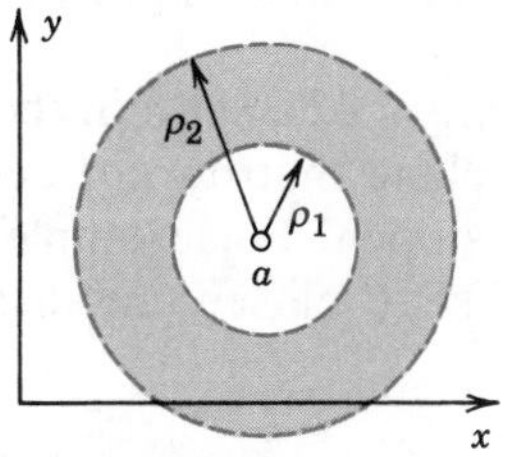

Fig. 296. Annulus in the complex plane

EXAMPLE 1 Circular disk

Determine the region in the complex plane given by $|z - 3 + i| \leqq 4$.

Solution. The inequality is valid precisely for all z whose distance from $a = 3 - i$ does not exceed 4. Hence this is a closed circular disk of radius 4 with center at $3 - i$. ∎

EXAMPLE 2 Unit circle and unit disk

Determine each of the regions

$$\text{(a)} \quad |z| < 1, \qquad \text{(b)} \quad |z| \leqq 1, \qquad \text{(c)} \quad |z| > 1.$$

Solution. (a) The interior of the unit circle. This is called the **open unit disk.**
(b) The unit circle and its interior. This is called the **closed unit disk.**
(c) The exterior of the unit circle. ∎

Half-planes. By the (open) *upper* **half-plane** we mean the set of all points $z = x + iy$ such that $y > 0$. Similarly, the condition $y < 0$ defines the *lower half-plane*, $x > 0$ the *right half-plane*, and $x < 0$ the *left half-plane*.

[8] More generally, any set that contains an open disk (2) is also called a *neighborhood of a*. For distinction, a disk (2) is often called an *open circular neighborhood of a*.

Concepts Related to Sets in the Complex Plane

We finally list a few concepts of general interest that will be used in our further work; this is for reference as needed.

The term **set of points** in the complex plane means any sort of collection of finitely or infinitely many points. For example, the solutions of a quadratic equation, the points on a line, and the points in the interior of a circle are sets.

A set S is called **open** if every point of S has a neighborhood consisting entirely of points that belong to S. For example, the points in the interior of a circle or a square form an open set, and so do the points of the right half-plane $\text{Re } z = x > 0$.

An open set S is called **connected** if any two of its points can be joined by a broken line of finitely many straight line segments all of whose points belong to S. An open connected set is called a **domain.** Thus an open disk (2) and an open annulus (3) are domains. An open square with a diagonal removed is not a domain since this set is not connected. (Why?)

The **complement** of a set S in the complex plane is the set of all points of the complex plane that do *not* belong to S. A set S is called **closed** if its complement is open. For example, the points on and inside the unit circle form a closed set ("closed unit disk"; see Example 2) since its complement $|z| > 1$ is open.

A **boundary point** of a set S is a point every neighborhood of which contains both points that belong to S and points that do not belong to S. For example, the boundary points of an annulus are the points on the two bounding circles. Clearly, if set S is open, then no boundary point belongs to S; if S is closed, then every boundary point belongs to S.

A **region** is a set consisting of a domain plus, perhaps, some or all of its boundary points. (The reader is warned that some authors use the term "region" for what we call a domain [following the modern standard terminology], and others make no distinction between the two terms.)

So far we have been concerned with complex numbers and the complex plane (just as at the beginning of calculus, one talks about real numbers and the real line). We shall start with complex calculus in the next section.

Problem Set 12.3

Determine and sketch the sets represented by

1. $|z - 4i| = 4$ **2.** $|z + 1| > 2$ **3.** $1/3 < |z - a| < 6$

4. $-\pi < \text{Im } z \leqq \pi$ **5.** $0 < \text{Re } z < \pi/2$ **6.** $\text{Re } z \geqq -1$

7. $|z - 1| \leqq |z + 1|$ **8.** $|\arg z| \leqq \pi/4$ **9.** $\text{Im } z^2 = 2$

10. $\left|\dfrac{z + i}{z - i}\right| = 1$ **11.** $\left|\dfrac{z + 1}{z - 1}\right| = 4$ **12.** $\left|\dfrac{2z - i}{-iz - 2}\right| \leqq 1$

12.4 Limit. Derivative. Analytic Function

Complex analysis is concerned with complex functions that are differentiable in some domain. Hence we should first say what we mean by a complex function and then define the concepts of limit and derivative in complex. This discussion will be similar to that in calculus.

Complex Function

Recall from calculus that a *real* function f defined on a set S of real numbers (usually an interval) is a rule that assigns to every x in S a real number $f(x)$, called the *value* of f at x.

Now in complex, S is a set of *complex* numbers. And a **function** f defined on S is a rule that assigns to every z in S a complex number w, called the *value* of f at z. We write

$$w = f(z).$$

Here z varies in S and is called a **complex variable.** The set S is called the **domain** *of definition*[9] *of* f.

Example: $w = f(z) = z^2 + 3z$ is a complex function defined for all z; that is, its domain S is the whole complex plane.

The set of all values of a function f is called the **range** *of* f.

w is complex, and we write $w = u + iv$, where u and v are the real and imaginary parts, respectively. Now w depends on $z = x + iy$. Hence u becomes a real function of x and y, and so does v. We may thus write

$$w = f(z) = u(x, y) + iv(x, y).$$

This shows that a *complex* function $f(z)$ is equivalent to a pair of *real* functions $u(x, y)$ and $v(x, y)$, each depending on the two real variables x and y.

EXAMPLE 1 Function of a complex variable

Let $w = f(z) = z^2 + 3z$. Find u and v and calculate the value of f at $z = 1 + 3i$.

Solution. $u = \text{Re } f(z) = x^2 - y^2 + 3x$ and $v = 2xy + 3y$. Also,

$$f(1 + 3i) = (1 + 3i)^2 + 3(1 + 3i) = 1 - 9 + 6i + 3 + 9i = -5 + 15i.$$

This shows that $u(1, 3) = -5$ and $v(1, 3) = 15$. ∎

---

[9]This is a standard term. In most cases, a domain of definition will be an open and connected set (a *domain* as defined in Sec. 12.3); exceptions rarely occur in applications.

In the literature on complex analysis, one sometimes uses relations such that to a value of z there may correspond more than one value of w, and it is customary to call such a relation a function (a "multivalued function"). We shall not adopt this convention, but assume that all occurring functions are ***single-valued*** relations, that is, functions in the usual sense: to each z in S corresponds only *one* value $w = f(z)$. (But, of course, several z may correspond to the same value $w = f(z)$, just as in calculus.)

Strictly speaking, $f(z)$ denotes the value of f at z, but it is a convenient abuse of language to talk about *the function* $f(z)$ (instead of *the function* f), thereby exhibiting the notation for the independent variable.

EXAMPLE 2 Function of a complex variable

Let $w = f(z) = 2iz + 6\bar{z}$. Find u and v and the value of f at $z = \frac{1}{2} + 4i$.

Solution. $f(z) = 2i(x + iy) + 6(x - iy)$ gives $u(x, y) = 6x - 2y$ and $v(x, y) = 2x - 6y$. Also,

$$f(\tfrac{1}{2} + 4i) = 2i(\tfrac{1}{2} + 4i) + 6(\overline{\tfrac{1}{2} - 4i}) = i - 8 + 3 - 24i = -5 - 23i.$$

Limit, Continuity

A function $f(z)$ is said to have the **limit** l as z approaches a point z_0, written

$$\lim_{z \to z_0} f(z) = l, \tag{1}$$

if f is defined in a neighborhood of z_0 (except perhaps at z_0 itself) and if the values of f are "close" to l for all z "close" to z_0; that is, in precise terms, for every positive real ϵ we can find a positive real δ such that for all $z \neq z_0$ in the disk $|z - z_0| < \delta$ (Fig. 297) we have

$$|f(z) - l| < \epsilon; \tag{2}$$

that is, for every $z \neq z_0$ in that δ-disk the value of f lies in the disk (2).

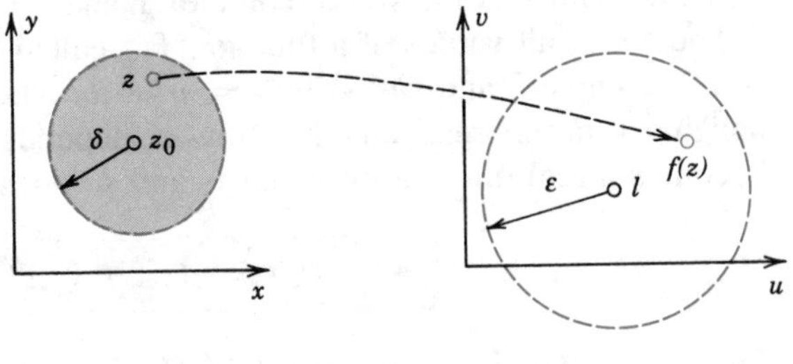

Fig. 297. Limit

Formally, this definition is similar to that in calculus, but there is a big difference. Whereas in the real case, x can approach an x_0 only along the real line, here, by definition, z may approach z_0 ***from any direction*** in the complex plane. This will be quite essential in what follows.

If a limit exists, it is unique. (See Prob. 28.)

A function $f(z)$ is said to be **continuous** at $z = z_0$ if $f(z_0)$ is defined and

$$\lim_{z \to z_0} f(z) = f(z_0). \tag{3}$$

Note that by the definition of a limit this implies that $f(z)$ is defined in some neighborhood of z_0.

$f(z)$ is said to be *continuous in a domain* if it is continuous at each point of this domain.

Derivative

The **derivative** of a complex function f at a point z_0 is written $f'(z_0)$ and is defined by

$$f'(z_0) = \lim_{\Delta z \to 0} \frac{f(z_0 + \Delta z) - f(z_0)}{\Delta z} \tag{4}$$

provided this limit exists. Then f is said to be **differentiable** at z_0. If we write $\Delta z = z - z_0$, we also have, since $z = z_0 + \Delta z$,

$$f'(z_0) = \lim_{z \to z_0} \frac{f(z) - f(z_0)}{z - z_0}. \tag{4'}$$

Now comes an important point. Remember that, by the definition of limit, $f(z)$ is defined in a neighborhood of z_0 and z in (4′) may approach z_0 from any direction in the complex plane. Hence differentiability at z_0 means that, along whatever path z approaches z_0, the quotient in (4′) always approaches a certain value and all these values are equal. This is important and should be kept in mind.

EXAMPLE 3 Differentiability. Derivative

The function $f(z) = z^2$ is differentiable for all z and has the derivative $f'(z) = 2z$ because

$$f'(z) = \lim_{\Delta z \to 0} \frac{(z + \Delta z)^2 - z^2}{\Delta z} = \lim_{\Delta z \to 0} (2z + \Delta z) = 2z.$$ ■

*The **differentiation rules** are the same as in real calculus,* since their proofs are literally the same. Thus,

$$(cf)' = cf', \quad (f + g)' = f' + g', \quad (fg)' = f'g + fg', \quad \left(\frac{f}{g}\right)' = \frac{f'g - fg'}{g^2}$$

as well as the chain rule and the power rule $(z^n)' = nz^{n-1}$ (n integer) hold.

Also, if $f(z)$ is differentiable at z_0, it is continuous at z_0. (See Prob. 30.)

EXAMPLE 4 $\bar{z}$ not differentiable

It is important to note that there are many simple functions that do not have a derivative at any point. For instance, $f(z) = \bar{z} = x - iy$ is such a function. Indeed, if we write $\Delta z = \Delta x + i\,\Delta y$, we have

$$\frac{f(z + \Delta z) - f(z)}{\Delta z} = \frac{\overline{(z + \Delta z)} - \bar{z}}{\Delta z} = \frac{\overline{\Delta z}}{\Delta z} = \frac{\Delta x - i\,\Delta y}{\Delta x + i\,\Delta y}. \tag{5}$$

If $\Delta y = 0$, this is $+1$. If $\Delta x = 0$, this is -1. Thus (5) approaches $+1$ along path I in Fig. 298 but -1 along path II. Hence, by definition, the limit of (5) as $\Delta z \to 0$ does not exist at any z. ■

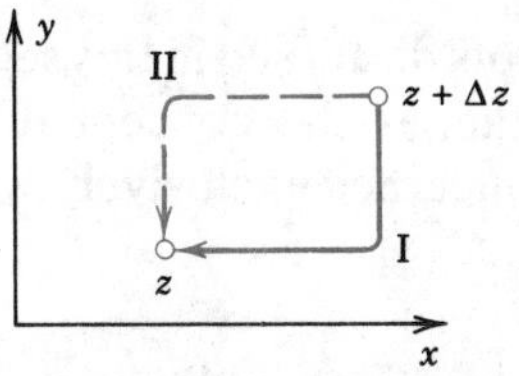

Fig. 298. Paths in (5)

The example just discussed may be surprising, but it merely illustrates that differentiability of a complex function is a rather severe requirement.

The idea of proof (approach from different directions) is basic and will be used again in the next section.

Analytic Functions

These are the functions that are differentiable in some domain, so that we can do "calculus in complex." They are the main concern of complex analysis. Their introduction is our main goal in this section.

Defintion (Analyticity)

A function $f(z)$ is said to be *analytic in a domain D* if $f(z)$ is defined and differentiable at all points of D. The function $f(z)$ is said to be *analytic at a point* $z = z_0$ in D if $f(z)$ is analytic in a neighborhood (see Sec. 12.3) of z_0.

Also, by an **analytic function** we mean a function that is analytic in *some* domain. ∎

Hence analyticity of $f(z)$ at z_0 means that $f(z)$ has a derivative at every point in some neighborhood of z_0 (including z_0 itself since, by definition, z_0 is a point of all its neighborhoods). This concept is *motivated* by the fact that it is of no practical interest if a function is differentiable merely at a single point z_0 but not throughout some neighborhood of z_0. Problem 26 gives an example.

A more modern term for *analytic in D* is *holomorphic in D*.

EXAMPLE 5 Polynomials, rational functions

The integer powers $1, z, z^2, \cdots$ and, more generally, **polynomials,** that is, functions of the form

$$f(z) = c_0 + c_1 z + c_2 z^2 + \cdots + c_n z^n$$

where $c_0, \cdots, c_n$ are complex constants, are analytic in the entire complex plane.

The quotient of two polynomials $g(z)$ and $h(z)$,

$$f(z) = \frac{g(z)}{h(z)},$$

is called a **rational function.** This f is analytic except at the points where $h(z) = 0$; here we assume that common factors of g and h have been canceled. **Partial fractions**

$$\frac{c}{(z - z_0)^m} \qquad (c \neq 0)$$

(c and z_0 complex, m a positive integer) are special rational functions; they are analytic except at z_0. It is proved in algebra that every rational function can be written as a sum of a polynomial (which may be 0) and finitely many partial fractions. ∎

The concepts discussed in this section extend familiar concepts of calculus. Most important is the concept of an analytic function. Indeed, complex analysis is concerned exclusively with analytic functions, and although many simple functions are not analytic, the large variety of remaining functions will yield a branch of mathematics that is most beautiful from the theoretical point of view and most useful for practical purposes.

Problem Set 12.4

Find $f(3 + i)$, $f(-i)$, $f(-4 + 2i)$ where $f(z)$ equals

1. $z^2 + 2z$ **2.** $1/(1 - z)$ **3.** $1/z^3$

Find the real and imaginary parts of the following functions.

4. $f(z) = z/(1 + z)$ **5.** $f(z) = 2z^3 - 3z$ **6.** $f(z) = z^2 + 4z - 1$

Suppose that z varies in a region R in the z-plane. Find the (precise) region in the w-plane in which the corresponding values of $w = f(z)$ lie, and show the two regions graphically.

7. $f(z) = z^2$, $|z| > 3$ **8.** $f(z) = 1/z$, $\text{Re}\, z > 0$ **9.** $f(z) = z^3$, $|\arg z| \leqq \frac{1}{4}\pi$

In each case, find whether $f(z)$ is continuous at the origin, assuming that $f(0) = 0$ and, for $z \neq 0$, the function $f(z)$ equals

10. $\text{Re}\, z/|z|$ **11.** $\text{Re}\,(z^2)/|z^2|$ **12.** $\text{Im}\, z/(1 + |z|)$

Differentiate

13. $(z^2 + i)^3$ **14.** $(z^2 - 4)/(z^2 + 1)$ **15.** $i/(1 - z)^2$

16. $(z + i)/(z - i)$ **17.** $(iz + 2)/(3z - 6i)$ **18.** $z^2/(z + i)^2$

Find the value of the derivative of

19. $(z + i)/(z - i)$ at $-i$ **20.** $(z^2 - i)^2$ at $3 - 2i$ **21.** $1/z^3$ at $3i$

22. $z^3 - 2z$ at $-i$ **23.** $(1 + i)/z^4$ at 2 **24.** $(2 + iz)^6$ at $2i$

25. Show that $f(z) = \text{Re}\, z = x$ is not differentiable at any z.

26. Show that $f(z) = |z|^2$ is differentiable only at $z = 0$; hence it is nowhere analytic. *Hint.* Use the relation $|z + \Delta z|^2 = (z + \Delta z)(\bar{z} + \overline{\Delta z})$.

27. Prove that (1) is equivalent to the pair of relations

$$\lim_{z \to z_0} \text{Re}\, f(z) = \text{Re}\, l, \qquad \lim_{z \to z_0} \text{Im}\, f(z) = \text{Im}\, l.$$

28. If $\lim_{z \to z_0} f(z)$ exists, show that this limit is unique.

29. If $z_1, z_2, \cdots$ are complex numbers for which $\lim_{n \to \infty} z_n = a$, and if $f(z)$ is continuous at $z = a$, show that

$$\lim_{n \to \infty} f(z_n) = f(a).$$

30. If $f(z)$ is differentiable at z_0, show that $f(z)$ is continuous at z_0.

12.5 Cauchy–Riemann Equations

We shall now derive a very important criterion (a test) for the analyticity of a complex function

$$w = f(z) = u(x, y) + iv(x, y).$$

Roughly, f is analytic in a domain D if and only if the first partial derivatives of u and v satisfy the two equations

$$u_x = v_y, \qquad u_y = -v_x \tag{1}$$

everywhere in D; here $u_x = \partial u/\partial x$ and $u_y = \partial u/\partial y$ (and similarly for v) are the usual notations for partial derivatives. The precise formulation of this statement is given in Theorems 1 and 2 below. The equations (1) are called the **Cauchy–Riemann equations.** They are the most important equations in the whole chapter.[10]

Example: $f(z) = z^2 = x^2 - y^2 + 2ixy$ is analytic for all z, and $u = x^2 - y^2$ and $v = 2xy$ satisfy (1), namely, $u_x = 2x = v_y$ as well as $u_y = -2y = -v_x$. More examples will follow.

Theorem 1 (Cauchy–Riemann equations)

Let $f(z) = u(x, y) + iv(x, y)$ be defined and continuous in some neighborhood of a point $z = x + iy$ and differentiable at z itself. Then at that point, the first-order partial derivatives of u and v exist and satisfy the Cauchy–Riemann equations (1).

Hence if $f(z)$ is analytic in a domain D, those partial derivatives exist and satisfy (1) *at all points of D.*

Proof. By assumption, the derivative $f'(z)$ at z exists. It is given by

$$f'(z) = \lim_{\Delta z \to 0} \frac{f(z + \Delta z) - f(z)}{\Delta z}. \tag{2}$$

The idea of the proof is very simple. By the definition of a limit in complex (Sec. 12.4) we can let Δz approach zero along any path in a neighborhood of z. Thus we may choose the two paths I and II in Fig. 299 and equate the

[10] AUGUSTIN-LOUIS CAUCHY (see Sec. 2.6) and the German mathematicians BERNHARD RIEMANN (1826—1866) and KARL WEIERSTRASS (1815—1897; see also Sec. 14.6) are the founders of complex analysis. Riemann received his Ph.D. (in 1851) under Gauss (Sec. 5.4) at Göttingen, where he also taught from 1854 until he died, when he was only 39 years old. He introduced the concept of the integral as it is used in basic calculus courses, in connection with his work on Fourier series. He created new methods in ordinary and partial differential equations and made basic contributions to number theory and mathematical physics. He also developed the so-called Riemannian geometry, which is the mathematical base of Einstein's theory of relativity. His important work gave the impetus to many ideas in modern mathematics (in particular, in topology and functional analysis). See N. Bourbaki, *Elements of Mathematics, General Topology,* Part 1, pp. 161–166 (Paris: Hermann, 1966).

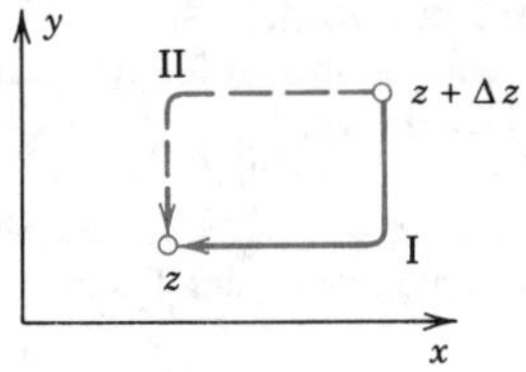

Fig. 299. Paths in (2)

results. By comparing the real parts we shall obtain the first Cauchy–Riemann equation and by comparing the imaginary parts the other equation in (1). The technical details are as follows.

We write $\Delta z = \Delta x + i\Delta y$. In terms of u and v, the derivative in (2) becomes

$$(3)\quad f'(z) = \lim_{\Delta z \to 0} \frac{[u(x + \Delta x, y + \Delta y) + iv(x + \Delta x, y + \Delta y)] - [u(x, y) + iv(x, y)]}{\Delta x + i\Delta y}.$$

We first choose path I in Fig. 299. Thus we let $\Delta y \to 0$ first and then $\Delta x \to 0$. After Δy is zero, $\Delta z = \Delta x$. Then (3) becomes, if we first write the two u-terms and then the two v-terms,

$$f'(z) = \lim_{\Delta x \to 0} \frac{u(x + \Delta x, y) - u(x, y)}{\Delta x} + i \lim_{\Delta x \to 0} \frac{v(x + \Delta x, y) - v(x, y)}{\Delta x}.$$

Since $f'(z)$ exists, the two real limits on the right exist. By definition, they are the partial derivatives of u and v with respect to x. Hence the derivative $f'(z)$ of $f(z)$ can be written

$$(4)\quad f'(z) = u_x + iv_x.$$

Similarly, if we choose path II in Fig. 299, we let $\Delta x \to 0$ first and then $\Delta y \to 0$. After Δx is zero, $\Delta z = i\Delta y$, so that from (3) we now obtain

$$f'(z) = \lim_{\Delta y \to 0} \frac{u(x, y + \Delta y) - u(x, y)}{i\,\Delta y} + i \lim_{\Delta y \to 0} \frac{v(x, y + \Delta y) - v(x, y)}{i\,\Delta y}.$$

Since $f'(z)$ exists, the limits on the right exist and give the partial derivatives of u and v with respect to y; noting that $1/i = -i$, we thus obtain

$$(5)\quad f'(z) = -iu_y + v_y.$$

The existence of the derivative $f'(z)$ thus implies the existence of the four partial derivatives in (4) and (5). By equating the real parts u_x and v_y in (4) and (5) we obtain the first Cauchy–Riemann equation (1). Equating the imaginary parts gives the other. This proves the first statement of the theorem and implies the second because of the definition of analyticity. ∎

Formulas (4) and (5) are also quite practical for calculating derivatives $f'(z)$, as we shall see.

EXAMPLE 1 **Cauchy–Riemann equations**

$f(z) = z^2$ is analytic for all z. It follows that the Cauchy–Riemann equations must be satisfied (as we have verified above).

For $f(z) = \bar{z} = x - iy$ we have $u = x$, $v = -y$ and see that the second Cauchy–Riemann equation is satisfied, $u_y = -v_x = 0$, but the first is not: $u_x = 1 \neq v_y = -1$. We conclude that $f(z) = \bar{z}$ is not analytic, confirming Example 4 of Sec. 12.4. Note the savings in calculation! ∎

The Cauchy–Riemann equations are fundamental because they are not only necessary but also sufficient for a function to be analytic. More precisely, the following theorem holds.

Theorem 2 **(Cauchy–Riemann equations)**

If two real-valued continuous functions $u(x, y)$ and $v(x, y)$ of two real variables x and y have ***continuous*** *first partial derivatives that satisfy the Cauchy–Riemann equations in some domain D, then the complex function $f(z) = u(x, y) + iv(x, y)$ is analytic in D.*

The proof is more involved than that of Theorem 1 and we leave it optional (see Appendix 4).

Theorems 1 and 2 are of great practical importance, since by using the Cauchy–Riemann equations we can now easily find out whether or not a given complex function is analytic.

EXAMPLE 2 **Cauchy–Riemann equations**

Is $f(z) = z^3$ analytic?

Solution. We find $u = x^3 - 3xy^2$ and $v = 3x^2y - y^3$. Next we calculate

$$u_x = 3x^2 - 3y^2, \qquad v_y = 3x^2 - 3y^2$$

$$u_y = -6xy, \qquad v_x = 6xy.$$

We see that the Cauchy–Riemann equations are satisfied for every z. Hence $f(z) = z^3$ is analytic for every z, by Theorem 2. ∎

EXAMPLE 3 **Determination of an analytic function with given real part**

We illustrate another class of problems that can be solved by the Cauchy–Riemann equations.

Find the most general analytic function $f(z)$ whose real part is $u = x^2 - y^2 - x$.

Solution. $u_x = 2x - 1 = v_y$ by the first Cauchy–Riemann equation. This we integrate with respect to y:

$$v = 2xy - y + k(x).$$

As an important point, since we integrated a *partial* derivative with respect to y, the "constant" of integration k may depend on the other variable, x. (To understand this, calculate v_y from this v.) From v and the second Cauchy–Riemann equation,

$$u_y = -v_x = -2y + \frac{dk}{dx}.$$

On the other hand, from the given $u = x^2 - y^2 - x$ we have $u_y = -2y$. By comparison, $dk/dx = 0$, hence $k = const$, which must be real. (Why?) The result is

$$f(z) = u + iv = x^2 - y^2 - x + i(2xy - y + k).$$

This we can express in terms of z, namely, $f(z) = z^2 - z + ik$. ∎

EXAMPLE 4 An analytic function of constant absolute value is constant

The Cauchy–Riemann equations also help in deriving general properties of analytic functions.

For example, show that if $f(z)$ is analytic in a domain D and $|f(z)| = k = const$ in D, then $f(z) = const$ in D.

Solution. By assumption, $u^2 + v^2 = k^2$. By differentiation,

$$uu_x + vv_x = 0, \qquad uu_y + vv_y = 0.$$

Now use $v_x = -u_y$ in the first equation and $v_y = u_x$ in the second, to get

$$\text{(6)} \qquad \text{(a)} \quad uu_x - vu_y = 0, \qquad \text{(b)} \quad uu_y + vu_x = 0.$$

To get rid of u_y, multiply (6a) by u and (6b) by v and add. Similarly, to eliminate u_x, multiply (6a) by $-v$ and (6b) by u and add. This yields

$$(u^2 + v^2)u_x = 0, \qquad (u^2 + v^2)u_y = 0.$$

If $k^2 = u^2 + v^2 = 0$, then $u = v = 0$, hence $f = 0$. If $k \neq 0$, then $u_x = u_y = 0$, hence, by the Cauchy–Riemann equations, also $v_x = v_y = 0$. Together, $u = const$ and $v = const$, hence $f = const$. ∎

We mention that if we use the polar form $z = r(\cos\theta + i \sin\theta)$ and set $f(z) = u(r, \theta) + iv(r, \theta)$, then the Cauchy–Riemann equations are

$$\text{(7)} \qquad u_r = \frac{1}{r} v_\theta, \qquad v_r = -\frac{1}{r} u_\theta \qquad (r > 0).$$

Laplace's Equation. Harmonic Functions

One of the main reasons for the great practical importance of complex analysis in engineering mathematics results from the fact that both the real part and the imaginary part of an analytic function satisfy the most important differential equation of physics, Laplace's equation, which occurs in gravitation, electrostatics, fluid flow, heat conduction, and so on (see Chaps. 11 and 17).

Theorem 3 (Laplace's equation)

If $f(z) = u(x, y) + iv(x, y)$ is analytic in a domain D, then u and v satisfy ***Laplace's equation***

$$\text{(8)} \qquad \nabla^2 u = u_{xx} + u_{yy} = 0$$

(∇^2 read "nabla squared") *and*

$$\text{(9)} \qquad \nabla^2 v = v_{xx} + v_{yy} = 0,$$

respectively, in D and have continuous second partial derivatives in D.

Proof. Differentiating $u_x = v_y$ with respect to x and $u_y = -v_x$ with respect to y, we have

$$(10) \qquad u_{xx} = v_{yx}, \qquad u_{yy} = -v_{xy}.$$

Now the derivative of an analytic function is itself analytic, as we shall prove later (in Sec. 13.6). This implies that u and v have continuous partial derivatives of all orders; in particular, the mixed second derivatives are equal: $v_{yx} = v_{xy}$. By adding (10) we thus obtain (8). Similarly, (9) is obtained by differentiating $u_x = v_y$ with respect to y and $u_y = -v_x$ with respect to x and subtraction, using $u_{xy} = u_{yx}$. ∎

Solutions of Laplace's equation having ***continuous*** second-order partial derivatives are called **harmonic functions** and their theory is called **potential theory** (see also Sec. 11.11). Hence the real and imaginary parts of an analytic function are harmonic functions.

If two harmonic functions u and v satisfy the Cauchy–Riemann equations in a domain D, they are the real and imaginary parts of an analytic function f in D. Then v is said to be a **conjugate harmonic function** of u in D. (Of course, this has nothing to do with the use of "conjugate" for $\bar{z}$.)

Let us illustrate that a conjugate harmonic function can be obtained by the Cauchy–Riemann equations.

EXAMPLE 5 Conjugate harmonic function

Verify that $u = x^2 - y^2 - y$ is harmonic in the whole complex plane and find a conjugate harmonic function v of u.

Solution. $\nabla^2 u = 0$ by direct calculation. Now $u_x = 2x$ and $u_y = -2y - 1$. Hence a conjugate v of u must satisfy

$$v_y = u_x = 2x, \qquad v_x = -u_y = 2y + 1.$$

Integrating the first equation with respect to y and differentiating the result with respect to x, we obtain

$$v = 2xy + h(x), \qquad v_x = 2y + \frac{dh}{dx}.$$

A comparison with the second equation shows that $dh/dx = 1$. This gives $h(x) = x + c$. Hence $v = 2xy + x + c$ (c any real constant) is the most general conjugate harmonic of the given u. The corresponding analytic function is

$$f(z) = u + iv = x^2 - y^2 - y + i(2xy + x + c) = z^2 + iz + ic.$$

Can you see that the present task was quite similar to that in Example 3? Explain. ∎

The Cauchy–Riemann equations are the most important equations in this chapter. Their relation to Laplace's equation opens wide ranges of engineering and physical applications, as we show in Chap. 17.

Problem Set 12.5

Are the following functions analytic? [Use (1) or (7).]

1. $f(z) = z^8$ **2.** $f(z) = \text{Re}\,(z^2)$ **3.** $f(z) = e^x(\cos y + i \sin y)$

4. $f(z) = i/z^4$ **5.** $f(z) = 1/(1 - z)$ **6.** $f(z) = z - \bar{z}$

7. $f(z) = \ln |z| + i \,\text{Arg}\, z$ **8.** $f(z) = 1/(1 - z^4)$ **9.** $f(z) = \text{Arg}\, z$

10. $f(z) = z + 1/z$ **11.** $f(z) = z^2 - \bar{z}^2$ **12.** $f(z) = e^x(\sin y - i \cos y)$

Are the following functions harmonic? If so, find a corresponding analytic function $f(z) = u(x, y) + iv(x, y)$.

13. $u = xy$ **14.** $v = xy$ **15.** $u = x/(x^2 + y^2)$

16. $v = 1/(x^2 + y^2)$ **17.** $u = x^3 - 3xy^2$ **18.** $u = \sin x \cosh y$

19. $u = e^x \cos 2y$ **20.** $v = i \ln |z|$ **21.** $v = (x^2 - y^2)^2$

Determine a, b, c such that the given functions are harmonic and find a conjugate harmonic.

22. $u = e^{2x} \cos ay$ **23.** $u = \cos bx \cosh y$ **24.** $u = \sin x \cosh cy$

25. Show that if u is harmonic and v a conjugate harmonic of u, then u is a conjugate harmonic of $-v$.

26. Show that, in addition to (4) and (5),

$$f'(z) = u_x - iu_y, \qquad f'(z) = v_y + iv_x. \tag{11}$$

27. Formulas (4), (5), and (11) are needed from time to time. Familiarize yourself with them by calculating $(z^3)'$ by one of them and verifying that the result is as expected.

28. Show that if $f(z)$ is analytic and Re $f(z)$ is constant, then $f(z)$ is constant.

29. **(Identically vanishing derivative)** Using (4), show that an analytic function whose derivative is identically zero is a constant.

30. Derive the Cauchy–Riemann equations in polar form (7) from (1).

12.6 Exponential Function

In the remaining sections of this chapter we discuss the most important elementary complex functions, the exponential function, trigonometric functions, logarithm, and so on. These will be defined so that for real $z = x$ they reduce to the familiar functions of calculus. These functions are indispensable throughout applications and some of them have interesting properties not apparent when $z = x$ is real. For these reasons, the student should follow the discussion with particular care.

We begin with the complex **exponential function**

$$e^z, \qquad \text{also written} \qquad \exp z.$$

e^z is one of the most important analytic functions. The definition of e^z in terms of the real functions e^x, $\cos y$, and $\sin y$ is

$$e^z = e^x(\cos y + i \sin y). \tag{1}$$

This definition[11] is motivated by requirements that make e^z a natural extension of the real exponential function e^x, namely,

(a) e^z should reduce to the latter when $z = x$ is real;

(b) e^z should be an **entire function,** that is, analytic for all z;

(c) similar to calculus, its derivative should be

$$(e^z)' = e^z. \tag{2}$$

From (1) we see that (a) holds, since $\cos 0 = 1$ and $\sin 0 = 0$. That e^z is entire is easily verified by the Cauchy–Riemann equations. Formula (2) then follows from (4) in Sec. 12.5:

$$(e^z)' = (e^x \cos y)_x + i(e^x \sin y)_x = e^x \cos y + ie^x \sin y = e^z.$$

Further Properties. e^z has further interesting properties. Let us first show that, as in real, we have the functional relation

$$e^{z_1+z_2} = e^{z_1}e^{z_2} \tag{3}$$

for any $z_1 = x_1 + iy_1$ and $z_2 = x_2 + iy_2$. Indeed, by (1),

$$e^{z_1}e^{z_2} = e^{x_1}(\cos y_1 + i \sin y_1)e^{x_2}(\cos y_2 + i \sin y_2).$$

Since $e^{x_1}e^{x_2} = e^{x_1+x_2}$ for these *real* functions, by an application of the addition formulas for the cosine and sine functions (similar to that in Sec. 12.2) we find that this equals

$$e^{z_1}e^{z_2} = e^{x_1+x_2}[\cos (y_1 + y_2) + i \sin (y_1 + y_2)] = e^{z_1+z_2},$$

as asserted. An interesting special case is $z_1 = x$, $z_2 = iy$:

$$e^z = e^x e^{iy}. \tag{4}$$

[11]This definition provides for a relatively simple discussion. We could also define e^z by the familiar series $e^x = \sum_{n=0}^{\infty} x^n/n!$ with x replaced by z, but then we would first have to discuss complex series at this early stage. We show the connection in Sec. 14.4.

Furthermore, for $z = iy$ we have from (1) the so-called **Euler formula**

$$e^{iy} = \cos y + i \sin y. \tag{5}$$

Hence the **polar form** of a complex number, $z = r(\cos \theta + i \sin \theta)$, may now be written

$$z = re^{i\theta}. \tag{6}$$

From (5) we also see that

$$|e^{iy}| = |\cos y + i \sin y| = \sqrt{\cos^2 y + \sin^2 y} = 1. \tag{7}$$

That is, for pure imaginary exponents the exponential function has absolute value one, a result the student should remember. From (7) and (1),

$$|e^z| = e^x. \qquad \text{Hence} \qquad \arg e^z = y \pm 2n\pi \qquad (n = 0, 1, 2, \cdots), \tag{8}$$

since $|e^z| = e^x$ shows that (1) is actually e^z in polar form.

EXAMPLE 1 Illustration of some properties of the exponential function

Computation of values from (1) provides no problem. For instance, verify that

$$e^{1.4-0.6i} = e^{1.4}(\cos 0.6 - i \sin 0.6) = 4.055(0.825 - 0.565i) = 3.347 - 2.290i,$$

$$|e^{1.4-0.6i}| = e^{1.4} = 4.055, \qquad \operatorname{Arg} e^{1.4-0.6i} = -0.6.$$

Since $\cos 2\pi = 1$ and $\sin 2\pi = 0$, we have from (5)

$$e^{2\pi i} = 1. \tag{9}$$

Furthermore, use (1), (5), or (6) to verify these important special values:

$$e^{\pi i/2} = i, \qquad e^{\pi i} = -1, \qquad e^{-\pi i/2} = -i, \qquad e^{-\pi i} = -1. \tag{10}$$

To illustrate (3), take the product of

$$e^{2+i} = e^2(\cos 1 + i \sin 1) \qquad \text{and} \qquad e^{4-i} = e^4(\cos 1 - i \sin 1)$$

and verify that it equals

$$e^2e^4(\cos^2 1 + \sin^2 1) = e^6 = e^{(2+i)+(4-i)}.$$

Finally, conclude from $|e^z| = e^x \neq 0$ in (8) that

$$e^z \neq 0 \qquad \text{for all } z. \tag{11}$$

So here we have an entire function that never vanishes, in contrast to (nonconstant) polynomials, which are also entire (Example 5 in Sec. 12.4) but always have a zero, as is proved in algebra. [Can you obtain (11) from (3)?] ∎

Periodicity of e^z with period $2\pi i$,

(12) $$e^{z+2\pi i} = e^z \qquad \text{for all } z$$

is a basic property that follows from (1) and the periodicity of cos y and sin y. [It also follows from (3) and (9).] Hence all the values that $w = e^z$ can assume are already assumed in the horizontal strip of width 2π

(13) $$-\pi < y \leqq \pi \qquad \text{(Fig. 300).}$$

This infinite strip is called a **fundamental region** of e^z.

EXAMPLE 2 **Solution of an equation**

Find all solutions of $e^z = 3 + 4i$.

Solution. $|e^z| = e^x = 5$, $x = \ln 5 = 1.609$ is the real part of all solutions. Now, since $e^x = 5$,

$$e^x \cos y = 3, \qquad e^x \sin y = 4, \qquad \cos y = 0.6, \qquad \sin y = 0.8, \qquad y = 0.927.$$

Ans. $z = 1.609 + 0.927i \pm 2n\pi i$ $(n = 0, 1, 2, \cdots)$. These are infinitely many solutions (due to the periodicity of e^z). They lie on the vertical line $x = 1.609$ at a distance 2π from their neighbors. ∎

To summarize: many properites of $e^z = \exp z$ parallel those of e^x; an exception is the periodicity of e^z with $2\pi i$, which suggested the concept of a fundamental region. Keep in mind that e^z is an *entire function.* (Do you still remember what that means?)

Problem Set 12.6

1. Using the Cauchy–Riemann equations, show that e^z is analytic for all z.

Compute e^z (in the form $u + iv$) and $|e^z|$ if z equals

2. $3 + \pi i$ **3.** $1 + i$ **4.** $2 + 5\pi i$ **5.** $\sqrt{2} - \frac{1}{2}i$

6. $7\pi i/2$ **7.** $(1 + i)\pi$ **8.** $-1 + 1.4i$ **9.** $-9\pi i/2$

Find the real and imaginary parts of

10. e^{-z^2} **11.** e^{z^3} **12.** $e^{-\pi z}$ **13.** e^{-2z}

Write the following expressions in the polar form (6).

14. $1 + i$ **15.** $\sqrt{i}$, $\sqrt{-i}$ **16.** $\sqrt[n]{z}$ **17.** $3 + 4i$

Find all values of z such that

18. e^z is real **19.** $|e^{-z}| < 1$ **20.** $e^{\bar{z}} = \overline{e^z}$ **21.** Re $e^{2z} = 0$

Find all solutions and plot some of them in the complex plane.

22. $e^{3z} = 3$ **23.** $e^z = -2$ **24.** $e^z = -3 + 4i$ **25.** $e^z = 0$

26. Show that $u = e^{xy} \cos(x^2/2 - y^2/2)$ is harmonic and find a conjugate.

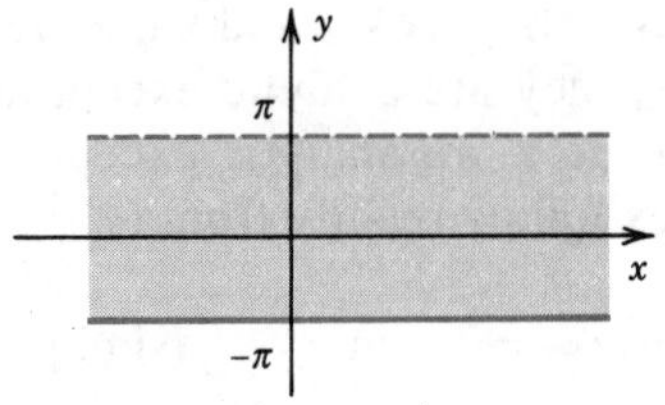

Fig. 300. Fundamental region of the exponential function e^z in the z-plane

27. Find all values of k such that $f(z) = e^x(\cos ky + i \sin ky)$ is analytic.

28. Show that $f(z) = e^{\bar{z}}$ is nowhere analytic.

29. It is interesting that $f(z) = e^z$ is *uniquely* determined by the two properties $f(x + i0) = e^x$ and $f'(z) = f(z)$, where f is assumed to be entire. Prove this. *Hint.* Let g be entire with these two properties and show that $(g/f)' = 0$.

30. Prove the statement in Prob. 29, using only the Cauchy–Riemann equations.

12.7 Trigonometric Functions, Hyperbolic Functions

Just as e^z extends e^x to complex, we want the *complex* trigonometric functions to extend the familiar *real* trigonometric functions. The idea of making the connection is the use of the Euler formulas (Sec. 12.6)

$$e^{ix} = \cos x + i \sin x, \qquad e^{-ix} = \cos x - i \sin x.$$

By addition and subtraction we obtain for the *real* cosine and sine

$$\cos x = \frac{1}{2}(e^{ix} + e^{-ix}), \qquad \sin x = \frac{1}{2i}(e^{ix} - e^{-ix}).$$

This suggests the following definitions for complex values $z = x + iy$:

$$\cos z = \frac{1}{2}(e^{iz} + e^{-iz}), \qquad \sin z = \frac{1}{2i}(e^{iz} - e^{-iz}). \tag{1}$$

It is quite remarkable that here in complex, functions come together that are unrelated in real. This is not an isolated incident but is typical of the general situation and shows the advantage of working in complex.

Furthermore, as in calculus we define

$$\tan z = \frac{\sin z}{\cos z}, \qquad \cot z = \frac{\cos z}{\sin z} \tag{2}$$

and

$$\sec z = \frac{1}{\cos z}, \qquad \csc z = \frac{1}{\sin z}. \tag{3}$$

Since e^z is entire, $\cos z$ and $\sin z$ are entire functions. $\tan z$ and $\sec z$ are not entire; they are analytic except at the points where $\cos z$ is zero; and $\cot z$ and $\csc z$ are analytic except where $\sin z$ is zero. Formulas for the derivatives follow readily from $(e^z)' = e^z$ and (1)–(3); as in calculus,

$$(4) \qquad (\cos z)' = -\sin z, \qquad (\sin z)' = \cos z, \qquad (\tan z)' = \sec^2 z,$$

etc. Equation (1) also shows that **Euler's formula** *is valid in complex:*

$$(5) \qquad \boxed{e^{iz} = \cos z + i \sin z} \qquad \text{for all } z.$$

Real and imaginary parts of $\cos z$ and $\sin z$ are needed in computing values, and they also help in displaying properties of our functions. We illustrate this by a typical example.

EXAMPLE 1 **Real and imaginary parts. Absolute value. Periodicity**

Show that

$$(6) \qquad \begin{array}{ll} \text{(a)} & \cos z = \cos x \cosh y - i \sin x \sinh y \\ \text{(b)} & \sin z = \sin x \cosh y + i \cos x \sinh y \end{array}$$

and

$$(7) \qquad \begin{array}{ll} \text{(a)} & |\cos z|^2 = \cos^2 x + \sinh^2 y \\ \text{(b)} & |\sin z|^2 = \sin^2 x + \sinh^2 y \end{array}$$

and give some applications of these formulas.

Solution. From (1),

$$\begin{aligned} \cos z &= \tfrac{1}{2}(e^{i(x+iy)} + e^{-i(x+iy)}) \\ &= \tfrac{1}{2}e^{-y}(\cos x + i \sin x) + \tfrac{1}{2}e^{y}(\cos x - i \sin x) \\ &= \tfrac{1}{2}(e^y + e^{-y})\cos x - \tfrac{1}{2}i(e^y - e^{-y})\sin x. \end{aligned}$$

This yields (6a) since, as is known from calculus,

$$(8) \qquad \cosh y = \tfrac{1}{2}(e^y + e^{-y}), \qquad \sinh y = \tfrac{1}{2}(e^y - e^{-y});$$

(6b) is obtained similarly. From (6a) and $\cosh^2 y = 1 + \sinh^2 y$ we obtain

$$|\cos z|^2 = \cos^2 x\,(1 + \sinh^2 y) + \sin^2 x \sinh^2 y.$$

Since $\sin^2 x + \cos^2 x = 1$, this gives (7a), and (7b) is obtained similarly.

For instance, $\cos(2 + 3i) = \cos 2 \cosh 3 - i \sin 2 \sinh 3 = -4.190 - 9.109i$.

From (6) we see that $\cos z$ and $\sin z$ are ***periodic with period 2π***, just as in real. Periodicity of $\tan z$ and $\cot z$ with period π now follows.

Formula (7) points to an essential difference between the real and the complex cosine and sine: whereas $|\cos x| \leqq 1$ and $|\sin x| \leqq 1$, the complex cosine and sine functions are no longer bounded but approach infinity in absolute value as $y \to \infty$, since then $\sinh y \to \infty$ in (7). ■

EXAMPLE 2 **Solution of equations. Zeros of cos z and sin z**

Solve (a) $\cos z = 5$ (which has no real solution!), (b) $\cos z = 0$, (c) $\sin z = 0$.

Solution. (a) $e^{2iz} - 10e^{iz} + 1 = 0$ from (1) by multiplication by e^{iz}. This is a quadratic equation in e^{iz}, with solutions (3D values)

$$e^{iz} = e^{-y+ix} = 5 \pm \sqrt{25 - 1} = 9.899 \quad \text{and} \quad 0.101.$$

Thus $e^{-y} = 9.899$ or 0.101, $e^{ix} = 1$, $y = \pm 2.292$, $x = 2n\pi$.

Ans. $z = \pm 2n\pi \pm 2.292i$ ($n = 0, 1, 2, \cdots$). Can you obtain this by using (6a)?

(b) $\cos x = 0$, $\sinh y = 0$ by (7a), $y = 0$. *Ans.* $z = \pm\frac{1}{2}(2n + 1)\pi$ ($n = 0, 1, 2, \cdots$).

(c) $\sin x = 0$, $\sinh y = 0$ by (7b). *Ans.* $z = 2n\pi$ ($n = 0, 1, 2, \cdots$). Hence the only zeros of $\cos z$ and $\sin z$ are those of the real cosine and sine functions. ∎

General formulas *for the real trigonometric functions continue to hold for complex values.* This follows immediately from the definitions. We mention in particular the addition rules

$$\begin{aligned} \cos(z_1 \pm z_2) &= \cos z_1 \cos z_2 \mp \sin z_1 \sin z_2 \\ \sin(z_1 \pm z_2) &= \sin z_1 \cos z_2 \pm \sin z_2 \cos z_1 \end{aligned} \tag{9}$$

and the formula

$$\cos^2 z + \sin^2 z = 1. \tag{10}$$

Some further useful formulas are included in the problem set.

Hyperbolic Functions

The complex **hyperbolic cosine** and **sine** are defined by the formulas

$$\cosh z = \tfrac{1}{2}(e^z + e^{-z}), \qquad \sinh z = \tfrac{1}{2}(e^z - e^{-z}). \tag{11}$$

This is suggested by the familiar definitions for a real variable [see (8)]. These functions are entire, with derivatives

$$(\cosh z)' = \sinh z, \qquad (\sinh z)' = \cosh z, \tag{12}$$

as in calculus. The other hyperbolic functions are defined by

$$\begin{aligned} \tanh z &= \frac{\sinh z}{\cosh z}, \qquad \coth z = \frac{\cosh z}{\sinh z}, \\ \operatorname{sech} z &= \frac{1}{\cosh z}, \qquad \operatorname{csch} z = \frac{1}{\sinh z}. \end{aligned} \tag{13}$$

Complex trigonometric and hyperbolic functions are related. If in (11), we replace z by iz and use (1), we obtain

$$\cosh iz = \cos z, \qquad \sinh iz = i \sin z. \tag{14}$$

From this, since cosh is even and sinh is odd, conversely,

$$\cos iz = \cosh z, \qquad \sin iz = i \sinh z. \tag{15}$$

Here we have another case of *unrelated* real functions that have *related* complex analogs, pointing again to the advantage of working in complex in order to get both a more unified formalism and a deeper understanding of special functions. This is one of three main reasons for the practical importance of complex analysis, mentioned at the beginning of the chapter.

Problem Set 12.7

1. Prove that $\cos z$, $\sin z$, $\cosh z$, $\sinh z$ are entire functions.

2. Verify by differentiation that $\operatorname{Re} \cos z$ and $\operatorname{Im} \sin z$ are harmonic.

Compute (in the form $u + iv$)

3. $\cos(1.7 + 1.5i)$ **4.** $\sin(1.7 + 1.5i)$ **5.** $\sin 10i$

6. $\cos 10i$ **7.** $\sin(\sqrt{2} - 4i)$ **8.** $\cos(\pi + \pi i)$

9. $\cos 3\pi i$ **10.** $\sin(3 + 2i)$ **11.** $\cos(2.1 - 0.2i)$

12. Show that

$$\cosh z = \cosh x \cos y + i \sinh x \sin y,$$
$$\sinh z = \sinh x \cos y + i \cosh x \sin y.$$

Compute (in the form $u + iv$)

13. $\cosh(-2 + 3i)$ **14.** $\sinh(4 - 3i)$ **15.** $\sinh(2 + i)$

Find all solutions of the following equations.

16. $\cosh z = 0$ **17.** $\cos z = 3i$ **18.** $\sin z = 1000$

19. $\sin z = i \sinh 1$ **20.** $\sin z = \cosh 3$ **21.** $\cosh z = \frac{1}{2}$

22. Find all values of z for which (a) $\cos z$, (b) $\sin z$ has real values.

23. Obtain $\cosh(-1.5 + 1.7i)$ from (15) and the answer to one of the above problems.

24. Find $\operatorname{Re} \tan z$ and $\operatorname{Im} \tan z$.

25. Prove that $\cos z$ is even, $\cos(-z) = \cos z$, and $\sin z$ is odd, $\sin(-z) = -\sin z$.

26. Show that $\cos z = \sin(z + \frac{1}{2}\pi)$ and $\sin(\pi - z) = \sin z$, as in real.

27. Show that $\sinh z$ and $\cosh z$ are periodic with period $2\pi i$.

28. From (9) and (15) derive the addition rules

$$\cosh(z_1 + z_2) = \cosh z_1 \cosh z_2 + \sinh z_1 \sinh z_2,$$
$$\sinh(z_1 + z_2) = \sinh z_1 \cosh z_2 + \cosh z_1 \sinh z_2.$$

29. Prove

$$\cos^2 z + \sin^2 z = 1, \qquad \cos^2 z - \sin^2 z = \cos 2z$$
$$\cosh^2 z - \sinh^2 z = 1, \qquad \cosh^2 z + \sinh^2 z = \cosh 2z.$$

30. Show that $|\sinh y| \leqq |\cos z| \leqq \cosh y$ and $|\sinh y| \leqq |\sin z| \leqq \cosh y$. Conclude that the complex cosine and sine are not bounded in the whole complex plane.

12.8 Logarithm. General Power

As the last of the functions we introduce the *complex logarithm,* which is more complicated than the real logarithm (which it includes as a special case) and historically puzzled mathematicians for some time (so if you first get puzzled—which need not happen!—be patient and work this section with extra care).

The **natural logarithm** of $z = x + iy$ is denoted by $\ln z$ (sometimes also by $\log z$) and is defined as the inverse of the exponential function; that is, $w = \ln z$ is defined for $z \neq 0$ by the relation

$$e^w = z.$$

(Note that $z = 0$ is impossible, since $e^w \neq 0$ for all w; see Sec. 12.6.) If we set $w = u + iv$ and $z = re^{i\theta}$, this becomes

$$e^w = e^{u+iv} = re^{i\theta}.$$

Now from Sec. 12.6 we know that e^{u+iv} has the absolute value e^u and the argument v. These must be equal to the absolute value and argument on the right:

$$e^u = r, \qquad v = \theta.$$

$e^u = r$ gives $u = \ln r$, where $\ln r$ is the familiar *real* natural logarithm of the positive number $r = |z|$. Hence $w = u + iv = \ln z$ is given by

$$\text{(1)} \qquad \boxed{\ln z = \ln r + i\theta} \qquad (r = |z| > 0, \quad \theta = \arg z).$$

Now comes an important point (without analog in real calculus): Since the argument of z is determined only up to integer multiples of 2π, *the complex natural logarithm* $\ln z$ $(z \neq 0)$ *is infinitely many-valued.*

The value of $\ln z$ corresponding to the principal value $\operatorname{Arg} z$ (see Sec. 12.2) is denoted by $\operatorname{Ln} z$ and is called the **principal value** of $\ln z$. Thus

$$\text{(2)} \qquad \boxed{\operatorname{Ln} z = \ln |z| + i \operatorname{Arg} z} \qquad (z \neq 0).$$

The uniqueness of $\operatorname{Arg} z$ for given z $(\neq 0)$ implies that $\operatorname{Ln} z$ is single-valued, that is, a function in the usual sense. Since the other values of $\arg z$ differ by integer multiples of 2π, the other values of $\ln z$ are given by

$$\text{(3)} \qquad \boxed{\ln z = \operatorname{Ln} z \pm 2n\pi i} \qquad (n = 1, 2, \cdots).$$

They all have the same real part, and their imaginary parts differ by integer multiples of 2π.

If z is positive real, then Arg $z = 0$, and Ln z becomes identical with the real natural logarithm known from calculus. If z is negative real (so that the natural logarithm of calculus is not defined!), then Arg $z = \pi$ and

$$\operatorname{Ln} z = \ln |z| + \pi i.$$

EXAMPLE 1 **Natural logarithm. Principal value**

$$\ln 1 = 0, \pm 2\pi i, \pm 4\pi i, \cdots \qquad \operatorname{Ln} 1 = 0$$

$$\ln 4 = 1.386\,294 \pm 2n\pi i \qquad \operatorname{Ln} 4 = 1.386\,294$$

$$\ln(-1) = \pm\pi i, \pm 3\pi i, \pm 5\pi i, \cdots \qquad \operatorname{Ln}(-1) = \pi i$$

$$\ln(-4) = 1.386\,294 \pm (2n+1)\pi i \qquad \operatorname{Ln}(-4) = 1.386\,294 + \pi i$$

$$\ln i = \pi i/2, -3\pi i/2, 5\pi i/2, \cdots \qquad \operatorname{Ln} i = \pi i/2$$

$$\ln 4i = 1.386\,294 + \pi i/2 \pm 2n\pi i \qquad \operatorname{Ln} 4i = 1.386\,294 + \pi i/2$$

$$\ln(-4i) = 1.386\,294 - \pi i/2 \pm 2n\pi i \qquad \operatorname{Ln}(-4i) = 1.386\,294 - \pi i/2$$

$$\ln(3-4i) = \ln 5 + i \arg(3-4i) \qquad \operatorname{Ln}(3-4i) = 1.609\,438 - 0.927\,295i$$

$$= 1.609\,438 - 0.927\,295i \pm 2n\pi i$$

∎

The familiar relations for the natural logarithm continue to hold for complex values, that is,

$$\text{(4)} \quad \text{(a)} \quad \ln(z_1 z_2) = \ln z_1 + \ln z_2, \qquad \text{(b)} \quad \ln(z_1/z_2) = \ln z_1 - \ln z_2$$

but these relations are to be understood in the sense that each value of one side is also contained among the values of the other side.

EXAMPLE 2 **Illustration of the functional relations (4) in complex**

Let

$$z_1 = z_2 = e^{\pi i} = -1.$$

If we take

$$\ln z_1 = \ln z_2 = \pi i,$$

then (4a) holds provided we write $\ln(z_1 z_2) = \ln 1 = 2\pi i$; it is not true for the principal value, $\operatorname{Ln}(z_1 z_2) = \ln 1 = 0$. ∎

From (1) and $e^{\ln r} = r$ for positive real r we obtain

$$\text{(5a)} \qquad e^{\ln z} = z$$

as expected, but, since $\arg(e^z) = y \pm 2n\pi$ is multivalued,

$$\text{(5b)} \qquad \ln(e^z) = z \pm 2n\pi i.$$

For every fixed nonnegative integer n, formula (3) defines a function. We prove that each of these functions, in particular the principal value Ln z, is analytic except at $z = 0$ and except on the negative real axis (where the imaginary part of such a function is not even continuous but has a jump of magnitude 2π). We do this by proving that

$$(6) \qquad (\ln z)' = \frac{1}{z} \qquad [n \text{ in (3) fixed, } z \text{ not negative real or zero}].$$

In (1) we have $\ln z = u + iv$, where

$$u = \ln r = \ln |z| = \tfrac{1}{2} \ln (x^2 + y^2), \qquad v = \arg z = \arctan \frac{y}{x} + c,$$

where c is a constant (a multiple of $2n\pi$). We calculate the partial derivatives of u and v and find that they satisfy the Cauchy–Riemann equations:

$$u_x = \frac{x}{x^2 + y^2} = v_y = \frac{1}{1 + (y/x)^2} \cdot \frac{1}{x},$$

$$u_y = \frac{y}{x^2 + y^2} = -v_x = -\frac{1}{1 + (y/x)^2}\left(-\frac{y}{x^2}\right).$$

Formula (4) in Sec. 12.5 now yields the desired result:

$$(\ln z)' = u_x + iv_x = \frac{x}{x^2 + y^2} + i\frac{1}{1 + (y/x)^2}\left(-\frac{y}{x^2}\right) = \frac{x - iy}{x^2 + y^2} = \frac{1}{z}. \quad \blacksquare$$

General Powers

General powers of a complex number $z = x + iy$ are defined by the formula

$$(7) \qquad z^c = e^{c \ln z} \qquad (c \text{ complex, } z \neq 0).$$

Since $\ln z$ is infinitely many-valued, z^c will, in general, be multivalued. The particular value

$$z^c = e^{c \operatorname{Ln} z}$$

is called the **principal value** *of* z^c.

If $c = n = 1, 2, \cdots$, then z^n is single-valued and identical with the usual nth power of z. If $c = -1, -2, \cdots$, the situation is similar.

If $c = 1/n$ where $n = 2, 3, \cdots$, then

$$z^c = \sqrt[n]{z} = e^{(1/n) \ln z} \qquad (z \neq 0),$$

the exponent is determined up to multiples of $2\pi i/n$ and we obtain the n distinct values of the nth root, in agreement with the result in Sec. 12.2. If $c = p/q$, the quotient of two positive integers, the situation if similar, and z^c has only finitely many distinct values. However, if c is real irrational or genuinely complex, then z^c is infinitely many-valued.

EXAMPLE 3 General power

$$i^i = e^{i \ln i} = \exp(i \ln i) = \exp\left[i\left(\frac{\pi}{2} i \pm 2n\pi i\right)\right] = e^{-(\pi/2) \mp 2n\pi}.$$

All these values are real, and the principal value ($n = 0$) is $e^{-\pi/2}$. Similarly, by direct calculation and multiplying out in the exponent,

$$(1 + i)^{2-i} = \exp[(2 - i) \ln(1 + i)] = \exp[(2 - i)\{\ln \sqrt{2} + \tfrac{1}{4}\pi i \pm 2n\pi i\}]$$
$$= 2e^{\pi/4 \pm 2n\pi}[\sin(\tfrac{1}{2} \ln 2) + i \cos(\tfrac{1}{2} \ln 2)].$$

It is a *convention* that for real positive $z = x$ the expression z^c means $e^{c \ln x}$ where $\ln x$ is the elementary real natural logarithm (that is, the principal value $\text{Ln } z$ ($z = x > 0$) in the sense of our definition). Also, if $z = e$, the base of the natural logarithm, $z^c = e^c$ is *conventionally* regarded as the unique value obtained from (1) in Sec. 12.6.

From (7) we see that for any complex number a,

$$a^z = e^{z \ln a}. \tag{8}$$

We have now introduced the complex functions needed in practical work, some of them (e^z, $\cos z$, $\sin z$, $\cosh z$, $\sinh z$) entire (Sec. 12.6), some of them ($\tan z$, $\cot z$, $\tanh z$, $\coth z$) analytic except at certain points, and one of them ($\ln z$) splitting up into infinitely many functions, each analytic except at 0 and on the negative real axis.

Problem Set 12.8

1. Verify the computations in Example 1. Plot some of the values of $\ln(3 - 4i)$ in the complex plane.
2. Verify (4) for $z_1 = -i$ and $z_2 = -1$.
3. Prove the analyticity of $\text{Ln } z$ by using the Cauchy–Riemann equations in polar form (see Sec. 12.5).
4. Prove (5a) and (5b).

Compute the principal value $\text{Ln } z$ if z equals

5. $1 + i$ **6.** -4 **7.** $-3 - 4i$ **8.** $2.5 + 3.8i$
9. -100 **10.** $-16.0 - 0.1i$ **11.** $-16.0 + 0.1i$ **12.** $0.6 + 0.8i$

Find all values of the given expressions and plot some of them in the complex plane.

13. $\ln 1$ **14.** $\ln e$ **15.** $\ln(-7)$ **16.** $\ln(-0.01i)$
17. $\ln(0.8 - 0.6i)$ **18.** $\ln(e^{2i})$ **19.** $\ln(-e^{-i})$ **20.** $\ln(1.3 + 2.8i)$

Solve the following equations for z.

21. $\ln z = 3 - i$ **22.** $\ln z = -2 - \frac{3}{2}i$ **23.** $\ln z = 2 + \frac{1}{4}\pi i$
24. $\ln z = (2 - \frac{1}{2}i)\pi$ **25.** $\ln z = 0.3 + 0.7i$ **26.** $\ln z = \sqrt{2} + \pi i$
27. $\ln z = 4 - 3i$ **28.** $\ln z = -5 + 0.01i$

Find the principal value of

29. $i^{1/2}$ **30.** $(1 + i)^i$ **31.** 3^{3-i} **32.** 2^{2i}

33. $(1 - i)^{1+i}$ **34.** $(1 + i)^{1-i}$ **35.** $(5 - 2i)^{3+\pi i}$ **36.** $(-5)^{2-4i}$

37. $(2 - i)^{1+i}$ **38.** $(3 + 4i)^{1/3}$

By definition, the **inverse sine** $w = \sin^{-1} z$ is the relation such that $\sin w = z$. The **inverse cosine** $w = \cos^{-1} z$ is the relation such that $\cos w = z$. The **inverse tangent, inverse cotangent, inverse hyperbolic sine,** etc., are defined and denoted in a similar fashion. (Note that all these relations are ***multivalued.***) Using $\sin w = (e^{iw} - e^{-iw})/2i$ and similar representations of $\cos w$, etc., show that

39. $\sin^{-1} z = -i \ln (iz + \sqrt{1 - z^2})$ **40.** $\cos^{-1} z = -i \ln (z + \sqrt{z^2 - 1})$

41. $\cosh^{-1} z = \ln (z + \sqrt{z^2 - 1})$ **42.** $\sinh^{-1} z = \ln (z + \sqrt{z^2 + 1})$

43. $\tan^{-1} z = \dfrac{i}{2} \ln \dfrac{i + z}{i - z}$ **44.** $\tanh^{-1} z = \dfrac{1}{2} \ln \dfrac{1 + z}{1 - z}$

45. Show that $w = \sin^{-1} z$ is infinitely many-valued, and if w_1 is one of these values, the others are of the form $w_1 \pm 2n\pi$ and $\pi - w_1 \pm 2n\pi$, $n = 0, 1, \cdots$. (The *principal value* of $w = u + iv = \sin^{-1} z$ is defined to be the value for which $-\pi/2 \leqq u \leqq \pi/2$ if $v \geqq 0$ and $-\pi/2 < u < \pi/2$ if $v < 0$.)

12.9 Mapping by Special Functions *Optional*

In this optional section we discuss elementary functions geometrically as mappings. This will supplement the student's understanding of these functions. Mapping properties of analytic functions in general will be discussed in Chap. 16, along with applications in Chap. 17.

From calculus we know that we can plot a *real* function $y = f(x)$ in the xy-plane. For a *complex* function

$$w = u + iv = f(z) \qquad (z = x + iy)$$

we need two planes, the **z-plane** in which we plot values of z, and the **w-plane** in which we plot the corresponding function values $w = f(z)$. In this way, a given function f assigns to each point z in its domain of definition D the corresponding point $w = f(z)$ in the w-plane. We say that f defines a **mapping** of D into the w-plane. For any point z_0 in D we call the point $w_0 = f(z_0)$ the **image** of z_0 with respect to f. More generally, for the points of a curve C in D, the image points form the **image** of C; similarly for other point sets in D.

The next idea is this. By plotting single points and their images we would not gain much insight into the mapping properties of a given function. For this purpose it is much better if we graph and study regions and their images, or families of simple curves (for instance, parallel straight lines or concentric circles) and their images. We shall explain this method for the elementary functions in terms of selected examples of practical importance.

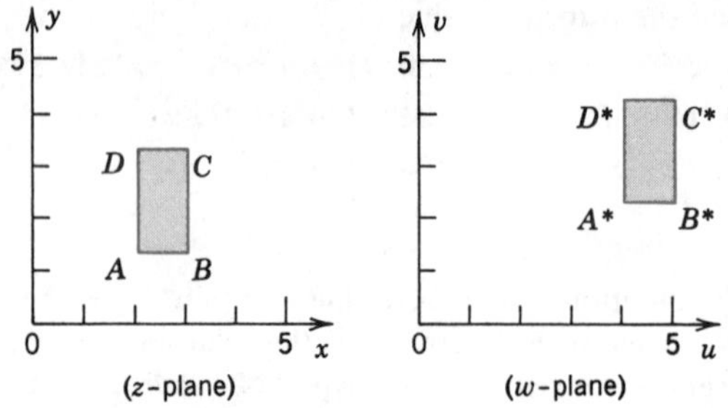

Fig. 301. Translation $w = z + 2 + i$

Linear Functions. Positive Integer Powers of z

EXAMPLE 1 **Mapping $w = az + b$**

Describe, in geometrical terms, the mapping properties of the linear function $w = az + b$.

Solution. If $a = 1$, this is a **translation** $w = z + b$. It moves (translates) every point in the direction of b through the same distance $|b|$. Hence every figure and its image are congruent. Figure 301 shows an example. $b = 0$ gives the **identity mapping** $w = z$, which leaves every point fixed.

If $b = 0$, we get $w = az$. For $|a| = 1$ we have $a = e^{i\alpha}$ and $w = e^{i\alpha}z$, which is a rotation through the angle α. Figure 302 shows an example. For real a with $0 < a < 1$ we get a uniform contraction and for real $a > 1$ a uniform expansion. (Explain! What about negative real a?)

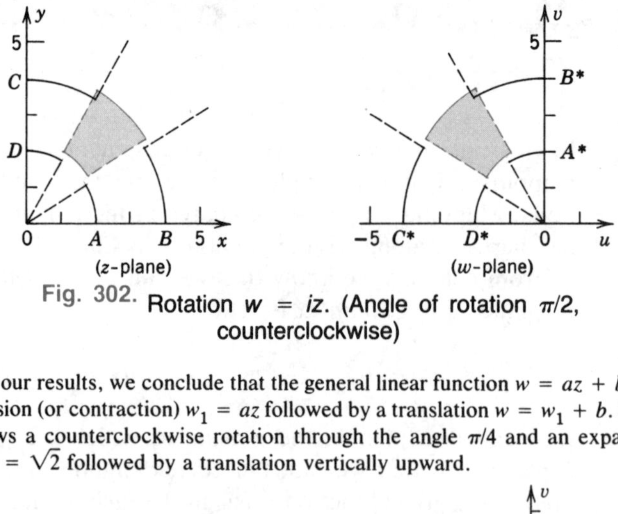

Fig. 302. Rotation $w = iz$. (Angle of rotation $\pi/2$, counterclockwise)

Combining our results, we conclude that the general linear function $w = az + b$ is a rotation and an expansion (or contraction) $w_1 = az$ followed by a translation $w = w_1 + b$. For instance, Fig. 303 shows a counterclockwise rotation through the angle $\pi/4$ and an expansion by the factor $|1 + i| = \sqrt{2}$ followed by a translation vertically upward.

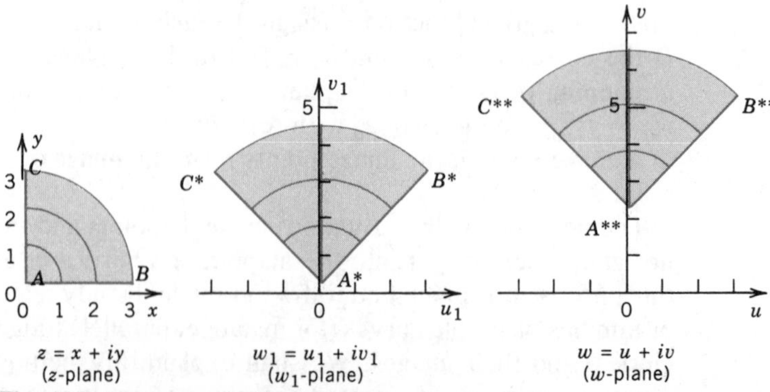

Fig. 303. Linear transformation $w = u + iv = (1 + i)z + 2i$, consisting of the rotation and expansion $w_1 = u_1 + iv_1 = (1 + i)z$ followed by the translation $w = w_1 + 2i$

EXAMPLE 2 Mapping $w = z^2$

Find the images of the circles $|z| = r_0 = const$ and of the rays $\theta = \theta_0 = const$ under the mapping $w = z^2$. Also find the image of the region $1 \leqq |z| \leqq 3/2$, $\pi/6 < \theta < \pi/3$.

Solution. In terms of polar coordinates

$$w = Re^{i\phi} = z^2 = r^2 e^{2i\theta} \qquad (z = re^{i\theta}).$$

This shows that any circle $|z| = r_0$ maps onto the circle $|w| = r_0^2$, and any ray $\theta = \theta_0$ onto the ray $\phi = 2\theta_0$. Hence that region is mapped onto the region $1 \leqq |w| \leqq 9/4$, $\pi/3 < \phi < 2\pi/3$ in the w-plane (see Fig. 304).

The circles and rays considered intersect at right angles, and the mapping $w = z^2$ preserves these angles: the corresponding images also intersect at right angles. This is typical of analytic functions f: they preserve angles between any intersecting curves whatsoever, as we shall see in Chap. 16. An exception occurs at points at which $f'(z) = 0$. For $w = z^2$ we have $w' = 2z = 0$ at $z = 0$; indeed, we see that images of rays do not make the same angle as the rays themselves, but make *twice* this angle. ∎

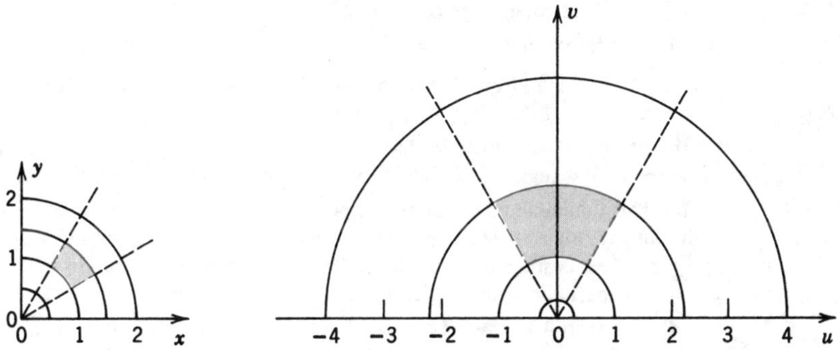

Fig. 304. Mapping $w = z^2$

Example 2 is a model case for all positive integer powers z^n. In terms of polar coordinates,

$$w = Re^{i\phi} = z^n = r^n e^{in\theta}.$$

This shows that $w = z^n$ maps circles $|z| = r_0$ onto circles $|w| = r_0^n$, and rays $\theta = \theta_0$ onto rays $\phi = n\theta_0$. For instance, $w = z^3$ triples angles at 0. Hence $w = z^n$ maps the angular region $0 \leqq \theta \leqq \pi/n$ onto the upper half of the w-plane (Fig. 305).

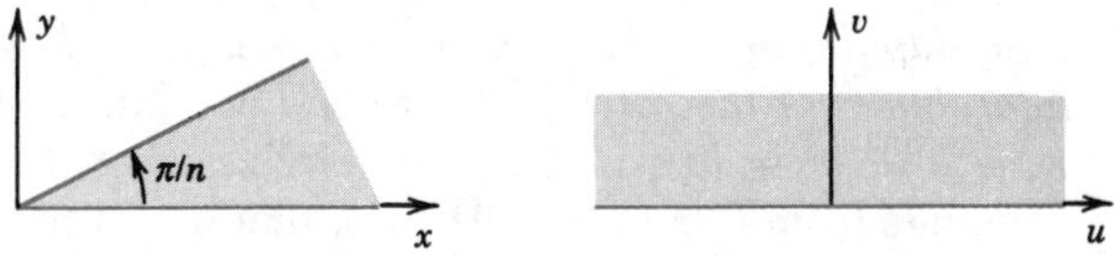

Fig. 305. Mapping defined by $w = z^n$

Exponential Function and Logarithm

As in the previous examples, we shall exhibit various properties of the mappings by considering curves and regions for which we can obtain the corresponding images without computations but simply by using properties of the functions under consideration. This is the general idea underlying our discussion in the whole section.

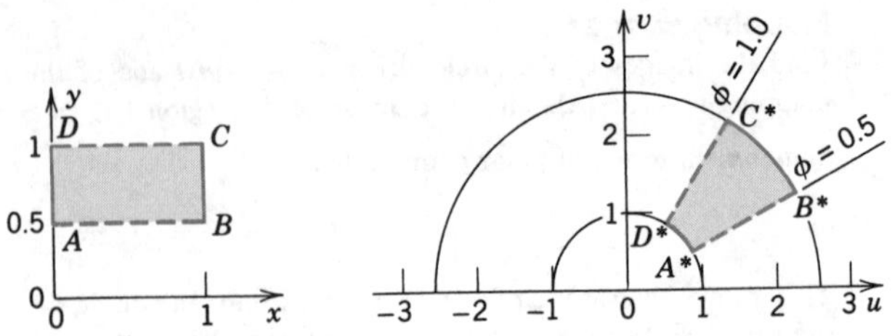

Fig. 306. Mapping by $w = e^z$ [Example 3, part (b)]

EXAMPLE 3 **Exponential function $w = e^z$**

For $w = e^z$, find the image of:

(a) the straight lines $x = x_0 = const$ and $y = y_0 = const$,

(b) the rectangle $0 \leqq x \leqq 1, \frac{1}{2} < y < 1$,

(c) the fundamental region $-\pi < y \leqq \pi$,

(d) the horizontal strip $0 \leqq y \leqq \pi$.

Solution. **(a)** Recall from Sec. 12.6 that $|w| = e^x$ and $\arg w = y$. Hence $x = x_0$ is mapped onto the circle $|w| = e^{x_0}$, and $y = y_0$ onto the ray $\arg w = y_0$.

(b) From (a) we conclude that any rectangle with sides parallel to the coordinate axes is mapped onto a region bounded by portions of rays and circles. See Fig. 306.

(c) This fundamental region is mapped onto the full w-plane (cut along the negative real axis) without the origin 0. More generally, every horizontal strip bounded by two lines $y = c$ and $y = c + 2\pi$ is mapped onto the full w-plane without the origin. This reflects the fact that $w = e^z$ is periodic with the period $2\pi i$.

(d) The strip $0 \leqq y \leqq \pi$ is mapped onto the upper half of the w-plane. Its right half ($x > 0$) is mapped onto the exterior of the unit circle $|w| = 1$ and its left half ($x < 0$) onto the interior without 0 (Fig. 307). ■

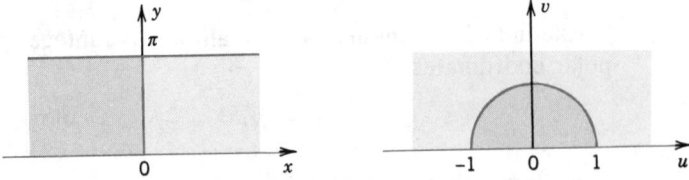

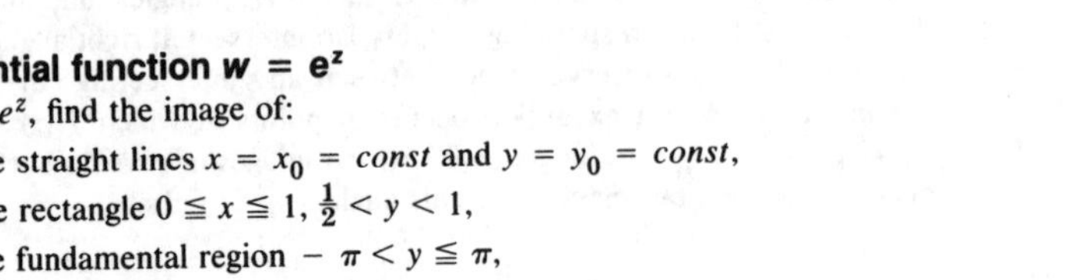

Fig. 307. Mapping by $w = e^z$ [Example 3, part (d)]

We now turn to the mapping given by the **natural logarithm.** In discussing it we use the following basic idea and principle. *The properties of the inverse of a mapping $w = f(z)$ can be obtained from those of the mapping itself by interchanging the roles of the z- and w-planes.* Since the natural logarithm $w = u + iv = \ln z$ is the inverse relation of the exponential function, we conclude from Example 3, part (c), that $w = \text{Ln } z, z \neq 0$ (the principal value of $\ln z$) maps the z-plane (cut along the negative real axis and with $z = 0$ omitted) onto the horizontal strip $-\pi < v \leqq \pi$ of the w-plane. Since the mapping

$$w = \text{Ln } z + 2\pi i$$

differs from $w = \text{Ln } z$ by the translation $2\pi i$ (vertically upward), this mapping maps the z-plane (cut as before and 0 omitted) onto the strip $\pi < v \leqq 3\pi$. Similarly for each of the infinitely many mappings

$$w = \ln z = \text{Ln } z \pm 2n\pi i \qquad (n = 0, 1, 2, \cdots).$$

The corresponding horizontal strips of width 2π (images of the z-plane under these mappings) together cover the whole w-plane without overlapping. We invite the reader to interpret the other parts of Example 3 in terms of the mapping $w = \text{Ln } z$, by interchanging the roles of z and w in the example.

Mapping properties of further functions (trigonometric, hyperbolic, fractional linear functions, etc.) will be discussed in Chap. 16.

Problem Set 12.9

Consider the mapping $w = (1 - i)z + 2$. Find and plot the images of the given curves or regions.

1. $x = 0, 1, 2, 3$ **2.** $y = -2, 0, 2, 4$ **3.** $|z + 1| \leqq 1$

Sketch the images of the following regions under the mapping $w = z^2$.

4. $|z| \leqq 3$ **5.** $1.5 \leqq |z| < 2.1$ **6.** $|\arg z| \leqq \pi/8$

7. $|z| > 4$ **8.** $\pi/2 \leqq \arg z \leqq \pi$ **9.** $0 < |z| \leqq 1/2$

10. **(Level curves)** Consider $w = u + iv = z^2$. Show that the *level curves* of u, that is, the curves $u = const$ in the z-plane, are hyperbolas with the asymptotes $y = \pm x$. Sketch some of them.

11. Find the level curves $v = const$ in Prob. 10 and sketch some of them.

12. Show that $w = u + iv = z^2$ maps $x = c$ onto the parabola $v^2 = 4c^2(c^2 - u)$ and graph it for $c = 2, 1, \frac{1}{2}, 0$.

13. Find the image of $y = k$ under the mapping $w = u + iv = z^2$.

Find and sketch the images of the following curves and regions under the mapping $w = 1/z$, where $z \neq 0$.

14. $|z| < 1$ **15.** $|z| = 1/2$ **16.** $y > 0$

Find and sketch the images of the given regions or curves under the mapping $w = e^z$.

17. $x \leqq 1, \quad |y| \leqq \pi$ **18.** $x = 0$ **19.** $x > 1/2, \quad 0 < y < \pi$

20. $|x| < a, \quad |y| < b$

Review Questions and Problems for Chapter 12

1. The definition of the multiplication of complex numbers looks strange at first sight. What is the motivation for it?
2. What does multiplication of a complex number by i mean geometrically?
3. Is a complex number $|z|e^{i\theta}$ uniquely determined if $|z|$ and arc tan (y/x) are given?
4. What do we mean by the principal value of the argument of a complex number?
5. What is the triangle inequality? What does it mean geometrically?
6. What is De Moivre's formula and what is it good for?

7. State from memory what it means that $f(z)$ is analytic in a domain D. That $f(z)$ is analytic at a point. What is a domain?

8. Are $|z|$, Re z, Im z analytic? Give a reason.

9. How would you test for analyticity?

10. The formula for the derivative looks the same as in calculus, but there is a big difference. Explain.

11. Can a function be differentiable at a point without being analytic there?

12. What is an entire function? Give examples and counterexamples.

13. How is e^z defined? What is a fundamental strip?

14. What properties of e^z are similar to those of e^x? Which one is different?

15. How are $\cos z$ and $\sin z$ related to e^z?

16. Does the equation $\sin z = -100$ have solutions?

17. In what respect is the complex logarithm much more complicated than the real logarithm?

18. How is Ln z defined?

19. What can you remember about $\sqrt[4]{1}$, $\sqrt[8]{1}$, $\sqrt[n]{1}$?

20. If you know the values of $\sqrt[8]{1}$, how do you get from them the values of $\sqrt[8]{z}$ for any z?

Compute, in the form $x + iy$,

21. $(-2 + 6i)^2$ **22.** $(1 + i)^4$ **23.** $(6 + i)/(2 - 3i)$ **24.** $25/(3 + 4i)^2$

25. $\dfrac{52.5 - 12.5i}{3 - i}$ **26.** $\dfrac{202}{10 - i}$ **27.** $\dfrac{(1 + i)^{10}}{(1 - i)^6}$ **28.** $\dfrac{23 + 2i}{2 - 2.5i}$

Compute

29. $\mathrm{Re}\,\dfrac{1}{2 + i}$ **30.** $\mathrm{Im}\,\dfrac{2 - i}{4 - 3i}$ **31.** $\mathrm{Re}\,\dfrac{16 + 2i}{3 + 2i}$ **32.** $\mathrm{Re}\,\dfrac{-4 - 6i}{(1 + i)^6}$

33. $\left|\dfrac{(3 + 4i)^4}{(3 - 4i)^3}\right|$ **34.** $\left|\dfrac{iz}{\bar{z}}\right|$ **35.** $\left|\dfrac{13 - 17i}{17 + 13i}\right|$ **36.** $\dfrac{4}{|7 - \pi i|}$

Find and plot all values of the roots

37. $\sqrt{-i}$ **38.** $\sqrt{3 + 4i}$ **39.** $\sqrt{-8 - 6i}$ **40.** $\sqrt[6]{-1}$

Represent in polar form

41. $-6 + 6i$ **42.** $17 + 4i$ **43.** $\sqrt{10}/(3 + i)$ **44.** $2.6 + 1.7i$

Sketch the following curves or regions in the complex plane.

45. $|z + 1 - 3i| = \frac{1}{2}$ **46.** $2 < |z + i| < 3$

47. $\mathrm{Re}\, z \geqq \mathrm{Im}\, z$ **48.** $\mathrm{Re}\,(z^2) \leqq 1$

Find an analytic function $f(z) = u(x, y) + iv(x, y)$ such that

49. $v = y/(x^2 + y^2)$ **50.** $u = e^{x^2-y^2} \cos 2xy$

51. $u = \cos x \cosh y$ **52.** $u = x^2 - 2xy - y^2$

Are the following functions harmonic? If so, find a conjugate harmonic.

53. x^2y^2 **54.** $e^{x/2} \cos \frac{1}{2}y$ **55.** $x^3 - 3xy^2$ **56.** $\sin x \cosh y$

Find the value of the derivative of

57. $2/z^4$ at $1 + i$ **58.** $e^{1/z}$ at i

59. $\sin z$ at $3 + 2i$ **60.** $\cosh z$ at $4 - 3i$

Summary of Chapter 12 Complex Numbers. Complex Analytic Functions

For arithmetic operations with **complex numbers**

$$z = x + iy = re^{i\theta} = r(\cos \theta + i \sin \theta), \tag{1}$$

$r = |z| = \sqrt{x^2 + y^2}$, $\theta = \arctan(y/x)$, and for their representation in the complex plane, see Secs. 12.1 and 12.2.

A complex function $f(z) = u(x, y) + iv(x, y)$ is **analytic** in a domain D if it has a **derivative** (Sec. 12.4)

$$f'(z) = \lim_{\Delta z \to 0} \frac{f(z + \Delta z) - f(z)}{\Delta z} \tag{2}$$

everywhere in D. Also, $f(z)$ is *analytic at a point* $z = z_0$ if it has a derivative in a neighborhood of z_0 (not merely at z_0 itself).

If $f(z)$ is analytic in D, then $u(x, y)$ and $v(x, y)$ satisfy the (very important!) **Cauchy–Riemann equations** (Sec. 12.5)

$$\frac{\partial u}{\partial x} = \frac{\partial v}{\partial y}, \qquad \frac{\partial u}{\partial y} = -\frac{\partial v}{\partial x} \tag{3}$$

everywhere in D. Then u and v also satisfy **Laplace's equation**

$$u_{xx} + u_{yy} = 0, \qquad v_{xx} + v_{yy} = 0 \tag{4}$$

everywhere in D. If $u(x, y)$ and $v(x, y)$ are continuous and have *continuous* partial derivatives in D that satisfy (3) in D, then $f(z) = u(x, y) + iv(x, y)$ is analytic in D. See Sec. 12.5. (More on Laplace's equation and complex analysis follows in Chap. 17.)

The complex **exponential function** (Sec. 12.6)

$$e^z = \exp z = e^x(\cos y + i \sin y) \tag{5}$$

is periodic with $2\pi i$, reduces to e^x if $z = x$ $(y = 0)$, and has the derivative e^z.

The **trigonometric functions** are (Sec. 12.7)

$$\begin{aligned} \cos z &= \frac{1}{2}(e^{iz} + e^{-iz}) = \cos x \cosh y - i \sin x \sinh y \\ \sin z &= \frac{1}{2i}(e^{iz} - e^{-iz}) = \sin x \cosh y + i \cos x \sinh y \end{aligned} \tag{6}$$

$\tan z = (\sin z)/\cos z$, $\cot z = 1/\tan z$, etc.

The **hyperbolic functions** are (Sec. 12.7)

$$\begin{aligned} \cosh z &= \frac{1}{2}(e^z + e^{-z}) = \cos iz, \\ \sinh z &= \frac{1}{2}(e^z - e^{-z}) = -i \sin iz, \end{aligned} \tag{7}$$

etc. An **entire function** is a function that is analytic everywhere in the complex plane. The functions in (5)–(7) are entire.

The **natural logarithm** is (Sec. 12.8)

$$\begin{aligned} \ln z &= \ln |z| + i \arg z & (\arg z = \theta, z \neq 0) \\ &= \ln |z| + i \operatorname{Arg} z \pm 2n\pi i & (n = 0, 1, \cdots), \end{aligned} \tag{8}$$

where $\operatorname{Arg} z$ is the **principal value** of $\arg z$, that is, $-\pi < \operatorname{Arg} z \leqq \pi$. We see that $\ln z$ is infinitely many-valued. Taking $n = 0$ gives the **principal value** $\operatorname{Ln} z$ of $\ln z$; thus

$$\operatorname{Ln} z = \ln |z| + i \operatorname{Arg} z. \tag{8*}$$

General powers are defined by (Sec. 12.8)

$$z^c = e^{c \ln z} \qquad (c \text{ complex}, z \neq 0). \tag{9}$$

A complex function $w = f(z)$ defines a mapping of its domain of definition in the z-plane into the w-plane. The mappings given by the functions (5), (8) and some others are discussed in Sec. 12.9. (More on mapping by analytic functions follows in Chap. 16.)

Chapter 13

Complex Integration

Integration in the complex plane is important for two reasons:

1. In applications there occur real integrals that can be evaluated by complex integration, whereas the usual methods of real integral calculus are not successful.

2. Some basic properties of analytic functions can be established by integration, but would be difficult to prove by other methods. The existence of higher derivatives of analytic functions is a striking property of this type.[1]

In this chapter we define and explain complex integrals. The most fundamental result in the whole chapter is Cauchy's integral theorem (Sec. 13.3). It implies the important Cauchy integral formula (Sec. 13.5). In Sec. 13.6 we prove that if a function is analytic, it has derivatives of all orders. Hence in this respect, complex analytic functions behave much more simply than real-valued functions of real variables.

(Integration by means of residues and applications to real integrals will be considered in Chap. 15.)

Prerequisite for this chapter: Chap. 12.
References: Appendix 1, Part D.
Answers to problems: Appendix 2.

13.1 Line Integral in the Complex Plane

As in real calculus, we distinguish between definite integrals and indefinite integrals or antiderivatives. An **indefinite integral** is a function whose derivative equals a given analytic function in a region. By inverting known differentiation formulas we may find many types of indefinite integrals.

The definition of definite integrals, or line integrals, of complex functions $f(z)$ of $z = x + iy$ needs a short preparation on curves in the complex plane, as follows.

Path of integration. In the case of a *real* definite integral in calculus we integrate over an interval (a segment) on the real line. In the case of a *complex* definite integral—a line integral—we integrate along a curve C in the complex

[1]Proved without integration or equivalent methods only relatively recently, in 1961 [by P. Porcelli and E. H. Connell (*Bulletin of the American Mathematical Society*, vol. 67, pp. 177–181), who make use of a topological theorem by G. T. Whyburn].

plane,[2] called the **path of integration.** Such a curve can be represented in the form

$$\text{(1)} \qquad z(t) = x(t) + iy(t) \qquad (a \leqq t \leqq b)$$

where t is a real parameter. For example,

$$z(t) = t + 3it \qquad (0 \leqq t \leqq 2)$$

represents a portion of the line $y = 3x$ (sketch it!),

$$z(t) = 4 \cos t + 4i \sin t \qquad (-\pi \leqq t \leqq \pi)$$

gives the circle $|z| = 4$, and so on. (More examples below.)

C is called a **smooth curve** if it has a continuous and nonzero derivative

$$\dot{z}(t) = \frac{dz}{dt} = \dot{x}(t) + i\dot{y}(t)$$

at each point. Geometrically this means that C has everywhere a continuously turning tangent, as follows directly from the definition

$$\dot{z}(t) = \lim_{\Delta t \to 0} \frac{z(t + \Delta t) - z(t)}{\Delta t} \qquad \text{(Fig. 308).}$$

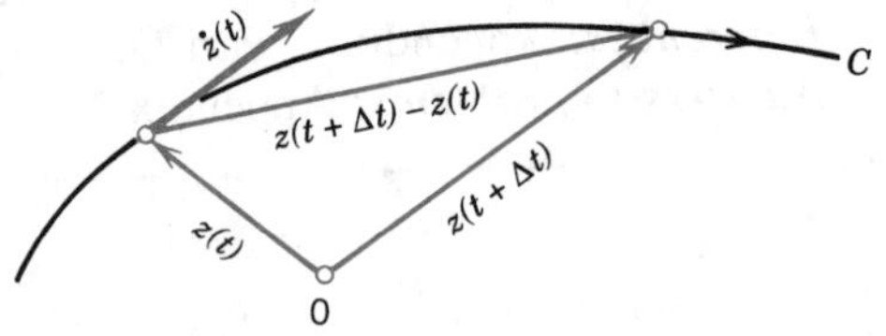

Fig. 308. Tangent vector $\dot{z}(t)$ of a curve C in the complex plane given by $z(t)$. The arrow on the curve indicates the positive sense (sense of increasing t).

Definition of the Complex Line Integral

This is similar to the method in calculus. We consider a smooth curve C in the complex plane given in the form (1) and a continuous function $f(z)$ defined (at least) at each point of C. We now subdivide ("partition") the interval $a \leqq t \leqq b$ in (1) by points

$$t_0 (= a), \quad t_1, \quad \cdots, \quad t_{n-1}, \quad t_n (= b)$$

where $t_0 < t_1 \cdots < t_n$. To this subdivision there corresponds a subdivision

[2] Actually, often only along a portion (an *arc*) of a curve, but we shall say **"curve"** in either case, for simplicity. Readers who have studied Sec. 9.1 will note that our present discussion is similar.

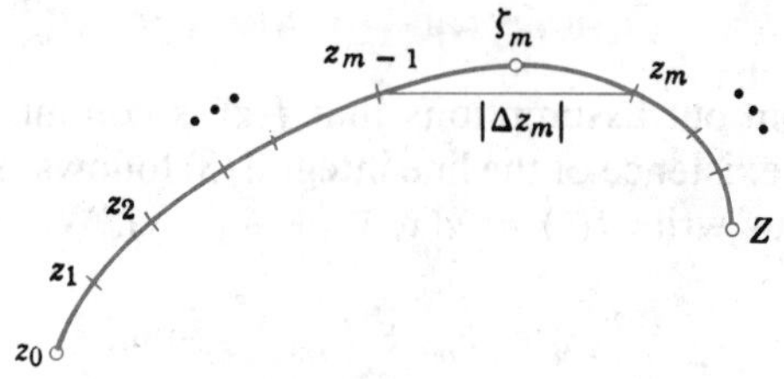

Fig. 309. Complex line integral

of C by points

$$z_0, \quad z_1, \quad \cdots, \quad z_{n-1}, \quad z_n \; (= Z) \qquad \text{(Fig. 309)},$$

where $z_j = z(t_j)$. On each portion of subdivision of C we choose an arbitrary point, say, a point ζ_1 between z_0 and z_1 (that is, $\zeta_1 = z(t)$ where t satisfies $t_0 \leqq t \leqq t_1$), a point ζ_2 between z_1 and z_2, etc. Then we form the sum

$$S_n = \sum_{m=1}^{n} f(\zeta_m)\, \Delta z_m \tag{2}$$

where

$$\Delta z_m = z_m - z_{m-1}.$$

This we do for each $n = 2, 3, \cdots$ in a completely independent manner, but so that the greatest $|\Delta t_m| = |t_m - t_{m-1}|$ approaches zero as $n \to \infty$; this implies that the greatest $|\Delta z_m|$ also approaches zero because it cannot exceed the length of the arc of C from z_{m-1} to z_m and the latter goes to zero since the arc length of the smooth curve C is a continuous function of t. The limit of the sequence of complex numbers $S_2, S_3, \cdots$ thus obtained is called the **line integral** (or simply the *integral*) of $f(z)$ over the oriented curve C (called the **path of integration**) and is denoted by

$$\int_C f(z)\, dz. \tag{3}$$

If C is a **closed path** (one whose terminal point Z coincides with its initial point z_0, as for a circle or an 8-shaped curve), we also use the (standard) notation

$$\oint_C f(z)\, dz.$$

Examples follow in the next section.

General Assumption. *All paths of integration for complex line integrals are* **piecewise smooth,** *that is, they consist of finitely many smooth curves joined end to end.*

Existence of the Complex Line Integral[3]

From our assumptions that $f(z)$ is continuous and C is piecewise smooth, the existence of the line integral (3) follows. In fact, as in the previous chapter let us write $f(z) = u(x, y) + iv(x, y)$. We also set

$$\xi_m = \xi_m + i\eta_m \qquad \text{and} \qquad \Delta z_m = \Delta x_m + i\Delta y_m.$$

Then (2) may be written

$$S_n = \sum (u + iv)(\Delta x_m + i\Delta y_m) \tag{4}$$

where $u = u(\xi_m, \eta_m)$, $v = v(\xi_m, \eta_m)$ and we sum over m from 1 to n. We may now split up S_n into four sums:

$$S_n = \sum u\,\Delta x_m - \sum v\,\Delta y_m + i\left[\sum u\,\Delta y_m + \sum v\,\Delta x_m\right].$$

These sums are real. Since f is continuous, u and v are continuous. Hence, if we let n approach infinity in the aforementioned way, then the greatest Δx_m and Δy_m will approach zero and each sum on the right becomes a real line integral:

$$\lim_{n\to\infty} S_n = \int_C f(z)\,dz = \int_C u\,dx - \int_C v\,dy + i\left[\int_C u\,dy + \int_C v\,dx\right]. \tag{5}$$

This shows that under our assumptions (f continuous on C and C piecewise smooth) the line integral (3) exists and its value is independent of the choice of subdivisions and intermediate points ζ_m. ∎

Three Basic Properties of Complex Line Integrals

We list three properties of complex line integrals that are quite similar to those of real definite integrals (and real line integrals) and follow immediately from the definition.

1. Integration is a **linear operation,** that is, a sum of two (or more) functions can be integrated term by term, and constant factors can be taken out from under the integral sign:

$$\int_C [k_1 f_1(z) + k_2 f_2(z)]\,dz = k_1 \int_C f_1(z)\,dz + k_2 \int_C f_2(z)\,dz. \tag{6}$$

2. Decomposing C into two portions C_1 and C_2 (Fig. 310), we get

$$\int_C f(z)\,dz = \int_{C_1} f(z)\,dz + \int_{C_2} f(z)\,dz. \tag{7}$$

[3]If one does not want to refer to real line integrals, one can take (1) in the next section as the ***definition*** of the complex line integral.

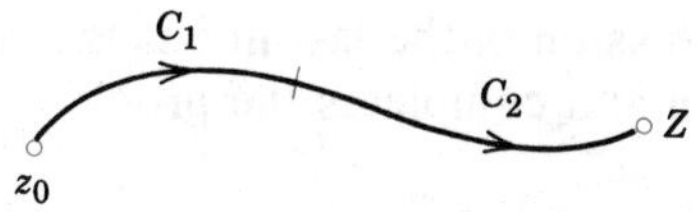

Fig. 310. Subdivision of path [formula (7)]

3. Reversing the sense of integration, we get the negative of the original value:

$$\int_{z_0}^{Z} f(z)\,dz = -\int_{Z}^{z_0} f(z)\,dz; \tag{8}$$

here ***the path C*** with endpoints z_0 and Z ***is the same;*** on the left we integrate from z_0 to Z, and on the right from Z to z_0.

Integration methods and worked examples of integrals follow in the next section and problems at the end of it.

13.2 Two Integration Methods. Examples

Complex integration is rich in methods for evaluating integrals. We discuss the first two of them, and others follow later in this chapter and in Chap. 15.

First Method: Use of a Representation of the Path

This method applies to any continuous complex function.

Theorem 1 **(Integration by the use of the path)**

Let C be a piecewise smooth path, represented by $z = z(t)$, where $a \leqq t \leqq b$. Let $f(z)$ be a continuous function on C. Then

$$\int_C f(z)\,dz = \int_a^b f[z(t)]\dot{z}(t)\,dt \qquad \left(\dot{z} = \frac{dz}{dt}\right). \tag{1}$$

Proof. The left side of (1) is given by (5), Sec. 13.1, in terms of real line integrals, and we show that the right side of (1) also equals (5). We have $z = x + iy$, hence $\dot{z} = \dot{x} + i\dot{y}$. We simply write u for $u[x(t), y(t)]$ and v for $v[x(t), y(t)]$. We also have $dx = \dot{x}\,dt$ *and* $dy = \dot{y}\,dt$. Consequently, in (1),

$$\int_a^b f[z(t)]\dot{z}(t)\,dt = \int_a^b (u + iv)(\dot{x} + i\dot{y})\,dt$$

$$= \int_C [u\,dx - v\,dy + i(u\,dy + v\,dx)].$$

The expression in the last line is the right side of (5) in Sec. 13.1. This was our claim and completes the proof. ∎

Steps in applying Theorem 1

(A) Represent the path C in the form $z(t)$ $(a \leqq t \leqq b)$.

(B) Calculate the derivative $\dot{z}(t) = dz/dt$.

(C) Substitute $z(t)$ for every z in $f(z)$ (hence $x(t)$ for x and $y(t)$ for y).

(D) Integrate $f[z(t)]\dot{z}(t)$ over t from a to b.

EXAMPLE 1 **A basic result: Integral of 1/z around the unit circle**

Show that

$$\oint_C \frac{dz}{z} = 2\pi i \qquad \text{(C the unit circle, counterclockwise).} \tag{2}$$

This is a very important result that we shall need quite often.

Solution. We may represent the unit circle C (see Sec. 12.3) in the form

$$z(t) = \cos t + i \sin t \qquad (0 \leqq t \leqq 2\pi),$$

so that the counterclockwise integration corresponds to an increase of t from 0 to 2π. By differentiation,

$$\dot{z}(t) = -\sin t + i \cos t.$$

Also $f[z(t)] = 1/z(t)$. Formula (1) now yields the desired result

$$\oint_C \frac{dz}{z} = \int_0^{2\pi} \frac{1}{\cos t + i \sin t}(-\sin t + i \cos t)\, dt = i \int_0^{2\pi} dt = 2\pi i.$$

The Euler formula (Sec. 12.6) helps to save work by representing the unit circle simply by

$$z(t) = e^{it}. \qquad \text{Then} \qquad 1/z(t) = e^{-it}, \qquad dz = ie^{it}\, dt.$$

As before, we now get more quickly

$$\oint_C \frac{dz}{z} = \int_0^{2\pi} e^{-it} ie^{it}\, dt = i \int_0^{2\pi} dt = 2\pi i.$$

∎

EXAMPLE 2 **Integral of integer powers**

Let $f(z) = (z - z_0)^m$ where m is an integer and z_0 a constant. Integrate counterclockwise around the circle C of radius ρ with center at z_0 (Fig. 311).

Solution. We may represent C in the form

$$z(t) = z_0 + \rho(\cos t + i \sin t) = z_0 + \rho e^{it} \qquad (0 \leqq t \leqq 2\pi).$$

Then we have

$$(z - z_0)^m = \rho^m e^{imt}, \qquad dz = i\rho e^{it}\, dt$$

and obtain

$$\oint_C (z - z_0)^m\, dz = \int_0^{2\pi} \rho^m e^{imt} i\rho e^{it}\, dt = i\rho^{m+1} \int_0^{2\pi} e^{i(m+1)t}\, dt.$$

By the Euler formula (5) in Sec. 12.6 the right side equals

$$i\rho^{m+1}\left[\int_0^{2\pi} \cos{(m+1)t}\,dt + i\int_0^{2\pi} \sin{(m+1)t}\,dt\right].$$

If $m = -1$, we have $\rho^{m+1} = 1$, $\cos 0 = 1$, $\sin 0 = 0$ and thus obtain $2\pi i$. For integer $m \neq 1$ each of the two integrals is zero because we integrate over an interval of length 2π, equal to a period of sine and cosine. Hence the result is

(3) $$\oint_C (z - z_0)^m\,dz = \begin{cases} 2\pi i & (m = -1), \\ 0 & (m \neq -1 \text{ and integer}). \end{cases}$$

Dependence on path. Now comes a very important fact. If we integrate a given function $f(z)$ from a point z_0 to a point z_1 along different paths, the integrals will in general have different values. In other words, ***a complex line integral depends not only on the endpoints of the path but in general also on the path itself.*** See the next example.

EXAMPLE 3 Integral of a nonanalytic function. Dependence on path

Integrate $f(z) = \text{Re}\, z = x$ from 0 to $1 + i$ (a) along C^* in Fig. 312, (b) along C consisting of C_1 and C_2.

Solution. (a) C^* can be represented by $z(t) = t + it$ $(0 \leqq t \leqq 1)$. Hence

$$\dot{z}(t) = 1 + i \qquad \text{and} \qquad f[z(t)] = x(t) = t \qquad \text{(on } C^*\text{).}$$

We now calculate

$$\int_{C^*} \text{Re}\, z\,dz = \int_0^1 t(1 + i)\,dt = \tfrac{1}{2}(1 + i).$$

(b) C_1 can be represented by $z(t) = t$ $(0 \leqq t \leqq 1)$. Hence

$$\dot{z}(t) = 1 \qquad \text{and} \qquad f[z(t)] = x(t) = t \qquad \text{(on } C_1\text{).}$$

C_2 can be represented by $z(t) = 1 + it$ $(0 \leqq t \leqq 1)$. Hence

$$\dot{z}(t) = i \qquad \text{and} \qquad f[z(t)] = x(t) = 1 \qquad \text{(on } C_2\text{).}$$

Using (7) in Sec. 13.1, we calculate

$$\int_C \text{Re}\, z\,dz = \int_{C_1} \text{Re}\, z\,dz + \int_{C_2} \text{Re}\, z\,dz = \int_0^1 t\,dt + \int_0^1 1 \cdot i\,dt = \tfrac{1}{2} + i.$$

Note that this result differs from the result in (a).

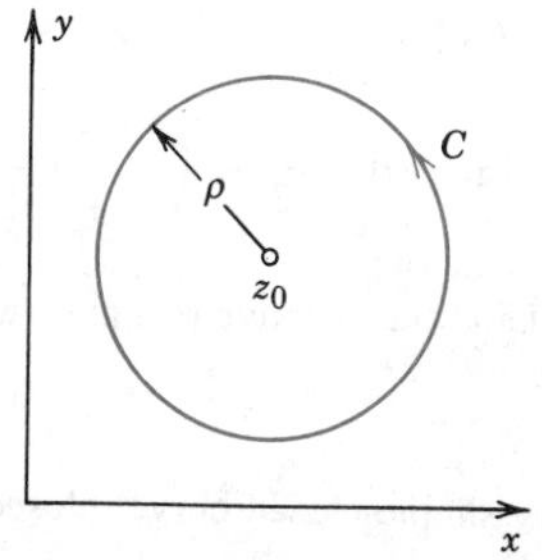

Fig. 311. Path in Example 2

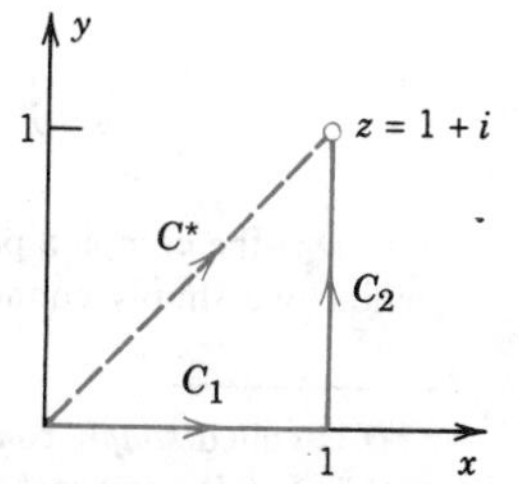

Fig. 312. Paths in Example 3

Second Method: Indefinite Integration

In real calculus, if for a given continuous $f(x)$ we know an $F(x)$ such that $F'(x) = f(x)$, then by the so-called first fundamental theorem of integral calculus,

$$\int_a^b f(x)\,dx = F(b) - F(a) \qquad [F'(x) = f(x)].$$

This method extends to complex analytic functions under the conditions of the next theorem, and is much simpler than the method just discussed, once we have found an $F(z)$ whose derivative $F'(z)$ equals the given function $f(z)$ to be integrated. Differentiation formulas will often help in finding $F(z)$, so that the method is of great practical importance.

Theorem 2 (Indefinite integration of analytic functions)
Let $f(z)$ be analytic in a simply connected[4] *domain D. Then there exists an indefinite integral of $f(z)$ in the domain D, that is, an analytic function $F(z)$ such that $F'(z) = f(z)$ in D, and for all paths in D joining two points z_0 and z_1 in D we have*

$$\int_{z_0}^{z_1} f(z)\,dz = F(z_1) - F(z_0) \qquad [F'(z) = f(z)]. \tag{4}$$

(Note that we can write z_0 and z_1 instead of C, since we get the same value for all those C from z_0 to z_1.)

This theorem will be proved in Sec. 13.4 (by using Cauchy's integral theorem, which we discuss in the next section).

If $f(z)$ is entire (Sec. 12.6), we can take for D the complex plane (which is certainly simply connected).

EXAMPLE 4

$$\int_0^{1+i} z^2\,dz = \frac{1}{3}z^3 \Big|_0^{1+i} = \frac{1}{3}(1+i)^3 = -\frac{2}{3} + \frac{2}{3}i$$ ■

EXAMPLE 5

$$\int_{-\pi i}^{\pi i} \cos z\,dz = \sin z \Big|_{-\pi i}^{\pi i} = 2\sin \pi i = 2i \sinh \pi = 23.097i$$ ■

EXAMPLE 6

$$\int_{8+\pi i}^{8-3\pi i} e^{z/2}\,dz = 2e^{z/2} \Big|_{8+\pi i}^{8-3\pi i} = 2(e^{4-3\pi i/2} - e^{4+\pi i/2}) = 0$$

since e^z is periodic with period $2\pi i$. ■

EXAMPLE 7

$$\int_{-i}^{i} \frac{dz}{z} = \operatorname{Ln} i - \operatorname{Ln}(-i) = \frac{i\pi}{2} - \left(-\frac{i\pi}{2}\right) = i\pi.$$

Here D is the complex plane without 0 and the negative real axis (where $\operatorname{Ln} z$ is not analytic), obviously a simply connected domain. ■

[4] D is called ***simply connected*** if every simple closed curve (closed curve without self-intersections) in D encloses only points of D.

EXAMPLE 8 Simple connectedness in Theorem 2 is essential

Equation (4) with $z_1 = z_0$ gives $F(z_0) - F(z_0) = 0$: the integral over a *closed* path is zero. But (see Example 1)

$$\oint \frac{dz}{z} = 2\pi i \qquad \text{(counterclockwise over the unit circle)}$$

although $1/z$ is analytic in the annulus $\frac{1}{2} < z < \frac{3}{2}$; this domain is not simply connected. ∎

Bound for the Absolute Value of Integrals

There will be a frequent need for estimating the absolute value of complex line integrals. The basic formula is

$$\left|\int_C f(z)\, dz\right| \leqq ML \qquad (5) \qquad \text{(}ML\text{-inequality);}$$

L is the length of C and M a constant such that $|f(z)| \leqq M$ everywhere on C.

Proof. We consider S_n as given by (2), Sec. 13.1. By the generalized triangle inequality (6) in Sec. 12.2 we obtain

$$|S_n| = \left|\sum_{m=1}^{n} f(\zeta_m)\, \Delta z_m\right| \leqq \sum_{m=1}^{n} |f(\zeta_m)|\, |\Delta z_m| \leqq M \sum_{m=1}^{n} |\Delta z_m|.$$

Now $|\Delta z_m|$ is the length of the chord whose endpoints are z_{m-1} and z_m, see Fig. 309 in Sec. 13.1. Hence the sum on the right represents the length L^* of the broken line of chords whose endpoints are $z_0, z_1, \cdots, z_n$ $(= Z)$. If n approaches infinity so that the greatest $|\Delta z_m|$ approaches zero as in Sec. 13.1, then L^* approaches the length L of the curve C, by the definition of the length of a curve. From this the inequality (5) follows. ∎

We cannot see from (5) how close to the bound ML the actual absolute value of the integral is, but this will be no handicap in applying (5), as we shall see. For the time being we explain the practical use of (5) by a simple example.

EXAMPLE 9 Estimation of an integral

Find an upper bound for the absolute value of the integral

$$\int_C z^2\, dz, \qquad C \text{ the straight-line segment from 0 to } 1 + i.$$

Solution. $L = \sqrt{2}$ and $|f(z)| = |z^2| \leqq 2$ on C gives by (5)

$$\left|\int_C z^2\, dz\right| \leqq 2\sqrt{2} = 2.8284.$$

The absolute value of the integral is $\left|-\frac{2}{3} + \frac{2}{3}i\right| = \frac{2}{3}\sqrt{2} = 0.9428$ (see Example 4). ∎

Problem Set 13.1–13.2

Find a representation $z = z(t)$ of the straight line segment with endpoints

1. $z = 0$ and $z = 1 + 2i$

2. $z = -3 + 2i$ and $z = -4 + 5i$

3. $z = 4 + 2i$ and $z = 3 + 5i$

4. $z = 0$ and $z = 5 + 10i$

5. $z = -4i$ and $z = -7 + 38i$

6. $z = 1 - i$ and $z = 9 - 5i$

What curves are represented by the following functions?

7. $(1 + 2i)t, \quad 0 \leqq t \leqq 3$

8. $3 - it, \quad -1 \leqq t \leqq 1$

9. $1 - i - 2e^{it}, \quad 0 \leqq t \leqq \pi$

10. $2 + i + 3e^{it} \quad 0 \leqq t < 2\pi$

11. $t + 3t^2 i, \quad -1 \leqq t \leqq 2$

12. $t + 2it^3, \quad -2 \leqq t \leqq 2$

13. $\cos t + 2i \sin t, \quad -\pi < t < \pi$

14. $t + t^{-1}i, \quad \frac{1}{2} \leqq t \leqq 5$

Represent the following curves in the form $z = z(t)$.

15. $|z - 3 + 4i| = 4$

16. $|z - i| = 2$

17. $y = 1/x$ from $(1, 1)$ to $(3, \frac{1}{3})$

18. $y = x^2$ from $(0, 0)$ to $(2, 4)$

19. $x^2 + 4y^2 = 4$

20. $4(x - 1)^2 + 9(y + 2)^2 = 36$

Evaluate $\int_C f(z)\, dz$ by the method in Theorem 1 and check the result by Theorem 2:

21. $f(z) = az + b, \quad C$ the line segment from $-1 - i$ to $1 + i$

22. $f(z) = e^{2z}, \quad C$ the segment in Prob. 1

23. $f(z) = z^3, \quad C$ the semicircle $|z| = 2$ from $-2i$ to $2i$ in the right half-plane

24. $f(z) = 5z^2, \quad C$ the boundary of the triangle with vertices $0, 1, i$ (clockwise)

Evaluate $\int_C f(z)\, dz$, where

25. $f(z) = 2z^4 - z^{-4}, \quad C$ the unit circle (counterclockwise)

26. $f(z) = \text{Re}\, z, \quad C$ the parabola $y = x^2$ from 0 to $1 + i$

27. $f(z) = \text{Im}\, z, \quad C$ the circle $|z| = r$ (counterclockwise)

28. $f(z) = 4z - 3, \quad C$ the straight line segment from i to $1 + i$

29. $f(z) = (z - 1)^{-1} + 2(z - 1)^{-2}, \quad C$ the circle $|z - 1| = 4$ (clockwise)

30. $f(z) = \sin z, \quad C$ the line segment from 0 to i

31. $f(z) = e^{2z}, \quad C$ the vertical segment from πi to $2\pi i$

32. $f(z) = z \cos z^2, \quad C$ any path from 0 to πi

33. $f(z) = \cosh 3z, \quad C$ any path from $\pi i/6$ to 0

34. $f(z) = e^z$, C the boundary of the square with vertices $0, 1, 1 + i, i$ (clockwise)

35. $f(z) = \text{Re}\,(z^2), \quad C$ the square in Prob. 34

36. $f(z) = \text{Im}\,(z^2), \quad C$ the square in Prob. 34

37. $f(z) = \bar{z}, \quad C$ the parabola $y = x^2$ from 0 to $1 + i$

38. $f(z) = (z - 1)^{-1} - (z - 1)^{-2}, \quad C$ the circle $|z - 1| = \frac{1}{2}$ (clockwise)

39. $f(z) = \sin^2 z, \quad C$ the semicircle $|z| = \pi$ from $-\pi i$ to πi in the right half-plane

40. $f(z) = \sec^2 z, \quad C$ any path from $\pi i/4$ to $\pi/4$ in the unit disk

41. Evaluate $\int_C \text{Im}\,(z^2)\,dz$ from 0 to $2 + 4i$ along (a) the line segment, (b) the x-axis to 2 and then vertically to $2 + 4i$, (c) the parabola $y = x^2$.

42. Evaluate $\int_C (z^{-5} + z^3)\,dz$ from 1 to -1 along (a) the upper arc of the unit circle, (b) the lower arc of the unit circle.

43. Evaluate $\int_C |z|\,dz$ from A: $z = -i$ to B: $z = i$ along (a) the line segment AB, (b) the unit circle in the left half-plane, (c) the unit circle in the right half-plane.

44. Prove (6) in Sec. 13.1.

45. Verify (6) in Sec. 13.1 for $k_1f_1 + k_2f_2 = 3z - z^2$, where C is the upper half of the unit circle from 1 to -1.

46. Verify (7) in Sec. 13.1 for $f(z) = 1/z$, where C is the unit circle, C_1 its upper half, and C_2 its lower half.

47. Verify (8) in Sec. 13.1 for $f(z) = z^2$, where C is the line segment from $-1 - i$ to $1 + i$.

Using the ML-inequality (5), find upper bounds for the following integrals, where C is the line segment from 0 to $3 + 4i$.

48. $\int_C \text{Ln}\,(z + 1)\,dz$

49. $\int_C e^z\,dz$

50. Find a better bound in Prob. 49 by decomposing C into two arcs.

13.3 Cauchy's Integral Theorem

A line integral of a function $f(z)$ depends not merely on the endpoints of the path, but also on the choice of the path itself (see Sec. 13.2). This dependence is awkward and one looks for situations when it does not occur. The answer will be simple: if $f(z)$ is analytic in a domain that is simply connected (definition below). This result (Theorem 2) follows from Cauchy's integral theorem, along with other basic consequences that make Cauchy's theorem the ***most important*** theorem in this chapter and fundamental throughout complex analysis. To state Cauchy's integral theorem we need the following two concepts.

1. A simple closed path is a closed path (Sec. 13.1) that does not intersect or touch itself (Fig. 313). For example, a circle is simple, an 8-shaped curve is not.

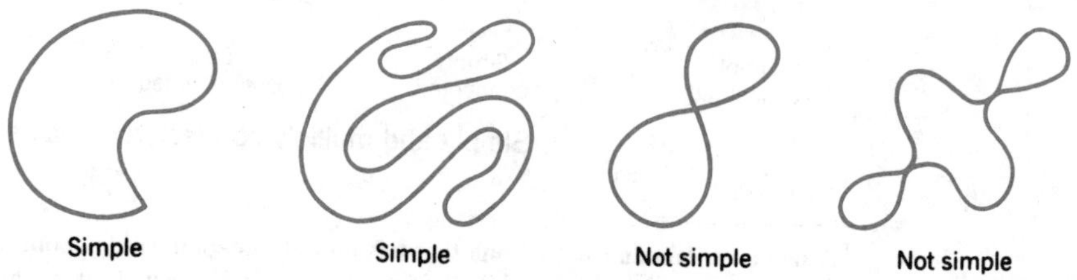

Fig. 313. Closed paths

2. A **simply connected domain** D in the complex plane is a domain (Sec. 12.3) such that every simple closed path in D encloses only points of D. A domain that is not simply connected is called **multiply connected.**

For instance, the interior of a circle ("circular disk"), ellipse, or square is simply connected. More generally, the interior of a simple closed curve is simply connected. An annulus (Sec. 12.3) is multiply connected (more precisely, double connected[5]). Figure 314 shows further examples.

Recalling that, by definition, a function is a ***single-valued*** relation (see Sec. 12.4), we can now state Cauchy's integral theorem as follows. This theorem is sometimes also called the **Cauchy–Goursat theorem.**

Theorem 1 Cauchy's integral theorem

If $f(z)$ is analytic in a simply connected domain D, then for every simple closed path C in D.

$$\oint_C f(z)\, dz = 0. \tag{1}$$

See Fig. 315.

Before we prove the theorem, let us consider some examples in order to really understand what is going on. A simple closed path is sometimes called a *contour* and an integral over such a path a **contour integral.** Thus, (1) and our examples involve contour integrals.

EXAMPLE 1

$$\oint_C e^z\, dz = 0, \qquad \oint_C \cos z\, dz = 0, \qquad \oint_C z^n\, dz = 0 \qquad (n = 0, 1, \cdots)$$

for any closed path, since these functions are entire (analytic for all z). ∎

EXAMPLE 2

$$\oint_C \sec z\, dz = 0, \qquad \oint_C \frac{dz}{z^2 + 4} = 0$$

where C is the unit circle, $\sec z = 1/\cos z$ is not analytic at $z = \pm\pi/2, \pm 3\pi/2, \cdots$, but all these points lie outside C; none lies on C or inside C. Similarly for the second integral, whose integrand is not analytic at $z = \pm 2i$ outside C. ∎

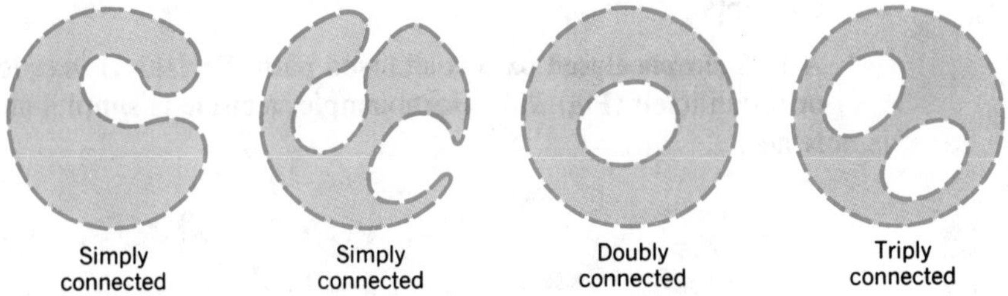

Fig. 314. Simply and multiply connected domains

[5] A **bounded domain** (that is, one that lies entirely in some circle about the origin) is said to be *p-fold connected* if its boundary consists of p closed connected sets without common points. For the annulus, $p = 2$, because the boundary consists of two circles having no points in common.

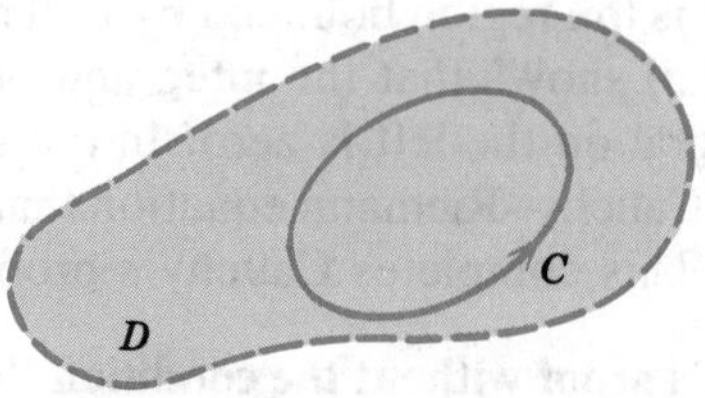

Fig. 315. Cauchy's integral theorem

EXAMPLE 3

$$\oint_C \bar{z}\, dz = 2\pi i$$

(C the unit circle, counterclockwise) does not contradict Cauchy's theorem, since $f(z) = \bar{z}$ is not analytic, so that the theorem does not apply. (Verify this result!) ■

EXAMPLE 4

$$\oint_C \frac{dz}{z^2} = 0,$$

where C is the unit circle. This result does ***not*** follow from Cauchy's theorem, because $f(z) = 1/z^2$ is not analytic at $z = 0$. Hence *the condition that f be analytic in D is sufficient rather than necessary for* (1) *to be true.* ■

EXAMPLE 5

$$\oint_C \frac{dz}{z} = 2\pi i$$

for counterclockwise integration around the unit circle (see Sec. 13.2). C lies in the annulus $\frac{1}{2} < |z| < \frac{3}{2}$ where $1/z$ is analytic, but this domain is not simply connected, so that Cauchy's theorem cannot be applied. Hence ***the condition that the domain D be simply connected is quite essential.***

In other words, by Cauchy's theorem, if $f(z)$ is analytic on a simple closed path C and everywhere in C, with no exception, not even a single point, then (1) holds. The point that causes trouble here is $z = 0$ where $1/z$ is not analytic. ■

EXAMPLE 6

$$\oint_C \frac{7z - 6}{z^2 - 2z}\, dz = \oint_C \frac{3}{z}\, dz + \oint_C \frac{4}{z - 2}\, dz = 3 \cdot 2\pi i + 0 = 6\pi i$$

(C the unit circle, counterclockwise) by partial fraction reduction, Example 1 in Sec. 13.2, and Cauchy's theorem. ■

Cauchy's proof under the condition that $f'(z)$ is continuous. From (5) in Sec. 13.1 we have

$$\oint_C f(z)\, dz = \oint_C (u\, dx - v\, dy) + i \oint_C (u\, dy + v\, dx).$$

Since $f(z)$ is analytic in D, its derivative $f'(z)$ exists in D. Since $f'(z)$ is assumed to be continuous, (4) and (5) in Sec. 12.5 imply that u and v have *continuous* partial derivatives in D. Hence Green's theorem (Sec. 9.4) (with u and $-v$ instead of F_1 and F_2) is applicable and gives

$$\oint_C (u\, dx - v\, dy) = \iint_R \left(-\frac{\partial v}{\partial x} - \frac{\partial u}{\partial y} \right) dx\, dy$$

where R is the region bounded by C. The second Cauchy–Riemann equation (Sec. 12.5) shows that the integrand on the right is identically zero. Hence the integral on the left is zero. In the same fashion it follows by the use of the first Cauchy–Riemann equation that the last integral in the above formula is zero. This completes Cauchy's proof. ∎

Goursat's proof without the condition that $f'(z)$ is continuous[6] is substantially more complicated. We leave it optional and include it in Appendix 4.

Independence of Path, Deformation of Path

We know from Sec. 13.1 that if we integrate a function $f(z)$ from a point z_1 to a point z_2, the result will in general depend on the path over which we integrate, not merely on z_1 and z_2. This is awkward, and one wants to find out when this is not the case. Accordingly, we call the integral of $f(z)$ **independent of path in a domain D** if its value depends only on z_1 and z_2 but not on the choice of the path C in D (so that any path C_1 in D from z_1 to z_2 gives the same value of the integral as any other path C_2 in D from z_1 to z_2). We prove that this happens if $f(z)$ is analytic in D and D is simply connected:

Theorem 2 (Independence of path)
If $f(z)$ is analytic in a simply connected domain D, then the integral of $f(z)$ is independent of path in D.

Proof. Let z_1 and z_2 be any points in D. Consider two paths C_1 and C_2 in D from z_1 to z_2 without further common points, as in Fig. 316. Denote by C_2^* the path C_2 with the orientation reversed (Fig. 317). Integrate from z_1 over C_1 to z_2 and over C_2^* back to z_1. This is a simple closed path, and Cauchy's theorem applies under our assumptions of the present theorem and gives zero:

$$(2') \qquad \int_{C_1} f\,dz + \int_{C_2^*} f\,dz = 0, \qquad \text{thus} \qquad \int_{C_1} f\,dz = -\int_{C_2^*} f\,dz.$$

But the minus sign on the right disappears if we integrate in the reverse

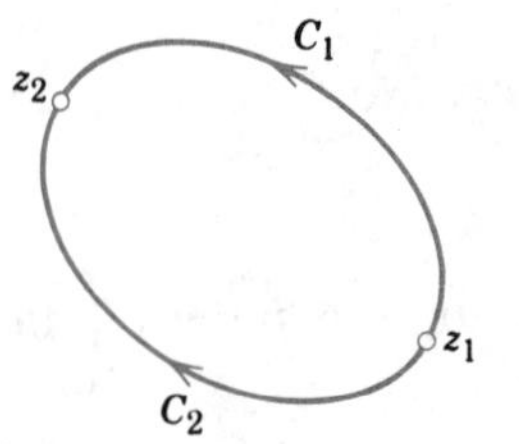

Fig. 316. Formula (2)

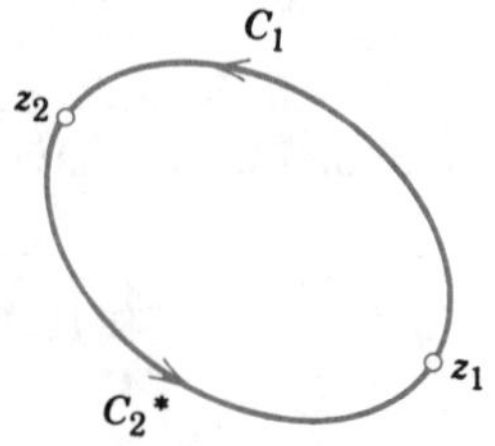

Fig. 317. Formula (2′)

[6] ÉDOUARD GOURSAT (1858—1936), French mathematician. Cauchy published the theorem in 1825. The removal of that condition by Goursat (see *Transactions Amer. Math. Soc.*, vol. 1, 1900) is quite important, for instance, in connection with the fact that derivatives of analytic functions are also analytic, as we shall prove soon. Goursat also made basic contributions to partial differential equations.

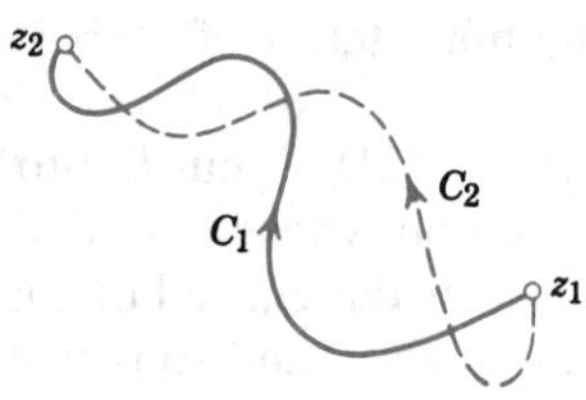

Fig. 318. Paths with more common points

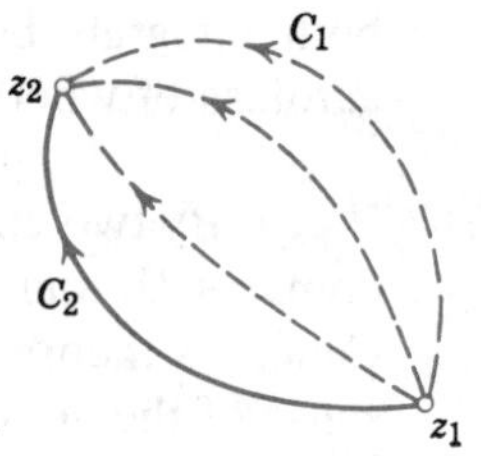

Fig. 319. Continuous deformation of path

direction, from z_1 to z_2, which shows that the integrals of $f(z)$ over C_1 and C_2 are equal,

$$\int_{C_1} f(z)\, dz = \int_{C_2} f(z)\, dz \qquad \text{(Fig. 316).} \tag{2}$$

This proves the theorem for paths that have only the endpoints in common. For paths that have finitely many further common points, apply the present argument to each "loop" (portions of C_1 and C_2 between consecutive common points; four in Fig. 318). For paths with infinitely many common points we would need additional argumentation not to be presented here. ∎

Principle of Deformation of Path. Related to independence of path is the following idea. We may imagine that the path C_2 in (2) was obtained from C_1 by a continuous deformation. (Figure 319 shows two of the infinitely many intermediate paths.) It follows that for a given integral of an analytic function we may impose a continuous deformation on the path of integration (keeping the endpoints fixed); as long as our deforming path never contains any points at which $f(z)$ is not analytic, the value of the line integral remains the same under the deformation. This is called the **principle of deformation of path.**

Cauchy's Theorem for Multiply Connected Domains

Cauchy's theorem applies to multiply connected domains. We first explain this for a **doubly connected domain** D with outer boundary curve C_1 and inner C_2 (Fig. 320). If a function $f(z)$ is analytic in any domain D^* that contains D as well as its boundary curves, we claim that

$$\int_{C_1} f(z)\, dz = \int_{C_2} f(z)\, dz \qquad \text{(Fig. 320),} \tag{3}$$

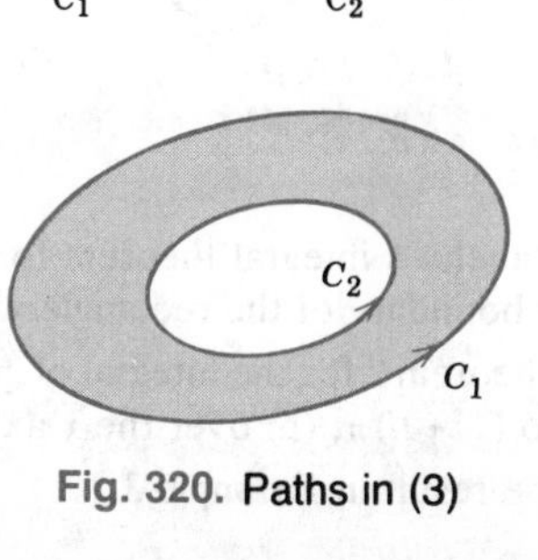

Fig. 320. Paths in (3)

both integrals being taken counterclockwise (or both clockwise, and regardless of whether or not the full interior of C_2 belongs to D^*).

Proof. By two cuts $\tilde{C}_1$ and $\tilde{C}_2$ (Fig. 321) we cut D into two simply connected domains D_1 and D_2 in which and on whose boundaries $f(z)$ is analytic. By Cauchy's theorem the integral over the entire boundary of D_1 (taken in the sense of the arrows in Fig. 321) is zero, and so is that over the boundary of D_2, and thus their sum. In this sum the integrals over the cuts $\tilde{C}_1$ and $\tilde{C}_2$ cancel because we integrate over them in both directions—this is the key—and we are left with the integrals over C_1 (counterclockwise) and C_2 (clockwise; see Fig. 321); hence by reversing the integration over C_2 (to counterclockwise) we have

$$\int_{C_1} f\,dz - \int_{C_2} f\,dz = 0$$

and (3) follows. ∎

For domains of higher connectivity the idea remains the same. Thus, for a **triply connected domain** we use three cuts $\tilde{C}_1$, $\tilde{C}_2$, $\tilde{C}_3$ (Fig. 322). Adding integrals as before, the integrals over the cuts cancel and the sum of the integrals over C_1 (counterclockwise) and C_2, C_3 (clockwise) is zero. Hence the integral over C_1 equals the sum of the integrals over C_2 and C_3, all three now taken counterclockwise, and so on.

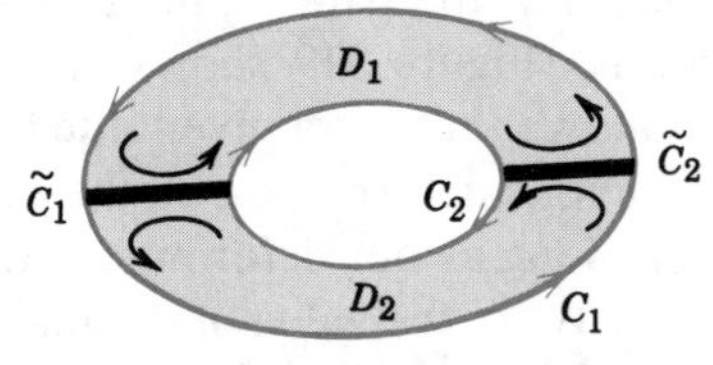

Fig. 321. Doubly connected domain

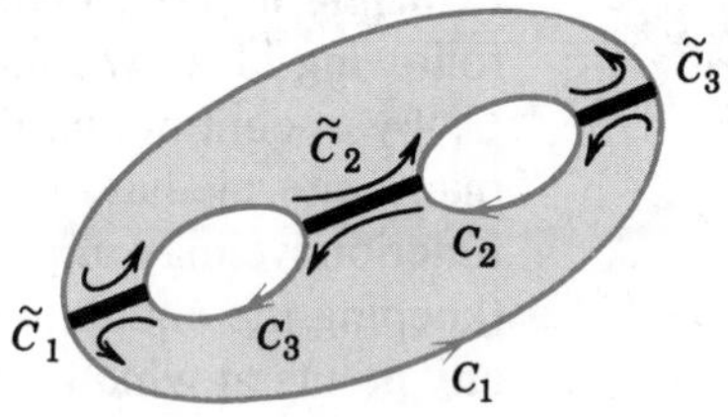

Fig. 322. Triply connected domain

EXAMPLE 7 **A basic result: Integral of integer powers**

From (3) and Sec. 13.2, Example 2, it now follows that

$$\oint_C (z - z_0)^m\,dz = \begin{cases} 2\pi i & (m = -1) \\ 0 & (m \neq -1 \text{ and integer}) \end{cases}$$

for counterclockwise integration around any simple closed path containing z_0 in its interior. ∎

Problem Set 13.3

1. Verify Cauchy's integral theorem for the integral of z^2 taken counterclockwise over the boundary of the rectangle with vertices $-1, 1, 1 + i, -1 + i$.
2. Verify Theorem 2 for the integral of $\sin z$ from 0 to $(1 + i)\pi$ (a) over the segment from 0 to $(1 + i)\pi$, (b) over the x-axis to π and then straight up to $(1 + i)\pi$.
3. Verify the result in Example 3.

4. For what contours C will it follow from Cauchy's theorem that

$$\text{(a)} \oint_C \frac{dz}{z} = 0, \qquad \text{(b)} \oint_C \frac{\cos z}{z^6 - z^2}\, dz = 0, \qquad \text{(c)} \oint_C \frac{e^{1/z}}{z^2 + 9}\, dz = 0?$$

5. The integral in Example 4 is zero. Can we conclude from this that it is zero over the contour in Prob. 1?

6. Can we conclude from Example 2 that the integral of $1/(z^2 + 4)$ taken over (a) $|z - 2| = 2$, (b) $|z - 2| = 3$ is zero? Give a reason.

Integrate $f(z)$ counterclockwise over the unit circle and indicate whether Cauchy's theorem may be applied.

7. $f(z) = |z|$ **8.** $f(z) = e^{z^2}$ **9.** $f(z) = \text{Im } z$
10. $f(z) = 1/(2z - 5)$ **11.** $f(z) = 1/\bar{z}$ **12.** $f(z) = 1/(\pi z - 3)$
13. $f(z) = \tan z$ **14.** $f(z) = \bar{z}$ **15.** $f(z) = \bar{z}^2$
16. $f(z) = 1/|z|^3$ **17.** $f(z) = 1/(z^2 + 2)$ **18.** $f(z) = z^2 \sec z$

Evaluate the following integrals. (*Hint.* If necessary, represent the integrand in terms of partial fractions.)

19. $\oint_C \frac{dz}{z - i}$, C the circle $|z| = 2$ (counterclockwise)

20. $\oint_C \frac{dz}{\sinh z}$, C the circle $|z - \frac{1}{2}\pi i| = 1$ (clockwise)

21. $\oint_C \frac{\cos z}{z}\, dz$, C consists of $|z| = 1$ (counterclockwise) and $|z| = 3$ (clockwise)

22. $\oint_C \frac{2z - 1}{z^2 - z}\, dz$, C the contour in Fig. 323

23. $\oint_C \frac{dz}{z^2 - 1}$, C the contour in Fig. 324

24. $\oint_C \text{Re } z\, dz$, C the contour in Fig. 325

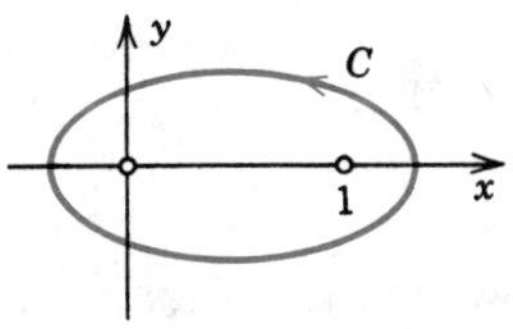

Fig. 323. Problem 22

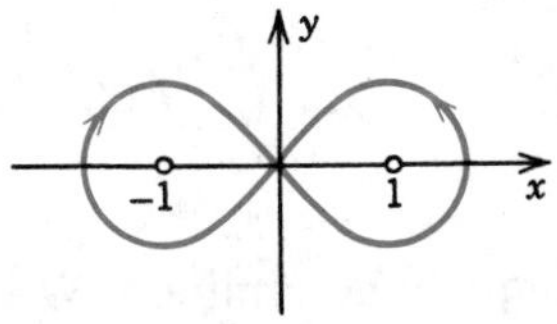

Fig. 324. Problem 23

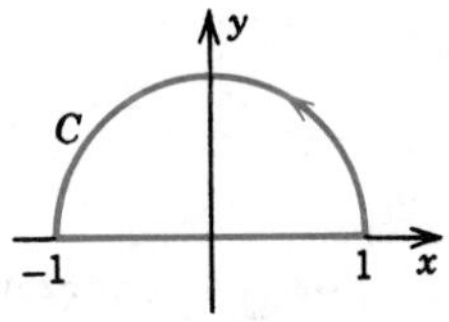

Fig. 325. Problem 24

25. $\oint_C \frac{dz}{z^2 + 1}$, C: (a) $|z + i| = 1$, (b) $|z - i| = 1$ (counterclockwise)

26. $\oint_C \frac{\sin z}{z + 3i}\, dz$, C: $|z - 2 + 3i| = 1$ (counterclockwise)

27. $\oint_C \frac{2z + 1}{z^2 + z}\, dz$, C: (a) $|z| = \frac{1}{4}$, (b) $|z - \frac{1}{2}| = \frac{1}{4}$, (c) $|z| = 2$ (clockwise)

Evaluate (*continued*)

28. $\oint_C \frac{dz}{1+z^3}$, C: $|z+1| = 1$ (counterclockwise)

29. $\oint_C \frac{3z+1}{z^3-z}\,dz$, C: (a) $|z| = 1/2$, (b) $|z| = 2$ (counterclockwise)

30. $\oint_C \text{Re}\,(z^2)\,dz$, C the boundary of the triangle with vertices at 0, 2, and $2+i$ (counterclockwise)

13.4 Existence of Indefinite Integral

In this short section we use Cauchy's integral theorem to establish the existence of an indefinite integral $F(z)$ of a given analytic function $f(z)$ and thereby justify the evaluation of line integrals by indefinite integration and substitution of the limits of integration (see Sec. 13.2):

$$\int_{z_0}^{z_1} f(z)\,dz = F(z_1) - F(z_0) \qquad [F'(z) = f(z)], \tag{1}$$

where $F(z)$ is an indefinite integral of $f(z)$, that is, $F'(z) = f(z)$, as indicated. In most applications, such an $F(z)$ can be found from differentiation formulas.

Theorem 1 **(Existence of an indefinite integral)**

If $f(z)$ is analytic in a simply connected domain D (see Sec. 13.3), *then there exists an indefinite integral $F(z)$ of $f(z)$ in D—thus, $F'(z) = f(z)$—which is analytic in D, and for all paths in D joining any two points z_0 and z_1 in D, the integral of $f(z)$ from z_0 to z_1 can be evaluated by formula* (1).

Proof. The conditions of Cauchy's integral theorem are satisfied. Hence the line integral of $f(z)$ from any z_0 in D to any z in D is independent of path in D. We keep z_0 fixed. Then this integral becomes a function of z, call it $F(z)$,

$$F(z) = \int_{z_0}^{z} f(z^*)\,dz^*, \tag{2}$$

which is uniquely determined. We show that this $F(z)$ is analytic in D and $F'(z) = f(z)$. The idea of doing this is as follows. We form the difference quotient

$$\begin{aligned} \frac{F(z+\Delta z) - F(z)}{\Delta z} &= \frac{1}{\Delta z}\left[\int_{z_0}^{z+\Delta z} f(z^*)\,dz^* - \int_{z_0}^{z} f(z^*)\,dz^*\right] \\ &= \frac{1}{\Delta z}\int_{z}^{z+\Delta z} f(z^*)\,dz^*, \end{aligned} \tag{3}$$

subtract $f(z)$ from it, and show that the resulting expression approaches zero as $\Delta z \to 0$. The details are as follows.

We keep z fixed. Then we choose $z + \Delta z$ in D so that the whole segment with endpoints z and $z + \Delta z$ is in D (Fig. 326). This can be done because D is a domain, hence it contains a neighborhood of z. We use this segment as the path of integration in (3). Now we subtract $f(z)$. This is a constant because z is kept fixed. Hence we can write

$$\int_z^{z+\Delta z} f(z)\, dz^* = f(z) \int_z^{z+\Delta z} dz^* = f(z)\,\Delta z. \text{ Thus } f(z) = \frac{1}{\Delta z} \int_z^{z+\Delta z} f(z)\, dz^*.$$

By this trick and from (3) we get a single integral:

$$\frac{F(z + \Delta z) - F(z)}{\Delta z} - f(z) = \frac{1}{\Delta z} \int_z^{z+\Delta z} [f(z^*) - f(z)]\, dz^*.$$

Since $f(z)$ is analytic, it is continuous. An $\epsilon > 0$ being given, we can thus find a $\delta > 0$ such that

$$|f(z^*) - f(z)| < \epsilon \qquad \text{if} \qquad |z^* - z| < \delta.$$

Hence, letting $|\Delta z| < \delta$, we see that the ML-inequality (Sec. 13.2) yields

$$\left| \frac{F(z + \Delta z) - F(z)}{\Delta z} - f(z) \right| = \frac{1}{|\Delta z|} \left| \int_z^{z+\Delta z} [f(z^*) - f(z)]\, dz^* \right| \leqq \frac{1}{|\Delta z|} \epsilon |\Delta z| = \epsilon.$$

By the definition of limit and derivative, this proves that

$$F'(z) = \lim_{\Delta z \to 0} \frac{F(z + \Delta z) - F(z)}{\Delta z} = f(z).$$

Since z is any point in D, this implies that $F(z)$ is analytic in D and is an indefinite integral or antiderivative of $f(z)$ in D, written

$$F(z) = \int f(z)\, dz.$$

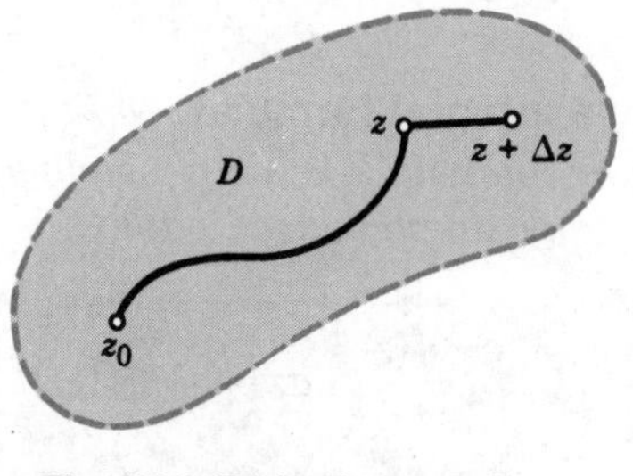

Fig. 326. Path of integration

Also, if $G'(z) = f(z)$, then $F'(z) - G'(z) \equiv 0$ in D; hence $F(z) - G(z)$ is constant in D (see Prob. 29 in Problem Set 12.5). That is, two indefinite integrals of $f(z)$ can differ only by a constant. The latter drops out in (1). Hence in (1) we can use any indefinite integral of $f(z)$. This proves the theorem. ∎

Examples on indefinite integration and additional problems are given in Sec. 13.2.

Problem Set 13.4

Evaluate the following integrals.

1. $\int_{i}^{2+i} z\,dz$ **2.** $\int_{0}^{1+i} z^2\,dz$ **3.** $\int_{0}^{\pi i} e^z\,dz$

4. $\int_{-\pi i}^{\pi i} \sin^2 z\,dz$ **5.** $\int_{-\pi i}^{3\pi i} e^{2z}\,dz$ **6.** $\int_{-i}^{i} \sin z\,dz$

7. $\int_{\pi i/6}^{0} \cosh 3z\,dz$ **8.** $\int_{-\pi i}^{\pi i} \cos z\,dz$ **9.** $\int_{0}^{i} \sinh \pi z\,dz$

10. $\int_{\pi}^{\pi i} \sin 2z\,dz$ **11.** $\int_{-1}^{1} z \cosh z^2\,dz$ **12.** $\int_{\pi i}^{0} z \cos z\,dz$

13. $\int_{1}^{i} ze^{z^2}\,dz$ **14.** $\int_{1+i}^{1} z^3 e^{z^4}\,dz$ **15.** $\int_{i}^{2i} (z^2 - 1)^3\,dz$

13.5 Cauchy's Integral Formula

The most important consequence of Cauchy's integral theorem is Cauchy's integral formula. This formula is useful for evaluating integrals (examples below). Equally important is its key role in proving the surprising fact that analytic functions have derivatives of all orders (Sec. 13.6), in establishing Taylor series representations (Sec. 14.4), and so on. Cauchy's integral formula and its conditions of validity may be stated as follows.

Theorem 1 **(Cauchy's integral formula)**

Let $f(z)$ be analytic in a simply connected domain D. Then for any point z_0 in D and any simple closed path C in D that encloses z_0 (Fig. 327),

$$\oint_C \frac{f(z)}{z - z_0}\,dz = 2\pi i f(z_0) \qquad \textbf{(Cauchy's integral formula)}, \tag{1}$$

the integration being taken counterclockwise.

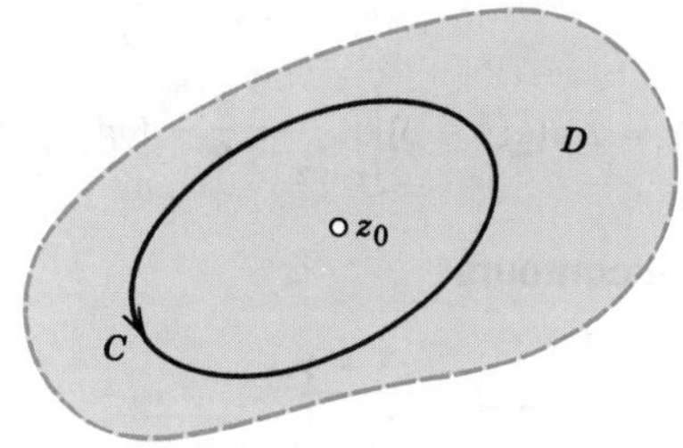

Fig. 327. Cauchy's integral formula

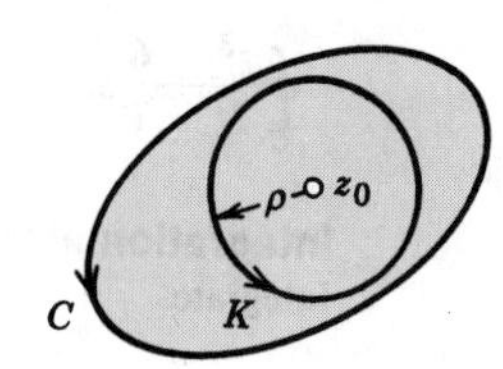

Fig. 328. Proof of Cauchy's integral formula

Proof. By addition and subtraction, $f(z) = f(z_0) + [f(z) - f(z_0)]$. Inserting this into (1) on the left and taking the constant factor $f(z_0)$ out from under the integral sign, we have

$$\oint_C \frac{f(z)}{z - z_0}\,dz = f(z_0) \oint_C \frac{dz}{z - z_0} + \oint_C \frac{f(z) - f(z_0)}{z - z_0}\,dz. \tag{2}$$

The first term on the right equals $f(z_0) \cdot 2\pi i$ (see Example 7 in Sec. 13.3, with $m = -1$). This proves the theorem, provided the second integral on the right is zero. This is what we are now going to show. Its integrand is analytic, except at z_0. Hence by the principle of deformation of path (Sec. 13.3) we can replace C by a small circle K of radius ρ and center z_0 (Fig. 328), without altering the value of the integral. Since $f(z)$ is analytic, it is continuous. Hence, an $\epsilon > 0$ being given, we can find a $\delta > 0$ such that

$$|f(z) - f(z_0)| < \epsilon \qquad \text{for all } z \text{ in the disk } |z - z_0| < \delta.$$

Choosing the radius ρ of K smaller than δ, we thus have the inequality

$$\left| \frac{f(z) - f(z_0)}{z - z_0} \right| < \frac{\epsilon}{\rho}$$

at each point of K. The length of K is $2\pi\rho$. Hence, by the *ML*-inequality in Sec. 13.2,

$$\left| \oint_K \frac{f(z) - f(z_0)}{z - z_0}\,dz \right| < \frac{\epsilon}{\rho} 2\pi\rho = 2\pi\epsilon.$$

Since ϵ (> 0) can be chosen arbitrarily small, it follows that the last integral in (2) has the value zero, and the theorem is proved. ∎

EXAMPLE 1 Cauchy's integral formula

$$\oint_C \frac{e^z}{z - 2}\,dz = 2\pi i e^z \Big|_{z=2} = 2\pi i e^2$$

for any contour enclosing $z_0 = 2$ (since e^z is entire), and zero for any contour for which $z_0 = 2$ lies outside (by Cauchy's integral theorem). ∎

EXAMPLE 2 **Cauchy's integral formula**

$$\oint_C \frac{z^3 - 6}{2z - i}\,dz = \oint_C \frac{\frac{1}{2}z^3 - 3}{z - \frac{1}{2}i}\,dz = 2\pi i[\tfrac{1}{2}z^3 - 3]\Big|_{z=i/2} = \frac{\pi}{8} - 6\pi i \qquad (z_0 = \tfrac{1}{2}i \text{ inside } C).$$ ■

EXAMPLE 3 **Integration around different contours**

Integrate

$$g(z) = \frac{z^2 + 1}{z^2 - 1}$$

in the counterclockwise sense around a circle of radius 1 with center at the point

(a) $z = 1$ (b) $z = \frac{1}{2}$ (c) $z = -1 + \frac{1}{2}i$ (d) $z = i$.

Solution. To see what is going on, locate the points where $g(z)$ is not analytic and sketch them along with the contours (Fig. 329). These points are -1 and 1. We see that (b) will give the same result as (a), by the principle of deformation of path. And (d) gives zero, by Cauchy's integral theorem. We consider (a) and afterward (c).

(a) Here $z_0 = 1$, so that $z - z_0 = z - 1$ in (1). Hence we must write

$$g(z) = \frac{z^2 + 1}{z^2 - 1} = \frac{z^2 + 1}{z + 1}\,\frac{1}{z - 1}; \qquad \text{thus} \qquad f(z) = \frac{z^2 + 1}{z + 1},$$

and (1) gives

$$\oint_C \frac{z^2 + 1}{z^2 - 1}\,dz = 2\pi i f(1) = 2\pi i \left[\frac{z^2 + 1}{z + 1}\right]_{z=1} = 2\pi i.$$

(c) $g(z)$ is the same as before, but $f(z)$ changes, because we must now take $z_0 = -1$ (instead of 1). This gives a factor $z - z_0 = z + 1$ in (1). Hence we must now write

$$g(z) = \frac{z^2 + 1}{z - 1}\,\frac{1}{z + 1}; \qquad \text{thus} \qquad f(z) = \frac{z^2 + 1}{z - 1}.$$

Compare this for a minute with the previous expression and then go on:

$$\oint_C \frac{z^2 + 1}{z^2 - 1}\,dz = 2\pi i f(-1) = 2\pi i \left[\frac{z^2 + 1}{z - 1}\right]_{z=-1} = -2\pi i.$$ ■

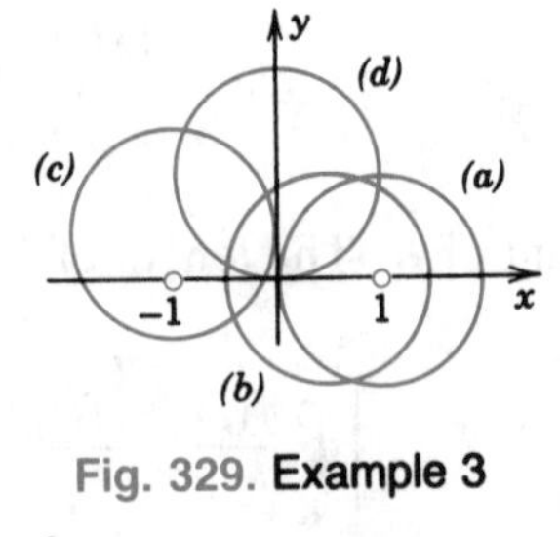

Fig. 329. Example 3

EXAMPLE 4 **Use of partial fractions**

Integrate $g(z) = (z^2 - 1)^{-1} \tan z$ around the circle C: $|z| = 3/2$ (counterclockwise).

Solution. $\tan z$ is not analytic at $\pm\pi/2, \pm 3\pi/2, \cdots$, but all these points lie outside the contour. $(z^2 - 1)^{-1} = 1/(z - 1)(z + 1)$ is not analytic at 1 and -1. To get integrals of the form (1), with only a *single* point inside C at which the integrand is not analytic, we use partial fractions:

$$\frac{1}{z^2 - 1} = \frac{1}{2}\left(\frac{1}{z - 1} - \frac{1}{z + 1}\right).$$

From this and (1) we obtain

$$\oint_C \frac{\tan z}{z^2 - 1}\, dz = \frac{1}{2}\left[\oint_C \frac{\tan z}{z - 1}\, dz - \oint_C \frac{\tan z}{z + 1}\, dz\right]$$

$$= \frac{2\pi i}{2}\,[\tan 1 - \tan(-1)] = 2\pi i \tan 1 \approx 9.785i.$$

Multiply connected domains may be handled as in Sec. 13.3. For instance, if $f(z)$ is analytic on C_1 and C_2 and in the ring shaped domain bounded by C_1 and C_2 (Fig. 330) and z_0 is any point in that domain, then

$$f(z_0) = \frac{1}{2\pi i}\oint_{C_1} \frac{f(z)}{z - z_0}\, dz + \frac{1}{2\pi i}\oint_{C_2} \frac{f(z)}{z - z_0}\, dz, \tag{3}$$

where the outer integral (over C_1) is taken counterclockwise and the inner clockwise, as indicated in Fig. 330.

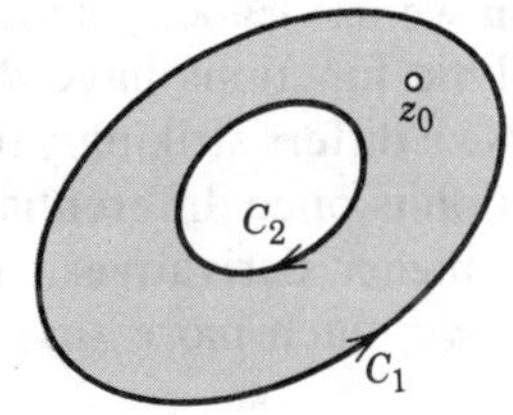

Fig. 330. Formula (3)

Problem Set 13.5

Integrate $1/(z^4 - 1)$ counterclockwise around the circle

1. $|z - i| = 1$ **2.** $|z - 1| = 1$ **3.** $|z + 1| = 1$ **4.** $|z + 3| = 1$

Integrate $(z^2 - 1)/(z^2 + 1)$ counterclockwise around the circle

5. $|z - 2i| = 2$ **6.** $|z - i| = 1$ **7.** $|z| = \frac{1}{2}$ **8.** $|z + i| = 1$

Integrate the following functions counterclockwise around the unit circle.

9. $(\cos z)/2z$ **10.** e^z/z **11.** $(z + 2)/(z - 2)$ **12.** $(e^z - 1)/z$

13. $z^3/(2z - i)$ **14.** $(\sin z)/2z$ **15.** $(z - \pi)^{-1}\cos z$ **16.** $e^{3z}/(3z - i)$

Integrate the given function over the given contour C (counterclockwise or as indicated).

17. $(z^3 - i)/\pi z$, C the circle $|z| = 3$

18. $e^{-3\pi z}/(2z + i)$, C the boundary of the triangle with vertices $-1, 1, -i$

19. $(\tan z)/(z - i)$, C the boundary of the triangle with vertices $-1, 1, 2i$

20. $\cosh(z^2 - \pi i)/(z - \pi i)$, C any ellipse with foci 0 and πi

21. $[\mathrm{Ln}(z - 1)]/(z - 5)$, C the circle $|z - 4| = 2$

22. $[\mathrm{Ln}(z + 1)]/(z^2 + 1)$, C consists of the boundary of the triangle with vertices $1 - \frac{1}{2}i, -1 - \frac{1}{2}i, 2i$ (counterclockwise) and $|z| = \frac{1}{4}$ (clockwise)

Integrate (*continued*)

23. $e^{z^3}/[(z - 1 - i)z^2]$, C consists of $|z| = 3$ (counterclockwise) and $|z| = 1$ (clockwise)

24. $(\sin z)/(4z^2 - 8iz)$, C consists of the boundaries of the squares with vertices ± 3, $\pm 3i$ (counterclockwise) and ± 1, $\pm i$ (clockwise)

25. Show that $\oint_C (z - z_1)^{-1}(z - z_2)^{-1}\, dz = 0$ for a simple closed path C enclosing z_1 and z_2, which are arbitrary.

13.6 Derivatives of Analytic Functions

In this section we use Cauchy's integral formula to show the basic fact that complex analytic functions have ***derivatives of all orders***. This is very surprising because it differs strikingly from the situation in real calculus. Indeed, if a *real* function is once differentiable, nothing follows about the existence of second or higher derivatives. Thus, in this respect, complex analytic functions behave much more simply than real functions that are once differentiable.

Theorem 1 **(Derivatives of an analytic function)**

If $f(z)$ is analytic in a domain D, then it has derivatives of all orders in D, which are then also analytic functions in D. The values of these derivatives at a point z_0 in D are given by the formulas

$$f'(z_0) = \frac{1}{2\pi i}\oint_C \frac{f(z)}{(z - z_0)^2}\, dz, \qquad (1')$$

$$f''(z_0) = \frac{2!}{2\pi i}\oint_C \frac{f(z)}{(z - z_0)^3}\, dz, \qquad (1'')$$

and in general

$$f^{(n)}(z)_0 = \frac{n!}{2\pi i}\oint_C \frac{f(z)}{(z - z_0)^{n+1}}\, dz \qquad (n = 1, 2, \cdots); \qquad (1)$$

here C is any simple closed path in D that encloses z_0 and whose full interior belongs to D; and we integrate counterclockwise around C (Fig. 331).

Comment. For memorizing (1), it is useful to observe that these formulas are obtained formally by differentiating the Cauchy formula (1), Sec. 13.5, under the integral sign *with respect to z_0*.

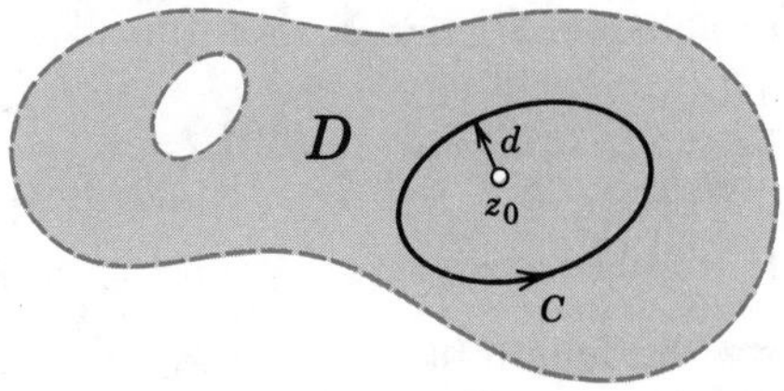

Fig. 331. Theorem 1 and its proof

Proof of Theorem 1. We prove (1′). We start from the definition

$$f'(z_0) = \lim_{\Delta z \to 0} \frac{f(z_0 + \Delta z) - f(z_0)}{\Delta z}.$$

On the right we represent $f(z_0 + \Delta z)$ and $f(z_0)$ by Cauchy's integral formula:

$$\frac{f(z_0 + \Delta z) - f(z_0)}{\Delta z} = \frac{1}{2\pi i \Delta z}\left[\oint_C \frac{f(z)}{z - (z_0 + \Delta z)}\, dz - \oint_C \frac{f(z)}{z - z_0}\, dz\right].$$

We write the two integrals as a single integral. Taking the common denominator gives the numerator $f(z)\{z - z_0 - [z - (z_0 + \Delta z)]\} = f(z)\, \Delta z$, so that Δz drops out and we get

$$\frac{f(z_0 + \Delta z) - f(z_0)}{\Delta z} = \frac{1}{2\pi i}\oint_C \frac{f(z)}{(z - z_0 - \Delta z)(z - z_0)}\, dz.$$

Clearly, we can now establish (1′) by showing that, as $\Delta z \to 0$, the integral on the right approaches the integral in (1′). To do this, we consider the difference between these two integrals. We can write this difference as a single integral by taking the common denominator and simplifying. This gives

$$\oint_C \frac{f(z)}{(z - z_0 - \Delta z)(z - z_0)}\, dz - \oint_C \frac{f(z)}{(z - z_0)^2}\, dz$$

$$= \oint_C \frac{f(z)\Delta z}{(z - z_0 - \Delta z)(z - z_0)^2}\, dz.$$

We show by the *ML*-inequality (Sec. 13.2) that the integral on the right approaches zero as $\Delta z \to 0$.

Being analytic, the function $f(z)$ is continuous on C, hence bounded in absolute value, say, $|f(z)| \leqq K$. Let d be the smallest distance from z_0 to the points of C (see Fig. 331). Then for all $|z|$ on C,

$$|z - z_0|^2 \geqq d^2, \qquad \text{hence} \qquad \frac{1}{|z - z_0|^2} \leqq \frac{1}{d^2}.$$

Furthermore, if $|\Delta z| \leqq d/2$, then for all z on C we also have

$$|z - z_0 - \Delta z| \geqq \frac{d}{2}, \qquad \text{hence} \qquad \frac{1}{|z - z_0 - \Delta z|} \leqq \frac{2}{d}.$$

Let L be the length of C. Then by the ML-inequality, if $|\Delta z| \leqq d/2$, then

$$\left| \oint_C \frac{f(z)\Delta z}{(z - z_0 - \Delta z)(z - z_0)^2}\, dz \right| \leqq K|\Delta z| \frac{2}{d} \cdot \frac{1}{d^2}.$$

This approaches zero as $\Delta z \to 0$. Formula (1′) is proved.

Note that we used Cauchy's integral formula (1), Sec. 13.5, but if all we had known about $f(z_0)$ is the fact that it can be represented by (1), Sec. 13.5, our argument would have established the existence of the derivative $f'(z_0)$ of $f(z)$. This is essential to the continuation and completion of this proof, because it implies that (1″) can be proved by a similar argument, with f replaced by f', and that the general formula (1) follows by induction. ∎

EXAMPLE 1 **Evaluation of line integrals**

From (1′), for any contour enclosing the point πi (counterclockwise)

$$\oint_C \frac{\cos z}{(z - \pi i)^2}\, dz = 2\pi i (\cos z)'\Big|_{z=\pi i} = -2\pi i \sin \pi i = 2\pi \sinh \pi.$$ ∎

EXAMPLE 2 From (1″), for any contour enclosing the point $-i$ (counterclockwise)

$$\oint_C \frac{z^4 - 3z^2 + 6}{(z + i)^3}\, dz = \pi i (z^4 - 3z^2 + 6)''\Big|_{z=-i} = \pi i [12z^2 - 6]_{z=-i} = -18\pi i.$$ ∎

EXAMPLE 3 By (1′), for any contour for which 1 lies inside and $\pm 2i$ lie outside (counterclockwise),

$$\oint_C \frac{e^z}{(z - 1)^2 (z^2 + 4)}\, dz = 2\pi i \left(\frac{e^z}{z^2 + 4} \right)'\Bigg|_{z=1} = 2\pi i \frac{e^z(z^2 + 4) - e^z 2z}{(z^2 + 4)^2}\Bigg|_{z=1} = \frac{6e\pi}{25} i \approx 2.050i.$$ ∎

Theorem 1 is also important in deriving general results on analytic functions. Let us show this by proving the converse of Cauchy's integral theorem:

Morera's[7] theorem

If $f(z)$ is continuous in a simply connected domain D and if

$$\oint_C f(z)\, dz = 0 \tag{2}$$

for every closed path in D, then $f(z)$ is analytic in D.

[7]GIACINTO MORERA (1856—1909), Italian mathematician, who worked in Genoa and Turin.

Proof. In Sec. 13.4 it was shown that if $f(z)$ is analytic in D, then

$$F(z) = \int_{z_0}^{z} f(z^*)\, dz^*$$

is analytic in D and $F'(z) = f(z)$. In the proof we used only the continuity of $f(z)$ and the property that its integral around every closed path in D is zero; from these assumptions we concluded that $F(z)$ is analytic. By Theorem 1, the derivative of $F(z)$ is analytic, that is, $f(z)$ is analytic in D, and Morera's theorem is proved. ∎

Cauchy's inequality. Theorem 1 also yields a basic inequality that has many applications. To get it, all we have to do is to choose for C in (1) a circle of radius r and center z_0 and apply the ML-inequality (Sec. 13.2); with $|f(z)| \leqq M$ on C we obtain from (1)

$$|f^{(n)}(z_0)| = \frac{n!}{2\pi}\left|\oint_C \frac{f(z)}{(z - z_0)^{n+1}}\, dz\right| \leqq \frac{n!}{2\pi} M \frac{1}{r^{n+1}} 2\pi r.$$

This gives **Cauchy's inequality**

$$|f^{(n)}(z_0)| \leqq \frac{n!M}{r^n}. \tag{3}$$

To gain a first impression of the importance of this inequality, let us prove a famous theorem on entire functions. (Definition in Sec. 12.6.)

Liouville's[8] theorem

If an entire function $f(z)$ is bounded in absolute value for all z, then $f(z)$ must be a constant.

Proof. By assumption, $|f(z)|$ is bounded, say, $|f(z)| < K$ for all z. Using (3), we see that $|f'(z_0)| < K/r$. Since $f(z)$ is entire, this is true for every r, so that we can take r as large as we please and conclude that $f'(z_0) = 0$. Since z_0 is arbitrary, $f'(z) = 0$ for all z, and $f(z)$ is constant (see Prob. 29 in Problem Set 12.5). This proves the theorem. ∎

This is the end of Chap. 13 on complex integration, which gave us a first impression of methods that have no counterpart in real integral calculus. We have seen that these methods result directly or indirectly from Cauchy's integral theorem (Sec. 13.3). More on integration follows in Chap. 15, based on material of Chap. 14.

[8]See Sec. 5.8, footnote 16.

Problem Set 13.6

Integrate the following functions counterclockwise around the circle $|z| = 2$. (n in Probs. 10 and 12–15 is a positive integer. a in Probs. 12 and 16 is any number.)

1. $z^4/(z - 3i)^2$ **2.** $z^2/(z - i)^2$ **3.** $e^{\pi z}/z^2$ **4.** $(\cos z)/z^3$
5. $(\cos z)/z^2$ **6.** $e^{z^2}/(z - 1)^2$ **7.** $z^3/(z + 1)^3$ **8.** $(\sin \pi z)/z^3$
9. $(e^z \sin z)/z^2$ **10.** e^z/z^n **11.** e^{z^3}/z^3 **12.** e^{az}/z^{n+1}
13. $z^n/(z + 1)^{n+1}$ **14.** $(\sin z)/z^{2n}$ **15.** $(\cos z)/z^{2n}$ **16.** $(\sinh az)/z^4$

Integrate $f(z)$ around the contour C (counterclockwise or as indicated).

17. $f(z) = z^{-2} \tan \pi z$, C any ellipse with foci $\pm i$

18. $f(z) = z^{-4}e^{z/2}$, C the circle $|z - 1 - i| = 2$

19. $f(z) = (z - \frac{1}{2}\pi)^{-2} \cot z$, C the boundary of the triangle with vertices $\pm i$ and 2

20. $f(z) = (z - 4)^{-3} \operatorname{Ln} z$, C the circle $|z - 5| = 3$

21. $f(z) = \dfrac{e^{z^2}}{z(z - 2i)^2}$, C consists of the boundary of the square with vertices $\pm 3 \pm 3i$ (counterclockwise) and $|z| = 1$ (clockwise)

22. $f(z) = \dfrac{\operatorname{Ln}(z + 3) + \cos z}{(z + 1)^2}$, C the boundary of the square with vertices ± 2, $\pm 2i$

23. If $f(z)$ is not a constant and is analytic for all (finite) z, and R and M are any positive real numbers (no matter how large), show that there exist values of z for which $|z| > R$ and $|f(z)| > M$. *Hint.* Use Liouville's theorem.

24. If $f(z)$ is a polynomial of degree $n > 0$ and M an arbitrary positive real number (no matter how large), show that there exists a positive real number R such that $|f(z)| > M$ for all $|z| > R$.

25. Show that $f(z) = e^z$ has the property characterized in Prob. 23, but does not have that characterized in Prob. 24.

26. Prove the **Fundamental theorem of algebra:** If $f(z)$ is a polynomial in z, not a constant, then $f(z) = 0$ for at least one value of z. *Hint.* Assume $f(z) \neq 0$ for all z and apply the result of Prob. 23 to $g = 1/f$.

Review Questions and Problems for Chapter 13

1. If a curve C is represented by $z = z(t)$, what is $\dot{z} = dz/dt$ geometrically?
2. What did we assume about paths of integration throughout this chapter?
3. What do you get if you integrate $1/z$ counterclockwise around the unit circle? (This is basic, so you should memorize it.) If you integrate $1/z^m$ ($m = 2, 3, \cdots$)?
4. Make a list of the integration methods in this chapter and illustrate each with a simple example.
5. Which integration methods in this chapter apply only to analytic functions and which to general continuous complex functions?

6. Which theorem in this chapter do you regard as the most important one? Why? State it from memory.
7. What do we mean by independence of path?
8. What is the principle of deformation of path? Give an example.
9. What is a doubly connected domain? How did we handle it in connection with Cauchy's theorem?
10. Don't confuse Cauchy's theorem and Cauchy's formula. State both. Which follows from which?
11. If the integral of a function $f(z)$ over the unit circle equals 2 and over the circle $|z| = 2$ equals 1, why can $f(z)$ not be analytic everywhere in the annulus bounded by these circles?
12. State the inequality that gives upper bounds for the absolute value of an integral. Does it apply only to analytic functions or more generally?
13. An analytic function $f(z)$ in a domain D is differentiable in D. Does it have higher derivatives in D? Answer the same question for a real function $f(x)$ that is differentiable on some interval of the real line.
14. Is $\operatorname{Re} \int_C f(z)\, dz = \int_C \operatorname{Re} f(z)\, dz$?
15. Using Liouville's theorem, show that e^z and $\cos z$ are not bounded in the complex plane.

Integrate:

16. $3z^2$ from $3i$ to $4 - i$ over any path
17. $z + 1/z$ clockwise around the unit circle
18. $\operatorname{Im} z$ counterclockwise around $|z| = r$
19. $|z|$ from $-i$ to i over the unit circle in the right half-plane
20. $(\operatorname{Ln} z)/(z - 2i)^2$ counterclockwise around $|z - 2i| = \pi/2$
21. e^{2z} from $14 + 3\pi i$ to $14 - \pi i$ along any path
22. e^z/z^4 counterclockwise around $|z| = 1/4$
23. $\bar{z}/|z|$ counterclockwise around $|z| = 4$
24. $z/(z^2 + 1)$ clockwise around $|z + i| = 1$
25. $z^{-2} \tan z$ clockwise around any contour enclosing 0
26. $5 \cos iz$ along the real axis from $-\pi$ to π
27. $\cosh z$ from 0 to i along the imaginary axis
28. $(z + i)^{-1}e^z \tan z$ counterclockwise around $|z - i| = 1$
29. $|z| + z$ clockwise around the unit circle
30. $(3z + 5)/(z^2 + z)$ counterclockwise around any ellipse with foci ± 1
31. $e^z \cos (e^z)$ from $-\pi i$ to $2\pi i$ along any path
32. e^z/z counterclockwise around $|z| = 1$ and clockwise around $|z| = 0.1$
33. $z \cos z^2$ from 0 to πi over any path
34. $(2z - 1)^{-2}(1 + z) \sin z$ counterclockwise around $|z - i| = 2$
35. $1/(z^2 + 4)$ clockwise around the ellipse $x^2 + 4(y - 2)^2 = 4$
36. $z^{-3} \cosh z$ counterclockwise around the unit circle
37. $(2z^3 - 3)/[z(z - 1 - i)^2]$ counterclockwise around $|z| = 2$ and clockwise around $|z| = 1$
38. $(z - i)^{-3}(z^3 + \sin z)$ clockwise around any ellipse with foci $\pm 2i$
39. $(\tan \pi z)/(z - 1)^2$ counterclockwise around the circle $|z - 1| = 0.1$
40. $(\sin 2\pi z)/(z - i)^3$ counterclockwise around $|z| = 2$

Summary of Chapter 13 Complex Integration

The **complex line integral** of a function $f(z)$ taken over a path C is denoted by (Sec. 13.1)

$$\int_C f(z)\,dz \qquad \text{or, if } C \text{ is closed, also by} \qquad \oint_C f(z)\,dz.$$

Such an integral can be evaluated by using the equation $z = z(t)$ of C, where $a \leqq t \leqq b$ (Sec. 13.2):

$$(1) \qquad \int_C f(z)\,dz = \int_a^b f(z(t))\dot{z}(t)\,dt \qquad \left(\dot{z} = \frac{dz}{dt}\right).$$

As another method, if $f(z)$ is analytic in a simply connected domain D (Sec. 13.3), then there exists an $F(z)$ in D such that $F'(z) = f(z)$ and for every path C in D from a point z_0 to a point z_1 we have (Secs. 13.2, 13.4)

$$(2) \qquad \int_C f(z)\,dz = F(z_1) - F(z_0) \qquad [F'(z) = f(z)].$$

Cauchy's integral theorem states that if $f(z)$ is analytic in a simply connected domain D, then for every closed path C in D (Sec. 13.3),

$$(3) \qquad \oint_C f(z)\,dz = 0.$$

Under the same assumptions and for any z_0 in D and closed path C in D containing z_0 in its interior we also have **Cauchy's integral formula**

$$(4) \qquad f(z_0) = \frac{1}{2\pi i}\oint_C \frac{f(z)}{z - z_0}\,dz.$$

Furthermore, then $f(z)$ has derivatives of all orders in D that are themselves analytic functions in D and (Sec. 13.6)

$$(5) \qquad f^{(n)}(z_0) = \frac{n!}{2\pi i}\oint_C \frac{f(z)}{(z - z_0)^{n+1}}\,dz \qquad (n = 1, 2, \cdots).$$

This implies *Morera's theorem* (the converse of Cauchy's integral theorem) and *Cauchy's inequality* (Sec. 13.6).

Chapter 14

Power Series, Taylor Series, Laurent Series

Section 14.1 contains the basic concepts and convergence tests for complex series, which are similar to those for real series usually discussed in calculus. Readers reasonably familiar with the latter may use Sec. 14.1 for reference and start with Sec. 14.2, in which we give a thorough discussion of **power series.**

Beginning in Sec. 14.3, we explain why power series play a fundamental role in complex analysis and its applications. The reason is that power series represent analytic functions (Sec. 14.3, Theorem 5) and, conversely, *every* analytic function can be represented by power series, called **Taylor series,** analogs of those in calculus; see Sec. 14.4). Section 14.5 concerns practical methods for obtaining them. In Sec. 14.6 we discuss uniform convergence of power and other series.

Laurent series (Sec. 14.7) are series of positive *and negative* integer powers of z (or $z - z_0$). These also help in classifying singularities (Sec. 14.8) and lead to an elegant general integration method (''residue integration,'' to be explained in the next chapter).

Prerequisites for this chapter: Chaps. 12, 13.
Sections that may be omitted in a shorter course: Secs. 14.5, 14.6, 14.8 (and use Sec. 14.1 only for reference).
References: Appendix 1, Part D.
Answers to Problems: Appendix 2.

14.1 Sequences, Series, Convergence Tests

In this section we define the basic concepts for ***complex*** sequences and series and discuss tests for convergence and divergence. This is very similar to *real* sequences and series in calculus. ***If you feel at home with the latter and want to take for granted that the ratio test also holds in complex, skip this section and go to Sec. 14.2.***

Sequences

The basic definitions are as in calculus. An *infinite sequence* or, briefly, a **sequence,** is obtained by assigning to each positive integer n a number z_n, called a **term** of the sequence, and is written

$$z_1, z_2, \cdots \qquad \text{or} \qquad \{z_1, z_2, \cdots\} \qquad \text{or briefly} \qquad \{z_n\}.$$

We may also write $z_0, z_1, \cdots$ or $z_2, z_3, \cdots$ or start with some other integer if convenient.

A *real sequence* is one whose terms are real.

Convergence. A **convergent sequence** $z_1, z_2, \cdots$ is one that has a limit c, written

$$\lim_{n\to\infty} z_n = c \qquad \text{or simply} \qquad z_n \to c.$$

By definition of limit this means that for every $\epsilon > 0$ we can find an N such that

$$(1) \qquad |z_n - c| < \epsilon \qquad \text{for all } n > N;$$

geometrically, all z_n with $n > N$ lie in the open disk of radius ϵ and center c (Fig. 332) and only finitely many do not lie in that disk. [For a *real* sequence, (1) gives an open interval of length 2ϵ and real midpoint c on the real line; see Fig. 333.]

A **divergent sequence** is one that does not converge.

EXAMPLE 1 Convergent and divergent sequences

The sequence $\{i^n/n\} = \{i, -1/2, -i/3, 1/4, \cdots\}$ is convergent with limit 0. The sequence $\{i^n\} = \{i, -1, -i, 1, \cdots\}$ is divergent, and so is $\{z_n\}$ with $z_n = (1 + i)^n$. ∎

EXAMPLE 2 Convergent sequence

The sequence $\{z_n\}$ with $z_n = x_n + iy_n = 2 - \frac{1}{n} + i\left(1 + \frac{2}{n}\right)$ is

$$1 + 3i, \quad \tfrac{3}{2} + 2i, \quad \tfrac{5}{3} + \tfrac{5}{3}i, \quad \tfrac{7}{4} + \tfrac{3}{2}i, \quad \cdots,$$

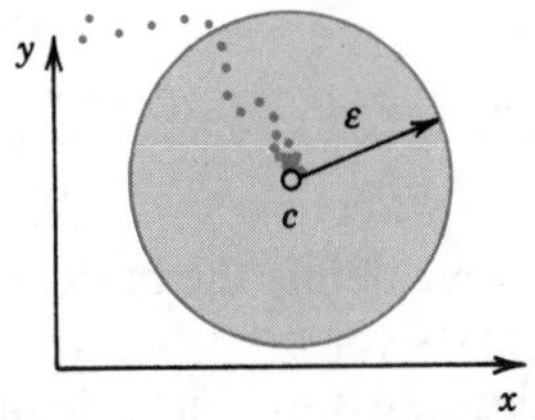

Fig. 332. Convergent complex sequence

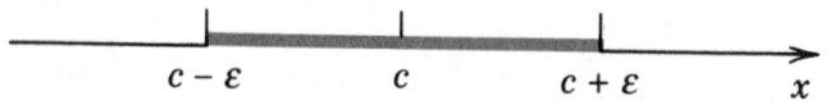

Fig. 333. Convergent real sequence

see Fig. 334. This sequence is convergent, the limit being $c = 2 + i$. Indeed, in (1) we have

$$|z_n - c| = \left|\frac{2n-1}{n} + i\frac{n+2}{n} - (2+i)\right| = \left|-\frac{1}{n} + \frac{2i}{n}\right| = \frac{\sqrt{5}}{n} < \epsilon \quad \text{if} \quad n > \frac{\sqrt{5}}{\epsilon}.$$

Furthermore, we see that the sequence $\{x_n\}$ of the real parts converges to $2 = \text{Re } c$, and $\{y_n\}$ converges to $1 = \text{Im } c$. This is typical. It illustrates the following theorem by which the convergence of a *complex* sequence can be referred back to that of the two *real* sequences of the real and the imaginary parts. ■

Theorem 1 **(Sequences of the real and the imaginary parts)**

A sequence $z_1, z_2, \cdots, z_n, \cdots$ of complex numbers $z_n = x_n + iy_n$ (where $n = 1, 2, \cdots$) converges to $c = a + ib$ if and only if the sequence of the real parts $x_1, x_2, \cdots$ converges to a and the sequence of the imaginary parts $y_1, y_2, \cdots$ converges to b.

Proof. If $|z_n - c| < \epsilon$, then $z_n = x_n + iy_n$ is within the circle of radius ϵ about $c = a + ib$ so that necessarily (Fig. 335*a*)

$$|x_n - a| < \epsilon, \qquad |y_n - b| < \epsilon.$$

Thus convergence $z_n \to c$ implies convergence $x_n \to a$ and $y_n \to b$.

Conversely, if $x_n \to a$ and $y_n \to b$ as $n \to \infty$, then for a given $\epsilon > 0$ we can choose N so large that, for every $n > N$,

$$|x_n - a| < \frac{\epsilon}{2}, \qquad |y_n - b| < \frac{\epsilon}{2}.$$

These two inequalities imply that $z_n = x_n + iy_n$ lies in a square with center c and side ϵ. Hence, z_n must lie within a circle of radius ϵ with center c (Fig. 335*b*). ■

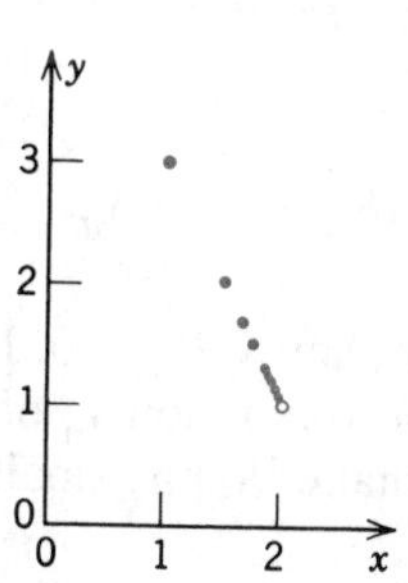

Fig. 334. Example 2

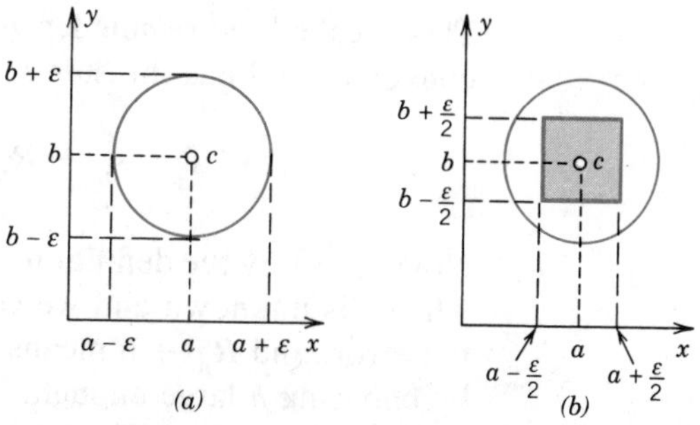

Fig. 335. Proof of Theorem 1

Series

Given a sequence $z_1, z_2, \cdots, z_m, \cdots$, we may form the sequence of sums

$$
\text{(2)} \qquad
\begin{aligned}
s_1 &= z_1 \\
s_2 &= z_1 + z_2 \\
&\cdots\cdots\cdots \\
s_n &= z_1 + z_2 + \cdots + z_n \\
&\cdots,
\end{aligned}
$$

called the sequence of **partial sums** of the *infinite series*, or **series**

$$
\text{(3)} \qquad \sum_{m=1}^{\infty} z_m = z_1 + z_2 + \cdots .
$$

The $z_1, z_2, \cdots$ are called the **terms** of the series. (Our usual ***summation letter*** is n, unless we need n for another purpose, as here.)

A **convergent series** is one whose sequence of partial sums converges, say,

$$\lim_{n\to\infty} s_n = s.$$

s is called the **sum** or *value* of the series and we write

$$s = \sum_{m=1}^{\infty} z_m = z_1 + z_2 + \cdots .$$

A **divergent series** is one that does not converge.

If we omit the terms of s_n from (3), there remains

$$
\text{(4)} \qquad R_n = z_{n+1} + z_{n+2} + z_{n+3} + \cdots .
$$

This is called the **remainder** *of the series* (3) *after the term* z_n. Clearly, if (3) converges and has the sum s, then

$$s = s_n + R_n, \qquad \text{thus} \qquad R_n = s - s_n.$$

Now $s_n \to s$ by the definition of convergence; hence $R_n \to 0$. In applications, when s is unknown and we compute an approximation s_n of s, then $|R_n|$ is the error, and $R_n \to 0$ means that we can make $|R_n|$ as small as we please, by choosing n large enough.

An application of Theorem 1 to the partial sums immediately relates the convergence of a complex series to that of the two series of its real parts and of its imaginary parts:

Theorem 2 **(Real and imaginary parts)**

A series (3) with $z_m = x_m + iy_m$ converges with sum $s = u + iv$ if and only if $x_1 + x_2 + \cdots$ converges with the sum u and $y_1 + y_2 + \cdots$ converges with the sum v.

Tests for Convergence and Divergence of Series

Convergence tests in complex are practically the same as in calculus. We apply them before we use a series, to make sure that the series converges.

Divergence can often be shown very simply as follows.

Theorem 3 **(Divergence)**

If a series $z_1 + z_2 + \cdots$ converges, then $\lim_{m\to\infty} z_m = 0$. Hence if this does not hold, the series diverges.

Proof. If $z_1 + z_2 + \cdots$ converges with the sum s, then, since $z_m = s_m - s_{m-1}$,

$$\lim_{m\to\infty} z_m = \lim_{m\to\infty} (s_m - s_{m-1}) = \lim_{m\to\infty} s_m - \lim_{m\to\infty} s_{m-1} = s - s = 0. \quad \blacksquare$$

Caution! $z_m \to 0$ is *necessary* for convergence but *not sufficient,* as we see from the harmonic series $1 + \frac{1}{2} + \frac{1}{3} + \frac{1}{4} + \cdots$, which satisfies this condition but diverges, as is shown in calculus (see, for example Ref. [12] listed in Appendix 1).

The practical difficulty in proving convergence is that in most cases the sum of a series is unknown. Cauchy overcame this by showing that a series converges if and only if its partial sums eventually get close to each other:

Theorem 4 **(Cauchy's convergence principle for series)**

A series $z_1 + z_2 + \cdots$ is convergent if and only if for every given $\epsilon > 0$ (no matter how small) we can find an N (which depends on ϵ, in general) such that

$$|z_{n+1} + z_{n+2} + \cdots + z_{n+p}| < \epsilon \qquad \text{for every } n > N \text{ and } p = 1, 2, \cdots. \tag{5}$$

The somewhat involved proof is left optional (see Appendix 4).

Absolute convergence. A series $z_1 + z_2 + \cdots$ is called **absolutely convergent** if the series of the absolute values of the terms

$$\sum_{m=1}^{\infty} |z_m| = |z_1| + |z_2| + \cdots$$

is convergent.

If $z_1 + z_2 + \cdots$ converges but $|z_1| + |z_2| + \cdots$ diverges, then the series $z_1 + z_2 + \cdots$ is called, more precisely, **conditionally convergent.**

EXAMPLE 3 A conditionally convergent series

The series $1 - \frac{1}{2} + \frac{1}{3} - \frac{1}{4} + - \cdots$ converges, but only conditionally since the harmonic series diverges, as mentioned above (after Theorem 3). ▮

If a series is absolutely convergent, it is convergent.

This follows readily from Cauchy's principle (Problem 14). This principle also yields the following general convergence test.

Theorem 5 (Comparison test)

If a series $z_1 + z_2 + \cdots$ is given and we can find a converging series $b_1 + b_2 + \cdots$ with nonnegative real terms such that

$$|z_n| \leqq b_n \qquad \text{for } n = 1, 2, \cdots,$$

then the given series converges, even absolutely.

Proof. By Cauchy's principle, since $b_1 + b_2 + \cdots$ converges, for any given $\epsilon > 0$ we can find an N such that

$$b_{n+1} + \cdots + b_{n+p} < \epsilon \qquad \text{for every } n > N \text{ and } p = 1, 2, \cdots.$$

From this and $|z_1| \leqq b_1$, $|z_2| \leqq b_2$, $\cdots$ we conclude that for those n and p,

$$|z_{n+1}| + \cdots + |z_{n+p}| \leqq b_{n+1} + \cdots + b_{n+p} < \epsilon.$$

Hence, again by Cauchy's principle, $|z_1| + |z_2| + \cdots$ converges, so that $z_1 + z_2 + \cdots$ is absolutely convergent. ▮

A good comparison series is the geometric series, which behaves as follows.

Theorem 6 (Geometric series)

*The **geometric series***

$$\sum_{m=0}^{\infty} q^m = 1 + q + q^2 + \cdots$$

converges with the sum $1/(1 - q)$ if $|q| < 1$ and diverges if $|q| \geqq 1$.

Proof. If $|q| \geqq 1$, then $|q^m| \geqq 1$ and Theorem 3 implies divergence.

Now let $|q| < 1$. The nth partial sum is

$$s_n = 1 + q + \cdots + q^n.$$

From this,

$$qs_n = \quad q + \cdots + q^n + q^{n+1}.$$

On subtraction, most terms on the right cancel in pairs, and we are left with

$$s_n - qs_n = (1 - q)s_n = 1 - q^{n+1}.$$

Now $1 - q \neq 0$ since $q \neq 1$, and we may solve for s_n, finding

(6) $$s_n = \frac{1 - q^{n+1}}{1 - q} = \frac{1}{1 - q} - \frac{q^{n+1}}{1 - q}.$$

Since $|q| < 1$, the last term approaches zero as $n \to \infty$. Hence the series is convergent and has the sum $1/(1 - q)$. This completes the proof. ∎

Ratio Test

This is the most important test in our further work. We get it by taking the geometric series as the comparison series $b_1 + b_2 + \cdots$ in Theorem 5:

Theorem 7 **(Ratio test)**

If a series $z_1 + z_2 + \cdots$ with $z_n \neq 0$ $(n = 1, 2, \cdots)$ has the property that for every n greater than some N,

(7) $$\left|\frac{z_{n+1}}{z_n}\right| \leqq q < 1 \qquad (n > N)$$

(where $q < 1$ is fixed), this series converges absolutely. If for every $n > N$,

(8) $$\left|\frac{z_{n+1}}{z_n}\right| \geqq 1 \qquad (n > N),$$

the series diverges.

Proof. If (8) holds, then $|z_{n+1}| \geqq |z_n|$ for those n, so that divergence of the series follows from Theorem 3.

If (7) holds, then $|z_{n+1}| \leqq |z_n|q$ for $n > N$, in particular,

$$|z_{N+2}| \leqq |z_{N+1}|q, \qquad |z_{N+3}| \leqq |z_{N+2}|q \leqq |z_{N+1}|q^2, \qquad \text{etc.},$$

and in general, $|z_{N+p}| \leqq |z_{N+1}|q^{p-1}$. Hence

$$|z_{N+1}| + |z_{N+2}| + |z_{N+3}| + \cdots \leqq |z_{N+1}| (1 + q + q^2 + \cdots).$$

Absolute convergence of $z_1 + z_2 + \cdots$ now follows from the comparison test and the geometric series, since $q < 1$ (Theorems 5 and 6). ∎

Caution! The inequality (7) implies $|z_{n+1}/z_n| < 1$, but this does ***not*** imply convergence, as we see from the harmonic series, which satisfies $z_{n+1}/z_n = n/(n + 1) < 1$ for all n but diverges.

If the sequence of the ratios in (7) and (8) converges, we get the more convenient

Theorem 8 **(Ratio test)**

If a series $z_1 + z_2 + \cdots$ *with* $z_n \neq 0$ $(n = 1, 2, \cdots)$ *is such that*

$$\lim_{n\to\infty} \left|\frac{z_{n+1}}{z_n}\right| = L, \tag{9}$$

then we have the following.

(a) *If* $L < 1$, *the series converges absolutely.*

(b) *If* $L > 1$, *it diverges.*

(c) *If* $L = 1$, *the test fails; that is, no conclusion is possible.*

Proof. **(a)** Write $k_n = |z_{n+1}/z_n|$ in (9). Let $L = 1 - b < 1$. Then, by the definition of a limit, the k_n must eventually get close to $1 - b$, say, $k_n \leqq q = 1 - \frac{1}{2}b < 1$ for all n greater than some N. Convergence of $z_1 + z_2 + \cdots$ now follows from Theorem 7.

(b) Similarly, for $L = 1 + c > 1$ we have $k_n \geqq 1 + \frac{1}{2}c > 1$ for all $n > N^*$ (sufficiently large), which implies divergence of $z_1 + z_2 + \cdots$ by Theorem 7.

(c) The harmonic series has $z_{n+1}/z_n = n/(n+1)$, hence $L = 1$ and diverges. The series

$$1 + \frac{1}{4} + \frac{1}{9} + \frac{1}{16} + \frac{1}{25} + \cdots \quad \text{has} \quad \frac{z_{n+1}}{z_n} = \frac{n^2}{(n+1)^2},$$

hence also $L = 1$, but converges. Convergence follows from (Fig. 336)

$$s_n = 1 + \frac{1}{4} + \cdots + \frac{1}{n^2} \leqq 1 + \int_1^n \frac{dx}{x^2} = 2 - \frac{1}{n},$$

so that $s_1, s_2, \cdots$ is a bounded sequence and is monotone increasing (since the terms of the series are all positive); both properties together are sufficient for the convergence of the real sequence $s_1, s_2, \cdots$. ∎

EXAMPLE 4 **Ratio test**

Is the following series convergent or divergent? (First guess, then calculate.)

$$\sum_{n=0}^{\infty} \frac{(100+75i)^n}{n!} = 1 + (100+75i) + \frac{1}{2!}(100+75i)^2 + \cdots$$

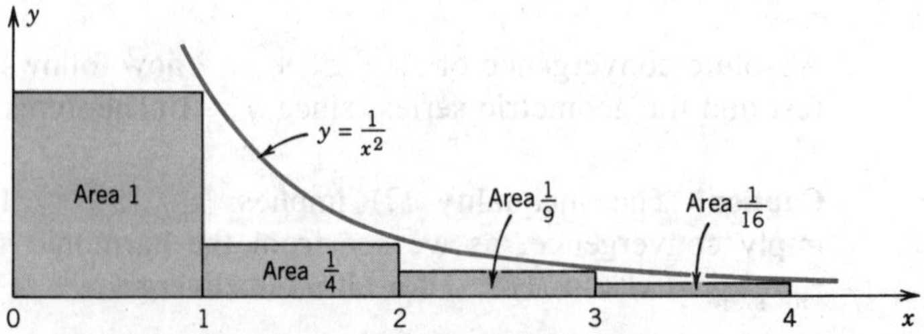

Fig. 336. Convergence of the series $1 + \frac{1}{4} + \frac{1}{9} + \frac{1}{16} + \cdots$

Solution. By Theorem 8, the series is convergent, since

$$\left|\frac{z_{n+1}}{z_n}\right| = \frac{|100 + 75i|^{n+1}/(n+1)!}{|100 + 75i|^n/n!} = \frac{|100 + 75i|}{n+1} = \frac{125}{n+1} \quad \rightarrow \quad L = 0.$$ ∎

EXAMPLE 5 **Theorem 7 more general than Theorem 8**

Is the following series convergent or divergent?

$$\sum_{n=0}^{\infty} \left(\frac{i}{2^{3n}} + \frac{1}{2^{3n+1}}\right) = i + \frac{1}{2} + \frac{i}{8} + \frac{1}{16} + \frac{i}{64} + \frac{1}{128} + \cdots .$$

Solution. The ratios of the absolute values of successive terms are $\frac{1}{2}, \frac{1}{4}, \frac{1}{2}, \frac{1}{4}, \cdots$. Hence convergence follows from Theorem 7. Since the sequence of these ratios has no limit, Theorem 8 is not applicable. ∎

Root Test

The two practically most important tests are the ratio test and the root test. The ratio test is generally simpler; the root test is somewhat more general.

Theorem 9 **(Root test)**

If a series $z_1 + z_2 + \cdots$ is such that for every n greater than some N,

$$\sqrt[n]{|z_n|} \leqq q < 1 \qquad (n > N) \tag{10}$$

(where $q < 1$ is fixed), this series converges absolutely. If for infinitely many n,

$$\sqrt[n]{|z_n|} \geqq 1, \tag{11}$$

the series diverges.

Proof. If (10) holds, then $|z_n| \leqq q^n < 1$ for all $n > N$. Hence the series $|z_1| + |z_2| + \cdots$ converges by comparison with the geometric series, so that $z_1 + z_2 + \cdots$ converges absolutely. If (11) holds, then $|z_n| \geqq 1$ for infinitely many n. Divergence of $z_1 + z_2 + \cdots$ now follows from Theorem 3. ∎

Caution! Equation (10) implies $\sqrt[n]{|z_n|} < 1$, but this does not imply convergence, as we see from the harmonic series, which satisfies $\sqrt[n]{1/n} < 1$ but diverges.

If the sequence of the roots in (10) and (11) converges, we more conveniently have

Theorem 10 **(Root test)**

If a series $z_1 + z_2 + \cdots$ is such that

$$\lim_{n \to \infty} \sqrt[n]{|z_n|} = L, \tag{12}$$

then we have the following.

(a) *If $L < 1$, the series converges absolutely.*

(b) *If $L > 1$, it diverges.*

(c) *If $L = 1$, the test fails; that is, no conclusion is possible.*

Proof. The proof parallels that of Theorem 8.

(a) Let $L = 1 - a^* < 1$. Then by the definition of a limit we have $\sqrt[n]{|z_n|} < q = 1 - \frac{1}{2}a^* < 1$ for all n greater than some (sufficiently large) N^*. Hence $|z_n| < q^n < 1$ for all $n > N^*$. Absolute convergence of the series $z_1 + z_2 + \cdots$ now follows by the comparison with the geometric series.

(b) If $L > 1$, also $\sqrt[n]{|z_n|} > 1$ for all sufficiently large n. Hence $|z_n| > 1$ for those n. Theorem 3 now implies that $z_1 + z_2 + \cdots$ diverges.

(c) Both the *divergent* harmonic series and the *convergent* series $1 + \frac{1}{4} + \frac{1}{9} + \frac{1}{16} + \frac{1}{25} + \cdots$ give $L = 1$. This can be seen from $(\ln n)/n \to 0$ and

$$\sqrt[n]{\frac{1}{n}} = \frac{1}{n^{1/n}} = \frac{1}{e^{(1/n)\ln n}} \to \frac{1}{e^0}, \qquad \sqrt[n]{\frac{1}{n^2}} = \frac{1}{n^{2/n}} = \frac{1}{e^{(2/n)\ln n}} \to \frac{1}{e^0}. \quad ■$$

EXAMPLE 6 Root test

Is the following series convergent or divergent?

$$\sum_{n=0}^{\infty} \frac{(-1)^n}{2^{2n} + 3}(4 - i)^n = \frac{1}{4} - \frac{1}{7}(4 - i) + \frac{1}{19}(4 - i)^2 - + \cdots.$$

Solution. By Theorem 10, the series diverges, since

$$\sqrt[n]{\frac{|(4 - i)^n|}{2^{2n} + 3}} = \frac{|4 - i|}{\sqrt[n]{4^n + 3}} = \frac{\sqrt{17}}{\sqrt[n]{4^n + 3}} \to L = \frac{\sqrt{17}}{4} > 1. \quad ■$$

This is the end of our discussion of basic concepts and facts on complex series and convergence tests. In the next section we begin our actual work.

Problem Set 14.1

Sequences

Are the following sequences $z_1, z_2, \cdots, z_n, \cdots$ bounded? Convergent? Find their limit points.

1. $z_n = (-1)^n + 2i$ **2.** $z_n = e^{in\pi/2}$ **3.** $z_n = (-1)^n/(n + i)$

4. $z_n = (3 + 4i)^n/n!$ **5.** $(3i)^n - (1 + i)^n$ **6.** $z_n = \frac{1}{2}\pi + e^{in\pi/4}/n\pi$

7. $z_n = (-1)^n + i/n$ **8.** $z_n = n\pi/(1 + 2in)$ **9.** $z_n = i^n \cos n\pi$

10. (Uniqueness of limit) Show that if a sequence converges, its limit is unique.

11. If $z_1, z_2, \cdots$ converges with the limit l and $z_1^*, z_2^*, \cdots$ converges with the limit l^*, show that $z_1 + z_1^*, z_2 + z_2^*, \cdots$ converges with the limit $l + l^*$.

12. Show that under the assumptions of Prob. 11 the sequence $z_1z_1^*, z_2z_2^*, \cdots$ converges with the limit ll^*.

13. Show that a complex sequence $z_1, z_2, \cdots$ is bounded if and only if the two corresponding sequences of the real parts and the imaginary parts are bounded.

Series

14. (Absolute convergence) Show that if a series converges absolutely, it is convergent.

Are the following series convergent or divergent?

15. $\sum_{n=0}^{\infty} \frac{(100 + 200i)^n}{n!}$ **16.** $\sum_{n=0}^{\infty} \frac{n - i}{3n + 2i}$ **17.** $\sum_{n=1}^{\infty} \frac{i^n}{n}$

18. $\sum_{n=0}^{\infty} n\left(\frac{1 + i}{2}\right)^n$ **19.** $\sum_{n=0}^{\infty} \left(\frac{8i}{9}\right)^n n^4$ **20.** $\sum_{n=1}^{\infty} \frac{n^{2n} + i^{2n}}{n!}$

21. $\sum_{n=1}^{\infty} \frac{(2i)^n n!}{n^n}$ **22.** $\sum_{n=0}^{\infty} \frac{(10 + 7i)^{8n}}{(2n)!}$ **23.** $\sum_{n=1}^{\infty} \frac{n + 1}{2^n n}$

24. Suppose that $|z_{n+1}/z_n| \leqq q < 1$, so that the series $z_1 + z_2 + \cdots$ converges by the ratio test (Theorem 7). Show that the remainder $R_n = z_{n+1} + z_{n+2} + \cdots$ satisfies $|R_n| \leqq |z_{n+1}|/(1 - q)$. (*Hint.* Use the fact that the ratio test is a comparison of the series $z_1 + z_2 + \cdots$ with the geometric series.)

25. Using Prob. 24, find how many terms suffice for computing the sum s of the series in Prob. 23 with an error not exceeding 0.05 and compute s to this accuracy.

14.2 Power Series

Power series are the most important series in complex analysis, as was mentioned at the beginning of the chapter and as we shall now see in detail.

A **power series** *in powers of* $z - z_0$ is a series of the form

$$\sum_{n=0}^{\infty} a_n(z - z_0)^n = a_0 + a_1(z - z_0) + a_2(z - z_0)^2 + \cdots \tag{1}$$

where z is a variable, $a_0, a_1, \cdots$ are constants, called the **coefficients** of the series, and z_0 is a constant, called the **center** of the series.

If $z_0 = 0$, we obtain as a particular case a *power series in powers of* z:

$$\sum_{n=0}^{\infty} a_n z^n = a_0 + a_1 z + a_2 z^2 + \cdots . \tag{2}$$

Convergence Behavior of Power Series

We have made the definitions in the last section for series of *constant terms*. If the terms of a series are *variable*, say, functions of a variable z (for example, powers of z, as in a power series), they assume definite values if we fix z, and then all those definitions apply. Clearly, for a series of functions of z, the partial sums, the remainders, and the sum will be functions of z. Usually such a series will converge for some z, for instance, throughout some region, and diverge for the other z. In general, such a region may be complicated. For a *power series* it is simple. Indeed, we shall see (in Theorem 1, below) that (1) converges throughout a disk with center z_0—this is the general case—or in the whole complex plane; or it converges merely at the center z_0 [in which case (1) is practically useless]. Let us first illustrate these three possibilities with three typical examples.

EXAMPLE 1 Convergence in a disk. Geometric series

The *geometric series*

$$\sum_{n=0}^{\infty} z^n = 1 + z + z^2 + \cdots$$

converges absolutely if $|z| < 1$ and diverges if $|z| \geqq 1$ (see Theorem 6 in Sec. 14.1). ∎

EXAMPLE 2 Convergence for every z

The power series

$$\sum_{n=0}^{\infty} \frac{z^n}{n!} = 1 + z + \frac{z^2}{2!} + \frac{z^3}{3!} + \cdots$$

is absolutely convergent for every z. In fact, by the ratio test, for any fixed z,

$$\left| \frac{z^{n+1}/(n+1)!}{z^n/n!} \right| = \frac{|z|}{n+1} \to 0 \quad \text{as} \quad n \to \infty.$$ ∎

EXAMPLE 3 Convergence only at the center. (Useless series)

The power series

$$\sum_{n=0}^{\infty} n!\, z^n = 1 + z + 2z^2 + 6z^3 + \cdots$$

converges only at $z = 0$, but diverges for every $z \neq 0$. In fact, from the ratio test we have

$$\left| \frac{(n+1)!\,z^{n+1}}{n!\,z^n} \right| = (n+1)|z| \to \infty \quad \text{as} \quad n \to \infty \quad (z \text{ fixed and } \neq 0).$$ ∎

Every power series (1) converges at the center because for $z = z_0$ the series reduces to the single term a_0. If this is not the only point of convergence, so that (1) also converges for some $z_1 \neq z_0$, then (1) converges for every z that is closer to z_0 than z_1:

Theorem 1 (Convergence of a power series)

If the power series (1) *converges at a point* $z = z_1 \neq z_0$, *it converges absolutely for every* z *closer to* z_0 *than* z_1, *that is,* $|z - z_0| < |z_1 - z_0|$. See Fig. 337.

If (1) *diverges at a* $z = z_2$, *it diverges for every* z *farther away from* z_0 *than* z_2. See Fig. 337.

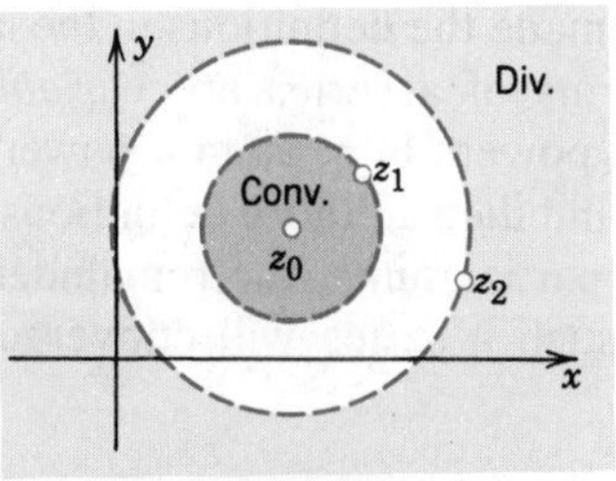

Fig. 337. Theorem 1

Proof. Since the series (1) converges for z_1. Theorem 3 in Sec. 14.1 gives

$$a_n(z_1 - z_0)^n \to 0 \quad \text{as} \quad n \to \infty.$$

This implies that for $z = z_1$ the terms of the series (1) are bounded, say,

$$|a_n(z_1 - z_0)^n| < M \qquad \text{for every } n = 0, 1, \cdots.$$

Multiplying and dividing by $(z_1 - z_0)^n$, we obtain from this

$$\text{(3)} \qquad |a_n(z - z_0)^n| = \left| a_n(z_1 - z_0)^n \left(\frac{z - z_0}{z_1 - z_0}\right)^n \right| \leqq M \left|\frac{z - z_0}{z_1 - z_0}\right|^n.$$

Now our assumption $|z - z_0| < |z_1 - z_0|$ implies that

$$\left|\frac{z - z_0}{z_1 - z_0}\right| < 1. \qquad \text{Hence the series} \qquad M \sum_{n=0}^{\infty} \left|\frac{z - z_0}{z_1 - z_0}\right|^n$$

is a converging geometric series (see Theorem 6 in Sec. 14.1). The convergence of (1) when $|z - z_0| < |z_1 - z_0|$ now follows from (3) and the comparison test in Sec. 14.1.

If the second statement of our theorem were false, we would have convergence at a z_3 such that $|z_3 - z_0| > |z_2 - z_0|$, implying convergence at z_2 by the statement just proved, contradicting the assumed divergence at z_2. ∎

Radius of Convergence of a Power Series

Examples 2 and 3 illustrate that a power series may converge for all z or only for $z = z_0$. Let us exclude these two cases for a moment. Then, if a power series (1) is given, we may consider all the points z in the complex plane for which the series converges. Let R be such that each of those points has distance from the center z_0 at most equal to R, and assume that R is the smallest possible number with this property; in other words, R is the radius of the *smallest* circle with center z_0 that includes all the points at which the series converges. Theorem 1 then implies convergence for all z within that circle, that is, for all z for which

$$\textbf{(4)} \qquad |z - z_0| < R.$$

And since R is as small as possible, the series diverges for all z for which

$$|z - z_0| > R.$$

The circle

$$|z - z_0| = R$$

is called the **circle of convergence** and its radius R the **radius of convergence** of (1). See Fig. 338 on the next page.

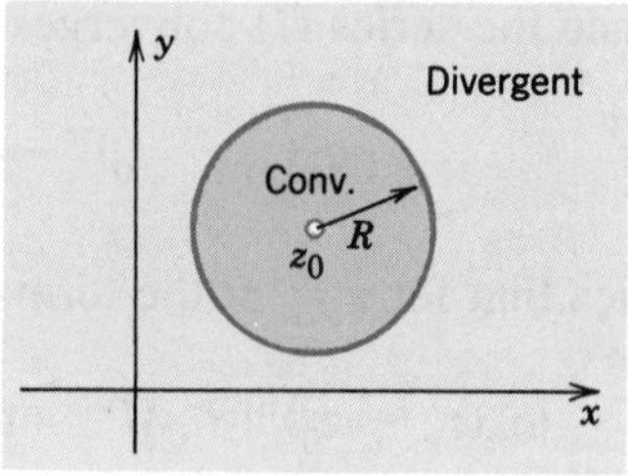

Fig. 338. Circle of convergence

To include those two excluded cases in the notation, we write

$R = \infty$ if the series (1) converges for all z (as in Example 2),

$R = 0$ if (1) converges only at the center $z = z_0$ (as in Example 3).

These are convenient notations, but nothing else.

For a *real* power series (1) in powers of $x - x_0$ with real coefficients and center, formula (4) gives the **convergence interval** $|x - x_0| < R$ of length $2R$ on the real line.

To avoid misunderstandings: no general statements can be made about the convergence of a power series (1) ***on the circle of convergence*** itself. The series (1) may converge at some or all or none of these points. Details will not be essential to us; hence a simple example may just give us the idea.

EXAMPLE 4 Behavior on the circle of convergence

On the circle of convergence (radius $R = 1$ in all three series),

$\Sigma z^n/n^2$ converges everywhere since $\Sigma 1/n^2$ converges,

$\Sigma z^n/n$ converges at -1 (by Leibniz's test) but diverges at 1,

Σz^n diverges everywhere. ∎

The radius of convergence R of the power series (1) may be determined from the coefficients of the series as follows.

Theorem 2 (Radius of convergence *R*)

Suppose that the sequence $|a_{n+1}/a_n|$, $n = 1, 2, \cdots$, *converges with limit* L^*. *If* $L^* = 0$, *then* $R = \infty$; *that is, the power series* (1) *converges for all* z. *If* $L^* \neq 0$ (*hence* $L^* > 0$), *then*

$$R = \frac{1}{L^*} \qquad \text{(5)} \qquad \textbf{(Cauchy–Hadamard formula[1]).}$$

If $|a_{n+1}/a_n| \to \infty$, *then* $R = 0$ (*convergence only at the center* z_0).

Proof. The series (1) has the terms $Z_n = a_n(z - z_0)^n$. Hence in the ratio test (Theorem 8 in Sec. 14.1),

$$L = \lim_{n\to\infty} \left|\frac{Z_{n+1}}{Z_n}\right| = \lim_{n\to\infty} \left|\frac{a_{n+1}(z - z_0)^{n+1}}{a_n(z - z_0)^n}\right| = \lim_{n\to\infty} \left|\frac{a_{n+1}}{a_n}\right| |z - z_0|,$$

[1]Named after the French mathematicians, A. L. CAUCHY (see the footnote in Sec. 2.6) and JACQUES HADAMARD (1865—1963). Hadamard made basic contributions to the theory of power series and devoted his lifework to partial differential equations.

that is,

$$L = L^*|z - z_0|.$$

If $L^* = 0$, then $L = 0$ for every z, and the ratio test gives convergence for all z, as asserted. Let $L^* > 0$. If $|z - z_0| < 1/L^*$, then $L = L^*|z - z_0| < 1$, and (1) converges by the ratio test. If $|z - z_0| > 1/L^*$, then $L > 1$, and (1) diverges by the ratio test. By definition, this shows that $1/L^*$ is the radius of convergence R of (1) and proves (5). If $|a_{n+1}/a_n| \to \infty$, then $|Z_{n+1}/Z_n| \geqq 1$ for every $z \neq z_0$ and all sufficiently large n, so that divergence of (1) for all $z \neq z_0$ now follows from Theorem 7 in Sec. 14.1. ∎

Comment. If $L^* = \lim |a_{n+1}/a_n| \neq 0$, then (5) gives

$$R = \lim_{n\to\infty} \left|\frac{a_n}{a_{n+1}}\right|. \tag{6}$$

This is plausible, because if $a_n \to 0$ rapidly, then $|a_n/a_{n+1}|$ will be large on the average (why?) and the series converges in a large disk. If $a_n \to 0$ slowly, then $|a_n/a_{n+1}|$ will be relatively small and the series converges in a relatively small disk.

Formulas (5) and (6) will not help if L^* does not exist, but extensions of Theorem 2 are still possible, as we discuss in Example 6 below.

EXAMPLE 5 Radius of convergence

Determine the radius of convergence R of the power series

$$\sum_{n=0}^{\infty} \frac{(2n)!}{(n!)^2}(z - 3i)^n.$$

Solution. From (5),

$$L^* = \lim_{n\to\infty} \frac{(2n+2)!/[(n+1)!]^2}{(2n)!/(n!)^2} = \lim_{n\to\infty} \frac{(2n+2)(2n+1)}{(n+1)^2} = 4, \qquad R = \frac{1}{L^*} = \frac{1}{4}.$$

The series converges in the open disk $|z - 3i| < 1/4$, of radius 1/4 and center $3i$. ∎

EXAMPLE 6 Extension of Theorem 2

Find the radius of convergence R of the power series

$$\sum_{n=0}^{\infty}\left[1 + (-1)^n + \frac{1}{2^n}\right]z^n = 3 + 2^{-1}z + (2 + 2^{-2})z^2 + 2^{-3}z^3 + (2 + 2^{-4})z^4 + \cdots.$$

Solution. Since the sequence of the ratios $1/6$, $2(2 + 2^{-2})$, $1/2^3(2 + 2^{-2})$, $\cdots$ does not converge, Theorem 2 is of no help. It can be shown that

$$R = 1/\tilde{L}, \qquad \tilde{L} = \lim_{n\to\infty} \sqrt[n]{|a_n|}. \tag{5*}$$

This does still not help here, since $\{\sqrt[n]{|a_n|}\}$ does not converge because inspection shows that it has two limit points 1/2 and 1. It can further be shown that

$$R = 1/\tilde{l}, \qquad \tilde{l} \text{ the greatest limit point of the sequence } \{\sqrt[n]{|a_n|}\}. \tag{5**}$$

Here $\tilde{l} = 1$, so that $R = 1$. *Answer.* The series converges for $|z| < 1$. ∎

Summary. Power series converge in an open circular disk or some even for every z (or some only at the center, but they are useless); for the radius of convergence, see (5), (6), or Example 6.

Except for the useless ones, power series have sums that are analytic functions (as we show in the next section); this accounts for their importance.

Problem Set 14.2

Find the center and the radius of convergence of the following power series.

1. $\sum_{n=0}^{\infty} (z + 4i)^n$ **2.** $\sum_{n=1}^{\infty} n\pi^n(z - i)^n$ **3.** $\sum_{n=0}^{\infty} \left(\frac{\pi}{4}\right)^n z^{2n}$

4. $\sum_{n=0}^{\infty} \frac{z^{2n}}{n!}$ **5.** $\sum_{n=0}^{\infty} \frac{(z - 2i)^n}{5^n}$ **6.** $\sum_{n=1}^{\infty} \frac{z^{2n}}{n^2}$

7. $\sum_{n=0}^{\infty} \frac{2^{10n}}{n!}(z + i)^n$ **8.** $\sum_{n=1}^{\infty} n^n(z + 1)^n$ **9.** $\sum_{n=0}^{\infty} (3z - 2i)^n$

10. $\sum_{n=0}^{\infty} \frac{(-1)^n}{2^{2n}(n!)^2} z^{2n}$ **11.** $\sum_{n=0}^{\infty} \frac{i^n n^3}{2^n} z^{2n}$ **12.** $\sum_{n=0}^{\infty} \frac{(3n)!}{(n!)^3}(z + \pi i)^n$

13. $\sum_{n=0}^{\infty} \frac{z^{2n+1}}{(2n + 1)!}$ **14.** $\sum_{n=1}^{\infty} \frac{(z - 1)^n}{n^n}$ **15.** $\sum_{n=0}^{\infty} \frac{z^{n+2}}{(n + 1)(n + 2)}$

16. $\sum_{n=1}^{\infty} \frac{n!}{n^n}(z + 2)^n$ **17.** $\sum_{n=1}^{\infty} \frac{3^n}{2^n - 1} z^{2n}$ **18.** $\sum_{n=1}^{\infty} \frac{n^n}{n!}(z - 1)^{2n}$

19. Show that if a power series $\Sigma a_n z^n$ has radius of convergence R (assumed finite), then $\Sigma a_n z^{2n}$ has the radius of convergence $\sqrt{R}$.

20. Does there exist a power series in powers of z that converges at $z = 30 + 10i$ and diverges at $z = 31 - 6i$? (Give a reason.)

14.3 Functions Given by Power Series

The main goal of this section is to show that power series represent analytic functions (Theorem 5). On the way we shall see that power series behave nicely under addition, multiplication, differentiation, and integration, which makes them very useful in complex analysis.

To simplify the formulas in this section, we take $z_0 = 0$ and write

$$\sum_{n=0}^{\infty} a_n z^n. \tag{1}$$

This is no restriction, since in a series in powers of $z^* - z_0$ with any center z_0 we can always set $z^* - z_0 = z$ to reduce it to the form (1).

If an arbitrary power series (1) has a nonzero radius of convergence, its sum is a function of z, say, $f(z)$. Then we write

$$f(z) = \sum_{n=0}^{\infty} a_n z^n = a_0 + a_1 z + a_2 z^2 + \cdots \qquad (|z| < R). \tag{2}$$

We say that $f(z)$ *is* **represented** *by the power series* or that *it is* **developed** *in the power series*. For instance, the geometric series represents the function $f(z) = 1/(1 - z)$ in the interior of the unit circle $|z| = 1$. (See Example 1 in Sec. 14.2.)

Our first goal is to show the **uniqueness** of such a representation; that is, *a function $f(z)$ cannot be represented by two different power series with the same center.* If $f(z)$ can at all be developed in a power series with center z_0, the development is unique. This important fact is frequently used in complex and real analysis. This result (Theorem 2, below) will follow from

Theorem 1 **(Continuity of the sum of a power series)**

The function $f(z)$ in (2) with $R > 0$ is continuous at $z = 0$.

Proof. By the definition of continuity we must show that

$$\lim_{z \to 0} f(z) = f(0) = a_0,$$

that is, we must show that for a given $\epsilon > 0$ there is a $\delta > 0$ such that $|z| < \delta$ implies $|f(z) - a_0| < \epsilon$. Now (2) converges absolutely for $|z| \leqq r < R$, by Theorem 1 in Sec. 14.2. Hence the series

$$\sum_{n=1}^{\infty} |a_n| r^{n-1} = \frac{1}{r} \sum_{n=1}^{\infty} |a_n| r^n \qquad (r > 0)$$

converges. Let S be its sum. Then, for $0 < |z| \leqq r$,

$$\begin{aligned} |f(z) - a_0| = \left| \sum_{n=1}^{\infty} a_n z^n \right| &\leqq |z| \sum_{n=1}^{\infty} |a_n| \, |z|^{n-1} \\ &\leqq |z| \sum_{n=1}^{\infty} |a_n| r^{n-1} = |z| S. \end{aligned}$$

This is less than ϵ for $|z| < \delta$, where $\delta > 0$ is less than both r and ϵ/S. ∎

From this theorem we can now readily obtain the desired uniqueness theorem (again assuming $z_0 = 0$ without loss of generality):

Theorem 2 **(Identity theorem for power series)**
Suppose that the power series

$$\sum_{n=0}^{\infty} a_n z^n \qquad \textit{and} \qquad \sum_{n=0}^{\infty} b_n z^n$$

both converge for $|z| < R$, *where* R *is positive, and have the same sum for all these* z. *Then these series are identical, that is,*

$$a_n = b_n \qquad \textit{for all } n = 0, 1, \cdots.$$

Proof. We proceed by induction. By assumption,

$$a_0 + a_1 z + a_2 z^2 + \cdots = b_0 + b_1 z + b_2 z^2 + \cdots \qquad (|z| < R).$$

The sums of these two power series are continuous at $z = 0$, by Theorem 1. Hence $a_0 = b_0$: the assertion is true for $n = 0$. Now assume that $a_n = b_n$ for $n = 0, 1, \cdots, m$. Then we may omit on both sides the terms that are equal and divide the result by z^{m+1} ($\neq 0$); this gives

$$a_{m+1} + a_{m+2} z + a_{m+3} z^2 + \cdots = b_{m+1} + b_{m+2} z + b_{m+3} z^2 + \cdots.$$

By Theorem 1, each of these power series represents a function that is continuous at $z = 0$. Hence $a_{m+1} = b_{m+1}$. This completes the proof. ∎

Power Series Represent Analytic Functions

This will be shown as the main goal in this section, after a short preparation. In the next section we shall see that, conversely, *every* analytic function can be represented by power series (called *Taylor series*). This is quite surprising and accounts for the great importance of power series in complex analysis.

Termwise addition or subtraction of two power series with radii of convergence R_1 and R_2 yields a power series with radius of convergence at least equal to the smaller of R_1 and R_2. *Proof.* Add (or subtract) the partial sums s_n and s_n^* term by term and use $\lim (s_n \pm s_n^*) = \lim s_n \pm \lim s_n^*$.

Termwise multiplication of two power series

$$f(z) = \sum_{k=0}^{\infty} a_k z^k = a_0 + a_1 z + \cdots$$

and

$$g(z) = \sum_{m=0}^{\infty} b_m z^m = b_0 + b_1 z + \cdots,$$

means the multiplication of each term of the first series by each term of the second series and the collection of like powers of z. This gives a power series, which is called the **Cauchy product** of the two series and is given by

$$a_0b_0 + (a_0b_1 + a_1b_0)z + (a_0b_2 + a_1b_1 + a_2b_0)z^2 + \cdots$$
$$= \sum_{n=0}^{\infty} (a_0b_n + a_1b_{n-1} + \cdots + a_nb_0)z^n.$$

We mention without proof that this power series converges absolutely for each z within the circle of convergence of each of the two given series and has the sum $s(z) = f(z)g(z)$. For a proof, see [D7] listed in Appendix 1.

Termwise differentiation and integration of power series is permissible, as we show next. We call **derived series** *of the power series* (1) the power series obtained from (1) by termwise differentiation, that is,

$$\sum_{n=1}^{\infty} na_nz^{n-1} = a_1 + 2a_2z + 3a_3z^2 + \cdots. \tag{3}$$

Theorem 3 (Termwise differentiation of a power series)
The derived series of a power series has the same radius of convergence as the original series.

Proof. This follows from (5) in Sec. 14.2 because

$$\lim_{n\to\infty} \frac{(n+1)a_{n+1}}{na_n} = \lim_{n\to\infty} \frac{n+1}{n} \lim_{n\to\infty} \frac{a_{n+1}}{a_n} = \lim_{n\to\infty} \frac{a_{n+1}}{a_n}$$

or, if the last limit does not exist, from (5**) in Sec. 14.2 by noting that $\sqrt[n]{n} \to 1$ as $n \to \infty$. ∎

EXAMPLE 1 An application of Theorem 3
Find the radius of convergence R of the following series by applying Theorem 3.

$$\sum_{n=2}^{\infty} \binom{n}{2} z^n = z^2 + 3z^3 + 6z^4 + 10z^5 + \cdots.$$

Solution. Differentiate the geometric series twice term by term and multiply the result by $z^2/2$. This yields the given series. Hence $R = 1$ by Theorem 3. ∎

Theorem 4 (Termwise integration of power series)
The power series

$$\sum_{n=0}^{\infty} \frac{a_n}{n+1} z^{n+1} = a_0z + \frac{a_1}{2} z^2 + \frac{a_2}{3} z^3 + \cdots$$

obtained by integrating the series $a_0 + a_1z + a_2z^2 + \cdots$ term by term has the same radius of convergence as the original series.

The proof is similar to that of Theorem 3.

With Theorem 3 as a tool, we are now ready to establish our main result in this section:

Theorem 5 **(Analytic functions. Their derivatives)**
A power series with a nonzero radius of convergence R represents an analytic function at every point interior to its circle of convergence. The derivatives of this function are obtained by differentiating the original series term by term. All the series thus obtained have the same radius of convergence as the original series. Hence, by the first statement, each of them represents an analytic function.

Proof. **(a)** We consider any power series (1) with positive radius of convergence R. Let $f(z)$ be its sum and $f_1(z)$ the sum of its derived series; thus

$$f(z) = \sum_{n=0}^{\infty} a_n z^n \qquad \text{and} \qquad f_1(z) = \sum_{n=1}^{\infty} n a_n z^{n-1}. \tag{4}$$

We show that $f(z)$ is analytic and has the derivative $f_1(z)$ in the interior of the circle of convergence. We do this by proving that for any fixed z with $|z| < R$ and $\Delta z \to 0$ the difference quotient $[f(z + \Delta z) - f(z)]/\Delta z$ approaches $f_1(z)$. By termwise addition we first have from (4)

$$\frac{f(z + \Delta z) - f(z)}{\Delta z} - f_1(z) = \sum_{n=2}^{\infty} a_n \left[\frac{(z + \Delta z)^n - z^n}{\Delta z} - nz^{n-1}\right]. \tag{5}$$

Note that the summation starts with 2, since the constant term drops out in taking the difference $f(z + \Delta z) - f(z)$, and so does the linear term when we subtract $f_1(z)$ from the difference quotient.

(b) We claim that the series in (5) can be written

$$\sum_{n=2}^{\infty} a_n \Delta z[(z + \Delta z)^{n-2} + 2z(z + \Delta z)^{n-3} + \cdots + (n - 1)z^{n-2}]. \tag{6}$$

The somewhat technical proof of this is given in Appendix 4.

(c) We consider (6). The brackets contain $n - 1$ terms, and the largest coefficient is $n - 1$. Since $(n - 1)^2 < n(n - 1)$, we see that for $|z| \leqq R_0$ and $|z + \Delta z| \leqq R_0$, $R_0 < R$, the absolute value of this series cannot exceed

$$|\Delta z| \sum_{n=2}^{\infty} |a_n| n(n - 1) R_0^{\,n-2}.$$

This series with a_n instead of $|a_n|$ is the second derived series of (2) at $z = R_0$ and converges absolutely by Theorem 3 and Sec. 14.2, Theorem 1. Hence our present series converges. Let $K(R_0)$ be its sum. Then we can write our present result

$$\left|\frac{f(z + \Delta z) - f(z)}{\Delta z} - f_1(z)\right| \leqq |\Delta z|\, K(R_0).$$

Letting $\Delta z \to 0$ and noting that R_0 ($< R$) is arbitrary, we conclude that $f(z)$ is analytic at any point interior to the circle of convergence and its derivative is represented by the derived series. From this the statements about the higher derivatives follow by induction. ∎

Summary. The results in this section show that power series are about as nice as we could hope for: we can differentiate and integrate them term by term (Theorems 3 and 4). Theorem 5 accounts for the great importance of power series in complex analysis: the sum of such a series (with a positive radius of convergence) is an analytic function and has derivatives of all orders, which thus are analytic functions. But this is only part of the story. In the next section we show that, conversely, *every* given analytic function $f(z)$ can be represented by power series.

Problem Set 14.3

Find the radius of convergence of the following series in two ways, (a) directly by the Cauchy–Hadamard formula (Sec. 14.2), (b) by Theorem 3 or Theorem 4 and a series with simpler coefficients.

1. $\displaystyle\sum_{n=2}^{\infty} \frac{n(n-1)}{2^n}(z-i)^n$ **2.** $\displaystyle\sum_{n=1}^{\infty} \frac{3^n}{n(n+1)} z^n$ **3.** $\displaystyle\sum_{n=1}^{\infty} \frac{n}{5^n}(z+1)^{2n}$

4. $\displaystyle\sum_{n=0}^{\infty} \frac{(-1)^n}{2n+1}\left(\frac{z}{\pi}\right)^{2n+1}$ **5.** $\displaystyle\sum_{n=1}^{\infty} \frac{2^n n(n+1)}{7^n} z^{2n}$ **6.** $\displaystyle\sum_{n=0}^{\infty} \binom{n+m}{m} z^n$

7. $\displaystyle\sum_{n=1}^{\infty} \frac{(-6)^n}{n(n+1)(n+2)} z^n$ **8.** $\displaystyle\sum_{n=0}^{\infty} \left[\binom{n+k}{n}\right]^{-1} z^{n+k}$ **9.** $\displaystyle\sum_{n=0}^{\infty} \frac{(2n)!}{(n+1)(n!)^2} z^{n+1}$

10. In the proof of Theorem 3, we claimed that $\sqrt[n]{n} \to 1$ as $n \to \infty$. Prove this. *Hint.* Set $\sqrt[n]{n} = 1 + c_n$, where $c_n > 0$, and show that $c_n \to 0$ as $n \to \infty$.

11. Show that $(1-z)^{-2} = \sum_{n=0}^{\infty} (n+1)z^n$ (a) by using the Cauchy product, (b) by differentiating a suitable series.

12. Applying Theorem 2 to $(1+z)^p(1+z)^q = (1+z)^{p+q}$ (p and q positive integers), show that

$$\sum_{n=0}^{r} \binom{p}{n}\binom{q}{r-n} = \binom{p+q}{r}.$$

13. If $f(z)$ in (1) is even, show that $a_n = 0$ for odd n. (Use Theorem 2.)

14. **(Fibonacci[2] numbers)** The *Fibonacci numbers* are recursively defined by $a_0 = a_1 = 1$, $a_n = a_{n-2} + a_{n-1}$ if $n \geqq 2$. Show that if a power series $a_0 + a_1 z + \cdots$ represents $f(z) = 1/(1 - z - z^2)$, it must have these numbers as coefficients and conversely. *Hint.* Start from $f(z)(1 - z - z^2) = 1$ and use Theorem 2.

15. Write out the proof on termwise addition and subtraction of power series indicated in the text.

---

[2]LEONARDO OF PISA, called FIBONACCI (= son of Bonaccio), about 1180—1250, Italian mathematician, credited with the first renaissance of mathematics on Christian soil.

14.4 Taylor Series

The sum of any power series (with positive radius of convergence) is an analytic function, as we have just seen (Theorem 5, Sec. 14.3). We shall now show that, conversely, ***every*** analytic function $f(z)$ can be represented by power series, which are called **Taylor series** of $f(z)$ and are of the same form as in calculus, with x replaced by complex z.

So let $f(z)$ be analytic in a neighborhood of a point $z = z_0$ and let us first derive **Taylor's formula** and from it the Taylor series of $f(z)$ with center z_0.

As the key tool in that derivation we use Cauchy's integral formula (1) in Sec. 13.5; writing z and z^* instead of z_0 and z, we have

$$f(z) = \frac{1}{2\pi i} \oint_C \frac{f(z^*)}{z^* - z}\, dz^*. \tag{1}$$

z lies inside C, for which we take a circle of radius r with center z_0 in that neighborhood. z^* is the complex variable of integration (Fig. 339). Our next idea is to develop $1/(z^* - z)$ in (1) in powers of $z - z_0$. By a standard algebraic manipulation (worth remembering!) we first have

$$\frac{1}{z^* - z} = \frac{1}{z^* - z_0 - (z - z_0)} = \frac{1}{(z^* - z_0)\left(1 - \dfrac{z - z_0}{z^* - z_0}\right)}. \tag{2}$$

For later use we note that since z^* is on C while z is inside C, we have

$$\left|\frac{z - z_0}{z^* - z_0}\right| < 1. \tag{2*}$$

To (2) we now apply the sum formula for a finite geometric sum

$$1 + q + \cdots + q^n = \frac{1 - q^{n+1}}{1 - q} = \frac{1}{1 - q} - \frac{q^{n+1}}{1 - q} \qquad (q \neq 1), \tag{3*}$$

which we use in the form

$$\frac{1}{1 - q} = 1 + q + \cdots + q^n + \frac{q^{n+1}}{1 - q}. \tag{3}$$

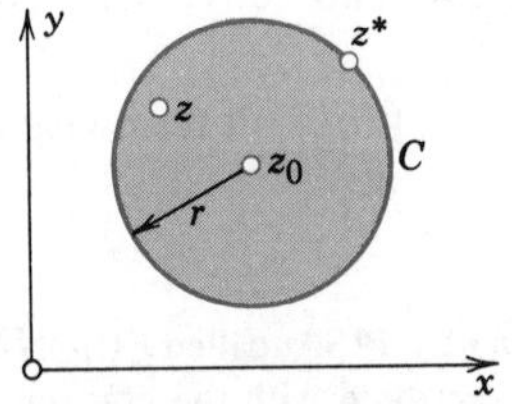

Fig. 339. Cauchy formula (1)

Applying (3) with $q = (z - z_0)/(z^* - z_0)$ to the right side of (2), we get

$$\frac{1}{z^* - z} = \frac{1}{z^* - z_0}\left[1 + \frac{z - z_0}{z^* - z_0} + \left(\frac{z - z_0}{z^* - z_0}\right)^2 + \cdots + \left(\frac{z - z_0}{z^* - z_0}\right)^n\right]$$
$$+ \frac{1}{z^* - z}\left(\frac{z - z_0}{z^* - z_0}\right)^{n+1}.$$

This we insert into (1). Powers of $z - z_0$ do not depend on the variable of integration z^*, so that we may take them out from under the integral sign, obtaining

$$f(z) = \frac{1}{2\pi i}\oint_C \frac{f(z^*)}{z^* - z_0}\,dz^* + \frac{z - z_0}{2\pi i}\oint_C \frac{f(z^*)}{(z^* - z_0)^2}\,dz^* + \cdots \tag{4}$$
$$\cdots + \frac{(z - z_0)^n}{2\pi i}\oint_C \frac{f(z^*)}{(z^* - z_0)^{n+1}}\,dz^* + R_n(z)$$

where the last term is given by the formula

$$R_n(z) = \frac{(z - z_0)^{n+1}}{2\pi i}\oint_C \frac{f(z^*)}{(z^* - z_0)^{n+1}(z^* - z)}\,dz^*. \tag{5}$$

Using the integral formulas (1) in Sec. 13.6 for the derivatives, we obtain

$$f(z) = f(z_0) + \frac{z - z_0}{1!} f'(z_0) + \frac{(z - z_0)^2}{2!} f''(z_0) + \cdots \tag{6}$$
$$\cdots + \frac{(z - z_0)^n}{n!} f^{(n)}(z_0) + R_n(z).$$

This representation is called **Taylor's[3] formula.** $R_n(z)$ is called the *remainder*. Since the analytic function $f(z)$ has derivatives of all orders, we may take n in (6) as large as we please. If we let n approach infinity, we obtain from (6) the power series

$$f(z) = \sum_{m=0}^{\infty} \frac{f^{(m)}(z_0)}{m!}(z - z_0)^m. \tag{7}$$

This series is called the **Taylor series** of $f(z)$ *with center* z_0. The particular case of (7) with center $z_0 = 0$ is called the **Maclaurin[4] series** of $f(z)$.

Clearly, the series (7) will converge and represent $f(z)$ if and only if

$$\lim_{n\to\infty} R_n(z) = 0. \tag{8}$$

[3] BROOK TAYLOR (1685—1731), English mathematician, who introduced this formula for functions of a real variable.

[4] COLIN MACLAURIN (1698—1746), Scots mathematician, professor at Edinburgh.

To prove (8), we consider (5). Since z^* is on C while z is inside C (Fig. 339), we have $|z^* - z| > 0$. From this and the analyticity of $f(z)$ inside and on C, it follows that the absolute value of $f(z^*)/(z^* - z)$ is bounded, say,

$$\left|\frac{f(z^*)}{z^* - z}\right| \leqq \tilde{M}$$

for all z^* on C. Also, $|z^* - z_0| = r$, the radius of C, and C has the length $2\pi r$. Hence by applying the ML-inequality (Sec. 13.2) to (5) we obtain

$$|R_n| = \frac{|z - z_0|^{n+1}}{2\pi}\left|\int_C \frac{f(z^*)}{(z^* - z_0)^{n+1}(z^* - z)}\,dz^*\right|$$

$$\leqq \frac{|z - z_0|^{n+1}}{2\pi}\tilde{M}\frac{1}{r^{n+1}}2\pi r = \tilde{M}r\left|\frac{z - z_0}{r}\right|^{n+1}.$$

On the right, $|z - z_0| < r$, since z lies *inside* the circle C. Thus $|z - z_0|/r < 1$, so that the right side approaches zero as $n \to \infty$. This proves (8) for all z inside C. By Theorem 2 in the last section, the representation of $f(z)$ in the form (7) is unique in the sense that (7) is the only power series with center at z_0 that represents the given function $f(z)$. Hence we may sum up our result as follows.

Theorem 1 **(Taylor's theorem)**

Let $f(z)$ be analytic in a domain D, and let $z = z_0$ be any point in D. Then there exists precisely one power series with center z_0 that represents $f(z)$. This series is of the form

$$f(z) = \sum_{n=0}^{\infty} a_n(z - z_0)^n \quad \text{where} \quad a_n = \frac{1}{n!}f^{(n)}(z_0). \tag{9}$$

This representation is valid in the largest open disk with center z_0 contained in D. The remainders $R_n(z)$ of (9) can be represented in the form (5). The coefficients satisfy the inequality

$$|a_n| \leqq \frac{M}{r^n} \tag{10}$$

where M is the maximum of $|f(z)|$ on the circle $|z - z_0| = r$.

Inequality (10) follows from Cauchy's inequality (3) in Sec. 13.6. Formula (1) in Sec. 13.6 also gives for the coefficients in (9)

$$a_n = \frac{1}{2\pi i}\oint_C \frac{f(z)}{(z - z_0)^{n+1}}\,dz, \tag{9*}$$

where we integrate counterclockwise around a simple closed path containing z_0 in its interior.

Practically, (8) means that for all z for which (9) converges, the nth partial sum of (9) will approximate $f(z)$ to any assigned degree of accuracy; we just have to choose n large enough.

Singular points of an analytic function $f(z)$ are points at which $f(z)$ ceases to be analytic. More precisely, $z = c$ is called a *singular point* of $f(z)$ if $f(z)$ is not differentiable at $z = c$, but every disk with center c contains points at which $f(z)$ is differentiable. We also say that $f(z)$ ***has a singularity** at* $z = c$. *Examples:* $1/(1 - z)$ at $z = 1$, $\tan z$ at $\pm\pi/2, \pm 3\pi/2, \cdots$.

Using this concept, we may say that there is at least one singular point of $f(z)$ on the circle of convergence[5] of the development (9).

Important Special Taylor Series

These are as in calculus, with x replaced by complex z. Can you see why? (*Answer.* The coefficient formulas are the same.)

EXAMPLE 1 Geometric series

Let $f(z) = 1/(1 - z)$. Then we have $f^{(n)}(z) = n!/(1 - z)^{n+1}$, $f^{(n)}(0) = n!$. Hence the Maclaurin expansion of $1/(1 - z)$ is the geometric series

$$\frac{1}{1 - z} = \sum_{n=0}^{\infty} z^n = 1 + z + z^2 + \cdots \qquad (|z| < 1). \tag{11}$$

$f(z)$ is singular at $z = 1$; this point lies on the circle of convergence. ■

EXAMPLE 2 Exponential function

We know that the exponential function e^z (Sec. 12.6) is analytic for all z, and $(e^z)' = e^z$. Hence from (9) with $z_0 = 0$ we obtain the Maclaurin series

$$e^z = \sum_{n=0}^{\infty} \frac{z^n}{n!} = 1 + z + \frac{z^2}{2!} + \cdots. \tag{12}$$

This series is also obtained if we replace x in the familiar Maclaurin series of e^x by z.

Furthermore, by setting $z = iy$ in (12) and separating the series into the real and imaginary parts (see Theorem 2, Sec. 14.1) we obtain

$$e^{iy} = \sum_{n=0}^{\infty} \frac{(iy)^n}{n!} = \sum_{k=0}^{\infty} (-1)^k \frac{y^{2k}}{(2k)!} + i \sum_{k=0}^{\infty} (-1)^k \frac{y^{2k+1}}{(2k+1)!}.$$

Since the series on the right are the familiar Maclaurin series of the real functions $\cos y$ and $\sin y$, this shows that we have rediscovered the **Euler formula**

$$e^{iy} = \cos y + i \sin y. \tag{13}$$

Indeed, one may use (12) for ***defining*** e^z and derive from (12) the basic properties of e^z. For instance, the differentiation formula $(e^z)' = e^z$ follows readily from (12) by termwise differentiation. ■

---

[5]The radius of convergence of (9) will in general be equal to the distance from z_0 to the nearest singular point of $f(z)$, but it may be larger; for example, Ln z is singular along the negative real axis, and the distance from $z_0 = -1 + i$ to that axis is 1, but the Taylor series of Ln z with center $z_0 = -1 + i$ has radius of convergence $\sqrt{2}$.

EXAMPLE 3 Trigonometric and hyperbolic functions

By substituting (12) into (1) of Sec. 12.7 we obtain

$$\text{(14)} \qquad \begin{aligned} \cos z &= \sum_{n=0}^{\infty} (-1)^n \frac{z^{2n}}{(2n)!} = 1 - \frac{z^2}{2!} + \frac{z^4}{4!} - + \cdots \\ \sin z &= \sum_{n=0}^{\infty} (-1)^n \frac{z^{2n+1}}{(2n+1)!} = z - \frac{z^3}{3!} + \frac{z^5}{5!} - + \cdots . \end{aligned}$$

When $z = x$ these are the familiar Maclaurin series of the real functions $\cos x$ and $\sin x$. Similarly, by substituting (12) into (11), Sec. 12.7, we obtain

$$\text{(15)} \qquad \begin{aligned} \cosh z &= \sum_{n=0}^{\infty} \frac{z^{2n}}{(2n)!} = 1 + \frac{z^2}{2!} + \frac{z^4}{4!} + \cdots \\ \sinh z &= \sum_{n=0}^{\infty} \frac{z^{2n+1}}{(2n+1)!} = z + \frac{z^3}{3!} + \frac{z^5}{5!} \cdots . \end{aligned}$$

■

EXAMPLE 4 Logarithm

From (9) it follows that

$$\text{(16)} \qquad \operatorname{Ln}(1 + z) = z - \frac{z^2}{2} + \frac{z^3}{3} - + \cdots \qquad (|z| < 1).$$

Replacing z by $-z$ and multiplying both sides by -1, we get

$$\text{(17)} \qquad -\operatorname{Ln}(1 - z) = \operatorname{Ln} \frac{1}{1 - z} = z + \frac{z^2}{2} + \frac{z^3}{3} + \cdots \qquad (|z| < 1).$$

By adding both series we obtain

$$\text{(18)} \qquad \operatorname{Ln} \frac{1 + z}{1 - z} = 2\left(z + \frac{z^3}{3} + \frac{z^5}{5} + \cdots\right) \qquad (|z| < 1).$$

■

In the next section we explain some practical methods of obtaining Taylor series that avoid the cumbersome calculations of the derivatives in (9).

Relation to Last Section

Our discussion in the last section can be nicely related to the present section:

Theorem 2 *Every power series with a nonzero radius of convergence is the Taylor series of the function represented by that power series* (more briefly: *is the Taylor series of its sum*).

Proof. Consider any power series with positive radius of convergence R and call its sum $f(z)$; thus,

$$f(z) = a_0 + a_1(z - z_0) + a_2(z - z_0)^2 + \cdots .$$

From Theorem 5 in the last section it follows that

$$f'(z) = a_1 + 2a_2(z - z_0) + \cdots$$

and more generally

$$f^{(n)}(z) = n!a_n + (n + 1)n \cdots 3 \cdot 2a_{n+1}(z - z_0) + \cdots;$$

all these series converge in the disk $|z - z_0| < R$ and represent analytic functions. Hence these functions are continuous at $z = z_0$, by Theorem 1 in the last section. If we set $z = z_0$, we thus obtain

$$f(z_0) = a_0, \qquad f'(z_0) = a_1, \qquad \cdots, \qquad f^{(n)}(z_0) = n!a_n, \qquad \cdots.$$

Since these formulas are identical with those in Taylor's theorem, the proof is complete. ∎

Comment. Comparison with real functions
One surprising property of complex analytic functions is that they have derivatives of all orders, and now we have discovered the other surprising property that they can always be represented by power series of the form (9). This is not true in general for ***real functions;*** there are real functions that have derivatives of all orders but cannot be represented by a power series. (Example: $f(x) = \exp(-1/x^2)$ if $x \neq 0$ and $f(0) = 0$; this function cannot be represented by a Maclaurin series since all its derivatives at 0 are zero.)

Problem Set 14.4

Find the Taylor series of the given function with the given point as center and determine the radius of convergence. (More problems of this kind follow in the next section, after the discussion of practical methods.)

1. e^{-z}, 0
2. e^{2z}, $2i$
3. $\sin \pi z$, 0
4. $\cos z$, $-\pi/2$
5. $\sin z$, $\pi/2$
6. $1/z$, 1
7. $1/(1 - z)$, -1
8. $1/(1 - z)$, i
9. $\operatorname{Ln} z$, 1
10. $\sinh(z - 2i)$, $2i$
11. z^5, -1
12. $z^4 - z^2 + 1$, 1
13. $\sin^2 z$, 0
14. $\cos^2 z$, 0
15. $\cos(z - \pi/2)$, $\pi/2$

Problems 16–26 illustrate how you can obtain properties of functions from their Maclaurin series.

16. Using (12), prove $(e^z)' = e^z$.

17. Derive (14) and (15) from (12). Obtain (16) from Taylor's theorem.

18. Using (14), show that $\cos z$ is even and $\sin z$ is odd.

19. Using (15), show that $\cosh z \neq 0$ for all real $z = x$.

20. Using (14), show that $\sin z \neq 0$ for all pure imaginary $z = iy \neq 0$.

21. Using (14), show that $(\sin z)' = \cos z$ and $(\cos z)' = -\sin z$.

22. Using (14), show that $\cos z + i \sin z$ yields the Maclaurin series for e^{iz} and thus a proof of the Euler formula for complex z.

23. $f(z) = (\sin z)/z$ is undefined at $z = 0$. Define $f(0)$ so that $f(z)$ becomes entire. (Give a reason.)

24. Find the derivative of $\operatorname{Ln}(1 + z)$ by differentiating (16).

25. Derive the relations (14) and (15), Sec. 12.7, between the trigonometric and hyperbolic sine and cosine from the present (14) and (15).

26. Obtain the Maclaurin series of $\cosh^2 z$ from Prob. 14.

Problems 27–30 concern the **error function** erf z (being related to the normal distribution in Chaps. 23, 24), the **sine integral** Si(z), and the **Fresnel integrals**[6] S(z) and C(z). These nonelementary special functions are defined by integrals that cannot be evaluated by the usual methods of calculus, which is good to know because they occur occasionally in applications. Find their Maclaurin series by termwise integration of the Maclaurin series of the integrand.

27. $\operatorname{erf} z = \dfrac{2}{\sqrt{\pi}} \displaystyle\int_0^z e^{-t^2}\, dt$

28. $\operatorname{Si}(z) = \displaystyle\int_0^z \frac{\sin t}{t}\, dt$

29. $S(z) = \displaystyle\int_0^z \sin t^2\, dt$

30. $C(z) = \displaystyle\int_0^z \cos t^2\, dt$

14.5 Power Series: Practical Methods

In most cases it would be complicated or time-consuming to determine the coefficients of Taylor series from the formula in Taylor's theorem, and there are better and quicker ways for that purpose, as we now illustrate in terms of typical examples. Regardless of the method used, one gets the ***same result,*** as follows from the uniqueness theorem (Theorem 2 in Sec. 14.3).

EXAMPLE 1 Substitution

Find the Maclaurin series of $f(z) = 1/(1 + z^2)$.

Solution. By substituting $-z^2$ for z in (11), Sec. 14.4, we obtain

$$\frac{1}{1+z^2} = \frac{1}{1-(-z^2)} = \sum_{n=0}^{\infty} (-z^2)^n = \sum_{n=0}^{\infty} (-1)^n z^{2n} \tag{1}$$

$$= 1 - z^2 + z^4 - z^6 + \cdots \qquad (|z| < 1). \quad ■$$

EXAMPLE 2 Integration

Find the Maclaurin series of $f(z) = \tan^{-1} z$.

Solution. We have $f'(z) = 1/(1 + z^2)$. Integrating (1) term by term and using $f(0) = 0$ we get

$$\tan^{-1} z = \sum_{n=0}^{\infty} \frac{(-1)^n}{2n+1} z^{2n+1} = z - \frac{z^3}{3} + \frac{z^5}{5} - + \cdots \qquad (|z| < 1);$$

this series represents the principal value of $w = u + iv = \tan^{-1} z$, defined as that value for which $|u| < \pi/2$. ■

[6]AUGUSTIN FRESNEL (1788—1827), French physicist, known for his work in optics.

EXAMPLE 3 Development by using the geometric series

Develop $1/(c - bz)$ in powers of $z - a$ where $c - ab \neq 0$ and $b \neq 0$.

Solution. To get powers of $z - a$ later on, we first use simple algebra:

$$\frac{1}{c - bz} = \frac{1}{c - ab - b(z - a)} = \frac{1}{(c - ab)\left[1 - \dfrac{b(z - a)}{c - ab}\right]}.$$

To the last expression we apply (11) in Sec. 14.4 with z replaced by $b(z - a)/(c - ab)$, finding

$$\frac{1}{c - bz} = \frac{1}{c - ab}\sum_{n=0}^{\infty}\left[\frac{b(z - a)}{c - ab}\right]^n = \sum_{n=0}^{\infty}\frac{b^n}{(c - ab)^{n+1}}(z - a)^n$$

$$= \frac{1}{c - ab} + \frac{b}{(c - ab)^2}(z - a) + \frac{b^2}{(c - ab)^3}(z - a)^2 + \cdots.$$

This series converges for

$$\left|\frac{b(z - a)}{c - ab}\right| < 1, \qquad \text{that is,} \qquad |z - a| < \left|\frac{c - ab}{b}\right| = \left|\frac{c}{b} - a\right|.$$ ∎

EXAMPLE 4 Binomial series, reduction by partial fractions

Find the Taylor series of the following function with center $z_0 = 1$.

$$f(z) = \frac{2z^2 + 9z + 5}{z^3 + z^2 - 8z - 12}$$

Solution. Given a rational function, we may first represent it as a sum of partial fractions and then apply the **binomial series**

$$\text{(2)} \qquad \begin{aligned} \frac{1}{(1 + z)^m} &= (1 + z)^{-m} = \sum_{n=0}^{\infty}\binom{-m}{n} z^n \\ &= 1 - mz + \frac{m(m + 1)}{2!} z^2 - \frac{m(m + 1)(m + 2)}{3!} z^3 + \cdots. \end{aligned}$$

Since the function on the left is singular at $z = -1$, the series converges in the disk $|z| < 1$. In our case we first have

$$f(z) = \frac{1}{(z + 2)^2} + \frac{2}{z - 3} = \frac{1}{[3 + (z - 1)]^2} - \frac{2}{2 - (z - 1)}.$$

This may be written in the form

$$f(z) = \frac{1}{9}\left(\frac{1}{[1 + \frac{1}{3}(z - 1)]^2}\right) - \frac{1}{1 - \frac{1}{2}(z - 1)}.$$

By using the binomial series we now obtain

$$f(z) = \frac{1}{9}\sum_{n=0}^{\infty}\binom{-2}{n}\left(\frac{z - 1}{3}\right)^n - \sum_{n=0}^{\infty}\left(\frac{z - 1}{2}\right)^n.$$

We may add the two series on the right term by term. Since the binomial coefficient in the first series equals $(-2)(-3) \cdots (-[n + 1])/n! = (-1)^n(n + 1)$, we find

$$f(z) = \sum_{n=0}^{\infty}\left[\frac{(-1)^n(n + 1)}{3^{n+2}} - \frac{1}{2^n}\right](z - 1)^n = -\frac{8}{9} - \frac{31}{54}(z - 1) - \frac{23}{108}(z - 1)^2 - \cdots.$$

Since $z = 3$ is the singular point of $f(z)$ that is nearest to the center $z = 1$, the series converges in the disk $|z - 1| < 2$. ∎

EXAMPLE 5 **Use of differential equations**

Find the Maclaurin series of $f(z) = \tan z$.

Solution. We have $f'(z) = \sec^2 z$ and, therefore, since $f(0) = 0$,

$$f'(z) = 1 + f^2(z), \qquad f'(0) = 1.$$

Observing that $f(0) = 0$, we obtain by successive differentiation

$$\begin{aligned} f'' &= 2ff', & f''(0) &= 0, & & \\ f''' &= 2f'^2 + 2ff'', & f'''(0) &= 2, & f'''(0)/3! &= 1/3, \\ f^{(4)} &= 6f'f'' + 2ff''', & f^{(4)}(0) &= 0, & & \\ f^{(5)} &= 6f''^2 + 8f'f''' + 2ff^{(4)}, & f^{(5)}(0) &= 16, & f^{(5)}(0)/5! &= 2/15, \quad \text{etc.} \end{aligned}$$

Hence the result is

$$\tan z = z + \frac{1}{3}z^2 + \frac{2}{15}z^5 + \frac{17}{315}z^7 + \cdots \qquad \left(|z| < \frac{\pi}{2}\right). \tag{3}$$ ■

EXAMPLE 6 **Undetermined coefficients**

Find the Maclaurin series of $\tan z$ by using those of $\cos z$ and $\sin z$ (Sec. 14.4).

Solution. Since $\tan z$ is odd, the desired expansion will be of the form

$$\tan z = a_1 z + a_3 z^3 + a_5 z^5 + \cdots .$$

Using $\sin z = \tan z \cos z$ and inserting those developments, we obtain

$$z - \frac{z^3}{3!} + \frac{z^5}{5!} - + \cdots = (a_1 z + a_3 z^3 + a_5 z^5 + \cdots)\left(1 - \frac{z^2}{2!} + \frac{z^4}{4!} - + \cdots\right).$$

Since $\tan z$ is analytic except at $z = \pm\pi/2, \pm 3\pi/2, \cdots$, its Maclaurin series converges in the disk $|z| < \pi/2$, and for these z we may form the Cauchy product of the two series on the right (see Sec. 14.3), that is, multiply the series term by term and arrange the resulting series in powers of z. By Theorem 2 in Sec. 14.3 the coefficient of each power of z is the same on both sides. This yields

$$1 = a_1, \qquad -\frac{1}{3!} = -\frac{a_1}{2!} + a_3, \qquad \frac{1}{5!} = \frac{a_1}{4!} - \frac{a_3}{2!} + a_5, \qquad \text{etc.}$$

Hence $a_1 = 1$, $a_3 = \frac{1}{3}$, $a_5 = \frac{2}{15}$, etc., as before. ■

Problem Set 14.5

Find the Maclaurin series of the following functions and determine the radius of convergence.

1. $\dfrac{1}{1 + z^4}$ **2.** $\dfrac{1}{1 - z^5}$ **3.** $\dfrac{z + 2}{1 - z^2}$

4. $\dfrac{4 - 3z}{(1 - z)^2}$ **5.** $\sin 2z^2$ **6.** $\dfrac{1}{(z + 3 - 4i)^2}$

7. $\dfrac{e^{z^4} - 1}{z^3}$ **8.** $e^{z^2}\displaystyle\int_0^z e^{-t^2}\,dt$ **9.** $\dfrac{2z^2 + 15z + 34}{(z + 4)^2(z - 2)}$

Find the Taylor series of the given function with the given point as center and determine the radius of convergence.

10. $\frac{1}{z}$, 1 **11.** $\frac{1}{z}$, $1+i$ **12.** $\frac{1}{(z+i)^2}$, $-2i$

13. $z^5 + z^3 - z$, i **14.** $(z+i)^3$, $1-i$ **15.** $\frac{1 + z - \sin(z+1)}{(z+1)^3}$, -1

16. e^z, $-\pi i$ **17.** $\cosh z$, $\pi i/2$ **18.** $\sin \pi z$, $1/2$

Find the first three nonzero terms of the Taylor series with the given point as center and determine the radius of convergence.

19. $e^{z^2} \sin z^2$, 0 **20.** $\frac{\cos 2z}{1 - 4z^2}$, 0 **21.** $\tan z$, $\frac{\pi}{4}$

22. $e^{z^2}/\cos z$, 0 **23.** $\cos\left(\frac{z}{3-z}\right)$, 0 **24.** $\frac{4 - 6z}{2z^2 - 3z + 1}$, -1

25. **(Euler numbers)** The Maclaurin series

$$\sec z = E_0 - \frac{E_2}{2!} z^2 + \frac{E_4}{4!} z^4 - + \cdots \tag{4}$$

defines the *Euler numbers* E_{2n}. Show that[7] $E_0 = 1$, $E_2 = -1$, $E_4 = 5$, $E_6 = -61$.

26. **(Bernoulli numbers)** The Maclaurin series

$$\frac{z}{e^z - 1} = 1 + B_1 z + \frac{B_2}{2!} z^2 + \frac{B_3}{3!} z^3 + \cdots \tag{5}$$

defines the *Bernoulli numbers* B_n. Using undetermined coefficients, show that[7]

$$B_1 = -\frac{1}{2}, \quad B_2 = \frac{1}{6}, \quad B_3 = 0, \quad B_4 = -\frac{1}{30}, \quad B_5 = 0, \quad B_6 = \frac{1}{42}, \quad \cdots. \tag{6}$$

27. Using (1), (2), Sec. 12.7, and (5), show that

$$\tan z = \frac{2i}{e^{2iz} - 1} - \frac{4i}{e^{4iz} - 1} - i = \sum_{n=1}^{\infty} (-1)^{n-1} \frac{2^{2n}(2^{2n} - 1)}{(2n)!} B_{2n} z^{2n-1}. \tag{7}$$

28. Developing $1/\sqrt{1 - z^2}$ and integrating, show that

$$\sin^{-1} z = z + \left(\frac{1}{2}\right)\frac{z^3}{3} + \left(\frac{1 \cdot 3}{2 \cdot 4}\right)\frac{z^5}{5} + \left(\frac{1 \cdot 3 \cdot 5}{2 \cdot 4 \cdot 6}\right)\frac{z^7}{7} + \cdots \qquad (|z| < 1).$$

Show that this series represents the principal value of $\sin^{-1} z$ (defined in Prob. 45, Sec. 12.8).

29. Was the radius of convergence in Example 3 to be expected from the form of the given function?

30. Find a Maclaurin series for which the corresponding function has more than one singularity on the circle of convergence.

[7] For tables, see Ref. [1], p. 810, in Appendix 1.

14.6 Uniform Convergence

We know that power series are *absolutely convergent* (Sec. 14.2, Theorem 1) and, as another basic property, we now show that they are *uniformly convergent*. Since uniform convergence is of great general importance, for instance, in connection with termwise integration of series, we shall discuss it quite thoroughly.

To define uniform convergence; we consider a series whose terms are functions $f_0(z), f_1(z), \cdots$:

$$\sum_{m=0}^{\infty} f_m(z) = f_0(z) + f_1(z) + f_2(z) + \cdots . \tag{1}$$

(For the special $f_m(z) = a_m(z - z_0)^m$ this is a power series.) We assume that this series converges for all z in some region G. We call its sum $s(z)$ and its nth partial sum $s_n(z)$; thus

$$s_n(z) = f_0(z) + f_1(z) + \cdots + f_n(z).$$

Convergence in G means the following. If we pick a $z = z_1$ in G, then, by the definition of convergence at z_1, for given $\epsilon > 0$ we can find an $N_1(\epsilon)$ such that

$$|s(z_1) - s_n(z_1)| < \epsilon \qquad \text{for all } n > N_1(\epsilon).$$

If we pick a z_2 in G, keeping ϵ as before, we can find an $N_2(\epsilon)$ such that

$$|s(z_2) - s_n(z_2)| < \epsilon \qquad \text{for all } n > N_2(\epsilon),$$

and so on. Hence, given an $\epsilon > 0$, to each z in G there corresponds a number $N_z(\epsilon)$. This number tells us how many terms we need (what s_n we need) at a z to make $|s(z) - s_n(z)|$ smaller than ϵ. It measures the speed of convergence.

Small $N_z(\epsilon)$ means rapid convergence, large $N_z(\epsilon)$ means slow convergence. Now, if we can find an $N(\epsilon)$ larger than *all* these $N_z(\epsilon)$, we say that the convergence of the series (1) in G is *uniform:*

Definition (Uniform convergence)
A series (1) with sum $s(z)$ is called **uniformly convergent** in a region G if for every $\epsilon > 0$ we can find an $N = N(\epsilon)$, ***not depending on z,*** such that

$$|s(z) - s_n(z)| < \epsilon \qquad \text{for all } n > N(\epsilon) \textbf{ and all } z \textbf{ in } G.$$

Uniformity of convergence is thus a property that always refers to an *infinite set* in the z-plane.

EXAMPLE 1 **Geometric series**

Show that the geometric series $1 + z + z^2 + \cdots$ is (a) uniformly convergent in any closed disk $|z| \leqq r < 1$, (b) not uniformly convergent in its whole disk of convergence $|z| < 1$.

Solution. **(a)** For z in that closed disk we have $|1 - z| \geqq 1 - r$ (sketch it). This implies that $1/|1 - z| \leqq 1/(1 - r)$. Hence (remember (3*) or (3) in Sec. 14.4 with $q = z$)

$$|s(z) - s_n(z)| = \left| \sum_{m=n+1}^{\infty} z^m \right| = \left| \frac{z^{n+1}}{1 - z} \right| \leqq \frac{r^{n+1}}{1 - r}.$$

Since $r < 1$, we can make the right side as small as we want by choosing n large enough, and since the right side does not depend on z (in the closed disk considered), this means uniform convergence.

(b) For given real K (no matter how large) and n we can always find a z in the disk $|z| < 1$ such that

$$\left| \frac{z^{n+1}}{1 - z} \right| = \frac{|z|^{n+1}}{|1 - z|} > K,$$

simply by taking z close enough to 1. Hence no single $N(\epsilon)$ will suffice to make $|s(z) - s_n(z)|$ smaller than a given $\epsilon > 0$ *throughout the whole disk.* By definition, this shows that the convergence of the geometric series in $|z| < 1$ is not uniform. ∎

This example suggests that for a power series, the uniformity of convergence may at most be disturbed near the circle of convergence. This is true:

Theorem 1 **(Uniform convergence of power series)**

A power series

$$\sum_{m=0}^{\infty} a_m(z - z_0)^m \tag{2}$$

with a nonzero radius of convergence R is uniformly convergent in every circular disk $|z - z_0| \leqq r$ of radius $r < R$.

Proof. For $|z - z_0| \leqq r$ and any positive integers n and p we have

$$\begin{aligned} &|a_{n+1}(z - z_0)^{n+1} + \cdots + a_{n+p}(z - z_0)^{n+p}| \\ &\qquad \leqq |a_{n+1}|r^{n+1} + \cdots + |a_{n+p}|r^{n+p}. \end{aligned} \tag{3}$$

Now (2) converges absolutely if $|z - z_0| = r < R$ (by Theorem 1 in Sec. 14.2). Hence it follows from the Cauchy convergence principle (Sec. 14.1) that, an $\epsilon > 0$ being given, we can find an $N(\epsilon)$ such that

$$|a_{n+1}|r^{n+1} + \cdots + |a_{n+p}|r^{n+p} < \epsilon \quad \text{for } n > N(\epsilon) \quad \text{and} \quad p = 1, 2, \cdots.$$

From this and (3) we obtain

$$|a_{n+1}(z - z_0)^{n+1} + \cdots + a_{n+p}(z - z_0)^{n+p}| < \epsilon$$

for all z in the disk $|z - z_0| \leqq r$, every $n > N(\epsilon)$, and every $p = 1, 2, \cdots$. Since $N(\epsilon)$ is independent of z, this shows uniform convergence, and the theorem is proved. ∎

Theorem 1 meets with our immediate need and concern, which is power series. The remainder of this section should provide a deeper understanding of the concept of uniform convergence as such.

Properties of Uniformly Convergent Series (*Optional*)

Uniform convergence derives its main importance from two facts:

1. If a series of *continuous* terms is uniformly convergent, its sum is also continuous (Theorem 2, below).

2. Under the same assumptions, termwise integration is permissible (Theorem 3).

This raises two questions:

1. How can a converging series of continuous terms manage to have a discontinuous sum? (Example 2)

2. How can something go wrong in termwise integration? (Example 3)

Another natural question is:

3. What is the relation between absolute convergence and uniform convergence? The surprising answer: none. (Example 4)

These are the ideas we shall discuss.

If we add *finitely many* continuous functions, we get a continuous function as their sum. Example 2 will show that this is no longer true for an infinite series, even if it converges absolutely. However, if it converges *uniformly,* this cannot happen, as follows.

Theorem 2 (Continuity of the sum)

Let the series

$$\sum_{m=0}^{\infty} f_m(z) = f_0(z) + f_1(z) + \cdots$$

be uniformly convergent in a region G. Let F(z) be its sum. Then if each term $f_m(z)$ is continuous at a point z_1 in G, the function F(z) is continuous at z_1.

Proof. Let $s_n(z)$ be the nth partial sum of the series and $R_n(z)$ the corresponding remainder:

$$s_n = f_0 + f_1 + \cdots + f_n, \qquad R_n = f_{n+1} + f_{n+2} + \cdots .$$

Since the series converges uniformly, for a given $\epsilon > 0$ we can find an $n = N(\epsilon)$ such that

$$|R_N(z)| < \frac{\epsilon}{3} \qquad \text{for all } z \text{ in } G.$$

Since $s_N(z)$ is a sum of finitely many functions that are continuous at z_1, this sum is continuous at z_1. Therefore we can find a $\delta > 0$ such that

$$|s_N(z) - s_N(z_1)| < \frac{\epsilon}{3} \qquad \text{for all } z \text{ in } G \text{ for which } |z - z_1| < \delta.$$

Using $F = s_N + R_N$ and the triangle inequality (Sec. 12.2), for these z we thus obtain

$$|F(z) - F(z_1)| = |s_N(z) + R_N(z) - [s_N(z_1) + R_N(z_1)]|$$

$$\leqq |s_N(z) - s_N(z_1)| + |R_N(z)| + |R_N(z_1)| < \frac{\epsilon}{3} + \frac{\epsilon}{3} + \frac{\epsilon}{3} = \epsilon.$$

This implies that $F(z)$ is continuous at z_1, and the theorem is proved. ∎

EXAMPLE 2 **Series of continuous terms with a discontinuous sum**

Consider the series

$$x^2 + \frac{x^2}{1 + x^2} + \frac{x^2}{(1 + x^2)^2} + \frac{x^2}{(1 + x^2)^3} + \cdots \qquad (x \text{ real}).$$

Using the formula (3*), Sec. 14.4, for a finite geometric sum with $q = 1/(1 + x^2)$, hence $1/(1 - q) = (1 + x^2)/x^2$, we get for the nth partial sum

$$s_n(z) = 1 + x^2 - \frac{1}{(1 + x^2)^n}.$$

The exciting Fig. 340 "explains" what is going on. We see that if $x \neq 0$, the sum is

$$s(x) = \lim s_n(x) = 1 + x^2,$$

but for $x = 0$ we have

$$s_n(0) = 1 - 1 = 0 \text{ for all } n,$$

hence

$$s(0) = 0.$$

So we have the surprising fact that the sum is discontinuous (at $x = 0$), although all the terms

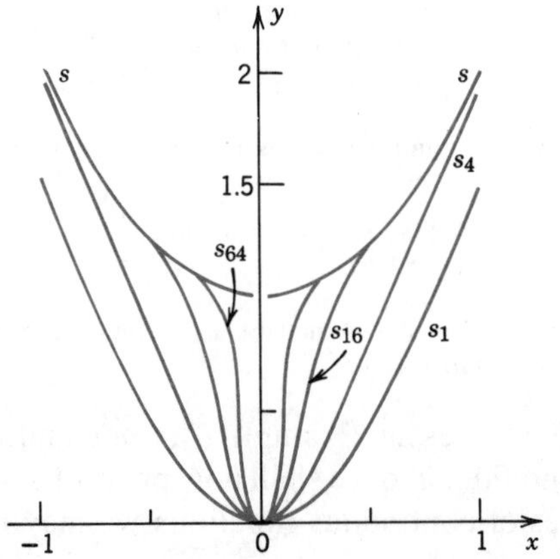

Fig. 340. Partial sums in Example 2

are continuous and the series converges even absolutely (its terms are nonnegative, thus equal to their absolute value!).

Theorem 2 now tells us that the convergence cannot be uniform in an interval containing $x = 0$. We can verify this also directly. Indeed, for $x \neq 0$ the remainder has the absolute value

$$|R_n(x)| = |s(x) - s_n(x)| = \frac{1}{(1 + x^2)^n}$$

and we see that for a given ϵ (< 1) we cannot find an N depending only on ϵ such that $|R_n| < \epsilon$ for all $n > N(\epsilon)$ and all x, say, in the interval $0 \leqq x \leqq 1$. ∎

Termwise Integration (*Optional*)

This is our second topic in connection with uniform convergence, and we begin with an example to become aware of the danger.

EXAMPLE 3 A series for which termwise integration is not permissible

Let

$$u_m(x) = mxe^{-mx^2}$$

and consider the series

$$\sum_{m=1}^{\infty} f_m(x) \qquad \text{where} \qquad f_m(x) = u_m(x) - u_{m-1}(x)$$

in the interval $0 \leqq x \leqq 1$. The nth partial sum is

$$s_n = u_1 - u_0 + u_2 - u_1 + \cdots + u_n - u_{n-1} = u_n - u_0 = u_n.$$

Hence the series has the sum

$$F(x) = \lim_{n\to\infty} s_n(x) = \lim_{n\to\infty} u_n(x) = 0 \qquad (0 \leqq x \leqq 1).$$

From this we obtain

$$\int_0^1 F(x)\,dx = 0.$$

On the other hand, by integrating term by term and using $f_1 + f_2 + \cdots + f_n = s_n$,

$$\sum_{m=1}^{\infty} \int_0^1 f_m(x)\,dx = \lim_{n\to\infty} \sum_{m=1}^{n} \int_0^1 f_m(x)\,dx = \lim_{n\to\infty} \int_0^1 s_n(x)\,dx.$$

Now $s_n = u_n$ and the expression on the right becomes

$$\lim_{n\to\infty} \int_0^1 u_n(x)\,dx = \lim_{n\to\infty} \int_0^1 nxe^{-nx^2}\,dx = \lim_{n\to\infty} \frac{1}{2}(1 - e^{-n}) = \frac{1}{2},$$

but not 0. This shows that the series under consideration cannot be integrated term by term from $x = 0$ to $x = 1$. ∎

The series in Example 3 is not uniformly convergent in the interval of integration, and we shall now prove that in the case of a uniformly convergent series of continuous functions we may integrate term by term.

Theorem 3 **(Termwise integration)**

Let

$$F(z) = \sum_{m=0}^{\infty} f_m(z) = f_0(z) + f_1(z) + \cdots$$

be a uniformly convergent series of continuous functions in a region G. Let C be any path in G. Then the series

$$\sum_{m=0}^{\infty} \int_C f_m(z)\,dz = \int_C f_0(z)\,dz + \int_C f_1(z)\,dz + \cdots \tag{4}$$

is convergent and has the sum $\int_C F(z)\,dz$.

Proof. From Theorem 2 it follows that $F(z)$ is continuous. Let $s_n(z)$ be the nth partial sum of the given series and $R_n(z)$ the corresponding remainder. Then $F = s_n + R_n$ and by integration,

$$\int_C F(z)\,dz = \int_C s_n(z)\,dz + \int_C R_n(z)\,dz.$$

Let L be the length of C. Since the given series converges uniformly, for every given $\epsilon > 0$ we can find a number N such that

$$|R_n(z)| < \frac{\epsilon}{L} \qquad \text{for all } n > N \text{ and all } z \text{ in } G.$$

By applying the ML-inequality (Sec. 13.2) we thus obtain

$$\left| \int_C R_n(z)\,dz \right| < \frac{\epsilon}{L} L = \epsilon \qquad \text{for all } n > N.$$

Since $R_n = F - s_n$, this means that

$$\left| \int_C F(z)\,dz - \int_C s_n(z)\,dz \right| < \epsilon \qquad \text{for all } n > N.$$

Hence, the series (4) converges and has the sum indicated in the theorem. This completes the proof. ∎

Theorems 2 and 3 characterize the two most important properties of uniformly convergent series.

Of course, since differentiation and integration are inverse processes, we readily conclude from Theorem 3 that a convergent series may be differentiated term by term, provided the terms of the given series have continuous derivatives and the resulting series is uniformly convergent; more precisely:

Theorem 4 **(Termwise differentiation)**
Let the series $f_0(z) + f_1(z) + f_2(z) + \cdots$ *be convergent in a region G and let F(z) be its sum. Suppose that the series* $f_0'(z) + f_1'(z) + f_2'(z) + \cdots$ *converges uniformly in G and its terms are continuous in G. Then*

$$F'(z) = f_0'(z) + f_1'(z) + f_2'(z) + \cdots \qquad \textit{for all } z \textit{ in } G.$$

The simple proof is left to the student (Prob. 18).

Test for Uniform Convergence (*Optional*)

Uniform convergence is usually proved by the following comparison test.

Theorem 5 **(Weierstrass[8] *M*-test for uniform convergence)**
Consider a series of the form (1) *in a region G of the z-plane. Suppose that one can find a convergent series of constant terms,*

$$M_0 + M_1 + M_2 + \cdots, \tag{5}$$

such that $|f_m(z)| \leqq M_m$ *for all z in G and every* $m = 0, 1, \cdots$. *Then* (1) *is uniformly convergent in G.*

The simple proof is left to the student (Prob. 17).

EXAMPLE 4 **Weierstrass *M*-test**
Does the following series converge uniformly in the disk $|z| \leqq 1$?

$$\sum_{m=0}^{\infty} \frac{z^m + 1}{m^2 + \cosh m|z|}.$$

Solution. Uniform convergence follows by the Weierstrass M-test and the convergence of $\Sigma 1/m^2$ (see Sec. 14.1, in the proof of Theorem 8) because

$$\left|\frac{z^m + 1}{m^2 + \cosh m|z|}\right| \leqq \frac{|z|^m + 1}{m^2} \leqq \frac{2}{m^2}. \qquad \blacksquare$$

[8]KARL WEIERSTRASS (1815—1897), great German mathematician, whose lifework was the development of complex analysis based on the concept of power series (see the footnote in Sec. 12.5). He also made basic contributions to the calculus, the calculus of variations, approximation theory, and differential geometry. He obtained the concept of uniform convergence in 1841 (published 1894); the first publications on the concept were by G. G. STOKES (see Sec. 9.9) in 1847 and PHILIPP LUDWIG VON SEIDEL (1821—1896) in 1848.

No Relation Between Absolute and Uniform Convergence (*Optional*)

We finally show the surprising fact that there are series that converge absolutely but not uniformly, and others that converge uniformly but not absolutely, so that there is no relation between the two concepts.

EXAMPLE 5 No relation between absolute and uniform convergence

The series in Example 2 converges absolutely but not uniformly, as we have shown. On the other hand, the series

$$\sum_{m=1}^{\infty} \frac{(-1)^{m-1}}{x^2+m} = \frac{1}{x^2+1} - \frac{1}{x^2+2} + \frac{1}{x^2+3} - + \cdots \qquad (x \text{ real})$$

converges uniformly on the whole real line but not absolutely.

Proof. By the familiar Leibniz test of calculus (see Appendix A3.3) the remainder R_n does not exceed its first term in absolute value, since we have a series of alternating terms whose absolute values form a monotone decreasing sequence with limit zero. Hence, given $\epsilon > 0$, for all x we have

$$|R_n(x)| \leqq \frac{1}{x^2+n+1} < \frac{1}{n} < \epsilon \qquad \text{if } n > N(\epsilon) = \frac{1}{\epsilon}.$$

This proves uniform convergence, since $N(\epsilon)$ does not depend on x.

The convergence is not absolute because

$$\left|\frac{(-1)^{m-1}}{x^2+m}\right| = \frac{1}{x^2+m} > \frac{k}{m}$$

(k a suitable constant) and $k\Sigma 1/m$ diverges. ∎

Problem Set 14.6

Prove that the following series converge uniformly in the given regions.

1. $\sum_{n=0}^{\infty} z^{2n}, \quad |z| \leqq 0.999$

2. $\sum_{n=0}^{\infty} \frac{z^{2n+1}}{(2n+1)!}, \quad |z| \leqq 10^{10}$

3. $\sum_{n=1}^{\infty} \frac{\pi^n}{n^4} z^{2n}, \quad |z| \leqq 0.56$

4. $\sum_{n=1}^{\infty} \frac{z^n}{n^2}, \quad |z| \leqq 1$

5. $\sum_{n=1}^{\infty} \frac{\sin n|z|}{n^2}, \quad \text{all } z$

6. $\sum_{n=1}^{\infty} \frac{\cos^n |\pi z|}{n(n+1)}, \quad \text{all } z$

7. $\sum_{n=1}^{\infty} \frac{nz^n}{n^3+|z|}, \quad |z| \leqq 1$

8. $\sum_{n=1}^{\infty} \frac{z^n}{n^2 \cosh n|z|}, \quad |z| \leqq 1$

9. $\sum_{n=1}^{\infty} \frac{\tanh^n |z|}{n(n+1)}, \quad \text{all } z$

10. $\sum_{n=0}^{\infty} \frac{z^n}{|z|^{2n}+2}, \quad 2 \leqq |z| \leqq 3$

Where do the following power series converge uniformly?

11. $\sum_{n=0}^{\infty} \frac{(z+2i)^{2n}}{3^n}$

12. $\sum_{n=0}^{\infty} \frac{(z-1+i)^n}{n!}$

Where do the following power series converge uniformly?

13. $\sum_{n=1}^{\infty} \frac{(-1)^n}{4^n n} z^n$

14. $\sum_{n=2}^{\infty} \binom{n}{2} (2z + i)^n$

15. $\sum_{n=1}^{\infty} \frac{n!}{n^2} (z - i)^n$

16. $\sum_{n=1}^{\infty} (2^n \tanh n) z^{2n}$

17. Give a proof of the Weierstrass M-test (Theorem 5).

18. Derive Theorem 4 from Theorem 3.

19. If the series (1) converges uniformly in a region G, show that it converges uniformly in any portion of G. Is the converse true?

20. Find the precise region of convergence of the series in Example 2 with x replaced by a complex variable z.

21. Determine the smallest integer n such that $|R_n| < 0.01$ in Example 1, if $z = x = 0.5, 0.6, 0.7, 0.8, 0.9$. What does the result mean from the viewpoint of computing $1/(1 - x)$ with an absolute error less than 0.01 by means of the geometric series?

22. Show that $x^2 \sum_{m=1}^{\infty} (1 + x^2)^{-m} = 1$ if $x \neq 0$ and 0 if $x = 0$. Verify by computation that the partial sums s_1, s_2, s_3 look as shown in Fig. 341 and compute and graph s_4, s_5, s_6.

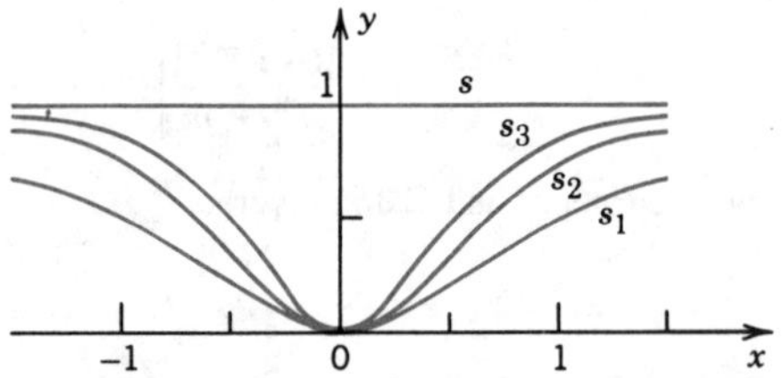

Fig. 341. Sum s and partial sums in Problem 22

Heat equation. Show that (10), Sec. 11.5, with coefficients (11) is a solution of the heat equation for $t > 0$, assuming that $f(x)$ is continuous on the interval $0 \leqq x \leqq L$ and has one-sided derivatives at all interior points of that interval. Proceed as follows.

23. Show that $|B_n|$ is bounded, say, $|B_n| < K$ for all n. Conclude that

$$|u_n| < Ke^{-\lambda_n^2 t_0} \quad \text{if} \quad t \geqq t_0 > 0$$

and, by the Weierstrass test, the series (10) converges uniformly with respect to x and t for $t \geqq t_0$, $0 \leqq x \leqq L$. Using Theorem 2, show that $u(x, t)$ is continuous for $t \geqq t_0$ and thus satisfies the boundary conditions (2) for $t \geqq t_0$.

24. Show that $|\partial u_n/\partial t| < \lambda_n^2 Ke^{-\lambda_n^2 t_0}$ if $t \geqq t_0$ and the series of the expressions on the right converges, by the ratio test. Conclude from this, the Weierstrass test, and Theorem 4 that the series (10) can be differentiated term by term with respect to t and the resulting series has the sum $\partial u/\partial t$. Show that (10) can be differentiated twice with respect to x and the resulting series has the sum $\partial^2 u/\partial x^2$. Conclude from this and the result of Prob. 23 that (10) is a solution of the heat equation for all $t \geqq t_0$. (The proof that (10) satisfies the given initial condition can be found in Ref. [C12] listed in Appendix 1.)

14.7 Laurent Series

In applications it will often be necessary to expand a function $f(z)$ around points at which it is no longer analytic, but is singular (as defined in Sec. 14.4). Then Taylor's theorem no longer applies, but we need a new type of series, known as *Laurent series,* consisting of positive *and negative* integer powers of $z - z_0$ and being convergent in some annulus (bounded by two circles with center z_0) in which $f(z)$ is analytic. $f(z)$ may have singular points not only outside the larger circle (as for a Taylor series) but also inside the smaller circle—this is a new feature.

Theorem 1 **Laurent's theorem**

If $f(z)$ is analytic on two concentric circles[9] *C_1 and C_2 with center z_0 and in the annulus between them, then $f(z)$ can be represented by the* ***Laurent series***

$$f(z) = \sum_{n=0}^{\infty} a_n(z - z_0)^n + \sum_{n=1}^{\infty} \frac{b_n}{(z - z_0)^n}$$

$$= a_0 + a_1(z - z_0) + a_2(z - z_0)^2 + \cdots \tag{1}$$

$$\cdots + \frac{b_1}{z - z_0} + \frac{b_2}{(z - z_0)^2} + \cdots.$$

The coefficients of this Laurent series are given by the integrals[10]

$$a_n = \frac{1}{2\pi i} \oint_C \frac{f(z^*)}{(z^* - z_0)^{n+1}}\, dz^*, \quad b_n = \frac{1}{2\pi i} \oint_C (z^* - z_0)^{n-1} f(z^*)\, dz^*, \tag{2}$$

each integral being taken counterclockwise around any simple closed path C that lies in the annulus and encircles the inner circle (Fig. 342, p. 822).

This series converges and represents $f(z)$ in the open annulus obtained from the given annulus by continuously increasing the circle C_1 and decreasing C_2 until each of the two circles reaches a point where $f(z)$ is singular.

In the important special case that z_0 is the only singular point of $f(z)$ inside C_2, this circle can be shrunk to the point z_0, giving convergence in a disk except at the center.

Comment. Obviously, instead of (1), (2) we may write (denoting b_n by a_{-n})

$$f(z) = \sum_{n=-\infty}^{\infty} a_n(z - z_0)^n \tag{1'}$$

---

[9]Recall that by the definition of analyticity, this means that $f(z)$ is analytic in some domain containing the annulus as well as its boundary circles.

The theorem is named after the French ingenieur and mathematician, PIERRE ALPHONSE LAURENT (1813—1854), who published it in 1843.

[10]We denote the variable of integration by z^* because z is used in $f(z)$.

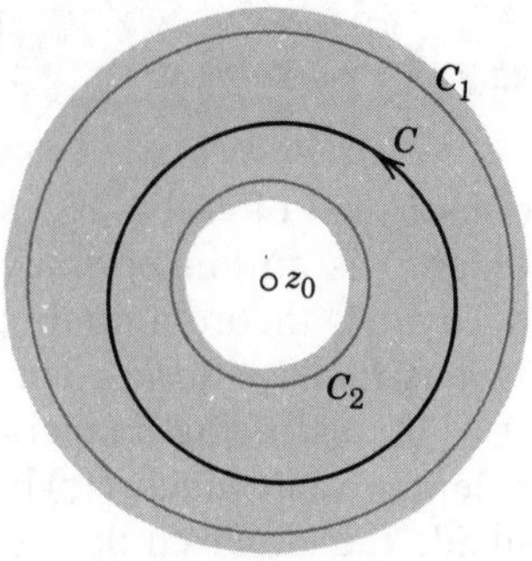

Fig. 342. Laurent's theorem

where

$$(2') \qquad a_n = \frac{1}{2\pi i} \oint_C \frac{f(z^*)}{(z^* - z_0)^{n+1}} \, dz^*.$$

Proof of Laurent's theorem. Let z be any point in the given annulus. Then from Cauchy's integral formula [(3) in Sec. 13.5] it follows that

$$(3) \qquad f(z) = \frac{1}{2\pi i} \oint_{C_1} \frac{f(z^*)}{z^* - z} \, dz^* - \frac{1}{2\pi i} \oint_{C_2} \frac{f(z^*)}{z^* - z} \, dz^*,$$

where we integrate counterclockwise over C_1 and also over C_2. We transform each of these two integrals as in Sec. 14.4. The first integral is *precisely* as in Sec. 14.4, so we get precisely the same result

$$(4) \qquad \frac{1}{2\pi i} \oint_{C_1} \frac{f(z^*)}{z^* - z} \, dz^* = \sum_{n=0}^{\infty} a_n (z - z_0)^n$$

with coefficients [see (9*), Sec. 14.4, counterclockwise integration]

$$(5) \qquad a_n = \frac{1}{2\pi i} \oint_{C_1} \frac{f(z^*)}{(z^* - z_0)^{n+1}} \, dz^*.$$

Here we can replace C_1 by C (see Fig. 342), by the principle of deformation of path, since z_0, the point where the integrand in (5) is not analytic, is not a point of the annulus. This proves the formula for the a_n in (2).

We turn to the second integral in (3) and shall get the formula for the b_n from it. Since z lies in the annulus, it lies *outside* the path C_2. Hence the situation differs from that for the first integral. The essential point is that instead of (2*) in Sec. 14.4 we now have

$$(6) \qquad \left| \frac{z^* - z_0}{z - z_0} \right| < 1;$$

consequently, we must develop $1/(z^* - z)$ in the integrand in powers of

$(z^* - z_0)/(z - z_0)$ [instead of the reciprocal of this] to get a *convergent* series. We find

$$\frac{1}{z^* - z} = \frac{1}{z^* - z_0 - (z - z_0)} = \frac{-1}{(z - z_0)\left(1 - \dfrac{z^* - z_0}{z - z_0}\right)}.$$

Compare this for a moment with (2) in Sec. 14.4, to really understand the difference. Then go on and apply formula (3), Sec. 14.4, for a finite geometric sum, obtaining

$$\frac{1}{z^* - z} = -\frac{1}{z - z_0}\left\{1 + \frac{z^* - z_0}{z - z_0} + \left(\frac{z^* - z_0}{z - z_0}\right)^2 + \cdots + \left(\frac{z^* - z_0}{z - z_0}\right)^n\right\}$$
$$- \frac{1}{z - z^*}\left(\frac{z^* - z_0}{z - z_0}\right)^{n+1}.$$

Multiplication by $-f(z^*)/2\pi i$ and integration over C_2 on both sides now yields

$$-\frac{1}{2\pi i}\oint_{C_2} \frac{f(z^*)}{z^* - z}\, dz^*$$
$$= \frac{1}{2\pi i}\left\{\frac{1}{z - z_0}\oint_{C_2} f(z^*)\, dz^* + \frac{1}{(z - z_0)^2}\oint_{C_2} (z^* - z_0) f(z^*)\, dz^*\right.$$
$$\left. + \cdots + \frac{1}{(z - z_0)^{n+1}}\oint_{C_2} (z^* - z_0)^n f(z^*)\, dz^*\right\} + R_n^*(z)$$

with the last term on the right given by

$$R_n^*(z) = \frac{1}{2\pi i (z - z_0)^{n+1}}\oint_{C_2} \frac{(z^* - z_0)^{n+1}}{z - z^*} f(z^*)\, dz^*. \tag{7}$$

As before, we can integrate over C instead of C_2 in the integrals on the right. We see that on the right, the power $1/(z - z_0)^n$ is multiplied by b_n as given in (2). This establishes Laurent's theorem, provided

$$\lim_{n \to \infty} R_n^*(z) = 0. \tag{8}$$

We prove (8). The expression $f(z^*)/(z - z^*)$ in (7) is bounded in absolute value, say,

$$\left|\frac{f(z^*)}{z - z^*}\right| < \tilde{M} \qquad \text{for all } z^* \text{ on } C_2$$

because $f(z^*)$ is analytic in the annulus and on C_2, and z^* lies on C_2 and z outside, so that $z - z^* \neq 0$. From this and the *ML*-inequality (Sec. 13.2) applied to (7) we get (L = length of C_2)

$$|R_n^*(z)| \leqq \frac{1}{2\pi|z - z_0|^{n+1}} |z^* - z_0|^{n+1} \widetilde{M}L = \frac{\widetilde{M}L}{2\pi} \left| \frac{z^* - z_0}{z - z_0} \right|^{n+1}.$$

From (6) we see that the expression on the right approaches zero as n approaches infinity. This proves (8). The representation (1) with coefficients (2) is now established in the given annulus.

Finally let us prove convergence of (1) in the open annulus characterized at the end of the theorem.

We denote the sums of the two series in (1) by $g(z)$ and $h(z)$, and the radii of C_1 and C_2 by r_1 and r_2, respectively. Then $f = g + h$. The first series is a power series. Since it converges in the annulus, it must converge in the entire disk bounded by C_1, and g is analytic in that disk.

If we set $Z = 1/(z - z_0)$, the last series becomes a power series in Z. The annulus $r_2 < |z - z_0| < r_1$ then corresponds to the annulus $1/r_1 < |Z| < 1/r_2$, the new series converges in this annulus and, therefore, in the entire disk $|Z| < 1/r_2$. Since this disk corresponds to $|z - z_0| > r_2$, the exterior of C_2, the given series converges for all z outside C_2, and h is analytic for all these z.

Since $f = g + h$, it follows that g must be singular at all those points outside C_1 where f is singular, and h must be singular at all those points inside C_2 where f is singular. Consequently, the first series converges for all z inside the circle about z_0 whose radius is equal to the distance of that singularity of f outside C_1 which is closest to z_0. Similarly, the second series converges for all z outside the circle about z_0 whose radius is equal to the maximum distance of the singularities of f inside C_2. The domain common to both of those domains of convergence is the open annulus characterized near the end of the theorem, and the proof of Laurent's theorem is complete. ∎

Uniqueness. *The Laurent series of a given analytic function $f(z)$ in its annulus of convergence is unique* (see Prob. 35). *However, $f(z)$ may have different Laurent series in two annuli with the same center;* see the examples below. The uniqueness is essential. As for a Taylor series, to obtain the coefficients of Laurent series, we do not generally use the integral formulas (2) but various other methods, some of which we shall illustrate by our examples. If a Laurent series has been found by any such process, the uniqueness guarantees that it must be *the* Laurent series of the given function in the given annulus.

EXAMPLE 1 **Use of Maclaurin series**

Find the Laurent series of $z^{-5} \sin z$ with center 0.

Solution. By (14), Sec. 14.4, we obtain

$$z^{-5} \sin z = \sum_{n=0}^{\infty} \frac{(-1)^n}{(2n+1)!} z^{2n-4} = \frac{1}{z^4} - \frac{1}{6z^2} + \frac{1}{120} - \frac{1}{5040} z^2 + - \cdots \qquad (|z| > 0).$$

Here the "annulus" of convergence is the whole complex plane without the origin. ∎

EXAMPLE 2 **Substitution**

Find the Laurent series of $z^2 e^{1/z}$ with center 0.

Solution. From (12) in Sec. 14.4 with z replaced by $1/z$ we obtain

$$z^2 e^{1/z} = z^2 \left(1 + \frac{1}{1!z} + \frac{1}{2!z^2} + \cdots \right) = z^2 + z + \frac{1}{2} + \frac{1}{3!z} + \frac{1}{4!z^2} + \cdots \quad (|z| > 0). \quad \blacksquare$$

EXAMPLE 3   Develop $1/(1 - z)$ (a) in nonnegative powers of z, (b) in negative powers of z.

Solution.

(a)
$$\frac{1}{1 - z} = \sum_{n=0}^{\infty} z^n \qquad \qquad \text{(valid if } |z| < 1).$$

(b)
$$\frac{1}{1 - z} = \frac{-1}{z(1 - z^{-1})} = - \sum_{n=0}^{\infty} \frac{1}{z^{n+1}} = - \frac{1}{z} - \frac{1}{z^2} - \cdots \qquad \text{(valid if } |z| > 1). \quad \blacksquare$$

EXAMPLE 4   **Laurent expansions in different concentric annuli**
Find all Laurent series of $1/(z^3 - z^4)$ with center 0.

Solution. Multiplying by $1/z^3$, we get from Example 3

(I)
$$\frac{1}{z^3 - z^4} = \sum_{n=0}^{\infty} z^{n-3} = \frac{1}{z^3} + \frac{1}{z^2} + \frac{1}{z} + 1 + z + \cdots \qquad (0 < |z| < 1),$$

(II)
$$\frac{1}{z^3 - z^4} = - \sum_{n=0}^{\infty} \frac{1}{z^{n+4}} = - \frac{1}{z^4} - \frac{1}{z^5} - \cdots \qquad (|z| > 1). \quad \blacksquare$$

EXAMPLE 5   **Use of partial fractions**
Find all Taylor and Laurent series of $f(z) = \dfrac{-2z + 3}{z^2 - 3z + 2}$ with center 0.

Solution. In terms of partial fractions,

$$f(z) = - \frac{1}{z - 1} - \frac{1}{z - 2}.$$

(a) and (b) in Example 3 take care of the first fraction. For the second fraction,

(c)
$$- \frac{1}{z - 2} = \frac{1}{2 \left(1 - \frac{1}{2} z \right)} = \sum_{n=0}^{\infty} \frac{1}{2^{n+1}} z^n \qquad (|z| < 2),$$

(d)
$$- \frac{1}{z - 2} = - \frac{1}{z \left(1 - \frac{2}{z} \right)} = - \sum_{n=0}^{\infty} \frac{2^n}{z^{n+1}} \qquad (|z| > 2).$$

(I) From (a) and (c), valid for $|z| < 1$ (see Fig. 343)

$$f(z) = \sum_{n=0}^{\infty} \left(1 + \frac{1}{2^{n+1}} \right) z^n = \frac{3}{2} + \frac{5}{4} z + \frac{9}{8} z^2 + \cdots .$$

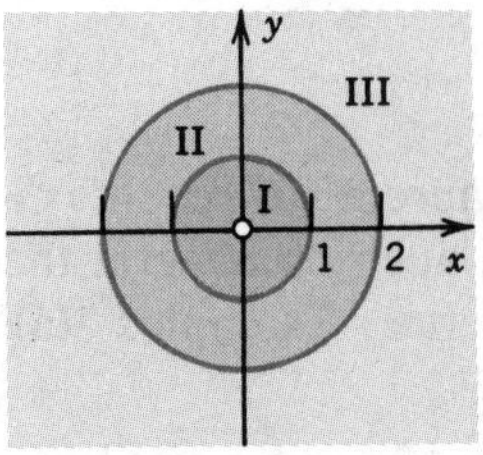

Fig. 343. Regions of convergence
in Example 5

(II) From (c) and (b), valid for $1 < |z| < 2$,

$$f(z) = \sum_{n=0}^{\infty} \frac{1}{2^{n+1}} z^n - \sum_{n=0}^{\infty} \frac{1}{z^{n+1}} = \frac{1}{2} + \frac{1}{4}z + \frac{1}{8}z^2 + \cdots - \frac{1}{z} - \frac{1}{z^2} - \cdots .$$

(III) From (d) and (b), valid for $|z| > 2$,

$$f(z) = -\sum_{n=0}^{\infty} (2^n + 1)\frac{1}{z^{n+1}} = -\frac{2}{z} - \frac{3}{z^2} - \frac{5}{z^3} - \frac{9}{z^4} - \cdots . \qquad \blacksquare$$

EXAMPLE 6   Find the Laurent series of $f(z) = 1/(1 - z^2)$ that converges in the annulus $1/4 < |z - 1| < 1/2$ and determine the precise region of convergence.

Solution. The annulus has center 1, so that we must develop

$$f(z) = \frac{-1}{(z - 1)(z + 1)}$$

in powers of $z - 1$. We calculate

$$\frac{1}{z + 1} = \frac{1}{2 + (z - 1)} = \frac{1}{2} \frac{1}{\left[1 - \left(-\dfrac{z - 1}{2}\right)\right]}$$

$$= \frac{1}{2} \sum_{n=0}^{\infty} \left(-\frac{z - 1}{2}\right)^n = \sum_{n=0}^{\infty} \frac{(-1)^n}{2^{n+1}} (z - 1)^n;$$

this series converges in the disk $|(z - 1)/2| < 1$, that is, $|z - 1| < 2$. Multiplication by $-1/(z - 1)$ now gives the desired series

$$f(z) = \sum_{n=0}^{\infty} \frac{(-1)^{n+1}}{2^{n+1}} (z - 1)^{n-1} = \frac{-1/2}{z - 1} + \frac{1}{4} - \frac{1}{8}(z - 1) + \frac{1}{16}(z - 1)^2 - + \cdots .$$

The precise region of convergence is $0 < |z - 1| < 2$; see Fig. 344. We confirm this by noting that $1/(z + 1)$ in $f(z)$ is singular at -1, at distance 2 from the center of the series. $\blacksquare$

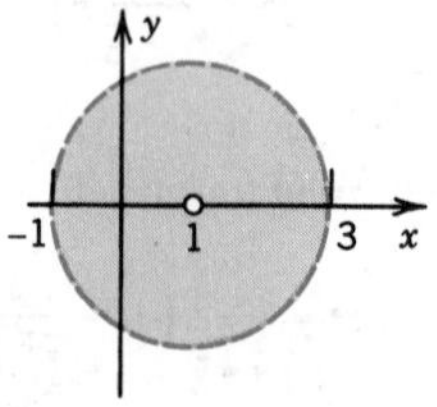

Fig. 344.   Region of convergence in Example 6

If $f(z)$ in Laurent's theorem is analytic inside C_2, the coefficients b_n in (2) are zero by Cauchy's integral theorem, so that the Laurent series reduces to a Taylor series. Examples 3(a) and 5(I) illustrate this.

Problem Set 14.7

Expand each of the following functions in a Laurent series that converges for $0 < |z| < R$ and determine the precise region of convergence.

1. $\dfrac{e^z}{z^2}$ **2.** $\dfrac{\sin 4z}{z^4}$ **3.** $\dfrac{\cosh 2z}{z}$

4. $\dfrac{1}{z^3(1 - z)}$ **5.** $\dfrac{1}{z(1 + z^2)}$ **6.** $\dfrac{8 - 2z}{4z - z^3}$

7. $z \cos \dfrac{1}{z}$ **8.** $\dfrac{e^{-1/z^2}}{z^5}$ **9.** $\dfrac{1}{z^6(1 + z)^2}$

Expand each of the following functions in a Laurent series that converges for $0 < |z - z_0| < R$ and determine the precise region of convergence.

10. $\dfrac{e^z}{z - 1}, \quad z_0 = 1$ **11.** $\dfrac{1}{z^2 + 1}, \quad z_0 = i$ **12.** $z^2 \sinh \dfrac{1}{z}, \quad z_0 = 0$

13. $\dfrac{\cos z}{(z - \pi)^3}, \quad z_0 = \pi$ **14.** $\dfrac{z^4}{(z + 2i)^2}, \quad z_0 = -2i$ **15.** $\dfrac{z^2 - 4}{z - 1}, \quad z_0 = 1$

16. $\dfrac{\sin z}{(z - \frac{1}{4}\pi)^3}, \quad z_0 = \dfrac{\pi}{4}$ **17.** $\dfrac{1}{(z + i)^2 - (z + i)}, \quad z_0 = -i$

18. $\dfrac{1}{1 - z^4}, \quad z_0 = -1$

Find the Taylor or Laurent series of $1/(1 - z^2)$ in the region

19. $0 \leqq |z| < 1$ **20.** $|z| > 1$ **21.** $0 < |z - 1| < 2$

Using partial fractions, find the Laurent series of $(3z^2 - 6z + 2)/(z^3 - 3z^2 + 2z)$ in the region

22. $0 < |z| < 1$ **23.** $1 < |z| < 2$ **24.** $|z| > 2$

Find all Taylor and Laurent series with center $z = z_0$ and determine the precise region of convergence.

25. $\dfrac{1}{1 - z^3}, \quad z_0 = 0$ **26.** $\dfrac{2}{1 - z^2}, \quad z_0 = 1$ **27.** $\dfrac{z^2}{1 - z^4}, \quad z_0 = 0$

28. $\dfrac{1}{z^2}, \quad z_0 = i$ **29.** $\dfrac{1}{z}, \quad z_0 = 1$ **30.** $\dfrac{\sinh z}{(z - 1)^2}, \quad z_0 = 1$

31. $\dfrac{\sin z}{z + \frac{1}{2}\pi}, \quad z_0 = -\dfrac{1}{2}\pi$ **32.** $\dfrac{z^3 - 2iz^2}{(z - i)^2}, \quad z_0 = i$ **33.** $\dfrac{4z - 1}{z^4 - 1}, \quad z_0 = 0$

34. Does $\tan (1/z)$ have a Laurent series convergent in a region $0 < |z| < R$?

35. Prove that the Laurent expansion of a given analytic function in a given annulus is unique.

14.8 Singularities and Zeros. Infinity

Roughly, a *singularity* of an analytic function $f(z)$ is a z at which $f(z)$ ceases to be analytic, and a *zero* is a z at which $f(z) = 0$. Precise definitions follow below. Singularities may be discussed and classified by means of Laurent series. This is what we do first. Then we show that zeros may be discussed by means of Taylor series.

We say that a function[11] $f(z)$ **is singular** or **has a singularity** at a point $z = z_0$ if $f(z)$ is not analytic (perhaps not even defined) at $z = z_0$, but every neighborhood of $z = z_0$ contains points at which $f(z)$ is analytic.

We call $z = z_0$ an **isolated singularity** of $f(z)$ if $z = z_0$ has a neighborhood without further singularities of $f(z)$. *Example:* tan z has isolated singularities at $\pm\pi/2$, $\pm 3\pi/2$, etc.; tan $(1/z)$ has a nonisolated singularity at 0. (Explain!) Isolated singularities of $f(z)$ at $z = z_0$ can be classified by the Laurent series

$$f(z) = \sum_{n=0}^{\infty} a_n(z - z_0)^n + \sum_{n=1}^{\infty} \frac{b_n}{(z - z_0)^n} \qquad \text{(Sec. 14.7)} \tag{1}$$

valid ***in the immediate neighborhood*** of the singular point $z = z_0$, except at z_0 itself, that is, in a region of the form

$$0 < |z - z_0| < R.$$

The sum of the first series is analytic at $z = z_0$, as we know from the last section. The second series, containing the negative powers, is called the **principal part** of (1). If it has only finitely many terms, it is of the form

$$\frac{b_1}{z - z_0} + \cdots + \frac{b_m}{(z - z_0)^m} \qquad (b_m \neq 0). \tag{2}$$

Then the singularity of $f(z)$ at $z = z_0$ is called a **pole,** and m is called its **order.** Poles of the first order are also known as *simple poles.*

If the principal part of (1) has infinitely many terms, we say that $f(z)$ has at $z = z_0$ an **isolated essential singularity.**

We leave aside nonisolated singularities (see the example above).

EXAMPLE 1 Poles. Essential singularities

The function

$$f(z) = \frac{1}{z(z - 2)^5} + \frac{3}{(z - 2)^2}$$

has a simple pole at $z = 0$ and a pole of fifth order at $z = 2$. Examples of functions having an isolated essential singularity at $z = 0$ are

[11]We recall that, by definition, a function is a ***single-valued*** relation. (See Sec. 12.4.)

$$e^{1/z} = \sum_{n=0}^{\infty} \frac{1}{n!\, z^n} = 1 + \frac{1}{z} + \frac{1}{2!\, z^2} + \cdots$$

and

$$\sin \frac{1}{z} = \sum_{n=0}^{\infty} \frac{(-1)^n}{(2n+1)!\, z^{2n+1}} = \frac{1}{z} - \frac{1}{3!\, z^3} + \frac{1}{5!\, z^5} - + \cdots .$$

Section 14.7 provides further examples. For instance, Example 1 shows that $z^{-5} \sin z$ has a fourth-order pole at 0. Example 4 shows that $1/(z^3 - z^4)$ has a third-pole at 0 and a Laurent series with infinitely many negative powers. This is no contradiction, since this series is valid for $|z| > 1$; it merely tells us that it is quite important to consider the Laurent series valid ***in the immediate neighborhood*** of a singularity. ∎

The classification of singularities into poles and essential singularities is not merely a formal matter, because the behavior of an analytic function in a neighborhood of an essential singularity is entirely different from that in the neighborhood of a pole.

EXAMPLE 2 **Behavior near a pole**

The function $f(z) = 1/z^2$ has a pole at $z = 0$, and $|f(z)| \to \infty$ as $z \to 0$ in any manner. ∎

This example illustrates

Theorem 1 **(Poles)**

If $f(z)$ is analytic and has a pole at $z = z_0$, then $|f(z)| \to \infty$ as $z \to z_0$ in any manner. (See Prob. 16.)

EXAMPLE 3 **Behavior near an essential singularity**

The function $f(z) = e^{1/z}$ has an essential singularity at $z = 0$. It has no limit for approach along the imaginary axis; it becomes infinite if $z \to 0$ through positive real values, but it approaches zero if $z \to 0$ through negative real values. It takes on any given value $c = c_0 e^{i\alpha} \neq 0$ in an arbitrarily small neighborhood of $z = 0$. In fact, setting $z = re^{i\theta}$, we must solve the equation

$$e^{1/z} = e^{(\cos\theta - i \sin\theta)/r} = c_0 e^{i\alpha}$$

for r and θ. Equating the absolute values and the arguments, we have $e^{(\cos\theta)/r} = c_0$, that is,

$$\cos\theta = r \ln c_0,$$

and

$$\sin\theta = -\alpha r.$$

From these two equations and $\cos^2\theta + \sin^2\theta = 1$ we obtain the formulas

$$r^2 = \frac{1}{(\ln c_0)^2 + \alpha^2}$$

and

$$\tan\theta = -\frac{\alpha}{\ln c_0}.$$

Hence r can be made arbitrarily small by adding multiples of 2π to α, leaving c unaltered. ∎

This example illustrates the following famous theorem.

Theorem 2 (Picard's[12] theorem)

If $f(z)$ is analytic and has an isolated essential singularity at a point z_0, it takes on every value, with at most one exceptional value, in an arbitrarily small neighborhood of z_0.

In Example 3, the exceptional value is $z = 0$. The proof of Picard's theorem is rather complicated; it can be found in Ref. [D10].

Removable singularities. We say that a function $f(z)$ has a *removable singularity* at $z = z_0$ if $f(z)$ is not analytic at $z = z_0$, but can be made analytic there by assigning a suitable value $f(z_0)$. Such singularities are of no interest since they can be removed as just indicated. *Example:* $f(z) = (\sin z)/z$ becomes analytic at $z = 0$ if we define $f(0) = 1$.

Zeros of Analytic Functions

We say that a function $f(z)$ that is analytic in some domain D has a **zero** at a point $z = z_0$ in D if $f(z_0) = 0$. We also say that this zero is of **order** n if not only f but also the derivatives $f', f'', \cdots, f^{(n-1)}$ are all zero at $z = z_0$ but $f^{(n)}(z_0) \neq 0$.

A zero of first order is also called a *simple zero;* for it, $f(z_0) = 0$ but $f'(z_0) \neq 0$. For a second-order zero, $f(z_0) = 0$, $f'(z_0) = 0$ but $f''(z_0) \neq 0$, and so on.

EXAMPLE 4 Zeros

The function $1 + z^2$ has simple zeros at $\pm i$. The function $(1 - z^4)^2$ has second-order zeros at ± 1 and $\pm i$. The function $(z - a)^3$ has a third-order zero at $z = a$. The function e^z has no zeros (see Sec. 12.6). The function $\sin z$ has simple zeros at $0, \pm\pi, \pm 2\pi, \cdots$, and $\sin^2 z$ has second-order zeros at these points. The function $1 - \cos z$ has second-order zeros at $0, \pm 2\pi, \pm 4\pi, \cdots$, and $(1 - \cos z)^2$ has fourth-order zeros at these points. ■

Taylor series at a zero. At an nth-order zero $z = z_0$ of $f(z)$, the derivatives $f'(z_0), \cdots, f^{(n-1)}(z_0)$ are zero, by definition. Hence the first few coefficients $a_0, \cdots, a_{n-1}$ of the Taylor series (9), Sec. 14.4, are zero, too, whereas $a_n \neq 0$, so that this series takes the form

$$\begin{aligned} f(z) &= a_n(z - z_0)^n + a_{n+1}(z - z_0)^{n+1} + \cdots \\ &= (z - z_0)^n[a_n + a_{n+1}(z - z_0) + a_{n+2}(z - z_0)^2 + \cdots] \end{aligned} \qquad (a_n \neq 0). \tag{3}$$

This is characteristic of such a zero, because if $f(z)$ has such a Taylor series, it has an nth-order zero at $z = z_0$, as follows by differentiation.

Whereas nonisolated singularities may occur, for zeros we have

Theorem 3 (Zeros)

The zeros of an analytic function $f(z)$ ($\not\equiv 0$) are isolated; that is, each of them has a neighborhood that contains no further zeros of $f(z)$.

---

[12]See the footnote in Sec. 1.10.

Proof. In (3), the factor $(z - z_0)^n$ is zero only at $z = z_0$. The power series in the brackets $[\cdot \cdot \cdot]$ represents an analytic function (by Theorem 5 in Sec. 14.3), call it $g(z)$. Now $g(z_0) = a_n \neq 0$, since an analytic function is continuous, and because of this continuity, also $g(z) \neq 0$ in some neighborhood of $z = z_0$. Hence the same holds for $f(z)$. ∎

This theorem is illustrated by the functions in Example 4.

Poles are often caused by zeros in the denominator. (*Example:* tan z has poles where cos z is zero.) This is a major reason for the importance of zeros. The key to the connection is

Theorem 4 **(Poles and zeros)**

Let $f(z)$ be analytic at $z = z_0$ and have a zero of nth order at $z = z_0$. Then $1/f(z)$ has a pole of nth order at $z = z_0$.

The same holds for $h(z)/f(z)$ if $h(z)$ is analytic at $z = z_0$ and $h(z_0) \neq 0$.

The proof follows from (3) (see Prob. 27).

Riemann Number Sphere. Infinity

The study of the behavior of analytic functions $f(z)$ for large $|z|$ is a natural task of practical importance. When $|z|$ is large, the complex plane becomes somewhat inconvenient. We may then prefer a representation of the complex numbers on a sphere, which was suggested by Riemann and is obtained as follows.

Let S be a sphere of diameter 1 that touches the complex z-plane at the origin (Fig. 345). Let N be the "North Pole" of S (the point diametrically opposite to the point of contact between the sphere and the plane). Let P be any point in the complex plane. Then the straight line segment with endpoints P and N intersects S at a point P^*. We let P and P^* correspond to each other. In this way we obtain a mapping of the complex plane into

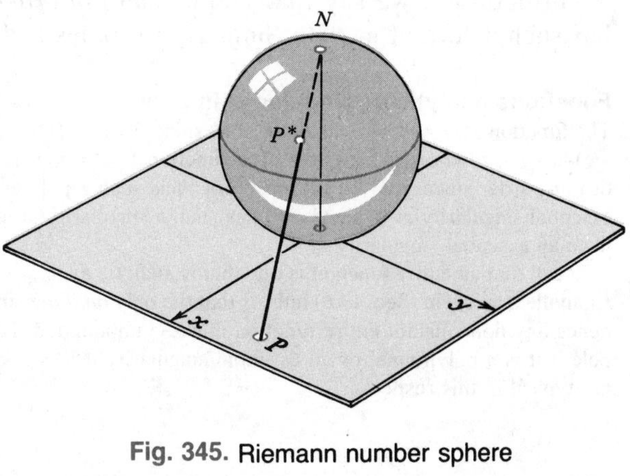

Fig. 345. Riemann number sphere

the sphere S, and P^* is the image point of P with respect to this mapping. The complex numbers, first represented in the plane, are now represented by points on S. To each z there corresponds a point on S.

Conversely, each point on S represents a complex number z, except for the point N, which does not correspond to any point in the complex plane. This suggests that we introduce an additional point, called the **point at infinity** and denoted by the symbol ∞ (*infinity*). The complex plane together with the point ∞ is called the **extended complex plane.** The complex plane without that point ∞ is often called the *finite complex plane,* for distinction, or simply the *complex plane,* as before.

Of course, we now let the point $z = \infty$ correspond to N. Then our mapping becomes a one-to-one mapping of the extended complex plane onto S. The sphere S is called the **Riemann number sphere.** The particular mapping we have used is called a **stereographic projection.**

Obviously, the unit circle is mapped onto the "equator" of S. The interior of the unit circle corresponds to the "Southern Hemisphere" and the exterior to the "Northern Hemisphere." Numbers z whose absolute values are large lie close to the North Pole N. The x and y axes (and, more generally, all the straight lines through the origin) are mapped onto "meridians," while circles with center at the origin are mapped onto "parallels." It can be shown that any circle or straight line in the z-plane is mapped onto a circle on S.

Analytic or Singular at Infinity

If we want to investigate a function $f(z)$ for large $|z|$, we may now set $z = 1/w$ and investigate $f(z) = f(1/w) \equiv g(w)$ in a neighborhood of $w = 0$. We define $f(z)$ to be **analytic or singular at infinity** if $g(w)$ is analytic or singular, respectively, at $w = 0$. We also define

$$g(0) = \lim_{w \to 0} g(w) \tag{4}$$

if this limit exists.

Furthermore, we say that $f(z)$ has an *nth-order zero at infinity* if $f(1/w)$ has such a zero at $w = 0$. Similarly for poles and essential singularities.

EXAMPLE 5 Functions analytic or singular at infinity

The function $f(z) = 1/z^2$ is analytic at ∞ since $g(w) = f(1/w) = w^2$ is analytic at $w = 0$, and $f(z)$ has a second-order zero at ∞. The function $f(z) = z^3$ is singular at ∞ and has there a pole of third order since $g(w) = f(1/w) = 1/w^3$ has such a pole at $w = 0$. The function e^z has an essential singularity at ∞ since $e^{1/w}$ has such a singularity at $w = 0$. Similarly, $\cos z$ and $\sin z$ have an essential singularity at ∞.

Recall that an **entire function** is one that is analytic everywhere in the (finite) complex plane. Liouville's theorem (Sec. 13.6) tells us that the only *bounded* entire functions are the constants, hence any nonconstant entire function must be unbounded. Hence it has a singularity at ∞, a pole if it is a polynomial or an essential singularity if it is not. The functions just considered are typical in this respect. ∎

A **meromorphic function** is an analytic function whose only singularities in the finite plane are poles.

EXAMPLE 6 Meromorphic functions

Rational functions with nonconstant denominator, $\tan z$, $\cot z$, $\sec z$, and $\csc z$ are meromorphic functions. ∎

This is the end of Chap. 14 on power series, particularly Taylor series (which play an even greater role here than in calculus), and on Laurent series. Interestingly enough, the latter will provide us with another powerful integration method in the next chapter.

Problem Set 14.8

Singularities. Determine the location and type of the singularities of the following functions, including those at infinity. (In the case of poles also state the order.)

1. $\cot z$
2. $1/(z + a)^4$
3. $z + 1/z$
4. $\frac{3}{z} - \frac{1}{z^2} - \frac{2}{z^3}$
5. $\frac{\cos 4z}{(z^4 - 1)^3}$
6. $\frac{\sin^2 z}{z^4 \cos 2z}$
7. $e^{\pi z}/(z^2 - iz + 2)^2$
8. $e^{1/(z+i)} + z^2$
9. $(e^z - 1 - z)/z^3$
10. $\cosh [1/(z^2 + 1)]$
11. $\tan 1/z$
12. $(\cos z - \sin z)^{-1}$
13. $\cos z - \sin z$
14. $1/\sinh \frac{1}{2}z$
15. $e^{1/(z-1)}/(e^z - 1)$

16. Verify Theorem 1 for $f(z) = z^{-3} - z^{-1}$. Prove Theorem 1.

Zeros. Determine the location and order of the zeros of the following functions.

17. $(z^4 - 16)^2$
18. $(z - 16)^8$
19. $z \sin^2 \pi z$
20. $e^z - e^{2z}$
21. $z^{-2} \cos^3 \pi z$
22. $\cosh^2 z$
23. $(3z^2 - 1)/(z^2 - 2iz + 3)^2$
24. $(z^2 - 1)^2(e^{z^2} - 1)$
25. $(1 - \cos z)^2$

26. If $f(z)$ has a zero of order n at $z = z_0$, show that $f^2(z)$ has a zero of order $2n$, and the derivative $f'(z)$ has a zero of order $n - 1$ at $z = z_0$ (provided $n > 1$).

27. Prove Theorem 4.

28. If $f_1(z)$ and $f_2(z)$ are analytic in a domain D and equal at a sequence of points z_n in D that converges in D, show that $f_1(z) \equiv f_2(z)$ in D.

29. Show that the points at which a nonconstant analytic function $f(z)$ assumes a given value k are isolated.

Riemann number sphere. Assuming that we let the image of the x-axis be the meridians 0° and 180°, describe and sketch the images of the following regions on the Riemann number sphere.

30. $|z| \leqq 1$
31. First quadrant
32. Second quadrant
33. $|z| > 100$
34. Lower half-plane
35. $\frac{1}{2} \leqq |z| \leqq 2$

Review Questions and Problems for Chapter 14

1. What is a power series? Why are these series very important in complex analysis?
2. State from memory the ratio test, the root test, and the Cauchy–Hadamard formula for the radius of convergence.
3. What is absolute convergence? Conditional convergence? Uniform convergence?
4. What do you know about the convergence of a power series?
5. Is termwise differentiation of a power series permissible? Termwise integration?
6. What is a Taylor series? What was the idea of getting it from Cauchy's integral formula?
7. Write down the Maclaurin series of e^z, $\cos z$, $\sin z$, $\cosh z$, $\sinh z$.
8. Does $\operatorname{Ln} z$ have a Maclaurin series? Explain.
9. Do we obtain an analytic function if we replace x by z in the Maclaurin series of a real function $f(x)$?
10. Give examples illustrating practical methods for getting Taylor series.
11. What is a Laurent series? What do you know about its convergence?
12. Is the Taylor series of a given function with a given center unique? Answer the same question for a Laurent series.
13. What is a singularity? What role did it play in connection with the convergence of Taylor and Laurent series?
14. How did we use Laurent series in classifying singularities?
15. What is a pole? An isolated essential singularity?
16. Does an entire function have poles? Essential singularities? (Think of typical examples.)
17. What is a zero of an analytic function? What do you know about zeros of analytic functions?
18. Do you know a nonconstant analytic function without zeros?
19. What is a meromorphic function? Give examples.
20. What is the Riemann number sphere? When would you use it?

Find the radius of convergence of the following power series. Can you identify the sum as a familiar function in some of the problems?

21. $\displaystyle\sum_{n=0}^{\infty} \frac{(-1)^n}{(2n+1)!}(z-2)^{2n+1}$
22. $\displaystyle\sum_{n=0}^{\infty} \frac{z^{2n+1}}{2n+1}$
23. $\displaystyle\sum_{n=1}^{\infty} \frac{(-2)^{n+1}}{2n} z^n$
24. $\displaystyle\sum_{n=0}^{\infty} \frac{n^{10}}{n!}(z-1+2i)^n$
25. $\displaystyle\sum_{n=1}^{\infty} \frac{3^n}{n^{100}} z^n$
26. $\displaystyle\sum_{n=0}^{\infty} \frac{(z-i)^n}{(3+4i)^n}$
27. $\displaystyle\sum_{n=0}^{\infty} \frac{(-1)^n}{n!} z^{2n}$
28. $\displaystyle\sum_{n=0}^{\infty} \pi^n (z-2i)^{2n}$
29. $\displaystyle\sum_{n=0}^{\infty} \frac{z^n}{(2n)!}$

Find the Taylor series of the following functions with the given point as center and determine the radius of convergence.

30. e^z, πi
31. e^{-2z}, 0
32. $\operatorname{Ln} z$, 2

33. $\frac{1}{1-z}$, -1 34. $\frac{1}{2z-i}$, -1 35. $\frac{1}{z}$, $2+3i$

36. $\sin \pi z^2$, 0 37. $\frac{1}{(1-z)^3}$, 0 38. $\frac{1}{4-3z}$, $1+i$

Expand the given functions in a Laurent series that converges for $0 < |z - z_0| < R$ and determine the precise region of convergence as well as the type of singularity of the function at z_0.

39. $\frac{\cos z}{(z - \frac{1}{2}\pi)^3}$, $z_0 = \frac{\pi}{2}$

40. $\frac{1 - z^2}{z^4}$, $z_0 = 0$

41. $\frac{1}{z^4 - z^5}$, $z_0 = 0$

42. $\frac{e^z}{(z-1)^2}$, $z_0 = 1$

43. $z^3 \cosh \frac{1}{z}$, $z_0 = 0$

44. $\frac{z + 1 + i}{(z+i)^2}$, $z_0 = -i$

45. $(z-1)^{-3} \operatorname{Ln} z$, $z_0 = 1$

46. $\frac{1}{z^4} \sin 2z^2$, $z_0 = 0$

47. $\frac{1}{z^2} \int_0^z \frac{e^t - 1}{t}\, dt$, $z_0 = 0$

48. $\frac{\sinh z + \sin z}{z^3}$, $z_0 = 0$

49. $\frac{e^z}{(z-i)^5}$, $z_0 = i$

50. $(z+i)^3 \sin \frac{1}{2z + 2i}$, $z_0 = -i$

Summary of Chapter 14 Power Series, Taylor Series, Laurent Series

Sequences, series, and convergence tests are discussed in Sec. 14.1. A **power series** is of the form (Sec. 14.2)

$$(1) \quad \sum_{n=0}^{\infty} a_n(z - z_0)^n = a_0 + a_1(z - z_0) + a_2(z - z_0)^2 + \cdots ;$$

z_0 is its *center*. The series (1) converges for $|z - z_0| < R$ and diverges for $|z - z_0| > R$. Some power series converge for all z (then we write $R = \infty$). R is the *radius of convergence*. Also, $R = \lim |a_n/a_{n+1}|$ if this limit exists. The series (1) converges absolutely (Sec. 14.2) and **uniformly** (Sec. 14.6) in every closed disk $|z - z_0| \leqq r < R$. If $R > 0$, it represents an analytic function $f(z)$ for $|z - z_0| < R$. The derivatives $f'(z)$, $f''(z)$, $\cdots$ are obtained by termwise differentiation of (1), and these series have the same radius of convergence R as (1). See Sec. 14.3.

Conversely, *every* analytic function $f(z)$ can be represented by power series. These **Taylor series** of $f(z)$ are of the form (Sec. 14.4)

$$(2) \qquad f(z) = \sum_{n=0}^{\infty} \frac{1}{n!} f^{(n)}(z_0)(z - z_0)^n \qquad (|z - z_0| < R),$$

as in calculus. They converge for all z if $f(z)$ is **entire** (= analytic for all z) or in the open disk with center z_0 and radius equal to the distance from z_0 to the nearest **singularity** of $f(z)$ (= point at which $f(z)$ ceases to be analytic, see Sec. 14.8). The functions e^z, $\cos z$, $\sin z$, etc. have Maclaurin and Taylor series similar to those in calculus (Sec. 14.4).

A **Laurent series** is of the form (Sec. 14.7)

$$(3) \qquad f(z) = \sum_{n=0}^{\infty} a_n(z - z_0)^n + \sum_{n=1}^{\infty} \frac{b_n}{(z - z_0)^n}$$

or, more briefly written [but this means the same as (3)!]

$$(3^*) \quad f(z) = \sum_{n=-\infty}^{\infty} a_n(z - z_0)^n, \quad a_n = \frac{1}{2\pi i} \oint_C \frac{f(z^*)}{(z^* - z_0)^{n+1}}\, dz^*.$$

This series converges in an open annulus (ring) A with center z_0. In A the function $f(z)$ is analytic. At points not in A it may have singularities. The first series in (3) is a power series. The second series is called the **principal part** of the Laurent series. In a given annulus, a Laurent series of $f(z)$ is unique, but $f(z)$ may have different Laurent series in different annuli with the same center.

If $f(z)$ has an isolated singularity (Sec. 14.8) at $z = z_0$, the Laurent series of $f(z)$ that converges for $0 < |z - z_0| < R$ (R suitable) can be used for classifying this singularity, which is called a **pole** if the principal part of this Laurent series is a finite sum, otherwise an (isolated) **essential singularity.**

A pole is said to be of **order** n if $1/(z - z_0)^n$ is the highest negative power of the principal part in (3). A first-order pole is also called a *simple pole*.

Similarly, the Taylor series (2) can be used for classifying **zeros** of $f(z)$. The function $f(z)$ given by (2) has a zero of ***n*th order** at z_0 if f and its derivatives f', f'', $\cdots$, $f^{(n-1)}$ are all zero at z_0, whereas $f^{(n)}(z_0) \neq 0$. A first-order zero is also called a *simple zero*. See Sec. 14.8.

Section 14.8 also includes a discussion of the **extended complex plane,** obtained from the complex plane by attaching an improper point ∞ ("*infinity*").

Chapter 15

Residue Integration Method

Since there are various methods for determining the coefficients of a Laurent series (1), Sec. 14.7, without using the integral formulas (2), Sec. 14.7, we may use the formula for b_1 for evaluating complex integrals in a very elegant and simple fashion. b_1 will be called the **residue** of $f(z)$ at $z = z_0$. This powerful method may also be applied for evaluating certain real integrals, as we shall see in Secs. 15.3 and 15.4.

Prerequisites for this chapter: Chaps. 12–14.
References: Appendix 1, Part D.
Answers to problems: Appendix 2.

15.1 Residues

Let us first explain what a residue is and how it can be used for evaluating integrals

$$\oint_C f(z)\, dz.$$

These will be contour integrals taken around a simple closed path C.

If $f(z)$ is analytic everywhere on C and inside C, such an integral is zero by Cauchy's integral theorem (Sec. 13.3), and we are done.

If $f(z)$ has a singularity at a point $z = z_0$ inside C, but is otherwise analytic on C and inside C, then $f(z)$ has a Laurent series

$$f(z) = \sum_{n=0}^{\infty} a_n(z - z_0)^n + \frac{b_1}{z - z_0} + \frac{b_2}{(z - z_0)^2} + \cdots$$

that converges for all points near $z = z_0$ (except at $z = z_0$ itself), in some domain of the form $0 < |z - z_0| < R$. Now comes the key idea. The coefficient b_1 of the first negative power $1/(z - z_0)$ of this Laurent series is given by the integral formula (2), Sec. 14.7, with $n = 1$, that is,

$$b_1 = \frac{1}{2\pi i} \oint_C f(z)\, dz,$$

but since we can obtain Laurent series by various methods, without using the integral formulas for the coefficients (see the examples in Sec. 14.7), we can find b_1 by one of those methods and then use the formula for b_1 for evaluating the integral:

$$\oint_C f(z)\, dz = 2\pi i b_1. \tag{1}$$

Here we integrate counterclockwise around the simple closed path C that contains $z = z_0$ in its interior.

The coefficient b_1 is called the **residue** of $f(z)$ at $z = z_0$ and we denote it by

$$b_1 = \operatorname*{Res}_{z=z_0} f(z). \tag{2}$$

EXAMPLE 1 **Evaluation of an integral by means of a residue**

Integrate the function $f(z) = z^{-4} \sin z$ counterclockwise around the unit circle C.

Solution. From (14) in Sec. 14.4 we obtain the Laurent series

$$f(z) = \frac{\sin z}{z^4} = \frac{1}{z^3} - \frac{1}{3!z} + \frac{z}{5!} - \frac{z^3}{7!} + - \cdots$$

which converges for $|z| > 0$ (that is, for all $z \neq 0$). This series shows that $f(z)$ has a pole of third order at $z = 0$ and the residue

$$b_1 = -1/3!.$$

From (1) we thus obtain the answer

$$\oint_C \frac{\sin z}{z^4}\, dz = 2\pi i b_1 = -\frac{\pi i}{3}.$$

■

EXAMPLE 2 **Be careful to use the *right* Laurent series!**

Integrate $f(z) = 1/(z^3 - z^4)$ clockwise around the circle C: $|z| = 1/2$.

Solution. $z^3 - z^4 = z^3(1 - z)$ shows that $f(z)$ is singular at $z = 0$ and $z = 1$. Now $z = 1$ lies outside C. Hence it is of no interest here. So we need the residue of $f(z)$ at 0. We find it from the Laurent series that converges for $0 < |z| < 1$. This is series (I) in Example 4, Sec. 14.7:

$$\frac{1}{z^3 - z^4} = \frac{1}{z^3} + \frac{1}{z^2} + \frac{1}{z} + 1 + z + \cdots \qquad (0 < |z| < 1).$$

We see from it that this residue is 1. Clockwise integration thus yields

$$\oint_C \frac{dz}{z^3 - z^4} = -2\pi i \operatorname*{Res}_{z=0} f(z) = -2\pi i.$$

Caution! Had we used the wrong series (II) in Example 4, Sec. 14.7,

$$\frac{1}{z^3 - z^4} = -\frac{1}{z^4} - \frac{1}{z^5} - \frac{1}{z^6} - \cdots \qquad (|z| > 1),$$

we would have obtained the wrong answer, 0, because this series has no power $1/z$. ■

Two Formulas for Residues at Simple Poles

Before we continue integration, we ask the following. To get a residue, a single coefficient of a Laurent series, must we derive the whole series or is there a more economical way? For poles, there is. We shall derive, once and for all, some formulas for residues at poles, so that in this case we no longer need the whole series.

Let $f(z)$ have a simple pole at $z = z_0$. Then the corresponding Laurent series is [see (2) with $m = 1$ in Sec. 14.8]

$$f(z) = \frac{b_1}{z - z_0} + a_0 + a_1(z - z_0) + a_2(z - z_0)^2 + \cdots \qquad (0 < |z - z_0| < R).$$

Here $b_1 \neq 0$. (Why?) Multiplying both sides by $z - z_0$ we have

$$(z - z_0)f(z) = b_1 + (z - z_0)[a_0 + a_1(z - z_0) + \cdots].$$

We now let $z \to z_0$. Then the right side approaches b_1. This gives

$$\text{(3)} \qquad \operatorname*{Res}_{z=z_0} f(z) = b_1 = \lim_{z \to z_0} (z - z_0)f(z).$$

EXAMPLE 3 **Residue at a simple pole**

$$\operatorname*{Res}_{z=i} \frac{9z + i}{z(z^2 + 1)} = \lim_{z \to i} (z - i) \frac{9z + i}{z(z + i)(z - i)} = \left[\frac{9z + i}{z(z + i)}\right]_{z=i} = \frac{10i}{-2} = -5i. \qquad \blacksquare$$

Another, sometimes simpler formula for the residue at a simple pole is obtained by starting from

$$f(z) = \frac{p(z)}{q(z)}$$

with analytic $p(z)$ and $q(z)$, where we assume that $p(z_0) \neq 0$ and $q(z)$ has a simple zero at $z = z_0$ (so that $f(z)$ has a simple pole at $z = z_0$ as wanted, by Theorem 4 in Sec. 14.8). By the definition of a simple zero, $q(z)$ has a Taylor series of the form

$$q(z) = (z - z_0)q'(z_0) + \frac{(z - z_0)^2}{2!} q''(z_0) + \cdots.$$

This we substitute into $f = p/q$ and then f into (3), finding

$$\begin{aligned} \operatorname*{Res}_{z=z_0} f(z) &= \lim_{z \to z_0} (z - z_0) \frac{p(z)}{q(z)} \\ &= \lim_{z \to z_0} \frac{(z - z_0)p(z)}{(z - z_0)[q'(z_0) + (z - z_0)q''(z_0)/2 + \cdots]}. \end{aligned}$$

We now see that on the right, a factor $z - z_0$ is canceled and the resulting denominator has the limit $q'(z_0)$. Hence our second formula for the residue at a simple pole is

$$\text{(4)} \qquad \operatorname*{Res}_{z=z_0} f(z) = \operatorname*{Res}_{z=z_0} \frac{p(z)}{q(z)} = \frac{p(z_0)}{q'(z_0)} .$$

EXAMPLE 4 Residue at a simple pole calculated by formula (4)

$$\operatorname*{Res}_{z=i} \frac{9z + i}{z(z^2 + 1)} = \left[\frac{9z + i}{3z^2 + 1}\right]_{z=i} = \frac{10i}{-2} = -5i \qquad \text{(see Example 3).} \quad ▮$$

EXAMPLE 5 Another application of formula (4)

Find all poles and the corresponding residues of the function

$$f(z) = \frac{\cosh \pi z}{z^4 - 1} .$$

Solution. $p(z) = \cosh \pi z$ is entire, and $q(z) = z^4 - 1$ has simple zeros at $1, i, -1, -i$. Hence $f(z)$ has simple poles at these points (and no further poles). Since $q'(z) = 4z^3$, we see from (4) that the residues equal the values of $(\cosh \pi z)/4z^3$ at those points, that is,

$$\frac{\cosh \pi}{4} \approx 2.8980, \qquad \frac{\cosh \pi i}{4i^3} = \frac{\cos \pi}{-4i} = -\frac{i}{4}, \qquad -\frac{\cosh \pi}{4}, \qquad \frac{\cosh(-\pi i)}{4(-i)^3} = \frac{i}{4} . \quad ▮$$

Formula for the Residue at a Pole of Any Order

Let $f(z)$ be an analytic function that has a pole of any order $m > 1$ at a point $z = z_0$. Then, by the definition of such a pole (Sec. 14.8), the Laurent series of $f(z)$ converging near $z = z_0$ (except at $z = z_0$ itself) is

$$f(z) = \frac{b_m}{(z - z_0)^m} + \frac{b_{m-1}}{(z - z_0)^{m-1}} + \cdots + \frac{b_2}{(z - z_0)^2} + \frac{b_1}{z - z_0}$$
$$+ a_0 + a_1(z - z_0) + \cdots ,$$

where $b_m \neq 0$. Multiplying both sides by $(z - z_0)^m$, we have

$$(z - z_0)^m f(z) = b_m + b_{m-1}(z - z_0) + \cdots + b_2(z - z_0)^{m-2} + b_1(z - z_0)^{m-1}$$
$$+ a_0(z - z_0)^m + a_1(z - z_0)^{m+1} + \cdots .$$

We see that the residue b_1 of $f(z)$ at $z = z_0$ is now the coefficient of the power $(z - z_0)^{m-1}$ in the Taylor series of the function

$$g(z) = (z - z_0)^m f(z)$$

on the left, with center $z = z_0$. Thus, by Taylor's theorem (Sec. 14.4),

$$b_1 = \frac{1}{(m-1)!} g^{(m-1)}(z_0).$$

Hence if $f(z)$ has a pole of mth order at $z = z_0$, the residue is given by

(5)
$$\operatorname*{Res}_{z=z_0} f(z) = \frac{1}{(m-1)!} \lim_{z\to z_0} \left\{ \frac{d^{m-1}}{dz^{m-1}} [(z-z_0)^m f(z)] \right\}.$$

In particular, for a second-order pole ($m = 2$),

(5*)
$$\operatorname*{Res}_{z=z_0} f(z) = \lim_{z\to z_0} \left\{ [(z-z_0)^2 f(z)]' \right\}.$$

EXAMPLE 6 **Residue at a pole of higher order**

The function

$$f(z) = \frac{50z}{(z+4)(z-1)^2}$$

has a pole of second order at $z = 1$, and from (5*) we obtain the corresponding residue

$$\operatorname*{Res}_{z=1} f(z) = \lim_{z\to 1} \frac{d}{dz} [(z-1)^2 f(z)] = \lim_{z\to 1} \frac{d}{dz} \left(\frac{50z}{z+4} \right) = 8.$$ ∎

EXAMPLE 7 **Residues from partial fractions**

If $f(z)$ is rational, we can also determine its residues from partial fractions. In Example 6,

$$f(z) = \frac{50z}{(z+4)(z-1)^2} = \frac{-8}{z+4} + \frac{8}{z-1} + \frac{10}{(z-1)^2}.$$

This shows that the residue at $z = 1$ is 8 (as before), and at $z = -4$ (simple pole) it is -8.

Why is this so? Consider $z = 1$. There the Laurent series has the last two fractions as its principal part and the first fraction as the sum of its other part. This first fraction is analytic at $z = 1$, so that it has a Taylor series with center $z = 1$, as it should be. Similarly, at $z = -4$ the first fraction is the principal part of the Laurent series. ∎

EXAMPLE 8 **Integration around a second-order pole**

Counterclockwise integration of $f(z)$ in Examples 6 and 7 around any simple closed path C such that $z = 1$ is inside C and $z = -4$ is outside C gives (see Example 6 or 7)

$$\oint_C \frac{50z}{(z+4)(z-1)^2}\, dz = 2\pi i \operatorname*{Res}_{z=1} \frac{50z}{(z+4)(z-1)^2} = 2\pi i \cdot 8 = 16\pi i \approx 50.27i.$$ ∎

Problem Set 15.1

Find the residues at the singular points of the following functions.

1. $\dfrac{3}{1-z}$ **2.** $\dfrac{4}{z^3} - \dfrac{1}{z^2}$ **3.** $\dfrac{\sin z}{z^4}$

4. $\dfrac{z^2+1}{z^2-z}$ **5.** $\dfrac{z^4}{z^2 - iz + 2}$ **6.** $\dfrac{e^z}{(z-\pi i)^5}$

7. $\cot z$ **8.** $\sec z$ **9.** $2/(z^2-1)^2$

Find the residues at those singular points which lie inside the circle $|z| = 2$.

10. $\dfrac{z^2}{z^4 - 1}$ **11.** $\dfrac{2z - 3}{z^3 + 3z^2}$ **12.** $\dfrac{z - 23}{z^2 - 4z - 5}$

13. $\dfrac{3}{(z^4 - 1)^2}$ **14.** $\dfrac{-z^2 - 22z + 8}{z^3 - 5z^2 + 4z}$ **15.** $\dfrac{3z + 6}{(z + 1)(z^2 + 16)}$

Evaluate the following integrals where C is the unit circle (counterclockwise).

16. $\oint_C e^{1/z}\, dz$ **17.** $\oint_C \tan z\, dz$ **18.** $\oint_C \csc 2z\, dz$

19. $\oint_C \cot z\, dz$ **20.** $\oint_C \dfrac{dz}{\sinh \frac{1}{2}\pi z}$ **21.** $\oint_C \dfrac{z^4 + 6}{z^2 - 2z}\, dz$

22. $\oint_C \dfrac{\sin \pi z}{z^6}\, dz$ **23.** $\oint_C \dfrac{z^3 + 2}{4z + \pi}\, dz$ **24.** $\oint_C \dfrac{\tanh (z + 1)}{e^z \sin z}\, dz$

25. Another derivation of (5). Obtain (5) without using Taylor's theorem, by $m - 1$ differentiations of the formula for $(z - z_0)^m f(z)$.

15.2 Residue Theorem

So far we can evaluate integrals of analytic functions $f(z)$ over closed curves C when $f(z)$ has only *one* singular point inside C. We shall now see that the residue integration method can be extended to the case of several singular points of $f(z)$ inside C. This extension is surprisingly simple, as follows.

Theorem 1 **Residue theorem**

Let $f(z)$ be a function that is analytic inside a simple closed path C and on C, except for finitely many singular points $z_1, z_2, \cdots, z_k$ inside C. Then

$$\oint_C f(z)\, dz = 2\pi i \sum_{j=1}^{k} \operatorname*{Res}_{z=z_j} f(z), \tag{1}$$

the integral being taken counterclockwise around the path C.

Proof. We enclose each of the singular points z_j in a circle C_j with radius small enough that those k circles and C are all separated (Fig. 346). Then $f(z)$ is analytic in the multiply connected domain D bounded by C and $C_1, \cdots, C_k$ and on the entire boundary of D. From Cauchy's integral theorem we thus have

$$\oint_C f(z)\, dz + \oint_{C_1} f(z)\, dz + \oint_{C_2} f(z)\, dz + \cdots + \oint_{C_k} f(z)\, dz = 0, \tag{2}$$

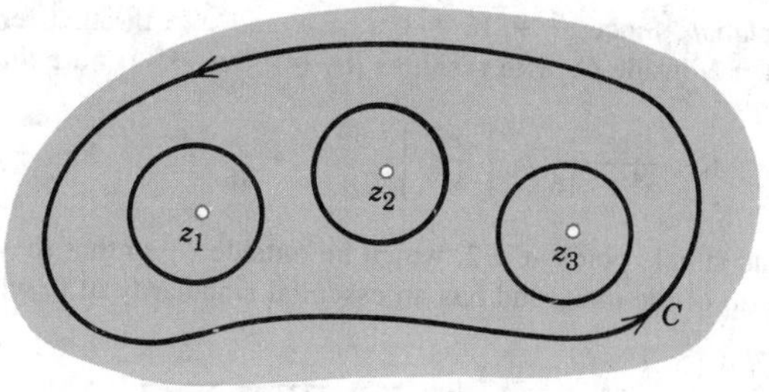

Fig. 346. Residue theorem

the integral along C being taken counterclockwise and the other integrals clockwise (see Sec. 13.3). We now reverse the sense of integration along $C_1, \cdots, C_k$. Then the signs of the values of these integrals change, and we obtain from (2)

$$\text{(3)} \qquad \oint_C f(z)\,dz = \oint_{C_1} f(z)\,dz + \oint_{C_2} f(z)\,dz + \cdots + \oint_{C_k} f(z)\,dz.$$

All these integrals are now taken counterclockwise. By (1) in the last section,

$$\oint_{C_j} f(z)\,dz = 2\pi i \operatorname*{Res}_{z=z_j} f(z),$$

so that (3) yields (1), and the theorem is proved. ∎

This important theorem has various applications in connection with complex and real integrals. We first consider some complex integrals.

EXAMPLE 1 Integration by the residue theorem

Evaluate the following integral counterclockwise around any simple closed path such that (a) 0 and 1 are inside C, (b) 0 is inside, 1 outside, (c) 1 is inside, 0 outside, (d) 0 and 1 are outside.

$$\oint_C \frac{4 - 3z}{z^2 - z}\,dz$$

Solution. The integrand has simple poles at 0 and 1, with residues [by (3), Sec. 15.1]

$$\operatorname*{Res}_{z=0} \frac{4 - 3z}{z(z-1)} = \left[\frac{4 - 3z}{z - 1}\right]_{z=0} = -4, \qquad \operatorname*{Res}_{z=1} \frac{4 - 3z}{z(z-1)} = \left[\frac{4 - 3z}{z}\right]_{z=1} = 1.$$

[Confirm this by (4), Sec. 15.1.] *Ans.* (a) $2\pi i(-4 + 1) = -6\pi i$, (b) $-8\pi i$, (c) $2\pi i$, (d) 0. ∎

EXAMPLE 2 Poles and essential singularities

Evaluate the following integral, where C is the ellipse $9x^2 + y^2 = 9$ (counterclockwise).

$$\oint_C \left(\frac{ze^{\pi z}}{z^4 - 16} + ze^{\pi/z}\right) dz$$

Solution. Since $z^4 - 16 = 0$ at $\pm 2i$ and ± 2, the first term of the integrand has simple poles at $\pm 2i$ inside C, with residues [by (4), Sec. 15.1; note that $e^{2\pi i} = 1$]

$$\operatorname*{Res}_{z=2i} \frac{ze^{\pi z}}{z^4 - 16} = \left[\frac{ze^{\pi z}}{4z^3}\right]_{z=2i} = -\frac{1}{16}, \qquad \operatorname*{Res}_{z=-2i} \frac{ze^{\pi z}}{z^4 - 16} = \left[\frac{ze^{\pi z}}{4z^3}\right]_{z=-2i} = -\frac{1}{16}$$

and simple poles at ± 2, which lie outside C, so that they are of no interest here. The second term of the integrand has an essential singularity at 0, with residue $\pi^2/2$ as obtained from

$$ze^{\pi/z} = z\left(1 + \frac{\pi}{z} + \frac{\pi^2}{2!z^2} + \frac{\pi^3}{3!z^3} + \cdots\right) = z + \pi + \frac{\pi^2}{2} \cdot \frac{1}{z} + \cdots.$$

Ans. $2\pi i(-1/16 - 1/16 + \pi^2/2) = \pi(\pi^2 - 1/4)i = 30.221i$ by the residue theorem. ∎

EXAMPLE 3 Confirmation of an earlier basic result

Integrate $1/(z - z_0)^m$ (m a positive integer) counterclockwise around any simple closed path C enclosing the point $z = z_0$.

Solution. $1/(z - z_0)^m$ is its own Laurent series with center $z = z_0$ consisting of this one-term principal part, and

$$\operatorname*{Res}_{z=z_0} \frac{1}{z - z_0} = 1,$$

$$\operatorname*{Res}_{z=z_0} \frac{1}{(z - z_0)^m} = 0 \qquad (m = 2, 3, \cdots).$$

In agreement with Example 2, Sec. 13.2, we thus obtain

$$\oint_C \frac{dz}{(z - z_0)^m} = \begin{cases} 2\pi i & \text{if } m = 1 \\ 0 & \text{if } m = 2, 3, \cdots. \end{cases}$$ ∎

Problem Set 15.2

Integrate $\dfrac{15z + 9}{z^3 - 9z}$ counterclockwise around the following paths C.

1. $|z| = 1$ **2.** $|z| = 4$ **3.** $|z + 2 + i| = 3$

4. $|z - 3| = 2$ **5.** $|z - \frac{3}{2} + 2i| = 2.4$ **6.** $|z - 1| = 3$

Evaluate the following integrals, where C is any simple closed path such that all the singularities lie inside C (counterclockwise).

7. $\oint_C \dfrac{5z}{z^2 + 4}\, dz$ **8.** $\oint_C \dfrac{z}{1 + 9z^2}\, dz$ **9.** $\oint_C \dfrac{z \cosh \pi z}{z^4 + 13z^2 + 36}\, dz$

10. $\oint_C \dfrac{\sinh z}{2z - i}\, dz$ **11.** $\oint_C \dfrac{z + e^z}{z^3 - z}\, dz$ **12.** $\oint_C \dfrac{z^2 \sin z}{4z^2 - 1}\, dz$

Evaluate the following integrals where C is the unit circle (counterclockwise).

13. $\oint_C \dfrac{z}{z^2 - \frac{1}{4}}\, dz$ **14.** $\oint_C \dfrac{7z}{z^2 + \frac{1}{9}}\, dz$ **15.** $\oint_C \dfrac{dz}{z^2 + 6iz}$

16. $\oint_C \frac{e^{-z^2}}{\sin 4z}\,dz$

17. $\oint_C \frac{e^{-z^2}}{\sin 2z}\,dz$

18. $\oint_C \frac{30z^2 - 23z + 5}{(2z - 1)^2(3z - 1)}\,dz$

19. $\oint_C \cot \frac{z}{4}\,dz$

20. $\oint_C e^z \cot 4z\,dz$

21. $\oint_C \frac{\sinh z}{4z^2 + 1}\,dz$

22. $\oint_C \frac{e^z}{z(z - \pi i/4)^2}\,dz$

23. $\oint_C \tan 2\pi z\,dz$

24. $\oint_C \frac{1 - 4z + 6z^2}{(z^2 + \frac{1}{4})(2 - z)}\,dz$

25. $\oint_C \tan \pi z\,dz$

26. $\oint_C \coth z\,dz$

27. $\oint_C \frac{\tan \pi z}{z^3}\,dz$

28. $\oint_C \frac{\cosh z}{z^2 - 3iz}\,dz$

29. $\oint_C \frac{e^z}{\cos \pi z}\,dz$

30. $\oint_C \frac{(z + 4)^3}{z^4 + 5z^3 + 6z^2}\,dz$

15.3 Evaluation of Real Integrals

We now show the very surprising fact that the residue theorem also yields a very elegant and simple method for evaluating certain classes of complicated *real* integrals.

Integrals of Rational Functions of $\cos\theta$ and $\sin\theta$

We first consider integrals of the type

$$I = \int_0^{2\pi} F(\cos\theta, \sin\theta)\,d\theta \tag{1}$$

where $F(\cos\theta, \sin\theta)$ is a real rational function of $\cos\theta$ and $\sin\theta$ [for example, $(\sin^2\theta)/(5 - 4\cos\theta)$] and is finite on the interval of integration. Setting $e^{i\theta} = z$, we obtain

$$\begin{aligned} \cos\theta &= \frac{1}{2}(e^{i\theta} + e^{-i\theta}) = \frac{1}{2}\left(z + \frac{1}{z}\right) \\ \sin\theta &= \frac{1}{2i}(e^{i\theta} - e^{-i\theta}) = \frac{1}{2i}\left(z - \frac{1}{z}\right) \end{aligned} \tag{2}$$

and we see that the integrand becomes a rational function of z, say, $f(z)$. As θ ranges from 0 to 2π, the variable z ranges once around the unit circle $|z| = 1$ in the counterclockwise sense. Since $dz/d\theta = ie^{i\theta}$, we have $d\theta = dz/iz$, and the given integral takes the form

$$I = \oint_C f(z)\,\frac{dz}{iz}, \tag{3}$$

the integration being taken counterclockwise around the unit circle.

EXAMPLE 1 **An integral of the type (1)**

Show by the present method that

$$\int_0^{2\pi} \frac{d\theta}{\sqrt{2} - \cos\theta} = 2\pi.$$

Solution. We use $\cos\theta = \frac{1}{2}(z + 1/z)$ and $d\theta = dz/iz$. Then the integral becomes

$$\oint_C \frac{dz/iz}{\sqrt{2} - \frac{1}{2}\left(z + \frac{1}{z}\right)} = \oint_C \frac{dz}{-\frac{i}{2}(z^2 - 2\sqrt{2}z + 1)} = -\frac{2}{i}\oint_C \frac{dz}{(z - \sqrt{2} - 1)(z - \sqrt{2} + 1)}.$$

We see that the integrand has two simple poles, one at $z_1 = \sqrt{2} + 1$, which lies outside the unit circle C: $|z| = 1$ and is thus of no interest, and the other at $z_2 = \sqrt{2} - 1$ inside C, where the residue is [by (3) in Sec. 15.1]

$$\operatorname*{Res}_{z=z_2} \frac{1}{(z - \sqrt{2} - 1)(z - \sqrt{2} + 1)} = \left[\frac{1}{z - \sqrt{2} - 1}\right]_{z=\sqrt{2}-1} = -\frac{1}{2}.$$

Together with the factor $-2/i$ in front of the integral this yields the desired result $2\pi i(-2/i)(-1/2) = 2\pi$. ∎

Improper Integrals of Rational Functions

We now consider real integrals of the type

$$\int_{-\infty}^{\infty} f(x)\,dx. \tag{4}$$

Such an integral, for which the interval of integration is not finite, is called an **improper integral,** and it has the meaning

$$\int_{-\infty}^{\infty} f(x)\,dx = \lim_{a\to-\infty}\int_a^0 f(x)\,dx + \lim_{b\to\infty}\int_0^b f(x)\,dx. \tag{5'}$$

If both limits exist, we may couple the two independent passages to $-\infty$ and ∞, and write[1]

$$\int_{-\infty}^{\infty} f(x)\,dx = \lim_{R\to\infty}\int_{-R}^{R} f(x)\,dx. \tag{5}$$

---

[1]The expression on the right side of (5) is called the **Cauchy principal value** of the integral; it may exist even if the limits in (5′) do not exist. For instance,

$$\lim_{R\to\infty}\int_{-R}^{R} x\,dx = \lim_{R\to\infty}\left(\frac{R^2}{2} - \frac{R^2}{2}\right) = 0, \quad \text{but} \quad \lim_{b\to\infty}\int_0^b x\,dx = \infty.$$

We assume that the function $f(x)$ in (4) is a real rational function whose denominator is different from zero for all real x and is of degree at least two units higher than the degree of the numerator. Then the limits in (5′) exist, and we may start from (5). We consider the corresponding contour integral

$$\oint_C f(z)\,dz \tag{5*}$$

around a path C in Fig. 347. Since $f(x)$ is rational, $f(z)$ has finitely many poles in the upper half-plane, and if we choose R large enough, then C encloses all these poles. By the residue theorem we then obtain

$$\oint_C f(z)\,dz = \int_S f(z)\,dz + \int_{-R}^{R} f(x)\,dx = 2\pi i \sum \operatorname{Res} f(z)$$

where the sum consists of all the residues of $f(z)$ at the points in the upper half-plane at which $f(z)$ has a pole. From this we have

$$\int_{-R}^{R} f(x)\,dx = 2\pi i \sum \operatorname{Res} f(z) - \int_S f(z)\,dz. \tag{6}$$

We prove that, if $R \to \infty$, the value of the integral over the semicircle S approaches zero. If we set $z = Re^{i\theta}$, then S is represented by $R = const$, and as z ranges along S, the variable θ ranges from 0 to π. Since, by assumption, the degree of the denominator of $f(z)$ is at least two units higher than the degree of the numerator, we have

$$|f(z)| < \frac{k}{|z|^2} \qquad (|z| = R > R_0)$$

for sufficiently large constants k and R_0. By the ML-inequality in Sec. 13.2,

$$\left|\int_S f(z)\,dz\right| < \frac{k}{R^2}\,\pi R = \frac{k\pi}{R} \qquad (R > R_0).$$

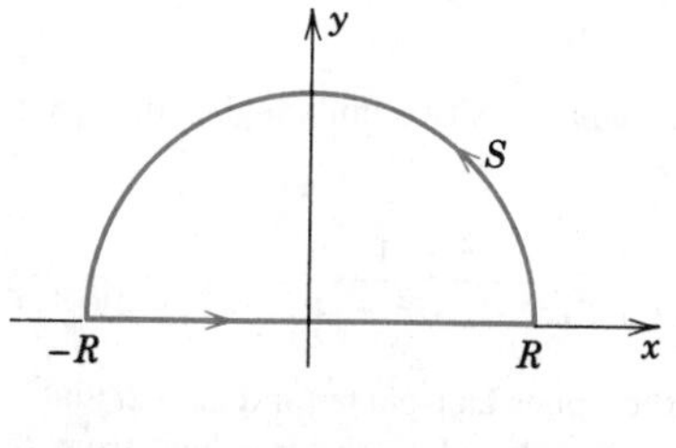

Fig. 347. Path C of the contour integral in (5*)

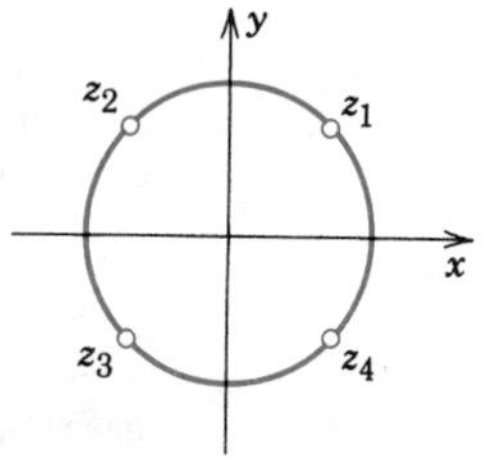

Fig. 348. Example 2, p. 848

Hence, as R approaches infinity, the value of the integral over S approaches zero, and (5) and (6) yield the result

$$\int_{-\infty}^{\infty} f(x)\,dx = 2\pi i \sum \operatorname{Res} f(z), \tag{7}$$

where we sum over all the residues of $f(z)$ corresponding to the poles of $f(z)$ in the upper half-plane.

EXAMPLE 2 An improper integral from 0 to ∞

Using (7), show that

$$\int_0^{\infty} \frac{dx}{1+x^4} = \frac{\pi}{2\sqrt{2}}.$$

Solution. Indeed, $f(z) = 1/(1+z^4)$ has four simple poles at the points

$$z_1 = e^{\pi i/4}, \qquad z_2 = e^{3\pi i/4}, \qquad z_3 = e^{-3\pi i/4}, \qquad z_4 = e^{-\pi i/4}.$$

The first two of the these poles lie in the upper half-plane (Fig. 348 on p. 847). From (4) in Sec. 15.1 we find

$$\operatorname*{Res}_{z=z_1} f(z) = \left[\frac{1}{(1+z^4)'}\right]_{z=z_1} = \left[\frac{1}{4z^3}\right]_{z=z_1} = \frac{1}{4}e^{-3\pi i/4} = -\frac{1}{4}e^{\pi i/4},$$

$$\operatorname*{Res}_{z=z_2} f(z) = \left[\frac{1}{(1+z^4)'}\right]_{z=z_2} = \left[\frac{1}{4z^3}\right]_{z=z_2} = \frac{1}{4}e^{-9\pi i/4} = \frac{1}{4}e^{-\pi i/4}.$$

By (1) in Sec. 12.7 and (7) in the current section,

$$\int_{-\infty}^{\infty} \frac{dx}{1+x^4} = \frac{2\pi i}{4}(-e^{\pi i/4} + e^{-\pi i/4}) = \pi \sin\frac{\pi}{4} = \frac{\pi}{\sqrt{2}}.$$

Since $1/(1+x^4)$ is an even function, we thus obtain, as asserted,

$$\int_0^{\infty} \frac{dx}{1+x^4} = \frac{1}{2}\int_{-\infty}^{\infty} \frac{dx}{1+x^4} = \frac{\pi}{2\sqrt{2}}.$$

■

EXAMPLE 3 Another improper integral

Using (7), show that

$$\int_{-\infty}^{\infty} \frac{x^2-1}{x^4+5x^2+4}\,dx = \frac{\pi}{6}.$$

Solution. The degree of the denominator is two units higher than that of the numerator, so that our method again applies. Now

$$f(z) = \frac{p(z)}{q(z)} = \frac{z^2-1}{z^4+5z^2+4} = \frac{z^2-1}{(z^2+4)(z^2+1)}$$

has simple poles at $2i$ and i in the upper half-plane (and at $-2i$ and $-i$ in the lower half-plane, which are of no interest here). We calculate the residues from (4), Sec. 15.1, noting that $q'(z) = 4z^3 + 10z$,

$$\operatorname*{Res}_{z=2i} f(z) = \left[\frac{z^2-1}{4z^3+10z}\right]_{z=2i} = \frac{5}{12i}, \qquad \operatorname*{Res}_{z=i} f(z) = \left[\frac{z^2-1}{4z^3+10z}\right]_{z=i} = \frac{-2}{6i}.$$

Ans. $2\pi i(5/12i - 1/3i) = \pi/6$, as asserted. ■

Looking back, we realize that the key ideas of our present methods were these. In the first method we mapped the interval of integration on the real axis onto a closed curve in the complex plane (the unit circle). In the second method we attached to an interval on the real axis a semicircle such that we got a closed curve in the complex plane, which we then "blew up." This second method can be applied to further types of integrals, as we show in the next section, the last in this chapter.

Problem Set 15.3

Evaluate the following integrals involving cosine and sine.

1. $\displaystyle\int_0^{2\pi} \frac{d\theta}{1 + \frac{1}{2}\cos\theta}$ **2.** $\displaystyle\int_0^{\pi} \frac{d\theta}{\pi + \cos\theta}$ **3.** $\displaystyle\int_0^{2\pi} \frac{d\theta}{37 - 12\cos\theta}$

4. $\displaystyle\int_0^{2\pi} \frac{d\theta}{5 - 3\sin\theta}$ **5.** $\displaystyle\int_0^{2\pi} \frac{d\theta}{5/4 - \sin\theta}$ **6.** $\displaystyle\int_0^{2\pi} \frac{\cos\theta}{3 + \sin\theta}\,d\theta$

7. $\displaystyle\int_0^{2\pi} \frac{\cos\theta}{17 - 8\cos\theta}\,d\theta$ **8.** $\displaystyle\int_0^{2\pi} \frac{\sin^2\theta}{5 - 4\cos\theta}\,d\theta$ **9.** $\displaystyle\int_0^{2\pi} \frac{\cos\theta}{13 - 12\cos 2\theta}\,d\theta$

10. $\displaystyle\int_0^{2\pi} \frac{1 + 4\cos\theta}{17 - 8\cos\theta}\,d\theta$

Hint. Use $\cos 2\theta = \frac{1}{2}(z^2 + z^{-2})$.

Evaluate the following improper integrals.

11. $\displaystyle\int_{-\infty}^{\infty} \frac{dx}{1 + x^2}$ **12.** $\displaystyle\int_{-\infty}^{\infty} \frac{dx}{(1 + x^2)^3}$ **13.** $\displaystyle\int_0^{\infty} \frac{1 + x^2}{1 + x^4}\,dx$

14. $\displaystyle\int_{-\infty}^{\infty} \frac{x}{x^4 + 1}\,dx$ **15.** $\displaystyle\int_{-\infty}^{\infty} \frac{dx}{1 + x^6}$ **16.** $\displaystyle\int_{-\infty}^{\infty} \frac{dx}{x^4 + 16}$

17. $\displaystyle\int_{-\infty}^{\infty} \frac{x}{(x^2 - 2x + 2)^2}\,dx$ **18.** $\displaystyle\int_{-\infty}^{\infty} \frac{dx}{(x^2 + 1)(x^2 + 9)}$ **19.** $\displaystyle\int_{-\infty}^{\infty} \frac{dx}{(4 + x^2)^2}$

20. $\displaystyle\int_{-\infty}^{\infty} \frac{dx}{(x^2 - 2x + 5)^2}$

15.4 Further Types of Real Integrals

There are further classes of real integrals that can be evaluated by applying the residue theorem to suitable complex integrals. In applications such integrals may arise in connection with integral transforms or representations of special functions. In the present section we consider two such classes of integrals. One is important in problems involving the Fourier integral representation (Sec. 10.9). The other consists of real integrals whose integrand is infinite at some point in the interval of integration.

Fourier Integrals

Real integrals of the form

$$\text{(1)} \qquad \int_{-\infty}^{\infty} f(x) \cos sx \, dx \quad \text{and} \quad \int_{-\infty}^{\infty} f(x) \sin sx \, dx \qquad (s \text{ real})$$

occur in connection with the Fourier integral (Sec. 10.9).

If $f(x)$ is a rational function satisfying the assumptions on the degree stated in connection with (4), Sec. 15.3, then the integrals (1) may be evaluated in a way similar to that used for the integral in (4) of the last section. In fact, we may then consider the corresponding integral

$$\oint_C f(z) e^{isz} \, dz \qquad (s \text{ real and positive})$$

over the contour C in Fig. 347 (Sec. 15.3). Instead of (7), Sec. 15.3, we get

$$\text{(2)} \qquad \int_{-\infty}^{\infty} f(x) e^{isx} \, dx = 2\pi i \sum \operatorname{Res} [f(z) e^{isz}] \qquad (s > 0)$$

where we sum the residues of $f(z)e^{isz}$ at its poles in the upper half-plane. Equating the real and the imaginary parts on both sides of (2), we have

$$\text{(3)} \qquad \begin{aligned} \int_{-\infty}^{\infty} f(x) \cos sx \, dx &= -2\pi \sum \operatorname{Im} \operatorname{Res} [f(z) e^{isz}], \\ \int_{-\infty}^{\infty} f(x) \sin sx \, dx &= 2\pi \sum \operatorname{Re} \operatorname{Res} [f(z) e^{isz}]. \end{aligned} \qquad (s > 0)$$

We remember that (7), Sec. 15.3, was established by proving that the value of the integral over the semicircle S in Fig. 347 approaches zero as $R \to \infty$. To establish (2), we must prove the same for our present contour integral, as follows. Since $s > 0$ and S lies in the upper half-plane $y \geqq 0$, we see that

$$|e^{isz}| = |e^{isx}| \, |e^{-sy}| = e^{-sy} \leqq 1 \qquad (s > 0, \quad y \geqq 0).$$

From this we obtain the inequality

$$|f(z) e^{isz}| = |f(z)| \, |e^{isz}| \leqq |f(z)| \qquad (s > 0, \quad y \geqq 0).$$

This reduces our present problem to that in the previous section. Continuing as before, we see that the value of the integral over S approaches zero as R approaches infinity. This establishes (2), which implies (3).

EXAMPLE 1 **An application of (3)**

Show that

$$\int_{-\infty}^{\infty} \frac{\cos sx}{k^2 + x^2}\,dx = \frac{\pi}{k}e^{-ks}, \qquad \int_{-\infty}^{\infty} \frac{\sin sx}{k^2 + x^2}\,dx = 0 \qquad (s > 0,\quad k > 0).$$

Solution. In fact, $e^{isz}/(k^2 + z^2)$ has only one pole in the upper half-plane, namely, a simple pole at $z = ik$, and from (4) in Sec. 15.1 we obtain

$$\operatorname*{Res}_{z=ik} \frac{e^{isz}}{k^2 + z^2} = \left[\frac{e^{isz}}{2z}\right]_{z=ik} = \frac{e^{-ks}}{2ik}.$$

Therefore,

$$\int_{-\infty}^{\infty} \frac{e^{isx}}{k^2 + x^2}\,dx = 2\pi i\,\frac{e^{-ks}}{2ik} = \frac{\pi}{k}e^{-ks}.$$

Since $e^{isx} = \cos sx + i \sin sx$, this yields the above results [see also (15) in Sec. 10.9.] ▮

Other Types of Real Improper Integrals

Another kind of improper integral is a definite integral

$$\int_A^B f(x)\,dx \tag{4}$$

whose integrand becomes infinite at a point a in the interval of integration,

$$\lim_{x \to a} |f(x)| = \infty.$$

Then the integral (4) means

$$\int_A^B f(x)\,dx = \lim_{\epsilon \to 0} \int_A^{a-\epsilon} f(x)\,dx + \lim_{\eta \to 0} \int_{a+\eta}^B f(x)\,dx \tag{5}$$

where both ϵ and η approach zero independently and through positive values. It may happen that neither limit exists if $\epsilon, \eta \to 0$ independently, but

$$\lim_{\epsilon \to 0} \left[\int_A^{a-\epsilon} f(x)\,dx + \int_{a+\epsilon}^B f(x)\,dx\right] \tag{6}$$

exists. This is called the **Cauchy principal value** of the integral. It is written

$$\text{pr. v.} \int_A^B f(x)\,dx.$$

For example,

$$\text{pr. v.} \int_{-1}^{1} \frac{dx}{x^3} = \lim_{\epsilon \to 0} \left[\int_{-1}^{-\epsilon} \frac{dx}{x^3} + \int_{\epsilon}^{1} \frac{dx}{x^3} \right] = 0;$$

the principal value exists, although the integral itself has no meaning. The whole situation is quite similar to that in the second part of Sec. 15.3.

To evaluate an improper integral whose integrand has poles on the real axis, we use a path that avoids these singularities by following small semicircles with centers at the singular points; this method may be illustrated by the following example.

EXAMPLE 2 Integrand having a pole on the real axis. Sine integral

Show that

$$\int_0^\infty \frac{\sin x}{x}\, dx = \frac{\pi}{2}.$$

(This is the limit of the sine integral Si(x) as $x \to \infty$; see Sec. 10.9.)

Solution. **(a)** We do not consider $(\sin z)/z$ because this function does not behave suitably at infinity. We consider e^{iz}/z, which has a simple pole at $z = 0$, and integrate around the contour in Fig. 349. Since e^{iz}/z is analytic inside and on C, Cauchy's integral theorem gives

$$\oint_C \frac{e^{iz}}{z}\, dz = 0. \tag{7}$$

(b) We prove that the value of the integral over the large semicircle C_1 approaches zero as R approaches infinity. Setting $z = Re^{i\theta}$, we have $dz = iRe^{i\theta}\, d\theta$, $dz/z = i\, d\theta$ and therefore

$$\left| \int_{C_1} \frac{e^{iz}}{z}\, dz \right| = \left| \int_0^\pi e^{iz}\, i\, d\theta \right| \leqq \int_0^\pi |e^{iz}|\, d\theta \qquad (z = Re^{i\theta}).$$

In the integrand on the right,

$$|e^{iz}| = |e^{iR(\cos\theta + i\sin\theta)}| = |e^{iR\cos\theta}|\, |e^{-R\sin\theta}| = e^{-R\sin\theta}.$$

We insert this into the integral and use $\sin(\pi - \theta) = \sin\theta$ to get an integral from 0 to $\pi/2$:

$$\int_0^\pi |e^{iz}|\, d\theta = \int_0^\pi e^{-R\sin\theta}\, d\theta = 2\int_0^{\pi/2} e^{-R\sin\theta}\, d\theta.$$

Now Fig. 350 shows that $\sin\theta \geqq 2\theta/\pi$ if $0 \leqq \theta \leqq \pi/2$. Hence $-\sin\theta \leqq -2\theta/\pi$. From this and by integration,

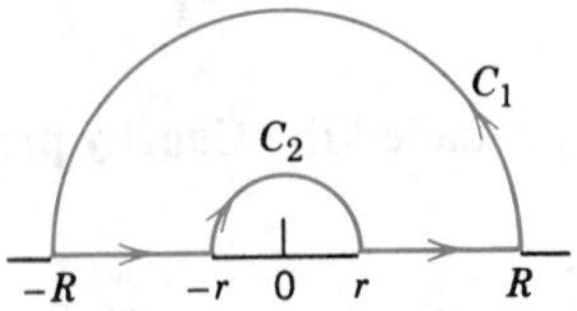

Fig. 349. Contour in Example 2

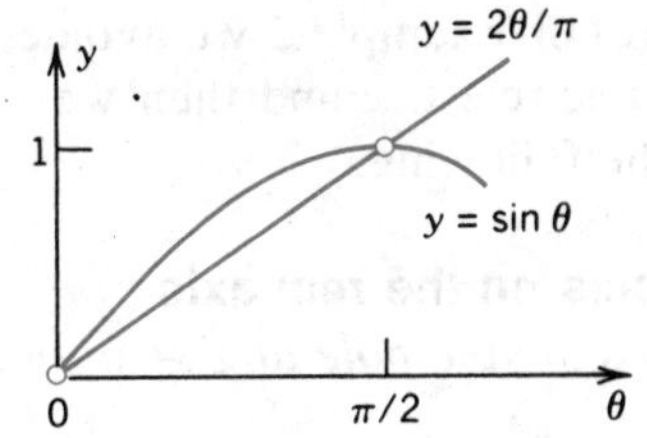

Fig. 350. Inequality in Example 2

$$2\int_0^{\pi/2} e^{-R\sin\theta}\,d\theta \leqq 2\int_0^{\pi/2} e^{-2R\theta/\pi}\,d\theta = \frac{\pi}{R}(1 - e^{-R}) \quad \to 0 \quad \text{as} \quad R \to \infty.$$

Hence the value of the integral over C_1 approaches 0 as $R \to \infty$.

(c) For the integral over the small semicircle C_2 in Fig. 349 we have

$$\int_{C_2} \frac{e^{iz}}{z}\,dz = \int_{C_2} \frac{dz}{z} + \int_{C_2} \frac{e^{iz} - 1}{z}\,dz.$$

The first integral on the right equals $-\pi i$. The integrand of the second integral is analytic and thus bounded, say, less than some constant M in absolute value for all z on C_2 and between C_2 and the x-axis. Hence by the ML-inequality (Sec. 13.2), the absolute value of this integral cannot exceed $M\pi r$. This approaches 0 as $r \to 0$. Because of part (b), from (7) we thus obtain

$$\int_C \frac{e^{iz}}{z}\,dz = \text{pr. v.} \int_{-\infty}^{\infty} \frac{e^{ix}}{x}\,dx + \lim_{r\to 0} \int_{C_2} \frac{e^{iz}}{z}\,dz = \text{pr. v.} \int_{-\infty}^{\infty} \frac{e^{ix}}{x}\,dx - \pi i = 0.$$

Hence this principal value equals πi; its real part is 0 and its imaginary part is

$$\text{(8)} \qquad \text{pr. v.} \int_{-\infty}^{\infty} \frac{\sin x}{x}\,dx = \pi.$$

(d) Now the integrand in (8) is not singular at $x = 0$. Furthermore, since for positive x the function $1/x$ decreases, the areas under the curve of the integrand between two consecutive positive zeros decrease in a monotone fashion, that is, the absolute values of the integrals

$$I_n = \int_{n\pi}^{n\pi+\pi} \frac{\sin x}{x}\,dx \qquad (n = 0, 1, \cdots)$$

form a monotone decreasing sequence $|I_1|, |I_2|, \cdots$, and $I_n \to 0$ as $n \to \infty$. Since these integrals have alternating sign (why?), it follows from the Leibniz test (in Appendix 3) that the infinite series $I_0 + I_1 + I_2 + \cdots$ converges. Clearly, the sum of the series is the integral

$$\int_0^{\infty} \frac{\sin x}{x}\,dx = \lim_{b\to\infty} \int_0^{b} \frac{\sin x}{x}\,dx$$

which therefore exists. Similarly, the integral from 0 to $-\infty$ exists. Hence we need not take the principal value in (8), and

$$\int_{-\infty}^{\infty} \frac{\sin x}{x}\,dx = \pi.$$

Since the integrand is an even function, the desired result follows. ∎

In part (c) of Example 2 we avoided the simple pole by integrating along a small semicircle C_2, and then we let C_2 shrink to a point. This process suggests the following.

Theorem 1 **Simple poles on the real axis**

If $f(z)$ has a simple pole at $z = a$ on the real axis, then (Fig. 351)

$$\lim_{r \to 0} \int_{C_2} f(z)\, dz = \pi i \operatorname*{Res}_{z=a} f(z).$$

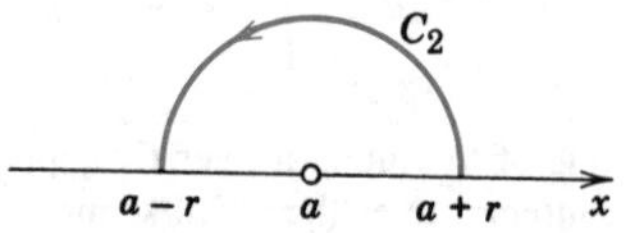

Fig. 351. Theorem 1

Proof. By the definition of a simple pole (Sec. 14.8) the integrand $f(z)$ has at $z = a$ the Laurent series

$$f(z) = \frac{b_1}{z - a} + g(z), \qquad b_1 = \operatorname*{Res}_{z=a} f(z)$$

where $g(z)$ is analytic on the semicircle of integration (Fig. 351)

$$C_2\colon \quad z = a + re^{i\theta}, \qquad 0 \leqq \theta \leqq \pi,$$

and for all z between C_2 and the x-axis. By integration,

$$\int_{C_2} f(z)\, dz = \int_0^{\pi} \frac{b_1}{re^{i\theta}}\, ire^{i\theta}\, d\theta + \int_{C_2} g(z)\, dz.$$

The first integral on the right equals $b_1 \pi i$. The second cannot exceed $M\pi r$ in absolute value, by the ML-inequality (Sec. 13.2), and $M\pi r \to 0$ as $r \to 0$. ∎

We may combine this theorem with (7) of Sec. 15.3 or (3) in this section. Thus [see (7), Sec. 15.3],

$$\text{(9)} \qquad \text{pr. v.} \int_{-\infty}^{\infty} f(x)\, dx = 2\pi i \sum \operatorname{Res} f(z) + \pi i \sum \operatorname{Res} f(z)$$

(summation over all poles in the upper half-plane in the first sum, and on the x-axis in the second), valid for rational $f(x) = p(x)/q(x)$ with degree $q \geqq$ degree $p + 2$, having simple poles on the x-axis.

Problem Set 15.4

1. Derive (3) from (2).

Evaluate the following real integrals.

2. $\int_{-\infty}^{\infty} \frac{\cos x}{x^4 + 1} dx$
3. $\int_{-\infty}^{\infty} \frac{\sin x}{1 + x^4} dx$
4. $\int_{-\infty}^{\infty} \frac{\sin x}{x^2 + x + 1} dx$
5. $\int_{-\infty}^{\infty} \frac{\cos x}{x^2 + x + 1} dx$
6. $\int_{-\infty}^{\infty} \frac{\sin 2x}{x^2 + x + 1} dx$
7. $\int_{-\infty}^{\infty} \frac{\cos x}{(x^2 + 1)^2} dx$
8. $\int_{-\infty}^{\infty} \frac{\cos 2x}{(x^2 + 1)^2} dx$
9. $\int_{-\infty}^{\infty} \frac{\sin nx}{1 + x^4} dx$
10. $\int_{-\infty}^{\infty} \frac{\cos 4x}{x^4 + 5x^2 + 4} dx$
11. $\int_{-\infty}^{\infty} \frac{\sin x}{x^2 + 2x + 4} dx$
12. $\int_{\infty}^{\infty} \frac{\cos 2x}{(x^2 + 4)^2} dx$
13. $\int_{0}^{\infty} \frac{\cos 2x}{4x^4 + 13x^2 + 9} dx$

14. Integrating e^{-z^2} around the boundary of the rectangle with vertices $-a$, a, $a + ib$, $-a + ib$, letting $a \to \infty$, and using

$$\int_0^{\infty} e^{-x^2}\, dx = \frac{\sqrt{\pi}}{2}, \quad \text{show that} \quad \int_0^{\infty} e^{-x^2} \cos 2bx\, dx = \frac{\sqrt{\pi}}{2} e^{-b^2}.$$

Poles on the real axis. Find the Cauchy principal value of the following integrals.

15. $\int_{-\infty}^{\infty} \frac{dx}{x^2 - ix}$
16. $\int_{-\infty}^{\infty} \frac{dx}{(x + 1)(x^2 + 2)}$
17. $\int_{-\infty}^{\infty} \frac{x}{8 - x^3} dx$
18. $\int_{-\infty}^{\infty} \frac{dx}{x^2 - 2ix}$
19. $\int_{-\infty}^{\infty} \frac{dx}{(x^2 + 1)(x - 1)}$
20. $\int_{-\infty}^{\infty} \frac{dx}{x^4 - 1}$
21. $\int_{-\infty}^{\infty} \frac{\sin \frac{1}{4}\pi x}{2x - x^2} dx$
22. $\int_{-\infty}^{\infty} \frac{\cos \frac{1}{2}\pi x}{x^2 - 1} dx$
23. $\int_{-\infty}^{\infty} \frac{\sin \pi x}{x - x^5} dx$

24. Show that the result in Example 2 remains the same if we replace the upper semicircle C_2 by the corresponding lower semicircle.

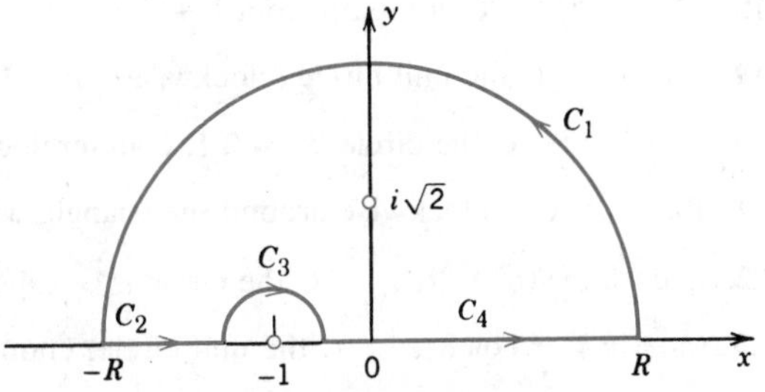

Fig. 352. Problem 16

Review Questions and Problems for Chapter 15

1. What is a singularity of an analytic function $f(z)$ and what is the residue of $f(z)$ at a singular point?
2. What is the idea of residue integration?
3. Why is it important in residue integration to use that Laurent series which converges in a neighborhood of the singular point z_0 (except at z_0 itself)?
4. Can you remember some of the methods by which we determined residues?
5. Can the residue at a singular point be zero? Give a reason.
6. Can the residue at a simple pole be zero?
7. Can we apply residue integration in the case of a function that is not analytic but merely continuous?
8. Can we use residue integration for evaluating the integral of tan $(1/z)$ around a closed contour containing $z = 0$ in its interior? Answer the same question for $e^{1/z}$.
9. In residue integration we need *closed* paths. How were we nevertheless able to evaluate (real) integrals over intervals?
10. What is the Cauchy principal value of an integral, and why did it occur in the present chapter?

Integrate the given function over the given path C, using residue integration or one of the methods discussed in Chap. 13, and indicate whether residue integration can be used.

11. $z^{-4} \cos 2z$, C any circle $|z| = const$, counterclockwise
12. $z^{-4} \sin 2z$, C the unit circle, clockwise
13. $5z/(2z + i)$, C the circle $|z - 2i| = 3$, counterclockwise
14. $(z^2 + 1)/(z^2 - 2z)$, C the ellipse $x^2 + 2y^2 = 2$, counterclockwise
15. $(9z - 8)/(z^2 + z - 6)$, C the circle $|z - i| = 4$, clockwise
16. $(iz + 1)/(z^2 - iz + 2)$, C the circle $|z - 1| = 3$, clockwise
17. $z/|z|^2$, C the unit circle, counterclockwise
18. $z^3 \exp z^4$, C any path from $1 + i$ to 1
19. $z^n e^{1/z}$, C the unit circle (clockwise), $n = 1, 2, \cdots$
20. $z/\sin^2 z$, C the circle $|z| = 0.1$, counterclockwise
21. Re z, C: clockwise around the triangle with vertices $0, 1, 1 + i$
22. $(z \cosh z^2)/(z - 2i)^3$, C the circle $|z - 2i| = 1$, counterclockwise
23. $(z - \pi/4)^{-3} \cos 8z$, C the unit circle, counterclockwise
24. $(4z^3 + 7z)/\cos z$, C the circle $|z + 1| = 1$, counterclockwise
25. $\cot 8z$, C the circle $|z| = 0.3$, counterclockwise

Evaluate by the methods in this chapter:

26. $\displaystyle\int_0^{\pi} \frac{d\theta}{k + \cos\theta}$ $(k > 1)$ 27. $\displaystyle\int_0^{2\pi} \frac{d\theta}{25 - 24\cos\theta}$ 28. $\displaystyle\int_0^{2\pi} \frac{d\theta}{13 - 5\sin\theta}$

29. $\displaystyle\int_0^{2\pi} \frac{d\theta}{1 - \frac{1}{2}\sin\theta}$ 30. $\displaystyle\int_0^{2\pi} \frac{\sin\theta}{3 + \cos\theta}$ 31. $\displaystyle\int_0^{2\pi} \frac{\sin\theta}{34 - 16\sin\theta}$

32. $\displaystyle\int_0^{2\pi} \frac{\frac{1}{2} + 2\sin\theta}{17 - 8\sin\theta}\,d\theta$ 33. $\displaystyle\int_{-\infty}^{\infty} \frac{dx}{1 + 4x^4}$ 34. $\displaystyle\int_{-\infty}^{\infty} \frac{dx}{1 + 4x^2}$

35. $\displaystyle\int_{-\infty}^{\infty} \frac{x}{(1 + x^2)^2}\,dx$ 36. $\displaystyle\int_0^{\infty} \frac{dx}{(1 + x^2)^2}$ 37. $\displaystyle\int_{-\infty}^{\infty} \frac{x^2}{x^4 + 5x^2 + 4}\,dx$

38. $\displaystyle\int_{-\infty}^{\infty} \frac{x^2}{x^4 + 13x^2 + 36}\,dx$ 39. $\displaystyle\int_0^{\infty} \frac{1 + 2x^2}{1 + 4x^4}\,dx$ 40. $\displaystyle\int_0^{\infty} \frac{dx}{x^4 + 10x^2 + 9}$

Summary of Chapter 15 Residue Integration Method

The **residue** of an analytic function $f(z)$ at a point $z = z_0$ is the coefficient b_1 of the power $1/(z - z_0)$ in that Laurent series

$$f(z) = a_0 + a_1(z - z_0) + \cdots + \frac{b_1}{z - z_0} + \frac{b_2}{(z - z_0)^2} + \cdots$$

of $f(z)$ which converges near z_0 (except at z_0 itself). This residue is given by the integral (Sec. 15.1)

$$(1) \qquad b_1 = \frac{1}{2\pi i} \oint_C f(z)\,dz$$

but can be obtained in various other ways, so that one can use (1) for evaluating integrals over closed curves. More generally, the **residue theorem** (Sec. 15.2) states that if $f(z)$ is analytic in a domain D except at finitely many points z_j and C is a simple closed path in D such that no z_j lies on C and the full interior of C belongs to D, then

$$(2) \qquad \oint_C f(z)\,dz = \frac{1}{2\pi i} \sum_j \operatorname*{Res}_{z=z_j} f(z)$$

(summation only over those z_j that lie *inside* C).

This integration method is elegant and powerful. Formulas for the residue at **poles** are (m = order of the pole (Sec. 15.1))

$$\operatorname*{Res}_{z=z_0} f(z) = \frac{1}{(m-1)!} \lim_{z\to z_0} \left(\frac{d^{m-1}}{dz^{m-1}} [(z - z_0)^m f(z)] \right), \tag{3}$$

where $m = 1, 2, \cdots$. Hence for a simple pole ($m = 1$),

$$\operatorname*{Res}_{z=z_0} f(z) = \lim_{z\to z_0} (z - z_0) f(z). \tag{3*}$$

Another formula for the residue at a simple pole of $f(z) = p(z)/q(z)$ is

$$\operatorname*{Res}_{z=z_0} f(z) = \frac{p(z_0)}{q'(z_0)}. \tag{3**}$$

Residue integration involves *closed* curves, but the real interval of integration $0 \leqq \theta \leqq 2\pi$ is transformed into the unit circle by setting $z = e^{i\theta}$, so that by residue integration we can integrate **real integrals** of the form (Sec. 15.3)

$$\int_0^{2\pi} F(\cos\theta, \sin\theta)\, d\theta,$$

where F is a rational function of $\cos\theta$ and $\sin\theta$, such as, for instance, $(\sin^2\theta)/(5 - 4\cos\theta)$, etc.

Another method of integrating *real* integrals by residues is the use of a closed contour consisting of an interval $-R \leqq x \leqq R$ on the real axis and a semicircle $|z| = R$. From the residue theorem, if we let $R \to \infty$, we obtain for rational $f(x) = p(x)/q(x)$ (with $q(x) \neq 0$ and degree $q \geqq$ degree $p + 2$)

$$\int_{-\infty}^{\infty} f(x)\, dx = 2\pi i \sum \operatorname{Res} f(z) \qquad \text{(Sec. 15.3)}$$

$$\int_{-\infty}^{\infty} f(x) \cos sx\, dx = -2\pi \sum \operatorname{Im} \operatorname{Res} [f(z) e^{isz}]$$

$$\int_{-\infty}^{\infty} f(x) \sin sx\, dx = 2\pi \sum \operatorname{Re} \operatorname{Res} [f(z) e^{isz}] \qquad \text{(Sec. 15.4)}$$

(sum of all residues at poles in the upper half-plane). In Sec. 15.4 we also extend this method to real integrals whose integrands become infinite at some point in the interval of integration.

Chapter 16

Conformal Mapping

If a complex function $w = f(z)$ is defined in a domain D of the z-plane, then to each point in D there corresponds a point in the w-plane. In this way we have a *mapping* of D onto the range of values of $f(z)$ in the w-plane. This "geometric approach" to complex analysis helps us to "visualize" the nature of a complex function by considering the manner in which the function maps certain curves and regions. We have seen this for special functions in Sec. 12.9, which the reader may want to review before going on.

This chapter concerns a systematic approach to mappings by general *analytic* functions $w = f(z)$. We show (in Sec. 16.1) that such a mapping is **conformal;** that is, it preserves angles in magnitude and sense, except at "*critical points*" (points at which the derivative $f'(z)$ is zero).

Conformal mapping is important in engineering mathematics, since it is a standard method for solving boundary value problems in two-dimensional potential theory by transforming a given complicated region into a simpler one (details in Chap. 17). For this task of mapping, linear fractional transformations (Secs. 16.2, 16.3) play a basic role, and other special functions (Sec. 16.4) can often be used. In the last section (Sec. 16.5) we discuss the idea of the Riemann surface.

Applications to potential theory follow in Chap. 17.

Prerequisites for this chapter: Chap. 12.
Sections that may be omitted in a shorter course: Secs. 16.3, 16.4.
References: Appendix 1, Part D.
Answers to problems: Appendix 2.

16.1 Conformal Mapping

A complex-valued function

$$w = f(z) = u(x, y) + iv(x, y) \qquad (z = x + iy)$$

of a complex variable z gives a **mapping** of its domain of definition in the complex z-plane onto[1] its range of values in the complex w-plane. Examples

[1]The general terminology is as follows. A mapping of a set A into a set B is called **surjective** or a mapping of A **onto** B if every element of B is the image of at least one element of A. It is called **injective** or **one-to-one** if different elements of A have different images in B. Finally, it is called **bijective** if it is both surjective and injective.

are shown in Sec. 12.9, at which the student may take a quick look before going on. If $f(z)$ is *analytic*, its most important mapping property is its *conformality*, which is defined as follows.

The mapping (1) is called **conformal** if it preserves angles between oriented curves in magnitude as well as in sense.

Figure 353 shows what this means: The images $C_1{}^*$ and $C_2{}^*$ of two oriented curves make the same **angle** α (this is the angle α $(0 \leqq \alpha \leqq \pi)$ between their oriented tangents at the intersection point) as the curves C_1 and C_2 themselves, in magnitude and in direction.

We prove the asserted conformality in the case of an *analytic* function $f(z)$, except at a **critical point;** this is a point at which the derivative $f'(z)$ is zero. *Example.* $f(z) = z^2$ has the critical point $z = 0$, where $f'(z) = 2z = 0$ and the angles are doubled (see Example 2, Sec. 12.9), so that the mapping is certainly not conformal there. $f(z) = \cos z$ has $0, \pm\pi, \pm 2\pi, \cdots$ as critical points, etc.

Theorem 1 (Conformal mapping)

The mapping defined by an analytic function $f(z)$ is conformal, except at critical points, that is, points at which the derivative $f'(z)$ is zero.

Proof. The idea is to consider a curve

$$C: \quad z(t) = x(t) + iy(t) \tag{1}$$

in the domain of $f(z)$ and to show that the mapping $w = f(z)$ rotates the tangent to C at any point z_0 of C [with $f'(z_0) \neq 0$] through an angle that is independent of C, so that the tangents of *two* curves C_1 and C_2 passing through z_0 (Fig. 353) are rotated through the same angle, so that the images of these curves make the same angle in size and sense as the curves themselves, which means conformality, by definition. The details are as follows.

We assume that C is a **smooth curve;** that is, $z(t)$ in (1) is differentiable and the derivative $\dot{z}(t) = dz/dt$ is continuous and nowhere zero. We claim that then C has a unique continuously turning tangent. In fact, by definition,

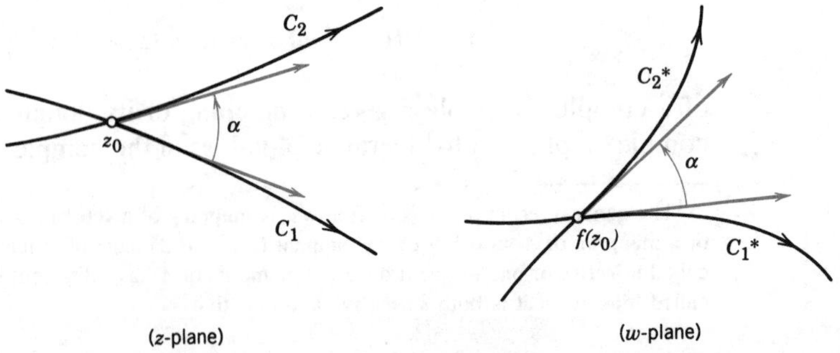

Fig. 353. Curves C_1 and C_2 and their respective images $C_1{}^*$ and $C_2{}^*$ under a conformal mapping

$$(2)\qquad \dot{z}(t_0) = \left.\frac{dz}{dt}\right|_{t_0} = \lim_{\Delta t \to 0} \frac{z_1 - z_0}{\Delta t} = \lim_{\Delta t \to 0} \frac{z(t_0 + \Delta t) - z(t_0)}{\Delta t} \qquad \text{(Fig. 354).}$$

Now the numerator $z_1 - z_0$ represents a chord of C (see Fig. 354), and $(z_1 - z_0)/\Delta t$ with positive Δt has the same direction. As $\Delta t \to 0$, the point z_1 approaches z_0 along the curve, and

$$(z_1 - z_0)/\Delta t \quad \to \quad \dot{z}(t_0).$$

Hence $\dot{z}(t_0)$ is tangent to C at z_0, briefly called a **tangent vector** of C at z_0, and our claim is proved.

Each tangent is now *oriented:* the *positive sense* on it and on C is the sense of increasing t in (1), in which also the tangent vector points.

We now turn to the image C^* of C under $w = f(z)$ (not a constant). C^* is a curve represented by

$$w = f[z(t)]$$

because $z(t)$ gives C and f maps it. To $z_0 = z(t_0)$ on C corresponds on C^* the point $w_0 = w(t_0)$ and a tangent vector of C^* at that point is $\dot{w}(t_0)$. Now

$$(3)\qquad \frac{dw}{dt} = \frac{df}{dz}\frac{dz}{dt}, \qquad \text{briefly,} \qquad \dot{w}(t) = f'[z(t)]\dot{z}(t),$$

by the chain rule. Let $f'(z_0) \neq 0$. Then $\arg f'(z_0)$ is defined, and taking arguments in (3) gives [by (9), Sec. 12.2]

$$(4)\qquad \arg \dot{w}(t_0) = \arg f'(z_0) + \arg \dot{z}(t_0).$$

Hence the angle between the tangent of C at z_0 and the tangent of C^* at w_0, that is, the angle α through which the mapping rotates the tangent, is

$$\alpha = \arg \dot{w}(t_0) - \arg \dot{z}(t_0) = \arg f'(z_0).$$

Now comes the key point: Since the right side is independent of the choice of C (it depeneds only on z_0), that angle is independent of C; it is the same for all curves passing through z_0 and their images, so that two images C_1^* and C_2^* make the same angle as the original curves C_1 and C_2. But this means conformality, by definition, and proves the theorem. ∎

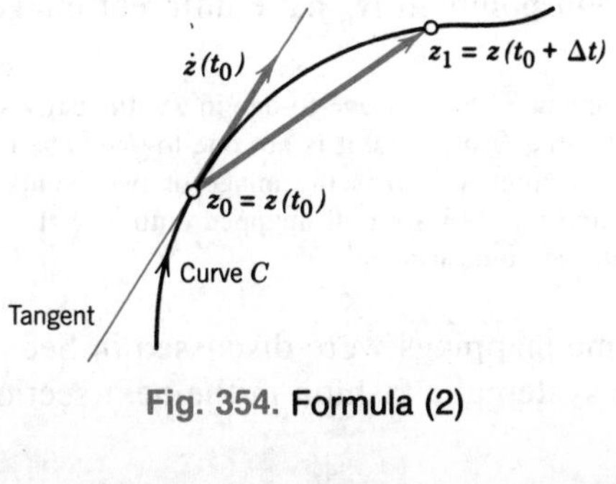

Fig. 354. Formula (2)

EXAMPLE 1 **Conformality of $w = z^n$ and $w = e^z$**

The mapping $w = z^n$, $n = 2, 3, \cdots$, is conformal except at $z = 0$, where $w' = nz^{n-1} = 0$. For $w = z^2$ this is shown in Fig. 304 (Sec. 12.9), in which the image curves intersect at right angles, except at $z = 0$ where the angles are doubled under the mapping, since every ray $\arg z = c = const$ transforms into a ray $\arg w = 2c$.

The mapping $w = e^z$ is conformal for all z since $w' = e^z$ is not 0 for any z. ∎

Magnification Ratio. By the definition of the derivative we have

$$\lim_{z \to z_0} \left| \frac{f(z) - f(z_0)}{z - z_0} \right| = |f'(z_0)|.$$

Therefore, the mapping $w = f(z)$ magnifies (or shortens) the lengths of short lines by approximately the factor $|f'(z_0)|$. The image of a small figure ***conforms*** to the original figure in the sense that it has approximately the same shape. However, since $f'(z)$ varies from point to point, a *large* figure may have an image whose shape is quite different from that of the original figure.

More on the Condition $f'(z) \neq 0$. From (4) in Sec. 12.5 and the Cauchy–Riemann equations we obtain

$$(5') \qquad |f'(z)|^2 = \left| \frac{\partial u}{\partial x} + i \frac{\partial v}{\partial x} \right|^2 = \left(\frac{\partial u}{\partial x} \right)^2 + \left(\frac{\partial v}{\partial x} \right)^2 = \frac{\partial u}{\partial x} \frac{\partial v}{\partial y} - \frac{\partial u}{\partial y} \frac{\partial v}{\partial x},$$

that is,

$$(5) \qquad |f'(z)|^2 = \begin{vmatrix} \dfrac{\partial u}{\partial x} & \dfrac{\partial u}{\partial y} \\ \dfrac{\partial v}{\partial x} & \dfrac{\partial v}{\partial y} \end{vmatrix} = \frac{\partial(u, v)}{\partial(x, y)}.$$

This determinant is the so-called **Jacobian** (see Sec. 9.3) of the transformation $w = f(z)$, written in real form

$$u = u(x, y), \qquad v = v(x, y).$$

Hence, the condition $f'(z_0) \neq 0$ implies that the Jacobian is not zero at z_0. This condition suffices that the restriction of the mapping $w = f(z)$ to a sufficiently small neighborhood N_0 of z_0 is ***one-to-one*** or injective, that is, different points in N_0 have different images; see Ref. [5] in Appendix 1.

EXAMPLE 2 The mapping $w = z^2$ is one-to-one in a sufficiently small neighborhood of any point $z \neq 0$. In a neighborhood of $z = 0$ it is not one-to-one. The full z-plane is mapped onto the w-plane so that each point $w \neq 0$ is the image of two points in the z-plane. For instance, the points $z = 1$ and $z = -1$ are both mapped onto $w = 1$, and, more generally, z_1 and $-z_1$ have the same image point $w = z_1^2$. ∎

Some mappings were discussed in Sec. 12.9 and further ones follow in a more systematic fashion in the next sections.

Problem Set 16.1

Find and sketch the images of the given curves under the mapping $w = u + iv = z^2$.

1. $x = 1, 2, 3, 4$ **2.** $y = 1, 2, 3, 4$ **3.** $xy = 10$

4. $y = x + 1$ **5.** $y = x, y = -x$ **6.** $y^2 = x^2 - 1$

Plot the images of the following regions under the mapping $w = z^2$.

7. $|z| < \frac{1}{3}$ **8.** $|z| \geqq 5$ **9.** $|\arg z| < \pi/3$

10. $\text{Re } z > 0$ **11.** $\pi/4 < \arg z < \pi/2$ **12.** $0 < y < 1$

Represent the following curves in the z-plane ($z = x + iy$) in the form $z = z(t)$ and determine the corresponding tangent vector $\dot{z}(t)$. Sketch the curve and some of the vectors.

13. $x^2 + y^2 = 16$ **14.** $(x - 1)^2 + (y + 2)^2 = 25$ **15.** $x^2 + 9y^2 = 9$

16. $x^2 - y^2 = 4$ **17.** $y = 1/x$ **18.** $y = 12 - 3x^2$

Determine the points in the z-plane at which the mapping $w = f(z)$ fails to be conformal, where $f(z)$ equals

19. $z^2 + az + b$ **20.** $\cos \pi z$ **21.** $\exp (z^5 - 80z)$

22. $\exp (z^9 - 9z)$ **23.** $\cosh 2z$ **24.** $z + z^{-1} \quad (z \neq 0)$

25. Why do the images of the curves $|z| = const$ and $\arg z = const$ under a mapping by an analytic function intersect at right angles?

26. Why do the level curves $u = const$ and $v = const$ of an analytic function $w = u + iv = f(z)$ intersect at right angles at each point at which $f'(z) \neq 0$?

27. Does the mapping $w = \bar{z} = x - iy$ preserve angles in size as well as in sense?

28. Verify (5) for $f(z) = e^z$.

29. **(Magnification of angles)** Let the function $f(z)$ be analytic at z_0. Suppose that $f'(z_0) = 0, \cdots, f^{(k-1)}(z_0) = 0$, whereas $f^{(k)}(z_0) \neq 0$. Then the mapping $w = f(z)$ magnifies angles with vertex at z_0 by a factor k. Illustrate this with examples for $k = 2, 3, 4$.

30. Prove the statement in Prob. 29 for arbitrary k. *Hint.* Use the Taylor series.

16.2 Linear Fractional Transformations

Conformal mapping has various physical applications, as we shall see in Chap. 17, but to make practical use of it, one must know what function to choose, for instance, in the frequent task of mapping a complicated domain onto a simple one, such as a disk. This requires a study of properties of mapping functions, and we begin with the following very important class.

Linear fractional transformations (or **Möbius transformations**) are mappings

$$w = \frac{az + b}{cz + d} \qquad (ad - bc \neq 0) \tag{1}$$

where a, b, c, d are complex or real numbers. These mappings are of practical importance in applications to boundary value problems since they are needed for mapping disks conformally onto half-planes or onto other disks, and conversely, as we shall see. They also provide further motivation of the extended complex plane (Sec. 14.8), which plays a role in fluid flow and other applications.

The condition $ad - bc \neq 0$ in (1) becomes clear if we differentiate:

$$w' = \frac{a(cz + d) - c(az + b)}{(cz + d)^2} = \frac{ad - bc}{(cz + d)^2}.$$

We see that $ad - bc \neq 0$ implies that w' is nowhere zero, hence the mapping (1) is everywhere conformal. From $ad - bc = 0$ we would obtain the uninteresting case that w' is identically zero, which we exclude once and for all. We begin our discussion with special cases of (1).

EXAMPLE 1 Translations, rotations, expansions, contractions

These are special cases of (1) of the form

$$w = z + b \qquad (2) \quad \textit{(translation)}$$

and

$$w = az, \qquad (3)$$

which is a *rotation* when $|a| = 1$, say, $a = e^{i\alpha}$ (α the angle of rotation), an *expansion* for real $a > 1$, and a *contraction* for $0 < a < 1$. These are also special cases of the **linear transformation**

$$w = az + b$$

obtained from (1) when $c = 0$ (except for the notation). ∎

EXAMPLE 2 Mapping $w = 1/z$. Inversion

The mapping

$$w = \frac{1}{z} \qquad (4)$$

is an important special case of (1). It is best studied in terms of polar forms $z = re^{i\theta}$ and $w = Re^{i\phi}$. Then $w = 1/z$ becomes

$$Re^{i\phi} = \frac{1}{re^{i\theta}} \qquad \text{and gives} \qquad R = \frac{1}{r}, \qquad \phi = -\theta$$

by equating absolute values and arguments on both sides. We see from this that the image $w = 1/z$ of a $z \neq 0$ lies on the ray from the origin through $\bar{z}$, at the distance $1/|z|$. In particular, $z = e^{i\theta}$ on the unit circle $|z| = 1$ is mapped onto $w = e^{i\phi} = e^{-i\theta}$ on the unit circle $|w| = 1$.

We mention that $w = 1/z$ can be obtained geometrically from z by an *inversion in the unit circle* (Fig. 355) followed by a reflection in the x-axis. The reader may prove this, using similar triangles.

Figure 356 shows that $w = 1/z$ maps horizontal and vertical straight lines onto circles or straight lines. Even the following is true.

$w = 1/z$ maps every straight line or circle onto a circle or straight line.

Proof. Every straight line or circle in the z-plane can be written

$$A(x^2 + y^2) + Bx + Cy + D = 0 \qquad (A, B, C, D \text{ real}).$$

$A = 0$ gives a straight line and $A \neq 0$ a circle. In terms of z and $\bar{z}$, this equation becomes

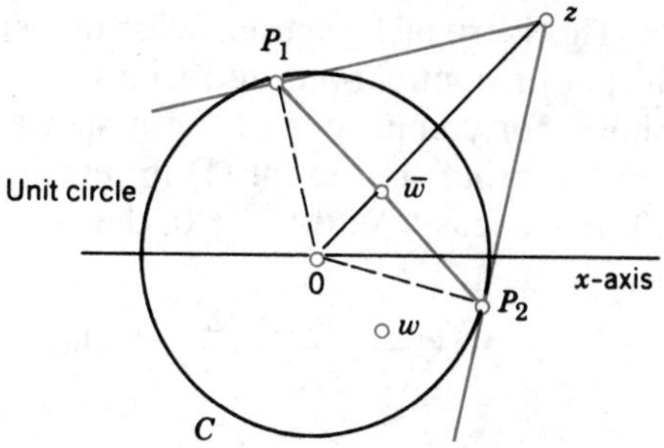

Fig. 355. Geometrical construction of $w = 1/z$. Here, $\overline{w}$ is the intersection of Oz and P_1P_2, where P_1 and P_2 are the points of contact of the tangents to the unit circle passing through z.

$$Az\bar{z} + B\frac{z + \bar{z}}{2} + C\frac{z - \bar{z}}{2i} + D = 0.$$

Now $w = 1/z$. Substitution of $z = 1/w$ and multiplication by $w\overline{w}$ gives the equation

$$A + B\frac{\overline{w} + w}{2} + C\frac{\overline{w} - w}{2i} + Dw\overline{w} = 0$$

or, in terms of u and v,

$$A + Bu - Cv + D(u^2 + v^2) = 0.$$

This represents a circle (if $D \neq 0$) or a straight line (if $D = 0$) in the w-plane. ∎

The proof in this example suggests the use of z and $\bar{z}$ instead of x and y, a general principle that is often quite useful in practice. Surprisingly, *every* linear fractional transformation has the property just proved:

Theorem 1 (Circles and straight lines)

Every linear fractional transformation (1) *maps the totality of circles and straight lines in the z-plane onto the totality of circles and straight lines in the w-plane.*

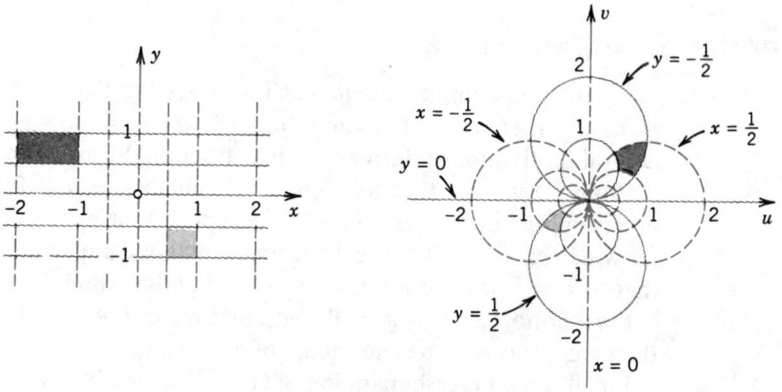

Fig. 356. Mapping $w = 1/z$

Proof. This is trivial for a translation or rotation, fairly obvious for a uniform expansion or contraction, and true for $w = 1/z$, as just proved. Hence it also holds for composites of these special mappings. Now comes the key idea of the proof: represent (1) in terms of these special mappings. When $c = 0$, this is easy. When $c \neq 0$, the representation is

$$w = K \frac{1}{cz + d} + \frac{a}{c} \qquad \text{where} \qquad K = -\frac{ad - bc}{c}.$$

This can be verified by substituting K, taking the common denominator and simplifying; this yields (1). We can now set

$$w_1 = cz, \qquad w_2 = w_1 + d, \qquad w_3 = \frac{1}{w_2}, \qquad w_4 = Kw_3,$$

and see from the previous formula that then $w = w_4 + a/c$. This tells us that (1) is indeed a composite of those special mappings and completes the proof. ∎

EXAMPLE 3 Image of a circle

Using Theorem 1, find the image of the unit circle $|z| = 1$ under the linear fractional transformation

$$w(z) = \frac{2iz - 2 - 2i}{(1 - i)z - 1}.$$

Solution. By Theorem 1, the image is a circle (or a straight line), hence determined by the images of three points $1, -1$, and i on $|z| = 1$. We calculate these images, writing $w(z)$ more simply as

$$w(z) = 2i \frac{z - 1 + i}{(1 - i)z - 1}; \qquad \text{thus} \qquad w(1) = 2i \frac{1 - 1 + i}{1 - i - 1} = -2i;$$

similarly, $w(-1) = 2i$, so that the center of the circle must lie on the u-axis. Finally, $w(i) = -2 + 4i$. The perpendicular bisector of $2i$ and $-2 + 4i$ intersects the u-axis at -4 (make a sketch); this is the center of the circle. *Ans.* $|w + 4| = \sqrt{20}$.

We will develop a more powerful method for problems of this type in the next section, along with other applications of Theorem 1. ∎

Extended Complex Plane

The extended complex plane was introduced in Sec. 14.8. It is the complex plane together with the point ∞ (*infinity*). We can now give it an even more natural motivation in terms of linear fractional transformations.

From (1) we see that to each z for which $cz + d \neq 0$ there corresponds precisely one complex number w. Suppose that $c \neq 0$. Then to $z = -d/c$, for which $cz + d = 0$, there does not correspond any number w. This suggests that to $z = -d/c$ we let $w = \infty$ correspond as the image.

Furthermore, when $c = 0$, we must have $a \neq 0$ and $d \neq 0$ (why?), and then we let $w = \infty$ be the image of $z = \infty$.

Finally, the inverse mapping of (1) is obtained by solving (1) for z; we find that it is again a linear fractional transformation:

(5) $$z = \frac{-dw + b}{cw - a}.$$

When $c \neq 0$, then $cw - a = 0$ for $w = a/c$, and we let a/c be the image of $z = \infty$. With these settings, the linear fractional transformation (1) is now a one-to-one conformal mapping of the extended z-plane onto the extended w-plane. We also say that every linear fractional transformation maps "the extended complex plane in a one-to-one and conformal manner onto itself."

Our present discussion suggests the following:

General Remark. If $z = \infty$, then the right-hand side of (1) becomes the meaningless expression $(a \cdot \infty + b)/(c \cdot \infty + d)$. We assign to it the value $w = a/c$ if $c \neq 0$ and $w = \infty$ if $c = 0$.

Fixed Points

Fixed points of a mapping $w = f(z)$ are points that are mapped onto themselves, are "kept fixed" under the mapping. Thus they are obtained from

$$w = f(z) = z.$$

The **identity mapping**

$$w = z$$

has every point as a fixed point. The mapping $w = \bar{z}$ has infinitely many fixed points, $w = 1/z$ has two, a rotation has one, and a translation none in the finite plane. (Find them in each case.) For (1), the fixed-point condition $w = z$ is

$$z = \frac{az + b}{cz + d}$$

or

(6) $$cz^2 - (a - d)z - b = 0.$$

This is a quadratic equation in z whose coefficients all vanish if and only if the mapping is the identity mapping $w = z$ (in this case, $a = d \neq 0$, $b = c = 0$). Hence we have

Theorem 2 (Fixed points)

A linear fractional transformation, not the identity, has at most two fixed points. If a linear fractional transformation is known to have three or more fixed points, it must be the identity mapping $w = z$.

To make our present general discussion of linear fractional transformations really useful from a practical point of view, we extend it by further facts and typical examples, in the problem set as well as in the next section.

Problem Set 16.2

Find the fixed points of the following mappings.

1. $w = (2 + i)z$
2. $w = z + 3i$
3. $w = z^5$
4. $w = (z - i)^2$
5. $w = (z + 1)^2$
6. $w = z^5 + 10z^3 + 10z$
7. $w = \dfrac{2iz - 1}{z + 2i}$
8. $w = \dfrac{3z + 2}{z - 1}$
9. $w = \dfrac{z - 1}{z + 1}$

Find a linear fractional transformation whose (only) fixed points are

10. 0, 1
11. $-2, 2$
12. 0

Find all linear fractional transformations whose (only) fixed points are

13. $0, \infty$
14. $-1, 1$
15. $-i, i$

16. Find all linear fractional transformations without fixed points in the finite plane.

17. Find the inverse of $w = (3z + 4i)/(z + 2i)$ directly, without using (5).

18. If z_1 is a fixed point of a linear fractional transformation $w = f(z)$, it is clear that z_1 must also be a fixed point of the inverse $z = g(w)$. Give a proof of this.

19. Show the calculations of deriving (5) from (1). Check (5) by substitution into (1).

20. Prove Theorem 1 by substituting (1) in its given form into the equation of a circle or straight line. *Hint.* Show that the latter can be written ($w = u + iv$; $A = 0$ gives a straight line)

$$Aw\bar{w} + Bw + \overline{B}\bar{w} + C = 0.$$

16.3 Special Linear Fractional Transformations

We shall now learn how we can determine linear fractional transformations

$$w = \frac{az + b}{cz + d} \qquad (ad - bc \neq 0) \tag{1}$$

for mapping certain simple domains onto others, and how we can discuss properties of (1).

Four given numbers a, b, c, d determine a unique mapping (1), but we can drop or introduce a common factor without altering a given mapping (1). That is, (1) depends on *three* essential constants, namely, the ratios of any three of the a, b, c, d to the fourth. Hence we should get a unique mapping (1) if we impose three conditions, say, the conditions that three distinct points in the z-plane have specified distinct images in the w-plane. We show that this is the case and, more important, we give a formula that yields the mapping:

Theorem 1 (Three points and their images given)
Three given distinct points z_1, z_2, z_3 can always be mapped onto three prescribed distinct points w_1, w_2, w_3 by one, and only one, linear fractional transformation $w = f(z)$. This mapping is given implicitly by the equation

$$\frac{w - w_1}{w - w_3} \cdot \frac{w_2 - w_3}{w_2 - w_1} = \frac{z - z_1}{z - z_3} \cdot \frac{z_2 - z_3}{z_2 - z_1}. \tag{2}$$

(If one of these points is the point ∞, the quotient of the two differences containing this point must be replaced by 1.)

Proof. Equation (2) is of the form $F(w) = G(z)$, where F and G denote linear fractional functions of the respective variables. From this we readily obtain $w = f(z) = F^{-1}[G(z)]$, where F^{-1} denotes the inverse function of F. Since the inverse of a linear fractional transformation and the composite of linear fractional transformations are linear fractional transformations [see (5) in Sec. 16.2 and Prob. 17 at the end of this section], $w = f(z)$ is a linear fractional transformation. Furthermore, from (2) we see by direct calculation that

$$F(w_1) = 0, \quad F(w_2) = 1, \quad F(w_3) = \infty$$

$$G(z_1) = 0, \quad G(z_2) = 1, \quad G(z_3) = \infty$$

[just set $w = w_1$ on the left, then $w = w_2$, then $w = w_3$, and then do the same on the right side of (2)]. Hence, $w_1 = f(z_1)$, $w_2 = f(z_2)$, $w_3 = f(z_3)$. This proves the existence of a linear fractional transformation $w = f(z)$ that maps z_1, z_2, z_3 onto w_1, w_2, w_3, respectively.

We prove that $w = f(z)$ is uniquely determined. Suppose that $w = g(z)$ is another linear fractional transformation that maps z_1, z_2, z_3 onto w_1, w_2, w_3, respectively. Then its inverse $g^{-1}(w)$ maps w_1 onto z_1, w_2 onto z_2, and w_3 onto z_3. Consequently, the composite mapping $H = g^{-1}[f(z)]$ maps each of the points z_1, z_2, z_3 onto itself; that is, it has three distinct fixed points z_1, z_2, z_3. From Theorem 2 in the preceding section it follows that H is the identity mapping, and, therefore, $g(z) \equiv f(z)$.

The last statement of the theorem follows from the general remark in the preceding section. This completes the proof. ▮

Mapping of Half-Planes onto Disks

This is a task of practical interest, for instance in potential problems. Without loss of generality, let us map the upper half-plane $y \geqq 0$ onto the unit disk $|w| \leqq 1$. The boundary of this half-plane is the x-axis; clearly, it must be mapped onto the unit circle $|w| = 1$. This gives us an idea: to find a mapping, choose three points on the x-axis, prescribe their images on that circle and apply Theorem 1. Make sure that the half-plane $y \geqq 0$ is mapped onto the interior but not onto the exterior of that circle.

EXAMPLE 1 **Mapping of a half-plane onto a disk**

Find the linear fractional transformation (1) that maps $z_1 = -1$, $z_2 = 0$, $z_3 = 1$ onto $w_1 = -1$, $w_2 = -i$, $w_3 = 1$, respectively.

Solution. From (2) we obtain

$$\frac{w-(-1)}{w-1}\cdot\frac{-i-1}{-i-(-1)} = \frac{z-(-1)}{z-1}\cdot\frac{0-1}{0-(-1)},$$

thus

$$w = \frac{z-i}{-iz+1}. \tag{3}$$

Let us show that we can determine the specific properties of such a mapping without difficult calculations. For $z = x$ we have $w = (x-i)/(-ix+1)$, thus $|w| = 1$, so that the x-axis maps onto the unit circle. Since $z = i$ gives $w = 0$, the upper half-plane maps onto the interior of that circle and the lower half-plane onto the exterior. $z = 0, i, \infty$ go onto $w = -i, 0, i$, so that the positive imaginary axis maps onto the segment S: $u = 0$, $-1 \lneqq v \leqq 1$. The vertical lines $x = const$ map onto circles (by Theorem 1, Sec. 16.2) through $w = i$ (the image of $z = \infty$) and perpendicular to $|w| = 1$ (by conformality; see Fig. 357). Similarly, the horizontal lines $y = const$ map onto circles through $w = i$ and perpendicular to S (by conformality). Figure 357 gives these circles for $y \geqq 0$, and for $y < 0$ they lie outside the unit disk shown. ∎

EXAMPLE 2 **Occurrence of ∞**

Determine the linear fractional transformation that maps $z_1 = 0$, $z_2 = 1$, $z_3 = \infty$ onto $w_1 = -1$, $w_2 = -i$, $w_3 = 1$, respectively.

Solution. From (2) we obtain the desired mapping

$$w = \frac{z-i}{z+i}. \tag{4}$$

This is sometimes called the *Cayley transformation*.[2] In this case, (2) gave at first the quotient $(1-\infty)/(z-\infty)$, which we had to replace by 1. ∎

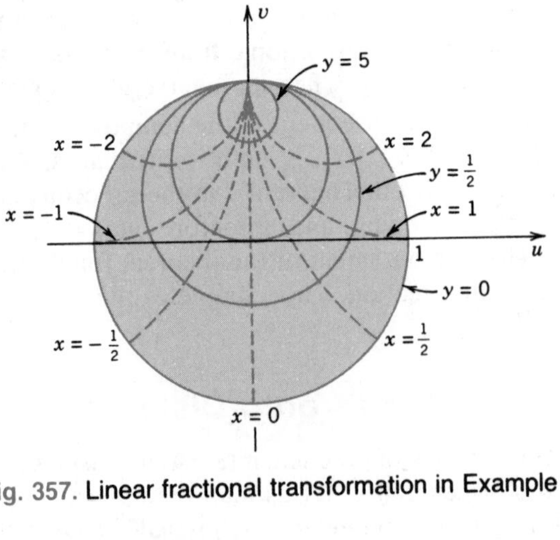

Fig. 357. Linear fractional transformation in Example 1

[2]ARTHUR CAYLEY (1821—1895), English mathematician and professor at Cambridge, is known for his important work in algebra, matrix theory, and differential equations.

Mappings of Disks onto Half-Planes. This is quite similar to the case just discussed.

EXAMPLE 3 **Mapping of the unit disk onto the right half-plane**

Find the linear fractional transformation that maps $z_1 = -1$, $z_2 = i$, $z_3 = 1$ onto $w_1 = 0$, $w_2 = i$, $w_3 = \infty$, respectively. (Make a sketch of the disk and the half-plane.)

Solution. From (2) we obtain, after replacing $(i - \infty)/(w - \infty)$ by 1,

$$w = -\frac{z + 1}{z - 1}.$$

■

Mappings of Half-Planes onto Half-Planes

This is another task of practical interest. We may map the upper half-plane $y \geqq 0$ onto the upper half-plane $v \geqq 0$, as a typical case. Then the x-axis must be mapped onto the u-axis.

EXAMPLE 4 **Mapping of a half-plane onto a half-plane**

Find the linear fractional transformation that maps the points $z_1 = -2$, $z_2 = 0$, $z_3 = 2$ onto the points $w_1 = \infty$, $w_2 = \frac{1}{4}$, $w_3 = \frac{3}{8}$, respectively.

Solution. From (2) we obtain

$$w = \frac{z + 1}{2z + 4}, \tag{5}$$

as the student may verify. What is the image of the x-axis? ■

Mapping of Disks onto Disks

This is a third class of practical problems. We may map the unit disk in the z-plane onto the unit disk in the w-plane. It can be readily verified that the function

$$w = \frac{z - z_0}{cz - 1}, \qquad c = \bar{z}_0, \qquad |z_0| < 1 \tag{6}$$

is of the desired type and maps the point z_0 onto the center $w = 0$ (see Prob. 14).

EXAMPLE 5 **Mapping of the unit disk onto the unit disk**

If we require that $z_0 = \frac{1}{2}$ be mapped onto $w = 0$, then (6) takes the form

$$w(z) = \frac{2z - 1}{z - 2}.$$

The real axes correspond to each other; in particular,

$$w(-1) = 1, \qquad w(0) = \tfrac{1}{2}, \qquad w(1) = -1.$$

Since the mapping is conformal and straight lines are mapped onto circles or straight lines and $w(\infty) = 2$, the images of the lines $x = const$ are circles through $w = 2$ with centers on the u-axis; the lines $y = const$ are mapped onto circles that are orthogonal to the aforementioned circles (see Fig. 358 on the next page). ■

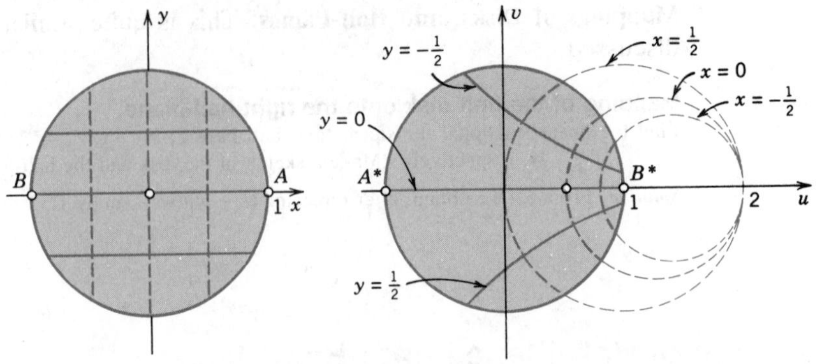

Fig. 358. Mapping in Example 5

Mappings of angular regions onto the unit disk may be obtained by combining linear fractional transformations and transformations of the form $w = z^n$, where n is an integer greater than 1.

EXAMPLE 6 Mapping of an angular region onto the unit disk

Map the angular region D: $-\pi/6 \leqq \arg z \leqq \pi/6$ onto the unit disk $|w| \leqq 1$.

Solution. We may proceed as follows. The mapping

$$Z = z^3$$

maps D onto the right half of the Z-plane. Then we may apply a linear fractional transformation that maps this half-plane onto the unit disk, for example, the transformation

$$w = i\frac{Z - 1}{Z + 1}.$$

By inserting $Z = z^3$ into this mapping we find

$$w = i\frac{z^3 - 1}{z^3 + 1};$$

this mapping has the required properties (see Fig. 359). ■

This is the end of our discussion of linear fractional transformations. In the next section we turn to conformal mappings by other analytic functions (sine, cosine, etc.).

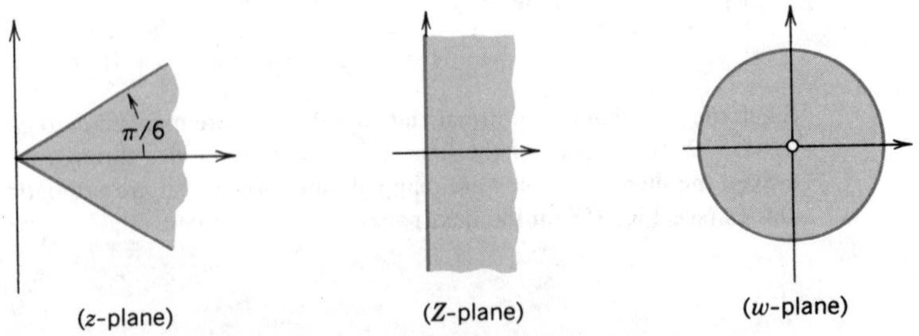

Fig. 359. Mapping in Example 6

Problem Set 16.3

Find the linear fractional transformation that maps

1. $0, 1, 2$ onto $2, 5, 8$, respectively

2. $0, -i, i$ onto $-1, 0, \infty$, respectively

3. $-1, 0, 1$ onto $0, 1, -1$, respectively

4. $\infty, 0, -1$ onto $1, 0, (1 + i)/2$, respectively

5. $0, 1, \infty$ onto $\infty, 1, 0$, respectively

6. $0, 2i, -2i$ onto $-1, 0, \infty$, respectively

7. $2i, i, 0$ onto $5i/2, 2i, \infty$, respectively

8. $i, 0, -i/2$ onto $\infty, i, 0$, respectively

9. $1, i, -1$ onto $i, -1, -i$, respectively

10. $-i, 0, 1$ onto $i, 2i, 1 + i$, respectively

11. Derive (4) from (2).

12. Find the inverse of (3). Show that (3) maps the lines $x = c = const$ onto circles with centers on the line $v = 1$.

13. Derive (5) from (2). Find the inverse of (5) and graph the curves corresponding to $u = const$ and $v = const$.

14. Prove the statement involving (6).

15. Find a linear fractional transformation that maps $|z| \leqq 1$ onto $|w| \leqq 1$ such that $z = i/2$ is mapped onto $w = 0$ and sketch the images of the lines $x = const$ and $y = const$.

16. Find all linear fractional transformations $w(z)$ that map the x-axis onto the u-axis.

17. Show that the composite of two linear fractional transformations is a linear fractional transformation.

18. Find an analytic function that maps the second quadrant of the z-plane onto the interior of the unit circle in the w-plane.

19. Find an analytic function $w = f(z)$ that maps the region $0 \leqq \arg z \leqq \pi/4$ onto the unit disk $|w| \leqq 1$.

20. Find an analytic function $w = f(z)$ that maps the region $2 \leqq y \leqq x + 1$ onto the unit disk $|w| \leqq 1$.

16.4 Mapping by Other Functions

Mappings by z^n, e^z, and $\ln z$ were discussed in Sec. 12.9, which the student may want to study or review before going on with the present section.

Sine Function

We now consider the mapping (Sec. 12.7)

$$w = u + iv = \sin z = \sin x \cosh y + i \cos x \sinh y \tag{1}$$

where

$$u = \sin x \cosh y, \qquad v = \cos x \sinh y. \tag{2}$$

Since $\sin z$ is periodic with period 2π, the mapping (1) is certainly not one-to-one if we consider it in the full z-plane. We restrict z to the vertical infinite strip S defined by $-\frac{1}{2}\pi \leqq x \leqq \frac{1}{2}\pi$ (Fig. 360). Since $f'(z) = \cos z = 0$ at $z = \pm\frac{1}{2}\pi$, the mapping is not conformal at these two critical points. We can explore the properties of the mapping by determining the images of the vertical lines $x = const$ and the horizontal lines $y = const$.

If $x = 0$, then $u = 0$ and $v = \sinh y$ from (2). Hence the y-axis ($x = 0$) is mapped onto the v-axis.

If $x = \pm\frac{1}{2}\pi$, then $u = \pm\cosh y$ and $v = 0$ from (2). Since $\cosh y \geqq 1$, the vertical boundaries $x = \pm\frac{1}{2}\pi$ of the strip S are thus mapped onto the portions $u \leqq -1$ and $u \geqq 1$ of the u-axis, the images being "folded up" as indicated in Fig. 360.

If $x \neq 0, \pm\frac{1}{2}\pi$, using $\cosh^2 x - \sinh^2 x = 1$, we get from (2)

$$\frac{u^2}{\sin^2 x} - \frac{v^2}{\cos^2 x} = 1. \tag{3}$$

For constant $x \neq 0$ these are hyperbolas. Hence these are the images of the vertical lines $x = const$.

If $y = 0$, then $\sinh y = 0$, hence $v = 0$ and $u = \sin x$ from (2). Hence the x-axis ($y = 0$) is mapped onto the segment $-1 \leqq u \leqq 1$ of the u-axis.

If $y \neq 0$, using $\cos^2 x + \sin^2 x = 1$, we get from (2)

$$\frac{u^2}{\cosh^2 y} + \frac{v^2}{\sinh^2 y} = 1. \tag{4}$$

For constant $y \neq 0$ these are confocal ellipses with foci ± 1. Hence these are the images of the horizontal lines $y = const$.

The image curves of $x = const$ and $y = const$ form an orthogonal net (intersect at right angles) because of conformality, except at the critical points $z = \pm\frac{1}{2}\pi$. See Fig. 360.

From our discussion we see that $w = \sin z$ *maps the infinite vertical strip* $-\frac{1}{2}\pi < x < \frac{1}{2}\pi$ *onto the* w*-plane cut along the rays* $u \leqq -1$ *and* $u \geqq 1$, $v = 0$, *and this mapping is one-to-one and conformal.*

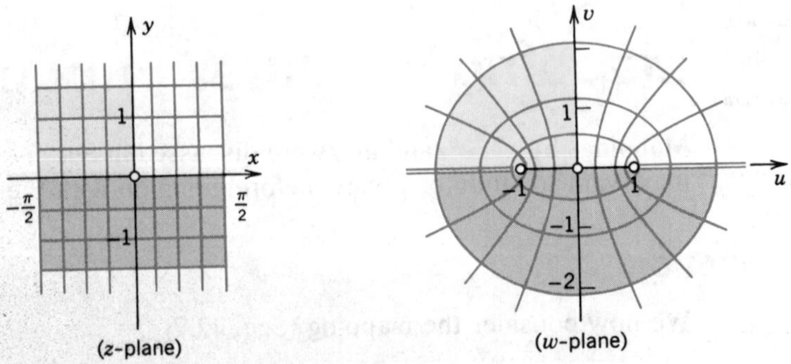

Fig. 360. Mapping $w = u + iv = \sin z$

EXAMPLE 1 **Mapping of a rectangle onto an elliptical disk.**
Find the image of the rectangle in Fig. 361 under $w = \sin z$.

Solution. The boundary of the rectangle corresponds to $x = \pm\pi/2$ and $y = \pm 1$. Hence the upper and lower edges are mapped onto the upper half and the lower half, respectively, of the ellipse (4) whose semiaxes are $\cosh 1 = 1.54$ and $\sinh 1 = 1.18$. The left edge is mapped onto the segment $-\cosh 1 \leqq u \leqq -1$ and the right edge onto the segment $1 \leqq u \leqq \cosh 1$ of the u-axis. Note that any two points $\pm y_0$ of those edges have the same image $-\cosh y_0$ (left edge) or $\cosh y_0$ (right edge) on the u-axis. See Fig. 361. ∎

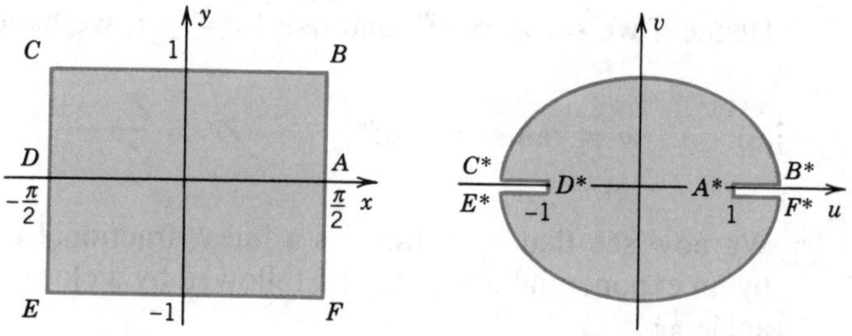

Fig. 361. Mapping by $w = \sin z$ in Example 1

EXAMPLE 2 **Mapping of a rectangle onto an elliptical ring**
Find the image of the rectangle

$$R: \quad -\pi < x < \pi, \qquad 1/2 < y < 1$$

under $w = \sin z$. See Fig. 362.

Solution. The upper edge of the rectangle is $y = 1$ and is mapped onto the ellipse (4) with semiaxes $\cosh 1$ and $\sinh 1$ (Fig. 362), the lower edge onto (4) with semiaxes $\cosh 1/2 = 1.13$ and $\sinh 1/2 = 0.52$. The lateral edges $x = \pm\pi$ are mapped onto the segment of the v-axis given by $-\sinh 1 \leqq v \leqq -\sinh 1/2$, as follows directly from (2) and $\sin(\pm\pi) = 0$, $\cos(\pm\pi) = -1$. Figure 362 shows the result. ∎

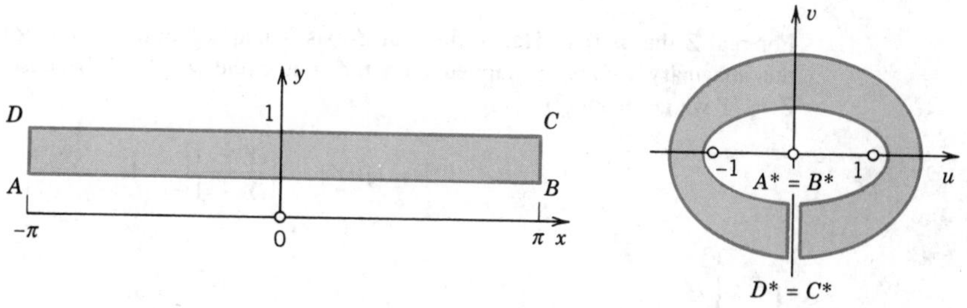

Fig. 362. Mapping by $w = \sin z$ in Example 2

Cosine and Tangent

Cosine. The mapping $w = \cos z$ could be discussed independently, but since

$$(5) \qquad w = \cos z = \sin(z + \tfrac{1}{2}\pi),$$

we see at once that this is the same mapping as $\sin z$ preceded by a translation to the right through $\frac{1}{2}\pi$ units.

Tangent. The idea of expressing mappings as composites of known mappings also helps in the case of $w = \tan z = \sin z \,/\, \cos z$. Expressing sine and cosine by exponential functions (Sec. 12.7) and multiplying both the numerator and the denominator by e^{iz}, we obtain

$$w = \tan z = \frac{\sin z}{\cos z} = \frac{(e^{iz} - e^{-iz})/i}{e^{iz} + e^{-iz}} = \frac{(e^{2iz} - 1)/i}{e^{2iz} + 1}.$$

Hence if we set $Z = e^{2iz}$ and use $1/i = -i$, we have

$$\text{(6)} \qquad w = \tan z = -iZ^*, \qquad Z^* = \frac{Z - 1}{Z + 1}, \qquad Z = e^{2iz}.$$

We now see that $w = \tan z$ is a linear fractional transformation preceded by an exponential mapping and followed by a clockwise rotation through an angle $\frac{1}{2}\pi$.

EXAMPLE 3 **Mapping of an infinite strip onto a circular disk**

Find the image of the infinite vertical strip S: $-\pi/4 < x < \pi/4$ (Fig. 363) under the mapping $w = \tan z$.

Solution. We start with the last mapping in (6). Since $Z = e^{2iz} = e^{-2y+2ix}$, we obtain

$$|Z| = e^{-2y}, \qquad \text{Arg } Z = 2x$$

[see (8), Sec. 12.6]. Hence the vertical lines $x = -\pi/4, 0, \pi/4$ are mapped onto the rays $\text{Arg } Z = -\pi/2, 0, \pi/2$, respectively. Hence S is mapped onto the right Z-half-plane. Also $|Z| = e^{-2y} < 1$ if $y > 0$ and $|Z| > 1$ if $y < 0$. Hence the upper half of S is mapped inside the unit circle $|Z| = 1$ and the lower half of S outside $|Z| = 1$ (Fig. 363).

Now comes the linear fractional transformation in (6), which we denote by $g(Z)$:

$$\text{(7)} \qquad Z^* = g(Z) = \frac{Z - 1}{Z + 1}.$$

For real Z this is real. Hence the real Z-axis is mapped onto the real Z^*-axis. Furthermore, the imaginary Z-axis is mapped onto the unit circle $|Z^*| = 1$ because for pure imaginary $Z = iY$ we get from (7)

$$|Z^*| = |g(iY)| = \left|\frac{iY - 1}{iY + 1}\right| = 1.$$

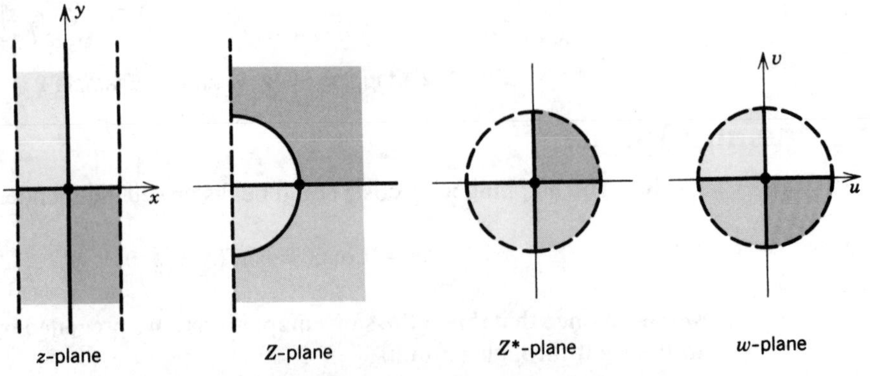

Fig. 363. Mappings in Example 3

The right Z-half-plane is mapped inside this unit circle $|Z^*| = 1$, not outside, because $Z = 1$ has its image $g(1) = 0$ inside that circle. Finally, the unit circle $|Z| = 1$ is mapped onto the imaginary Z^*-axis, because this circle is $Z = e^{i\phi}$, so that (7) gives a pure imaginary expression, namely,

$$g(e^{i\phi}) = \frac{e^{i\phi} - 1}{e^{i\phi} + 1} = \frac{e^{i\phi/2} - e^{-i\phi/2}}{e^{i\phi/2} + e^{-i\phi/2}} = \frac{i \sin (\phi/2)}{\cos (\phi/2)}.$$

From the Z^*-plane we get to the w-plane simply by a clockwise rotation through $\pi/2$; see (6).

Ans. $w = \tan z$ maps S: $-\pi/4 < \text{Re } z < \pi/4$ onto the unit disk $|w| = 1$, with the four quarters of S mapped as indicated in Fig. 363. This mapping is conformal and one-to-one. ∎

Hyperbolic Functions

Hyperbolic functions can be treated independently in a similar fashion or reduced to the trigonometric functions just discussed. In particular, the **hyperbolic sine**

$$w = \sinh z = -i \sin (iz) \tag{8}$$

defines a mapping that is a rotation $Z = iz$ followed by the mapping $Z^* = \sin Z$ and another rotation $w = -iZ^*$.

Similarly, the **hyperbolic cosine**

$$w = \cosh z = \cos (iz) \tag{9}$$

defines a mapping that is a rotation $Z = iz$ followed by the mapping $w = \cos Z$.

EXAMPLE 4 Mapping of a semi-infinite strip onto a half-plane

Find the image of the semi-infinite strip $x \geqq 0$, $0 \leqq y \leqq \pi$ (Fig. 364) under the mapping (9).

Solution. We set $w = u + iv$. Since $\cosh 0 = 1$, the point $z = 0$ is mapped onto $w = 1$. For real $z = x \geqq 0$, $\cosh z$ is real and increases with increasing x in a monotone fashion, starting from 1. Hence the positive x-axis is mapped onto the portion $u \geqq 1$ of the u-axis.

For pure imaginary $z = iy$ we have $\cosh iy = \cos y$. Hence the left boundary of the strip is mapped onto the segment $1 \geqq u \geqq -1$ of the u-axis, the point $z = \pi i$ corresponding to

$$w = \cosh i\pi = \cos \pi = -1.$$

On the upper boundary of the strip, $y = \pi$, and since $\sin \pi = 0$, $\cos \pi = -1$, it follows that this part of the boundary is mapped onto the portion $u \leqq -1$ of the u-axis. Hence the boundary of the strip is mapped onto the u-axis. It is not difficult to see that the interior of the strip is mapped onto the upper half of the w-plane, and the mapping is one-to-one. ∎

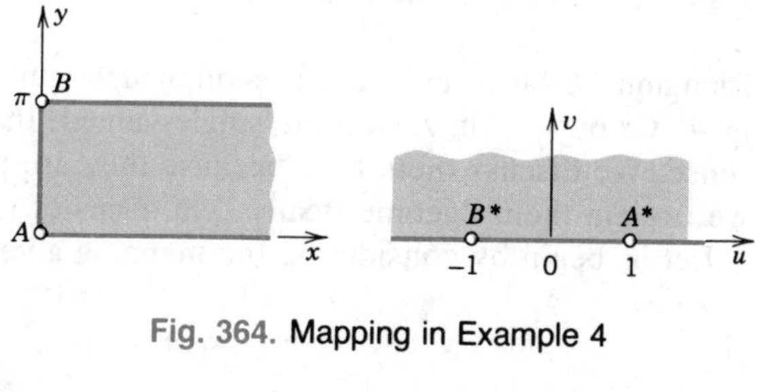

Fig. 364. Mapping in Example 4

Problem Set 16.4

Find and sketch the images of the following regions under the mapping $w = e^z$.

1. $0 \leqq x \leqq 1, \quad 0 \leqq y \leqq \pi/2$
2. $-1 < x < 1, \quad \pi/4 < y < 3\pi/4$
3. $1 < x < 2, \quad -\pi/2 < y < \pi/2$
4. $-2 < x < -1, \quad 0 < y < \pi$

Find and graph the images of the following regions under the mapping $w = \sin z$.

5. $0 < x < \pi/2, \quad 0 < y < 2$
6. $0 < x < \pi/6, \quad y$ arbitrary
7. $0 < x < 2\pi, \quad 1 < y < 2$
8. $-\pi/4 < x < \pi/4, \quad 0 < y < 3$

9. Find an analytic function that maps the region bounded by the positive x and y axes and the hyperbola $xy = \pi/2$ in the first quadrant onto the upper half-plane. *Hint.* First map that region onto a horizontal strip.
10. Find and plot the images of the lines $x = 0, \pm\pi/6, \pm\pi/3, \pm\pi/2$ under the mapping $w = \sin z$.
11. Determine all points at which the mapping $w = \sin z$ is not conformal.
12. Describe the transformation $w = \cosh z$ in terms of the transformation $w = \sin z$ and rotations and translations.
13. Find all points at which the mapping $w = \cosh z$ is not conformal.
14. Find the images of the lines $y = const$ under the mapping $w = \cos z$.

Find and graph the images of the following regions under the mapping $w = \cos z$.

15. $0 < x < \pi, \quad y < 0$
16. $0 < x < \pi/2, \quad 0 < y < 1$
17. $\pi < x < 2\pi, \quad y < 0$
18. $0 < x < 2\pi, \quad 1/2 < y < 1$

19. Find the image of $2 \leqq |z| \leqq 3, \ \pi/4 \leqq \theta \leqq \pi/2$ under the mapping $w = \text{Ln } z$.
20. Show that $w = \text{Ln } \dfrac{z-1}{z+1}$ maps the upper half-plane onto the horizontal strip $0 \leqq \text{Im } w \leqq \pi$ as shown in Fig. 365.

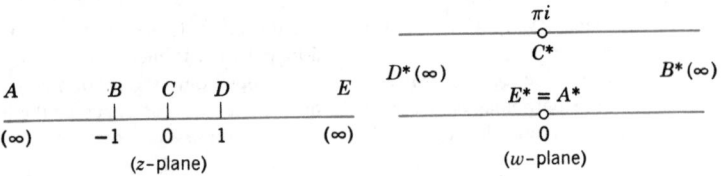

Fig. 365. Problem 20

16.5 Riemann Surfaces

Riemann surfaces are surfaces on which multivalued relations, such as $w = \sqrt{z}$ or $w = \ln z$, become single-valued, that is, functions in the usual sense. We discuss them here because they are probably best understood if we explain them "geometrically," in terms of mappings.

Let us begin by considering the mapping given by

$$w = u + iv = z^2 \qquad \text{(Sec. 12.9)}, \tag{1}$$

which is conformal, except at the critical point $z = 0$ where $w' = 2z = 0$.

At $z = 0$ the angles are doubled under the mapping. The right half of the z-plane (including the positive y-axis) is mapped onto the full w-plane cut along the negative half of the u-axis; the mapping is one-to-one. Similarly, the left half of the z-plane (including the negative y-axis) is mapped onto the cut w-plane in a one-to-one manner.

Obviously, the mapping of the full z-plane is not one-to-one, because every point $w \neq 0$ corresponds to precisely two points z. In fact, if z_1 is one of these points, then the other is $-z_1$. For example, $z = i$ and $z = -i$ have the same image, namely, $w = -1$, etc. Hence the w-plane is "covered twice" by the image of the z-plane. We say that the full z-plane is mapped onto the *doubly covered* w-plane. We can still give our imagination the necessary support as follows.

We imagine one of the two previously obtained copies of the cut w-plane to be placed on the other so that the upper sheet is the image of the right half of the z-plane and the lower sheet is the image of the left half of the z-plane; we denote these half-planes by R and L, respectively. When we pass from R to L, the corresponding image point should pass from the upper to the lower sheet. For this reason we join the two sheets crosswise along the cut, that is, along the negative real axis. (This construction can be carried out only in imagination, since the penetration of the two sheets of a material model can be realized only imperfectly.) The two origins are fastened together. The configuration thus obtained is called a **Riemann surface.** On it every point $w \neq 0$ appears twice, at superposed positions, and the origin appears precisely once. The function $w = z^2$ now maps the full z-plane onto this Riemann surface in a one-to-one manner, and the mapping is conformal, except for the "winding point" or **branch point** at $w = 0$ (Fig. 366). Such a branch point connecting two sheets is said to be of the *first order*. (More generally, a branch point connecting n sheets is said to be of **order** $n - 1$.)

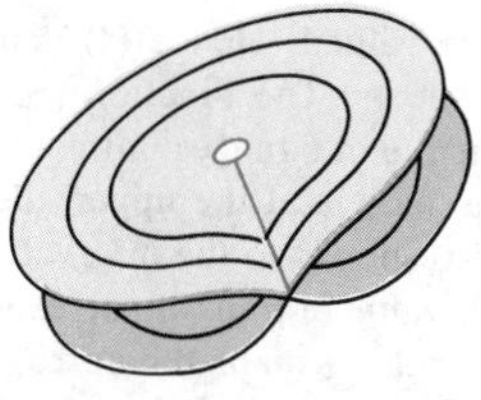

Fig. 366. Riemann surface of $\sqrt{z}$

We now consider the double-valued relation

$$w = \sqrt{z} \qquad (2) \quad \text{(Sec. 12.2).}$$

To each $z \neq 0$ there correspond two values w, one of which is the principal value. If we replace the z-plane by the two-sheeted Riemann surface just considered, then each complex number $z \neq 0$ is represented by two points of the surface at superposed positions. We let one of these points correspond to the principal value—say, the point in the upper sheet—and the other to the other value. Then (2) becomes single-valued, that is, (2) is a function of the points of the Riemann surface, and to any continuous motion of z on the surface there corresponds a continuous motion of the corresponding point in the w-plane. The function maps the sheet corresponding to the principal value onto the right half of the w-plane and the other sheet onto the left half of the w-plane.

Let us consider some further important examples.

EXAMPLE 1 Riemann surface of $\sqrt[n]{z}$

In the case of the relation

$$w = \sqrt[n]{z} \qquad (n = 3, 4, \cdots), \tag{3}$$

we need a Riemann surface consisting of n sheets and having a branch point of order $n - 1$ at $z = 0$. One of the sheets corresponds to the principal value and the other $n - 1$ sheets to the other $n - 1$ values. Figure 367 shows the Riemann surface of $w = \sqrt[3]{z}$. ∎

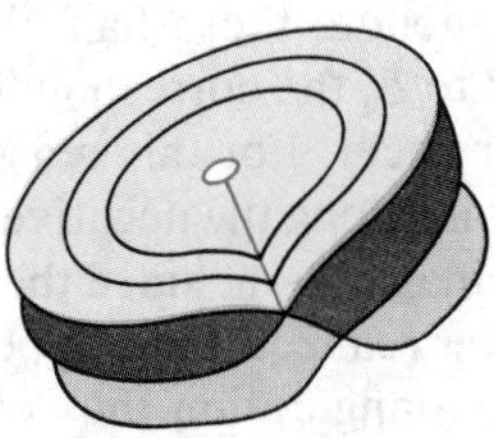

Fig. 367. Riemann surface of $\sqrt[3]{z}$

EXAMPLE 2 Riemann surface of the natural logarithm

For every $z \neq 0$ the relation

$$w = \ln z = \text{Ln } z + 2n\pi i \qquad (n = 0, \pm 1, \pm 2, \cdots, z \neq 0) \tag{4}$$

is infinitely many-valued. Hence (4) defines a function on a Riemann surface consisting of infinitely many sheets. The function $w = \text{Ln } z$ corresponds to one of these sheets. On this sheet the argument θ of z ranges in the interval $-\pi < \theta \leqq \pi$ (Sec. 12.8). The sheet is cut along the negative real axis, and the upper edge of the slit is joined to the lower edge of the next sheet, which corresponds to the interval $\pi < \theta \leqq 3\pi$, that is, to the function $w = \text{Ln } z + 2\pi i$. In this way each value of n in (4) corresponds to precisely one of these infinitely many sheets. The function $w = \text{Ln } z$ maps the corresponding sheet onto the horizontal strip $-\pi < v \leqq \pi$ in the w-plane. The next sheet is mapped onto the neighboring strip $\pi < v \leqq 3\pi$, etc. The function $w = \ln z$ thus maps all the sheets of the corresponding Riemann surface onto the entire w-plane, the correspondence between the points $z \neq 0$ of the Riemann surface and those of the w-plane being one-to-one. ∎

EXAMPLE 3 Mapping $w = z + z^{-1}$. Airfoils

Let us consider the mapping defined by

$$w = z + \frac{1}{z} \qquad (z \neq 0) \tag{5}$$

which is important in aerodynamics (see below). Since this function has the derivative

$$w' = 1 - \frac{1}{z^2} = \frac{(z+1)(z-1)}{z^2},$$

the mapping is conformal except at the points $z = 1$ and $z = -1$; these points correspond to $w = 2$ and $w = -2$, respectively. From (5) we find

$$z = \frac{w}{2} \pm \sqrt{\frac{w^2}{4} - 1} = \frac{w}{2} \pm \frac{1}{2}\sqrt{(w + 2)(w - 2)}. \tag{6}$$

Hence, the points $w = 2$ and $w = -2$ are branch points of the first order of $z = z(w)$. To any value w ($\neq 2, \neq -2$) there correspond two values of z. Consequently, (5) maps the z-plane onto a two-sheeted Riemann surface, the two sheets being connected crosswise from $w = -2$ to $w = 2$ (Fig. 368), and this mapping is one-to-one. We set $z = re^{i\theta}$ and determine the images of the curves $r = const$ and $\theta = const$. From (5) we obtain

$$w = u + iv = re^{i\theta} + \frac{1}{r}e^{-i\theta} = \left(r + \frac{1}{r}\right)\cos\theta + i\left(r - \frac{1}{r}\right)\sin\theta.$$

By equating the real and imaginary parts on both sides we have

$$u = \left(r + \frac{1}{r}\right)\cos\theta, \qquad v = \left(r - \frac{1}{r}\right)\sin\theta. \tag{7}$$

From this we find

$$\frac{u^2}{a^2} + \frac{v^2}{b^2} = 1, \qquad \text{where} \qquad a = r + \frac{1}{r}, \qquad b = \left|r - \frac{1}{r}\right|.$$

The circles $r = const$ are thus mapped onto ellipses whose principal axes lie in the u and v axes and have the lengths $2a$ and $2b$, respectively. Since $a^2 - b^2 = 4$, independent of r, these ellipses are confocal, with foci at $w = -2$ and $w = 2$. The unit circle $r = 1$ is mapped onto the line segment from $w = -2$ to $w = 2$. For every $r \neq 1$ the two circles with radii r and $1/r$ are mapped onto the same ellipse in the w-plane, corresponding to the two sheets of the Riemann surface. Hence, the interior of the unit circle $|z| = 1$ corresponds to one sheet, and the exterior to the other.

Furthermore, from (7) we obtain

$$\frac{u^2}{\cos^2\theta} - \frac{v^2}{\sin^2\theta} = 4. \tag{8}$$

The lines $\theta = const$ are thus mapped onto the hyperbolas that are the orthogonal trajectories of those ellipses. The real axis (that is, the rays $\theta = 0$ and $\theta = \pi$) is mapped onto the part of the real axis from $w = 2$ via ∞ to $w = -2$. The y-axis is mapped onto the v-axis. Any other pair of rays $\theta = \theta_0$ and $\theta = \theta_0 + \pi$ is mapped onto the two branches of the same hyperbola.

An exterior region of one of the above ellipses is free of branch points and corresponds to either the interior or the exterior of the corresponding circle in the z-plane, depending on the sheet of the Riemann surface to which the region belongs. In particular, the full w-plane corresponds to the interior or the exterior of the unit circle $|z| = 1$, as we mentioned before.

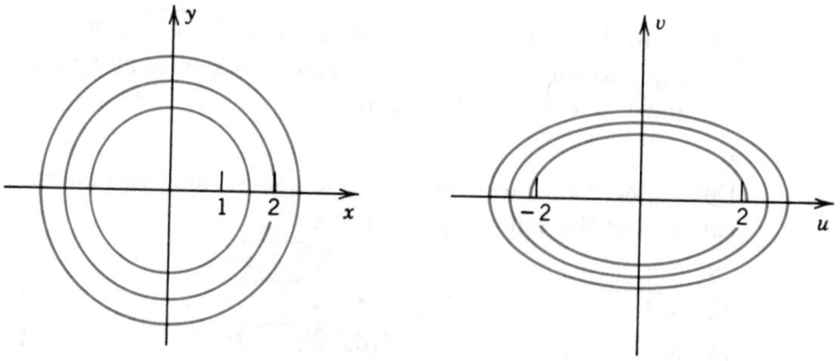

Fig. 368. Example 3

The mapping (5) transforms suitable circles into airfoils with a sharp trailing edge whose interior angle is zero; these airfoils are known as **Joukowski airfoils.**[3] Since the airfoil to be obtained has a sharp edge, it is clear that the circle to be mapped must pass through one of the points $z = \pm 1$ at which the mapping is not conformal.

In Fig. 369 the larger circle through -1 is mapped onto the contour of the airfoil and the dashed circle (which is *not* the unit circle!) onto the dashed arc inside the profile. Reference [D8] in Appendix 1 discusses still more details.

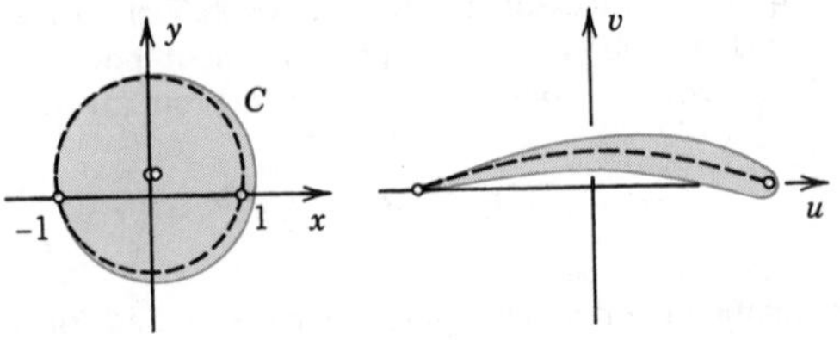

Fig. 369. Joukowski airfoil

This is the end of Chap. 16, but some further applications of conformal mapping will occur in Chap. 17.

Problem Set 16.5

1. Consider $w = \sqrt{z}$. Find the path of the image point w of a point z that moves twice around the unit circle, starting from the initial position $z = 1$.
2. Show that the Riemann surface of $w = \sqrt[3]{z}$ consists of three sheets and has a branch point of second order at $z = 0$. Find the path of the image point w of a point z that moves three times around the unit circle starting from the initial position $z = 1$.
3. Make a sketch, similar to Fig. 366, of the Riemann surface of $\sqrt[4]{z}$.
4. Consider the Riemann surfaces of $w = \sqrt[4]{z}$ and $w = \sqrt[5]{z}$ in a fashion similar to that in Prob. 2.
5. Determine the path of the image of a point z under the mapping $w = \ln z$ as z moves several times counterclockwise around the unit circle.
6. Show that the Riemann surface of $w = \sqrt{(z-1)(z-2)}$ has branch points at $z = 1$ and $z = 2$ and consists of two sheets that may be cut along the line segment from 1 to 2 and joined crosswise. *Hint*. Introduce polar coordinates $z - 1 = r_1 e^{i\theta_1}$, $z - 2 = r_2 e^{i\theta_2}$.
7. Find the branch points and the number of sheets of the Riemann surface of $w = \sqrt{z^2 - 1}$.
8. Show that the Riemann surface of $w = \sqrt{(1 - z^2)(4 - z^2)}$ has four branch points and two sheets that may be joined crosswise along the segments $-2 \leqq x \leqq -1$ and $1 \leqq x \leqq 2$ of the x-axis.

Determine the location of the branch points and the number of sheets of the Riemann surfaces of the following functions.

9. $i + \sqrt{z}$
10. $\sqrt{i + z}$
11. $\sqrt[3]{z - a}$
12. $\sqrt{3z + 4}$
13. $\sqrt{z^3 - z}$
14. $3 + 2z - \sqrt{2z}$
15. $\ln(z + 1)$
16. $\sqrt[4]{z - 16}$
17. $3 + \sqrt[3]{2z + i}$
18. $\ln(4z - 3i)$
19. $e^{\sqrt{z}}$
20. $\sqrt{e^z}$

[3]NIKOLAI JEGOROVICH JOUKOWSKI (1847—1921), Russian mathematician.

Review Questions and Problems for Chapter 16

1. How did we define the angle of intersection of two oriented curves, and what does it mean to say that a mapping is conformal?
2. At what points is a mapping $w = f(z)$ by an analytic function not conformal? Give examples.
3. What happens to angles at z_0 under a mapping $w = f(z)$ if $f'(z_0) = 0$, $f''(z_0) = 0$, $f'''(z_0) \neq 0$?
4. What is a linear fractional transformation? Why are these mappings of great practical interest?
5. Why did we require that the coefficients of a linear fractional transformation satisfy $ad - bc \neq 0$?
6. What is the extended complex plane, and why did it occur in this chapter?
7. What is a fixed point of a mapping? Give examples of mappings by analytic functions that have no fixed points, precisely one fixed point, infinitely many fixed points.
8. What is a Riemann surface, and what purpose does it serve?
9. How many sheets does the Riemann surface of $w = \sqrt[4]{z}$ have? Where is its branch point?
10. Why is the mapping $w = z + 1/z$ of practical interest? How many sheets does the Riemann surface of the inverse of this mapping have?

Find and sketch the images of the following curves or regions under the mapping $w = u + iv = z^2$.

11. $y = -1, 1$
12. $x = -1, 1, 2$
13. $|z| = 2.5, |\arg z| < \pi/8$
14. $xy = -4$
15. $y = x/2$
16. $\text{Im } z > 0$
17. $\pi/8 < \arg z \leqq \pi/4$
18. $0 < y < 2$
19. $\frac{1}{2} < x < 1$

Find and sketch the following curves or regions under the mapping $w = 1/z$.

20. $|z| < 1/2, y < 0$
21. $|\arg z| < \pi/8$
22. $x < 0, y > 0, |z| < 1$
23. $|z - 1/2| = 1/2$
24. $y = 1$
25. $x = -1$

Find all points at which the following mappings $w = f(z)$ fail to be conformal, where $f(z)$ equals

26. $\cos \pi z^2$
27. $3z^5 + 5z^3$
28. $\cosh 2z$
29. $\exp(z^3 + z)$
30. $z + 1/z \quad (z \neq 0)$
31. $\sin z + \cos z$

Find the linear fractional transformation that maps

32. $-1, 1, i$ onto $-1, 1, -i$, respectively
33. $0, 1, 2$ onto $0, i, 2i$, respectively
34. $0, i, \infty$ onto $1, 1 + i, \infty$, respectively
35. $0, \infty, -2$ onto $0, 1, \infty$, respectively
36. $0, \infty, 1$ onto $-2i, 0, 1 - i$, respectively
37. $i, 1, \infty$ onto $5i, 5, \infty$, respectively
38. $-1, 1, 2$ onto $0, 2, 3/2$, respectively
39. $0, 1, -1$ onto $0, \infty, 1$, respectively
40. $0, i, 2i$ onto $0, \infty, 2i$, respectively

Find an analytic function $w = u + iv = f(z)$ that maps

41. The region $0 < \arg z < \pi/3$ onto the region $u < 1$

42. The half-plane $x \geqq 0$ onto the region $u \geqq 2$ so that 0 has the image $2 + i$

43. The right half-plane onto the upper half-plane

44. The interior of the unit circle $|z| = 1$ onto the exterior of the circle $|w + 1| = 5$

45. The infinite strip $0 < y < \pi/3$ onto the upper half-plane $v > 0$

46. The semidisk $|z| < 1$, $x > 0$ onto the exterior of the unit circle $|w| = 1$

47. The region $x > 0$, $y > 0$, $xy < k$ onto the strip $0 < v < 1$

Find all fixed points of the mappings $w = f(z)$, where $f(z)$ equals

48. $z^4 + z - 81$ **49.** $(3z + 2)/(z - 1)$ **50.** $(2iz - 1)/(z + 2i)$

Summary of Chapter 16 Conformal Mapping

A complex function $w = f(z)$ gives a **mapping** of its domain of definition in the complex z-plane onto its range of values in the complex w-plane. If $f(z)$ is *analytic,* this mapping is **conformal,** that is, angle-preserving: the images of any two intersecting curves make the same angle of intersection, in both magnitude and sense, as the curves themselves (Sec. 16.1).

For mapping properties of e^z, $\cos z$, $\sin z$, etc., see Secs. 12.9 and 16.4.

Linear fractional transformations, also called *Möbius transformations* (Secs. 16.2, 16.3)

$$w = \frac{az + b}{cz + d} \qquad (ad - bc \neq 0) \tag{1}$$

map the extended complex plane (Sec. 16.2) conformally onto itself. They solve the problems of mapping half-planes onto half-planes or disks, and disks onto disks or half-planes. Prescribing the images of three points determines (1) uniquely (Sec. 16.3).

Riemann surfaces (Sec. 16.5) consist of several sheets connected at certain points called *branch points*. On them, multivalued relations become single-valued, that is, functions in the usual sense. *Examples.* For $w = \sqrt{z}$ we need two sheets since this relation is doubly valued. For $w = \ln z$ we need infinitely many sheets since this relation is infinitely many-valued (see Sec. 12.8).

Chapter 17

Complex Analysis Applied to Potential Theory

Laplace's equation $\nabla^2\Phi = 0$ is one of the most important partial differential equations in engineering mathematics, because it occurs in connection with gravitational fields (Sec. 8.9), electrostatic fields (Sec. 11.11), steady-state heat conduction (Secs. 9.8, 11.5), incompressible fluid flow, etc. The theory of the solutions of this equation is called **potential theory,** and solutions whose second partial derivatives are continuous are called **harmonic functions.**

In the "two-dimensional case" when Φ depends only on two Cartesian coordinates x and y, Laplace's equation becomes

$$\nabla^2\Phi = \Phi_{xx} + \Phi_{yy} = 0.$$

We know that then its solutions are closely related to complex analytic functions (see Sec. 12.5).[1] In this chapter we discuss this connection and its consequences in more detail and explain it in terms of practical problems taken from electrostatics (Secs. 17.1, 17.2), heat conduction (Sec. 17.3), and hydrodynamics (Sec. 17.4). In most cases this requires the solution of a **boundary value problem,** called the **Dirichlet problem** (or *first boundary value problem*) for the Laplace equation, that is, to find the solution of the Laplace equation in a given domain D assuming given values on the boundary of D. Such a problem can often be solved by the method of conformal mapping (Sec. 17.2), that is, by mapping D conformally onto a simpler domain (disk, half-plane, etc.) for which the solution is known or can be obtained from a formula, such as Poisson's integral formula (Sec. 17.5). For *mixed boundary value problems* the situation is similar.

In the last section (Sec. 17.6) we show that various general properties of harmonic functions follow from results on analytic fuctions.

Prerequisites for this chapter: Chaps. 12, 13, 16.
References: Appendix 1, Part D.
Answers to problems: Appendix 2.

[1]No such close relation exists in the three-dimensional case.

On notation. We write Φ and later $\Phi + i\Psi$ since u and $u + iv$ will be needed in conformal mapping from Sec. 17.2 on.

17.1 Electrostatic Fields

The electrical force of attraction or repulsion between charged particles is governed by Coulomb's law. This force is the gradient of a function Φ, called the **electrostatic potential.** At any points free of charges, Φ is a solution of Laplace's equation

$$\nabla^2\Phi = 0.$$

The surfaces $\Phi = const$ are called **equipotential surfaces.** At each point P the gradient of Φ is perpendicular to the surface $\Phi = const$ through P; that is, the electrical force has the direction perpendicular to the equipotential surface. (See also Secs. 8.9 and 11.11.)

The problems we shall discuss in this whole chapter are **two-dimensional** (for the reason just given in the chapter opening), that is, they concern physical systems that lie in three-dimensional space (of course!), but are such that the potential Φ is independent of one of the space coordinates, so that Φ depends only on two coordinates, which we call x and y. Then **Laplace's equation** becomes

$$\nabla^2\Phi = \frac{\partial^2\Phi}{\partial x^2} + \frac{\partial^2\Phi}{\partial y^2} = 0. \tag{1}$$

Equipotential surfaces now appear as **equipotential lines** (curves) in the xy-plane.

EXAMPLE 1 Potential between parallel plates

Find the potential Φ of the field between two parallel conducting plates extending to infinity (Fig. 370), which are kept at potentials Φ_1 and Φ_2, respectively.

Solution. From the shape of the plates it follows that Φ depends only on x, and Laplace's equation becomes $\Phi'' = 0$. By integrating twice we obtain $\Phi = ax + b$, where the constants a and b are determined by the given boundary values of Φ on the plates. For example, if the plates correspond to $x = -1$ and $x = 1$, the solution is

$$\Phi(x) = \tfrac{1}{2}(\Phi_2 - \Phi_1)x + \tfrac{1}{2}(\Phi_2 + \Phi_1).$$

The equipotential surfaces are parallel planes. ∎

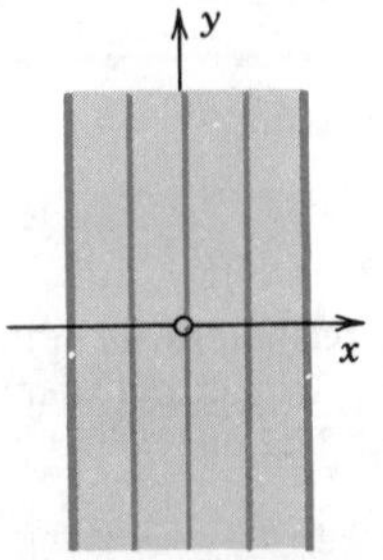

Fig. 370. Potential in Example 1

EXAMPLE 2 **Potential between coaxial cylinders**

Find the potential Φ between two coaxial conducting cylinders extending to infinity on both ends (Fig. 371) and kept at potentials Φ_1 and Φ_2, respectively.

Solution. Here Φ depends only on $r = \sqrt{x^2 + y^2}$, for reasons of symmetry, and Laplace's equation becomes

$$r\Phi'' + \Phi' = 0 \qquad \text{(see (4) in Sec. 11.9).}$$

By separating variables and integrating we obtain

$$\frac{\Phi''}{\Phi'} = -\frac{1}{r}, \qquad \ln \Phi' = -\ln r + \tilde{a}, \qquad \Phi' = \frac{a}{r}, \qquad \Phi = a \ln r + b$$

and a and b are determined by the given values of Φ on the cylinders. Although no infinitely extended conductors exist, the field in our idealized conductor will approximate the field in a long finite conductor in that part which is far away from the ends of the two cylinders. ∎

EXAMPLE 3 **Potential in an angular region**

Find the potential Φ between the conducting plates in Fig. 372, which are kept at potentials Φ_1 (the lower plate) and Φ_2, and make an angle α, where $0 < \alpha \leqq \pi$. (In the figure we have $\alpha = 120° = 2\pi/3$.)

Solution. $\theta = \text{Arg } z$ ($z = x + iy \neq 0$) is constant on rays $\theta = const.$ It is harmonic since it is the imaginary part of an analytic function, Ln z (Sec. 12.8). Hence the solution is

$$\Phi(x, y) = a + b \text{ Arg } z$$

with a and b determined from the two boundary conditions (given values on the plates)

$$a + b(-\tfrac{1}{2}\alpha) = \Phi_1, \qquad a + b(\tfrac{1}{2}\alpha) = \Phi_2.$$

Thus $a = (\Phi_2 + \Phi_1)/2$, $b = (\Phi_2 - \Phi_1)/\alpha$. The answer is

$$\Phi(x, y) = \frac{1}{2}(\Phi_2 + \Phi_1) + \frac{1}{\alpha}(\Phi_2 - \Phi_1)\theta, \qquad \theta = \text{arc tan } \frac{y}{x}. \quad ∎$$

Fig. 371. Potential in Example 2

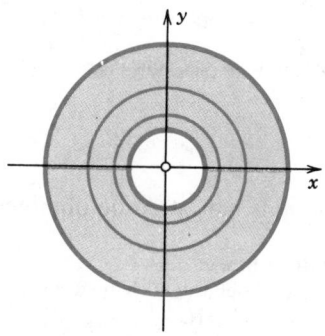

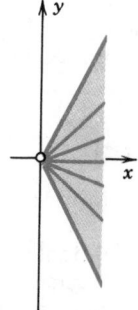

Fig. 372. Potential in Example 3

Complex Potential

Let $\Phi(x, y)$ be harmonic in some domain D and $\Psi(x, y)$ a conjugate harmonic of Φ in D (Sec. 12.5). Then[2]

$$F(z) = \Phi(x, y) + i\Psi(x, y)$$

[2]We write $F = \Phi + i\Psi$, reserving $f = u + iv$ for conformal mapping, as needed from the next section on.

is an analytic function of $z = x + iy$. This function F is called the **complex potential** corresponding to the real potential Φ. Recall from Sec. 12.5 that for given Φ, a conjugate Ψ is uniquely determined except for an additive real constant. Hence we may say ***the*** complex potential, without causing misunderstandings.

The use of F has two advantages, a technical one and a physical one. Technically, F is easier to handle than real or imaginary parts, in connection with methods of complex analysis. Physically, Ψ has a meaning. By conformality, the curves $\Psi = const$ intersect the equipotential lines $\Phi = const$ at right angles [except where $F'(z) = 0$]. Hence they have the direction of the electrical force and, therefore, are called **lines of force.** They are the paths of moving charged particles (electrons in an electron microscope, etc.).

EXAMPLE 4 Complex potential

In Example 1, a conjugate is $\Psi = ay$. It follows that the complex potential is

$$F(z) = az + b = ax + b + iay,$$

and the lines of force are straight lines $y = const$ parallel to the x-axis. ∎

EXAMPLE 5 Complex potential

In Example 2 we have $\Phi = a \ln r + b = a \ln |z| + b$. A conjugate is $\Psi = a \operatorname{Arg} z$. Hence the complex potential is

$$F(z) = a \operatorname{Ln} z + b$$

and the lines of force are straight lines through the origin. $F(z)$ may also be interpreted as the complex potential of a source line whose trace in the xy-plane is the origin. ∎

EXAMPLE 6 Complex potential

In Example 3 we get $F(z)$ by noting that $i \operatorname{Ln} z = i \ln |z| - \operatorname{Arg} z$, multiplying this by $-b$, and adding a:

$$F(z) = a - ib \operatorname{Ln} z = a + b \operatorname{Arg} z - ib \ln |z|.$$

We see from this that the lines of force are concentric circles $|z| = const$. ∎

Superposition

More complicated potentials can often be obtained by superposition.

EXAMPLE 7 Potential of a pair of source lines

Determine the potential of a pair of oppositely charged source lines of the same strength at the points $z = c$ and $z = -c$ on the real axis.

Solution. From Examples 2 and 5 it follows that the potential of each of the source lines is

$$\Phi_1 = K \ln |z - c| \qquad \text{and} \qquad \Phi_2 = -K \ln |z + c|,$$

respectively. Here the real constant K measures the strength (amount of charge). These are the real parts of the complex potentials

$$F_1(z) = K \operatorname{Ln} (z - c) \qquad \text{and} \qquad F_2(z) = -K \operatorname{Ln} (z + c).$$

Hence the complex potential of the combination of the two source lines is

$$F(z) = F_1(z) + F_2(z) = K[\operatorname{Ln} (z - c) - \operatorname{Ln} (z + c)]. \tag{2}$$

The equipotential lines are the curves

$$\Phi = \operatorname{Re} F(z) = K \ln \left|\frac{z-c}{z+c}\right| = const, \qquad \text{thus} \qquad \left|\frac{z-c}{z+c}\right| = const.$$

These are circles, as the student may show by direct calculation. The lines of force are

$$\Psi = \operatorname{Im} F(z) = K[\operatorname{Arg}(z-c) - \operatorname{Arg}(z+c)] = const.$$

We write this briefly (Fig. 373)

$$\Psi = K(\theta_1 - \theta_2) = const.$$

Now $\theta_1 - \theta_2$ is the angle between the line segments from z to c and $-c$ (Fig. 373). Hence the lines of force are the curves along each of which the line segment S: $-c \leqq x \leqq c$ appears under a constant angle. These curves are the totality of circular arcs over S, as is well known from elementary geometry. Hence the lines of force are circles. Figure 374 shows some of them together with some equipotential lines.

In addition to the interpretation as the potential of two source lines, this potential could also be thought of as the potential between two circular cylinders whose axes are parallel but do not coincide, or as the potential between two equal cylinders that lie outside each other, or as the potential between a cylinder and a plane wall. Explain this, using Fig. 374. ▮

The idea of the complex potential as just explained is a key to a close relation of potential theory to complex analysis and will recur in heat flow and fluid flow.

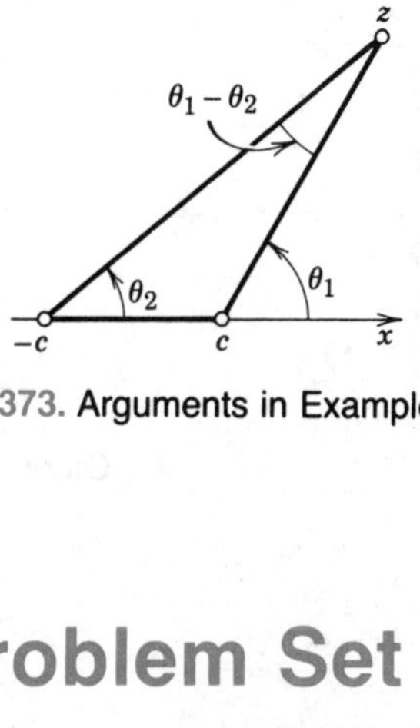

Fig. 373. Arguments in Example 7

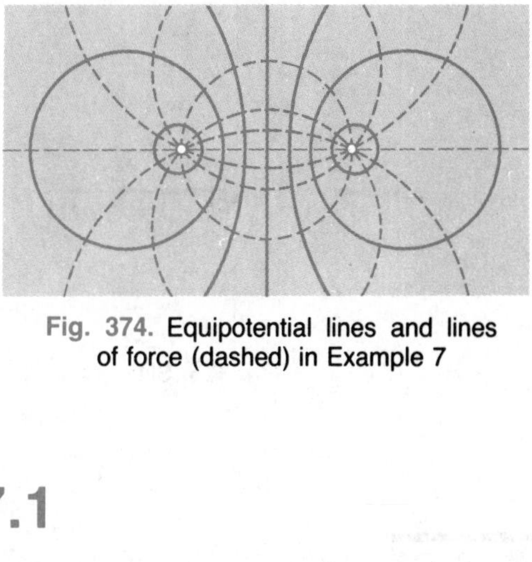
Fig. 374. Equipotential lines and lines of force (dashed) in Example 7

Problem Set 17.1

1. Find and sketch the equipotential surfaces between two parallel plates at $x = -3$ and $x = 3$ having potential 140 and 260 volts, respectively.
2. Find the complex potential in Prob. 1.
3. Find and sketch the equipotential surfaces between two parallel plates at $y = 0$ and $y = d$ having potentials $U_1 = 20$ kV ($= 20\,000$ volts) and $U_2 = 0$, respectively.
4. Find the potential Φ and the complex potential F between the plates $y = x$ and $y = x + k$ having potentials 0 and 100 volts, respectively.
5. Find the potential in the first quadrant of the xy-plane between the axes (having potential 110 volts) and the hyperbola $xy = 1$ (having potential 60 volts).

Find the potential Φ between two infinite coaxial cylinders of radii r_1 and r_2 $(> r_1)$ having potentials U_1 and U_2, respectively, where

6. $r_1 = 1, r_2 = 10, U_1 = 100$ volts, $U_2 = 1000$ volts

7. $r_1 = 0.5, r_2 = 2, U_1 = -110$ volts, $U_2 = 110$ volts

8. $r_1 = 2, r_2 = 8, U_1 = 1$ kV, $U_2 = 2$ kV

9. $r_1 = 1, r_2 = 4, U_1 = 200$ volts, $U_2 = 0$

10. $r_1 = 0.1, r_2 = 10, U_1 = 50$ volts, $U_2 = 150$ volts

Find the equipotential lines of the complex potential $F(z)$ and show these lines graphically, where

11. $F(z) = (1 + 2i)z$ **12.** $F(z) = -iz^2$ **13.** $F(z) = z^2$

14. $F(z) = (1 + i)/z$ **15.** $F(z) = 1/z$ **16.** $F(z) = i \operatorname{Ln} z$

17. Find the potential of two source lines at $z = a$ and $z = -a$ having the same charge.

18. Near each of the two source lines in Fig. 374 the circles $\Phi = const$ are almost concentric. Is this physically understandable?

19. Verify that the equipotential lines in Example 7 are circles.

20. Show that $F(z) = \cos^{-1} z$ may be interpreted as the complex potential of the configurations in Figs. 375 and 376.

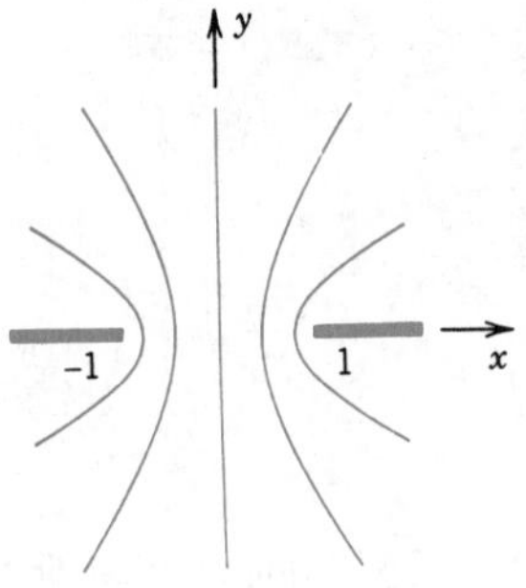

Fig. 375. Slit

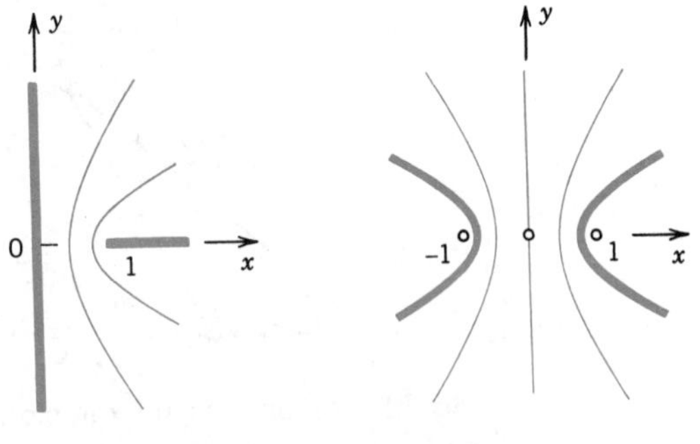

Fig. 376. Other apertures

17.2 Use of Conformal Mapping

Complex potentials relate potential theory closely to complex analysis, as we have just seen. Another close relation results from the use of conformal mapping in solving **boundary value problems** for the Laplace equation, that is, in finding a solution of the equation in some domain assuming given values on the boundary (**"Dirichlet problem"**; see also Sec. 11.5). Then conformal mapping is used to map a given complicated domain onto a simpler one where the solution is known or can be found more easily. This solution is then mapped back to the given domain. This is the idea. That it works is due to the fact that harmonic functions remain harmonic under conformal mapping:

Theorem 1 **(Harmonic functions under conformal mapping)**
If $\Phi^*(u, v)$ is harmonic in a domain D^* in the w-plane, and if an analytic function $w = u + iv = f(z)$ maps a domain D in the z-plane conformally onto D^*, then

$$\Phi(x, y) = \Phi^*(u(x, y), v(x, y)) \tag{1}$$

is harmonic in D.

Proof. The composite of analytic functions is analytic, as follows from the chain rule. Hence, taking a harmonic conjugate[3] $\Psi^*(u, v)$ of Φ^* and forming the analytic function $F^*(w) = \Phi^*(u, v) + i\Psi^*(u, v)$, we conclude that $F(z) = F^*(f(z))$ is analytic in D and its real part, $\Phi(x, y) = \text{Re } F(z)$ is harmonic in D. ∎

EXAMPLE 1 **Potential between noncoaxial cylinders**

Find the potential between the cylinders C_1: $|z| = 1$ (grounded, that is, having potential $U_1 = 0$) and C_2: $|z - 2/5| = 2/5$ (having potential $U_2 = 110$ volts).

Solution. We map the unit disk $|z| = 1$ onto the unit disk $|w| = 1$ in such a way that C_2 is mapped onto some cylinder C_2^*: $|w| = r_0$. By (6), Sec. 16.3, a linear fractional transformation mapping the unit disk onto the unit disk is

$$w = \frac{z - b}{bz - 1} \tag{2}$$

where we have chosen $b = z_0$ real without restriction. z_0 is of no immediate help here because centers of circles do not map onto centers of the images, in general. However, we now have two free constants b and r_0 and shall succeed by imposing two reasonable conditions, namely, that 0 and 4/5 (Fig. 377) should be mapped onto r_0 and $-r_0$, respectively. This gives by (2)

$$r_0 = \frac{-b}{-1}, \qquad -r_0 = \frac{4/5 - b}{4b/5 - 1} = \frac{4/5 - r_0}{4r_0/5 - 1},$$

a quadratic equation in r_0 with solutions 2 (no good because $r_0 < 1$) and 1/2. Hence our mapping function (2) with $b = 1/2$ becomes

$$w = f(z) = \frac{2z - 1}{z - 2}. \tag{3}$$

From Example 5 in Sec. 17.1, writing w for z we have as the complex potential in the w-plane

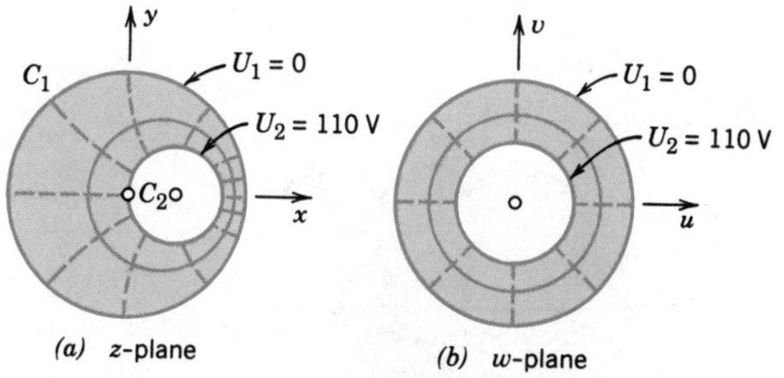

Fig. 377. Example 1

[3]See Sec. 12.5. We mention without proof that if D^* is simply connected (Sec. 13.3), then a conjugate harmonic of Φ^* exists. Another proof without the use of a harmonic conjugate is given in Appendix 4.

and from this the real potential

$$\Phi^*(u, v) = \text{Re } F^*(w) = a \ln |w| + k.$$

We determine a and k from the boundary conditions. If $|w| = 1$, then $\Phi^* = a \ln 1 + k = 0$, hence $k = 0$. If $|w| = 1/2$, then $\Phi^* = a \ln (1/2) = 110$, hence $a = 110/\ln (1/2) = -158.7$. Substitution of (3) now gives the desired solution in the given domain in the z-plane

$$F(z) = F^*(f(z)) = a \,\text{Ln}\, \frac{2z - 1}{z - 2}.$$

The real potential is

$$\Phi(x, y) = \text{Re } F(z) = a \ln \left|\frac{2z - 1}{z - 2}\right|, \qquad a = -158.7.$$

Can we "see" this result? Well, $\Phi(x, y) = const$ if and only if $|(2z - 1)/(z - 2)| = const$, that is, $|w| = const$ by (5). These circles correspond to circles in the z-plane, as follows from Theorem 1, Sec. 16.2, together with the fact that the inverse of a linear fractional transformation is linear fractional [see (5), Sec. 16.2]. Similarly for the rays $\arg w = const$. Hence the equipotential lines $\Phi(x, y) = const$ are circles, and the lines of force are circular arcs (dashed in Fig. 377). These two families of curves intersect orthogonally, that is, at right angles, as shown in Fig. 377. ∎

EXAMPLE 2 Potential between two semicircular plates

Find the potential between two semicircular plates P_1 and P_2 in Fig. 378*a* having potentials -3000 and 3000 volts, respectively. Use Example 3 in Sec. 17.1 and conformal mapping.

Solution. First step. We map the unit disk in Fig. 378*a* onto the right half of the w-plane (Fig. 378*b*) by using the linear fractional transformation in Example 3, Sec. 16.3:

$$w = f(z) = \frac{1 + z}{1 - z}.$$

The boundary $|z| = 1$ is mapped onto the boundary $u = 0$ (the v-axis), with $z = -1, i, 1$ going onto $w = 0, i, \infty$, respectively, and $z = -i$ onto $w = -i$. Hence the upper semicircle of $|z| = 1$ is mapped onto the upper half, and the lower semicircle onto the lower half of the v-axis, so that the boundary conditions in the w-plane are as indicated in Fig. 378b.

Second step. We determine the potential $\Phi^*(u, v)$ in the right half-plane of the w-plane. Example 3 in Sec. 17.1 with $\alpha = \pi$, $U_1 = -3000$, and $U_2 = 3000$ [with $\Phi^*(u, v)$ instead of $\Phi(x, y)$] yields

$$\Phi^*(u, v) = \frac{6000}{\pi}\varphi, \qquad \varphi = \arctan \frac{v}{u}.$$

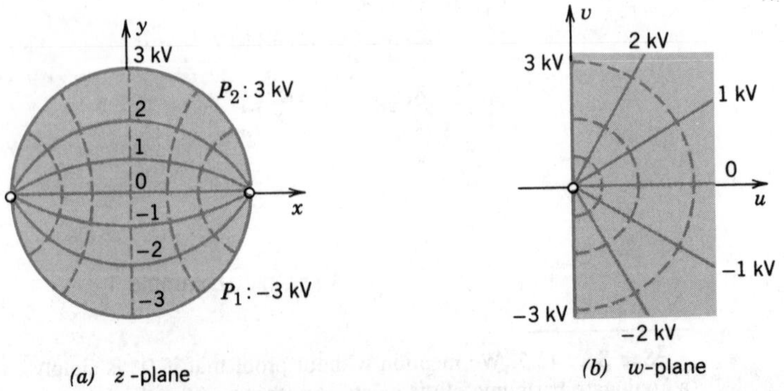

Fig. 378. Example 2

On the positive half of the imaginary axis ($\varphi = \pi/2$), this equals 3000 and on the negative half -3000, as it should be. Φ^* is the real part of the complex potential

$$F^*(w) = -\frac{6000\,i}{\pi}\,\mathrm{Ln}\,w.$$

Third step. We substitute the mapping function into F^* to get the complex potential $F(z)$ in Fig. 378*a* in the form

$$F(z) = F^*(f(z)) = -\frac{6000\,i}{\pi}\,\mathrm{Ln}\,\frac{1+z}{1-z}.$$

The real part of this is the potential we wanted to determine:

$$\Phi(x, y) = \mathrm{Re}\,F(z) = \frac{6000}{\pi}\,\mathrm{Im}\,\mathrm{Ln}\,\frac{1+z}{1-z} = \frac{6000}{\pi}\,\mathrm{Arg}\,\frac{1+z}{1-z}.$$

As in Example 1 we conclude that the equipotential lines $\Phi(x, y) = const$ are circular arcs because they correspond to $\mathrm{Arg}\,[(1+z)/(1-z)] = const$, hence to $\mathrm{Arg}\,w = const$. Also, $\mathrm{Arg}\,w = const$ are rays from 0 to ∞, the images of $z = -1$ and $z = 1$, respectively. Hence the equipotential lines all have -1 and 1 (the points where the boundary potential jumps) as their endpoints (Fig. 378*a*). The lines of force are circular arcs, too, and since they must be orthogonal to the equipotential lines, their centers can be obtained as intersections of tangents to the unit circle with the x-axis. (Explain!) ∎

Basic comment. We formulated the examples in this section in terms of the electrostatic potential. It is quite important to realize that this is accidental. We could equally well have phrased everything in terms of (time-independent) heat flow; then instead of voltages we would have had temperatures, the equipotential lines would have become isotherms (= lines of constant temperature), and the lines of the electrical force would have become lines along which heat flows from higher to lower temperatures (more on this in the next section). Or we could have talked about fluid flow; then the electrostatic equipotential lines would have become streamlines (more on this in Sec. 17.4). What we again see here is the **unifying power of mathematics:** different phenomena and systems from different areas in physics having the same types of model can be treated by the same mathematical methods.

Problem Set 17.2

1. Verify Theorem 1 for $\Phi^*(u, v) = u^2 - v^2$ and $w = f(z) = e^z$.
2. Write down the formulas in the second proof of Theorem 1 (given in Appendix 4) and carry out all the steps in detail.
3. Find the potential Φ in the region R in the first quadrant of the z-plane bounded by the axes (having potential 0) and the hyperbola $y = 1/x$ (having potential U_2) in two ways: (i) directly, (ii) by mapping R onto a suitable infinite strip.
4. Verify the steps of the derivation of (3) from (2).
5. Applying a suitable conformal mapping, obtain from Fig. 378*b* the potential Φ in the angular region $-\pi/4 < \mathrm{Arg}\,z < \pi/4$, $\Phi = -3$ kV when $\mathrm{Arg}\,z = -\pi/4$ and $\Phi = 3$ kV when $\mathrm{Arg}\,z = \pi/4$.
6. Show that in Example 2, the y-axis is mapped onto the unit circle in the w-plane.
7. At $z = \pm 1$ in Fig. 378*a* the tangents to the equipotential lines shown make equal angles ($\pi/6$). Why?

8. In Example 2, set $z = Z^2$ $(Z = X + iY)$ and show that the resulting potential $\Phi(X, Y) = \text{Re } F(Z^2)$ is the potential in the portion of the unit disk $|Z| \leqq 1$ in the first quadrant having the boundary values 0 on the axes and 3 on $|Z| = 1$.

9. Find the linear fractional transformation $z = g(Z)$ that maps $|Z| \leqq 1$ onto $|z| \leqq 1$ with $Z = i/2$ being mapped onto $z = 0$. Show that $Z_1 = (3 + 4i)/5$ is mapped onto $z = -1$, and $Z_2 = (-3 + 4i)/5$ onto $z = 1$, so that the equipotential lines of Example 2 look in $|Z| \leqq 1$ as shown in Fig. 379.

10. Why are the equipotential lines in Prob. 9 circles?

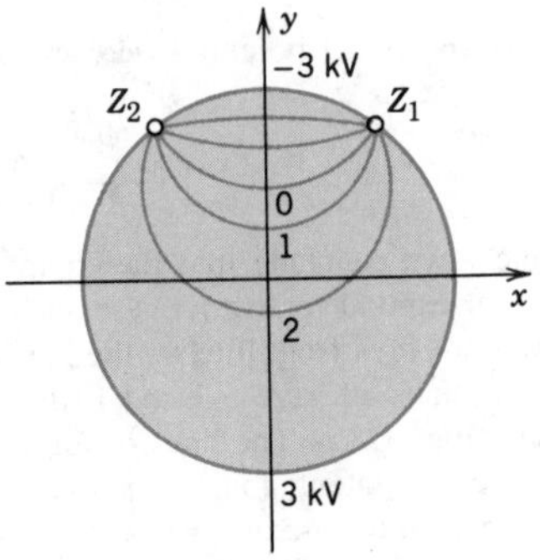

Fig. 379. Problem 9

17.3 Heat Problems

Laplace's equation also governs heat flow problems that are **steady,** that is, time-independent. Indeed, heat conduction in a body of homogeneous material is modeled by the **heat equation**

$$T_t = c^2\nabla^2 T$$

where the function T is temperature, $T_t = \partial T/\partial t$, t is time, and c^2 is a positive constant (depending on the material of the body). Hence if a problem is ***steady,*** so that $T_t = 0$, and two-dimensional, then the heat equation reduces to the two-dimensional Laplace equation

$$\nabla^2 T = T_{xx} + T_{yy} = 0, \tag{1}$$

so that the problem can be treated by our present methods.

$T(x, y)$, called the **heat potential,** is the real part of the **complex heat potential**

$$F(z) = T(x, y) + i\Psi(x, y).$$

The curves $T(x, y) = const$ are called **isotherms** (= lines of constant temperature) and the curves $\Psi(x, y) = const$ **heat flow lines,** because along them, heat flows from higher to lower temperatures.

It follows that all the examples considered so far (Secs. 17.1, 17.2) can now be reinterpreted as problems on heat flow. The electrostatic equipotential lines $\Phi(x, y) = const$ now become isotherms $T(x, y) = const$, and the lines of electrical force become lines of heat flow. Mathematically, the calculations remain the same. New problems may arise, involving boundary conditions that would make no sense physically in electrostatics or would be of no practical interest there. Examples 3 and 4 (below) illustrate this.

Physically, to have a steady problem, the boundary of the domain of heat flow must be kept at constant temperature, by heating or cooling.

EXAMPLE 1 **Temperature between parallel plates**

Find the temperature between two parallel plates $x = 0$ and $x = d$ in Fig. 380 having temperatures 0 and 100°C, respectively.

Solution. As in Sec. 17.1 we conclude that $T(x, y) = ax + b$. From the boundary conditions, $b = 0$ and $a = 100/d$. The answer is

$$T(x, y) = \frac{100}{d} x \text{ [°C]}.$$

The corresponding complex potential is $F(z) = (100/d)z$. Heat flows horizontally, in the negative x-direction, along the lines $y = const.$ ∎

EXAMPLE 2 **Temperature distribution between a wire and a cylinder**

Find the temperature field around a long thin wire of radius $r_1 = 1$ mm that is electrically heated to $T_1 = 500$°F and is surrounded by a circular cylinder of radius $r_2 = 100$ mm, which is kept at temperature $T_2 = 60$°F by cooling it with air. See Fig. 381.

Solution. T depends only on r, for reasons of symmetry. Hence, as in Sec. 17.1 (Example 2),

$$T(x, y) = a \ln r + b.$$

The boundary conditions are

$$T_1 = 500 = a \ln 1 + b, \qquad T_2 = 60 = a \ln 100 + b.$$

Hence $b = 500$ (since $\ln 1 = 0$) and $a = (60 - b)/\ln 100 = -95.54$. The answer is

$$T(x, y) = 500 - 95.54 \ln r \text{ [°F]}.$$

The isotherms are concentric circles. Heat flows from the wire radially outward to the cylinder. ∎

EXAMPLE 3 **A mixed boundary value problem**

Find the temperature distribution in the region in Fig. 382 (cross section of a solid cylinder), whose vertical portion of the boundary is at 20°C, the horizontal portion at 50°C, and the circular portion is insulated.

Solution. The insulated portion of the boundary must be a heat flow line, since by the insulation, heat is prevented from crossing such a curve, hence it must flow along. Thus the isotherms must meet such a curve at right angles. Since T is constant along an isotherm, this means that

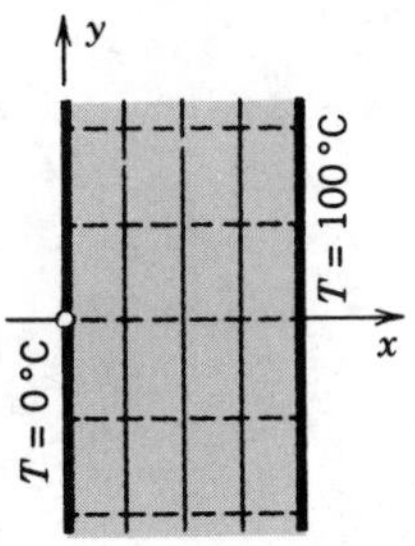

Fig. 380. Example 1

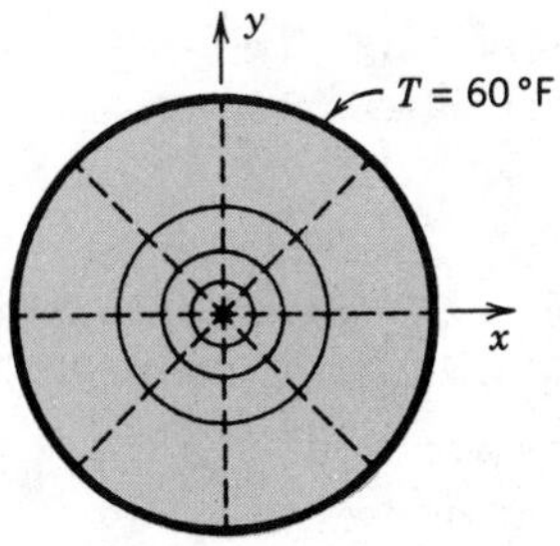

Fig. 381. Example 2

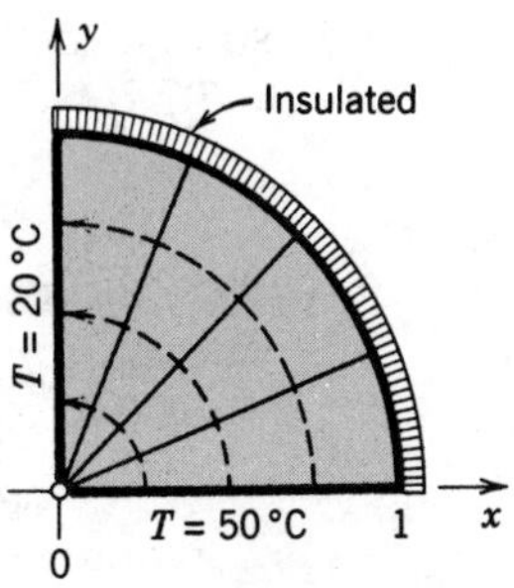

Fig. 382. Example 3

(2) $$\frac{\partial T}{\partial n} = 0 \quad \text{along an isolated portion of the boundary,}$$

where $\partial T/\partial n$ is the **normal derivative** of T, that is, the directional derivative (Sec. 8.9) of T in the **normal direction** to the insulated boundary (that is, the direction perpendicular to this boundary). Such a problem in which T is prescribed on one portion of the boundary and $\partial T/\partial n$ on the other portion is called a **mixed boundary value problem.**

In our case, the normal direction to the insulated circular boundary curve is the radial direction toward the origin. Hence (2) becomes $\partial T/\partial r = 0$, meaning that along this curve the solution must not depend on r. Now Arg $z = \theta$ satisfies (1) as well as this condition, and is constant (0 and $\pi/2$) on the straight portions of the boundary. Hence the solution is of the form

$$T(x, y) = a\theta + b.$$

The boundary conditions yield $a(\pi/2) + b = 20$ and $a \cdot 0 + b = 50$. This gives

$$T(x, y) = 50 - \frac{60}{\pi}\theta, \qquad \theta = \arctan\frac{y}{x}.$$

The isotherms are portions of rays $\theta = const.$ Heat flows from the x-axis along circles $r = const$ (dashed in Fig. 382) to the y-axis. ∎

EXAMPLE 4 **Another mixed boundary value problem in heat conduction**

Find the temperature field in the upper half-plane when the x-axis is at $T = 0$°C for $x < -1$, insulated for $-1 < x < 1$, and at $T = 20$°C for $x > 1$ (Fig. 383*a*).

Solution. We map the half-plane in Fig. 383*a* onto the vertical strip in Fig. 383*b*, find the temperature $T^*(u, v)$ there, and map it back to get the temperature $T(x, y)$ in the half-plane.

The idea of using that strip is suggested by Fig. 360 in Sec. 16.4 with the roles of $z = x + iy$ and $w = u + iv$ interchanged, showing that $z = \sin w$ maps our present strip onto our half-plane in Fig. 383*a*. Hence the inverse function

$$w = f(z) = \sin^{-1} z$$

maps that half-plane onto the strip in the w-plane. This is the mapping function that we need according to Theorem 1 in Sec. 17.2.

The insulated segment $-1 < x < 1$ on the x-axis maps onto the segment $-\pi/2 < u < \pi/2$ on the u-axis. The rest of the x-axis maps onto the two vertical boundary portions $u = -\pi/2$ and $\pi/2$, $v > 0$, of the strip. This gives the transformed boundary conditions in Fig. 383*b* for $T^*(u, v)$, where on the insulated horizontal boundary, $\partial T^*/\partial n = \partial T^*/\partial v = 0$ because v is a coordinate normal to that segment.

Similar to Example 1 we obtain

$$T^*(u, v) = 10 + \frac{20}{\pi}u$$

which satisfies all the boundary conditions. This is the real part of the complex potential $F^*(w) = 10 + (20/\pi)w$. Hence the complex potential in the z-plane is

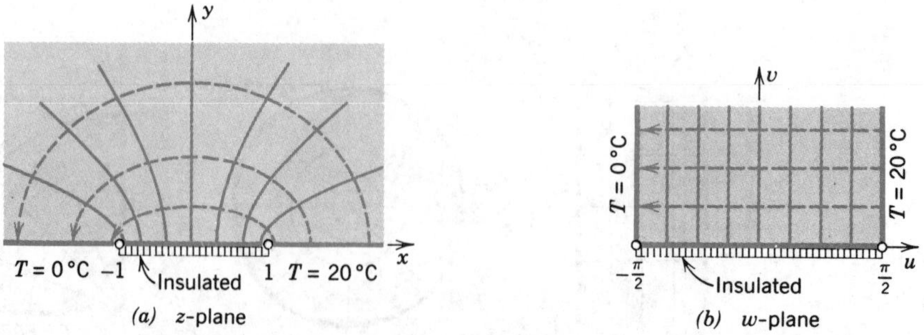

Fig. 383. Example 4

$$F(z) = F^*(f(z)) = 10 + \frac{20}{\pi}\sin^{-1} z,$$

and $T(x, y) = \operatorname{Re} F(z)$ is the solution. The isotherms are $u = const$ in the strip and the hyperbolas in the z-plane, perpendicular to which heat flows along the dashed ellipses from the 20°-portion to the cooler 0°-portion of the boundary, a physically very reasonable result. ∎

These examples have shown the usefulness of conformal mappings and complex potentials. The latter will also play a role in the next section on fluid flow.

Problem Set 17.3

1. Find the temperature between two parallel plates $y = 0$ and $y = d$ that are kept at temperatures 0 and 100°C, respectively. (i) Proceed directly. (ii) Use Example 1 and a suitable mapping.
2. Find the temperature and the complex potential in an infinite plate with edges $y = x - 2$ and $y = x + 2$ kept at -10°C and 20°C, respectively.
3. Find the temperature in Fig. 382 if $T = -20$°C on the y-axis, $T = 100$°C on the x-axis, and the circular portion of the boundary is insulated as before.
4. Find the temperature T in the sector $0 \leqq \operatorname{Arg} z \leqq \pi/3$, $|z| \leqq 1$, if $T = 20$°C on the x-axis, $T = 50$°C on $y = \sqrt{3}x$, and the curved portion is insulated.
5. Find the temperature and the complex potential in the first quadrant of the z-plane when the y-axis is kept at 100°C, the segment $0 < x < 1$ of the x-axis is insulated, and the portion $x > 1$ of the x-axis is kept at 200°C. *Hint.* Use Example 4.
6. Interpret Problem 9 in Problem Set 17.2 as a heat flow problem (with boundary temperatures, say, 20°C and 300°C). Along what curves does the heat flow?
7. Find the temperature in the upper half-plane in Fig. 384 satisfying the given boundary conditions.
8. Find the complex potential in Prob. 7.
9. Find the temperature T^* in the upper half-plane subject to the boundary conditions shown in Fig. 385.
10. Find the complex potential in Prob. 9.
11. Obtain the result of Prob. 9 from Prob. 7 by superposition.
12. What temperature in the first quadrant of the z-plane is obtained from Prob. 7 by the mapping $w = a + z^2$, and what are the transformed boundary conditions?
13. Using Prob. 9 and the mapping $w = \cosh z$ (see Example 4 in Sec. 16.4), show that the temperature in the infinite strip in Fig. 386 is

$$T(x, y) = \frac{T_0}{\pi} \operatorname{Arg} \frac{\cosh z - 1}{\cosh z + 1} = \frac{2T_0}{\pi} \operatorname{Arg}\left(\tanh \frac{z}{2}\right).$$

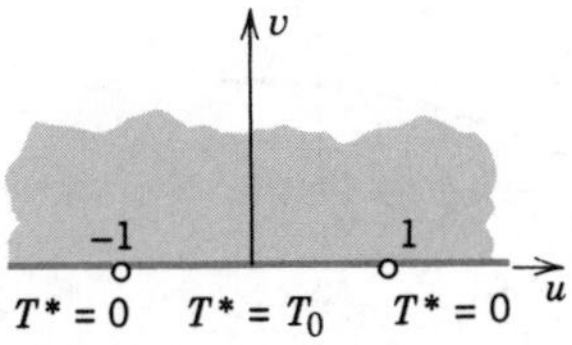

Fig. 384. Problem 7

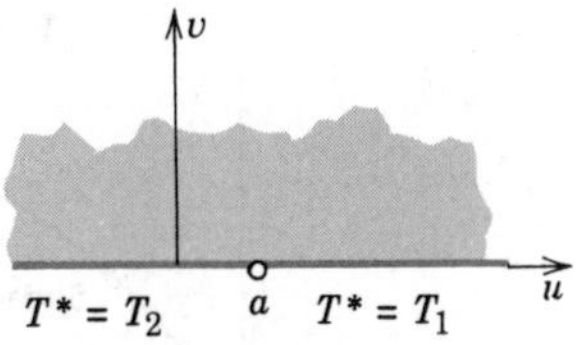

Fig. 385. Problem 9

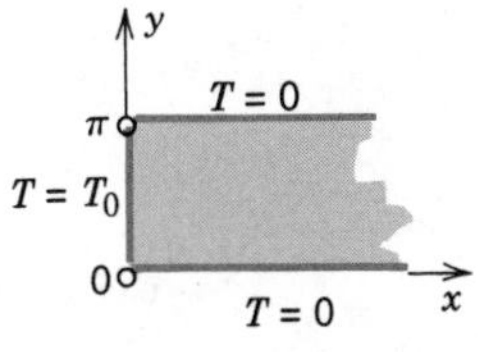

Fig. 386. Problem 13

14. Show by direct calculation that $(\text{Im} \tanh \frac{1}{2}z)/(\text{Re} \tanh \frac{1}{2}z) = (\sin y)/(\sinh x)$, so that $T(x, y) = (2T_0/\pi) \arc\tan [(\sin y)/(\sinh x)]$ in Prob. 13.

Find the temperature $T(x, y)$ in the given thin metal plate whose faces are insulated and whose edges are kept at the temperatures shown in the figure.

15.

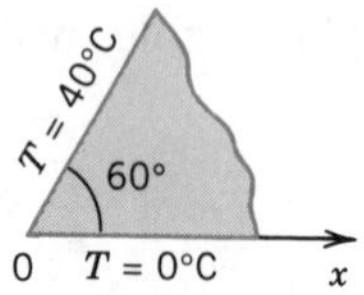

16.

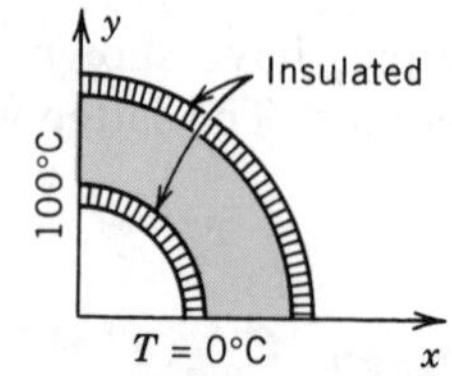

17.

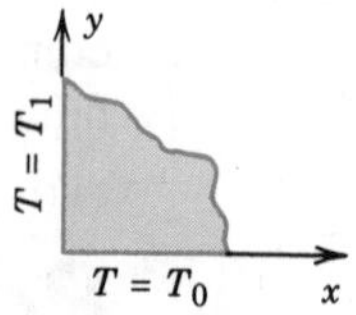

18.

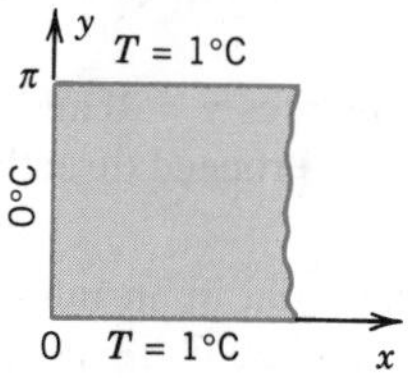

19.

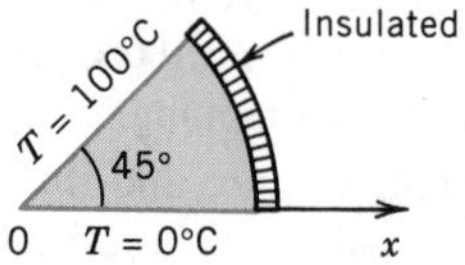

20.

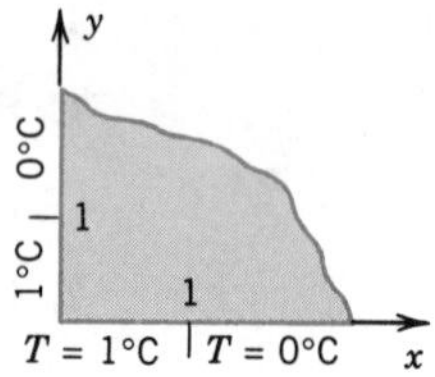

17.4 Fluid Flow

Laplace's equation also plays a basic role in hydrodynamics, in steady nonviscous fluid flow under physical conditions discussed later in this section. To keep in touch with complex analysis, our problems will be ***two-dimensional,*** so that the velocity vector V by which the motion of the fluid can be given depends only on two space variables x and y and the motion is the same in all planes parallel to the xy-plane.

Then we can use for V a complex vector function

$$V = V_1 + iV_2 \tag{1}$$

giving the magnitude and direction of the velocity at each point $z = x + iy$. Here V_1 and V_2 are the components of the velocity in the x and y directions. V is tangential to the path of the moving particles, called a **streamline** of the motion (Fig. 387).

We show that under suitable assumptions (explained in detail following the examples), for a given flow there exists an analytic function

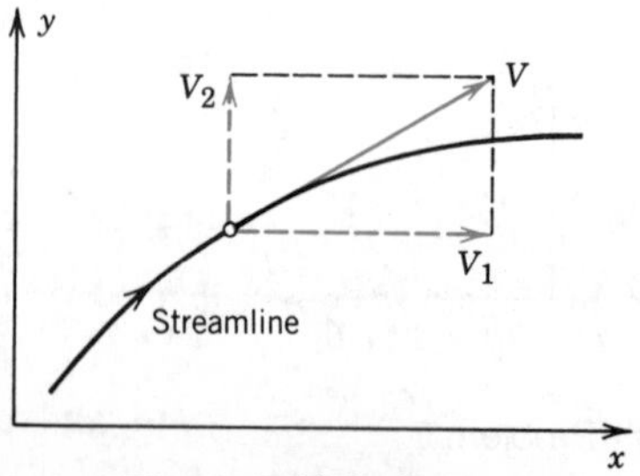

Fig. 387. Velocity

$$F(z) = \Phi(x, y) + i\Psi(x, y), \tag{2}$$

called the **complex potential** of the flow, such that the streamlines are given by $\Psi(x, y) = const$, and the velocity is given by

$$V = V_1 + iV_2 = \overline{F'(z)} \tag{3}$$

where the bar denotes the complex conjugate. Ψ is called the **stream function.** The function Φ is called the **velocity potential.**[4] The curves $\Phi(x, y) = const$ are called **equipotential lines.** V is the **gradient** of Φ; by definition, this means that

$$V_1 = \frac{\partial \Phi}{\partial x}, \qquad V_2 = \frac{\partial \Phi}{\partial y}. \tag{4}$$

Indeed, by (4), Sec. 12.5, and the second Cauchy–Riemann equation, with $f = F$, $u = \Phi$, $v = \Psi$,

$$\overline{F'(z)} = \Phi_x - i\Psi_x = \Phi_x + i\Phi_y = V_1 + iV_2 = V.$$

Furthermore, since $F(z)$ is analytic, Φ and Ψ satisfy Laplace's equation

$$\nabla^2\Phi = \frac{\partial^2 \Phi}{\partial x^2} + \frac{\partial^2 \Phi}{\partial y^2} = 0, \qquad \nabla^2\Psi = \frac{\partial^2 \Psi}{\partial x^2} + \frac{\partial^2 \Psi}{\partial y^2} = 0. \tag{5}$$

Whereas in electrostatics the boundaries (conducting plates) are equipotential lines, in fluid flow a boundary across which fluid cannot flow must be a streamline. Hence in fluid flow the stream function is of particular importance.

Before discussing the conditions for the validity of the statements involving (2)–(5), let us consider two flows of practical interest, so that we first see what is going on from a practical point of view. Further flows follow in the problem set.

EXAMPLE 1 **Flow around a corner**

The complex potential

$$F(z) = z^2 = x^2 - y^2 + 2ixy$$

describes a flow whose equipotential lines are the hyperbolas

$$\Phi = x^2 - y^2 = const$$

and whose streamlines are the hyperbolas

$$\Psi = 2xy = const.$$

From (3) we obtain the velocity vector

$$V = 2\bar{z} = 2(x - iy), \qquad \text{that is,} \qquad V_1 = 2x, \qquad V_2 = -2y.$$

The speed (magnitude of the velocity) is

[4] Some authors use $-\Phi$ (instead of Φ) as the velocity potential.

$$|V| = \sqrt{V_1^2 + V_2^2} = 2\sqrt{x^2 + y^2}.$$

The flow may be interpreted as the flow in a channel bounded by the positive coordinates axes and a hyperbola, say, $xy = 1$ (Fig. 388). We note that the speed along a streamline S has a minimum at the point P where the cross section of the channel is large. ∎

EXAMPLE 2 **Flow around a cylinder**

Consider the complex potential

$$F(z) = \Phi(x, y) + i\,\Psi(x, y) = z + \frac{1}{z}.$$

Using the polar form $z = re^{i\theta}$, we obtain

$$F(z) = re^{i\theta} + \frac{1}{r}e^{-i\theta} = \left(r + \frac{1}{r}\right)\cos\theta + i\left(r - \frac{1}{r}\right)\sin\theta.$$

Hence the streamlines are

$$\Psi(x, y) = \left(r - \frac{1}{r}\right)\sin\theta = const.$$

In particular, $\Psi(x, y) = 0$ gives $r - 1/r = 0$ or $\sin\theta = 0$. Hence this streamline consists of the unit circle ($r = 1/r$ gives $r = 1$) and the x-axis ($\theta = 0$ and $\theta = \pi$). For large $|z|$ the term $1/z$ in $F(z)$ is small in absolute value, so that for these z the flow is nearly uniform and parallel to the x-axis. Hence we can interpret this as a flow around a long circular cylinder of unit radius. The flow has two **stagnation points** (that is, points at which the velocity $V = 0$), at $z = \pm 1$. This follows from

$$F'(z) = 1 - \frac{1}{z^2}$$

and (3). See Fig. 389. ∎

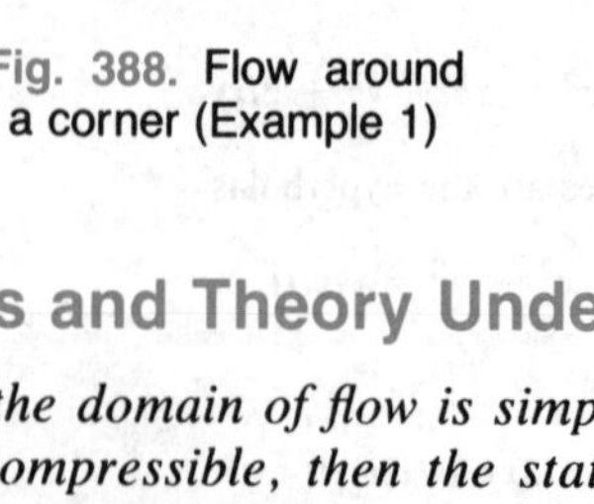

Fig. 388. Flow around a corner (Example 1)

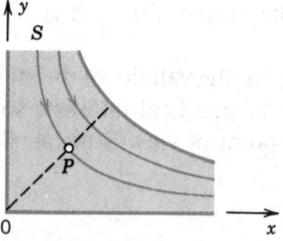

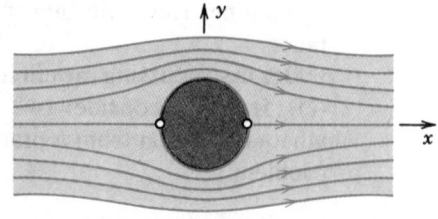

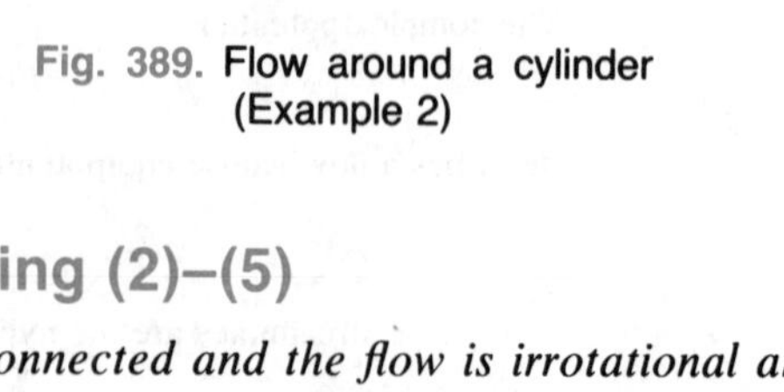

Fig. 389. Flow around a cylinder (Example 2)

Assumptions and Theory Underlying (2)–(5)

If the domain of flow is simply connected and the flow is irrotational and incompressible, then the statements involving (2)–(5) *hold. In particular, then the flow has a complex potential* $F(z)$, *which is an analytic function.* (Explanation of terms below.)

We prove this, along with a discussion of basic concepts related to fluid flow. Consider any smooth curve C in the z-plane, given by $z(s) = x(s) + iy(s)$, where s is the arc length of C. Let the real variable V_t be the component of the velocity V tangent to C (Fig. 390). Then the value of the real line integral

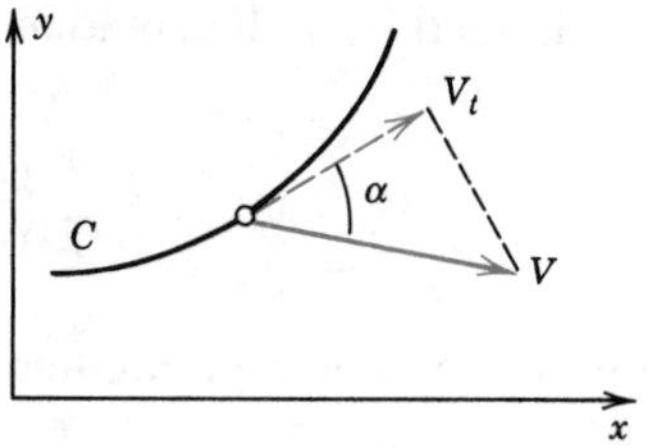

Fig. 390. Tangential component of the velocity with respect to a curve C

$$\int_C V_t \, ds \tag{6}$$

taken along C in the sense of increasing values of s is called the **circulation** of the fluid along C. Dividing the circulation by the length of C, we obtain the mean velocity[5] of the flow along the curve C. Now

$$V_t = |V| \cos \alpha \qquad \text{(Fig. 390)}.$$

Hence V_t is the dot product (Sec. 8.2) of V and the tangent vector dz/ds of C (Sec. 16.1); thus in (6),

$$V_t \, ds = \left(V_1 \frac{dx}{ds} + V_2 \frac{dy}{ds} \right) ds = V_1 \, dx + V_2 \, dy.$$

The circulation (6) along C now becomes

$$\int_C V_t \, ds = \int_C (V_1 \, dx + V_2 \, dy). \tag{7}$$

As the next idea, let C be a *closed* curve, namely, the boundary of a simply connected domain D, and suppose that V has continuous partial derivatives in a domain containing D and C. Then we can use Green's theorem (Sec. 9.4) to represent the circulation around C by a double integral,

$$\oint_C (V_1 \, dx + V_2 \, dy) = \iint_D \left(\frac{\partial V_2}{\partial x} - \frac{\partial V_1}{\partial y} \right) dx \, dy. \tag{8}$$

The integrand of this double integral is called the **vorticity** of the flow, and

[5]*Definitions:* $\displaystyle\frac{1}{b-a}\int_a^b f(x)\,dx$ = mean value of f on the interval $a \leqq x \leqq b$,

$\displaystyle\frac{1}{L}\int_C f(s)\,ds$ = mean value of f on C (L = length of C),

$\displaystyle\frac{1}{A}\iint_D f(x, y)\,dx\,dy$ = mean value of f on D (A = area of D).

the vorticity divided by 2 the **rotation**

(9) $$\omega(x, y) = \frac{1}{2}\left(\frac{\partial V_2}{\partial x} - \frac{\partial V_1}{\partial y}\right).$$

We assume the flow to be **irrotational,** that is, $\omega(x, y) \equiv 0$ throughout the flow; thus,

(10) $$\frac{\partial V_2}{\partial x} - \frac{\partial V_1}{\partial y} = 0.$$

To understand the physical meaning of vorticity and rotation, take for C in (8) a circle of radius r. Then the circulation divided by the length $2\pi r$ of C is the mean velocity of the fluid along C. Hence by dividing this by r we obtain the mean *angular* velocity ω_0 of the fluid about the axis of the circle:

$$\omega_0 = \frac{1}{2\pi r^2}\iint_D \left(\frac{\partial V_2}{\partial x} - \frac{\partial V_1}{\partial y}\right) dx\, dy = \frac{1}{\pi r^2}\iint_D \omega(x, y)\, dx\, dy.$$

If we now let $r \to 0$, the limit of ω_0 is the value of ω at the center of C. Hence, $\omega(x, y)$ is the limiting angular velocity of a circular element of the fluid as the circle shrinks to the point (x, y). Roughly speaking, if a spherical element of the fluid were suddenly solidified and the surrounding fluid simultaneously annihilated, the element would rotate with the angular velocity ω.

Our second assumption is that the fluid is ***incompressible.*** (Examples are water and oil, whereas air is compressible.) Then

(11) $$\frac{\partial V_1}{\partial x} + \frac{\partial V_2}{\partial y} = 0 \qquad \text{[see (7), Sec. 8.10]}$$

in every region that is free of **sources** or **sinks,** that is, points at which fluid is produced or disappears. [The expression in (11) is called the ***divergence*** of V and is denoted by div V.]

If the domain D of the flow is simply connected and the flow is irrotational, then the line integral (7) is independent of path in D (by Theorem 3 in Sec. 9.2, where $F_1 = V_1$, $F_2 = V_2$, $F_3 = 0$, and z is the third coordinate in space and has nothing to do with our present z). Hence if we integrate from a fixed point (a, b) in D to a variable point (x, y) in D, the integral becomes a function of the point (x, y), say, $\Phi(x, y)$:

(12) $$\Phi(x, y) = \int_{(a,\, b)}^{(x,\, y)} (V_1\, dx + V_2\, dy).$$

We claim that Φ is the desired velocity potential. To prove this, all we have to do is to show that (4) holds. Now since the integral (7) is independent of path, $V_1\, dx + V_2\, dy$ is an exact differential (Sec. 9.2), namely, the differential of Φ, that is,

$$V_1\,dx + V_2\,dy = \frac{\partial\Phi}{\partial x}\,dx + \frac{\partial\Phi}{\partial y}\,dy.$$

From this we see that $V_1 = \partial\Phi/\partial x$ and $V_2 = \partial\Phi/\partial y$, which gives (4).

That Φ is harmonic follows at once by substituting (4) into (11), which gives the first Laplace's equation in (5).

We finally take a conjugate harmonic Ψ of Φ. Then the other equation in (5) holds. Also, assuming that the second partial derivatives of Φ and Ψ are continuous, we have that the complex function

$$F(z) = \Phi(x, y) + i\Psi(x, y)$$

is analytic in D. Since the curves $\Psi(x, y) = const$ are perpendicular to the equipotential curves $\Phi(x, y) = const$ (except where $F'(z) = 0$), we conclude that they are the streamlines. Hence Ψ is the stream function and $F(z)$ is the complex potential of the flow. ∎

Problem Set 17.4

1. **(Parallel flow)** Show that $F(z) = Kz$ (K positive real) describes a uniform flow to the right, which can be interpreted as a uniform flow between two parallel lines (between two parallel planes in three-dimensional space). See Fig. 391. Find the velocity vector, the streamlines, and the equipotential lines.
2. Show that $F(z) = iz^2$ describes a flow around a corner. Find and sketch the streamlines and the equipotential lines. Find the velocity vector V.
3. Obtain the flow in Example 1 from that in Prob. 1 by a conformal mapping of the first quadrant onto the upper half-plane.
4. Make a slight change of $F(z)$ in Example 2 so that you obtain a flow around a cylinder of radius r_0, that yields the flow in Prob. 1 as $r_0 \to 1$.

Consider the flow corresponding to the given complex potential $F(z)$. Show some streamlines and equipotential lines graphically. Find the velocity vector V. Determine all points at which V is parallel to the x-axis.

5. iz **6.** $(1 - 2i)z$ **7.** $(1 + i)z$ **8.** $z^2 + 2z$ **9.** z^3 **10.** iz^3

11. **(Source and sink)** Consider the complex potential $F(z) = (c/2\pi)\ln z$, where c is positive real. Show that $V = (c/2\pi r^2)(x + iy)$, where $r = \sqrt{x^2 + y^2}$, and this implies that the flow is directed radially outward (see Fig. 392), so that this potential corresponds to a **point source** at $z = 0$ (that is, a **source line** $x = 0$, $y = 0$ in space). (The constant c is called the **strength** or **discharge** of the source. If c is negative real, the flow is said to have a **sink** at $z = 0$, it is directed radially inward, and fluid disappears at the singular point $z = 0$ of the complex potential.)

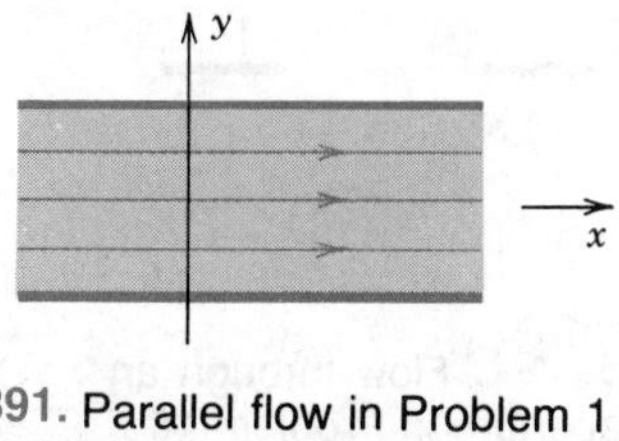

Fig. 391. Parallel flow in Problem 1

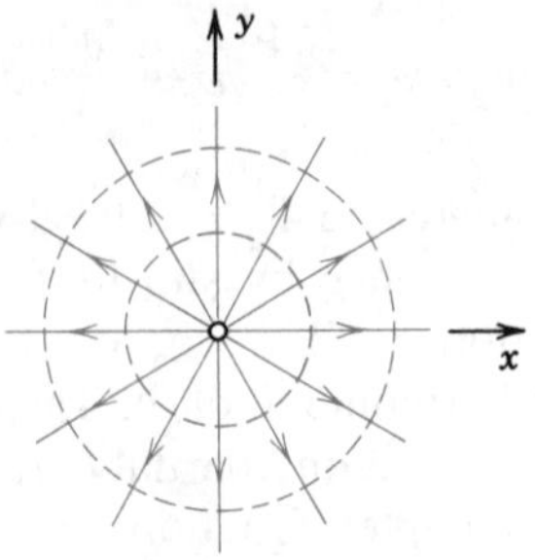

Fig. 392. Point source in Problem 11

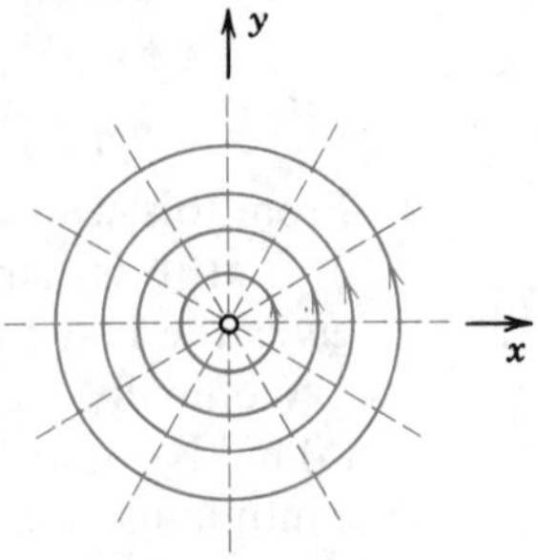

Fig. 393. Vortex flow in Problem 12

12. **(Vortex line)** Show that $F(z) = -(iK/2\pi) \ln z$ with real positive K describes a flow circulating around the origin in the counterclockwise sense; see Fig. 393. (The point $z = 0$ is a **vortex;** the potential increases by K each time we travel around the vortex.)
13. Find the complex potential of a flow that has a point source of strength 1 at $z = -a$.
14. Find the complex potential of a flow that has a point sink of strength 1 at $z = a$.
15. Show that vector addition of the velocity vectors of two flows leads to a flow whose complex potential is obtained by adding those of the two flows.
16. Add the potentials in Probs. 13 and 14 and show the streamlines graphically.
17. Find the streamlines of the flow corresponding to $F(z) = 1/z$. Show that for small $|a|$ the streamlines in Prob. 16 look similar to those in the present problem.
18. Show that $F(z) = \cosh^{-1} z$ corresponds to a flow whose streamlines are confocal hyperbolas with foci at $z = \pm 1$, and the flow may be interpreted as a flow through an aperture. See Fig. 394.
19. Show that $F(z) = \cos^{-1} z$ can be interpreted as the complex potential of a flow circulating around an elliptic cylinder or a plate (the straight segment from $z = -1$ to $z = 1$). Show that the streamlines are confocal ellipses with foci at $z = \pm 1$. See Fig. 395.

20. **(Flow with circulation around a cylinder)** Show that addition of the potentials in Prob. 12 and Example 2 gives a flow such that the cylinder wall $|z| = 1$ is a streamline. Find the speed and show that the stagnation points are

$$z = \frac{iK}{4\pi} \pm \sqrt{\frac{-K^2}{16\pi^2} + 1};$$

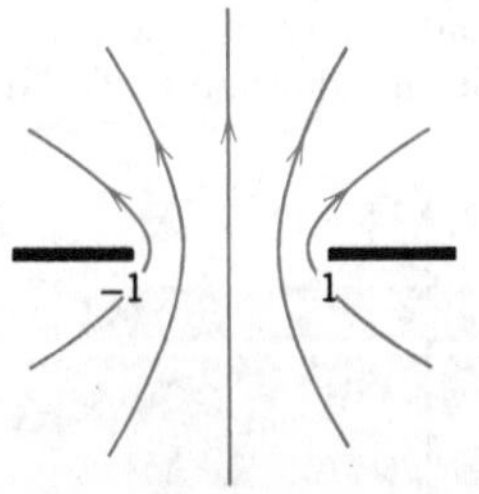

Fig. 394. Flow through an aperture in Problem 18

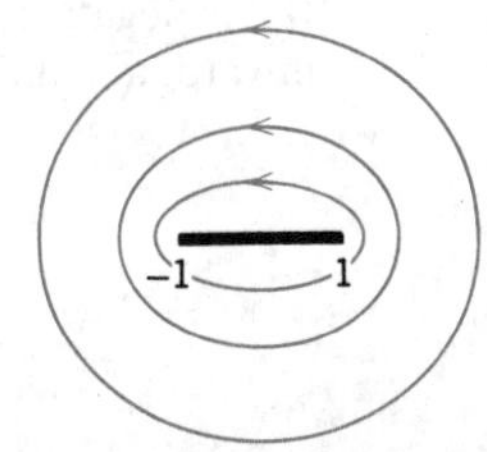

Fig. 395. Flow around a plate in Problem 19

if $K = 0$ they are at ± 1, as K increases they move up on the unit circle until they unite at $z = i$ ($K = 4\pi$, see Fig. 396), and if $K > 4\pi$ they lie on the imaginary axis (one lies in the field of flow and the other one lies inside the cylinder and has no physical meaning).

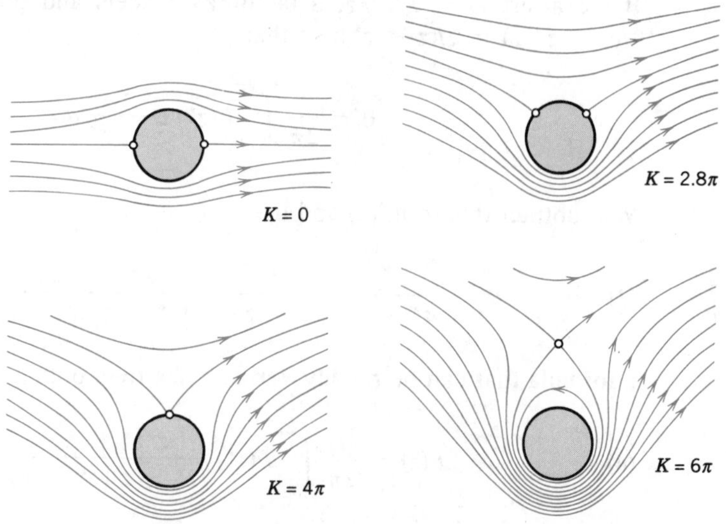

Fig. 396. Flow around a cylinder without circulation ($K = 0$) and with circulation (Problem 20)

17.5 Poisson's Integral Formula

So far in this chapter we have seen that complex analysis offers powerful methods for discussing and solving two-dimensional potential problems based on conformal mappings and complex potentials. A further method results from complex integration. From Cauchy's integral formula we shall derive Poisson's integral formula for the solution of the Dirichlet problem (see Sec. 17.2) in a disk (from which solutions in complicated domains are obtained by conformal mapping, as we have seen).

To derive Poisson's formula, we start from Cauchy's integral formula

$$F(z) = \frac{1}{2\pi i} \oint_C \frac{F(z^*)}{z^* - z}\, dz^*, \tag{1}$$

integrating counterclockwise over the circle C: $z^* = Re^{i\alpha}$ ($0 \leqq \alpha \leqq 2\pi$) and assuming that $F(z^*)$ is analytic in a domain containing C and its full interior.

Since $dz^* = iRe^{i\alpha}\, d\alpha = iz^*\, d\alpha$, we obtain from (1)

$$F(z) = \frac{1}{2\pi} \int_0^{2\pi} F(z^*) \frac{z^*}{z^* - z}\, d\alpha \qquad (z^* = Re^{i\alpha},\ z = re^{i\theta}). \tag{2}$$

On the other hand, if instead of z we take a point Z outside C, say, the point $Z = z^*\bar{z}^*/\bar{z}$ (whose absolute value is $R^2/r > R$), then the integrand in (1) is analytic in the disk $|z| \leqq R$, so that the integral is zero by Cauchy's theorem:

$$0 = \frac{1}{2\pi i} \int_C \frac{F(z^*)}{z^* - Z}\, dz^* = \frac{1}{2\pi} \int_0^{2\pi} F(z^*) \frac{z^*}{z^* - Z}\, d\alpha.$$

If we insert $Z = z^*\bar{z}^*/\bar{z}$, a factor z^* cancels and the fraction becomes $1/(1 - \bar{z}^*/\bar{z}) = \bar{z}/(\bar{z} - \bar{z}^*)$, so that

$$0 = \frac{1}{2\pi} \int_0^{2\pi} F(z^*) \frac{\bar{z}}{\bar{z} - \bar{z}^*}\, d\alpha.$$

We subtract this from (2) and use

$$\text{(3)} \qquad \frac{z^*}{z^* - z} - \frac{\bar{z}}{\bar{z} - \bar{z}^*} = \frac{z^*\bar{z}^* - z\,\bar{z}}{(z^* - z)(\bar{z}^* - \bar{z})},$$

a formula that can be readily verified. We then obtain

$$\text{(4)} \qquad F(z) = \frac{1}{2\pi} \int_0^{2\pi} F(z^*) \frac{z^*\bar{z}^* - z\,\bar{z}}{(z^* - z)(\bar{z}^* - \bar{z})}\, d\alpha.$$

From the polar representations of z and z^* we see that the quotient in the integrand is equal to

$$\frac{R^2 - r^2}{(Re^{i\alpha} - re^{i\theta})(Re^{-i\alpha} - re^{-i\theta})} = \frac{R^2 - r^2}{R^2 - 2Rr\cos(\theta - \alpha) + r^2}.$$

Consequently, writing

$$F(z = \Phi(r, \theta) + i\Psi(r, \theta)$$

and taking the real part on both sides of formula (4), we obtain **Poisson's integral formula**[6]

$$\text{(5)} \qquad \Phi(r, \theta) = \frac{1}{2\pi} \int_0^{2\pi} \Phi(R, \alpha) \frac{R^2 - r^2}{R^2 - 2Rr\cos(\theta - \alpha) + r^2}\, d\alpha$$

which represents the harmonic function Φ in the disk $|z| \leqq R$ in terms of its values $\Phi(R, \alpha)$ on the boundary (the circle) $|z| = R$.

Formula (5) is still valid if the boundary function $\Phi(R, \alpha)$ is merely piecewise continuous (as is practically often the case; see Fig. 399, p. 909, for an example). Then (5) gives a function harmonic in the open disk, and on the circle $|z| = R$ equal to the given boundary function, except at points where the latter is discontinuous. A proof can be found in Ref. [D1].

---

[6]SIMÉON DENIS POISSON (1781—1840), French mathematician and physicist, professor in Paris from 1809. His work includes potential theory, partial differential equations (Poisson equation, Sec. 11.1), and probability.

Series Representation of Potential in a Disk

From (5) we may obtain an important series development of Φ in terms of simple harmonic functions. We remember that the quotient in the integrand of (5) was derived from (3). We claim that the right side of (3) is the real part of

$$\frac{z^* + z}{z^* - z} = \frac{(z^* + z)(\bar{z}^* - \bar{z})}{(z^* - z)(\bar{z}^* - \bar{z})} = \frac{z^*\bar{z}^* - z\bar{z} - z^*\bar{z} + z\bar{z}^*}{|z^* - z|^2}.$$

Indeed, the denominator is real and so is $z^*\bar{z}^* - z\bar{z}$ in the numerator, whereas $-z^*\bar{z} + z\bar{z}^* = 2i \operatorname{Im}(z\bar{z}^*)$ in the numerator is pure imaginary. This verifies our claim. Now by the use of the geometric series we obtain

$$\text{(6)} \qquad \frac{z^* + z}{z^* - z} = \frac{1 + (z/z^*)}{1 - (z/z^*)} = \left(1 + \frac{z}{z^*}\right) \sum_{n=0}^{\infty} \left(\frac{z}{z^*}\right)^n = 1 + 2 \sum_{n=1}^{\infty} \left(\frac{z}{z^*}\right)^n.$$

Since $z = re^{i\theta}$ and $z^* = Re^{i\alpha}$, we have

$$\operatorname{Re} \left(\frac{z}{z^*}\right)^n = \operatorname{Re} \left[\frac{r^n}{R^n} e^{in\theta} e^{-in\alpha}\right] = \left(\frac{r}{R}\right)^n \cos (n\theta - n\alpha).$$

On the right, $\cos (n\theta - n\alpha) = \cos n\theta \cos n\alpha + \sin n\theta \sin n\alpha$, so that from (6) we obtain

$$\operatorname{Re} \frac{z^* + z}{z^* - z} = 1 + 2 \sum_{n=1}^{\infty} \left(\frac{r}{R}\right)^n (\cos n\theta \cos n\alpha + \sin n\theta \sin n\alpha).$$

This expression is equal to the quotient in (5), as we have mentioned before, and by inserting the series into (5) and integrating term by term we find

$$\text{(7)} \qquad \Phi(r, \theta) = a_0 + \sum_{n=1}^{\infty} \left(\frac{r}{R}\right)^n (a_n \cos n\theta + b_n \sin n\theta)$$

where the coefficients are

$$\text{(8)} \qquad \begin{aligned} a_0 &= \frac{1}{2\pi} \int_0^{2\pi} \Phi(R, \alpha)\, d\alpha, \qquad a_n = \frac{1}{\pi} \int_0^{2\pi} \Phi(R, \alpha) \cos n\alpha\, d\alpha, \\ b_n &= \frac{1}{\pi} \int_0^{2\pi} \Phi(R, \alpha) \sin n\alpha\, d\alpha, \qquad n = 1, 2, \cdots, \end{aligned}$$

the Fourier coefficients of $\Phi(R, \alpha)$; see Sec. 10.2. Now for $r = R$ the series (7) becomes the Fourier series of $\Phi(R, \alpha)$, and therefore the representation (7) will be valid whenever $\Phi(R, \alpha)$ can be represented by a Fourier series.

EXAMPLE 1 **Dirichlet problem for the unit disk**

Find the electrostatic potential $\Phi(r, \theta)$ in the unit disk $r < 1$ having the boundary values

$$\Phi(1, \alpha) = \begin{cases} -\alpha/\pi & \text{if } -\pi < \alpha < 0 \\ \alpha/\pi & \text{if } 0 < \alpha < \pi \end{cases} \qquad \text{(Fig. 397).}$$

Solution. Since $\Phi(1, \alpha)$ is even, $b_n = 0$, and from (8) we obtain $a_0 = \frac{1}{2}$ and

$$a_n = \frac{1}{\pi}\left[-\int_{-\pi}^{0} \frac{\alpha}{\pi} \cos n\alpha \, d\alpha + \int_{0}^{\pi} \frac{\alpha}{\pi} \cos n\alpha \, d\alpha\right] = \frac{2}{n^2\pi^2}(\cos n\pi - 1).$$

Hence, $a_n = -4/n^2\pi^2$ if n is odd, $a_n = 0$ if $n = 2, 4, \cdots$, and the potential is

$$\Phi(r, \theta) = \frac{1}{2} - \frac{4}{\pi^2}\left[r \cos \theta + \frac{r^3}{3^2} \cos 3\theta + \frac{r^5}{5^2} \cos 5\theta + \cdots\right].$$

Figure 398 shows the unit disk and some of the equipotential lines (curves $\Phi = const$). ∎

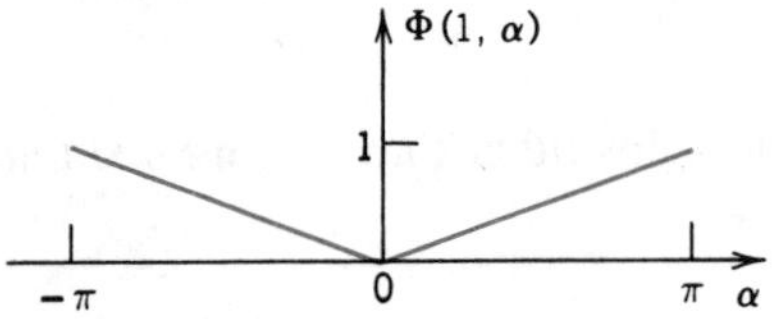

Fig. 397. Boundary values in Example 1

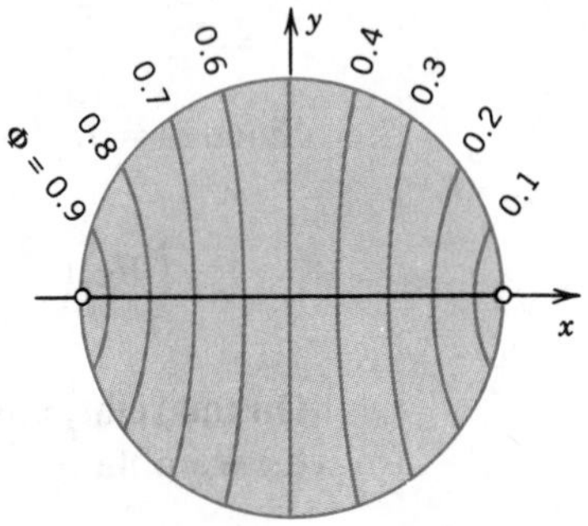

Fig. 398. Potential in Example 1

Problem Set 17.5

1. Verify (3).
2. Show that each term in (7) is a harmonic function in the disk $r^2 < R^2$.

Using (7), find the potential $\Phi(r, \theta)$ in the unit disk $r < 1$ having the given boundary values $\Phi(1, \theta)$. Using the first few terms of the series, compute some values of Φ and sketch a figure of the equipotential lines.

3. $\Phi(1, \theta) = \sin \theta$
4. $\Phi(1, \theta) = \sin 3\theta$
5. $\Phi(1, \theta) = 2 - \cos \theta$
6. $\Phi(1, \theta) = 1 + \cos 2\theta$
7. $\Phi(1, \theta) = \frac{1}{2} \sin 5\theta$
8. $\Phi(1, \theta) = \cos 2\theta - \cos 4\theta$
9. $\Phi(1, \theta) = \sin^2 \theta$
10. $\Phi(1, \theta) = \cos^4 \theta$
11. $\Phi(1, \theta) = \theta$ if $-\pi < \theta < \pi$
12. $\Phi(1, \theta) = \theta$ if $0 < \theta < 2\pi$
13. $\Phi(1, \theta) = \theta$ if $-\pi/2 < \theta < \pi/2$, $\Phi(1, \theta) = \pi - \theta$ if $\pi/2 < \theta < 3\pi/2$
14. $\Phi(1, \theta) = 1$ if $0 < \theta < \pi/2$, $\Phi(1, \theta) = -1$ if $\pi/2 < \theta < \pi$ and 0 otherwise

15. Using (18) in Sec. 14.4, show that the result of Prob. 14 may be written

$$\Phi(r, \theta) = \frac{1}{\pi} \operatorname{Im} \operatorname{Ln} \frac{(1 + iz)(1 + z^2)}{(1 - iz)(1 - z^2)}.$$

16. Show that the potential in Prob. 11 may be written $\Phi(r, \theta) = 2 \operatorname{Im} \operatorname{Ln} (1 + z)$.

17. Using (7) and (8), show that the potential $\Phi(r, \theta)$ in the unit disk $r < 1$ having the boundary values -1 if $-\pi < \theta < 0$ and 1 if $0 < \theta < \pi$ is given by the series

$$\Phi(r, \theta) = \frac{4}{\pi}\left(r \sin \theta + \frac{r^3}{3} \sin 3\theta + \frac{r^5}{5} \sin 5\theta + \cdots\right).$$

Compute some values of Φ by the use of the first few terms of this series and sketch some of the equipotential lines. Compare the result with Fig. 399. Sketch the lines of force (orthogonal trajectories).

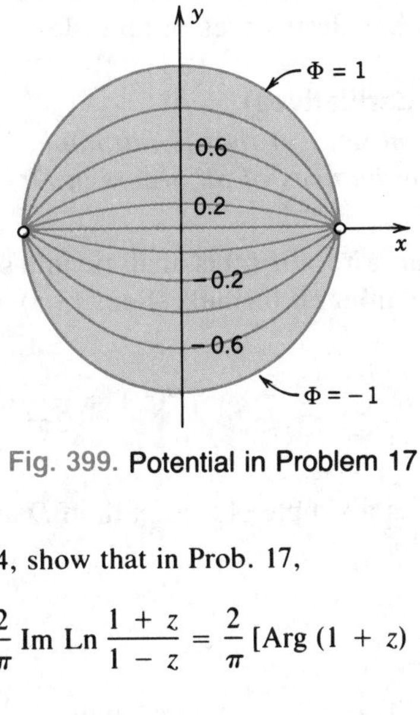

Fig. 399. Potential in Problem 17

18. Using (18), Sec. 14.4, show that in Prob. 17,

$$\Phi(r, \theta) = \frac{2}{\pi} \operatorname{Im} \operatorname{Ln} \frac{1 + z}{1 - z} = \frac{2}{\pi} [\operatorname{Arg} (1 + z) - \operatorname{Arg} (1 - z)].$$

19. Show that

$$\Phi^*(w) = 1 + \frac{2}{\pi} \operatorname{Im} \operatorname{Ln} \frac{w + 1}{w - 1} \qquad (w = u + iv)$$

is harmonic in the upper half-plane $v > 0$ and has the values -1 for $-1 < u < 1$ and $+1$ on the remaining part of the u-axis.

20. Show that the linear fractional transformation which maps $w_1 = -1$, $w_2 = 0$, $w_3 = 1$ onto $z_1 = -1$, $z_2 = -i$, $z_3 = 1$, respectively, is

$$z = \frac{w - i}{-iw + 1}.$$

Find the inverse $w = w(z)$, insert it into Φ^* in Prob. 19, and show that the resulting harmonic function is that in Prob. 18. Compare also with Example 2 in Sec. 17.2.

17.6 General Properties of Harmonic Functions

Complex analysis helps not only in solving two-dimensional potential problems, as we have seen, but also in deriving general properties of harmonic functions, as we show in this last section of the chapter.

If $\Phi(x, y)$ is harmonic in a domain D and D is simply connected, then Φ has a conjugate harmonic Ψ in D and

$$F(z) = \Phi(x, y) + i\Psi(x, y)$$

is analytic in D. (See Sec. 12.5 and footnote 3 in Sec. 17.2.) Since an analytic function has derivatives of all orders, our first result is

Theorem 1 (Partial derivatives)
A function $\Phi(x, y)$ that is harmonic in a simply connected domain D has partial derivatives of all orders in D.

Furthermore, if $F(z)$ is analytic in a simply connected domain D, then, by Cauchy's integral formula (Sec. 13.5),

$$F(z_0) = \frac{1}{2\pi i}\oint_C \frac{F(z)}{z - z_0}\,dz \tag{1}$$

where C is a simple closed path in D and z_0 lies inside C. Choosing for C a circle

$$z = z_0 + re^{i\alpha}$$

in D, we have $z - z_0 = re^{i\alpha}$ and $dz = ire^{i\alpha}\,d\alpha$, so that (1) becomes

$$F(z_0) = \frac{1}{2\pi}\int_0^{2\pi} F(z_0 + re^{i\alpha})\,d\alpha. \tag{2}$$

The right side is the mean value of F on the circle (= value of the integral divided by the length of the interval of integration). This proves the following theorem.

Theorem 2 (Mean value property of analytic functions)
Let $F(z)$ be analytic in a simply connected domain D. Then the value of $F(z)$ at a point z_0 in D is equal to the mean value of $F(z)$ on any circle in D with center at z_0.

This theorem and the next one, both characterizing important properties of analytic functions, will subsequently be used to derive basic properties of harmonic functions.

Theorem 3 **(Maximum modulus theorem for analytic functions)**
Let $F(z)$ be analytic and nonconstant in a domain containing a bounded[7] region D and its boundary. Then the absolute value $|F(z)|$ cannot have a maximum at an interior point of D. Consequently, the maximum of $|F(z)|$ is taken on the boundary of D. If $F(z) \neq 0$ in D, the same is true with respect to the minimum of $|F(z)|$.

Proof. We assume that $|F(z)|$ has a maximum at an interior point z_0 of D and show that this leads to a contradiction. Let $|F(z_0)| = M$ be this maximum. Since $F(z)$ is not constant, $|F(z)|$ is not constant, as follows from Example 4 in Sec. 12.5. Consequently, we can find a circle C of radius r with center at z_0 such that the interior of C is in D and $|F(z)|$ is smaller than M at some point P of C. Since $|F(z)|$ is continuous, it will be smaller than M on an arc C_1 of C that contains P, say,

$$|F(z)| \leqq M - k \quad (k > 0) \qquad \text{for all } z \text{ on } C_1 \qquad \text{(Fig. 400).}$$

If C_1 has the length L_1, the complementary arc C_2 of C has the length $2\pi r - L_1$. By applying the ML-inequality (Sec. 13.2) to (1) and noting that $|z - z_0| = r$, we thus obtain

$$M = |F(z_0)| \leqq \frac{1}{2\pi}\left|\int_{C_1} \frac{F(z)}{z - z_0}\,dz\right| + \frac{1}{2\pi}\left|\int_{C_2} \frac{F(z)}{z - z_0}\,dz\right|$$

$$\leqq \frac{1}{2\pi}\left(\frac{M - k}{r}\right) L_1 + \frac{1}{2\pi}\left(\frac{M}{r}\right)(2\pi r - L_1) = M - \frac{kL_1}{2\pi r} < M,$$

that is,

$$M < M,$$

which is impossible. Hence, our assumption is false and the first statement of the theorem is proved.

We prove the second statement. If $F(z) \neq 0$ in D, then $1/F(z)$ is analytic in D. From the statement already proved it follows that the maximum of $1/|F(z)|$ lies on the boundary of D. But this maximum corresponds to a minimum of $|F(z)|$. This completes the proof. ∎

Basic consequences of Theorems 2 and 3 for harmonic functions can now be stated and derived as follows.

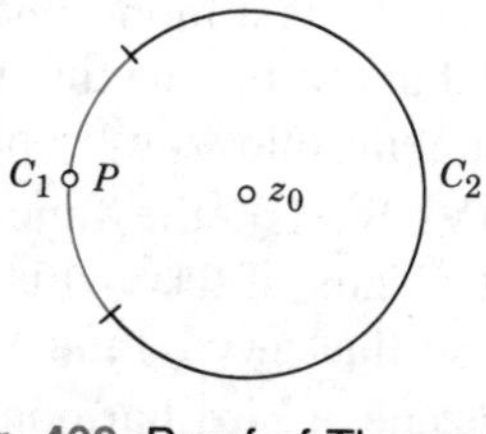

Fig. 400. Proof of Theorem 3

[7]See Sec. 9.3.

Theorem 4 **(Harmonic functions)**
Let $\Phi(x, y)$ be harmonic in a domain containing a simply connected bounded domain D and its boundary curve C. Then $\Phi(x, y)$ has the following properties.

I. The value of $\Phi(x, y)$ at a point (x_0, y_0) in the domain D is equal to the mean valuue of $\Phi(x, y)$ on any circle in D with center at (x_0, y_0).

II. The value of $\Phi(x, y)$ at the point (x_0, y_0) is equal to the mean value of $\Phi(x, y)$ in any circular disk in D with center at (x_0, y_0) [See footnote 5 in Sec. 17.4.]

III. **(Maximum principle)** *If $\Phi(x, y)$ is not constant, it has neither a maximum nor a minimum in D. Consequently, the maximum and the minimum are taken on the boundary of D.*

IV. If $\Phi(x, y)$ is constant on C, then $\Phi(x, y)$ is a constant.

V. If $h(x, y)$ is harmonic in D and on C and if $h(x, y) = \Phi(x, y)$ on C, then $h(x, y) = \Phi(x, y)$ everywhere in D.

Proof. Statement I follows from (2) by taking the real parts on both sides:

$$\Phi(x_0, y_0) = \text{Re } F(x_0 + iy_0) = \frac{1}{2\pi} \int_0^{2\pi} \Phi(x_0 + r \cos \alpha, y_0 + r \sin \alpha) \, d\alpha.$$

If we multiply both sides of this by r and integrate over r from 0 to r_0, where r_0 is the radius of a circular disk in D with center at (x_0, y_0), then we obtain on the left side $\frac{1}{2}r_0^2\Phi(x_0, y_0)$ and therefore

$$\Phi(x_0, y_0) = \frac{1}{\pi r_0^2} \int_0^{r_0} \int_0^{2\pi} \Phi(x_0 + r \cos \alpha, y_0 + r \sin \alpha) r \, d\alpha \, dr.$$

This proves the second statement.

We prove statement III. Let $\Psi(x, y)$ be a conjugate harmonic function of $\Phi(x, y)$ in D. Then $F(z) = \Phi(x, y) + i\Psi(x, y)$ is analytic in D, and so is

$$G(z) = e^{F(z)}.$$

The absolute value is

$$|G(z)| = e^{\text{Re } F(z)} = e^{\Phi(x, y)}.$$

From Theorem 3 it follows that $|G(z)|$ cannot have a maximum at an interior point of D. Since e^{Φ} is a monotone increasing function of the real variable Φ, statement III about the maximum of Φ follows. From this, the statement about the minimum follows by replacing Φ by $-\Phi$.

We prove IV. By III, the function $\Phi(x, y)$ takes its maximum and its minimum on C. Thus, if $\Phi(x, y)$ is constant on C, its minimum must equal its maximum, so that $\Phi(x, y)$ must be a constant.

Finally, if h and Φ are harmonic in D and on C, then $h - \Phi$ is also harmonic in D and on C, and by assumption, $h - \Phi = 0$ everywhere on C. Hence, by IV, we have $h - \Phi = 0$ everywhere in D, and statement V is proved. This completes the proof of Theorem 4. ∎

The last statement of Theorem 4 is very important. It means that a *harmonic function is uniquely determined in D by its values on the boundary of D*. Usually, $\Phi(x, y)$ is required to be harmonic in D and continuous on the boundary[8] of D. Under these circumstances the maximum principle (Theorem 4, III) is still applicable. The problem of determining $\Phi(x, y)$ when the boundary values are given is called the Dirichlet problem for the Laplace equation in two variables, as we know. From Theorem 4, V we thus have

Theorem 5 **(Dirichlet problem)**
If for a given region and given boundary values the Dirichlet problem for the Laplace equation in two variables has a solution, the solution is unique.

This is the end of Chap. 17 and Part D on complex analysis. We hope that the reader has gained an impression of the usefulness of complex analysis to the engineer and physicist, and of the mathematical beauty of this area.

Problem Set 17.6

1. Verify Theorem 2 for $F(z) = (z + 2)^2$, $z_0 = 1$, and a circle of radius 1 with center at z_0.
2. Verify Theorem 2 for $F(z) = 5z^4$, $z_0 = 0$, and a circle of radius 1 about 0.
3. Integrate $F(z) = |z|$ around the unit circle. Does the result contradict Theorem 2?
4. Verify Theorem 3 for $F(z) = z^2$ and the rectangle $1 < x < 9$, $3 < y < 5$.
5. Verify Theorem 3 for $F(z) = e^z$ and any bounded domain.

Find the maximum of $|F(z)|$ in the disk $|z| \leqq 1$ and the corresponding z, where

6. $F(z) = az + b$ 7. $F(z) = z^2 - 1$ 8. $F(z) = \cos z$

9. The function $F(z) = 1 + 3|z|^2$ is not zero in the disk $|z| \leqq 2$ and has a minimum at an interior point of that disk. Does this contradict Theorem 3?
10. The *real* function $F(x) = \sin x$ has a maximum at $x = \pi/2$. Using Theorem 3, conclude that this cannot be a maximum of the absolute value of the *complex* function $F(z) = \sin z$ in a domain containing $z = \pi/2$.
11. Let $F(z)$ be analytic (not a constant) in the closed disk $|z| \leqq 1$ and suppose that $|F(z)| = c = const$ on $|z| = 1$. Show that then $F(z)$ must have a zero in that disk.
12. If $F(z)$ is analytic (not a constant) in a simply connected domain D, and the curve given by $|F(z)| = c$ (c any fixed constant) lies in D and is closed, show that $F(z) = 0$ at a point in the interior of that curve. Give examples.
13. Verify statement I in Theorem 4 for $\Phi(x, y) = x^2 - y^2$ and a circle of radius 1 about $(x_0, y_0) = (1, 0)$.
14. Verify the maximum principle in Theorem 4 for $\Phi(x, y) = xy$ and the disk $x^2 + y^2 \leqq 8$. Determine the maximum and minimum values of Φ in that disk and the points at which Φ takes those values.
15. Prove statement I of Theorem 4 by applying Poisson's integral formula.

[8]That is, $\lim_{\substack{x \to x_0 \\ y \to y_0}} \Phi(x, y) = \Phi(x_0, y_0)$, where (x_0, y_0) is on the boundary and (x, y) is in D.

Review Questions and Problems for Chapter 17

1. Why can complex analytic functions be used in solving two-dimensional potential problems? What do we mean by a "two-dimensional" problem?
2. What is a complex potential, and what advantage does it have?
3. What do we mean by the Dirichlet problem?
4. How is conformal mapping applied in solving the Dirichlet problem?
5. What is a mixed boundary value problem, and where in this chapter did it occur?
6. What is a streamline?
7. State the conditions a flow must satisfy if it is to be treated by methods of complex analysis.
8. Explain why steady-state heat flow is related to potential theory.
9. What do we mean by superposition of potentials?
10. List some of the remarkable properties common to all harmonic functions.

11. Find the potential between the cylinders $|z| = 1$, having potential zero, and $|z| = 5$, having potential 100 volts.
12. Find the complex potential in Prob. 11.
13. Find the potential between the cylinders $|z| = 10$ and $|z| = 100$ kept at the potentials 10 kV and 0, respectively.
14. Find the equipotential line $U = 0$ between the cylinders $|z| = 0.25$ and $|z| = 4$ kept at -220 V and 220 V, respectively. (Guess first.)
15. Find the potential Φ in the first quadrant of the xy-plane if the x-axis has potential 110 V and the y-axis is grounded (0 volts).
16. Find the potential between the plates $\operatorname{Arg} z = \pi/6$, kept at 4 kV, and $\operatorname{Arg} z = \pi/3$, kept at 3 kV.
17. What is the complex potential in the upper half-plane if the negative x-axis has potential 400 V and the positive x-axis is grounded?
18. Find the potential on the ray $y = x$, $x > 0$, and on the positive x-axis if the positive y-axis is at 1200 V and the negative y-axis at 0.
19. Find the potential between the plates $y = x/2$, kept at 100 volts, and $y = x/2 + 4$, kept at 420 volts.
20. In Prob. 18, find a formula for the potential on the line $x = 1$. Find the potential at the point $x = 1$, $y = 2$.
21. Find the isotherms of $F(z) = 10(1 + i)z$ and show that $F(z)$ can be interpreted as the complex potential of heat flow between two parallel plates.
22. Find the temperature in the upper half-plane if the portion $x > 4$ of the x-axis is kept at 100°C and the other portion at 0°C.
23. Show that the isotherms of $F(z) = -iz^2 + 2z$ are hyperbolas.
24. If the region between two concentric cylinders of radii 2 cm and 10 cm contains water and the outer cylinder is kept at 20°C, to what temperature must we heat the inner cylinder in order to have 30°C at distance 5 cm from the axis?
25. Find the isotherms of $F(z) = k \operatorname{Ln}(z - 1 - i)$, where k is real, and show that this is the complex potential between coaxial cylinders with axis at $1 + i$.
26. Find the temperature in the unit disk $|z| \leqq 1$ in the form of an infinite series if the left semicircle of $|z| = 1$ has the temperature of 100°C and the right semicircle the temperature zero.
27. Same task as in Prob. 26 if the upper semicircle is at 20°C and the lower at zero.

28. Find the complex potential in Prob. 27 in the form of a series. What is the sum of this series? *Hint.* Use Example 4 in Sec. 14.4.

29. Same task as in Prob. 26 if $T(1, \theta) = 30\theta^2$ $(-\pi < \theta \leqq \pi)$ on the unit circle.

30. Find and sketch the equipotential lines of $F(z) = 1/(1 - z)$.

Find the streamlines and the velocity of the flow with complex potential

31. $-ikz$ (k real) **32.** $z^2 + z$ **33.** $z + 4/z$ **34.** $z^2 + 1/z^2$

35. Show that the flow in Prob. 34 has stagnation points at ± 1 and $\pm i$. Can you see the relation to Example 2, Sec. 17.4?

Summary of Chapter 17 Complex Analysis Applied to Potential Theory

Potential theory is the theory of solutions of **Laplace's equation**

$$\nabla^2\Phi = 0. \tag{1}$$

Solutions whose second partial derivatives are ***continuous*** are called **harmonic functions.** Equation (1) is the most important partial differential equation in physics, where it is of interest in two and three dimensions. It appears in electrostatics (Sec. 17.1), steady-state heat problems (Sec. 17.3), fluid flow (Sec. 17.4), gravity, etc. Whereas the three-dimensional case requires other methods (see Chap. 11), two-dimensional potential theory can be handled by complex analysis, since the real and imaginary parts of an analytic function are harmonic (Sec. 12.5). They remain harmonic under conformal mapping (Sec. 17.2), so that **conformal mapping** becomes a powerful tool in solving boundary value problems for (1), as is illustrated in this chapter. With a real potential Φ we can associate a **complex potential** (Sec. 17.1)

$$F(z) = \Phi + i\Psi. \tag{2}$$

Then both families of curves $\Phi = \mathit{const}$ and $\Psi = \mathit{const}$ have a physical meaning. In electrostatics, they are equipotential lines and lines of electrical force of attraction or repulsion (Sec. 17.1). In heat problems, they are isotherms (curves of constant temperature) and lines of heat flow (Sec. 17.3). In fluid flow, they are equipotential lines of the velocity potential and streamlines (Sec. 17.4).

For the disk, the solution of the Dirichlet problem is given by the **Poisson formula** (Sec. 17.5) or by a series that on the boundary circle becomes the Fourier series of the given boundary values (Sec. 17.5).

Harmonic functions, like analytic functions, have a number of general properties; particularly important are the **mean value property** and the **maximum modulus property** (Sec. 17.6), which implies the uniqueness of the solution of the Dirichlet problem (Theorem 5 in Sec. 17.6).

PART E

NUMERICAL METHODS

No other field of mathematics has shown a recent increase in importance to the engineer comparable to that of numerical methods, nor has any other field developed as rapidly. Of course, the main reason for this evolution is the development of various computers, from the personal computer to the Cray, CM-5, and beyond, and we can see no end to it. Indeed, each new generation of computers invites new tasks in numerical analysis; in this connection, even small improvements in algorithms may have a great impact on time, storage demand, accuracy, and stability. This opens up wide areas of research, a main goal being the development of well-structured software.

Chapters 18–20 concern the study and application of **numerical methods,** which provide the transition from the mathematical model of a problem (the equations or functions obtained in calculus or algebra, etc.) to an **algorithm** that we can program (or use directly on a pocket calculator) to obtain the solution of the problem in the form of numbers. This includes the investigation of the range of applicability of numerical methods, and their error analysis, stability, and properties in general.

We begin with numerical methods of a general nature in Chap. 18. In Chap. 19 we discuss numerical methods for problems in linear algebra, in particular, the solution of linear systems of equations and algebraic eigenvalue problems. Chapter 20 is devoted to the numerical solution of ordinary and partial differential equations.

We give the algorithms in a form that seems best for showing how a method works and how to program it, even with little experience. The student is encouraged to program the given algorithms and try them out on the computer.[1] We also recommend strongly that the student makes use of programs

[1]No actual programs, FORTRAN or other, are given, because, in our experience, this could encourage some students to generate results without fully understanding the underlying numerical method.

in high-quality commercial program libraries. The two major ones of them for mainframe computers are those of IMSL (NBC Building, 7500 Bellaire Boulevard, Houston, TX 77036-5085), written in FORTRAN 77, and NAG (Mayfield House, 256 Banbury Road, Oxford OX2 7DE, United Kingdom), available in both FORTRAN and ALGOL. Other packages can also be obtained from National Energy Software Center (Argonne National Laboratory, 9700 South Cass Avenue, Argonne, IL 60439). See also EISPACK and LINPACK (Refs. [E6], [E9], and [E19] in Appendix 1).

Chapter 18

Numerical Methods in General

Numerical methods are methods for solving problems on a **computer** (or on a pocket calculator if the problem is simple). The computer has become very important in engineering work. It provides access to problems so large that they were out of reach in precomputer times. Much computing today is *"real-time"*: it is done almost simultaneously with the process of generating data, for instance, in controlling ongoing chemical processes or guiding airplanes. Questions of speed, storage demand, and timing of portions of long programs then become very crucial.

Computers have changed, almost revolutionized, numerical methods—the field as a whole as well as many individual methods—and that development is continuing. Much research work is going on in creating new methods, adapting existing methods to new computer generations, improving methods—in large-scale work even small improvements bring large savings in time or storage space—and investigating stability and accuracy of methods.

The purpose of this chapter is twofold. First, for the most important practical tasks, the student should become familiar with the most basic (but not too complicated) solution methods.[2] Such methods are needed because for many problems there is no solution formula (think of a complicated integral or of the roots of a polynomial of high degree) or in other cases a solution formula may be practically useless.

Second, the student should learn to understand some basic ideas and concepts that are important throughout the field; this includes the idea of an algorithm, rounding errors, error estimation in general, ill-conditioning, order of convergence, and stability.

In the first section we explain some concepts that are basic in numerical work; this includes general remarks on computing and computers. Each of the other sections of the chapter is devoted to methods for a specific task that is important throughout applied mathematics, regardless of the particular field of application.

Prerequisite for this chapter: elementary calculus.
References: Appendix 1, Part E.
Answers to problems: Appendix 2.

[2]This includes those for numerical linear algebra and differential equations in Chaps. 19 and 20.

18.1 Introduction

Numerical methods are methods for solving problems on (large or small) computers (or on calculators if the problem is simple). The steps from the given situation (in physics, economics, etc.) to the final answer are usually as follows.

1. Modeling. Setting up a mathematical model, that is, formulating the problem in mathematical terms.

2. Choice of mathematical methods, together with a preliminary error analysis (estimation of error, determination of step size, etc.; see below).

3. Programming. Writing a program, say, in FORTRAN, resulting from an algorithm or a flow chart (showing a block diagram of the procedures to be performed by the computer).

4. Running the program.

5. Interpretation of results. This may also include decisions to rerun if further results are needed.

Steps 1 and 2 are related: a slight change in the model may often permit the use of a more efficient method.

In Step 3 the **program** consists of all relevant given data and a sequence of instructions that the computer will execute in a certain order, thereby eventually producing a numerical answer (or a graph, etc.) to the problem. The program is usually written in FORTRAN or some other high-level language. A compiler then translates this program into a sequence of machine instructions that performs the desired task.

Beginning in the next section and continuing until the end of Chap. 20, our main concern will be Step 2, the discussion and application of basic numerical methods for the most important classes of problems that arise in practice. To create a good understanding of the nature of numerical work, we continue in this section with some simple remarks on computing and computers.

Floating-Point Form of Numbers

In the decimal notation, every real number is represented by a finite or infinite sequence of decimal digits. For machine computation the number must be replaced by a number of finitely many digits. Most digital computers have two ways of representing numbers, called *fixed point* and *floating point*. In a **fixed-point** system all numbers are given with a *fixed number of decimal places,* for example, 62.358, 0.013, 1.000. In a **floating-point** system numbers are given with a *fixed number of significant digits,* for instance,

$$0.6238 \times 10^3 \qquad 0.1714 \times 10^{-13} \qquad -0.2000 \times 10^1$$

also written[3]

$$0.6238\text{E}03 \qquad 0.1714\text{E}-13 \qquad -0.2000\text{E}01.$$

Significant digit of a number c is any given digit of c, except possibly for zeros to the left of the first nonzero digit that serve only to fix the position

[3] One also uses a representation of the form 6.238×10^2, 1.1714×10^{-14}, etc.

of the decimal point. (Thus, any other zero is a significant digit of c.) For instance, each of the numbers 1360, 1.360, 0.001 360 has 4 significant digits.[4]

Most computers use (internally) the binary number system, whose base is 2 (see Probs. 3–5). A binary digit is briefly called a **bit.** By grouping bits, one can obtain octal (base 8) or hexadecimal (base 16) representations. In the computer, a number, represented in floating point, consists of the sign, a fractional part, called the *mantissa,*[5] and the exponential part, called the *characteristic*. For example, in the IBM series, a (single-precision) floating-point number consists of 1 bit for the sign, 24 bits for the mantissa (corresponding to 6–7 decimal digits), and 7 bits for the (base 16) exponent, allowing values from -64 to 63. If in a computation a number greater than 16^{63} ($\approx 10^{76}$) occurs (which does happen!), this is called **overflow.** Then the computer halts. If a number less than 16^{-64} occurs, this is called **underflow.** This is also similar for a computer with another range of the size of numbers. In many computers, numbers causing underflow are set to 0.

The fixed-point form of numbers is impractical in physics, chemistry, etc., because of its limited range (explain!) and will not be of concern to us.

Round-Off

An error is caused by **chopping** (= discarding all decimals from some decimal on) or **rounding.** The rule for rounding off a number to k decimals is as follows. (The rule for rounding off to k significant figures is the same, with "decimal" replaced by "significant figure.")

Round-off rule. Discard the $(k + 1)$th and all subsequent decimals. (a) If the number thus discarded is less than half a unit in the kth place, leave the kth decimal unchanged ("*rounding down*"). (b) If it is greater than half a unit in the kth place, add one to the kth decimal ("*rounding up*"). (c) If it is exactly half a unit, round off to the nearest *even* decimal. (Example: Rounding off 3.45 and 3.55 to 1 decimal gives 3.4 and 3.6, respectively.)

The last part of the rule is supposed to ensure that in discarding exactly half a decimal, rounding up and rounding down happens about equally often, on the average.

If we round off 1.2535 to 3, 2, 1 decimals, we get 1.254, 1.25, 1.3, but if 1.25 is rounded off to one decimal, without further information, we get 1.2.

Chopping is not recommended since it introduces an error that is systematic and can be larger than an error in rounding off. Nevertheless, surprisingly

[4]In tables of functions showing k significant digits, it is conventionally assumed that any given value $\tilde{a}$ deviates from the corresponding exact value a by at most ± 0.5 unit of the last given digit, unless otherwise stated; for example, if $a = 1.1996$, then a table with 4 significant digits should show $\tilde{a} = 1.200$. Correspondingly, if 12 000 is correct to three digits only, we should write 120×10^2, etc. "Decimal" is abbreviated by D and "significant digit" by S. For example, 5D means 5 decimals, and 8S means 8 significant digits.

[5]This has nothing to do with "mantissa" as used in connection with logarithms. **"Single precision"** means the number of bits normally used in computations by the computer; **"double precision"** means twice as many bits, etc.

many computers use chopping! A reason is that rounding is time-consuming, and manufacturers try all sorts of shortcuts to make the basic arithmetical operations as fast as possible. Most computers that use rounding off always round *up* in case (c) of the rule (or in the corresponding case for a base other than 10), since this is easier to realize technically.

Rounding errors may ruin a computation completely, even a small computation. In general, they are the more dangerous the more arithmetic operations (perhaps several millions!) we have to perform. It is therefore important to analyze computational programs for rounding errors to be expected and to find an arrangement of the computations such that the effect of rounding errors is as small as possible.

Algorithm. Stability

An **algorithm** is a finite sequence of rules for performing computations on a computer such that at each instant the rules determine exactly what the computer has to do next. These rules include a *"stopping rule"* that makes the computer stop, so it cannot run on indefinitely. Important algorithms follow in the next sections.

Stability. To be useful, an algorithm should be **stable;** that is, small changes in the initial data should give only correspondingly small changes in the final results. However, if small changes in the initial data produce large changes in the final results, we call the algorithm **unstable.**

This *"numerical instability,"* which can be avoided by choosing a better algorithm, must be distinguished from *"mathematical instability"* of a problem, which is called *"ill-conditioning,"* a concept we discuss in the next section.

Some algorithms are stable only for certain initial data, so that one must be careful in such a case.

Programming Errors

A common name for all kinds of errors in a computer program is **bugs,** and **debugging** means locating and removing bugs. Programming needs experience. Experience cannot be *taught* but must be *gained.* Nevertheless, a few general hints may be in order.

Prepare your program with the greatest care—it is easier to avoid errors this way than it is to discover them later. There are no general rules guaranteeing the detection of all bugs in all programs. Compilers provide *diagnostics* indicating all errors in a source program, except for errors in logic. Perhaps the best way to determine whether a program has bugs is to run it with data for which the answers are known or can be easily obtained in some other way.

If you are convinced that bugs exist because of nonsensical results but tests fail to actually find the bugs, apply selective (or even full) **tracing,** that is, printing out intermediate results and checking them over step by step.

Errors of Numerical Results

Final results of computations of unknown quantities generally are **approximations;** that is, they are not exact but involve errors. Such an error may result from a combination of the following effects. **Round-off errors** result from rounding, as discussed above. **Experimental errors** are errors of given data (probably arising from measurements). **Truncating errors** result from truncating (prematurely breaking off), for instance, if we replace a Taylor series by the sum of its first few terms. These errors depend on the computational method used and must be dealt with individually for each method. ["Truncating" is sometimes used as a term for chopping off (see before), a terminology that is not recommended.]

Formulas for errors. If $\tilde{a}$ is an approximate value of a quantity whose exact value is a, we call the difference

$$\epsilon = a - \tilde{a} \tag{1}$$

the **error** of $\tilde{a}$. Hence[6]

$$a = \tilde{a} + \epsilon, \qquad \text{True value} = \text{Approximation} + \text{Error}.$$

For instance, if $\tilde{a} = 10.5$ is an approximation of $a = 10.2$, its error is $\epsilon = -0.3$. The error of an approximation $\tilde{a} = 1.60$ of $a = 1.82$ is $\epsilon = 0.22$.

The **relative error** ϵ_r of $\tilde{a}$ is defined by

$$\epsilon_r = \frac{\epsilon}{a} = \frac{a - \tilde{a}}{a} = \frac{\text{Error}}{\text{True value}} \qquad (a \neq 0). \tag{2}$$

This looks useless because a is unknown. But if $|\epsilon|$ is much less than $|\tilde{a}|$, then we can use $\tilde{a}$ instead of a and get

$$\epsilon_r \approx \frac{\epsilon}{\tilde{a}}. \tag{2'}$$

This still looks problematic because ϵ is unknown—if it were known, we could get $a = \tilde{a} + \epsilon$ from (1) and we would be done. But what one can get in practice is an **error bound** for $\tilde{a}$, that is, a number β such that

$$|\epsilon| \leqq \beta, \qquad \text{hence} \qquad |a - \tilde{a}| \leqq \beta.$$

This tells us how far away from our computed $\tilde{a}$ the unknown a can at most lie. Similarly, for the relative error, an error bound is a number β_r such that

$$|\epsilon_r| \leqq \beta_r, \qquad \text{hence} \qquad \left|\frac{a - \tilde{a}}{a}\right| \leqq \beta_r.$$

[6]***Attention!*** Users of the last edition of this book will notice that we have changed our definition from $\epsilon = \tilde{a} - a$ to $\epsilon = a - \tilde{a}$ because the latter seems slightly preferable, resembling standard situations in analysis, such as series = partial sum + remainder, integral = Riemann sum + error, etc.

Both definitions are used in the literature.

Error Propagation

This is an important matter. It refers to how errors at the beginning and in later steps (round-off, for example) propagate into the computation and affect accuracy, sometimes very drastically. We state here what happens to error bounds in addition and subtraction: they add, whereas error bounds for the relative errors add under multiplication and division.

Theorem 1 (Error propagation)

(a) *In addition and subtraction, an error bound for the results is given by the sum of the error bounds for the terms.*

(b) *In multiplication and division, an error bound for the relative error of the results is given (approximately) by the sum of error bounds for the relative errors of the given numbers.*

Proof. **(a)** We use the notations $x = \tilde{x} + \epsilon_1$, $y = \tilde{y} + \epsilon_2$, $|\epsilon_1| \leqq \beta_1$, $|\epsilon_2| \leqq \beta_2$. Then for the error ϵ of the *difference* we get

$$\begin{aligned} |\epsilon| &= |x - y - (\tilde{x} - \tilde{y})| \\ &= |x - \tilde{x} - (y - \tilde{y})| \\ &= |\epsilon_1 - \epsilon_2| \leqq |\epsilon_1| + |\epsilon_2| \leqq \beta_1 + \beta_2. \end{aligned}$$

The proof for the *sum* is similar and is left to the student.

(b) For the relative error ϵ_r of $\tilde{x}\tilde{y}$ we get from the relative errors ϵ_{r1} and ϵ_{r2} of $\tilde{x}$, $\tilde{y}$ and bounds β_{r1}, β_{r2}

$$\begin{aligned} |\epsilon_r| &= \left|\frac{xy - \tilde{x}\tilde{y}}{xy}\right| = \left|\frac{xy - (x - \epsilon_1)(y - \epsilon_2)}{xy}\right| = \left|\frac{\epsilon_1 y + \epsilon_2 x - \epsilon_1\epsilon_2}{xy}\right| \\ &\approx \left|\frac{\epsilon_1 y + \epsilon_2 x}{xy}\right| \leqq |\epsilon_{r1}| + |\epsilon_{r2}| \leqq \beta_{r1} + \beta_{r2}. \end{aligned}$$

This proof shows what "approximately" means: we neglected $\epsilon_1\epsilon_2$ as small in absolute value compared to $|\epsilon_1|$ and $|\epsilon_2|$. The proof for the quotient is similar but slightly more tricky (see Prob. 21). ∎

Comment on Selection of Method

The more powerful the computer is and the bigger your problem is, the more important is the choice of a best possible method—because the greater would be the loss caused by inferior mathematics or inferior computing methods. Brilliant programs cannot make up for poor choices of methods; on the other hand, poor programs can spoil good methods, giving inaccurate results after much too much time. If you use a package program from the library, you should know exactly what its purpose and limitations are. In many cases you should be prepared to modify or exchange the program until you reach the optimum for your problem.

The computer does not reduce the need for a good understanding of the problem area and the related mathematics. Even in very simple cases you can use numerically good or numerically poor mathematics, as we illustrate next.

EXAMPLE 1 Quadratic equation

Find the roots of the equation

$$x^2 - 40x + 2 = 0,$$

using 4 significant digits in the computation.

Solution. A formula for the roots x_1, x_2 of a quadratic equation $ax^2 + bx + c = 0$ is

$$(3) \qquad x_1 = \frac{1}{2a}(-b + \sqrt{b^2 - 4ac}), \qquad x_2 = \frac{1}{2a}(-b - \sqrt{b^2 - 4ac}).$$

Furthermore, since $x_1 x_2 = c/a$, another formula for those roots is

$$(4) \qquad x_1 \text{ as before}, \qquad x_2 = \frac{c}{ax_1}.$$

From (3) we get $x = 20 \pm \sqrt{398} = 20.00 \pm 19.95$, $x_1 = 39.95$, $x_2 = 0.05$, which is poor, whereas (4) gives $x_1 = 39.95$, $x_2 = 2.000/39.95 = 0.05006$, in error by less than one unit of the last digit, as a computation with more figures shows.

Comment. To avoid misunderstandings: 4S was for convenience; (4) is better than (3) regardless of the number of figures used. For instance, the 8S-computation by (3) is $x_1 = 39.949\,937$, $x_2 = 0.050\,063$, which is poor, and by (4) it is x_1 as before, $x_2 = 2/x_1 = 0.050\,062\,657$. ∎

Problem Set 18.1

1. **(Floating point)** Write 98.17, −100.988, 0.004 7869, −13 800 in floating-point form, with 4 significant digits.
2. Write −0.016 8409, 10.27845, −30 681.55 in floating-point form with 6 significant digits.
3. **(Binary representation)** Most computers use the binary number system or a variant of it, such as base 8 or base 16. Writing the familiar base 10 representation in the form $(81.5)_{10}$, verify that

$$(81.5)_{10} = 8 \cdot 10^1 + 1 \cdot 10^0 + 5 \cdot 10^{-1} = 2^6 + 2^4 + 2^0 + 2^{-1} = (1010001.1)_2.$$

   Convert $(100)_{10}$, $(29.25)_{10}$, and $(3.75)_{10}$ to binary form.
4. Convert $(1000)_{10}$, $(13.78125)_{10}$ and $(-22.0625)_{10}$ to binary form.
5. Convert $(11100.1)_2$, $(0.00101)_2$, $(1.10111)_2$, and $(-11.01101)_2$ to base 10 form.
6. **(Small differences of large numbers)** Rounding errors can become particularly disadvantageous in expressions $a - b$ when a and b are close together. To illustrate this, calculate $0.36443/(17.862 - 17.798)$, first using the numbers as given, and then rounding stepwise to 4, 3, and 2 significant digits.
7. The quotient in Prob. 6 is of the form $a/(b - c)$. Write it as $a(b + c)/(b^2 - c^2)$ and compute it first with 5 significant digits, then rounding numerator 12.996 and denominator 2.2822 stepwise as in Prob. 6.

8. Solve $x^2 - 20x + 1 = 0$ by (3) and by (4), using 6 significant digits (6S) in the computation. Compare and comment.

9. Do the computations in Prob. 8 with 4S and 2S.

10. What is a good way to compute $\cos a - \cos b$ if a and b are nearly equal?

11. Indicate how $\log a - \log b = \log (a/b)$ and $e^{a-b} = e^a/e^b$ may be used in computations to avoid loss of significant digits.

12. Approximations to $\pi = 3.141\,592\,653\,589\,79 \cdots$ are 22/7 and 355/113. Determine the corresponding errors and relative errors to 3 significant digits.

13. Compute π by Machin's approximation $16 \arctan (1/5) - 4 \arctan (1/239)$ to 10 significant digits. [Note that these digits are correct! For the first 100 000 digits of π, see D. Shanks and J. W. Wrench, *Mathematics of Computation* **16** (1962), pp. 76–99.]

14. Let 32.03 and 13.2381 be correctly rounded to the number of digits shown. Compute the smallest interval in which the exact sum s of the numbers must lie.

15. Answer the same question as in Prob. 14, for the difference $32.03 - 13.2381$.

16. Illustrate with an example that in computations with a fixed number of significant digits the result of adding numbers depends on the order in which they are added.

17. We are given n numbers $a_1, \cdots, a_n$, where a_j is correctly rounded to D_j decimals. In calculating the sum $a_1 + \cdots + a_n$, retaining $D = \min D_j$ decimals, is it essential that we add first and then round the result or that we first round each number to D decimals and then add?

18. Prove Theorem 1(a) for addition.

19. Compute a two-decimal table[7] of $f(x) = x/16$, $x = 0(1)20$ and find out how the rounding error is distributed.

20. Show that in Example 1 the absolute value of the error of $x_2 = 2.000/39.95 = 0.05006$ is less than 0.00001.

21. Prove Theorem 1(b) for division.

18.2 Solution of Equations by Iteration

From here on, each section will be devoted to some basic kind of problem and corresponding solution methods. We begin with methods of finding solutions of a single equation

$$f(x) = 0, \tag{1}$$

a task for which there are practically no formulas (except in a few simple cases), so that one depends almost entirely on numerical algorithms. f in (1) is a given function. A **solution** of (1) is a number $x = s$ such that $f(s) = 0$. Here, s suggests "solution," but we shall also use other letters.

Examples are $x^2 - 3x + 2 = 0$, $x^3 + x = 1$, $\sin x = 0.5x$, $\tan x = x$, $\cosh x = \sec x$, $\cosh x \cos x = -1$, which can all be written in the form (1). The first two are **algebraic equations** because the corresponding f is a

[7] The notation $x = a(h)b$ means $x = a, a + h, a + 2h, \cdots, b$.

polynomial, and in this case the solutions are also called **roots** of the equations. The other equations are **transcendental equations** because they involve transcendental functions. Solving equations (1) is a task of prime importance because engineering applications abound: some occur in Chaps. 2, 4, 7 (characteristic equations), 6 (partial fractions), 11 (eigenvalues, zeros of Bessel functions), and 15 (integration), but there are many, many others.

Formulas that give exact numerical values of the solution will exist only in very simple situations. In most cases we have to use an approximation method, in particular an **iteration method,** that is, a method in which we start from an initial guess x_0 (which may be poor) and compute step by step (in general better and better) approximations $x_1, x_2, \cdots$ of an unknown solution of (1). We discuss three such methods that are of particular practical importance and mention two others in the problem set.

In general, iteration methods are easy to program because the computational operations are the same in each step—just the data change from step to step—and, more important, if in a concrete case a method converges, it is stable (see Sec. 18.1) in general.

Fixed-Point Iteration[8] for Solving Equations $f(x) = 0$

We transform (1) *algebraically* into the form

$$x = g(x). \tag{2}$$

Then we choose an x_0 and compute $x_1 = g(x_0)$, $x_2 = g(x_1)$, and in general

$$\boxed{x_{n+1} = g(x_n)} \qquad (n = 0, 1, \cdots). \tag{3}$$

A solution of (2) is called a **fixed point** of g, motivating the name of the method, and this is a solution of (1), since from $x = g(x)$ we can return to the original form $f(x) = 0$. From (1) we may get several different forms of (2), and the behavior of corresponding iterative sequences $x_0, x_1, \cdots$ may differ accordingly, in particular, with respect to their speed of convergence. Let us illustrate this with a simple example.

EXAMPLE 1 An iteration process (Fixed-point iteration)

Set up an iteration process for the equation $f(x) = x^2 - 3x + 1 = 0$. Since we know the solutions

$$x = 1.5 \pm \sqrt{1.25}, \qquad \text{thus} \qquad 2.618\,034 \qquad \text{and} \qquad 0.381\,966,$$

we can watch the behavior of the error as the iteration proceeds.

Solution. The equation may be written

$$x = g_1(x) = \tfrac{1}{3}(x^2 + 1), \qquad \text{thus} \qquad x_{n+1} = \tfrac{1}{3}(x_n^{\,2} + 1), \tag{4a}$$

---

[8]Our present use of the word "fixed point" has absolutely nothing to do with that in the last section.

and if we choose $x_0 = 1$, we obtain the sequence (Fig. 401*a*)

$$x_0 = 1.000, \quad x_1 = 0.667, \quad x_2 = 0.481, \quad x_3 = 0.411, \quad x_4 = 0.390, \cdots$$

which seems to approach the smaller solution. If we choose $x_0 = 2$, the situation is similar. If we choose $x_0 = 3$, we obtain the sequence (Fig. 401*a*, upper part)

$$x_0 = 3.000, \quad x_1 = 3.333, \quad x_2 = 4.037, \quad x_3 = 5.766, \quad x_4 = 11.415, \cdots$$

which seems to diverge.

Our equation may also be written

$$\text{(4b)} \qquad x = g_2(x) = 3 - \frac{1}{x}, \qquad \text{thus} \qquad x_{n+1} = 3 - \frac{1}{x_n},$$

and if we choose $x_0 = 1$, we obtain the sequence (Fig. 401*b*)

$$x_0 = 1.000, \quad x_1 = 2.000, \quad x_2 = 2.500, \quad x_3 = 2.600, \quad x_4 = 2.615, \cdots$$

which seems to approach the larger solution. Similarly, if we choose $x_0 = 3$, we obtain the sequence (Fig. 401*b*)

$$x_0 = 3.000, \quad x_1 = 2.667, \quad x_2 = 2.625, \quad x_3 = 2.619, \quad x_4 = 2.618, \cdots.$$

Our figures show the following. In the lower part of Fig. 401*a* the slope of $g_1(x)$ is less than the slope of $y = x$, which is 1, thus $|g_1'(x)| < 1$, and we seem to have convergence. In the upper part, $g_1(x)$ is steeper ($g_1'(x) > 1$) and we have divergence. In Fig. 401*b* the slope of $g_2(x)$ is less near the intersection point ($x = 2.618$, fixed point of g_2, solution of $f(x) = 0$), and both sequences seem to converge. From all this we conclude that convergence seems to depend on the fact that in a neighborhood of a solution the curve of $g(x)$ is less steep than the straight line $y = x$, and we shall now see that this condition $|g'(x)| < 1$ (= slope of $y = x$) is sufficient for convergence. ■

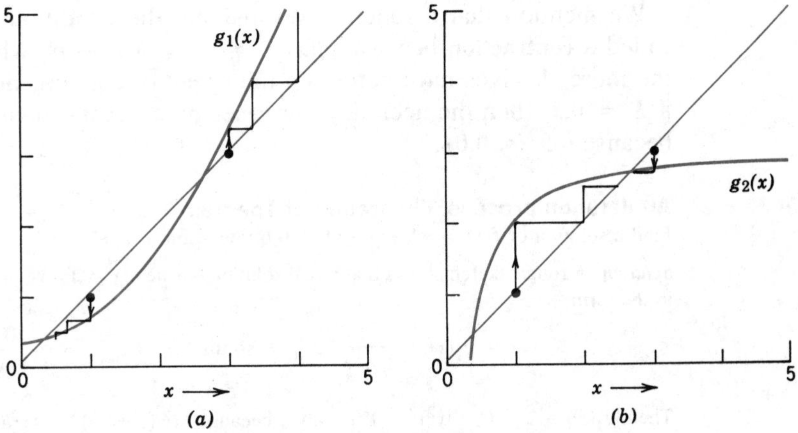

Fig. 401. Example 1, iterations (4*a*) and (4*b*)

An iteration process defined by (3) is called **convergent** for an x_0 if the corresponding sequence $x_0, x_1, \cdots$ is convergent.

A sufficient condition for convergence is given in the following theorem, which has various practical applications.

Theorem 1 **(Convergence of fixed-point iteration)**
Let $x = s$ be a solution of $x = g(x)$ and suppose that g has a continuous derivative in some interval J containing s. Then if $|g'(x)| \leqq K < 1$ in J, the iteration process defined by (3) *converges for any x_0 in J.*

Proof. By the mean value theorem of differential calculus there is a t between x and s such that

$$g(x) - g(s) = g'(t)(x - s) \qquad (x \text{ in } J).$$

Since $g(s) = s$ and $x_1 = g(x_0)$, $x_2 = g(x_1)$, $\cdots$, we obtain from this

$$\begin{aligned} |x_n - s| &= |g(x_{n-1}) - g(s)| \\ &= |g'(t)|\,|x_{n-1} - s| \\ &\leqq K|x_{n-1} - s| \\ &= K|g(x_{n-2}) - g(s)| \\ &= K|g'(\tilde{t})|\,|x_{n-2} - s| \\ &\leqq K^2|x_{n-2} - s| \\ \cdots &\leqq K^n|x_0 - s|. \end{aligned}$$

Since $K < 1$, we have $K^n \to 0$; hence $|x_n - s| \to 0$ as $n \to \infty$. ∎

We mention that a function g satisfying the condition in Theorem 1 is called a **contraction** because $|g(x) - g(v)| \leqq K|x - v|$, where $K < 1$. Furthermore, K gives information on the speed of convergence. For instance, if $K = 0.5$, then the accuracy increases by at least 2 digits in only 7 steps because $0.5^7 < 0.01$.

EXAMPLE 2 **An iteration process. Illustration of Theorem 1**
Find a solution of $f(x) = x^3 + x - 1 = 0$ by iteration.

Solution. A rough sketch shows that a real solution lies near $x = 1$. We may write the equation in the form

$$x = g_1(x) = \frac{1}{1 + x^2}, \qquad \text{so that} \qquad x_{n+1} = \frac{1}{1 + x_n^{\,2}}.$$

Then $|g_1'(x)| = 2\,|x|/(1 + x^2)^2 < 1$ for any x because $4x^2/(1 + x^2)^4 = 4x^2/(1 + 4x^2 + \cdots) < 1$, so that we have convergence for any x_0. Choosing $x_0 = 1$, we obtain (Fig. 402)

$$x_1 = 0.500, \quad x_2 = 0.800, \quad x_3 = 0.610, \quad x_4 = 0.729, \quad x_5 = 0.653, \quad x_6 = 0.701, \cdots.$$

The solution exact to 6D is $s = 0.682\,328$. The equation may also be written

$$x = g_2(x) = 1 - x^3. \qquad \text{Then} \qquad |g_2'(x)| = 3x^2$$

and this is greater than 1 near the solution, so that we cannot expect convergence. Try $x_0 = 1$, $x_0 = 0.5$, $x_0 = 2$ and see what happens. ∎

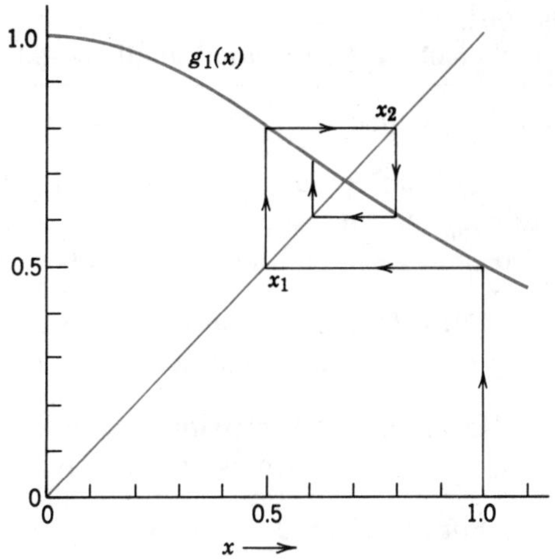

Fig. 402. Iteration in Example 2

Newton's Method for Solving Equations $f(x) = 0$

The **Newton method** or **Newton–Raphson[9] method** is another iteration method for solving equations $f(x) = 0$, where f is assumed to have a continuous derivative f'. The method is commonly used because of its simplicity and great speed. The underlying idea is that we approximate the graph of f by suitable tangents. Using an approximate value x_0 obtained from the graph of f, we let x_1 be the point of intersection of the x-axis and the tangent to the curve of f at x_0 (see Fig. 403). Then

$$\tan \beta = f'(x_0) = \frac{f(x_0)}{x_0 - x_1}, \qquad \text{hence} \qquad x_1 = x_0 - \frac{f(x_0)}{f'(x_0)}.$$

In the second step we compute $x_2 = x_1 - f(x_1)/f'(x_1)$, in the third step x_3 from x_2 again by the same formula, and so on. We thus have the algorithm shown in Table 18.1. Formula (5) in this algorithm can also be obtained from Taylor's formula (see Prob. 20).

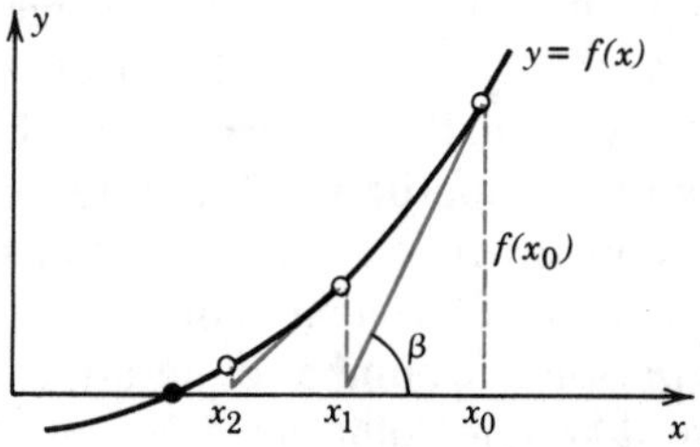

Fig. 403. Newton's method

[9]JOSEPH RAPHSON (1648—1715), English mathematician, who published a method similar to Newton's method. For historical details, see Ref. [2], p. 203, listed in Appendix 1.

Table 18.1
Newton's Method for Solving Equations $f(x) = 0$

ALGORITHM NEWTON (f, f', x_0, ϵ, N)

This algorithm computes a solution of $f(x) = 0$ given an initial approximation x_0 (starting value of the iteration). Here the function $f(x)$ is continuous and has a continuous derivative $f'(x)$.

INPUT: Initial approximation x_0, tolerance $\epsilon > 0$, maximum number of iterations N.

OUTPUT: Approximate solution x_n ($n \leqq N$) or message of failure.

For $n = 0, 1, 2, \cdots, N - 1$ do:

1 Compute $f'(x_n)$.

2 If $f'(x_n) = 0$ then OUTPUT "Failure". Stop.
[*Procedure completed unsuccessfully*]

3 Else compute

$$x_{n+1} = x_n - \frac{f(x_n)}{f'(x_n)}. \qquad (5)$$

4 If $|x_{n+1} - x_n| \leqq \epsilon$ then OUTPUT x_{n+1}. Stop.
[*Procedure completed successfully*]

End

5 OUTPUT "Failure". Stop.

[*Procedure completed unsuccessfully after N iterations*]

End NEWTON

If it happens that $f'(x_n) = 0$ for some n (see line 2 of the algorithm), then try another x_0. Line 3 is the heart of Newton's method. The inequality in line 4 is a **termination criterion;** if it holds, we have reached the desired accuracy of the approximation and stop. Line 5 gives another *termination criterion* and is needed since Newton's method may diverge or our initial guess x_0 may have been so poor that we will not reach the desired accuracy within a reasonable number of iterations. Then we try another x_0. If $f(x) = 0$ has more than one solution, different choices of x_0 may produce different solutions. Also, an iterative sequence may sometimes converge to a solution different from the expected one.

EXAMPLE 3 **Square root**

Set up a Newton iteration for computing the square root x of a given positive number c and apply it to $c = 2$.

Solution. We have $x = \sqrt{c}$, hence $f(x) = x^2 - c = 0$, $f'(x) = 2x$, and (5) takes the form

$$x_{n+1} = x_n - \frac{x_n^2 - c}{2x_n} = \frac{1}{2}\left(x_n + \frac{c}{x_n}\right).$$

For $c = 2$, choosing $x_0 = 1$, we obtain

$$x_1 = 1.500\,000, \quad x_2 = 1.416\,667, \quad x_3 = 1.414\,216, \quad x_4 = 1.414\,214, \cdots.$$

x_4 is exact to 6D. ∎

EXAMPLE 4 **Iteration for a transcendental equation**

Find the positive solution of $2 \sin x = x$.

Solution. Setting $f(x) = x - 2 \sin x$, we have $f'(x) = 1 - 2 \cos x$, and (5) gives

$$x_{n+1} = x_n - \frac{x_n - 2 \sin x_n}{1 - 2 \cos x_n} = \frac{2(\sin x_n - x_n \cos x_n)}{1 - 2 \cos x_n} = \frac{N_n}{D_n}.$$

From the graph of f we conclude that the solution is near $x_0 = 2$. We compute:

n	x_n	N_n	D_n	x_{n+1}
0	2.00000	3.48318	1.83229	1.90100
1	1.90100	3.12470	1.64847	1.89552
2	1.89552	3.10500	1.63809	1.89550
3	1.89550	3.10493	1.63806	1.89549

$x_4 = 1.89549$ is exact to 5D since the solution to 6D is 1.895 494. ∎

EXAMPLE 5 **Newton's method applied to an algebraic equation**

Apply Newton's method to the equation $f(x) = x^3 + x - 1 = 0$.

Solution. From (5) we have

$$x_{n+1} = x_n - \frac{x_n^3 + x_n - 1}{3x_n^2 + 1} = \frac{2x_n^3 + 1}{3x_n^2 + 1}.$$

Starting from $x_0 = 1$, we obtain

$$x_1 = 0.750\,000, \quad x_2 = 0.686\,047, \quad x_3 = 0.682\,340, \quad x_4 = 0.682\,328, \cdots$$

where x_4 is exact to 6D. A comparison with Example 2 shows that the present convergence is much more rapid. This may motivate the concept of the *order of an iteration process,* to be discussed next. ∎

Order of an Iteration Method. Speed of Convergence

We shall now see how we can characterize the quality of an iteration method by judging the speed of convergence, as follows.

Let $x_{n+1} = g(x_n)$ define an iteration method, and let x_n approximate a solution s of $x = g(x)$. Then $x_n = s - \epsilon_n$, where ϵ_n is the error of x_n. Suppose that g is differentiable a number of times, so that the Taylor formula gives

$$(6)\qquad \begin{aligned} x_{n+1} = g(x_n) &= g(s) + g'(s)(x_n - s) + \tfrac{1}{2}g''(s)(x_n - s)^2 + \cdots \\ &= g(s) - g'(s)\epsilon_n + \tfrac{1}{2}g''(s)\epsilon_n{}^2 + \cdots . \end{aligned}$$

The exponent of ϵ_n in the first nonvanishing term after $g(s)$ is called the **order** of the iteration process defined by g. The order measures the speed of convergence.

To see this, subtract $g(s) = s$ on both sides of (6). Then on the left you get $x_{n+1} - s = -\epsilon_{n+1}$, where ϵ_{n+1} is the error of x_{n+1}. And on the right the remaining expression equals about its first nonzero term because $|\epsilon_n|$ is small in the case of convergence. Thus

$$(7)\qquad \begin{aligned} &\text{(a)}\quad \epsilon_{n+1} \approx +g'(s)\epsilon_n && \text{in the case of first order,} \\ &\text{(b)}\quad \epsilon_{n+1} \approx -\tfrac{1}{2}g''(s)\epsilon_n{}^2 && \text{in the case of second order,} \qquad \text{etc.} \end{aligned}$$

Thus if $\epsilon_n = 10^{-k}$ in some step, then for second order, $\epsilon_{n+1} = const \cdot 10^{-2k}$, so that the number of significant digits is about doubled in each step.

For example, in **Newton's method,** $g(x) = x - f(x)/f'(x)$ and by differentiation,

$$g'(x) = 1 - \frac{f'(x)^2 - f(x)f''(x)}{f'(x)^2} = \frac{f(x)f''(x)}{f'(x)^2}.$$

Since $f(s) = 0$, this shows that also $g'(s) = 0$. Hence Newton's method is at least of second order. If we differentiate again and set $x = s$, we find that

$$(8)\qquad g''(s) = \frac{f''(s)}{f'(s)},$$

which will not be zero in general. This proves

Theorem 2 (Second-order convergence of Newton's method)
If $f(x)$ is three times differentiable and f' and f'' are not zero at a solution s of $f(x) = 0$, then for x_0 sufficiently close to s, Newton's method is of second order.

Comment. For Newton's method, (7b) becomes, by (8),

$$(9)\qquad \epsilon_{n+1} \approx -\frac{f''(s)}{2f'(s)}\epsilon_n{}^2.$$

EXAMPLE 6 Prior error estimate

Use $x_0 = 2$ and $x_1 = 1.901$ in Example 4 for estimating how many iterations we need to produce the solution to 5D accuracy. This is an **a priori estimate** or **prior estimate** because we can compute it after only one iteration, prior to further iterations.

Solution. We have

$$\frac{f''(s)}{2f'(s)} \approx \frac{f''(x_1)}{2f'(x_1)} = \frac{2\sin x_1}{2(1 - 2\cos x_1)} \approx 0.57.$$

Hence (9) gives

$$|\epsilon_{n+1}| \approx 0.57\epsilon_n^{\ 2} \approx 0.57^3\epsilon_{n-1}^4 \approx \cdots \approx 0.57^M\epsilon_0^{\ 2n+2} \leqq 5 \cdot 10^{-6}$$

where $M = 2^n + 2^{n-1} + \cdots + 2 + 1 = 2^{n+1} - 1$. We show below that $\epsilon_0 \approx -0.11$. Consequently, our condition becomes

$$0.57^M 0.11^{2n+2} = 5 \cdot 10^{-6}.$$

Hence $n = 2$ is the smallest possible n, according to this crude estimate, in good agreement with Example 4.

$\epsilon_0 \approx -0.11$ is obtained from $\epsilon_1 - \epsilon_0 = (\epsilon_1 - s) - (\epsilon_0 - s) = -x_1 + x_0 \approx 0.10$, hence $\epsilon_1 = \epsilon_0 + 0.10 \approx -0.57\epsilon_0^{\ 2}$ or $0.57\epsilon_0^{\ 2} + \epsilon_0 + 0.10 \approx 0$, which gives $\epsilon_0 \approx -0.11$. ∎

Difficulties in Newton's method may arise if $|f'(x)|$ is very small near a solution of $f(x) = 0$, or, geometrically, if the tangent of $f(x)$ at the intersection point almost coincides with the x-axis (so that double precision may be needed to get $f(x)$ and $f'(x)$ accurately enough). Then for values $x = \tilde{s}$ far away from a solution s we can still have small function values

$$R(\tilde{s}) = f(\tilde{s}).$$

In this case we call the equation $f(x) = 0$ **ill-conditioned.** $R(\tilde{s})$ is called the **residual** of $f(x) = 0$ at $\tilde{s}$. Thus a small residual guarantees a small error of $\tilde{s}$ only if the equation is ***not*** ill-conditioned.

EXAMPLE 7 An ill-conditioned equation.

$f(x) = x^5 + 10^{-4}x = 0$ is ill-conditioned. $x = 0$ is a solution. $f'(0) = 10^{-4}$ is small. At $\tilde{s} = 0.1$ the residual $f(0.1) = 2 \cdot 10^{-5}$ is small, but the error -0.1 is larger in absolute value by a factor 5000. Invent a more drastic example of your own. ∎

Secant Method for Solving Equations $f(x) = 0$

We obtain the **secant method** from Newton's method if we replace the derivative $f'(x)$ by the difference quotient,

$$f'(x_n) \approx \frac{f(x_n) - f(x_{n-1})}{x_n - x_{n-1}}.$$

Then instead of (5) we have

$$x_{n+1} = x_n - f(x_n)\,\frac{x_n - x_{n-1}}{f(x_n) - f(x_{n-1})}. \tag{10}$$

Geometrically, we intersect the x-axis at x_{n+1} with the secant of $f(x)$ passing through P_{n-1} and P_n in Fig. 404. We need two starting values x_0 and x_1, but

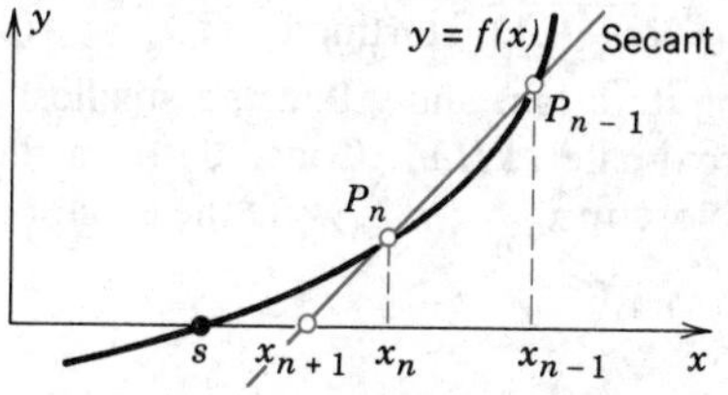

Fig. 404. Secant method

avoid the evaluation of derivatives. This method is not quite as good as Newton's (see Ref. [E16] in Appendix 1), but is still preferable if computing an $f'(x)$ takes more operations than 1/2 times that of computing an $f(x)$. The algorithm is similar to that of Newton's method, as the student may show. It is ***not*** good to write (10) as

$$x_{n+1} = \frac{x_{n-1}f(x_n) - x_n f(x_{n-1})}{f(x_n) - f(x_{n-1})},$$

because this may lead to loss of significant digits if x_n and x_{n-1} are about equal. (Can you see this from the formula?)

EXAMPLE 8 Secant method

Find the positive solution of $f(x) = x - 2 \sin x = 0$ by the secant method, starting from $x_0 = 2$, $x_1 = 1.9$.

Solution. Here, (10) is

$$x_{n+1} = x_n - \frac{(x_n - 2 \sin x_n)(x_n - x_{n-1})}{x_n - x_{n-1} + 2(\sin x_{n-1} - \sin x_n)} = x_n - \frac{N_n}{D_n}.$$

Numerical values are:

n	x_{n-1}	x_n	N_n	D_n	$x_{x+1} - x_n$
1	2.000 000	1.900 000	−0.000 740	−0.174 005	−0.004 253
2	1.900 000	1.895 747	−0.000 002	−0.006 986	−0.000 252
3	1.895 747	1.895 494	0		0

$x_3 = 1.895\ 494$ is exact to 6D. See Example 4. ∎

The (rather poor) **bisection method** and the **method of false position (regula falsi)** will be considered in the problem set.

Problem Set 18.2

Fixed-Point Iteration

1. Why do we obtain a monotone sequence in Example 1, but not in Example 2?
2. Perform the iterations indicated at the end of Example 2. Sketch a figure similar to Fig. 402.
3. Compute the solution of $x^4 = x + 0.12$ near $x = 0$ by transforming the equation algebraically into the form (2) and starting from $x_0 = 0$ (8S = 8 significant digits).
4. The equation in Prob. 3 has a solution near $x = 1$. Compute it from $x = \sqrt[4]{x + 0.12}$, starting from $x_0 = 1$ (8S).
5. Using iteration, show that the smallest positive solution of $x = \tan x$ is 4.49, approximately. *Hint.* Conclude from the graphs of x and $\tan x$ that a solution lies close to $x_0 = 3\pi/2$; write the equation in the form $x = \pi + \arctan x$. (Why?)

6. Solve $x = \cos x$ by iteration ($x_0 = 1$, 20 steps, 6S)

7. Show that $x = \cos x$ can be transformed to $x = 1 - (\sin^2 x)/(1 + x)$ and 15 steps with $x_0 = 1$ give $x = 0.739\,085$ (exact to 6S).

8. Show that $x = \cos x$ can be transformed to $(x \cos x)^{1/2} = x$ and this gives the result of Prob. 7 with 6 steps.

9. Show that if g is continuous in a closed interval I and its range lies in I, then the equation $x = g(x)$ has at least one solution in that interval. Illustrate that it may have more than one solution.

10. Of what orders are the processes in Example 1?

Newton's Method

Compute a solution (6D-accuracy) by Newton's method, starting from the given x_0. (First try to sketch the function, to see what is going on.)

11. $x^3 - 5x + 3 = 0, \quad x_0 = 2$

12. $x = \cos x, \quad x_0 = 1$

13. $x^5 + 0.85x^4 + 0.7x^3 - 3.45x^2 - 1.1x + 1.265 = 0, \quad x_0 = 1$

14. $e^{-x} - \tan x = 0, \quad x_0 = 1$

15. $\sin x = \cot x, \quad x_0 = 1$

16. $x + \ln x = 2, \quad x_0 = 2$

17. $e^{-x^2} - x, \quad x_0 = 0.5$

18. Design a Newton iteration for $\sqrt[k]{c}$ $(c > 0)$. Use it to compute $\sqrt{2}$, $\sqrt[3]{2}$, $\sqrt[4]{2}$, $\sqrt[5]{2}$ (6D, $x_0 = 1$).

19. Design a Newton iteration for cube roots and compute $\sqrt[3]{7}$ (6D, $x_0 = 2$).

20. Obtain the formula for Newton's method by truncating the Taylor series.

Secant Method

Solve the given problem by the secant method, using x_0 and x_1 as indicated.

21. Prob. 11; $x_0 = 1.5, x_1 = 2$

22. Prob. 14; $x_0 = 1, x_1 = 0.7$

23. Prob. 15; $x_0 = 1, x_1 = 0.5$

24. Prob. 12; $x_0 = 0.5, x_1 = 1$

Bisection Method and Method of False Position

25. (Bisection Method) This simple but slowly convergent method for finding a solution of $f(x) = 0$ with continuous f is based on the **intermediate value theorem,** which states that if a continuous function f has opposite signs at some $x = a$ and $x = b$ $(> a)$, that is, either $f(a) < 0$, $f(b) > 0$ or $f(a) > 0$, $f(b) < 0$, then f must be 0 somewhere on $[a, b]$. The solution is found by repeated bisection of the interval and in each iteration picking that half which also satisfies that sign condition. Write down an algorithm for the bisection method, solve $\cos x = x$, taking $a = 0$, $b = 1$, and compare the speed with that in Prob. 12.

Find a solution of the following equations by the bisection method (5 iterations), using the given a and b.

26. $e^{-x} = \ln x, \quad a = 1, b = 2$

27. $e^x + x^4 + x = 2, \quad a = 0, b = 1$

28. $x^3 - 2x^2 + 6x = 10, a = 1.7, b = 1.8$

29. $x^2 = \ln x + 3, \quad a = 1, b = 2$

30. (Method of false position; regula falsi) Figure 405 shows the idea. We assume that f is continuous. Compute the x-intercept c_0 of the line through $(a_0, f(a_0))$, $(b_0, f(b_0))$. If $f(c_0) = 0$, we are done. If $f(a_0)f(c_0) < 0$ (as in Fig. 405), set $a_1 = a_0$, $b_1 = c_0$ and repeat to get c_1, etc. If $f(a_0)f(c_0) > 0$, then $f(c_0)f(b_0) < 0$ and we set $a_1 = c_0$, $b_1 = b_0$, etc. Show that

$$c_0 = \frac{a_0 f(b_0) - b_0 f(a_0)}{f(b_0) - f(a_0)}$$

and write an algorithm for the method. [The method is generally of first order and is good for starting, but it should not be used near a solution.]

Find all real solutions of the following equations by the method of false position.

31. $x^3 = 5x + 6$ **32.** $\cos x = \sqrt{x}$ **33.** $x^4 = 2$

34. In Prob. 33, the approximate values of the positive solution will always be somewhat smaller than the exact value of the solution. Why?

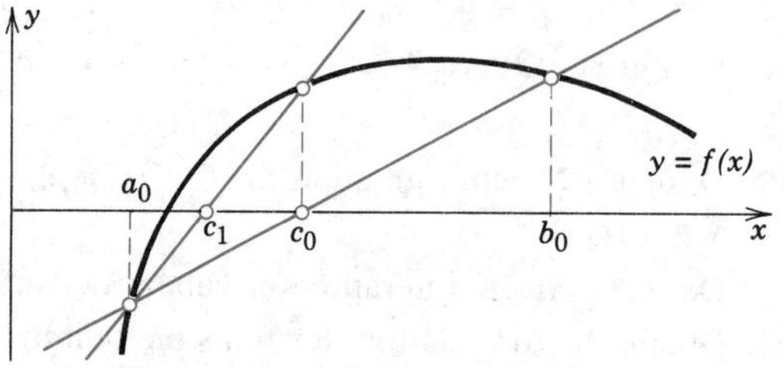

Fig. 405. Method of false position

18.3 Interpolation

Interpolation means to find (approximate) values of a function $f(x)$ for an x between ***different*** x-values $x_0, x_1, \cdots, x_n$ at which the values of $f(x)$ are given. You know this for a table of logarithms, or you may think of recorded values (temperatures, results of censuses as in Prob. 54 of Problem Set 1.7, etc.). So the given values

$$f_0 = f(x_0), \quad f_1 = f(x_1), \quad \cdots, \quad f_n = f(x_n)$$

may come from a *"mathematical function"* given by a formula or from an *"empirical function"* resulting from observations or experiments.

A standard idea in interpolation now is to find a polynomial $p_n(x)$ of degree n (or less) that assumes the given values; thus

$$p_n(x_0) = f_0, \quad p_n(x_1) = f_1, \quad \cdots, \quad p_n(x_n) = f_n. \tag{1}$$

We call this p_n an **interpolation polynomial** and $x_0, \cdots, x_n$ the **nodes.** And if $f(x)$ is a mathematical function, we call p_n an **approximation** of f (or a

polynomial approximation, because there are other kinds of approximations, as we shall see later). We use p_n to get (approximate) values of f for x's between x_0 and x_n (**"interpolation"**) or sometimes outside that interval (**"extrapolation"**).

Existence and uniqueness. p_n satisfying (1) for given data exists—we give formulas for it below. p_n is unique. Indeed, if a polynomial q_n also satisfies $q_n(x_0) = f_0, \cdots, q_n(x_n) = f_n$, then $p_n(x) - q_n(x) = 0$ at $x_0, \cdots, x_n$, but a polynomial $p_n - q_n$ of degree n (or less) with $n + 1$ roots must be identically zero, as we know from algebra; thus $p_n \equiv q_n$, the uniqueness. ■

How to find p_n? This is the important practical question. We answer it by explaining several standard methods. For given data, these methods give the same polynomial, by the uniqueness just proved (which is thus of practical interest!), but in several forms, which differ in the amount of computation.

Lagrange Interpolation

Given $(x_0, f_0), \cdots, (x_n, f_n)$ with arbitrarily spaced x_j, Lagrange had the idea of multiplying each f_j by a polynomial that is 1 at x_j and 0 at the other n nodes, and then to take the sum of these $n + 1$ polynomials to get the unique interpolation polynomial of degree n or less. Beginning with the simplest case, let us see how this works.

Linear interpolation is interpolation by the straight line through (x_0, f_0), (x_1, f_1); see Fig. 406. Thus by that idea, the linear Lagrange polynomial p_1 is a sum $p_1 = L_0 f_0 + L_1 f_1$ with L_0 the linear polynomial that is 1 at x_0 and 0 at x_1; similarly, L_1 is 0 at x_0 and 1 at x_1. Obviously,

$$L_0(x) = \frac{x - x_1}{x_0 - x_1}, \qquad L_1(x) = \frac{x - x_0}{x_1 - x_0}.$$

This gives the linear Lagrange polynomial

$$p_1(x) = L_0(x)f_0 + L_1(x)f_1 = \frac{x - x_1}{x_0 - x_1} \cdot f_0 + \frac{x - x_0}{x_1 - x_0} \cdot f_1. \tag{2}$$

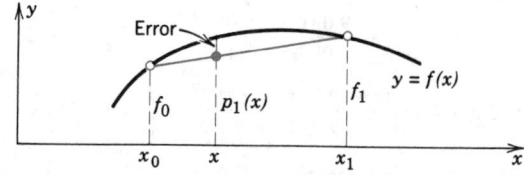

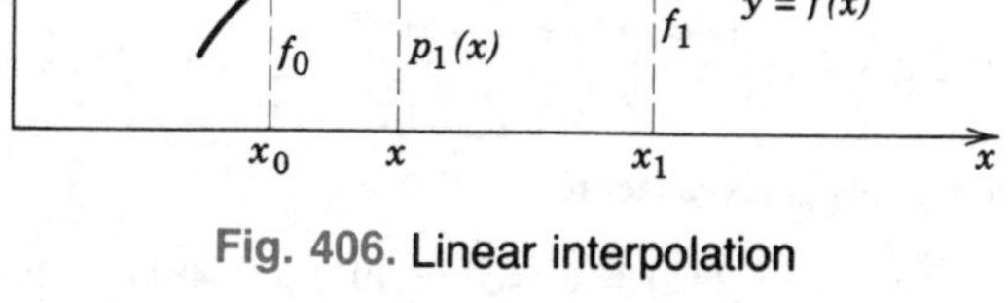

Fig. 406. Linear interpolation

EXAMPLE 1 **Linear Lagrange interpolation**

Compute ln 9.2 from ln 9.0 = 2.1972, ln 9.5 = 2.2513 by linear Lagrange interpolation and determine the error from ln 9.2 = 2.2192 (4D).

Solution. $x_0 = 9.0,\quad x_1 = 9.5,\quad f_0 = \ln 9.0,\quad f_1 = \ln 9.5.$ In (2) we need

$$L_0(9.2) = \frac{9.2 - 9.5}{9.0 - 9.5} = 0.6, \qquad L_1(9.2) = \frac{9.2 - 9.0}{9.5 - 9.0} = 0.4$$

and we get the answer

$$\ln 9.2 \approx p_1(9.2) = L_0(9.2)f_0 + L_1(9.2)f_1 = 0.6 \cdot 2.1972 + 0.4 \cdot 2.2513 = 2.2188.$$

The error is $\epsilon = a - \tilde{a} = 2.2192 - 2.2188 = 0.0004$. Hence linear interpolation is not sufficient here to get 4D-accuracy; it would suffice for 3D-accuracy. ∎

Quadratic interpolation is interpolation of given (x_0, f_0), (x_1, f_1), (x_2, f_2) by a second-degree polynomial $p_2(x)$, which by Lagrange's idea is

$$p_2(x) = L_0(x)f_0 + L_1(x)f_1 + L_2(x)f_2 \qquad \textbf{(3a)}$$

with $L_0(x_0) = 1$, $L_1(x_1) = 1$, $L_2(x_2) = 1$, and $L_0(x_1) = L_0(x_2) = 0$, etc. We claim that

$$\textbf{(3b)} \qquad \begin{aligned} L_0(x) &= \frac{l_0(x)}{l_0(x_0)} = \frac{(x - x_1)(x - x_2)}{(x_0 - x_1)(x_0 - x_2)} \\ L_1(x) &= \frac{l_1(x)}{l_1(x_1)} = \frac{(x - x_0)(x - x_2)}{(x_1 - x_0)(x_1 - x_2)} \\ L_2(x) &= \frac{l_2(x)}{l_2(x_2)} = \frac{(x - x_0)(x - x_1)}{(x_2 - x_0)(x_2 - x_1)}. \end{aligned}$$

How did we get this? Well, the numerator makes $L_k(x_j) = 0$ if $j \neq k$. And the denominator makes $L_k(x_k) = 1$ because it equals the numerator at $x = x_k$.

EXAMPLE 2 **Quadratic Lagrange interpolation**

Compute ln 9.2 by (3) from the data in Example 1 and ln 11.0 = 2.3979.

Solution. In (3),

$$L_0(x) = \frac{(x - 9.5)(x - 11.0)}{(9.0 - 9.5)(9.0 - 11.0)} = x^2 - 20.5x + 104.5, \qquad L_0(9.2) = 0.5400,$$

$$L_1(x) = \frac{(x - 9.0)(x - 11.0)}{(9.5 - 9.0)(9.5 - 11.0)} = -\frac{1}{0.75}(x^2 - 20x + 99), \qquad L_1(9.2) = 0.4800,$$

$$L_2(x) = \frac{(x - 9.0)(x - 9.5)}{(11.0 - 9.0)(11.0 - 9.5)} = \frac{1}{3}(x^2 - 18.5x + 85.5), \qquad L_2(9.2) = -0.0200,$$

so that (3a) gives, exact to 4D,

$$\ln 9.2 \approx p_2(9.2) = 0.5400 \cdot 2.1972 + 0.4800 \cdot 2.2513 - 0.0200 \cdot 2.3979 = 2.2192. \qquad ∎$$

General Lagrange interpolation polynomial. For general n we obtain

$$f(x) \approx p_n(x) = \sum_{k=0}^{n} L_k(x) f_k = \sum_{k=0}^{n} \frac{l_k(x)}{l_k(x_k)} f_k \tag{4a}$$

where $L_k(x_k) = 1$ and 0 at the other nodes. We get this if we take

$$l_0(x) = (x - x_1)(x - x_2) \cdots (x - x_n),$$

$$l_k(x) = (x - x_0) \cdots (x - x_{k-1})(x - x_{k+1}) \cdots (x - x_n), \quad 0 < k < n, \tag{4b}$$

$$l_n(x) = (x - x_0)(x - x_1) \cdots (x - x_{n-1}).$$

We can easily see that $p_n(x_k) = f_k$. Indeed, inspection of (4b) shows that $l_k(x_j) = 0$ if $j \neq k$, so that for $x = x_k$, the sum in (4a) reduces to the single term $(l_k(x_k)/l_k(x_k))f_k = f_k$.

Error estimate. If f is itself a polynomial of degree n (or less), it must coincide with p_n because the $n + 1$ data $(x_0, f_0), \cdots, (x_n, f_n)$ determine a polynomial uniquely, so the error is zero. Now the special f has its $(n + 1)$st derivative identically zero. This makes it plausible that for a *general* f its $(n + 1)$st derivative $f^{(n+1)}$ should measure the error $\epsilon_n(x) = f(x) - p_n(x)$. It can be shown that this is true if $f^{(n+1)}$ exists and is continuous, and that, with a suitable t between x_0 and x_n (or between x_0, x_n and x if we extrapolate),

$$\epsilon_n(x) = f(x) - p_n(x) = (x - x_0)(x - x_1) \cdots (x - x_n) \frac{f^{(n+1)}(t)}{(n + 1)!}. \tag{5}$$

Thus, $\epsilon_n(x) = 0$ at the nodes. Also, we get error bounds by taking the smallest and the largest value of $f^{(n+1)}(t)$ on the interval $x_0 \leqq t \leqq x_n$ (or on the interval also containing x if we *extra*polate). Most important: since p_n is unique, as we have shown, we have

Theorem 1 (Error of interpolation)

*Formula (5) gives the error for **any** polynomial interpolation method if $f(x)$ has a continuous $(n + 1)$st derivative.*

EXAMPLE 3 Error estimate of linear interpolation. Damage by round-off

Estimate the error in Example 1 by (5).

Solution. $n = 1, \quad f(t) = \ln t, \quad f'(t) = 1/t, \quad f''(t) = -1/t^2,$ hence

$$\epsilon_1(x) = (x - 9.0)(x - 9.5)\frac{(-1)}{2t^2}, \qquad \text{thus} \qquad \epsilon_1(9.2) = \frac{0.03}{t^2}.$$

$t = 9.0$ gives the maximum $0.03/9^2 = 0.00037$ and $t = 9.5$ gives the minimum $0.03/9.5^2 = 0.00033$, so that we get $0.00033 \leqq \epsilon_1(9.2) \leqq 0.00037$, or better, 0.00038 because $0.3/81 = 0.003\,703 \cdots$.

But the error 0.0004 in Example 1 disagrees, and we can learn something! Repetition of the computation there with 5D instead of 4D gives

$$\ln 9.2 \approx p_1(9.2) = 0.6 \cdot 2.19722 + 0.4 \cdot 2.25129 = 2.21885$$

with an actual error $\epsilon = 2.21920 - 2.21885 = 0.00035$, which lies nicely near the middle between our two error bounds.

This shows that the discrepancy (0.0004 vs. 0.00038) was caused by round-off, which is not taken into account in (5). ∎

Newton's Divided Difference Interpolation

In practice we often do not know the degree of the interpolation polynomial that will give the required accuracy, so we should be prepared to increase the degree if necessary. Whereas in Lagrange interpolation we would need an entirely new polynomial, in Newton's interpolation we can use our previous work and simply add another term, as we now show. Let $p_{n-1}(x)$ be the $(n - 1)$st Newton polynomial (whose form we shall determine); thus, $p_{n-1}(x_0) = f_0, \cdots, p_{n-1}(x_{n-1}) = f_{n-1}$. Furthermore, let us write the nth Newton polynomial as

$$p_n(x) = p_{n-1}(x) + g_n(x), \tag{6}$$

with

$$g_n(x) = p_n(x) - p_{n-1}(x) \tag{6'}$$

to be determined so that $p_n(x_0) = f_0, \cdots, p_n(x_n) = f_n$.

Since p_n and p_{n-1} agree at $x_0, \cdots, x_{n-1}$, we see that g_n is zero there. Also, g_n will generally be a polynomial of nth degree because so is p_n, whereas p_{n-1} can be of degree $n - 1$ at most. Hence g_n must be of the form

$$g_n(x) = a_n(x - x_0)(x - x_1) \cdots (x - x_{n-1}). \tag{6''}$$

We determine the constant a_n. For this we set $x = x_n$ and solve (6″) algebraically for a_n. Replacing $g_n(x_n)$ according to (6′) and using $p_n(x_n) = f_n$, we see that this gives

$$a_n = \frac{f_n - p_{n-1}(x_n)}{(x_n - x_0)(x_n - x_1) \cdots (x_n - x_{n-1})}. \tag{7}$$

We show that a_k equals the **kth divided difference,** recursively denoted and defined as follows:

$$a_1 = f[x_0, x_1] = \frac{f_1 - f_0}{x_1 - x_0},$$

$$a_2 = f[x_0, x_1, x_2] = \frac{f[x_1, x_2] - f[x_0, x_1]}{x_2 - x_0},$$

and in general

$$\text{(8)} \qquad a_k = f[x_0, \cdots, x_k] = \frac{f[x_1, \cdots, x_k] - f[x_0, \cdots, x_{k-1}]}{x_k - x_0}.$$

Proof. If $n = 1$, then $p_{n-1}(x_n) = p_0(x_1) = f_0$, so that (7) gives

$$a_1 = \frac{f_1 - p_0(x_1)}{x_1 - x_0} = \frac{f_1 - f_0}{x_1 - x_0} = f[x_0, x_1],$$

and (6) and (6″) give the Newton interpolation polynomial of the first degree

$$p_1(x) = f_0 + (x - x_0)f[x_0, x_1].$$

If $n = 2$, then this p_1 and (7) give

$$a_2 = \frac{f_2 - p_1(x_2)}{(x_2 - x_0)(x_2 - x_1)} = \frac{f_2 - f_0 - (x_2 - x_0)f[x_0, x_1]}{(x_2 - x_0)(x_2 - x_1)} = f[x_0, x_1, x_2]$$

where the last equality follows by straightforward calculation and comparison with the definition of the right side. (Verify it.) From (6) and (6″) we thus obtain the second Newton polynomial

$$p_2(x) = f_0 + (x - x_0)f[x_0, x_1] + (x - x_0)(x - x_1)f[x_0, x_1, x_2].$$

For $n = k$, formula (6) gives

$$\text{(9)} \qquad p_k(x) = p_{k-1}(x) + (x - x_0)(x - x_1) \cdots (x - x_{k-1})f[x_0, \cdots, x_k].$$

With $p_0(x) = f_0$ by repeated application with $k = 1, \cdots, n$ this finally gives **Newton's divided difference interpolation formula**

$$\text{(10)} \qquad \begin{aligned} f(x) \approx f_0 &+ (x - x_0)f[x_0, x_1] + (x - x_0)(x - x_1)f[x_0, x_1, x_2] \\ &+ \cdots + (x - x_0) \cdots (x - x_{n-1})f[x_0, \cdots, x_n]. \end{aligned}$$

An algorithm is shown in Table 18.2 on the next page. The first do-loop computes the divided differences and the second the desired value $p_n(\hat{x})$.

Example 4 shows how to arrange differences near the values from which they are obtained; the latter always stand a half-line above and a half-line below in the previous column. Such an arrangement is called a (divided) **difference table.** ∎

Table 18.2
Newton's Divided Difference Interpolation

ALGORITHM INTERPOL ($x_0, \cdots, x_n; f_0, \cdots, f_n; \hat{x}$)

This algorithm computes an approximation $p_n(\hat{x})$ of $f(\hat{x})$ at $\hat{x}$.

INPUT: Data $(x_0, f_0), (x_1, f_1), \cdots, (x_n, f_n)$; $\hat{x}$

OUTPUT: Approximation $p_n(\hat{x})$ of $f(\hat{x})$

Set $f[x_j] = f_j \quad (j = 0, \cdots, n)$.

For $m = 1, \cdots, n - 1$ do:

For $j = 0, \cdots, n - m$ do:

$$f[x_j, \cdots, x_{j+m}] = \frac{f[x_{j+1}, \cdots, x_{j+m}] - f[x_j, \cdots, x_{j+m-1}]}{x_{j+m} - x_j}$$

End

End

Set $p_0(x) = f_0$.

For $k = 1, \cdots, n$ do:

$$p_k(\hat{x}) = p_{k-1}(\hat{x}) + (\hat{x} - x_0) \cdots (\hat{x} - x_{k-1}) f[x_0, \cdots, x_k]$$

End

OUTPUT $p_n(\hat{x})$

End INTERPOL

EXAMPLE 4 Newton's Divided Difference Interpolation Formula

Compute $f(9.2)$ from the given values.

x_j	$f_j = f(x_j)$	$f[x_j, x_{j+1}]$	$f[x_j, x_{j+1}, x_{j+2}]$	$f[x_j, \cdots, x_{j+3}]$
8.0	2.079 442			
		0.117 783		
9.0	2.197 225		−0.0006 433	
		0.108 134		0.000 411
9.5	2.251 292		−0.005 200	
		0.097 735		
11.0	2.397 895			

Solution. We compute the divided differences as shown. Sample computation: (0.097 735 − 0.108 134)/(11 − 9) = −0.005 200. The values we need in (10) are circled. We have

$$f(x) \approx p_3(x) = 2.079\ 442 + 0.117\ 783(x - 8.0) - 0.006\ 433(x - 8.0)(x - 9.0)$$
$$+ 0.000\ 411(x - 8.0)(x - 9.0)(x - 9.5).$$

At $x = 9.2$,

$$f(9.2) \approx 2.079\ 442 + 0.141\ 340 - 0.001\ 544 - 0.000\ 030 = 2.219\ 208.$$

The value exact to 6D is $f(9.2) = \ln 9.2 = 2.219\ 203$. Note that we can nicely see how the accuracy increases from term to term:

$$p_1(9.2) = 2.220\ 782, \quad p_2(9.2) = 2.219\ 238, \quad p_3(9.2) = 2.219\ 203.$$ ■

Equal Spacing: Newton's Forward Difference Formula

Newton's formula (10) is valid for *arbitrarily spaced* nodes as they may occur in practice in experiments or observations. However, in many applications the x_j's are *regularly spaced*—for instance, in function tables or in measurements at regular intervals of time. Then we can write

$$x_0, \quad x_1 = x_0 + h, \quad x_2 = x_0 + 2h, \quad \cdots, \quad x_n = x_0 + nh. \tag{11}$$

We show how (8) and (10) simplify in this case.

To get started, let us define the *first forward difference* of f at x_j by

$$\Delta f_j = f_{j+1} - f_j,$$

the *second forward difference* of f at x_j by

$$\Delta^2 f_j = \Delta f_{j+1} - \Delta f_j,$$

and, continuing in this way, the **kth forward difference** of f at x_j by

$$\Delta^k f_j = \Delta^{k-1} f_{j+1} - \Delta^{k-1} f_j \qquad (k = 1, 2, \cdots). \tag{12}$$

Examples and an explanation of the name "forward" follow below. What is the point of this? We show that if we have regular spacing (11), then

$$f[x_0, \cdots, x_k] = \frac{1}{k!h^k} \Delta^k f_0. \tag{13}$$

We prove (13) by induction. It is true for $k = 1$ because $x_1 = x_0 + h$, so that

$$f[x_0, x_1] = \frac{f_1 - f_0}{x_1 - x_0} = \frac{1}{h}(f_1 - f_0) = \frac{1}{1!h} \Delta f_0.$$

Assuming (13) to be true for all forward differences of order k, we show that it holds for $k + 1$. We use (8) with $k + 1$ instead of k, then

$x_{k+1} = x_0 + (k+1)h$, resulting from (11), and finally (12) with $j = 0$. This gives

$$f[x_0, \cdots, x_{k+1}] = \frac{f[x_1, \cdots, x_{k+1}] - f[x_0, \cdots, x_k]}{(k+1)h}$$

$$= \frac{1}{(k+1)h}\left[\frac{1}{k!h^k}\Delta^k f_1 - \frac{1}{k!h^k}\Delta^k f_0\right]$$

$$= \frac{1}{(k+1)!h^{k+1}}\Delta^{k+1} f_0$$

which is (13) with $k+1$ instead of k. Formula (13) is proved. ∎

In (10) we finally set $x = x_0 + rh$. Then $x - x_0 = rh$, $x - x_1 = (r-1)h$ since $x_1 - x_0 = h$, and so on. With this and (13), formula (10) becomes **Newton's** (or *Gregory*[10]*–Newton's*) **forward difference interpolation formula**

$$f(x) \approx p_n(x) = \sum_{s=0}^{n} \binom{r}{s} \Delta^s f_0 \qquad (x = x_0 + rh,\quad r = (x - x_0)/h)$$

(14)

$$= f_0 + r\Delta f_0 + \frac{r(r-1)}{2!}\Delta^2 f_0 + \cdots + \frac{r(r-1)\cdots(r-n+1)}{n!}\Delta^n f_0$$

where the **binomial coefficients** in the first line are defined by

(15) $$\binom{r}{0} = 1, \quad \binom{r}{s} = \frac{r(r-1)(r-2)\cdots(r-s+1)}{s!} \qquad (s > 0, \text{integer})$$

and $s! = 1 \cdot 2 \cdots s$.

Error. From (5) we get, with $x - x_0 = rh$, $x - x_1 = (r-1)h$, etc.,

(16) $$\epsilon_n(x) = f(x) - p_n(x) = \frac{h^{n+1}}{(n+1)!} r(r-1)\cdots(r-n) f^{(n+1)}(t)$$

with t as in (5). The next example explains the application.

Comments on accuracy. **(A)** The error $\epsilon_n(x)$ is about of the order of magnitude of the next difference not used in $p_n(x)$.

(B) One should choose $x_0, \cdots, x_n$ such that the x at which one interpolates is as well centered between $x_0, \cdots, x_n$ as possible.

The reason for (A) is that in (16),

$$f^{n+1}(t) \approx \frac{\Delta^{n+1} f(t)}{h^{n+1}}, \qquad \frac{|r(r-1)\cdots(r-n)|}{1 \cdot 2 \cdots (n+1)} \leqq 1 \quad \text{if} \quad |r| \leqq 1$$

(and actually for any r as long as we do not *extrapolate*). The reason for (B) is that $|r(r-1)\cdots(r-n)|$ becomes smallest for that choice.

[10]JAMES GREGORY (1638—1675), Scots mathematician, professor at St. Andrews and Edinburgh.

EXAMPLE 5 Newton's forward difference formula. Error estimation

Compute cosh 0.56 from (14) and the four values in the following table and estimate the error.

j	x_j	$f_j = \cosh x_j$	Δf_j	$\Delta^2 f_j$	$\Delta^3 f_j$
0	0.5	1.127 626			
			0.057 839		
1	0.6	1.185 465		0.011 865	
			0.069 704		0.000 697
2	0.7	1.255 169		0.012 562	
			0.082 266		
3	0.8	1.337 435			

Solution. We compute the forward differences as shown in the table. The values we need are circled. In (14) we have $r = (0.56 - 0.50)/0.1 = 0.6$, so that (14) gives

$$\cosh 0.56 \approx 1.127\,626 + 0.6 \cdot 0.057\,839 + \frac{0.6(-0.4)}{2} \cdot 0.011\,865 + \frac{0.6(-0.4)(-1.4)}{6} \cdot 0.000\,697$$

$$= 1.127\,626 + 0.034\,703 - 0.001\,424 + 0.000\,039 = 1.160\,944.$$

Error estimate. From (16), since $\cosh^{(4)} t = \cosh t$,

$$\epsilon_3(0.56) = \frac{0.1^4}{4!} \cdot 0.6(-0.4)(-1.4)(-2.4) \cosh t = A \cosh t,$$

where $A = -0.000\,003\,36$ and $0.5 \leqq t \leqq 0.8$. We do not know t, but we get an inequality by taking the largest and smallest $\cosh t$ in that interval:

$$A \cosh 0.8 \leqq \epsilon_3(0.62) \leqq A \cosh 0.5.$$

Since $f(x) = p_3(x) + \epsilon_3(x)$, this gives

$$p_3(0.56) + A \cosh 0.8 \leqq \cosh 0.56 \leqq p_3(0.56) + A \cosh 0.5.$$

Numerical values are

$$1.160\,939 \leqq \cosh 0.56 \leqq 1.160\,941.$$

The exact 6D-value is $\cosh 0.56 = 1.160\,941$. It lies within these bounds. Such bounds are not always so tight. Also, we did not consider round-off errors, which will depend on the number of operations. ∎

This example also explains the name "*forward* difference formula": we see that the differences in the formula slope forward in the difference table.

Equal Spacing: Newton's Backward Difference Formula

Instead of forward-sloping differences we may also employ backward-sloping differences. The difference table remains the same as before (same numbers, in the same positions), except for a very harmless change of the running subscript j (which we explain in Example 6, below). Nevertheless, purely for reasons of convenience it is standard to introduce a second name and notation for differences as follows. We define the *first backward difference* of f at x_j by

$$\nabla f_j = f_j - f_{j-1},$$

the *second backward difference* of f at x_j by

$$\nabla^2 f_j = \nabla f_j - \nabla f_{j-1},$$

and, continuing in this way, the **kth backward difference** of f at x_j by

$$\nabla^k f_j = \nabla^{k-1} f_j - \nabla^{k-1} f_{j-1} \qquad (k = 1, 2, \cdots). \tag{17}$$

A formula similar to (14) but involving backward differences is **Newton's** (or *Gregory–Newton's*) **backward difference interpolation formula**

$$\begin{aligned} f(x) \approx p_n(x) &= \sum_{s=0}^{n} \binom{r+s-1}{s} \nabla^s f_0 \qquad (x = x_0 + rh, r = (x - x_0)/h) \\ &= f_0 + r\nabla f_0 + \frac{r(r+1)}{2!}\nabla^2 f_0 + \cdots + \frac{r(r+1)\cdots(r+n-1)}{n!}\nabla^n f_0. \end{aligned} \tag{18}$$

EXAMPLE 6 Newton's forward and backward interpolations

Compute a 7D-value of the Bessel function $J_0(x)$ for $x = 1.72$ from the four values in the following table, using (a) Newton's forward formula (14), (b) Newton's backward formula (18).

j_{for}	j_{back}	x_j	$J_0(x_j)$	1st Diff.	2nd Diff.	3rd Diff.
0	−3	1.7	0.397 9849			
				−0.057 9985		
1	−2	1.8	0.339 9864		−0.000 1693	
				−0.058 1678		0.000 4093
2	−1	1.9	0.281 8186		0.000 2400	
				−0.057 9278		
3	0	2.0	0.223 8908			

Solution. The computation of the differences is the same in both cases. Only their notation differs.

(a) Forward. In (14) we have $r = (1.72 - 1.70)/0.1 = 0.2$, and j goes from 0 to 3 (see first column). In each column we need the first given number, and (14) thus gives

$$\begin{aligned} J_0(1.72) &\approx 0.397\ 9849 + 0.2(-0.057\ 9985) + \frac{0.2(-0.8)}{2}(-0.000\ 1693) \\ &\quad + \frac{0.2(-0.8)(-1.8)}{6} \cdot 0.000\ 4093 \\ &= 0.397\ 9849 - 0.011\ 5997 + 0.000\ 0135 + 0.000\ 0196 \\ &= 0.386\ 4183, \end{aligned}$$

which is exact to 6D, the exact 7D-value being 0.386 4185.

(b) Backward. For (18) we use j shown in the second column, and in each column the last number. Since $r = (1.72 - 2.00)/0.1 = -2.8$, we thus get from (18)

$$J_0(1.72) \approx 0.223\ 8908 - 2.8(-0.057\ 9278) + \frac{-2.8(-1.8)}{2} \cdot 0.000\ 2400$$
$$+ \frac{-2.8(-1.8)(-0.8)}{6} \cdot 0.000\ 4093$$
$$= 0.223\ 8908 + 0.162\ 1978 + 0.000\ 6048 - 0.000\ 2750$$
$$= 0.386\ 4184.$$

Central Difference Notation

There is a third notation for differences, which is useful in certain interpolation formulas (see, for instance, Prob. 21), in numerical differentiation to be discussed in Sec. 18.5, and in connection with differential equations (Chap. 20). This is the central difference notation. The *first central difference* of $f(x)$ at x_j is defined by

$$\delta f_j = f_{j+1/2} - f_{j-1/2}$$

and the ***k*th central difference** of $f(x)$ at x_j by

$$\textbf{(19)} \qquad \delta^k f_j = \delta^{k-1} f_{j+1/2} - \delta^{k-1} f_{j-1/2} \qquad (j = 2, 3, \cdots).$$

Thus in this notation a difference table, for example, for f_{-1}, f_0, f_1, f_2, looks as follows:

x_{-1}	f_{-1}			
		$\delta f_{-1/2}$		
x_0	f_0		$\delta^2 f_0$	
		$\delta f_{1/2}$		$\delta^3 f_{1/2}$
x_1	f_1		$\delta^2 f_1$	
		$\delta f_{3/2}$		
x_2	f_2			

Inverse Interpolation

The problem of finding x for given $f(x)$ is known as **inverse interpolation.** If f is differentiable and df/dx is not zero near the point at which the inverse interpolation is to be effected, the inverse $x = F(y)$ of $y = f(x)$ exists locally near the given value of f and it may happen that F can be approximated in that neighborhood by a polynomial of moderately low degree. Then we may effect the inverse interpolation by tabulating F as a function of y and applying methods of direct interpolation to F. If $df/dx = 0$ near or at the desired point, it may be useful to solve $p(x) = \tilde{f}$ by iteration; here $p(x)$ is a polynomial that approximates $f(x)$ and $\tilde{f}$ is the given value.

Problem Set 18.3

1. Calculate $p_1(x)$ in Example 1. Compute from it $\ln 9.4 \approx p_1(9.4)$.

2. Estimate the error in Prob. 1 by (5).

3. Calculate the Lagrange polynomial $p_2(x)$ for the 4D-values of the gamma function [(24), Appendix 3.1], namely, $\Gamma(1.00) = 1.0000$, $\Gamma(1.02) = 0.9888$, $\Gamma(1.04) = 0.9784$, and from it approximations of $\Gamma(1.01)$ and $\Gamma(1.03)$.

4. Calculate $p_2(x)$ in Example 2. Compute from it approximations of ln 9.4, ln 10, ln 10.5, ln 11.5, ln 12, compute the errors by using exact 4D-values, and comment.

5. Calculate the Lagrange polynomial $p_2(x)$ for the 4D-values of the sine integral $\mathrm{Si}(x)$ [(40), Appendix 3.1], namely, $\mathrm{Si}(0) = 0$, $\mathrm{Si}(1) = 0.9461$, $\mathrm{Si}(2) = 1.6054$, and from p_2 approximations of $\mathrm{Si}(0.5)$ (= 0.4931, 4D) and $\mathrm{Si}(1.5)$ (= 1.3247, 4D).

6. Derive error bounds for $p_2(9.2)$ in Example 2 from (5).

7. Calculate the Lagrange polynomial $p_2(x)$ for the 5D-values of the error function $f(x) = \operatorname{erf} x = (2/\sqrt{\pi}) \int_0^x e^{-w^2}\, dw$, namely, $f(0.25) = 0.27633$, $f(0.5) = 0.52050$, $f(1) = 0.84270$, and from p_2 an approximation of $f(0.75)$ (= 0.71116, 5D).

8. Derive an error bound in Prob. 7 from (5).

9. Set up Newton's forward difference formula for the data in Prob. 3 and compute from it $\Gamma(1.01)$, $\Gamma(1.03)$, $\Gamma(1.05)$.

10. Set up Newton's divided difference formula for the data in Prob. 4 and derive from it $p_2(x)$ in Prob. 4.

11. Do the same task as in Prob. 10, for the data in Prob. 7.

12. In Example 5, write down the backward difference formula with general x and then use it to verify cosh 0.56 in Example 5.

13. Set up the Newton forward difference formula for the data in Prob. 5 and derive from it $p_2(x)$ in Prob. 5.

14. Compute $f(6.5)$ from $f(6.0) = 0.1506$, $f(7.0) = 0.3001$, $f(7.5) = 0.2663$, $f(7.7) = 0.2346$ by cubic interpolation, using (10).

15. Compute $f(0.8)$ and $f(0.9)$ from $f(0.5) = 0.479$, $f(1.0) = 0.841$, $f(2.0) = 0.909$ by quadratic interpolation.

16. Using a difference table, compute the Bessel function $J_1(x)$ for[11] $x = 0.1(0.2)0.9$ from $J_1(x)$ for $x = 0(0.2)1.0$ given by 0, 0.09950, 0.19603, 0.28670, 0.36884, 0.44005, respectively.

17. Write the differences in the table of Example 5 in central difference notation.

18. Compute a difference table of $f(x) = x^3$ for[11] $x = 0(1)5$. Choose $x_0 = 2$ and write all occurring numbers in terms of the notations (a) for central differences, (b) for forward differences, (c) for backward differences.

19. Show that $\delta^n f_m = \Delta^n f_{m-n/2} = \nabla^n f_{m+n/2}$.

20. Show that $\delta^2 f_m = f_{m+1} - 2f_m + f_{m-1}$, $\delta^3 f_{m+1/2} = f_{m+2} - 3f_{m+1} + 3f_m - f_{m-1}$.

21. **(Everett interpolation formula)** There are formulas involving only even-order differences. A particularly useful such formula is the **Everett formula**

$$\text{(20)}\quad f(x) \approx (1 - r)f_0 + rf_1 + \frac{(2 - r)(1 - r)(-r)}{3!}\,\delta^2 f_0 + \frac{(r + 1)r(r - 1)}{3!}\,\delta^2 f_1,$$

where $r = (x - x_0)/h$. Applying this formula, compute $e^{1.24}$ from $e^{1.1} = 3.004\ 166$, $e^{1.2} = 3.320\ 117$, $e^{1.3} = 3.669\ 297$, $e^{1.4} = 4.055\ 200$.

[11]The standard notation $x = a(h)b$ means $x = a, a + h, a + 2h, \cdots, b$.

22. The error in the Everett formula is

$$\epsilon(x) = f(x) - p(x) = h^4 \binom{r+1}{4} f^{(4)}(t), \tag{21}$$

where $p(x)$ denotes the right side of (20), and $x_0 - h < t < x_0 + 2h$. Use this formula to get error bounds in Prob. 21.

23. Using (20), compute $J_0(1.72)$ from $J_0(1.60) = 0.455\ 4022$ and $J_0(1.7)$, $J_0(1.8)$, $J_0(1.9)$ in Example 6. Note that the accuracy is higher than in Example 6, although the work is less. Can you explain why?

24. Compute $J_1(x)$ for $x = 0.1(0.2)0.9$ from $J_1(x)$ for $x = 0.2(0.2)0.8$ [given in Prob. 16] by the Everett formula.

18.4 Splines

One might expect the quality of interpolation to increase with increasing degree n of the polynomials used. Unfortunately, this is not generally true. Indeed, for various functions f the corresponding interpolation polynomials may tend to oscillate more and more between nodes as n increases. Hence we must be prepared for possible ***numerical instability.*** Figure 407 on p. 950 shows a famous example for which C. Runge[12] has proved that for equidistant nodes the maximum error approaches infinity as $n \to \infty$. (See Refs. [E13], pp. 275–279, [E24], p. 148.) Figure 408 shows how the error becomes larger with increasing n.

In the present section we discuss a popular and important method, the method of splines, which avoids that disadvantage. This approach was initiated in 1946 by I. J. Schoenberg (*Quarterly of Applied Mathematics* **4** (1946), pp. 45–99, 112–141) and has found various applications.

Spline interpolation is piecewise polynomial interpolation. This means that a function $f(x)$ is given on an interval $a \leqq x \leqq b$, and we want to approximate $f(x)$ on this interval by a function $g(x)$ that is obtained as follows. We *partition* $a \leqq x \leqq b$; that is, we subdivide it into subintervals with common endpoints, again called **nodes,**

$$a = x_0 < x_1 < \cdots < x_n = b. \tag{1}$$

We now require that $g(x_0) = f(x_0), \cdots, g(x_n) = f(x_n)$, and that in these subintervals $g(x)$ be given by polynomials, one polynomial per subinterval, such that at the nodes, $g(x)$ is several times differentiable. Hence, instead of approximating $f(x)$ by a single polynomial on the entire interval $a \leqq x \leqq b$, we now approximate $f(x)$ by n polynomials. In this way we may obtain an approximating function $g(x)$ that is more suitable in many problems of interpolation and approximation. In particular, this method will in general be numerically ***stable;*** these $g(x)$ will not be as oscillatory between nodes as a single polynomial on $a \leqq x \leqq b$ often is. Functions $g(x)$ thus obtained are called **splines.** This name is derived from thin elastic rods, called *splines,*

[12]See Prob. 20 in Problem Set 19.7.

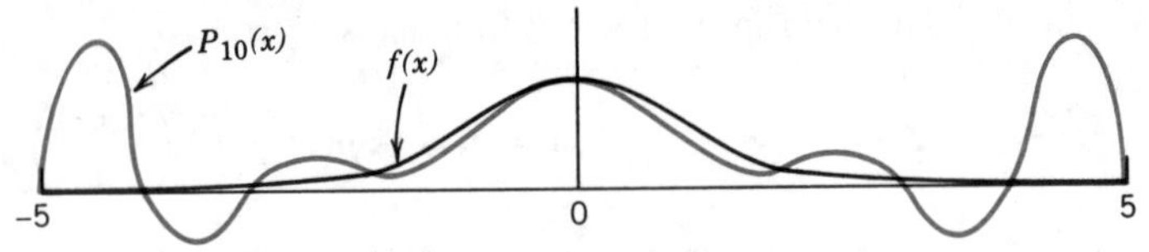

Fig. 407. Runge's example $f(x) = 1/(1 + x^2)$ and interpolating polynomial $P_{10}(x)$

which engineers have used for a long time to fit curves through given points.

The simplest continuous piecewise polynomial approximation would be by piecewise *linear* functions. But the graph of such a function has corners and would be of little practical interest—think of designing the body of a ship or a car. Hence it is preferable to use functions that have a certain number of derivatives *everywhere* on the interval $a \leqq x \leqq b$.

We shall consider cubic splines, which are perhaps the most important ones from a practical point of view. By definition, a **cubic spline** $g(x)$ on $a \leqq x \leqq b$ corresponding to the partition (1) is a continuous function $g(x)$ that has continuous first and second derivatives everywhere in that interval and, in each subinterval of that partition (1), is represented by a polynomial of degree not exceeding three. Hence $g(x)$ consists of n such polynomials, one in each subinterval.

If $f(x)$ is given on $a \leqq x \leqq b$ and a partition (1) has been chosen, we obtain a cubic spline $g(x)$ that approximates $f(x)$ by requiring that

$$(2) \quad g(x_0) = f(x_0) = f_0, \quad g(x_1) = f(x_1) = f_1, \quad \cdots, \quad g(x_n) = f(x_n) = f_n,$$

as in the classical interpolation problem of the previous section. We claim that there is a cubic spline $g(x)$ satisfying these conditions (2). And if we also require that

$$(3) \qquad g'(x_0) = k_0, \qquad g'(x_n) = k_n$$

(k_0 and k_n given numbers), then we have a uniquely determined cubic spline. This is the content of the following existence and uniqueness theorem, whose proof will also pave the way for the practical determination of splines.

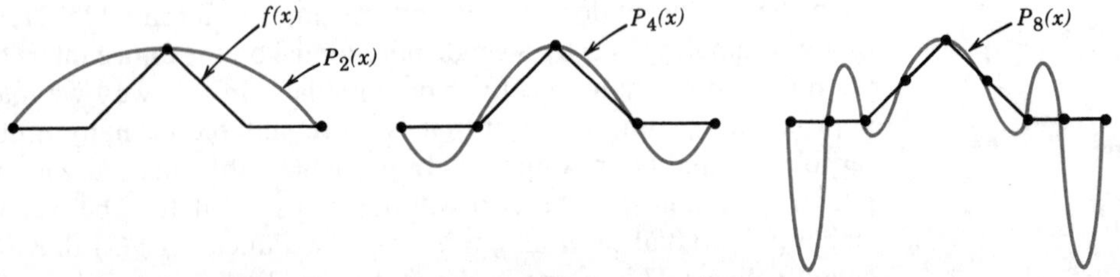

Fig. 408. Piecewise linear function $f(x)$ and interpolation polynomials of increasing degrees

Theorem 1 **(Cubic splines)**

Let $f(x)$ be defined on the interval $a \leqq x \leqq b$, let a partition (1) be given, and let k_0 and k_n be any two given numbers. Then there exists one and only one cubic spline $g(x)$ corresponding to (1) and satisfying (2) and (3).

Proof. By definition, on every subinterval I_j given by $x_j \leqq x \leqq x_{j+1}$, the spline $g(x)$ must agree with a polynomial $p_j(x)$ of degree not exceeding three, such that

$$p_j(x_j) = f(x_j), \qquad p_j(x_{j+1}) = f(x_{j+1}). \tag{4}$$

We write $1/(x_{j+1} - x_j) = c_j$ and

$$p_j'(x_j) = k_j, \qquad p_j'(x_{j+1}) = k_{j+1}, \tag{5}$$

where k_0 and k_n are given, and $k_1, \cdots, k_{n-1}$ will be determined later. Equations (4) and (5) are four conditions for $p_j(x)$. By direct calculation we can verify that the unique cubic polynomial $p_j(x)$ satisfying (4) and (5) is

$$\begin{aligned} p_j(x) = {} & f(x_j)c_j^2(x - x_{j+1})^2[1 + 2c_j(x - x_j)] \\ & + f(x_{j+1})c_j^2(x - x_j)^2[1 - 2c_j(x - x_{j+1})] \\ & + k_jc_j^2(x - x_j)(x - x_{j+1})^2 \\ & + k_{j+1}c_j^2(x - x_j)^2(x - x_{j+1}). \end{aligned} \tag{6}$$

Differentiating twice, we obtain

$$p_j''(x_j) = -6c_j^2f(x_j) + 6c_j^2f(x_{j+1}) - 4c_jk_j - 2c_jk_{j+1} \tag{7}$$

$$p_j''(x_{j+1}) = 6c_j^2f(x_j) - 6c_j^2f(x_{j+1}) + 2c_jk_j + 4c_jk_{j+1}. \tag{8}$$

By definition, $g(x)$ has continuous second derivatives. This gives the conditions

$$p_{j-1}''(x_j) = p_j''(x_j) \qquad (j = 1, \cdots, n - 1).$$

If we use (8) with j replaced by $j - 1$, and (7), these $n - 1$ equations become

$$c_{j-1}k_{j-1} + 2(c_{j-1} + c_j)k_j + c_jk_{j+1} = 3[c_{j-1}^2 \nabla f_j + c_j^2 \nabla f_{j+1}]. \tag{9}$$

In these equations, $\nabla f_j = f(x_j) - f(x_{j-1})$ and $\nabla f_{j+1} = f(x_{j+1}) - f(x_j)$ and $j = 1, \cdots, n - 1$, as before. This linear system of $n - 1$ equations has a unique solution $k_1, \cdots, k_{n-1}$ because all the coefficients of the system are nonnegative, and each entry on the main diagonal is greater than the sum of the other entries in the corresponding row, so that the coefficient determinant cannot be zero. (See Prob. 20 in Problem Set 19.7.) Hence we are able to determine unique values $k_1, \cdots, k_{n-1}$ of the first derivative of $g(x)$ at the nodes. This completes the proof. ∎

We show how the formulas in this proof can be used for the actual determination of a spline. For simplicity we consider the case of equidistant nodes x_0, $x_1 = x_0 + h, \cdots, x_n = x_0 + nh$, the idea for nonequidistant nodes being the same (just the formulas would look somewhat more complicated). We then have $c_j = 1/(x_{j+1} - x_j) = 1/h$, so that (9), after multiplication by h and with our previous notation $f(x_j) = f_j$, becomes simply

$$\textbf{(10)}\quad k_{j-1} + 4k_j + k_{j+1} = \frac{3}{h}(f_{j+1} - f_{j-1}), \qquad j = 1, \cdots, n-1,$$

k_0 and k_n are given, for instance, $k_0 = f'(a)$, $k_n = f'(b)$ or otherwise, and $k_1, \cdots, k_{n-1}$ are obtained by solving the linear system (10) consisting of $n - 1$ equations. This was the first step.

Second step. We determine the coefficients of the spline $g(x)$. On the interval $x_j \leqq x \leqq x_{j+1} = x_j + h$ the spline $g(x)$ is given by a cubic polynomial, which we write in the form

$$(11)\qquad p_j(x) = a_{j0} + a_{j1}(x - x_j) + a_{j2}(x - x_j)^2 + a_{j3}(x - x_j)^3.$$

Using Taylor's formula, we obtain

$$\textbf{(12)}\quad \begin{aligned} a_{j0} &= p(x_j) = f_j && \text{by (2)},\\ a_{j1} &= p_j'(x_j) = k_j && \text{by (5)},\\ a_{j2} &= \frac{1}{2}p_j''(x_j) = \frac{3}{h^2}(f_{j+1} - f_j) - \frac{1}{h}(k_{j+1} + 2k_j) && \text{by (7)},\\ a_{j3} &= \frac{1}{6}p_j'''(x_j) = \frac{2}{h^3}(f_j - f_{j+1}) + \frac{1}{h^2}(k_{j+1} + k_j) \end{aligned}$$

with a_{j3} obtained by calculating $p_j''(x_{j+1})$ from (11) and equating the result to (8), that is,

$$p_j''(x_{j+1}) = 2a_{j2} + 6a_{j3}h = \frac{6}{h^2}(f_j - f_{j+1}) + \frac{2}{h}(k_j + 2k_{j+1}),$$

and now subtracting from this $2a_{j2}$ as given in (12) and simplifying.

EXAMPLE 1 **Spline interpolation**

Interpolate $f(x) = x^4$ on the interval $-1 \leqq x \leqq 1$ by the cubic spline $g(x)$ corresponding to the partition $x_0 = -1$, $x_1 = 0$, $x_2 = 1$ and satisfying $g'(-1) = f'(-1)$, $g'(1) = f'(1)$.

Solution. We first write the given data in our standard notation, $f_0 = f(-1) = 1$, $f_1 = f(0) = 0$, $f_2 = f(1) = 1$. The given interval is partitioned into $n = 2$ parts of length $h = 1$. Hence our spline g consists of $n = 2$ polynomials (11),

$$p_0(x) = a_{00} + a_{01}(x+1) + a_{02}(x+1)^2 + a_{03}(x+1)^3 \qquad (-1 \leqq x \leqq 0),$$

$$p_1(x) = a_{10} + a_{11}x + a_{12}x^2 + a_{13}x^3 \qquad (0 \leqq x \leqq 1).$$

Step 1. Our goal is to determine the coefficients of p_0 and p_1, but (12) shows that they involve the k_j's, which we must thus first determine from (10). Since $n = 2$, we have $j = 1$ in (10) and get a single equation,

$$k_0 + 4k_1 + k_2 = \frac{3}{1}(f_2 - f_0) = 3(1 - 1) = 0.$$

Now (5) shows that $p_0'(x_0) = k_0$ and $p_1'(x_2) = k_2$. But $g = p_0$ at $x_0 = -1$ and $g = p_1$ at $x_2 = 1$. Hence, as given ($g' = f'$ at ± 1),

$$f'(-1) = -4 = g'(-1) = p_0(-1) = k_0, \qquad f'(1) = 4 = g'(1) = p_1(1) = k_2.$$

Substitution of $k_0 = -4$ and $k_2 = 4$ into (10) gives $k_1 = 0$.

Step 2. From (12) we can now obtain the coefficients of p_0,

$$a_{00} = f_0 = 1, \qquad a_{01} = k_0 = -4$$

$$a_{02} = \frac{3}{1^2}(f_1 - f_0) - \frac{1}{1}(k_1 + 2k_0) = 3(0 - 1) - (0 - 8) = 5$$

$$a_{03} = \frac{2}{1^3}(f_0 - f_1) + \frac{1}{1^2}(k_1 + k_0) = 2(1 - 0) + (0 - 4) = -2.$$

Similarly, for the coefficients of p_1 we obtain from (12)

$$a_{10} = f_1 = 0, \qquad a_{11} = k_1 = 0$$

$$a_{12} = 3(f_2 - f_1) - (k_2 + 2k_1) = 3(1 - 0) - (4 + 0) = -1$$

$$a_{13} = 2(f_1 - f_2) + (k_2 + k_1) = 2(0 - 1) + (4 + 0) = 2.$$

This gives the polynomials

$$p_0(x) = 1 - 4(x + 1) + 5(x + 1)^2 - 2(x + 1)^3 = -x^2 - 2x^3,$$

$$p_1(x) = -x^2 + 2x^3$$

of which the spline $g(x)$ consists. Figure 409 shows $f(x)$ and this spline

$$g(x) = \begin{cases} -x^2 - 2x^3 & \text{if } -1 \leqq x \leqq 0 \\ -x^2 + 2x^3 & \text{if } \quad 0 \leqq x \leqq 1. \end{cases}$$

Do you see that we could have saved over half of our work by using symmetry? ■

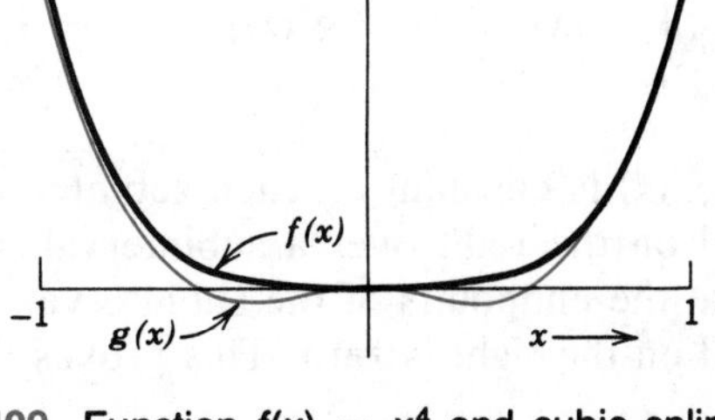

Fig. 409. Function $f(x) = x^4$ and cubic spline $g(x)$ in Example 1

EXAMPLE 2 **Spline Interpolation**

Interpolate $f_0 = f(0) = 1$, $f_1 = f(2) = 9$, $f_2 = f(4)$, $= 41$, $f_3 = f(6) = 41$ by the cubic spline satisfying $k_0 = 0$, $k_3 = -12$.

Solution. $n = 3$, $h = 2$, so that (10) is

$$\begin{aligned} k_0 + 4k_1 + k_2 \qquad &= \tfrac{3}{2}(f_2 - f_0) = 60 \\ k_1 + 4k_2 + k_3 &= \tfrac{3}{2}(f_3 - f_1) = 48. \end{aligned}$$

Since $k_0 = 0$ and $k_3 = -12$, the solution is $k_1 = 12$, $k_2 = 12$.

In (12) with $j = 0$ we have $a_{00} = f_0 = 1$, $a_{01} = k_0 = 0$,

$$a_{02} = \tfrac{3}{4}(9 - 1) - \tfrac{1}{2}(12 + 0) = 0$$

$$a_{03} = \tfrac{2}{8}(1 - 9) + \tfrac{1}{4}(12 + 0) = 1.$$

From this and, similarly, from (12) with $j = 1$ and $j = 2$ we get the spline $g(x)$ consisting of the three polynomials (see Fig. 410)

$$p_0(x) = 1 + x^3 \qquad (0 \leqq x \leqq 2)$$

$$p_1(x) = 9 + 12(x - 2) + 6(x - 2)^2 - 2(x - 2)^3 = 25 - 36x + 18x^2 - 2x^3 \qquad (2 \leqq x \leqq 4)$$

$$p_2(x) = 41 + 12(x - 4) - 6(x - 4)^2 = -103 + 60x - 6x^2 \qquad (4 \leqq x \leqq 6).$$ ∎

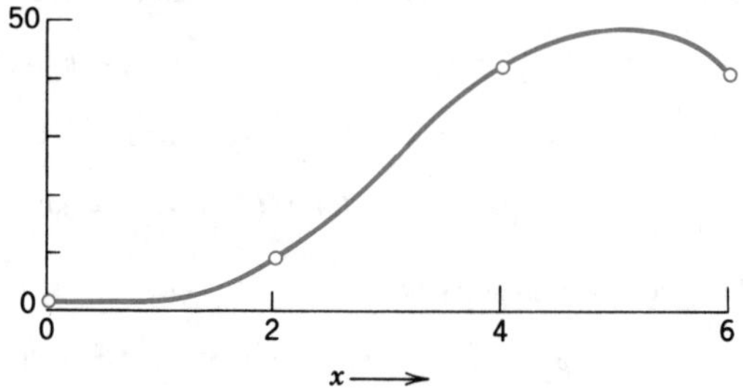

Fig. 410. Spline in Example 2

Splines have an interesting minimum property, which we shall now derive. Suppose that $f(x)$ in Theorem 1 is continuous and has continuous first and second derivatives on $a \leqq x \leqq b$. Assume that (3) is of the form

$$g'(a) = f'(a), \qquad g'(b) = f'(b) \tag{13}$$

(as in Example 1). Then $f' - g'$ is zero at a and b. Thus, by integration by parts,

$$\int_a^b g''(x)[f''(x) - g''(x)]\, dx = -\int_a^b g'''(x)[f'(x) - g'(x)]\, dx.$$

Since $g'''(x)$ is constant on each subinterval of the partition, evaluating the integral on the right over a subinterval, we obtain $const \cdot [f(x) - g(x)]$, taken at the endpoints of the subinterval, which is zero by (2). Hence that integral on the right is zero. This proves the following formula:

$$\int_a^b f''(x)g''(x)\,dx = \int_a^b g''(x)^2\,dx.$$

Consequently,

$$\int_a^b [f''(x) - g''(x)]^2\,dx = \int_a^b f''(x)^2\,dx - 2\int_a^b f''(x)g''(x)\,dx + \int_a^b g''(x)^2\,dx$$

$$= \int_a^b f''(x)^2\,dx - \int_a^b g''(x)^2\,dx.$$

Since the integrand on the left is nonnegative, so is the integral. This yields

$$\textbf{(14)} \qquad \int_a^b f''(x)^2\,dx \geqq \int_a^b g''(x)^2\,dx.$$

Our result can be stated as follows.

Theorem 2 **(Minimum property of cubic splines)**

Let $f(x)$ be continuous and have continuous first and second derivatives on some interval $a \leqq x \leqq b$. Let $g(x)$ be the cubic spline corresponding to some partition (1) *of this interval and satisfying* (2) *and* (13). *Then $f(x)$ and $g(x)$ satisfy the inequality* (14), *and equality holds if and only if $f(x)$ is the cubic spline $g(x)$.*

Engineers use thin rods called *splines* to fit curves through given points, as was mentioned before, and the strain energy minimized by such splines is proportional, approximately, to the integral of the square of the second derivative of the spline. Hence inequality (14) explains the use of the term *splines* for the functions $g(x)$ considered in this section.

Cubic splines $g(x)$ that instead of (13) satisfy

$$(15) \qquad g''(a) = 0, \qquad g''(b) = 0$$

are called **natural splines** because they have the following interesting minimum property. Among all twice differentiable functions $g(x)$ on $a \leqq x \leqq b$ that satisfy (2) and whose second derivative is continuous, natural splines are the functions that give the integral

$$\int_a^b g''(x)^2\,dx$$

the smallest possible value. Condition (15) means that $g(x)$ has an almost linear graph near the ends a and b.

Problem Set 18.4

1. Verify the computations in Example 1.
2. Compare the spline $g(x)$ in Example 1 with the quadratic interpolation polynomial $p(x)$ over the whole interval. What are the maximum deviations of $g(x)$ and $p(x)$ from $f(x)$? Comment.
3. Derive $p_1(x)$ and $p_2(x)$ in Example 2 and verify that $g(x)$ in Example 2 has continuous first and second derivatives.
4. Verify that (6) satisfies (4) and (5).
5. Obtain (7) and (8) from (6) as indicated in the text.
6. Verify the derivation of (9) indicated in the text.
7. Derive (10) from (9).
8. Give the details of the derivation of a_{j2} and a_{j3} in (12).

Find the cubic spline to the given data, with k_0 and k_n as indicated.

9. $f_0 = f(0) = 0,\ f_1 = f(1) = 0,\ f_2 = f(2) = 4,\quad k_0 = -1,\ k_2 = 5$
10. $f_0 = f(-2) = 1,\ f_1 = f(0) = 5,\ f_2 = f(2) = 17,\quad k_0 = -2,\ k_2 = -14$
11. $f_0 = f(1) = 1,\ f_1 = f(4) = 46,\ f_2 = f(9) = 86,\quad k_0 = 0,\ k_2 = -157$
12. $f_0 = f(0) = 0,\ f_1 = f(1) = 1,\ f_2 = f(2) = 6,\ f_3 = f(3) = 10,\quad k_0 = 0,\ k_3 = 0$
13. $f_0 = f(0) = 1,\ f_1 = f(1) = 0,\ f_2 = f(2) = -1,\ f_3 = f(3) = 0,\quad k_0 = 0,$ $k_3 = -6$
14. $f_0 = f(0) = 4,\ f_1 = f(2) = 0,\ f_2 = f(4) = 4,\ f_3 = f(6) = 80,\quad k_0 = 0,\ k_3 = 0$
15. $f(-2) = f(-1) = f(1) = f(2) = 0,\ f(0) = 1,\quad k_0 = k_4 = 0$

16. Rewrite the polynomials in the answer to Prob. 15 in powers of x and show that $g(x)$ is an even function of x, that is, $g(-x) = g(x)$. Was this to be expected?
17. If we started from the piecewise linear function $f(x)$ in Fig. 411, we would obtain $g(x)$ in Prob. 15 as the spline satisfying $g'(-2) = f'(-2),\ g'(2) = f'(2)$. Find and sketch the corresponding interpolation polynomial of fourth degree and compare it with that spline (see Fig. 411). Comment.

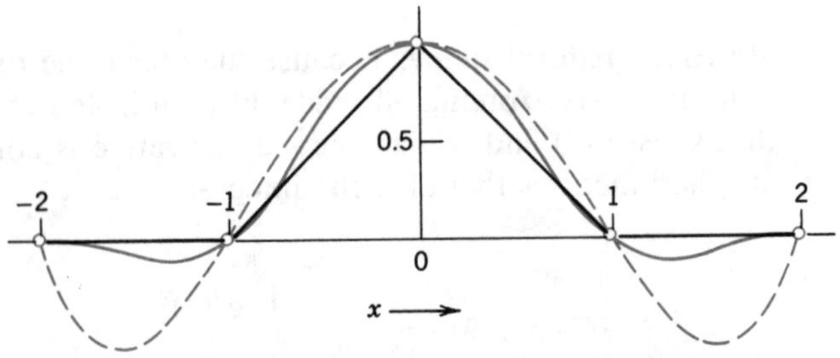

Fig. 411. Spline and interpolation polynomial in Problems 15–17

18. It may sometimes happen that a spline is represented by the same polynomial in adjacent subintervals of $a \leqq x \leqq b$. To illustrate this, find the cubic spline $g(x)$ for $f(x) = \sin x$ corresponding to the partition $x_0 = -\pi/2$, $x_1 = 0$, $x_2 = \pi/2$ of the interval $-\pi/2 \leqq x \leqq \pi/2$ and satisfying $g'(-\pi/2) = f'(-\pi/2)$ and $g'(\pi/2) = f'(\pi/2)$.

19. A possible geometric interpretation of (14) is that a cubic spline function minimizes the integral of the square of the curvature, at least approximately. Explain.

20. If a cubic spline is three times continuously differentiable (that is, it has continuous first, second, and third derivatives), show that it must be a polynomial.

18.5 Numerical Integration and Differentiation

The problem of **numerical integration** is the numerical evaluation of integrals

$$J = \int_a^b f(x)\,dx$$

where a and b are given and f is a function given analytically by a formula or empirically by a table of values. Geometrically, J is the area under the curve of f between a and b (Fig. 412).

We know that if f is such that we can find a differentiable function F whose derivative is f, then we can evaluate J by applying the familiar formula

$$J = \int_a^b f(x)\,dx = F(b) - F(a) \qquad [F'(x) = f(x)],$$

and tables of integrals, such as those in Ref. [3] in Appendix 1, may be helpful for that purpose.

However, in applications there frequently occur integrals whose integrand is an empirical function given by measured values or the integral cannot be integrated by the usual methods of calculus. Then we may use a numerical method of approximate integration.

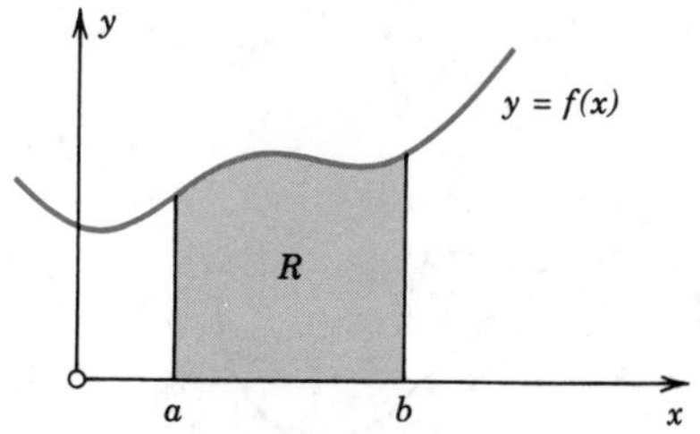

Fig. 412. Geometrical interpretation of a definite integral

Rectangular Rule. Trapezoidal Rule

Numerical integration methods are obtained by approximating the integrand f by functions that can easily be integrated. The simplest formula, the **rectangular rule,** is obtained if we subdivide the interval of integration $a \leqq x \leqq b$ into n subintervals of equal length $h = (b - a)/n$ and in each subinterval approximate f by the constant $f(x_j^*)$, the value of f at the midpoint x_j^* of the jth subinterval (Fig. 413). Then f is approximated by a **step function** (piecewise constant function), the n rectangles in Fig. 413 have the areas $f(x_1^*)h, \cdots, f(x_n^*)h$, and the **rectangular rule** is

$$J = \int_a^b f(x)\,dx \approx h[f(x_1^*) + f(x_2^*) + \cdots + f(x_n^*)] \tag{1}$$

where $h = (b - a)/n$.

The **trapezoidal rule** is generally more accurate, and we obtain it if we take the same subdivision as before and approximate f by a broken line of segments (chords) with endpoints $[a, f(a)]$, $[x_1, f(x_1)]$, $\cdots$, $[b, f(b)]$ on the curve of f (Fig. 414). Then the area under the curve of f between a and b is approximated by n trapezoids of areas

$$\tfrac{1}{2}[f(a) + f(x_1)]h, \quad \tfrac{1}{2}[f(x_1) + f(x_2)]h, \quad \cdots, \quad \tfrac{1}{2}[f(x_{n-1}) + f(b)]h,$$

and by taking their sum we obtain the **trapezoidal rule**

$$J = \int_a^b f(x)\,dx \approx h[\tfrac{1}{2}f(a) + f(x_1) + f(x_2) + \cdots + f(x_{n-1}) + \tfrac{1}{2}f(b)] \tag{2}$$

where $h = (b - a)/n$, as in (1).

EXAMPLE 1 Trapezoidal rule

Evaluate $J = \int_0^1 e^{-x^2}\,dx$ by means of (2) with $n = 10$.

Solution. $J \approx 0.1(0.5 \cdot 1.367\ 879 + 6.778\ 167) = 0.746\ 211$ from Table 18.3. ∎

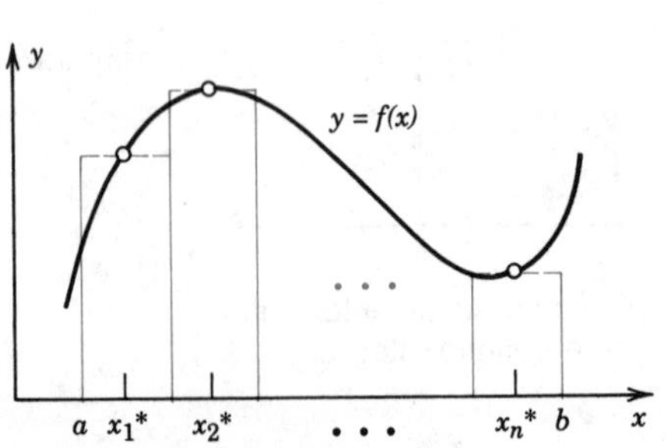

Fig. 413. Rectangular rule

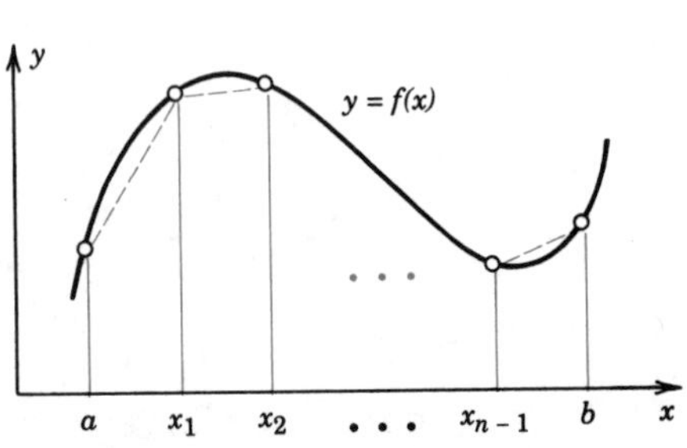

Fig. 414. Trapezoidal rule

Table 18.3
Computation in Example 1

j	x_j	x_j^2	$e^{-x_j^2}$	
0	0	0	1.000 000	
1	0.1	0.01		0.990 050
2	0.2	0.04		0.960 789
3	0.3	0.09		0.913 931
4	0.4	0.16		0.852 144
5	0.5	0.25		0.778 801
6	0.6	0.36		0.697 676
7	0.7	0.49		0.612 626
8	0.8	0.64		0.527 292
9	0.9	0.81		0.444 858
10	1.0	1.00	0.367 879	
Sums			1.367 879	6.778 167

Error Bounds for the Trapezoidal Rule

This error estimate can be derived from (5), Sec. 18.3, with $n = 1$, as follows. Beginning with a single subinterval, we have

$$f(x) - p_1(x) = (x - x_0)(x - x_1)\frac{f''(t)}{2}$$

with a suitable t depending on x, between x_0 and x_1. Integration over x from $a = x_0$ to $x_1 = x_0 + h$ gives

$$\int_{x_0}^{x_0+h} f(x)\,dx - \frac{h}{2}[f(x_0) + f(x_1)] = \int_{x_0}^{x_0+h} (x - x_0)(x - x_0 - h)\frac{f''(t(x))}{2}\,dx.$$

Setting $x - x_0 = v$ and applying the mean value theorem of integral calculus, which we can use because $(x - x_0)(x - x_0 - h)$ does not change sign, we find that the right side equals

$$\int_0^h v(v - h)\,dv\,\frac{f''(\tilde{t})}{2} = \left(\frac{h^3}{3} - \frac{h^3}{2}\right)\frac{f''(\tilde{t})}{2} = -\frac{h^3}{12}f''(\tilde{t}),$$

where $\tilde{t}$ is a suitable value between x_0 and x_1. Hence the **error** ϵ of (2) is the sum of n such contributions from the n subintervals; since $h = (b - a)/n$, we thus obtain

$$\epsilon = -\frac{(b - a)^3}{12n^2}f''(\hat{t})$$

with $\hat{t}$ suitable between a and b. Finally, **error bounds** are now obtained by taking for f'' the largest value, say, M_2, and the smallest value, M_2^*, in the interval of integration, obtaining

(3) $$KM_2 \leqq \epsilon \leqq KM_2^* \quad \text{where} \quad K = -\frac{(b-a)^3}{12n^2}.$$

EXAMPLE 2 Error estimate for the trapezoidal rule

Estimate the error of the approximate value in Example 1.

Solution. We use (3). By differentiation, $f''(x) = 2(2x^2 - 1)e^{-x^2}$. Also, $f'''(x) > 0$ if $0 < x < 1$, so that the minimum and maximum occur at the ends of the interval. We compute $M_2 = f''(1) = 0.735\ 759$ and $M_2^* = f''(0) = -2$. Furthermore, $K = -1200$, and (3) gives

$$-0.000\ 614 \leqq \epsilon \leqq 0.001\ 667.$$

Hence the exact value of J must lie between

$$0.746\ 211 - 0.000\ 614 = 0.745\ 597 \quad \text{and} \quad 0.746\ 211 + 0.001\ 667 = 0.747\ 878.$$

(Actually, $J = 0.746\ 824$, exact to 6D.) ∎

Simpson's Rule of Integration

Piecewise constant approximation of f led to the rectangular rule (1), piecewise linear approximation to the trapezoidal rule (2), and piecewise quadratic approximation will give Simpson's rule, which is of great practical importance because it is sufficiently accurate for most problems, but still sufficiently simple.

To derive Simpson's rule, we divide the interval of integration $a \leqq x \leqq b$ into an ***even*** number of equal subintervals, say, into $n = 2m$ subintervals of length $h = (b - a)/2m$, with endpoints $x_0 (= a), x_1, \cdots, x_{2m-1}, x_{2m} (= b)$; see Fig. 415. We now take the first two subintervals and approximate $f(x)$ in the interval $x_0 \leqq x \leqq x_2 = x_0 + 2h$ by the Lagrange polynomial $p_2(x)$ through $(x_0, f_0), (x_1, f_1), (x_2, f_2)$, where $f_j = f(x_j)$. From (4a) in Sec. 18.3 we obtain

(4*) $$p_2(x) = \frac{(x - x_1)(x - x_2)}{(x_0 - x_1)(x_0 - x_2)} f_0 + \frac{(x - x_0)(x - x_2)}{(x_1 - x_0)(x_1 - x_2)} f_1 + \frac{(x - x_0)(x - x_1)}{(x_2 - x_0)(x_2 - x_1)} f_2.$$

The denominators are $2h^2$, $-h^2$, and $2h^2$, respectively. Setting

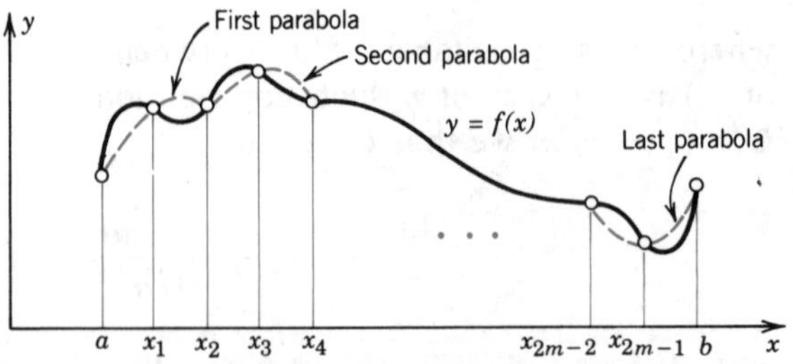

Fig. 415. Simpson's rule

$s = (x - x_1)/h$, we have $x - x_0 = (s + 1)h$, $x - x_1 = sh$, $x - x_2 = (s - 1)h$, and obtain

$$p_2(x) = \tfrac{1}{2}s(s - 1)f_0 - (s + 1)(s - 1)f_1 + \tfrac{1}{2}(s + 1)sf_2.$$

We now integrate with respect to x from x_0 to x_2. This corresponds to integrating with respect to s from -1 to 1. Since $dx = h\,ds$, the result is

$$\int_{x_0}^{x_2} f(x)\,dx \approx \int_{x_0}^{x_2} p_2(x)\,dx = h\left(\frac{1}{3}f_0 + \frac{4}{3}f_1 + \frac{1}{3}f_2\right).$$

A similar formula holds for the next two subintervals from x_2 to x_4, and so on. By summing all these m formulas we obtain **Simpson's rule**[13]

$$\text{(4)} \qquad \int_a^b f(x)\,dx \approx \frac{h}{3}(f_0 + 4f_1 + 2f_2 + 4f_3 + \cdots + 2f_{2m-2} + 4f_{2m-1} + f_{2m}),$$

where $h = (b - a)/2m$ and $f_j = f(x_j)$. Table 18.4 shows an algorithm for Simpson's rule.

Table 18.4
Simpson's Rule of Integration

ALGORITHM SIMPSON ($x_j, f_j, j = 0, 1, \cdots, 2m$)

This algorithm computes the integral $J = \int_a^b f(x)\,dx$ from given values $f_j = f(x_j)$ at equidistant $x_0 = a, x_1 = x_0 + h, \cdots, x_{2m} = x_0 + 2mh = b$ by Simpson's rule (4), where $h = (b - a)/2m$.

INPUT: $a, b, m, f_0, \cdots, f_{2m}$

OUTPUT: Approximate value $\tilde{J}$ of J

Compute $s_0 = f_0 + f_{2m}$

$s_1 = f_1 + f_3 + \cdots + f_{2m-1}$

$s_2 = f_2 + f_4 + \cdots + f_{2m-2}$

$h = (b - a)/2m$

$\tilde{J} = \dfrac{h}{3}(s_0 + 4s_1 + 2s_2)$

OUTPUT $\tilde{J}$. Stop.

End SIMPSON

[13]THOMAS SIMPSON (1710—1761), self-taught English mathematician, author of several popular textbooks. Simpson's rule was used much earlier by Torricelli, Gregory (in 1668), and Newton (in 1676).

Bounds for the error ϵ_S in (4) can be obtained by a method similar to that in the case of the trapezoidal rule (2), assuming that the fourth derivative of f exists and is continuous in the interval of integration. The result is

$$CM_4 \leqq \epsilon_S \leqq CM_4^*, \qquad \text{where} \qquad C = -\frac{(b-a)^5}{180(2m)^4} \tag{5}$$

and M_4 and M_4^* are the largest and the smallest value of the fourth derivative of f in the interval of integration. It is trivial that $\epsilon_S = 0$ for a quadratic polynomial f, but (5) shows that $\epsilon_S = 0$ even for a cubic polynomial, because the fourth derivative of such a polynomial is zero.

Simpson's rule is sufficiently accurate for most practical purposes and is preferred to more complicated formulas of higher precision, which can be obtained by the use of approximating polynomials of higher degree. A list of these so-called **Newton–Cotes formulas** is given on p. 316 of Ref. [E13] listed in Appendix 1.

EXAMPLE 3 Simpson's rule. Error estimate

Evaluate $J = \int_0^1 e^{-x^2}\,dx$ by Simpson's rule with $2m = 10$ and estimate the error.

Solution. Since $h = 0.1$, Table 18.5 gives

$$J \approx \frac{0.1}{3}(1.367\ 879 + 4 \cdot 3.740\ 266 + 2 \cdot 3.037\ 901) = 0.746\ 825.$$

Table 18.5
Computations in Example 3

j	x_j	x_j^2	$e^{-x_j^2}$		
0	0	0	1.000 000		
1	0.1	0.01		0.990 050	
2	0.2	0.04			0.960 789
3	0.3	0.09		0.913 931	
4	0.4	0.16			0.852 144
5	0.5	0.25		0.778 801	
6	0.6	0.36			0.697 676
7	0.7	0.49		0.612 626	
8	0.8	0.64			0.527 292
9	0.9	0.81		0.444 858	
10	1.0	1.00	0.367 879		
Sums			1.367 879	3.740 266	3.037 901

Estimate of error. Differentiation gives $f^{\text{IV}}(x) = 4(4x^4 - 12x^2 + 3)e^{-x^2}$. By considering the derivative f^{V} of f^{IV} we find that the largest value of f^{IV} in the interval of integration occurs at $x = 0$ and the smallest value at $x = x^* = 2.5 + 0.5\sqrt{10}$. Computation gives the values

$M_4 = f^{IV}(0) = 12$ and $M_4{}^* = f^{IV}(x^*) = -7.359$. Since $2m = 10$ and $b - a = 1$, we obtain $C = -1/1\ 800\ 000 = -0.000\ 000\ 56$. Therefore, from (5),

$$-0.000\ 007 \leqq \epsilon_S \leqq 0.000\ 005.$$

Hence J must lie between the values $0.746\ 825 - 0.000\ 007 = 0.746\ 818$ and $0.746\ 825 + 0.000\ 005 = 0.746\ 830$, so that at least four digits of our approximate value are exact. Actually, 0.746 825 is exact to 5D because $J = 0.746\ 824$ (exact to 6D).

Thus our result is much better than that in Example 1 obtained by the trapezoidal rule, whereas the number of operations is nearly the same in both cases. ∎

In our example we used some $n = 2m$ and then (5) for estimating the resulting error. More realistically, a certain accuracy will be required and $n = 2m$ determined from it. This simply means that given ϵ_S, we solve (5) for n, as the following example shows.

EXAMPLE 4 Determination of *n* from the required accuracy

What n should we choose in Example 3 to get 6D-accuracy?

Solution. Using $M_4 = 12$ (which is bigger in absolute value than $M_4{}^*$), we get from (5), with $b - a = 1$ and the required accuracy,

$$|CM_4| = \frac{12}{180(2m)^4} = \frac{1}{2} \cdot 10^{-6}, \qquad \text{thus} \qquad m = \left[\frac{2 \cdot 10^6 \cdot 12}{180 \cdot 2^4}\right]^{1/4} = 9.55.$$

Hence we should choose $n = 2m = 20$. The student may do the computation, which parallels that in Example 3.

Note that the error bounds in (3) or (5) may sometimes be loose, so that in such a case a smaller $n = 2m$ may already suffice. ∎

Numerical Stability. *All Newton–Cotes formulas are stable with respect to round-off error.*

For Simpson's rule this can be proved as follows (and for other Newton–Cotes formulas by practically the same idea). The round-off error ϵ_j of a value $f(x_j)$ does not exceed the **rounding unit** u, $|\epsilon_j| \leqq u$ (for example, $u = \frac{1}{2} \cdot 10^{-6}$ if we round to 6D). Each of the $2m + 1$ function values contributes an ϵ_j times its coefficient 1, 4, or 2, so that for the sum of these errors we have

$$|\epsilon_0 + 4\epsilon_1 + 2\epsilon_2 + \cdots + \epsilon_{2m}| \leqq (1 + 4 + \cdots + 4 + 1)u = 6mu.$$

In (4) this is multiplied by $h/3 = (b - a)/6m$, thus giving the bound

$$6mu \cdot (b - a)/6m = (b - a)u.$$

This bound is independent of m. Hence the method is stable as h approaches zero, or, practically, as we increase accuracy by choosing smaller and smaller values of h. ∎

Gaussian Integration Formulas

Our integration formulas discussed so far involve function values for equidistant x-values and give exact results for polynomials not exceeding a certain degree. More generally, we may set

$$\int_{-1}^{1} f(x)\,dx \approx \sum_{j=1}^{n} A_j f_j \qquad [f_j = f(x_j)] \tag{6}$$

(choosing $a = -1$ and $b = 1$, which may be accomplished by a linear transformation of scale) and determine the $2n$ constants $A_1, \cdots, A_n$, $x_1, \cdots, x_n$ so that (6) gives exact results for polynomials of degree k as high as possible. Since $2n$ is the number of coefficients of a polynomial of degree $2n - 1$, it follows that $k \leqq 2n - 1$. Gauss has shown that exactness for polynomials of degree not exceeding $2n - 1$ (instead of $n - 1$ or n) can be attained if and only if the values $x_1, \cdots, x_n$ are the n zeros of the *Legendre polynomial* $P_n(x)$, where

$$P_0 = 1, \quad P_1(x) = x, \quad P_2(x) = \tfrac{1}{2}(3x^2 - 1), \quad P_3(x) = \tfrac{1}{2}(5x^3 - 3x), \cdots,$$

[see (11) in Sec. 5.3 for the general formula] and the coefficients A_j are suitably chosen. Then (6) is called a **Gaussian integration formula.** Some numerical values are (more values in Ref. [1], listed in Appendix 1)

n	Zeros		Coefficients
2	$\pm 1/\sqrt{3}$	$= \pm 0.57735\ 02692$	1
3	0		8/9
	$\pm\sqrt{3/5}$	$= \pm 0.77459\ 66692$	5/9
4	$\pm\sqrt{(15 - \sqrt{120})/35}$	$= \pm 0.33998\ 10436$	0.65214 51549
	$\pm\sqrt{(15 + \sqrt{120})/35}$	$= \pm 0.86113\ 63116$	0.34785 48451

EXAMPLE 5 Gaussian integration formula with $n = 3$

Evaluate the integral in Example 3 by the Gaussian formula (6) with $n = 3$.

Solution. We have to convert our integral from 0 to 1 into an integral from -1 to 1. We set $x = \frac{1}{2}(t + 1)$. Then $dx = \frac{1}{2}dt$, and (6) with $n = 3$ and the above values of the zeros and the coefficients yields

$$\int_0^1 \exp(-x^2)\,dx = \frac{1}{2}\int_{-1}^{1} \exp\left(-\frac{1}{4}(t+1)^2\right) dt$$

$$\approx \frac{1}{2}\left[\frac{5}{9}\exp\left(-\frac{1}{4}\left(1 - \sqrt{\frac{3}{5}}\right)^2\right) + \frac{8}{9}\exp\left(-\frac{1}{4}\right) + \frac{5}{9}\exp\left(-\frac{1}{4}\left(1 + \sqrt{\frac{3}{5}}\right)^2\right)\right] = 0.746\ 815$$

(exact to 6D: 0.746 825), which is almost as accurate as the Simpson result obtained in Example 3 with a much larger number of arithmetical operations. ∎

The Gaussian integration formula (6) has the advantage of high accuracy. Its disadvantage is the irregular spacing of $x_1, \cdots, x_n$ and the somewhat inconvenient values of the coefficients, but this is not essential if the $f(x_j)$ are also computed (not taken from tables) or obtained from an experiment in which the x_j can be set once and for all. Formula (6) is hardly practical if $f(x_j)$ results from interpolation in a table giving f at equal intervals, because such an interpolation may outweigh any gain due to greater accuracy.

Since the endpoints -1 and 1 of the interval of integration in (6) are not zeros of P_n, they do not occur among $x_0, \cdots, x_n$, and the Gauss formula (6) is called, therefore, an **open formula,** in contrast to a **closed formula,** in which the endpoints of the interval of integration are x_0 and x_n. [For example, (2) and (4) are closed formulas.]

We mention that, just as in the case of interpolation, there are numerical integration methods based on differences. A very efficient method uses the **central difference formula by Gauss,**

$$\textbf{(7)} \qquad \int_{x_0}^{x_1} f(x)\,dx \approx \frac{h}{2}\left(f_0 + f_1 - \frac{\delta^2 f_0 + \delta^2 f_1}{12} + \frac{11(\delta^4 f_0 + \delta^4 f_1)}{720}\right).$$

For more details see Ref. [E24] in Appendix 1.

Numerical Differentiation

Numerical differentiation is the computation of values of the derivative of a function f from given values of f. Numerical differentiation should be avoided whenever possible, because, whereas *integration* is a *smoothing* process and not affected much by small inaccuracies in function values, *differentiation* tends to make matters *rough* and generally give values of f' much less accurate than those of f—remember that the derivative is the limit of the difference quotient, and in the latter you usually have a small difference of large quantities that you then divide by a small quantity. However, the formulas to be obtained will be basic in the numerical solution of differential equations.

We use the notations $f_j' = f'(x_j)$, $f_j'' = f''(x_j)$, etc., and may obtain rough approximation formulas for derivatives by remembering that

$$f'(x) = \lim_{h\to 0} \frac{f(x+h) - f(x)}{h}.$$

This suggests

$$\textbf{(8)} \qquad f_{1/2}' \approx \frac{\delta f_{1/2}}{h} = \frac{f_1 - f_0}{h}.$$

Similarly, for the second derivative we obtain

$$f_1'' \approx \frac{\delta^2 f_1}{h^2} = \frac{f_2 - 2f_1 + f_0}{h^2}, \tag{9}$$

and so on.

More accurate approximations are obtained by differentiating suitable Lagrange polynomials. Differentiating (4*) and remembering that the denominators in (4*) are $2h^2$, $-h^2$, $2h^2$, we have

$$f'(x) \approx p_2'(x) = \frac{2x - x_1 - x_2}{2h^2} f_0 - \frac{2x - x_0 - x_2}{h^2} f_1 + \frac{2x - x_0 - x_1}{2h^2} f_2.$$

Evaluating this at x_0, x_1, x_2, we obtain the "three-point formulas"

(10)

$$\text{(a)} \quad f_0' \approx \frac{1}{2h}(-3f_0 + 4f_1 - f_2),$$

$$\text{(b)} \quad f_1' \approx \frac{1}{2h}(-f_0 + f_2),$$

$$\text{(c)} \quad f_2' \approx \frac{1}{2h}(f_0 - 4f_1 + 3f_2).$$

Applying the same idea to the Lagrange polynomial $p_4(x)$, we obtain similar formulas, in particular,

$$f_2' \approx \frac{1}{12h}(f_0 - 8f_1 + 8f_3 - f_4). \tag{11}$$

Further details and formulas are included in Ref. [E11] listed in Appendix 1.

Problem Set 18.5

1. Show that (1) with $n = 5$ gives $J \approx 0.748\ 053$ for the integral in Example 1 [exact to 6D: 0.746 824].
2. To get a feeling for the increase in accuracy, compute $\int_0^1 x^2\,dx$ by (1) with $h = 1$, $h = 0.5$, $h = 0.25$.
3. Perform the task in Prob. 2, using the trapezoidal rule (2) instead of (1). Also graph x^2 and the trapezoids.
4. Derive a formula that gives lower and upper bounds for J in connection with the rectangular rule (1). Apply it to Prob. 1.

Evaluate the following integrals numerically as indicated, and compare the result with the exact value obtained by a formula known from calculus.

$$A(x) = \int_0^x \frac{dx^*}{1 + x^{*2}}, \qquad B(x) = \int_1^x \frac{dx^*}{x^*}, \qquad D(x) = \int_0^x x^* e^{-x^{*2}} \, dx^*.$$

5. $A(1)$ by (2), $n = 4$

6. $A(1)$ by (2), $n = 20$

7. $A(1)$ by (4), $2m = 4$

8. $A(1)$ by (4), $2m = 10$

9. $A(2)$ by (4), $2m = 10$

10. $A(10)$ by (2), $n = 10$

11. $B(2)$ by (4), $2m = 4$

12. $B(2)$ by (4), $2m = 10$

13. $B(2)$ by (2), $n = 10$

14. $D(0.4)$ by (4), $2m = 4$

Compute error bounds in

15. Prob. 5 **16.** Prob. 6 **17.** Prob. 11

18. What would be the smallest n in computing B(2) for which 5D-accuracy is guaranteed (a) by (3) in the use of (2), (b) by (5) in the use of (4). Compare and comment.

Evaluate the following integrals numerically as indicated.

$$\text{Si}(x) = \int_0^x \frac{\sin x^*}{x^*} \, dx^*, \qquad \text{C}(x) = \int_0^x \cos(x^{*2}) \, dx^*, \qquad J(x) = \int_0^x J_0(x^*) \, dx^*.$$

(These are nonelementary integrals. Si(x) is called the *sine integral,* C(x) the *Fresnel integral,* and $J_0(x)$ the *Bessel function of order zero.*)

19. Si(1) by (2), $n = 5$, $n = 10$

20. Si(1) by (4), $2m = 2$ and $2m = 10$

21. C(2) by (4), $2m = 10$

22. C(2) by (2), $n = 10$

23. $J(1)$ by (2), $n = 10$, and values from Table 1, Appendix 5

24. $J(1)$ by (4), $2m = 10$

25. Prove that the trapezoidal rule is stable with respect to round-off.

26. Carry out the details of the derivation of (10).

27. Consider $f(x) = x^4$ for $x_0 = 0$, $x_1 = 0.2$, $x_2 = 0.4$, $x_3 = 0.6$, $x_4 = 0.8$. Calculate f_2' from (10a), (10b), (10c), (11). Determine the errors. Compare and comment.

28. A "four-point formula" for the derivative is

$$f_2' \approx \frac{1}{6h}(-2f_1 - 3f_2 + 6f_3 - f_4).$$

Apply it to $f(x) = x^4$ with $x_1, \cdots, x_4$ as in Prob. 27, determine the error, and compare it with that in the case of (11).

29. The derivative $f'(x)$ can also be approximated in terms of first-order and higher order differences:

$$f'(x_0) \approx \frac{1}{h}\left(\Delta f_0 - \frac{1}{2}\Delta^2 f_0 + \frac{1}{3}\Delta^3 f_0 - \frac{1}{4}\Delta^4 f_0 + - \cdots\right).$$

Compute $f'(0.4)$ in Prob. 27 from this formula, using differences up to and including first order, second order, third order, fourth order.

30. Derive the formula in Prob. 29 from (14) in Sec. 18.3.

Review Questions and Problems for Chapter 18

1. How are numbers represented in a computer?
2. What do we mean by floating point and fixed point?
3. What are the rules for rounding?
4. Name and explain some of the sources of errors in computations.
5. What do we mean by stability of an algorithm? Why is this important?
6. State the definitions of error and relative error of an approximate value.
7. How do error and relative error behave under addition? Under multiplication?
8. Why is the selection of a good method at least as important on a large computer as it is on a pocket calculator?
9. What is fixed-point iteration? State a condition sufficient for convergence.
10. What do we mean by the order of convergence of an iteration? Why is this important?

11. Make a difference table of $f(x) = 1/x^2$, $x = 1.0(0.2)2.0$, 4D, and write the differences in all three notations.
12. What is the purpose of the Newton (–Raphson) method? What is its idea?
13. What is the bisection method? Can it diverge? Is it fast?
14. What is the regula falsi? When would you apply it?
15. What is the advantage of Newton's interpolation formulas over Lagrange's?

16. In deriving Simpson's rule we used a Newton interpolation polynomial. Explain how.
17. What is spline interpolation? Why can it be better than the usual polynomial interpolation?
18. What do you remember about the error of polynomial interpolation?
19. How did we obtain formulas for numerical differentiation?
20. Why is numerical differentiation generally more delicate than numerical integration?

21. Write -0.53678, 1186.699, -0.00604, 23.9481, $1/3$, $85/7$ in floating-point form, with 5 significant digits.
22. Convert $(1992)_{10}$, $(5.25)_{10}$, $(15.664\ 0625)_{10}$, and $(-24.875)_{10}$ to binary form.
23. Convert $(111.101)_2$, $(100010.1)_2$, and $(11.1011)_2$ to base 10 form.
24. Compute $0.29731/(4.1232 - 4.0872)$ with the numbers as given and then rounded stepwise to 4, 3, 2 significant digits. Comment.

25. Solve $x^2 - 50x + 1 = 0$ by (3) and by (4), Sec. 18.1, using 5 significant digits in the computation. Compare and comment.
26. What would be a good way to compute $\sqrt{x^2 + 9} - 3$ for small $|x|$?

27. Let 4.81 and 12.752 be correctly rounded to the number of digits shown. Determine the smallest interval in which the sum $s = 4.81 + 12.752$ (using true instead of rounded values of the quantities) must lie.
28. Answer the question in Prob. 27, for the difference $d = 4.81 - 12.752$.
29. What is the relative error of $n\tilde{a}$ in terms of that of $\tilde{a}$?
30. Show that the relative error of $\tilde{a}^2$ is twice that of $\tilde{a}$.

31. Find the solution of $x^5 = x + 0.2$ near $x = 0$ by transforming the equation algebraically into the form (2), Sec. 18.2, and starting from $x_0 = 0$.
32. The equation in Prob. 31 has a solution near $x = 1$. Find this solution by writing the equation in the form $x = \sqrt[5]{x + 0.2}$ and iterating, starting with $x_0 = 1$.
33. Solve $x^6 + 3x^4 - 6x^2 - 8 = 0$ by iteration with $x_0 = 1$, 10 steps, 6S.
34. Solve $\cos x = x$ by iteration (6S, $x_0 = 1$), writing it as $x = (0.74x + \cos x)/1.74$, obtaining $x_4 = 0.739\,085$ (exact to 6S!). Why does this converge so rapidly?

35. Solve $x^4 - 4.00322x^3 + x^2 + 6.43105 = 0$ by Newton's method with $x_0 = 3.5$, 6S-accuracy.
36. Solve $x^4 - x^3 - 2x - 34 = 0$ by Newton's method, with $x_0 = 3$, 6S-accuracy.
37. Solve $e^x - 1/x = 0$ by Newton's method with $x_0 = 0.5$, 6S-accuracy.
38. Solve $\cos x - x = 0$ by the method of false position.

39. Compute sinh 0.3 from $\sinh(-0.5) = -0.521$, $\sinh 0 = 0$, $\sinh 1 = 1.175$ by quadratic interpolation.
40. Compute $f(1.75)$ from $f(1.0) = 3.00000$, $f(1.2) = 2.98007$, $f(1.4) = 2.92106$, $f(1.6) = 2.82534$, $f(1.8) = 2.69671$, $f(2.0) = 2.54030$ by linear interpolation. By quadratic interpolation, using the last three values.
41. Do the same task as in Prob. 40, for $f(1.28)$.
42. In Prob. 40, compute $f(1.75)$ by the Everett formula (see Problem Set 18.3).
43. In Prob. 40, compute $f(1.28)$ by the usual cubic interpolation.
44. Compute $f(0.3)$ from $f(0) = 0.50000$, $f(0.2) = 0.69867$, $f(0.4) = 0.88942$, $f(0.6) = 1.06464$ by the Everett formula in Problem Set 18.3.

45. Find the cubic spline for the data $f(0) = 1$, $f(1) = 0$, $f(2) = -3$, $k_0 = k_2 = 0$.
46. Find the cubic spline for the data $f(-1) = 3$, $f(1) = 1$, $f(3) = 23$, $f(5) = 45$, $k_0 = k_3 = 3$.

47. Using the rectangular rule (1) in Sec. 18.5 with $n = 5$, compute $\int_0^1 x^3\,dx$. What is the error?
48. Compute the integral in Prob. 47 by the trapezoidal rule (2) in Sec. 18.5 with $n = 5$. What error bounds are obtained from (3)? What is the actual error of the result? Why is this result larger than the exact value?
49. Compute the Fresnel integral $C(x) = \int_0^x \cos(t^2)\,dt$ for $x = 1$ by Simpson's rule with $2m = 2$.
50. Compute the integral in Prob. 49 by Simpson's rule with $2m = 10$.

Summary of Chapter 18 Numerical Methods in General

In this chapter we discussed concepts that are relevant throughout numerical work as a whole and methods of a general nature, as opposed to methods for problems in linear algebra (Chap. 19) or in differential equations (Chap. 20).

In scientific computations we use the **floating-point** representation of numbers (Sec. 18.1); fixed-point representation is less suitable in most cases.

Numerical methods give approximate values $\tilde{a}$ of quantities. The **error** ϵ of $\tilde{a}$ is

$$\epsilon = a - \tilde{a} \qquad \text{(Sec. 18.1)} \tag{1}$$

where a is the exact value. The *relative error* of $\tilde{a}$ is ϵ/a. Errors arise from rounding off, inaccuracies of measured values, truncation (that is, replacement of integrals by sums, derivatives by difference quotients, series by partial sums), and so on.

An algorithm is called **numerically stable** if small changes in the initial data give only correspondingly small changes in the final results. Unstable algorithms are generally useless because errors may become so large that results will be very inaccurate. Numerical instability of algorithms must not be confused with mathematical instability of problems ("*ill-conditioned problems*," Sec. 18.2).

Fixed-point iteration is a method for solving equations $f(x) = 0$ in which the equation is first transformed algebraically to $x = g(x)$, an initial guess x_0 for the solution is made, and then approximations x_1, x_2, $\cdots$, are successively computed by iteration from (see Sec. 18.2)

$$x_{n+1} = g(x_n) \qquad (n = 0, 1, \cdots). \tag{2}$$

Newton's method for solving equations $f(x) = 0$ is an iteration

$$x_{n+1} = x_n - \frac{f(x_n)}{f'(x_n)} \qquad \text{(Sec. 18.2).} \tag{3}$$

x_{n+1} is the x-intercept of the tangent of the curve $y = f(x)$ at the point x_n. This method is of second order (Theorem 2, Sec. 18.2). If we replace

f' in (3) by a difference quotient (geometrically: we replace the tangent by a secant), we obtain the **secant method;** see (10) in Sec. 18.2. For the *bisection method* (which is slow) and the *method of false position,* see Problem Set 18.2.

Polynomial interpolation means the determination of an *interpolating polynomial,* that is, a polynomial $p_n(x)$ such that $p_n(x_j) = f_j$, where $j = 0, \cdots, n$ and $(x_0, f_0), \cdots, (x_n, f_n)$ are measured or observed values, values of a function, etc. For given data, $p_n(x)$ of degree n (or less) is unique. However, it can be written in different forms, notably in **Lagrange's form** (4), Sec. 18.3, or in **Newton's divided difference form** (10), Sec. 18.3, which requires fewer operations and for regularly spaced $x_0, x_1 = x_0 + h, \cdots, x_n = x_0 + nh$, becomes **Newton's forward difference formula** (Sec. 18.3)

$$\begin{aligned} f(x) \approx p_n(x) &= f_0 + r\Delta f_0 + \frac{r(r-1)}{2!}\Delta^2 f_0 + \cdots \\ &\quad \cdots + \frac{r(r-1)\cdots(r-n+1)}{n!}\Delta^n f_0 \end{aligned} \tag{4}$$

where $r = (x - x_0)/h$ and the forward differences are $\Delta f_j = f_{j+1} - f_j$ and

$$\Delta^k f_j = \Delta^{k-1} f_{j+1} - \Delta^{k-1} f_j.$$

A similar formula is *Newton's backward difference interpolation formula* (Sec. 18.3).

Interpolation polynomials may become numerically unstable as n increases, and instead of interpolating and approximating by a single high-degree polynomial it is preferable to use a cubic **spline** $g(x)$, that is, a twice continuously differentiable interpolation function [thus, $g(x_j) = f_j$], which in each subinterval $x_j \leqq x \leqq x_{j+1}$ consists of a cubic polynomial $p_j(x)$; see Sec. 18.4.

Simpson's rule of numerical integration is (Sec. 18.5)

$$\begin{aligned} \int_a^b f(x)\,dx \approx \frac{h}{3}(f_0 &+ 4f_1 + 2f_2 + 4f_3 + \cdots \\ &+ 2f_{2m-2} + 4f_{2m-1} + f_{2m}) \end{aligned} \tag{5}$$

where $h = (b - a)/2m$ and $f_j = f(x_j)$. For error estimates and other integration methods, see Sec. 18.5.

Chapter 19

Numerical Methods in Linear Algebra

In this chapter we consider some of the most important numerical methods for solving linear systems of equations (Secs. 19.1–19.4), for fitting straight lines (Sec. 19.5), and for matrix eigenvalue problems (Secs. 19.6–19.10). These and similar methods are of great practical importance. Indeed, many engineering or other (for instance, statistical) problems lead to mathematical models whose solution requires methods of numerical linear algebra.

The present chapter is independent of Chap. 18 and can be studied immediately after Chap. 7.

Prerequisite for this chapter: Secs. 7.1–7.3, 7.10.
Sections that may be omitted in a shorter course: Secs. 19.4–19.6, 19.9, 19.10.
References: Appendix 1, Part E.
Answers to Problems: Appendix 2.

19.1 Linear Systems: Gauss Elimination

A **linear system of n equations** in n *unknowns* $x_1, \cdots, x_n$ is a set of equations $E_1, \cdots, E_n$ of the form

$$
\text{(1)} \qquad
\begin{aligned}
E_1{:}\quad & a_{11}x_1 + \cdots + a_{1n}x_n = b_1 \\
E_2{:}\quad & a_{21}x_1 + \cdots + a_{2n}x_n = b_2 \\
& \cdots\cdots\cdots\cdots\cdots\cdots \\
E_n{:}\quad & a_{n1}x_1 + \cdots + a_{nn}x_n = b_n
\end{aligned}
$$

where the **coefficients** a_{jk} and the b_j are given numbers. The system is called **homogeneous** if all the b_j are zero; otherwise **nonhomogeneous.** Using matrix multiplication (Sec. 7.3), we can write (1) as a single vector equation

$$\mathbf{Ax} = \mathbf{b} \tag{2}$$

where the **coefficient matrix** $\mathbf{A} = [a_{ik}]$ is the $n \times n$ matrix

$$\mathbf{A} = \begin{bmatrix} a_{11} & a_{12} & \cdots & a_{1n} \\ a_{21} & a_{22} & \cdots & a_{2n} \\ \cdot & \cdot & \cdots & \cdot \\ a_{n1} & a_{n2} & \cdots & a_{nn} \end{bmatrix}, \quad \text{and} \quad \mathbf{x} = \begin{bmatrix} x_1 \\ \cdot \\ \cdot \\ \cdot \\ x_n \end{bmatrix} \quad \text{and} \quad \mathbf{b} = \begin{bmatrix} b_1 \\ \cdot \\ \cdot \\ \cdot \\ b_n \end{bmatrix}$$

are column vectors. The **augmented matrix** $\tilde{\mathbf{A}}$ of the system (1) is

$$\tilde{\mathbf{A}} = [\mathbf{A} \quad \mathbf{b}] = \begin{bmatrix} a_{11} & \cdots & a_{1n} & b_1 \\ a_{21} & \cdots & a_{2n} & b_2 \\ \cdot & \cdots & \cdot & \cdot \\ a_{n1} & \cdots & a_{nn} & b_n \end{bmatrix}.$$

A **solution** of (1) is a set of numbers $x_1, \cdots, x_n$ that satisfy all the n equations, and a **solution vector** of (1) is a vector $\mathbf{x}$ whose components constitute a solution of (1).

The method of solving such a system by determinants (Cramer's rule in Sec. 7.9) is not practical, even with efficient methods for evaluating the determinants.

A practical method for the solution of a linear system is the so-called *Gauss elimination,* which we shall now discuss (proceeding independently of Sec. 7.4).

Gauss Elimination

This standard method for solving linear systems (1) is a systematic process of elimination that reduces (1) to "**triangular form**" because then the system can be easily solved by "**back substitution.**" For instance, a triangular system is

$$\begin{aligned} 3x_1 + 5x_2 + 2x_3 &= 8 \\ 8x_2 + 2x_3 &= -7 \\ 6x_3 &= 3 \end{aligned}$$

and back substitution gives $x_3 = 3/6 = 1/2$ from the third equation, then

$$x_2 = \tfrac{1}{8}(-7 - 2x_3) = -1$$

from the second equation, and finally from the first equation

$$x_1 = \tfrac{1}{3}(8 - 5x_2 - 2x_3) = 4.$$

How do we reduce a given system (1) to triangular form? In the first step we eliminate x_1 from equations E_2 to E_n in (1). This we do by subtracting suitable multiples of the first equation from the other equations. This first equation is called the **pivot equation** in this step, and a_{11} is called the **pivot.** This equation is left unaltered. In the second step we take the *new* second equation (which no longer contains x_1) as the pivot equation and use it to eliminate x_2 from the new equations E_3^* to E_n^* obtained in the first step, and so on. This gives a triangular system that can be solved by back substitution as just shown. In this way we obtain precisely all solutions of the *given* system (proof in Sec. 7.4). The pivots must be different from zero. To achieve this, we may have to change the order of equations. This is called **partial pivoting.**[1] Also, pivots should not be too small in absolute value, because of round-off errors, and this may be another reason for partial pivoting, as we shall discuss later. Let us first consider a simple example.

EXAMPLE 1 Gauss elimination. Pivoting

Solve the system

$$\begin{aligned} 8x_2 + 2x_3 &= -7 \\ 3x_1 + 5x_2 + 2x_3 &= 8 \\ 6x_1 + 2x_2 + 8x_3 &= 26. \end{aligned}$$

Solution. To get a pivot equation containing x_1, we have to reorder the equations, say, by interchanging the first equation and the first equation containing x_1 (that is, the second equation of our present system):

$$\begin{aligned} 3x_1 + 5x_2 + 2x_3 &= 8 \\ 8x_2 + 2x_3 &= -7 \\ 6x_1 + 2x_2 + 8x_3 &= 26. \end{aligned}$$

First Step. Elimination of x_1

It would suffice to show the augmented matrix and operate on it. We show both the equations and the augmented matrix. In the first step, the first equation is the pivot equation. Thus

$$\begin{aligned} \text{Pivot} \rightarrow \boxed{3x_1} + 5x_2 + 2x_3 &= 8 \\ 8x_2 + 2x_3 &= -7 \\ \text{Eliminate} \rightarrow \boxed{6x_1} + 2x_2 + 8x_3 &= 26 \end{aligned} \qquad \begin{bmatrix} 3 & 5 & 2 & 8 \\ 0 & 8 & 2 & -7 \\ 6 & 2 & 8 & 26 \end{bmatrix}$$

To eliminate x_1 from the other equations (here, from the third equation), do:

Subtract $6/3 = 2$ times the pivot equation from the third equation.

The result is shown below.

Second Step. Elimination of x_2

We leave the first equation untouched and take the *new* second equation as the pivot equation. (In the present case, this is still the old second equation because there was no x_1-term to eliminate in the first step.)

$$\begin{aligned} 3x_1 + 5x_2 + 2x_3 &= 8 \\ \text{Pivot} \rightarrow \boxed{8x_2} + 2x_3 &= -7 \\ \text{Eliminate} \rightarrow \boxed{-8x_2} + 4x_3 &= 10 \end{aligned} \qquad \begin{bmatrix} 3 & 5 & 2 & 8 \\ 0 & 8 & 2 & -7 \\ 0 & -8 & 4 & 10 \end{bmatrix}$$

[1] As opposed to **total pivoting,** which we discuss later in this section. For the Gauss–Jordan elimination, see Sec. 19.2.

To eliminate x_2 from the equations below the pivot equation (here, from the third equation), do:

Subtract $-8/8 = -1$ times the pivot equation from the third equation.

The resulting triangular system is shown below. This is the end of the forward elimination. Now comes the back substitution.

Back Substitution. Determination of x_3, x_2, x_1
The triangular system obtained in Step 2 is

$$\begin{aligned} 3x_1 + 5x_2 + 2x_3 &= 8 \\ 8x_2 + 2x_3 &= -7 \\ 6x_3 &= 3 \end{aligned} \qquad \begin{bmatrix} 3 & 5 & 2 & 8 \\ 0 & 8 & 2 & -7 \\ 0 & 0 & 6 & 3 \end{bmatrix}$$

From it, taking first the last equation, then the second equation and then the first equation, we compute the solution

$$\begin{aligned} x_3 &= \tfrac{3}{6} = \tfrac{1}{2} \\ x_2 &= \tfrac{1}{8}(-7 - 2x_3) = -1 \\ x_1 &= \tfrac{1}{3}(8 - 5x_2 - 2x_3) = 4. \end{aligned}$$

This agrees with the values given above. ∎

The general algorithm is shown in Table 19.1 on p. 976. To explain it, we have numbered some of its lines. b_j is denoted by $a_{j,n+1}$, for uniformity. m_{jk} in line 3 suggests *multiplier,* since these are the factors by which we have to multiply the pivot equation $E_k{}^*$ in Step k before subtracting it from an equation $E_j{}^*$ below $E_k{}^*$ from which we want to eliminate x_k. Here we have written $E_k{}^*$ and $E_j{}^*$ to indicate that after Step 1 these are no longer the given equations in (1), but these underwent a change in each step, as indicated in line 4. Accordingly, the possible pivoting in line 1 always refers to the most recent equations, and $j \geqq k$ indicates that we leave untouched all the equations that have served as pivot equations in previous steps. For $p = k$ in line 4 we get 0 on the right, as it should be in the elimination,

$$a_{jk} - m_j a_{kk} = a_{jk} - \frac{a_{jk}}{a_{kk}} a_{kk} = 0.$$

In line 5, if the last equation in the *triangular* system is $0 = b_n{}^* \neq 0$, we have no solution. If it is $0 = b_n{}^* = 0$, we have no unique solution because we then have fewer equations than unknowns.

EXAMPLE 2 **Gauss elimination in Table 19.1, sample computation**

In Example 1 we had $a_{11} = 0$ and $a_{21} \neq 0$, thus $j = 2$ and, by line 2 of the algorithm, interchanged the equations E_1 and E_2. Then we computed $m_{21} = 0/3 = 0$ and $m_{31} = 6/3 = 2$, so that line 4 with $p = 2, 3, 4$ gave no change in E_2, and in E_3 (labeling the new elements with an asterisk)

$$\begin{aligned} &(p = 2) \quad a_{32}^* = a_{32} - m_{31}a_{12} = 2 - 2 \cdot 5 = -8 \\ &(p = 3) \quad a_{33}^* = a_{33} - m_{31}a_{13} = 8 - 2 \cdot 2 = 4 \\ &(p = 4) \quad a_{34}^* = a_{34} - m_{31}a_{14} = b_3 - m_{31}b_1 = 26 - 2 \cdot 8 = 10 \end{aligned}$$

as the coefficients of $E_3{}^*$ in Step 2. ∎

Table 19.1
Gauss Elimination

ALGORITHM GAUSS ($\tilde{\mathbf{A}} = [a_{jk}] = [\mathbf{A} \quad \mathbf{b}]$)

This algorithm computes a unique solution $\mathbf{x} = [x_j]$ of the system (1) or indicates that (1) has no unique solution.

INPUT: Augmented $n \times (n + 1)$ matrix $\tilde{\mathbf{A}} = [a_{jk}]$, where $a_{j,n+1} = b_j$

OUTPUT: Solution $\mathbf{x} = [x_j]$ of (1) or message that the system (1) has no unique solution

For $k = 1, \cdots, n - 1$, do:

1 Find the smallest $j \geqq k$ such that $a_{jk} \neq 0$.
If no such j exists then OUTPUT "No unique solution exists." Stop.
[*Procedure completed unsuccessfully;* **A** *is singular*]

2 Else exchange the contents of rows j and k of $\tilde{\mathbf{A}}$.
For $j = k + 1, \cdots, n$, do:

3
$$m_{jk} := \frac{a_{jk}}{a_{kk}}$$

For $p = k + 1, \cdots, n + 1$, do:

4
$$a_{jp} := a_{jp} - m_{jk}a_{kp}$$

End

End

End

5 If $a_{nn} = 0$ then OUTPUT "No unique solution exists."
Stop.
Else,

6
$$x_n = \frac{a_{n,n+1}}{a_{nn}}$$
[*Start back substitution*]

For $i = n - 1, \cdots, 1$, do:

7
$$x_i = \frac{1}{a_{ii}}\left(a_{i,n+1} - \sum_{j=i+1}^{n} a_{ij}x_j\right)$$

End

OUTPUT $\mathbf{x} = [x_j]$. Stop

End GAUSS

Operations Count

Quite generally, the quality of a numerical method is judged in terms of:

Amount of storage
Amount of time ($\equiv$ number of operations)
Effect of round-off error.

For the Gauss elimination, the operations count is as follows. In Step k we eliminate x_k from $n - k$ equations. This needs $n - k$ divisions in computing the m_{jk} (line 3) and $(n - k)(n - k + 1)$ multiplications and as many subtractions (both in line 4). Since we do $n - 1$ steps, k goes from 1 to $n - 1$ and thus the total number of operations in this forward elimination is

$$f(n) = \sum_{k=1}^{n-1} (n - k) + 2 \sum_{k=1}^{n-1} (n - k)(n - k + 1) \qquad (\text{write } n - k = s)$$

$$= \sum_{s=1}^{n-1} s + 2 \sum_{s=1}^{n-1} s(s + 1) = \tfrac{1}{2}(n - 1)n + \tfrac{2}{3}(n^2 - 1)n \approx \tfrac{2}{3}n^3$$

where $2n^3/3$ is obtained by dropping lower powers of n. We see that $f(n)$ grows about proportional to n^3. We say that $f(n)$ is of *order* n^3 and write

$$f(n) = O(n^3),$$

where O suggests **order.** The general definition of O is as follows. We write

$$f(n) = O(h(n))$$

if the quotient $|f(n)/h(n)|$ remains bounded (does not trail off to infinity) as $n \to \infty$. In our present case, $h(n) = n^3$ and, indeed, $f(n)/n^3 \to 2/3$ because the omitted terms divided by n^3 go to zero as $n \to \infty$.

In the back substitution of x_i we make $n - i$ multiplications and as many subtractions, as well as 1 division. Hence the number of operations in the back substitution is

$$b(n) = \sum_{i=1}^{n} (n - i) + n = \sum_{s=1}^{n} s + n = \tfrac{1}{2}n(n + 1) + n \approx \tfrac{1}{2}n^2 = O(n^2).$$

We see that it grows more slowly than the number of operations in the forward elimination of the Gauss algorithm, so that it is negligible for large systems because it is smaller by a factor n, approximately. For instance, if an operation takes 10^{-6} sec, then the times needed are:

n	Elimination	Back Substitution
100	0.7 sec	0.005 sec
1000	11 min	0.5 sec

More on Pivoting. Scaling

We know that a pivot entry a_{kk}^* must be different from zero. If it is zero, we *must* pivot (interchange equations). But this is not all: if a_{kk}^* is small in absolute value, we *should* pivot, because then we would have to subtract large multiples of the pivot equation from the other equations, thereby amplifying round-off errors. This may affect the accuracy of the result. Before discussing how to overcome this difficulty, let us illustrate it with a simple example.

EXAMPLE 3 Difficulty with small pivot entries

The solution of the system

$$0.0004x_1 + 1.402x_2 = 1.406$$

$$0.4003x_1 - 1.502x_2 = 2.501$$

is $x_1 = 10$, $x_2 = 1$. We solve this system by the Gauss elimination, using four-digit floating-point arithmetic.

(a) Picking the first equation as the pivot equation, we have to multiply this equation by $m = 0.4003/0.0004 = 1001$ and subtract the result from the second equation, obtaining

$$-1405x_2 = -1404.$$

Hence $x_2 = -1404/(-1405) = 0.9993$, and from the first equation, instead of $x_1 = 10$, we get

$$x_1 = \frac{1}{0.0004}(1.406 - 1.402 \cdot 0.9993) = \frac{0.005}{0.0004} = 12.5.$$

This failure occurs because $|a_{11}|$ is small compared to $|a_{12}|$, so that a small round-off error in x_2 led to a large error in x_1.

(b) Picking the second equation as the pivot equation, we have to multiply this equation by $0.0004/0.4003 = 0.000\,9993$ and subtract the result from the first equation, obtaining

$$1.404x_2 = 1.404.$$

Hence $x_2 = 1$, and from the second equation $x_1 = 10$. This success occurs because $|a_{21}|$ is not very small compared to $|a_{22}|$, so that a small round-off error in x_2 would not lead to a large error in x_1. Indeed, for instance, if we had $x_2 = 1.002$, we would still have from the second equation the good value $x_1 = (2.501 + 1.505)/0.4003 = 10.01$. ■

In *partial pivoting,* in the first step we usually pick as the pivot equation an equation in which the coefficient of x_1 is largest in absolute value; similarly for x_2 in the second step, and so on. **Total pivoting** would mean that we look for a coefficient that is largest in absolute value *in the entire system* and start the elimination with the corresponding variable, using this coefficient as the pivot entry; similarly in the further steps. This is hardly done in practice, since it is much more expensive than partial pivoting.

There is a catch: we could magnify a coefficient by multiplying a whole equation, *but this would not change the computed solution.* Multiplication of an equation by a factor is called **row scaling;** here, one normally uses a power of 10 (or of the machine basis β) such that afterward the absolutely largest coefficients of the equation have absolute value between 0.1 and 1 (or β^{-1} and 1, respectively).

In practice, one uses scaled partial pivoting, that is, in the kth step ($k = 1, 2, \cdots$) of the elimination one chooses as pivot equation an equation from the $n - k + 1$ available equations such that the quotient of the coefficient of x_k and the absolutely largest coefficient in the equation is maximum in absolute value. If there are several such equations, take the first of them.

EXAMPLE 4 Choice of pivot equations

Apply the method just mentioned to the selection of pivot equations in the following system and compute the solution.

$$\begin{aligned} 3x_1 - 4x_2 + 5x_3 &= -1 \\ -3x_1 + 2x_2 + x_3 &= 1 \\ 6x_1 + 8x_2 - x_3 &= 35. \end{aligned}$$

Solution. We have

$$\frac{|a_{11}|}{\max |a_{1k}|} = \frac{3}{5}, \qquad \frac{|a_{21}|}{\max |a_{2k}|} = \frac{3}{3}, \qquad \frac{|a_{31}|}{\max |a_{3k}|} = \frac{6}{8}.$$

Accordingly, we select the second equation as the pivot equation and eliminate x_1, obtaining

$$\begin{aligned} -3x_1 + 2x_2 + x_3 &= 1 \\ -2x_2 + 6x_3 &= 0 \\ 12x_2 + x_3 &= 37. \end{aligned}$$

We now have

$$\frac{|a^*_{22}|}{\max |a^*_{2k}|} = \frac{2}{6}, \qquad \frac{|a^*_{32}|}{\max |a^*_{3k}|} = \frac{12}{12}.$$

Hence we select the last equation as the pivot equation and get the triangular system

$$\begin{aligned} -3x_1 + 2x_2 + x_3 &= 1 \\ 12x_2 + x_3 &= 37 \\ \tfrac{37}{6} x_3 &= \tfrac{37}{6}. \end{aligned}$$

Back substitution now gives the solution $x_3 = 1$, $x_2 = 3$, $x_1 = 2$, in this order. ∎

No pivoting is needed for two important classes of matrices, namely, the **diagonally dominant matrices,** that is,

$$|a_{jj}| > \sum_{\substack{k=1 \\ k \neq j}}^{n} |a_{jk}| \qquad (j = 1, \cdots, n)$$

and the symmetric and positive definite matrices, that is,

$$\mathbf{A}^\mathsf{T} = \mathbf{A}, \qquad \mathbf{x}^\mathsf{T}\mathbf{A}\mathbf{x} > 0 \quad \text{for all } \mathbf{x} \neq \mathbf{0}.$$

Error estimates for the Gauss elimination are discussed in Ref. [E13], p. 37, listed in Appendix 1.

Problem Set 19.1

Solve the following linear systems by the Gauss elimination (with partial pivoting if necessary).

1. $x_1 - 4x_2 = -2$
$3x_1 + x_2 = 7$

2. $6x_1 + x_2 = -3$
$4x_1 - 2x_2 = 6$

3. $5x_1 + 2x_2 = -1.5$
$12x_1 - 3x_2 = 12.0$

4. $4x_1 + 2x_2 - 2x_3 = 0$
$7x_2 + 4x_3 = 5$
$3x_2 + 2x_3 = 3$

5. $x_2 + 3x_3 = 9$
$2x_1 + 2x_2 - x_3 = 8$
$-x_1 + 5x_3 = 8$

6. $x_2 - 4x_3 = 2$
$x_1 + 3x_2 - 2x_3 = 4$
$2x_1 + 5x_2 - 7x_3 = 6$

7. $2x_1 + 2x_2 + 2x_3 = 4$
$-x_1 + 2x_2 - 3x_3 = 32$
$3x_1 - 4x_3 = 17$

8. $-2.0x_2 + 12.0x_3 = 12.00$
$2.5x_1 + x_2 - 3.2x_3 = -1.88$
$0.5x_1 - 1.4x_2 + 0.5x_3 = 5.84$

9. $-0.8x_1 + 1.6x_2 - 4.3x_3 = 5.28$
$x_1 - 7.3x_2 + 0.5x_3 = -32.04$
$4.5x_1 - 3.5x_2 = -3.30$

10. $5x_1 + 3x_2 + x_3 = 2$
$-4x_2 + 8x_3 = -3$
$10x_1 - 6x_2 + 26x_3 = 0$

11. $3.0x_1 + x_2 + 2.0x_3 = 0$
$-2.5x_1 - 0.5x_2 - 11.0x_3 = 0$
$1.8x_1 + 2.8x_2 + 4.0x_3 = 16.1$

12. $2x_1 - x_2 + 5x_3 = 4$
$-6x_1 + 3x_2 - 9x_3 = -6$
$4x_1 - 2x_2 = -2$

13. $4x_1 + 10x_2 - 2x_3 = -20$
$-x_1 - 15x_2 + 3x_3 = 30$
$25x_2 - 5x_3 = -50$

Solve the following linear systems by the Gauss elimination with selection of pivot equations by the method in Example 4.

14. $x_1 - x_2 + 2x_3 = 3.8$
$4x_1 + 3x_2 - x_3 = -5.7$
$5x_1 + x_2 + 3x_3 = 2.8$

15. $5x_1 + 10x_2 - 2x_3 = -0.30$
$2x_1 - x_2 + x_3 = 1.91$
$3x_1 + 4x_2 = 1.16$

16. $x_1 + 2x_2 - 3x_3 = -11$
$10x_1 + x_2 + x_3 = 8$
$10x_2 + 2x_3 = 2$

17. Solve the following system by the Gauss elimination without pivoting.

$$\epsilon x_1 + x_2 = 1$$

$$x_1 + x_2 = 2.$$

Show that, for any fixed machine word length and sufficiently small $\epsilon > 0$, the computer gives $x_2 = 1$ and then $x_1 = 0$. Solve the system exactly and show that for the exact solution, $x_1 \to 1$ and $x_2 \to 1$ as $\epsilon \to 0$. Comment.

18. Solve the system in Prob. 17 by the Gauss elimination with pivoting. Compare and comment.

19. Try to solve each of the following two systems by the Gauss elimination. Explain why the Gauss elimination fails if no solution exists.

$$\begin{aligned} x_1 + x_2 + x_3 &= 3 \\ 4x_1 + 2x_2 - x_3 &= 5 \\ 9x_1 + 5x_2 - x_3 &= 13 \end{aligned} \qquad \begin{aligned} x_1 + x_2 + x_3 &= 3 \\ 4x_1 + 2x_2 - x_3 &= 5 \\ 9x_1 + 5x_2 - x_3 &= 12. \end{aligned}$$

20. Verify the calculation of the operations count of the Gauss method in the text.

19.2 Linear Systems: LU-Factorization, Matrix Inversion

We continue our discussion of numerical methods for solving linear systems of n equations in n unknowns $x_1, \cdots, x_n$,

$$\mathbf{Ax} = \mathbf{b}, \tag{1}$$

where $\mathbf{A} = [a_{jk}]$ is the $n \times n$ coefficient matrix and $\mathbf{x}^\mathsf{T} = [x_1 \cdots x_n]$ and $\mathbf{b}^\mathsf{T} = [b_1 \cdots b_n]$. We present three related methods that are modifications of the Gauss elimination. They are named after Doolittle, Crout, and Cholesky and use the idea of the LU-factorization of **A**, which we explain first.

An **LU-factorization** of a given square matrix **A** is of the form

$$\mathbf{A} = \mathbf{LU} \tag{2}$$

where **L** is lower triangular and **U** is upper triangular. For example,

$$\mathbf{A} = \begin{bmatrix} 2 & 3 \\ 8 & 5 \end{bmatrix} = \mathbf{LU} = \begin{bmatrix} 1 & 0 \\ 4 & 1 \end{bmatrix} \begin{bmatrix} 2 & 3 \\ 0 & -7 \end{bmatrix}.$$

It can be proved that for any nonsingular matrix (see Sec. 7.7) the rows can be reordered so that the resulting matrix **A** has an LU-factorization (2) in which **L** turns out to be the matrix of the multipliers m_{jk} of the Gauss elimination, with main diagonal $1, \cdots, 1$, and **U** is the matrix of the triangular system at the end of the Gauss elimination. (See Ref. [E3], pp. 155–156.)

The ***crucial idea*** now is that **L** and **U** in (2) can be computed directly, without solving simultaneous equations (thus, without using the Gauss elimination). As a count shows, this needs about $n^3/3$ operations, about half as many as the Gauss elimination, which needs about $2n^3/3$ (see Sec. 19.1). And once we have (2), we can use it for solving $\mathbf{Ax} = \mathbf{b}$ in two steps, involving only about n^2 operations, simply by noting that $\mathbf{Ax} = \mathbf{LUx} = \mathbf{b}$ may be written

$$\text{(3)} \qquad \text{(a)} \quad \mathbf{Ly} = \mathbf{b} \qquad \text{where} \qquad \text{(b)} \quad \mathbf{Ux} = \mathbf{y}$$

and solving first (3a) for **y** and then (3b) for **x**. This is called **Doolittle's method.** Both systems (3a) and (3b) are triangular, so their solution is the same as back substitution in the Gauss elimination.

A similar method, **Crout's method,** is obtained from (2) if **U** (instead of **L**) is required to have main diagonal $1, \cdots, 1$. In either case the factorization (2) is unique.

EXAMPLE 1 Doolittle's method

Solve the system in Example 1 of Sec. 19.1 by Doolittle's method.

Solution. The decomposition (2) is obtained from

$$\mathbf{A} = [a_{jk}] = \begin{bmatrix} 3 & 5 & 2 \\ 0 & 8 & 2 \\ 6 & 2 & 8 \end{bmatrix} = \begin{bmatrix} 1 & 0 & 0 \\ m_{21} & 1 & 0 \\ m_{31} & m_{32} & 1 \end{bmatrix} \begin{bmatrix} u_{11} & u_{12} & u_{13} \\ 0 & u_{22} & u_{23} \\ 0 & 0 & u_{33} \end{bmatrix}$$

by determining the m_{jk} and u_{jk}, using matrix multiplication. By going through **A** row by row we get successively

$a_{11} = 3 = u_{11}$	$a_{12} = 5 = u_{12}$	$a_{13} = 2 = u_{13}$
$a_{21} = 0 = m_{21}u_{11}$	$a_{22} = 8 = m_{21}u_{12} + u_{22}$	$a_{23} = 2 = m_{21}u_{13} + u_{23}$
$m_{21} = 0$	$u_{22} = 8$	$u_{23} = 2$
$a_{31} = 6 = m_{31}u_{11}$	$a_{32} = 2 = m_{31}u_{12} + m_{32}u_{22}$	$a_{33} = 8 = m_{31}u_{13} + m_{32}u_{23} + u_{33}$
$= m_{31} \cdot 3$	$= 2 \cdot 5 + m_{32} \cdot 8$	$= 2 \cdot 2 - 1 \cdot 2 + u_{33}$
$m_{31} = 2$	$m_{32} = -1$	$u_{33} = 6$

Thus the factorization (2) is

$$\begin{bmatrix} 3 & 5 & 2 \\ 0 & 8 & 2 \\ 6 & 2 & 8 \end{bmatrix} = \begin{bmatrix} 1 & 0 & 0 \\ 0 & 1 & 0 \\ 2 & -1 & 1 \end{bmatrix} \begin{bmatrix} 3 & 5 & 2 \\ 0 & 8 & 2 \\ 0 & 0 & 6 \end{bmatrix}.$$

We first solve $\mathbf{Ly} = \mathbf{b}$, determining y_1, then y_2, then y_3, that is,

$$\begin{bmatrix} 1 & 0 & 0 \\ 0 & 1 & 0 \\ 2 & -1 & 1 \end{bmatrix} \begin{bmatrix} y_1 \\ y_2 \\ y_3 \end{bmatrix} = \begin{bmatrix} 8 \\ -7 \\ 26 \end{bmatrix}. \qquad \text{Solution} \qquad \mathbf{y} = \begin{bmatrix} 8 \\ -7 \\ 3 \end{bmatrix}.$$

Then we solve $\mathbf{Ux} = \mathbf{y}$, determining x_3, then x_2, then x_1, that is,

$$\begin{bmatrix} 3 & 5 & 2 \\ 0 & 8 & 2 \\ 0 & 0 & 6 \end{bmatrix} \begin{bmatrix} x_1 \\ x_2 \\ x_3 \end{bmatrix} = \begin{bmatrix} 8 \\ -7 \\ 3 \end{bmatrix}. \qquad \text{Solution} \qquad \mathbf{x} = \begin{bmatrix} 4 \\ -1 \\ 1/2 \end{bmatrix}.$$

This agrees with the solution in Example 1 of Sec. 19.1. ∎

Our formulas in Example 1 suggest that for general n the elements of the matrices $\mathbf{L} = [m_{jk}]$ (with main diagonal $1, \cdots, 1$) and $\mathbf{U} = [u_{jk}]$ in the ***Doolittle method*** are computed from

$$
\textbf{(4)}\qquad
\begin{aligned}
u_{1k} &= a_{1k} && k = 1, \cdots, n \\
u_{jk} &= a_{jk} - \sum_{s=1}^{j-1} m_{js}u_{sk} && k = j, \cdots, n; \quad j \geqq 2 \\
m_{j1} &= \frac{a_{j1}}{u_{11}} && j = 2, \cdots, n \\
m_{jk} &= \frac{1}{u_{kk}}\left(a_{jk} - \sum_{s=1}^{k-1} m_{js}u_{sk}\right) && j = k+1, \cdots, n; \quad k \geqq 2.
\end{aligned}
$$

The corresponding formulas for the **LU**-factorization in ***Crout's method*** are quite similar:

$$
\textbf{(5)}\qquad
\begin{aligned}
m_{j1} &= a_{j1} && j = 1, \cdots, n \\
m_{jk} &= a_{jk} - \sum_{s=1}^{k-1} m_{js}u_{sk} && j = k, \cdots, n; \quad k \geqq 2 \\
u_{1k} &= \frac{a_{1k}}{m_{11}} && k = 2, \cdots, n \\
u_{jk} &= \frac{1}{m_{jj}}\left(a_{jk} - \sum_{s=1}^{j-1} m_{js}u_{sk}\right) && k = j+1, \cdots, n; \quad j \geqq 2.
\end{aligned}
$$

Cholesky's Method

For a ***symmetric, positive definite*** matrix $\mathbf{A}$ (thus $\mathbf{A} = \mathbf{A}^\mathsf{T}$, $\mathbf{x}^\mathsf{T}\mathbf{A}\mathbf{x} > 0$ for all $\mathbf{x} \neq \mathbf{0}$) we can in (2) even choose $\mathbf{U} = \mathbf{L}^\mathsf{T}$, thus $u_{jk} = m_{kj}$ (but impose no conditions on the main diagonal entries).

For example,

$$
\text{(6)}\qquad
\begin{bmatrix} 4 & 2 & 14 \\ 2 & 17 & -5 \\ 14 & -5 & 83 \end{bmatrix}
=
\begin{bmatrix} 2 & 0 & 0 \\ 1 & 4 & 0 \\ 7 & -3 & 5 \end{bmatrix}
\begin{bmatrix} 2 & 1 & 7 \\ 0 & 4 & -3 \\ 0 & 0 & 5 \end{bmatrix}.
$$

The popular method of solving $\mathbf{A}\mathbf{x} = \mathbf{b}$ based on this factorization $\mathbf{A} = \mathbf{L}\mathbf{L}^\mathsf{T}$ is called **Cholesky's method.**

The formulas for the factorization are

$$
(7)\qquad
\begin{aligned}
m_{11} &= \sqrt{a_{11}} \\
m_{jj} &= \sqrt{a_{jj} - \sum_{s=1}^{j-1} m_{js}^2} && j = 2, \cdots, n \\
m_{j1} &= \frac{a_{j1}}{m_{11}} && j = 2, \cdots, n \\
m_{jk} &= \frac{1}{m_{kk}}\left(a_{jk} - \sum_{s=1}^{k-1} m_{js}m_{ks}\right) && j = k+1, \cdots, n; \quad k \geqq 2.
\end{aligned}
$$

If A is symmetric but not positive definite, this method could still be applied, but then leads to a *complex* matrix L, so that it becomes impractical.

EXAMPLE 2 Cholesky's method

Solve by Cholesky's method:

$$
\begin{aligned}
4x_1 + 2x_2 + 14x_3 &= 14 \\
2x_1 + 17x_2 - 5x_3 &= -101 \\
14x_1 - 5x_2 + 83x_3 &= 155.
\end{aligned}
$$

Solution. From (7) or from the form of the factorization

$$
\begin{bmatrix} 4 & 2 & 14 \\ 2 & 17 & -5 \\ 14 & -5 & 83 \end{bmatrix} = \begin{bmatrix} m_{11} & 0 & 0 \\ m_{21} & m_{22} & 0 \\ m_{31} & m_{32} & m_{33} \end{bmatrix} \begin{bmatrix} m_{11} & m_{21} & m_{31} \\ 0 & m_{22} & m_{32} \\ 0 & 0 & m_{33} \end{bmatrix}
$$

we compute, in the given order,

$$
m_{11} = \sqrt{a_{11}} = 2 \qquad m_{21} = \frac{a_{21}}{m_{11}} = \frac{2}{2} = 1 \qquad m_{31} = \frac{a_{31}}{m_{11}} = \frac{14}{2} = 7
$$

$$
m_{22} = \sqrt{a_{22} - m_{21}^2} = \sqrt{17 - 1} = 4
$$

$$
m_{32} = \frac{1}{m_{22}}(a_{32} - m_{31}m_{21}) = \frac{1}{4}(-5 - 7 \cdot 1) = -3
$$

$$
m_{33} = \sqrt{a_{33} - m_{31}^2 - m_{32}^2} = \sqrt{83 - 7^2 - (-3)^2} = 5.
$$

This agrees with (6). We now have to solve $\mathbf{Ly} = \mathbf{b}$, that is,

$$
\begin{bmatrix} 2 & 0 & 0 \\ 1 & 4 & 0 \\ 7 & -3 & 5 \end{bmatrix} \begin{bmatrix} y_1 \\ y_2 \\ y_3 \end{bmatrix} = \begin{bmatrix} 14 \\ -101 \\ 155 \end{bmatrix}. \qquad \text{Solution} \qquad \mathbf{y} = \begin{bmatrix} 7 \\ -27 \\ 5 \end{bmatrix}.
$$

As the second step, we have to solve $\mathbf{Ux} = \mathbf{L}^\mathsf{T}\mathbf{x} = \mathbf{y}$, that is,

$$
\begin{bmatrix} 2 & 1 & 7 \\ 0 & 4 & -3 \\ 0 & 0 & 5 \end{bmatrix} \begin{bmatrix} x_1 \\ x_2 \\ x_3 \end{bmatrix} = \begin{bmatrix} 7 \\ -27 \\ 5 \end{bmatrix}. \qquad \text{Solution} \qquad \mathbf{x} = \begin{bmatrix} 3 \\ -6 \\ 1 \end{bmatrix}.
$$

■

Gauss–Jordan Elimination. Matrix Inversion

Another variant of the Gauss elimination is the **Gauss–Jordan elimination,** introduced by W. Jordan in 1920, in which back substitution is avoided by additional computations that reduce the matrix to diagonal form, instead of the triangular form in the Gauss elimination. But this reduction from the Gauss triangular to diagonal form requires more operations than back substitution does, so that the method is *disadvantageous* for solving systems $\mathbf{Ax} = \mathbf{b}$. But it may be used for matrix inversion, where the situation is as follows.

The **inverse** of a nonsingular square matrix **A** may be determined in principle by solving the n systems

$$\mathbf{Ax} = \mathbf{b}_j \qquad (j = 1, \cdots, n) \tag{8}$$

where $\mathbf{b}_j$ is the jth column of the $n \times n$ unit matrix.

However, it is preferable to produce $\mathbf{A}^{-1}$ by operating on the unit matrix **I** in the same way as the Gauss–Jordan algorithm, reducing **A** to **I**. A typical illustrative example of this method is given in Sec. 7.7.

Problem Set 19.2

Solve the following linear systems by Doolittle's method. Show also the LU-factorization.

1. $3x_1 + 2x_2 = 18$
$18x_1 + 17x_2 = 123$

2. $2x_1 - 4x_2 = 0$
$8x_1 - 15x_2 = 0.5$

3. $4x_1 - 6x_2 = -17.0$
$8x_1 - 7x_2 = -26.5$

4. $x_1 + x_2 + x_3 = 5$
$x_1 + 2x_2 + 2x_3 = 6$
$x_1 + 2x_2 + 3x_3 = 8$

5. $x_1 + 2x_2 + 5x_3 = -11$
$x_2 = 1$
$2x_1 + 9x_2 + 11x_3 = -20$

6. $3x_1 + 2x_2 = 14$
$12x_1 + 13x_2 + 6x_3 = 40$
$-3x_1 + 8x_2 + 9x_3 = -28$

7. $5x_1 + 4x_2 + x_3 = 3.4$
$10x_1 + 9x_2 + 4x_3 = 8.8$
$10x_1 + 13x_2 + 15x_3 = 19.2$

8. $x_1 - x_2 + 2.6x_3 = -4.94$
$0.5x_1 - 3.0x_2 + 3.3x_3 = -8.27$
$-1.5x_1 - 3.5x_2 - 10.4x_3 = 10.51$

9. $3x_1 + 9x_2 + 6x_3 = 23$
$18x_1 + 48x_2 + 39x_3 = 136$
$9x_1 - 27x_2 + 42x_3 = 45$

Solve by Crout's method:

10. Prob. 1 **11.** Prob. 5 **12.** Prob. 6

Solve the following systems of linear equations by Cholesky's method.

13. $9x_1 + 6x_2 + 12x_3 = 174$
$6x_1 + 13x_2 + 11x_3 = 236$
$12x_1 + 11x_2 + 26x_3 = 308$

14. $0.01x_1 + 0.03x_3 = 0.14$
$0.16x_2 + 0.08x_3 = 0.16$
$0.03x_1 + 0.08x_2 + 0.14x_3 = 0.54$

15. $\begin{aligned} 4x_1 + 6x_2 + 8x_3 &= 0 \\ 6x_1 + 34x_2 + 52x_3 &= -160 \\ 8x_1 + 52x_2 + 129x_3 &= -452 \end{aligned}$

16. $\begin{aligned} 4x_1 + 10x_2 + 8x_3 &= 44 \\ 10x_1 + 26x_2 + 26x_3 &= 128 \\ 8x_1 + 26x_2 + 61x_3 &= 214 \end{aligned}$

Find the inverse of the given matrix by the Gauss–Jordan method.

17. $\begin{bmatrix} 2 & -0.5 \\ -1 & 1.5 \end{bmatrix}$

18. $\begin{bmatrix} 3 & -5 \\ -2 & -1 \end{bmatrix}$

19. $\begin{bmatrix} 0.5 & -0.5 \\ -0.7 & 0.2 \end{bmatrix}$

20. $\begin{bmatrix} 6 & 4 & 3 \\ 4 & 3 & 2 \\ 3 & 4 & 2 \end{bmatrix}$

21. $\begin{bmatrix} 0 & -\frac{2}{3} & \frac{1}{3} \\ \frac{1}{6} & \frac{1}{6} & -\frac{1}{6} \\ -\frac{1}{3} & \frac{2}{3} & 0 \end{bmatrix}$

22. $\begin{bmatrix} 5 & 7 & 6 & 5 \\ 7 & 10 & 8 & 7 \\ 6 & 8 & 10 & 9 \\ 5 & 7 & 9 & 10 \end{bmatrix}$

19.3 Linear Systems: Solution by Iteration

The Gauss elimination and its variants in the last two sections belong to the **direct methods** for solving linear systems of equations; these are methods that give solutions after an amount of computation that can be specified in advance. In contrast, in an **indirect** or **iterative method** we start from an approximation to the true solution and, if successful, obtain better and better approximations from a computational cycle repeated as often as may be necessary for achieving a required accuracy, so that the amount of arithmetic depends upon the accuracy required and varies from case to case.

We apply iterative methods if the convergence is rapid (if matrices have large main diagonal entries, as we shall see), so that we save operations compared to a direct method, or if a large system is **sparse,** that is, has very many zero coefficients, so that one would waste space in storing zeros, for instance, 9995 zeros per equation in a potential problem of 10^4 equations in 10^4 unknowns with typically only 5 nonzero terms per equation (more on this in Sec. 20.4).

Gauss–Seidel Iteration Method

This is an iterative method of great practical importance, which we can simply explain in terms of an example.

EXAMPLE 1 Gauss–Seidel iteration

We consider the linear system

$$\begin{aligned} x_1 - 0.25x_2 - 0.25x_3 \qquad\qquad &= 50 \\ -0.25x_1 + x_2 \qquad\qquad - 0.25x_4 &= 50 \\ -0.25x_1 \qquad\qquad + x_3 - 0.25x_4 &= 25 \\ - 0.25x_2 - 0.25x_3 + x_4 &= 25. \end{aligned} \tag{1}$$

(Equations of this form arise in the numerical solution of partial differential equations and in spline interpolation.) We write the system in the form

$$
(2) \qquad
\begin{aligned}
x_1 &= \phantom{0.25x_1 +{}} 0.25x_2 + 0.25x_3 \phantom{{}+ 0.25x_4} + 50\\
x_2 &= 0.25x_1 \phantom{{}+ 0.25x_2 + 0.25x_3} + 0.25x_4 + 50\\
x_3 &= 0.25x_1 \phantom{{}+ 0.25x_2 + 0.25x_3} + 0.25x_4 + 25\\
x_4 &= \phantom{0.25x_1 +{}} 0.25x_2 + 0.25x_3 \phantom{{}+ 0.25x_4} + 25.
\end{aligned}
$$

We use these equations for iteration, that is, we start from a (possibly poor) approximation to the solution, say, $x_1^{(0)} = 100$, $x_2^{(0)} = 100$, $x_3^{(0)} = 100$, $x_4^{(0)} = 100$, and compute from (2) a presumably better approximation

Use "old" values
("New" values here not yet available)

$$
(3) \qquad
\begin{aligned}
x_1^{(1)} &= \phantom{0.25x_1^{(1)} +{}} 0.25x_2^{(0)} + 0.25x_3^{(0)} \phantom{{}+ 0.25x_4^{(0)}} + 50.00 = 100.00\\
x_2^{(1)} &= 0.25x_1^{(1)} \phantom{{}+ 0.25x_2^{(0)} + 0.25x_3^{(0)}} + 0.25x_4^{(0)} + 50.00 = 100.00\\
x_3^{(1)} &= 0.25x_1^{(1)} \phantom{{}+ 0.25x_2^{(0)} + 0.25x_3^{(0)}} + 0.25x_4^{(0)} + 25.00 = 75.00\\
x_4^{(1)} &= \phantom{0.25x_1^{(1)} +{}} 0.25x_2^{(1)} + 0.25x_3^{(1)} \phantom{{}+ 0.25x_4^{(0)}} + 25.00 = 68.75.
\end{aligned}
$$

Use "new" values

We see that these equations are obtained from (2) by substituting on the right the ***most recent*** approximations. In fact, corresponding elements replace previous ones as soon as they have been computed, so that in the second and third equations we use $x_1^{(1)}$ (not $x_1^{(0)}$), and in the last equation of (3) we use $x_2^{(1)}$ and $x_3^{(1)}$ (not $x_2^{(0)}$ and $x_3^{(0)}$). The next step yields

$$
\begin{aligned}
x_1^{(2)} &= \phantom{0.25x_1^{(2)} +{}} 0.25x_2^{(1)} + 0.25x_3^{(1)} \phantom{{}+ 0.25x_4^{(1)}} + 50.00 = 93.75\\
x_2^{(2)} &= 0.25x_1^{(2)} \phantom{{}+ 0.25x_2^{(1)} + 0.25x_3^{(1)}} + 0.25x_4^{(1)} + 50.00 = 90.62\\
x_3^{(2)} &= 0.25x_1^{(2)} \phantom{{}+ 0.25x_2^{(1)} + 0.25x_3^{(1)}} + 0.25x_4^{(1)} + 25.00 = 65.62\\
x_4^{(2)} &= \phantom{0.25x_1^{(2)} +{}} 0.25x_2^{(2)} + 0.25x_3^{(2)} \phantom{{}+ 0.25x_4^{(1)}} + 25.00 = 64.06.
\end{aligned}
$$

In practice, one would do further steps and obtain a more accurate approximate solution.
The reader may show that the exact solution is $x_1 = x_2 = 87.5$, $x_3 = x_4 = 62.5$. ▮

To obtain an algorithm for the Gauss–Seidel iteration, let us derive the general formulas for this iteration.

We assume that $a_{jj} = 1$ for $j = 1, \cdots, n$. (Note that this can be achieved if we can rearrange the equations so that no diagonal coefficient is zero; then we may divide each equation by the corresponding diagonal coefficient.) We now write

$$
\textbf{(4)} \qquad \mathbf{A} = \mathbf{I} + \mathbf{L} + \mathbf{U}
$$

where $\mathbf{I}$ is the $n \times n$ unit matrix and $\mathbf{L}$ and $\mathbf{U}$ are respectively lower and upper triangular matrices with zero main diagonals. If we substitute (4) into $\mathbf{Ax} = \mathbf{b}$, we have

$$\mathbf{Ax} = (\mathbf{I} + \mathbf{L} + \mathbf{U})\mathbf{x} = \mathbf{b}.$$

Taking $\mathbf{Lx}$ and $\mathbf{Ux}$ to the right, we obtain, since $\mathbf{Ix} = \mathbf{x}$,

$$\mathbf{x} = \mathbf{b} - \mathbf{Lx} - \mathbf{Ux}. \tag{5}$$

Remembering from our computation in Example 1 that below the main diagonal we took "new" approximations and above the main diagonal "old" approximations, we obtain from (5) the desired iteration formulas

$$\mathbf{x}^{(m+1)} = \mathbf{b} - \mathbf{Lx}^{(m+1)} - \mathbf{Ux}^{(m)} \tag{6}$$

where $\mathbf{x}^{(m)} = [x_j^{(m)}]$ is the mth approximation and $\mathbf{x}^{(m+1)} = [x_j^{(m+1)}]$ is the $(m + 1)$st approximation. In components this gives the formula in line 1 of the algorithm in Table 19.2, where the factor $1/a_{jj}$ accomplishes our assumption in (4) and (5) that the main diagonal entries of $\mathbf{A}$ are all 1.

Table 19.2
Gauss–Seidel Iteration

ALGORITHM GAUSS–SEIDEL ($\mathbf{A}$, $\mathbf{b}$, $\mathbf{x}^{(0)}$, ϵ, N)

This algorithm computes a solution $\mathbf{x}$ of the system $\mathbf{Ax} = \mathbf{b}$ given an initial approximation $\mathbf{x}^{(0)}$, where $\mathbf{A} = [a_{jk}]$ is an $n \times n$ matrix with $a_{jj} \neq 0$, $j = 1, \cdots, n$.

INPUT: $\mathbf{A}$, $\mathbf{b}$, initial approximation $\mathbf{x}^{(0)}$, tolerance $\epsilon > 0$, maximum number of iterations N

OUTPUT: Approximate solution $\mathbf{x}^{(m)} = [x_j^{(m)}]$ or failure message that $\mathbf{x}^{(N)}$ does not satisfy the tolerance condition

For $m = 0, \cdots, N - 1$, do:

For $j = 1, \cdots, n$, do:

1 $$x_j^{(m+1)} = -\frac{1}{a_{jj}}\left(\sum_{k=1}^{j-1} a_{jk}x_k^{(m+1)} + \sum_{k=j+1}^{n} a_{jk}x_k^{(m)} - b_j\right)$$

End

2 If $\max_j |x_j^{(m+1)} - x_j^{(m)}| < \epsilon$ then OUTPUT $\mathbf{x}^{(m+1)}$. Stop

[*Procedure completed successfully*]

End

OUTPUT: "No solution satisfying the tolerance condition obtained after N iteration steps." Stop

[*Procedure completed unsuccessfully*]

End GAUSS–SEIDEL

Convergence. Matrix Norms

An iteration method for solving $\mathbf{Ax} = \mathbf{b}$ is said to **converge** for an initial $\mathbf{x}^{(0)}$ if the corresponding iterative sequence $\mathbf{x}^{(0)}, \mathbf{x}^{(1)}, \mathbf{x}^{(2)}, \cdots$ converges to a solution of the given system. Convergence depends on the relation between $\mathbf{x}^{(m)}$ and $\mathbf{x}^{(m+1)}$. To get this relation for the Gauss–Seidel method, we use (6). We first have

$$(\mathbf{I} + \mathbf{L})\mathbf{x}^{(m+1)} = \mathbf{b} - \mathbf{U}\mathbf{x}^{(m)}$$

and by multiplying by $(\mathbf{I} + \mathbf{L})^{-1}$ from the left,

$$\mathbf{x}^{(m+1)} = \mathbf{C}\mathbf{x}^{(m)} + (\mathbf{I} + \mathbf{L})^{-1}\mathbf{b} \qquad \text{where} \qquad \mathbf{C} = -(\mathbf{I} + \mathbf{L})^{-1}\mathbf{U}. \tag{7}$$

The Gauss–Seidel iteration converges for every $\mathbf{x}^{(0)}$ if and only if all the eigenvalues (Sec. 7.10) of the "iteration matrix" $\mathbf{C} = [c_{jk}]$ have absolute value less than 1. (Proof in Ref. [E3], p. 191, listed in Appendix 1.) ***Caution!*** If you want to get $\mathbf{C}$, first divide the rows of $\mathbf{A}$ by a_{jj} to have main diagonal $1, \cdots, 1$. If the **spectral radius** of $\mathbf{C}$ (= maximum of those absolute values) is small, then the convergence is rapid.

Sufficient convergence condition. A sufficient condition for convergence is

$$\|\mathbf{C}\| < 1. \tag{8}$$

Here $\|\mathbf{C}\|$ is some **matrix norm**, such as

$$\|\mathbf{C}\| = \sqrt{\sum_{j=1}^{n}\sum_{k=1}^{n} c_{jk}^{\,2}} \qquad \textbf{(Frobenius norm)} \tag{9}$$

or

$$\|\mathbf{C}\| = \max_{k} \sum_{j=1}^{n} |c_{jk}| \qquad \textbf{(Column "sum" norm)} \tag{10}$$

or

$$\|\mathbf{C}\| = \max_{j} \sum_{k=1}^{n} |c_{jk}| \qquad \textbf{(Row "sum" norm).} \tag{11}$$

These are the most frequently used matrix norms in numerical work. In (9) we take the sum of the squares of all the entries and then the root of it. In (10) we take the sum of the $|c_{jk}|$ in column k, where $k = 1, \cdots, n$, and then the largest of these n sums. In (11) we sum the $|a_{jk}|$ in each row and then take the largest of these n sums.

EXAMPLE 2 Test of convergence of the Gauss–Seidel iteration

Test whether the Gauss–Seidel iteration converges for the system

$$\begin{aligned} 2x + y + z &= 4 \\ x + 2y + z &= 4 \\ x + y + 2z &= 4 \end{aligned} \qquad \text{written} \qquad \begin{aligned} x &= 2 - \tfrac{1}{2}y - \tfrac{1}{2}z \\ y &= 2 - \tfrac{1}{2}x - \tfrac{1}{2}z \\ z &= 2 - \tfrac{1}{2}x - \tfrac{1}{2}y. \end{aligned}$$

Solution. The decomposition

$$\begin{bmatrix} 1 & 1/2 & 1/2 \\ 1/2 & 1 & 1/2 \\ 1/2 & 1/2 & 1 \end{bmatrix} = \mathbf{I} + \mathbf{L} + \mathbf{U} = \mathbf{I} + \begin{bmatrix} 0 & 0 & 0 \\ 1/2 & 0 & 0 \\ 1/2 & 1/2 & 0 \end{bmatrix} + \begin{bmatrix} 0 & 1/2 & 1/2 \\ 0 & 0 & 1/2 \\ 0 & 0 & 0 \end{bmatrix}$$

shows that

$$\mathbf{C} = -(\mathbf{I} + \mathbf{L})^{-1}\mathbf{U} = -\begin{bmatrix} 1 & 0 & 0 \\ -1/2 & 1 & 0 \\ -1/4 & -1/2 & 1 \end{bmatrix}\begin{bmatrix} 0 & 1/2 & 1/2 \\ 0 & 0 & 1/2 \\ 0 & 0 & 0 \end{bmatrix} = \begin{bmatrix} 0 & -1/2 & -1/2 \\ 0 & 1/4 & -1/4 \\ 0 & 1/8 & 3/8 \end{bmatrix}.$$

We compute the Frobenius norm of $\mathbf{C}$

$$\|\mathbf{C}\| = \left(\frac{1}{4} + \frac{1}{4} + \frac{1}{16} + \frac{1}{16} + \frac{1}{64} + \frac{9}{64}\right)^{1/2} = \left(\frac{50}{64}\right)^{1/2} = 0.884 < 1$$

and conclude from (8) that this Gauss–Seidel iteration converges. It is interesting that the other two norms would permit no conclusion, as the student should verify. ∎

Residual. Given a system $\mathbf{Ax} = \mathbf{b}$, the **residual r** is defined by

$$\mathbf{r} = \mathbf{b} - \mathbf{Ax}.$$

Clearly, $\mathbf{r} = \mathbf{0}$ if and only if $\mathbf{x}$ is a solution. Hence $\mathbf{r} \neq \mathbf{0}$ for an approximate solution. In the Gauss–Seidel iteration, at each stage we modify or *relax* a component of an approximate solution in order to reduce a component of $\mathbf{r}$ to zero. Hence the Gauss–Seidel iteration belongs to a class of methods often called **relaxation methods.** More about the residual follows in the next section.

Jacobi Iteration

The Gauss–Seidel iteration is a method of **successive corrections** because we replace approximations by corresponding new ones as soon as the latter have been computed. A method is called a method of **simultaneous corrections** if no component of an approximation $\mathbf{x}^{(m+1)}$ is used until *all* the components of $\mathbf{x}^{(m+1)}$ have been computed. A method of this type is the **Jacobi iteration,** which is similar to the Gauss–Seidel iteration but consists in *not* using improved values until a step has been completed and then replacing $\mathbf{x}^{(m)}$ by $\mathbf{x}^{(m+1)}$ at once, directly before the beginning of the next cycle. Hence, if we write $\mathbf{Ax} = \mathbf{b}$ in the form $\mathbf{x} = \mathbf{b} + (\mathbf{I} - \mathbf{A})\mathbf{x}$, the Jacobi iteration in matrix notation is

$$\mathbf{x}^{(m+1)} = \mathbf{b} + (\mathbf{I} - \mathbf{A})\mathbf{x}^{(m)}. \tag{12}$$

This method is largely of theoretical interest. It converges for every choice of $\mathbf{x}^{(0)}$ if and only if the spectral radius of $\mathbf{I} - \mathbf{A}$ is less than 1; here we again assume $a_{jj} = 1$ for $j = 1, \cdots, n$.

Iteration for Computing the Inverse

The **inverse** of a nonsingular square matrix $\mathbf{A}$ may also be determined by an iteration method suggested by the following idea. The reciprocal x of a given number a satisfies $xa = 1$, and if we want to determine x without division, we may apply Newton's method (Sec. 18.2) to the function $f(x) = x^{-1} - a$. Since we have $f'(x) = -1/x^2$, the Newton iteration is

$$x_{m+1} = x_m - (x_m^{-1} - a)(-x_m^2) = x_m(2 - ax_m).$$

This suggests an analogous iteration formula for determining the inverse $\mathbf{X} = \mathbf{A}^{-1}$ of $\mathbf{A}$, namely,

$$\mathbf{X}^{(m+1)} = \mathbf{X}^{(m)}(2\mathbf{I} - \mathbf{A}\mathbf{X}^{(m)}). \tag{13}$$

This process converges (produces $\mathbf{A}^{-1}$ as $m \to \infty$) if and only if an $\mathbf{X}^{(0)}$ is chosen so that every eigenvalue of $\mathbf{I} - \mathbf{A}\mathbf{X}^{(0)}$ is of absolute value less than 1 (see Ref. [E24], p. 225, listed in Appendix 1). The method is of practical interest if the occurring multiplications are simple, for instance, if $\mathbf{A}$ has many zeros. However, a suitable choice of $\mathbf{X}^{(0)}$ is generally difficult, and the method is mostly used for improving an inaccurate inverse obtained by another method.

Problem Set 19.3

Apply the Gauss–Seidel iteration to the following systems. Perform five steps, starting from 1, 1, 1 (from 1, 1, 1, 1 in Probs. 7 and 8), using 6 significant digits in the computation. *Hint*. Make sure that at the beginning you solve each equation for the variable that has the largest coefficient. (Why?)

1. $$\begin{aligned} 4x_1 + 2x_2 + x_3 &= 14 \\ x_1 + 5x_2 - x_3 &= 10 \\ x_1 + x_2 + 8x_3 &= 20 \end{aligned}$$

2. $$\begin{aligned} 10x_1 - x_2 - x_3 &= 13 \\ x_1 + 10x_2 + x_3 &= 36 \\ -x_1 - x_2 + 10x_3 &= 35 \end{aligned}$$

3. $$\begin{aligned} 6x_1 + x_2 + x_3 &= 107 \\ x_1 + 9x_2 - 2x_3 &= 36 \\ 2x_1 - x_2 + 8x_3 &= 121 \end{aligned}$$

4. $$\begin{aligned} 2x_1 + 10x_2 - x_3 &= -32 \\ -x_1 + 2x_2 + 15x_3 &= 17 \\ 10x_1 - x_2 + 2x_3 &= 58 \end{aligned}$$

Gauss-Seidel iteration (*continued;* see previous page for details)

5. $$\begin{aligned} 10x_1 + x_2 + x_3 &= 6 \\ x_1 + 10x_2 + x_3 &= 6 \\ x_1 + x_2 + 10x_3 &= 6 \end{aligned}$$

6. $$\begin{aligned} 6x_1 + x_2 - x_3 &= 3 \\ -x_1 + x_2 + 7x_3 &= -17 \\ x_1 + 5x_2 + x_3 &= 0 \end{aligned}$$

7. $$\begin{aligned} 8x_1 + x_2 + x_3 - x_4 &= 18 \\ -2x_1 + 12x_2 - x_3 \phantom{{}+2x_4} &= -7 \\ 2x_1 + 16x_3 + 2x_4 &= 54 \\ x_2 + 2x_3 - 20x_4 &= -14 \end{aligned}$$

8. $$\begin{aligned} 7x_1 + x_3 - x_4 &= 1.2 \\ -x_1 + x_2 + 8x_3 &= -8.3 \\ 10x_2 - x_3 + x_4 &= 2.6 \\ 10x_1 + x_2 + 30x_4 &= 22.1 \end{aligned}$$

9. Apply the Gauss–Seidel iteration (3 steps) to the system in Prob. 5, starting from (a) 0, 0, 0, (b) 10, 10, 10. Compare and comment.

10. Apply the Gauss–Seidel and Jacobi iterations (3 steps each) to the system in Prob. 5, starting from 1, 1, 1. Compare and comment.

11. Starting from 0, 0, 0, show that for the system in Example 2 the Jacobi iteration diverges.

12. In Prob. 5, compute **C** (a) if you solve the first equation for x_1, the second for x_2, the third for x_3, proving convergence; (b) if you nonsensically solve the third equation for x_1, the first for x_2, the second for x_3, proving divergence.

13. Consider an approximation $\mathbf{X}^{(0)}$ to the inverse of a matrix **A**, where

$$\mathbf{X}^{(0)} = \begin{bmatrix} 0.5 & -0.1 & 0.4 \\ 0 & 0.2 & 0 \\ -0.4 & 0.3 & -1.5 \end{bmatrix} \quad \text{and} \quad \mathbf{A} = \begin{bmatrix} 3 & 0 & 1 \\ 0 & 5 & 0 \\ -1 & 1 & -1 \end{bmatrix}.$$

Calculate $\mathbf{X}^{(1)}$ by (13). Determine $\mathbf{A}^{-1}$ and show that each element of $\mathbf{X}^{(0)}$ deviates from the corresponding element of $\mathbf{A}^{-1}$ by at most 0.1, whereas for $\mathbf{X}^{(1)}$ that maximum deviation is 0.03.

14. Apply (13) with

$$\mathbf{X}^{(0)} = \begin{bmatrix} 2.9 & -0.9 \\ -4.9 & 1.9 \end{bmatrix} \quad \text{to the matrix} \quad \mathbf{A} = \begin{bmatrix} 2 & 1 \\ 5 & 3 \end{bmatrix},$$

verify that the condition for convergence is satisfied, perform two steps, and compare the results with the exact inverse.

Compute each of the norms (9), (10), (11) of the following matrices.

15. $\begin{bmatrix} -1 & -3 \\ 0 & 3 \end{bmatrix}$ **16.** $\begin{bmatrix} 0.8 & 0.6 \\ -0.6 & 0.8 \end{bmatrix}$ **17.** $\begin{bmatrix} 1 & 0 \\ 0 & 1 \end{bmatrix}$

18. $\begin{bmatrix} 4 & -8 & 12 \\ 0 & 1 & -3 \\ 0 & 0 & 5 \end{bmatrix}$ **19.** $\begin{bmatrix} -2 & 0 & -5 \\ 8 & -8 & 0 \\ 22 & 0 & 7 \end{bmatrix}$ **20.** $\begin{bmatrix} 0.3 & -1.2 & 2.8 \\ -0.3 & 1.2 & -2.8 \\ 0 & 0 & 0 \end{bmatrix}$

19.4 Linear Systems: Ill-Conditioning, Norms

One does not need much experience to observe that some systems $\mathbf{Ax} = \mathbf{b}$ are good, giving accurate solutions even under round-off or inaccuracies of coefficients, whereas others are bad, so that those factors affect the solution strongly. We want to see what is going on and how we can find out whether or not we can "trust" a system.

We call $\mathbf{Ax} = \mathbf{b}$ **well-conditioned** if small errors in the coefficients or in the solution process have only a small effect on the solution, and **ill-conditioned** if the effect on the solution is large. These are basic concepts.

For two equations in two unknowns, Fig. 416 explains these concepts. Ill-conditioning occurs if and only if the equations give two nearly parallel lines, so that their intersection point (the solution of the system) moves very much, for example, if we lift one of the lines just a little.

EXAMPLE 1 An ill-conditioned system

The reader may verify that the system

$$\begin{aligned} 0.9999x - 1.0001y &= 1 \\ x - \quad\quad y &= 1 \end{aligned}$$

has the solution $x = 0.5$, $y = -0.5$, whereas the system

$$\begin{aligned} 0.9999x - 1.0001y &= 1 \\ x - \quad\quad y &= 1 + \epsilon \end{aligned}$$

has the solution $x = 0.5 + 5000.5\epsilon$, $y = -0.5 + 4999.5\epsilon$. This shows that the system is ill-conditioned because a change on the right of magnitude ϵ produces a change in the solution of magnitude 5000ϵ, approximately. We see that the lines given by the equations have nearly the same slope. ∎

For larger systems, the situation is similar in principle, but geometry no longer helps. We may regard ill-conditioning as an approach toward singularity. We may encounter loss of significant digits, making it difficult to get accurate solutions or inverses. Double or triple precision may help if we get the coefficients from formulas, but if they are measured, with limited

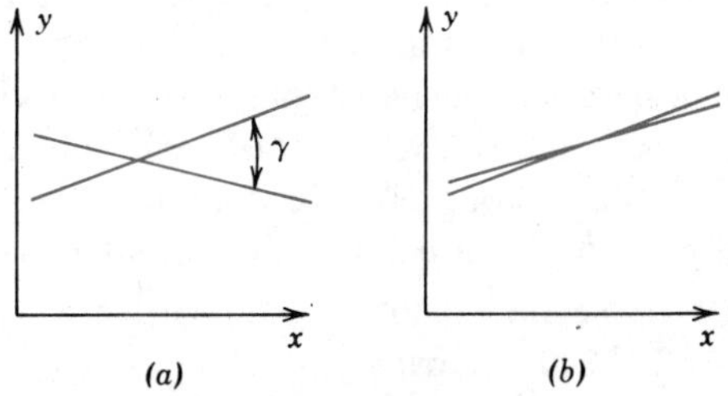

Fig. 416. (*a*) Well-conditioned and (*b*) ill-conditioned linear system of two equations in two unknowns

accuracy, then there is no help and one should try to ***reformulate the problem*** in terms of a well-conditioned system.

Well-conditioning can be asserted if the main diagonal entries of **A** are large in absolute value compared to the off-diagonal entries. Similarly if $\mathbf{A}^{-1}$ and **A** have maximum entries of about the same absolute value.

Symptoms of ill-conditioning are that $|\det \mathbf{A}|$ is small compared to the maximum of the $|a_{jk}|$'s, that $\mathbf{A}^{-1}$ has large entries in absolute value compared to the absolute values of the components of the solution, and that poor approximate solutions may still produce small residuals.

Residual. The *residual* **r** of an approximate solution $\tilde{\mathbf{x}}$ of $\mathbf{Ax} = \mathbf{b}$ is defined as

$$\mathbf{r} = \mathbf{b} - \mathbf{A}\tilde{\mathbf{x}}. \tag{1}$$

Now $\mathbf{b} = \mathbf{Ax}$, so that

$$\mathbf{r} = \mathbf{A}(\mathbf{x} - \tilde{\mathbf{x}}). \tag{2}$$

Hence **r** is small if $\tilde{\mathbf{x}}$ has high accuracy, but the converse may be false:

EXAMPLE 2 Poor approximate solution with a small residual

The system

$$\begin{aligned} 1.0001x_1 + \quad x_2 &= 2.0001 \\ x_1 + 1.0001x_2 &= 2.0001 \end{aligned}$$

has the exact solution $x_1 = 1$, $x_2 = 1$. Can you see this by inspection? The very poor approximation $\tilde{x}_1 = 2.0000$, $\tilde{x}_2 = 0.0001$ has the very small residual (to 4 decimals)

$$\mathbf{r} = \begin{bmatrix} 2.0001 \\ 2.0001 \end{bmatrix} - \begin{bmatrix} 1.0001 & 1.0000 \\ 1.0000 & 1.0001 \end{bmatrix} \begin{bmatrix} 2.0000 \\ 0.0001 \end{bmatrix} = \begin{bmatrix} 2.0001 \\ 2.0001 \end{bmatrix} - \begin{bmatrix} 2.0003 \\ 2.0001 \end{bmatrix} = \begin{bmatrix} -0.0002 \\ 0.0000 \end{bmatrix}.$$

From this, a naive person might draw the false conclusion that the approximation should be accurate to 3 or 4 decimals. ∎

Vector Norms

The probably unexpected result in Example 2 has to do with ill-conditioning and requires further investigation. Several measures for ill-conditioning have been proposed. We shall explain the most widely used measure, the *condition number* of a matrix, which is of general basic importance in numerical mathematics (far beyond our present purpose). This needs some preparation: we shall proceed in three steps, discussing

1. **Vector norms,**
2. **Matrix norms,**
3. **Condition number** κ of a square matrix.

Vector norm. A *vector norm* for all column vectors $\mathbf{x} = [x_j]$ with n components (n fixed) is a generalized length, is denoted by $\|\mathbf{x}\|$ and is defined by four properties of the usual length of vectors in three-dimensional space, namely,

(3)
(a) $\|\mathbf{x}\|$ is a nonnegative real number.

(b) $\|\mathbf{x}\| = 0$ if and only if $\mathbf{x} = \mathbf{0}$.

(c) $\|k\mathbf{x}\| = |k|\,\|\mathbf{x}\|$ for all k.

(d) $\|\mathbf{x} + \mathbf{y}\| \leqq \|\mathbf{x}\| + \|\mathbf{y}\|$ (Triangle inequality).

If we use several norms, we label them by a subscript. Most important in connection with computations is the *p-norm* defined by

$$\|\mathbf{x}\|_p = (|x_1|^p + |x_2|^p + \cdots + |x_n|^p)^{1/p} \tag{4}$$

where p is a fixed number and $p \geqq 1$. In practice, one usually takes $p = 1$ or 2 and, as a third norm, $\|\mathbf{x}\|_\infty$ (the latter as defined below), that is,

$$\|\mathbf{x}\|_1 = |x_1| + \cdots + |x_n| \qquad (\text{"}l_1\text{-norm"}) \tag{5}$$

$$\|\mathbf{x}\|_2 = \sqrt{x_1^{\,2} + \cdots + x_n^{\,2}} \qquad (\text{"Euclidean" or "}l_2\text{-norm"}) \tag{6}$$

$$\|\mathbf{x}\|_\infty = \max_j |x_j| \qquad (\text{"}l_\infty\text{-norm"}). \tag{7}$$

For $n = 3$ the l_2-norm is the usual length of a vector in three-dimensional space. The l_1-norm and l_∞-norm are usually more convenient in computation.

EXAMPLE 3 Vector norms

If $\mathbf{x}^T = [2 \quad -3 \quad 0 \quad 1 \quad -4]$, then

$$\|\mathbf{x}\|_1 = 10, \qquad \|\mathbf{x}\|_2 = \sqrt{30}, \qquad \|\mathbf{x}\|_\infty = 4. \qquad \blacksquare$$

In three-dimensional space, two points with position vectors $\mathbf{x}$ and $\tilde{\mathbf{x}}$ have distance $|\mathbf{x} - \tilde{\mathbf{x}}|$ from each other. This suggests, for a linear system $\mathbf{Ax} = \mathbf{b}$, that we take $\|\mathbf{x} - \tilde{\mathbf{x}}\|$ as a measure of accuracy and call it the **distance** between an exact and an approximate solution, or the **error** of $\tilde{\mathbf{x}}$.

Matrix Norms

If $\mathbf{A}$ is an $n \times n$ matrix and $\mathbf{x}$ any vector with n components, then $\mathbf{Ax}$ is a vector with n components. We now take a vector norm and consider $\|\mathbf{x}\|$ and $\|\mathbf{Ax}\|$. One can prove (see Ref. [9], p. 95, listed in Appendix 1) that there is a number c (depending on $\mathbf{A}$) such that

$$\|\mathbf{Ax}\| \leqq c\|\mathbf{x}\| \qquad \text{for all } \mathbf{x}. \tag{8}$$

Let $\mathbf{x} \neq \mathbf{0}$. Then $\|\mathbf{x}\| > 0$ by (3b) and division gives

$$\frac{\|\mathbf{Ax}\|}{\|\mathbf{x}\|} \leqq c.$$

We obtain the smallest possible c valid for *all* $\mathbf{x}$ ($\neq \mathbf{0}$) by taking the maximum on the left. This smallest c is called the **matrix norm of A** *resulting from the vector norm we picked* and is denoted by $\|\mathbf{A}\|$. Thus

(9) $$\|\mathbf{A}\| = \max \frac{\|\mathbf{Ax}\|}{\|\mathbf{x}\|} \qquad (\mathbf{x} \neq \mathbf{0}),$$

the maximum being taken over all $\mathbf{x} \neq \mathbf{0}$. Alternatively (Prob. 24),

(10) $$\|\mathbf{A}\| = \max_{\|\mathbf{x}\|=1} \|\mathbf{Ax}\|.$$

Note well that $\|\mathbf{A}\|$ depends on the vector norm that we picked.[2] In particular, one can show that we get

for the l_1-norm (5) the column "sum" norm (10), Sec. 19.3,

for the l_∞-norm (7) the row "sum" norm (11), Sec. 19.3.

By taking our best possible (our smallest) $c = \|\mathbf{A}\|$ we have from (8)

(11) $$\|\mathbf{Ax}\| \leqq \|\mathbf{A}\| \, \|\mathbf{x}\|.$$

This is the formula we need. Formula (9) also implies for two $n \times n$ matrices (see Ref. [9], p. 98)

(12) $$\|\mathbf{AB}\| \leqq \|\mathbf{A}\| \, \|\mathbf{B}\|, \qquad \text{thus} \qquad \|\mathbf{A}^n\| \leqq \|\mathbf{A}\|^n.$$

See Refs. [9] and [E10] for other useful formulas on norms.

Before we go on, let us do a simple illustrative computation.

EXAMPLE 4 Matrix norms

Compute the matrix norms of the coefficient matrix $\mathbf{A}$ in Example 1 and of its inverse $\mathbf{A}^{-1}$ assuming that we use (a) the l_1-vector norm, (b) the l_∞-vector norm.

Solution. We use (4), Sec. 7.7, and then (10) and (11) in Sec. 19.3. Thus

$$\mathbf{A} = \begin{bmatrix} 0.9999 & -1.0001 \\ 1.0000 & -1.0000 \end{bmatrix}, \qquad \mathbf{A}^{-1} = \begin{bmatrix} -5000.0 & 5000.5 \\ -5000.0 & 4999.5 \end{bmatrix}.$$

Hence (a) $\|\mathbf{A}\| = 2.0001$, $\|\mathbf{A}^{-1}\| = 10\,000$; (b) $\|\mathbf{A}\| = 2$, $\|\mathbf{A}^{-1}\| = 10\,000.5$. We notice that $\|\mathbf{A}^{-1}\|$ is surprisingly large, which makes $\|\mathbf{A}\| \, \|\mathbf{A}^{-1}\|$ large (20 001). We shall see below that (and why) this is typical of an ill-conditioned system. ∎

Condition Number of a Matrix

We are now ready to introduce the key concept in our discussion of ill-conditioning, the **condition number** $\kappa(\mathbf{A})$ of a (nonsingular) square matrix $\mathbf{A}$, defined by

(13) $$\kappa(\mathbf{A}) = \|\mathbf{A}\| \, \|\mathbf{A}^{-1}\|.$$

[2]The maximum in (10) [hence in (9)] exists by Theorem 2.5-3 in Ref. [9], p. 77, since the norm is continuous (see Ref. [9], p. 60). The name "matrix *norm*" is justified since $\|\mathbf{A}\|$ satisfies (3) with $\mathbf{x}$ and $\mathbf{y}$ replaced by $\mathbf{A}$ and $\mathbf{B}$ (see Ref. [9], pp. 92–93).

The role of the condition number is seen from the following theorem.

Theorem 1 **(Condition number)**
A linear system of equations $\mathbf{Ax} = \mathbf{b}$ *whose condition number* (13) *is small is well-conditioned. A large condition number indicates ill-conditioning.*

Proof. $\mathbf{b} = \mathbf{Ax}$ and (11) give $\|\mathbf{b}\| \leqq \|\mathbf{A}\| \, \|\mathbf{x}\|$. Let $\mathbf{b} \neq \mathbf{0}$ and $\mathbf{x} \neq \mathbf{0}$. Then division by $\|\mathbf{b}\| \, \|\mathbf{x}\|$ gives

$$\frac{1}{\|\mathbf{x}\|} \leqq \frac{\|\mathbf{A}\|}{\|\mathbf{b}\|}. \tag{14}$$

Multiplying (2) by $\mathbf{A}^{-1}$ from the left and interchanging sides, we have $\mathbf{x} - \tilde{\mathbf{x}} = \mathbf{A}^{-1}\mathbf{r}$. Now (11) with $\mathbf{A}^{-1}$ and $\mathbf{r}$ instead of $\mathbf{A}$ and $\mathbf{x}$ yields

$$\|\mathbf{x} - \tilde{\mathbf{x}}\| = \|\mathbf{A}^{-1}\mathbf{r}\| \leqq \|\mathbf{A}^{-1}\| \, \|\mathbf{r}\|.$$

Division by $\|\mathbf{x}\|$ and use of (14) finally gives

$$\frac{\|\mathbf{x} - \tilde{\mathbf{x}}\|}{\|\mathbf{x}\|} \leqq \frac{1}{\|\mathbf{x}\|} \|\mathbf{A}^{-1}\| \, \|\mathbf{r}\| \leqq \frac{\|\mathbf{A}\|}{\|\mathbf{b}\|} \|\mathbf{A}^{-1}\| \, \|\mathbf{r}\| = \kappa(\mathbf{A}) \frac{\|\mathbf{r}\|}{\|\mathbf{b}\|}. \tag{15}$$

Hence if $\kappa(\mathbf{A})$ is small, a small $\|\mathbf{r}\|/\|\mathbf{b}\|$ implies a small relative error $\|\mathbf{x} - \tilde{\mathbf{x}}\|/\|\mathbf{x}\|$, so that the system is well-conditioned. However, this does not hold if $\kappa(\mathbf{A})$ is large. ∎

EXAMPLE 5 **Condition number for an ill-conditioned system**
From Example 4 we see that the condition number for the system in Example 1 is very large, $\kappa(\mathbf{A}) = \|\mathbf{A}\| \, \|\mathbf{A}^{-1}\| = 20\ 001$. This confirms that this system is very ill-conditioned. ∎

EXAMPLE 6 **Condition number for an ill-conditioned system**
By (4) in Sec. 7.7 the inverse of the matrix $\mathbf{A}$ in Example 2 is

$$\mathbf{A}^{-1} = \frac{1}{0.0002}\begin{bmatrix} 1.0001 & -1.0000 \\ -1.0000 & 1.0001 \end{bmatrix} = \begin{bmatrix} 5000.5 & -5000.0 \\ -5000.0 & 5.000.5 \end{bmatrix}.$$

From (10), Sec. 19.3, we thus get the very large value

$$\kappa(\mathbf{A}) = (1.0001 + 1.0000)(5000.5 + 5000.0) \approx 20\ 002;$$

similarly from (9) or (11) in Sec. 19.3. This shows that the system in Example 2 is very ill-conditioned and explains the surprising result in Example 2. ∎

In practice, $\mathbf{A}^{-1}$ will not be known, so that in computing the condition number $\kappa(\mathbf{A})$, one must estimate $\|\mathbf{A}^{-1}\|$. A method for this proposed in 1979 is explained in Ref. [E10] listed in Appendix 1.

Inaccurate matrix entries. $\kappa(\mathbf{A})$ can also be used for estimating the effect $\delta\mathbf{x}$ of an inaccuracy $\delta\mathbf{A}$ of $\mathbf{A}$ (errors of measurements of the a_{jk}, for instance), so that instead of $\mathbf{Ax} = \mathbf{b}$ we have

$$(\mathbf{A} + \delta\mathbf{A})(\mathbf{x} + \delta\mathbf{x}) = \mathbf{b}.$$

Multiplying out and subtracting $\mathbf{Ax} = \mathbf{b}$ on both sides, we have

$$\mathbf{A}\delta\mathbf{x} + \delta\mathbf{A}(\mathbf{x} + \delta\mathbf{x}) = \mathbf{0}.$$

Multiplication by $\mathbf{A}^{-1}$ from the left gives

$$\delta\mathbf{x} = -\mathbf{A}^{-1}\delta\mathbf{A}(\mathbf{x} + \delta\mathbf{x}).$$

Applying (11) with $\mathbf{A}^{-1}$ and vector $\delta\mathbf{A}(\mathbf{x} + \delta\mathbf{x})$ instead of $\mathbf{A}$ and $\mathbf{x}$, we get

$$\|\delta\mathbf{x}\| = \|\mathbf{A}^{-1}\delta\mathbf{A}(\mathbf{x} + \delta\mathbf{x})\| \leqq \|\mathbf{A}^{-1}\| \, \|\delta\mathbf{A}(\mathbf{x} + \delta\mathbf{x})\|.$$

Again applying (11), with $\delta\mathbf{A}$ and $\mathbf{x} - \delta\mathbf{x}$ instead of $\mathbf{A}$ and $\mathbf{x}$, we obtain

$$\|\delta\mathbf{x}\| \leqq \|\mathbf{A}^{-1}\| \, \|\delta\mathbf{A}\| \, \|\mathbf{x} + \delta\mathbf{x}\|.$$

Now $\|\mathbf{A}^{-1}\| = \kappa(\mathbf{A})/\|\mathbf{A}\|$, so that division by $\|\mathbf{x} + \delta\mathbf{x}\|$ shows that the relative inaccuracy of $\mathbf{x}$ is related to that of $\mathbf{A}$ via the condition number by the inequality

$$\text{(16)} \qquad \frac{\|\delta\mathbf{x}\|}{\|\mathbf{x}\|} \approx \frac{\|\delta\mathbf{x}\|}{\|\mathbf{x} + \delta\mathbf{x}\|} \leqq \|\mathbf{A}^{-1}\| \, \|\delta\mathbf{A}\| = \kappa(\mathbf{A}) \frac{\|\delta\mathbf{A}\|}{\|\mathbf{A}\|}.$$

Conclusion. If the system is well-conditioned, small inaccuracies $\|\delta\mathbf{A}\|/\|\mathbf{A}\|$ can have only a small effect on the solution. However, in the case of ill-conditioning, if $\|\delta\mathbf{A}\|/\|\mathbf{A}\|$ is small, $\|\delta\mathbf{x}\|/\|\mathbf{x}\|$ *may* be large.

Inaccurate right side. The student may show that, similarly, when $\mathbf{A}$ is accurate, an inaccuracy $\delta\mathbf{b}$ of $\mathbf{b}$ causes an inaccuracy $\delta\mathbf{x}$ satisfying

$$\text{(17)} \qquad \frac{\|\delta\mathbf{x}\|}{\|\mathbf{x}\|} \leqq \kappa(\mathbf{A}) \frac{\|\delta\mathbf{b}\|}{\|\mathbf{b}\|}.$$

Hence $\|\delta\mathbf{x}\|/\|\mathbf{x}\|$ must remain relatively small whenever $\kappa(\mathbf{A})$ is small.

Further Comments on Condition Numbers. The following additional explanations may be helpful.

1. There is no sharp dividing line between "well conditioned" and "ill conditioned", but generally the situation will get worse as we go from systems with small $\kappa(\mathbf{A})$ to systems with larger $\kappa(\mathbf{A})$. Now always $\kappa(\mathbf{A}) \geqq 1$, so that values of 10 or 20 or so give no reason for concern, whereas $\kappa(\mathbf{A}) = 100$, say, calls for caution, and systems such as those in Examples 1 and 2 are extremely ill-conditioned.

2. If $\kappa(\mathbf{A})$ is large (or small) in one norm, it will be large (or small, respectively) in any other norm.

3. The literature on ill-conditioning is rather extensive. For an introduction to it, see [E10], W. Kahan [*Canad. Math. Bull.* **9** (1966), 757-801] and J. Rice [*SIAM Journal Num. Analysis* **3** (1966), 287-310].

This is the end of our discussion of numerical methods for solving linear systems. An important application follows in the next section.

Problem Set 19.4

Compute each of the vector norms (5), (6), (7) of the following vectors.

1. $[-1 \quad -3 \quad 4]$ **2.** $[0.3 \quad 1.2 \quad 0]$ **3.** $[4 \quad -3 \quad -1 \quad 5 \quad -5]$

4. $[0 \quad 0 \quad 0 \quad 1]$ **5.** $[3 \quad -4]$ **6.** $[3 \quad 4 \quad -7 \quad 0]$

7. $[1 \quad -1 \quad 1 \quad -1]$ **8.** $[5 \quad 4 \quad 3 \quad 3 \quad 2 \quad 1]$

For the following matrices compute the matrix norm and the condition number corresponding to the l_∞-vector norm.

9. $\begin{bmatrix} -3 & 1 \\ 4 & 2 \end{bmatrix}$ **10.** $\begin{bmatrix} 5 & 3 \\ 3 & 5 \end{bmatrix}$ **11.** $\begin{bmatrix} 10 & 7 \\ 7 & 5 \end{bmatrix}$

12. $\begin{bmatrix} -1 & 1 & 2 \\ 3 & -1 & 1 \\ -1 & 3 & 4 \end{bmatrix}$ **13.** $\begin{bmatrix} 100 & 0 & 0 \\ 0 & 100 & 0 \\ 0 & 0 & 0.01 \end{bmatrix}$ **14.** $\begin{bmatrix} 3.0 & 1.5 & 1.0 \\ 1.5 & 1.0 & 0.75 \\ 1.0 & 0.75 & 0.60 \end{bmatrix}$

15. Verify (11) for $\mathbf{x}^\mathsf{T} = [3 \quad 15 \quad -4]$ taken with the l_∞-norm and the matrix in Prob. 12.

16. Solve each of the following two systems, compare the solutions, and comment.

$$\begin{aligned} 2x_1 + 1.4x_2 &= 1.4 \\ 1.4x_1 + x_2 &= 1 \end{aligned} \qquad \begin{aligned} 2x_1 + 1.4x_2 &= 1.44 \\ 1.4x_1 + x_2 &= 1 \end{aligned}$$

17. Compute the condition number in Prob. 16 with respect to the l_1-vector norm and comment.

18. Solve each of the following two systems, compare the solutions, and comment.

$$\begin{aligned} 5x_1 - 7x_2 &= -2 \\ -7x_1 + 10x_2 &= 3 \end{aligned} \qquad \begin{aligned} 5x_1 - 7x_2 &= -2 \\ -7x_1 + 10x_2 &= 3.1 \end{aligned}$$

19. Compute the condition numbers in Prob. 18 with respect to the l_1- and l_∞-vector norms.

20. Show that the solution of the system

$$\begin{aligned} 6x_1 + 7x_2 + 8x_3 &= 21 \\ 7x_1 + 8x_2 + 9x_3 &= 24 \\ 8x_1 + 9x_2 + 9x_3 &= 26 \end{aligned}$$

is $x_1 = 1$, $x_2 = 1$, $x_3 = 1$. Compute the determinant of the system and the residual corresponding to $x_1 = -0.8$, $x_2 = 2.9$, $x_3 = 0.7$. Comment.

21. **(Hilbert matrix)** Using the Gauss elimination, solve the system

$$\begin{aligned} x_1 + \tfrac{1}{2}x_2 + \tfrac{1}{3}x_3 &= 1 \\ \tfrac{1}{2}x_1 + \tfrac{1}{3}x_2 + \tfrac{1}{4}x_3 &= 0 \\ \tfrac{1}{3}x_1 + \tfrac{1}{4}x_2 + \tfrac{1}{5}x_3 &= 0. \end{aligned}$$

Computing the condition number (in the column sum norm), show that the system is ill-conditioned. (The coefficient matrix is called the 3×3 *Hilbert matrix*.)

22. The $n \times n$ Hilbert matrix $\mathbf{H}_n = [h_{jk}]$ has the entries $h_{jk} = 1/(j + k - 1)$ by definition. The elements of the inverse $\mathbf{H}_n^{-1}$ grow rapidly in absolute value as n increases. Illustrate this by computing $\mathbf{H}_2^{-1}$, $\mathbf{H}_3^{-1}$, $\mathbf{H}_4^{-1}$. ($\mathbf{H}_n$ is not an academic curiosity, but comparable matrices occur in connection with curve fitting by least squares.)

23. Show that $\kappa(\mathbf{A}) \geqq 1$ for the matrix norms (10), (11), Sec. 19.3, and $\kappa(\mathbf{A}) \geqq \sqrt{n}$ for the Frobenius norm (9), Sec. 19.3.

24. Derive (10) from (9).

25. Show that $\|\mathbf{x}\|_\infty \leqq \|\mathbf{x}\|_1 \leqq n\,\|\mathbf{x}\|_\infty$.

19.5 Method of Least Squares

In **curve fitting** we are given n points (pairs of numbers)

$$(x_1, y_1), \cdots, (x_n, y_n)$$

and we want to determine a function $f(x)$ such that $f(x_j) \approx y_j$, $j = 1, \cdots, n$. The type of function (for example, polynomials, exponential functions, sine and cosine functions) may be suggested by the nature of the problem (the underlying physical law, for instance), and in many cases a polynomial of a certain degree will be appropriate.

If we require strict equality $f(x_1) = y_1, \cdots, f(x_n) = y_n$ and use polynomials of sufficiently high degree, we may apply one of the methods discussed in Sec. 18.3 in connection with interpolation. However, in certain situations this would not be the appropriate solution of the actual problem. For instance, to the four points

$$(1) \quad (-1.0,\ 1.000), \qquad (-0.1,\ 1.099), \qquad (0.2,\ 0.808), \qquad (1.0,\ 1.000)$$

there corresponds the Lagrange polynomial $f(x) = x^3 - x + 1$ (Fig. 417), but if we graph the points, we see that they lie nearly on a straight line. Hence if these values are obtained in an experiment and thus involve an experimental error, and if the nature of the experiment suggests a linear relation, we better fit a straight line through the points (Fig. 417). Such a

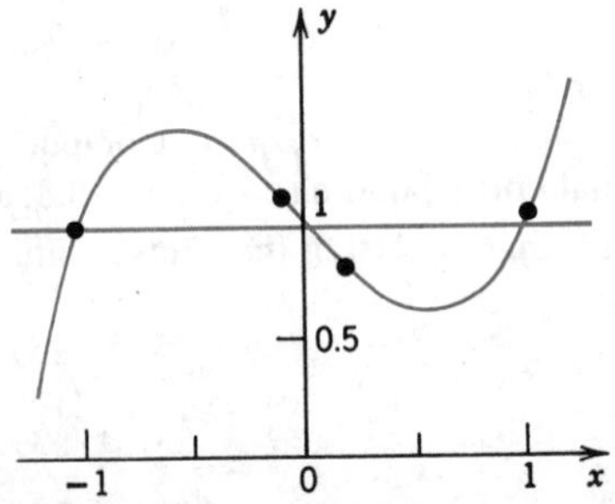

Fig. 417. The approximate fitting of a straight line

line may be useful for predicting values to be expected for other values of x. In simple cases a straight line may be fitted by eye, but if the points are scattered, this becomes unreliable and we better use a mathematical principle. A widely used procedure of this type is the **method of least squares** by Gauss. In the present situation it may be formulated as follows.

Method of least squares. *The straight line*

$$y = a + bx$$

should be fitted through the given points $(x_1, y_1), \cdots, (x_n, y_n)$ *so that the sum of the squares of the distances of those points from the straight line is minimum, where the distance is measured in the vertical direction (the y-direction).*

The point on the line with abscissa x_j has the ordinate $a + bx_j$. Hence its distance from (x_j, y_j) is $|y_j - a - bx_j|$ (Fig. 418) and that sum of squares is

$$q = \sum_{j=1}^{n} (y_j - a - bx_j)^2.$$

q depends on a and b. A necessary condition for q to be minimum is

$$\begin{aligned} \frac{\partial q}{\partial a} &= -2 \sum (y_j - a - bx_j) = 0 \\ \frac{\partial q}{\partial b} &= -2 \sum x_j(y_j - a - bx_j) = 0 \end{aligned} \tag{2}$$

(where we sum over j from 1 to n). Writing each sum as three sums and taking one of them to the right, we obtain the result

$$\begin{aligned} an \quad + b \sum x_j &= \sum y_j \\ a \sum x_j + b \sum x_j^2 &= \sum x_j y_j. \end{aligned} \tag{3}$$

These equations are called the **normal equations** of our problem.

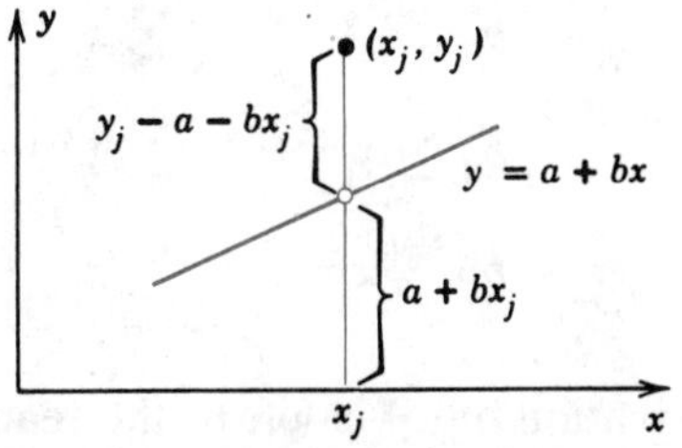

Fig. 418. Vertical distance of a point (x_j, y_j) from a straight line $y = a + bx$

EXAMPLE 1 Straight line

Using the method of least squares, fit a straight line to the four points given in formula (1).

Solution. We obtain

$$n = 4, \quad \sum x_j = 0.1, \quad \sum x_j^2 = 2.05, \quad \sum y_j = 3.907, \quad \sum x_j y_j = 0.0517.$$

Hence the normal equations are

$$\begin{aligned} 4a + 0.10b &= 3.9070 \\ 0.1a + 2.05b &= 0.0517. \end{aligned}$$

The solution is $a = 0.9773$, $b = -0.0224$, and we obtain the straight line (Fig. 417)

$$y = 0.9773 - 0.0224x.$$

Our method of curve fitting can be generalized from a polynomial $y = a + bx$ to a polynomial degree m

$$p(x) = b_0 + b_1 x + \cdots + b_m x^m$$

where $m \leqq n - 1$. Then q takes the form

$$q = \sum_{j=1}^{n} (y_j - p(x_j))^2$$

and depends on $m + 1$ parameters $b_0, \cdots, b_m$. Instead of (2) we then have $m + 1$ conditions

$$\frac{\partial q}{\partial b_0} = 0, \quad \cdots, \quad \frac{\partial q}{\partial b_m} = 0$$

which give a system of $m + 1$ normal equations.

In the case of a quadratic polynomial

$$p(x) = b_0 + b_1 x + b_2 x^2 \tag{4}$$

the normal equations are (summation from 1 to n)

$$\begin{aligned} b_0 n \quad\;\; + b_1 \sum x_j + b_2 \sum x_j^2 &= \sum y_j \\ b_0 \sum x_j + b_1 \sum x_j^2 + b_2 \sum x_j^3 &= \sum x_j y_j \\ b_0 \sum x_j^2 + b_1 \sum x_j^3 + b_2 \sum x_j^4 &= \sum x_j^2 y_j. \end{aligned} \tag{5}$$

The derivation of (5) is left to the reader.

Note that this system is symmetric. To solve it for the unknowns b_0, b_1, b_2, we may apply one of the methods discussed in Secs. 19.1–19.3.

Various applications are included in the problem set.

Problem Set 19.5

Using the method of least squares, fit a straight line to the given points (x, y). Check the result by sketching the points and the line.

1. (2, 0), (3, 4), (4, 10), (5, 16)

2. (0, 0), (2, 1.8), (4, 3.4), (6, 4.6)

3. (5, 8.0), (10, 6.9), (15, 6.2), (20, 5.0)

4. (4, −20), (15, −7), (30, −10), (100, 47), (200, 67)

5. (9.6, 1), (8.5, 2.3), (7.0, 2.0), (0, 7.7), (−100, 9.7)

6.

Revolutions per minute x	400	500	600	700	750
Power of a Diesel engine y [hp]	580	1030	1420	1880	2100

7.

Density of ore x [grams/cm^3]	2.8	2.9	3.0	3.1	3.2	3.2	3.2	3.3	3.4
Iron content y [percent]	30	26	33	31	33	35	37	36	33

8. If a car is traveling along a straight road with constant speed $v = b_1$ [m/sec], its position y[m] at time t[sec] is $y = b_0 + b_1 t$. Suppose that measurements are

t	0	3	5	8	10
y	200	230	240	270	290

Fit a straight line to these data by least squares and estimate from it the speed.

9. Fit a straight line by least squares to the data (I, U) = (0.8, 20), (1.7, 40), (2.3, 60), I [amperes] the current, U [volts] the voltage, and estimate from it the resistance $R = U/I$ [ohms] (Ohm's law).

10. Fit a straight line by least squares to the data (s, F) = (0.9, 10), (0.5, 5), (1.6, 15), (2.1, 20), s the elongation of an elastic spring under a force F, and estimate from it the spring modulus $k = F/s$. ($F = ks$ is called *Hooke's law*.)

Using the method of least squares, fit a parabola (4) to the given data.

11. (−1, 2), (0, 0), (0, 1), (1, 2)

12. (0, 3), (1, 1), (2, 0), (4, 1), (6, 4)

13. (−1, 0), (0, −2), (0, −1), (1, 0)

14. (1.09, 1.35), (1.28, 1.58), (1.36, 1.68), (1.44, 1.85), (1.60, 2.23), (1.65, 2.38)

15.

Speed of a tractor x [km/hr]	1.4	1.8	2.3	3.0	4.0
Power of a tractor y [kg]	7400	7500	7600	7500	7200

16.

Worker's time on duty x [hours]	1	2	3	4	5	6
Worker's reaction time y [sec]	1.50	1.48	1.75	1.65	1.72	1.55

17. Solve Prob. 13 by the Cholesky method (Sec. 19.2).

18. Determine the normal equations in the case of a polynomial of the third degree.

19. In the least-squares principle involving a polynomial we seek to satisfy

$$b_0 + b_1 x_j + b_2 x_j^2 + \cdots + b_m x_j^m = y_j \qquad (j = 1, \cdots, n)$$

as best as possible. Introduce a matrix $\mathbf{C}$ such that this can be written $\mathbf{Cb} = \mathbf{y}$ and show that then the normal equations can be written $\mathbf{C}^\mathsf{T}\mathbf{Cb} = \mathbf{C}^\mathsf{T}\mathbf{y}$.

20. In problems of growth it is often required to fit an exponential function $y = b_0 e^{bx}$ by the method of least squares. Show that by taking logarithms this task can be reduced to that of determining a straight line.

19.6 Matrix Eigenvalue Problems: Introduction

In the remaining sections of this chapter we discuss some of the most important ideas and numerical methods for matrix eigenvalue problems. This very extensive part of numerical linear algebra is of great practical importance, with much research going on, particularly since 1945, and hundreds, if not thousands of papers published in various mathematical journals (see the references in [E9], [E10], [E12], [E19], [E22], [E26], [E27]). We begin with the concepts and general results we shall need in explaining and applying numerical methods for eigenvalue problems. (For typical applications, see Secs. 7.11–7.15.)

An **eigenvalue** or **characteristic value** (or *latent root*) of a given $n \times n$ matrix $\mathbf{A} = [a_{jk}]$ is a real or complex number λ such that the vector equation

$$\mathbf{A}\mathbf{x} = \lambda\mathbf{x} \tag{1}$$

has a nontrivial solution, that is, a solution $\mathbf{x} \neq \mathbf{0}$, which is then called an **eigenvector** or **characteristic vector** of $\mathbf{A}$ corresponding to that eigenvalue λ. The set of all eigenvalues of $\mathbf{A}$ is called the **spectrum** of $\mathbf{A}$. Equation (1) can be written

$$(\mathbf{A} - \lambda\mathbf{I})\mathbf{x} = \mathbf{0} \tag{2}$$

where $\mathbf{I}$ is the $n \times n$ unit matrix. This homogeneous system has a nontrivial solution if and only if the **characteristic determinant** det $(\mathbf{A} - \lambda\mathbf{I})$ is 0 (see Theorem 2 in Sec. 7.6). This gives (see Sec. 7.10)

Theorem 1 **(Eigenvalues)**

The eigenvalues of $\mathbf{A}$ *are the solutions of the* **characteristic equation**

$$\det(\mathbf{A} - \lambda\mathbf{I}) = \begin{vmatrix} a_{11} - \lambda & a_{12} & \cdots & a_{1n} \\ a_{21} & a_{22} - \lambda & \cdots & a_{2n} \\ \cdot & \cdot & \cdots & \cdot \\ a_{n1} & a_{n2} & \cdots & a_{nn} - \lambda \end{vmatrix} = 0. \tag{3}$$

Developing the characteristic determinant, we obtain the **characteristic polynomial** of $\mathbf{A}$, which is of degree n in λ. Hence $\mathbf{A}$ has at least one and at most n numerically different eigenvalues. If $\mathbf{A}$ is real, so are the coefficients of the characteristic polynomial. By familiar algebra it follows that then the roots (the eigenvalues of $\mathbf{A}$) are real or complex conjugates in pairs.

We shall usually denote the eigenvalues of $\mathbf{A}$ by

$$\lambda_1, \lambda_2, \cdots, \lambda_n$$

with the understanding that some (or all) of them may be numerically equal.

The sum of these n eigenvalues equals the sum of the entries on the main diagonal of **A**, called the **trace of A**; thus

$$\text{trace } \mathbf{A} = \sum_{j=1}^{n} a_{jj} = \sum_{k=1}^{n} \lambda_k. \tag{4}$$

Also,

$$\det \mathbf{A} = \lambda_1 \lambda_2 \cdots \lambda_n. \tag{5}$$

Both formulas follow from the product representation of the characteristic polynomial $f(\lambda)$,

$$f(\lambda) = (-1)^n(\lambda - \lambda_1)(\lambda - \lambda_2) \cdots (\lambda - \lambda_n).$$

If we take equal factors together and denote the *numerically distinct* eigenvalues of **A** by $\lambda_1, \cdots, \lambda_r$ ($r \leqq n$), then the product becomes

$$f(\lambda) = (-1)^n (\lambda - \lambda_1)^{m_1} (\lambda - \lambda_2)^{m_2} \cdots (\lambda - \lambda_r)^{m_r}. \tag{6}$$

The exponent m_j is called the **algebraic multiplicity** of λ_j. The maximum number of linearly independent eigenvectors corresponding to λ_j is called the **geometric multiplicity** of λ_j. It is equal to or smaller than m_j.

Similarity. Spectral Shift. Special matrices

An $n \times n$ matrix **B** is called **similar** to **A** if there is a matrix **T** such that

$$\mathbf{B} = \mathbf{T}^{-1}\mathbf{A}\mathbf{T}. \tag{7}$$

Similarity is important for the following reason.

Theorem 2 (Similar matrices)
Similar matrices have the same eigenvalues. If **x** *is an eigenvector of* **A**, *then* $\mathbf{y} = \mathbf{T}^{-1}\mathbf{x}$ *is an eigenvector of* **B** *in* (7) *corresponding to the same eigenvalue*. (Proof in Sec. 7.14.)

Another theorem that has various applications in numerical work is as follows.

Theorem 3 (Spectral shift)
If **A** *has the eigenvalues* $\lambda_1, \cdots, \lambda_n$, *then* $\mathbf{A} - k\mathbf{I}$ *has the eigenvalues* $\lambda_1 - k, \cdots, \lambda_n - k$.

This theorem is a special case of the following **spectral mapping theorem.**

Theorem 4 **(Polynomial matrices)**
If λ *is an eigenvalue of* $\mathbf{A}$, *then*

$$q(\lambda) = \alpha_s \lambda^s + \alpha_{s-1}\lambda^{s-1} + \cdots + \alpha_1 \lambda + \alpha_0$$

is an eigenvalue of the **polynomial matrix**

$$q(\mathbf{A}) = \alpha_s \mathbf{A}^s + \alpha_{s-1}\mathbf{A}^{s-1} + \cdots + \alpha_1 \mathbf{A} + \alpha_0 \mathbf{I}.$$

Proof. $\mathbf{Ax} = \lambda\mathbf{x}$ implies $\mathbf{A}^2\mathbf{x} = \mathbf{A}\lambda\mathbf{x} = \lambda\mathbf{A}\mathbf{x} = \lambda^2\mathbf{x}$, $\mathbf{A}^3\mathbf{x} = \lambda^3\mathbf{x}$, etc. Thus

$$\begin{aligned} q(\mathbf{A})\mathbf{x} &= (\alpha_s \mathbf{A}^s + \alpha_{s-1}\mathbf{A}^{s-1} + \cdots)\mathbf{x} \\ &= \alpha_s \mathbf{A}^s\mathbf{x} + \alpha_{s-1}\mathbf{A}^{s-1}\mathbf{x} + \cdots \\ &= \alpha_s \lambda^s\mathbf{x} + \alpha_{s-1}\lambda^{s-1}\mathbf{x} + \cdots = q(\lambda)\mathbf{x}. \end{aligned}$$ ∎

The eigenvalues of important special matrices can be characterized by

Theorem 5 **(Special matrices)**
The eigenvalues of Hermitian matrices (i.e., $\overline{\mathbf{A}}^{\mathsf{T}} = \mathbf{A}$), *hence of real symmetric matrices* (i.e., $\mathbf{A}^{\mathsf{T}} = \mathbf{A}$), *are real. The eigenvalues of skew-Hermitian matrices* (i.e., $\overline{\mathbf{A}}^{\mathsf{T}} = -\mathbf{A}$), *hence of real skew-symmetric matrices* (i.e., $\mathbf{A}^{\mathsf{T}} = -\mathbf{A}$) *are pure imaginary or* 0. *The eigenvalues of unitary matrices* (i.e., $\overline{\mathbf{A}}^{\mathsf{T}} = \mathbf{A}^{-1}$), *hence of orthogonal matrices* (i.e., $\mathbf{A}^{\mathsf{T}} = \mathbf{A}^{-1}$), *have absolute value* 1. (Proof in Sec. 7.13.)

The **choice of a numerical method** for matrix eigenvalue problems depends essentially on two circumstances, on the kind of matrix (real symmetric, real general, complex, sparse, or full) and on the kind of information to be obtained, that is, whether one wants to know all eigenvalues or merely specific ones, for instance, the largest eigenvalue, whether eigenvalues *and* eigenvectors are wanted, and so on. It is clear that we cannot enter into a systematic discussion of all these and further possibilities that arise in practice—look quickly into Ref. [E27] to get an idea—but we shall concentrate on some basic aspects and methods that will give us a good feeling and general understanding of this fascinating field.

19.7 Inclusion of Matrix Eigenvalues

By "**inclusion**" we mean the determination of approximate values of eigenvalues and corresponding error bounds. As a first important "inclusion theorem," the theorem by Gerschgorin gives a region consisting of closed circular disks in the complex plane and including all the eigenvalues of a given matrix. Indeed, for each $j = 1, \cdots, n$ the inequality (1) in the theorem determines a closed circular disk in the complex λ-plane with center a_{jj} and radius given by the right side of (1); and Theorem 1 states that each of the eigenvalues of $\mathbf{A}$ lies in one of these n disks.

Theorem 1 (Gerschgorin's theorem)
Let λ be an eigenvalue of an arbitrary $n \times n$ matrix $\mathbf{A} = [a_{jk}]$. Then for some integer j $(1 \leqq j \leqq n)$ we have

$$(1) \quad |a_{jj} - \lambda| \leqq |a_{j1}| + |a_{j2}| + \cdots + |a_{j,j-1}| + |a_{j,j+1}| + \cdots + |a_{jn}|.$$

Proof. Let $\mathbf{x}$ be an eigenvector corresponding to that eigenvalue λ. Then

$$(2) \qquad \mathbf{Ax} = \lambda\mathbf{x} \qquad \text{or} \qquad (\mathbf{A} - \lambda\mathbf{I})\mathbf{x} = \mathbf{0}.$$

Let x_j be a component of $\mathbf{x}$ that is largest in absolute value. Then we have $|x_m/x_j| \leqq 1$ for $m = 1, \cdots, n$. The vector equation (2) is equivalent to a system of n equations for the n components of the vectors on both sides, and the jth of these n equations is

$$a_{j1}x_1 + \cdots + a_{j,j-1}x_{j-1} + (a_{jj} - \lambda)x_j + a_{j,j+1}x_{j+1} + \cdots + a_{jn}x_n = 0.$$

Division by x_j and reshuffling terms gives

$$a_{jj} - \lambda = -a_{j1}\frac{x_1}{x_j} - \cdots - a_{j,j-1}\frac{x_{j-1}}{x_j} - a_{j,j+1}\frac{x_{j+1}}{x_j} - \cdots - a_{jn}\frac{x_n}{x_j}.$$

By taking absolute values on both sides of this equation, applying the triangle inequality $|a + b| \leqq |a| + |b|$ (where a and b are any complex numbers), and observing that because of the choice of j (which is crucial!),

$$\left|\frac{x_1}{x_j}\right| \leqq 1, \qquad \cdots, \qquad \left|\frac{x_n}{x_j}\right| \leqq 1,$$

we obtain (1), and the theorem is proved. ∎

EXAMPLE 1 Gerschgorin's theorem
For the eigenvalues of the matrix

$$\mathbf{A} = \begin{bmatrix} 0 & 1/2 & 1/2 \\ 1/2 & 5 & 1 \\ 1/2 & 1 & 1 \end{bmatrix}$$

we get the Gerschgorin disks (Fig. 419 on the next page)

D_1: Center 0, radius 1

D_2: Center 5, radius 1.5

D_3: Center 1, radius 1.5.

Since $\mathbf{A}$ is symmetric, it follows from this and Theorem 5, Sec. 19.6, that the spectrum of $\mathbf{A}$ must actually lie in the intervals $[-1, 2.5]$ and $[3.5, 6.5]$ on the real axis. The student may confirm this by showing that, to 3 decimals, $\lambda_1 = -0.209$, $\lambda_2 = 5.305$, $\lambda_3 = 0.904$.

It is interesting that here the Gerschgorin disks form two disjoint sets, namely, $D_1 \cup D_3$, which contains two eigenvalues, and D_2, which contains one eigenvalue. This is typical, as the following theorem shows. ∎

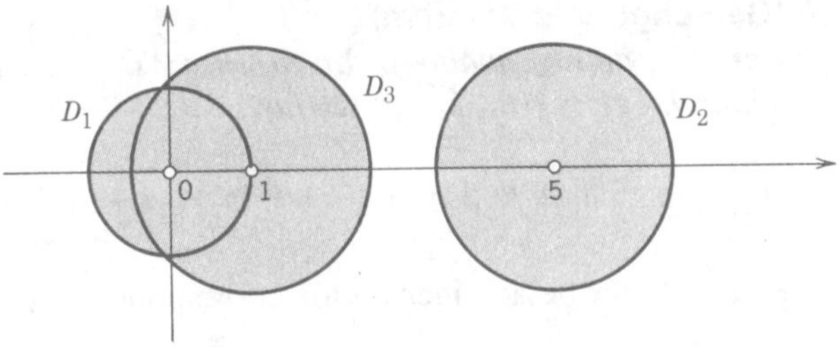

Fig. 419. Gerschgorin disks in Example 1

Theorem 2 **(Extension of Gerschgorin's theorem)**
If p Gerschgorin disks form a set S that is disjoint from the $n - p$ other disks of a given matrix **A**, *then S contains precisely p eigenvalues of* **A** (*each counted with its algebraic multiplicity, as defined in Sec.* 19.6).

Proof. This is a "*continuity proof.*" Let $S = D_1 \cup D_2 \cup \cdots \cup D_p$ without restriction, where D_j is the Gerschgorin disk with center a_{jj}. We write $\mathbf{A} = \mathbf{B} + \mathbf{C}$, where $\mathbf{B} = \text{diag}\,(a_{jj})$ is the diagonal matrix with the main diagonal of **A** as its diagonal. We now consider

$$\mathbf{A}_t = \mathbf{B} + t\mathbf{C} \qquad \text{for } 0 \leqq t \leqq 1.$$

Then $\mathbf{A}_0 = \mathbf{B}$ and $\mathbf{A}_1 = \mathbf{A}$. Now by algebra, the roots of the characteristic polynomial $f_t(\lambda)$ of $\mathbf{A}_t$ (that is, the eigenvalues of $\mathbf{A}_t$) depend continuously on the coefficients of $f_t(\lambda)$, which in turn depend continuously on t. For $t = 0$, the eigenvalues are $a_{11}, \cdots, a_{nn}$. If we let t increase continuously from 0 to 1, the eigenvalues move continuously and, by Theorem 1, for each t lie in the Gerschgorin disks with centers a_{jj} and radii

$$tr_j \qquad \text{where} \qquad r_j = \sum_{k \neq j} |a_{jk}|.$$

Since at the end, S is disjoint from the other disks, the assertion follows. ∎

EXAMPLE 2 **Another application of Gerschgorin's theorem. Similarity**

Suppose that we have diagonalized a matrix by some numerical method that left us with some off-diagonal entries of size 10^{-5}, say,

$$\mathbf{A} = \begin{bmatrix} 2 & 10^{-5} & 10^{-5} \\ 10^{-5} & 2 & 10^{-5} \\ 10^{-5} & 10^{-5} & 4 \end{bmatrix}.$$

What can we conclude about deviations of the eigenvalues from the main diagonal entries?

Solution. By Theorem 2, one eigenvalue must lie in the disk of radius $2 \cdot 10^{-5}$ centered at 4 and two eigenvalues (or an eigenvalue of algebraic multiplicity 2) in the disk of radius $2 \cdot 10^{-5}$ centered at 2. Actually, since the matrix is symmetric, these eigenvalues must lie in the intersections of these disks and the real axis, by Theorem 5 in Sec. 19.6.

We show that and how an isolated disk can always be reduced in size by a similarity transformation. The matrix

$$\mathbf{B} = \mathbf{T}^{-1}\mathbf{A}\mathbf{T} = \begin{bmatrix} 1 & 0 & 0 \\ 0 & 1 & 0 \\ 0 & 0 & 10^{-5} \end{bmatrix} \begin{bmatrix} 2 & 10^{-5} & 10^{-5} \\ 10^{-5} & 2 & 10^{-5} \\ 10^{-5} & 10^{-5} & 4 \end{bmatrix} \begin{bmatrix} 1 & 0 & 0 \\ 0 & 1 & 0 \\ 0 & 0 & 10^{5} \end{bmatrix}$$

$$= \begin{bmatrix} 2 & 10^{-5} & 1 \\ 10^{-5} & 2 & 1 \\ 10^{-10} & 10^{-10} & 4 \end{bmatrix}$$

is similar to **A**. Hence by Theorem 2, Sec. 19.6, it has the same eigenvalues as **A**. From row 3 we get the smaller disk of radius $2 \cdot 10^{-10}$. Note that the other disks got bigger, approximately by a factor 10^5. And in choosing **T** we have to watch that the new disks do not overlap with the disk whose size we want to decrease. ∎

Further Inclusion Theorems

An **inclusion theorem** is a theorem that specifies a set which contains at least one eigenvalue of a given matrix. Thus, Theorems 1 and 2 are inclusion theorems; they even include the whole spectrum. We now discuss some famous theorems that yield inclusions of eigenvalues.

Theorem 3 **(Schur's theorem[3])**

Let $\mathbf{A} = [a_{jk}]$ *be an* $n \times n$ *matrix. Let* $\lambda_1, \cdots, \lambda_n$ *be its eigenvalues. Then*

$$\sum_{i=1}^{n} |\lambda_i|^2 \leqq \sum_{j=1}^{n} \sum_{k=1}^{n} |a_{jk}|^2 \qquad \textbf{(Schur's inequality)}. \tag{3}$$

In (3) *the equality sign holds if and only if* **A** *is such that*

$$\bar{\mathbf{A}}^{\mathsf{T}}\mathbf{A} = \mathbf{A}\bar{\mathbf{A}}^{\mathsf{T}}. \tag{4}$$

Matrices that satisfy (4) are called **normal matrices.** It is not difficult to see that Hermitian, skew-Hermitian, and unitary matrices are normal, and so are real symmetric, skew-symmetric, and orthogonal matrices.

Let λ_m be any eigenvalue of the matrix **A** in Theorem 3. Then $|\lambda_m|^2$ is less than or equal to the sum on the left side of (3), and by taking square roots we obtain from (3)

$$|\lambda_m| \leqq \sqrt{\sum_{j=1}^{n} \sum_{k=1}^{n} |a_{jk}|^2}. \tag{5}$$

Note that the right side is the Frobenius norm of **A** (see Sec. 19.3).

---

[3]ISSAI SCHUR (1875—1941), German mathematician, professor in Berlin, also known by his important work in group theory.

EXAMPLE 3 Bounds for eigenvalues obtained from Schur's inequality

For the matrix

$$\mathbf{A} = \begin{bmatrix} 26 & -2 & 2 \\ 2 & 21 & 4 \\ 4 & 2 & 28 \end{bmatrix}$$

we obtain from Schur's inequality

$$|\lambda| \leqq \sqrt{1949} < 44.2.$$

(The eigenvalues of $\mathbf{A}$ are 30, 25, and 20; thus $30^2 + 25^2 + 20^2 = 1925 < 1949$; in fact, $\mathbf{A}$ is not normal.) ▪

The preceding theorems are valid for every real or complex square matrix. There are other theorems that hold for special classes of matrices only. The following theorem by Frobenius,[4] which we state without proof, is of this type.

Theorem 4 (Perron–Frobenius's theorem)

Let $\mathbf{A}$ *be a real square matrix whose elements are all positive. Then* $\mathbf{A}$ *has at least one real positive eigenvalue* λ, *and the corresponding eigenvector can be chosen real and such that all its components are positive.*

From this theorem we may derive the following useful result by Collatz.[5]

Theorem 5 (Collatz's theorem)

Let $\mathbf{A} = [a_{jk}]$ *be a real* $n \times n$ *matrix whose elements are all positive. Let* $\mathbf{x}$ *be any real vector whose components* $x_1, \cdots, x_n$ *are positive, and let* $y_1, \cdots, y_n$ *be the components of the vector* $\mathbf{y} = \mathbf{Ax}$. *Then the closed interval on the real axis bounded by the smallest and the largest of the* n *quotients* $q_j = y_j/x_j$ *contains at least one eigenvalue of* $\mathbf{A}$.

Proof. We have $\mathbf{Ax} = \mathbf{y}$ or

$$\mathbf{y} - \mathbf{Ax} = \mathbf{0}. \tag{6}$$

The transpose $\mathbf{A}^\mathsf{T}$ satisfies the conditions of Theorem 4. Hence $\mathbf{A}^\mathsf{T}$ has a positive eigenvalue λ and, corresponding to this eigenvalue, an eigenvector $\mathbf{u}$ whose components u_j are all positive. Thus $\mathbf{A}^\mathsf{T}\mathbf{u} = \lambda\mathbf{u}$, and by taking the transpose we obtain $\mathbf{u}^\mathsf{T}\mathbf{A} = \lambda\mathbf{u}^\mathsf{T}$. From this and (6),

$$\mathbf{u}^\mathsf{T}(\mathbf{y} - \mathbf{Ax}) = \mathbf{u}^\mathsf{T}\mathbf{y} - \mathbf{u}^\mathsf{T}\mathbf{Ax} = \mathbf{u}^\mathsf{T}(\mathbf{y} - \lambda\mathbf{x}) = 0$$

or written out

$$\sum_{j=1}^{n} u_j(y_j - \lambda x_j) = 0.$$

---

[4] See the footnote in Sec. 5.4. For a proof of the theorem, see Ref. [B2], vol. II, pp. 53–62.

[5] LOTHAR COLLATZ (1910—1990), German mathematician, known for his work in numerical analysis.

Since all the components u_j are positive, it follows that

$$(7)\qquad \begin{aligned} &y_j - \lambda x_j \geqq 0, \quad \text{that is,} \quad q_j \geqq \lambda \quad \text{for at least one } j, \quad \text{and} \\ &y_j - \lambda x_j \leqq 0, \quad \text{that is,} \quad q_j \leqq \lambda \quad \text{for at least one } j. \end{aligned}$$

Since $\mathbf{A}$ and $\mathbf{A}^\mathsf{T}$ have the same eigenvalues, λ is an eigenvalue of $\mathbf{A}$, and from (7) the statement of the theorem follows. ▮

EXAMPLE 4 Bounds for eigenvalues from Collatz's theorem

Let

$$\mathbf{A} = \begin{bmatrix} 8 & 1 & 1 \\ 1 & 5 & 2 \\ 1 & 2 & 5 \end{bmatrix}. \qquad \text{Choose} \qquad \mathbf{x} = \begin{bmatrix} 1 \\ 1 \\ 1 \end{bmatrix}. \qquad \text{Then} \qquad \mathbf{y} = \begin{bmatrix} 10 \\ 8 \\ 8 \end{bmatrix}.$$

Hence $q_1 = 10$, $q_2 = 8$, $q_3 = 8$, and Theorem 5 implies that one of the eigenvalues of $\mathbf{A}$ must lie in the interval $8 \leqq \lambda \leqq 10$. Of course, the length of such an interval depends on the choice of $\mathbf{x}$. The student may show that $\lambda = 9$ is an eigenvalue of $\mathbf{A}$. ▮

Problem Set 19.7

Using Theorem 1, determine and sketch disks that contain the eigenvalues of the following matrices.

1. $\begin{bmatrix} 1 & 2 & 3 \\ 2 & 4 & 6 \\ 3 & 6 & 1 \end{bmatrix}$ **2.** $\begin{bmatrix} 2.2 & 0.1 & 0 \\ 0.1 & 3.0 & -0.1 \\ 0 & -0.1 & 4.5 \end{bmatrix}$ **3.** $\begin{bmatrix} 5 & -2 & 2 \\ 2 & 0 & 4 \\ 4 & 2 & 7 \end{bmatrix}$

4. $\begin{bmatrix} 4 & 1 & 1 \\ 2 & 4 & 1 \\ 0 & 1 & 4 \end{bmatrix}$ **5.** $\begin{bmatrix} 5 & 10^{-2} & 10^{-2} \\ 10^{-2} & 8 & 10^{-2} \\ 10^{-2} & 10^{-2} & 9 \end{bmatrix}$ **6.** $\begin{bmatrix} 0 & 0.5i & -i \\ 1-i & 1+i & 0 \\ 0.1i & 1 & -i \end{bmatrix}$

7. Show that in Probs. 1 and 4 the eigenvalues are 10, 0, -4 and 6, 3, 3, respectively.

8. Find a 2×2 matrix that illustrates that an eigenvalue may very well lie on a Gerschgorin circle (so that Gerschgorin disks can generally not be replaced by smaller disks without losing the inclusion property).

9. Find a similarity transformation $\mathbf{T}^{-1}\mathbf{AT}$ that reduces the radius of the Gerschgorin circle with center 5 in Prob. 5 to 1/100 of its original value.

Using (5), obtain an upper bound for the absolute value of the eigenvalues of the matrix:

10. In Prob. 2 **11.** In Prob. 3 **12.** In Prob. 6 **13.** In Prob. 5

Apply Theorem 5 to the following matrices, choosing the given vectors as vectors $\mathbf{x}$.

14. $\begin{bmatrix} 20 & 8 & 1 \\ 8 & 21 & 8 \\ 1 & 8 & 20 \end{bmatrix}$, $\begin{bmatrix} 1 \\ 1 \\ 1 \end{bmatrix}$, $\begin{bmatrix} 1 \\ 2 \\ 1 \end{bmatrix}$, $\begin{bmatrix} 2 \\ 3 \\ 2 \end{bmatrix}$ **15.** $\begin{bmatrix} 5 & 1 & 1 \\ 1 & 5 & 1 \\ 1 & 1 & 5 \end{bmatrix}$, $\begin{bmatrix} 1 \\ 2 \\ 1 \end{bmatrix}$, $\begin{bmatrix} 1 \\ 1 \\ 1 \end{bmatrix}$

16. Can you conjecture from Prob. 15 that if a matrix $\mathbf{A}$ has equal row sums or equal column sums, say, equal to k, then k is an eigenvalue of $\mathbf{A}$? Prove it.

17. Show that the matrix in Example 4 has the eigenvalues 9, 6, 3, and (3) holds with the equality sign. Conclude that the matrix is normal. Verify it.

18. Show that Hermitian, skew-Hermitian, and unitary matrices are normal.

19. Show that the matrix in Prob. 3 is not normal and has the eigenvalues $-1, 4, 9$.

20. **(Nonzero determinant)** If in each row of a determinant the entry on the main diagonal is larger in absolute value than the sum of the absolute values of the other entries in that row, show that the value of the determinant cannot be zero. What does this imply with respect to the solvability of linear systems of equations? With respect to the spectrum of the corresponding matrix?

19.8 Eigenvalues by Iteration (Power Method)

A simple standard procedure for computing approximate values of the eigenvalues of an $n \times n$ matrix $\mathbf{A} = [a_{jk}]$ is the **power method.** In this method we start from any vector $\mathbf{x}_0$ ($\neq \mathbf{0}$) with n components and compute successively

$$\mathbf{x}_1 = \mathbf{A}\mathbf{x}_0, \qquad \mathbf{x}_2 = \mathbf{A}\mathbf{x}_1, \qquad \cdots, \qquad \mathbf{x}_s = \mathbf{A}\mathbf{x}_{s-1}.$$

For simplifying notation, we denote $\mathbf{x}_{s-1}$ by $\mathbf{x}$ and $\mathbf{x}_s$ by $\mathbf{y}$, so that $\mathbf{y} = \mathbf{A}\mathbf{x}$. If $\mathbf{A}$ is real symmetric, the following theorem gives an approximation and error bounds.

Theorem 1 *Let $\mathbf{A}$ be an $n \times n$ real symmetric matrix. Let $\mathbf{x}$ ($\neq \mathbf{0}$) be any real vector with n components. Furthermore, let*

$$\mathbf{y} = \mathbf{A}\mathbf{x}, \qquad m_0 = \mathbf{x}^\mathsf{T}\mathbf{x}, \qquad m_1 = \mathbf{x}^\mathsf{T}\mathbf{y}, \qquad m_2 = \mathbf{y}^\mathsf{T}\mathbf{y}.$$

Then the quotient

$$q = \frac{m_1}{m_0} \qquad \textbf{(Rayleigh quotient[6])}$$

is an approximation for an eigenvalue[7] λ of $\mathbf{A}$, and if we set $q = \lambda - \epsilon$, so that ϵ is the error of q, then

$$\textbf{(1)} \qquad |\epsilon| \leqq \sqrt{\frac{m_2}{m_0} - q^2}.$$

[6]LORD RAYLEIGH (JOHN WILLIAM STRUTT) (1842—1919), English physicist and mathematician, professor at Cambridge and London, known by his important contributions to various branches of applied mathematics and theoretical physics, in particular, the theory of waves, elasticity, and hydrodynamics.

[7]Ordinarily that λ which is greatest in absolute value, but no general statements are possible.

Proof. Let δ^2 denote the radicand in (1). Then, since $m_1 = qm_0$, we have

$$(2) \quad (\mathbf{y} - q\mathbf{x})^\mathsf{T}(\mathbf{y} - q\mathbf{x}) = m_2 - 2qm_1 + q^2m_0 = m_2 - q^2m_0 = \delta^2 m_0.$$

Since $\mathbf{A}$ is real symmetric, it has an orthogonal set of n real unit eigenvectors $\mathbf{z}_1, \cdots, \mathbf{z}_n$ corresponding to the eigenvalues $\lambda_1, \cdots, \lambda_n$, respectively (some of which may be equal). (Proof in Ref. [B2], vol. 1, pp. 270–272, listed in Appendix 1.) Then $\mathbf{x}$ has a representation of the form

$$\mathbf{x} = a_1\mathbf{z}_1 + \cdots + a_n\mathbf{z}_n.$$

Now $\mathbf{A}\mathbf{z}_1 = \lambda_1\mathbf{z}_1$, etc., and we obtain

$$\mathbf{y} = \mathbf{A}\mathbf{x} = a_1\lambda_1\mathbf{z}_1 + \cdots + a_n\lambda_n\mathbf{z}_n$$

and, since the $\mathbf{z}_j$ are orthogonal unit vectors.

$$(3) \qquad m_0 = \mathbf{x}^\mathsf{T}\mathbf{x} = a_1^{\,2} + \cdots + a_n^{\,2}.$$

It follows that in (2),

$$\mathbf{y} - q\mathbf{x} = a_1(\lambda_1 - q)\mathbf{z}_1 + \cdots + a_n(\lambda_n - q)\mathbf{z}_n.$$

Since the $\mathbf{z}_j$ are orthogonal unit vectors, we thus obtain from (2)

$$\delta^2 m_0 = a_1^{\,2}(\lambda_1 - q)^2 + \cdots + a_n^{\,2}(\lambda_n - q)^2.$$

Replacing each $(\lambda_j - q)^2$ by the smallest of these terms, we have from (3)

$$\delta^2 m_0 \geqq (\lambda_c - q)^2(a_1^{\,2} + \cdots + a_n^{\,2}) = (\lambda_c - q)^2 m_0$$

where λ_c is an eigenvalue to which q is closest. Dividing this inequality by m_0 and taking square roots, we obtain (1), and the theorem is proved. ∎

EXAMPLE 1 **An application of Theorem 1**

As in Example 4 of the preceding section, let us consider the real symmetric matrix

$$\mathbf{A} = \begin{bmatrix} 8 & -2 & 2 \\ -2 & 6 & -4 \\ 2 & -4 & 6 \end{bmatrix} \quad \text{and choose} \quad \mathbf{x}_0 = \begin{bmatrix} 1 \\ 1 \\ 1 \end{bmatrix}.$$

Then we obtain successively

$$\mathbf{x}_1 = \begin{bmatrix} 8 \\ 0 \\ 4 \end{bmatrix}, \quad \mathbf{x}_2 = \begin{bmatrix} 72 \\ -32 \\ 40 \end{bmatrix}, \quad \mathbf{x}_3 = \begin{bmatrix} 720 \\ -496 \\ 512 \end{bmatrix}, \quad \mathbf{x}_4 = \begin{bmatrix} 7776 \\ -6464 \\ 6496 \end{bmatrix}.$$

Taking $\mathbf{x} = \mathbf{x}_3$ and $\mathbf{y} = \mathbf{x}_4$, we have

$$m_0 = \mathbf{x}^\mathsf{T}\mathbf{x} = 1\ 026\ 560, \qquad m_1 = \mathbf{x}^\mathsf{T}\mathbf{y} = 12\ 130\ 816, \qquad m_2 = \mathbf{y}^\mathsf{T}\mathbf{y} = 144\ 447\ 488.$$

From this we calculate

$$q = \frac{m_1}{m_0} = 11.817, \qquad |\epsilon| \leqq \sqrt{\frac{m_2}{m_0} - q^2} = 1.034.$$

This shows that $q = 11.817$ is an approximation for an eigenvalue that must lie between 10.783 and 12.851. The student may show that an eigenvalue is $\lambda = 12$. ∎

Spectral shift, the transition from $\mathbf{A}$ to $\mathbf{A} - k\mathbf{I}$ (Sec. 19.6), may help to improve convergence of the power method, which is simple, but sometimes rather slowly convergent. Information on the spectrum resulting from the problem or bounds obtained by another method or small preliminary computational experiments can help to find a good k. In Example 1 the matrix is symmetric and Gerschgorin's theorem gives $0 \leqq \lambda \leqq 12$. If we shift too much, say, $k = -6$, getting $-6 \leqq \tilde{\lambda} \leqq 6$, we may get large negative eigenvalues of the new matrix, so let us settle for less, say, $k = 4$:

EXAMPLE 2 Spectral shift

For $\mathbf{A}$ and $\mathbf{x}_0$ in Example 1 and $\tilde{\mathbf{A}} = \mathbf{A} - 4\mathbf{I}$ we get

$$\tilde{\mathbf{A}} = \begin{bmatrix} 4 & -2 & 2 \\ -2 & 2 & -4 \\ 2 & -4 & 2 \end{bmatrix}, \ \tilde{\mathbf{x}}_1 = \begin{bmatrix} 4 \\ -4 \\ 0 \end{bmatrix}, \ \tilde{\mathbf{x}}_2 = \begin{bmatrix} 24 \\ -16 \\ 24 \end{bmatrix}, \ \tilde{\mathbf{x}}_3 = \begin{bmatrix} 176 \\ -176 \\ 160 \end{bmatrix}, \ \tilde{\mathbf{x}}_4 = \begin{bmatrix} 1376 \\ -1344 \\ 1376 \end{bmatrix}.$$

From this we obtain

$$\tilde{m}_0 = 87\,552, \qquad \tilde{m}_1 = 698\,880, \qquad \tilde{m}_2 = 5\,593\,088, \qquad \tilde{q} = 7.982$$

so that $q = \tilde{q} + 4 = 11.982$, in error by 0.15%, as opposed to q in Example 1, whose error is 1.5%; hence the improvement is by a factor ten!. Also, $|\epsilon| \leqq 0.404$ gives $11.578 \leqq \lambda \leqq 12.386$. It is typical that $|\epsilon|$ is much larger than the actual error, but the resulting bounds cannot be improved, as examples show. ∎

Problem Set 19.8

Choosing $\mathbf{x}_0 = [1 \quad 1]^\mathsf{T}$ or $\mathbf{x}_0 = [1 \quad 1 \quad 1]^\mathsf{T}$, respectively, apply the power method (3 steps) to the following matrices, computing in each step the Rayleigh quotient and an error bound.

1. $\begin{bmatrix} 10 & 4 \\ 4 & 2 \end{bmatrix}$ **2.** $\begin{bmatrix} 3 & 2 \\ 2 & 5 \end{bmatrix}$ **3.** $\begin{bmatrix} 0.6 & 0.8 \\ 0.8 & -0.6 \end{bmatrix}$

4. $\begin{bmatrix} 2 & -1 & 1 \\ -1 & 3 & 2 \\ 1 & 2 & 3 \end{bmatrix}$ **5.** $\begin{bmatrix} -2 & 2 & 3 \\ 2 & 1 & 6 \\ 3 & 6 & -2 \end{bmatrix}$ **6.** $\begin{bmatrix} 3 & 2 & 3 \\ 2 & 6 & 6 \\ 3 & 6 & 3 \end{bmatrix}$

7. Compute the exact values of the eigenvalues in Prob. 1. Verify that the actual errors of the approximations are much smaller than the error bounds in Prob. 1. This is typical.

To see how the power method behaves for different choices of $\mathbf{x}_0$, apply it to the matrix in Prob. 4, call it $\mathbf{A}$; do 3 steps and compute the Rayleigh quotient and error bound corresponding to the third step. Verify that $\mathbf{A}$ has the eigenvalues 5, 3, 0 and indicate which eigenvalue is approximated by the result.

8. $[0 \quad 1 \quad -1]^\mathsf{T}$ **9.** $[1 \quad 1 \quad 0]^\mathsf{T}$ **10.** $[0 \quad 0 \quad 1]^\mathsf{T}$

Choosing $\mathbf{x}_0 = [1 \quad 1 \quad 1 \quad 1]^\mathsf{T}$, compute $\mathbf{x}_1$, $\mathbf{x}_2$ and approximations $q = \mathbf{x}_1{}^\mathsf{T}\mathbf{x}_0/\mathbf{x}_0{}^\mathsf{T}\mathbf{x}_0$, $q = \mathbf{x}_2{}^\mathsf{T}\mathbf{x}_1/\mathbf{x}_1{}^\mathsf{T}\mathbf{x}_1$ and corresponding error bounds for an eigenvalue of each of the following matrices.

11. $\begin{bmatrix} 2 & 0 & 1 & 0 \\ 0 & 0 & 3 & 1 \\ 1 & 3 & 4 & -2 \\ 0 & 1 & -2 & 0 \end{bmatrix}$ **12.** $\begin{bmatrix} 3 & 2 & 0 & 1 \\ 2 & 0 & 5 & -1 \\ 0 & 5 & 2 & 1 \\ 1 & -1 & 1 & 4 \end{bmatrix}$ **13.** $\begin{bmatrix} 1 & 0 & 0 & 1 \\ 0 & 2 & -1 & 0 \\ 0 & -1 & 3 & 0 \\ 1 & 0 & 0 & -1 \end{bmatrix}$

14. To understand the importance of the error bound (1), consider the matrix

$$\mathbf{A} = \begin{bmatrix} 3 & 4 \\ 4 & -3 \end{bmatrix}, \quad \text{choose} \quad \mathbf{x}_0 = \begin{bmatrix} 3 \\ -1 \end{bmatrix},$$

and show that $q = 0$ for all s. Find the eigenvalues and explain what happened. Start again, choosing another $\mathbf{x}_0$.

15. **(Spectral shift)** Show that the matrix in Prob. 6 is $\mathbf{A} + 5\mathbf{I}$, where $\mathbf{A}$ is the matrix in Prob. 5. Verify that $\mathbf{A}$ has the eigenvalues 7, -3, -7. (a) What are the eigenvalues of $\mathbf{A} + 5\mathbf{I}$? (b) Why can we expect an improvement of convergence by the "spectral shift" (see Sec. 19.6) from $\mathbf{A}$ to $\mathbf{A} + 5\mathbf{I}$?

16. Why is the Rayleigh quotient $q = m_1/m_0$ in general an approximation of that eigenvalue, λ_1, which is largest in absolute value? *Hint.* Let $\mathbf{z}_1, \cdots, \mathbf{z}_n$ be as in the proof of Theorem 1. Show that if

$$\mathbf{x}_0 = \sum c_j\mathbf{z}_j, \quad \text{then} \quad \mathbf{x} = \mathbf{x}_{s-1} = \sum c_j\lambda_j^{s-1}\mathbf{z}_j, \quad \mathbf{y} = \mathbf{x}_s = \sum c_j\lambda_j^s\mathbf{z}_j,$$

$$q = m_1/m_0 = (c_1{}^2\lambda_1^{2s-1} + \cdots)/(c_1{}^2\lambda_1^{2s-2} + \cdots) \approx \lambda_1.$$

Under what conditions will this be a good approximation?

17. Show that if $\mathbf{x}$ is an eigenvector, then $\epsilon = 0$ in (1).

18. Let $\mathbf{A}$ be symmetric, with eigenvalues $\lambda_1 > \lambda_2 \geqq \cdots \geqq \lambda_{n-1} > \lambda_n$, assuming $|\lambda_1| > |\lambda_n|$, and let α be a good estimate of λ_1. Then the power method applied to $\mathbf{B} = \mathbf{A} - \alpha\mathbf{I}$ will in general yield approximations to λ_n. Why is this so, and what does "in general" mean?

19. Apply the method in Prob. 18 to the matrix in Prob. 4 with $\alpha = 4.9$ (as suggested by the answer to Prob. 4), $\mathbf{x}_0 = [1 \quad 1 \quad 1]^\mathsf{T}$ and $\mathbf{x}_2$, $\mathbf{x}_3$ as $\mathbf{x}$, $\mathbf{y}$.

20. **Collatz's theorem** (Sec. 19.7) applies to the matrix in Prob. 6. Use it with $\mathbf{x}_0 = [1 \quad 1 \quad 1]^\mathsf{T}$, do 2 steps, and compare the results with those obtained from our Theorem 1.

19.9 Deflation of a Matrix

In practice, it often happens that we have computed or guessed one of the eigenvalues of a given matrix **A**, call it λ_1, and we want to find the further eigenvalues $\lambda_2, \cdots, \lambda_n$ of **A**. We show that then we can obtain from **A** a matrix $\mathbf{A}_1$ whose spectrum consists of 0 and the (still unknown) eigenvalues $\lambda_2, \cdots, \lambda_n$ of **A**. The process of replacing **A** by $\mathbf{A}_1$ in the further determination of those eigenvalues is called **deflation of A.** We discuss the popular Wielandt's method. A simpler but numerically poor method is mentioned in Prob. 16 below.

Wielandt's Deflation[8]

Let **A** be any real $n \times n$ matrix and λ_1 a known eigenvalue of **A**, as before. Let $\mathbf{x}_1$ be an eigenvector of **A** corresponding to λ_1. Let **u** be any vector (to be determined later) such that

$$\mathbf{u}^\mathsf{T}\mathbf{x}_1 = \lambda_1. \tag{1}$$

Consider

$$\mathbf{A}_1 = \mathbf{A} - \mathbf{x}_1\mathbf{u}^\mathsf{T}. \tag{2}$$

We first show that 0 is an eigenvalue of $\mathbf{A}_1$ and $\mathbf{x}_1$ is a corresponding eigenvector. Indeed, by (1),

$$\mathbf{A}_1\mathbf{x}_1 = \mathbf{A}\mathbf{x}_1 - \mathbf{x}_1\mathbf{u}^\mathsf{T}\mathbf{x}_1 = \lambda_1\mathbf{x}_1 - \lambda_1\mathbf{x}_1 = \mathbf{0}.$$

Let λ_j, $j = 2, \cdots, n$, be the other eigenvalues of **A**, with corresponding eigenvectors $\mathbf{x}_j$. We show that $\lambda_j \neq 0$ is an eigenvalue of $\mathbf{A}_1$, with corresponding eigenvector

$$\mathbf{y}_j = \mathbf{x}_j - \frac{\mathbf{u}^\mathsf{T}\mathbf{x}_j}{\lambda_j}\mathbf{x}_1. \tag{3}$$

Using the definitions, we have

$$\mathbf{A}_1\mathbf{y}_j = (\mathbf{A} - \mathbf{x}_1\mathbf{u}^\mathsf{T})\left(\mathbf{x}_j - \frac{\mathbf{u}^\mathsf{T}\mathbf{x}_j}{\lambda_j}\mathbf{x}_1\right).$$

Multiplying out and noting that $\mathbf{u}^\mathsf{T}\mathbf{x}_j/\lambda_j$ is a scalar, so that we can move it around in products, we obtain

$$\mathbf{A}_1\mathbf{y}_j = \mathbf{A}\mathbf{x}_j - \frac{\mathbf{u}^\mathsf{T}\mathbf{x}_j}{\lambda_j}\mathbf{A}\mathbf{x}_1 - \mathbf{x}_1(\mathbf{u}^\mathsf{T}\mathbf{x}_j) + \frac{\mathbf{u}^\mathsf{T}\mathbf{x}_j}{\lambda_j}\mathbf{x}_1\mathbf{u}^\mathsf{T}\mathbf{x}_1.$$

[8]HELMUT WIELANDT (born 1910), German mathematician.

Now on the right, $\mathbf{A}\mathbf{x}_1 = \lambda_1\mathbf{x}_1$ in the second term, and $\mathbf{x}_1\mathbf{u}^\mathsf{T}\mathbf{x}_1 = \mathbf{x}_1\lambda_1$ in the fourth term [by (1)], so that these two terms cancel. Hence we are left with

$$\mathbf{A}_1\mathbf{y}_j = \lambda_j\mathbf{x}_j - \mathbf{x}_1(\mathbf{u}^\mathsf{T}\mathbf{x}_j) = \lambda_j\mathbf{y}_j.$$

This shows that λ_j is an eigenvalue of $\mathbf{A}_1$.

We show how to get $\mathbf{u}$ in practice, and we shall also see that our assumption $\lambda_j \neq 0$ is no handicap to the method. We have $\mathbf{A}\mathbf{x}_1 = \lambda_1\mathbf{x}_1$. We can easily find $\mathbf{v}$ such that $\mathbf{v}^\mathsf{T}\mathbf{x}_1 = 1$. Then

$$\mathbf{v}^\mathsf{T}\mathbf{A}\mathbf{x}_1 = \mathbf{v}^\mathsf{T}\lambda_1\mathbf{x}_1 = \lambda_1\mathbf{v}^\mathsf{T}\mathbf{x}_1 = \lambda_1.$$

Because of (1) this shows that we can take

$$\mathbf{u}^\mathsf{T} = \mathbf{v}^\mathsf{T}\mathbf{A}.$$

Then (2) becomes

$$\mathbf{A}_1 = \mathbf{A} - \mathbf{x}_1\mathbf{v}^\mathsf{T}\mathbf{A} \tag{4}$$

and in (3) we now get rid of λ_j, since $\mathbf{A}\mathbf{x}_j = \lambda_j\mathbf{x}_j$ and thus

$$\mathbf{y}_j = \mathbf{x}_j - \frac{\mathbf{v}^\mathsf{T}\mathbf{A}\mathbf{x}_j}{\lambda_j}\mathbf{x}_j = \mathbf{x}_j - \mathbf{v}^\mathsf{T}\mathbf{x}_j\mathbf{x}_1.$$

Furthermore, if $\mathbf{x}_1$ has a nonzero first component, we can normalize $\mathbf{x}_1$ to make its first component 1. Then we can simply take the vector $\mathbf{v}^\mathsf{T} = \mathbf{e}_1{}^\mathsf{T} = [1 \quad 0 \quad \cdots \quad 0]$. For this choice,

$$\boxed{\mathbf{A}_1 = (\mathbf{I} - \mathbf{x}_1\mathbf{e}_1{}^\mathsf{T})\mathbf{A}.} \tag{5}$$

EXAMPLE 1 Wielandt's deflation

Find the spectrum of the matrix

$$\mathbf{A} = \begin{bmatrix} 8 & -2 & 2 \\ -2 & 6 & -4 \\ 2 & -4 & 6 \end{bmatrix}$$

in Example 1 of the last section, assuming that one of the eigenvalues, $\lambda_1 = 12$, is known.

Solution. We first compute an eigenvector $\mathbf{x}_1$ of $\mathbf{A}$ corresponding to $\lambda_1 = 12$ by the Gauss elimination, finding $\mathbf{x}_1 = [1 \quad -1 \quad 1]^\mathsf{T}$. In Wielandt's method we can now take the vector $\mathbf{v}_1{}^\mathsf{T} = \mathbf{e}_1{}^\mathsf{T} = [1 \quad 0 \quad 0]$ since then $\mathbf{v}_1{}^\mathsf{T}\mathbf{x}_1 = 1$. We compute

$$\mathbf{x}_1\mathbf{e}_1{}^\mathsf{T} = \begin{bmatrix} 1 \\ -1 \\ 1 \end{bmatrix} [1 \quad 0 \quad 0] = \begin{bmatrix} 1 & 0 & 0 \\ -1 & 0 & 0 \\ 1 & 0 & 0 \end{bmatrix}$$

and from this, by (5),

$$\mathbf{A}_1 = (\mathbf{I} - \mathbf{x}_1\mathbf{e}_1^{\mathsf{T}})\mathbf{A} = \begin{bmatrix} 0 & 0 & 0 \\ 1 & 1 & 0 \\ -1 & 0 & 1 \end{bmatrix} \begin{bmatrix} 8 & -2 & 2 \\ -2 & 6 & -4 \\ 2 & -4 & 6 \end{bmatrix} = \begin{bmatrix} 0 & 0 & 0 \\ 6 & 4 & -2 \\ -6 & -2 & 4 \end{bmatrix}.$$

We can now determine the other two eigenvalues directly from $\det(\mathbf{A}_1 - \lambda\mathbf{I}) = 0$, where

$$\det(\mathbf{A}_1 - \lambda\mathbf{I}) = \begin{vmatrix} -\lambda & 0 & 0 \\ 6 & 4-\lambda & -2 \\ -6 & -2 & 4-\lambda \end{vmatrix} = -\lambda \begin{vmatrix} 4-\lambda & -2 \\ -2 & 4-\lambda \end{vmatrix} = -\lambda(\lambda^2 - 8\lambda + 12).$$

We obtain $\lambda_2 = 6$, $\lambda_3 = 2$.

In the case of a larger matrix, the next step would have been the computation of another eigenvalue (for instance, by the power method) followed by another application of Wielandt's method, and so on. ∎

In the next section we show that for determining the whole spectrum of a matrix there are better methods than the successive application of deflations.

Problem Set 19.9

1. Verify that $\mathbf{A}_1\mathbf{x}_1 = \mathbf{0}$ in Example 1 in the text.

Apply Wielandt's method of deflation to the following matrices. Show the deflated matrix and determine the further eigenvalues.

2. $\begin{bmatrix} -3 & 1 & 1 \\ 1 & -3 & 1 \\ 1 & 1 & -3 \end{bmatrix}$, $\mathbf{x}_1 = \begin{bmatrix} 1 \\ 1 \\ 1 \end{bmatrix}$

3. $\begin{bmatrix} 31 & 16 & 72 \\ -24 & -12 & -57 \\ -8 & -4 & -19 \end{bmatrix}$, $\mathbf{x}_1 = \begin{bmatrix} 4 \\ -3 \\ -1 \end{bmatrix}$

4. $\begin{bmatrix} 6 & -2 & -2 \\ 0 & -1 & 1 \\ 3 & -6 & 2 \end{bmatrix}$, $\mathbf{x}_1 = \begin{bmatrix} 2 \\ 1 \\ 3 \end{bmatrix}$

5. $\begin{bmatrix} 0 & 2 & -1 \\ -2 & 2 & 1 \\ -3 & 8 & 0 \end{bmatrix}$, $\mathbf{x}_1 = \begin{bmatrix} 5 \\ 1 \\ 7 \end{bmatrix}$

6. $\begin{bmatrix} 6 & -1 & -5 \\ -4 & 2 & -2 \\ 18 & -5 & -9 \end{bmatrix}$, $\mathbf{x}_1 = \begin{bmatrix} 1 \\ 2 \\ 1 \end{bmatrix}$

7. $\begin{bmatrix} -5 & 11 & -13 \\ -92 & 39 & 22 \\ 74 & -13 & -44 \end{bmatrix}$, $\mathbf{x}_1 = \begin{bmatrix} 1 \\ 2 \\ 1 \end{bmatrix}$

8. $\begin{bmatrix} 1.2 & -0.4 & 0.4 \\ 0.4 & 0.2 & 0.8 \\ 0.8 & 0.4 & 1.6 \end{bmatrix}$, $\mathbf{x}_1 = \begin{bmatrix} 1 \\ 2 \\ 4 \end{bmatrix}$

9. $\begin{bmatrix} -1 & 2 & 3 \\ 2 & 2 & 6 \\ 3 & 6 & -1 \end{bmatrix}$, $\mathbf{x}_1 = \begin{bmatrix} 3 \\ 6 \\ 5 \end{bmatrix}$

10. $\begin{bmatrix} 8 & 1 & 0 \\ 1 & 8 & 2 \\ 1 & 1 & 8 \end{bmatrix}$, $\mathbf{x}_1 = \begin{bmatrix} 2 \\ 4 \\ 3 \end{bmatrix}$

11. $\begin{bmatrix} 6 & 7 & 7 \\ 7 & 6 & 7 \\ 7 & 7 & 6 \end{bmatrix}$, $\mathbf{x}_1 = \begin{bmatrix} 1 \\ 1 \\ 1 \end{bmatrix}$

12. $\begin{bmatrix} -6 & 4 & 4 & 4 \\ -6 & 4 & 4 & 4 \\ -4 & 0 & 6 & 4 \\ -2 & 0 & 0 & 8 \end{bmatrix}, \quad \mathbf{x}_1 = \begin{bmatrix} 1 \\ 1 \\ 1 \\ 1 \end{bmatrix}$ **13.** $\begin{bmatrix} 4 & 4 & 1 & 1 \\ 4 & 4 & 1 & 1 \\ 1 & 1 & 3 & 2 \\ 1 & 1 & 2 & 3 \end{bmatrix}, \quad \mathbf{x}_1 = \begin{bmatrix} 2 \\ 2 \\ 1 \\ 1 \end{bmatrix}$

14. $\begin{bmatrix} 2 & -5 & 0 & 3 \\ 0 & 2 & -3 & -5 \\ 5 & -3 & 2 & 0 \\ 3 & 0 & 5 & 2 \end{bmatrix}, \quad \mathbf{x}_1 = \begin{bmatrix} 1 \\ -1 \\ 1 \\ 1 \end{bmatrix}$ **15.** $\begin{bmatrix} 3.4 & 3.4 & 1.3 & 1.3 \\ 3.4 & 3.4 & 1.3 & 1.3 \\ 1.3 & 1.3 & 1.5 & 1.4 \\ 1.3 & 1.3 & 1.4 & 1.5 \end{bmatrix}, \quad \mathbf{x}_1 = \begin{bmatrix} 2 \\ 2 \\ 1 \\ 1 \end{bmatrix}$

16. **(Hotelling's deflation)** Let $\mathbf{A}$ be symmetric, λ_1 a known eigenvalue of $\mathbf{A}$, and $\mathbf{x}_1$ a corresponding normalized eigenvector, $\mathbf{x}_1^{\mathsf{T}}\mathbf{x}_1 = 1$. Show that

$$\mathbf{A}_1 = \mathbf{A} - \lambda_1 \mathbf{x}_1 \mathbf{x}_1^{\mathsf{T}} \qquad \text{(Hotelling's deflation)}$$

has the eigenvalues $\lambda_1 = 0$ and the others equal to those of $\mathbf{A}$. (This method is numerically poor with respect to round-off). *Hint.* Use Theorem 4, Sec. 7.14.

19.10 Householder Tridiagonalization and QR-Factorization

We consider the problem of computing *all* the eigenvalues of a ***real symmetric*** matrix $\mathbf{A} = [a_{jk}]$. Successive deflations would not be good, because of round-off error growth. We discuss a method widely used in practice. In the first stage we apply Householder's method,[9] which reduces the given matrix to a **tridiagonal matrix,** that is, a matrix having all its nonzero entries on the main diagonal and in the positions immediately adjacent to the main diagonal (such as $\mathbf{A}_3$ in Fig. 420). In the second stage, the tridiagonal matrix is factorized in the form **QR,** where **Q** is orthogonal and **R** upper triangular, and the eigenvalues are actually determined (approximately); we discuss this afterward. (For extensions to general matrices, see Ref. [E27] listed in Appendix 1.)

Householder's Method

This method reduces a given real ***symmetric*** $n \times n$ matrix $\mathbf{A} = [a_{jk}]$ by $n - 2$ successive similarity transformations (see Sec. 19.6) to tridiagonal form. The matrices $\mathbf{P}_1, \mathbf{P}_2, \cdots, \mathbf{P}_{n-2}$ are orthogonal and symmetric matrices. Hence $\mathbf{P}_1^{-1} = \mathbf{P}_1^{\mathsf{T}} = \mathbf{P}_1$ and similarly for the others. The $n - 2$ similarity transformations that produce from the given $\mathbf{A}_0 = \mathbf{A} = [a_{jk}]$ successively the matrices $\mathbf{A}_1 = [a_{jk}^{(1)}]$, $\mathbf{A}_2 = [a_{jk}^{(2)}]$, etc. look as follows.

[9] *Journal of the Association for Computing Machinery* **5** (1958), 335–342. See also Refs. [E25], [E26], etc. in Appendix 1.

$$
\text{(1)} \qquad
\begin{aligned}
\mathbf{A}_1 &= \mathbf{P}_1\mathbf{A}_0\mathbf{P}_1 \\
\mathbf{A}_2 &= \mathbf{P}_2\mathbf{A}_1\mathbf{P}_2 \\
&\cdots\cdots\cdots\cdots \\
\mathbf{B} = \mathbf{A}_{n-2} &= \mathbf{P}_{n-2}\mathbf{A}_{n-3}\mathbf{P}_{n-2}.
\end{aligned}
$$

These transformations create the necessary zeros, in the first step in row 1 and column 1, in the second step in row 2 and column 2, etc., as Fig. 420 illustrates for a 5 × 5 matrix. **B** is tridiagonal.

How to determine $\mathbf{P}_1, \mathbf{P}_2, \cdots$? All these $\mathbf{P}_r$ are of the form

$$
\text{(2)} \qquad \mathbf{P}_r = \mathbf{I} - 2\mathbf{v}_r\mathbf{v}_r^{\mathsf{T}} \qquad (r = 1, \cdots, n-2)
$$

where $\mathbf{v}_r = [v_{jr}]$ is a unit vector with its first r components 0; thus

$$
\text{(3)} \qquad \mathbf{v}_1 = \begin{bmatrix} 0 \\ * \\ * \\ \vdots \\ * \end{bmatrix}, \quad \mathbf{v}_2 = \begin{bmatrix} 0 \\ 0 \\ * \\ \vdots \\ * \end{bmatrix}, \quad \cdots, \quad \mathbf{v}_{n-2} = \begin{bmatrix} 0 \\ 0 \\ \vdots \\ * \\ * \end{bmatrix}
$$

where the asterisks denote any other components.

First Step. $\mathbf{v}_1$ has the components

$$
\begin{aligned}
&& v_{11} &= 0 \\
&\text{(a)} & v_{21} &= \sqrt{\frac{1}{2}\left(1 + \frac{|a_{21}|}{S_1}\right)} \\
\text{(4)}\ &\text{(b)} & v_{j1} &= \frac{a_{j1}\ \operatorname{sgn} a_{21}}{2v_{21}S_1} \qquad j = 3, 4, \cdots, n \\
&& &\text{where} \\
&\text{(c)} & S_1 &= \sqrt{a_{21}^2 + a_{31}^2 + \cdots + a_{n1}^2}
\end{aligned}
$$

$$
\begin{bmatrix} * & * & & & \\ * & * & * & * & * \\ & * & * & * & * \\ & * & * & * & * \\ & * & * & * & * \end{bmatrix}
\qquad
\begin{bmatrix} * & * & & & \\ * & * & * & & \\ & * & * & * & * \\ & & * & * & * \\ & & * & * & * \end{bmatrix}
\qquad
\begin{bmatrix} * & * & & & \\ * & * & * & & \\ & * & * & * & \\ & & * & * & * \\ & & & * & * \end{bmatrix}
$$

First Step $\mathbf{A}_1 = \mathbf{P}_1\mathbf{A}\mathbf{P}_1$ — Second Step $\mathbf{A}_2 = \mathbf{P}_2\mathbf{A}_1\mathbf{P}_2$ — Third Step $\mathbf{A}_3 = \mathbf{P}_3\mathbf{A}_2\mathbf{P}_3$

Fig. 420. Householder's method for a 5 × 5 matrix. Positions left blank are zeros created by the method.

where $S_1 > 0$, and sgn $a_{21} = +1$ if $a_{21} \geqq 0$ and sgn $a_{21} = -1$ if $a_{21} < 0$. With this we compute $\mathbf{P}_1$ by (2) and then $\mathbf{A}_1$ by (1). This was the first step.

Second Step. We compute $\mathbf{v}_2$ by (4) with all subscripts increased by 1 and the a_{jk} replaced by $a_{jk}^{(1)}$, the entries of $\mathbf{A}_1$ just computed. Thus [see also (3)]

$$\begin{aligned} v_{12} &= v_{22} = 0 \\ v_{32} &= \sqrt{\frac{1}{2}\left(1 + \frac{|a_{32}^{(1)}|}{S_2}\right)} \qquad (4^*) \\ v_{j2} &= \frac{a_{j2}^{(1)} \operatorname{sgn} a_{32}^{(1)}}{2 v_{32} S_2} \qquad j = 4, 5, \cdots, n \end{aligned}$$

where

$$S_2 = \sqrt{a_{32}^{(1)2} + a_{42}^{(1)2} + \cdots + a_{n2}^{(1)2}}.$$

With this we compute $\mathbf{P}_2$ by (2) and then $\mathbf{A}_2$ by (1).

Third Step. We compute $\mathbf{v}_3$ by (4*) with all subscripts increased by 1 and the $a_{jk}^{(1)}$ replaced by the entries $a_{jk}^{(2)}$ of $\mathbf{A}_2$, and so on.

EXAMPLE 1 Householder's method

Tridiagonalize the real symmetric matrix

$$\mathbf{A} = \mathbf{A}_0 = \begin{bmatrix} 6 & 4 & 1 & 1 \\ 4 & 6 & 1 & 1 \\ 1 & 1 & 5 & 2 \\ 1 & 1 & 2 & 5 \end{bmatrix}.$$

Solution. First Step. We compute $S_1^2 = 4^2 + 1^2 + 1^2 = 18$ from (4c). Since $a_{21} = 4 > 0$, we have sgn $a_{21} = +1$ in (4b) and get from (4) by straightforward computation

$$\mathbf{v}_1 = \begin{bmatrix} 0 \\ v_{21} \\ v_{31} \\ v_{41} \end{bmatrix} = \begin{bmatrix} 0 \\ 0.985\,598\,56 \\ 0.119\,573\,16 \\ 0.119\,573\,16 \end{bmatrix}.$$

From this and (2),

$$\mathbf{P}_1 = \begin{bmatrix} 1 & 0 & 0 & 0 \\ 0 & -0.942\,809\,04 & -0.235\,702\,26 & -0.235\,702\,26 \\ 0 & -0.235\,702\,26 & 0.971\,404\,52 & -0.028\,595\,48 \\ 0 & -0.235\,702\,26 & -0.028\,595\,48 & 0.971\,404\,52 \end{bmatrix}.$$

From the first line in (1) we now get

$$\mathbf{A}_1 = \mathbf{P}_1\mathbf{A}_0\mathbf{P}_1 = \begin{bmatrix} 6 & -\sqrt{18} & 0 & 0 \\ -\sqrt{18} & 7 & -1 & -1 \\ 0 & -1 & 9/2 & 3/2 \\ 0 & -1 & 3/2 & 9/2 \end{bmatrix}.$$

Second Step. From (4*) we compute $S_2^{\,2} = 2$ and

$$\mathbf{v}_2 = \begin{bmatrix} 0 \\ 0 \\ v_{32} \\ v_{42} \end{bmatrix} = \begin{bmatrix} 0 \\ 0 \\ 0.923\,879\,53 \\ 0.382\,683\,43 \end{bmatrix}.$$

From this and (2),

$$\mathbf{P}_2 = \begin{bmatrix} 1 & 0 & 0 & 0 \\ 0 & 1 & 0 & 0 \\ 0 & 0 & -1/\sqrt{2} & -1/\sqrt{2} \\ 0 & 0 & -1/\sqrt{2} & 1/\sqrt{2} \end{bmatrix}.$$

The second line in (1) now gives

$$\mathbf{B} = \mathbf{A}_2 = \mathbf{P}_2\mathbf{A}_1\mathbf{P}_2 = \begin{bmatrix} 6 & -\sqrt{18} & 0 & 0 \\ -\sqrt{18} & 7 & \sqrt{2} & 0 \\ 0 & \sqrt{2} & 6 & 0 \\ 0 & 0 & 0 & 3 \end{bmatrix}.$$

This matrix **B** is tridiagonal. Since our given matrix has order $n = 4$, we needed $n - 2 = 2$ steps to accomplish this reduction, as claimed. (Do you see that we got more zeros than we can expect in general?) **B** is similar to **A**, as we now show in general. This is essential because **B** thus has the same spectrum as **A**, by Theorem 2 in Sec. 19.6. ∎

We show that **B** in (1) is similar to $\mathbf{A} = \mathbf{A}_0$.

The matrix $\mathbf{P}_r$ is symmetric,

$$\mathbf{P}_r^{\mathsf{T}} = (\mathbf{I} - 2\mathbf{v}_r\mathbf{v}_r^{\mathsf{T}})^{\mathsf{T}} = \mathbf{I}^{\mathsf{T}} - 2(\mathbf{v}_r\mathbf{v}_r^{\mathsf{T}})^{\mathsf{T}}$$
$$= \mathbf{I} - 2\mathbf{v}_r\mathbf{v}_r^{\mathsf{T}} = \mathbf{P}_r.$$

Also, $\mathbf{P}_r$ is orthogonal because

$$\mathbf{P}_r\mathbf{P}_r^{\mathsf{T}} = \mathbf{P}_r^{\,2} = (\mathbf{I} - 2\mathbf{v}_r\mathbf{v}_r^{\mathsf{T}})^2 = \mathbf{I} - 4\mathbf{v}_r\mathbf{v}_r^{\mathsf{T}} + 4\mathbf{v}_r\mathbf{v}_r^{\mathsf{T}}\mathbf{v}_r\mathbf{v}_r^{\mathsf{T}}$$

and $\mathbf{v}_r^{\mathsf{T}}\mathbf{v}_r = 1$ in the last term because $\mathbf{v}_r$ is a unit vector (see above), so that the right side reduces to **I**. This gives $\mathbf{P}_r^{-1} = \mathbf{P}_r^{\mathsf{T}} = \mathbf{P}_r$. Hence from (1) we now obtain

$$\mathbf{P} = \mathbf{P}_{n-2}\mathbf{A}_{n-3}\mathbf{P}_{n-2} = \cdots$$
$$\cdots = \mathbf{P}_{n-2}\mathbf{P}_{n-3}\cdots\mathbf{P}_1\mathbf{A}\mathbf{P}_1\cdots\mathbf{P}_{n-3}\mathbf{P}_{n-2}$$
$$= \mathbf{P}_{n-2}^{-1}\mathbf{P}_{n-3}^{-1}\cdots\mathbf{P}_1^{-1}\mathbf{A}\mathbf{P}_1\cdots\mathbf{P}_{n-3}\mathbf{P}_{n-2}$$
$$= \mathbf{T}^{-1}\mathbf{A}\mathbf{T}$$

where $\mathbf{T} = \mathbf{P}_1\mathbf{P}_2\cdots\mathbf{P}_{n-2}$. This proves our assertion. ∎

QR-Factorization Method

In 1958 H. Rutishauser proposed the idea of using the LU-factorization (see Sec. 19.2; he called it LR-factorization) in solving eigenvalue problems. An improved version of Rutishauser's method (avoiding breakdown if certain submatrices become singular, etc.; see Ref. [E26]) is the **QR-method,**[10] which is based on the factorization **QR**, where **R** is upper triangular as before but **Q** is orthogonal (instead of lower triangular). We discuss the QR-method for a real symmetric matrix. (For extensions to general real or complex matrices, see Refs. [E26] and [E27] in Appendix 1.) In this method we start from a real ***symmetric*** tridiagonal matrix $\mathbf{B}_0 = \mathbf{B}$ (as obtained from a real symmetric matrix **A** by Householder's method). We compute stepwise $\mathbf{B}_1, \mathbf{B}_2, \cdots$ according to this rule:

First Step. Factor

$$\mathbf{B}_0 = \mathbf{Q}_0\mathbf{R}_0$$

where $\mathbf{Q}_0$ is orthogonal and $\mathbf{R}_0$ is upper triangular. Then compute

$$\mathbf{B}_1 = \mathbf{R}_0\mathbf{Q}_0.$$

Second Step. Factor $\mathbf{B}_1 = \mathbf{Q}_1\mathbf{R}_1$. Then compute $\mathbf{B}_2 = \mathbf{R}_1\mathbf{Q}_1$.

General Step. Factor

$$\mathbf{B}_s = \mathbf{Q}_s\mathbf{R}_s \tag{5}$$

where $\mathbf{Q}_s$ is orthogonal and $\mathbf{R}_s$ is upper triangular. Then compute

$$\mathbf{B}_{s+1} = \mathbf{R}_s\mathbf{Q}_s. \tag{6}$$

The method of obtaining the factorization (5) will be explained below.

From (5) we have $\mathbf{R}_s = \mathbf{Q}_s^{-1}\mathbf{B}_s$. Substitution into (6) gives

$$\mathbf{B}_{s+1} = \mathbf{R}_s\mathbf{Q}_s = \mathbf{Q}_s^{-1}\mathbf{B}_s\mathbf{Q}_s. \tag{7}$$

Thus $\mathbf{B}_{s+1}$ is similar to $\mathbf{B}_s$. Hence $\mathbf{B}_{s+1}$ is similar to $\mathbf{B}_0 = \mathbf{B}$ for all s. By Theorem 2, Sec. 19.6, this implies that $\mathbf{B}_{s+1}$ has the same eigenvalues as **B.**

Also, $\mathbf{B}_{s+1}$ is symmetric. This follows by induction. Indeed, $\mathbf{B}_0 = \mathbf{B}$ is symmetric. Assuming $\mathbf{B}_s$ to be symmetric and using $\mathbf{Q}_s^{-1} = \mathbf{Q}_s^{\mathsf{T}}$ (since $\mathbf{Q}_s$ is orthogonal), we get from (7)

$$\mathbf{B}_{s+1}^{\mathsf{T}} = (\mathbf{Q}_s^{\mathsf{T}}\mathbf{B}_s\mathbf{Q}_s)^{\mathsf{T}} = \mathbf{Q}_s^{\mathsf{T}}\mathbf{B}_s^{\mathsf{T}}\mathbf{Q}_s = \mathbf{Q}_s^{\mathsf{T}}\mathbf{B}_s\mathbf{Q}_s = \mathbf{B}_{s+1}$$

as claimed.

---

[10] Proposed independently by J. G. F. Francis, *Computer Journal* **4** (1961–62), 265–271, 332–345, and V. N. Kublanovskaya, *Zhurnal Vych. Mat. i Mat. Fiz.* **1** (1961), 555–570.

If the eigenvalues of $\mathbf{B}$ are different in absolute value, say, if we have $|\lambda_1| > |\lambda_2| > \cdots > |\lambda_n|$, then

$$\lim_{s\to\infty} \mathbf{B}_s = \mathbf{D}$$

where $\mathbf{D}$ is diagonal, with main diagonal entries $\lambda_1, \lambda_2, \cdots, \lambda_n$. (Proof in Ref. [E26] listed in Appendix 1.)

How to get the QR-factorization, say, $\mathbf{B} = \mathbf{B}_0 = [b_{jk}] = \mathbf{Q}_0\mathbf{R}_0$? The tridiagonal matrix $\mathbf{B}$ has $n - 1$ generally nonzero entries below the main diagonal. These are $b_{21}, b_{32}, \cdots, b_{n,n-1}$. We multiply $\mathbf{B}$ from the left by a matrix $\mathbf{C}_2$ such that $\mathbf{C}_2\mathbf{B} = [b_{jk}^{(2)}]$ has $b_{21}^{(2)} = 0$. This we multiply by a matrix $\mathbf{C}_3$ such that $\mathbf{C}_3\mathbf{C}_2\mathbf{B} = [b_{jk}^{(3)}]$ has $b_{32}^{(3)} = 0$, etc. After $n - 1$ such multiplications we are left with an upper triangular matrix $\mathbf{R}_0$, namely,

$$\mathbf{C}_n\mathbf{C}_{n-1} \cdots \mathbf{C}_3\mathbf{C}_2\mathbf{B}_0 = \mathbf{R}_0. \tag{8}$$

These $\mathbf{C}_j$ are very simple. $\mathbf{C}_j$ has the 2×2 submatrix

$$\begin{bmatrix} \cos\theta_j & \sin\theta_j \\ -\sin\theta_j & \cos\theta_j \end{bmatrix} \qquad (\theta_j \text{ suitable})$$

in rows $j - 1$ and j and columns $j - 1$ and j, entries 1 everywhere else on the main diagonal and all other entries 0. (This submatrix is the matrix of a plane rotation through the angle θ_j see Prob. 17, Sec. 7.3.) For instance, if $n = 4$, writing $c_j = \cos\theta_j$, $s_j = \sin\theta_j$, we have

$$\mathbf{C}_2 = \begin{bmatrix} c_2 & s_2 & 0 & 0 \\ -s_2 & c_2 & 0 & 0 \\ 0 & 0 & 1 & 0 \\ 0 & 0 & 0 & 1 \end{bmatrix}, \qquad \mathbf{C}_3 = \begin{bmatrix} 1 & 0 & 0 & 0 \\ 0 & c_3 & s_3 & 0 \\ 0 & -s_3 & c_3 & 0 \\ 0 & 0 & 0 & 1 \end{bmatrix},$$

$$\mathbf{C}_4 = \begin{bmatrix} 1 & 0 & 0 & 0 \\ 0 & 1 & 0 & 0 \\ 0 & 0 & c_4 & s_4 \\ 0 & 0 & -s_4 & c_4 \end{bmatrix}.$$

These $\mathbf{C}_j$ are orthogonal. Hence their product in (8) is orthogonal, and so is the inverse of this product. We call this inverse $\mathbf{Q}_0$. Then from (8),

$$\mathbf{B}_0 = \mathbf{Q}_0\mathbf{R}_0 \tag{9a}$$

where, with $\mathbf{C}_j^{-1} = \mathbf{C}_j^{\mathsf{T}}$,

$$\mathbf{Q}_0 = (\mathbf{C}_n\mathbf{C}_{n-1} \cdots \mathbf{C}_3\mathbf{C}_2)^{-1} = \mathbf{C}_2^{\mathsf{T}}\mathbf{C}_3^{\mathsf{T}} \cdots \mathbf{C}_{n-1}^{\mathsf{T}}\mathbf{C}_n^{\mathsf{T}}. \tag{9b}$$

This is our QR-factorization of $\mathbf{B}_0$. From it we have by (6) with $s = 0$

$$\mathbf{B}_1 = \mathbf{R}_0\mathbf{Q}_0 = \mathbf{R}_0\mathbf{C}_2^{\mathsf{T}}\mathbf{C}_3^{\mathsf{T}} \cdots \mathbf{C}_{n-1}^{\mathsf{T}}\mathbf{C}_n^{\mathsf{T}}. \tag{10}$$

We do not need $\mathbf{Q}_0$ explicitly, but to get $\mathbf{B}_1$ from (10), we first compute $\mathbf{R}_0\mathbf{C}_2^{\mathsf{T}}$, then $(\mathbf{R}_0\mathbf{C}_2^{\mathsf{T}})\mathbf{C}_3^{\mathsf{T}}$, etc. Similarly in the further steps that produce $\mathbf{B}_2, \mathbf{B}_3, \cdots$.

We show next how to determine $\cos\theta_2$ and $\sin\theta_2$ in $\mathbf{C}_2$ such that $b_{21}^{(2)} = 0$ in the product

$$\mathbf{C}_2\mathbf{B} = \begin{bmatrix} c_2 & s_2 & 0 & \cdots \\ -s_2 & c_2 & 0 & \cdots \\ \cdot & \cdot & \cdot & \cdots \\ \cdot & \cdot & \cdot & \cdots \end{bmatrix} \begin{bmatrix} b_{11} & b_{12} & \cdots \\ b_{21} & b_{22} & \cdots \\ \cdot & \cdot & \cdots \\ \cdot & \cdot & \cdots \end{bmatrix}.$$

Now $b_{21}^{(2)}$ is obtained by multiplying the second row of $\mathbf{C}_2$ by the first column of $\mathbf{B}$, that is,

$$b_{21}^{(2)} = -s_2b_{11} + c_2b_{21} = 0.$$

Hence $\tan\theta_2 = s_2/c_2 = b_{21}/b_{11}$, and

$$\cos\theta_2 = \frac{1}{\sqrt{1 + (b_{21}/b_{11})^2}}, \qquad \sin\theta_2 = \frac{b_{21}/b_{11}}{\sqrt{1 + (b_{21}/b_{11})^2}}. \tag{11}$$

Similarly for $\theta_3, \theta_4, \cdots$.

The next example illustrates all this.

EXAMPLE 2 **QR-factorization method**

Compute all eigenvalues of the matrix

$$\mathbf{A} = \begin{bmatrix} 6 & 4 & 1 & 1 \\ 4 & 6 & 1 & 1 \\ 1 & 1 & 5 & 2 \\ 1 & 1 & 2 & 5 \end{bmatrix}.$$

Solution. We first reduce $\mathbf{A}$ to tridiagonal form. Applying Householder's method, we obtain (see Example 1)

$$\mathbf{A}_2 = \begin{bmatrix} 6 & -\sqrt{18} & 0 & 0 \\ -\sqrt{18} & 7 & \sqrt{2} & 0 \\ 0 & \sqrt{2} & 6 & 0 \\ 0 & 0 & 0 & 3 \end{bmatrix}.$$

From the characteristic determinant we see that $\mathbf{A}_2$, hence $\mathbf{A}$, has the eigenvalue 3. Hence it suffices to apply the QR-method to the 3×3 matrix

$$\mathbf{B}_0 = \mathbf{B} = \begin{bmatrix} 6 & -\sqrt{18} & 0 \\ -\sqrt{18} & 7 & \sqrt{2} \\ 0 & \sqrt{2} & 6 \end{bmatrix}.$$

First step. We multiply **B** by

$$\mathbf{C}_2 = \begin{bmatrix} \cos\theta_2 & \sin\theta_2 & 0 \\ -\sin\theta_2 & \cos\theta_2 & 0 \\ 0 & 0 & 1 \end{bmatrix} \quad \text{and then } \mathbf{C}_2\mathbf{B} \text{ by} \quad \mathbf{C}_3 = \begin{bmatrix} 1 & 0 & 0 \\ 0 & \cos\theta_3 & \sin\theta_3 \\ 0 & -\sin\theta_3 & \cos\theta_3 \end{bmatrix}.$$

Here $(-\sin\theta_2)\cdot 6 + (\cos\theta_2)(-\sqrt{18}) = 0$ gives [see (11)]

$$\cos\theta_2 = 0.816\,496\,58, \qquad \sin\theta_2 = -0.577\,350\,27.$$

With these values, we compute

$$\mathbf{C}_2\mathbf{B} = \begin{bmatrix} 7.348\,469\,23 & -7.505\,553\,50 & -0.816\,496\,58 \\ 0 & 3.265\,986\,32 & 1.154\,700\,54 \\ 0 & 1.414\,213\,56 & 6.000\,000\,00 \end{bmatrix}.$$

In $\mathbf{C}_3$ we get from $(-\sin\theta_3)\cdot 3.265\,986\,32 + (\cos\theta_3)\cdot 1.414\,213\,56 = 0$ the values

$$\cos\theta_3 = 0.917\,662\,94, \qquad \sin\theta_3 = 0.397\,359\,71.$$

This gives

$$\mathbf{R}_0 = \mathbf{C}_3\mathbf{C}_2\mathbf{B} = \begin{bmatrix} 7.348\,469\,23 & -7.505\,553\,50 & -0.816\,496\,58 \\ 0 & 3.559\,026\,08 & 3.443\,784\,13 \\ 0 & 0 & 5.047\,146\,15 \end{bmatrix}.$$

From this we compute

$$\mathbf{B}_1 = \mathbf{R}_0\mathbf{C}_2^{\mathsf{T}}\mathbf{C}_3^{\mathsf{T}} = \begin{bmatrix} 10.333\,333\,33 & -2.054\,804\,67 & 0 \\ -2.054\,804\,67 & 4.035\,087\,72 & 2.005\,532\,51 \\ 0 & 2.005\,532\,51 & 4.631\,578\,95 \end{bmatrix}$$

which is symmetric and tridiagonal. The off-diagonal entries in $\mathbf{B}_1$ are large in absolute value. Hence we have to go on.

Second Step. We do the same computations as in the first step, with $\mathbf{B}_0 = \mathbf{B}$ replaced by $\mathbf{B}_1$. We obtain

$$\mathbf{R}_1 = \begin{bmatrix} 10.535\,653\,75 & -2.802\,322\,42 & -0.391\,145\,88 \\ 0 & 4.083\,295\,83 & 3.988\,240\,28 \\ 0 & 0 & 3.068\,326\,68 \end{bmatrix}$$

and from this

$$\mathbf{B}_2 = \begin{bmatrix} 10.879\,879\,88 & -0.796\,379\,19 & 0 \\ -0.796\,379\,19 & 5.447\,386\,64 & 1.507\,025\,00 \\ 0 & 1.507\,025\,00 & 2.672\,733\,48 \end{bmatrix}.$$

We see that the off-diagonal entries are somewhat smaller in absolute value than those of $\mathbf{B}_1$.

Third Step. We now obtain

$$\mathbf{R}_2 = \begin{bmatrix} 10.908\,987\,40 & -1.191\,925\,04 & -0.110\,016\,02 \\ 0 & 5.581\,996\,06 & 2.168\,774\,94 \\ 0 & 0 & 2.167\,703\,93 \end{bmatrix}$$

and from this

$$\mathbf{B}_3 = \begin{bmatrix} 10.966\,892\,94 & -0.407\,497\,54 & 0 \\ -0.407\,497\,54 & 5.945\,898\,56 & 0.585\,235\,81 \\ 0 & 0.585\,235\,82 & 2.987\,208\,51 \end{bmatrix}.$$

The reader may verify that the eigenvalues of **B** are 11, 6, 2. Hence the given matrix **A** has the spectrum 11, 6, 3, 2. ∎

Looking back at our discussion, we recognize that the purpose of applying Householder's method before the QR-factorization method is a substantial reduction of cost in each QR-factorization.

Convergence acceleration can be achieved by a **spectral shift** (see Sec. 19.6) in each step, that is, instead of $\mathbf{B}_s$ we take $\mathbf{B}_s - k_s\mathbf{I}$ with a suitable k_s. For instance, if we take $k_s = b_{nn}^{(s)}$ (the entry in the lower right corner of $\mathbf{B}_s$), this will generally result in a more rapid decrease to 0 of the other entry in the nth row of $\mathbf{B}_s$, so that we can soon determine an eigenvalue with sufficient accuracy and reduce the size of the matrix in our further work. (See Ref. [E26], p. 510, for a discussion of other choices of k_s.) It is fair to say that the QR-method, preceded by the Householder tridiagonalization, is a general-purpose method for symmetric matrices that is better than any other presently known method.

EXAMPLE 3 Spectral shift in the QR-method

In Example 2 we have $b_{nn} = b_{33} = 6$, so that the spectral shift with $k_0 = b_{nn} = 6$ gives the matrix

$$\hat{\mathbf{B}}_0 = \mathbf{B} - b_{33}\mathbf{I} = \mathbf{B} - 6\mathbf{I} = \begin{bmatrix} 0 & -\sqrt{18} & 0 \\ -\sqrt{18} & 1 & \sqrt{2} \\ 0 & \sqrt{2} & 0 \end{bmatrix}.$$

From this we get $(-\sin\theta_2)\cdot 0 + (\cos\theta_2)(-\sqrt{18}) = 0$, $\cos\theta_2 = 0$, $\sin\theta_2 = 1$ and

$$\mathbf{C}_2\hat{\mathbf{B}}_0 = \begin{bmatrix} -\sqrt{18} & 1 & \sqrt{2} \\ 0 & \sqrt{18} & 0 \\ 0 & \sqrt{2} & 0 \end{bmatrix}.$$

From this we get $(-\sin\theta_3)\sqrt{18} + (\cos\theta_3)\sqrt{2} = 0$, hence

$$\cos\theta_3 = 0.948\,683\,30, \qquad \sin\theta_3 = 0.316\,227\,77$$

and

$$\hat{\mathbf{R}}_0 = \mathbf{C}_3\mathbf{C}_2\hat{\mathbf{B}}_0 = \begin{bmatrix} -4.242\,640\,69 & 1.000\,000\,00 & 1.414\,213\,56 \\ 0 & 4.472\,135\,96 & 0 \\ 0 & 0 & 0 \end{bmatrix}.$$

Hence

$$\hat{\mathbf{B}}_1 = \hat{\mathbf{R}}_0\mathbf{C}_2{}^{\mathsf{T}}\mathbf{C}_3{}^{\mathsf{T}} = \begin{bmatrix} 1.000\,000\,00 & 4.472\,135\,96 & 0 \\ 4.472\,135\,96 & 0 & 0 \\ 0 & 0 & 0 \end{bmatrix}.$$

This shows that 0 is an eigenvalue of $\hat{\mathbf{B}}_0$. Hence 6 is an eigenvalue of **B**, by Theorem 3 in Sec. 19.6.

The result is ***not*** typical. Typically we could only expect that the off-diagonal entry in the nth row becomes rapidly smaller within the next few steps.

That we can now find the other eigenvalues of **B** from a quadratic equation is a consequence of the small size of our matrix, which we picked on purpose—we would not learn more from a monster. In practice, for a larger matrix we would now have to go on with QR (and shifts) as applied to a matrix of order $n - 1$. ∎

Nonsymmetric matrices **A** can also be treated by the QR-method. Instead of reducing **A** to tridiagonal form, we first reduce it to an **upper Hessenberg matrix B,** that is, **B** may have nonzero entries only where an upper triangular matrix or a tridiagonal matrix has them (see Fig. 421). For details, see Refs. [E26] or [E27] listed in Appendix 1.

$$\begin{bmatrix} * & * & * & * & * \\ * & * & * & * & * \\ & * & * & * & * \\ & & * & * & * \\ & & & * & * \end{bmatrix}$$

Fig. 421. A 5 × 5 upper Hessenberg matrix. Positions left blank are zeros.

Problem Set 19.10

Tridiagonalize the following matrices by Householder's method.

1. $\begin{bmatrix} 0 & 1 & 1 \\ 1 & 0 & 1 \\ 1 & 1 & 0 \end{bmatrix}$ **2.** $\begin{bmatrix} 4 & 5 & 9 \\ 5 & 1 & 6 \\ 9 & 6 & 15 \end{bmatrix}$ **3.** $\begin{bmatrix} 1 & 2 & 3 \\ 2 & 4 & 6 \\ 3 & 6 & 9 \end{bmatrix}$

4. $\begin{bmatrix} 2 & -1 & 1 \\ -1 & 3 & 2 \\ 1 & 2 & 3 \end{bmatrix}$ **5.** $\begin{bmatrix} -1 & 2 & 3 \\ 2 & 2 & 6 \\ 3 & 6 & -1 \end{bmatrix}$ **6.** $\begin{bmatrix} 4 & 4 & 1 & 1 \\ 4 & 4 & 1 & 1 \\ 1 & 1 & 3 & 2 \\ 1 & 1 & 2 & 3 \end{bmatrix}$

7. Perform 1 step of the QR-method for the tridiagonal matrix in the answer to Prob. 5.

8. Determine the eigenvalues of the matrix in Prob. 2 by applying the QR-method to the tridiagonalized matrix in the answer to Prob. 2.

9. Perform 1 step of the QR-method (with spectral shift) for the matrix $\hat{\mathbf{B}}_0 = \mathbf{B}_0 - b_{33}\mathbf{I} = \mathbf{B}_0 + 4.461\,538\,46\mathbf{I}$, where $\mathbf{B}_0$ is the matrix in the answer to Prob. 5. Compare with the result in Prob. 7 and comment.

10. Apply another QR-step with spectral shift to the matrix in the answer to Prob. 9. Applying Gerschgorin's theorem to row 3 of the answer and using the symmetry of the matrix, find an inclusion interval and deduce from it that the given matrix **A** in Prob. 5 has an eigenvalue in the interval $-6.08 \leqq \lambda \leqq -5.92$. Verify directly that $\lambda = -6$ is an eigenvalue of **A**.

Review Questions and Problems for Chapter 19

1. What is pivoting? When would you apply it?
2. What happens if you apply Gauss elimination to a system without solution?
3. How would you compute the inverse of a matrix? Is the method good for solving systems of equations? Explain.
4. Is the Gauss–Seidel method a direct or an indirect method? Explain.
5. For what systems would you apply an indirect method of solution?
6. What is the difference between the Gauss–Seidel method and the Jacobi method? Which method is better?
7. What do you know about the convergence of the Gauss–Seidel method?
8. What are vector and matrix norms? How did we use them?
9. What is an ill-conditioned system? What would you do to solve it?
10. Does a small residual indicate great accuracy of an approximate solution? Explain.
11. What is the idea common to Doolittle's and Crout's methods? The difference?
12. When would you apply Cholesky's method?
13. What is the method of least squares? Give a typical example.
14. State Gerschgorin's theorem from memory. Can you remember its proof?
15. Can you remember how we decreased the size of Gerschgorin circles?
16. What is a similarity transformation of a matrix and why is it important in numerical work?
17. What is the power method for eigenvalues? What are its advantages and disadvantages?
18. State Schur's inequality. What is it good for?
19. What is tridiagonalization? When would you apply it?
20. What is the basic idea of the QR-method? When would you apply it?

Solve the following linear systems by the Gauss elimination.

21.
$$\begin{aligned} 2x_1 \phantom{{}-x_2} + 3x_3 &= 15 \\ 4x_2 - x_3 &= -13 \\ 3x_1 - x_2 + 5x_3 &= 26 \end{aligned}$$

22.
$$\begin{aligned} 0.1x_1 + x_2 - 2x_3 &= -0.3 \\ 10x_1 - 3x_2 + 0.5x_3 &= 149.9 \\ -3x_1 + 4x_2 \phantom{{}+0.5x_3} &= -44.2 \end{aligned}$$

23.
$$\begin{aligned} x_1 + x_2 + x_3 &= 5 \\ x_1 + 2x_2 + 2x_3 &= 6 \\ x_1 + 2x_2 + 3x_3 &= 8 \end{aligned}$$

24.
$$\begin{aligned} x_1 + 2x_2 + 4x_3 &= 3.5 \\ 2x_1 + 13x_2 + 23x_3 &= 13.0 \\ 4x_1 + 23x_2 + 77x_3 &= 6.0 \end{aligned}$$

25.
$$\begin{aligned} 2x_1 + x_2 + 2x_3 &= 5.6 \\ 8x_1 + 5x_2 + 13x_3 &= 20.9 \\ 6x_1 + 3x_2 + 12x_3 &= 11.4 \end{aligned}$$

26.
$$\begin{aligned} 4x_2 - 3x_3 &= 11.8 \\ 5x_1 + 3x_2 + x_3 &= 34.2 \\ 6x_1 - 7x_2 + 2x_3 &= -3.1 \end{aligned}$$

27. Solve Prob. 23 by Doolittle's method.
28. Solve Prob. 25 by Doolittle's method.
29. Solve Prob. 23 by Cholesky's method.
30. Solve Prob. 24 by Cholesky's method.

Compute the inverse:

31. $\begin{bmatrix} -0.1 & 0.1 & 0.2 \\ 0.3 & -0.1 & 0.1 \\ -0.1 & 0.3 & 0.4 \end{bmatrix}$ **32.** $\begin{bmatrix} 1 & 2 & 0.5 \\ 0.5 & 1 & 0.5 \\ 1.5 & 2 & 1 \end{bmatrix}$ **33.** $\begin{bmatrix} 1.4 & -0.4 & -0.6 \\ 2.6 & 0.4 & -1.4 \\ -1.6 & -0.4 & 0.4 \end{bmatrix}$

Apply the Gauss–Seidel iteration to the following systems. Perform three steps, starting from 1, 1, 1.

34. $$\begin{aligned} 10x_1 - x_2 + 2x_3 &= 38 \\ -x_1 + 15x_2 + 3x_3 &= 26 \\ 2x_1 + 3x_2 + 12x_3 &= 14 \end{aligned}$$

35. $$\begin{aligned} x_1 + 15x_2 - x_3 &= 11 \\ 10x_1 + 3x_2 &= -17 \\ 2x_1 - x_2 + 5x_3 &= 5 \end{aligned}$$

36. $$\begin{aligned} 4x_1 - x_2 &= 5.5 \\ 4x_2 - x_3 &= 0.4 \\ -x_1 + 4x_3 &= 11.2 \end{aligned}$$

Compute the l_1-, l_2-, and l_∞-norms of the vectors

37. $[0 \quad 4 \quad -8 \quad 3]^T$ **38.** $[3 \quad 8 \quad 11]^T$ **39.** $[-4 \quad 1 \quad 0 \quad 2]^T$

40. $[0 \quad 0 \quad 1 \quad 0]^T$ **41.** $[5 \quad -2 \quad 7 \quad 0 \quad -8]^T$ **42.** $[0.3 \; 1.4 \; 0.2 \; -0.6]^T$

Compute the matrix norm corresponding to the l_∞-vector norm, for the coefficient matrix of the system of equations:

43. In Prob. 24 **44.** In Prob. 25 **45.** In Prob. 26

Compute the condition number (corresponding to the l_1-vector norm) of the matrix:

46. In Prob. 32 **47.** In Prob. 31 **48.** In Prob. 33

49. Fit a straight line by least squares to the data (−2, 0.1), (0, 1.9), (2, 3.8), (4, 6.1), (6, 7.8).

50. Fit a quadratic parabola by least squares to the data (1, 9), (2, 5), (3, 4) (4, 5) (5, 7).

For each matrix find three circular disks that must contain all the eigenvalues.

51. $\begin{bmatrix} 11.4 & 2.0 & 0.6 \\ 2.0 & 14.4 & 1.2 \\ 0.6 & 1.2 & 14.6 \end{bmatrix}$ **52.** $\begin{bmatrix} 5 & 1 & 1 \\ 1 & 6 & 0 \\ 1 & 0 & 8 \end{bmatrix}$ **53.** $\begin{bmatrix} 10 & 0.1 & 0.1 \\ 0.1 & 4 & 0.2 \\ 0.1 & 0.2 & 3 \end{bmatrix}$

54. Apply Wielandt's method of deflation to the matrix in Prob. 51, using that 16.4 is an eigenvalue.

55. Apply 4 steps of the power method to the matrix in Prob. 52, starting from $[1 \quad 1 \quad 1]^T$ and computing the Rayleigh quotients and error bounds.

Summary of Chapter 19 Numerical Methods in Linear Algebra

This chapter deals with three large classes of numerical problems arising from linear algebra, namely, the numerical solution of linear systems (Secs. 19.1–19.4), the fitting of straight lines or parabolas through given data (represented as points in the plane; Sec. 19.5), and the numerical solution of eigenvalue problems (Secs. 19.6–19.10).

Thus, in Secs. 19.1–19.4 we solve systems $\mathbf{Ax} = \mathbf{b}$, where $\mathbf{A} = [a_{jk}]$ is an $n \times n$ matrix, written out

$$
\begin{array}{ll}
E_1: & a_{11}x_1 + \cdots + a_{1n}x_n = b_1 \\
E_2: & a_{21}x_1 + \cdots + a_{2n}x_n = b_2 \\
\multicolumn{2}{c}{\cdots\cdots\cdots\cdots\cdots\cdots} \\
E_n: & a_{n1}x_1 + \cdots + a_{nn}x_n = b_n.
\end{array}
\tag{1}
$$

There are two types of numerical methods for this task, **direct methods,** such as the Gauss elimination, in which the amount of computation to get a solution can be specified in advance, and **indirect** or **iterative methods,** such as the Gauss–Seidel method, in which we start from a (possibly crude) approximation and improve it stepwise by repeatedly performing the same cycle of computation, with changing data (see below).

The **Gauss elimination** (Sec. 19.1) is a systematic elimination process that reduces (1) stepwise to triangular form. In step 1 we eliminate x_1 from equations E_2 to E_n by subtracting $(a_{21}/a_{11})E_1$ from E_2, then $(a_{31}/a_{11})E_1$ from E_3, etc. Equation E_1 is called the **pivot equation** in this step and a_{11} the **pivot.** In step 2 we take the new second equation as pivot equation and eliminate x_2, etc. If the triangular form is reached, we get x_n from the last equation, then x_{n-1} from the second last, etc. **Partial pivoting** (= interchange of equations) is *necessary* if pivots are zero, and *advisable* if they are small.

Variants of the Gauss elimination (**Doolittle's, Crout's, Cholesky's methods,** Sec. 19.2) use the idea of factoring

$$\mathbf{A} = \mathbf{LU} \tag{2}$$

(**L** lower triangular, **U** upper triangular), setting $\mathbf{Ux} = \mathbf{y}$ and then solving $\mathbf{Ax} = \mathbf{LUx} = \mathbf{Ly} = \mathbf{b}$ by first solving the triangular system $\mathbf{Ly} = \mathbf{b}$ for $\mathbf{y}$ and then the triangular system $\mathbf{Ux} = \mathbf{y}$ for $\mathbf{x}$.

Iterative methods are obtained by writing $\mathbf{Ax} = \mathbf{b}$ as

$$\mathbf{x} = \mathbf{b} - (\mathbf{A} - \mathbf{I})\mathbf{x} \tag{3}$$

and then substituting approximations on the right to get new approximations on the left. In particular, in the **Gauss–Seidel method** we first divide each equation to make $a_{11} = a_{22} = \cdots = a_{nn} = 1$, then write $\mathbf{A} = \mathbf{I} + \mathbf{L} + \mathbf{U}$, so that (3) becomes

$$\mathbf{x} = \mathbf{b} - \mathbf{Lx} - \mathbf{Ux}$$

and always take the most recent approximate x_j's on the right. This gives the iteration formula (Sec. 19.3)

$$\mathbf{x}^{(m+1)} = \mathbf{b} - \mathbf{L}\mathbf{x}^{(m+1)} - \mathbf{U}\mathbf{x}^{(m)}. \tag{4}$$

If $\|\mathbf{C}\| < 1$, where

$$\mathbf{C} = -(\mathbf{I} + \mathbf{L})^{-1}\mathbf{U},$$

then this process converges. Here, $\|\mathbf{C}\|$ denotes any matrix norm (Sec. 19.3).

If the **condition number**

$$\kappa(\mathbf{A}) = \|\mathbf{A}\|\,\|\mathbf{A}^{-1}\|$$

of $\mathbf{A}$ is large, then the system $\mathbf{Ax} = \mathbf{b}$ is **ill-conditioned** (Sec. 19.4), and a small **residual**

$$\mathbf{r} = \mathbf{b} - \mathbf{A}\tilde{\mathbf{x}}$$

does not imply that $\tilde{\mathbf{x}}$ is close to the exact solution.

The fitting of a polynomial $p(x) = b_0 + b_1x + \cdots + b_mx^m$ through given data (points in the xy-plane) $(x_1, y_1), \cdots, (x_n, y_n)$ by the method of **least squares** is discussed in Sec. 19.5.

An **eigenvalue** of an $n \times n$ matrix $\mathbf{A} = [a_{jk}]$ is a number λ such that

$$\mathbf{Ax} = \lambda\mathbf{x}, \qquad \text{thus} \qquad (\mathbf{A} - \lambda\mathbf{I})\mathbf{x} = \mathbf{0} \tag{5}$$

has a solution $\mathbf{x} \neq \mathbf{0}$, called an **eigenvector** of **A** corresponding to that λ. In Sec. 19.6 we list basic facts about the eigenvalue problem needed in numerical methods.

Section 19.7 concerns theorems that give **inclusion sets** (sets in the complex λ-plane that contain one or several eigenvalues of **A**), notably the famous **Gerschgorin theorem,** which states that the **spectrum** (= set of all eigenvalues) of **A** lies in the n disks

$$|a_{jj} - \lambda| \leqq \sum_{\substack{k=1 \\ k \neq j}}^{n} |a_{jk}| \qquad (j = 1, \cdots, n) \tag{6}$$

whose centers are $a_{11}, a_{22}, \cdots, a_{nn}$ and whose radii are given by the sums on the right.

The **power method** (Sec. 19.8) gives approximations

$$q = \frac{(\mathbf{Ax})^{\mathsf{T}}\mathbf{x}}{\mathbf{x}^{\mathsf{T}}\mathbf{x}} \qquad \textit{(Rayleigh quotient)}, \tag{7}$$

usually to the eigenvalue that is largest in absolute value, and, if **A** is symmetric, error bounds

$$|\epsilon| \leqq \sqrt{\frac{(\mathbf{Ax})^{\mathsf{T}}\mathbf{Ax}}{\mathbf{x}^{\mathsf{T}}\mathbf{x}} - q^2}. \tag{8}$$

Practically, one chooses any vector $\mathbf{x}_0 \neq \mathbf{0}$, computes the vectors

$$\mathbf{x}_1 = \mathbf{Ax}_0, \quad \mathbf{x}_2 = \mathbf{Ax}_1, \quad \cdots, \quad \mathbf{x}_s = \mathbf{Ax}_{s-1},$$

and takes

$$\mathbf{x} = \mathbf{x}_{s-1} \qquad \text{and} \qquad \mathbf{Ax} = \mathbf{x}_s$$

in (7) and (8).

If we know an eigenvalue λ_1 of **A**, we can **deflate A,** that is, compute a matrix $\mathbf{A}_1$ with eigenvalues $0, \lambda_2, \cdots, \lambda_n$ (Sec. 19.9).

If we want to know all eigenvalues of a symmetric matrix, then better than repeated deflation is the **QR-method** preceded by **Householder's tridiagonalization** (Sec. 19.10). The QR-method is based on a factorization $\mathbf{A} = \mathbf{QR}$, where **Q** is orthogonal and **R** is upper triangular, and similarity transformations.

Chapter 20

Numerical Methods for Differential Equations

Numerical methods for differential equations are of great importance to the engineer and physicist because practical problems often lead to differential equations that cannot be solved by one of the methods in Chaps. 1–6 or similar methods, or to equations for which the solutions in terms of formulas are so complicated that one often prefers to compute a table of values by applying a numerical method to such an equation.

The present chapter includes basic methods for the numerical solution of ordinary differential equations (Secs. 20.1–20.3) and partial differential equations (Secs. 20.4–20.7).

Sections 20.1, 20.2, and 20.3 may also be studied immediately after Chaps. 1 and 2, respectively, since they are independent of Chaps. 18 and 19.

Sections 20.4–20.7 may also be studied immediately after Chap. 11, provided the reader has some knowledge of linear systems of algebraic equations.

Prerequisite for Secs. 20.1–20.3: Secs. 1.1–1.7, 2.1–2.3.
Prerequisite for Secs. 20.4–20.7: Secs. 11.1–11.3, 11.5, 11.11.
References: Appendix 1, Part E (see also Parts A and C).
Answers to problems: Appendix 2.

20.1 Methods for First-Order Differential Equations

From Chap. 1 we know that a *differential equation of the first order* is of the form $F(x, y, y') = 0$, and often it will be possible to write the equation in the *explicit form* $y' = f(x, y)$. An **initial value problem** consists of a differential equation and a condition the solution must satisfy (or several conditions referring to the same value of x if the equation is of higher order). In this section we shall consider initial value problems of the form

$$y' = f(x, y), \qquad y(x_0) = y_0 \tag{1}$$

assuming f to be such that the problem has a unique solution on some interval containing x_0.

We shall discuss methods for computing numerical values of the solutions, which are needed if a formula for the solution of an equation is not available or is too complicated to be of practical use.

These methods are **step-by-step methods,** that is, we start from the given $y_0 = y(x_0)$ and proceed stepwise, computing approximate values of the solution $y(x)$ at the *"mesh points"*

$$x_1 = x_0 + h, \qquad x_2 = x_0 + 2h, \qquad x_3 = x_0 + 3h, \qquad \cdots,$$

where the **step size** h is a fixed number, for instance 0.2 or 0.1 or 0.01, whose choice we discuss later in this section.

The computation in each step is done by the same formula. Such formulas are suggested by the Taylor series

$$y(x + h) = y(x) + hy'(x) + \frac{h^2}{2}y''(x) + \cdots. \tag{2}$$

Now for a small value of h, the higher powers h^2, h^3, $\cdots$ are very small. This suggests the crude approximation

$$y(x + h) \approx y(x) + hy'(x) = y(x) + hf(x, y)$$

(with the right side obtained from the given differential equation) and the following iteration process. In the first step we compute

$$y_1 = y_0 + hf(x_0, y_0)$$

which approximates $y(x_1) = y(x_0 + h)$. In the second step we compute

$$y_2 = y_1 + hf(x_1, y_1)$$

which approximates $y(x_2) = y(x_0 + 2h)$, etc., and in general

(3)
$$\boxed{y_{n+1} = y_n + hf(x_n, y_n)} \qquad (n = 0, 1, \cdots).$$

This is called the **Euler method** or **Euler–Cauchy method.** Geometrically it is an approximation of the curve of $y(x)$ by a polygon whose first side is tangent to the curve at x_0 (see Fig. 422).

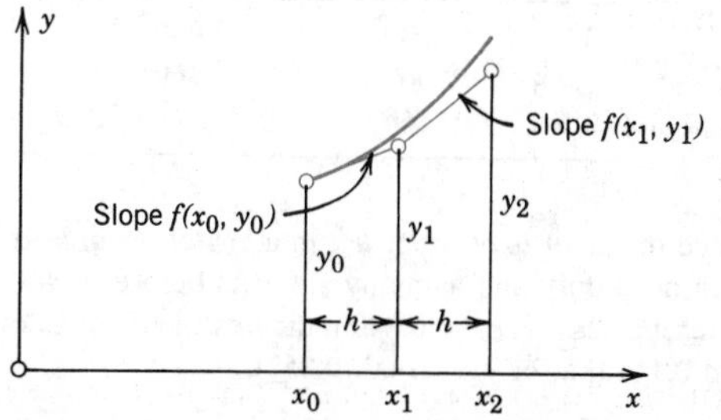

Fig. 422. Euler method

This crude method is hardly ever used in practice, but since it is simple, it nicely explains the principle of methods based on the Taylor series.

Euler's method is called a **first-order method,** because in (2) we take only the constant terms and the term containing the first power of h. The omission of the further terms in (2) causes an error, which is called the **truncation error** of the method. For small h, the third and higher powers of h will be small compared with h^2 in the first neglected term in (2), and we therefore say that the **truncation error per step** (or *local truncation error*) is of **order** h^2. In addition there are **round-off errors** in this and other methods, which may affect the accuracy of the values $y_1, y_2, \cdots$ more and more as n increases; we shall return to this point in the next section.

Table 20.1
Euler Method Applied to (4) in Example 1 and Error

n	x_n	y_n	$0.2(x_n + y_n)$	Exact Values	Error
0	0.0	0.000	0.000	0.000	0.000
1	0.2	0.000	0.040	0.021	0.021
2	0.4	0.040	0.088	0.092	0.052
3	0.6	0.128	0.146	0.222	0.094
4	0.8	0.274	0.215	0.426	0.152
5	1.0	0.489		0.718	0.229

EXAMPLE 1 Euler method

Apply the Euler method to the following initial value problem, choosing $h = 0.2$ and computing $y_1, \cdots, y_5$:

$$y' = x + y, \qquad y(0) = 0. \tag{4}$$

Solution. Here $f(x, y) = x + y$, and we see that (3) becomes

$$y_{n+1} = y_n + 0.2(x_n + y_n).$$

Table 20.1 shows the computations, the values of the exact solution

$$y(x) = e^x - x - 1$$

obtained from (4) in Sec. 1.7, and the error. In practice the exact solution is unknown, but an indication of the accuracy of the values can be obtained by applying the Euler method once more with step $2h = 0.4$ and comparing corresponding approximations. This computation is:

x_n	y_n	$0.4(x_n + y_n)$	y_n in Table 20.1	Difference
0.0	0.000	0.000	0.000	0.000
0.4	0.000	0.160	0.040	0.040
0.8	0.160		0.274	0.114

Since the error is of order h^2, in a switch from h to $2h$ it is multiplied by $2^2 = 4$, but since we then need only half as many steps as before, it will only be multiplied by $4/2 = 2$. Hence the difference $2\epsilon_2 - \epsilon_2 = 0.040$ indicates the error ϵ_2 of y_2 in Table 20.1 (which actually is 0.052), and 0.114 that of y_4 (actual: 0.152). ∎

Improved Euler Method (Heun's Method)

By taking more terms in (2) into account we obtain numerical methods of higher order and precision. But there is a practical problem. If we substitute $y' = f(x, y(x))$ into (2), we have

$$y(x + h) = y(x) + hf + \tfrac{1}{2}h^2f' + \tfrac{1}{6}h^3f'' + \cdots \tag{2*}$$

where, since y in f depends on x,

$$f' = f_x + f_y y' = f_x + f_y f$$

and the further derivatives f'', f''' become even much more cumbersome. The ***general strategy*** now is to avoid their computation and replace it by computing f for one or several suitably chosen auxiliary values of (x, y), where "suitably" means that they are chosen to make the order of the method as high as possible (to have high accuracy). Let us discuss two such methods that are of practical importance.

The first method is the so-called **improved Euler method** or **improved Euler–Cauchy method** (sometimes also called **Heun's method**). In each step of this method we compute first the auxiliary value

(5a) $$y^*_{n+1} = y_n + hf(x_n, y_n)$$

and then the new value

(5b) $$y_{n+1} = y_n + \tfrac{1}{2}h[f(x_n, y_n) + f(x_{n+1}, y^*_{n+1})].$$

This method has a simple geometric interpretation. In fact, we may say that in the interval from x_n to $x_n + \frac{1}{2}h$ we approximate the solution y by the straight line through (x_n, y_n) with slope $f(x_n, y_n)$, and then we continue along the straight line with slope $f(x_{n+1}, y^*_{n+1})$ until x reaches x_{n+1} (see Fig. 423, where $n = 0$).

The improved Euler–Cauchy method is a **predictor–corrector method,** because in each step we first predict a value by (5a) and then correct it by (5b).

In algorithmic form, using the notations $k_1 = hf(x_n, y_n)$ in (5a) and $k_2 = hf(x_{n+1}, y^*_{n+1})$ in (5b) we can write this method as shown in Table 20.2 on the next page.

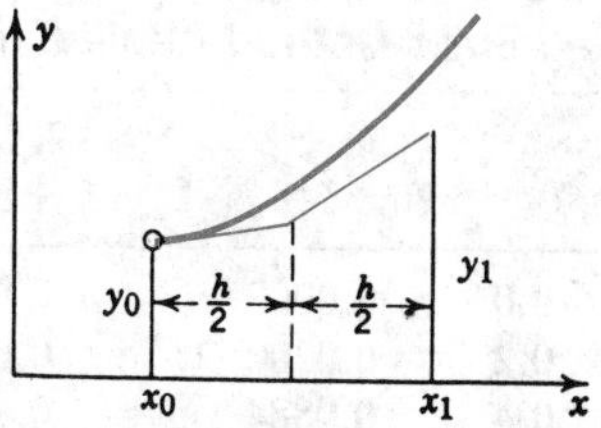

Fig. 423. Improved Euler method

Table 20.2
Improved Euler Method (Heun's Method)

ALGORITHM EULER (f, x_0, y_0, h, N)

This algorithm computes the solution of the initial value problem $y' = f(x, y)$, $y(x_0) = y_0$ at equidistant points $x_1 = x_0 + h$, $x_2 = x_0 + 2h, \cdots, x_N = x_0 + Nh$; here f is such that this problem has a unique solution on the interval $[x_0, x_N]$ (see Sec. 1.11).

INPUT: Initial values x_0, y_0, step size h, number of steps N

OUTPUT: Approximation y_{n+1} to the solution $y(x_{n+1})$ at $x_{n+1} = x_0 + (n + 1)h$, where $n = 0, \cdots, N - 1$

For $n = 0, 1, \cdots, N - 1$ do:

$x_{n+1} = x_n + h$

$k_1 = hf(x_n, y_n)$

$k_2 = hf(x_{n+1}, y_n + k_1)$

$y_{n+1} = y_n + \frac{1}{2}(k_1 + k_2)$

OUTPUT x_{n+1}, y_{n+1}

End

Stop

End EULER

EXAMPLE 2 **Improved Euler method**

Apply the improved Euler method to the initial value problem (4), choosing $h = 0.2$, as before.

Solution. For the present problem,

$$k_1 = 0.2(x_n + y_n)$$

$$k_2 = 0.2(x_n + 0.2 + y_n + 0.2(x_n + y_n))$$

$$y_{n+1} = y_n + \frac{0.2}{2}(2.2x_n + 2.2y_n + 0.2)$$

$$= y_n + 0.22(x_n + y_n) + 0.02.$$

Table 20.3 shows that our present results are more accurate than those in Example 1; see also Table 20.6 on p. 1041. ∎

Table 20.3
Improved Euler Method Applied to (4) and Error

n	x_n	y_n	$0.22(x_n + y_n) + 0.02$	Exact Values	Error
0	0.0	0.0000	0.0200	0.0000	0.0000
1	0.2	0.0200	0.0684	0.0214	0.0014
2	0.4	0.0884	0.1274	0.0918	0.0034
3	0.6	0.2158	0.1995	0.2221	0.0063
4	0.8	0.4153	0.2874	0.4255	0.0102
5	1.0	0.7027		0.7183	0.0156

The improved Euler method is a **second-order method,** *because the truncation error per step is of order* h^3.

Proof. Setting $\tilde{f}_n = f(x_n, y(x_n))$ and using (2*), we have

$$\text{(6a)} \qquad y(x_n + h) - y(x_n) = h\tilde{f}_n + \tfrac{1}{2}h^2\tilde{f}'_n + \tfrac{1}{6}h^3\tilde{f}''_n + \cdots .$$

Approximating the expression in the brackets in (5b) by $\tilde{f}_n + \tilde{f}_{n+1}$ and again using the Taylor expansion, we obtain from (5b)

$$\text{(6b)} \qquad \begin{aligned} y_{n+1} - y_n &\approx \tfrac{1}{2}h[\tilde{f}_n + \tilde{f}_{n+1}] \\ &= \tfrac{1}{2}h[\tilde{f}_n + (\tilde{f}_n + h\tilde{f}'_n + \tfrac{1}{2}h^2\tilde{f}''_n + \cdots)]. \end{aligned}$$

Subtraction of (6a) from (6b) gives the truncation error per step

$$\frac{h^3}{4}\tilde{f}''_n - \frac{h^3}{6}\tilde{f}''_n + \cdots = \frac{h^3}{12}\tilde{f}''_n + \cdots .$$

This proves the assertion. ∎

Choice of step size. This is an important matter in any step-by-step method. h should not be too small, to avoid excessively many steps and corresponding round-off error accumulation. But h should not be too large either, to avoid a large truncation error per step and an additional error, call it φ_n, caused by the evaluation of f at (x_n, y_n) instead of $(x_n, y(x_n))$. Now φ_n would be zero if f were independent of y; thus it will matter the more the faster f varies with y, that is, the larger the absolute value of the partial derivative $f_y = \partial f/\partial y$ is. More precisely, by the definition of φ_n and the mean value theorem we get

$$\varphi_n = f(x_n, y(x_n)) - f(x_n, y_n) = f_y(x_n, \tilde{y})\eta_n$$

where $\eta_n = y(x_n) - y_n$ is the error of y_n and $\tilde{y}$ lies between $y(x_n)$ and y_n. Hence the contribution of φ_n to the error of y_{n+1} is approximately $h\varphi_n = hf_y(x_n, \tilde{y})\eta_n$. This suggests to take a close upper bound K of $|f_y|$ in the region of interest and to choose h such that

$$\kappa = hK$$

is not too large. We see that if $|f_y|$ is large (strong dependence of f on y), then K is large and h must be small, which is understandable. (In Examples 1 and 2, $f_y = 1$, $K = 1$, $hK = 0.2$.) If f_y varies very much, we may choose a close upper bound K_n of $|f_y(x_n, \tilde{y})|$ and choose two or even three different values of h in different regions, to keep

$$\kappa_n = hK_n$$

within a certain interval (for example, $0.05 \leqq \kappa_n \leqq 0.1$), which depends on the desired accuracy; of course, because of the truncation error per step, we cannot let h increase beyond a certain value.

Runge–Kutta Method

A still more accurate method of great practical importance is the **Runge–Kutta method,**[1] shown in Table 20.4. We see that in each step we first compute four auxiliary quantities k_1, k_2, k_3, k_4 and then the new value y_{n+1}. These formulas look complicated at first sight, but they are in fact very easy to program.

It can be shown that the truncation error per step is of the order h^5 (see Ref. [E2] in Appendix 1) and the method is, therefore, a fourth-order method.

Table 20.4
Runge–Kutta Method (of Fourth Order)

ALGORITHM RUNGE–KUTTA (f, x_0, y_0, h, N).

This algorithm computes the solution of the initial value problem $y' = f(x, y)$, $y(x_0) = y_0$ at equidistant points

$$x_1 = x_0 + h, \ x_2 = x_0 + 2h, \cdots, x_N = x_0 + Nh;$$

here f is such that this problem has a unique solution on the interval $[x_0, x_N]$ (see Sec. 1.11).

INPUT: Initial values x_0, y_0, step size h, number of steps N

OUTPUT: Approximation y_{n+1} to the solution $y(x_{n+1})$ at $x_{n+1} = x_0 + (n + 1)h$, where $n = 0, 1, \cdots, N - 1$

For $n = 0, 1, \cdots, N - 1$ do:

$$k_1 = hf(x_n, y_n)$$

$$k_2 = hf(x_n + \tfrac{1}{2}h, y_n + \tfrac{1}{2}k_1)$$

$$k_3 = hf(x_n + \tfrac{1}{2}h, y_n + \tfrac{1}{2}k_2)$$

$$k_4 = hf(x_n + h, y_n + k_3)$$

$$x_{n+1} = x_n + h$$

$$y_{n+1} = y_n + \tfrac{1}{6}(k_1 + 2k_2 + 2k_3 + k_4)$$

OUTPUT x_{n+1}, y_{n+1}

End

Stop

End RUNGE–KUTTA

---

[1]Named after the German mathematicians CARL DAVID TOLMÉ RUNGE (1856—1927), professor of applied mathematics at Göttingen, and WILHELM KUTTA (1867—1944).

Without much historical justification, the improved Euler method is sometimes called a **second-order Runge–Kutta method.** The Runge–Kutta method discussed here is often called the **fourth-order Runge–Kutta method** because there are Runge–Kutta methods of still higher order based on the same principle of replacing the computation of derivatives by the computation of auxiliary values (values of f at certain points). For details, see Ref. [E21] in Appendix 1.

Note that if f depends only on x, this method reduces to Simpson's rule of integration (Sec. 18.5).

In hand calculations, the frequent calculation of $f(x, y)$ is laborious. On a computer this does not matter too much, and the method is well suited because it needs no special starting procedure, makes light demands on storage, requires no estimation, and repeatedly uses the same straightforward computational procedure.

EXAMPLE 3 Runge–Kutta method

Apply the Runge–Kutta method to the initial value problem (4) in Example 1, choosing $h = 0.2$, as before, and computing five steps.

Solution. For the present problem we have $f(x, y) = x + y$. Hence

$$k_1 = 0.2(x_n + y_n), \qquad k_2 = 0.2(x_n + 0.1 + y_n + 0.5k_1),$$

$$k_3 = 0.2(x_n + 0.1 + y_n + 0.5k_2), \qquad k_4 = 0.2(x_n + 0.2 + y_n + k_3).$$

Since these expressions are so simple, we find it convenient to insert k_1 into k_2, obtaining $k_2 = 0.22(x_n + y_n) + 0.02$, insert this into k_3, finding $k_3 = 0.222(x_n + y_n) + 0.022$, and finally insert this into k_4, finding $k_4 = 0.2444(x_n + y_n) + 0.0444$. If we use these expressions, the formula for y_{n+1} in Table 20.4 becomes

$$y_{n+1} = y_n + 0.2214(x_n + y_n) + 0.0214. \tag{7}$$

Of course, our present inserting process is ***not typical*** of the Runge–Kutta method and should not be tried in general. Table 20.5 shows the computations. From Table 20.6 we see that the values are much more accurate than those in Examples 1 and 2. ■

Table 20.5
Runge–Kutta Method Applied to (4); Computations by the Use of (7)

n	x_n	y_n	$0.2214(x_n + y_n) + 0.0214$	Exact Values $y = e^x - x - 1$	$10^6 \times$ Error of y_n
0	0.0	0	0.021 400	0.000 000	0
1	0.2	0.021 400	0.070 418	0.021 403	3
2	0.4	0.091 818	0.130 289	0.091 825	7
3	0.6	0.222 107	0.203 414	0.222 119	11
4	0.8	0.425 521	0.292 730	0.425 541	20
5	1.0	0.718 251		0.718 282	31

Table 20.6
Comparison of the Accuracy of the Three Methods Under Consideration in the Case of the Initial Value Problem (4), with h = 0.2

x	$y = e^x - x - 1$	Error		
		Euler Method (Table 20.1)	Improved Euler (Table 20.3)	Runge–Kutta (Table 20.5)
0.2	0.021 403	0.021	0.0014	0.000 003
0.4	0.091 825	0.052	0.0034	0.000 007
0.6	0.222 119	0.094	0.0063	0.000 011
0.8	0.425 541	0.152	0.0102	0.000 020
1.0	0.718 282	0.229	0.0156	0.000 031

Step size h. The step size h should not be greater than a certain value H, which depends on the desired accuracy, and should otherwise be such that

$$\kappa = hK \qquad (K \text{ a close upper bound for } |\partial f/\partial y|)$$

lies about between 0.01 and 0.05; this is similar to the case of the improved Euler method discussed before.

It is an advantage of the Runge–Kutta method that we may **control *h*** by means of k_1, k_2, k_3, because from the definition of f_y we have

$$\kappa = hK \approx h|f_y| \approx h\left|\frac{f(x, y^*) - f(x, y^{**})}{y^* - y^{**}}\right|,$$

and if in the numerator we choose

$$hf(x, y^*) = k_3 = hf(x_n + \tfrac{1}{2}h, y_n + \tfrac{1}{2}k_2), \quad \text{thus,} \quad y^* = y_n + \tfrac{1}{2}k_2,$$

$$hf(x, y^{**}) = k_2 = hf(x_n + \tfrac{1}{2}h, y_n + \tfrac{1}{2}k_1), \quad \text{thus,} \quad y^{**} = y_n + \tfrac{1}{2}k_1,$$

we have in the denominator $y^* - y^{**} = \frac{1}{2}(k_2 - k_1)$ and get the desired formula for κ in terms of computed quantities,

(8) $$\kappa \approx 2\left|\frac{k_3 - k_2}{k_2 - k_1}\right|.$$

We may now make provision to leave h unchanged if, say, $0.01 \leqq \kappa_n \leqq 0.05$, to decrease h by 50% if $\kappa_n > 0.05$, and to double h if $\kappa_n < 0.01$ (if doubling is possible without increasing h beyond a suitably chosen number H, which depends on the desired accuracy).

Another control of h results from performing the computation simultaneously with step $2h$, which corresponds to increasing the truncation error per step by a factor $2^5 = 32$, but since the number of steps decreases, the actual increase is by a factor $2^5/2 = 16$. Hence the error ϵ of an approximation $\tilde{y}$ obtained with step h equals about 1/15 times the difference $\delta = \tilde{y} - \tilde{\tilde{y}}$ of corresponding approximations obtained with steps h and $2h$, respectively,

(9) $$\epsilon \approx \frac{1}{15}(\tilde{y} - \tilde{\tilde{y}}).$$

We may now choose a number ϵ (for example, 1 unit of the last digit that is supposed to be significant) and leave h unchanged if $0.2\epsilon \leqq |\delta| \leqq 10\epsilon$, decrease h by 50% if $|\delta| > 10\epsilon$, and double h if $|\delta| < 0.2\epsilon$; of course, in doubling we must take care that the step does not become larger than a suitable number H; this is as before.

Let us illustrate the error estimate (9) by a simple example.

EXAMPLE 4 Runge–Kutta method (of fourth order), error estimate
Solve the initial value problem

$$y' = (y - x - 1)^2 + 2, \qquad y(0) = 1$$

by the Runge–Kutta method for $0 \leqq x \leqq 0.4$ with step $h = 0.1$ and estimate the error by (9).

Solution. The numerical results are shown in Table 20.7. They also illustrate how the accuracy increases with decreasing step (from $2h = 0.2$ to $h = 0.1$). The error estimates (9) are close to the actual error. Although we cannot always expect this, formula (9) will certainly give information about the order of magnitude of the error. ∎

It can be shown that the methods discussed in this section are *numerically stable* (definition in Sec. 18.1). They are **one-step methods** because in each step we use the data of just one preceding step, in contrast to **multistep methods,** which in each step use data from several preceding steps, as we shall see in the next section.

Table 20.7
Runge–Kutta Method Applied to the Initial Value Problem in Example 4 and Error Estimate

x	$\tilde{y}$ (Step h)	$\tilde{\tilde{y}}$ (Step $2h$)	Error Estimate (9)	Actual Error	Exact Solution (9D)
0.0	1.000 000 000	1.000 000 000	0.000 000 000	0.000 000 000	1.000 000 000
0.1	1.200 334 589			0.000 000 083	1.200 334 672
0.2	1.402 709 878	1.402 707 341	0.000 000 181	0.000 000 157	1.402 710 036
0.3	1.609 336 039			0.000 000 210	1.609 336 250
0.4	1.822 792 993	1.822 788 917	0.000 000 291	0.000 000 226	1.822 793 219

Problem Set 20.1

Apply the Euler method to the following initial value problems. Do 10 steps. Solve the problem exactly. Compute the errors.

1. $y' = y,\ y(0) = 1, \quad h = 0.1$
2. $y' = y,\ y(0) = 1, \quad h = 0.01$
3. $y' + 5x^4y^2 = 0,\ y(0) = 1, \quad h = 0.1$
4. $y' = (y + x)^2,\ y(0) = 0, \quad h = 0.1$

Apply the improved Euler method to the following initial value problems. Do 10 steps with $h = 0.1$. Solve the problem exactly. Compute the errors.

5. $y' = y, \quad y(0) = 1$
6. $y' = 1 + y^2, \quad y(0) = 0$
7. $y' + y \tan x = \sin 2x, \quad y(0) = 1$
8. $y' = y - y^2, \quad y(0) = 0.5$

Apply the Runge–Kutta method (of fourth order) to the following initial value problems. Do 5 steps with $h = 0.2$. Solve the problem exactly. Compute the errors.

9. $y' = xy, \quad y(0) = 1$
10. $y' = y - y^2, \quad y(0) = 0.5$
11. $y' = (1 + x^{-1})y, \quad y(1) = e$
12. $y' = \frac{1}{2}(y/x - x/y), \quad y(2) = 2$

13. Apply the Euler method and the improved Euler method with $h = 0.1$ and 10 steps to the initial value problem $y' = 2 - 2y$, $y(0) = 0$, and determine and compare the errors.

14. Apply the Runge–Kutta method with $h = 0.1$ and 10 steps to Prob. 13 and determine and compare the errors with those in Prob. 13.

15. Solve $y' = 2x^{-1}\sqrt{y - \ln x} + x^{-1}$, $y(1) = 0$ for $1 \leqq x \leqq 1.8$ by Euler's method with $h = 0.1$. Verify that the exact solution is $y = (\ln x)^2 + \ln x$ and compute the error.

16. Solve Prob. 15 by the improved Euler method with $h = 0.2$, determine the error, compare with Prob. 15, and comment. Note that this is a fair comparison because here we evaluate $f(x, y)$ eight times (4 steps with 2 evaluations each), just as in Prob. 15.

17. Solve Prob. 15 by the Runge–Kutta method with $h = 0.4$, determine the error, and compare with Prob. 15. (Note that these 2 Runge–Kutta steps require 8 evaluations of $f(x, y)$, just as many as in Prob. 15.)

18. Solve Prob. 15 by the Runge–Kutta method with $h = 0.1$ and compare the error with that in Prob. 15.

19. Another Euler–Cauchy type method is given by

$$y_{n+1} = y_n + hf(x_n + \tfrac{1}{2}h, y^*_{n+1}),$$

where $y^*_{n+1} = y_n + \frac{1}{2}hf(x_n, y_n)$. Give a geometric motivation of the method. Apply it to (4), choosing $h = 0.2$ and calculating 5 steps.

20. **Kutta's third-order method** is defined by

$$y_{n+1} = y_n + \tfrac{1}{6}(k_1 + 4k_2 + k_3^*)$$

where k_1 and k_2 are as in Table 20.4 and $k_3^* = hf(x_{n+1}, y_n - k_1 + 2k_2)$. Apply this method to (4) in Example 1. Choose $h = 0.2$ and do 5 steps. Compare with Table 20.6.

20.2 Multistep Methods

A **one-step method** is a method that in each step uses only values obtained in a single step, namely, in the preceding step. Examples are the Runge–Kutta method and all the other methods in the last section. In contrast, a method that uses values from more than one preceding step is called a **multistep method.** We shall explain the idea of obtaining such methods in terms of the derivation of the **Adams–Moulton method,** which is of great practical importance. The initial value problem is as before,

$$y' = f(x, y), \qquad y(x_0) = y_0 \tag{1}$$

where f is assumed to be such that the problem has a unique solution in some interval containing x_0 as well as all the x-values at which we shall compute approximate values of the solution.

Adams–Bashford Method

Integrating the differential equation (1) from x_n to $x_{n+1} = x_n + h$, we have

$$\int_{x_n}^{x_{n+1}} f(x, y(x))\,dx = \int_{x_n}^{x_{n+1}} y'(x)\,dx = y(x_{n+1}) - y(x_n). \tag{2}$$

Our notations are as before; in particular, $x_n = x_0 + nh$, and y_n denotes an approximate value of $y(x_n)$. In (2) we replace f by an interpolation polynomial $p_3(x)$ of third degree, so that we can later integrate. For $p_3(x)$ we take the polynomial that at $x_n, x_{n-1}, x_{n-2}, x_{n-3}$ has the values

$$\begin{gathered} f_n = f(x_n, y_n), \quad f_{n-1} = f(x_{n-1}, y_{n-1}), \quad f_{n-2} = f(x_{n-2}, y_{n-2}), \\ f_{n-3} = f(x_{n-3}, y_{n-3}), \end{gathered} \tag{3}$$

respectively. (Actually, we could take a polynomial of higher degree, but a cubic polynomial is commonly used in practice.) We can obtain $p_3(x)$, for instance, from the Newton backward difference formula (18), Sec. 18.3:

$$p_3(x) = f_n + r\nabla f_n + \tfrac{1}{2}r(r + 1)\nabla^2 f_n + \tfrac{1}{6}r(r + 1)(r + 2)\nabla^3 f_n$$

where $r = (x - x_n)/h$. We integrate $p_3(x)$ over x from x_n to $x_{n+1} = x_n + h$, thus over r from 0 to 1. Since $x = x_n + hr$, we have $dx = h\,dr$. The integral of $\frac{1}{2}r(r + 1)$ is 5/12 and that of $\frac{1}{6}r(r + 1)(r + 2)$ is 3/8. We thus obtain

$$\int_{x_n}^{x_{n+1}} p_3\,dx = h\int_0^1 p_3\,dr = h\left(f_n + \frac{1}{2}\nabla f_n + \frac{5}{12}\nabla^2 f_n + \frac{3}{8}\nabla^3 f_n\right). \tag{4}$$

It is practical to replace these differences by their expressions in terms of f:

$$\begin{aligned} \nabla f_n &= f_n - f_{n-1} \\ \nabla^2 f_n &= f_n - 2f_{n-1} + f_{n-2} \\ \nabla^3 f_n &= f_n - 3f_{n-1} + 3f_{n-2} - f_{n-3}. \end{aligned}$$

We substitute this into (4) and collect terms. By (2) we then obtain the multistep formula of the **Adams–Bashford method,**

$$y^*_{n+1} = y_n + \frac{h}{24}(55f_n - 59f_{n-1} + 37f_{n-2} - 9f_{n-3}). \tag{5}$$

It expresses the new value y^*_{n+1} [approximation of the solution y of (1) at x_{n+1}] in terms of 4 values of f computed from the y-values obtained in the preceding 4 steps. What about the beginning? Wait and see.

Adams–Moulton method

We wrote y^*_{n+1} in (5), instead of the usual y_{n+1}, because we want to extend the method, using (5) as a predictor and another formula [(6), below] as a corrector. A corrector is obtained by the same idea of integrating a cubic Newton backward polynomial $\tilde{p}_3(x)$, which at $x_{n+1}, x_n, x_{n-1}, x_{n-2}$ equals $f_{n+1}, f_n, f_{n-1}, f_{n-2}$, respectively; here,

$$f_{n+1} = f(x_{n+1}, y^*_{n+1})$$

and the other f's are as in (3). The derivation is quite similar to that of (5). Indeed,

$$\tilde{p}_3(x) = f_{n+1} + r\nabla f_{n+1} + \tfrac{1}{2}r(r+1)\nabla^2 f_{n+1} + \tfrac{1}{6}r(r+1)(r+2)\nabla^3 f_{n+1}$$

where $r = (x - x_{n+1})/h$. We integrate over x from x_n to x_{n+1} as before. This now corresponds to integrating over r from -1 to 0. We obtain

$$\int_{x_n}^{x_{n+1}} \tilde{p}_3(x)\,dx = h\left(f_{n+1} - \frac{1}{2}\nabla f_{n+1} - \frac{1}{12}\nabla^2 f_{n+1} - \frac{1}{24}\nabla^3 f_{n+1}\right).$$

Expressing differences in terms of values of f as before, we finally arrive at the corrector formula

$$\textbf{(6)} \qquad y_{n+1} = y_n + \frac{h}{24}(9f^*_{n+1} + 19f_n - 5f_{n-1} + f_{n-2})$$

where $f^*_{n+1} = f(x_{n+1}, y^*_{n+1})$ and the other f's are as in (3). The predictor–corrector method (5), (6) is called the **Adams–Moulton method.** The program can provide for repeated application of the corrector for a fixed n, say, until the relative difference of successive values (for the same n) in absolute value becomes less than a small preassigned positive number.

Getting started. Whereas one-step methods are *"self-starting"* (need no starting data beyond the given initial condition), multistep methods are not. This is an important point—and a disadvantage! In (5) we need f_0, f_1, f_2, f_3; thus, by (3), we must first compute y_1, y_2, y_3 by some other method, say, by the Runge–Kutta method (to have great accuracy). But the advantage of (5), (6) is that it is faster than the Runge–Kutta method because we now need only two new values of f per step, in contrast to the four values in each Runge-Kutta step. It can be shown that the present method is of fourth order, like the Runge–Kutta method, and that it is ***numerically stable.***

Furthermore, predictor–corrector methods have the advantage that they provide an estimate of the error. Specifically, large $|y_n - y_n^*|$ indicates that the error of y_n is probably large in absolute value and calls for a reduction of h. On the other hand, if $|y_n - y_n^*|$ is very small, then h may be increased, say, by doubling it. (For more details, see Ref. [E13], p. 388–393, listed in Appendix 1.)

EXAMPLE 1 Adams–Moulton method

Solve the initial value problem

$$y' = x + y, \quad y(0) = 0 \tag{7}$$

by the Adams–Moulton method on the interval $0 \leqq x \leqq 2$, choosing $h = 0.2$.

Solution. The problem is the same as in Examples 1–3, Sec. 20.1, so that we can compare the results. We compute starting values y_1, y_2, y_3 by the Runge–Kutta method. Then in each step we predict by (5) and make one correction by (6) before we execute the next step. The results are shown and compared with the exact values in Table 20.8. We see that the corrections improve the accuracy considerably. This is typical. ∎

Table 20.8
Adams–Moulton Method Applied to the Initial Value Problem (7); Predicted Values Computed by (5) and Corrected Values by (6)

n	x_n	Starting y_n	Predicted y_n^*	Corrected y_n	Exact Values	$10^6 \times$ Error of y_n
0	0.0	0.000 000			0.000 000	0
1	0.2	0.021 400			0.021 403	3
2	0.4	0.091 818			0.091 825	7
3	0.6	0.222 107			0.222 119	12
4	0.8		0.425 361	0.425 529	0.425 541	12
5	1.0		0.718 066	0.718 270	0.718 282	12
6	1.2		1.119 855	1.120 106	1.120 117	11
7	1.4		1.654 885	1.655 191	1.655 200	9
8	1.6		2.352 653	2.353 026	2.353 032	6
9	1.8		3.249 190	3.249 646	3.249 647	1
10	2.0		4.388 505	4.389 062	4.389 056	−6

This is the end of our discussion of methods for first-order differential equations. In the next section we turn to differential equations of **second order.**

Problem Set 20.2

Solve the following initial value problems by the Adams–Moulton method (10 steps, 1 correction per step). Compute the errors by using the exact solution. (The given starting values should help you to spend your time entirely on the new method. Use the Runge–Kutta method where no such values are given.)

1. $y' = 2xy$, $y(0) = 1$, $h = 0.1$; (1, 1.010 050, 1.040 811, 1.094 174)

2. $y' = y$, $y(0) = 1$, $h = 0.1$; (1, 1.105 171, 1.221 403, 1.349 859)

3. $y' = 1 + y^2$, $y(0) = 0$, $h = 0.1$; (0, 0.100 335, 0.202 710, 0.309 336)

4. $y' = x + y$, $y(0) = 0$, $h = 0.1$

5. $y' = (x + y - 4)^2$, $y(0) = 4$, $h = 0.2$, 7 steps; (4, 4.002 707, 4.022 789, 4.084 133)

6. Solve the equation in Prob. 5 exactly and find the reason for the restriction to 7 steps.

7. Carry out the details of the calculations that lead to (5) and (6).

8. Accurate starting values are important to the Adams–Moulton method. Illustrate this by replacing the starting values in Example 1 with the corresponding values obtained by the improved Euler–Cauchy method, computing y for $x = 0.8, 1.0$ with 1 correction per step and comparing the results with those in Table 20.8.
9. Show that by applying the method in the text to a polynomial of second degree we obtain the predictor and corrector formulas

$$y^*_{n+1} = y_n + \frac{h}{12}(23f_n - 16f_{n-1} + 5f_{n-2}),$$

$$y_{n+1} = y_n + \frac{h}{12}(5f_{n+1} + 8f_n - f_{n-1}).$$

10. Apply the method in Prob. 9 with $h = 0.2$ to the initial value problem $y' = x + y$, $y(0) = 0$; perform 5 steps and compare with the exact values.
11. Apply the method in Prob. 9 with $h = 0.1$, 10 steps, to the initial value problem $y' = 2xy$, $y(0) = 1$, using Runge–Kutta starting values. Compare with the exact solution.
12. Apply the method in Prob. 9 with $h = 0.2$, 5 steps, to the initial value problem in Prob. 5. Compare with the exact solution.

20.3 Methods for Second-Order Differential Equations

This section may also be studied immediately after Chap. 2, since it is independent of the preceding sections on numerical methods in Chaps. 18–20.

An **initial value problem** for a second-order differential equation consists of that equation and two conditions (*initial conditions*) referring to the same point. In this section we shall consider two numerical methods for solving initial value problems of the form

$$y'' = f(x, y, y'), \qquad y(x_0) = y_0, \qquad y'(x_0) = y'_0 \tag{1}$$

assuming f to be such that the problem has a unique solution on some interval containing x_0 as well as the x-values at which we wish to compute approximate values of the solution. The first method is simple (but inaccurate) and serves to illustrate the principle, whereas the second method is of great precision and practical importance.

In both methods we shall obtain approximate values of the solution $y(x)$ of (1) at equidistant points $x_1 = x_0 + h$, $x_2 = x_0 + 2h$, $\cdots$; these values will be denoted by $y_1, y_2, \cdots$, respectively. Similarly, approximate values of the derivative $y'(x)$ at those points will be denoted by $y'_1, y'_2, \cdots$, respectively.

The methods in Sec. 20.1 were suggested by the Taylor expansion

$$\text{(2)} \qquad y(x + h) = y(x) + hy'(x) + \frac{h^2}{2} y''(x) + \frac{h^3}{3!} y'''(x) + \cdots$$

which we shall now use for the same purpose, together with the expansion for the derivative

$$\text{(3)} \qquad y'(x + h) = y'(x) + hy''(x) + \frac{h^2}{2} y'''(x) + \cdots .$$

Roughest Method, to Explain the Principle

The roughest numerical method is obtained by neglecting the terms containing y''' and the further terms in (2) and (3); this yields the approximations for $y(x + h)$ and the derivative $y'(x + h)$ given by the formulas

$$y(x + h) \approx y(x) + hy'(x) + \frac{h^2}{2} y''(x),$$

$$y'(x + h) \approx y'(x) + hy''(x).$$

In the first step of the method we compute

$$y_0'' = f(x_0, y_0, y_0')$$

from (1), then

$$y_1 = y_0 + hy_0' + \frac{h^2}{2} y_0''$$

which approximates $y(x_1) = y(x_0 + h)$, and furthermore

$$y_1' = y_0' + hy_0''$$

which will be needed in the next step. In the second step we compute

$$y_1'' = f(x_1, y_1, y_1')$$

from (1), then

$$y_2 = y_1 + hy_1' + \frac{h^2}{2} y_1''$$

which approximates $y(x_2) = y(x_0 + 2h)$, and furthermore

$$y_2' = y_1' + hy_1''.$$

In the $(n + 1)$th step we compute

$$y''_n = f(x_n, y_n, y'_n)$$

from (1), then the new value

$$y_{n+1} = y_n + hy'_n + \frac{h^2}{2} y''_n \tag{4a}$$

which is an approximation for $y(x_{n+1})$, and furthermore

$$y'_{n+1} = y'_n + hy''_n \tag{4b}$$

which is an approximation for $y'(x_{n+1})$ needed in the next step.

Note that, geometrically speaking, this method is an approximation of the curve of $y(x)$ by portions of parabolas.

EXAMPLE 1 **An application of the method defined by (4)**

Apply (4) to the following initial value problem, choosing $h = 0.2$:

$$y'' = \frac{1}{2}(x + y + y' + 2), \qquad y(0) = 0, \qquad y'(0) = 0. \tag{5}$$

Solution. For the present problem, formulas (4) become

$$y_{n+1} = y_n + 0.2y'_n + 0.02y''_n$$

$$y'_{n+1} = y'_n + 0.2y''_n$$

where

$$y''_n = \frac{1}{2}(x_n + y_n + y'_n + 2).$$

The computations are shown in Table 20.9. The student may verify that the exact solution is

$$y = e^x - x - 1.$$

We see that the errors of our approximate values are very large. This is typical, because in most practical cases our present method would be too inaccurate. ∎

Table 20.9
Computations in Example 1

n	x_n	y_n	y'_n	y''_n	Exact (4D)	Error
0	0.0	0.0000	0.0000	1.0000	0.0000	0.0000
1	0.2	0.0200	0.2000	1.2100	0.0214	0.0014
2	0.4	0.0842	0.4420	1.4631	0.0918	0.0076
3	0.6	0.2019	0.7346	1.7682	0.2221	0.0202
4	0.8	0.3842	1.0883	2.1362	0.4255	0.0413
5	1.0	0.6446			0.7183	0.0737

Runge–Kutta–Nyström Method

Much more accurate than (4) is the **Runge–Kutta–Nyström method,** which generalizes the Runge–Kutta method in Sec. 20.1. We mention without proof that this is a *fourth-order method,* which means that in the Taylor formulas for y and y' the first terms up to and including the term that contains h^4 are given exactly.

The computation in this method can be done as shown in Table 20.10. We see that in each step we compute four auxiliary quantities k_1, k_2, k_3, k_4 and then from them the new approximate value y_{n+1} of the solution y as well as an approximation of the derivative y' needed in the next step.

Table 20.10
Runge–Kutta–Nyström Method

ALGORITHM R–K–N (f, x_0, y_0, y'_0, h, N).

This algorithm computes the solution of the initial value problem $y'' = f(x, y, y')$, $y(x_0) = y_0$, $y'(x_0) = y'_0$ at equidistant points $x_1 = x_0 + h$, $x_2 = x_0 + 2h$, $\cdots$, $x_N = x_0 + Nh$; here f is such that this problem has a unique solution on the interval $[x_0, x_N]$.

INPUT: Initial values x_0, y_0, y'_0, step size h, number of steps N

OUTPUT: Approximation y_{n+1} to the solution $y(x_{n+1})$ at $x_{n+1} = x_0 + (n + 1)h$, where $n = 0, 1, \cdots, N - 1$

For $n = 0, 1, \cdots, N - 1$ do:

$$k_1 = \tfrac{1}{2}hf(x_n, y_n, y'_n)$$

$$k_2 = \tfrac{1}{2}hf(x_n + \tfrac{1}{2}h, y_n + K, y'_n + k_1)$$

where $K = \tfrac{1}{2}h(y'_n + \tfrac{1}{2}k_1)$

$$k_3 = \tfrac{1}{2}hf(x_n + \tfrac{1}{2}h, y_n + K, y'_n + k_2)$$

$$k_4 = \tfrac{1}{2}hf(x_n + h, y_n + L, y'_n + 2k_3)$$

where $L = h(y'_n + k_3)$

$$x_{n+1} = x_n + h$$

$$y_{n+1} = y_n + h(y'_n + \tfrac{1}{3}(k_1 + k_2 + k_3))$$

OUTPUT x_{n+1}, y_{n+1}

[*Approximation to the solution at* x_{n+1}]

$$y'_{n+1} = y'_n + \tfrac{1}{3}(k_1 + 2k_2 + 2k_3 + k_4)$$

[*Auxiliary value needed in the next step*]

End

Stop

End R–K–N

h can be controlled as described near the end of Sec. 20.1, where we now take for δ the larger of δ^* and δ^{**}; here δ^* is $\frac{1}{15}$ times the difference of corresponding values of y, and δ^{**} is $\frac{1}{15}$ times the difference of corresponding values of y'.

EXAMPLE 2 Runge–Kutta–Nyström method

Apply the Runge–Kutta–Nyström method to the initial value problem (5), choosing $h = 0.2$.

Solution. Here

$$f = 0.5(x + y + y' + 2).$$

Hence

$$k_1 = 0.05(x_n + y_n + y_n' + 2)$$

$$k_2 = 0.05(x_n + 0.1 + y_n + K + y_n' + k_1 + 2) \qquad K = 0.1(y_n' + \tfrac{1}{2}k_1)$$

$$k_3 = 0.05(x_n + 0.1 + y_n + K + y_n' + k_2 + 2)$$

$$k_4 = 0.05(x_n + 0.2 + y_n + L + y_n' + 2k_3 + 2) \qquad L = 0.2(y_n' + k_3).$$

In the present case the differential equation is simple, and so are the expressions for k_1, k_2, k_3, k_4. Hence we may insert k_1 into k_2, then k_2 into k_3, and, finally, k_3 into k_4. The result of this simple calculation is

$$k_2 = 0.05[1.0525(x_n + y_n) + 1.152\,5y_n' + 2.205]$$

$$k_3 = 0.05[1.055\,125(x_n + y_n) + 1.160\,125y_n' + 2.215\,25]$$

$$k_4 = 0.05[1.116\,063\,75(x_n + y_n) + 1.327\,613\,75y_n' + 2.443\,677\,5].$$

From this we obtain

$$\begin{aligned} y_{n+1} &= y_n + a(x_n + y_n) + by_n' + c \\ y_{n+1}' &= y_n' + a^*(x_n + y_n) + b^*y_n' + c^* \end{aligned} \tag{6}$$

where

$$a = 0.010\,3588 \qquad b = 0.211\,0421 \qquad c = 0.021\,4008$$

$$a^* = 0.105\,5219 \qquad b^* = 0.115\,8811 \qquad c^* = 0.221\,4030.$$

Table 20.11 shows the corresponding computations. The errors of the approximate values for $y(x)$ are much smaller than those in Example 1 (see Table 20.12). ∎

Table 20.11
Runge–Kutta–Nyström Method (with $h = 0.2$) Applied to the Initial Value Problem (5); Five Steps Computed by the Use of (6)

n	x_n	y_n	y_n'	$a(x_n + y_n) + by_n' + c$	$a^*(x_n + y_n) + b^*y_n' + c^*$
0	0.0	0.000 0000	0.000 0000	0.021 4008	0.221 4030
1	0.2	0.021 4008	0.021 4030	0.070 4196	0.270 4220
2	0.4	0.091 8204	0.491 8250	0.130 2913	0.330 2940
3	0.6	0.222 1117	0.822 1190	0.203 4186	0.403 4219
4	0.8	0.425 5303	1.225 5409	0.292 7365	0.492 7403
5	1.0	0.718 2668	1.718 2812		

Table 20.12
Comparison of Accuracy of the Two Methods Under Consideration in the Case of the Initial Value Problem (5), with $h = 0.2$

x	$y = e^x - x - 1$	Error: Example 1	Error: Table 20.11
0.2	0.021 4028	0.0014	0.000 0020
0.4	0.091 8247	0.0076	0.000 0043
0.6	0.222 1188	0.0202	0.000 0071
0.8	0.425 5409	0.0413	0.000 0106
1.0	0.718 2818	0.0737	0.000 0150

EXAMPLE 3 Runge–Kutta–Nyström method. Airy's equation. Airy function Ai(x)

Solve the initial value problem

$$y'' = xy, \qquad y(0) = 3^{-2/3}/\Gamma(2/3) = 0.355\ 028\ 05, \qquad y'(0) = -3^{-1/3}/\Gamma(1/3) = -0.258\ 819\ 40$$

by the Runge–Kutta–Nyström method with $h = 0.2$; do 5 steps. This is **Airy's equation,**[2] which arose in optics (see Ref. [A9], p. 188, listed in Appendix 1). Γ is the gamma function (see Appendix 3). The initial conditions are such that we obtain a standard solution, the **Airy function** Ai(x), which has been tabulated and investigated (see Ref. [1], pp. 446, 475).

Solution. For this equation and $h = 0.2$ we obtain from the general formulas in the algorithm

$$k_1 = 0.1x_n y_n,$$

$$k_2 = k_3 = 0.1(x_n + 0.1)(y_n + 0.1y_n' + 0.05k_1),$$

$$k_4 = 0.1(x_n + 0.2)(y_n + 0.2y_n' + 0.2k_2).$$

Table 20.13 shows the results. The error was determined by comparing with the values in Ref. [1], p. 475, listed in Appendix 1. ∎

Our work in Example 3 also illustrates that methods for differential equations are often useful in the tabulation of "**higher transcendental functions**" given by formulas that are less practical in numerical work [in our example: a power series or integral representation of Ai(x)].

Table 20.13
Runge–Kutta–Nyström Method Applied to Airy's Equation, Computation of the Airy Function y = Ai(x)

x_n	y_n	y_n'	$y(x)$ exact (8D)	$10^8 \times$ Error of y_n
0.0	0.355 028 05	−0.258 819 40	0.355 028 05	0
0.2	0.303 703 04	−0.252 404 64	0.303 703 15	11
0.4	0.254 742 11	−0.235 830 70	0.254 742 35	24
0.6	0.209 799 74	−0.212 791 72	0.209 800 06	32
0.8	0.169 845 99	−0.186 411 34	0.169 846 32	33
1.0	0.135 292 18	−0.159 146 09	0.135 292 42	24

[2]Named after Sir GEORGE BIDELL AIRY (1801—1892), English mathematician, who is known for his work in elasticity and in partial differential equations.

The methods discussed in this section involve a truncation error. In addition, there are **round-off errors,** and we want to warn the reader that round-off errors may affect results to a large extent. For instance, the solution of the problem $y'' = y$, $y(0) = 1$, $y'(0) = -1$ is $y = e^{-x}$, but the round-off error will introduce a small multiple of the unwanted solution e^{x}, which may eventually (after sufficiently many steps) swamp the required solution. This is known as **building-up error.** In our simple example we may avoid it by starting from known values of e^{-x} and its derivative for a large x and compute in the reverse direction, but in more complicated cases considerable experience is needed to avoid that phenomenon.

In the remaining sections of this chapter we consider numerical methods for **partial differential equations.**

Problem Set 20.3

1. Repeat the computation in Example 1, choosing $h = 0.1$, and compare the errors of the values thus obtained with those in Example 1.
2. Perform the same task as in Prob. 1, choosing $h = 0.05$.

Apply (4) to the following initial value problems (5 steps).

3. $y'' = -y, \quad y(0) = 0, \quad y'(0) = 1, \quad h = 0.1$
4. $y'' = -y, \quad y(0) = 1, \quad y'(0) = 0, \quad h = 0.1$
5. $y'' = y, \quad y(0) = 1, \quad y'(0) = 1, \quad h = 0.1$
6. $y'' = y, \quad y(0) = 1, \quad y'(0) = -1, \quad h = 0.1$

7. Verify the formulas and computations for the Airy equation in Example 3.
8. Verify, to 3D-accuracy, the values y_0 and y_0' in Example 3, using (25) in Appendix A3.1 and Table A2 in Appendix 5.

Apply (4) to the following initial value problems ($h = 0.1$, 5 steps). Verify the given solution. Compute the error of y_n.

9. $y'' = xy' - 4y$, $y(0) = 3$, $y'(0) = 0$. Exact solution $y = x^4 - 6x^2 + 3$.
10. $y'' = xy' - 3y$, $y(0) = 0$, $y'(0) = -3$. Exact solution $y = x^3 - 3x$.
11. $(1 - x^2)y'' - 2xy' + 6y = 0$, $y(0) = -\frac{1}{2}$, $y'(0) = 0$. Exact: $y = \frac{1}{2}(3x^2 - 1)$.

Apply the Runge–Kutta–Nyström method to the following initial value problems (5 steps). Compute the error.

12. $xy'' + y' + xy = 0$, $y(1) = 0.765\,198$, $y'(1) = -0.440\,051$, $h = 0.5$. (These initial conditions lead to the standard solution $y = J_0(x)$, the Bessel function of the first kind of order 0, whose 6D-values at $x = 1, 1.5, \cdots$ are 0.765 198, 0.511 828, 0.223 891, −0.048 384, −0.260 052, −0.380 128.)
13. Prob. 9, $h = 0.2$

14. (Boundary value problem) Numerical methods for boundary value problems will not be considered in any generality, but the reader may show the following. To solve

$$y'' + f(x)y' + g(x)y = r(x), \qquad y(0) = 0, \quad y(b) = k$$

numerically, apply one of the previous methods to get a solution $Y(x)$ of the equation satisfying the initial conditions $Y(0) = 0$, $Y'(0) = 1$. Then determine by that method a solution $z(x)$ of the homogeneous equation satisfying the initial conditions $z(0) = 0$, $z'(0) = 1$. Show that

$$y(x) = Y(x) + cz(x)$$

with a suitable c satisfies the given problem. What is the condition for c?

15. To solve the boundary value problem

$$y'' + f(x)\,y' + g(x)\,y = r(x), \qquad y(a) = k_1, \quad y(b) = k_2$$

numerically, show that one may determine a solution $y_1(x)$ satisfying $y_1(a) = k_1$, $y_1'(a) = c_1$ and a solution $y_2(x)$ satisfying $y_2(a) = k_1$, $y_2'(a) = c_2$ and such that $y_1(b) \neq y_2(b)$ and then obtain the solution of the given problem in the form

$$y(x) = \frac{1}{y_1(b) - y_2(b)}\,[(k_2 - y_2(b))\,y_1(x) + (y_1(b) - k_2)\,y_2(x)].$$

20.4 Numerical Methods for Elliptic Partial Differential Equations

The remaining sections of this chapter are devoted to numerical methods for partial differential equations, particularly for the Laplace, Poisson, heat, and wave equations, which are basic in applications and, at the same time, are model cases of elliptic, parabolic, and hyperbolic equations. The definitions are as follows.

A partial differential equation is called **quasilinear** if it is linear in the highest derivatives. Hence a second-order quasilinear equation in two independent variables x, y can be written

$$au_{xx} + 2bu_{xy} + cu_{yy} = F(x, y, u, u_x, u_y). \tag{1}$$

u is the unknown function. This equation is said to be of

elliptic type if $ac - b^2 > 0$ (example: *Laplace equation*)

parabolic type if $ac - b^2 = 0$ (example: *heat equation*)

hyperbolic type if $ac - b^2 < 0$ (example: *wave equation*)

(where in the heat and wave equations, y is time t). Here, the *coefficients* a, b, c may be functions of x, y, so that the type of (1) may be different in different regions of the xy-plane. This classification is not merely a formal matter but is of great practical importance because the general behavior of solutions differs from type to type and so do the additional conditions (boundary and initial conditions) that must be taken into account.

Applications involving *elliptic* equations usually lead to boundary value problems in a region R, called a *first boundary value problem* or **Dirichlet problem** if u is prescribed on the boundary curve C of R, a *second boundary value problem* or **Neumann problem** if $u_n = \partial u/\partial n$ (normal derivative of u) is prescribed on C, and a *third* or **mixed problem** if u is prescribed on a part of C and u_n on the remaining part. C usually is a closed curve (or sometimes consists of two or more such curves).

Difference Equations for the Laplace and Poisson Equations

In this section we consider the **Laplace equation**

$$\nabla^2 u = u_{xx} + u_{yy} = 0 \tag{2}$$

and the **Poisson equation**

$$\nabla^2 u = u_{xx} + u_{yy} = f(x, y) \tag{3}$$

which are the most important elliptic equations in applications. To obtain methods of numerical solution, we replace the partial derivatives in a given equation by corresponding difference quotients, as follows. By the Taylor formula,

$$\text{(4)}\quad \begin{aligned} &\text{(a)}\ u(x+h,y) = u(x,y) + hu_x(x,y) + \tfrac{1}{2}h^2u_{xx}(x,y) + \tfrac{1}{6}h^3u_{xxx}(x,y) + \cdots \\ &\text{(b)}\ u(x-h,y) = u(x,y) - hu_x(x,y) + \tfrac{1}{2}h^2u_{xx}(x,y) - \tfrac{1}{6}h^3u_{xxx}(x,y) + \cdots . \end{aligned}$$

Subtracting (4b) from (4a) and neglecting terms in $h^3, h^4, \cdots$, we obtain

$$u_x(x, y) \approx \frac{1}{2h}[u(x + h, y) - u(x - h, y)]. \tag{5a}$$

Similarly,

$$u_y(x, y) \approx \frac{1}{2k}[u(x, y + k) - u(x, y - k)]. \tag{5b}$$

We turn to second derivatives. Adding (4a) and (4b) and neglecting terms in $h^4, h^5, \cdots$, we obtain

$$u(x + h, y) + u(x - h, y) \approx 2u(x, y) + h^2u_{xx}(x, y).$$

Solving for u_{xx}, we have

$$u_{xx}(x, y) \approx \frac{1}{h^2}[u(x+h, y) - 2u(x, y) + u(x-h, y)]. \tag{6a}$$

Similarly,

$$u_{yy}(x, y) \approx \frac{1}{k^2}[u(x, y+k) - 2u(x, y) + u(x, y-k)]. \tag{6b}$$

We shall not need (Prob. 3)

$$\begin{aligned} u_{xy}(x, y) \approx \frac{1}{4hk}[&u(x+h, y+k) - u(x-h, y+k) \\ &- u(x+h, y-k) + u(x-h, y-k)]. \end{aligned} \tag{6c}$$

Figure 424 shows the points $(x+h, y)$, $(x-h, y)$, $\cdots$ in (5) and (6).

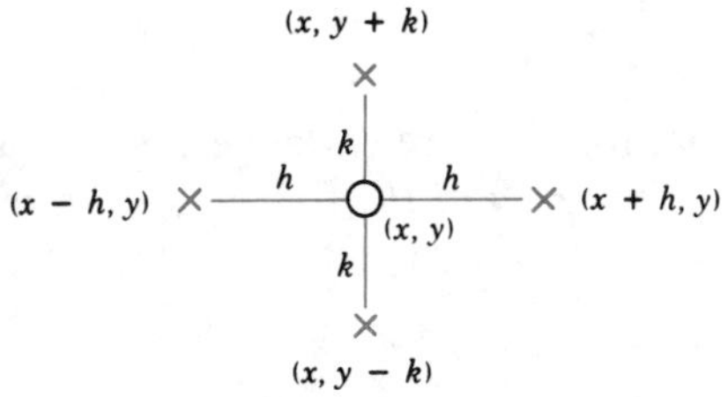

Fig. 424. Points in (5) and (6)

We now substitute (6a) and (6b) into the ***Poisson equation*** (3), choosing $k = h$ to obtain a simple formula:

$$\begin{aligned} u(x+h, y) + u(x, y+h) + u(x-h, y) + u(x, y-h) - 4u(x, y) \\ = h^2 f(x, y). \end{aligned} \tag{7}$$

This is a **difference equation** corresponding to (3). Hence for the ***Laplace equation*** (2) the corresponding difference equation is

$$u(x+h, y) + u(x, y+h) + u(x-h, y) + u(x, y-h) - 4u(x, y) = 0. \tag{8}$$

h is called the **mesh size.** Equation (8) relates u at (x, y) to u at the four neighboring points shown in Fig. 425 on p. 1058. For convenience, these neighboring points are often called E (East). N (North), W (West), S (South). Then Fig. 425 takes the form of Fig. 426, and (7) becomes

$$u(E) + u(N) + u(W) + u(S) - 4u(x, y) = h^2 f(x, y). \tag{7*}$$

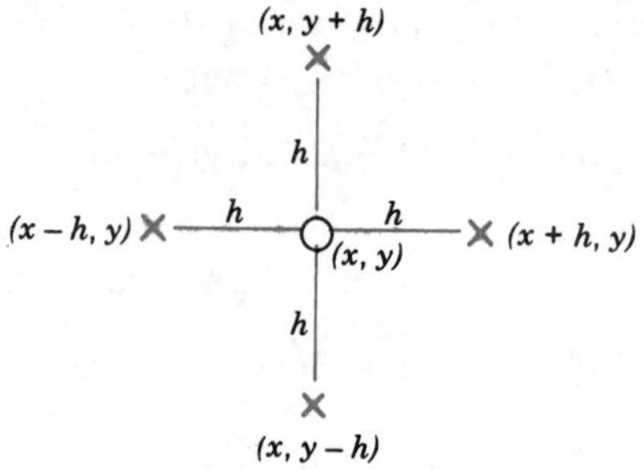

Fig. 425. Points in (7) and (8)

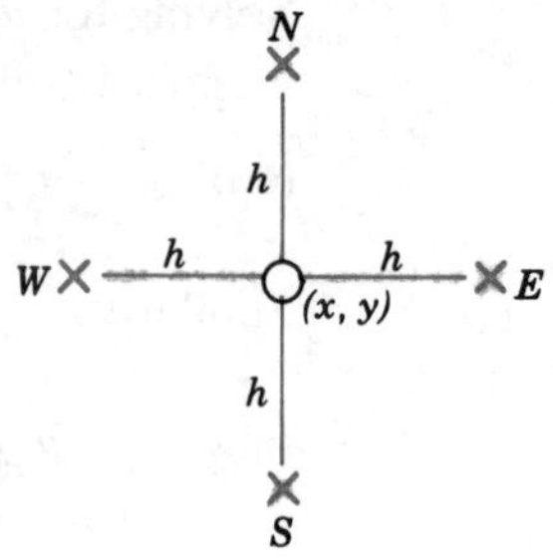

Fig. 426. Notation in (7*)

Our approximation of $h^2\nabla^2 u$ in (7) and (8) is a 5-point approximation with the coefficient scheme or **pattern**

$$\textbf{(9)}\qquad \left\{\begin{matrix} & 1 & \\ 1 & -4 & 1 \\ & 1 & \end{matrix}\right\}$$

Note that (8) has a remarkable interpretation: *u at* (x, y) *equals the mean of the values of u at the four neighboring mesh points.*

To grasp (7) at first sight, we may very conveniently represent it by that pattern:

$$\left\{\begin{matrix} & 1 & \\ 1 & -4 & 1 \\ & 1 & \end{matrix}\right\} u = h^2 f(x, y).$$

Dirichlet Problem

In the numerical solution of the Dirichlet problem (definition above) in a region R we first choose h and introduce in R a grid consisting of equidistant horizontal and vertical straight lines of distance h. Their intersections are called **mesh points** (or *nodes* or *lattice points*). See Fig. 427. Then we use a

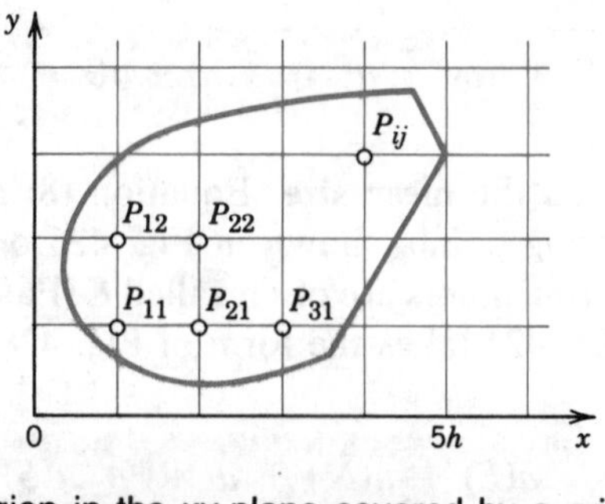

Fig. 427. Region in the xy-plane covered by a grid of mesh h, also showing mesh points $P_{11} = (h, h), \cdots, P_{ij} = (ih, jh), \cdots$

difference equation approximating the given partial differential equation—formula (8) in the case of the Laplace equation—by which we relate the unknown values of u at the mesh points in R to each other and to the given boundary values, as will be discussed below. This yields a linear system of *algebraic* equations. By solving it we obtain approximations to the unknown values of u at the mesh points in R. We shall see that the number of equations equals the number of unknowns, that is, the number of mesh points in R. Since at each mesh point, u is only related to the values at the neighboring mesh points, the coefficients of the system form a **sparse matrix,** that is, a matrix with relatively few nonzero entries. In practice, this matrix will be large, since for obtaining high accuracy one needs many mesh points, and a 500×500 or larger matrix may cause a storage problem.[3] Hence an indirect method (see Sec. 19.3) is preferable to a direct one. In particular, we may use the **Gauss–Seidel method,** which in the present context is also called **Liebmann's method.** We illustrate the whole approach by an example, in which we keep the number of equations small, for simplicity. As convenient ***notations for mesh points and corresponding values of the solution*** (and of approximate solutions) we use (see also Fig. 427)

$$P_{ij} = (ih, jh), \qquad u_{ij} = u(ih, jh). \tag{10}$$

With this notation we can write (8) for any mesh point P_{ij} in the form

$$u_{i+1,j} + u_{i,j+1} + u_{i-1,j} + u_{i,j-1} - 4u_{ij} = 0. \tag{11}$$

EXAMPLE 1 Laplace equation. Liebmann's method

The four sides of a square plate of side 12 cm made of homogeneous material are kept at constant temperature 0°C and 100°C as shown in Fig. 428*a*. Using a (very wide) grid of mesh 4 cm and applying Liebmann's method (that is, Gauss–Seidel iteration), find the (steady-state) temperature at the mesh points.

Solution. In the case of independence of time, the heat equation (see Sec. 9.8)

$$u_t = c^2(u_{xx} + u_{yy})$$

reduces to the Laplace equation. Hence our problem is a Dirichlet problem for this equation.

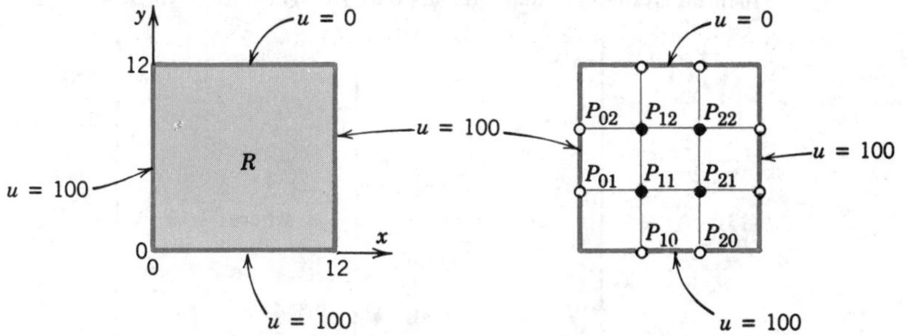

(*a*) Given problem (*b*) Grid and mesh points

Fig. 428. Example 1

[3]The present matrix is *not* tridiagonal! (Definition below.) If it were, we could apply the Gauss elimination without causing that storage problem.

We choose the grid shown in Fig. 428*b* and consider the mesh points in the order P_{11}, P_{21}, P_{12}, P_{22}. We use (11) and, in each equation, take to the right all the terms resulting from the given boundary values. Then we obtain the system

$$
\begin{aligned}
-4u_{11} + u_{21} + u_{12} \qquad\qquad &= -200 \\
u_{11} - 4u_{21} \qquad\quad + u_{22} &= -200 \\
u_{11} \qquad\quad - 4u_{12} + u_{22} &= -100 \\
u_{21} + u_{12} - 4u_{22} &= -100
\end{aligned}
\tag{12}
$$

In practice, one would solve such a small system by the Gauss elimination, finding

$$u_{11} = u_{21} = 87.5, \qquad u_{12} = u_{22} = 62.5.$$

More exact values (exact to 1D) of the solution of our problem are 88.1 and 61.9, respectively. (These were obtained by using Fourier series.) Hence the error is about 1%, which is surprisingly accurate for a grid of such a large mesh size h. If the system of equations were large, one would solve it by an indirect method, such as Liebmann's method. For (12) this is as follows. We write (12) in the form

$$
\begin{aligned}
u_{11} &= \qquad\quad 0.25u_{21} + 0.25u_{12} \qquad\qquad + 50 \\
u_{21} &= 0.25u_{11} \qquad\qquad\qquad + 0.25u_{22} + 50 \\
u_{12} &= 0.25u_{11} \qquad\qquad\qquad + 0.25u_{22} + 25 \\
u_{22} &= \qquad\quad 0.25u_{21} + 0.25u_{12} \qquad\qquad + 25
\end{aligned}
$$

These equations are now used for the Gauss–Seidel iteration. They are identical with (2) in Sec. 19.3, where $u_{11} = x_1$, $u_{21} = x_2$, $u_{12} = x_3$, $u_{22} = x_4$, and the iteration is explained there, where 100, 100, 100, 100 are chosen as starting values. Using physical intuition on what the values at the mesh points might approximately be, one can save some work by choosing better starting values. The reader may verify that the exact solution of the system is $u_{11} = u_{21} = 87.5$, $u_{12} = u_{22} = 62.5$.

For a further idea that leads to substantial simplification, see Prob. 6.

Remark. It is interesting to note that if we choose mesh $h = L/n$ (L = side of R) and consider the $(n-1)^2$ inner mesh points (i.e., mesh points not on the boundary) row by row in the order

$$P_{11}, P_{21}, \cdots, P_{n-1,1}, P_{12}, P_{22}, \cdots, P_{n-1,2}, \cdots,$$

then the system of equations has the $(n-1)^2 \times (n-1)^2$ coefficient matrix

$$
\mathbf{A} = \begin{bmatrix} \mathbf{B} & \mathbf{I} & & & \\ \mathbf{I} & \mathbf{B} & \mathbf{I} & & \\ & & \cdot & & \\ & & & \cdot & \\ & & & & \cdot \\ & & \mathbf{I} & \mathbf{B} & \mathbf{I} \\ & & & \mathbf{I} & \mathbf{B} \end{bmatrix} \quad \text{where} \quad \mathbf{B} = \begin{bmatrix} -4 & 1 & & & \\ 1 & -4 & 1 & & \\ & & \cdot & & \\ & & & \cdot & \\ & & & & \cdot \\ & & 1 & -4 & 1 \\ & & & 1 & -4 \end{bmatrix}
\tag{13}
$$

is an $(n-1) \times (n-1)$ matrix, and it can be shown that $\mathbf{A}$ is nonsingular. ∎

A matrix is called a **band matrix** if it has all its nonzero entries on the main diagonal and on sloping lines parallel to it (separated by sloping lines of zeros or not). For example, **A** in (13) is a band matrix. Although the Gauss elimination does not preserve zeros between bands, it does not introduce nonzero elements outside the limits defined by the original bands. Hence a band structure is advantageous. In (13) it has been achieved by carefully ordering the mesh points.

ADI Method

A matrix is called a **tridiagonal matrix** if it has all its nonzero entries on the main diagonal and in the positions immediately adjacent to the main diagonal. In this case the Gauss elimination is particularly simple.

This raises the question of whether in the solution of the Dirichlet problem for the Laplace or Poisson equations one could obtain a system of equations whose coefficient matrix is tridiagonal. The answer is yes, and a popular method of that kind, called the **ADI method** (*alternating direction implicit method*) was developed by Peaceman and Rachford. The idea is as follows. The pattern in (9) shows that we could obtain a tridiagonal matrix if there were only the three points in a row (or only the three points in a column). This suggests to write (11) in the form

$$u_{i-1,j} - 4u_{ij} + u_{i+1,j} = -u_{i,j-1} - u_{i,j+1} \tag{14a}$$

so that the left side belongs to y-row j and the right side to x-column i. Of course, we can also write (11) in the form

$$u_{i,j-1} - 4u_{ij} + u_{i,j+1} = -u_{i-1,j} - u_{i+1,j} \tag{14b}$$

so that the left side belongs to column i and the right side to row j. In the ADI method we proceed by iteration. At every mesh point we choose an arbitrary starting value $u_{ij}^{(0)}$. In each step we compute new values at all mesh points. In one step we use an iteration formula resulting from (14a) and in the next step an iteration formula resulting from (14b), and so on in alternating order.

In detail: suppose approximations $u_{ij}^{(m)}$ have been computed. Then, to obtain the next approximations $u_{ij}^{(m+1)}$, we substitute the $u_{ij}^{(m)}$ on the right side of (14a) and solve for the $u_{ij}^{(m+1)}$ on the left side; that is, we use

$$u_{i-1,j}^{(m+1)} - 4u_{ij}^{(m+1)} + u_{i+1,j}^{(m+1)} = -u_{i,j-1}^{(m)} - u_{i,j+1}^{(m)}. \tag{15a}$$

We use this for a fixed j, that is, ***for a fixed row*** j, and for all internal mesh points in this row. This gives a linear system of N algebraic equations (N = number of internal mesh points per row) in N unknowns, the new approximations of u at these mesh points. Note that (15a) involves not only approximations computed in the previous step but also given boundary values. We solve the system (15a) (j fixed!) by the Gauss elimination. Then we go to the next row, obtain another system of N equations and solve it by Gauss, and so on, until all rows are done. In the next step we ***alternate direction,*** that is, we compute the next approximations $u_{ij}^{(m+2)}$ column by column from the $u_{ij}^{(m+1)}$ and the given boundary values, using a formula obtained from (14b) by substituting the $u_{ij}^{(m+1)}$ on the right:

$$\textbf{(15b)}\qquad u_{i,j-1}^{(m+2)} - 4u_{ij}^{(m+2)} + u_{i,j+1}^{(m+2)} = -u_{i-1,j}^{(m+1)} - u_{i+1,j}^{(m+1)}.$$

For each fixed i, that is, ***for each column,*** this is a system of M equations (M = number of internal mesh points per column) in M unknowns, which we solve by the Gauss elimination. Then we go to the next column, and so on, until all columns are done.

Let us consider an example that merely serves to explain the whole method. (In practice one would solve this problem directly by the Gauss elimination.)

EXAMPLE 2 Dirichlet problem. ADI method

Explain the procedure and formulas of the ADI method in terms of the problem in Example 1, using the same grid and starting values 100, 100, 100, 100.

Solution. While working, we keep an eye on Fig. 428*b*, p. 1059, and the given boundary values. We obtain first approximations $u_{11}^{(1)}, u_{21}^{(1)}, u_{12}^{(1)}, u_{22}^{(1)}$ from (15a) with $m = 0$. We write boundary values contained in (15a) without an upper index, for better identification and to indicate that these given values remain the same during the iteration. From (15a) with $m = 0$ we have for $j = 1$ (first row) the system

$$\begin{aligned} &(i = 1) \qquad u_{01} - 4u_{11}^{(1)} + u_{21}^{(1)} &&= -u_{10} - u_{12}^{(0)} \\ &(i = 2) \qquad u_{11}^{(1)} - 4u_{21}^{(1)} + u_{31} &&= -u_{20} - u_{22}^{(0)}. \end{aligned}$$

The solution is $u_{11}^{(1)} = u_{21}^{(1)} = 100$. For $j = 2$ (second row) we obtain from (15a) the system

$$\begin{aligned} &(i = 1) \qquad u_{02} - 4u_{12}^{(1)} + u_{22}^{(1)} &&= -u_{11}^{(0)} - u_{13} \\ &(i = 2) \qquad u_{12}^{(1)} - 4u_{22}^{(1)} + u_{32} &&= -u_{21}^{(0)} - u_{23}. \end{aligned}$$

The solution is $u_{12}^{(1)} = u_{22}^{(1)} = 66.667$.

Second approximations $u_{11}^{(2)}, u_{21}^{(2)}, u_{12}^{(2)}, u_{22}^{(2)}$ are now obtained from (15b) with $m = 1$ by using the first approximations just computed and the boundary values. For $i = 1$ (first column) we obtain from (15b) the system

$$\begin{aligned} &(j = 1) \qquad u_{10} - 4u_{11}^{(2)} + u_{12}^{(2)} &&= -u_{01} - u_{21}^{(1)} \\ &(j = 2) \qquad u_{11}^{(2)} - 4u_{12}^{(2)} + u_{13} &&= -u_{02} - u_{22}^{(1)}. \end{aligned}$$

The solution is $u_{11}^{(2)} = 91.11$, $u_{12}^{(2)} = 64.44$. For $i = 2$ (second column) we obtain from (15b) the system

$$\begin{aligned} &(j = 1) \qquad u_{20} - 4u_{21}^{(2)} + u_{22}^{(2)} &&= -u_{11}^{(1)} - u_{31} \\ &(j = 2) \qquad u_{21}^{(2)} - 4u_{22}^{(2)} + u_{23} &&= -u_{12}^{(1)} - u_{32}. \end{aligned}$$

The solution is $u_{21}^{(2)} = 91.11$, $u_{22}^{(2)} = 64.44$.

In this example, which merely serves to explain the practical procedure in the ADI method, the accuracy of the second approximations is about the same as of those of two Gauss–Seidel steps in Sec. 19.3 (where $u_{11} = x_1$, $u_{21} = x_2$, $u_{12} = x_3$, $u_{22} = x_4$), as the following table shows.

	u_{11}	u_{21}	u_{12}	u_{22}
ADI, 2nd approximations	91.11	91.11	64.44	64.44
Gauss–Seidel, 2nd approximations	93.75	90.62	65.62	64.06
Exact solution of (12)	87.50	87.50	62.50	62.50

Improving convergence. Additional improvement of the convergence of the ADI method results from the following interesting idea. Introducing a parameter p, we can also write (11) in the form

$$u_{i-1,j} - (2 + p)u_{ij} + u_{i+1,j} = -u_{i,j-1} + (2 - p)u_{ij} - u_{i,j+1} \tag{16a}$$

and

$$u_{i,j-1} - (2 + p)u_{ij} + u_{i,j+1} = -u_{i-1,j} + (2 - p)u_{ij} - u_{i+1,j}. \tag{16b}$$

This gives the more general ADI iteration formulas

$$u_{i-1,j}^{(m+1)} - (2 + p)u_{ij}^{(m+1)} + u_{i+1,j}^{(m+1)} = -u_{i,j-1}^{(m)} + (2 - p)u_{ij}^{(m)} - u_{i,j+1}^{(m)} \tag{17a}$$

and

$$u_{i,j-1}^{(m+2)} - (2 + p)u_{ij}^{(m+2)} + u_{i,j+1}^{(m+2)} = -u_{i-1,j}^{(m+1)} + (2 - p)u_{ij}^{(m+1)} - u_{i+1,j}^{(m+1)}. \tag{17b}$$

For $p = 2$, this is (15). The parameter p may be used for improving convergence. Indeed, one can show that the ADI method converges for positive p, and that the optimum value for maximum rate of convergence is

$$p_0 = 2 \sin \frac{\pi}{K} \tag{18}$$

where K is the larger of $M + 1$ and $N + 1$ (see above). Even better results can be achieved by letting p vary from step to step. More details of the ADI method and variants are discussed in Ref. [E8] listed in Appendix 1.

Problem Set 20.4

1. Derive (5b). **2.** Derive (6b). **3.** Derive (6c).

4. Solve Example 1, choosing $h = 3$.

5. In Example 1, take $h = 6$, compute u_{11}, and compare it with the exact value 75.

6. **(Use of symmetry)** Conclude from the boundary values in Example 1 that $u_{21} = u_{11}$ and $u_{22} = u_{12}$. Show that this leads to a system of two equations and solve it.

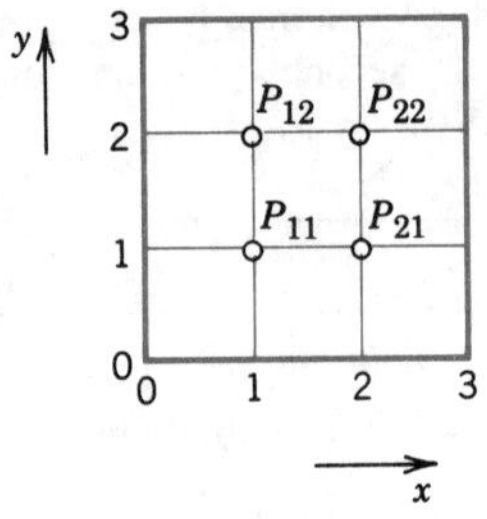

Fig. 429. Problems 7–13

Using the grid in Fig. 429, compute the potential at the four interior points for the following boundary values.

7. $u(1, 0) = 60$, $u(2, 0) = 300$, $u = 100$ on the three other edges of the boundary

8. $u = 220$ on the upper and lower edges, $u = 110$ on the left edge, $u = -10$ on the right edge

9. $u = x^4$ on the lower edge, $81 - 54y^2 + y^4$ on the right edge, $x^4 - 54x^2 + 81$ on the upper edge, y^4 on the left edge. (Verify that the exact solution is $x^4 - 6x^2y^2 + y^4$ and determine the error.)

10. $u = 0$ on the left edge, x^3 on the lower edge, $27 - 9y^2$ on the right edge, $x^3 - 27x$ on the upper edge

11. $u = \sin \frac{1}{3}\pi x$ on the upper edge, 0 on the other edges

12. Apply the ADI method (2 steps) to the Dirichlet problem in Prob. 11, using the grid in Fig. 429, as before, and starting values zero.

13. What p_0 in (18) should we choose for Prob. 12? Apply the ADI formulas (17) with $p_0 = 1.7$ to Prob. 12, performing 1 step. Illustrate the improved convergence by comparing with the corresponding values 0.077, 0.308 after the first step in Prob. 12. (Use starting values zero.)

14. Find the potential in Fig. 430 using (a) the coarse grid, (b) the fine grid, and Gauss elimination. *Hint.* In (b), use symmetry; take $u = 0$ as the boundary value at the two points at which the potential has a jump.

15. How many Gauss–Seidel steps would be needed to obtain the answer to Prob. 14, coarse grid, to 5S (5 significant figures) if one started from 0, 0? In the case of the fine grid in Prob. 14, the Gauss–Seidel method converges more slowly. Can you see the reason by inspecting the system of equations?

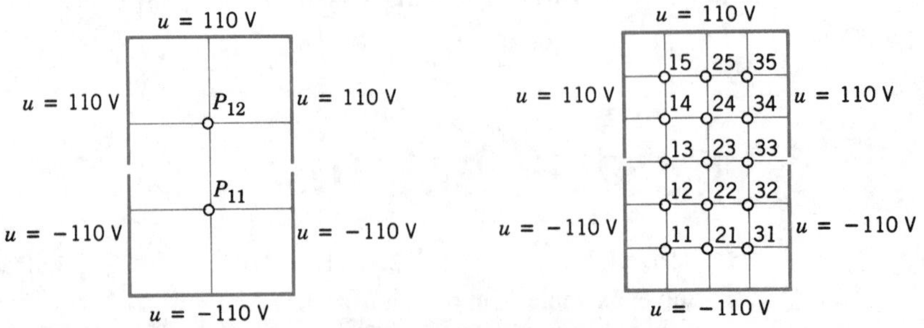

Fig. 430. Region and grids in Problem 14

20.5 Neumann and Mixed Problems. Irregular Boundary

We continue our discussion of the numerical solution of boundary value problems for elliptic equations in a region R in the xy-plane. The Dirichlet problem was studied in the last section. In **Neumann** and **mixed problems** (defined in the last section) we are confronted with a new situation, because there are boundary points at which the (outer) **normal derivative** $u_n = \partial u/\partial n$ of the solution is given, but u itself is unknown since it is not given. To handle such points we need a new idea. This idea is the same for Neumann and mixed problems. Hence we may explain it in connection with one of these two types of problem. We shall do so and consider a typical example as follows.

EXAMPLE 1 **Mixed boundary value problem for a Poisson equation**

Solve the mixed boundary value problem for the Poisson equation

$$\nabla^2 u = u_{xx} + u_{yy} = 12xy$$

shown in Fig. 431*a*.

Solution. We use the grid shown in Fig. 431*b*, where $h = 0.5$. From the formulas $u = 3y^3$ and $u_n = 6x$ given on the boundary we compute the boundary data

$$\text{(1)} \qquad u_{31} = 0.375, \qquad u_{32} = 3, \qquad \frac{\partial u_{12}}{\partial n} = \frac{\partial u_{12}}{\partial y} = 3, \qquad \frac{\partial u_{22}}{\partial n} = \frac{\partial u_{22}}{\partial y} = 6.$$

P_{11} and P_{21} are interior mesh points and can be handled as in the previous section. Indeed, from (7), Sec. 20.4, with $h^2 = 0.25$ and $f(x, y) = 12xy$, and from the given boundary values we obtain two equations corresponding to P_{11} and P_{21}:

$$\text{(2a)} \qquad \begin{aligned} -4u_{11} + u_{21} + u_{12} \qquad &= 0.75 \\ u_{11} - 4u_{21} \qquad + u_{22} &= 1.5 - 0.375 = 1.125. \end{aligned}$$

The only difficulty with these equations seems to be that they involve the unknown values u_{12} and u_{22} of u at P_{12} and P_{22} on the boundary, where the normal derivative $u_n = \partial u/\partial n = \partial u/\partial y$ is given, instead of u; but we shall overcome this difficulty as we go on.

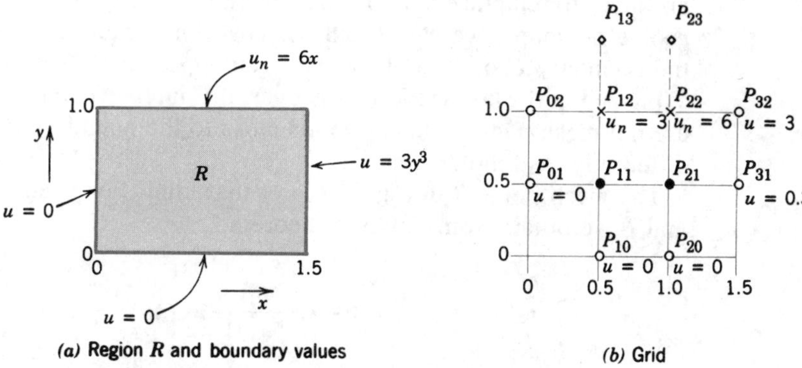

Fig. 431. Mixed boundary value problem in Example 1

We consider P_{12} and P_{22}. The idea that will help us here is as follows. We imagine the region R to be extended above to the first row of external mesh points (corresponding to $y = 1.5$), and we assume that the differential equation also holds in the extended region. Then we can write down two more equations as before (Fig. 431*b*)

$$\text{(2b)} \qquad \begin{aligned} u_{11} \qquad - 4u_{12} + u_{22} + u_{13} \qquad &= 1.5 \\ u_{21} + u_{12} - 4u_{22} \qquad + u_{23} &= 3 - 3 = 0. \end{aligned}$$

We remember that we have not yet used the boundary condition on the upper part of the boundary of R, and we also notice that in (2b) we have introduced two more unknowns u_{13}, u_{23}. But we can now use that condition and get rid of u_{13}, u_{23} by applying the central difference formula for u_y. From (1) we then obtain (see Fig. 431*b*)

$$3 = \frac{\partial u_{12}}{\partial y} \approx \frac{u_{13} - u_{11}}{2h} = u_{13} - u_{11}, \qquad \text{hence} \qquad u_{13} = u_{11} + 3$$

$$6 = \frac{\partial u_{22}}{\partial y} \approx \frac{u_{23} - u_{21}}{2h} = u_{23} - u_{21}, \qquad \text{hence} \qquad u_{23} = u_{21} + 6.$$

Substituting these results into (2b) and simplifying, we have

$$\begin{aligned} 2u_{11} \qquad - 4u_{12} + u_{22} &= 1.5 - 3 = -1.5 \\ 2u_{21} + u_{12} - 4u_{22} &= 0 - 6 \quad = -6. \end{aligned}$$

Together with (2a) this yields, written in matrix form,

$$\text{(3)} \qquad \begin{bmatrix} -4 & 1 & 1 & 0 \\ 1 & -4 & 0 & 1 \\ 2 & 0 & -4 & 1 \\ 0 & 2 & 1 & -4 \end{bmatrix} \begin{bmatrix} u_{11} \\ u_{21} \\ u_{12} \\ u_{22} \end{bmatrix} = \begin{bmatrix} 0.750 \\ 1.125 \\ -1.500 \\ -6.000 \end{bmatrix}.$$

The solution is as follows; the exact values of the problem are given in parentheses.

$$u_{12} = 0.866 \quad \text{(exact 1)} \qquad u_{22} = 1.812 \quad \text{(exact 2)}$$

$$u_{11} = 0.077 \quad \text{(exact 0.125)} \qquad u_{21} = 0.191 \quad \text{(exact 0.25)}.$$

■

Irregular Boundary

We continue our discussion of the numerical solution of boundary value problems for elliptic equations in a region R in the xy-plane. If R has a simple geometric shape, we can usually arrange for certain mesh points to lie on the boundary C of R, and then we can approximate partial derivatives as explained in the last section. However, if C intersects the grid at points that are not mesh points, then at points close to the boundary we must proceed differently, as follows.

The mesh point O in Fig. 432 is of that kind. For O and its neighbors A and P we obtain from Taylor's theorem

$$\text{(4)} \qquad \begin{aligned} &\text{(a)} \quad u_A = u_O + ah\frac{\partial u_O}{\partial x} + \frac{1}{2}(ah)^2\frac{\partial^2 u_O}{\partial x^2} + \cdots \\ &\text{(b)} \quad u_P = u_O - h\frac{\partial u_O}{\partial x} + \frac{1}{2}h^2\frac{\partial^2 u_O}{\partial x^2} + \cdots . \end{aligned}$$

We disregard the terms marked by dots and eliminate $\partial u_O/\partial x$. Equation (4b) times a plus equation (4a) gives

$$u_A + au_P \approx (1 + a)u_O + \frac{1}{2}a(a + 1)h^2 \frac{\partial^2 u_O}{\partial x^2}.$$

We solve this algebraically for the derivative, obtaining

$$\frac{\partial^2 u_O}{\partial x^2} \approx \frac{2}{h^2}\left[\frac{1}{a(1 + a)} u_A + \frac{1}{1 + a} u_P - \frac{1}{a} u_O\right].$$

Similarly, by considering the points O, B, and Q,

$$\frac{\partial^2 u_O}{\partial y^2} \approx \frac{2}{h^2}\left[\frac{1}{b(1 + b)} u_B + \frac{1}{1 + b} u_Q - \frac{1}{b} u_O\right].$$

By addition,

$$\text{(5)} \qquad \nabla^2 u_O \approx \frac{2}{h^2}\left[\frac{u_A}{a(1+a)} + \frac{u_B}{b(1+b)} + \frac{u_P}{1+a} + \frac{u_Q}{1+b} - \frac{(a+b)u_O}{ab}\right].$$

For example, if $a = \frac{1}{2}$, $b = \frac{1}{2}$, instead of the pattern (see Sec. 20.4)

$$\left\{\begin{matrix} & 1 & \\ 1 & -4 & 1 \\ & 1 & \end{matrix}\right\} \quad \text{we now have} \quad \left\{\begin{matrix} & \frac{4}{3} & \\ \frac{2}{3} & -4 & \frac{4}{3} \\ & \frac{2}{3} & \end{matrix}\right\},$$

the sum of all five terms still being zero (which is useful for checking).

Using the same ideas, the student may show that in the case of Fig. 433,

$$\text{(6)} \qquad \nabla^2 u_O \approx \frac{2}{h^2}\left[\frac{u_A}{a(a+p)} + \frac{u_B}{b(b+q)} + \frac{u_P}{p(p+a)} + \frac{u_Q}{q(q+b)} - \frac{ap+bq}{abpq} u_O\right],$$

a formula that takes care of all conceivable cases.

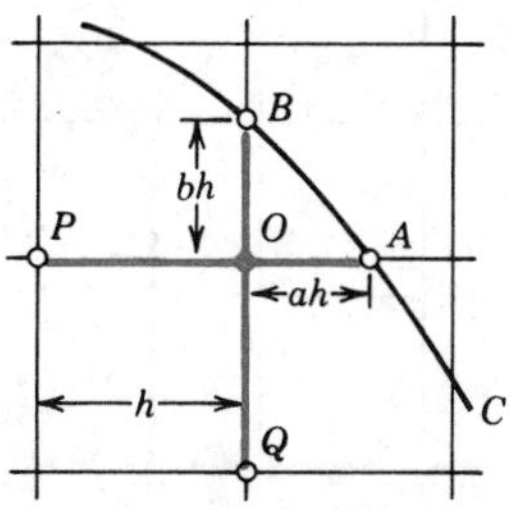

Fig. 432. Curved boundary C of a region R, a mesh point O near C, and neighbors A, B, P, Q

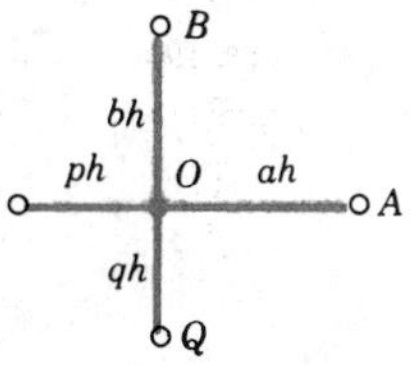

Fig. 433. Neighboring points A, B, P, Q of a mesh point O and notations in formula (6)

EXAMPLE 2 **Dirichlet problem for the Laplace equation. Curved boundary**

Find the potential u in the region in Fig. 434 that has the boundary values given in that figure; here the curved portion of the boundary is an arc of the circle of radius 10 about (0, 0). Use the grid in the figure.

Solution. u is a solution of the Laplace equation. From the given formulas for the boundary values $u = x^3$, $u = 512 - 24y^2, \cdots$ we compute the values at the points where we need them; the result is shown in the figure. For P_{11} and P_{12} we have the usual regular pattern, and for P_{21} and P_{22} we use (6), obtaining

$$(7)\quad P_{11}, P_{12}: \begin{Bmatrix} & 1 & \\ 1 & -4 & 1 \\ & 1 & \end{Bmatrix}, \quad P_{21}: \begin{Bmatrix} & 0.5 & \\ 0.6 & -2.5 & 0.9 \\ & 0.5 & \end{Bmatrix}, \quad P_{22}: \begin{Bmatrix} & 0.9 & \\ 0.6 & -3 & 0.9 \\ & 0.6 & \end{Bmatrix}.$$

We use this and the boundary values and take the mesh points in the order $P_{11}, P_{21}, P_{12}, P_{22}$. Then we obtain the system

$$\begin{aligned} -4u_{11} + u_{21} + u_{12} \qquad\qquad &= 0 - 27 &&= -27 \\ 0.6u_{11} - 2.5u_{21} \qquad + 0.5u_{22} &= -0.9 \cdot 296 - 0.5 \cdot 216 &&= -374.4 \\ u_{11} \qquad - 4u_{12} + u_{22} &= 702 + 0 &&= 702 \\ 0.6u_{21} + 0.6u_{12} - 3u_{22} &= 0.9 \cdot 352 + 0.9 \cdot 936 &&= 1159.2. \end{aligned}$$

In matrix form,

$$(8)\quad \begin{bmatrix} -4 & 1 & 1 & 0 \\ 0.6 & -2.5 & 0 & 0.5 \\ 1 & 0 & -4 & 1 \\ 0 & 0.6 & 0.6 & -3 \end{bmatrix} \begin{bmatrix} u_{11} \\ u_{21} \\ u_{12} \\ u_{22} \end{bmatrix} = \begin{bmatrix} -27 \\ -374.4 \\ 702 \\ 1159.2 \end{bmatrix}.$$

The Gauss elimination yields the (rounded) values

$$u_{11} = -55.6 \qquad u_{21} = 49.2, \qquad u_{12} = -298.5, \qquad u_{22} = -436.3.$$

Clearly, from a grid with so few mesh points we cannot expect great accuracy. The exact values are

$$u_{11} = -54, \qquad u_{21} = 54, \qquad u_{12} = -297, \qquad u_{22} = -432.$$

In practice one would use a much finer grid and solve the resulting large system by an indirect method. ∎

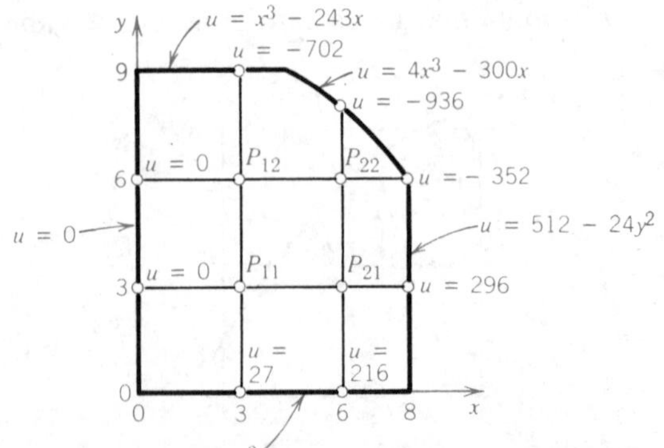

Fig. 434. Region, boundary values of the potential, and grid in Example 2

Problem Set 20.5

1. Check the values given at the end of Example 1 by solving the system (3) by the Gauss elimination.

2. Solve the mixed boundary value problem for the Laplace equation $\nabla^2 u = 0$ in the rectangle in Fig. 431 (using the grid in Fig. 431*b*) and the boundary conditions $u_x = 0$ on the left edge, $u_x = 3$ on the right edge, $u = x^2$ on the lower edge, and $u = x^2 - 1$ on the upper edge.

3. Solve the mixed boundary value problem for the Poisson equation $\nabla^2 u = 2(x^2 + y^2)$ in the region and for the boundary conditions shown in Fig. 435, using the indicated grid.

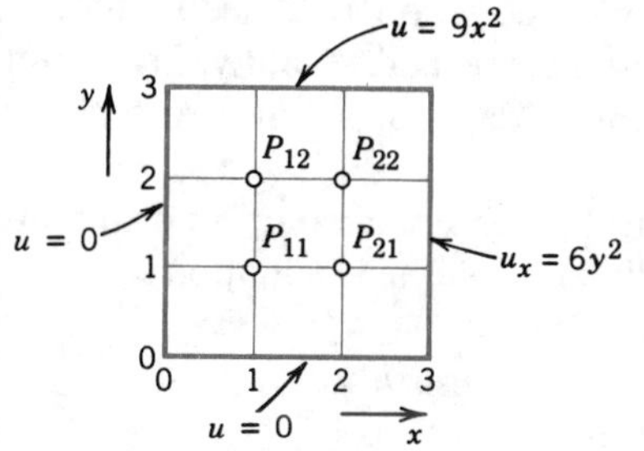

Fig. 435. Problem 3

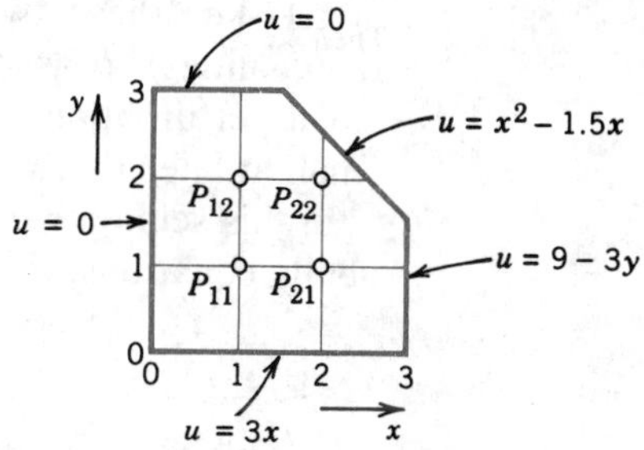

Fig. 436. Problem 9

4. Eliminating $\partial^2 u_O / \partial x^2$ from (4), show that

$$\frac{\partial u_O}{\partial x} \approx \frac{1}{h}\left[\frac{1}{a(1+a)} u_A - \frac{1-a}{a} u_O - \frac{a}{1+a} u_P\right].$$

5. Verify the sample calculation right after (5).

6. Give a detailed derivation of (6).

7. Verify (7).

8. Solve (8) by the Gauss elimination.

9. Solve the Laplace equation in the region and for the boundary values shown in Fig. 436, using the indicated grid. (The sloping portion of the boundary is $y = 4.5 - x$.)

10. Solve the Poisson equation $\nabla^2 u = 2$ in the region and for the boundary values shown in Fig. 437, using the grid also shown in the figure.

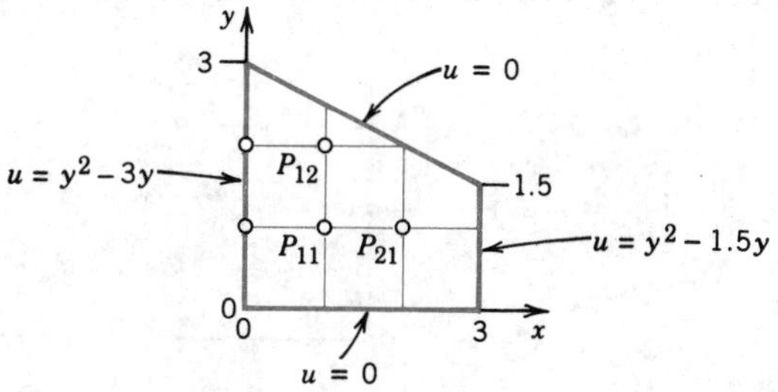

Fig. 437. Problem 10

20.6 Methods for Parabolic Equations

The last two sections concerned elliptic equations, and we now turn to parabolic equations. The definitions of elliptic, parabolic, and hyperbolic equations were given in Sec. 20.4. There it was also mentioned that the general behavior of solutions differs from type to type, and so do the problems of practical interest. This reflects on numerical methods as follows. For all three types, one replaces the equation by a corresponding difference equation, but for *parabolic* and *hyperbolic* equations this does not automatically guarantee the **convergence** of the approximate solution to the exact solution as the mesh $h \to 0$; in fact, it does not even guarantee convergence at all. For these two types of equation one needs additional conditions (inequalities) to assure convergence and **stability,** the latter meaning that small perturbations in the initial data (or small errors at any time) remain small at later times.

In this section we explain the numerical solution of the prototype of parabolic equations, the one-dimensional heat equation

$$u_t = c^2 u_{xx} \qquad (c \text{ constant}).$$

This equation is usually considered for x in some fixed interval, say, $0 \leqq x \leqq L$, and time $t \geqq 0$, and one prescribes the initial temperature $u(x, 0) = f(x)$ (f given) and boundary conditions at $x = 0$ and $x = L$ for all $t \geqq 0$, for instance $u(0, t) = 0$, $u(L, t) = 0$. We may assume $c = 1$ and $L = 1$; this can always be accomplished by a linear transformation of x and t (Prob. 11). Then the **heat equation** and those conditions are

$$(1) \qquad u_t = u_{xx} \qquad 0 \leqq x \leqq 1,\ t \geqq 0$$

$$(2) \qquad u(x, 0) = f(x) \qquad \text{(initial condition)}$$

$$(3) \qquad u(0, t) = u(1, t) = 0 \qquad \text{(boundary conditions).}$$

A simple finite difference approximation of (1) is [see (6a) in Sec. 20.4]

$$(4) \qquad \frac{1}{k}(u_{i,j+1} - u_{ij}) = \frac{1}{h^2}(u_{i+1,j} - 2u_{ij} + u_{i-1,j}).$$

Figure 438 shows a corresponding grid and mesh points. The mesh size is

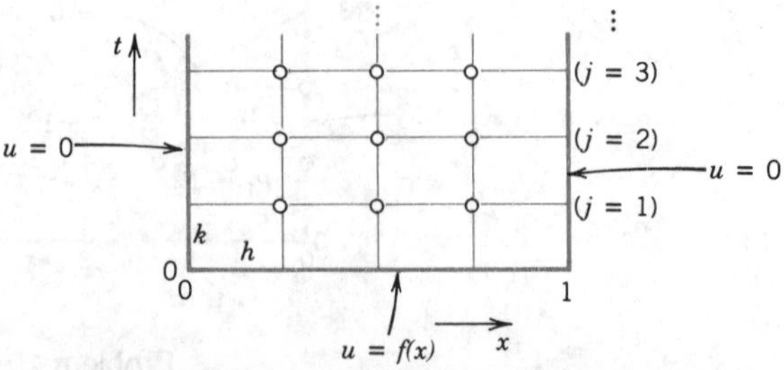

Fig. 438. Grid and mesh points corresponding to (4), (5)

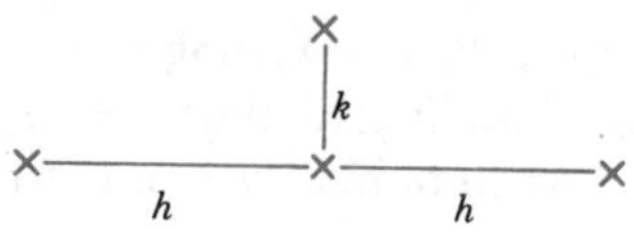

Fig. 439. The four points in (4) and (5)

h in the x-direction and k in the t-direction. Formula (4) involves the four points shown in Fig. 439. On the left we have used a *forward* difference quotient since we have no information for negative t at the start. From (4) we calculate $u_{i,j+1}$, which corresponds to time row $j + 1$, in terms of the three other u that correspond to time row j; solving (4) for $u_{i,j+1}$, we have

$$u_{i,j+1} = (1 - 2r)u_{ij} + r(u_{i+1,j} + u_{i-1,j}), \qquad r = \frac{k}{h^2}. \tag{5}$$

Computations by means of this formula are simple and straightforward. However, it can be shown that crucial to the convergence of this method is the condition

$$r = \frac{k}{h^2} \leqq \frac{1}{2}, \tag{6}$$

that is, that u_{ij} have a positive coefficient in (5) or (for $r = \frac{1}{2}$) be absent from (5). Intuitively, (6) means that we should not move too fast in the t-direction. An example is given below.

Crank–Nicolson Method

Condition (6) is a handicap in practice. Indeed, to attain sufficient accuracy, we have to choose h small, which makes k very small by (6). For example, if $h = 0.1$, then $k \leqq 0.005$. And a switch to $\frac{1}{2}h$ quadruples the number of time steps needed to reach a certain t-value. Accordingly, we should look for a method based on a more satisfactory discretization of the heat equation.

Such a method that imposes no restriction on $r = k/h^2$ is the **Crank–Nicholson method,** which uses values of u at the six points in Fig. 440. The idea of the method is the replacement of the difference quotient on the right side of (4) by $\frac{1}{2}$ times the sum of two such difference quotients at two time rows (see Fig. 440). Instead of (4) we then have

$$\begin{aligned} \frac{1}{k}(u_{i,j+1} - u_{ij}) &= \frac{1}{2h^2}(u_{i+1,j} - 2u_{ij} + u_{i-1,j}) \\ &\quad + \frac{1}{2h^2}(u_{i+1,j+1} - 2u_{i,j+1} + u_{i-1,j+1}). \end{aligned} \tag{7}$$

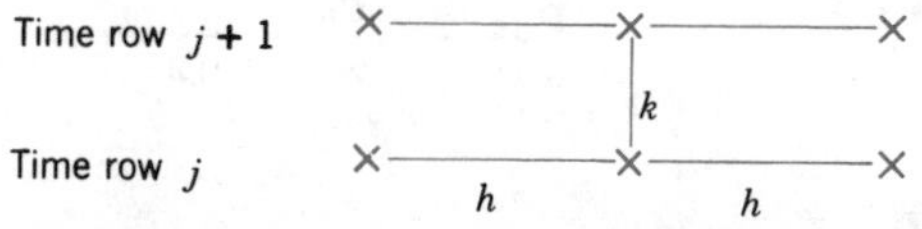

Fig. 440. The six points in the Crank–Nicolson formulas (7) and (8)

Multiplying by $2k$ and writing $r = k/h^2$ as before, we can collect the three terms corresponding to time row $j + 1$ on the left and the three terms corresponding to time row j on the right; this gives

$$(8) \quad (2 + 2r)u_{i,j+1} - r(u_{i+1,j+1} + u_{i-1,j+1}) = (2 - 2r)u_{ij} + r(u_{i+1,j} + u_{i-1,j}).$$

How to use (8)? In general, the three values on the left are unknown, whereas the three values on the right are known. If we divide the x-interval $0 \leqq x \leqq 1$ in (1) into n equal intervals, we have $n - 1$ internal mesh points per time row (see Fig. 438, where $n = 4$). Then for $j = 0$ and $i = 1, \cdots, n - 1$, formula (8) gives a linear system of $n - 1$ equations for the $n - 1$ unknown values $u_{11}, u_{21}, \cdots, u_{n-1,1}$ in the first time row in terms of the initial values $u_{00}, u_{10}, \cdots, u_{n0}$ and the boundary values u_{01}, u_{n1} $(= 0)$. Similarly for $j = 1, j = 2$, and so on; that is, for each time row we have to solve such a system of $n - 1$ linear equations resulting from (8).

Although $r = k/h^2$ is no longer restricted, smaller r will still give better results. In practice, one chooses a k by which one can save a considerable amount of work, without making r too large. For instance, often a good choice is $r = 1$ (which would be impossible in the previous direct method). Then (8) becomes simply

$$(9) \qquad 4u_{i,j+1} - u_{i+1,j+1} - u_{i-1,j+1} = u_{i+1,j} + u_{i-1,j}.$$

EXAMPLE 1 **Temperature in a bar. Crank–Nicolson method, direct method**

Consider a laterally insulated metal bar of length 1 and such that $c^2 = 1$ in the heat equation. Suppose that the ends of the bar are kept at temperature $u = 0°C$ and the temperature in the bar at some instant—call it $t = 0$—is $f(x) = \sin \pi x$. Applying the Crank-Nicolson method with $h = 0.2$ and $r = 1$, find the temperature $u(x, t)$ in the bar for $0 \leqq t \leqq 0.2$. Compare the results with the exact solution. Also apply (5) with an r satisfying (6), say, $r = 0.25$, and with values not satisfying (6), say, $r = 1$ and $r = 2.5$.

Solution by Crank–Nicolson. Since $r = 1$, formula (8) takes the form (9). Since $h = 0.2$ and $r = k/h^2 = 1$, we have $k = h^2 = 0.04$. Hence we have to do 5 steps. Figure 441 shows the grid. We shall need the initial values

$$u_{10} = \sin 0.2\pi = 0.587\ 785, \qquad u_{20} = \sin 0.4\pi = 0.951\ 057.$$

Also, $u_{30} = u_{20}$ and $u_{40} = u_{10}$. (Recall that u_{10} means u at P_{10} in Fig. 441, etc.) In each time row in Fig. 441 there are 4 internal mesh points. Hence in each time step we would have to

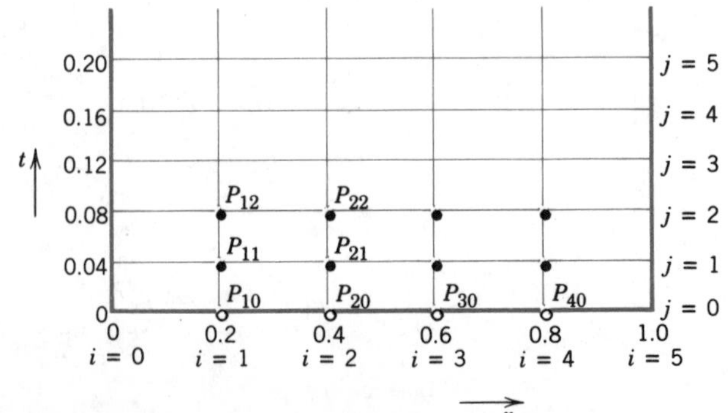

Fig. 441. Grid in Example 1

solve 4 equations in 4 unknowns. But since the initial temperature distribution is symmetric with respect to $x = 0.5$ and $u = 0$ at both ends for all t, we have $u_{31} = u_{21}$, $u_{41} = u_{11}$ in the first time row and similarly for the other rows. This reduces each system to 2 equations in 2 unknowns. By (9), since $u_{31} = u_{21}$, for $j = 0$ these equations are

$$4u_{11} - u_{21} = u_{00} + u_{20} = 0.951\ 057$$

$$-u_{11} + 4u_{21} - u_{21} = u_{10} + u_{20} = 1.538\ 842.$$

The solution is $u_{11} = 0.399\ 274$, $u_{21} = 0.646\ 039$. Similarly, for $j = 1$ we have the system

$$4u_{12} - u_{22} = u_{01} + u_{21} = 0.646\ 039$$

$$-u_{12} + 3u_{22} = u_{11} + u_{21} = 1.045\ 313.$$

The solution is $u_{12} = 0.271\ 221$, $u_{22} = 0.438\ 845$, and so on. This gives the temperature distribution (Fig. 442):

t	$x = 0$	$x = 0.2$	$x = 0.4$	$x = 0.6$	$x = 0.8$	$x = 1$
0.00	0	0.588	0.951	0.951	0.588	0
0.04	0	0.399	0.646	0.646	0.399	0
0.08	0	0.271	0.439	0.439	0.271	0
0.12	0	0.184	0.298	0.298	0.184	0
0.16	0	0.125	0.202	0.202	0.125	0
0.20	0	0.085	0.138	0.138	0.085	0

Comparison with the exact solution. The present problem can be solved exactly by separating variables (Sec. 11.5); the result is

$$u(x, t) = \sin \pi x\ e^{-\pi^2 t}. \tag{10}$$

Solution by the direct method (5) with $r = 0.25$. For $h = 0.2$ and $r = k/h^2 = 0.25$ we have $k = rh^2 = 0.25 \cdot 0.04 = 0.01$. Hence we have to perform 4 times as many steps as with the Crank–Nicholson method! Formula (5) with $r = 0.25$ is

$$u_{i,j+1} = 0.25(u_{i-1,j} + 2u_{ij} + u_{i+1,j}). \tag{11}$$

We can make use of the symmetry. For $j = 0$ we need $u_{00} = 0$, $u_{10} = 0.587\ 785$, $u_{20} = u_{30} = 0.951\ 057$ and compute

$$u_{11} = 0.25(u_{00} + 2u_{10} + u_{20}) = 0.531\ 657$$

$$u_{21} = 0.25(u_{10} + 2u_{20} + u_{30}) = 0.25(u_{10} + 3u_{20}) = 0.860\ 239.$$

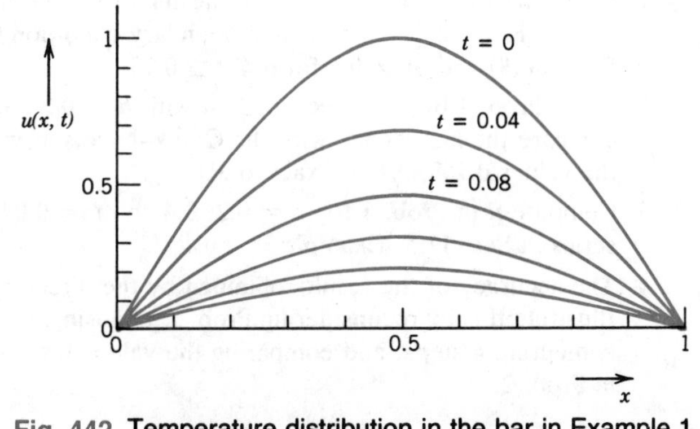

Fig. 442. Temperature distribution in the bar in Example 1

Of course we can omit the boundary terms $u_{01} = 0$, $u_{02} = 0, \cdots$ from the formulas. For $j = 1$ we compute

$$u_{12} = 0.25(2u_{11} + u_{21}) = 0.480\ 888$$

$$u_{22} = 0.25(\ u_{11} + 3u_{21}) = 0.778\ 094$$

and so on. We have to perform 20 steps, but the numerical values show that the accuracy is only about the same as that of the Crank–Nicolson values CN (exact values given with 3 decimals):

t	$x = 0.2$			$x = 0.4$		
	CN	By (11)	Exact	CN	By (11)	Exact
0.04	0.399	0.393	0.396	0.646	0.637	0.641
0.08	0.271	0.263	0.267	0.439	0.426	0.432
0.12	0.184	0.176	0.180	0.298	0.285	0.291
0.16	0.125	0.118	0.121	0.202	0.191	0.196
0.20	0.085	0.079	0.082	0.138	0.128	0.132

Failure of (5) with r violating (6). Formula (5) with $h = 0.2$ and $r = 1$—which violates (6)—is

$$u_{i,j+1} = u_{i-1,j} - u_{ij} + u_{i+1,j}$$

and gives very poor values; some of them are as follows:

t	$x = 0.2$	Exact	$x = 0.4$	Exact
0.4	0.363	0.396	0.588	0.641
0.12	0.139	0.180	0.225	0.291
0.20	0.053	0.082	0.086	0.132

Formula (5) with an even larger $r = 2.5$ (and $h = 0.2$ as before) gives completely nonsensical results; some of them are

t	$x = 0.2$	Exact	$x = 0.4$	Exact
0.1	0.0265	0.2191	0.0429	0.3545
0.3	0.0001	0.0304	0.0001	0.0492
0.5	0.0018	0.0042	−0.0011	0.0068

■

Problem Set 20.6

1. Solve the heat equation (1) for the initial condition $f(x) = x$ if $0 \leqq x \leqq \frac{1}{2}$, $f(x) = 1 - x$ if $\frac{1}{2} < x \leqq 1$, and boundary condition (3) by the Crank–Nicolson formula (9) with $h = 0.2$ for $0 \leqq t \leqq 0.20$.
2. Solve Prob. 1 by the direct method with $h = 0.2$ and $r = 0.25$; do 8 steps and compare the last values with the Crank–Nicolson values 0.107, 0.175 (3D) and the values 0.108, 0.175 (exact to 3D).
3. Compute u in Prob. 1 for $x = 0.2$, 0.4 and $t = 0.04, 0.08, \cdots, 0.20$ from the series in Sec. 11.5, Example 3.
4. The accuracy of the results obtained by the direct method depends on r ($\leqq \frac{1}{2}$). Illustrate this by reconsidering Prob. 2, choosing $r = \frac{1}{2}$ (and $h = 0.2$ as before), computing 4 steps, and comparing the values for $t = 0.04$ and 0.08 with those in Prob. 2.

5. Using the direct method given by (5), with $h = 1$ and $k = 0.5$, find the temperature at $t = 2$ in a laterally insulated bar of length 10 whose ends at $x = 0$ and $x = 10$ are kept at zero temperature and whose initial temperature is $f(x) = 0.1x - 0.01x^2$.
6. Solve Prob. 5, choosing $k = 0.25$ (instead of 0.5), leaving everything else as before. To how many digits do the two solutions agree?
7. Extend the computation in Prob. 5 to $t = 4$. Then find the temperature at $t = 4$ and $x = 2, 4, 6, 8$ by the Crank–Nicolson formula (9) with $h = 2$ (thus $k = 4$) and compare the values.
8. Find the exact solution in Prob. 5 by the method in Sec. 11.5. Note that the series thus obtained converges rapidly, so that we can expect the sum of the first 2 terms to give 3D-values of the solution for $t = 2$ and $t = 4$. Compute $u(2, 2)$ and $u(4, 2)$ in this way and compare with the values in Prob. 5. Compute $u(2, 4)$, $u(4, 4)$ and compare with Prob. 7.
9. If the left end of a laterally insulated bar extending from $x = 0$ to $x = 1$ is insulated, the boundary condition at $x = 0$ is $u_n(0, t) = u_x(0, t) = 0$. Show that in the application of the direct method given by (5), we can compute $u_{0,j+1}$ by the formula

$$u_{0,j+1} = (1 - 2r)u_{0j} + 2ru_{1j}.$$

10. Applying the direct method with $h = 0.2$ and $r = 0.25$, determine the temperature $u(x, t)$, $0 \leqq t \leqq 0.12$ in a laterally insulated bar extending from $x = 0$ to $x = 1$ if $u(x, 0) = 0$, the left end is insulated and the right end is kept at temperature $g(t) = \sin \frac{50}{3} \pi t$. *Hint.* See Prob. 9.
11. **(Nondimensional form)** Show that the heat equation $\tilde{u}_{\tilde{t}} = c^2 \tilde{u}_{\tilde{x}\tilde{x}}$, $0 \leqq \tilde{x} \leqq L$, can be transformed to the "nondimensional" standard form $u_t = u_{xx}$, $0 \leqq x \leqq 1$, by setting $x = \tilde{x}/L$, $t = c^2\tilde{t}/L^2$, $u = \tilde{u}/u_0$, where u_0 is any constant temperature.

20.7 Methods for Hyperbolic Equations

In this section we consider the numerical solution of problems involving hyperbolic equations. We explain a standard method in terms of a typical setting for the prototype of a hyperbolic equation, the **wave equation:**

(1)	$u_{tt} = u_{xx}$	$0 \leqq x \leqq 1,\ t \geqq 0$
(2)	$u(x, 0) = f(x)$	(Given initial displacement)
(3)	$u_t(x, 0) = g(x)$	(Given initial velocity)
(4)	$u(0, t) = u(1, t) = 0$	(Boundary conditions).

Note that an equation $u_{tt} = c^2 u_{xx}$ and another x-interval can be reduced to the form (1) by a linear transformation of x and t. (This is similar to Sec. 20.6, Prob. 11.)

For instance, (1)–(4) is the model of a vibrating elastic string with fixed ends at $x = 0$ and $x = 1$ (see Sec. 11.2). Although an analytic solution of the problem is given in (7), Sec. 11.4, we use the problem for explaining basic ideas of the numerical approach that are also relevant for more complicated hyperbolic equations.

Replacing the derivatives by difference quotients as before, we obtain from (1) [see (6) in Sec. 20.4 with $y = t$]

$$\frac{1}{k^2}(u_{i,j+1} - 2u_{ij} + u_{i,j-1}) = \frac{1}{h^2}(u_{i+1,j} - 2u_{ij} + u_{i-1,j}) \tag{5}$$

where h is the mesh size in x, and k is the mesh size in t. This difference equation relates 5 points as shown in Fig. 443*a*. It suggests a rectangular grid similar to those for parabolic equations in the preceding section. We choose $r^* = k^2/h^2 = 1$. Then u_{ij} drops out and we have

$$u_{i,j+1} = u_{i-1,j} + u_{i+1,j} - u_{i,j-1} \qquad \text{(Fig. 443}b\text{)}. \tag{6}$$

It can be shown that for $0 < r^* \leqq 1$ the present method is stable, so that from (6) we may expect reasonable results for initial data that have no discontinuities. (For a hyperbolic equation, the latter would propagate into the solution domain—a phenomenon that would be difficult to deal with on our present grid; for details, see Ref. [E8] in Appendix 1.)

Equation (6) still involves 3 time steps $j - 1, j, j + 1$, whereas the formulas in the parabolic case involved only 2 time steps. Furthermore, we now have 2 initial conditions. So we ask how we get started and how we can use the initial condition (3). This can be done as follows. From $u_t(x, 0) = g(x)$ we derive the difference formula

$$\frac{1}{2k}(u_{i1} - u_{i,-1}) = g_i, \qquad \text{hence} \qquad u_{i,-1} = u_{i1} - 2kg_i \tag{7}$$

where $g_i = g(ih)$. For $t = 0$, that is, $j = 0$, equation (6) is

$$u_{i1} = u_{i-1,0} + u_{i+1,0} - u_{i,-1}.$$

Into this we substitute (7), then take $-u_{i1}$ from the right to the left and divide by 2, finding

$$u_{i1} = \tfrac{1}{2}(u_{i-1,0} + u_{i+1,0}) + kg_i \tag{8}$$

which expresses u_{i1} in terms of the initial data.

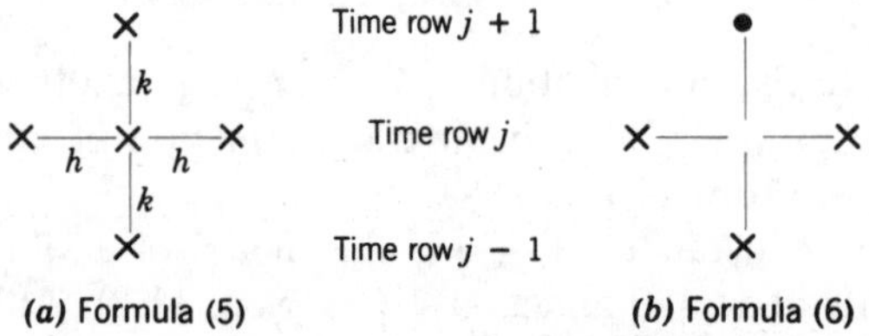

Fig. 443. Mesh points used in (5) and (6)

EXAMPLE 1 Vibrating string

Apply the present method of numerical solution with $h = k = 0.2$ to the problem (1)–(4), where

$$f(x) = \sin \pi x, \qquad g(x) = 0.$$

Solution. The grid is the same as in Fig. 441, Sec. 20.6, except for the values of t, which now are 0.2, 0.4, $\cdots$ (instead of 0.04, 0.08, $\cdots$). The initial values u_{00}, u_{10}, $\cdots$ are the same as in Example 1, Sec. 20.6. From (8) and $g(x) = 0$ we have

$$u_{i1} = \tfrac{1}{2}(u_{i-1,0} + u_{i+1,0}).$$

From this we compute

$$u_{11} = \tfrac{1}{2}(u_{00} + u_{20}) = \tfrac{1}{2} \cdot 0.951\ 057 = 0.475\ 528$$

$$u_{21} = \tfrac{1}{2}(u_{10} + u_{30}) = \tfrac{1}{2} \cdot 1.538\ 842 = 0.769\ 421,$$

and $u_{31} = u_{21}$, $u_{41} = u_{11}$ by symmetry as in Sec. 20.6, Example 1. From (6) with $j = 1$, using $u_{01} = u_{02} = \cdots = 0$, we now compute

$$u_{12} = u_{01} + u_{21} - u_{10} = 0.769\ 421 - 0.587\ 785 = 0.181\ 636$$

$$u_{22} = u_{11} + u_{31} - u_{20} = 0.475\ 528 + 0.769\ 421 - 0.951\ 057 = 0.293\ 892,$$

and $u_{32} = u_{22}$, $u_{42} = u_{12}$ by symmetry; and so on. We thus obtain the following values of the displacement $u(x, t)$ of the string over the first half-cycle:

t	$x = 0$	$x = 0.2$	$x = 0.4$	$x = 0.6$	$x = 0.8$	$x = 1$
0.0	0	0.588	0.951	0.951	0.588	0
0.2	0	0.476	0.769	0.769	0.476	0
0.4	0	0.182	0.294	0.294	0.182	0
0.6	0	−0.182	−0.294	−0.294	−0.182	0
0.8	0	−0.476	−0.769	−0.769	−0.476	0
1.0	0	−0.588	−0.951	−0.951	−0.588	0

These values are exact, the exact solution of the problem being (see Sec. 11.3)

$$u(x, t) = \sin \pi x \cos \pi t.$$

The reason for the exactness follows from d'Alembert's solution (6), Sec. 11.4. (See Prob. 4, below.) ∎

This is the end of Chap. 20 on numerical methods for ordinary and partial differential equations. We have concentrated on some basic ideas and methods in this field, which is full of interesting open questions and is developing quite rapidly. This is also the end of Part E on numerical methods.

Problem Set 20.7

Solve the vibrating string problem (1)–(4) by the present numerical method with $h = k = 0.2$ on the given t-interval for initial velocity 0 and given initial deflection $f(x)$.

1. $0 \leqq t \leqq 2$, $f(x) = 0.1x(1 - x)$

2. $0 \leqq t \leqq 1$, $f(x) = x^2(1 - x^2)$

3. $0 \leqq t \leqq 1$, $f(x) = x$ if $0 \leqq x \leqq 0.2$, $f(x) = 0.25(1 - x)$ if $0.2 < x \leqq 1$

4. Show that from d'Alembert's solution (14) in Sec. 11.4 with $c = 1$ it follows that (6) in the present section gives the exact value $u_{i,j+1} = u(ih, (j + 1)h)$.
5. If the string governed by (1) starts from its equilibrium position with initial velocity $g(x) = \sin \pi x$, what is its displacement at time $t = 0.4$ and $x = 0.2$, 0.4, 0.6, 0.8? (Use the present method with $h = 0.2$, $k = 0.2$. Compare with the exact values obtained from (13) in Sec. 11.4.)
6. Compute approximate values in Prob. 5, using a finer grid ($h = 0.1$, $k = 0.1$), and notice the increase in accuracy.
7. Illustrate the starting procedure for the present method in the case where both f and g are not identically zero, say,

$$f(x) = 1 - \cos 2\pi x, \qquad g(x) = x - x^2;$$

choose $h = k = 0.1$ and compute 2 time steps.
8. Show that (13) in Sec. 11.4 gives as another starting formula

$$u_{i1} = \frac{1}{2}(u_{i+1,0} + u_{i-1,0}) + \frac{1}{2}\int_{x_i-k}^{x_i+k} g(s)\, ds$$

(where one can evaluate the integral numerically if necessary). In what case is this identical with (8)?
9. Compute u in Prob. 7 for $t = 0.1$ and $x = 0.1, 0.2, \cdots, 0.9$, using the formula in Prob. 8, and compare the values.
10. Solve (1) numerically, subject to the conditions

$$u(x, 0) = x^2, \qquad u_t(x, 0) = 2x, \qquad u_x(0, t) = 2t, \qquad u(1, t) = (1 + t)^2,$$

choosing $h = k = 0.2$ (5 time steps).

Review Questions and Problems for Chapter 20

1. Explain Euler's method and the idea by which we got it.
2. Is Euler's method practically useful? How can it be improved?
3. In each Runge–Kutta step we computed auxiliary quantities. Why?
4. Explain the similarities and differences of the Runge–Kutta and Runge–Kutta–Nyström methods.
5. We got some of the methods from the Taylor series. Can you remember how?
6. What do we mean by one-step and multistep methods?
7. What are the advantages and disadvantages of multistep methods?
8. What do you know about the choice of a step size? About the advantages and disadvantages of a very small step size?
9. Can you recall the idea by which we got the Adams–Bashford method?
10. Explain the difference between the Adams–Bashford and Adams–Moulton methods.
11. Explain why and how we used finite differences in this chapter.
12. In numerical methods for the Laplace equation $\nabla^2 u = 0$ in two variables, by what formula did we relate u at a point P to u at its neighbors E, N, W, S?

13. Derive the formula in Prob. 12 from memory.

14. If we take a grid and solution of Laplace's equation, say, $u = e^x \cos y$, can we expect the formula in Prob. 12 to hold exactly?

15. How did we handle boundary value problems on domains of irregular shape? Write down what you remember; only then look up more details in the text.

16. What was the idea in using given data for the normal derivative on the boundary of a region?

17. What do we mean by an elliptic, parabolic, or hyperbolic equation? Name and write down one equation of each type.

18. Why do we need different numerical methods for different types of partial differential equations?

19. How many initial conditions did we prescribe for the wave equation? For the heat equation?

20. Can some of our methods diverge sometimes? Explain.

21. Solve $y' = 2xy$, $y(0) = 1$ by Euler's method (10 steps, $h = 0.1$). Compute the errors.

22. Solve $y' = 2xy + 1$, $y(0) = 0$ by Euler's method (10 steps, $h = 0.1$). Write the exact solution as an integral.

23. Apply the improved Euler method to the initial value problem $y' = 2x$, $y(0) = 0$, choosing $h = 0.1$. Why are the errors zero?

24. Compute $y = e^x$ for $x = 0, 0.1, 0.2, \cdots, 1.0$ by applying the Runge–Kutta method to $y' = y$, $y(0) = 1$ ($h = 0.1$). Show that the first five decimals of the result are correct.

25. Apply the Runge–Kutta method to the problem $y' = 1 + y^2$, $y(0) = 0$. Choose $h = 0.1$, compute 2 steps, and compare with the exact 8D-values 0.100 334 67 and 0.202 710 04. Solve the problem analytically.

26. Solve $y' = (x + y - 4)^2$, $y(0) = 4$, by the Runge–Kutta method with step $h = 0.2$ for $x = 0, 0.2, \cdots, 1.4$.

27. Show that by applying the method in Sec. 20.2 to a polynomial of first degree we obtain the multistep predictor and corrector formulas

$$y^*_{n+1} = y_n + \frac{h}{2}(3f_n - f_{n-1}), \qquad y_{n+1} = y_n + \frac{h}{2}(f_{n+1} + f_n).$$

28. Apply the multistep method in Prob. 27 to the initial value problem $y' = x + y$, $y(0) = 0$, choosing $h = 0.2$ and doing 5 steps. Compare with the exact values.

29. Solve $y' = (y - x - 1)^2 + 2$, $y(0) = 1$ for $0 \leqq x \leqq 1$ by the Adams–Moulton method (Sec. 20.2) with $h = 0.1$ and starting values from Example 4 in Sec. 20.1.

Apply the method given by (4), Sec. 20.3, with $h = 0.1$ to the given initial value problem (5 steps) and compare with the exact solution.

30. $y'' = y$, $y(0) = 3$, $y'(0) = 1$

31. $y'' = y$, $y(0) = 1$, $y'(0) = 0$

32. $y'' = xy' - 3y$, $y(0) = 0$, $y'(0) = -6$. Exact solution $y = 2x^3 - 6x$

33. Apply the Runge–Kutta–Nyström method to the initial value problem $y'' = -y$, $y(0) = 0$, $y'(0) = 1$; choose $h = 0.2$ and do 5 steps. Compare with the exact solution.

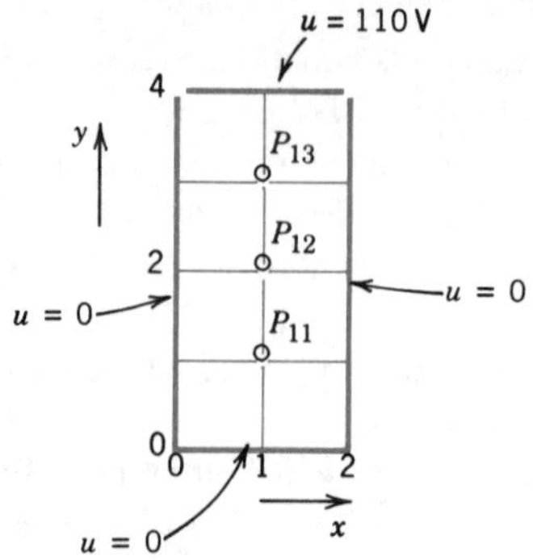

Fig. 444. Problem 35

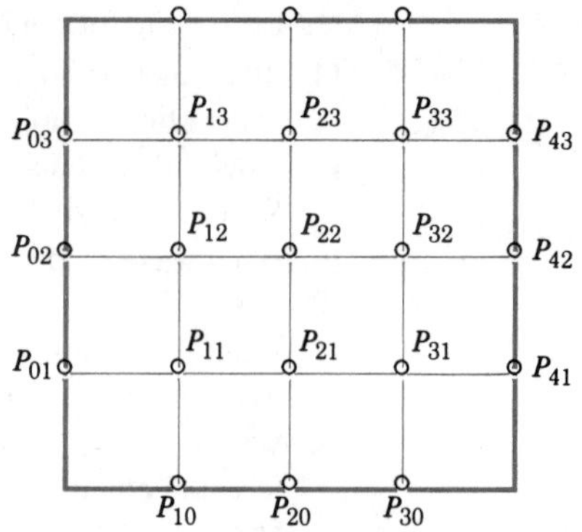

Fig. 445. Problems 36–38

34. In Prob. 33, replace $h = 0.2$ by $h = 0.1$, do 4 steps, and compare the results with those in Prob. 33 and with the exact 9D-values 0.099 833 417, 0.198 669 331, 0.295 520 207, 0.389 418 342.

35. Find rough approximate values of the electrostatic potential at P_{11}, P_{12}, P_{13} in Fig. 444 that lie in a field between conducting plates (in Fig. 444 appearing as sides of a rectangle) kept at potentials 0 and 220 volts as shown. (Use the indicated grid.)

Find the potential in Fig. 445, using the given grid and the following boundary values:

36. $u = 70$ on the upper and left sides, $u = 0$ on the lower and right sides

37. $u(P_{10}) = u(P_{30}) = 960$, $u(P_{20}) = -480$, $u = 0$ elsewhere on the boundary

38. $u(P_{01}) = u(P_{03}) = u(P_{41}) = u(P_{43}) = 200$, $u(P_{10}) = u(P_{30}) = -400$, $u(P_{20}) = 1600$, $u(P_{02}) = u(P_{42}) = u(P_{14}) = u(P_{24}) = u(P_{34}) = 0$

39. Verify (13) in Sec. 20.4 for the system (12) and show that **A** in (12) is nonsingular.

40. Formula (17) in Sec. 20.4 with $p = 0$ would not work. Show this by applying it to the initial value problem in Example 2 of Sec. 20.4, with grid and starting values as before; perform 2 steps. What happens?

41. Solve the heat equation (1), Sec. 20.6, for the initial condition $f(x) = x$ if $0 \leqq x \leqq 0.2$, $f(x) = 0.25(1 - x)$ if $0.2 < x \leqq 1$ and boundary condition (3), Sec. 20.6, by the direct method [formula (5) in Sec. 20.6] with $h = 0.2$ and $k = 0.01$ so that you get values of the temperature at $t = 0.05$ as the answer.

42. A laterally insulated homogeneous bar with ends at $x = 0$ and $x = 1$ has initial temperature 0. Its left end is kept at 0, whereas the temperature at the right end varies sinusoidally according to

$$u(t, 1) = g(t) = \sin \tfrac{25}{3}\pi t.$$

Find the temperature $u(x, t)$ in the bar [solution of (1) in Sec. 20.6] by the direct method with $h = 0.2$ and $r = 0.5$ (one period, that is, $0 \leqq t \leqq 0.24$).

43. Find $u(x, 0.12)$ and $u(x, 0.24)$ in Prob. 42 if the left end of the bar is kept at $-g(t)$ (instead of 0), all the other data being as before.

44. Find out how the results of Prob. 42 can be used for obtaining the results in Prob. 43. Use the values 0.054, 0.172, 0.325, 0.406 ($t = 0.12$, $x = 0.2, 0.4, 0.6, 0.8$) and -0.009, -0.086, -0.252, -0.353 ($t = 0.24$) from the answer to Prob. 42 to check your answer to Prob. 43.

45. Derive the difference approximation of the heat equation.

Summary of Chapter 20 Numerical Methods for Differential Equations

In this chapter we discussed numerical methods for ordinary differential equations (Secs. 20.1–20.3) and partial differential equations (Secs. 20.4–20.7). For first-order equations we considered initial value problems of the form (Sec. 20.1)

$$y' = f(x, y), \qquad y(x_0) = y_0. \tag{1}$$

Numerical methods for solving such a problem can be obtained by truncating the Taylor series

$$y(x + h) = y(x) + hy'(x) + \frac{h^2}{2} y''(x) + \cdots$$

where, by (1), $y' = f$, $y'' = f' = \partial f/\partial x + (\partial f/\partial y)y'$, etc. Truncating after the term hy', we get the *Euler method,* in which we compute step by step

$$y_{n+1} = y_n + hf(x_n, y_n) \qquad (n = 0, 1, \cdots). \tag{2}$$

Taking one more term into account, we obtain the *improved Euler* or *Heun's method.* Both methods show the basic idea in a simple form, but are too inaccurate for most practical purposes. Truncating after the term in h^4, we get the important **Runge–Kutta method** (of fourth order). The crucial idea in this and similar methods is to replace the cumbersome evaluation of derivatives by the evaluation of $f(x, y)$ at suitable points (x, y); thus in each step we first compute four auxiliary quantities (Sec. 20.1)

$$\begin{aligned} k_1 &= hf(x_n, y_n) \\ k_2 &= hf(x_n + \tfrac{1}{2}h,\ y_n + \tfrac{1}{2}k_1) \\ k_3 &= hf(x_n + \tfrac{1}{2}h,\ y_n + \tfrac{1}{2}k_2) \\ k_4 &= hf(x_n + h,\ y_n + k_3) \end{aligned} \tag{3a}$$

and then from them the new value

$$y_{n+1} = y_n + \tfrac{1}{6}(k_1 + 2k_2 + 2k_3 + k_4). \tag{3b}$$

This is a **one-step method,** since y_{n+1} is computed from data of a single step.

A **multistep method** uses data from several preceding steps, thereby avoiding computations such as (3a). By integrating cubic interpolation polynomials we obtained as a multistep method the important **Adams–Moulton method** in which we first compute the **predictor** (Sec. 20.2)

$$\text{(4a)} \qquad y^*_{n+1} = y_n + \frac{1}{24} h(55f_n - 59f_{n-1} + 37f_{n-2} - 9f_{n-3})$$

where $f_j = f(x_j, y_j)$, and then from it the **corrector** (the actual new value)

$$\text{(4b)} \qquad y_{n+1} = y_n + \frac{1}{24} h(9f^*_{n+1} + 19f_n - 5f_{n-1} + f_{n-2})$$

where $f^*_{n+1} = f(x_{n+1}, y^*_{n+1})$. Here, to get started, y_1, y_2, y_3 must be computed by the Runge–Kutta method or some other accurate method.

Section 20.3 concerned second-order ordinary differential equations, in particular, the **Runge–Kutta–Nyström method,** an extension of the Runge–Kutta method to these equations.

Numerical methods for **partial differential equations** are obtained by replacing partial derivatives by difference quotients. This leads to approximating difference equations, for the **Laplace equation** to (Sec. 20.4)

$$\text{(5)} \qquad u_{i+1,j} + u_{i,j+1} + u_{i-1,j} + u_{i,j-1} - 4u_{ij} = 0,$$

for the **heat equation** to (Sec. 20.6)

$$\text{(6)} \qquad \frac{1}{k}(u_{i,j+1} - u_{ij}) = \frac{1}{h^2}(u_{i+1,j} - 2u_{ij} + u_{i-1,j})$$

and for the **wave equation** to (Sec. 20.7)

$$\text{(7)} \qquad \frac{1}{k^2}(u_{i,j+1} - 2u_{ij} + u_{i,j-1}) = \frac{1}{h^2}(u_{i+1,j} - 2u_{ij} + u_{i-1,j});$$

here h and k are the mesh sizes of a grid in the x- and y-directions, respectively, where in (6) and (7) the variable y is time t.

These equations are *elliptic, parabolic,* and *hyperbolic*, respectively. Corresponding numerical methods differ, for the following reason. For elliptic equations we have boundary value problems, and we discussed for them the **Gauss–Seidel** or **Liebmann method** and the **ADI method** (Secs. 20.4, 20.5). For parabolic equations we are given one initial condition and boundary conditions, and we discussed a *direct method* and the **Crank–Nicolson method** (Sec. 20.6). For hyperbolic equations, the problems are similar but we are given a second initial condition; in Sec. 20.7 we explained how to handle such problems numerically.

PART F

OPTIMIZATION, GRAPHS

The ideas of optimization and the application of graphs and digraphs (directed graphs) play an increasing role in engineering, computer science, systems theory, economics, and other areas. In this part we explain some basic concepts, methods, and results in unconstrained optimization (Sec. 21.1), linear programming (Secs. 21.1–21.4), graphs and digraphs, and their application in combinatorial optimization (Chap. 22). Combinatorial methods involving graphs and digraphs are relatively new and an interesting area of ongoing applied and theoretical research.

Chapter 21

Unconstrained Optimization, Linear Programming

This chapter provides an introduction to the more important concepts, methods, and results of optimization. Optimization principles are of increasing importance in modern design and systems operation in various areas. The recent development has been affected by the modern computer capable of solving large-scale problems, and by the corresponding creation of new optimization techniques, so that the whole field is in the process of becoming a large area of its own.

Prerequisite: A modest knowledge of linear systems of equations.
References: Appendix 1, Part F.
Answers to problems: Appendix 2.

21.1 Basic Concepts. Unconstrained Optimization

In an **optimization problem,** the objective is to *optimize* (*maximize* or *minimize*) some function f. This function f is called the **objective function.**

For example, an objective function f to be *maximized* may be the revenue in a production of TV sets, the yield per minute in a chemical process, the mileage per gallon of a certain type of car, the hourly number of customers served in some office, the hardness of steel, or the tensile strength of a rope.

Similarly, we may want to *minimize* f if f is the cost per unit of producing certain cameras, the operating cost of some power plant, the daily loss of heat in a heating system, the idling time of some lathe, or the time needed to produce a fender.

In most optimization problems the objective function f depends on several variables

$$x_1, \cdots, x_n.$$

These are called **control variables** because we can "control" them, that is, choose their values.

For example, the yield of a chemical process may depend on pressure x_1 and temperature x_2. The efficiency of a certain air conditioning system may depend on temperature x_1, air pressure x_2, moisture content x_3, cross-sectional area of outlet x_4, and so on.

Optimization theory develops methods for optimal choices of $x_1, \cdots, x_n$, which maximize (or minimize) the objective function f, that is, methods for finding optimal values of $x_1, \cdots, x_n$.

In many problems the choice of values of $x_1, \cdots, x_n$ is not entirely free but is subject to some **constraints,** that is, additional conditions arising from the nature of the problem and the variables.

For example, if x_1 is production cost, then $x_1 \geqq 0$, and there are many other variables (time, weight, distance traveled by a salesman, etc.) that can take nonnegative values only. Constraints can also have the form of equations (instead of inequalities).

Let us first consider **unconstrained optimization** in the case of a real-valued function $f(x_1, \cdots, x_n)$. We also write $\mathbf{x} = (x_1, \cdots, x_n)$ and $f(\mathbf{x})$, for convenience.

By definition, f has a **minimum** at a point $\mathbf{x} = \mathbf{X}_0$ in a region R where f is defined if $f(\mathbf{x}) \geqq f(\mathbf{X}_0)$ for all $\mathbf{x}$ in R. Similarly, f has a **maximum** at $\mathbf{X}_0$ if $f(\mathbf{x}) \leqq f(\mathbf{X}_0)$ for all $\mathbf{x}$ in R. Minima and maxima together are called **extrema.**

Furthermore, f is said to have a **local minimum** at $\mathbf{X}_0$ if $f(\mathbf{x}) \geqq f(\mathbf{X}_0)$ for all $\mathbf{x}$ in a neighborhood of $\mathbf{X}_0$, say, for all $\mathbf{x}$ satisfying

$$|\mathbf{x} - \mathbf{X}_0| = [(x_1 - X_1)^2 + \cdots + (x_n - X_n)^2]^{1/2} < r,$$

where $\mathbf{X}_0 = (X_1, \cdots, X_n)$ and $r > 0$ is sufficiently small. A *local maximum* is defined similarly.

If f is differentiable and has an extremum at a point $\mathbf{X}_0$ in the *interior of* R (that is, not on the boundary), then the partial derivatives $\partial f/\partial x_1, \cdots, \partial f/\partial x_n$ must be zero at $\mathbf{X}_0$. These are the components of a vector that is called the **gradient** of f and denoted by grad f or ∇f. (For $n = 3$ this agrees with Sec. 8.9.) Thus

$$\nabla f(\mathbf{X}_0) = \mathbf{0}. \tag{1}$$

A point $\mathbf{X}_0$ at which (1) holds is called a **stationary point** of f.

Condition (1) is necessary for an extremum of f at $\mathbf{X}_0$ in the interior of R, but is not sufficient. Indeed, if $n = 1$, then for $y = f(x)$, condition (1) is $y' = f'(x) = 0$; and, for instance, $y = x^3$ satisfies $y' = 3x^2 = 0$ at $x = X_0 = 0$ where f has no extremum but a point of inflection. Similarly, for $f(\mathbf{x}) = x_1x_2$ we have $\nabla f(\mathbf{0}) = \mathbf{0}$, and f does not have an extremum but a saddle point at $\mathbf{0}$. Hence after solving (1), one must still find out whether one has obtained an extremum. In the case $n = 1$ the conditions $y'(X_0) = 0$, $y''(X_0) > 0$ guarantee a local minimum at X_0 and the conditions $y'(X_0) = 0$, $y''(X_0) < 0$ a local maximum, as is known from calculus. For $n > 1$ there exist similar criteria. However, in practice even solving (1) will often be difficult. For this reason, one generally prefers solution by iteration, by search processes that start at some point and move stepwise to points at which f is smaller (if a minimum of f is wanted) or bigger (in the case of a maximum).

Cauchy's **method of steepest descent** or **gradient method** is of this type. It was introduced in 1847 and has enjoyed popularity ever since.[1] The idea is to find a minimum of $f(\mathbf{x})$ by repeatedly computing minima of a function $g(t)$ of a single variable t, as follows. Suppose that f has a minimum at $\mathbf{X}_0$ and we start at a point $\mathbf{x}$. Then we look for a minimum of f closest to $\mathbf{x}$ along the straight line in the direction of

$$-\nabla f(\mathbf{x}),$$

the direction of steepest descent (= direction of maximum decrease) of f at $\mathbf{x}$; that is, we determine the value of t and the corresponding point

$$\mathbf{z}(t) = \mathbf{x} - t\nabla f(\mathbf{x}) \tag{2}$$

at which the function

$$g(t) = f(\mathbf{z}(t)) \tag{3}$$

has a minimum. We take this $\mathbf{z}(t)$ as our next approximation to $\mathbf{X}_0$.

EXAMPLE 1 Method of steepest descent

Determine a minimum of

$$f(\mathbf{x}) = x_1^2 + 3x_2^2, \tag{4}$$

starting from $\mathbf{x}_0 = 6\mathbf{i} + 3\mathbf{j}$ and applying the method of steepest descent.

Solution. Clearly, inspection shows that $f(\mathbf{x})$ has a minimum at $\mathbf{0}$. Knowing the solution gives us a better feeling of how the method works.

We obtain $\nabla f(\mathbf{x}) = 2x_1\mathbf{i} + 6x_2\mathbf{j}$ and from this

$$\mathbf{z}(t) = \mathbf{x} - t\nabla f(\mathbf{x}) = (1 - 2t)x_1\mathbf{i} + (1 - 6t)x_2\mathbf{j}$$

$$g(t) = f(\mathbf{z}(t)) = (1 - 2t)^2x_1^2 + 3(1 - 6t)^2x_2^2.$$

We now calculate

$$g'(t) = 2(1 - 2t)x_1^2(-2) + 6(1 - 6t)x_2^2(-6),$$

set $g'(t) = 0$, and solve for t, finding

$$t = \frac{x_1^2 + 9x_2^2}{2x_1^2 + 54x_2^2}.$$

Starting from $\mathbf{x}_0 = 6\mathbf{i} + 3\mathbf{j}$, we compute the values in Table 21.1, which are shown in Fig. 446. ∎

Figure 446 suggests that in the case of slimmer ellipses ("a long narrow valley"), convergence would be poor. The reader may confirm this by replacing in (4) the coefficient 3 with a large coefficient. For more sophisticated descent and other methods, some of them also applicable to vector functions of vector variables, we refer to the references listed in Part F of Appendix 1.

[1]Convergence can sometimes be slow. For more refined methods and recent developments, see Ref. [F7] in Appendix 1.

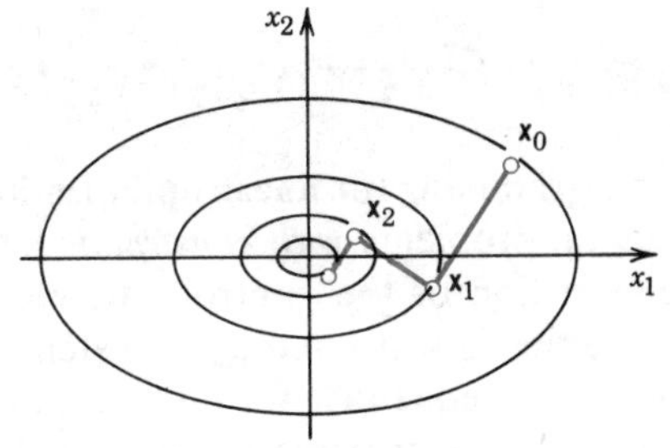

Fig. 446. Method of steepest descent in Example 1

Table 21.1
Method of Steepest Descent, Computations in Example 1

n	$\mathbf{x}$		t	$1 - 2t$	$1 - 6t$
0	6.000	3.000	0.210	0.581	−0.258
1	3.484	−0.774	0.310	0.381	−0.857
2	1.327	0.664	0.210	0.581	−0.258
3	0.771	−0.171	0.310	0.381	−0.857
4	0.294	0.147	0.210	0.581	−0.258
5	0.170	−0.038	0.310	0.381	−0.857
6	0.065	0.032			

Problem Set 21.1

1. What happens if you apply the method of steepest descent to $f(\mathbf{x}) = x_1^2 + x_2^2$? (Explain!)

2. Apply the method of steepest descent to $f(\mathbf{x}) = x_1^2 + 1.1x_2^2$, starting from $\mathbf{x}_0 = 6\mathbf{i} + 3\mathbf{j}$ (3 steps). Why is the convergence faster than in Example 1?

3. In Prob. 2, start from $\mathbf{x}_0 = 1.1\mathbf{i} + \mathbf{j}$. Show that the next approximations are $\mathbf{x}_1 = k(1.1\mathbf{i} - \mathbf{j})$, $\mathbf{x}_2 = k^2\mathbf{x}_0$, etc., where $k = 0.1/2.1$.

4. Verify that in Example 1, successive gradients are orthogonal (perpendicular). Is this just by chance?

5. Apply the method of steepest descent to $f(\mathbf{x}) = x_1^2 + cx_2^2$. Starting from $\mathbf{x}_0 = c\mathbf{i} + \mathbf{j}$, show that subsequent approximations are $\mathbf{x}_m = a_m(c\mathbf{i} + (-1)^m\mathbf{j})$, where $a_m = (c - 1)^m/(c + 1)^m$. Conclude that for large c (slim ellipses $f(\mathbf{x}) = const$) the convergence becomes poor.

6. Apply the method of steepest descent to $f(\mathbf{x}) = x_1^4 + x_2^4 - 12$, starting from $2\mathbf{i} + 2\mathbf{j}$.

7. Sketch some curves $f(\mathbf{x}) = const$ in Prob. 6 for obtaining a qualitative picture of the behavior of the method of steepest descent for different choices of the starting point.

8. Apply the method of steepest descent to $f(\mathbf{x}) = x_1^2 - x_2$, which has no minimum, and see what will happen. Start from $\mathbf{x}_0 = \mathbf{i} + \mathbf{j}$.

9. Apply the method of steepest descent to $f(\mathbf{x}) = x_1^4 + 16x_2^4$. Start from $\mathbf{x}_0 = \mathbf{i} + \mathbf{j}$. Do 1 step.

10. What happens if you apply the method of steepest descent to $f(\mathbf{x}) = ax_1 + bx_2$? First guess, then calculate.

21.2 Linear Programming

Linear programming (or **linear optimization**) consists of methods for solving optimization problems ***with constraints*** in which the objective function f is a ***linear*** function of the control variables $x_1, \cdots, x_n$, and the domain of these variables is restricted by a system of linear inequalities. Problems of this type arise frequently, for instance, in production, distribution of goods, economics, and approximation theory. Let us illustrate this with a simple example.

EXAMPLE 1 Production plan

Suppose that in producing two types of containers K and L one uses two machines M_1, M_2. To produce a container K, one needs M_1 two minutes and M_2 four minutes. Similarly, L occupies M_1 eight minutes and M_2 four minutes. The net profit for a container K is \$29 and for L it is \$45. Determine the production plan that maximizes the net profit.

Solution. If we produce x_1 containers K and x_2 containers L per hour, the profit per hour is

$$f(x_1, x_2) = 29x_1 + 45x_2.$$

The constraints are

$$\begin{array}{llr} & 2x_1 + 8x_2 \leqq 60 & \text{(resulting from machine } M_1) \\ (1) & 4x_1 + 4x_2 \leqq 60 & \text{(resulting from machine } M_2) \\ & x_1 \qquad \geqq 0 & \\ & \qquad x_2 \geqq 0. & \end{array}$$

Figure 447 shows these constraints. (x_1, x_2) must lie in the first quadrant and below or on the straight line $2x_1 + 8x_2 = 60$ as well as below or on the line $4x_1 + 4x_2 = 60$. Thus (x_1, x_2) is restricted to the quadrangle $OABC$. We have to find (x_1, x_2) in $OABC$ such that $f(x_1, x_2)$ is maximum. Now $f(x_1, x_2) = 0$ gives $x_2 = -(29/45)x_1$ (see Fig. 447). The lines $f(x_1, x_2) = const$ are parallel to that line. We see that B, that is, $x_1 = 10$, $x_2 = 5$, gives the optimum $f(10, 5) = 515$. Hence the answer is that the optimal production plan that maximizes the profit is achieved by producing containers K and L in the ratio 2:1, the maximum profit being \$515 per hour. ∎

Note that the problem in Example 1 or similar optimization problems

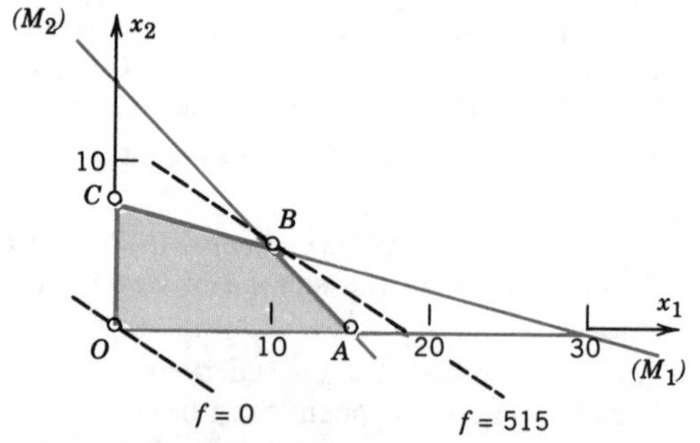

Fig. 447. Linear programming in Example 1

cannot be solved by setting certain partial derivatives equal to zero, because crucial to such problems is the region in which the control variables are allowed to vary.

Furthermore, our "geometrical" or graphical method illustrated in Example 1 is confined to two variables x_1, x_2. However, most practical problems involve more variables, so that we need other methods of solution.

Normal Form of a Linear Programming Problem

To prepare for those methods, we show that constraints can be written more uniformly. Let us explain the idea in terms of the first inequality in (1),

$$2x_1 + 8x_2 \leqq 60.$$

This inequality implies $60 - 2x_1 - 8x_2 \geqq 0$ (and conversely), that is, the quantity

$$x_3 = 60 - 2x_1 - 8x_2$$

is nonnegative. Accordingly, our original inequality can now be written

$$2x_1 + 8x_2 + x_3 = 60,$$

$$x_3 \geqq 0.$$

x_3 is a nonnegative auxiliary variable introduced for the purpose of converting inequalities to equations. Such a variable is called a **slack variable.**

EXAMPLE 2 Conversion of inequalities to equations by using slack variables

With the help of two slack variables x_3, x_4 we can write the linear programming problem in Example 1 in the following form. ***Maximize***

$$f = 29x_1 + 45x_2$$

under the constraints

$$\begin{aligned} &2x_1 + 8x_2 + x_3 = 60 \\ &4x_1 + 4x_2 + x_4 = 60 \\ &x_i \geqq 0 \quad (i = 1, \cdots, 4). \end{aligned} \tag{2}$$

We now have $n = 4$ variables and $m = 2$ (linearly independent) equations, so that two of the four variables, for example, x_1, x_2, determine the others. Also note that each of the four sides of the quadrangle in Fig. 447 now has an equation of the form $x_i = 0$:

$$OA: x_2 = 0, \qquad AB: x_4 = 0, \qquad BC: x_3 = 0, \qquad CO: x_1 = 0.$$

A vertex of the quadrangle is the intersection of two sides. Hence at a vertex, $n - m = 4 - 2 = 2$ of the variables are zero and the others are nonnegative. Thus at A we have $x_2 = 0$, $x_4 = 0$, and so on. ∎

Our example suggests that a general linear optimization problem can be brought to the following **normal form.** *Maximize*

$$f = c_1x_1 + c_2x_2 + \cdots + c_nx_n \tag{3}$$

subject to the constraints

$$\begin{aligned} a_{11}x_1 + \cdots + a_{1n}\,x_n &= b_1 \\ a_{21}x_1 + \cdots + a_{2n}\,x_n &= b_2 \\ \cdots\cdots\cdots&\cdots\cdots\cdots \\ a_{m1}x_1 + \cdots + a_{mn}x_n &= b_m \\ x_i \geqq 0 \qquad (i = 1&, \cdots, n). \end{aligned} \tag{4}$$

Here $x_1, \cdots, x_n$ include the slack variables (for which the c's in f are zero). We assume that the equations in (4) are linearly independent. Then, if we choose values for $n - m$ of the variables, the system uniquely determines the others.

Recall that our problem also includes the **minimization** of an objective function f since this corresponds to maximizing $-f$ and thus needs no separate consideration.

An n-tuple $(x_1, \cdots, x_n)$ that satisfies all the constraints in (4) is called a *feasible point* or **feasible solution.** A feasible solution is called an **optimal solution** if for it the objective function f becomes maximum, compared with the values of f at all feasible solutions.

Finally, by a **basic feasible solution** we mean a feasible solution for which at least $n - m$ of the variables $x_1, \cdots, x_n$ are zero. For instance, in Example 2 we have $n = 4$, $m = 2$, and the basic feasible solutions are the four vertices O, A, B, C in Fig. 447. Here B is an optimal solution (the only one in this example).

The following theorem is fundamental.

Theorem 1 (Optimal solution)

Some optimal solution of a linear programming problem (3), (4) *is also a basic feasible solution of* (3), (4).

For a proof, see Ref. [F10], Chap. 3 (listed in Appendix 1). A problem can have many optimal solutions and not all of them may be *basic* feasible solutions; but the theorem guarantees that we can find an optimal solution by searching through the basic feasible solutions. This is a great simplification; but since there are $\binom{n}{n-m}$ different ways of equating $n - m$ of the n variables to zero, considering all these possibilities, dropping those which are not feasible and then searching through the rest would still involve very much work, even when n and m are relatively small. Hence a systematic search is needed. We shall explain an important method of this type in the next section.

Problem Set 21.2

In each case, describe and graph the region in the first quadrant of the x_1x_2-plane determined by the given set of linear inequalities.

1. $x_1 + 2x_2 \leqq 8$, $x_1 - x_2 \leqq 0$, $x_2 \leqq 1.5$

2. $x_1 - 2x_2 \leqq -2$, $0.8x_1 + x_2 \leqq 6$

3. $-0.5x_1 + x_2 \leqq 2$, $x_1 + x_2 \geqq 2$, $-x_1 + 5x_2 \geqq 5$

4. $-x_1 + x_2 \geqq 0$, $x_1 + x_2 \leqq 4$, $-x_1 + x_2 \leqq 3.2$

5. $x_1 + 3x_2 \leqq 6$, $2x_1 + x_2 \geqq 14$

6. $x_1 + 2x_2 \leqq 4$, $2x_1 - 3x_2 \leqq 3$, $0.4x_1 - x_2 \geqq 2$

7. $3x_1 - 7x_2 \geqq -28$, $x_1 + x_2 \geqq 6$, $9x_1 - 2x_2 \leqq 36$, $5x_1 + 6x_2 \leqq 48$

8. $x_1 + x_2 \geqq 3$, $x_1 + x_2 \leqq 9$, $-x_1 + x_2 \geqq -3$, $-x_1 + x_2 \leqq 3$

9. $-2x_1 + 3x_2 \leqq 9$, $5x_1 + x_2 \leqq 25$, $2x_1 + 3x_2 \geq 3$, $x_1 - 4x_2 \leqq 4$

Maximize the given objective function f subject to the given constraints.

10. $f = 2x_1 + 3x_2$, $4x_1 + 3x_2 \geqq 12$, $x_1 - x_2 \geqq -3$, $x_2 \leqq 6$, $2x_1 - 3x_2 \leqq 0$

11. $f = -10x_1 + 2x_2$, $x_1 \geqq 0$, $x_2 \geqq 0$, $-x_1 + x_2 \geqq -1$, $x_1 + x_2 \leqq 6$, $x_2 \leqq 5$

12. $f = 3x_1 - 6x_2$, $4x_1 + x_2 \geqq 4$, $-x_1 + 2x_2 \geqq 6$, $x_1 + 2x_2 \leqq 14$

13. Minimize f in Prob. 10.

14. Minimize f in Prob. 11.

15. Maximize the daily profit in manufacturing two alloys A_1, A_2 that are different mixtures of two metals M_1, M_2 as shown.

Metal	Proportion of Metal In Alloy A_1	Proportion of Metal In Alloy A_2	Daily Supply (tons)
M_1	0.5	0.25	10
M_2	0.5	0.75	15
Net profit ($ per ton)	60	50	

16. Maximize the total daily output $f = x_1 + x_2$ of a production in which one can choose from two production processes subject to the constraints

$$2x_1 + 4x_2 \leqq 200 \text{ (labor hours)}, \qquad 5x_1 + 2x_2 \leqq 150 \text{ (machine hours)}.$$

17. A company manufactures and sells two models of lamps L_1, L_2, the profit being \$15 and \$10, respectively. The process involves two workers W_1 and W_2 who are available for this kind of work 100 and 80 hours per month, respectively. W_1 assembles L_1 in 20 and L_2 in 30 minutes. W_2 paints L_1 in 20 and L_2 in 10 minutes. Assuming that all lamps made can be sold without difficulty, determine production figures that maximize the profit.

18. What is the meaning of the slack variables x_3, x_4 in Example 2 in terms of the problem in Example 1?

19. Could one find a profit $f(x_1, x_2) = a_1x_1 + a_2x_2$ whose maximum is at an interior point of the quadrangle in Fig. 447? (Give a reason for your answer.)

20. In Example 1, the solution is unique. Can we always expect uniqueness? (Give a reason.)

21.3 Simplex Method

From the last section we recall the following. A linear optimization problem (linear programming problem) can be written in normal form; that is, ***maximize***

$$f = c_1x_1 + \cdots + c_nx_n \tag{1}$$

subject to the constraints

$$\begin{aligned} a_{11}x_1 + \cdots + a_{1n}x_n &= b_1 \\ a_{21}x_1 + \cdots + a_{2n}x_n &= b_2 \\ \cdots\cdots&\cdots\cdots \\ a_{m1}x_1 + \cdots + a_{mn}x_n &= b_m \\ x_i \geqq 0 \qquad (i = 1, \cdots, n)&. \end{aligned} \tag{2}$$

For finding an optimal solution of this problem, we need to consider only the **basic feasible solutions** (defined in Sec. 21.2), but these are still so many that we have to follow a systematic search procedure. In 1948 G. B. Dantzig published an iterative method, called the **simplex method,** for that purpose. In this method, one proceeds stepwise from one basic feasible solution to another in such a way that the objective function f always increases its value. The method begins with an initial *Operation* I_0 in which we find a basic feasible solution to start from. Each further step then consists of three operations

Operation O_1*:* Test for optimality,
Operation O_2*:* Location of a better basic feasible solution,
Operation O_3*:* Transition to that better solution.

We describe the details and simultaneously illustrate them by a simple example (the example from Sec. 21.2, which can be solved by an elementary method, as we have seen).

Initial Operation I_0

Find any basic feasible solution (needed to begin the iteration). That is, divide the variables $x_1, \cdots, x_n$ into two classes by selecting m variables—call them **basic variables;** then the other $n - m$ variables are those which must be zero at a basic feasible solution; these are called **nonbasic variables** or **right-hand variables** because we shall write them on the right side of our system (2), which we solve for the basic variables. For illustration:

Maximize

$$f = 29x_1 + 45x_2 \tag{3a}$$

subject to the constraints

$$
\begin{aligned}
2x_1 + 8x_2 + x_3 \qquad &= 60 \\
4x_1 + 4x_2 \qquad + x_4 &= 60 \\
x_i \geqq 0 \qquad (i = 1, \cdots, 4).&
\end{aligned} \tag{3b}
$$

For instance, let us take x_3, x_4 as basic variables. We solve (3b) for these variables:

$$
\begin{aligned}
x_3 &= 60 - 2x_1 - 8x_2 \\
x_4 &= 60 - 4x_1 - 4x_2.
\end{aligned} \tag{4}
$$

In favorable cases, such as the present, we get the values of the variables of a basic feasible solution by setting the right-hand variables equal to zero. Indeed, $x_3 = 60$, $x_4 = 60$, and our basic feasible solution is the point O, the origin $x_1 = 0$, $x_2 = 0$ in Fig. 447 in the last section.

If this method should give a negative value for some of the basic variables, we must try another set of basic variables.

1st Step. Operation O_1: Test for optimality
Find out whether all coefficients of f, expressed as a function of the present right-hand variables, are negative or zero.

If this **optimality criterion** is satisfied, then our basic feasible solution is optimal.

Indeed, then f cannot increase if we assign positive values (instead of zero) to the right-hand variables—remember that negative values are not allowed. Hence this condition is sufficient for optimality, and it can be shown that it is also necessary.

For the example, since by our choice the right-hand variables are x_1, x_2, we must consider f *as function of these,* that is, $f = 29x_1 + 45x_2$. The criterion shows that $x_1 = 0$, $x_2 = 0$ is not optimal, because 29 and 45 are positive.

1st Step. Operation O_2: Search for a better basic feasible solution
If the basic feasible solution just tested is not optimal, go to a neighboring basic feasible solution for which f is larger, as follows.

To go to a neighboring basic feasible solution means to go to a point at which another x_i is zero; that is, we have to make an **exchange:** the variable x_i which will now be zero leaves the set of basic variables, and one other variable becomes a basic variable instead. We explain this **method for exchange.**

Consider a right-hand variable x_R that has a positive coefficient in f. Keep the other right-hand variables at zero. Determine the largest increase Δx_R of x_R such that *all* the (old) basic variables are still nonnegative; also list the corresponding increase Δf of f.

In (4), for x_1 this looks as follows.

$$
(5)\qquad \begin{aligned} &x_3 = 60 - 2x_1 - 8x_2, \quad 0 = 60 - 2x_1, \quad \Delta x_1 = 30, \quad \Delta f = 29\Delta x_1 = 870.\\ &x_4 = 60 - \underline{4x_1} - 4x_2, \quad 0 = 60 - 4x_1, \quad \Delta x_1 = 15, \quad \Delta f = 29\Delta x_1 = 435.\end{aligned}
$$

Underline $x_R = x_1$ in the equation that blocked any further increase in x_R [as shown in (5)]. Do this for every right-hand variable x_R whose coefficient in f, *expressed as a function of the right-hand variables,* is positive.

Hence we must also consider $x_R = x_2$, underlining as before:

$$
(6)\qquad \begin{aligned} &x_3 = 60 - 2x_1 - \underline{8x_2}, \quad 0 = 60 - 8x_2, \quad \Delta x_2 = 7.5, \quad \Delta f = 45\Delta x_2 = 337.5\\ &x_4 = 60 - 4x_1 - 4x_2, \quad 0 = 60 - 4x_2, \quad \Delta x_2 = 15, \quad \Delta f = 45\Delta x_2 = 675.\end{aligned}
$$

1st Step. Operation O_3: Exchange variable

Consider the equations in which a variable is underlined. Thus, in (5), (6),

$$
(7)\qquad \begin{aligned} x_3 &= 60 - 2x_1 - \underline{8x_2}, \qquad \Delta f = 337.5\\ x_4 &= 60 - \underline{4x_1} - 4x_2, \qquad \Delta f = 435.\end{aligned}
$$

Exchange that right-hand variable x_R which gave the greatest Δf with the basic variable in the equation where x_R is underlined. Thus, by (7), we have to exchange x_1 and x_4, so that x_1 now becomes basic and x_4 right-hand. Solve the system for the new basic variables. Thus, by solving the second equation in (7) for x_1 and inserting this x_1 into the first equation in (7) we get

$$
(8)\qquad \begin{aligned} x_1 &= 15 - x_2 - \tfrac{1}{4}x_4\\ x_3 &= 30 - 6x_2 + \tfrac{1}{2}x_4.\end{aligned}
$$

2nd Step. Perform Operations O_1, O_2, O_3,
using (8). Thus, in Operation O_1 we must now express $f = 29x_1 + 45x_2$ in terms of the new right-hand variables x_2, x_4. Using the first equation in (8), we find

$$
(9)\qquad f = 435 + 16x_2 - 7.25x_4.
$$

By the optimality criterion, the present basic feasible solution (point A in Fig. 447, Sec. 21.2) is not optimal because x_2 has a positive coefficient in (9).

In Operation O_2 we now consider $x_R = x_2$. From (8),

$$
(10)\qquad \begin{aligned} &x_1 = 15 - x_2 - \tfrac{1}{4}x_4, \quad 0 = 15 - x_2, \quad \Delta x_2 = 15, \quad \Delta f = 16\Delta x_2 = 240\\ &x_3 = 30 - \underline{6x_2} + \tfrac{1}{2}x_4, \quad 0 = 30 - 6x_2, \quad \Delta x_2 = 5, \quad \Delta f = 16\Delta x_2 = 80.\end{aligned}
$$

x_4 need not be considered because its coefficient in (9) is negative.

In Operation O_3 we exchange x_2 and x_3 and solve the system for the new basic variables x_1, x_2 by solving the second equation in (10) for x_2 and then inserting this x_2 into the first equation in (10), obtaining

$$
\begin{aligned}
x_1 &= 10 + \tfrac{1}{6}x_3 - \tfrac{1}{3}x_4 \\
x_2 &= 5 - \tfrac{1}{6}x_3 + \tfrac{1}{12}x_4.
\end{aligned}
\tag{11}
$$

3rd Step. Perform Operations O_1, O_2, O_3, using (11). Thus, in Operation O_1 we must express $f = 29x_1 + 45x_2$ in terms of x_3, x_4. By (11),

$$
f = 515 - 2.667x_3 - 5.917x_4.
\tag{12}
$$

By the optimality criterion, this basic feasible solution is optimal because the right-hand variables x_3 and x_4 have negative coefficients. $f_{\text{opt}} = 515$ is obtained from (12) with $x_3 = 0$, $x_4 = 0$. The point (x_1, x_2) at which the optimum occurs follows from (11) with $x_3 = 0$, $x_4 = 0$, namely, $x_1 = 10$, $x_2 = 5$. This agrees with our result in Sec. 21.2.

Table 21.2 shows a tabular arrangement of calculations. In each step, the right-hand variables are listed in the top line. Arrows point to variables that enter or leave the basic variables. ***Caution!*** In the literature one sometimes lists the *negative* of the right-hand variables ($-x_1$, $-x_2$, etc.); this introduces a factor -1 in all coefficients. Since, however, this is a convention that adds neither to the idea of the method as such nor to a better understanding of it, we do not adopt it.

Table 21.2
Simplex Method, Calculations in Tabular Form

	Basic Variables	Constants	⬇ x_1	x_2
1st Step	x_3	60	-2	-8
	⬅ x_4	60	$\underline{-4}$	-4
	f	0	29	45
	Basic Variables	**Constants**	**⬇ x_2**	**x_4**
2nd Step	x_1	15	-1	-0.25
	⬅ x_3	30	$\underline{-6}$	0.5
	f	435	16	-7.25
	Basic Variables	**Constants**	**x_3**	**x_4**
3rd Step	x_1	10	0.167	-0.333
	x_2	5	-0.167	0.083
	f	515	-2.667	-5.917

Problem Set 21.3

1. In the example discussed in the text, apply the simplex method, starting from x_2, x_4 as basic variables.

2. In the example in the text, replace f by $f = 15x_1 + 30x_2$ and apply the simplex method, starting from x_3, x_4 as basic variables. In step 1, operation O_3, you will have a choice in exchanging variables. Explain why.

3. Maximize $f = 3x_1 + 2x_2$ subject to the constraints $x_1 \geqq 0$, $x_2 \geqq 0$,

$$3x_1 + 4x_2 \leqq 60, \qquad 4x_1 + 3x_2 \leqq 60, \qquad 10x_1 + 2x_2 \leqq 120;$$

write the problem in normal form and use the simplex method, taking the slack variables as the initial basic variables.

4. Using the simplex method, maximize the daily output $f = x_1 + x_2$ in producing chairs by two different processes subject to the constraints

$$3x_1 + 4x_2 \leqq 550 \quad \text{(machine hours)}, \qquad 5x_1 + 4x_2 \leqq 650 \quad \text{(labor)}.$$

5. Using the simplex method, maximize the profit in the daily production of two kinds of metal frames F_1 (profit per frame \$90) and F_2 (profit per frame \$50) subject to the constraints (x_1, x_2 = numbers of F_1, F_2 produced per day)

$$\begin{aligned} x_1 + 3x_2 &\leqq 18 \quad &\text{(material)} \\ x_1 + x_2 &\leqq 10 \quad &\text{(machine hours)} \\ 3x_1 + x_2 &\leqq 24 \quad &\text{(labor)}. \end{aligned}$$

6. Maximize $f = 2x_1 + 3x_2 + x_3$ subject to

$$x_1 \geqq 0,\ x_2 \geqq 0,\ x_3 \geqq 0,\ x_1 + x_2 + x_3 \leqq 4.8,\ 10x_1 + x_3 \leqq 9.9,\ x_2 - x_3 \leqq 0.2.$$

7. Maximize $f = x_1 + 4x_2 - x_3$ subject to the same constraints as in Prob. 6.

8. Maximize $f = 2x_1 + x_2 + 3x_3$ subject to

$$x_1 \geqq 0,\ x_2 \geqq 0,\ x_3 \geqq 0,\ 4x_1 + 3x_2 + 6x_3 \leqq 12.$$

9. A company produces batteries I, II, choosing from four production processes P_1, P_2 (for I) and P_3, P_4 (for II). The profit per battery is \$10 for I and \$20 for II. Maximize the total profit $f = 10x_1 + 10x_2 + 20x_3 + 20x_4$ subject to the constraints

$$\begin{aligned} 12x_1 + 8x_2 + 6x_3 + 4x_4 &\leqq 120 \quad &\text{(machine hours)} \\ 3x_1 + 6x_2 + 12x_3 + 24x_4 &\leqq 180 \quad &\text{(labor hours)}. \end{aligned}$$

10. Suppose that for certain fuses the profit is \$3 per carton and one can choose from six processes of production $P_1, \cdots, P_6$. Maximize the total profit subject to the constraints (x_j = number of cartons of fuses produced by P_j)

$$\begin{aligned} 10x_1 + 4x_2 + 16x_3 + 7.5x_4 + 5.6x_5 + 4.4x_6 &\leqq 400 \quad &\text{(machine } M_1) \\ 2x_1 + 4x_2 + 1.6x_3 + 2.4x_4 + 2.8x_5 + 3.3x_6 &\leqq 120 \quad &\text{(machine } M_2). \end{aligned}$$

21.4 Simplex Method: Degeneracy, Difficulties in Starting

We recall from the last section that in the simplex method we proceed stepwise from one basic feasible solution to another, thereby increasing the value of the objective function f until we reach an optimal solution. Each such transition is accomplished by an exchange in which a right-hand variable becomes a basic variable and vice versa.

It can happen, however, that a basic feasible solution is not optimal but one cannot increase f by exchanging a *single* variable without coming into conflict with the constraints. This situation can occur only for a basic feasible solution for which more than $n - m$ variables are zero (with m and n defined as in Sec. 21.3). Such a solution is called a **degenerate feasible solution.** In such a case one exchanges a right-hand variable with a basic variable which is zero for that solution, and then one proceeds as usual. In more complicated cases, one may even have to perform several such additional exchanges.

Theoretically, in the case of degeneracies there is a possibility that the simplex algorithm will get caught in an infinite loop, that is, will not terminate. Practically, the occurrence of such a loop has been extremely rare, so that it is sufficient to mention that techniques for handling degeneracies are further described in Refs. [F6] and [F10] listed in Appendix 1.

EXAMPLE 1 Simplex method, degenerate feasible solution

A foundry produces two kinds of iron I_1, I_2 by using three kinds of raw material R_1, R_2, R_3 (scrap iron and two kinds of ore) as shown. Maximize the daily net profit.

Raw Material	Raw Material Needed per Ton		Raw Material Available per Day (tons)
	Iron I_1	Iron I_2	
R_1	2	1	16
R_2	1	1	8
R_3	0	1	3.5
Net profit per ton	\$150	\$300	

Solution. Let x_1 and x_2 denote the amount (in tons) of iron I_1 and I_2, respectively, produced per day. Then our problem is as follows. Maximize

$$f = 150x_1 + 300x_2 \tag{1}$$

subject to the constraints $x_1 \geqq 0$, $x_2 \geqq 0$ and

$$2x_1 + x_2 \leqq 16$$

$$x_1 + x_2 \leqq 8$$

$$x_2 \leqq 3.5.$$

By introducing slack variables x_3, x_4, x_5 we obtain the normal form

$$
\begin{aligned}
2x_1 + x_2 + x_3 \qquad\quad &= 16 \\
x_1 + x_2 \qquad + x_4 \qquad &= 8 \\
x_2 \qquad\qquad + x_5 &= 3.5
\end{aligned}
\tag{2}
$$

$$x_i \geqq 0 \qquad (i = 1, \cdots, 5).$$

Initial operation. To get started, we choose x_3, x_4, x_5 as basic variables and solve (2) for these:

$$
\begin{aligned}
x_3 &= 16 - 2x_1 - x_2 \\
x_4 &= 8 - x_1 - x_2 \\
x_5 &= 3.5 \qquad - x_2.
\end{aligned}
\tag{3}
$$

Setting $x_1 = 0$, $x_2 = 0$, we get the positive values $x_3 = 16$, $x_4 = 8$, $x_5 = 3.5$, so that all the constraints are satisfied, that is, we have a basic feasible solution (point O in Fig. 448) and may begin with the iterative process.

1st Step. We test for optimality. In (3), the right-hand variables are x_1, x_2. In terms of these, f is given by (1). The optimality criterion (Sec. 21.3) shows that f is not optimal.

We prepare for an exchange of a variable. We have to consider both x_1 and x_2 since both x_1, x_2 have positive coefficients in f; see (1). We consider x_1. From (3) and (1) we obtain

$$
\begin{aligned}
x_3 &= 16 - \underline{2x_1} - x_2, \qquad && \Delta x_1 = 8, \qquad && \Delta f = 150\Delta x_1 = 1200 \\
x_4 &= 8 - \underline{x_1} - x_2. && \Delta x_1 = 8, && \Delta f = 150\Delta x_1 = 1200 \\
x_5 &= 3.5 \qquad - x_2,
\end{aligned}
$$

Similarly, for x_2 we obtain from (3) and (1)

$$
\begin{aligned}
x_3 &= 16 - 2x_1 - x_2, \qquad && \Delta x_2 = 16 \\
x_4 &= 8 - x_1 - x_2, && \Delta x_2 = 8 \\
x_5 &= 3.5 \qquad \underline{- x_2}, && \Delta x_2 = 3.5, \qquad && \Delta f = 300\Delta x_2 = 1050.
\end{aligned}
$$

We see that we have to exchange x_1 since then we get the maximum possible increase, $\Delta f = 1200$. From the underlining we see that we can exchange x_1 and x_3 or x_1 and x_4. We exchange x_1 and x_3. Solving for the new basic variables x_1, x_4, x_5, we obtain

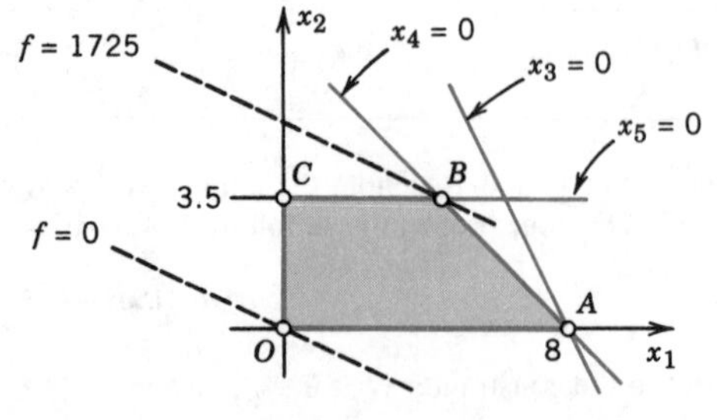

Fig. 448. Example 1, where A is degenerate

$$
\begin{aligned}
x_1 &= 8 \quad - 0.5x_2 - 0.5x_3 \\
x_4 &= \quad\;\; - 0.5x_2 + 0.5x_3 \\
x_5 &= 3.5 - \quad x_2.
\end{aligned} \tag{4}
$$

Hence the present basic feasible solution is

$$x_1 = 8, \qquad x_2 = 0, \qquad x_3 = 0, \qquad x_4 = 0, \qquad x_5 = 3.5$$

(point A in Fig. 448), and we note that it is ***degenerate***, since it has more than $n - m = 5 - 3 = 2$ zeros, namely, 3 zeros.

2nd Step. We test for optimality. For this we have to express f in terms of the new right-hand variables x_2, x_3. From (1) and the first equation in (4) we obtain

$$f = 1200 + 225x_2 - 75x_3. \tag{5}$$

The optimality criterion shows that the present basic feasible solution is not optimal.

We consider x_2 for an exchange. We do not consider x_3 since its coefficient in (5) is negative, for reasons explained in Sec. 21.3. From (4) we obtain

$$
\begin{aligned}
x_1 &= 8 \quad - 0.5x_2 - 0.5x_3, & \Delta x_2 &= 16, & & \\
x_4 &= \quad\;\; - \underline{0.5x_2} + 0.5x_3, & \Delta x_2 &= 0, & \Delta f &= 0, \\
x_5 &= 3.5 - \quad x_2, & \Delta x_2 &= 3.5. & &
\end{aligned}
$$

Because of the degeneracy we exchange x_2 and x_4 (which are both zero at A) without increasing f and without leaving A. We solve for the new basic variables x_1, x_2, x_5:

$$
\begin{aligned}
x_1 &= 8 \quad - x_3 + x_4 \\
x_2 &= \qquad\;\; x_3 - 2x_4 \\
x_5 &= 3.5 - x_3 + 2x_4.
\end{aligned} \tag{6}
$$

3rd Step. From (1) and (6) we obtain f in terms of the new right-hand variables x_3, x_4:

$$f = 1200 + 150x_3 - 450x_4. \tag{7}$$

By the optimality criterion, f is not optimal at A. To prepare for an exchange we consider x_3. We do not consider x_4 since its coefficient is negative. From (6) and (7),

$$
\begin{aligned}
x_1 &= 8 \quad - x_3 + x_4, & \Delta x_3 &= 8 & & \\
x_2 &= \qquad\;\; x_3 - 2x_4 & & & & \\
x_5 &= 3.5 - \underline{x_3} + 2x_4, & \Delta x_3 &= 3.5, & \Delta f &= 150\,\Delta x_3 = 525.
\end{aligned}
$$

We exchange x_3 and x_5 and solve for the new basic variables x_1, x_2, x_3:

$$
\begin{aligned}
x_1 &= 4.5 - x_4 + x_5 \\
x_2 &= 3.5 \qquad\;\; - x_5 \\
x_3 &= 3.5 + 2x_4 - x_5.
\end{aligned} \tag{8}
$$

4th Step. From (1) and (8) we obtain f in terms of the new right-hand variables x_4, x_5:

$$f = 1725 - 150x_4 - 150x_5. \tag{9}$$

The optimality criterion shows that our present basic feasible solution is optimal. From (8) we see that this solution is

$$x_1 = 4.5, \qquad x_2 = 3.5, \qquad x_3 = 3.5, \qquad x_4 = 0, \qquad x_5 = 0$$

(point B in Fig. 448), and $f_{\text{opt}} = 1725$ from (9).

Answer. The net profit is maximized if we produce $x_1 = 4.5$ tons of iron I_1 and $x_2 = 3.5$ tons of iron I_2 per day. This maximum amount is \$1725. Also $x_3 = 3.5$ tons of raw material R_1 remains unused in this plan. ■

Difficulties in Starting

It may sometimes be difficult to find a basic feasible solution to start from. In such a case the idea of an *artificial variable* (or several such variables) is helpful. We explain this method in terms of a typical example.

EXAMPLE 2 Simplex method: difficult start, artificial variable

Maximize

$$f = 2x_1 + x_2 \tag{10}$$

subject to the constraints $x_1 \geqq 0$, $x_2 \geqq 0$ and

$$\begin{aligned} x_1 - \tfrac{1}{2}x_2 &\geqq 1 \\ x_1 - \phantom{\tfrac{1}{2}}x_2 &\leqq 2 \\ x_1 + \phantom{\tfrac{1}{2}}x_2 &\leqq 4. \end{aligned}$$

Solution. By means of slack variables we achieve the normal form

$$\begin{aligned} -x_1 + \tfrac{1}{2}x_2 + x_3 \qquad\qquad &= -1 \\ x_1 - x_2 \qquad + x_4 \qquad &= 2 \\ x_1 + x_2 \qquad\qquad + x_5 &= 4 \\ x_i \geqq 0 \quad (i = 1, \cdots, 5). \end{aligned} \tag{11}$$

We want to use x_3, x_4, x_5 as basic variables and start by solving accordingly:

$$\begin{aligned} x_3 &= -1 + x_1 - \tfrac{1}{2}x_2 \\ x_4 &= 2 - x_1 + x_2 \\ x_5 &= 4 - x_1 - x_2. \end{aligned} \tag{12}$$

But for $x_1 = 0$, $x_2 = 0$ we have the negative value $x_3 = -1$, so that we cannot proceed immediately. Now, instead of searching for other basic variables we can do the following. We introduce a new variable x_6, called an **artificial variable** and defined by

(13) $$x_3 = -1 + x_1 - \tfrac{1}{2}x_2 + x_6$$

with the constraint $x_6 \geqq 0$, and we take x_4, x_5, x_6 as basic variables:

(14) $$\begin{aligned} x_4 &= 2 - x_1 + x_2 \\ x_5 &= 4 - x_1 - x_2 \\ x_6 &= 1 - x_1 + \tfrac{1}{2}x_2 + x_3. \end{aligned}$$

These are positive when the new right-hand variables x_1, x_2, x_3 are zero. Hence we have found a basic feasible solution $x_1 = x_2 = x_3 = 0$, $x_4 = 2$, $x_5 = 4$, $x_6 = 1$ for an **extended problem** involving 6 variables and 3 equations. For this new problem to yield a solution of the given problem we must take care that x_6 will eventually disappear. We accomplish this by the idea of modifying the objective function by adding a term $-Mx_6$, where M is very large; that is, we replace f in (10) by [see (10) and (14)]

(15) $$\hat{f} = f - Mx_6 = 2x_1 + x_2 - Mx_6 = (M + 2)x_1 - (\tfrac{1}{2}M - 1)x_2 - Mx_3 - M.$$

Then a positive x_6 will make $\hat{f}$ very small, so that $x_6 = 0$ eventually.

1st Step. $\hat{f}$ is not maximum at our present basic feasible solution. For large M, the only positive coefficient in (15) is that of x_1; hence, by (14) and (15), our preparation for exchange is

$$\begin{aligned} x_4 &= 2 - x_1 + x_2, & \Delta x_1 &= 2, & & \\ x_5 &= 4 - x_1 - x_2, & \Delta x_1 &= 4, & & \\ x_6 &= 1 - \underline{x_1} + \tfrac{1}{2}x_2 + x_3, & \Delta x_1 &= 1, & \Delta\hat{f} &= M + 2. \end{aligned}$$

We exchange x_1 and x_6 and solve (14) and (15) accordingly:

(16) $$\begin{aligned} x_1 &= 1 + \tfrac{1}{2}x_2 + x_3 - x_6 \\ x_4 &= 1 + \tfrac{1}{2}x_2 - x_3 + x_6 \\ x_5 &= 3 - \tfrac{3}{2}x_2 - x_3 + x_6 \end{aligned}$$

and

$$\hat{f} = 2 + 2x_2 + 2x_3 - (M + 2)x_6.$$

We see that the artificial variable x_6 has become a right-hand variable. We can now set $x_6 = 0$ since (16) then gives $x_1 = 1$, $x_2 = 0$, $x_3 = 0$, $x_4 = 1$, $x_5 = 3$ as a basic feasible solution of the *original* problem.

The further steps follow the usual pattern (see Sec. 21.3) and may be left to the student. ∎

Problem Set 21.4

1. Maximize $f = 300x_1 + 500x_2$ subject to $x_i \geqq 0$ $(i = 1, \cdots, 5)$,

$$2x_1 + 8x_2 + x_3 = 60, \qquad 4x_1 + 4x_2 + x_4 = 60, \qquad 2x_1 + x_2 + x_5 = 30,$$

choosing x_3, x_4, x_5 as basic variables and exchanging x_1 and x_5 in the first step.

2. Maximize the daily output $f = x_1 + x_2$ in producing x_1 glass plates by a process P_1 and x_2 glass plates by a process P_2 subject to the constraints

$$2x_1 + 3x_2 \leqq 130 \quad \text{(labor hours)}$$

$$3x_1 + 8x_2 \leqq 300 \quad \text{(machine hours)}$$

$$4x_1 + 2x_2 \leqq 140 \quad \text{(raw material supply)}$$

3. Maximize $f = 6x_1 + 6x_2 + 9x_3$ subject to $x_j \geqq 0$ $(j = 1, \cdots, 5)$, and

$$x_1 + x_3 + x_4 = 1, \qquad x_2 + x_3 + x_5 = 1.$$

4. Maximize the total output $f = x_1 + x_2 + x_3$ (production figures of three different production processes) subject to input constraints (limitation of machine time)

$$4x_1 + 5x_2 + 8x_3 \leqq 12, \qquad 8x_1 + 5x_2 + 4x_3 \leqq 12.$$

5. Continue and finish the solution process in Example 2.
6. Maximize $f = 4x_1 + x_2 + 2x_3$ subject to

$$x_1 \geqq 0, x_2 \geqq 0, x_3 \geqq 0, x_1 + x_2 + x_3 \leqq 1, x_1 + x_2 - x_3 \leqq 0.$$

7. Maximize the total daily profit f of producing x_1, x_2, x_3 lamps of types I, II, III, respectively, if the profits per lamp are \$10, \$8, \$5, respectively, and the constraints are (daily metal supply, plastic and paper supply, glass supply)

$$3x_1 + x_2 \leqq 450, \qquad 2x_2 + 3x_3 \leqq 900, \qquad 2x_1 + x_2 \leqq 350.$$

8. Using an artificial variable, maximize $f = 2x_1 + x_2$ subject to

$$x_1 \geqq 0, \qquad x_2 \geqq 0, \qquad x_1 + 2x_2 \geqq 4, \qquad x_1 + x_2 \leqq 3.$$

9. Using an artificial variable, minimize $f = 2x_1 - x_2$ subject to

$$x_1 \geqq 0, \qquad x_2 \geqq 0, \qquad x_1 + x_2 \geqq 5, \qquad -x_1 + x_2 \leqq 1, \qquad 5x_1 + 4x_2 \leqq 40.$$

10. If one uses the method of artificial variables in a problem without solution, this nonexistence will become apparent by the fact that one cannot get rid of the artificial variable. Illustrate this by trying to maximize $f = 2x_1 + x_2$ subject to $x_1 \geqq 0$, $x_2 \geqq 0$ and

$$2x_1 + x_2 \leqq 2, \qquad x_1 + 2x_2 \geqq 6, \qquad x_1 + x_2 \leqq 4.$$

Review Questions and Problems for Chapter 21

1. What is the difference between constrained and unconstrained optimization?
2. Explain the idea of the method of steepest descent. On what will the speed of convergence depend?
3. Why does the function value decrease in each step of the method of steepest descent (unless we reach a point at which the gradient is the zero vector)?
4. What differentiability condition must we make in the method of steepest descent?

5. Write an algorithm for the method of steepest descent.
6. What assumptions about the objective function do we make in linear programming?
7. Why can we not use methods of calculus for determining maxima or minima in a linear programming problem?
8. Explain the basic idea of the systematic search for a maximum or minimum in linear programming.
9. What are slack variables? Artificial variables? Why did we use them?
10. What do you know about difficulties that may arise in a linear programming problem? Discuss ways to overcome them.

11. Apply the method of steepest descent to $f(\mathbf{x}) = x_1^2 + 1.5x_2^2$, starting from $\mathbf{x}_0 = 6\mathbf{i} + 3\mathbf{j}$ (3 steps). Why is the convergence faster than in Example 1, Sec. 21.1?
12. In Prob. 11, start from $\mathbf{x}_0 = 1.5\mathbf{i} + \mathbf{j}$. Show that the next approximations are $\mathbf{x}_1 = k(1.5\mathbf{i} - \mathbf{j})$, $\mathbf{x}_2 = k^2\mathbf{x}_0$, etc., where $k = 0.2$.
13. Verify that in Prob. 12, successive gradients are orthogonal. What is the reason?

14. What result will the method of steepest descent give when you apply it to $f(\mathbf{x}) = x_1x_2 + 1$, starting (a) from $4\mathbf{i} + 4\mathbf{j}$, (b) from $2\mathbf{i} + \mathbf{j}$ (1 step)? Compare and explain.
15. What does the method of steepest descent amount to in the case of a single variable?
16. Design a "method of steepest ascent" for determining maxima.

Describe and sketch the region in the first quadrant of the x_1x_2-plane given by the following linear inequalities.

17. $x_1 - 2x_2 \geqq -4$, $2x_1 + x_2 \leqq 12$, $x_1 + x_2 \leqq 8$

18. $x_1 + x_2 \leqq 5$, $x_2 \leqq 3$, $-x_1 + x_2 \leqq 2$

19. $x_1 + x_2 \geqq 2$, $2x_1 - 3x_2 \geqq -12$, $x_1 \leqq 15$

20. $x_1 + 2x_2 \geqq 2$, $x_1 \leqq 6$, $x_1 - 2x_2 \geqq 0$, $x_1 + 2x_2 \leqq 6$

21. $3x_1 - 5x_2 \leqq 0$, $4x_1 + 3x_2 \leqq 28$, $5x_1 - 6x_2 \geqq -12$, $4x_1 + x_2 \geqq 4$

22. $x_1 - 3x_2 \geqq -3$, $x_1 + x_2 \leqq 9$, $2x_1 + 7x_2 \leqq 14$, $3x_1 - x_2 \geqq 24$

Minimize the given function f subject to $x_1 \geqq 0$, $x_2 \geqq 0$ and the given further constraints.

23. $f = 3x_1 + 2x_2$, $x_1 + x_2 \geqq 2$, $x_1 - x_2 \geqq -1$, $5x_1 + 3x_2 \leqq 15$

24. $f = 2x_1 + 8x_2$, $2x_1 + 4x_2 \geqq 8$, $2x_1 - 5x_2 \leqq 0$, $-x_1 + 5x_2 \leqq 5$

25. $f = -2x_1 + 10x_2$, $4 \leqq x_1 + x_2 \leqq 10$, $-6 \leqq x_1 - x_2 \leqq 0$

26. Maximize $f = x_1 + x_2$ subject to $x_1 \geqq 0$, $x_2 \geqq 0$ and

$$x_1 + 2x_2 \leqq 10, \quad 2x_1 + x_2 \leqq 10, \quad x_2 \leqq 4.$$

27. A factory produces three kinds of gaskets G_1, G_2, G_3 with net profits of \$4, \$6, \$8, respectively. Maxmize the total daily profit subject to the constraints (x_j = number of gaskets G_j produced per day)

$$x_1 + 2x_2 + 4x_3 \leqq 100 \quad \text{(machine hours)}, \qquad 2x_1 + x_2 \leqq 40 \quad \text{(labor)}.$$

28. Graph the region of feasible solutions in Prob. 27 and mark the optimal solution.

29. Maximize $f = 12x_1 + 6x_2 + 4x_3$ subject to the constraints $x_1 \geqq 0$, $x_2 \geqq 0$, $x_3 \geqq 0$,

$$4x_1 + 2x_2 + x_3 \leqq 60, \qquad 2x_1 + 3x_2 + 3x_3 \leqq 50, \qquad x_1 + 3x_2 + x_3 \leqq 45.$$

30. Is the solution of a minimization problem unique? Explain.

Summary of Chapter 21 Unconstrained Optimization, Linear Programming

In optimization we maximize or minimize an **objective function** f depending on control variables x_j whose domain is either unrestricted ("**unconstrained optimization,**" Sec. 21.1) or restricted by constraints in the form of inequalities or equations or both ("**constrained optimization,**" Sec. 21.2).

If the objective function is *linear* and the constraints are *linear inequalities* in the control variables $x_1, \cdots, x_m$, by introducing **slack variables** $x_{m+1}, \cdots, x_n$ we can write the optimization problem in **normal form** with the objective function given by

$$f_1 = c_1x_1 + \cdots + c_nx_n \tag{1}$$

(where $c_{m+1} = \cdots = c_n = 0$) and the constraints given by

$$\begin{aligned} a_{11}x_1 + a_{12}x_2 + \cdots + a_{1n}x_n &= b_1 \\ a_{21}x_1 + a_{22}x_2 + \cdots + a_{2n}x_n &= b_2 \\ \cdots\cdots\cdots\cdots\cdots\cdots&\cdots\cdots \\ a_{m1}x_1 + a_{m2}x_2 + \cdots + a_{mn}x_n &= b_m \\ x_1 \geqq 0, \cdots, x_n &\geqq 0. \end{aligned} \tag{2}$$

In this case we can then apply the widely used **simplex method** (Sec. 21.3), a systematic stepwise search through a very much reduced subset of all feasible solutions. Section 21.4 shows how to overcome difficulties that may arise in connection with this method.

Chapter 22

Graphs and Combinatorial Optimization

Graphs and **digraphs** (= directed graphs) are presently developing into more and more powerful tools in electrical and civil engineering, communication networks, industrial management, operations research, computer science, marketing, economics, sociology, and so on. An accelerating factor of this growth is the impact of fast large computers and their use in large-scale optimization problems that can be modeled in terms of graphs and solved by algorithms provided by graph theory. This approach yields models of general applicability and economic importance. It lies in the center of **"combinatorial optimization,"** a rather recently introduced term denoting optimization problems that have pronounced discrete or combinatorial structures.

The present chapter gives an introduction to this wide new area, which is full of new ideas as well as unsolved problems—in connection, for instance, with efficient computer algorithms and ideas of computational complexity. The classes of problems we shall consider often form the core of larger and more complex practical problems.

Prerequisites: None.
References: Appendix 1, Part F.
Answers to problems: Appendix 2.

22.1 Graphs and Digraphs

Roughly, a *graph* consists of points, called *vertices,* and lines connecting them, called *edges*. For example, these may be four cities and five highways connecting them, as in Fig. 449 on the next page. Or the points may represent some people, and we connect by an edge those who do business with each other. Or the vertices may represent computers and the edges connections between them. Let us now give a formal definition.

Definition of a graph

A **graph** G consists of two finite sets (sets having finitely many elements), a set V of points, called **vertices,** and a set E of connecting lines, called **edges,** such that each edge connects two vertices, called the *endpoints* of the edge. We write

$$G = (V, E).$$

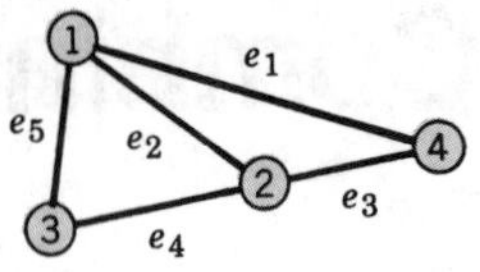

Fig. 449. Graph consisting of 4 vertices and 5 edges

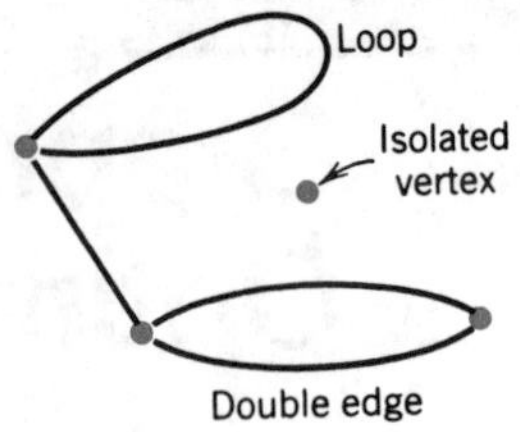

Fig. 450. Isolated vertex, loop, double edge. (Excluded by definition.)

For simplicity we exclude[1] isolated vertices (vertices that are not endpoints of any edge), *loops* (edges whose endpoints coincide), and *multiple edges* (edges that have both endpoints in common). See Fig. 450. This is practical for our purpose. ∎

We denote vertices by letters, $u, v, \cdots$ or $v_1, v_2, \cdots$ or simply by numbers $1, 2, \cdots$ (as in Fig. 449). We denote edges by $e_1, e_2, \cdots$ or by their two endpoints; for instance, $e_1 = (1, 4)$, $e_2 = (1, 2)$ in Fig. 449.

We say that a vertex v_i is **incident** with an edge (v_i, v_j); similarly for v_j. The number of edges incident with a vertex v is called the **degree** of v. We call two vertices **adjacent** in G if they are connected by an edge in G (that is, if they are the two endpoints of some edge in G).

We meet graphs in different fields under different names: as "networks" in electrical engineering, "structures" in civil engineering, "molecular structures" in chemistry, "organizational structures" in economics, "sociograms," "road maps," "telecommunication networks," and so on.

Nets of one-way streets, pipeline networks, sequences of jobs in construction work, flows of computation in a computer, producer–consumer relations, and many other applications suggest the idea of a "digraph" (= directed graph) in which each edge has as direction (indicated by an arrow, as in Fig. 451).

Definition of a digraph (directed graph)

A **digraph** $G = (V, E)$ is a graph in which each edge $e = (i, j)$ has a direction from its *"initial point"* i to its *"terminal point"* j.

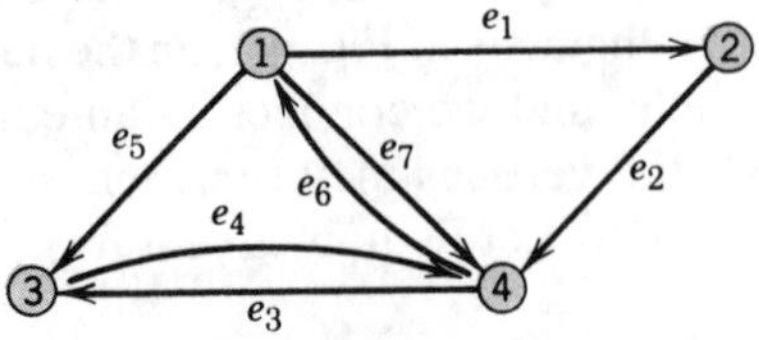

Fig. 451. Digraph

[1] As many authors do, but there is no uniformity, and one must be careful. Some authors permit multiple edges and call graphs without them **"simple graphs."** Others permit loops or isolated vertices, depending on the purpose.

Two edges connecting the same two points i, j are now permitted, provided they have opposite directions, that is, they are (i, j) and (j, i). *Example.* (1, 4) and (4, 1) in Fig. 451. ∎

A **subgraph** or subdigraph of a given graph or digraph $G = (V, E)$, respectively, is a graph or digraph obtained by deleting some of the edges and vertices of G, retaining the other edges of G (together with their pairs of endpoints). For instance, e_1, e_3 (together with the vertices 1, 2, 4) form a subgraph in Fig. 449, and e_3, e_4, e_5 (together with the vertices 1, 3, 4) form a subdigraph in Fig. 451.

Computer Representation of Graphs and Digraphs

Drawings of graphs are useful to people in explaining or illustrating specific situations. Here one should be aware that a graph may be sketched in various ways (see Fig. 452). For handling graphs and digraphs in computers, one uses matrices or lists, as follows.

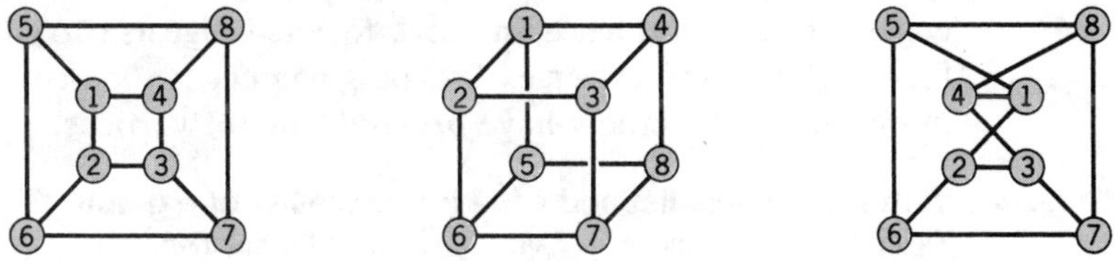

(a) *(b)* *(c)*

Fig. 452. Different sketches of the same graph

Adjacency matrix A $= [a_{ij}]$ **of a graph** G

$$a_{ij} = \begin{cases} 1 & \text{if } G \text{ has an edge } (i, j), \\ 0 & \text{else.} \end{cases}$$

Here, by definition, no vertex is considered to be adjacent to itself. **A** is symmetric, $a_{ij} = a_{ji}$. (Why?)

The adjacency matrix of a graph is generally much smaller than the so-called *incidence matrix* (see Probs. 21–24) and is preferred over the latter if one decides to store a graph in a computer in matrix form.

EXAMPLE 1 **Adjacency matrix of a graph**

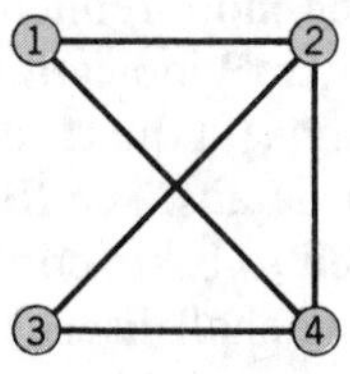

Vertex	1	2	3	4
Vertex 1	0	1	0	1
2	1	0	1	1
3	0	1	0	1
4	1	1	1	0

∎

Adjacency matrix $\mathbf{A} = [a_{ij}]$ of a digraph G

$$a_{ij} = \begin{cases} 1 & \text{if } G \text{ has a directed edge } (i, j), \\ 0 & \text{else.} \end{cases}$$

This matrix **A** is not symmetric. (Why?)

EXAMPLE 2 Adjacency matrix of a digraph

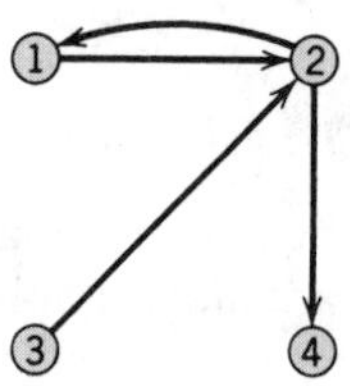

	To vertex 1	2	3	4
From vertex 1	0	1	0	0
2	1	0	0	1
3	0	1	0	0
4	0	0	0	0

Lists. The **vertex incidence list** of a graph shows for each vertex the incident edges. The **edge incidence list** shows for each edge its two endpoints. Similarly for a *digraph;* in the vertex list, outgoing edges then get a minus sign, and in the edge list we now have *ordered* pairs of vertices.

EXAMPLE 3 Vertex incidence list and edge incidence list of a graph

This graph is the same as in Example 1, except for notation.

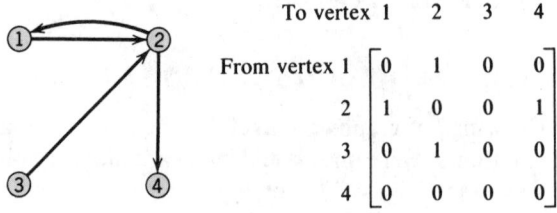

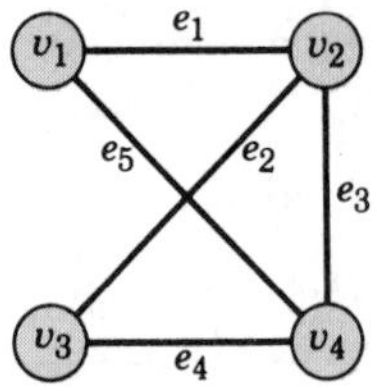

Vertex	Incident edges
v_1	e_1, e_5
v_2	e_1, e_2, e_3
v_3	e_2, e_4
v_4	e_3, e_4, e_5

Edge	Endpoints
e_1	v_1, v_2
e_2	v_2, v_3
e_3	v_2, v_4
e_4	v_3, v_4
e_5	v_1, v_4

"Sparse graphs" are graphs with few edges (far fewer than the maximum possible number $n(n - 1)/2$, where n is the number of vertices). For these graphs, matrices are not efficient. *Lists* then have the advantage of needing much less storage and being easier to handle; they can be ordered, sorted, or manipulated in various other ways directly within the computer. For instance, in tracing a "walk" (a connected sequence of edges with pairwise common endpoints), one can easily go back and forth between the two lists just discussed, instead of scanning a large column of a matrix for a single 1.

Computer science has developed more refined lists, which, in addition to the actual content, contain "pointers" indicating the preceding item or the next item to be scanned or both items (in the case of a "walk": the preceding edge or the subsequent one). For details, see Ref. [F12] in Appendix 1.

The present section was devoted to basic concepts and notations needed throughout this chapter, in which we shall discuss some of the most important classes of combinatorial optimization problems. This will at the same time help us to become more and more familiar with graphs and digraphs.

Problem Set 22.1

1. Sketch the graph consisting of the vertices and edges of a tetrahedron.
2. Worker W_1 can do jobs J_1 and J_3, worker W_2 job J_4, worker W_3 jobs J_2 and J_3. Represent this by a graph.
3. Explain how the following may be regarded as graphs or digraphs: air connections between given cities; memberships of some persons in some committees; relations between chapters of a book; a tennis tournament; countries on a map.
4. How would you represent a net consisting of one-way and two-way streets by a digraph?
5. Give further examples that could be represented by a graph or a digraph.
6. Find the adjacency matrix of the graph in Fig. 449.

In each case find the adjacency matrix of the given graph or digraph.

7.

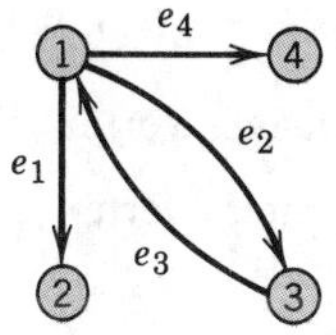

8.

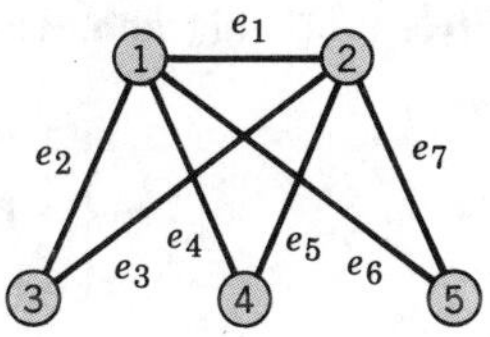

9.

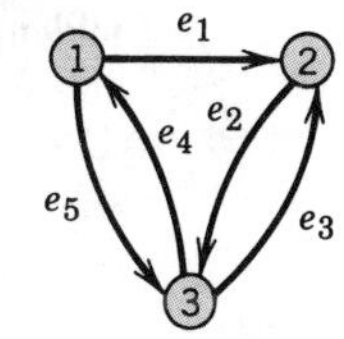

10.

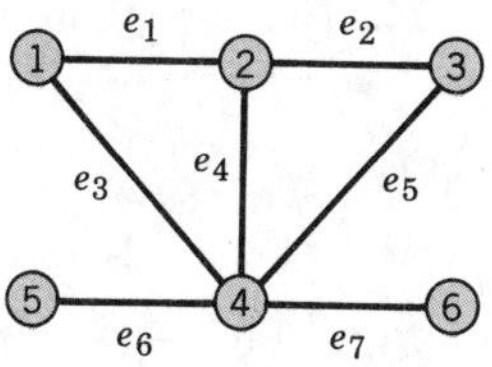

11.

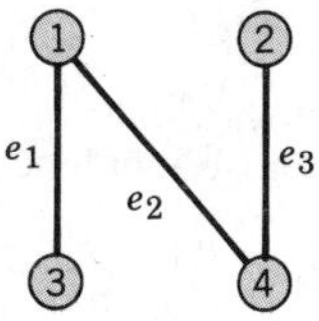

12.

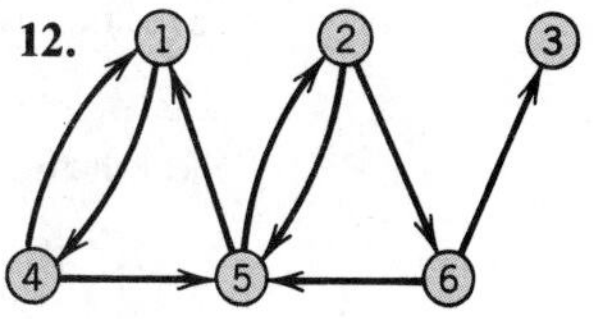

13. Show that the adjacency matrix of a graph is symmetric.

Sketch the graph whose adjacency matrix is

14. $\begin{bmatrix} 0 & 1 & 1 & 1 \\ 1 & 0 & 0 & 1 \\ 1 & 0 & 0 & 1 \\ 1 & 1 & 1 & 0 \end{bmatrix}$ **15.** $\begin{bmatrix} 0 & 1 & 0 & 0 \\ 1 & 0 & 0 & 0 \\ 0 & 0 & 0 & 1 \\ 0 & 0 & 1 & 0 \end{bmatrix}$ **16.** $\begin{bmatrix} 0 & 1 & 1 & 1 \\ 1 & 0 & 1 & 1 \\ 1 & 1 & 0 & 1 \\ 1 & 1 & 1 & 0 \end{bmatrix}$

Sketch the digraph whose adjacency matrix is:

17. The matrix in Prob. 14.
18. The matrix in Prob. 16.

19. **(Complete graph)** Show that a graph G with n vertices can have at most $n(n-1)/2$ edges, and G has exactly $n(n-1)/2$ edges if G is *complete,* that is, if every pair of vertices of G is joined by an edge. (Recall that loops and multiple edges are excluded.)
20. In what case are all the off-diagonal elements of the adjacency matrix of a graph G equal to 1?

Incidence matrix B of a graph. The definition is $\mathbf{B} = [b_{jk}]$, where

$$b_{jk} = \begin{cases} 1 & \text{if vertex } j \text{ is an endpoint of edge } e_k, \\ 0 & \text{otherwise.} \end{cases}$$

Find the incidence matrix of

21. The graph in Prob. 11.

22. The graph in Prob. 10.

Sketch the graph whose incidence matrix is

23. $$\begin{bmatrix} 0 & 1 & 1 & 0 \\ 1 & 1 & 0 & 1 \\ 1 & 0 & 1 & 0 \\ 0 & 0 & 0 & 1 \end{bmatrix}$$

24. $$\begin{bmatrix} 0 & 1 & 0 & 0 & 1 \\ 0 & 0 & 1 & 1 & 1 \\ 1 & 1 & 0 & 1 & 0 \\ 1 & 0 & 1 & 0 & 0 \end{bmatrix}$$

Incidence matrix $\widetilde{\mathbf{B}}$ of a digraph. The definition is $\widetilde{\mathbf{B}} = (\widetilde{b}_{jk})$, where

$$\widetilde{b}_{jk} = \begin{cases} -1 & \text{if edge } e_k \text{ leaves vertex } j, \\ 1 & \text{if edge } e_k \text{ enters vertex } j, \\ 0 & \text{otherwise.} \end{cases}$$

Find the incidence matrix of

25. The digraph in Prob. 7.

26. The digraph in Prob. 9.

Sketch the digraph whose incidence matrix is

27. $$\begin{bmatrix} 1 & -1 & 0 & 0 \\ -1 & 1 & -1 & 1 \\ 0 & 0 & 1 & -1 \end{bmatrix}$$

28. $$\begin{bmatrix} 1 & 1 & -1 & 0 \\ 0 & -1 & 0 & 1 \\ -1 & 0 & 1 & -1 \end{bmatrix}$$

29. Make the vertex incidence list of the digraph in Prob. 7.

30. Make the edge incidence list of the digraph in Fig. 451.

22.2 Shortest Path Problems. Complexity

Beginning in this section, we shall discuss some of the most important classes of optimization problems that concern graphs and digraphs as they arise in applications. Basic ideas and algorithms will be explained and illustrated by small graphs, but you should keep in mind that real-life problems may often involve many thousands or even millions of vertices and edges (think of telephone networks, worldwide air travel, companies that have offices and stores in all larger cities), and then reliable and efficient systematic methods are an absolute necessity—solution by inspection or by trial and error would no longer work, even if "nearly optimal" solutions were acceptable.

We begin with **shortest path problems,** as they arise, for instance, in designing shortest (or least expensive, or fastest) routes for a traveling salesman, for a cargo ship, etc. Let us first explain what we mean by a path.

In a graph $G = (V, E)$ we can walk from a vertex v_1 along some edges to some other vertex v_k. Here we can

(A) make no restrictions, or

(B) require that each *edge* of G be traversed at most once, or

(C) require that each *vertex* be visited at most once.

In case (A) we call this a **walk.** Thus a walk from v_1 to v_k is of the form

$$(1) \qquad (v_1, v_2), (v_2, v_3), \cdots, (v_{k-1}, v_k),$$

where some of these edges or vertices may be the same. In case (B), where each *edge* may occur at most once, we call the walk a **trail.** Finally, in case (C), where each *vertex* may occur at most once (and thus each edge automatically occurs at most once), we call the trail a **path.**

We admit that a walk, trail, or path may end at the vertex it started from, in which case we call it **closed;** then $v_k = v_1$ in (1).

A closed path is called a **cycle.** *A cycle has at least three edges* (because we do not have double edges; see Sec. 22.1). Figure 453 illustrates all these concepts.

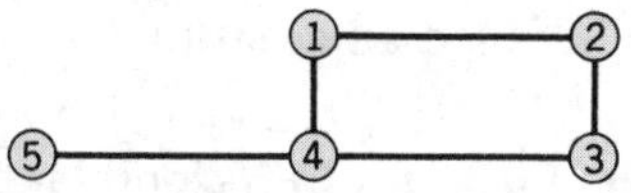

Fig. 453. Walk, trail, path, cycle

1 – 2 – 3 – 2 is a walk (not a trail).
4 – 1 – 2 – 3 – 4 – 5 is a trail (not a path).
1 – 2 – 3 – 4 – 5 is a path (not a cycle).
1 – 2 – 3 – 4 – 1 is a cycle.

Shortest path. Having said what a path is, we should say next what we mean by a shortest path in a graph $G = (V, E)$. For this, each edge (v_i, v_j) in G must have a given "length" $l_{ij} > 0$. Then a **shortest path** $v_1 \to v_k$ (with fixed v_1 and v_k) is a path (1) such that the sum of the lengths of its edges

$$l_{12} + l_{23} + l_{34} + \cdots + l_{k-1,k}$$

is minimum (as small as possible among all paths from v_1 to v_k). Similarly, a **longest path** $v_1 \to v_k$ is one for which that sum is maximum.

Shortest (and longest) path problems are among the most important optimization problems. Here, "length" l_{ij} (often also called "cost" or "weight") can be an actual length measured in miles or travel time or gasoline expenses, but it may also be something entirely different.

For instance, the "*traveling salesman problem*" requires the determination of a shortest **Hamiltonian**[2] **cycle** in a given graph, that is, a cycle that contains all the vertices of the graph.

As another example, by choosing the "most profitable" route $v_1 \to v_k$, a salesman may want to maximize Σl_{ij}, where l_{ij} is his expected commission minus his travel expenses for going from town i to town j.

In an investment problem, i may be the day an investment is made, j the day it matures, and l_{ij} the resulting profit, and one gets a graph by considering the various possibilities of investing and reinvesting over a given period of time.

Shortest Path If All Edges Have Length 1

Obviously, if all edges have length 1, then a shortest path $v_1 \to v_k$ is one that has the smallest number of edges among all paths $v_1 \to v_k$ in a given graph G. For this problem we discuss a BFS algorithm. BFS stands for **Breadth First Search.** This means that in each step the algorithm visits *all neighboring* (all adjacent) vertices of a vertex reached, as opposed to a DFS algorithm (**Depth First Search** algorithm), which makes a long trail (as in a maze). This widely used algorithm is shown in Table 22.1.

We want to find a shortest path in G from a vertex s (***start***) to a vertex t (***terminal***). To guarantee that there is a path from s to t, we make sure that G does not consist of separate portions. Thus we assume that G is **connected,** that is, for any two vertices v and w there is a path $v \to w$ in G. (Recall that a vertex v is called **adjacent** to a vertex u if there is an edge (u, v) in G.)

EXAMPLE 1 **Application of Moore's BFS algorithm**

Find a shortest path $s \to t$ in the graph G shown in Fig. 454.

Solution. Figure 454 shows the labels. The heavy edges form a shortest path (length 4). There is another shortest path $s \to t$. (Can you find it?) Hence in the program we must introduce a rule that makes backtracking unique because otherwise the computer would not know what to do next if at some step there is a choice (for instance, in Fig. 454 when it got back to the vertex labeled 2). The following rule seems to be natural.

Backtracking rule. Using the numbering of the vertices from 1 to n (not the labeling!), at each step, if a vertex labeled i is reached, take as the next vertex that with the smallest number (not label!) among all the vertices labeled $i - 1$. ∎

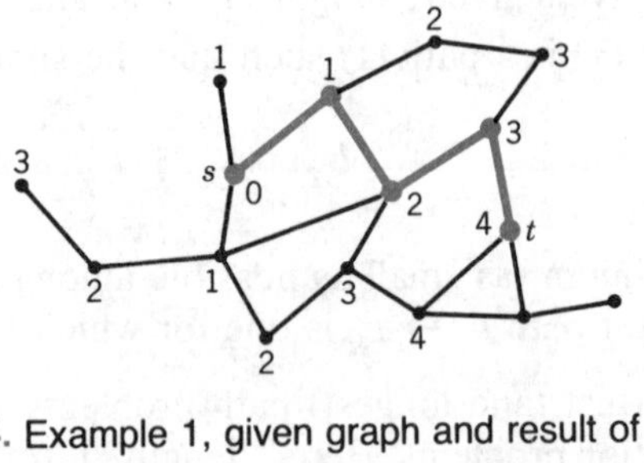

Fig. 454. Example 1, given graph and result of labeling

---

[2]WILLIAM ROWAN HAMILTON (1805—1865), Irish mathematician, known for his work in dynamics.

Table 22.1
Moore's BFS for Shortest Path (All Lengths One)[3]

ALGORITHM MOORE [$G = (V, E), s, t$]
This algorithm determines a shortest path in a connected graph $G = (V, E)$ from a vertex s to a vertex t.

INPUT: Connected graph $G = (V, E)$, in which one vertex is denoted by s and one by t, and each edge (i, j) has length $l_{ij} = 1$. Initially all vertices are unlabeled.

OUTPUT: A shortest path $s \rightarrow t$ in $G = (V, E)$

1. Label s with 0.

2. Set $i = 0$.

3. Find all *unlabeled* vertices adjacent to a vertex labeled i.

4. Label the vertices just found with $i + 1$.

5. If vertex t is labeled, then "backtracking" gives the shortest path

$$k \text{ (= label of } t), k - 1, k - 2, \cdots, 0$$

OUTPUT $k, k - 1, k - 2, \cdots, 0$. Stop

Else increase i by 1. Go to Step 3.

End MOORE

Complexity of an Algorithm

Complexity *of Moore's algorithm.* To find the vertices to be labeled 1, we have to scan all edges incident with s. Next, when $i = 1$, we have to scan all edges incident with vertices labeled 1, etc. Hence each edge is scanned twice. These are $2m$ operations (m = number of edges of G). This is a function $c(m)$. Whether it is $2m$ or $5m + 3$ or $12m$ is not so essential; it *is* essential that $c(m)$ is proportional to m (not m^2, for example); it is of the "order" m. We write for any function $am + b$ simply $O(m)$, for any function $am^2 + bm + d$ simply $O(m^2)$, and so on; here, O suggests "order." The underlying idea and practical aspect are as follows.

In judging an algorithm, we are mostly interested in its behavior for very large problems (large m in the present case), since these are going to determine the limits of the applicability of the algorithm. Thus, the essential item is the fastest growing term (am^2 in $am^2 + bm + d$, etc.) since it will overwhelm the others when m is large enough. Also, a constant factor in this term is not very essential; for instance, the difference between two algorithms of orders, say, $5m^2$ and $8m^2$ is generally not very essential and can be made irrelevant by a modest increase in the speed of computers. However, it does make a great practical difference whether an algorithm is of order m or m^2 or of a still higher power m^p. And the biggest difference occurs between these "polynomial orders" and "exponential orders," such as 2^m.

[3] *Proceedings of the International Symposium for Switching Theory,* Part II, pp. 285–292. Cambridge: Harvard University Press, 1959.

For instance, on a computer that does 10^9 operations per second, a problem of size $m = 50$ will take 0.3 seconds with an algorithm that requires m^5 operations, but 13 days with an algorithm that requires 2^m operations. But this is not our only reason for regarding polynomial orders as good and exponential orders as bad. Another reason is the gain in using a faster computer. For example, since $1000 = 31.6^2 = 2^{9.97}$, an increase in speed by a factor 1000 will permit us to do per hour problems 1000 and 31.6 times as big if the algorithms are $O(m)$ and $O(m^2)$, respectively, but when an algorithm is $O(2^m)$, all we gain is a relatively modest increase of 10 in problem size since $2^{9.97} \cdot 2^m = 2^{m+9.97}$.

The **symbol *O*** is quite practical and commonly used whenever the order of growth is essential, but not the specific form of a function. Thus if a function $g(m)$ is of the form

$$g(m) = kh(m) + \text{more slowly growing terms} \qquad (k \text{ constant}),$$

we say that $g(m)$ is of the *order* $h(m)$ and write

$$g(m) = O(h(m)).$$

For instance,

$$am + b = O(m), \quad am^2 + bm + d = O(m^2), \quad 5 \cdot 2^m + 3m^2 = O(2^m),$$

and so on.

We want an algorithm $\mathcal{A}$ to be "efficient," that is, "good" with respect to

(i) *Time* (number $c_{\mathcal{A}}(m)$ of computer operations)

or

(ii) *Space* (storage needed in the internal memory)

or both. Here $c_{\mathcal{A}}$ suggests **"complexity"** of $\mathcal{A}$. Two popular choices for $c_{\mathcal{A}}$ are

(*Worst case*) $c_{\mathcal{A}}(m)$ = longest time $\mathcal{A}$ takes for a problem of size m,

(*Average case*) $c_{\mathcal{A}}(m)$ = average time $\mathcal{A}$ takes for a problem of size m.

In problems on graphs, the "size" will often be m (number of edges) or n (number of vertices). For our present simple algorithm, $c_{\mathcal{A}}(m) = 2m$ in both cases.

For a "good" algorithm $\mathcal{A}$, we want that $c_{\mathcal{A}}(m)$ does not grow too fast. Accordingly, we call $\mathcal{A}$ **efficient** if $c_{\mathcal{A}}(m) = O(m^k)$ for some integer $k \geqq 0$; that is, $c_{\mathcal{A}}$ may contain only powers of m (or functions that grow even more slowly, such as $\ln m$), but no exponential functions. Furthermore, we call $\mathcal{A}$ **polynomially bounded** if $\mathcal{A}$ is efficient when we choose the "worst case" $c_{\mathcal{A}}(m)$. These conventional concepts have intuitive appeal, as our discussion shows.

Complexity should be investigated for every algorithm, so that one can also compare different algorithms for the same task. This may often exceed the level in this chapter; accordingly, we shall confine ourselves to a few occasional comments in this direction.

Problem Set 22.2

In each graph find a shortest path P: $s \rightarrow t$ and its length by Moore's BFS algorithm; sketch the graph with the labels and indicate P by heavier lines (as in Fig. 454).

1.

2.

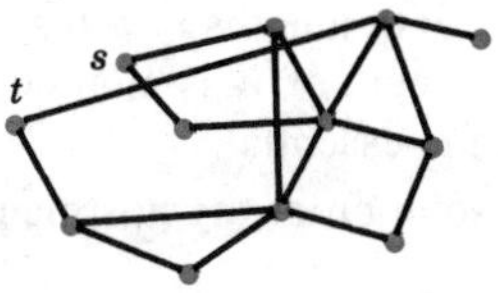

3.

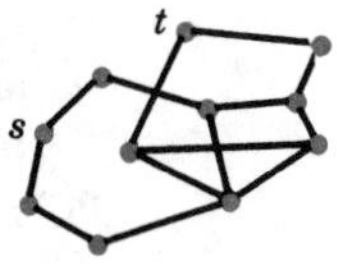

4.

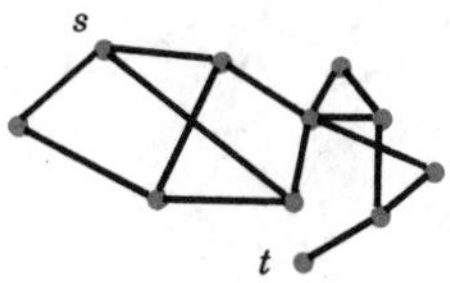

5.

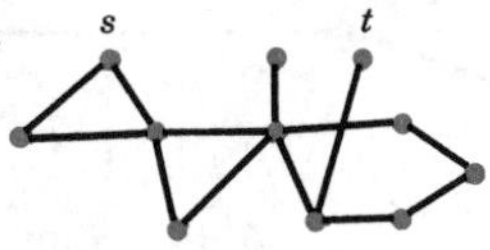

6.

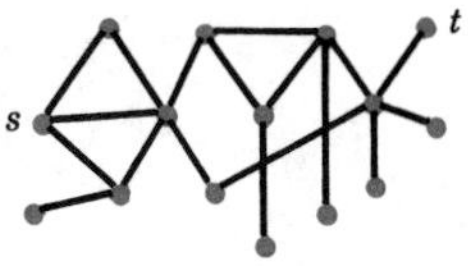

7. A shortest path $s \rightarrow t$ for given s and t need not be unique. Illustrate this by finding another shortest path $s \rightarrow t$ in Example 1.

8. How many edges can a shortest path between any two vertices in a graph with n vertices at most have? Give a reason. In a complete graph with all edges of length 1?

9. **(Moore's algorithm)** Show that if vertex v has label $\lambda(v) = k$, then there is a path $s \rightarrow v$ of length k.

10. **(Moore's algorithm)** Call the length of a shortest path $s \rightarrow v$ briefly the *distance* of v from s. Show that if v has distance l, it has label $\lambda(v) = l$.

11. **(Hamiltonian cycle)** Find and sketch a Hamiltonian cycle in the graph in Prob. 3.

12. Does the graph in Prob. 2 have a Hamiltonian cycle? (Give a reason.)

13. Find and sketch a Hamiltonian cycle in Fig. 452, Sec. 22.1.

14. Find and sketch a Hamiltonian cycle in the graph of a dodecahedron, which has 12 pentagonal faces and 20 vertices (Fig. 455). This is a problem Hamilton considered.

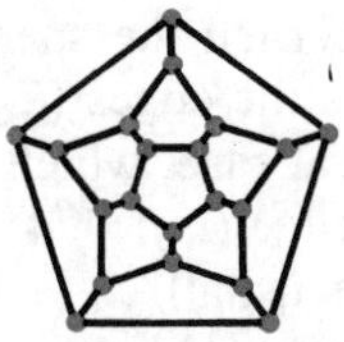

Fig. 455. Problem 14

15. Divide a square into 5 × 5 congruent squares and find in the resulting graph a Hamiltonian cycle with 36 vertices.

16. **(Euler graph)** An *Euler graph* G is a graph that has a closed Euler trail. An **Euler trail** is a trail that contains every edge of G exactly once. Which subgraph with four edges of the graph in Example 1, Sec. 22.1, is an Euler graph?

17. Is the graph in Fig. 456 an Euler graph? (Give a reason.)

18. Find four different closed Euler trails in Fig. 457.

19. **(Postman problem)** The *postman problem* (or *Chinese postman problem*[4]) is the problem of finding a closed walk $W: s \to s$ (s is the post office) in a graph G with edges (i, j) of lengths $l_{ij} > 0$ such that every edge of G is traversed at least once and the length of W is minimum. Explain some other applications in which this problem is essential.

20. Find a solution of the postman problem in Fig. 456 by inspection.

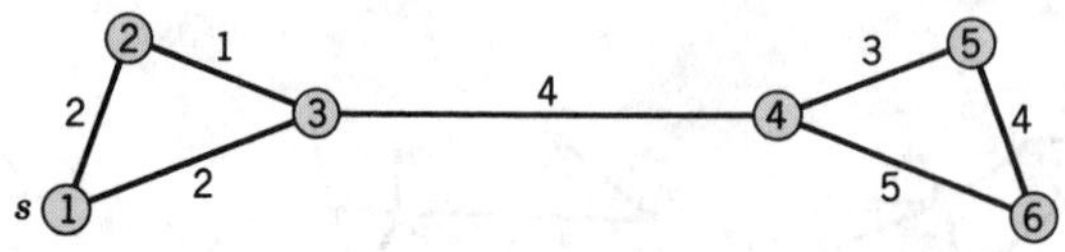

Fig. 456. Problems 17, 20

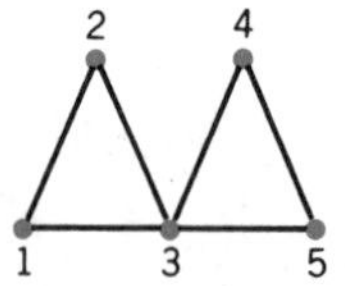

Fig. 457. Problem 18

21. What is the difference between the postman problem and the traveling salesman problem mentioned in the text?

22. Show that the length of a shortest postman trail is the same for every starting vertex.

23. **(Order)** Show that $O(m^3) + O(m^3) = O(m^3)$ and $kO(m^p) = O(m^p)$.

24. Show that $\sqrt{1 + m^2} = O(m)$, $0.02e^m + 100m^2 = O(e^m)$.

25. If we switch from one computer to another that is 100 times as fast, what is our gain in problem size per hour in the use of an algorithm that is $O(m)$, $O(m^2)$, $O(m^5)$, $O(e^m)$?

22.3 Bellman's Optimality Principle. Dijkstra's Algorithm

We continue our discussion of the shortest path problem in a graph G. The last section concerned the special case that all edges had length 1. But in most applications the edges (i, j) will have any lengths $l_{ij} > 0$, and we now turn to this general case, which is of greater practical importance. We write $l_{ij} = \infty$ for any edge (i, j) that does not exist in G (setting $\infty + a = \infty$ for any number a, as usual).

We consider the problem of finding shortest paths from a given vertex, denoted by 1 and called the **origin,** to *all* other vertices $2, 3, \cdots, n$ of G. We let L_j denote the length of a shortest path $1 \to j$ in G.

---

[4]Since it was first considered in the journal *Chinese Mathematics* **1** (1962), 273–77.

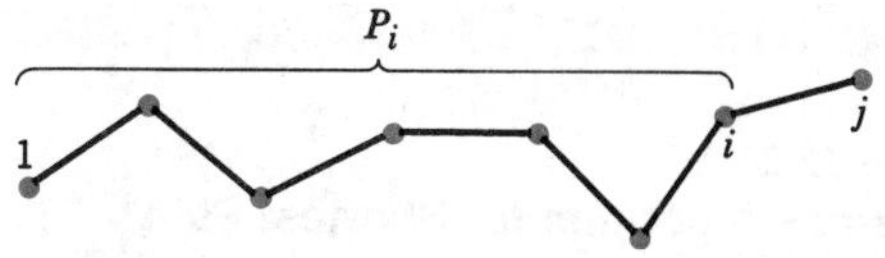

Fig. 458. Paths P and P_i in Bellman's minimality principle

Bellman's minimality principle (or optimality principle)

If P: $1 \to j$ is a shortest path from 1 to j in G and (i, j) is the last edge of P (Fig. 458), *then P_i: $1 \to i$ [obtained by dropping (i, j) from P] is a shortest path $1 \to i$.*

Proof. Suppose that the conclusion is false. Then there is a path P_i^*: $1 \to i$ that is shorter than P_i. Hence if we now add (i, j) to P_i^*, we get a path $1 \to j$ that is shorter than P. This contradicts our assumption that P is shortest. ∎

From Bellman's principle we can derive basic equations as follows. For fixed j we may obtain various paths $1 \to j$ by taking shortest paths P_i for various i for which there is in G an edge (i, j), and add (i, j) to the corresponding P_i. These paths obviously have lengths $L_i + l_{ij}$ (L_i = length of P_i). We can now take the minimum over i, that is, pick an i for which $L_i + l_{ij}$ is smallest. By the Bellman principle, this gives a shortest path $1 \to j$. It has the length

$$\begin{aligned} L_1 &= 0 \\ L_j &= \min_{i \neq j} (L_i + l_{ij}), \qquad j = 2, \cdots, n. \end{aligned}$$

These are the **Bellman equations.** Since $l_{ii} = 0$ by definition, instead of $\min_{i \neq j}$ we can simply write $\min_i$. These equations suggest the idea of one of the best known algorithms for the shortest path problem, as follows.

Dijkstra's algorithm is shown in Table 22.2 on p. 1118. It is a labeling procedure. At each stage of the computation, each vertex v gets a label, either

(PL) a *permanent label* = length L_v of a shortest path $1 \to v$

or

(TL) a *temporary label* = upper bound $\tilde{L}_v$ for the length of a shortest path $1 \to v$.

We denote by $\mathcal{PL}$ and $\mathcal{TL}$ the sets of vertices with a permanent label and with a temporary label, respectively. The algorithm has an initial step in which vertex 1 gets the permanent label $L_1 = 0$ and the other vertices get temporary labels, and then the algorithm alternates between Steps 2 and 3. In Step 2 the idea is to pick k "minimally." In Step 3 the idea is that the upper bounds will in general improve (decrease) and must be updated accordingly.

Dijkstra's Algorithm for Shortest Paths

Table 22.2
Dijkstra's Algorithm for Shortest Path[5]

ALGORITHM DIJKSTRA [$G = (V, E)$, $V = \{1, \cdots, n\}$, l_{ij} for all (i, j) in E]

Given a connected graph $G = (V, E)$ with vertices $1, \cdots, n$ and edges (i, j) having lengths $l_{ij} > 0$, this algorithm determines the lengths of shortest paths from vertex 1 to the vertices $2, \cdots, n$.

INPUT: Number of vertices n, edges (i, j), and lengths l_{ij}
OUTPUT: Lengths L_j of shortest paths $1 \to j$, $j = 2, \cdots, n$

1. *Initial step*
   Vertex 1 gets PL: $L_1 = 0$.
   Vertex j $(= 2, \cdots, n)$ gets TL: $\tilde{L}_j = l_{1j}$ ($= \infty$ if there is no edge $(1, j)$ in G).
   Set $\mathcal{PL} = \{1\}$, $\mathcal{TL} = \{2, 3, \cdots, n\}$.

2. *Fixing a permanent label*
   Find a k in $\mathcal{TL}$ for which $\tilde{L}_k$ is minimum, set $L_k = \tilde{L}_k$. Take the smallest k if there are several. Delete k from $\mathcal{TL}$ and include it in $\mathcal{PL}$.
   If $\mathcal{TL} = \varnothing$ (that is, $\mathcal{TL}$ is empty) then
   OUTPUT $L_2, \cdots, L_n$. Stop
   Else continue (that is, go to Step 3).

3. *Updating tentative labels*
   For all j in $\mathcal{TL}$, set[6] $\tilde{L}_j = \min_k \{\tilde{L}_j, L_k + l_{kj}\}$.
   Go to Step 2.

End DIJKSTRA

EXAMPLE 1 Application of Dijkstra's algorithm

Applying Dijkstra's algorithm to the graph in Fig. 459, find shortest paths from vertex 1 to vertices 2, 3, 4.

Solution. We list the steps and computations.

1. $L_1 = 0, \tilde{L}_2 = 8, \tilde{L}_3 = 5, \tilde{L}_4 = 7,$ $\quad \mathcal{PL} = \{1\},$ $\quad \mathcal{TL} = \{2, 3, 4\}$
2. $L_3 = \min \{\tilde{L}_2, \tilde{L}_3, \tilde{L}_4\} = 5, k = 3,$ $\quad \mathcal{PL} = \{1, 3\},$ $\quad \mathcal{TL} = \{2, 4\}$
3. $\tilde{L}_2 = \min \{8, L_3 + l_{32}\} = \min \{8, 5 + 1\} = 6$
   $\tilde{L}_4 = \min \{7, L_3 + l_{34}\} = \min \{7, \infty\} = 7$
2. $L_2 = \min \{\tilde{L}_2, \tilde{L}_4\} = 6, k = 2,$ $\quad \mathcal{PL} = \{1, 2, 3\},$ $\quad \mathcal{TL} = \{4\}$
3. $\tilde{L}_4 = \min \{7, L_2 + l_{24}\} = \min \{7, 6 + 2\} = 7$
2. $L_4 = 7, k = 4$ $\quad \mathcal{PL} = \{1, 2, 3, 4\},$ $\quad \mathcal{TL} = \varnothing.$

Figure 459b shows the resulting shortest paths, of lengths $L_2 = 6$, $L_3 = 5$, $L_4 = 7$. ∎

[5] *Numerische Mathematik* **1** (1959), 269–271.
[6] That is, take the number $\min_k \{\tilde{L}_j, L_k + l_{kj}\}$ as your new $\tilde{L}_j$.

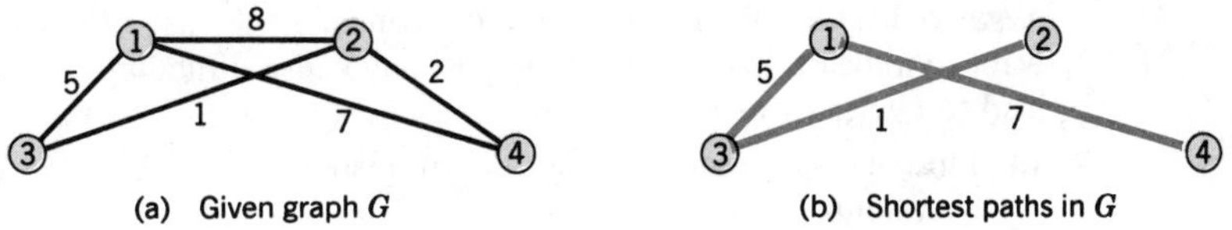

Fig. 459. Example 1

Complexity. *Dijkstra's algorithm is $O(n^2)$.*

Proof. Step 2 requires comparison of elements, first $n - 2$, the next time $n - 3$, etc., a total of $(n - 2)(n - 1)/2$. Step 3 requires the same number of comparisons, a total of $(n - 2)(n - 1)/2$, as well as additions, first $n - 2$, the next time $n - 3$, etc., again a total of $(n - 2)(n - 1)/2$. Hence the total number of operations is $3(n - 2)(n - 1)/2 = O(n^2)$. ∎

Problem Set 22.3

1. The net of roads in Fig. 460 connecting four villages is to be reduced to minimum length, but so that one can still reach every village from every other village. Which of the roads should be retained? Find the solution (a) by inspection, (b) by Dijkstra's algorithm.

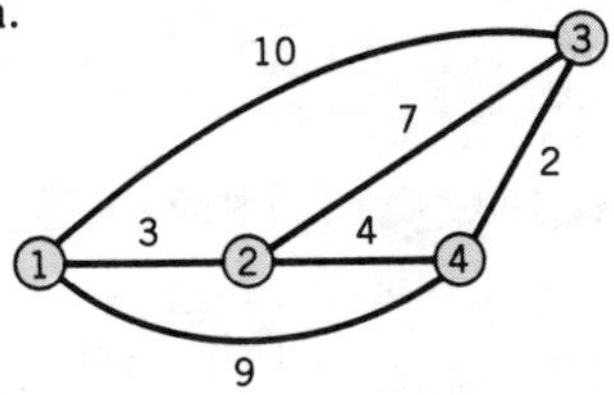

Fig. 460. Problem 1

Applying Dijkstra's algorithm, find shortest paths for the following graphs.

2.

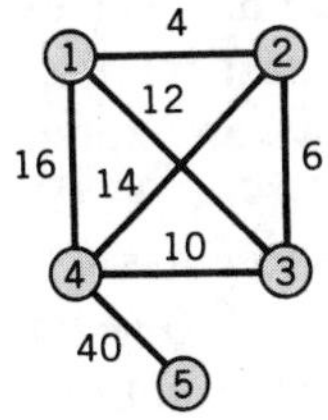

3.

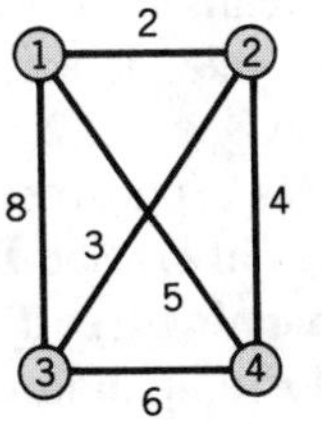

4.

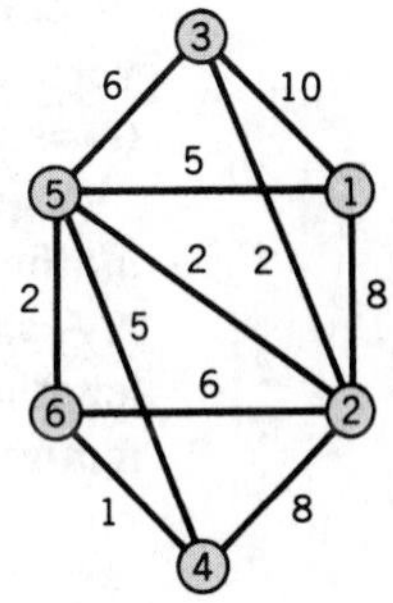

5.

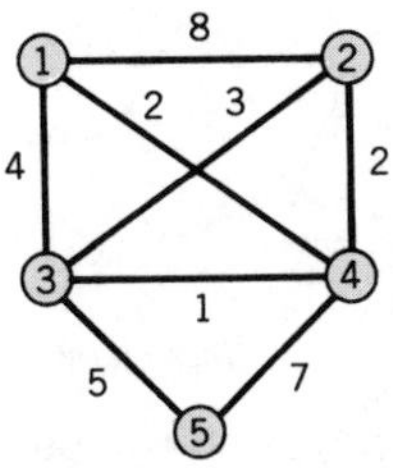

6.

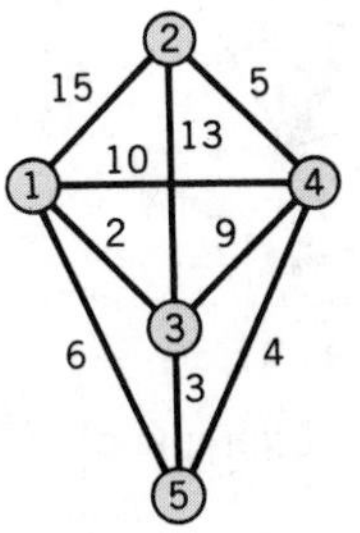

7.

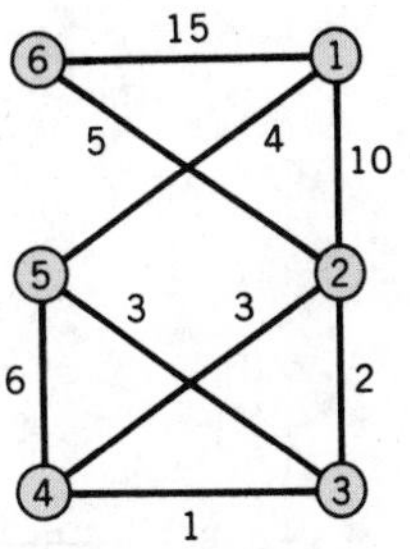

8. Show that in Dijkstra's algorithm, for L_k there is a path P: $1 \to k$ of length L_k.

9. Show that in Dijkstra's algorithm, at each instant the demand on storage is light (data for less than n edges).

Negative lengths. We do not generally consider the case of negative lengths, but we want to indicate that negative lengths may cause difficulties, particularly when they lead to cycles of negative length.

10. Find shortest paths in Fig. 461 by inspection and confirm them by Bellman's equations.

11. Find shortest paths in Fig. 462 by inspection and show that Bellman's equations do not have a unique solution.

12. Show by inspection that there are no unique shortest paths in Fig. 463.

Fig. 461. Problem 10

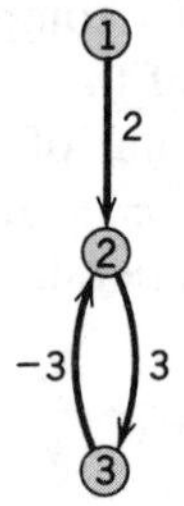

Fig. 462. Problem 11

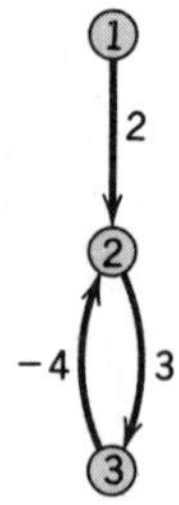

Fig. 463. Problem 12

22.4 Shortest Spanning Trees. Kruskal's Greedy Algorithm

So far we have discussed shortest path problems. We now turn to a particularly important kind of graph, called a tree, along with related optimization problems that arise quite often in practice.

By definition, a **tree** T is a graph that is connected and has no cycles. **"Connected"** means that there is a path from any vertex in T to any other vertex in T. A **cycle**[7] is a path $s \to t$ of at least three edges that is closed ($t = s$); see also Sec. 22.2. Figure 464a shows an example.

A **spanning tree** T in a given connected graph $G = (V, E)$ is a tree containing *all* the n vertices of G. See Fig. 464b. Such a tree has $n - 1$ edges. (Proof?)

A **shortest spanning tree** T in a connected graph G (whose edges (i, j) have lengths $l_{ij} > 0$) is a spanning tree for which Σl_{ij} (sum over all edges of T) is minimum compared to Σl_{ij} for any other spanning tree in G.

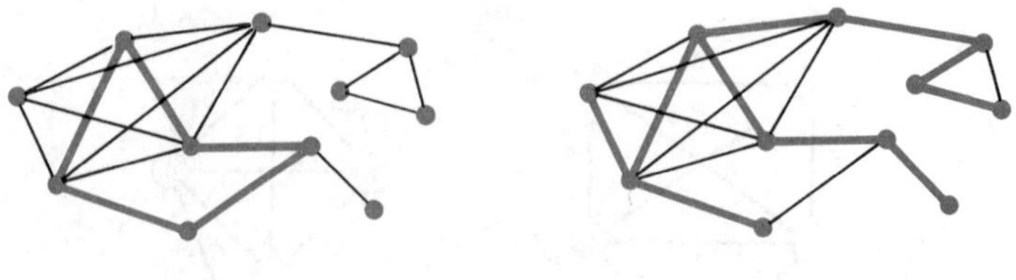

(a) A cycle *(b)* A spanning tree

Fig. 464. Example of (a) a cycle, (b) a spanning tree in a graph

[7]Or **circuit.** Caution! The terminology varies considerably.

Trees are among the most important types of graphs, and they occur in various applications. Familiar examples are family trees and organization charts. Trees can be used to exhibit, organize, or analyze electrical networks, producer–consumer and other business relations, information in database systems, syntactic structure of computer programs, etc. We mention a few specific applications that need no lengthy additional explanations.

The set of shortest paths from vertex 1 to the vertices $2, \cdots, n$ in the last section forms a spanning tree.

Railway lines connecting a number of cities (the vertices) can be set up in the form of a spanning tree, the "length" of a line (edge) being the construction cost, and one wants to minimize the total construction cost. Similarly for bus lines, where "length" may be the average annual operating cost. Or for steamship lines (freight lines), where "length" may be profit and the goal is the maximization of total profit. Or in a network of telephone lines between some cities, a shortest spanning tree may simply represent a selection of lines that connect all the cities at minimal cost.

As a somewhat more sophisticated example, consider a private communication network G, let p_{ij} be the probability of intercepting line (i, j) by an outsider, and suppose that one wants to communicate a confidential message to all participants (vertices) along a spanning tree T in G that minimizes the product of the p_{ij} of all the edges (lines) of T, that is, assuming stochastic independence, the total probability of interception; equivalently, a tree T that minimizes the sum of the logarithms of the p_{ij} or, better, the sum of $l_{ij} = \ln \tilde{p}_{ij}$, where $\tilde{p}_{ij} = Kp_{ij}$ with K so large that $\tilde{p}_{ij} > 1$ for each edge of the network G, thus $l_{ij} > 0$.

In addition to these examples from transportation and communication networks, one could mention others from distribution networks, and so on.

We shall now discuss a simple algorithm for the shortest spanning tree problem, which is particularly suitable for sparse graphs (graphs with very few edges; see Sec. 22.1). This algorithm is shown in Table 22.3.

Table 22.3
Kruskal's Greedy Algorithm for Shortest Spanning Trees[8]

ALGORITHM KRUSKAL [$G = (V, E)$, l_{ij} for all (i, j) in E]

Given a connected graph $G = (V, E)$ with edges (i, j) having length $l_{ij} > 0$, the algorithm determines a shortest spanning tree T in G.

INPUT: Edges (i, j) of G and their lengths l_{ij}

OUTPUT: Shortest spanning tree T in G

1. Order the edges of G in ascending order of length.
2. Choose them in this order as edges of T, rejecting an edge only if it forms a cycle with edges already chosen.

If $n - 1$ edges have been chosen, then

OUTPUT T (= the set of edges chosen). Stop

End KRUSKAL

[8] *Proceedings of the American Mathematical Society* 7 (1956), 48–50.

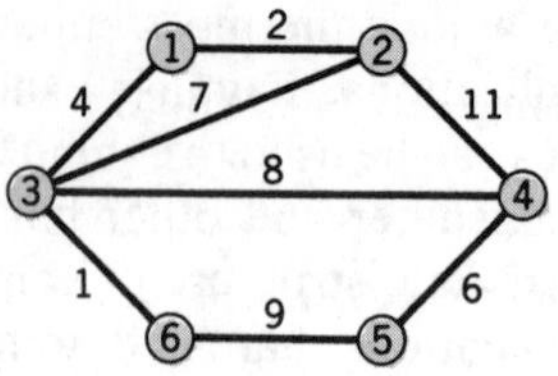

Fig. 465. Graph in Example 1

EXAMPLE 1 **Application of Kruskal's algorithm**

Find a shortest spanning tree in the graph in Fig. 465.

Solution. See Table 22.4. In some of the intermediate stages the edges chosen form a *disconnected* graph (see Fig. 466); this is typical. We stop after $n - 1 = 5$ choices since a spanning tree has $n - 1$ edges. In our problem the edges chosen are in the upper part of the list. This is typical of problems of any size; in general, edges farther down in the list have a smaller chance of being chosen. ∎

Table 22.4
Solution in Example 1

Edge	Length	Choice
(3, 6)	1	1st
(1, 2)	2	2nd
(1, 3)	4	3rd
(4, 5)	6	4th
(2, 3)	7	Reject
(3, 4)	8	5th
(5, 6)	9	
(2, 4)	11	

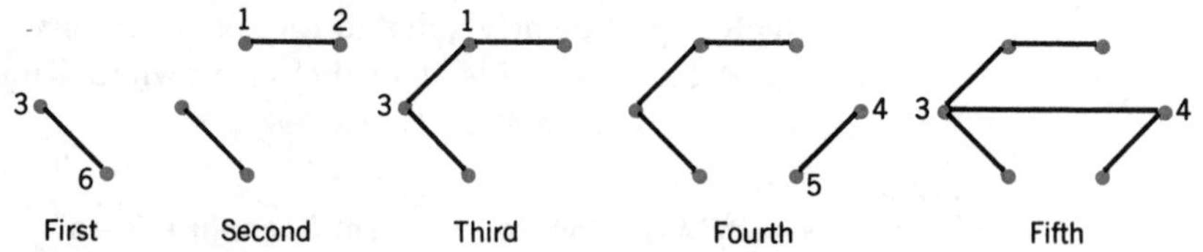

Fig. 466. Choice process in Example 1

The efficiency of Kruskal's method is greatly increased by

Double labeling of vertices

Each vertex i carries a double label (r_i, p_i), *where*

r_i = *Root of the subtree to which i belongs,*
p_i = *Predecessor of i in its subtree,*
p_i = 0 *for roots.*

This simplifies

Rejecting. *If* (i, j) *is next in the list to be considered, reject* (i, j) *if* $r_i = r_j$ (that is, i and j are in the same subtree, so that they are already joined by edges and (i, j) would thus create a cycle). *If* $r_i \neq r_j$, *include* (i, j) *in T.*

If there are several choices for r_i, choose the smallest. If subtrees merge (become a single tree), retain the smallest root as the root of the new subtree.

For Example 1, the double-label list is shown in Table 22.5. In storing it, at each instant one may retain only the latest double label. We show all double labels in order to exhibit the process in all its stages. Labels that remain unchanged are not listed again. Underscored are the two 1's that are the common root of vertices 2 and 3, the reason for rejecting the edge (2, 3). By reading for each vertex the latest label we can read from this list that 1 is the vertex we have chosen as a root and the tree is as shown in the last part of Fig. 466. This is made possible by the predecessor label that each vertex carries. Also, for accepting or rejecting an edge we have to make only one comparison (the roots of the two endpoints of the edge).

Table 22.5
List of Double Labels in Example 1

Vertex	Choice 1 (3, 6)	Choice 2 (1, 2)	Choice 3 (1, 3)	Choice 4 (4, 5)	Choice 5 (3, 4)
1		(1, 0)			
2		(1, 1)			
3	(3, 0)		(1, 1)		
4				(4, 0)	(1, 3)
5				(4, 4)	(1, 4)
6	(3, 3)		(1, 3)		

Ordering is the more expensive part of the algorithm. It is a standard process in data processing for which various methods have been suggested (see Ref. [E14] in Appendix 1). For a complete list of m edges, an algorithm would be $O(m \log_2 m)$, but since the $n - 1$ edges of the tree are most likely to be found earlier, by inspecting the q $(< m)$ topmost edges, for such a list of q edges one would have $O(q \log_2 m)$.

Problem Set 22.4

Applying Kruskal's algorithm, find a shortest spanning tree for the following graphs.

1.

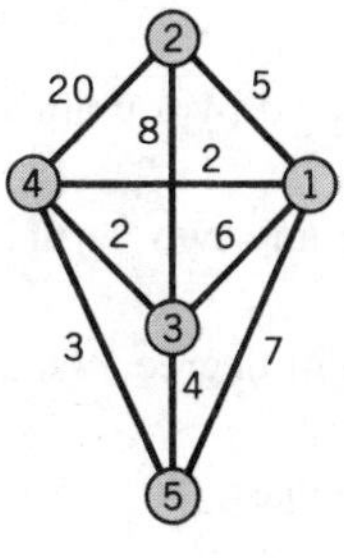

2.

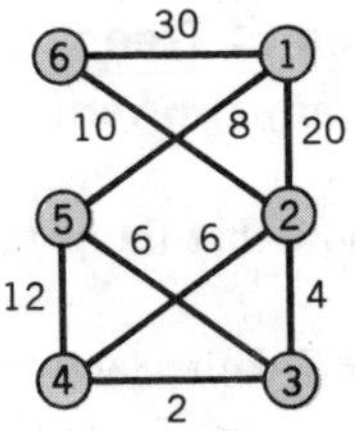

3.

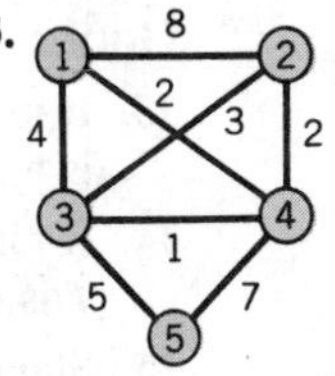

Kruskal's algorithm (*continued*)

4.

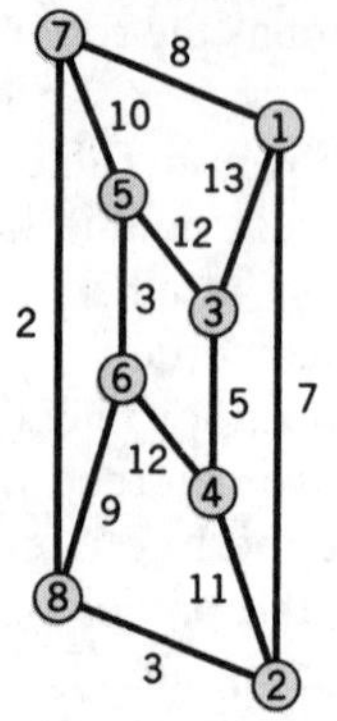

5.

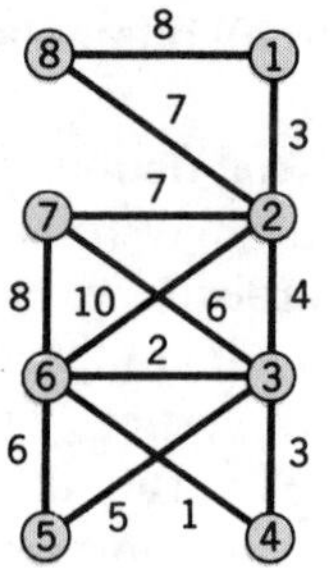

6.

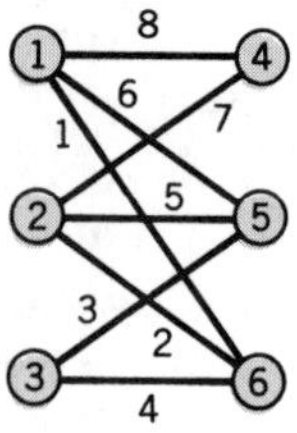

7. In the graph in Fig. 467 find (a) shortest paths from 1 to 2 and from 1 to 3, (b) a shortest spanning tree. Compare the sum of the lengths of the edges used in (a) with that in (b) and comment.

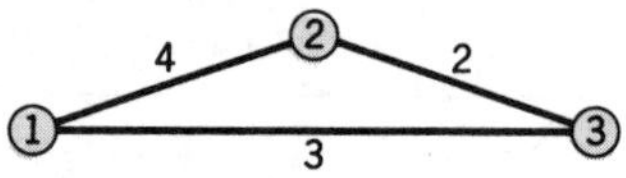

Fig. 467. Problem 7

8. Design an algorithm for obtaining longest spanning trees.

9. Apply the algorithm in Prob. 8 to the graph in Example 1. Compare with the result in Example 1.

10. To get a minimum spanning tree, instead of adding shortest edges, one could think of deleting longest edges. For what graphs would this be feasible? Describe an algorithm for this.

11. Apply the method suggested in Prob. 10 to the graph in Example 1. Do you get the same tree?

12. Find a shortest spanning tree in the complete graph of all possible 30 air connections between the six cities given (distances in miles, rounded). Can you think of a practical application of the result?

	Dallas	Denver	Los Angeles	New York	Washington, DC
Chicago	800	900	1800	700	650
Dallas		650	1300	1350	1200
Denver			850	1650	1500
Los Angeles				2500	2350
New York					200

General Properties of Trees

13. If in a graph G, any two vertices are connected by a unique path, show that G is a tree.

14. **(Uniqueness)** Show that the path connecting any two vertices u and v in a tree T is unique.

15. Show that a tree T with exactly two vertices of degree 1 (see Sec. 22.1) must be a path.

16. Show that a tree with n vertices has $n - 1$ edges.

17. **(Forest)** A (not necessarily connected) graph without cycles is called a *forest*. Give typical examples of applications in which graphs occur that are forests or trees.
18. Show that if a graph G has no cycles, then G must have at least 2 vertices of degree 1.
19. Show that if one joins two vertices in a tree T by a new edge, then a cycle is formed.
20. Show that a graph G with n vertices is a tree if and only if G has $n - 1$ edges and has no cycles.

22.5 Prim's Algorithm for Shortest Spanning Trees

Table 22.6
Prim's Algorithm for Shortest Spanning Tree[9]

ALGORITHM PRIM [$G = (V, E)$, $V = \{1, \cdots, n\}$, l_{ij} for all (i, j) in E]

Given a connected graph $G = (V, E)$ with vertices $1, 2, \cdots, n$ and edges (i, j) having length $l_{ij} > 0$, this algorithm determines a shortest spanning tree T in G and its length $L(T)$.

INPUT: n, edges (i, j) of G and their lengths l_{ij}

OUTPUT: Edge set S of a shortest spanning tree T in G; $L(T)$

[*Initially, all vertices are unlabeled.*]

1. *Initial step*

   Set $i(k) = 1$, $U = \{1\}$, $S = \varnothing$.

   Label vertex k $(= 2, \cdots, n)$ with $\lambda_k = l_{ik}$ [$= \infty$ if G has no edge $(1, k)$].

2. *Addition of an edge to the tree T*

   Let λ_j be the smallest λ_k for k not in U. Include vertex j in U and edge $(i(j), j)$ in S.

   If $U = V$ then compute

   $L(T) = \Sigma l_{ij}$ (sum over all edges in S)

   OUTPUT S, $L(T)$. Stop

   [*S is the edge set of a shortest spanning tree T in G.*]

   Else continue (that is, go to Step 3).

3. *Label updating*

   For every k not in U, if $l_{jk} < \lambda_k$, then set $\lambda_k = l_{jk}$ and $i(k) = j$.

   Go to Step 2.

End PRIM

[9] *Bell System Technical Journal* **36** (1957), 1389–1401. For an improved version of the algorithm, see Cheriton and Tarjan, *SIAM Journal on Computation* **5** (1976), 724–742.

Prim's algorithm shown in Table 22.6 is another popular algorithm for the shortest spanning tree problem (see Sec. 22.4). This algorithm gives a tree T at each stage, a property that Kruskal's algorithm in the last section did not have (look back at Fig. 466 if you did not notice it).

In Prim's algorithm, starting from any single vertex, which we call 1, we "grow" the tree T by adding edges to it, one at a time, according to some rule (below) until T finally becomes a *spanning* tree, which is shortest.

We denote by U the set of vertices of the growing tree T and by S the set of its edges. Thus, initially $U = \{1\}$ and $S = \varnothing$; at the end, $U = V$, the vertex set of the given graph $G = (V, E)$, whose edges (i, j) have length $l_{ij} > 0$, as before.

Thus at the beginning (Step 1) the labels $\lambda_2, \cdots, \lambda_n$ of the vertices $2, \cdots, n$ are the lengths of the edges connecting them to vertex 1 (or ∞ if there is no such edge in G). And we pick (Step 2) the shortest of these as the first edge of the growing tree T and include its other end j in U (choosing the smallest j if there are several, to make the process unique). Updating labels in Step 3 (at this stage and at any later stage) concerns each vertex k not yet in U. Vertex k has label $\lambda_k = l_{i(k),k}$ from before. If $l_{jk} < \lambda_k$, this means that k is closer to the new member j just included in U than k is to its old "closest neighbor" $i(k)$ in U. Then we update the label of k, replacing $\lambda_k = l_{i(k),k}$ by $\lambda_k = l_{jk}$ and setting $i(k) = j$. If, however, $l_{jk} \geqq \lambda_k$ (the *old* label of k), we don't touch the old label. Thus the label λ_k always identifies the closest neighbor of k in U, and this is updated in Step 3 as U and the tree T grow. From the final labels we can backtrack the final tree, and from their numerical values we compute the total length (sum of the lengths of the edges) of this tree.

EXAMPLE 1 Application of Prim's algorithm

Find a shortest spanning tree in the graph in Fig. 468 (which is the same as in Example 1, Sec. 22.4, so that we can compare).

Solution. The steps are as follows.

1. $i(k) = 1$, $U = \{1\}$, $S = \varnothing$, initial labels see Table 22.7.

2. $\lambda_2 = l_{12} = 2$ is smallest, $U = \{1, 2\}$, $S = \{(1, 2)\}$

3. Update labels as shown in Table 22.7, column (I).

2. $\lambda_3 = l_{13} = 4$ is smallest, $U = \{1, 2, 3\}$, $S = \{(1, 2), (1, 3)\}$

3. Update labels as shown in Table 22.7, column (II).

2. $\lambda_6 = l_{36} = 1$ is smallest, $U = \{1, 2, 3, 6\}$, $S = \{(1, 2), (1, 3), (3, 6)\}$

3. Update labels as shown in Table 22.7, column (III).

2. $\lambda_4 = l_{34} = 8$ is smallest, $U = \{1, 2, 3, 4, 6\}$, $S = \{(1, 2), (1, 3), (3, 4), (3, 6)\}$

3. Update labels as shown in Table 22.7, column (IV).

2. $\lambda_5 = l_{45} = 6$ is smallest, $U = V$, $S = (1, 2), (1, 3)\ (3, 4), (3, 6), (4, 5)$. Stop.

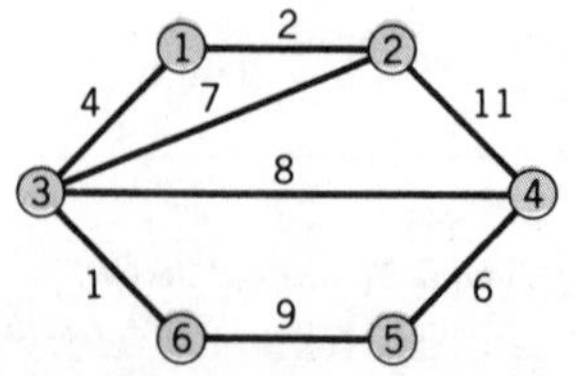

Fig. 468. Graph in Example 1

Table 22.7
Labeling of Vertices in Example 1

Vertex	Initial Label	Relabeling			
		(I)	(II)	(III)	(IV)
2	$l_{12} = 2$	—	—	—	—
3	$l_{13} = 4$	$l_{13} = 4$	—	—	—
4	∞	$l_{24} = 11$	$l_{34} = 8$	$l_{34} = 8$	—
5	∞	∞	∞	$l_{65} = 9$	$l_{45} = 6$
6	∞	∞	$l_{36} = 1$	—	—

The tree is the same as in Example 1, Sec. 22.4. Its length is 21. You will find it interesting to compare the growth process of the present tree with that in Sec. 22.4. ∎

Problem Set 22.5

Applying Prim's algorithm, find a shortest spanning tree T in the following graphs and draw a sketch of T.

1. The graph in Prob. 1, Sec. 22.4

2. The graph in Prob. 2, Sec. 22.4

3.

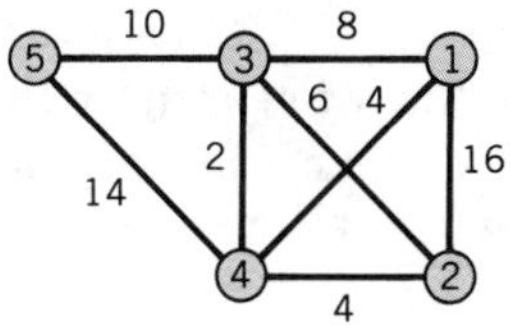

4.

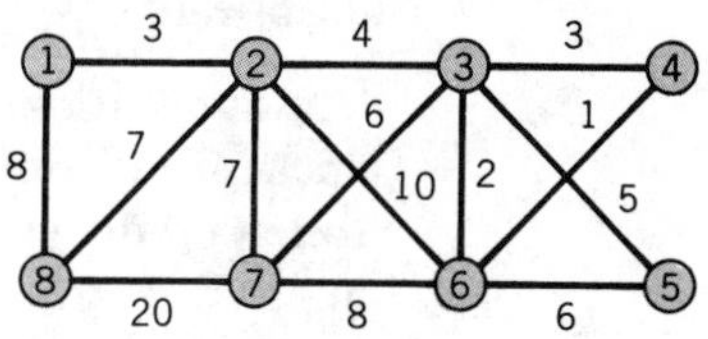

5.

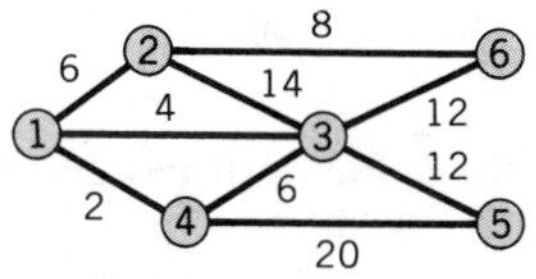

6.

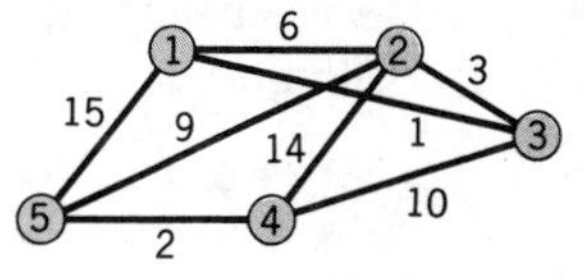

7. How does Prim's algorithm prevent the formation of cycles as one grows T?

8. **(Complexity)** Show that Prim's algorithm has complexity $O(n^2)$.

9. In what case will at the end $S = E$ in Prim's algorithm?

10. For a complete graph (or one that is almost complete), if our data is an $n \times n$ distance table (as in Prob. 12, Sec. 22.4), show that the present algorithm [which is $O(n^2)$] cannot easily be replaced by an algorithm of order less than $O(n^2)$.

11. What would the result be if one applied Prim's algorithm to a graph that is not connected?

12. **(Distance, eccentricity)** Call the length of a shortest path $u \to v$ in a graph $G = (V, E)$ the *distance* $d(u, v)$ from u to v. For fixed u, call the greatest $d(u, v)$ as v ranges over V the *eccentricity* $\epsilon(u)$ of u. Find the eccentricity of vertices 1, 2, 3 in the graph in Prob. 5.

13. **(Diameter, radius, center)** The *diameter* $d(G)$ of a graph $G = (V, E)$ is the maximum of $d(u, v)$ (see Prob. 12) as u and v vary over V, and the *radius* $r(G)$ is the smallest eccentricity $\epsilon(v)$ of the vertices v. A vertex v with $\epsilon(v) = r(G)$ is called a *central vertex*. The set of all central vertices is called the *center* of G. Find $d(G)$, $r(G)$ and the center of the graph in Prob. 5.

14. What are diameter, radius, and center of the spanning tree in Example 1?

15. Explain how the idea of a center can be used in setting up an emergency service facility on a transportation network. In setting up a fire station, a shopping center. How would you generalize the concepts in the case of two or more such facilities?

16. Show that a tree T whose edges all have length 1 has a center consisting of either one vertex or two adjacent vertices.

17. Set up an algorithm of complexity $O(n)$ for finding the center of a tree T.

22.6 Networks. Flow Augmenting Paths

After shortest path problems and problems for trees, as a third large area in combinatorial optimization we discuss **flow problems in networks** (electrical, water, communication, traffic, business connections, etc.), turning from graphs to digraphs (directed graphs; see Sec. 22.1).

By definition, a **network** is a digraph $G = (V, E)$ in which each edge (i, j) has assigned to it a **capacity** $c_{ij} > 0$ [= maximum possible flow along (i, j)], and at one vertex, s, called the **source,** a flow is produced that flows along the edges to another vertex, t, called the **target** or **sink,** where the flow disappears.

In applications, this may be the flow of electricity in wires, of water in pipes, of cars on roads, of people in a public transportation system, of goods from a producer to consumers, of letters from senders to recipients, and so on.

We denote the flow along a (directed!) edge (i, j) by f_{ij} and impose two conditions:

1. For each edge (i, j) in G the flow does not exceed the capacity c_{ij},

$$0 \leqq f_{ij} \leqq c_{ij} \quad \text{(``\textit{Edge condition}'')} \tag{1}$$

2. For each vertex i, not s or t,

$$\text{Inflow} = \text{Outflow} \qquad \text{(``\textit{Vertex condition},'' ``\textit{Kirchhoff's law}'');}$$

in a formula,

$$\underbrace{\sum_k f_{ki}}_{\text{Inflow}} - \underbrace{\sum_j f_{ij}}_{\text{Outflow}} = \begin{cases} 0 \text{ if vertex } i \neq s,\ i \neq t, \\ -f \text{ at the source } s, \\ f \text{ at the target (sink) } t, \end{cases} \tag{2}$$

where f is the total flow (and at s the inflow is zero, whereas at t the outflow is zero). Figure 469 illustrates the notation (for some hypothetical figures).

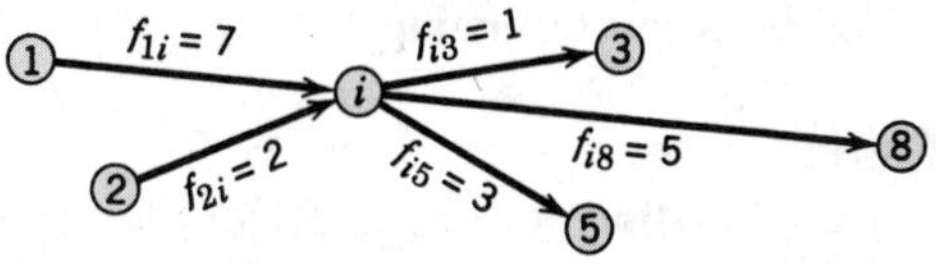

Fig. 469. Notation in (2): inflow and outflow for a vertex i (not s or t)

By a **path** $v_1 \to v_k$ from a vertex v_1 to a vertex v_k in a digraph G we mean a sequence of ***undirected*** *edges*

$$(v_1, v_2), \quad (v_2, v_3), \quad \cdots, \quad (v_{k-1}, v_k)$$

that forms a path as defined in Sec. 22.2. Thus, when we travel from v_1 to v_k we may travel through some edge *in* its given direction—then we call it a **forward edge**—or *opposite to* its given direction—then we call it a **backward edge.** Figure 470 shows a forward edge (u, v) and a backward edge (w, v) of a path $v_1 \to v_k$.

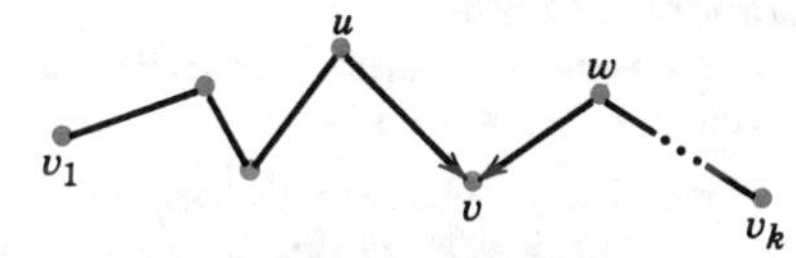

Fig. 470. Forward edge (u, v) and backward edge (w, v) of a path $v_1 \to v_k$

Caution! Each edge in a network has a given direction, *which we cannot change*. Accordingly, if (u, v) is a forward edge in a path $v_1 \to v_k$, then (u, v) can become a backward edge only in another path $x_1 \to x_j$ in which it is an edge and is traversed in the opposite direction as one goes from x_1 to x_j; see Fig. 471. This should be kept in mind, to avoid misunderstandings.

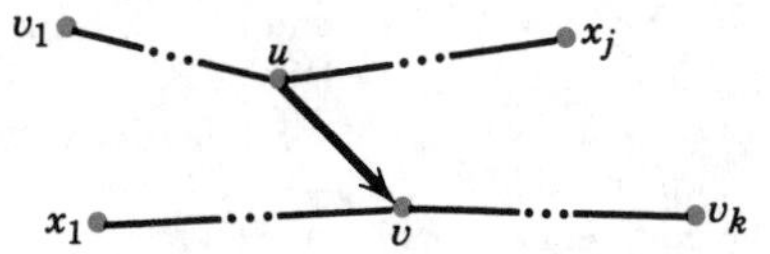

Fig. 471. Edge (u, v) as forward edge in the path $v_1 \to v_k$ and as backward edge in the path $x_1 \to x_j$

Flow Augmenting Paths

Our goal will be to **maximize the flow** from the source s to the target t of a given network. We shall do this by developing methods for increasing an existing flow (including the special case in which the latter is zero). The idea then is to find a path P: $s \to t$ all of whose edges are not fully used, so that we can push additional flow through P by

(i) increasing the flow on forward edges, and

(ii) decreasing the flow on backward edges.

This suggests the following

Definition

A **flow augmenting path** in a network with a given flow f_{ij} on each edge (i, j) is a path $P: s \to t$ such that

(i) no forward edge is used to capacity; thus $f_{ij} < c_{ij}$ for these;

(ii) no backward edge has flow 0; thus $f_{ij} > 0$ for these.

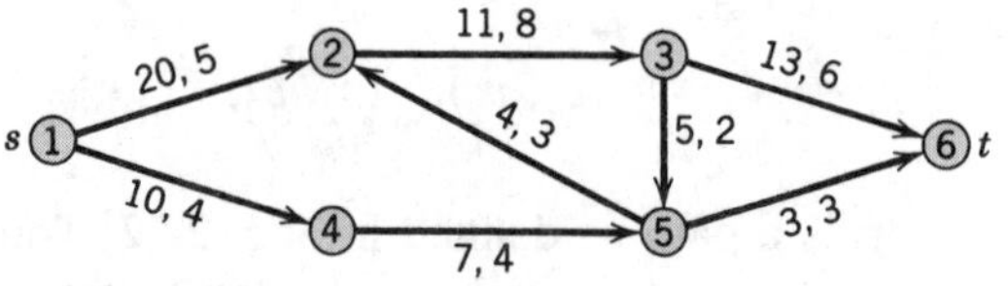

Fig. 472. Network in Example 1
First number = Capacity, Second number = Given flow

EXAMPLE 1 Flow augmenting paths

Find flow augmenting paths in the network in Fig. 472, where the first number is the capacity and the second number a given flow.

Solution. In practical problems, networks are large and one needs a ***systematic method for augmenting flows, which we discuss in the next section.*** In our small network, which should help to illustrate and clarify the concepts and ideas, we can find flow augmenting paths by inspection and augment the existing flow $f = 9$ in Fig. 472. (The outflow from s is $5 + 4 = 9$, which equals the inflow $6 + 3$ into t.)

We use the notation

$$\Delta_{ij} = c_{ij} - f_{ij} \quad \text{for forward edges}$$

$$\Delta_{ij} = f_{ij} \quad \text{for backward edges}$$

$$\Delta = \min \Delta_{ij} \quad \text{taken over all edges of a path.}$$

From Fig. 472 we see that a flow augmenting path $P_1: s \to t$ is P_1: $1 - 2 - 3 - 6$ (Fig. 473), with $\Delta_{12} = 20 - 5 = 15$, etc., and $\Delta = 3$. Hence we can use P_1 to increase the given flow 9 to $f = 9 + 3 = 12$. All three edges of P_1 are forward edges. We augment the flow by 3. Then the flow in each of the edges of P_1 is increased by 3, so that we now have $f_{12} = 8$ (instead of 5), $f_{23} = 11$ (instead of 8) and $f_{36} = 9$ (instead of 6). Edge (2, 3) is now used to capacity. The flow in the other edges remains as before.

We shall now try to increase the flow in this network (Fig. 472) beyond $f = 12$.

There is another flow augmenting path P_2: $s \to t$, namely, P_2: $1 - 4 - 5 - 3 - 6$ (Fig. 473). It shows how a backward edge comes in and how it is handled. Edge (3, 5) is a backward

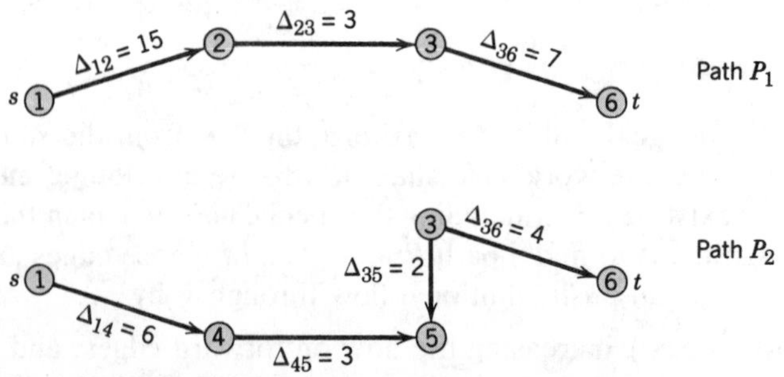

Fig. 473. Flow augmenting paths in Example 1

edge. It has flow 2, so that $\Delta_{35} = 2$. We compute $\Delta_{14} = 10 - 4 = 6$, etc. (Fig. 473) and $\Delta = 2$. Hence we can use P_2 for another augmentation to get $f = 12 + 2 = 14$. The new flow is shown in Fig. 474. No further augmentation is possible. We shall confirm later that $f = 14$ is maximum. (The "cut" in Fig. 474 will be explained below.) ∎

Cut Sets

A "cut set" is a set of edges in a network. The underlying idea is simple and natural. If we want to find out what is flowing from s to t in a network, we may cut the network somewhere between s and t (Fig. 474 shows an example) and see what is flowing in the edges hit by the cut, because any flow from s to t must sometimes pass through some of these edges. These form what is called a **cut set.** [In Fig. 474, the cut set consists of the edges (2, 3), (5, 2), (4, 5).] We denote this cut set by (S, T). Here S is the set of vertices on that side of the cut on which s lies ($S = \{s, 2, 4\}$ for the cut in Fig. 474) and T is the set of the other vertices ($T = \{3, 5, t\}$ in Fig. 474). We say that a cut ***"partitions"*** the vertex set V into two parts S and T. Obviously, the corresponding cut set (S, T) consists of all the edges in the network with one end in S and the other end in T.

By definition, the **capacity** cap (S, T) of a cut set (S, T) is the sum of the capacities of all **forward edges** in (S, T) (forward edges only!), that is, the edges that are directed *from S to T*,

$$\text{cap } (S, T) = \Sigma c_{ij} \qquad \text{[sum over the forward edges of } (S, T)]. \tag{3}$$

Thus, cap $(S, T) = 11 + 7 = 18$ in Fig. 474.

The other edges (directed *from T to S*) are called **backward edges** of the cut set (S, T), and by the **net flow** through a cut set we mean the sum of the flows in the forward edges minus the sum of the flows in the backward edges of the cut set.

Caution! Distinguish well between forward and backward edges in a cut set and in a path: (5, 2) in Fig. 474 is a backward edge for the cut shown but a forward edge in the path $1 - 4 - 5 - 2 - 3 - 6$.

For the cut in Fig. 474 the net flow is $11 + 6 - 3 = 14$. For the same cut in Fig. 472 (not indicated there), the net flow is $8 + 4 - 3 = 9$. In both cases it equals the flow f. We claim that this is not just by chance, but cuts do serve the purpose for which we have introduced them:

Theorem 1 (Net flow in cut sets)

Any given flow in a network G is the net flow through any cut set (S, T) of G.

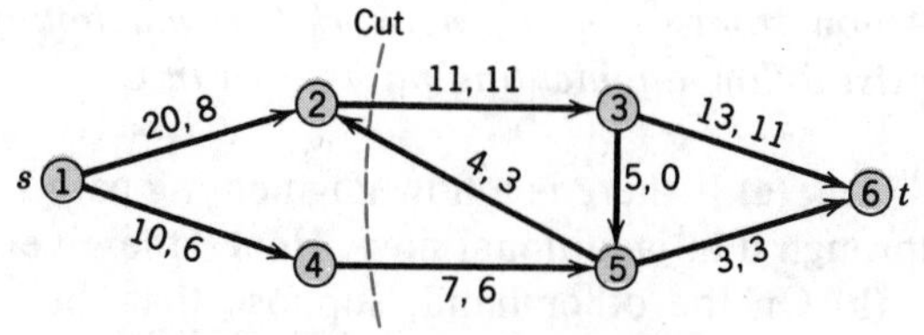

Fig. 474. Maximum flow in Example 1
(The "cut" is explained after Example 1.)

Proof. By Kirchhoff's law (2), at a vertex i,

$$(4) \qquad \underbrace{\sum_j f_{ij}}_{\text{Outflow}} - \underbrace{\sum_l f_{li}}_{\text{Inflow}} = \begin{cases} 0 & \text{if } i \neq s, t, \\ f & \text{if } i = s. \end{cases}$$

Here we can sum over j and l from 1 to n (= number of vertices) by putting $f_{ij} = 0$ for $j = i$ and also for edges without flow or nonexisting edges; hence we can write the two sums as one,

$$\sum_j (f_{ij} - f_{ji}) = \begin{cases} 0 & \text{if } i \neq s, t, \\ f & \text{if } i = s. \end{cases}$$

We now sum over all i in S. Since $s \in S$, this sum equals f:

$$(5) \qquad \sum_{i \in S} \sum_{j \in V} (f_{ij} - f_{ji}) = f.$$

We claim that in this sum, only the edges belonging to the cut set contribute. Indeed, edges with both ends in T cannot contribute, since we sum only over i in S; but edges (i, j) with both ends in S contribute $+f_{ij}$ at one end and $-f_{ij}$ at the other, a total contribution of 0. Hence the left side of (5) equals the net flow through the cut set. By (5), this is equal to the flow f and proves the theorem. ∎

This theorem has the following consequence, which we shall need below.

Theorem 2 **(Upper bound for flows)**
A flow f in a network G cannot exceed the capacity of any cut set (S, T) in G.

Proof. By Theorem 1, the flow f equals the net flow through the cut set, $f = f_1 - f_2$, where f_1 is the sum of the flows through the forward edges and f_2 ($\geqq 0$) is the sum of the flows through the backward edges of the cut set. Thus $f \leqq f_1$. Now f_1 cannot exceed the sum of the capacities of the forward edges; but this sum equals the capacity of the cut set, by definition. Together, $f \leqq \text{cap}\,(S, T)$, as asserted. ∎

Cut sets will now bring out the full importance of augmenting paths:

Theorem 3 **Main Theorem (Augmenting path theorem for flows)**
A flow from s to t in a network G is maximum if and only if there does not exist a flow augmenting path $s \to t$ in G.

Proof. **(a)** If there is a flow augmenting path P: $s \to t$, we can use it to push through it an additional flow. Hence the given flow cannot be maximum.

(b) On the other hand, suppose that there is no flow augmenting path $s \to t$ in G. Let S_0 be the set of all vertices i (including s) such that there is a flow augmenting path $s \to i$, and let T_0 be the set of the other vertices in G. Consider any edge (i, j) with i in S_0 and j in T_0. Then we have a flow augmenting path $s \to i$ since i is in S_0, but $s \to i \to j$ is not flow augmenting because j is not in S_0. Hence we must have

$$(6)\quad f_{ij} = \begin{cases} c_{ij} \\ 0 \end{cases} \text{ if } (i, j) \text{ is a } \begin{cases} \text{forward} \\ \text{backward} \end{cases} \text{ edge of the path } s \to i \to j.$$

Otherwise we could use (i, j) to get a flow augmenting path $s \to i \to j$. Now (S_0, T_0) defines a cut set (since t is in T_0; why?). Since by (6), forward edges are used to capacity and backward edges carry no flow, the net flow through the cut set (S_0, T_0) equals the sum of the capacities of the forward edges, which is cap (S_0, T_0) by definition. This net flow equals the given flow f by Theorem 1. Thus f = cap (S_0, T_0). Also, $f \leqq$ cap (S_0, T_0) by Theorem 2. Hence f must be maximum since we have reached equality. ∎

The end of this proof yields another basic result (by Ford and Fulkerson, *Canadian Journal of Mathematics* **8** (1956), 399–404), namely, the so-called

Theorem 4 **Max-flow min-cut theorem**

The maximum flow[10] *in any network G equals the capacity of a "**minimum cut set**" (= a cut set of minimum capacity) in G.*

Proof. We have just seen that f = cap (S_0, T_0) for a maximum flow f and a suitable cut set (S_0, T_0). Now by Theorem 2 we also have $f \leqq$ cap (S, T) for this f and any cut set (S, T) in G. Together, cap $(S_0, T_0) \leqq$ cap (S, T). Hence (S_0, T_0) is a minimum cut set, and the theorem is proved. ∎

The two basic tools in connection with networks are flow augmenting paths and cut sets. In the next section we show how flow augmenting paths can be used as the basic tool in an algorithm for maximum flows.

Problem Set 22.6

Find flow augmenting paths:

1.

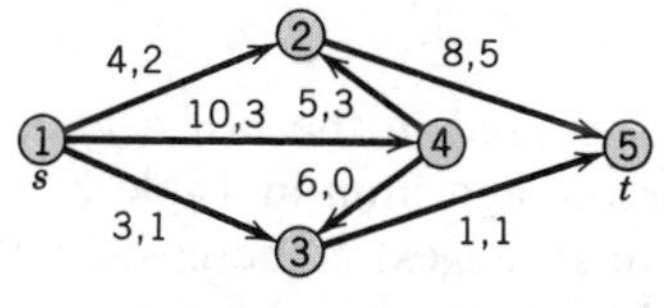

2.

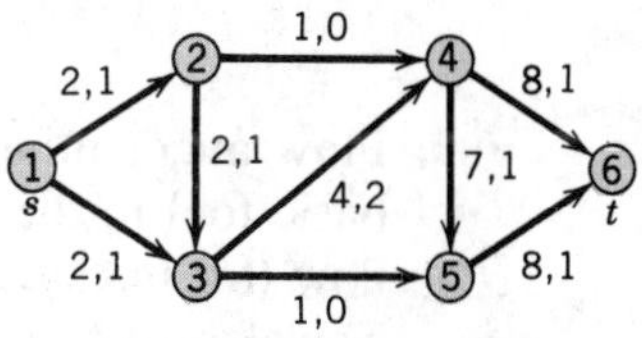

3.

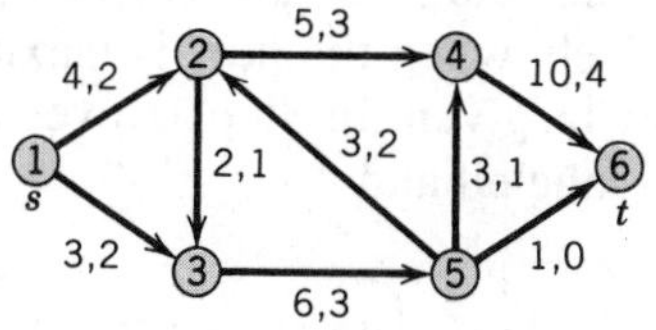

4.

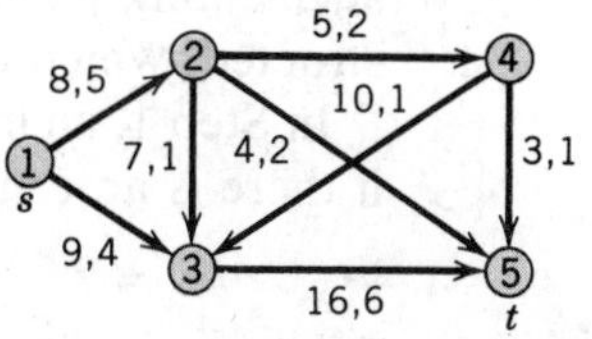

[10]The existence of a maximum flow follows in the case of rational capacities from the algorithm in the next section and in the case of arbitrary capacities from a modification of this algorithm mentioned in footnote 11 in the next section.

Find the maximum flow by inspection:

5. In Prob. 1. **6.** In Prob. 2.

7. In Prob. 3. **8.** In Prob. 4.

In Fig. 472, find T and cap (S, T) if S equals

9. {1, 2, 3} **10.** {1, 2, 4, 5} **11.** {1, 3, 5}

12. Find a minimum cut set in Fig. 472 and verify that its capacity equals the maximum flow $f = 14$.

13. Find examples of flow augmenting paths and the maximum flow in the network in Fig. 475.

In Fig. 475, find T and cap (S, T) if S equals

14. {1, 2, 4} **15.** {1, 2, 4, 6} **16.** {1, 2, 3, 4, 5}

17. In Fig. 475, find a minimum cut set and its capacity.

18. Why are backward edges not considered in the definition of the capacity of a cut set?

19. In which case can an edge (i, j) be used as a forward as well as a backward edge of a path in a network with a given flow?

20. **(Incremental network)** Sketch the network in Fig. 475, and on each edge (i, j) write $c_{ij} - f_{ij}$ and f_{ij}. Do you recognize that from this "incremental network" one can more easily see flow augmenting paths?

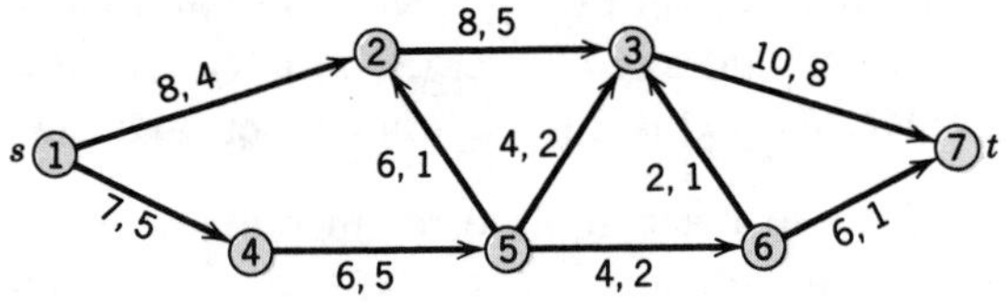

Fig. 475. Problems 13–17

22.7 Ford–Fulkerson Algorithm for Maximum Flow

Flow augmenting paths, as discussed in the last section, are used as the basic tool in the Ford–Fulkerson algorithm in Table 22.8 in which a given flow (for instance, zero flow in all edges) is increased until it is maximum. The algorithm accomplishes the increase by a stepwise construction of flow augmenting paths, one at a time, until no further such paths can be constructed, which happens precisely when the flow is maximum.

In Step 1, an initial flow may be given. In Step 3, a vertex j can be labeled if there is an edge (i, j) with i labeled and

$$c_{ij} > f_{ij} \qquad (\text{"forward edge"})$$

or if there is an edge (j, i) with i labeled and

$$f_{ji} > 0 \qquad (\text{"backward edge"}).$$

Table 22.8
Ford–Fulkerson Algorithm for Maximum Flow

ALGORITHM FORD–FULKERSON

$[G = (V, E)$, vertices $1\ (= s), \cdots, n\ (= t)$, edges (i, j), $c_{ij}]$

This algorithm computes the maximum flow in a network G with source s, sink t, and capacities $c_{ij} > 0$ of the edges (i, j).

INPUT: n, $s = 1$, $t = n$, edges (i, j) of G, c_{ij}

OUTPUT: Maximum flow f in G

1. Assign an initial flow f_{ij} (for instance, $f_{ij} = 0$ for all edges), compute f.
2. Label s by $\varnothing$. Mark the other vertices *"unlabeled."*
3. Find a labeled vertex i that has not yet been scanned. Scan i as follows.

   For every unlabeled adjacent vertex j, if $c_{ij} > f_{ij}$, compute

   $$\Delta_{ij} = c_{ij} - f_{ij} \quad \text{and} \quad \Delta_j = \begin{cases} \Delta_{1j} & \text{if } i = 1 \\ \min(\Delta_i, \Delta_{ij}) & \text{if } i > 1 \end{cases}$$

   and label j with a *"forward label"* (i^+, Δ_j); or if $f_{ji} > 0$, compute

   $$\Delta_j = \min(\Delta_i, f_{ji})$$

   and label j by a "backward label" (i^-, Δ_j).

   If no such j exists then OUTPUT f. Stop

   [*f is the maximum flow.*]

   Else continue (that is, go to Step 4).
4. Repeat Step 3 until t is reached.

   [*This gives a flow augmenting path P: $s \rightarrow t$.*]

   If it is impossible to reach t then OUTPUT f. Stop

   [*f is the maximum flow.*]

   Else continue (that is, go to Step 5).
5. Backtrack the path P, using the labels.
6. Using P, augment the existing flow by Δ_t. Set $f = f + \Delta_t$.
7. Remove all labels from vertices $2, \cdots, n$. Go to Step 3.

End FORD–FULKERSON

To **scan** a labeled vertex i means to label every unlabeled vertex j adjacent to i that can be labeled. Before scanning a labeled vertex i, scan all the vertices that got labeled before i. This BFS (Breadth First Search) strategy was suggested by Edmonds and Karp[11] in 1972. It has the effect that one gets shortest possible augmenting paths. The computational advantage of this is illustrated in Prob. 13.

[11] *Journal of the Association for Computing Machinery* **19** (1972), 248–64. The Ford–Fulkerson algorithm was first published in *Canadian Journal of Mathematics* **9** (1957), 210–218.

EXAMPLE 1 Ford–Fulkerson algorithm

Applying the Ford–Fulkerson algorithm, determine the maximum flow for the network in Fig. 476 (which is the same as that in Example 1, Sec. 22.6, so that we can compare).

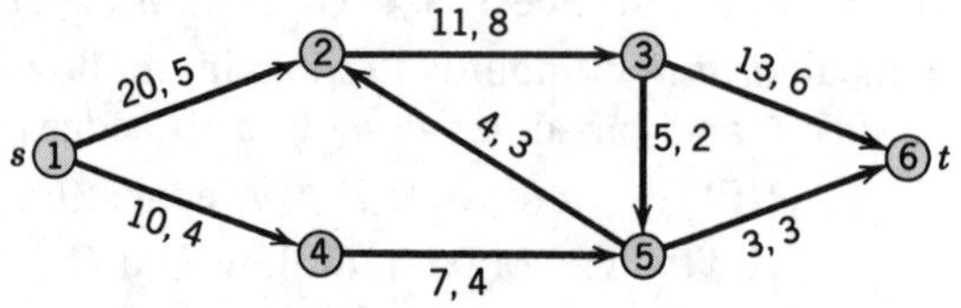

Figure 476. Network in Example 1 with capacities (first numbers) and given flow

Solution. The algorithm proceeds as follows.

1. An initial flow $f = 9$ is given.

2. Label s (= 1) by $\varnothing$. Mark 2, 3, 4, 5, 6 "unlabeled."

3. Scan 1.

Compute $\Delta_{12} = 20 - 5 = 15 = \Delta_2$. Label 2 by $(1^+, 15)$.

Compute $\Delta_{14} = 10 - 4 = 6 = \Delta_4$. Label 4 by $(1^+, 6)$.

4. Scan 2.

Compute $\Delta_{23} = 11 - 8 = 3$, $\Delta_3 = \min(\Delta_2, 3) = 3$. Label 3 by $(2^+, 3)$.

Compute $\Delta_5 = \min(\Delta_2, 3) = 3$. Label 5 by $(2^-, 3)$.

Scan 3.

Compute $\Delta_{36} = 13 - 6 = 7$, $\Delta_6 = \Delta_t = \min(\Delta_3, 7) = 3$. Label 6 by $(3^+, 3)$.

5. P: $1 - 2 - 3 - 6\ (= t)$ is a flow augmenting path.

6. $\Delta_t = 3$. Augmentation gives $f_{12} = 8$, $f_{23} = 11$, $f_{36} = 9$, other f_{ij} unchanged. Augmented flow $f = 9 + 3 = 12$.

7. Remove labels on vertices $2, \cdots, 6$. Go to Step 3.

3. Scan 1.

Compute $\Delta_{12} = 20 - 8 = 12 = \Delta_2$. Label 2 by $(1^+, 12)$.

Compute $\Delta_{14} = 10 - 4 = 6 = \Delta_4$. Label 4 by $(1^+, 6)$.

4. Scan 2.

Compute $\Delta_5 = \min(\Delta_2, 3) = 3$. Label 5 by $(2^-, 3)$.

Scan 4. [*No vertex left for labeling.*]

Scan 5.

Compute $\Delta_3 = \min(\Delta_5, 2) = 2$. Label 3 by $(5^-, 2)$.

Scan 3.

Compute $\Delta_{36} = 13 - 9 = 4$, $\Delta_6 = \min(\Delta_3, 4) = 2$. Label 6 by $(3^+, 2)$.

5. P: $1 - 2 - 5 - 3 - 6\ (= t)$ is a flow augmenting path.

6. $\Delta_t = 2$. Augmentation gives $f_{12} = 10$, $f_{52} = 1$, $f_{35} = 0$, $f_{36} = 11$, other f_{ij} unchanged. Augmented flow $f = 12 + 2 = 14$.

7. Remove labels on vertices $2, \cdots, 6$. Go to Step 3.

One can now scan 1 and then scan 2, as before, but in scanning 4 and then 5 one finds that no vertex is left for labeling. Thus one can no longer reach t. Hence the flow obtained (Fig. 477) is maximum, in agreement with our result in the last section. ∎

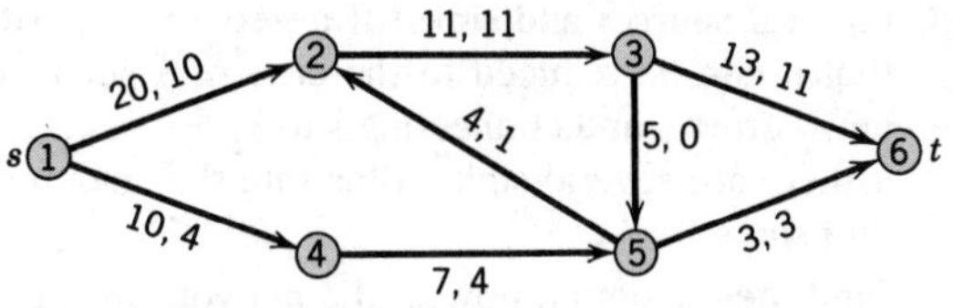

Fig. 477. Maximum flow in Example 1

Problem Set 22.7

1. Carry out the details of the further computations indicated in Example 1.

2. Apply the Ford–Fulkerson algorithm to Example 1 with initial flow 0. Comment on the amount of work compared to that in Example 1.

3. Which are the "bottleneck" edges by which the flow in Example 1 is actually limited? Hence which capacities could be decreased without affecting the maximum flow?

Applying the Ford–Fulkerson algorithm, find the maximum flow:

4. In Prob. 2, Sec. 22.6.

5. In Prob. 1, Sec. 22.6.

6. In Prob. 4, Sec. 22.6.

7. In Prob. 3, Sec. 22.6.

8. What is the (simple) reason that in the augmentation of a flow by the use of a flow augmenting path, Kirchhoff's law is preserved?

9. How does the Ford–Fulkerson algorithm prevent the formation of cycles?

10. How can one see from the algorithm that Ford and Fulkerson follow a BFS technique?

11. Are the consecutive flow augmenting paths produced by the Ford–Fulkerson algorithm unique?

12. **(Integer flow theorem)** Show that if the capacities in a network G are integers, there is a maximum flow that is an integer.

13. How many augmentations would you need to obtain the maximum flow in the network in Fig. 478 if you started from zero flow and alternatingly used the paths P_1: $s - 2 - 3 - t$ and P_2: $s - 3 - 2 - t$? How does the Ford–Fulkerson algorithm prevent this poor choice?

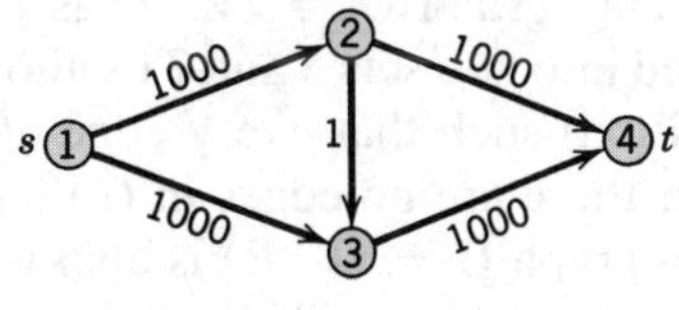

Fig. 478. Problem 13

14. If the Ford–Fulkerson algorithm stops without reaching t, show that the edges with one end labeled and the other end unlabeled form a cut set (S, T) whose capacity equals the maximum flow.

15. **(Several sources and sinks)** If a network has several sources $s_1, \cdots, s_k$, show that it can be reduced to the case of a single-source network by introducing a new vertex s and connecting s to $s_1, \cdots, s_k$ by k edges of capacity ∞. Similarly if there are several sinks. Illustrate this idea by a network with two sources and two sinks.
16. Find the maximum flow in the network in Fig. 479 with two sources (factories) and two sinks (consumers).
17. Find a minimum cut set in Fig. 476 and its capacity.
18. Show that in a network G with all $c_{ij} = 1$, the maximum flow equals the number of edge-disjoint paths $s \to t$.
19. In Prob. 17, the cut set contains precisely all forward edges used to capacity by the maximum flow (Fig. 477). Is this just by chance?
20. Show that in a network G with capacities all equal to 1, the capacity of a minimum cut set (S, T) equals the minimum number q of edges whose deletion destroys all directed paths $s \to t$. (A **directed path** $v \to w$ is a path in which each edge has the direction in which it is traversed in going from v to w.)

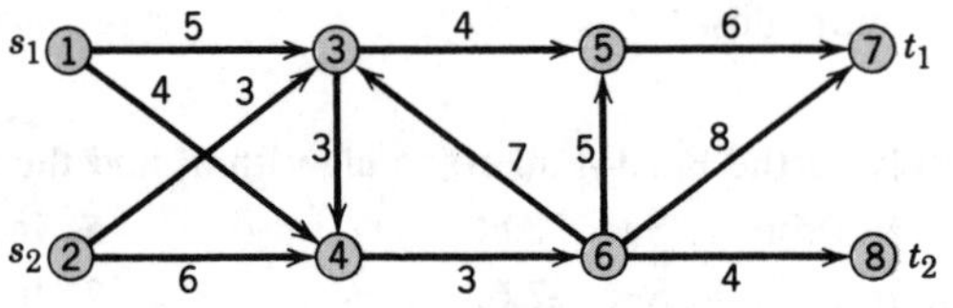

Fig. 479. Problem 16

22.8 Assignment Problems. Bipartite Matching

From digraphs we return to graphs and discuss another important class of combinatorial optimization problems that arises in **assignment problems** of workers to jobs, jobs to machines, goods to storage, ships to piers, classes to classrooms, exams to time periods, and so on. To explain the problem, we need the following concepts.

A **bipartite graph** $G = (V, E)$ is a graph in which the vertex set V is partitioned into two sets S and T (without common elements, by the definition of a partition) such that every edge of G has one end in S and the other in T, so that there are no edges in G that have both ends in S or both ends in T. Such a graph $G = (V, E)$ is also written $G = (S, T; E)$.

Figure 480 shows an illustration, in which V consists of seven elements, three workers a, b, c, making up the set S, and four jobs 1, 2, 3, 4, making up the set T. The edges indicate that worker a can do the jobs 1 and 2, worker b the jobs 1, 2, 3, etc., and the problem is to assign one job to each worker so that every worker gets one job to do. This suggests the next concept:

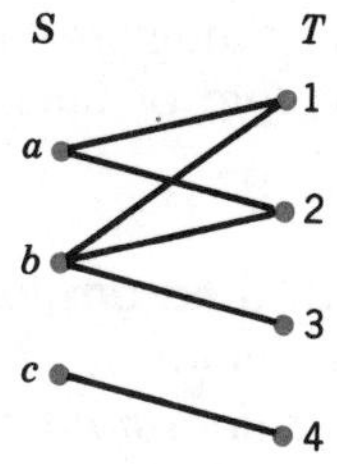

Fig. 480. Bipartite graph in the assignment of a set $S = \{a, b, c\}$ of workers to a set $T = \{1, 2, 3, 4\}$ of jobs

A **matching** in $G = (S, T; E)$ is a set M of edges of G such that no two of them have a vertex in common. If M consists of the greatest possible number of edges, we call it a **maximum cardinality matching**[12] in G.

For instance, a matching in Fig. 480 is $M_1 = \{(a, 2), (b, 1)\}$. Another is $M_2 = \{(a, 1), (b, 3), (c, 4)\}$; obviously, this is of maximum cardinality.

A vertex v is **exposed** (or *not covered*) by a matching M if v is not an endpoint of an edge of M. This concept, which always refers to some matching, will be of interest when we begin to augment given matchings (below). If a matching leaves no vertex exposed, we call it a **complete matching.** Obviously, a complete matching can exist only if S and T consist of the same number of vertices.

We now want to show how one can stepwise increase the cardinality of a matching M until it becomes maximum. Central in this task is the concept of an augmenting path:

An **alternating path** is a path that consists alternately of edges in M and not in M (Fig. 481A). An **augmenting path** is an alternating path both of whose endpoints (a and b in Fig. 481B) are exposed. By dropping from the matching M the edges that are on an augmenting path P (two edges in Fig. 481B) and adding to M the other edges of P (three in the figure), we get a new matching, with one more edge than M. This is how we use an augmenting path in augmenting a given matching by one edge. We assert that this will

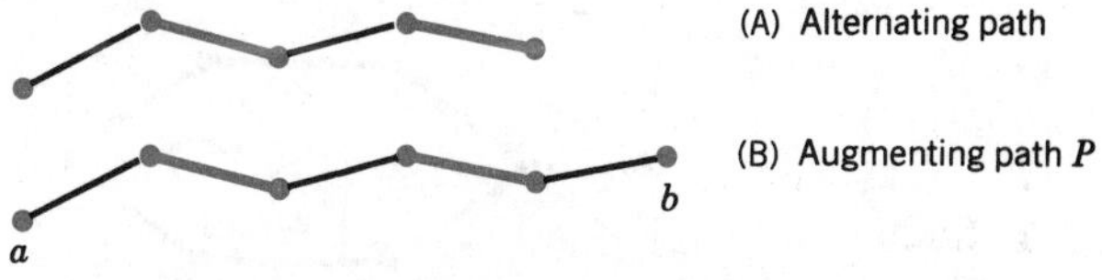

Fig. 481. Alternating and augmenting paths. Heavy edges are those belonging to a matching M.

[12]Or simply a **maximum matching,** but this term is sometimes also used in a different sense, which does not interest us here.

always lead, after a number of steps, to a maximum cardinality matching; indeed the basic role of augmenting paths is expressed in the following theorem.

Theorem 1 **Augmenting path theorem for bipartite matching**

A matching M in a bipartite graph $G = (S, T; E)$ is of maximum cardinality if and only if there does not exist an augmenting path P with respect to M.

Proof. **(a)** We show that if such a path P exists, then M is not of maximum cardinality. Let P have q edges belonging to M. Then P has $q + 1$ edges not belonging to M. (In Fig. 481B we have $q = 2$.) The endpoints a and b of P are exposed, and all the other vertices on P are endpoints of edges in M, by the definition of an alternating path. Hence if an edge of M is not an edge of P, it cannot have an endpoint on P since then M would not be a matching. Consequently, the edges of M not on P, together with the $q + 1$ edges of P not belonging to M form a matching of cardinality one more than the cardinality of M because we omitted q edges from M and added $q + 1$ instead. Hence M cannot be of maximum cardinality.

(b) We now show that if there is no augmenting path for M, then M is of maximum cardinality. Let M^* be a maximum cardinality matching and consider the graph H consisting of all edges that belong to M or to M^*, but not to both. Then it is possible that two edges of H have a vertex in common, but three edges cannot have a vertex in common since then two of the three would have to belong to M (or to M^*), violating that M and M^* are matchings. So every v in V can be in common with two edges of H or with one or none. Hence we can characterize each "component" (= maximal *connected* subset) of H; a component can be:

(A) A closed path with an *even* number of edges (in the case of an *odd* number, two edges from M or two from M^* would meet, violating the matching property). See (A) in Fig. 482.

(B) An open path P with the same number of edges from M and edges from M^*, for the following reason. P must be alternating, that is, an edge of M is followed by an edge of M^*, etc. (since M and M^* are matchings). Now if P had an edge more from M^*, then P would be augmenting for M [see (B2) in Fig. 482], contradicting our assumption that there is no aug-

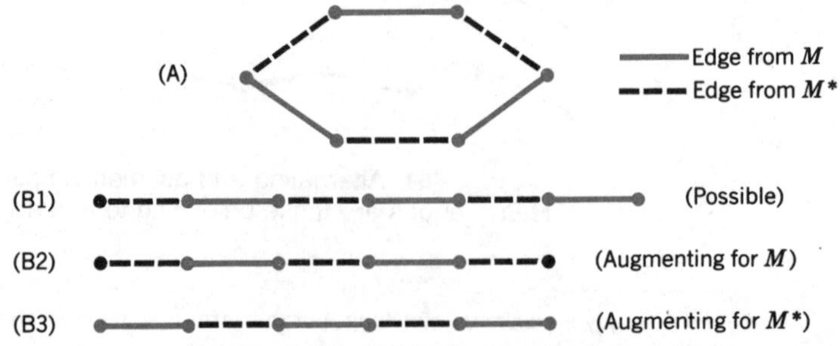

Fig. 482. Proof of the augmenting path theorem for bipartite matching

menting path for M. If P had an edge more from M, it would be augmenting for M^* [see (B3) in Fig. 482], violating the maximum cardinality of M^*, by part (a) of this proof. Hence in each component of H, the two matchings have the same number of edges. Adding to this the number of edges that belong to both M and M^* (which we left aside when we made up H), we conclude that M and M^* must have the same number of edges. M^* being of maximum cardinality, this shows that the same holds for M, as we wanted to prove. ∎

This theorem suggests an algorithm for obtaining augmenting paths in which vertices are labeled for the purpose of backtracking paths. Such a label is *in addition* to the number of the vertex, which is also retained. Clearly, to get an augmenting path, one must start from an *exposed* vertex, and then trace an alternating path until one arrives at another *exposed* vertex. Table 22.9 shows such an algorithm. After Step 3 all vertices in S are labeled. In Step 4, the set T contains at least one exposed vertex, since otherwise we would have stopped at Step 1.

Table 22.9
Bipartite Maximum Cardinality Matching

ALGORITHM MATCHING [$G = (S, T; E), M, n$]

This algorithm determines a maximum cardinality matching M in a bipartite graph G by augmenting a given matching in G.

INPUT: Bipartite graph $G = (S, T; E)$ with vertices $1, \cdots, n$, matching M in G (for instance, $M = \varnothing$)

OUTPUT: Maximum cardinality matching M in G

1. If there is no exposed vertex in S then
   OUTPUT M. Stop
   [*M is of maximum cardinality in G.*]
   Else label all *exposed* vertices *in S* with $\varnothing$.
2. For each i in S and edge (i, j) *not* in M, label j with i, unless already labeled.
3. For each *nonexposed* j in T, label i with j, where i is the other end of the unique edge (i, j) in M.
4. Backtrack the alternating paths P ending on an exposed vertex in T by using the labels on the vertices.
5. If no P in Step 4 is augmenting then
   OUTPUT M. [*M is of maximum cardinality in G.*]
   Else augment M by using an augmenting path P. Remove all labels. Go to Step 1.

End MATCHING

EXAMPLE 1 **Maximum cardinality matching**

Is the matching M_1 in Fig. 483a of maximum cardinality? If not, augment it until maximum cardinality will be reached.

Solution. We apply the algorithm.

1. Label 1 and 4 with ∅.

2. Label 7 with 1. Label 5, 6, 8 with 3.

3. Label 2 with 6, and 3 with 7.

[*All vertices are now labeled as shown in Fig. 483a.*]

4. P_1: 1 – 7 ▬ 3 – 5. [*By backtracking. P_1 is augmenting.*]

P_2: 1 – 7 ▬ 3 – 8. [*P_2 is augmenting.*]

5. Augment M_1 by using P_1, dropping (3, 7) from M_1 and including (1, 7) and (3, 5). Remove all labels. Go to Step 1.

Figure 483b shows the resulting matching M_2 = {(1, 7), (2, 6), (3, 5)}.

1. Label 4 with ∅.

2. Label 7 with 2. Label 6 and 8 with 3.

3. Label 1 with 7, and 2 with 6, and 3 with 5.

4. P_3: 5 ▬ 3 – 8. [*P_3 is alternating but not augmenting.*]

5. Stop. M_2 *is of maximum cardinality* (*namely*, 3). ∎

This is the end of Chap. 22 and of Part F on optimization, an area that is new, is full of unsolved problems, has various applications (many of which have not yet been explored!), and is of rapidly increasing interest to the engineer and computer scientist, as well as in management, micro- and macroeconomics, and other fields.

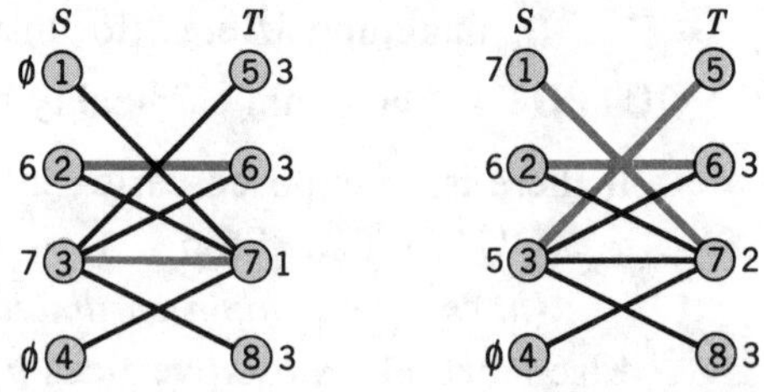

(*a*) Given graph and matching M_1 (*b*) Matching M_2 and new labels

Fig. 483. Example 1

Problem Set 22.8

Are the following graphs bipartite? If your answer is yes, find S and T.

1. 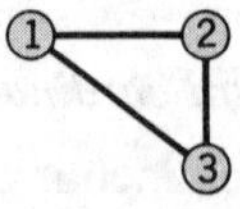

2. 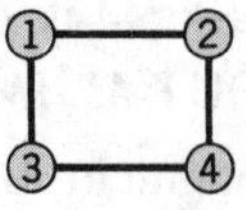

3. 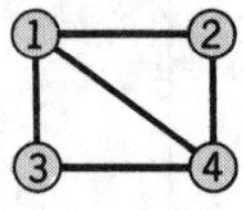

4. 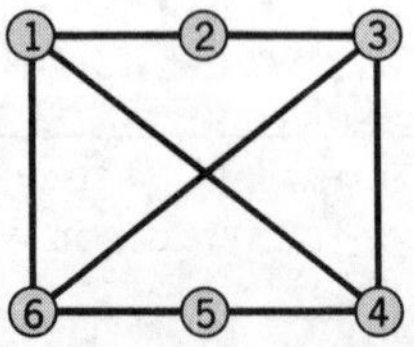

5. 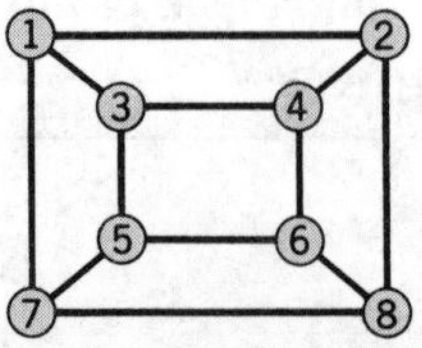

6. 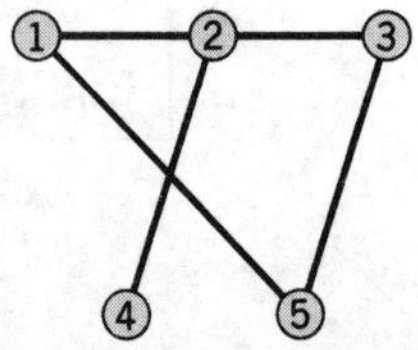

7. Can you obtain the answer to Prob. 3 from the answer to Prob. 1?

Find an augmenting path in the following graphs.

8. 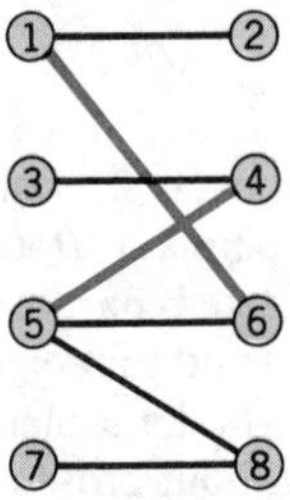

9. 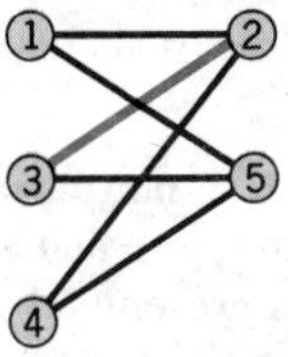

10. 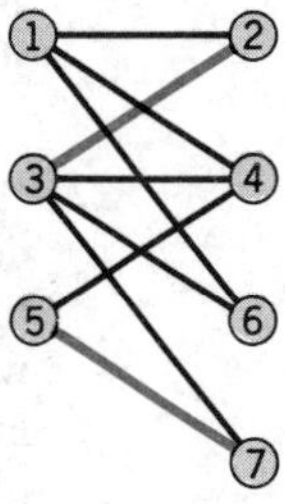

By augmenting the given matching, find a maximum cardinality matching of the graph:

11. In Prob. 9. 12. In Prob. 8. 13. In Prob. 10.

14. **(Timetabling and matching)** Three teachers x_1, x_2, x_3 teach four classes y_1, y_2, y_3, y_4 for these numbers of periods:

	y_1	y_2	y_3	y_4
x_1	1	0	1	1
x_2	1	1	1	1
x_3	0	1	1	1

Show that this arrangement can be represented by a bipartite graph G and that a teaching schedule for one period corresponds to a matching in G. Set up a teaching schedule with the smallest possible number of periods.

15. **(Vertex coloring and exam scheduling)** What is the smallest number of exam periods for six subjects a, b, c, d, e, f if some of the students take a, b, f, some c, d, e, some a, c, e, and some c, e? Solve this as follows. Sketch a graph with six vertices $a, \cdots, f$ and join vertices if they represent subjects simultaneously taken by some students. Color the vertices so that adjacent vertices receive different colors. (Use numbers 1, 2, $\cdots$ instead of actual colors if you want.) What is the minimum number of colors you need? For any graph G, this minimum number is called the (vertex) **chromatic number** $\chi_v(G)$. Why is this the answer to the problem? Write down a possible schedule.

16. How many colors do you need in vertex coloring the graph in Prob. 5?

17. Show that all trees can be vertex colored with two colors.

18. In some computation, temporary storage of frequently used variables $v_1, \cdots, v_6$ during overlapping time intervals (0, 3), (2, 4), (3, 6), (1, 4), (5, 7), (3, 6), respectively, is required. How many index registers (storage locations) are needed? *Hint.* Join v_i and v_j by an edge if their intervals overlap. Then color vertices.

19. What would be the answer to Prob. 18 if not only overlapping but also common endpoints are excluded?

20. **(Complete bipartite graphs)** A bipartite graph $G = (S, T; E)$ is called *complete* if every vertex in S is joined to every vertex in T by an edge, and is denoted by K_{n_1,n_2}, where n_1 and n_2 are the numbers of vertices in S and T, respectively. How many edges does this graph have?

21. **(Planar graph)** A *planar graph* is a graph that can be drawn on a sheet of paper so that no two edges cross. Show that the complete graph K_4 with four vertices is planar. The complete graph K_5 with five vertices is not planar. Make this plausible by attempting to draw K_5 so that no edges cross. Interpret the result in terms of a net of roads between five cities.

22. **(Bipartite graph $K_{3,3}$ not planar)** Three factories 1, 2, 3 are each supplied underground by water, gas, and electricity, from points A, B, C, respectively. Show that this can be represented by $K_{3,3}$ (the complete bipartite graph $G = (S, T; E)$ with S and T consisting of three vertices each) and that eight of the nine supply lines (edges) can be laid out without crossing. Make it plausible that $K_{3,3}$ is not planar by attempting to draw the ninth line without crossing the others.

23. **(Four- (vertex) color theorem)** The famous *four-color theorem* states that one can color the vertices of any *planar* graph (so that adjacent vertices get different colors) with at most four colors. It had been conjectured for a long time and was eventually proved in 1976 by Appel and Haken [*Bulletin of the American Mathematical Society* **82** (1976), 711–712]. Can you color the complete graph K_5 with four colors? Does the result contradict the four-color theorem? (For more details, see Chap. 12 of Ref. [F13] in Appendix 1.)

24. **(Edge coloring)** The *edge chromatic number* $\chi_e(G)$ of a graph G is the minimum number of colors needed for coloring the edges of G so that incident edges get different colors. Clearly, $\chi_e(G) \geqq \max d(u)$, where $d(u)$ is the degree of vertex u. If $G = (S, T; E)$ is bipartite, the equality sign holds. Prove this for $K_{n,n}$.

25. **Vizing's theorem** states that for any graph G (without multiple edges!), $\max d(u) \leqq \chi_e(G) \leqq \max d(u) + 1$. Give an example of a graph for which $\chi_e(G)$ does exceed $\max d(u)$.

26. **(Vertex cover of edges)** A *vertex cover K_v of a set L of edges* in a graph G is a set of vertices of G such that at least one endpoint of each edge of L is in K_v. How many police officers (at least) are needed to cover all blocks (edges) in the graph in Fig. 484? Where should they be placed?

27. How many police officers (at least) are needed to keep every corner in Fig. 484 under surveillance (that is, an officer being at most one block away from every corner)? Where should they be placed?

28. **(Edge cover of vertices)** An *edge cover K_e* of a set M of vertices in a graph G is a set of edges such that each vertex in M is endpoint of at least one edge in K_e. Find a minimum edge cover of the graph in Fig. 484 (consisting of the smallest possible number of edges).

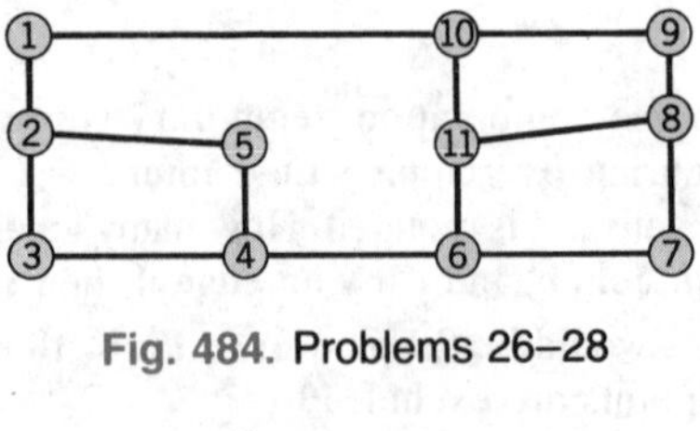

Fig. 484. Problems 26–28

29. **(König's theorem)** A famous theorem of D. König states that in a bipartite graph G, the number of edges in a maximum cardinality matching equals the number of vertices in a minimum vertex cover of G. Illustrate this by an example. Show that it is not true for a nonbipartite graph (with "*matching*" defined as in the bipartite case).

30. **Euler's polyhedron formula** is $n - m + f = 2$, where n, m, f are the numbers of vertices, edges, and faces, respectively, of a polyhedron. Verify this for the cube. This holds for a connected planar graph (drawn so that no edges cross), where "face" now means a plane region bounded by edges and such that any two points in the region can be joined by a continuous curve that meets no edges or vertices; here the exterior region that extends to infinity must be counted. Verify this for Fig. 452a in Sec. 22.1.

Use the Euler formula to prove that the complete graph K_5 with five vertices is not planar. *Hint.* Use that each region is bounded by at least three edges, but each edge is a part of the boundary of at most two faces.

Review Questions and Problems for Chapter 22

1. What is a graph? A digraph? A tree?
2. What matrices and lists did we use in representing graphs?
3. What is a path? What do we mean by a shortest path problem?
4. What is BFS? DFS? In what connection did these concepts occur in this chapter?
5. What is the "traveling salesman problem"?
6. What is the intuitive idea of Bellman's optimality principle, and how is the principle used in Dijkstra's algorithm?
7. Name some applications in which spanning trees play a role.
8. What is the basic idea of Kruskal's greedy algorithm?
9. What is a network? What kind of optimization problems are connected with it?
10. There is a famous theorem on cut sets. Can you remember it?
11. What is a flow augmenting path and why is this concept important?
12. Can a forward edge in one path be a backward edge in another path? In a cut set? Explain.
13. What is a bipartite graph? Give some typical applications that motivate this concept.
14. What kind of optimization problems did we consider in connection with bipartite graphs?
15. What is an augmenting path in bipartite matching? How is it used in obtaining a maximum cardinality matching?

Find the adjacency matrices of the following graphs and digraphs.

16.

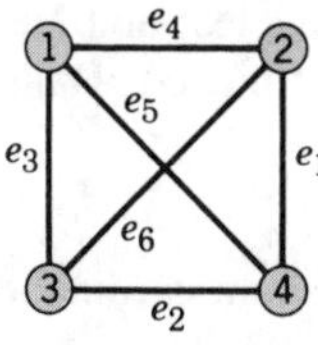

17.

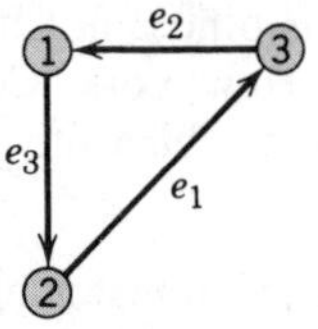

18.

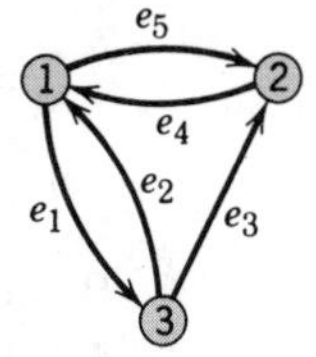

19.

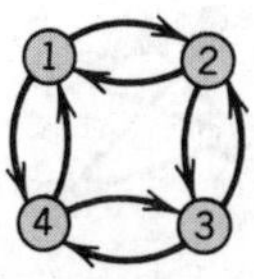

20.

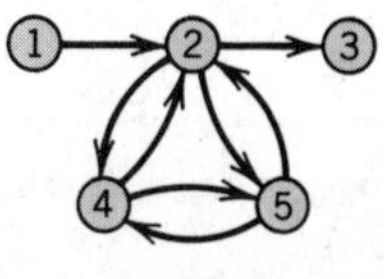

21.

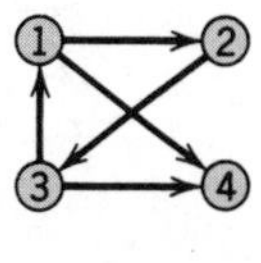

Sketch the graph whose adjacency matrix is

22. $\begin{bmatrix} 0 & 1 & 0 & 1 \\ 1 & 0 & 1 & 0 \\ 0 & 1 & 0 & 0 \\ 1 & 0 & 0 & 0 \end{bmatrix}$ **23.** $\begin{bmatrix} 0 & 1 & 1 \\ 1 & 0 & 1 \\ 1 & 1 & 0 \end{bmatrix}$ **24.** $\begin{bmatrix} 0 & 1 & 0 & 1 \\ 1 & 0 & 0 & 1 \\ 0 & 0 & 0 & 1 \\ 1 & 1 & 1 & 0 \end{bmatrix}$

25. Under what condition will the adjacency matrix of a digraph be symmetric?

Make a vertex incidence list of the graph or digraph:

26. In Prob. 16. **27.** In Prob. 17. **28.** In Prob. 18.

Find a shortest path and its length by Moore's BFS algorithm, assuming that all the edges have length 1:

29.

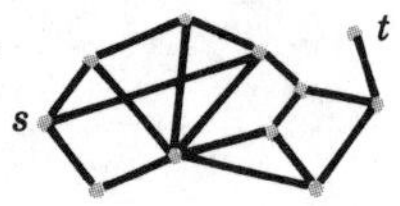

30.

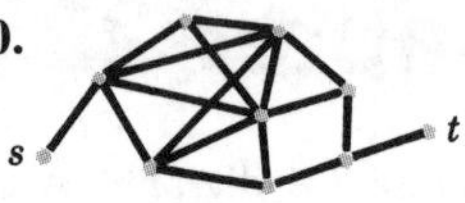

31.

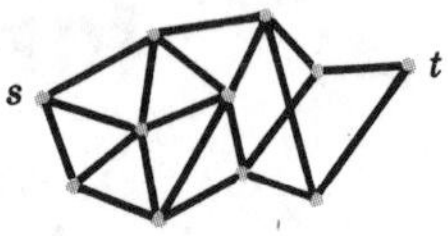

Find shortest paths by Dijkstra's algorithm:

32.

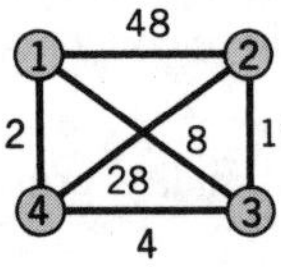

33.

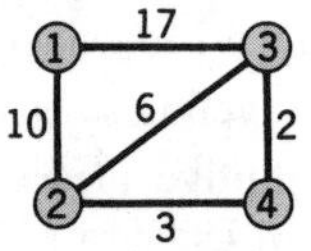

34.

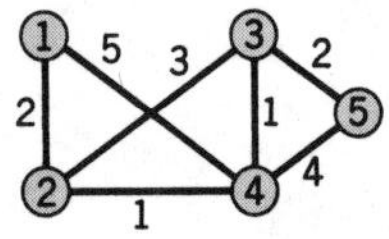

Find a shortest spanning tree for the graph:

35. In Prob. 32. **36.** In Prob. 33. **37.** In Prob. 34.

38. Show that a connected graph G with n vertices and $n - 1$ edges is a tree.

39. **Cayley's theorem** states that the number of spanning trees in a complete graph with n vertices is n^{n-2}. Verify this for $n = 2, 3, 4$.

40. Show that $O(m^3) + O(m^2) = O(m^3)$.

Find the maximum flow in the following networks, where the given numbers are capacities.

41.

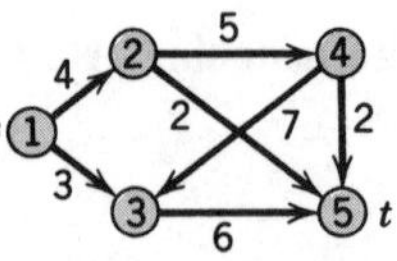

42.

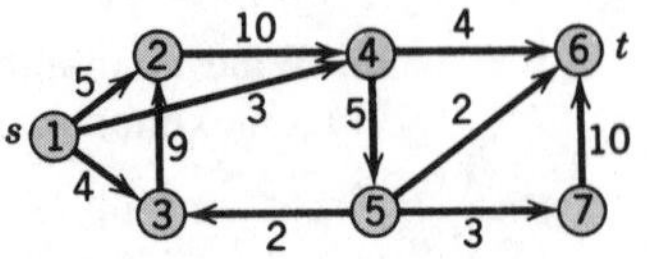

43. Company A has offices in Chicago, Los Angeles, and New York, Company B in Boston and New York, Company C in Chicago, Dallas, and Los Angeles. Represent this by a bipartite graph.

By augmenting the given matching, find a maximum cardinality matching:

44.

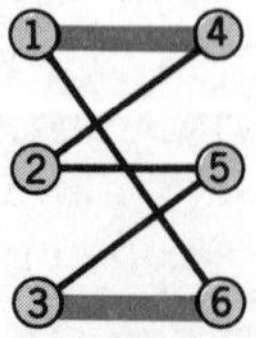

45.

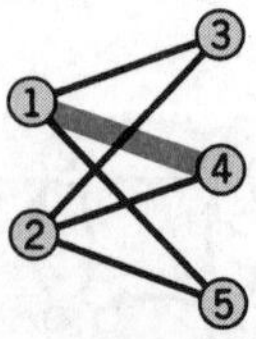

Summary of Chapter 22 Graphs and Combinatorial Optimization

Combinatorial optimization concerns optimization problems of a discrete or combinatorial structure. It uses graphs and digraphs (Sec. 22.1) as basic tools.

A **graph** $G = (V, E)$ consists of a set V of **vertices** $v_1, v_2, \cdots$ (often simply denoted by $1, 2, \cdots, n$) and a set E of **edges** $e_1, e_2, \cdots$ each of which connects two vertices. We also write (i, j) for an edge with vertices i and j as endpoints. A **digraph** (= directed graph) is a graph in which each edge has a direction (indicated by an arrow). For handling graphs and digraphs in computers, one can use *matrices* or *lists* (Sec. 22.1).

This chapter concerns important classes of optimization problems for graphs that all arise from practical applications, and corresponding algorithms, as follows.

In a **shortest path problem** (Sec. 22.2) we determine a path of minimum length (consisting of edges) from a vertex s to a vertex t in a graph whose edges (i, j) have a "length" $l_{ij} > 0$, which may be an actual length or a travel time or cost or an electrical resistance [if (i, j) is a wire in a net], and so on. **Dijkstra's algorithm** (Sec. 22.3) or, when all $l_{ij} = 1$, **Moore's algorithm** (Sec. 22.2) are suitable for these problems.

A **tree** is a graph that is connected and has no **cycles** (no closed paths). Trees are very important in practice. A *spanning tree* in a graph G is a tree containing *all* the vertices of G. If the edges of G have lengths, we can determine a **shortest spanning tree,** for which the sum of the lengths of all its edges is minimum. Corresponding algorithms are those by **Kruskal** (Sec. 22.4) and by **Prim** (Sec. 22.5).

A **network** (Sec. 22.6) is a digraph in which each edge (i, j) has a *capacity* $c_{ij} > 0$ [= maximum possible flow along (i, j)] and at one vertex, the *source* s, a flow is produced that flows along the edges to a vertex t, the *sink* or *target,* where the flow disappears. The problem is to maximize the flow, for instance, by applying the **Ford–Fulkerson algorithm** (Sec. 22.7), which uses **flow augmenting paths** (Sec. 22.6). Another related concept is that of a **cut set,** as defined in Sec. 22.6.

A **bipartite graph** $G = (V, E)$ (Sec. 22.8) is a graph whose vertex set V consists of two parts S and T such that every edge of G has one end in S and the other in T, so that there are no edges connecting vertices in S or vertices in T. A **matching** in G is a set of edges, no two of which have an endpoint in common. The problem then is to find a **maximum cardinality matching** in G, that is, a matching M that has a maximum number of edges. For an algorithm, see Sec. 22.8.

PART G

PROBABILITY AND STATISTICS

This last part of the book is devoted to **mathematical statistics** and its foundation, which is **probability theory.** Both areas are important to various tasks arising in practice, such as testing materials, performance tests of systems, robotics and automatization in general, control of production processes, optimization based on statistical methods, quality control, marketing problems, and so on. In recent decades, probability and statistics have broadened their range of engineering applications and have also entered computer science, for instance, in computer vision, performance analysis of algorithms, and data flow in computer networks. To this we could add a long list of applications in agriculture, biology, demography, economics, geography, management of natural resources, medicine, meteorology, politics, psychology, sociology, traffic control and planning, etc.

It is fortunate that statistics is ***uniform*** in the sense that many of its basic methods can be used in various different areas of application. This suggests our general plan, as follows.

In Chap. 23 on probability we provide the basis of statistics by considering the concept of probability and apply it to develop mathematical models of processes governed or affected by "chance effects," that is, effects which we cannot control or predict with certainty. In order to create a good understanding necessary for successful work in statistics, our emphasis will be on concepts as well as on typical applications.

In Chap. 24 we use the laws and probability models of Chap. 23 in developing methods of statistical inference, that is, conclusions from *samples* to the corresponding entities, called *populations*. Here we shall concentrate on principles and most important standard methods of estimation and testing that are universally applicable in any of the many fields just mentioned.

Chapter 23

Probability Theory

The concept of probability (Sec. 23.2) had its origin in connection with games of chance, such as flipping coins, rolling dice, or playing cards. Nowadays it yields mathematical models of chance processes (briefly called "experiments"; Sec. 23.1). In any such experiment we observe a "random variable" X (a function whose values in the experiment occur "by chance"; Sec. 23.4), which is characterized by a probability distribution (Secs. 23.5–23.7). Or we observe more than one random variable, for example, height and weight of persons, hardness and tensile strength of steel; this is discussed in Sec. 23.8, which will also provide the basis for the mathematical justification of the statistical methods in Chap. 24.

Prerequisite for this chapter: Calculus.
References: Appendix 1, Part G.
Answers to problems: Appendix 2.

23.1 Experiments, Outcomes, Events

In probability and statistics we are concerned with **data** that result either from controlled experimentation in the laboratory or from observation in nature. In either case we use the word "experiment":

An **experiment** is a process by which a measurement or observation is obtained. A single performance of an experiment is briefly called a **trial.** ***Examples*** are

(1) Inspecting a light bulb whether or not it is defective.
(2) Rolling a die and observing what number appears.
(3) Making a measurement of daily rainfall.
(4) Measuring the tensile strength of some steel wire.
(5) Randomly selecting a person and asking whether he or she likes some new car model.

Thus "experiment" is used in a very broad sense. Our interest is in experiments that involve **randomness,** chance effects, so that we cannot predict the result exactly. That result is called an **outcome** or **sample point,** and the set of all possible outcomes is called the **sample space** S of the experiment.

In our examples,

(1) $S = \{D, N\}$, D = defective, N = nondefective
(2) $S = \{1, 2, 3, 4, 5, 6\}$
(3) S the nonnegative numbers in some interval $0 \leqq x \leqq K$.
(4) S the numbers in some interval $a \leqq x \leqq b$
(5) $S = \{L, D, U\}$, L = like, D = dislike, U = undecided.

The subsets of S are called **events** and the outcomes, which obviously are special subsets, **simple events.** In (2), events are $A = \{1, 3, 5\}$ ("*odd number*"), $B = \{2, 4, 6\}$ ("*even number*"), $C = \{5, 6\}$, the six simple events $\{1\}, \{2\}, \cdots, \{6\}$, etc. In practical problems it is events rather than outcomes that interest us.

EXAMPLE 6 Events

Given four gaskets numbered 1, 2, 3, 4, two of which are defective, say, 1, 2. The experiment is that we draw two gaskets at random, so that the sample space S consists of 6 outcomes (unordered pairs)

$$(1, 2), \quad (1, 3), \quad (1, 4), \quad (2, 3), \quad (2, 4), \quad (3, 4)$$

and we are interested in the number of defectives to be obtained in drawing, that is, in the events (subsets of S)

$A = \{(3, 4)\}$ "No defectives"

$B = \{(1, 3), (1, 4), (2, 3), (2, 4)\}$ "1 defective"

$C = \{(1, 2)\}$ "2 defectives." ∎

Subsets. In a trial, if an outcome happens that is a point of an event A, we say that A happens. For example, if we get a 3 on a die, we say that the event A: *Odd number* happens. This is quite natural.

Similarly, $A \subseteq B$ ("A is a subset of B") means that all the points of A are also points of B. And if A happens, we say that B also happens. For example, in a trial, if $A = \{4, 5\}$ happens (meaning that the die shows a 4 or 5), then $B = \{4, 5, 6\}$ also happens in that trial.

Unions, Intersections, Complements

From events $A, B, C, \cdots$ of a given sample space S we can derive further events for practical or theoretical reasons as follows.

The **union** $A \cup B$ of A and B consists of all points in A or B or both.

The **intersection** $A \cap B$ of A and B consists of all points that are in both A and B.

If A and B have no common points [as $A = \{1, 3, 5\}$ and $B = \{2, 4, 6\}$ in (2)], we call them **mutually exclusive,** because the occurrence of one of those events excludes the simultaneous occurrence of the other—if you get an odd number, you cannot get an even number in the same trial. We then write

$$A \cap B = \varnothing,$$

where $\varnothing$ is the *empty set* (set with no elements).

The **complement**[1] A^C of an event A consists of all the points of S *not* in A. In (2), the complement of $A = \{1, 3, 5\}$ is $A^C = \{2, 4, 6\} = B$. Note that an event and its complement are always mutually exclusive, and their union is the whole space S,

$$A \cap A^C = \varnothing, \qquad A \cup A^C = S.$$

Unions and intersections of more events are defined similarly. The **union**

$$\bigcup_{j=1}^{m} A_j = A_1 \cup A_2 \cup \cdots \cup A_m$$

of events $A_1, \cdots, A_m$ consists of all the points that are in at least one A_j, and the **intersection**

$$\bigcap_{j=1}^{m} A_j = A_1 \cap A_2 \cap \cdots \cap A_m$$

consists of the points of S that are in each of the events $A_1, \cdots, A_m$.

Venn diagrams[2] are graphical representations of events in a sample space and are quite useful in working with events, in particular, if one has several of them. Figures 485 and 486 show typical examples.

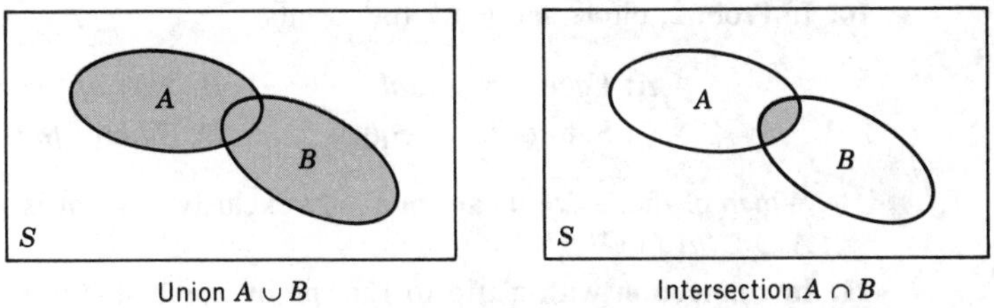

Fig. 485. Venn diagrams showing two events A and B in a sample space S and their union $A \cup B$ (colored) and intersection $A \cap B$ (colored)

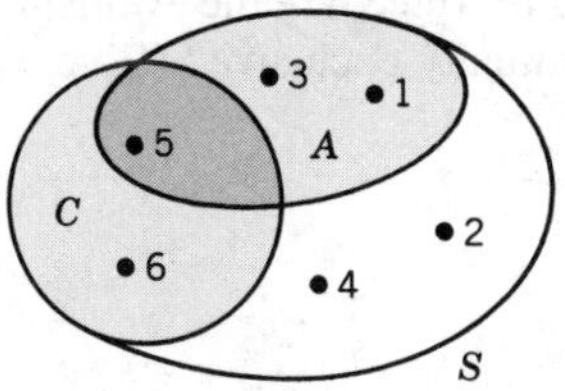

Fig. 486. Venn diagram for the experiment of rolling a die, showing S, $A = \{1, 3, 5\}$, $C = \{5, 6\}$, $A \cup C = \{1, 3, 5, 6\}$, $A \cap C = \{5\}$

[1]Or $\bar{A}$, but we shall not use this because in set theory it is used for another purpose (to denote the closure of A).

[2]JOHN VENN (1834—1923), English mathematician.

EXAMPLE 7 Unions and intersections of 3 events

In rolling a die, consider the events

A: *Number greater than* 3, B: *Number less than* 6, C: *Even number.*

Then $A \cap B = \{4, 5\}$, $B \cap C = \{2, 4\}$, $C \cap A = \{4, 6\}$, $A \cap B \cap C = \{4\}$. Can you sketch a Venn diagram of this? Furthermore, $A \cup B = S$, hence $A \cup B \cup C = S$ (why?), etc. ∎

Problem Set 23.1

Graph a sample space for the following experiments.

1. Tossing two coins

2. Rolling two dice

3. Drawing three screws from a lot containing right-handed and left-handed screws

4. Interviewing two persons about whether they like a certain movie, distinguishing between Yes, No, Undecided

5. Drawing balls from a box, containing 9 blue balls and 1 red ball, until the red ball is drawn, assuming **"sampling without replacement,"** that is, balls drawn are *not* returned to the box

6. Rolling a die until the first 6 appears

7. Recording the lifetime of each of three electronic components

8. Choosing a committee of three from a group of five people

9. Recording the daily rainfall X and maximum temperature Y in a city

10. In Prob. 2, circle and mark the events:

A: *Faces are equal.* B: *Sum of faces exceeds* 9.
C: Sum of faces equals 7. D: *Sum of faces is even.*

Which of these events are mutually exclusive? What is A^C, B^C, $A \cup B$, $A \cap B$, $A \cup C$, $A \cap C$?

11. In connection with a trip to Europe by some students, consider the events P that they see Paris, G that they have a good time, and M that they run out of money, and describe in words the events 1, · · · , 7 in the diagram.

12. In Prob. 4, what is the complement of the event consisting of the three outcomes YY, NN, UU?

13. In rolling two dice, are the events A: "*Sum divisible by* 3" and B: "*Sum divisible by* 4" mutually exclusive? (Give a reason.)

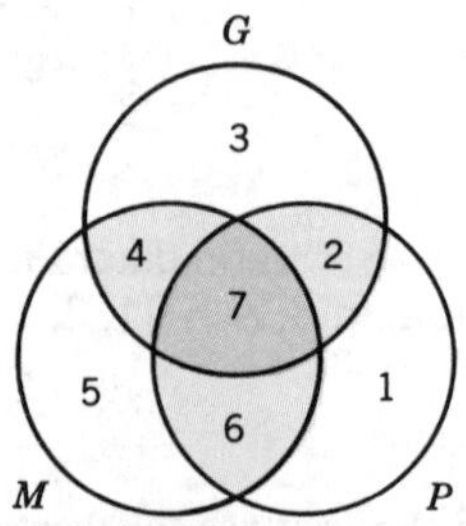

Problem 11

14. In Prob. 6, list the outcomes that make up the event E: *First "Six" in rolling at most* 4 *times*. Describe E^C.

15. List all eight subsets of the sample space $S = \{a, b, c\}$.

16. Using Venn diagrams, graph and check the rules

$$A \cup (B \cap C) = (A \cup B) \cap (A \cup C)$$

$$A \cap (B \cup C) = (A \cap B) \cup (A \cap C).$$

17. (De Morgan's laws) Using Venn diagrams, graph and check *De Morgan's laws*

$$(A \cup B)^C = A^C \cap B^C$$

$$(A \cap B)^C = A^C \cup B^C.$$

18. Using a Venn diagram, show that $A \subseteq B$ if and only if $A \cap B = A$.

19. Show that, by the definition of complement, for any subset A of a sample space S,

$$(A^C)^C = A, \qquad S^C = \varnothing, \qquad \varnothing^C = S, \qquad A \cup A^C = S, \qquad A \cap A^C = \varnothing.$$

20. Using a Venn diagram, show that $A \subseteq B$ if and only if $A \cup B = B$.

23.2 Probability

The "probability" of an event A in an experiment is supposed to measure how frequently A is about to occur if we make many trials. If we flip a coin, then heads H and tails T will appear *about* equally often—we say that H and T are **"equally likely."** Table 23.1 on p. 1154 confirms this. Similarly for a regularly shaped die (**"fair die"**) each of the six outcomes $1, \cdots, 6$ will be equally likely. These are examples of experiments in which the sample space S consists of finitely many outcomes (points) that for reasons of some symmetry can be regarded as equally likely. This suggests the following

Definition 1. Probability

If the sample space S of an experiment consists of finitely many outcomes (points) that are equally likely, then the probability $P(A)$ of an event A is

$$P(A) = \frac{\text{Number of points in } A}{\text{Number of points in } S}. \tag{1}$$

Thus, in particular,

$$P(S) = 1. \tag{2}$$

EXAMPLE 1 Probability

In rolling a fair die, what is the probability $P(A)$ of A of obtaining at least a 5? The probability of B: "*Even number*"?

Solution. The six outcomes are equally likely, so that each has probability 1/6. Thus $P(A) = 2/6 = 1/3$ because $A = \{5, 6\}$ has 2 points, and $P(B) = 3/6 = 1/2$. ∎

Table 23.1
Coin Tossing

Experiments by	Number of Throws	Number of Heads	Relative Frequency of Heads
BUFFON	4,040	2,048	0.5069
K. PEARSON	12,000	6,019	0.5016
K. PEARSON	24,000	12,012	0.5005

Definition 1 takes care of many games as well as some practical applications, as we shall see, but certainly not of all experiments, simply because in many problems we do not have finitely many equally likely outcomes. For instance, if we draw screws, in batches of 3 at a time, from a lot just produced and inspect them for defectives, how can we assign a probability to A: "*At most one defective*"? Well, we can make a larger number n of trials (draw 3 screws many times at random from the lot and inspect them) and take the **relative frequency** $f_{\text{rel}}(A)$ as an "approximation" of the unknown probability $P(A)$. Here, by definition,

$$f_{\text{rel}}(A) = \frac{f(A)}{n} = \frac{\text{Number of times } A \text{ occurs}}{\text{Number of trials}} \tag{3}$$

and $f(A)$ is called the **frequency** of A. Clearly,

$$0 \leqq f_{\text{rel}}(A) \leqq 1 \tag{4*}$$

where $f_{\text{rel}}(A) = 0$ if A does not occur in a sequence of trials and $f_{\text{rel}}(A) = 1$ if A occurs in each trial of the sequence.

Also, in each of the n trials, one of the outcomes $D_0, \cdots, D_3$ (D_j = *Number of defectives in a batch*) must happen; thus their frequencies must add up to n. For the sample space $S = \{D_0, D_1, D_2, D_3\}$ this gives

$$f_{\text{rel}}(S) = \frac{f(D_0) + \cdots + f(D_3)}{n} = \frac{n}{n} = 1. \tag{5*}$$

Furthermore, if A and B are mutually exclusive events ($A \cap B = \varnothing$, see Sec. 23.1), at most A or B can occur in a trial, so that we get

$$f_{\text{rel}}(A \cup B) = \frac{f(A) + f(B)}{n} = f_{\text{rel}}(A) + f_{\text{rel}}(B). \tag{6*}$$

Probability as the counterpart of relative frequency. We are now ready to extend the definition of probability to experiments in which equally likely outcomes are not available. Of course, the extended definition should include Definition 1. Since probabilities are supposed to be the theoretical counterpart of relative frequencies, we choose the properties in (4*), (5*), (6*) as axioms. (Historically, such a choice is the result of a long process of gaining experience on what might be best and most practical.)

Definition 2. Probability
Given a sample space S, with each[3] event A of S (subset of S) there is associated a number $P(A)$, called the **probability** of A, such that the following **axioms of probability** are satisfied.

1. For every A in S,

$$0 \leqq P(A) \leqq 1. \tag{4}$$

2. The entire sample space S has the probability

$$P(S) = 1. \tag{5}$$

3. For mutually exclusive events A and B ($A \cap B = \varnothing$; see Sec. 23.1),

$$P(A \cup B) = P(A) + P(B) \qquad (A \cap B = \varnothing). \tag{6}$$

If S is infinite (has infinitely many points), Axiom 3 has to be replaced by
3′. For mutually exclusive events $A_1, A_2, \cdots$,

$$P(A_1 \cup A_2 \cup \cdots) = P(A_1) + P(A_2) + \cdots. \tag{6'}$$

Basic Theorems for Probability

We shall see that the axioms of probability will enable us to build up probability theory and its application to statistics. We begin with three basic theorems. The first of them is useful if we can get the probability of the complement A^C more easily than $P(A)$ itself.

Theorem 1 (Complementation rule)
For an event A and its complement A^C in a sample space S,

$$P(A^C) = 1 - P(A). \tag{7}$$

Proof. By the definition of complement (Sec. 23.1) we have $S = A \cup A^C$ and $A \cap A^C = \varnothing$. Hence by Axioms 2 and 3,

$$1 = P(S) = P(A) + P(A^C), \qquad \text{thus} \qquad P(A^C) = 1 - P(A). \quad \blacksquare$$

EXAMPLE 2 Coin tossing
Five coins are tossed simultaneously. Find the probability of the event A: *At least one head turns up.* Assume that the coins are fair.

Solution. Since each coin can turn up heads or tails, the sample space consists of $2^5 = 32$ outcomes. Since the coins are fair, we may assign the same probability (1/32) to each outcome. Then the event A^C (*No heads turn up*) consists of only 1 outcome. Hence $P(A^C) = 1/32$, and the answer is $P(A) = 1 - P(A^C) = 31/32$. ■

---

[3]In the infinite case, for a *theoretical* restriction of no *practical* consequence to us, see "σ-algebra," for example, in Ref. [8] listed in Appendix 1.

The next theorem is a simple extension of Axiom 3, which you can readily prove by induction:

Theorem 2 **(Addition rule for mutually exclusive events)**
For mutually exclusive events $A_1, \cdots, A_m$ in a sample space S,

$$(8) \qquad P(A_1 \cup A_2 \cup \cdots \cup A_m) = P(A_1) + P(A_2) + \cdots + P(A_m).$$

EXAMPLE 3 **Mutually exclusive events**

If the probability that on any workday a garage will get 10–20, 21–30, 31–40, over 40 cars to service is 0.20, 0.35, 0.25, 0.12, respectively, what is the probability that on a given day the garage gets at least 21 cars to service?

Solution. Since these are mutually exclusive events, Theorem 2 gives the answer $0.35 + 0.25 + 0.12 = 0.72$. ∎

In many cases, events will not be mutually exclusive. Then we have

Theorem 3 **(Addition rule for arbitrary events)**
For events A and B in a sample space,

$$(9) \qquad P(A \cup B) = P(A) + P(B) - P(A \cap B).$$

Proof. C, D, E in Fig. 487 are disjoint, and $A = C \cup D$, $B = D \cup E$. Hence by Axiom 3,

$$P(A) = P(C) + P(D), \qquad P(B) = P(D) + P(E).$$

By addition,

$$P(A) + P(B) = P(C) + P(D) + P(\mathrm{D}) + P(\mathrm{E}).$$

Now subtract $P(D)$ on both sides,

$$P(A) + P(B) - P(D) = P(C) + P(D) + P(E).$$

The left side is $P(A) + P(B) - P(A \cap B)$ because $D = A \cap B$. The right side is $P(A \cup B)$ by Theorem 1 because $A \cup B = C \cup D \cup E$ and C, D, E are disjoint. This proves (9). ∎

Note that for mutually exclusive events A and B we have $A \cap B = \varnothing$ by definition and, by comparing (9) and (6),

$$(10) \qquad \boxed{P(\varnothing) = 0.}$$

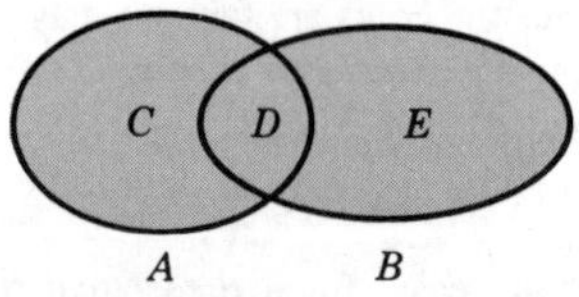

Fig. 487. Proof of Theorem 3

EXAMPLE 4 **Union of arbitrary events**

In tossing a fair die, what is the probability of getting an odd number or a number less than 4?

Solution. Let A be the event "*Odd number*" and B the event "*Number less than* 4." Then Theorem 3 gives the answer

$$P(A \cup B) = \tfrac{3}{6} + \tfrac{3}{6} - \tfrac{2}{6} = \tfrac{2}{3}$$

because $A \cap B$ = "*Odd number less than* 4" = $\{1, 3\}$. ∎

Conditional Probability. Independent Events

Often it is required to find the probability of an event B under the condition that an event A occurs. This probability is called the **conditional probability** *of B given A* and is denoted by $P(B|A)$. In this case A serves as a new (reduced) sample space, and that probability is the fraction of $P(A)$ which corresponds to $A \cap B$. Thus

$$P(B|A) = \frac{P(A \cap B)}{P(A)} \qquad [P(A) \neq 0]. \tag{11}$$

Similarly, the *conditional probability of A given B* is

$$P(A|B) = \frac{P(A \cap B)}{P(B)} \qquad [P(B) \neq 0]. \tag{12}$$

Solving (11) and (12) for $P(A \cap B)$, we obtain

Theorem 4 **(Multiplication rule)**

If A and B are events in a sample space S and $P(A) \neq 0$, $P(B) \neq 0$, *then*

$$P(A \cap B) = P(A)P(B|A) = P(B)P(A|B). \tag{13}$$

EXAMPLE 5 **Multiplication rule**

In producing screws, let A mean "screw too slim" and B "screw too short." Let $P(A) = 0.1$ and let the conditional probability that a slim screw is also too short be $P(B|A) = 0.2$. What is the probability that a screw that we pick randomly from the lot produced will be both too slim and too short?

Solution. $P(A \cap B) = P(A)P(B|A) = 0.1 \cdot 0.2 = 0.02 = 2\%$, by Theorem 4. ∎

Independent events. If events A and B are such that

$$P(A \cap B) = P(A)P(B), \tag{14}$$

they are called **independent events.** Assuming $P(A) \neq 0$, $P(B) \neq 0$, we see from (11)–(13) that in this case

$$P(A|B) = P(A), \qquad P(B|A) = P(B).$$

This means that the probability of A does not depend on the occurrence or nonoccurrence of B, and conversely. This justifies the term "independent."

Independence of *m* events. Similarly, m events $A_1, \cdots, A_m$ are called **independent** if

$$P(A_1 \cap \cdots \cap A_m) = P(A_1) \cdots P(A_m) \tag{15a}$$

as well as for every k different events $A_{j_1}, A_{j_2}, \cdots, A_{j_k}$

$$P(A_{j_1} \cap A_{j_2} \cap \cdots \cap A_{j_k}) = P(A_{j_1})P(A_{j_2}) \cdots P(A_{j_k}) \tag{15b}$$

where $k = 2, 3, \cdots, m - 1$.

Accordingly, three events A, B, C are independent if

$$\begin{aligned} P(A \cap B) &= P(A)P(B), \\ P(B \cap C) &= P(B)P(C), \\ P(C \cap A) &= P(C)P(A), \\ P(A \cap B \cap C) &= P(A)P(B)P(C). \end{aligned} \tag{16}$$

Sampling. Our next example has to do with randomly drawing objects, *one at a time,* from a given set of objects. This is called **sampling from a population,** and there are two ways of sampling, as follows.

1. In **sampling with replacement,** the object that was drawn at random is placed back to the given set and the set is mixed thoroughly. Then we draw the next object at random.

2. In **sampling without replacement** the object that was drawn is put aside.

EXAMPLE 6 Sampling with and without replacement

A box contains 10 screws, three of which are defective. Two screws are drawn at random. Find the probability that none of the two screws is defective.

Solution. We consider the events

A: First drawn screw nondefective.

B: Second drawn screw nondefective.

Clearly, $P(A) = \frac{7}{10}$ because 7 of the 10 screws are nondefective and we sample at random, so that each screw has the same probability ($\frac{1}{10}$) of being picked. If we sample with replacement, the situation before the second drawing is the same as at the beginning, and $P(B) = \frac{7}{10}$. The events are independent, and the answer is

$$P(A \cap B) = P(A)P(B) = 0.7 \cdot 0.7 = 0.49 = 49\%.$$

If we sample without replacement, then $P(A) = \frac{7}{10}$, as before. If A has occurred, then there are 9 screws left in the box, 3 of which are defective. Thus $P(B|A) = \frac{6}{9} = \frac{2}{3}$, and Theorem 4 yields the answer

$$P(A \cap B) = \frac{7}{10} \cdot \frac{2}{3} \approx 47\%.$$ ∎

Problem Set 23.2

1. What is the probability of obtaining at least one head in tossing six fair coins?
2. In rolling two fair dice, what is the probability of obtaining a sum greater than 10 or a sum divisible by 6?
3. Three screws are drawn at random from a lot of 100 screws, 10 of which are defective. Find the probability of the event that all 3 screws drawn are nondefective, assuming that we draw (a) with replacement, (b) without replacement.
4. Under what conditions will it make *practically* no difference whether we sample with or without replacement?
5. Three boxes contain five chips each, numbered from 1 to 5, and one chip is drawn from each box. Find the probability of the event E that the sum of the numbers on the drawn chips is greater than 4.

6. A batch of 100 iron rods consists of 25 oversized rods, 25 undersized rods, and 50 rods of the desired length. If two rods are drawn at random without replacement, what is the probability of obtaining (a) two rods of the desired length, (b) one of the desired length, (c) none of the desired length, (d) two undersized rods?
7. If a certain kind of tire has a life exceeding 25 000 miles with probability 0.95, what is the probability that a set of these tires on a car will last longer than 25 000 miles?
8. In Prob. 7, what is the probability that at least one of the tires will not last for 25 000 miles?
9. A pressure control apparatus contains 4 electronic tubes. The apparatus will not work unless all tubes are operative. If the probability of failure of each tube during some interval of time is 0.03, what is the corresponding probability of failure of the apparatus?
10. If a circuit contains three automatic switches and we want that, with a probability of 95%, during a given time interval they are all working, what probability of failure per time interval can we admit for a single switch?

11. If we inspect sheets of paper by drawing 3 sheets without replacement from every lot of 100 sheets, what is the probability of getting 3 clean sheets although 8% of the sheets contain impurities?
12. What gives the greater probability of hitting at least once: (a) hitting with probability 1/2 and firing 1 shot, or (b) hitting with probability 1/4 and firing 2 shots?
13. In rolling two fair dice, what is the probability of getting equal numbers or numbers with an even product?
14. Suppose that we draw cards repeatedly and with replacement from a file of 200 cards, 100 of which refer to male and 100 to female persons. What is the probability of obtaining the second "female" card before the third "male" card?
15. What is the complementary event of the event considered in Prob. 14? Calculate its probability and use it to check your result in Prob. 14.
16. Suppose that in a production of fuses the fraction of defective fuses has been constant at 2% over a long time, and that this process is controlled every half hour by drawing and inspecting two fuses just produced. Find the probabilities of getting (a) no defectives, (b) 1 defective, (c) 2 defectives. What is the sum of these probabilities?

17. A motor drives an electric generator. During a 30-day period, the motor needs repair with probability 8% and the generator needs repair with probability 4%. What is the probability that during a given period, the entire apparatus will need repair?

18. Show that if B is a subset of A, then $P(B) \leqq P(A)$.

19. Extending Theorem 4, show that $P(A \cap B \cap C) = P(A)P(B|A)P(C|A \cap B)$.

20. You may wonder whether in (16) the last relation follows from the others, but the answer is no. To see this, imagine that a chip is drawn from a box containing 4 chips numbered 000, 011, 101, 110, and let A, B, C be the events that the first, second, and third digit, respectively, on the drawn chip is 1. Show that then the first three formulas in (16) hold but the last one does not hold.

23.3 Permutations and Combinations

In this section we get help in **systematic counting** of sample points in events A. This is necessary for computing the probability $P(A)$ in experiments with a finite sample space S consisting of k equally likely outcomes. Then each outcome has probability $1/k$, and if A consists of m outcomes, then

$$P(A) = \frac{m}{k}.$$

And in practice, m or k can be so large that forgetting points or counting them twice is almost inevitable. For instance, the number of seating orders of 10 people is 3 628 800 (giving 1/3 628 800 as the probability that randomly seating those persons will result in a certain given order)—can you see that it would not be very practical to figure that out by making a list of those orders? It is here that "permutations" and "combinations" are indispensable in giving help.

Permutations. Given n different things (*elements* or *objects*), we may arrange them in a row *in any order*. Each such arrangement is called a **permutation** of the given things. For example, we have 6 permutations of the three letters a, b, c, namely, abc, acb, bac, bca, cab, cba. This illustrates

Theorem 1 (Permutations)

The number of permutations of n different things taken all at a time is

$$n! = 1 \cdot 2 \cdot 3 \cdots n \qquad \text{(read "}n\text{ factorial").} \tag{1}$$

In fact, there are n possibilities for filling the first place in the row; then $n - 1$ objects are still available for filling the second place, etc. Similarly, if not all given things are different, we obtain for their number of permutations the following theorem.

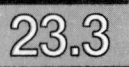

Theorem 2 (Permutations)

If n given things can be divided into c classes such that things belonging to the same class are alike while things belonging to different classes are different, then the number of permutations of these things taken all at a time is

$$\frac{n!}{n_1!\,n_2!\cdots n_c!} \qquad (n_1 + n_2 + \cdots + n_c = n) \tag{2}$$

where n_j is the number of things in the jth class.

EXAMPLE 1 Illustration of Theorem 1

If there are 10 different screws in a box that are needed in a certain order for assembling a certain product, and these screws are drawn at random from the box, the probability P of picking them in the desired order is very small, namely (see Theorem 1),

$$P = 1/10! = 1/3\,628\,800 \approx 0.000\,03\%.$$ ∎

EXAMPLE 2 Illustration of Theorem 2

If a box contains 6 red and 4 blue balls, the probability of drawing first the red and then the blue balls is (see Theorem 2)

$$P = 6!4!/10! = 1/210 \approx 0.5\%.$$ ∎

A **permutation of *n* things taken *k* at a time** is a permutation containing only k of the n given things. Two such permutations consisting of the same k elements, in a different order, are different, by definition. For example, there are 6 different permutations of the three letters a, b, c, taken two letters at a time, namely,

$$ab, \quad ac, \quad bc, \quad ba, \quad ca, \quad cb.$$

A **permutation of *n* things taken *k* at a time with repetitions** is an arrangement obtained by putting any given thing in the first position, any given thing, including a repetition of the one just used, in the second, and continuing until k positions are filled. For example, there are $3^2 = 9$ different such permutations of a, b, c taken 2 letters at a time, namely, the preceding 6 permutations and aa, bb, cc. The student may prove the following theorem (Prob. 15).

Theorem 3 (Permutations)

The number of different permutations of n different things taken k at a time without repetitions is

$$n(n-1)(n-2)\cdots(n-k+1) = \frac{n!}{(n-k)!} \tag{3a}$$

and with repetitions is

$$n^k. \tag{3b}$$

EXAMPLE 3   **Illustration of Theorem 3**

In a coded telegram the letters are arranged in groups of five letters, called *words*. From (3b) we see that the number of different such words is

$$26^5 = 11\ 881\ 376.$$

From (3a) it follows that the number of different such words containing each letter no more than once is

$$26!/(26 - 5)! = 26 \cdot 25 \cdot 24 \cdot 23 \cdot 22 = 7\ 893\ 600. \qquad ■$$

Combinations

In a permutation, the order of the selected things is essential. In contrast, a **combination** of given things means any selection of one or more things *without regard to order*. There are two kinds of combinations, as follows.

The number of **combinations of n different things, taken k at a time, without repetitions** is the number of sets that can be made up from the n given things, each set containing k different things and no two sets containing exactly the same k things.

The number of **combinations of n different things, taken k at a time, with repetitions** is the number of sets that can be made up of k things chosen from the given n things, each being used as often as desired.

For example, there are three combinations of the three letters a, b, c, taken two letters at a time, without repetitions, namely, ab, ac, bc, and six such combinations with repetitions, namely, ab, ac, bc, aa, bb, cc.

Theorem 4   **(Combinations)**

The number of different combinations of n different things, k at a time, without repetitions, is

(4a)
$$\binom{n}{k} = \frac{n!}{k!(n-k)!} = \frac{n(n-1)\cdots(n-k+1)}{1 \cdot 2 \cdots k},$$

and the number of those combinations with repetitions is

(4b)
$$\binom{n+k-1}{k}.$$

The statement involving (4a) follows from the first part of Theorem 3 by noting that there are $k!$ permutations of k things from the given n things that differ by the order of the elements (see Theorem 1), but there is only a single combination of those k things of the type characterized in the first statement of Theorem 4. The last statement of Theorem 4 can be proved by induction (see Prob. 16).

EXAMPLE 4   **Illustration of Theorem 4**

The number of samples of five light bulbs that can be selected from a lot of 500 bulbs is [see (4a)]

$$\binom{500}{5} = \frac{500!}{5!495!} = \frac{500 \cdot 499 \cdot 498 \cdot 497 \cdot 496}{1 \cdot 2 \cdot 3 \cdot 4 \cdot 5} = 255\ 244\ 687\ 600. \qquad ■$$

Factorial Function

In (1)–(4) the **factorial function** is basic. By definition,

$$0! = 1. \tag{5}$$

Values may be computed recursively from given values by

$$(n + 1)! = (n + 1)n!. \tag{6}$$

For large n the function is very large (see Table A3 in Appendix 5). A convenient approximation for large n is the **Stirling formula**[4]

$$n! \sim \sqrt{2\pi n}\left(\frac{n}{e}\right)^n \qquad (e = 2.718 \cdots) \tag{7}$$

where $\sim$ is read "*asymptotically equal*" and means that the ratio of the two sides of (7) approaches 1 as n approaches infinity.

EXAMPLE 5 Stirling formula

To get a feeling for the Stirling formula, verify the following approximations:

$n!$	By (7)	Exact Value	Relative Error
4!	23.5	24	1.4%
10!	3 598 696	3 628 800	0.8%
20!	$2.422\ 79 \cdot 10^{18}$	2 432 902 008 176 640 000	0.4%

■

Binomial Coefficients

The binomial coefficients are defined by the formula

$$\binom{a}{k} = \frac{a(a-1)(a-2)\cdots(a-k+1)}{k!} \qquad (k \geqq 0, \text{ integer}). \tag{8}$$

The numerator has k factors. Furthermore, we define

$$\binom{a}{0} = 1, \quad \text{in particular,} \quad \binom{0}{0} = 1. \tag{9}$$

For integer $a = n$ we obtain from (8)

$$\binom{n}{k} = \binom{n}{n-k} \qquad (n \geqq 0,\ 0 \leqq k \leqq n). \tag{10}$$

Binomial coefficients may be computed recursively, because

[4]JAMES STIRLING (1692—1770), Scots mathematician.

$$\binom{a}{k} + \binom{a}{k+1} = \binom{a+1}{k+1} \qquad (k \geqq 0, \text{ integer}). \tag{11}$$

Formula (8) also yields

$$\binom{-m}{k} = (-1)^k \binom{m+k-1}{k} \qquad (k \geqq 0, \text{ integer}) \ (m > 0). \tag{12}$$

There are numerous further relations; we mention

$$\sum_{s=0}^{n-1} \binom{k+s}{k} = \binom{n+k}{k+1} \qquad (k \geqq 0, n \geqq 1, \text{ both integer}) \tag{13}$$

and

$$\sum_{k=0}^{r} \binom{p}{k}\binom{q}{r-k} = \binom{p+q}{r}. \tag{14}$$

Problem Set 23.3

1. List all permutations of four digits 1, 2, 3, 4, taken all at a time.

2. List (a) all permutations, (b) all combinations without repetitions, (c) all combinations with repetitions, of 5 letters a, e, i, o, u taken 2 at a time.

3. In how many ways can we assign 8 workers to 8 jobs (one worker to each job and conversely)?

4. In how many ways can we choose a committee of 3 from 6 persons?

5. How many different samples of 3 objects can be drawn from a lot of 50 objects?

6. If a cage contains 100 mice, two of which are male, what is the probability that the two male mice will be included if 12 mice are randomly selected?

7. Of a lot of 10 items, 2 are defective. (a) Find the number of different samples of 4. Find the number of samples of 4 containing (b) no defectives, (c) 1 defective, (d) 2 defectives.

8. An urn contains 2 blue, 3 green, and 4 red balls. We draw 1 ball at random and put it aside. Then we draw the next ball, and so on. Find the probability of drawing at first the 2 blue balls, then the 3 green ones, and finally the red ones.

9. Determine the number of different bridge hands. (A bridge hand consists of 13 cards selected from a full deck of 52 cards.)

10. In how many different ways can 5 people be seated at a round table?

11. If 3 suspects who committed a burglary and 6 innocent persons are lined up, what is the probability that a witness who is not sure and has to pick three persons will pick the three suspects by chance? That the witness picks 3 innocent persons by chance?

12. In how many different ways can we select a committee consisting of 4 engineers, 2 chemists, and 2 mathematicians from 10 engineers, 5 chemists, and 7 mathematicians? (First guess, then compute.)

13. How many different license plates showing 5 symbols, namely, 2 letters followed by 3 digits, could be made?

14. What is the probability that in a group of 20 people (that includes no twins) at least two have the same birthday, if we assume that the probability of having birthday at a given day is 1/365 for every day. First guess, then calculate.

15. Prove Theorem 3.

16. Prove the last statement of Theorem 4. *Hint.* Use (13).

17. Using (7), compute approximate values of 4! and 8! and determine the absolute and relative errors.

18. **(Binomial theorem)** By the binomial theorem,

$$(a + b)^n = \sum_{k=0}^{n} \binom{n}{k} a^k b^{n-k},$$

so that $a^k b^{n-k}$ has the coefficient $\binom{n}{k}$. Can you conclude this from Theorem 4 or is this merely a coincidence?

19. Derive (11) from (8).

20. Prove (14) by applying the binomial theorem (Prob. 18) to

$$(1 + b)^p(1 + b)^q = (1 + b)^{p+q}.$$

23.4 Random Variables, Probability Distributions

If we roll two dice, we know that the sum X of the two numbers that turn up must be an integer between 2 and 12, but we cannot predict which value of X will occur in the next trial, and we may say that X depends on "chance." Similarly, if we want to draw 5 bolts from a lot of bolts and measure their diameter, we cannot predict how many will be defective, that is, will not meet given requirements; hence X = *number of defectives* is again a function that depends on "chance." The lifetime X of a light bulb to be drawn at random from a lot of bulbs also depends on "chance," and so does the content X of a bcttle of lemonade filled by a filling machine and selected at random from a given lot.

Roughly speaking, a **random variable** X (also called **stochastic variable** or **variate**) is a function associated with an experiment whose values are real numbers and their occurrence in the trials (the performances of the experiment) depends on "chance." More precisely, we have the following definition.

Definition (Random variable, its probability distribution)
A **random variable** X is a function with the following properties.

1. *X is defined on the sample space S of an experiment, and its values are real numbers.*
2. *For every real number a the probability*

$$P(X = a)$$

that X assume the value a in a trial is well defined; likewise, for every interval I the probability

$$P(X \in I)$$

that X assume **any** *value in I in a trial is well defined.*

These probabilities form the **probability distribution** or, briefly, the **distribution** *of* X, given by the **distribution function**[5]

$$F(x) = P(X \leqq x), \tag{1}$$

for every x giving the probability that X assume ***any*** value not exceeding x.

Although the definition of a distribution is very general, only a very small number of distributions will occur over and over again in applications. Most important is a classification of the practically important distributions into two large classes, the *discrete distributions,* which occur in experiments in which we ***count*** (cars on a road, deaths by cancer, tosses until the first Six turns up, etc.), and *continuous distributions,* which occur in experiments in which we ***measure*** (electric voltage, blood pressure, rainfall, etc.). Let us discuss these two kinds of distributions one after another.

Discrete Random Variables and Distributions

Definition (Discrete random variable and distribution)
A random variable X and its distribution are called **discrete** if X can assume only finitely many or at most countably infinitely many values, called the **possible values** of X, say,

$$x_1, \quad x_2, \quad x_3, \quad \cdots$$

with positive probabilities

$$p_1, \quad p_2, \quad p_3, \quad \cdots,$$

respectively, whereas the probability for any interval containing no possible values of X is zero. ∎

---

[5] ***Caution!*** The terminology is not uniform! $F(x)$ is sometimes also called the **cumulative distribution function.**

Obviously, that discrete distribution is given by the so-called **probability function** *of X*, defined by

$$f(x) = \begin{cases} p_j & \text{if } x = x_j \qquad (j = 1, 2, \cdots), \\ 0 & \text{otherwise.} \end{cases} \tag{2}$$

From this we get the values of the distribution function $F(x)$ by taking sums,

$$F(x) = \sum_{x_j \leqq x} f(x_j) = \sum_{x_j \leqq x} p_j, \tag{3}$$

where for any given x we sum all the probabilities p_j for which x_j is smaller than or equal to that x. This is a **step function** with upward jumps of size p_j at the possible values x_j of X and constant in between.

EXAMPLE 1 **Probability function and distribution function**

Figure 488 shows the probability function $f(x)$ and the distribution function $F(x)$ of the discrete random variable

$$X = \textit{Number a fair die turns up.}$$

X has the possible values $x = 1, 2, 3, 4, 5, 6$ with probability 1/6 each, and precisely at these x the distribution function has upward jumps of magnitude 1/6. Hence from the graph of $f(x)$ we can construct the graph of $F(x)$, and conversely.

In Figure 488 (and the next one) the ***fat dot*** indicates the ***function value at the jump!*** ∎

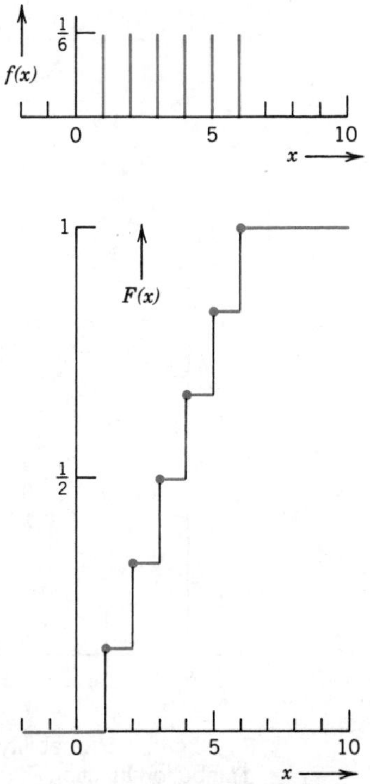

Fig. 488. Probability function $f(x)$ and distribution function $F(x)$ of the random variable X = *Number obtained in tossing a fair die once*

EXAMPLE 2 Probability function and distribution function

The random variable

$$X = \textit{Sum of the two numbers two fair dice turn up}$$

is discrete and has the possible values 2 (= 1 + 1), 3, 4, $\cdots$, 12 (= 6 + 6). There are $6 \cdot 6 = 36$ equally likely outcomes

$$(1, 1), \quad (1, 2), \quad \cdots, \quad (6, 6),$$

where the first number is that shown on the first die and the second number that on the other die. Each such outcome has probability 1/36. Now $X = 2$ occurs in the case of the outcome (1, 1); $X = 3$ in the case of the two outcomes (1, 2) and (2, 1); $X = 4$ in the case of the three outcomes (1, 3), (2, 2), (3, 1), and so on. Hence $f(x) = P(X = x)$ and $F(x) = P(X \leqq x)$ have the values

x	2	3	4	5	6	7	8	9	10	11	12
$f(x)$	1/36	2/36	3/36	4/36	5/36	6/36	5/36	4/36	3/36	2/36	1/36
$F(x)$	1/36	3/36	6/36	10/36	15/36	21/36	26/36	30/36	33/36	35/36	36/36

Figure 489 shows a bar chart of this function and the graph of the distribution function, which is again a step function, with jumps at the possible values of X. ∎

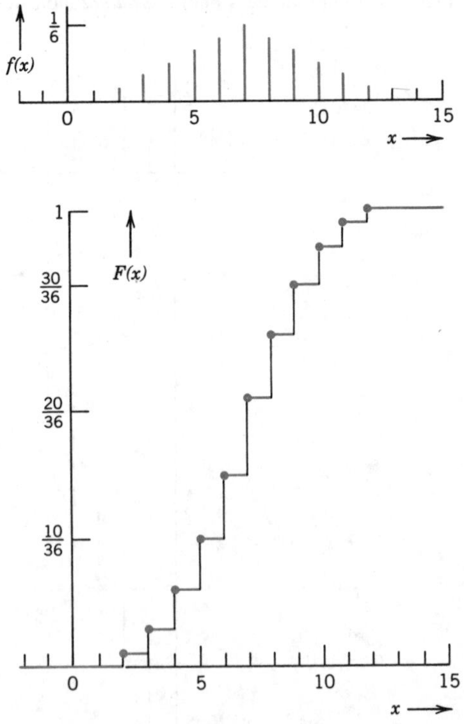

Fig. 489. Probability function $f(x)$ and distribution function $F(x)$ of the random variable X = *Sum of the two numbers obtained in tossing two fair dice once*

Probability corresponding to intervals. In applications, one is often interested in the probability $P(a < X \leqq b)$ that X assume any value in some given interval $a < x \leqq b$. From (1) we obtain the basic formula

$$P(a < X \leqq b) = F(b) - F(a), \tag{4}$$

valid for any distribution. Indeed, the events $X \leqq a$ ("*X assumes any value not exceeding a*") and $a < X \leqq b$ ("*X assumes any value in the interval* $a < x \leqq b$") are mutually exclusive, so that Axiom 3, Sec. 23.2, together with (1), gives

$$F(b) = P(x \leqq b) = P(X \leqq a) + P(a < X \leqq b) = F(a) + P(a < X \leqq b).$$

To get (4), subtract $F(a)$ on both sides. ∎

If X is discrete, (4) and (3) give

$$P(a < X \leqq b) = \sum_{a<x_j\leqq b} p_j, \tag{5}$$

the sum of all probabilities p_j for which x_j satisfies $a < x_j \leqq b$. (Be careful about $<$ and $\leqq$!)

EXAMPLE 3 Illustration of formula (5)

In Example 2, compute the probability of a sum of at least 4 and at most 8.

Solution. $P(3 < X \leqq 8) = F(8) - F(3) = \frac{26}{36} - \frac{3}{36} = \frac{23}{36}$. ∎

Another useful formula to keep in mind results from $P(S) = 1$ and (5):

$$\sum_j p_j = 1 \qquad \text{(sum of all probabilities).} \tag{6}$$

EXAMPLE 4 Waiting problem. Countably infinite sample space

In tossing a fair coin, let X = *Number of trials until the first head appears*. Then, by independence of events (Sec. 23.2),

$$\begin{aligned} P(X = 1) &= P(H) = \tfrac{1}{2} && (H = \text{Head}) \\ P(X = 2) &= P(TH) = \tfrac{1}{2}\cdot\tfrac{1}{2} = \tfrac{1}{4} && (T = \text{Tail}) \\ P(X = 3) &= P(TTH) = \tfrac{1}{2}\cdot\tfrac{1}{2}\cdot\tfrac{1}{2} = \tfrac{1}{8}, \text{ etc.} \end{aligned}$$

and in general $P(X = n) = \left(\frac{1}{2}\right)^n$, $n = 1, 2, \cdots$. Also, (6) can be confirmed by the sum formula for the geometric series,

$$\tfrac{1}{2} + \tfrac{1}{4} + \tfrac{1}{8} + \cdots = -1 + \frac{1}{1 - \frac{1}{2}} = -1 + 2 = 1.$$ ∎

Continuous Random Variables and Distributions

Experiments in which we *count* involve *discrete* random variables and those in which we *measure* involve *continuous* random variables:

Definition (Continuous random variable and distribution)
A random variable X and its distribution are called *of continuous type* or, briefly, **continuous,** if the corresponding distribution function $F(x)$ can be given by an integral in the form[6]

$$F(x) = \int_{-\infty}^{x} f(v)\, dv \tag{7}$$

with the "**density**" f of the distribution being continuous (possibly except for finitely many values of v) and nonnegative. ∎

By differentiation of (7) we get

$$F'(x) = f(x)$$

for every x at which $f(x)$ is continuous.

From (7) and Axiom 2, Sec. 23.2, we also have

$$\int_{-\infty}^{\infty} f(v)\, dv = 1. \tag{8}$$

Furthermore, from (4) and (7) we obtain the important formula

$$P(a < X \leqq b) = F(b) - F(a) = \int_{a}^{b} f(v)\, dv. \tag{9}$$

Hence this probability equals the area under the curve of the density $f(x)$ between $x = a$ and $x = b$, as shown in Fig. 490.

Obviously, for any fixed a and b ($> a$), in the case of a *continuous* random variable X the probabilities corresponding to the intervals $a < X \leqq b$, $a < X < b$, $a \leqq X < b$, and $a \leqq X \leqq b$ are all the same. This is different from the case of a discrete distribution. (Explain why!)

The next example illustrates notations and typical applications of our present formulas.

---

[6] $F(x)$ is continuous, but continuity of $F(x)$ does not imply the existence of a representation of the form (7). Since continuous distribution functions that cannot be represented in the form (7) are rare in practice, the widely accepted terms "continuous random variable" and "continuous distribution" should not be confusing.

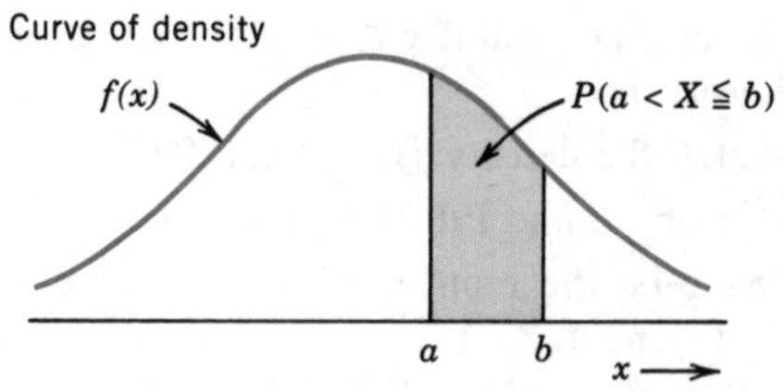

Fig. 490. Example illustrating formula (9)

EXAMPLE 5 A continuous distribution

Let X have the density function $f(x) = 0.75(1 - x^2)$ if $-1 \leqq x \leqq 1$ and zero otherwise. Find the distribution function. Find the probabilities $P(-\frac{1}{2} \leqq X \leqq \frac{1}{2})$ and $P(\frac{1}{4} \leqq X \leqq 2)$. Find x such that $P(X \leqq x) = 0.95$.

Solution. From (7) we obtain $F(x) = 0$ if $x \leqq -1$,

$$F(x) = 0.75 \int_{-1}^{x} (1 - v^2)\, dv = 0.5 + 0.75x - 0.25x^3 \qquad \text{if } -1 < x \leqq 1,$$

and $F(x) = 1$ if $x > 1$. From this and (9) we get

$$P(-\tfrac{1}{2} \leqq X \leqq \tfrac{1}{2}) = F(\tfrac{1}{2}) - F(-\tfrac{1}{2}) = 0.75 \int_{-1/2}^{1/2} (1 - v^2)\, dv = 68.75\%$$

(because $P(-\frac{1}{2} \leqq X \leqq \frac{1}{2}) = P(-\frac{1}{2} < X \leqq \frac{1}{2})$ for a continuous distribution) and

$$P(\tfrac{1}{4} \leqq X \leqq 2) = F(2) - F(\tfrac{1}{4}) = 0.75 \int_{1/4}^{1} (1 - v^2)\, dv = 31.64\%.$$

(Note that the upper limit of integration is 1, not 2. Why?) Finally,

$$P(X \leqq x) = F(x) = 0.5 + 0.75x - 0.25x^3 = 0.95.$$

Algebraic simplification gives $3x - x^3 = 1.8$. A solution is $x = 0.73$, approximately.

Sketch $f(x)$ and mark $x = -\frac{1}{2}, \frac{1}{2}, \frac{1}{4}$, and 0.73, so that you can see the results (the probabilities) as areas under the curve. Sketch also $F(x)$. ∎

Further examples of continuous distributions are included in the next problem set and in later sections.

Problem Set 23.4

1. Sketch the probability function $f(x) = x^2/30$ ($x = 1, 2, 3, 4$) and the distribution function.
2. Sketch f and F when $f(0) = f(3) = \frac{1}{8}$, $f(1) = f(2) = \frac{3}{8}$. Can f have further positive values?
3. Sketch the density $f(x) = x^2/9$ if $0 \leqq x \leqq 3$ and zero otherwise, and the distribution function $F(x)$.
4. Sketch the distribution function $F(x) = 0$ if $x \leqq 0$, $F(x) = 1 - e^{-2x}$ if $x > 0$, and the density $f(x)$.

5. Sketch $F(x) = 0$ if $x \leqq 0$, $F(x) = 0.2x$ if $0 < x \leqq 5$, $F(x) = 1$ if $x > 5$, and its density $f(x)$.
6. Sketch the density $f(x) = 0.4$ if $3 < x < 5.5$ and the distribution function.
7. In Prob. 6, find $P(0 \leqq X \leqq 4)$. Find x such that $P(X \leqq x) = \frac{1}{2}$.
8. Let X be the number of years before a particular type of machine will need replacement. Assume that X has the probability function $f(1) = 0.1$, $f(2) = 0.2$, $f(3) = 0.2$, $f(4) = 0.2$, $f(5) = 0.3$. Sketch f and F. Find the probability that the machine needs no replacement during the first 3 years.
9. Find the probability that none of three bulbs in a traffic signal will have to be replaced during the first 1200 hours of operation if the lifetime X of a bulb is a random variable with the density $f(x) = 6[0.25 - (x - 1.5)^2]$ when $1 \leqq x \leqq 2$ and $f(x) = 0$ otherwise, where x is measured in multiples of 1000 hours.

10. Suppose that certain bolts have length $L = 200 + X$ mm, where X is a random variable with density $f(x) = \frac{3}{4}(1 - x^2)$ if $-1 \leqq x \leqq 1$ and 0 otherwise. Determine c so that with a probability of 95% a bolt will have any length between $200 - c$ and $200 + c$. *Hint.* See also Example 5.
11. Let $f(x) = kx^2$ if $0 \leqq x \leqq 1$ and 0 otherwise. Find k. Find c_1 and c_2 such that $P(X \leqq c_1) = 0.1$ and $P(X \leqq c_2) = 0.9$.

12. Suppose that in an automatic process of filling oil into cans, the content of a can (in gallons) is $Y = 50 + X$, where X is a random variable with density $f(x) = 1 - |x|$ when $|x| \leqq 1$ and 0 when $|x| > 1$. Sketch $f(x)$ and $F(x)$. In a lot of 100 cans, about how many will contain 50 gallons or more? What is the probability that a can will contain less than 49.5 gallons? Less than 49 gallons?
13. Let X [millimeters] be the thickness of washers a machine turns out. Assume that X has the density $f(x) = kx$ if $1.9 < x < 2.1$ and 0 otherwise. Find k. What is the probability that a washer will have thickness between 1.95 mm and 2.05 mm?
14. Consider the random variable X = *Number of times a fair die is rolled until the first Six appears*. Find the probability function of X; show that it satisfies (6).

15. Show that $b < c$ implies $P(X \leqq b) \leqq P(X \leqq c)$.
16. Let X be the ratio of sales to profits of some firm. Assume that X has the distribution function $F(x) = 0$ if $x < 2$, $F(x) = (x^2 - 4)/5$ if $2 \leqq x < 3$, $F(x) = 1$ if $x \geqq 3$. Find and sketch the density. What is the probability that X is between 2.5 (40% profit) and 5 (20% profit)?
17. If the life of ball bearings has the density $f(x) = ke^{-x}$ if $0 \leqq x \leqq 2$ and 0 otherwise, what is k? What is the probability $P(X \geqq 1)$?
18. If the diameter X of axles has the density $f(x) = k$ if $119.9 \leqq x \leqq 120.1$ and 0 otherwise, how many defectives will a lot of 500 axles approximately contain if defectives are axles slimmer than 119.92 or thicker than 120.08?

19. Let X be a random variable that can assume every real value. What are the complements of the events $X \leqq b$, $X < b$, $X \geqq c$, $X > c$, $b \leqq X \leqq c$, $b < X \leqq c$?
20. A box contains 4 right-handed and 6 left-handed screws. Two screws are drawn at random without replacement. Let X be the number of left-handed screws drawn. Find the probabilities $P(X = 0)$, $P(X = 1)$, $P(X = 2)$, $P(1 < X < 2)$, $P(X \leqq 1)$, $P(X \geqq 1)$, $P(X > 1)$, and $P(0.5 < X < 10)$.

23.5 Mean and Variance of a Distribution

The distribution function (and equally well the probability function or the density, respectively) determines all the properties of a random variable completely. In addition, we may compute from the distribution function certain quantities, called **parameters,** that express important properties of the distribution, such as *central location, dispersion* (spread), asymmetry, and so on. In this section we discuss the mean and the variance of a distribution, the two most important parameters in practice.

The *mean value* or **mean** of a distribution is denoted by μ and is defined by

(1)

(a) $$\mu = \sum_j x_j f(x_j) \qquad \text{(Discrete distribution)}$$

(b) $$\mu = \int_{-\infty}^{\infty} x f(x)\, dx \qquad \text{(Continuous distribution).}$$

In (1a) the function $f(x)$ is the probability function of the random variable X considered, and we sum over all possible values (see Sec. 23.4). In (1b) the function $f(x)$ is the density of X. The mean is also known as the *mathematical expectation of* X and is sometimes denoted by $E(X)$. By definition it is assumed that the series in (1a) converges absolutely and the integral of $|x| f(x)$ from $-\infty$ to ∞ exists. If this does not hold, we say that the distribution does not have a mean; this case will rarely occur in applications.

A distribution is said to be **symmetric** with respect to a number $x = c$ if for every real x,

$$f(c + x) = f(c - x). \tag{2}$$

The student may prove

Theorem 1 **(Mean of a symmetric distribution)**
If a distribution is symmetric with respect to $x = c$ *and has a mean* μ, *then* $\mu = c$.

The **variance** of a distribution is denoted by σ^2 and is defined by the formula

(3)

(a) $$\sigma^2 = \sum_j (x_j - \mu)^2 f(x_j) \qquad \text{(Discrete distribution)}$$

(b) $$\sigma^2 = \int_{-\infty}^{\infty} (x - \mu)^2 f(x)\, dx \qquad \text{(Continuous distribution)}$$

where, by definition, it is assumed that the series in (3a) converges and the integral in (3b) exists (has a finite value).

In the case of a discrete distribution with $f(x) = 1$ at a point and $f = 0$ otherwise, we have $\sigma^2 = 0$. This case is of no practical interest. In any other case

$$\sigma^2 > 0. \tag{4}$$

The positive square root of the variance is called the **standard deviation** and is denoted by σ.

Variance and standard deviation measure the spread of a distribution, as Example 2 (below) illustrates.

EXAMPLE 1 Mean and variance

The random variable

$$X = \textit{Number of heads in a single toss of a fair coin}$$

has the possible values $X = 0$ and $X = 1$ with probabilities $P(X = 0) = \frac{1}{2}$ and $P(X = 1) = \frac{1}{2}$. From (1a) we thus obtain the mean value $\mu = 0 \cdot \frac{1}{2} + 1 \cdot \frac{1}{2} = \frac{1}{2}$, and (3a) yields

$$\sigma^2 = (0 - \tfrac{1}{2})^2 \cdot \tfrac{1}{2} + (1 - \tfrac{1}{2})^2 \cdot \tfrac{1}{2} = \tfrac{1}{4}.$$ ∎

EXAMPLE 2 Uniform distribution

The distribution with the density

$$f(x) = \frac{1}{b - a} \quad \text{if} \quad a < x < b$$

and $f = 0$ otherwise is called the **uniform distribution** on the interval $a < x < b$. From Theorem 1 or from (1b) we find that $\mu = (a + b)/2$, and (3b) yields the variance

$$\sigma^2 = \int_a^b \left(x - \frac{a + b}{2}\right)^2 \frac{1}{b - a}\, dx = \frac{(b - a)^2}{12}.$$

Figure 491 shows special cases illustrating that the spread is large if and only if σ^2 (and equally well σ) is large. ∎

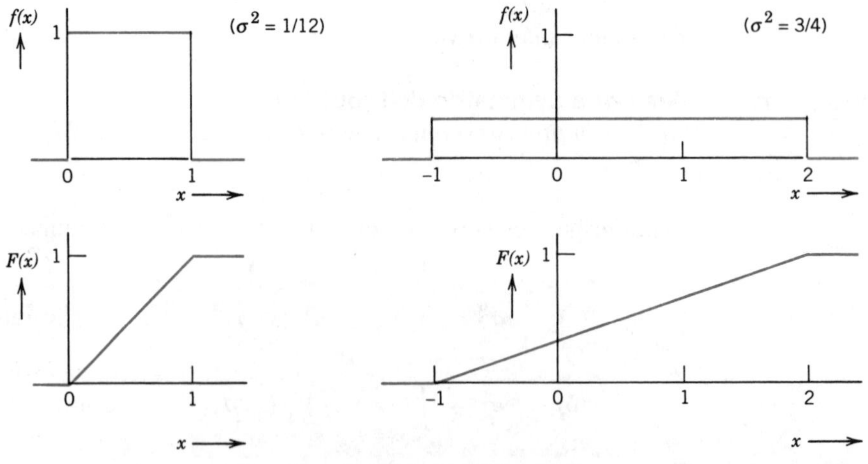

Fig. 491. Uniform distributions having the same mean (0.5) but different variances σ^2

Transformation of mean and variance. From a given random variable X we often want to obtain a new random variable of the form

$$X^* = c_1 X + c_2 \qquad (c_1 \neq 0), \tag{5}$$

in particular, one that has mean 0 and variance 1. (A basic case of this follows in Sec. 23.7.) Then we must know the mean and the variance of X^*:

Theorem 2 (Transformation of mean and variance)

(a) *If a random variable X has mean μ and variance σ^2, then X^* given by (5) has the mean*

$$\mu^* = c_1\mu + c_2 \tag{6}$$

and the variance

$$\sigma^{*2} = c_1{}^2\sigma^2. \tag{7}$$

(b) *In particular, the* **standardized random variable** *Z corresponding to X, given by*

$$\boxed{Z = \frac{X - \mu}{\sigma}} \tag{8}$$

has the mean 0 *and the variance* 1.

Proof. We prove (6) in the continuous case, first assuming that $c_1 > 0$. For corresponding x and $x^* = c_1 x + c_2$ the densities $f(x)$ of X and $f^*(x^*)$ of X^* satisfy the relation $f^*(x^*) = f(x)/c_1$, because to a small interval of length Δx on the X-axis there corresponds the probability $f(x)\,\Delta x$ (approximately) and this must equal $f^*(x^*)\,\Delta x^*$, where $\Delta x^* = c_1 \Delta x$ is the length of the corresponding interval on the X^*-axis. Since $dx^*/dx = c_1$, $dx^* = c_1\,dx$, we thus have $f^*(x^*)\,dx^* = f(x)\,dx$. Hence

$$\mu^* = \int_{-\infty}^{\infty} x^* f^*(x^*)\,dx^* = \int_{-\infty}^{\infty} (c_1 x + c_2) f(x)\,dx$$

$$= c_1 \int_{-\infty}^{\infty} x f(x)\,dx + c_2 \int_{-\infty}^{\infty} f(x)\,dx.$$

The last integral equals 1 [see (8), Sec. 23.4] and formula (6) is proved. Since

$$x^* - \mu^* = (c_1 x + c_2) - (c_1\mu + c_2) = c_1 x - c_1\mu,$$

the definition of the variance yields

$$\sigma^{*2} = \int_{-\infty}^{\infty} (x^* - \mu^*)^2 f^*(x^*)\, dx^* = \int_{-\infty}^{\infty} (c_1 x - c_1\mu)^2 f(x)\, dx = c_1^2\sigma^2,$$

which proves (7).

If $c_1 < 0$, the results remain the same, because we get two additional minus signs, one from changing the direction of integration in x (note that $x^* = -\infty$ corresponds to $x = \infty$) and the other one from $f^*(x^*) = f(x)/(-c_1)$; here $-c_1 > 0$ is needed since densities are nonnegative.

For a discrete variable X the proof is similar.

(b) is obtained by choosing $c_1 = 1/\sigma$ and $c_2 = -\mu/\sigma$ in (5)–(7). ∎

Expectation, moments. For any random variable X and any continuous function $g(X)$ defined for all real X, the **mathematical expectation** of $g(X)$ is defined by

(9)

$$\text{(a)}\quad E(g(X)) = \sum_j g(x_j) f(x_j) \qquad (X \text{ discrete})$$

$$\text{(b)}\quad E(g(X)) = \int_{-\infty}^{\infty} g(x) f(x)\, dx \qquad (X \text{ continuous})$$

where f in (9a) is the probability function and in (9b) the density of X. Note that for $g(X) = X$ this gives the mean of X,

(10) $$\mu = E(X).$$

Taking $g(X) = X^k$ $(k = 1, 2, \cdots)$ in (9), we get the **kth moment** of X given by

(11) $$E(X^k) = \sum_j x_j^k f(x_j) \quad \text{and} \quad E(X^k) = \int_{-\infty}^{\infty} x^k f(x)\, dx,$$

respectively. Finally, taking $g(X) = (X - \mu)^k$ in (9) gives the **kth central moment**

(12) $$E([X - \mu]^k) = \sum_j (x_j - \mu)^k f(x_j) \quad \text{and} \quad \int_{-\infty}^{\infty} (x - \mu)^k f(x)\, dx,$$

respectively. We see that the second central moment ($k = 2$) is the variance,

(13) $$\sigma^2 = E([X - \mu]^2).$$

For later use, the student may show that

(14) $$E(1) = 1.$$

Problem Set 23.5

1. Find the mean and variance of a discrete random variable X having the probability function $f(0) = \frac{1}{4}$, $f(1) = \frac{1}{2}$, $f(2) = \frac{1}{4}$.

2. Do the same task as in Prob. 1, when $f(0) = 0.512$, $f(1) = 0.384$, $f(2) = 0.096$, $f(3) = 0.008$.

3. Let X have the density $f(x) = 2x$ if $0 \leqq x \leqq 1$ and 0 otherwise. Show that X has the mean 2/3 and the variance 1/18.

4. Find the mean and the variance of $Y = -4X + 5$, where X is the random variable in Prob. 3.

5. Find the mean and the variance of X: *Number that shows up on a fair die.*

6. If in rolling a fair die, Jack wins as many dimes as the die shows, how much per game should Jack pay to make the game fair?

7. What is the expected daily profit if a small grocery store sells X turkeys per day with probabilities $f(5) = 0.1$, $f(6) = 0.3$, $f(7) = 0.4$, $f(8) = 0.2$ and the profit per turkey is \$3.50?

8. If the life X (in hours) of certain light bulbs has the density $f(x) = 0.001e^{-0.001x}$ if $x > 0$ and 0 otherwise, what is the mean life of such a bulb?

9. Let X [cm] be the diameter of bolts in a production. Assume that X has the density $f(x) = k(x - 0.9)(1.1 - x)$ if $0.9 < x < 1.1$ and 0 otherwise. Determine k, sketch $f(x)$, and find μ and σ^2.

10. Suppose that in Prob. 9, a bolt is regarded as being defective if its diameter deviates from 1.00 cm by more than 0.09 cm. What percentage of defective bolts should we then expect?

11. A small filling station is supplied with gasoline every Saturday afternoon. Assume that its volume X of sales in ten thousands of gallons has the probability density $f(x) = 6x(1 - x)$ if $0 \leqq x \leqq 1$ and 0 otherwise. Determine the mean, the variance, and the standardized variable.

12. What capacity must the tank in Prob. 11 have in order that the probability that the tank will be emptied in a given week be 5%?

13. Prove (10), (13), and (14).

14. Show that $E(X - \mu) = 0$ and $\sigma^2 = E(X^2) - \mu^2$.

15. Let X have the density $f(x) = 2x$ if $0 < x < 1$ and 0 otherwise. Find all the moments. Calculate σ^2 by the formula in Prob. 14.

16. Show that $E(ag(X) + bh(X)) = aE(g(X)) + bE(h(X))$; [$a$ and b are constants].

17. Find all the moments of the uniform distribution on an interval $a \leqq x \leqq b$.

18. The **skewness** γ of a random variable X is defined by

$$\gamma = \frac{1}{\sigma^3} E([X - \mu]^3).$$

Show that for a symmetric distribution (whose third central moment exists) the skewness is zero.

19. Find the skewness of the distribution with density $f(x) = xe^{-x}$ when $x > 0$ and $f(x) = 0$ otherwise. Sketch $f(x)$.

20. The **moment generating function** $G(t)$ of a discrete or continuous random variable X is defined by

$$G(t) = E(e^{tX}) = \sum_j e^{tx_j} f(x_j) \quad \text{and} \quad G(t) = E(e^{tX}) = \int_{-\infty}^{\infty} e^{tx} f(x)\, dx,$$

respectively. Assuming that differentiation under the summation sign and the integral sign is permissible, show that $E(X^k) = G^{(k)}(0)$, in particular $\mu = G'(0)$, where $G^{(k)}(t)$ is the kth derivative of G with respect to t.

23.6 Binomial, Poisson, and Hypergeometric Distributions

We have said before (in Sec. 23.4) that in practice we need only a handful of special random variables and their distributions. In this section we discuss three particularly important discrete distributions, and in the next section the most important continuous distribution. (Three more continuous distributions follow in Chap. 24, when we shall set up tests.)

Binomial Distribution

We start with the binomial distribution, which is obtained if we are interested in the number of times an event A occurs in n independent performances of an experiment, assuming that A has probability $P(A) = p$ in a single trial. Then $q = 1 - p$ is the probability that in a single trial the event A does not occur. We assume that the experiment is performed n times and consider the random variable

$$X = \textit{Number of times A occurs.}$$

Then X can assume the values $0, 1, \cdots, n$, and we want to determine the corresponding probabilities. For this purpose we consider any of these values, say, $X = x$, which means that in x of the n trials A occurs and in $n - x$ trials A does not occur. This may look as follows:

$$\text{(1)} \qquad \underbrace{AA \cdots A}_{x \text{ times}} \ \underbrace{BB \cdots B}_{n - x \text{ times}}.$$

Here $B = A^c$, that is, A does not occur. We assume that the trials are *independent*, that is, do not influence each other. Then, since $P(A) = p$ and $P(B) = q$, we see that to (1) there corresponds the probability

$$\underbrace{pp \cdots p}_{x \text{ times}} \ \underbrace{qq \cdots q}_{n - x \text{ times}} = p^x q^{n-x}.$$

Clearly, (1) is merely *one* order of arranging x A's and $n - x$ B's, and the probability $P(X = x)$ thus equals $p^x q^{n-x}$ times the number of different arrangements of x A's and $n - x$ B's, as follows from Theorem 1 in Sec. 23.2. We may number the n trials from 1 to n and pick x of these numbers corresponding to those trials in which A happens. Since the order in which we pick the x numbers does not matter, we see from (4a) in Sec. 23.3 that we can pick those x numbers from the n numbers in $\binom{n}{x}$ different ways. Hence the probability $P(X = x)$ corresponding to $X = x$ equals

$$f(x) = \binom{n}{x} p^x q^{n-x} \qquad (x = 0, 1, \cdots, n). \tag{2}$$

and $f(x) = 0$ for any other value of x. This is the probability that in n independent trials an event A occurs precisely x times, where p is the probability of A in a single trial and $q = 1 - p$. The distribution with the probability function (2) is called the **binomial distribution** or *Bernoulli distribution.* The occurrence of A is called *success,* and the nonoccurrence is called *failure.* p is called the *probability of success in a single trial.* Figure 492 shows illustrative examples of $f(x)$.

If $p = q = 1/2$, then from (2) we simply have

$$f(x) = \binom{n}{x} \left(\frac{1}{2}\right)^n \qquad (x = 0, 1, \cdots, n). \tag{2*}$$

The binomial distribution has the mean (see Prob. 17)

$$\mu = np \tag{3}$$

and the variance (see Prob. 18)

$$\sigma^2 = npq. \tag{4}$$

For tables of the binomial distribution see Appendix 5 and Ref. [G14].

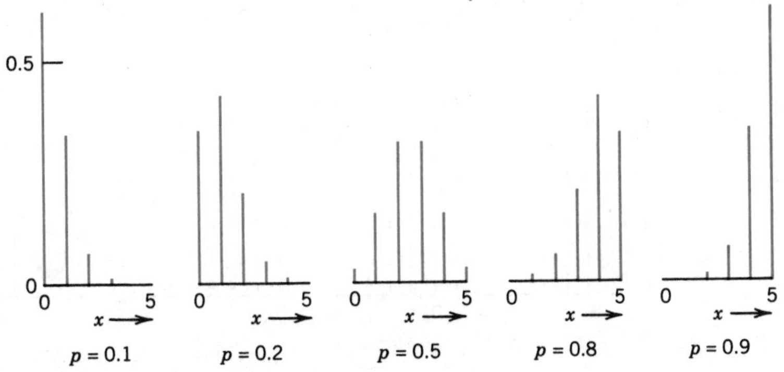

Fig. 492. Probability function (2) of the binomial distribution for $n = 5$ and various values of p

EXAMPLE 1 **Binomial distribution**

Compute the probability of obtaining at least two "*Six*" in rolling a fair die 4 times.

Solution. $p = P(A) = P(\text{"}Six\text{"}) = 1/6$, $q = 5/6$, $n = 4$. *Answer:*

$$P = f(2) + f(3) + f(4) = \binom{4}{2}\left(\frac{1}{6}\right)^2\left(\frac{5}{6}\right)^2 + \binom{4}{3}\left(\frac{1}{6}\right)^3\left(\frac{5}{6}\right) + \binom{4}{4}\left(\frac{1}{6}\right)^4$$

$$= \frac{1}{6^4}(6 \cdot 25 + 4 \cdot 5 + 1) = \frac{171}{1296} = 13.2\%.$$

Poisson Distribution

The discrete distribution with the probability function

$$f(x) = \frac{\mu^x}{x!} e^{-\mu} \qquad (x = 0, 1, \cdots) \tag{5}$$

is called the **Poisson distribution,** named after S. D. Poisson (Sec. 17.5). Figure 493 shows (5) for some values of μ. It can be proved that this distribution is obtained as a limiting case of the binomial distribution, if we let $p \to 0$ and $n \to \infty$ so that the mean $\mu = np$ approaches a finite value. The Poisson distribution has the mean μ and the variance (see Prob. 19)

$$\sigma^2 = \mu. \tag{6}$$

EXAMPLE 2 **Poisson distribution**

If the probability of producing a defective screw is $p = 0.01$, what is the probability that a lot of 100 screws will contain more than 2 defectives?

Solution. For the complementary event A^C: *Not more than 2 defectives* we compute the Poisson approximation of the binomial distribution with $\mu = np = 1$

$$P(A^C) = \binom{100}{0} 0.99^{100} + \binom{100}{1} 0.01 \cdot 0.99^{99} + \binom{100}{2} 0.01^2 \cdot 0.99^{98}$$

$$\approx e^{-1}\left(1 + 1 + \frac{1}{2}\right) = 91.97\%.$$

Thus $P(A) = 8.03\%$. Show that the binomial distribution gives $P(A) = 7.94\%$, so that the Poisson approximation is quite good.

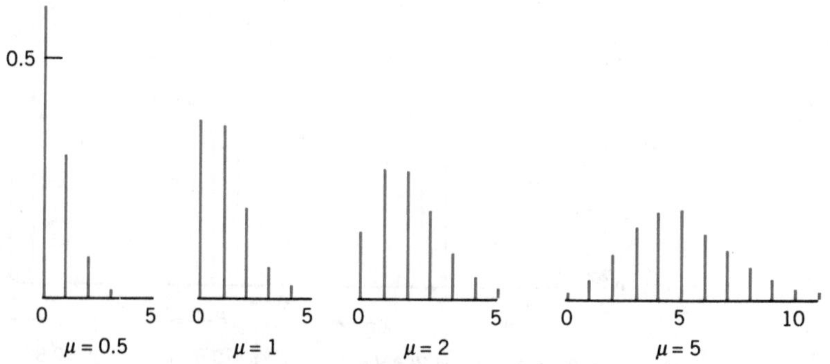

Fig. 493. Probability function (5) of the Poisson distribution for various values of μ

EXAMPLE 3 Poisson distribution

If on the average, 2 cars enter a certain parking lot per minute, what is the probability that during any given minute 4 or more cars will enter the lot?

Solution. To understand that the Poisson distribution is a model of the situation, we imagine the minute to be divided into very many short time intervals, let p be the (constant) probability that a car will enter the lot during any such short interval, and assume independence of the events that happen during those intervals. Then we are dealing with a binomial distribution with very large n and very small p, which we can approximate by the Poisson distribution with $\mu = np = 2$. Thus the complementary event "3 *cars or fewer enter the lot*" has the probability

$$f(0) + f(1) + f(2) + f(3) = e^{-2}\left(\frac{2^0}{0!} + \frac{2^1}{1!} + \frac{2^2}{2!} + \frac{2^3}{3!}\right) = 0.857.$$

Hence the answer is 14%. ∎

Sampling With and Without Replacement: Hypergeometric Distribution

Sampling with replacement means that we draw one thing at a time and put it back to the given set, mixing the latter before we draw the next thing (see also Sec. 23.2). This guarantees independence of trials and leads to the **binomial distribution.** Indeed, if a box contains N things, for example, screws, M of which are defective, then the probability of drawing a defective screw in a trial is

$$p = \frac{M}{N}.$$

Hence in drawing a sample of n screws *with replacement,* the probability of obtaining precisely x defective screws is [see (2)]

$$f(x) = \binom{n}{x}\left(\frac{M}{N}\right)^x\left(1 - \frac{M}{N}\right)^{n-x} \qquad (x = 0, 1, \cdots, n). \tag{7}$$

In **sampling without replacement,** in which we return no screws to the box, that probability is

$$f(x) = \frac{\binom{M}{x}\binom{N-M}{n-x}}{\binom{N}{n}} \qquad (x = 0, 1, \cdots, n). \tag{8}$$

This distribution with the probability function (8) is called the **hypergeometric distribution.**[7]

[7]Because the moment generating function (see Prob. 20, Sec. 23.5) of this distribution can be represented in terms of the hypergeometric function.

To verify our statement we first note that, by (4a) in Sec. 23.3, there are

(a) $\binom{N}{n}$ different ways of picking n things from N,

(b) $\binom{M}{x}$ different ways of picking x defectives from M,

(c) $\binom{N-M}{n-x}$ different ways of picking $n - x$ nondefectives from $N - M$,

and each way in (b) combined with each way in (c) gives the total number of mutually exclusive ways of obtaining x defectives in n drawings without replacement. Since (a) is the total number of outcomes and we draw at random, each such way has the probability $1\Big/\binom{N}{n}$. From this, (8) follows. ∎

The hypergeometric distribution has the mean (Prob. 20)

$$\mu = n\frac{M}{N} \tag{9}$$

and the variance

$$\sigma^2 = \frac{nM(N - M)(N - n)}{N^2(N - 1)}. \tag{10}$$

EXAMPLE 4 Sampling with and without replacement

We want to draw random samples of two gaskets from a box containing 10 gaskets, three of which are defective. Find the probability function of the random variable

X = *Number of defectives in the sample.*

Solution. We have $N = 10$, $M = 3$, $N - M = 7$, $n = 2$. For sampling with replacement, (7) yields

$$f(x) = \binom{2}{x}\left(\frac{3}{10}\right)^x\left(\frac{7}{10}\right)^{2-x}, \qquad f(0) = 0.49, \quad f(1) = 0.42, \quad f(2) = 0.09.$$

For sampling without replacement we have to use (8), finding

$$f(x) = \binom{3}{x}\binom{7}{2-x}\Big/\binom{10}{2}, \qquad f(0) = f(1) = \frac{21}{45} \approx 0.47, \quad f(2) = \frac{3}{45} \approx 0.07.$$ ∎

If N, M, and $N - M$ are large compared with n, then it does not matter too much whether we sample with or without replacement, and in this case the hypergeometric distribution may be approximated by the binomial distribution (with $p = M/N$), *which is somewhat simpler.*

Hence in sampling from an indefinitely large population (**"infinite population"**) *we may use the binomial distribution, regardless of whether we sample with or without replacement.*

Problem Set 23.6

1. Four fair coins are tossed simultaneously. Find the probability function of the random variable X = *Number of heads* and compute the probabilities of obtaining no heads, precisely 1 head, at least 1 head, not more than 3 heads.
2. If the probability of hitting a target is 10% and 10 shots are fired independently, what is the probability that the target will be hit at least once?
3. In Prob. 2, if the probability of hitting would be 5% and we fired 20 shots, would the probability of hitting at least once be less than, equal to, or greater than in Prob. 2? Guess first, then compute.
4. Suppose that a telephone switchboard handles 300 calls on the average during a rush hour, and that the board can make at most 10 connections per minute. Using the Poisson distribution, estimate the probability that the board will be overtaxed during a given minute. (Use Table 6 in Appendix 5.)
5. Let p = 1% be the probability that a certain type of light bulb will fail in a 24-hour test. Find the probability that a sign consisting of 10 such bulbs will burn 24 hours with no bulb failures.
6. Suppose that 3% of bolts made by a machine are defective, the defectives occurring at random during production. If the bolts are packaged 50 per box, what is the Poisson approximation of the probability that a given box will contain $x = 0, 1, \cdots, 5$ defectives?
7. Suppose that in the production of 50-ohm radio resistors, nondefective items are those that have a resistance between 45 and 55 ohms and the probability of a resistor's being defective is 0.2%. The resistors are sold in lots of 100, with the guarantee that all resistors are nondefective. What is the probability that a given lot will violate this guarantee? (Use the Poisson distribution.)
8. Suppose that a certain type of magnetic tape contains, on the average, 2 defects per 100 meters. What is the probability that a roll of tape 300 meters long will contain (a) x defects, (b) no defects?
9. Classical experiments by E. Rutherford and H. Geiger in 1910 showed that the number of alpha particles emitted per second in a radioactive process is a random variable X having a Poisson distribution. If X has mean 0.5, what is the probability of observing two or more particles during any given second?
10. A process of manufacturing screws is checked every hour by inspecting n screws selected at random from that hour's production. If one or more screws are defective, the process is halted and carefully examined. How large should n be if the manufacturer wants the probability to be about 95% that the process will be halted when 10% of the screws being produced are defective? (Assume independence of the quality of any item of that of the other items.)
11. Let X be the number of cars per minute passing a certain point of some road between 8 A.M. and 10 A.M. on a Sunday. Assume that X has a Poisson distribution with mean 5. Find the probability of observing 3 or fewer cars during any given minute.
12. A carton contains 20 fuses, 5 of which are defective. Find the probability that, if a sample of 3 fuses is chosen from the carton by random drawing without replacement, x fuses in the sample will be defective.
13. Suppose that a test for extrasensory perception consists of naming (in any order) 3 cards randomly drawn from a deck of 13 cards. Find the probability that by chance alone, the person will correctly name (a) no cards, (b) 1 card, (c) 2 cards, (d) 3 cards.

14. A distributor sells rubber bands in packages of 100 and guarantees that at most 10% are defective. A consumer controls each package by drawing 10 bands without replacement. If the sample contains no defective rubber bands, he accepts the package. Otherwise he rejects it. Find the probability that in this process the consumer rejects a package that contains 10 defective bands (so that it still satisfies the guarantee).

15. **(Multinomial distribution)** Suppose a trial can result in precisely one of k mutually exclusive events $A_1, \cdots, A_k$ with probabilities $p_1, \cdots, p_k$, respectively, where $p_1 + \cdots + p_k = 1$. Suppose that n independent trials are performed. Show that the probability of getting x_1 A_1's, $\cdots$, x_k A_k's is

$$f(x_1, \cdots, x_k) = \frac{n!}{x_1! \cdots x_k!} p_1^{x_1} \cdots p_k^{x_k}$$

where $0 \leqq x_j \leqq n$, $j = 1, \cdots, k$, and $x_1 + \cdots + x_k = n$. The distribution having this probability function is called the *multinomial distribution*.

16. Using the binomial theorem, show that the binomial distribution has the moment generating function (see Prob. 20, Sec. 23.5)

$$G(t) = \sum_{x=0}^{n} e^{tx} \binom{n}{x} p^x q^{n-x} = \sum_{x=0}^{n} \binom{n}{x} (pe^t)^x q^{n-x} = (pe^t + q)^n.$$

17. Using Prob. 16 and $p + q = 1$, prove (3).

18. Prove (4). *Hint.* Use the moment generating function in Prob. 16.

19. Show that the Poisson distribution has the moment generating function

$$G(t) = e^{-\mu} e^{\mu e^t}$$

and prove (6).

20. Prove (9). *Hint.* Prove that $x\binom{M}{x} = M\binom{M-1}{x-1}$. Use (14) in Sec. 23.3.

23.7 Normal Distribution

Turning from discrete to continuous distributions, in this section we discuss the normal distribution. This is the most important continuous distribution because in applications many random variables are **normal random variables,** (that is, they have a normal distribution) or they are approximately normal or can be transformed into normal random variables in a relatively simple fashion. Furthermore, the normal distribution is a useful approximation of more complicated distributions, and it also occurs in the proofs of various statistical tests.

The **normal distribution** or *Gauss distribution* is defined as the distribution with the density

$$\text{(1)} \qquad f(x) = \frac{1}{\sigma\sqrt{2\pi}} e^{-\frac{1}{2}\left(\frac{x-\mu}{\sigma}\right)^2} \qquad (\sigma > 0).$$

Here, μ is the mean. The curve of $f(x)$ is symmetric with respect to $x = \mu$ because the exponent contains $(x - \mu)^2$. Changing μ corresponds to moving the curve to another position (translating it), and for $\mu = 0$ it is symmetric with respect to the ordinate, as shown in Fig. 494. The curves in Fig. 494 are called *bell-shaped curves*. They have a peak at $x = 0$ (or $x = \mu$ when translated). σ^2 is the variance, and we see that for small σ^2 we get a high peak and steep slopes, whereas with increasing σ^2 the curve gets flatter and flatter, the density is "spread out" more and more, in agreement with the fact that the variance measures the spread.

Distribution function. From the density (1) we get the distribution function $F(x)$ of the normal distribution by integrating from $-\infty$ to x [see (7), Sec. 23.4], that is,

$$F(x) = \frac{1}{\sigma\sqrt{2\pi}} \int_{-\infty}^{x} e^{-\frac{1}{2}\left(\frac{v-\mu}{\sigma}\right)^2} dv. \tag{2}$$

Hence, by (9) in Sec. 23.4, the probability that a normal random variable X assume any value in some interval $a < x \leqq b$ is

$$P(a < X \leqq b) = F(b) - F(a) = \frac{1}{\sigma\sqrt{2\pi}} \int_{a}^{b} e^{-\frac{1}{2}\left(\frac{v-\mu}{\sigma}\right)^2} dv. \tag{3}$$

The integral (2) cannot be integrated by one of the methods of calculus, but has been tabulated because one needs it quite frequently. Now if this were done for many μ and σ, the table would be bulky and impractical. Fortunately, it is enough to tabulate the distribution function of the standardized normal random variable $Z = (X - \mu)/\sigma$ with mean 0 and variance 1 [see (8), Sec. 23.5], which is

$$\Phi(z) = \frac{1}{\sqrt{2\pi}} \int_{-\infty}^{z} e^{-u^2/2}\, du \qquad \text{(Fig. 495).} \tag{4}$$

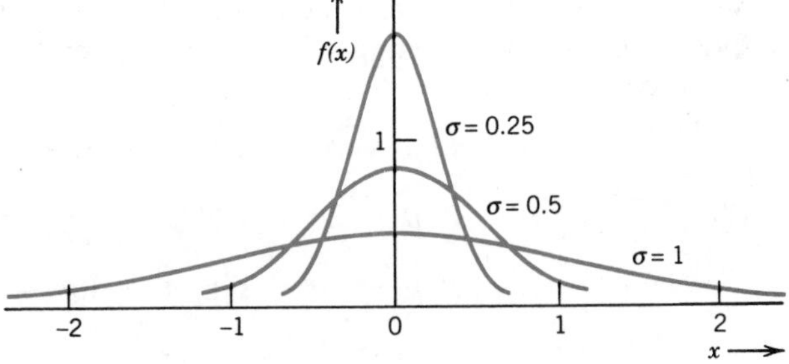

Fig. 494. Density (1) of the normal distribution with $\mu = 0$ for various values of σ

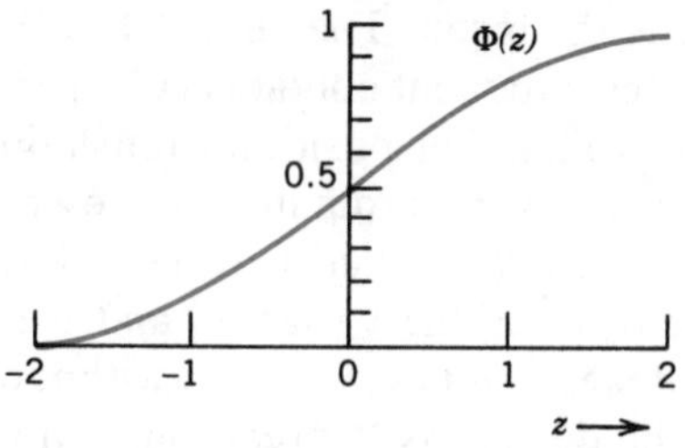

Fig. 495. Distribution function Φ(z) of the normal distribution with mean 0 and variance 1

Values of $\Phi(z)$ are given in Table A7, Appendix 5.

Comparing the exponents in (2) and (4) shows that for getting $F(x)$ in terms of $\Phi(z)$ we should put

$$\frac{v-\mu}{\sigma} = u. \qquad \text{Then} \qquad \frac{dv}{\sigma} = du$$

and $v = x$ gives the new upper limit of integration $u = (x - \mu)/\sigma$:

$$F(x) = \frac{1}{\sigma\sqrt{2\pi}} \int_{-\infty}^{(x-\mu)/\sigma} e^{-\mu^2/2}\,\sigma\,du.$$

σ drops out, and the expression on the right equals (4) where $z = (x - \mu)/\sigma$:

(5)
$$F(x) = \Phi\left(\frac{x-\mu}{\sigma}\right).$$

From this important formula and (3) we obtain another important formula:

(6)
$$P(a < X \leqq b) = F(b) - F(a) = \Phi\left(\frac{b-\mu}{\sigma}\right) - \Phi\left(\frac{a-\mu}{\sigma}\right).$$

In particular, if $a = \mu - \sigma$ and $b = \mu + \sigma$, the right side equals $\Phi(1) - \Phi(-1)$; to $a = \mu - 2\sigma$ and $b = \mu + 2\sigma$ there corresponds the value $\Phi(2) - \Phi(-2)$, etc. Using Table A7 in Appendix 5 we thus find (Fig. 496)

(7)

(a) $P(\mu - \sigma < X \leqq \mu + \sigma) \approx 68\%$,

(b) $P(\mu - 2\sigma < X \leqq \mu + 2\sigma) \approx 95.5\%$,

(c) $P(\mu - 3\sigma < X \leqq \mu + 3\sigma) \approx 99.7\%$.

Hence we may expect that a large number of observed values of a normal random variable X will be distributed as follows:

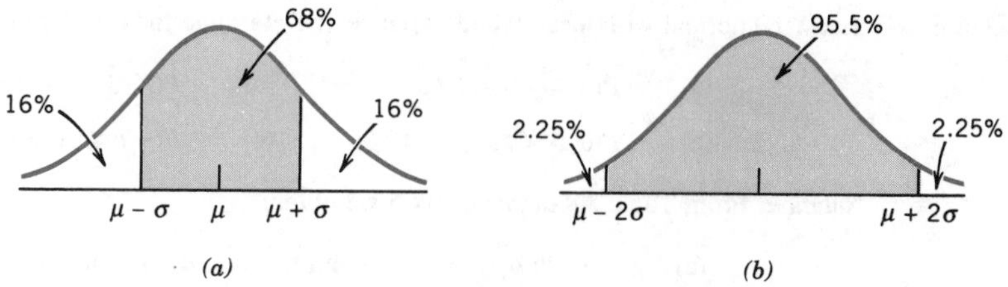

Fig. 496. Illustration of formula (7)

(a) *About $\frac{2}{3}$ of the values will lie between $\mu - \sigma$ and $\mu + \sigma$.*

(b) *About 95% of the values will lie between $\mu - 2\sigma$ and $\mu + 2\sigma$.*

(c) *About $99\frac{3}{4}$% of the values will lie between $\mu - 3\sigma$ and $\mu + 3\sigma$.*

This may also be expressed as follows.

A value that deviates more than σ from μ will occur about once in 3 trials. A value that deviates more than 2σ or 3σ from μ will occur about once in 20 or 400 trials, respectively. Practically speaking, this means that all the values will lie between $\mu - 3\sigma$ and $\mu + 3\sigma$; these two numbers are called **three-sigma limits.** In a similar fashion we obtain

$$\text{(8)}\quad \begin{aligned} &\textbf{(a)}\quad P(\mu - 1.96\sigma < X \leqq \mu + 1.96\sigma) = 95\%,\\ &\textbf{(b)}\quad P(\mu - 2.58\sigma < X \leqq \mu + 2.58\sigma) = 99\%,\\ &\textbf{(c)}\quad P(\mu - 3.29\sigma < X \leqq \mu + 3.29\sigma) = 99.9\%. \end{aligned}$$

Use of normal tables. The following typical examples should help the student to understand the practical use of Tables A7 and A8 in Appendix 5.

EXAMPLE 1 For a normal random variable X with mean 0 and variance 1 find the probabilities

(a) $P(X \leqq 2.44)$, (b) $P(X \leqq -1.16)$, (c) $P(X \geqq 1)$, (d) $P(2 \leqq X \leqq 10)$.

Solution. Since $\mu = 0$ and $\sigma^2 = 1$, we may obtain the desired values directly from Table A7:

(a) 0.9927, (b) 0.1230, (c) $1 - P(X \leqq 1) = 1 - 0.8413 = 0.1587$ [see (7), Sec. 23.2],

(d) $\Phi(10) = 1.0000$ (why?), $\Phi(2) = 0.9772$, $\Phi(10) - \Phi(2) = 0.0228$. ∎

EXAMPLE 2 Compute the probabilities in Example 1 for a normal X with mean 0.8 and variance 4.

Solution. From (6) and Table A7 we obtain

(a) $$F(2.44) = \Phi\left(\frac{2.44 - 0.80}{2}\right) = \Phi(0.82) = 0.7939$$

(b) $$F(-1.16) = \Phi(-0.98) = 0.1635$$

(c) $$1 - P(X \leqq 1) = 1 - F(1) = 1 - \Phi(0.1) = 0.4602$$

(d) $$F(10) - F(2) = \Phi(4.6) - \Phi(0.6) = 1 - 0.7257 = 0.2743.$$ ∎

EXAMPLE 3 Let X be normal with mean 0 and variance 1. Determine the constant c such that

(a) $P(X \geqq c) = 10\%$, (b) $P(X \leqq c) = 5\%$,

(c) $P(0 \leqq X \leqq c) = 45\%$, (d) $P(-c \leqq X \leqq c) = 99\%$.

Solution. From Table A8 in Appendix 5 we obtain

(a) $1 - P(X \leqq c) = 1 - \Phi(c) = 0.1, \Phi(c) = 0.9, c = 1.282$

(b) $c = -1.645$

(c) $\Phi(c) - \Phi(0) = \Phi(c) - 0.5 = 0.45, \Phi(c) = 0.95, c = 1.645$

(d) $c = 2.576.$ ∎

EXAMPLE 4 Let X be normal with mean -2 and variance 0.25. Determine c such that

(a) $P(X \geqq c) = 0.2$ (b) $P(-c \leqq X \leqq -1) = 0.5$

(c) $P(-2 - c \leqq X \leqq -2 + c) = 0.9$ (d) $P(-2 - c \leqq X \leqq -2 + c) = 99.6\%$.

Solution. Using Table A8 in Appendix 5, we obtain the following results.

(a) $$1 - P(X \leqq c) = 1 - \Phi\left(\frac{c + 2}{0.5}\right) = 0.2,$$

$$\Phi(2c + 4) = 0.8,\ 2c + 4 = 0.842,\ c = -1.579$$

(b) $$\Phi\left(\frac{-1 + 2}{0.5}\right) - \Phi\left(\frac{-c + 2}{0.5}\right) = 0.9772 - \Phi(4 - 2c) = 0.5,$$

$$\Phi(4 - 2c) = 0.4772,\ 4 - 2c = -0.057,\ c = 2.03$$

(c) $$\Phi\left(\frac{-2 + c + 2}{0.5}\right) - \Phi\left(\frac{-2 - c + 2}{0.5}\right)$$

$$= \Phi(2c) - \Phi(-2c) = 0.9,\ 2c = 1.645,\ c = 0.823$$

(d) $$\Phi(2c) - \Phi(-2c) = 99.6\%,\ 2c = 2.878,\ c = 1.439.$$ ∎

EXAMPLE 5 Suppose that iron plates in production are required to have a certain thickness, and the machine work is done by a shaper. In any case industrial products will differ slightly from each other because the properties of the material and the behavior of the machines and tools used show slight random variations caused by small disturbances we cannot predict. We may therefore regard the thickness X [mm] of the plates as a random variable. We assume that for a certain setting the variable X is normal with mean $\mu = 10$ mm and standard deviation $\sigma = 0.02$ mm. We want to determine the percentage of defective plates to be expected, assuming that defective plates are (a) plates thinner than 9.97 mm, (b) plates thicker than 10.05 mm, (c) plates that deviate more than 0.03 mm from 10 mm. (d) How should we choose numbers $10 - c$ and $10 + c$ to ensure that the expected percentage of defectives will not be greater than 5%? (e) How does the percentage of defectives in part (d) change if μ is shifted from 10 mm to 10.01 mm?

Solution. Using Table A7 in Appendix 5 and (6), we obtain the following solutions.

(a) $$P(X \leqq 9.97) = \Phi\left(\frac{9.97 - 10.00}{0.02}\right) = \Phi(-1.5) = 0.0668 \approx 6.7\%$$

(b) $$P(X \geqq 10.05) = 1 - P(X \leqq 10.05) = 1 - \Phi\left(\frac{10.05 - 10.00}{0.02}\right)$$

$$= 1 - \Phi(2.5) = 1 - 0.9938 \approx 0.6\%$$

(c) $$P(9.97 \leqq X \leqq 10.03) = \Phi\left(\frac{10.03 - 10.00}{0.02}\right) - \Phi\left(\frac{9.97 - 10.00}{0.02}\right)$$

$$= \Phi(1.5) - \Phi(-1.5) = 0.8664.$$ *Answer.* $1 - 0.8664 \approx 13\%$.

(d) From (8a) we obtain $c = 1.96\sigma = 0.039$. *Answer.* 9.961 and 10.039 mm.

(e) $$P(9.961 \leqq X \leqq 10.039) = \Phi\left(\frac{10.039 - 10.010}{0.02}\right) - \Phi\left(\frac{9.961 - 10.010}{0.02}\right)$$

$$= \Phi(1.45) - \Phi(-2.45) = 0.9265 - 0.0071 \approx 92\%;$$

hence the answer is 8%, and we see that this slight change in the adjustment of the tool bit causes a considerable increase of the percentage of defectives. ▮

Transformation of normal random variables. These variables remain normal under a translation and change of scale. In fact, using (5), the reader may prove

Theorem 1 (Transformation)

If X is normal with mean μ and variance σ^2, then $X^* = c_1X + c_2$ $(c_1 \neq 0)$ is normal with mean $\mu^* = c_1\mu + c_2$ and variance $\sigma^{*2} = c_1{}^2\sigma^2$.

Approximation of the binomial distribution with large *n*. If n is large, the probability function of the binomial distribution (Sec. 23.6)

$$f(x) = \binom{n}{x} p^x q^{n-x} \qquad (x = 0, 1, \cdots, n) \tag{9}$$

involves large binomial coefficients and powers, so that it becomes very inconvenient. It is of great practical (and theoretical) importance that in this case the normal distribution provides a good approximation of the binomial distribution, according to the following theorem, one of the most important theorems in whole probability theory.

Theorem 2 (Limit theorem of De Moivre and Laplace)

For large n,

$$f(x) \sim f^*(x) \qquad (x = 0, 1, \cdots, n)$$

where f is given by (9),

$$f^*(x) = \frac{1}{\sqrt{2\pi}\sqrt{npq}}\, e^{-z^2/2}, \qquad z = \frac{x - np}{\sqrt{npq}} \tag{10}$$

is the density of the normal distribution with mean $\mu = np$ and variance $\sigma^2 = npq$ (the mean and variance of the binomial distribution), and the symbol $\sim$ (read **asymptotically equal**) *means that the ratio of both sides approaches 1 as n approaches ∞. Furthermore, for any nonnegative integers a and b $(> a)$,*

$$\begin{gathered} P(a \leqq X \leqq b) = \sum_{x=a}^{b} \binom{n}{x} p^x q^{n-x} \sim \Phi(\beta) - \Phi(\alpha), \\ \alpha = \frac{a - np - 0.5}{\sqrt{npq}}, \qquad \beta = \frac{b - np + 0.5}{\sqrt{npq}}. \end{gathered} \tag{11}$$

A proof of this theorem can be found in [G3] listed in Appendix 1. The proof shows that the term 0.5 in α and β is a correction caused by the change from a discrete to a continuous distribution.

Problem Set 23.7

1. Let X be normal with mean 80 and variance 9. Find $P(X > 83)$, $P(X < 81)$, $P(X < 80)$, and $P(78 < X < 82)$.

2. Let X be normal with mean 105 and variance 25. Find $P(X \leqq 112.5)$, $P(X > 100)$, $P(110.5 < X < 111.25)$.

3. Let X be normal with mean 14 and variance 4. Determine c such that $P(X \leqq c) = 95\%$, $P(X \leqq c) = 5\%$, $P(X \leqq c) = 99.5\%$.

4. Let X be normal with mean 3.6 and variance 0.01. Find c such that $P(X \leqq c) = 50\%$, $P(X > c) = 10\%$, $P(-c < X - 3.6 \leqq c) = 99.9\%$.

5. What percentage of defective iron plates can we expect in part (c) of Example 5 if we use a better shaper so that $\sigma = 0.01$ mm?

6. In Example 5, part (c), what value should σ have in order that the percentage of defective plates reduces to 6%?

7. If the lifetime X of a certain kind of automobile battery is normally distributed with a mean of 4 years and a standard deviation of 1 year, and the manufacturer wishes to guarantee the battery for 3 years, what percentage of the batteries will he have to replace under the guarantee?

8. What is the probability of obtaining at least 2048 heads if a coin is tossed 4040 times and heads and tails are equally likely? (See Table 23.1 in Sec. 23.2.)

9. Specifications for a certain job call for bolts with a diameter of 0.280 ± 0.002 cm. If the diameters of the bolts made by some manufacturer are normally distributed with $\mu = 0.279$ and $\sigma = 0.001$, what percentage of these bolts will meet specifications?

10. A producer sells electric bulbs in cartons of 1000 bulbs. Using (11), find the probability that any given carton contains not more than 1% defective bulbs, assuming the production process to be a Bernoulli experiment with $p = 1\%$ (= probability that any given bulb will be defective).

11. A manufacturer knows from experience that the resistance of resistors he produces is normal with mean $\mu = 150$ ohms and standard deviation $\sigma = 5$ ohms. What percentage of the resistors will have resistance between 148 ohms and 152 ohms? Between 140 ohms and 160 ohms?

12. The breaking strength X [kg] of a certain type of plastic block is normally distributed with a mean of 1250 kg and a standard deviation of 55 kg. What is the maximum load such that we can expect no more than 5% of the blocks to break?

13. A manufacturer produces airmail envelopes whose weight is normal with mean $\mu = 1.950$ grams and standard deviation $\sigma = 0.025$ grams. The envelopes are sold in lots of 1000. How many envelopes in a lot will be heavier than 2 grams?

14. Prove Theorem 1.

15. Derive the formulas in (8), using the normal table.

16. **(Bernoulli's law of large numbers)** In an experiment let an event A have probability p $(0 < p < 1)$, and let X be the number of times A happens in n independent trials. Show that for any given $\epsilon > 0$,

$$P\left(\left|\frac{X}{n} - p\right| \leqq \epsilon\right) \rightarrow 1 \qquad \text{as } n \rightarrow \infty.$$

17. Show that $\Phi(-z) = 1 - \Phi(z)$.

18. Considering $\Phi^2(\infty)$ and introducing polar coordinates in the double integral, prove

$$\Phi(\infty) = \frac{1}{\sqrt{2\pi}} \int_{-\infty}^{\infty} e^{-u^2/2}\, du = 1. \tag{12}$$

19. Show that the curve of (1) has two points of inflection corresponding to $x = \mu \pm \sigma$.

20. Using (12) and integration by parts, show that σ in (1) is the standard deviation of the normal distribution.

23.8 Distributions of Several Random Variables

In this last section of the chapter we discuss probability distributions of several random variables, as they occur in experiments in which we observe several quantities, such as carbon content X and hardness Y of steel; amount of fertilizer X and yield Y; height X_1, weight X_2, and blood pressure X_3 of persons; etc. Another reason for our discussion is that those distributions also play a basic role in the justification of statistical methods in Chap. 24.

If we observe two quantities X and Y, each trial gives a pair of values $X = x$, $Y = y$, briefly $(X, Y) = (x, y)$, which we can plot as a point in the XY-plane. Now for a single random variable X, the probability distribution is determined by the distribution function $F(x) = P(X \leqq x)$. Similarly, for two random variables X, Y, which we also call a **two-dimensional random variable** (X, Y), the probability distribution is determined by the **distribution function**

$$F(x, y) = P(X \leqq x, Y \leqq y). \tag{1}$$

This is the probability that (X, Y) assume ***any*** pair of values in the blue region (extending to $-\infty$ to the left and below) in Fig. 497. It determines the probability distribution uniquely because in analogy to the formula $P(a < X \leqq b) = F(b) - F(a)$ we now have for any rectangle (see Prob. 16)

$$\begin{aligned} &P(a_1 < X \leqq b_1, a_2 < Y \leqq b_2) \\ &\quad = F(b_1, b_2) - F(a_1, b_2) - F(b_1, a_2) + F(a_1, a_2). \end{aligned} \tag{2}$$

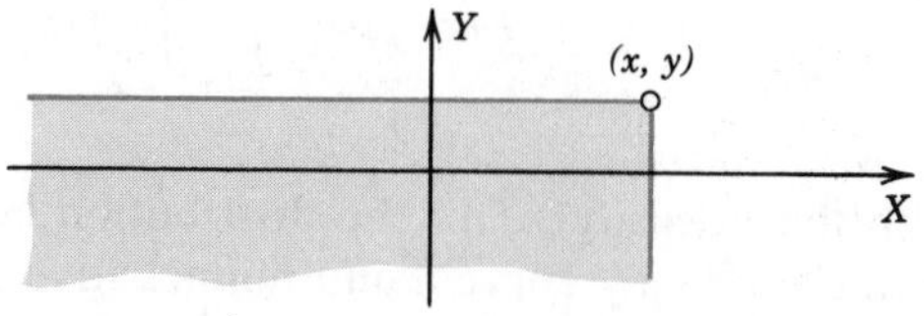

Fig. 497. Formula (1)

Discrete Two-Dimensional Distributions

Definition (Discrete two-dimensional random variable and distribution)
A random variable (X, Y) and its distribution are called **discrete** if (X, Y) can assume only finitely many or at most countably infinitely many pairs of values $(x_1, y_1), (x_2, y_2), \cdots$ with positive probabilities, whereas the probability for any domain containing none of those values of (X, Y) is zero. ∎

Let (x_i, y_j) be any of those pairs and let $P(X = x_i, Y = y_j) = p_{ij}$ (where we admit that p_{ij} may be 0 for certain pairs of subscripts i, j). Then we define the **probability function** $f(x, y)$ of (X, Y) by

$$f(x, y) = p_{ij} \quad \text{if} \quad x = x_i, y = y_j \quad \text{and} \quad f(x, y) = 0 \text{ otherwise;} \tag{3}$$

here, $i = 1, 2, \cdots$ and $j = 1, 2, \cdots$ independently. In analogy to (3), Sec. 23.4, we now have for the distribution function the formula

$$F(x, y) = \sum_{x_i \leqq x} \sum_{y_j \leqq y} f(x_i, y_j), \tag{4}$$

and instead of (6), Sec. 23.4, the condition

$$\sum_i \sum_j f(x_i, y_j) = 1. \tag{5}$$

EXAMPLE 1 **Two-dimensional discrete distribution**

If we simultaneously toss a dime and a nickel and consider

X = *Number of heads the dime turns up,*

Y = *Number of heads the nickel turns up,*

then X and Y can have the values 0 or 1, and the probability function is

$$f(0, 0) = f(1, 0) = f(0, 1) = f(1, 1) = \tfrac{1}{4}, \quad f(x, y) = 0 \text{ otherwise.}$$ ∎

Continuous Two-Dimensional Distributions

Definition
(Continuous two-dimensional random variable and distribution)
A random variable (X, Y) and its distribution are called **continuous** if the corresponding distribution function $F(x, y)$ can be given by a double integral

$$F(x, y) = \int_{-\infty}^{y} \int_{-\infty}^{x} f(x^*, y^*)\, dx^*\, dy^* \tag{6}$$

with the "**density**" f of the distribution being continuous (possibly except on finitely many curves) and nonnegative. ∎

From (6) we get for the probability that (X, Y) assume any value in a rectangle (Fig. 498) the formula

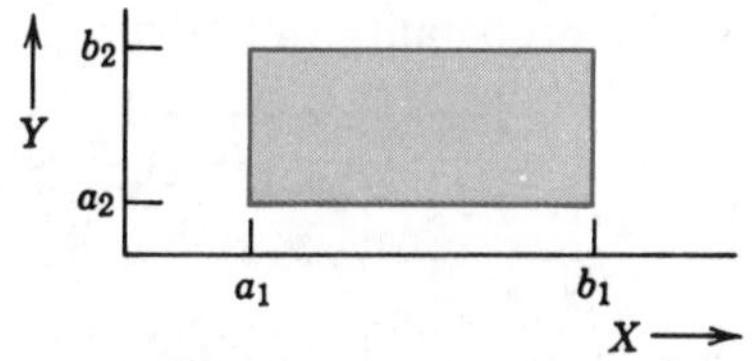

Fig. 498. Notion of a two-dimensional distribution

(7) $$P(a_1 < X \leqq b_1, a_2 < Y \leqq b_2) = \int_{a_2}^{b_2} \int_{a_1}^{b_1} f(x, y)\, dx\, dy.$$

EXAMPLE 2 **Two-dimensional uniform distribution in a rectangle**

Let R be the rectangle $\alpha_1 < x \leqq \beta_1$, $\alpha_2 < y \leqq \beta_2$. The density (see Fig. 499)

(8) $$f(x, y) = 1/k \quad \text{if } (x, y) \text{ is in } R, \qquad f(x, y) = 0 \text{ otherwise}$$

defines the so-called **uniform distribution** *in the rectangle R*; here k is the area of R, that is, $k = (\beta_1 - \alpha_1)(\beta_2 - \alpha_2)$. The distribution function is shown in Fig. 500. ∎

Marginal Distributions of a Discrete Distribution

In the case of a discrete random variable (X, Y) with probability function $f(x, y)$ we may ask for the probability $P(X = x, Y \text{ arbitrary})$ that X assume a value x, while Y may assume any value, which we ignore. This probability is a function of x, say, $f_1(x)$, and we have the formula

(9) $$f_1(x) = P(X = x, Y \text{ arbitrary}) = \sum_y f(x, y)$$

where we sum all the values of $f(x, y)$ that are not 0 for that x. Clearly, $f_1(x)$ is the probability function of the probability distribution of a single random variable. This distribution is called the **marginal distribution** *of X with respect to the given two-dimensional distribution*. It has the distribution function

(10) $$F_1(x) = P(X \leqq x, Y \text{ arbitrary}) = \sum_{x^* \leqq x} f_1(x^*).$$

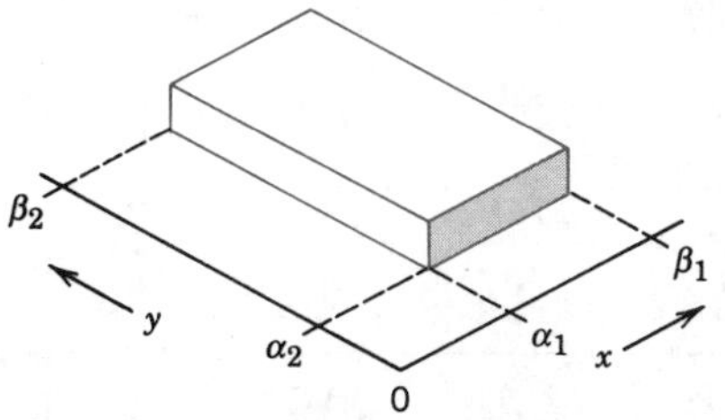

Fig. 499. Probability density function (8) of the uniform distribution

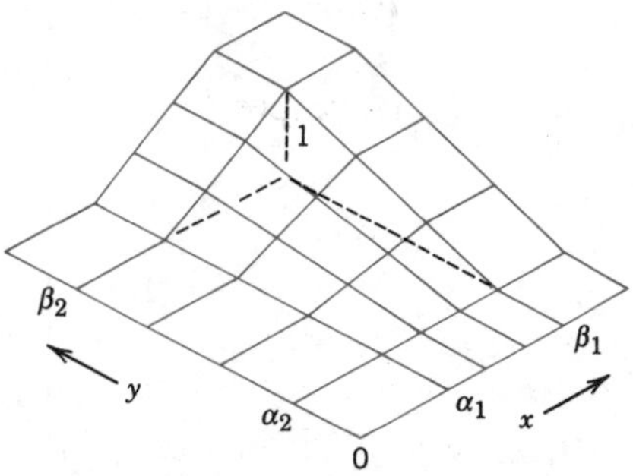

Fig. 500. Distribution function of the uniform distribution defined by (8)

Similarly, the probability function

$$f_2(y) = P(X \text{ arbitrary}, Y = y) = \sum_x f(x, y) \tag{11}$$

determines the so-called **marginal distribution** *of Y with respect to the given two-dimensional distribution.* In (11) we sum all the values of $f(x, y)$ that are not 0 for the corresponding y. The distribution function of this distribution is

$$F_2(y) = P(X \text{ arbitrary}, Y \leqq y) = \sum_{y^* \leqq y} f_2(y^*). \tag{12}$$

Obviously, both marginal distributions of a discrete random variable (X, Y) are discrete.

EXAMPLE 3 Marginal distributions of a discrete two-dimensional random variable

Suppose that we want to draw three cards with replacement from a bridge deck and observe the two-dimensional discrete random variable (X, Y), where

$$X = \textit{Number of queens},$$

$$Y = \textit{Number of kings or aces}.$$

Since the 52 cards of the deck contain four queens, four kings, and four aces, the probability of obtaining a queen in drawing a single card is $4/52 = 1/13$, and a king or ace is obtained with probability $8/52 = 2/13$. Hence (X, Y) has the probability function

$$f(x, y) = \frac{3!}{x!\, y!\, (3 - x - y)!} \left(\frac{1}{13}\right)^x \left(\frac{2}{13}\right)^y \left(\frac{10}{13}\right)^{3-x-y} \qquad (x + y \leqq 3)$$

and $f(x, y) = 0$ otherwise. Table 23.2 shows the values of $f(x, y)$ in the center and, on the right and lower margins, the values of the probability functions $f_1(x)$ and $f_2(y)$ of the marginal distributions of X and Y, respectively. ■

Table 23.2
Values of the Probability Functions $f(x, y)$, $f_1(x)$, $f_2(y)$ in Drawing Three Cards with Replacement from a Bridge Deck, where X is the Number of Queens Drawn and Y is the Number of Kings or Aces Drawn

x \ y	0	1	2	3	$f_1(x)$
0	$\frac{1000}{2197}$	$\frac{600}{2197}$	$\frac{120}{2197}$	$\frac{8}{2197}$	$\frac{1728}{2197}$
1	$\frac{300}{2197}$	$\frac{120}{2197}$	$\frac{12}{2197}$	0	$\frac{432}{2197}$
2	$\frac{30}{2197}$	$\frac{6}{2197}$	0	0	$\frac{36}{2197}$
3	$\frac{1}{2197}$	0	0	0	$\frac{1}{2197}$
$f_2(y)$	$\frac{1331}{2197}$	$\frac{726}{2197}$	$\frac{132}{2197}$	$\frac{8}{2197}$	

Marginal Distributions of a Continuous Distribution

Similarly, in the case of a continuous random variable (X, Y) with density $f(x, y)$ we may consider

$$(X \leqq x,\ Y \text{ arbitrary}) \quad \text{or} \quad (X \leqq x,\ -\infty < Y < \infty).$$

Obviously, the corresponding probability is

$$F_1(x) = P(X \leqq x,\ -\infty < Y < \infty) = \int_{-\infty}^{x} \left(\int_{-\infty}^{\infty} f(x^*, y)\, dy \right) dx^*.$$

Setting

$$f_1(x) = \int_{-\infty}^{\infty} f(x, y)\, dy, \tag{13}$$

we may write

$$F_1(x) = \int_{-\infty}^{x} f_1(x^*)\, dx^*. \tag{14}$$

$f_1(x)$ is called the *density* and $F_1(x)$ the *distribution function* of the **marginal distribution** *of X with respect to the given continuous distribution.*

Similarly, the function

$$f_2(y) = \int_{-\infty}^{\infty} f(x, y)\, dx \tag{15}$$

is called the *density* and

$$F_2(y) = \int_{-\infty}^{y} f_2(y^*)\, dy^* = \int_{-\infty}^{y} \int_{-\infty}^{\infty} f(x, y^*)\, dx\, dy^* \tag{16}$$

is called the *distribution function* of the **marginal distribution** *of Y with respect to the given two-dimensional distribution*. We see that both marginal distributions of a continuous distribution are continuous.

Independence of Random Variables

The two random variables X and Y of a two-dimensional (X, Y)-distribution with distribution function $F(x, y)$ are said to be **independent** if

$$F(x, y) = F_1(x)F_2(y) \tag{17}$$

holds for all (x, y). Otherwise these variables are said to be **dependent.**

Suppose that X and Y are either both discrete or both continuous. Then X and Y are independent if and only if the corresponding probability functions or densities $f_1(x)$ and $f_2(y)$ satisfy

$$f(x, y) = f_1(x)f_2(y) \tag{18}$$

for all (x, y); see Prob. 18. For example, the variables in Table 23.2 are dependent. The variables X = *Number of heads on a dime*, Y = *Number of heads on a nickel* in tossing a dime and a nickel may assume the values 0 or 1 and are independent.

The notions of independence and dependence may be extended to the n random variables of an n-dimensional $(X_1, \cdots, X_n)$-distribution with distribution function

$$F(x_1, \cdots, x_n) = P(X_1 \leqq x_1, \cdots, X_n \leqq x_n).$$

These random variables are said to be **independent** if for all $(x_1, \cdots, x_n)$,

$$F(x_1, \cdots, x_n) = F_1(x_1)F_2(x_2) \cdots F_n(x_n), \tag{19}$$

where $F_j(x_j)$ is the distribution function of the marginal distribution of X_j, that is,

$$F_j(x_j) = P(X_j \leqq x_j, X_k \text{ arbitrary}, k \neq j).$$

Otherwise these random variables are said to be **dependent.**

Functions of Random Variables

Let (X, Y) be a random variable with probability function or density $f(x, y)$ and distribution function $F(x, y)$, and let $g(x, y)$ be any continuous function that is defined for all (x, y) and is not constant. Then $Z = g(X, Y)$ is a random variable, too. For example, if we roll two dice and X is the number that the first die turns up whereas Y is the number that the second die turns up, then $Z = X + Y$ is the sum of those two numbers (see Fig. 489 in Sec. 23.4).

If $(X_1, \cdots, X_n)$ is an n-dimensional random variable and $g(x_1, \cdots, x_n)$ is a continuous function that is defined for all $(x_1, \cdots, x_n)$ and is not constant, then $Z = g(X_1, \cdots, X_n)$ is a random variable, too.

In the case of a *discrete* random variable (X, Y) we may obtain the probability function $f(z)$ of $Z = g(X, Y)$ by summing all $f(x, y)$ for which $g(x, y)$ equals the value of z considered; thus

$$f(z) = P(Z = z) = \sum_{g(x,y)=z}\sum f(x, y). \tag{20}$$

The distribution function of Z is

$$F(z) = P(Z \leqq z) = \sum_{g(x,y) \leqq z} \sum f(x, y) \tag{21}$$

where we sum all values of $f(x, y)$ for which $g(x, y) \leqq z$.

In the case of a *continuous* random variable (X, Y) we similarly have

$$F(z) = P(Z \leqq z) = \iint_{g(x,y) \leqq z} f(x, y)\, dx\, dy \tag{22}$$

where for each z we integrate over the region $g(x, y) \leqq z$ in the xy-plane.

Addition of Means and Variances

The number

$$E(g(X, Y)) = \begin{cases} \displaystyle\sum_x \sum_y g(x, y) f(x, y) & [(X, Y) \text{ discrete}] \\ \displaystyle\int_{-\infty}^{\infty} \int_{-\infty}^{\infty} g(x, y) f(x, y)\, dx\, dy & [(X, Y) \text{ continuous}] \end{cases} \tag{23}$$

is called the **mathematical expectation** or, briefly, the **expectation of** $g(X, Y)$. Here it is assumed that the double series converges absolutely and the integral of $|g(x, y)|\, f(x, y)$ over the xy-plane exists (is finite). Since summation and integration are linear processes, we have from (23)

$$E(ag(X, Y) + bh(X, Y)) = aE(g(X, Y)) + bE(h(X, Y)). \tag{24}$$

An important special case is $E(X + Y) = E(X) + E(Y)$, and by induction we have the following result.

Theorem 1 (Addition of means)

The mean (expectation) of a sum of random variables equals the sum of the means (expectations), that is,

$$E(X_1 + X_2 + \cdots + X_n) = E(X_1) + E(X_2) + \cdots + E(X_n). \tag{25}$$

Furthermore we readily obtain

Theorem 2 (Multiplication of means)

The mean (expectation) of the product of ***independent*** *random variables equals the product of the means (expectations), that is,*

$$E(X_1 X_2 \cdots X_n) = E(X_1) E(X_2) \cdots E(X_n). \tag{26}$$

Proof. If X and Y are independent random variables (both discrete or both continuous), then $E(XY) = E(X)E(Y)$. In fact, in the discrete case we have

$$E(XY) = \sum_x \sum_y xyf(x, y) = \sum_x xf_1(x) \sum_y yf_2(y) = E(X)E(Y),$$

and in the continuous case the proof of the relation is similar. Extension to n independent random variables gives (26), and Theorem 2 is proved. ∎

We shall now discuss the **addition of variances.** Let $Z = X + Y$ and let μ and σ^2 denote the mean and variance of Z. We first have (see Prob. 14 in Problem Set 23.5)

$$\sigma^2 = E([Z - \mu]^2) = E(Z^2) - [E(Z)]^2.$$

From (24) we see that the first term on the right equals

$$E(Z^2) = E(X^2 + 2XY + Y^2) = E(X^2) + 2E(XY) + E(Y^2),$$

and for the second term on the right, we obtain from Theorem 1

$$[E(Z)]^2 = [E(X) + E(Y)]^2 = [E(X)]^2 + 2E(X)E(Y) + [E(Y)]^2.$$

By substituting these expressions in the formula for σ^2 we have

$$\sigma^2 = E(X^2) - [E(X)]^2 + E(Y^2) - [E(Y)]^2$$
$$+ 2[E(XY) - E(X)E(Y)].$$

From Prob. 14, Sec. 23.5, we see that the expression in the first line on the right is the sum of the variances of X and Y, which we denote by σ_1^2 and σ_2^2, respectively. The quantity

(27) $$\sigma_{XY} = E(XY) - E(X)E(Y)$$

is called the **covariance** of X and Y. Consequently, our result is

(28) $$\sigma^2 = \sigma_1^2 + \sigma_2^2 + 2\sigma_{XY}.$$

If X and Y are independent, then $E(XY) = E(X)E(Y)$; hence $\sigma_{XY} = 0$, and

(29) $$\sigma^2 = \sigma_1^2 + \sigma_2^2.$$

Extension to more than two variables yields

Theorem 3 **(Addition of variances)**

The variance of the sum of ***independent*** *random variables equals the sum of the variances of these variables.*

Caution! In the numerous applications of Theorems 1 and 3 we must always remember that Theorem 3 holds only for ***independent*** variables.

This is the end of Chap. 23 on probability theory. Most of the concepts, methods, and special distributions discussed in this chapter will play a fundamental role in the next chapter, which deals with methods of **statistical inference,** that is, conclusions from samples to populations, whose unknown properties we want to know, and try to discover by looking at suitable properties of samples that we have obtained.

Problem Set 23.8

1. Let $f(x, y) = k$ when $8 \leqq x \leqq 12$ and $0 \leqq y \leqq 2$ and zero elsewhere. Find k. Find $P(X \leqq 11, 1 \leqq Y \leqq 1.5)$ and $P(9 \leqq X \leqq 13, Y \leqq 1)$.

2. Find $P(X > 2, Y > 2)$ and $P(X \leqq 1, Y \leqq 1)$ if (X, Y) has the density $f(x, y) = 1/8$ if $x \geqq 0, y \geqq 0, x + y \leqq 4$.

3. Let (X, Y) have the density $f(x, y) = kxy$ if $0 \leqq x \leqq 1, 0 \leqq y \leqq 1$ and 0 otherwise. Find k and $P(X > 0.5, Y > 0.5)$.

4. Find the density of the marginal distribution of X in Prob. 2.

5. Find the density of the marginal distribution of Y in Fig. 499.

6. If certain sheets of wrapping paper have a mean weight of 10 g each, with a standard deviation of 0.05 g, what are the mean weight and standard deviation of a pack of 10 000 sheets?

7. A four-gear assembly is put together with spacers between the gears. The mean thickness of the gears is 5.020 cm with a standard deviation of 0.003 cm. The mean thickness of the spacers is 0.040 cm with a standard deviation of 0.002 cm. Find the mean and standard deviation of the assembled units consisting of 4 randomly selected gears and 3 randomly selected spacers.

8. If the weight of certain (empty) containers has mean 2 lb and standard deviation 0.1 lb, and if the filling of the containers has mean weight 75 lb and standard deviation 0.8 lb, what are the mean weight and standard deviation of filled containers?

9. What are the mean thickness and the standard deviation of transformer cores each consisting of 50 layers of sheet metal and 49 insulating paper layers if the metal sheets have mean thickness 0.5 mm each with a standard deviation of 0.05 mm and the paper layers have mean 0.05 mm each with a standard deviation of 0.02 mm?

10. Let X [cm] and Y [cm] be the diameter of a pin and hole, respectively. Suppose that (X, Y) has the density

$$f(x, y) = 2500 \quad \text{if} \quad 0.99 < x < 1.01, \; 1.00 < y < 1.02$$

and 0 otherwise. (a) Find the marginal distributions. (b) What is the probability that a pin chosen at random will fit a hole whose diameter is 1.00?

11. An electronic device consists of two components. Let X and Y [months] be the length of time until failure of the first and second component, respectively. Assume that (X, Y) has the probability density

$$f(x, y) = 0.01e^{-0.1(x+y)} \qquad \text{if } x > 0 \text{ and } y > 0$$

and 0 otherwise. (a) Are X and Y dependent or independent? (b) Find the densities of the marginal distributions. (c) What is the probability that the first component has a lifetime of 10 months or longer?

12. Give an example of two different discrete distributions that have the same marginal distributions.

13. Show that the random variables with the densities $f(x, y) = x + y$ and $g(x, y) = (x + \frac{1}{2})(y + \frac{1}{2})$ if $0 \leqq x \leqq 1$, $0 \leqq y \leqq 1$, have the same marginal distributions.

14. Let (X, Y) have the density $f(x, y) = k$ if $x^2 + y^2 < 1$ and 0 otherwise. Determine k. Find the densities of the marginal distributions. Find the probability $P(X^2 + Y^2 < 1/2)$.

15. Find $P(X > Y)$ when (X, Y) has the density $f(x, y) = e^{-(x+y)}$ if $x \geqq 0$, $y \geqq 0$ and 0 otherwise.

16. Prove (2).

17. Let (X, Y) have the probability function $f(0, 0) = f(1, 1) = 1/8$, $f(0, 1) = f(1, 0) = 3/8$. Are X and Y independent?

18. Prove the statement involving (18).

19. Using Theorems 1 and 3, obtain the formulas for the mean and the variance of the binomial distribution.

20. Using Theorem 1, obtain the formula for the mean of the hypergeometric distribution. Can you use Theorem 3 to obtain the variance of that distribution?

Review Questions and Problems for Chapter 23

1. Give examples of random experiments in which one has equally likely cases and others in which that is not so.
2. What is the difference between mutually exclusive events and independent events?
3. What is the difference between the concepts of a permutation and a combination?
4. List all permutations and all combinations of the letters a, b, c, d taken two at a time.
5. What do we mean by conditional probability? Give a typical example.
6. If $P(A) = P(B)$ and $A \subseteq B$, can $A \neq B$?
7. If $E \neq S$ (= the sample space), can $P(E) = 1$?
8. What distributions correspond to sampling with replacement and without replacement?

9. Under what conditions will it practically make no difference whether we sample with or without replacement?
10. In what cases will the Poisson distribution be a good approximation of the binomial distribution?
11. What do you know about the approximation of the binomial distribution by the normal distribution?
12. What do we mean by "three-sigma limits"?
13. Write down the formula for the distribution function $F(x)$ of an arbitrary normal random variable in terms of the distribution function of the standardized normal distribution. Why is this practically important?
14. How is the density of a continuous random variable related to the distribution function?
15. What property of a distribution does the variance characterize? The mean?

16. What is the sum of the probabilities of all the possible values of a discrete random variable? What is the analog of this for a continuous random variable?
17. Can the probability function of a discrete random variable have infinitely many possible values? (Give a reason for your answer.)
18. If X is a continuous random variable, what is the value of $P(X = a)$, where a is any given number?
19. In what kind of experiment do we have a discrete random variable and in what kind a continuous random variable?
20. Under what conditions does the addition formula for variances hold? The addition formula for means?

21. Three screws are drawn at random from a lot of 50 screws, 10 of which are defective. Find the probability of the event that all three screws drawn are nondefective, assuming that we draw (a) with replacement, (b) without replacement.
22. What is the probability of a sum 17 or 18 in rolling three fair dice?
23. What is the probability of obtaining a sum divisible by 4 if we roll two fair dice?
24. In drawing 4 screws without replacement from a box containing 10 left-handed and 20 right-handed screws, what is the probability of obtaining at least 1 right-handed screw?
25. A box contains 50 screws, five of which are defective. Find the probability function of the random variable X = *Number of defective screws in drawing two screws without replacement* and compute its values.
26. Find the values of the distribution function in Prob. 25.

27. Find the probability function of X = *Number of times of rolling a fair die until the first even number appears.*
28. In Prob. 27, find the mean.
29. If X has the probability function $f(4) = 0.2$, $f(6) = 0.5$, $f(8) = 0.1$, $f(10) = 0.2$, what is the probability that X will assume any value greater than 6? Less than 6? Less than 4?
30. Of a lot of 12 items, 3 are defective. (a) Find the number of different samples of 3 items. Find the number of samples of 3 items containing (b) no defectives, (c) 1 defective, (d) 2 defectives, (e) 3 defectives.

31. How many automobile registrations may the police have to check in a hit-and-run accident if a witness reports KDP7 and cannot remember the last two digits on the license plate but is certain that all three were different?

32. If 6 different inks are available, in how many ways can we select two colors for a printing job? Four colors?

33. Compute 5! by the Stirling formula and find the absolute and relative errors.

34. In how many ways can a committee of 6 persons be chosen from 20 persons?

35. If we know that a process of producing bolts is running smoothly, with a fraction defective of 3%, and we inspect the process by taking samples of two bolts, what is the probability of no defectives? One defective? Two defectives?

36. Let $f(x) = ke^{-x}$ when $x > 0$, $f(x) = 0$ when $x < 0$. Find k. Sketch f and F.

37. Let $f(x) = kx^2$ when $0 \leqq x \leqq 1$ and 0 otherwise. Find k. Find the number c such that $P(X \leqq c) = 95\%$.

38. Find μ and σ^2 of the distribution with the probability function $f(1) = 0.3$, $f(2) = 0.4$, $f(3) = 0.3$.

39. Find the mean and the variance of the distribution having the density $f(x) = \frac{1}{2}e^{-|x|}$.

40. If the mileage X (in thousands of miles) of a certain kind of tire has the density $f(x) = 0.05e^{-0.05x}$ $(x > 0)$, what mileage can one expect to get? What is the probability that the tire will last at least 30 000 miles?

41. If X has the density $f(x) = 0.5x$ $(0 \leqq x \leqq 2)$ and 0 otherwise, what are the mean and the variance of $X^* = -2X + 5$?

42. Find the skewness of the distribution with density $f(x) = 2(1 - x)$ if $0 < x < 1$, $f(x) = 0$ otherwise.

43. Let $p = 2\%$ be the probability that a certain type of light bulb will fail in a 24-hour test. Find the probability that a sign consisting of 15 such bulbs will burn for 24 hours without bulb failures.

44. Suppose that 4% of screws made by a machine are defective, the defectives occurring at random during production. If the screws are packaged 100 per box, what is the Poisson approximation of the probability that a given box will contain x defectives?

45. Show that the distribution function of the Poisson distribution satisfies $F(\infty) = 1$.

46. Find the probability that a random variable X assume any value in the interval $\mu - \sigma \leqq X \leqq \mu + \sigma$ if X has (a) a binomial distribution with $n = 8$ and $p = 0.5$, (b) the normal distribution.

47. If the breaking strength of certain ropes is normal with mean 100 kg and standard deviation 10 kg, what is the maximum load such that we can expect at most 1% of the ropes to break?

48. What is the probability of obtaining at least 12 012 heads if a coin is tossed 24 000 times and heads and tails are equally likely? (See Table 23.1 in Sec. 23.2.)

49. If the life of tires is normal with mean 25 000 km and variance 25 000 000 km^2, what is the probability that a given one of those tires will last at least 30 000 km? At least 35 000 km?

50. If the weight of bags of cement is normal with mean 50 kg and standard deviation 1 kg, what is the probability that 100 bags will be heavier than 5030 kg?

Summary of Chapter 23
Probability Theory

A *random experiment,* briefly called **experiment,** is a process in which the result ("**outcome**") depends on "chance" (effects of factors unknown to us). Examples are games of chance with dice or cards, measuring the hardness of steel, observing weather conditions, counting the number of telephone calls in an office, or recording the number of accidents in a city. (Thus the word "experiment" is used here in a much wider sense than in common language.) The outcomes are regarded as points (elements) of a set S, called the **sample space,** whose subsets are called **events.** For events E we define a **probability** $P(E)$ by the axioms (Sec. 23.2)

$$
\begin{aligned}
&0 \leqq P(E) \leqq 1 \\
(1)\qquad &P(S) = 1 \\
&P(E_1 \cup E_2 \cup \cdots) = P(E_1) + P(E_2) + \cdots \qquad (E_j \cap E_k = \varnothing).
\end{aligned}
$$

The complement E^C of E then has the probability

$$
(2)\qquad P(E^C) = 1 - P(E).
$$

The **conditional probability** of an event B under the condition that an event A happens is

$$
(3)\qquad P(B|A) = \frac{P(A \cap B)}{P(A)}.
$$

A and B are called **independent** if

$$
(4)\qquad P(A \cap B) = P(A)P(B).
$$

With an experiment we associate a **random variable** X; this is a function defined on S whose values are real numbers, and X is such that the probability $P(X = a)$ that X assume any value a, and the probability $P(a < X \leqq b)$ that X assume any value in any interval are defined (Sec. 23.4). The **probability distribution** of X is determined by the distribution function (Sec. 23.4)

$$
(5)\qquad F(x) = P(X \leqq x).
$$

There are two practically important kinds of random variables: those of **discrete** type, which appear if we count (defective items, customers in a bank, etc.) and those of **continuous** type, which appear if we measure (length, speed, temperature, weight, etc.).

A discrete random variable has a **probability function**

$$f(x) = P(X = x). \tag{6}$$

Its **mean** μ and **variance** σ^2 are (Sec. 23.5)

$$\mu = \sum_j x_j f(x_j) \quad \text{and} \quad \sigma^2 = \sum_j (x_j - \mu)^2 f(x_j) \tag{7}$$

where the x_j are the values for which X has a positive probability. Important discrete random variables and distributions are the **binomial, Poisson,** and **hypergeometric distributions** (Sec. 23.6).

A continuous random variable has a **density**

$$f(x) = F'(x) \qquad \text{[see (5)].} \tag{8}$$

Its mean and variance are (Sec. 23.5)

$$\mu = \int_{-\infty}^{\infty} x f(x)\, dx \quad \text{and} \quad \sigma^2 = \int_{-\infty}^{\infty} (x - \mu)^2 f(x)\, dx. \tag{9}$$

Very important is the **normal distribution** whose density is (Sec. 23.7)

$$f(x) = \frac{1}{\sigma\sqrt{2\pi}} \exp\left[-\frac{1}{2}\left(\frac{x-\mu}{\sigma}\right)^2\right] \tag{10}$$

and whose distribution function is (Sec. 23.7; Tables A7, A8 in Appendix 5)

$$F(x) = \Phi\left(\frac{x-\mu}{\sigma}\right). \tag{11}$$

A **two-dimensional random variable** (X, Y) occurs if we simultaneously observe two quantities (for example, height X and weight Y of persons). Its distribution function is (Sec. 23.8)

$$F(x, y) = P(X \leqq x, Y \leqq y). \tag{12}$$

X and Y have the distribution functions (Sec. 23.8)

$$F_1(x) = P(X \leqq x, Y \text{ arbitrary}), \quad F_2(y) = P(x \text{ arbitrary}, Y \leqq y); \tag{13}$$

their distributions are called **marginal distributions.** If both X and Y are discrete, then (X, Y) has a probability function

$$f(x, y) = P(X = x, Y = y).$$

If both X and Y are continuous, then (X, Y) has a density $f(x, y)$.

Chapter 24

Mathematical Statistics

In **probability theory** we set up mathematical models of processes and systems that are affected by "chance." In mathematical statistics or, briefly, **statistics,** we check these models against the reality, to determine whether they are faithful and accurate enough for practical purposes. This is done by drawing **samples,** inspecting their properties, and then making conclusions about properties of the population from which the samples were drawn. These conclusions are true, not absolutely, but with some (usually high) probability that we can choose (95%, for instance) or at least compute. Such conclusions result from methods of statistical inference that we shall develop in this chapter, using probability theory as the foundation of our approach.

Those conclusions are then used as bases for making predictions, decisions, and actions, for instance, in weather forecast or market analysis or traffic planning, in developing economical production plans, in buying raw materials or machines or equipment, in methods of advertising and selling products, in taking actions against dangerous diseases, and so on.

We talk about the whole approach in general terms in Sec. 24.1, and then we discuss samples and their means (Secs. 24.2–24.4). **Statistical methods** begin in Sec. 24.5 with point and interval **estimation of parameters** (Secs. 24.5, 24.6) and continue with **hypothesis testing** and applications (Secs. 24.7–24.9), **goodness of fit** (test for distribution functions, Sec. 24.10), and some **nonparametric tests** (Sec. 24.11). The last section deals with *pairs of measurements* and **regression analysis.**

Prerequisites for this chapter: Chap. 23.
Sections that may be omitted in a shorter course: 24.3, 24.8, 24.9, 24.11.
References: Appendix 1, Part G.
Answers to problems: Appendix 2.
Statistical tables: Appendix 5.

24.1 Nature and Purpose of Statistics

In statistics we are concerned with methods for designing and evaluating experiments to obtain information about practical problems, for example, the inspection of quality of raw material or manufactured products, the comparison of machines and tools or methods used in production, the output of workers, the reaction of consumers to new products, the yield of a chemical process under various conditions, the relation between the iron content of iron ore and the density of the ore, the efficiency of air-conditioning systems under various temperatures, the relation between the Rockwell hardness and the carbon content of steel, and so on.

For instance, in a process of mass production (of screws, bolts, light bulbs, transistors, etc.) we ordinarily have *nondefective items*, that is, items which satisfy the quality specifications, and *defective items*, which do not satisfy those specifications. Such specifications may include, for instance, maximum and minimum diameters of axle shafts, minimum lifetimes of light bulbs, limiting values of the resistance of resistors in TV sets, maximum weights of airmail envelopes, minimum contents of automatically filled bottles, maximum reaction times of switches and minimum values of strength of yarn.

The reason for the differences in the quality of products is *variation* due to numerous factors (in the raw material, function of automatic machinery, workmanship, etc.) whose influence cannot be predicted, so that the variation must be regarded as a ***random variation***. In the case of evaluating the efficiency of production methods and in the other preceding examples the situation is similar.

In most cases the inspection of *each* item of a production would be too expensive and time-consuming. It may even be impossible if it leads to the destruction of the item. Hence instead of inspecting all the manufactured items, just a few of them (a "**sample**") are inspected and from this inspection conclusions can be drawn about the totality (the "**population**"). If we draw 100 screws from a lot of 10,000 screws and find that 4 of the 100 screws are defective, we are inclined to conclude that *about* 4% of the lot are defective, provided the screws were drawn "**at random,**" that is, so that each screw of the lot had the same "chance" of being drawn. It is clear that such a conclusion is not absolutely certain; that is, we cannot say that the lot contains *precisely* 4% defectives, but in most cases such a precise statement would not be of particular practical interest anyway. We also feel that the conclusion is the more dependable the more (randomly selected!) screws we inspect. Using the theory of mathematical probability we shall see that these somewhat vague intuitive notions and ideas can be made precise. Furthermore, this theory will also yield measures for the reliability of conclusions about populations obtained from samples by statistical methods. Hence probability theory forms the basis of statistical methods.

Similarly, to obtain information about the iron content μ of iron ore, we may pick a certain number n of specimens at random and measure their iron content. This yields a sample of n numbers $x_1, \cdots, x_n$ (the results of the n measurements) whose average $\bar{x} = (x_1 + x_2 + \cdots + x_n)/n$ is an approximate value for μ.

Problems of differing natures may require different methods and techniques, but the steps leading from the formulation of a problem to its solution are the same in almost all cases. They are as follows.

1. ***Formulation of the problem.*** It is important to formulate the problem in a precise fashion and to limit the investigation, so that we can expect a useful answer within a prescribed interval of time, taking into account the cost of the statistical investigation, the skill of the investigators, and the facilities available. This step must also include the creation of a mathematical model based on clear concepts. (For example, we must define what we mean by a defective item in a specific situation.)

2. ***Design of experiment.*** This step includes the choice of the statistical method to be used in the last step, the sample size n (number of items to be drawn and inspected or number of experiments to be made, etc.), and the physical methods and techniques to be used in the experiment. The goal is to obtain a maximum of information using a minimum of cost and time.

3. ***Experimentation or collection of data.*** This step should adhere to strict rules.

4. ***Data processing.*** In this step we arrange the data in a clear and simple tabular form, and we may represent them graphically by diagrams, bar charts, and so on. We also compute sample parameters (numbers), in particular, the sample mean $\bar{x}$ (average size of the sample values) and the sample variance s^2 (which measures the spread of the sample values).

5. ***Statistical inference.*** In this step we use the sample and apply a suitable statistical method for drawing conclusions about the unknown properties of the population so that we obtain the answer to our problem.

It seems clear that **samples** will play a crucial role in the whole approach. Accordingly, we should make this intuitive notion rigorous. We do this in the next section.

24.2 Random Sampling. Random Numbers

From the last section we know that samples are selections from populations and that they are basic in statistics because we need them to make conclusions from their properties about properties of the population we want to know. For this, it is quite essential that a sample be a **random selection,** that is, each element of the population must have a known probability of being taken into the sample. This condition must be satisfied (at least approximately); otherwise statistical methods may yield misleading results.

If the sample space is infinite, the sample values will be **independent;** that is, the results of the n performances of an experiment made for obtaining n sample values will not influence each other. This certainly applies to samples from a normal population. If the sample space is finite, the sample values will still be independent if we sample with replacement; they will be *practically* independent if we sample without replacement but keep the size of the sample small compared to the size of the population (for instance, samples of 5 or 10 values from a population of 1000 values). However, if we sample without replacement and take large samples from a finite population, the dependence will matter considerably.

Random Numbers

It is not so easy to satisfy the requirement that a sample be a random selection, because there are many and subtle factors of various types that can bias the result of sampling. For example, if we want to draw and inspect a sample of 10 ball bearings from a lot of 80 before deciding whether to purchase the lot, how should we select physically those 10 bearings in order to be reasonably sure that all

$$\binom{80}{10} \approx 1.6 \cdot 10^{12}$$

samples of size 10 would be equally probable?

Methods to solve this problem have been developed, and we shall now describe such a procedure, which is frequently used.

We number the items of that lot from 1 to 80. Then we select 10 items using Table A9 in Appendix 5, which contains *random numbers* made up of sets of **random digits,** as follows. We first select a row number from 0 to 99 at random. This can be done by tossing a fair coin seven times, denoting heads by 1 and tails by 0, thus generating a seven-place binary number that will represent $0, 1, \cdots, 127$ with equal probabilities. We use this number if it is $0, 1, \cdots$, or 99. Otherwise we ignore it and repeat this process. Then we select a column number from 0 to 9 by generating a four-place binary number in a similar fashion. Suppose that we obtained 0011010 (= 26) and 0111 (= 7), respectively. In row 26 and column 7 of Table A9 we find 44973. We keep the first two digits, that is, 44. We move down the column, starting with 44973 and keeping only the first two digits. In this way we find

44 44 83 91 55 etc.

We omit numbers that are greater than 80 or occur for the second time and continue until 10 numbers are obtained. This gives

44 55 53 03 52 61 67 78 39 54.

The 10 items having these numbers represent the desired selection.

For a larger table of random digits see Ref. [G15], Appendix 1.

However, for large samples such tables may become cumbersome. For this reason, procedures for generating numbers with similar properties have been developed and are available in a so-called **random number generator** in many computer languages and subroutine libraries.

We shall say more about samples from the data processing point of view in the next section.

Problem Set 24.2

1. Suppose that in the example explained in the text we started from row 83 and column 2 of Table A9 in Appendix 5, and moved upward. What items would then be included in a sample of size 10?
2. Using Table A9 in Appendix 5, select a sample of size 20 from a lot of 250 given items.
3. How can fair dice be used in connection with random selection?

4. Consider a random variable Y having the uniform density $f(y) = 1$ if $0 < y < 1$, and 0 otherwise. We can easily simulate Y (that is, sample the values of Y) with the use of random digits. For instance, to obtain 20 values rounded to 2 decimals, consider any of the 10 columns in Table A9, start with some randomly chosen row and move downward, taking the first two digits of the five digits given in the column and put a decimal point to the left of the first digit. Suppose your choice was column 8 and row 12. Show that you get the following sample and graph its values as dots on a scale between 0 and 1.

0.91 0.42 0.64 0.55 0.66 0.50 0.34 0.28 0.07 0.64
0.77 0.03 0.88 0.11 0.37 0.54 0.92 0.96 0.97 0.48

5. How would you modify the method in Prob. 4 if you want a sample of size 20 from the uniform distribution on the interval $2 < y < 6$?

6. Random digits can also be used to simulate *any* continuous random variable X. For this end we graph the distribution function of X, use random digits to obtain values of the variable Y described in Prob. 4, plot these values on the vertical axis, and read off the corresponding values of X. Illustrate this procedure for a normal random variable X with mean 0 and variance 1, using the sample in Prob. 4. (Of course, in practice a computer would do this by the use of a random number generator and the distribution function of X.)
7. Write the method in Prob. 6 in terms of $\Phi(z)$, so that you could program it on a computer.
8. Extend the method in Prob. 6 to the normal distribution with mean 3 and variance 4. What do you then get from 0.91, 0.42, 0.64 in Prob. 4?
9. What sample from a random variable with density $f(x) = e^{-x}$ if $x > 0$, $f(x) = 0$ if $x < 0$, would you get from the sample in Prob. 4? Compute the first three values.
10. The simulation technique described in Prob. 6 applies also to *discrete* random variables. Explain how you would proceed if X is the sum of the two numbers obtained in rolling two fair dice (see Fig. 489 in Sec. 23.4).

24.3 Processing of Samples

In this section we discuss methods for handling samples and representing them by diagrams (graphical representations); more fundamentally, we also introduce concepts related to samples that we shall need throughout this chapter.

In the course of a statistical experiment we usually obtain a sequence of observations (numbers in most cases). These should always be recorded in the order in which they occur. They are called **sample values;** their number is called the *sample* **size** and is denoted by n. As a typical example, Table 24.1 on the next page shows a sample of size $n = 100$ values. These data were obtained by making standard test cylinders (diameter 6 inches, length 12 inches) of concrete and splitting them 28 days later.

Table 24.1
Sample of 100 Values of the Splitting Tensile Strength (lb/in.^{2}) of Concrete Cylinders

320	380	340	410	380	340	360	350	320	370
350	340	350	360	370	350	380	370	300	420
370	390	390	440	330	390	330	360	400	370
320	350	360	340	340	350	350	390	380	340
400	360	350	390	400	350	360	340	370	420
420	400	350	370	330	320	390	380	400	370
390	330	360	380	350	330	360	300	360	360
360	390	350	370	370	350	390	370	370	340
370	400	360	350	380	380	360	340	330	370
340	360	390	400	370	410	360	400	340	360

D. L. IVEY, Splitting tensile tests on structural lightweight aggregate concrete. Texas Transportation Institute, College Station, Texas.

Frequency Table of a Sample. Frequency Function

From a table such as Table 24.1, in which the sample values $x_1, \cdots, x_n$ are recorded in their order of occurrence during the experimentation, we cannot see much. Accordingly, we next use a sorting subroutine that will produce a **frequency table,** such as Table 24.2, in which column 1 lists the occurring numerical values and column 2 their **absolute frequencies** (= numbers of times a value occurs in the sample), also simply called **frequencies.** (Thus, $x = 330$ has the absolute frequency 6, etc.) Division by the sample size n gives the **relative frequency** $\tilde{f}(x)$ in column 3. (For instance, $\tilde{f}(300) = 2/100 = 0.02$, etc.) Clearly, the extreme possible values are $\tilde{f}(x) = 0$ if x does not occur in the sample, and $\tilde{f}(x) = n/n = 1$ if all the sample values are equal (equal to that x). This gives

Theorem 1 (Relative frequency)
The relative frequency satisfies

$$0 \leqq \tilde{f}(x) \leqq 1. \tag{1}$$

For a given sample we call $\tilde{f}(x)$, with $\tilde{f}(x) = 0$ for all x not in the sample, the **frequency function** of the sample and say that it determines the **frequency distribution** of the sample, because it shows how the sample values are distributed.

Theorem 1 is the empirical counterpart of Axiom 1 for probability in Sec. 23.2. Furthermore, the sum of all absolute frequencies in a sample must equal n (why?), so that division by n gives

Theorem 2 (Sum of all relative frequencies)
The sum of all relative frequencies in a sample equals 1,

$$\sum_x \tilde{f}(x) = 1. \tag{2}$$

Table 24.2
Frequency Table of the Sample in Table 24.1

1 Tensile Strength x [lb/in.^{2}]	2 Absolute Frequency	3 Relative Frequency $\tilde{f}(x)$	4 Cumulative Absolute Frequency	5 Cumulative Relative Frequency $\tilde{F}(x)$
300	2	0.02	2	0.02
310	0	0.00	2	0.02
320	4	0.04	6	0.06
330	6	0.06	12	0.12
340	11	0.11	23	0.23
350	14	0.14	37	0.37
360	16	0.16	53	0.53
370	15	0.15	68	0.68
380	8	0.08	76	0.76
390	10	0.10	86	0.86
400	8	0.08	94	0.94
410	2	0.02	96	0.96
420	3	0.03	99	0.99
430	0	0.00	99	0.99
440	1	0.01	100	1.00

Formula (2) is the empirical counterpart of Axiom 2 in Sec. 23.2. (The analog of Axiom 3 would be of no great interest.) In (2) we have a finite sum, which we can write in more orderly form if we first denote the ***numerically different*** values in a sample of size n by

$$x_1, \quad x_2, \quad \cdots, \quad x_m \qquad (m \leqq n).$$

(Thus $n = 100$ but $m = 15$ in Table 24.2, and $x_1 = 330$, $x_2 = 340$, $\cdots$, $x_{15} = 440$.) Then (2) becomes

$$\sum_{j=1}^{m} \tilde{f}(x_j) = 1 \tag{2*}$$

(summation from 1 to m, not n). Verify (2*) for Table 24.2.

The frequency function $\tilde{f}(x)$ of a sample is an empirical counterpart or analog of the probability function or density $f(x)$ of the corresponding population—although these functions are conceptually quite different; most obviously: a population has *one* $f(x)$, but if we take 10 samples from the same population, we shall generally get 10 different sample frequency functions $\tilde{f}(x)$. (Why?) In our further work we shall also need an analog of the distribution function $F(x)$ of a population. We define it as follows.

Sample Distribution Function

Let

(3) $\tilde{F}(x)$ = *Sum of the relative frequencies of all the values that are smaller than x or equal to x.*

This function is called the **cumulative frequency function of the sample** or the **sample distribution function.** An example is shown in Fig. 501.

$\tilde{F}(x)$ is a *step function* (piecewise constant function) having jumps of magnitude $\tilde{f}(x)$ precisely at those x at which $\tilde{f}(x) \neq 0$. Before the first jump, $\tilde{F}(x) = 0$. The first jump is at the smallest sample value and the last at the largest. Afterward, $\tilde{F}(x) = 1$.

The relation between $\tilde{f}(x)$ and $\tilde{F}(x)$ is

$$\tilde{F}(x) = \sum_{t \leqq x} \tilde{f}(t) \tag{4}$$

where $t \leqq x$ means that for an x we have to sum all those $\tilde{f}(t)$ for which t is less than x or equal to x.

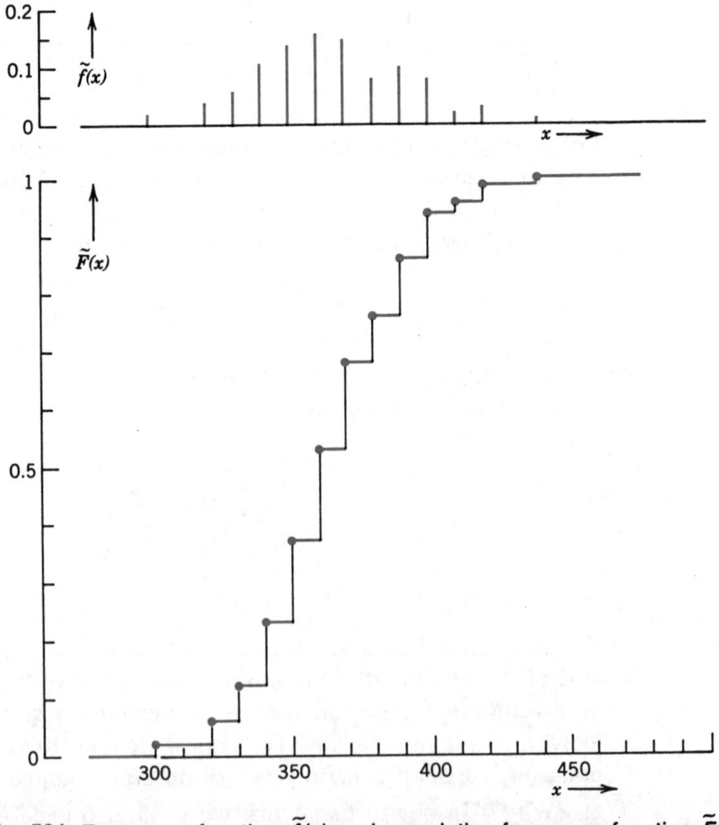

Fig. 501. Frequency function $\tilde{f}(x)$ and cumulative frequency function $\tilde{F}(x)$ of the sample in Table 24.2. (The dots mark the values of $\tilde{F}(x)$ at the jumps.)

Graphical Representations of Samples

Figures 502–505 show possibilities of representing samples by graphs. These are self-explanatory, but the following comment is of practical interest.

In Fig. 504 the area of each rectangle is equal to the corresponding relative frequency. Hence the ordinate should be labeled "*relative frequency per unit interval.*" Since in the present case the rectangles are equally wide, those values on the ordinate are proportional to $\tilde{f}(x)$ and we may label the ordinate in terms of $\tilde{f}(x)$. However, this would no longer be true if the rectangles were of different width.

In Fig. 505 the situation is similar.

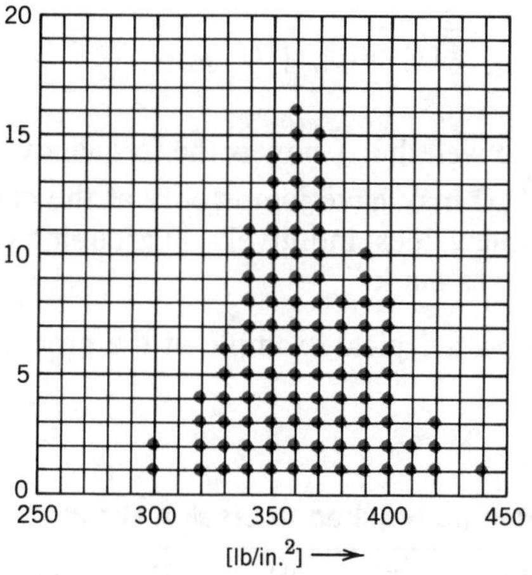

Fig. 502. Dot frequency diagram of the sample in Table 24.2

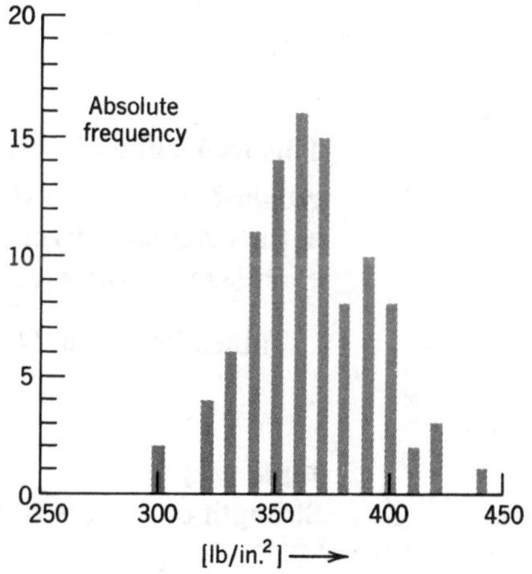

Fig. 503. Bar chart of the sample in Table 24.2

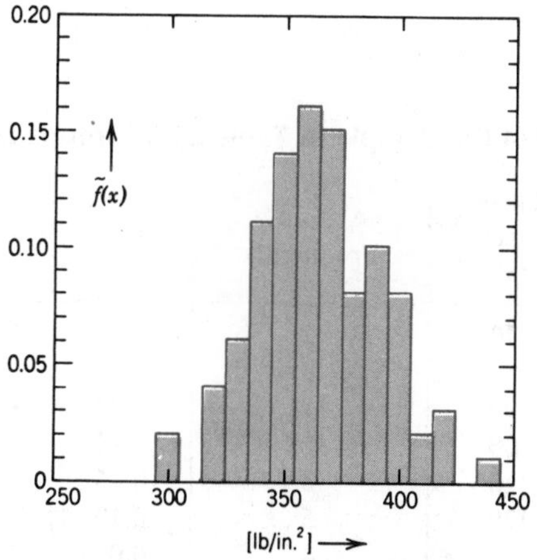

Fig. 504. Frequency histogram of the sample in Table 24.2

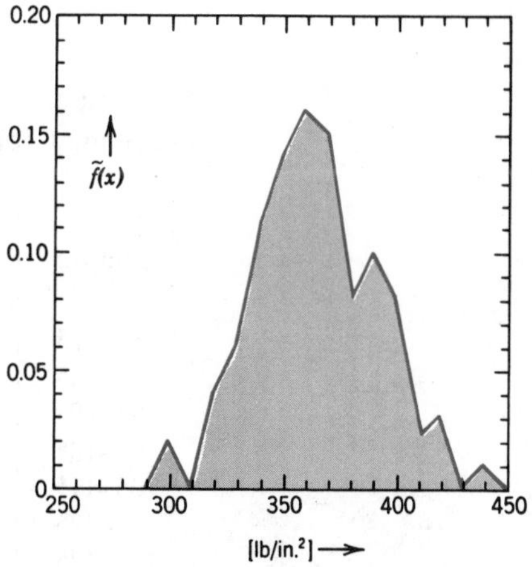

Fig. 505. Frequency polygon of the sample in Table 24.2

Grouping of Samples

If a sample consists of too many numerically different sample values, then its tabular and graphical representations are unnecessarily complicated but may be simplified by the process of **grouping,** as follows.

A sample being given, we choose a convenient interval I that contains all the sample values and subdivide I into subintervals. We call these **class intervals** and their midpoints **class midpoints** or **class marks.** The set of sample values in a class interval is called a **class** and their number the corresponding **absolute class frequency.** Division by n gives the **relative class frequency** $\tilde{f}(x)$, and $\tilde{f}(x)$, with $\tilde{f}(x) = 0$ if x is not a class mark, is called the **frequency function of the grouped sample.** From it we get the **distribution function** $\tilde{F}(x)$ of the grouped sample by summation,

$$\tilde{F}(x) = \sum_{t \leq x} \tilde{f}(t). \tag{4*}$$

This looks like (4), but note well that $\tilde{f}$ now is the frequency function of the ***grouped*** sample, so that $\tilde{F}(x)$ may have jumps only at the class marks, not at the originally given sample values; intuitively, in grouping we have moved each sample value to its class mark.

Tables 24.3 and 24.4 show a typical example of the process of grouping.

Table 24.3
Strength of 50 Lots of Cotton (lb required to break a skein)

114	118	86	107	87	94	82	81	98	84
120	126	98	89	114	83	94	106	96	111
123	110	83	118	83	96	96	74	91	81
102	107	103	80	109	71	96	91	86	129
130	104	86	121	96	96	127	94	102	87

Table 24.4
Frequency Table of the Sample in Table 24.3 (Grouped)

Class Interval	Class Mark x	Absolute Frequency	$\tilde{f}(x)$	$\tilde{F}(x)$
65– 75	70	2	0.04	0.04
75– 85	80	8	0.16	0.20
85– 95	90	11	0.22	0.42
95–105	100	12	0.24	0.66
105–115	110	8	0.16	0.82
115–125	120	5	0.10	0.92
125–135	130	4	0.08	1.00
		Sum 50	1.00	

Grouping rules. The fewer classes we choose, the simpler the distribution of the grouped sample becomes but the more information we lose, because the original sample values no longer appear explicitly. Grouping should be done so that only unessential details are eliminated. Unnecessary complications in the later use of a grouped sample are avoided by obeying the following rules.

1. All the class intervals should have the same length.

2. The class intervals should be chosen so that the class marks correspond to simple numbers (numbers with few nonzero digits).

3. If a sample value x_j coincides with the common endpoint of two class intervals, take it into the class interval that extends from x_j to the right.

Problem Set 24.3

In each case make a frequency table of the given sample and represent the sample by a dot frequency diagram, a bar chart, and a histogram.

1. Resistance [ohms] of resistors

99 100 102 101 98 103 100 102 99 101
100 100 99 101 100 102 99 101 98 100

2. Numbers that turned up on a die

3 3 2 1 6 5 6 3 4 6 2 4 1 2 4

3. Release time [sec] of a relay

1.3 1.2 1.4 1.5 1.3 1.3 1.4 1.1 1.5 1.4
1.6 1.3 1.5 1.1 1.4 1.2 1.3 1.5 1.4 1.4

4. Carbon content [%] of coal

86 87 86 81 77 85 87 86 85 87
82 84 83 79 82 73 86 84 83 83

5. Tensile strength [kg/mm^2] of sheet steel

43 44 45 46 44 43 41 41 44 44 43 44 42 45 43
45 42 44 44 42 45 41 44 44 43 44 46 41 43 45

6. Number of sheets of paper over and under the desired number of 100 sheets per package in a packaging process

1 2 0 1 0 0 −1 0 0 1

7. Miles per gallon of gasoline required by six cars of the same make

15.0 15.5 14.5 15.0 15.5 15.0

8. Weight of filled bags [grams] in an automatic filling process

200 203 199 198 201 200 201 201

9. Waiting time [min, rounded] of a commuter for a train in a certain subway

3 1 5 3 3 4 1 0 2 2

10. Graph the cumulative frequency function of the sample in Prob. 3.

11. Graph the bar chart, the histogram, and the frequency polygon of the grouped sample in Table 24.4.

12. Graph the histogram of the following sample of lifetimes [hours] of light bulbs.

Lifetime	Absolute Frequency	Lifetime	Absolute Frequency	Lifetime	Absolute Frequency
950–1050	4	1350–1450	51	1750–1850	20
1050–1150	9	1450–1550	58	1850–1950	9
1150–1250	19	1550–1650	53	1950–2050	3
1250–1350	36	1650–1750	37	2050–2150	1

13. Group the sample in Table 24.1, using class intervals with midpoints 300, 320, 340, · · · . Make up the corresponding frequency table. Graph the histogram and compare it with Fig. 504. Graph the cumulative frequency function.

14. Group the sample shown in Table 24.3, using class intervals with midpoints 75, 85, 95, · · · . Make up the corresponding frequency table. Graph the histogram and compare it with that in Prob. 11.

15. The smallest of 1500 measurements was 10.8 cm and the largest was 11.9 cm. Suggest class intervals for grouping these data.

24.4 Sample Mean, Sample Variance

For populations we have defined numbers, called **parameters** (the mean μ and variance σ^2 above all; see Sec. 23.5) that characterize important properties of the distribution. For samples we can do the same, as follows.

Sample mean. The *sample mean* $\bar{x}$ of a sample $x_1, x_2, \cdots, x_n$ is defined by

$$\bar{x} = \frac{1}{n}\sum_{j=1}^{n} x_j = \frac{1}{n}(x_1 + x_2 + \cdots + x_n). \tag{1}$$

It is the sum of all the sample values, divided by the size n of the sample. Obviously, it measures the average size of the sample values.

Sample variance. The *sample variance* s^2 of a sample $x_1, \cdots, x_n$ is defined by

$$s^2 = \frac{1}{n-1}\sum_{j=1}^{n} (x_j - \bar{x})^2 = \frac{1}{n-1}[(x_1 - \bar{x})^2 + \cdots + (x_n - \bar{x})^2]. \tag{2}$$

It is the sum of the squares of the deviations of the sample values from the mean $\bar{x}$, divided by $n - 1$. It measures the spread or dispersion of the sample values and is positive, except for the rare case when all the sample values are equal (and are then equal to $\bar{x}$). The positive square root of the sample variance s^2 is called the **standard deviation** of the sample and is denoted by s.

EXAMPLE 1 Sample mean and sample variance

Ten randomly selected nails had the lengths [in.]

0.80 0.81 0.81 0.82 0.81 0.82 0.80 0.82 0.81 0.81.

Find the mean and the variance of this sample.

Solution. From (1) we see that the sample mean is

$$\bar{x} = \tfrac{1}{10}(0.80 + 0.81 + 0.81 + 0.82 + \cdots + 0.81) = 0.811 \text{ [in.]}.$$

Applying (2), we thus obtain the sample variance

$$s^2 = \tfrac{1}{9}[(0.800 - 0.811)^2 + \cdots + (0.810 - 0.811)^2] = 0.000\,054 \text{ [in.}^2\text{]}.$$

This computation becomes simpler if we take equal sample values together. Then

$$\bar{x} = \tfrac{1}{10}(2 \cdot 0.80 + 5 \cdot 0.81 + 3 \cdot 0.82) = 0.811.$$

In the parentheses we have the sum of the three ***numerically different*** sample values $x_1 = 0.80$, $x_2 = 0.81$, $x_3 = 0.82$, each multiplied by its absolute frequency. Similarly,

$$s^2 = \tfrac{1}{9}[2(0.800 - 0.811)^2 + 5(0.810 - 0.811)^2 + 3(0.820 - 0.811)^2] = 0.000\,054. \quad \blacksquare$$

The example illustrates how we may compute $\bar{x}$ and s^2 from the frequency function $\tilde{f}(x)$ of the sample. If a sample of n values contains precisely m ***numerically different*** sample values

$$x_1, \qquad x_2, \qquad \cdots, \qquad x_m$$

(where $m \leqq n$), the corresponding relative frequencies are

$$\tilde{f}(x_1), \qquad \tilde{f}(x_2), \qquad \cdots, \qquad \tilde{f}(x_m)$$

and the corresponding absolute frequencies needed in the computation are

$$a(x_1) = n\tilde{f}(x_1), \qquad a(x_2) = n\tilde{f}(x_2), \qquad \cdots, \qquad a(x_m) = n\tilde{f}(x_m).$$

We see that (1) now takes the form

$$\textbf{(3)} \qquad \bar{x} = \frac{1}{n}\sum_{j=1}^{m} x_j a(x_j), \qquad a(x_j) = n\tilde{f}(x_j),$$

and (2) takes the form

$$\textbf{(4)} \qquad s^2 = \frac{1}{n-1}\sum_{j=1}^{m} (x_j - \bar{x})^2 a(x_j).$$

Note that in (1) and (2) we sum over *all* the sample values, whereas now we sum over the *numerically different* sample values. The absolute frequencies $a(x_j)$ are integers, while the relative frequencies $\tilde{f}(x_j)$ may be messy, for example, if $n = 23$ or $n = 84$, etc.

This is the end of our discussion on getting and handling samples, in preparation for the further sections, which will concern statistical inference from samples to populations by statistical methods.

Problem Set 24.4

1. Compute the mean and variance of the sample in Prob. 2, Sec. 24.3, using (1), (2) or (3), (4).

2. Compute the mean and variance of the sample 3, 4, 10, 4, 4. Notice the large contribution of 10 to s^2 and comment.

3. Graph a histogram of the sample 8, 2, 4, 10 and guess $\bar{x}$ and s by inspecting the histogram. Then calculate $\bar{x}$, s^2, and s.

4. Compute the mean and variance of the sample in Prob. 4, Sec. 24.3.

5. To illustrate that s^2 measures the spread, compute s^2 for the two samples 109, 110, 111 and 105, 110, 115, and compare.

6. Prove that $\bar{x}$ lies between the smallest and the largest sample values.

7. Prove that $s^2 = 0$ if and only if all sample values are equal.

8. (Working origin) If $x_j = x_j^* + c$, where $j = 1, \cdots, n$ and c is any constant, show that

$$\bar{x} = c + \bar{x}^* \qquad \left(\bar{x}^* = \frac{1}{n}\sum_{j=1}^{n} x_j^*\right), \qquad s^2 = s^{*2}$$

where s^{*2} is the variance of the x_j^*'s. (In practice, c is chosen so that the x_j^*'s are small in absolute value. Geometrically this is a shift of the origin and is called the *method of working origin.*)

9. Apply the method of working origin to the sample in Example 1.

10. (Full coding) If $x_j = c_1x_j^* + c_2$, where $j = 1, \cdots, n$ and c_1 and c_2 are constants, show that

$$\bar{x} = c_1\bar{x}^* + c_2, \qquad s^2 = c_1^2 s^{*2}$$

where $\bar{x}^*$ and s^{*2} have the same meaning as in Prob. 8. (This is called the *method of full coding*. Obviously it is valuable in computations with a pocket calculator, for instance, for quick checks.)

11. Using the method of working origin, compute the mean of the sample in Prob. 1 of Sec. 24.3.

12. Apply full coding to the sample in Example 1.

13. (Range of a sample) The difference between the largest and the smallest values in a sample is called the *range of the sample*. Find the range of the sample in Example 1.

14. An advantage of the range is that it can be computed more easily than s^2. Can you think of a disadvantage?

15. (Percentile, median) The pth **percentile** of a sample is a number Q_p such that at least p% of the sample values are smaller than or equal to Q_p and also at least $(100 - p)$% of those values are larger than or equal to Q_p. If there is more than one such number (in which case there will be an interval of them), the pth percentile is defined as the average of the numbers (midpoint of that interval). In particular, Q_{50} is called the **middle quartile** or **median** and is denoted by $\tilde{x}$. Find $\tilde{x}$ for the sample in Table 24.2 (Sec. 24.3).

16. The percentiles Q_{25} and Q_{75} of a sample are called the **lower** and **upper quartiles** of the sample, and $Q_{75} - Q_{25}$, which is a measure for the spread, is called the **interquartile range.** Find Q_{25}, Q_{75}, and $Q_{75} - Q_{25}$ for the sample in Table 24.2.

17. Do the same tasks as in Probs. 15 and 16 for the sample in Table 24.3.

18. **(Mode)** A *mode* of a sample is a sample value that occurs most frequently in the sample. Find the mean, median, and mode of the following sample. Comment.

Total market value of stocks owned ($):	100	1000	100,000
Absolute frequency (= number of owners):	100	90	20

19. Compute the mean and the variance of the ungrouped sample in Table 24.3 (Sec. 24.3) and the grouped sample in Table 24.4 and compare the results.

20. If a sample is being grouped, its mean will change, in general. Show that the change cannot exceed $L/2$ where L is the length of each class interval.

24.5 Estimation of Parameters

Beginning in this section, we shall discuss the most basic practical tasks in statistics and corresponding statistical methods. The first of them is point estimation of **parameters,** that is, quantities appearing in distributions, such as p in the binomial distribution and μ and σ in the normal distribution.

A **point estimate** of a parameter is a number (point on the real line), which is computed from a given sample and serves as an approximation of the unknown exact value of the parameter. An **interval estimate** is an interval ("*confidence interval*") obtained from a sample; such estimates will be considered in the next section. Estimation of parameters is of great practical importance in many applications.

As an approximation of the mean μ of a population we may take the mean $\bar{x}$ of a corresponding sample. This gives the estimate $\hat{\mu} = \bar{x}$ for μ, that is,

$$\hat{\mu} = \bar{x} = \frac{1}{n}(x_1 + \cdots + x_n) \tag{1}$$

where n is the size of the sample. Similarly, an estimate $\hat{\sigma}^2$ for the variance of a population is the variance s^2 of a corresponding sample, that is,

$$\hat{\sigma}^2 = s^2 = \frac{1}{n-1}\sum_{j=1}^{n}(x_j - \bar{x})^2. \tag{2}$$

Clearly, (1) and (2) are estimates of parameters for distributions in which μ or σ^2 appear explicitly as parameters, such as the normal and Poisson distributions. For the binomial distribution, $p = \mu/n$ [see (3) in Sec. 23.6]. From (1) we thus obtain for p the estimate

$$\hat{p} = \frac{\bar{x}}{n}. \tag{3}$$

We mention that (1) is a special case of the so-called **method of moments.** In this method the parameters to be estimated are expressed in terms of the moments of the distribution (see Sec. 23.5). In the resulting formulas those moments are replaced by the corresponding moments of the sample. This gives the estimates. Here the ***k*th moment of a sample** $x_1, \cdots, x_n$ is

$$m_k = \frac{1}{n} \sum_{j=1}^{n} x_j^{\,k}.$$

Maximum Likelihood Method

Another method for obtaining estimates is the so-called **maximum likelihood method** of R. A. Fisher [*Messenger Math.* **41** (1912), 155–160]. To explain it, we consider a discrete (or continuous) random variable X whose probability function (or density) $f(x)$ depends on a single parameter θ and take a corresponding sample of n independent values $x_1, \cdots, x_n$. Then in the discrete case the probability that a sample of size n consists precisely of those n values is

$$l = f(x_1)f(x_2) \cdots f(x_n). \tag{4}$$

In the continuous case the probability that the sample consists of values in the small intervals $x_i \leqq x \leqq x_i + \Delta x$ $(i = 1, 2, \cdots, n)$ is

$$f(x_1)\Delta x\, f(x_2)\Delta x \cdots f(x_n)\Delta x = l(\Delta x)^n. \tag{5}$$

Since $f(x_i)$ depends on θ, the function l depends on $x_1, \cdots, x_n$ and θ. We imagine $x_1, \cdots, x_n$ to be given and fixed. Then l is a function of θ, which is called the **likelihood function.** The basic idea of the maximum likelihood method is very simple, as follows. We choose that approximation for the unknown value of θ for which l is as large as possible. If l is a differentiable function of θ, a necessary condition for l to have a maximum (not at the boundary) is

$$\frac{\partial l}{\partial \theta} = 0. \tag{6}$$

(We write a *partial* derivative, because l depends also on $x_1, \cdots, x_n$.) A solution of (6) depending on $x_1, \cdots, x_n$ is called a **maximum likelihood estimate** for θ. We may replace (6) by

$$\frac{\partial \ln l}{\partial \theta} = 0, \tag{7}$$

because $f(x_j) > 0$, a maximum of l is in general positive, and $\ln l$ is a monotone increasing function of l. This often simplifies calculations.

Several parameters. If the distribution of X involves r parameters $\theta_1, \cdots, \theta_r$, then instead of (6) we have the r conditions $\partial l/\partial\theta_1 = 0, \cdots, \partial l/\partial\theta_r = 0$, and instead of (7) we have

$$(8) \qquad \frac{\partial \ln l}{\partial \theta_1} = 0, \qquad \cdots, \qquad \frac{\partial \ln l}{\partial \theta_r} = 0.$$

EXAMPLE 1 **Normal distribution**

Find maximum likelihood estimates for μ and σ in the case of the normal distribution.

Solution. From (1), Sec. 23.7, and (4) we obtain

$$l = \left(\frac{1}{\sqrt{2\pi}}\right)^n \left(\frac{1}{\sigma}\right)^n e^{-h} \qquad \text{where} \qquad h = \frac{1}{2\sigma^2}\sum_{i=1}^{n}(x_i - \mu)^2.$$

Taking logarithms, we have

$$\ln l = -n \ln \sqrt{2\pi} - n \ln \sigma - h.$$

The first equation in (8) is $\partial \ln l/\partial\mu = 0$, written out

$$\frac{\partial \ln l}{\partial \mu} = -\frac{\partial h}{\partial \mu} = \frac{1}{\sigma^2}\sum_{i=1}^{n}(x_i - \mu) = 0, \qquad \text{hence} \qquad \sum_{i=1}^{n} x_i - n\mu = 0.$$

The solution is the desired estimate $\hat{\mu}$ for μ; we find

$$\hat{\mu} = \frac{1}{n}\sum_{i=1}^{n} x_i = \bar{x}.$$

The second equation in (8) is $\partial \ln l/\partial\sigma = 0$, written out

$$\frac{\partial \ln l}{\partial \sigma} = -\frac{n}{\sigma} - \frac{\partial h}{\partial \sigma} = -\frac{n}{\sigma} + \frac{1}{\sigma^3}\sum_{i=1}^{n}(x_i - \mu)^2 = 0.$$

Replacing μ by $\hat{\mu}$ and solving for σ^2, we obtain the estimate

$$\tilde{\sigma}^2 = \frac{1}{n}\sum_{i=1}^{n}(x_i - \bar{x})^2$$

which we shall use in Sec. 24.10. Note that this differs from (2). We cannot discuss criteria for the goodness of estimates, but want to mention that for small n, formula (2) is preferable. ▮

Problem Set 24.5

1. Find the maximum likelihood estimate for the parameter μ of a normal distribution with known variance $\sigma^2 = \sigma_0^2$.
2. Apply the maximum likelihood method to the normal distribution with $\mu = 0$.
3. **(Binomial distribution)** Derive a maximum likelihood estimate for p.
4. Extend Prob. 3 as follows. Suppose that m times n trials were made and in the first n trials A happened k_1 times, in the second n trials A happened k_2 times, $\cdots$, in the mth n trials A happened k_m times. Find a maximum likelihood estimate of p based on this information.

5. Suppose that in Prob. 4 we made 3 times 4 trials and A happened 2, 3, 2 times, respectively. Estimate p.
6. Consider X = *Number of independent trials until an event A occurs*. Show that X has the probability function $f(x) = pq^{x-1}$, $x = 1, 2, \cdots$, where p is the probability of A in a single trial and $q = 1 - p$. Find the maximum likelihood estimate of p corresponding to a single observed value x of X.
7. In Prob. 6, find the maximum likelihood estimate of p resulting from a sample $x_1, \cdots, x_n$.
8. In rolling a die, suppose that we get the first Six in the 7th trial and in doing it again we get it in the 6th trial. Estimate the probability p of getting a Six in rolling that die once.

9. **(Poisson distribution)** Apply the maximum likelihood method to the Poisson distribution.
10. **(Uniform distribution)** Show that in the case of the parameters a and b of the uniform distribution (see Sec. 23.5), the maximum likelihood estimate cannot be obtained by equating the first derivative to zero. How can we obtain maximum likelihood estimates in this case?
11. Find the maximum likelihood estimate of θ in the density $f(x) = \theta e^{-\theta x}$ if $x \geqq 0$ and $f(x) = 0$ if $x < 0$.
12. In Prob. 11, find the mean μ, substitute it in $f(x)$, find the maximum likelihood estimate of μ, and show that it is identical with the estimate for μ which can be obtained from that for θ in Prob. 11.
13. Compute $\hat{\theta}$ in Prob. 11 from the sample 1.9, 0.4, 0.7, 0.6, 1.4. Graph the sample distribution function $\hat{F}(x)$ and the distribution function $F(x)$ of the random variable, with $\theta = \hat{\theta}$, on the same axes. Do they agree reasonably well? (We consider goodness of fit systematically in Sec. 24.10.)
14. Do the same task as in Prob. 13 if the given sample is 0.4, 0.7, 0.2, 1.1, 0.1.
15. Using Table A9 in Appendix 5 and the method explained in Prob. 6 of Sec. 24.2, obtain a sample of size 20 from the distribution with density $f(x) = 0.5e^{-0.5x}$ if $x > 0$, $f(x) = 0$ if $x < 0$, and compare the distribution functions of the sample and of the population (as in Prob. 13).

24.6 Confidence Intervals

The last section was devoted to point estimates of parameters, and we shall now discuss **interval estimates,** starting with a general motivation.

Whenever we use mathematical approximation formulas, we should try to find out how much the approximate value can at most deviate from the unknown true value. For example, in the case of numerical integration methods there exist "error formulas" from which we can compute the maximum possible error (that is, the difference between the true value and the approximate value). Suppose that in a certain case we obtain 2.47 as an approximate value of a given integral and ± 0.02 as the maximum possible deviation from the unknown exact value. Then we are sure that the values $2.47 - 0.02 = 2.45$ and $2.47 + 0.02 = 2.49$ "*include*" the unknown exact value; that is, 2.45 is smaller than or equal to that value and 2.49 is larger than or equal to that value.

In estimating a parameter θ, the corresponding problem would be the determination of two numerical quantities that depend on the sample values and include the unknown value of the parameter with certainty. However, we already know that from a sample we cannot draw conclusions about the corresponding population that are 100% certain. So we have to be more modest and modify our task, as follows.

Choose a probability γ close to 1 (for example, $\gamma = 95\%, 99\%$, or the like). Then determine two quantities Θ_1 and Θ_2 such that the probability that Θ_1 and Θ_2 include the exact unknown value of the parameter θ is equal to γ.

Here the idea is that we replace the impossible requirement "with certainty" by the attainable requirement "with a preassigned probability close to 1."

Numerical values of those two quantities should be computed from a given sample $x_1, \cdots, x_n$. For this we regard the n sample values as observed values of n random variables $X_1, \cdots, X_n$ (with the same distribution). This is the crucial idea. Then Θ_1 and Θ_2 are functions of these random variables and therefore random variables, too. Our requirement may thus be written

$$P(\Theta_1 \leqq \theta \leqq \Theta_2) = \gamma.$$

If we know such functions Θ_1 and Θ_2 and a sample is given, we may compute from it a value θ_1 of Θ_1 and a value θ_2 of Θ_2. The interval with endpoints θ_1 and θ_2 is called a **confidence interval**[1] or *interval estimate* for the unknown parameter θ, and we shall denote it by

$$\text{CONF}\,\{\theta_1 \leqq \theta \leqq \theta_2\}.$$

The values θ_1 and θ_2 are called **lower** and **upper confidence limits** for θ. The number γ is called the **confidence level.** One chooses $\gamma = 95\%, 99\%$, sometimes 99.9%.

Clearly, if we intend to obtain a sample and determine a corresponding confidence interval, then γ is the probability of getting an interval that will include the unknown exact value of the parameter.

For example, if we choose $\gamma = 95\%$, then we can expect that *about* 95% of the samples that we may obtain will yield confidence intervals that do include the value θ, whereas the remaining 5% do not. Hence the statement "the confidence interval includes θ" will be correct in *about* 19 out of 20 cases, while in the remaining case it will be false.

Choosing $\gamma = 99\%$ instead of 95%, we may expect that statement to be correct even in *about* 99 out of 100 cases. But we shall see later that the intervals corresponding to $\gamma = 99\%$ are longer than those corresponding to $\gamma = 95\%$. This is the price to pay for an increase of γ.

---

[1]The modern theory and terminology of confidence intervals were developed by J. Neyman [*Annals of Mathematical Statistics* **6** (1935), 111–116]. See also footnote 3 in the next section.
In mathematics, $\theta_1 \leqq \theta \leqq \theta_2$ means that θ lies between θ_1 and θ_2, and, to avoid misunderstandings, it seems worthwhile to characterize a confidence interval by a special symbol, such as CONF.

Choice of γ. What γ should we choose in a concrete case? This is not a mathematical question but one to be answered from the viewpoint of the application by considering the affordable risk of making a false decision—is it a "problem" of getting wet or taking an umbrella, or is it a medical matter of life or death?

It is clear that the uncertainty involved in the present method as well as in the methods yet to be discussed comes from the sampling process, so that the statistician should be prepared for his share of mistakes. However, he is no worse off than a judge or a banker, who is subject to the laws of chance. Quite the contrary, he has the advantage that he can *measure* his chances of making a mistake.

Normal Distribution: Confidence Intervals for μ and σ^2

We shall now consider methods for obtaining confidence intervals for the mean (Tables 24.5, 24.6) and the variance (Table 24.7) of the **normal distribution.** The corresponding theory will be explained in the last part of this section.

EXAMPLE 1 **Confidence interval for μ of the normal distribution with known σ^2**

Determine a 95% confidence interval for the mean of a normal distribution with variance $\sigma^2 = 9$, using a sample of $n = 100$ values with mean $\bar{x} = 5$.

Solution. 1st Step. $\gamma = 0.95$ is required.

2nd Step. The corresponding c equals 1.960; see Table 24.5.

3rd Step. $\bar{x} = 5$ is given.

4th Step. We need $k = 1.960 \cdot 3/\sqrt{100} = 0.588$. Hence $\bar{x} - k = 4.412$, $\bar{x} + k = 5.588$ and the confidence interval is

$$\text{CONF}\,\{4.412 \leqq \mu \leqq 5.588\}.$$

Sometimes this is written $\mu = 5 \pm 0.588$, but we shall not use this notation, which can be misleading. ■

Table 24.5
Determination of a Confidence Interval for the Mean μ of a Normal Distribution with Known Variance σ^2

1st Step. Choose a confidence level γ (95%, 99%, or the like).

2nd Step. Determine the corresponding c:

γ	0.90	0.95	0.99	0.999
c	1.645	1.960	2.576	3.291

3rd Step. Compute the mean $\bar{x}$ of the sample $x_1, \cdots, x_n$.

4th Step. Compute $k = c\sigma/\sqrt{n}$. The confidence interval for μ is

$$\text{CONF}\,\{\bar{x} - k \leqq \mu \leqq \bar{x} + k\}. \tag{1}$$

EXAMPLE 2 Sample size needed for a confidence interval of prescribed length

How large must n in the last example be if we want to obtain a 95% confidence interval of length $L = 0.4$?

Solution. The interval (1) has the length $L = 2k = 2c\sigma/\sqrt{n}$. Solving for n, we obtain

$$n = (2c\sigma/L)^2.$$

In the present case the answer is $n = (2 \cdot 1.960 \cdot 3/0.4)^2 \approx 870$.

Figure 506 shows how L decreases as n increases and that for $\gamma = 99\%$ the confidence interval is substantially longer than for $\gamma = 95\%$ (and the same sample size n). ▮

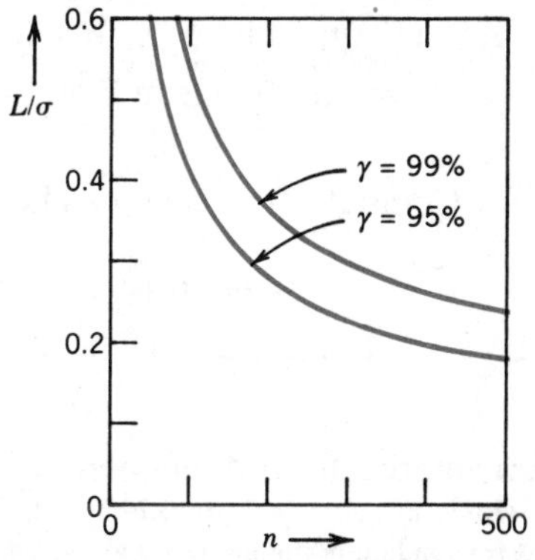

Fig. 506. Length of the confidence interval (1) (measured in multiples of σ) as a function of the sample size n for $\gamma = 95\%$ and $\gamma = 99\%$

Confidence interval for the mean if the variance is unknown. So far we have assumed the variances to be known. If this is no longer the case, as in most applications, the whole theory changes, although the practical technicalities remain quite similar. Now using less information (σ^2 is not known), we should expect somewhat longer intervals (which is true). We may hope that the sample variance will help us out (which is also true). Table 24.6 on p. 1226 shows the steps. k differs from that in Table 24.5 and c now depends on n and must be determined from Table A10 in Appendix 5. That table contains values z corresponding to given values of the distribution function

$$F(z) = K_m \int_{-\infty}^{z} \left(1 + \frac{u^2}{m}\right)^{-(m+1)/2} du$$

of the so-called ***t*-distribution** of Student.[2] Here, m $(= 1, 2, \cdots)$ is a parameter, called the **number of degrees of freedom** of the distribution. The constant K_m is such that $F(\infty) = 1$. By integration it turns out that $K_m = \Gamma(\frac{1}{2}m + \frac{1}{2})/[\sqrt{m\pi}\,\Gamma(\frac{1}{2}m)]$, where Γ is the gamma function (see (24) in Appendix 3).

[2]Pseudonym for WILLIAM SEALY GOSSET (1876—1937), English statistician, who discovered the t-distribution in 1907–1908.

Table 24.6
Determination of a Confidence Interval for the Mean μ of a Normal Distribution with Unknown Variance σ^2

1st Step. Choose a confidence level γ (95%, 99%, or the like).

2nd Step. Determine the solution c of the equation

$$F(c) = \tfrac{1}{2}(1 + \gamma) \tag{2}$$

from the table of the t-distribution with $n - 1$ degrees of freedom (Table A10 in Appendix 5; n = sample size).

3rd Step. Compute the mean $\bar{x}$ and the variance s^2 of the sample $x_1, \cdots, x_n$.

4th Step. Compute $k = sc/\sqrt{n}$. The confidence interval is

$$\text{CONF}\ \{\bar{x} - k \leqq \mu \leqq \bar{x} + k\}. \tag{3}$$

EXAMPLE 3 Confidence interval for μ of the normal distribution with unknown σ^2

Using the sample in Table 24.2, Sec. 24.3, determine a 99% confidence interval for the mean μ of the corresponding population, assuming that the population is normal. (This assumption will be justified in Sec. 24.10.)

Solution. 1st Step. $\gamma = 0.99$ is required.

2nd Step. Since $n = 100$, we obtain $c = 2.63$.

3rd Step. Computation gives $\bar{x} = 364.70$ and $s = \sqrt{720.1} = 26.83$.

4th Step. We find $k = 26.83 \cdot 2.63/10 = 7.06$. Hence the confidence interval is

$$\text{CONF}\ \{357.64 \leqq \mu \leqq 371.76\}.$$

For comparison, if σ were known and equal to 26.83, Table 24.5 would give the value $k = 2.576 \cdot 26.83/\sqrt{100} = 6.91$, and CONF $\{357.79 \leqq \mu \leqq 371.61\}$. This differs but little from the preceding result because n is large. For smaller n the difference would be considerable, as Fig. 507 illustrates. ∎

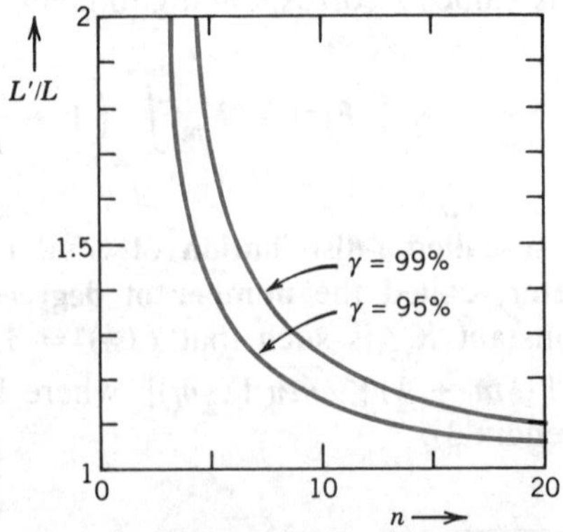

Fig. 507. Ratio of the lengths L' and L of the confidence intervals (3) and (1) with γ = 95% and γ = 99% as a function of the sample size n for equal s and σ

Confidence interval for the variance of the normal distribution. This is our third task. Table 24.7 shows the steps, which are similar to those in Tables 24.5 and 24.6, but we now have to determine two numbers c_1 and c_2. Both are obtained from Table A11 in Appendix 5, which contains values z corresponding to given values of the distribution function $F(z) = 0$ if $z < 0$ and

$$F(z) = C_m \int_0^z e^{-u/2} u^{(m-2)/2} \, du \quad \text{if} \quad z \geqq 0.$$

This is the distribution function of the so-called **χ^2-distribution** (*chi-square distribution*); here, m $(= 1, 2, \cdots)$ is a parameter, called the **number of degrees of freedom** of the distribution, and $C_m = 1/[2^{m/2}\Gamma(\frac{1}{2}m)]$.

EXAMPLE 4 Confidence interval for the variance of the normal distribution

Using the sample in Table 24.2, Sec. 24.3, determine a 95% confidence interval for the variance of the corresponding population.

Solution. 1st Step. $\gamma = 0.95$ is required.

2nd Step. For $n - 1 = 99$ we find by linear interpolation $c_1 = 73.3$ and $c_2 = 128$.

3rd Step. From Table 24.2 we compute $99s^2 = 71\,291$.

4th Step. The confidence interval is

$$\text{CONF}\,\{556 \leqq \sigma^2 \leqq 973\}.$$

Other distributions. Confidence intervals for the mean and variance of other distributions may be obtained by using the previous methods and ***sufficiently large samples.*** *Practically* speaking, if the sample indicates that the skewness of the unknown distribution is small, one should take samples of size $n = 20$ at least for obtaining confidence intervals for μ and samples of size $n = 50$ at least for obtaining confidence intervals for σ^2. The reason for this method will be explained at the end of this section.

Table 24.7
Determination of a Confidence Interval for the Variance σ^2 of a Normal Distribution, Whose Mean Need Not Be Known

1st Step. Choose a confidence level γ (95%, 99%, or the like).

2nd Step. Determine solutions c_1 and c_2 of the equations

$$F(c_1) = \tfrac{1}{2}(1 - \gamma), \qquad F(c_2) = \tfrac{1}{2}(1 + \gamma) \tag{4}$$

from the table of the chi-square distribution with $n - 1$ degrees of freedom (Table A11 in Appendix 5; n = sample size).

3rd Step. Compute $(n - 1)s^2$, where s^2 is the variance of the sample $x_1, \cdots, x_n$.

4th Step. Compute $k_1 = (n - 1)s^2/c_1$ and $k_2 = (n - 1)s^2/c_2$. The confidence interval is

$$\text{CONF}\,\{k_2 \leqq \sigma^2 \leqq k_1\}. \tag{5}$$

Theory for Table 24.5

We shall now discuss the theory that justifies our methods for obtaining confidence intervals, using the following simple but very important idea.

So far we have regarded the values $x_1, \cdots, x_n$ of a sample as n observed values of a single random variable X. We may equally well regard these n values as single observations of n random variables $X_1, \cdots, X_n$ that have the same distribution (the distribution of X) and are independent because the sample values are assumed to be independent.

For deriving (1) in Table 24.5 we need

Theorem 1 **(Sum of independent normal random variables)**
Suppose that $X_1, X_2, \cdots, X_n$ are independent normal random variables with means $\mu_1, \mu_2, \cdots, \mu_n$ and variances $\sigma_1^2, \sigma_2^2, \cdots, \sigma_n^2$, respectively. Then the random variable

$$X = X_1 + X_2 + \cdots + X_n$$

is normal with the mean

$$\mu = \mu_1 + \mu_2 + \cdots + \mu_n$$

and the variance

$$\sigma^2 = \sigma_1^2 + \sigma_2^2 + \cdots + \sigma_n^2.$$

The statements about μ and σ follow directly from Theorems 1 and 3 in Sec. 23.8. The proof that X is normal can be found in Ref. [G3] listed in Appendix 1.

From this theorem, Theorem 1 in Sec. 23.7, and Theorem 2(b) in Sec. 23.5 we obtain the following theorem about the distribution of the sum of independent identically distributed normal random variables.

Theorem 2 *If $X_1, \cdots, X_n$ are independent normal random variables each of which has mean μ and variance σ^2, then the random variable*

$$\bar{X} = \frac{1}{n}(X_1 + \cdots + X_n) \tag{6}$$

is normal with the mean μ and the variance σ^2/n, and the random variable

$$Z = \sqrt{n}\,\frac{\bar{X} - \mu}{\sigma} \tag{7}$$

is normal with the mean 0 and the variance 1.

Derivation of (1). From the general motivation at the beginning of this section we recall that our goal is to find two random variables Θ_1 and Θ_2 such that

$$P(\Theta_1 \leqq \mu \leqq \Theta_2) = \gamma, \tag{8}$$

where γ is chosen, and the sample gives observed values θ_1 of Θ_1 and θ_2 of Θ_2, which then yield a confidence interval CONF $\{\theta_1 \leqq \mu \leqq \theta_2\}$. In the present case this can be done as follows. We choose a number γ between 0 and 1 and determine c from Table A8, Appendix 5, such that $P(-c \leqq Z \leqq c) = \gamma$. (If $\gamma = 0.90$, etc., we obtain the values c in Table 24.5.) The inequality $-c \leqq Z \leqq c$ with Z given by (7) is

$$-c \leqq \sqrt{n}\,(\bar{X} - \mu)/\sigma \leqq c$$

and can be transformed into an inequality for μ. In fact, multiplication by $\sigma/\sqrt{n}$ yields $-k \leqq \bar{X} - \mu \leqq k$, where $k = c\sigma/\sqrt{n}$. Multiplying by -1 and adding $\bar{X}$, we get

$$\bar{X} + k \geqq \mu \geqq \bar{X} - k. \tag{9}$$

Thus $P(-c \leqq Z \leqq c) = \gamma$ is equivalent to $P(\bar{X} - k \leqq \mu \leqq \bar{X} + k) = \gamma$. This is of the form (8) with $\Theta_1 = \bar{X} - k$ and $\Theta_2 = \bar{X} + k$. Under our assumptions it means that with probability γ the random variables $\bar{X} - k$ and $\bar{X} + k$ will assume values that include the unknown mean μ. Regarding the sample values $x_1, \cdots, x_n$ in Table 24.5 as observed values of n independent normal random variables $X_1, \cdots, X_n$, we see that the sample mean $\bar{x}$ is an observed value of (6), and by inserting this value into (9) we obtain (1). ∎

Theory for Table 24.6

For deriving (3) in Table 24.6 we need

Theorem 3 *Let $X_1, \cdots, X_n$ be independent normal random variables with the same mean μ and the same variance σ^2. Then the random variable*

$$T = \sqrt{n}\,\frac{\bar{X} - \mu}{S} \tag{10}$$

has a t-distribution (see p. 1225) *with $n - 1$ degrees of freedom; here $\bar{X}$ is given by* (6) *and*

$$S^2 = \frac{1}{n-1}\sum_{j=1}^{n}(X_j - \bar{X})^2. \tag{11}$$

Proof in Ref. [G3] in Appendix 1.

Derivation of (3). This is similar to the derivation of (1). We choose a number γ between 0 and 1 and determine a number c from Table A10, Appendix 5, with $n - 1$ degrees of freedom such that

$$P(-c \leqq T \leqq c) = F(c) - F(-c) = \gamma. \tag{12}$$

Since the t-distribution is symmetric, we have $F(-c) = 1 - F(c)$, and (12) assumes the form (2). Transforming $-c \leqq T \leqq c$ in (12) as before, we get

$$\bar{X} - K \leqq \mu \leqq \bar{X} + K \qquad \text{where} \qquad K = cS/\sqrt{n}, \tag{13}$$

and (12) becomes $P(\bar{X} - K \leqq \mu \leqq \bar{X} + K) = \gamma$. By inserting the observed values $\bar{x}$ of $\bar{X}$ and s^2 of S^2 into (13) we obtain (3). ∎

Theory for Table 24.7

For deriving (5) in Table 24.7 we need

Theorem 4 *Under the assumptions in Theorem 3 the random variable*

$$Y = (n - 1)\frac{S^2}{\sigma^2} \tag{14}$$

with S^2 given by (11) *has a chi-square distribution* (see p. 1227) *with $n - 1$ degrees of freedom.*

Proof in Ref. [G3] in Appendix 1.

Derivation of (5). This is similar to the derivation of (1) and (3). We choose a number γ between 0 and 1 and determine c_1 and c_2 from Table A11, Appendix 5, such that [see (4)]

$$P(Y \leqq c_1) = F(c_1) = \tfrac{1}{2}(1 - \gamma), \qquad P(Y \leqq c_2) = F(c_2) = \tfrac{1}{2}(1 + \gamma).$$

Subtraction yields

$$P(c_1 \leqq Y \leqq c_2) = P(Y \leqq c_2) - P(Y \leqq c_1) = \gamma.$$

Transforming $c_1 \leqq Y \leqq c_2$ with Y given by (14) into an inequality for σ^2, we obtain

$$\frac{n-1}{c_2} S^2 \leqq \sigma^2 \leqq \frac{n-1}{c_1} S^2.$$

By inserting the observed value s^2 of S^2 we obtain (5). ∎

Confidence Intervals for Other Distributions

In the case of other distributions we may also obtain confidence intervals by the methods in Tables 24.5 and 24.7, but in this case we must use large samples. This follows from the basic

Theorem 5 (Central limit theorem)

Let $X_1, \cdots, X_n, \cdots$ be independent random variables that have the same distribution function and therefore the same mean μ and the same variance σ^2. Let $Y_n = X_1 + \cdots + X_n$. Then the random variable

$$Z_n = \frac{Y_n - n\mu}{\sigma\sqrt{n}} \tag{15}$$

*is **asymptotically normal** with mean 0 and variance 1; that is, the distribution function $F_n(x)$ of Z_n satisfies*

$$\lim_{n\to\infty} F_n(x) = \Phi(x) = \frac{1}{\sqrt{2\pi}} \int_{-\infty}^{x} e^{-u^2/2}\, du.$$

A proof can be found in Ref. [G3] listed in Appendix 1.

We know that if $X_1, \cdots, X_n$ are independent random variables with the same mean μ and the same variance σ^2, then their sum $X = X_1 + \cdots + X_n$ has the following properties.

(A) X has the mean $n\mu$ and the variance $n\sigma^2$ (by Theorems 1 and 3 in Sec. 23.8).

(B) If those variables are normal, then X is normal (by Theorem 1).

If those variables are not normal, then (B) fails to hold, but if n is large, then X is approximately normal (see Theorem 5) and this justifies the application of methods for the normal distribution to other distributions, but in such a case we have to use large samples.

Problem Set 24.6

1. Why are interval estimates in most cases more useful than point estimates?
2. Find a 95% confidence interval for the mean μ of a normal population with standard deviation 5.00, using the sample 32, 24, 20, 38, 30.
3. Determine a 99% confidence interval for the mean of a normal population with standard deviation 2.2, using the sample 28, 24, 31, 27, 22.
4. Determine a 95% confidence interval for the mean μ of a normal population with variance $\sigma^2 = 9$, using a sample of size 100 with mean 38.25.
5. What will happen to the length of the interval in Prob. 4 if we reduce the sample size to 25?

6. Obtain a 99% confidence interval for the mean of a normal population with variance $\sigma^2 = 0.36$ from Fig. 506, using a sample of size 290 with mean 16.30. (This problem should merely help you to understand the meaning of the figure.)

7. What sample size would be needed to produce a 95% confidence interval (1) of length (a) 2σ, (b) σ?

Assuming that the populations from which the following samples are taken are normal, determine a 99% confidence interval for the mean μ of the population.

8. 325, 320, 325, 335

9. A sample of lengths of 20 bolts with mean 10.20 cm and variance 0.04 cm^2.

10. Flash point (°F) of Diesel oil (2-D) 124, 127, 126, 122, 124

11. Find a 95% confidence interval for the percentage of cars on a certain highway that have poorly adjusted brakes, using a random sample of 500 cars stopped at a roadblock on that highway, 87 of which had poorly adjusted brakes.

12. Find a 99% confidence interval for the parameter p of the binomial distribution, using Pearson's result in the last row of Table 23.1 in Sec. 23.2.

Assuming that the populations from which the following samples are taken are normal, determine a 95% confidence interval for the variance σ^2 of the population.

13. A sample of size 30 with variance 0.0007

14. Ultimate tensile strength (kpsi) of alloy steel (Maraging H) at room temperature: 251, 255, 258, 253, 253, 252, 250, 252, 255, 256

15. Mean energy (keV) of delayed neutron group (Group 3, half-life 6.2 sec.) for uranium U^{235} fission: 435, 451, 430, 444, 438

16. Carbon monoxide emission (grams per mile) of a certain type of passenger car (cruising at 55 mph): 17.3, 17.8, 18.0, 17.7, 18.2, 17.4, 17.6, 18.1

17. If X is normal with mean 27 and variance 16, what distributions do $-X$, $3X$, and $5X - 2$ have?

18. If X_1 and X_2 are independent normal random variables with mean 23 and 4 and variance 3 and 1, respectively, what distribution does $4X_1 - X_2$ have?

19. A machine fills boxes weighing Y lb with X lb of salt, where X and Y are normal with mean 100 lb and 5 lb and standard deviation 1 lb and 0.5 lb, respectively. What percent of filled boxes weighing between 104 lb and 106 lb are to be expected?

20. If the weight X of bags of cement is normally distributed with a mean of 40 kg and a standard deviation of 2 kg, how many bags can a delivery truck carry so that the probability of the total load exceeding 2000 kg will be 5%?

24.7 Testing of Hypotheses, Decisions

The idea of confidence intervals that we have just discussed and applied is characteristic of modern statistics, and the other equally (or even more) important idea is that of tests of hypotheses. Here, a statistical **hypothesis** is an assumption about the distribution of a random variable, for example, that a certain distribution has mean 20.3, etc. A statistical **test** of a hypothesis is a procedure in which a sample is used to find out whether we may **"not reject"** (**"accept"**) the hypothesis, that is, act as though it is true, or whether we should **"reject"** it, that is, act as though it is false.

Tests are applied quite frequently, and we may first ask why they are important. Well, we often have to make decisions in situations where chance variation plays a role. If we have a choice between, say, two possibilities, the decision may often be based on the result of some test.[3]

For example, if we want to use a certain lathe for producing bolts whose diameter should lie between given limits and we allow at most 2% defective bolts, we may take a sample of 100 bolts produced on that lathe and use it for testing the hypothesis $\sigma^2 = \sigma_0^2$ that the variance σ^2 of the corresponding population has a certain value σ_0^2, which we choose such that we can expect to obtain not more than 2% defectives. A meaningful **alternative** in this case is $\sigma^2 > \sigma_0^2$. Depending on the result of the test, we either do not reject the hypothesis $\sigma^2 = \sigma_0^2$ (and then use that lathe) or we reject is, assert that $\sigma^2 > \sigma_0^2$, and use a better lathe. In the latter case we say that the test indicates a **significant deviation** of σ^2 from σ_0^2, that is, a deviation that is not caused merely by the unavoidable influence of chance factors but by the lack of precision of the lathe.

In other cases we may want to compare two things, for example, two different medications, two methods of performing a certain work, the accuracy of two methods of measurement, the quality of products produced by two different tools, etc. Depending on the result of a suitable test, we decide to use one of the two medications, to introduce the better method of working, etc.

Typical sources for hypotheses are as follows.

1. The hypothesis comes from a quality requirement. (Experience about attainable quality may be gained by producing a larger number of items with special care.)

2. The hypothesis is based on values known from previous experience.

3. The hypothesis results from a theory one wants to verify.

4. The hypothesis is a pure guess caused by occasional observations.

Let us start with a simple introductory example, which will illustrate the basic ideas.

EXAMPLE 1 Test of a hypothesis

The birth of a single child may be regarded as an experiment with two possible outcomes, namely, B: *Birth of a boy* and G: *Birth of a girl*. Intuitively, we should feel that both outcomes are about equally likely. However, in the literature it is often claimed that births of boys are somewhat more frequent than births of girls. On the basis of this situation we want to test the hypothesis that the two outcomes B and G have the same probability. If we let p denote the probability of the outcome B, our hypothesis to be tested is $p = 50\% = 0.5$. Because of those claims we choose the alternative $p > 0.5$.

For the test we use a sample of $n = 3000$ babies from a city of about 250 000 inhabitants; 1578 of these babies were boys.

If the hypothesis is true, we expect that in a sample of $n = 3000$ births there will be *about* 1500 boys. If the alternative holds, then we expect more than 1500 boys, on the average. Hence if the number of boys actually observed is much larger than 1500, we can use this as an indication that the hypothesis may be false, and we reject it.

---

[3]A systematic theory of tests was developed by the American statistician JERZY NEYMAN (1894—1981) and the English statistician EGON SHARPE PEARSON (1895—1980), the son of Karl Pearson (see footnote 6), beginning around 1930.

To perform our test, we proceed as follows. We first determine a critical value c. Because of the alternative, c will be greater than 1500. (A method for determining c will be given below.) Then, if the observed number of boys is greater than c, we reject the hypothesis. If that number is not greater than c, we do not reject the hypothesis.

The basic question now is how we should choose c, that is, where we should draw the line between small random deviations and large significant deviations. Different people may have different opinions, and to answer the question we must use mathematical arguments. In the present case these are very simple, as we now show.

We determine c such that if the hypothesis is true, the probability of observing more than c boys in a sample of 3000 single births is a very small number, call it α. It is customary to choose $\alpha = 1\%$ or 5%. Choosing $\alpha = 1\%$ (or 5%) we risk about once in 100 cases (in 20 cases, respectively) rejecting a hypothesis even though it is true. We shall return to this point later. Let us choose $\alpha = 1\%$ and consider the random variable

$$X = \textit{Number of boys in 3000 births}.$$

Assuming that the hypothesis is true, we obtain the critical value c from the equation

$$P(X > c)_{p=0.5} = \alpha = 0.01. \tag{1}$$

(That assumption is indicated by the subscript $p = 0.5$.) If the observed value 1578 is greater than c we reject the hypothesis. If $1578 \leqq c$, we do not reject the hypothesis.

To determine c from (1) we must know the distribution of X. For our purpose the binomial distribution is a sufficiently accurate model. Hence if the hypothesis is true, X has a binomial distribution with $p = 0.5$ and $n = 3000$. This distribution can be approximated by the normal distribution with mean $\mu = np = 1500$ and variance $\sigma^2 = npq = 750$; see Sec. 23.7. [For the sake of simplicity we shall disregard the term 0.5 in (11), Sec. 23.7.] The curve of the density is shown in Fig. 508. Using (1), we thus obtain

$$P(X > c) = 1 - P(X \leqq c) \approx 1 - \Phi\left(\frac{c - 1500}{\sqrt{750}}\right) = 0.01.$$

Table A8 in Appendix 5 gives $(c - 1500)/\sqrt{750} = 2.326$. Hence $c = 1564$. Since $1578 > c$, we reject the hypothesis and assert that $p > 0.5$, that is, we assert that births of boys are more frequent than births of girls. This completes the test. ∎

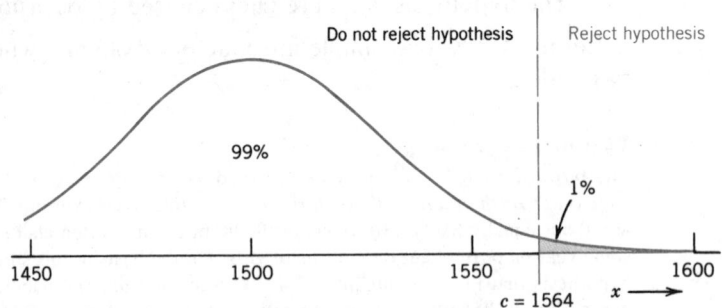

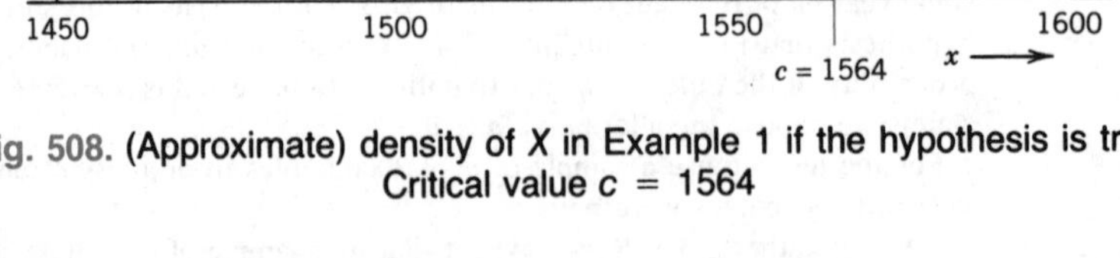

Fig. 508. (Approximate) density of X in Example 1 if the hypothesis is true. Critical value $c = 1564$

Standard Terms

The **hypothesis** to be tested is sometimes called the **null hypothesis,** and a counterassumption (such as $p > 0.5$ in Example 1) is called an *alternative hypothesis* or, briefly, an **alternative.** The number α (or $100\alpha\%$) is called the **significance level** of the test and c the **critical value.** The region containing the values for which we reject the hypothesis is called the **rejection region** or **critical region.** The region of values for which we do not reject the hypothesis is called the **acceptance region.** A frequent choice for α is 5%.

Kinds of Alternatives (Fig. 509)

Let θ be an unknown parameter in a distribution, and suppose that we want to test the hypothesis $\theta = \theta_0$. Then there are three main kinds of alternatives, namely,

(2) $$\theta > \theta_0$$

(3) $$\theta < \theta_0$$

(4) $$\theta \neq \theta_0.$$

(2) and (3) are called **one-sided alternatives,** and (4) is called a **two-sided alternative.** (2) is of the type considered in Example 1 (where $\theta_0 = p = 0.5$ and $\theta = p > 0.5$); c lies to the right of θ_0 and the rejection region extends from c to ∞ (Fig. 509, upper part). This test is called a **right-sided test.** In the case of (3) the number c lies to the left of θ_0, the rejection region extends from c to $-\infty$ (Fig. 509, middle part) and the test is called a **left-sided test.** Tests of both kinds are called **one-sided tests.** In the case of (4) we have two critical values c_1 and c_2 ($> c_1$), the rejection region extends from c_1 to $-\infty$ and from c_2 to ∞, and the test is called a **two-sided test.**

All three kinds of alternatives are of practical importance. For example, (3) may appear in connection with testing strength of material. θ_0 may then be the required strength, and the alternative characterizes an undesirable weakness. The case that the material may be stronger than required is acceptable, of course, and therefore needs no special attention. (4) may be important, for example, in connection with the diameter of an axle shaft. Then θ_0 is the required diameter, and slimmer axle-shafts are as bad as thicker ones, so that one has to watch for undesirable deviations from θ_0 in both directions.

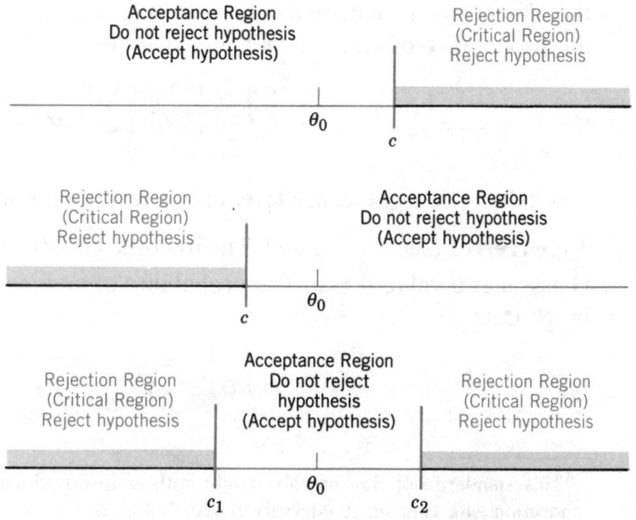

Fig. 509. Test in the case of alternative (2) (upper part of the figure), alternative (3) (middle part), and alternative (4)

Types of Errors in Tests

Tests always involve **risks of making false decisions:**

(I) Rejecting a true hypothesis **(Type I error),**

(II) Accepting a false hypothesis **(Type II error).**

This is clear because no absolutely certain conclusions about populations can be drawn from samples. We cannot *avoid* those errors, but we show next that we can calculate the probabilities α and β, respectively, of their occurrence; more important, there are ways and means to choose suitable levels of risks (that is, values of α and β) that we can afford depending on the nature of the problem. (Recall from Sec. 24.6: Is it a question of getting wet or not, or is it a problem of life or death?) Let us discuss this systematically for a test of a hypothesis $\theta = \theta_0$ against an alternative that is a single number[4] θ_1, for simplicity. We let $\theta_1 > \theta_0$, so that we have a right-sided test. For a left-sided or a two-sided test the discussion is quite similar.

We choose a critical $c > \theta_0$ (as in the upper part of Fig. 509, by methods discussed below). From a given sample $x_1, \cdots, x_n$ we then compute a value

$$\hat{\theta} = g(x_1, \cdots, x_n)$$

with a suitable g (whose choice will be a main point of our further discussion; for instance, take $g = (x_1 + \cdots + x_n)/n$ in the case in which θ is the mean). If $\hat{\theta} > c$, we reject the hypothesis. If $\hat{\theta} \leqq c$, we accept it. Here, the value $\hat{\theta}$ can be regarded as an observed value of the random variable

$$\hat{\Theta} = g(X_1, \cdots, X_n)$$

because x_j may be regarded as an observed value of X_j, $j = 1, \cdots, n$. In this test there are two possibilities of making an error, as follows.

Type I error (see Table 24.8). The hypothesis is true but is rejected (hence the alternative is accepted) because Θ assumes a value $\hat{\theta} > c$. Obviously, the probability of making such an error equals

$$P(\hat{\Theta} > c)_{\theta=\theta_0} = \alpha. \tag{5}$$

α is called the **significance level** of the test, as mentioned before.

Type II error (see Table 24.8). The hypothesis is false but is accepted because $\hat{\Theta}$ assumes a value $\hat{\theta} \leqq c$. The probability of making such an error is denoted by β; thus

$$P(\hat{\Theta} \leqq c)_{\theta=\theta_1} = \beta. \tag{6}$$

---

[4]This standard notation has absolutely nothing to do with the use of the notation θ_1 in connection with confidence intervals in Sec. 24.6.

Table 24.8
Type I and Type II Errors in Testing a Hypothesis $\theta = \theta_0$
Against an Alternative $\theta = \theta_1$

		Unknown Truth	
		$\theta = \theta_0$	$\theta = \theta_1$
Accepted	$\theta = \theta_0$	True decision $P = 1 - \alpha$	Type II error $P = \beta$
	$\theta = \theta_1$	Type I error $P = \alpha$	True decision $P = 1 - \beta$

$\eta = 1 - \beta$ is called the **power** of the test. Obviously, this is the probability of avoiding a Type II error.

Formulas (5) and (6) show that both α and β depend on c, and we would like to choose c so that these probabilities of making errors are as small as possible. But Fig. 510 shows that these are conflicting requirements because, to let α decrease we must shift c to the right, but then β increases. In practice we first choose α (5%, sometimes 1%), then determine c, and finally compute β. If β is large so that the power $\eta = 1 - \beta$ is small, we should repeat the test, choosing a larger sample, for reasons that will appear shortly.

If the alternative is not a single number but is of the form (2)–(4), then β becomes a function of θ. This function $\beta(\theta)$ is called the **operating characteristic** (OC) of the test and its curve the **OC curve.** Clearly, in this case $\eta = 1 - \beta$ also depends on θ, and this function $\eta(\theta)$ is called the **power function** of the test.

Of course, from a test that leads to the acceptance of a certain hypothesis θ_0, it does *not* follow that this is the only possible hypothesis or the best possible hypothesis. Hence the terms **"not reject"** or **"fail to reject"** are perhaps better than the term **"accept."**

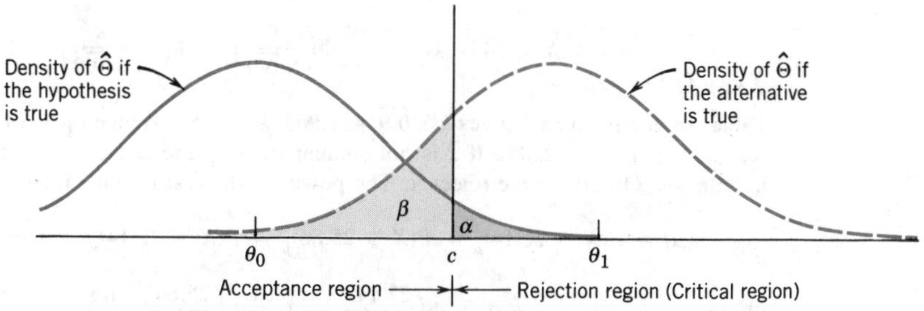

Fig. 510. Illustration of Type I and II errors in testing a hypothesis $\theta = \theta_0$ against an alternative $\theta = \theta_1$ ($> \theta_0$)

Tests in the Case of the Normal Distribution

The following examples will explain tests of practically important hypotheses.

EXAMPLE 2 Test for the mean of the normal distribution with known variance

Let X be a normal random variable with variance $\sigma^2 = 9$. Using a sample of size $n = 10$ with mean $\bar{x}$, test the hypothesis $\mu = \mu_0 = 24$ against the three kinds of alternatives, namely,

$$(a)\quad \mu > \mu_0 \qquad (b)\quad \mu < \mu_0 \qquad (c)\quad \mu \neq \mu_0.$$

Solution. We choose the significance level $\alpha = 0.05$. An estimate of the mean will be obtained from

$$\overline{X} = \frac{1}{n}(X_1 + \cdots + X_n).$$

If the hypothesis is true, $\overline{X}$ is normal with mean $\mu = 24$ and variance $\sigma^2/n = 0.9$, see Theorem 2, Sec. 24.6. Hence we may obtain the critical value c from Table A8 in Appendix 5.

Case (a). We determine c from $P(\overline{X} > c)_{\mu=24} = \alpha = 0.05$, that is,

$$P(\overline{X} \leqq c)_{\mu=24} = \Phi\left(\frac{c - 24}{\sqrt{0.9}}\right) = 1 - \alpha = 0.95.$$

Table A8 in Appendix 5 gives $(c - 24)/\sqrt{0.9} = 1.645$, and $c = 25.56$, which is greater than μ_0, as in the upper part of Fig. 509. If $\bar{x} \leqq 25.56$, the hypothesis is accepted. If $\bar{x} > 25.56$, it is rejected. The power of the test is (Fig. 511)

$$\eta(\mu) = P(\overline{X} > 25.56)_\mu = 1 - P(\overline{X} \leqq 25.56)_\mu$$

(7)
$$= 1 - \Phi\left(\frac{25.56 - \mu}{\sqrt{0.9}}\right) = 1 - \Phi(26.94 - 1.05\mu).$$

Case (b). The critical value c is obtained from the equation

$$P(\overline{X} \leqq c)_{\mu=24} = \Phi\left(\frac{c - 24}{\sqrt{0.9}}\right) = \alpha = 0.05.$$

Table A8 in Appendix 5 yields $c = 24 - 1.56 = 22.44$. If $\bar{x} \geqq 22.44$, we accept the hypothesis. If $\bar{x} < 22.44$, we reject it. The power of the test is

(8)
$$\eta(\mu) = P(\overline{X} \leqq 22.44)_\mu = \Phi\left(\frac{22.44 - \mu}{\sqrt{0.9}}\right) = \Phi(23.65 - 1.05\mu).$$

Case (c). Since the normal distribution is symmetric, we choose c_1 and c_2 equidistant from $\mu = 24$, say, $c_1 = 24 - k$ and $c_2 = 24 + k$, and determine k from

$$P(24 - k \leqq \overline{X} \leqq 24 + k)_{\mu=24} = \Phi\left(\frac{k}{\sqrt{0.9}}\right) - \Phi\left(-\frac{k}{\sqrt{0.9}}\right) = 1 - \alpha = 0.95.$$

Table A8 in Appendix 5 gives $k/\sqrt{0.9} = 1.960$, $k = 1.86$. Hence $c_1 = 24 - 1.86 = 22.14$ and $c_2 = 24 + 1.86 = 25.86$. If $\bar{x}$ is not smaller than c_1 and not greater than c_2, we accept the hypothesis. Otherwise we reject it. The power of the test is (Fig. 511)

$$\eta(\mu) = P(\overline{X} < 22.14)_\mu + P(\overline{X} > 25.86)_\mu = P(\overline{X} < 22.14)_\mu + 1 - P(\overline{X} \leqq 25.86)_\mu$$

(9)
$$= 1 + \Phi\left(\frac{22.14 - \mu}{\sqrt{0.9}}\right) - \Phi\left(\frac{25.86 - \mu}{\sqrt{0.9}}\right)$$

$$= 1 + \Phi(23.34 - 1.05\mu) - \Phi(27.26 - 1.05\mu).$$

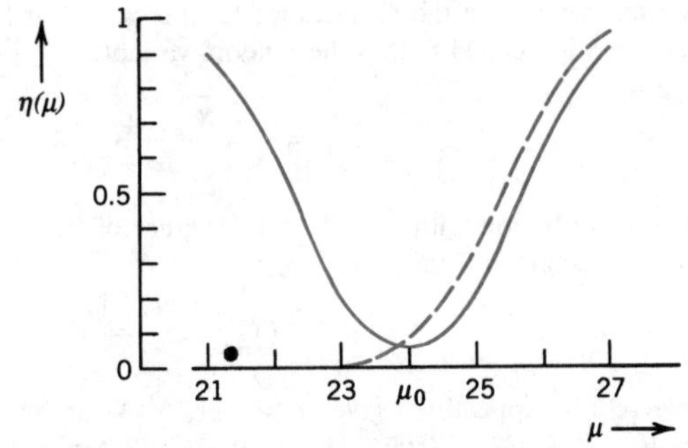

Fig. 511. Power $\eta(\mu)$ in Example 2, case (*a*) (dashed) and case (*c*)

Consequently, the operating characteristic $\beta(\mu) = 1 - \eta(\mu)$ (see before) is (Fig. 512)

$$\beta(\mu) = \Phi(27.26 - 1.05\mu) - \Phi(23.34 - 1.05\mu).$$

If we take a larger sample, say, of size $n = 100$ (instead of 10), then $\sigma^2/n = 0.09$ (instead of 0.9) and the critical values are $c_1 = 23.41$ and $c_2 = 24.59$, as can be readily verified. Then the operating characteristic of the test is

$$\beta(\mu) = \Phi\left(\frac{24.59 - \mu}{\sqrt{0.09}}\right) - \Phi\left(\frac{23.41 - \mu}{\sqrt{0.09}}\right)$$
$$= \Phi(81.97 - 3.33\mu) - \Phi(78.03 - 3.33\mu).$$

Figure 512 shows that the corresponding OC curve is steeper than that for $n = 10$. This means that the increase of n has led to an improvement of the test. In any practical case, n is chosen as small as possible but so large that the test brings out deviations between μ and μ_0 that are of practical interest. For instance, if deviations of ± 2 units are of interest, we see from Fig. 512 that $n = 10$ is much too small because when $\mu = 24 - 2 = 22$ or $\mu = 24 + 2 = 26$, then β is almost 50%. On the other hand, we see that $n = 100$ is sufficient for that purpose. ∎

EXAMPLE 3 Test for the mean of the normal distribution with unknown variance

The tensile strength of a sample of $n = 16$ manila ropes (diameter 3 in.) was measured. The sample mean was $\bar{x} = 4482$ kg, and the sample standard deviation was $s = 115$ kg (N. C. Wiley, 41st Annual Meeting of the American Society for Testing Materials). Assuming that the tensile strength is a normal random variable, test the hypothesis $\mu_0 = 4500$ kg against the alternative $\mu_1 = 4400$ kg. Here μ_0 may be a value given by the manufacturer, while μ_1 may result from previous experience.

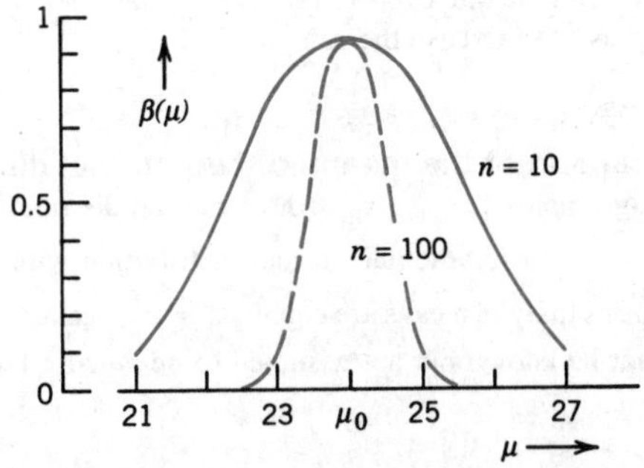

Fig. 512. Curves of the operating characteristic (OC curves) in Example 2, case (*c*), for two different sample sizes n

Solution. We choose the significance level $\alpha = 5\%$. If the hypothesis is true, it follows from Theorem 3 in Sec. 24.6, that the random variable

$$T = \sqrt{n}\,\frac{\bar{X} - \mu_0}{S} = 4\,\frac{\bar{X} - 4500}{S}$$

has a t-distribution with $n - 1 = 15$ degrees of freedom. The test is left-sided. The critical value c is obtained from

$$P(T < c)_{\mu_0} = \alpha = 0.05.$$

Table A10 in Appendix 5 gives $c = -1.75$. As an observed value of T we obtain from the sample $t = 4(4482 - 4500)/115 = -0.626$. We see that $t > c$ and accept the hypothesis. For obtaining numerical values of the power of the test, we would need tables called noncentral Student t-tables; we shall not discuss this question here. ∎

EXAMPLE 4 **Test for the variance of the normal distribution**

Using a sample of size $n = 15$ and sample variance $s^2 = 13$ from a normal population, test the hypothesis $\sigma^2 = \sigma_0^2 = 10$ against the alternative $\sigma^2 = \sigma_1^2 = 20$.

Solution. We choose the significance level $\alpha = 5\%$. If the hypothesis is true, then

$$Y = (n - 1)\frac{S^2}{\sigma_0^2} = 14\,\frac{S^2}{10} = 1.4S^2$$

has a chi-square distribution with $n - 1 = 14$ degrees of freedom, by Theorem 4, Sec. 24.6. From

$$P(Y > c) = \alpha = 0.05, \qquad \text{that is,} \qquad P(Y \leqq c) = 0.95,$$

and Table A11 in Appendix 5 with 14 degrees of freedom we obtain $c = 23.68$. This is the critical value of Y. Hence to $S^2 = \sigma_0^2 Y/(n - 1) = 0.714Y$ there corresponds the critical value $c^* = 0.714 \cdot 23.68 = 16.91$. Since $s^2 < c^*$, we accept the hypothesis.

If the alternative is true, the variable

$$Y_1 = 14\,\frac{S^2}{\sigma_1^2} = 0.7S^2$$

has a chi-square distribution with 14 degrees of freedom. Hence our test has the power

$$\eta = P(S^2 > c^*)_{\sigma^2=20} = P(Y_1 > 0.7c^*)_{\sigma^2=20} = 1 - P(Y_1 \leqq 11.84)_{\sigma^2=20} \approx 62\%$$

and we see that the Type II risk is very large, namely, 38%. To make this risk smaller, we would have to increase the sample size. ∎

EXAMPLE 5 **Comparison of the means of two normal distributions**

Using a sample $x_1, \cdots, x_{n_1}$ from a normal distribution with unknown mean μ_1 and a sample $y_1, \cdots, y_{n_2}$ from another normal distribution with unknown mean μ_2, we want to test the hypothesis that the means are equal, $\mu_1 = \mu_2$, against an alternative, say, $\mu_1 > \mu_2$. The variances need not be known but are assumed to be equal.[5] Two cases are of practical importance:

[5]If the test in the next example shows that the variances differ significantly, then choose two not too small samples of the same size $n_1 = n_2 = n$ (> 30, say), use the fact that (12) is an observed value of an approximately normal random variable with mean 0 and variance 1, and proceed as in Example 2.

Case A. The samples have the same size. Furthermore, each value of the first sample corresponds to precisely one value of the other, because corresponding values result from the same person or thing **(paired comparison)**; for example, two measurements of the same thing by two different methods or two measurements from the two eyes of the same animal; more generally, they may result from pairs of *similar* individuals or things, for example, identical twins, pairs of used front tires from the same car, etc. Then we should form the differences of corresponding values and test the hypothesis that the population corresponding to the differences has mean 0, using the method in Example 3. If we have a choice, this method is better than the following.

Case B. *The two samples are independent and not necessarily of the same size.* Then we may proceed as follows. Suppose that the alternative is $\mu_1 > \mu_2$. We choose a significance level α. Then we compute the sample means $\bar{x}$ and $\bar{y}$ as well as $(n_1 - 1)s_1^2$ and $(n_2 - 1)s_2^2$, where s_1^2 and s_2^2 are the sample variances. Using Table A10 in Appendix 5 with $n_1 + n_2 - 2$ degrees of freedom, we now determine c from

$$P(T \leqq c) = 1 - \alpha. \tag{10}$$

We finally compute

$$t_0 = \sqrt{\frac{n_1 n_2(n_1 + n_2 - 2)}{n_1 + n_2}} \; \frac{\bar{x} - \bar{y}}{\sqrt{(n_1 - 1)s_1^2 + (n_2 - 1)s_2^2}} . \tag{11}$$

It can be shown that this is an observed value of a random variable that has a t-distribution with $n_1 + n_2 - 2$ degrees of freedom, provided the hypothesis is true. If $t_0 \leqq c$, the hypothesis is accepted. If $t_0 > c$, it is rejected.

If the alternative is $\mu_1 \neq \mu_2$, then (10) must be replaced by

$$P(T \leqq c_1) = 0.5\alpha, \qquad P(T \leqq c_2) = 1 - 0.5\alpha. \tag{10*}$$

Note that for samples of equal size $n_1 = n_2 = n$, formula (11) reduces to

$$t_0 = \sqrt{n} \, \frac{\bar{x} - \bar{y}}{\sqrt{s_1^2 + s_2^2}} . \tag{12}$$

To illustrate the computations, let us consider the two samples

	105	108	86	103	103	107	124	105
and	89	92	84	97	103	107	111	97

showing the relative output of tin plate workers under two different working conditions [J.J.B. Worth, *Journal of Industrial Engineering* **9** (1958), 249–253). Assuming that the corresponding populations are normal and have the same variance, let us test the hypothesis $\mu_1 = \mu_2$ against the alternative $\mu_1 \neq \mu_2$. (Equality of variances will be tested in the next example.)

Solution. We find

$$\bar{x} = 105.125, \qquad \bar{y} = 97.500, \qquad s_1^2 = 106.125, \qquad s_2^2 = 84.000.$$

We choose the significance level $\alpha = 5\%$. From (10*) with $0.5\alpha = 2.5\%$, $1 - 0.5\alpha = 97.5\%$ and Table A10 in Appendix 5 with 14 degrees of freedom we obtain $c_1 = -2.15$ and $c_2 = 2.15$. Formula (12) with $n = 8$ gives the value

$$t_0 = \sqrt{8} \cdot 7.625/\sqrt{190.125} = 1.56.$$

Since $c_1 \leqq t_0 \leqq c_2$, we accept the hypothesis $\mu_1 = \mu_2$ that under both conditions the mean output is the same.

Case A applies to the example, because the two first sample values correspond to a certain type of work, the next two were obtained in another kind of work, etc. So we may use the differences

$$16 \quad 16 \quad 2 \quad 6 \quad 0 \quad 0 \quad 13 \quad 8$$

of corresponding sample values and the method in Example 3 to test the hypothesis $\mu = 0$, where μ is the mean of the population corresponding to the differences. As a logical alternative we take $\mu \neq 0$. The sample mean is $\bar{d} = 7.625$, and the sample variance is $s^2 = 45.696$. Hence

$$t = \sqrt{8}\,(7.625 - 0)/\sqrt{45.696} = 3.19.$$

From $P(T \leqq c_1) = 2.5\%$, $P(T \leqq c_2) = 97.5\%$ and Table A10 in Appendix 5 with $n - 1 = 7$ degrees of freedom we obtain $c_1 = -2.37$, $c_2 = 2.37$ and reject the hypothesis because $t = 3.19$ does not lie between c_1 and c_2. Hence our present test, in which we used more information (but the same samples), shows that the difference in output is significant. ■

EXAMPLE 6 **Comparison of the variances of two normal distributions**

Using the two samples in the last example, test the hypothesis $\sigma_1^2 = \sigma_2^2$; assume that the corresponding populations are normal and the nature of the experiment suggests the alternative $\sigma_1^2 > \sigma_2^2$.

Solution. We find $s_1^2 = 106.125$, $s_2^2 = 84.000$. We choose the significance level $\alpha = 5\%$. Using $P(V \leqq c) = 1 - \alpha = 95\%$ and Table A12 in Appendix 5, with $(n_1 - 1, n_2 - 1) = (7, 7)$ degrees of freedom, we determine $c = 3.79$. We finally compute $v_0 = s_1^2/s_2^2 = 1.26$. Since $v_0 \leqq c$, we accept the hypothesis. If $v_0 > c$, we would reject it.

This test is justified by the fact that v_0 is an observed value of a random variable which has a so-called ***F*-distribution** with $(n_1 - 1, n_2 - 1)$ degrees of freedom, provided the hypothesis is true. (Proof in Ref. [G3] listed in Appendix 1.) The F-distribution with (m, n) degrees of freedom was introduced by R. A. Fisher[6] and has the distribution function $F(z) = 0$ if $z < 0$ and

$$F(z) = K_{mn} \int_0^z t^{(m-2)/2}(mt + n)^{-(m+n)/2}\, dt \qquad (z \geqq 0), \tag{13}$$

where $K_{mn} = m^{m/2}n^{n/2}\Gamma(\frac{1}{2}m + \frac{1}{2}n)/\Gamma(\frac{1}{2}m)\Gamma(\frac{1}{2}n)$. (For Γ see Appendix 3.) ■

This long section contained the basic ideas and concepts of testing, along with typical applications, and the student may perhaps want to review it quickly before going on, because the next sections concern an adaption of these ideas to tasks of great practical importance and resulting tests in connection with quality control, acceptance (or rejection) of goods produced, and so on.

Problem Set 24.7

1. Test $\mu = 0$ against $\mu > 0$, assuming normality and using the sample 1, −1, 1, 3, −8, 6, 0 (deviations of the azimuth [multiples of 0.01 radian] in some revolution of a satellite). Choose $\alpha = 5\%$.
2. Using the data of Buffon in Table 23.1 (Sec. 23.2), test the hypothesis that the coin is fair, that is, that heads and tails have the same probability of occurrence, against the alternative that heads are more likely than tails. (Choose $\alpha = 5\%$.)
3. Do the same test as in Prob. 2, using the data of Pearson in Table 23.1.

[6] After the pioneering work of the English statistician and biologist, KARL PEARSON (1857—1936), the founder of the English school of statistics, and W. S. GOSSET (see the footnote in Sec. 24.6), the English statistician Sir RONALD AYLMER FISHER (1890—1962), professor of eugenics in London (1933–1943) and professor of genetics in Cambridge, England (1943–1957), had great influence on the further development of modern statistics.

4. Assuming normality and known variance $\sigma^2 = 4$, test the hypothesis $\mu = 30.0$ against the alternative (a) $\mu = 28.5$, (b) $\mu = 30.7$, using a sample of size 10 with mean $\bar{x} = 28.5$ and choosing $\alpha = 5\%$.

5. How does the result in Prob. 4(a) change if we use a smaller sample, say, of size 4, the other data ($\bar{x} = 28.5$, $\alpha = 5\%$, etc.) remaining as before?

6. Determine the power of the test in Prob. 4(a).

7. What is the rejection region in Prob. 4 in the case of a two-sided test with $\alpha = 5\%$?

8. Using the sample in Example 1, Sec. 24.4, test the hypothesis $\mu = 0.80$ in. (the length indicated on the box) against the alternative $\mu \neq 0.80$ in. (Assume normality, choose $\alpha = 5\%$.)

9. A firm sells oil in cans containing 1000 g oil per can and is interested to know whether the mean weight differs significantly from 1000 g at the 5% level, in which case the filling machine has to be adjusted. Set up a hypothesis and an alternative and perform the test, assuming normality and using a sample of 20 fillings have a mean of 996 g and a standard deviation of 5 g.

10. If a sample of 50 tires of a certain kind has a mean life of 32 000 miles and a standard deviation of 4000 miles, can the manufacturer claim that the true mean life of such tires is greater than 30 000 miles? Set up and test a corresponding hypothesis at a 5% level, assuming normality.

11. If simultaneous measurements of electric voltage by two different types of voltmeter yield the differences (in volts) 0.8, 0.2, −0.3, 0.1, 0.0, 0.5, 0.7, 0.2, can we assert at the 5% level that there is no significant difference in the calibration of the two types of instruments? (Assume normality.)

12. If a standard medication cures about 70% of patients with a certain disease and a new medication cured 148 of the first 200 patients on whom it was tried, can we conclude that the new medication is better? (Choose $\alpha = 5\%$.)

13. Suppose that in the past the standard deviation of weights of certain 25.0-oz packages filled by a machine was 0.4 oz. Test the hypothesis H_0: $\sigma = 0.4$ against the alternative H_1: $\sigma > 0.4$ (an undesirable increase), using a sample of 10 packages with standard deviation 0.5 oz and assuming normality. (Choose $\alpha = 5\%$.)

14. Suppose that in operating battery-powered electrical equipment, it is less expensive to replace all batteries at fixed intervals than to replace each battery individually when it breaks down, provided the standard deviation of the lifetime is less than a certain limit, say, less than 5 hours. Set up and apply a suitable test, using a sample of 28 values of lifetimes with standard deviation $s = 3.5$ hours and assuming normality; choose $\alpha = 5\%$.

15. Brand *A* gasoline was used in 9 similar automobiles under identical conditions. The corresponding sample of 9 values (miles per gallon) had mean 20.2 and standard deviation 0.5. Under the same conditions, high-power brand *B* gasoline gave a sample of 10 values with mean 21.8 and standard deviation 0.6. Is the mileage of *B* significantly better than that of *A*? (Test at the 5% level; assume normality.)

16. The two samples 50, 90, 100, 90, 110, 80 and 110, 110, 120, 110, 130, 110, 120 are values of the differences of temperatures (°C) of iron at two stages of casting, taken from two different crucibles. Is the variance of the first population larger than that of the second? (Assume normality. Choose $\alpha = 5\%$.)

17. Using samples of sizes 10 and 16 with variances $s_1^2 = 50$ and $s_2^2 = 30$ and assuming normality of the corresponding populations, test the hypothesis H_0: $\sigma_1^2 = \sigma_2^2$ against the alternative $\sigma_1^2 > \sigma_2^2$. Choose $\alpha = 5\%$.

18. Assuming normality and equal variance and using independent samples with $n_1 = 9$, $\bar{x} = 12$, $s_1 = 2$, $n_2 = 9$, $\bar{y} = 15$, $s_2 = 2$, test H_0: $\mu_1 = \mu_2$ against $\mu_1 \neq \mu_2$; choose $\alpha = 5\%$.

19. Show that for a normal distribution the two types of errors in a test of a hypothesis H_0: $\mu = \mu_0$ against an alternative H_1: $\mu = \mu_1$ can be made as small as one pleases (not zero) by taking the sample sufficiently large.

20. Graph the OC curves in Example 2, cases (*a*) and (*b*).

24.8 Quality Control

The ideas on testing can be adapted and extended in various ways to serve basic practical needs in engineering and other fields. We show this in the remaining sections for some of the most important tasks solvable by statistical methods. As a first such area of problems, we discuss industrial quality control.

No production process is so perfect that all the products are completely alike. There is always a small variation that is caused by a great number of small, uncontrollable factors and must therefore be regarded as a chance variation. It is important to make sure that the products have required values (for example, length, strength, or whatever property may be of importance in a particular case). For this purpose one makes a test of the hypothesis that the products have the required property, say, $\mu = \mu_0$, where μ_0 is a required value. If this is done after an entire lot has been produced (for example, a lot of 100,000 screws), the test will tell us how good or how bad the products are, but it is obviously too late to alter undesirable results. It is much better to test during the production run. This is done at regular intervals of time (for example, every hour or half-hour) and is called **quality control.** Each time a sample of the same size is taken, in practice 3 to 10 items. If the hypothesis is rejected, we stop the production process and look for the trouble that causes the deviation.

If we stop the production process even though it is progressing properly, we make a Type I error. If we do not stop the process even though something is not in order, we make a Type II error (see Sec. 24.7).

The result of each test is marked in graphical form on what is called a **control chart.** This was proposed by W. A. Shewhart in 1924 and makes quality control particularly effective.

Control Chart for the Mean

An illustration and example of a control chart is given in the upper part of Fig. 513. This control chart for the mean shows the **lower control limit** LCL, the **center control line** CL, and the **upper control limit** UCL. The two **control limits** correspond to the critical values c_1 and c_2 in case (*c*) of Example 2 in Sec. 24.7. As soon as a sample mean falls outside the range between the control limits, we reject the hypothesis and assert that the production process

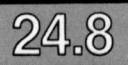

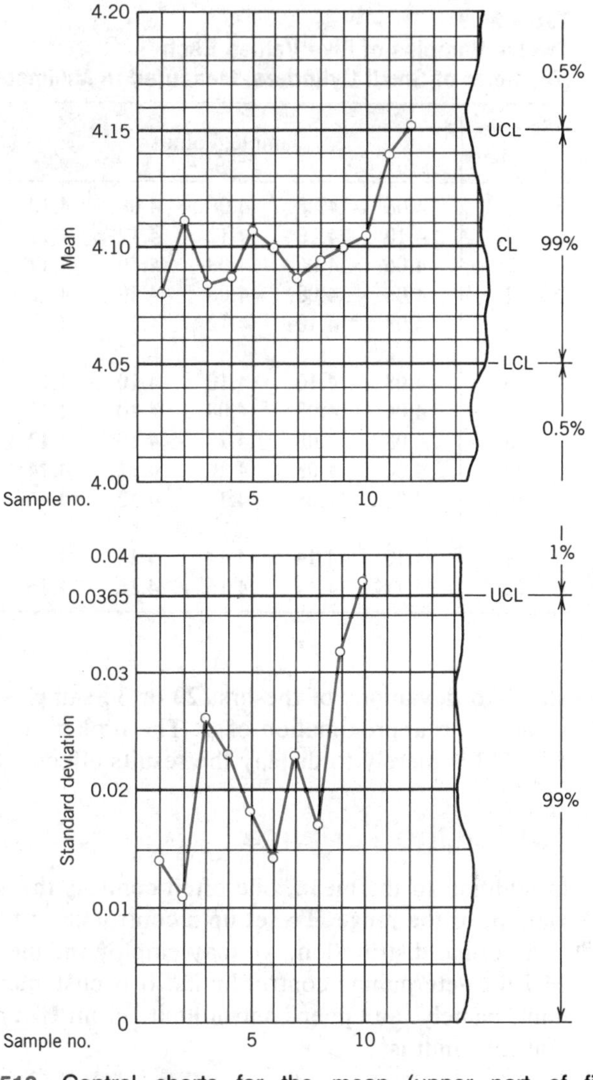

Fig. 513. Control charts for the mean (upper part of figure) and the standard deviation in the case of the sample in Table 24.9, p. 1246

is "out of control"; that is, we assert that there has been a shift in process level. Action is called for whenever a point exceeds the limits.

If we choose control limits that are too loose, we shall not detect process shifts. On the other hand, if we choose control limits that are too tight, we shall be unable to run the process because of frequent searches for nonexistent trouble. The usual significance level is $\alpha = 1\%$. From Theorem 2 in Sec. 24.6 and Table A8 in Appendix 5 we see that in the case of the normal distribution the corresponding control limits for the mean are

$$\text{(1)} \qquad \text{LCL} = \mu_0 - 2.58\frac{\sigma}{\sqrt{n}}, \qquad \text{UCL} = \mu_0 + 2.58\frac{\sigma}{\sqrt{n}}.$$

Here σ is assumed to be known. If σ is unknown, we may compute the

Table 24.9
Twelve Samples of Five Values Each
(Diameter of Small Cylinders, Measured in Millimeters)

Sample Number	Sample Values					$\bar{x}$	s	R
1	4.06	4.08	4.08	4.08	4.10	4.080	0.014	0.04
2	4.10	4.10	4.12	4.12	4.12	4.112	0.011	0.02
3	4.06	4.06	4.08	4.10	4.12	4.084	0.026	0.06
4	4.06	4.08	4.08	4.10	4.12	4.088	0.023	0.06
5	4.08	4.10	4.12	4.12	4.12	4.108	0.018	0.04
6	4.08	4.10	4.10	4.10	4.12	4.100	0.014	0.04
7	4.06	4.08	4.08	4.10	4.12	4.088	0.023	0.06
8	4.08	4.08	4.10	4.10	4.12	4.096	0.017	0.04
9	4.06	4.08	4.10	4.12	4.14	4.100	0.032	0.08
10	4.06	4.08	4.10	4.12	4.16	4.104	0.038	0.10
11	4.12	4.14	4.14	4.14	4.16	4.140	0.014	0.04
12	4.14	4.14	4.16	4.16	4.16	4.152	0.011	0.02

standard deviations of the first 20 or 30 samples and take their arithmetic mean as an approximation of σ. The broken line connecting the means in Fig. 513 is merely to display the results effectively.

Control Chart for the Variance

In addition to the mean, one often controls the variance, the standard deviation, or the range. To set up a control chart for the variance in the case of a normal distribution, we may employ the method in Example 4 of Sec. 24.7 for determining control limits. It is customary to use only one control limit, namely, an upper control limit. From Example 4 of Sec. 24.7 we see that this limit is

$$\text{UCL} = \frac{\sigma^2 c}{n-1} \tag{2}$$

where c is obtained from the equation

$$P(Y > c) = \alpha, \qquad \text{that is,} \qquad P(Y \leqq c) = 1 - \alpha$$

and the table of the chi-square distribution (Table A11 in Appendix 5) with $n - 1$ degrees of freedom; here α (5% or 1%, say) is the probability that an observed value s^2 of S^2 in a sample is greater than the upper control limit.

If we wanted a control chart for the variance with both an upper control limit UCL and a lower control limit LCL, these limits would be

$$\text{LCL} = \frac{\sigma^2 c_1}{n-1} \qquad \text{and} \qquad \text{UCL} = \frac{\sigma^2 c_2}{n-1} \tag{3}$$

where c_1 and c_2 are obtained from the equations

(4) $$P(Y \leqq c_1) = \frac{\alpha}{2} \quad \text{and} \quad P(Y \leqq c_2) = 1 - \frac{\alpha}{2}$$

and Table A11 in Appendix 5 with $n - 1$ degrees of freedom.

Control Chart for the Standard Deviation

Similarly, to set up a control chart for the standard deviation, we need an upper control limit

(5) $$\text{UCL} = \frac{\sigma\sqrt{c}}{\sqrt{n-1}}$$

obtained from (2). For example, in Table 24.9 we have $n = 5$. Assuming that the corresponding population is normal with standard deviation $\sigma = 0.02$ and choosing $\alpha = 1\%$, we obtain from the equation

$$P(Y \leqq c) = 1 - \alpha = 99\%$$

and Table A11 in Appendix 5 with 4 degrees of freedom the critical value $c = 13.28$ and from (5) the corresponding value

$$\text{UCL} = \frac{0.02\sqrt{13.28}}{\sqrt{4}} = 0.0365,$$

which is shown in the lower part of Fig. 513.

A control chart for the standard deviation with both an upper and a lower control limit is obtained from (3).

Control Chart for the Range

Instead of the variance or standard deviation, one often controls the **range** R (= largest sample value minus smallest sample value). It can be shown that in the case of the normal distribution, the standard deviation σ is proportional to the expectation of the random variable R^* for which R is an observed value, say, $\sigma = \lambda_n E(R^*)$, where the factor of proportionality λ_n depends on the sample size n and has the values

n	2	3	4	5	6	7	8	9	10
$\lambda_n = \sigma/E(R^*)$	0.89	0.59	0.49	0.43	0.40	0.37	0.35	0.34	0.32

n	12	14	16	18	20	30	40	50
$\lambda_n = \sigma/E(R^*)$	0.31	0.29	0.28	0.28	0.27	0.25	0.23	0.22

Since R depends on two sample values only, it gives less information about a sample than s does. Clearly, the larger the sample size n is, the more information we lose in using R instead of s. A practical rule is to use s when n is larger than 10.

Problem Set 24.8

1. Suppose a machine for filling cans with lubricating oil is set so that it will generate fillings which form a normal population with mean 1 gal and standard deviation 0.03 gal. Set up a control chart of the type shown in Fig. 513 for controlling the mean (that is, find LCL and UCL), assuming that the sample size is 6.
2. **(Three-sigma control chart)** Show that in Prob. 1, the requirement of the significance level $\alpha = 0.3\%$ leads to $\text{LCL} = \mu - 3\sigma/\sqrt{n}$ and $\text{UCL} = \mu + 3\sigma/\sqrt{n}$, and find the corresponding numerical values.
3. What sample size should we choose in Prob. 1 if we want LCL and UCL somewhat closer together, say, $\text{UCL} - \text{LCL} = 0.05$, without changing the significance level?
4. How does the meaning of the control limits (1) change if we apply a control chart with these limits in the case of a population that is not normal?
5. How should we change the sample size in controlling the mean of a normal population if we want UCL − LCL to decrease to half its original value?
6. What LCL and UCL should we use instead of (1) if instead of $\bar{x}$ we use the sum $x_1 + \cdots + x_n$ of the sample values? Determine these limits in the case of Fig. 513.
7. Ten samples of size 2 were taken from a production lot of bolts. The values (length in millimeters) are

Sample No.	1	2	3	4	5	6	7	8	9	10
Length	27.4	27.4	27.5	27.3	27.9	27.6	27.6	27.8	27.5	27.3
	27.6	27.4	27.7	27.4	27.5	27.5	27.4	27.3	27.4	27.7

Assuming that the population is normal with mean 27.5 and variance 0.024 and using (1), set up a control chart for the mean and graph the sample means on the chart.

8. Graph the means of the following 10 samples (thickness of washers, coded values) on a control chart for means, assuming that the population is normal with mean 5 and standard deviation 1.55.

Time	8:00	8:30	9:00	9:30	10:00	10:30	11:00	11:30	12:00	12:30
	3	3	5	7	7	4	5	6	5	5
Sample	4	6	2	5	3	4	6	4	5	2
Values	8	6	5	4	6	3	4	6	6	5
	4	8	6	4	5	6	6	4	4	3

9. Graph the ranges of the samples in Prob. 8 on a control chart for ranges.
10. Graph $\lambda_n = \sigma/E(R^*)$ as a function of n. What is the reason that λ_n is a monotone decreasing function of n?
11. Since the presence of a point outside control limits for the mean indicates trouble, how often would we be making the mistake of looking for nonexistent trouble if we used (a) 1-sigma limits, (b) 2-sigma limits? (Assume normality.)
12. How would progressive tool wear in an automatic lathe operation be indicated by a control chart for the mean? Answer the same question for a sudden change in the position of the tool in that operation.
13. **(Number of defectives)** Find formulas for the UCL, CL, and LCL (corresponding to 3σ-limits) in the case of a control chart for the number of defectives, assuming that in a state of statistical control the fraction of defectives is p.

14. (Attribute control charts). Twenty samples of size 100 were taken from a production of containers. The numbers of defectives (leaking containers) in those samples (in the order observed) were

3 7 6 1 4 5 4 9 7 0 5 6 13 4 9 0 2 1 12 8

From previous experience it was known that the average fraction defective is $p = 5\%$ provided that the process of production is running properly. Using the binomial distribution, set up a *fraction defective chart* (also called a **p-chart**), that is, choose the LCL = 0 and determine the UCL for the fraction defective (in percent) by the use of 3-sigma limits, where σ^2 is the variance of the random variable $\bar{X}$ = *Fraction defective in a sample of size* 100. Is the process in control?

15. (Number of defects per unit) A so-called *c-chart* or *defects-per-unit chart* is used for the control of the number X of defects per unit (for instance, the number of defects per 10 meters of paper, the number of missing rivets in an airplane wing, etc.) (a) Set up formulas for CL and LCL, UCL corresponding to $\mu \pm 3\sigma$, assuming that X has a Poisson distribution. (b) Compute CL, LCL, and UCL in a control process of the number of imperfections in sheet glass; assume that this number is 2.5 per sheet on the average when the process is under control.

24.9 Acceptance Sampling

Acceptance sampling is another testing procedure of practical importance. It is applied in mass production if a **producer** supplies to a **consumer** lots consisting of N items. Then one has to decide whether to accept or reject an individual lot (for instance, a box of screws), depending on the quality of the lot. This decision is often made by taking and inspecting a sample of size n from the lot and determining the number of **defectives** (short for "defective items") in the sample, that is, items that do not meet the specifications (size, color, strength, or whatever may be important). If the number of defectives x in the sample is not greater than a specified number c $(< n)$, the lot is accepted. If $x > c$, the lot is rejected. c is called the *allowable number of defectives* or the **acceptance number.** It is clear that the producer and the consumer must agree on a certain **sampling plan,** that is, on a certain sample size n and an acceptance number c. Such a plan is called a **single sampling plan** because it is based on a single sample. *Double sampling plans,* in which two samples are used, will be mentioned later.

Let A be the event that a lot is accepted. It is clear that the corresponding probability $P(A)$ depends not only on n and c but also on the number of defectives in the lot. Let M denote this number. Furthermore, the number x of defectives in a given sample is an observed value of a random variable X. Now suppose that we sample without replacement. Then X has a hypergeometric distribution (see Sec. 23.6), so that

$$\textbf{(1)} \qquad P(A) = P(X \leqq c) = \sum_{x=0}^{c} \binom{M}{x}\binom{N-M}{n-x} \bigg/ \binom{N}{n}.$$

If $M = 0$ (no defectives in the lot), then X must assume the value 0, and

$$P(A) = \binom{0}{0}\binom{N}{n} \bigg/ \binom{N}{n} = 1.$$

For fixed n and c and increasing M the probability $P(A)$ decreases. If $M = N$ (all items of the lot defective), then X must assume the value n, and we have $P(A) = P(X \leqq c) = 0$ because $c < n$.

The ratio $\theta = M/N$ is called the **fraction defective** in the lot. Note that $M = N\theta$, so that (1) may be written

$$P(A;\,\theta) = \sum_{x=0}^{c} \binom{N\theta}{x}\binom{N - N\theta}{n - x} \bigg/ \binom{N}{n}. \tag{2}$$

Since θ can have one of the $N + 1$ values $0, 1/N, 2/N, \cdots, N/N$, the probability $P(A)$ is defined for these values only. For fixed n and c we may plot $P(A)$ as a function of θ. These are $N + 1$ points. Through these points we may then draw a smooth curve, which is called the **operating characteristic curve** (OC curve) of the sampling plan considered.

EXAMPLE 1 Single sampling plan

Suppose that certain tool bits are packaged 20 to a box, and the following single sampling plan is used. A sample of two tool bits is drawn, and the corresponding box is accepted if and only if both bits in the sample are good. In this case, $N = 20$, $n = 2$, $c = 0$, and (2) takes the form

$$P(A;\,\theta) = \binom{20\theta}{0}\binom{20 - 20\theta}{2} \bigg/ \binom{20}{2} = \frac{(20 - 20\theta)(19 - 20\theta)}{380}.$$

Numerical values are

θ	0.00	0.05	0.10	0.15	0.20	$\cdots$
$P(A;\,\theta)$	1.00	0.90	0.81	0.72	0.63	$\cdots$

The OC curve is shown in Fig. 514. ■

In most practical cases θ will be small (less than 10%). In many cases the lot size N will be very large (1000, 10,000, etc.), so that we may approximate the hypergeometric distribution in (1) and (2) by the binomial distribution with $p = \theta$. Then if n is such that $n\theta$ is moderate (say, less than 20), we may approximate that distribution by the Poisson distribution with mean $\mu = n\theta$. From (2) we then have

$$P(A;\,\theta) \sim e^{-\mu} \sum_{x=0}^{c} \frac{\mu^x}{x!} \qquad (\mu = n\theta). \tag{3}$$

EXAMPLE 2 Single sampling plan. Poisson distribution

Suppose that for large lots the following single sampling plan is used. A sample of size $n = 20$ is taken. If it contains not more than one defective, the lot is accepted. If the sample contains two or more defectives, the lot is rejected. In this plan, we obtain from (3)

$$P(A;\,\theta) \sim e^{-20\theta}(1 + 20\theta).$$

The corresponding OC curve is shown in Fig. 515. ■

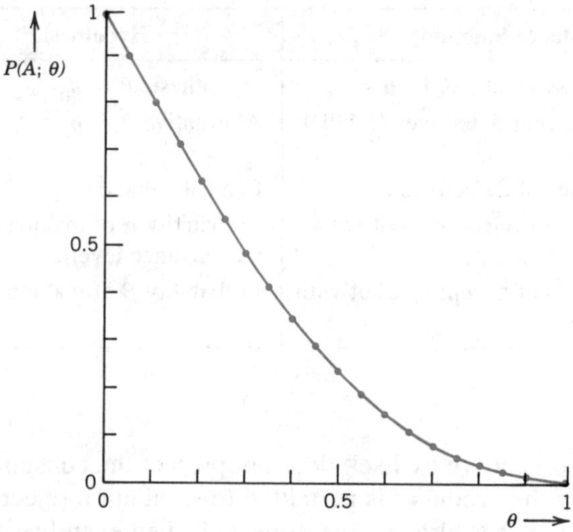

Fig. 514. OC curve of the single sampling plan with $n = 2$ and $c = 0$ for lots of size $N = 20$

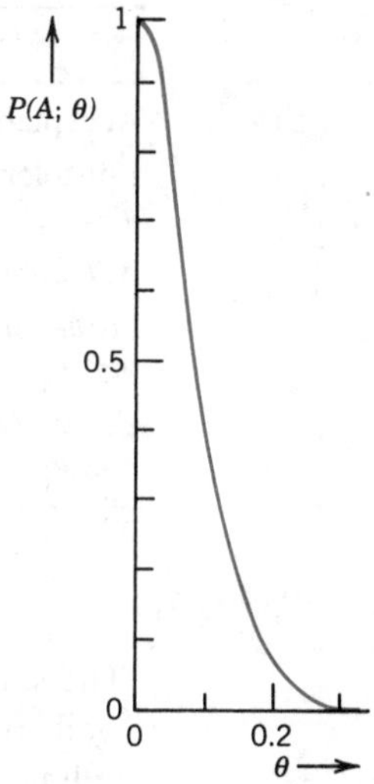

Fig. 515. OC curve in Example 2

Errors in Acceptance Sampling

Since acceptance sampling is a test, as we have just seen, we should ask how it fits into the circle of ideas on tests; in particular, we should find out what is going on with respect to errors and discuss the related practical problem of choosing n and c.

In acceptance sampling the producer and the consumer have different interests. The producer may require the probability of rejecting a "good" or "acceptable" lot to be a small number, call it α. The consumer (buyer) may want the probability of accepting a "bad" or "unacceptable" lot to be a small number β. More precisely, suppose that the two parties agree that a lot for which θ does not exceed a certain number θ_0 is an **acceptable lot,** whereas a lot for which θ is greater than or equal to a certain number θ_1 is an **unacceptable lot.** Then α is the probability of rejecting a lot with $\theta \leqq \theta_0$ and is called **producer's risk.** This corresponds to a Type I error in testing a hypothesis (Sec. 24.7). β is the probability of accepting a lot with $\theta \geqq \theta_1$ and is called **consumer's risk.** This corresponds to a Type II error in Sec. 24.7. Figure 516 on p. 1252 shows an illustrative example. θ_0 is called the **acceptable quality level** (AQL), and θ_1 is called the **lot tolerance percent defective** (LTPD) or the **rejectable quality level** (RQL). A lot with $\theta_0 < \theta < \theta_1$ may be called an *indifferent lot.*

From Fig. 516 we see that the points $(\theta_0, 1 - \alpha)$ and (θ_1, β) lie on the OC curve. It can be shown that for large lots we can choose θ_0, θ_1 $(> \theta_0)$, α, β and then determine n and c such that the OC curve runs very close to those prescribed points. Sampling plans for specified α, β, θ_0, and θ_1 have been published; see Ref. [G5] in Appendix 1.

Our present discussion of the conceptual relations between acceptance sampling and tests in general can be completed and summarized in the following table:

Acceptance Sampling	Hypothesis Testing
Acceptable quality level (AQL) $\theta = \theta_0$	Hypothesis $\theta = \theta_0$
Lot tolerance percent defectives (LTPD) $\theta = \theta_1$	Alternative $\theta = \theta_1$
Allowable number of defectives c	Critical value c
Producer's risk α of rejecting a lot with $\theta \leqq \theta_0$	Probability α of making a Type I error (significance level)
Consumer's risk β of accepting a lot with $\theta \geqq \theta_1$	Probability β of making a Type II error

Rectification

The sampling procedure by itself does not protect the consumer sufficiently well. In fact, if the producer is permitted to resubmit a rejected lot without telling that the lot has already been rejected, then even bad lots will eventually be accepted. To protect the consumer against this and other possibilities, the producer may agree with the consumer that a rejected lot is **rectified,** that is, is inspected 100%, item by item, and all defective items in the lot are removed and replaced by nondefective items.[7] Suppose that a plant produces $100\theta\%$ defective items and rejected lots are rectified. Then K lots of size N contain KN items, $KN\theta$ of which are defective. $KP(A; \theta)$

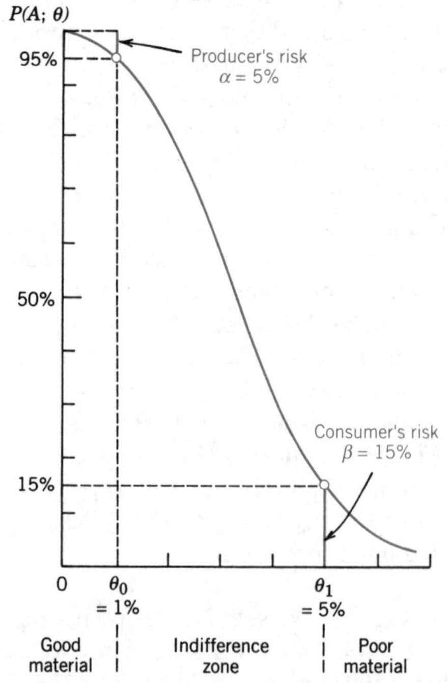

Fig. 516. OC curve, producer's and consumer's risks

[7]Of course, rectification is impossible if the inspection is destructive, or is not worthwhile if it is too expensive, compared with the value of the lot. The rejected lot may then be sold at a cut-rate price or scrapped.

of the lots are accepted; these contain a total of $KPN\theta$ defective items. The rejected and rectified lots contain no defective items. Hence after the rectification the fraction defective in the K lots equals $KPN\theta/KN = \theta P(A;\, \theta)$. This function of θ is called the **average outgoing quality** (AOQ) and is denoted by AOQ(θ). Thus

$$\text{AOQ}(\theta) = \theta P(A;\, \theta). \tag{4}$$

A sampling plan being given, this function and its graph, the *average outgoing quality curve* (AOQ curve), can readily be obtained from $P(A;\, \theta)$ and the OC curve. An example is shown in Fig. 517.

Clearly AOQ(0) = 0. Also, AOQ(1) = 0 because $P(A;\, 1) = 0$. From this and AOQ $\geqq 0$ we conclude that this function must have a maximum at some $\theta = \theta^*$. The corresponding value AOQ(θ^*) is called the **average outgoing quality limit** (AOQL). This is the worst average quality that may be expected to be accepted under the rectifying procedure.

It turns out that several single sampling plans may correspond to the same AOQL; see Ref. [G5] in Appendix 1. Hence if the AOQL is all the consumer cares about, the producer has some freedom in choosing a sampling plan and may select a plan that minimizes the amount of sampling, that is, the number of inspected items per lot. This number is

$$nP(A;\, \theta) + N(1 - P(A;\, \theta))$$

where the first term corresponds to the accepted lots and the last term to the rejected and rectified lots; in fact, rectification requires the inspection of all N items of the lot, and $1 - P(A;\, \theta)$ is the probability of rejecting a lot.

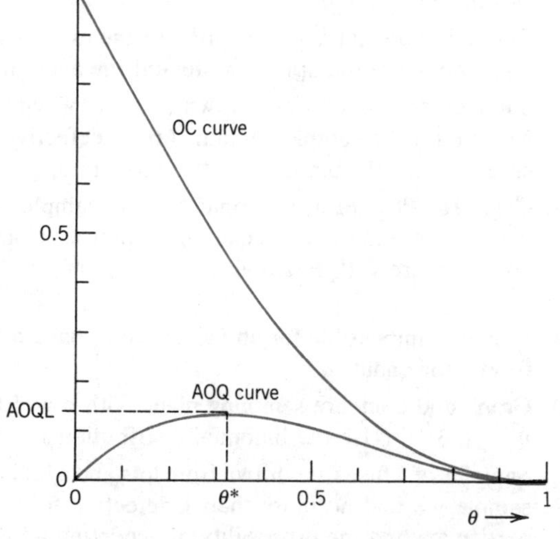

Fig. 517. OC curve and AOQ curve for the sampling plan in Fig. 514

Double sampling plans. We finally mention that inspection work can be saved by using a *double sampling plan,* in which the sample of size n is broken up into two samples of sizes n_1 and n_2 (where $n_1 + n_2 = n$). If the lot is very good or very bad, it may then be possible to decide about acceptance or rejection, using one sample only, so that the other sample will be necessary in the case of a lot of intermediate quality only. Reference [G5] contains double sampling plans using rectifying inspection of the following type (where x_1 and x_2 are the numbers of defectives in those two samples).

1. If $x_1 \leqq c_1$, accept the lot. If $x_1 > c_2$, reject the lot.

2. If $c_1 < x_1 \leqq c_2$, use the second sample, too. If $x_1 + x_2 \leqq c_2$, accept the lot. If $x_1 + x_2 > c_2$, reject the lot.

Problem Set 24.9

1. Large lots of razor blades are inspected by a single sampling plan that uses a sample of size 20 and the acceptance number $c = 1$. What are the probabilities of accepting a lot with 1%, 2%, 10% defectives (dull blades)? Use Table A6 in Appendix 5. Graph the OC curve.

2. Large lots of batteries for pocket calculators are inspected according to the following plan. $n = 30$ batteries are randomly drawn from a lot and tested. If this sample contains at most $c = 1$ defective battery, the lot is accepted. Otherwise it is rejected. Graph the OC curve of the plan, using the Poisson approximation.

3. Graph the AOQ curve in Prob. 2. Determine the AOQL, assuming that rectification is applied.

4. Do the work required in Prob. 2 if $n = 50$ and $c = 0$.

5. In Example 1, what are the producer's and consumer's risks if the AQL is 0.1 and the RQL is 0.6?

6. Find the binomial approximation of the hypergeometric distribution in Example 1 and compare the approximate and the accurate values.

7. Samples of 5 screws are drawn from a lot with fraction defective θ. The lot is accepted if the sample contains (a) no defective screws, (b) at most 1 defective screw. Using the binomial distribution, find, graph, and compare the OC curves.

8. Calculate $P(A; \theta)$ in Example 1 if the sample size is increased from $n = 2$ to $n = 3$, the other data remaining as before. Compute $P(A; 0.10)$ and $P(A; 0.20)$ and compare with Example 1.

9. Why is it impossible for an OC curve to have a vertical portion separating good from poor quality?

10. Graph and compare sampling plans with $c = 1$ and increasing values of n, say, $n = 2, 3, 4$. (Use the binomial distribution.)

11. Samples of 3 fuses are drawn from lots and a lot is accepted if in the corresponding sample we find no more than 1 defective fuse. Criticize this sampling plan. In particular, find the probability of accepting a lot that is 50% defective. (Use the binomial distribution.)

12. If in a single sampling plan for large lots of spark plugs, the sample size is 100 and we want the AQL to be 5% and the producer's risk 2%, what acceptance number c should we choose? (Use the normal approximation.)

13. What is the consumer's risk in Prob. 12 if we want the RQL to be 12%?

14. Graph the OC curve and the AOQ curve for the single sampling plan for large lots with $n = 5$ and $c = 0$, and find the AOQL.

15. Find the risks in the single sampling plan with $n = 5$ and $c = 0$, assuming that the AQL is $\theta_0 = 1\%$ and the RQL is $\theta_1 = 15\%$.

24.10 Goodness of Fit. χ^2-Test

So far we have discussed tests for unknown parameters (μ, σ, etc.) in distributions whose kind we knew or did not care about because we had a large sample, so that by the central limit theorem (Sec. 24.6) we could also apply methods designed for the normal distribution to other distributions (with proper caution!). However, what can we do if we want to test whether a distribution is of a certain kind, for instance, normal? This means we wish to test that a certain function $F(x)$ is the distribution function of a distribution from which we have a sample $x_1, \cdots, x_n$. This important practical task is called testing for **goodness of fit:** How well does the sample distribution function $\tilde{F}(x)$ fit the hypothetical distribution function $F(x)$?

Clearly, the sample distribution function $\tilde{F}(x)$ is an approximation of $F(x)$, and if it approximates $F(x)$ "sufficiently well," we shall accept the hypothesis that $F(x)$ is the distribution function of the population. If $\tilde{F}(x)$ deviates "too much" from $F(x)$, we shall reject the hypothesis.

To decide in this fashion, we have to know how much $\tilde{F}(x)$ can differ from $F(x)$ if the hypothesis is true. Hence we must first introduce a quantity that measures the deviation of $\tilde{F}(x)$, and we must know the probability distribution of this quantity under the assumption that the hypothesis is true. Then we proceed as follows. We determine a number c such that if the hypothesis is true, a deviation greater than c has a small preassigned probability. If, nevertheless, a deviation larger than c occurs, we have reason to doubt that the hypothesis is true and we reject it. On the other hand, if the deviation does not exceed c, so that $\tilde{F}(x)$ approximates $F(x)$ sufficiently well, we accept the hypothesis. Of course, if we accept the hypothesis, this means that we have insufficient evidence to reject it and does not exclude the possibility that there are other functions that would not be rejected in the test. In this respect the situation is quite similar to that in Sec. 24.7.

Table 24.10 on p. 1256 shows a test of that type, which was introduced by R. A. Fisher. This test is justified by the fact that if the hypothesis is true, then χ_0^2 is an observed value of a random variable whose distribution function approaches that of the chi-square distribution with $K - 1$ degrees of freedom (or $K - r - 1$ degrees of freedom if r parameters are estimated) as n approaches infinity. The requirement that at least five sample values lie in each interval in Table 24.10 results from the fact that for finite n that

Table 24.10
Chi-square Test for the Hypothesis That $F(x)$ is the Distribution Function of a Population from Which a Sample $x_1, \cdots, x_n$ Is Taken

1st Step. Subdivide the x-axis into K intervals $I_1, I_2, \cdots, I_K$ such that each interval contains at least 5 values of the given sample $x_1, \cdots, x_n$. Determine the number b_j of sample values in the interval I_j, where $j = 1, \cdots, K$. If a sample value lies at a common boundary point of two inervals, add 0.5 to each of the two corresponding b_j.

2nd Step. Using $F(x)$, compute the probability p_j that the random variable X under consideration assumes any value in the interval I_j, where $j = 1, \cdots, K$. Compute

$$e_j = np_j.$$

(This is the number of sample values theoretically expected in I_j if the hypothesis is true.)

3rd Step. Compute the deviation

$$\chi_0^2 = \sum_{j=1}^{K} \frac{(b_j - e_j)^2}{e_j}. \tag{1}$$

4th Step. Choose a significance level (5%, 1%, or the like).

5th Step. Determine the solution c of the equation

$$P(\chi^2 \leqq c) = 1 - \alpha$$

from the table of the chi-square distribution with $K - 1$ degrees of freedom (Table A11, Appendix 5). If r parameters of $F(x)$ are unknown and their maximum likelihood estimates (Sec. 24.5) are used, then use $K - r - 1$ degrees of freedom (instead of $K - 1$). If $\chi_0^2 \leqq c$, accept the hypothesis. If $\chi_0^2 > c$, reject the hypothesis.

random variable has only approximately a chi-square distribution. A proof can be found in Ref. [G3] listed in Appendix 1. If the sample is so small that the requirement cannot be satisfied, one may continue with the test, but the result should then be used with great caution.

EXAMPLE 1 **Test of normality**

Test whether the population from which the sample in Table 24.2, Sec. 24.3, was taken is normal.

Solution. The maximum likelihood estimates for μ and σ^2 are $\hat{\mu} = \bar{x} = 364.7$ and $\tilde{\sigma}^2 = 712.9$. The computation in Table 24.11 yields $\chi_0^2 = 2.790$. We choose $\alpha = 5\%$. Since $K = 10$ and we estimated $r = 2$ parameters, we have to use Table A11, Appendix 5, with $K - r - 1 = 7$ degrees of freedom, finding $c = 14.07$ as the solution of $P(\chi^2 \leqq c) = 95\%$. Since $\chi_0^2 < c$, we accept the hypothesis that the population is normal. ∎

Table 24.11
Computations in Example 1

x_j	$\frac{x_j - 364.7}{26.7}$	$\Phi\left(\frac{x_j - 364.7}{26.7}\right)$	e_j	b_j	Terms in (1)
$-\infty \cdots 325$	$-\infty \cdots -1.49$	$0.0000 \cdots 0.0681$	6.81	6	0.096
$325 \cdots 335$	$-1.49 \cdots -1.11$	$0.0681 \cdots 0.1335$	6.54	6	0.045
$335 \cdots 345$	$-1.11 \cdots -0.74$	$0.1335 \cdots 0.2296$	9.61	11	0.201
$345 \cdots 355$	$-0.74 \cdots -0.36$	$0.2296 \cdots 0.3594$	12.98	14	0.080
$355 \cdots 365$	$-0.36 \cdots 0.00$	$0.3594 \cdots 0.5000$	14.06	16	0.268
$365 \cdots 375$	$0.00 \cdots 0.39$	$0.5000 \cdots 0.6517$	15.17	15	0.002
$375 \cdots 385$	$0.39 \cdots 0.76$	$0.6517 \cdots 0.7764$	12.47	8	1.602
$385 \cdots 395$	$0.76 \cdots 1.13$	$0.7764 \cdots 0.8708$	9.44	10	0.033
$395 \cdots 405$	$1.13 \cdots 1.51$	$0.8708 \cdots 0.9345$	6.37	8	0.417
$405 \cdots \infty$	$1.51 \cdots \infty$	$0.9345 \cdots 1.0000$	6.55	6	0.046
					$\chi_0^2 = 2.790$

Problem Set 24.10

1. If in rolling a die 180 times we get $1, \cdots, 6$ with the absolute frequencies 25, 31, 33, 27, 29, 35, can we claim on the 5% level that the die is fair?

2. Solve Prob. 1 if the sample is 39, 22, 41, 26, 20, 32.

3. In a classical experiment, R. Wolf rolled a die 20 000 times and obtained $1, \cdots, 6$ with the absolute frequencies 3407, 3631, 3176, 2916, 3448, 3422. Test whether the die was fair, using $\alpha = 5\%$.

4. If 100 flips of a coin result in 40 heads and 60 tails, can we assert on the 5% level that the coin is fair?

5. If in 10 flips of a coin we get the same ratio as in Prob. 4 (4 heads and 6 tails), is the conclusion the same as in Prob. 4? First conjecture, then compute.

6. A manufacturer claims that in a process of producing razor blades, only 2.5% of the blades are dull. Test the claim against the alternative that more than 2.5% of the blades are dull, using a sample of 400 blades containing 17 dull blades. (Use $\alpha = 5\%$.)

7. In a table of random digits, even and odd digits should be *about* equally frequent. Test this, using as a sample the 50 digits in row 0 of Table A9, Appendix 5. (Use $\alpha = 5\%$.)

8. In tossing a coin 50 times, what would be the minimum number of heads (greater than 25) that would lead to the rejection of the hypothesis that the coin is fair, at the 5% level?

9. Between 1 P.M. and 2 P.M. on five consecutive days (Monday through Friday) a certain service station has 92, 60, 66, 62, and 90 customers, respectively. Test the hypothesis that the expected number of customers during that hour is the same on those days. (Use $\alpha = 5\%$)

10. Three samples of 200 rivets each were taken from a large production of each of three machines. The numbers of defective rivets in the samples were 7, 8, and 12. Is this difference significant? (Use $\alpha = 5\%$.)

11. Using the given sample, test that the corresponding population has a Poisson distribution. x is the number of alpha particles per 7.5-second intervals observed by E. Rutherford and H. Geiger in one of their classical experiments in 1910, and $a(x)$ is the absolute frequency (= number of time periods during which exactly x particles were observed).

x	$a(x)$	x	$a(x)$	x	$a(x)$
0	57	5	408	10	10
1	203	6	273	11	4
2	383	7	139	12	2
3	525	8	45	$\geqq 13$	0
4	532	9	27		

12. Can we assert that the traffic on the three lanes of an expressway (in one direction) is about the same on each lane if a count gives 910, 850, 720 cars on the right, middle, and left lanes, respectively, during the same interval of time?

13. If it is known that 25% of certain steel rods produced by a standard process will break when subjected to a load of 5000 lb, can we claim that a new process yields the same breakage rate if we find that in a sample of 80 rods produced by the new process, 27 rods broke when subjected to that load?

14. Was the sample in Table 24.4, Sec. 24.3, taken from a normal population?

15. Verify the computations in Example 1.

24.11 Nonparametric Tests

Nonparametric tests, also called *distribution-free tests,* are valid for any distribution; hence they are practical in cases when the kind of distribution is unknown, or is known but such that no tests specifically designed for it are available. In the present section we shall explain the basic idea of these tests, which are based on *"order statistics"* and are very simple. If there is a choice, then tests designed for a specific distribution generally give better results than do nonparametric tests; for instance, this applies to the tests in Sec. 24.7 for the normal distribution.

We shall discuss two tests in terms of typical examples. In deriving the distributions used in the test, it is essential that the distributions from which we sample are continuous. (Nonparametric tests can also be derived for discrete distributions, but this is slightly more complicated.)

EXAMPLE 1 Sign test for the median

A **median** is a solution $x = \tilde{\mu}$ of the equation $F(x) = 0.5$, where F is the distribution function.

Suppose that eight radio operators were tested, first in rooms without air-conditioning and then in air-conditioned rooms over the same period of time, and the difference of errors (unconditioned minus conditioned) were

$$9 \quad 4 \quad 0 \quad 6 \quad 4 \quad 0 \quad 7 \quad 11.$$

Test the hypothesis $\tilde{\mu} = 0$ (that is, air-conditioning has no effect) against the alternative $\tilde{\mu} > 0$ (that is, inferior performance in unconditioned rooms).

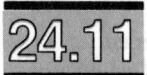

Solution. We choose the significance level $\alpha = 5\%$. If the hypothesis is true, the probability p of a positive difference is the same as that of a negative difference. Hence in this case, $p = 0.5$, and the random variable

$$X = \textit{Number of positive values among } n \textit{ values}$$

has a binomial distribution with $p = 0.5$. Our sample has eight values. We omit the values 0, which do not contribute to the decision. Then six values are left, all of which are positive. Since

$$P(X = 6) = \binom{6}{6}(0.5)^6(0.5)^0 = 0.0156 = 1.56\% < \alpha,$$

we reject the hypothesis and assert that the number of errors made in unconditioned rooms is significantly higher, so that installing air-conditioning should be considered. ∎

EXAMPLE 2 **Test for arbitrary trend**

A certain machine is used for cutting lengths of wire. Five successive pieces had the lengths

29 31 28 30 32.

Using this sample, test the hypothesis that there is **no trend,** that is, the machine does not have the tendency to produce longer and longer pieces or shorter and shorter pieces. Assume that the type of machine suggests the alternative that there is *positive trend,* that is, there is the tendency of successive pieces to get longer.

Solution. We count the number of **transpositions** in the sample, that is, the number of times a larger value precedes a smaller value:

29 precedes 28 (1 transposition),

31 precedes 28 and 30 (2 transpositions).

The remaining three sample values follow in ascending order. Hence in the sample there are $1 + 2 = 3$ transpositions. We now consider the random variable

$$T = \textit{Number of transpositions.}$$

If the hypothesis is true (no trend), then each of the $5! = 120$ permutations of five elements 1 2 3 4 5 has the same probability (1/120). We arrange these permutations according to their number of transpositions:

$T = 0$	$T = 1$	$T = 2$	$T = 3$	
1 2 3 4 5	1 2 3 5 4	1 2 4 5 3	1 2 5 4 3	
	1 2 4 3 5	1 2 5 3 4	1 3 4 5 2	
	1 3 2 4 5	1 3 2 5 4	1 3 5 2 4	
	2 1 3 4 5	1 3 4 2 5	1 4 2 5 3	
		1 4 2 3 5	1 4 3 2 5	
		2 1 3 5 4	1 5 2 3 4	
		2 1 4 3 5	2 1 4 5 3	
		2 3 1 4 5	2 1 5 3 4	etc.
		3 1 2 4 5	2 3 1 5 4	
			2 3 4 1 5	
			2 4 1 3 5	
			3 1 2 5 4	
			3 1 4 2 5	
			3 2 1 4 5	
			4 1 2 3 5	

From this we obtain

$$P(T \leqq 3) = \tfrac{1}{120} + \tfrac{4}{120} + \tfrac{9}{120} + \tfrac{15}{120} = \tfrac{29}{120} = 24\%.$$

Hence we accept the hypothesis.

Values of the distribution function of T in the case of no trend are shown in Table A13, Appendix 5. Our method and those values refer to *continuous* distributions. Theoretically, we may then expect that all the values of a sample are different. Practically, some sample values may still be equal, because of rounding off. If m values are equal, add $m(m-1)/4$ (= mean value of the transpositions in the case of the permutations of m elements), that is, $\frac{1}{2}$ for each pair of equal values, $\frac{3}{2}$ for each triple, etc. ∎

Problem Set 24.11

1. Are air filters of type A better than type B filters if in 10 trials, A gave cleaner air than B in 7 cases, B gave cleaner air than A in 1 case, whereas in 2 of the trials the results for A and B were practically the same?

2. Test whether a thermostatic switch is properly set to 20°C against the alternative that its setting is too low. Use a sample of 9 values, 8 of which are less than 20°C and 1 is greater than 20°C.

3. Test $\tilde{\mu} = 0$ against $\tilde{\mu} > 0$, using $1, -1, 1, 3, -8, 6, 0$ (deviations of the azimuth [multiples of 0.01 radian] in some revolution of a satellite).

4. Each of 10 patients were given two different sedatives A and B. The following table shows the effect (increase of sleeping time, measured in hours). Using the sign test, find out whether the difference is significant.

A	1.9	0.8	1.1	0.1	−0.1	4.4	5.5	1.6	4.6	3.4
B	0.7	−1.6	−0.2	−1.2	−0.1	3.4	3.7	0.8	0.0	2.0
Difference	1.2	2.4	1.3	1.3	0.0	1.0	1.8	0.8	4.6	1.4

5. Assuming that the populations corresponding to the samples in Prob. 4 are normal, apply the test explained in Example 3, Sec. 24.7.

6. Does a process of producing plastic pipes of length $\tilde{\mu} = 2$ meters need adjustment if in a sample, 4 pipes have the exact length and 15 are shorter and 3 longer than 2 meters? (Use the normal approximation of the binomial distribution; see Sec. 23.7.)

7. Do the computations in Prob. 6 without the use of the DeMoivre–Laplace limit theorem.

8. Set up a sign test for the lower quartile q_{25} (defined by the condition $F(q_{25}) = 0.25$).

9. How would you proceed in the sign test if the hypothesis is $\tilde{\mu} = \tilde{\mu}_0$ (any number) instead of $\tilde{\mu} = 0$?

10. Do increases in the price of milk lead to a decrease in sales if daily sales figures [gallons] (ordered according to increasing daily prices) are as follows?

79 76 42 60 65 49 34 56 28

11. Does the yield of some chemical process depend on temperature? (Test against positive trend.)

Temperature [°C]	10	20	30	40	60	80
Yield [kg/min]	0.6	1.1	0.9	1.6	1.2	2.0

12. Test the hypothesis that for a certain type of voltmeter, readings are independent of temperature T [°C] against the alternative that they tend to increase with T. Use a sample of values obtained by applying a constant voltage:

Temperature T [°C]	10	20	30	40	50
Reading V [volts]	99.5	101.1	100.4	100.8	101.6

13. In a swine feeding experiment, the following gains in weight [kg] of 10 animals (ordered according to increasing amounts of food given per day) were recorded:

$$20 \quad 17 \quad 19 \quad 18 \quad 23 \quad 16 \quad 25 \quad 28 \quad 24 \quad 22.$$

Test for no trend against positive trend.

14. Does the amount of fertilizer increase the yield of wheat X [kg/plot]? Use a sample of values ordered according to increasing amounts of fertilizer:

$$41.4 \quad 43.3 \quad 39.6 \quad 43.0 \quad 44.1 \quad 45.6 \quad 44.5 \quad 46.7.$$

15. Apply the test in Example 2 to the following data (x = disulfide content of a certain type of wool, measured in percent of the content in unreduced fibers; y = saturation water content of the wool, measured in percent).

x	10	15	30	40	50	55	80	100
y	50	46	43	42	36	39	37	33

24.12 Pairs of Measurements. Fitting Straight Lines

So far we have been concerned with random experiments in which we observed a single quantity (random variable) and got samples whose values were single numbers. In this last section of the chapter we discuss experiments in which we observe or measure two quantities simultaneously, so that we get samples of *pairs* of values $(x_1, y_1), (x_2, y_2), \cdots, (x_n, y_n)$. In practice we may distinguish between two kinds of experiments, as follows.

1. In **correlation analysis** both quantities are random variables and we are interested in relations between them. Examples are the relation (one says "correlation") between wear X and wear Y of the front tires of cars, between grades X and Y of students in mathematics and in physics, respectively, between the hardness X of steel plates in the center and the hardness Y near the edges of the plates, etc. (We shall not discuss this branch of statistics.)

2. In **regression analysis** one of the two variables, call it x, can be regarded as an ordinary variable because we can measure it without substantial error or we can even give it values we want. x is called the **independent variable,** or sometimes the **controlled variable** because we can control it (set it at values we choose). The other variable, Y, is a random variable, and we are interested in the dependence of Y on x. Typical examples are the dependence of the blood pressure Y on the age x of a person or, as we shall now say, the regression of Y on x, the regression of the gain of weight Y of certain animals on the daily ration of food x, the regression of the heat conductivity Y of cork on the specific weight x of the cork, etc.

In general, in an experiment we usually choose $x_1, \cdots, x_n$ and then observe corresponding values $y_1, \cdots, y_n$ of Y, so that we get a sample $(x_1, y_1), \cdots, (x_n, y_n)$. Now in regression analysis the dependence of Y on x is a dependence of the mean μ of Y on x, so that $\mu = \mu(x)$ is a function in the ordinary sense. The curve of $\mu(x)$ is called the **regression curve** of Y on x.

In this section we discuss the simplest case, namely, that of a straight line

$$\mu(x) = \kappa_0 + \kappa_1 x. \tag{1}$$

Then we may want to plot the sample values as n points in the xY-plane, fit a straight line through them, and use it for estimating $\mu(x)$ at values of x that interest us, so that we know what values of Y we can expect for those x. Fitting that line by eye would not be good because it would be subjective; that is, different persons' results would come out differently, particularly if the points are scattered. So we need a mathematical method that gives a unique result depending only on the n points. A widely used procedure is the **method of least squares** by Gauss. For our present task we may formulate it as follows.

Least squares principle. *The straight line should be fitted through the given points so that the sum of the squares of the distances of those points from the straight line is minimum, where the distance is measured in the vertical direction (the y-direction).*

To get uniqueness, we need some extra condition. To see this, take the sample (0, 1), (0, -1). Then all the lines $y = k_1 x$ with any k_1 satisfy the principle. (Can you see it?) The following assumption will imply uniqueness, as we shall find out.

General assumption (A1)

The x-values $x_1, \cdots, x_n$ in our sample $(x_1, y_1), \cdots, (x_n, y_n)$ are not all equal.

From a given sample $(x_1, y_1), \cdots, (x_n, y_n)$ we shall now determine a straight line by least squares. We write it as

$$y = k_0 + k_1 x \tag{2}$$

and call it the **sample regression line** because it will be the counterpart of the population regression line.

Now a sample point (x_j, y_j) has the vertical distance (distance measured in the y-direction) from (2) given by

$$|y_j - k_0 - k_1 x_j| \qquad \text{(see Fig. 518).}$$

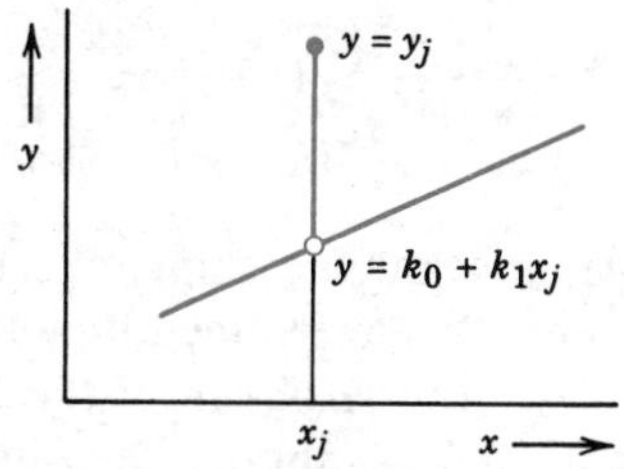

Fig. 518. Vertical distance of a point (x_j, y_j) from a straight line $y = k_0 + k_1x$

Hence the sum of the squares of these distances is

$$q = \sum_{j=1}^{n} (y_j - k_0 - k_1x_j)^2. \tag{3}$$

In the method of least squares we now have to determine k_0 and k_1 such that q is minimum. From calculus we know that a necessary condition for this is

$$\frac{\partial q}{\partial k_0} = 0 \quad \text{and} \quad \frac{\partial q}{\partial k_1} = 0. \tag{4}$$

We shall see that from this condition we obtain for the **sample regression line** the formula

$$\boxed{y - \bar{y} = k_1(x - \bar{x})} \tag{5}$$

where $\bar{x}$ and $\bar{y}$ are the means of the x- and the y-values in our sample,

$$\bar{x} = \frac{1}{n}(x_1 + \cdots + x_n) \quad \text{and} \quad \bar{y} = \frac{1}{n}(y_1 + \cdots + y_n), \tag{6}$$

and the slope k_1, called the **regression coefficient**, is given by

$$\boxed{k_1 = \frac{s_{xy}}{s_1^{\,2}}} \tag{7}$$

with the "**sample covariance**" s_{xy} given by

$$s_{xy} = \frac{1}{n-1}\sum_{j=1}^{n}(x_j - \bar{x})(y_j - \bar{y}) = \frac{1}{n-1}\left[\sum_{j=1}^{n} x_jy_j - \frac{1}{n}\left(\sum_{i=1}^{n} x_i\right)\left(\sum_{j=1}^{n} y_j\right)\right] \tag{8}$$

and

(9) $$s_1^2 = \frac{1}{n-1}\sum_{j=1}^{n}(x_j - \bar{x})^2 = \frac{1}{n-1}\left[\sum_{j=1}^{n} x_j^2 - \frac{1}{n}\left(\sum_{j=1}^{n} x_j\right)^2\right].$$

From (5) we see that the sample regression line passes through the point $(\bar{x}, \bar{y})$, by which it is determined, together with the regression coefficient (7). We may call s_1^2 the *variance* of the x-values in our sample [analogous to (2), Sec. 24.4], but we should keep in mind that x is not a random variable but an ordinary variable.

Derivation of (5) and (7). Differentiating (3) and using (4), we first obtain

$$\frac{\partial q}{\partial k_0} = -2\sum(y_j - k_0 - k_1 x_j) = 0$$

$$\frac{\partial q}{\partial k_1} = -2\sum x_j(y_j - k_0 - k_1 x_j) = 0$$

where we sum over j from 1 to n. We now divide by 2, write each of the two sums as three sums, and take those containing y_j and $x_j y_j$ over to the right. Then we get the **"normal equations"**

(10) $$\begin{aligned} nk_0 + k_1\sum x_j &= \sum y_j \\ k_0\sum x_j + k_1\sum x_j^2 &= \sum x_j y_j. \end{aligned}$$

This is a linear system of two equations in the two unknowns k_0 and k_1. Its coefficient determinant is [see (9)]

$$\begin{vmatrix} n & \sum x_j \\ \sum x_j & \sum x_j^2 \end{vmatrix} = n\sum x_j^2 - \left(\sum x_j\right)^2 = n(n-1)s_1^2 = n\sum(x_j - \bar{x})^2$$

and is not zero because of Assumption (A1), so that the system has a unique solution. Dividing the first equation by n and using (6), we get $k_0 = \bar{y} - k_1\bar{x}$. Together with $y = k_0 + k_1 x$ this gives (5). To get (7), we solve the system by Cramer's rule or elimination, finding

(11) $$k_1 = \frac{n\sum x_j y_j - \sum x_i \sum y_j}{n(n-1)s_1^2}.$$

This gives (7)–(9) and completes the derivation.

[The equality of the two expressions in (8) and in (9) may be shown by the student; see Prob. 10.] ∎

Table 24.12
Regression of the Decrease of Volume y [%] of Leather on the Pressure x [Atmospheres]

Given Values		Auxiliary Values	
x_j	y_j	x_j^2	$x_j y_j$
4,000	2.3	16,000,000	9,200
6,000	4.1	36,000,000	24,600
8,000	5.7	64,000,000	45,600
10,000	6.9	100,000,000	69,000
28,000	19.0	216,000,000	148,400

(C. E. Weir, *Journ. Res. Nat. Bureau of Standards* **45** (1950), 468.)

EXAMPLE 1 **Regression line**

The decrease of volume y [%] of leather for certain fixed values of high pressure x [atmospheres] was measured. The results are shown in the first two columns of Table 24.12. Find the regression line of y on x.

Solution. We see that $n = 4$ and obtain the values $\bar{x} = 28\,000/4 = 7000$, $\bar{y} = 19.0/4 = 4.75$,

$$s_1^2 = \frac{1}{3}\left(216\,000\,000 - \frac{28\,000^2}{4}\right) = \frac{20\,000\,000}{3}$$

$$s_{xy} = \frac{1}{3}\left(148\,400 - \frac{28\,000 \cdot 19}{4}\right) = \frac{15\,400}{3}.$$

Hence $k_1 = 15\,400/20\,000\,000 = 0.000\,77$, and the regression line is

$$y - 4.75 = 0.000\,77(x - 7000) \qquad \text{or} \qquad y = 0.000\,77x - 0.64.$$

Confidence Intervals in Regression Analysis

If we want to get confidence intervals, we have to make assumptions about the distribution of Y (which we have not made so far; least squares is a "geometric principle," nowhere involving probabilities!). We assume normality and independence in sampling:

Assumption (A2)

For each fixed x the random variable Y is normal with mean

$$\mu(x) = \kappa_0 + \kappa_1 x \tag{12}$$

and variance σ^2 independent of x.

Assumption (A3)

The n performances of the experiment by which we obtain a sample $(x_1, y_1), \cdots, (x_n, y_n)$ are independent (p. 1207).

κ_1 in (12) is called the *population* **regression coefficient,** because it can be shown that under Assumptions (A1)–(A3) the maximum likelihood estimate of κ_1 is the sample regression coefficient k_1 given by (11).

Under Assumptions (A1)–(A3) we may now obtain a confidence interval for κ_1, as shown in Table 24.13.

Table 24.13
Determination of a Confidence Interval for κ_1 in (1) under Assumptions (A1)–(A3)

1st Step. Choose a confidence level γ (95%, 99%, or the like).

2nd Step. Determine the solution c of the equation

(13) $$F(c) = \tfrac{1}{2}(1 + \gamma)$$

from the table of the t-distribution with $n - 2$ degrees of freedom (Table A10 in Appendix 5; n = sample size).

3rd Step. Using a sample $(x_1, y_1), \cdots, (x_n, y_n)$, compute $(n - 1)s_1^2$ from (9), $(n - 1)s_{xy}$ from (8), k_1 from (7),

(14) $$(n - 1)s_2^2 = \sum_{j=1}^{n} y_j^2 - \frac{1}{n}\left(\sum_{j=1}^{n} y_j\right)^2,$$

and

(15) $$q_0 = (n - 1)(s_2^2 - k_1^2 s_1^2).$$

4th Step. Compute $K = c\sqrt{q_0/(n - 2)(n - 1)s_1^2}$. The confidence interval is

(16) $$\text{CONF}\,\{k_1 - K \leqq \kappa_1 \leqq k_1 + K\}.$$

EXAMPLE 2 **Confidence interval for the regression coefficient**

Using the sample in Table 24.12, determine a confidence interval for κ_1 by the method in Table 24.13.

Solution. 1st Step. We choose $\gamma = 0.95$.

2nd Step. Equation (13) takes the form $F(c) = 0.975$, and Table A10 in Appendix 5 with $n - 2 = 2$ degrees of freedom gives $c = 4.30$.

3rd Step. From Example 1 we have $3s_1^2 = 20\,000\,000$ and $k_1 = 0.00077$. From Table 24.12 we compute

$$3s_2^2 = 102.2 - \frac{19^2}{4} = 11.95, \qquad q_0 = 11.95 - 20\,000\,000 \cdot 0.00077^2 = 0.092.$$

4th Step. We thus obtain $K = 4.30\sqrt{0.092/2 \cdot 20\,000\,000} = 0.000\,206$ and

$$\text{CONF}\,\{0.00056 \leqq \kappa_1 \leqq 0.00098\}.$$

■

Problem Set 24.12

1. Fit a straight line by eye. Estimate the stopping distance of a car traveling at 35 miles per hour.

x = Speed (mph)	20	30	40	50
y = Stopping distance (ft)	50	95	150	210

In each case find and sketch the sample regression line of y on x.

2. (3, 3.5), (5, 2), (7, 4.5), (9, 3) **3.** (−1, 1), (0, 1.7), (1, 3)

4. (11, 22), (15, 18), (17, 16), (20, 9), (22, 10)

5. Number of revolutions (per minute) x and power y [hp] of a Diesel engine

x	400	500	600	700	750
y	580	1030	1420	1880	2100

6. Deformation x [mm] and Brinell hardness y [kg/mm^2] of a certain type of steel

x	6	9	11	13	22	26	28	33	35
y	68	67	65	53	44	40	37	34	32

In each case find a 95% confidence interval for the regression coefficient κ_1, using the given sample and supposing that Assumptions (A2) and (A3) are satisfied.

7. x = humidity of air [%], y = expansion of gelatin [%]

x	10	20	30	40
y	0.8	1.6	2.3	2.8

8. The sample in Prob. 5.

9. The sample in Prob. 6.

10. Derive the second expression for s_1^2 in (9) from the first one.

Review Questions and Problems for Chapter 24

1. Why and how did we use probability theory in statistics?
2. What are point estimates? Interval estimates?
3. What do we mean by a statistical test? What errors occur in testing?
4. What will generally happen with respect to the errors in testing if we take larger samples?
5. In what cases did we use the t-distribution?
6. What is the χ^2-test? Give a simple example from memory.
7. What are nonparametric tests? When would you apply them?
8. What do we mean by one-sided and two-sided tests? Give examples.
9. What do we mean by "goodness of fit"?

10. Quality control is a kind of test. Explain.
11. Acceptance sampling uses principles of testing. Explain.
12. What is the significance level of a test? How would you choose it?
13. What is the power of a test? What do you do if it is low?
14. Can you also apply tests for the normal distribution to other distributions? Explain.
15. Explain the idea of a maximum likelihood estimate from memory.
16. Why can we generally obtain better results in testing by taking larger samples?
17. In what way does the length of a confidence interval depend on the sample size? On the confidence level?
18. What is Gauss's least squares principle (which he found at age 18)?
19. Why do we not simply fit a straight line through given points by eye?
20. Why do we take samples?

21. To illustrate that s^2 measures the spread, compute s^2 for the two samples 108, 110, 112 and 100, 110, 120, and compare.
22. Graph a histogram of the sample 2, 1, 4, 5 and guess $\bar{x}$ and s by inspecting the histogram. Then compute $\bar{x}$, s^2, and s.

Find the mean, variance, and range of the samples:

23. 110 120 100 90 100 110 130 110 90 100
24. 43 51 50 47 46
25. 1.6 1.8 1.7 1.5 1.4 1.2 1.7 1.6

Find the maximum likelihood estimates of mean and variance of a normal distribution, using the sample:

26. 5 4 6 5 3 5 7 4 6 5 8 6
27. 12 16 15 17 16 15 16 14 17 10 18 15 16 14 20 18

28. Determine a 95% confidence interval for the mean μ of a normal population with variance $\sigma^2 = 16$, using a sample of size 400 with mean 53.
29. What will happen to the length of the interval in Prob. 28 if we reduce the sample size to 100?
30. Determine a 99% confidence interval for the mean of a normal population with standard deviation 2.2, using the sample 28, 24, 31, 27, 22.
31. What confidence interval do we obtain in Prob. 30 if we assume the variance to be unknown?
32. Assuming normality, find a 95% confidence interval for the variance from the sample 145.3, 145.1, 145.4, 146.2.

Assuming that the populations from which the following samples are taken are normal, determine a 95% confidence interval for the mean μ of the population.

33. Nitrogen content [%] of steel 0.74, 0.75, 0.73, 0.75, 0.74, 0.72
34. A sample of diameters of 10 gaskets with mean 4.37 cm and standard deviation 0.157 cm
35. Density [g/cm³] of coke 1.40, 1.45, 1.39, 1.44, 1.38
36. What sample size should we use in Prob. 34 if we want to obtain a confidence interval of length 0.1?

37. The specific heat of iron was measured 41 times at a temperature of 25°C. The sample had the mean 0.106 [cal/g °C] and the standard deviation 0.002 [cal/g °C]. What can we assert with a probability of 99% about the possible size of the error if we use that sample mean to estimate the true specific heat of iron?

38. Find a 95% confidence interval for the percentage of cars on a certain highway that have poorly adjusted brakes, using a random sample of 1000 cars stopped at a roadblock on that highway, 188 of which had poorly adjusted brakes.

Assuming that the populations from which the following samples are taken are normal, determine a 99% confidence interval for the variance σ^2 of the population.

39. Rockwell hardness of tool bits 64.9, 64.1, 63.8, 64.0

40. A sample of size $n = 128$ with variance $s^2 = 1.921$

41. Using a sample of 10 values with mean 14.5 from a normal population with variance $\sigma^2 = 0.25$, test the hypothesis $\mu_0 = 15.0$ against the alternative $\mu_1 = 14.5$ on the 5% level.

42. In Prob. 41, change the alternative to $\mu \neq 15.0$ and test as before.

43. Find the power in Prob. 41.

44. Using a sample of 15 values with mean 36.2 and variance 0.9, test the hypothesis $\mu_0 = 35.0$ against the alternative $\mu_1 = 37.0$, assuming normality and taking $\alpha = 1\%$.

45. Using a sample of 20 values with variance 8.25 from a normal population, test the hypothesis $\sigma_0{}^2 = 5.0$ against the alternative $\sigma_1{}^2 = 8.1$, choosing $\alpha = 5\%$.

46. A firm sells paint in cans containing 1 kg of paint per can and is interested to know whether the mean weight differs significantly from 1 kg, in which case the filling machine must be adjusted. Set up a hypothesis and an alternative and perform the test, assuming normality and using a sample of 20 fillings having a mean of 991 g and a standard deviation of 8 g. (Choose $\alpha = 5\%$.)

47. If a sample of 100 tires of a certain kind has a mean life of 26 000 km and a standard deviation of 2000 km, can the manufacturer claim that the true mean life of such tires is greater than 25 000 km? Set up and test a corresponding hypothesis, assuming normality and choosing $\alpha = 1\%$.

48. Three specimens of high-quality concrete had compressive strength 357, 359, 413 [kg/cm²], and for three specimens of ordinary concrete the values were 346, 358, 302. Test for equality of the population means, $\mu_1 = \mu_2$, against the alternative $\mu_1 > \mu_2$. (Assume normality and equality of variances. Choose $\alpha = 5\%$.)

49. Using samples of sizes 10 and 5 with variances $s_1{}^2 = 50$ and $s_2{}^2 = 20$ and assuming normality of the corresponding populations, test the hypothesis H_0: $\sigma_1{}^1 = \sigma_2{}^2$ against the alternative $\sigma_1{}^2 > \sigma_2{}^2$. Choose $\alpha = 5\%$.

50. Set up a control chart (find LCL and UCL) for the mean of a normal distribution (length of screws) with mean 5.00 cm and standard deviation 0.03 cm, for the sample size $n = 8$.

51. What effect on UCL − LCL in a control chart for the mean does it have if we double the sample size? If we switch from $\alpha = 1\%$ to $\alpha = 5\%$?

52. The following samples of screws (length in inches) were taken from an ongoing production:

Sample No.	1	2	3	4	5	6	7	8
Length	3.49 3.50	3.48 3.47	3.52 3.49	3.50 3.51	3.51 3.48	3.49 3.50	3.52 3.50	3.53 3.49

Assuming that the population is normal with mean 3.500 and variance 0.0004, set up a control chart for the mean and graph the sample means on the chart.

53. A purchaser checks gaskets by a single sampling plan that uses a sample size of 40 and an acceptance number of 1. Use Table A6 in Appendix 5 to compute the probability of acceptance of lots containing the following percentages of defective gaskets $\frac{1}{4}\%$, $\frac{1}{2}\%$, 1%, 2%, 5%. 10%. Graph the OC curve. (Use the Poisson approximation.)

54. Lots of copper pipes are inspected according to a single sample plan that uses sample size 30 and acceptance number 1. Graph the OC curve of the plan, using the Poisson approximation.

55. Graph the AOQ curve in Prob. 54. Determine the AOQL, assuming that rectification is applied.

56. Do the work required in Prob. 54 if $n = 35$ and $c = 0$.

57. If in some week the numbers of customers in a bank from Monday to Friday were 2680, 1600, 2020, 2250, 3650, can we assert on the 1% level that the random variable X = *Number of customers per day* has the same probability (0.2) for each weekday?

58. Repeat the test in Prob. 57 for the more reasonable model that the five probabilities are 0.22, 0.14, 0.16, 0.18, 0.30, respectively, choosing $\alpha = 1\%$.

59. Does an automatic cutter have the tendency of cutting longer and longer pieces of wire if the lengths of subsequent pieces [in.] were 10.1, 9.8, 9.9, 10.2, 10.6, 10.5?

60. Does the sample 4.8, 4.7, 4.9, 5.1, 4.6, 5.0 (ordered by increasing amounts of irrigation) indicate that the yield of corn per square meter increases with increasing irrigation?

Summary of Chapter 24 Mathematical Statistics

With an experiment in which we observe some quantity (number of defectives, height of persons, etc.) there is associated a random variable X whose probability distribution is given by a distribution function

$$F(x) = P(X \leqq x) \qquad \text{(1)} \quad \text{(Sec. 23.4)}$$

which for each x gives the probability that X assumes any value not exceeding x. In statistics we take random samples $x_1, \cdots, x_n$ of size n by performing that experiment n times (Secs. 24.2, 24.3). The purpose is to draw conclusions from properties of samples about properties of the distribution of the corresponding X (Sec. 24.1). We do this by calculating *point estimates* or *confidence intervals* or by performing a *test* for **parameters** (μ and σ^2 in the normal distribution, p in the binomial distribution, etc.) or by a test for distribution functions.

A **point estimate** (Sec. 24.5) is an approximate value for a parameter in the distribution of X obtained from a sample. For instance, the **sample mean** (Sec. 24.4)

$$\bar{x} = \frac{1}{n}\sum_{j=1}^{n} x_j = \frac{1}{n}(x_1 + \cdots + x_n) \qquad \text{(2)}$$

is an estimate of the mean μ of X, and the **sample variance** (Sec. 24.4)

$$s^2 = \frac{1}{n-1}\sum_{j=1}^{n}(x_j - \bar{x})^2 \tag{3}$$

$$= \frac{1}{n-1}[(x_1 - \bar{x})^2 + \cdots + (x_n - \bar{x})^2]$$

is an estimate of the variance σ^2 of X. Point estimation can be done by the basic **maximum likelihood method** (Sec. 24.5).

Confidence intervals (Sec. 24.6) are intervals $\theta_1 \leqq \theta \leqq \theta_2$ with endpoints calculated from a sample such that with a high probability γ, which we can choose (95%, for instance), we obtain an interval that contains the unknown true value of the parameter θ in the distribution of X. We denote such an interval by CONF $\{\theta_1 \leqq \theta \leqq \theta_2\}$.

In a **test** for a parameter we test a *hypothesis* $\theta = \theta_0$ against an *alternative* $\theta = \theta_1$ [this notation has nothing to do with the θ_1 above] and then, on the basis of a sample, reject (or do not reject) the hypothesis in favor of the alternative (Sec. 24.7). Like any conclusion about X from samples, this may involve errors leading to a false decision, as follows. There is a small probability α (which we can choose, 5% or 1%, for instance) that we reject a true hypothesis, and there is a probability β (which we can compute and decrease by taking larger samples) that we accept a false hypothesis. α is called the **significance level** and $1 - \beta$ the **power** of the test. Among many applications, testing is used in **quality control** (Sec. 24.8) and **acceptance sampling** (Sec. 24.9).

If not merely a parameter but the kind of distribution of X is unknown, we can use the **chi-square test** (Sec. 24.10) for testing the hypothesis that some function $F(x)$ is the unknown distribution function of X. This is done by determining the discrepancy between $F(x)$ and the distribution function $\tilde{F}(x)$ of a given sample.

"Distribution-free" or **nonparametric tests** are tests that apply to any distribution, since they are based on combinatorial ideas. Two such tests are discussed in Sec. 24.11.

Finally, Sec. 24.12 deals with samples consisting of pairs of values, as they arise if we simultaneously consider two quantities, and gives an introduction to **regression analysis.** One of these two quantities is an ordinary variable x, the other a random variable Y whose mean depends on x,

$$\mu(x) = \kappa_0 + \kappa_1 x.$$

From a given sample $(x_1, y_1), \cdots, (x_n, y_n)$ one can obtain as an analog the **sample regression line**

$$y - \bar{y} = k_1(x - \bar{x}) \tag{4}$$

with $\bar{x} = (\Sigma\, x_j)/n$, $\bar{y} = (\Sigma\, y_j)/n$, by Gauss's **least squares principle.**

[illegible]

is an estimate of the mean μ of X, and the sample variance (Sec. 24.4)

$$(3) \quad s^2 = \frac{1}{n-1}\sum_{j=1}^{n} (x_j - \bar{x})^2$$

is an estimate of the variance σ^2 of X. Point estimation can be done by the basic maximum likelihood method (Sec. 24.5).

Confidence intervals (Sec. 24.6) are intervals $\theta_1 \leq \theta \leq \theta_2$ with endpoints calculated from a sample such that with a high probability γ [illegible] which we can choose (95%, for instance), we obtain an interval that contains the unknown true value of the parameter θ of the distribution of interest. [illegible]

In a test for a parameter we test a hypothesis $\theta = \theta_0$ against an alternative $\theta = \theta_1$ and then, on the basis of a sample, accept or reject the hypothesis [illegible] about X [illegible] false conclusion [illegible] is small (5% or 1%, for instance) [illegible] the probability which we can [illegible] the latter [illegible] that we accept [illegible] is called the significance level [illegible] is the power of the test. [illegible] testing is used in quality control (Sec. 24.7) and acceptance sampling (Sec. [illegible]).

If [illegible] the distribution [illegible] of a distribution [illegible] we can use the chi-square test (Sec. 24.10) [illegible] the function $F(x)$ [illegible] the distribution function [illegible] of X [illegible] and the distribution function $\tilde{F}(x)$ of the given sample.

Distribution-free or nonparametric tests are tests that apply to any distribution, since they are based on combinatorial [illegible] such tests are discussed in Sec. 24.11.

Finally, Sec. 24.12 deals with samples consisting of pairs of values [illegible] if they arise if we simultaneously consider two quantities, and distinguish correlation and regression analysis. One of these two quantities is an ordinary variable [illegible] variable [illegible] whose mean depends on

$$[illegible]$$

From a given sample $(x_1, y_1), \cdots, (x_n, y_n)$ one can obtain an estimate [illegible] sample regression line

$$(4) \quad [illegible]$$

[illegible] $= (\Sigma x_j y_j - n\bar{x}\bar{y})/[illegible]$, by Gauss's least squares principle.

Appendix 1

References

General References

[1] Abramowitz, M. and I. A. Stegun (eds.), *Handbook of Mathematical Functions*. 10th printing, with corrections. Washington, DC: National Bureau of Standards, 1972. (Also New York: Dover, 1965.)

[2] Cajori, F., *A History of Mathematics*. 3rd ed. New York: Chelsea, 1980.

[3] *CRC Handbook of Mathematical Sciences*. 6th ed. Boca Raton, FL: CRC Press, 1987.

[4] Courant, R. and D. Hilbert, *Methods of Mathematical Physics*. 2 vols. New York: Wiley-Interscience, 1989.

[5] Courant, R. and F. John, *Introduction to Calculus and Analysis*. 2 vols. New York: Wiley-Interscience, 1965, 1974.

[6] Erdélyi, A., W. Magnus, F. Oberhettinger and F. G. Tricomi, *Higher Transcendental Functions*. 3 vols. New York: McGraw-Hill, 1953, 1955.

[7] Hildebrand, F. B., *Advanced Calculus for Applications*. 2nd ed. Englewood Cliffs, NJ: Prentice-Hall, 1976.

[8] Itô, K. (ed.), *Encyclopedic Dictionary of Mathematics*. 4 vols. 2nd ed. Cambridge, MA: MIT Press, 1987.

[9] Kreyszig, E., *Introductory Functional Analysis with Applications*. New York: Wiley, 1989.

[10] Magnus, W., F. Oberhettinger and R. P. Soni, *Formulas and Theorems for the Special Functions of Mathematical Physics*. 3rd ed. New York: Springer, 1966.

[11] Sneddon, I. N. (ed.), *Encyclopedic Dictionary of Mathematics*. Elmsford, NY: Pergamon, 1976.

[12] Thomas, G. B. and R. L. Finney, *Calculus and Analytic Geometry*. 8th ed. Reading, MA: Addison-Wesley, 1992.

Part A. Ordinary Differential Equations (Chaps. 1–6)

See also Part E: Numerical Analysis.

[A1] Birkhoff, G. and G.-C. Rota, *Ordinary Differential Equations*. 4th ed. New York: Wiley, 1989.

[A2] Cesari, L., *Asymptotic Behavior and Stability Problems in Ordinary Differential Equations*. 3rd ed. New York: Springer, 1971.

[A3] Churchill, R. V., *Operational Mathematics*. 3rd ed. New York: McGraw-Hill, 1972.

[A4] Coddington, E. A. and N. Levinson, *Theory of Ordinary Differential Equations*. New York: McGraw-Hill, 1955.

[A5] Erdélyi, A., W. Magnus, F. Oberhettinger and F. Tricomi, *Tables of Integral Transforms*. 2 vols. New York: McGraw-Hill, 1954.

[A6] Ince, E. L., *Ordinary Differential Equations*. New York: Dover, 1956.

[A7] Kamke, E., *Differentialgleichungen, Lösungsmethoden und Lösungen. I. Gewöhnliche Differentialgleichungen*. 3rd ed. New York: Chelsea, 1948. (This extremely useful book contains a systematic list of more than 1500 differential equations and their solutions.)

[A8] Oberhettinger, F. and L. Badii, *Tables of Laplace Transforms*. New York: Springer, 1973.

[A9] Watson, G. N., *A Treatise on the Theory of Bessel Functions*. 2nd ed. Cambridge: University Press, 1944. (Reprinted 1966.)

[A10] Widder, D. V., *The Laplace Transform*. Princeton, NJ: Princeton University Press, 1941.

Part B. Linear Algebra, Vector Calculus (Chaps. 7–9)

For books on *numerical* linear algebra see also Part E: Numerical Analysis.

[B1] Bellman, R., *Introduction to Matrix Analysis*. 2nd ed. New York: McGraw-Hill, 1970.

[B2] Gantmacher, F. R., *The Theory of Matrices*. 2 vols., 2nd ed. New York: Chelsea, 1990, 1988.

[B3] Gohberg, I., P. Lancaster and L. Rodman, *Invariant Subspaces of Matrices with Applications*. New York: Wiley, 1986.

[B4] Herstein, I. N., *Topics in Algebra*. 2nd ed. Lexington, MA: Xerox College Publishing, 1975.

[B5] Kreyszig, E., *Differential Geometry*. Mineola, NY: Dover, 1991.

[B6] Lang, S., *Linear Algebra*. 3rd ed. New York: Springer, 1987.

[B7] MacDuffee, C. C., *The Theory of Matrices*. New York: Chelsea, 1946.

[B8] Nef, W., *Linear Algebra*. 2nd ed. New York: Dover, 1988.

[B9] Wilkinson, J. H., *The Algebraic Eigenvalue Problem*. Oxford: Clarendon, 1965.

[B10] Wilkinson, J. H. and C. Reinsch, *Linear Algebra*. 2nd ed. New York: Springer, 1986.

Part C. Fourier Analysis and Partial Differential Equations (Chaps. 10, 11)

For books on *numerical* methods for partial differential equations, see also Part E: Numerical Analysis.

[C1] Carslaw, H. S. and J. C. Jaeger, *Conduction of Heat in Solids*. 2nd ed. Oxford: Clarendon, 1959. (Reprinted 1978.)

[C2] Churchill, R. V. and J. W. Brown, *Fourier Series and Boundary Value Problems*. 4th ed. New York: McGraw-Hill, 1987.

[C3] Epstein, B., *Partial Differential Equations*. Malabar, FL: Krieger, 1983.

[C4] Erdélyi, A., W. Magnus, F. Oberhettinger and F. Tricomi, *Tables of Integral Transforms*. 2 vols. New York: McGraw-Hill, 1954.

[C5] Gilbarg, D. and N. S. Trudinger, *Elliptic Partial Differential Equations of Second Order*. 2nd ed. New York: Springer, 1983.

[C6] Gustafson, K. E., *Introduction to Partial Differential Equations and Hilbert Space Methods*. New York: Wiley, 1980.

[C7] John, F., *Partial Differential Equations*. 4th ed. New York: Springer, 1982.

[C8] Sneddon, I. N., *Elements of Partial Differential Equations*. New York: McGraw-Hill, 1957.

[C9] Sommerfeld, A., *Partial Differential Equations in Physics*. New York: Academic Press, 1949.

[C10] Szegö, G., *Orthogonal Polynomials*. 4th ed. New York: American Mathematical Society, 1975.

[C11] Tolstov, G. P., *Fourier Series*. New York: Dover, 1976.

[C12] Widder, D. V., *The Heat Equation*. New York: Academic Press, 1975.

[C13] Zauderer, E., *Partial Differential Equations of Applied Mathematics*. 2nd ed. New York: Wiley, 1989.

[C14] Zygmund, A., *Trigonometric Series*. 2nd ed., reprinted with corrections. Cambridge: University Press, 1988.

Part D. Complex Analysis (Chaps. 12–17)

[D1] Ahlfors, L. V., *Complex Analysis*. 3rd ed. New York: McGraw-Hill, 1979.

[D2] Ahlfors, L. V. and L. Sario, *Riemann Surfaces*. Princeton, NJ: Princeton University Press, 1971.

[D3] Bieberbach, L., *Conformal Mapping*. New York: Chelsea, 1964.

[D4] Henrici, P., *Applied and Computational Complex Analysis*. 3 vols. New York: Wiley, 1988, 1991, 1977.

[D5] Hille, E., *Analytic Function Theory*. 2 vols. 2nd ed. New York: Chelsea, 1990, 1987.

[D6] Knopp, K., *Theory of Functions*. 2 parts. New York: Dover, 1945, 1947.

[D7] Knopp, K., *Elements of the Theory of Functions*. New York: Dover, 1952.

[D8] Rothe, R., F. Ollendorf and K. Pohlhausen, *Theory of Functions as Applied to Engineering Problems*. New York: Dover, 1961.

[D9] Rudin, W., *Real and Complex Analysis*. 3rd ed. New York: McGraw-Hill, 1987.

[D10] Titchmarsh, E. C., *The Theory of Functions*. 2nd ed. London: Oxford University Press, 1939. (Reprinted 1975.)

[D11] Weyl, H., *The Concept of a Riemann Surface*. Reading, MA: Addison-Wesley, 1964.

Part E. Numerical Analysis (Chaps. 18–20)

[E1] Aho, A. V., J. E. Hopcroft and J. D. Ullman, *The Design and Analysis of Computer Algorithms*. Reading, MA: Addison-Wesley, 1974.

[E2] Collatz, L., *The Numerical Treatment of Differential Equations*. 3rd ed. New York: Springer, 1966.

[E3] Dahlquist, G. and Å. Björck, *Numerical Methods*. Englewood Cliffs, NJ: Prentice-Hall, 1974.

[E4] Davis, P. and P. Rabinowitz, *Methods of Numerical Integration*. 2nd ed. New York: Academic Press, 1984.

[E5] DeBoor, C., *A Practical Guide to Splines*. New York: Springer, 1978. (Reprinted 1991.)

[E6] Dongerra, J. J. et al., *LINPACK Users Guide*. Philadelphia: SIAM Publications, 1978.

[E7] Dongerra, J. J. et al., *Solving Linear Systems on Vector and Shared Memory Computers*. Philadelphia: SIAM Publications, 1990.

EISPACK, see [E19].

EISPACK 2, see [E9].

[E8] Forsythe, G. E. and W. R. Wasow, *Finite-Difference Methods for Partial Differential Equations*. New York: Wiley, 1960.

[E9] Garbow, B. S. et al., *Matrix Eigensystem Routines: EISPACK Guide Extension*. New York: Springer, 1972. (Reprinted 1990.)

[E10] Golub, G. H. and C. F. Van Loan, *Matrix Computations*. 2nd ed. Baltimore: Johns Hopkins University Press, 1989.

[E11] Hildebrand, F. B., *Introduction to Numerical Analysis*. 2nd ed. New York: McGraw-Hill, 1974.

[E12] IMSL (International Mathematical and Statistical Libraries), *FORTRAN Subroutines for Mathematics and Statistics. User's Manuals, Version 2.0.* 6 vols. Houston, TX: IMSL, 1991.

[E13] Isaacson, E. and H. B. Keller, *Analysis of Numerical Methods*. New York: Wiley, 1966.

[E14] Knuth, D. E., *The Art of Computer Programming*. 3 vols. 2nd ed. Reading, MA: Addison-Wesley, 1981.

[E15] Lambert, J. D., *Computational Methods in Ordinary Differential Equations*. New York: Wiley, 1983.

LINPACK, see [E6].

[E16] Ostrowski, A., *Solution of Equations in Euclidean and Banach Spaces*. 3rd ed. New York: Academic Press, 1973.

[E17] Prenter, P. M., *Splines and Variational Methods*. New York: Wiley, 1989.

[E18] Press, W. H. et al., *Numerical Recipes in C: The Art of Scientific Computing*. Cambridge: University Press, 1988.

[E19] Smith, B. T. et al., *Matrix Eigensystems Routines-EISPACK Guide*. 2nd ed. New York: Springer, 1976. (Reprinted 1990.)

[E20] Smith, G. D., *Numerical Solution of Partial Differential Equations: Finite Difference Methods*. 3rd ed. London: Oxford University Press, 1986.

[E21] Stetter, H. J., *Analysis of Discretization Methods for Ordinary Differential Equations*. New York: Springer, 1973.

[E22] Stewart, G. W., *Introduction to Matrix Computations*. New York: Academic Press, 1973.

[E23] Strang, G. and G. J. Fix, *An Analysis of the Finite Element Method*. Englewood Cliffs, NJ: Prentice-Hall, 1973.

[E24] Todd, J. (ed.), *Survey of Numerical Analysis*. New York: McGraw-Hill, 1962.

[E25] Todd, J., *Basic Numerical Mathematics*. 2 vols. New York: Academic Press, 1978–1980.

[E26] Wilkinson, J. H., *The Algebraic Eigenvalue Problem*. Oxford: Clarendon, 1965.

[E27] Wilkinson, J. H. and C. Reinsch, *Linear Algebra*. New York: Springer, 1971.

Part F. Optimization, Graphs (Chaps. 21, 22)

[F1] Berge, C., *Graphs*. 3rd ed. New York: North-Holland, 1985.

[F2] Bondy, J. A. and U. S. R. Murty, *Graph Theory with Applications*. New York: North-Holland, 1976.

[F3] Christofides, N. et al. (eds.), *Combinatorial Optimization*. New York: Wiley, 1979.

[F4] Chvátal, V., *Linear Programming*. San Francisco: Freeman, 1983.

[F5] Cook, W. J., W. H. Cunningham, W. R. Pulleyblank and A. Schrijver, *Combinatorial Optimization*. New York: Wiley, 1993.

[F6] Dantzig, G. B., *Linear Programming and Extensions*. 2nd ed. Princeton, NJ: Princeton University Press, 1963.

[F7] Dixon, L. C. W., *Nonlinear Optimization*. London: English Universities Press, 1972.

[F8] Fletcher, R., *Practical Methods of Optimization*. 2nd ed. New York: Wiley, 1988.

[F9] Ford, L. R., Jr., and D. R. Fulkerson, *Flows in Networks*. Princeton, NJ: Princeton University Press, 1962.

[F10] Gass, S. I., *Linear Programming, Methods and Applications*. 3rd ed. New York: McGraw-Hill, 1969.

[F11] Gondran, M. and M. Minoux, *Graphs and Algorithms*. New York: Wiley, 1984.

[F12] Gotlieb, C. C. and L. R. Gotlieb, *Data Types and Structures*. Englewood Cliffs, NJ: Prentice-Hall, 1978.

[F13] Harari, F., *Graph Theory*. Reading, MA: Addison-Wesley, 1969.

[F14] Papadimitriou, C. H. and K. Steiglitz, *Combinatorial Optimization: Algorithms and Complexity*. Englewood Cliffs, NJ: Prentice-Hall, 1982.

Part G. Probability and Statistics (Chaps. 23, 24)

[G1] Anderson, T. W., *An Introduction to Multivariate Statistical Analysis*. 2nd ed. New York: Wiley, 1984.

[G2] Cochran, W. G., *Sampling Techniques*. 3rd ed. New York: Wiley, 1977.

[G3] Cramér, H., *Mathematical Methods of Statistics*. Princeton: Princeton University Press, 1946. (Reprinted 1961.)

[G4] David, H. A., *Order Statistics*. 2nd ed. New York: Wiley, 1981.

[G5] Dodge, H. F. and H. G. Romig, *Sampling Inspection Tables*. 2nd ed. New York: Wiley, 1959.

[G6] Feller, W., *An Introduction to Probability Theory and Its Applications*. Vol. I. 3rd ed. New York: Wiley, 1968.

[G7] Gibbons, J. D., *Nonparametric Statistical Inference*. 2nd ed. New York: Dekker, 1985.

[G8] Hooke, R., *How to Tell the Liars from the Statisticians*. New York: Dekker, 1983.

[G9] Huber, P. J., *Robust Statistics*. New York: Wiley, 1981.

[G10] IMSL, *FORTRAN Subroutines for Mathematics and Statistics. User's Manuals, Verson 2.0*. 6 vols. Houston, TX: IMSL, 1991.

[G11] Kendall, M. and A. Stuart, *The Advanced Theory of Statistics*. 3 vols. 4th ed. New York: Macmillan, 1977, 1979, 1983. (Vol. 3 coauthored by K. Ord.) (*See also* [G16].)

[G12] Kreyszig, E., *Introductory Mathematical Statistics. Principles and Methods*. New York: Wiley, 1970.

[G13] Maindonald, J. H., *Statistical Computation*. New York: Wiley, 1984.

[G14] Pearson, E. S. and H. O. Hartley, *Biometrika Tables for Statisticians*. Vol. I. 3rd ed. Cambridge: University Press, 1966.

[G15] Rand Corporation, *A Million Random Digits with 100,000 Normal Deviates*. Glencoe, IL: Free Press, 1955.

[G16] Stuart, A. and J. K. Orth, *Kendall's Advanced Theory of Statistics*. Vols. 1, 2. 5th ed. Kent, U.K.: Arnold, 1987, 1991.

Appendix 2

Answers to Odd-Numbered Problems

PROBLEM SET 1.1, page 8

1. First order
3. First order
5. First order
7. $-\frac{1}{2}e^{-2x} + c$
9. $-2 \sin \frac{1}{2}x + c$
11. $1.5e^{-x} + 1$
13. $3x^2$
15. $-2 \sin x$
17. $y' = 2y/x$
19. $y' = y^2 + 1$
23. $y'' = k$, $y = \frac{1}{2}kt^2 + 5t$, $y(40) = 800k + 200 = 1800$, $k = 2$, $y'(40) = 85$ m/sec $= 190$ mi/hr
25. (a) $3e^x$, (b) $y(2) = ce^2 = 100$, $y(0) = c = 13.53$
27. $y = 5.3 \exp(0.030t) = 32$ (1860), 78 (1890), the other values being much too large. A better model is the "logistic law" in Sec. 1.7.
29. About 1600 years

PROBLEM SET 1.2, page 13

1. Suppose you forgot c in Example 2, wrote arc tan $y = x$, transformed this to $y = \tan x$ and afterward added a constant c. Then $y = \tan x + c$, which, for $c \neq 0$, is not a solution of $y' = 1 + y^2$.
3. $y = ce^{3x} - 1$
5. $y = \tan(x + \frac{1}{2}x^2 + c)$
7. $y = c(\sin 2x)^{1/2}$
9. $y = c \cosh x$
11. $\tan y = \frac{1}{2}e^{2x} + c$
13. $y = c \ln |x|$
15. $y = \sin(x + c)$
17. $y = -\sqrt{4 - x^2}$
19. $\cot y = \tanh x + 1$
21. $(x - 1)^2 + y^2 = 1$
23. $v^2 = v_0^2 + 2g(t - t_0)$
25. $r = -0.3 \sin \theta$
27. $I = I_0 e^{-(R/L)t}$

PROBLEM SET 1.3, page 18

1. $-v_0$
3. 78.5% of y_0
5. 70.7%, 25%
7. 10 hours
9. $\Delta A = -kA\Delta x$ (A = amount of incident light, ΔA = absorbed light, Δx = thickness, $-k$ = constant of proportionality). Let $\Delta x \to 0$. Then $A' = -kA$. Hence $A(x) = A_0e^{-kx}$ is the amount of light in a thick layer at depth x from the surface of incidence.
11. 11.3, 19.2, 38.4
15. 1.21 meter
17. $pV = c = const$, the law empirically found by Boyle (1662) and Mariotte (1676).
19. $y(t) = 0.01e^{-0.275t}$

PROBLEM SET 1.4, page 22

1. $y = x(\ln|x| + c)$
3. $y^2 = x^2 - cx$
5. $y = x \tan(\ln|x| + c)$
7. $y = -x/(\ln|x| + c)$
9. $y^2 - 2xy - x^2 = c$
11. $y = x(\ln|x| - 7.4)$
13. $y = x(e^x - e)^{1/4}$
15. $y = x(x^2 + 5)^{1/2}$
17. $y = -x + \tan(x + c)$
19. $\sin(x^2y^2) = ce^x$
21. $(y - x)^2 + 10y - 2x = c$
23. $u' = a + by' = a + bf(u),\ du/(a + bf(u)) = dx$
25. $y = ce^{2x} - 4x - 2$
27. $y = [(c - 15x)^{-1/3} - 2]/5$
29. $y = ax,\ y/x = a = const,\ y' = g(a) = const$

PROBLEM SET 1.5, page 26

1. $2x\,dx + 2y\,dy = 0$
3. $(b - y)\,dx - (x - a)\,dy = 0$
5. $e^{xy^2}(y^2\,dx + 2xy\,dy) = 0$
7. $2(y\,dx + x\,dy)/xy = 0$
9. $2(y - x + 1)(-dx + dy) = 0$
11. $xy = c$
13. $xy^3 = c$
15. $re^{-\theta} = c$
17. $\sinh x \cos y = c$
19. $x(x + e^y) = c$
21. No, $y = 1/(5 + \ln|x|)$
23. Yes, $(x - 3)(y - 1) = 1$
25. Yes, $e^{y/x} = e$ or $y = x$
27. No, $y = \sqrt{x^2 + 3x}$. Use $u = y/x$.
29. No, $y = \pi/\cosh x$
31. $y = cx$
33. $b = k,\ ax^2 + 2kxy + ly^2 = c$
35. $\partial f/\partial y = g'(x)h(y)$
37. $x^2 + y/x = c$
39. $y = cx^3$

PROBLEM SET 1.6, page 29

Comment. Since integrating factors are not unique, your F's may differ from those given here.

7. $x^2y = 2$
9. $xe^{xy} = 1$
11. $x^2y^3 - x^3y^2 = -2$
13. $y\cosh(x - y) = 3$
15. $F = e^{2x},\quad e^{2x}\cos \pi y = c$
17. $F = 1/xy,\quad x^2y^2e^x = c$
19. $F = e^{x^2},\quad (x + 2y)e^{x^2} = c$
21. $F = 1/(x + 1)(y + 1),\quad y + 1 = c(x + 1)$
23. $F = x^{a-1}y^{b-1},\quad x^ay^b = c$
25. $\partial P/\partial y = \partial Q/\partial x + (a/x)Q$
27. $(b/y)P + \partial P/\partial y = (a/x)Q + \partial Q/\partial x$
29. $\dfrac{b}{y}\left(\dfrac{2y}{x} - 3\right) + \dfrac{2}{x} = \dfrac{a}{x}\left(3 - \dfrac{2x}{y}\right) - \dfrac{2}{y}$. Now equate the terms in $1/x$ and in $1/y$ to get $a = b = 2$.

PROBLEM SET 1.7, page 35

5. $y = ce^{4x} - 0.2$
7. $y = ce^{-2x} + 2e^x$
9. $y = ce^{4x} + x^2$
11. $y = cx^{-2} + 3x$
13. $y = (c + x)e^{-3x}$
15. $y = (c + 2e^{x^2})x^{-2}$
17. $y = cx^2 + x^2e^x$
19. $y = 2e^{-3x} + 4$
21. $y = 2e^{-x} + x^2 + 1$
23. $y = e^{-x^2} + 2$
25. $y = 1 - \cosh 2x$
27. $y = (0.7 + x)e^{-kx}$
37. $y = 1/(1 + ce^x)$
39. $\sin y = 2 + ce^{-x^2/2}$
41. $x = (c + y)e^y$

43. $y^{-3} = ce^x - 2x - 1$ **45.** $y^2 = cxe^{-x} + \frac{1}{2}xe^x$ **47.** 327.5 min

49. $y' + 0.025y = 0$, $y = \tilde{c}e^{-0.025t}$ if $t \geqq 10$, $y(10) = \tilde{c}e^{-0.25} = 119.6$ from Example 2, $\tilde{c} = 153.6$, $y(t) = 0.01$ if $t = 385.6$ min.

51. $T(t) = T_1 + (T_0 - T_1)e^{-kt}$, where T_1 is the temperature of the water

53. Eq. (7) gives $y' = Ay[1 - (B/A)y] > 0$ if $0 < y < A/B$ and $y' < 0$ if $y > A/B$.

55. Eq. (7) gives $y'' = Ay' - 2Byy' = 0$ if $A - 2By = 0$.

57. Substitute $y = w + v$, $y' = w' + v'$. Since v is a solution, there remains $w' = -pw + g(w^2 + 2wv)$, a Bernoulli equation.

59. $y' = y' + xy'' + (dg/ds)y''$, $y''(x + dg/ds) = 0$, etc.

PROBLEM SET 1.8, page 42

5. It will decrease. **7.** $L = 0.348$ henry

11. $I(0) = -E_0(R^2 + \omega^2L^2)^{-1/2} \sin \delta = -E_0\omega L/(R^2 + \omega^2L^2)$

13. (a) $I(t) = ce^{-(R/L)t} + e^{-t}/(R - L)$, (b) $I(t) = (c + t/L)e^{-t}$

15. $I = I_1 = 1 - \frac{1}{2}e^{-t}$ if $0 \leqq t \leqq 3$, $I = I_2 = c_2e^{-(t-3)}$,
$c_2 = I_2(3) = I_1(3) = 1 - \frac{1}{2}e^{-3} = 0.975$

21. $RQ' + Q/C = E_0$, $Q(t) = E_0C(1 - e^{-t/RC})$, $V(t) = Q(t)/C = 12(1 - e^{-0.05t})$

23. $Q = ce^{-2t} + 0.001(1 - 2e^{-t})$ **25.** $Q = \sin \pi t - 2\pi \cos \pi t$

27. Since the charge on the capacitor cannot change abruptly, we must have J = *jump of RI*, hence *jump of I* = J/R.

29. $I = 1$ if $0 < t < a$ and $I = -ae^{a-t}$ if $t > a$.

PROBLEM SET 1.9, page 46

1. $xy' + y = 0$ **3.** $y' = 2y \cot 2x$ **5.** $xy' = 2y \ln |y|$

7. $xy' - 2y = x^3e^x$ **9.** $y' + y = x/y$ **11.** $y = c^*x$

13. $y = c^* - \frac{1}{2}x^2$ **15.** $y = c^*e^{-2x}$ **17.** $y = c^*e^{x^2}$

19. $y = c^*x^4$ **21.** $y = \frac{1}{2}\sqrt{c - x}$ **23.** $y = c^*x$

25. $x^2 - y^2 = c^*$ **29.** $e^x \sin y = c^*$

PROBLEM SET 1.10, page 51

1. $y = cx^2$ **3.** $y = ce^{-x^2/2}$ **5.** $x^2 + 9y^2 = c$

7. $v(t) = 5(1 - e^{-4t})/(1 + e^{-4t})$, $5(1 + \frac{1}{6}e^{-4t})/(1 - \frac{1}{6}e^{-4t})$

13. $y_1(1) = 2$, $y_2(1) = 2.5$, $y_3(1) = 2.667$

15. $y_n = 1 + 2\int_0^x y_{n-1}(t)\, dt$, $y_0 = 1$, $y_1 = 1 + 2x$, $y_2 = 1 + 2x + (2x)^2/2!$, etc., $y = e^{2x}$

17. $y_0 = -1$, $y_n = -1 - x + x^{n+1}/(n + 1)!$, $y = -1 - x$

19. $y_0 = 0$, $y_1 = x^2 - \frac{1}{4}x^4$, $y_2 = x^2 - \frac{1}{4 \cdot 6}x^6$, etc., $y = x^2$

PROBLEM SET 1.11, page 57

1. $y = cx^3$, c arbitrary. No; $f(x, y) = 3y/x$ is not defined when $x = 0$.

3. $y = (x - 1)^2$, $y \equiv 0$

5. $y = cx^2$, (a) $y(0) = k \neq 0$, (b) $y(0) = 0$, (c) $y(x_0) = y_0$, $x_0 \neq 0$. No, since $f(x, y) = 2y/x$ is not defined when $x = 0$.

7. The xy-plane without the origin

9. This would violate the uniqueness.

11. $\frac{1}{2}$, the largest value of $b/(1 + b^2)$, where $1 + b^2$ is the smallest K.

13. $\alpha = \frac{1}{4}$, the largest value of $b/(b^2 + 4)$, where $b^2 + 4$ is the smallest K. The solution $y = 2 \tan 2x$ exists for $|x| < \pi/4$.

15. $y = (2x - 4)^2$, $y \equiv 0$. Not on the x-axis ($y = 0$).

17. Yes, with $M = 1$. Not on the x-axis ($y = 0$).

19. $f(x, y) = r(x) - p(x)y$, $\partial f/\partial y = p(x)$

CHAPTER 1 (REVIEW QUESTIONS AND PROBLEMS), page 58

11. $e^{2xy} + \sin y = c$

13. $y = ce^{x^2} - x^2$

15. $y = \tan(\frac{1}{2}x^2 + x + c)$

17. $y^2 = 1 + ce^{-x^2}$

19. $x^4 + x^3y^2 = c$

21. $e^{x^2} \tan y = c$

23. $y = x + x(c - x^2)^{-1/2}$

25. $u = y/x$, $\ln(x^2 + y^2) + 2 \arctan(y/x) = c$

27. $4x^2 + 9y^2 = 36$

29. $y = \frac{1}{2} - x^{-1} + \frac{1}{2}x^{-2}$

31. $y = \sin x + x^2$

33. $x^2 + xe^y = 2$

35. $x = \frac{1}{2}y^4 + 4y^2$

37. $y_n = 1 + 2\int_0^x t y_{n-1}(t)\, dt$, $y_0 = 1$, $y_1 = 1 + x^2$, $y_2 = 1 + x^2 + \frac{1}{2}x^4$, etc., $y = e^{x^2}$, $y_1(1) = 2$, $y_2(1) = 2.5$, $y_3(1) = 2.667$

39. $(y - 1)^2 - (x - 1)^2 = c^*$

41. 0.93%, 8.94%, 29.3%

43. 7.27 days, 11.5 days

45. $y = cx^k$

47. 46 ohms

49. $m\, dv/dt = mg(\sin\alpha - 0.2\cos\alpha) = 9.80(0.500 - 0.2 \cdot 0.866)m = 3.203m$, $v = 3.203t$, $s = 3.203t^2/2 = 10$. *Ans.* 8.00 meters/sec.

51. $dy/(y - a)(y - b) = k\, dx$. Now use partial fractions. *Ans.* $y(t) = (b - acf)/(1 - cf)$, where $f(t) = \exp(b - a)kt$

53. $r^2 = a^2 \cos 2\theta$

55. $y' = y$, $y = ce^x$

PROBLEM SET 2.1, page 67

3. Since $y'' + py' + qy = cr$

5. Since $y'' + py' + qy = r - r$

7. $y = c_1x^3 + c_2$

9. $y = c_1 \cosh 2x + c_2$

11. $y = c_1^{-1} \ln|c_1x + 1| + c_2$

15. $x = \frac{1}{4}e^{2y} + c_1y + c_2$

17. $x = -\cos y + c_1y + c_2$

19. $(y - 1)e^y = c_1x + c_2$

21. $xy = 1$

23. $y = e^{3x} + e^{-3x} = 2\cosh 3x$

25. $y = (4 - x)e^x$

27. $y = e^x - 2e^{3x}$

29. $y = (2 + \ln|x|)x^2$

31. Linearly independent

33. Linearly dependent

35. Linearly dependent

37. Linearly independent

39. Linearly independent

PROBLEM SET 2.2, page 72

1. $y = c_1e^{-x} + c_2e^{-2x}$ **3.** $y = (c_1 + c_2x)e^{-5x}$ **5.** $y = (c_1 + c_2x)e^{3x}$
7. $y'' + y' - 6y = 0$ **9.** $y'' + 2.5y' + y = 0$ **11.** $y'' - y = 0$
15. $(\lambda - \lambda_1)(\lambda - \lambda_2) = \lambda^2 - (\lambda_1 + \lambda_2)\lambda + \lambda_1\lambda_2 = \lambda^2 + a\lambda + b$. Now compare.
17. $y = 3e^{4x} - 2e^{-4x}$ **19.** $y = (2x - 4)e^{-3x}$ **21.** $y = 3e^{-0.2x} + 0.3e^{-2x}$
23. $y = e^{-0.5x}$ **25.** $y = e^{3x} + 3e^{-3x} = 4 \cosh 3x - 2 \sinh 3x$

PROBLEM SET 2.3, page 76

1. $y = A \cos 3x + B \sin 3x$ **3.** $y = e^{-\alpha x}(A \cos x + B \sin x)$
5. $y = e^{-\alpha x}(A \cos \omega x + B \sin \omega x)$ **7.** I, $y = c_1e^{5x} + c_2e^{-5x}$
9. II, $y = (c_1 + c_2x)e^{-3x}$ **11.** I, $y = c_1 + c_2e^{-2x}$
13. III, $e^{-0.3x}(A \cos x + B \sin x)$ **15.** III, $y = e^{-x}(A \cos \omega x + B \sin \omega x)$
17. $y = 2 \cos 3x - \sin 3x$ **19.** $y = (3 + 4x)e^{2x}$
21. $y = (3 + 2x)e^{-10x}$ **23.** $y = e^{0.5x} - e^{-x}$
25. $y = 5e^{4x}$ **27.** $y = e^{2x} - 2$
29. $y = -3e^{x} \cos x$

PROBLEM SET 2.4, page 79

1. $3x^2 + 20x,\ 0,\ (6 + \pi) \cos \pi x + (3 - 2\pi) \sin \pi x$
3. $0,\ -5e^{-4x},\ 0$ **5.** $y = e^{-x}(A \cos x + B \sin x)$
7. $y = (c_1 + c_2x)e^{3x/2}$ **9.** $y = (c_1 + c_2x)e^{2x/\pi}$
11. $y = (c_1 + c_2x)e^{-0.6x}$ **13.** $y = -3e^{-2x} \sin x$
15. $y = e^{x}(\cos \pi x - \sin \pi x)$ **17.** $y = (1 - 3x)e^{-x/3}$
19. $y = (1 - x)e^{-x}$

PROBLEM SET 2.5, page 87

1. $y = (y_0^2 + v_0^2\omega_0^{-2})^{1/2} \cos (\omega_0 t - \arctan (v_0/y_0\omega_0))$
3. Lower by $\sqrt{2}$, higher, unchanged
7. $1/k = 1/k_1 + 1/k_2,\ k = k_1k_2/(k_1 + k_2) = 21/10 = 2.1$
9. $mL\theta'' = -mg \sin \theta \approx -mg\theta$ (the tangential component of $w = mg$),
$\theta'' + \omega_0^2\theta = 0,\ \omega_0^2 = g/L$. *Ans.* $\sqrt{g/L}/2\pi$.
11. $T = 2\pi \sqrt{\dfrac{mL}{mg + (k_1 + k_2)L}}$
13. $\theta(t) = 0.2618 \cos 3.7t + 0.0472 \sin 3.7t$ [rad]
17. If c_1 and c_2 have the same sign
19. $y = [(v_0 + \alpha y_0)t + y_0]e^{-\alpha t}$
21. $y = c_2e^{-v_0/c_2}/\alpha$, where $c_2 = v_0 + \alpha y_0$
25. $\omega^* = \omega_0[1 - (c^2/4mk)]^{1/2} \approx \omega_0[1 - \frac{1}{2}(c^2/4mk)]$
27. The positive solutions of $\tan t = 1$, that is, $\pi/4$ (max), $5\pi/4$ (min), etc.
29. 0.046 kg/sec, from $\exp(-20 \cdot 3c/2m) = \frac{1}{2}$.

PROBLEM SET 2.6, page 93

3. $y = c_1 + c_2x^{-3}$ **5.** $y = (c_1 + c_2 \ln x)x^{-4}$ **7.** $y = (c_1 + c_2 \ln x)/x$
9. $y = (c_1 + c_2 \ln x)x^{-2.6}$ **11.** $y = A \cos(\ln x) + B \sin(\ln x)$
13. $y = c_1x^{1.5} + c_2x^{-2.5}$ **15.** $y = 4/\sqrt{x}$ **17.** $y = -x \cos(\ln x)$
19. $y = x^{0.1}$
21. $x = e^t$, $t = \ln x$; by the chain rule, $y' = \dot{y}t' = \dot{y}/x$, $y'' = \ddot{y}/x^2 - \dot{y}/x^2$, hence $\ddot{y} - \dot{y} + a\dot{y} + by = 0$.
25. $z - 2 = x$, $y = c_1(z - 2)^{-3} + c_2(z - 2)^{-1}$

PROBLEM SET 2.7, page 98

1. $W = (\lambda_2 - \lambda_1) \exp(\lambda_1 x + \lambda_2 x)$ **3.** $W = e^{-\alpha x}\omega$
5. $W = \nu x^{2\mu - 1}$ **7.** $y'' - 2y' + y = 0$, $W = e^{2x}$
9. $y'' - 2y' + 2y = 0$, $W = e^{2x}$ **11.** $y'' + \pi^2 y = 0$, $W = \pi$
13. $xy'' - 2y' = 0$, $W = 3x^2$ **15.** $x^2y'' - xy' + 0.75y = 0$, $W = x$
17. $W = 0$, since $y_1' = 0$, $y_2' = 0$ at such an x, by calculus.
19. $z_1 = (y_1 + y_2)/2$, $z_2 = (y_1 - y_2)/2$, $\det [a_{jk}] = -1/2$
21. $y_2 = x^2 + x$ **23.** $y_2 = (1 - x)^{-2}$
25. $y_2 = x^{-1} \cos x$

PROBLEM SET 2.8, page 102

1. $c_1e^x + c_2e^{-x} + e^{2x}$ **3.** $A \cos x + B \sin x + \sin 2x$
5. $c_1e^x + c_2e^{-4x} + xe^x$ **7.** $c_1x + c_2x^2 - 5x \cos x$
9. $(c_1 + c_2 \ln x)x^{1/2} + \cos \frac{1}{2}x$ **11.** $(x + x^2)e^{3x}$
13. $(1 + 3x) \cos 2x$ **15.** $(1 - \sin x)e^x$
19. $A \cos x + B \sin x + e^x + x \sin x$

PROBLEM SET 2.9, page 105

1. $A \cos x + B \sin x + 3x^2 - 6$ **3.** $(c_1 + c_2x)e^{-3x} + \sin 3x$
5. $c_1e^{-x} + c_2e^{2x} - \frac{1}{2}e^x - \frac{1}{2}x + \frac{1}{4}$ **7.** $e^{-x}(A \cos 2x + B \sin 2x) + x^2$
9. $c_1e^x + c_2e^{3x} + \sin x$ **11.** $(c_1 + c_2x)e^x + x^2e^x$
13. $A \cos 3x + B \sin 3x + \frac{1}{6}x \sin 3x$ **15.** $(c_1 + \frac{1}{4}x)e^{2x} + (c_2 + \frac{1}{4}x)e^{-2x} - \frac{1}{4}x$
17. $e^{-x} - e^{2x} + xe^{2x}$ **19.** $4e^x + 2e^{-2x} + x^2 - 6$
21. $\cos x - 3 \sin x$ **23.** $3 \cos 2x - e^{-2x}$
25. $c_1e^{-2x} + \frac{1}{4} \cos 2x + \frac{1}{4} \sin 2x$

PROBLEM SET 2.10, page 109

1. $(c_1 + c_2x + \frac{4}{35}x^{7/2})e^x$ **3.** $(c_1 + c_2x)e^x + 6e^x/x$ **5.** $(c_1 + c_2x - \ln|x|)e^{-2x}$
7. $(c_1 + c_2x + 4x^{7/2})e^x + x^2 + 4x + 6$
9. $(c_1 + c_2x - \sin x)e^x$ **11.** $(c_1 + c_2x + x^{-2})e^{2x}$
13. $(c_1 + c_2x + (2x)^{-1})e^x$ **15.** $c_1x^2 + c_2x^3 + x^{-4}$

17. $c_1x + c_2x^2 - 5x\cos x$

19. $c_1 + c_2x^2 + (x - 1)e^x$

21. $c_1x + c_2x^2 + 2x^{-2}$

23. $c_1x^2 + c_2x^3 + \frac{1}{42}x^{-4}$

25. $c_1e^{-x} + c_2e^{-3x} + 8\sin 2x - \cos 2x$

PROBLEM SET 2.11, page 115

1. $A\cos 2t + B\sin 2t + 5\sin t$. Note that the general solution corresponding to the free motion will not die out since there is no damping.
3. $4\sin t - 12\cos t$
5. $y = -0.4\cos\frac{1}{2}t + 0.92\sin\frac{1}{2}t$
7. $-0.025\cos 0.2t + 0.252\sin 0.2t$
9. $c_1\cos 5t + c_2\sin 5t + 2\sin t$
11. $e^{-t}(A\cos t + B\sin t) + 0.2\cos t + 0.4\sin t$
13. $e^{-t/4}[A\cos(\sqrt{31}/4)t + B\sin(\sqrt{31}/4)t] + 4\cos t + 2\sin t$
15. $y = A\cos t + B\sin t + (1 - \omega^2)^{-1}\cos\omega t$
17. $\cos 3t + \sin t$
19. $[y_0 - (\omega_0^2 - \omega^2)^{-1}]\cos\omega_0t + (v_0/\omega_0)\sin\omega_0t + (\omega_0^2 - \omega^2)^{-1}\cos\omega t$
21. $y = 3te^{-t/2} + 2\sin t - \sin 2t$
25. $y = (1 - \omega^2)^{-1}(\cos\omega t - \cos t)$

PROBLEM SET 2.12, page 120

3. $R > R_{crit} = 2\sqrt{L/C}$ is Case I, etc.
5. $I = 0$
7. $I = (4\cos t - 198\sin t)/1961$
9. $I = e^{-t}(A\cos t + B\sin t) + \cos t + 2\sin t$
11. $I = 5e^{-2t}\sin t$
13. $I = e^{-3t}(3\sin 4t - 4\cos 4t) + 4\cos 5t$
15. $I = A\cos 5t + B\sin 5t + [275\omega/(25 - \omega^2)]\cos\omega t$
17. $I = A\cos 20t + B\sin 20t + 0.02t$
19. $I = 10(\cos t - \cos 2t)$
21. $I = 1 - \cos t$
23. $I = \sin t$ if $0 < t < 1$, $I = \sin t - \sin(t - 1)$ if $t > 1$
25. $I = \frac{1}{2}(e^{-t} - \cos t + \sin t)$ if $0 < t < \pi$,
    $I = -\frac{1}{2}(1 + e^{-\pi})\cos t + \frac{1}{2}(3 - e^{-\pi})\sin t$ if $t > \pi$

PROBLEM SET 2.13, page 123

1. $\frac{1}{73}(50\sin 4t - 110\cos 4t)$
3. $-0.9704\cos 5t + 0.4621\sin 5t$
5. $-0.1\cos 2t + 0.2\sin 2t$
9. $y_p = -4\cos 2t$
11. $0.2\cos t + 0.4\sin t$
13. $8\sin 2t - \cos 2t$
15. $-3\cos 3t - 4\sin 3t$

CHAPTER 2 (REVIEW QUESTIONS AND PROBLEMS), page 124

17. $c_1e^x + c_2e^{-2x} + xe^x$
19. $(c_1 + c_2 \ln x)x^{0.7}$
21. $(c_1 + c_2x + x \ln x - x)e^{2x}$
23. $(c_1 + c_2x - \cos x)e^{-x}$
25. $c_1x^2 + c_2x^3 - x^2 \sin x$
27. $c_1e^{-x} + c_2e^{-3x} + \frac{1}{2} \sin x$
29. $(c_1 + c_2x - \ln x)e^{-2x}$
31. $0.3e^x + 2x^2$
33. $x^{-1/2} + 2x^{-3/2}$
35. $x - 4x^{-1} + 2x^3$
37. $3e^{-\alpha x} \cos \pi x$
39. $2e^{-x} \cos x$
41. $e^{-x} \cos x$
43. $\sin x$
45. $e^{x/2} - 2 \cos x + e^x$
47. $c_1c^{-1999.87t} + c_2e^{-0.125\,008t}$
49. $\frac{1}{73}(50 \sin 4t - 110 \cos 4t)$
51. $\frac{1}{2} \sin 2t$
53. $e^{-t} \cos \sqrt{5}t + \frac{1}{2} \sin 2t$
55. $c_1e^{-4x} + c_2e^{-2x} - 3 \cos 2x + \sin 2x$

PROBLEM SET 3.1, page 134

17. $W = 12$, $y = 6x - 8x^3$
19. $W = 2x^{-3}$, $y = 2 - 3x + 5x^{-1}$
21. Linearly independent
23. Linearly dependent
25. Linearly independent
27. Linearly independent
29. Linearly dependent
31. Linearly dependent
33. $I = [-1, 1]$. Linearly dependent on $[0, 1]$ or $[-1, 0]$, for instance.
39. We would have $W = 0$ at such a point.

PROBLEM SET 3.2, page 140

1. $y''' - 6y'' + 11y' - 6y = 0$
3. $y''' - 3y'' + 3y' - y = 0$
5. $y^{IV} - 2y'' + y = 0$
7. $y^{IV} + 4y'' = 0$
9. $c_1 + c_2e^{-x} + c_3e^x$
11. $(c_1 + c_2x)e^x + c_3e^{-x}$
13. $(A + Cx) \cos x + (B + Dx) \sin x$
15. $c_1e^{3x} + c_2e^{-3x} + A \cos 3x + B \sin 3x$
17. $4x^2 - 4$
19. $(2 - x)e^x$
21. $3 \cosh x$
23. $c_1e^x + c_2e^{-2x} + c_3e^{3x}$
25. $c_1e^{2x} + A \cos 2x + B \sin 2x$
27. $c_1e^{4x} + c_2e^{-x/2} + c_2e^{-3x/2}$
33. $c_1x + c_2x^2 + c_3x^{-2}$
35. $c_1x^{-1} + c_2x^{-2} + c_3x$

PROBLEM SET 3.4, page 146

1. $c_1e^{-x} + c_2e^x + c_3e^{-2x} + e^{2x}$
3. $c_1 + c_2e^{-x} + c_3e^x - \sin 2x$
5. $(c_1 + c_2x + c_3x^2)e^x + 2x^3e^x$
7. $(c_1 + c_2x + c_3x^2)e^{-x} + 2e^x + x$
9. $\cosh x + \cos x$
11. $x - 3x^2 + x^3/3 + x^4/6$
13. $e^{-x} \sin x - x^2 - x - 1$
15. $\sin x + \sin 3x + 0.1 \sinh x$

PROBLEM SET 3.5, page 149

1. $[c_1 + c_2x + c_3x^2 + (8/105)x^{7/2}]e^x$
3. $c_1e^x + c_2e^{2x} + c_3e^{3x} + \frac{1}{2}e^{2x}\cos x$
5. $c_1 + c_2\cos x + c_3\sin x + \ln|\sec x + \tan x| - x\cos x + (\sin x)\ln|\cos x|$
7. $c_1x^{-1} + c_2 + c_3x + x^{-1}e^x$
9. $c_1x^{-1} + c_2x + c_3x^2 - 1/12x^2$

CHAPTER 3 (REVIEW QUESTIONS AND PROBLEMS), page 150

11. $c_1e^{-x} + c_2e^x + c_3xe^x$
13. $c_1x^{1/2} + c_2x^{-1/2} + c_3$
15. $c_1 + c_2x^2 + c_3x^4$
17. $c_1\cosh x + c_2\sinh x + c_3\cos x + c_4\sin x + 2e^{2x} + x^2$
19. $c_1e^x + c_2e^{-x} + c_3e^{2x} + c_4e^{-2x} + \cos 2x$
21. $4 - 3e^x + 2.4e^{-x}$
23. $(2 - x)e^x$
25. $(1 + x)e^x$
27. $3x^2 - 0.5x^{-1}$
29. $(2 + 4x^2)e^x$

PROBLEM SET 4.1, page 163

3. $y_1 = (2/\sqrt{6})\sin\sqrt{6}t,\ y_2 = -(1/\sqrt{6})\sin\sqrt{6}t$
5. $y_1 = -2\cos t + 8\sin\sqrt{6}t,\ y_2 = -4\cos t - 4\sin\sqrt{6}t$
7. $I_1 = 16e^{-2t} - 10e^{-0.8t} + 3,\ I_2 = 8e^{-2t} - 8e^{-0.8t}$

PROBLEM SET 4.3, page 174

1. $y_1 = c_1e^t + c_2e^{-t},\ y_2 = c_1e^t - c_2e^{-t}$
3. $y_1 = A\cos 2t + B\sin 2t,\ y_2 = A\sin 2t - B\cos 2t$
5. $y_1 = c_1e^t + c_2e^{3t},\ y_2 = (-c_1e^t + c_2e^{3t})/3$
7. $y_1 = -2c_1e^{-t} - 3c_2e^{-2t},\ y_2 = c_1e^{-t} + c_2e^{-2t}$
9. $y_1 = e^t(A\cos t + B\sin t),\ y_2 = e^t(A\sin t - B\cos t)$
11. $y_1 = 3e^{2t} - 12e^{-2t},\ y_2 = 3e^{2t} + 12e^{-2t}$
13. $y_1 = -6e^{4t} + 2,\ y_2 = -3e^{4t} - 1$
15. $y_1 = 4e^{2t} - e^{-5t},\ y_2 = 3e^{2t} + e^{-5t}$
17. $y_2 = y_1' + y_1,\ y_2' = y_1'' + y_1' = -y_1 - (y_1' + y_1),\ y_1 = e^{-t}(A\cos t + B\sin t)$, $y_2 = y_1' + y_1 = e^{-t}(B\cos t - A\sin t)$
19. $I_1 = c_1e^{-t} + 3c_2e^{-3t},\ I_2 = -3c_1e^{-t} - c_2e^{-3t}$

PROBLEM SET 4.4, page 179

1. $y_1 = c_1e^{2t} + c_2e^{4t},\ y_2 = c_1e^{2t} - c_2e^{4t}$; unstable node
3. $y_1 = c_1e^t + c_2e^{-2t},\ y_2 = 4c_1e^t + c_2e^{-2t}$; saddle point
5. $y_1 = 3c_1e^{5t} + c_2e^{-2t},\ y_2 = 4c_1e^{5t} - c_2e^{-2t}$; saddle point
7. $y_1 = e^{-2t}(A\cos 2t + B\sin 2t),\ y_2 = e^{-2t}(-A\sin 2t + B\cos 2t)$; stable and attractive spiral point

9. $y_1 = 3c_1e^{3t} + 3c_2e^{9t}$, $y_2 = -c_1e^{3t} + c_2e^{9t}$; unstable node
13. $\Delta = -16$ for any k, so that the answer is no.
15. $y = c_1e^{kt} + c_2e^{-kt}$; hyperbolas $k^2y_1^2 - y_2^2 = const$
17. $y = A \cos \omega_0 t + B \sin \omega_0 t$; ellipses $\omega_0^2 y_1^2 + y_2^2 = const$
19. Ellipses

PROBLEM SET 4.5, page 186

1. At (0, 0), $y_1' = y_2$, $y_2' = y_1$; saddle point. At (1, 0), set $y_1 = 1 + \tilde{y}_1$. Then $y_1 - y_1^2 = -\tilde{y}_1 - \tilde{y}_1^2 \approx -\tilde{y}_1$; thus $\tilde{y}_1' = y_2$, $y_2' = -\tilde{y}_1$; center.
3. (0, 0) saddle point; (−3, 0) and (3, 0) centers
5. $(\frac{1}{2}\pi \pm 2n\pi, 0)$ saddle points; $(-\frac{1}{2}\pi \pm 2n\pi, 0)$ centers
7. $(n\pi, 0)$ saddle points (n even) or centers (n odd)
11. Since the paths in the phase plane then approach circles
13. Consider $dy_2/dy_1 = 0$.
15. Hyperbolas $y_1y_2 = const$

PROBLEM SET 4.6, page 192

1. $y_1 = c_1e^t + c_2e^{-t} + 3e^{2t}$, $y_2 = c_1e^t - c_2e^{-t}$
3. $y_1 = c_1e^{-2t} + c_2e^{-4t} + \cos t$, $y_2 = c_1e^{-2t} - c_2e^{-4t} - \sin t$
5. $y_1 = A \cos 2t + B \sin 2t + 4 \cos t$, $y_2 = -2A \sin 2t + 2B \cos 2t + \sin t$
7. $y_1 = \cos 5t + 2 \sin 5t + 3t$, $y_2 = -\sin 5t + 2 \cos 5t - 4$
9. $y_1 = e^{10t} - e^{-4t} + 2 \sin t$, $y_2 = 0.75e^{10t} + e^{-4t} + \sin t$
11. $y_1 = 4e^{6t} + e^t + t^2 - 5$, $y_2 = e^{6t} - e^t - t$
13. $I_1 = 2c_1e^{\lambda_1 t} + 2c_2e^{\lambda_2 t} + 100$,
$I_2 = (1.1 + \sqrt{0.41})c_1e^{\lambda_1 t} + (1.1 - \sqrt{0.41})c_2e^{\lambda_2 t}$,
$\lambda_1 = -0.9 + \sqrt{0.41}$, $\lambda_2 = -0.9 - \sqrt{0.41}$
15. $c_1 = 17.948$, $c_2 = -67.948$

CHAPTER 4 (REVIEW QUESTIONS AND PROBLEMS), page 193

11. $y_1 = 4c_1e^{5t} + c_2e^{-2t}$, $y_2 = 3c_1e^{5t} - c_2e^{-2t}$; saddle point
13. $y_1 = e^{-t}(2A \cos t + 2B \sin t)$,
$y_2 = e^{-t}[A(5 \cos t + \sin t) + B(5 \sin t - \cos t)]$; spiral point
15. (0, 0) saddle point; (−1, 0), (1, 0) centers
17. $y_2 = \frac{1}{4}(y_1' - 2y_1)$, $y_2' = \frac{1}{4}(y_1'' - 2y_1') = 3y_1 + \frac{1}{4}(y_1' - 2y_1)$,
$y_1 = c_1e^{5t} + c_2e^{-2t}$, $y_2 = \frac{1}{4}(y_1' - 2y_1) = \frac{3}{4}c_1e^{5t} - c_2e^{-2t}$
19. Saddle point
21. $I_1 = (-19 - 62.5t)e^{-5t} - 19 \cos t + 62.5 \sin t$,
$I_2 = (-6 + 62.5t)e^{-5t} + 6 \cos t + 2.5 \sin t$
23. $y_1 = 30e^{-0.03t} - 30e^{-0.15t}$, $y_2 = 45e^{-0.03t} + 45e^{-0.15t}$
27. $y_1 = c_1e^{-t} + 3c_2e^t + te^{-t} + e^{-t}$, $y_2 = -c_1e^{-t} - c_2e^t - te^{-t}$
29. $y_1 = 4e^{2t} - e^{-5t} + e^t - t$, $y_2 = 3e^{2t} + e^{-5t} - e^t - 3$

PROBLEM SET 5.2, page 208

1. $y = -1 + c_3x^3$

3. $y = a_0e^{x^2}$

5. $y = a_0(1 + 3x + 4x^2 + \frac{10}{3}x^3 + 2x^4 + \frac{14}{15}x^5 + \cdots) = a_0(x + 1)e^{2x}$

7. $y = c_0 + c_1x + (\frac{3}{2}c_1 - c_0)x^2 + (\frac{7}{6}c_1 - c_0)x^3 + \cdots$. Setting $c_0 = A + B$ and $c_1 = A + 2B$, we obtain $y = Ae^x + Be^{2x}$. This illustrates the fact that even if the solution of an equation is a known function, the power series method may not yield it immediately in the usual form. Practically, this does not matter because the main interest concerns equations for which the power series solutions define *new* functions.

9. $y = a_1x + a_0(1 - x^2 - \frac{1}{3}x^4 - \frac{1}{5}x^6 - \frac{1}{7}x^8 - \cdots)$. [This is a particular case of Legendre's equation ($n = 1$), which we consider in Sec. 5.3.]

11. $(t + 1)\dot{y} - y = t + 1, y = c_0(1 + t) + (1 + t)(t - \frac{1}{2}t^2 + - \cdots)$
$= c_0x + x \ln x \qquad (\dot{y} = dy/dt)$

13. $y = a_0(1 - t^2/2! + t^4/4! - + \cdots) + a_1(t - t^3/3! + - \cdots)$
$= a_0 \cos (x - 1) + a_1 \sin (x - 1)$

15. 3 17. ∞ 19. $\sqrt{|k|}$ 21. $\sqrt{5/7}$ 23. 0 25. $\sqrt{2}$

29. $\sum_{m=1}^{\infty} \frac{(m + 1)(m + 2)}{(m + 1)^2 + 1} x^m$; 1

PROBLEM SET 5.3, page 213

9. Set $(x^2 - 1)^n = v(x)$. Since $v', v'', \cdots, v^{(n-1)}$ are zero at $x \pm 1$ and $v^{(2n)} = (2n)!$, we obtain from (12)

$$(2^n n!)^2 \int_{-1}^{1} P_n^2 \, dx = \int_{-1}^{1} v^{(n)}v^{(n)} \, dx = [v^{(n-1)}v^{(n)}]|_{-1}^{1} - \int_{-1}^{1} v^{(n-1)}v^{(n+1)} \, dx$$

$$= \cdots = (-1)^n \int_{-1}^{1} vv^{(2n)} \, dx = (-1)^n(2n)! \int_{-1}^{1} (x^2 - 1)^n \, dx$$

$$= 2(2n)! \int_{0}^{1} (1 - x^2)^n \, dx = 2(2n)! \int_{0}^{\pi/2} \sin^{2n+1} \beta \, d\beta$$

$$= 2(2n)! \frac{2 \cdot 4 \cdots (2n)}{1 \cdot 3 \cdot 5 \cdots (2n + 1)} = \frac{2}{2n + 1}.$$

$(x = \cos \beta)$.

13. Set $x = -1$ and use the formula for the sum of a geometric series.

15. Set $x = 0$ and use $(1 + u^2)^{-1/2} = \sum \binom{-1/2}{m} u^{2m}$.

PROBLEM SET 5.4, page 223

1. $y_1 = x^{-1} \cos x, y_2 = x^{-1} \sin x$

3. $y_1 = \frac{1}{x^2}\left(1 + \frac{x^2}{3!} + \frac{x^4}{5!} + \cdots\right) = \frac{\sinh x}{x^3}, y_2 = \frac{\cosh x}{x^3}$

5. $y_1 = (\sin x)/x^3, y_2 = (\cos x)/x^3$

9. $y_1 = \frac{1}{x^2}\left(1 - \frac{x^2}{2!} + \frac{x^4}{4!} - + \cdots\right) = \frac{\text{cox } x}{x^2}, y_2 = \frac{\sin x}{x^2}$

11. $y_1 = e^x, y_2 = e^x \ln x$

13. $y_1 = x^{-2} \sin x^2, y_2 = x^{-2} \cos x^2$

15. $y_1 = (x - 1)^2, y_2 = (x - 1)^{-2}$

17. $y_1 = x, y_2 = x \ln x + x^2$

19. $y_1 = \sqrt{x}\, e^x, y_2 = (x + 1)e^x$

35. $y = AF(\frac{1}{2}, \frac{1}{4}, \frac{1}{2}; x) + B\sqrt{x}\, F(1, \frac{3}{4}, \frac{3}{2}; x)$

37. $y = AF(\frac{1}{5}, -1, \frac{4}{5}; x) + Bx^{1/5}F(\frac{2}{5}, -\frac{4}{5}, \frac{6}{5}; x)$

39. $y = A(1 - 8x + \frac{32}{5}x^2) + Bx^{3/4}F(\frac{7}{4}, -\frac{5}{4}, \frac{7}{4}; x)$

41. $y = AF(-\frac{3}{2}, -\frac{5}{2}, \frac{1}{2}; x) + Bx^{1/2}(1 + \frac{4}{3}x)$

43. $y = C_1F(-\frac{1}{2}, -\frac{1}{2}, \frac{1}{2}; t - 1) + C_2(t - 1)^{1/2}$

45. $y = c_1F(-1, \frac{1}{3}, \frac{1}{3}; t + 1) + c_2(t + 1)^{2/3}F(-\frac{1}{3}, 1, \frac{5}{3}; t + 1)$

PROBLEM SET 5.5, page 231

1. Use (7b), Sec. 5.2, and $\frac{x^{2m+2}}{2^{2m+2+n}(m + 1)!(n + m + 1)!} \Big/ \frac{x^{2m}}{2^{2m+n}m!(n + m)!} \to 0$

as $m \to \infty$ (x fixed).

5. Approximate values $\frac{3}{4}\pi + k\pi = 2.356, 5.498, 8.639, 11.781, 14.923$.

7. $\frac{1}{4}\pi + k\pi$, $k = 1, 2, 3, 4$; 2.5%, 0.8%, 0.4%, 0.2%

11. $AJ_{1/3}(x) + BJ_{-1/3}(x)$

13. $AJ_{3/2}(5x) + BJ_{-3/2}(5x)$

15. $AJ_{2/3}(x^2) + BJ_{-2/3}(x^2)$

17. $x^{-2}J_2(x)$; see Theorem 2.

19. $x^{1/3}(AJ_{1/3}(x^{1/3}) + BJ_{-1/3}(x^{1/3}))$

PROBLEM SET 5.6, page 235

5. $J_2(x) = 2x^{-1}J_1(x) - J_0(x)$; see (3) with $\nu = 1$.

7. Start from (2), with ν replaced by $\nu - 1$, and replace $J_{\nu-1}$ by using (1).

9. Let $J_n(x_1) = J_n(x_2) = 0$. Then $x_1^{-n}J_n(x_1) = x_2^{-n}J_n(x_2) = 0$, and $[x^{-n}J_n(x)]' = 0$ somewhere between x_1 and x_2 by Rolle's theorem. Now use (2). Then use (1) with $\nu = n + 1$.

13. $-2J_2(x) - J_0(x) + c$ by (4)

15. $-2J_4 - 2J_2 - J_0 + c$ by (4)

17. 0.886 227, 1.329 340, 3.323 351

PROBLEM SET 5.7, page 240

1. $AJ_2(x) + BY_2(x)$

3. $AJ_0(\sqrt{x}) + BY_0(\sqrt{x})$

5. $AJ_{1/4}(x^2) + BY_{1/4}(x^2)$

7. $x^2[AJ_2(x^2) + BY_2(x^2)]$

9. $\sqrt{x}\,[AJ_{1/2}(\sqrt{x}) + BJ_{-1/2}(\sqrt{x})] = x^{1/4}[\tilde{A} \sin \sqrt{x} + \tilde{B} \cos \sqrt{x}]$

11. $\sqrt{x}\,[AJ_{1/4}(\frac{1}{2}x^2) + BY_{1/4}(\frac{1}{2}x^2)]$

13. $\sqrt{x}[AJ_{1/4}(\frac{1}{2}kx^2) + BY_{1/4}(\frac{1}{2}kx^2)]$

15. 1.1

17. Set $H_\nu^{(1)} = kH_\nu^{(2)}$, use (10), obtain a contradiction.

19. For $x \neq 0$ all the terms of the series (13) are real and positive.

PROBLEM SET 5.8, page 248

3. $\lambda = [(2n+1)\pi/2]^2,\ n = 0, 1, \cdots; y_n(x) = \sin(\frac{1}{2}(2n+1)\pi x)$

5. $\lambda = [(2n+1)\pi/2L]^2,\ n = 0, 1, \cdots; y_n(x) = \sin((2n+1)\pi x/2L)$

7. $\lambda = n^2,\ n = 0, 1, \cdots; y_n(x) = \cos nx$

9. $\lambda = (n\pi/L)^2,\ n = 0, 1, \cdots; y_n(x) = \cos(n\pi x/L)$

11. $\lambda = ((2n+1)\pi/2)^2,\ n = 0, 1, \cdots; y_n(x) = \sin\left((2n+1)\frac{\pi}{2}\ln|x|\right)$

13. $\lambda = n^2\pi^2,\ n = 1, 2, \cdots; y_n(x) = x\sin(n\pi\ln|x|)$; Euler-Cauchy equation

15. $P_0/\sqrt{2},\ \sqrt{\frac{3}{2}}P_1(x),\ \sqrt{\frac{5}{2}}P_2(x)$

17. Set $x = ct + k$.

PROBLEM SET 5.9, page 255

1. $4P_1 + 2P_3$

3. $x^2 = \frac{1}{3}P_0 + \frac{2}{3}P_2, x^3 = \frac{3}{5}P_1 + \frac{2}{5}P_3$

5. $\frac{1}{4}P_0 + \frac{1}{2}P_1 + \frac{5}{16}P_2 + \cdots$

7. $\frac{1}{2}P_0 + \frac{5}{8}P_2 + \cdots$

15. Since the highest power in L_m is x^m, it suffices to show that $\int e^{-x}x^kL_n\,dx = 0$ for $k < n$:

$$\int_0^\infty e^{-x}x^kL_n(x)\,dx = \frac{1}{n!}\int_0^\infty x^k\frac{d^n}{dx^n}(x^ne^{-x})\,dx = -\frac{k}{n!}\int_0^\infty x^{k-1}\frac{d^{n-1}}{dx^{n-1}}(x^ne^{-x})\,dx$$

$$= \cdots = (-1)^k\frac{k!}{n!}\int_0^\infty\frac{d^{n-k}}{dx^{n-k}}(x^ne^{-x})\,dx = 0.$$

21. $G_x = \sum a_n'(x)t^n = \sum He_n'(x)t^n/n! = tG = \sum He_{n-1}(x)t^n/(n-1)!$, etc.

23. Write $e^{-x^2/2} = v$, $v^{(n)} = d^nv/dx^n$, etc., integrate by parts, use the formula $He_n' = nHe_{n-1}$ (Prob. 21); then, for $n > m$,

$$\int_{-\infty}^{\infty} vHe_mHe_n\,dx = (-1)^n\int_{-\infty}^{\infty} He_mv^{(n)}\,dx = (-1)^{n-1}\int_{-\infty}^{\infty} He_m'v^{(n-1)}\,dx$$

$$= (-1)^{n-1}m\int_{-\infty}^{\infty} He_{m-1}v^{(n-1)}\,dx = \cdots$$

$$= (-1)^{n-m}m!\int_{-\infty}^{\infty} He_0\,v^{(n-m)}\,dx = 0.$$

25. By (1) in Sec. 5.6, with $\nu = 1$,

$$a_m = \frac{2}{R^2J_1^{\,2}(\alpha_{m0})}\int_0^R xJ_0\left(\frac{\alpha_{m0}}{R}x\right)dx = \frac{2}{\alpha_{m0}^{\,2}J_1^{\,2}(\alpha_{m0})}\int_0^{\alpha_{m0}} wJ_0(w)\,dw$$

$$= \frac{2}{\alpha_{m0}J_1(\alpha_{m0})},\qquad f = 2\left(\frac{J_0(\lambda_{10}x)}{\alpha_{10}J_1(\alpha_{10})} + \frac{J_0(\lambda_{20}x)}{\alpha_{20}J_1(\alpha_{20})} + \cdots\right).$$

27. $a_m = \dfrac{2akJ_1(\alpha_{m0}a/R)}{\alpha_{m0}RJ_1^{\,2}(\alpha_{m0})}$

29. $a_m = \dfrac{4J_2(\alpha_{m0})}{\alpha_{m0}^{\,2}J_1^{\,2}(\alpha_{m0})}$

31. $a_m = \dfrac{2R^2}{\alpha_{m0}J_1(\alpha_{m0})}\left[1 - \dfrac{2J_2(\alpha_{m0})}{\alpha_{m0}J_1(\alpha_{m0})}\right]$

35. $x^3 = 16\left[\dfrac{J_3(\alpha_{13}x/2)}{\alpha_{13}J_4(\alpha_{13})} + \dfrac{J_3(\alpha_{23}x/2)}{\alpha_{23}J_4(\alpha_{23})} + \cdots\right]$

CHAPTER 5 (REVIEW QUESTIONS AND PROBLEMS), page 258

17. $(1 - x)^{-1}, x^{-1}$
19. $e^x, e^x \ln x$
21. $\sqrt{x}, 1 + x$
23. $x + 1, (x + 1) \ln x$
25. $x^2 + x^3, (x^2 + x^3) \ln x$
27. $[AJ_1(x) + BY_1(x)]/x$
29. $x[AJ_1(x) + BY_1(x)]$
31. $1/\sqrt{2}, \cos \pi x, \cos 2\pi x, \cdots$
33. $\sqrt{2/\pi}, (2/\sqrt{\pi}) \cos 4mx, (2/\sqrt{\pi}) \sin 4mx$
35. $\frac{1}{2}P_0(\frac{1}{2}x), \frac{1}{2}\sqrt{3}\,P_1(\frac{1}{2}x), \frac{1}{2}\sqrt{5}\,P_2(\frac{1}{2}x)$
37. $\lambda = n^2, y_n(x) = \sin(n \ln |x|)$
39. $\lambda = n^2\pi^2, y_n(x) = \cos(n\pi \ln |x|)$
41. $2(-P_0 + P_1 - P_2 + P_3)$
43. $-4P_0 + 10P_2 + 16P_4$
45. $-30P_0 + \frac{138}{5}P_1 - 6P_2 + \frac{2}{5}P_3$

PROBLEM SET 6.1, page 267

1. $3/s^2 + 4/s$
3. $2/s^3 + a/s^2 + b/s$
5. $(s \cos \theta - \omega \sin \theta)/(s^2 + \omega^2)$
7. $2n\pi[T(s^2 + (2n\pi/T)^2)]^{-1}$
9. $\cos^2 \omega t = \frac{1}{2} + \frac{1}{2} \cos 2\omega t$. *Ans.* $1/2s + s/(2s^2 + 8\omega^2)$.
11. $1/2s + s/(2s^2 + 8)$
13. $e^b/(s - a)$
15. $\frac{1}{2}[s/(s^2 - 16) - 1/s]$
17. $k(1 - e^{-cs})/s$
19. $k/s - k(1 - e^{-cs})/cs^2$
27. $5e^{-3t}$
29. $\frac{1}{5} \sin 5t$
31. $t^3/6$
33. $1 - e^{-t}$
35. $3 - 3e^{-3t}$
37. $2 + e^{-2t}$

PROBLEM SET 6.2, page 274

1. $(s^2 + 2)/s(s^2 + 4)$
3. $(s - a)^{-2}$
5. $(s^2 - 8)/s(s^2 - 16)$
7. $(s^2 + 18)/s(s^2 + 36)$
15. $3 - 3e^{-t}$
17. $1 - \cos 2t$
19. $\sinh 2t - 2t$
21. $(e^{2t} - 1 - 2t - 2t^2)/8$
23. $t^2/2 - t/\pi + (e^{\pi t} - 1)/\pi^2$
25. $1 + t - \cos t - \sin t$
27. $2 \cos 2t - 4 \sin 2t$
29. $y = A \cos \omega t + (B/\omega) \sin \omega t$
31. $e^{-2t} - e^{-3t}$
33. $\cos 5t + \frac{1}{25}t$
35. $y = 2e^{kt}$

PROBLEM SET 6.3, page 281

1. $4.5/(s - 3.5)^2$
3. $1/(s^2 - 2s + 2)$
5. $(\omega \cos \theta + (s + 1) \sin \theta)/[(s + 1)^2 + \omega^2]$
7. $12/(s + \frac{1}{2})^4$
9. $[A(s + \alpha) + B\beta]/[(s + \alpha)^2 + \beta^2]$
11. $\pi t e^{-\pi t}$
13. $e^{2t} \cos t$
15. $e^{-3t}(\cos t - 3 \sin t)$
17. $2e^{2t} \sinh 3t$
23. $k(e^{-as} - e^{-bs})/s$
25. $u(t) - u(t - 1) + u(t - 2) - + \cdots,$
$s^{-1}(1 - e^{-s} + e^{-2s} - + \cdots) = s^{-1}(1 + e^{-s})^{-1}$ (geometric series)

27. e^{-s}/s^2

29. $2e^{-s}/s^3$

31. $e^{1/2}e^{-s/2}/(s - 1)$

33. $-se^{-\pi s}/(s^2 + 1)$

35. $t - tu(t - 1) = t - (t - 1)u(t - 1) - u(t - 1),\ \ s^{-2} - s^{-2}e^{-s} - s^{-1}e^{-s}$

37. $(1 - e^{1-s})/(s - 1)$

39. $s^{-2} - e^{-as}(s^{-2} + as^{-1})$

41. $-2s(e^{-s} + e^{-2s})/(s^2 + \pi^2)$

43. $t - 3$ if $t > 3$; 0 if $t < 3$

45. -3 if $1 < t < 4$; 0 otherwise

47. $\cos 2t$ if $t > \pi$; 0 otherwise

49. $(t - 1)^3/6$ if $t > 1$; 0 otherwise

51. $e^{-t} \sin t$

53. $3e^{t/2} \cos 3t$

55. $e^{-2t} \sin 3t + 9 \cos 2t + 8 \sin 2t$

57. $e^{-2t} + e^{-3t} + e^{-4t} + e^{-5t}$

59. $t - \sin t$ if $0 < t < 1$; $\cos (t - 1) + \sin (t - 1) - \sin t$ if $t > 1$

61. $\sin 3t + \sin t$ if $0 < t < \pi$; $\frac{4}{3} \sin 3t$ if $t > \pi$

63. $\sin 2t + \sin t$ if $0 < t < \pi$; $-\sin t$ if $t > \pi$

65. $e^t - \sin t$ if $0 < t < 2\pi$; $e^t - \frac{1}{2} \sin 2t$ if $t > 2\pi$

67. $i(t) = V_0(1 - e^{-Rt/L})/R$

69. $sI + I/s = (1 - e^{-s})/s^2$. *Ans.* $1 - \cos t$ if $0 < t < 1$, $\cos (t - 1) - \cos t$ if $t > 1$.

PROBLEM SET 6.4, page 288

1. $i = 0\ (t < 1),\ \ i = e^{-0.1(t-1)}\ (1 < t < 2),\ \ i = e^{-0.1(t-1)} - e^{-0.1(t-2)}\ (t > 2)$

3. $i = 0\ (t < 2),\ \ i = (10e^{-(t-2)} - e^{-0.1(t-2)})/900e^2\ (t > 2)$

5. $i = 20(1 - e^{-0.1t})\ (0 < t < 1),\ \ i = 20(e^{0.1} - 1)e^{-0.1t}\ (t > 1)$

7. $y = 0\ (t < 1),\ \ y = \frac{1}{3} \sin (3t - 3)\ (t > 1)$

9. $y = \sin t\ (0 < t < \pi),\ \ y = 0\ (\pi < t < 2\pi),\ \ y = -\sin t\ (t > 2\pi)$

11. $y = e^{-t} \cos t\ (0 < t < 2\pi),\ \ y = e^{-t}(\cos t + e^2 \sin t)\ (t > 2\pi)$

13. $y = 3e^{-2t} \sin t\ (0 < t < 1),\ \ y = e^{-2t}[3 \sin t + e^2 \sin (t - 1)]\ (t > 1)$

15. $y = 2e^{-3t} + e^{-t}\ (0 < t < \frac{1}{2}),\ \ y = 2e^{-3t} + e^{-t} + \frac{1}{4}[e^{-3(t-1/2)} - e^{t-1/2}]\ (t > \frac{1}{2})$

17. $y = e^{-t} \cos 2t + e^t\ (0 < t < 1)$, $y = e^{-t} \cos 2t + e^t + e^{-(t-1)} \sin 2(t - 1)\ (t > 1)$

19. $y = 5t - 2\ (0 < t < \pi),\ \ y = 5t - 2 - \frac{1}{2}e^{-(t-\pi)} \sin 2t\ (t > \pi)$

PROBLEM SET 6.5, page 293

1. $\dfrac{2s^2 - 8}{(s^2 + 4)^2}$

3. $\dfrac{2}{(s - 1)^3}$

5. $\dfrac{6s}{(s^2 - 9)^2}$

7. $\dfrac{2\omega s}{(s^2 + \omega^2)^2}$

9. $\dfrac{2s(s^2 - 3)}{(s^2 + 1)^3}$

11. $\dfrac{s + 2}{(s^2 + 4s + 5)^2}$

13. $4e^{-t}t$

15. $t \sinh 2t$

17. $\frac{1}{2}t \sin t$

19. $(e^t - 1)/t$

21. $(\sin \omega t)/t$

PROBLEM SET 6.6, page 297

1. t

3. te^t

5. $t^5/30$

7. $\frac{1}{2}t \sin \omega t$

9. $\frac{2}{3} \sin t - \frac{1}{3} \sin 2t$

11. $0\ (t < \pi),\ -\sin t\ (t > \pi)$

13. $e^t * e^t = te^t$

15. $(e^{3t} - 1)/9 - t/3$

17. $(1 - \cos \omega t)/\omega^2$

19. $(t \sin \omega t)/2\omega$

21. $t \cos \omega t$

23. $e^{-t} - e^{-2t}$

25. Set $t - \tau = \sigma$. Then $\int_0^t f(\tau)g(t - \tau)\, d\tau = \int_t^0 g(\sigma)f(t - \sigma)(-d\sigma)$.

29. Let $t > k$. Then $(f_k * f)(t) = \int_0^k \frac{1}{k} f(t - \tau)\, d\tau = f(t - \tilde{t})$ for some $\tilde{t}$ between 0 and k. Now let $k \to 0$. Then $\tilde{t} \to 0$ and $f_k(t - \tilde{t}) \to \delta(t)$, so that the formula follows.

31. $\frac{3}{8} \sin t - \frac{1}{8} \sin 3t$

33. $t - \sin t$

35. $0.2e^{-t} + \cos 5t - 2 \sin 5t$

39. 0 if $0 < t < 1$; $\frac{1}{4} - \frac{1}{4} \cos (2t - 2)$ if $t > 1$

41. $\sin t + \sin 3t$ if $0 < t < \pi$; $\frac{4}{3} \sin 3t$ if $t > \pi$

43. $\sin t + \sin 2t$ if $0 < t < \pi$; $-\sin t$ if $t > \pi$

45. e^t

47. $\cos t$

49. $\sinh t$

PROBLEM SET 6.7, page 308

1. $(e^{4t} - e^t)/3$

3. $2e^{-4t} + e^{2t}$

5. $e^{-t}(\cos 3t + 4 \sin 3t)$

7. $2 + e^t - 2e^{2t}$

9. $e^{2t}(2t - 4)$

11. $e^{-t}(1 - t - t^2)$

13. $\cosh 3t - 3 \cosh t$

15. $e^{-t}(\cos 2t - 2t \sin 2t)$

17. $(1 + t^2 - 2t^3)e^{-2t}$

19. $2e^{2t} + te^t$

27. $y_1 = \cos t, \quad y_2 = \sin t$

29. $y_1 = e^{-t} \cos t, \quad y_2 = -e^{-t} \sin t$

31. $y_1 = -6e^{4t} + 2, \quad y_2 = -3e^{4t} - 1$

33. $y_1 = 4e^t - e^{-2t}, \quad y_2 = e^t - e^{-2t}$

35. $y_1 = 3e^{2t} + e^{-5t}, \quad y_2 = 4e^{2t} - e^{-5t}$

37. $y_1 = e^t + e^{2t}, \quad y_2 = e^{2t}$

39. $y_1 = e^t, \quad y_2 = e^{-t}, \quad y_3 = e^t - e^{-t}$

PROBLEM SET 6.8, page 313

1. $\dfrac{\pi s - 1 + (\pi s + 1)e^{-2\pi s}}{s^2(1 - e^{-2\pi s})} = \dfrac{\pi}{s} \coth \pi s - \dfrac{1}{s^2}$

3. $\dfrac{(4\pi^2 s^2 - 2)e^{2\pi s} + 4\pi s + 2}{s^3(e^{2\pi s} - 1)}$

5. $\dfrac{e^{2(1-s)\pi} - 1}{(1 - s)(1 - e^{-2\pi s})}$

7. $\left[\dfrac{1}{s^2}(1 - e^{-\pi s}) - \dfrac{\pi}{s} e^{-\pi s}\right] \Big/ (1 - e^{-2\pi s})$

9. $\left[\dfrac{\pi}{s} e^{-\pi s}(e^{-\pi s} - 1) + \dfrac{1}{s^2}(e^{-\pi s} - 1)^2\right] \Big/ (1 - e^{-2\pi s})$

15. $\dfrac{\omega}{s^2 + \omega^2} \coth \dfrac{\pi s}{2\omega}$

21. $i(t) = (V_0/\omega^* L)e^{-\alpha t} \sin \omega^* t$ if $0 < t < a$, $\alpha = R/2L$, $\omega^{*2} = 1/LC - \alpha^2$,
$i(t) = (V_0/\omega^* L)[e^{-\alpha t} \sin \omega^* t - e^{-\alpha(t-a)} \sin \{\omega^*(t - a)\}]$ if $t > a$

25. $i(t) = \dfrac{1}{R}\left(\dfrac{e^{-kt}}{1 - e^{-k}} + t - \dfrac{L}{R}\right)$ if $0 < t < 1$, and $i(t + 1) = i(t)$, $k = \dfrac{R}{L}$.

27. $Ri + q/C = v(t)$, and since the charge q on the capacitor cannot change abruptly, i must have a jump.

29. $i(t) = \begin{cases} a \cos \omega t + b \sin \omega t - ae^{-t/(RC)} & \text{if } 0 < t < \pi/\omega \\ -a(1 + e^{\pi/(\omega RC)})e^{-t/(RC)} & \text{if } t > \pi/\omega \end{cases}$
where $a = \omega CK$, $b = \omega^2 RC^2 K$, $K = 1/[1 + (\omega RC)^2]$.

31. $i(0) = 0$, $i'(0) = -\frac{100}{16}$, $i = \frac{20}{3}(e^{-t} - e^{-t/16})$, $q = \frac{20}{3}(16e^{-t/16} - e^{-t})$

33. $y = \dfrac{P}{K}\left[1 + \dfrac{1}{2\beta}(\alpha - \beta)e^{-(\alpha+\beta)t} - \dfrac{1}{2\beta}(\alpha + \beta)e^{(-\alpha+\beta)t}\right]$, where $\beta^2 = -\omega^{*2}$,
$y = P(1 - (1 + \alpha t)e^{-\alpha t})/K$

35. For $0 < t < 1$ the solution is as in Prob. 34. If $t > 1$,
$$v = \omega - \frac{P}{M_1 + M_2} + \frac{P}{k(M_1 + M_2)}[\sin kt - \sin (k(t - 1))],$$
$$k^2 = c\left(\frac{1}{M_1} + \frac{1}{M_2}\right), \quad c = \text{stiffness of shaft.}$$

39. $i_1 = V_0\left[e^{-\alpha t}\left(-\dfrac{1}{R_1}\cos \omega^* t + \dfrac{1}{\omega^*}\left(\dfrac{L_2}{A} - \dfrac{\alpha}{R_1}\right)\sin \omega^* t\right) + \dfrac{1}{R_1}\right]$

CHAPTER 6 (REVIEW QUESTIONS AND PROBLEMS), page 320

17. $2\pi/[(s + 1)^2 + 4\pi^2]$

19. $(s^2 - 2)/(s^3 - 4s)$

21. $2s^2/(s^2 + 1)^2$

23. $1/2s + s/(2s^2 + 8)$

25. arc tan $(1/s)$

27. $e^{-t}(\cos 3t - \sin 3t)$

29. $(1 - e^{-3t})/3$

31. $e^{2t} - 1 - 2t - 2t^2$

33. $2 + e^t - 2e^{2t}$

35. $2t + 3 + \frac{1}{2}e^{3t} - 2e^{2t} - \frac{1}{2}e^t$

37. 0 if $0 < t < 3$, $0.2[1 - \cos (3t - 9)]$ if $t > 3$

39. 0 if $t < 2\pi$, $-\frac{1}{2}\cos t + \frac{1}{2}\sin t + \frac{1}{2}e^{-(t-2\pi)}$ if $t > 2\pi$

41. $1.5e^t \cos t + te^t \sin t$

43. $2te^{2t} + (20 + 12t + 3t^2)e^t$

45. $\frac{3}{2}\sin 2t$ if $\pi < t < 2\pi$, $\sin 2t$ otherwise

47. e^{-3t}

51. $y_1 = \cos \sqrt{2}t$, $y_2 = \cos \sqrt{2}t$

53. $1 - e^{-t}$ if $0 < t < 4$, $(e^4 - 1)e^{-t}$ if $t > 4$

55. $i(t) = e^{-4t}(\frac{3}{26}\cos 3t - \frac{10}{39}\sin 3t) - \frac{3}{26}\cos 10t + \frac{8}{65}\sin 10t$

57. $i_1 = 2(1 - e^{-t})$, $i_2 = 2e^{-t}$

59. $i_1 = e^{-4t} + t$, $i_2 = -4e^{-4t} + \frac{1}{5}$

PROBLEM SET 7.1–7.2, page 332

1. $\begin{bmatrix} 3 & 0 \\ 8 & 6 \end{bmatrix}$

3. $\begin{bmatrix} 18 & 2 & -6 \\ 18 & -4 & 28 \end{bmatrix}$

5. All undefined

7. $\begin{bmatrix} 0 & 24 \\ 0 & 60 \end{bmatrix}$

9. $\begin{bmatrix} 4 & 3 & -8 \\ -1 & 3 & 8 \\ 8 & -4 & 5 \end{bmatrix}$, $\begin{bmatrix} 4 & -1 & 8 \\ 3 & 3 & -4 \\ -8 & 8 & 5 \end{bmatrix}$

11. 0, 0, [11 1 9]

13. $\begin{bmatrix} 8 & 6 & -16 \\ -2 & 6 & 16 \\ 16 & -8 & 10 \end{bmatrix}$

19. $\begin{bmatrix} 0 & 1 \\ 1 & 5 \end{bmatrix} + \begin{bmatrix} 0 & -3 \\ 3 & 0 \end{bmatrix}$

23. $\begin{bmatrix} -1 & 1 & 0 & 0 & -1 & -1 \\ 1 & -1 & 0 & 0 & 0 & 0 \\ 0 & 0 & 1 & -1 & 1 & 0 \end{bmatrix}$

25.

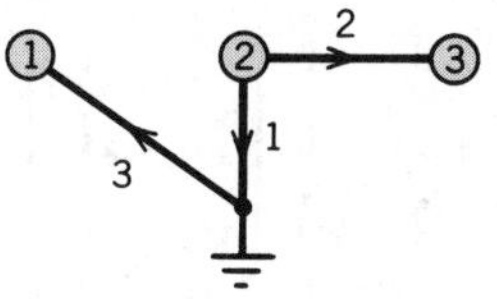

27.

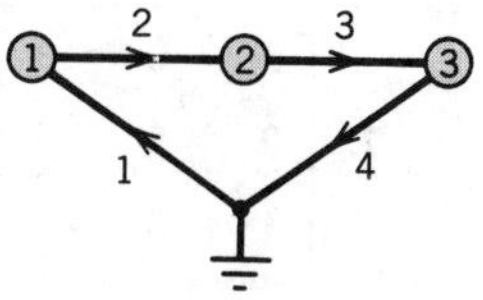

PROBLEM SET 7.3, page 343

1. $\begin{bmatrix} -2 & -2 \\ 0 & -4 \\ 8 & 6 \end{bmatrix}, \begin{bmatrix} -2 & 0 & 8 \\ -2 & -4 & 6 \end{bmatrix}$

3. $\begin{bmatrix} -1 \\ 9 \\ 19 \end{bmatrix}, \begin{bmatrix} -1 \\ 2 \\ 8 \end{bmatrix}, \begin{bmatrix} 1 \\ 6 \\ 7 \end{bmatrix}$

5. $\begin{bmatrix} 0 & 6 & 8 \\ -3 & 3 & 11 \end{bmatrix}, \begin{bmatrix} 4 & 6 \\ 6 & 14 \end{bmatrix}$

7. [26], [82], [15]

11. $\begin{bmatrix} 0 & 1 \\ 0 & 0 \end{bmatrix}, \begin{bmatrix} 0 & 0 \\ 1 & 0 \end{bmatrix}$, etc.

13. $\begin{bmatrix} 1 & 0 \\ 0 & 0 \end{bmatrix}, \begin{bmatrix} 1 & 1 \\ 0 & 0 \end{bmatrix}$, etc.

15. **0**

19. Uniform stretch or contraction, same in all directions

21. (a) C_1 to J_2, C_3 to J_1, J_3; (b) C_1 to J_2, C_2 to J_1, C_3 to J_3

23. [0.7 0.7], [0.49 0.91], [0.427 0.973], [0.4081 0.9919]

PROBLEM SET 7.4, page 352

1. $x = -4, y = 4$

3. $x = 6, y = 5$

5. $x = -2, y = 0, z = 4$

7. $x = -2, y = 0, z = 1$

9. $x = 2, y = 1, z = 3$

11. $x = 3y + 2, z = 0$

13. $w = 2x + 1, y = 1, z = 2$

15. $w = 4, x = 0, y = 2, z = 6$

17. $I_1 = (R_1 + R_2)E_0/R_1R_2$, $I_2 = E_0/R_1$, $I_3 = E_0/R_2$ (amperes)

19. $I_1 = 2, I_2 = 6, I_3 = 8$

23. $P_1 = 6, P_2 = 10, D_1 = S_1 = 18, D_2 = S_2 = 26$

PROBLEM SET 7.5, page 360

1. Yes, 2, $[1 \quad -\frac{1}{2} \quad 0]^T$, $[0 \quad 0 \quad 1]^T$

3. No

5. Yes, 1, 1

7. No

9. Yes, 1, $[0 \quad \cdots \quad 0 \quad 1]^T$

11. 2

13. 3 **15.** 2

17. 2 **19.** 4

21. Use Theorem 1.

23. Use Theorem 1.

25. $[1 \quad 1]$, $[1 \quad -1]$ and $[7 \quad \frac{1}{2} \quad 0]^T$, $[7 \quad -\frac{1}{2} \quad 0]^T$

27. Linearly independent

29. Linearly dependent

31. Linearly dependent

33. Linearly independent

35. Linearly dependent

PROBLEM SET 7.7, page 369

1. $\begin{bmatrix} -2 & 1 \\ \frac{3}{2} & -\frac{1}{2} \end{bmatrix}$ **3.** $\begin{bmatrix} 0.6 & 0.8 \\ 0.8 & -0.6 \end{bmatrix}$ **5.** No inverse

7. $\begin{bmatrix} 3 & 0 & -1 \\ 0 & 5 & 1 \\ 1 & 0 & -1 \end{bmatrix}$ **9.** $\begin{bmatrix} 1 & 0 & 0 \\ -\frac{1}{2} & 1 & 0 \\ \frac{3}{4} & -\frac{5}{2} & \frac{1}{2} \end{bmatrix}$ **11.** $\begin{bmatrix} 1 & -0.8 & -0.64 \\ 0 & 0.2 & -0.04 \\ 0 & 0 & 0.10 \end{bmatrix}$

13. $\begin{bmatrix} 2 & 0 & -1 \\ 5 & 1 & 0 \\ 0 & 1 & 3 \end{bmatrix}$ **15.** $\begin{bmatrix} 11 & 4 & 24 \\ 20 & 9 & 44 \\ 64 & 20 & 139 \end{bmatrix}$ **19.** Use Prob. 18.

PROBLEM SET 7.8, page 379

1. 257 **3.** -40.12 **5.** 180 **7.** 0 **9.** 45.045 **11.** 72
13. -24 **15.** 72

17. $\begin{vmatrix} 20 & 31 & 66 \\ 60 & 93 & 90 \\ 40 & 64 & 34 \end{vmatrix}$

PROBLEM SET 7.9, page 385

1. 1 **3.** 2 **5.** 3

7. $\begin{bmatrix} 14 & -5 \\ -25 & 9 \end{bmatrix}$ **9.** $\begin{bmatrix} \cos 3\theta & -\sin 3\theta \\ \sin 3\theta & \cos 3\theta \end{bmatrix}$ **11.** $\begin{bmatrix} 1 & 0 & 0 \\ 2 & 1 & 0 \\ 5 & 4 & 1 \end{bmatrix}$

13. $\begin{bmatrix} 10 & 20 & 5 \\ 5 & 10 & 5 \\ 15 & 20 & 10 \end{bmatrix}$ **15.** $\begin{bmatrix} 1 & 2 & 5 \\ 0 & -1 & 2 \\ 2 & 4 & 11 \end{bmatrix}$ **17.** $x = 2, y = 1, z = -3$

21. Use Theorem 4. **25.** $(x - 2)^2 + (y - 1)^2 = 25$

PROBLEM SET 7.10, page 391

1. $1, \begin{bmatrix} 1 \\ 0 \end{bmatrix}, -4, \begin{bmatrix} 0 \\ 1 \end{bmatrix}$ **3.** $2, \begin{bmatrix} 3 \\ -4 \end{bmatrix}, 5, \begin{bmatrix} 0 \\ 1 \end{bmatrix}$ **5.** 0, any $\mathbf{x} \neq \mathbf{0}$

7. $1, \begin{bmatrix} 1 \\ 1 \end{bmatrix}, -1, \begin{bmatrix} 1 \\ -1 \end{bmatrix}$ **9.** $4, \begin{bmatrix} 2 \\ 9 \end{bmatrix}, -3, \begin{bmatrix} 1 \\ 1 \end{bmatrix}$ **11.** $i, \begin{bmatrix} 1 \\ i \end{bmatrix}, -i, \begin{bmatrix} 1 \\ -i \end{bmatrix}$

13. $4, 8, 6, \begin{bmatrix} 1 \\ 0 \\ 0 \end{bmatrix}, \begin{bmatrix} 0 \\ 1 \\ 0 \end{bmatrix}, \begin{bmatrix} 0 \\ 0 \\ 1 \end{bmatrix}$ **15.** $0, \begin{bmatrix} 0 \\ 0 \\ 1 \end{bmatrix}, \begin{bmatrix} 1 \\ 1 \\ 0 \end{bmatrix}, -2, \begin{bmatrix} 1 \\ -1 \\ 0 \end{bmatrix}$

17. 6, $\begin{bmatrix}1\\0\\0\end{bmatrix}$, 8, $\begin{bmatrix}5\\1\\0\end{bmatrix}$, 2, $\begin{bmatrix}7\\-4\\2\end{bmatrix}$

19. 9, $\begin{bmatrix}3\\1\\1\end{bmatrix}$, 2, $\begin{bmatrix}0\\1\\0\end{bmatrix}$, $\begin{bmatrix}1\\0\\-2\end{bmatrix}$

21. 0, $\begin{bmatrix}-2\\1\\0\end{bmatrix}$, $\begin{bmatrix}3\\0\\1\end{bmatrix}$, 8, $\begin{bmatrix}1\\2\\-1\end{bmatrix}$

23. This follows from the fact that the zeros of a polynomial with real coefficients are real or complex conjugates in pairs.

27. $\mathbf{Ax}_j = \lambda_j\mathbf{x}_j$ $(\mathbf{x}_j \neq \mathbf{0})$, $(\mathbf{A} - k\mathbf{I})\mathbf{x}_j = \lambda_j\mathbf{x}_j - k\mathbf{x}_j = (\lambda_j - k)\mathbf{x}_j$

29. Use induction; premultiply $\mathbf{A}^k\mathbf{x}_j = \lambda_j^k\mathbf{x}_j$ by $\mathbf{A}$.

PROBLEM SET 7.11, page 396

1. 35.3°, 4; 125.3°, 1
3. 45°, 1.2; 135°, 0.8
5. 45°, 3; 135°, 1
7. $[5 \quad 7]^\mathsf{T}$
9. $[1 \quad 4 \quad 3.2]^\mathsf{T}$
11. 2
13. $c[1 \quad 1.6 \quad 2.5]$, $c > 0$
15. $\mathbf{x} = (\mathbf{I} - \mathbf{A})^{-1}\mathbf{y} = [0.55 \quad 0.64375 \quad 0.6875]^\mathsf{T}$

PROBLEM SET 7.12, page 400

1. $\begin{bmatrix}4 & 2\\2 & -1\end{bmatrix} + \begin{bmatrix}0 & -5\\5 & 0\end{bmatrix}$

3. $\begin{bmatrix}0 & 4 & -3\\4 & 1 & 2\\-3 & 2 & -1\end{bmatrix} + \begin{bmatrix}0 & -5 & 7\\5 & 0 & -9\\-7 & 9 & 0\end{bmatrix}$

5. Symmetric, 10, -10
7. Skew-symmetric, $3i$, $-3i$
9. Orthogonal, 1, i, $-i$
11. Symmetric, 3, 0
13. Skew-symmetric, 0, $50i$, $-50i$
15. No
17. $|\mathbf{v}| = |\mathbf{w}| = \sqrt{20}$, $|\mathbf{x}| = |\mathbf{y}| = \sqrt{5}$; Theorem 2
19. Rotation through an angle θ about the x_3-axis
21. No, yes
23. $(\mathbf{AB})^{-1} = \mathbf{B}^\mathsf{T}\mathbf{A}^\mathsf{T} = (\mathbf{AB})^\mathsf{T}$
25. $(\mathbf{A}^{-1})^\mathsf{T} = (\mathbf{A}^\mathsf{T})^{-1} = (-\mathbf{A})^{-1} = -\mathbf{A}^{-1}$ (see Prob. 18, Sec. 7.7).

PROBLEM SET 7.13, page 406

3. $[2 + i \quad i]^\mathsf{T}$, $[2 + i \quad -5i]^\mathsf{T}$
5. Skew-Hermitian; i, $[1 \quad 1]^\mathsf{T}$; $-i$, $[1 \quad -1]^\mathsf{T}$
7. Hermitian; $3 + \sqrt{2}$, $[-i \quad 1 - \sqrt{2}]$; $3 - \sqrt{2}$, $[-i \quad 1 + \sqrt{2}]$
9. Hermitian, unitary; 1, $[1 \quad i - i\sqrt{2}]$; -1, $[1 \quad i + i\sqrt{2}]$
11. For $\mathbf{A}$, $\mathbf{B}$ unitary, $(\mathbf{AB})^{-1} = \mathbf{B}^{-1}\mathbf{A}^{-1} = \overline{\mathbf{B}}^\mathsf{T}\overline{\mathbf{A}}^\mathsf{T} = \overline{(\mathbf{AB})}^\mathsf{T}$.

17. $\begin{bmatrix}1 & -3\\-3 & 9\end{bmatrix}$

19. $\begin{bmatrix}-3 & 2 & 1\\2 & -1 & 0\\1 & 0 & -5\end{bmatrix}$

21. $\begin{bmatrix}1 & 1 & 0 & 0\\1 & 1 & 0 & 0\\0 & 0 & 1 & 1\\0 & 0 & 1 & 1\end{bmatrix}$

23. -2 **25.** $5i$ **27.** 1

29. $3|x_1|^2 - 2\,\mathrm{Im}\,(x_1\bar{x}_2 + 2x_3\bar{x}_2) + 4|x_3|^2$

PROBLEM SET 7.14, page 413

1. $\begin{bmatrix} 14 & -6 \\ 36 & -16 \end{bmatrix}$, 2, $\begin{bmatrix} 1 \\ 2 \end{bmatrix}$, $\mathbf{x} = \begin{bmatrix} 0 \\ -1 \end{bmatrix}$; -4, $\begin{bmatrix} 1 \\ 3 \end{bmatrix}$, $\mathbf{x} = \begin{bmatrix} -2 \\ 0 \end{bmatrix}$

3. $\begin{bmatrix} 2 & 0 \\ 3 & 3 \end{bmatrix}$, 3, $\begin{bmatrix} 0 \\ 1 \end{bmatrix}$, $\mathbf{x} = \begin{bmatrix} 2 \\ 1 \end{bmatrix}$; 2, $\begin{bmatrix} 1 \\ -3 \end{bmatrix}$, $\mathbf{x} = \begin{bmatrix} -1 \\ -1 \end{bmatrix}$

5. $\begin{bmatrix} 9 & 0 \\ 0 & -5 \end{bmatrix}$, 9, $\begin{bmatrix} 1 \\ 0 \end{bmatrix}$, $\mathbf{x} = \begin{bmatrix} 5 \\ 2 \end{bmatrix}$; -5, $\begin{bmatrix} 0 \\ 1 \end{bmatrix}$, $\mathbf{x} = \begin{bmatrix} 1 \\ -1 \end{bmatrix}$

7. $\begin{bmatrix} 2 & 2 & 1 \\ 0 & 8 & 3 \\ 0 & 2 & 3 \end{bmatrix}$, 9, $\begin{bmatrix} 1 \\ 3 \\ 1 \end{bmatrix}$, $\mathbf{x} = \begin{bmatrix} 3 \\ 1 \\ 1 \end{bmatrix}$; 2, $\begin{bmatrix} 1 \\ 0 \\ 0 \end{bmatrix}$, $\begin{bmatrix} 0 \\ 1 \\ -2 \end{bmatrix}$, $\mathbf{x} = \begin{bmatrix} 0 \\ 1 \\ 0 \end{bmatrix}$, $\begin{bmatrix} 1 \\ 0 \\ -2 \end{bmatrix}$

11. Compare the coefficient of λ^{n-1} in

$$(-1)^n(\lambda - \lambda_1) \cdots (\lambda - \lambda_n) = (-1)^n\lambda^n + (-1)^{n-1}(\lambda_1 + \cdots + \lambda_n)\lambda^{n-1} + \cdots$$

with that of λ^{n-1} in the development of the characteristic determinant.

13. $[1/\sqrt{2} \quad -i/\sqrt{2}]^\mathsf{T}$, $[1/\sqrt{2} \quad i/\sqrt{2}]^\mathsf{T}$

15. $[3/\sqrt{10} \quad 1/\sqrt{10}]^\mathsf{T}$, $[1/\sqrt{10} \quad -3/\sqrt{10}]^\mathsf{T}$

17. $[1/\sqrt{2} \quad i/\sqrt{2}]^\mathsf{T}$, $[1/\sqrt{2} \quad -i/\sqrt{2}]^\mathsf{T}$

19. $\begin{bmatrix} 2 \\ 1 \end{bmatrix}$, $\begin{bmatrix} 2 \\ -1 \end{bmatrix}$, $\begin{bmatrix} 8 & 0 \\ 0 & -8 \end{bmatrix}$

21. $\begin{bmatrix} 1 \\ i \end{bmatrix}$, $\begin{bmatrix} 1 \\ -i \end{bmatrix}$, $\begin{bmatrix} i & 0 \\ 0 & -i \end{bmatrix}$

23. $\begin{bmatrix} 0 \\ 1 \\ 0 \end{bmatrix}$, $\begin{bmatrix} 1 \\ -2 \\ 1 \end{bmatrix}$, $\begin{bmatrix} 1 \\ 0 \\ 0 \end{bmatrix}$, $\begin{bmatrix} 3 & 0 & 0 \\ 0 & 2 & 0 \\ 0 & 0 & 1 \end{bmatrix}$

25. Hyperbola $2y_1^2 - 3y_2^2 = 1$; $x_1 = 0.8y_1 - 0.6y_2$, $x_2 = 0.6y_1 + 0.8y_2$

27. Ellipse $9y_1^2 + y_2^2 = 9$; $x_1 = \frac{1}{2}y_1 - \frac{1}{2}\sqrt{3}\,y_2$, $x_2 = \frac{1}{2}\sqrt{3}\,y_1 + \frac{1}{2}y_2$

29. Parallel straight lines $y_1 = \pm 1$; $x_1 = (y_1 - 3y_2)/\sqrt{10}$, $x_2 = (3y_1 + y_2)/\sqrt{10}$

PROBLEM SET 7.15, page 421

1. 5; $1, x, x^2, x^3, x^4$

3. 1; $\begin{bmatrix} 0 & 1 \\ -1 & 0 \end{bmatrix}$

5. No

7. 2; xe^{-x}, e^{-x}

9. If $S_0 = \{\mathbf{a}_{(1)}, \cdots, \mathbf{a}_{(k)}\}$, then $c_1\mathbf{a}_{(1)} + \cdots + c_k\mathbf{a}_{(k)} = \mathbf{0}$ for $c_1, \cdots, c_k$ not all 0, and $c_1\mathbf{a}_{(1)} + \cdots + c_k\mathbf{a}_{(k)} + \cdots + c_m\mathbf{a}_{(m)} = \mathbf{0}$ for those $c_1, \cdots, c_k$ and $c_{k+1} = \cdots = c_m = 0$.

11. For instance, $[1, \ 0]$, $[0, \ 1]$ and $[1, \ 1]$, $[1, \ -1]$ and $[1, \ 0]$, $[0, \ -1]$

13. $\sqrt{11}$ **15.** 5 **17.** $\sqrt{62}$

19. $10 \leqq \sqrt{11}\sqrt{20} = 14.83$

21. $\|[8 \quad 1 \quad 0 \quad 10]^T\| = \sqrt{165} = 12.85 \leqq 5 + \sqrt{62} = 12.87$

23. $[v_1, -\frac{1}{2}v_1, v_3]$; v_1, v_3 arbitrary. Yes.

25. $x_1 = 2y_1 + y_2, x_2 = 5y_1 + 3y_2$

27. $x_1 = -2y_2 + y_3$, $x_2 = \frac{1}{2}y_1 + \frac{1}{2}y_2 - \frac{1}{2}y_3$, $x_3 = -y_1 + 2y_2$

29. $x_1 = 4y_1 - 2y_2 + 2y_3$, $x_2 = -2y_1 - 4y_2 + 4y_3$, $x_3 = -4y_1 + 2y_2 + 8y_3$

CHAPTER 7 (REVIEW QUESTIONS AND PROBLEMS), page 422

17. $\begin{bmatrix} 26 \\ -16 \\ 11 \end{bmatrix}$ **19.** $\begin{bmatrix} 15 \\ -22 \\ 0 \end{bmatrix}$ **21.** $\begin{bmatrix} -0.7 & -0.75 & 1 \\ 0.1 & 0.25 & 0 \\ 0.2 & 0 & 0 \end{bmatrix}$ **23.** 1

25. $\begin{bmatrix} 25 & -10 & 10 \\ -10 & 20 & 8 \\ 10 & 8 & 14 \end{bmatrix}$ **27.** $\begin{bmatrix} 52 \\ 42 \\ -42 \end{bmatrix}$ **29.** $\begin{bmatrix} -15 \\ -12 \\ -11 \end{bmatrix}$

31. $\begin{bmatrix} 4 & 0 & -2 \\ 0 & 0 & 0 \\ -2 & 0 & 1 \end{bmatrix}$, [5] **33.** $x = 2, y = 1, z = 3$ **35.** $x = \frac{1}{2}, y = \frac{1}{4}, z = -\frac{1}{2}$

37. $x = z = y + 3$ **39.** $x = -2z, y = \frac{1}{4}$ **41.** 2 **43.** 3

45. $\begin{bmatrix} -1.6 & 1 \\ 0.2 & 0 \end{bmatrix}$ **47.** $\begin{bmatrix} 0 & -2 & 1 \\ \frac{1}{2} & \frac{1}{2} & -\frac{1}{2} \\ -1 & 2 & 0 \end{bmatrix}$ **49.** $\begin{bmatrix} 5 & -1 & -6 \\ 0 & 2 & -8 \\ 0 & 0 & 10 \end{bmatrix}$

51. $1, \begin{bmatrix} 1 \\ 1 \end{bmatrix}, -1, \begin{bmatrix} 0 \\ 1 \end{bmatrix}$ **53.** $10, \begin{bmatrix} 3 \\ 0 \\ 1 \end{bmatrix}; 6, \begin{bmatrix} 0 \\ 1 \\ 0 \end{bmatrix}; 0, \begin{bmatrix} 1 \\ -1 \\ 1 \end{bmatrix}$

55. $5, \begin{bmatrix} 1 \\ 0 \\ 1 \end{bmatrix}; 0, \begin{bmatrix} 2 \\ -1 \\ 0 \end{bmatrix}; -5, \begin{bmatrix} 0 \\ 2 \\ -1 \end{bmatrix}$ **57.** $3, \begin{bmatrix} 2 \\ 3 \\ 3 \end{bmatrix}; 0, \begin{bmatrix} 1 \\ 0 \\ -1 \end{bmatrix}; -5, \begin{bmatrix} 2 \\ -5 \\ 3 \end{bmatrix}$

59. $\begin{bmatrix} 2 \\ 3 \end{bmatrix}, \begin{bmatrix} 2 \\ -3 \end{bmatrix}, \begin{bmatrix} 6 & 0 \\ 0 & -6 \end{bmatrix}$ **61.** $\begin{bmatrix} 0 \\ 0 \\ 1 \end{bmatrix}, \begin{bmatrix} 0 \\ 1 \\ 0 \end{bmatrix}, \begin{bmatrix} 1 \\ 1 \\ 0 \end{bmatrix}, \begin{bmatrix} 2 & 0 & 0 \\ 0 & 0 & 0 \\ 0 & 0 & -2 \end{bmatrix}$

63. Hyperbola $y_1^2 - y_2^2 = 1$; $x_1 = 0.8y_1 - 0.6y_2, x_2 = 0.6y_1 + 0.8y_2$

65. Ellipse $4y_1^2 + 9y_2^2 = 1$; $x_1 = (12y_1 - 5y_2)/13, x_2 = (5y_1 + 12y_2)/13$

67. $2i|x_1|^2 + 8i \operatorname{Im} \bar{x}_1 x_2$ **69.** 2

71. $I_1 = 15, I_2 = 10, I_3 = 5$ [amps] **73.** $I_1 = 4, I_2 = 5, I_3 = 1$ [amps]

PROBLEM SET 8.1, page 435

1. 3, 2, 0, $|\mathbf{v}| = \sqrt{13}$
3. -3, 0, 3, $|\mathbf{v}| = 3\sqrt{2}$
5. 4, 1, 1, $|\mathbf{v}| = 3\sqrt{2}$
7. 0, 0, 0, $|\mathbf{v}| = 0$
9. Q: (3, 0, 0), $|\mathbf{v}| = \sqrt{2}$
11. Q: (0, 0, 0), $|\mathbf{v}| = \sqrt{41}$
13. Q: (1, -1, -2), $|\mathbf{v}| = 0$
15. Q: (6, -6, 12), $|\mathbf{v}| = \sqrt{56}$
17. [3, 0, 2]
19. [4, -5, 27]
21. [3, 3, -27]
23. $\sqrt{13}$, $\sqrt{14} + \sqrt{3}$
27. [6, 2, 0], $\sqrt{40}$
29. [14, 4, 2], $\sqrt{216}$
31. [-1, -2, 4]
33. $1 \leqq |\mathbf{p} + \mathbf{q}| \leqq 3$, nothing about the direction.
35. The vector $\mathbf{r}$ from 0 to P is of the form $\mathbf{r} = k(\mathbf{a} + \mathbf{b})$. We also have $\mathbf{r} = \mathbf{a} + l(\mathbf{b} - \mathbf{a})$. Thus, $k\mathbf{a} + k\mathbf{b} = (1 - l)\mathbf{a} + l\mathbf{b}$. Form this, $k = 1 - l$, $k = l$; hence $k = \frac{1}{2}$, $l = \frac{1}{2}$, and the proof is complete.

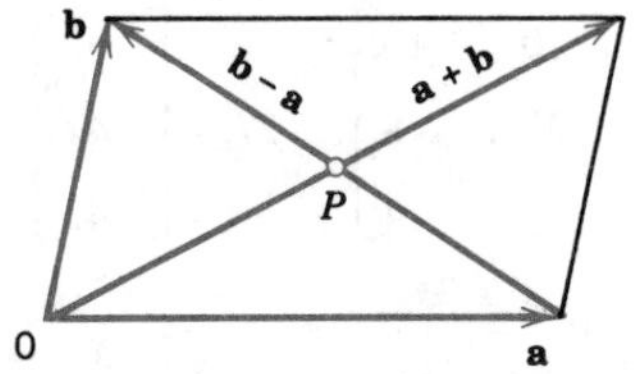

Problem 35

PROBLEM SET 8.2, page 441

1. -10
3. 6
5. $\sqrt{37}$
7. -21, 21
9. $\sqrt{122}$
11. $\sqrt{6}$, 0
15. 2, 0, -2
17. 0
19. 2
23. $-8/3$
27. $\pm(1/3)$ [2, 1, -2]
29. $c = -2$
31. 0
33. $1/\sqrt{50}$
35. 90°
37. $\mathbf{a} = \mathbf{i} + 2\mathbf{j} + \mathbf{k}$ and $\mathbf{b} = 2\mathbf{i} - \mathbf{j} + 3\mathbf{k}$ are normal; (3) yields the value $71° = \arccos \sqrt{3/28}$.
39. 55°, 79°, 46°, approximately
41. $|\mathbf{a} - \mathbf{b}|^2 = (\mathbf{a} - \mathbf{b})\cdot(\mathbf{a} - \mathbf{b}) = |\mathbf{a}|^2 + |\mathbf{b}|^2 - 2|\mathbf{a}||\mathbf{b}| \cos \gamma$, etc.
43. 2
45. 0
47. $-\sqrt{13}$
49. $|\mathbf{a} + \mathbf{b}|^2 = (\mathbf{a} + \mathbf{b})\cdot(\mathbf{a} + \mathbf{b}) \leqq |\mathbf{a}|^2 + 2|\mathbf{a}||\mathbf{b}| + |\mathbf{b}|^2 = (|\mathbf{a}| + |\mathbf{b}|)^2$

PROBLEM SET 8.3, page 449

1. [0, 0, 3], [0, 0, -3]
3. [1, -1, 1], $\sqrt{3}$, 5
5. [3, 0, -6]
7. [1, -1, 1]
9. **0**
11. [-9, -9, 45]
13. [2, 3, 1], 3
15. 1, -1
17. [-3, 39, 0]
19. -87
21. 3
23. 180
27. **0**
29. [13, -10, 14]

31. $[-12, -12, -6]$ **33.** 10 **35.** 9
37. $\sqrt{34}$ **39.** $\frac{1}{2}\sqrt{53}$ **41.** $3x - 2y + 4z = 0$
43. $4x + z = 5$ **45.** 8 **47.** 4
49. 1/6 **51.** 7/6 **53.** Yes **55.** No
59. This follows from (15) with **b** replaced by $\mathbf{a} \times \mathbf{b}$.

PROBLEM SET 8.4, page 456

9. $f(x, 2x^2) = x^2 + 16x^4$ **11.** Parallel planes
13. Coaxial cylinders **15.** Cones of revolution
17. Ellipsoids of revolution **19.** Ellipses $3x^2 + y^2 = const$
27. $x^2 + y^2 = const,\ x/y = const$ **29.** $y^2 + x^4 = const,\ y = cx^2$
31. $[1,\ 2t,\ 0], [0,\ 2,\ 0], \sqrt{1 + 4t^2}, 2$
33. $[-3 \sin t,\ 2 \cos t,\ 0], [-3 \cos t,\ -2 \sin t,\ 0], \sqrt{9 \sin^2 t + 4 \cos^2 t}$, $\sqrt{9 \cos^2 t + 4 \sin^2 t}$
35. $[-3 \sin t,\ 3 \cos t,\ 4], [-3 \cos t,\ -3 \sin t,\ 0], 5, 3$
37. $[-\sin t,\ 2 \cos 2t,\ 0], [-\cos t,\ -4 \sin 2t,\ 0], \sqrt{\sin^2 t + 4 \cos^2 2t}$, $\sqrt{\cos^2 t + 16 \sin^2 2t}$
39. $[1,\ 0,\ 0], [0,\ 1,\ 0], [0,\ 0,\ 1]$ **41.** $[y,\ 0,\ 0], [x,\ z,\ 0], [0,\ y,\ 0]$
43. $[0,\ 0,\ 2x], [2y,\ 0,\ 0], [0,\ 2z,\ 0]$
45. $-6t - 3$ **47.** $32t^7 - 216t^5 - 4t^3 - 3t^2$

PROBLEM SET 8.5, page 463

1. $\mathbf{r}(t) = \mathbf{i} + \mathbf{j} + t\mathbf{k}$ **3.** $\mathbf{r}(t) = (-3 + 3t)\mathbf{i} + \mathbf{j} - (2 + t)\mathbf{k}$
5. $\mathbf{r}(t) = t\mathbf{i} + t\mathbf{j} + t\mathbf{k}$ **7.** $\mathbf{r}(t) = (3 + 2t)\mathbf{i} - (1 + 4t)\mathbf{j} + 5\mathbf{k}$
9. $\mathbf{r}(t) = (1 + t)\mathbf{i} + (4 - 2t)\mathbf{j} + (-2 + 5t)\mathbf{k}$
11. $\mathbf{r}(t) = t\mathbf{i} + t\mathbf{j}$ **13.** $\mathbf{r}(t) = (1 - 2t)\mathbf{i} + t\mathbf{j} + t\mathbf{k}$
15. $xy = 1, z = 0$ **17.** $4y^2 + (z - 2)^2 = 4, x = 0$
19. $x^2 - \frac{1}{4}y^2 = 1, z = 0$ **21.** $\mathbf{r}(t) = \cos t\,\mathbf{i} + \sin t\,\mathbf{j}$
23. $\mathbf{r}(t) = 2 \cos t\,\mathbf{i} + 4 \sin t\,\mathbf{j} + \mathbf{k}$ **25.** $\mathbf{r}(t) = 3 \cosh t\,\mathbf{i} + 2 \sinh t\,\mathbf{j}$
27. $\mathbf{r}(t) = (1 + 2 \cos t)\mathbf{i} + (-2 + \sin t)\mathbf{j} + \frac{1}{2}\mathbf{k}$
31. (a) $\mathbf{u} = (1 + 4t^2)^{-1/2}(\mathbf{i} + 2t\mathbf{j})$, (b) $\mathbf{u}(P) = 5^{-1/2}(\mathbf{i} + 2\mathbf{j})$, (c) $(1 + w)\mathbf{i} + (1 + 2w)\mathbf{j}$
33. (a) $\mathbf{u} = (1 + 9t^4)^{-1/2}(\mathbf{i} + 3t^2\mathbf{j})$, (b) $\mathbf{u}(P) = 145^{-1/2}(\mathbf{i} + 12\mathbf{j})$, (c) $(2 + w)\mathbf{i} + (8 + 12w)\mathbf{j}$
35. (a) $\mathbf{u} = (4 \sin^2 t + \cos^2 t)^{-1/2}(-2 \sin t\,\mathbf{i} + \cos t\,\mathbf{j})$, (b) $\mathbf{u}(P) = (2/5)^{1/2}(-2^{1/2}\mathbf{i} + 2^{-1/2}\mathbf{j})$, (c) $2^{1/2}(1 - w)\mathbf{i} + 2^{-1/2}(1 + w)\mathbf{j}$
37. (a) $\mathbf{u} = (1/5)(-3 \sin t\,\mathbf{i} + 3 \cos t\,\mathbf{j} + 4\mathbf{k})$, (b) $t = \frac{1}{2}\pi$, $\mathbf{u}(P) = (1/5)(-3\mathbf{i} + 4\mathbf{k})$, (c) $-3w\mathbf{i} + 3\mathbf{j} + (2\pi + 4w)\mathbf{k}$
39. $\sinh 1 = 1.175$ **41.** $8(\sqrt{1000} - 1)/27 = 9.073$
43. Start from $\mathbf{r}(t) = t\mathbf{i} + f(t)\mathbf{j}$.

PROBLEM SET 8.6, page 468

1. Positive x-axis, $3\mathbf{i}$, 3, $\mathbf{0}$

3. Positive z-axis, $8t\mathbf{k}$, $8t$, $8\mathbf{k}$

5. Circle, $-6 \sin 2t\,\mathbf{i} + 6 \cos 2t\,\mathbf{j}$, 6, $-12 \cos 2t\,\mathbf{i} - 12 \sin 2t\,\mathbf{j}$

7. Hyperbola $xy = 1$, $x > 0$, $e^t\mathbf{i} - e^{-t}\mathbf{j}$, $(2 \cosh 2t)^{1/2}$, $e^t\mathbf{i} + e^{-t}\mathbf{j}$

9. Segment $-1 \leqq y \leqq 1$ on the y-axis, $\cos t\,\mathbf{j}$, $|\cos t|$, $-\sin t\,\mathbf{j}$

11. $\mathbf{r}(t) = \frac{1}{2}\mathbf{a}_0 t^2 + \mathbf{v}_0 t + \mathbf{r}_0$ ($\mathbf{a}_0$, $\mathbf{v}_0$, $\mathbf{r}_0$ constant vectors)

13. (a) $\cos t\,\mathbf{j}$, $-\sin t\,\mathbf{j}$, (b) at $\pm\mathbf{j}$, (c) at $\pm\mathbf{j}$

15. $[3 \sin t \cos t/(\sin^2 t + 4 \cos^2 t)](\sin t\,\mathbf{i} - 2 \cos t\,\mathbf{j})$

17. $\mathbf{v}(0) = 2\omega R\mathbf{i}$, $\mathbf{a}(0) = -\omega^2 R\mathbf{j}$, $\mathbf{v}(\pi/\omega) = \mathbf{0}$, $\mathbf{a}(\pi/\omega) = \omega^2 R\mathbf{j}$

19. $|\mathbf{a}| = \omega^2 R = |\mathbf{v}|^2/R = 30^2/(30 \cdot 86400 \cdot 365/2\pi) = 5.98 \cdot 10^{-6}$ [km/sec²]

21. $R = 3960 + 80$ mi $= 2.133 \cdot 10^7$ ft, $g = |\mathbf{a}| = \omega^2 R = |\mathbf{v}|^2/R$,
$|\mathbf{v}| = \sqrt{gR} = \sqrt{6.61 \cdot 10^8} = 25700$ [ft/sec] $= 17500$ [mi/h]

PROBLEM SET 8.7, page 471

3. $\mathbf{r}(t) = [t, \ y(t), \ 0]$, $\mathbf{r}' = [1, \ y', \ 0]$, $\mathbf{r}' \bullet \mathbf{r}' = 1 + y'^2$, $\mathbf{r}'' = [0, \ y'', \ 0]$, etc.

5. $2/(1 + 4x^2)^{3/2}$

7. $a/(a^2 + c^2)$

9. $6/\sqrt{t}\,(4 + 9t)^{3/2}$

15. $3/(1 + 9t^2 + 9t^4)$

PROBLEM SET 8.8, page 474

1. $(e^{2t} + \sin t \cos t)/\sqrt{e^{2t} + \sin^2 t}$

3. $(\cos t)^{(\sin t)-1}[\cos^2 t \ln (\cos t) - \sin^2 t]$

5. $(2 + 3e^t - e^{3t})/(e^{2t} + e^{-t})^2$

7. $e^{2u} \sin 2v$, $e^{2u} \cos 2v$

9. $-(u^2 + v^2)^{-3}u$, $-(u^2 + v^2)^{-3}v$

PROBLEM SET 8.9, page 480

1. $y\mathbf{i} + x\mathbf{j}$

3. $2x\mathbf{i} - 2y\mathbf{j}$

5. $y^{-1}\mathbf{i} - xy^{-2}\mathbf{j}$

7. $e^x \sin y\,\mathbf{i} + e^x \cos y\,\mathbf{j}$

9. $(x\mathbf{i} + y\mathbf{j})/(x^2 + y^2)$

11. $-(x^2 + y^2 + z^2)^{-3/2}(x\mathbf{i} + y\mathbf{j} + z\mathbf{k})$

13. $-2z(x^2 + y^2)^{-2}(x\mathbf{i} + y\mathbf{j}) + (x^2 + y^2)^{-1}\mathbf{k}$

15. $4yz(y^2 + z^2)^{-2}(z\mathbf{i} - y\mathbf{j})$

17. $x - y + z$

19. e^{xy}

21. $\frac{1}{2} \ln (x^2 + y^2)$

23. $x/y + \frac{1}{2}z^2$

29. $1/\sqrt{5}$

31. $-\sqrt{2}$

33. 0

35. $-2e^2/\sqrt{13}$

37. 7/3

39. $0.8\mathbf{i} - 0.6\mathbf{j}$

41. $0.6\mathbf{i} + 0.8\mathbf{j}$

43. $\mathbf{j}$

45. $(1/5\sqrt{2})(3\mathbf{i} + 4\mathbf{j} - 5\mathbf{k})$

47. The direction of $-3\mathbf{i} + 4\mathbf{j}$

49. The direction of $6\mathbf{i} - 5\mathbf{j}$

PROBLEM SET 8.10, page 485

1. 3

3. $2x^2z$

5. 0

7. $(x + y)^2$

9. 0

11. 28

13. 0

15. $2xz/y^3$

17. $6x\mathbf{i} + 2\mathbf{j}$

19. $2\mathbf{i} + 2\mathbf{j} - 8\mathbf{k}$

25. $(a + b + c)(xyz + 1)e^{xyz}$

27. $2(x^2 + x - y^2 - y)e^{x+y}$

PROBLEM SET 8.11, page 487

1. $\mathbf{k}$ **3.** $\mathbf{0}$ **5.** 0

7. $e^x[\sin z\,(\mathbf{i} - \mathbf{j}) + \cos z\,\mathbf{k}]$ **9.** $-3y(x^2 + y^2)^{-1}\mathbf{k}$

11. curl $\mathbf{v} = -2z\mathbf{i}$, incompressible, $\mathbf{r} = c_1\mathbf{i} + (c_3{}^2t + c_2)\mathbf{j} + c_3\mathbf{k}$

13. curl $\mathbf{v} = -2\mathbf{k}$, incompressible, $x' = y,\ y' = -x,\ x^2 + y^2 = c_1,\ z = c_2$

15. curl $\mathbf{v} = \mathbf{0}$, div $\mathbf{v} = 2$, compressible, $\mathbf{r} = c_1e^t\mathbf{i} + c_2e^t\mathbf{j} + c_3\mathbf{k}$

23. $(1 - y)\mathbf{i} + (1 - z)\mathbf{j} + (1 - x)\mathbf{k}$ **25.** $2(xz + xy + yz)$

27. $-yz - zx - xy,\ (yz - x^2)\mathbf{i} + (xz - y^2)\mathbf{j} + (xy - z^2)\mathbf{k}$

29. $z(y - x)\mathbf{i} + x(z - y)\mathbf{j} + y(x - z)\mathbf{k}$

CHAPTER 8 (REVIEW QUESTIONS AND PROBLEMS), page 494

21. $[1, 1, -3]$ **23.** $[16, -20, 44]$ **25.** $\sqrt{33},\ 3 - \sqrt{14}$

27. 19, 19, 4, 4 **29.** $\pm[7, -1, -3]$ **31.** 0, 0

33. -15, 30 **35.** $[-3, 5, -6]$ **37.** 55.5°

39. 25 **41.** 0 **43.** -27 **45.** 1 **47.** $-6/\sqrt{35}$

49. If $|\mathbf{a}| = |\mathbf{b}|$ or $\mathbf{a}$ and $\mathbf{b}$ are orthogonal **51.** $-17\mathbf{k}$

53. $[21, -3, -1]$ **55.** $[1, -2, 1]$ **57.** 57

59. 7/6 **61.** $7x - 8y + 6z = 5$ **63.** $[256, -4, 1]$

65. -2 **67.** $2x^2 + 4(y - z)^2$ **69.** $4xy(4z(x^2 + y^2) + 1)$

71. 0 **73.** 0 **75.** $4\sqrt{2}$

77. $84/\sqrt{481}$ **79.** Hyperbola, $\mathbf{i} - \frac{1}{4}\mathbf{j},\ \frac{1}{4}\sqrt{17},\ \frac{1}{4}\mathbf{j}$

PROBLEM SET 9.1, page 506

1. 2/3 **3.** 82/81 **5.** -38.4 **7.** 6π **9.** 3/2

11. $-e^{-3} + \frac{1}{2}e^{-1} + \frac{26}{3}$ **13.** -60 **15.** 0

17. $-32/3$ **19.** $80\sqrt{10}$ **21.** 5 **23.** $3\pi/2$

PROBLEM SET 9.2, page 514

1. $f = 4x - 3y^2 + 5z;\ 6$ **3.** $f = x^2 + y^3z;\ 20$

5. $f = x^3y^2z^4;\ -324$ **7.** $f = -\cos xy;\ 2$

9. $f = x^2yz^2/2;\ a^2bc^2/2$ **11.** No

13. $f = xe^y - ze^x;\ ae^b - ce^a$ **15.** $f = \sin xy - z;\ \sin ab - c$

17. $f = x^2e^{2z}/2;\ a^2e^{2c}/2$ **19.** No

PROBLEM SET 9.3, page 521

1. 11/6 **3.** 1/8 **5.** $\pi/4$ **7.** $(1 - e^{-4})/2$ **9.** 1/12

11. 2/3 **13.** $a^3/3$ **15.** 8/15 **17.** $\pi/2$

19. $\pi(e^{-1} - e^{-4})$ **21.** 1 **23.** 1

25. $2b/3,\ h/3$ **27.** $I_x = bh^3/12,\quad I_y = b^3h/4$

29. $I_x = h^3(b + a)/24,\quad I_y = h(a^4 - b^4)/48(a - b)$

PROBLEM SET 9.4, page 528

1. $3a^2$ 3. 24π 5. 1 7. $2 \cdot 5^7/7$ 9. 10/3

11. $-\frac{1}{6} + e^{-1} + \frac{1}{2}e^2 - \frac{1}{3}e^3$ 13. $-1/3$ 15. $9a^4\pi/16$

17. $a^2(3\pi/8 - 1)$ 19. $4e^2 - 4$ 21. 80π

23. 2 25. $2e - 15/4$ 27. $4e^4 - 4$ 29. 16π

PROBLEM SET 9.5, page 533

1. Straight lines, **k**
3. $x^2/16 + y^2/4 = 1$, straight lines, ellipses, $2 \cos u\, \mathbf{i} + 4 \sin u\, \mathbf{j}$
5. $z = c\sqrt{x^2 + y^2}$, circles, straight lines, $-cu \cos v\, \mathbf{i} - cu \sin v\, \mathbf{j} + u\mathbf{k}$
7. $z = x^2 + y^2$, circles, parabolas, $-2u^2 \cos v\, \mathbf{i} - 2u^2 \sin v\, \mathbf{j} + u\mathbf{k}$
9. $z = x^2/a^2 + y^2/b^2$, ellipses, parabolas, $-2bu^2 \cos v\, \mathbf{i} - 2au^2 \sin v\, \mathbf{j} + abu\mathbf{k}$
11. $x^2/a^2 + y^2/b^2 - z^2/c^2 + 1 = 0$, ellipses, hyperbolas, $-bc \sinh^2 u \cos v\, \mathbf{i} - ac \sinh^2 u \sin v\, \mathbf{j} + ab \cosh u \sinh u\, \mathbf{k}$
13. $u\mathbf{i} + v\mathbf{j} + v\mathbf{k}$, $-\mathbf{j} + \mathbf{k}$
15. $u\mathbf{i} + v\mathbf{j} + (1 - u - v)\mathbf{k}$, $\mathbf{i} + \mathbf{j} + \mathbf{k}$
17. $\cosh u\, \mathbf{i} + \sinh u\, \mathbf{j} + v\mathbf{k}$, $\cosh u\, \mathbf{i} - \sinh u\, \mathbf{j}$
19. $u \cos v\, \mathbf{i} + u \sin v\, \mathbf{j} + 16u^2\mathbf{k}$, $-32u^2 \cos v\, \mathbf{i} - 32u^2 \sin v\, \mathbf{j} + u\mathbf{k}$
21. $u\mathbf{i} + v\mathbf{j} + 16(u^2 + v^2)\mathbf{k}$, $-32u\mathbf{i} - 32v\mathbf{j} + \mathbf{k}$
25. $\mathbf{r}_u$ is tangent to the curves $v = const$, and $\mathbf{r}_v$ is tangent to $u = const$.
27. $(1/\sqrt{38})(2\mathbf{i} + 3\mathbf{j} - 5\mathbf{k})$ 29. $(1/3)(x\mathbf{i} + y\mathbf{j} + z\mathbf{k})$
31. $(1/a)(x\mathbf{i} + y\mathbf{j})$ 33. $(x^2 + y^2 + 1)^{-1/2}(-y\mathbf{i} - x\mathbf{j} + \mathbf{k})$
37. $z^* = 2\sqrt{2} + y^*$ 39. $3x^* + 4y^* = 25$

PROBLEM SET 9.6, page 542

1. -16 3. $2e^3 - 3e^2 + 1$ 5. 7π 7. -1250π

9. $17h/4$ 11. $-59/180$ 13. $-53/2$

15. $2 \cosh^3 2 - 2 \approx 104.5$ 17. $3\sqrt{3}$ 19. $\pi a^3h^3/6$

21. $17^{5/2}\pi/6$ 23. $10^{3/2} - 1$ 25. $3^{5/2} - 2^{7/2} + 1$

29. $2\pi h$ 31. $\pi h^4/\sqrt{2}$ 33. $h\pi + 2h^3\pi/3$

35. Use $d\mathbf{r} = \mathbf{r}_u\, du + \mathbf{r}_v\, dv$.

PROBLEM SET 9.7, page 549

1. 8/27 3. 96π 5. 1/120 7. $8\pi^2$ 9. 19/180

11. 2/3 13. $\pi ha^4/2$ 15. $8\pi a^5/15$ 17. 16 19. $6\pi a^2h$

21. 2π 23. 1/6 25. $40\pi/3$ 27. $384\pi/5$ 29. $-\pi/8$

PROBLEM SET 9.8, page 555

3. Put $f = g$ in (10). 5. Use (11). 9. $r = a$, $\cos \phi = 1$

PROBLEM SET 9.9, page 561

1. ± 3
3. ± 6
5. $\pm(e^2 - 1)$
7. $\pm 2(1 - e^2)$
11. 24π
13. 2π
15. $-1/3$
17. $-2/3$
19. $-18\pi\sqrt{2}$

CHAPTER 9 (REVIEW QUESTIONS AND PROBLEMS), page 562

17. $2/3$
19. $e^2/2$
21. $\pm 25\pi$
23. 0
25. $e^4 + 15$
27. $\pm 12\pi$
29. 0
31. $8e^{-24}$
33. 0
35. $\pm(1 - \cosh 1)$
37. $(e^2 - 1)^2/2$
39. $1/8$
41. $189/20$
43. $4/5, 8/15$
45. $8a/5\pi, 8a/5\pi$
47. $3 \cos u\, \mathbf{i} + 3 \sin u\, \mathbf{j} + v\mathbf{k}, \quad 3 \cos u\, \mathbf{i} + 3 \sin u\, \mathbf{j}$
49. $u \cos v\, \mathbf{i} + u \sin v\, \mathbf{j} + u\mathbf{k}, \quad -u \cos v\, \mathbf{i} - u \sin v\, \mathbf{j} + u\mathbf{k}$
51. $2u \cos v\, \mathbf{i} + u \sin v\, \mathbf{j} + u^2\mathbf{k}, \quad -2u^2 \cos v\, \mathbf{i} - 4u^2 \sin v\, \mathbf{j} + 2u\mathbf{k}$
53. $5/216$
55. $4/9$
57. $-4/3 + 8\pi/3$
59. 40

PROBLEM SET 10.1, page 568

1. $2\pi, 2\pi, \pi, \pi, 2, 2, 1, 1$
17. 0 (n even), $2/n$ (n odd)
19. 0
21. $n[(-1)^n e^{-\pi} - 1]/(1 + n^2)$
23. $[(-1)^n e^{\pi} - 1]/(1 + n^2)$
25. $0\ (n = 0),\ \pi/n\ (n = 1, 3, \cdots),\ -\pi/n\ (n = 2, 4, \cdots)$

PROBLEM SET 10.2, page 576

1. $\frac{1}{2} + \frac{2}{\pi}\left(\cos x - \frac{1}{3}\cos 3x + \frac{1}{5}\cos 5x - + \cdots\right)$

3. $\frac{1}{2} + \frac{2}{\pi}\left(\sin x + \frac{1}{3}\sin 3x + \frac{1}{5}\sin 5x + \cdots\right)$

5. $\frac{4}{\pi}\left(\cos x - \frac{1}{3}\cos 3x + \frac{1}{5}\cos 5x - + \cdots\right)$

7. $2\left(\sin x - \frac{1}{2}\sin 2x + \frac{1}{3}\sin 3x - \frac{1}{4}\sin 4x + - \cdots\right)$

9. $\frac{\pi^2}{3} - 4\left(\cos x - \frac{1}{4}\cos 2x + \frac{1}{9}\cos 3x - \frac{1}{16}\cos 4x + - \cdots\right)$

11. $2\left[\left(\frac{\pi^2}{1} - \frac{6}{1^3}\right)\sin x - \left(\frac{\pi^2}{2} - \frac{6}{2^3}\right)\sin 2x + \left(\frac{\pi^2}{3} - \frac{6}{3^3}\right)\sin 3x - + \cdots\right]$

13. $\frac{2}{\pi}\sin x + \frac{1}{2}\sin 2x - \frac{2}{9\pi}\sin 3x - \frac{1}{4}\sin 4x + \frac{2}{25\pi}\sin 5x + \cdots$

15. $\frac{4}{\pi}\left(\sin x - \frac{1}{9}\sin 3x + \frac{1}{25}\sin 5x - + \cdots\right)$

PROBLEM SET 10.3, page 580

1. $\frac{4}{\pi}\left(\sin \pi x + \frac{1}{3}\sin 3\pi x + \frac{1}{5}\sin 5\pi x + \cdots\right)$

3. $1 + \frac{4}{\pi}\left(\sin\frac{\pi x}{2} + \frac{1}{3}\sin\frac{3\pi x}{2} + \frac{1}{5}\sin\frac{5\pi x}{2} + \cdots\right)$

5. $\frac{2}{3} + \frac{4}{\pi^2}\left(\cos \pi x - \frac{1}{4}\cos 2\pi x + \frac{1}{9}\cos 3\pi x - + \cdots\right)$

7. $\frac{1}{4} - \frac{2}{\pi^2}\left(\cos \pi x + \frac{1}{9}\cos 3\pi x + \cdots\right) + \frac{1}{\pi}\left(\sin \pi x - \frac{1}{2}\sin 2\pi x + - \cdots\right)$

9. $-\frac{4}{\pi^2}\left(\cos \pi x + \frac{1}{9}\cos 3\pi x + \cdots\right) + \frac{2}{\pi}\left(2\sin \pi x - \frac{1}{2}\sin 2\pi x + - \cdots\right)$

11. $1 - \frac{12}{\pi^2}\left(\cos \pi x - \frac{1}{4}\cos 2\pi x + \frac{1}{9}\cos 3\pi x - \frac{1}{16}\cos 4\pi x + - \cdots\right)$

13. $4\left(\frac{1}{2} - \frac{1}{1\cdot 3}\cos 2\pi x - \frac{1}{3\cdot 5}\cos 4\pi x - \frac{1}{5\cdot 7}\cos 6\pi x - \cdots\right)$

17. Write τ for x in Example 1 and set $\tau = x - 1$.

PROBLEM SET 10.4, page 584

1. Even: $|x^3|$, x^2 cox nx, $\cosh x$. Odd: x cox nx, $\sinh x$, $x|x|$.

3. Odd 5. Odd 7. Even 9. Neither 11. Neither

13. $1/(1 - x^2) + x/(1 - x^2)$ 15. $\cosh kx + \sinh kx$

25. $\frac{k}{2} + \frac{2k}{\pi}\left(\cos x - \frac{1}{3}\cos 3x + \frac{1}{5}\cos 5x - + \cdots\right)$

27. $\frac{4}{\pi}\left(\sin x - \frac{1}{9}\sin 3x + \frac{1}{25}\sin 5x - + \cdots\right)$

29. $\frac{\pi^2}{6} - \frac{4}{\pi}\cos x - \frac{2}{2^2}\cos 2x + \frac{4}{3^3\pi}\cos 3x + \frac{2}{4^2}\cos 4x - \frac{4}{5^3\pi}\cos 5x + \cdots$

31. $\frac{\pi^2}{12} - \cos x + \frac{1}{4}\cos 2x - \frac{1}{9}\cos 3x + \frac{1}{16}\cos 4x - + \cdots$

PROBLEM SET 10.5, page 588

1. $\frac{4k}{\pi}\left(\sin\frac{\pi x}{L} + \frac{1}{3}\sin\frac{3\pi x}{L} + \frac{1}{5}\sin\frac{5\pi x}{L} + \cdots\right)$

3. $\frac{2L^2}{\pi}\left[\left(1 - \frac{4}{\pi^2}\right)\sin\frac{\pi x}{L} - \frac{1}{2}\sin\frac{2\pi x}{L} + \left(\frac{1}{3} - \frac{4}{3^3\pi^2}\right)\sin\frac{3\pi x}{L} - \frac{1}{4}\sin\frac{4\pi x}{L} + \cdots\right]$

5. $\frac{2L}{\pi}\left(\sin\frac{\pi x}{L} + \frac{1}{2}\sin\frac{2\pi x}{L} + \frac{1}{3}\sin\frac{3\pi x}{L} + \cdots\right)$

7. $\left(1 + \frac{2}{\pi}\right)\sin x - \frac{1}{2}\sin 2x + \left(\frac{1}{3} - \frac{2}{9\pi}\right)\sin 3x - \frac{1}{4}\sin 4x + \cdots$

9. $\frac{4L}{\pi^2}\left(\frac{1}{1^2}\sin\frac{\pi x}{L} - \frac{1}{3^2}\sin\frac{3\pi x}{L} + \frac{1}{5^2}\sin\frac{5\pi x}{L} - + \cdots\right)$

11. $\frac{L}{2} - \frac{4L}{\pi^2}\left(\cos\frac{\pi x}{L} + \frac{1}{9}\cos\frac{3\pi x}{L} + \frac{1}{25}\cos\frac{5\pi x}{L} + \cdots\right)$

13. $\frac{L^2}{3} - \frac{4L^2}{\pi^2}\left(\cos\frac{\pi x}{L} - \frac{1}{4}\cos\frac{2\pi x}{L} + \frac{1}{9}\cos\frac{3\pi x}{L} - \frac{1}{16}\cos\frac{4\pi x}{L} + - \cdots\right)$

15. $\frac{1}{2} - \frac{2}{\pi}\left(\cos\frac{\pi x}{L} - \frac{1}{3}\cos\frac{3\pi x}{L} + \frac{1}{5}\cos\frac{5\pi x}{L} - + \cdots\right)$

17. $a_0 = \frac{1}{L}(e^L - 1),\quad a_n = \frac{2L}{L^2 + n^2\pi^2}[(-1)^n e^L - 1]$

19. $\frac{2}{\pi} - \frac{4}{\pi}\left(\frac{1}{1\cdot 3}\cos\frac{2\pi x}{L} + \frac{1}{3\cdot 5}\cos\frac{4\pi x}{L} + \frac{1}{5\cdot 7}\cos\frac{6\pi x}{L} + \cdots\right)$

PROBLEM SET 10.6, page 591

1. Use (7).

3. $-\frac{2i}{\pi}\sum_{n=-\infty}^{\infty}\frac{1}{2n+1}e^{(2n+1)ix}$

5. $i\sum_{\substack{n=-\infty\\ n\neq 0}}^{\infty}\frac{(-1)^n}{n}e^{inx}$

7. $\pi + i\sum_{\substack{n=-\infty\\ n\neq 0}}^{\infty}\frac{1}{n}e^{inx}$

9. $\frac{\pi^2}{3} + 2\sum_{\substack{n=-\infty\\ n\neq 0}}^{\infty}\frac{(-1)^n}{n^2}e^{inx}$

PROBLEM SET 10.7, page 594

3. $y = C_1\cos\omega t + C_2\sin\omega t + A(\omega)\sin t,\quad A(\omega) = 1/(\omega^2 - 1),$
$A(0.5) = -1.33,\quad A(0.7) = -1.96,\quad A(0.9) = -5.3,\quad A(1.1) = 4.8,$
$A(1.5) = 0.8,\quad A(2) = 0.33,\quad A(10) = 0.01$

5. $y = C_1\cos\omega t + C_2\sin\omega t + \sum_{n=1}^{N}\frac{a_n}{\omega^2 - n^2}\cos nt$

7. $y = C_1\cos\omega t + C_2\sin\omega t$
$$+ \frac{\pi}{2\omega^2} + \frac{4}{\pi}\left(\frac{1}{\omega^2 - 1}\cos t + \frac{1}{9(\omega^2 - 9)}\cos 3t + \cdots\right)$$

9. $y = \frac{1 - n^2}{D}a_n\cos nt + \frac{nc}{D}a_n\sin nt,\quad D = (1 - n^2)^2 + n^2c^2$

11. $y = -\frac{3c}{64 + 9c^2}\cos 3t - \frac{8}{64 + 9c^2}\sin 3t$

13. $y = \sum_{n=1}^{\infty}\left[\frac{(-1)^n c}{n^2 D_n}\cos nt - \frac{(-1)^n(1 - n^2)}{n^3 D_n}\sin nt\right],\ D_n = (1 - n^2)^2 + n^2c^2$

15. $I = \sum_{n=1}^{\infty}(A_n\cos nt + B_n\sin nt),\quad A_n = (-1)^{n+1}\frac{240(10 - n^2)}{n^2 D_n},$
$B_n = \frac{(-1)^{n+1}2400}{nD_n},\quad D_n = (10 - n^2)^2 + 100n^2$

PROBLEM SET 10.8, page 598

1. $F = \frac{4}{\pi}\left[\sin x + \frac{1}{3}\sin 3x + \cdots + \frac{1}{N}\sin Nx\right]$ (N odd)

5. $F = 2\left(\sin x - \frac{1}{2}\sin 2x + \cdots + \frac{(-1)^{N+1}}{N}\sin Nx\right)$,
$E^* \approx 8, 5, 3.6, 2.8, 2.3$

7. $F = \frac{\pi^2}{3} - 4\left(\cos x - \frac{1}{4}\cos 2x + \frac{1}{9}\cos 3x - \cdots + \frac{(-1)^{N+1}}{N^2}\cos Nx\right)$,
$E^* \approx 4.14, 1.00, 0.38, 0.18, 0.10$

9. $F = \frac{2}{\pi}\sin x + \frac{1}{2}\sin 2x - \frac{2}{9\pi}\sin 3x - \frac{1}{4}\sin 4x + \frac{2}{25\pi}\sin 5x + \cdots$,
$E^* = \frac{\pi^3}{12} - \pi\left[\frac{4}{\pi^2} + \frac{1}{4} + \frac{4}{81\pi^2} - \frac{1}{16} + \frac{4}{625\pi^2} + \cdots\right]$; 1.311, 0.525, 0.509, 0.313, 0.311

15. Use the Fourier series $\cos^3 x = \frac{3}{4}\cos x + \frac{1}{4}\cos 3x$.

PROBLEM SET 10.9, page 605

9. $\frac{2}{\pi}\int_0^\infty \left[\left(1 - \frac{2}{w^2}\right)\sin w + \frac{2}{w}\cos w\right]\frac{\cos wx}{w}\,dw$

11. $\frac{2}{\pi}\int_0^\infty \left[\frac{a\sin aw}{w} + \frac{\cos aw - 1}{w^2}\right]\cos xw\,dw$

13. $A = \frac{2}{\pi}\int_0^\infty \frac{\cos wv}{1 + v^2}\,dv = e^{-w}\ (w > 0)$, $f(x) = \int_0^\infty e^{-w}\cos wx\,dw$

15. $f(ax) = \int_0^\infty A(w)\cos axw\,dw = \int_0^\infty A\left(\frac{p}{a}\right)\cos xp\,\frac{dp}{a}$, where $wa = p$.
If we write again w instead of p, the result follows.

17. Differentiating (10) we have $\frac{d^2A}{dw^2} = -\frac{2}{\pi}\int_0^\infty f^*(v)\cos wv\,dv$, $f^*(v) = v^2 f(v)$, and the result follows.

PROBLEM SET 10.10, page 610

1. $\sqrt{2/\pi}\,(\sin 2w - 2\sin w)/w$

3. $\sqrt{2/\pi}\,(aw\sin aw + \cos aw - 1)/w^2$

7. $e^{-w}\sqrt{\pi/2}$

9. $\sqrt{2/\pi}\,[(2 - w^2)\cos w + 2w\sin w - 2]/w^3$

11. $\sqrt{\pi/2}$ if $0 < w < \pi$, 0 if $w > \pi$

13. $\sqrt{\pi/2}\cos w$ if $|w| < \pi/2$, 0 if $|w| > \pi/2$

17. $\sqrt{\pi/2}\,e^{+w}\cos w$

19. No

PROBLEM SET 10.11, page 618

1. $1/(1 + iw)\sqrt{2\pi}$

3. $\sqrt{2/\pi}\,(2 - w)^{-1}\sin(2 - w)$

5. $[-1 + (1 + iaw)e^{-iaw}]/w^2\sqrt{2\pi}$

7. $i\sqrt{2/\pi}\,(\cos w - 1)/w$

CHAPTER 10 (REVIEW QUESTIONS AND PROBLEMS), page 622

17. $\frac{1}{2} - \frac{2}{\pi}\left(\sin x + \frac{1}{3}\sin 3x + \frac{1}{5}\sin 5x + \cdots\right)$

19. $\frac{2}{\pi}\left(\sin x - \frac{2}{2}\sin 2x + \frac{1}{3}\sin 3x + \frac{1}{5}\sin 5x - \frac{2}{6}\sin 6x + \cdots\right)$

21. $\frac{\pi}{2} - \frac{4}{\pi}\left(\cos x + \frac{1}{9}\cos 3x + \frac{1}{25}\cos 5x + \cdots\right)$

23. $-\sin x + \frac{1}{2}\sin 2x - \frac{1}{3}\sin 3x + \frac{1}{4}\sin 4x - + \cdots$

25. $\frac{8}{\pi}\left(\cos x + \frac{1}{9}\cos 3x + \frac{1}{25}\cos 5x + \cdots\right)$

27. $\frac{\pi^2}{12} - \frac{2}{\pi}\cos x - \frac{1}{2^2}\cos 2x + \frac{2}{3^3\pi}\cos 3x + \frac{1}{4^2}\cos 4x - \cdots$

29. $\frac{8}{\pi}\left(\sin x + \frac{1}{3^3}\sin 3x + \frac{1}{5^3}\sin 5x + \cdots\right)$

31. $-\frac{4}{\pi}\left(\sin \pi x + \frac{1}{3}\sin 3\pi x + \frac{1}{5}5\pi x + \cdots\right)$

33. $\frac{4}{\pi}\left(\sin\frac{\pi}{2}x - \frac{1}{2}\sin \pi x + \frac{1}{3}\sin\frac{3\pi}{2}x - + \cdots\right)$

35. $\frac{1}{4} - \frac{2}{\pi^2}\left(\cos \pi x + \frac{1}{9}\cos 3\pi x + \cdots\right) - \frac{1}{\pi}\left(\sin \pi x - \frac{1}{2}\sin 2\pi x + - \cdots\right)$

37. $-\frac{4}{\pi^2}\left(\cos \pi x + \frac{1}{9}\cos 3\pi x + \cdots\right) + \frac{2}{\pi}\left(2\sin \pi x - \frac{1}{2}\sin 2\pi x + - \cdots\right)$

39. $-\frac{1}{3} + \frac{4}{\pi^2}\left(\cos \pi x - \frac{1}{4}\cos 2\pi x + \frac{1}{9}\cos 3\pi x - + \cdots\right) + \frac{2}{\pi}\left(\sin \pi x - \frac{1}{2}\sin 2\pi x + - \cdots\right)$

41. $\pi/4$ **43.** $\pi^3/32$ **47.** 5.168, 0.075, 0.075, 0.012, 0.012, 0.004

49. $y = C_1\cos\omega t + C_2\sin\omega t + \frac{\pi^2}{12\omega^2} - \frac{1}{\omega^2 - 1}\cos t + \frac{1}{4(\omega^2 - 4)}\cos 2t - + \cdots$

PROBLEM SET 11.1, page 628

25. $u = f(x)$ **27.** $u_x = f(y)$, $u = xf(y) + g(y)$

29. $u = c(y)e^{x^2y}$ **31.** $u = v(x) + w(y)$

33. $u = c = const$ **35.** $u = cx + g(y)$

PROBLEM SET 11.3, page 637

1. $u = 0.02\cos t\sin x$ **3.** $u = k(\cos t\sin x - \cos 2t\sin 2x)$

5. $u = \frac{4}{5\pi}\left(\frac{1}{4}\cos 2t\sin 2x - \frac{1}{36}\cos 6t\sin 6x + \frac{1}{100}\cos 10t\sin 10x - + \cdots\right)$

7. $u = \frac{8k}{\pi}\left(\cos t\sin x + \frac{1}{3^3}\cos 3t\sin 3x + \frac{1}{5^3}\cos 5t\sin 5x + \cdots\right)$

9. $u = 12k\left[\left(\frac{\pi}{1^3} - \frac{8}{1^5\pi}\right)\cos t \sin x + \left(\frac{\pi}{3^3} - \frac{8}{3^5\pi}\right)\cos 3t \sin 3x + \cdots\right]$

11. $u = 0.1 \sin x (\cos t - 2 \sin t)$

15. $27,\ 960/\pi^6 \approx 0.9986$

17. $u = ke^{c(x+y)}$

19. $u = k \exp [c(x^2 + y^2)]$

21. $u = ky^c e^{cx}$

23. $u = k \exp (cx + y/c)$

PROBLEM SET 11.4, page 642

9. $17.5n$ cycles/sec

13. $u = f_1(x) + f_2(xy)$

15. $u = xf_1(x - y) + f_2(x - y)$

17. $u = f_1(x + y) + f_2(3x + y)$

23. $u = \frac{8L^2}{\pi^3}\left(\cos c\left(\frac{\pi}{L}\right)^2 t \sin \frac{\pi x}{L} + \frac{1}{3^3}\cos c\left(\frac{3\pi}{L}\right)^2 t \sin \frac{3\pi x}{L} + \cdots\right)$

25. $u(0, t) = 0,\ u(L, t) = 0,\ u_x(0, t) = 0,\ u_x(L, t) = 0$

27. $\beta L \approx \frac{3}{2}\pi, \frac{5}{2}\pi, \frac{7}{2}\pi, \cdots$ (more exactly 4.730, 7.853, 10.996, $\cdots$)

29. $\beta L \approx \frac{1}{2}\pi, \frac{3}{2}\pi, \frac{5}{2}\pi, \cdots$ (more exactly 1.875, 4.694, 7.855, $\cdots$)

PROBLEM SET 11.5, page 654

3. $\lambda_1^2 = (\ln 2)/10,\ c^2 = 0.00702L^2$

5. $u = \sin 0.1\pi x\, e^{-1.752\pi^2 t/100}$

7. $u = \frac{40}{\pi^2}\left(\sin 0.1\pi x\, e^{-0.01752\pi^2 t} - \frac{1}{9}\sin 0.3\pi x\, e^{-0.01752(3\pi)^2 t} + - \cdots\right)$

9. $u = \frac{8}{\pi^3}\left(\sin 0.1\pi x\, e^{-0.01752\pi^2 t} + \frac{1}{3^3}\sin 0.3\pi x\, e^{-0.01752(3\pi)^2 t} + \cdots\right)$

11. Since the temperatures at the ends are kept constant, the temperature will approach a steady-state (time-independent) distribution $u_I(x)$ as $t \to \infty$, and $u_I = U_1 + (U_2 - U_1)x/L$, the solution of (1) with $\partial u/\partial t = 0$ satisfying the boundary conditions.

15. $u = 1$

17. $u = 0.5 \cos 2x\, e^{-4t}$

19. $u = \frac{\pi}{4} - \frac{8}{\pi}\left(\frac{1}{4}\cos 2x\, e^{-4t} + \frac{1}{36}\cos 6x\, e^{-36t} + \cdots\right)$

21. $u = \frac{\pi}{8} + \left(1 - \frac{2}{\pi}\right)\cos x\, e^{-t} - \frac{1}{\pi}\cos 2x\, e^{-4t} - \left(\frac{1}{3} + \frac{2}{9\pi}\right)\cos 3x\, e^{-9t} + \cdots$

25. $w = e^{-\beta t}$

27. $-\frac{K\pi}{L}\sum_{n=1}^{\infty} nB_n e^{-\lambda_n^2 t}$

29. 2.57, 0.52, 0.10°C

31. $u = (\sin \frac{1}{2}\pi x \sinh \frac{1}{2}\pi y)/\sinh \pi$

33. $u = \frac{80}{\pi}\sum_{n=1}^{\infty}\frac{1}{2n - 1}\sin\frac{(2n - 1)\pi x}{24}\frac{\sinh [(2n - 1)\pi y/24]}{\sinh (2n - 1)\pi}$

35. $u(x, y) = \sum_{n=1}^{\infty} A_n \sin \frac{n\pi x}{a} \sinh \frac{n\pi(b - y)}{a}$,

$$A_n = \frac{2}{a \sinh (n\pi b/a)}\int_0^a f(x) \sin \frac{n\pi x}{a}\, dx$$

37. $u(x, y) = \frac{A_0}{24} x + \sum_{n=1}^{\infty} A_n \frac{\sinh (n\pi x/24)}{\sinh n\pi} \cos \frac{n\pi y}{24}$,

$$A_0 = \frac{1}{24} \int_0^{24} f(y)\, dy, \quad A_n = \frac{1}{12} \int_0^{24} f(y) \cos \frac{n\pi y}{24}\, dy$$

39. $u = \sum_{n=0}^{\infty} A_n \cos nx\, e^{-ny}, \quad A_0 = \frac{1}{\pi} \int_0^{\pi} f(x)\, dx,$

$A_n = \frac{2}{\pi} \int_0^{\pi} f(x) \cos nx\, dx, \quad n = 1, 2, \cdots$

PROBLEM SET 11.6, page 660

7. $\frac{1}{\sqrt{\pi}} \int_{(a-x)/\tau}^{(b-x)/\tau} e^{-w^2}\, dw - \frac{1}{\sqrt{\pi}} \int_{(a+x)/\tau}^{(b+x)/\tau} e^{-w^2}\, dw$

11. $A(p) = \frac{2}{\pi(1 + p^2)}, \; B(p) = 0, \; u = \frac{2}{\pi} \int_0^{\infty} \frac{1}{1 + p^2} \cos px\, e^{-c^2p^2t}\, dp$

13. $A(p) = \frac{2 \sin p}{\pi p}, \; B(p) = 0, \; u = \frac{2}{\pi} \int_0^{\infty} \frac{\sin p}{p} \cos px\, e^{-c^2p^2t}\, dp$

PROBLEM SET 11.8, page 671

1. c increases and so does the frequency.

5. $c\pi \sqrt{260}$ (corresponding eigenfunctions $F_{4,16}$, $F_{16,14}$), etc.

7. $A = ab$, $b = A/a$, $(ma^{-2} + na^2A^{-2})' = 0$ gives $a^2/b^2 = m/n$.

9. $f_1(x) = 2(4x - x^2)$, $f_2(y) = 2y - y^2$

11. $B_{mn} = (-1)^{m+1} 8/mn\pi$ (n odd), 0 (n even)

13. $4(\cos m\pi/2 - (-1)^m)(\cos n\pi/2 - (-1)^n)/mn\pi^2$

15. $B_{mn} = (-1)^{m+n} ab/mn\pi^2$

17. $B_{mn} = 4[1 - (-1)^n(b + 1)][1 - (-1)^m(a + 1)]/mn\pi^2$

19. $B_{mn} = (-1)^{m+n} \frac{144a^3b^3}{m^3n^3\pi^6}$

21. $u = k \cos \pi\sqrt{5}t \sin \pi x \sin 2\pi y$

23. $u = k \cos 5\pi t \sin 3\pi x \sin 4\pi y$

PROBLEM SET 11.9, page 673

9. $u = 30r \sin \theta - 10r^3 \sin 3\theta$

11. $u = \frac{400}{\pi} \left(r \sin \theta + \frac{1}{3} r^3 \sin 3\theta + \frac{1}{5} r^5 \sin 5\theta + \cdots \right)$

13. $u = \frac{2}{\pi} r \sin \theta + \frac{1}{2} r^2 \sin 2\theta - \frac{2}{9\pi} r^3 \sin 3\theta - \frac{1}{4} r^4 \sin 4\theta + \cdots$

15. $u = \frac{\pi}{2} - \frac{4}{\pi} \left(r \cos \theta + \frac{1}{9} r^3 \cos 3\theta + \frac{1}{25} r^5 \cos 5\theta + \cdots \right)$

17. $u = \pi/2$

19. $u = \frac{4u_0}{\pi}\left(\frac{r}{a}\sin\theta + \frac{1}{3a^3}r^3\sin 3\theta + \frac{1}{5a^5}r^5\sin 5\theta + \cdots\right)$

21. $a^2u_{x^*x^*} + c^2u_{y^*y^*}$

23. $4x^*u_{x^*x^*} + 4y^*u_{y^*y^*} + 2u_{x^*} + 2u_{y^*}$

25. $u_{x^*x^*} + u_{y^*y^*}$

27. Use $\nabla^2 u = 0$ and $u_n = u_r$.

PROBLEM SET 11.10, page 680

5. $T = 6.828\rho R^2 f_1^{\,2}$, f_1 the fundamental frequency

9. No

15. $u = 4k \sum_{m=1}^{\infty} \frac{J_2(\alpha_m)}{\alpha_m^{\,2} J_1^{\,2}(\alpha_m)} \cos \alpha_m t\, J_0(\alpha_m r)$

23. $\alpha_{11}/2\pi \approx 0.6099$ (see Table A1 in Appendix 5)

PROBLEM SET 11.11, page 684

3. $u = 160/r + 30$

5. $u = -40 \ln r/(\ln 2) + 150$

17. $u = (u_1 - u_0)(\ln r)/\ln(r_1/r_0) + (u_0 \ln r_1 - u_1 \ln r_0) / \ln(r_1/r_0)$

PROBLEM SET 11.12, page 690

5. $u = 1$

7. $\cos 2\phi = 2\cos^2\phi - 1$, $2x^2 - 1 = \frac{4}{3}P_2(x) - \frac{1}{3}$, $u = \frac{4}{3}r^2P_2(\cos\phi) - \frac{1}{3}$

9. $x^3 = \frac{2}{5}P_3(x) + \frac{3}{5}P_1(x)$, $u = \frac{2}{5}r^3P_3(\cos\phi) + \frac{3}{5}rP_1(\cos\phi)$

11. $u = 4r^3P_3(\cos\phi) - 2r^2P_2(\cos\phi) + rP_1(\cos\phi) - 2$

17. This is the analog of Example 1 with 55 replaced by 10.

19. $55(3/8 + 5/16) \approx 37.8$

25. $u = U_0 \cos(\pi t/l\sqrt{LC}) \sin(\pi x/l)$

PROBLEM SET 11.13, page 695

5. $U(x, s) = \frac{c(s)}{x^s} + \frac{x}{s^2(s+1)}$, $U(0, s) = 0$, $c(s) = 0$,
$u(x, t) = x(t - 1 + e^{-t})$

9. Set $x^2/4c^2\tau = z^2$. Use z as a new variable of integration. Use $\mathrm{erf}(\infty) = 1$.

CHAPTER 11 (REVIEW QUESTIONS AND PROBLEMS), page 700

21. $u = A(x)\cos 4y + B(x)\sin 4y$

23. $u = A(y)e^{-2x} + B(y)e^{x} - 5$

25. $u = g(x)(1 - e^{-y}) + f(x)$

27. $u = f_1(y) + f_2(x + y)$

29. $u = yf_1(x + y) + f_2(x + y)$

31. $u = f_1(x + y) + f_2(2y - x)$

33. $u = f_1(2x - y) + f_2(2y - x)$

35. $-0.1 \cos 3t \sin 3x$

37. $\frac{3}{4}\cos t \sin x - \frac{1}{4}\cos 3t \sin 3x$

39. $u = k$

41. $u = \frac{1}{2} + \frac{1}{2}\cos 2x\, e^{-4t}$

43. $u = \sin 0.02\pi x\, e^{-0.004572t}$

45. $u = \frac{3}{4}\sin\frac{\pi x}{10} e^{-0.1143t} - \frac{1}{4}\sin\frac{3\pi x}{10} e^{-1.029t}$

47. $u = \dfrac{200}{\pi^2}\left(\sin\dfrac{\pi x}{50}\,e^{-0.004572t} - \dfrac{1}{9}\sin\dfrac{3\pi x}{50}\,e^{-0.04115t} + \cdots\right)$

49. $u = 95\cos 2x\, e^{-4t}$ **59.** $u = 275/r - 27.5$

PROBLEM SET 12.1, page 711

3. $32 - 24i$ **5.** $-\frac{7}{41} + \frac{22}{41}i$ **7.** $-47.2 - 23i$ **9.** $-10 - 24i$
11. $31/50$ **13.** $2xy/(x^2 + y^2)$ **15.** $x^2 - y^2, x^2$ **17.** 16

PROBLEM SET 12.2, page 717

3. 2.5 **5.** 1 **7.** 1 **9.** 8/17
11. $\sqrt{2}(\cos\frac{1}{4}\pi + i\sin\frac{1}{4}\pi)$ **13.** $10(\cos 0.927 + i\sin 0.927)$
15. $\frac{1}{4}(\cos\frac{1}{4}\pi + i\sin\frac{1}{4}\pi)$ **17.** $0.563(\cos 0.308 + i\sin 0.308)$
19. -3.042 **21.** $\pi/4$ **23.** $-2 + 2i$
25. $-0.227 - 0.974i$ **27.** $\pm(2 - 2i)$
29. $\pm 1, \pm i, \pm(1 \pm i)/\sqrt{2}$ **31.** $\pm(1 \pm i)/\sqrt{2}$
33. $\sqrt[6]{2}\left(\cos\dfrac{k\pi}{12} + i\sin\dfrac{k\pi}{12}\right),\ k = 1, 9, 17$
35. $3 + 2i, 2 - i$ **37.** $|z| = \sqrt{x^2 + y^2} \geqq |x|$, etc.
39. Equation (5) holds when $z_1 + z_2 = 0$. Let $z_1 + z_2 \neq 0$ and $c = a + ib = z_1/(z_1 + z_2)$. By (19) in Prob. 37, $|a| \leqq |c|$, $|a - 1| \leqq |c - 1|$. Thus $|a| + |a - 1| \leqq |c| + |c - 1|$. Clearly $|a| + |a - 1| \geqq 1$. Together we have the inequality below; multiply by $|z_1 + z_2|$ to get (5).

$$1 \leqq |c| + |c - 1| = \left|\frac{z_1}{z_1 + z_2}\right| + \left|\frac{z_2}{z_1 + z_2}\right|$$

PROBLEM SET 12.3, page 720

1. Circle, radius 4, center $4i$ **3.** Annulus with center a
5. Vertical infinite strip **7.** Right half-plane
9. Region between the two branches of the hyperbola $xy = 1$
11. Circle $(x - 17/15)^2 + y^2 = (8/15)^2$

PROBLEM SET 12.4, page 725

1. $14 + 8i, -1 - 2i, 4 - 12i$ **3.** $(9 - 13i)/500, -i, (-2 - 11i)/1000$
5. $2(x^3 - 3xy^2) - 3x, 2(3x^2y - y^3) - 3y$
7. $|w| > 9$ **9.** $|\arg w| \leqq 3\pi/4$
11. $\text{Re}\,(z^2)/|z^2| = (x^2 - y^2)/(x^2 + y^2) = 1$ if $y = 0$ and -1 if $x = 0$. *Ans.* No.
13. $6z(z^2 + i)^2$ **15.** $2i/(1 - z)^3$ **17.** 0 **19.** $i/2$
21. $-1/27$ **23.** $-\frac{1}{8}(1 + i)$
25. The quotient in (4) is $\Delta x/\Delta z$, which is 0 if $\Delta x = 0$ but 1 if $\Delta y = 0$, so that it has no limit as $\Delta z \to 0$.
27. Use $\text{Re}\, f(z) = [f(z) + \overline{f(z)}]/2$, $\text{Im}\, f(z) = [f(z) - \overline{f(z)}]/2i$.
29. By continuity, for any $\epsilon > 0$ there is a $\delta > 0$ so that $|f(z) - f(a)| < \epsilon$ when $|z - a| < \delta$. Now $|z_n - a| < \delta$ for all sufficiently large n since $\lim z_n = a$. Thus $|f(z_n) - f(a)| < \epsilon$ for these n.

PROBLEM SET 12.5, page 731

1. Yes **3.** Yes **5.** For $z \neq 1$ **7.** Yes **9.** No

11. No **13.** $f(z) = -iz^2/2$ **15.** $f(z) = 1/z$

17. $f(z) = z^3$ **19.** No **21.** No

23. $b = 1$, $v = -\sin x \sinh y$

29. $f'(z) = u_x = iv_x = 0$, $u_x = v_x = 0$, hence $v_y = u_y = 0$ by (1), $u = const$, $v = const$, $f = u + iv = const$.

PROBLEM SET 12.6, page 734

3. $1.469 + 2.287i$, 2.718 **5.** $3.610 - 1.972i$, 4.113

7. -23.141, 23.141 **9.** $-i$, 1

11. $\exp(x^3 - 3xy^2)\cos(3x^2y - y^3)$, $\exp(x^3 - 3xy^2)\sin(3x^2y - y^3)$

13. $e^{-2x}\cos 2y$, $-e^{-2x}\sin 2y$ **15.** $e^{\pi i/4}$, $e^{-3\pi i/4}$, $e^{-\pi i/4}$, $e^{3\pi i/4}$

17. $5\exp(i \arctan \frac{4}{3})$ **19.** $x > 0$

21. $y = (2n + 1)\pi/4$, $n = 0, \pm 1, \cdots$

23. $z = \ln 2 + (2n + 1)\pi i$, $n = 0, \pm 1, \cdots$

25. No solutions **27.** $k = 1$

29. $(g/f)' = (g'f - gf')/g^2 = 0$, since $g' = g$, $f' = f$; $g/f = k = const$ by Prob. 29, Sec. 12.5. $g(0) = f(0) = e^0 = 1$ gives $k = 1$, $g = f$.

PROBLEM SET 12.7, page 738

3. $-0.303 - 2.112i$ **5.** $11013i$

7. $26.974 - 4.256i$ **9.** $\cosh 3\pi = 6195.8$

11. $-0.5150 + 0.1738i$ **13.** $-3.725 - 0.512i$

15. $1.960 + 3.166i$ **17.** $\frac{1}{2}(2n + 1)\pi - (-1)^n 1.818i$

19. $\pm 2n\pi + i$, $\pm(2n + 1)\pi - i$ **21.** $\pm(\pi/3)i \pm 2n\pi i$, $n = 0, 1, \cdots$

23. Use Prob. 3.

PROBLEM SET 12.8, page 742

5. $\frac{1}{2}\ln 2 + \frac{1}{4}\pi i = 0.347 + 0.785i$ **7.** $1.609 - 2.214i$

9. $4.605 + 3.142i$ **11.** $2.773 + 3.135i$

13. $\pm 2n\pi i$, $n = 0, 1, \cdots$ **15.** $1.946 + (1 \pm 2n)\pi i$, $n = 0, 1, \cdots$

17. $(-0.644 \pm 2n\pi)i$, $n = 0, 1, \cdots$ **19.** $(\pi - 1 \pm 2n\pi)i$, $n = 0, 1, \cdots$

21. $10.85 - 16.90i$ **23.** $(1 + i)e^2/\sqrt{2}$

25. $1.032 + 0.870i$ **27.** $-54.05 - 7.70i$

29. $(1 + i)/\sqrt{2}$ **31.** $27[\cos(\ln 3) - i \sin(\ln 3)]$

33. $\sqrt{2}\, e^{\pi/4}[\cos(\ln\sqrt{2} - \frac{1}{4}\pi) + i \sin(\ln\sqrt{2} - \frac{1}{4}\pi)]$

35. $\exp[(3 + \pi i)(\ln\sqrt{29} - i \arctan 0.4)] = -276.2 - 436.0i$

37. $3.350 + 1.189i$

41. $\cosh w = \frac{1}{2}(e^w + e^{-w}) = z$, $(e^w)^2 - 2ze^w + 1 = 0$, $e^w = z + \sqrt{z^2 - 1}$

PROBLEM SET 12.9, page 747

1. $v = u - 2x - 2$ **3.** $|w - 1 - i| \leqq \sqrt{2}$ **5.** $2.25 \leqq |w| < 4.41$
7. $|w| > 16$ **9.** $0 \leqq |w| \leqq 1/4$ **11.** $2xy = const$
13. $v^2 = 4k^2(k^2 + u)$ **15.** $|w| = 2$ **17.** $|w| \leqq e, w \neq 0$
19. $|w| > \sqrt{e}, 0 < \arg w < \pi$

CHAPTER 12 (REVIEW QUESTIONS AND PROBLEMS), page 747

21. $-32 - 24i$ **23.** $\frac{9}{13} + \frac{20}{13}i$ **25.** $17 + 1.5i$ **27.** 4
29. 2/5 **31.** 4 **33.** 5 **35.** 1
37. $\pm(1 - i)/\sqrt{2}$ **39.** $\pm(1 - 3i)$ **41.** $6\sqrt{2}\, e^{3\pi i/4}$ **43.** $e^{-0.3218i}$
45. Circle, radius $\frac{1}{2}$, center $-1 + 3i$ **47.** Half-plane below $y = x$
49. $-1/z$ **51.** $\cos z$ **53.** No **55.** $3x^2y - y^3$
57. $1 - i$ **59.** $-3.725 - 0.512i$

PROBLEM SET 13.1–13.2, page 760

1. $z = (1 + 2i)t, 0 \leqq t \leqq 1$
3. $z = 4 + 2i + (-1 + 3i)t, 0 \leqq t \leqq 1$
5. $z = -4i + (-1 + 6i)t, 0 \leqq t \leqq 7$ **7.** Straight segment from 0 to $3 + 6i$
9. Lower semicircle (radius 2, center $1 - i$)
11. Parabola $y = 3x^2$ from $(-1, 3)$ to $(2, 12)$
13. Ellipse $4x^2 + y^2 = 4$ **15.** $3 - 4i + 4e^{it}, 0 \leqq t \leqq 2\pi$
17. $t + i/t, 1 \leqq t \leqq 3$ **19.** $2 \cos t + i \sin t, 0 \leqq t \leqq 2\pi$
21. $2b(1 + i)$ **23.** 0 **25.** 0 **27.** $-\pi r^2$
29. $-2\pi i$ **31.** 0 **33.** $-i/3$ **35.** $-1 - i$
37. $1 + i/3$ **39.** $(\pi - \frac{1}{2} \sinh 2\pi)i$
41. $32/3 + 64i/3, 32i, 8 + 128i/5$ **43.** $i, 2i, 2i$ **49.** $5e^3$

PROBLEM SET 13.3, page 766

5. Yes, by the deformation principle
7. 0, no **9.** $-\pi$, no **11.** 0, no **13.** 0, yes **15.** 0, no
17. 0, yes **19.** $2\pi i$ **21.** 0 **23.** $2\pi i$ **25.** $-\pi, \pi$
27. $-2\pi i, 0, -4\pi i$ **29.** $-2\pi i, 0$

PROBLEM SET 13.4, page 770

1. $2 + 2i$ **3.** -2 **5.** 0 **7.** $-i/3$ **9.** $-2/\pi$
11. 0 **13.** $-\sinh 1$ **15.** $-1566i/35$

PROBLEM SET 13.5, page 773

1. $-\pi/2$ 3. $-\pi i/2$ 5. -2π 7. 0 9. πi
11. 0 13. $\pi/8$ 15. 0 17. 2 19. $-2\pi \tanh 1$
21. $2\pi i \operatorname{Ln} 4 = 8.710i$ 23. $\pi e^{-2+2i} = -0.1769 + 0.3866i$
25. Use partial fractions.

PROBLEM SET 13.6, page 778

1. 0 3. $2\pi^2 i$ 5. 0 7. $-6\pi i$ 9. $2\pi i$ 11. 0
13. $2\pi i$ 15. 0 17. $2\pi^2 i$ 19. $-2\pi i$ 21. $\frac{9}{2}\pi e^{-4}i$

CHAPTER 13 (REVIEW QUESTIONS AND PROBLEMS), page 778

17. $-2\pi i$ 19. $2i$ 21. 0 23. $2\pi i$
25. $-2\pi i$ 27. $i \sin 1$ 29. 0 31. $2 \sin 1$
33. $-\frac{1}{2}\sin \pi^2$ 35. $-\pi/2$ 37. $-5 + 8\pi i$ 39. $2\pi^2 i$

PROBLEM SET 14.1, page 790

1. Yes, no, $\pm 1 + 2i$ 3. Yes, yes, 0 5. No, no, none
7. Yes, no, ± 1 9. Yes, no, ± 1, $\pm i$ 15. Convergent
17. Convergent 19. Convergent 21. Divergent
23. Convergent
25. $\dfrac{(n+2)n}{2(n+1)^2} < \dfrac{1}{2}$, $|R_n| \leqq \dfrac{|w_{n+1}|}{1-q} = \dfrac{n+2}{2^n(n+1)} < 0.05$, $n = 5$, $s \approx 1.657$

PROBLEM SET 14.2, page 796

1. $-4i$, 1 3. 0, $2/\sqrt{\pi}$ 5. $2i$, 5 7. $-i$, ∞
9. $2i/3$, $1/3$ 11. 0, $\sqrt{2}$ 13. 0, ∞ 15. 0, 1
17. 0, $\sqrt{2/3}$ 19. $\Sigma\, a_n z^{2n} = \Sigma\, a_n (z^2)^n$, $|z^2| < R$, etc.

PROBLEM SET 14.3, page 801

1. 2 3. $\sqrt{5}$ 5. $\sqrt{7/2}$ 7. 1/6 9. 1/4

PROBLEM SET 14.4, page 807

1. $1 - z + z^2/2! - z^3/3! + - \cdots$, $R = \infty$
3. $\pi z - \pi^3 z^3/3! + \pi^5 z^5/5! - + \cdots$, $R = \infty$
5. $1 - (z - \frac{1}{2}\pi)^2/2! + (z - \frac{1}{2}\pi)^4/4! - + \cdots$, $R = \infty$
7. $\frac{1}{2} + \frac{1}{4}(z+1) + \frac{1}{8}(z+1)^2 + \frac{1}{16}(z+1)^3 + \cdots$, $R = 2$
9. $(z-1) - \frac{1}{2}(z-1)^2 + \frac{1}{3}(z-1)^3 - + \cdots$, $R = 1$
11. $-1 + 5(z+1) - 10(z+1)^2 + 10(z+1)^3 - 5(z+1)^4 + (z+1)^5$
13. $\sin^2 z = \frac{1}{2} - \frac{1}{2}\cos 2z = z^2 - 2^3 z^4/4! + 2^5 z^6/6! - + \cdots$, $R = \infty$
15. $1 - (z - \frac{1}{2}\pi)^2/2! + (z - \frac{1}{2}\pi)^4/4! - + \cdots$, $R = \infty$

27. $(2/\sqrt{\pi})(z - z^3/3 + z^5/2!5 - z^7/3!7 + \cdots), \quad R = \infty$

29. $z^3/1!3 - z^7/3!7 + z^{11}/5!11 - + \cdots, \quad R = \infty$

PROBLEM SET 14.5, page 810

1. $1 - z^4 + z^8 - z^{12} + - \cdots, \quad R = 1$

3. $2 + z + 2z^2 + z^3 + 2z^4 + \cdots, \quad R = 1$

5. $2z^2 - 2^3z^6/3! + 2^5z^{10}/5! - + \cdots, \quad R = \infty$

7. $z + z^5/2! + z^9/3! + \cdots, \quad R = \infty$

9. $2/(z - 2) - 1/(z + 4)^2 = -17/16 - (15/32)z - (67/256)z^2 + \cdots, \quad R = 2$

11. $(1 - i)/2 - [(1 - i)^2/4](z - 1 - i) + [(1 - i)^3/8](z - 1 - i)^2 - + \cdots, \quad R = \sqrt{2}$

13. $-i + (z - i) - 7i(z - i)^2 - 9(z - i)^3 + 5i(z - i)^4 + (z - i)^5$

15. $1/3! - (z + 1)^2/5! + (z + 1)^4/7! - + \cdots, \quad R = \infty$

17. $i(z - \frac{1}{2}\pi i) + i(z - \frac{1}{2}\pi i)^3/3! + i(z - \frac{1}{2}\pi i)^5/5! + \cdots, \quad R = \infty$

19. $z^2 + z^4 + z^6/3 + \cdots, \quad R = \infty$

21. $1 + 2(z - \frac{1}{4}\pi) + 2(z - \frac{1}{4}\pi)^2 + \frac{8}{3}(z - \frac{1}{4}\pi)^3 + \cdots, \quad R = \frac{1}{4}\pi$

23. $1 - z^2/18 - z^3/27 + \cdots, \quad R = 3$

PROBLEM SET 14.6, page 819

1. Use Theorem 1.

3. $R = 1/\sqrt{\pi} > 0.56$

5. $|\sin n|z|| \leqq 1$; $\Sigma\, 1/n^2$ converges.

7. $|z^n| \leqq 1$; $\Sigma\, n/(n^3 + |z|) \leqq \Sigma\, 1/n^2$

9. $|\tanh^n x| \leqq 1, \quad 1/n(n + 1) < 1/n^2$

11. $|z + 2i| \leqq \sqrt{3} - \delta, \quad \delta > 0$

13. $|z| \leqq 4 - \delta, \quad \delta > 0$

15. Nowhere

17. Convergence follows from the comparison test (Sec. 14.1). Let $R_n(z)$ and R_n^* be the remainders of (1) and (5), respectively. Since (5) converges, for given $\epsilon > 0$ we can find an $N(\epsilon)$ such that $R_n^* < \epsilon$ for all $n > N(\epsilon)$. Since $|f_n(z)| \leqq M_n$ for all z in the region G, we also have $|R_n(z)| \leqq R_n^*$ and therefore $|R_n(z)| < \epsilon$ for all $n > N(\epsilon)$ and all z in the region G. This proves that the convergence of (1) in G is uniform.

19. No. Why?

21. $n = 7, 10, 16, 27, 65$

PROBLEM SET 14.7, page 827

1. $\displaystyle\sum_{n=0}^{\infty} \frac{z^{n-2}}{n!} = \frac{1}{z^2} + \frac{1}{z} + \frac{1}{2!} + \frac{z}{3!} + \cdots, \quad R = \infty$

3. $\displaystyle\sum_{n=0}^{\infty} \frac{2^{2n}}{(2n)!} z^{2n-1} = \frac{1}{z} + 2z + \frac{2}{3} z^3 + \cdots, \quad R = \infty$

5. $\displaystyle\sum_{n=0}^{\infty} (-1)^n z^{2n-1} = \frac{1}{z} - z + z^3 - z^5 + - \cdots, \quad R = 1$

7. $\displaystyle\sum_{n=0}^{\infty} \frac{(-1)^n}{(2n)!} z^{1-2n} = z - \frac{1}{2z} + \frac{1}{24z^3} - + \cdots, \quad R = \infty$

9. $\displaystyle\sum_{n=1}^{\infty} (-1)^{n+1} n z^{n-7} = \frac{1}{z^6} - \frac{2}{z^5} + \frac{3}{z^4} - \frac{4}{z^3} + - \cdots, \quad R = 1$

11. $-\sum_{n=0}^{\infty}\left(\frac{i}{2}\right)^{n+1}(z-i)^{n-1} = -\frac{i/2}{z-i}+\frac{1}{4}+\frac{i}{8}(z-i)+\cdots, \quad R=2$

13. $\sum_{n=0}^{\infty}\frac{(-1)^{n+1}}{(2n)!}(z-\pi)^{2n-3} = -\frac{1}{(z-\pi)^3}+\frac{1}{2!(z-\pi)}-\frac{1}{4!}(z-\pi)+-\cdots,$

$R=\infty$

15. $(z-1)+2-3/(z-1)$

17. $-\sum_{n=0}^{\infty}(z+i)^{n-1} = -(z+i)^{-1}-1-(z+i)-\cdots, \quad R=1$

19. $\sum_{n=0}^{\infty} z^{2n}$

21. $\sum_{n=0}^{\infty}\frac{(-1)^{n+1}}{2^{n+1}}(z-1)^{n-1}$

23. $\frac{1}{z}+\sum_{n=0}^{\infty}\frac{1}{z^{n+1}}-\sum_{n=0}^{\infty}\frac{z^n}{2^{n+1}}$

25. $\sum_{n=0}^{\infty} z^{3n}, \quad |z|<1; \quad -\sum_{n=0}^{\infty}\frac{1}{z^{3n+3}}, \quad |z|>1$

27. $\sum_{n=0}^{\infty} z^{4n+2}, \quad |z|<1, \quad -\sum_{n=0}^{\infty}\frac{1}{z^{4n+2}}, \quad |z|>1$

29. $\sum_{n=0}^{\infty}(-1)^n(z-1)^n, \quad 0<|z-1|<1; \quad \sum_{n=0}^{\infty}\frac{(-1)^n}{(z-1)^{n+1}}, \quad |z-1|>1$

31. $\sum_{n=0}^{\infty}\frac{(-1)^{n+1}(z+\frac{1}{2}\pi)^{2n-1}}{(2n)!}, \quad |z+\frac{1}{2}\pi|>0$

33. $(1-4z)\sum_{n=0}^{\infty} z^{4n}, \quad |z|<1, \quad \left(\frac{4}{z^3}-\frac{1}{z^4}\right)\sum_{n=0}^{\infty}\frac{1}{z^{4n}}, \quad |z|>1$

35. Let $\sum_{-\infty}^{\infty} a_n(z-z_0)^n$ and $\sum_{-\infty}^{\infty} c_n(z-z_0)^n$ be two Laurent series of the same function $f(z)$ in the same annulus. We multiply both series by $(z-z_0)^{-k-1}$ and integrate along a circle with center at z_0 in the interior of the annulus. Since the series converge uniformly, we may integrate term by term. This yields $2\pi i a_k = 2\pi i c_k$. Thus, $a_k = c_k$ for all $k = 0, \pm 1, \cdots$.

PROBLEM SET 14.8, page 833

1. $0, \pm\pi, \pm 2\pi, \cdots$ (simple poles), ∞ (essential singularity)
3. $0, \infty$ (simple poles)
5. $\pm 1, \pm i$ (third-order poles), ∞ (essential singularity)
7. $-i, 2i$ (second-order poles), ∞ (essential singularity)
9. 0 (simple pole), ∞ (essential singularity)
11. 0 (essential singularity), $\pm 2/\pi, \pm 2/3\pi, \cdots$ (simple poles)
13. ∞ (essential singularity)
15. $1, \infty$ (essential singularities), $\pm 2n\pi i$ ($n = 0, 1, \cdots$, simple poles)
17. $\pm 2, \pm 2i$ (second order)

19. 0 (third order), $\pm 1, \pm 2, \cdots$ (second order)

21. $(2n + 1)/2$ (third order) **23.** $\pm 1/\sqrt{3}$ (simple)

25. $0, \pm 2\pi, \pm 4\pi, \cdots$ (fourth order), by (6), Sec. 12.7

27. $f(z) = (z - z_0)^n g(z)$ by (3), and $g(z_0) \neq 0$. Hence $1/g(z_0)$ is analytic at $z = z_0$. Let its Taylor series be
$$\frac{1}{g(z)} = c_0 + c_1(z - z_0) + \cdots. \text{ Then } \frac{1}{f(z)} = \frac{c_0 + c_1(z - z_0) + \cdots}{(z - z_0)^n},$$
which proves the first statement. Multiplication of $h(z)$ does not change this.

29. Apply Theorem 3 to $f(z) - k$.

31. Region between the 0° and 90° meridians

33. Small spherical disk centered at the North Pole

35. Belt between two parallels that includes the equator

CHAPTER 14 (REVIEW QUESTONS AND PROBLEMS), page 834

21. $\infty, \quad \sin(z - 2)$ **23.** $1/2, \quad \text{Ln}(1 + 2z)$ **25.** $1/3$

27. $\infty, \quad e^{-z^2}$ **29.** $\infty, \quad \cosh \sqrt{z}$

31. $1 - 2z + (2z)^2/2! - (2z)^3/3! + - \cdots, \quad R = \infty$

33. $1/2 + (z + 1)/4 + (z + 1)^2/8 + (z + 1)^3/16 + \cdots, \quad R = 2$

35. $k - k^2(z - 2 - 3i) + k^3(z - 2 - 3i)^2 - + \cdots, k = (2 - 3i)/13, R = \sqrt{13}$

37. $1 + 3z + 6z^2 + 10z^3 + \cdots, \quad R = 1$

39. $\displaystyle\sum_{n=0}^{\infty} \frac{(-1)^{n+1}}{(2n + 1)!}\left(z - \frac{\pi}{2}\right)^{2n-2}, \quad \left|z - \frac{\pi}{2}\right| > 0,$ pole of second order

41. $\displaystyle\sum_{n=0}^{\infty} z^{n-4}, \quad 0 < |z| < 1,$ pole of fourth order

43. $\displaystyle\sum_{n=0}^{\infty} \frac{1}{(2n)!z^{2n-3}}, \quad |z| > 0,$ essential singularity

45. $\displaystyle\sum_{n=1}^{\infty} \frac{(-1)^{n+1}}{n}(z - 1)^{n-3}, \quad 0 < |z - 1| < 1,$ pole of second order

47. $\displaystyle\sum_{n=1}^{\infty} \frac{1}{n!n} z^{n-2}, \quad |z| > 0,$ simple pole

49. $\displaystyle\sum_{n=0}^{\infty} \frac{e^i}{n!}(z - i)^{n-5}, \quad |z - i| > 0,$ pole of fifth order

PROBLEM SET 15.1, page 841

1. -3 (at $z = 1$) **3.** $-1/3!$ (at $z = 0$)

5. $i/3$ (at $-i$), $-16i/3$ (at $2i$) **7.** 1 (at $z = \pm n\pi$)

9. $\mp\frac{1}{2}$ (at ± 1) by (5) **11.** 1 (at $z = 0$)

13. $-9/16, 9/16, -9i/16, 9i/16$ (poles of second order at $z = 1, -1, i, -i$)

15. $3/17$ (at $z = -1$) **17.** 0 **19.** $2\pi i$

21. $-6\pi i$ **23.** $(1 - \pi^3/128)\pi i$

PROBLEM SET 15.2, page 844

1. $-2\pi i$ **3.** $-6\pi i$ **5.** 0 **7.** $10\pi i$ **9.** $4\pi i/5$
11. $2\pi i(\cosh 1 - 1) = 3.412i$ **13.** $2\pi i$ **15.** $\pi/3$
17. πi **19.** $8\pi i$ **21.** $\pi i \sin \frac{1}{2} = 1.506i$ **23.** $-4i$
25. $-4i$ **27.** 0 **29.** $-4i \sinh \frac{1}{2} = -2.084i$

PROBLEM SET 15.3, page 849

1. $4\pi/\sqrt{3}$ **3.** $2\pi/35$ **5.** $8\pi/3$ **7.** $\pi/30$ **9.** 0
11. π **13.** $\pi/\sqrt{2}$ **15.** $2\pi/3$ **17.** $\pi/2$ **19.** $\pi/16$

PROBLEM SET 15.4, page 855

3. 0 **5.** $(2\pi/\sqrt{3})e^{-\sqrt{3}/2} \cos \frac{1}{2} = 1.339$ **7.** π/e
9. 0 **11.** $-\pi(\sin 1)/\sqrt{3}e^{\sqrt{3}} = -0.2700$
13. $\pi/10e^2 - 2\pi/30e^3$ **15.** π **17.** $-\sqrt{3}\pi/6$
19. $-\pi/2$ **21.** $\pi/2$ **23.** $(3 - e^{-\pi})\pi/2 = 4.645$

CHAPTER 15 (REVIEW QUESTIONS AND PROBLEMS), page 856

11. 0, yes **13.** $5\pi/2$, yes **15.** $-18\pi i$, yes **17.** 0, no
19. $-2\pi i/(n + 1)!$, yes **21.** $-i/2$, no **23.** $-64\pi i$, yes
25. $\pi i/4$, yes **27.** $2\pi/7$ **29.** $4\pi/\sqrt{3}$ **31.** $\pi/60$
33. $\pi/2$ **35.** 0 **37.** $\pi/3$ **39.** $\pi/2$

PROBLEM SET 16.1, page 863

1. $u = 1 - \frac{1}{4}v^2, 4 - \frac{1}{16}v^2, 9 - \frac{1}{36}v^2, 16 - \frac{1}{64}v^2$ **3.** $v = 20$
5. The positive and the negative v-axis, respectively **7.** $|w| < \frac{1}{9}$
9. $|\arg w| < 2\pi/3$ **11.** $\pi/2 < \arg w < \pi$ **13.** $4e^{it}$, $4ie^{it}$
15. $3 \cos t + i \sin t,\quad -3 \sin t + i \cos t$
17. $t + it^{-1}, 1 - it^2$ **19.** $-a/2$ **21.** $\pm 2, \pm 2i$
23. $0, \pm\pi i/2, \pm\pi i, \cdots$ **25.** By conformality **27.** Only in size

PROBLEM SET 16.2, page 868

1. $z = 0$ **3.** $0, \pm 1, \pm i$ **5.** $-\frac{1}{2} \pm \frac{1}{2}\sqrt{3}i$ **7.** $\pm i$
9. $\pm i$ **11.** $w = 4/z$, $w = (z + 4)/(z + 1)$, etc. **13.** $w = az/d$
15. $w = (az + b)/(a - bz)$ **17.** $z = (2iw - 4i)/(-w + 3)$

PROBLEM SET 16.3, page 873

1. $w = 3z + 2$ **3.** $w = (z + 1)/(-3z + 1)$
5. $w = 1/z$ **7.** $w = (3iz + 1)/z$
9. $w = iz$ **13.** $z = (-4w + 1)/(2w - 1)$
15. $w = (2z - i)/(-iz - 2)$ **19.** $w = (z^4 - i)/(-iz^4 + 1)$

PROBLEM SET 16.4, page 878

1. $1 \leqq |w| \leqq e$, $0 \leqq \arg w \leqq \frac{1}{2}\pi$
3. $e < |w| < e^2$, $-\frac{1}{2}\pi < \arg w < \frac{1}{2}\pi$
5. Interior of the ellipse $u^2/(\cosh^2 2) + v^2/(\sinh^2 2) = 1$ in the first quadrant
7. Elliptical annulus bounded by $u^2/\cosh^2 1 + v^2/\sinh^2 1 = 1$ and $u^2/\cosh^2 2 + v^2/\sinh^2 2 = 1$ and cut along the positive imaginary axis
9. $t = z^2$ maps the given region onto the strip $0 < \text{Im } t < \pi$, and $w = e^t$ maps this strip onto the upper half-plane. *Ans.* $w = e^{z^2}$.
11. $w' = \cos z = 0$ at $z = \pm(2n + 1)\pi/2$, $n = 0, 1, \cdots$
13. $\pm n\pi i$, $n = 0, 1, \cdots$
15. Upper half-plane $v > 0$
17. Lower half-plane $v < 0$
19. $\ln 2 \leqq u \leqq \ln 3$, $\pi/4 \leqq v \leqq \pi/2$

PROBLEM SET 16.5, page 882

1. w moves once around the unit circle $|w| = 1$.
5. $|z| = 1$; $\ln z = \ln |z| + i\theta = i\theta$ moves up the v-axis by 2π each time.
7. ± 1 (first order), 2 sheets
9. 0, 2 sheets
11. a, 3 sheets
13. 0, ± 1, 2 sheets
15. -1, infinitely many sheets
17. $-\frac{1}{2}i$, 3 sheets
19. 0, 2 sheets

CHAPTER 16 (REVIEW QUESTIONS AND PROBLEMS), page 883

11. $u = \frac{1}{4}v^2 - 1, \frac{1}{4}v^2 - 1$
13. $|w| = 6.25$, $|\arg w| < \pi/4$
15. $v = 4u/3$
17. $\pi/4 < \arg w \leqq \pi/2$
19. The domain between the parabolas $u = \frac{1}{4} - v^2$ and $u = 1 - \frac{1}{4}v^2$
21. $|\arg w| < \pi/8$
23. $u = 1$
25. $|w + \frac{1}{2}| = \frac{1}{2}$
27. $0, \pm i$
29. $\pm i/\sqrt{3}$
31. $(\pm n + \frac{1}{4})\pi$
33. iz
35. $z/(z + 2)$
37. $5z$
39. $2z/(z - 1)$
41. $w = iz^3 + 1$
43. $w = iz$
45. $w = e^{3z}$
47. $w = z^2/2k$
49. $2 \pm \sqrt{6}$

PROBLEM SET 17.1, page 889

1. $\Phi = 20x + 200$
3. $\Phi = 20(1 - y/d)$
5. $\Phi = 110 - 50xy$
7. $(110 \ln r)/\ln 2$
9. $200 - (100 \ln r)/\ln 2$
11. $y = x/2 + c$
13. $x^2 - y^2 = const$
15. $(x - 1/2c)^2 + y^2 = 1/4c^2$
17. $u = c \text{ Re } [\text{Ln } (z - a) + \text{Ln } (z + a)] = c \ln |z^2 - a^2|$

PROBLEM SET 17.2, page 893

3. $\Phi(x, y) = U_2 xy$. $w = u + iv = iz^2$ maps R onto $-2 \leqq u \leqq 0$.
5. Apply $w = z^2$.
7. Corresponding rays in the w-plane make equal angles, and the mapping is conformal.
9. $z = (2Z - i)/(-iZ - 2)$

PROBLEM SET 17.3, page 897

1. $(100/d)y$. Rotate through $\pi/2$.
3. $100 - 240\theta/\pi$
5. $\text{Re}\, F(z) = 100 + (200/\pi)\, \text{Re}\,(\sin^{-1} z)$
7. $T_1 + \pi^{-1}(T_2 - T_1)\, \text{Arg}\,(w - a)$
9. $T_0[\text{Arg}\,(w - 1) - \text{Arg}\,(w + 1)]/\pi$
15. $120\pi^{-1}\, \text{Arg}\, z$
17. $T_0 + 2\pi^{-1}(T_1 - T_0)\, \text{Arg}\, z$
19. $400(\text{Arg}\, z)/\pi$

PROBLEM SET 17.4, page 903

1. $V = V_1 = K$, $Ky = const$, $Kx = const$
3. $F^*(w) = w$ by Prob. 1; $w = f(z) = z^2$ maps that quadrant on the half-plane; $F(z) = F^*(f(z)) = z^2$.
5. Parallel flow in the negative y-direction, $V = -i$
7. Parallel flow in the direction of $y = -x$, $V = 1 - i$
9. $V = 3(x^2 - y^2) - 6xyi$, $V_2 = 0$ on the coordinate axes
13. $F(z) = [\ln\,(z + a)]/2\pi$
17. The streamlines are circles $y/(x^2 + y^2) = c$ or $x^2 + (y - k)^2 = k^2$.
19. Use that $w = \cos^{-1} z$ is the mapping $w = \cos z$ with the roles of the z- and w-planes interchanged.

PROBLEM SET 17.5, page 908

3. $\Phi = r \sin \theta$
5. $\Phi = 2 - r \cos \theta$
7. $\Phi = \frac{1}{2}r^5 \sin 5\theta$
9. $\Phi = \frac{1}{2} - \frac{1}{2}r^2 \cos 2\theta$
11. $\Phi = 2(r \sin \theta - \frac{1}{2}r^2 \sin 2\theta + \frac{1}{3}r^3 \sin 3\theta - + \cdots)$
13. $\Phi = (4/\pi)(r \sin \theta - \frac{1}{9}r^3 \sin 3\theta + \frac{1}{25}r^5 \sin 5\theta - + \cdots)$

PROBLEM SET 17.6, page 913

1. Use (2); $F(1) = 9$.
3. $|z|$ is not analytic.
5. Use that $|e^z| = e^x$ is monotone.
7. 2, at $z = \pm i$
11. Use Theorem 3.
13. Use $x = 1 + \cos \theta$, $y = \sin \theta$.
15. Set $r = 0$ in that formula.

CHAPTER 17 (REVIEW QUESTIONS AND PROBLEMS), page 914

11. $\Phi = (100 \ln r)/\ln 5 = 62.13 \ln r$
13. $\Phi = (10/\ln 10)(\ln 100 - \ln r) = 20.00 - 4.34 \ln r$
15. $\Phi = 110[1 - (2/\pi)\, \text{Arg}\, z]$
17. $\Phi = -(400i/\pi)\, \text{Ln}\, z$
19. $\Phi = 100 + 80(y - \frac{1}{2}x)$
21. $x + y = const$
23. $2x(y + 1) = const$
25. $|z - 1 - i| = const$
27. $10 + (40/\pi)(r \sin \theta + \frac{1}{3}r^3 \sin 3\theta + \frac{1}{5}r^5 \sin 5\theta + \cdots)$
29. $10\pi^2 - 120(r \cos \theta - \frac{1}{4}r^2 \cos 2\theta + \frac{1}{9}r^3 \cos 3\theta - + \cdots)$
31. Flow in the y-direction, $V = ik$
33. $(r - 4/r) \sin \theta = const$, $V = 1 - 4/\bar{z}^2$

PROBLEM SET 18.1, page 924

1. 0.9817E02, −0.1010E03, 0.4787E−02, −0.1380E05

3. $(1100100)_2$, $(11101.01)_2$, $(11.11)_2$

5. 28.5, 0.15625, 1.71875, −3.40625

7. 5.6945, 5.697, 5.70, 5.7

9. 19.95, 0.05, 0.05013; 20, 0, 0.05

15. $18.78685 \leqq d \leqq 18.79695$

17. Add first, then round.

21. $\frac{a_1}{a_2} = \frac{\tilde{a}_1 + \epsilon_1}{\tilde{a}_2 + \epsilon_2} = \frac{\tilde{a}_1 + \epsilon_1}{\tilde{a}_2}\left(1 - \frac{\epsilon_2}{\tilde{a}_2} + \frac{\epsilon_2^{\,2}}{\tilde{a}_2^{\,2}} - + \cdots\right) \approx \frac{\tilde{a}_1}{\tilde{a}_2} + \frac{\epsilon_1}{\tilde{a}_2} - \frac{\epsilon_2}{\tilde{a}_2} \cdot \frac{\tilde{a}_1}{\tilde{a}_2}$,

$$\left|\left(\frac{a_1}{a_2} - \frac{\tilde{a}_1}{\tilde{a}_2}\right) \Big/ \frac{a_1}{a_2}\right| \approx \left|\frac{\epsilon_1}{a_1} - \frac{\epsilon_2}{a_2}\right| \leqq |\epsilon_{r1}| + |\epsilon_{r2}| \leqq \beta_{r1} + \beta_{r2}$$

PROBLEM SET 18.2, page 934

3. −0.12, −0.119 792 64, −0.119 794 07

7. $x_5 = 0.736\ 058$, $x_{10} = 0.739\ 126$

9. This follows from the intermediate value theorem of calculus.

11. 1.834 243 ($= x_4$)

13. 1.150 000 ($= x_5$)

15. 0.904 557 ($= x_3$)

17. 0.652 919 ($= x_3$)

19. $x_{n+1} = (2x_n + 7/x_n^{\,2})/3$, 1.912 931 ($= x_3$)

21. 1.834 243

23. 0.904 557

25. **ALGORITHM BISECT** (f, a_0, b_0, N) **Bisection Method**

This algorithm computes an interval $[a_n, b_n]$ containing a solution of $f(x) = 0$ (f continuous), or it computes a solution c_n, given an initial interval $[a_0, b_0]$ such that $f(a_0)f(b_0) < 0$.

INPUT: Initial interval $[a_0, b_0]$, maximum number of iterations N.
OUTPUT: Interval $[a_N, b_N]$ containing a solution, or a solution c_n.
For $n = 0, 1, \cdots, N - 1$ do:
  Compute $c_n = \frac{1}{2}(a_n + b_n)$.
  If $f(c_n) = 0$ then OUTPUT c_n. Stop. [*Procedure completed*]
  Else continue.
  If $f(a_n)f(c_n) < 0$ then $a_{n+1} = a_n$ and $b_{n+1} = c_n$.
  Else set $a_{n+1} = c_n$ and $b_{n+1} = b_n$.
End
OUTPUT $[a_N, b_N]$. Stop.
[*Procedure completed*]
End BISECT

27. [0.40625, 0.43750]

29. [1.90625, 1.93750]

31. 2.689

33. ±1.189

PROBLEM SET 18.3, page 948

1. $L_0(x) = -2x + 19$, $L_1(x) = 2x - 18$, $p_1(x) = 0.1082x + 1.2234$, $p_1(9.4) = 2.2405$

3. $p_2(x) = x^2 - 2.580x + 2.580$, $\Gamma(1.01) = 0.9943$, $\Gamma(1.03) = 0.9835$ (exact to 4D)

5. $p_2(x) = -0.1434x^2 + 1.0895x$, $p_2(0.5) = 0.5089$, $p_2(1.5) = 1.3116$

7. $p_2(x) = -0.44304x^2 + 1.30896x - 0.02322$, $p_2(0.75) = 0.70929$

9. $p_2(x) = 1.0000 - 0.0112r + 0.0008r(r - 1)/2 = x^2 - 2.580x + 2.580$, $r = (x - 1)/0.02$; 0.9943, 0.9835, 0.9735

11. $p_2(x) = 0.27633 + 0.97678(x - 0.25) - 0.44304(x - 0.25)(x - 0.5)$
$= -0.44304x^2 + 1.30896x - 0.02322$

13. $p_2(x) = 0.9461x - 0.2868x(x - 1)/2 = -0.1434x^2 + 1.0895x$

15. 0.722, 0.786

17. $\delta f_{1/2} = 0.057\,839$, $\delta f_{3/2} = 0.069\,704$, etc.

21. 3.455606, exact to 5D

23. 0.386 4185, exact to 7D

PROBLEM SET 18.4, page 956

9. $-x + x^3$, $2(x - 1) + 3(x - 1)^2 - (x - 1)^3$

11. $1 - (x - 1)^2 + 2(x - 1)^3$, $46 + 48(x - 4) + 17(x - 4)^2 - 5(x - 4)^3$

13. $1 - x^2$, $-2(x - 1) - (x - 1)^2 + 2(x - 1)^3$,
$-1 + 2(x - 2) + 5(x - 2)^2 - 6(x - 2)^3$

15. $-\frac{3}{4}(x + 2)^2 + \frac{3}{4}(x + 2)^3$, $\frac{3}{4}(x + 1) + \frac{3}{2}(x + 1)^2 - \frac{5}{4}(x + 1)^3$,
$1 - \frac{9}{4}x^2 + \frac{5}{4}x^3$, $-\frac{3}{4}(x - 1) + \frac{3}{2}(x - 1)^2 - \frac{3}{4}(x - 1)^3$

17. $1 - 5x^2/4 + x^4/4$

19. The curvature is $f''(x)/(1 + f'(x)^2)^{3/2}$, which equals $f''(x)$, approximately, if $|f'(x)|$ is small.

PROBLEM SET 18.5, page 966

3. 0.5, 0.375, 0.34375

5. 0.782 794 (exact 0.785 398)

7. 0.785 392 (exact 0.785 398)

9. 1.107 147 (exact 1.107 149)

11. 0.693 254 (exact 0.693 147)

13. 0.693 771 (exact 0.693 147)

15. $M_2 = 1/2$, $M_2^* = -2$, $-0.002\,604 \leqq \epsilon \leqq 0.010\,417$, $0.780\,190 \leqq A(1) \leqq 0.793\,211$

17. $(1/x)^{iv} = 24/x^5$, $M_4 = 24$, $M_4^* = 0.75$, $-0.000\,521 \leqq \epsilon \leqq -0.000\,017$

19. 0.94508, 0.94583 (exact 0.94608)

21. 0.4612 (exact 0.4615)

23. 0.91936 (exact 0.91973)

25. $h|\frac{1}{2}\epsilon_0 + \epsilon_1 + \cdots + \epsilon_{n-1} + \frac{1}{2}\epsilon_n| \leqq [(b - a)/n]nu = (b - a)u$

27. 0.08, 0.32, 0.176, 0.256 (exact)

29. 0.52, 0.080, 0.304, 0.256

CHAPTER 18 (REVIEW QUESTIONS AND PROBLEMS), page 968

21. −0.53678E00, 0.11867E04, −0.60400E−02, 0.23948E02, 0.33333E00, 0.12143E02

23. $(7.625)_{10}$, $(34.5)_{10}$, $(3.6875)_{10}$

25. 49.980, 0.020; 49.980, 0.020008

27. $17.5565 \leqq s \leqq 17.5675$

29. The same as that of $\tilde{a}$

31. −0.2, −0.200 32, −0.200 323

33. $x = (8 + 6x^2 - 3x^4)^{1/6}$, $x_5 = 1.403\,007$, $x_{10} = 1.414\,557$, exact $\sqrt{2}$

35. 3.584 631

37. 0.567 143

39. 0.334 [exact 0.305 (3D)]

41. 2.95647, 2.96087

43. 2.96111

45. $1 - x^3$, $-3(x - 1) - 3(x - 1)^2 + 3(x - 1)^3$

47. 0.245, $\epsilon = 0.005$

49. 0.9027 (0.9045 exact to 4D)

PROBLEM SET 19.1, page 980

1. $x_1 = 2, \quad x_2 = 1$ **3.** $x_1 = \frac{1}{2}, \quad x_2 = -2$

5. $x_1 = 2, \quad x_2 = 3, \quad x_3 = 2$ **7.** $x_1 = -1, \quad x_2 = 8, \quad x_3 = -5$

9. $x_1 = 3.0, \quad x_2 = 4.8, \quad x_3 = 0$ **11.** $x_1 = -2.5, \quad x_2 = 7.0, \quad x_3 = 0.25$

13. $x_1 = 0$, x_2 arbitrary, $x_3 = 5x_2 + 10$ **15.** $x_1 = 0.4, \quad x_2 = -0.01, \quad x_3 = 1.1$

17. $(1 - 1/\epsilon)x_2 = 2 - 1/\epsilon$ eventually becomes $x_2/\epsilon \approx 1/\epsilon$, $x_2 \approx 1$, $x_1 = (1 - x_2)/\epsilon \approx 0$.

19. $x_1 = x_2 = x_3 = 1$; no solution

PROBLEM SET 19.2, page 985

1. $x_1 = 4, \ x_2 = 3, \quad \begin{bmatrix} 1 & 0 \\ 6 & 1 \end{bmatrix} \begin{bmatrix} 3 & 2 \\ 0 & 5 \end{bmatrix}$

3. $x_1 = -2, \ x_2 = 1.5 \quad \begin{bmatrix} 1 & 0 \\ 2 & 1 \end{bmatrix} \quad \begin{bmatrix} 4 & -6 \\ 0 & 5 \end{bmatrix}$

5. $x_1 = 2, \ x_2 = 1, \ x_3 = -3, \quad \begin{bmatrix} 1 & 0 & 0 \\ 0 & 1 & 0 \\ 2 & 5 & 1 \end{bmatrix} \quad \begin{bmatrix} 1 & 2 & 5 \\ 0 & 1 & 0 \\ 0 & 0 & 1 \end{bmatrix}$

7. $x_1 = 0.2, \ x_2 = 0.4, \ x_3 = 0.8, \quad \begin{bmatrix} 1 & 0 & 0 \\ 2 & 1 & 0 \\ 2 & 5 & 1 \end{bmatrix} \quad \begin{bmatrix} 5 & 4 & 1 \\ 0 & 1 & 2 \\ 0 & 0 & 3 \end{bmatrix}$

9. $x_1 = -\frac{1}{3}, \ x_2 = \frac{4}{3}, \ x_3 = 2, \quad \begin{bmatrix} 1 & 0 & 0 \\ 6 & 1 & 0 \\ 3 & 9 & 1 \end{bmatrix} \begin{bmatrix} 3 & 9 & 6 \\ 0 & -6 & 3 \\ 0 & 0 & -3 \end{bmatrix}$

11. **L** and **U** as in Prob. 5 (why?)

13. $x_1 = 6, \ x_2 = 12, \ x_3 = 4, \quad \begin{bmatrix} 3 & 0 & 0 \\ 2 & 3 & 0 \\ 4 & 1 & 3 \end{bmatrix} \quad \begin{bmatrix} 3 & 2 & 4 \\ 0 & 3 & 1 \\ 0 & 0 & 3 \end{bmatrix}$

15. $x_1 = 8, \ x_2 = 0, \ x_3 = -4, \quad \begin{bmatrix} 2 & 0 & 0 \\ 3 & 5 & 0 \\ 4 & 8 & 7 \end{bmatrix} \quad \begin{bmatrix} 2 & 3 & 4 \\ 0 & 5 & 8 \\ 0 & 0 & 7 \end{bmatrix}$

17. $\begin{bmatrix} 0.6 & 0.2 \\ 0.4 & 0.8 \end{bmatrix}$ **19.** $\begin{bmatrix} -0.8 & -2.0 \\ -2.8 & -2.0 \end{bmatrix}$ **21.** $\begin{bmatrix} 6 & 12 & 3 \\ 3 & 6 & 3 \\ 9 & 12 & 6 \end{bmatrix}$

PROBLEM SET 19.3, page 991

1. Exact 2, 2, 2 **3.** Exact 15, 5, 12 **5.** Exact 0.5, 0.5, 0.5

7. Exact 2, 0, 3, 1

9. (a) $\mathbf{x}^{(3)\mathsf{T}} = [0.49982 \quad 0.50001 \quad 0.50002]$, (b) $\mathbf{x}^{(3)\mathsf{T}} = [0.50333 \quad 0.49985 \quad 0.49968]$

13. $\mathbf{X}^{(1)} = \begin{bmatrix} 0.49 & -0.1 & 0.51 \\ 0 & 0.2 & 0 \\ -0.51 & 0.3 & -1.47 \end{bmatrix}, \quad \mathbf{A}^{-1} = \begin{bmatrix} 0.5 & -0.1 & 0.5 \\ 0 & 0.2 & 0 \\ -0.5 & 0.3 & -1.5 \end{bmatrix}$

15. $\sqrt{19}$, 6, 4 **17.** $\sqrt{2}$, 1, 1 **19.** $\sqrt{690}$, 32, 29

PROBLEM SET 19.4, page 999

1. 8, $\sqrt{26}$, 4 **3.** 18, $\sqrt{76}$, 5 **5.** 7, 5, 4 **7.** 4, 2, 1

9. 6, 4.2 **11.** 17, 289 **13.** 100, 10^4

15. $\|\mathbf{Ax}\| = 26 < 8 \cdot 15$ **17.** 289 **19.** 289, 289

21. $x_1 = 9,\ x_2 = -36,\ x_3 = 30$; $\kappa(\mathbf{A}) = \frac{11}{6} \cdot 408 = 748$

23. By (12), $1 = \|\mathbf{I}\| = \|\mathbf{AA}^{-1}\| \leqq \|\mathbf{A}\|\,\|\mathbf{A}^{-1}\| = \kappa(\mathbf{A})$. For the Frobenius norm, $\sqrt{n} = \|\mathbf{I}\| \leqq \kappa(\mathbf{A})$.

25. $\|\mathbf{x}\|_\infty = |x_1| \leqq \Sigma\, |x_j| = \|\mathbf{x}\|_1 \leqq n|x_1| = n\|\mathbf{x}\|_\infty$, where x_1 is largest in absolute value.

PROBLEM SET 19.5, page 1003

1. $y = -11.4 + 5.4x$ **3.** $y = 8.96 - 0.194x$

5. $y = 3.573 - 0.065x$ **7.** $y = -5.05 + 12.1x$

9. $U = -2.1 + 26.3I,\ R = 26.3$ **11.** $y = 0.5 + 1.5x^2$

13. $y = -1.5 + 1.5x^2$ **15.** $y = 6642 + 762.3x - 156.1x^2$

19. $\mathbf{C} = [c_{jk}],\ c_{jk} = x_j^{k-1},\ \mathbf{b}^\mathsf{T} = [b_0 \cdots b_m]$

PROBLEM SET 19.7, page 1011

1. 1, 4, 1, radii 5, 8, 9 **3.** 5, 0, 7, radii 4, 6, 6

5. 5, 8, 9, radii 0.02 **9.** $t_{11} = 100,\ t_{22} = t_{33} = 1$

11. $\sqrt{122}$ **13.** $\sqrt{170.0006}$ **15.** $6 \leqq \lambda \leqq 8$; 7 (an eigenvalue)

PROBLEM SET 19.8, page 1014

1. $q = 10,\ 11.655,\ 11.656\,853$; $|\epsilon| \leqq 4,\ 0.138,\ 0.004\,061$

3. $q = 0.8,\ 0.8, \cdots$; $|\epsilon| \leqq 0.6,\ 0.6, \cdots$

5. $q = 6.333,\ 6.871,\ 6.976$; $|\epsilon| \leqq 2.5,\ 1.2,\ 0.49$

7. 11.656 854 $(6 \pm \sqrt{32})$; errors 1.66, 0.0019, 10^{-6}

9. $q = 4.917$, $|\epsilon| \leqq 0.399$ **11.** $q = 3,\ 5.774$; $|\epsilon| \leqq 2.55,\ 1.337$

13. $q = 5/4,\ 14/9$; $|\epsilon| \leqq \sqrt{11}/4,\ \sqrt{101}/9$

15. (a) 12, 2, -2; (b) The ratio 12/2 is large.

17. $\mathbf{y} = \mathbf{Ax} = \lambda\mathbf{x},\ m_1 - \lambda\mathbf{x}^\mathsf{T}\mathbf{x} = \lambda m_0,\ q = m_1/m_0 = \lambda,$ $m_2 = \lambda^2 m_0,\ m_2/m_0 - q^2 = 0$

19. $-4.770 + 4.900 = 0.130$ (exact 0)

PROBLEM SET 19.9, page 1018

3. $\begin{bmatrix} 0 & 0 & 0 \\ -0.75 & 0 & -3 \\ -0.25 & 0 & -1 \end{bmatrix}$ 1, 0, −1

5. $\begin{bmatrix} 0 & 0 & 0 \\ -2 & 1.6 & 1.2 \\ -3 & 5.2 & 1.4 \end{bmatrix}$ −1, −1, 4

7. $\begin{bmatrix} 0 & 0 & 0 \\ -82 & 17 & 48 \\ 79 & -24 & -31 \end{bmatrix}$ $4, -7 \pm 24i$

9. $\begin{bmatrix} 0 & 0 & 0 \\ 4 & -2 & 0 \\ 14/3 & 8/3 & -6 \end{bmatrix}$ 8, −2, −6

11. $\begin{bmatrix} 0 & 0 & 0 \\ 1 & -1 & 0 \\ 1 & 0 & -1 \end{bmatrix}$ 20, −1, −1

13. $\begin{bmatrix} 0 & 0 & 0 & 0 \\ 0 & 0 & 0 & 0 \\ -1 & -1 & 2.5 & 1.5 \\ -1 & -1 & 1.5 & 2.5 \end{bmatrix}$ 9, 4, 1, 0

15. $\begin{bmatrix} 0 & 0 & 0 & 0 \\ 0 & 0 & 0 & 0 \\ -0.4 & -0.4 & 0.85 & 0.75 \\ -0.4 & -0.4 & 0.75 & 0.85 \end{bmatrix}$ 8.1, 1.6, 0.1, 0

PROBLEM SET 19.10, page 1028

1. $\begin{bmatrix} 0 & -\sqrt{2} & 0 \\ -\sqrt{2} & 1 & 0 \\ 0 & 0 & -1 \end{bmatrix}$

3. $\begin{bmatrix} 1 & -\sqrt{13} & 0 \\ -\sqrt{13} & 13 & 0 \\ 0 & 0 & 0 \end{bmatrix}$

5. $\begin{bmatrix} -1 & -3.605\,551\,28 & 0 \\ -3.605\,551\,28 & 5.461\,538\,46 & 3.692\,307\,69 \\ 0 & 3.692\,307\,69 & -4.461\,538\,46 \end{bmatrix}$

7. $\begin{bmatrix} 3.142\,857\,14 & 5.938\,459\,91 & 0 \\ 5.938\,459\,91 & 0.190\,476\,19 & -2.494\,438\,26 \\ 0 & -2.494\,438\,26 & -3.333\,333\,33 \end{bmatrix}$

9. $\begin{bmatrix} 10.426\,483\,98 & -4.072\,844\,87 & 0 \\ -4.072\,844\,87 & 4.223\,244\,41 & -1.093\,611\,69 \\ 0 & -1.093\,611\,69 & -1.265\,113\,01 \end{bmatrix}$

CHAPTER 19 (REVIEW QUESTIONS AND PROBLEMS), page 1029

21. $[6 \quad -3 \quad 1]^T$ **23.** $[4 \quad -1 \quad 2]^T$ **25.** $[2.2 \quad 3.0 \quad -0.9]^T$

27. All entries of the triangular matrices are 1.

31. $\begin{bmatrix} -7 & 2 & 3 \\ -13 & -2 & 7 \\ 8 & 2 & -2 \end{bmatrix}$ **33.** $\begin{bmatrix} 0.5 & -0.5 & -1 \\ -1.5 & 0.5 & -0.5 \\ 0.5 & -1.5 & -2 \end{bmatrix}$ **35.** Exact $-2, 1, 2$

37. $15, \sqrt{89}, 8$ **39.** $7, \sqrt{21}, 4$ **41.** $22, \sqrt{142}, 8$

43. 104 **45.** 15 **47.** $5.6 \cdot 3.5 = 19.6$

49. $1.98 + 0.98x$ **51.** Centers 11.4, 14.4, 14.6; radii 2.6, 3.2, 1.8

53. Centers 10, 4, 3; radii 0.2, 0.3, 0.3

55. $q_1 = 23/3, |\epsilon_1| \leqq 0.95;$ $q_2 = 7.88, |\epsilon_2| \leqq 0.82;$ $q_3 = 8.04,$ $|\epsilon_3| \leqq 0.69;$ $q_4 = 8.15, |\epsilon_4| \leqq 0.57$

PROBLEM SET 20.1, page 1043

1. $y = e^x$; 0.051 521, 0.145 975 (errors of x_5, x_{10})

3. $y = 1/(x^5 + 1)$; -0.012 735, -0.021 255 (errors of x_5, x_{10})

5. $y = e^x$; 0.001 275, 0.004 201 (errors of x_5, x_{10})

7. See Example 4, Sec. 1.7; 0.001 047, 0.004 935 (errors of x_5, x_{10})

9. $y = e^{x^2/2}$; $0, 7 \cdot 10^{-9}, 7 \cdot 10^{-8}, 4 \cdot 10^{-7}, 2 \cdot 10^{-6}, 5 \cdot 10^{-6}$

11. $y = xe^x$; errors 0, 0.000 190, 0.000 461, 0.000 846, 0.001 385, 0.002 131

13. For instance, $x = 0.5$, $y = 0.672$ (error $-$ 0.040), 0.6293 (error 0.0029); $x = 1$, $y = 0.893$ (error $-$ 0.028), 0.8626 (error 0.0021)

15. $y = 0$, 0.1000, 0.2034, 0.3109, 0.4217, 0.5348, 0.6494, 0.7649, 0.8806; error 0, 0.0044, 0.0122, $\cdots$, 0.0527

17. $y = 0$, 04352, 0.9074; error 0.0145, 0.0258 (about 50% of that in Prob. 15)

PROBLEM SET 20.2, page 1047

1. $y_4 = 1.173\,518$, $y_5 = 1.284\,044$, $y_6 = 1.433\,364$, $y_7 = 1.632\,374$, $y_8 = 1.896\,572$, $y_9 = 2.248\,046$, $y_{10} = 2.718\,486$

3. $y_4 = 0.422\,798$, $y_5 = 0.546\,315$, $y_6 = 0.684\,161$, $y_7 = 0.842\,332$, $y_8 = 1.029\,714$, $y_9 = 1.260\,288$, $y_{10} = 1.555\,763$

5. $y_4 = 4.229\,690$, $y_5 = 4.556\,859$, $y_6 = 5.360\,657$, $y_7 = 8.082\,563$

11. $y_1 = 1.010\,050$, $y_2 = 1.040\,811$, $y_3 = 1.094\,224$, $y_4 = 1.173\,622$, $y_5 = 1.284\,219$, $y_6 = 1.433\,636$, $y_7 = 1.632\,782$, $y_8 = 1.897\,074$, $y_9 = 2.248\,543$, $y_{10} = 2.718\,780$

PROBLEM SET 20.3, page 1054

9. $y \approx 3$, 2.94, 2.7606, 2.4642, 2.055 606, 1.542 042

11. You get the exact solution. Can you see why?

13. You get the exact solution, except for a round-off error [e.g., $y_1 = 2.761\,608$, $y(0.2) = 2.7616$ (exact), etc.]. Why?

PROBLEM SET 20.4, page 1063

3. $u_{xy}(x, y) \approx \frac{1}{2k}[u_x(x, y + k) - u_x(x, y - k)]$. Now substitute

$$u_x(x, y \pm k) \approx \frac{1}{2h}[u(x + h, y \pm k) - u(x - h, y \pm k)].$$

5. 75

7. $u(P_{11}) = u(P_{12}) = 105$, $u(P_{21}) = 155$, $u(P_{22}) = 115$

9. $-2, -5, -5, -62$

11. $u_{11} = u_{21} = 0.1083$, $u_{12} = u_{22} = 0.3248$; exact to 4D: 0.0937, 0.2999

13. $\sqrt{3}$. $u_{11} = u_{21} = 0.0859$, $u_{12} = u_{22} = 0.3180$. (0.1083, 0.3248 are 4D-values of the solution of the linear equations of the problem.)

15. Only 5 steps

PROBLEM SET 20.5, page 1069

3. $u_{11} = 1$, $u_{21} = u_{12} = 4$, $u_{22} = 16$

9. $-4u_{11} + u_{21} + u_{12} = -3$, $u_{11} - 4u_{21} + u_{22} = -12$, $u_{11} - 4u_{12} + u_{22} = 0$, $2u_{21} + 2u_{12} - 12u_{22} = -14$, $u_{11} = u_{22} = 2$, $u_{21} = 4$, $u_{12} = 1$

PROBLEM SET 20.6, page 1074

1. Calculations to 6D, values rounded off to 3D

t	$x = 0.2$	Exact	$x = 0.4$	Exact
0.04	0.164	0.159	0.255	0.260
0.08	0.107	0.108	0.175	0.175
0.12	0.073	0.073	0.119	0.118
0.16	0.050	0.049	0.081	0.079
0.20	0.034	0.033	0.055	0.054

3. The values are given as exact values in the answer to Prob. 1. The series converges rapidly.

5. 0, 0.06625, 0.12500, 0.17125, 0.20000, 0.21000, 0.20000, etc.

7. 0, 0.05352, 0.10156 [0.10182 by (9)], 0.13984, 0.16406 [0.16727 by (9)], 0.17266, etc.

9. Use (5) and $0 = \partial u_{0j}/\partial x = (u_{1j} - u_{-1,j})/2h$.

PROBLEM SET 20.7, page 1077

1. For $x = 0.2, 0.4$ we obtain 0.012, 0.02 ($t = 0.2$), 0.004, 0.008 ($t = 0.4$), -0.004, -0.008 ($t = 0.6$), etc.

3. $u(x, 1) = 0, -0.05, -0.10, -0.15, -0.20, 0$

5. 0.190, 0.308, 0.308, 0.190 (0.178, 0.288, 0.288, 0.178 exact to 3D)

7. 0, 0.354, 0.766, 1.271, 1.679, 1.834, $\cdots$ ($t = 0.1$); 0, 0.575, 0.935, 1.135, 1.296, 1.357, $\cdots$ ($t = 0.2$)

CHAPTER 20 (REVIEW QUESTIONS AND PROBLEMS), page 1078

21. 1, 1, 1.02, 1.0608, etc. Errors 0, 0.01, 0.021, 0.0334, etc.

25. $y = \tan x$

29. $y(0.4) = 1.822\,798$, $y(0.5) = 2.046\,315$, $y(0.6) = 2.284\,155$, $y(0.7) = 2.542\,325$, $y(0.8) = 2.829\,706$, $y(0.9) = 3.160\,277$, $y(1.0) = 3.557\,611$

35. 1.96, 7.86, 29.46

37. $u(P_{11}) = u(P_{31}) = 270$, $u(P_{21}) = u(P_{13}) = u(P_{23}) = u(P_{33}) = 30$, $u(P_{12}) = u(P_{32}) = 90$, $u(P_{22}) = 60$

41. 0.06279, 0.09336, 0.08364, 0.04707

43. 0, −0.352, −0.153, 0.153, 0.352, 0 if $t = 0.12$ and 0, 0.344, 0.166, −0.166, −0.344, 0 if $t = 0.24$

PROBLEM SET 21.1, page 1087

9. $g'(t) = -16x_1^3(x_1 - 4tx_1^3)^3 - 4096x_2^3(x_2 - 64tx_2^3)^3 = 0$. For $\mathbf{x}_0 = \mathbf{i} + \mathbf{j}$ the solution is $t_1 = 0.017\,909\,5$ (7D), and $\mathbf{z}(t_1) = 0.928\,362\mathbf{i} - 0.146\,202\mathbf{j}$.

PROBLEM SET 21.2, page 1091

5. Contradictory, no solution

11. $f_{\max} = f(0, 5) = 10$

13. $f_{\min} = f(2, 4/3) = 8$

15. $x_1 = 15$ tons of A_1, $x_2 = 10$ tons of A_2, daily profit \$1400

17. $f_{\max} = f(210, 60) = 3750$

19. No

PROBLEM SET 21.3, page 1096

1. You will start at C (Fig. 447) and then exchange x_1 and x_4.

3. $f_{\max} = f(\frac{120}{11}, \frac{60}{11}) = \frac{480}{11}$

5. $x_3 = 18 - x_1 - 3x_2$, $x_4 = 10 - x_1 - x_2$, $x_5 = 24 - 3x_1 - x_2$; $x_1 \leftrightarrow x_5$; $x_2 \leftrightarrow x_4$; $f_{\max} = f(7, 3) = 780$

7. $f_{\max} = f(0, 2.5, 2.3) = 7.7$

9. Take x_5, x_6 as basic. Exchange $x_3 \leftrightarrow x_6$, then $x_1 \leftrightarrow x_5$. $f_{\max} = f(2.857, 0, 14.286, 0) = 314$.

PROBLEM SET 21.4, page 1101

1. $x_1 \leftrightarrow x_5$, then $x_2 \leftrightarrow x_4$ (without gain), then $x_3 \leftrightarrow x_5$. $f_{\max} = f(10, 5) = 5500$.

3. $f_{\max} = f(1, 1, 0) = 12$. The basic feasible solution $x_1 = x_2 = x_4 = x_5 = 0$, $x_3 = 1$ is degenerate.

5. Exchange $x_2 \leftrightarrow x_5$, then $x_3 \leftrightarrow x_4$. $f_{\max} = f(3, 1) = 7$.

7. Exchange $x_2 \leftrightarrow x_6$, then $x_3 \leftrightarrow x_5$, then $x_1 \leftrightarrow x_4$. $f_{\max} = f(100, 150, 200) = 3200$.

9. Maximize $\tilde{f} = -f$. $x_4 = 1 + x_1 - x_2$, $x_5 = 40 - 5x_1 - 4x_2$, $x_6 = 5 - x_1 - x_2 + x_3$; $x_1 \leftrightarrow x_6$, $x_2 \leftrightarrow x_4$; $-f_{\min} = \tilde{f}_{\max} = \tilde{f}(2, 3) = -4 + 3 = -1$; $f_{\min} = 1$.

CHAPTER 21 (REVIEW QUESTIONS AND PROBLEMS), page 1102

11. 6i + 3j, 0.9153i − 0.8136j, 0.2219i + 0.1109j, 0.0338i − 0.0301j

23. $f_{\min} = f(0.5, 1.5) = 4.5$ **25.** $f_{\min} = f(2, 2) = 16$

27. $f_{\max} = f(0, 40, 5) = 280$ **29.** $f_{\max} = f(13, 0, 8) = 188$

PROBLEM SET 22.1, page 1109

7. $\begin{bmatrix} 0 & 1 & 1 & 1 \\ 0 & 0 & 0 & 0 \\ 1 & 0 & 0 & 0 \\ 0 & 0 & 0 & 0 \end{bmatrix}$ **9.** $\begin{bmatrix} 0 & 1 & 1 \\ 0 & 0 & 1 \\ 1 & 1 & 0 \end{bmatrix}$ **11.** $\begin{bmatrix} 0 & 0 & 1 & 1 \\ 0 & 0 & 0 & 1 \\ 1 & 0 & 0 & 0 \\ 1 & 1 & 0 & 0 \end{bmatrix}$

15.

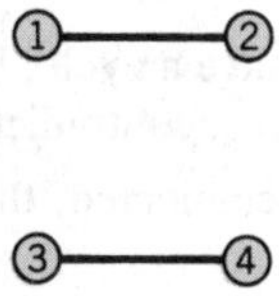

17.

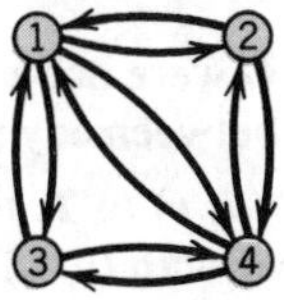

21. Vertex $\begin{matrix} 1 \\ 2 \\ 3 \\ 4 \end{matrix} \begin{bmatrix} 1 & 1 & 0 \\ 0 & 0 & 1 \\ 1 & 0 & 0 \\ 0 & 1 & 1 \end{bmatrix}$

23.

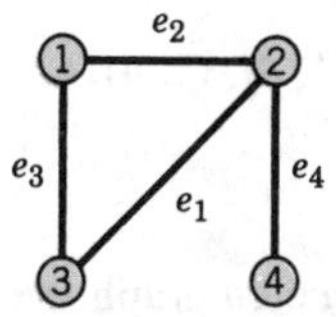

25. Vertex $\begin{matrix} 1 \\ 2 \\ 3 \\ 4 \end{matrix} \begin{bmatrix} -1 & -1 & 1 & -1 \\ 1 & 0 & 0 & 0 \\ 0 & 1 & -1 & 0 \\ 0 & 0 & 0 & 1 \end{bmatrix}$

27.

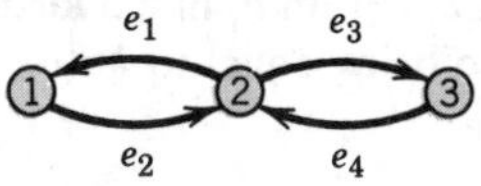

29.

Vertex	Incident Edges
1	$-e_1, -e_2, e_3, -e_4$
2	e_1
3	$e_2, -e_3$
4	e_4

PROBLEM SET 22.2, page 1115

1. 4 **3.** 5 **5.** 4

9. The idea is to go backward. There is a v_{k-1} adjacent to v_k and labeled $k - 1$, etc. Now the only vertex labeled 0 is s. Hence $\lambda(v_0) = 0$ implies $v_0 = s$, so that $v_0 - v_1 - \cdots - v_{k-1} - v_k$ is a path $s \to v_k$ that has length k.

17. No; there is no way of traveling along (3, 4) only once.

19. Police patrol, track repair crew, farmer's best route for seeding his fields.

25. From m to $100m$, $10m$, $2.5m$, $m + 4.6$

PROBLEM SET 22.3, page 1119

1. (1, 2), (2, 4), (4, 3); $L_2 = 3$, $L_3 = 9$, $L_4 = 7$

3. (1, 2), (1, 4), (2, 3); $L_2 = 2$, $L_3 = 5$, $L_4 = 5$

5. (1, 4), (2, 4), (3, 4), (3, 5); $L_2 = 4, L_3 = 3, L_4 = 2, L_5 = 8$

7. (1, 5), (2, 3), (2, 6), (3, 4), (3, 5); $L_2 = 9, L_3 = 7, L_4 = 8, L_5 = 4, L_6 = 14$

11. $L_2 = 2, L_3 = 5$

PROBLEM SET 22.4, page 1123

1. 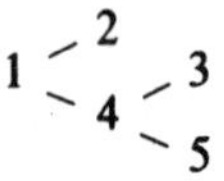

3. 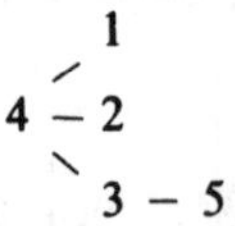

5. 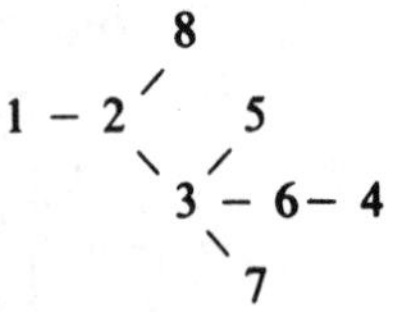

7. (a) $L_2 = 4, L_3 = 3$; (b) $L = 5$

9. 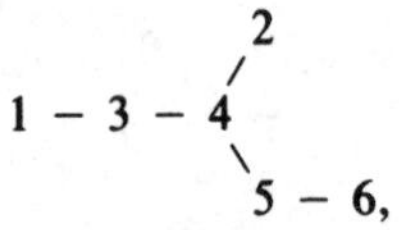

$L = 38$

11. Yes

13. G is connected. If G were not a tree, it would have a cycle, but this cycle would provide two paths between any pair of its vertices, contradicting the uniqueness.

19. If we add an edge (u, v) to T, then since T is connected, there is a path $u \to v$ in T which, together with (u, v), forms a cycle.

PROBLEM SET 22.5, page 1127

1. (1, 2), (1, 4), (3, 4), (4, 5), $L = 12$

3. (1, 4), (3, 4), (2, 4), (3, 5), $L = 20$

5. (1, 2), (1, 3), (1, 4), (2, 6), (3, 5), $L = 32$

9. If G is a tree

11. A shortest spanning tree of the largest connected graph that contains vertex 1

13. $d(G) = 24$, $r(G) = 12 = \epsilon(3)$, center {3}

17. Choose a vertex u and find a farthest v_1. From v_1 find a farthest v_2. Find w such that $d(w, v_1)$ is as close as possible to being equal to $\frac{1}{2}d(v_1, v_2)$.

PROBLEM SET 22.6, page 1133

1. $1 - 2 - 5, \Delta f = 2; 1 - 4 - 2 - 5, \Delta f = 2$, etc.

3. $1 - 2 - 4 - 6, \Delta f = 2; 1 - 2 - 3 - 5 - 6, \Delta f = 1$, etc.

5. $f_{12} = 4, f_{13} = 1, f_{14} = 4, f_{42} = 4, f_{43} = 0, f_{25} = 8, f_{35} = 1, f = 9$

7. $f_{12} = 4, f_{13} = 3, f_{24} = 4, f_{35} = 3, f_{54} = 2, f_{46} = 6, f_{56} = 1, f = 7$

9. {4, 5, 6}, 28

11. {2, 4, 6}, 50

13. $1 - 2 - 3 - 7, \Delta f = 2; 1 - 4 - 5 - 6 - 7, \Delta f = 1;$
$1 - 2 - 3 - 6 - 7, \Delta f = 1; f_{max} = 14$

15. {3, 5, 7}, 22

17. $S = \{1,4\}, \text{cap}(S,T) = 6 + 8 = 14$

19. If $f_{ij} < c_{ij}$ as well as $f_{ij} > 0$

PROBLEM SET 22.7, page 1137

3. (2, 3) and (5, 6)

5. $1 - 2 - 5, \quad \Delta_t = 2; \quad 1 - 4 - 2 - 5, \quad \Delta_t = 1; \quad f = 6 + 2 + 1 = 9$

7. $1 - 2 - 4 - 6, \quad \Delta_t = 2; \quad 1 - 3 - 5 - 6, \quad \Delta_t = 1; \quad f = 4 + 2 + 1 = 7$

9. By considering only edges with one labeled end and one unlabeled

13. 2000

17. $S = \{1, 2, 4, 5\}$, $T = \{3, 6\}$, $\operatorname{cap}(S, T) = 14$

PROBLEM SET 22.8, page 1142

1. No **3.** No **5.** Yes, $S = \{1, 4, 5, 8\}$

7. Yes; a graph is not bipartite if it has a nonbipartite subgraph.

9. 1 − 2 ▬ 3 − 5 **11.** (1, 5), (2, 3)

13. (1, 4), (2, 3), (5, 7) **15.** 3 **19.** 5

23. No; K_5 is not planar. **25.** K_3

27. 3; at 1, 4, 8 **29.** K_3

CHAPTER 22 (REVIEW QUESTIONS AND PROBLEMS), page 1145

17. $\begin{bmatrix} 0 & 1 & 0 \\ 0 & 0 & 1 \\ 1 & 0 & 0 \end{bmatrix}$ **19.** $\begin{bmatrix} 0 & 1 & 0 & 1 \\ 1 & 0 & 1 & 0 \\ 0 & 1 & 0 & 1 \\ 1 & 0 & 1 & 0 \end{bmatrix}$ **21.** $\begin{bmatrix} 0 & 1 & 0 & 1 \\ 0 & 0 & 1 & 0 \\ 1 & 0 & 0 & 1 \\ 0 & 0 & 0 & 0 \end{bmatrix}$

23.

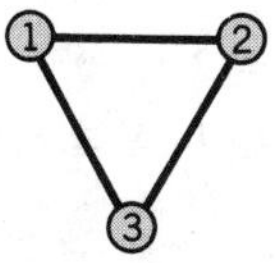

27.

Vertex	Incident Edges
1	$e_2, -e_3$
2	$-e_1, e_3$
3	$e_1, -e_2$

29. 4 **31.** 4

33. $L_2 = 10, L_3 = 15, L_4 = 13$ **35.** 1 − 4 − 3 − 2

37. 1 − 2 − 4 − 3 − 5 **41.** $f = 7$ **45.** (1, 3), (2, 4)

PROBLEM SET 23.1, page 1152

1. 4 outcomes: *HH*, *HT*, *TH*, *TT* (H = head, T = tail)

3. 2^3 outcomes: *RRR*, *RRL*, *RLR*, *LRR*, *RLL*, *LRL*, *LLR*, *LLL*

5. 10 outcomes: *R*, *BR*, *BBR*, *BBBR*, etc.

7. The space of ordered triples of nonnegative numbers

9. The space of ordered pairs of numbers $X = x \geqq 0$, $Y = y$

13. No

15. Do not forget to list S and the empty set.

PROBLEM SET 23.2, page 1159

1. $63/64 \approx 0.98$

3. (a) $0.9^3 = 72.9\%$, (b) $\frac{90}{100} \cdot \frac{89}{99} \cdot \frac{88}{98} = 72.65\%$

5. $1 - 4/125$ **7.** $0.95^4 \approx 81.5\%$

9. $1 - 0.97^4 = 11.5\%$ **11.** 77.66%

13. $\frac{6}{36} + \frac{27}{36} - \frac{3}{36} = \frac{30}{36}$

15. $P(MMM) + P(MMFM) + P(MFMM) + P(FMMM) = \frac{1}{8} + 3 \cdot \frac{1}{16} = \frac{5}{16}$

17. 11.68%

PROBLEM SET 23.3, page 1164

3. In 40320 ways **5.** 19600 **7.** 210, 70, 112, 28

9. $\binom{52}{13} = 635\ 013\ 559\ 600$ **11.** 1/84, 5/21 **13.** 676 000

15. The idea of proof is the same as that of Theorem 1, but instead of filling n places we now have to fill only k places. If repetitions are permitted, we have n elements for filling each of the k places.

17. 23.5, 0.5, 2%; 39 902, 400, 1%

PROBLEM SET 23.4, page 1171

3. $F(x) = 0$ if $x \leqq 0$, $F(x) = x^3/27$ if $0 < x \leqq 3$, $F(x) = 1$ if $x > 3$

5. Special uniform distribution, $f(x) = 0.2$ if $0 < x < 5$ and 0 otherwise.

7. 40%, $x = 4.25$

9. $P(X > 1200) = \int_{1.2}^{2} 6[0.25 - (x - 1.5)^2]\, dx = 0.896$. *Ans.* $0.896^3 = 72\%$.

11. $k = 3$, $c_1 = 0.464$, $c_2 = 0.965$ **13.** $k = 2.5$; 50%

17. $k = 1.1565$; 26.9% **19.** $X > b$, $X \geqq b$, $X < c$, $X \leqq c$, etc.

PROBLEM SET 23.5, page 1177

1. $1, \frac{1}{2}$ **5.** 3.5, 2.917 **7.** \$23.45

9. 750, 1, 0.002 **11.** $\frac{1}{2}, \frac{1}{20}, (X - \frac{1}{2})\sqrt{20}$

15. $E(X^k) = 2/(k + 2)$, $\sigma^2 = \frac{1}{18}$

17. $E(X^k) = (b^{k+1} - a^{k+1})/[(b - a)(k + 1)]$

19. $\gamma = 4/2\sqrt{2} = \sqrt{2}$

PROBLEM SET 23.6, page 1183

1. 0.0625, 0.25, 0.9375, 0.9375 **3.** 64%

5. $0.99^{10} \approx 90.4\%$ **7.** $1 - e^{-0.2} = 18\%$

9. $f(x) = 0.5^x e^{-0.5}/x!$, $f(0) + f(1) = e^{-0.5}(1.0 + 0.5) = 0.91$. *Ans.* 9%.

11. 0.265 **13.** $\frac{120}{286}, \frac{135}{286}, \frac{30}{286}, \frac{1}{286}$

19. $G(0) = 1$, $G'(t) = e^{-\mu} \exp[\mu e^t]\, \mu e^t = \mu e^t G(t)$, $G''(t) = \mu e^t[G(t) + G'(t)]$, $E(X^2) = G''(0) = \mu + \mu^2$, $\sigma^2 = E(X^2) - \mu^2 = \mu$

PROBLEM SET 23.7, page 1190

1. 0.1587, 0.6306, 0.5, 0.4950 **3.** 17.29, 10.71, 19.152

5. 0.27% **7.** 16% **9.** 84% **11.** 31.1%, 95.5%

13. About 22 **19.** Use $f''(x) = 0$.

PROBLEM SET 23.8, page 1199

1. 1/8, 3/16, 3/8 **3.** $k = 4$, $P = 9/16$
5. $f_2(y) = 1/(\beta_2 - \alpha_2)$ if $\alpha_2 < y < \beta_2$ **7.** 20.200, about 0.007
9. 27.45 mm, 0.38 mm
11. Independent, $f_1(x) = 0.1e^{-0.1x}$ if $x > 0$, $f_2(y) = 0.1e^{-0.1y}$ if $y > 0$, 36.8%
15. 50% **17.** No

CHAPTER 23 (REVIEW QUESTIONS AND PROBLEMS), page 1200

21. (a) 0.512, (b) About 0.504 **23.** 1/4
25. $f(0) = 0.80816$, $f(1) = 0.18367$, $f(2) = 0.00816$
27. $f(x) = 2^{-x}$, $x = 1, 2, \cdots$ **29.** 0.3, 0.2, 0
31. 72 **33.** 118.019, 1.98, 1.65% **35.** 94.09%, 5.82%, 0.09%
37. $k = 3$, $c = 0.983$ **39.** 0, 2 **41.** 7/3, 8/9 **43.** 0.98^{15}
47. 76.74 kg **49.** 16%, 2.25% (see Fig. 496)

PROBLEM SET 24.2, page 1208

1. The items 38, 69, 02, 49, 23, 52, 73, 29, 09, 05
5. Take $\tilde{y}_j = 2 + 4y_j$ with y_j selected as before.
7. $x = 0.91 = \Phi(z)$ gives $z = 1.341$ (Table A8), etc.
9. 2.408, 0.545, 1.022

PROBLEM SET 24.4, page 1218

1. $\bar{x} = 3.47$, $s^2 = 2.98$ **3.** $\bar{x} = 6$, $s^2 = 13.3$
5. 1, 25 **11.** 100.25 **13.** 0.02 **15.** 360
17. 96, 86, 110, 24 **19.** 99.2, 234.7; grouped: 99.4, 254.7

PROBLEM SET 24.5, page 1221

3. $l = p^k(1 - p)^{n-k}$, $\hat{p} = k/n$, k = number of successes in n trials
5. 7/12 **7.** $\hat{p} = 1/\bar{x}$ **9.** $\hat{\mu} = \bar{x}$
11. $\hat{\theta} = n/\Sigma x_j = 1/\bar{x}$ **13.** $\hat{\theta} = 1$

PROBLEM SET 24.6, page 1231

3. CONF $\{23.86 \leqq \mu \leqq 28.93\}$ **5.** It will double.
7. About 4, 16 **9.** CONF $\{10.07 \leqq \mu \leqq 10.33\}$
11. $c = 1.96$, $\bar{x} = 87$, $s^2 = 71.86$, $k \approx cs/\sqrt{n} = 0.742$, CONF $\{86 \leqq \mu \leqq 88\}$, CONF $\{0.17 \leqq p \leqq 0.18\}$
13. CONF $\{0.00044 \leqq \sigma^2 \leqq 0.00127\}$ **15.** CONF $\{23 \leqq \sigma^2 \leqq 553\}$
17. Normal distributions, means -27, 81, 133, variances 16, 144, 400

19. $Z = X + Y$ is normal with mean 105 and variance 1.25.
Ans. $P(104 \leqq Z \leqq 106) = 63\%$.

PROBLEM SET 24.7, page 1242

1. $t = \sqrt{7}(0.286 - 0)/4.31 = 0.18 < c = 1.94$; do not reject the hypothesis.

3. $c = 6090 > 6019$, $c = 12\,127 > 12\,012$; do not reject the hypothesis.

5. $\sigma^2/n = 1$, $c = 28.36$; do not reject the hypothesis.

7. $\mu < 28.76$ or $\mu > 31.24$

9. Alternative $\mu \neq 1000$, $t = \sqrt{20}(996 - 1000)/5 = -3.58 < c = -2.09$ (Table A10, 19 degrees of freedom). Reject the hypothesis $\mu = 1000$ g.

11. Test $\mu = 0$ against $\mu \neq 0$. $t = 2.11 < c = 2.37$ (7 degrees of freedom). Do not reject the hypothesis.

13. $\alpha = 5\%$, $c = 16.92 > 9 \cdot 0.5^2/0.4^2 = 14.06$; do not reject the hypothesis.

15. $t_0 = \sqrt{10 \cdot 9 \cdot 17/19}\,(21.8 - 20.2)/\sqrt{9 \cdot 0.6^2 + 8 \cdot 0.5^2} = 6.3 > c = 1.74$ (17 degrees of freedom). Reject the hypothesis and assert that B is better.

17. $v_0 = 50/30 = 1.67 < c = 2.59$ [(9, 15) degrees of freedom]; do not reject the hypothesis.

PROBLEM SET 24.8, page 1248

1. LCL $= 1 - 2.58 \cdot 0.03/\sqrt{6} = 0.968$, UCL $= 1.032$

3. $n = 10$

5. Choose 4 times the original sample size.

7. $2.58\sqrt{0.024}/\sqrt{2} = 0.283$, UCL $= 27.783$, LCL $= 27.217$

11. In 30% (5%) of the cases, approximately

13. UCL $= np + 3\sqrt{np(1 - p)}$, CL $= np$, LCL $= np - 3\sqrt{np(1 - p)}$

15. CL $= \mu = 2.5$, UCL $= \mu + 3\sqrt{\mu} \approx 7$, LCL $= \mu - 3\sqrt{\mu}$ is negative in (b) and we set LCL $= 0$.

PROBLEM SET 24.9, page 1254

1. 0.9825, 0.9384, 0.4060

3. 0.028 (at $\theta = 0.054$)

5. 19%, 15%

7. $(1 - \theta)^5$, $(1 - \theta)^5 + 5\theta(1 - \theta)^4$

9. Because n is finite

11. $(1 - \frac{1}{2})^3 + 3 \cdot \frac{1}{2}(1 - \frac{1}{2})^2 = \frac{1}{2}$

13. 22% (if $c = 9$)

15. $\alpha = 5\%$, $\beta = 44\%$

PROBLEM SET 24.10, page 1257

1. $\chi_0^2 = 2.33 < c = 11.07$. Yes.

3. $\chi_0^2 = 94.19 > c = 11.07$. Reject.

5. $\chi_0^2 = 0.4 < c = 3.84$. Assert that the coin is fair.

7. 20 odd and 30 even digits, $\chi_0^2 = 2 < 3.84$. Do not reject the hypothesis.

9. $e_j = np_j = 370/5 = 74$, $\chi_0^2 = 984/74 = 13.3$, $c = 9.49$. Reject the hypothesis.

11. Combining the last three rows, we have $K - r - 1 = 9$ ($r = 1$ since we estimated the mean, $\frac{10094}{2608} \approx 3.87$). $\chi_0^2 = 12.8 < c = 16.92$. Do not reject.

13. $\chi_0^2 = 49/20 + 49/60 = 3.27 < c = 3.84$ (1 degree of freedom, $\alpha = 5\%$), which supports the claim.

PROBLEM SET 24.11, page 1260

1. Hypothesis: A and B are equally good. Then the probability of at least 7 trials favorable to A is $\frac{1}{2}^8 + 8 \cdot \frac{1}{2}^8 = 3.5\%$. Reject the hypothesis.

3. $\frac{1}{2}^6 + 6 \cdot (\frac{1}{2})^6 + 15 \cdot (\frac{1}{2})^6 = 34\%$ is the probability of at most 2 negative values if $\tilde{\mu} = 0$, which we do not reject.

5. Hypothesis $\mu = 0$. Alternative $\mu > 0$, $\bar{x} = 1.58$,
$t = \sqrt{10} \cdot 1.58/1.23 = 4.06 > c = 1.83$ ($\alpha = 5\%$). Hypothesis rejected.

7. $(\frac{1}{2})^{18}(1 + 18 + 153 + 816) = 0.0038$

9. Consider $y_j = x_j - \tilde{\mu}_0$.

11. $P(T \leqq 2) = 2.8\%$. Reject.

13. $P(T \leqq 14) = 7.8\%$. Do not reject.

15. $P(T \leqq 2) = 0.1\%$. Reject.

PROBLEM SET 24.12, page 1267

1. About 120 feet

3. $y - 1.9 = x$

5. $y - 1402 = 4.32(x - 590)$

7. $3s_1^2 = 500$, $3s_{xy} = 33.5$, $k_1 = 0.067$, $3s_2^2 = 2.268$, $q_0 = 0.023$,
$K = 0.021$, CONF $\{0.046 \leqq \kappa_1 \leqq 0.088\}$

9. $q_0 = 76$, $K = 2.37\sqrt{76/(7 \cdot 944)} = 0.254$, CONF $\{-1.58 \leqq \kappa_1 \leqq -1.06\}$

CHAPTER 24 (REVIEW QUESTIONS AND PROBLEMS), page 1267

21. 4, 100

23. 106, 160, 40

25. 1.5625, 0.03696, 0.6

27. $\hat{\mu} = 15.5625$, $\tilde{\sigma}^2 = 5.3711$

29. It will double.

31. CONF $\{19.2 \leqq \mu \leqq 33.6\}$

33. CONF $\{0.726 \leqq \mu \leqq 0.751\}$

35. CONF $\{1.373 \leqq \mu \leqq 1.451\}$

37. CONF $\{0.105 \leqq \mu \leqq 0.107\}$; that error is 0.001, approximately.

39. CONF $\{0.05 \leqq \sigma^2 \leqq 10\}$

41. $c = 14.74 > 14.5$; reject μ_0.

43. $\Phi\left(\dfrac{14.74 - 14.50}{\sqrt{0.025}}\right) = 0.9357$

45. $30.14/3.8 = 7.93 < 8.25$. Reject.

47. $t = \sqrt{100}(26\,000 - 25\,000)/2000 = 5 > c = 2.37$ (Table A10). Reject the hypothesis $\mu_0 = 25\,000$ and assert that the manufacturer's claim is justified.

49. $v_0 = 2.5 < 6.0$ [(9, 4) degrees of freedom]; do not reject the hypothesis.

51. Decrease by a factor $\sqrt{2}$. By a factor $2.58/1.96 = 1.32$.

53. 0.9953, 0.9825, 0.9384, etc.

55. 0.028 (at $\theta = 0.054$)

57. $\chi_0^2 = 999.9 > c = 13.28$. No.

59. $P(T \leqq 3) = 6.8\%$ (Table A13). No.

Appendix 3

Auxiliary Material

A3.1 Formulas for Special Functions

For tables of numerical values, see Appendix 5.

Exponential function e^x (Fig. 519)

$$e = 2.71828\ 18284\ 59045\ 23536\ 02874\ 71353$$

$$e^x e^y = e^{x+y}, \qquad e^x/e^y = e^{x-y}, \qquad (e^x)^y = e^{xy} \tag{1}$$

Natural logarithm (Fig. 520)

$$\ln(xy) = \ln x + \ln y, \qquad \ln(x/y) = \ln x - \ln y, \qquad \ln(x^a) = a \ln x \tag{2}$$

$\ln x$ is the inverse of e^x, and $e^{\ln x} = x$, $e^{-\ln x} = e^{\ln(1/x)} = 1/x$.

Logarithm of base ten $\log_{10} x$ or simply $\log x$

$$\log x = M \ln x, \quad M = \log e = 0.43429\ 44819\ 03251\ 82765\ 11289\ 18917 \tag{3}$$

$$\ln x = \frac{1}{M} \log x, \quad \frac{1}{M} = 2.30258\ 50929\ 94045\ 68401\ 79914\ 54684 \tag{4}$$

$\log x$ is the inverse of 10^x, and $10^{\log x} = x$, $10^{-\log x} = 1/x$.

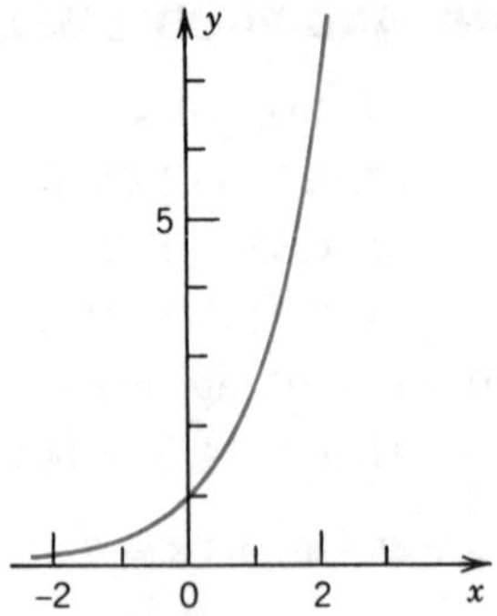

Fig. 519. Exponential function e^x

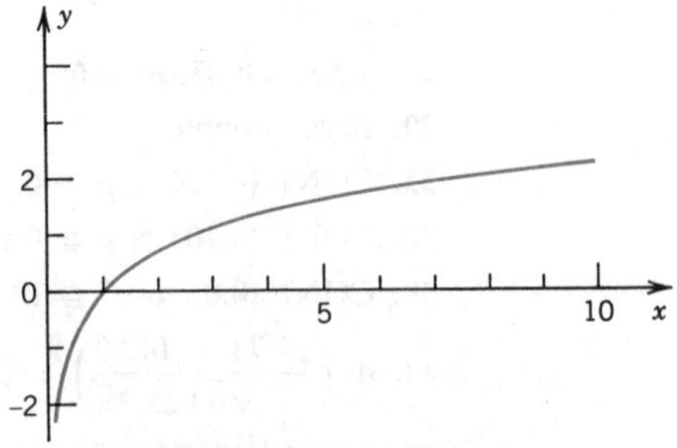

Fig. 520. Natural logarithm ln x

Sine and cosine functions (Figs. 521, 522). In calculus, angles are measured in radians, so that $\sin x$ and $\cos x$ have period 2π.

$\sin x$ is odd, $\sin(-x) = -\sin x$, and $\cos x$ is even, $\cos(-x) = \cos x$.

$$1° = 0.01745\ 32925\ 19943 \text{ radian}$$

$$1 \text{ radian} = 57°\ 17'\ 44.80625''$$

$$= 57.29577\ 95131°$$

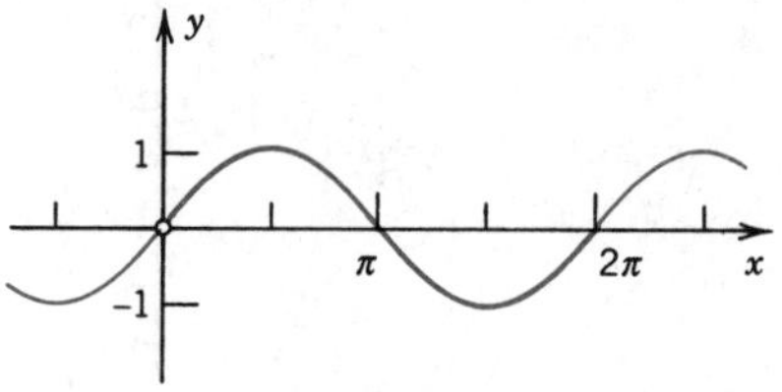

Fig. 521. sin *x*

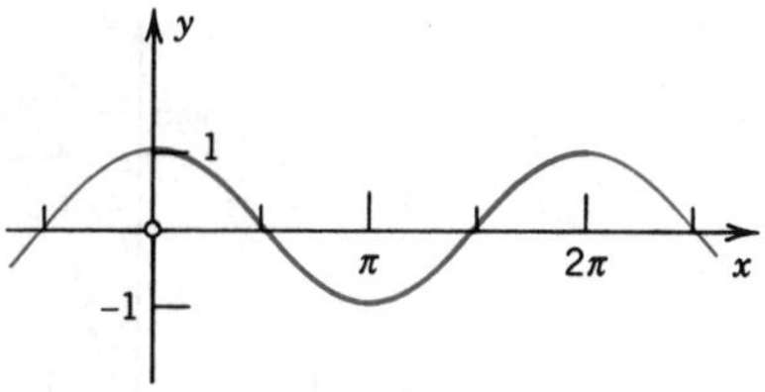

Fig. 522. cos *x*

$$\sin^2 x + \cos^2 x = 1 \tag{5}$$

$$\begin{cases} \sin(x+y) = \sin x \cos y + \cos x \sin y \\ \sin(x-y) = \sin x \cos y - \cos x \sin y \\ \cos(x+y) = \cos x \cos y - \sin x \sin y \\ \cos(x-y) = \cos x \cos y + \sin x \sin y \end{cases} \tag{6}$$

$$\sin 2x = 2 \sin x \cos x, \qquad \cos 2x = \cos^2 x - \sin^2 x \tag{7}$$

$$\begin{cases} \sin x = \cos\left(x - \dfrac{\pi}{2}\right) = \cos\left(\dfrac{\pi}{2} - x\right) \\ \cos x = \sin\left(x + \dfrac{\pi}{2}\right) = \sin\left(\dfrac{\pi}{2} - x\right) \end{cases} \tag{8}$$

$$\sin(\pi - x) = \sin x, \qquad \cos(\pi - x) = -\cos x \tag{9}$$

$$\cos^2 x = \tfrac{1}{2}(1 + \cos 2x), \qquad \sin^2 x = \tfrac{1}{2}(1 - \cos 2x) \tag{10}$$

$$\begin{cases} \sin x \sin y = \tfrac{1}{2}[-\cos(x+y) + \cos(x-y)] \\ \cos x \cos y = \tfrac{1}{2}[\cos(x+y) + \cos(x-y)] \\ \sin x \cos y = \tfrac{1}{2}[\sin(x+y) + \sin(x-y)] \end{cases} \tag{11}$$

$$\begin{cases} \sin u + \sin v = 2 \sin \dfrac{u+v}{2} \cos \dfrac{u-v}{2} \\ \cos u + \cos v = 2 \cos \dfrac{u+v}{2} \cos \dfrac{u-v}{2} \\ \cos v - \cos u = 2 \sin \dfrac{u+v}{2} \sin \dfrac{u-v}{2} \end{cases} \tag{12}$$

$$A \cos x + B \sin x = \sqrt{A^2 + B^2} \cos(x \pm \delta), \quad \tan \delta = \frac{\sin \delta}{\cos \delta} = \mp \frac{B}{A} \tag{13}$$

$$A \cos x + B \sin x = \sqrt{A^2 + B^2} \sin(x \pm \delta), \quad \tan \delta = \frac{\sin \delta}{\cos \delta} = \pm \frac{A}{B} \tag{14}$$

Tangent, cotangent, secant, cosecant (Figs. 523, 524, p. A68)

$$\tan x = \frac{\sin x}{\cos x}, \quad \cot x = \frac{\cos x}{\sin x}, \quad \sec x = \frac{1}{\cos x}, \quad \csc x = \frac{1}{\sin x} \tag{15}$$

$$\tan(x+y) = \frac{\tan x + \tan y}{1 - \tan x \tan y}, \quad \tan(x-y) = \frac{\tan x - \tan y}{1 + \tan x \tan y} \tag{16}$$

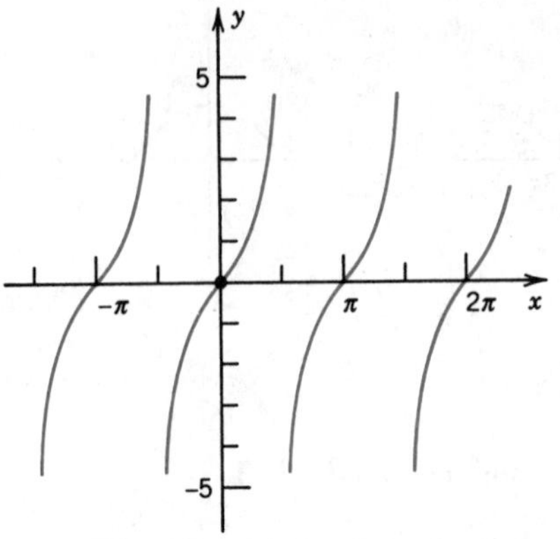

Fig. 523. tan *x*

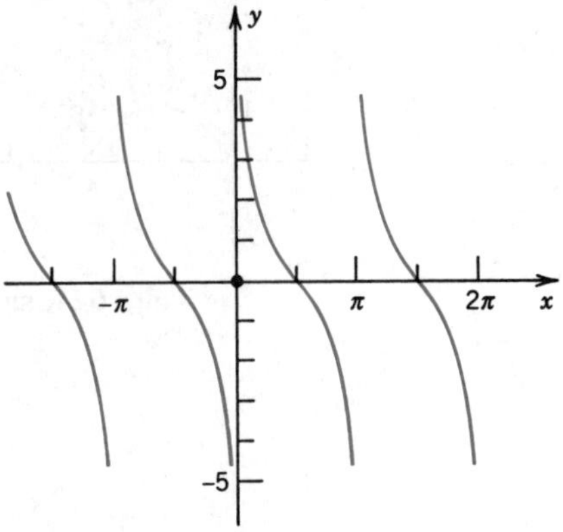

Fig. 524. cot *x*

Hyperbolic functions (hyperbolic sine sinh x, etc.; Figs. 525, 526)

$$\sinh x = \tfrac{1}{2}(e^x - e^{-x}), \qquad \cosh x = \tfrac{1}{2}(e^x + e^{-x}) \tag{17}$$

$$\tanh x = \frac{\sinh x}{\cosh x}, \qquad \coth x = \frac{\cosh x}{\sinh x} \tag{18}$$

$$\cosh x + \sinh x = e^x, \qquad \cosh x - \sinh x = e^{-x} \tag{19}$$

$$\cosh^2 x - \sinh^2 x = 1 \tag{20}$$

$$\sinh^2 x = \tfrac{1}{2}(\cosh 2x - 1), \qquad \cosh^2 x = \tfrac{1}{2}(\cosh 2x + 1) \tag{21}$$

$$\begin{cases} \sinh(x \pm y) = \sinh x \cosh y \pm \cosh x \sinh y \\ \cosh(x \pm y) = \cosh x \cosh y \pm \sinh x \sinh y \end{cases} \tag{22}$$

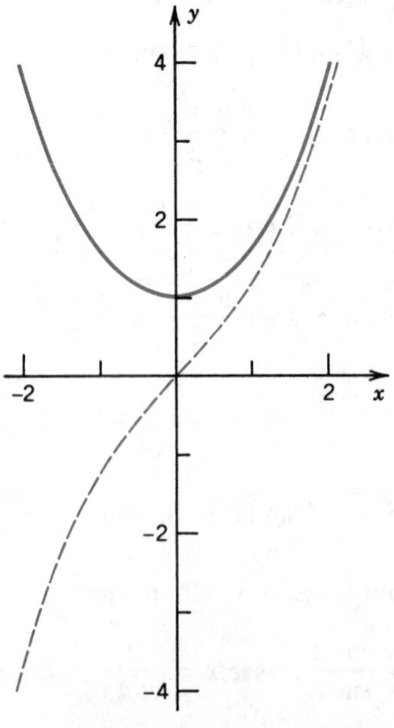

Fig. 525. sinh *x* (dashed) and cosh *x*

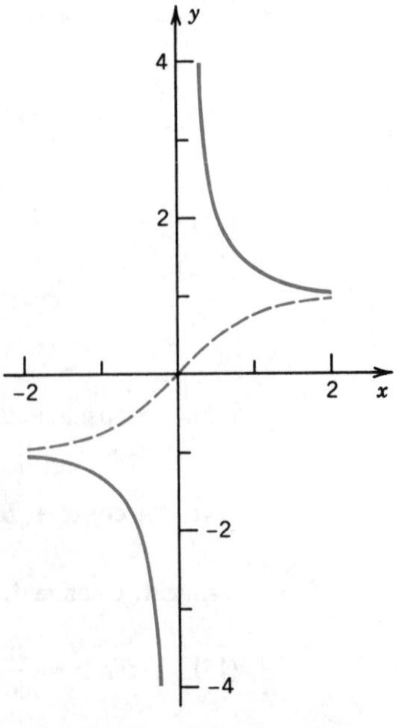

Fig. 526. tanh *x* (dashed) and coth *x*

(23) $$\tanh (x \pm y) = \frac{\tanh x \pm \tanh y}{1 \pm \tanh x \tanh y}$$

Gamma function (Fig. 527 and Table A2 in Appendix 5). The gamma function $\Gamma(\alpha)$ is defined by the integral

(24) $$\Gamma(\alpha) = \int_0^\infty e^{-t} t^{\alpha-1}\, dt \qquad (\alpha > 0)$$

which is meaningful only if $\alpha > 0$ (or, if we consider complex α, for those α whose real part is positive). Integration by parts gives the important *functional relation of the gamma function,*

(25) $$\Gamma(\alpha + 1) = \alpha\Gamma(\alpha).$$

From (24) we readily have $\Gamma(1) = 1$; hence if α is a positive integer, say k, then by repeated application of (25) we obtain

(26) $$\Gamma(k + 1) = k! \qquad (k = 0, 1, \cdots).$$

This shows that *the gamma function can be regarded as a generalization of the elementary factorial function.* [Sometimes the notation $(\alpha - 1)!$ is used for $\Gamma(\alpha)$, even for noninteger values of α, and the gamma function is also known as the **factorial function.**]

By repeated application of (25) we obtain

$$\Gamma(\alpha) = \frac{\Gamma(\alpha + 1)}{\alpha} = \frac{\Gamma(\alpha + 2)}{\alpha(\alpha + 1)} = \cdots = \frac{\Gamma(\alpha + k + 1)}{\alpha(\alpha + 1)(\alpha + 2)\cdots(\alpha + k)}$$

and we may use this relation

(27) $$\Gamma(\alpha) = \frac{\Gamma(\alpha + k + 1)}{\alpha(\alpha + 1)\cdots(\alpha + k)} \qquad (\alpha \neq 0, -1, -2, \cdots)$$

for defining the gamma function for negative α ($\neq -1, -2, \cdots$), choosing for k the

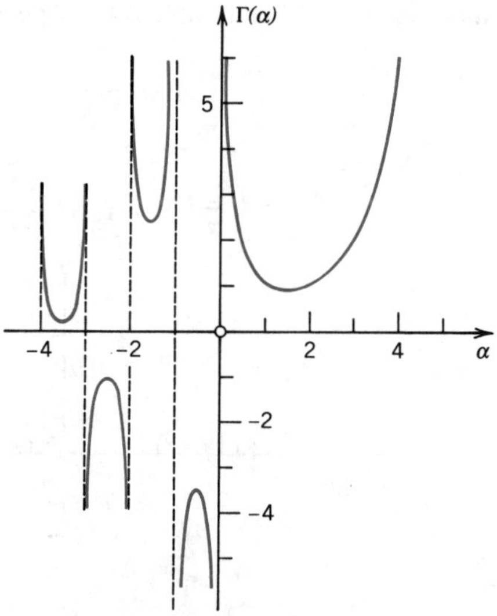

Fig. 527. Gamma function

smallest integer such that $\alpha + k + 1 > 0$. *Together with* (24), *this then gives a definition of* $\Gamma(\alpha)$ *for all* α *not equal to zero or a negative integer* (Fig. 527).

It can be shown that the gamma function may also be represented as the limit of a product, namely, by the formula

(28) $$\Gamma(\alpha) = \lim_{n\to\infty} \frac{n!\, n^{\alpha}}{\alpha(\alpha+1)(\alpha+2)\cdots(\alpha+n)} \qquad (\alpha \neq 0, -1, \cdots).$$

From (27) or (28) we see that, for complex α, the gamma function $\Gamma(\alpha)$ is a meromorphic function that has simple poles at $\alpha = 0, -1, -2, \cdots$.

An approximation of the gamma function for large positive α is given by the **Stirling formula**

(29) $$\Gamma(\alpha + 1) \approx \sqrt{2\pi\alpha}\left(\frac{\alpha}{e}\right)^{\alpha}$$

where e is the base of the natural logarithm. We finally mention the special value

(30) $$\Gamma(\tfrac{1}{2}) = \sqrt{\pi}.$$

Incomplete gamma functions

(31) $$P(\alpha, x) = \int_0^x e^{-t} t^{\alpha-1}\, dt, \qquad Q(\alpha, x) = \int_x^{\infty} e^{-t} t^{\alpha-1}\, dt \qquad (\alpha > 0)$$

(32) $$\Gamma(\alpha) = P(\alpha, x) + Q(\alpha, x)$$

Beta function

(33) $$B(x, y) = \int_0^1 t^{x-1}(1-t)^{y-1}\, dt \qquad (x > 0,\ y > 0)$$

Representation in terms of gamma functions:

(34) $$B(x, y) = \frac{\Gamma(x)\Gamma(y)}{\Gamma(x+y)}$$

Error function (Fig. 528 and Table A4 in Appendix 5)

(35) $$\operatorname{erf} x = \frac{2}{\sqrt{\pi}} \int_0^x e^{-t^2}\, dt$$

(36) $$\operatorname{erf} x = \frac{2}{\sqrt{\pi}}\left(x - \frac{x^3}{1!3} + \frac{x^5}{2!5} - \frac{x^7}{3!7} + - \cdots\right)$$

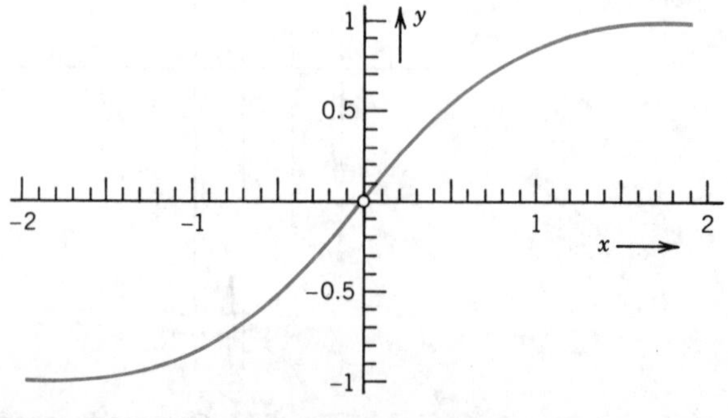

Fig. 528. Error function

erf (∞) = 1, *complementary error function*

(37)
$$\operatorname{erfc} x = 1 - \operatorname{erf} x = \frac{2}{\sqrt{\pi}} \int_x^\infty e^{-t^2}\, dt$$

Fresnel integrals[1] (Fig. 529)

(38)
$$C(x) = \int_0^x \cos(t^2)\, dt, \qquad S(x) = \int_0^x \sin(t^2)\, dt$$

$C(\infty) = \sqrt{\pi/8}$, $S(\infty) = \sqrt{\pi/8}$, *complementary functions*

(39)
$$c(x) = \sqrt{\frac{\pi}{8}} - C(x) = \int_x^\infty \cos(t^2)\, dt$$
$$s(x) = \sqrt{\frac{\pi}{8}} - S(x) = \int_x^\infty \sin(t^2)\, dt$$

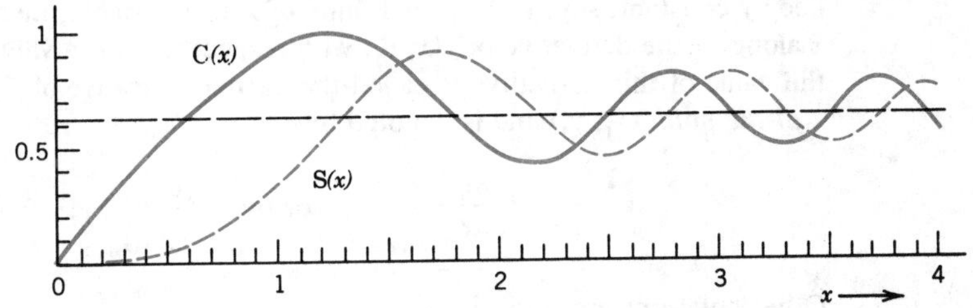

Fig. 529. Fresnel integrals

Sine integral (Fig. 530 and Table A4 in Appendix 5)

(40)
$$\operatorname{Si}(x) = \int_0^x \frac{\sin t}{t}\, dt$$

$\operatorname{Si}(\infty) = \pi/2$, *complementary function*

(41)
$$\operatorname{si}(x) = \frac{\pi}{2} - \operatorname{Si}(x) = \int_x^\infty \frac{\sin t}{t}\, dt$$

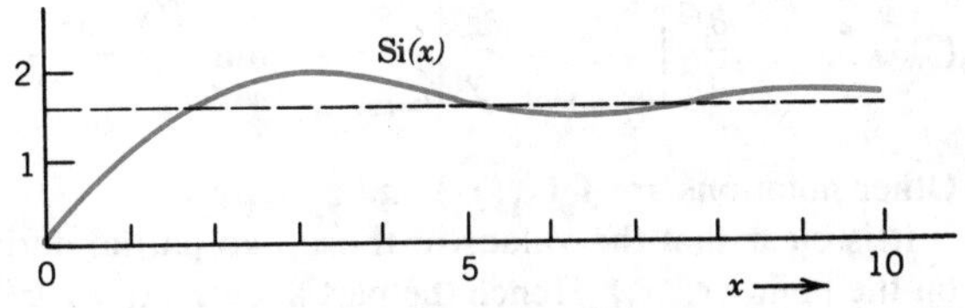

Fig. 530. Sine integral

Cosine integral (Table A4 in Appendix 5)

(42)
$$\operatorname{ci}(x) = \int_x^\infty \frac{\cos t}{t}\, dt \qquad (x > 0)$$

[1]AUGUSTIN FRESNEL (1788—1827), French physicist and mathematician. For tables see Ref. [1].

Exponential integral

(43) $$\mathrm{Ei}(x) = \int_x^\infty \frac{e^{-t}}{t}\,dt \qquad (x > 0)$$

Logarithmic integral

(44) $$\mathrm{li}(x) = \int_0^x \frac{dt}{\ln t}$$

A3.2 Partial Derivatives

For differentiation formulas, see inside of front cover

Let $z = f(x, y)$ be a real function of two independent real variables, x and y. If we keep y constant, say, $y = y_1$, and think of x as a variable, then $f(x, y_1)$ depends on x alone. If the derivative of $f(x, y_1)$ with respect to x for a value $x = x_1$ exists, then the value of this derivative is called the **partial derivative** of $f(x, y)$ *with respect to x at the point* (x_1, y_1) and is denoted by

$$\left.\frac{\partial f}{\partial x}\right|_{(x_1, y_1)} \quad \text{or by} \quad \left.\frac{\partial z}{\partial x}\right|_{(x_1, y_1)}.$$

Other notations are

$$f_x(x_1, y_1) \quad \text{and} \quad z_x(x_1, y_1);$$

these may be used when subscripts are not used for another purpose and there is no danger of confusion.

We thus have, by the definition of the derivative,

(1) $$\left.\frac{\partial f}{\partial x}\right|_{(x_1, y_1)} = \lim_{\Delta x \to 0} \frac{f(x_1 + \Delta x, y_1) - f(x_1, y_1)}{\Delta x}.$$

The partial derivative of $z = f(x, y)$ with respect to y is defined similarly; we now keep x constant, say, equal to x_1, and differentiate $f(x_1, y)$ with respect to y. Thus

(2) $$\left.\frac{\partial f}{\partial y}\right|_{(x_1, y_1)} = \left.\frac{\partial z}{\partial y}\right|_{(x_1, y_1)} = \lim_{\Delta y \to 0} \frac{f(x_1, y_1 + \Delta y) - f(x_1, y_1)}{\Delta y}.$$

Other notations are $f_y(x_1, y_1)$ and $z_y(x_1, y_1)$.

It is clear that the values of those two partial derivatives will in general depend on the point (x_1, y_1). Hence the partial derivatives $\partial z/\partial x$ and $\partial z/\partial y$ at a variable point (x, y) are functions of x and y. The function $\partial z/\partial x$ is obtained as in ordinary calculus by differentiating $z = f(x, y)$ with respect to x, *treating y as a constant*, and $\partial z/\partial y$ is obtained by differentiating z with respect to y, *treating x as a constant*.

EXAMPLE 1 Let $z = f(x, y) = x^2y + x \sin y$. Then

$$\frac{\partial f}{\partial x} = 2xy + \sin y, \qquad \frac{\partial f}{\partial y} = x^2 + x \cos y.$$ ■

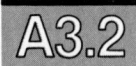

The partial derivatives $\partial z/\partial x$ and $\partial z/\partial y$ of a function $z = f(x, y)$ have a very simple **geometric interpretation.** The function $z = f(x, y)$ can be represented by a surface in space. The equation $y = y_1$ then represents a vertical plane intersecting the surface in a curve, and the partial derivative $\partial z/\partial x$ at a point (x_1, y_1) is the slope of the tangent (that is, $\tan \alpha$ where α is the angle shown in Fig. 531) to the curve. Similarly, the partial derivative $\partial z/\partial y$ at (x_1, y_1) is the slope of the tangent to the curve $x = x_1$ on the surface $z = f(x, y)$ at (x_1, y_1).

The partial derivatives $\partial z/\partial x$ and $\partial z/\partial y$ are called *first partial derivatives* or *partial derivatives of first order*. By differentiating these derivatives once more, we obtain the four *second partial derivatives* (or *partial derivatives of second order*)[2]

$$
\begin{aligned}
\frac{\partial^2 f}{\partial x^2} &= \frac{\partial}{\partial x}\left(\frac{\partial f}{\partial x}\right) = f_{xx} \\
\frac{\partial^2 f}{\partial x\, \partial y} &= \frac{\partial}{\partial x}\left(\frac{\partial f}{\partial y}\right) = f_{yx} \\
\frac{\partial^2 f}{\partial y\, \partial x} &= \frac{\partial}{\partial y}\left(\frac{\partial f}{\partial x}\right) = f_{xy} \\
\frac{\partial^2 f}{\partial y^2} &= \frac{\partial}{\partial y}\left(\frac{\partial f}{\partial y}\right) = f_{yy}.
\end{aligned}
\tag{3}
$$

It can be shown that if all the derivatives concerned are continuous, then the two mixed partial derivatives are equal, so that then the order of differentiation does not matter (see Ref. [5] in Appendix 1), that is,

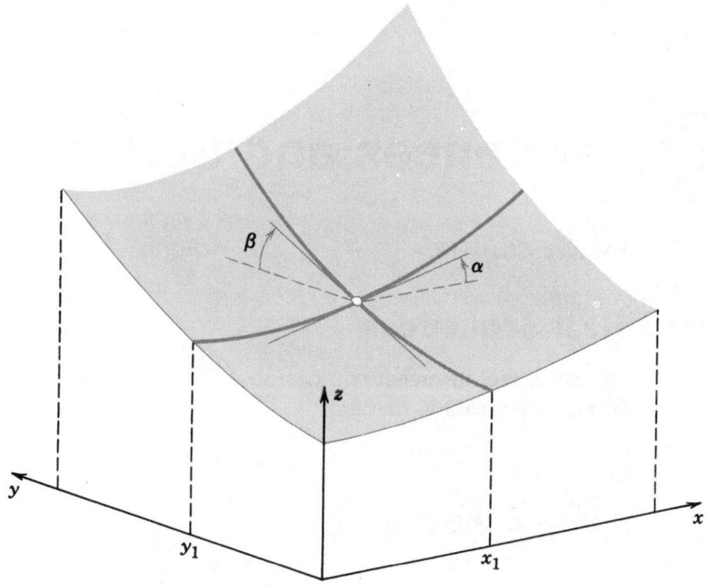

Fig. 531. Geometrical interpretation of first partial derivatives

[2]Caution! In the subscript notation the subscripts are written in the order in which we differentiate, whereas in the "∂" notation the order is opposite.

$$\frac{\partial^2 z}{\partial x\, \partial y} = \frac{\partial^2 z}{\partial y\, \partial x}. \tag{4}$$

EXAMPLE 2 For the function in Example 1,

$$f_{xx} = 2y, \quad f_{xy} = 2x + \cos y = f_{yx}, \quad f_{yy} = -x \sin y.$$ ∎

By differentiating the second partial derivatives again with respect to x and y, respectively, we obtain the *third partial derivatives* or *partial derivatives of the third order* of f, etc.

If we consider a function $f(x, y, z)$ of **three independent variables,** then we have the three first partial derivatives $f_x(x, y, z)$, $f_y(x, y, z)$, and $f_z(x, y, z)$. Here f_x is obtained by differentiating f with respect to x, *treating both y and z as constants.* Thus, in analogy to (1), we now have

$$\left.\frac{\partial f}{\partial x}\right|_{(x_1, y_1, z_1)} = \lim_{\Delta x \to 0} \frac{f(x_1 + \Delta x, y_1, z_1) - f(x_1, y_1, z_1)}{\Delta x},$$

etc. By differentiating f_x, f_y, f_z again in this fashion we obtain the second partial derivatives of f, etc.

EXAMPLE 3 Let $f(x, y, z) = x^2 + y^2 + z^2 + xy\, e^z$. Then

$$f_x = 2x + y\, e^z, \qquad f_y = 2y + x\, e^z, \qquad f_z = 2z + xy\, e^z,$$

$$f_{xx} = 2, \qquad f_{xy} = f_{yx} = e^z, \qquad f_{xz} = f_{zx} = y\, e^z,$$

$$f_{yy} = 2, \qquad f_{yz} = f_{zy} = x\, e^z, \qquad f_{zz} = 2 + xy\, e^z.$$ ∎

A3.3 Sequences and Series

See also Chap. 14.

Monotone Real Sequences

We call a real sequence $x_1, x_2, \cdots, x_n, \cdots$ a **monotone sequence** if it is either **monotone increasing,** that is,

$$x_1 \leqq x_2 \leqq x_3 \leqq \cdots$$

or **monotone decreasing,** that is,

$$x_1 \geqq x_2 \geqq x_3 \geqq \cdots.$$

We call $x_1, x_2, \cdots$ a **bounded sequence** if there is a positive constant K such that $|x_n| < K$ for all n.

Theorem 1 *If a real sequence is bounded and monotone, it converges.*

Proof. Let $x_1, x_2, \cdots$ be a bounded monotone increasing sequence. Then its terms are smaller than some number B and, since $x_1 \leqq x_n$ for all n, they lie in the interval

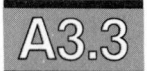

$x_1 \leqq x_n \leqq B$, which will be denoted by I_0. We bisect I_0; that is, we subdivide it into two parts of equal length. If the right half (together with its endpoints) contains terms of the sequence, we denote it by I_1. If it does not contain terms of the sequence, then the left half of I_0 (together with its endpoints) is called I_1. This is the first step.

In the second step we bisect I_1, select one half by the same rule, and call it I_2, and so on (see Fig. 532).

In this way we obtain shorter and shorter intervals $I_0, I_1, I_2, \cdots$ with the following properties. Each I_m contains all I_n for $n > m$. No term of the sequence lies to the right of I_m, and, since the sequence is monotone increasing, all x_n with n greater than some number N lie in I_m; of course, N will depend on m, in general. The lengths of the I_m approach zero as m approaches infinity. Hence there is precisely one number, call it L, that lies in all those intervals,[3] and we may now easily prove that the sequence is convergent with the limit L.

In fact, given an $\epsilon > 0$, we choose an m such that the length of I_m is less than ϵ. Then L and all the x_n with $n > N(m)$ lie in I_m, and, therefore, $|x_n - L| < \epsilon$ for all those n. This completes the proof for an increasing sequence. For a decreasing sequence the proof is the same, except for a suitable interchange of "left" and "right" in the construction of those intervals. ∎

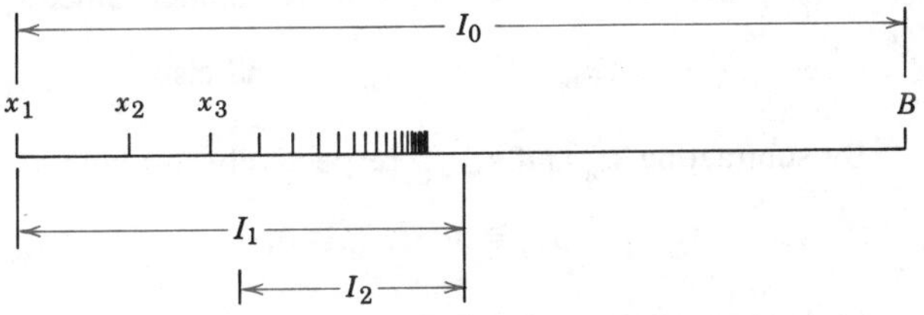

Fig. 532. Proof of Theorem 1

Real Series

Theorem 2 **Leibniz test for real series**

Let $x_1, x_2, \cdots$ be real and monotone decreasing to zero, that is,

$$(1) \qquad \text{(a) } x_1 \geqq x_2 \geqq x_3 \geqq \cdots, \qquad \text{(b) } \lim_{m\to\infty} x_m = 0.$$

Then the series with terms of alternating signs

$$x_1 - x_2 + x_3 - x_4 + - \cdots$$

converges, and for the remainder R_n after the nth term we have the estimate

$$(2) \qquad |R_n| \leqq x_{n+1}.$$

---

[3]This statement seems to be obvious, but actually it is not; it may be regarded as an axiom of the real number system in the following form. Let $J_1, J_2, \cdots$ be closed intervals such that each J_m contains all J_n with $n > m$, and the lengths of the J_m approach zero as m approaches infinity. Then there is precisely one real number that is contained in all those intervals. This is the so-called **Cantor–Dedekind axiom,** named after the German mathematicians GEORG CANTOR (1845—1918), the creator of set theory, and RICHARD DEDEKIND (1831—1916), known for his fundamental work in number theory. For further details see Ref. [2] in Appendix 1. (An interval I is said to be **closed** if its two endpoints are regarded as points belonging to I. It is said to be **open** if the endpoints are not regarded as points of I.)

Proof. Let s_n be the nth partial sum of the series. Then, because of (1a),

$$s_1 = x_1, \qquad s_2 = x_1 - x_2 \leqq s_1,$$

$$s_3 = s_2 + x_3 \geqq s_2, \qquad s_3 = s_1 - (x_2 - x_3) \leqq s_1,$$

so that $s_2 \leqq s_3 \leqq s_1$. Proceeding in this fashion, we conclude that (Fig. 533)

$$s_1 \geqq s_3 \geqq s_5 \geqq \cdots \geqq s_6 \geqq s_4 \geqq s_2 \tag{3}$$

which shows that the odd partial sums form a bounded monotone sequence, and so do the even partial sums. Hence, by Theorem 1, both sequences converge, say,

$$\lim_{n\to\infty} s_{2n+1} = s, \qquad \lim_{n\to\infty} s_{2n} = s^*.$$

Now, since $s_{2n+1} - s_{2n} = x_{2n+1}$, we readily see that (1b) implies

$$s - s^* = \lim_{n\to\infty} s_{2n+1} - \lim_{n\to\infty} s_{2n} = \lim_{n\to\infty} (s_{2n+1} - s_{2n}) = \lim_{n\to\infty} x_{2n+1} = 0.$$

Hence $s^* = s$, and the series converges with the sum s.

We prove the estimate (2) for the remainder. Since $s_n \to s$, it follows from (3) that

$$s_{2n+1} \geqq s \geqq s_{2n} \qquad \text{and also} \qquad s_{2n-1} \geqq s \geqq s_{2n}.$$

By subtracting s_{2n} and s_{2n-1}, respectively, we obtain

$$s_{2n+1} - s_{2n} \geqq s - s_{2n} \geqq 0, \qquad 0 \geqq s - s_{2n-1} \geqq s_{2n} - s_{2n-1}.$$

In these inequalities, the first expression is equal to x_{2n+1}, the last is equal to $-x_{2n}$, and the expressions between the inequality signs are the remainders R_{2n} and R_{2n-1}. Thus the inequalities may be written

$$x_{2n+1} \geqq R_{2n} \geqq 0, \qquad 0 \geqq R_{2n-1} \geqq -x_{2n}$$

and we see that they imply (2). This completes the proof. ∎

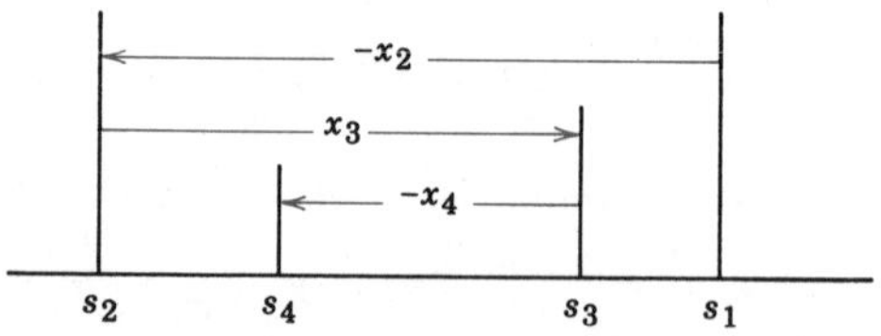

Fig. 533. Proof of the Leibniz test

Appendix 4

Additional Proofs

SECTION 2.7, page 94

Proof of Theorem 1 (Uniqueness)[1]

Assuming that the problem consisting of the differential equation

$$y'' + p(x)y' + q(x)y = 0 \tag{1}$$

and the two initial conditions

$$y(x_0) = K_0, \qquad y'(x_0) = K_1 \tag{3}$$

has two solutions $y_1(x)$ and $y_2(x)$ on the interval I in the theorem, we show that their difference

$$y(x) = y_1(x) - y_2(x)$$

is identically zero on I; then $y_1 \equiv y_2$ on I, which implies uniqueness.

Since (1) is homogeneous and linear, y is a solution of that equation on I, and since y_1 and y_2 satisfy the same initial conditions, y satisfies the conditions

$$y(x_0) = 0, \qquad y'(x_0) = 0. \tag{11}$$

We consider the function

$$z(x) = y(x)^2 + y'(x)^2$$

and its derivative

$$z' = 2yy' + 2y'y''.$$

From the differential equation we have

$$y'' = -py' - qy.$$

By substituting this in the expression for z' we obtain

$$z' = 2yy' - 2py'^2 - 2qyy'. \tag{12}$$

Now, since y and y' are real,

$$(y \pm y')^2 = y^2 \pm 2yy' + y'^2 \geqq 0.$$

---

[1]This proof was suggested by my colleague, Prof. A. D. Ziebur. In this proof we use formula numbers that have not yet been used in Sec. 2.7.

From this we immediately obtain the two inequalities

(13) (a) $2yy' \leqq y^2 + y'^2 = z,$ (b) $-2yy' \leqq y^2 + y'^2 = z.$

From (13b) we have $2yy' \geqq -z$. Together, $|2yy'| \leqq z$. For the last term in (12) we now obtain

$$-2qyy' \leqq |-2qyy'| = |q||2yy'| \leqq |q|z.$$

Using this result as well as $-p \leqq |p|$ and applying (13a) to the term $2yy'$ in (12), we find

$$z' \leqq z + 2|p|y'^2 + |q|z.$$

Since $y'^2 \leqq y^2 + y'^2 = z$, we obtain from this

$$z' \leqq (1 + 2|p| + |q|)z$$

or, denoting the function in parentheses by h,

(14a) $$z' \leqq hz \qquad \text{for all } x \text{ on } I.$$

Similarly, from (12) and (13) it follows that

(14b) $$\begin{aligned} -z' &= -2yy' + 2py'^2 + 2qyy' \\ &\leqq z + 2|p|z + |q|z = hz. \end{aligned}$$

The inequalities (14a) and (14b) are equivalent to the inequalities

(15) $$z' - hz \leqq 0, \qquad z' + hz \geqq 0.$$

Integrating factors for the two expressions on the left are

$$F_1 = e^{-\int h(x)\,dx} \qquad \text{and} \qquad F_2 = e^{\int h(x)\,dx}.$$

The integrals in the exponents exist because h is continuous. Since F_1 and F_2 are positive, we thus have from (15)

$$F_1(z' - hz) = (F_1 z)' \leqq 0 \qquad \text{and} \qquad F_2(z' + hz) = (F_2 z)' \geqq 0,$$

which means that $F_1 z$ is nonincreasing and $F_2 z$ is nondecreasing on I. Since $z(x_0) = 0$ by (11), we thus obtain when $x \leqq x_0$

$$F_1 z \geqq (F_1 z)_{x_0} = 0, \qquad F_2 z \leqq (F_2 z)_{x_0} = 0$$

and similarly, when $x \geqq x_0$,

$$F_1 z \leqq 0, \qquad F_2 z \geqq 0.$$

Dividing by F_1 and F_2 and noting that these functions are positive, we altogether have

$$z \leqq 0, \qquad z \geqq 0 \qquad \text{for all } x \text{ on } I.$$

This implies $z = y^2 + y'^2 \equiv 0$ on I. Hence $y \equiv 0$ or $y_1 \equiv y_2$ on I. ∎

SECTION 5.4, page 218

Proof of Theorem 2 (***Frobenius method. Basis of solutions***)
The formula numbers in this proof are the same as in the text of Sec. 5.4. An additional formula not appearing in Sec. 5.4 will be called (A) (see below).

The differential equation in Theorem 2 is

$$y'' + \frac{b(x)}{x} y' + \frac{c(x)}{x^2} y = 0, \tag{1}$$

where $b(x)$ and $c(x)$ are analytic functions. We can write it

$$x^2y'' + xb(x)y' + c(x)y = 0. \tag{1'}$$

The indicial equation of (1) is

$$r(r - 1) + b_0 r + c_0 = 0. \tag{4}$$

The roots r_1, r_2 of this quadratic equation determine the general form of a basis of solutions of (1), and there are three possible cases as follows.

Case 1 (Roots not differing by an integer). A first solution of (1) of the form

$$y_1(x) = x^{r_1}(a_0 + a_1x + a_2x^2 + \cdots) \tag{5}$$

can be determined as in the power series method. For a proof that in this case, equation (1) has a second independent solution of the form

$$y_2(x) = x^{r_2}(A_0 + A_1x + A_2x^2 + \cdots), \tag{6}$$

see Ref. [A6] listed in Appendix 1.

Case 2 (Double root). The indicial equation (4) has a double root r if and only if $(b_0 - 1)^2 - 4c_0 = 0$, and then $r = \frac{1}{2}(1 - b_0)$. A first solution

$$y_1(x) = x^r(a_0 + a_1x + a_2x^2 + \cdots), \qquad r = \tfrac{1}{2}(1 - b_0), \tag{7}$$

can be determined as in Case 1. We show that a second independent solution is of the form

$$y_2(x) = y_1(x) \ln x + x^r(A_1x + A_2x^2 + \cdots) \qquad (x > 0). \tag{8}$$

We use the method of reduction of order (see Sec. 2.7), that is, we determine $u(x)$ such that $y_2(x) = u(x)y_1(x)$ is a solution of (1). By inserting this and the derivatives

$$y_2' = u'y_1 + uy_1', \qquad y_2'' = u''y_1 + 2u'y_1' + uy_1''$$

into the differential equation (1′) we obtain

$$x^2(u''y_1 + 2u'y_1' + uy_1'') + xb(u'y_1 + uy_1') + cuy_1 = 0.$$

Since y_1 is a solution of (1′), the sum of the terms involving u is zero, and this equation reduces to

$$x^2 y_1 u'' + 2x^2 y_1' u' + xby_1 u' = 0.$$

By dividing by $x^2 y_1$ and inserting the power series for b we obtain

$$u'' + \left(2\frac{y_1'}{y_1} + \frac{b_0}{x} + \cdots\right) u' = 0.$$

Here and in the following the dots designate terms that are constant or involve positive powers of x. Now from (7) it follows that

$$\frac{y_1'}{y_1} = \frac{x^{r-1}[ra_0 + (r+1)a_1 x + \cdots]}{x^r[a_0 + a_1 x + \cdots]}$$

$$= \frac{1}{x}\left(\frac{ra_0 + (r+1)a_1 x + \cdots}{a_0 + a_1 x + \cdots}\right) = \frac{r}{x} + \cdots.$$

Hence the previous equation can be written

(A) $$u'' + \left(\frac{2r + b_0}{x} + \cdots\right) u' = 0.$$

Since $r = (1 - b_0)/2$, the term $(2r + b_0)/x$ equals $1/x$, and by dividing by u' we thus have

$$\frac{u''}{u'} = -\frac{1}{x} + \cdots.$$

By integration we obtain $\ln u' = -\ln x + \cdots$, hence $u' = (1/x)e^{(\cdots)}$. Expanding the exponential function in powers of x and integrating once more, we see that u is of the form

$$u = \ln x + k_1 x + k_2 x^2 + \cdots.$$

Inserting this into $y_2 = uy_1$, we obtain for y_2 a representation of the form (8).

Case 3 (Roots differing by an integer). We write $r_1 = r$ and $r_2 = r - p$ where p is a ***positive*** integer. A first solution

(9) $$y_1(x) = x^{r_1}(a_0 + a_1 x + a_2 x^2 + \cdots)$$

can be determined as in Cases 1 and 2. We show that a second independent solution is of the form

(10) $$y_2(x) = ky_1(x) \ln x + x^{r_2}(A_0 + A_1 x + A_2 x^2 + \cdots)$$

where we may have $k \neq 0$ or $k = 0$. As in Case 2 we set $y_2 = uy_1$. The first steps are literally as in Case 2 and give equation (A),

$$u'' + \left(\frac{2r + b_0}{x} + \cdots\right) u' = 0.$$

Now by elementary algebra, the coefficient $b_0 - 1$ of r in (4) equals minus the sum of the roots,

$$b_0 - 1 = -(r_1 + r_2) = -(r + r - p) = -2r + p.$$

Hence $2r + b_0 = p + 1$, and division by u' gives

$$\frac{u''}{u'} = -\left(\frac{p+1}{x} + \cdots\right).$$

The further steps are as in Case 2. Integrating, we find

$$\ln u' = -(p+1)\ln x + \cdots, \qquad \text{thus} \qquad u' = x^{-(p+1)}e^{(\cdots)}$$

where dots stand for some series of nonnegative integer powers of x. By expanding the exponential function as before we obtain a series of the form

$$u' = \frac{1}{x^{p+1}} + \frac{k_1}{x^p} + \cdots + \frac{k_{p-1}}{x^2} + \frac{k_p}{x} + k_{p+1} + k_{p+2}x + \cdots.$$

We inegrate once more. Writing the resulting logarithmic term first, we get

$$u = k_p \ln x + \left(-\frac{1}{px^p} - \cdots - \frac{k_{p-1}}{x} + k_{p+1}x + \cdots\right).$$

Hence, by (9) we get for $y_2 = uy_1$ the formula

$$y_2 = k_p y_1 \ln x + x^{r_1 - p}\left(-\frac{1}{p} - \cdots - k_{p-1}x^{p-1} + \cdots\right)(a_0 + a_1 x + \cdots).$$

But this is of the form (10) with $k = k_p$ since $r_1 - p = r_2$ and the product of the two series involves nonnegative integer powers of x only. ∎

SECTION 5.8, page 243

Theorem (Reality of eigenvalues)

If p, q, r, and r' in the Sturm–Liouville equation (1) of Sec. 5.8 are real-valued and continuous on the interval $a \leqq x \leqq b$ and $p(x) > 0$ throughout that interval (or $p(x) < 0$ throughout that interval), then all the eigenvalues of the Sturm–Liouville problem (1), (2), Sec. 5.8, are real.

Proof. Let $\lambda = \alpha + i\beta$ be an eigenvalue of the problem and let

$$y(x) = u(x) + iv(x)$$

be a corresponding eigenfunction; here α, β, u, and v are real. Substituting this into (1), Sec. 5.8, we have

$$(ru' + irv')' + (q + \alpha p + i\beta p)(u + iv) = 0.$$

This complex equation is equivalent to the following pair of equations for the real and imaginary parts:

$$(ru')' + (q + \alpha p)u - \beta p v = 0,$$
$$(rv')' + (q + \alpha p)v + \beta p u = 0.$$

Multiplying the first equation by v, the second by $-u$ and adding, we get

$$-\beta(u^2 + v^2)p = u(rv')' - v(ru')'$$
$$= [(rv')u - (ru')v]'.$$

The expression in brackets is continuous on $a \leqq x \leqq b$, for reasons similar to those in the proof of Theorem 1, Sec. 5.8. Integrating over x from a to b, we thus obtain

$$-\beta \int_a^b (u^2 + v^2)p\,dx = \Big[r(uv' - u'v)\Big]_a^b .$$

Because of the boundary conditions the right side is zero; this is as in that proof. Since y is an eigenfunction, $u^2 + v^2 \not\equiv 0$. Since y and p are continuous and $p > 0$ (or $p < 0$) on the interval $a \leqq x \leqq b$, the integral on the left is not zero. Hence, $\beta = 0$, which means that $\lambda = \alpha$ is real. This completes the proof. ∎

SECTION 7.8, page 373

Proof that the definition of a determinant in Sec. 7.8 is unambiguous

We show that the definition of a determinant

$$D = \det \mathbf{A} = \begin{vmatrix} a_{11} & a_{12} & \cdots & a_{1n} \\ a_{21} & a_{22} & \cdots & a_{2n} \\ \cdot & \cdot & \cdots & \cdot \\ \cdot & \cdot & \cdots & \cdot \\ a_{n1} & a_{n2} & \cdots & a_{nn} \end{vmatrix} \tag{1}$$

as given in Sec. 7.8 is unambiguous, that is, yields the same value of D no matter which rows or columns we choose. (Here we shall use formula numbers not yet used in Sec. 7.8.)

We shall prove first that *the same value is obtained no matter which row is chosen.*

The proof is by induction. The statement is true for a second-order determinant (see Example 2). Assuming that it is true for a determinant of order $n - 1$, we prove that it is true for a determinant D of order n.

For this purpose we expand D, given in the definition, in terms of each of two arbitrary rows, say the ith and the jth, and compare the results. Without loss of generality let us assume $i < j$.

First expansion. We expand D by the ith row. A typical term in this expansion is

$$a_{ik}C_{ik} = a_{ik} \cdot (-1)^{i+k} M_{ik}. \tag{14}$$

The minor M_{ik} of a_{ik} in D is an $(n - 1)$th order determinant. By the induction hypothesis we may expand it by any row. We expand it by the row corresponding to the jth row of D. This row contains the entries a_{jl} $(l \neq k)$. It is the $(j - 1)$th row of M_{ik}, because M_{ik} does not contain entries of the ith row of D, and $i < j$. We have to distinguish between two cases as follows.

Case I. If $l < k$, then the entry a_{jl} belongs to the lth column of M_{ik} (see Fig. 534). Hence the term involving a_{jl} in this expansion is

$$a_{jl} \cdot (\text{cofactor of } a_{jl} \text{ in } M_{ik}) = a_{jl} \cdot (-1)^{(j-1)+l} M_{ikjl} \tag{15}$$

where M_{ikjl} is the minor of a_{jl} in M_{ik}. Since this minor is obtained from M_{ik} by deleting the row and column of a_{jl}, it is obtained from D by deleting the ith and jth rows and the kth and lth columns of D. We insert the expansions of the M_{ik} into that of D. Then it follows from (14) and (15) that the terms of the resulting representation of D are of the form

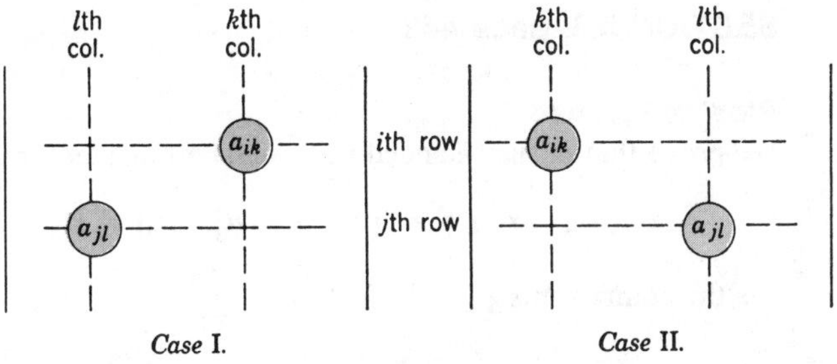

Fig. 534. Cases I and II of the two expansions of D

(16a) $$a_{ik}a_{jl} \cdot (-1)^b M_{ikjl} \qquad (l < k)$$

where

$$b = i + k + j + l - 1.$$

Case II. If $l > k$, the only difference is that then a_{jl} belongs to the $(l - 1)$th column of M_{ik}, because M_{ik} does not contain entries of the kth column of D, and $k < l$. This causes an additional minus sign in (15), and, instead of (16a), we therefore obtain

(16b) $$-a_{ik}a_{jl} \cdot (-1)^b M_{ikjl} \qquad (l > k)$$

where b is the same as before.

Second expansion. We now expand D at first by the jth row. A typical term in this expansion is

(17) $$a_{jl}C_{jl} = a_{jl} \cdot (-1)^{j+l} M_{jl}.$$

By the induction hypothesis we may expand the minor M_{jl} of a_{jl} in D by its ith row, which corresponds to the ith row of D, since $j > i$.

Case I. If $k > l$, the entry a_{ik} in that row belongs to the $(k - 1)$th column of M_{jl}, because M_{jl} does not contain entries of the lth column of D, and $l < k$ (see Fig. 534). Hence the term involving a_{ik} in this expansion is

(18) $$a_{ik} \cdot (\text{cofactor of } a_{ik} \text{ in } M_{jl}) = a_{ik} \cdot (-1)^{i+(k-1)} M_{ikjl},$$

where the minor M_{ikjl} of a_{ik} in M_{jl} is obtained by deleting the ith and jth rows and the kth and lth columns of D [and is, therefore, identical with M_{ikjl} in (15), so that our notation is consistent]. We insert the expansions of the M_{jl} into that of D. It follows from (17) and (18) that this yields a representation whose terms are identical with those given by (16a) when $l < k$.

Case II. If $k < l$, then a_{ik} belongs to the kth column of M_{jl}, we obtain an additional minus sign, and the result agrees with that characterized by (16b).

We have shown that the two expansions of D consist of the same terms, and this proves our statement concerning rows.

The proof of the statement concerning *columns* is quite similar; if we expand D in terms of two arbitrary columns, say, the kth and the lth, we find that the general term involving $a_{jl}a_{ik}$ is exactly the same as before. This proves that not only all column expansions of D yield the same value, but also that their common value is equal to the common value of the row expansions of D.

This completes the proof and shows that *our definition of an nth-order determinant is unambiguous.* ∎

SECTION 8.3, page 443

Proof of Formula (2)

We prove that in right-handed Cartesian coordinates, the vector product

$$\mathbf{v} = \mathbf{a} \times \mathbf{b} = [a_1, \quad a_2, \quad a_3] \times [b_1, \quad b_2, \quad b_3]$$

has the components

$$(2) \qquad v_1 = a_2b_3 - a_3b_2, \qquad v_2 = a_3b_1 - a_1b_3, \qquad v_3 = a_1b_2 - a_2b_1.$$

We need only consider the case $\mathbf{v} \neq \mathbf{0}$. Since $\mathbf{v}$ is perpendicular to both $\mathbf{a}$ and $\mathbf{b}$, Theorem 1 in Sec. 8.2 gives $\mathbf{a} \cdot \mathbf{v} = 0$ and $\mathbf{b} \cdot \mathbf{v} = 0$; in components [see (2), Sec. 8.2],

$$(3) \qquad \begin{aligned} a_1v_1 + a_2v_2 + a_3v_3 &= 0, \\ b_1v_1 + b_2v_2 + b_3v_3 &= 0. \end{aligned}$$

Multiplying the first equation by b_3, the last by a_3, and subtracting, we obtain

$$(a_3b_1 - a_1b_3)v_1 = (a_2b_3 - a_3b_2)v_2.$$

Multiplying the first equation by b_1, the last by a_1, and subtracting, we obtain

$$(a_1b_2 - a_2b_1)v_2 = (a_3b_1 - a_1b_3)v_3.$$

We can easily verify that these two equations are satisfied by

$$(4) \qquad v_1 = c(a_2b_3 - a_3b_2), \qquad v_2 = c(a_3b_1 - a_1b_3), \qquad v_3 = c(a_1b_2 - a_2b_1),$$

where c is a constant. The reader may verify by inserting that (4) also satisfies (3). Now each of the equations in (3) represents a plane through the origin in $v_1v_2v_3$-space. The vectors $\mathbf{a}$ and $\mathbf{b}$ are normal vectors of these planes (see Example 6 in Sec. 8.2). Since $\mathbf{v} \neq \mathbf{0}$, these vectors are not parallel and the two planes do not coincide. Hence their intersection is a straight line L through the origin. Since (4) is a solution of (3) and, for varying c, represents a straight line, we conclude that (4) represents L, and every solution of (3) must be of the form (4). In particular, the components of $\mathbf{v}$ must be of this form, where c is to be determined. From (4) we obtain

$$|\mathbf{v}|^2 = v_1^2 + v_2^2 + v_3^2 = c^2[(a_2b_3 - a_3b_2)^2 + (a_3b_1 - a_1b_3)^2 + (a_1b_2 - a_2b_1)^2].$$

This can be written

$$|\mathbf{v}|^2 = c^2[(a_1^2 + a_2^2 + a_3^2)(b_1^2 + b_2^2 + b_3^2) - (a_1b_1 + a_2b_2 + a_3b_3)^2],$$

as can be verified by performing the indicated multiplications in both formulas and comparing. Using (2) in Sec. 8.2, we thus have

$$|\mathbf{v}|^2 = c^2[(\mathbf{a} \cdot \mathbf{a})(\mathbf{b} \cdot \mathbf{b}) - (\mathbf{a} \cdot \mathbf{b})^2].$$

By comparing this with (13) in the problem set for Sec. 8.3 we conclude that $c = \pm 1$.

We show that $c = +1$. This can be done as follows.

If we change the lengths and directions of **a** and **b** continuously and so that at the end **a** = **i** and **b** = **j** (Fig. 163*a* in Sec. 8.3), then **v** will change its length and direction continuously, and at the end, **v** = **i** × **j** = **k**. Obviously we may effect the change so that both **a** and **b** remain different from the zero vector and are not parallel at any instant. Then **v** is never equal to the zero vector, and since the change is continuous and c can only assume the values $+1$ or -1, it follows that at the end c must have the same value as before. Now at the end **a** = **i**, **b** = **j**, **v** = **k** and, therefore, $a_1 = 1$, $b_2 = 1$, $v_3 = 1$, and the other components in (4) are zero. Hence from (4) we see that $v_3 = c = +1$. This proves Theorem 1.

For a left-handed coordinate system, **i** × **j** = −**k** (see Fig. 163*b* in Sec. 8.3), resulting in $c = -1$. This proves the statement right after formula (2). ∎

SECTION 8.10, page 482

Proof of Theorem 1 ***(Invariance of the divergence)***
This proof will follow from two theorems (A and B), which we prove first.

Theorem A **(Transformation law for vector components)**
For any vector **v**, *the components* v_1, v_2, v_3 *and* v_1^*, v_2^*, v_3^* *in any two systems of Cartesian coordinates* x_1, x_2, x_3 *and* x_1^*, x_2^*, x_3^*, *respectively, are related by*

$$\begin{aligned} v_1^* &= c_{11}v_1 + c_{12}v_2 + c_{13}v_3 \\ v_2^* &= c_{21}v_1 + c_{22}v_2 + c_{23}v_3 \\ v_3^* &= c_{31}v_1 + c_{32}v_2 + c_{33}v_3, \end{aligned} \tag{1}$$

and conversely

$$\begin{aligned} v_1 &= c_{11}v_1^* + c_{21}v_2^* + c_{31}v_3^* \\ v_2 &= c_{12}v_1^* + c_{22}v_2^* + c_{32}v_3^* \\ v_3 &= c_{13}v_1^* + c_{23}v_2^* + c_{33}v_3^* \end{aligned} \tag{2}$$

with coefficients

$$\begin{aligned} c_{11} &= \mathbf{i}^*\cdot\mathbf{i} & c_{12} &= \mathbf{i}^*\cdot\mathbf{j} & c_{13} &= \mathbf{i}^*\cdot\mathbf{k} \\ c_{21} &= \mathbf{j}^*\cdot\mathbf{i} & c_{22} &= \mathbf{j}^*\cdot\mathbf{j} & c_{23} &= \mathbf{j}^*\cdot\mathbf{k} \\ c_{31} &= \mathbf{k}^*\cdot\mathbf{i} & c_{32} &= \mathbf{k}^*\cdot\mathbf{j} & c_{33} &= \mathbf{k}^*\cdot\mathbf{k} \end{aligned} \tag{3}$$

satisfying

$$\sum_{j=1}^{3} c_{kj}c_{mj} = \delta_{km} \qquad (k, m = 1, 2, 3), \tag{4}$$

where the **Kronecker delta**[2] *is given by*

[2]LEOPOLD KRONECKER (1823—1891), German mathematician at Berlin, who made important contributions to algebra, group theory, and number theory.

We shall keep our discussion completely independent of Chap. 7, but readers familiar with matrices should recognize that we are dealing with **orthogonal transformations and matrices,** and that our present theorem follows from Theorem 2 in Sec. 7.12.

$$\delta_{km} = \begin{cases} 0 & (k \neq m) \\ 1 & (k = m). \end{cases}$$

and **i**, **j**, **k** *and* **i***, **j***, **k*** *denote the unit vectors in the positive* x_1-, x_2-, x_3- *and* x_1^*-, x_2^*-, x_3^*-*directions, respectively.*

Proof. The representations of **v** in the two systems are

$$\text{(5)} \qquad \text{(a)} \quad \mathbf{v} = v_1\mathbf{i} + v_2\mathbf{j} + v_3\mathbf{k} \qquad \text{(b)} \quad \mathbf{v} = v_1^*\mathbf{i}^* + v_2^*\mathbf{j}^* + v_3^*\mathbf{k}^*.$$

Since $\mathbf{i}^*\bullet\mathbf{i}^* = 1$, $\mathbf{i}^*\bullet\mathbf{j}^* = 0$, $\mathbf{i}^*\bullet\mathbf{k}^* = 0$, we get from (5b) simply $\mathbf{i}^*\bullet\mathbf{v} = v_1^*$ and from this and (5a)

$$v_1^* = \mathbf{i}^*\bullet\mathbf{v} = \mathbf{i}^*\bullet v_1\mathbf{i} + \mathbf{i}^*\bullet v_2\mathbf{j} + \mathbf{i}^*\bullet v_3\mathbf{k} = v_1\mathbf{i}^*\bullet\mathbf{i} + v_2\mathbf{i}^*\bullet\mathbf{j} + v_3\mathbf{i}^*\bullet\mathbf{k}.$$

Because of (3), this is the first formula in (1), and the other two formulas are obtained similarly, by considering $\mathbf{j}^*\bullet\mathbf{v}$ and then $\mathbf{k}^*\bullet\mathbf{v}$. Formula (2) follows by the same idea, taking $\mathbf{i}\bullet\mathbf{v} = v_1$ from (5a) and then from (5b) and (3)

$$v_1 = \mathbf{i}\bullet\mathbf{v} = v_1^*\mathbf{i}\bullet\mathbf{i}^* + v_2^*\mathbf{i}\bullet\mathbf{j}^* + v_3^*\mathbf{i}\bullet\mathbf{k}^* = c_{11}v_1^* + c_{21}v_2^* + c_{31}v_3^*,$$

and similarly for the other two components.

We prove (4). We can write (1) and (2) briefly as

$$\text{(6)} \qquad \text{(a)} \quad v_j = \sum_{m=1}^{3} c_{mj}v_m^*, \qquad \text{(b)} \quad v_k^* = \sum_{j=1}^{3} c_{kj}v_j.$$

Substituting v_j into v_k^*, we get

$$v_k^* = \sum_{j=1}^{3} c_{kj} \sum_{m=1}^{3} c_{mj}v_m^* = \sum_{m=1}^{3} v_m^* \left(\sum_{j=1}^{3} c_{kj}c_{mj}\right),$$

where $k = 1, 2, 3$. Taking $k = 1$, we have

$$v_1^* = v_1^*\left(\sum_{j=1}^{3} c_{1j}c_{1j}\right) + v_2^*\left(\sum_{j=1}^{3} c_{1j}c_{2j}\right) + v_3^*\left(\sum_{j=1}^{3} c_{1j}c_{3j}\right).$$

For this to hold for *every* vector **v**, the first sum must be 1 and the other two sums 0. This proves (4) with $k = 1$ for $m = 1, 2, 3$. Taking $k = 2$ and then $k = 3$, we obtain (4) with $k = 2$ and 3, for $m = 1, 2, 3$. ∎

The most general transformation of a Cartesian coordinate system into another such system may be decomposed into a transformation of the type just considered and a translation. Under a translation, corresponding coordinates differ merely by a constant. We thus obtain

Theorem B **(Transformation law for Cartesian coordinates)**
The transformation of any Cartesian $x_1x_2x_3$-*coordinate system into any other Cartesian* $x_1^*x_2^*x_3^*$-*coordinate system is of the form*

$$\text{(7)} \qquad x_m^* = \sum_{j=1}^{3} c_{mj}x_j + b_m, \qquad m = 1, 2, 3,$$

with coefficients (3) *and constants* b_1, b_2, b_3; *conversely,*

$$\text{(8)} \qquad x_k = \sum_{n=1}^{3} c_{nk}x_n^* + \tilde{b}_k, \qquad k = 1, 2, 3.$$

***Proof of Theorem* 1 *in Sec.* 8.10 (*Invariance of the divergence*)**

We write x_1, x_2, x_3 instead of x, y, z. Then by definition,

$$\text{(9)} \qquad \operatorname{div} \mathbf{v} = \sum_{j=1}^{3} \frac{\partial v_j}{\partial x_j} = \frac{\partial v_1}{\partial x_1} + \frac{\partial v_2}{\partial x_2} + \frac{\partial v_3}{\partial x_3}.$$

We also write x_1^*, x_2^*, x_3^* instead of x^*, y^*, z^*. We have to show that in these coordinates.

$$\text{(10)} \qquad \operatorname{div} \mathbf{v} = \sum_{k=1}^{3} \frac{\partial v_k^*}{\partial x_k^*} = \frac{\partial v_1^*}{\partial x_1^*} + \frac{\partial v_2^*}{\partial x_2^*} + \frac{\partial v_3^*}{\partial x_3^*}.$$

From the chain rule for functions of several variables (Sec. 8.8) we obtain

$$\text{(11)} \qquad \frac{\partial v_j}{\partial x_j} = \sum_{m=1}^{3} \frac{\partial v_j}{\partial x_m^*} \frac{\partial x_m^*}{\partial x_j}.$$

In this formula, $\partial x_m^*/\partial x_j = c_{mj}$ by (7); and by (2),

$$\frac{\partial v_j}{\partial x_m^*} = \sum_{k=1}^{3} c_{kj} \frac{\partial v_k^*}{\partial x_m^*}.$$

Substituting all this into (11), summing over j from 1 to 3, and noting (4), we get

$$\operatorname{div} \mathbf{v} = \sum_{j=1}^{3} \frac{\partial v_j}{\partial x_j} = \sum_{j=1}^{3} \sum_{k=1}^{3} \sum_{m=1}^{3} c_{kj} \frac{\partial v_k^*}{\partial x_m^*} c_{mj}$$

$$= \sum_{k=1}^{3} \sum_{m=1}^{3} \delta_{km} \frac{\partial v_k^*}{\partial x_m^*} = \sum_{k=1}^{3} \frac{\partial v_k^*}{\partial x_k^*}.$$

This proves (10). ∎

SECTION 8.11, page 487

***Proof of Theorem* 1 (*Invariance of the curl*)**

We write again x_1, x_2, x_3 instead of x, y, z, and similarly x_1^*, x_2^*, x_3^* for other Cartesian coordinates, assuming that both systems are right-handed. Let a_1, a_2, a_3 denote the components of curl **v** in the $x_1x_2x_3$-coordinates, as given by (1), Sec. 8.11, with

$$x = x_1, \qquad y = x_2, \qquad z = x_3.$$

Similarly, let a_1^*, a_2^*, a_3^* denote the components of curl **v** in the $x_1^*x_2^*x_3^*$-coordinate system. We prove that the length and direction of curl **v** are independent of the particular choice of Cartesian coordinates, as asserted in the theorem. We do this by showing that the components of curl **v** satisfy the transformation law (2), p. A85, which is characteristic of vector components. We consider a_1. We use (6a), p. A86, and then the chain rule for functions of several variables (Sec. 8.8). This gives

$$a_1 = \frac{\partial v_3}{\partial x_2} - \frac{\partial v_2}{\partial x_3} = \sum_{m=1}^{3} \left(c_{m3} \frac{\partial v_m^*}{\partial x_2} - c_{m2} \frac{\partial v_m^*}{\partial x_3} \right)$$

$$= \sum_{m=1}^{3} \sum_{j=1}^{3} \left(c_{m3} \frac{\partial v_m^*}{\partial x_j^*} \frac{\partial x_j^*}{\partial x_2} - c_{m2} \frac{\partial v_m^*}{\partial x_j^*} \frac{\partial x_j^*}{\partial x_3} \right).$$

From this and (7), p. A86, we obtain

$$a_1 = \sum_{m=1}^{3} \sum_{j=1}^{3} (c_{m3}c_{j2} - c_{m2}c_{j3}) \frac{\partial v_m^*}{\partial x_j^*}$$

$$= (c_{33}c_{22} - c_{32}c_{23})\left(\frac{\partial v_3^*}{\partial x_2^*} - \frac{\partial v_2^*}{\partial x_3^*}\right) + \cdots$$

$$= (c_{33}c_{22} - c_{32}c_{23})a_1^* + (c_{13}c_{32} - c_{12}c_{33})a_2^* + (c_{23}c_{12} - c_{22}c_{13})a_3^*.$$

Note what we did. The double sum had $3 \times 3 = 9$ terms, 3 of which were zero (when $m = j$), and the remaining 6 terms we combined in pairs as we needed them in getting a_1^*, a_2^*, a_3^*.

We now use (3), p. A85, Lagrange's identity (Prob. 60, Sec. 8.3) and $\mathbf{k}^* \times \mathbf{j}^* = -\mathbf{i}^*$ and $\mathbf{k} \times \mathbf{j} = -\mathbf{i}$. Then

$$c_{33}c_{22} - c_{32}c_{23} = (\mathbf{k}^* \cdot \mathbf{k})(\mathbf{j}^* \cdot \mathbf{j}) - (\mathbf{k}^* \cdot \mathbf{j})(\mathbf{j}^* \cdot \mathbf{k})$$

$$= (\mathbf{k}^* \times \mathbf{j}^*) \cdot (\mathbf{k} \times \mathbf{j}) = \mathbf{i}^* \cdot \mathbf{i} = c_{11}, \qquad \text{etc.}$$

Hence $a_1 = c_{11}a_1^* + c_{21}a_2^* + c_{31}a_3^*$. This is of the form of the first formula in (2), p. A85, and the other two formulas of the form (2) are obtained similarly. This proves the theorem for right-handed systems. If the $x_1x_2x_3$-coordinates are left-handed, then $\mathbf{k} \times \mathbf{j} = +\mathbf{i}$, but then there is a minus sign in front of the determinant in (1), Sec. 8.11. ■

SECTION 9.2, page 509

Theorem 1 (Independence of path)

A line integral

$$\int_C \mathbf{F}(\mathbf{r}) \cdot d\mathbf{r} = \int_C (F_1\,dx + F_2\,dy + F_3\,dz) \tag{1}$$

with continuous F_1, F_2, F_3 in a domain D is independent of path in D if and only if $\mathbf{F}$ = grad f in D for some f; in components,

$$F_1 = \frac{\partial f}{\partial x}, \qquad F_2 = \frac{\partial f}{\partial y}, \qquad F_3 = \frac{\partial f}{\partial z}. \tag{2'}$$

Proof. **(a)** That (2′) implies independence of path was proved in the text.

(b) Conversely, assume that (1) is independent of path in D. Choose any fixed A: (x_0, y_0, z_0) in D and any B: (x, y, z) in D and define f by

$$f(x, y, z) = f_0 + \int_A^B (F_1\,dx^* + F_2\,dy^* + F_3\,dz^*), \tag{3}$$

with any constant f_0 and any path from A to B in D. Since A is fixed and we have independence of path, the integral depends only on the coordinates x, y, z, so that (3) defines a function $f(x, y, z)$ in D. We show that $\mathbf{F}$ = grad f with this f, beginning with the first of the three relations (2′). Because of independence of path, we may integrate from A to B_1: (x_1, y, z) and then parallel to the x-axis along the segment B_1B in Fig. 535 with B_1 chosen so that the whole segment lies in D. Then

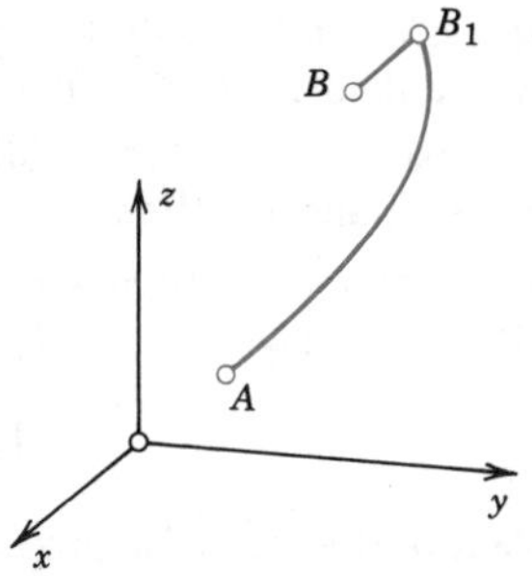

Fig. 535. Proof of Theorem 1

$$f(x, y, z) = f_0 + \int_A^{B_1} (F_1\, dx^* + F_2\, dy^* + F_3\, dz^*)$$

$$+ \int_{B_1}^{B} (F_1\, dx^* + F_2\, dy^* + F_3\, dz^*).$$

We now take the partial derivative with respect to x on both sides. On the left we get $\partial f/\partial x$. We show that on the right we get F_1. The derivative of the first integral is zero because A: (x_0, y_0, z_0) and B_1: (x_1, y, z) do not depend on x. We consider the second integral. Since on the segment B_1B, both y and z are constant, the terms $F_2\, dy^*$ and $F_3\, dz^*$ do not contribute to the derivative of the integral. The remaining part can be written as a definite integral,

$$\int_{B_1}^{B} F_1\, dx^* = \int_{x_1}^{x} F_1(x^*, y, z)\, dx^*.$$

Hence its partial derivative with respect to x is $F_1(x, y, z)$, and the first of the relations (2′) is proved. The other two formulas in (2′) follow by the same argument. ∎

SECTION 12.5, page 728

Proof of Theorem 2 (*Cauchy–Riemann equations*)
We prove that the Cauchy–Riemann equations

$$(1) \qquad u_x = v_y, \qquad u_y = -v_x$$

are sufficient for a complex function

$$f(z) = u(x, y) + iv(x, y)$$

to be analytic; precisely, *if the real part u and the imaginary part v of f(z) satisfy* (1) *in a domain D in the complex plane and if the partial derivatives in* (1) *are* ***continuous*** *in D, then f(z) is analytic in D.*

In this proof we write $\Delta z = \Delta x + i\Delta y$ and $\Delta f = f(z + \Delta z) - f(z)$. The idea of proof is as follows.

(a) We express Δf in terms of first partial derivatives of u and v, by applying the mean value theorem of Sec. 8.8.

(b) We get rid of partial derivatives with respect to y, by applying the Cauchy–Riemann equations.

(c) We let Δz approach zero and show that then $\Delta f/\Delta z$ as obtained approaches a limit, which is equal to $u_x + iv_x$, the right side of (4) in Sec. 12.5,

$$f'(z) = u_x + iv_x$$

regardless of the way of approach to zero.

The details are as follows.

(a) Let P: (x, y) be any fixed point in D. Since D is a domain, it contains a neighborhood of P. We can choose a point Q: $(x + \Delta x, y + \Delta y)$ in this neighborhood such that the straight line segment PQ is in D. Because of our continuity assumptions we may apply the mean value theorem in Sec. 8.8. This yields

$$u(x + \Delta x, y + \Delta y) - u(x, y) = (\Delta x)u_x(M_1) + (\Delta y)u_y(M_1)$$

$$v(x + \Delta x, y + \Delta y) - v(x, y) = (\Delta x)v_x(M_2) + (\Delta y)v_y(M_2).$$

where M_1 and M_2 ($\neq M_1$ in general!) are suitable points on that segment. The first line is Re Δf and the second is Im Δf, so that

$$\Delta f = (\Delta x)u_x(M_1) + (\Delta y)u_y(M_1) + i[(\Delta x)v_x(M_2) + (\Delta y)v_y(M_2)].$$

(b) $u_y = -v_x$ and $v_y = u_x$ by the Cauchy–Riemann equations, so that

$$\Delta f = (\Delta x)u_x(M_1) - (\Delta y)v_x(M_1) + i[(\Delta x)v_x(M_2) + (\Delta y)u_x(M_2)].$$

Also $\Delta z = \Delta x + i\Delta y$, so that we can write $\Delta x = \Delta z - i\Delta y$ in the first term and $\Delta y = (\Delta z - \Delta x)/i = -i(\Delta z - \Delta x)$ in the second term. This gives

$$\Delta f = (\Delta z - i\Delta y)u_x(M_1) + i(\Delta z - \Delta x)v_x(M_1) + i[(\Delta x)v_x(M_2) + (\Delta y)u_x(M_2)].$$

By performing the multiplications and reordering we obtain

$$\Delta f = (\Delta z)u_x(M_1) - i\Delta y\{u_x(M_1) - u_x(M_2)\}$$

$$+ i[(\Delta z)v_x(M_1) - \Delta x\{v_x(M_1) - v_x(M_2)\}].$$

Division by Δz now yields

$$\text{(A)} \quad \frac{\Delta f}{\Delta z} = u_x(M_1) + iv_x(M_1) - \frac{i\Delta y}{\Delta z}\{u_x(M_1) - u_x(M_2)\} - \frac{i\Delta x}{\Delta z}\{v_x(M_1) - v_x(M_2)\}.$$

(c) We finally let Δz approach zero and note that $|\Delta y/\Delta z| \leqq 1$ and $|\Delta x/\Delta z| \leqq 1$ in (A). Then Q: $(x + \Delta x, y + \Delta y)$ approaches P: (x, y), so that M_1 and M_2 must approach P. Also, since the partial derivatives in (A) are assumed to be continuous, they approach their value at P. In particular, the differences in the braces $\{\cdots\}$ in (A) approach zero. Hence the limit of the right side of (A) exists and is independent of the path along which $\Delta z \to 0$. We see that this limit equals the right side of (4) in Sec. 12.5. This means that $f(z)$ is analytic at every point z in D, and the proof is complete. ∎

SECTION 13.3, pages 762 and 764

Goursat's proof of Cauchy's integral theorem without the condition that $f'(z)$ is continuous

We start with the case when C is the boundary of a triangle. We orient C counterclockwise. By joining the midpoints of the sides we subdivide the triangle into four congruent triangles (Fig. 536). Let C_{I}, C_{II}, C_{III}, C_{IV} denote their boundaries. We claim that (see Fig. 536)

$$\oint_C f\,dz = \oint_{C_{\text{I}}} f\,dz + \oint_{C_{\text{II}}} f\,dz + \oint_{C_{\text{III}}} f\,dz + \oint_{C_{\text{IV}}} f\,dz. \tag{1}$$

Indeed, on the right we integrate along each of the three segments of subdivision in both possible directions (Fig. 536), so that the corresponding integrals cancel out in pairs, and the sum of the integrals on the right equals the integral on the left. We now pick an integral on the right that is biggest in absolute value and call its path C_1. Then, by the triangle inequality (Sec. 12.2),

$$\left|\oint_C f\,dz\right| \leqq \left|\oint_{C_{\text{I}}} f\,dz\right| + \left|\oint_{C_{\text{II}}} f\,dz\right| + \left|\oint_{C_{\text{III}}} f\,dz\right| + \left|\oint_{C_{\text{IV}}} f\,dz\right|$$

$$\leqq 4\left|\oint_{C_1} f\,dz\right|.$$

We now subdivide the triangle bounded by C_1 as before and select a triangle of subdivision with boundary C_2 for which

$$\left|\oint_{C_1} f\,dz\right| \leqq 4\left|\oint_{C_2} f\,dz\right|. \qquad \text{Then} \qquad \left|\oint_C f\,dz\right| \leqq 4^2\left|\oint_{C_2} f\,dz\right|.$$

Continuing in this fashion, we obtain a sequence of triangles $T_1, T_2, \cdots$ with boundaries $C_1, C_2, \cdots$ that are similar and such that T_n lies in T_m when $n > m$, and

$$\left|\oint_C f\,dz\right| \leqq 4^n\left|\oint_{C_n} f\,dz\right|, \qquad n = 1, 2, \cdots. \tag{2}$$

Let z_0 be the point that belongs to all these triangles. Since f is differentiable at $z = z_0$, the derivative $f'(z_0)$ exists. Let $h(z)$ denote the difference between the difference quotient and the derivative:

$$h(z) = \frac{f(z) - f(z_0)}{z - z_0} - f'(z_0). \tag{3}$$

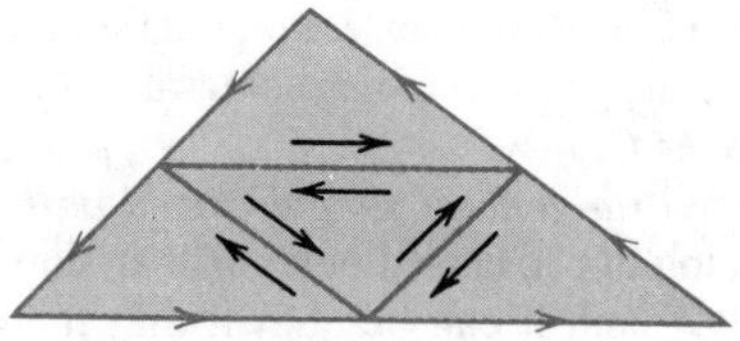

Fig. 536. Proof of Cauchy's integral theorem

Since $f'(z_0)$ is the limit of this difference quotient, we see that $h(z)$ can be made as small in absolute value as we please; for a given positive number ϵ we can find a positive number δ such that

$$|h(z)| < \epsilon \qquad \text{when} \qquad |z - z_0| < \delta. \tag{4}$$

We shall need this in a minute. Solving (3) algebraically for $f(z)$, we have

$$f(z) = f(z_0) + (z - z_0)f'(z_0) + h(z)(z - z_0).$$

Integrating this over the boundary C_n of the triangle T_n gives

$$\oint_{C_n} f(z)\,dz = \oint_{C_n} f(z_0)\,dz + \oint_{C_n} (z - z_0)f'(z_0)\,dz + \oint_{C_n} h(z)(z - z_0)\,dz.$$

Since $f(z_0)$ and $f'(z_0)$ are constants and C_n is a closed path, the first two integrals on the right are zero, as follows from Cauchy's proof, which is applicable because the integrands do have continuous derivatives (0 and *const*, respectively). We thus have

$$\oint_{C_n} f(z)\,dz = \oint_{C_n} h(z)(z - z_0)\,dz.$$

We may now take n so large that the triangle T_n lies in the disk $|z - z_0| < \delta$. Let L_n be the length of C_n. Then $|z - z_0| < L_n$ for all z on C_n and z_0 in T_n. From this and (4) we have $|h(z)(z - z_0)| < \epsilon L_n$. The *ML*-inequality in Sec. 13.2 now gives

$$\left|\oint_{C_n} f(z)\,dz\right| = \left|\oint_{C_n} h(z)(z - z_0)\,dz\right| \leqq \epsilon L_n \cdot L_n = \epsilon L_n^{\,2}. \tag{5}$$

Now let L be the length of C. Then the path C_1 has the length $L_1 = L/2$, the path C_2 has the length $L_2 = L_1/2 = L/4$, etc., and C_n has the length

$$L_n = \frac{L}{2^n}. \qquad \text{Hence} \qquad L_n^{\,2} = \frac{L^2}{4^n}.$$

From (2) and (5) we thus obtain

$$\left|\oint_C f\,dz\right| \leqq 4^n \left|\oint_{C_n} f\,dz\right| \leqq 4^n \epsilon L_n^{\,2} = 4^n \epsilon \frac{L^2}{4^n} = \epsilon L^2.$$

By choosing ϵ (> 0) sufficiently small we can make the expression on the right as small as we please, while the expression on the left is the definite value of an integral. Consequently, this value must be zero, and the proof is complete.

The proof for *the case in which C is the boundary of a polygon* follows from the previous proof by subdividing the polygon into triangles (Fig. 537). The integral corresponding to each such triangle is zero. The sum of these integrals is equal to the integral over C, because we integrate along each segment of subdivision in both directions, the corresponding integrals cancel out in pairs, and we are left with the integral over C.

The case of a general simple closed path C can be reduced to the preceding one by inscribing in C a closed polygon P of chords, which approximates C "sufficiently accurately," and it can be shown that there is a polygon P such that the integral over P differs from that over C by less than any preassigned positive real number $\tilde{\epsilon}$, no matter how small. The details of this proof are somewhat involved and can be found in Ref. [D6] listed in Appendix 1. ∎

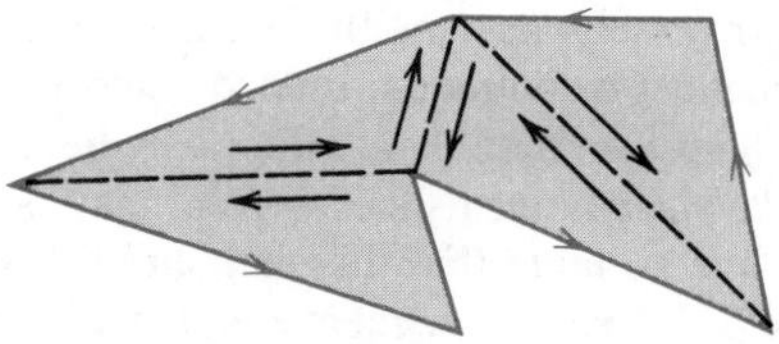

Fig. 537. Proof of Cauchy's integral theorem for a polygon

SECTION 14.1, page 785

Proof of Theorem 4 (***Cauchy's convergence principle***)

(a) In this proof we need two concepts and a theorem, which we list first.

1. A **bounded sequence** $s_1, s_2, \cdots$ is a sequence whose terms all lie in a disk of (sufficiently large, finite) radius K with center at the origin; thus $|s_n| < K$ for all n.

2. A **limit point** a of a sequence $s_1, s_2, \cdots$ is a point such that, given an $\epsilon > 0$, there are infinitely many terms satisfying $|s_n - a| < \epsilon$. (Note that this does *not* imply convergence, since there may still be infinitely many terms that do not lie within that circle of radius ϵ and center a.)

Example: $\frac{1}{4}, \frac{3}{4}, \frac{1}{8}, \frac{7}{8}, \frac{1}{16}, \frac{15}{16}, \cdots$ has the limit points 0 and 1 and diverges.

3. A bounded sequence in the complex plane has at least one limit point. (Bolzano-Weierstrass theorem; proof below. Recall that "sequence" always means *infinite* sequence.)

(b) We now turn to the actual proof that $z_1 + z_2 + \cdots$ converges if and only if for every $\epsilon > 0$ we can find an N such that

$$(1) \qquad |z_{n+1} + \cdots + z_{n+p}| < \epsilon \qquad \text{for every } n > N \text{ and } p = 1, 2, \cdots.$$

Here, by the definition of partial sums,

$$s_{n+p} - s_n = z_{n+1} + \cdots + z_{n+p}.$$

Writing $n + p = r$, we see from this that (1) is equivalent to

$$(1^*) \qquad |s_r - s_n| < \epsilon \qquad \text{for all } r > N \text{ and } n > N.$$

Suppose that $s_1, s_2, \cdots$ converges. Denote its limit by s. Then for a given $\epsilon > 0$ we can find an N such that

$$|s_n - s| < \frac{\epsilon}{2} \qquad \text{for every } n > N.$$

Hence, if $r > N$ and $n > N$, then by the triangle inequality (Sec. 12.2),

$$|s_r - s_n| = |(s_r - s) - (s_n - s)| \leqq |s_r - s| + |s_n - s| < \frac{\epsilon}{2} + \frac{\epsilon}{2} = \epsilon,$$

that is, (1*) holds.

(c) Conversely, assume that $s_1, s_2, \cdots$ satisfies (1*). We first prove that then the sequence must be bounded. Indeed, choose a fixed ϵ and a fixed $n = n_0 > N$ in (1*). Then (1*) implies that all s_r with $r > N$ lie in the disk of radius ϵ and center s_{n_0} and only *finitely many* terms $s_1, \cdots, s_N$ may not lie in this disk. Clearly, we can now find a circle so large that this disk and these finitely many terms all lie within this new circle. Hence the sequence is bounded. By the Bolzano–Weierstrass theorem, it has at least one limit point, call it s.

We now show that the sequence is convergent with the limit s. Let $\epsilon > 0$ be given. Then there is an N^* such that $|s_r - s_n| < \epsilon/2$ for all $r > N^*$ and $n > N^*$, by (1*). Also, by the definition of a limit point, $|s_n - s| < \epsilon/2$ for *infinitely many n*, so that we can find and fix an $n > N^*$ such that $|s_n - s| < \epsilon/2$. Together, for *every* $r > N^*$,

$$|s_r - s| = |(s_r - s_n) + (s_n - s)| \leqq |s_r - s_n| + |s_n - s| < \frac{\epsilon}{2} + \frac{\epsilon}{2} = \epsilon;$$

that is, the sequence $s_1, s_2, \cdots$ is convergent with the limit s. ∎

Bolzano–Weierstrass theorem[3]

A bounded infinite sequence $z_1, z_2, z_3, \cdots$ *in the complex plane has at least one limit point.*

Proof. It is obvious that we need both conditions: a finite sequence cannot have a limit point, and the sequence $1, 2, 3, \cdots$, which is infinite but not bounded, has no limit point. To prove the theorem, consider a bounded infinite sequence $z_1, z_2, \cdots$ and let K be such that $|z_n| < K$ for all n. If only finitely many values of the z_n are different, then, since the sequence is infinite, some number z must occur infinitely many times in the sequence, and, by definition, this number is a limit point of the sequence.

We may now turn to the case when the sequence contains infinitely many different terms. We draw a large square Q_0 that contains all z_n. We subdivide Q_0 into four congruent squares, which we number 1, 2, 3, 4. Clearly, at least one of these squares (each taken with its complete boundary) must contain infinitely many terms of the sequence. The square of this type with the lowest number (1, 2, 3, or 4) will be denoted by Q_1. This is the first step. In the next step we subdivide Q_1 into four congruent squares and select a square Q_2 according to the same rule, and so on. This yields an infinite sequence of squares $Q_0, Q_1, Q_2, \cdots, Q_n, \cdots$ with the property that the side of Q_n approaches zero as n approaches infinity, and Q_m contains all Q_n with $n > m$. It is not difficult to see that the number which belongs to all these squares,[4] call it $z = a$, is a limit point of the sequence. In fact, given an $\epsilon > 0$, we can choose an N so large that the side of the square Q_N is less than ϵ and, since Q_N contains infinitely many z_n we have $|z_n - a| < \epsilon$ for infinitely many n. This completes the proof. ∎

---

[3]BERNARD BOLZANO (1781—1848), Austrian mathematician and professor of religious studies, is a pioneer in the study of point sets, the foundations of analysis, and mathematical logic.

For Weierstrass, see footnote 8 in Sec. 14.6.

[4]The fact that such a unique number $z = a$ exists seems to be obvious, but it actually follows from an axiom of the real number system, the so-called *Cantor–Dedekind axiom;* see footnote 3 in Appendix 3.3.

SECTION 14.3, page 800

Part (b) of the proof of Theorem 5

We have to show that

$$\sum_{n=2}^{\infty} a_n \left[\frac{(z+\Delta z)^n - z^n}{\Delta z} - nz^{n-1}\right]$$

$$= \sum_{n=2}^{\infty} a_n\, \Delta z[(z+\Delta z)^{n-2} + 2z(z+\Delta z)^{n-3} + \cdots + (n-1)z^{n-2}],$$

thus,

$$\frac{(z+\Delta z)^n - z^n}{\Delta z} - nz^{n-1}$$

$$= \Delta z[(z+\Delta z)^{n-2} + 2z(z+\Delta z)^{n-3} + \cdots + (n-1)z^{n-2}]$$

or, setting $z + \Delta z = b$ and $z = a$, thus $\Delta z = b - a$,

(7a) $$\frac{b^n - a^n}{b-a} - na^{n-1} = (b-a)A_n \qquad (n = 2, 3, \cdots),$$

where A_n is the expression in the brackets on the right,

(7b) $$A_n = b^{n-2} + 2ab^{n-3} + 3a^2b^{n-4} + \cdots + (n-1)a^{n-2};$$

thus, $A_2 = 1$, $A_3 = b + 2a$, etc. We use induction. When $n = 2$, then (7) holds, since then

$$\frac{b^2 - a^2}{b-a} - 2a = \frac{b^2 - a^2 - 2a(b-a)}{b-a} = \frac{(b-a)^2}{b-a} = b - a = (b-a)A_2.$$

Assuming that (7) holds for $n = k$, we show that it holds for $n = k + 1$. By adding and subtracting a term in the numerator and then dividing we get

$$\frac{b^{k+1} - a^{k+1}}{b-a} = \frac{b^{k+1} - ba^k + ba^k - a^{k+1}}{b-a} = b\,\frac{b^k - a^k}{b-a} + a^k.$$

By the induction hypothesis, the right side equals

$$b[(b-a)A_k + ka^{k-1}] + a^k.$$

Direct calculation shows that this is equal to

$$(b-a)\{bA_k + ka^{k-1}\} + aka^{k-1} + a^k.$$

From (7b) with $n = k$ we see that the expression in the braces $\{\cdots\}$ equals

$$b^{k-1} + 2ab^{k-2} + \cdots + (k-1)ba^{k-2} + ka^{k-1} = A_{k+1}.$$

Hence our result is

$$\frac{b^{k+1} - a^{k+1}}{b-a} = (b-a)A_{k+1} + (k+1)a^k.$$

Taking the last term to the left, we obtain (7) with $n = k + 1$. This proves (7) for any integer $n \geqq 2$ and completes the proof. ∎

SECTION 17.2, page 891

Another proof of Theorem 1, without the use of a harmonic conjugate

We show that if $w = u + iv = f(z)$ is analytic and maps a domain D conformally onto a domain D^* and $\Phi^*(u, v)$ is harmonic in D^*, then

$$\Phi(x, y) = \Phi^*(u(x, y), v(x, y)) \tag{1}$$

is harmonic in D, that is, $\nabla^2\Phi = 0$ in D. We make no use of a harmonic conjugate of Φ^*, but use straightforward differentiation. By the chain rule,

$$\Phi_x = \Phi^*_u u_x + \Phi^*_v v_x.$$

We apply the chain rule again, underscoring the terms that will drop out when we form $\nabla^2\Phi$:

$$\begin{aligned}\Phi_{xx} &= \underline{\Phi^*_u u_{xx}} + (\Phi^*_{uu} u_x + \underline{\Phi^*_{uv} v_x}) u_x \\ &\quad + \underline{\Phi^*_v v_{xx}} + (\underline{\Phi^*_{vu} u_x} + \Phi^*_{vv} v_x) v_x.\end{aligned}$$

Φ_{yy} is the same with each x replaced by y. We form the sum $\nabla^2\Phi$. In it, $\Phi^*_{vu} = \Phi^*_{uv}$ is multiplied by

$$u_x v_x + u_y v_y,$$

which is 0 by the Cauchy–Riemann equations. Also $\nabla^2 u = 0$ and $\nabla^2 v = 0$. There remains

$$\nabla^2\Phi = \Phi^*_{uu}(u_x^2 + u_y^2) + \Phi^*_{vv}(v_x^2 + v_y^2).$$

By the Cauchy–Riemann equations this becomes

$$\nabla^2\Phi = (\Phi^*_{uu} + \Phi^*_{vv})(u_x^2 + v_x^2)$$

and is 0 since Φ^* is harmonic. ∎

Appendix 5

Tables

Tables of Laplace transforms in Secs. 6.9 and 6.10
Tables of Fourier transforms in Sec. 10.12

Table A1
Bessel Functions

For more extensive tables see Ref. [1] in Appendix 1.

x	$J_0(x)$	$J_1(x)$	x	$J_0(x)$	$J_1(x)$	x	$J_0(x)$	$J_1(x)$
0.0	1.0000	0.0000	3.0	−0.2601	0.3991	6.0	0.1506	−0.2767
0.1	0.9975	0.0499	3.1	−0.2921	0.3009	6.1	0.1773	−0.2559
0.2	0.9900	0.0995	3.2	−0.3202	0.2613	6.2	0.2017	−0.2329
0.3	0.9776	0.1483	3.3	−0.3443	0.2207	6.3	0.2238	−0.2081
0.4	0.9604	0.1960	3.4	−0.3643	0.1792	6.4	0.2433	−0.1816
0.5	0.9385	0.2423	3.5	−0.3801	0.1374	6.5	0.2601	−0.1538
0.6	0.9120	0.2867	3.6	−0.3918	0.0955	6.6	0.2740	−0.1250
0.7	0.8812	0.3290	3.7	−0.3992	0.0538	6.7	0.2851	−0.0953
0.8	0.8463	0.3688	3.8	−0.4026	0.0128	6.8	0.2931	−0.0652
0.9	0.8075	0.4059	3.9	−0.4018	−0.0272	6.9	0.2981	−0.0349
1.0	0.7652	0.4401	4.0	−0.3971	−0.0660	7.0	0.3001	−0.0047
1.1	0.7196	0.4709	4.1	−0.3887	−0.1033	7.1	0.2991	0.0252
1.2	0.6711	0.4983	4.2	−0.3766	−0.1386	7.2	0.2951	0.0543
1.3	0.6201	0.5220	4.3	−0.3610	−0.1719	7.3	0.2882	0.0826
1.4	0.5669	0.5419	4.4	−0.3423	−0.2028	7.4	0.2786	0.1096
1.5	0.5118	0.5579	4.5	−0.3205	−0.2311	7.5	0.2663	0.1352
1.6	0.4554	0.5699	4.6	−0.2961	−0.2566	7.6	0.2516	0.1592
1.7	0.3980	0.5778	4.7	−0.2693	−0.2791	7.7	0.2346	0.1813
1.8	0.3400	0.5815	4.8	−0.2404	−0.2985	7.8	0.2154	0.2014
1.9	0.2818	0.5812	4.9	−0.2097	−0.3147	7.9	0.1944	0.2192
2.0	0.2239	0.5767	5.0	−0.1776	−0.3276	8.0	0.1717	0.2346
2.1	0.1666	0.5683	5.1	−0.1443	−0.3371	8.1	0.1475	0.2476
2.2	0.1104	0.5560	5.2	−0.1103	−0.3432	8.2	0.1222	0.2580
2.3	0.0555	0.5399	5.3	−0.0758	−0.3460	8.3	0.0960	0.2657
2.4	0.0025	0.5202	5.4	−0.0412	−0.3453	8.4	0.0692	0.2708
2.5	−0.0484	0.4971	5.5	−0.0068	−0.3414	8.5	0.0419	0.2731
2.6	−0.0968	0.4708	5.6	0.0270	−0.3343	8.6	0.0146	0.2728
2.7	−0.1424	0.4416	5.7	0.0599	−0.3241	8.7	−0.0125	0.2697
2.8	−0.1850	0.4097	5.8	0.0917	−0.3110	8.8	−0.0392	0.2641
2.9	−0.2243	0.3754	5.9	0.1220	−0.2951	8.9	−0.0653	0.2559

$J_0(x) = 0$ for $x = 2.405, 5.520, 8.654, 11.792, 14.931, \cdots$
$J_1(x) = 0$ for $x = 0, 3.832, 7.016, 10.173, 13.324, \cdots$

Table A1 (continued)

x	$Y_0(x)$	$Y_1(x)$	x	$Y_0(x)$	$Y_1(x)$	x	$Y_0(x)$	$Y_1(x)$
0.0	$(-\infty)$	$(-\infty)$	2.5	0.498	0.146	5.0	−0.309	0.148
0.5	−0.445	−1.471	3.0	0.377	0.325	5.5	−0.339	−0.024
1.0	0.088	−0.781	3.5	0.189	0.410	6.0	−0.288	−0.175
1.5	0.382	−0.412	4.0	−0.017	0.398	6.5	−0.173	−0.274
2.0	0.510	−0.107	4.5	−0.195	0.301	7.0	−0.026	−0.303

Table A2
Gamma Function [see (24) in Appendix 3]

α	$\Gamma(\alpha)$	α	$\Gamma(\alpha)$	α	$\Gamma(\alpha)$	α	$\Gamma(\alpha)$	α	$\Gamma(\alpha)$
1.00	1.000 000	1.20	0.918 169	1.40	0.887 264	1.60	0.893 515	1.80	0.931 384
1.02	0.988 844	1.22	0.913 106	1.42	0.886 356	1.62	0.895 924	1.82	0.936 845
1.04	0.978 438	1.24	0.908 521	1.44	0.885 805	1.64	0.898 642	1.84	0.942 612
1.06	0.968 744	1.26	0.904 397	1.46	0.885 604	1.66	0.901 668	1.86	0.948 687
1.08	0.959 725	1.28	0.900 718	1.48	0.885 747	1.68	0.905 001	1.88	0.955 071
1.10	0.951 351	1.30	0.897 471	1.50	0.886 227	1.70	0.908 639	1.90	0.961 766
1.12	0.943 590	1.32	0.894 640	1.52	0.887 039	1.72	0.912 581	1.92	0.968 774
1.14	0.936 416	1.34	0.892 216	1.54	0.888 178	1.74	0.916 826	1.94	0.976 099
1.16	0.929 803	1.36	0.890 185	1.56	0.889 639	1.76	0.921 375	1.96	0.983 743
1.18	0.923 728	1.38	0.888 537	1.58	0.891 420	1.78	0.926 227	1.98	0.991 708
1.20	0.918 169	1.40	0.887 264	1.60	0.893 515	1.80	0.931 384	2.00	1.000 000

Table A3
Factorial Function

n	$n!$	$\log(n!)$	n	$n!$	$\log(n!)$	n	$n!$	$\log(n!)$
1	1	0.000 000	6	720	2.857 332	11	39 916 800	7.601 156
2	2	0.301 030	7	5 040	3.702 431	12	479 001 600	8.680 337
3	6	0.778 151	8	40 320	4.605 521	13	6 227 020 800	9.794 280
4	24	1.380 211	9	362 880	5.559 763	14	87 178 291 200	10.940 408
5	120	2.079 181	10	3 628 800	6.559 763	15	1 307 674 368 000	12.116 500

Table A4
Error Function, Sine and Cosine Integrals [see (35), (40), (42) in Appendix 3]

x	erf x	Si(x)	ci(x)	x	erf x	Si(x)	ci(x)
0.0	0.0000	0.0000	∞	2.0	0.9953	1.6054	−0.4230
0.2	0.2227	0.1996	1.0422	2.2	0.9981	1.6876	−0.3751
0.4	0.4284	0.3965	0.3788	2.4	0.9993	1.7525	−0.3173
0.6	0.6039	0.5881	0.0223	2.6	0.9998	1.8004	−0.2533
0.8	0.7421	0.7721	−0.1983	2.8	0.9999	1.8321	−0.1865
1.0	0.8427	0.9461	−0.3374	3.0	1.0000	1.8487	−0.1196
1.2	0.9103	1.1080	−0.4205	3.2	1.0000	1.8514	−0.0553
1.4	0.9523	1.2562	−0.4620	3.4	1.0000	1.8419	0.0045
1.6	0.9763	1.3892	−0.4717	3.6	1.0000	1.8219	0.0580
1.8	0.9891	1.5058	−0.4568	3.8	1.0000	1.7934	0.1038
2.0	0.9953	1.6054	−0.4230	4.0	1.0000	1.7582	0.1410

Table A5
Binomial Distribution

Probability function $f(x)$ [see (2), Sec. 23.6] and distribution function $F(x)$

n	x	$p = 0.1$		$p = 0.2$		$p = 0.3$		$p = 0.4$		$p = 0.5$	
		$f(x)$	$F(x)$	$f(x)$	$F(x)$	$f(x)$	$F(x)$	$f(x)$	$F(x)$	$f(x)$	$F(x)$
		0.		**0.**		**0.**		**0.**		**0.**	
1	0	9000	0.9000	8000	0.8000	7000	0.7000	6000	0.6000	5000	0.5000
	1	1000	1.0000	2000	1.0000	3000	1.0000	4000	1.0000	5000	1.0000
2	0	8100	0.8100	6400	0.6400	4900	0.4900	3600	0.3600	2500	0.2500
	1	1800	0.9900	3200	0.9600	4200	0.9100	4800	0.8400	5000	0.7500
	2	0100	1.0000	0400	1.0000	0900	1.0000	1600	1.0000	2500	1.0000
3	0	7290	0.7290	5120	0.5120	3430	0.3430	2160	0.2160	1250	0.1250
	1	2430	0.9720	3840	0.8960	4410	0.7840	4320	0.6480	3750	0.5000
	2	0270	0.9990	0960	0.9920	1890	0.9730	2880	0.9360	3750	0.8750
	3	0010	1.0000	0080	1.0000	0270	1.0000	0640	1.0000	1250	1.0000
4	0	6561	0.6561	4096	0.4096	2401	0.2401	1296	0.1296	0625	0.0625
	1	2916	0.9477	4096	0.8192	4116	0.6517	3456	0.4752	2500	0.3125
	2	0486	0.9963	1536	0.9728	2646	0.9163	3456	0.8208	3750	0.6875
	3	0036	0.9999	0256	0.9984	0756	0.9919	1536	0.9744	2500	0.9375
	4	0001	1.0000	0016	1.0000	0081	1.0000	0256	1.0000	0625	1.0000
5	0	5905	0.5905	3277	0.3277	1681	0.1681	0778	0.0778	0313	0.0313
	1	3281	0.9185	4096	0.7373	3602	0.5282	2592	0.3370	1563	0.1875
	2	0729	0.9914	2048	0.9421	3087	0.8369	3456	0.6826	3125	0.5000
	3	0081	0.9995	0512	0.9933	1323	0.9692	2304	0.9130	3125	0.8125
	4	0005	1.0000	0064	0.9997	0284	0.9976	0768	0.9898	1563	0.9688
	5	0000	1.0000	0003	1.0000	0024	1.0000	0102	1.0000	0313	1.0000
6	0	5314	0.5314	2621	0.2621	1176	0.1176	0467	0.0467	0156	0.0156
	1	3543	0.8857	3932	0.6554	3025	0.4202	1866	0.2333	0938	0.1094
	2	0984	0.9841	2458	0.9011	3241	0.7443	3110	0.5443	2344	0.3438
	3	0146	0.9987	0819	0.9830	1852	0.9295	2765	0.8208	3125	0.6563
	4	0012	0.9999	0154	0.9984	0595	0.9891	1382	0.9590	2344	0.8906
	5	0001	1.0000	0015	0.9999	0102	0.9993	0369	0.9959	0938	0.9844
	6	0000	1.0000	0001	1.0000	0007	1.0000	0041	1.0000	0156	1.0000
7	0	4783	0.4783	2097	0.2097	0824	0.0824	0280	0.0280	0078	0.0078
	1	3720	0.8503	3670	0.5767	2471	0.3294	1306	0.1586	0547	0.0625
	2	1240	0.9743	2753	0.8520	3177	0.6471	2613	0.4199	1641	0.2266
	3	0230	0.9973	1147	0.9667	2269	0.8740	2903	0.7102	2734	0.5000
	4	0026	0.9998	0287	0.9953	0972	0.9712	1935	0.9037	2734	0.7734
	5	0002	1.0000	0043	0.9996	0250	0.9962	0774	0.9812	1641	0.9375
	6	0000	1.0000	0004	1.0000	0036	0.9998	0172	0.9984	0547	0.9922
	7	0000	1.0000	0000	1.0000	0002	1.0000	0016	1.0000	0078	1.0000
8	0	4305	0.4305	1678	0.1678	0576	0.0576	0168	0.0168	0039	0.0039
	1	3826	0.8131	3355	0.5033	1977	0.2553	0896	0.1064	0313	0.0352
	2	1488	0.9619	2936	0.7969	2965	0.5518	2090	0.3154	1094	0.1445
	3	0331	0.9950	1468	0.9437	2541	0.8059	2787	0.5941	2188	0.3633
	4	0046	0.9996	0459	0.9896	1361	0.9420	2322	0.8263	2734	0.6367
	5	0004	1.0000	0092	0.9988	0467	0.9887	1239	0.9502	2188	0.8555
	6	0000	1.0000	0011	0.9999	0100	0.9987	0413	0.9915	1094	0.9648
	7	0000	1.0000	0001	1.0000	0012	0.9999	0079	0.9993	0313	0.9961
	8	0000	1.0000	0000	1.0000	0001	1.0000	0007	1.0000	0039	1.0000

Table A6
Poisson Distribution

Probability function $f(x)$ [see (5), Sec. 23.6] and distribution function $F(x)$

	$\mu = 0.1$		$\mu = 0.2$		$\mu = 0.3$		$\mu = 0.4$		$\mu = 0.5$	
x	$f(x)$	$F(x)$	$f(x)$	$F(x)$	$f(x)$	$F(x)$	$f(x)$	$F(x)$	$f(x)$	$F(x)$
	0.		0.		0.		0.		0.	
0	9048	0.9048	8187	0.8187	7408	0.7408	6703	0.6703	6065	0.6065
1	0905	0.9953	1637	0.9825	2222	0.9631	2681	0.9384	3033	0.9098
2	0045	0.9998	0164	0.9989	0333	0.9964	0536	0.9921	0758	0.9856
3	0002	1.0000	0011	0.9999	0033	0.9997	0072	0.9992	0126	0.9982
4	0000	1.0000	0001	1.0000	0003	1.0000	0007	0.9999	0016	0.9998
5							0001	1.0000	0002	1.0000

	$\mu = 0.6$		$\mu = 0.7$		$\mu = 0.8$		$\mu = 0.9$		$\mu = 1$	
x	$f(x)$	$F(x)$	$f(x)$	$F(x)$	$f(x)$	$F(x)$	$f(x)$	$F(x)$	$f(x)$	$F(x)$
	0.		0.		0.		0.		0.	
0	5488	0.5488	4966	0.4966	4493	0.4493	4066	0.4066	3679	0.3679
1	3293	0.8781	3476	0.8442	3595	0.8088	3659	0.7725	3679	0.7358
2	0988	0.9769	1217	0.9659	1438	0.9526	1647	0.9371	1839	0.9197
3	0198	0.9966	0284	0.9942	0383	0.9909	0494	0.9865	0613	0.9810
4	0030	0.9996	0050	0.9992	0077	0.9986	0111	0.9977	0153	0.9963
5	0004	1.0000	0007	0.9999	0012	0.9998	0020	0.9997	0031	0.9994
6			0001	1.0000	0002	1.0000	0003	1.0000	0005	0.9999
7									0001	1.0000

	$\mu = 1.5$		$\mu = 2$		$\mu = 3$		$\mu = 4$		$\mu = 5$	
x	$f(x)$	$F(x)$	$f(x)$	$F(x)$	$f(x)$	$F(x)$	$f(x)$	$F(x)$	$f(x)$	$F(x)$
	0.		0.		0.		0.		0.	
0	2231	0.2231	1353	0.1353	0498	0.0498	0183	0.0183	0067	0.0067
1	3347	0.5578	2707	0.4060	1494	0.1991	0733	0.0916	0337	0.0404
2	2510	0.8088	2707	0.6767	2240	0.4232	1465	0.2381	0842	0.1247
3	1255	0.9344	1804	0.8571	2240	0.6472	1954	0.4335	1404	0.2650
4	0471	0.9814	0902	0.9473	1680	0.8153	1954	0.6288	1755	0.4405
5	0141	0.9955	0361	0.9834	1008	0.9161	1563	0.7851	1755	0.6160
6	0035	0.9991	0120	0.9955	0504	0.9665	1042	0.8893	1462	0.7622
7	0008	0.9998	0034	0.9989	0216	0.9881	0595	0.9489	1044	0.8666
8	0001	1.0000	0009	0.9998	0081	0.9962	0298	0.9786	0653	0.9319
9			0002	1.0000	0027	0.9989	0132	0.9919	0363	0.9682
10					0008	0.9997	0053	0.9972	0181	0.9863
11					0002	0.9999	0019	0.9991	0082	0.9945
12					0001	1.0000	0006	0.9997	0034	0.9980
13							0002	0.9999	0013	0.9993
14							0001	1.0000	0005	0.9998
15									0002	0.9999
16									0000	1.0000

Table A7
Normal Distribution

Values of the distribution function $\Phi(z)$ [see (4), Sec. 23.7]
$\Phi(-z) = 1 - \Phi(z)$, $\Phi(0) = 0.5000$

z	$\Phi(z)$	z	$\Phi(z)$	z	$\Phi(z)$	z	$\Phi(z)$	z	$\Phi(z)$	z	$\Phi(z)$
	0.		0.		0.		0.		0.		0.
0.01	5040	0.51	6950	1.01	8438	1.51	9345	2.01	9778	2.51	9940
0.02	5080	0.52	6985	1.02	8461	1.52	9357	2.02	9783	2.52	9941
0.03	5120	0.53	7019	1.03	8485	1.53	9370	2.03	9788	2.53	9943
0.04	5160	0.54	7054	1.04	8508	1.54	9382	2.04	9793	2.54	9945
0.05	5199	0.55	7088	1.05	8531	1.55	9394	2.05	9798	2.55	9946
0.06	5239	0.56	7123	1.06	8554	1.56	9406	2.06	9803	2.56	9948
0.07	5279	0.57	7157	1.07	8577	1.57	9418	2.07	9808	2.57	9949
0.08	5319	0.58	7190	1.08	8599	1.58	9429	2.08	9812	2.58	9951
0.09	5359	0.59	7224	1.09	8621	1.59	9441	2.09	9817	2.59	9952
0.10	5398	0.60	7257	1.10	8643	1.60	9452	2.10	9821	2.60	9953
0.11	5438	0.61	7291	1.11	8665	1.61	9463	2.11	9826	2.61	9955
0.12	5478	0.62	7324	1.12	8686	1.62	9474	2.12	9830	2.62	9956
0.13	5517	0.63	7357	1.13	8708	1.63	9484	2.13	9834	2.63	9957
0.14	5557	0.64	7389	1.14	8729	1.64	9495	2.14	9838	2.64	9959
0.15	5596	0.65	7422	1.15	8749	1.65	9505	2.15	9842	2.65	9960
0.16	5636	0.66	7454	1.16	8770	1.66	9515	2.16	9846	2.66	9961
0.17	5675	0.67	7486	1.17	8790	1.67	9525	2.17	9850	2.67	9962
0.18	5714	0.68	7517	1.18	8810	1.68	9535	2.18	9854	2.68	9963
0.19	5753	0.69	7549	1.19	8830	1.69	9545	2.19	9857	2.69	9964
0.20	5793	0.70	7580	1.20	8849	1.70	9554	2.20	9861	2.70	9965
0.21	5832	0.71	7611	1.21	8869	1.71	9564	2.21	9864	2.71	9966
0.22	5871	0.72	7642	1.22	8888	1.72	9573	2.22	9868	2.72	9967
0.23	5910	0.73	7673	1.23	8907	1.73	9582	2.23	9871	2.73	9968
0.24	5948	0.74	7704	1.24	8925	1.74	9591	2.24	9875	2.74	9969
0.25	5987	0.75	7734	1.25	8944	1.75	9599	2.25	9878	2.75	9970
0.26	6026	0.76	7764	1.26	8962	1.76	9608	2.26	9881	2.76	9971
0.27	6064	0.77	7794	1.27	8980	1.77	9616	2.27	9884	2.77	9972
0.28	6103	0.78	7823	1.28	8997	1.78	9625	2.28	9887	2.78	9973
0.29	6141	0.79	7852	1.29	9015	1.79	9633	2.29	9890	2.79	9974
0.30	6179	0.80	7881	1.30	9032	1.80	9641	2.30	9893	2.80	9974
0.31	6217	0.81	7910	1.31	9049	1.81	9649	2.31	9896	2.81	9975
0.32	6255	0.82	7939	1.32	9066	1.82	9656	2.32	9898	2.82	9976
0.33	6293	0.83	7967	1.33	9082	1.83	9664	2.33	9901	2.83	9977
0.34	6331	0.84	7995	1.34	9099	1.84	9671	2.34	9904	2.84	9977
0.35	6368	0.85	8023	1.35	9115	1.85	9678	2.35	9906	2.85	9978
0.36	6406	0.86	8051	1.36	9131	1.86	9686	2.36	9909	2.86	9979
0.37	6443	0.87	8078	1.37	9147	1.87	9693	2.37	9911	2.87	9979
0.38	6480	0.88	8106	1.38	9162	1.88	9699	2.38	9913	2.88	9980
0.39	6517	0.89	8133	1.39	9177	1.89	9706	2.39	9916	2.89	9981
0.40	6554	0.90	8159	1.40	9192	1.90	9713	2.40	9918	2.90	9981
0.41	6591	0.91	8186	1.41	9207	1.91	9719	2.41	9920	2.91	9982
0.42	6628	0.92	8212	1.42	9222	1.92	9726	2.42	9922	2.92	9982
0.43	6664	0.93	8238	1.43	9236	1.93	9732	2.43	9925	2.93	9983
0.44	6700	0.94	8264	1.44	9251	1.94	9738	2.44	9927	2.94	9984
0.45	6736	0.95	8289	1.45	9265	1.95	9744	2.45	9929	2.95	9984
0.46	6772	0.96	8315	1.46	9279	1.96	9750	2.46	9931	2.96	9985
0.47	6808	0.97	8340	1.47	9292	1.97	9756	2.47	9932	2.97	9985
0.48	6844	0.98	8365	1.48	9306	1.98	9761	2.48	9934	2.98	9986
0.49	6879	0.99	8389	1.49	9319	1.99	9767	2.49	9936	2.99	9986
0.50	6915	1.00	8413	1.50	9332	2.00	9772	2.50	9938	3.00	9987

Table A8
Normal Distribution

Values of z for given values of $\Phi(z)$ [see (4), Sec. 23.7] and $D(z) = \Phi(z) - \Phi(-z)$
Example: $z = 0.279$ if $\Phi(z) = 61\%$; $z = 0.860$ if $D(z) = 61\%$.

%	$z(\Phi)$	$z(D)$	%	$z(\Phi)$	$z(D)$	%	$z(\Phi)$	$z(D)$
1	−2.326	0.013	41	−0.228	0.539	81	0.878	1.311
2	−2.054	0.025	42	−0.202	0.553	82	0.915	1.341
3	−1.881	0.038	43	−0.176	0.568	83	0.954	1.372
4	−1.751	0.050	44	−0.151	0.583	84	0.994	1.405
5	−1.645	0.063	45	−0.126	0.598	85	1.036	1.440
6	−1.555	0.075	46	−0.100	0.613	86	1.080	1.476
7	−1.476	0.088	47	−0.075	0.628	87	1.126	1.514
8	−1.405	0.100	48	−0.050	0.643	88	1.175	1.555
9	−1.341	0.113	49	−0.025	0.659	89	1.227	1.598
10	−1.282	0.126	50	0.000	0.674	90	1.282	1.645
11	−1.227	0.138	51	0.025	0.690	91	1.341	1.695
12	−1.175	0.151	52	0.050	0.706	92	1.405	1.751
13	−1.126	0.164	53	0.075	0.722	93	1.476	1.812
14	−1.080	0.176	54	0.100	0.739	94	1.555	1.881
15	−1.036	0.189	55	0.126	0.755	95	1.645	1.960
16	−0.994	0.202	56	0.151	0.772	96	1.751	2.054
17	−0.954	0.215	57	0.176	0.789	97	1.881	2.170
18	−0.915	0.228	58	0.202	0.806	97.5	1.960	2.241
19	−0.878	0.240	59	0.228	0.824	98	2.054	2.326
20	−0.842	0.253	60	0.253	0.842	99	2.326	2.576
21	−0.806	0.266	61	0.279	0.860	99.1	2.366	2.612
22	−0.772	0.279	62	0.305	0.878	99.2	2.409	2.652
23	−0.739	0.292	63	0.332	0.896	99.3	2.457	2.697
24	−0.706	0.305	64	0.358	0.915	99.4	2.512	2.748
25	−0.674	0.319	65	0.385	0.935	99.5	2.576	2.807
26	−0.643	0.332	66	0.412	0.954	99.6	2.652	2.878
27	−0.613	0.345	67	0.440	0.974	99.7	2.748	2.968
28	−0.583	0.358	68	0.468	0.994	99.8	2.878	3.090
29	−0.553	0.372	69	0.496	1.015	99.9	3.090	3.291
30	−0.524	0.385	70	0.524	1.036			
31	−0.496	0.399	71	0.553	1.058	99.91	3.121	3.320
32	−0.468	0.412	72	0.583	1.080	99.92	3.156	3.353
33	−0.440	0.426	73	0.613	1.103	99.93	3.195	3.390
34	−0.412	0.440	74	0.643	1.126	99.94	3.239	3.432
35	−0.385	0.454	75	0.674	1.150	99.95	3.291	3.481
36	−0.358	0.468	76	0.706	1.175	99.96	3.353	3.540
37	−0.332	0.482	77	0.739	1.200	99.97	3.432	3.615
38	−0.305	0.496	78	0.772	1.227	99.98	3.540	3.719
39	−0.279	0.510	79	0.806	1.254	99.99	3.719	3.891
40	−0.253	0.524	80	0.842	1.282			

Table A9
Random Digits

See Sec. 24.2.

Row No.	Column Number 0	1	2	3	4	5	6	7	8	9
0	87331	82442	28104	26432	83640	17323	68764	84728	37995	96106
1	33628	17364	01409	87803	65641	33433	48944	64299	79066	31777
2	54680	13427	72496	16967	16195	96593	55040	53729	62035	66717
3	51199	49794	49407	10774	98140	83891	37195	24066	61140	65144
4	78702	98067	61313	91661	59861	54437	77739	19892	54817	88645
5	55672	16014	24892	13089	00410	81458	76156	28189	40595	21500
6	18880	58497	03862	32368	59320	24807	63392	79793	63043	09425
7	10242	62548	62330	05703	33535	49128	66298	16193	55301	01306
8	54993	17182	94618	23228	83895	73251	68199	64639	83178	70521
9	22686	50885	16006	04041	08077	33065	35237	02502	94755	72062
10	42349	03145	15770	70665	53291	32288	41568	66079	98705	31029
11	18093	09553	39428	75464	71329	86344	80729	40916	18860	51780
12	11535	03924	84252	74795	40193	84597	42497	21918	91384	84721
13	35066	73848	65351	53270	67341	70177	92373	17604	42204	60476
14	57477	22809	73558	96182	96779	01604	25748	59553	64876	94611
15	48647	33850	52956	45410	88212	05120	99391	32276	55961	41775
16	86857	81154	22223	74950	53296	67767	55866	49061	66937	81818
17	20182	36907	94644	99122	09774	29189	27212	79000	50217	71077
18	83687	31231	01133	41432	54542	60204	81618	09586	34481	87683
19	81315	12390	46074	47810	90171	36313	95440	77583	28506	38808
20	87026	52826	58341	76549	04105	66191	12914	55348	07907	06978
21	34301	76733	07251	90524	21931	83695	41340	53581	64582	60210
22	70734	24337	32674	49508	49751	90489	63202	24380	77943	09942
23	94710	31527	73445	32839	68176	53580	85250	53243	03350	00128
24	76462	16987	07775	43162	11777	16810	75158	13894	88945	15539
25	14348	28403	79245	69023	34196	46398	05964	64715	11330	17515
26	74618	89317	30146	25606	94507	98104	04239	44973	37636	88866
27	99442	19200	85406	45358	86253	60638	38858	44964	54103	57287
28	26869	44399	89452	06652	31271	00647	46551	83050	92058	83814
29	80988	08149	50499	98584	28385	63680	44638	91864	96002	87802
30	07511	79047	89289	17774	67194	37362	85684	55505	97809	67056
31	49779	12138	05048	03535	27502	63308	10218	53296	48687	61340
32	47938	55945	24003	19635	17471	65997	85906	98694	56420	78357
33	15604	06626	14360	79542	13512	87595	08542	03800	35443	52823
34	12307	27726	21864	00045	16075	03770	86978	52718	02693	09096
35	02450	28053	66134	99445	91316	25727	89399	85272	67148	78358
36	57623	54382	35236	89244	27245	90500	75430	96762	71968	65838
37	91762	78849	93105	40481	99431	03304	21079	86459	21287	76566
38	87373	31137	31128	67050	34309	44914	80711	61738	61498	24288
39	67094	41485	54149	86088	10192	21174	39948	67268	29938	32476
40	94456	66747	76922	87627	71834	57688	04878	78348	68970	60048
41	68359	75292	27710	86889	81678	79798	58360	39175	75667	65782
42	52393	31404	32584	06837	79762	13168	76055	54833	22841	98889
43	59565	91254	11847	20672	37625	41454	86861	55824	79793	74575
44	48185	11066	20162	38230	16043	48409	47421	21195	98008	57305
45	19230	12187	86659	12971	52204	76546	63272	19312	81662	96557
46	84327	21942	81727	68735	89190	58491	55329	96875	19465	89687
47	77430	71210	00591	50124	12030	50280	12358	76174	48353	09682
48	12462	19108	70512	53926	25595	97085	03833	59806	12351	64253
49	11684	06644	57816	10078	45021	47751	38285	73520	08434	65627

Table A9
Random Digits (*continued*)

Row No.	Column Number 0	1	2	3	4	5	6	7	8	9
50	12896	36576	68686	08462	65652	76571	70891	09007	04581	01684
51	59090	05111	27587	90349	30789	50304	70650	06646	70126	15284
52	42486	67483	65282	19037	80588	73076	41820	46651	40442	40718
53	88662	03928	03249	85910	97533	88643	29829	21557	47328	36724
54	69403	03626	92678	53460	15465	83516	54012	80509	55976	46115
55	56434	70543	38696	98502	32092	95505	62091	39549	30117	98209
56	58227	62694	42837	29183	11393	68463	25150	86338	95620	39836
57	41272	94927	15413	40505	33123	63218	72940	98349	57249	40170
58	36819	01162	30425	15546	16065	68459	35776	64276	92868	07372
59	31700	66711	26115	55755	33584	18091	38709	57276	74660	90392
60	69855	63699	36839	90531	97125	87875	62824	03889	12538	24740
61	44322	17569	45439	41455	34324	90902	07978	26268	04279	76816
62	62226	36661	87011	66267	78777	78044	40819	49496	39814	73867
63	27284	19737	98741	72531	52741	26699	98755	19657	08665	16818
64	88341	21652	94743	77268	79525	44769	66583	30621	90534	62050
65	53266	18783	51903	56711	38060	69513	61963	80470	88018	86510
66	50527	49330	24839	42529	03944	95219	88724	37247	84166	23023
67	15655	07852	77206	35944	71446	30573	19405	57824	23576	23301
68	62057	22206	03314	83465	57466	10465	19891	32308	01900	67484
69	41769	56091	19892	96253	92808	45785	52774	49674	68103	65032
70	25993	72416	44473	41299	93095	17338	69802	98548	02429	85238
71	22842	57871	04470	37373	34516	04042	04078	35336	34393	97573
72	55704	31982	05234	22664	22181	40358	28089	15790	33340	18852
73	94258	18706	09437	96041	90052	80862	20420	24323	11635	91677
74	74145	20453	29657	98868	56695	53483	87449	35060	98942	62697
75	88881	12673	73961	89884	73247	97670	69570	88888	58560	72580
76	01508	56780	52223	35632	73347	71317	46541	88023	36656	76332
77	92069	43000	23233	06058	82527	25250	27555	20426	60361	63525
78	53366	35249	02117	68620	39388	69795	73215	01846	16983	78560
79	88057	54097	49511	74867	32192	90071	04147	46094	63519	07199
80	85492	82238	02668	91854	86149	28590	77853	81035	45561	16032
81	39453	62123	69611	53017	34964	09786	24614	49514	01056	18700
82	82627	98111	93870	56969	69566	62662	07353	84838	14570	14508
83	61142	51743	38209	31474	96095	15163	54380	77849	20465	03142
84	12031	32528	61311	53730	89032	16124	58844	35386	45521	59368
85	31313	59838	29147	76882	74328	09955	63673	96651	53264	29871
86	50767	41056	97409	44376	62219	35439	70102	99248	71179	26052
87	30522	95699	84966	26554	24768	72247	84993	85375	92518	16334
88	74176	19870	89874	64799	03792	57006	57225	36677	46825	14087
89	17114	93248	37065	91346	04657	93763	92210	43676	44944	75798
90	53005	11825	64608	87587	05742	31914	55044	41818	29667	77424
91	31985	81539	79942	49471	46200	27639	94099	42085	79231	03932
92	63499	60508	77522	15624	15088	78519	52279	79214	43623	69166
93	30506	42444	99047	66010	91657	37160	37408	85714	21420	80996
94	78248	16841	92357	10130	68990	38307	61022	56806	81016	38511
95	64996	84789	50185	32200	64382	29752	11876	00664	54547	62597
96	11963	13157	09136	01769	30117	71486	80111	09161	08371	71749
97	44335	91450	43456	90449	18338	19787	31339	60473	06606	89788
98	42277	11868	44520	01113	11341	11743	97949	49718	99176	42006
99	77562	18863	58515	90166	78508	14864	19111	57183	85808	59385

Table A10
t-Distribution

Values of z for given values of the distribution function $F(z)$ (see p. 1225)
Example: For 9 degrees of freedom, $z = 1.83$ when $F(z) = 0.95$.

$F(z)$	Number of Degrees of Freedom									
	1	2	3	4	5	6	7	8	9	10
0.5	0.00	0.00	0.00	0.00	0.00	0.00	0.00	0.00	0.00	0.00
0.6	0.33	0.29	0.28	0.27	0.27	0.27	0.26	0.26	0.26	0.26
0.7	0.73	0.62	0.58	0.57	0.56	0.55	0.55	0.55	0.54	0.54
0.8	1.38	1.06	0.98	0.94	0.92	0.91	0.90	0.89	0.88	0.88
0.9	3.08	1.89	1.64	1.53	1.48	1.44	1.42	1.40	1.38	1.37
0.95	6.31	2.92	2.35	2.13	2.02	1.94	1.90	1.86	1.83	1.81
0.975	12.7	4.30	3.18	2.78	2.57	2.45	2.37	2.31	2.26	2.23
0.99	31.8	6.97	4.54	3.75	3.37	3.14	3.00	2.90	2.82	2.76
0.995	63.7	9.93	5.84	4.60	4.03	3.71	3.50	3.36	3.25	3.17
0.999	318.3	22.3	10.2	7.17	5.89	5.21	4.79	4.50	4.30	4.14

$F(z)$	Number of Degrees of Freedom									
	11	12	13	14	15	16	17	18	19	20
0.5	0.00	0.00	0.00	0.00	0.00	0.00	0.00	0.00	0.00	0.00
0.6	0.26	0.26	0.26	0.26	0.26	0.26	0.26	0.26	0.26	0.26
0.7	0.54	0.54	0.54	0.54	0.54	0.54	0.53	0.53	0.53	0.53
0.8	0.88	0.87	0.87	0.87	0.87	0.87	0.86	0.86	0.86	0.86
0.9	1.36	1.36	1.35	1.35	1.34	1.34	1.33	1.33	1.33	1.33
0.95	1.80	1.78	1.77	1.76	1.75	1.75	1.74	1.73	1.73	1.73
0.975	2.20	2.18	2.16	2.15	2.13	2.12	2.11	2.10	2.09	2.09
0.99	2.72	2.68	2.65	2.62	2.60	2.58	2.57	2.55	2.54	2.53
0.995	3.11	3.06	3.01	2.98	2.95	2.92	2.90	2.88	2.86	2.85
0.999	4.03	3.93	3.85	3.79	3.73	3.69	3.65	3.61	3.58	3.55

$F(z)$	Number of Degrees of Freedom									
	22	24	26	28	30	40	50	100	200	∞
0.5	0.00	0.00	0.00	0.00	0.00	0.00	0.00	0.00	0.00	0.00
0.6	0.26	0.26	0.26	0.26	0.26	0.26	0.26	0.25	0.25	0.25
0.7	0.53	0.53	0.53	0.53	0.53	0.53	0.53	0.53	0.53	0.52
0.8	0.86	0.86	0.86	0.86	0.85	0.85	0.85	0.85	0.84	0.84
0.9	1.32	1.32	1.32	1.31	1.31	1.30	1.30	1.29	1.29	1.28
0.95	1.72	1.71	1.71	1.70	1.70	1.68	1.68	1.66	1.65	1.65
0.975	2.07	2.06	2.06	2.05	2.04	2.02	2.01	1.98	1.97	1.96
0.99	2.51	2.49	2.48	2.47	2.46	2.42	2.40	2.37	2.35	2.33
0.995	2.82	2.80	2.78	2.76	2.75	2.70	2.68	2.63	2.60	2.58
0.999	3.51	3.47	3.44	3.41	3.39	3.31	3.26	3.17	3.13	3.09

Table A11
Chi-square Distribution

Values of x for given values of the distribution function $F(z)$ (see p. 1227)
Example: For 3 degrees of freedom, $z = 11.34$ when $F(z) = 0.99$.

$F(z)$	Number of Degrees of Freedom									
	1	2	3	4	5	6	7	8	9	10
0.005	0.00	0.01	0.07	0.21	0.41	0.68	0.99	1.34	1.73	2.16
0.01	0.00	0.02	0.11	0.30	0.55	0.87	1.24	1.65	2.09	2.56
0.025	0.00	0.05	0.22	0.48	0.83	1.24	1.69	2.18	2.70	3.25
0.05	0.00	0.10	0.35	0.71	1.15	1.64	2.17	2.73	3.33	3.94
0.95	3.84	5.99	7.81	9.49	11.07	12.59	14.07	15.51	16.92	18.31
0.975	5.02	7.38	9.35	11.14	12.83	14.45	16.01	17.53	19.02	20.48
0.99	6.63	9.21	11.34	13.28	15.09	16.81	18.48	20.09	21.67	23.21
0.995	7.88	10.60	12.84	14.86	16.75	18.55	20.28	21.96	23.59	25.19

$F(z)$	Number of Degrees of Freedom									
	11	12	13	14	15	16	17	18	19	20
0.005	2.60	3.07	3.57	4.07	4.60	5.14	5.70	6.26	6.84	7.43
0.01	3.05	3.57	4.11	4.66	5.23	5.81	6.41	7.01	7.63	8.26
0.025	3.82	4.40	5.01	5.63	6.26	6.91	7.56	8.23	8.91	9.59
0.05	4.57	5.23	5.89	6.57	7.26	7.96	8.67	9.39	10.12	10.85
0.95	19.68	21.03	22.36	23.68	25.00	26.30	27.59	28.87	30.14	31.41
0.975	21.92	23.34	24.74	26.12	27.49	28.85	30.19	31.53	32.85	34.17
0.99	24.73	26.22	27.69	29.14	30.58	32.00	33.41	34.81	36.19	37.57
0.995	26.76	28.30	29.82	31.32	32.80	34.27	35.72	37.16	38.58	40.00

$F(z)$	Number of Degrees of Freedom									
	21	22	23	24	25	26	27	28	29	30
0.005	8.0	8.6	9.3	9.9	10.5	11.2	11.8	12.5	13.1	13.8
0.01	8.9	9.5	10.2	10.9	11.5	12.2	12.9	13.6	14.3	15.0
0.025	10.3	11.0	11.7	12.4	13.1	13.8	14.6	15.3	16.0	16.8
0.05	11.6	12.3	13.1	13.8	14.6	15.4	16.2	16.9	17.7	18.5
0.95	32.7	33.9	35.2	36.4	37.7	38.9	40.1	41.3	42.6	43.8
0.975	35.5	36.8	38.1	39.4	40.6	41.9	43.2	44.5	45.7	47.0
0.99	38.9	40.3	41.6	43.0	44.3	45.6	47.0	48.3	49.6	50.9
0.995	41.4	42.8	44.2	45.6	46.9	48.3	49.6	51.0	52.3	53.7

$F(z)$	Number of Degrees of Freedom							
	40	50	60	70	80	90	100	>100 (Approximation)
0.005	20.7	28.0	35.5	43.3	51.2	59.2	67.3	$\frac{1}{2}(h - 2.58)^2$
0.01	22.2	29.7	37.5	45.4	53.5	61.8	70.1	$\frac{1}{2}(h - 2.33)^2$
0.025	24.4	32.4	40.5	48.8	57.2	65.6	74.2	$\frac{1}{2}(h - 1.96)^2$
0.05	26.5	34.8	43.2	51.7	60.4	69.1	77.9	$\frac{1}{2}(h - 1.64)^2$
0.95	55.8	67.5	79.1	90.5	101.9	113.1	124.3	$\frac{1}{2}(h + 1.64)^2$
0.975	59.3	71.4	83.3	95.0	106.6	118.1	129.6	$\frac{1}{2}(h + 1.96)^2$
0.99	63.7	76.2	88.4	100.4	112.3	124.1	135.8	$\frac{1}{2}(h + 2.33)^2$
0.995	66.8	79.5	92.0	104.2	116.3	128.3	140.2	$\frac{1}{2}(h + 2.58)^2$

In the last column, $h = \sqrt{2m - 1}$ where m is the number of degrees of freedom.

Table A12
F-Distribution with (*m*, *n*) Degrees of Freedom

Values of z for which the distribution function $F(z)$ [see (13), Sec. 24.7] has the value **0.95**

Example: For (7, 4) degrees of freedom, $z = 6.09$ if $F(z) = 0.95$.

n	$m = 1$	$m = 2$	$m = 3$	$m = 4$	$m = 5$	$m = 6$	$m = 7$	$m = 8$	$m = 9$
1	161	200	216	225	230	234	237	239	241
2	18.5	19.0	19.2	19.2	19.3	19.3	19.4	19.4	19.4
3	10.1	9.55	9.28	9.12	9.01	8.94	8.89	8.85	8.81
4	7.71	6.94	6.59	6.39	6.26	6.16	6.09	6.04	6.00
5	6.61	5.79	5.41	5.19	5.05	4.95	4.88	4.82	4.77
6	5.99	5.14	4.76	4.53	4.39	4.28	4.21	4.15	4.10
7	5.59	4.74	4.35	4.12	3.97	3.87	3.79	3.73	3.68
8	5.32	4.46	4.07	3.84	3.69	3.58	3.50	3.44	3.39
9	5.12	4.26	3.86	3.63	3.48	3.37	3.29	3.23	3.18
10	4.96	4.10	3.71	3.48	3.33	3.22	3.14	3.07	3.02
11	4.84	3.98	3.59	3.36	3.20	3.09	3.01	2.95	2.90
12	4.75	3.89	3.49	3.26	3.11	3.00	2.91	2.85	2.80
13	4.67	3.81	3.41	3.18	3.03	2.92	2.83	2.77	2.71
14	4.60	3.74	3.34	3.11	2.96	2.85	2.76	2.70	2.65
15	4.54	3.68	3.29	3.06	2.90	2.79	2.71	2.64	2.59
16	4.49	3.63	3.24	3.01	2.85	2.74	2.66	2.59	2.54
17	4.45	3.59	3.20	2.96	2.81	2.70	2.61	2.55	2.49
18	4.41	3.55	3.16	2.93	2.77	2.66	2.58	2.51	2.46
19	4.38	3.52	3.13	2.90	2.74	2.63	2.54	2.48	2.42
20	4.35	3.49	3.10	2.87	2.71	2.60	2.51	2.45	2.39
22	4.30	3.44	3.05	2.82	2.66	2.55	2.46	2.40	2.34
24	4.26	3.40	3.01	2.78	2.62	2.51	2.42	2.36	2.30
26	4.23	3.37	2.98	2.74	2.59	2.47	2.39	2.32	2.27
28	4.20	3.34	2.95	2.71	2.56	2.45	2.36	2.29	2.24
30	4.17	3.32	2.92	2.69	2.53	2.42	2.33	2.27	2.21
32	4.15	3.30	2.90	2.67	2.51	2.40	2.31	2.24	2.19
34	4.13	3.28	2.88	2.65	2.49	2.38	2.29	2.23	2.17
36	4.11	3.26	2.87	2.63	2.48	2.36	2.28	2.21	2.15
38	4.10	3.24	2.85	2.62	2.46	2.35	2.26	2.19	2.14
40	4.08	3.23	2.84	2.61	2.45	2.34	2.25	2.18	2.12
50	4.03	3.18	2.79	2.56	2.40	2.29	2.20	2.13	2.07
60	4.00	3.15	2.76	2.53	2.37	2.25	2.17	2.10	2.04
70	3.98	3.13	2.74	2.50	2.35	2.23	2.14	2.07	2.02
80	3.96	3.11	2.72	2.49	2.33	2.21	2.13	2.06	2.00
90	3.95	3.10	2.71	2.47	2.32	2.20	2.11	2.04	1.99
100	3.94	3.09	2.70	2.46	2.31	2.19	2.10	2.03	1.97
150	3.90	3.06	2.66	2.43	2.27	2.16	2.07	2.00	1.94
200	3.89	3.04	2.65	2.42	2.26	2.14	2.06	1.98	1.93
1000	3.85	3.00	2.61	2.38	2.22	2.11	2.02	1.95	1.89
∞	3.84	3.00	2.60	2.37	2.21	2.10	2.01	1.94	1.88

Table A12
F-Distribution with (*m*, *n*) Degrees of Freedom (*continued*)

Values of z for which the distribution function $F(z)$ [see (13), Sec. 24.7) has the value **0.95**

n	$m = 10$	$m = 15$	$m = 20$	$m = 30$	$m = 40$	$m = 50$	$m = 100$	∞
1	242	246	248	250	251	252	253	254
2	19.4	19.4	19.4	19.5	19.5	19.5	19.5	19.5
3	8.79	8.70	8.66	8.62	8.59	8.58	8.55	8.53
4	5.96	5.86	5.80	5.75	5.72	5.70	5.66	5.63
5	4.74	4.62	4.56	4.50	4.46	4.44	4.41	4.37
6	4.06	3.94	3.87	3.81	3.77	3.75	3.71	3.67
7	3.64	3.51	3.44	3.38	3.34	3.32	3.27	3.23
8	3.35	3.22	3.15	3.08	3.04	3.02	2.97	2.93
9	3.14	3.01	2.94	2.86	2.83	2.80	2.76	2.71
10	2.98	2.85	2.77	2.70	2.66	2.64	2.59	2.54
11	2.85	2.72	2.65	2.57	2.53	2.51	2.46	2.40
12	2.75	2.62	2.54	2.47	2.43	2.40	2.35	2.30
13	2.67	2.53	2.46	2.38	2.34	2.31	2.26	2.21
14	2.60	2.46	2.39	2.31	2.27	2.24	2.19	2.13
15	2.54	2.40	2.33	2.25	2.20	2.18	2.12	2.07
16	2.49	2.35	2.28	2.19	2.15	2.12	2.07	2.01
17	2.45	2.31	2.23	2.15	2.10	2.08	2.02	1.96
18	2.41	2.27	2.19	2.11	2.06	2.04	1.98	1.92
19	2.38	2.23	2.16	2.07	2.03	2.00	1.94	1.88
20	2.35	2.20	2.12	2.04	1.99	1.97	1.91	1.84
22	2.30	2.15	2.07	1.98	1.94	1.91	1.85	1.78
24	2.25	2.11	2.03	1.94	1.89	1.86	1.80	1.73
26	2.22	2.07	1.99	1.90	1.85	1.82	1.76	1.69
28	2.19	2.04	1.96	1.87	1.82	1.79	1.73	1.65
30	2.16	2.01	1.93	1.84	1.79	1.76	1.70	1.62
32	2.14	1.99	1.91	1.82	1.77	1.74	1.67	1.59
34	2.12	1.97	1.89	1.80	1.75	1.71	1.65	1.57
36	2.11	1.95	1.87	1.78	1.73	1.69	1.62	1.55
38	2.09	1.94	1.85	1.76	1.71	1.68	1.61	1.53
40	2.08	1.92	1.84	1.74	1.69	1.66	1.59	1.51
50	2.03	1.87	1.78	1.69	1.63	1.60	1.52	1.44
60	1.99	1.84	1.75	1.65	1.59	1.56	1.48	1.39
70	1.97	1.81	1.72	1.62	1.57	1.53	1.45	1.35
80	1.95	1.79	1.70	1.60	1.54	1.51	1.43	1.32
90	1.94	1.78	1.69	1.59	1.53	1.49	1.41	1.30
100	1.93	1.77	1.68	1.57	1.52	1.48	1.39	1.28
150	1.89	1.73	1.64	1.53	1.48	1.44	1.34	1.22
200	1.88	1.72	1.62	1.52	1.46	1.41	1.32	1.19
1000	1.84	1.68	1.58	1.47	1.41	1.36	1.26	1.08
∞	1.83	1.67	1.57	1.46	1.39	1.35	1.24	1.00

Table A12
F-Distribution with (*m*, *n*) Degrees of Freedom (*continued*)

Values of z for which the distribution function $F(z)$ [see (13), Sec. 24.7] has the value **0.99**

n	$m = 1$	$m = 2$	$m = 3$	$m = 4$	$m = 5$	$m = 6$	$m = 7$	$m = 8$	$m = 9$
1	4052.	4999	5403	5625	5764	5859	5928	5982	6022
2	98.5	99.0	99.2	99.3	99.3	99.3	99.4	99.4	99.4
3	34.1	30.8	29.5	28.7	28.2	27.9	27.7	27.5	27.3
4	21.2	18.0	16.7	16.0	15.5	15.2	15.0	14.8	14.7
5	16.3	13.3	12.1	11.4	11.0	10.7	10.5	10.3	10.2
6	13.7	10.9	9.78	9.15	8.75	8.47	8.26	8.10	7.98
7	12.2	9.55	8.45	7.85	7.46	7.19	6.99	6.84	6.72
8	11.3	8.65	7.59	7.01	6.63	6.37	6.18	6.03	5.91
9	10.6	8.02	6.99	6.42	6.06	5.80	5.61	5.47	5.35
10	10.0	7.56	6.55	5.99	5.64	5.39	5.20	5.06	4.94
11	9.65	7.21	6.22	5.67	5.32	5.07	4.89	4.74	4.63
12	9.33	6.93	5.95	5.41	5.06	4.82	4.64	4.50	4.39
13	9.07	6.70	5.74	5.21	4.86	4.62	4.44	4.30	4.19
14	8.86	6.51	5.56	5.04	4.70	4.46	4.28	4.14	4.03
15	8.68	6.36	5.42	4.89	4.56	4.32	4.14	4.00	3.89
16	8.53	6.23	5.29	4.77	4.44	4.20	4.03	3.89	3.78
17	8.40	6.11	5.18	4.67	4.34	4.10	3.93	3.79	3.68
18	8.29	6.01	5.09	4.58	4.25	4.01	3.84	3.71	3.60
19	8.18	5.93	5.01	4.50	4.17	3.94	3.77	3.63	3.52
20	8.10	5.85	4.94	4.43	4.10	3.87	3.70	3.56	3.46
22	7.95	5.72	4.82	4.31	3.99	3.76	3.59	3.45	3.35
24	7.82	5.61	4.72	4.22	3.90	3.67	3.50	3.36	3.26
26	7.72	5.53	4.64	4.14	3.82	3.59	3.42	3.29	3.18
28	7.64	5.45	4.57	4.07	3.75	3.53	3.36	3.23	3.12
30	7.56	5.39	4.51	4.02	3.70	3.47	3.30	3.17	3.07
32	7.50	5.34	4.46	3.97	3.65	3.43	3.26	3.13	3.02
34	7.44	5.29	4.42	3.93	3.61	3.39	3.22	3.09	2.98
36	7.40	5.25	4.38	3.89	3.57	3.35	3.18	3.05	2.95
38	7.35	5.21	4.34	3.86	3.54	3.32	3.15	3.02	2.92
40	7.31	5.18	4.31	3.83	3.51	3.29	3.12	2.99	2.89
50	7.17	5.06	4.20	3.72	3.41	3.19	3.02	2.89	2.79
60	7.08	4.98	4.13	3.65	3.34	3.12	2.95	2.82	2.72
70	7.01	4.92	4.08	3.60	3.29	3.07	2.91	2.78	2.67
80	6.96	4.88	4.04	3.56	3.26	3.04	2.87	2.74	2.64
90	6.93	4.85	4.01	3.54	3.23	3.01	2.84	2.72	2.61
100	6.90	4.82	3.98	3.51	3.21	2.99	2.82	2.69	2.59
150	6.81	4.75	3.92	3.45	3.14	2.92	2.76	2.63	2.53
200	6.76	4.71	3.88	3.41	3.11	2.89	2.73	2.60	2.50
1000	6.66	4.63	3.80	3.34	3.04	2.82	2.66	2.53	2.43
∞	6.63	4.61	3.78	3.32	3.02	2.80	2.64	2.51	2.41

Table A12
F-Distribution with (*m*, *n*) Degrees of Freedom (*continued*)

Values of z for which the distribution function $F(z)$ [see (13), Sec. 24.7] has the value **0.99**

n	$m = 10$	$m = 15$	$m = 20$	$m = 30$	$m = 40$	$m = 50$	$m = 100$	∞
1	6056	6157	6209	6261	6287	6300	6330	6366
2	99.4	99.4	99.4	99.5	99.5	99.5	99.5	99.5
3	27.2	26.9	26.7	26.5	26.4	26.4	26.2	26.1
4	14.5	14.2	14.0	13.8	13.7	13.7	13.6	13.5
5	10.1	9.72	9.55	9.38	9.29	9.24	9.13	9.02
6	7.87	7.56	7.40	7.23	7.14	7.09	6.99	6.88
7	6.62	6.31	6.16	5.99	5.91	5.86	5.75	5.65
8	5.81	5.52	5.36	5.20	5.12	5.07	4.96	4.86
9	5.26	4.96	4.81	4.65	4.57	4.52	4.42	4.31
10	4.85	4.56	4.41	4.25	4.17	4.12	4.01	3.91
11	4.54	4.25	4.10	3.94	3.86	3.81	3.71	3.60
12	4.30	4.01	3.86	3.70	3.62	3.57	3.47	3.36
13	4.10	3.82	3.66	3.51	3.43	3.38	3.27	3.17
14	3.94	3.66	3.51	3.35	3.27	3.22	3.11	3.00
15	3.80	3.52	3.37	3.21	3.13	3.08	2.98	2.87
16	3.69	3.41	3.26	3.10	3.02	2.97	2.86	2.75
17	3.59	3.31	3.16	3.00	2.92	2.87	2.76	2.65
18	3.51	3.23	3.08	2.92	2.84	2.78	2.68	2.57
19	3.43	3.15	3.00	2.84	2.76	2.71	2.60	2.49
20	3.37	3.09	2.94	2.78	2.69	2.64	2.54	2.42
22	3.26	2.98	2.83	2.67	2.58	2.53	2.42	2.31
24	3.17	2.89	2.74	2.58	2.49	2.44	2.33	2.21
26	3.09	2.82	2.66	2.50	2.42	2.36	2.25	2.13
28	3.03	2.75	2.60	2.44	2.35	2.30	2.19	2.06
30	2.98	2.70	2.55	2.39	2.30	2.25	2.13	2.01
32	2.93	2.66	2.50	2.34	2.25	2.20	2.08	1.96
34	2.89	2.62	2.46	2.30	2.21	2.16	2.04	1.91
36	2.86	2.58	2.43	2.26	2.17	2.12	2.00	1.87
38	2.83	2.55	2.40	2.23	2.14	2.09	1.97	1.84
40	2.80	2.52	2.37	2.20	2.11	2.06	1.94	1.80
50	2.70	2.42	2.27	2.10	2.01	1.95	1.82	1.68
60	2.63	2.35	2.20	2.03	1.94	1.88	1.75	1.60
70	2.59	2.31	2.15	1.98	1.89	1.83	1.70	1.54
80	2.55	2.27	2.12	1.94	1.85	1.79	1.66	1.49
90	2.52	2.24	2.09	1.92	1.82	1.76	1.62	1.46
100	2.50	2.22	2.07	1.89	1.80	1.73	1.60	1.43
150	2.44	2.16	2.00	1.83	1.73	1.66	1.52	1.33
200	2.41	2.13	1.97	1.79	1.69	1.63	1.48	1.28
1000	2.34	2.06	1.90	1.72	1.61	1.54	1.38	1.11
∞	2.32	2.04	1.88	1.70	1.59	1.52	1.36	1.00

Table A13
Distribution Function $F(x) = P(T \leqq x)$ of the Random Variable T in Section 24.11

If $n = 3$, then $F(2) = 1 - 0.167 = 0.833$.
If $n = 4$, then $F(3) = 1 - 0.375 = 0.625$, $F(4) = 1 - 0.167 = 0.833$, etc.

x	$n = 3$
	0.
0	167
1	500

x	$n = 4$
	0.
0	042
1	167
2	375

x	$n = 5$
	0.
0	008
1	042
2	117
3	242
4	408

x	$n = 6$
	0.
0	001
1	008
2	028
3	068
4	136
5	235
6	360
7	500

x	$n = 7$
	0.
1	001
2	005
3	015
4	035
5	068
6	119
7	191
8	281
9	386
10	500

x	$n = 8$
	0.
2	001
3	003
4	007
5	016
6	031
7	054
8	089
9	138
10	199
11	274
12	360
13	452

x	$n = 9$
	0.
4	001
5	003
6	006
7	012
8	022
9	038
10	060
11	090
12	130
13	179
14	238
15	306
16	381
17	460

x	$n = 10$
	0.
6	001
7	002
8	005
9	008
10	014
11	023
12	036
13	054
14	078
15	108
16	146
17	190
18	242
19	300
20	364
21	431
22	500

x	$n = 11$
	0.
8	001
9	002
10	003
11	005
12	008
13	013
14	020
15	030
16	043
17	060
18	082
19	109
20	141
21	179
22	223
23	271
24	324
25	381
26	440
27	500

x	$n = 12$
	0.
11	001
12	002
13	003
14	004
15	007
16	010
17	016
18	022
19	031
20	043
21	058
22	076
23	098
24	125
25	155
26	190
27	230
28	273
29	319
30	369
31	420
32	473

x	$n = 13$
	0.
14	001
15	001
16	002
17	003
18	005
19	007
20	011
21	015
22	021
23	029
24	038
25	050
26	064
27	082
28	102
29	126
30	153
31	184
32	218
33	255
34	295
35	338
36	383
37	429
38	476

x	$n = 14$
	0.
18	001
19	002
20	002
21	003
22	005
23	007
24	010
25	013
26	018
27	024
28	031
29	040
30	051
31	063
32	079
33	096
34	117
35	140
36	165
37	194
38	225
39	259
40	295
41	334
42	374
43	415
44	457
45	500

x	$n = 15$
	0.
23	001
24	002
25	003
26	004
27	006
28	008
29	010
30	014
31	018
32	023
33	029
34	037
35	046
36	057
37	070
38	084
39	101
40	120
41	141
42	164
43	190
44	218
45	248
46	279
47	313
48	349
49	385
50	423
51	461
52	500

x	$n = 16$
	0.
27	001
28	002
29	002
30	003
31	004
32	006
33	008
34	010
35	013
36	016
37	021
38	026
39	032
40	039
41	048
42	058
43	070
44	083
45	097
46	114
47	133
48	153
49	175
50	199
51	225
52	253
53	282
54	313
55	345
56	378
57	412
58	447
59	482

x	$n = 17$
	0.
32	001
33	002
34	002
35	003
36	004
37	005
38	007
39	009
40	011
41	014
42	017
43	021
44	026
45	032
46	038
47	046
48	054
49	064
50	076
51	088
52	102
53	118
54	135
55	154
56	174
57	196
58	220
59	245
60	271
61	299
62	328
63	358
64	388
65	420
66	452
67	484

x	$n = 18$
	0.
38	001
39	002
40	003
41	003
42	004
43	005
44	007
45	009
46	011
47	013
48	016
49	020
50	024
51	029
52	034
53	041
54	048
55	056
56	066
57	076
58	088
59	100
60	115
61	130
62	147
63	165
64	184
65	205
66	227
67	250
68	275
69	300
70	327
71	354
72	383
73	411
74	441
75	470
76	500

x	$n = 19$
	0.
43	001
44	002
45	002
46	003
47	003
48	004
49	005
50	006
51	008
52	010
53	012
54	014
55	017
56	021
57	025
58	029
59	034
60	040
61	047
62	054
63	062
64	072
65	082
66	093
67	105
68	119
69	133
70	149
71	166
72	184
73	203
74	223
75	245
76	267
77	290
78	314
79	339
80	365
81	391
82	418
83	445
84	473
85	500

x	$n = 20$
	0.
50	001
51	002
52	002
53	003
54	004
55	005
56	006
57	007
58	008
59	010
60	012
61	014
62	017
63	020
64	023
65	027
66	032
67	037
68	043
69	049
70	056
71	064
72	073
73	082
74	093
75	104
76	117
77	130
78	144
79	159
80	176
81	193
82	211
83	230
84	250
85	271
86	293
87	315
88	339
89	362
90	387
91	411
92	436
93	462
94	487

INDEX

Page numbers A1, A2, A3, · · · refer to Appendix 1 to Appendix 5 at the end of the book.

A

B

C

D

E

F

G

N

O

P

S

T

U

V

W

Z

Typeset in Australia by Essay Composition, Alexandria NSW 2015
Printed by Times Printers Pte Ltd

Some Constants

$$
\begin{aligned}
e &= 2.71828\ 18284\ 59045\ 23536 \\
\sqrt{e} &= 1.64872\ 12707\ 00128\ 14685 \\
e^2 &= 7.38905\ 60989\ 30650\ 22723
\end{aligned}
$$

$$
\begin{aligned}
\pi &= 3.14159\ 26535\ 89793\ 23846 \\
\pi^2 &= 9.86960\ 44010\ 89358\ 61883 \\
\sqrt{\pi} &= 1.77245\ 38509\ 05516\ 02730
\end{aligned}
$$

$$
\begin{aligned}
\log_{10} \pi &= 0.49714\ 98726\ 94133\ 85435 \\
\ln \pi &= 1.14472\ 98858\ 49400\ 17414 \\
\log_{10} e &= 0.43429\ 44819\ 03251\ 82765 \\
\ln 10 &= 2.30258\ 50929\ 94045\ 68402
\end{aligned}
$$

$$
\begin{aligned}
\sqrt{2} &= 1.41421\ 35623\ 73095\ 04880 \\
\sqrt[3]{2} &= 1.25992\ 10498\ 94873\ 16477 \\
\sqrt{3} &= 1.73205\ 08075\ 68877\ 29353 \\
\sqrt[3]{3} &= 1.44224\ 95703\ 07408\ 38232 \\
\ln 2 &= 0.69314\ 71805\ 59945\ 30942 \\
\ln 3 &= 1.09861\ 22886\ 68109\ 69140
\end{aligned}
$$

$$
\begin{aligned}
\gamma &= 0.57721\ 56649\ 01532\ 86061 \\
\ln \gamma &= -0.54953\ 93129\ 81644\ 82234
\end{aligned}
$$

(see Sec. 5.7)

$$
\begin{aligned}
1^\circ &= 0.01745\ 32925\ 19943\ 29577 \text{ rad} \\
1 \text{ rad} &= 57.29577\ 95130\ 82320\ 87680^\circ \\
&= 57^\circ 17' 44.806''
\end{aligned}
$$

Polar Coordinates

$$x = r \cos \theta \qquad r = \sqrt{x^2 + y^2}$$

$$y = r \sin \theta \qquad \theta = \arctan \frac{y}{x}$$

$$dx\, dy = r\, dr\, d\theta$$

Series

$$\frac{1}{1-x} = \sum_{m=0}^{\infty} x^m \qquad (|x| < 1)$$

$$e^x = \sum_{m=0}^{\infty} \frac{x^m}{m!}$$

$$\sin x = \sum_{m=0}^{\infty} \frac{(-1)^m x^{2m+1}}{(2m+1)!}$$

$$\cos x = \sum_{m=0}^{\infty} \frac{(-1)^m x^{2m}}{(2m)!}$$

$$\ln (1 - x) = -\sum_{m=0}^{\infty} \frac{x^m}{m} \qquad (|x| < 1)$$

$$\arctan x = \sum_{m=0}^{\infty} \frac{(-1)^m x^{2m+1}}{2m+1} \qquad (|x| < 1)$$

Greek Alphabet

α	Alpha	ν	Nu
β	Beta	ξ	Xi
γ, Γ	Gamma	o	Omicron
δ, Δ	Delta	π	Pi
ϵ	Epsilon	ρ	Rho
ζ	Zeta	σ, Σ	Sigma
η	Eta	τ	Tau
θ, ϑ, Θ	Theta	υ, Υ	Upsilon
ι	Iota	ϕ, φ, Φ	Phi
κ	Kappa	χ	Chi
λ, Λ	Lambda	ψ, Ψ	Psi
μ	Mu	ω, Ω	Omega

Vectors

$$\mathbf{a} \cdot \mathbf{b} = a_1 b_1 + a_2 b_2 + a_3 b_3$$

$$\mathbf{a} \times \mathbf{b} = \begin{vmatrix} \mathbf{i} & \mathbf{j} & \mathbf{k} \\ a_1 & a_2 & a_3 \\ b_1 & b_2 & b_3 \end{vmatrix}$$

$$\operatorname{grad} f = \nabla f = \frac{\partial f}{\partial x}\mathbf{i} + \frac{\partial f}{\partial y}\mathbf{j} + \frac{\partial f}{\partial z}\mathbf{k}$$

$$\operatorname{div} \mathbf{v} = \nabla \cdot \mathbf{v} = \frac{\partial v_1}{\partial x} + \frac{\partial v_2}{\partial y} + \frac{\partial v_3}{\partial z}$$

$$\operatorname{curl} \mathbf{v} = \nabla \times \mathbf{v} = \begin{vmatrix} \mathbf{i} & \mathbf{j} & \mathbf{k} \\ \dfrac{\partial}{\partial x} & \dfrac{\partial}{\partial y} & \dfrac{\partial}{\partial z} \\ v_1 & v_2 & v_3 \end{vmatrix}$$